of the Elements

18

2
4.2
0.95
0.176¤ **He**
Helium
4.002 602 ±2

Atomic Mass Interval

H	[1.007 84; 1.008 11]
Li	[6.938; 6.997]
B	[10.806; 10.821]
C	[12.009 6; 12.011 6]
N	[14.006 43; 14.007 28]
O	[15.999 03; 15.999 77]
Si	[28.084; 28.086]
S	[32.059; 32.076]
Cl	[35.446; 35.457]
Tl	[204.382; 204.385]

13 14 15 16 17

5 +3	6 ±4,2	7 ±3,5,4,2	8 -2	9 -1	10
4275	4470	77	90	85	27
2300	4100	63	50	53	25
2.34 **B**	2.62 **C**	1.234¤ **N**	1.410¤ **O**	1.674¤ **F**	0.889¤ **Ne**
Boron	Carbon	Nitrogen	Oxygen	Fluorine	Neon
10.814 ±8	12.010 6 ±10	14.006 8 ±4	15.999 4 ±4	18.998 403 2 ±5	20.179 7 ±6

13 +3	14 +4	15 ±3,5,4	16 ±2,4,6	17 ±1,3,5,7	18
2793	3540	550	718	239	87
933	1685	317	388	172	84
2.70 **Al**	2.33 **Si**	1.82 **P**	2.07 **S**	3.12¤ **Cl**	1.760¤ **Ar**
Aluminum	Silicon	Phosphorus	Sulfur	Chlorine	Argon
26.981 538 6 ±8	28.085	30.973 762 ±2	32.068 ±9	35.452 ±6	39.948

10 11 12

28 +2,3	29 +2,1	30 +2	31 +3	32 +4	33 ±3,5	34 -2,4,6	35 ±1,5	36
3187	2836	1180	2478	3107	876	958	332	120
1726	1358	693	303	1210	—	494	266	116
8.90 **Ni**	8.96 **Cu**	7.14 **Zn**	5.91 **Ga**	5.32 **Ge**	5.72 **As**	4.80 **Se**	3.12 **Br**	3.69¤ **Kr**
Nickel	Copper	Zinc	Gallium	Germanium	Arsenic	Selenium	Bromine	Krypton
58.693 4 ±4	63.546 ±3	65.38 ±2	69.723	72.63	74.921 60 ±2	78.96 ±3	79.904	83.798 ±2

46 +2,4	47 +1	48 +2	49 +3	50 +4,2	51 ±3,5	52 -2,4,6	53 ±1,5,7	54
3237	2436	1040	2346	2876	1860	1261	458	165
1825	1234	594	430	505	904	723	387	161
12.0 **Pd**	10.5 **Ag**	8.65 **Cd**	7.31 **In**	7.30 **Sn**	6.68 **Sb**	6.24 **Te**	4.92 **I**	5.78¤ **Xe**
Palladium	Silver	Cadmium	Indium	Tin	Antimony	Tellurium	Iodine	Xenon
106.42	107.868 2 ±2	112.411 ±8	114.818 ±3	118.710 ±7	121.760	127.60 ±3	126.904 47 ±3	131.293 ±6

78 +2,4	79 +3,1	80 +2,1	81 +3,1	82 +4,2	83 +3,5	84 +4,2	85 ±1,3,5,7	86
4100	3130	630	1746	2023	1837	1235	610	211
2045	1338	234	577	601	545	527	575	202
21.4 **Pt**	19.3 **Au**	13.5 **Hg**	11.85 **Tl**	11.4 **Pb**	9.8 **Bi**	9.4 **Po**	9.78¤ **At**	**Rn**
Platinum	Gold	Mercury	Thallium	Lead	Bismuth	Polonium	Astatine	Radon
195.084 ±9	196.966 569 ±4	200.59 ±2	204.384 ±2	207.2	208.980 40	(209)	(210)	(222)

110	111	112	113	114	115	116	117	118
Ds	**Rg**	**Cn**		**Fl**		**Lv**		
Darmstadtium	Roentgenium	Copernicium		Flerovium		Livermorium		
(281)	(280)	(285)	(284)	(289)	(288)	(293)	(294)	(294)

64 +3	65 +3,4	66 +3	67 +3	68 +3	69 +3,2	70 +3,2	71 +3
3539	3496	2835	2968	3136	2220	1467	3668
1585	1630	1682	1743	1795	1818	1097	1936
7.89 **Gd**	8.27 **Tb**	8.54 **Dy**	8.80 **Ho**	9.05 **Er**	9.33 **Tm**	6.98 **Yb**	9.84 **Lu**
Gadolinium	Terbium	Dysprosium	Holmium	Erbium	Thulium	Ytterbium	Lutetium
157.25 ±3	158.925 35 ±2	162.500	164.930 32 ±2	167.259 ±3	168.934 21 ±2	173.054 ±5	174.966 8

96 +3	97 +4,3	98 +3	99	100	101	102	103
1340		900					
13.5 **Cm**	**Bk**	**Cf**	**Es**	**Fm**	**Md**	**No**	**Lr**
Curium	Berkelium	Californium	Einsteinium	Fermium	Mendelevium	Nobelium	Lawrencium
(247)	(247)	(251)	(252)	(257)	(260)	(259)	(262)

Quantitative Chemical Analysis

Quantitative
Chemical Analysis

Ninth Edition

Daniel C. Harris
Michelson Laboratory, China Lake, California

Charles A. Lucy
Contributing Author
University of Alberta, Edmonton, Alberta

 W. H. FREEMAN
& COMPANY

A Macmillan Education Imprint

Publisher: Kate Parker

Senior Acquisitions Editor: Lauren Schultz

Development Editors: Brittany Murphy, Anna Bristow

Editorial Assistant: Shannon Moloney

Photo Editor: Cecilia Varas

Photo Researcher: Richard Fox

Cover and Text Designer: Vicki Tomaselli

Project Editor: J. Carey Publishing Service

Manuscript Editor: Marjorie Anderson

Illustrations: Network Graphics, Precision Graphics

Illustration Coordinators: Matthew McAdams, Janice Donnola

Production Coordinator: Julia DeRosa

Composition and Text Layout: Aptara®, Inc.

Printing and Binding: LSC Communications

Front Cover/Title Page Photo Credit: © The Natural History Museum/The Image Works

Back Cover Photo Credit: Pascal Goetgheluck/Science Source

Library of Congress Control Number: 2014950382
ISBN-13: 978-1-4641-3538-5
ISBN-10: 1-4641-3538-X

Printed in the United States of America
Third Printing

W. H. Freeman and Company
41 Madison Avenue
New York, NY 10010

www.whfreeman.com

BRIEF CONTENTS

v

CONTENTS

EXPERIMENTS

Experiments are found at the website
www.whfreeman.com/qca/

0. Green Analytical Chemistry
1. Calibration of Volumetric Glassware
2. Gravimetric Determination of Calcium as $CaC_2O_4 \cdot H_2O$
3. Gravimetric Determination of Iron as Fe_2O_3
4. Penny Statistics
5. Statistical Evaluation of Acid-Base Indicators
6. Preparing Standard Acid and Base
7. Using a pH Electrode for an Acid-Base Titration
8. Analysis of a Mixture of Carbonate and Bicarbonate
9. Analysis of an Acid-Base Titration Curve: The Gran Plot
10. Fitting a Titration Curve with Excel Solver
11. Kjeldahl Nitrogen Analysis
12. EDTA Titration of Ca^{2+} and Mg^{2+} in Natural Waters
13. Synthesis and Analysis of Ammonium Decavanadate
14. Iodimetric Titration of Vitamin C
15. Preparation and Iodometric Analysis of High-Temperature Superconductor
16. Potentiometric Halide Titration with Ag^+
17. Electrogravimetric Analysis of Copper
18. Polarographic Measurement of an Equilibrium Constant
19. Coulometric Titration of Cyclohexene with Bromine
20. Spectrophotometric Determination of Iron in Vitamin Tablets
21. Microscale Spectrophotometric Measurement of Iron in Foods by Standard Addition
22. Spectrophotometric Measurement of an Equilibrium Constant
23. Spectrophotometric Analysis of a Mixture: Caffeine and Benzoic Acid in a Soft Drink
24. Mn^{2+} Standardization by EDTA Titration
25. Measuring Manganese in Steel by Spectrophotometry with Standard Addition
26. Measuring Manganese in Steel by Atomic Absorption Using a Calibration Curve
27. Properties of an Ion-Exchange Resin
28. Analysis of Sulfur in Coal by Ion Chromatography
29. Measuring Carbon Monoxide in Automobile Exhaust by Gas Chromatography
30. Amino Acid Analysis by Capillary Electrophoresis
31. DNA Composition by High-Performance Liquid Chromatography
32. Analysis of Analgesic Tablets by High Performance Liquid Chromatography
33. Anion Content of Drinking Water by Capillary Electrophoresis
34. Green Chemistry: Liquid Carbon Dioxide Extraction of Lemon Peel Oil

SPREADSHEET TOPICS

SPREADSHEETS AT WEBSITE

[Emilio Segre Visual Archives/ Science Source.]

Maria Goeppert Mayer (1906–1972) was the second and, so far, last woman (after Marie Curie) to receive the Nobel Prize in Physics. She shared half of the 1963 prize with Hans Jensen for their independent theories of atomic nuclear shell structure published in 1949.

What does she have to do with this book? The back cover shows evidence that the body temperature of certain dinosaurs was similar to that of warm blooded animals. In 1947, she and Jacob Bigeleisen published a paper, "Calculation of Equilibrium Constants for Isotopic Exchange Reactions."[*] This paper was one of the foundational studies for paleothermometry—the use of isotopes to deduce the temperature at which objects such as dinosaur teeth were formed. From mathematical physics to analytical chemistry to dinosaurs, there is a thread of connection.

Maria was born to a sixth-generation university professor in Göttingen, Germany.[†] From early childhood, she knew that she would acquire a university education, but there were few avenues for girls' education. She attended a small, private girls' school, which closed before her studies were complete. Against all advice, she took and passed the University of Göttingen entrance examination to be admitted in 1924. Her first exposure to quantum mechanics by Max Born hooked her. She received a Ph.D. in 1930, with three Nobel Prize winners on her committee.

Maria married Joe Mayer, a Caltech- and Berkeley-educated physical chemist who was a postdoctoral boarder in the Goeppert household. They moved to the U.S., where Joe began a distinguished career at Johns Hopkins University, Columbia University, and the University of Chicago. In 1940 they coauthored *Statistical Mechanics*, a textbook used for more than 40 years. Maria was regarded as at least equally gifted, but she was not offered a paid position at any university despite teaching courses, advising graduate students, serving on committees, and writing graduate examinations—all as a volunteer! Her first paid appointment as a professor at the University of California at San Diego came in 1960, four years after her election to the National Academy of Sciences.

---

[*] J. Bigeleisen and M. G. Mayer, *J. Chem. Phys.* **1947**, *15*, 261.

[†] S. B. McGrayne, *Nobel Prize Women in Science* (Washington DC: Joseph Henry Press, 1998).

Goals of This Book

My goals are to provide a sound physical understanding of the principles of analytical chemistry and to show how these principles are applied in chemistry and related disciplines—especially in life sciences and environmental science. I have attempted to present the subject in a rigorous, readable, and interesting manner, lucid enough for nonchemistry majors, but containing the depth required by advanced undergraduates. This book grew out of an introductory analytical chemistry course that I taught mainly for nonmajors at the University of California at Davis and from a course for third-year chemistry students at Franklin and Marshall College in Lancaster, Pennsylvania.

What's New?

Beginning with dinosaur body temperature on the back cover of this book, analytical chemistry addresses interesting questions in the wider world. The facing page draws a connection between the back cover and underlying human achievement in physics that enables us to deduce body temperature from the isotopic composition of teeth. The story of Maria Goeppert Mayer is a lesson for us all in how women in science were so poorly treated not so long ago.

In this edition, the introduction to titrations has been consolidated in Chapter 7. Acid-base, EDTA, redox, and spectrophotometric titrations are still treated in other chapters. The power of the spreadsheet is unleashed in Chapter 8 to reach numerical solutions to equilibrium problems and in Chapter 19 to compute equilibrium constants from spectrophotometric data. Atomic spectroscopy Chapter 21 has a new section on X-ray fluorescence as a routine analytical tool. Mass spectrometry Chapter 22 has been expanded to increase the level of detail and to help keep up with new developments. Chapter 27 has an extraordinary sequence of micrographs showing the onset of crystallization of a precipitate. Three new methods in sample preparation were added to Chapter 28. Appendix B takes a deeper look at propagation of uncertainty and Appendix C treats analysis of variance.

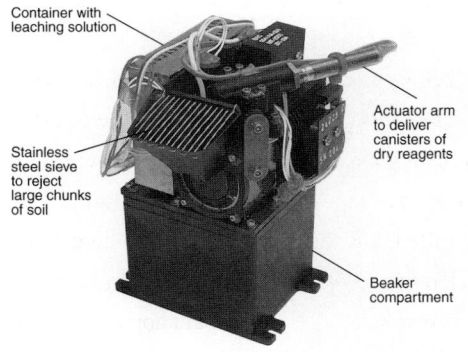

BOX 15-3 Measuring sulfate on Mars by titration with barium [Mars Lander: NASA/JPL-Caltech/University of Arizona/Max Planck Institute.]

FIGURE FROM PROBLEM 7-21 Barium sulfate precipitation titration from Phoenix Mars Lander [Data courtesy S. Kounaves, Tufts University.]

For the first time since I began work on this book in 1978, I have taken on a contributing author for part of this revision. Professor Chuck Lucy of the University of Alberta shares his expertise and teaching experience with us in Chapters 23–26 on chromatography and capillary electrophoresis. He improved the discussion of the efficiency of separation and mechanisms of band spreading. Emphasis is placed on types of interactions between solutes and the stationary phase. Types of solvent polarity are distinguished in liquid chromatography. Examples are given for the selection of stationary phase and pH for liquid chromatography separations. Electrophoresis has more emphasis on the effects of ion size and pH on mobility. Chuck contributes the views of a specialist in separation science to these chapters.

New boxed applications include a home pregnancy test (Chapter 0 opener), observing the addition of one base to DNA with a quartz crystal microbalance (Chapter 2 opener), medical implications of false positive results (Box 5-1), a titration on Mars (Chapter 7 opener),

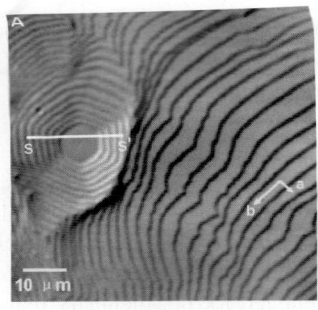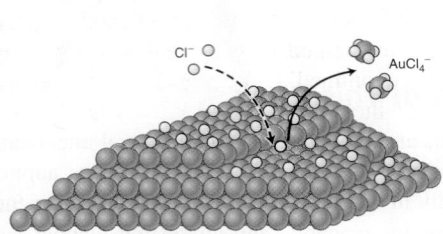

FIGURE FROM BOX 17-1 Anodic dissolution of gold at atomic steps [R. Wen, A. Lahiri, M. Azhagurajan, S. Kobayashi, K. Itaya, "A New in situ Optical Microscope with Single Atomic Layer Resolution for Observation of Electrochemical Dissolution of Au (111)," *J Am Chem Soc* **2010,** *132,*13657, Figure 2. Reprinted with permission © 2010, American Chemical Society.]

microequilibrium constants (Box 10-3), acid-base titration of RNA to provide evidence for the mechanism of RNA catalysis (Chapter 11 opener), the hydrogen-oxygen fuel cell and the *Apollo* 13 accident (Box 14-2), the lead-acid battery (Box 14-3), high-throughput DNA sequencing by counting protons (Chapter 15 opener), how perchlorate was discovered on Mars (Box 15-3), ion-selective electrode with a conductive polymer for a sandwich immuno-assay (Box 15-4), metal reaction at atomic steps (Box 17-1), an aptamer biosensor for clinical use (Box 17-5), Bunsen burner flame photometer (Box 21-2), atomic emission spectroscopy on Mars (Box 21-3), making elephants fly (mechanism of protein electrospray, Box 22-5), chromatographic analysis of breast milk (Chapter 23 opener), doping in sports (Chapter 24 opener), two-dimensional gas chromatography (Box 24-3), million-plate separation by slip flow chromatography (Box 25-1), forensic DNA profiling (Chapter 26 opener and Section 26-8), and measuring van der Waals attraction (Box 27-1). New Color Plates illustrate the effect of ionic strength on ion dissociation (Color Plate 4), the mechanism of chromatography by partitioning of analyte between phases (Color Plate 30), and separation of dyes by solid-phase extraction (Color Plate 36).

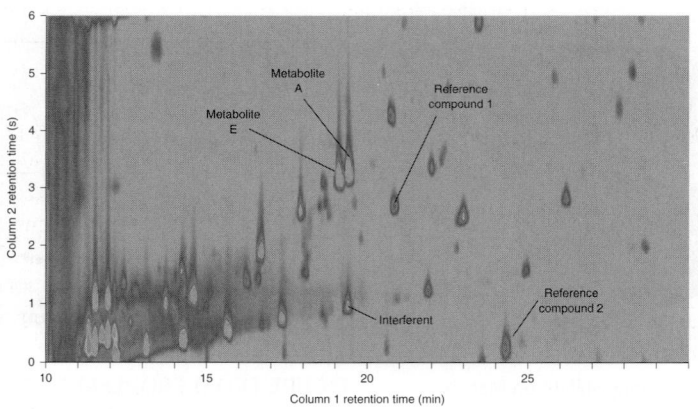

CHAPTER 24 OPENING IMAGE Two-dimensional gas chromatography–combustion isotope ratio mass spectrometry to detect doping in athletes [H. J. Tobias, Y. Zhang, R. J. Auchus, J. T. Brenna, "Detection of Synthetic Testosterone Use by Novel Comprehensive Two-Dimensional Gas Chromatography Combustion Isotope Ratio Mass Spectrometry," *Anal Chem* **2011,** *83,* 7158, Figure 4A. Reprinted with permission © 2011, American Chemical Society.]

Pedagogical changes in this edition include more discussion of serial dilution to prepare standards in Chapters 2, 3, and 18, distinction between standard uncertainty and standard deviation in statistics, more discussion of hypothesis testing in statistics, employing the *F* test before the *t* test for comparison of means, using a graphical treatment for internal standards, emphasis on electron flow toward the more positive electrode in electrochemical cells, using nanoscale observations to probe phenomena such as van der Waals forces and the amorphous structure of glass in a pH electrode, polynomial smoothing of noisy data, expanded discussion of the time-of-flight mass spectrometer and ion mobility separations, enhanced discussion of intermolecular forces in chromatography, enhanced discussion of method development in liquid chromatography, use of a free, online liquid chromatography simulator, introduction of two literature search questions in chromatography, and taking more advantage of the power of Excel for numerical analysis. Box 3-3 explains how I have chosen to handle atomic weight intervals in the latest periodic table of the elements.

Features

Topics are introduced and illustrated with concrete, interesting examples. In addition to their pedagogic value, Chapter Openers, Boxes, Demonstrations, and Color Plates are intended to help lighten the load of a very dense subject. **Chapter Openers** show the relevance of analytical chemistry to the real world and to other disciplines of science. I can't come to your classroom to present **Chemical Demonstrations,** but I can tell you about some of my favorites and show how they look with the **Color Plates** located near the center of the book. **Boxes** discuss interesting topics related to what you are studying or amplify points in the text.

Problem Solving

Nobody can do your learning for you. The two most important ways to master this course are to work problems and to gain experience in the laboratory. **Worked Examples** are a principal pedagogic tool to teach problem solving and to illustrate how to apply what you have just read. Each worked example ends with a **Test Yourself** question that you are encouraged to answer to apply what you learned in the example. There are Exercises and Problems at the end of each chapter. **Exercises** are the minimum set of problems that apply most major concepts of each chapter. Please struggle mightily with an Exercise before consulting the solution at the back of the book. **Problems** at the end of the chapter cover the entire content of the book. **Short Answers** are at the back of the book and complete solutions appear in the **Solutions Manual.**

Spreadsheets are indispensable for science and engineering and uses far beyond this course. You can cover this book without using spreadsheets, but you will never regret taking the time to learn to use them. A few of the powerful features of Microsoft Excel are described as they are needed, including graphing in Chapters 2 and 4, statistical functions and regression in Chapter 4, solving equations with Goal Seek, Solver, and circular definitions in Chapters 7, 8, 13, and 19, and some matrix operations in Chapter 19. The text teaches you how to construct spreadsheets to simulate many types of titrations, to solve chemical equilibrium problems, and to simulate chromatographic separations.

Other Features of This Book

Terms to Understand Essential vocabulary, highlighted in **bold** in the text, is collected at the end of the chapter. Other unfamiliar or new terms are *italic* in the text.

Glossary **Bold** vocabulary terms and many of the italic terms are defined in the glossary.

Appendixes Tables of solubility products, acid dissociation constants, redox potentials, and formation constants appear at the back of the book. You will also find discussions of logarithms and exponents, propagation of error, analysis of variance, balancing redox equations, normality, analytical standards, and a little bit about DNA.

Notes and References Citations in the chapters appear at the end of the book.

Inside Cover Here is your trusty periodic table, as well as tables of physical constants and other information.

EXAMPLE A Weak-Acid Problem

Find the pH of 0.050 M trimethylammonium chloride.

Cl^- Trimethylammonium chloride

Solution We assume that ammonium halide salts are completely dissociated to give $(CH_3)_3NH^+$ and Cl^-.* We then recognize that trimethylammonium ion is a weak acid, being the conjugate acid of trimethylamine, $(CH_3)_3N$, a weak base. Cl^- has no basic or

CHAPTER 9 EXAMPLE PAGE 193

	A	B	C	D	E	F
1	Thallium azide equilibria					
2	1. *Estimate* values of pC = –log[C] for N_3^- and OH^- in cells B6 and B7					
3	2. Use Solver to adjust the values of pC to minimize the sum in cell F8					
4						
5	Species	pC	C (= 10^-pC)		Mass and charge balances	b_i
6	N_3^-	2	0.01	C6 = 10^-B6	$b_1 = 0 = [Tl^+] - [N_3^-] - [HN_3] =$	1.19E-02
7	OH^-	4	0.0001	C7 = 10^-B7	$b_2 = 0 = [Tl^+] + [H^+] - [N_3^-] - [OH^-] =$	1.18E-02
8	Tl^+		0.021877616	C8 = D12/C6	$\Sigma b_i^2 =$	2.80E-04
9	HN_3		4.46684E-08	C9 = D13*C6/C7	F6 = C8-C6-C9	
10	H^+		1E-10	C10 = D14/C7	F7 = C8+C10-C6-C7	
11					F8 = F6^2+F7^2	
12	pK_{sp}	3.66	$K_{sp} =$	0.000218776	= 10^-B12	
13	pK_b	9.35	$K_b =$	4.46684E-10	= 10^-B13	
14	pK_w	14.00	$K_w =$	1E-14	= 10^-B14	

FIGURE 8-9 Thallium azide solubility spreadsheet without activity coefficients. Initial estimates $pN_3^- = 2$ and $pOH^- = 4$ appear in cells B6 and B7. From these two numbers, the spreadsheet computes concentrations in cells C6:C10. Solver then varies pN_3^- and pOH^- in cells B6 and B7 until the charge and mass balances in cell F8 are satisfied.

Media and Supplements

The *Solutions Manual for Quantitative Chemical Analysis* contains complete solutions to all problems.

New Clicker Questions allow instructors to integrate active learning in the classroom and to assess students' understanding of key concepts during lectures. Available in Microsoft Word and PowerPoint (PPT).

New Lecture PowerPoints have been developed to minimize preparation time for new users of the book. These files offer suggested lectures including key illustrations and summaries that instructors can adapt to their teaching styles.

New Test Bank offers questions in editable Microsoft Word format.

Premium WebAssign with e-Book www.webassign.com features time-tested, secure, online environment already used by millions of students worldwide. Featuring algorithmic problem generation, students receive homework problems containing unique values for computation, encouraging them to work out the problems on their own. Additionally, there is complete access to the e-Book, from a live table of contents.

Sapling Learning with e-Book www.sapling.com provides highly effective interactive homework and instruction that improve student learning outcomes for the problem-solving disciplines. Sapling Learning offers an enjoyable teaching and effective learning experience that is distinctive in three important ways: (1) ease of use: Sapling Learning's easy-to-use interface keeps students engaged in problem-solving, not struggling with the software; (2) targeted instructional content: Sapling Learning increases student engagement and comprehension by delivering immediate feedback and targeted instructional content; (3) unsurpassed service and support: Sapling Learning makes teaching more enjoyable by providing a dedicated Masters- and Ph.D.-level colleague to service instructors' unique needs throughout the course, including content customization.

The **student website** www.whfreeman.com/qca has **directions for experiments** which may be reproduced for your use. You will also find **lists of experiments** from the *Journal of Chemical Education*. **Supplementary topics** at the website include spreadsheets for precipitation and redox titrations, discussion of microequilibrium constants, a spreadsheet simulation of gradient liquid chromatography, and Fourier transformation of an interferogram into an infrared spectrum. You will also find 24 **selected Excel spreadsheets** from the textbook ready to use at the student website.

The **instructors' website**, www.whfreeman.com/qca, has all **artwork** and **tables** from the book in preformatted PowerPoint slides.

The People

My wife Sally works on every aspect of this book and the Solutions Manual. She contributes mightily to whatever clarity and accuracy we have achieved.

Solutions to problems and exercises were meticulously checked by Heather Audesirk, a graduate student at Caltech, and by Julia Lee, a senior at Harvey Mudd College.

A book of this size and complexity is the work of many people. Brittany Murphy, Anna Bristow, and Lauren Schultz provided editorial and market guidance. Jennifer Carey was the Project Editor responsible for making sure that all pieces of this book fell into the right place. Marjorie Anderson attended to the challenging details of copyediting. Photo research and permissions were ably handled by Cecilia Varas and Richard Fox. Matthew McAdams, Janice Donnola, and Tracey Kuehn coordinated the illustration program. Anna Skiba-Crafts was the courageous proofreader.

In Closing

This book is dedicated to the students who use it, who occasionally smile when they read it, who gain new insight, and who feel satisfaction after struggling to solve a problem. I have been successful if this book helps you develop critical, independent reasoning that you can apply to new problems in or out of chemistry. I truly relish your comments, criticisms, suggestions, and corrections. Please address correspondence to me at the Chemistry Division (Mail Stop 6303), Research Department, Michelson Laboratory, China Lake, CA 93555.

Dan Harris
March 2015

Acknowledgements

I am indebted to many people who provided new information for this edition, asked probing questions, and made good suggestions. Pete Palmer of San Francisco State University graciously shared his instructional material for X-ray fluorescence and provided a detailed critique of my draft, as well as suggestions for mass spectrometry. Karyn Usher of Metropolitan State University, Saint Paul, Minnesota, photographed her solid-phase extraction experiment that appears in Color Plate 36. Martin Mirenda of the Universidad de Buenos Aires provided Color Plate 4 showing the instructive effect of ionic strength on the color of bromocresol green. Jim De Yoreo and Mike Nielsen of Battelle Pacific Northwest National Laboratory provided the exquisite time-lapse calcium carbonate nucleation transmission electron micrographs in Figure 27-2.

Barbara Belmont of California State University, Dominguez Hills asked a seemingly simple question in 2011 about the propagation of uncertainty that required the knowledge of my statistician colleague, Dr. Ding Huang, to answer. This question led to the expanded Appendix B. D. Brynn Hibbert of the University of New South Wales, Australia, was also a resource for statistics. Jürgen Gross of Heidelberg University and David Sparkman of the University of the Pacific in California were resources for mass spectrometry. Dale Lecaptain of Central Michigan University requested more emphasis on serial dilutions, which has been added. Brian K. Niece of Assumption College, Worcester, Massachusetts, corrected my procedure for using hydroxynaphthol blue indicator for EDTA titrations. Micha Enevoldsen of Frederiksberg, Denmark, taught me that Kjeldahl was a Danish chemist, not a Dutch chemist. He also taught me that Kjeldahl was one of the "three great pH's," who also include S. P. L. Sørensen and K. U. Linderstrøm-Lang. Chan Kang of Chonbuk National University, Korea, pointed out that I had been using the letter n to mean more than one thing in electrochemistry, which I have attempted to correct in this edition. Alena Kubatova of the University of North Dakota provided some of her teaching materials for mass spectrometry. Other helpful corrections and suggestions came from Richard Gregor (Rollins College, Florida), Franco Basile (University of Wyoming), Jeffrey Smith (Carleton University, Ottawa), Kris Varazo (Francis Marion University, Florence, South Carolina), Doo Soo Chung (Seoul National University), Ron Cooke (California State University, Chico), David D. Weiss (Kansas University), Steven Brown (University of Delaware), Athula Attygalle (Stevens Institute of Technology, Hoboken, New Jersey), and Peter Liddel (Glass Expansion, West Melbourne, Australia).

People who reviewed the 8th edition of Quantitative Chemical Analysis and parts of the manuscript for the 9th edition include Truis Smith-Palmer (St. Francis Xavier University), William Lammela (Nazareth College), Nelly Mateeva (Florida A&M University), Alena Kubatova (University of North Dakota), Barry Ryan (Emory University), Neil Jespersen (St. John's University), David Kreller (Georgia Southern University), Darcey Wayment (Nicholls State University), Karla McCain (Austin College), Grant Wangila (University of Arkansas), James Rybarczyk (Ball State University), Frederick Northrup (Northwestern University), Mark Even (Kent State University), Jill Robinson (Indiana University), Pete Palmer (San Francisco State University), Cindy Burkhardt (Radford University), Nathanael Fackler (Nebraska Weslyan University), Stuart Chalk (University of North Florida), Reynaldo Barreto (Purdue University North Central), Susan Varnum (Temple University), Wendy Cory (College of Charleston), Eric D. Dodds (University of Nebraska, Lincoln), Troy D. Wood (University of Buffalo), Roy Cohen (Xavier University), Christopher Easley (Auburn University), Leslie Sombers (North Carolina State University), Victor Hugo Vilchiz (Virginia State University), Yehia Mechref (Texas Tech University), Lenuta Cires Gonzales (California State University, San Marcos), Wendell Griffith (University of Toledo), Anahita Izadyar (Arkansas State University), Leslie Hiatt (Austin Peay State University), David Carter (Angelo State University), Andre Venter (Western Michigan University), Rosemarie Chinni (Alvernia University), Mary Sohn (Florida Technical College), Christopher Babayco (Columbia College), Razi Hassan (Alabama A&M University), Chris Milojevich (University of Tampa), Steven Brown (University of Delaware), Anne Falke (Worcester State University), Julio Alvarez (Virginia Commonwealth University), Keith Kuwata (Macalaster College), Levi Mielke (University of Indianapolis), Simon Mwongela (Georgia Gwinnett College), Omowunmi Sadik (State University of New York at Binghamton), Jingdong Mao (Old Dominion University), Jani Ingram (Northern Arizona University), Matthew Mongelli (Kean University), Vince Cammarata (Auburn University), Ed Segstro (University of Winnipeg), Tiffany Mathews (Villanova University), Andrea Matti (Wayne State University), Rebecca Barlag (Ohio University), Barbara Munk (Wayne State University), John Berry (Florida International University), Patricia Cleary (University of Wisconsin, Eau Claire), and Sandra Barnes (Alcorn State University).

0 | The Analytical Process

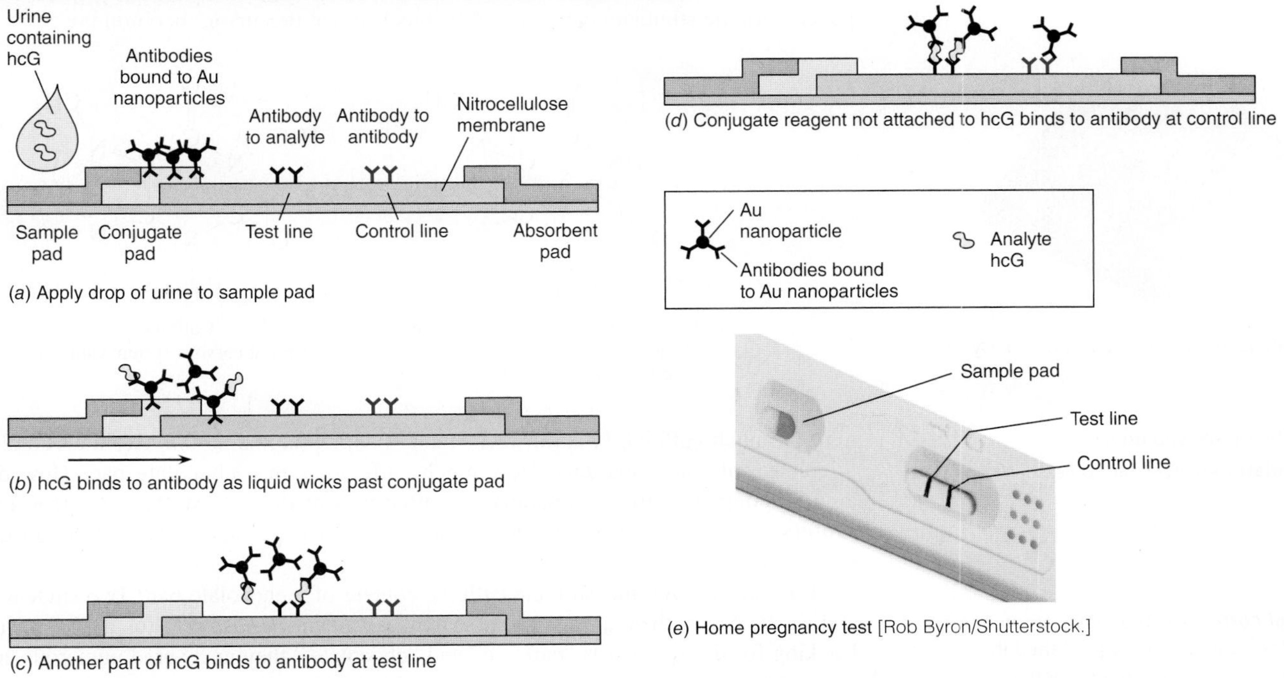

(a) Apply drop of urine to sample pad

(b) hcG binds to antibody as liquid wicks past conjugate pad

(c) Another part of hcG binds to antibody at test line

(d) Conjugate reagent not attached to hcG binds to antibody at control line

(e) Home pregnancy test [Rob Byron/Shutterstock.]

A common home pregnancy test detects a hormone called hcG in urine. This hormone begins to be secreted shortly after conception.

An *antibody* is a protein secreted by white blood cells to bind to a foreign molecule called an *antigen*. Antibody-antigen binding is the first step in the immune response that eventually removes a foreign substance or an invading cell from your body. Antibodies to human proteins such as hcG can be cultivated in animals.

In the *lateral flow home pregnancy immunoassay* shown in the diagram, urine is applied to the sample pad at the left end of a horizontal test strip made of nitrocellulose that serves as a wick. Liquid flows from left to right by capillary action. Liquid first encounters detection reagent on the conjugate pad. The reagent is called a conjugate because it consists of hcG antibody attached to red-colored gold nanoparticles. The antibody binds to one site on hcG.

As liquid flows to the right, hcG bound to the conjugate is trapped at the test line, which contains an antibody that binds to another site on hcG. Gold nanoparticles trapped with hcG at the test line create a visible red line. As liquid continues to the right, it encounters the control line with antibodies that bind to the conjugate reagent. A second red line forms at the control line. At the far right is an absorbent pad that soaks up liquid containing anything that was not retained at the test or control lines.

In a positive pregnancy test, both lines turn red. The test is negative if only the control line turns red. If the control line fails to turn red, the test is invalid.

Bold terms should be learned. *Italicized* terms are less important. A glossary of terms is found at the back of the book.

Quantitative analysis: How much is present?
Qualitative analysis: What is present?

Quantitative chemical analysis is the measurement of *how much* of a chemical substance is present. The purpose of quantitative analysis is usually to answer a question such as "Does this mineral contain enough copper to be an economical source of copper?" The home pregnancy test above is a **qualitative chemical analysis,** which looks for the presence of a hormone that is produced during pregnancy. This test answers the even more important question, "Am I pregnant?" Qualitative analysis tells us *what* is present and quantitative analysis tells us *how much* is present. In quantitative analysis, the chemical measurement is only part of a process that includes asking a meaningful question, collecting a relevant sample, treating the sample so that the chemical of interest can be measured, making the measurement, interpreting the results, and providing a report.

0-1 The Analytical Chemist's Job

My favorite chocolate bar,[1] jammed with 33% fat and 47% sugar, propels me over mountains in California's Sierra Nevada. In addition to its high energy content, chocolate packs an extra punch with the stimulant caffeine and its biochemical precursor, theobromine.

Chocolate is great to eat, but not so easy to analyze. [Dima Sobko/Shutterstock.]

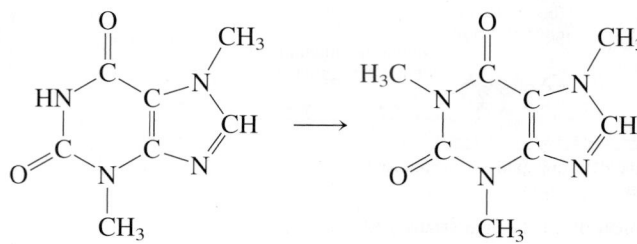

Theobromine (from Greek "food of the gods")
A diuretic, smooth muscle relaxant, cardiac stimulant, and vasodilator

Caffeine
A central nervous system stimulant

Notes and references appear after the last chapter of the book.

A **diuretic** makes you urinate.
A **vasodilator** enlarges blood vessels.

Chemical Abstracts is the most comprehensive source for locating articles published in chemistry journals. *SciFinder* is software that accesses *Chemical Abstracts.*

Too much caffeine is harmful for many people, and some unlucky individuals cannot tolerate even small amounts. How much caffeine is in a chocolate bar? How does that amount compare with the quantity in coffee or soft drinks? At Bates College in Maine, Professor Tom Wenzel teaches his students chemical problem solving through questions such as these.[2]

But, how *do* you measure the caffeine content of a chocolate bar? Two students, Denby and Scott, began their quest with a search of *Chemical Abstracts* for analytical methods. Looking for the key words "caffeine" and "chocolate," they uncovered numerous articles in chemistry journals. Two reports, both entitled "High-Pressure Liquid Chromatographic Determination of Theobromine and Caffeine in Cocoa and Chocolate Products,"[3] described a procedure suitable for the equipment in their laboratory.[4]

Sampling

The first step in any chemical analysis is procuring a representative sample to measure—a process called **sampling.** Is all chocolate the same? Of course not. Denby and Scott bought one chocolate bar and analyzed pieces of it. If you wanted to make broad statements about "caffeine in chocolate," you would need to analyze a variety of chocolates. You would also need to measure multiple samples of each type to determine the range of caffeine in each kind of chocolate.

A pure chocolate bar is fairly **homogeneous,** which means that its composition is the same everywhere. It might be safe to assume that a piece from one end has the same caffeine content as a piece from the other end. Chocolate with a macadamia nut in the middle is an example of a **heterogeneous** material—one whose composition differs from place to place. The nut is different from the chocolate. To sample a heterogeneous material, you need to use a strategy different from that used to sample a homogeneous material. You would need to know the average mass of chocolate and the average mass of nuts in many candies. You would need to know the average caffeine content of the chocolate and of the macadamia nut (if it has any caffeine). Only then could you make a statement about the average caffeine content of macadamia chocolate.

Homogeneous: same throughout
Heterogeneous: differs from region to region

Sample Preparation

The first step in the procedure calls for weighing out some chocolate and extracting fat from it by dissolving the fat in a hydrocarbon solvent. Fat needs to be removed because it would interfere with chromatography later in the analysis. Unfortunately, if you just shake a chunk of chocolate with solvent, extraction is not very effective because the solvent has no access to the inside of the chocolate. So, our resourceful students sliced the chocolate into small bits and placed the pieces into a mortar and pestle (Figure 0-1), thinking they would grind the solid into small particles.

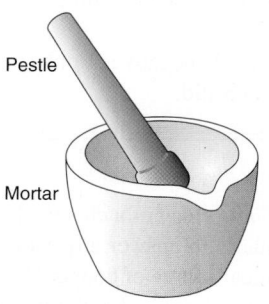

Pestle

Mortar

FIGURE 0-1 Ceramic mortar and pestle used to grind solids into fine powders.

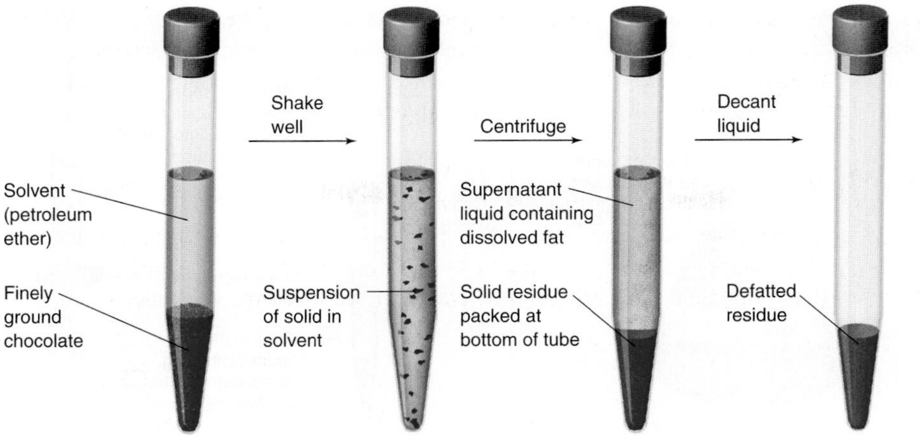

FIGURE 0-2 Extracting fat from chocolate to leave defatted solid residue for analysis.

Imagine trying to grind chocolate! The solid is too soft to grind. So Denby and Scott froze the mortar and pestle with its load of sliced chocolate. Once the chocolate was cold, it was brittle enough to grind. Small pieces were placed in a preweighed 15-milliliter (mL) centrifuge tube, and their mass was noted.

Figure 0-2 shows the next part of the procedure, which is to remove fat that would interfere with subsequent chromatography. A 10-mL portion of the solvent, petroleum ether, was added to the tube, and the top was capped with a stopper. The tube was shaken vigorously to dissolve fat from the solid chocolate into the solvent. Caffeine and theobromine are insoluble in this solvent. The mixture of liquid and fine particles was then spun in a centrifuge to pack the chocolate at the bottom of the tube. The clear liquid, containing dissolved fat, could now be **decanted** (poured off) and discarded. Extraction with fresh portions of solvent was repeated twice more to remove more fat from the chocolate. Residual solvent in the chocolate was then removed by heating the centrifuge tube in a beaker of boiling water. The mass of chocolate residue could be calculated by weighing the tube plus its content of defatted chocolate residue and subtracting the known mass of the empty tube.

Substances being measured—caffeine and theobromine in this case—are called **analytes.** The next step in the sample preparation procedure was to make a **quantitative transfer** (a complete transfer) of the fat-free chocolate residue to an Erlenmeyer flask and to dissolve the analytes in water for the chemical analysis. If any residue were not transferred from the tube to the flask, then the final analysis would be in error because not all of the analyte would be present. To perform the quantitative transfer, Denby and Scott added a few milliliters of pure water to the centrifuge tube and used stirring and heating to dissolve or suspend as much of the chocolate as possible. Then they poured the **slurry** (a suspension of solid in a liquid) into a 50-mL flask. They repeated the procedure several times with fresh portions of water to ensure that every bit of chocolate was transferred from the centrifuge tube to the flask.

To complete the dissolution of analytes, Denby and Scott added water to bring the volume up to about 30 mL. They heated the flask in a boiling water bath to extract all the caffeine and theobromine from the chocolate into the water. To compute the quantity of analyte later, the total mass of water must be known. Denby and Scott knew the mass of chocolate residue in the centrifuge tube and they knew the mass of the empty Erlenmeyer flask. So they put the flask on a balance and added water drop by drop until there were 33.3 g of water in the flask. Later, they would compare known solutions of pure analyte in water with the unknown solution containing 33.3 g of water.

Before Denby and Scott could inject the unknown solution into a chromatograph for the chemical analysis, they had to clean up the unknown even further (Figure 0-3). The chocolate residue in water contained tiny solid particles that would surely clog their expensive chromatography column and ruin it. So they transferred a portion of the slurry to a centrifuge tube and centrifuged the mixture to pack as much of the solid as possible at the bottom of the tube. The cloudy, tan, **supernatant liquid** (liquid above the packed solid) was then filtered in a further attempt to remove tiny particles of solid from the liquid.

It is critical to avoid injecting solids into a chromatography column, but the tan liquid still looked cloudy. So Denby and Scott took turns between classes to repeat the centrifugation and filtration five times. After each cycle in which the supernatant liquid was filtered and centrifuged, it became a little cleaner. But the liquid was never completely clear. Given enough time, more solid always seemed to precipitate from the filtered solution.

A solution of anything in water is called an **aqueous** solution.

Real-life samples rarely cooperate with you!

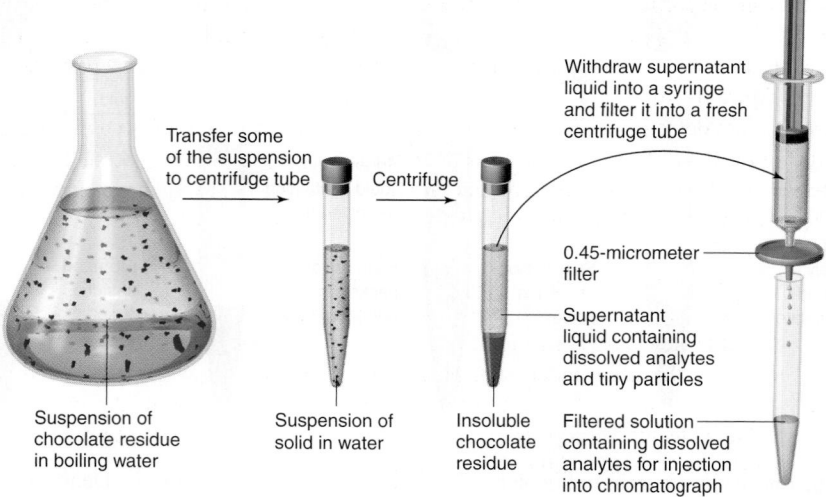

Transfer some of the suspension to centrifuge tube

Centrifuge

Withdraw supernatant liquid into a syringe and filter it into a fresh centrifuge tube

0.45-micrometer filter

Supernatant liquid containing dissolved analytes and tiny particles

Suspension of chocolate residue in boiling water

Suspension of solid in water

Insoluble chocolate residue

Filtered solution containing dissolved analytes for injection into chromatograph

The tedious procedure described so far is called **sample preparation**—transforming a sample into a state that is suitable for analysis. In this case, fat had to be removed from the chocolate, analytes had to be extracted into water, and residual solid had to be separated from the water.

Chemical Analysis (At Last!)

Denby and Scott finally decided that the solution of analytes was as clean as they could make it in the time available. The next step was to inject solution into a *chromatography* column, which would separate the analytes and measure the quantity of each. The column in Figure 0-4a is packed with tiny particles of silica (SiO_2) to which are attached long hydrocarbon molecules. Twenty microliters (20.0×10^{-6} liters) of the chocolate extract were injected into the column and washed through with a solvent made by mixing 79 mL of pure water, 20 mL of methanol, and 1 mL of acetic acid. Caffeine has greater affinity

Chromatography solvent is selected by a systematic trial-and-error process described in Chapter 25. Acetic acid reacts with negative oxygen atoms on the silica surface. When not neutralized, these oxygen atoms tightly bind a small fraction of caffeine and theobromine.

$$\text{silica-O}^- \xrightarrow{\text{Acetic acid}} \text{silica-OH}$$

Binds analytes very tightly

Does not bind analytes strongly

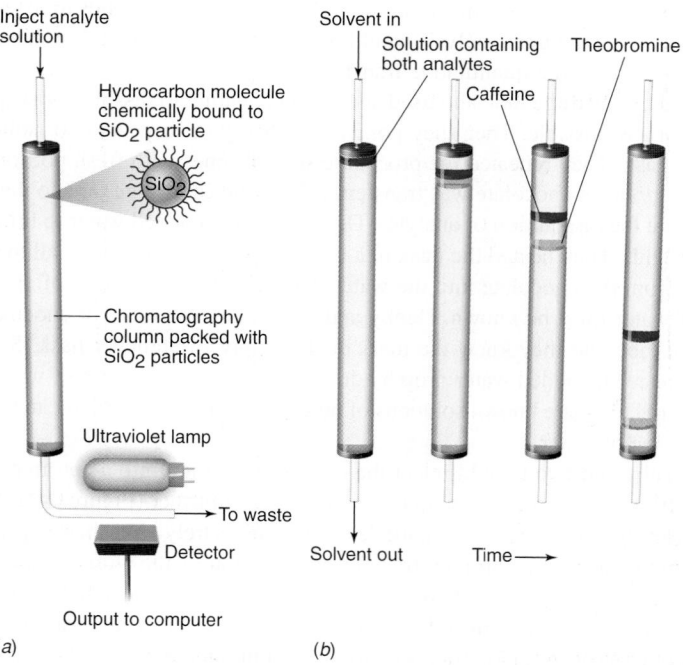

Inject analyte solution

Hydrocarbon molecule chemically bound to SiO_2 particle

Chromatography column packed with SiO_2 particles

Ultraviolet lamp

To waste

Detector

Output to computer

(a)

Solvent in

Solution containing both analytes

Caffeine

Theobromine

Solvent out

Time →

(b)

FIGURE 0-4 Principle of liquid chromatography. (*a*) Chromatography apparatus with an ultraviolet absorbance monitor to detect analytes at the column outlet. (*b*) Separation of caffeine and theobromine by chromatography. Caffeine has greater affinity than theobromine for the hydrocarbon layer on the particles in the column. Therefore, caffeine is retained more strongly and moves through the column more slowly than theobromine.

CHAPTER 0 The Analytical Process

than theobromine for the hydrocarbon on the silica surface. Therefore, caffeine "sticks" to the coated silica particles in the column more strongly than theobromine does. When both analytes are flushed through the column by solvent, theobromine reaches the outlet before caffeine (Figure 0-4b).

Analytes are detected at the outlet by their ability to absorb ultraviolet radiation from the lamp in Figure 0-4a. The graph of detector response versus time in Figure 0-5 is called a *chromatogram*. Theobromine and caffeine are the major peaks in the chromatogram. Small peaks arise from other substances extracted from the chocolate.

The chromatogram alone does not tell us what compounds are present. One way to identify individual peaks is to measure spectral characteristics of each one as it emerges from the column. Another way is to add an authentic sample of either caffeine or theobromine to the unknown and see whether one of the peaks grows in magnitude.

In Figure 0-5, the *area* under each peak is proportional to the quantity of compound passing through the detector. The best way to measure area is with a computer attached to the chromatography detector. Denby and Scott did not have a computer linked to their chromatograph, so they measured the *height* of each peak instead.

Only substances that absorb ultraviolet radiation at a wavelength of 254 nanometers are observed in Figure 0-5. The major components in the aqueous extract are sugars, but they are not detected in this experiment.

Calibration Curves

In general, analytes with equal concentrations give different detector responses. Therefore, the response must be measured for known concentrations of each analyte. A graph of detector response as a function of analyte concentration is called a **calibration curve** or a *standard curve*. To construct such a curve, **standard solutions** containing known concentrations of pure theobromine or caffeine were prepared and injected into the column, and the resulting peak heights were measured. Figure 0-6 is a chromatogram of one of the standard solutions, and Figure 0-7 shows calibration curves made by injecting solutions containing 10.0, 25.0, 50.0, or 100.0 micrograms of each analyte per gram of solution.

Straight lines drawn through the calibration points could then be used to find the concentrations of theobromine and caffeine in an unknown. From the equation of the theobromine

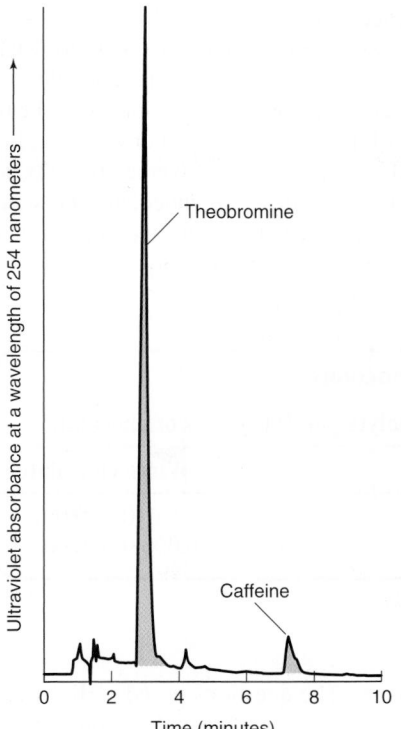

FIGURE 0-5 Chromatogram of 20.0 microliters of dark chocolate extract. A 150-mm-long × 4.6-mm-diameter column, packed with 5-micrometer-diameter particles of Hypersil ODS, was eluted (washed) with water:methanol:acetic acid (79:20:1 by volume) at a rate of 1.0 mL per minute.

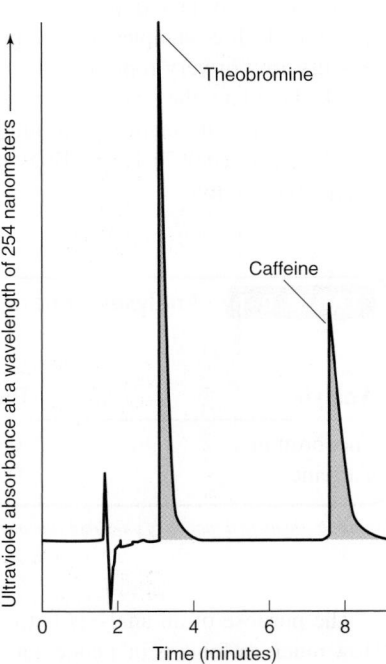

FIGURE 0-6 Chromatogram of 20.0 microliters of a standard solution containing 50.0 micrograms of theobromine and 50.0 micrograms of caffeine per gram of solution.

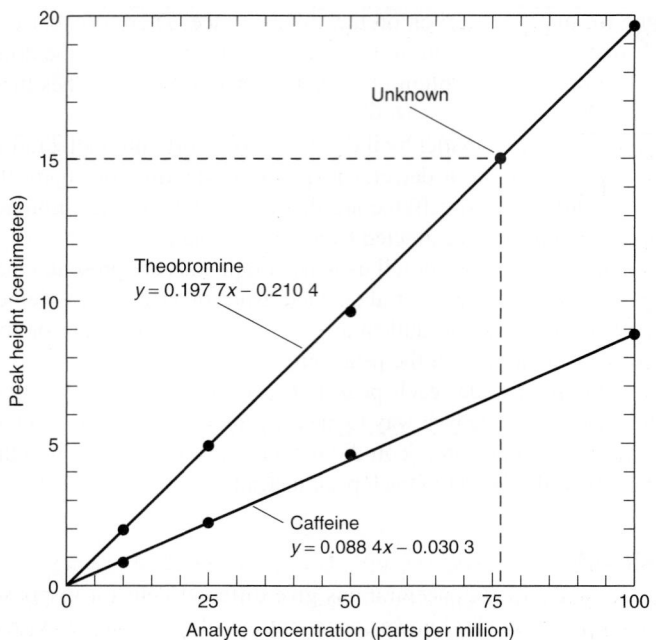

FIGURE 0-7 Calibration curves show observed peak heights for known concentrations of pure compounds. One *part per million* is one microgram of analyte per gram of solution. Equations of the straight lines drawn through the experimental data points were determined by the *method of least squares* described in Chapter 4.

Theobromine
$y = 0.197\ 7x - 0.210\ 4$

Caffeine
$y = 0.088\ 4x - 0.030\ 3$

Unknown

Peak height (centimeters)

Analyte concentration (parts per million)

line in Figure 0-7, we can say that, if the observed peak height of theobromine from an unknown solution is 15.0 cm, then the concentration is 76.9 micrograms per gram of solution.

Interpreting the Results

Knowing how much analyte is in the aqueous extract of the chocolate, Denby and Scott could calculate how much theobromine and caffeine were in the original chocolate. Results for dark and white chocolates are shown in Table 0-1. The quantities found in white chocolate are only about 2% as great as the quantities in dark chocolate.

The table also reports the *standard deviation* of three replicate measurements for each sample. Standard deviation, discussed in Chapter 4, is a measure of the reproducibility of the results. If three samples were to give identical results, the standard deviation would be 0. If results are not very reproducible, then the standard deviation is large. For theobromine in dark chocolate, the standard deviation (0.002) is less than 1% of the average (0.392), so we say the measurement is reproducible. For theobromine in white chocolate, the standard deviation (0.007) is nearly as great as the average (0.010), so the measurement is poorly reproducible.

TABLE 0-1 Analyses of dark and white chocolate

	Grams of analyte per 100 grams of chocolate	
Analyte	**Dark chocolate**	**White chocolate**
Theobromine	0.392 ± 0.002	0.010 ± 0.007
Caffeine	0.050 ± 0.003	$0.000\ 9 \pm 0.001\ 4$

Average ± standard deviation of three replicate injections of each extract.

The purpose of an analysis is to reach a conclusion. The questions posed earlier were "How much caffeine is in a chocolate bar?" and "How does it compare with the quantity in coffee or soft drinks?" After all this work, Denby and Scott discovered how much caffeine was in *one* particular chocolate bar that they analyzed. It would take a great deal more work to sample many chocolate bars of the same type and many different types of chocolate to gain a broad view. Table 0-2 compares results from analyses of different sources of caffeine. A can of soft drink or a cup of tea contains less than one-half of the caffeine in a small cup of coffee. Chocolate contains even less caffeine, but a hungry backpacker eating enough baking chocolate can get a pretty good jolt!

TABLE 0-2	Caffeine content of beverages and foods	
Source	Caffeine (milligrams per serving)	Serving size[a] (ounces)
Regular coffee	106–164	5
Decaffeinated coffee	2–5	5
Tea	21–50	5
Cocoa beverage	2–8	6
Baking chocolate	35	1
Sweet chocolate	20	1
Milk chocolate	6	1
Caffeinated soft drinks	36–57	12
Red Bull	80	8.2

a. 1 ounce = 28.35 grams
DATA SOURCES: http://www.holymtn.com/tea/caffeine_content.htm. Red Bull from http://wilstar.com/caffeine.htm.

Simplifying Sample Preparation with Solid-Phase Extraction

The procedure followed by Denby and Scott in the mid-1990s was developed before *solid-phase extraction* (page 785) came into use. Today, solid-phase extraction simplifies sample preparation by separating some major interfering components of the mixture from the desired analytes.[5] The procedure shown in Figure 0-8 features a short, disposable column containing a chromatography solid phase that can clean the sample enough prior to performing chromatography on an expensive analytical column.

Denby and Scott extracted fat with organic solvent. Then they extracted caffeine and theobromine with hot water and laboriously removed fine particles by repeated centrifugation and filtration. Solid-phase extraction in Figure 0-8 removes sugars, fats, and fine

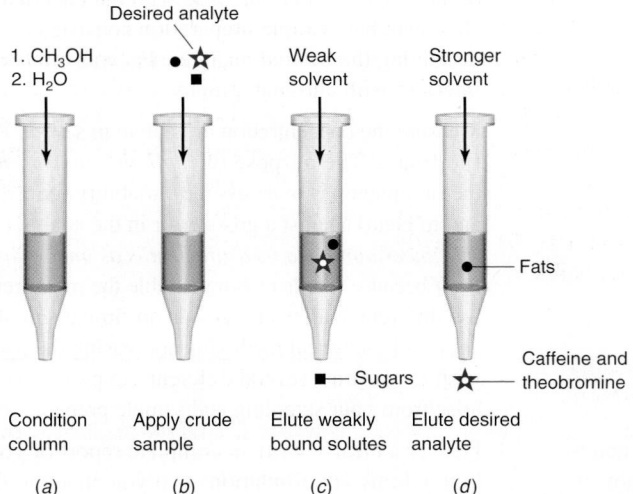

FIGURE 0-8 Solid-phase extraction separates caffeine and theobromine from sugars and fats found in chocolate. Sugars wash right through the column because they are not attracted to the hydrocarbon that is covalently attached to the particles on the column. Fats are so soluble in the hydrocarbon that they are not washed off the column by methanol. Caffeine and theobromine are soluble in the hydrocarbon but are washed off the column with methanol.

particles from the aqueous sample, replacing the extraction with organic solvent, centrifugation, and filtration. Crushed whole chocolate (0.5 gram) is suspended in 20 mL of water at 80°C for 15 minutes to extract caffeine, theobromine, and other water-soluble components. A solid-phase extraction column containing 0.5 gram of silica particles with covalently attached hydrocarbons (like the particles on the column in Figure 0-4) is cleaned with 1 mL of methanol followed by 1 mL of water. When 0.5 mL of aqueous extract is applied to the column, theobromine and caffeine adhere to the hydrocarbon on the silica particles in the column. Many water-soluble components such as sugars are washed through with 1 mL of water. Caffeine and theobromine are then washed from the column with 2.5 mL of methanol. Fats remain on the column. After evaporating the methanol to dryness, the residue is dissolved in 1 mL of water and is ready for chromatography. See Color Plate 36 near the center of this book for an example of solid-phase extraction.

0-2 General Steps in a Chemical Analysis

The analytical process often begins with a question that is not phrased in terms of a chemical analysis. The question could be "Is this water safe to drink?" or "Does emission testing of automobiles reduce air pollution?" A scientist translates such questions into the need for particular measurements. An analytical chemist then chooses or invents a procedure to carry out those measurements.

When the analysis is complete, the analyst must translate the results into terms that can be understood by others—preferably by the general public. A critical feature of any result is its reliability. What is the statistical uncertainty in reported results? If you took samples in a different manner, would you obtain the same results? Is a tiny amount (a *trace*) of analyte found in a sample really there or is it contamination from the analytical procedure? Only after we understand the results and their limitations can we draw conclusions.

Here are the general steps in the analytical process:

Formulating the question	Translate general questions into specific questions to be answered through chemical measurements.
Selecting analytical procedures	Search the chemical literature to find appropriate procedures or, if necessary, devise new procedures to make the required measurements.
Sampling	*Sampling* is the process of selecting representative material to analyze. Box 0-1 provides some ideas on how to do so. If you begin with a poorly chosen sample or if the sample changes between the time it is collected and the time it is analyzed, results are meaningless. "Garbage in—garbage out!"
Sample preparation	Converting a representative sample into a form suitable for analysis is called *sample preparation,* which usually means dissolving the sample. Samples with a low concentration of analyte may need to be concentrated prior to analysis. It may be necessary to remove or *mask* species that interfere with the chemical analysis. For a chocolate bar, sample preparation consisted of removing fat and dissolving the desired analytes. Fat was removed because it would interfere with chromatography.
Analysis	Measure the concentration of analyte in several identical **aliquots** (portions). The purpose of *replicate measurements* (repeated measurements) is to assess the variability (uncertainty) in the analysis and to guard against a gross error in the analysis of a single aliquot. *The uncertainty of a measurement is as important as the measurement itself* because it tells us how reliable the measurement is. If necessary, use different analytical methods on similar samples to show that the choice of analytical method is not biasing the result. You may also wish to construct several different samples to see what variations arise from your sampling and sample preparation procedure.
Reporting and interpretation	Deliver a clearly written, complete report of your results, highlighting any limitations that you attach to them. Your report might be written to be read only by a specialist (such as your instructor), or it might be written for a general audience (such as a legislator or newspaper reporter). Be sure the report is appropriate for its intended audience.
Drawing conclusions	Once a report is written, the analyst might not be involved in what is done with the information, such as modifying the raw material supply for a factory or creating new laws to regulate food additives. The more clearly a report is written, the less likely it is to be misinterpreted by those who use it.

Most of this book deals with measuring chemical concentrations in homogeneous aliquots of an unknown. Analysis is meaningless unless you have collected the sample properly, you have taken measures to ensure the reliability of the analytical method, and you communicate your results clearly and completely. The chemical analysis is only the middle portion of a process that begins with a question and ends with a conclusion.

Chemists use the term **species** to refer to any chemical of interest. Species is both singular and plural.

Interference occurs when a species other than analyte increases or decreases the response of the analytical method and makes it appear that there is more or less analyte than is actually present.

Masking is the transformation of an interfering species into a form that is not detected. For example, Ca^{2+} in lake water can be measured with a reagent called EDTA. Al^{3+} interferes with this analysis because it also reacts with EDTA. Al^{3+} can be masked with excess F^- to form AlF_6^{3-}, which does not react with EDTA.

BOX 0-1 Constructing a Representative Sample

In a **random heterogeneous material,** differences in composition occur randomly and on a fine scale. When you collect a portion of the material for analysis, you obtain some of each of the different compositions. To construct a representative sample from a heterogeneous material, you can first visually divide the material into segments. A **random sample** is collected by taking portions from the desired number of segments chosen at random. If you wanted to measure the magnesium content of the grass in the 10-meter × 20-meter field in panel *a,* you could divide the field into 20 000 small patches that are 10 centimeters on a side. After assigning a number to each small patch, you could use a computer program to pick 100 numbers at random from 1 to 20 000. Then harvest and combine the grass from each of these 100 patches to construct a representative bulk sample for analysis.

For a **segregated heterogeneous material** (in which large regions have obviously different compositions), a representative **composite sample** must be constructed. For example, the field in panel *b* has three different types of grass segregated into regions A, B, and C. You could draw a map of the field on graph paper and measure the area in each region. In this case, 66% of the area lies in region A, 14% lies in region B, and 20% lies in region C. To construct a representative bulk sample from this segregated material, take 66 of the small patches from region A, 14 from region B, and 20 from region C. You could do so by drawing random numbers from 1 to 20 000 to select patches until you have the desired number from each region.

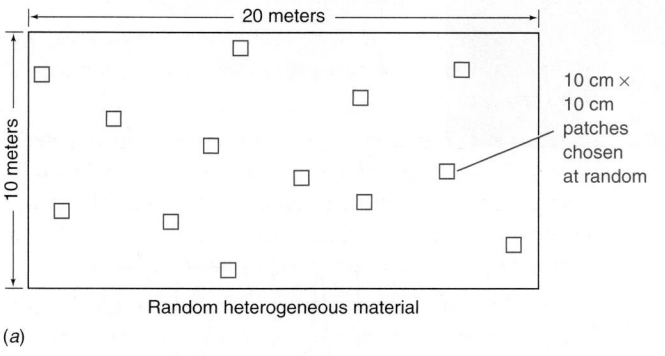

(a)

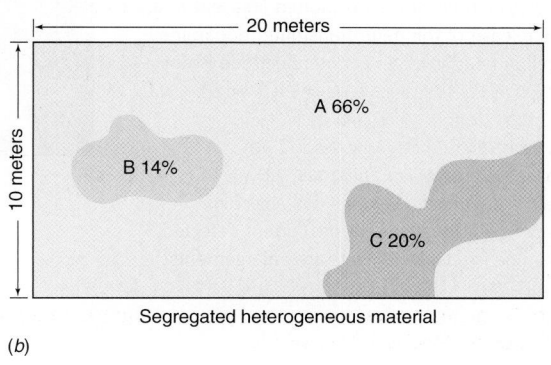

(b)

Terms to Understand

Terms are introduced in **bold** type in the chapter and are also defined in the Glossary.

aliquot	heterogeneous	quantitative transfer	segregated heterogeneous
analyte	homogeneous	random heterogeneous	material
aqueous	interference	material	slurry
calibration curve	masking	random sample	species
composite sample	qualitative chemical analysis	sample preparation	standard solution
decant	quantitative chemical analysis	sampling	supernatant liquid

Problems

Complete solutions to Problems can be found in the *Solutions Manual.* Short answers to numerical problems are at the back of the book.

0-1. What is the difference between *qualitative* and *quantitative* analysis?

0-2. List the steps in a chemical analysis.

0-3. What does it mean to *mask* an interfering species?

0-4. What is the purpose of a calibration curve?

0-5. (a) What is the difference between a homogeneous material and a heterogeneous material?

(b) After reading Box 0-1, state the difference between a segregated heterogeneous material and a random heterogeneous material.

(c) How would you construct a representative sample from each type of material?

0-6. The iodide (I^-) content of a commercial mineral water was measured by two methods that produced wildly different results.[6] Method A found 0.23 milligrams of I^- per liter (mg/L) and method B found 0.009 mg/L. When Mn^{2+} was added to the water, the I^- content found by method A increased each time that more Mn^{2+} was added, but results from method B were unchanged. Which of the *Terms to Understand* describes what is occurring in these measurements? Explain your answer. Which result is more reliable?

1 | Chemical Measurements

(*a*) Carbon-fiber electrode with a 100-nanometer-diameter (100×10^{-9} meter) tip extending from glass capillary. The marker bar is 200 micrometers (200×10^{-6} meter). [W.-H. Huang, D.-W. Pang, H. Tong, Z.-L. Wang, and J.-K. Cheng, "A Method for the Fabrication of Low-Noise Carbon Fiber Nanoelectrodes," *Anal. Chem.* **2001**, *73*, 1048. Reprinted with permission © 2001 American Chemical Society.] (*b*) Electrode positioned adjacent to a cell detects release of the neurotransmitter dopamine from the cell. A nearby, larger counterelectrode is not shown. [W.-Z. Wu, W.-H. Huang, W. Wang, Z.-L. Wang, J.-K. Cheng, T. Xu, R.-Y. Zhang, Y. Chen, and J. Liu, "Monitoring Dopamine Release from Single Living Vesicles with Nanoelectrodes," *J. Amer. Chem. Soc.* **2005**, *127*, 8914, Figure 1. Reprinted with permission © 2005 American Chemical Society.] (*c*) Bursts of electric current detected when dopamine is released. Insets are enlargements. [Data from W.-Z. Wu, ibid.]

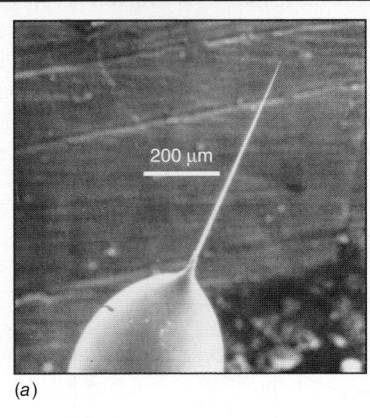
(*a*)

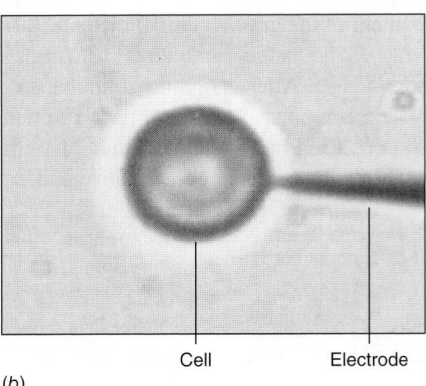
(*b*)

Cell   Electrode

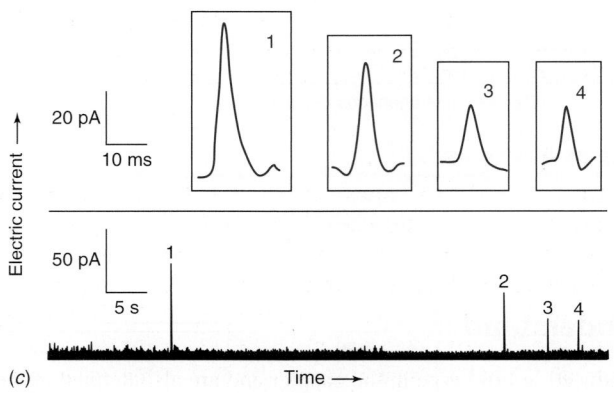
(*c*)

An electrode with a tip smaller than a single cell allows us to measure neurotransmitter molecules released by a nerve cell in response to a chemical stimulus. We call the electrode a *nanoelectrode* because its active region has dimensions of nanometers (10^{-9} meters). Neurotransmitter molecules released from one *vesicle* (a small compartment) of a nerve cell diffuse to the electrode, where they donate or accept electrons, thereby generating an electric current measured in picoamperes (10^{-12} amperes) for a period of milliseconds (10^{-3} seconds). This chapter discusses units that describe chemical and physical measurements of objects ranging in size from atoms to galaxies.

N eurotransmitter measurements illustrate the need for units of measurement covering many *orders of magnitude* (powers of 10) in range. This chapter introduces those units and reviews chemical concentrations, solution preparation, and stoichiometry of chemical reactions.

1-1 SI Units

SI units of measurement, used by scientists around the world, derive their name from the French *Système International d'Unités*. *Fundamental units* (base units) from which all others are derived are listed in Table 1-1. Standards of length, mass, and time are the *meter* (m), *kilogram* (kg), and *second* (s), respectively. Temperature is measured in *kelvins* (K), amount of substance in *moles* (mol), and electric current in *amperes* (A).

Table 1-2 lists some quantities that are defined in terms of the fundamental quantities. For example, force is measured in *newtons* (N), pressure in *pascals* (Pa), and energy in *joules* (J), each of which can be expressed in terms of length, time, and mass.

For readability, we insert a space after every third digit on either side of the decimal point. Commas are not used, because in some parts of the world a comma has the same meaning as a decimal point. Examples:

speed of light: 299 792 458 m/s

Avogadro's number: $6.022\ 141\ 29 \times 10^{23}\ \text{mol}^{-1}$

Pressure is force per unit area: 1 pascal (Pa) = 1 N/m^2. The pressure of the atmosphere is approximately 100 000 Pa.

Quantity	Unit (symbol)	Definition
Length	meter (m)	One meter is the distance light travels in a vacuum during $\frac{1}{299\ 792\ 458}$ of a second.
Mass	kilogram (kg)	One kilogram is the mass of the Pt-Ir alloy prototype kilogram made in 1885 and kept under an inert atmosphere at Sèvres, France. This object has been removed from its protective enclosure only in 1890, 1948, and 1992 to weigh secondary standards kept in several countries. Unfortunately, the mass of the prototype kilogram can change slowly over time by chemical reaction with the atmosphere or from mechanical wear. Work in progress will replace the prototype kilogram with a standard based on unchanging properties of nature that can be measured with high precision.[a]
Time	second (s)	One second is the duration of 9 192 631 770 periods of the radiation corresponding to a certain atomic transition of ^{133}Cs.
Electric current	ampere (A)	One ampere of current produces a force of 2×10^{-7} newtons per meter of length when maintained in two straight, parallel conductors of infinite length and negligible cross section, separated by 1 meter in a vacuum.
Temperature	kelvin (K)	Temperature is defined such that the triple point of water (at which solid, liquid, and gaseous water are in equilibrium) is 273.16 K, and the temperature of absolute zero is 0 K.
Luminous intensity	candela (cd)	Candela is a measure of luminous intensity visible to the human eye.
Amount of substance	mole (mol)	One mole is the number of particles equal to the number of atoms in exactly 0.012 kg of ^{12}C (approximately 6.022×10^{23}).
Plane angle	radian (rad)	There are 2π radians in a circle.
Solid angle	steradian (sr)	There are 4π steradians in a sphere.

TABLE 1-1 Fundamental SI units

a. P.F. Rusch, "Redefining the Kilogram and Mole," Chem. Eng. News, 30 May 2011, p. 58.

Using Prefixes as Multipliers

We use prefixes from Table 1-3 to express large or small quantities. For example, consider the pressure of ozone (O_3) in the stratosphere (Figure 1-1). Upper atmospheric ozone absorbs ultraviolet radiation from the sun that damages living organisms and causes skin cancer. Each Antarctic spring, much ozone disappears from the Antarctic stratosphere, creating an ozone "hole." The opening of Chapter 18 discusses the chemistry of this process. By contrast, ozone in the lower atmosphere harms animals and plants because it oxidizes sensitive cells.

At an altitude of 1.7×10^4 meters above Earth's surface, the ozone pressure over Antarctica reaches a peak of 0.019 Pa. Let's express these numbers with prefixes from Table 1-3. We use prefixes for every third power of ten (10^{-9}, 10^{-6}, 10^{-3}, 10^3, 10^6, 10^9). The number 1.7×10^4 is $> 10^3$ and $< 10^6$, so we use a multiple of 10^3 m (= kilometers, km):

$$1.7 \times 10^4\ \text{m} \times \frac{1\ \text{km}}{10^3\ \text{m}} = 17\ \text{km}$$

TABLE 1-2 SI-derived units with special names

Quantity	Unit	Symbol	Expression in terms of other units	Expression in terms of SI base units
Frequency	hertz	Hz		1/s
Force	newton	N		$m \cdot kg/s^2$
Pressure	pascal	Pa	N/m^2	$kg/(m \cdot s^2)$
Energy, work, quantity of heat	joule	J	$N \cdot m$	$m^2 \cdot kg/s^2$
Power, radiant flux	watt	W	J/s	$m^2 \cdot kg/s^3$
Quantity of electricity, electric charge	coulomb	C		$s \cdot A$
Electric potential, potential difference, electromotive force	volt	V	W/A	$m^2 \cdot kg/(s^3 \cdot A)$
Electric resistance	ohm	Ω	V/A	$m^2 \cdot kg/(s^3 \cdot A^2)$
Electric capacitance	farad	F	C/V	$s^4 \cdot A^2/(m^2 \cdot kg)$

Frequency *is the number of cycles per unit time for a repetitive event.* Force *is the product mass × acceleration.* Pressure *is force per unit area.* Energy *or* work *is force × distance = mass × acceleration × distance.* Power *is energy per unit time. The* electric potential difference *between two points is the work required to move a unit of positive charge between the two points.* Electric resistance *is the potential difference required to move one unit of charge per unit time between two points. The* electric capacitance *of two parallel surfaces is the quantity of electric charge on each surface when there is a unit of electric potential difference between the two surfaces.*

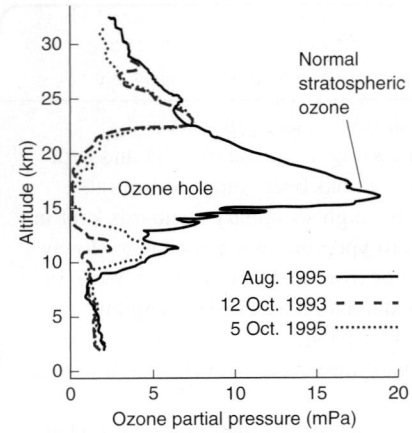

FIGURE 1-1 An ozone "hole" forms each year in the stratosphere over the South Pole at the beginning of spring in October. The graph compares ozone pressure in August, when there is no hole, with the pressure in October, when the hole is deepest. Less severe ozone loss is observed at the North Pole. [Data from National Oceanic and Atmospheric Administration.]

Of course you recall that $10^0 = 1$.

TABLE 1-3		**Prefixes**			
Prefix	**Symbol**	**Factor**	**Prefix**	**Symbol**	**Factor**
yotta	Y	10^{24}	deci	d	10^{-1}
zetta	Z	10^{21}	centi	c	10^{-2}
exa	E	10^{18}	milli	m	10^{-3}
peta	P	10^{15}	micro	μ	10^{-6}
tera	T	10^{12}	nano	n	10^{-9}
giga	G	10^{9}	pico	p	10^{-12}
mega	M	10^{6}	femto	f	10^{-15}
kilo	k	10^{3}	atto	a	10^{-18}
hecto	h	10^{2}	zepto	z	10^{-21}
deca	da	10^{1}	yocto	y	10^{-24}

The number 0.019 Pa is more than 10^{-3} Pa and less than 10^0 Pa, so we use a multiple of 10^{-3} Pa (= millipascals, mPa):

$$0.019 \text{ Pa} \times \frac{1 \text{ mPa}}{10^{-3} \text{ Pa}} = 19 \text{ mPa}$$

Figure 1-1 is labeled with km on the y-axis and mPa on the x-axis. The y-axis of a graph is called the **ordinate** and the x-axis is called the **abscissa.**

It is a fabulous idea to write units beside each number in a calculation and to cancel identical units in the numerator and denominator. This practice ensures that you know the units for your answer. If you intend to calculate pressure and your answer comes out with units other than pascals (N/m^2 or $kg/[m \cdot s^2]$ or other units of force/area), then you have made a mistake.

Converting Between Units

Although SI is the internationally accepted system of measurement in science, other units are encountered. Useful conversion factors are found in Table 1-4. For example, common non-SI

Oops! In 1999, the $125 million Mars Climate Orbiter spacecraft was lost when it entered the Martian atmosphere 100 km lower than planned. *This navigation error would have been avoided if people had written their units of measurement.* Engineers who built the spacecraft calculated thrust in the English unit, pounds of force. Jet Propulsion Laboratory engineers thought they were receiving the information in the metric unit, newtons. Nobody caught the error.

[JPL/NASA image.]

TABLE 1-4	**Conversion factors**		
Quantity	**Unit**	**Symbol**	**SI equivalent**[a]
Volume	liter	L	*10^{-3} m^3
	milliliter	mL	*10^{-6} m^3
Length	angstrom	Å	*10^{-10} m
	inch	in.	*0.025 4 m
Mass	pound	lb	*0.453 592 37 kg
	metric ton		*1 000 kg
Force	dyne	dyn	*10^{-5} N
Pressure	bar	bar	*10^5 Pa
	atmosphere	atm	*101 325 Pa
	atmosphere	atm	*1.013 25 bar
	atmosphere	atm	760 mm Hg = 760 torr
	torr (= 1 mm Hg)	Torr	133.322 Pa
	pound/in.^{2}	psi	6 894.76 Pa
Energy	erg	erg	*10^{-7} J
	electron volt	eV	1.602 176 655 × 10^{-19} J
	calorie, thermochemical	cal	*4.184 J
	Calorie (with a capital C)	Cal	*1 000 cal = 4.184 kJ
	British thermal unit	Btu	1 055.06 J
Power	horsepower		745.700 W
Temperature	centigrade (= Celsius)	°C	*K − 273.15
	Fahrenheit	°F	*1.8(K − 273.15) + 32

a. An asterisk (*) indicates that the conversion is exact (by definition).

CHAPTER 1 Chemical Measurements

units for energy are the *calorie* (cal) and the *Calorie* (written with a capital C and standing for 1 000 calories, or 1 kcal). Table 1-4 states that 1 cal is exactly 4.184 J (joules).

Your *basal metabolism* uses approximately 46 Calories per hour (h) per 100 pounds (lb) of body mass just to carry out basic functions for life, apart from exercise. A person walking at 2 miles per hour on a level path uses approximately 45 Calories per hour per 100 pounds of body mass beyond basal metabolism. The same person swimming at 2 miles per hour consumes 360 Calories per hour per 100 pounds beyond basal metabolism.

One **calorie** is the energy required to heat 1 gram of water from 14.5° to 15.5°C.

One **joule** is the energy expended when a force of 1 newton acts over a distance of 1 meter. This much energy can raise 102 g (the mass of a hamburger) by 1 meter.

$$1 \text{ cal} = 4.184 \text{ J}$$

1 **pound** (mass) $\approx$ 0.453 6 kg
1 **mile** $\approx$ 1.609 km
The symbol $\approx$ is read "**is approximately equal to.**"

EXAMPLE Unit Conversions

Express the rate of energy used by a person walking 2 miles per hour (46 + 45 = 91 Calories per hour per 100 pounds of body mass) in kilojoules per hour per kilogram of body mass.

Solution We will convert each non-SI unit separately. First, note that 91 Calories = 91 kcal. Table 1-4 states that 1 cal = 4.184 J; so 1 kcal = 4.184 kJ, and

$$91 \text{ kcal} \times 4.184 \frac{\text{kJ}}{\text{kcal}} = 3.8 \times 10^2 \text{ kJ}$$

Table 1-4 also says that 1 lb is 0.453 6 kg; so 100 lbs = 45.36 kg. The rate of energy consumption is therefore

$$\frac{91 \text{ kcal/h}}{100 \text{ lb}} = \frac{3.8 \times 10^2 \text{ kJ/h}}{45.36 \text{ kg}} = 8.4 \frac{\text{kJ/h}}{\text{kg}}$$

We could have written one long calculation:

$$\text{Rate} = \frac{91 \text{ kcal/h}}{100 \text{ lb}} \times 4.184 \frac{\text{kJ}}{\text{kcal}} \times \frac{1 \text{ lb}}{0.453 \text{ 6 kg}} = 8.4 \frac{\text{kJ/h}}{\text{kg}}$$

TEST YOURSELF A person swimming at 2 miles per hour requires 360 + 46 Calories per hour per 100 pounds of body mass. Express the energy use in kJ/h per kg of body mass. (*Answer:* 37 kJ/h per kg)

Significant figures are discussed in Chapter 3. For multiplication and division, the number with the fewest digits determines how many digits should be in the answer. The number 91 kcal at the beginning of this problem limits the answer to 2 digits.

1-2 Chemical Concentrations

A *solution* is a *homogeneous* mixture of two or more substances. A minor species in a solution is called **solute** and the major species is the **solvent.** In this book, most discussions concern *aqueous solutions*, in which the solvent is water. **Concentration** states how much solute is contained in a given volume or mass of solution or solvent.

A **homogeneous** substance has a uniform composition. Sugar dissolved in water is homogeneous. A mixture that is not the same everywhere (such as orange juice, which has suspended solids) is **heterogeneous.**

Molarity and Molality

A **mole** (mol) is *Avogadro's number* of particles (atoms, molecules, ions, or anything else). **Molarity** (M) is the number of moles of a substance per liter of solution. A **liter** (L) is the volume of a cube that is 10 cm on each edge. Because 10 cm = 0.1 m, 1 L = $(0.1 \text{ m})^3$ = 10^{-3} m^3. Chemical concentrations, denoted with square brackets, are usually expressed in moles per liter (M). Thus "[H^+]" means "the concentration of H^+."

The **atomic mass** of an element is the number of grams containing Avogadro's number of atoms. The **molecular mass** of a compound is the sum of atomic masses of the atoms in the molecule. It is the number of grams containing Avogadro's number of molecules.

An **electrolyte** is a substance that dissociates into ions in solution. In general, electrolytes are more dissociated in water than in other solvents. We refer to a compound that is mostly dissociated into ions as a *strong electrolyte*. One that is partially dissociated is called a *weak electrolyte*.

Magnesium chloride is a strong electrolyte. In 0.44 M $MgCl_2$ solution, 70% of the magnesium is free Mg^{2+} and 30% is $MgCl^+$. The concentration of $MgCl_2$ molecules is close to 0. Sometimes the molarity of a strong electrolyte is called the **formal concentration** (F), which is a description of how the solution was made by dissolving F moles per liter, even if the

Avogadro's number = number of atoms in 12 g of ^{12}C

$$\text{Molarity (M)} = \frac{\text{moles of solute}}{\text{liters of solution}}$$

Atomic masses are shown in the periodic table inside the cover of this book. See Box 3-3 for more on atomic mass. Physical constants such as Avogadro's number are also listed inside the cover.

Strong electrolyte: mostly dissociated into ions in solution

Weak electrolyte: partially dissociated into ions in solution

$MgCl^+$ is called an **ion pair.** See Box 8-1.

substance is converted into other species in solution. When we say that the "concentration" of $MgCl_2$ is 0.054 M in seawater, we are really speaking of its formal concentration (0.054 F). The "molecular mass" of a strong electrolyte is called the **formula mass** (FM), because it is the sum of atomic masses of atoms in the formula, even though there are very few molecules with that formula. *We are going to use the abbreviation FM for both formula mass and molecular mass.*

EXAMPLE Molarity of Salts in the Sea

(a) Typical seawater contains 2.7 g of salt (sodium chloride, NaCl) per 100 mL ($= 100 \times 10^{-3}$ L). What is the molarity of NaCl in the ocean? (b) $MgCl_2$ has a concentration of 0.054 M in the ocean. How many grams of $MgCl_2$ are present in 25 mL of seawater?

Solution (a) The molecular mass of NaCl is 22.99 g/mol (Na) + 35.45 g/mol (Cl) = 58.44 g/mol. The moles of salt in 2.7 g are (2.7 g)/(58.44 g/mol) = 0.046 mol, so the molarity is

$$\text{Molarity of NaCl} = \frac{\text{mol NaCl}}{\text{L of seawater}} = \frac{0.046 \text{ mol}}{100 \times 10^{-3} \text{ L}} = 0.46 \text{ M}$$

(b) The molecular mass of $MgCl_2$ is 24.30 g/mol (Mg) + 2×35.45 g/mol (Cl) = 95.20 g/mol. The number of grams in 25 mL is

$$\text{Garms of } MgCl_2 = \left(0.054\frac{\text{mol}}{\text{L}}\right)\left(95.20\frac{\text{g}}{\text{mol}}\right)(25 \times 10^{-3} \text{ L}) = 0.13 \text{ g}$$

TEST YOURSELF Calculate the formula mass of $CaSO_4$. What is the molarity of $CaSO_4$ in a solution containing 1.2 g of $CaSO_4$ in a volume of 50 mL? How many grams of $CaSO_4$ are in 50 mL of 0.086 M $CaSO_4$? (*Answer:* 136.13 g/mol, 0.18 M, 0.59 g)

For a *weak electrolyte* such as acetic acid, CH_3CO_2H, some of the molecules dissociate into ions in solution:

Formal concentration	Percent dissociated
0.10 F	1.3%
0.010 F	4.1%
0.001 0 F	12%

Molality (m) is concentration expressed as moles of substance per kilogram of solvent (not total solution). Molality is independent of temperature. Molarity changes with temperature because the volume of a solution usually increases when it is heated.

Percent Composition

The percentage of a component in a mixture or solution is usually expressed as a **weight percent** (wt%):

$$\text{Weight percent} = \frac{\text{mass of solute}}{\text{mass of total solution or mixture}} \times 100 \qquad (1\text{-}1)$$

Ethanol (CH_3CH_2OH) is often purchased as a 95 wt% solution containing 95 g of ethanol per 100 g of total solution. The remainder is water. **Volume percent** (vol%) is defined as

$$\text{Volume percent} = \frac{\text{volume of solute}}{\text{volume of total solution}} \times 100 \qquad (1\text{-}2)$$

Although units of mass or volume should always be expressed to avoid ambiguity, mass is usually implied when units are absent.

Confusing abbreviations:

mol = moles

$\mathbf{M} = \text{molarity} = \dfrac{\text{mol solute}}{\text{L solution}}$

$\boldsymbol{m} = \text{molality} = \dfrac{\text{mol solute}}{\text{kg solvent}}$

EXAMPLE **Converting Weight Percent into Molarity and Molality**

Find the molarity and molality of 37.0 wt% HCl. The **density** of a substance is the mass per unit volume. The table inside the back cover of this book tells us that the density of the reagent is 1.19 g/mL.

Density $= \dfrac{mass}{volume} = \dfrac{g}{mL}$

A closely related dimensionless quantity is

Specific gravity $= \dfrac{\text{density of a substance}}{\text{density of water at } 4^\circ C}$

The density of water at 4°C is close to 1 g/mL, so specific gravity is nearly the same as density.

Solution For *molarity*, we need to find the moles of HCl per liter of solution. The mass of a liter of solution is $(1.19\ \text{g/mL})(1\,000\ \text{mL}) = 1.19 \times 10^3$ g. The mass of HCl in a liter is

$$\text{Mass of HCl per liter} = \left(1.19 \times 10^3 \frac{\text{g solution}}{\text{L}}\right)\left(0.370 \underbrace{\frac{\text{g HCl}}{\text{g solution}}}_{\substack{\text{This is what} \\ \text{37.0 wt\% means}}}\right) = 4.40 \times 10^2 \frac{\text{g HCl}}{\text{L}}$$

The molecular mass of HCl is 36.46 g/mol, so the molarity is

$$\text{Molarity} = \frac{\text{mol HCl}}{\text{L solution}} = \frac{4.40 \times 10^2\ \text{g HCl/L}}{36.46\ \text{g HCl/mol}} = 12.1 \frac{\text{mol}}{\text{L}} = 12.1\ \text{M}$$

For *molality*, we need to find the moles of HCl per kilogram of solvent (which is H_2O). The solution is 37.0 wt% HCl, so we know that 100.0 g of solution contains 37.0 g of HCl and $100.0 - 37.0 = 63.0$ g of H_2O ($= 0.063\,0$ kg). But 37.0 g of HCl contains 37.0 g/(36.46 g/mol) $= 1.01$ mol. The molality is therefore

$$\text{Molality} = \frac{\text{mol HCl}}{\text{kg of solvent}} = \frac{1.01\ \text{mol HCl}}{0.063\,0\ \text{kg } H_2O} = 16.1\ m$$

If you divide 1.01/0.063 0, you get 16.0. I got 16.1 because I kept all the digits in my calculator and did not round off until the end. The number 1.01 was really 1.014 8 and (1.014 8)/(0.063 0) = 16.1.

TEST YOURSELF Calculate the molarity and molality of 49.0 wt% HF using the density given inside the back cover of this book. (*Answer:* 28.4 M, 48.0 *m*)

Figure 1-2 illustrates a weight percent measurement in the application of analytical chemistry to archaeology.[1] Gold and silver are found together in nature. Dots in Figure 1-2 show the weight percent of gold in more than 1300 silver coins minted over a 500-year period. Prior to A.D. 500, it was rare for the gold content to be below 0.3 wt%. By A.D. 600, people developed techniques for removing more gold from the silver, so some coins had as little as 0.02 wt% gold. Colored squares in Figure 1-2 represent known, modern forgeries made from silver whose gold content is always less than the prevailing gold content in the years A.D. 200 to 500. Chemical analysis makes it easy to detect the forgeries.

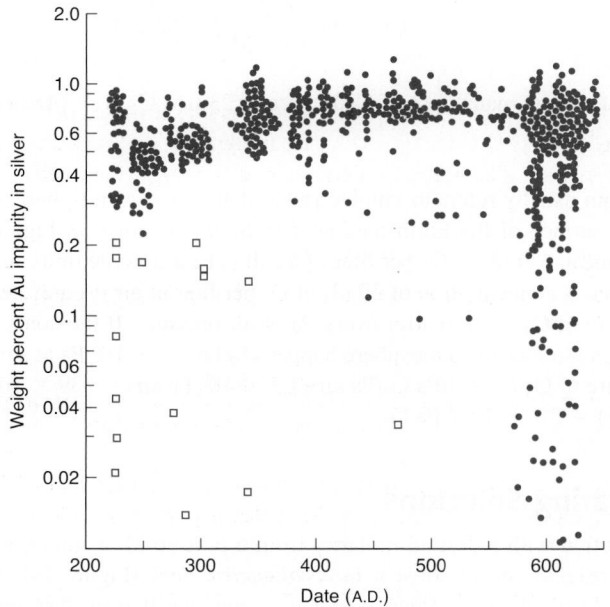

FIGURE 1-2 Weight percent of gold impurity in silver coins from Persia. Colored squares are known, modern forgeries. Note that the ordinate scale is logarithmic. [Data from A. A. Gordus and J. P. Gordus, *Archaeological Chemistry*, Adv. Chem. No. 138, American Chemical Society, Washington, DC, 1974, pp. 124–147.]

Parts per Million and Parts per Billion

Sometimes composition is expressed as **parts per million (ppm)** or **parts per billion (ppb)**, which mean grams of substance per million or billion grams of total solution or mixture.

$$\text{ppm} = \frac{\text{mass of substance}}{\text{mass of sample}} \times 10^6 \qquad (1\text{-}3)$$

$$\text{ppb} = \frac{\text{mass of substance}}{\text{mass of sample}} \times 10^9 \qquad (1\text{-}4)$$

A solution concentration of 1 ppm means 1 μg of solute per gram of solution. Because the density of a dilute aqueous solution is close to 1.00 g/mL, *we frequently equate 1 g of water with 1 mL of water.* Therefore, 1 ppm in dilute aqueous solution corresponds *approximately* to 1 μg/mL (= 1 mg/L) and 1 ppb is 1 ng/mL (= 1 μg/L).

EXAMPLE Converting Parts per Billion into Molarity

Normal alkanes are hydrocarbons with the formula C_nH_{2n+2}. Plants selectively synthesize alkanes with an odd number of carbon atoms. The concentration of $C_{29}H_{60}$ in summer rainwater collected in Hannover, Germany, is 34 ppb. Find the molarity of $C_{29}H_{60}$ and express the answer with a prefix from Table 1-3.

Solution A concentration of 34 ppb means there are 34 ng of $C_{29}H_{60}$ per gram of rainwater, which is nearly the same as 34 ng/mL because the density of rainwater is close to 1.00 g/mL. To find the molarity, we need to know how many grams of $C_{29}H_{60}$ are contained in a liter. Multiplying nanograms and milliliters by 1 000 gives 34 μg of $C_{29}H_{60}$ per liter of rainwater:

$$\frac{34 \text{ ng } C_{29}H_{60}}{\text{mL}}\left(\frac{1\,000 \text{ mL/L}}{1\,000 \text{ ng/μg}}\right) = \frac{34 \text{ μg } C_{29}H_{60}}{\text{L}}$$

The molecular mass of $C_{29}H_{60}$ is $29 \times 12.011 + 60 \times 1.008 = 408.8$ g/mol, so the molarity is

$$\text{Molarity of } C_{29}H_{60} \text{ in rainwater} = \frac{34 \times 10^{-6} \text{ g/L}}{408.8 \text{ g/mol}} = 8.3 \times 10^{-8} \text{ M}$$

An appropriate prefix from Table 1-3 would be nano (n), which is a multiple of 10^{-9}:

$$8.3 \times 10^{-8} \text{ M} \left(\frac{1 \text{ nM}}{10^{-9} \text{ M}}\right) = 83 \text{ nM}$$

nM = nanomoles per liter

TEST YOURSELF How many ppm of $C_{29}H_{60}$ are in 23 μM $C_{29}H_{60}$? (*Answer:* 9.4 ppm)

For gases, ppm usually refers to volume rather than mass. Atmospheric ozone (O_3) concentration at the surface of the Earth measured in Spain is shown in Figure 1-3. The peak value of 39 ppb means 39 nL of O_3 per liter of air. It is best to write units such as "nL O_3/L" to avoid confusion. A concentration of 39 nL of O_3 per liter of air is equivalent to saying that the partial pressure of O_3 is 39 nPa for every Pa of air pressure. If the concentration of O_3 is 39 ppm and the pressure of the atmosphere happens to be 1.3×10^4 Pa at some altitude, then the partial pressure of O_3 is (39 nPa O_3/Pa air)(1.3×10^4 Pa air) = (39×10^{-9} Pa O_3/Pa air) (1.3×10^4 Pa air) = 5.1×10^{-4} Pa O_3.

1-3 Preparing Solutions

To prepare a solution with a desired molarity from a pure solid or liquid, we weigh out the correct mass of reagent and dissolve it in a *volumetric flask* (Figure 1-4) with *distilled* or *deionized* water. In distillation, water is boiled to separate it from less volatile impurities and the vapor is condensed to liquid that is collected in a clean container. In deionization (page 719), water is passed through a column that removes ionic impurities. Nonionic impurities remain in the water. Distilled and deionized water are used almost interchangeably.

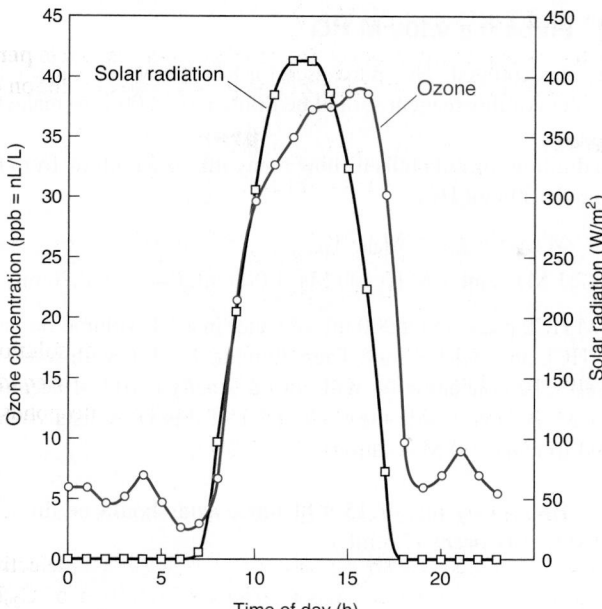

FIGURE 1-3 Ozone concentration (ppb by volume = nL/L) and solar radiation (W/m^2) measured by students in Argamasilla de Calatrava, Spain, on 6 February 2008. Ozone at Earth's surface arises largely from the reactions NO_2 + sunlight → NO + O followed by $O + O_2 → O_3$. The data show O_3 concentration peaking shortly after the peak in solar radiation. [Data from Y. T. Diaz-de-Mera, A. Notario, A. Aranda, J. A. Adame, A. Parra, E. Romero, J. Parra, and F. Muñoz, "Research Study of Tropospheric Ozone and Meteorological Parameters to Introduce High School Students to Scientific Procedures," *J. Chem. Ed.* **2011**, *88*, 392.]

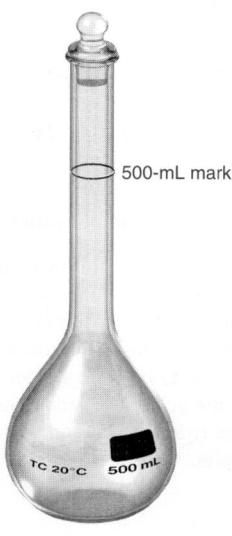

FIGURE 1-4 A *volumetric flask* contains a specified volume when the liquid level is adjusted to the middle of the mark in the thin neck of the flask. Use of this flask is described in Section 2-5.

EXAMPLE **Preparing a Solution with a Desired Molarity**

Copper(II) sulfate pentahydrate, $CuSO_4 \cdot 5H_2O$, has 5 moles of H_2O for each mole of $CuSO_4$ in the solid crystal. The formula mass of $CuSO_4 \cdot 5H_2O$ (= $CuSO_9H_{10}$) is 249.68 g/mol. (Copper(II) sulfate without water in the crystal has the formula $CuSO_4$ and is said to be **anhydrous.**) How many grams of $CuSO_4 \cdot 5H_2O$ should be dissolved in a volume of 500.0 mL to make 8.00 mM Cu^{2+}?

Solution An 8.00 mM solution contains 8.00×10^{-3} mol/L. We need

$$8.00 \times 10^{-3} \frac{mol}{L} \times 0.500\,0\,L = 4.00 \times 10^{-3} \text{ mol } CuSO_4 \cdot 5H_2O$$

The mass of reagent is $(4.00 \times 10^{-3} \text{ mol}) \times \left(249.68 \frac{g}{mol}\right) = 0.999$ g.

Using a volumetric flask: The procedure is to place 0.999 g of solid $CuSO_4 \cdot 5H_2O$ into a 500-mL volumetric flask, add about 400 mL of distilled water, and swirl to dissolve the reagent. Then dilute with distilled water up to the 500-mL mark and invert the flask several times to ensure complete mixing.

TEST YOURSELF Find the formula mass of anhydrous $CuSO_4$. How many grams should be dissolved in 250.0 mL to make a 16.0 mM solution? (***Answer:*** 159.60 g/mol, 0.638 g)

Dilution

You can prepare a dilute solution from a more concentrated solution. Transfer a calculated volume of the concentrated solution to a volumetric flask and dilute to the final volume. The number of moles of reagent in V liters containing M moles per liter is the product $M \cdot V = mol/L \cdot L$. Equating the number of moles taken from the concentrated (conc) solution with the number of moles delivered to the dilute (dil) solution gives us the *dilution formula:*

Dilution formula:
$$M_{conc} \cdot V_{conc} = M_{dil} \cdot V_{dil} \qquad (1-5)$$

Moles taken from concentrated solution   Moles placed in dilute solution

You can use any units for concentration (such as mmol/L or g/mL) and any units for volume (such as mL or μL), as long as you use the same units on both sides. We frequently use mL for volume.

EXAMPLE Preparing 0.100 M HCl

The molarity of "concentrated" HCl purchased for laboratory use is approximately 12.1 M. How many milliliters of this reagent should be diluted to 1.000 L to make 0.100 M HCl?

Solution The dilution formula tells us how many mL to withdraw from the concentrated solution to obtain 0.100 mol HCl:

$$M_{conc} \cdot V_{conc} = M_{dil} \cdot V_{dil}$$
$$(12.1\ M)(x\ mL) = (0.100\ M)(1\ 000\ mL) \Rightarrow x = 8.26\ mL$$

To make 0.100 M HCl, place about 900 mL of water in a 1-L volumetric flask, add 8.26 mL of concentrated HCl, and swirl to mix. Then dilute to 1.000 L with water and invert many times to mix well. The concentration will not be exactly 0.100 M because the reagent is not exactly 12.1 M. A table inside the back cover of this book lists volumes of common reagents required to make 1.0 M solutions.

TEST YOURSELF How many mL of 15.8 M nitric acid should be diluted to 0.250 L to make 3.00 M HNO_3? (*Answer:* 47.5 mL)

The symbol ⇒ is read **implies that.**

In a chemical reaction, species on the left side are called **reactants** and species on the right are called **products.** NH_3 is a reactant and NH_4^+ is a product in Reaction 1-6.

EXAMPLE A More Complicated Dilution Calculation

A solution of ammonia in water is called "ammonium hydroxide" because of the equilibrium

$$NH_3\ +\ H_2O\ \rightleftharpoons\ NH_4^+\ +\ OH^- \qquad (1\text{-}6)$$
$$\text{Ammonia} \qquad\qquad\qquad \text{Ammonium}\quad \text{Hydroxide}$$

The density of concentrated ammonium hydroxide, which contains 28.0 wt% NH_3, is 0.899 g/mL. What volume of this reagent should be diluted to 500.0 mL to make 0.250 M NH_3?

Solution To use Equation 1-5, we need to know the molarity of the concentrated reagent. The solution contains 0.899 g of solution per milliliter and there is 0.280 g of NH_3 per gram of solution (28.0 wt%), so we can write

$$\text{Molarity of } NH_3 = \frac{899\dfrac{\text{g solution}}{L} \times 0.280\dfrac{\text{g } NH_3}{\text{g solution}}}{17.03\dfrac{\text{g } NH_3}{\text{mol } NH_3}} = 14.8\ M$$

Now we find the volume of 14.8 M NH_3 required to prepare 500.0 mL of 0.250 M NH_3:

$$M_{conc} \cdot V_{conc} = M_{dil} \cdot V_{dil}$$
$$14.8\ M \times V_{conc} = 0.250\ M \times 500.0\ mL \Rightarrow V_{conc} = 8.46\ mL$$

The procedure is to place about 400 mL of water in a 500-mL volumetric flask, add 8.46 mL of concentrated reagent, and swirl to mix. Then dilute to 500.0 mL with water and invert the flask many times to mix well.

TEST YOURSELF From the density of 70.4 wt% HNO_3 given on the inside cover, calculate the molarity of HNO_3. (*Answer:* 15.8 M)

1-4 Stoichiometry Calculations for Gravimetric Analysis

Chemical analysis based on weighing a final product is called **gravimetric analysis.** Gravimetric analysis and titrations (*volumetric analysis*) were practiced long before electronic instruments were available to make chemical measurements. We call gravimetric and volumetric analysis "classical" or "wet chemical" methods to distinguish them from instrumental methods of analysis that have been added to the arsenal of analytical chemistry in the last century. Classical methods still have a place in modern analytical chemistry. They can be more accurate than instrumental methods and they can be used to prepare standards for instrumental methods of analysis.

Iron from a dietary supplement tablet can be measured gravimetrically by dissolving the tablet and then converting the dissolved iron into solid Fe_2O_3. The mass of Fe_2O_3 tells us the mass of iron in the original tablet.

Here are the steps in the procedure:

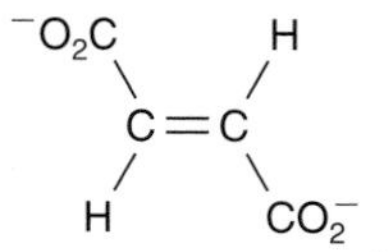

Step 1 Tablets containing iron(II) fumarate ($Fe^{2+}C_4H_2O_4^{2-}$) and inert binder are mixed with 150 mL of 0.100 M HCl to dissolve the Fe^{2+}. The mixture is filtered to remove insoluble binder.

Step 2 Iron(II) in the clear liquid is oxidized to iron(III) with excess hydrogen peroxide:

$$2Fe^{2+} \;+\; H_2O_2 \;+\; 2H^+ \;\rightarrow\; 2Fe^{3+} \;+\; 2H_2O \qquad (1\text{-}7)$$

Iron(II)   Hydrogen peroxide   Iron(III)
(ferrous ion)   FM 34.01   (ferric ion)

Step 3 Ammonium hydroxide is added to precipitate hydrous iron(III) oxide, which is a gel. The gel is filtered and heated in a furnace to convert it to pure solid Fe_2O_3.

$$Fe^{3+} + 3OH^- + (x-1)H_2O \longrightarrow FeOOH \cdot xH_2O(s) \xrightarrow{900°C} Fe_2O_3(s) \qquad (1\text{-}8)$$

Hydroxide   Hydrous iron(III) oxide   Iron(III) oxide
                                                               FM 159.69

We now work through some practical laboratory calculations for this analysis.

EXAMPLE   **How Many Tablets Should We Analyze?**

In a gravimetric analysis, we need enough product to weigh accurately. Each tablet provides ~15 mg of iron. How many tablets should we analyze to provide 0.25 g of Fe_2O_3 product?

Solution   We can answer the question if we know how many grams of iron are in 0.25 g of Fe_2O_3. The formula mass of Fe_2O_3 is 159.69 g/mol, so 0.25 g is equal to

$$\text{mol } Fe_2O_3 = \frac{0.25 \text{ g}}{159.69 \text{ g/mol}} = 1.6 \times 10^{-3} \text{ mol}$$

Each mol of Fe_2O_3 has 2 mol of Fe, so 0.25 g of Fe_2O_3 contains

$$1.6 \times 10^{-3} \text{ mol } Fe_2O_3 \times \frac{2 \text{ mol Fe}}{1 \text{ mol } Fe_2O_3} = 3.2 \times 10^{-3} \text{ mol Fe}$$

The mass of Fe is

$$3.2 \times 10^{-3} \text{ mol Fe} \times \frac{55.845 \text{ g Fe}}{\text{mol Fe}} = 0.18 \text{ g Fe}$$

If each tablet contains 15 mg Fe, the number of tablets required is

$$\text{Number of tablets} = \frac{0.18 \text{ g Fe}}{0.015 \text{ g Fe/tablet}} = 12 \text{ tablets}$$

TEST YOURSELF   If each tablet provides ~20 mg of iron, how many tablets should we analyze to provide ~0.50 g of Fe_2O_3? (*Answer:* 18)

EXAMPLE   **How Much H_2O_2 Is Required?**

What mass of 3.0 wt% H_2O_2 solution is required to provide a 50% excess of reagent for Reaction 1-7 with 12 dietary iron tablets?

Solution   Twelve tablets provide 12 tablets × (0.015 g Fe^{2+}/tablet) = 0.18 g Fe^{2+}, or (0.18 g Fe^{2+})/(55.845 g Fe^{2+}/mol Fe^{2+}) = 3.2 × 10⁻³ mol Fe^{2+}. Reaction 1-7 requires 1 mol of H_2O_2 for every 2 mol of Fe^{2+}. Therefore, 3.2 × 10⁻³ mol Fe^{2+} requires (3.2 × 10⁻³ mol Fe^{2+})(1 mol H_2O_2/2 mol Fe^{2+}) = 1.6 × 10⁻³ mol H_2O_2. A 50% excess means that we want to use 1.50 times the stoichiometric quantity: (1.50)(1.6 × 10⁻³ mol H_2O_2) = 2.4 × 10⁻³ mol H_2O_2. The formula mass of H_2O_2 is 34.01 g/mol, so the required

Side notes (right margin):

Fumarate anion, $C_4H_2O_4^{2-}$

The units of formula mass (FM) are g/mol

$Fe_2O_3(s)$ means that Fe_2O_3 is a *solid*. Other abbreviations for phases are (*l*) for *liquid*, (*g*) for *gas*, and (*aq*) for *aqueous* (meaning dissolved in water).

The symbol ~ is read **approximately.**

$$\text{Moles} = \frac{\text{grams}}{\text{grams per mol}} = \frac{\text{grams}}{\text{formula mass}}$$

The atomic mass of Fe, 55.845 g/mol, is in the periodic table inside the cover.

3.0 wt% means 3.0 g H_2O_2 per 100 g solution or 0.030 g H_2O_2 per gram of solution.

$$\text{Moles} = \frac{\text{grams}}{\text{formula mass}} = \frac{g}{g/mol}$$

You should be able to use this relationship in your sleep.

mass of pure H_2O_2 is $(2.4 \times 10^{-3} \text{ mol})(34.01 \text{ g/mol}) = 0.082$ g. But hydrogen peroxide is available as a 3.0 wt% solution, so the required mass of solution is

$$\text{Mass of } H_2O_2 \text{ solution} = \frac{0.082 \text{ g } H_2O_2}{0.030 \text{ g } H_2O_2/\text{g solution}} = 2.7 \text{ g solution}$$

TEST YOURSELF What mass of 3.0 wt% H_2O_2 solution is required to provide a 25% excess of reagent for Reaction 1-7 with 12 dietary iron tablets? (**Answer:** 2.3 g)

EXAMPLE **The Gravimetric Calculation**

The mass of Fe_2O_3 isolated from analysis of 12 tablets was 0.277 g. What is the average mass of iron per dietary tablet?

Solution The moles of isolated Fe_2O_3 are $(0.277 \text{ g})/(159.69 \text{ g/mol}) = 1.73 \times 10^{-3}$ mol. There are 2 mol Fe per formula unit, so the moles of Fe in the product are

$$(1.73 \times 10^{-3} \text{ mol } Fe_2O_3)\left(\frac{2 \text{ mol Fe}}{1 \text{ mol } Fe_2O_3}\right) = 3.47 \times 10^{-3} \text{ mol Fe}$$

The mass of Fe is $(3.47 \times 10^{-3} \text{ mol Fe})(55.845 \text{ g Fe/mol Fe}) = 0.194$ g Fe. Each of the 12 tablets therefore contains an average of $(0.194 \text{ g Fe})/12 = 0.016\ 1 \text{ g} = 16.1$ *mg*.

Retain all digits in your calculator during a series of calculations. The product 1.73×2 is not 3.47; but, with extra digits in the calculator, the answer is 3.47.

TEST YOURSELF If the mass of isolated Fe_2O_3 were 0.300 g, what would be the average mass of iron per tablet? (**Answer:** 17.5 mg)

Limiting Reagent

The **limiting reagent** in a chemical reaction is the one that is consumed first. Once the limiting reagent is gone, the reaction ceases. To decide which reagent is limiting, find the number of moles of each reactant that is available. Compare the number of moles present to the number of moles required for complete reaction.

EXAMPLE **Limiting Reagent**

Reaction 1-9 requires one mole of oxalate for each mole of calcium.

$$\underset{\text{Oxalate}}{Ca^{2+} + C_2O_4^{2-}} \rightarrow \underset{\text{Calcium oxalate}}{Ca(C_2O_4) \cdot H_2O(s)} \qquad (1\text{-}9)$$

If you mix 1.00 g of $CaCl_2$ (FM 110.98) with 1.15 g of $Na_2C_2O_4$ (FM 134.00) in water, which is the limiting reagent? What fraction of the nonlimiting reagent is left over?

Solution The available moles of each reagent are

$$\frac{1.00 \text{ g } CaCl_2}{110.98 \text{ g/mol}} = 9.01 \text{ mmol } Ca^{2+} \qquad \frac{1.15 \text{ g } Na_2C_2O_4}{134.00 \text{ g/mol}} = 8.58 \text{ mmol } C_2O_4^{2-}$$

The reaction requires 1 mol Ca^{2+} for 1 mol $C_2O_4^{2-}$, so oxalate will be used up first. The Ca^{2+} remaining is $9.01 - 8.58 = 0.43$ mmol. The fraction of unreacted Ca^{2+} is $(0.43 \text{ mmol}/9.01 \text{ mmol}) = 4.8\%$

TEST YOURSELF The reaction $5H_2C_2O_4 + 2MnO_4^- + 6H^+ \rightarrow 10CO_2 + 2Mn^{2+} + 8H_2O$ requires 5 mol $H_2C_2O_4$ for 2 mol MnO_4^-. If you mix 1.15 g $Na_2C_2O_4$ (FM 134.00) with 0.60 g $KMnO_4$ (FM 158.03) and excess aqueous acid, which reactant is limiting? How much CO_2 is produced?

(**Answer:** 8.58 mmol $C_2O_4^{2-}$ require $\left(\dfrac{2 \text{ mol } MnO_4^-}{5 \text{ mol } C_2O_4^{2-}}\right)(8.58 \text{ mmol } C_2O_4^{2-}) = 3.43$ mmol MnO_4^-. $KMnO_4$ available is 0.60 g/(158.03 g/mol) = 3.80 mmol, which is more than enough. So, $Na_2C_2O_4$ is the limiting reagent. Reaction of 8.58 mmol $C_2O_4^{2-}$ produces (10 mol CO_2/5 mol $C_2O_4^{2-}$)(8.58 mmol $C_2O_4^{2-}$) = 17.16 mmol CO_2.)

Terms to Understand

Terms are introduced in **bold** type in the chapter and are also defined in the Glossary.

abscissa	formula mass	molecular mass	solute
anhydrous	gravimetric analysis	ordinate	solvent
atomic mass	limiting reagent	ppb (parts per billion)	stoichiometry
concentration	liter	ppm (parts per million)	volume percent
density	molality	product	weight percent
electrolyte	molarity	reactant	
formal concentration	mole	SI units	

Summary

SI base units include the meter (m), kilogram (kg), second (s), ampere (A), kelvin (K), and mole (mol). Derived quantities such as force (newton, N), pressure (pascal, Pa), and energy (joule, J) can be expressed in terms of base units. In calculations, units should be carried along with the numbers. Prefixes such as kilo- and milli- are used to denote multiples of units. Common expressions of concentration are molarity (moles of solute per liter of solution), molality (moles of solute per kilogram of solvent), formal concentration (formula units per liter), percent composition, and parts per million.

To calculate quantities of reagents needed to prepare solutions, the relation $M_{conc} \cdot V_{conc} = M_{dil} \cdot V_{dil}$ is useful because it equates moles of reagent removed from a stock solution to moles delivered into a new solution. You should be able to use stoichiometry relationships to calculate required masses or volumes of reagents for chemical reactions. From the mass of product of a reaction, you should be able to compute how much reactant was consumed. The limiting reagent in a chemical reaction is the one that is consumed first. Once the limiting reagent is gone, the reaction ceases.

Exercises

Complete solutions to *Exercises* are provided at the back of the book, whereas only short answers to *Problems* are provided. Complete solutions to *Problems* are in the *Solutions Manual*. *Exercises* cover many of the major ideas in each chapter.

1-A. A solution with a final volume of 500.0 mL was prepared by dissolving 25.00 mL of methanol (CH_3OH, density = 0.791 4 g/mL) in chloroform.

(a) Calculate the *molarity* of methanol in the solution.

(b) The solution has a density of 1.454 g/mL. Find the *molality* of methanol.

1-B. A 48.0 wt% solution of HBr in water has a density of 1.50 g/mL.

(a) Find the formal concentration of HBr.

(b) What mass of solution contains 36.0 g of HBr?

(c) What volume of solution contains 233 mmol of HBr?

(d) How much solution is required to prepare 0.250 L of 0.160 M HBr?

1-C. (a) An aqueous solution contains 12.6 ppm of dissolved $Ca(NO_3)_2$ (which gives Ca^{2+} + $2NO_3^-$). Find the concentration of NO_3^- in parts per million.

(b) How many ppm of $Ca(NO_3)_2$ are in 0.144 mM $Ca(NO_3)_2$?

(c) How many ppm of nitrate (NO_3^-) are in 0.144 mM $Ca(NO_3)_2$?

1-D. Ammonia reacts with hypobromite, OBr^-, by the reaction $2NH_3 + 3OBr^- \rightarrow N_2 + 3Br^- + 3H_2O$. Which is the limiting reagent if 5.00 mL of 0.623 M NaOBr solution are added to 183 µL of 28 wt% NH_3 (14.8 M NH_3 on the inside back cover of the book)? How much of the excess reagent is left over?

Problems

Units and Conversions

1-1. (a) List the SI units of length, mass, time, electric current, temperature, and amount of substance; write the abbreviation for each.

(b) Write the units and symbols for frequency, force, pressure, energy, and power.

1-2. Write the names and abbreviations for each of the prefixes from 10^{-24} to 10^{24}. Which abbreviations are capitalized?

1-3. Write the name and number represented by each symbol. For example, for kW you should write kW = kilowatt = 10^3 watts.

(a) mW	**(c)** kΩ	**(e)** TJ	**(g)** fg
(b) pm	**(d)** µF	**(f)** ns	**(h)** dPa

1-4. Express the following quantities with abbreviations for units and prefixes from Tables 1-1 through 1-3:

(a) 10^{-13} joules	**(d)** 10^{-10} meters
(b) 4.317 28 × 10^{-8} farads	**(e)** 2.1 × 10^{13} watts
(c) 2.997 9 × 10^{14} hertz	**(f)** 48.3 × 10^{-20} moles

1-5. The burning of fossil fuel by humans in 2012 introduced approximately 8 petagrams (Pg) of carbon per year into the atmosphere in the form of CO_2.

(a) How many kg of C are placed in the atmosphere each year?

(b) How many kg of CO_2 are placed in the atmosphere each year?

(c) A metric ton is 1 000 kg. How many metric tons of CO_2 are placed in the atmosphere each year? There are 7 billion people on

Earth. Find the per capita rate of CO_2 production (tons of CO_2 per person per year).

1-6. I have always enjoyed eating tuna fish. Unfortunately, a study of the mercury content of canned tuna in 2010 found that chunk *white* tuna contains 0.6 ppm Hg and chunk *light* tuna contains 0.14 ppm.[2] The U.S. Environmental Protection Agency recommends no more than 0.1 μg Hg/kg body weight per day. I weigh 68 kg. How often may I eat a can containing 6 ounces (1 lb = 16 oz) of chunk *white* tuna so that I do not average more than 0.1 μg Hg/kg body weight per day? If I switch to chunk *light* tuna, how often can I eat one can?

1-7. How many joules per second and how many calories per hour are produced by a 100.0-horsepower engine?

1-8. A 120-pound woman working in an office consumes about 2.2×10^3 kcal/day, whereas the same woman climbing a mountain needs 3.4×10^3 kcal/day.

(a) Express these numbers in terms of joules per second per kilogram of body mass (= watts per kilogram).

(b) Which consumes more power (watts), the office worker or a 100-W light bulb?

1-9. How many joules per second (J/s) are used by a device that requires 5.00×10^3 British thermal units per hour (Btu/h)? How many watts (W) does this device use?

1-10. The table shows fuel efficiency for several automobiles.

Car model	Fuel consumption (L/100 km)	CO_2 emission (g CO_2/km)
Gasoline engine		
Peugeot 107	4.6	109
Audi Cabriolet	11.1	266
Chevrolet Tahoe	14.6	346
Diesel engine		
Peugeot 107	4.1	109
Audi Cabriolet	8.4	223

SOURCE: M. T. Oliver-Hoyo and G. Pinto, "Using the Relationship Between Vehicle Fuel Consumption and CO₂ Emissions to Illustrate Chemical Principles," J. Chem. Ed. **2008**, 85, 218.

(a) A mile is 5 280 feet and a foot is 12 inches. Use Table 1-4 to find how many miles are in 1 km.

(b) The gasoline-engine Peugeot 107 consumes 4.6 L of fuel per 100 km. Express the fuel efficiency in miles per gallon. A U.S. liquid gallon is 3.785 4 L.

(c) The diesel Cabriolet is more efficient than the gasoline Cabriolet. How many metric tons of CO_2 are produced by the diesel and gasoline Cabriolets in 15 000 miles of driving? A metric ton is 1 000 kg.

1-11. Newton's law states that force = mass × acceleration. Acceleration has the units m/s². You also know that energy = force × distance and pressure = force/area. From these relations, derive the dimensions of newtons, joules, and pascals in terms of the fundamental SI units in Table 1-1. Check your answers in Table 1-2.

1-12. Dust falls on Chicago at a rate of 65 mg m^{-2} day^{-1}. Major metallic elements in the dust include Al, Mg, Cu, Zn, Mn, and Pb.[3] Pb accumulates at a rate of 0.03 mg m^{-2} day^{-1}. How many metric tons (1 metric ton = 1 000 kg) of Pb fall on the 535 square kilometers of Chicago in 1 year?

Chemical Concentrations

1-13. Define the following terms:

(a) molarity
(b) molality
(c) density
(d) weight percent
(e) volume percent
(f) parts per million
(g) parts per billion
(h) formal concentration

1-14. Why is it more accurate to say that the concentration of a solution of acetic acid is 0.01 F rather than 0.01 M? (Despite this distinction, we will usually write 0.01 M.)

1-15. What is the formal concentration (expressed as mol/L = M) of NaCl when 32.0 g are dissolved in water and diluted to 0.500 L?

1-16. How many grams of methanol (CH_3OH, FM 32.04) are contained in 0.100 L of 1.71 M aqueous methanol (i.e., 1.71 mol CH_3OH/L solution)?

1-17. (a) Figure 1-1 shows a peak O_3 concentration of 19 mPa in the stratosphere. Figure 1-3 shows a peak concentration of O_3 of 39 ppb at ground level at one particular location. To compare these concentrations, convert 39 ppb to pressure in mPa. Which concentration is higher? To convert ppb to mPa, assume that atmospheric pressure at the surface of the Earth is 1 bar ≡ 10^5 Pa. If atmospheric pressure is 1 bar, then a concentration of 1 ppb is 10^{-9} bar.

(b) The pressure of the atmosphere at 16 km altitude in the stratosphere is 9.6 kPa. The peak pressure of O_3 at this altitude in the stratosphere in Figure 1-1 is 19 mPa. Convert the pressure of 19 mPa to ppb when the atmospheric pressure is 9.6 kPa.

1-18. The concentration of a gas is related to its pressure by the *ideal gas law:*

$$\text{Concentration}\left(\frac{\text{mol}}{\text{L}}\right) = \frac{n}{V} = \frac{P}{RT}, \quad R = \text{gas constant} = 0.083\,14\frac{\text{L} \cdot \text{bar}}{\text{mol} \cdot \text{K}}$$

where n is the number of moles, V is volume (L), P is pressure (bar), and T is temperature (K).

(a) The maximum pressure of ozone in the Antarctic stratosphere in Figure 1-1 is 19 mPa. Convert this pressure into bars.

(b) Find the molar concentration of ozone in part **(a)** if the temperature is −70°C.

1-19. Noble gases (Group 18 in the periodic table) have the following volume concentrations in dry air: He, 5.24 ppm; Ne, 18.2 ppm; Ar, 0.934 vol%; Kr, 1.14 ppm; Xe, 87 ppb.

(a) A concentration of 5.24 ppm He means 5.24 μL of He per liter of air. Using the ideal gas law in Problem 1-18, find how many moles of He are contained in 5.24 μL of He at 25.00°C (298.15 K) and 1.000 bar. This number is the molarity of He in the air.

(b) Find the molar concentrations of Ar, Kr, and Xe in air at 25°C and 1 bar.

1-20. Any dilute aqueous solution has a density near 1.00 g/mL. Suppose the solution contains 1 ppm of solute; express the concentration of solute in g/L, μg/L, μg/mL, and mg/L.

1-21. The concentration of the alkane $C_{20}H_{42}$ (FM 282.56) in a particular sample of rainwater is 0.2 ppb. Assume that the density of rainwater is close to 1.00 g/mL and find the molar concentration of $C_{20}H_{42}$.

1-22. How many grams of perchloric acid, $HClO_4$, are contained in 37.6 g of 70.5 wt% aqueous perchloric acid? How many grams of water are in the same solution?

1-23. The density of 70.5 wt% aqueous perchloric acid is 1.67 g/mL. Recall that grams refer to grams of *solution* ($= g \ HClO_4 + g \ H_2O$).

(a) How many grams of solution are in 1.000 L?

(b) How many grams of $HClO_4$ are in 1.000 L?

(c) How many moles of $HClO_4$ are in 1.000 L?

1-24. An aqueous solution containing 20.0 wt% KI has a density of 1.168 g/mL. Find the molality (m, not M) of the KI solution.

1-25. A cell in your adrenal gland has about 2.5×10^4 tiny compartments called *vesicles* that contain the hormone epinephrine (also called adrenaline).

(a) An entire cell has about 150 fmol of epinephrine. How many attomoles (amol) of epinephrine are in each vesicle?

(b) How many molecules of epinephrine are in each vesicle?

(c) The volume of a sphere of radius r is $\frac{4}{3}\pi r^3$. Find the volume of a spherical vesicle of radius 200 nm. Express your answer in cubic meters (m^3) and liters, remembering that $1 \ L = 10^{-3} \ m^3$.

(d) Find the molar concentration of epinephrine in the vesicle if it contains 10 amol of epinephrine.

1-26. The concentration of sugar (glucose, $C_6H_{12}O_6$) in human blood ranges from about 80 mg/dL before meals to 120 mg/dL after eating. The abbreviation dL stands for deciliter $= 0.1$ L. Find the molarity of glucose in blood before and after eating.

1-27. An aqueous solution of antifreeze contains 6.067 M ethylene glycol ($HOCH_2CH_2OH$, FM 62.07) and has a density of 1.046 g/mL.

(a) Find the mass of 1.000 L of this solution and the number of grams of ethylene glycol per liter.

(b) Find the molality of ethylene glycol in this solution.

1-28. Protein and carbohydrates provide 4.0 Cal/g, whereas fat gives 9.0 Cal/g. (Remember that 1 Calorie, with a capital C, is really 1 kcal.) The weight percentages of these components in some foods are

Food	wt% protein	wt% carbohydrate	wt% fat
Shredded wheat	9.9	79.9	—
Doughnut	4.6	51.4	18.6
Hamburger (cooked)	24.2	—	20.3
Apple	—	12.0	—

Calculate the number of calories per gram and calories per ounce in each of these foods. (Use Table 1-4 to convert grams into ounces, remembering that there are 16 ounces in 1 pound.)

1-29. It is recommended that drinking water contain 1.6 ppm fluoride (F^-) to prevent tooth decay. Consider a cylindrical reservoir with a diameter of 4.50×10^2 m and a depth of 10.0 m. (The volume is $\pi r^2 h$, where r is the radius and h is the height.) How many grams of F^- should be added to give 1.6 ppm? Fluoride is provided by hydrogen hexafluorosilicate, H_2SiF_6. How many grams of H_2SiF_6 contain this much F^-?

Preparing Solutions

1-30. How many grams of boric acid, $B(OH)_3$ (FM 61.83), should be used to make 2.00 L of 0.050 0 M solution? What kind of flask is used to prepare this solution?

1-31. Describe how you would prepare approximately 2 L of 0.050 0 *m* boric acid, $B(OH)_3$.

1-32. What is the maximum volume of 0.25 M sodium hypochlorite solution (NaOCl, laundry bleach) that can be prepared by dilution of 1.00 L of 0.80 M NaOCl?

1-33. How many grams of 50 wt% NaOH (FM 40.00) should be diluted to 1.00 L to make 0.10 M NaOH? (Answer with two digits.)

1-34. A bottle of concentrated aqueous sulfuric acid, labeled 98.0 wt% H_2SO_4, has a concentration of 18.0 M.

(a) How many milliliters of reagent should be diluted to 1.000 L to give 1.00 M H_2SO_4?

(b) Calculate the density of 98.0 wt% H_2SO_4.

1-35. What is the density of 53.4 wt% aqueous NaOH (FM 40.00) if 16.7 mL of the solution diluted to 2.00 L give 0.169 M NaOH?

Stoichiometry Calculations

1-36. How many milliliters of 3.00 M H_2SO_4 are required to react with 4.35 g of solid containing 23.2 wt% $Ba(NO_3)_2$ if the reaction is $Ba^{2+} + SO_4^{2-} \rightarrow BaSO_4(s)$?

1-37. How many grams of 0.491 wt% aqueous HF are required to provide a 50% excess to react with 25.0 mL of 0.023 6 M Th^{4+} by the reaction $Th^{4+} + 4F^- \rightarrow ThF_4(s)$?

1-38. To entertain children between the ages of 2 and 90, I enjoy popping corks from bottles containing vinegar and baking soda. I pour about 50 mL of vinegar into a 500-mL plastic bottle. Then I wrap about 5 g of baking soda (which is sodium bicarbonate, $NaHCO_3$) in one layer of tissue and drop the tissue into the bottle. I place a cork tightly in the mouth of the bottle and step back. The chemical reaction generates $CO_2(g)$ that pressurizes the bottle and eventually bursts the cork into the air. Everyone smiles.

$$CH_3CO_2H + NaHCO_3 \rightarrow CH_3CO_2^- + Na^+ + CO_2(g) + H_2O$$

Acetic acid   Sodium bicarbonate
in vinegar   in baking soda

(a) Find the formula mass of acetic acid and of sodium bicarbonate.

(b) How many grams of acetic acid are required to react with 5 g of $NaHCO_3$?

(c) Vinegar contains ~5 wt% acetic acid. How many grams of vinegar are required to react with 5 g of $NaHCO_3$? The density of vinegar is close to 1.0 g/mL. How many mL of vinegar are required to react with 5 g of $NaHCO_3$?

(d) Which is the limiting reagent when you mix 50 mL of vinegar with 5 g of $NaHCO_3$?

(e) Use the ideal gas law (Problem 1-18) to calculate how many L of $CO_2(g)$ are generated if $P = 1$ bar and $T = 300$ K. If there is 0.5 L of air space in the bottle, what pressure can be generated to pop the cork?

2 | Tools of the Trade

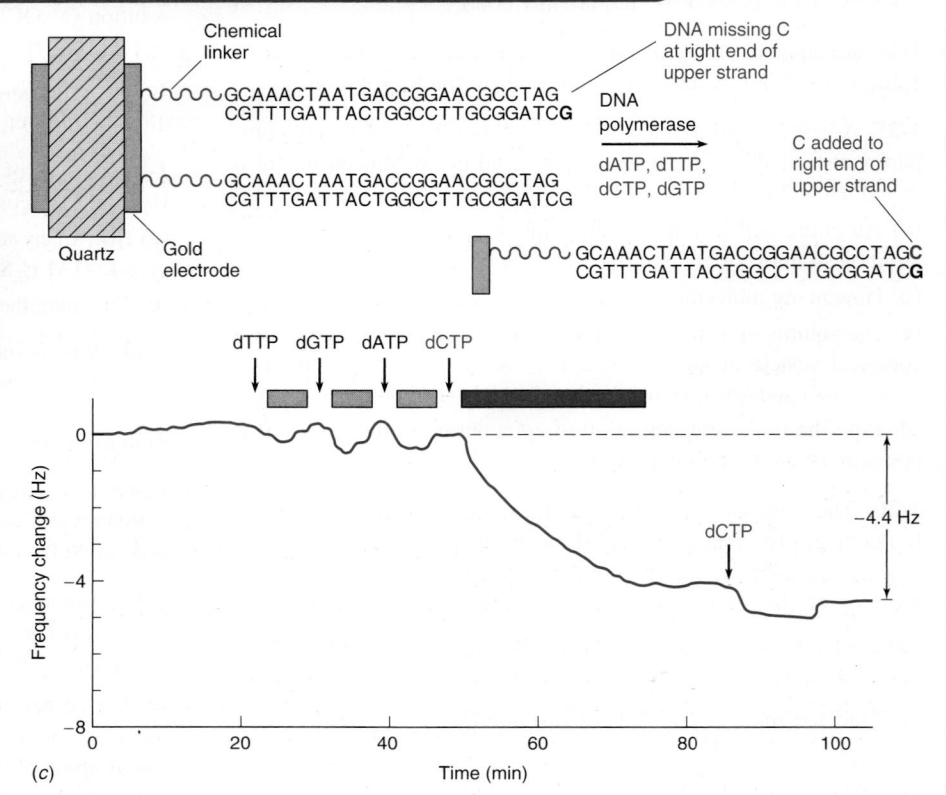

QUARTZ CRYSTAL MICROBALANCE MEASURES ONE BASE ADDED TO DNA

(a) Slice of quartz used to make (b) microbalance. [Photo courtesy LapTech Precision Inc.] (c) Response of piezoelectric quartz crystal to addition of nucleotides to a DNA template that can accommodate just one cytosine (C) on the end of the growing strand. DNA coverage on the gold is 20 ng/cm^2 = 1.2 pmol/cm^2. Stabilizing the temperature of the crystal and reagent solutions to ±0.001°C gives a noise of ±0.05 Hz. [Data from H. Yoshimine, T. Kojima, H. Furusawa, and Y. Okahata, "Small Mass-Change Detectable Quartz Crystal Microbalance and Its Application to Enzymatic One-Base Elongation on DNA," *Anal. Chem.* **2011,** *83*, 8741.]

A substance, such as quartz, whose dimensions change when an electric field is applied is said to be **piezoelectric.**

A quartz crystal stimulated to vibrate at its resonant frequency by an oscillating electric field keeps accurate time in your wristwatch. A quartz crystal microbalance, capable of weighing nanograms, consists of a slice of quartz sandwiched between two thin gold electrodes.[1, 2] When additional mass is attached to the surface of the gold electrodes, the oscillation frequency decreases in proportion to the bound mass.[3] Binding of 0.62 ng (nanograms, 10^{-9} g) to a 1 cm^2 area of a gold electrode lowers the 27-MHz resonant frequency of the crystal by an observable 1 Hz.

Deoxyribonucleic acid (DNA), whose structure is described in Appendix L, carries genetic information encoded in the sequence of four nucleotide bases designated A, T, C, and G. In the double-helical structure of DNA, A and T are always hydrogen bonded to each other and C and G are always hydrogen bonded to each other. To replicate DNA, the enzyme *DNA polymerase* uses a single DNA template strand and the building blocks dATP, dTTP, dCTP, and dGTP to stitch together a new, complementary DNA strand. T on the template specifies that A will be used in the corresponding position of the new strand. Similarly, A specifies T, C specifies G, and G specifies C.

With a quartz crystal microbalance, it is possible to measure the addition of one nucleotide to a growing chain of DNA. DNA is chemically anchored to a gold electrode of the piezoelectric crystal. When the correct nucleotide is added, the mass of the DNA increases and the oscillation frequency of the quartz crystal decreases. In panel *c*, double-stranded DNA is missing C at the right end of the upper chain. The graph shows changes in the crystal oscillation frequency when nucleotide building blocks are added in the presence of DNA polymerase. Only small changes in frequency are observed upon addition of dTTP, dGTP, and dATP. These changes are reversed when the reagents are washed away. However, addition of dCTP near 50 minutes causes a larger, irreversible change when C is covalently attached to the terminal position. Addition of more dCTP near 85 minutes has no further effect. The observed frequency change of −4.4 Hz is consistent with the additional mass of a single nucleotide to the DNA.

This chapter describes basic laboratory apparatus and manipulations associated with gravimetric and volumetric "wet" chemical measurements.[4] We also introduce spreadsheets, which are essential to everyone who manipulates quantitative data.

2-1 Safe, Ethical Handling of Chemicals and Waste

Chemical experimentation, like driving a car or operating a household, creates hazards. *The primary safety rule is to evaluate the hazards and then avoid what you (or your instructor or supervisor) deem to be dangerous.* If you believe that an operation is hazardous, discuss it first and do not proceed until sensible precautions are in place.

Before working, familiarize yourself with safety features of your laboratory.[5] You should wear goggles (Figure 2-1) or safety glasses with side shields at all times in the lab to protect your eyes from liquids and glass, which fly around when least expected. Contact lenses are not recommended in the lab because vapors can be trapped between the lens and your eye. Protect your skin from spills and flames by wearing a flame-resistant lab coat. Use rubber gloves when pouring concentrated acids. Do not eat or drink in the lab.

Organic solvents, concentrated acids, and concentrated ammonia should be handled in a fume hood. Air flowing into the hood keeps fumes out of the lab and dilutes the fumes before expelling them from the roof. Never generate large quantities of toxic fumes that are allowed to escape through the hood. Wear a respirator when handling fine powders, which could produce a cloud of dust that might be inhaled.

Clean up spills immediately to prevent accidental contact by the next person who comes along. Treat spills on your skin first by flooding with water. In anticipation of splashes on your body or in your eyes, know where to find and how to operate the emergency shower and eyewash. If the sink is closer than an eyewash, use the sink first for splashes in your eyes. Know how to operate the fire extinguisher and how to use an emergency blanket to extinguish burning clothes. A first-aid kit should be available, and you should know how and where to seek emergency medical assistance.

Label all vessels to indicate what they contain. An unlabeled bottle left and forgotten in a refrigerator or cabinet presents an expensive disposal problem, because the contents must be analyzed before they can be legally discarded. A Material Safety Data Sheet provided with each chemical sold in the United States lists hazards and precautions for that chemical. It gives first-aid procedures and instructions for handling spills.

If we want our grandchildren to inherit a habitable planet, we need to minimize waste production and dispose of chemical waste responsibly. When it is economically feasible, recycling of chemicals is preferable to waste disposal.[6] Carcinogenic dichromate ($Cr_2O_7^{2-}$) waste provides an example of an accepted disposal strategy. Cr(VI) from dichromate should be reduced to the less toxic Cr(III) with sodium hydrogen sulfite ($NaHSO_3$) and precipitated with hydroxide as insoluble $Cr(OH)_3$. The solution is evaporated to dryness and the solid is discarded in an approved landfill that is lined to prevent escape of the chemicals. Wastes such as silver and gold that can be economically recycled should be chemically treated to recover the metal.[7]

Green chemistry is a set of principles intended to change our behavior to help sustain a habitable planet.[8] Examples of unsustainable behavior are consumption of limited resources and careless disposal of waste. Green chemistry seeks to design chemical products and processes to reduce the use of resources and energy and the generation of hazardous waste. It is better to design a process to prevent waste than to dispose of waste. For example, NH_3 can be measured with an ion-selective electrode instead of using the spectrophotometric Nessler procedure, which generates HgI_2 waste. "Microscale" classroom experiments are encouraged to reduce the cost of reagents and the generation of waste.

2-2 The Lab Notebook

The critical functions of your lab notebook are to state *what you did* and *what you observed*, and it should be *understandable by a stranger.* The greatest error, made even by experienced scientists, is writing incomplete or unintelligible notes. Using *complete sentences* is an excellent way to prevent incomplete descriptions.

Beginning students often find it useful to write a complete description of an experiment, with sections dealing with purpose, methods, results, and conclusions. Arranging a notebook to accept numerical data prior to coming to the lab is an excellent way to prepare for an experiment. It is good practice to write a balanced chemical equation for every reaction you use.

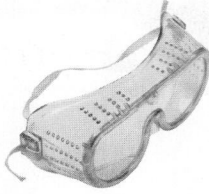

FIGURE 2-1 Goggles or safety glasses with side shields should be worn at all times in every lab. [©Stockbyte/Getty Images.]

Why we wear lab coats

In 2008, 23-year-old University of California Los Angeles research assistant Sheharbano Sangji was withdrawing *t*-butyllithium solution with a syringe from a bottle. She was not wearing a lab coat. The plunger came out of the syringe and the pyrophoric liquid burst into flames, igniting her sweater and gloves. Burns on 40% of her body proved fatal. A flame-resistant lab coat might have protected her. In 2014, a plea bargain was reached in the criminal case brought against the professor in whose lab the accident occurred for "willfully violating health and safety standards."

Limitations of gloves

In 1997, popular Dartmouth College chemistry professor Karen Wetterhahn, age 48, died from a drop of dimethylmercury absorbed through the latex rubber gloves she was wearing. Many organic compounds readily penetrate rubber. Wetterhahn was an expert in the biochemistry of metals and the first female professor of chemistry at Dartmouth. She was a mother of two children and played a major role in bringing more women into science and engineering.

Electronics should be taken to a collection center for recycling, rather than be sent to a landfill for burial. *Fluorescent bulbs must not be discarded as ordinary waste because they contain mercury.* Light-emitting diodes (LEDs) are even more efficient than fluorescent lamps, contain no mercury, and will eventually replace fluorescent lights.

The **lab notebook** should
1. State what was done
2. State what was observed
3. Be understandable to someone else

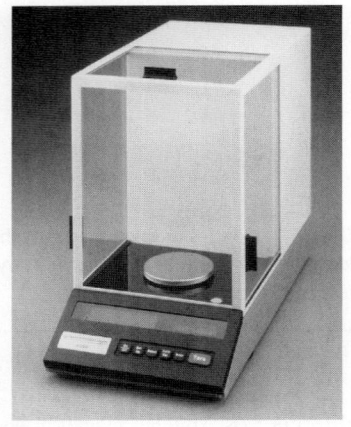

FIGURE 2-2 Electronic analytical balance measures mass down to 0.1 mg. [Courtesy Thermo Fisher Scientific Inc.]

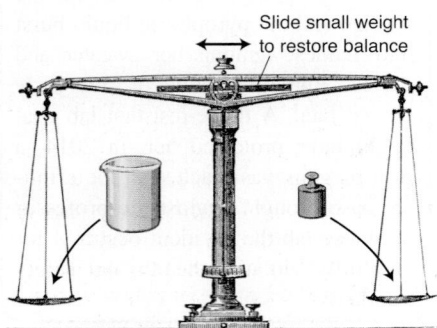

Slide small weight to restore balance

FIGURE 2-3 Equal-long-arm nineteenth-century balance. [Reproduced from *Fresenius' Quantitative Chemical Analysis,* 2nd American ed., 1881.]

This practice helps you understand what you are doing and may point out what you do not understand about what you are doing.

The measure of scientific "truth" is the ability of different people to reproduce an experiment. A good lab notebook will state everything that was done and what you observed and will allow you or anyone else to repeat the experiment.

Record in your notebook the names of computer files where programs and data are stored. Paste hard copies of important data into your notebook. The lifetime of a printed page is 10 to 100 times longer than that of a computer file.

2-3 Analytical Balance

An *electronic balance* uses electromagnetic force compensation to balance the load on the pan. Figure 2-2 shows a typical analytical balance with a capacity of 100–200 g and a readability of 0.01–0.1 mg. *Readability* is the smallest increment of mass that can be indicated. A *microbalance* weighs milligram quantities with a readability of 1μg.

To weigh a chemical, first place a clean receiving vessel on the balance pan. The mass of the empty vessel is called the **tare.** On most balances, you can press a button to reset the tare to 0. Add the chemical to the vessel and read its mass. If there is no automatic tare operation, subtract the tare mass from that of the filled vessel. To protect the balance from corrosion, *chemicals should never be placed directly on the weighing pan.* Also, be careful not to spill chemicals into the mechanism below the balance pan.

An alternative procedure, called *weighing by difference,* is necessary for **hygroscopic** reagents, which rapidly absorb moisture from the air. First weigh a capped bottle containing dry reagent. Then quickly pour some reagent from the weighing bottle into a receiver. Cap the weighing bottle and weigh it again. The difference is the mass of reagent delivered from the weighing bottle. With an electronic balance, set the initial mass of the weighing bottle to zero with the tare button. Then deliver reagent from the bottle and reweigh the bottle. The negative reading on the balance is the mass of reagent delivered from the bottle.[9]

The classic *mechanical balance* in Figure 2-3 has two pans suspended on opposite ends of an equal-arm lever balanced at its center on a knife edge. An unknown mass is placed on the left hand pan and standard masses are placed on the right until the balance is nearly restored to its level position. Then a small mass is moved along the horizontal slide until balance is perfectly restored. The sum of the standard mass plus the mass on the slide is equal to the unknown mass. A mechanical balance should be in its arrested position (in which the balance is prevented from moving) when you load or unload the pans and in the half-arrested position when you use the sliding mass. Arresting the balance minimizes wear on the knife edge that supports the balance beam.

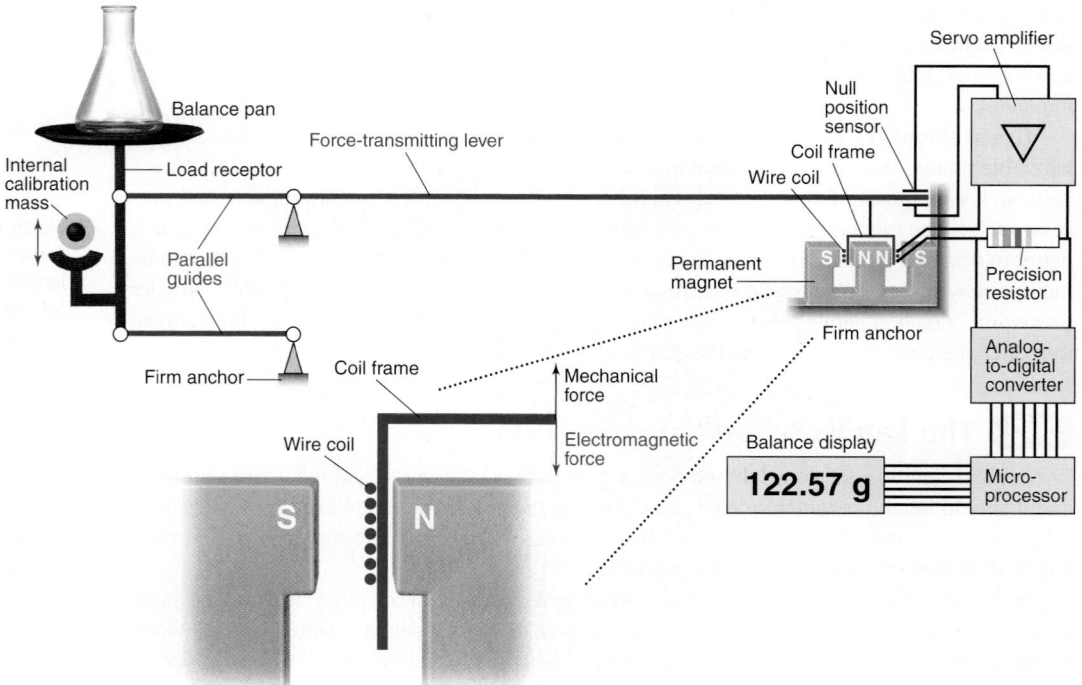

FIGURE 2-4 Schematic diagram of electronic balance. [Information from C. Berg, *The Fundamentals of Weighing Technology* (Göttingen, Germany: Sartorius AG, 1996).]

How an Electronic Balance Works

An object placed on the electronic balance in Figure 2-2 pushes the pan down with a force $m \times g$, where m is the mass of the object and g is the acceleration of gravity. The balance generates an electric current to exactly cancel the motion of the pan. The magnitude of the current tells us how much mass was placed on the pan.

Figure 2-4 shows how the balance works. The pan is located on the short side of a lever. The weight of the sample pushes the left side of the lever down and moves the right side of the lever up. The null position sensor on the far right detects the smallest movement of the lever arm away from its equilibrium (null) position. When the null sensor detects displacement of the lever arm, the servo amplifier sends electric current through the force compensation wire coil in the field of a permanent magnet. The enlargement at the lower left shows part of the coil and magnet. Electric current in the coil interacts with the permanent magnet to produce a downward force. The servo amplifier provides current that exactly compensates for the upward force on the lever arm to maintain a null position. Current flowing through the coil creates a voltage across the precision resistor, which is converted to a digital signal and ultimately to a readout in grams. The conversion between current and mass is accomplished by measuring the current required to balance an internal calibration mass. Figure 2-5 shows the layout of components inside a balance.

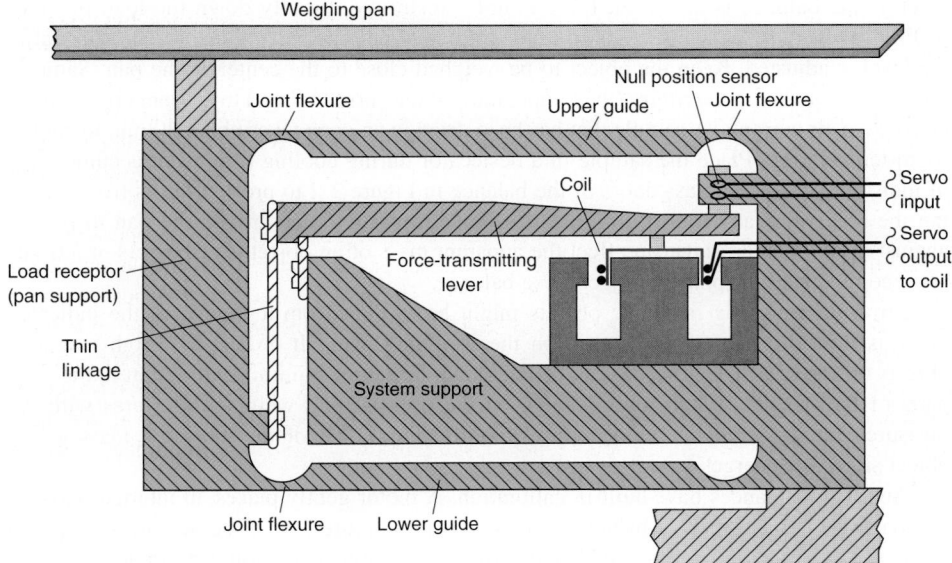

(a)

(b)

FIGURE 2-5 (*a*) Mechanical layout of an electronic balance. The lever ratio is such that the electromagnetic force must be only ~10% of the load on the pan. [Information from C. Berg, *The Fundamentals of Weighing Technology* (Göttingen, Germany: Sartorius AG, 1996).] (*b*) Internal components of Sartorius analytical balance with a capacity of 300 g and readability of 0.1 mg. The monolithic (one-piece) metal weighing system has nonmagnetic calibration weights, which are placed gently on the load receptor by a microprocessor-activated motor. Calibration is automatically activated by temperature changes. [Courtesy J. Barankewitz, Sartorius AG, Göttingen, Germany.]

TABLE 2-1	Tolerances for laboratory balance weights[a]					
Denomination	Tolerance (mg)			Denomination	Tolerance (mg)	
Grams	Class 1	Class 2		Milligrams	Class 1	Class 2
500	1.2	2.5		500	0.010	0.025
200	0.50	1.0		200	0.010	0.025
100	0.25	0.50		100	0.010	0.025
50	0.12	0.25		50	0.010	0.014
20	0.074	0.10		20	0.010	0.014
10	0.050	0.074		10	0.010	0.014
5	0.034	0.054		5	0.010	0.014
2	0.034	0.054		2	0.010	0.014
1	0.034	0.054		1	0.010	0.014

a. Tolerances are defined in ASTM (American Society for Testing and Materials) Standard E 617. Classes 1 and 2 are the most accurate. Larger tolerances exist for Classes 3–6, which are not given in this table.

Weighing Errors

An analytical balance should be located on a heavy table, such as a marble slab, to minimize vibrations. The balance has adjustable feet and a bubble meter that allow you to keep it level. If the balance is not level, force is not transmitted directly down the load receptor in Figure 2-5 and an error results. Press the calibrate button to recalibrate the balance after the level is adjusted. Keep the object to be weighed close to the center of the pan. Samples must be at *ambient temperature* (the temperature of the surroundings) to prevent errors due to convective air currents. A sample that has been dried in an oven takes about 30 min to cool to room temperature. Place the sample in a desiccator during cooling to prevent accumulation of moisture. Close the glass doors of the balance in Figure 2-2 to prevent drafts from affecting the reading. Many top-loading balances have a plastic fence around the pan to protect against drafts. Fingerprints can affect the apparent mass of an object, so tweezers or a tissue are recommended for placing objects on a balance.

Errors in weighing magnetic objects might be evident from a change of the indicated mass as the object is moved around on the weighing pan.[10] It is best to weigh magnetic objects on top of a spacer such as an upside-down beaker to minimize attraction to the steel parts of the balance. Electrostatic charge on the object being weighed interferes with the measurement and could be evident from unidirectional drift of the indicated mass as the object slowly discharges.

Analytical balances have built-in calibration. A motor gently places an internal mass on the load receptor beneath the balance pan, as shown in Figure 2-5b. Electric current required to balance this mass is measured. For external calibration, you should periodically weigh standard masses and verify that the reading is within allowed limits. Table 2-1 lists *tolerances* (allowable deviations) for standard masses. Another test for a balance is to weigh a standard mass six times and calculate the standard deviation (Section 4-1). Variations are partly from the balance but also reflect factors such as drafts and vibrations.

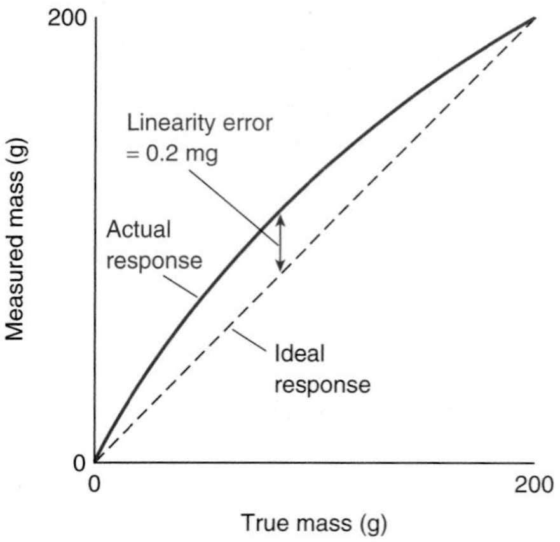

Linearity error (or *linearity*) of a balance is the maximum error that can occur as a result of nonlinear response of the system to added mass after the balance has been calibrated (Figure 2-6). A balance with a capacity of 220 g and a readability of 0.1 mg might have a linearity of ± 0.2 mg. Even though the scale can be read to 0.1 mg, the error in mass can be as large as ± 0.2 mg in some part of the allowed range.

After a balance is calibrated, the reading might drift if room temperature changes. If a balance has a temperature coefficient of sensitivity of 2 ppm/°C, and temperature changes by 4°C, the apparent mass will change by (4°C)(2 ppm/°C) = 8 ppm. For a mass of 100 g, 8 ppm is $(100 \text{ g})(8 \times 10^{-6}) = 0.8$ mg. You can recalibrate the balance at its current temperature by pressing the calibrate button. For temperature stability, leave a balance in standby mode when not in use.

Buoyancy

You can float in water because your weight when swimming is nearly zero. **Buoyancy** is the upward force exerted on an object in a liquid or gaseous fluid.[11] An object weighed in air appears lighter than its actual mass by an amount equal to the mass of air that it displaces. True mass is the mass measured in vacuum. A standard mass in a balance is also affected by

FIGURE 2-6 *Linearity error.* Dashed line is ideal response proportional to mass on the balance, which has been calibrated at 0 and 200 g. Actual response deviates from the straight line. Linearity error is the maximum deviation, which is greatly exaggerated in this drawing.

buoyancy, so it weighs less in air than in vacuum. A buoyancy error occurs whenever the density of the object being weighed is not equal to the density of the standard mass.

If mass m' is read on a balance, the true mass m of the object weighed in vacuum is[12]

Buoyancy equation:
$$m = \frac{m'\left(1 - \dfrac{d_a}{d_w}\right)}{\left(1 - \dfrac{d_a}{d}\right)}$$
(2-1)

Equation 2-1 applies to mechanical and electronic balances.

where d_a is the density of air (0.001 2 g/mL near 1 bar and 25°C),[13] d_w is the density of the calibration weights (8.0 g/mL), and d is the density of the object being weighed.

EXAMPLE Buoyancy Correction

A pure compound called "tris" is used as a *primary standard* to measure concentrations of acids. The volume of acid that reacts with a known mass of tris tells us the concentration of acid. Find the true mass of tris (density = 1.33 g/mL) if the apparent mass weighed in air is 100.00 g.

Solution If the density of air is 0.001 2 g/mL, we find the true mass by using Equation 2-1:

$$m = \frac{100.00 \text{ g}\left(1 - \dfrac{0.001\ 2 \text{ g/mL}}{8.0 \text{ g/mL}}\right)}{1 - \dfrac{0.001\ 2 \text{ g/mL}}{1.33 \text{ g/mL}}} = 100.08 \text{ g}$$

Unless we correct for buoyancy, we would think that the mass of tris is 0.08% less than its actual mass and we would think that the molarity of acid reacting with the tris is 0.08% less than the actual molarity.

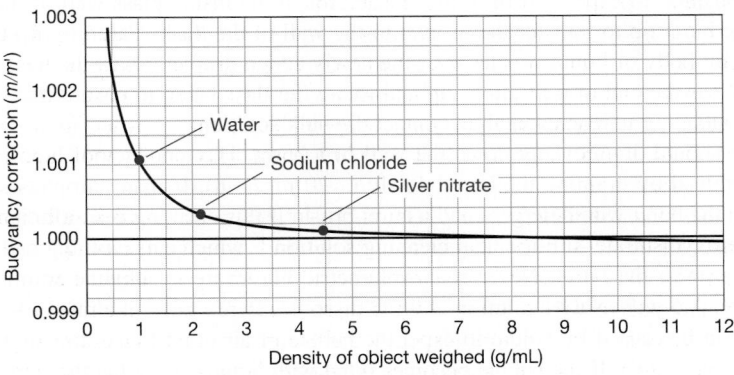

FIGURE 2-7 Buoyancy correction, assuming d_a = 0.001 2 g/mL and d_w = 8.0 g/mL. The apparent mass measured in air (1.000 0 g) is multiplied by the buoyancy correction to find the true mass.

Figure 2-7 shows buoyancy corrections for several substances. When you weigh water with a density of 1.00 g/mL, the true mass is 1.001 1 g when the balance reads 1.000 0 g. The error is 0.11%. For NaCl with a density of 2.16 g/mL, the error is 0.04%; for $AgNO_3$ with a density of 4.45 g/mL, the error is only 0.01%.

2-4 Burets

The **buret** in Figure 2-8 is a precisely manufactured glass tube with graduations enabling you to measure the volume of liquid delivered through the stopcock (the valve) at the bottom. The 0-mL mark is near the top. If the initial liquid level is 0.83 mL and the final level is 27.16 mL, then you have delivered 27.16 − 0.83 = 26.33 mL. Class A burets (the most accurate grade) are certified to meet tolerances in Table 2-2. If the reading of a 50-mL buret is 27.16 mL, the true volume can be anywhere in the range 27.21 to 27.11 mL and still be within the tolerance of ±0.05 mL.

Even though the tolerance of the buret may be ±0.05 mL, you should interpolate readings to 0.01 mL. Never throw away precision in reading an instrument.

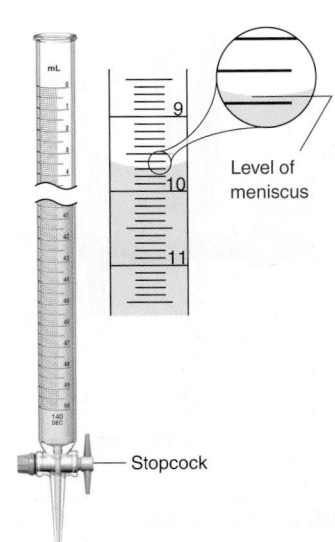

FIGURE 2-8 Glass buret with Teflon® stopcock. Enlargement shows meniscus at 9.68 mL. Estimate the reading of any scale to the nearest tenth of a division. This buret has 0.1-mL divisions, so we estimate the reading to the nearest 0.01 mL.

TABLE 2-2	Tolerances of Class A burets	
Buret volume (mL)	Smallest graduation (mL)	Tolerance (mL)
5	0.01	±0.01
10	0.05 *or* 0.02	±0.02
25	0.1	±0.03
50	0.1	±0.05
100	0.2	±0.10

Operating a buret:
- Wash buret with new solution
- Eliminate air bubble before use
- Drain liquid slowly
- Deliver a fraction of a drop near end point
- Read bottom of concave meniscus
- Estimate reading to 1/10 of a division
- Avoid parallax
- Account for graduation thickness in readings

In a **titration,** increments of reagent in the buret are added to analyte until reaction is complete. From the volume delivered, we calculate the quantity of analyte.

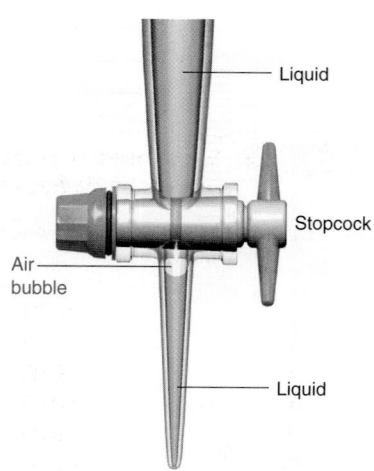

FIGURE 2-9 An air bubble trapped beneath the stopcock should be expelled before you use the buret.

Precision refers to reproducibility.

Relative uncertainty is the uncertainty in a quantity divided by the magnitude of the quantity. We usually express relative uncertainty as a percentage:

Relative uncertainty: $$\text{Relative uncertainty (\%)} = \frac{\text{uncertainty in quantity}}{\text{magnitude of quantity}} \times 100 \qquad (2\text{-}2)$$

If you deliver 20 mL from a 50-mL buret, the relative uncertainty is (0.05 mL/20 mL) × 100 = 0.25%. If you deliver 40 mL from the same buret, the relative uncertainty is (0.05 mL/40 mL) × 100 = 0.12%. We reduce relative uncertainty by delivering a larger volume from the buret. You can reduce the uncertainty for a given buret by calibrating it as described on page 45.

When reading the liquid level in a buret, your eye should be at the same height as the top of the liquid. If your eye is too high, the liquid seems to be higher than it really is. If your eye is too low, the liquid appears too low. The error that occurs when your eye is not at the same height as the liquid is called **parallax.**

The surface of most liquids forms a concave **meniscus** like that shown at the right side of Figure 2-8.[14] It is helpful to use black tape on a white card as a background for locating the precise position of the meniscus. Move the black strip up the buret to approach the meniscus. The bottom of the meniscus turns dark as the black strip approaches, thus making the meniscus more easily readable. Highly colored solutions may appear to have two meniscuses; either one may be used. Because volumes are determined by subtracting one reading from another, the important point is to read the level of the meniscus reproducibly. Estimate the reading to the nearest tenth of a division (such as 0.01 mL) between marks.

The thickness of the markings on a 50-mL buret corresponds to about 0.02 mL. For best accuracy, select one part of the marking as zero. For example, you can say that the liquid level is *at* the mark when the bottom of the meniscus just touches the top of the mark. When the meniscus is at the *bottom* of the same mark, the reading is 0.02 mL greater.

The solution in the buret is called the **titrant.** For precise location of the end of a *titration,* deliver less than one drop of titrant at a time near the end point. (A drop from a 50-mL buret is about 0.05 mL.) To deliver a fraction of a drop, carefully open the stopcock until part of a drop is hanging from the buret tip. (Some people prefer to rotate the stopcock rapidly through the open position to expel part of a drop.) Then touch the inside glass wall of the receiving flask to the buret tip to transfer the droplet to the wall of the flask. Carefully tip the flask so that the main body of liquid washes over the newly added droplet. Swirl the flask to mix the contents. Near the end of a titration, tip and rotate the flask often to ensure that droplets on the wall containing unreacted analyte contact the bulk solution.

Liquid should drain evenly down the wall of a buret. The tendency of liquid to stick to glass is reduced by draining the buret slowly (<20 mL/min). If many droplets stick to the wall, clean the buret with detergent and a buret brush. If this cleaning is insufficient, soak the buret in peroxydisulfate–sulfuric acid cleaning solution,[15] which eats clothing and people, as well as grease in the buret. Never soak volumetric glassware in alkaline solutions, which attack glass. A 5 wt% NaOH solution at 95°C dissolves Pyrex glass at a rate of 9 μm/h.

Error can be caused by failure to expel the bubble of air often found directly beneath the stopcock (Figure 2-9). If the bubble becomes filled with liquid during the titration, then some volume that drained from the graduated portion of the buret did not reach the titration vessel. The bubble can be dislodged by draining the buret for a second or two with the stopcock wide open. You can expel a tenacious bubble by abruptly shaking the buret while draining it into a sink.

Before you fill a buret with fresh solution, it is a wonderful idea to rinse the buret several times with small portions of the new solution, discarding each wash. It is not necessary to fill the buret with wash solution. Add a small volume of wash solution to the buret and tilt the buret to allow all surfaces to contact the liquid. This same technique should be used with any vessel (such as a spectrophotometer cuvet or a pipet) that is reused without drying.

The labor of conducting a titration is greatly reduced by using an autotitrator (Figure 2-10) instead of a buret. This device delivers reagent from a reservoir and records the volume of reagent and the response of an electrode immersed in the solution being titrated. Output can go directly to a computer for manipulation in a spreadsheet.

Mass Titrations and Microscale Titrations

Mass can be measured more precisely than volume. For best precision in a titration, measure the *mass* of solution, instead of the volume, delivered from a buret or syringe or pipet.[16] Exercise 7-B provides an example.

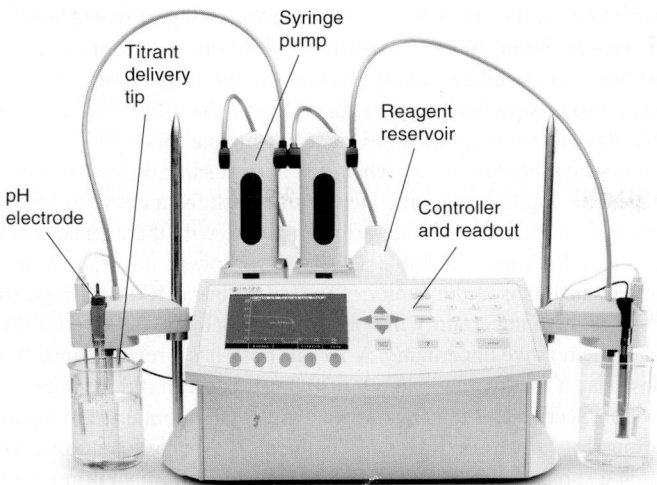

Titrant delivery tip

Syringe pump

pH electrode

Reagent reservoir

Controller and readout

FIGURE 2-10 Autotitrator delivers reagent from a reservoir bottle into a beaker containing analyte. The electrode immersed in the beaker monitors pH or concentrations of specific ions. Volume and pH readings can be exported to a spreadsheet. [Courtesy Hanna Instruments.]

For procedures that can tolerate poor precision, "microscale" student experiments reduce consumption of reagents and generation of waste. An inexpensive student buret can be constructed from a 2-mL pipet graduated in 0.01-mL intervals.[17] Volume can be read to 0.001 mL, and titrations can be carried out with a precision of 1%.

2-5 Volumetric Flasks

A **volumetric flask** is calibrated to contain a particular volume of solution at 20°C when the bottom of the meniscus is adjusted to the center of the mark on the neck of the flask (Figure 2-11, Table 2-3). Most flasks bear the label "TC 20°C," which means *to contain* at 20°C. (Pipets and burets are calibrated *to deliver*, "TD," their indicated volume.) The temperature of the container is relevant because both liquid and glass expand when heated.

To use a volumetric flask, dissolve the desired mass of reagent in the flask by swirling with *less* than the final volume of liquid. Then add more liquid and swirl the solution again. Adjust the final volume with as much well-mixed liquid in the flask as possible. (When two different liquids are mixed, there is generally a small volume change. The total volume is *not* the sum of the two volumes that were mixed. By swirling the liquid in a nearly full volumetric flask before liquid reaches the thin neck, you minimize the change in volume when the

Volumetric glassware made of Pyrex, Kimax, or other low-expansion glass can be safely dried in an oven heated to at least 320°C without harm,[18] although there is rarely reason to go above 150°C.

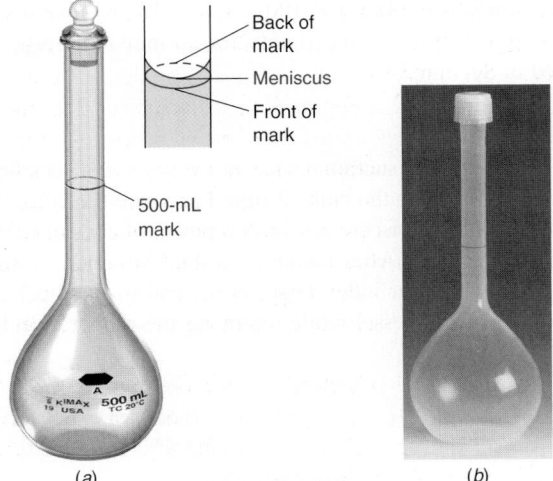

Back of mark

Meniscus

Front of mark

500-mL mark

(a)

(b)

FIGURE 2-11 (a) Class A glass volumetric flask showing proper position of the meniscus—at the center of the ellipse formed by the front and back of the calibration mark when viewed from above or below. Volumetric flasks and transfer pipets are calibrated to this position. (b) VITLAB® Class A perfluoroalkoxy copolymer (PFA) plastic volumetric flask for trace analysis. PFA is stable from −200° to +260°C, resists common acids, and has low levels of leachable metals. [Courtesy BrandTech® Scientific, Essex, CT.]

TABLE 2-3	Tolerances of Class A volumetric flasks[a]
Flask capacity (mL)	**Tolerance (mL)**
1	±0.02
2	±0.02
5	±0.02
10	±0.02
25	±0.03
50	±0.05
100	±0.08
200	±0.10
250	±0.12
500	±0.20
1 000	±0.30
2 000	±0.50

a. Class B glassware tolerances are twice as large as for Class A.

last liquid is added.) For good control, add the final drops of liquid with a pipet, *not a squirt bottle*. After adjusting the liquid to the correct level, hold the cap firmly in place and invert the flask about 10 times to complete mixing. Before the liquid is homogeneous, we observe streaks (called *schlieren*) arising from regions that refract light differently. After the schlieren are gone, invert the flask a few more times to ensure complete mixing.

Figure 2-11 shows how liquid appears when it is at the *center* of the mark of a volumetric flask or a pipet. Adjust the liquid level while viewing the flask from above or below the level of the mark. The front and back of the mark describe an ellipse with the meniscus at the center.

The manufacturer's tolerance in Table 2-3 is the allowed uncertainty in the volume contained when the meniscus is at the center of the mark. For a 100-mL flask, the volume is 100 ± 0.08 mL. The relative uncertainty in volume is $(0.08/100) \times 100 = 0.08\%$. The larger the flask, the smaller the relative uncertainty. A 10-mL flask has a relative uncertainty of 0.2%, but a 1000-mL flask has a relative uncertainty of 0.03%. You can reduce uncertainty by calibration as described in Section 2-9 to measure what is actually contained in a particular flask.

Glass is notorious for *adsorbing* traces of chemicals—especially cations. **Adsorption** is the process in which a substance sticks to a surface. (In contrast, **absorption** is the process in which a substance is taken inside another, as water is taken into a sponge.) For critical work, you should **acid wash** glassware to replace low concentrations of cations on the surface with H^+. To do this, soak already thoroughly cleaned glassware in 3–6 M HCl or HNO_3 (in a fume hood) for >1 h. Then rinse it well with distilled water and, finally, soak it in distilled water. Acid can be reused many times, as long as it is only used for clean glassware. Acid washing is *especially* appropriate for new glassware, which you should assume is not clean. A high-density polyethylene or polypropylene or perfluoralkoxy (PFA) plastic volumetric flask (Figure 2-11b) is preferred for trace analysis (parts per billion concentrations) in which cations might be lost by adsorption on the walls of a glass flask.

When collecting and storing samples such as natural waters for trace analysis, bottles made of plastics such as high-density polyethylene are recommended for ionic analytes so that traces of analyte are not lost by adsorption on the glass surface or contaminated by metal leached from the glass surface. By contrast, aqueous samples to be analyzed for part-per-trillion (pg/g) traces of organic materials, such as pharmaceuticals, personal care products, and steroids, are best collected and stored in amber (dark) glass bottles, not plastic bottles.[20]

2-6 Pipets and Syringes

Pipets deliver known volumes of liquid. The *transfer pipet* in Figure 2-12a is calibrated to deliver one fixed volume. The last drop does not drain out of the pipet and *should not be blown out*. The *measuring pipet* in Figure 2-12b is calibrated like a buret. It is used to deliver a variable volume such as 5.6 mL by starting delivery at the 1.0-mL mark and terminating at the 6.6-mL mark. The transfer pipet is more accurate, with tolerances listed in Table 2-4. The larger the transfer pipet, the smaller is its relative uncertainty. The relative uncertainty in a 1-mL pipet is $(0.006/1) \times 100 = 0.6\%$. The relative uncertainty in a 25 mL pipet is $(0.03/25) \times 100 = 0.12\%$. Uncertainty for an individual pipet can be reduced by calibration described in Section 2-9.

Using a Transfer Pipet

Using a rubber bulb or other pipet suction device, *not your mouth*, suck liquid up slightly past the calibration mark. (But not into the bulb! If liquid gets into the bulb, start over again with fresh solution and a fresh bulb.) Discard one or two pipet volumes of liquid to rinse traces of previous reagents from the pipet. After taking up a third volume past the calibration mark, quickly replace the bulb with your index finger at the end of the pipet. Gently pressing the pipet against the bottom of the vessel while removing the rubber bulb helps prevent liquid

Example of acid washing: High purity HNO_3 delivered from an acid-washed glass pipet had no detectable level of Ti, Cr, Mn, Fe, Co, Ni, Cu, and Zn (<0.01 ppb). The same acid delivered from a clean—but not acid washed—pipet contained each metal at a level of 0.5 to 9 ppb.[19]

Do not blow the last drop out of a transfer pipet.

TABLE 2-4	Tolerances of Class A transfer pipets
Volume (mL)	Tolerance (mL)
0.5	±0.006
1	±0.006
2	±0.006
3	±0.01
4	±0.01
5	±0.01
10	±0.02
15	±0.03
20	±0.03
25	±0.03
50	±0.05
100	±0.08

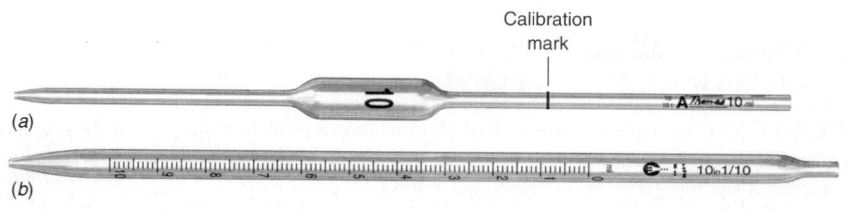

FIGURE 2-12 (*a*) Transfer pipet and (*b*) measuring (Mohr) pipet. [Courtesy A. H. Thomas Co., Philadelphia, PA.]

from draining below the mark while you put your finger in place. (A better alternative is to use an automatic suction device such as that in Figure 2-13 that remains attached to the pipet.) Wipe the excess liquid off the outside of the pipet with a clean tissue. *Touch the tip of the pipet to the side of a beaker* and drain the liquid until the bottom of the meniscus just reaches the center of the mark, as in Figure 2-11. Touching the beaker draws liquid from the pipet without leaving part of a drop hanging when the liquid reaches the calibration mark.

Transfer the pipet to a receiving vessel and drain it by gravity *while holding the tip against the wall of the vessel.* After liquid stops draining, hold the pipet to the wall for a few more seconds to complete draining. *Do not blow out the last drop.* The pipet should be nearly vertical at the end of delivery. When you finish using a pipet, rinse it with distilled water or soak it until you are ready to clean it. Solutions should never be allowed to dry inside a pipet because it is hard to remove dried internal deposits.

Serial Dilution

Serial dilution is the process of making successive dilutions to obtain a desired concentration of reagent. The purpose is to transfer accurately small amounts of material that are too little to weigh accurately. Here is an example of a process that you could use to prepare standards for instrumental analysis.

EXAMPLE Serial Dilution

You wish to prepare a solution containing 2.00 μg Cs/mL (really Cs$^+$) as a standard for atomic emission analysis. You have available 250-, 500-, and 1 000-mL volumetric flasks and 5-, 10-, and 25-mL transfer pipets. To obtain 4-digit weighing accuracy, you want to weigh out at least 1 g of pure CsCl with a balance that is accurate to the milligram decimal place. Design a procedure to use pure CsCl to prepare a concentrated stock solution from which you can make a series of dilutions to obtain 2.00 μg Cs/mL.

Solution One possible strategy is to weigh out enough CsCl to contain a convenient multiple of 2.00 μg of Cs and dissolve it in a volumetric flask. Then make a series of dilutions to bring the concentration down to 2.00 μg Cs/mL. For example, you could make a stock solution containing 1 000 μg Cs/mL. This solution is 500 times more concentrated than we want. If you make two successive dilutions by factors of 50 and 10, you will obtain the desired 500-fold dilution.

To make a liter of stock solution, you need (1 000 μg Cs/mL)(1 000 mL) = 1.000 g Cs. The atomic mass of Cs is 132.91 and the formula mass of CsCl is 168.36, so the mass of CsCl containing 1.000 g Cs is

$$(1.000 \text{ g Cs})\left(\frac{168.36 \text{ g CsCl/mol}}{132.91 \text{ g Cs/mol}}\right) = 1.267 \text{ g CsCl}$$

To make the stock solution containing 1 000 μg Cs/mL, weight out 1.267 g of CsCl (corrected for buoyancy) and dissolve it in a 1.000 L volumetric flask. In general, it is easier to get near the desired mass and measure the actual mass, such as 1.284 g instead of 1.267 g. With 1.284 g, the concentration will be 1 014 μg Cs/mL instead of 1 000 μg Cs/mL.

For a 50-fold dilution, you could transfer 10.00 mL of stock solution with a 10-mL transfer pipet into a 500-mL volumetric flask and dilute to volume. Call this solution B. Its concentration is (1 000 μg Cs/mL)/50 = 20.0 μg Cs/mL. For a 10-fold dilution of solution B, you could transfer 25.00 mL with a 25-mL transfer pipet into a 250-mL volumetric flask and dilute to volume. This final solution has the desired concentration of 2.00 μg Cs/mL. In Chapter 3 we will see how to use uncertainties in each measurement to find the number of significant digits in the final concentration.

You could accomplish the same dilutions in multiple ways. For example, you could make a 50-fold dilution by transferring 5.00 mL of solution into a 250-mL volumetric flask.

TEST YOURSELF How can you dilute solution B containing 20.0 mg Cs/mL to give three new solutions with 4, 3, and 1 μg Cs/mL?

(Answer: For 1 μg/mL, you could dilute 25 mL of solution B up to 500 mL. For 3 μg/mL, dilute (25 + 25 + 25) mL of solution B up to 500 mL. For 4 μg/mL, dilute (25+25) mL of solution B up to 250 mL.)

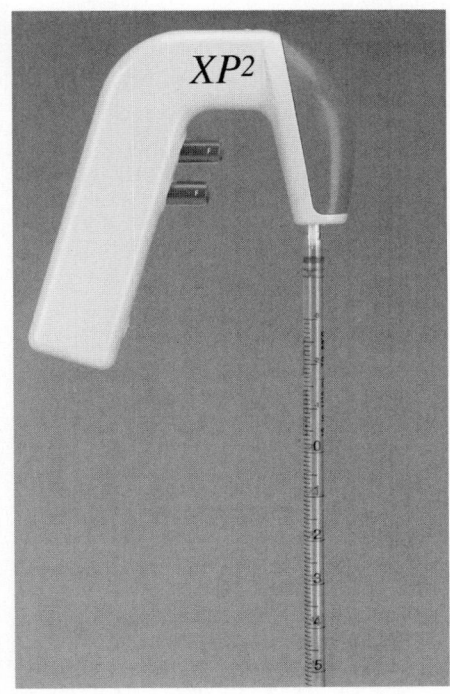

FIGURE 2-13 Electronic Pipet-Aid® allows you to fill the pipet by pressing the top button and to drain the pipet by pressing the lower button. To protect the device and to avoid contamination of solutions, a filter in the nosepiece immediately above the glass pipet ceases to pass air in either direction if the filter gets wet. [Courtesy Drummond Scientific Co., Broomall, PA.]

1 μg = 1 microgram = 10^{-6} g

Relative uncertainty in serial dilution is improved by using larger pipets and larger flasks. If you have a choice of transferring 1 mL into a 100-mL volumetric flask or transferring 10 mL into a 1-L volumetric flask, the result will be more accurate if you use the larger glassware. Accuracy is improved if each item of glassware has been individually calibrated. Large glassware generates extra waste that might be hazardous or expensive to discard. You might have to choose between how much accuracy you need and how much waste you will generate.

Micropipets

Micropipets (Figure 2-14) deliver volumes of 1 to 1 000 μL (1 $\mu L = 10^{-6}$ L). Liquid is contained in the disposable polypropylene tip, which is stable to most aqueous solutions and many organic solvents except chloroform ($CHCl_3$). The tip also is not resistant to concentrated nitric or sulfuric acids. To prevent *aerosols* from entering the pipet shaft, tips are available with polyethylene filters. Aerosols can corrode mechanical parts of the pipet or cross contaminate biological experiments.

To use a micropipet, place a fresh tip tightly on the barrel. Keep tips in their package or dispenser so that you do not contaminate the tips with your fingers. Set the desired volume with the knob at the top of the pipet. Depress the plunger to the first stop, which corresponds to the selected volume. Hold the pipet *vertically*, dip it 3–5 mm into the reagent solution, and *slowly* release the plunger to suck up liquid. Leave the tip in the liquid for a few seconds to allow the aspiration of liquid into the tip to go to completion. Withdraw the pipet vertically from the liquid without touching the tip to the side of the vessel. The volume of liquid taken into the tip depends on the angle at which the pipet is held and how far beneath the liquid surface the tip is held during uptake. To dispense liquid, touch the tip to the wall of the receiver and gently depress the plunger to the first stop. Wait a few seconds to allow liquid to drain down the tip, and then depress the plunger further to squirt out the last liquid. Prior to dispensing liquid, clean and wet a fresh tip by taking up and discarding three squirts of reagent. The tip can be discarded or rinsed well with a squirt bottle and reused. A tip with a filter (Figure 2-14b) cannot be cleaned for reuse.

The procedure we just described for *aspirating* (sucking in) and delivering liquids is called "forward mode." The plunger is depressed to the first stop and liquid is then taken up. To expel liquid, the plunger is depressed beyond the first stop. In "reverse mode," the plunger is depressed beyond the first stop and *excess* liquid is taken in. To deliver the correct volume, depress the plunger to the first stop and *not beyond*. Reverse mode with slow operation of the plunger improves precision for foamy solutions (proteins or surfactants) and viscous (syrupy) liquids.[22] Reverse pipetting is also good for volatile liquids such as methanol and hexane. For volatile liquids, pipet rapidly to minimize evaporation.

Table 2-5 lists tolerances for micropipets from one manufacturer. As internal parts wear out, both precision and accuracy can decline by an order of magnitude. In a study[23] of

An **aerosol** is a suspension of fine liquid droplets or solid particles in the gas phase.

Avoiding errors with micropipets:[21]
- Use tip recommended by manufacturer. Other tips might make inadequate seal.
- To wet the pipet tip and equilibrate the inside with vapor, take up and expel liquid three times before delivery.
- Unnecessary wiping of the tip can cause loss of sample.
- Liquid must be at same temperature as pipet. Less than the indicated volume of cold liquid is delivered and more than the indicated volume of warm liquid is delivered. Errors are greatest for smallest volumes.
- Micropipets are calibrated at sea level pressure. They are out of calibration at higher elevations. Errors are greatest for smallest volumes. Calibrate your pipet by weighing the water it delivers.

TABLE 2-5 Manufacturer's tolerances for micropipets				
	At 10% of pipet volume		At 100% of pipet volume	
Pipet volume (μL)	Accuracy (%)	Precision (%)	Accuracy (%)	Precision (%)
Adjustable Pipets				
0.2–2	±8	±4	±1.2	±0.6
1–10	±2.5	±1.2	±0.8	±0.4
2.5–25	±4.5	±1.5	±0.8	±0.2
10–100	±1.8	±0.7	±0.6	±0.15
30–300	±1.2	±0.4	±0.4	±0.15
100–1 000	±1.6	±0.5	±0.3	±0.12
Fixed Pipets				
10			±0.8	±0.4
25			±0.8	±0.3
100			±0.5	±0.2
500			±0.4	±0.18
1 000			±0.3	±0.12

SOURCE: *Data from Hamilton Co., Reno, NV.*

54 micropipets in use at a biomedical lab, 12 were accurate and precise to ≤1%. Five of 54 had errors ≥10%. When 54 quality control technicians at four pharmaceutical companies used a properly functioning micropipet, 10 people were accurate and precise to ≤1%. Six were inaccurate by ≥10%. Micropipets require periodic calibration and maintenance (cleaning, seal replacement, and lubrication), and operators (people) require certification. If the mean time between falling out of tolerance for micropipets is 2 years, calibration is required every *2 months* to be confident that 95% of the micropipets in a laboratory are operating within specifications.[24] You can calibrate a micropipet by measuring the mass of water it delivers, as described in Section 2-9, or with a commercial colorimetric kit.[25]

Accuracy refers to nearness to the true value. **Precision** refers to reproducibility.

Syringes

Microliter *syringes*, such as that in Figure 2-15, come in sizes from 1 to 500 μL and have an accuracy and precision near 1%. When using a syringe, take up and discard several volumes of liquid to wash the glass walls and to remove air bubbles from the barrel. The steel needle is attacked by strong acid and will contaminate strongly acidic solutions with iron. A syringe is more reliable than a micropipet, but the syringe requires more care in handling and cleaning. Figure 2-16 is an example of a programmable dual syringe diluter that automatically dispenses microliter volumes and can create reproducible mixtures of two solutions from the two syringes.

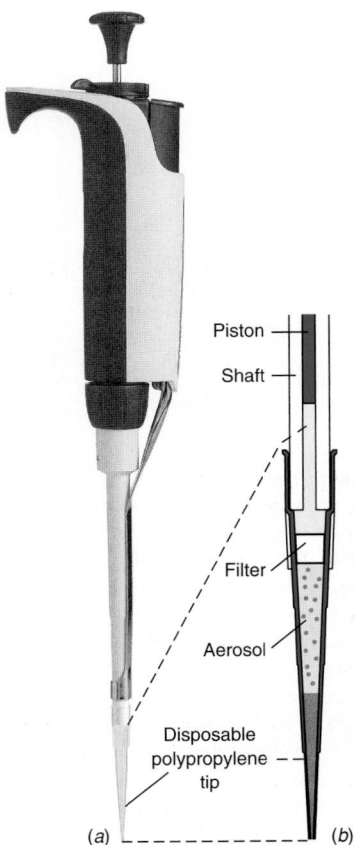

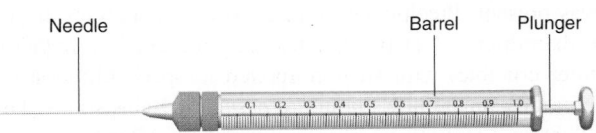

FIGURE 2-15 Hamilton syringe with a volume of 1 μL and divisions of 0.02 μL on the glass barrel. [Courtesy Hamilton Co., Reno, NV.]

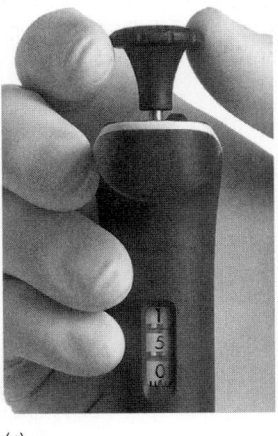

(c)

FIGURE 2-14 (*a*) Micropipet with disposable plastic tip. (*b*) Enlarged view of disposable tip containing polyethylene filter to prevent aerosol from contaminating the shaft of the pipet. (*c*) Volume selection dial set to 150 μL. [Courtesy of Rainin Instrument, LLC, Oakland, CA.]

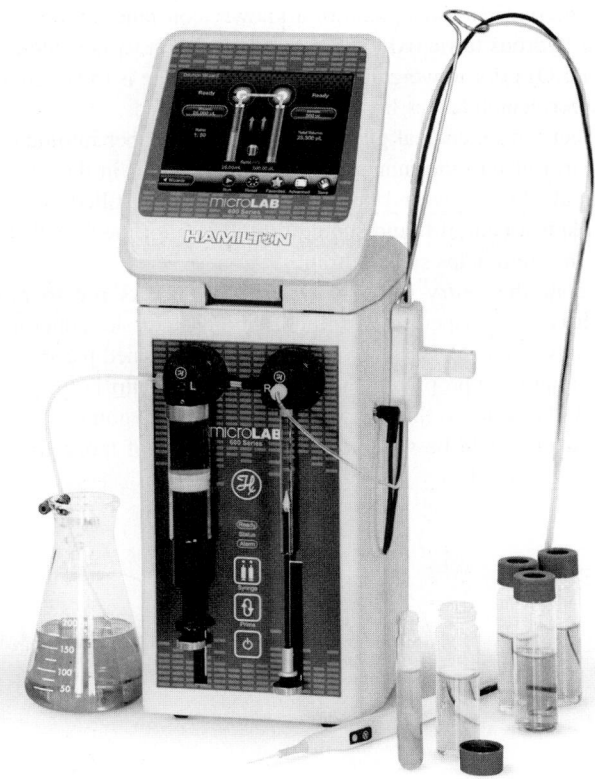

FIGURE 2-16 Microlab 600 Dual Syringe Diluter is a programmable dispenser of microliter quantities from two syringes shown at the front of the instrument. It reproducibly dispenses a single liquid or mixtures of two liquids. [Courtesy Hamilton Co., Reno, NV.]

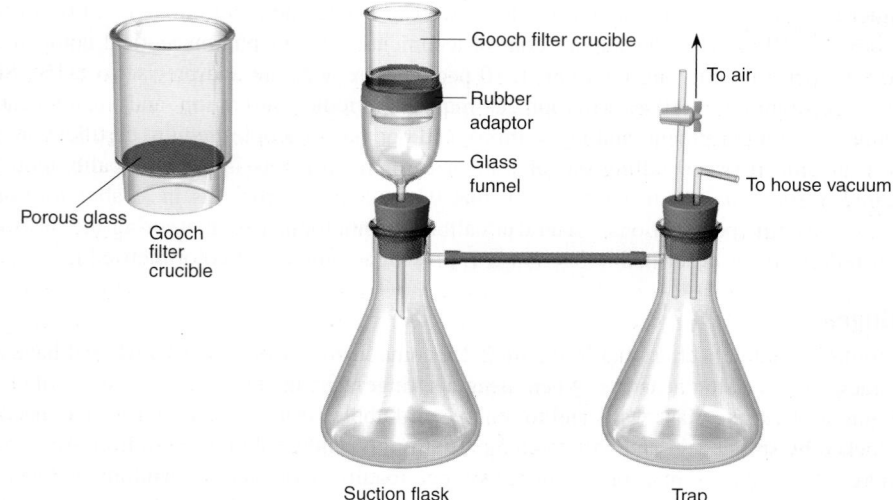

Gooch filter crucible

Porous glass

Gooch
filter
crucible

Rubber
adaptor

Glass
funnel

To air

To house vacuum

Suction flask

Trap

FIGURE 2-17 Filtration with a Gooch crucible that has a porous (*fritted*) glass disk through which liquid can pass. The trap prevents liquid from being accidentally sucked into the vacuum system.

2-7 Filtration

In *gravimetric analysis*, the mass of product from a reaction is measured to determine how much unknown was present. Precipitates from gravimetric analyses are collected by filtration, washed, and then dried. Most precipitates are collected in a *fritted-glass funnel* (also called a Gooch filter crucible) with suction applied to speed filtration (Figure 2-17). The porous glass plate in the funnel allows liquid to pass but retains solids. The empty funnel is first dried at 110°C or in a microwave oven and weighed. After collecting solid and drying again, the funnel and its contents are weighed a second time to determine the mass of collected solid. Liquid from which a substance precipitates or crystallizes is called the **mother liquor.** Liquid that passes through the filter is called **filtrate.**

In some gravimetric procedures, **ignition** (heating at high temperature over a burner or in a furnace) is used to convert a precipitate to a known, constant composition. For example, Fe^{3+} precipitates as hydrous ferric oxide, $FeOOH \cdot xH_2O$, with variable composition. Ignition converts it to pure Fe_2O_3 prior to weighing. When a precipitate is to be ignited, it is collected in **ashless filter paper,** which leaves little residue when burned.

To use filter paper with a conical glass funnel, fold the paper into quarters, tear off one corner (to allow a firm fit into the funnel), and place the paper in the funnel (Figure 2-18). The filter paper should fit snugly and be seated with some distilled water. When liquid is poured in, an unbroken stream of liquid should fill the stem of the funnel (Figure 2-19). The weight of liquid in the stem helps speed filtration.

For filtration, pour the slurry of precipitate down a glass rod to prevent splattering (Figure 2-19). (A **slurry** is a suspension of solid in liquid.) Particles adhering to the beaker or rod can be dislodged with a *rubber policeman*, which is a flattened piece of rubber at the end of a glass rod. Use a jet of appropriate wash liquid from a squirt bottle to transfer particles from the rubber and glassware to the filter. If the precipitate is going to be ignited, particles remaining in the beaker should be wiped onto a small piece of moist filter paper. Add that paper to the filter to be ignited.

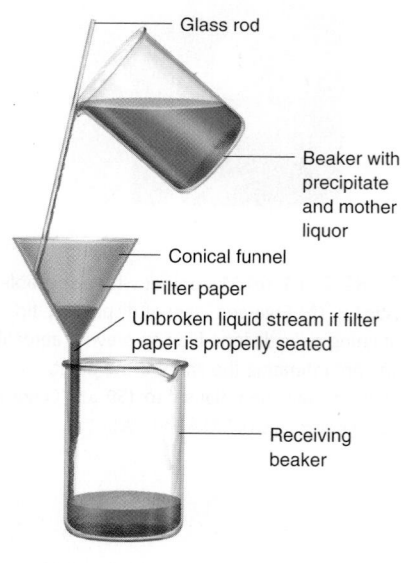

Glass rod

Beaker with
precipitate
and mother
liquor

Conical funnel

Filter paper

Unbroken liquid stream if filter
paper is properly seated

Receiving
beaker

FIGURE 2-19 Filtering a precipitate. The conical funnel is supported by a metal ring attached to a ring stand, neither of which is shown.

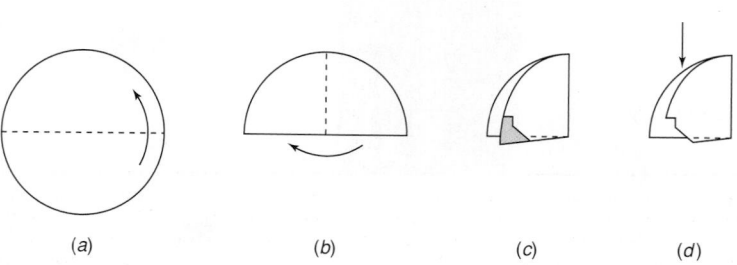

(a)   (b)   (c)   (d)

FIGURE 2-18 Folding filter paper for a conical funnel. (*a*) Fold the paper in half. (*b*) Then fold it in half again. (*c*) Tear off a corner to allow better seating of the paper in the funnel. (*d*) Open the side that was not torn when fitting the paper in the funnel.

2-8 Drying

Reagents, precipitates, and glassware are conveniently dried in an oven at 110°C. (Some chemicals require other temperatures.) *Anything that you put in the oven should be labeled.* Use a beaker and watchglass (Figure 2-20) to minimize contamination by dust during drying. It is good practice to cover all vessels on the benchtop to prevent dust contamination.

The mass of a gravimetric precipitate is measured by weighing a dry, empty filter crucible before the procedure and reweighing the same crucible filled with dry product after the procedure. To weigh the empty crucible, first bring it to "constant mass" by drying it in the oven for 1 h or longer and then cooling it for 30 min in a desiccator. Weigh the crucible and then heat it again for about 30 min. Cool it and reweigh it. When successive weighings agree to ±0.3 mg, the filter has reached "constant mass." If the crucible is warm when it is weighed, it creates convection currents that give a false weight. Do not touch the crucible with your fingers because fingerprints can change the mass. You can use a microwave oven instead of an electric oven for drying reagents and crucibles. Try an initial heating time of 4 min, with subsequent 2-min heatings. Use a 15-min cooldown before weighing.

A **desiccator** (Figure 2-21) is a closed chamber containing a drying agent called a **desiccant** (Table 2-6). The lid is greased to make an airtight seal, and desiccant is placed in the bottom beneath the perforated disk. Another useful desiccant, which is not listed in the table, is 98 wt% sulfuric acid. After placing a hot object in the desiccator, leave the lid cracked open

Dust is a source of contamination in all experiments, so...

> Cover all vessels whenever possible.

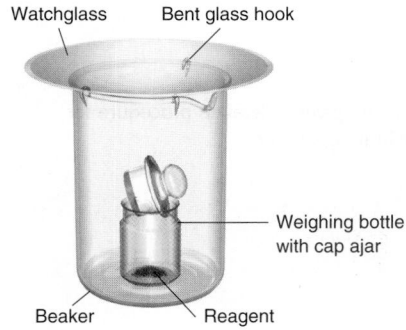

FIGURE 2-20 Use a watchglass as a dust cover while drying reagents or crucibles in the oven.

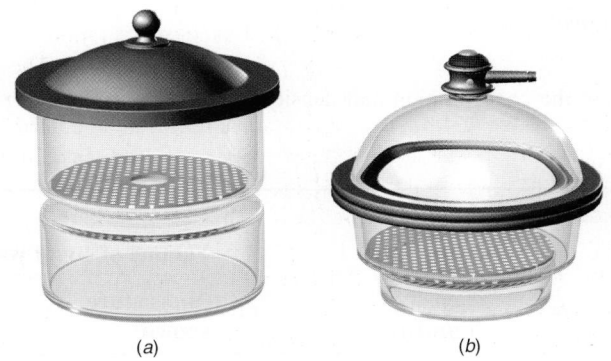

(a)   (b)

FIGURE 2-21 (a) Ordinary desiccator. (b) Vacuum desiccator that can be evacuated through the side arm at the top and then sealed by rotating the joint containing the side arm. Drying is more efficient at low pressure. [Information from A. H. Thomas Co., Philadelphia, PA.]

TABLE 2-6	Efficiencies of drying agents	
Agent	**Formula**	**Water left in atmosphere (μg H_2O/L)^a**
Magnesium perchlorate, anhydrous	$Mg(ClO_4)_2$	0.2
"Anhydrone"	$Mg(ClO_4)_2 \cdot 1-1.5H_2O$	1.5
Barium oxide	BaO	2.8
Alumina	Al_2O_3	2.9
Phosphorus pentoxide	P_4O_{10}	3.6
Calcium sulfate (Drierite)^b	$CaSO_4$	67
Silica gel	SiO_2	70

a. *Moist nitrogen was passed over each desiccant, and the water remaining in the gas was condensed and weighed.* [A. I. Vogel, A Textbook of Quantitative Inorganic Analysis, 3rd ed. (New York: Wiley. 1961), p. 178.] *For drying gases, the gas can be passed through a 60-cm-long Nafion tube. At 25°C, the residual moisture is 10 μg/L. If the drier is held at 0°C, the residual moisture is 0.8 μg/L.* [K. J. Leckrone and J. M. Hayes, "Efficiency and Temperature Dependence of Water Removal by Membrane Dryers," Anal. Chem. **1997,** 69, 911.]

b. *Used Drierite can be regenerated by irradiating 1.5-kg batches in a 100 × 190 mm Pyrex crystallizing dish in a microwave oven for 15 min. Stir the solid, heat a second time for 15 min. and place the hot, dry material back in its original container. Use small glass spacers between the crystallizing dish and the glass tray of the oven to protect the tray.* [J. A. Green and R. W. Goetz, "Recycling Drierite," J. Chem. Ed. **1991,** 68, 429.]

for a minute until the object has cooled slightly. This practice prevents the lid from popping open when the air inside warms up. To open a desiccator, slide the lid sideways rather than trying to pull it straight up.

2-9 Calibration of Volumetric Glassware

Page 45 gives a detailed procedure for calibrating a buret.

Each instrument that we use has a scale of some sort to measure a quantity such as mass, volume, force, or electric current. Manufacturers usually certify that the indicated quantity lies within a certain *tolerance* from the true quantity. For example, a Class A transfer pipet is certified to deliver 10.00 ± 0.02 mL when you use it properly. Your individual pipet might always deliver 10.016 ± 0.004 mL in a series of trials. That is, your pipet delivers an average of 0.016 mL more than the indicated volume in repeated trials. **Calibration** is the process of measuring the actual quantity that corresponds to an indicated quantity on the scale of an instrument.

For greatest accuracy, we calibrate volumetric glassware to measure the volume actually contained in or delivered by a particular piece of equipment. We do this by measuring the mass of water contained or delivered by the vessel and using the density of water to convert mass into volume.

In the most careful work, it is necessary to account for thermal expansion of solutions and glassware with changing temperature. For this purpose, you should know the lab temperature when a solution was prepared and when it is used. Table 2-7 shows that water expands 0.02% per degree near 20°C. Because the concentration of a solution is proportional to its density, we can write

Concentration decreases when the temperature increases.

$$\text{Correction for thermal expansion:} \qquad \frac{c'}{d'} = \frac{c}{d} \qquad\qquad (2\text{-}3)$$

where c' and d' are the concentration and density at temperature T', and c and d apply at temperature T.

TABLE 2-7	Density of water		
		Volume of 1 g of water (mL)	
Temperature (°C)	**Density (g/mL)**	**At temperature shown**[a]	**Corrected to 20°C**[b]
10	0.999 702 6	1.001 4	1.001 5
11	0.999 608 4	1.001 5	1.001 6
12	0.999 500 4	1.001 6	1.001 7
13	0.999 380 1	1.001 7	1.001 8
14	0.999 247 4	1.001 8	1.001 9
15	0.999 102 6	1.002 0	1.002 0
16	0.998 946 0	1.002 1	1.002 1
17	0.998 777 9	1.002 3	1.002 3
18	0.998 598 6	1.002 5	1.002 5
19	0.998 408 2	1.002 7	1.002 7
20	0.998 207 1	1.002 9	1.002 9
21	0.997 995 5	1.003 1	1.003 1
22	0.997 773 5	1.003 3	1.003 3
23	0.997 541 5	1.003 5	1.003 5
24	0.997 299 5	1.003 8	1.003 8
25	0.997 047 9	1.004 0	1.004 0
26	0.996 786 7	1.004 3	1.004 2
27	0.996 516 2	1.004 6	1.004 5
28	0.996 236 5	1.004 8	1.004 7
29	0.995 947 8	1.005 1	1.005 0
30	0.995 650 2	1.005 4	1.005 3

a. Corrected for buoyancy with Equation 2-1.
b. Corrected for buoyancy and expansion of borosilicate glass ($0.001\ 0\%\ K^{-1}$).

EXAMPLE **Effect of Temperature on Solution Concentration**

A 0.031 46 M aqueous solution was prepared in winter when the lab temperature was 17°C. What is the molarity of the solution on a warm day when the temperature is 25°C?

Solution We assume that the thermal expansion of a dilute solution is equal to the thermal expansion of pure water. Then, using Equation 2-3 and densities from Table 2-7, we write

$$\frac{c' \text{ at } 25°C}{0.997\ 05 \text{ g/mL}} = \frac{0.031\ 46 \text{ M}}{0.998\ 78 \text{ g/mL}} \implies c' = 0.031\ 41 \text{ M}$$

The concentration has decreased by 0.16% on the warm day.

Pyrex and other borosilicate glasses expand by 0.001 0% per degree near room temperature. If the temperature increases by 10°C, the volume of a piece of glassware increases by (10°C)(0.001 0%/°C) = 0.010%. For most work, this expansion is insignificant.

To calibrate a 25-mL transfer pipet, first weigh an empty weighing bottle like the one in Figure 2-20. Then fill the pipet to the mark with distilled water, drain it into the weighing bottle, and cap the bottle to prevent evaporation. Weigh the bottle again to find the mass of water delivered from the pipet. Finally, convert mass into volume.

$$\text{True volume} = (\text{grams of water}) \times (\text{volume of 1 g of } H_2O \text{ in Table 2-7}) \qquad \textbf{(2-4)}$$

> Small or odd-shaped vessels can be calibrated with Hg, which is easier than water to pour out of glass and is 13.6 times denser than water. This procedure is for researchers, not students. Mercury spills are a health hazard.

EXAMPLE **Calibration of a Pipet**

An empty weighing bottle had a mass of 10.313 g. After the addition of water from a 25-mL pipet, the mass was 35.225 g. If the lab temperature was 27°C, find the volume of water delivered by the pipet.

Solution The mass of water is 35.225 − 10.313 = 24.912 g. From Equation 2-4 and the next-to-last column of Table 2-7, the volume of water is (24.912 g)(1.004 6 mL/g) = 25.027 mL at 27°C. The last column in Table 2-7 tells us what the volume would be if the pipet were at 20°C. This pipet would deliver (24.912 g)(1.004 5 mL/g) = 25.024 mL at 20°C.

> The pipet delivers less volume at 20°C than at 27°C because glass contracts slightly as the temperature is lowered. Volumetric glassware is usually calibrated at 20°C.

The value of calibration is illustrated by a manufacturer's bulletin for digitally controlled, electronically operated glass microliter syringes. The manufacturer's tolerance for accuracy of an uncalibrated syringe that can deliver variable volumes of 2 to 50 μL is ±1.0% (±0.5 μL) when delivering 50 μL. When the syringe is calibrated by weighing the water it delivers, the accuracy becomes ±0.2% (±0.1 μL).

2-10 Introduction to Microsoft Excel®

If you already use a spreadsheet, you can skip this section. The computer spreadsheet is an essential tool for manipulating quantitative information. In analytical chemistry, spreadsheets can help us with calibration curves, statistical analysis, titration curves, and equilibrium problems. Spreadsheets allow us to conduct "what if" experiments such as investigating the effect of a stronger acid or a different ionic strength on a titration curve. We use Microsoft Excel in this book as a tool for solving problems in analytical chemistry.[26] Although you can skip over spreadsheets with no loss of continuity, spreadsheets will enrich your understanding of chemistry and provide a valuable tool for use outside this course.

Getting Started: Calculating the Density of Water

Let's prepare a spreadsheet to compute the density of water from the equation

$$\text{Density (g/mL)} = a_0 + a_1 * T + a_2 * T^2 + a_3 * T^3 \qquad \textbf{(2-5)}$$

where T is temperature (°C) and $a_0 = 0.999\ 89$, $a_1 = 5.332\ 2 \times 10^{-5}$, $a_2 = -7.589\ 9 \times 10^{-6}$, and $a_3 = 3.671\ 9 \times 10^{-8}$.

> This equation is accurate to five decimal places over the range 4° to 40°C.

Columns

(a)

	A	B	C
1			
2			
3			
4		cell B4	
5			
6			
7			
8			
9			
10			
11			
12			

Rows

(b)

	A	B	C
1	Calculating Density of H2O with Equation 2-5		
2	(from the delightful book by Dan Harris)		
3			
4	Constants:		
5	a0 =		
6	0.99989		
7	a1 =		
8	5.3322E-05		
9	a2 =		
10	-7.5899E-06		
11	a3 =		
12	3.6719E-08		

(c)

	A	B	C
1	Calculating Density of H2O with Equation 2-5		
2	(from the delightful book by Dan Harris)		
3			
4	Constants:	Temp (°C)	Density (g/mL)
5	a0 =	5	0.99997
6	0.99989	10	
7	a1 =	15	
8	5.3322E-05	20	
9	a2 =	25	
10	-7.5899E-06	30	
11	a3 =	35	
12	3.6719E-08	40	

(d)

	A	B	C
1	Calculating Density of H2O with Equation 2-5		
2	(from the delightful book by Dan Harris)		
3			
4	Constants:	Temp (°C)	Density (g/mL)
5	a0 =	5	0.99997
6	0.99989	10	0.99970
7	a1 =	15	0.99911
8	5.3322E-05	20	0.99821
9	a2 =	25	0.99705
10	-7.5899E-06	30	0.99565
11	a3 =	35	0.99403
12	3.6719E-08	40	0.99223
13			
14	Formula:		
15	C5 = $A$6+$A$8*B5+$A$10*B5^2+$A$12*B5^3		

FIGURE 2-22 Evolution of a spreadsheet for computing the density of water.

The blank spreadsheet in Figure 2-22a has columns labeled A, B, C and rows numbered 1, 2, 3, . . . , 12. The box in column B, row 4 is called *cell* B4.

Begin each spreadsheet with a title to help make the spreadsheet more readable. In Figure 2-22b, we click in cell A1 and type "Calculating Density of H2O with Equation 2-5". Then we click in cell A2 and write "(from the delightful book by Dan Harris)" without quotation marks. The computer automatically spreads the text to adjoining cells. To save your worksheet, click on the File menu at the upper left in Excel 2010 and select Save As. In Excel 2007, click the Office button at the upper left to find Save As. Give your spreadsheet a descriptive name that will tell you what is in it long after you have forgotten about it. File it in a location that you can find in the future. Information in your computer is only as good as your ability to retrieve it.

We adopt a convention in this book in which constants are collected in column A. Type "Constants:" in cell A4. Then select cell A5 and type "a0 =". Now select cell A6 and type the number 0.99989 (without extra spaces). In cells A7 to A12, enter the remaining constants. Powers of 10 are written, for example, as E-5 for 10^{-5}.

After you type "5.3322E-5" in cell A8, the spreadsheet probably displays "5.33E-5" even though additional digits are kept in memory. To display a desired number of digits in scientific format, click in cell A8 and select the Home ribbon at the top of the spreadsheet. Go to the part of the ribbon that says Number and click on the little arrow at the lower right. The Format Cells window should appear. Select Scientific and 4 decimal places. When you click OK, the entry in cell A8 will be "5.3322E-5". If you need more room for the number of digits, grab the vertical line at the top of the column with your mouse and resize the column. You can format all numbers in column A by clicking on the top of the column before setting the format. Your spreadsheet should now look like Figure 2-22b.

In cell B4, write the heading "Temp (°C)". You can find the degree sign on the Insert ribbon by clicking Symbol. Now enter temperatures from 5 through 40 in cells B5 through B12. This is our *input* to the spreadsheet. The *output* will be computed values of density in column C.

In cell C4, enter the heading "Density (g/mL)". Cell C5 is the most important one in the table. In this one, you will write the *formula*

$$= \$A\$6 + \$A\$8*B5 + \$A\$10*B5^2 + \$A\$12*B5^3$$

Formulas begin with an equal sign. Arithmetic operations in a spreadsheet are
+ addition
− subtraction
* multiplication
/ division
^ exponentiation

It doesn't matter whether or not you use spaces around the arithmetic operators. When you hit Return, the number 0.99997 appears in cell C5. The formula above is the spreadsheet translation of Equation 2-5. $\$A\$6$ refers to the constant in cell A6. We will explain the dollar signs shortly. B5 refers to the temperature in cell B5. The times sign is * and the exponentiation sign is ^. For example, the term "$\$A\$12*B5^3$" means "(contents of cell A12) × (contents of cell B5)^3."

Now comes the most magical property of a spreadsheet. Highlight cell C5, which has the formula that you wish to copy into cells C6 to C12. Grab the small dark square in the lower right corner of cell C5 and drag it down into cells C6 through C12. Excel copies the formula from C5 into the cells below it and evaluates the number in each cell. The density of water at each temperature now appears in column C in Figure 2-22d. You can make numbers appear as decimals, rather than scientific notation, by clicking on the arrow at the lower right in Number in the Home ribbon. In Format Cells, select Number and 5 decimal places.

In this example, we made three types of entries. *Labels* such as "a0 =" were typed in as text. An entry that does not begin with a digit or an equal sign is treated as text. *Numbers*, such as 25, were typed in some cells. The spreadsheet treats a number differently from text. In cell C5, we entered a *formula* that necessarily begins with an equal sign.

Three kinds of entries:
label   a3 =
number 4.4E-05
formula = $A$8*B5

Arithmetic Operations and Functions

Addition, subtraction, multiplication, division, and exponentiation have the symbols +, −, *, /, and ^. *Functions* such as Exp(·) can be typed or can be selected from the Formula ribbon. Exp(·) raises e to the power in parentheses. Other functions such as Ln(·), Log(·), Sin(·), and Cos(·) are also available.

The order of arithmetic operations in formulas is negation first, followed by ^, followed by * and / (evaluated in order from left to right as they appear), finally followed by + and − (also evaluated from left to right). Make liberal use of parentheses to be sure that the computer does what you intend. The contents of parentheses are evaluated first, before carrying out operations outside the parentheses. Here are some examples:

Order of operations:
1. Negation (a minus sign before a term)
2. Exponentiation
3. Multiplication and division (in order from left to right)
4. Addition and subtraction (in order from left to right)

Operations within parentheses are evaluated first, from the innermost set.

$$9/5*100+32 = (9/5)*100+32 = (1.8)*100+32 = (1.8*100)+32 = (180)+32 = 212$$

$$9/5*(100+32) = 9/5*(132) = (1.8)*(132) = 237.6$$

$$9+5*100/32 = 9+(5*100)/32 = 9+(500)/32 = 9+(15.625) = 24.625$$

$$9/5^2+32 = 9/(5^2)+32 = (9/25)+32 = (0.36)+32 = 32.36$$

$$-2^2 = 4 \quad \text{but} \quad -(2^2) = -4$$

When in doubt about how an expression will be evaluated, use parentheses to force what you intend.

Documentation and Readability

The first important *documentation* in the spreadsheet is the name of the file. A name such as "Expt 10 Gran Plot" is more meaningful than "Chem Lab". The next important feature is a title at the top of the spreadsheet, which tells its purpose. To tell what formulas were used in the spreadsheet, we added text (labels) at the bottom. In cell A14, write "Formula:" and in cell A15 write "C5 = $\$A\$6+\$A\$8*B5+\$A\$10*B5^2+\$A\$12*B5^3$". The surest way to document a formula is to copy the text from the formula bar for cell C5. Go to cell A15, type "C5," and then paste in the text you copied.

Documentation means labeling. If your spreadsheet cannot be read by another person without your help, it needs better documentation. (The same is true of your lab notebook!)

We improve the *readability* of data in a spreadsheet by selecting the number (decimal) or scientific format and specifying how many decimal places will be shown. The spreadsheet retains more digits in its memory, even though just five might be displayed.

Absolute and Relative References

The formula "$= \$A\$8*B5$" refers to cells A8 and B5 in different manners. $\$A\$8$ is an *absolute reference* to the contents of cell A8. No matter where cell $\$A\$8$ is called from in the spreadsheet,

Absolute reference: $A$8
Relative reference: B5

Save your files frequently while you are working and make a backup file of anything that you don't want to lose.

the computer goes to cell A8 to look for a number. "B5" is a *relative reference* in the formula in cell C5. When called from cell C5, the computer goes to cell B5 to find a number. When called from cell C6, the computer goes to cell B6 to look for a number. If called from cell C19, the computer would look in cell B19. This is why the cell written without dollar signs is called a relative reference. If you want the computer to always look only in cell B5, then you should write "$B$5".

2-11 Graphing with Microsoft Excel

Graphs are critical to understanding quantitative relations. To make a graph in Excel 2010 or 2007 from the spreadsheet in Figure 2-22d, go to the Insert ribbon and select Chart. Click on Scatter and select the icon for Scatter with Smooth Lines and Markers. The other most common graph we will make is Scatter with only Markers. Grab the blank chart with your mouse and move it to the right of the data. In Chart Tools, select Design and click on Select Data. Click on Add. For Series name, write "Density" (without quotation marks). For X values, highlight cells B5:B12. For Y values, delete what was in the box and highlight cells C5:C12. Click OK twice. Click inside the plot area and select the Chart Tools Format ribbon. In Plot Area, Format Selection provides options for the border and fill color of the graph. For Fill, select Solid fill and Color white. For Border Color, select solid line and Color black. We now have a white graph surrounded by a black border.

To add an X axis title, select Chart Tools Layout. Click on Axis Titles and Primary Horizontal Axis Title. Click on Title Below Axis. A generic axis title appears on the graph. Highlight it and type "Temperature (°C)" over the title. Get the degree sign from Insert Symbol. To put a title on the Y axis, select Chart Tools Layout again. Click on Axis Titles and Primary Vertical Axis Title. Click on Rotated Title. Then type "Density (g/mL)" for the title. Select the title that appears above the graph and remove it with the delete key. Your graph probably looks like the one in Figure 2-23 now.

Let's change the graph so that it looks like Figure 2-24. Click on the curve on the graph to highlight all data points. If only one point is highlighted, click elsewhere on the line. Select Chart Tools Format. In Current Selection, choose Format Selection. A Format Data Series window appears. For Marker Options, choose Built-in. Select the Type circle and Size 6. For Marker Fill, select Solid fill and a Color of your choice. Select Marker Line Color then Solid line then the same Color as the marker. To change the appearance of the curve on the graph, use Line Color and Line Style. Create a solid black line with a width of 1.5 points.

To change the appearance of the Y axis, click any number on the Y axis and they will all be highlighted. Select Chart Tools and Format and Format Selection. The Format Axis box appears. For Axis Options, Minimum, click on Fixed and set the value to 0.992. For Axis Options, Maximum, click on Fixed and set the value to 1.000. For Major unit, click on Fixed and set the value to 0.002. For Minor unit, click on Fixed and set the value to 0.0004.

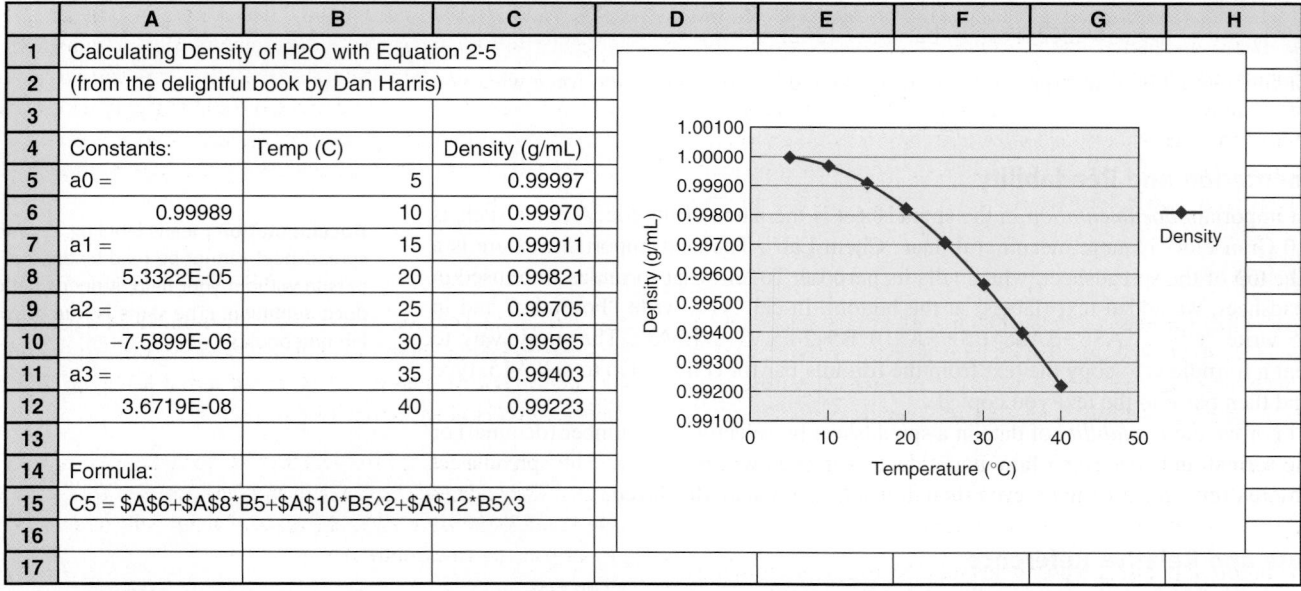

	A	B	C	D	E	F	G	H
1	Calculating Density of H2O with Equation 2-5							
2	(from the delightful book by Dan Harris)							
3								
4	Constants:	Temp (C)	Density (g/mL)					
5	a0 =	5	0.99997					
6	0.99989	10	0.99970					
7	a1 =	15	0.99911					
8	5.3322E-05	20	0.99821					
9	a2 =	25	0.99705					
10	−7.5899E-06	30	0.99565					
11	a3 =	35	0.99403					
12	3.6719E-08	40	0.99223					
13								
14	Formula:							
15	C5 = $A$6+$A$8*B5+$A$10*B5^2+$A$12*B5^3							
16								
17								

FIGURE 2-23 Initial density chart drawn by Excel.

Set Minor tick mark type to Outside. In the Format Axis window, select Number and set a display of 3 decimal places. Close the Format Axis window to finish with the vertical axis.

In a similar manner, select a number on the X axis and change the appearance so that it looks like Figure 2-24 with a Minimum of 0, Maximum of 40, Major unit of 10, and Minor unit of 5. Place Minor tick marks Outside. To add vertical grid lines, go to Chart Tools and select Layout and Grid Lines. Select Primary Vertical Gridlines and Major Gridlines.

Add a title back to the chart. In Chart Tools Layout, select Chart Title and highlight Above Chart. Type "Density of Water". In the Home ribbon, select a font size of 10 points. Your chart ought to look much like Figure 2-24 now. You can resize the chart from its lower right corner. To draw on an Excel worksheet, select Insert and then Shapes.

To write on the chart, go to the Insert ribbon and select Text Box. Click in the chart and begin typing. Drag the text box where you want it to be. To format the box, click on its border. Go to the Format ribbon and use Shape Fill and Shape Outline. To add arrows or lines, go to the Insert ribbon and select Shapes. To change the data point symbol, click on one point. On the Format Ribbon, click on Format Selection. The box that appears lets you change the appearance of the points and the line.

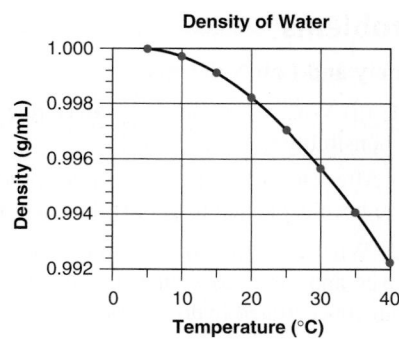

FIGURE 2-24 Density chart after reformatting.

Terms to Understand

Terms are introduced in **bold** type in the chapter and are also defined in the Glossary.

absorption	calibration	ignition	serial dilution
acid wash	desiccant	meniscus	slurry
adsorption	desiccator	mother liquor	tare
ashless filter paper	filtrate	parallax	titrant
buoyancy	green chemistry	pipet	volumetric flask
buret	hygroscopic	relative uncertainty	

Summary

Safety requires you to plan in advance about what you will do and to consider the hazards of each operation before executing it. Do not carry out a procedure until adequate safety precautions are in place. Know how to use safety equipment such as goggles, fume hood, lab coat, gloves, emergency shower, eyewash, and fire extinguisher. Chemicals should be stored and used in a manner that minimizes human contact. Environmentally acceptable disposal procedures should be established in advance for every chemical that you use. Your lab notebook tells what you did and what you observed; it should be understandable to other people. It also should allow you to repeat an experiment in the same manner in the future.

You should understand the principle of operation of an electronic balance and treat it as delicate equipment. Buoyancy corrections are required in accurate work. Burets should be read in a reproducible manner and drained slowly for best results. Interpolate between markings to obtain accuracy one decimal place beyond the graduations. Volumetric flasks are used to prepare solutions with known volume. Transfer pipets deliver fixed volumes; less accurate measuring pipets deliver variable volumes. The larger the volumetric glassware, the smaller is its relative uncertainty. You should understand how to design a serial dilution to make a less concentrated solution from a more concentrated solution by using transfer pipets and volumetric flasks. Do not be lulled into complacency by the nice digital reading on a micropipet. Unless your pipet has been calibrated recently and your personal technique tested, micropipets can have gross errors. Filtration and collection of precipitates require careful technique, as does the drying of reagents, precipitates, and glassware in ovens and desiccators. Volumetric glassware is calibrated by weighing water contained in or delivered by the vessel. In the most careful work, solution concentrations and volumes of vessels should be corrected for changes in temperature.

If you plan to use spreadsheets in this course, you should know how to enter formulas in a spreadsheet and how to draw a graph of data from a spreadsheet.

Exercises

2-A. What is the true mass of water if the measured mass in the atmosphere is 5.397 4 g? When you look up the density of water, assume that the lab temperature is **(a)** 15°C and **(b)** 25°C. Take the density of air to be 0.001 2 g/mL and the density of balance weights to be 8.0 g/mL.

2-B. A sample of ferric oxide (Fe_2O_3, density = 5.24 g/mL) obtained from ignition of a gravimetric precipitate weighed 0.296 1 g in the atmosphere. What is the true mass in vacuum?

2-C. A solution of potassium permanganate ($KMnO_4$) was found by titration to be 0.051 38 M at 24°C. What is the molarity when the lab temperature drops to 16°C?

2-D. A stock solution contains 51.38 mmol $KMnO_4$/L. How can you use pipets in Table 2-4 plus a 100- or 250-mL volumetric flask to obtain approximately 1, 2, 3, and 4 mmol $KMnO_4$/L? What will be the exact concentrations of the solutions?

2-E. Water was drained from a buret between the 0.12- and 15.78-mL marks. The apparent volume delivered was $15.78 - 0.12 = 15.66$ mL. Measured in the air at 22°C, the mass of water delivered was 15.569 g. What was the true volume?

2-F. [img] Reproduce the spreadsheet in Figure 2-23 and the graph in Figure 2-24.

Problems

Safety and Lab Notebook

2-1. (a) What is the primary safety rule and what is your implied responsibility to make it work?

(b) After safety features and safety procedures in your laboratory have been explained to you, make a list of them.

2-2. What class of liquids might easily penetrate through rubber gloves and get onto your skin? Would rubber gloves protect you from concentrated hydrochloric acid?

2-3. For chemical disposal, why is dichromate converted to $Cr(OH)_3(s)$?

2-4. What does "green chemistry" mean?

2-5. State three essential attributes of a lab notebook.

Analytical Balance

2-6. Explain the principle of operation of an electronic balance.

2-7. Why is the buoyancy correction equal to 1 in Figure 2-7 when the density of the object being weighed is 8.0 g/mL?

2-8. Pentane (C_5H_{12}) is a liquid with a density of 0.626 g/mL near 25°C. Find the true mass of pentane when the mass in air is 14.82 g. Assume air density = 0.001 2 g/mL.

2-9. The densities (g/mL) of several substances are: acetic acid, 1.05; CCl_4, 1.59; S, 2.07; Li, 0.53; Hg, 13.5; PbO_2, 9.4; Pb, 11.4; Ir, 22.5. From Figure 2-8, predict which substances will have the smallest and largest buoyancy corrections.

2-10. Potassium hydrogen phthalate is a primary standard used to measure the concentration of NaOH solutions. Find the true mass of potassium hydrogen phthalate (density = 1.636 g/mL) if the mass weighed in air is 4.236 6 g. If you did not correct the mass for buoyancy, would the calculated molarity of NaOH be too high or too low? By what percentage?

2-11. Accounting for buoyancy, what apparent mass of CsCl (density = 3.988 g/mL) in air should you weigh out to obtain a true mass of 1.267 g?

2-12. (a) Use the ideal gas law (Problem 1-18) to calculate the density (g/mL) of helium at 20°C and 1.00 bar.

(b) Find the true mass of Na (density = 0.97 g/mL) weighed in a glove box with a He atmosphere, if the apparent mass is 0.823 g.

2-13. (a) The equilibrium vapor pressure of water at 20°C is 2 330 Pa. What is the vapor pressure of water in the air at 20°C if the relative humidity is 42%? (*Relative humidity* is the percentage of the equilibrium water vapor pressure in the air.)

(b) Use note 13 for Chapter 2 at the end of the book to find the air density (g/mL, not g/L) under the conditions of part **(a)** if the barometric pressure is 94.0 kPa.

(c) What is the true mass of water in part **(b)** if the mass in air is 1.000 0 g?

2-14. (a) *Effect of altitude on electronic balance.* If an object weighs m_a grams at distance r_a from the center of the Earth, it will weigh $m_b = m_a(r_a^2/r_b^2)$ when raised to r_b. An object weighs 100.000 0 g on the first floor of a building at r_a = 6 370 km. How much will it weigh on the tenth floor, which is 30 m higher?

(b) If you press the "calibrate" button of the electronic balance on the tenth floor before weighing the object, the observed mass will be 100.000 0 g. Why?

2-15. *Quartz crystal microbalance.* The area of the gold electrodes on the quartz crystal microbalance at the opening of Chapter 2 is 3.3 mm². One gold electrode is covered with DNA at a surface density of 1.2 pmol/cm².

(a) How much mass of the nucleotide cytosine (C) is bound to the surface of the electrode when each bound DNA is elongated by one unit of C? The formula mass of the bound nucleotide is cytosine + deoxyribose + phosphate = $C_9H_{10}N_3O_6P$ = 287.2 g/mol.

(b) The shift in quartz crystal oscillator frequency for binding of DNA to the gold electrode was found to be -10 Hz for each ng/cm² bound to the electrode. Calculate how many ng of cytosine per square centimeter of electrode area are bound when the observed frequency change is -4.4 Hz. Is the frequency change consistent with the extension of DNA by one unit of C?

Glassware and Thermal Expansion

2-16. What do the symbols "TD" and "TC" mean on volumetric glassware?

2-17. Describe how to prepare 250.0 mL of 0.150 0 M K_2SO_4 with a volumetric flask.

2-18. When is it preferable to use a plastic volumetric flask instead of a glass flask?

2-19. (a) Describe how to deliver 5.00 mL of liquid by using a transfer pipet.

(b) Which is more accurate, a transfer pipet or a measuring pipet?

2-20. (a) Describe how to deliver 50.0 μL by using a 100-μL adjustable micropipet.

(b) What would you do differently in **(a)** if the liquid foams?

2-21. What is the purpose of the trap in Figure 2-17 and the watch-glass in Figure 2-20?

2-22. Which drying agent is more efficient, Drierite or phosphorus pentoxide?

2-23. (a) How much of the primary standard benzoic acid (FM 122.12, density = 1.27 g/mL) should you weigh out to obtain a 100.0 mM aqueous solution in a volume of 250 mL?

(b) What apparent mass in air will give you the true mass in **(a)**?

(c) *Serial dilution.* You have available 5- and 10-mL transfer pipets plus volumetric flasks of the following sizes: 100, 250, 500, and 1 000 mL. Devise a serial dilution that will give 50.0 μM benzoic acid.

2-24. An empty 10-mL volumetric flask weighs 10.263 4 g. When the flask is filled to the mark with distilled water and weighed again in the air at 20°C, the mass is 20.214 4 g. What is the true volume of the flask at 20°C?

2-25. By what percentage does a dilute aqueous solution expand when heated from 15° to 25°C? If a 0.500 0 M solution is prepared at 15°C, what would its molarity be at 25°C?

2-26. The true volume of a 50-mL volumetric flask is 50.037 mL at 20°C. What mass of water measured **(a)** in vacuum and **(b)** in air at 20°C would be contained in the flask?

2-27. You want to prepare 500.0 mL of 1.000 M KNO_3 at 20°C, but the lab (and water) temperature is 24°C at the time of preparation. How many grams of solid KNO_3 (density = 2.109 g/mL) should be dissolved in a volume of 500.0 mL at 24°C to give a concentration of 1.000 M at 20°C? What apparent mass of KNO_3 weighed in air is required?

2-28. *Accuracy of serial dilution.* To make a 1/100 dilution of a solution, which, if either, procedure provides more accuracy: (i) transfer 1 mL with a pipet to a 100-mL volumetric flask or (ii) transfer 10 mL with a pipet to a 1-L volumetric flask? How can you improve the accuracy of either procedure?

2-29. A simple model for the fraction of micropipets that operate within specifications after time *t* is

$$\text{Fraction within specifications} = e^{-t(\ln 2)/t_m}$$

where t_m is the mean time between failure (the time when the fraction meeting specifications is reduced to 50%). Suppose that $t_m = 2.00$ years.

(a) Show that the equation predicts that the time at which 50% of micropipets remain within specifications is 2 yr if $t_m = 2.00$ yr.

(b) Find the time *t* at which pipets should be recalibrated (and repaired, if necessary) so that 95% of all pipets will operate within specifications.

Reference Procedure: Calibrating a 50-mL Buret

This procedure tells how to construct a graph such as Figure 3-3 to convert the measured volume delivered by a buret to the true volume delivered at 20°C.

0. Measure the temperature in the laboratory. Distilled water for this experiment must be at laboratory temperature.

1. Fill the buret with distilled water and force any air bubbles out the tip. See whether the buret drains without leaving drops on the walls. If drops are left, clean the buret with soap and water or soak it with cleaning solution.[15] Adjust the meniscus to be at or slightly below 0.00 mL, and touch the buret tip to a beaker to remove the suspended drop of water. Allow the buret to stand for 5 min while you weigh a 125-mL flask fitted with a rubber stopper. (Hold the flask with a tissue or paper towel, not with your hands, to prevent fingerprint residue from changing its mass.) If the level of the liquid in the buret has changed, tighten the stopcock and repeat the procedure. Record the level of the liquid.

2. Drain approximately 10 mL of water at a rate < 20 mL/min into the weighed flask, and cap it tightly to prevent evaporation. Allow about 30 s for the film of liquid on the walls to descend before you read the buret. Estimate all readings to the nearest 0.01 mL. Weigh the flask again to determine the mass of water delivered.

3. Now drain the buret from 10 to 20 mL, and measure the mass of water delivered. Repeat the procedure for 30, 40, and 50 mL. Then do the entire procedure (10, 20, 30, 40, 50 mL) a second time.

4. Use Table 2-7 to convert the mass of water into the volume delivered. Repeat any set of duplicate buret corrections that do not agree to within 0.04 mL. Prepare a calibration graph like that in Figure 3-3, showing the correction factor at each 10-mL interval.

EXAMPLE Buret Calibration

When draining the buret at 24°C, you observe the following values:

Final reading	10.01	10.08 mL
Initial reading	0.03	0.04
Difference	9.98	10.04 mL
Mass	9.984	10.056 g
Actual volume delivered	10.02	10.09 mL
Correction	+0.04	+0.05 mL
Average correction		+0.045 mL

2-30. Glass is a notorious source of metal ion contamination. Three glass bottles were crushed and sieved to collect 1-mm pieces.[27] To see how much Al^{3+} could be extracted, 200 mL of a 0.05 M solution of the metal-binding compound EDTA were stirred with 0.50 g of ~1-mm glass particles in a polyethylene flask. The Al content of the solution after 2 months was 5.2 μM. The total Al content of the glass, measured after completely dissolving some glass in 48 wt% HF with microwave heating, was 0.80 wt%. What fraction of the Al was extracted from glass by EDTA?

2-31. [icon] The efficiency of a gas chromatography column is measured by a parameter called plate height (*H*, mm), which is related to the gas flow rate (*u*, mL/min) by the van Deemter equation: $H = A + B/u + Cu$, where *A*, *B*, and *C* are constants. Prepare a spreadsheet with a graph showing values of *H* as a function of *u* for $u = 4, 6, 8, 10, 20, 30, 40, 50, 60, 70, 80, 90,$ and 100 mL/min. Use the values $A = 1.65$ mm, $B = 25.8$ mm·mL/min, and $C = 0.023\ 6$ mm·min/mL.

To calculate the actual volume delivered when 9.984 g of water are delivered at 24°C, look at the column of Table 2-7 headed "Corrected to 20°C." In the row for 24°C, you find that 1.000 0 g of water occupies 1.003 8 mL. Therefore, 9.984 g occupy (9.984 g)(1.003 8 mL/g) = 10.02 mL. The average correction for both sets of data is +0.045 mL.

To obtain the correction for a volume greater than 10 mL, add successive masses of water collected in the flask. Suppose that the following masses were measured:

Volume interval (mL)	Mass delivered (g)
0.03–10.01	9.984
10.01–19.90	9.835
19.90–30.06	10.071
Sum 30.03 mL	29.890 g

The total volume of water delivered is (29.890 g)(1.003 8 mL/g) = 30.00 mL. Because the indicated volume is 30.03 mL, the buret correction at 30 mL is −0.03 mL.

What does this mean? Suppose that Figure 3-3 applies to your buret. If you begin a titration at 0.04 mL and end at 29.00 mL, you would deliver 28.96 mL if the buret were perfect. Figure 3-3 tells you that the buret delivers 0.03 mL less than the indicated amount, so only 28.93 mL were actually delivered. To use the calibration curve, either begin all titrations near 0.00 mL or correct both the initial and the final readings. Use the calibration curve whenever you use your buret.

3 Experimental Error

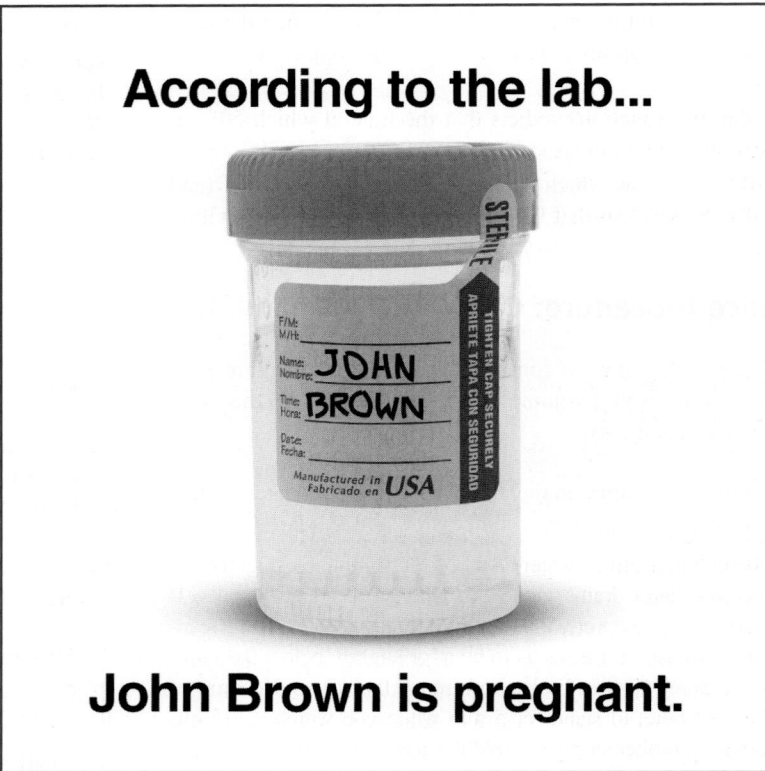

According to the lab...

John Brown is pregnant.

[Photo © bhathaway/Shutterstock.]

Some laboratory errors are more obvious than others, but there is error associated with every measurement. There is no way to measure the "true" value of anything. The best we can do in a chemical analysis is to carefully apply a technique that experience tells us is reliable. Repetition of one method of measurement several times tells us the *precision* (reproducibility) of the measurement. If the results of measuring the same quantity by different methods agree with one another, then we become confident that the results are *accurate,* which means they are near the "true" value.

Suppose that you determine the density of a mineral by measuring its mass (4.635 ± 0.002 g) and volume (1.13 ± 0.05 mL). Density is mass per unit volume: 4.635 g/1.13 mL = 4.101 8 g/mL. The uncertainties in measured mass and volume are ± 0.002 g and ± 0.05 mL, but what is the uncertainty in the computed density? And how many significant figures should be used for the density? This chapter discusses the propagation of uncertainty in lab calculations.

3-1 Significant Figures

Significant figures: minimum number of digits required to express a value in scientific notation without loss of precision.

The number of **significant figures** is the minimum number of digits needed to write a given value in scientific notation without loss of precision. The number 142.7 has four significant figures, because it can be written 1.427×10^2. If you write $1.427\,0 \times 10^2$, you imply that you know the value of the digit after 7, which is not the case for the number 142.7. The number $1.427\,0 \times 10^2$ has five significant figures.

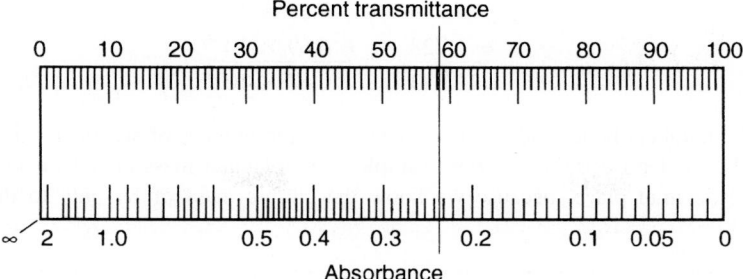

Percent transmittance

Absorbance

The number 6.302×10^{-6} has four significant figures, because all four digits are necessary. You could write the same number as 0.000 006 302, which also has just *four* significant figures. The zeros to the left of the 6 are merely holding decimal places. The number 92 500 is ambiguous. It could mean any of the following:

$$9.25 \times 10^4 \qquad \text{3 significant figures}$$
$$9.250 \times 10^4 \qquad \text{4 significant figures}$$
$$9.250\ 0 \times 10^4 \qquad \text{5 significant figures}$$

You should write one of the three numbers above, instead of 92 500, to indicate how many figures are actually known.

Zeros are significant when they occur (1) in the middle of a number or (2) at the end of a number on the right-hand side of a decimal point.

The last significant digit (farthest to the right) in a measured quantity always has some associated uncertainty. The minimum uncertainty is ± 1 in the last digit. The scale of an analog Spectronic 20 spectrophotometer is drawn in Figure 3-1. The needle in the figure appears to be at an absorbance of 0.234. We say that this number has three significant figures because the numbers 2 and 3 are completely certain and the number 4 is an estimate. The value might be read 0.233 or 0.235 by other people. The percent transmittance is near 58.3. The transmittance scale is smaller than the absorbance scale at this point, so there is more uncertainty in the last digit of transmittance. A reasonable estimate of uncertainty might be 58.3 ± 0.2. There are three significant figures in the number 58.3.

When reading the scale of any apparatus, try to estimate to the nearest tenth of a division. On a 50-mL buret, which is graduated to 0.1 mL, read the level to the nearest 0.01 mL. For a ruler calibrated in millimeters, estimate distances to the nearest 0.1 mm.

There is uncertainty in any *measured* quantity, even if the measuring instrument has a digital readout that does not fluctuate. When a digital pH meter indicates a pH of 3.51, there is uncertainty in the digit 1 (and maybe even in the digit 5). By contrast, integers are exact. To calculate the average height of four people, you would divide the sum of heights (which is a measured quantity with some uncertainty) by the integer 4. There are exactly 4 people, not 4.000 ± 0.002 people!

Significant zeros below are **bold**:

10**6**   0.010 **6**   0.10**6**   0.106 **0**

Interpolation: Estimate all readings to the nearest tenth of the distance between scale divisions.

3-2 Significant Figures in Arithmetic

We now consider how many digits to retain in the answer after you have performed arithmetic operations with your data. Rounding should only be done on the *final answer* (not intermediate results), to avoid accumulating round-off errors. Retain all of the digits for intermediate results in your calculator or spreadsheet.

Addition and Subtraction

If the numbers to be added or subtracted have equal numbers of digits, the answer goes to the *same decimal place* as in any of the individual numbers:

$$
\begin{aligned}
&1.362 \times 10^{-4} \\
+\ &3.111 \times 10^{-4} \\
\hline
&4.473 \times 10^{-4}
\end{aligned}
$$

The number of significant figures in the answer may exceed or be less than that in the original data.

$$\begin{array}{r} 5.345 \\ + 6.728 \\ \hline 12.073 \end{array} \qquad \begin{array}{r} 7.26 \times 10^{14} \\ - 6.69 \times 10^{14} \\ \hline 0.57 \times 10^{14} \end{array}$$

The periodic table inside the cover of this book gives uncertainty in the last digit of atomic mass:

F: $18.998\ 403\ 2 \pm 0.000\ 000\ 5$

Kr: 83.798 ± 0.002

If the numbers being added do not have the same number of significant figures, we are limited by the least certain one. For example, the molecular mass of KrF_2 is known only to the third decimal place, because we know the atomic mass of Kr only to three decimal places:

$$\begin{array}{rl} 18.998\ 403\ 2 & \text{(F)} \\ + 18.998\ 403\ 2 & \text{(F)} \\ + 83.798 & \text{(Kr)} \\ \hline 121.794\ \underbrace{806\ 4}_{\text{Not significant}} \end{array}$$

The number $121.794\ 806\ 4$ should be rounded to 121.795 as the final answer.

Rules for rounding off numbers

When rounding off, look at *all* the digits *beyond* the last place desired. In the preceding example, the digits 806 4 lie beyond the last significant decimal place. Because this number is more than halfway to the next higher digit, we round the 4 up to 5 (that is, we round up to 121.795 instead of down to 121.794). If the insignificant figures were less than halfway, we would round down. For example, 121.794 3 is rounded to 121.794.

In the special case where the number is exactly halfway, round to the nearest *even* digit. Thus, 43.55 is rounded to 43.6, if we can only have three significant figures. If we are retaining only three figures, 1.425×10^{-9} becomes 1.42×10^{-9}. The number $1.425\ 01 \times 10^{-9}$ would become 1.43×10^{-9}, because 501 is more than halfway to the next digit. The rationale for rounding to an even digit is to avoid systematically increasing or decreasing results through successive round-off errors. Half the round-offs will be up and half down.

Addition and subtraction: Express all numbers with the same exponent and align all numbers with respect to the decimal point. Round off the answer according to the number of decimal places in the number with the fewest decimal places.

In the addition or subtraction of numbers expressed in scientific notation, all numbers should first be expressed with the same exponent:

$$\begin{array}{r} 1.632 \times 10^5 \\ + 4.107 \times 10^3 \\ + 0.984 \times 10^6 \end{array} \rightarrow \begin{array}{r} 1.632 \times 10^5 \\ + 0.041\ 07 \times 10^5 \\ + 9.84 \times 10^5 \\ \hline 11.51 \times 10^5 \end{array}$$

The sum $11.513\ 07 \times 10^5$ is rounded to 11.51×10^5 because the number 9.84×10^5 limits us to two decimal places when all numbers are expressed as multiples of 10^5.

Multiplication and Division

Multiplication and division: Answer is limited to number of digits in the number with the fewest significant figures.

In multiplication and division, we are normally limited to the number of digits contained in the number with the fewest significant figures:

$$\begin{array}{r} 3.26 \times 10^{-5} \\ \times 1.78\phantom{\times 10^{-5}} \\ \hline 5.80 \times 10^{-5} \end{array} \qquad \begin{array}{r} 4.317\ 9 \times 10^{12} \\ \times 3.6 \times 10^{-19} \\ \hline 1.6 \times 10^{-6} \end{array} \qquad \begin{array}{r} 34.60 \\ \div 2.462\ 87 \\ \hline 14.05 \end{array}$$

The power of 10 has no influence on the number of figures that should be retained. Page 54 explains why it is reasonable to keep an extra digit when the first digit of the answer is 1. The middle product above could be expressed as 1.55×10^{-6} instead of 1.6×10^{-6} to avoid throwing away some of the precision of the factor 3.6 in the multiplication.

Logarithms and Antilogarithms

If $n = 10^a$, then we say that a is the base 10 **logarithm** of n:

Logarithm of n: $\qquad n = 10^a$ means that $\log n = a \qquad$ **(3-1)**

$10^{-3} = \dfrac{1}{10^3} = \dfrac{1}{1\ 000} = 0.001$

For example, 2 is the logarithm of 100 because $100 = 10^2$. The logarithm of 0.001 is -3 because $0.001 = 10^{-3}$. Most calculators have a *log* key to find the logarithm of a number.

In Equation 3-1, the number n is said to be the **antilogarithm** of a. That is, the antilogarithm of 2 is 100 because $10^2 = 100$, and the antilogarithm of -3 is 0.001 because $10^{-3} = 0.001$. Your calculator probably has either a 10^x key or an *antilog* key that you can use to find the antilogarithm of a number.

A logarithm is composed of a **characteristic** and a **mantissa.** The characteristic is the integer part and the mantissa is the decimal part:

$$\log 339 = \underbrace{2}_{\text{Characteristic}}.\underbrace{530}_{\text{Mantissa}} \qquad \log 3.39 \times 10^{-5} = -\underbrace{4}_{\text{Characteristic}}.\underbrace{470}_{\text{Mantissa}}$$

$$= 2 \qquad = 0.530 \qquad\qquad = -4 \qquad = 0.470$$

The number 339 can be written 3.39×10^2. *The number of digits in the mantissa of log 339 should equal the number of significant figures in 339.* The logarithm of 339 is properly expressed as 2.530. The *characteristic*, 2, corresponds to the exponent in 3.39×10^2.

To see that the third decimal place is the last significant place, consider the following results:

$$10^{2.531} = 340 \ (339.6)$$
$$10^{2.530} = 339 \ (338.8)$$
$$10^{2.529} = 338 \ (338.1)$$

The numbers in parentheses are the results prior to rounding to three figures. Changing the exponent in the third decimal place changes the answer in the third place of 339.

In the conversion of a logarithm into its antilogarithm, *the number of significant figures in the antilogarithm should equal the number of digits in the mantissa.* Thus,

$$\text{antilog}(-3.\underbrace{42}_{\text{2 digits}}) = 10^{-3.\underbrace{42}_{\text{2 digits}}} = \underbrace{3.8}_{\text{2 digits}} \times 10^{-4}$$

Here are several examples showing the proper use of significant figures:

$$\log 0.001\ 237 = -2.907\ 6 \qquad \text{antilog } 4.37 = 2.3 \times 10^4$$
$$\log 1\ 237 = 3.092\ 4 \qquad\qquad 10^{4.37} = 2.3 \times 10^4$$
$$\log 3.2 = 0.51 \qquad\qquad\qquad 10^{-2.600} = 2.51 \times 10^{-3}$$

Significant Figures and Graphs

When drawing a graph on a computer, consider whether the graph is meant to display qualitative behavior of the data (Figure 3-2) or precise values that must be read with several significant figures. If someone will use the graph (such as Figure 3-3) to read points, it should at least have tick marks on both sides of the horizontal and vertical scales. Better still is a fine grid superimposed on the graph.

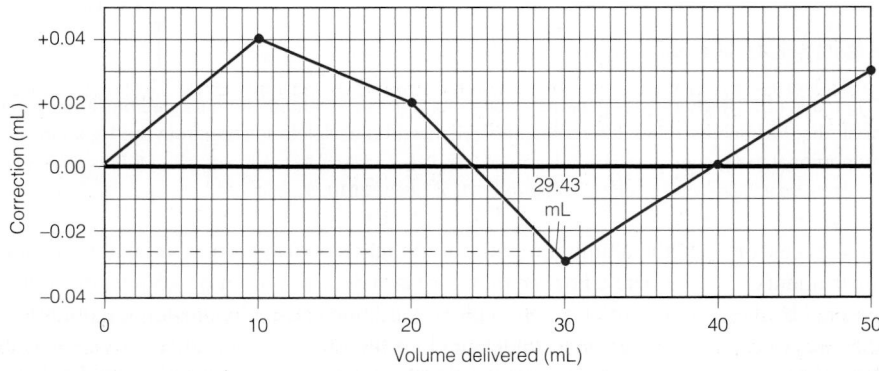

FIGURE 3-3 Calibration curve for a 50-mL buret. The volume delivered can be read to the nearest 0.1 mL. If your buret reading is 29.43 mL, you can find the correction factor accurately enough by locating 29.4 mL on the graph. The correction factor on the ordinate (y-axis) for 29.4 mL on the abscissa (x-axis) is −0.03 mL (to the nearest 0.01 mL).

Number of digits in **mantissa** of log x = number of significant figures in x:

$$\log(5.\underbrace{403}_{\text{4 digits}} \times 10^{-8}) = -7.\underbrace{267\ 4}_{\text{4 digits}}$$

Number of digits in antilog x $(= 10^x)$ = number of significant figures in **mantissa** of x:

$$10^{6.\underbrace{142}_{\text{3 digits}}} = 1.\underbrace{39}_{\text{3 digits}} \times 10^6$$

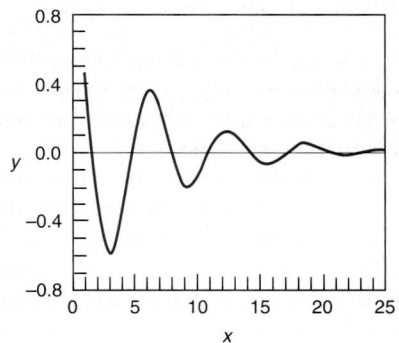

FIGURE 3-2 Example of a graph intended to show the qualitative behavior of the function $y = e^{-x/6} \cos x$. You are not expected to be able to read coordinates accurately on this graph.

Problem 3-8 shows you how to control gridlines in an Excel graph.

3-3 Types of Error

Every measurement has some uncertainty, which is called *experimental error*. Conclusions can be expressed with a high or a low degree of confidence, but never with complete certainty. Experimental error is classified as either *systematic* or *random*.

Systematic Error

Systematic error is a consistent error that can be detected and corrected. Box 3-1 provides an example from environmental analysis.

Systematic error, also called **determinate error,** arises from a flaw in equipment or the design of an experiment. If you conduct the experiment again in exactly the same manner, the error is reproducible. In principle, systematic error can be discovered and corrected, although this may not be easy.

For example, a pH meter that has been standardized incorrectly produces a systematic error. Suppose you think that the pH of the buffer used to standardize the meter is 7.00, but it is really 7.08. Then all your pH readings will be 0.08 pH unit too low. When you read a pH of 5.60, the actual pH of the sample is 5.68. This systematic error could be discovered by using a second buffer of known pH to test the meter.

Another systematic error arises from an uncalibrated buret. The manufacturer's tolerance for a Class A 50-mL buret is ±0.05 mL. When you think you have delivered 29.43 mL, the real volume could be anywhere from 29.38 to 29.48 mL and still be within tolerance. One way to correct for an error of this type is to construct a calibration curve, such as

---

BOX 3-1   **Case Study in Ethics: Systematic Error in Ozone Measurement[1]**

Ozone (O_3) is an oxidizing, corrosive gas that harms your lungs and all forms of life. It is formed near the surface of the Earth by the action of sunlight on NO_2 largely derived from automobile exhaust (Figure 1-3). The U.S. Environmental Protection Agency sets an 8-h average O_3 limit of 75 ppb (75 nL/L by volume) in air. Regions that fail to meet this standard can be required to reduce sources of pollution that contribute to O_3 formation. Error in ozone measurement can have serious consequences for the health and economy of a region.

A variety of instruments can be used to monitor compliance. The instrument in the diagram pumps air through a cell whose three sections have a total pathlength of 15 cm. Ultraviolet radiation from a mercury lamp is partially absorbed by O_3. The more O_3 in the air, the less radiation reaches the photodiode detector. From the measured absorbance, the instrument computes O_3 concentration. In routine use, the operator only adjusts the zero control, which sets the meter to read zero when O_3-free air is drawn through the instrument. Periodically, the instrument is recalibrated with a source of known O_3.

A study of commercial O_3 monitors found that controlled changes in humidity led to *systematic errors in the apparent O_3 concentration of tens to hundreds of ppb* (errors several times greater than the O_3 being measured). Increasing humidity produced systematic *positive* errors in some types of instruments and systematic *negative* errors in other instruments.

Water does not absorb the ultraviolet wavelength measured by the detector, so humidity is not interfering by absorbing radiation. A perceptive analysis of the problem led to the hypothesis that *adsorption* of moisture on the inside surface of the measurement cell changed the reflectivity of that surface. Depending on the type of instrument, adsorbed water could decrease or increase the amount of light reaching the detector, thereby creating false high or low O_3 readings.

The problem was solved by two means. The first step was to install a length of water-permeable Nafion® tubing just before the absorption cell to equalize the humidity in air being measured and that used to zero the instrument. The second step was to eliminate water adsorption on aluminum surfaces in the optical path by lining

Schematic of 2B Technologies personal ozone monitor. The solenoid alternately admits ambient air or air that has been scrubbed free of O_3. Absorbance of ultraviolet radiation from the Hg lamp is proportional to O_3 concentration. [Information from P. C. Andersen, C. J. Williford, and J. W. Birks, "Miniature Personal Ozone Monitor Based on UV Absorbance," *Anal. Chem.* **2010,** *82,* 7924.]

the aluminum with fused quartz (SiO_2 glass). A miniature, battery-operated ozone monitor has a detection limit of 4.5 ppb and a precision of 1.5 ppb, with no variation caused by changing humidity.

Prior to understanding the effect of humidity on O_3 measurement, it was known that O_3 monitors often exhibit erratic behavior on hot, humid days. Some people conjectured that half of the regions deemed to be out of compliance with the O_3 standard might actually have been under the legal limit. This error could force expensive remediation measures when none were required. Conversely, there were rumors that some unscrupulous operators of O_3 monitors were aware that zeroing their instrument at night when the humidity is higher produced lower O_3 readings the next day, thereby reducing the number of days when a region is deemed out of compliance.

BOX 3-2 **Certified Reference Materials**

Inaccurate laboratory measurements can cause wrong medical diagnoses and treatments, lost production time, wasted energy and materials, manufacturing rejects, and product liability. National standards laboratories around the world distribute **certified reference materials,** such as metals, chemicals, rubber, plastics, engineering materials, radioactive substances, and environmental and clinical standards that can be used to test the accuracy of analytical procedures.[2] The U.S. National Institute of Standards and Technology calls its certified materials *Standard Reference Materials*. The quantity of analyte in a reference material is certified—with painstaking care—to lie in a stated range.

For example, in treating patients with epilepsy, physicians depend on laboratory tests to measure concentrations of anticonvulsant drugs in blood serum. Drug levels that are too low lead to seizures; high levels are toxic. Because tests of identical serum specimens at different laboratories were giving an unacceptably wide range of results, the National Institute of Standards and Technology developed a standard reference material containing known levels of antiepilepsy drugs in serum. The reference material now enables different laboratories to detect and correct errors in their assay procedures.

Before the introduction of this reference material, five laboratories analyzing identical samples reported a range of results with relative errors of 40% to 110% of the expected value. After distribution of the reference material, the error was reduced to 20% to 40%.

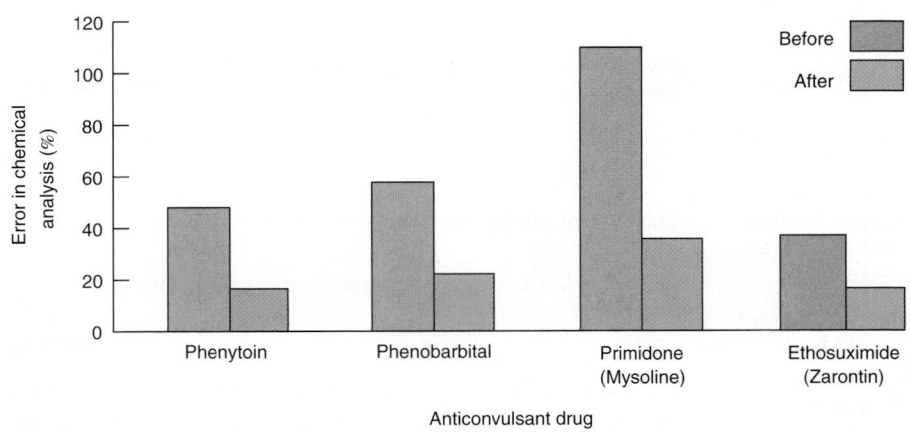

that in Figure 3-3, by the procedure on page 45. To do this, deliver distilled water from the buret into a flask and weigh it. Determine the volume of water from its mass by using Table 2-7. Figure 3-3 tells us to apply a correction factor of -0.03 mL to the measured value of 29.43 mL. The actual volume delivered is $29.43 - 0.03 = 29.40$ mL.

A key feature of systematic error is that it is reproducible. For the buret just discussed, the error is always -0.03 mL when the buret reading is 29.43 mL. Systematic error may always be positive in some regions and always negative in others. With care and cleverness, you can detect and correct a systematic error.

Random Error

Random error, also called **indeterminate error,** arises from uncontrolled (and maybe uncontrollable) variables in the measurement. Random error has an equal chance of being positive or negative. It is always present and cannot be corrected. There is random error associated with reading a scale. Different people reading the scale in Figure 3-1 report a range of values representing their subjective interpolation between the markings. One person reading the same instrument several times might report several different readings. Another random error results from electrical noise in an instrument. Positive and negative fluctuations occur with approximately equal frequency and cannot be completely eliminated.

Precision and Accuracy

Precision describes the reproducibility of a result. If you measure a quantity several times and the values agree closely with one another, your measurement is precise. If the values vary widely, your measurement is not precise. **Accuracy** describes how close a measured value is to the "true" value. If a known standard is available, accuracy is how close your value is to the known value.

A measurement might be reproducible, but wrong. If you made a mistake preparing a solution for a titration, you might do a series of reproducible titrations but report an incorrect result because the concentration of the titrating solution was not what you intended. In this case, precision is good but accuracy is poor. Conversely, it is possible to make poorly reproducible measurements clustered around the correct value. For this case, precision is poor but accuracy is good. An ideal procedure is both precise and accurate.

Ways to detect systematic error:

1. Analyze a known sample, such as a certified reference material (Box 3-2). Your method should reproduce the known answer.

2. Analyze blank samples containing no analyte being sought. If you observe a nonzero result, your method responds to more than you intend. Section 5-1 discusses different kinds of blanks.

3. Use different analytical methods to measure the same quantity. If results do not agree, there is error in one (or more) of the methods.

4. Round robin experiment: Different people in several laboratories analyze identical samples by the same or different methods. Disagreement beyond the estimated random error is systematic error (see Box 15-1).

Random error cannot be eliminated, but it might be reduced by a better experiment.

Precision: reproducibility
Accuracy: nearness to the "truth"

Accuracy is defined as nearness to the "true" value. *True* is in quotes because somebody must *measure* the "true" value, and there is error associated with *every* measurement. The "true" value is best obtained by an experienced person using a well-tested procedure. It is desirable to test the result by using different procedures, because systematic error could lead to poor agreement between methods. Good agreement among several methods affords us confidence, but never proof, that results are accurate.

Absolute and Relative Uncertainty

Absolute uncertainty expresses the margin of uncertainty associated with a measurement. If the estimated uncertainty in reading a calibrated buret is ±0.02 mL, we say that ±0.02 mL is the absolute uncertainty associated with the reading.

Relative uncertainty compares the size of the absolute uncertainty with the size of its associated measurement. The relative uncertainty of a buret reading of 12.35 ± 0.02 mL is a dimensionless quotient:

An uncertainty of ±0.02 mL means that, when the reading is 13.33 mL, the true value could be in the range 13.31 to 13.35 mL.

Relative uncertainty:

$$\text{Relative uncertainty} = \frac{\text{absolute uncertainty}}{\text{magnitude of measurement}} \qquad (3\text{-}2)$$

$$= \frac{0.02 \text{ mL}}{12.35 \text{ mL}} = 0.002$$

The percent relative uncertainty is simply

If you use a 50-mL buret, design your titration to require 20–40 mL of reagent to produce a small relative uncertainty of 0.1–0.05%.

Percent relative uncertainty:

$$\text{Percent relative uncertainty} = 100 \times \text{relative uncertainty} \qquad (3\text{-}3)$$

$$= 100 \times 0.002 = 0.2\%$$

In a gravimetric analysis, plan to have enough precipitate for a low relative uncertainty. If weighing precision is ±0.3 mg, a 100-mg precipitate has a relative weighing error of 0.3% and a 300-mg precipitate has an uncertainty of 0.1%.

If the absolute uncertainty in reading a buret is constant at ±0.02 mL, the percent relative uncertainty is 0.2% for a volume of 10 mL and 0.1% for a volume of 20 mL.

3-4 Propagation of Uncertainty from Random Error[3]

We can usually estimate or measure the random error associated with a measurement, such as the length of an object or the temperature of a solution. Uncertainty might be based on how well we can read an instrument or on our experience with a particular method. If possible, uncertainty will be expressed as the *standard deviation, standard deviation of the mean*, or a *confidence interval*, which we discuss in Chapter 4. This section applies only to random error. We assume that systematic error has been detected and corrected.

Most propagation of uncertainty computations that you will encounter deal with random error, not systematic error. Our goal is always to eliminate systematic error.

For most experiments, we need to perform arithmetic operations on several numbers, each of which has a random error. The most likely uncertainty in the result is not the sum of individual errors, because some of them are likely to be positive and some negative. We expect some cancellation of errors.

Addition and Subtraction

Suppose you wish to perform the following arithmetic, in which the experimental uncertainties, designated e_1, e_2, and e_3, are given in parentheses.

$$
\begin{array}{r}
1.76\ (\pm0.03) \leftarrow e_1 \\
+\ 1.89\ (\pm0.02) \leftarrow e_2 \\
-\ 0.59\ (\pm0.02) \leftarrow e_3 \\
\hline
3.06\ (\pm e_4)
\end{array}
\qquad (3\text{-}4)
$$

The arithmetic answer is 3.06. But what is the uncertainty associated with this result?

For addition and subtraction, the uncertainty in the answer is obtained from the *absolute uncertainties* of the individual terms as follows:

For **addition and subtraction,** use **absolute uncertainty.**

Uncertainty in addition and subtraction:

$$e_4 = \sqrt{e_1^2 + e_2^2 + e_3^2} \qquad (3\text{-}5)$$

For the sum in Equation 3-4, we can write

$$e_4 = \sqrt{(0.03)^2 + (0.02)^2 + (0.02)^2} = 0.04_1$$

CHAPTER 3 Experimental Error

The absolute uncertainty e_4 is ± 0.04, and we express the answer as 3.06 ± 0.04. Although there is only one significant figure in the uncertainty, we wrote it initially as 0.04_1, with the first insignificant figure subscripted. We retain one or more insignificant figures to avoid introducing round-off errors into later calculations through the number 0.04_1. The insignificant figure was subscripted to remind us where the last significant figure should be when we conclude the calculations.

To find the percent relative uncertainty in the sum of Equation 3-4, we write

$$\text{Percent relative uncertainty} = \frac{0.04_1}{3.06} \times 100 = 1._3\%$$

The uncertainty, 0.04_1, is $1._3\%$ of the result, 3.06. When the first digit of uncertainty is 1, it is reasonable to retain an extra digit to avoid throwing away information. We express the final result as

3.06 (± 0.04)   (absolute uncertainty)

3.06 ($\pm 1.3\%$)   (relative uncertainty)

> When the first digit of uncertainty is 1, retain an extra digit to avoid throwing away information.

> For addition and subtraction, use absolute uncertainty. Relative uncertainty can be found at the end of the calculation.

EXAMPLE Uncertainty in a Buret Reading

The volume delivered by a buret is the difference between final and initial readings. If the uncertainty in each reading is ± 0.02 mL, what is the uncertainty in the volume delivered?

Solution Suppose that the initial reading is 0.05 (± 0.02) mL and the final reading is 17.88 (± 0.02) mL. The volume delivered is the difference:

$$
\begin{array}{r}
17.88\,(\pm 0.02) \\
-\ 0.05\,(\pm 0.02) \\
\hline
17.83\,(\pm e)
\end{array}
\qquad e = \sqrt{0.02^2 + 0.02^2} = 0.02_8 \approx 0.03
$$

Regardless of the initial and final readings, if the uncertainty in each one is ± 0.02 mL, the uncertainty in volume delivered is ± 0.03 mL.

TEST YOURSELF What would be the uncertainty in volume delivered if the uncertainty in each reading were 0.03 mL? (***Answer:*** ± 0.04 mL)

Multiplication and Division

For multiplication and division, first convert all uncertainties into percent relative uncertainties. Then calculate the error of the product or quotient as follows:

Uncertainty in multiplication and division:

$$\%e_4 = \sqrt{(\%e_1)^2 + (\%e_2)^2 + (\%e_3)^2} \tag{3-6}$$

> For **multiplication and division**, use **percent relative uncertainty.**

For example, consider the following operations:

$$\frac{1.76\,(\pm 0.03) \times 1.89\,(\pm 0.02)}{0.59\,(\pm 0.02)} = 5.64 \pm e_4$$

First convert absolute uncertainties into percent relative uncertainties.

$$\frac{1.76\,(\pm 1._7\%) \times 1.89\,(\pm 1._1\%)}{0.59\,(\pm 3._4\%)} = 5.64 \pm e_4$$

> **Advice:** Retain one or more extra insignificant figures until you have finished your entire calculation. Then round to the correct number of digits. When storing intermediate results in a calculator, keep all digits without rounding.

Then find the percent relative uncertainty of the answer by using Equation 3-6.

$$\%e_4 = \sqrt{(1._7)^2 + (1._1)^2 + (3._4)^2} = 4._0\%$$

The answer is $5.6_4\,(\pm 4._0\%)$.

To convert relative uncertainty into absolute uncertainty, find $4._0\%$ of the answer.

$$4._0\% \times 5.6_4 = 0.04_0 \times 5.6_4 = 0.2_2$$

The answer is $5.6_4\,(\pm 0.2_2)$. Finally, drop the insignificant digits.

5.6 (± 0.2)   (absolute uncertainty)

5.6 ($\pm 4\%$)   (relative uncertainty)

> For multiplication and division, use percent relative uncertainty. Absolute uncertainty can be found at the end of the calculation.

The denominator of the original problem, 0.59, limits the answer to two digits.

Mixed Operations

Now consider a computation containing subtraction and division:

$$\frac{[1.76\,(\pm 0.03) - 0.59\,(\pm 0.02)]}{1.89\,(\pm 0.02)} = 0.619_0 \pm \,?$$

First work out the difference in the numerator, using absolute uncertainties. Thus,

$$1.76\,(\pm 0.03) - 0.59\,(\pm 0.02) = 1.17\,(\pm 0.03_6)$$

because $\sqrt{(0.03)^2 + (0.02)^2} = 0.03_6$.

Then convert into percent relative uncertainties. Thus,

$$\frac{1.17\,(\pm 0.03_6)}{1.89\,(\pm 0.02)} = \frac{1.17\,(\pm 3._1\%)}{1.89\,(\pm 1._1\%)} = 0.619_0\,(\pm 3._3\%)$$

because $\sqrt{(3._1\%)^2 + (1._1\%)^2} = 3._3\%$.

The percent relative uncertainty is $3._3\%$, so the absolute uncertainty is $0.03_3 \times 0.619_0 = 0.02_0$. The final answer can be written as

$$0.619\,(\pm 0.02_0) \qquad \text{(absolute uncertainty)}$$
$$0.619\,(\pm 3._3\%) \qquad \text{(relative uncertainty)}$$

Because the uncertainty begins in the 0.01 decimal place, it is reasonable to round the result to the 0.01 decimal place:

$$0.62\,(\pm 0.02) \qquad \text{(absolute uncertainty)}$$
$$0.62\,(\pm 3\%) \qquad \text{(relative uncertainty)}$$

> The result of a calculation ought to be written in a manner consistent with its uncertainty.

The *Real* Rule for Significant Figures

> The real rule: The first uncertain figure is the last significant figure.

The first digit of the absolute uncertainty is the last significant digit in the answer. For example, in the quotient

$$\frac{0.002\,364\,(\pm 0.000\,003)}{0.025\,00\,(\pm 0.000\,05)} = 0.094\,6\,(\pm 0.000\,2)$$

uncertainty ($\pm 0.000\,2$) begins in the fourth decimal place. Therefore, the answer $0.094\,6$ is properly expressed with *three* significant figures, even though there are four figures in the original data. The first uncertain figure of the answer is the last significant figure. The quotient

$$\frac{0.002\,664\,(\pm 0.000\,003)}{0.025\,00\,(\pm 0.000\,05)} = 0.106\,6\,(\pm 0.000\,2)$$

is expressed with *four* significant figures because the uncertainty occurs in the fourth place. The quotient

$$\frac{0.821\,(\pm 0.002)}{0.803\,(\pm 0.002)} = 1.022\,(\pm 0.004)$$

> In multiplication and division, keep an extra digit when the answer lies between 1 and 2.

is expressed with *four* figures even though the dividend and divisor each have *three* figures.

Now you can appreciate why *it is all right to keep one extra digit when the first digit of an answer lies between 1 and 2.* The quotient 82/80 is better written as 1.02 than 1.0. If the uncertainties in 82 and 80 are in the ones place, the uncertainty is of the order of 1%, which is in the second decimal place of 1.02. If I write 1.0, you can surmise that the uncertainty is at least $1.0 \pm 0.1 = \pm 10\%$, which is much larger than the actual uncertainty.

EXAMPLE Significant Figures in Laboratory Work

You prepared a 0.250 M NH_3 solution by diluting 8.46 (± 0.04) mL of 28.0 (± 0.5) wt% NH_3 [density = 0.899 (± 0.003) g/mL] up to 500.0 (± 0.2) mL. Find the uncertainty in 0.250 M. The molecular mass of NH_3, 17.031 g/mol, has negligible uncertainty relative to other uncertainties in this problem.

Solution To find the uncertainty in molarity, we need the uncertainty in moles delivered to the 500-mL flask. The concentrated reagent contains 0.899 (± 0.003) g of solution per mL. Weight percent tells us that the reagent contains 0.280 ($\pm$0.005) g of NH_3 per gram of solution. In our calculations, we retain extra insignificant digits and round off only at the end.

$$\text{Grams of } NH_3 \text{ per mL} \atop \text{in concentrated reagent} = 0.899 \,(\pm0.003) \, \frac{\text{g solution}}{\text{mL}} \times 0.280 \,(\pm0.005) \, \frac{\text{g } NH_3}{\text{g solution}}$$

$$= 0.899 \,(\pm0.334\%) \, \frac{\text{g solution}}{\text{mL}} \times 0.280 \,(\pm1.79\%) \, \frac{\text{g } NH_3}{\text{g solution}}$$

$$= 0.251\,7 \,(\pm1.82\%) \, \frac{\text{g } NH_3}{\text{mL}}$$

Convert absolute uncertainty into percent relative uncertainty for multiplication.

because $\sqrt{(0.334\%)^2 + (1.79\%)^2} = 1.82\%$.

Next, we find the moles of ammonia contained in 8.46 ($\pm$0.04) mL of concentrated reagent. The relative uncertainty in volume is 0.04/8.46 = 0.473%.

$$\text{mol } NH_3 = \frac{0.251\,7 \,(\pm1.82\%) \, \dfrac{\text{g } NH_3}{\text{mL}} \times 8.46 \,(\pm0.473\%) \text{ mL}}{17.031 \,(\pm0\%) \, \dfrac{\text{g } NH_3}{\text{mol}}}$$

$$= 0.125\,0 \,(\pm1.88\%) \text{ mol}$$

because $\sqrt{(1.82\%)^2 + (0.473\%)^2 + (0\%)^2} = 1.88\%$.

This much ammonia was diluted to 0.500 0 ($\pm$0.000 2) L. The relative uncertainty in the final volume is 0.000 2/0.500 0 = 0.04%. The molarity is

$$\frac{\text{mol } NH_3}{\text{L}} = \frac{0.125\,0 \,(\pm1.88\%) \text{ mol}}{0.500\,0 \,(\pm0.04\%) \text{ L}}$$

$$= 0.250\,1 \,(\pm1.88\%) \text{ M}$$

because $\sqrt{(1.88\%)^2 + (0.04\%)^2} = 1.88\%$. The absolute uncertainty is 1.88% of 0.250 1 M = 0.004 7 M. The uncertainty in molarity is in the third decimal place, so our final, rounded answer is

$$[NH_3] = 0.250 \,(\pm0.005) \text{ M}$$

TEST YOURSELF Suppose that you used smaller volumetric apparatus to prepare 0.250 M NH_3 solution by diluting 84.6 ($\pm$0.8) μL of 28.0 ($\pm$0.5) wt% NH_3 up to 5.00 ($\pm$0.02) mL. Find the uncertainty in 0.250 M. (*Answer:* 0.250 ($\pm$0.005) M. The uncertainty in density of concentrated NH_3 overwhelms all other smaller uncertainties in this procedure.)

EXAMPLE Volumetric Versus Gravimetric Dilution

Let's compare the uncertainty resulting from a 10-fold volumetric dilution with a 10-fold gravimetric dilution. **(a)** For the volumetric dilution, suppose that you have a standard reagent with a concentration of 0.046 80 M. For the sake of this example, we suppose that there is negligible uncertainty in the initial concentration. To dilute by a factor of 10, you use a micropipet to deliver 1 000 μL (= 1.000 mL) into a 10-mL volumetric flask and dilute to volume. **(b)** For the gravimetric dilution, suppose that you have a standard reagent with a concentration of 0.046 80 mol reagent/kg solution. To dilute it by a factor of close to 10, you weight out 983.2 mg (= 0.983 2 g) of solution ($\approx$ 1 mL) and add 9.026 6 g of water ($\approx$ 9 mL). For each procedure, find the resulting concentration and its relative uncertainty.

mol reagent/kg solution is a convenient unit that is *not* molality, which is mol reagent/kg solvent.

The factor V_{dil}/V_{conc} comes from Equation 1-5:

$$M_{conc} \cdot V_{conc} = M_{dil} \cdot V_{dil}$$

$$M_{conc} \cdot \frac{V_{conc}}{V_{dil}} = M_{dil}$$

$$\frac{M_{conc}}{V_{dil}/V_{conc}} = M_{dil}$$

where M is concentration, V is volume, "dil" is dilute, and "conc" is concentrated

Solution (a) We will use the tolerance for the volumetric flask from Table 2-3 (10.00 $\pm$ 0.02 mL = 10.00 mL $\pm$ 0.2%) and for the micropipet from Table 2-5 (1 000 μL $\pm$ 0.3%). The *dilution factor* is

$$\text{Dilution factor} = \frac{V_{final}}{V_{initial}} = \frac{10.00 \,(\pm0.2\%) \text{ mL}}{1.000 \,(\pm0.3\%) \text{ mL}} = 10.00 \pm 0.3_6 \%$$

because $\sqrt{(0.2\%)^2 + (0.3\%)^2} = 0.3_6\%$. The concentration of the dilute solution is

$$\frac{0.046\ 80\ M}{10.00\ \pm\ 0.3_6\%} = 0.004\ 680\ M\ \pm\ 0.3_6\% = 0.004\ 680\ \pm\ 0.000\ 017\ M$$

(b) In the gravimetric procedure, we dilute 0.983 2 g of concentrated solution up to (0.983 2 g + 9.026 6 g) = 10.009 8 g. The dilution factor is (10.009 8 g)/(0.983 2 g) = 10.180 8. Suppose that the uncertainty in each mass is ±0.3 mg. The uncertainty in the mass of concentrated solution is $0.983\ 2\ \pm\ 0.000\ 3$ g = $0.983\ 2\ (\pm0.03_{05}\%)$ g. The absolute uncertainty in the sum (0.983 2 g + 9.026 6 g) is $\sqrt{(0.000\ 3)^2 + (0.000\ 3)^2} = 0.000\ 4_2$ g, which is $0.004_2\%$. The uncertainty in the dilution factor is

$$\text{Dilution factor} = \frac{10.009\ 8\ (\pm0.004_2\%)\ g}{0.983\ 2\ (\pm0.03_{05}\%)\ g} = 10.180\ 8\ \pm\ 0.03_{08}\%$$

because $\sqrt{(0.004_2\%)^2 + (0.03_{05}\%)^2} = 0.03_{08}\%$. The concentration of the dilute solution is

$$\frac{0.046\ 80\ \text{mol reagent/kg solution}}{10.180\ 8\ \pm\ 0.03_{08}\%} = 0.004\ 596_9\ M\ \pm\ 0.03_{08}\%$$

$$= 0.004\ 596_9\ \pm\ 0.000\ 001_4\ \text{mol reagent/kg solution}$$

In this example, the gravimetric dilution is 10 times more precise than the volumetric dilution (0.03% versus 0.4%). Increased precision is the reason why gravimetric titrations are recommended over volumetric titrations, though the latter are less tedious.

TEST YOURSELF Describe a volumetric dilution procedure that would be more precise than using a 1 000-μL micropipet and a 10-mL volumetric flask. What would be the relative uncertainty in the dilution?

(*Answer:* Using a 10-mL transfer pipet and a 100-mL volumetric flask gives an uncertainty of $\sqrt{(0.2\%)^2 + (0.08\%)^2} = 0.2\%$. A 100-mL transfer pipet with a 1-L volumetric flask gives an uncertainty of $\sqrt{(0.08\%)^2 + (0.03\%)^2} = 0.09\%$.)

Exponents and Logarithms

For the function $y = x^a$, the relative uncertainty in y ($\%e_y$) is a times the relative uncertainty in x ($\%e_x$):

Uncertainty for powers and roots:
$$y = x^a \quad \Rightarrow \quad \%e_y = a(\%e_x) \qquad (3\text{-}7)$$

For example, if $y = \sqrt{x} = x^{1/2}$, a 2% uncertainty in x will yield a $(\frac{1}{2})(2\%) = 1\%$ uncertainty in y. If $y = x^2$, a 3% uncertainty in x leads to a $(2)(3\%) = 6\%$ uncertainty in y.

To calculate a power or root on your calculator, use the y^x button. For example, to find a cube root ($y^{1/3}$), raise y to the 0.333 333 333... power with the y^x button. In Excel, y^x is y^x. The cube root is y^(1/3).

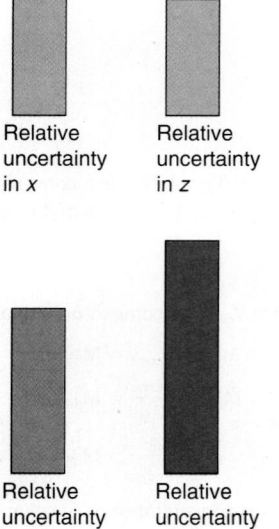

Relative uncertainty in x Relative uncertainty in z

Relative uncertainty in $x \cdot z$ Relative uncertainty in x^2

EXAMPLE Propagation of Uncertainty in the Product $x \cdot x$

If an object falls for t seconds, the distance it travels is $\frac{1}{2}gt^2$, where g is the acceleration of gravity (9.8 m/s²) at the surface of the Earth. (This equation ignores the effect of drag produced by air, which slows the falling object.) If the object falls for 2.34 s, the distance traveled is $\frac{1}{2}(9.8\ \text{m/s}^2)(2.34\ \text{s})^2 = 26.8$ m. If the relative uncertainty in time is $\pm1.0\%$, what is the relative uncertainty in distance?

Solution Equation 3-7 tells us that for $y = x^a$, the relative uncertainty in y is a times the relative uncertainty in x:

$$y = x^a \Rightarrow \%e_y = a(\%e_x)$$
$$\text{Distance} = \tfrac{1}{2}gt^2 \Rightarrow \%e_{\text{distance}} = 2(\%e_t) = 2(1.0\%) = 2.0\%$$

If you write distance = $\frac{1}{2}gt \cdot t$, you might be tempted to say that the relative uncertainty in distance is $\sqrt{1.0^2 + 1.0^2} = 1.4\%$. This answer is wrong because the error in a single, measured value of t is always positive or always negative. If t is 1.0% high, then t^2 is 2% high because we are multiplying a high value by a high value: $(1.01)^2 = 1.02$.

Equation 3-6 presumes that the uncertainty in each factor of the product $x \cdot z$ is random and independent of the other. In the product $x \cdot z$, the measured value of x could be high sometimes and the measured value of z could be low sometimes. In the majority of cases, the uncertainty in the product $x \cdot z$ is not as great as the uncertainty in x^2.

TEST YOURSELF   You can calculate the time it will take for an object to fall from the top of a building to the ground if you know the height of the building. If the height has an uncertainty of 1.0%, what is the uncertainty in time? (*Answer:* 0.5%)

If y is the base 10 logarithm of x, then the absolute uncertainty in $y(e_y)$ is proportional to the relative uncertainty in x, which is e_x/x:

Uncertainty for logarithm:
$$y = \log x \quad \Rightarrow \quad e_y = \frac{1}{\ln 10} \frac{e_x}{x} \approx 0.434\,29 \frac{e_x}{x} \tag{3-8}$$

You should not work with percent relative uncertainty [$100 \times (e_x/x)$] in calculations with logs and antilogs because one side of Equation 3-8 has relative uncertainty and the other has absolute uncertainty.

The **natural logarithm** (ln) of x is the number y, whose value is such that $x = e^y$, where $e\ (= 2.718\,28\ldots)$ is called the base of the natural logarithm. The absolute uncertainty in y is equal to the relative uncertainty in x.

Uncertainty for natural logarithm:
$$y = \ln x \quad \Rightarrow \quad e_y = \frac{e_x}{x} \tag{3-9}$$

Now consider $y =$ antilog x, which is the same as saying $y = 10^x$. In this case, the relative uncertainty in y is proportional to the absolute uncertainty in x.

Uncertainty for 10^x:
$$y = 10^x \quad \Rightarrow \quad \frac{e_y}{y} = (\ln 10)e_x \approx 2.302\,6\,e_x \tag{3-10}$$

If $y = e^x$, the relative uncertainty in y equals the absolute uncertainty in x.

Uncertainty for e^x:
$$y = e^x \quad \Rightarrow \quad \frac{e_y}{y} = e_x \tag{3-11}$$

Table 3-1 summarizes rules for propagation of uncertainty. You need not memorize the rules for exponents, logs, and antilogs, but you should be able to use them.

> Use relative uncertainty (e_x/x), not percent relative uncertainty [$100 \times (e_x/x)$], in calculations involving log x, ln x, 10^x, and e^x.
>
> In Excel, the base 10 logarithm is log(x). The natural logarithm is ln(x). The expression 10^x is 10^x and the expression e^x is exp(x).

> Appendix B gives a general rule for propagation of random uncertainty for any function.

TABLE 3-1	**Summary of rules for propagation of uncertainty**			
Function	**Uncertainty**		**Function**[a]	**Uncertainty**[b]
$y = x_1 + x_2$	$e_y = \sqrt{e_{x_1}^2 + e_{x_2}^2}$		$y = x^a$	$\%e_y = a\%e_x$
$y = x_1 - x_2$	$e_y = \sqrt{e_{x_1}^2 + e_{x_2}^2}$		$y = \log x$	$e_y = \frac{1}{\ln 10}\frac{e_x}{x} \approx 0.434\,29\frac{e_x}{x}$
$y = x_1 \cdot x_2$	$\%e_y = \sqrt{\%e_{x_1}^2 + \%e_{x_2}^2}$		$y = \ln x$	$e_y = \frac{e_x}{x}$
$y = \dfrac{x_1}{x_2}$	$\%e_y = \sqrt{\%e_{x_1}^2 + \%e_{x_2}^2}$		$y = 10^x$	$\frac{e_y}{y} = (\ln 10)\,e_x \approx 2.302\,6\,e_x$
			$y = e^x$	$\frac{e_y}{y} = e_x$

a. x represents a variable and a represents a constant that has no uncertainty.
b. e_x/x is the relative error in x and $\%e_x$ is $100 \times e_x/x$.

EXAMPLE   **Uncertainty in H$^+$ Concentration**

Consider the function pH $= -\log$ [H$^+$], where [H$^+$] is the molarity of H$^+$. For pH $= 5.21 \pm 0.03$, find [H$^+$] and its uncertainty.

Solution First solve the equation pH = −log[H⁺] for [H⁺]: If $a = b$, then $10^a = 10^b$. If pH = −log[H⁺], then log[H⁺] = −pH and $10^{\log[H^+]} = 10^{-pH}$. But $10^{\log[H^+]} = [H^+]$. We therefore need to find the uncertainty in the equation

$$[H^+] = 10^{-pH} = 10^{-(5.21 \pm 0.03)}$$

In Table 3-1, the relevant function is $y = 10^x$, in which $y = [H^+]$ and $x = -(5.21 \pm 0.03)$. For $y = 10^x$, the table tells us that

$$\frac{e_y}{y} = 2.302\ 6\ e_x$$

$$\frac{e_{[H^+]}}{[H^+]} = 2.302\ 6\ e_{pH} = (2.302\ 6)(0.03) = 0.069\ 1 \qquad (3\text{-}12)$$

The relative uncertainty in [H⁺] is 0.069 1. For $[H^+] = 10^{-pH} = 10^{-5.21} = 6.17 \times 10^{-6}$ M, we find

$$\frac{e_{[H^+]}}{[H^+]} = \frac{e_{[H^+]}}{6.17 \times 10^{-6}\ M} = 0.069\ 1 \Rightarrow e_{[H^+]} = 4.26 \times 10^{-7}\ M$$

The concentration of H⁺ is 6.17 (±0.426) × 10⁻⁶ M = 6.2 (±0.4) × 10⁻⁶ M. An uncertainty of 0.03 in pH gives an uncertainty of 7% in [H⁺]. Notice that extra digits were retained in the intermediate results and were not rounded off until the final answer.

TEST YOURSELF If uncertainty in pH is doubled to ±0.06, what is the relative uncertainty in [H⁺]? (***Answer:*** 14%)

3-5 Propagation of Uncertainty from Systematic Error

Systematic error occurs in some common situations and is treated differently from random error in arithmetic operations. We will look at examples of systematic error in molecular mass and volumetric glassware.

Uncertainty in Atomic Mass

The periodic table inside the cover of this book gives the atomic mass of oxygen as 15.999 4 ± 0.000 4 g/mol. The key for the table states that, if no uncertainty is listed, the uncertainty is ±1 in the last decimal place. For most of the elements, which have more than one isotope, the uncertainty is *not* from random error in measurement. The uncertainty is from isotopic variation in different materials (Box 3-3).[4] The atomic mass of oxygen in a particular substance has a *systematic* uncertainty. It could be, say, 15.999 7, with only a small random variation around that value.

Uncertainty in Molecular Mass

What is the uncertainty in molecular mass of O_2? If the mass of each oxygen atom were at the upper limit of the uncertainty range in the periodic table (15.999 8), the mass of O_2 is 2 × 15.999 8 = 31.999 6 g/mol. If the mass of each oxygen atom were at the lower limit of the standard uncertainty range (15.999 0), the mass of O_2 is 2 × 15.999 0 = 31.998 0 g/mol. The mass of O_2 is somewhere in the range 31.998 8 ± 0.000 8. The uncertainty in the mass of n atoms is $n \times$ (uncertainty of one atom) = 2 × (±0.000 8) = ±0.001 6. The uncertainty is not $\pm\sqrt{0.000\ 8^2 + 0.000\ 8^2} = \pm 0.001\ 1$. *For systematic uncertainty, we add the uncertainties of each term in a sum or difference.*

Now let's find the uncertainty in molecular mass of C_2H_4.

$$\text{Atomic mass of C} = 12.010\ 6 \pm 0.001\ 0$$

$$\text{Atomic mass of H} = 1.007\ 98 \pm 0.000\ 14$$

Uncertainties in the masses of atoms in C_2H_4 are obtained by multiplying the uncertainty for each atom by the number of atoms of each type:

2C:	2(12.010 6 ± 0.001 0)	= 24.021 2 ± **0.002 0**	← 2 × 0.001 0
4H:	4(1.007 98 ± 0.000 14)	= 4.031 92 ± **0.000 56**	← 4 × 0.000 14
		28.053 12 ± ?	(3-13)

Propagation of systematic uncertainty:
Uncertainty in mass of *n* identical atoms
= *n* × (uncertainty in atomic mass)

BOX 3-3 **Atomic Masses of the Elements**

Every four years, a commission of the International Union for Pure and Applied Chemistry uses the best available measurements to reevaluate atomic masses. (The commission calls these atomic *weights*.) The **atomic mass** of an element is a weighted average of the masses of its isotopes found in terrestrial sources. If an element has just one stable isotope, its mass can be measured precisely by mass spectrometry. For the more than 80% of elements with several isotopes, the average atomic mass depends on the mole fraction of each isotope in the material. Different materials have different fractions of isotopes, so the average atomic mass of an element varies from one specimen to another. The atomic mass of sodium, which has just one stable isotope, is $22.989\,769\,28 \pm 0.000\,000\,02$. The atomic mass of lead, which has varying proportions of several isotopes, is listed as 207.2 ± 0.1 because of that variation.

In 2009, a significant change was made in the way atomic masses are listed. For elements with multiple stable isotopes, the atomic mass is now given as an *interval* covering the range found nature. In 2005, the atomic mass of oxygen was listed as $15.999\,4 \pm 0.000\,3$. As of 2009, the atomic mass is shown as an *interval* $[15.999\,03;\ 15.999\,77]$ which, you can see in the chart, covers the range of atomic mass of oxygen from many sources.

Nonetheless, you and I need "an atomic mass" to use for everyday calculations. For this purpose, I choose to use the midpoint of the atomic mass interval and to add an uncertainty that covers the full range of the interval. So, for example the atomic mass interval for H is $[1.007\,84;\ 1.008\,11]$. The midpoint is $1.007\,98$ and the uncertainty $\pm0.000\,14$ covers the interval. The atomic mass listed in the periodic table inside the cover of this book is $1.007\,98 \pm 14$, where 14 is the uncertainty in the last two decimal places. Atomic masses of H, Li, B, C, N, O, Si, S, Cl, and Tl shown in the periodic table in this book are similarly derived from their atomic mass intervals.

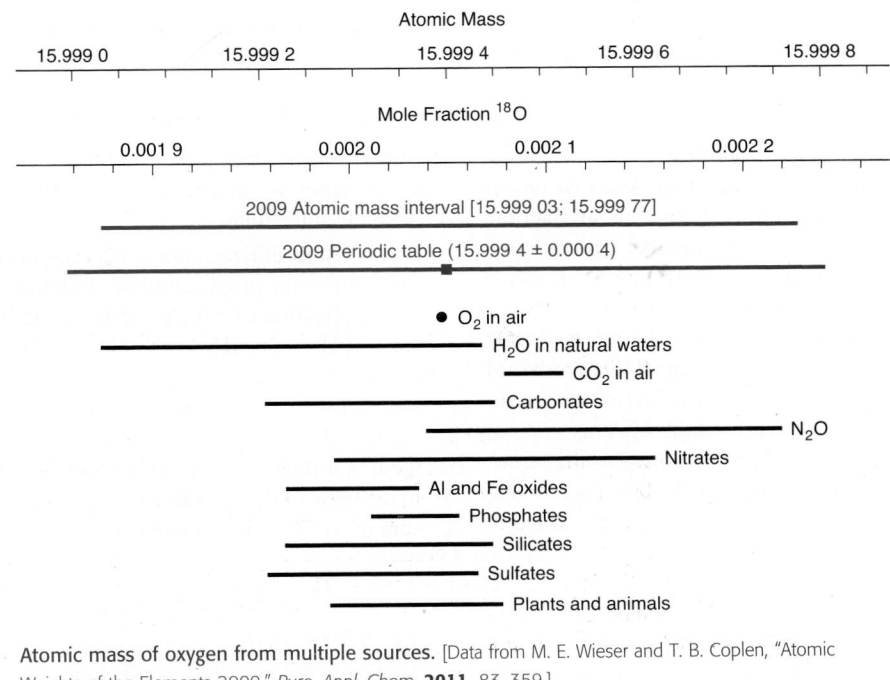

Atomic mass of oxygen from multiple sources. [Data from M. E. Wieser and T. B. Coplen, "Atomic Weights of the Elements 2009," *Pure. Appl. Chem.* **2011,** *83*, 359.]

For the uncertainty in the sum of the masses of 2C + 4H, we use Equation 3-5, which applies to random error, because the uncertainties in the masses of C and H are independent of each other. One might be positive and one might be negative. So the molecular mass of C_2H_4 is

$$28.053\,12 \pm \sqrt{0.002\,0^2 + 0.000\,56^2}$$
$$28.053_1 \quad \pm 0.002_1 \text{ g/mol}$$

The relative uncertainty is $100 \times (0.002_1/28.053_1) = 0.007_4\%$.

> Use the rule for propagation of random uncertainty for the sum of atomic masses of different elements because uncertainties for different elements are independent.

Multiple Deliveries from One Pipet: The Virtue of Calibration

A 25-mL Class A volumetric pipet is certified by the manufacturer to deliver 25.00 ± 0.03 mL. The volume delivered by a given pipet is reproducible but can be in the range 24.97 to 25.03 mL. If you use an uncalibrated 25-mL Class A volumetric pipet four times to deliver a total of 100 mL, what is the uncertainty in 100 mL? The uncertainty is a systematic error, so the uncertainty in four pipet volumes is like the uncertainty in the mass of 4 mol of oxygen: the uncertainty is $\pm4 \times 0.03 = \pm0.12$ mL, not $\pm\sqrt{0.03^2 + 0.03^2 + 0.03^2 + 0.03^2} = \pm0.06$ mL.

The difference between 25.00 mL and the actual volume delivered by a particular pipet is a *systematic* error. It is always the same, within a small random error. You could calibrate a pipet by weighing the water it delivers, as in Section 2-9. Calibration eliminates systematic error, because we would know that the pipet always delivers, say, 24.991 ± 0.006 mL. The uncertainty ± 0.006 mL is a *random* error.

Calibration improves certainty by removing systematic error. If a calibrated pipet delivers a mean volume of 24.991 mL with an uncertainty of ± 0.006 mL, and you deliver four aliquots, the volume delivered is $4 \times 24.991 = 99.964$ mL and the uncertainty is $\pm \sqrt{0.006^2 + 0.006^2 + 0.006^2 + 0.006^2} = \pm 0.012$ mL. For an uncalibrated pipet, the uncertainty is $\pm 4 \times 0.03 = \pm 0.12$ mL.

$$\text{Calibrated pipet volume} = 99.964 \pm 0.012 \text{ mL}$$

$$\text{Uncalibrated pipet volume} = 100.00 \pm 0.12 \text{ mL}$$

Terms to Understand

absolute uncertainty	certified reference material	logarithm	random error
accuracy	characteristic	mantissa	relative uncertainty
antilogarithm	determinate error	natural logarithm	significant figure
atomic mass	indeterminate error	precision	systematic error

Summary

The number of significant digits in a number is the minimum required to write the number in scientific notation. The first uncertain digit is the last significant figure. In addition and subtraction, the last significant figure is determined by the number with the fewest decimal places (when all exponents are equal). In multiplication and division, the number of figures is usually limited by the factor with the smallest number of digits. The number of figures in the mantissa of the logarithm of a quantity should equal the number of significant figures in the quantity. Random (indeterminate) error affects the precision (reproducibility) of a result, whereas systematic (determinate) error affects the accuracy (nearness to the "true" value). With diligence, systematic error can be discovered and eliminated, but some random error is always present. We strive to eliminate systematic errors in all measurements. For random errors, propagation of uncertainty in addition and subtraction requires absolute uncertainties ($e_3 = \sqrt{e_1^2 + e_2^2}$), whereas multiplication and division utilize relative uncertainties ($\%e_3 = \sqrt{\%e_1^2 + \%e_2^2}$). Other rules for propagation of random error are found in Table 3-1, one application of which is calculating hydrogen ion concentration from pH. For $[H^+] = 10^{-pH}$, the uncertainty in $[H^+]$ is $e_{[H^+]}/[H^+] = 2.302\,6\,e_{pH}$. Always retain more digits than necessary during a calculation and round off to the appropriate number of digits at the end. Systematic error in the mass on n atoms of one element is n times the uncertainty in mass of that element. Uncertainty in the mass of a molecule with several elements is computed from the sum of squares of the systematic uncertainty for each element.

Exercises

3-A. An empty crucible weighs 12.437 2 g and the same crucible containing a precipitate from a gravimetric analysis weighs 12.529 6 g.

(a) Each mass has six significant digits. What is the mass of precipitate contained in the crucible and how many significant digits are in that mass?

(b) The manufacturer states that the balance has an uncertainty of ± 0.3 mg. Find the absolute and relative uncertainty of the mass of the precipitate and write the mass with a reasonable number of digits.

3-B. Write each answer with a reasonable number of figures. Find the absolute and percent relative uncertainty for each answer.

(a) $[12.41\,(\pm 0.09) \div 4.16\,(\pm 0.01)] \times 7.068\,2\,(\pm 0.000\,4) = ?$

(b) $[3.26\,(\pm 0.10) \times 8.47\,(\pm 0.05)] - 0.18\,(\pm 0.06) = ?$

(c) $6.843\,(\pm 0.008) \times 10^4 \div [2.09\,(\pm 0.04) - 1.63\,(\pm 0.01)] = ?$

(d) $\sqrt{3.24 \pm 0.08} = ?$

(e) $(3.24 \pm 0.08)^4 = ?$

(f) $\log\,(3.24 \pm 0.08) = ?$

(g) $10^{3.24 \pm 0.08} = ?$

3-C. (a) How many milliliters of 53.4 (± 0.4) wt% NaOH with a density of 1.52 (± 0.01) g/mL will you need to prepare 2.000 L of 0.169 M NaOH?

(b) If the uncertainty in delivering NaOH is ± 0.01 mL, calculate the absolute uncertainty in the molarity (0.169 M). Assume there is negligible uncertainty in the formula mass of NaOH and in the final volume (2.000 L).

3-D. The pH of a solution is 4.44 ± 0.04. Find $[H^+]$ and its absolute uncertainty.

3-E. We have a 37.0 (± 0.5) wt% HCl solution with a density of 1.18 (± 0.01) g/mL. To deliver 0.050 0 mol of HCl requires 4.18 mL of solution. If the uncertainty that can be tolerated in 0.050 0 mol is $\pm 2\%$, how big can the absolute uncertainty in 4.18 mL be? (*Caution:* In this problem, you have to work backward. You would normally compute the uncertainty in mol HCl from the uncertainty in volume:

$$\text{mol HCl} = \frac{\text{mL solution} \times \dfrac{\text{g solution}}{\text{mL solution}} \times \dfrac{\text{g HCl}}{\text{g solution}}}{\dfrac{\text{g HCl}}{\text{mol HCl}}}$$

But, in this case, we know the uncertainty in mol HCl (2%) and we need to find what uncertainty in mL solution leads to that 2% uncertainty. The arithmetic has the form $a = b \times c \times d$, for which $\%e_a^2 = \%e_b^2 + \%e_c^2 + \%e_d^2$. If we know $\%e_a$, $\%e_c$, and $\%e_d$, we can find $\%e_b$ by subtraction: $\%e_b^2 = \%e_a^2 - \%e_c^2 - \%e_d^2$.)

Problems

Significant Figures

3-1. How many significant figures are there in the following numbers?

(a) 1.903 0   **(b)** 0.039 10   **(c)** 1.40×10^4

3-2. Round each number as indicated:

(a) 1.236 7 to 4 significant figures

(b) 1.238 4 to 4 significant figures

(c) 0.135 2 to 3 significant figures

(d) 2.051 to 2 significant figures

(e) 2.005 0 to 3 significant figures

3-3. Round each number to three significant figures:

(a) 0.216 74   **(b)** 0.216 5   **(c)** 0.216 500 3

3-4. *Vernier scale.* The figure below shows a scale found on instruments such as a micrometer caliper used for accurately measuring dimensions of objects. The lower scale slides along the upper scale and is used to interpolate between the markings on the upper scale. In (*a*), the reading (at the left-hand 0 of the

lower scale) is between 1.4 and 1.5 on the upper scale. To find the exact reading, observe which mark on the lower scale is aligned with a mark on the upper scale. Because the 6 on the lower scale is aligned with the upper scale, the correct reading is 1.46. Write the correct readings in (*b*) and (*c*) and indicate how many significant figures are in each reading.

3-5. Write each answer with the correct number of digits.

(a) $1.021 + 2.69 = 3.711$

(b) $12.3 - 1.63 = 10.67$

(c) $4.34 \times 9.2 = 39.928$

(d) $0.060\ 2 \div (2.113 \times 10^4) = 2.849\ 03 \times 10^{-6}$

(e) $\log (4.218 \times 10^{12}) = ?$

(f) $\text{antilog}(-3.22) = ?$

(g) $10^{2.384} = ?$

3-6. Write the formula mass of **(a)** BaF_2 and **(b)** $C_6H_4O_4$ with a reasonable number of digits. Use the periodic table inside the cover of this book to find atomic masses.

3-F. Compute the molecular mass and its uncertainty for NH_3. What is the percent relative uncertainty in molecular mass?

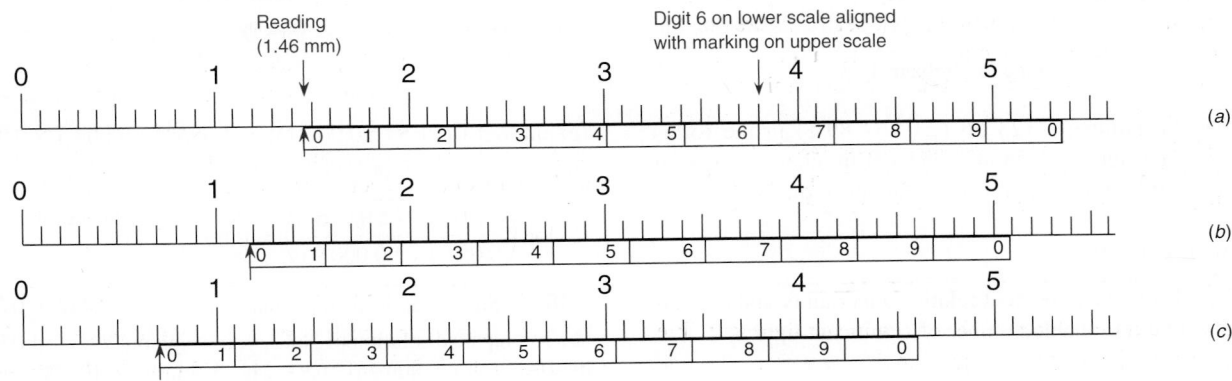

Figure for Problem 3-4.

3-7. Write each answer with the correct number of significant figures.

(a) $1.0 + 2.1 + 3.4 + 5.8 = 12.300\ 0$

(b) $106.9 - 31.4 = 75.500\ 0$

(c) $107.868 - (2.113 \times 10^2) + (5.623 \times 10^3) = 5\ 519.568$

(d) $(26.14/37.62) \times 4.38 = 3.043\ 413$

(e) $(26.14/(37.62 \times 10^8)) \times (4.38 \times 10^{-2}) = 3.043\ 413 \times 10^{-10}$

(f) $(26.14/3.38) + 4.2 = 11.933\ 7$

(g) $\log(3.98 \times 10^4) = 4.599\ 9$

(h) $10^{-6.31} = 4.897\ 79 \times 10^{-7}$

3-8. 🖼 *Controlling the appearance of a graph.* Figure 3-3 requires gridlines to read buret corrections. In this exercise, you will format a graph so that it looks like Figure 3-3. Follow the procedure in Section 2-11 to graph the data in the following table. For Excel

2007 or 2010, insert a Chart of the type Scatter with data points connected by straight lines. Delete the legend and title. With Chart Tools, Layout, Axis Titles, add labels for both axes. Click any number on the abscissa (*x*-axis) and go to Chart Tools, Format. In Format Selection, Axis Options, choose Minimum = 0, Maximum = 50, Major unit = 10, and Minor unit = 1. For Major tick mark type, select Outside. In Format Selection, Number, choose Number and set Decimal places = 0. In a similar manner, set the ordinate (*y*-axis) to run from −0.04 to +0.05 with a Major unit of 0.02 and a Minor unit of 0.01 and with Major tick marks Outside. To display gridlines, go to Chart Tools, Layout, and Gridlines. For Primary Horizontal Gridlines, select Major & Minor Gridlines. For Primary Vertical Gridlines, select Major & Minor Gridlines. To move *x*-axis labels from the middle of the chart to the bottom, click a number on the *y*-axis (not the *x*-axis) and select Chart Tools, Layout, Format Selection. In Axis Options, choose Horizontal axis crosses Axis value and type in −0.04. Close the Format Axis window and your graph should look like Figure 3-3.

Volume (mL)	Correction (mL)
0.03	0.00
10.04	0.04
20.03	0.02
29.98	−0.03
40.00	0.00
49.97	0.03

Types of Error

3-9. Why do we use quotation marks around the word *true* in the statement that accuracy refers to how close a measured value is to the "true" value?

3-10. Explain the difference between systematic and random error.

3-11. Suppose that in a gravimetric analysis, you forget to dry the filter crucibles before collecting precipitate. After filtering the product, you dry the product and crucible thoroughly before weighing them. Is the apparent mass of product always high or always low? Is the error in mass systematic or random?

3-12. State whether the errors in **(a)**–**(d)** are random or systematic:

(a) A 25-mL transfer pipet consistently delivers 25.031 ± 0.009 mL.

(b) A 10-mL buret consistently delivers 1.98 ± 0.01 mL when drained from exactly 0 to exactly 2 mL and consistently delivers 2.03 mL ± 0.02 mL when drained from 2 to 4 mL.

(c) A 10-mL buret delivered 1.983 9 g of water when drained from exactly 0.00 to 2.00 mL. The next time I delivered water from the 0.00 to the 2.00 mL mark, the delivered mass was 1.990 0 g.

(d) Four consecutive 20.0-µL injections of a solution into a chromatograph were made and the area of a particular peak was 4 383, 4 410, 4 401, and 4 390 units.

3-13. Cheryl, Cynthia, Carmen, and Chastity shot the following targets at Girl Scout camp. Match each target with the proper description.

(a) accurate and precise

(b) accurate but not precise

(c) precise but not accurate

(d) neither precise nor accurate

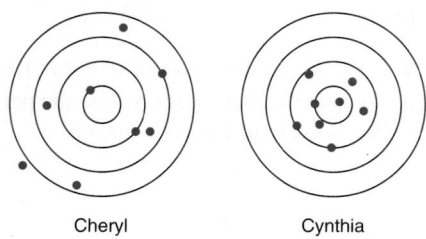
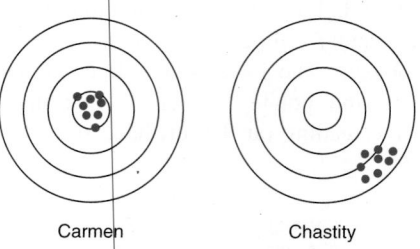

Figure for Problem 3-13.

3-14. Rewrite the number 3.123 56 (± 0.167 89%) in the forms **(a)** number ($\pm$ absolute uncertainty) and **(b)** number ($\pm$ percent relative uncertainty) with an appropriate number of digits.

Propagation of Uncertainty

3-15. Find the absolute and percent relative uncertainty and express each answer with a reasonable number of significant figures.

(a) $6.2\ (\pm 0.2) - 4.1\ (\pm 0.1) = ?$

(b) $9.43\ (\pm 0.05) \times 0.016\ (\pm 0.001) = ?$

(c) $[6.2\ (\pm 0.2) - 4.1\ (\pm 0.1)] \div 9.43\ (\pm 0.05) = ?$

(d) $9.43\ (\pm 0.05) \times \{[6.2\ (\pm 0.2) \times 10^{-3}] + [4.1\ (\pm 0.1) \times 10^{-3}]\} = ?$

3-16. Find the absolute and percent relative uncertainty and express each answer with a reasonable number of significant figures.

(a) $9.23\ (\pm 0.03) + 4.21\ (\pm 0.02) - 3.26\ (\pm 0.06) = ?$

(b) $91.3\ (\pm 1.0) \times 40.3\ (\pm 0.2)/21.1\ (\pm 0.2) = ?$

(c) $[4.97\ (\pm 0.05) - 1.86\ (\pm 0.01)]/21.1\ (\pm 0.2) = ?$

(d) $2.016\ 4\ (\pm 0.000\ 8) + 1.233\ (\pm 0.002) + 4.61\ (\pm 0.01) = ?$

(e) $2.016\ 4\ (\pm 0.000\ 8) \times 10^3 + 1.233\ (\pm 0.002) \times 10^2 + 4.61\ (\pm 0.01) \times 10^1 = ?$

(f) $[3.14\ (\pm 0.05)]^{1/3} = ?$

(g) $\log[3.14\ (\pm 0.05)] = ?$

3-17. Verify the following calculations:

(a) $\sqrt{3.141\ 5\ (\pm 0.001\ 1)} = 1.772\ 4_3\ (\pm 0.000\ 3_1)$

(b) $\log[3.141\ 5\ (\pm 0.001\ 1)] = 0.497\ 1_4\ (\pm 0.000\ 1_5)$

(c) $\text{antilog}[3.141\ 5\ (\pm 0.001\ 1)] = 1.385_2\ (\pm 0.003_5) \times 10^3$

(d) $\ln[3.141\ 5\ (\pm 0.001\ 1)] = 1.144\ 7_0\ (\pm 0.000\ 3_5)$

(e) $\log\left(\dfrac{\sqrt{0.104\ (\pm 0.006)}}{0.051\ 1\ (\pm 0.000\ 9)}\right) = 0.80_0\ (\pm 0.01_5)$

3-18. **(a)** Show that the formula mass of NaCl is 58.442 ± 0.006 g/mol.

(b) To prepare a solution of NaCl, you weigh out $2.634\ (\pm 0.002)$ g and dissolve it in a volumetric flask whose volume is $100.00\ (\pm 0.08)$ mL. Express the molarity of the solution, along with its uncertainty, with an appropriate number of digits.

3-19. What is the true mass of water in vacuum if the apparent mass weighed in air at 24°C is $1.034\ 6 \pm 0.000\ 2$ g? The density of air is $0.001\ 2 \pm 0.000\ 1$ g/mL and the density of balance weights is 8.0 ± 0.5 g/mL. The uncertainty in the density of water in Table 2-7 is negligible in comparison to the uncertainty in the density of air.

3-20. We can measure the concentration of HCl solution by reaction with pure sodium carbonate: $2H^+ + Na_2CO_3 \rightarrow 2Na^+ + H_2O + CO_2$. Complete reaction with $0.967\ 4 \pm 0.000\ 9$ g of Na_2CO_3 (FM $105.988\ 4 \pm 0.000\ 7$) required 27.35 ± 0.04 mL of HCl.

(a) Find the formula mass (and its uncertainty) for Na_2CO_3.

(b) Find the molarity of the HCl and its absolute uncertainty.

(c) The purity of primary standard Na_2CO_3 is stated to be 99.95 to 100.05 wt%, which means that it can react with $(100.00 \pm 0.05)\%$ of the theoretical amount of H^+. Recalculate your answer to **(b)** with this additional uncertainty.

3-21. Express the molecular mass ($\pm$uncertainty) of $C_9H_9O_6N_3$ with the correct number of significant figures.

3-22. You have a stock solution certified by a manufacturer to contain 150.0 ± 0.3 μg SO_4^{2-}/mL. You would like to dilute it by a factor of 100 to obtain 1.500 μg/mL. Two possible methods of dilution are stated below. For each method, calculate the resulting uncertainty in concentration. Use manufacturer's tolerances in Tables 2-3 and 2-4 for uncertainties. Explain why one method is more precise than the other.

(a) Dilute 10.00 mL up to 100 mL with a transfer pipet and volumetric flask. Then take 10.00 mL of the dilute solution and dilute it again to 100 mL.

(b) Dilute 1.000 mL up to 100 mL with a transfer pipet and volumetric flask.

3-23. Avogadro's number can be computed from the following measured properties of pure crystalline silicon:[5] (1) atomic mass (obtained from the mass and abundance of each isotope), (2) density of the crystal, (3) size of the unit cell (the smallest repeating unit in the crystal), and (4) number of atoms in the unit cell. For the material that was used, the average atomic mass of Si is $m_{Si} = 28.085\ 384\ 2\ (35)$ g/mol, where 35 is the uncertainty (standard deviation) in the last two digits. The density is $\rho = 2.329\ 031\ 9\ (18)$ g/cm^3, the size of the cubic unit cell is $c_o = 5.431\ 020\ 36\ (33) \times 10^{-8}$ cm, and there are 8 atoms per unit cell. Avogadro's number is computed from the equation

$$N_A = \frac{m_{Si}}{(\rho c_o^3)/8}$$

From the measured properties and their uncertainties, compute Avogadro's number and its uncertainty. To find the uncertainty of c_o^3, use the function $y = x^a$ in Table 3-1.

4 Statistics

IS MY RED BLOOD CELL COUNT HIGH TODAY?

Red blood cells, also called erythrocytes.
[Susumu Nishinaga/Science Source.]

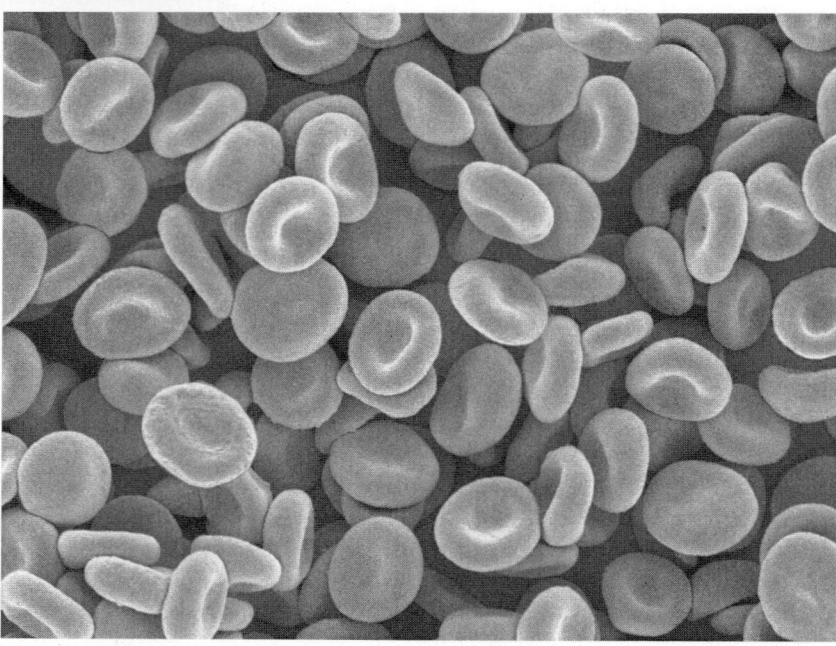

All measurements contain experimental error, so it is never possible to be completely certain of a result. Nevertheless, we often seek the answers to questions such as "Is my red blood cell count today higher than usual?" If today's count is twice as high as usual, it is probably truly higher than normal. But what if the "high" count is not excessively above "normal" counts?

Count on "normal" days	Today's count
5.1 ⎫	
5.3 ⎪	
4.8 ⎬ $\times 10^6$ cells/μL	5.6×10^6 cells/μL
5.4 ⎪	
5.2 ⎭	
Average = 5.16	

The number 5.6 is higher than the five normal values, but the random variation in normal values might lead us to expect that 5.6 will be observed on some "normal" days.

We will learn on page 79 that there is only a 1.3% random chance of observing a value as far from the average as 5.6 on a "normal" day. It is still up to you to decide what to do with this information.

Experimental measurements always contain some variability, so no conclusion can be drawn with certainty. Statistics gives us tools to accept conclusions that have a high probability of being correct and to reject conclusions that do not.[1] In this chapter we apply statistics for the purpose of accepting or rejecting conclusions based on their probability of being correct. We learn the method of least squares for fitting experimental data to a straight line and estimating uncertainties associated with that line. After you have studied this chapter, you will have the tools to appreciate a more general treatment of propagation of uncertainty in Appendix B that assigns a confidence level to uncertainty computed from experimental data.

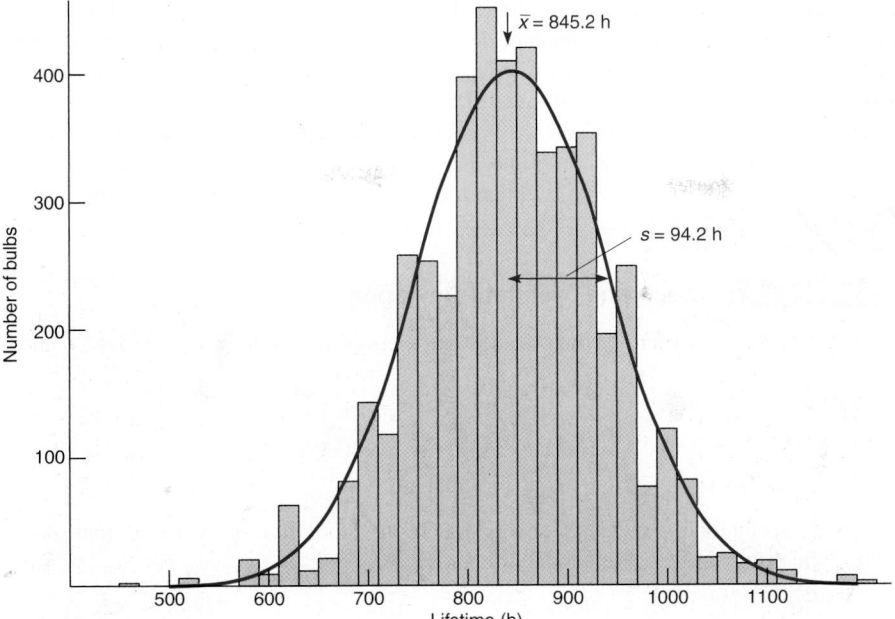

FIGURE 4-1 Bar graph and Gaussian curve describing the lifetimes of a hypothetical set of incandescent light bulbs. The smooth curve has the same mean, standard deviation, and area as the bar graph. Any finite set of data, however, will differ from the bell-shaped curve. The more measurements we make, the closer the results will come to the smooth curve.

4-1 Gaussian Distribution

If an experiment is repeated a great many times and if the errors are purely random, then the results tend to cluster symmetrically about the average value (Figure 4-1). The more times the experiment is repeated, the more closely the results approach an ideal bell-shaped curve called the **Gaussian distribution.** In general, we cannot make so many measurements in a lab experiment. We are more likely to repeat an experiment 3 to 5 times than 2 000 times. From the small set of results, we can estimate the properties of the hypothetical large set.

Mean Value and Standard Deviation

In the hypothetical case in Figure 4-1, a manufacturer tested the lifetimes of 4 768 incandescent light bulbs. The bar graph shows the number of bulbs with a lifetime in each 20-h interval. Lifetimes approximate a Gaussian distribution because variations in the construction of light bulbs, such as filament thickness and quality of attachments, are random. The smooth curve is the Gaussian distribution that best fits the data. Any finite set of data will vary somewhat from the Gaussian curve.

Light bulb lifetimes, and the corresponding Gaussian curve, are characterized by two parameters. The arithmetic **mean,** $\bar{x}$—also called the **average**—is the sum of the measured values divided by n, the number of measurements:

Mean:
$$\bar{x} = \frac{\sum_i x_i}{n} \qquad (4\text{-}1)$$

where x_i is the lifetime of an individual bulb. The Greek capital sigma, Σ, means summation: $\sum_i x_i = x_1 + x_2 + x_3 + \cdots + x_n$. In Figure 4-1, the mean value is 845.2 h.

The **standard deviation,** s, measures how closely the data are clustered about the mean. *The smaller the standard deviation, the more closely the data are clustered about the mean* (Figure 4-2).

Standard deviation:
$$s = \sqrt{\frac{\sum_i (x_i - \bar{x})^2}{n - 1}} \qquad (4\text{-}2)$$

In Figure 4-1, $s = 94.2$ h. A set of light bulbs having a small standard deviation in lifetime is more uniformly manufactured than a set with a large standard deviation.

We say that experimental data are **normally distributed** when replicate measurements exhibit the bell-shaped distribution in Figure 4-1. It is equally probable that a measurement will be higher or lower than the mean. The probability of observing any value decreases as its distance from the mean increases.

The mean gives the center of the distribution. The standard deviation measures the width of the distribution.

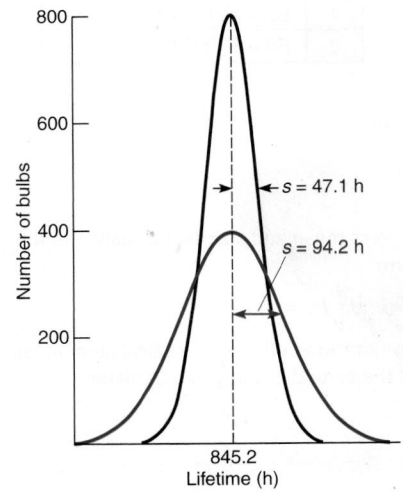

FIGURE 4-2 Gaussian curves for two sets of light bulbs, one having a standard deviation half as great as the other. The number of bulbs described by each curve is the same.

An experiment that produces a small standard deviation is more **precise** than one that produces a large standard deviation. Greater precision does not necessarily imply greater **accuracy,** which is nearness to the "truth."

	A	B
1		821
2		783
3		834
4		855
5	Average =	823.25
6	Std dev =	30.27
7	B5 = AVERAGE(B1:B4)	
8	B6 = STDEV(B1:B4)	

For an *infinite* set of data, the mean is designated by the lowercase Greek letter mu, μ (the population mean), and the standard deviation is written as a lowercase Greek sigma, σ (the population standard deviation). We can never measure μ and σ, but the values of $\bar{x}$ and s approach μ and σ as the number of measurements increases.

The quantity $n - 1$ in Equation 4-2 is called the **degrees of freedom.** The square of the standard deviation is called the **variance.** The standard deviation expressed as a percentage of the mean value ($=100 \times s/\bar{x}$) is called the *relative standard deviation* or the *coefficient of variation.*

EXAMPLE Mean and Standard Deviation

Find the average, standard deviation, and coefficient of variation for 821, 783, 834, and 855.

Solution The average is

$$\bar{x} = \frac{821 + 783 + 834 + 855}{4} = 823._2$$

To avoid accumulating round-off errors, retain one more digit in the mean than was present in the original data. The standard deviation is

$$s = \sqrt{\frac{(821 - 823.2)^2 + (783 - 823.2)^2 + (834 - 823.2)^2 + (855 - 823.2)^2}{(4 - 1)}} = 30._3$$

The average and the standard deviation should both end at the *same decimal place.* For $\bar{x} = 823._2$, we will write $s = 30._3$. The coefficient of variation is the percent relative uncertainty:

$$\text{Coefficient of variation} = 100 \times \frac{s}{\bar{x}} = 100 \times \frac{30._3}{823._2} = 3.7\%$$

TEST YOURSELF If each of the four numbers 821, 783, 834, and 855 in the example is divided by 2, how will the mean, standard deviation, and coefficient of variation be affected? (*Answer:* $\bar{x}$ and s will be divided by 2, but the coefficient of variation is unchanged)

Spreadsheets have built-in functions for the average and standard deviation. In the adjacent spreadsheet, data points are entered in cells B1 through B4. The average in cell B5 is computed with the statement "=AVERAGE(B1:B4)". B1:B4 means cells B1, B2, B3, and B4. The standard deviation in cell B6 is computed with "=STDEV(B1:B4)".

For ease of reading, cells B5 and B6 were set to display two decimal places. A heavy line was placed beneath cell B4 in Excel 2007 or 2010 by highlighting the cell, going to Home, Font, and selecting the Border icon.

Significant Figures in Mean and Standard Deviation

We commonly express experimental results in the form $\bar{x} \pm s(n = \_\,)$, where n is the number of data points. It is sensible to write the preceding result as 823 ± 30 ($n = 4$) or even $8.2\ (\pm 0.3) \times 10^2$ ($n = 4$) to indicate that the mean has just two significant figures. The expressions 823 ± 30 and $8.2\ (\pm 0.3) \times 10^2$ are not suitable for continued calculations in which $\bar{x}$ and s are intermediate results. Retain one or more insignificant digits to avoid introducing round-off errors into subsequent work. Try not to go into cardiac arrest over significant figures when you see $823._2 \pm 30._3$ as the answer to a problem in this book.

Standard Deviation and Probability

The formula for a Gaussian curve is

Gaussian curve:
$$y = \frac{1}{\sigma\sqrt{2\pi}}\, e^{-(x-\mu)^2/2\sigma^2} \tag{4-3}$$

where e ($= 2.718\ 28\ldots$) is the base of the natural logarithm. For a finite set of data, we approximate μ by $\bar{x}$ and σ by s. A graph of Equation 4-3 is shown in Figure 4-3, in which the values $\sigma = 1$ and $\mu = 0$ are used for simplicity. The maximum value of y is at $x = \mu$ and the curve is symmetric about $x = \mu$.

TABLE 4-1	Ordinate and area for the normal (Gaussian) error curve, $y = \dfrac{1}{\sqrt{2\pi}} e^{-z^2/2}$							
$|z|^a$	y	$Area^b$	$|z|$	y	$Area$	$|z|$	y	$Area$
0.0	0.398 9	0.000 0	1.4	0.149 7	0.419 2	2.8	0.007 9	0.497 4
0.1	0.397 0	0.039 8	1.5	0.129 5	0.433 2	2.9	0.006 0	0.498 1
0.2	0.391 0	0.079 3	1.6	0.110 9	0.445 2	3.0	0.004 4	0.498 650
0.3	0.381 4	0.117 9	1.7	0.094 1	0.455 4	3.1	0.003 3	0.499 032
0.4	0.368 3	0.155 4	1.8	0.079 0	0.464 1	3.2	0.002 4	0.499 313
0.5	0.352 1	0.191 5	1.9	0.065 6	0.471 3	3.3	0.001 7	0.499 517
0.6	0.333 2	0.225 8	2.0	0.054 0	0.477 3	3.4	0.001 2	0.499 663
0.7	0.312 3	0.258 0	2.1	0.044 0	0.482 1	3.5	0.000 9	0.499 767
0.8	0.289 7	0.288 1	2.2	0.035 5	0.486 1	3.6	0.000 6	0.499 841
0.9	0.266 1	0.315 9	2.3	0.028 3	0.489 3	3.7	0.000 4	0.499 904
1.0	0.242 0	0.341 3	2.4	0.022 4	0.491 8	3.8	0.000 3	0.499 928
1.1	0.217 9	0.364 3	2.5	0.017 5	0.493 8	3.9	0.000 2	0.499 952
1.2	0.194 2	0.384 9	2.6	0.013 6	0.495 3	4.0	0.000 1	0.499 968
1.3	0.171 4	0.403 2	2.7	0.010 4	0.496 5	∞	0	0.5

a. $z = (x - \mu)/\sigma$
b. The area refers to the area between $z = 0$ and $z =$ the value in the table. Thus the area from $z = 0$ to $z = 1.4$ is 0.419 2.
The area from $z = -0.7$ to $z = 0$ is the same as from $z = 0$ to $z = 0.7$. The area from $z = -0.5$ to $z = +0.3$ is $(0.191\ 5 + 0.117\ 9) = 0.309\ 4$. The total area between $z = -\infty$ and $z = +\infty$ is unity.

It is useful to express deviations from the mean value in multiples, z, of the standard deviation. That is, we transform x into z, given by

$$z = \frac{x - \mu}{\sigma} \approx \frac{x - \bar{x}}{s} \qquad (4\text{-}4)$$

The probability of measuring z in a certain *range* is equal to the *area* of that range. For example, the probability of observing z between -2 and -1 is 0.136. This probability corresponds to the shaded area in Figure 4-3. The area under each portion of the Gaussian curve is given in Table 4-1. Because the sum of the probabilities of all the measurements must be unity, the area under the whole curve from $z = -\infty$ to $z = +\infty$ must be unity. The number $1/(\sigma\sqrt{2\pi})$ in Equation 4-3 is called the *normalization factor*. It makes the area under the entire curve unity. A Gaussian curve with unit area is called a *normal error curve*.

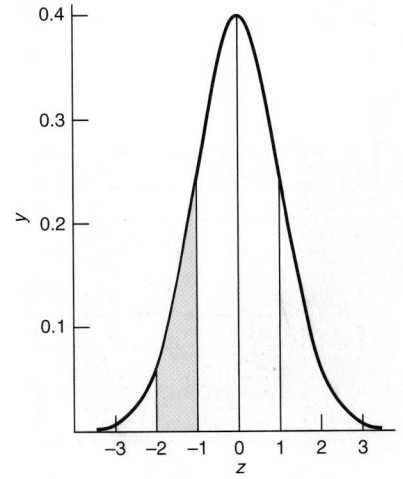

FIGURE 4-3 A Gaussian curve in which $\mu = 0$ and $\sigma = 1$. A Gaussian curve whose area is unity is called a normal error curve. The abscissa $z = (x - \mu)/\sigma$ is the distance away from the mean, measured in units of the standard deviation. When $z = 2$, we are two standard deviations away from the mean.

When $z = +1$, x is one standard deviation above the mean. When $z = -2$, x is two standard deviations below the mean.

EXAMPLE Area Under a Gaussian Curve

Suppose the manufacturer of the bulbs used for Figure 4-1 offers to replace free of charge any bulb that burns out in less than 600 hours. If she plans to sell a million bulbs, how many extra bulbs should she keep available as replacements?

Solution We need to express the desired interval in multiples of the standard deviation and then find the area of the interval in Table 4-1. Because $\bar{x} = 845.2$ and $s = 94.2$, $z = (600 - 845.2)/94.2 = -2.60$. The area under the curve between the mean value and $z = -2.60$ is 0.495 3 in Table 4-1. The entire area from $-\infty$ to the mean value is 0.500 0, so the area from $-\infty$ to -2.60 must be $0.500\ 0 - 0.495\ 3 = 0.004\ 7$. The area to the left of 600 hours in Figure 4-1 is only 0.47% of the entire area under the curve. Only 0.47% of the bulbs are expected to fail in fewer than 600 h. If the manufacturer sells 1 million bulbs a year, she should make 4 700 extra bulbs to meet the replacement demand.

TEST YOURSELF If the manufacturer will replace bulbs that burn out in less than 620 hours, how many extras should she make? (*Answer:* $z \approx -2.4$, area $\approx 0.008\ 2 = 8\ 200$ bulbs)

EXAMPLE Using a Spreadsheet to Find Area Under a Gaussian Curve

What fraction of bulbs is expected to have a lifetime between 900 and 1 000 h?

Solution *We need to find the fraction of the area of the Gaussian curve between $x = 900$ and $x = 1\ 000$ h. The function NORMDIST in Excel gives the area of the curve from $-\infty$ up to a specified point, x. Here is the strategy: We will find the area from $-\infty$ to 900 h, which is the shaded area to the left of 900 h in Figure 4-4. Then we will find the area*

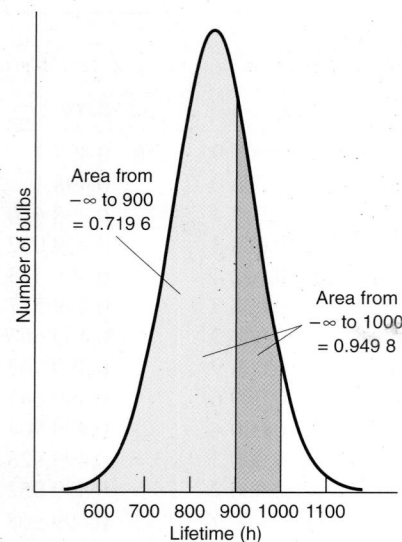

FIGURE 4-4 Use of the Gaussian curve to find the fraction of bulbs with a lifetime between 900 and 1 000 h. We find the area between −∞ and 1 000 h and subtract the area between −∞ and 900 h.

Range	Percentage of measurements
$\mu \pm 1\sigma$	68.3
$\mu \pm 2\sigma$	95.5
$\mu \pm 3\sigma$	99.7

from −∞ to 1 000 h, which is all the shaded area to the left of 1 000 h in Figure 4-4. The difference between the two is the area from 900 to 1 000 h:

Area from 900 to 1 000 = (area from −∞ to 1 000) − (area from −∞ to 900)

In a spreadsheet, enter the mean in cell A2 and the standard deviation in cell B2. To find the area under the Gaussian curve from −∞ to 900 h in cell C4, select cell C4 and, in Excel 2007 or 2010, go to Formulas, Insert Function. In the window that appears, select Statistical functions and find NORMDIST from the list of possibilities. (Excel 2010 has both NORMDIST and NORM.DIST. Both give the same result.) Double click on NORMDIST and another window appears asking for four values that will be used by NORMDIST. (If you click on Help, you will find a cryptic explanation of how to use NORMDIST.)

	A	B	C
1	Mean =	Std dev =	
2	845.2	94.2	
3			
4	Area from −∞ to 900 =		0.7196
5	Area from −∞ to 1000 =		0.9498
6	Area from 900 to 1000		0.2302
7			
8	C4 = NORMDIST(900,$A$2,$B$2,TRUE)		
9	C5 = NORMDIST(1000,$A$2,$B$2,TRUE)		
10	C6 = C5-C4		

Values provided to the function NORMDIST(x,mean,standard_dev,cumulative) are called *arguments* of the function. The first argument is x, which is 900. The second argument is the mean, which is 845.2. You can enter 845.2 for the mean or enter A2, which is the cell containing 845.2. We will enter $A2 so that we can move the formula to other cells and still always refer to cell A2. The third argument is the standard deviation, for which we enter $B$2. The last argument is called "cumulative." When it has the value TRUE, NORMDIST gives the area under the Gaussian curve. When cumulative is FALSE, NORMDIST gives the ordinate (the y-value) of the Gaussian curve. We want area, so enter TRUE. The formula "= NORMDIST(900,$A$2,$B$2,TRUE)" in cell C4 returns 0.719 6. This is the area under the Gaussian curve from −∞ to 900 h. To get the area from −∞ to 1 000 h, write "= NORMDIST(1000,$A$2,$B$2,TRUE)" in cell C5. The value returned is 0.949 8. Then subtract the areas (C5 − C4) to obtain 0.230 2, which is the area from 900 to 1 000. That is, 23.02% of the area lies in the range 900 to 1 000 h. We expect 23% of the bulbs to have a lifetime of 900 to 1 000 h.

TEST YOURSELF Find the area from 800 to 1 000 hours. (*Answer:* 0.634 2)

The standard deviation measures the width of the Gaussian curve. The larger the value of σ, the broader the curve. In any Gaussian curve, 68.3% of the area is in the range from $\mu - 1\sigma$ to $\mu + 1\sigma$. That is, more than two-thirds of the measurements are expected to lie within one standard deviation from the mean. Also, 95.5% of the area lies within $\mu \pm 2\sigma$, and 99.7% of the area lies within $\mu \pm 3\sigma$. Suppose that you use two different techniques to measure sulfur in coal: Method A has a standard deviation of 0.4%, and method B has a standard deviation of 1.1%. You can expect that approximately two-thirds of measurements from method A will lie within 0.4% of the mean. For method B, two-thirds will lie within 1.1% of the mean.

Standard Deviation of the Mean

To measure the mean lifetime of a large number of light bulbs, we could select one at a time and measure its lifetime. Alternatively, we could select, say, four at a time, measure the life of each, and compute the average of the four. We repeat this process of measuring four at a time many times and compute a mean μ and a standard deviation, which we label σ_4 because it is based on sets of four bulbs. The mean of many sets of four bulbs is the same as the population mean. However, the standard deviation of the means of sets of four bulbs is smaller than the population standard deviation, σ. The relation is $\sigma_4 = \sigma/\sqrt{4}$. We call σ_4 the

standard deviation of the mean of sets of four samples. In general, the standard deviation of the mean for sets of n samples is:

Standard deviation of the mean of sets of n values:
$$\sigma_n = \frac{\sigma}{\sqrt{n}} \qquad (4\text{-}5)$$

The more times you measure a quantity, the more confident you can be that the mean is close to the population mean. Uncertainty decreases in proportion to $1/\sqrt{n}$, where n is the number of measurements. You can decrease uncertainty by a factor of $2(=\sqrt{4})$ by making four times as many measurements and by a factor of $10(=\sqrt{100})$ by making 100 times as many measurements.

- σ measures uncertainty in x. σ approaches a constant value as n approaches ∞.
- σ_n measures uncertainty in the mean, $\bar{x}$. σ_n approaches 0 as n approaches ∞.

Instruments with rapid data acquisition allow us to average many experiments in a short time to improve precision.

4-2 Comparison of Standard Deviations with the *F* Test

An important question in statistics is "Are the mean values of two sets of measurements 'statistically different' from each other when experimental uncertainty is considered?" To compare mean values in the next section, we must first decide whether the standard deviations of the two sets are "statistically different."

Consider the measurement of bicarbonate (HCO_3^-) in the blood of racehorses. Some trainers injected $NaHCO_3$ into a horse prior to a race to neutralize lactic acid that accumulates during strenuous activity. To enforce a ban on this practice, HCO_3^- in horse blood is measured after a race. When a manufacturer stopped making an instrument that was certified for such measurements, authorities needed to certify a new instrument.

Table 4-2 shows results from two instruments. The averages of 36.14 and 36.20 mM are similar, but the standard deviation (s) from the substitute instrument is almost twice as great as that from the original instrument (0.47 versus 0.28 mM). Is s from the substitute instrument "significantly" greater than s from the original instrument?

TABLE 4-2 Measurement of HCO_3^- in horse blood

	Original instrument	Substitute instrument
Mean ($\bar{x}$, mM)	36.14	36.20
Standard deviation (s, mM)	0.28	0.47
Number of measurements (n)	10	4

*Data from M. Jarrett, D. B. Hibbert, R. Osborne, and E. B. Young, Anal. Bioanal. Chem. **2010**, 397, 717.*

To answer this question, we use the **F test,** with the quotient F defined as

$$F_{\text{calculated}} = \frac{s_1^2}{s_2^2} \qquad (4\text{-}6)$$

The square of the standard deviation is the **variance.**

Put the larger standard deviation in the numerator so that $F \geq 1$. We test if the difference between s_1 and s_2 is significant by applying the F test in Table 4-3. If $F_{\text{calculated}} > F_{\text{table}}$, then the difference is significant. In Table 4-3, the *degrees of freedom* for n measurements are $n - 1$. If there are five measurements in one set, there are four degrees of freedom.

EXAMPLE **Is the Standard Deviation from the Substitute Instrument Significantly Greater Than That of the Original Instrument?**

In Table 4-2, the standard deviation from the substitute instrument is $s_1 = 0.47$ ($n_1 = 4$ measurements) and the standard deviation from the original instrument is $s_2 = 0.28$ ($n_2 = 10$).

Solution To answer the question, find F with Equation 4-6:

$$F_{\text{calculated}} = \frac{s_1^2}{s_2^2} = \frac{(0.47)^2}{(0.28)^2} = 2.8_2$$

In Table 4-3, we find $F_{\text{table}} = 3.86$ in the column with 3 degrees of freedom for s_1 (degrees of freedom $= n - 1$) and the row with 9 degrees of freedom for s_2. *Because $F_{\text{calculated}}$ (= 2.8_2) < F_{table} (= 3.86), we reject the hypothesis that s_1 is significantly larger* *than* s_2. You will see in the next section on hypothesis testing that there is more than a 5% chance that the two sets of data are drawn from populations with the same population standard deviation.

> **TEST YOURSELF** If there had been $n = 13$ replications in both data sets, would the difference in standard deviations be significant? (***Answer:*** Yes. $F_{calculated} = 2.8_2 > F_{table} = 2.69$)

Hypothesis Testing

Hypothesis test: A decision about measured data is made on the basis of the probability of observing that data if a stated hypothesis is true.

Null hypothesis: The statement that two sets of data are drawn from populations with the same properties such as standard deviation σ (F test) or mean μ (t test in Section 4-4).

The F test is an example of a **hypothesis test.** The **null hypothesis** for the F test is that the two sets of measurements are drawn from populations with the same population standard deviation (σ); observed differences arise only from random variation in the measurements. We test this null hypothesis by evaluating the probability of observing the quotient $F = s_1^2/s_2^2$ if the two sets of data are selected at random from populations with the same standard deviation. If there is less than a 5% probability of finding the observed value of F, we reject the null hypothesis and conclude that the two sets of data probably do not come from populations with the same standard deviation. If there is more than a 5% probability of finding the observed value of F, we accept the null hypothesis. The selection of 5% as the confidence level is customary. You could choose higher or lower levels to suit your needs.

Values of F in Table 4-3 are chosen such that there is a probability (p) of 5% to find the observed quotient s_1^2/s_2^2 if all measurements come from populations with the same population standard deviation. When $F_{calculated} > F_{table}$, there is *less* than a $p = 5\%$ chance that the two sets of measurements come from populations with the same population standard deviation. *We choose to reject the null hypothesis if there is less than a 5% chance that it is true.* Now read Box 4-1 to really appreciate what we mean by the null hypothesis.

TABLE 4-3 Critical values of $F = s_1^2/s_2^2$ at 95% confidence level

Degrees of freedom for s_2	Degrees of freedom for s_1													
	2	3	4	5	6	7	8	9	10	12	15	20	30	∞
2	19.0	19.2	19.2	19.3	19.3	19.4	19.4	19.4	19.4	19.4	19.4	19.4	19.5	19.5
3	9.55	9.28	9.12	9.01	8.94	8.89	8.84	8.81	8.79	8.74	8.70	8.66	8.62	8.53
4	6.94	6.59	6.39	6.26	6.16	6.09	6.04	6.00	5.96	5.91	5.86	5.80	5.75	5.63
5	5.79	5.41	5.19	5.05	4.95	4.88	4.82	4.77	4.74	4.68	4.62	4.56	4.50	4.36
6	5.14	4.76	4.53	4.39	4.28	4.21	4.15	4.10	4.06	4.00	3.94	3.87	3.81	3.67
7	4.74	4.35	4.12	3.97	3.87	3.79	3.73	3.68	3.64	3.58	3.51	3.44	3.38	3.23
8	4.46	4.07	3.84	3.69	3.58	3.50	3.44	3.39	3.35	3.28	3.22	3.15	3.08	2.93
9	4.26	3.86	3.63	3.48	3.37	3.29	3.23	3.18	3.14	3.07	3.01	2.94	2.86	2.71
10	4.10	3.71	3.48	3.33	3.22	3.14	3.07	3.02	2.98	2.91	2.84	2.77	2.70	2.54
11	3.98	3.59	3.36	3.20	3.10	3.01	2.95	2.90	2.85	2.79	2.72	2.65	2.57	2.40
12	3.88	3.49	3.26	3.11	3.00	2.91	2.85	2.80	2.75	2.69	2.62	2.54	2.47	2.30
13	3.81	3.41	3.18	3.02	2.92	2.83	2.77	2.71	2.67	2.60	2.53	2.46	2.38	2.21
14	3.74	3.34	3.11	2.96	2.85	2.76	2.70	2.65	2.60	2.53	2.46	2.39	2.31	2.13
15	3.68	3.29	3.06	2.90	2.79	2.71	2.64	2.59	2.54	2.48	2.40	2.33	2.25	2.07
16	3.63	3.24	3.01	2.85	2.74	2.66	2.59	2.54	2.49	2.42	2.35	2.28	2.19	2.01
17	3.59	3.20	2.96	2.81	2.70	2.61	2.55	2.49	2.45	2.38	2.31	2.23	2.15	1.96
18	3.56	3.16	2.93	2.77	2.66	2.58	2.51	2.46	2.41	2.34	2.27	2.19	2.11	1.92
19	3.52	3.13	2.90	2.74	2.63	2.54	2.48	2.42	2.38	2.31	2.23	2.16	2.07	1.88
20	3.49	3.10	2.87	2.71	2.60	2.51	2.45	2.39	2.35	2.28	2.20	2.12	2.04	1.84
30	3.32	2.92	2.69	2.53	2.42	2.33	2.27	2.21	2.16	2.09	2.01	1.93	1.84	1.62
∞	3.00	2.60	2.37	2.21	2.10	2.01	1.94	1.88	1.83	1.75	1.67	1.57	1.46	1.00

Critical values of F for a one-tailed test of the hypothesis that $s_1 > s_2$. There is a 5% probability of observing F above the tabulated value if the two sets of data come from populations with the same population standard deviation.

You can compute F for a chosen level of confidence with the Excel function FINV(Probability, deg_freedom1, deg_freedom2). The statement "= FINV(0.05,7,6)" reproduces the value $F = 4.21$ in this table. The statement "= FINV(0.1,7,6)" gives $F = 3.01$ for 90% confidence.

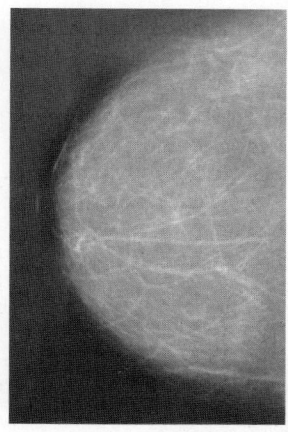
4-3 Confidence Intervals

Student's *t* is a statistical tool used most frequently to express confidence intervals and to compare results from different experiments. It is the tool you could use to evaluate the probability that your red blood cell count will be found in a certain range on "normal" days.

Calculating Confidence Intervals

From a limited number of measurements (*n*), we cannot find the true population mean, μ, or the true standard deviation, σ. What we determine are $\bar{x}$ and s, the sample mean and the sample standard deviation. The **confidence interval** is computed from the equation

$$\text{Confidence interval} = \bar{x} \pm \frac{ts}{\sqrt{n}} = \bar{x} \pm tu_x \qquad (4\text{-}7)$$

where t is Student's t, taken from Table 4-4 for a desired level of confidence, such as 95%. The expression on the right replaces the standard deviation of the mean $s/\sqrt{n}$ with the equivalent quantity u_x, called the **standard uncertainty** ($u_x = s/\sqrt{n}$).

The meaning of the confidence interval is this: If we were to repeat the *n* measurements many times to compute the mean and standard deviation, the 95% confidence interval would include the true population mean (whose value we do not know) in 95% of the sets of *n* measurements. We say (somewhat imprecisely) that "we are 95% confident that the true mean lies within the confidence interval."

"Student" was the pseudonym of W. S. Gosset, whose employer, the Guinness Breweries of Ireland, restricted publications for proprietary reasons. Because of the importance of Gosset's work, he was allowed to publish it under an assumed name (*Biometrika* **1908,** 6, 1).

Standard uncertainty = standard deviation of the mean

$$u_x = s/\sqrt{n}$$

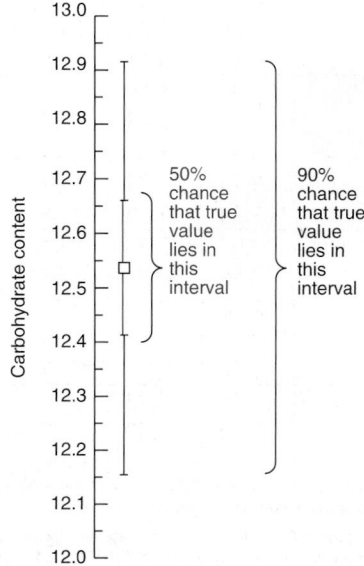

EXAMPLE   Calculating Confidence Intervals

The carbohydrate content of a glycoprotein (a protein with sugars attached to it) is found to be 12.6, 11.9, 13.0, 12.7, and 12.5 wt% (g carbohydrate/100 g glycoprotein) in replicate analyses. Find the 50% and 90% confidence intervals for the carbohydrate content.

Solution First calculate $\bar{x}$ (= 12.5_4) and s (= 0.4_0) for the five measurements. For the 50% confidence interval, look up t in Table 4-4 under 50 and across from *four* degrees of freedom (degrees of freedom = $n - 1$). The value of t is 0.741, so the 50% confidence interval is

$$50\% \text{ confidence interval} = \bar{x} \pm \frac{ts}{\sqrt{n}} = 12.5_4 \pm \frac{(0.741)(0.4_0)}{\sqrt{5}} = 12.5_4 \pm 0.1_3 \text{ wt\%}$$

The 90% confidence interval is

$$90\% \text{ confidence interval} = \bar{x} \pm \frac{ts}{\sqrt{n}} = 12.5_4 \pm \frac{(2.132)(0.4_0)}{\sqrt{5}} = 12.5_4 \pm 0.3_8 \text{ wt\%}$$

If we repeat sets of five measurements many times, half of the 50% confidence intervals are expected to include the true mean, μ. Nine-tenths of the 90% confidence intervals are expected to include the true mean, μ.

TEST YOURSELF Carbohydrate measured on one more sample was 12.3 wt%. Using six results, find the 90% confidence interval. (**Answer:** $12.5_0 \pm (2.015)(0.3_7)/\sqrt{6} = 12.5_0 \pm 0.3_1$ wt%)

TABLE 4-4 **Values of Student's *t***

Degrees of freedom	Confidence level (%)						
	50	90	95	98	99	99.5	99.9
1	1.000	6.314	12.706	31.821	63.656	127.321	636.578
2	0.816	2.920	4.303	6.965	9.925	14.089	31.598
3	0.765	2.353	3.182	4.541	5.841	7.453	12.924
4	0.741	2.132	2.776	3.747	4.604	5.598	8.610
5	0.727	2.015	2.571	3.365	4.032	4.773	6.869
6	0.718	1.943	2.447	3.143	3.707	4.317	5.959
7	0.711	1.895	2.365	2.998	3.500	4.029	5.408
8	0.706	1.860	2.306	2.896	3.355	3.832	5.041
9	0.703	1.833	2.262	2.821	3.250	3.690	4.781
10	0.700	1.812	2.228	2.764	3.169	3.581	4.587
15	0.691	1.753	2.131	2.602	2.947	3.252	4.073
20	0.687	1.725	2.086	2.528	2.845	3.153	3.850
25	0.684	1.708	2.060	2.485	2.787	3.078	3.725
30	0.683	1.697	2.042	2.457	2.750	3.030	3.646
40	0.681	1.684	2.021	2.423	2.704	2.971	3.551
60	0.679	1.671	2.000	2.390	2.660	2.915	3.460
120	0.677	1.658	1.980	2.358	2.617	2.860	3.373
∞	0.674	1.645	1.960	2.326	2.576	2.807	3.291

In calculating confidence intervals, σ may be substituted for s in Equation 4-7 if you have a great deal of experience with a particular method and have therefore determined its "true" population standard deviation. If σ is used instead of s, the value of t to use in Equation 4-7 comes from the bottom row of this table.

Values of t in this table apply to two-tailed tests illustrated in Figure 4-9a. The 95% confidence level specifies the regions containing 2.5% of the area in each wing of the curve. For a one-tailed test, we use values of t listed for 90% confidence. Each wing outside of t for 90% confidence contains 5% of the area of the curve.

The Meaning of a Confidence Interval

Figure 4-5 illustrates the meaning of confidence intervals. A computer chose numbers at random from a Gaussian population with a population mean (μ) of 10 000 and a population standard deviation (σ) of 1 000 in Equation 4-3. In trial 1, four numbers were chosen, and their mean and standard deviation were calculated with Equations 4-1 and 4-2. The 50% confidence interval was then calculated with Equation 4-7, using $t = 0.765$ from Table 4-4 (50% confidence, 3 degrees of freedom). This trial is plotted as the first point at the left in Figure 4-5a; the square is centered at the mean value of 9 526, and the error bar extends from the lower limit to the upper limit of the 50% confidence interval ($\pm$290). The experiment was repeated 100 times to produce the points in Figure 4-5a.

The 50% confidence interval is defined such that, if we repeated this experiment an infinite number of times, 50% of the error bars in Figure 4-5a would include the true population mean of 10 000. In fact, I did the experiment 100 times, and 45 of the error bars in Figure 4-5a pass through the horizontal line at 10 000.

Figure 4-5b shows the same experiment with the same set of random numbers, but this time the 90% confidence interval was calculated. For an infinite number of experiments, we would expect 90% of the confidence intervals to include the population mean of 10 000. In Figure 4-5b, 89 of the 100 error bars cross the horizontal line at 10 000.

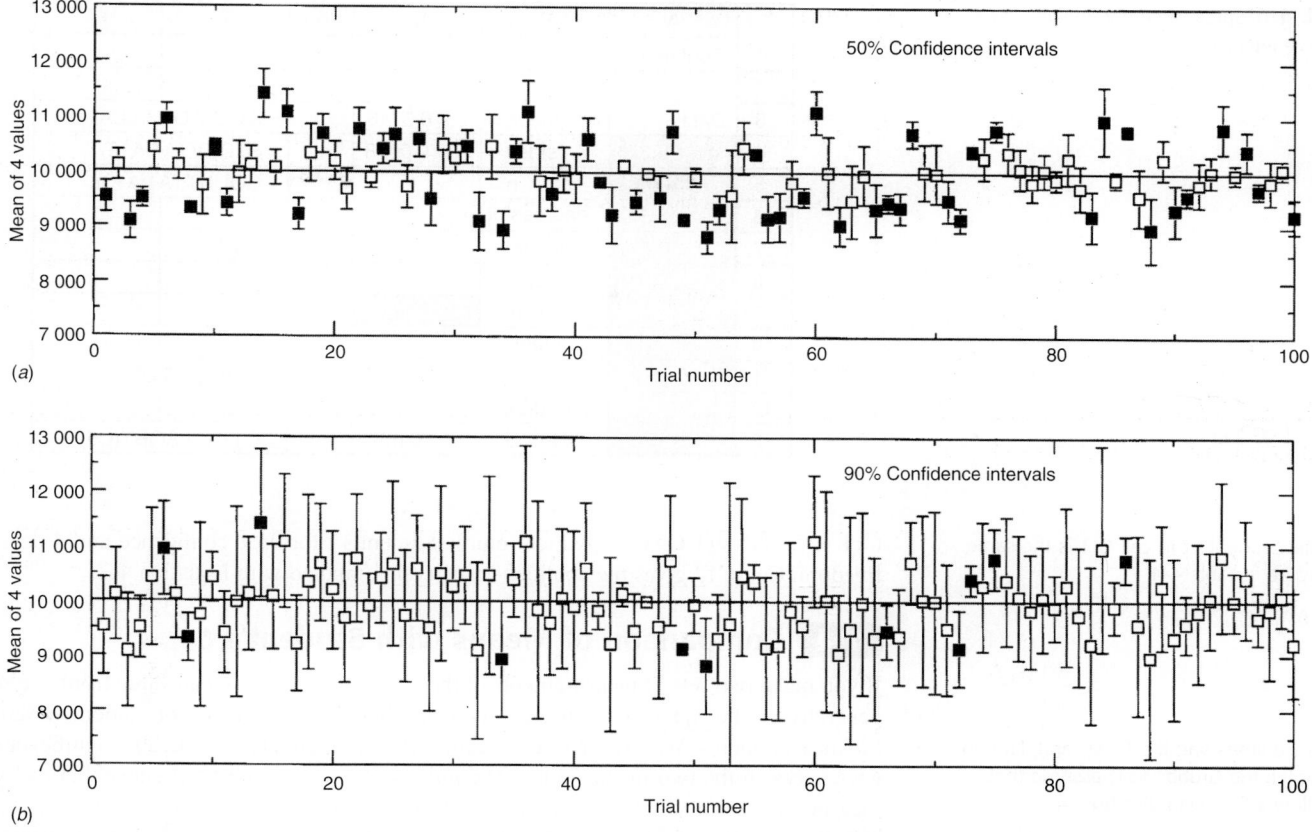

FIGURE 4-5 Confidence intervals of 50% and 90% for the same set of random data. Filled squares are the data points whose confidence interval does not include the true population mean of 10 000.

Confidence Interval as a Measure of Uncertainty

Suppose you measure the volume of a vessel five times and observe values of 6.375, 6.372, 6.374, 6.377, and 6.375 mL. The average is $\bar{x} = 6.374_6$ mL and the standard deviation is $s = 0.001_8$ mL. You could report a volume of $\bar{x} \pm s = 6.374_6 \pm 0.001_8$ mL ($n = 5$), where n is the number of measurements. Alternatively, you could report the standard deviation of the mean as the uncertainty: $\bar{x} \pm s/\sqrt{n} = 6.374_6 \pm 0.000_8$ mL ($n = 5$). Recall that the standard deviation of the mean is also called the *standard uncertainty*, u_x.

Alternatively, you could choose a confidence interval (such as 95%) as the uncertainty. Using Equation 4-7 with 4 degrees of freedom, you find that the 95% confidence interval is $\pm ts/\sqrt{n} = \pm(2.776)(0.001_8)/\sqrt{5} = \pm0.002_3$. By this criterion, the uncertainty in volume is $\pm0.002_3$ mL. *Always state what kind of uncertainty you are reporting*, such as the standard deviation for n measurements, the standard deviation of the mean for n measurements, or the 95% confidence interval for n measurements.

We reduce uncertainty by making more measurements. If we make 21 measurements and have the same standard deviation, the 95% confidence interval is reduced from $\pm0.002_3$ mL to $\pm(2.086)(0.001_8$ mL$)/\sqrt{21} = \pm0.000\ 8$ mL.

Finding Confidence Intervals with Excel

Excel has a built-in function to compute Student's t. In Figure 4-6, we enter data in the block of cells A4:A13. We reserved 10 cells for data, but you could modify a spreadsheet to accept more data. For the five data points entered in Figure 4-6, the mean is computed in cell C3 with the statement "= AVERAGE(A4:A13)" even though some of the cells in the range A4:A13 are empty. Excel ignores empty cells and does not consider them to be 0, which would produce an incorrect average. Standard deviation is computed in cell C4. We find the number of data points, n, with the statement in cell C5 "=COUNT(A4:A13)". Degrees of freedom are computed as $n - 1$ in cell C7. Cell C9 has the desired confidence level (0.95), which is our only input other than the data.

The function to find Student's t in cell C11 is "=TINV(probability,deg_freedom)". Probability in this function is $1 -$ confidence level $= 1 - 0.95 = 0.05$. So, the statement in cell

"Analytical chemists must always emphasize to the public that *the single most important characteristic of any result obtained from one or more analytical measurements is an adequate statement of its uncertainty interval.*"[3]

Always state what kind of uncertainty you are reporting.

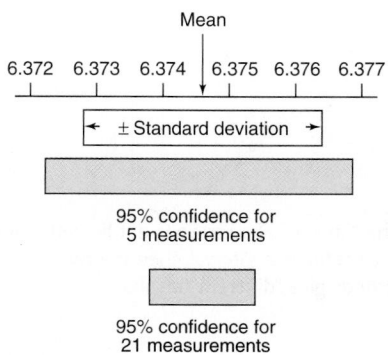

FIGURE 4-6 Spreadsheet for finding confidence interval.

	A	B	C	D	E	F
1	Confidence Interval					
2						
3	Data	mean =	6.3746		= AVERAGE (A4:A13)	
4	6.375	stdev =	0.0018		= STDEV (A4:A13)	
5	6.372	n =	5		= COUNT (A4:A13)	
6	6.374	degrees of				
7	6.377	freedom =	4		= C5-1	
8	6.375	confidence				
9		level =	0.95			
10		Student's				
11		t =	2.776		= TINV-(1-C9,C7)	
12		confidence				
13		interval =	0.0023		= C11*C4/SQRT(C5)	

Verify that Student's t in cell C11 is the same as the value in Table 4-4.

C11 is "=TINV(1-C9,C7)", which returns Student's t for 95% confidence and 4 degrees of freedom. Cell C13 gives the confidence interval computed with Equation 4-7.

4-4 Comparison of Means with Student's t

If you make two sets of measurements of the same quantity, the mean value from one set will generally not be equal to the mean value from the other set because of random variations in the measurements. We use a **t test** to decide whether there is a statistically significant difference between the two mean values. The *null hypothesis* for the t test states that two sets of measurements come from populations with the same population mean. We reject the null hypothesis if there is less than a $p = 5\%$ chance that the two sets of measurements come from populations with the same population mean. Statistics gives us a probability that the observed difference between two means arises from random measurement uncertainty.

Confidence limits and the t test (and, later in this chapter, the Grubbs test) assume that data follow a Gaussian distribution.

Null hypothesis for t test: Data come from sets with the same population mean. Reject the null hypothesis if there is less than a 5% probability that it is true.

Here are three cases that are handled in slightly different manners:

Case 1   We measure a quantity several times, obtaining an average value and standard deviation. We need to compare our answer with an accepted answer. The average is not exactly the same as the accepted answer. Does our measured answer agree with the accepted answer "within experimental error"?

Case 2   We measure a quantity multiple times by two different methods that give two different answers, each with its own standard deviation. Do the two results agree with each other "within experimental error"?

Case 3   Sample A is measured once by method 1 and once by method 2; the two measurements do not give exactly the same result. Then a different sample, designated B, is measured once by method 1 and once by method 2; and, again, the results are not exactly equal. The procedure is repeated for n different samples. Do the two methods agree with each other "within experimental error"?

Case 1. Comparing a Measured Result with a "Known" Value

You purchased a Standard Reference Material coal sample certified by the National Institute of Standards and Technology to contain 3.19 wt% sulfur. You are testing a new analytical method to see whether it can reproduce the known value. The measured values are 3.29, 3.22, 3.30, and 3.23 wt% sulfur, giving a mean of $\bar{x} = 3.26_0$ and a standard deviation of $s = 0.04_1$. Does your answer agree with the known answer? To find out, *compute the 95% confidence interval for your answer and see if that range includes the known answer*. If the known answer is not within your 95% confidence interval, then the results do not agree.

If the "known" answer does not lie within the 95% confidence interval, then the two methods give "different" results.

So here we go. For four measurements, there are 3 degrees of freedom and $t_{95\%} = 3.182$ in Table 4-4. The 95% confidence interval is

Retain many digits in this calculation.

$$95\% \text{ confidence interval} = \bar{x} \pm \frac{ts}{\sqrt{n}} = 3.26_0 \pm \frac{(3.182)(0.04_1)}{\sqrt{4}} = 3.26_0 \pm 0.06_5 \quad \textbf{(4-8)}$$

$$95\% \text{ confidence interval} = 3.19_5 \text{ to } 3.32_5 \text{ wt\%}$$

The known answer (3.19 wt%) is just outside the 95% confidence interval. Therefore we conclude that there is less than a 5% chance that our method agrees with the known answer.

We conclude that our method gives a "different" result from the known result. However, in this case, the 95% confidence interval is so close to including the known result that it would be prudent to make more measurements before concluding that our new method is not accurate.

Case 2. Comparing Replicate Measurements

Do the results of two different sets of measurements agree "within experimental error"?[4] To answer this question, we first compare the standard deviations with the F test (Equation 4-6). If the standard deviations are not significantly different, we conduct a *t* **test** with Equations 4-9a and 4-10a to see if the two mean values are significantly different. If the standard deviations are significantly different, we use Equations 4-9b and 4-10b to see if the means differ significantly.

Case 2a: Standard Deviations Are Not Significantly Different

Let's look again at Table 4-2 and ask whether the two mean values of 36.14 and 36.20 mM are significantly different from each other. We answer this question with the t *test*. If the F test tells us that the two standard deviations are not significantly different, then for data sets consisting of n_1 and n_2 measurements (with averages $\bar{x}_1$ and $\bar{x}_2$), calculate t from the formula

t test for comparison of means:
$$t = \frac{|\bar{x}_1 - \bar{x}_2|}{s_{pooled}} \sqrt{\frac{n_1 n_2}{n_1 + n_2}}$$
(4-9a)

t test when the standard deviations are not significantly different.

where

$$s_{pooled} = \sqrt{\frac{\sum_{set\ 1}(x_i - \bar{x}_1)^2 + \sum_{set\ 2}(x_j - \bar{x}_2)^2}{n_1 + n_2 - 2}} = \sqrt{\frac{s_1^2(n_1 - 1) + s_2^2(n_2 - 1)}{n_1 + n_2 - 2}}$$
(4-10a)

Here s_{pooled} is a *pooled* standard deviation making use of both sets of data. The absolute value of $\bar{x}_1 - \bar{x}_2$ is used in Equation 4-9a so that t is always positive. The value of t from Equation 4-9a is to be compared with the value of t in Table 4-4 for $(n_1 + n_2 - 2)$ degrees of freedom. *If the calculated t is greater than the tabulated t at the 95% confidence level, the two results are considered to be significantly different.*

If $t_{calculated} > t_{table}$ (95%), the difference is significant.

In Table 4-2, the means are $\bar{x}_1 = 36.14$ and $\bar{x}_2 = 36.20$ mM with $n_1 = 10$ and $n_2 = 4$ measurements. The standard deviations are $s_1 = 0.28$ and $s_2 = 0.47$ mM, which we found with the F test in Equation 4-6 not to differ significantly from each other. Therefore we use Equations 4-9a and 4-10a to compare the means. The pooled standard deviation is

$$s_{pooled} = \sqrt{\frac{s_1^2(n_1 - 1) + s_2^2(n_2 - 1)}{n_1 + n_2 - 2}} = \sqrt{\frac{0.28^2(10 - 1) + 0.47^2(4 - 1)}{10 + 4 - 2}} = 0.33_8$$

Retain at least one extra insignificant digit at this point to avoid introducing round-off error into subsequent calculations.

To compare the means, we calculate the value of t with Equation 4-9a:

$$t_{calculated} = \frac{|\bar{x}_1 - \bar{x}_2|}{s_{pooled}} \sqrt{\frac{n_1 n_2}{n_1 + n_2}} = \frac{|36.14 - 36.20|}{0.33_8} \sqrt{\frac{10 \cdot 4}{10 + 4}} = 0.30_0$$

The calculated value of t is 0.30_0. The critical value of t in Table 4-4 for $(n_1 + n_2 - 2) = 12$ degrees of freedom lies between 2.228 and 2.131 listed for 10 and 15 degrees of freedom in the column for 95% confidence. *Because $t_{calculated} < t_{table}$, the difference in mean values is not significant.* You could have expected this conclusion because the difference is less than the standard deviation of either measurement.

$t_{calculated} < t_{table}$ (95%), so difference is *not* significant.

Case 2b: Standard Deviations Are Significantly Different

An example comes from the work of Lord Rayleigh (John W. Strutt), who is remembered today for landmark investigations of light scattering, blackbody radiation, and elastic waves in solids. His Nobel Prize in 1904 was received for discovering the inert gas argon. This discovery occurred when he noticed a small discrepancy between two sets of measurements of the density of nitrogen gas.

In Rayleigh's time, it was known that dry air was composed of about one-fifth oxygen and four-fifths nitrogen. Rayleigh removed O_2 from air by passing the air through red-hot copper to make CuO. He then measured the density of the remaining gas by collecting it in a fixed volume at constant temperature and pressure. He also prepared the same volume of

Removing O_2 from air:
$Cu(s) + \frac{1}{2}O_2(g) \rightarrow CuO(s)$

TABLE 4-5	Masses of gas isolated by Lord Rayleigh	
From air (g)	**From chemical decomposition (g)**	
2.310 17	2.301 43	
2.309 86	2.298 90	
2.310 10	2.298 16	
2.310 01	2.301 82	
2.310 24	2.298 69	
2.310 10	2.299 40	
2.310 28	2.298 49	
—	2.298 89	
Average		
2.310 11	2.299 47	
Standard deviation		
$0.000\ 14_3$	0.001 38	

SOURCE: R. D. Larsen, J. Chem. Ed. **1990**, 67, 925; see also C. J. Giunta, J. Chem. Ed. **1998**, 75, 1322.

If two variances were $s_1^2 = (0.002\ 00)^2$ (7 degrees of freedom) and $s_2^2 = (0.001\ 00)^2$ (6 degrees of freedom), is the difference significant? (Answer: No. $F_{calculated} = 4.00 < F_{table} = 4.21$)

If $F_{calculated} < F_{table}$, the standard deviations do not differ significantly and we use Equations 4-9a and 4-10a to compare means with the t test.
If $F_{calculated} > F_{table}$, the standard deviations differ significantly and we use Equations 4-9b and 4-10b to compare means with the t test.

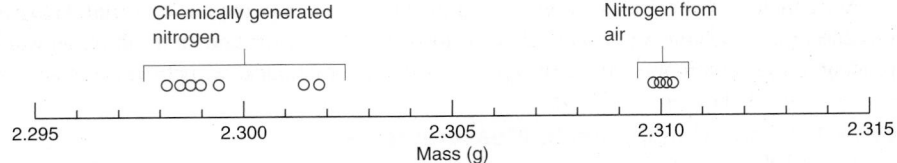

FIGURE 4-7 Rayleigh's measurements of the mass of constant volumes of gas (at constant temperature and pressure) isolated by removing oxygen from air or generated by decomposition of nitrogen compounds. Rayleigh recognized that the difference between the two clusters was outside of his experimental error and deduced that a heavier component, which turned out to be argon, was present in gas isolated from air.

pure N_2 by chemical decomposition of nitrous oxide (N_2O), nitric oxide (NO), or ammonium nitrite ($NH_4^+NO_2^-$). Table 4-5 and Figure 4-7 show the mass of gas collected in each experiment. The average mass collected from air (2.310 11 g) is 0.46% greater than the average mass of the same volume of gas from chemical sources (2.299 47 g).

If Rayleigh's measurements had not been performed with care, this difference might have been attributed to experimental error. Instead, Rayleigh understood that the discrepancy was outside his margin of error, and he postulated that gas collected from air was a mixture of nitrogen with a small amount of a heavier gas, which turned out to be argon.

In Figure 4-7, the two sets of data are clustered in different regions. The range of results for chemically generated nitrogen is larger than the range for nitrogen from air. Are the two standard deviations in Table 4-5 statistically different from each other? We answer this question with the F test (Equation 4-6):

$$F_{calculated} = \frac{s_1^2}{s_2^2} = \frac{(0.001\ 37_9)^2}{(0.000\ 14_3)^2} = 93._1$$

The critical value of F in Table 4-3 for $n - 1 = 7$ degrees of freedom for the numerator (s_1) and 6 degrees of freedom for the denominator (s_2) is 4.21. Because $F_{calculated} > F_{table}$, the difference in standard deviations is significant.

When the standard deviations of two sets of measurements are significantly different, the equations for the t test are

$$t_{calculated} = \frac{|\bar{x}_1 - \bar{x}_2|}{\sqrt{(s_1^2/n_1) + (s_2^2/n_2)}} = \frac{|\bar{x}_1 - \bar{x}_2|}{\sqrt{(u_1^2) + (u_2^2)}} \qquad \text{(4-9b)}$$

$$\text{Degrees of freedom} = \frac{(s_1^2/n_1 + s_2^2/n_2)^2}{\dfrac{(s_1^2/n_1)^2}{n_1 - 1} + \dfrac{(s_2^2/n_2)^2}{n_2 - 1}} = \frac{(u_1^2 + u_2^2)^2}{\dfrac{u_1^4}{n_1 - 1} + \dfrac{u_2^4}{n_2 - 1}} \qquad \text{(4-10b)}$$

where the standard uncertainty (u_i) of each variable is the standard deviation of the mean ($u_i = s_i/\sqrt{n_i}$). Round the degrees of freedom from Equation 4-10b to the nearest integer.

EXAMPLE Is Rayleigh's N_2 from Air Denser than N_2 from Chemicals?

The average mass of nitrogen from air in Table 4-5 is $\bar{x}_1 = 2.310\ 10_9$ g, with a standard deviation of $s_1 = 0.000\ 14_3$ (for $n_1 = 7$ measurements). The average mass from chemical sources is $\bar{x}_2 = 2.299\ 47_2$ g, with a standard deviation of $s_2 = 0.001\ 37_9$ (for $n_2 = 8$ measurements). Are the two masses significantly different?

Solution The F test told us that the standard deviations are significantly different, so we use Equations 4-9b and 4-10b:

$$t_{calculated} = \frac{|\bar{x}_1 - \bar{x}_2|}{\sqrt{(s_1^2/n_1) + (s_2^2/n_2)}} = \frac{|2.310\ 10_9 - 2.299\ 47_2|}{\sqrt{0.000\ 14_3^2/7 + 0.001\ 37_9^2/8}} = 21.7$$

$$\text{Degrees of freedom} = \frac{(s_1^2/n_1 + s_2^2/n_2)^2}{\dfrac{(s_1^2/n_1)^2}{n_1 - 1} + \dfrac{(s_2^2/n_2)^2}{n_2 - 1}} = \frac{(0.000\ 14_3^2/7 + 0.001\ 37_9^2/8)^2}{\dfrac{(0.000\ 14_3^2/7)^2}{7 - 1} + \dfrac{(0.001\ 37_9^2/8)^2}{8 - 1}} = 7.17$$

Equation 4-10b gives us 7.17 degrees of freedom, which we round to 7. For 7 degrees of freedom, the critical value of t in Table 4-4 for 95% confidence is 2.365. The observed value $t_{calculated} = 21.7$ far exceeds t_{table}. The obvious difference between the two data sets in Figure 4-7 is highly significant.

TEST YOURSELF If the difference between the two mean values were half as great as Rayleigh found, but the standard deviations were unchanged, would the difference still be significant? (*Answer:* $t_{calculated} = 10.8 > t_{table} = 2.365$—still a highly significant difference)

Case 3. Paired t Test for Comparing Individual Differences

In this case, we use two methods to make single measurements on several different samples. No measurement has been duplicated. Do the two methods give the same answer "within experimental error"? Figure 4-8 shows measurements of nitrate in 8 different plant extracts. Results from a spectrophotometric method in column B and from an electrochemical *biosensor* in column C are similar, but not identical.

To see if there is a significant difference between the methods, we use the paired t test. First, column D computes the difference (d_i) between the two results for each sample. The mean of the 8 differences ($\bar{d} = 0.11_4$) is computed in cell D14 and the standard deviation of the 8 differences (s_d) is computed in cell D15.

> **Biosensor:** a device that uses biological components such as enzymes, antibodies, or DNA, in combination with electric, optical, or other signals, to achieve selective response to one analyte.

$$s_d = \sqrt{\frac{\sum (d_i - \bar{d})^2}{n - 1}} \tag{4-11}$$

$$s_d = \sqrt{\frac{(0.01 - \bar{d})^2 + (0.37 - \bar{d})^2 + (-0.14 - \bar{d})^2 + \ldots + (0.09 - \bar{d})^2}{8 - 1}} = 0.40_1$$

Once you have the mean and standard deviation, compute $t_{calculated}$ in cell D16:

$$t_{calculated} = \frac{|\bar{d}|}{s_d}\sqrt{n} \tag{4-12}$$

where $|\bar{d}|$ is the absolute value of the mean difference, so that $t_{calculated}$ is always positive. Inserting numbers into Equation 4-12 gives

$$t_{calculated} = \frac{0.11_4}{0.40_1}\sqrt{8 - 1} = 0.80_3$$

	A	B	C	D
1	Comparison of methods for measuring nitrate			
2				
3		Nitrate (ppm) in plant extract		
4		Spectrophotometry		
5	Sample	with Cd reduction	Experimental biosensor	Difference (d_i)
6	1	1.22	1.23	0.01
7	2	1.21	1.58	0.37
8	3	4.18	4.04	−0.14
9	4	3.96	4.92	0.96
10	5	1.18	0.96	−0.22
11	6	3.65	3.37	−0.28
12	7	4.36	4.48	0.12
13	8	1.61	1.70	0.09
14			Mean difference =	0.114
15			Standard deviation of differences =	0.401
16			$t_{calculated}$ =	0.803
17			t_{table} =	2.365
18	D6 = C6–B6			
19	D14 = AVERAGE(D6:D13)			
20	D15 = STDEV(D6:D13)			
21	D16 = ABS(D14)*SQRT(A13)/D15 (ABS = absolute value)			
22	D17 = TINV(0.05,A13-1)			

Figure 4-8 uses the Excel formula =TINV (0.05,A13-1) to compute $t_{table} = 2.365$ in cell D17 for $(1 - 0.05) = 95\%$ confidence and $8 - 1 = 7$ degrees of freedom.

FIGURE 4-8 Measurement of nitrate in plant extract by two methods. [Data from N. Plumeré, J. Henig, and W. H. Campbell, "Enzyme-Catalyzed O_2 Removal System for Electrochemical Analysis Under Ambient Air: Application in an Amperometric Nitrate Biosensor," *Anal. Chem.* **2012,** *84,* 2141.]

We find that $t_{calculated}$ (0.80_3) is less than t_{table} (2.365) listed in Table 4-4 for 95% confidence and 7 degrees of freedom. *There is more than a 5% chance that the two sets of results come from populations with the same mean, so we conclude that the results are not significantly different.* (The paired *t* test presumes that the two sets of measurements have similar standard deviations. We have no way to test this assumption without replicate results from each method.)

One-Tailed and Two-Tailed Significance Tests

In Equation 4-8, we sought to compare the mean from four replicate measurements with a certified value. The curve in Figure 4-9a is the *t* distribution for 3 degrees of freedom. If the certified value lies in the outer 5% of the area under the curve, we reject the null hypothesis and conclude with 95% confidence that the measured mean is not equivalent to the certified value. The critical value of *t* for rejecting the null hypothesis is 3.182 for 3 degrees of freedom in Table 4-4. In Figure 4-9a, *2.5% of the area* beneath the curve lies above *t* = 3.182 and *2.5% of the area* lies below *t* = −3.182. We call this a *two-tailed test* because we reject the null hypothesis if the certified value lies in the low-probability region on either side of the mean.

If we have a preconceived reason to believe that our method gives systematically low values, we could use the *one-tailed t test* in Figure 4-9b. In this case, we reject the null hypothesis (which states that there is no significant difference between the measured and certified values) if $t_{calculated}$ is greater than 2.353. Figure 4-9b shows that *5% of the area* beneath the curve lies above *t* = 2.353. We did not consider area at the left side of the curve because we had a reason to believe that our method produces low results, not high results.

How can you find the value of *t* that bounds the upper 5% of the area of the curve? Because the *t* distribution is symmetric, the two-tailed value *t* = 2.353 for 90% confidence in Table 4-2 must be the value we seek, because 5% of the area lies above *t* = 2.353 and 5% of the area lies below *t* = −2.353.

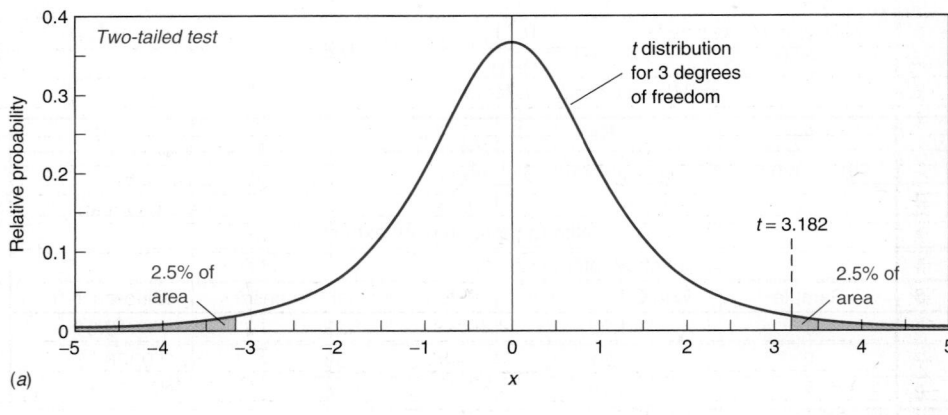

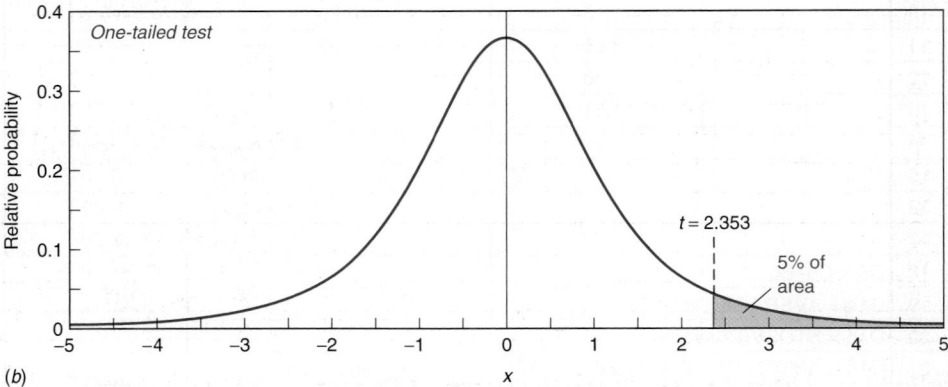

FIGURE 4-9 Student's *t* distribution for 3 degrees of freedom. In panel *a*, each shaded tail contains 2.5% of the area under the curve. In panel *b*, the single shaded tail contains 5% of the area. The fewer the degrees of freedom, the broader the distribution. As the degrees of freedom increase, the curve approaches a Gaussian curve.

The purpose of this discussion was to inform you of the distinction between one- and two-tailed tests. All t tests in this book will be two-tailed.

Is My Red Blood Cell Count High Today?

At the opening of this chapter, red cell counts on five "normal" days were 5.1, 5.3, 4.8, 5.4, and 5.2×10^6 cells/μL. The question was whether today's count of 5.6×10^6 cells/μL is "significantly" higher than normal? Disregarding the factor of 10^6, the mean of the normal values is $\bar{x} = 5.16$ and the standard deviation is $s = 0.23$. For today's value of 5.6,

$$t_{calculated} = \frac{|\text{today's count} - \bar{x}|}{s}\sqrt{n} = \frac{|5.16 - 5.6|}{0.23}\sqrt{5} = 4.28$$

What is the probability of finding $t = 4.28$ for 4 degrees of freedom?

In Table 4-4, looking across the row for 4 degrees of freedom, we see that 4.28 lies between the 98% ($t = 3.747$) and 99% ($t = 4.604$) confidence levels. Today's red cell count lies in the upper tail of the curve containing less than 2% of the area of the curve. There is less than a 2% probability of observing a count of 5.6×10^6 cells/μL on "normal" days. It is reasonable to conclude that today's count is elevated.

Table 4-2 places the probability of today's red cell count between 1 and 2%. Excel provides a probability with the function TDIST(x,deg_freedom,tails), where x is $t_{calculated}$, deg_freedom = 4, and tails = 2. The function TDIST(4.28,4,2) returns the value 0.013. Today's red cell count lies in the upper 1.3% of the area of the t-distribution.

Find probability with Excel:
TDIST(x,deg_freedom,tails)

4-5 t Tests with a Spreadsheet

To compare Rayleigh's two data sets in Table 4-5, enter his data in columns B and C of a spreadsheet (Figure 4-10). In rows 13 and 14, we computed the averages and standard deviations, but we did not need to do this.

	A	B	C	D	E	F	G
1	Analysis of Rayleigh's Data				t-Test: Two-Sample Assuming Equal Variances		
2						*Variable 1*	*Variable 2*
3		Mass of gas (g) collected from			Mean	2.310109	2.299473
4		air	chemical		Variance	2.03E-08	1.9E-06
5		2.31017	2.30143		Observations	7	8
6		2.30986	2.29890		Pooled Variance	1.03E-06	
7		2.31010	2.29816		Hypothesized Mean Diff	0	
8		2.31001	2.30182		df	13	
9		2.31024	2.29869		t Stat	20.21372	
10		2.31010	2.29940		P(T<=t) one-tail	1.66E-11	
11		2.31028	2.29849		t Critical one-tail	1.770932	
12			2.29889		P(T<=t) two-tail	3.32E-11	
13	Average	2.31011	2.29947		t Critical two-tail	2.160368	
14	Std Dev	0.00014	0.00138				
15					t-Test: Two-Sample Assuming Unequal Variances		
16	B13 = AVERAGE(B5:B12)					*Variable 1*	*Variable 2*
17	B14 = STDEV(B5:B12)				Mean	2.310109	2.299473
18					Variance	2.03E-08	1.9E-06
19					Observations	7	8
20					Hypothesized Mean Diff	0	
21					df	7	
22					t Stat	21.68022	
23					P(T<=t) one-tail	5.6E-08	
24					t Critical one-tail	1.894578	
25					P(T<=t) two-tail	1.12E-07	
26					t Critical two-tail	2.364623	

FIGURE 4-10 Spreadsheet for comparing mean values of Rayleigh's measurements in Table 4-5.

In the Excel Data ribbon, you might find Data Analysis as an option. If not, click the File menu in Excel 2010. Select Excel Options and Add-Ins. Select Analysis ToolPak, click GO, and then OK to load the Analysis ToolPak. For future use, follow the same steps to load Solver Add-In.

Returning to Figure 4-10, we want to test the null hypothesis that the two sets of data are drawn from populations with the same population standard deviation, μ. Excel gives us the choice to compare the means with Equations 4-9a and 4-10a when the standard deviations are not significantly different or with Equations 4-9b and 4-10b when the standard deviations are significantly different. We illustrate both cases in Figure 4-10. In the Data ribbon, select Data Analysis. In the window that appears, select t-Test: Two-Sample Assuming Equal Variances. Click OK. The next window asks you in which cells the data are located. Write B5:B12 for Variable 1 and C5:C12 for Variable 2. The routine ignores the blank space in cell B12. For the Hypothesized Mean Difference enter 0 and for Alpha enter 0.05. Alpha is the level of probability to which we are testing the difference in the means. With Alpha = 0.05, we are at the 95% confidence level. For Output Range, select cell E1 and click OK.

Excel goes to work and prints results in cells E1 to G13 of Figure 4-10. Mean values are in cells F3 and G3. Cells F4 and G4 give *variance*, which is the square of the standard deviation. Cell F6 gives *pooled variance*, which is the square of s_{pooled} from Equation 4-10a. That equation was painful to use by hand. Cell F8 shows degrees of freedom (df) = 13 and $t_{calculated}$ (t Stat) = 20.2 from Equation 4-9a appears in cell F9.

At this point in Section 4-4, we consulted Table 4-4 to find that t_{table} lies between 2.228 and 2.131 for 95% confidence and 13 degrees of freedom. Excel gives us the critical value of t (2.160) in cell F13 of Figure 4-10. Because $t_{calculated}$ (= 20.2) > t_{table} (= 2.160), we conclude that the two means are significantly different. Cell F12 states that the probability of observing these two mean values by random chance if the data came from sets with the same population mean (μ) is $p = 3 \times 10^{-11}$. The difference is *highly* significant. For any value of $p < 0.05$ in cell F12, we reject the *null hypothesis* and conclude that the means *are different*.

The F test told us that the standard deviations of Rayleigh's two experiments are different. Therefore, we can select the other t test found in the Analysis ToolPak. Select t-Test: Two-Sample Assuming Unequal Variances and fill in the blanks as before. Results based on Equations 4-9b and 4-10b are displayed in cells E15 to G26 of Figure 4-10. Just as we found in Section 4-4, the degrees of freedom are 7 (cell F21) and $t_{calculated}$ = 21.7 (cell F22). Because $t_{calculated}$ is greater than the critical value of t (2.36 in cell F26), we reject the null hypothesis and conclude that the two means *are* significantly different. Cell F25 states that the probability of observing these two mean values by random chance if the data came from sets with the same population mean (μ) is $p = 1 \times 10^{-7}$. Biologists often state conclusions in terms of p values. *The smaller the value of p, the more confident we can be in rejecting the null hypothesis that the two sets of data come from populations with the same population mean.*

4-6 Grubbs Test for an Outlier

Students dissolved zinc from a galvanized nail and measured the mass lost by the nail to tell how much of the nail was zinc. Here are 12 results:

Mass loss (%): 10.2, 10.8, 11.6, 9.9, 9.4, 7.8, 10.0, 9.2, 11.3, 9.5, 10.6, 11.6

The value 7.8 appears unusually low. A datum that is far from other points is called an **outlier.** Should 7.8 be discarded before averaging the rest of the data or should 7.8 be retained?

We answer this question with the **Grubbs test.** First compute the average ($\bar{x} = 10.16$) and the standard deviation ($s = 1.11$) of the complete data set (all 12 points in this example). Then compute the Grubbs statistic G, defined as

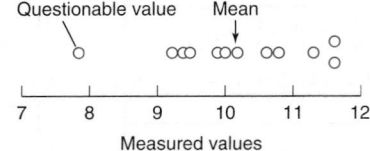

Questionable value   Mean

Measured values

The Grubbs test is recommended by the International Standards Organization and the American Society for Testing and Materials in place of the Q test, which was formerly used in this book.

$$\text{Grubbs test:} \qquad G_{calculated} = \frac{|\text{questionable value} - \bar{x}|}{s} \qquad (4\text{-}13)$$

where the numerator is the absolute value of the difference between the suspected outlier and the mean value. *If $G_{calculated}$ is greater than G in Table 4-6, the questionable point should be discarded.*

In our example, $G_{calculated} = |7.8 - 10.16|/1.11 = 2.13$. In Table 4-6, $G_{table} = 2.285$ for 12 observations. *Because $G_{calculated} < G_{table}$, the questionable point should be retained.* There is more than a 5% chance that the value 7.8 is a member of the same population as the other measurements.

Common sense must prevail. If you know that a result is low because you spilled part of a solution, then the probability that the result is wrong is 100% and the datum should be discarded. Any datum based on a faulty procedure should be discarded, no matter how well it fits the rest of the data.

4-7 The Method of Least Squares

For most chemical analyses, the response of the procedure must be evaluated for known quantities of analyte (called *standards*) so that the response to an unknown quantity can be interpreted. For this purpose, we commonly prepare a **calibration curve,** such as the one for caffeine in Figure 0-7. Most often, we work in a region where the calibration curve is a straight line.

We use the **method of least squares** to draw the "best" straight line through experimental data points that have some scatter and do not lie perfectly on a straight line.[5] The best line will be such that some of the points lie above and some lie below the line. We will learn to estimate the uncertainty in a chemical analysis from the uncertainties in the calibration curve and in the measured response to replicate samples of unknown.

Finding the Equation of the Line

The procedure we use assumes that the errors in the y values are substantially greater than the errors in the x values.[6] This condition is often true in a calibration curve in which the experimental response (y values) is less certain than the quantity of analyte (x values). A second assumption is that uncertainties (standard deviations) in all y values are similar.

We seek to draw the best straight line through the points in Figure 4-11 by minimizing the vertical deviations between the points and the line. We minimize only the vertical deviations because we assume that uncertainties in y values are much greater than uncertainties in x values.

Let the equation of the line be

Equation of straight line: $$y = mx + b \qquad \text{(4-14)}$$

in which m is the **slope** and b is the **y-intercept.** The vertical deviation for the point (x_i, y_i) in Figure 4-11 is $y_i - y$, where y is the ordinate of the straight line when $x = x_i$.

$$\text{Vertical deviation} = d_i = y_i - y = y_i - (mx_i + b) \qquad \text{(4-15)}$$

Some of the deviations are positive and some are negative. Because we wish to minimize the magnitude of the deviations irrespective of their signs, we square all the deviations so that we are dealing only with positive numbers:

$$d_i^2 = (y_i - y)^2 = (y_i - mx_i - b)^2$$

Because we minimize the squares of the deviations, this is called the *method of least squares.* It can be shown that minimizing the squares of the deviations (rather than simply their magnitudes) corresponds to assuming that the set of y values is the most probable set.

If uncertainties in x and y are comparable, it is appropriate to minimize a combination of vertical and horizontal deviations of points from the line, instead of the vertical distance shown in Figure 4-11. References 7 and 8 provide equations to handle uncertainty in both x and y.

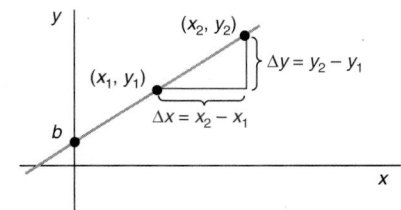

Equation for a straight line: $y = mx + b$

Slope (m) $= \dfrac{\Delta y}{\Delta x} = \dfrac{y_2 - y_1}{x_2 - x_1}$

y-intercept (b) = crossing point on y-axis

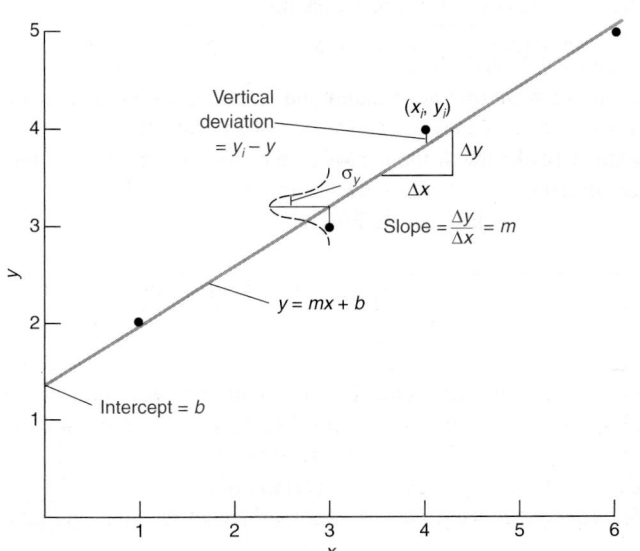

FIGURE 4-11 Least-squares curve fitting. The Gaussian curve drawn over the point (3,3) indicates schematically how each value of y_i is normally distributed about the straight line. That is, the most probable value of y will fall on the line, but there is a finite probability of measuring y some distance from the line.

To evaluate the determinant, multiply the diagonal elements $e \times h$ and then subtract the product of the other diagonal elements $f \times g$.

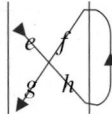

Translation of least-squares equations:

$$m = \frac{n\Sigma(x_i y_i) - \Sigma x_i \Sigma y_i}{n\Sigma(x_i^2) - (\Sigma x_i)^2}$$

$$b = \frac{\Sigma(x_i^2)\Sigma y_i - \Sigma(x_i y_i)\Sigma x_i}{n\Sigma(x_i^2) - (\Sigma x_i)^2}$$

Finding values of m and b that minimize the sum of the squares of the vertical deviations involves some calculus, which we omit. We express the final solution for slope and intercept in terms of *determinants*, which summarize certain arithmetic operations. The **determinant** $\begin{vmatrix} e & f \\ g & h \end{vmatrix}$ represents the value $eh - fg$. So, for example,

$$\begin{vmatrix} 6 & 5 \\ 4 & 3 \end{vmatrix} = (6 \times 3) - (5 \times 4) = -2$$

The slope and the intercept of the "best" straight line are found to be

Least-squares "best" line

$$Slope: \quad m = \begin{vmatrix} \Sigma(x_i y_i) & \Sigma x_i \\ \Sigma y_i & n \end{vmatrix} \div D \qquad (4\text{-}16)$$

$$Intercept: \quad b = \begin{vmatrix} \Sigma(x_i^2) & \Sigma(x_i y_i) \\ \Sigma x_i & \Sigma y_i \end{vmatrix} \div D \qquad (4\text{-}17)$$

where D is

$$D = \begin{vmatrix} \Sigma(x_i^2) & \Sigma x_i \\ \Sigma x_i & n \end{vmatrix} \qquad (4\text{-}18)$$

and n is the number of points.

Let's use these equations to find the slope and intercept of the best straight line through the four points in Figure 4-11. The work is set out in Table 4-7. Noting that $n = 4$ and putting the various sums into the determinants in Equations 4-16, 4-17, and 4-18 gives

$$m = \begin{vmatrix} 57 & 14 \\ 14 & 4 \end{vmatrix} \div \begin{vmatrix} 62 & 14 \\ 14 & 4 \end{vmatrix} = \frac{(57 \times 4) - (14 \times 14)}{(62 \times 4) - (14 \times 14)} = \frac{32}{52} = 0.615\,38$$

$$b = \begin{vmatrix} 62 & 57 \\ 14 & 14 \end{vmatrix} \div \begin{vmatrix} 62 & 14 \\ 14 & 4 \end{vmatrix} = \frac{(62 \times 14) - (57 \times 14)}{(62 \times 4) - (14 \times 14)} = \frac{70}{52} = 1.346\,15$$

The equation of the best straight line through the points in Figure 4-11 is therefore

$$y = 0.615\,38x + 1.346\,15$$

We tackle the question of significant figures for m and b in the next section.

EXAMPLE   **Finding Slope and Intercept with a Spreadsheet**

Excel has functions called SLOPE and INTERCEPT, whose use is illustrated here:

	A	B	C	D	E	F
1	x	y			Formulas:	
2	1	2		slope =		
3	3	3		0.61538	D3 = SLOPE(B2:B5,A2:A5)	
4	4	4		intercept =		
5	6	5		1.34615	D5 = INTERCEPT(B2:B5,A2:A5)	

The slope in cell D3 is computed with the formula "=SLOPE(B2:B5,A2:A5)", where B2:B5 is the range containing the y values and A2:A5 is the range containing x values.

TEST YOURSELF   Change the second value of x from 3 to 3.5 and find the slope and intercept. (*Answer:* 0.610 84, 1.285 71)

TABLE 4-7   **Calculations for least-squares analysis**

x_i	y_i	$x_i y_i$	x_i^2	$d_i (= y_i - mx_i - b)$	d_i^2
1	2	2	1	0.038 46	0.001 479 3
3	3	9	9	−0.192 31	0.036 982
4	4	16	16	0.192 31	0.036 982
6	5	30	36	−0.038 46	0.001 479 3
$\Sigma x_i = 14$	$\Sigma y_i = 14$	$\Sigma(x_i y_i) = 57$	$\Sigma(x_i^2) = 62$		$\Sigma(d_i^2) = 0.076\,923$

How Reliable Are Least-Squares Parameters?

To estimate uncertainties in the slope and intercept, an uncertainty analysis must be performed on Equations 4-16 and 4-17. Because the uncertainties in m and b are related to the uncertainty in measuring each value of y, we first estimate the standard deviation that describes the population of y values. This standard deviation, σ_y, characterizes the little Gaussian curve inscribed in Figure 4-11.

We estimate σ_y, the population standard deviation of all y values, by calculating s_y, the standard deviation, for the four measured values of y. The deviation of each value of y_i from the center of its Gaussian curve is $d_i = y_i - y = y_i - (mx_i + b)$. The standard deviation of these vertical deviations is

$$\sigma_y \approx s_y = \sqrt{\frac{\sum (d_i - \bar{d})^2}{(\text{degrees of freedom})}} \qquad \textbf{(4-19)}$$

Equation 4-19 is analogous to Equation 4-2.

But the average deviation, $\bar{d}$, is 0 for the best straight line, so the numerator of Equation 4-19 reduces to $\sum (d_i^2)$.

The *degrees of freedom* is the number of independent pieces of information available. For n data points, there are n degrees of freedom. If you were calculating the standard deviation of n points, you would first find the average to use in Equation 4-2. This calculation leaves $n - 1$ degrees of freedom in Equation 4-2 because only $n - 1$ pieces of information are available in addition to the average. If you know $n - 1$ values and you also know their average, then the nth value is fixed and you can calculate it.

For Equation 4-19, we began with n points. Two degrees of freedom were lost in determining the slope and the intercept. Therefore, $n - 2$ degrees of freedom remain. Equation 4-19 becomes

Standard deviation of y:
$$s_y = \sqrt{\frac{\sum (d_i^2)}{n - 2}} \qquad \textbf{(4-20)}$$

where d_i is given by Equation 4-15.

Uncertainty analysis for Equations 4-16 and 4-17 leads to the following results:

$$\textit{Standard uncertainty} \begin{cases} u_m^2 = \dfrac{s_y^2 n}{D} & \textbf{(4-21)} \\[2ex] u_b^2 = \dfrac{s_y^2 \sum (x_i^2)}{D} & \textbf{(4-22)} \end{cases}$$
of slope and intercept

where u_m is the *standard uncertainty* of the slope, u_b is the *standard uncertainty* of the intercept, s_y is the *standard deviation* of y given by Equation 4-20, and D is given by Equation 4-18. *Standard uncertainty* (u_m and u_b) is the standard deviation of the mean. If you double the number of calibration points, u_m and u_b decrease by $\sim 1/\sqrt{2}$. The standard deviation s_y is a characteristic of the population of measurements and is independent of the number of calibration points. If you double the number of points, s_y is nearly constant.

Standard uncertainty (u) = standard deviation of the mean
• u decreases if you measure more points
• Standard deviation (s) is approximately constant if you measure more points

At last, we can assign significant figures to the slope and the intercept in Figure 4-11. In Table 4-7, we see that $\sum (d_i^2) = 0.076\ 923$. Putting this number into Equation 4-20 gives

$$s_y^2 = \frac{0.076\ 923}{4 - 2} = 0.038\ 462$$

Now, we can plug numbers into Equations 4-21 and 4-22 to find

$$u_m^2 = \frac{s_y^2 n}{D} = \frac{(0.038\ 462)(4)}{52} = 0.002\ 958\ 6 \implies u_m = 0.054\ 39$$

$$u_b^2 = \frac{s_y^2 \sum (x_i^2)}{D} = \frac{(0.038\ 462)(62)}{52} = 0.045\ 859 \implies u_b = 0.214\ 15$$

4-7 The Method of Least Squares

83

Combining the results for m, u_m, b, and u_b, we write

<table>
<tr><td>The first digit of the uncertainty is the last significant figure. We often retain extra, insignificant digits to prevent round-off errors in further calculations. ──────</td></tr>
</table>

Slope: $\dfrac{0.615\,38}{\pm 0.054\,39} = 0.62 \pm 0.05 \quad \text{or} \quad 0.61_5 \pm 0.05_4$ (4-23)

Intercept: $\dfrac{1.346\,15}{\pm 0.214\,15} = 1.3 \pm 0.2 \quad \text{or} \quad 1.3_5 \pm 0.2_1$ (4-24)

where the uncertainties are u_m and u_b. *The first decimal place of the uncertainty is the last significant figure of the slope or intercept.* Many scientists write results such as 1.35 ± 0.21 to avoid excessive round-off.

To express uncertainty as a confidence interval, Equation 4-7 tells us to multiply the standard uncertainties in Equations 4-23 and 4-24 by Student's t from Table 4-2 for $n - 2$ degrees of freedom.

95% confidence interval for slope

$= \pm t u_m = \pm (4.303)(0.054) = \pm 0.23$

based on $n - 2 = 2$ degrees of freedom.

EXAMPLE   Finding s_y, u_m, and u_b with a Spreadsheet

The Excel function LINEST returns the slope and intercept and their standard uncertainties in a table (a *matrix*). As an example, enter x and y values in columns A and B. Then highlight the 3-row $\times$ 2-column region E3:F5 with your mouse. This block of cells is selected for the output of LINEST. On the Formulas ribbon, go to Insert Function. In the window that appears, go to Statistical and double click on LINEST. The new window asks for four inputs to the function. For y values, enter B2:B5. Then enter A2:A5 for x values. The next two entries are both "TRUE". The first TRUE tells Excel that we want to compute the y-intercept of the line and not force the intercept to be 0. The second TRUE tells Excel to print out the uncertainties as well as the slope and intercept. The formula you have just entered is "=LINEST(B2:B5,A2:A5,TRUE,TRUE)". Now press CONTROL+SHIFT+ENTER on a PC or COMMAND(⌘)+RETURN on a Mac. Excel prints out a matrix in cells E3:F5. Write labels around the block to indicate what is in each cell. The slope and intercept are on the top line. The second line contains u_m and u_b. Cell F5 contains s_y and cell E5 contains a quantity called R^2, which is defined in Equation 5-2 and is a measure of the goodness of fit of the data to the line. The closer R^2 is to unity, the better the fit.

	A	B	C	D	E	F	G
1	x	y			Output from LINEST		
2	1	2			Slope	Intercept	
3	3	3		Parameter	0.61538	1.34615	
4	4	4		u_m	0.05439	0.21414	u_b
5	6	5		R^2	0.98462	0.19612	s_y
6							
7	Highlight cells E3:F5						
8	Type "= LINEST(B2:B5,A2:A5,TRUE,TRUE)"						
9	Press CTRL+SHIFT+ENTER (on PC)						
10	Press COMMAND(⌘)+RETURN (on Mac)						

TEST YOURSELF   Change the second value of x from 3 to 3.5 and apply LINEST. What is the value of s_y from LINEST? (*Answer:* 0.364 70)

4-8 Calibration Curves

Sections 18-1 and 18-2 discuss absorption of light and define *absorbance,* which we use throughout this book. You may want to read these sections for background.

A *calibration curve* shows the response of an analytical method to known quantities of analyte.[9] Table 4-8 gives real data from a protein analysis that produces a colored product. A *spectrophotometer* measures the absorbance of light, which is proportional to the quantity of protein analyzed. Solutions containing known concentrations of analyte are called **standard solutions.** Solutions containing all reagents and solvents used in the analysis, but no deliberately added analyte, are called **blank solutions.** Blanks measure the response of the analytical procedure to impurities or interfering species in the reagents.

TABLE 4-8 Spectrophotometer data used to construct calibration curve

Amount of protein (μg)	Absorbance of independent samples			Range	Corrected absorbance		
0	0.099	0.099	0.100	0.001	-0.000_3	-0.000_3	0.000_7
5.0	0.185	0.187	0.188	0.003	0.085_7	0.087_7	0.088_7
10.0	0.282	0.272	0.272	0.010	0.182_7	0.172_7	0.172_7
15.0	0.345	0.347	(0.392)	0.047	0.245_7	0.247_7	—
20.0	0.425	0.425	0.430	0.005	0.325_7	0.325_7	0.330_7
25.0	0.483	0.488	0.496	0.013	0.383_7	0.388_7	0.396_7

When we scan across the three absorbance values in each row of Table 4-8, the number 0.392 seems out of line: It is inconsistent with the other values for 15.0 μg, and the range of values for the 15.0-μg samples is much bigger than the range for the other samples. The linear relation between the average values of absorbance up to the 20.0-μg sample also indicates that the value 0.392 is in error (Figure 4-12). We choose to omit 0.392 from subsequent calculations.

It is reasonable to ask whether all three absorbances for the 25.0-μg samples are low for some unknown reason, because this point falls below the straight line in Figure 4-12. Repetition of this analysis shows that the 25.0-μg point is consistently below the straight line and there is nothing "wrong" with the data in Table 4-8.

Constructing a Calibration Curve

We adopt the following procedure for constructing a calibration curve:

Step 1 Prepare known samples of analyte covering a range of concentrations expected for unknowns. Measure the response of the analytical procedure to these standards to generate data like the left half of Table 4-8.

Step 2 Subtract the average absorbance (0.099_3) of the *blank* samples from each measured absorbance to obtain *corrected absorbance*. The blank measures the response of the procedure when no protein is present.

Step 3 Make a graph of corrected absorbance versus quantity of protein analyzed (Figure 4-13). Use the least-squares procedure to find the best straight line through the linear portion of the data, up to and including 20.0 μg of protein (14 points, including the 3 corrected blanks, in the shaded portion of Table 4-8). Find the slope and intercept and their standard uncertainties with Equations 4-16, 4-17, 4-20, 4-21, and 4-22:

$$m = 0.016\,3_0 \qquad u_m = 0.000\,2_2 \qquad u_y = 0.005_9$$
$$b = 0.004_7 \qquad s_b = 0.002_6$$

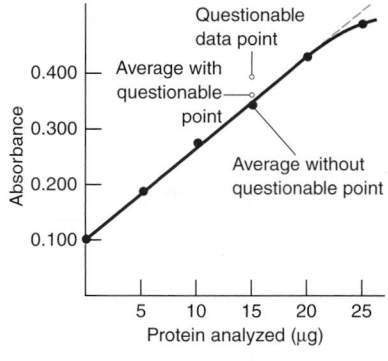

FIGURE 4-12 Average absorbance values in Table 4-8 versus micrograms of protein analyzed. Averages for 0 to 20 μg of protein lie on a straight line if the questionable datum 0.392 at 15 μg is omitted.

Absorbance of the blank can arise from the color of starting reagents, reactions of impurities, and reactions of interfering species. Blank values can vary from one set of reagents to another, but corrected absorbance should not.

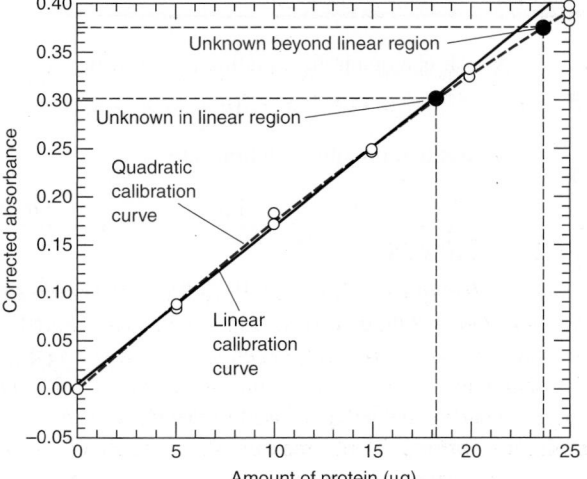

FIGURE 4-13 Calibration curve for protein analysis in Table 4-8. The equation of the solid straight line fitting the 14 data points (open circles) from 0 to 20 μg, derived by the method of least squares, is $y = 0.016\,3_0\ (\pm 0.000\,2_2)x + 0.004_7\ (\pm 0.002_6)$ with $s_y = 0.005_9$. The equation of the dashed quadratic curve that fits all 17 data points from 0 to 25 μg, determined by a nonlinear least squares procedure[5] is $y = -1.1_7\ (\pm 0.2_1) \times 10^{-4}\ x^2 + 0.018\,5_8\ (\pm 0.000\,4_6)\,x - 0.000\,7\ (\pm 0.001\,0)$ with $s_y = 0.004_6$.

Equation of calibration line:

$$y(\pm s_y) = [m(\pm u_m)]x + [b(\pm u_b)]$$

The equation of the linear calibration line is

$$\underbrace{\text{absorbance}}_{y} = m \times \underbrace{(\mu\text{g of protein})}_{x} + b$$

$$= (0.016\,3_0)(\mu\text{g of protein}) + 0.004_7 \qquad \textbf{(4-25)}$$

where y is the corrected absorbance (= observed absorbance − blank absorbance).

Step 4 If you analyze an unknown at a future time, run a blank at that time. Subtract the new blank absorbance from the unknown absorbance to obtain corrected absorbance.

> **EXAMPLE** **Using a Linear Calibration Curve**
>
> An unknown protein sample gave an absorbance of 0.406 and a blank had an absorbance of 0.104. How many micrograms of protein are in the unknown?
>
> **Solution** The corrected absorbance is $0.406 - 0.104 = 0.302$, which lies on the linear portion of the calibration curve in Figure 4-13. Rearranging Equation 4-25 gives
>
> $$\mu\text{g of protein} = \frac{\text{absorbance} - 0.004_7}{0.016\,3_0} = \frac{0.302 - 0.004_7}{0.016\,3_0} = 18.2_4 \ \mu\text{g} \qquad \textbf{(4-26)}$$
>
> **TEST YOURSELF** What mass of protein gives a corrected absorbance of 0.250? (**Answer:** $15.0_5 \ \mu\text{g}$)

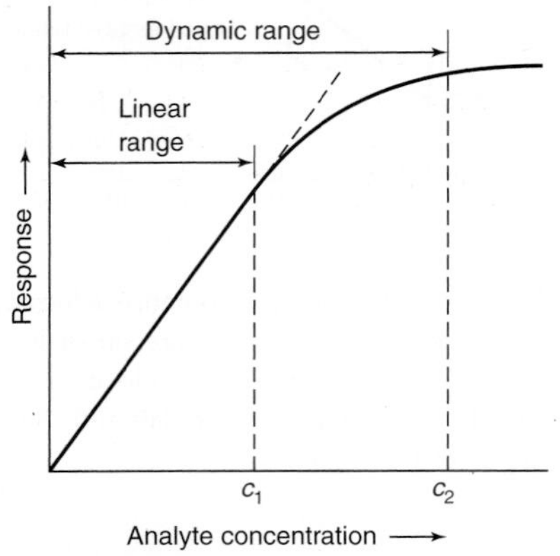

FIGURE 4-14 Calibration curve illustrating linear and dynamic ranges.

Examine your data for sensibility.

We prefer calibration procedures with a **linear response,** in which the corrected analytical signal (= signal from sample − signal from blank) is proportional to the quantity of analyte. Although we try to work in the linear range, you can obtain valid results beyond the linear region ($>20 \ \mu\text{g}$) in Figure 4-13. The dashed curve that goes up to 25 μg of protein comes from a least-squares fit of the data to the equation $y = ax^2 + bx + c$ (Box 4-2).

The **linear range** of an analytical method is the analyte concentration range over which response is proportional to concentration. A related quantity in Figure 4-14 is **dynamic range**—the concentration range over which there is a measurable response to analyte, even if the response is not linear.

Good Practice

Always make a graph of your data. The graph gives you an opportunity to reject bad data or the stimulus to repeat a measurement or decide that a straight line is not appropriate.

It is not reliable to extrapolate any calibration curve, linear or nonlinear, beyond the measured range of standards. Measure standards in the entire concentration range of interest.

At least six calibration concentrations and two replicate measurements of unknown are recommended. The most rigorous procedure is to make each calibration solution independently

---

BOX 4-2 Using a Nonlinear Calibration Curve

Consider an unknown whose corrected absorbance of 0.375 lies beyond the linear region in Figure 4-13. We can fit all the data points with the quadratic equation[5]

$$y = -1.17 \times 10^{-4} x^2 + 0.018\,58\,x - 0.000\,7 \qquad \textbf{(A)}$$

To find the quantity of protein, substitute the corrected absorbance into Equation A:

$$0.375 = -1.17 \times 10^{-4} x^2 + 0.018\,58\,x - 0.000\,7$$

This equation can be rearranged to

$$1.17 \times 10^{-4} x^2 - 0.018\,58\,x + 0.375\,7 = 0$$

which is a quadratic equation of the form

$$ax^2 + bx + c = 0$$

whose two possible solutions are

$$x = \frac{-b + \sqrt{b^2 - 4ac}}{2a} \qquad x = \frac{-b - \sqrt{b^2 - 4ac}}{2a}$$

Substituting $a = 1.17 \times 10^{-4}$, $b = -0.018\,58$, and $c = 0.375\,7$ into these equations gives

$$x = 135 \ \mu\text{g} \qquad x = 23.8 \ \mu\text{g}$$

Figure 4-13 tells us that the correct choice is 23.8 μg, not 135 μg.

from a certified material. Avoid serial dilution of a single stock solution. Serial dilution propagates systematic error in the stock solution. Measure calibration solutions in random order, not in consecutive order by increasing concentration.

Propagation of Uncertainty with a Calibration Curve

In the preceding example, an unknown with a corrected absorbance of $y = 0.302$ had a protein content of $x = 18.2_4$ µg. What is the uncertainty in the number 18.2_4? Propagation of uncertainty for the equation $y = mx + b$ (but not $y = mx$) gives the following result:[1,10]

Standard uncertainty in x = standard devation of the mean =

$$u_x = \frac{s_y}{|m|} \sqrt{\frac{1}{k} + \frac{1}{n} + \frac{(y - \bar{y})^2}{m^2 \Sigma (x_i - \bar{x})^2}} \qquad (4\text{-}27)$$

where s_y is the standard deviation of y (Equation 4-20), $|m|$ is the absolute value of the slope (= ABS(m) in Excel), k is the number of replicate measurements of the unknown, n is the number of data points for the calibration line (14 in Table 4-8), $\bar{y}$ is the mean value of y for the points on the calibration line, x_i are the individual values of x for the points on the calibration line, and $\bar{x}$ is the mean value of x for the points on the calibration line. For a single measurement of the unknown, $k = 1$ and Equation 4-27 gives $u_x = \pm 0.3_9$ µg. Four replicate unknowns ($k = 4$) with an average corrected absorbance of 0.302 reduce the uncertainty to $u_x = \pm 0.2_3$ µg.

The confidence interval for x is $\pm t u_x$, where t is Student's t (Table 4-4) for $n - 2$ degrees of freedom. If $u_x = 0.2_3$ µg and $n = 14$ points (12 degrees of freedom), the 95% confidence interval for x is $\pm t u_x = \pm (2.179)(0.2_3) = \pm 0.5_0$ µg. There is no $1/\sqrt{n}$ in the expression for confidence interval because u_x is the standard deviation of the mean.

u_x = standard uncertainty for x

y = corrected absorbance of unknown = 0.302

x_i = µg of protein in standards in Table 4-8
  = (0, 0, 0, 5.0, 5.0, 5.0, 10.0, 10.0, 10.0, 15.0, 15.0, 20.0, 20.0, 20.0,

$\bar{y}$ = average of 14 y values = 0.161_8

$\bar{x}$ = average of 14 x values = 9.64_3 µg

To find values of t that are not in Table 4-4, use the Excel function TINV. For 12 degrees of freedom and 95% confidence, the function TINV(0.05,12) returns $t = 2.179$.

Propagation of Uncertainty

You now have all the tools required for a more rigorous discussion of propagation of uncertainty than we had in Chapter 3. If you are so inclined, you will find that discussion in Appendix B.

4-9   A Spreadsheet for Least Squares

Figure 4-15 implements least-squares analysis, including propagation of error with Equation 4-27. Enter values of x and y in columns B and C. Then select cells B10:C12. Enter the formula "=LINEST(C4:C7,B4:B7,TRUE,TRUE)" and press CONTROL+SHIFT+ENTER on a PC or COMMAND(⌘)+RETURN on a Mac. LINEST returns m, b, u_m, u_b, R^2, and s_y in

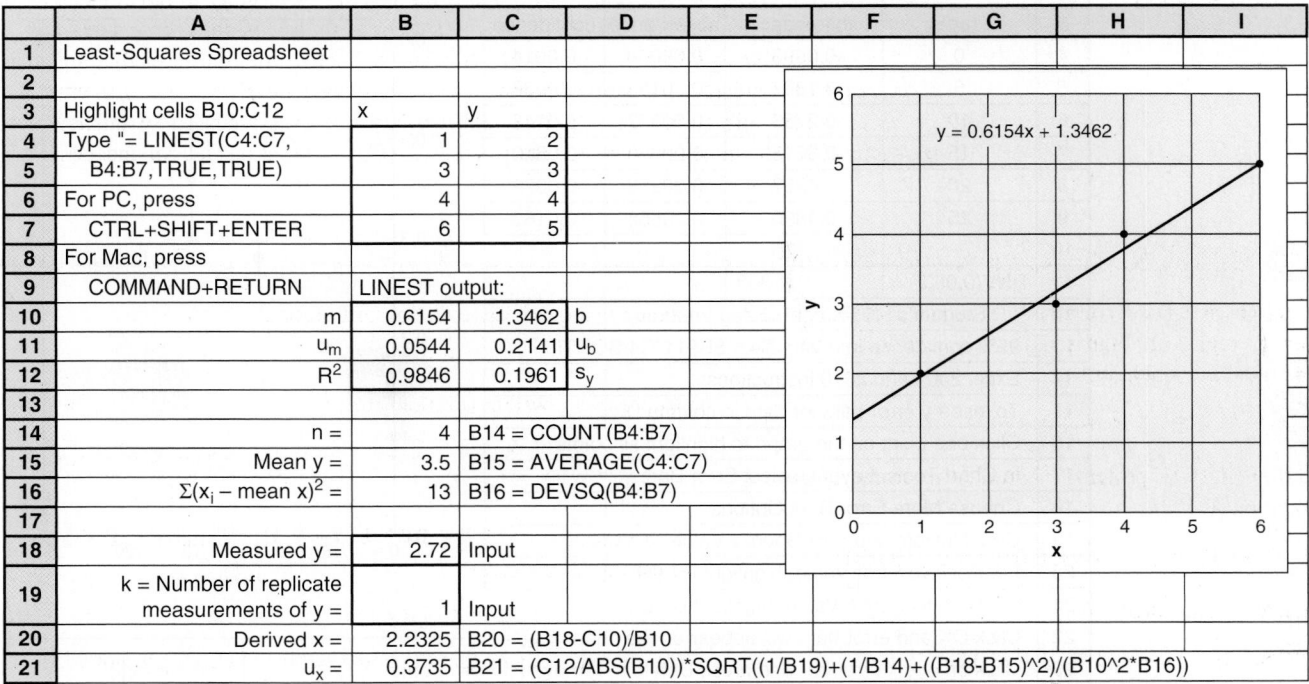

	A	B	C	D	E	F	G	H	I
1	Least-Squares Spreadsheet								
2									
3	Highlight cells B10:C12	x	y						
4	Type "= LINEST(C4:C7,	1	2						
5	B4:B7,TRUE,TRUE)	3	3						
6	For PC, press	4	4						
7	CTRL+SHIFT+ENTER	6	5						
8	For Mac, press								
9	COMMAND+RETURN	LINEST output:							
10	m	0.6154	1.3462	b					
11	u_m	0.0544	0.2141	u_b					
12	R^2	0.9846	0.1961	s_y					
13									
14	n =	4	B14 = COUNT(B4:B7)						
15	Mean y =	3.5	B15 = AVERAGE(C4:C7)						
16	$\Sigma(x_i - \text{mean } x)^2$ =	13	B16 = DEVSQ(B4:B7)						
17									
18	Measured y =	2.72	Input						
19	k = Number of replicate measurements of y =	1	Input						
20	Derived x =	2.2325	B20 = (B18-C10)/B10						
21	u_x =	0.3735	B21 = (C12/ABS(B10))*SQRT((1/B19)+(1/B14)+((B18-B15)^2)/(B10^2*B16))						

FIGURE 4-15 Spreadsheet for linear least-squares analysis.

cells B10:C12. Write labels in cells A10:A12 and D10:D12 so you know what the numbers in cells B10:C12 mean.

Cell B14 gives the number of data points with the formula "=COUNT(B4:B7)". Cell B15 computes the mean value of y. Cell B16 computes the sum $\Sigma(x_i - \bar{x})^2$ that we need for Equation 4-27. This sum is common enough that Excel has a built in function called DEVSQ that you can find in the Statistics menu of the Insert Function window.

Enter the measured mean value of y for replicate measurements of the unknown in cell B18. In cell B19, enter the number of replicate measurements of the unknown. Cell B20 computes the value of x corresponding to the measured mean value of y. Cell B21 uses Equation 4-27 to find the uncertainty u_x (the standard deviation of the mean) in the value of x for the unknown. If you want a confidence interval for x, multiply u_x times Student's t from Table 4-4 for $n - 2$ degrees of freedom and the desired confidence level.

We always want a graph to see if the calibration points lie on a straight line. Follow the instructions in Section 2-11 to plot the calibration data as a scatter plot with only markers (no line yet). To add a straight line in Excel 2007 or 2010, click on the chart to obtain the Chart Tools ribbon. Select Layout, then Trendline and choose More Trendline Options. Select Linear and Display Equation on chart. The least-squares straight line and its equation appear on the graph. Use the Forecast section of the Format Trendline box to extend the line forward or backward beyond the range of data. The Format Trendline box also allows you to select the color and style of the line.

Adding Error Bars to a Graph

Error bars on a graph help us judge the quality of the data and the fit of a curve to the data. Consider the data in Table 4-8. Let's plot the mean absorbance of columns 2 to 4 versus sample mass in column 1. Then we will add error bars corresponding to the 95% confidence interval for each point. Figure 4-16 lists mass in column A and mean absorbance in column B. The standard deviation of absorbance is given in column C. The 95% confidence interval for absorbance is computed in column D with the formula in the margin. Student's $t = 4.303$ can be found for 95% confidence and $3 - 1 = 2$ degrees of freedom in Table 4-4. Alternatively, compute Student's t with the function TINV(0.05,2) in cell B11. The parameters for TINV are 0.05 for 95% confidence and 2 for degrees of freedom. The 95% confidence interval in cell D4 is computed with "=$B$11*C4/SQRT(3)". You should be able to plot mean absorbance (y) in column B versus protein mass (x) in column A.

95% confidence interval for x in Figure 4-15:

$x \pm tu_x = 2.232\,5 \pm (4.303)(0.373\,5)$
$= 2.2 \pm 1.6$

(degrees of freedom $= n - 2 = 2$)

Confidence interval $= \pm ts/\sqrt{n}$

t = Student's t for 95% confidence and $n - 1 = 2$ degrees of freedom

s = standard deviation

n = number of values in average = 3

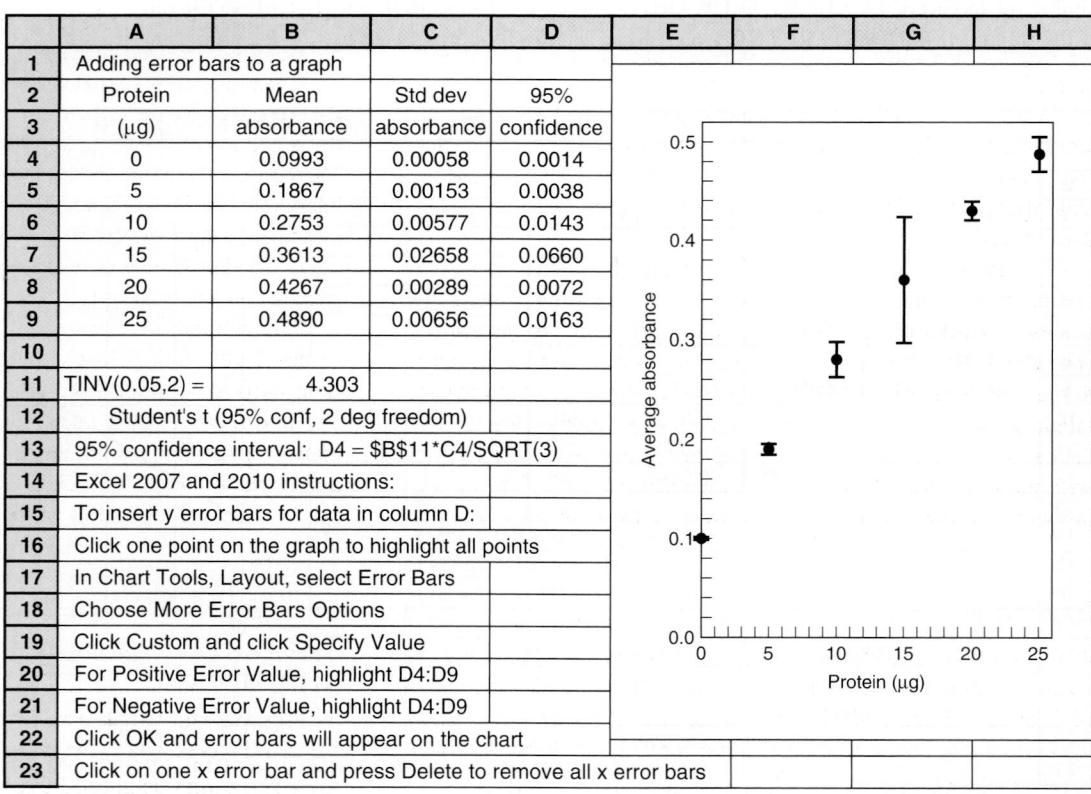

	A	B	C	D	E	F	G	H
1	Adding error bars to a graph							
2	Protein	Mean	Std dev	95%				
3	(µg)	absorbance	absorbance	confidence				
4	0	0.0993	0.00058	0.0014				
5	5	0.1867	0.00153	0.0038				
6	10	0.2753	0.00577	0.0143				
7	15	0.3613	0.02658	0.0660				
8	20	0.4267	0.00289	0.0072				
9	25	0.4890	0.00656	0.0163				
10								
11	TINV(0.05,2) =	4.303						
12	Student's t (95% conf, 2 deg freedom)							
13	95% confidence interval: D4 = $B$11*C4/SQRT(3)							
14	Excel 2007 and 2010 instructions:							
15	To insert y error bars for data in column D:							
16	Click one point on the graph to highlight all points							
17	In Chart Tools, Layout, select Error Bars							
18	Choose More Error Bars Options							
19	Click Custom and click Specify Value							
20	For Positive Error Value, highlight D4:D9							
21	For Negative Error Value, highlight D4:D9							
22	Click OK and error bars will appear on the chart							
23	Click on one x error bar and press Delete to remove all x error bars							

FIGURE 4-16 Adding 95% confidence error bars to a graph.

To add error bars in Excel 2007 or 2010, click on one of the points to highlight all points on the graph. In Chart Tools, Layout, select Error Bars and choose More Error Bars Options. For Error Amount, choose Custom and Specify Value. For both Positive Value and Negative Value, enter D4:D9. You just told the spreadsheet to use 95% confidence intervals for error bars. When you click OK, the graph has both x and y error bars. Click on any x error bar and press Delete to remove all x error bars.

To add error bars in earlier versions of Excel, click on one of the points to highlight all points on the graph. In the Format menu, choose Selected Data Series. Select Y Error Bars and a window appears. Click Custom. Click in the positive error bars box and select cells D4:D9. Click the negative error bars box and select cells D4:D9 again. You just told Excel to use values in cells D4:D9 for the lengths of the error bars. Click OK and error bars appear in the graph.

Terms to Understand

average	F test	mean	standard solution
blank solution	Gaussian distribution	method of least squares	standard uncertainty
calibration curve	Grubbs test	null hypothesis	Student's t
confidence interval	hypothesis test	outlier	t test
degrees of freedom	intercept	slope	variance
determinant	linear range	standard deviation	
dynamic range	linear response	standard deviation of the mean	

Summary

The results of many measurements of an experimental quantity follow a Gaussian distribution. The measured mean, $\bar{x}$, approaches the true mean, μ, as the number of measurements becomes very large. The broader the distribution, the greater is σ, the standard deviation. For n measurements, an estimate of the standard deviation is $s = \sqrt{[\Sigma(x_i - \bar{x})^2]/(n - 1)}$. About two-thirds of all measurements lie within $\pm 1\sigma$ and 95% lie within $\pm 2\sigma$. The probability of observing a value within a certain interval is proportional to the area of that interval. The standard deviation s is a measure of the uncertainty of individual measurements. The standard deviation of the mean, $s/\sqrt{n}$ (also called the standard uncertainty, u), is a measure of the uncertainty of the mean of n measurements.

The F test is used to decide whether two standard deviations are significantly different from each other. If F ($= s_1^2/s_2^2$) is greater than the tabulated value, then the two data sets have less than a 5% chance of coming from distributions with the same population standard deviation.

Student's t is used to find confidence intervals ($\mu = \bar{x} \pm ts/\sqrt{n}$) and to compare mean values measured by different methods. If the standard deviations are not significantly different (as determined with the F test), find the pooled standard deviation with Equation 4-10a and compute t with Equation 4-9a. If t is greater than the tabulated value for $n_1 + n_2 - 2$ degrees of freedom, then the two data sets have less than a 5% chance of coming from distributions with the same population mean. If the standard deviations are significantly different, compute the degrees of freedom with Equation

4-10b and compute t with Equation 4-9b. Other applications of the t test are (1) comparing measured values to a "known" value and (2) comparing results from two analytical methods applied to identical samples with no replication (paired t test).

The Grubbs test helps you to decide whether or not a questionable datum should be discarded. It is best to repeat the measurement several times to increase the probability that your decision to accept or reject a datum is correct.

A calibration curve shows the response of a chemical analysis to known quantities (standard solutions) of analyte. When there is a linear response, the corrected analytical signal (= signal from sample − signal from blank) is proportional to the quantity of analyte. Blank solutions are prepared from the same reagents and solvents used to prepare standards and unknowns, but blanks have no intentionally added analyte. The blank tells us the response of the procedure to impurities or to interfering species in the reagents. The blank value is subtracted from measured values of standards prior to constructing the calibration curve. The blank value is subtracted from the response of an unknown prior to computing the quantity of analyte in the unknown.

The method of least squares is used to determine the equation of the "best" straight line through experimental data points. Equations 4-16 to 4-18 and 4-20 to 4-22 provide the least-squares slope and intercept and their standard uncertainties. Equation 4-27 estimates the standard uncertainty in x from a measured value of y with a calibration curve. A spreadsheet simplifies least-squares calculations and graphical display of the results.

Exercises

4-A. For the numbers 116.0, 97.9, 114.2, 106.8, and 108.3, find the mean, standard deviation, standard uncertainty (= standard deviation of the mean), range, and 90% confidence interval for the mean. Using the Grubbs test, decide whether the number 97.9 should be discarded.

4-B. 🖾 *Spreadsheet for standard deviation.* Let's create a spreadsheet to compute the mean and standard deviation of a column of

numbers in two different ways. The spreadsheet here is a template for this exercise.

(a) Reproduce the template on your spreadsheet. Cells B4 to B8 contain the data (x values) whose mean and standard deviation we will compute.

(b) Write a formula in cell B9 to compute the sum of numbers in B4 to B8.

(c) Write a formula in cell B10 to compute the mean value.

(d) Write a formula in cell C4 to compute (x − mean), where x is in cell B4 and the mean is in cell B10. Use Fill Down to compute values in cells C5 to C8.

(e) Write a formula in cell D4 to compute the square of the value in cell C4. Use Fill Down to compute values in cells D5 to D8.

(f) Write a formula in cell D9 to compute the sum of the numbers in cells D4 to D8.

(g) Write a formula in cell B11 to compute the standard deviation.

(h) Use cells B13 to B18 to document your formulas.

(i) Now we are going to simplify life by using formulas built into the spreadsheet. In cell B21 type "=SUM(B4:B8)", which means find the sum of numbers in cells B4 to B8. Cell B21 should display the same number as cell B9. In general, you will not know what functions are available and how to write them. In Excel 2010, use the Formulas ribbon and Insert Function to find SUM.

(j) Select cell B22. Go to Insert Function and find AVERAGE. When you type "=AVERAGE(B4:B8)" in cell B22, its value should be the same as B10.

(k) For cell B23, find the standard deviation function ("=STDEV(B4:B8)") and check that the value agrees with cell B11.

	A	B	C	D
1	Computing standard deviation			
2				
3		Data = x	x-mean	(x-mean)^2
4		17.4		
5		18.1		
6		18.2		
7		17.9		
8		17.6		
9	sum =			
10	mean =			
11	std dev =			
12				
13	Formulas:	B9 =		
14		B10 =		
15		B11 =		
16		C4 =		
17		D4 =		
18		D9 =		
19				
20	Calculations using built-in functions:			
21	sum =			
22	mean =			
23	std dev =			

4-C. Use Table 4-1 for this exercise. Suppose that the mileage at which 10 000 sets of automobile brakes had been 80% worn through was recorded. The average was 62 700, and the standard deviation was 10 400 miles.

(a) What fraction of brakes is expected to be 80% worn in less than 40 860 miles?

(b) What fraction is expected to be 80% worn at a mileage between 57 500 and 71 020 miles?

4-D. [image] Use the NORMDIST spreadsheet function to answer these questions about the brakes described in Exercise 4-C:

(a) What fraction of brakes is expected to be 80% worn in less than 45 800 miles?

(b) What fraction is expected to be 80% worn at a mileage between 60 000 and 70 000 miles?

4-E. Bicarbonate in replicate samples of horse blood was measured four times by each of two methods with the following results:

Method 1: 31.40, 31.24, 31.18, 31.43 mM
Method 2: 30.70, 29.49, 30.01, 30.15 mM

(a) Find the mean, standard deviation, and standard uncertainty (= standard deviation of the mean) for each analysis.

(b) Are the standard deviations significantly different at the 95% confidence level?

4-F. A reliable assay shows that the ATP (adenosine triphosphate) content of a certain cell type is 111 μmol/100 mL. You developed a new assay, which gave the values 117, 119, 111, 115, 120 μmol/100 mL (average = $116._4$) for replicate analyses. Do your results agree with the known value at the 95% confidence level?

4-G. Traces of toxic, man-made hexachlorohexanes in North Sea sediments were extracted by a known process and by two new procedures, and measured by chromatography.

Method	Concentration found (pg/g)	Standard deviation (pg/g)	Number of replications
Conventional	34.4	3.6	6
Procedure A	42.9	1.2	6
Procedure B	51.1	4.6	6

SOURCE: D. Sterzenbach, B. W. Wenclawiak, and V. Weigelt, Anal. Chem. **1997**, 69, 831.

(a) Is the concentration pg/g parts per million, parts per billion, or something else?

(b) Is the standard deviation for procedure B significantly different from that of the conventional procedure?

(c) Is the mean concentration found by procedure B significantly different from that of the conventional procedure?

(d) Answer the same two questions as parts (b) and (c) to compare procedure A to the conventional procedure.

4-H. [image] Calibration curve. (You can do this exercise with your calculator, but it is more easily done by the spreadsheet in Figure 4-15.) In the Bradford protein determination, the color of a dye changes from brown to blue when it binds to protein. Absorbance of light is measured.

Protein (μg):	0.00	9.36	18.72	28.08	37.44
Absorbance at 595 nm:	0.466	0.676	0.883	1.086	1.280

(a) Find the equation of the least-squares straight line through these points in the form $y = [m(\pm u_m)]x + [b(\pm u_b)]$ with a reasonable number of significant figures.

(b) Make a graph showing the experimental data and the calculated straight line.

(c) An unknown protein sample gave an absorbance of 0.973. Calculate the number of micrograms of protein in the unknown and estimate its uncertainty.

Problems

Gaussian Distribution

4-1. What is the relation between the standard deviation and the precision of a procedure? What is the relation between standard deviation and accuracy?

4-2. Use Table 4-1 to state what fraction of a Gaussian population lies within the following intervals:

(a) $\mu \pm \sigma$ (c) μ to $+\sigma$ (e) $-\sigma$ to -0.5σ

(b) $\mu \pm 2\sigma$ (d) μ to $+0.5\sigma$

4-3. The ratio of the number of atoms of the isotopes ^{69}Ga and ^{71}Ga in eight samples from different sources was measured in an effort to understand differences in reported values of the atomic mass of gallium:

Sample	^{69}Ga/^{71}Ga	Sample	^{69}Ga/^{71}Ga
1	1.526 60	5	1.528 94
2	1.529 74	6	1.528 04
3	1.525 92	7	1.526 85
4	1.527 31	8	1.527 93

SOURCE: Data from J. W. Gramlich and L. A. Machlan, Anal. Chem. **1985**, 57, 1788.

Find the (a) mean, (b) standard deviation, (c) variance, and (d) standard deviation of the mean. (e) Write the mean and standard deviation together with an appropriate number of significant digits.

4-4. Students at Francis Marion University measured the mass of each M&M candy in 16 sets of 4 candies and in 16 sets of 16 candies

Mean mass of each candy in sets of 4 candies		Mean mass of each candy in sets of 16 candies	
0.879 9	0.866 7	0.900 4	0.892 5
0.935 6	0.890 2	0.915 2	0.895 8
0.887 6	0.919 5	0.905 6	0.899 6
0.855 3	0.946 9	0.886 7	0.870 7
0.912 2	0.865 0	0.892 6	0.910 5
0.857 5	0.875 5	0.909 7	0.901 9
0.892 8	0.870 1	0.885 5	0.880 3
0.874 6	0.913 8	0.891 3	0.905 5

Mean =
Standard deviation =

SOURCE: Data from K. Varazo, Francis Marion University.

(a) Find the mean of the 16 values on the left side of the table and the mean of the 16 values on the right side of the table

(b) Find the standard deviation of the 16 mean values at the left side of the table and of the 16 mean values at the right side of the table.

(c) From the standard deviation of the mean values for sets of 4 candies, predict what the standard deviation is expected to be for the sets of 16 candies. Compare your prediction to the measured standard deviation in (b).

4-5. (a) Calculate the fraction of bulbs in Figure 4-1 expected to have a lifetime greater than 1 005.3 h.

(b) What fraction of bulbs is expected to have a lifetime between 798.1 and 901.7 h?

(c) ▦ Use the Excel NORMDIST function to find the fraction of bulbs expected to have a lifetime between 800 and 900 h.

4-6. ▦ Blood plasma proteins of patients with malignant breast tumors differ from proteins of healthy people in their solubility in the presence of various polymers. When the polymers dextran and poly(ethylene glycol) are mixed with water, a two-phase mixture is formed. When plasma proteins of tumor patients are added, the distribution of proteins between the two phases is different from that of plasma proteins of a healthy person. The distribution coefficient (K) for any substance is defined as K = [concentration of the substance in phase A]/[concentration of the substance in phase B]. Proteins of healthy people have a mean distribution coefficient of 0.75 with a standard deviation of 0.07. For the proteins of people with cancer, the mean is 0.92 with a standard deviation of 0.11.

(a) Suppose that K were used as a diagnostic tool and that a positive indication of cancer is taken as $K \geq 0.92$. What fraction of people with tumors would have a false negative indication of cancer because $K < 0.92$?

(b) What fraction of healthy people would have a false positive indication of cancer? This number is the fraction of healthy people with $K \geq 0.92$, shown by the shaded area in the graph below. Estimate an answer with Table 4-1 and obtain a more exact result with the NORMDIST function in Excel.

(c) Vary the first argument of the NORMDIST function to select a distribution coefficient that would identify 75% of people with tumors. That is, 75% of patients with tumors would have K above the selected distribution coefficient. With this value of K, what fraction of healthy people would have a false positive result indicating they have a tumor?

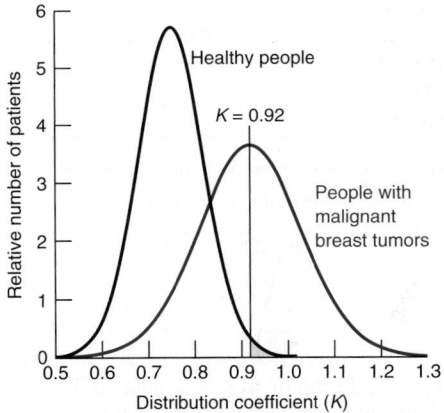

Distribution coefficients of plasma proteins from healthy people and from people with malignant breast tumors. [Data from B. Y. Zaslavsky, "Bioanalytical Applications of Partitioning in Aqueous Polymer Two-Phase Systems," *Anal. Chem.* **1992**, 64, 765A.]

4-7. ▦ The equation for the Gaussian curve in Figure 4-1 is

$$y = \frac{(\text{total bulbs})(\text{hours per bar})}{s\sqrt{2\pi}} e^{-(x-\bar{x})^2/2s^2}$$

where $\bar{x}$ is the mean value (845.2 h), s is the standard deviation (94.2 h), total bulbs = 4 768, and hours per bar (= 20) is the width of each bar in Figure 4-1. Set up a spreadsheet like the one with this problem to calculate the coordinates of the Gaussian curve in

Figure 4-1 from 500 to 1 200 h in 25-h intervals. Note the heavy use of parentheses in the formula at the bottom of the spreadsheet to force the computer to do the arithmetic as intended. Use Excel to graph your results.

	A	B	C
1	Gaussian curve for light bulbs (Fig 4-1)		
2			
3	mean =	x (hours)	y (bulbs)
4	845.2	500	0.49
5	std dev =	525	1.25
6	94.2	550	2.98
7	total bulbs =	600	13.64
8	4768	700	123.11
9	hours per bar =	800	359.94
10	20	845.2	403.85
11	sqrt(2 pi) =	900	340.99
12	2.506628	1000	104.67
13		1100	10.41
14		1200	0.34
15	Formula for cell C4 =		
16	($A$8*$A$10/($A$6*$A$12))*		
17	EXP(-((B4-$A$4)^2)/(2*$A$6^2))		

4-8. 🖳 Repeat Problem 4-7 but use the values 50, 100, and 150 for the standard deviation. Superimpose all three curves on a single graph.

F test, Confidence Intervals, t Test, and Grubbs Test

4-9. What is the meaning of a confidence interval?

4-10. What fraction of vertical bars in Figure 4-5a is expected to include the population mean (10 000) if many experiments are carried out? Why are the 90% confidence interval bars longer than the 50% bars in Figure 4-5?

4-11. List the three different cases that we studied for comparison of means, and write the equations used in each case.

4-12. The percentage of an additive in gasoline was measured six times with the following results: 0.13, 0.12, 0.16, 0.17, 0.20, 0.11%. Find the 90% and 99% confidence intervals for the percentage of the additive.

4-13. Sample 8 of Problem 4-3 was analyzed seven times, with $\bar{x} = 1.527\ 93$ and $s = 0.000\ 07$. Find the 99% confidence interval for sample 8.

4-14. A trainee in a medical lab will be released to work on her own when her results agree with those of an experienced worker at the 95% confidence level. Results for a blood urea nitrogen analysis are shown below.

Trainee: $\bar{x} = 14.5_7$ mg/dL $s = 0.5_3$ mg/dL $n = 6$ samples

Experienced
worker: $\bar{x} = 13.9_5$ mg/dL $s = 0.4_2$ mg/dL $n = 5$ samples

(a) What does the abbreviation dL stand for?

(b) Should the trainee be released to work alone?

4-15. The CdSe content (g/L) of six different samples of nanocrystals was measured by two methods. Do the two methods differ significantly at the 95% confidence level?

Sample	Method 1 Anodic stripping	Method 2 Atomic absorption
A	0.88	0.83
B	1.15	1.04
C	1.22	1.39
D	0.93	0.91
E	1.17	1.08
F	1.51	1.31

SOURCE: Data from E. Kuçur, F. M. Boldt, S. Cavaliere-Jaricont, J. Ziegler, and T. Nann, Anal. Chem. **2007**, 79, 8987.

4-16. 🖳 Now we use a built-in routine in Excel for the paired t test to see if the two methods in Problem 4-15 produce significantly different results. Enter the data for Methods 1 and 2 into two columns of a spreadsheet. For Excel 2007 and 2010, find Data Analysis in the Data ribbon. If Data Analysis does not appear, follow the instructions at the beginning of Section 4-5 to load this software. Select Data Analysis and then select t-Test: Paired Two Sample for Means. Follow the instructions of Section 4-5 and the routine will print out information including $t_{calculated}$ (labeled "t Stat") and t_{table} (labeled "t Critical two-tail"). You should reproduce the results of Problem 4-15.

4-17. Two methods were used to measure fluorescence lifetime of a dye. Are the standard deviations significantly different? Are the means significantly different?

Quantity	Method 1	Method 2
Mean lifetime (ns)	1.382	1.346
Standard deviation (ns)	0.025	0.039
Number of measurements	4	4

SOURCE: Data from N. Boens et al., Anal. Chem. **2007**, 79, 2137.

4-18. 🖳 Do the following two sets of measurements of $^6Li/^7Li$ in a Standard Reference Material give statistically equivalent results?

Method 1	Method 2
0.082 601	0.081 83
0.082 621	0.081 86
0.082 589	0.082 05
0.082 617	0.082 06
0.082 598	0.082 15
	0.082 08

SOURCE: Data from S. Ahmed, N. Jabeen, and E. ur Rehman, Anal. Chem. **2002**, 74, 4133; L. W. Green, J. J. Leppinen, and N. L. Elliot, Anal. Chem. **1988**, 60, 34.

4-19. If you measure a quantity four times and the standard deviation is 1.0% of the average, can you be 90% confident that the true value is within 1.2% of the measured average?

4-20. Students measured the concentration of HCl in a solution by titrating with different indicators to find the end point.

Indicator	Mean HCl concentration (M) (± standard deviation)	Number of measurements
1. Bromothymol blue	0.095 65 ± 0.002 25	28
2. Methyl red	0.086 86 ± 0.000 98	18
3. Bromocresol green	0.086 41 ± 0.001 13	29

SOURCE: Data from D. T. Harvey, J. Chem. Ed. **1991**, 68, 329.

Is the difference between indicators 1 and 2 significant at the 95% confidence level? Answer the same question for indicators 2 and 3.

4-21. Hydrocarbons in the cab of an automobile were measured during trips on the New Jersey Turnpike and trips through the Lincoln Tunnel connecting New York and New Jersey.[11] The concentrations (± standard deviations) of m- and p-xylene were

Turnpike:	$31.4 \pm 30.0 \ \mu g/m^3$	(32 measurements)
Tunnel:	$52.9 \pm 29.8 \ \mu g/m^3$	(32 measurements)

Do these results differ at the 95% confidence level? At the 99% confidence level?

4-22. A Standard Reference Material is certified to contain 94.6 ppm of an organic contaminant in soil. Your analysis gives values of 98.6, 98.4, 97.2, 94.6, and 96.2 ppm. Do your results differ from the expected result at the 95% confidence level? If you made one more measurement and found 94.5, would your conclusion change?

4-23. Nitrite (NO_2^-) was measured by two methods in rainwater and unchlorinated drinking water. The results ± standard deviation (number of samples) are

Sample source	Gas chromatography	Spectrophotometry
Rainwater	0.069 ± 0.005 mg/L ($n = 7$)	0.063 ± 0.008 mg/L ($n = 5$)
Drinking water	0.078 ± 0.007 mg/L ($n = 5$)	0.087 ± 0.008 mg/L ($n = 5$)

SOURCE: Data from I. Sarudi and I. Nagy, Talanta **1995**, *42, 1099.*

(a) Do the two methods agree with each other at the 95% confidence level for both rainwater and drinking water?

(b) For each method, does the drinking water contain significantly more nitrite than the rainwater (at the 95% confidence level)?

4-24. Should the value 216 be rejected from the set of results 192, 216, 202, 195, and 204?

4-25. Which statement about the *F* test is true? Explain your answer.

(i) If $F_{calculated} < F_{table}$, there is more than a 5% chance that the two sets of data are drawn from populations with the same population standard deviation.

(ii) If $F_{calculated} < F_{table}$, there is at least 95% probability that the two sets of data are drawn from populations with the same population standard deviation.

Linear Least Squares

4-26. A straight line is drawn through the points $(3.0, -3.87 \times 10^4)$, $(10.0, -12.99 \times 10^4)$, $(20.0, -25.93 \times 10^4)$, $(30.0, -38.89 \times 10^4)$, and $(40.0, -51.96 \times 10^4)$ to give $m = -1.298\ 72 \times 10^4$, $b = 256.695$, $u_m = 13.190$, $u_b = 323.57$, and $s_y = 392.9$. Express the slope and intercept and their uncertainties with reasonable significant figures.

4-27. Here is a least-squares problem that you can do by hand with a calculator. Find the slope and intercept and their standard deviations for the straight line drawn through the points $(x,y) = (0,1)$, $(2,2)$, and $(3,3)$. Make a graph showing the three points and the line. Place error bars ($\pm s_y$) on the points.

4-28. Set up a spreadsheet to reproduce Figure 4-15. *Add error bars:* Follow the procedure on pages 87–88. Use s_y for the + and − error.

4-29. *Excel LINEST function.* Enter the following data in a spreadsheet and use LINEST to find slope, intercept, and standard errors. Use Excel to draw a graph of the data and add a trendline. Draw error bars of $\pm s_y$ on the points.

x:	3.0	10.0	20.0	30.0	40.0
y:	-0.074	-1.411	-2.584	-3.750	-5.407

Calibration Curves

4-30. Explain the following statement: "The validity of a chemical analysis ultimately depends on measuring the response of the analytical procedure to known standards."

4-31. Suppose that you carry out an analytical procedure to generate a linear calibration curve like that shown in Figure 4-13. Then you analyze an unknown and find an absorbance that gives a negative concentration for the analyte. What might this mean?

4-32. A calibration curve based on $n = 10$ known points was used to measure the protein in an unknown. The results were protein $= 15.2_2 \ (\pm 0.4_6) \ \mu g$, where the standard uncertainty is $u_x = 0.4_6 \ \mu g$. Find the 90% and 99% confidence intervals for protein in the unknown.

4-33. Consider the least-squares problem in Figure 4-11.

(a) Suppose that a single new measurement produces a y value of 2.58. Find the corresponding x value and its standard uncertainty, u_x.

(b) Suppose you measure y four times and the average is 2.58. Calculate u_x based on four measurements, not one.

(c) Find the 95% confidence intervals for **(a)** and **(b)**.

4-34. (a) The linear calibration curve in Figure 4-13 is $y = 0.016\ 3_0$ $(\pm 0.000\ 2_2)\ x + 0.004_7\ (\pm 0.002_6)$ with $s_y = 0.005_9$. Find the quantity of unknown protein that gives a measured absorbance of 0.264 when a blank has an absorbance of 0.095.

(b) Figure 4-13 has $n = 14$ calibration points in the linear portion. You measure $k = 4$ replicate samples of unknown and find a mean corrected absorbance of 0.169. Find the standard uncertainty and 95% confidence interval for protein in the unknown.

4-35. Here are mass spectrometric signals for methane in H_2:

CH_4 (vol%):	0	0.062	0.122	0.245	0.486	0.971	1.921
Signal (mV):	9.1	47.5	95.6	193.8	387.5	812.5	1 671.9

(a) Subtract the blank value (9.1) from all other values. Then use the method of least squares to find the slope and intercept and their uncertainties. Construct a calibration curve.

(b) Replicate measurements of an unknown gave 152.1, 154.9, 153.9, and 155.1 mV, and a blank gave 8.2, 9.4, 10.6, and 7.8 mV. Subtract the average blank from the average unknown to find the average corrected signal for the unknown.

(c) Find the concentration of the unknown, its standard uncertainty (u_x), and the 95% confidence interval.

4-36. *Nonlinear calibration curve.* Following the procedure in Box 4-2, find how many micrograms (μg) of protein are contained in a sample with a corrected absorbance of 0.350 in Figure 4-13.

4-37. *Logarithmic calibration curve.* Calibration data spanning five orders of magnitude for an electrochemical determination of p-nitrophenol are given in the table. (The blank has already been subtracted from the measured current.) If you try to plot these data on a linear graph extending from 0 to 310 μg/mL and from 0 to 5 260 nA, most of the points will be bunched up near the origin. To handle data with such a large range, a logarithmic plot is helpful.

p-Nitrophenol (μg/mL)	Current (nA)	p-Nitrophenol (μg/mL)	Current (nA)
0.010 0	0.215	3.00	66.7
0.029 9	0.846	10.4	224
0.117	2.65	31.2	621
0.311	7.41	107	2 020
1.02	20.8	310	5 260

Data from Figure 4 of L. R. Taylor, Am. Lab., February 1993, p. 44.

(a) Make a graph of log(current) versus log(concentration). Over what range is the log-log calibration linear?

(b) Find the equation of the line in the form log(current) = $m \times$ log(concentration) + b.

(c) Find the concentration of p-nitrophenol corresponding to a signal of 99.9 nA.

(d) *Propagation of uncertainty with logarithm.* For a signal of 99.9 nA, log(concentration) and its standard uncertainty turn out to be 0.683 15 ± 0.045 22. With rules for propagation of uncertainty from Chapter 3, find the uncertainty in concentration.

5 Quality Assurance and Calibration Methods

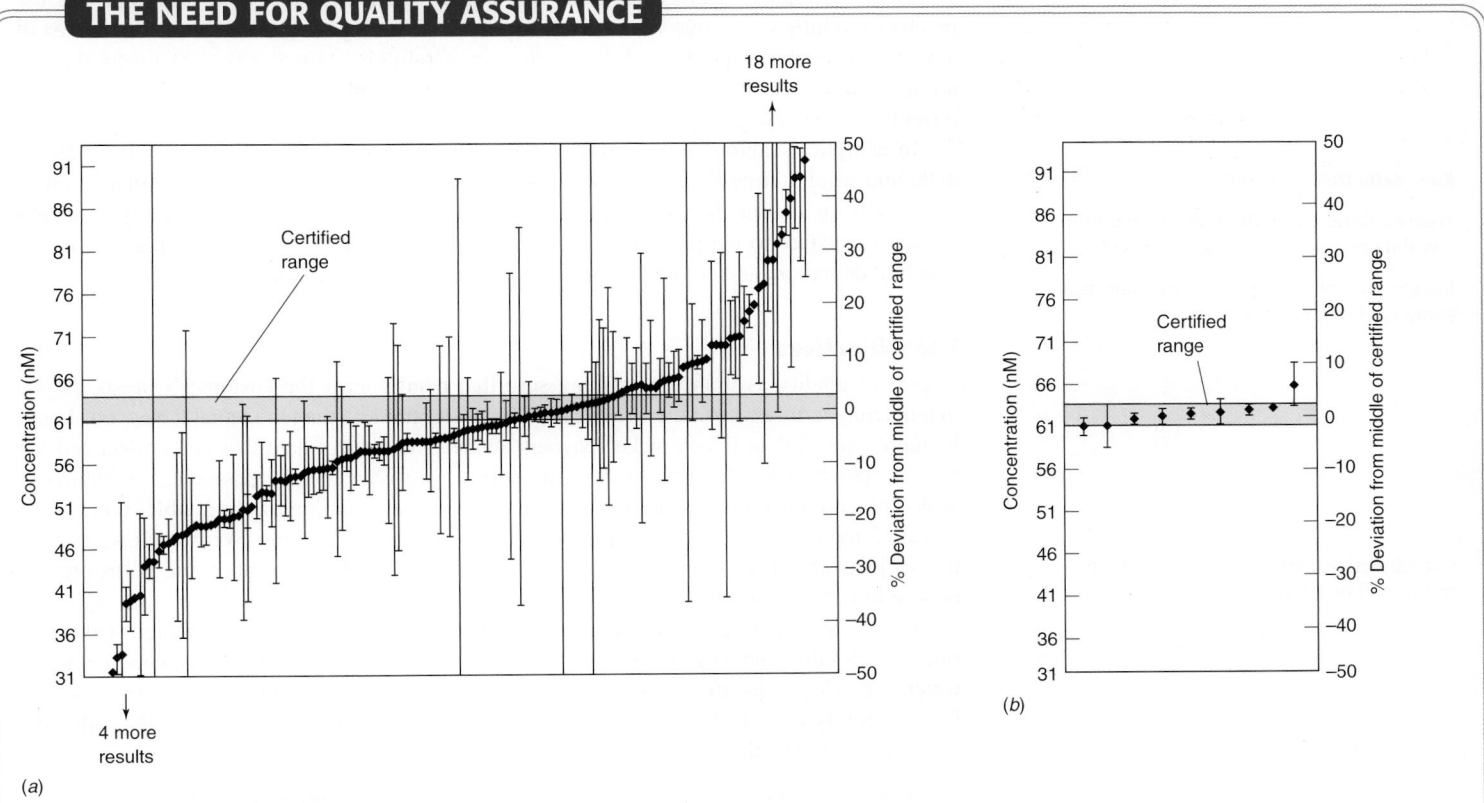

(a) Scattered measurements of lead in river water by different laboratories, each of which employed a recognized quality management system. (b) Reproducible results from national measurement institutes. [Data from P. De Bièvre and P. D. P. Taylor, "'Demonstration' vs. 'Designation' of Measurement Competence: The Need to Link Accreditation to Metrology," *Fresenius J. Anal. Chem.* **2000,** *368,* 567.]

The Institute for Reference Materials and Measurements in Belgium conducts an International Measurement Evaluation Program to allow laboratories to assess the reliability of their analyses. Panel *a* shows results for lead in river water. Of 181 labs, 18 reported results more than 50% above and 4 reported results more than 50% below the certified level of 62.3 ± 1.3 nM. Even though most labs in the study employed recognized quality management procedures, a large fraction of results did not include the certified range. Panel *b* shows that when this same river water was analyzed by nine different national measurement institutes, where the most care is taken, all results were close to the certified range.

This example illustrates that there is no guarantee that results are reliable, even if they are obtained by "accredited" laboratories using accepted procedures. A good way to assess the reliability of a lab working for you is to provide the lab with "blind" samples—similar to your unknowns—for which you know the "right" answer but the analyst does not. If the lab does not find the known result, there is a problem. Periodic blind check samples are required to demonstrate continuing reliability.

Data quality standards:
• Get the right data
• Get the data right
• Keep the data right
[Nancy W. Wentworth, U.S. Environmental Protection Agency[1]]

Quality assurance is what we do to get the right answer for our purpose. The answer should have sufficient accuracy and precision to support subsequent decisions. There is no point in spending extra money to obtain a more accurate or more precise answer if it is not necessary. This chapter describes basic issues and procedures in quality assurance[2] and introduces two more calibration methods. In Chapter 4, we discussed a *calibration curve*, such as Figure 4-13, made by preparing a series of known solutions of analyte and making a graph of instrument response versus analyte concentration. Known solutions of analyte that do not involve the unknown solution are called **external standards.** In Chapter 5, we describe *standard addition* and *internal standards*, both of which are made from the unknown solution.

"Suppose you are cooking for some friends. While making spaghetti sauce, you taste it, season it, taste it some more. Each tasting is a sampling event with a quality control test. You can taste the whole batch because there is only one batch. Now suppose you run a spaghetti sauce plant that makes 1 000 jars a day. You can't taste each one, so you decide to taste three a day, one each at 11 A.M., 2 P.M., and 5 P.M. If the three jars all taste OK, you conclude all 1 000 are OK. Unfortunately, that may not be true, but the relative risk—that a jar has too much or too little seasoning—is not very important because you agree to refund the money of any customer who is dissatisfied. If the number of refunds is small, say, 100 a year, there is no apparent benefit in tasting four jars a day." There would be 365 additional tests to avoid refunds on 100 jars, giving a net loss of 265 jars worth of profit.

In analytical chemistry, the product is not spaghetti sauce but, rather, raw data, treated data, and results. *Raw data* are measurements, such as peak areas from a chromatogram or volumes from a buret. *Treated data* are concentrations or amounts found by applying a calibration procedure to the raw data. *Results* are what we ultimately report, such as the mean, standard deviation, and confidence interval, after applying statistics to treated data.

Raw data: measurements

Treated data: concentrations derived from raw data by use of calibration methods

Results: quantities reported after statistical analysis of treated data

Use Objectives

A goal of quality assurance is making sure that results meet the customer's needs. If you manufacture a drug whose therapeutic dose is just a little less than the lethal dose, you should be more careful than if you make spaghetti sauce. The kind of data that you collect and the way in which you collect them depend on how you plan to use those data. A bathroom scale does not have to measure mass to the nearest milligram, but a drug tablet required to contain 2 mg of active ingredient probably cannot contain 2 ± 1 mg. Writing clear, concise **use objectives** for data and results is a critical step in quality assurance and helps prevent misuse of data and results.

Here is an example of a use objective. Drinking water is usually disinfected with chlorine, which kills microorganisms. Unfortunately, chlorine also reacts with organic matter in water to produce "disinfection by-products"—compounds that might harm humans. A disinfection facility was planning to introduce a new chlorination process and wrote the following analytical use objective:

Use objective: states purpose for which results will be used

> Analytical data and results shall be used to determine whether the modified chlorination process results in at least a 10% reduction of formation of selected disinfection by-products.

The new process was expected to decrease the disinfection by-products. The use objective says that uncertainty in the analysis must be small enough that a 10% decrease in selected by-products is clearly distinguishable from experimental error. In other words, is an observed decrease of 10% real?

Specifications

Once you have use objectives, you are ready to write **specifications** stating how good the numbers need to be and what precautions are required in the analytical procedure. How shall samples be taken and how many are needed? Are special precautions required to protect samples and ensure that they are not degraded? Within practical restraints, such as cost, time, and limited amounts of material available for analysis, what level of accuracy and precision will satisfy the use objectives? What rate of false positives or false negatives is acceptable? These questions need to be answered in detailed specifications.

Quality assurance begins with sampling. We must collect representative samples, and analyte must be preserved after sample is collected. If our sample is not representative or if analyte is lost after collection, then even the most accurate analysis is meaningless. Samples for trace metal analysis are usually collected in plastic or Teflon containers—not glass—because metal ions found on glass surfaces leach out into the sample over time. Samples for organic analysis are usually collected in glass containers—not plastic—because organic plasticizers leached from plastic containers can contaminate the sample. Samples are often stored in the dark in a refrigerator to minimize degradation of organic analytes.

What do we mean by *false positives* and *false negatives*? Suppose you must certify that a contaminant in drinking water is below a legal limit. A **false positive** says that the concentration

Specifications might include:
• sampling requirements
• accuracy and precision
• rate of false results
• selectivity
• sensitivity
• acceptable blank values
• recovery of fortification
• calibration checks
• quality control samples

Box 5-1 describes implications of false positive tests in medicine.

BOX 5-1 **Medical Implication of False Positive Results[3]**

A seemingly low rate of false positive results can have surprising consequences in medicine. Suppose that 0.2% of people have a particular type of cancer and suppose that a test for that cancer has a 99% probability of detecting the cancer when it is present. Suppose that the same test has a false positive rate of 1%. That is, the test indicates that 1% of healthy people have the cancer.

In a population of a million people, 0.2% or 2 000 people are likely to have the cancer. If a million people are screened for this cancer, the test will indicate that 99% of those 2 000 people (1 980 people) have the cancer and 1% (20 people) are free of the cancer, even though they are not. Of the remaining 998 000 people who are free of cancer, the 1% false positive rate identifies cancer in 9 980 of those people. Of the 1 980 + 9 980 = 11 960 positive test results, only 1 980/11 960 = 17% are true positives. The remaining 83% of positive tests falsely indicate cancer in healthy people. If 9 980 healthy people were subjected to dangerous treatments such as radiation, chemotherapy, or surgery, the test for cancer could do more harm than good. This arithmetic explains why a positive test result *must be confirmed* with a biopsy before commencing with treatment.

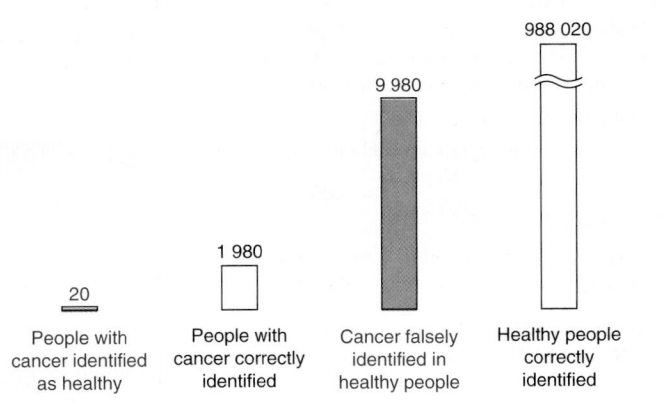

exceeds the legal limit when, in fact, the concentration is below the limit. A **false negative** says that the concentration is below the limit when it is actually above the limit. Even well-executed procedures produce some false conclusions because of the statistical nature of sampling and measurement. For drinking water, it is more important to have a low rate of false negatives than a low rate of false positives. It would be worse to certify that contaminated water is safe than to certify that safe water is contaminated. Drug testing of athletes is designed to minimize false positives so that an innocent athlete is not falsely accused of doping. In Section 5-2, we will see that there is a trade-off between false positives, false negatives, and the *detection limit* of an analytical method.

In choosing a method, we also consider selectivity and sensitivity. **Selectivity** (also called *specificity*) means being able to distinguish analyte from other species in the sample (avoiding interference). **Sensitivity** is the capability of responding reliably and measurably to changes in analyte concentration. The *detection limit* of an analytical method must be lower than the concentrations to be measured.

Sensitivity

$$= \text{slope of calibration curve}$$

$$= \frac{\text{change in signal}}{\text{change in analyte concentration}}$$

Specifications could include required accuracy and precision, reagent purity, apparatus tolerances, the use of certified reference materials, and acceptable values for blanks. *Certified reference materials* contain certified levels of analyte in realistic materials that you might be analyzing, such as blood, coal, or metal alloys. Your analytical method should produce an answer acceptably close to the certified level or there is insufficient accuracy in your method.

Blanks account for interference by other species in the sample and for traces of analyte found in reagents used for sample preservation, preparation, and analysis. Frequent measurements of blanks detect whether analyte from previous samples is carried into subsequent analyses by adhering to vessels or instruments.

A **method blank** is a sample containing all components except analyte, and it is taken through all steps of the analytical procedure. We subtract the response of the method blank from the response of a real sample prior to calculating the quantity of analyte in the sample. A **reagent blank** is similar to a method blank, but it has not been subjected to all sample preparation procedures. The method blank is a more complete estimate of the blank contribution to the analytical response.

A **field blank** is similar to a method blank, but it has been exposed to the site of sampling. For example, to analyze particulates in air, a certain volume of air could be sucked through a filter, which is then dissolved and analyzed. A field blank would be a filter carried to the collection site in the same package with the collection filters. The filter for the blank would be taken out of its package in the field and placed in the same kind of sealed container used for collection filters. The difference between the blank and the collection filters is that air was not sucked through the blank filter. Volatile organic compounds encountered during transportation or in the field are conceivable contaminants of a field blank.

Add a small volume of concentrated standard to avoid changing the volume of the sample significantly. For example, add 50.5 μL of 500 μg/L standard to 5.00 mL (= 5 000 μL) of sample to increase analyte by 5.00 μg/L.

Final concentration

= initial concentration × dilution factor

$$= \left(500\frac{\mu g}{L}\right)\left(\frac{50.5\ \mu L}{5\ 050.5\ \mu L}\right) = 5.00\frac{\mu g}{L}$$

Matrix is everything in the unknown other than analyte. The matrix can *decrease* (Figure 5-4 and Problem 5-25) or *increase* (Problem 5-33) response to analyte.

To gauge accuracy:
• calibration checks
• fortification recoveries
• quality control samples
• blanks

To gauge precision:
• replicate samples
• replicate portions of same sample

Another performance requirement often specified is *spike recovery*. Sometimes, response to analyte can be decreased or increased by something else in the sample. **Matrix** refers to everything in the sample other than analyte. A **spike,** also called a *fortification*, is a known quantity of analyte added to a sample to test whether the response to the spike is the same as that expected from a calibration curve. Spiked samples are analyzed in the same manner as unknowns. For example, if drinking water is found to contain 10.0 μg/L of nitrate, a spike of 5.0 μg/L could be added. Ideally, the concentration in the spiked portion found by analysis will be 15.0 μg/L. If a number other than 15.0 μg/L is found, then the matrix could be interfering with the analysis.

EXAMPLE Spike Recovery

Let C stand for concentration. One definition of spike recovery is

$$\% \text{ recovery} = \frac{C_{\text{spiked sample}} - C_{\text{unspiked sample}}}{C_{\text{added}}} \times 100 \tag{5-1}$$

An unknown was found to contain 10.0 μg of analyte per liter. A spike of 5.0 μg/L was added to a replicate portion of unknown. Analysis of the spiked sample gave a concentration of 14.6 μg/L. Find the percent recovery of the spike.

Solution The percent of the spike found by analysis is

$$\% \text{ recovery} = \frac{14.6\ \mu g/L - 10.0\ \mu g/L}{5.0\ \mu g/L} \times 100 = 92\%$$

If the acceptable recovery is specified to be in the range from 96% to 104%, then 92% is unacceptable. Something in your method or techniques needs improvement.

TEST YOURSELF Find % recovery if the spiked sample gave a concentration of 15.3 μg/L. (*Answer:* 106%)

When dealing with large numbers of samples and replicates, we perform periodic calibration checks to make sure that our instrument continues to work properly and the calibration remains valid. In a **calibration check,** we analyze solutions with known concentrations of analyte. A specification might, for example, call for one calibration check for every 10 samples. Calibration check solutions should be different from the ones used to prepare the original calibration curve. This practice helps verify that the initial calibration standards were made properly.

Performance test samples (also called *quality control samples* or *blind samples*) are a quality control measure to help eliminate bias introduced by an analyst who knows the concentration of the calibration check sample. These samples of known composition are provided to the analyst as unknowns. Results are then compared with the known values, usually by a quality assurance manager. For example, the U.S. Department of Agriculture maintains a bank of quality control homogenized food samples for distribution as blind samples to test laboratories that measure nutrients in foods.[4]

Together, raw data and results from calibration checks, spike recoveries, quality control samples, and blanks are used to gauge accuracy. Analytical performance on replicate samples and on replicate portions of the same sample measures precision. Fortification also helps ensure that qualitative identification of analyte is correct. If you spike the unknown in Figure 0-5 with extra caffeine and the area of a chromatographic peak not thought to be caffeine increases, then you have misidentified the caffeine peak.

Standard operating procedures stating what steps will be taken and how they will be carried out are the bulwark of quality assurance. For example, if a reagent has "gone bad" for some reason, control experiments built into your normal procedures should detect that something is wrong and your results should not be reported. It is implicit that everyone follows the standard operating procedures. Adhering to these procedures guards against the normal human desire to take shortcuts based on assumptions that could be false.

A meaningful analysis requires a meaningful sample that represents what is to be analyzed. The sample must be stored in containers and under conditions that do not allow relevant chemical characteristics to change. Protection might be needed to prevent oxidation, photodecomposition, or growth of organisms. The *chain of custody* is the trail followed by a sample from the time it is collected to the time it is analyzed and, possibly, archived. Documents are signed each time the material changes hands to indicate who is responsible for the sample. Each person in the chain of custody follows a written procedure telling how the sample is to be handled and stored. Each person receiving a sample should inspect it to see that it is in the expected condition in an appropriate container. If the original sample was a homogeneous liquid, but it contains a precipitate when you receive it, the standard operating procedure might dictate that you reject that sample.

Standard operating procedures specify how instruments are to be maintained and calibrated to ensure their reliability. Many labs have their own standard practices, such as recording temperatures of refrigerators, calibrating balances, conducting routine instrument maintenance, or replacing reagents. These practices are part of the overall quality management plan. The rationale behind standard practices is that some equipment is used by many people for different analyses. We save money by having one program to ensure that the most rigorous needs are met.

In drug testing of athletes, the chain of custody necessarily uses different people to collect samples and to analyze samples. The identity of the athlete is known to the collector, but not to the analyst, so that the analyst will not deliberately falsify a result to favor or incriminate a particular person or team.

Assessment

Assessment is the process of (1) collecting data to show that analytical procedures are operating within specified limits and (2) verifying that final results meet use objectives.

Documentation is critical for assessment. Standard *protocols* provide directions for what must be documented and how the documentation is to be done, including how to record information in notebooks. For labs that rely on manuals of standard practices, it is imperative that tasks done to comply with the manuals be monitored and recorded. *Control charts* (Box 5-2) can be used to monitor performance on blanks, calibration checks, and spiked samples to see if results are stable over time or to compare the work of different employees. Control charts can also monitor sensitivity or selectivity, especially if a laboratory encounters a wide variety of matrixes.

BOX 5-2 Control Charts

A **control chart** is a visual representation of confidence intervals for a Gaussian distribution. A control chart warns us when a property being monitored strays dangerously far from an intended *target value*.

Consider a laboratory measuring perchlorate (ClO_4^-) in human urine. For quality assurance, $n = 5$ replicate quality control samples made from synthetic urine spiked with perchlorate are measured every day. The control chart shows the mean value of the five samples observed each day over a series of days. The spike contains $\mu = 4.92$ ng/mL and the population standard deviation from many analyses over a long time is $\sigma = 0.40$ ng/mL.

For a Gaussian distribution, 95.5% of all observations are within $\pm 2\sigma \sqrt{n}$ from the mean and 99.7% are within $\pm 3\sigma/\sqrt{n}$. In these expressions, n is the number of replicate measurements ($= 5$) that are averaged each day. The $\pm 2\sigma \sqrt{n}$ limits are designated *warning lines* and the $\pm 3\sigma/\sqrt{n}$ limits are designated *action lines*. We expect ~4.5% of measurements to be outside the warning lines and ~0.3% to be outside the action lines. It is unlikely that we would observe two consecutive measurements at the warning line (probability = $0.045 \times 0.045 = 0.002\ 0$).

The following conditions are considered to be so unlikely that if they occur the process should be shut down for troubleshooting:

- one observation outside the action lines
- two out of three consecutive measurements between the warning and the action lines
- seven consecutive measurements all above or all below the center line

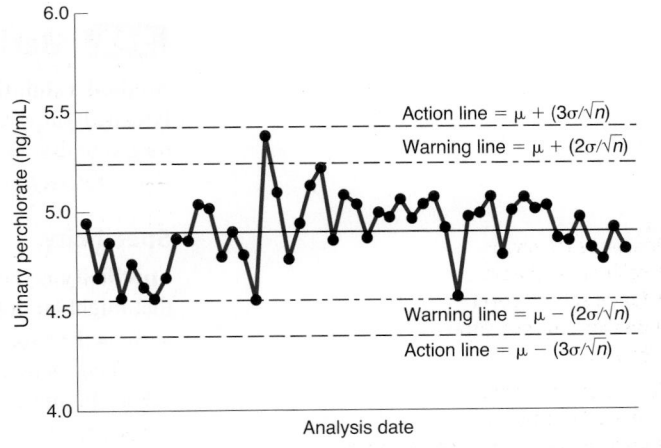

Control chart for ClO_4^- in urine. [Data from L. Valentin-Blasini, J. P. Mauldin, D. Maple, and B. C. Blount, "Analysis of Perchlorate in Human Urine Using Ion Chromatography and Electrospray Tandem Mass Spectrometry," *Anal. Chem.* **2005,** *77,* 2475.]

- six consecutive measurements all steadily increasing or all steadily decreasing, wherever they are located
- 14 consecutive points alternating up and down, regardless of where they are located
- an obvious nonrandom pattern

For quality assessment of an analytical process, a control chart might show the mean value of quality control samples or the precision of replicate analyses of unknowns or standards as a function of time.

TABLE 5-1	Quality assurance process
Question	**Actions**

Use Objectives

Why do you want the data and results and how will you use the results?
- Write use objectives

Specifications

How good do the numbers have to be?
- Write specifications
- Pick methods to meet specifications
- Consider sampling, precision, accuracy, selectivity, sensitivity, detection limit, robustness, rate of false results
- Employ blanks, fortification, calibration checks, quality control samples, and control charts to monitor performance
- Write and follow standard operating procedures

Assessment

Were the specifications achieved?
- Compare data and results with specifications
- Document procedures and keep records suitable to meet use objectives
- Verify that use objectives were met

Government agencies such as the U.S. Environmental Protection Agency set requirements for quality assurance for their own labs and for certification of other labs. Published standard methods specify precision, accuracy, numbers of blanks, replicates, and calibration checks. To monitor drinking water, regulations state how often and how many samples are to be taken. Documentation is necessary to demonstrate that all requirements have been met. Table 5-1 summarizes the quality assurance process.

5-2 Method Validation

Method validation is the process of proving that an analytical method is acceptable for its intended purpose.[5] In pharmaceutical chemistry, method validation requirements for regulatory submission include studies of *method specificity, linearity, accuracy, precision, range, limit of detection, limit of quantitation,* and *robustness.*

Specificity

Specificity is the ability of an analytical method to distinguish analyte from everything else that might be in the sample. *Electrophoresis* is an analytical method in which substances are separated from one another by their differing rates of migration in a strong electric field. An *electropherogram* is a graph of detector response versus time in an electrophoretic separation. Figure 5-1 shows an electropherogram of the drug cefotaxime (peak 4) spiked with

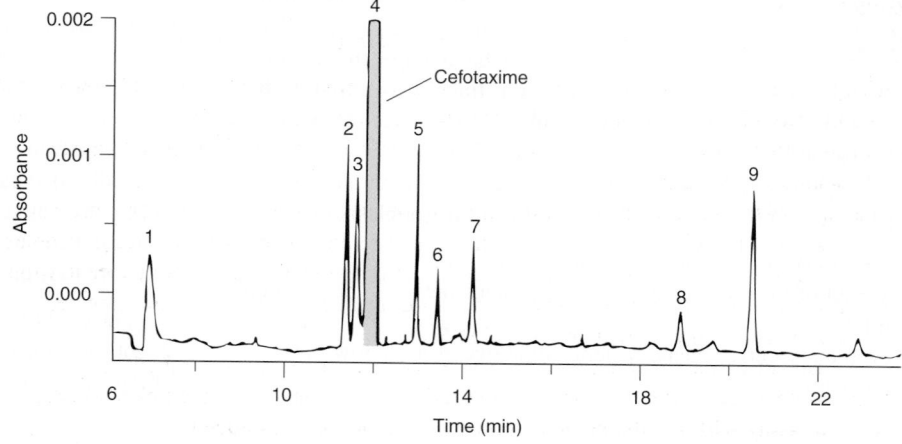

FIGURE 5-1 Electropherogram of the drug cefotaxime (peak 4) spiked with known impurities (peaks 2, 3, 5–9) from synthesis of the drug. Peak 1 is a marker for electroosmotic flow. Smaller peaks from unknown impurities are also observed. Separation was performed by micellar electrophoretic capillary chromatography (Section 26-7). [Data from H. Fabre and K. D. Altria, "Key Points for Validating CE Methods, Particularly in Pharmaceutical Analysis," *LCGC North Am.* **2001,** *19,* 498.]

0.2 wt% of known impurities normally present from the synthesis. A reasonable requirement for specificity might be that there is baseline separation of analyte (cefotaxime) from all impurities that might be present. *Baseline separation* means that the detector signal returns to its baseline before the next compound reaches the detector.

In Figure 5-1, impurity peak 3 is not completely resolved from cefotaxime. In this case, another reasonable criterion for specificity might be that unresolved impurities at their maximum expected concentration do not affect the assay of cefotaxime by more than 0.5%. If we were trying to measure impurities, as opposed to assaying cefotaxime, a reasonable criterion for specificity is that all impurities having >0.1% of the area in the electropherogram are baseline separated from cefotaxime. Figure 5-1 does not meet this criterion.

When developing an assay, we need to decide what impurities to deliberately add to test for specificity. For analysis of a drug formulation, we would want to compare the pure drug with one containing additions of all possible synthetic by-products and intermediates, degradation products, and *excipients* (substances added to give desirable form or consistency). Degradation products might be introduced by exposing pure material to heat, light, humidity, acid, base, and oxidants to decompose ~20% of the original material.

Linearity

Linearity measures how well a calibration curve follows a straight line, showing that response is proportional to the quantity of analyte. If you know the target concentration of analyte in a drug preparation, for example, test the calibration curve for linearity with five standard solutions spanning the range from 0.5 to 1.5 times the expected analyte concentration. Each standard should be prepared and analyzed three times. (This procedure requires $3 \times 5 = 15$ standards plus three blanks.) To prepare a calibration curve for an impurity that might be present at, say, 0.1 to 1 wt%, you might prepare a calibration curve with five standards spanning the range 0.05 to 2 wt%.

A superficial, but common, measure of linearity is the *square of the correlation coefficient*, R^2:

Square of correlation coefficient:
$$R^2 = \frac{[\Sigma(x_i - \bar{x})(y_i - \bar{y})]^2}{\Sigma(x_i - \bar{x})^2 \Sigma(y_i - \bar{y})^2}$$ (5-2)

where $\bar{x}$ is the mean of all the x values and $\bar{y}$ is the mean of all the y values. An easy way to find R^2 is with the LINEST function in Excel. In the example on page 84, values of x and y are entered in columns A and B. LINEST produces a table in cells E3:F5 that contains R^2 in cell E5.

R^2 is the fraction of observed variance that can be attributed to the mathematical model that was chosen (such as a straight line). If R^2 is not very close to 1, the mathematical model does not account for all sources of variance. For a major component of an unknown, a value of R^2 above 0.995 or, perhaps, 0.999, is deemed a good fit for many purposes.[6] For data in Figure 4-11, which do not lie very close to the straight line, $R^2 = 0.985$.

Another criterion for linearity is that the y-intercept of the calibration curve (after the response of the blank has been subtracted from each standard) should be close to 0. An acceptable degree of "closeness to 0" might be 2% of the response for the target value of analyte. For the assay of impurities, which are present at concentrations lower than that of the major component, an acceptable value of R^2 might be ≥0.98 for the range 0.1 to 2 wt% and the y-intercept should be ≤10% of the response for the 2 wt% standard.

R^2 can be used as a diagnostic. If R^2 decreases significantly after a method is established, something has gone wrong with the procedure.

Accuracy

Accuracy is "nearness to the truth." Ways to demonstrate accuracy include

1. Analyze a *certified reference material* in a matrix similar to that of your unknown. Your method should find the certified value for analyte, within the precision of your method.
2. Compare results from two or more different analytical methods. They should agree within their expected precision.
3. Analyze a blank sample spiked with a known addition of analyte. The matrix must be the same as your unknown. When assaying a major component, three replicate samples at each of three levels ranging from 0.5 to 1.5 times the expected sample concentration are customary. For impurities, spikes could cover three levels spanning an expected range of concentrations, such as 0.1 to 2 wt%.
4. If you cannot prepare a blank with the same matrix as the unknown, then it is appropriate to make *standard additions* (Section 5-3) of analyte to the unknown. An accurate assay will find the known amount of analyte that was added.

Spiking is the most common method to evaluate accuracy because reference materials are not usually available and a second analytical method may not be readily available. Spiking with a small volume of concentrated analyte ensures that the matrix remains nearly constant.

An example of a specification for accuracy is that the analysis will recover $100 \pm 2\%$ of the spike of a major constituent. For an impurity, the specification might be that recovery is within 0.1 wt% absolute or $\pm 10\%$ relative.

Precision

Precision is how well replicate measurements agree with one another, usually expressed as a standard deviation or standard uncertainty (standard deviation of the mean) or confidence interval. When an experienced analyst replicates his or her measurement by the same procedure using the same instrument, results could be highly repeatable. The 95% confidence interval could be small. When different people in different labs with different instruments perform the analysis, each person's confidence interval could be small, but it might not overlap the confidence intervals of others who did the same analysis. What are the sources of error? There could be differences in samples, differences in sample preparation, differences in technique among analysts, uncontrolled changes that occur in each lab from day to day, uncontrolled differences among laboratories, and differences among instruments.

Repeatability: describes how well one person can obtain the same results when analyzing the same sample by the same procedure with the same equipment in the same laboratory

Reproducibility: describes how well different people in different laboratories with different equipment can get the same results when analyzing similar samples by the same procedure

Two broad categories of precision are *repeatability* and *reproducibility*. **Repeatability** describes the spread in results when one person uses one procedure to analyze the same sample by the same method multiple times. **Reproducibility** describes the spread in results when different people in different labs using different instruments each try to follow the same procedure. Statistical procedures such as *analysis of variance* described in Appendix C help us find which factors account for the variability.

Some specific types of precision are defined below:

Instrument precision is the repeatability observed when the same quantity of one sample is repeatedly introduced (≥ 10 times) into an instrument. Variability could arise from variation in the injected quantity and variation of instrument response.

Autosamplers in chromatography and graphite furnace atomic spectroscopy, for example, have improved injection precision by a factor of 3–10 compared with that attained by humans.

Intra-assay precision is evaluated by analyzing aliquots of a homogeneous material several times by one person on one day with the same equipment. Each analysis is independent, so the intra-assay precision is telling us how reproducible the analytical method can be. Intra-assay variability is greater than instrument variability, because more steps are involved. Examples of specifications might be that instrument precision is $\leq 1\%$ and intra-assay precision is $\leq 2\%$.

Intermediate precision, formerly called <u>ruggedness</u>, is the variation observed when an assay is performed by <u>different people</u> on <u>different instruments</u> on <u>different days</u> in the <u>same lab</u>. Each analysis might incorporate fresh reagents and different chromatography columns.

Interlaboratory precision, which is the same as *reproducibility*, is the most general measure of reproducibility observed when aliquots of the same sample are analyzed by different people in different laboratories. Interlaboratory precision can be significantly poorer than intermediate precision. For example, a 13-laboratory study was conducted to validate a new method for measuring bisphenol A and related phenolic compounds in water. Intermediate precision (*within* laboratories) was 1.9 to 5.5% for several compounds. Interlaboratory precision with all labs following the same instructions was 10.8 to 22.5%.[7] Interlaboratory precision becomes poorer as analyte concentration decreases (Box 5-3).

Range

Confusing terms:

Linear range: concentration range over which calibration curve is linear (Figure 4-14)

Dynamic range: concentration range over which there is measurable response

Range: concentration range over which linearity, accuracy, and precision meet specifications for analytical method

Range is the concentration interval over which linearity, accuracy, and precision are all acceptable. An example of a specification for range for a major component of a mixture is the concentration interval providing a correlation coefficient of $R^2 \geq 0.995$ (a measure of linearity), spike recovery of $100 \pm 2\%$ (a measure of accuracy), and interlaboratory precision of $\pm 3\%$. For an impurity, an acceptable range might provide a correlation coefficient of $R^2 \geq 0.98$, spike recovery of $100 \pm 10\%$, and interlaboratory precision of $\pm 15\%$.

Limits of Detection and Quantitation

A good definition for you: The **detection limit** in Equation 5-5 is the concentration of analyte that gives a signal equal to three times the standard deviation of signal from a blank.

The **detection limit** (also called the *lower limit of detection*) is the smallest quantity of analyte that is "significantly different" from the blank.[9] We will describe a procedure that provides ~99% confidence that a signal above the detection limit arises from a sample that really does contain analyte. That is, only ~1% of samples containing no analyte will give a signal

CHAPTER 5 Quality Assurance and Calibration Methods

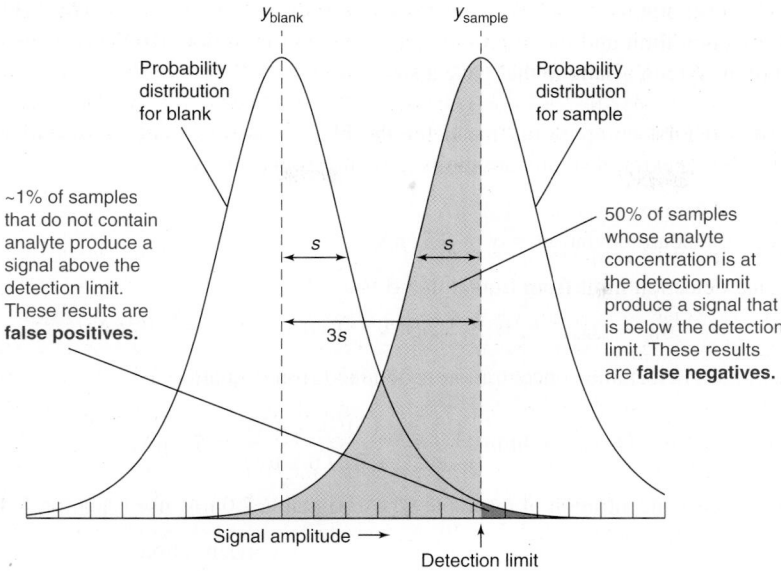

FIGURE 5-2 Detection limit. Distribution of measurements for a blank and a sample whose concentration is at the detection limit. The area of any region is proportional to the number of measurements in that region. Only ~1% of measurements for a blank are expected to exceed the detection limit. However, 50% of measurements for a sample containing analyte at the detection limit will be below the detection limit. There is a 1% chance of concluding that a blank has analyte above the detection limit (*false positive*). If a sample contains analyte at the detection limit, there is a 50% chance of concluding that analyte is *absent* because its signal is below the detection limit (*false negative*). Curves are Student's *t* distributions for six degrees of freedom and are broader than the corresponding Gaussian distributions.

greater than the detection limit (Figure 5-2). We say that there is a ~1% rate of *false positives* in Figure 5-2. This same definition of detection limit provides only 50% confidence that we can identify a sample that does contain analyte, if the analyte concentration is at the detection limit. That is, half of the samples whose analyte concentration is at the detection limit give *false negative* results below the detection limit in Figure 5-2. In the following procedure, we assume that the standard deviation of the signal from samples near the detection limit is similar to the standard deviation from blanks.

1. After estimating the detection limit from previous experience with the method, prepare a sample whose concentration is ~1 to 5 times the detection limit.
2. Measure the signal from *n* replicate samples ($n \geq 7$).
3. Compute the standard deviation (*s*) of the *n* measurements.
4. Measure the signal from *n* blanks (containing no analyte) and find the mean value, y_{blank}.
5. The minimum detectable signal, y_{dl}, is defined as

    Signal detection limit: $y_{dl} = y_{blank} + 3s$ (5-3)

6. The corrected signal, $y_{sample} - y_{blank}$, is proportional to sample concentration:

    Calibration line: $y_{sample} - y_{blank} = m \times$ sample concentration   (5-4)

    where y_{sample} is the signal observed for the sample and *m* is the slope of the linear calibration curve. The *minimum detectable concentration*, also called the detection limit, is obtained by substituting y_{dl} from Equation 5-3 for y_{sample} in Equation 5-4 to get

    Detection limit: Minimum detectable concentration $\equiv \dfrac{3s}{m}$ (5-5)

EXAMPLE **Detection Limit**

From previous measurements of a low concentration of analyte, the signal detection limit was estimated to be in the low nanoampere range. Signals from seven replicate samples with a concentration about three times the detection limit were 5.0, 5.0, 5.2, 4.2, 4.6, 6.0, and 4.9 nA. Reagent blanks gave values of 1.4, 2.2, 1.7, 0.9, 0.4, 1.5, and 0.7 nA. The

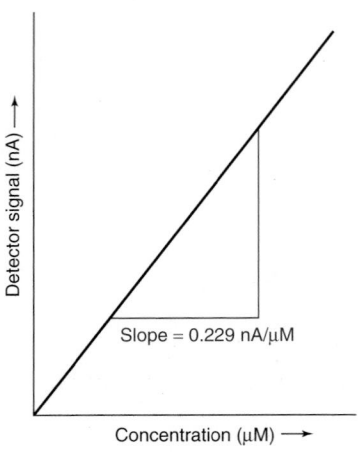

Slope = 0.229 nA/μM

Detector signal (nA) →

Concentration (μM) →

slope of the calibration curve for higher concentrations is $m = 0.229$ nA/μM. **(a)** Find the signal detection limit and the minimum detectable concentration. **(b)** What is the concentration of analyte in a sample that gave a signal of 7.0 nA?

Solution **(a)** First compute the mean for the blanks and the standard deviation of the samples. Retain extra, insignificant digits to reduce round-off errors.

Blank: Average $= y_{blank} = 1.2_6$ nA
Sample: Standard deviation $= s = 0.5_6$ nA

The signal detection limit from Equation 5-3 is

$$y_{dl} = y_{blank} + 3s = 1.2_6 \text{ nA} + (3)(0.5_6 \text{ nA}) = 2.9_4 \text{ nA}$$

The minimum detectable concentration is obtained from Equation 5-5:

$$\text{Detection limit} = \frac{3s}{m} = \frac{(3)(0.5_6 \text{ nA})}{0.229 \text{ nA/μM}} = 7._3 \text{ μM}$$

(b) To find the concentration of a sample whose signal is 7.0 nA, use Equation 5-4:

$$y_{sample} - y_{blank} = m \times \text{concentration}$$

$$\Rightarrow \text{Concentration} = \frac{y_{sample} - y_{blank}}{m} = \frac{7.0 \text{ nA} - 1.2_6 \text{ nA}}{0.229 \text{ nA/μM}} = 25._1 \text{ μM}$$

TEST YOURSELF Find the minimum detectable concentration if the average of the blanks is 1.0_5 nA and $s = 0.6_3$ nA. (**Answer:** $8._3$ μM)

BOX 5-3 The Horwitz Trumpet: Variation in Interlaboratory Precision

Interlaboratory tests are routinely used to validate new analytical procedures—especially those intended for regulatory use. Typically, five to 10 laboratories are given identical samples and the same written procedure. If all results are "similar," and there is no serious systematic error, then the method is considered "reliable."

The **coefficient of variation** is the standard deviation divided by the mean, usually expressed as a percentage: $CV(\%) = 100 \times s/\bar{x}$, where s is the standard deviation and $\bar{x}$ is the mean. The smaller the coefficient of variation, the more precise is a set of measurements.

In reviewing more than 150 interlaboratory studies with different analytes measured by different techniques, it was observed that the coefficient of variation of mean values reported by different laboratories increased as analyte concentration decreased. At best, the coefficient of variation never seemed to be better than[8]

Horwitz curve: $$CV(\%) \approx 2^{(1 - 0.5 \log C)}$$

where C is g analyte/g sample. The coefficient of variation within a laboratory is about one-half to two-thirds of the between-laboratory variation. Experimental results varied from the idealized curve by about a factor of 2 in the vertical direction and a factor of 10 in the horizontal direction. About 5–15% of all interlaboratory results were "outliers"—clearly outside the cluster of other results. This incidence of outliers is above the statistical expectation.

The Horwitz curve predicts that when the concentration of analyte is 1 ppm, the coefficient of variation between laboratories is ~16%. When the concentration is 1 ppb, the coefficient of variation is ~45%. If, perchance, you become a regulation writer one day, acceptable analyte levels should allow for variation among laboratories. The Gaussian distribution tells us that ~5% of measurements lie above $\bar{x} + 1.65s$ (Section 4-1). If the target allowable level of analyte is 1.0 ppb, the allowed observed amount might be

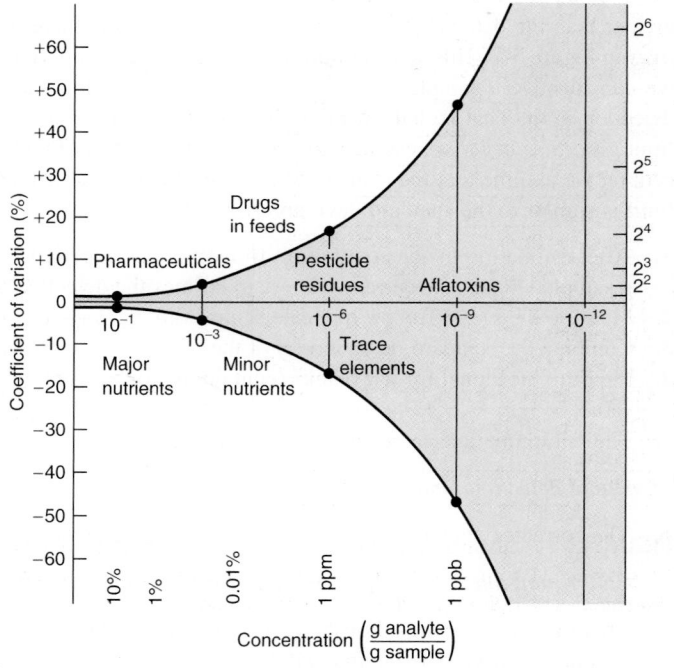

Coefficient of variation of interlaboratory results as a function of sample concentration (expressed as g analyte/g sample). The shaded region has been referred to as the "Horwitz trumpet" because of the way it flares open. [Data from W. Horwitz, "Evaluation of Analytical Methods Used for Regulation of Foods and Drugs," *Anal. Chem.* **1982,** *54,* 67A.]

set at $1 + 1.65 \times 0.45$ ppb, or about 1.7 ppb. This level gives a 5% rate of false positives that exceed the allowed value even though the true value is below 1.0 ppb.

Another common way to define detection limit is from the least-squares equation of a calibration curve: signal detection limit $= b + 3s_y$, where b is the y-intercept and s_y is given by Equation 4-20. A more rigorous procedure is described in the notes for this chapter.[10]

The lower limit of detection in Equation 5-5 is $3s/m$, where s is the standard deviation of a low-concentration sample and m is the slope of the calibration curve. The standard deviation is a measure of the *noise* (random variation) in a blank or a small signal. When the signal is 3 times greater than the noise, it is detectable, but still too small for accurate measurement. A signal that is 10 times greater than the noise is defined as the **lower limit of quantitation,** or the smallest amount that can be measured with reasonable accuracy.

$$\text{Lower limit of quantitation} \equiv \frac{10s}{m} \tag{5-6}$$

The *instrument detection limit* is obtained by replicate measurements ($n \geq 7$) of aliquots from one sample. The *method detection limit*, which is greater than the instrument detection limit, is obtained by preparing $n \geq 7$ individual samples and analyzing each one once.

The **reporting limit** is the concentration below which regulations say that a given analyte is reported as "not detected," which *does not mean* that analyte is not observed. It means that analyte is below a prescribed level. Reporting limits are set at least 5 to 10 times higher than the detection limit, so that detecting analyte at the reporting limit is not ambiguous.

Labels on U.S. packaged foods must state how much *trans* fat is present. This type of fat is derived mainly from partial hydrogenation of vegetable oil and is a major component of margarine and shortening. *Trans* fat is thought to increase risk of heart disease, stroke, and some cancers. However, the *reporting limit* for *trans* fat is 0.5 g per serving. If the concentration is <0.5 g/serving, it is reported as 0, as in Figure 5-3. By reducing the serving size, a manufacturer can state that *trans* fat content is 0. If your favorite snack food is made with partially hydrogenated oil, it contains *trans* fat even if the label says otherwise.

Detection limit $\equiv \dfrac{3s}{m}$

Quantitation limit $\equiv \dfrac{10s}{m}$

The symbol $\equiv$ means **"is defined as."**

Question A snack food contains 2.5 wt% *trans* fat. What is the largest serving size that could be set so that the manufacturer can list 0 *trans* fat on the package? (*Answer:* 20 g per serving)

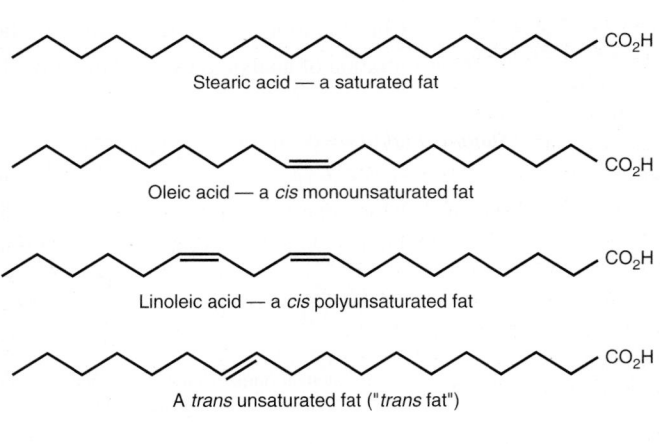

FIGURE 5-3 Nutritional label from a package of crackers. The *reporting limit* for *trans* fat is 0.5 g/serving. Any amount less than this is reported as 0. The end of Chapter 6 explains the shorthand used to draw these 18-carbon compounds.

Robustness

Robustness is the ability of an analytical method to be unaffected by small, deliberate changes in operating parameters. For example, a chromatographic method is robust if it gives acceptable results when small changes are made in solvent composition, pH, buffer concentration, temperature, injection volume, and detector wavelength. In tests for robustness, the organic solvent content in the mobile phase could be varied by, say, $\pm 2\%$, the eluent pH varied by ± 0.1, and column temperature varied by $\pm 5°C$. If acceptable results are obtained, the written procedure should state that these variations are tolerable. Capillary electrophoresis requires such small volumes that a given solution could conceivably be used for months before it is used up. Therefore, solution stability (shelf life) should be evaluated for robustness.

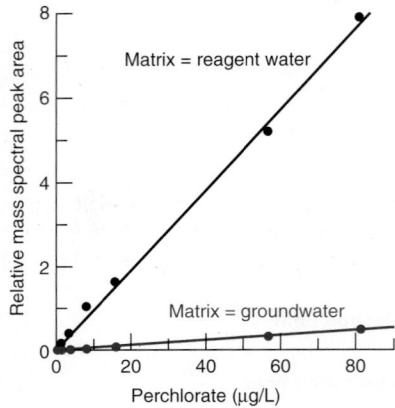

FIGURE 5-4 Calibration curves for perchlorate in pure water and in groundwater. [Data from C. J. Koester, H. R. Beller, and R. U. Halden, "Analysis of Perchlorate in Groundwater by Electrospray Ionization Mass Spectrometry/Mass Spectrometry," *Environ. Sci. Technol.* **2000,** *34,* 1862.]

The matrix affects the magnitude of the analytical signal. In standard addition, all samples are in the same matrix.

Derivation of Equation 5-7:

$I_X = k[X]_i$, where k is a constant of proportionality

$I_{S+X} = k([S]_f + [X]_f)$, where k is the same constant

Dividing one equation by the other gives

$$\frac{I_X}{I_{S+X}} = \frac{k[X]_i}{k([S]_f + [X]_f)} = \frac{[X]_i}{[S]_f + [X]_f}$$

5-3 Standard Addition[11,12]

In **standard addition,** known quantities of analyte are added to the unknown. From the increase in signal, we deduce how much analyte was in the original unknown. This method requires a linear response to analyte. As in titrations, higher precision can be achieved when standards are added by mass instead of volume.[13]

Standard addition is especially appropriate when the sample composition is unknown or complex and affects the analytical signal. In such case, it is impossible or difficult to create standards and blanks whose composition matches that of the sample. If standards and blanks do not match the composition of the unknown sample, a calibration curve is not reliable. The *matrix* is everything in the unknown, other than analyte. A **matrix effect** is a change in the analytical signal caused by anything in the sample other than analyte.

Figure 5-4 shows a strong matrix effect in the analysis of perchlorate (ClO_4^-) by mass spectrometry. Perchlorate at a level above 18 μg/L in drinking water is of concern because it can reduce thyroid hormone production. Standard solutions of ClO_4^- in pure water gave the upper calibration curve in Figure 5-4. The slope in the lower curve for standard solutions in groundwater was 15 times less. Reduction of the ClO_4^- signal is a *matrix effect* attributed to other anions present in the groundwater.

Different groundwaters have different concentrations of many anions, so there is no way to construct a calibration curve for this analysis that would apply to more than one specific groundwater. Hence, the method of standard addition is required. When we add a small volume of concentrated standard to an existing unknown, we do not change the concentration of the matrix very much.

Consider a standard addition in which a sample with unknown initial concentration of analyte $[X]_i$ gives a signal intensity I_X. Then a known concentration of standard, S, is added to an aliquot of the sample and a signal I_{S+X} is observed for this second solution. Addition of standard to the unknown changes the concentration of the original analyte because of dilution. Let's call the diluted concentration of analyte $[X]_f$, where "f" stands for "final." We designate the concentration of standard in the final solution as $[S]_f$. (Bear in mind that the chemical species X and S are the same.)

Signal is directly proportional to analyte concentration, so

$$\frac{\text{Concentration of analyte in initial solution}}{\text{Concentration of analyte plus standard in final solution}} = \frac{\text{signal from initial solution}}{\text{signal from final solution}}$$

Standard addition equation:
$$\frac{[X]_i}{[S]_f + [X]_f} = \frac{I_X}{I_{S+X}} \qquad (5\text{-}7)$$

For an initial volume V_o of unknown and added volume V_S of standard with concentration $[S]_i$, the total volume is $V = V_o + V_S$ and the concentrations in Equation 5-7 are

$$[X]_f = [X]_i\left(\frac{V_o}{V}\right) \qquad [S]_f = [S]_i\left(\frac{V_S}{V}\right) \qquad (5\text{-}8)$$

The quotient (initial volume/final volume), which relates final concentration to initial concentration, is the **dilution factor.** It comes directly from Equation 1-5.

By expressing the diluted concentration of analyte, $[X]_f$, in terms of the initial concentration of analyte, $[X]_i$, we can solve for $[X]_i$, because everything else in Equation 5-7 is known.

EXAMPLE Standard Addition

Serum containing Na^+ gave a signal of 4.27 mV in an atomic emission analysis. Then 5.00 mL of 2.08 M NaCl were added to 95.0 mL of serum. This spiked serum gave a signal of 7.98 mV. Find the original concentration of Na^+ in the serum.

Solution From Equation 5-8, the final concentration of Na^+ after dilution with the standard is $[X]_f = [X]_i(V_o/V) = [X]_i(95.0 \text{ mL}/100.0 \text{ mL})$. The final concentration of added standard is $[S]_f = [S]_i(V_S/V) = (2.08 \text{ M})(5.00 \text{ mL}/100.0 \text{ mL}) = 0.104 \text{ M}$. Equation 5-7 becomes

$$\frac{[Na^+]_i}{[0.104 \text{ M}] + 0.950[Na^+]_i} = \frac{4.27 \text{ mV}}{7.98 \text{ mV}} \Rightarrow [Na^+]_i = 0.113 \text{ M}$$

TEST YOURSELF If spiked serum gave a signal of 6.50 mV, what was the original concentration of Na^+? (*Answer:* 0.182 M)

Graphical Procedure for Standard Addition to a Single Solution

There are two common methods to perform standard addition. If the analysis does not consume solution, we begin with an unknown solution and measure the analytical signal. Then we add a small volume of concentrated standard and measure the signal again. We add several more small volumes of standard and measure the signal after each addition. Standard should be concentrated so that only small volumes are added and the sample matrix is not appreciably altered. Added standards should increase the analytical signal by a factor of 1.5 to 3. The other common procedure is described in the next section.

Figure 5-5 shows data for an experiment in which ascorbic acid (vitamin C) was measured in orange juice by an electrochemical method. The current between a pair of electrodes immersed in the juice is proportional to the concentration of ascorbic acid. Eight standard additions increased current from 1.78 to 5.82 μA (column C), which is at the upper end of the desired range of 1.5- to 3-fold increase in analytical signal.

A common mistake in standard addition is to increase the original signal by more than a factor of 3, which reduces the accuracy of the result.

	A	B	C	D	E
1	Vitamin C standard addition experiment				
2	Add 0.279 M ascorbic acid to 50.0 mL orange juice				
3					
4		Vs =			
5	Vo (mL) =	mL ascorbic	I(s+x) =	x-axis function	y-axis function
6	50	acid added	signal (μA)	Si*Vs/Vo	I(s+x)*V/Vo
7	[S]i (mM) =	0.000	1.78	0.000	1.780
8	279	0.050	2.00	0.279	2.002
9		0.250	2.81	1.395	2.824
10		0.400	3.35	2.232	3.377
11		0.550	3.88	3.069	3.923
12		0.700	4.37	3.906	4.431
13		0.850	4.86	4.743	4.943
14		1.000	5.33	5.580	5.437
15		1.150	5.82	6.417	5.954
16					
17	D7 = $A$8*B7/$A$6		E7 = C7*($A$6+B7)/$A$6		

FIGURE 5-5 Data for standard addition experiment with *variable total volume*.

Figure 5-6 allows us to find the original concentration of unknown. The theoretical response is derived by substituting expressions for $[X]_f$ and $[S]_f$ from Equations 5-8 into Equation 5-7. Following a little rearrangement, we find

For successive standard additions to one solution

$$\underbrace{I_{S+X}\left(\frac{V}{V_o}\right)}_{\substack{\text{Function to plot} \\ \text{on } y\text{-axis}}} = I_X + \underbrace{\frac{I_X}{[X]_i}[S]_i\left(\frac{V_S}{V_o}\right)}_{\substack{\text{Function to plot} \\ \text{on } x\text{-axis}}}$$

(5-9)

Successive standard additions to one solution:

Plot $I_{S+X}\left(\dfrac{V}{V_o}\right)$ versus $[S]_i\left(\dfrac{V_S}{V_o}\right)$

x-intercept is $[X]_i$

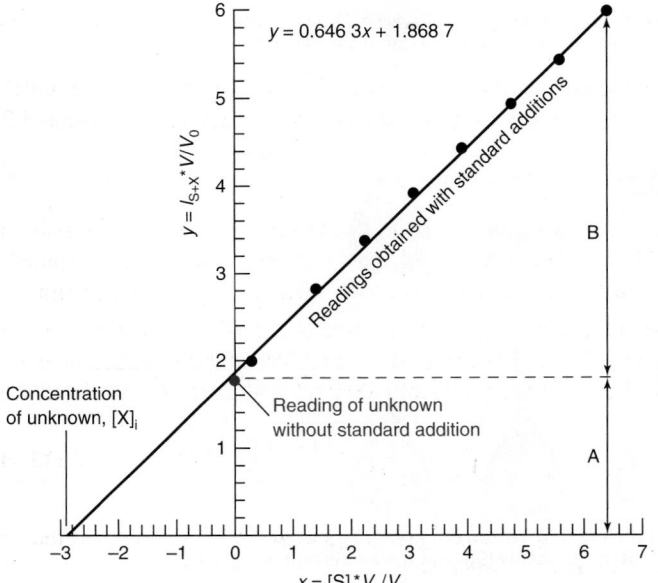

$y = 0.646\,3x + 1.868\,7$

Readings obtained with standard additions

Concentration of unknown, $[X]_i$

Reading of unknown without standard addition

$y = I_{S+X}*V/V_0$

$x = [S]_i*V_s/V_0$

FIGURE 5-6 Graphical treatment of standard additions to a single solution *with variable total volume*. Data from Figure 5-5. Standard additions should increase the analytical signal to between 1.5 and 3 times its original value (that is, B = 0.5A to 2A).

The equation of a line is $y = mx + b$. The x-intercept is obtained by setting $y = 0$:

$$0 = mx + b$$
$$x = -b/m$$

A graph of $I_{S+X}(V/V_o)$ (the *corrected response*) on the y-axis versus $[S]_i(V_S/V_o)$ on the x-axis should be a straight line. The data plotted in Figure 5-6 are computed in columns D and E of Figure 5-5. The right side of Equation 5-9 is 0 when $[S]_i (V_S/V_o) = -[X]_i$. The magnitude of the intercept on the x-axis is the *original* concentration of unknown, $[X]_i = 2.89$ mM in Figure 5-6.

The *standard uncertainty* in the x-intercept is[14]

Standard uncertainty = standard deviation of the mean, as in Equation 4-27.

Standard uncertainty
of x-intercept:
$$u_x = \frac{s_y}{|m|} \sqrt{\frac{1}{n} + \frac{\bar{y}^2}{m^2 \sum (x_i - \bar{x})^2}} \qquad (5\text{-}10)$$

where s_y is the standard deviation of y (Equation 4-20), $|m|$ is the absolute value of the slope of the least-squares line (Equation 4-16), n is the number of data points (nine in Figure 5-6), $\bar{y}$ is the mean value of y for the nine points, x_i are the individual values of x for the nine points, and $\bar{x}$ is the mean value of x for the nine points. For the points in Figure 5-6, the standard uncertainty in the x-intercept is $u_x = 0.09_8$ mM.

The confidence interval is $\pm t u_x$, where t is Student's t (Table 4-4) for $n - 2$ degrees of freedom. The 95% confidence interval for the intercept in Figure 5-6 is $\pm (2.365)(0.09_8$ mM) = ± 0.23 mM. The value $t = 2.365$ was taken from Table 4-4 for $9 - 2 = 7$ degrees of freedom.

Graphical Procedure for Multiple Solutions with Constant Volume

The second common standard addition procedure is shown in Figure 5-7. Equal volumes of unknown are pipetted into several volumetric flasks. Increasing volumes of standard are added to each flask. After addition of a constant volume of reagents for the chemical analysis, each sample is diluted to the *same final volume*. Every flask contains the same concentration of unknown and differing concentrations of standard. For each flask, a measurement of analytical signal, I_{S+X}, is then made. The method in Figure 5-7 is necessary when the analysis consumes some of the solution.

If all standard additions are made to a constant final volume, plot the signal I_{S+X} versus the concentration of diluted standard, $[S]_f$ (Figure 5-7). In this case, the x-intercept is the *final* concentration of unknown, $[X]_f$, after dilution to the final sample volume. Equation 5-10 still

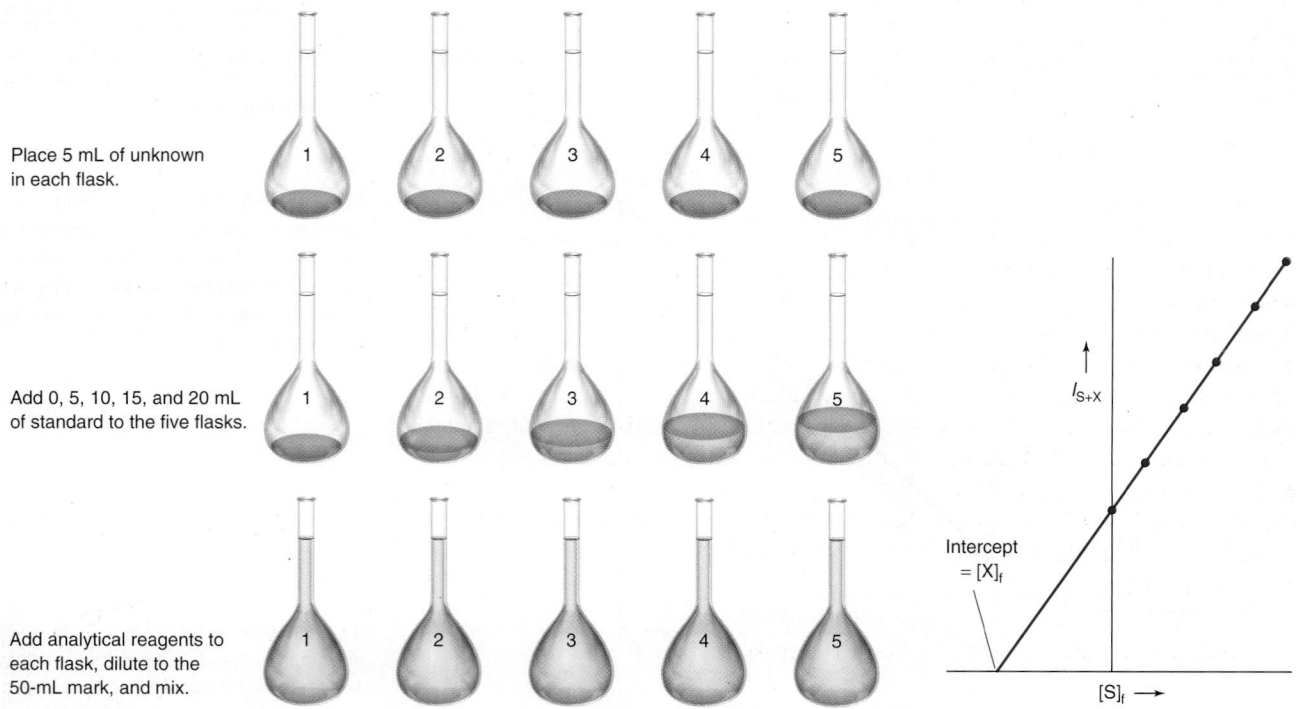

Place 5 mL of unknown in each flask.

Add 0, 5, 10, 15, and 20 mL of standard to the five flasks.

Add analytical reagents to each flask, dilute to the 50-mL mark, and mix.

Intercept = $[X]_f$

I_{S+X}

$[S]_f \longrightarrow$

FIGURE 5-7 Standard addition experiment with *constant total volume*. Plot I_{S+X} versus $[S]_f$, and the x-intercept is $[X]_f$. Lines in Figures 5-6 and 5-7 are both derived from Equation 5-9.

CHAPTER 5 Quality Assurance and Calibration Methods

applies to the uncertainty. The initial concentration of unknown, $[X]_i$, is calculated from the dilution that was applied to make the final sample.

> **EXAMPLE** **Standard Addition with Constant Total Volume**
>
> In Figure 5-7, 5.00 mL of unknown in each flask are diluted to 50.00 mL. If the x-intercept is 0.235 mM, what is the original concentration of analyte in the unknown?
>
> **Solution** Analyte was diluted by a factor of 5.00 mL/50.00 mL $=$ 10.00 in each flask. The x-intercept is the final concentration of diluted analyte, $[X]_f$. The original concentration was 10.00 times greater $=$ 2.35 mM.
>
> **TEST YOURSELF** 1.00 mL of blood serum was diluted to 25.00 mL in each flask of a standard addition experiment like Figure 5-7 to measure a hormone with a molecular mass of 373 g/mol. The x-intercept of the graph was 4.2 ppb (parts per billion). Find the concentration of hormone in the serum and express your answer in ppb and molarity. Assume that the density of serum and all solutions is close to 1.00 g/mL (*Answer:* 105 ppb, 0.28 μM)

5-4 Internal Standards

An **internal standard** is a known amount of a compound—*different from analyte*—that is added to the unknown. Signal from analyte is compared with signal from the internal standard to find out how much analyte is present. A carefully chosen internal standard will give an analytical signal (such as a chromatographic peak or spectrophotometric absorption) that is well separated from those of the analyte and other species in an unknown. The internal standard should be chemically stable and not react with components of the unknown. It is helpful for the internal standard to be chemically similar to analyte so that uncontrolled effects of the matrix that increase or decrease the analyte signal might have a similar effect on the signal from the standard.

Internal standards are especially useful for analyses in which the quantity of sample analyzed or the instrument response varies slightly from run to run. For example, gas or liquid flow rates that vary by a few percent in a chromatography experiment could change the detector response. A calibration curve is accurate only for the one set of conditions under which it was obtained. However, the *relative* response of the detector to the analyte and standard is usually constant over a range of conditions. If signal from the standard increases by 8.4% because of a change in flow rate, signal from the analyte usually increases by 8.4% also. As long as the concentration of standard is known, the correct concentration of analyte can be derived. Internal standards are used in chromatography because the small quantity of sample injected into the chromatograph is not reproducible.

Internal standards are desirable when sample loss can occur during sample preparation steps prior to analysis. If a known quantity of standard is added to the unknown prior to any manipulations, the ratio of standard to analyte remains constant because the same fraction of each is lost in any operation.

To use an internal standard, we prepare a known mixture of standard and analyte to measure the relative response of the detector to the two species. In Figure 5-8, the area, A, under each peak is proportional to the concentration of the species injected into a chromatography column. However, the detector generally has a different response to each component. For example, if both analyte (X) and internal standard (S) have concentrations of 10.0 mM, the area under the analyte peak might be 2.30 times greater than the area under the standard peak. We say that the **response factor,** F, is 2.30 times greater for X than for S.

$$\text{Response factor:} \quad \frac{\text{Signal from analyte}}{\text{Concentration of analyte}} = F\left(\frac{\text{signal from standard}}{\text{concentration of standard}}\right) \tag{5-11}$$

$$\frac{A_X}{[X]} = F\left(\frac{A_S}{[S]}\right)$$

Standard addition: Added standard is *the same* substance as analyte.

Internal standard: Added standard is *different* from analyte.

External standard: Solutions with known concentrations of analyte are used to prepare a calibration curve.

The assumption that relative response to analyte and standard remains constant over a range of concentrations must be verified.

If the detector responds equally to standard and analyte, $F = 1$. If the detector responds twice as much to analyte as to standard, $F = 2$. If the detector responds half as much to analyte as to standard, $F = 0.5$.

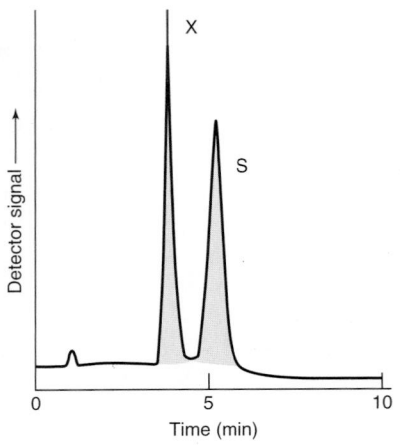

FIGURE 5-8 Chromatographic separation of unknown (X) and internal standard (S). A known amount of S was added to the unknown. The relative areas of the signals from X and S allow us to find out how much X is in the mixture. It is necessary first to measure the relative response of the detector to each compound.

The **dilution factor** $\dfrac{\text{initial volume}}{\text{final volume}}$ converts initial concentration into final concentration.

Sometimes a signal is an area (as in chromatography) and other times the signal might be a height rather than an area. [X] and [S] are the concentrations of analyte and standard *after they have been mixed together*. Equation 5-11 is predicated on linear response to analyte and standard.

EXAMPLE   **Using an Internal Standard**

In a preliminary experiment, a solution containing 0.083 7 M X and 0.066 6 M S gave peak areas of $A_X = 423$ and $A_S = 347$. (Areas are measured in arbitrary units by the instrument's computer.) To analyze the unknown, 10.0 mL of 0.146 M S were added to 10.0 mL of unknown, and the mixture was diluted to 25.0 mL in a volumetric flask. This mixture gave the chromatogram in Figure 5-8, for which $A_X = 553$ and $A_S = 582$. Find the concentration of X in the unknown.

Solution First use the standard mixture to find the response factor in Equation 5-11:

Standard mixture:
$$\frac{A_X}{[X]} = F\left(\frac{A_S}{[S]}\right)$$

$$\frac{423}{0.083\ 7} = F\left(\frac{347}{0.066\ 6}\right) \Rightarrow F = 0.970_0$$

In the mixture of unknown plus standard, the concentration of S is

$$[S] = \underbrace{(0.146\ \text{M})}_{\substack{\text{Initial}\\\text{concentration}}}\underbrace{\left(\frac{10.0}{25.0}\right)}_{\substack{\text{Dilution}\\\text{factor}}} = 0.058\ 4\ \text{M}$$

Using the known response factor, substitute back into Equation 5-11 to find the concentration of unknown in the mixture:

Unknown mixture:
$$\frac{A_X}{[X]} = F\left(\frac{A_S}{[S]}\right)$$

$$\frac{553}{[X]} = 0.970_0\left(\frac{582}{0.058\ 4}\right) \Rightarrow [X] = 0.057\ 2_1\ \text{M}$$

Because X was diluted from 10.0 to 25.0 mL when the mixture with S was prepared, the original concentration of X in the unknown was (25.0 mL/10.0 mL)(0.057 2$_1$ M) = 0.143 M.

TEST YOURSELF Suppose that peak areas for the known mixture were $A_X = 423$ and $A_S = 447$. Find [X] in the unknown. (***Answer:*** $F = 0.753_0$, [X] = 0.184 M)

Multipoint Calibration Curve for an Internal Standard

The example above uses a single mixture to find the response factor. If there were no experimental error, this "one-point calibration curve" would be sufficient to give an accurate response factor. There is always experimental error, so a multipoint calibration curve is preferred to average out some experimental variability. For this purpose, we rearrange Equation 5-11 so that the signals are on one side and the concentrations are on the other side:

Equation for internal standard calibration curve.

$$\frac{\text{Signal from analyte}}{\text{Signal from standard}} = F\left(\frac{\text{concentration of analyte}}{\text{concentration of standard}}\right) \tag{5-12}$$

$$\frac{A_X}{A_S} = F\left(\frac{[X]}{[S]}\right)$$

Then construct a graph in which the signal ratio on the left side of Equation 5-12 is plotted as a function of the concentration ratio on the right side. The graph should be linear with a zero intercept. The slope of this graph is the response factor. Let's look at an example.

CHAPTER 5 Quality Assurance and Calibration Methods

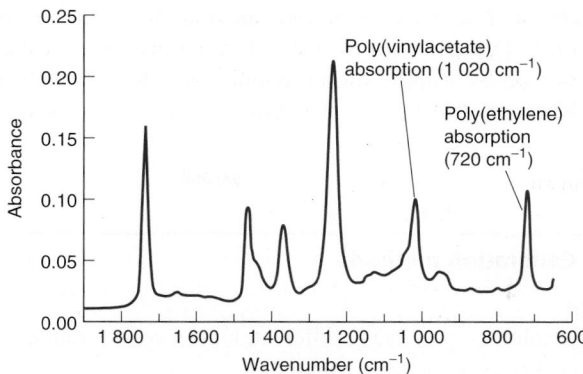

FIGURE 5-9 Infrared spectrum of poly(ethylene-co-vinyl acetate) containing 18 mol% vinyl acetate. The spectrum shows the absorption of infrared radiation as a function of wavenumber (= 1/wavelength). We define absorbance, wavelength, and wavenumber in Chapter 18. [Data from M. K. Bellamy, "Using FTIR-ATR Spectroscopy to Teach the Internal Standard Method," *J. Chem. Ed.* **2010,** *87,* 1399.]

Figure 5-9 shows the infrared absorbance spectrum of a polymer made from ethylene and vinyl acetate.

$$p\ H_2C{=}CH_2 + q\ H_2C{=}C \overset{\displaystyle \overset{O}{\underset{\displaystyle \|}{}}}{\underset{\displaystyle H}{\diagdown OCCH_3}} \xrightarrow[\text{catalyst}]{\text{Polymerization}} \quad (5\text{-}13)$$

Ethylene   Vinyl acetate   Poly(ethylene-co-vinyl acetate)   (OAc = acetate)

The polymer contains a mole ratio $p:q$ of ethylene and vinyl acetate units bonded in a random order. In Figure 5-9, the mole ratio is $p:q = 82:18$. The absorption peak at a wavenumber of 1 020 cm^{-1} arises from vinyl acetate units and the peak at 720 cm^{-1} arises from ethylene units. Our objective is to measure the quotient p/q from the relative absorbance of the two peaks in a polymer of unknown composition.

Figure 5-10 is an internal standard calibration curve obtained from infrared absorbance of polymers containing six different known mole ratios $p:q$. We arbitrarily choose to call ethylene the internal standard (S) and vinyl acetate the unknown (X). The ordinate (*y*-axis) is the quotient $A_X/A_S =$ (absorbance of vinyl acetate units at 1 020 cm^{-1})/(absorbance of ethylene units at 720 cm^{-1}). The abscissa (*x*-axis) is the quotient of concentrations [X]/[S]. In a quotient, we can use any concentration unit that we wish, so long as the same units are used in the numerator and denominator. It is sensible in this example to use (mol% vinyl acetate)/(mol% ethylene) $= q/p$ as the quotient [X]/[S] because the polymer is made by mixing known mole ratios of the two components.

Points in Figure 5-10 lie on a straight line whose slope is the response factor, F. The theoretical intercept for Equation 5-12 is 0. Problem 5-32 demonstrates that the observed intercept is within statistical uncertainty of 0. If polymer of unknown composition exhibits a

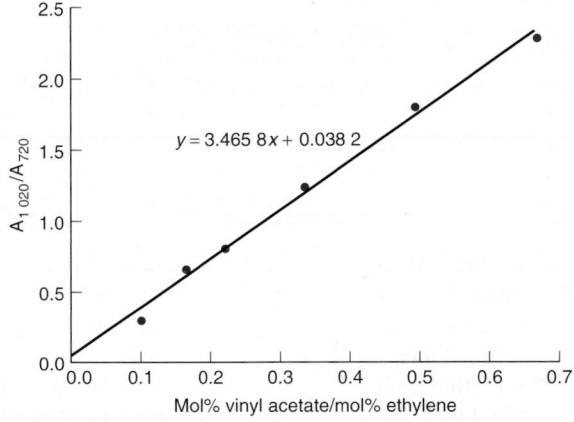

FIGURE 5-10 Infrared absorbance calibration curve for polymer in which ethylene is treated as the internal standard and vinyl acetate is treated as the unknown. [Data from M. K. Bellamy, "Using FTIR-ATR Spectroscopy to Teach the Internal Standard Method," *J. Chem. Ed.* **2010,** *87,* 1399.]

quotient of, say $A_X/A_S = 1.98$, we can use the equation of the line in Figure 5-10 to find that the composition is $[X]/[S] = q/p = 0.56$. The graph improves the accuracy of the internal standard method by using multiple points instead of a single point to find the response factor. The graph also verifies that Equation 5-12 is obeyed over the required range of composition.

TABLE 5-2 **Calibration methods**

External standard
- Prepare multiple solutions containing different, known concentrations of analyte
- Measure analytical response to each solution
- Prepare a calibration curve showing analytical response versus analyte concentration
- From observed analytical response to unknown solution, use calibration curve to find concentration of analyte in unknown

Internal standard
- Internal standard is a known quantity of a compound chemically similar to analyte that is mixed with the unknown solution
- Internal standard is especially useful if uncontrolled loss of sample occurs during sample preparation or analytical procedure
- Prepare a series of known mixtures of analyte and internal standard. Prepare a calibration curve showing (signal from analyte/signal from standard) versus (concentration of analyte/concentration of standard). The slope of this curve is the response factor
- Add known quantity of internal standard to unknown solution and measure signal from analyte/signal from standard. Use calibration curve to find concentration of analyte/concentration of standard in the unknown

Standard addition
- Method is based on addition of known quantities of analyte to solution of the unknown
- Standard addition is most useful when sample matrix is complex or unknown and effect of matrix on signal is not known
- Standard addition to a single solution of unknown
  - Measure analytical response to unknown
  - Make a series of small volume, known additions of concentrated analyte to the unknown. Measure analytical response after each addition. Total addition should increase analytical signal by a factor of 1.5 to 3
  - Prepare a graph of corrected analytical response versus known concentration of added standard. Corrected response = observed signal × (total volume/initial volume). Extrapolate line back to *x*-axis to find original concentration of analyte in unknown
- Multiple solutions with constant total volume
  - Place equal quantities of unknown solution into several volumetric flasks. Add an increasing, known amount of analyte to each flask. Dilute each flask to same final volume
  - Measure analytical signal for each solution
  - Prepare a graph of analytical signal versus concentration of added standard. Extrapolate line back to *x*-axis to find concentration of analyte in volumetric flask. From known dilution factor, compute concentration of analyte in original unknown

---

Terms to Understand

assessment	field blank	quality assurance	sensitivity
calibration check	internal standard	range	specifications
coefficient of variation	linearity	reagent blank	specificity
control chart	lower limit of quantitation	repeatability	spike
detection limit	matrix	reporting limit	standard addition
dilution factor	matrix effect	reproducibility	standard operating procedure
external standard	method blank	response factor	use objectives
false negative	method validation	robustness	
false positive	performance test sample	selectivity	

Summary

Quality assurance is what we do to get the right answer for our purpose. We begin by writing use objectives, from which specifications for data quality can be derived. Specifications could include requirements for sampling, accuracy, precision, specificity, detection limit, standards, and blank values. For any meaningful analysis, we must first collect a representative sample. A method blank containing all components except analyte is taken through all steps of the analytical procedure. We subtract the response of the method blank from the response of a real sample prior to calculating the quantity of analyte in the sample. A field blank tells us if analyte is inadvertently picked up by exposure to field conditions. Accuracy can be assessed by analyzing certified standards, by calibration checks performed by the analyst, with spikes made by the analyst, and by analyzing blind quality control samples. Written standard operating procedures must be followed rigorously to avoid inadvertent changes in procedure that could affect the outcome. Assessment is the process of (1) collecting data to show that analytical procedures are operating within specified limits and (2) verifying that final results meet use objectives. Control charts can be used to monitor accuracy, precision, or instrument performance as a function of time.

Method validation is the process of proving that an analytical method is acceptable for its intended purpose. In validating a method, we typically demonstrate that requirements are met for specificity, linearity, accuracy, precision, range, limit of detection, limit of quantitation, and robustness. Specificity is the ability to distinguish analyte from anything else. Linearity is usually measured by the square of the correlation coefficient for the calibration curve. Types of precision include instrument precision, intra-assay precision, intermediate precision, and, most generally, interlaboratory precision. The "Horwitz trumpet" is an empirical statement that precision becomes poorer as analyte concentration decreases.

Range is the concentration interval over which linearity, accuracy, and precision are acceptable. The detection limit is usually taken as 3 times the standard deviation of the blank. The lower limit of quantitation is 10 times the standard deviation of the blank. The reporting limit is the concentration below which regulations say that analyte is reported as "not detected," even when it is observed. Robustness is the ability of an analytical method to be unaffected by small changes in operating parameters.

A standard addition is a known quantity of analyte added to an unknown to increase the concentration of analyte. Standard additions are especially useful when matrix effects are important. A matrix effect is a change in the analytical signal caused by anything in the sample other than analyte. Use Equation 5-7 to compute the quantity of analyte after a single standard addition. For multiple standard additions to a single solution, use Equation 5-9 to construct the graph in Figure 5-6, in which the x-intercept gives us the concentration of analyte. For multiple solutions made up to the same final volume, the slightly different graph in Figure 5-7 is used. Equation 5-10 gives the x-intercept uncertainty in either graph.

An internal standard is a known amount of a compound, different from analyte, that is added to the unknown. Signal from analyte is compared with signal from the internal standard to find out how much analyte is present. Internal standards are useful when the quantity of sample analyzed is not reproducible, when instrument response varies from run to run, or when sample losses occur in sample preparation. The response factor is the relative response to analyte and standard. For the sake of accuracy, it is best to prepare a series of standard mixtures of analyte plus internal standard and prepare a graph such as Figure 5-10. The response factor is the slope of the graph and the intercept should be within statistical limits of zero. The graph should confirm linear response over the desired analytical range.

Exercises

5-A. *Detection limit.* In spectrophotometry, we measure the concentration of analyte by its absorbance of light. A low-concentration sample was prepared and nine replicate measurements gave absorbances of 0.004 7, 0.005 4, 0.006 2, 0.006 0, 0.004 6, 0.005 6, 0.005 2, 0.004 4, and 0.005 8. Nine reagent blanks gave values of 0.000 6, 0.001 2, 0.002 2, 0.000 5, 0.001 6, 0.000 8, 0.001 7, 0.001 0, and 0.001 1.

(a) Find the absorbance detection limit with Equation 5-3.

(b) The calibration curve is a graph of absorbance versus concentration. Absorbance is a dimensionless quantity. The slope of the calibration curve is $m = 2.24 \times 10^4 \ M^{-1}$. Find the concentration detection limit with Equation 5-5.

(c) Find the lower limit of quantitation with Equation 5-6.

5-B. *Standard addition.* An unknown sample of Ni^{2+} gave a current of 2.36 μA in an electrochemical analysis. When 0.500 mL of solution containing 0.028 7 M Ni^{2+} was added to 25.0 mL of unknown, the current increased to 3.79 μA.

(a) Denoting the initial, unknown concentration as $[Ni^{2+}]_i$, write an expression for the final concentration, $[Ni^{2+}]_f$, after 25.0 mL of unknown were mixed with 0.500 mL of standard. Use the dilution factor for this calculation.

(b) In a similar manner, write the final concentration of added standard Ni^{2+}, designated as $[S]_f$.

(c) Find $[Ni^{2+}]_i$ in the unknown.

5-C. In Figure 5-6, the x-intercept is -2.89 mM and its standard uncertainty is 0.09_8 mM. Find the 90% and 99% confidence intervals for the intercept.

5-D. *Internal standard.* A solution was prepared by mixing 5.00 mL of unknown element X with 2.00 mL of solution containing 4.13 μg of standard element S per milliliter, and diluting to 10.0 mL. The signal ratio in atomic absorption spectrometry was (signal from X)/(signal from S) = 0.808. In a separate experiment, with equal concentrations of X and S, (signal from X)/(signal from S) = 1.31. Find the concentration of X in the unknown.

5-E. 📖 *Internal standard graph.* Data are shown below for chromatographic analysis of naphthalene ($C_{10}H_8$), using deuterated naphthalene ($C_{10}D_8$, in which D is the isotope ^2H) as an internal standard. The two compounds emerge from the column at almost identical times and are measured by a mass spectrometer.

Sample	$C_{10}H_8$ (ppm)	$C_{10}D_8$ (ppm)	$C_{10}H_8$ peak area	$C_{10}D_8$ peak area
1	1.0	10.0	303	2 992
2	5.0	10.0	3 519	6 141
3	10.0	10.0	3 023	2 819

The volume of solution injected into the column was different in all three runs.

(a) Using a spreadsheet like Figure 4-15, prepare a graph of Equation 5-12 showing peak area ratio ($C_{10}H_8/C_{10}D_8$) versus concentration ratio ($[C_{10}H_8]/[C_{10}D_8]$). Find the least-squares slope and intercept and their standard uncertainties. What is the theoretical value of the intercept? Is the observed value of the intercept within experimental uncertainty of the theoretical value?

(b) Find the quotient $[C_{10}H_8]/[C_{10}D_8]$ for an unknown whose peak area ratio ($C_{10}H_8/C_{10}D_8$) is 0.652. Find the standard uncertainty u_x for the peak area ratio.

(c) Here is why we try not to use 3-point calibration curves. For $n = 3$ data points, there is $n - 2 = 1$ degree of freedom because 2 degrees of freedom are lost in computing the slope and intercept. Find the value of Student's t for 95% confidence and 1 degree of freedom. From the standard uncertainty in (b), compute the 95% confidence interval for the quotient $[C_{10}H_8]/[C_{10}D_8]$. What is the percent relative uncertainty in the quotient $[C_{10}H_8]/[C_{10}D_8]$? Why do we avoid 3-point calibration curves?

5-F. *Control chart.* Volatile compounds in human blood serum were measured by purge and trap gas chromatography/mass spectrometry. For quality control, serum was periodically spiked with a constant amount of 1,2-dichlorobenzene and the concentration (ng/g = ppb) was measured. Find the mean and standard deviation for the following spike data and prepare a control chart. State whether or not the observations (Obs.) meet each criterion for stability in a control chart.

Day	Obs. (ppb)	Day	Obs. (ppb)	Day	Obs. (ppb)	Day	Obs. (ppb)	Day	Obs. (ppb)
0	1.05	91	1.13	147	0.83	212	1.03	290	1.04
1	0.70	101	1.64	149	0.88	218	0.90	294	0.85
3	0.42	104	0.79	154	0.89	220	0.86	296	0.59
6	0.95	106	0.66	156	0.72	237	1.05	300	0.83
7	0.55	112	0.88	161	1.18	251	0.79	302	0.67
30	0.68	113	0.79	167	0.75	259	0.94	304	0.66
70	0.83	115	1.07	175	0.76	262	0.77	308	1.04
72	0.97	119	0.60	182	0.93	277	0.85	311	0.86
76	0.60	125	0.80	185	0.72	282	0.72	317	0.88
80	0.87	128	0.81	189	0.87	286	0.68	321	0.67
84	1.03	134	0.84	199	0.85	288	0.86	323	0.68

SOURCE: *Data from D. L. Ashley, M. A. Bonin, F. L. Cardinali, J. M. McCraw, J. S. Holler, L. L. Needham, and D. G. Patterson, Jr., "Determining Volatile Organic Compounds in Blood by Using Purge and Trap Gas Chromatography/Mass Spectrometry,"* Anal. Chem. **1992**, 64, 1021.

Problems

Quality Assurance and Method Validation

5-1. Explain the meaning of the quotation at the beginning of this chapter: "Get the right data. Get the data right. Keep the data right."

5-2. What are the three parts of quality assurance? What questions are asked in each part and what actions are taken in each part?

5-3. How can you validate precision and accuracy?

5-4. Distinguish *raw data*, *treated data*, and *results*.

5-5. What is the difference between a *calibration check* and a *performance test sample*?

5-6. What is a blank and what is its purpose? Distinguish *method blank*, *reagent blank,* and *field blank*.

5-7. Distinguish *linear range, dynamic range*, and *range*.

5-8. What is the difference between a *false positive* and a *false negative*?

5-9. Consider a sample that contains analyte at the detection limit defined in Figure 5-2. Explain the following statements: There is approximately a 1% chance of falsely concluding that a sample containing no analyte contains analyte above the detection limit. There is a 50% chance of concluding that a sample that really contains analyte at the detection limit does not contain analyte above the detection limit.

5-10. How is a control chart used? State six indications that a process is going out of control.

5-11. Here is a use objective for a chemical analysis to be performed at a drinking water purification plant: "Data and results collected quarterly shall be used to determine whether the concentrations of haloacetates in the treated water demonstrate compliance with the levels set by the Stage 1 Disinfection By-products Rule using Method 552.2" (a specification that sets precision, accuracy, and other requirements). Which one of the following questions best summarizes the meaning of the use objective?

(i) Are haloacetate concentrations known within specified precision and accuracy?

(ii) Are any haloacetates detectable in the water?

(iii) Do any haloacetate concentrations exceed the regulatory limit?

5-12. What is the difference between an instrument detection limit and a method detection limit? What is the difference between robustness and intermediate precision?

5-13. What is the difference between repeatability and reproducibility? Define the following terms: instrument precision, intra-assay precision, intermediate precision, and interlaboratory precision. Which type of precision is a synonym for reproducibility?

5-14. *Control chart.* A laboratory monitoring perchlorate (ClO_4^-) in urine measured quality control samples made from synthetic urine spiked with ClO_4^-. The graph in Box 5-2 shows consecutive quality control measurements. Are any troubleshooting conditions from Box 5-2 observed in these data?

5-15. *Correlation coefficient and Excel graphing.* Synthetic data are given below for a calibration curve in which random Gaussian noise with a standard deviation of 80 was superimposed on y values for the equation $y = 26.4x + 1.37$. This exercise shows that a high value of R^2 does not guarantee that data quality is excellent.

(a) Enter concentration in column A and signal in column B of a spreadsheet. Prepare an XY Scatter chart of signal versus concentration without a line as described in Section 2-11. Use LINEST (Section 4-7) to find the least-squares parameters including R^2.

(b) Now insert the Trendline by following instructions on page 88. In the Options window used to select the Trendline, select Display Equation and Display R-Squared. Verify that Trendline and LINEST give identical results.

(c) Add 95% confidence interval y error bars following the instructions at the end of Section 4-9. The 95% confidence interval is $\pm ts_y$, where s_y comes from LINEST and Student's t comes from Table 4-4 for 95% confidence and $11 - 2 = 9$ degrees of freedom. Also, compute t with the statement "=TINV(0.05,9)".

Concentration (x)	Signal (y)	Concentration (x)	Signal (y)
0	14	60	1 573
10	350	70	1 732
20	566	80	2 180
30	957	90	2 330
40	1 067	100	2 508
50	1 354		

5-16. In a murder trial in the 1990s, the defendant's blood was found at the crime scene. The prosecutor argued that blood was left by the defendant during the crime. The defense argued that police "planted" the defendant's blood from a sample collected later. Blood is normally collected in a vial containing the metal-binding compound EDTA (as an anticoagulant) at a concentration of ~4.5 mM after the vial is filled with blood. At the time of the trial, procedures to measure EDTA in blood were not well established. Even though the amount of EDTA found in the crime-scene blood was orders of magnitude below 4.5 mM, the jury acquitted the defendant. This trial motivated the development of a new method to measure EDTA in blood.

(a) *Precision and accuracy.* To measure accuracy and precision of the method, blood was fortified with EDTA to known levels.

$$\text{Accuracy} = 100 \times \frac{\text{mean value found} - \text{known value}}{\text{known value}}$$

$$\text{Precision} = 100 \times \frac{\text{standard deviation}}{\text{mean}} \equiv coefficient\ of\ variation$$

For each of the three spike levels in the table, find the precision and accuracy of the quality control samples.

EDTA measurements (ng/mL) at three fortification levels

Spike:	22.2 ng/mL	88.2 ng/mL	314 ng/mL
Found:	33.3	83.6	322
	19.5	69.0	305
	23.9	83.4	282
	20.8	100.0	329
	20.8	76.4	276

SOURCE: Data from R. L. Sheppard and J. Henion, Anal. Chem. **1997,** *69, 477A, 2901.*

(b) *Detection and quantitation limits.* Low concentrations of EDTA near the detection limit gave the following dimensionless instrument readings: 175, 104, 164, 193, 131, 189, 155, 133, 151, and 176. Ten blanks had a mean reading of $45._0$. The slope of the calibration curve is 1.75×10^9 M^{-1}. Estimate the signal and concentration detection limits and the lower limit of quantitation for EDTA.

5-17. (a) From Box 5-3, estimate the minimum expected coefficient of variation, CV(%), for interlaboratory results when the analyte concentration is (i) 1 wt% or (ii) 1 part per trillion.

(b) The coefficient of variation within a laboratory is typically ~0.5–0.7 of the between-laboratory variation. If your class analyzes an unknown containing 10 wt% NH_3, what is the minimum expected coefficient of variation for the class?

5-18. *Spike recovery and detection limit.* Species of arsenic found in drinking water include AsO_3^{3-} (arsenite), AsO_4^{3-} (arsenate), $(CH_3)_2AsO_2^-$ (dimethylarsinate), and $(CH_3)AsO_3^{2-}$ (methylarsonate). Pure water containing no arsenic was spiked with 0.40 μg arsenate/L. Seven replicate determinations gave 0.39, 0.40, 0.38, 0.41, 0.36, 0.35, and 0.39 μg/L.[15] Find the mean percent recovery of the spike and the concentration detection limit (μg/L).

5-19. *Detection limit.* A sensitive chromatographic method was developed to measure sub-part-per-billion levels of the disinfectant by-products iodate (IO_3^-), chlorite (ClO_2^-), and bromate (BrO_3^-) in drinking water. As the oxyhalides emerge from the column, they react with Br^- to make Br_3^-, which is measured by its strong absorption at 267 nm. For example, each mole of bromate makes 3 mol of Br_3^- by the reaction

$$BrO_3^- + 8Br^- + 6H^+ \rightarrow 3Br_3^- + 3H_2O$$

Bromate near its detection limit gave the following chromatographic peak heights and standard deviations (s). For each concentration, estimate the detection limit. Find the mean detection limit. The blank is 0 because chromatographic peak height is measured from the baseline adjacent to the peak. Because blank = 0, relative standard deviation applies to both peak height and concentration, which are proportional to each other. Detection limit is $3s$ for peak height or concentration.

Bromate concentration (μg/L)	Peak height (arbitrary units)	Relative standard deviation (%)	Number of measurements
0.2	17	14.4	8
0.5	31	6.8	7
1.0	56	3.2	7
2.0	111	1.9	7

SOURCE: Data from H. S. Weinberg and H. Yamada, "Post-Ion-Chromatography Derivatization for the Determination of Oxyhalides at Sub-PPB Levels in Drinking Water," Anal. Chem. **1998,** *70, 1.*

5-20. Olympic athletes are tested to see if they are using illegal performance-enhancing drugs. Suppose that urine samples are taken and analyzed and the rate of false positive results is 1%. Suppose also that it is too expensive to refine the method to reduce the rate of false positive results. We do not want to accuse innocent people of using illegal drugs. What can you do to reduce the rate of false accusations even though the test always has a false positive rate of 1%?

5-21. *Blind samples: interpreting statistics.* The U.S. Department of Agriculture provided homogenized beef baby food samples to three labs for analysis.[4] Results from the labs agreed well for protein, fat, zinc, riboflavin, and palmitic acid. Results for iron were questionable: Lab A, 1.59 ± 0.14 (13); Lab B, 1.65 ± 0.56 (8); Lab C, 2.68 ± 0.78 (3) mg/100 g. Uncertainty is the standard deviation with the number of replicate analyses in parentheses. Use two separate t tests to compare results from Lab C with those from Lab A and Lab

B at the 95% confidence level. Comment on the sensibility of the t test results and offer your own conclusions.

Standard Addition

5-22. Why is it desirable in the method of standard addition to add a small volume of concentrated standard rather than a large volume of dilute standard?

5-23. An unknown sample of Cu^{2+} gave an absorbance of 0.262 in an atomic absorption analysis. Then 1.00 mL of solution containing 100.0 ppm (= μg/mL) Cu^{2+} was mixed with 95.0 mL of unknown, and the mixture was diluted to 100.0 mL in a volumetric flask. The absorbance of the new solution was 0.500.

(a) Denoting the initial, unknown concentration as $[Cu^{2+}]_i$, write an expression for the final concentration, $[Cu^{2+}]_f$, after dilution. Units of concentration are ppm.

(b) In a similar manner, write the final concentration of added standard Cu^{2+}, designated as $[S]_f$.

(c) Find $[Cu^{2+}]_i$ in the unknown.

5-24. *Standard addition graph.* Tooth enamel consists mainly of the mineral calcium hydroxyapatite, $Ca_{10}(PO_4)_6(OH)_2$. Trace elements in teeth of archeological specimens provide anthropologists with clues about diet and diseases of ancient people. Students at Hamline University used atomic absorption spectroscopy to measure strontium in enamel from extracted wisdom teeth. Solutions were prepared with a constant total volume of 10.0 mL containing 0.750 mg of dissolved tooth enamel plus variable concentrations of added Sr.

Added Sr (ng/mL = ppb)	Signal (arbitrary units)
0	28.0
2.50	34.3
5.00	42.8
7.50	51.5
10.00	58.6

SOURCE: Data from V. J. Porter, P. M. Sanft, J. C. Dempich, D. D. Dettmer, A. E. Erickson, N. A. Dubauskie, S. T. Myster, E. H. Matts, and E. T. Smith, "Elemental Analysis of Wisdom Teeth by Atomic Spectroscopy Using Standard Addition," J. Chem. Ed. 2002, 79, 1114.

(a) Find the concentration of Sr and its uncertainty in the 10-mL sample solution in parts per billion = ng/mL.

(b) Find the concentration of Sr in tooth enamel in parts per million = μg/g.

(c) If the standard addition intercept is the major source of uncertainty, find the uncertainty in the concentration of Sr in tooth enamel in parts per million.

(d) Find the 95% confidence interval for Sr in tooth enamel.

5-25. Europium is a lanthanide element found at parts per billion levels in natural waters. It can be measured from the intensity of orange light emitted when a solution is illuminated with ultraviolet radiation. Certain organic compounds that bind Eu(III) are required to enhance the emission. The figure shows standard addition experiments in which 10.00 mL of sample and 20.00 mL containing a large excess of organic additive were placed in 50-mL volumetric flasks. Then Eu(III) standards (0, 5.00, 10.00, or 15.00 mL) were added and the flasks were diluted to 50.0 mL with H_2O. Standards added to tap water contained 0.152 ng/mL

(ppb) of Eu(III), but those added to pond water were 100 times more concentrated (15.2 ng/mL).

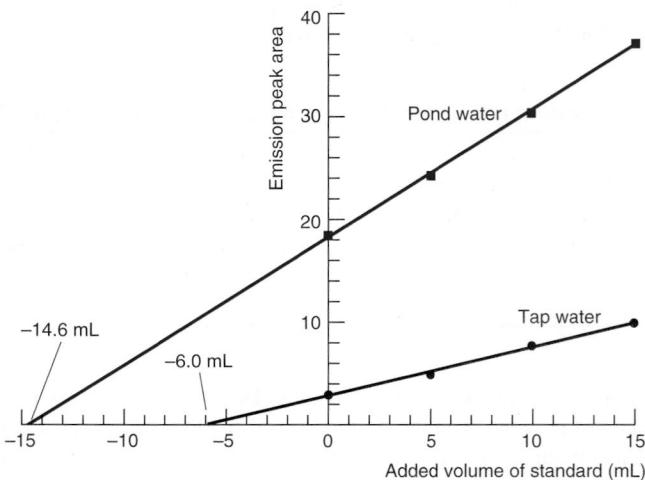

Standard addition of Eu(III) to pond water or tap water. [Data from A. L. Jenkins and G. M. Murray, "Enhanced Luminescence of Lanthanides," *J. Chem. Ed.* **1998**, *75*, 227.]

(a) Calculate the concentration of Eu(III) (ng/mL) in pond water and tap water.

(b) For tap water, emission peak area increases by 4.61 units when 10.00 mL of 0.152 ng/mL standard are added. This response is 4.61 units/1.52 ng = 3.03 units per ng of Eu(III). For pond water, the response is 12.5 units when 10.00 mL of 15.2 ng/mL standard are added, or 0.082 2 units per ng. How would you explain these observations? Why was standard addition necessary for this analysis?

5-26. 🖿 *Standard addition graph.* Students performed an experiment like that in Figure 5-7 in which each flask contained 25.00 mL of serum, varying additions of 2.640 M NaCl standard, and a total volume of 50.00 mL.

Flask	Volume of standard (mL)	Na^+ atomic emission signal (mV)
1	0	3.13
2	1.000	5.40
3	2.000	7.89
4	3.000	10.30
5	4.000	12.48

(a) Prepare a standard addition graph and find $[Na^+]$ in the serum.

(b) Find the standard deviation and 95% confidence interval for $[Na^+]$.

5-27. 🖿 *Standard addition graph.* Allicin is a ~0.4 wt% component in garlic with antimicrobial and possibly anticancer and antioxidant activity. It is unstable and therefore difficult to measure. An assay was developed in which the stable precursor alliin is added to freshly crushed garlic and converted to allicin by the enzyme alliinase found in garlic. Components of the garlic are extracted and measured by chromatography. The chromatogram shows standard additions reported as mg alliin added per gram of garlic. The chromatographic peak is allicin from the conversion of alliin.

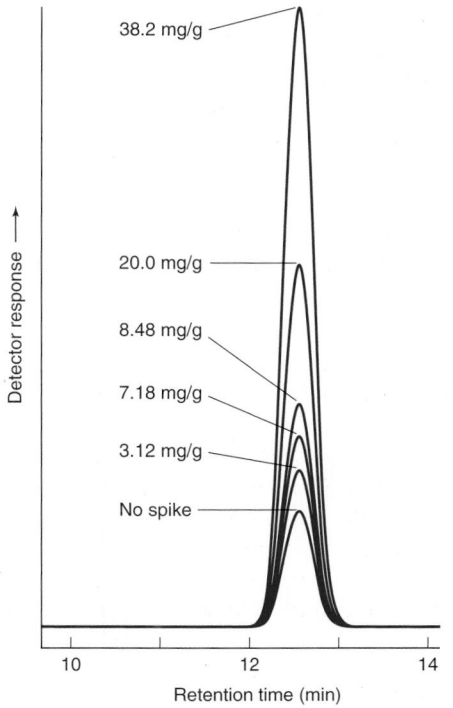

(a) The standard addition procedure has a constant total volume. Measure the responses in the figure and prepare a graph to find how much alliin equivalent was in the unspiked garlic. The units of your answer will be mg alliin/g garlic. Find the 95% confidence interval, as well.

Chromatographic measurement of alliin after standard addition to garlic.
[Data from M. E. Rybak, E. M. Calvey, and J. M. Harnly, "Quantitative Determination of Allicin in Garlic: Supercritical Fluid Extraction and Standard Addition of Alliin," *J. Agric. Food Chem.* **2004,** *52,* 682.]

(b) Given that 2 mol of alliin are converted to 1 mol of allicin, find the allicin content of garlic (mg allicin/g garlic) including the 95% confidence interval.

5-28. 🖥 *Standard addition.* Lead in dry river sediment was extracted with 25 wt% HNO_3 at 35°C for 1 h. Then 1.00 mL of filtered extract was mixed with other reagents to bring the total volume to $V_0 = 4.60$ mL. Pb(II) was measured electrochemically with a series of standard additions of 2.50 ppm Pb(II).

Pb(II) added (mL)	Signal (arbitrary units)
0	1.10
0.025	1.66
0.050	2.20
0.075	2.81

SOURCE: Data from M. J. Goldcamp, M. N. Underwood, J. L. Cloud, S. Harshman, and K. Ashley, "An Environmentally Friendly, Cost-Effective Determination of Lead in Environmental Samples Using Anodic Stripping Voltammetry," J. Chem. Ed. **2008,** 85, 976.

(a) Volume is not constant, so follow the procedure of Figures 5-5 and 5-6 to find ppm Pb(II) in the 1.00-mL extract.

(b) Find the standard uncertainty and 95% confidence interval for the x-intercept of the graph. Assuming that uncertainty in intercept is larger than other uncertainties, estimate the uncertainty in ppm Pb(II) in the 1.00-mL extract.

Internal Standards

5-29. State when standard additions and internal standards, instead of a calibration curve, are desirable, and why.

5-30. A solution containing 3.47 mM X (analyte) and 1.72 mM S (standard) gave peak areas of 3 473 and 10 222, respectively, in a chromatographic analysis. Then 1.00 mL of 8.47 mM S was added to 5.00 mL of unknown X, and the mixture was diluted to 10.0 mL. This solution gave peak areas of 5 428 and 4 431 for X and S, respectively.

(a) Calculate the response factor for the analyte.

(b) Find the concentration of S (mM) in the 10.0-mL mixture.

(c) Find the concentration of X (mM) in the 10.0-mL mixture.

(d) Find the concentration of X in the original unknown.

5-31. Chloroform is an internal standard in the determination of the pesticide DDT in a polarographic analysis in which each compound is reduced at an electrode surface. A mixture containing 0.500 mM chloroform and 0.800 mM DDT gave signals of 15.3 μA for chloroform and 10.1 μA for DDT. An unknown solution (10.0 mL) containing DDT was placed in a 100-mL volumetric flask and 10.2 μL of chloroform (FM 119.39, density = 1.484 g/mL) were added. After dilution to the mark with solvent, polarographic signals of 29.4 and 8.7 μA were observed for the chloroform and DDT, respectively. Find the concentration of DDT in the unknown.

5-32. 🖥 *Internal standard calibration curve.* Figure 5-10 is a graph of A_X/A_S versus [X]/[S] = (mol% vinyl acetate units)/(mol% ethylene units) = q/p in Reaction 5-13.

(a) Construct the graph in Figure 5-10 from the following data:

[X]/[S]	A_X/A_S	[X]/[S]	A_X/A_S
0.099	0.291	0.333	1.235
0.163	0.656	0.493	1.808
0.220	0.800	0.667	2.284

(b) Use a least-squares spreadsheet such as that in Figure 4-15 to find the slope and intercept and uncertainties s_y, u_m, and u_b of the calibration line in (a).

(c) From the equation of the line, find [X]/[S] for a measured value $A_X/A_S = 1.98$. Use Equation 4-27 in your spreadsheet to find the standard uncertainty u_x in the quotient [X]/[S]. The graph has 6 points with 4 degrees of freedom. Find the 95% confidence interval ($= \pm tu_x$) for [X]/[S].

(d) From the uncertainty u_b of the intercept, find the 95% confidence interval for the intercept. Does this interval include the theoretical value of zero?

5-33. *Correcting for matrix effects with an internal standard.* The appearance of pharmaceuticals in municipal wastewater (sewage) is an increasing problem that is likely to have adverse effects on our drinking water supply. Sewage is a complex matrix. The drug carbamazepine can be measured at low levels in sewage by liquid chromatography with mass spectral detection. When carbamazepine was spiked into sewage at a concentration of 5 ppb, chromatographic analysis gave an apparent spike recovery of 154%.[16] Deuterium (D) is the hydrogen isotope ^2H. Deuterated carbamazepine can be used as an internal standard for carbamazepine. The deuterated compound has the same retention time as undeuterated material in chromatography, but is distinguished by its higher mass in the mass spectrum. When deuterated carbamazepine was used as an internal standard for the analysis, the apparent recovery was 98%. Explain how the internal standard is used in this analysis and rationalize why it works so well to correct for matrix effects.

Carbamazepine   Carbamazepine-d_{10}

6 | Chemical Equilibrium

Paper mill on the Potomac River near Westernport, Maryland, neutralizes acid mine drainage in the water. Upstream of the mill, the river is acidic and lifeless; below the mill, the river teems with life. [Courtesy of Interstate Commission on the Potomac River Basin.]

Great Barrier Reef and other coral reefs are threatened with extinction by rising atmospheric CO_2. [© Sunpix Travel/Alamy.]

Part of the North Branch of the Potomac River runs crystal clear through the scenic Appalachian Mountains, but it is lifeless—a victim of acid drainage from abandoned coal mines. As the river passes a paper mill and a wastewater treatment plant near Westernport, Maryland, the pH rises from an acidic, lethal value of 4.5 to a neutral value of 7.2, at which fish and plants thrive. This happy "accident" comes about because calcium carbonate exiting the paper mill equilibrates with massive quantities of carbon dioxide from bacterial respiration at the sewage treatment plant. The resulting soluble bicarbonate neutralizes the acidic river and restores life downstream of the plant.[1] In the absence of CO_2, solid $CaCO_3$ would be trapped at the treatment plant and would never enter the river.

$$CaCO_3(s) + CO_2(aq) + H_2O(l) \rightleftharpoons Ca^{2+}(aq) + 2HCO_3^-(aq)$$

Calcium carbonate   Dissolved calcium bicarbonate
trapped at treatment plant   enters river and neutralizes acid

$$HCO_3^-(aq) + H^+(aq) \xrightarrow{\text{neutralization}} CO_2(g) + H_2O(l)$$

Bicarbonate   Acid in river

The chemistry that helps the Potomac River also endangers coral reefs, which are largely $CaCO_3$. Burning of fossil fuel has increased CO_2 in the atmosphere from 280 ppm when Captain Cook first sighted the Great Barrier Reef in 1770 to 400 ppm today (page 211). CO_2 in the atmosphere adds CO_2 to the ocean, dissolving $CaCO_3$ from coral. Rising CO_2 and rising atmospheric temperature from the greenhouse effect threaten coral reefs with extinction.[2] CO_2 has lowered the average pH of the ocean from its preindustrial value of 8.16 to 8.04 today.[3] Without changes in our activities, the pH could be 7.7 by 2100.

hemical equilibrium provides a foundation not only for chemical analysis, but also for other subjects such as biochemistry, geology, and oceanography. This chapter introduces equilibria for the solubility of ionic compounds, complex formation, and acid-base reactions.

6-1 The Equilibrium Constant

For the reaction

$$aA + bB \rightleftharpoons cC + dD \tag{6-1}$$

we write the **equilibrium constant,** K, in the form

Equilibrium constant:
$$K = \frac{[C]^c[D]^d}{[A]^a[B]^b} \tag{6-2}$$

where the lowercase superscript letters denote stoichiometry coefficients and each capital letter stands for a chemical species. The symbol [A] stands for the concentration of A relative to its standard state (defined next). By definition, *a reaction is favored whenever* $K > 1$.

In the thermodynamic derivation of the equilibrium constant, each quantity in Equation 6-2 is expressed as the *ratio* of the concentration of a species to its concentration in its **standard state.** For solutes, the standard state is 1 M. For gases, the standard state is 1 bar ($\equiv 10^5$ Pa; 1 atm $\equiv$ 1.013 25 bar), and for solids and liquids, the standard states are the pure solid or liquid. It is understood that [A] in Equation 6-2 really means [A]/(1 M) if A is a solute. If D is a gas, [D] really means (pressure of D in bars)/(1 bar). To emphasize that [D] means pressure of D, we usually write P_D in place of [D]. The terms in Equation 6-2 are actually dimensionless; therefore, all equilibrium constants are dimensionless.

For the ratios [A]/(1 M) and [D]/(1 bar) to be dimensionless, [A] *must* be expressed in moles per liter (M), and [D] *must* be expressed in bars. If C were a pure liquid or solid, the ratio [C]/(concentration of C in its standard state) would be unity (1) because the standard state is the pure liquid or solid. If C is a solvent, the concentration is so close to that of pure liquid C that the value of [C] is still essentially 1.

The take-home lesson is this: When you evaluate an equilibrium constant:

1. Concentrations of solutes should be expressed as moles per liter.
2. Concentrations of gases should be expressed in bars.
3. Concentrations of pure solids, pure liquids, and solvents are omitted because they are unity.

These conventions are arbitrary, but you must use them if you wish to use tabulated values of equilibrium constants, standard reduction potentials, and free energies.

Manipulating Equilibrium Constants

Consider the reaction

$$HA \rightleftharpoons H^+ + A^- \qquad K_1 = \frac{[H^+][A^-]}{[HA]}$$

If the direction of a reaction is reversed, the new value of K is simply the reciprocal of the original value of K.

Equilibrium constant for reverse reaction:
$$H^+ + A^- \rightleftharpoons HA \qquad K_1' = \frac{[HA]}{[H^+][A^-]} = 1/K_1$$

If two reactions are added, the new K is the product of the two individual values:

$$\begin{array}{lll} HA & \rightleftharpoons \cancel{[H^+]} + A^- & K_1 \\ \cancel{[H^+]} + C & \rightleftharpoons CH^+ & K_2 \\ \hline HA + C & \rightleftharpoons A^- + CH^+ & K_3 \end{array}$$

Equilibrium constant for sum of reactions:
$$K_3 = K_1 K_2 = \frac{\cancel{[H^+]}[A^-]}{[HA]} \cdot \frac{[CH^+]}{\cancel{[H^+]}[C]} = \frac{[A^-][CH^+]}{[HA][C]}$$

If n reactions are added, the overall equilibrium constant is the product of n individual equilibrium constants.

---

*Equation 6-2, the **law of mass action,** was formulated by the Norwegians C. M. Guldenberg and P. Waage in 1864. Their derivation was based on the idea that the forward and reverse rates of a reaction at equilibrium must be equal.[4]*

The equilibrium constant is more correctly expressed as a ratio of activities *rather than of concentrations. We reserve the discussion of activity for Chapter 8.*

Equilibrium constants are dimensionless.

Equilibrium constants are dimensionless but, when specifying concentrations, you must use units of molarity (M) for solutes and bars for gases.

Throughout this book, assume that all species in chemical equations are in aqueous solution, unless otherwise specified.

If a reaction is reversed, then $K' = 1/K$. If two reactions are added, then $K_3 = K_1 K_2$.

EXAMPLE **Combining Equilibrium Constants**

The equilibrium constant for the reaction $H_2O \rightleftharpoons H^+ + OH^-$ is called K_w ($= [H^+][OH^-]$) and has the value 1.0×10^{-14} at 25°C. Given that $K_{NH_3} = 1.8 \times 10^{-5}$ for the reaction $NH_3(aq) + H_2O \rightleftharpoons NH_4^+ + OH^-$, find K for the reaction $NH_4^+ \rightleftharpoons NH_3(aq) + H^+$.

Solution The third reaction can be obtained by reversing the second reaction and adding it to the first reaction:

$$H_2O \rightleftharpoons H^+ + OH^- \qquad K = K_w$$
$$\underline{NH_4^+ + OH^- \rightleftharpoons NH_3(aq) + H_2O} \qquad K = 1/K_{NH_3}$$
$$NH_4^+ \rightleftharpoons H^+ + NH_3(aq) \qquad K = K_w \cdot \frac{1}{K_{NH_3}} = 5.6 \times 10^{-10}$$

TEST YOURSELF For the reaction $Li^+ + H_2O \rightleftharpoons Li(OH)(aq) + H^+$, $K_{Li} = 2.3 \times 10^{-14}$. Combine this reaction with the K_w reaction to find the equilibrium constant for the reaction $Li^+ + OH^- \rightleftharpoons Li(OH)(aq)$. (**Answer:** 2.3)

6-2 Equilibrium and Thermodynamics

Equilibrium is controlled by the thermodynamics of a chemical reaction. The heat absorbed or released (*enthalpy*) and the dispersal of energy into molecular motions (*entropy*) independently contribute to the degree to which the reaction is favored or disfavored.

Enthalpy

The **enthalpy change**, ΔH, for a reaction is the heat absorbed or released when the reaction takes place under constant applied pressure.[5] The *standard enthalpy change*, $\Delta H°$, refers to the heat absorbed when all reactants and products are in their standard states:[*]

$$HCl(g) \rightleftharpoons H^+(aq) + Cl^-(aq) \qquad \Delta H° = -74.85 \text{ kJ/mol at } 25°C \qquad \textbf{(6-3)}$$

$\Delta H = (+)$
Heat is absorbed
Endothermic

$\Delta H = (-)$
Heat is liberated
Exothermic

The negative sign of $\Delta H°$ indicates that Reaction 6-3 releases heat—the solution becomes warmer. For other reactions, ΔH is positive, which means that heat is absorbed. Consequently, the solution gets colder during the reaction. A reaction for which ΔH is positive is said to be **endothermic.** Whenever ΔH is negative, the reaction is **exothermic.**

Entropy

When a chemical or physical change occurs in a reversible manner[†] at a constant temperature, the **entropy change**, ΔS, is equal to the heat absorbed (q_{rev}) divided by the temperature (T):

$$\Delta S = \frac{q_{rev}}{T} \qquad \textbf{(6-4)}$$

Positive q means that heat is absorbed by the system

Negative q means that heat is removed from the system

Consider a closed vessel at equilibrium containing liquid water, solid ice, and water vapor at 273.16 K. If a little heat enters the vessel from warmer fluid surrounding the vessel, a little ice melts and the temperature remains at 273.16 K. The heat absorbed breaks some hydrogen bonds between adjacent water molecules in the crystal and increases translational, rotational, and vibrational kinetic energy of the molecules that change from solid to liquid. (*Translation* means motion of the whole molecule through space.) The change in entropy of the contents of the vessel, given by Equation 6-4, is equal to the heat absorbed divided by the temperature. If the heat absorbed is 0.10 J, the change in entropy is $\Delta S = q_{rev}/T = (0.10 \text{ J})/(273.16 \text{ K}) = 0.000\ 37$ J/K. The entropy change is the amount of energy, at a given

---

[*]The definition of the standard state contains subtleties beyond the scope of this book. For Reaction 6-3, the standard state of H^+ or Cl^- is the hypothetical state in which each ion is present at a concentration of 1 M but behaves as if it were in an infinitely dilute solution. That is, the standard concentration is 1 M, but the standard behavior is what would be observed in a very dilute solution in which each ion is unaffected by surrounding ions.

[†]"Reversible" means that a small physical or chemical change can be reversed by a small external action. For example, a small addition of heat causes a small amount of melting. An equally small removal of heat freezes the same small amount of liquid. An example of an irreversible change is the explosion of a mixture of $H_2(g) + O_2(g)$ to make $H_2O(l)$ initiated by a spark in a closed vessel. No small external action can make the water dissociate back to $H_2(g) + O_2(g)$.

temperature, that is dispersed into motions of the molecules in the system.[*] For an irreversible change, ΔS can be found from a reversible path between the same initial and final states. The initial and final entropy only depend on the state of the system, and not how it got from one state to the other.

In principle, the *standard entropy*, $S°$, of a mole of substance can be measured by slowly heating it from absolute zero temperature (at which it has no entropy). For each small addition of heat q_{rev} at temperature T, a small change ΔT is produced. For each small step in this process, the entropy change q_{rev}/T is computed. The sum of these small entropy changes required to heat the substance from 0 to 298.15 K at 1 bar pressure is called the standard entropy, $S°$, of the substance.

A liquid has higher entropy than the same solid material because heat is required to separate molecules that are attracted to each other in the solid and to increase the kinetic energy of translation, rotation, and vibration of molecules in the liquid. A gas has higher entropy than a liquid because heat is required to separate molecules that are attracted to each other in the liquid and to increase the kinetic energy of translation, rotation, and vibration of molecules in the gas.

Ions in aqueous solution usually have greater entropy than their solid salt:

$$KCl(s) \rightleftharpoons K^+(aq) + Cl^-(aq) \qquad \Delta S° = +76.4 \text{ J}/(K \cdot mol) \text{ at } 25°C \qquad \text{(6-5)}$$

$\Delta S°$ is the change in entropy (entropy of products minus entropy of reactants) when all species are in their standard states. The positive value of $\Delta S°$ indicates that a mole of $K^+(aq)$ plus a mole of $Cl^-(aq)$ has more energy dispersed into translation and rotation and vibration of species in solution than there is energy dispersed into translation and rotation and vibration of ions in a crystal of $KCl(s)$. For Reaction 6-3, $HCl(g) \rightleftharpoons H^+(aq) + Cl^-(aq)$, $\Delta S°$ is *negative* $[-130.4 \text{ J}/(K \cdot mol)]$ at 25°C. The aqueous ions have *less* energy dispersed among translations and rotations and vibrations than does gaseous HCl.

Free Energy

Systems at constant temperature and pressure, which are common laboratory conditions, have a tendency toward lower enthalpy and higher entropy. A chemical reaction is driven toward the formation of products by a *negative* value of ΔH (heat given off) or a *positive* value of ΔS or both. When ΔH is negative and ΔS is positive, the reaction is clearly favored. When ΔH is positive and ΔS is negative, the reaction is clearly disfavored.

When ΔH and ΔS are both positive or both negative, what decides whether a reaction will be favored? The change in **Gibbs free energy,** ΔG, is the arbiter between opposing tendencies of ΔH and ΔS. At constant temperature, T,

Free energy:
$$\Delta G = \Delta H - T\Delta S \qquad \text{(6-6)}$$

A reaction is favored if ΔG is negative.

For the dissolution of $HCl(g)$ and dissociation into ions (Reaction 6-3), when all species are in their standard states, $\Delta H°$ favors the reaction and $\Delta S°$ disfavors it. To find the net result, we evaluate $\Delta G°$:

$$\Delta G° = \Delta H° - T\Delta S°$$
$$= (-74.85 \times 10^3 \text{ J/mol}) - (298.15 \text{ K})(-130.4 \text{ J/K} \cdot mol)$$
$$= -35.97 \text{ kJ/mol}$$

$\Delta G°$ is negative, so the reaction is favored when all species are in their standard states. The favorable influence of $\Delta H°$ is greater than the unfavorable influence of $\Delta S°$ in this case. To reach equilibrium, the reaction begins moving from its initial condition in the direction of negative ΔG until it reaches the minimum free energy of the system, at which point equilibrium has been reached.[6]

---

[*]Another definition of entropy change, which is equivalent to Equation 6-4 (but not obviously so), is the change in the amount of information needed to specify the distribution of positions, velocities, rotations, and vibrations of a collection of molecules under specified conditions. The more information required, the greater the entropy. Several highly readable and enjoyable books explain the relation of information and entropy: A. Ben-Naim, *Entropy and the Second Law: Interpretation and Misss-Interpretationsss* (Singapore: World Scientific, 2010); A. Ben-Naim, *Discover Entropy and the Second Law of Thermodynamics* (Singapore: World Scientific, 2010); A. Ben-Naim, *Entropy Demystified: The Second Law Reduced to Plain Common Sense* (Singapore: World Scientific, 2008).

Margin notes:

Note that 298.15 K = 25.00°C.

$\Delta S = (+)$
Products have greater entropy than reactants
$\Delta S = (-)$
Products have lower entropy than reactants

The point of discussing free energy is to relate the equilibrium constant to the energetics ($\Delta H°$ and $\Delta S°$) of a reaction. The equilibrium constant depends on $\Delta G°$ in the following manner:

Free energy and equilibrium: $$K = e^{-\Delta G°/RT}$$ (6-7)

Challenge Satisfy yourself that $K > 1$ if $\Delta G°$ is negative.

where R is the gas constant [$= 8.314\,5$ J/(K · mol)] and T is temperature (Kelvin). The more negative the value of $\Delta G°$, the larger is the equilibrium constant. For Reaction 6-3,

$$K = e^{-(-35.97 \times 10^3 \text{ J/mol})/[8.314\,5 \text{ J/(K·mol)}](298.15 \text{ K})} = 2.00 \times 10^6$$

Because the equilibrium constant is large, $HCl(g)$ is very soluble in water and is nearly completely ionized to H^+ and Cl^- when it dissolves.

To summarize, a chemical reaction is favored by the liberation of heat (ΔH negative) and by an increase in entropy (ΔS positive). ΔG takes both effects into account to determine whether or not a reaction is favorable. We say that a reaction is *spontaneous* under standard conditions if $\Delta G°$ is negative or, equivalently, if $K > 1$. The reaction is not spontaneous if $\Delta G°$ is positive ($K < 1$). You should be able to calculate K from $\Delta G°$ and vice versa.

$\Delta G = (+)$ $K < 1$
Reaction is disfavored

$\Delta G = (-)$ $K > 1$
Reaction is favored

Le Châtelier's Principle

Suppose that a system at equilibrium is subjected to a change that disturbs the system. **Le Châtelier's principle** states that the direction in which the system proceeds back to equilibrium is such that the change is partially offset.

To see what this statement means, let's see what happens when we change the concentration of one species of the following reaction in water:

$$BrO_3^- + 2Cr^{3+} + 4H_2O \rightleftharpoons Br^- + Cr_2O_7^{2-} + 8H^+$$ (6-8)
Bromate   Chromium(III)   Dichromate

$$K = \frac{[Br^-][Cr_2O_7^{2-}][H^+]^8}{[BrO_3^-][Cr^{3+}]^2} = 1 \times 10^{11} \text{ at } 25°C$$

Notice that H_2O is omitted from K because it is the solvent.

In one particular equilibrium state of this system, the concentrations are $[H^+] = 5.0$ M, $[Cr_2O_7^{2-}] = 0.10$ M, $[Cr^{3+}] = 0.003\,0$ M, $[Br^-] = 1.0$ M, and $[BrO_3^-] = 0.043$ M. Suppose that the equilibrium is disturbed by adding dichromate to the solution to increase $[Cr_2O_7^{2-}]$ from 0.10 to 0.20 M. In what direction will the reaction proceed to reach equilibrium?

According to the principle of Le Châtelier, the reaction should go back to the left to partially offset the increase in dichromate, which appears on the right side of Reaction 6-8. We can verify this algebraically by setting up the **reaction quotient,** Q, which has the same form as the equilibrium constant. The only difference is that Q is evaluated with whatever concentrations happen to exist, even though the solution is not at equilibrium. When the system reaches equilibrium, $Q = K$. For Reaction 6-8,

The reaction quotient has the same form as the equilibrium constant, but the concentrations are generally not the equilibrium concentrations.

$$Q = \frac{(1.0)(0.20)(5.0)^8}{(0.043)(0.003\,0)^2} = 2 \times 10^{11} > K$$

Because $Q > K$, the reaction must go to the left to decrease the numerator and increase the denominator, until $Q = K$.

If $Q < K$, then the reaction must proceed to the right to reach equilibrium. If $Q > K$, then the reaction must proceed to the left to reach equilibrium.

1. If a reaction is at equilibrium and products are added (or reactants are removed), the reaction goes to the left.
2. If a reaction is at equilibrium and reactants are added (or products are removed), the reaction goes to the right.

When the temperature of a system is changed, so is the equilibrium constant. Equations 6-6 and 6-7 can be combined to predict the effect of temperature on K:

$$K = e^{-\Delta G°/RT} = e^{-(\Delta H° - T\Delta S°)/RT} = e^{(-\Delta H°/RT + \Delta S°/R)}$$

$$= e^{-\Delta H°/RT} \cdot e^{\Delta S°/R}$$ (6-9)

$e^{(a+b)} = e^a \cdot e^b$

The term $e^{\Delta S°/R}$ is independent of T (at least over a limited temperature range in which $\Delta S°$ is constant). The term $e^{-\Delta H°/RT}$ increases with increasing temperature if $\Delta H°$ is positive and decreases if $\Delta H°$ is negative. Therefore:

1. The equilibrium constant of an endothermic reaction ($\Delta H° = +$) increases if the temperature is raised.

2. The equilibrium constant of an exothermic reaction ($\Delta H° = -$) decreases if the temperature is raised.

These statements can be understood in terms of Le Châtelier's principle as follows. Consider an endothermic reaction:

$$\text{Heat + reactants} \rightleftharpoons \text{products}$$

If the temperature is raised, then heat is added to the system. The reaction proceeds to the right to partially offset this change.[7]

In dealing with equilibrium problems, we are making *thermodynamic* predictions, not *kinetic* predictions. We are calculating what must happen for a system to reach equilibrium, but not how long it will take. Some reactions are over in an instant; others will not reach equilibrium in a million years. For example, dynamite remains unchanged indefinitely, until a spark sets off the spontaneous, explosive decomposition. The size of an equilibrium constant tells us nothing about the rate (the kinetics) of the reaction. A large equilibrium constant does not imply that a reaction is fast.

> Heat can be treated as if it were a reactant in an endothermic reaction and a product in an exothermic reaction.

6-3 Solubility Product

In chemical analysis, we encounter solubility in precipitation titrations, electrochemical reference cells, and gravimetric analysis. The effect of acid on the solubility of minerals and the effect of atmospheric CO_2 on the solubility (and death) of coral reefs and plankton are important in environmental science.

The **solubility product** is the equilibrium constant for the reaction in which a solid salt dissolves to give its constituent ions in solution. Solid is omitted from the equilibrium constant because it is in its standard state. Appendix F lists solubility products.

As an example, consider the dissolution of mercury(I) chloride (Hg_2Cl_2, also called mercurous chloride) in water. The reaction is

$$Hg_2Cl_2(s) \rightleftharpoons Hg_2^{2+} + 2Cl^- \tag{6-10}$$

for which the solubility product, K_{sp}, is

$$K_{sp} = [Hg_2^{2+}][Cl^-]^2 = 1.2 \times 10^{-18} \tag{6-11}$$

A **saturated solution** is one that contains excess, undissolved solid. The solution contains all the solid capable of being dissolved under the prevailing conditions.

The physical meaning of the solubility product is this: If an aqueous solution is left in contact with excess solid Hg_2Cl_2, the solid will dissolve until the condition $[Hg_2^{2+}][Cl^-]^2 = K_{sp}$ is satisfied. Thereafter, the amount of undissolved solid remains constant. Unless excess solid remains, there is no guarantee that $[Hg_2^{2+}][Cl^-]^2 = K_{sp}$. If Hg_2^{2+} and Cl^- are mixed together (with appropriate counterions) such that the product $[Hg_2^{2+}][Cl^-]^2$ exceeds K_{sp}, then Hg_2Cl_2 will precipitate.

We most commonly use the solubility product to find the concentration of one ion when the concentration of the other is known or fixed by some means. For example, what is the concentration of Hg_2^{2+} in equilibrium with 0.10 M Cl^- in a solution of KCl containing excess, undissolved $Hg_2Cl_2(s)$? To answer this question, we rearrange Equation 6-11 to find

$$[Hg_2^{2+}] = \frac{K_{sp}}{[Cl^-]^2} = \frac{1.2 \times 10^{-18}}{0.10^2} = 1.2 \times 10^{-16} \text{ M}$$

Because Hg_2Cl_2 is so slightly soluble, additional Cl^- obtained from Hg_2Cl_2 is negligible compared with 0.10 M Cl^-.

The solubility product does not tell the entire story of solubility. In addition to complications described in Box 6-1, most salts form soluble *ion pairs* to some extent. That is, MX(s) can give MX(aq) as well as $M^+(aq)$ and $X^-(aq)$. In a saturated solution of $CaSO_4$, for example, two-thirds of the dissolved calcium is Ca^{2+} and one third is $CaSO_4(aq)$.[8] The $CaSO_4(aq)$ **ion pair** is a closely associated pair of ions that behaves as one species in solution. Appendix J and Box 8-1 provide information on ion pairs.[9]

Mercurous ion, Hg_2^{2+}, is a *dimer* (pronounced DIE mer), which means that it consists of two identical units bound together:

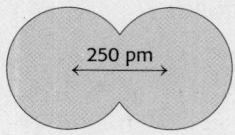

[Hg—Hg]$^{2+}$
+1 oxidation state of mercury

OH^-, S^{2-}, and CN^- stabilize Hg(II), thereby converting Hg(I) into Hg(0) and Hg(II):

$$Hg_2^{2+} + 2CN^- \rightarrow Hg(CN)_2(aq) + Hg(l)$$
Hg(I)   Hg(II)   Hg(0)

Disproportionation is the process in which an element in an intermediate oxidation state gives products in both higher and lower oxidation states.

> A *salt* is any ionic solid, such as Hg_2Cl_2 or $CaSO_4$.

If we want to know how much Hg_2^{2+} is dissolved in a saturated solution of Hg_2Cl_2, we are tempted to look at Reaction 6-10 and to note that two Cl^- are created for each Hg_2^{2+}. If we let x be the concentration of Hg_2^{2+}, we could say that the concentration of Cl^- is $2x$. Substituting these values into the solubility product 6-11, we could write $K_{sp} = [Hg_2^{2+}][Cl^-]^2 = (x)(2x)^2$, and we would find $[Hg_2^{2+}] = x = 6.7 \times 10^{-7}$ M.

However, this answer is incorrect because we have not accounted for other reactions, such as

Hydrolysis:	$Hg_2^{2+} + H_2O \rightleftharpoons Hg_2OH^+ + H^+$	$K = 10^{-5.3}$
Disproportionation:	$2Hg_2^{2+} \rightleftharpoons Hg^{2+} + Hg(l)$	$K = 10^{-2.1}$
$Hg^{2+} - Cl^-$ ion pairing:	$Hg^{2+} + Cl^- \rightleftharpoons HgCl^+$	$K = 10^{7.3}$
$Hg^{2+} - Cl^-$ ion pairing:	$Hg^{2+} + 2Cl^- \rightleftharpoons HgCl_2(aq)$	$K = 10^{14.00}$

These reactions lead to dissolution of more Hg_2Cl_2 than we predicted from K_{sp} alone. We need to know all significant chemical reactions to compute the solubility of a compound.

Common Ion Effect

For the ionic solubility reaction

$$CaSO_4(s) \rightleftharpoons Ca^{2+} + SO_4^{2-} \qquad K_{sp} = 2.4 \times 10^{-5}$$

the product $[Ca^{2+}][SO_4^{2-}]$ is constant at equilibrium in the presence of excess solid $CaSO_4$. If the concentration of Ca^{2+} were increased by adding another source of Ca^{2+}, such as $CaCl_2$, then the concentration of SO_4^{2-} must decrease so that the product $[Ca^{2+}][SO_4^{2-}]$ remains constant. In other words, less $CaSO_4(s)$ will dissolve if Ca^{2+} or SO_4^{2-} is already present from some other source. Figure 6-1 shows how the solubility of $CaSO_4$ decreases in the presence of dissolved $CaCl_2$.

This application of Le Châtelier's principle is called the **common ion effect**. *A salt will be less soluble if one of its constituent ions is already present in the solution.*

Common ion effect: A salt is less soluble if one of its ions is already present in the solution. Demonstration 6-1 illustrates the common ion effect.

Separation by Precipitation

Precipitations can sometimes be used to separate ions from each other.[12] For example, consider a solution containing lead(II) (Pb^{2+}) and mercury(I) (Hg_2^{2+}) ions, each at a concentration of 0.010 M. Each forms an insoluble iodide (Figure 6-2), but the mercury(I) iodide is considerably less soluble, as indicated by the smaller value of K_{sp}:

$$PbI_2(s) \rightleftharpoons Pb^{2+} + 2I^- \qquad K_{sp} = 7.9 \times 10^{-9}$$
$$Hg_2I_2(s) \rightleftharpoons Hg_2^{2+} + 2I^- \qquad K_{sp} = 4.6 \times 10^{-29}$$

The smaller K_{sp} implies a lower solubility for Hg_2I_2 because the stoichiometries of the two reactions are the same. If the stoichiometries were different, it does not follow that the smaller K_{sp} would imply lower solubility.

Is it possible to lower the concentration of Hg_2^{2+} by 99.990% by selective precipitation with I^-, without precipitating Pb^{2+}?

We are asking if we can lower $[Hg_2^{2+}]$ to 0.010% of 0.010 M $= 1.0 \times 10^{-6}$ M without precipitating Pb^{2+}. Here is the experiment: We add enough I^- to precipitate 99.990% of Hg_2^{2+} if all the I^- reacts with Hg_2^{2+} and none reacts with Pb^{2+}. To see if any Pb^{2+} should precipitate,

Fill two large test tubes about one-third full with saturated aqueous KCl containing no excess solid. The solubility of KCl is approximately 3.7 M, so the solubility product (ignoring activity effects introduced later) is

$$K_{sp} \approx [K^+][Cl^-] = (3.7)(3.7) = 13.7$$

Now add one-third of a test tube of 6 M HCl to one test tube and an equal volume of 12 M HCl to the other. Even though a common ion, Cl^-, is added in each case, KCl precipitates only in one tube.

To understand your observations, calculate the concentrations of K^+ and Cl^- in each tube after HCl addition. Then evaluate the reaction quotient, $Q = [K^+][Cl^-]$, for each tube. Explain your observations.

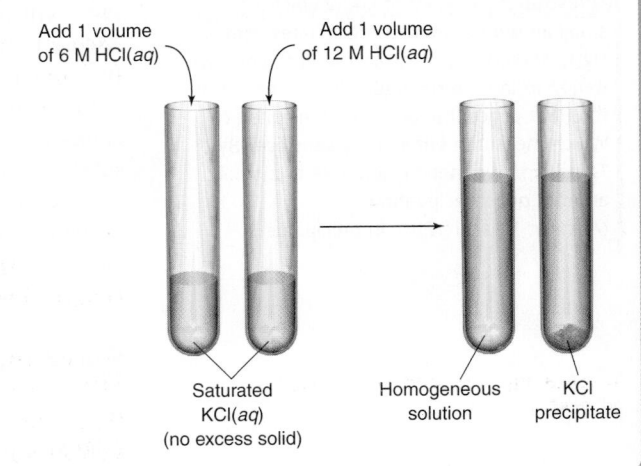

Add 1 volume of 6 M HCl(aq)
Add 1 volume of 12 M HCl(aq)

Saturated KCl(aq) (no excess solid)
Homogeneous solution
KCl precipitate

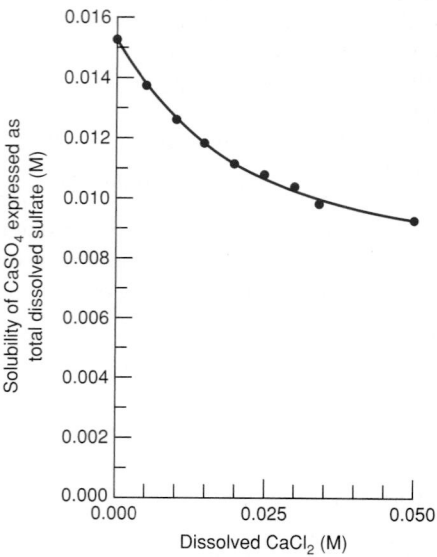

FIGURE 6-1 Solubility of $CaSO_4$ in solutions containing dissolved $CaCl_2$. Solubility is expressed as total dissolved sulfate, which is present as free SO_4^{2-} and as the ion pair, $CaSO_4(aq)$. [Data from W. B. Guenther, *Unified Equilibrium Calculations* (New York: Wiley, 1991).]

FIGURE 6-2 A yellow solid, lead(II) iodide (PbI_2), precipitates when a colorless solution of lead nitrate ($Pb(NO_3)_2$) is added to a colorless solution of potassium iodide (KI). [© 1989 Chip Clark – Fundamental Photographs.]

we need to know the concentration of I^- in equilibrium with precipitated $Hg_2I_2(s)$ plus the remaining 1.0×10^{-6} M Hg_2^{2+}.

$$Hg_2I_2(s) \xrightleftharpoons{K_{sp}} Hg_2^{2+} + 2I^-$$

$$[Hg_2^{2+}][I^-]^2 = K_{sp}$$

$$(1.0 \times 10^{-6})[I^-]^2 = 4.6 \times 10^{-29}$$

$$[I^-] = \sqrt{\frac{4.6 \times 10^{-29}}{1.0 \times 10^{-6}}} = 6.8 \times 10^{-12} \text{ M}$$

Will this concentration of I^- cause 0.010 M Pb^{2+} to precipitate? That is, is the solubility product of PbI_2 exceeded?

$$Q = [Pb^{2+}][I^-]^2 = (0.010)(6.8 \times 10^{-12})^2$$

$$= 4.6 \times 10^{-25} < K_{sp} \qquad \text{(for } PbI_2\text{)}$$

The reaction quotient, $Q = 4.6 \times 10^{-25}$, is less than K_{sp} for $PbI_2 = 7.9 \times 10^{-9}$. Therefore, Pb^{2+} will not precipitate. The separation of Pb^{2+} and Hg_2^{2+} is feasible. We predict that adding I^- to a solution of Pb^{2+} and Hg_2^{2+} will precipitate virtually all Hg_2^{2+} before any Pb^{2+} precipitates.

Life should be so easy! We have just made a thermodynamic prediction. If the system comes to equilibrium, we can achieve the desired separation. However, occasionally one substance *coprecipitates* with the other. In **coprecipitation,** a substance whose solubility is not exceeded precipitates along with another substance whose solubility is exceeded. For example, some Pb^{2+} might become adsorbed on the surface of the Hg_2I_2 crystal or might even occupy sites within the crystal. Our calculation says that the separation is worth trying. *Only an experiment can show whether the separation actually works.*

Question If you want to know whether a small amount of Pb^{2+} coprecipitates with Hg_2I_2, should you measure the Pb^{2+} concentration in the mother liquor (the solution) or the Pb^{2+} concentration in the precipitate? Which measurement is more sensitive? By "sensitive," we mean responsive to a small amount of coprecipitation.
(*Answer:* Measure Pb^{2+} in precipitate.)

6-4 Complex Formation

If anion X^- precipitates metal M^+, it is often observed that a high concentration of X^- causes solid MX to redissolve. The increased solubility arises from formation of **complex ions,** such as MX_2^-, which consist of two or more simple ions bonded to one another.

Lewis Acids and Bases

In complex ions such as PbI^+, PbI_3^-, and PbI_4^{2-}, iodide is said to be the *ligand* of Pb^{2+}. A **ligand** is any atom or group of atoms attached to the species of interest. We say that Pb^{2+} acts as a *Lewis acid* and I^- acts as a *Lewis base* in these complexes. A **Lewis acid** accepts a pair of electrons from a **Lewis base** when the two form a bond:

$$^{++}Pb\,\square + \ddot{:}\,\ddot{\underset{..}{I}}:^- \longrightarrow [Pb-\ddot{\underset{..}{I}}:]^+$$

Room to accept electrons

A pair of electrons to be donated

Lewis acid + Lewis base $\rightleftharpoons$ adduct

Electron pair acceptor

Electron pair donor

The product of the reaction between a Lewis acid and a Lewis base is called an *adduct*. The bond between a Lewis acid and a Lewis base is called a *dative* or *coordinate covalent* bond.

Effect of Complex Ion Formation on Solubility[13]

If Pb^{2+} and I^- only reacted to form solid PbI_2, then the solubility of Pb^{2+} would always be very low in the presence of excess I^-:

$$PbI_2(s) \xrightarrow{K_{sp}} Pb^{2+} + 2I^- \qquad K_{sp} = [Pb^{2+}][I^-]^2 = 7.9 \times 10^{-9} \qquad \textbf{(6-12)}$$

We observe, however, that high concentrations of I^- cause solid PbI_2 to dissolve. We explain this by the formation of a series of complex ions:

$$Pb^{2+} + I^- \xrightarrow{K_1} PbI^+ \qquad K_1 = [PbI^+]/[Pb^{2+}][I^-] = 1.0 \times 10^2 \qquad \textbf{(6-13)}$$

$$Pb^{2+} + 2I^- \xrightarrow{\beta_2} PbI_2(aq) \qquad \beta_2 = [PbI_2(aq)]/[Pb^{2+}][I^-]^2 = 1.4 \times 10^3 \qquad \textbf{(6-14)}$$

$$Pb^{2+} + 3I^- \xrightarrow{\beta_3} PbI_3^- \qquad \beta_3 = [PbI_3^-]/[Pb^{2+}][I^-]^3 = 8.3 \times 10^3 \qquad \textbf{(6-15)}$$

$$Pb^{2+} + 4I^- \xrightarrow{\beta_4} PbI_4^{2-} \qquad \beta_4 = [PbI_4^{2-}]/[Pb^{2+}][I^-]^4 = 3.0 \times 10^4 \qquad \textbf{(6-16)}$$

Notation for these equilibrium constants is discussed in Box 6-2.

The species $PbI_2(aq)$ in Reaction 6-14 is *dissolved* PbI_2, containing two iodine atoms bound to a lead atom. Reaction 6-14 is *not* the reverse of Reaction 6-12, in which the species is solid PbI_2.

BOX 6-2 **Notation for Formation Constants**

Formation constants are the equilibrium constants for complex ion formation. The **stepwise formation constants,** designated K_i, are defined as follows:

$$M + X \xrightarrow{K_1} MX \qquad K_1 = [MX]/[M][X]$$

$$MX + X \xrightarrow{K_2} MX_2 \qquad K_2 = [MX_2]/[MX][X]$$

$$MX_{n-1} + X \xrightarrow{K_n} MX_n \qquad K_n = [MX_n]/[MX_{n-1}][X]$$

The **overall,** or **cumulative, formation constants** are denoted β_i:

$$M + 2X \xrightarrow{\beta_2} MX_2 \qquad \beta_2 = [MX_2]/[M][X]^2$$

$$M + nX \xrightarrow{\beta_n} MX_n \qquad \beta_n = [MX_n]/[M][X]^n$$

A useful relation is that $\beta_n = K_1 K_2 \cdots K_n$. Some formation constants can be found in Appendix I.

At low I^- concentrations, the solubility of lead is governed by precipitation of $PbI_2(s)$. At high I^- concentrations, Reactions 6-13 through 6-16 are driven to the right (Le Châtelier's principle), and the total concentration of dissolved lead is considerably greater than that of Pb^{2+} alone (Figure 6-3).

A most useful characteristic of chemical equilibrium is that *all equilibria are satisfied simultaneously.* If we know $[I^-]$, we can calculate $[Pb^{2+}]$ by substituting the value of $[I^-]$ into the equilibrium constant expression for Reaction 6-12, regardless of whether there

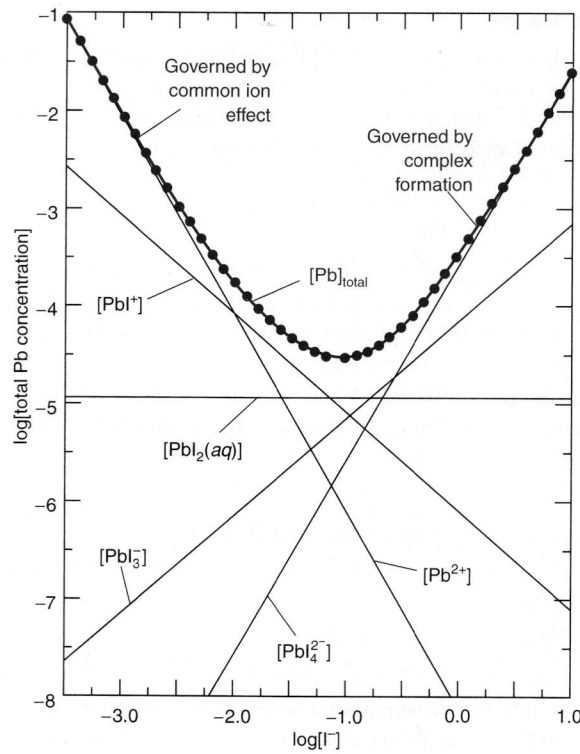

FIGURE 6-3 Total solubility of lead(II) (curve with circles) and solubilities of individual species (straight lines) as a function of the concentration of free iodide. To the left of the minimum, $[Pb]_{total}$ is governed by the solubility product for $PbI_2(s)$. As $[I^-]$ increases, $[Pb]_{total}$ decreases because of the common ion effect. At high values of $[I^-]$, $PbI_2(s)$ redissolves because it reacts with I^- to form soluble complex ions, such as PbI_4^{2-}. Note logarithmic scales. The solution is made slightly acidic so that $[PbOH^+]$ is negligible.

are other reactions involving Pb^{2+}. *The concentration of Pb^{2+} that satisfies any one equilibrium must satisfy all equilibria. There can be only one concentration of Pb^{2+} in the solution.*

EXAMPLE Effect of I^- on the Solubility of Pb^{2+}

Find the concentrations of PbI^+, $PbI_2(aq)$, PbI_3^-, and PbI_4^{2-} in a solution saturated with $PbI_2(s)$ and containing dissolved I^- with a concentration of **(a)** 0.001 0 M and **(b)** 1.0 M.

Solution **(a)** From K_{sp} for Reaction 6-12, we calculate

$$[Pb^{2+}] = K_{sp}/[I^-]^2 = (7.9 \times 10^{-9})/(0.001\ 0)^2 = 7.9 \times 10^{-3}\ M$$

From Reactions 6-13 through 6-16, we then calculate the concentrations of the other Pb[II] species:

$$[PbI^+] = K_1[Pb^{2+}][I^-] = (1.0 \times 10^2)(7.9 \times 10^{-3})(1.0 \times 10^{-3})$$
$$= 7.9 \times 10^{-4}\ M$$
$$[PbI_2(aq)] = \beta_2[Pb^{2+}][I^-]^2 = 1.1 \times 10^{-5}\ M$$
$$[PbI_3^-] = \beta_3[Pb^{2+}][I^-]^3 = 6.6 \times 10^{-8}\ M$$
$$[PbI_4^{2-}] = \beta_4[Pb^{2+}][I^-]^4 = 2.4 \times 10^{-10}\ M$$

(b) If, instead, we take $[I^-] = 1.0\ M$, then analogous computations show that

$$[Pb^{2+}] = 7.9 \times 10^{-9}\ M \qquad [PbI_3^-] = 6.6 \times 10^{-5}\ M$$
$$[PbI^+] = 7.9 \times 10^{-7}\ M \qquad [PbI_4^{2-}] = 2.4 \times 10^{-4}\ M$$
$$[PbI_2(aq)] = 1.1 \times 10^{-5}\ M$$

TEST YOURSELF Find $[Pb^{2+}]$, $PbI_2(aq)$, and $[PbI_3^-]$, in a saturated solution of $PbI_2(s)$ with $[I^-] = 0.10\ M$. (*Answer:* 7.9×10^{-7}, 1.1×10^{-5}, $6.6 \times 10^{-6}\ M$)

Pb^{2+} is dominant when $I^- = 0.001\ 0\ M$.
Pb_4^{2-} is dominant when $I^- = 1.0\ M$.

The total concentration of dissolved lead in the preceding example is

$$[Pb]_{total} = [Pb^{2+}] + [PbI^+] + [PbI_2(aq)] + [PbI_3^-] + [PbI_4^{2-}]$$

When $[I^-] = 10^{-3}$ M, $[Pb]_{total} = 8.7 \times 10^{-3}$ M, of which 91% is Pb^{2+}. As $[I^-]$ increases, $[Pb]_{total}$ decreases by the common ion effect operating in Reaction 6-12. However, at sufficiently high $[I^-]$, complex formation takes over and $[Pb]_{total}$ increases in Figure 6-3. When $[I^-] = 1.0$ M, $[Pb]_{total} = 3.2 \times 10^{-4}$ M, of which 76% is PbI_4^{2-}.

Tabulated Equilibrium Constants Are Usually Not "Constant"

If you look up the equilibrium constant of a chemical reaction in two different books, there is an excellent chance that the values will be different (sometimes by a factor of 10 or more).[14] This discrepancy occurs because the constant may have been determined under different conditions and, perhaps, by using different techniques.

A common source of variation in the reported value of K is the ionic composition of the solution. Note whether K is reported for a particular ionic composition (for example, 1 M $NaClO_4$) or whether K has been extrapolated to zero ion concentration. If you need an equilibrium constant for your own work, choose a value of K measured under conditions as close as possible to those you will employ.

The effect of dissolved ions on chemical equilibria is the subject of Chapter 8.

6-5 Protic Acids and Bases

Understanding the behavior of acids and bases is essential to every branch of science having anything to do with chemistry. In analytical chemistry, we almost always need to account for the effect of pH on analytical reactions involving complex formation or oxidation-reduction. pH can affect molecular charge and shape—factors that help determine which molecules can be separated from others in chromatography and electrophoresis and which molecules will be detected in some types of mass spectrometry.

In aqueous chemistry, an **acid** is a substance that increases the concentration of H_3O^+ **(hydronium ion)** when added to water. Conversely, a **base** decreases the concentration of H_3O^+. We will see that a decrease in H_3O^+ concentration necessarily requires an increase in OH^- concentration. Therefore, a base increases the concentration of OH^- in aqueous solution.

The word *protic* refers to chemistry involving transfer of H^+ from one molecule to another. The species H^+ is also called a *proton* because it is what remains when a hydrogen atom loses its electron. Hydronium ion, H_3O^+, is a combination of H^+ with H_2O. Although H_3O^+ is a more accurate representation than H^+ for the hydrogen ion in aqueous solution, we will use H_3O^+ and H^+ interchangeably in this book.

We will write H^+ when we really mean H_3O^+.

Brønsted-Lowry Acids and Bases

Brønsted and Lowry classified *acids* as *proton donors* and *bases* as *proton acceptors*. HCl is an acid (a proton donor) and it increases the concentration of H_3O^+ in water:

$$HCl + H_2O \rightleftharpoons H_3O^+ + Cl^-$$

The Brønsted-Lowry definition does not require that H_3O^+ be formed. This definition can therefore be extended to nonaqueous solvents and to the gas phase:

$$\underset{\substack{\text{Hydrochloric acid} \\ \text{(acid)}}}{HCl(g)} + \underset{\substack{\text{Ammonia} \\ \text{(base)}}}{NH_3(g)} \rightleftharpoons \underset{\substack{\text{Ammonium chloride} \\ \text{(salt)}}}{NH_4^+Cl^-(s)}$$

For the remainder of this book, when we speak of acids and bases, we are speaking of Brønsted-Lowry acids and bases.

Brønsted-Lowry acid: proton donor

Brønsted-Lowry base: proton acceptor

J. N. Brønsted (1879–1947) of the University of Copenhagen and T. M. Lowry (1874–1936) of Cambridge University independently published their definitions of acids and bases in 1923.

Salts

Any ionic solid, such as ammonium chloride, is called a **salt.** In a formal sense, a salt can be thought of as the product of an acid-base reaction. When an acid and a base react, they are said to **neutralize** each other. Most salts containing cations and anions with single positive and negative charges are *strong electrolytes*—they dissociate nearly completely into ions in dilute aqueous solution. Thus, ammonium chloride gives NH_4^+ and Cl^- in water:

$$NH_4^+Cl^-(s) \rightarrow NH_4^+(aq) + Cl^-(aq)$$

Conjugate Acids and Bases

The products of a reaction between an acid and a base are also classified as acids and bases:

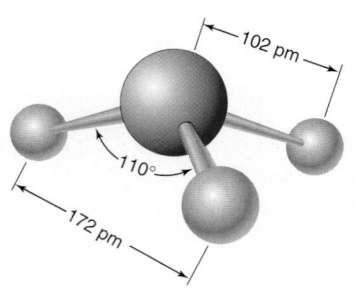

Eigen structure of H_3O^+

FIGURE 6-4 Structure of hydronium ion, H_3O^+, proposed by M. Eigen and found in many crystals.[15] The bond enthalpy (heat needed to break the OH bond) of H_3O^+ is 544 kJ/mol, about 84 kJ/mol greater than the bond enthalpy in H_2O.

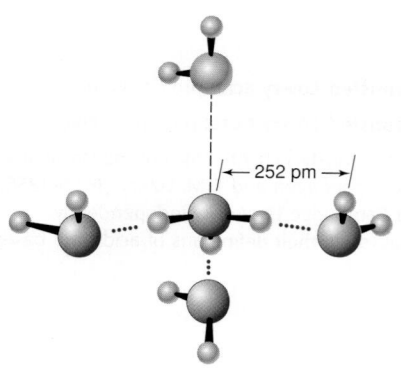

FIGURE 6-5 Environment of aqueous H_3O^+.[15] Three H_2O molecules are bound to H_3O^+ by strong hydrogen bonds (dotted lines), and one H_2O (at the top) is held by weaker ion-dipole attraction (dashed line). The O—H⋯O hydrogen-bonded distance of 252 pm (picometers, 10^{-12} m) compares with an O—H⋯O distance of 283 pm between hydrogen-bonded water molecules. The discrete cation $(H_2O)_3H_3O^+$ found in some crystals is similar in structure to $(H_2O)_4H_3O^+$, with the weakly bonded H_2O at the top removed.[16]

Acetate is a base because it can accept a proton to make acetic acid. Methylammonium ion is an acid because it can donate a proton and become methylamine. Acetic acid and the acetate ion are said to be a **conjugate acid-base pair.** Methylamine and methylammonium ion are likewise conjugate. *Conjugate acids and bases are related to each other by the gain or loss of one* H^+.

The Nature of H^+ and OH^-

The proton does not exist by itself in water. The simplest formula found in some crystalline salts is H_3O^+. For example, crystals of perchloric acid monohydrate contain pyramidal hydronium (also called *hydroxonium*) ions:

$$HClO_4 \cdot H_2O \text{ is really}$$

Hydronium   Perchlorate

The formula $HClO_4 \cdot H_2O$ is a way of specifying the composition of the substance when we are ignorant of its structure. A more accurate formula would be $H_3O^+ClO_4^-$.

Average dimensions of the H_3O^+ cation in many crystals are shown in Figure 6-4. In aqueous solution, H_3O^+ is tightly associated with three molecules of water through exceptionally strong hydrogen bonds (Figure 6-5). The $H_5O_2^+$ cation is another simple species in which a hydrogen ion is shared by two water molecules.[17, 18]

Zundel structure of $(H_3O^+ \cdot H_2O)$

← 243 pm →

In the gas phase, H_3O^+ can be surrounded by 20 molecules of H_2O in a regular dodecahedron held together by 30 hydrogen bonds.[19] In a salt containing the discrete cation $(C_6H_6)_3H_3O^+$, and in benzene solution, hydrogen atoms of the pyramidal H_3O^+ ion are each attracted toward the center of the pi electron cloud of a benzene ring (Figure 6-6).

The ion $H_3O_2^-$ ($OH^- \cdot H_2O$) has been observed by X-ray crystallography.[20] The central O⋯H⋯O linkage contains the shortest hydrogen bond involving H_2O that has ever been observed.

← 229 pm →

CHAPTER 6 Chemical Equilibrium

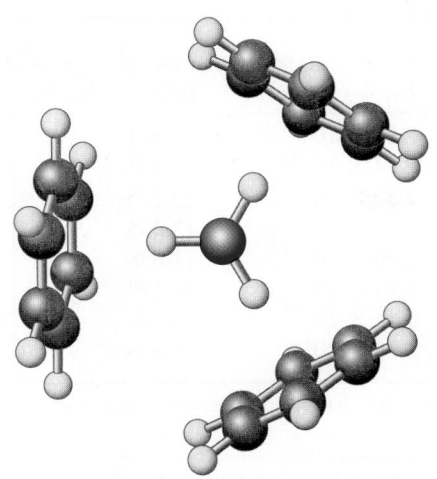

FIGURE 6-6 $H_3O^+ \cdot 3C_6H_6$ cation found in the crystal structure of $[(C_6H_6)_3H_3O^+][CHB_{11}Cl_{11}^-]$. [Data from E. S. Stoyanov, K.-C. Kim, and C. A. Reed, "The Nature of the H_3O^+ Hydronium Ion in Benzene and Chlorinated Hydrocarbon Solvents," *J. Am. Chem. Soc.* **2006,** *128,* 1948.]

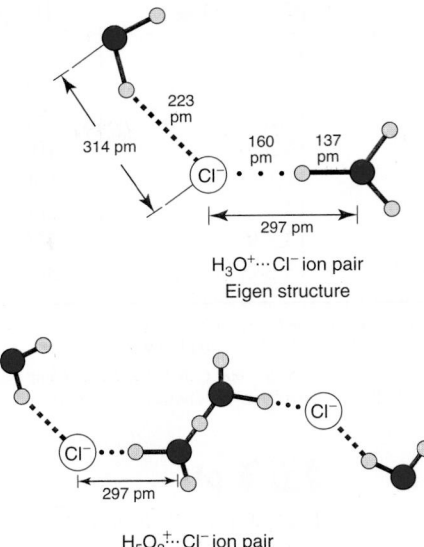

$H_3O^+ \cdots Cl^-$ ion pair
Eigen structure

$H_5O_2^+ \cdots Cl^-$ ion pair
Zundel structure

FIGURE 6-7 $Cl^- \cdots H_3O^+$ and $Cl^- \cdots H_5O_2^+$ ion pair structures in concentrated aqueous HCl. Each Cl^- is surrounded by a combination of approximately six ($H_2O + H_3O^+ + H_5O_2^+$) molecules, only some of which are shown. [Data from J. L. Fulton and M. Balasubramanian, "Structure of Hydronium (H_3O^+)/Chloride (Cl^-) Contact Ion Pairs in Aqueous Hydrochloric Acid Solution," *J. Am. Chem. Soc.* **2010,** *132,* 12597.]

In aqueous HCl, ion pairing of H_3O^+ with Cl^- is detectable at a concentration of ~6 *m* (molal). In 16.1 *m* HCl, all H_3O^+ is paired with Cl^- (Figure 6-7). There is a combination of approximately six Eigen (H_3O^+) and Zundel ($H_5O_2^+$) structures plus H_2O around each Cl^-. The $Cl^- \cdots H^+$ hydrogen bond length is reduced from 223 pm in $Cl^- \cdots H_2O$ to 160 pm in $Cl^- \cdots H_3O^+$.

We will ordinarily write H^+ in most chemical equations, although we really mean H_3O^+. To emphasize the chemistry of water, we will write H_3O^+. For example, water can be either an acid or a base. Water is an acid with respect to methoxide:

$$H-\overset{..}{\underset{..}{O}}-H \;+\; CH_3-\overset{..}{\underset{..}{O}}{}^- \;\rightleftharpoons\; H-\overset{..}{\underset{..}{O}}{}^- \;+\; CH_3-\overset{..}{\underset{..}{O}}-H$$
$$\text{Water} \qquad\qquad \text{Methoxide} \qquad\qquad \text{Hydroxide} \qquad\qquad \text{Methanol}$$

But with respect to hydrogen bromide, water is a base:

$$H_2O \;+\; HBr \;\rightleftharpoons\; H_3O^+ \;+\; Br^-$$
$$\text{Water} \quad\;\; \text{Hydrogen} \quad\;\; \text{Hydronium} \quad\;\; \text{Bromide}$$
$$\text{bromide} \qquad\;\; \text{ion}$$

Autoprotolysis

Water undergoes self-ionization, called **autoprotolysis** (also called *autoionization*), in which it acts as both an acid and a base:

$$H_2O + H_2O \rightleftharpoons H_3O^+ + OH^- \tag{6-17a}$$

or

$$H_2O \rightleftharpoons H^+ + OH^- \tag{6-17b}$$

Reactions 6-17a and 6-17b mean the same thing.

Protic solvents have a reactive H^+, and all protic solvents undergo autoprotolysis. An example is acetic acid:

$$2CH_3\overset{O}{\overset{\|}{C}}OH \;\rightleftharpoons\; CH_3\overset{OH}{\underset{OH}{C^+}} \;+\; H_3CC\overset{O}{\underset{O}{\diagup}}{}^- \qquad \text{(in acetic acid)} \tag{6-18}$$

The extent of these reactions is very small. The *autoprotolysis constants* (equilibrium constants) for Reactions 6-17 and 6-18 are 1.0×10^{-14} and 3.5×10^{-15}, respectively, at 25°C.

16.1 *m* HCl contains 16.1 mol HCl per kg H_2O. What is the molar ratio HCl:H_2O? (*Answer:* 16.1 : 55.5 = 1 : 3.45)

Examples of **protic solvents** (acidic proton **bold**):

H_2O	CH_3CH_2OH
Water	Ethanol

Examples of **aprotic solvents** (no acidic protons):

$CH_3CH_2OCH_2CH_3$	CH_3CN
Diethyl ether	Acetonitrile

TABLE 6-1 Temperature dependence of $K_w{}^a$

Temperature (°C)	K_w	$pK_w = -\log K_w$	Temperature (°C)	K_w	$pK_w = -\log K_w$
0	1.15×10^{-15}	14.938	40	2.88×10^{-14}	13.541
5	1.88×10^{-15}	14.726	45	3.94×10^{-14}	13.405
10	2.97×10^{-15}	14.527	50	5.31×10^{-14}	13.275
15	4.57×10^{-15}	14.340	100	5.43×10^{-13}	12.265
20	6.88×10^{-15}	14.163	150	2.30×10^{-12}	11.638
25	1.01×10^{-14}	13.995	200	5.14×10^{-12}	11.289
30	1.46×10^{-14}	13.836	250	6.44×10^{-12}	11.191
35	2.07×10^{-14}	13.685	300	3.93×10^{-12}	11.406

a. Concentrations in the product $[H^+][OH^-]$ in this table are expressed in molality rather than in molarity. Accuracy of $\log K_w$ is ± 0.01. To convert molality (mol/kg) into molarity (mol/L), multiply by the density of H_2O at each temperature. At 25°C, $K_w = 10^{-13.995}$ $(mol/kg)^2(0.997\,05\,kg/L)^2 = 10^{-13.998}$ $(mol/L)^2$.

SOURCE: W. L. Marshall and E. U. Franck, "Ion Product of Water Substance, 0–1 000°C, 1–10,000 Bars," J. Phys. Chem. Ref. Data **1981**, 10, 295. For values of K_w over a temperature range of 0–800°C and a density range of 0–1.2 g/cm³, see A. V. Bandura and S. N. Lvov, "The Ionization Constant of Water over Wide Ranges of Temperature and Pressure," J. Phys. Chem. Ref. Data **2006**, 35, 15.

6-6 pH

The autoprotolysis constant for H_2O has the special symbol K_w, where "w" stands for water:

Recall that H_2O (the solvent) is omitted from the equilibrium constant. The value $K_w = 1.0 \times 10^{-14}$ at 25°C is accurate enough for problems in this book.

Autoprotolysis of water:
$$H_2O \xrightarrow{\ K_w\ } H^+ + OH^- \qquad K_w = [H^+][OH^-] \qquad (6\text{-}19)$$

Table 6-1 shows how K_w varies with temperature. Its value at 25.0°C is 1.01×10^{-14}.

EXAMPLE Concentration of H^+ and OH^- in Pure Water at 25°C

Calculate the concentrations of H^+ and OH^- in pure water at 25°C.

Solution The stoichiometry of Reaction 6-19 tells us that H^+ and OH^- are produced in a 1:1 molar ratio. Their concentrations must be equal. Calling each concentration x, we can write

$$K_w = 1.0 \times 10^{-14} = [H^+][OH^-] = [x][x] \Rightarrow x = 1.0 \times 10^{-7}\ M$$

The concentrations of H^+ and OH^- are both $1.0 \times 10^{-7}\ M$ in pure water.

TEST YOURSELF Use Table 6-1 to find the $[H^+]$ in water at 100°C and at 0°C. (**Answer:** 7.4×10^{-7} and $3.4 \times 10^{-8}\ M$)

EXAMPLE Concentration of OH^- When $[H^+]$ Is Known

What is the concentration of OH^- if $[H^+] = 1.0 \times 10^{-3}\ M$? (From now on, assume that the temperature is 25°C unless otherwise stated.)

Solution Putting $[H^+] = 1.0 \times 10^{-3}\ M$ into the K_w expression gives

$$K_w = 1.0 \times 10^{-14} = (1.0 \times 10^{-3})[OH^-] \Rightarrow [OH^-] = 1.0 \times 10^{-11}\ M$$

A concentration of $[H^+] = 1.0 \times 10^{-3}\ M$ gives $[OH^-] = 1.0 \times 10^{-11}\ M$. *As the concentration of H^+ increases, the concentration of OH^- necessarily decreases, and vice versa.* A concentration of $[OH^-] = 1.0 \times 10^{-3}\ M$ gives $[H^+] = 1.0 \times 10^{-11}\ M$.

TEST YOURSELF Find $[OH^-]$ if $[H^+] = 1.0 \times 10^{-4}\ M$. (**Answer:** $1.0 \times 10^{-10}\ M$)

$pH \approx -\log[H^+]$. The term "pH" was introduced in 1909 by the Danish biochemist S. P. L. Sørensen, who called it the "hydrogen ion exponent."[21]

An approximate definition of **pH** is the negative logarithm of the H^+ concentration.

Approximate definition of pH: $\qquad pH \approx -\log[H^+] \qquad (6\text{-}20)$

Chapter 8 defines pH more accurately in terms of *activities*, but, for many purposes, Equation 6-20 is a good definition. The measurement of pH with glass electrodes and buffers used by the U.S. National Institute of Standards and Technology to define the pH scale are described in Chapter 15.

In pure water at 25°C with $[H^+] = 1.0 \times 10^{-7}\ M$, the pH is $-\log(1.0 \times 10^{-7}) = 7.00$. If $[OH^-] = 1.0 \times 10^{-3}\ M$, then $[H^+] = 1.0 \times 10^{-11}\ M$ and the pH is 11.00. A useful relation between $[H^+]$ and $[OH^-]$ is

Take the log of both sides of the K_w expression to derive Equation 6-21:

$$K_w = [H^+][OH^-]$$
$$\log K_w = \log[H^+] + \log[OH^-]$$
$$-\log K_w = -\log[H^+] - \log[OH^-]$$
$$14.00 = pH + pOH \quad (\text{at } 25°C)$$

$$pH + pOH = -\log(K_w) = 14.00 \text{ at } 25°C \qquad (6\text{-}21)$$

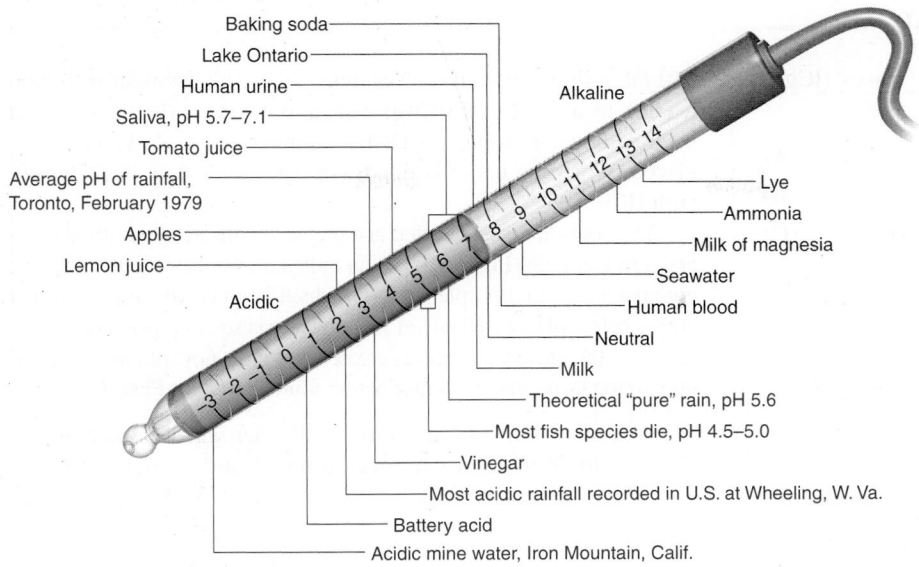

Baking soda
Lake Ontario
Human urine
Saliva, pH 5.7–7.1
Tomato juice
Average pH of rainfall,
Toronto, February 1979
Apples
Lemon juice

Acidic

Alkaline

Lye
Ammonia
Milk of magnesia
Seawater
Human blood
Neutral
Milk
Theoretical "pure" rain, pH 5.6
Most fish species die, pH 4.5–5.0
Vinegar
Most acidic rainfall recorded in U.S. at Wheeling, W. Va.
Battery acid
Acidic mine water, Iron Mountain, Calif.

FIGURE 6-8 pH of various substances.
[Data from *Chem. Eng. News*, 14 September 1981.]
The most acidic rainfall (Box 15-1) is a stronger acid than lemon juice. The most acidic natural waters known are mine waters, with total dissolved metal concentrations of 200 g/L and sulfate concentrations of 760 g/L.[23] The pH of this water, −3.6, does not mean that $[H^+] = 10^{3.6}$ M = 4 000 M! It means that the *activity* of H^+ (discussed in Chapter 8) is $10^{3.6}$.

where $pOH = -\log[OH^-]$, just as $pH = -\log[H^+]$. Equation 6-21 is a fancy way of saying that if pH = 3.58, then pOH = 14.00 − 3.58 = 10.42, or $[OH^-] = 10^{-10.42} = 3.8 \times 10^{-11}$ M.

A solution is **acidic** if $[H^+] > [OH^-]$. A solution is **basic** if $[H^+] < [OH^-]$. At 25°C, an acidic solution has a pH below 7 and a basic solution has a pH above 7.

$$\text{pH} \quad \begin{array}{ccccccccccccccccc} -1 & 0 & 1 & 2 & 3 & 4 & 5 & 6 & 7 & 8 & 9 & 10 & 11 & 12 & 13 & 14 & 15 \end{array}$$

$\longleftarrow$ Acidic $\longrightarrow$ $\uparrow$ $\longleftarrow$ Basic $\longrightarrow$

Neutral

pH is measured with a *glass electrode* described in Chapter 15.

The surface of water or ice is ~2 pH units more acidic than the bulk because H_3O^+ is more stable on the surface. Surface acidity could be important to the chemistry of atmospheric clouds.[22]

pH values for various common substances are shown in Figure 6-8.

Although pH generally falls in the range 0 to 14, these are not the limits of pH. A pH of −1, for example, means $-\log[H^+] = -1$; or $[H^+] = 10$ M. This concentration is attained in a concentrated solution of a strong acid such as HCl.

Is There Such a Thing as Pure Water?

In most labs, the answer is "No." Pure water at 25°C should have a pH of 7.00. Distilled water from the tap in most labs is acidic because it contains CO_2 from the atmosphere. CO_2 is an acid by virtue of the reaction

$$CO_2 + H_2O \rightleftharpoons HCO_3^- + H^+ \qquad \text{(6-22)}$$
$$\text{Bicarbonate}$$

CO_2 can be largely removed by boiling water and then protecting it from the atmosphere.

More than a century ago, careful measurements of the conductivity of water were made by F. Kohlrausch and his students. To remove impurities, they found it necessary to distill water *42 consecutive times* under vacuum to reduce conductivity to a limiting value.

6-7 Strengths of Acids and Bases

Acids and bases are commonly classified as strong or weak, depending on whether they react nearly "completely" or only "partially" to produce H^+ or OH^-. Although there is no sharp distinction between weak and strong, some acids or bases react so completely that they are easily classified as strong—and, by convention, everything else is termed weak.

Strong Acids and Bases

Common strong acids and bases are listed in Table 6-2, which you need to memorize. By definition, a strong acid or base is completely dissociated in aqueous solution. That is, the equilibrium constants for the following reactions are large.

$$HCl(aq) \rightleftharpoons H^+ + Cl^-$$
$$KOH(aq) \rightleftharpoons K^+ + OH^-$$

TABLE 6-2	Common strong acids and bases
Formula	**Name**
Acids	
HC1	Hydrochloric acid (hydrogen chloride)
HBr	Hydrogen bromide
HI	Hydrogen iodide
$H_2SO_4^a$	Sulfuric acid
HNO_3	Nitric acid
$HClO_4$	Perchloric acid
Bases	
LiOH	Lithium hydroxide
NaOH	Sodium hydroxide
KOH	Potassium hydroxide
RbOH	Rubidium hydroxide
CsOH	Cesium hydroxide
R_4NOH^b	Quaternary ammonium hydroxide

a. For H_2SO_4, only the first proton ionization is complete. Dissociation of the second proton has an equilibrium constant of 1.0×10^{-2}.

b. This is a general formula for any hydroxide salt of an ammonium cation containing four organic groups. An example is tetrabutylammonium hydroxide: $(CH_3CH_2CH_2CH_2)_4N^+OH^-$.

The complete dissociation of HCl into H^+ and Cl^- makes $HCl(g)$ extremely soluble in water.

$$HCl(g) \rightleftharpoons HCl(aq) \qquad \textbf{(A)}$$

$$\underline{HCl(aq) \rightleftharpoons H^+(aq) + Cl^-(aq)} \qquad \textbf{(B)}$$

Net reaction: $\quad HCl(g) \rightleftharpoons H^+(aq) + Cl^-(aq) \qquad \textbf{(C)}$

Because the equilibrium of Reaction B lies far to the right, it pulls Reaction A to the right as well.

Challenge The standard free energy change ($\Delta G°$) for Reaction C is -36.0 kJ/mol. Show that the equilibrium constant is 2.0×10^6.

The extreme solubility of $HCl(g)$ in water is the basis for the HCl fountain,[24] assembled as shown below. In Figure *a*, an inverted

250-mL round-bottom flask containing air is set up with its inlet tube leading to a source of $HCl(g)$ and its outlet tube directed into an inverted bottle of water. As HCl is admitted to the flask, air is displaced. When the bottle is filled with air, the flask is filled mostly with $HCl(g)$.

The hoses are disconnected and replaced with a beaker of indicator and a rubber bulb (Figure *b*). For an indicator, we use slightly alkaline, commercial methyl purple solution, which is green above pH 5.4 and purple below pH 4.8. When ~1 mL of water is squirted from the rubber bulb into the flask, a vacuum is created and indicator solution is drawn up into the flask, making a fascinating fountain (Color Plate 1).

Question Why is a vacuum created when water is squirted into the flask, and why does the indicator change color when it enters the flask?

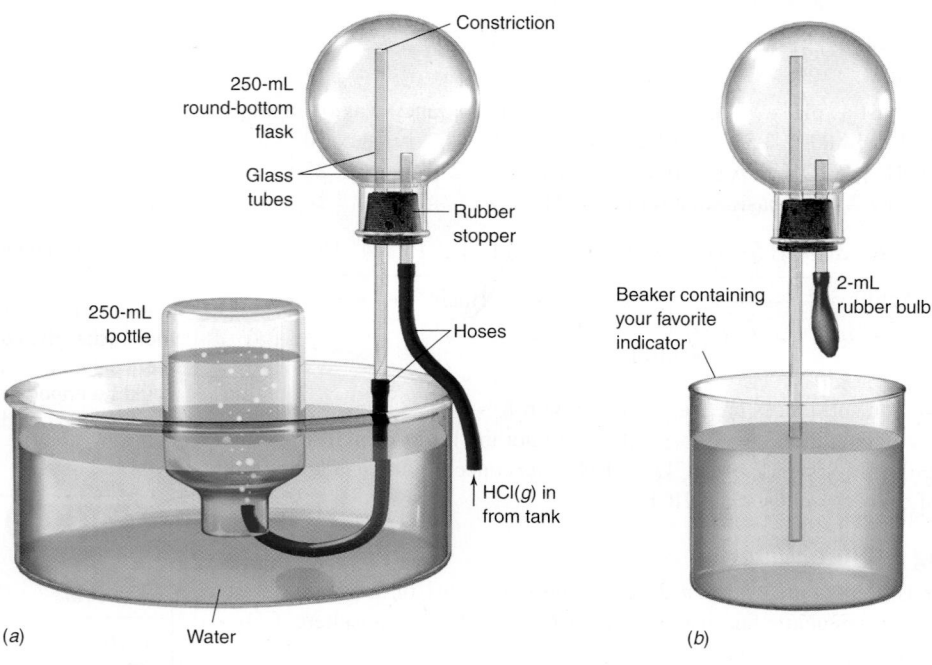

(a) Water

(b)

TABLE 6-3	Equilibria of alkaline earth metal hydroxides

$$M(OH)_2(s) \rightleftharpoons M^{2+} + 2OH^-$$
$$K_{sp} = [M^{2+}][OH^-]^2$$

$$M^{2+} + OH^- \rightleftharpoons MOH^+$$
$$K_1 = [MOH^+]/[M^{2+}][OH^-]$$

Metal	$\log K_{sp}$	$\log K_1$
Mg^{2+}	-11.15	2.58
Ca^{2+}	-5.19	1.30
Sr^{2+}	—	0.82
Ba^{2+}	—	0.64

NOTE: 25°C and ionic strength = 0.

Virtually no undissociated HCl or KOH exists in aqueous solution. Demonstration 6-2 shows a consequence of the strong-acid behavior of HCl.

Even though the hydrogen halides HCl, HBr, and HI are strong acids, HF is *not* a strong acid, as explained in Box 6-3. For most purposes, the hydroxides of the alkaline earth metals (Mg^{2+}, Ca^{2+}, Sr^{2+}, and Ba^{2+}) can be considered to be strong bases, although they are far less soluble than alkali metal hydroxides and have some tendency to form MOH^+ complexes (Table 6-3). The strongest known base is the gas phase molecule LiO^-.[27]

Weak Acids and Bases

All weak acids, denoted HA, react with water by donating a proton to H_2O:

Dissociation of weak acid: $\quad HA + H_2O \overset{K_a}{\rightleftharpoons} H_3O^+ + A^- \qquad \textbf{(6-23)}$

which means exactly the same as

Dissociation of weak acid:

$$HA \overset{K_a}{\rightleftharpoons} H^+ + A^- \qquad K_a = \frac{[H^+][A^-]}{[HA]} \qquad \textbf{(6-24)}$$

Acid dissociation constant: $K_a = \dfrac{[H^+][A^-]}{[HA]}$

The equilibrium constant is called K_a, the **acid dissociation constant.** By definition, a weak acid is one that is only partially dissociated in water, so K_a is "small."

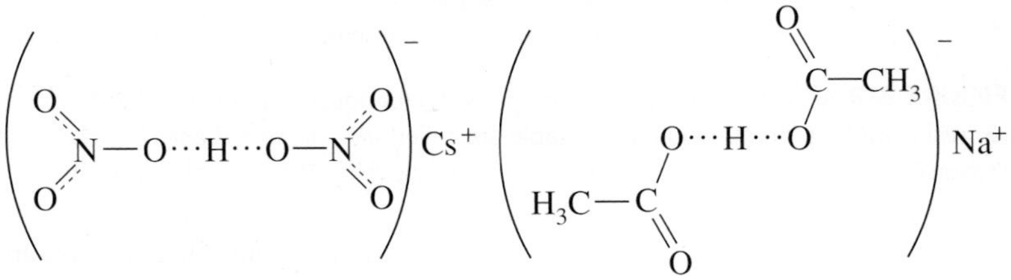

A weak base, B, reacts with water by abstracting a proton from H_2O:

Base hydrolysis: $\quad B + H_2O \xrightarrow{K_b} BH^+ + OH^- \qquad K_b = \dfrac{[BH^+][OH^-]}{[B]}$ **(6-25)**

Base hydrolysis constant: $K_b = \dfrac{[BH^+][OH^-]}{[B]}$

Hydrolysis refers to any reaction with water.

The equilibrium constant K_b is the **base hydrolysis constant,** which is "small" for a weak base.

Common Classes of Weak Acids and Bases

Acetic acid is a typical weak acid.

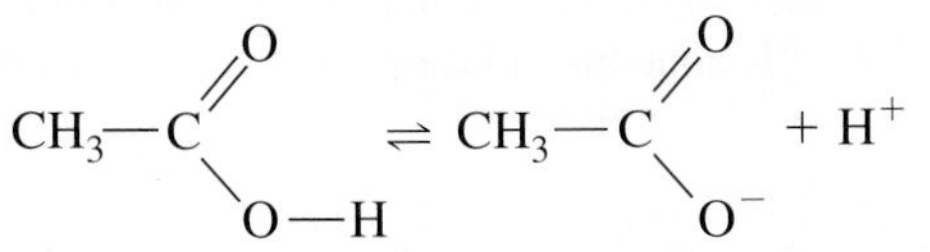

$$CH_3-C \rightleftharpoons CH_3-C + H^+ \qquad K_a = 1.75 \times 10^{-5}$$

(6-26)

Acetic acid
(HA)

Acetate
(A^-)

Acetic acid is a representative carboxylic acid, which has the general formula RCO_2H, where R is an organic substituent. *Most* **carboxylic acids** *are weak acids, and most* **carboxylate anions** *are weak bases.*

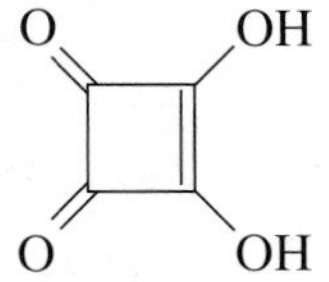

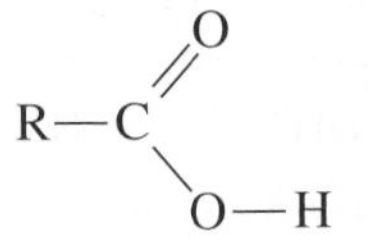

A carboxylic acid
(weak acid, HA)

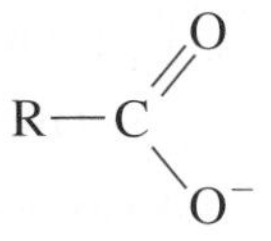

A carboxylate anion
(weak base, A^-)

Methylamine is a typical weak base.

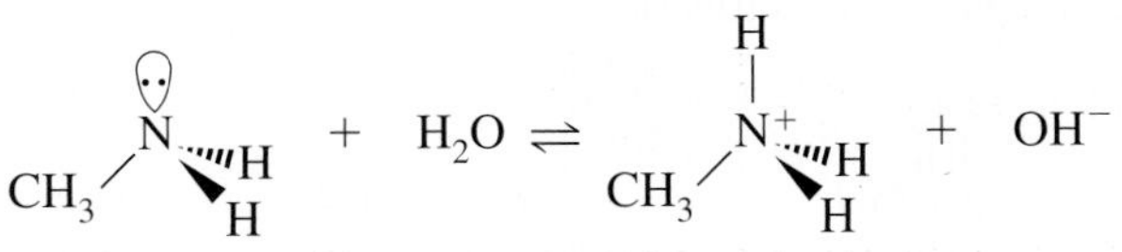

$$\text{Methylamine} + H_2O \rightleftharpoons \text{Methylammonium ion} + OH^- \qquad K_b = 4.47 \times 10^{-4}$$ **(6-27)**

Methylamine
B

Methylammonium ion
BH^+

Carboxylic acids (RCO_2H) and ammonium ions (R_3NH^+) are weak acids.

Carboxylate anions (RCO_2^-) and amines (R_3N) are weak bases.

Stronger acid →

Li^+ 13.64	Be													
Na^+ 13.9	Mg^{2+} 11.4											Al^{3+} 5.00		
K	Ca^{2+} 12.70	Sc^{3+} 4.3	Ti^{3+} 1.3	VO^{2+} 5.7	Cr^{2+} 5.5[a] / Cr^{3+} 3.66	Mn^{2+} 10.6	Fe^{2+} 9.4 / Fe^{3+} 2.19	Co^{2+} 9.7 / Co^{3+} 0.5[b]	Ni^{2+} 9.9	Cu^{2+} 7.5	Zn^{2+} 9.0	Ga^{3+} 2.6	Ge	
Rb	Sr 13.18	Y^{3+} 7.7	Zr^{4+} −0.3	Nb	Mo	Tc	Ru	Rh^{3+} 3.33[c]	Pd^{2+} 1.0	Ag^+ 12.0	Cd^{2+} 10.1	In^{3+} 3.9	Sn^{2+} 3.4	Sb
Cs	Ba^{2+} 13.36	La^{3+} 8.5	Hf	Ta	W	Re	Os	Ir	Pt	Au	Hg_2^{2+} 5.3[d] / Hg^{2+} 3.40	Tl^+ 13.21	Pb^{2+} 7.6	Bi^{3+} 1.1

Ce^{3+} 9.1[b]	Pr^{3+} 9.4[b]	Nd^{3+} 8.7[b]	Pm	Sm^{3+} 8.6[b]	Eu^{3+} 8.6[d]	Gd^{3+} 9.1[b]	Tb^{3+} 8.4[d]	Dy^{3+} 8.4[d]	Ho^{3+} 8.3	Er^{3+} 9.1[b]	Tm^{3+} 8.2[d]	Yb^{3+} 8.4[b]	Lu^{3+} 8.2[d]

Ionic strength = 0 unless noted by superscript
a. Ionic strength = 1 M *b.* Ionic strength = 3 M *c.* Ionic strength = 2.5 M *d.* Ionic strength = 0.5 M

FIGURE 6-9 Acid dissociation constants ($-\log K_a$) for aqueous metal ions: $M^{n+} + H_2O \overset{K_a}{\rightleftharpoons} MOH^{(n-1)+} + H^+$. For example, for Li^+, $K_a = 10^{-13.64}$. In Chapter 9 we will learn that the numbers in this table are called pK_a. Darkest shades are strongest acids. [Data from R. M. Smith, A. E. Martell, and R. J. Motekaitis, *NIST Critical Stability Constants of Metal Complexes Database 46* (Gaithersburg, MD: National Institute of Standards and Technology, 2001).]

Amines are nitrogen-containing compounds:

$R\ddot{N}H_2$	a primary amine	RNH_3^+
$R_2\ddot{N}H$	a secondary amine	$R_2NH_2^+$
$R_3\ddot{N}$	a tertiary amine	R_3NH^+

(ammonium ions)

Amines *are weak bases, and* **ammonium ions** *are weak acids.* The "parent" of all amines is ammonia, NH_3. When a base such as methylamine reacts with water, the product is the conjugate acid. That is, methylammonium ion produced in Reaction 6-27 is a weak acid:

Although we write a **base** as **B** and an **acid** as **HA**, it is important to realize that **BH^+** is also an **acid** and **A^-** is also a **base**.

$$\underset{BH^+}{CH_3\overset{+}{N}H_3} \overset{K_a}{\rightleftharpoons} \underset{B}{CH_3\ddot{N}H_2} + H^+ \qquad K_a = 2.33 \times 10^{-11} \tag{6-28}$$

The methylammonium ion is the conjugate acid of methylamine.

You should learn to recognize whether a compound is acidic or basic. The salt methylammonium chloride, for example, dissociates in aqueous solution to give methylammonium cation and chloride:

Methylammonium chloride is a weak acid because

1. It dissociates into $CH_3NH_3^+$ and Cl^-.

2. $CH_3NH_3^+$ is a weak acid, being conjugate to CH_3NH_2, a weak base.

3. Cl^- has no basic properties. It is conjugate to HCl, a strong acid. That is, HCl dissociates completely.

$$\underset{\text{Methylammonium chloride}}{CH_3\overset{+}{N}H_3Cl^-(s)} \rightarrow \underset{\text{Methylammonium ion}}{CH_3\overset{+}{N}H_3(aq)} + Cl^-(aq) \tag{6-29}$$

Methylammonium ion, being the conjugate acid of methylamine, is a weak acid (Reaction 6-28). Chloride is the conjugate base of HCl, a strong acid. In other words, Cl^- *has virtually no tendency to associate with* H^+, or else HCl would not be a strong acid. Methylammonium chloride is acidic because methylammonium ion is an acid and Cl^- is not a base.

Challenge Phenol (C_6H_5OH) is a weak acid. Explain why a solution of the ionic compound potassium phenolate ($C_6H_5O^-K^+$) is basic.

Metal cations, M^{n+}, act as weak acids by *acid hydrolysis* to form $M(OH)^{(n-1)+}$.[28] Figure 6-9 shows acid dissociation constants for the reaction

$$M^{n+} + H_2O \overset{K_a}{\rightleftharpoons} MOH^{(n-1)+} + H^+$$

Monovalent metal ions are very weak acids (Na^+, $K_a = 10^{-13.9}$). Divalent ions tend to be stronger (Fe^{2+}, $K_a = 10^{-9.4}$) and trivalent ions are stronger yet (Fe^{3+}, $K_a = 10^{-2.19}$).

Aqueous metal ions are associated with (*hydrated* by) several H_2O molecules, so a more accurate way to write the acid dissociation reaction is

$$M(H_2O)_x^{n+} \rightleftharpoons M(H_2O)_{x-1}(OH)^{(n-1)+} + H^+$$

Polyprotic Acids and Bases

Polyprotic acids and bases are compounds that can donate or accept more than one proton. For example, oxalic acid is diprotic and phenol is tribasic:

$$\underset{\text{Oxalic acid}}{HOC(=O)C(=O)OH} \rightleftharpoons H^+ + \underset{\text{Monohydrogen oxalate}}{^-OC(=O)C(=O)OH} \qquad K_{a1} = 5.62 \times 10^{-2} \tag{6-30}$$

$$\overset{\overset{\displaystyle O}{\|}\;\overset{\displaystyle O}{\|}}{{}^-\!OCCOH} \;\rightleftharpoons\; H^+ + \overset{\overset{\displaystyle O}{\|}\;\overset{\displaystyle O}{\|}}{{}^-\!OCCO^-} \qquad\qquad K_{a2} = 5.42 \times 10^{-5} \quad (6\text{-}31)$$

Oxalate

$$\text{Phosphate} + H_2O \;\rightleftharpoons\; \text{Monohydrogen phosphate} + OH^- \qquad K_{b1} = 2.3 \times 10^{-2} \quad (6\text{-}32)$$

$$\text{Dihydrogen phosphate} + H_2O \;\rightleftharpoons\; + OH^- \qquad K_{b2} = 1.60 \times 10^{-7} \quad (6\text{-}33)$$

$$+ H_2O \;\rightleftharpoons\; \text{Phosphoric acid} + OH^- \qquad K_{b3} = 1.42 \times 10^{-12} \quad (6\text{-}34)$$

Notation for acid and base equilibrium constants: K_{a1} refers to the acidic species with the most protons and K_{b1} refers to the basic species with the least protons. The subscript "a" in acid dissociation constants will usually be omitted.

Standard notation for successive acid dissociation constants of a polyprotic acid is K_1, K_2, K_3, and so on, with the subscript "a" usually omitted. We retain or omit the subscript as dictated by clarity. For successive base hydrolysis constants, we retain the subscript "b." The preceding examples illustrate that K_{a1} (or K_1) *refers to the acidic species with the most protons, and* K_{b1} *refers to the basic species with the least number of protons.* Carbonic acid, a very important diprotic carboxylic acid derived from CO_2, is described in Box 6-4.

BOX 6-4   Carbonic Acid[29]

Carbonic acid (H_2CO_3) is formed by the reaction of carbon dioxide with water:

$$CO_2(g) \rightleftharpoons CO_2(aq) \qquad K = \frac{[CO_2(aq)]}{P_{CO_2}} = 0.034\,4$$

$$CO_2(aq) + H_2O \rightleftharpoons \underset{\text{Carbonic acid}}{\overset{\overset{\displaystyle O}{\|}}{\underset{HO\quad OH}{C}}} \qquad K = \frac{[H_2CO_3]}{[CO_2(aq)]} \approx 0.002$$

$$H_2CO_3 \rightleftharpoons \underset{\text{Bicarbonate}}{HCO_3^-} + H^+ \qquad K_{a1} = 4.46 \times 10^{-7}$$

$$HCO_3^- \rightleftharpoons \underset{\text{Carbonate}}{CO_3^{2-}} + H^+ \qquad K_{a2} = 4.69 \times 10^{-11}$$

Its behavior as a diprotic acid appears anomalous at first, because the value of K_{a1} is about 10^2 to 10^4 times smaller than K_a for other carboxylic acids.

CH_3CO_2H	HCO_2H
Acetic acid	Formic acid
$K_a = 1.75 \times 10^{-5}$	$K_a = 1.80 \times 10^{-4}$
$N{\equiv}CCH_2CO_2H$	$HOCH_2CO_2H$
Cyanoacetic acid	Glycolic acid
$K_a = 3.37 \times 10^{-3}$	$K_a = 1.48 \times 10^{-4}$

The reason for this anomaly is not that H_2CO_3 is unusual but, rather, that the value commonly given for K_{a1} applies to the equation

$$\underset{(= CO_2(aq) + H_2CO_3)}{\text{All dissolved } CO_2} \rightleftharpoons HCO_3^- + H^+$$

$$K_{a1} = \frac{[HCO_3^-][H^+]}{[CO_2(aq) + H_2CO_3]} = 4.46 \times 10^{-7}$$

Only about 0.2% of dissolved CO_2 is in the form H_2CO_3. When the true value of $[H_2CO_3]$ is used instead of the value $[H_2CO_3 + CO_2(aq)]$, the value of the equilibrium constant becomes

$$K_{a1} = \frac{[HCO_3^-][H^+]}{[H_2CO_3]} = 2 \times 10^{-4}$$

The hydration of CO_2 (reaction of CO_2 with H_2O) and dehydration of H_2CO_3 are slow reactions that can be demonstrated in a classroom.[29] Living cells use the enzyme *carbonic anhydrase* to speed the rate at which H_2CO_3 and CO_2 equilibrate in order to process this key metabolite. The enzyme provides an environment just right for the reaction of CO_2 with OH^-, lowering the *activation energy* (the energy barrier for the reaction) from 50 to 26 kJ/mol and increasing the rate of reaction by more than a factor of 10^6. One molecule of carbonic anhydrase can catalyze the conversion of 600 000 molecules of CO_2 each second.

Carbonic acid is estimated to have a half life of 200 000 years in the gas phase at 300 K in the absence of water.[30] The presence of just two molecules of H_2O per H_2CO_3 is calculated to reduce the half life to 2 min. The dimer $(H_2CO_3)_2$ or oligomers $(H_2CO_3)_n$[31] and a crystalline form of H_2CO_3[32] are known in the solid state at cryogenic temperatures.

Relation Between K_a and K_b

A most important relation exists between K_a and K_b of a conjugate acid-base pair in aqueous solution. We can derive this result with the acid HA and its conjugate base A$^-$.

$$HA \rightleftharpoons H^+ + A^- \qquad K_a = \frac{[H^+][A^-]}{[HA]}$$

$$\underline{A^- + H_2O \rightleftharpoons HA + OH^-} \qquad K_b = \frac{[HA][OH^-]}{[A^-]}$$

$$H_2O \rightleftharpoons H^+ + OH^- \qquad K_w = K_a \cdot K_b$$

$$= \frac{[H^+][A^-]}{[HA]} \cdot \frac{[HA][OH^-]}{[A^-]}$$

When the reactions are added, their equilibrium constants are multiplied to give

$K_a \cdot K_b = K_w$ for a conjugate acid-base pair in aqueous solution.

Relation between K_a and K_b for a conjugate pair:

$$K_a \cdot K_b = K_w \qquad (6\text{-}35)$$

Equation 6-35 applies to any acid and its conjugate base in aqueous solution.

EXAMPLE   **Finding K_b for the Conjugate Base**

K_a for acetic acid is 1.75×10^{-5} (Reaction 6-26). Find K_b for acetate ion.

Solution This one is trivial:*

$$K_b = \frac{K_w}{K_a} = \frac{1.0 \times 10^{-14}}{1.75 \times 10^{-5}} = 5.7 \times 10^{-10}$$

TEST YOURSELF K_a for chloroacetic acid is 1.36×10^{-3}. Find K_b for chloroacetate ion. (*Answer:* 7.4×10^{-12})

EXAMPLE   **Finding K_a for the Conjugate Acid**

K_b for methylamine is 4.47×10^{-4} (Reaction 6-27). Find K_a for methylammonium ion.

Solution Once again,

$$K_a = \frac{K_w}{K_b} = 2.2 \times 10^{-11}$$

TEST YOURSELF K_b for dimethylamine is 5.9×10^{-4}. Find K_a for dimethylammonium ion. (*Answer:* 1.7×10^{-11})

For a diprotic acid, we can derive relationships between each of two acids and their conjugate bases:

$$H_2A \rightleftharpoons H^+ + HA^- \quad K_{a1} \qquad\qquad HA^- \rightleftharpoons H^+ + A^{2-} \quad K_{a2}$$

$$\underline{HA^- + H_2O \rightleftharpoons H_2A + OH^-} \quad K_{b2} \qquad \underline{A^{2-} + H_2O \rightleftharpoons HA^- + OH^-} \quad K_{b1}$$

$$H_2O \rightleftharpoons H^+ + OH^- \quad K_w \qquad\qquad H_2O \rightleftharpoons H^+ + OH^- \quad K_w$$

The final results are

General relation between K_a and K_b:

$$K_{a1} \cdot K_{b2} = K_w \qquad (6\text{-}36)$$

$$K_{a2} \cdot K_{b1} = K_w \qquad (6\text{-}37)$$

---

*In this text, we use $K_w = 10^{-14.00} = 1.0 \times 10^{-14}$ at 25°C. The more accurate value from Table 6-1 is $K_w = 10^{-13.995}$. For acetic acid with $K_a = 10^{-4.756}$, the accurate value of K_b is $10^{-(13.995-4.756)} = 10^{-9.239} = 5.77 \times 10^{-10}$.

Challenge Derive the following results for a triprotic acid:

$$K_{a1} \cdot K_{b3} = K_w \qquad \text{(6-38)}$$

$$K_{a2} \cdot K_{b2} = K_w \qquad \text{(6-39)}$$

$$K_{a3} \cdot K_{b1} = K_w \qquad \text{(6-40)}$$

Shorthand for Organic Structures

We are beginning to encounter organic (carbon-containing) compounds in this book. Chemists and biochemists use simple conventions for drawing molecules to avoid writing every atom. Each vertex of a structure is understood to be a carbon atom, unless otherwise labeled. In the shorthand, we usually omit bonds from carbon to hydrogen. Carbon forms four chemical bonds. If you see carbon forming fewer than four bonds, the remaining bonds are assumed to go to hydrogen atoms that are not written. Here is an example:

Epinephrine (also called adrenaline)

Shorthand drawing of epinephrine

Benzene, C_6H_6, has two equivalent resonance structures, so all C–C bonds are equivalent. We often draw benzene rings with a circle in place of three double bonds.

Benzene
C_6H_6

The shorthand shows that the carbon atom at the top right of the six-membered benzene ring forms three bonds to other carbon atoms (one single bond and one double bond), so there must be a hydrogen atom attached to this carbon atom. The carbon atom at the left side of the benzene ring forms three bonds to other carbon atoms and one bond to an oxygen atom. There is no hidden hydrogen atom attached to this carbon. In the CH_2 group adjacent to nitrogen, both hydrogen atoms are omitted in the shorthand structure.

Terms to Understand

acid	Brønsted-Lowry base	entropy change	overall formation constant
acid dissociation constant (K_a)	carboxylate anion	equilibrium constant	pH
acidic solution	carboxylic acid	exothermic	polyprotic acids
amine	common ion effect	Gibbs free energy	polyprotic bases
ammonium ion	complex ion	hydronium ion	protic solvent
aprotic solvent	conjugate acid-base pair	ion pair	reaction quotient
autoprotolysis	coprecipitation	Le Châtelier's principle	salt
base	cumulative formation constant	Lewis acid	saturated solution
base hydrolysis constant (K_b)	disproportionation	Lewis base	solubility product
basic solution	endothermic	ligand	standard state
Brønsted-Lowry acid	enthalpy change	neutralization	stepwise formation constant

Summary

For the reaction $aA + bB \rightleftharpoons cC + dD$, the equilibrium constant is $K = [C]^c[D]^d/[A]^a[B]^b$. Solute concentrations should be expressed in moles per liter; gas concentrations should be in bars; and the concentrations of pure solids, liquids, and solvents are omitted. If the direction of a reaction is changed, $K' = 1/K$. If two reactions are added, $K_3 = K_1K_2$. The equilibrium constant can be calculated from the free-energy change for a chemical reaction: $K = e^{-\Delta G°/RT}$. The equation $\Delta G = \Delta H - T\Delta S$ summarizes the observations that a reaction is favored if it liberates heat (exothermic, negative ΔH) or

increases entropy (positive ΔS). Entropy change is the amount of energy, at a given temperature, that is dispersed into motions of the molecules in the system. Le Châtelier's principle predicts the effect on a chemical reaction when reactants or products are added or temperature is changed. The reaction quotient, Q, tells how a system must change to reach equilibrium.

The solubility product is the equilibrium constant for the dissolution of a solid salt into its constituent ions in aqueous solution. The common ion effect is the observation that, if one of the ions of that salt is already present in the solution, the solubility of a salt is decreased. Sometimes, we can selectively precipitate one ion from a solution containing other ions by adding a suitable counterion. At high concentration of ligand, a precipitated metal ion may redissolve by forming soluble complex ions. In a metal-ion complex, the metal is a Lewis acid (electron pair acceptor) and the ligand is a Lewis base (electron pair donor).

Brønsted-Lowry acids are proton donors, and Brønsted-Lowry bases are proton acceptors. An acid increases the concentration of H_3O^+ in aqueous solution, and a base increases the concentration of OH^-. An acid-base pair related through the gain or loss of a single proton is described as conjugate. When a proton is transferred from one molecule to another molecule of a protic solvent, the reaction is called autoprotolysis.

The definition pH = $-\log[H^+]$ will be modified to include activity later. K_a is the equilibrium constant for the dissociation of an acid: HA + $H_2O \rightleftharpoons H_3O^+ + A^-$. The base hydrolysis constant for the reaction B + $H_2O \rightleftharpoons BH^+ + OH^-$ is K_b. When either K_a or K_b is large, the acid or base is said to be strong; otherwise, the acid or base is weak. Common strong acids and bases are listed in Table 6-2, which you should memorize. The most common weak acids are carboxylic acids (RCO_2H), and the most common weak bases are amines (R_3N:). Carboxylate anions (RCO_2^-) are weak bases, and ammonium ions (R_3NH^+) are weak acids. Metal cations are weak acids. For a conjugate acid-base pair in water, $K_a \cdot K_b = K_w$. For polyprotic acids, we denote successive acid dissociation constants as $K_{a1}, K_{a2}, K_{a3}, \ldots$, or just $K_1, K_2, K_3, \ldots$. For polybasic species, we denote successive hydrolysis constants $K_{b1}, K_{b2}, K_{b3}, \ldots$. For a diprotic system, the relations between successive acid and base equilibrium constants are $K_{a1} \cdot K_{b2} = K_w$ and $K_{a2} \cdot K_{b1} = K_w$. For a triprotic system the relations are $K_{a1} \cdot K_{b3} = K_w$, $K_{a2} \cdot K_{b2} = K_w$, and $K_{a3} \cdot K_{b1} = K_w$.

In the chemists' shorthand for organic structures, each vertex is a carbon atom. If fewer than four bonds to that carbon are shown, it is understood that H atoms are attached to the carbon so that it makes four bonds.

Exercises

6-A. Consider the following equilibria in aqueous solution:

(1) $Ag^+ + Cl^- \rightleftharpoons AgCl(aq)$ $K = 2.0 \times 10^3$
(2) $AgCl(aq) + Cl^- \rightleftharpoons AgCl_2^-$ $K = 9.3 \times 10^1$
(3) $AgCl(s) \rightleftharpoons Ag^+ + Cl^-$ $K = 1.8 \times 10^{-10}$

(a) Calculate the numerical value of the equilibrium constant for the reaction $AgCl(s) \rightleftharpoons AgCl(aq)$.

(b) Calculate the concentration of $AgCl(aq)$ in equilibrium with excess undissolved solid AgCl.

(c) Find the numerical value of K for the reaction $AgCl_2^- \rightleftharpoons AgCl(s) + Cl^-$.

6-B. Reaction 6-8 is allowed to come to equilibrium in a solution initially containing 0.010 0 M BrO_3^-, 0.010 0 M Cr^{3+}, and 1.00 M H^+. To find the concentrations at equilibrium, we construct the table at the bottom of the page showing initial and final concentrations. We use the stoichiometry coefficients of the reaction to say that if x mol of Br^- are created, then we must also make x mol of $Cr_2O_7^{2-}$ and $8x$ mol of H^+. To produce x mol of Br^-, we must have consumed x mol of BrO_3^- and $2x$ mol of Cr^{3+}.

(a) Write the equilibrium constant expression that you would use to solve for x to find the concentrations at equilibrium. Do not try to solve the equation.

(b) Because $K = 1 \times 10^{11}$, we suppose that the reaction will go nearly "to completion." That is, we expect both the concentration of Br^- and $Cr_2O_7^{2-}$ to be close to 0.005 00 M at equilibrium. (Why?) That is, $x \approx 0.005\ 00$ M. With this value of x, $[H^+] = 1.00 + 8x = 1.04$ M and $[BrO_3^-] = 0.010\ 0 - x = 0.005\ 0$ M. However, we cannot say $[Cr^{3+}] = 0.010\ 0 - 2x = 0$, because there must be some small concentration of Cr^{3+} at equilibrium. Write $[Cr^{3+}]$ for the concentration of Cr^{3+} and solve for $[Cr^{3+}]$. The *limiting reagent* in this example is Cr^{3+}. The reaction uses up Cr^{3+} before consuming BrO_3^-.

6-C. Find $[La^{3+}]$ in the solution when excess solid lanthanum iodate, $La(IO_3)_3$, is stirred with 0.050 M $LiIO_3$ until the system reaches equilibrium. Assume that IO_3^- from $La(IO_3)_3$ is negligible compared with IO_3^- from $LiIO_3$.

6-D. Which will be more soluble (moles of metal dissolved per liter of solution), $Ba(IO_3)_2$ ($K_{sp} = 1.5 \times 10^{-9}$) or $Ca(IO_3)_2$ ($K_{sp} = 7.1 \times 10^{-7}$)? Give an example of a chemical reaction that might occur that would reverse the predicted solubilities.

6-E. Fe(III) precipitates from acidic solution by addition of OH^- to form $Fe(OH)_3(s)$. At what concentration of OH^- will [Fe(III)] be reduced to 1.0×10^{-10} M? If Fe(II) is used instead, what concentration of OH^- will reduce [Fe(II)] to 1.0×10^{-10} M?

6-F. Is it possible to precipitate 99.0% of 0.010 M Ce^{3+} by adding oxalate ($C_2O_4^{2-}$) without precipitating 0.010 M Ca^{2+}?

$$CaC_2O_4 \qquad K_{sp} = 1.3 \times 10^{-8}$$
$$Ce_2(C_2O_4)_3 \qquad K_{sp} = 5.9 \times 10^{-30}$$

6-G. For a solution of Ni^{2+} and ethylenediamine, the following equilibrium constants apply at 20°C:

$$Ni^{2+} + H_2NCH_2CH_2NH_2 \rightleftharpoons Ni(en)^{2+} \qquad \log K_1 = 7.52$$
$$\text{Ethylenediamine}$$
$$\text{(abbreviated en)}$$
$$Ni(en)^{2+} + en \rightleftharpoons Ni(en)_2^{2+} \qquad\qquad \log K_2 = 6.32$$
$$Ni(en)_2^{2+} + en \rightleftharpoons Ni(en)_3^{2+} \qquad\qquad \log K_3 = 4.49$$

Calculate the concentration of free Ni^{2+} in a solution prepared by mixing 0.100 mol of en plus 1.00 mL of 0.010 0 M Ni^{2+} and diluting to 1.00 L with dilute base (which keeps all the en in its unprotonated form). Assume that nearly all nickel is in the form $Ni(en)_3^{2+}$ so $[Ni(en)_3^{2+}] = 1.00 \times 10^{-5}$ M. Calculate the concentrations of $Ni(en)^{2+}$ and $Ni(en)_2^{2+}$ to verify that they are negligible in comparison with $Ni(en)_3^{2+}$.

	BrO_3^-	+	$2Cr^{3+}$	+ $4H_2O$	$\rightleftharpoons$	Br^-	+	$Cr_2O_7^{2-}$	+ $8H^+$
Initial concentration	0.010 0		0.010 0						1.00
Final concentration	0.010 0 − x		0.010 0 − $2x$			x		x	1.00 + 8x

6-H. If each compound is dissolved in water, will the solution be acidic, basic, or neutral?

(a) Na^+Br^-

(b) $Na^+CH_3CO_2^-$

(c) $NH_4^+Cl^-$

(d) K_3PO_4

(e) $(CH_3)_4N^+Cl^-$

(f) $(CH_3)_4N^+$ ⟨benzene ring⟩$-CO_2^-$

(g) $Fe(NO_3)_3$

6-I. Succinic acid dissociates in two steps:

$$HOCCH_2CH_2COH \overset{K_1}{\rightleftharpoons} HOCCH_2CH_2CO^- + H^+$$
$$K_1 = 6.2 \times 10^{-5}$$

$$HOCCH_2CH_2CO^- \overset{K_2}{\rightleftharpoons} {}^-OCCH_2CH_2CO^- + H^+$$
$$K_2 = 2.3 \times 10^{-6}$$

Calculate K_{b1} and K_{b2} for the following reactions:

$${}^-OCCH_2CH_2CO^- + H_2O \overset{K_{b1}}{\rightleftharpoons} HOCCH_2CH_2CO^- + OH^-$$

$$HOCCH_2CH_2CO^- + H_2O \overset{K_{b2}}{\rightleftharpoons} HOCCH_2CH_2COH + OH^-$$

6-J. Histidine is a triprotic amino acid:

$K_1 = 3 \times 10^{-2}$

$K_2 = 8.5 \times 10^{-7}$

$K_3 = 4.6 \times 10^{-10}$

Find the value of the equilibrium constant for the reaction

6-K. (a) From K_w in Table 6-1, calculate the pH of pure water at 0°, 20°, and 40°C.

(b) For the reaction $D_2O \rightleftharpoons D^+ + OD^-$, $K = [D^+][OD^-] = 1.35 \times 10^{-15}$ at 25°C. In this equation, D stands for deuterium, which is the isotope ^2H. What is the pD $(= -\log[D^+])$ for neutral D_2O?

Problems

Equilibrium and Thermodynamics

6-1. To evaluate the equilibrium constant in Equation 6-2, we must express concentrations of solutes in mol/L, gases in bars, and omit solids, liquids, and solvents. Explain why.

6-2. Why do we say that the equilibrium constant for the reaction $H_2O \rightleftharpoons H^+ + OH^-$ (or any other reaction) is dimensionless?

6-3. Explain the statement that predictions about the direction of a reaction based on Gibbs free energy or Le Châtelier's principle are *thermodynamic*, not *kinetic*.

6-4. Write the expression for the equilibrium constant for each of the following reactions. Write the pressure of a gas, X, as P_X.

(a) $3Ag^+(aq) + PO_4^{3-}(aq) \rightleftharpoons Ag_3PO_4(s)$

(b) $C_6H_6(l) + \frac{15}{2}O_2(g) \rightleftharpoons 3H_2O(l) + 6CO_2(g)$

6-5. For the reaction $2A(g) + B(aq) + 3C(l) \rightleftharpoons D(s) + 3E(g)$, the concentrations at equilibrium are found to be

A: 2.8×10^3 Pa   C: 12.8 M   E: 3.6×10^4 Torr
B: 1.2×10^{-2} M   D: 16.5 M

Find the numerical value of the equilibrium constant that would appear in a conventional table of equilibrium constants.

6-6. From the equations

$$HOCl \rightleftharpoons H^+ + OCl^- \qquad K = 3.0 \times 10^{-8}$$
$$HOCl + OBr^- \rightleftharpoons HOBr + OCl^- \qquad K = 15$$

find the value of K for the reaction $HOBr \rightleftharpoons H^+ + OBr^-$.

6-7. (a) A favorable entropy change occurs when ΔS is positive. Does the order of the system increase or decrease when ΔS is positive?

(b) A favorable enthalpy change occurs when ΔH is negative. Does the system absorb heat or give off heat when ΔH is negative?

(c) Write the relation between ΔG, ΔH, and ΔS. Use the results of parts (a) and (b) to state whether ΔG must be positive or negative for a spontaneous change.

6-8. For the reaction $HCO_3^- \rightleftharpoons H^+ + CO_3^{2-}$, $\Delta G° = +59.0$ kJ/mol at 298.15 K. Find the value of K for the reaction.

6-9. The formation of tetrafluoroethylene from its elements is highly exothermic:

$$2F_2(g) + 2C(s) \rightleftharpoons F_2C{=}CF_2(g)$$
Fluorine   Graphite   Tetrafluoroethylene

(a) If a mixture of F_2, graphite, and C_2F_4 is at equilibrium in a closed container, will the reaction go to the right or to the left if F_2 is added?

(b) Rare bacteria from the planet Teflon eat C_2F_4 and make Teflon for their cell walls. Will the reaction go to the right or to the left if these bacteria are added?

Teflon

(c) Will the reaction go right or left if solid graphite is added? (Neglect any effect of increased pressure due to the decreased volume in the vessel when solid is added.)

(d) Will the reaction go right or left if the container is crushed to one-eighth of its original volume?

(e) Does the equilibrium constant become larger or smaller if the container is heated?

6-10. $BaCl_2 \cdot H_2O(s)$ loses water when it is heated in an oven:

$$BaCl_2 \cdot H_2O(s) \rightleftharpoons BaCl_2(s) + H_2O(g)$$

$$\Delta H° = 63.11 \text{ kJ/mol at } 25°C$$

$$\Delta S° = +148 \text{ J/(K} \cdot \text{mol) at } 25°C$$

(a) Write the equilibrium constant for this reaction. Calculate the vapor pressure of gaseous H_2O (P_{H_2O}) above $BaCl_2 \cdot H_2O$ at 298 K.

(b) If $\Delta H°$ and $\Delta S°$ are not temperature dependent (a poor assumption), estimate the temperature at which P_{H_2O} above $BaCl_2 \cdot H_2O(s)$ will be 1 bar.

6-11. The equilibrium constant for the reaction of $NH_3(aq)$ + $H_2O \rightleftharpoons NH_4^+ + OH^-$ is $K_b = 1.479 \times 10^{-5}$ at 5°C and 1.570×10^{-5} at 10°C.

(a) Assuming $\Delta H°$ and $\Delta S°$ are constant in the interval 5°−10°C (probably a good assumption for small ΔT), use Equation 6-9 to find $\Delta H°$ for the reaction in this temperature range.

(b) Describe how Equation 6-9 could be used to make a linear graph to determine $\Delta H°$, if $\Delta H°$ and $\Delta S°$ were constant over some temperature range.

6-12. For $H_2(g) + Br_2(g) \rightleftharpoons 2HBr(g)$, $K = 7.2 \times 10^{-4}$ at 1 362 K and $\Delta H°$ is positive. A vessel is charged with 48.0 Pa HBr, 1 370 Pa H_2, and 3 310 Pa Br_2 at 1 362 K.

(a) Will the reaction proceed to the left or the right to reach equilibrium?

(b) Calculate the pressure (Pa) of each species at equilibrium.

(c) If the mixture at equilibrium is compressed to half of its original volume, will the reaction proceed to the left or the right to reestablish equilibrium?

(d) If the mixture at equilibrium is heated from 1 362 to 1 407 K, will HBr be formed or consumed in order to reestablish equilibrium?

6-13. *Henry's law* states that the concentration of a gas dissolved in a liquid is proportional to the pressure of the gas. This law is a consequence of the equilibrium

$$X(g) \xrightarrow{K_h} X(aq) \qquad K_h = \frac{[X]}{P_X}$$

where K_h is called the Henry's law constant. For the gasoline additive MTBE, $K_h = 1.71$ M/bar. Suppose we have a closed container with aqueous solution and air in equilibrium. If the concentration of MTBE in the water is 1.00×10^2 ppm ($= 100 \mu g$ MTBE/g solution $\approx 100 \mu g/mL$), what is the pressure of MTBE in the air?

$$CH_3\text{—}O\text{—}C(CH_3)_3 \qquad \text{Methyl-\textit{t}-butylether (MTBE, FM 88.15)}$$

Solubility Product

6-14. Find $[Cu^+]$ in equilibrium with $CuBr(s)$ and 0.10 M Br^-.

6-15. What concentration of $Fe(CN)_6^{4-}$ (ferrocyanide) is in equilibrium with 1.0 μM Ag^+ and $Ag_4Fe(CN)_6(s)$? Express your answer with a prefix from Table 1-3.

6-16. Find $[Cu^{2+}]$ in a solution saturated with $Cu_4(OH)_6(SO_4)$ if $[OH^-]$ is *fixed* at 1.0×10^{-6} M. Note that $Cu_4(OH)_6(SO_4)$ gives 1 mol of SO_4^{2-} for 4 mol of Cu^{2+}.

$$Cu_4(OH)_6(SO_4)(s) \rightleftharpoons 4Cu^{2+} + 6OH^- + SO_4^{2-}$$

$$K_{sp} = 2.3 \times 10^{-69}$$

6-17. (a) From the solubility product of zinc ferrocyanide, $Zn_2Fe(CN)_6$, calculate the concentration of $Fe(CN)_6^{4-}$ in 0.10 mM $ZnSO_4$ saturated with $Zn_2Fe(CN)_6$. Assume that $Zn_2Fe(CN)_6$ is a negligible source of Zn^{2+}.

(b) What concentration of $K_4Fe(CN)_6$ should be in a suspension of solid $Zn_2Fe(CN)_6$ in water to give $[Zn^{2+}] = 5.0 \times 10^{-7}$ M?

6-18. Solubility products predict that cation A^{3+} can be 99.999% separated from cation B^{2+} by precipitation with anion X^-. When the separation is tried, we find 0.2% contamination of $AX_3(s)$ with B^{2+}. Explain what might be happening.

6-19. A solution contains 0.050 0 M Ca^{2+} and 0.030 0 M Ag^+. Can 99% of Ca^{2+} be precipitated by sulfate without precipitating Ag^+? What will be the concentration of Ca^{2+} when Ag_2SO_4 begins to precipitate?

6-20. A solution contains 0.010 M Ba^{2+} and 0.010 M Ag^+. Can 99.90% of either ion be precipitated by chromate (CrO_4^{2-}) without precipitating the other metal ion?

6-21. If a solution containing 0.10 M Cl^-, Br^-, I^-, and CrO_4^{2-} is treated with Ag^+, in what order will the anions precipitate?

Complex Formation

6-22. Explain why the total solubility of lead in Figure 6-3 first decreases and then increases as $[I^-]$ increases. Give an example of the chemistry in each of the two domains.

6-23. Identify the Lewis acids in the following reactions:
(a) $BF_3 + NH_3 \rightleftharpoons F_3\overset{-}{B}\text{—}\overset{+}{N}H_3$
(b) $F^- + AsF_5 \rightleftharpoons AsF_6^-$

6-24. The cumulative formation constant for $SnCl_2(aq)$ in 1.0 M $NaNO_3$ is $\beta_2 = 12$. Find the concentration of $SnCl_2(aq)$ for a solution in which the concentrations of Sn^{2+} and Cl^- are both somehow fixed at 0.20 M.

6-25. Given the following equilibria, calculate the concentration of each zinc species in a solution saturated with $Zn(OH)_2(s)$ and containing $[OH^-]$ at a fixed concentration of 3.2×10^{-7} M.

$Zn(OH)_2(s)$	$K_{sp} = 3 \times 10^{-16}$
$Zn(OH)^+$	$\beta_1 = 1 \times 10^4$
$Zn(OH)_2(aq)$	$\beta_2 = 2 \times 10^{10}$
$Zn(OH)_3^-$	$\beta_3 = 8 \times 10^{13}$
$Zn(OH)_4^{2-}$	$\beta_4 = 3 \times 10^{15}$

6-26. Although KOH, RbOH, and CsOH have little association between metal and hydroxide in aqueous solution, Li^+ and Na^+ do form complexes with OH^-:

$$Li^+ + OH^- \rightleftharpoons LiOH(aq) \qquad K_1 = \frac{[LiOH(aq)]}{[Li^+][OH^-]} = 0.83$$

$$Na^+ + OH^- \rightleftharpoons NaOH(aq) \qquad K_1 = 0.20$$

Prepare a table like the one in Exercise 6-B showing initial and final concentrations of Na^+, OH^-, and $NaOH(aq)$ in 1 F NaOH solution. Calculate the fraction of sodium in the form NaOH(*aq*) at equilibrium.

6-27. In Figure 6-3, the concentration of $PbI_2(aq)$ is independent of $[I^-]$. Use any of the equilibrium constants for Reactions 6-12 through 6-16 to find the equilibrium constant for the reaction $PbI_2(s) \rightleftharpoons PbI_2(aq)$, which is equal to the concentration of $PbI_2(aq)$.

Acids and Bases

6-28. Distinguish Lewis acids and bases from Brønsted-Lowry acids and bases. Give an example of each.

6-29. Fill in the blanks:

(a) The product of a reaction between a Lewis acid and a Lewis base is called _____.

(b) The bond between a Lewis acid and a Lewis base is called _____ or _____.

(c) Brønsted-Lowry acids and bases related by gain or loss of one proton are said to be _____.

(d) A solution is *acidic* if _____. A solution is *basic* if _____.

6-30. Why is the pH of distilled water usually <7? How can you prevent this from happening?

6-31. Gaseous SO_2 is created by combustion of sulfur-containing fuels, especially coal. Explain how SO_2 in the atmosphere makes acidic rain.

6-32. Use electron dot structures to show why tetramethylammonium hydroxide, $(CH_3)_4N^+OH^-$, is an ionic compound. That is, show why hydroxide is not covalently bound to the rest of the molecule.

6-33. Identify the Brønsted-Lowry acids among the reactants in the following reactions:

(a) $KCN + HI \rightleftharpoons HCN + KI$

(b) $PO_4^{3-} + H_2O \rightleftharpoons HPO_4^{2-} + OH^-$

6-34. Write the autoprotolysis reaction of H_2SO_4.

6-35. Identify the conjugate acid-base pairs in the following reactions:

(a) $H_3\overset{+}{N}CH_2CH_2\overset{+}{N}H_3 + H_2O \rightleftharpoons H_3\overset{+}{N}CH_2CH_2NH_2 + H_3O^+$

(b)

Benzoic acid $\quad$ Pyridine $\quad$ Benzoate $\quad$ Pyridinium

pH

6-36. Calculate $[H^+]$ and pH for the following solutions:

(a) 0.010 M HNO_3 **(d)** 3.0 M HCl

(b) 0.035 M KOH   **(e)** 0.010 M $[(CH_3)_4N^+]OH^-$
                                              Tetramethylammonium hydroxide

(c) 0.030 M HCl

6-37. Use Table 6-1 to calculate the pH of pure water at **(a)** 25°C and **(b)** 100°C.

6-38. The equilibrium constant for the reaction $H_2O \rightleftharpoons H^+ + OH^-$ is 1.0×10^{-14} at 25°C. What is the value of K for the reaction $4H_2O \rightleftharpoons 4H^+ + 4OH^-$?

6-39. An acidic solution containing 0.010 M La^{3+} is treated with NaOH until $La(OH)_3$ precipitates. At what pH does this occur?

6-40. Use Le Châtelier's principle and K_w in Table 6-1 to decide whether the autoprotolysis of water is endothermic or exothermic at **(a)** 25°C; **(b)** 100°C; **(c)** 300°C.

Strengths of Acids and Bases

6-41. Make a list of the common strong acids and strong bases. Memorize this list.

6-42. Write the formulas and names for three classes of weak acids and two classes of weak bases.

6-43. Explain why hydrated metal ions such as $(H_2O)_6Fe^{3+}$ hydrolyze to give H^+, but hydrated anions such as $(H_2O)_6Cl^-$ do not hydrolyze to give H^+.

6-44. Write the K_a reaction for trichloroacetic acid, Cl_3CCO_2H, for anilinium ion, $\langle\!\!\bigcirc\!\!\rangle\!\!-\overset{+}{N}H_3$, and for lanthanum ion, La^{3+}.

6-45. Write the K_b reaction for pyridine and for sodium 2-mercaptoethanol.

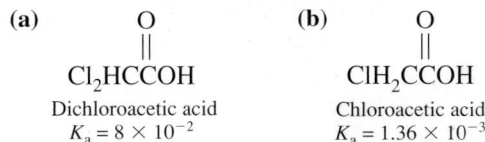

Pyridine   Sodium 2-mercaptoethanol

6-46. Write the K_a and K_b reactions of $NaHCO_3$.

6-47. Write the stepwise acid-base reactions for the following ions in water. Write the correct symbol (for example, K_{b1}) for the equilibrium constant for each reaction.

(a) $H_3\overset{+}{N}CH_2CH_2\overset{+}{N}H_3$ **(b)**
Ethylenediammonium ion
$$\underset{\text{Malonate ion}}{^-OCCH_2CO^-}$$
(with two C=O groups shown above)

6-48. Which is a stronger acid, **(a)** or **(b)**?

(a) Cl_2HCCOH (with C=O)
Dichloroacetic acid
$K_a = 8 \times 10^{-2}$

(b) ClH_2CCOH (with C=O)
Chloroacetic acid
$K_a = 1.36 \times 10^{-3}$

Which is a stronger base, **(c)** or **(d)**?

(c) H_2NNH_2
Hydrazine
$K_b = 1.1 \times 10^{-6}$

(d) H_2NCNH_2 (with C=O)
Urea
$K_b = 1.5 \times 10^{-14}$

6-49. Write the K_b reaction of CN^-. Given that the K_a value for HCN is 6.2×10^{-10}, calculate K_b for CN^-.

6-50. Write the K_{a2} reaction of phosphoric acid (H_3PO_4) and the K_{b2} reaction of disodium oxalate ($Na_2C_2O_4$).

6-51. From the K_b values for phosphate in Equations 6-32 through 6-34, calculate the three K_a values of phosphoric acid.

6-52. From the following equilibrium constants, calculate the equilibrium constant for the reaction $HO_2CCO_2H \rightleftharpoons 2H^+ + C_2O_4^{2-}$.

$HOCCOH \rightleftharpoons H^+ + HOCCO^-$ $K_1 = 5.6 \times 10^{-2}$
Oxalic acid

$HOCCO^- \rightleftharpoons H^+ + {}^-OCCO^-$ $K_2 = 5.4 \times 10^{-5}$
Oxalate

6-53. (a) Using only K_{sp} from Table 6-3, calculate how many moles of $Ca(OH)_2$ will dissolve in 1.00 L of water.

(b) How will the solubility calculated in part **(a)** be affected by the K_1 reaction in Table 6-3?

6-54. The planet Aragonose (which is made mostly of the mineral aragonite, or $CaCO_3$) has an atmosphere containing methane and carbon dioxide, each at a pressure of 0.10 bar. The oceans are saturated with aragonite and have a concentration of H^+ equal to 1.8×10^{-7} M. Given the following equilibria, calculate how many grams of calcium are contained in 2.00 L of Aragonose seawater.

$$CaCO_3(s, \text{aragonite}) \rightleftharpoons Ca^{2+}(aq) + CO_3^{2-}(aq) \quad K_{sp} = 6.0 \times 10^{-9}$$
$$CO_2(g) \rightleftharpoons CO_2(aq) \quad K_{CO_2} = 3.4 \times 10^{-2}$$
$$CO_2(aq) + H_2O(l) \rightleftharpoons HCO_3^-(aq) + H^+(aq) \quad K_1 = 4.5 \times 10^{-7}$$
$$HCO_3^-(aq) \rightleftharpoons H^+(aq) + CO_3^{2-}(aq) \quad K_2 = 4.7 \times 10^{-11}$$

Don't panic! Reverse the first reaction, add all the reactions together, and see what cancels.

7 Let the Titrations Begin

Robotic arm of *Phoenix Mars Lander* scoops up soil for chemical analysis on Mars. [NASA/JPL-Caltech/University of Arizona/Texas A&M University.]

In 2008, Professor Sam Kounaves and his students at Tufts University had the thrill of a lifetime as their Wet Chemistry Laboratory aboard the *Phoenix Mars Lander* returned a stream of information about the ionic composition of Martian soil scooped up by a robotic arm. The arm delivered ~1 gram of soil through a sieve into a "beaker" fitted with a suite of electrochemical sensors described in Chapter 15. Aqueous solution added to the beaker shown in Box 15-3 leached soluble salts from the soil, while sensors measured ions appearing in the liquid. Unlike other ions, sulfate was measured by a *precipitation titration* with Ba^{2+}:

$$BaCl_2(s) \rightarrow Ba^{2+} + 2Cl^-$$
$$SO_4^{2-} + Ba^{2+} \rightarrow BaSO_4(s)$$

As solid $BaCl_2$ from a reagent canister slowly dissolved in the aqueous liquid, $BaSO_4$ precipitated. In Problem 7-21, we see that one sensor showed a low level of Ba^{2+} until enough reagent had been added to react with all SO_4^{2-}. Another sensor found steadily increasing Cl^- as $BaCl_2$ dissolved. The end point of the titration is marked by a sudden increase of Ba^{2+} when the last SO_4^{2-} has precipitated and $BaCl_2$ continues to dissolve. The increase in Cl^- from the beginning of the titration up to the end point tells how much $BaCl_2$ was required to consume SO_4^{2-}. Titrations of two soil samples in two cells found ~1.3 (±0.5) wt% sulfate in the soil.[1] Other evidence suggests that the sulfate is mostly from $MgSO_4$.

Procedures in which we measure the volume of reagent needed to react with analyte are called **volumetric analysis.** In this chapter, we discuss principles that apply to all volumetric procedures and then focus on precipitation titrations. Acid-base, oxidation-reduction, complex formation, and spectrophotometric titrations are discussed later in the book in their respective chapters.

7-1 Titrations

In a **titration,** increments of reagent solution—the **titrant**—are added to analyte until their reaction is complete. From the quantity of titrant required, we can calculate the quantity of analyte that must have been present. Titrant is usually delivered from a buret (Figure 7-1).

The principal requirements for a titration reaction are that it have a large equilibrium constant and proceed rapidly. That is, each increment of titrant should be completely and quickly consumed by analyte until the analyte is used up. Common titrations rely on acid-base, oxidation-reduction, complex formation, or precipitation reactions.

The **equivalence point** occurs when the quantity of added titrant is the exact amount necessary for stoichiometric reaction with the analyte. For example, 5 mol of oxalic acid react with 2 mol of permanganate in hot acidic solution:

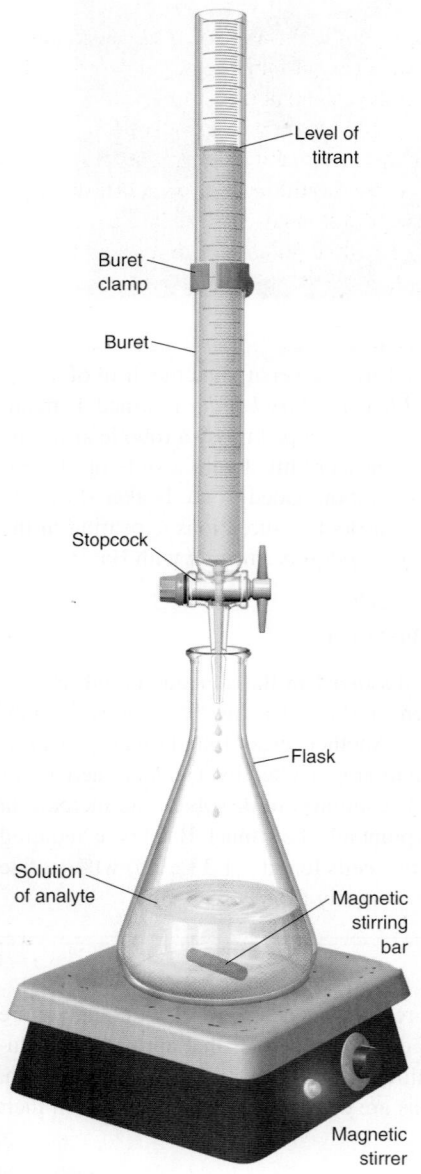

Reaction 7-1 is an oxidation-reduction reaction. Please study Appendix D if you need to relearn how to balance Reaction 7-1.

$$5HO-\overset{\overset{O}{\|}}{C}-\overset{\overset{O}{\|}}{C}-OH + 2MnO_4^- + 6H^+ \rightarrow 10CO_2 + 2Mn^{2+} + 8H_2O \qquad (7\text{-}1)$$

Analyte	**Titrant**		
Oxalic acid colorless	Permanganate purple	colorless	colorless

If the unknown contains 5.000 mmol of oxalic acid, the equivalence point is reached when 2.000 mmol of MnO_4^- have been added.

The equivalence point is the ideal (theoretical) result we seek in a titration. What we actually measure is the **end point,** which is marked by a sudden change in a physical property of the solution. In Reaction 7-1, a convenient end point is the abrupt appearance of the purple color of permanganate in the flask. Prior to the equivalence point, all permanganate is consumed by oxalic acid, and the titration solution remains colorless. After the equivalence point, unreacted MnO_4^- accumulates until there is enough to see. The *first trace* of purple color is the end point. The better your eyes, the closer will be your measured end point to the true equivalence point. Here, the end point cannot exactly equal the equivalence point, because extra MnO_4^-, beyond that needed to react with oxalic acid, is required to exhibit purple color.

Methods for determining when the analyte has been consumed include (1) detecting a sudden change in the voltage or current between a pair of electrodes (Figure 7-5), (2) observing an indicator color change (Color Plate 2), and (3) monitoring absorption of light (Figure 18-11). An **indicator** is a compound with a physical property (usually color) that changes abruptly near the equivalence point. The change is caused by the disappearance of analyte or the appearance of excess titrant.

The difference between the end point and the equivalence point is an inescapable **titration error.** By choosing a physical property whose change is easily observed (such as pH or the color of an indicator), the end point can be very close to the equivalence point. We estimate the titration error with a **blank titration,** in which we carry out the same procedure *without* analyte. For example, we can titrate a solution containing no oxalic acid to see how much MnO_4^- is needed to produce observable purple color. We then subtract this volume of MnO_4^- from the volume observed in the analytical titration.

The validity of an analytical result depends on knowing the amount of one of the reactants used. If a titrant is prepared by dissolving a weighed amount of pure reagent in a known volume of solution, its concentration can be calculated. We call such a reagent a **primary standard,** because it is pure enough to be weighed and used directly. A primary standard should be 99.9% pure, or better. It should not decompose under ordinary storage, and it should be stable when dried by heat or vacuum, because drying is required to remove traces of water adsorbed from the atmosphere. The highest quality, commercially available, primary standard pure materials, such as As_2O_3 (arsenic(III) oxide), $CaCO_3$, Hg metal, and Ni metal, have purities certified to within $\pm 0.05\%$. Primary standards for many elements are given in Appendix K. Box 7-1 discusses reagent purity. Box 3-2 describes certified reference materials that allow laboratories to test the accuracy of their procedures.

Many reagents used as titrants, such as HCl, are not available as primary standards. Instead, we prepare titrant with approximately the desired concentration and use it to titrate a primary standard. By this procedure, called **standardization,** we determine the concentration of titrant. We then say that the titrant is a **standard solution.** The validity of the analytical result ultimately depends on knowing the composition of a primary standard. Sodium oxalate ($Na_2C_2O_4$) is a commercially available primary standard for generating oxalic acid to standardize permanganate in Reaction 7-1.

In a **direct titration,** we add titrant to analyte until the reaction is complete. Occasionally, we perform a **back titration,** in which we add a known *excess* of one standard reagent to the analyte. Then we titrate the excess reagent with a second standard reagent. A back titration is useful when its end point is clearer than the end point of the direct titration, or

FIGURE 7-1 Typical setup for a titration. The analyte is contained in the flask, and the titrant is in the buret. The stirring bar is a magnet coated with Teflon, which is inert to most solutions. The bar is spun by a rotating magnet inside the stirring motor.

Labels on figure: Level of titrant; Buret clamp; Buret; Stopcock; Flask; Solution of analyte; Magnetic stirring bar; Magnetic stirrer

Chemicals are sold in many grades of purity. For analytical chemistry, we usually use **reagent-grade chemicals** meeting purity requirements set by the American Chemical Society (ACS) Committee on Analytical Reagents.[2] Sometimes "reagent grade" simply meets purity standards set by the manufacturer. An actual lot analysis for specified impurities should appear on the reagent bottle. For example, here is a lot analysis of zinc sulfate:

$ZnSO_4$	ACS Reagent	Lot Analysis:
Assay: 100.6%	Fe: 0.000 5%	Ca: 0.001%
Insoluble matter:		
0.002%	Pb: 0.002 8%	Mg: 0.000 3%
pH of 5% solution		
at 25°C: 5.6	Mn: 0.6 ppm	K: 0.002%
Ammonium:		
0.000 8%	Nitrate: 0.000 4%	Na: 0.003%
Chloride: 1.5 ppm		

The assay value of 100.6% means that a specified analysis for one of the major components produced 100.6% of the theoretical value. For example, if $ZnSO_4$ is contaminated with the lower-molecular-mass $Zn(OH)_2$, the assay for Zn^{2+} will be higher than the value for pure $ZnSO_4$. Less pure chemicals, generally unsuitable for

analytical chemistry, carry designations such as "chemically pure" (CP), "practical," "purified," or "technical."

A few chemicals are sold in high enough purity to be *primary standard grade*. Whereas reagent-grade potassium dichromate has a lot assay of $\geq99.0\%$, primary standard grade $K_2Cr_2O_7$ must be in the range 99.95–100.05%. Besides high purity, a key quality of primary standards is that they are indefinitely stable.

For **trace analysis** (analysis of species at ppm and lower levels), impurities in reagent chemicals must be extremely low. For this purpose, we use very-high-purity, expensive grades of acids such as "trace metal grade" HNO_3 or HCl to dissolve samples. We must pay careful attention to reagents and vessels whose impurity levels could be greater than the quantity of analyte we seek to measure.

The following steps help you to protect the purity of chemical reagents:

- Avoid putting a spatula into a bottle. Instead, pour chemical out of the bottle into a clean container (or onto weighing paper) and dispense the chemical from the clean container.
- Never pour unused chemical back into the reagent bottle.
- Replace the cap on the bottle immediately to keep dust out.
- Never put a glass stopper from a liquid-reagent container down on the lab bench. Either hold the stopper or place it in a clean place (such as a clean beaker) while you dispense reagent.
- Store chemicals in a cool, dark place. Do not expose them unnecessarily to sunlight.

when an excess of the first reagent is required for complete reaction with analyte. To appreciate the difference between direct and back titrations, consider first the addition of permanganate titrant to oxalic acid analyte in Reaction 7-1; this reaction is a direct titration. Alternatively, to perform a back titration, we could add a known *excess* of permanganate to consume oxalic acid. Then we could back titrate the excess permanganate with standard Fe^{2+} to measure how much permanganate was left after reaction with the oxalic acid.

An alternative to measuring titrant by volume is to measure the *mass* of titrant solution delivered in each increment. We call this procedure a **gravimetric titration.** Titrant can be delivered from a pipet. Titrant concentration is expressed as moles of reagent per kilogram of solution. Precision is improved from 0.3% attainable with a buret to 0.1% with a balance (see the dilution example on page 55). Experiments by Guenther and by Butler and Swift provide examples.[3] "Gravimetric titrations should become the gold standard, and volumetric glassware should be seen in museums only."[4]

7-2 Titration Calculations

Here are some examples to illustrate stoichiometry calculations in volumetric analysis. The key step is to *relate moles of titrant to moles of analyte.*

EXAMPLE Standardization of Titrant Followed by Analysis of Unknown

The calcium content of urine can be determined by the following procedure:

Step 1 Precipitate Ca^{2+} with oxalate in basic solution:

$$Ca^{2+} + \underset{\text{Oxalate}}{C_2O_4^{2-}} \rightarrow \underset{\text{Calcium oxalate}}{Ca(C_2O_4) \cdot H_2O(s)}$$

Step 2 Wash the precipitate with ice-cold water to remove free oxalate, and dissolve the solid in acid to obtain Ca^{2+} and $H_2C_2O_4$ in solution.

Step 3 Heat the solution to 60°C and titrate the oxalate with standardized potassium permanganate until the purple end point of Reaction 7-1 is observed.

Standardization Suppose that 0.356 2 g of $Na_2C_2O_4$ is dissolved in a 250.0-mL volumetric flask. If 10.00 mL of this solution require 48.36 mL of $KMnO_4$ solution for titration, what is the molarity of the permanganate solution?

Solution The concentration of the oxalate solution is

$$\frac{0.356\,2 \text{ g } Na_2C_2O_4/(134.00 \text{ g } Na_2C_2O_4/mol)}{0.250\,0 \text{ L}} = 0.010\,63_3 \text{ M}$$

The moles of $C_2O_4^{2-}$ in 10.00 mL are $(0.010\,63_3 \text{ mol/L})(0.010\,00 \text{ L}) = 1.063_3 \times 10^{-4} \text{ mol} = 0.106\,3_3$ mmol. Reaction 7-1 requires 2 mol of permanganate for 5 mol of oxalate, so the MnO_4^- delivered must have been

$$\text{Moles of } MnO_4^- = \left(\frac{2 \text{ mol } MnO_4^-}{5 \text{ mol } C_2O_4^{2-}}\right)(\text{mol } C_2O_4^{2-}) = 0.042\,53_1 \text{ mmol}$$

The concentration of MnO_4^- in the titrant is therefore

$$\text{Molarity of } MnO_4^- = \frac{0.042\,53_1 \text{ mmol}}{48.36 \text{ mL}} = 8.794_7 \times 10^{-4} \text{ M}$$

Analysis of Unknown Calcium in a 5.00-mL urine sample was precipitated with $C_2O_4^{2-}$ and redissolved; the sample then required 16.17 mL of standard MnO_4^- solution. Find the concentration of Ca^{2+} in the urine.

Solution In 16.17 mL of MnO_4^-, there are $(0.016\,17 \text{ L})(8.794_7 \times 10^{-4} \text{ mol/L}) = 1.422_1 \times 10^{-5}$ mol MnO_4^-. This quantity will react with

$$\text{Moles of } C_2O_4^{2-} = \left(\frac{5 \text{ mol } C_2O_4^{2-}}{2 \text{ mol } MnO_4^-}\right)(\text{mol } MnO_4^-) = 0.035\,55_3 \text{ mmol}$$

Because there is one oxalate ion for each calcium ion in $Ca(C_2O_4)\cdot H_2O$, there must have been $0.035\,55_3$ mmol of Ca^{2+} in 5.00 mL of urine:

$$[Ca^{2+}] = \frac{0.035\,55_3 \text{ mmol}}{5.00 \text{ mL}} = 0.007\,11_1 \text{ M}$$

TEST YOURSELF In standardization, 10.00 mL of $Na_2C_2O_4$ solution required 39.17 mL of $KMnO_4$. Find the molarity of $KMnO_4$. The unknown required 14.44 mL of MnO_4^-. Find $[Ca^{2+}]$ in the urine. (***Answer:*** 1.086×10^{-3} M, 7.840×10^{-3} M)

EXAMPLE **Titration of a Mixture**

A solid mixture weighing 1.372 g containing only sodium carbonate and sodium bicarbonate required 29.11 mL of 0.734 4 M HCl for complete titration:

$$Na_2CO_3 + 2HCl \rightarrow 2NaCl(aq) + H_2O + CO_2$$
FM 105.99

$$NaHCO_3 + HCl \rightarrow NaCl(aq) + H_2O + CO_2$$
FM 84.01

Find the mass of each component of the mixture.

Solution Let's denote the grams of Na_2CO_3 by x and grams of $NaHCO_3$ by $1.372 - x$. The moles of each component must be

$$\text{Moles of } Na_2CO_3 = \frac{x \text{ g}}{105.99 \text{ g/mol}} \qquad \text{Moles of } NaHCO_3 = \frac{(1.372 - x) \text{ g}}{84.01 \text{ g/mol}}$$

We know that the total number of moles of HCl used was $(0.029\,11 \text{ L})(0.734\,4 \text{ M}) = 0.021\,38$ mol. From the stoichiometry of the two reactions, we can say that

$$2(\text{mol } Na_2CO_3) + \text{mol } NaHCO_3 = 0.021\,38$$

$$2\left(\frac{x}{105.99}\right) + \frac{1.372 - x}{84.01} = 0.021\,38 \implies x = 0.724 \text{ g}$$

The mixture contains 0.724 g of Na_2CO_3 and $1.372 - 0.724 = 0.648$ g of $NaHCO_3$.

TEST YOURSELF A 2.000-g mixture containing only K_2CO_3 (FM 138.21) and $KHCO_3$ (FM 100.12) required 15.00 mL of 1.000 M HCl for complete titration. Find the mass of each component of the mixture. (**Answer:** K_2CO_3 = 1.811 g, $KHCO_3$ = 0.189 g)

7-3 Precipitation Titration Curves

In gravimetric analysis, we could measure an unknown concentration of I^- by adding excess Ag^+ and weighing the AgI precipitate $[I^- + Ag^+ \rightarrow AgI(s)]$. In a *precipitation titration*, we monitor the course of the reaction between analyte (I^-) and titrant (Ag^+) to locate the *equivalence point* at which there is exactly enough titrant for stoichiometric reaction with the analyte. Knowing how much titrant was added tells us how much analyte was present. We seek the equivalence point in a titration, but we observe the *end point* at which there is an abrupt change in a physical property (such as an electrode potential) that is being measured. The physical property is chosen to make the end point as close as possible to the equivalence point.

The **titration curve** is a graph showing how the concentration of a reactant varies as titrant is added. We will derive equations that can be used to predict precipitation titration curves. One reason to calculate titration curves is to understand the chemistry that occurs during titrations. A second reason is to learn how experimental control can be exerted to influence the quality of an analytical titration. Concentrations of analyte and titrant and the size of the *solubility product* (K_{sp}) influence the sharpness of the end point.

Concentration varies over orders of magnitude, so it is useful to plot the p function:

p function:
$$pX = -\log_{10}[X] \tag{7-2}$$

where [X] is the concentration of X.

Consider the titration of 25.00 mL of 0.100 0 M I^- with 0.050 00 M Ag^+,

Titration reaction:
$$I^- + Ag^+ \rightarrow AgI(s) \tag{7-3}$$

and suppose that we are monitoring $[Ag^+]$ with an electrode. Reaction 7-3 is the reverse of the dissolution of AgI(s), whose solubility product is rather small:

$$AgI(s) \rightleftharpoons Ag^+ + I^- \qquad K_{sp} = [Ag^+][I^-] = 8.3 \times 10^{-17} \tag{7-4}$$

The equilibrium constant for the *titration reaction* 7-3 is large ($K = 1/K_{sp} = 1.2 \times 10^{16}$), so the equilibrium lies far to the right. Each aliquot of Ag^+ reacts nearly completely with I^-, leaving only a tiny amount of Ag^+ in solution. At the equivalence point, there will be a sudden increase in $[Ag^+]$ because there is no I^- left to consume the added Ag^+.

What volume of Ag^+ titrant is needed to reach the equivalence point? We calculate this volume, designated V_e, with the fact that 1 mol of Ag^+ reacts with 1 mol of I^-.

$$\underbrace{(0.025\ 00\ L)(0.100\ 0\ mol\ I^-/L)}_{mol\ I^-} = \underbrace{(V_e)(0.050\ 00\ mol\ Ag^+/L)}_{mol\ Ag^+}$$

$$\Rightarrow V_e = 0.050\ 00\ L = 50.00\ mL$$

The titration curve has three distinct regions, depending on whether we are before, at, or after the equivalence point. Let's consider each region separately.

Before the Equivalence Point

Suppose that 10.00 mL of Ag^+ have been added. There are more moles of I^- than Ag^+ at this point, so virtually all Ag^+ is "used up" to make AgI(s). We want to find the small concentration of Ag^+ remaining in solution after reaction with I^-. Imagine that Reaction 7-3 has gone to completion and that some AgI redissolves (Reaction 7-4). The solubility of Ag^+ is determined by the concentration of free I^- remaining in the solution:

$$[Ag^+] = \frac{K_{sp}}{[I^-]} \tag{7-5}$$

Free I^- is overwhelmingly from the I^- that has not been precipitated by 10.00 mL of Ag^+. By comparison, I^- from dissolution of AgI(s) is negligible.

Please review Section 6-3 on the solubility product prior to studying precipitation titration curves.

Equivalence point: Point at which stoichiometric amounts of reactants have been mixed.

End point: Point near the equivalence point at which an abrupt change in a physical property is observed.

In Chapter 8, we write the p function more correctly in terms of activity instead of concentration. For now, we use pX = −log[X].

V_e = volume of titrant at equivalence point

Eventually, we will derive a single, unified equation for a spreadsheet that treats all regions of the titration curve. To understand the chemistry, we break the curve into three regions described by approximate equations that are easy to use.

When $V < V_e$, the concentration of unreacted I^- regulates the solubility of AgI.

So let's find the concentration of unprecipitated I^-:

$$\text{Moles of } I^- = \text{original moles of } I^- - \text{moles of } Ag^+ \text{ added}$$
$$= (0.025\ 00\ L)(0.100\ mol/L) - (0.010\ 00\ L)(0.050\ 00\ mol/L)$$
$$= 0.002\ 000\ mol\ I^-$$

The volume is 0.035 00 L (25.00 mL + 10.00 mL), so the concentration is

$$[I^-] = \frac{0.002\ 000\ mol\ I^-}{0.035\ 00\ L} = 0.057\ 14\ M \tag{7-6}$$

The concentration of Ag^+ in equilibrium with this much I^- is

$$[Ag^+] = \frac{K_{sp}}{[I^-]} = \frac{8.3 \times 10^{-17}}{0.057\ 14} = 1.4_5 \times 10^{-15}\ M \tag{7-7}$$

Finally, the p function we seek is

$$pAg^+ = -\log[Ag^+] = -\log(1.4_5 \times 10^{-15}) = 14.84 \tag{7-8}$$

Two significant figures in K_{sp} provide two significant figures in $[Ag^+]$. The two figures in $[Ag^+]$ translate into two figures in the *mantissa* of the p function, which is written as 14.84.

The preceding step-by-step calculation is a tedious way to find the concentration of I^-. Here is a streamlined procedure that is well worth learning. Bear in mind that $V_e = 50.00$ mL. When 10.00 mL of Ag^+ have been added, the reaction is one-fifth complete because 10.00 mL out of the 50.00 mL of Ag^+ needed for complete reaction have been added. Therefore, four-fifths of the I^- is unreacted. If there were no dilution, $[I^-]$ would be four-fifths of its original value. However, the original volume of 25.00 mL has been increased to 35.00 mL. If no I^- had been consumed, the concentration would be the original value of $[I^-]$ times (25.00/35.00). Accounting for both the reaction and the dilution, we can write

$$[I^-] = \underbrace{\left(\frac{4.000}{5.000}\right)}_{\substack{\text{Fraction} \\ \text{remaining}}} \underbrace{(0.100\ 0\ M)}_{\substack{\text{Original} \\ \text{concentration}}} \underbrace{\left(\frac{25.00}{35.00}\right)}_{\substack{\text{Dilution} \\ \text{factor}}} = 0.057\ 14\ M$$

(Original volume of I^- / Total volume of solution)

This is the same result found in Equation 7-6.

EXAMPLE Using the Streamlined Calculation

Let's calculate pAg^+ when V_{Ag^+} (the volume added from the buret) is 49.00 mL.

Solution Because $V_e = 50.00$ mL, the fraction of I^- reacted is 49.00/50.00, and the fraction remaining is 1.00/50.00. The total volume is 25.00 + 49.00 = 74.00 mL.

$$[I^-] = \underbrace{\left(\frac{1.00}{50.00}\right)}_{\substack{\text{Fraction} \\ \text{remaining}}} \underbrace{(0.100\ 0\ M)}_{\substack{\text{Original} \\ \text{concentration}}} \underbrace{\left(\frac{25.00}{74.00}\right)}_{\substack{\text{Dilution} \\ \text{factor}}} = 6.76 \times 10^{-4}\ M$$

$$[Ag^+] = K_{sp}/[I^-] = (8.3 \times 10^{-17})/(6.76 \times 10^{-4}) = 1.2_3 \times 10^{-13}\ M$$
$$pAg^+ = -\log[Ag^+] = 12.91$$

The concentration of Ag^+ is negligible compared with the concentration of unreacted I^-, even though the titration is 98% complete.

TEST YOURSELF Find pAg^+ at 49.1 mL. (*Answer:* 12.86)

At the Equivalence Point

Now we have added exactly enough Ag^+ to react with all the I^-. Imagine that all AgI precipitates and some redissolves to give equal concentrations of Ag^+ and I^-. The value of pAg^+ is found by setting $[Ag^+] = [I^-] = x$ in the solubility product:

$$[Ag^+][I^-] = K_{sp}$$
$$(x)(x) = 8.3 \times 10^{-17} \Rightarrow x = 9.1 \times 10^{-9} \Rightarrow pAg^+ = -\log x = 8.04$$

Margin notes:

$\log(1.4_5 \times 10^{-15}) = 14.84$

Two significant figures | Two digits in mantissa

Significant figures in logarithms were discussed in Section 3-2.

Streamlined calculation *well worth using.*

When $V = V_e$, $[Ag^+]$ is determined by the solubility of pure AgI. *This problem is the same as if we had just added AgI(s) to water.*

This value of pAg^+ is independent of the original concentrations or volumes.

After the Equivalence Point

Virtually all Ag^+ added *before* the equivalence point has precipitated. The solution contains all of the Ag^+ added *after* the equivalence point. Suppose that $V_{Ag^+} = 52.00$ mL. The volume past the equivalence point is 2.00 mL. The calculation proceeds as follows:

$$\text{Moles of } Ag^+ = (0.002\ 00\ L)(0.050\ 00\ mol\ Ag^+/L) = 0.000\ 100\ mol$$

$$[Ag^+] = (0.000\ 100\ mol)/(0.077\ 00\ L) = 1.30 \times 10^{-3}\ M \Rightarrow pAg^+ = 2.89$$

$$\text{Total volume} = 77.00\ mL$$

When $V > V_e$, $[Ag^+]$ is determined by the excess Ag^+ added from the buret.

We could justify three significant figures for the mantissa of pAg^+ because there are three significant figures in $[Ag^+]$. For consistency with earlier results, we retain only two figures.

For a streamlined calculation, the concentration of Ag^+ in the buret is 0.050 00 M, and 2.00 mL of titrant are being diluted to $(25.00 + 52.00) = 77.00$ mL. Hence, $[Ag^+]$ is

Volume of excess Ag^+

$$[Ag^+] = (0.050\ 00\ M)\left(\frac{2.00}{77.00}\right) = 1.30 \times 10^{-3}\ M$$

Total volume of solution

Original concentration of Ag^+ | Dilution factor

The Shape of the Titration Curve

Titration curves in Figure 7-2 illustrate the effect of reactant concentration. The equivalence point is the steepest point of the curve. It is the point of maximum slope (a negative slope in this case) and is therefore an inflection point (at which the second derivative is 0):

Steepest slope: $\dfrac{dy}{dx}$ reaches its greatest value

Inflection point: $\dfrac{d^2y}{dx^2} = 0$

In titrations involving 1 : 1 stoichiometry of reactants, the equivalence point is the steepest point of the titration curve. For stoichiometries other than 1 : 1, such as $2Ag^+ + CrO_4^{2-} \rightarrow Ag_2CrO_4(s)$, the curve is not symmetric. The equivalence point is not at the center of the steepest section of the curve, and it is not an inflection point. In practice, conditions are chosen

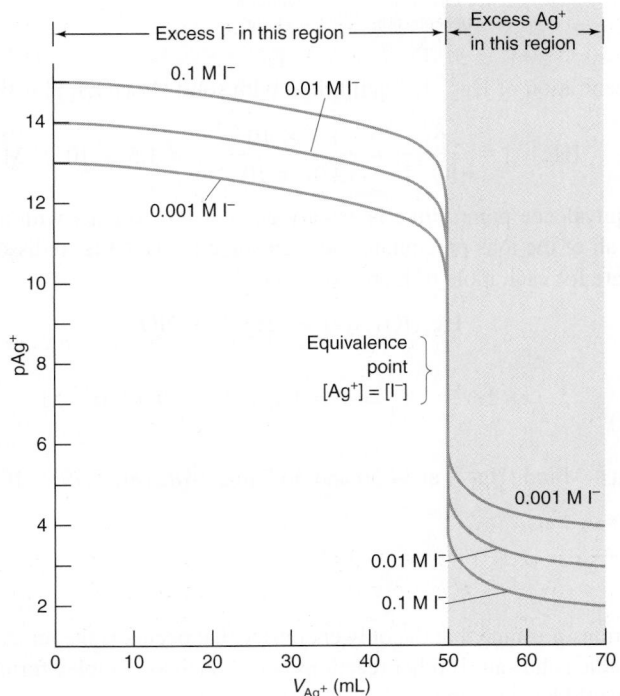

FIGURE 7-2 Titration curves showing the effect of diluting the reactants.

Outer curve: 25.00 mL of 0.100 0 M I^- titrated with 0.050 00 M Ag^+

Middle curve: 25.00 mL of 0.010 00 M I^- titrated with 0.005 000 M Ag^+

Inner curve: 25.00 mL of 0.001 000 M I^- titrated with 0.000 500 0 M Ag^+

At the equivalence point, the titration curve is steepest for the least soluble precipitate.

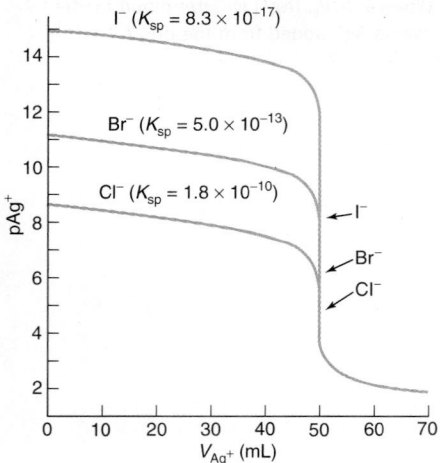

FIGURE 7-3 Titration curves showing the effect of K_{sp}. Each curve is calculated for 25.00 mL of 0.100 0 M halide titrated with 0.050 00 M Ag^+. Equivalence points are marked by arrows.

such that titration curves are steep enough for the steepest point to be a good estimate of the equivalence point, regardless of the stoichiometry.

Figure 7-3 illustrates how K_{sp} affects the titration of halide ions. The least soluble product, AgI, gives the sharpest change at the equivalence point. However, even for AgCl, the curve is steep enough to locate the equivalence point accurately. The larger the equilibrium constant for a titration reaction, the more pronounced will be the change in concentration near the equivalence point.

EXAMPLE Calculating Concentrations during a Precipitation Titration

25.00 mL of 0.041 32 M $Hg_2(NO_3)_2$ were titrated with 0.057 89 M KIO_3.

$$Hg_2^{2+} + 2IO_3^- \rightarrow Hg_2(IO_3)_2(s)$$
$$\text{Iodate}$$

For $Hg_2(IO_3)_2$, $K_{sp} = 1.3 \times 10^{-18}$. Find $[Hg_2^{2+}]$ in the solution after addition of **(a)** 34.00 mL of KIO_3; **(b)** 36.00 mL of KIO_3; and **(c)** at the equivalence point.

Solution The volume of iodate needed to reach the equivalence point is found as follows:

$$\text{Moles of } IO_3^- = \left(\frac{2 \text{ mol } IO_3^-}{1 \text{ mol } Hg_2^{2+}}\right)(\text{moles of } Hg_2^{2+})$$

$$\underbrace{(V_e)(0.057\ 89\ M)}_{\text{Moles of } IO_3^-} = \underbrace{2(25.00\ mL)(0.041\ 32\ M)}_{\text{Moles of } Hg_2^{2+}} \Rightarrow V_e = 35.69\ mL$$

(a) When $V = 34.00$ mL, the precipitation of Hg_2^{2+} is not yet complete.

$$[Hg_2^{2+}] = \underbrace{\left(\frac{35.69 - 34.00}{35.69}\right)}_{\substack{\text{Fraction} \\ \text{remaining}}}\underbrace{(0.041\ 32\ M)}_{\substack{\text{Original} \\ \text{concentration} \\ \text{of } Hg_2^{2+}}}\underbrace{\left(\frac{25.00}{25.00 + 34.00}\right)}_{\substack{\text{Dilution} \\ \text{factor}}} = 8.29 \times 10^{-4}\ M$$

Original volume of Hg_2^{2+}

Total volume of solution

(b) When $V = 36.00$ mL, the precipitation is complete. We have gone $(36.00 - 35.69) = 0.31$ mL *past* the equivalence point. The concentration of excess IO_3^- is

$$[IO_3^-] = \underbrace{(0.057\ 89\ M)}_{\substack{\text{Original} \\ \text{concentration} \\ \text{of } IO_3^-}}\underbrace{\left(\frac{0.31}{25.00 + 36.00}\right)}_{\substack{\text{Dilution} \\ \text{factor}}} = 2.9_4 \times 10^{-4}\ M$$

Volume of excess IO_3^-

Total volume of solution

The concentration of Hg_2^{2+} in equilibrium with solid $Hg_2(IO_3)_2$ plus this much IO_3^- is

$$[Hg_2^{2+}] = \frac{K_{sp}}{[IO_3^-]^2} = \frac{1.3 \times 10^{-18}}{(2.9_4 \times 10^{-4})^2} = 1.5 \times 10^{-11}\ M$$

(c) At the equivalence point, there is exactly enough IO_3^- to react with all Hg_2^{2+}. We can imagine that all of the ions precipitate and then some $Hg_2(IO_3)_2(s)$ redissolves, giving two moles of iodate for each mole of mercurous ion:

$$Hg_2(IO_3)_2(s) \rightleftharpoons \underset{x}{Hg_2^{2+}} + \underset{2x}{2IO_3^-}$$

$$(x)(2x)^2 = K_{sp} \Rightarrow x = [Hg_2^{2+}] = 6.9 \times 10^{-7}\ M$$

TEST YOURSELF Find $[Hg_2^{2+}]$ at 34.50 and 36.5 mL. (***Answer:*** 5.79×10^{-4} M, 2.2×10^{-12} M)

Our calculations presume that the only chemistry that occurs is the reaction of anion with cation to precipitate solid salt. If other reactions occur, such as complex formation or ion-pair formation, we would have to modify the calculations.

152

CHAPTER 7 Let the Titrations Begin

7-4 Titration of a Mixture

If a mixture of two ions is titrated, the less soluble precipitate forms first. If the solubilities are sufficiently different, the first precipitation is nearly complete before the second commences.

Consider the addition of $AgNO_3$ to a solution containing KI and KCl. Because $K_{sp}(AgI)$ $\ll K_{sp}(AgCl)$, AgI precipitates first. When precipitation of I^- is almost complete, the concentration of Ag^+ abruptly increases and AgCl begins to precipitate. When Cl^- is consumed, another abrupt increase in $[Ag^+]$ occurs. We expect two breaks in the titration curve, first at V_e for AgI and then at V_e for AgCl.

Figure 7-4 shows an experimental curve for this titration. The apparatus used to measure the curve is shown in Figure 7-5, and the theory of how this system measures Ag^+ concentration is discussed in Section 15-2.

The I^- end point is taken as the intersection of the steep and nearly horizontal curves at 23.85 mL shown in the inset of Figure 7-4. Precipitation of I^- is not quite complete when Cl^- begins to precipitate. (The way we know that I^- precipitation is not complete is by a calculation. That's what these obnoxious calculations are for!) Therefore, the end of the steep portion (the intersection) is a better approximation of the equivalence point than is the middle of the steep section. The Cl^- end point is taken as the midpoint of the second steep section, at 47.41 mL. The moles of Cl^- in the sample equal the moles of Ag^+ delivered between the first and second end points. That is, it requires 23.85 mL of Ag^+ to precipitate I^-, and $(47.41 - 23.85) = 23.56$ mL of Ag^+ to precipitate Cl^-.

Comparison of the I^-/Cl^- and pure I^- titration curves in Figure 7-4 shows that the I^- end point is 0.38% too high in the I^-/Cl^- titration. We expect the first end point at 23.76 mL, but it is observed at 23.85 mL. Two factors contribute to this high value. One is experimental error, which is always present. This discrepancy is as likely to be positive as negative. However, the end point in some titrations, especially Br^-/Cl^- titrations, is systematically 0 to 3% high, depending on conditions. This error is attributed to *coprecipitation* of AgCl with AgBr. Even though the solubility of AgCl has not been exceeded, some Cl^- becomes attached to AgBr crystallites (small crystals) as they precipitate. Each Cl^- carries down an equivalent amount of Ag^+. A high concentration of nitrate reduces coprecipitation, perhaps because NO_3^- competes with Cl^- for binding sites on AgBr(s).

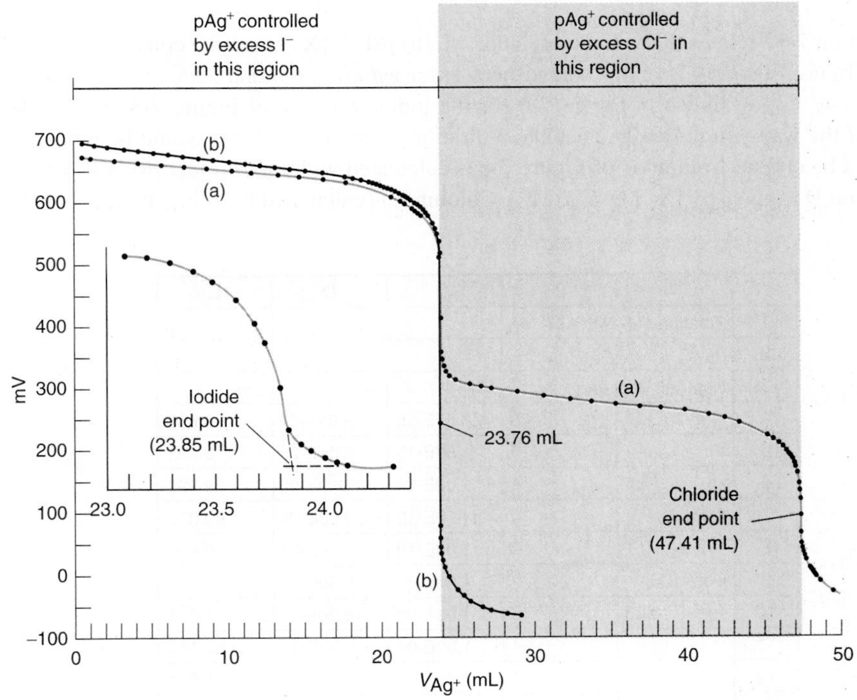

A liquid containing suspended particles is said to be **turbid** because the particles scatter light.

The product with the smaller K_{sp} precipitates first, if the stoichiometry of the precipitates is the same. Precipitation of I^- and Cl^- with Ag^+ produces two breaks in the titration curve, first for I^- and then for Cl^-.

FIGURE 7-4 Experimental titration curves. (*a*) Titration curve for 40.00 mL of 0.050 2 M KI plus 0.050 0 M KCl titrated with 0.084 5 M $AgNO_3$. The inset is an expanded view of the region near the first equivalence point. (*b*) Titration curve for 20.00 mL of 0.100 4 M I^- titrated with 0.084 5 M Ag^+.

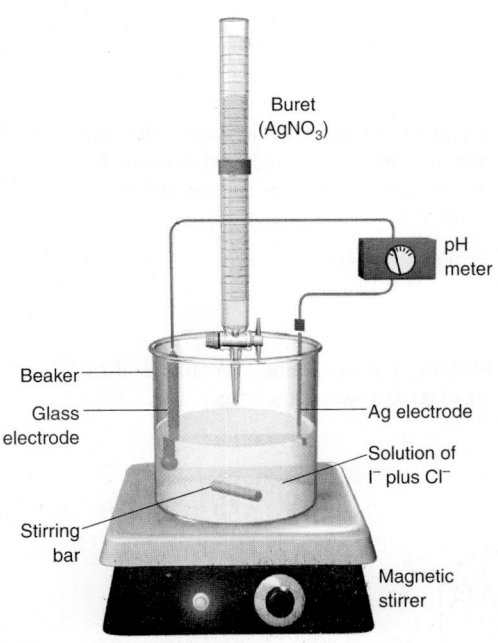

FIGURE 7-5 Apparatus for measuring the titration curves in Figure 7-4. The silver electrode responds to changes in Ag^+ concentration, and the glass electrode provides a constant reference potential in this experiment. The measured voltage changes by approximately 59 mV for each factor-of-10 change in $[Ag^+]$. All solutions, including $AgNO_3$, were maintained at pH 2.0 by using 0.010 M sulfate buffer prepared from H_2SO_4 and KOH.

The second end point in Figure 7-4 corresponds to complete precipitation of both halides. It is observed at the expected value of V_{Ag^+}. The concentration of Cl^-, found from the *difference* between the two end points, will be slightly low in Figure 7-4, because the first end point is slightly high.

7-5 📊 Calculating Titration Curves with a Spreadsheet

Now you understand the chemistry that occurs at different stages of a precipitation titration, and you should know how to calculate the shape of a titration curve. We now introduce spreadsheet calculations that are more powerful than hand calculations and less prone to error. If you are not interested in spreadsheets at this time, you can skip this section.

Consider the addition of V_M liters of cation M^+ (whose initial concentration is C_M^0) to V_X^0 liters of solution containing anion X^- with a concentration C_X^0.

$$M^+ + X^- \xrightleftharpoons{K_{sp}} MX(s) \tag{7-9}$$
$$\underset{\substack{\text{Titrant} \\ C_M^0, V_M}}{} \quad \underset{\substack{\text{Analyte} \\ C_X^0, V_X^0}}{}$$

The total moles of added M $(= C_M^0 \cdot V_M)$ must equal the moles of M^+ in solution $(= [M^+](V_M + V_X^0)$ plus the moles of precipitated $MX(s)$. (This equality is called a *mass balance*, even though it is really a *mole balance*.) In a similar manner, we can write a mass balance for X.

The **mass balance** states that the moles of an element in all species in a mixture equal the total moles of that element delivered to the solution.

Mass balance for M:
$$C_M^0 \cdot V_M = \underbrace{[M^+](V_M + V_X^0)}_{} + \text{mol } MX(s) \tag{7-10}$$

Total mol	Moles of M	Moles of M in
of added M	in solution	precipitate

Mass balance for X:
$$C_X^0 \cdot V_X^0 = \underbrace{[X^-](V_M + V_X^0)}_{} + \text{mol } MX(s) \tag{7-11}$$

Total mol	Moles of X	Moles of X in
of added X	in solution	precipitate

Now equate mol $MX(s)$ from Equation 7-10 with mol $MX(s)$ from Equation 7-11:

$$C_M^0 \cdot V_M - [M^+](V_M + V_X^0) = C_X^0 \cdot V_X^0 - [X^-](V_M + V_X^0)$$

which can be rearranged to

Precipitation of X^- with M^+:
$$V_M = V_X^0 \left(\frac{C_X^0 + [M^+] - [X^-]}{C_M^0 - [M^+] + [X^-]} \right) \tag{7-12}$$

A supplementary section at www.whfreeman.com/qca derives a spreadsheet equation for the titration of a mixture, such as that in Figure 7-4.

Equation 7-12 relates the volume of added M^+ to $[M^+]$, $[X^-]$, and the constants V_X^0, C_X^0, and C_M^0. To use Equation 7-12 in a spreadsheet, *enter values of pM and compute corresponding values of V_M*, as shown in Figure 7-6 for the iodide titration of Figure 7-3. This is the reverse of the way you normally calculate a titration curve in which V_M would be input and pM would be output. Column C of Figure 7-6 is calculated with the formula $[M^+] = 10^{-pM}$, and column D is given by $[X^-] = K_{sp}/[M^+]$. Column E is calculated from Equation 7-12. The

FIGURE 7-6 Spreadsheet for titration of 25 mL of 0.1 M I^- with 0.05 M Ag^+.

	A	B	C	D	E
1	Titration of I- with Ag+				
2					
3	Ksp(AgI) =	pAg	[Ag+]	[I-]	Vm
4	8.30E-17	15.08	8.32E-16	9.98E-02	0.035
5	Vo =	15	1.00E-15	8.30E-02	3.195
6	25	14	1.00E-14	8.30E-03	39.322
7	Co(I) =	12	1.00E-12	8.30E-05	49.876
8	0.1	10	1.00E-10	8.30E-07	49.999
9	Co(Ag) =	8	1.00E-08	8.30E-09	50.000
10	0.05	6	1.00E-06	8.30E-11	50.001
11		4	1.00E-04	8.30E-13	50.150
12		3	1.00E-03	8.30E-14	51.531
13		2	1.00E-02	8.30E-15	68.750
14	C4 = 10^-B4				
15	D4 = $A$4/C4				
16	E4 = $A$6*($A$8+C4-D4)/($A$10-C4+D4)				

first input value of pM (15.08) was selected by trial and error to produce a small V_M. You can start wherever you like. If your initial value of pM is before the true starting point, then V_M in column E will be negative. In practice, you will want more points than we have shown so that you can plot an accurate titration curve.

7-6 End-Point Detection

Precipitation titration end points are commonly found with electrodes (Figure 7-5) or indicators. We now describe two indicator methods for the titration of Cl^- with Ag^+:

Volhard titration: formation of a soluble, colored complex at the end point
Fajans titration: adsorption of a colored indicator on the precipitate at the end point

Volhard Titration

The Volhard method is a titration of Ag^+ in ~0.5 M HNO_3 with standard KSCN (potassium thiocyanate). To measure Cl^-, a back titration is necessary. First, Cl^- in ~0.5 M HNO_3 is precipitated in a vigorously stirred solution by a known, small excess of standard $AgNO_3$.

$$Ag^+ + Cl^- \rightarrow AgCl(s)$$

Vigorous stirring minimizes trapping of excess Ag^+ in the precipitate. $AgCl(s)$ is filtered and washed with dilute (~0.16 M) HNO_3 to collect excess Ag^+ from the precipitate. Then $Fe(NO_3)_3$ (ferric nitrate) or $Fe(NH_4)(SO_4)_2 \cdot 12H_2O$ (ferric ammonium sulfate also called ferric alum) solution is added to the combined filtrate to give ~0.02 M Fe^{3+}. Ag^+ is then titrated with standard KSCN:

$$Ag^+ + SCN^- \rightarrow AgSCN(s)$$

When Ag^+ has been consumed, the next drop of SCN^- reacts with Fe^{3+} to form a red complex.

$$Fe^{3+} + SCN^- \rightarrow \underset{\text{Red}}{FeSCN^{2+}}$$

The appearance of red color is the end point. Knowing how much SCN^- was required for the back titration tells us how much Ag^+ was left over from the reaction with Cl^-. The total amount of Ag^+ is known, so the amount consumed by Cl^- can be calculated.

The purpose of HNO_3 in the titration solution is to prevent hydrolysis of Fe^{3+} to form $Fe(OH)^{2+}$. Concentrated (~70 wt%) nitric acid is prepared for use by mixing with an equal volume of water and boiling for a few minutes (in a hood!) to remove NO_2, which has a red color that would make the color change at the end point harder to see. The Volhard analysis should not be conducted above room temperature to prevent oxidation of SCN^- by warm HNO_3. The ferric nitrate or ferric alum solutions used as indicators are stabilized by a few drops of concentrated HNO_3 to prevent precipitation of ferric hydroxide.

In the analysis of Cl^- by the Volhard method, the end point would slowly fade if AgCl were not filtered off, because AgCl is more soluble than AgSCN. AgCl slowly dissolves and is replaced by AgSCN. To eliminate this secondary reaction, we filter the AgCl, and titrate Ag^+ in the filtrate. An alternative to filtration is to add a few mL of nitrobenzene and stir vigorously to coat AgCl with nitrobenzene, which retards access to SCN^-. In the analysis of Br^- and I^-, whose silver salts are *less* soluble than AgSCN, it is not necessary to isolate the silver halide precipitate.

Fajans Titration

The Fajans titration uses an **adsorption indicator**. To see how this works, consider the electric charge at the surface of a precipitate. When Ag^+ is added to Cl^-, there is excess Cl^- in solution prior to the equivalence point. Some Cl^- is adsorbed onto the AgCl surface, imparting a negative charge to the crystal (Figure 7-7a). After the equivalence point, there is excess Ag^+ in solution. Adsorption of Ag^+ onto the AgCl surface places positive charge on the precipitate (Figure 7-7b). The abrupt change from negative to positive occurs at the equivalence point.

Common adsorption indicators are anionic dyes that are attracted to positively charged particles produced immediately after the equivalence point. Adsorption of the negatively charged dye onto the positively charged surface changes the color of the dye. The color change is the end point in the titration. Because the indicator reacts with the precipitate surface, we want as much surface area as possible. To attain maximum surface area, we use conditions that keep the particles as small as possible, because small particles have more

Titrations with Ag^+ are called **argentometric titrations.**

Because the Volhard method is a titration of Ag^+, it can be adapted for the determination of many anions that form insoluble silver salts.

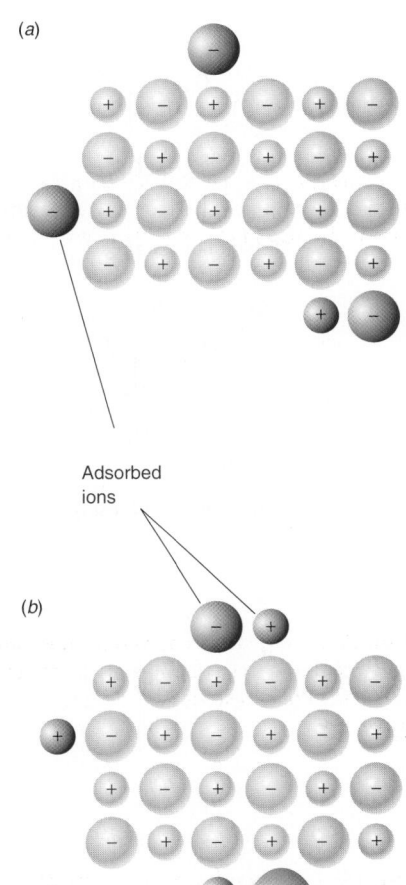

FIGURE 7-7 Ions from a solution are adsorbed on the surface of a growing crystallite. (*a*) A crystal growing in the presence of excess lattice anions (anions that belong in the crystal) will have a slight negative charge because anions are predominantly adsorbed. (*b*) A crystal growing in the presence of excess lattice cations will have a slight positive charge and can therefore adsorb a negative indicator ion. Anions and cations in the solution that do not belong in the crystal lattice are less likely to be adsorbed than are ions belonging to the lattice. These diagrams omit other ions in solution. Overall, each solution plus its growing crystallites must have zero total charge.

The Fajans titration of Cl$^-$ with Ag$^+$ convincingly demonstrates indicator end points in precipitation titrations. Dissolve 0.5 g of NaCl plus 0.15 g of dextrin in 400 mL of water. The purpose of the dextrin is to retard coagulation of the AgCl precipitate. Add 1 mL of dichlorofluorescein indicator solution containing either 1 mg/mL of dichlorofluorescein in 95% aqueous ethanol or 1 mg/mL of the sodium salt in water. Titrate the NaCl solution with a solution containing 2 g of AgNO$_3$ in 30 mL H$_2$O. About 20 mL are required to reach the end point.

Color Plate 2a shows the yellow color of the indicator in the NaCl solution prior to the titration. Color Plate 2b shows the milky white appearance of the AgCl suspension during titration, before the end point is reached. The pink suspension in Color Plate 2c appears at the end point, when the anionic indicator becomes adsorbed on the cationic particles of precipitate.

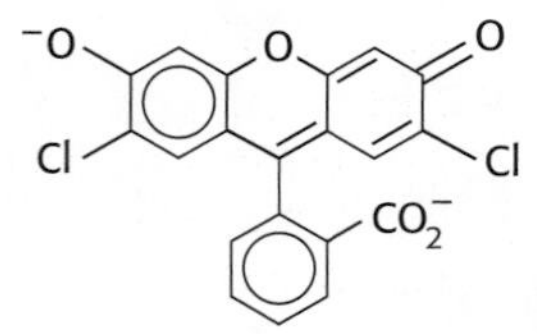

Dichlorofluorescein

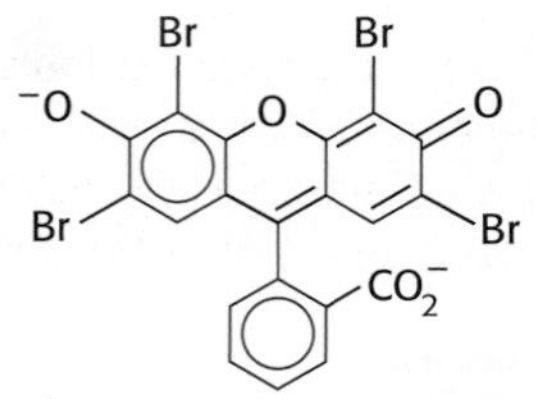

Tetrabromofluorescein (eosin)

surface area than an equal volume of large particles. Low electrolyte concentration helps prevent coagulation of the precipitate and maintains small particle size.

The most common indicator for AgCl is dichlorofluorescein. This dye is greenish yellow in solution but turns pink when adsorbed onto AgCl (Demonstration 7-1 and Color Plate 2). The pH of the reaction must be controlled because the indicator is a weak acid and must be present in its anionic form. The dye eosin is useful in the titration of Br$^-$, I$^-$, and SCN$^-$. It gives a sharper end point than dichlorofluorescein and is more sensitive (that is, less halide can be titrated). It cannot be used for AgCl, because eosin is more strongly bound than Cl$^-$ to AgCl. Eosin binds to AgCl crystallites even before the particles become positively charged.

In all argentometric titrations, but especially with adsorption indicators, strong light (such as daylight through a window) should be avoided. Light decomposes silver salts, and adsorbed indicators are especially light sensitive.

Applications of precipitation titrations are listed in Table 7-1. Whereas the Volhard method is an argentometric titration, the Fajans method has wider applications. Because the Volhard titration is carried out in acidic solution (typically 0.2 M HNO$_3$), it avoids certain interferences that affect other titrations. Silver salts of CO$_3^{2-}$, C$_2$O$_4^{2-}$, and AsO$_4^{3-}$ are soluble in acidic solution, so these anions do not interfere.

TABLE 7-1 Applications of precipitation titrations

Species analyzed	Notes
	Volhard Method
Br$^-$, I$^-$, SCN$^-$, CNO$^-$, AsO$_4^{3-}$,	Precipitate removal is unnecessary.
Cl$^-$, PO$_4^{3-}$, CN$^-$, C$_2$O$_4^{2-}$, CO$_3^{2-}$, S^{2-}, CrO$_4^{2-}$	Precipitate removal required.
BH$_4^-$	Back titration of Ag$^+$ left after reaction with BH$_4^-$:
	$$BH_4^- + 8Ag^+ + 8OH^- \rightarrow 8Ag(s) + H_2BO_3^- + 5H_2O$$
	Fajans Method
Cl$^-$, Br$^-$, I$^-$, SCN$^-$, Fe(CN)$_6^{4-}$	Titration with Ag$^+$. Detection with dyes such as fluorescein, dichlorofluorescein, eosin, bromophenol blue.
Zn^{2+}	Titration with K$_4$Fe(CN)$_6$ to produce K$_2$Zn$_3$[Fe(CN)$_6$]$_2$. End-point detection with diphenylamine.
SO$_4^{2-}$	Titration with Ba(OH)$_2$ in 50 vol% aqueous methanol using alizarin red S as indicator.
Hg$_2^{2+}$	Titration with NaCl to produce Hg$_2$Cl$_2$. End point detected with bromophenol blue.
PO$_4^{3-}$, C$_2$O$_4^{2-}$	Titration with Pb(CH$_3$CO$_2$)$_2$ to give Pb$_3$(PO$_4$)$_2$ or PbC$_2$O$_4$. End point detected with dibromofluorescein (PO$_4^{3-}$) or fluorescein (C$_2$O$_4^{2-}$).

Terms to Understand

adsorption indicator
argentometric titration
back titration
blank titration
direct titration
end point

equivalence point
Fajans titration
gravimetric titration
indicator
primary standard
reagent-grade chemical

standardization
standard solution
titrant
titration
titration curve
titration error

trace analysis
Volhard titration
volumetric analysis

Summary

The volume of reagent (titrant) required for stoichiometric reaction of analyte is measured in volumetric analysis. The stoichiometric point of the reaction is called the equivalence point. What we measure by an abrupt change in a physical property (such as the color of an indicator or the potential of an electrode) is the end point. The difference between the end point and the equivalence point is a titration error. This error can be reduced by subtracting results of a blank titration, in which the same procedure is carried out in the absence of analyte, or by standardizing the titrant, using the same reaction and a similar volume as that used for analyte.

The validity of an analytical result depends on knowing the amount of a primary standard. A solution with an approximately desired concentration can be standardized by titrating a primary standard. In a direct titration, titrant is added to analyte until the reaction is complete. In a back titration, a known excess of reagent is added to analyte, and the excess is titrated with a second standard reagent. Calculations of volumetric analysis relate the known moles of titrant to the unknown moles of analyte.

Concentrations of reactants and products during a precipitation titration are calculated in three regions. Before the equivalence point, there is excess analyte. Its concentration is the product (fraction remaining) $\times$ (original concentration) $\times$ (dilution factor). The concentration of titrant can be found from the solubility product of the precipitate and the known concentration of excess analyte. At the equivalence point, concentrations of both reactants are governed by the solubility product. After the equivalence point, the concentration of analyte can be determined from the known concentration of excess titrant and the solubility product.

The end points of two common argentometric titrations of anions that precipitate with with Ag^+ are marked by color changes. In the Volhard titration, excess standard $AgNO_3$ is added to the anion and the resulting precipitate is filtered off. Excess Ag^+ in the filtrate is back titrated with standard KSCN in the presence of Fe^{3+}. When Ag^+ has been consumed, SCN^- reacts with Fe^{3+} to form a red complex. The Fajans titration uses an adsorption indicator to find the end point of a direct titration of anion with standard $AgNO_3$. The indicator color changes right after the equivalence point when the charged indicator is adsorbed onto the oppositely charged surface of the precipitate.

Exercises

7-A. Ascorbic acid (vitamin C) reacts with I_3^- according to the equation

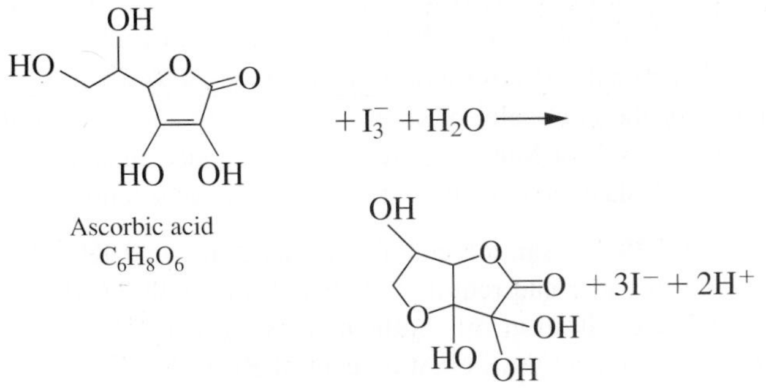

Ascorbic acid
$C_6H_8O_6$

$+ I_3^- + H_2O \longrightarrow$

Dehydroascorbic acid
$C_6H_8O_7$

$+ 3I^- + 2H^+$

Starch is used as an indicator in the reaction. The end point is marked by the appearance of a deep blue starch-iodine complex when the first fraction of a drop of unreacted I_3^- remains in the solution.

(a) Verify that the structures above have the chemical formulas written beneath them. You must be able to locate every atom in the formula. Use atomic masses from the periodic table on the inside cover of this book to find the formula mass of ascorbic acid.

(b) If 29.41 mL of I_3^- solution are required to react with 0.197 0 g of pure ascorbic acid, what is the molarity of the I_3^- solution?

(c) A vitamin C tablet containing ascorbic acid plus inert binder was ground to a powder, and 0.424 2 g was titrated by 31.63 mL of I_3^-. Find the weight percent of ascorbic acid in the tablet.

7-B. A solution of NaOH was standardized by gravimetric titration of a known quantity of the primary standard, potassium hydrogen phthalate:

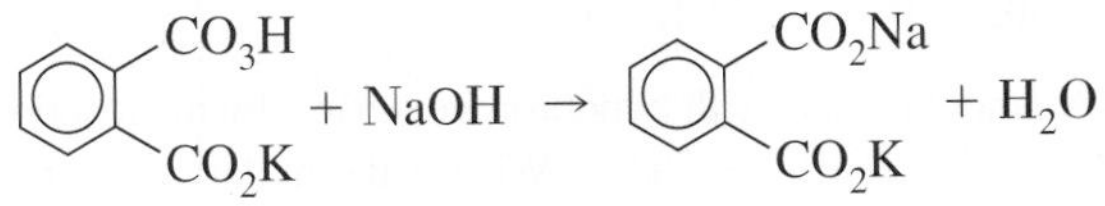

Potassium hydrogen phthalate
$C_8H_5O_4K$, FM 204.22

The NaOH was then used to find the concentration of an unknown solution of H_2SO_4:

$$H_2SO_4 + 2NaOH \rightarrow Na_2SO_4 + 2H_2O$$

(a) Verify from the structure of potassium hydrogen phthalate that its formula is $C_8H_5O_4K$.

(b) Titration of 0.824 g of potassium hydrogen phthalate required 38.314 g of NaOH solution to reach the end point detected by phenolphthalein indicator. Find the concentration of NaOH (mol NaOH/kg solution).

(c) A 10.00-mL aliquot of H_2SO_4 solution required 57.911 g of NaOH solution to reach the phenolphthalein end point. Find the molarity of H_2SO_4.

7-C. A solid sample weighing 0.237 6 g contained only malonic acid and aniline hydrochloride. It required 34.02 mL of 0.087 71 M NaOH to neutralize the sample. Find the weight percent of each component in the solid mixture. The reactions are

$$CH_2(CO_2H)_2 + 2OH^- \rightarrow CH_2(CO_2^-)_2 + 2H_2O$$

Malonic acid FM 104.06   Malonate

Aniline hydrochloride   Aniline
FM 129.59

7-D. A 50.0-mL sample of 0.080 0 M KSCN is titrated with 0.040 0 M Cu^+. The solubility product of CuSCN is 4.8×10^{-15}. At each of the following volumes of titrant, calculate pCu^+, and construct a graph of pCu^+ versus milliliters of Cu^+ added: 0.10, 10.0, 25.0, 50.0, 75.0, 95.0, 99.0, 100.0, 100.1, 101.0, 110.0 mL.

7-E. Construct a graph of pAg^+ versus milliliters of Ag^+ for the titration of 40.00 mL of solution containing 0.050 00 M Br^- and 0.050 00 M Cl^-. The titrant is 0.084 54 M $AgNO_3$. Calculate pAg^+ at the following volumes: 2.00, 10.00, 22.00, 23.00, 24.00, 30.00, 40.00 mL, second equivalence point, 50.00 mL.

7-F. Consider the titration of 50.00 (±0.05) mL of a mixture of I^- and SCN^- with 0.068 3 (±0.000 1) M Ag^+. The first equivalence point is observed at 12.6 (±0.4) mL, and the second occurs at 27.7 (±0.3) mL.

(a) Find the molarity and the uncertainty in molarity of thiocyanate in the original mixture.

(b) Suppose that the uncertainties are all the same, except that the uncertainty of the first equivalence point (12.6 $\pm$? mL) is variable. What is the maximum uncertainty (milliliters) of the first equivalence point if the uncertainty in SCN^- molarity is to be $\leq4.0\%$?

Problems

Volumetric Procedures and Calculations

7-1. Explain the following statement: "The validity of an analytical result ultimately depends on knowing the composition of some primary standard."

7-2. Distinguish between the terms *equivalence point* and *end point*.

7-3. How does a blank titration reduce titration error?

7-4. What is the difference between a direct titration and a back titration?

7-5. What is the difference between a reagent-grade chemical and a primary standard?

7-6. Why are ultrapure acid solvents required to dissolve samples for trace analysis?

7-7. How many milliliters of 0.100 M KI are needed to react with 40.0 mL of 0.040 0 M $Hg_2(NO_3)_2$ if the reaction is $Hg_2^{2+} + 2I^- \rightarrow Hg_2I_2(s)$?

7-8. For Reaction 7-1, how many milliliters of 0.165 0 M $KMnO_4$ are needed to react with 108.0 mL of 0.165 0 M oxalic acid? How many milliliters of 0.165 0 M oxalic acid are required to react with 108.0 mL of 0.165 0 M $KMnO_4$?

7-9. Ammonia reacts with hypobromite, OBr^-, by the reaction $2NH_3 + 3OBr^- \rightarrow N_2 + 3Br^- + 3H_2O$. What is the molarity of a hypobromite solution if 1.00 mL of the OBr^- solution reacts with 1.69 mg of NH_3?

7-10. Sulfamic acid is a primary standard that can be used to standardize NaOH.

$$^+H_3NSO_3^- + OH^- \longrightarrow H_2NSO_3^- + H_2O$$

Sulfamic acid
FM 97.094

What is the molarity of a sodium hydroxide solution if 34.26 mL react with 0.333 7 g of sulfamic acid?

7-11. Limestone consists mainly of the mineral calcite, $CaCO_3$. The carbonate content of 0.541 3 g of powdered limestone was measured by suspending the powder in water, adding 10.00 mL of 1.396 M HCl, and heating to dissolve the solid and expel CO_2:

$$CaCO_3(s) + 2H^+ \longrightarrow Ca^{2+} + CO_2\uparrow + H_2O$$

Calcium carbonate
FM 100.087

The excess acid required 39.96 mL of 0.100 4 M NaOH for complete titration to a phenolphthalein end point. Find the weight percent of calcite in the limestone.

7-12. Arsenic(III) oxide (As_2O_3) is available in pure form and is a useful (but carcinogenic) primary standard for oxidizing agents such as MnO_4^-. The As_2O_3 is dissolved in base and then titrated with MnO_4^- in acidic solution. A small amount of iodide (I^-) or iodate (IO_3^-) is used to catalyze the reaction between H_3AsO_3 and MnO_4^-.

$$As_2O_3 + 4OH^- \rightleftharpoons 2HAsO_3^{2-} + H_2O$$
$$HAsO_3^{2-} + 2H^+ \rightleftharpoons H_3AsO_3$$
$$5H_3AsO_3 + 2MnO_4^- + 6H^+ \rightarrow 5H_3AsO_4 + 2Mn^{2+} + 3H_2O$$

(a) A 3.214-g aliquot of $KMnO_4$ (FM 158.034) was dissolved in 1.000 L of water, heated to cause any reactions with impurities to occur, cooled, and filtered. What is the theoretical molarity of this solution if no MnO_4^- was consumed by impurities?

(b) What mass of As_2O_3 (FM 197.84) would be just sufficient to react with 25.00 mL of the $KMnO_4$ solution in part **(a)**?

(c) It was found that 0.146 8 g of As_2O_3 required 29.98 mL of $KMnO_4$ solution for the faint color of unreacted MnO_4^- to appear. In a blank titration, 0.03 mL of MnO_4^- was required to produce enough color to be seen. Calculate the molarity of the permanganate solution.

7-13. A 0.238 6-g sample contained only NaCl and KBr. It was dissolved in water and required 48.40 mL of 0.048 37 M $AgNO_3$ for complete titration of both halides [giving AgCl(s) and AgBr(s)]. Calculate the weight percent of Br in the solid sample.

7-14. A solid mixture weighing 0.054 85 g contained only ferrous ammonium sulfate and ferrous chloride. The sample was dissolved in 1 M H_2SO_4, and the Fe^{2+} required 13.39 mL of 0.012 34 M Ce^{4+} for complete oxidation to Fe^{3+} ($Ce^{4+} + Fe^{2+} \rightarrow Ce^{3+} + Fe^{3+}$). Calculate the weight percent of Cl in the original sample.

$$FeSO_4 \cdot (NH_4)_2SO_4 \cdot 6H_2O \qquad FeCl_2 \cdot 6H_2O$$

Ferrous ammonium sulfate   Ferrous chloride
FM 392.13   FM 234.84

7-15. A cyanide solution with a volume of 12.73 mL was treated with 25.00 mL of Ni^{2+} solution (containing excess Ni^{2+}) to convert the cyanide into tetracyanonickelate(II):

$$4CN^- + Ni^{2+} \rightarrow Ni(CN)_4^{2-}$$

The excess Ni^{2+} was then titrated with 10.15 mL of 0.013 07 M ethylenediaminetetraacetic acid (EDTA):

$$Ni^{2+} + EDTA^{4-} \rightarrow Ni(EDTA)^{2-}$$

$Ni(CN)_4^{2-}$ does not react with EDTA. If 39.35 mL of EDTA were required to react with 30.10 mL of the original Ni^{2+} solution, calculate the molarity of CN^- in the 12.73-mL cyanide sample.

7-16. *Managing a salt-water aquarium.* A tank at the New Jersey State Aquarium has a volume of 2.9 million liters.[5] Bacteria are used to remove nitrate that would otherwise build up to toxic levels. Aquarium water is first pumped into a 2 700-L deaeration tank containing bacteria that consume O_2 in the presence of added methanol:

$$2CH_3OH + 3O_2 \xrightarrow{\text{Bacteria}} 2CO_2 + 4H_2O \qquad (1)$$
Methanol

Anoxic (deoxygenated) water from the deaeration tank flows into a 1 500-L denitrification reactor containing colonies of *Pseudomonas* bacteria in a porous medium. Methanol is injected continuously and nitrate is converted into nitrite and then into nitrogen:

$$3NO_3^- + CH_3OH \xrightarrow{\text{Bacteria}} 3NO_2^- + CO_2 + 2H_2O \qquad (2)$$
Nitrate $\qquad\qquad\qquad$ nitrite

$$2NO_2^- + CH_3OH \xrightarrow{\text{Bacteria}} N_2 + CO_2 + H_2O + 2OH^- \qquad (3)$$

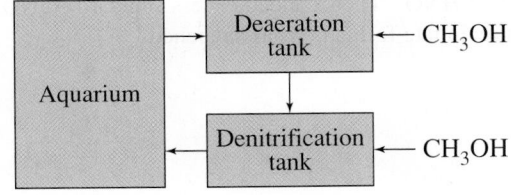

(a) Deaeration can be thought of as a slow, bacteria-mediated titration of O_2 by CH_3OH. The concentration of O_2 in seawater at 24°C is 220 μM. How many liters of CH_3OH (FM 32.04, density = 0.791 g/mL) are required by Reaction 1 for 2.9 million liters of aquarium water?

(b) Write the net reaction showing nitrate plus methanol going to nitrogen. How many liters of CH_3OH are required by the net reaction for 2.9 million liters of aquarium water with a nitrate concentration of 8 100 μM?

(c) In addition to consuming methanol for Reactions 1 through 3, the bacteria require 30% more methanol for their own growth. What is the total volume of methanol required to denitrify 2.9 million liters of aquarium water?

Shape of a Precipitation Curve

7-17. Describe the chemistry that occurs in each of the following regions in Figure 7-2: (i) before the equivalence point; (ii) at the equivalence point; and (iii) past the equivalence point. Write the equation to find $[Ag^+]$ in each region.

7-18. Consider the titration of 25.00 mL of 0.082 30 M KI with 0.051 10 M $AgNO_3$. Calculate pAg^+ at the following volumes of added $AgNO_3$: **(a)** 39.00 mL; **(b)** V_e; **(c)** 44.30 mL.

7-19. A 25.00-mL solution containing 0.031 10 M $Na_2C_2O_4$ was titrated with 0.025 70 M $Ca(NO_3)_2$ to precipitate calcium oxalate: $Ca^{2+} + C_2O_4^{2-} \rightarrow CaC_2O_4(s)$. Find pCa^{2+} at the following volumes of $Ca(NO_3)_2$: **(a)** 10.00, **(b)** V_e, **(c)** 35.00 mL.

7-20. In precipitation titrations of halides by Ag^+, the ion pair $AgX(aq)$ (X = Cl, Br, I) is in equilibrium with the precipitate. Use Appendix J to find the concentrations of $AgCl(aq)$, $AgBr(aq)$, and $AgI(aq)$ during the precipitations.

7-21. *Sulfate in soil on Mars.* A barium sulfate precipitation titration described at the opening of this chapter is shown in the figure. The initial concentration of Cl^- before adding $BaCl_2$ was 0.000 19 M in 25 mL of aqueous extract of Martian soil. At the end point, when there is a sudden rise in Ba^{2+}, $[Cl^-] = 0.009\ 6$ M.

(a) Write the titration reaction.

(b) How many mmol of $BaCl_2$ were required to reach the end point?

(c) How many mmol of SO_4^{2-} were contained in the 25 mL?

(d) If SO_4^{2-} is derived from 1.0 g of soil, what is the wt% of SO_4^{2-} in the soil?

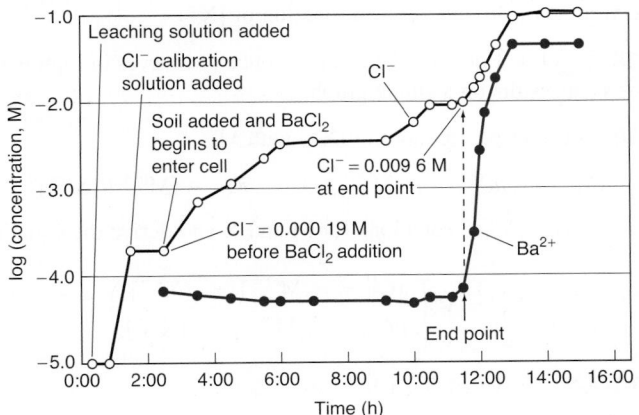

Barium sulfate precipitation titration from *Phoenix Mars Lander.* [Data from Reference 1. S. Kounaves, Tufts University.]

Titration of a Mixture

7-22. Describe the chemistry that occurs in each of the following regions in curve (a) in Figure 7-4: (i) before the first equivalence point; (ii) at the first equivalence point; (iii) between the first and second equivalence points; (iv) at the second equivalence point; and (v) past the second equivalence point. For each region except (ii), write the equation that you would use to calculate $[Ag^+]$.

7-23. The text claims that precipitation of I^- is not complete before Cl^- begins to precipitate in the titration in Figure 7-4. Calculate the concentration of Ag^+ at the equivalence point in the titration of I^- alone. Show that this concentration of Ag^+ will precipitate Cl^-.

7-24. A procedure[6] for determining halogens in organic compounds uses an argentometric titration. To 50 mL of anhydrous ether is added a carefully weighed sample (10–100 mg) of unknown, plus 2 mL of sodium dispersion and 1 mL of methanol. (Sodium dispersion is finely divided solid sodium suspended in oil. With methanol, it makes sodium methoxide, $CH_3O^-Na^+$, which attacks the organic compound, liberating halides.) Excess sodium is destroyed by slow addition of 2-propanol, after which 100 mL of water are added. (Sodium should not be treated directly with water, because the H_2 produced can explode in the presence of O_2: $2Na + 2H_2O \rightarrow 2NaOH + H_2$.) This procedure gives a two-phase mixture, with an ether layer floating on top of the aqueous layer that contains the halide salts. The aqueous layer is adjusted to pH 4 and titrated with Ag^+, using the electrodes in Figure 7-5. How much 0.025 70 M $AgNO_3$ solution will be required to reach each equivalence point when 82.67 mg of 1-bromo-4-chlorobutane ($BrCH_2CH_2CH_2CH_2Cl$; FM 171.46) are analyzed?

7-25. Calculate pAg^+ at the following points in titration (a) in Figure 7-4: **(a)** 10.00 mL; **(b)** 20.00 mL; **(c)** 30.00 mL; **(d)** second equivalence point; **(e)** 50.00 mL.

7-26. A mixture having a volume of 10.00 mL and containing 0.100 0 M Ag^+ and 0.100 0 M Hg_2^{2+} was titrated with 0.100 0 M KCN to precipitate $Hg_2(CN)_2$ and AgCN.

(a) Calculate pCN^- at each of the following volumes of added KCN: 5.00, 10.00, 15.00, 19.90, 20.10, 25.00, 30.00, 35.00 mL.

(b) Should any AgCN be precipitated at 19.90 mL?

Using Spreadsheets

7-27. Derive an expression analogous to Equation 7-12 for the titration of M^+ (concentration = C_M^0, volume = V_M^0) with X^- (titrant concentration = C_X^0). Your equation should allow you to compute the volume of titrant (V_X) as a function of $[X^-]$.

7-28. ▦ Use Equation 7-12 to reproduce the curves in Figure 7-3. Plot your results on a single graph.

7-29. Consider precipitation of X^{x-} with M^{m+}:

$$xM^{m+} + mX^{x-} \rightleftharpoons M_xX_m(s) \qquad K_{sp} = [M^{m+}]^x[X^{x-}]^m$$

Write mass balance equations for M and X and derive the equation

$$V_M = V_X^0\left(\frac{xC_X^0 + m[M^{m+}] - x[X^{x-}]}{mC_M^0 - m[M^{m+}] + x[X^{x-}]}\right)$$

where $[X^{x-}] = (K_{sp}/[M^{m+}]^x)^{1/m}$.

7-30. Use the equation in Problem 7-29 to calculate the titration curve for 10.0 mL of 0.100 M CrO_4^{2-} titrated with 0.100 M Ag^+ to produce $Ag_2CrO_4(s)$.

End-Point Detection

7-31. Why does the surface charge of a precipitate change sign at the equivalence point?

7-32. Examine the procedure in Table 7-1 for the Fajans titration of Zn^{2+}. Do you expect the charge on the precipitate to be positive or negative after the equivalence point?

7-33. Describe how to analyze a solution of NaI by using the Volhard titration.

7-34. A 30.00-mL solution of HBr was treated with 5 mL of freshly boiled and cooled 8 M HNO_3, and then with 50.00 mL of 0.365 0 M $AgNO_3$ with vigorous stirring. Then 1 mL of saturated ferric alum was added and the solution was titrated with 0.287 0 M KSCN. When 3.60 mL had been added, the solution turned red. What was the concentration of HBr in the original solution? How many milligrams of Br^- were in the original solution?

7-35. *What is wrong with this procedure?* According to Table 7-1, carbonate can be measured by a Volhard titration. Removal of the precipitate is required. To analyze an unknown solution of Na_2CO_3, I acidified the solution with freshly boiled and cooled HNO_3 to give ~0.5 M HNO_3. Then I added excess standard $AgNO_3$, but no Ag_2CO_3 precipitate formed. What happened?

8 Activity and the Systematic Treatment of Equilibrium

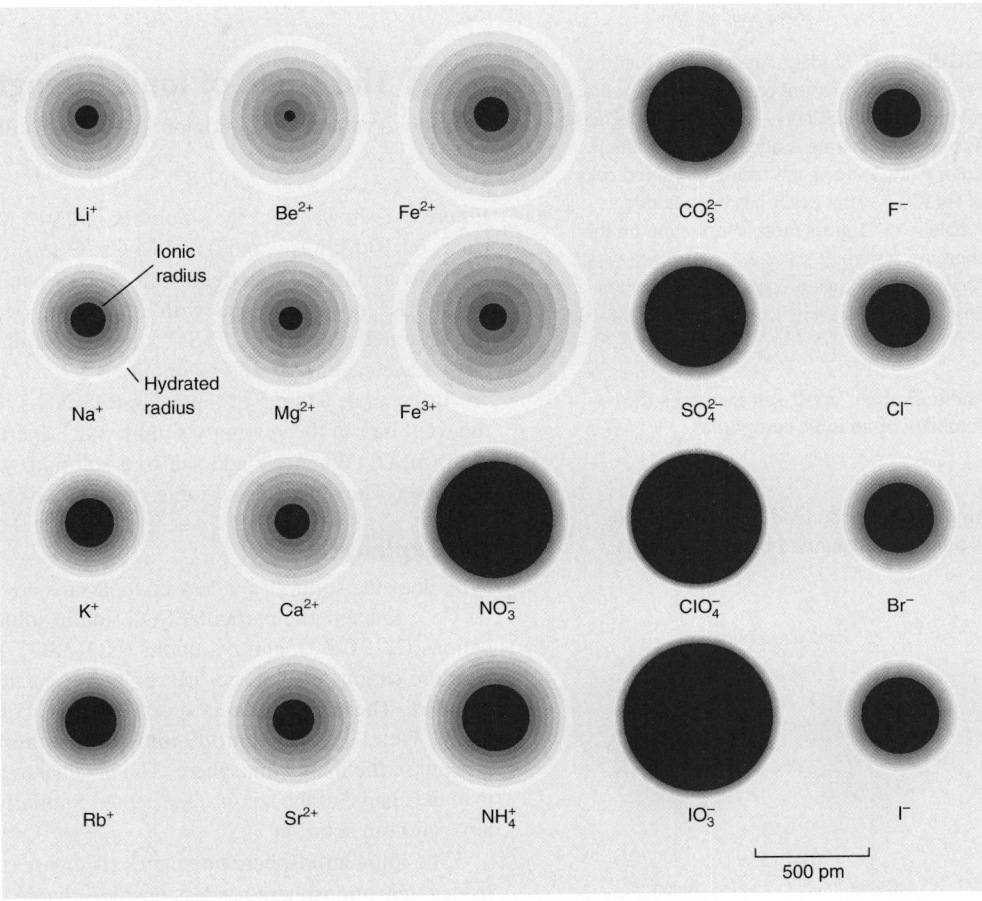

Ionic and hydrated radii of several ions. Smaller, more highly charged ions bind water molecules more tightly and behave as larger hydrated species.

Estimated number of waters of hydration

Molecule	Tightly bound H_2O
$CH_3CH_2CH_3$	0
C_6H_6	0
CH_3CH_2Cl	0
CH_3CH_2SH	0
$CH_3{-}O{-}CH_3$	1
CH_3CH_2OH	1
$(CH_3)_2C{=}O$	1.5
$CH_3CH{=}O$	1.5
CH_3CO_2H	2
$CH_3C{\equiv}N$	3
$CH_3\overset{\overset{\displaystyle O}{\|}}{C}NHCH_3$	4
CH_3NO_2	5
$CH_3CO_2^-$	5
CH_3NH_2	6
CH_3SO_3H	7
NH_3	9
$CH_3SO_3^-$	10
NH_4^+	12

S. Fu and C. A. Lucy, "Prediction of Electrophoretic Mobilities," Anal. Chem. *1998*, 70, 173.

Ions and molecules in solution are surrounded by an organized sheath of solvent molecules. The oxygen atom of H_2O has a partial negative charge and each hydrogen atom has half as much positive charge.

Water binds to cations through the oxygen atom. The first coordination sphere of the small cation Li^+ is composed of $\sim$4 H_2O molecules at the corners of a tetrahedron.[1] The first coordination sphere of the larger ion Cf^{3+} (element 98, californium) appears to be a square antiprism with $\sim$8 H_2O molecules.[2] Cl^- binds $\sim$6 H_2O molecules through hydrogen atoms.[1,3] H_2O exchanges rapidly between bulk solvent and ion-coordination sites.

Ionic radii in the chart above are measured by X-ray diffraction of ions in crystals. Hydrated radii are estimated from diffusion coefficients of ions in solution and from the mobilities of aqueous ions in an electric field.[4,5] Smaller, more highly charged ions bind more water molecules and behave as larger species in solution. The *activity* of aqueous ions, which we study in this chapter, is related to the size of the hydrated species.

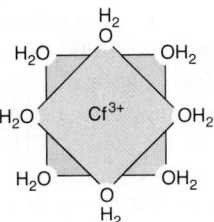

Cf^{3+} resides in the center between planes of 4 H_2O in a square antiprism geometry

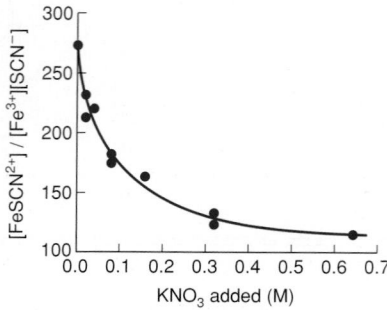

FIGURE 8-1 Student data showing that the equilibrium quotient of concentrations for the reaction $Fe^{3+} + SCN^- \rightleftharpoons Fe(SCN)^{2+}$ decreases as potassium nitrate is added to the solution. Color Plate 3 shows the fading of the red color of $Fe(SCN)^{2+}$ after KNO_3 has been added. Problem 13-11 gives more information on this chemical system. [Data from R. J. Stolzberg, "Discovering a Change in Equilibrium Constant with Change in Ionic Strength," *J. Chem. Ed.* **1999**, 76, 640.]

Addition of an "inert" salt increases the solubility of an ionic compound.

An anion is surrounded by excess cations. A cation is surrounded by excess anions.

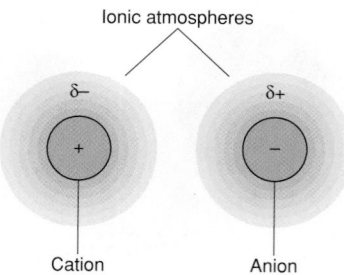

FIGURE 8-2 An ionic atmosphere, shown as a spherical cloud of charge $\delta+$ or $\delta-$, surrounds ions in solution. The charge of the atmosphere is less than the charge of the central ion. The greater the ionic strength of the solution, the greater the charge in each ionic atmosphere.

In Chapter 6, we wrote the equilibrium constant for a reaction in the form

$$Fe^{3+} + SCN^- \rightleftharpoons Fe(SCN)^{2+} \qquad K = \frac{[Fe(SCN)^{2+}]}{[Fe^{3+}][SCN^-]} \qquad (8\text{-}1)$$

Pale yellow   Colorless   Red

Figure 8-1, Demonstration 8-1, and Color Plate 3 show that the concentration quotient in Equation 8-1 decreases if you add the "inert" salt KNO_3 to the solution. That is, the equilibrium "constant" is not really constant. This chapter explains why concentrations are replaced by *activities* in the equilibrium constant and how activities are used.

8-1 The Effect of Ionic Strength on Solubility of Salts

Consider a saturated solution of $CaSO_4$ in distilled water.

$$CaSO_4(s) \rightleftharpoons Ca^{2+} + SO_4^{2-} \qquad K_{sp} = 2.4 \times 10^{-5} \qquad (8\text{-}2)$$

Figure 6-1 showed that the solubility is 0.015 M. The dissolved species are mainly 0.010 M Ca^{2+}, 0.010 M SO_4^{2-}, and 0.005 M $CaSO_4(aq)$ (an ion pair).

Now an interesting effect is observed when a salt such as KNO_3 is added to the solution. Neither K^+ nor NO_3^- reacts with either Ca^{2+} or SO_4^{2-}. Yet, when 0.050 M KNO_3 is added to the saturated solution of $CaSO_4$, more solid dissolves until the concentrations of Ca^{2+} and SO_4^{2-} have each increased by about 30%.

In general, adding an "inert" salt (KNO_3) to a sparingly soluble salt ($CaSO_4$) increases the solubility of the sparingly soluble salt. "Inert" means that KNO_3 has no chemical reaction with $CaSO_4$. When we add salt to a solution, we say that the *ionic strength* of the solution increases. The definition of ionic strength will be given shortly.

The Explanation

Why does the solubility increase when salts are added to the solution? Consider one particular Ca^{2+} ion and one particular SO_4^{2-} ion in solution. The SO_4^{2-} ion is surrounded by H_2O, by cations (K^+, Ca^{2+}), and by anions (NO_3^-, SO_4^{2-}). However, for the average anion, there will be more cations than anions nearby because cations are attracted to the anion, but anions are repelled. These interactions create a region of net positive charge around any particular anion. We call this region the **ionic atmosphere** (Figure 8-2). Ions continually diffuse into and out of the ionic atmosphere. The net charge in the atmosphere, averaged over time, is less than the charge of the anion at the center. Similarly, an atmosphere of negative charge surrounds any cation in solution.

The ionic atmosphere attenuates (decreases) the attraction between ions. The cation plus its negative atmosphere has less positive charge than the cation alone. The anion plus its ionic atmosphere has less negative charge than the anion alone. The net attraction between the cation with its ionic atmosphere and the anion with its ionic atmosphere is smaller than it would be between pure cation and anion in the absence of ionic atmospheres. *The greater the ionic strength of a solution, the higher the charge in the ionic atmosphere. Each ion-plus-atmosphere contains less net charge and there is less attraction between any particular cation and anion.*

DEMONSTRATION 8-1 **Effect of Ionic Strength on Ion Dissociation[6]**

This experiment demonstrates the effect of ionic strength on the dissociation of the red iron(III) thiocyanate complex:

$$Fe(SCN)^{2+} \rightleftharpoons Fe^{3+} + SCN^-$$

Red   Pale yellow   Colorless

Prepare a solution of 1 mM $FeCl_3$ by dissolving 0.27 g of $FeCl_3 \cdot 6H_2O$ in 1 L of water containing 3 drops of 15 M (concentrated) HNO_3. The acid slows the precipitation of $Fe(OH)_3$, which occurs within a few days and necessitates the preparation of fresh solution for this demonstration.

To demonstrate the effect of ionic strength on the dissociation reaction, mix 300 mL of 1 mM $FeCl_3$ with 300 mL of 1.5 mM NH_4SCN or KSCN. Divide the pale red solution into two equal portions and add 12 g of KNO_3 to one of them to increase the ionic strength to 0.4 M. As KNO_3 dissolves, the red $Fe(SCN)^{2+}$ complex dissociates and the color fades (Color Plate 3).

Add a few crystals of NH_4SCN or KSCN to either solution to drive the reaction toward formation of $Fe(SCN)^{2+}$, thereby intensifying the red color. This reaction demonstrates Le Châtelier's principle—adding a product creates more reactant.

CHAPTER 8 Activity and the Systematic Treatment of Equilibrium

Increasing ionic strength therefore reduces the attraction between any particular Ca^{2+} ion and any SO_4^{2-} ion, relative to their attraction for each other in distilled water. The effect is to reduce their tendency to come together, thereby *increasing* the solubility of $CaSO_4$.

Increasing ionic strength promotes dissociation into ions. Thus, each of the following reactions is driven to the right if the ionic strength is raised from, say, 0.01 to 0.1 M:

$$Fe(SCN)^{2+} \rightleftharpoons Fe^{3+} + SCN^-$$
Thiocyanate

Phenol $\rightleftharpoons$ Phenolate

Potassiumn hydrogen tartrate

$$HO_2CCHCHCO_2K(s) \rightleftharpoons HO_2CCHCHCO_2^- + K^+$$

Figure 8-3 shows the effect of added salt on the solubility of potassium hydrogen tartrate.

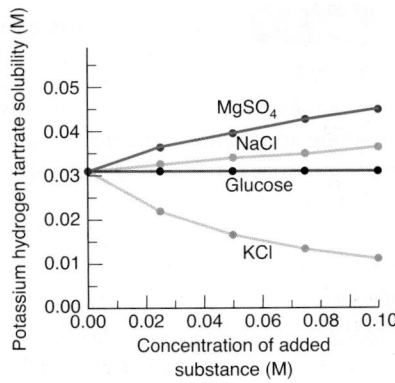

FIGURE 8-3 Solubility of potassium hydrogen tartrate increases when the salts $MgSO_4$ or NaCl are added. There is no effect when the neutral compound glucose is added. Addition of KCl decreases the solubility. (*Why?*) [Data from C. J. Marzzacco, "Effect of Salts and Nonelectrolytes on the Solubility of Potassium Bitartrate," *J. Chem. Ed.* **1998,** *75,* 1628.]

What Do We Mean by "Ionic Strength"?

Ionic strength, μ, is a measure of the total concentration of ions in solution. The more highly charged an ion, the more it is counted.

Ionic strength:
$$\mu = \frac{1}{2}(c_1 z_1^2 + c_2 z_2^2 + \cdots) = \frac{1}{2}\sum_i c_i z_i^2 \qquad (8\text{-}3)$$

where c_i is the concentration of the ith species and z_i is its charge. The sum extends over *all* ions in solution.

EXAMPLE Calculation of Ionic Strength

Find the ionic strength of **(a)** 0.10 M $NaNO_3$; **(b)** 0.010 M Na_2SO_4; and **(c)** 0.020 M KBr plus 0.010 M Na_2SO_4.

Solution

(a) $\mu = \frac{1}{2}\{[Na^+] \cdot (+1)^2 + [NO_3^-] \cdot (-1)^2\}$

$= \frac{1}{2}\{0.10 \cdot 1 + 0.10 \cdot 1\} = 0.10\ M$

(b) $\mu = \frac{1}{2}\{[Na^+] \cdot (+1)^2 + [SO_4^{2-}] \cdot (-2)^2\}$

$= \frac{1}{2}\{(0.020 \cdot 1) + (0.010 \cdot 4)\} = 0.030\ M$

Note that $[Na^+] = 0.020\ M$ because there are two moles of Na^+ per mole of Na_2SO_4.

(c) $\mu = \frac{1}{2}\{[K^+] \cdot (+1)^2 + [Br^-] \cdot (-1)^2 + [Na^+] \cdot (+1)^2 + [SO_4^{2-}] \cdot (-2)^2\}$

$= \frac{1}{2}\{(0.020 \cdot 1) + (0.020 \cdot 1) + (0.020 \cdot 1) + (0.010 \cdot 4)\} = 0.050\ M$

TEST YOURSELF What is the ionic strength of 1 mM $CaCl_2$? (*Answer:* 3 mM)

$NaNO_3$ is called a 1:1 electrolyte because the cation and the anion both have a charge of 1. For 1:1 electrolytes, ionic strength equals molarity. For other stoichiometries (such as the 2:1 electrolyte Na_2SO_4), ionic strength is greater than molarity.

Computing the ionic strength of any but the most dilute solutions is complicated because salts with ions of charge ≥ 2 are not fully dissociated. In Box 8-1 we find that, at a formal concentration of 0.025 M $MgSO_4$, 35% of Mg^{2+} is bound in the ion pair, $MgSO_4(aq)$. The higher the concentration and the higher the ionic charge, the more the ion pairing. There is no simple way to find the ionic strength of 0.025 M $MgSO_4$.

Electrolyte	Molarity	Ionic strength
1:1	M	M
2:1	M	3M
3:1	M	6M
2:2	M	4M

BOX 8-1 **Salts with Ions of Charge ≥|2| Do Not Fully Dissociate**[7]

Salts composed of cations and anions with charges of ± 1 dissociate almost completely at concentrations <0.1 M in water. Salts containing ions with a charge ≥ 2 are less dissociated, even in dilute solution. Appendix J gives formation constants for *ion pairing*:

Ion-pair formation constant:

$$M^{n+}(aq) + L^{m-}(aq) \rightleftharpoons \underset{\text{Ion pair}}{M^{n+}L^{m-}(aq)}$$

$$K = \frac{[ML]\gamma_{ML}}{[M]\gamma_M[L]\gamma_L}$$

where the γ_i are activity coefficients. With constants from Appendix J, activity coefficients from Equation 8-6, and the spreadsheet approach in Section 8-5, we calculate the following percentages of ion pairing in 0.025 F solutions:

Percentage of metal ion bound as ion pair in 0.025 F M_xL_y solution[a]

M \ L	Cl^-	SO_4^{2-}
Na^+	0.6%	9%
Mg^{2+}	8%	35%

a. The size of ML was taken as 500 pm to compute its activity coefficient.

The table tells us that 0.025 F NaCl is only 0.6% associated as NaCl(*aq*) and Na_2SO_4 is 9% associated as $NaSO_4^-$ (*aq*). For $MgSO_4$, 35% is ion paired. A solution of 0.025 F $MgSO_4$ contains 0.016 M Mg^{2+}, 0.016 M SO_4^{2-}, and 0.009 M $MgSO_4$(*aq*). The ionic strength of 0.025 F $MgSO_4$ is not 0.10 M, but just 0.065 M. Problem 8-30 teaches you how to compute the fraction of ion pairing.

Ion pairing is not a precisely defined phenomenon. One definition is that two ions are paired if they remain within a given distance for a time longer than that needed to diffuse over that same distance.

8-2 Activity Coefficients

Equation 8-1 does not predict any effect of ionic strength on a chemical reaction. To account for the effect of ionic strength, concentrations are replaced by **activities:**

Activity of C:

$$\mathcal{A}_C = [C]\gamma_C \qquad \text{(8-4)}$$

Activity of C   Concentration of C   Activity coefficient of C

Do not confuse the terms **activity** and **activity coefficient.**

The activity of species C is its concentration multiplied by its **activity coefficient**. The activity coefficient measures the deviation of behavior from ideality. If the activity coefficient were 1, then the behavior would be ideal and the form of the equilibrium constant in Equation 8-1 would be correct. Activity is a dimensionless quantity. Recall from Section 6-1 that [C] is really the dimensionless ratio of concentration divided by the concentration of the standard state. [C] in Equation 8-4 really means [C]/(1 M) if C is a solute or (pressure of C in bars)/(1 bar) if C is a gas. The activity of a pure solid or liquid is, by definition, unity.

The correct form of the equilibrium constant is

Equation 8-5 is the "real" equilibrium constant. Equation 6-2 gave the concentration quotient, K_c, which did not include activity coefficients:

$$K_c = \frac{[C]^c[D]^d}{[A]^a[B]^b} \qquad \text{(6-2)}$$

General form of equilibrium constant:

$$K = \frac{\mathcal{A}_C^c \mathcal{A}_D^d}{\mathcal{A}_A^a \mathcal{A}_B^b} = \frac{[C]^c\gamma_C^c[D]^d\gamma_D^d}{[A]^a\gamma_A^a[B]^b\gamma_B^b} \qquad \text{(8-5)}$$

Equation 8-5 allows for the effect of ionic strength on a chemical equilibrium because the activity coefficients depend on ionic strength.

For Reaction 8-2, the equilibrium constant is

$$K_{sp} = \mathcal{A}_{Ca^{2+}}\mathcal{A}_{SO_4^{2-}} = [Ca^{2+}]\gamma_{Ca^{2+}}[SO_4^{2-}]\gamma_{SO_4^{2-}}$$

If the concentrations of Ca^{2+} and SO_4^{2-} are to *increase* when a second salt is added to increase ionic strength, the activity coefficients must *decrease* with increasing ionic strength.

At low ionic strength, activity coefficients approach unity, and the thermodynamic equilibrium constant (8-5) approaches the "concentration" equilibrium constant (6-2). One way to measure a thermodynamic equilibrium constant is to measure the concentration ratio (6-2) at successively lower ionic strengths and extrapolate to zero ionic strength. Commonly, tabulated equilibrium constants are not thermodynamic constants but just the concentration ratio (6-2) measured under a particular set of conditions.

EXAMPLE **Exponents of Activity Coefficients**

Write the solubility product expression for $La_2(SO_4)_3(s) \rightleftharpoons 2La^{3+} + 3SO_4^{2-}$ with activity coefficients.

Solution Exponents of activity coefficients are the same as exponents of concentrations:

$$K_{sp} = \mathscr{A}_{La^{3+}}^2 \mathscr{A}_{SO_4^{2-}}^3 = [La^{3+}]^2\gamma_{La^{3+}}^2[SO_4^{2-}]^3\gamma_{SO_4^{2-}}^3$$

TEST YOURSELF Write the equilibrium expression for $Ca^{2+} + 2Cl^- \rightleftharpoons CaCl_2(aq)$ with activity coefficients. $\left(\textbf{\textit{Answer:}}\ K = \dfrac{\mathscr{A}_{CaCl_2}}{\mathscr{A}_{Ca^{2+}}\mathscr{A}_{Cl^-}^2} = \dfrac{[CaCl_2]\gamma_{CaCl_2}}{[Ca^{2+}]\gamma_{Ca^{2+}}[Cl^-]^2\gamma_{Cl^-}^2}\right)$

Activity Coefficients of Ions

The ionic atmosphere model leads to the **extended Debye-Hückel equation,** relating activity coefficients to ionic strength:

Extended Debye-Hückel equation:
$$\log \gamma = \frac{-0.51z^2\sqrt{\mu}}{1 + (\alpha\sqrt{\mu}/305)} \qquad \text{(at 25°C)} \qquad \textbf{(8-6)}$$

In Equation 8-6, γ is the activity coefficient of an ion of charge $\pm z$ and size α (picometers, pm) in an aqueous solution of ionic strength μ. The equation works fairly well for $\mu \leq 0.1$ M. To find activity coefficients for ionic strengths above 0.1 M (up to molalities of 2–6 mol/kg for many salts), more complicated *Pitzer equations* are usually used.[8]

Table 8-1 lists sizes (α) and activity coefficients of many ions. All ions of the same size and charge appear in the same group and have the same activity coefficients. For example, Ba^{2+} and succinate ion $[^-O_2CCH_2CH_2CO_2^-$, listed as $(CH_2CO_2^-)_2]$ each have a size of 500 pm and are listed among the charge ± 2 ions. At an ionic strength of 0.001 M, both of these ions have an activity coefficient of 0.868.

The ion size α in Equation 8-6 is an empirical parameter that provides agreement between measured activity coefficients and ionic strength up to $\mu \approx 0.1$ M. In theory, α is the diameter of the hydrated ion.[9] However, sizes in Table 8-1 cannot be taken literally. For example, the diameter of Cs^+ ion in crystals is 340 pm. The hydrated Cs^+ ion must be larger than the ion in the crystal, but the size of Cs^+ in Table 8-1 is only 250 pm.

Even though ion sizes in Table 8-1 are empirical parameters, trends among sizes are sensible. Small, highly charged ions bind solvent more tightly and have larger effective sizes than do larger or less highly charged ions. For example, the order of sizes in Table 8-1 is $Li^+ > Na^+ > K^+ > Rb^+$, even though crystallographic radii are $Li^+ < Na^+ < K^+ < Rb^+$.

Effect of Ionic Strength, Ion Charge, and Ion Size on the Activity Coefficient

Over the range of ionic strengths from 0 to 0.1 M, the effect of each variable on activity coefficients is as follows:

1. As ionic strength increases, the activity coefficient decreases (Figure 8-4). The activity coefficient (γ) approaches unity as the ionic strength (μ) approaches 0.
2. As the magnitude of the charge of the ion increases, the departure of its activity coefficient from unity increases. Activity corrections are more important for ions with a charge of ± 3 than for ions with a charge of ± 1 (Figure 8-4).
3. The smaller the ion size (α), the more important activity effects become.

1 pm (picometer) = 10^{-12} m

Ionic and hydrated ion sizes are shown at the opening of this chapter.

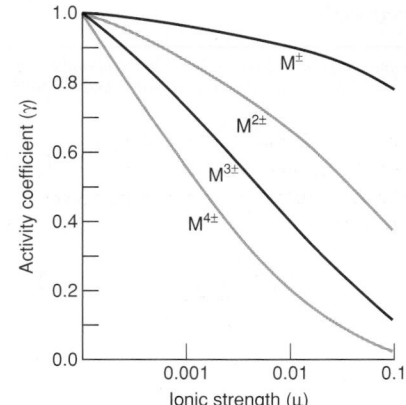

FIGURE 8-4 Activity coefficients for differently charged ions with a constant ionic size (α) of 500 pm. At zero ionic strength, $\gamma = 1$. The greater the charge of the ion, the more rapidly γ decreases as ionic strength increases. Note that the abscissa is logarithmic.

TABLE 8-1 Activity coefficients for aqueous solutions at 25°C

Ion	Ion size (α, pm)	Ionic strength (μ, M)				
		0.001	0.005	0.01	0.05	0.1
Charge = ±1						
H^+	900	0.967	0.933	0.914	0.86	0.83
$(C_6H_5)_2CHCO_2^-$, $(C_3H_7)_4N^+$	800	0.966	0.931	0.912	0.85	0.82
$(O_2N)_3C_6H_2O^-$, $(C_3H_7)_3NH^+$, $CH_3OC_6H_4CO_2^-$	700	0.965	0.930	0.909	0.845	0.81
Li^+, $C_6H_5CO_2^-$, $HOC_6H_4CO_2^-$, $ClC_6H_4CO_2^-$, $C_6H_5CH_2CO_2^-$, $CH_2{=}CHCH_2CO_2^-$, $(CH_3)_2CHCH_2CO_2^-$, $(CH_3CH_2)_4N^+$, $(C_3H_7)_2NH_2^+$	600	0.965	0.929	0.907	0.835	0.80
$Cl_2CHCO_2^-$, $Cl_3CCO_2^-$, $(CH_3CH_2)_3NH^+$, $(C_3H_7)NH_3^+$	500	0.964	0.928	0.904	0.83	0.79
Na^+, $CdCl^+$, ClO_2^-, IO_3^-, HCO_3^-, $H_2PO_4^-$, HSO_3^-, $H_2AsO_4^-$, $Co(NH_3)_4(NO_2)_2^+$, $CH_3CO_2^-$, $ClCH_2CO_2^-$, $(CH_3)_4N^+$, $(CH_3CH_2)_2NH_2^+$, $H_2NCH_2CO_2^-$	450	0.964	0.928	0.902	0.82	0.775
$^+H_3NCH_2CO_2H$, $(CH_3)_3NH^+$, $CH_3CH_2NH_3^+$	400	0.964	0.927	0.901	0.815	0.77
OH^-, F^-, SCN^-, OCN^-, HS^-, ClO_3^-, ClO_4^-, BrO_3^-, IO_4^-, MnO_4^-, HCO_2^-, $H_2citrate^-$, $CH_3NH_3^+$, $(CH_3)_2NH_2^+$	350	0.964	0.926	0.900	0.81	0.76
K^+, Cl^-, Br^-, I^-, CN^-, NO_2^-, NO_3^-	300	0.964	0.925	0.899	0.805	0.755
Rb^+, Cs^+, NH_4^+, Tl^+, Ag^+	250	0.964	0.924	0.898	0.80	0.75
Charge = ±2						
Mg^{2+}, Be^{2+}	800	0.872	0.755	0.69	0.52	0.45
$CH_2(CH_2CH_2CO_2^-)_2$, $(CH_2CH_2CH_2CO_2^-)_2$	700	0.872	0.755	0.685	0.50	0.425
Ca^{2+}, Cu^{2+}, Zn^{2+}, Sn^{2+}, Mn^{2+}, Fe^{2+}, Ni^{2+}, Co^{2+}, $C_6H_4(CO_2^-)_2$, $H_2C(CH_2CO_2^-)_2$, $(CH_2CH_2CO_2^-)_2$	600	0.870	0.749	0.675	0.485	0.405
Sr^{2+}, Ba^{2+}, Cd^{2+}, Hg^{2+}, S^{2-}, $S_2O_4^{2-}$, WO_4^{2-}, $H_2C(CO_2^-)_2$, $(CH_2CO_2^-)_2$, $(CHOHCO_2^-)_2$	500	0.868	0.744	0.67	0.465	0.38
Pb^{2+}, CO_3^{2-}, SO_3^{2-}, MoO_4^{2-}, $Co(NH_3)_5Cl^{2+}$, $Fe(CN)_5NO^{2-}$, $C_2O_4^{2-}$, $Hcitrate^{2-}$	450	0.867	0.742	0.665	0.455	0.37
Hg_2^{2+}, SO_4^{2-}, $S_2O_3^{2-}$, $S_2O_6^{2-}$, $S_2O_8^{2-}$, SeO_4^{2-}, CrO_4^{2-}, HPO_4^{2-}	400	0.867	0.740	0.660	0.445	0.355
Charge = ±3						
Al^{3+}, Fe^{3+}, Cr^{3+}, Sc^{3+}, Y^{3+}, In^{3+}, lanthanides[a]	900	0.738	0.54	0.445	0.245	0.18
$citrate^{3-}$	500	0.728	0.51	0.405	0.18	0.115
PO_4^{3-}, $Fe(CN)_6^{3-}$, $Cr(NH_3)_6^{3+}$, $Co(NH_3)_6^{3+}$, $Co(NH_3)_5H_2O^{3+}$	400	0.725	0.505	0.395	0.16	0.095
Charge = ±4						
Th^{4+}, Zr^{4+}, Ce^{4+}, Sn^{4+}	1 100	0.588	0.35	0.255	0.10	0.065
$Fe(CN)_6^{4-}$	500	0.57	0.31	0.20	0.048	0.021

a. Lanthanides are elements 57–71 in the periodic table.
SOURCE: J. Kielland, J. Am. Chem. Soc. **1937,** 59, 1675.

EXAMPLE Using Table 8-1

Find the activity coefficient of Ca^{2+} in a solution of 3.3 mM $CaCl_2$.

Solution The ionic strength is

$$\mu = \frac{1}{2}\{[Ca^{2+}] \cdot 2^2 + [Cl^-] \cdot (-1)^2\}$$

$$= \frac{1}{2}\{(0.003\ 3) \cdot 4 + (0.006\ 6) \cdot 1\} = 0.010\ M$$

In Table 8-1, Ca^{2+} is listed under the charge ±2 and has a size of 600 pm. Thus $\gamma = 0.675$ when $\mu = 0.010$ M.

TEST YOURSELF Find γ for Cl^- in 0.33 mM $CaCl_2$. (*Answer:* 0.964)

How to Interpolate

If you need to find an activity coefficient for an ionic strength that is between values in Table 8-1, you can use Equation 8-6. Alternatively, in the absence of a spreadsheet, it is not hard to interpolate in Table 8-1. In *linear interpolation*, we assume that values between two entries of a table lie on a straight line. For example, consider a table in which $y = 0.67$ when $x = 10$ and $y = 0.83$ when $x = 20$. What is the value of y when $x = 16$?

> **Interpolation** is the estimation of a number that lies *between* two values in a table. Estimating a number that lies *beyond* values in a table is called **extrapolation**.

x value	y value
10	0.67
16	?
20	0.83

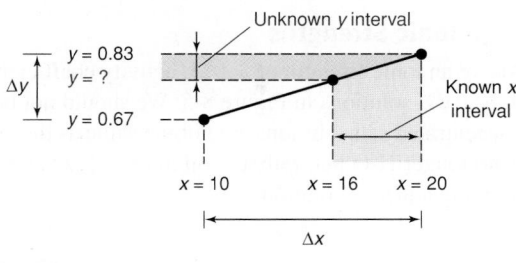

To interpolate a value of y, we can set up a proportion:

Interpolation:
$$\frac{\text{Unknown } y \text{ interval}}{\Delta y} = \frac{\text{known } x \text{ interval}}{\Delta x}$$
(8-7)

$$\frac{0.83 - y}{0.83 - 0.67} = \frac{20 - 16}{20 - 10} \Rightarrow y = 0.76_6$$

For $x = 16$, our estimate of y is 0.76_6.

> This calculation is equivalent to saying:
>
> "16 is 60% of the way from 10 to 20. Therefore the y value will be 60% of the way from 0.67 to 0.83."

EXAMPLE **Interpolating Activity Coefficients**

Calculate the activity coefficient of H^+ when $\mu = 0.025$ M.

Solution H^+ is the first entry in Table 8-1.

μ	γ for H^+
0.01	0.914
0.025	?
0.05	0.86

The linear interpolation is set up as follows:

$$\frac{\text{Unknown } \gamma \text{ interval}}{\Delta\gamma} = \frac{\text{known } \mu \text{ interval}}{\Delta\mu}$$

$$\frac{0.86 - \gamma}{0.86 - 0.914} = \frac{0.05 - 0.025}{0.05 - 0.01}$$

$$\gamma = 0.89_4$$

Solution A better and slightly more tedious calculation uses Equation 8-6 with the ion size $\alpha = 900$ pm listed for H^+ in Table 8-1:

$$\log\gamma_{H^+} = \frac{(-0.51)(1^2)\sqrt{0.025}}{1 + (900\sqrt{0.025}/305)} = -0.054_{98}$$

$$\gamma_{H^+} = 10^{-0.054_{98}} = 0.88_1$$

TEST YOURSELF By interpolation, find γ for H^+ when $\mu = 0.06$ M. (*Answer:* 0.85_4)

Activity Coefficients of Nonionic Compounds

Neutral molecules, such as benzene and acetic acid, have no ionic atmosphere because they have no charge. To a good approximation, their activity coefficients are unity when the ionic strength is less than 0.1 M. In this book, we set $\gamma = 1$ for neutral molecules. That is, *the activity of a neutral molecule will be assumed to be equal to its concentration.*

> For neutral species, $\mathcal{A}_C \approx$ [C]. A more accurate relation is $\log\gamma = k\mu$, where $k \approx 0$ for ion pairs, $k \approx 0.11$ for NH_3 and CO_2, and $k \approx 0.2$ for organic molecules. For an ionic strength of $\mu = 0.1$ M, $\gamma \approx 1.00$ for ion pairs, $\gamma \approx 1.03$ for NH_3, and $\gamma \approx 1.05$ for organic molecules.

For gases such as H_2, the activity is written

$$\mathcal{A}_{H_2} = P_{H_2}\gamma_{H_2}$$

where P_{H_2} is pressure in bars. The activity of a gas is called its *fugacity*, and the activity coefficient is called the *fugacity coefficient*. Deviation of gas behavior from the ideal gas law results in deviation of the fugacity coefficient from unity. For gases at or below 1 bar, $\gamma \approx 1$. Therefore, for gases, *we will set $\mathcal{A} = P$ (bar)*.

For gases, $\mathcal{A} \approx P$ (bar).

At high ionic strength, γ increases with increasing μ.

High Ionic Strengths

Above an ionic strength of ~1 M, activity coefficients of most ions increase, as shown for H^+ in $NaClO_4$ solutions in Figure 8-5. We should not be too surprised that activity coefficients in concentrated salt solutions are not the same as those in dilute aqueous solution. The "solvent" is no longer H_2O but, rather, a mixture of H_2O and $NaClO_4$. Hereafter, we limit our attention to dilute aqueous solutions.

EXAMPLE Using Activity Coefficients

Find the concentration of Ca^{2+} in equilibrium with 0.050 M NaF saturated with CaF_2. The solubility of CaF_2 is small, so the concentration of F^- is 0.050 M from NaF.

Solution We find $[Ca^{2+}]$ from the solubility product expression, including activity coefficients. The ionic strength of 0.050 M NaF is 0.050 M. At $\mu = 0.050\,0$ M in Table 8-1, we find $\gamma_{Ca^{2+}} = 0.485$ and $\gamma_{F^-} = 0.81$.

K_{sp} comes from Appendix F. Note that γ_{F^-} is squared.

$$K_{sp} = [Ca^{2+}]\gamma_{Ca^{2+}}[F^-]^2\gamma_{F^-}^2$$
$$3.2 \times 10^{-11} = [Ca^{2+}](0.485)(0.050)^2(0.81)^2$$
$$[Ca^{2+}] = 4.0 \times 10^{-8}\text{ M}$$

TEST YOURSELF Find $[Hg_2^{2+}]$ in equilibrium with 0.010 M KCl saturated with Hg_2Cl_2. (*Answer:* 2.2×10^{-14} M)

8-3 pH Revisited

For a deeper discussion of what pH really means and how the pH of primary standard solutions is measured, see B. Lunelli and F. Scagnolari, "pH Basics," *J. Chem. Ed.* **2009**, *86*, 246.

The definition pH $\approx -\log[H^+]$ in Chapter 6 is not exact. A better definition is

$$\mathbf{pH} = -\log\mathcal{A}_{H^+} = -\log[H^+]\gamma_{H^+} \tag{8-8}$$

When we measure pH with a pH meter, we are attempting to measure the negative logarithm of the hydrogen ion *activity*, not its concentration.

EXAMPLE pH of Pure Water at 25°C

Let's calculate the pH of pure water by using activity coefficients.

Solution The relevant equilibrium is

$$H_2O \xrightarrow{K_w} H^+ + OH^- \tag{8-9}$$
$$K_w = \mathcal{A}_{H^+}\mathcal{A}_{OH^-} = [H^+]\gamma_{H^+}[OH^-]\gamma_{OH^-} \tag{8-10}$$

H^+ and OH^- are produced in a $1:1$ mole ratio, so their concentrations must be equal. Calling each concentration x, we write

$$K_w = 1.0 \times 10^{-14} = (x)\gamma_{H^+}(x)\gamma_{OH^-}$$

But the ionic strength of pure water is so small that it is reasonable to guess that $\gamma_{H^+} = \gamma_{OH^-} = 1$. Using these values in the preceding equation gives

$$1.0 \times 10^{-14} = (x)(1)(x)(1) = x^2 \Rightarrow x = 1.0 \times 10^{-7}\text{ M}$$

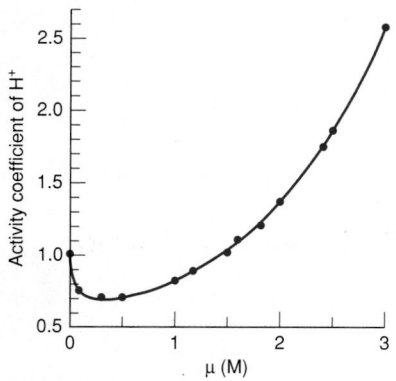

FIGURE 8-5 Activity coefficient of H^+ in solutions containing 0.010 0 M $HClO_4$ and varying amounts of $NaClO_4$. [Information from L. Pezza, M. Molina, M. de Moraes, C. B. Melios, and J. O. Tognolli, *Talanta* **1996**, *43*, 1689.] The authoritative source on electrolyte solutions is H. S. Harned and B. B. Owen, *The Physical Chemistry of Electrolyte Solutions* (New York: Reinhold, 1958 ed.).

The concentrations of H^+ and OH^- are both 1.0×10^{-7} M. The ionic strength is 1.0×10^{-7} M, so each activity coefficient is very close to 1.00. The pH is

$$pH = -\log[H^+]\gamma_{H^+} = -\log(1.0 \times 10^{-7})(1.00) = 7.00$$

EXAMPLE pH of Water Containing a Salt

Now let's calculate the pH of water containing 0.10 M KCl at 25°C.

Solution Reaction 8-9 tells us that $[H^+] = [OH^-]$. However, the ionic strength of 0.10 M KCl is 0.10 M. The activity coefficients of H^+ and OH^- in Table 8-1 are 0.83 and 0.76, respectively, when $\mu = 0.10$ M. Putting these values into Equation 8-10 gives

$$K_w = [H^+]\gamma_{H^+}[OH^-]\gamma_{OH^-}$$
$$1.0 \times 10^{-14} = (x)(0.83)(x)(0.76)$$
$$x = 1.26 \times 10^{-7}\ M$$

The concentrations of H^+ and OH^- are equal and are both greater than 1.0×10^{-7} M. The activities of H^+ and OH^- are not equal in this solution:

$$\mathscr{A}_{H^+} = [H^+]\gamma_{H^+} = (1.26 \times 10^{-7})(0.83) = 1.05 \times 10^{-7}$$
$$\mathscr{A}_{OH^-} = [OH^-]\gamma_{OH^-} = (1.26 \times 10^{-7})(0.76) = 0.96 \times 10^{-7}$$

Finally, we calculate $pH = -\log \mathscr{A}_{H^+} = -\log(1.05 \times 10^{-7}) = 6.98$.

TEST YOURSELF Find $[H^+]$ and the pH of 0.05 M $LiNO_3$. (*Answer:* $1.2_0 \times 10^{-7}$ M, 6.99)

The pH of water changes from 7.00 to 6.98 when we add 0.10 M KCl. KCl is not an acid or a base. The pH changes because KCl affects the activities of H^+ and OH^-. The pH change of 0.02 units lies at the limit of accuracy of pH measurements and is hardly important. However, the *concentration* of H^+ in 0.10 M KCl (1.26×10^{-7} M) is 26% greater than the concentration of H^+ in pure water (1.00×10^{-7} M).

8-4 Systematic Treatment of Equilibrium

The *systematic treatment of equilibrium* is a way to deal with all types of chemical equilibria, regardless of their complexity. After setting up general equations, we often introduce specific conditions or judicious approximations that allow simplification. Even simplified calculations are usually tedious, so we make liberal use of spreadsheets for numerical solutions. When you have mastered the systematic treatment of equilibrium, you should be able to explore the behavior of complex systems.

The systematic procedure is to write as many independent algebraic equations as there are unknowns (species) in the problem. The equations are generated by writing all the chemical equilibrium conditions plus two more: the balances of charge and of mass. There is only one charge balance in a given system, but there could be several mass balances.

Charge Balance

The **charge balance** is an algebraic statement of electroneutrality: *The sum of the positive charges in solution equals the sum of the negative charges in solution.*

Suppose that a solution contains the following ionic species: H^+, OH^-, K^+, $H_2PO_4^-$, HPO_4^{2-}, and PO_4^{3-}. The charge balance is

$$[H^+] + [K^+] = [OH^-] + [H_2PO_4^-] + 2[HPO_4^{2-}] + 3[PO_4^{3-}] \qquad (8\text{-}11)$$

This statement says that the total charge contributed by H^+ and K^+ equals the magnitude of the charge contributed by all of the anions on the right side of the equation. *The coefficient in front of each species always equals the magnitude of the charge on the ion.* This statement is true because a mole of, say, PO_4^{3-} contributes three moles of negative charge. If $[PO_4^{3-}] = 0.01$ M, the negative charge is $3[PO_4^{3-}] = 3(0.01) = 0.03$ M.

Equation 8-11 appears unbalanced to many people. "The right side of the equation has much more charge than the left side!" you might think. But you would be wrong.

Solutions must have zero total charge.

The coefficient of each term in the charge balance equals the magnitude of the charge on each ion.

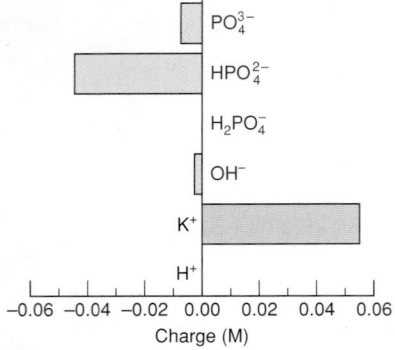

FIGURE 8-6 Charge contributed by each ion in 1.00 L of solution containing 0.025 0 mol KH_2PO_4 plus 0.030 0 mol KOH. The total positive charge equals the total negative charge.

Σ[positive charges] = Σ[negative charges]

Activity coefficients do not appear in the charge balance. The charge contributed by 0.1 M H^+ is exactly 0.1 M. Think about this.

For example, consider a solution prepared by weighing out 0.025 0 mol of KH_2PO_4 plus 0.030 0 mol of KOH and diluting to 1.00 L. The concentrations of the species at equilibrium are calculated to be

$$[H^+] = 5.1 \times 10^{-12} \text{ M} \qquad [H_2PO_4^-] = 1.3 \times 10^{-6} \text{ M}$$
$$[K^+] = 0.055\ 0 \text{ M} \qquad [HPO_4^{2-}] = 0.022\ 0 \text{ M}$$
$$[OH^-] = 0.002\ 0 \text{ M} \qquad [PO_4^{3-}] = 0.003\ 0 \text{ M}$$

This calculation, which you should be able to do when you have finished studying acids and bases, takes into account the reaction of OH^- with $H_2PO_4^-$ to produce HPO_4^{2-} and PO_4^{3-}.

Are the charges balanced? Yes, indeed. Plugging into Equation 8-11, we find

$$[H^+] + [K^+] = [OH^-] + [H_2PO_4^-] + 2[HPO_4^{2-}] + 3[PO_4^{3-}]$$
$$5.1 \times 10^{-12} + 0.055\ 0 = 0.002\ 0 + 1.3 \times 10^{-6} + 2(0.022\ 0) + 3(0.003\ 0)$$
$$0.055\ 0 \text{ M} = 0.055\ 0 \text{ M}$$

The total positive charge is 0.055 0 M, and the total negative charge also is 0.055 0 M (Figure 8-6). Charges must balance in every solution. Otherwise, a beaker with excess positive charge would glide across the lab bench and smash into a beaker with excess negative charge.

The general form of the charge balance for any solution is

Charge balance: $\qquad n_1[C_1] + n_2[C_2] + \cdots = m_1[A_1] + m_2[A_2] + \cdots$ (8-12)

where [C] is the concentration of a cation, n is the charge of the cation, [A] is the concentration of an anion, and m is the magnitude of the charge of the anion.

EXAMPLE **Writing a Charge Balance**

Write the charge balance for a solution containing H_2O, H^+, OH^-, ClO_4^-, $Fe(CN)_6^{3-}$, CN^-, Fe^{3+}, Mg^{2+}, CH_3OH, HCN, NH_3, and NH_4^+.

Solution Neutral species (H_2O, CH_3OH, HCN, and NH_3) contribute no charge, so the charge balance is

$$[H^+] + 3[Fe^{3+}] + 2[Mg^{2+}] + [NH_4^+] = [OH^-] + [ClO_4^-] + 3[Fe(CN)_6^{3-}] + [CN^-]$$

TEST YOURSELF What would be the charge balance if you add $MgCl_2$ to the solution and it dissociates into $Mg^{2+} + 2Cl^-$? (*Answer:* $[H^+] + 3[Fe^{3+}] + 2[Mg^{2+}] + [NH_4^+] = [OH^-] + [ClO_4^-] + 3[Fe(CN)_6^{3-}] + [CN^-] + [Cl^-]$)

Mass Balance

The mass balance is a statement of the conservation of matter. It really refers to conservation of atoms, not to mass.

The **mass balance**, also called the *material balance*, is a statement of the conservation of matter. The mass balance states that *the quantity of all species in a solution containing a particular atom (or group of atoms) must equal the amount of that atom (or group) delivered to the solution.* It is easier to see this relation through examples than by a general statement.

Suppose that a solution is prepared by dissolving 0.050 mol of acetic acid in water to give a total volume of 1.00 L. Acetic acid partially dissociates into acetate:

$$CH_3CO_2H \rightleftharpoons CH_3CO_2^- + H^+$$
$$\text{Acetic acid} \qquad \text{Acetate}$$

The mass balance states that the quantity of dissociated and undissociated acetic acid in the solution must equal the amount of acetic acid put into the solution.

Mass balance for acetic acid in water:
$$0.050 \text{ M} = [CH_3CO_2H] + [CH_3CO_2^-]$$
$$\underset{\substack{\text{What we put into} \\ \text{the solution}}}{} \quad \underset{\substack{\text{Undissociated} \\ \text{product}}}{} \quad \underset{\substack{\text{Dissociated} \\ \text{product}}}{}$$

Activity coefficients do not appear in the mass balance. The concentration of each species counts exactly the number of atoms of that species.

When a compound dissociates in several ways, the mass balance must include all the products. Phosphoric acid (H_3PO_4), for example, dissociates to $H_2PO_4^-$, HPO_4^{2-}, and PO_4^{3-}. The mass balance for phosphorus atoms in a solution prepared by dissolving 0.025 0 mol of H_3PO_4 in 1.00 L is

$$0.025\ 0 \text{ M} = [H_3PO_4] + [H_2PO_4^-] + [HPO_4^{2-}] + [PO_4^{3-}]$$

EXAMPLE **Mass Balance When the Total Concentration Is Known**

Write the mass balances for K^+ and for phosphate in a solution prepared by mixing 0.025 0 mol KH_2PO_4 plus 0.030 0 mol KOH and diluting to 1.00 L.

Solution The total K^+ is 0.025 0 M + 0.030 0 M, so one mass balance is

$$[K^+] = 0.055\ 0\ M$$

The total of *all forms* of phosphate is 0.025 0 M, so the mass balance for phosphate is

$$[H_3PO_4] + [H_2PO_4^-] + [HPO_4^{2-}] + [PO_4^{3-}] = 0.025\ 0\ M$$

TEST YOURSELF Write two mass balances for a 1.00-L solution containing 0.100 mol of sodium acetate. (*Answer:* $[Na^+] = 0.100\ M$: $[CH_3CO_2H] + [CH_3CO_2^-] = 0.100\ M$)

Now consider a solution prepared by dissolving $La(IO_3)_3$ in water.

$$La(IO_3)_3(s) \overset{K_{sp}}{\rightleftharpoons} La^{3+} + 3IO_3^-$$
$$\text{Iodate}$$

We do not know how much La^{3+} or IO_3^- is dissolved, but we do know that there must be three iodate ions for each lanthanum ion dissolved. That is, the iodate concentration must be three times the lanthanum concentration. If La^{3+} and IO_3^- are the only species derived from $La(IO_3)_3$, then the mass balance is

$$[IO_3^-] = 3[La^{3+}]$$

If the solution also contains the ion pair $LaIO_3^{2+}$ and the hydrolysis product $LaOH^{2+}$, the mass balance would be

$$[\text{Total iodate}] = 3[\text{total lanthanum}]$$
$$[IO_3^-] + [LaIO_3^{2+}] = 3\{[La^{3+}] + [LaIO_3^{2+}] + [LaOH^{2+}]\}$$

EXAMPLE **Mass Balance When the Total Concentration Is Unknown**

Write the mass balance for a saturated solution of the slightly soluble salt Ag_3PO_4, which produces PO_4^{3-} and $3Ag^+$ when it dissolves.

Solution If the phosphate in solution remained as PO_4^{3-}, we could write

$$[Ag^+] = 3[PO_4^{3-}]$$

because three silver ions are produced for each phosphate ion. However, phosphate reacts with water to give HPO_4^{2-}, $H_2PO_4^-$, and H_3PO_4, so the mass balance is

$$[Ag^+] = 3\{[PO_4^{3-}] + [HPO_4^{2-}] + [H_2PO_4^-] + [H_3PO_4]\}$$

That is, the number of atoms of Ag^+ must equal three times the total number of atoms of phosphorus, regardless of how many species contain phosphorus.

Atoms of Ag = 3(atoms of P)

Box 8-2 illustrates the operation of a mass balance in natural waters.

TEST YOURSELF Write the mass balance for a saturated solution of $Ba(HSO_4)_2$ if the species in solution are Ba^{2+}, $BaSO_4(aq)$, HSO_4^-, SO_4^{2-}, and $BaOH^+$. (*Answer:* 2 × total barium = total sulfate, or $2\{[Ba^{2+}] + [BaSO_4(aq)] + [BaOH^+]\} = [SO_4^{2-}] + [HSO_4^-] + [BaSO_4(aq)]$)

Systematic Treatment of Equilibrium

Now that we have considered the charge and mass balances, we are ready for the systematic treatment of equilibrium.[12] Here is the general prescription:

Step 1 Write the *pertinent reactions*.
Step 2 Write the *charge balance* equation.
Step 3 Write *mass balance* equations. There may be more than one.
Step 4 Write the *equilibrium constant expression* for each chemical reaction. This step is the only one in which activity coefficients appear.

BOX 8-2 **Calcium Carbonate Mass Balance in Rivers**

Ca^{2+} is the most common cation in rivers and lakes. It comes from dissolution of the mineral calcite by the action of CO_2 to produce 2 moles of HCO_3^- for each mole of Ca^{2+}:

$$CaCO_3(s) + CO_2(aq) + H_2O \rightleftharpoons Ca^{2+} + 2HCO_3^- \quad \text{(A)}$$

$\underset{\text{Calcite}}{} \qquad \qquad \qquad \underset{\text{Bicarbonate}}{}$

Near neutral pH, most of the product is bicarbonate, not CO_3^{2-} or H_2CO_3. The mass balance for the dissolution of calcite is therefore $[HCO_3^-] \approx 2[Ca^{2+}]$. Indeed, measurements of Ca^{2+} and HCO_3^- in many rivers conform to this mass balance, shown by the straight line on the graph. Rivers such as the Danube, the Mississippi, and the Congo, which lie on the line $[HCO_3^-] = 2[Ca^{2+}]$, appear to be saturated with calcium carbonate. If the river water were in equilibrium with atmospheric CO_2 ($P_{CO_2} = 10^{-3.4}$ bar), the concentration of Ca^{2+} would be 21 mg/L (see Problem 8-34). Rivers with more than 21 mg of Ca^{2+} per liter have a higher concentration of dissolved CO_2 produced by respiration or from inflow of groundwaters with a high CO_2 content. Rivers such as the Nile, the Niger, and the Amazon, for which $2[Ca^{2+}] < [HCO_3^-]$, are not saturated with $CaCO_3$.

Between 1960 and 2013, atmospheric CO_2 increased by 27%—mostly from our burning of fossil fuel. This increase drives Reaction A to the right and threatens the existence of coral reefs,[10] which are living structures consisting largely of $CaCO_3$. Coral reefs are a unique habitat for many aquatic species. Continued buildup of atmospheric CO_2 threatens certain plankton and other forms of sea life with $CaCO_3$ shells,[11] a loss that, in turn, threatens higher members of the food chain.

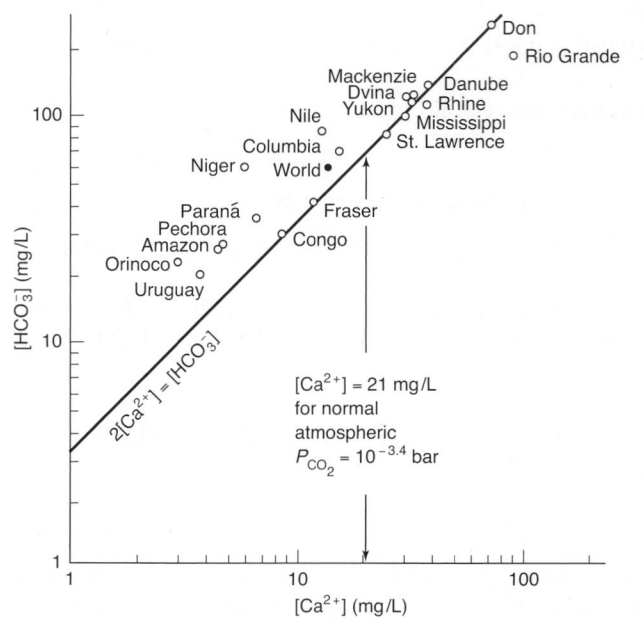

Concentrations of bicarbonate and calcium in many rivers conform to the mass balance $[HCO_3^-] \approx 2[Ca^{2+}]$. [Data from W. Stumm and J. J. Morgan, *Aquatic Chemistry*, 3rd ed. (New York: Wiley-Interscience, 1996), p. 189; and H. D. Holland, *The Chemistry of the Atmosphere and Oceans* (New York: Wiley-Interscience, 1978).]

Step 5 *Count the equations and unknowns.* There should be as many equations as unknowns (chemical species). If not, you must either find more equilibria or fix some concentrations at known values.

Step 6 *Solve* for all unknowns.

Steps 1 and 6 are the heart of the problem. Guessing what chemical equilibria exist in a given solution requires a fair degree of chemical intuition. In this book, you will usually be given help with step 1. Unless we know all the relevant equilibria, it is not possible to calculate the composition of a solution correctly. Because we do not know all the chemical reactions, we undoubtedly oversimplify many equilibrium problems. Step 6—solving the equations—is likely to be your biggest challenge. With n equations and n unknowns, the problem can be solved, at least in principle. In the simplest cases, you can do this by hand, but, for most problems, approximations are made or a spreadsheet is employed.

8-5 Applying the Systematic Treatment of Equilibrium

Now we examine some problems to learn to apply the systematic treatment of equilibrium and to illustrate what can be done by hand and when a spreadsheet is really helpful.

A Solution of Ammonia

Let's find the concentrations of species in an aqueous solution containing 0.010 0 mol NH_3 in 1.000 L. The primary equilibrium is

$$NH_3 + H_2O \overset{K_b}{\rightleftharpoons} NH_4^+ + OH^- \qquad K_b = 1.76 \times 10^{-5} \text{ at } 25°C \qquad \text{(8-13)}$$

A second equilibrium in every aqueous solution is

$$H_2O \overset{K_w}{\rightleftharpoons} H^+ + OH^- \qquad K_w = 1.0 \times 10^{-14} \text{ at } 25°C \qquad \text{(8-14)}$$

Our goal is to find $[NH_3]$, $[NH_4^+]$, $[H^+]$, and $[OH^-]$.

Step 1 Pertinent reactions. They are 8-13 and 8-14.

Step 2 Charge balance. The sum of positive charge equals the sum of negative charge:

$$[NH_4^+] + [H^+] = [OH^-] \qquad (8\text{-}15)$$

Step 3 Mass balance. All of the ammonia delivered to the solution is either in the form NH_3 or NH_4^+. These two must add up to 0.010 0 M.

$$[NH_3] + [NH_4^+] = 0.010\ 0\ M \equiv F \qquad (8\text{-}16)$$

The symbol ≡ means "is defined as."

where F stands for formal concentration.

Step 4 Equilibrium expressions

$$K_b = \frac{[NH_4^+]\gamma_{NH_4^+}[OH^-]\gamma_{OH^-}}{[NH_3]\gamma_{NH_3}} = 10^{-4.755} \qquad (8\text{-}17)$$

$$K_w = [H^+]\gamma_{H^+}[OH^-]\gamma_{OH^-} = 10^{-14.00} \qquad (8\text{-}18)$$

This is the only step in which activity coefficients enter the problem.

We use pK in spreadsheet 8-7. pK is the negative logarithm of an equilibrium constant. For $K_b = 10^{-4.755}$, pK_b = 4.755.

Step 5 Count equations and unknowns. We have four equations, 8-15 to 8-18, and four unknowns ($[NH_3]$, $[NH_4^+]$, $[H^+]$, and $[OH^-]$). We have enough information to solve the problem.

We need n equations to solve for n unknowns.

Step 6 Solve.

This "simple" problem is complicated. Let's start by ignoring activity coefficients; we return to them later in the $Mg(OH)_2$ example. Our approach is to eliminate one variable at a time until only one unknown is left. For an acid-base problem, I choose to express each concentration in terms of $[H^+]$. A substitution we can always make is $[OH^-] = K_w/[H^+]$. Putting this expression for $[OH^-]$ into the charge balance 8-15 gives

$$[NH_4^+] + [H^+] = \frac{K_w}{[H^+]}$$

from which we can solve for $[NH_4^+]$:

$$[NH_4^+] = \frac{K_w}{[H^+]} - [H^+] \qquad (8\text{-}19)$$

The mass balance tells us that $[NH_3] = F - [NH_4^+]$. We can substitute the expression for $[NH_4^+]$ from Equation 8-19 into the mass balance to express $[NH_3]$ in terms of $[H^+]$:

$$[NH_3] = F - [NH_4^+] = F - \left(\frac{K_w}{[H^+]} - [H^+]\right) \qquad (8\text{-}20)$$

Equation 8-19 gives $[NH_4^+]$ in terms of $[H^+]$. Equation 8-20 gives $[NH_3]$ in terms of $[H^+]$.

We can generate an equation in which the only unknown is $[H^+]$ by substituting our expressions for $[NH_4^+]$, $[NH_3]$, and $[OH^-]$ into the K_b equilibrium (still ignoring activity coefficients):

$$K_b = \frac{[NH_4^+][OH^-]}{[NH_3]} = \frac{\left(\dfrac{K_w}{[H^+]} - [H^+]\right)\left(\dfrac{K_w}{[H^+]}\right)}{\left(F - \dfrac{K_w}{[H^+]} + [H^+]\right)} \qquad (8\text{-}21)$$

Equation 8-21 is horrible, but $[H^+]$ is the only unknown. Excel has a built-in procedure called Goal Seek that solves equations with one unknown. Set up the spreadsheet in Figure 8-7, which uses Equations 8-19 and 8-20 for $[NH_4^+]$ and $[NH_3]$, plus $[OH^-] = K_w/[H^+]$ in cells B9, B11, and B10. Cell B5 contains the formal concentration of ammonia, F = 0.01 M. Cell B7 contains an estimate (a guess in this case) for pH from which $[H^+]$ is computed in cell B8. Ammonia is a base, so we guess a basic pH = 9. Cell B12 evaluates the reaction quotient $Q = [NH_4^+][OH^-]/[NH_3]$ with whatever (incorrect) concentrations are in cells B9:B11. Cell B13 shows the difference $K_b - [NH_4^+][OH^-]/[NH_3]$. If concentrations in cells B9:B11 are correct, then $K_b - [NH_4^+][OH^-]/[NH_3]$ would be 0. Before calling Goal Seek in Excel 2010, in the File Tab, click Options and select Formulas. In Calculation Options, set Maximum Change to 1E-14 and click OK. You just set the precision for the calculation. Back in the spreadsheet, select the Data tab and click on What If Analysis in Data Tools. Select Goal Seek. In the Goal Seek window in Figure 8-8a, Set cell <u>B13</u>, To value <u>0</u>, By changing cell <u>B7</u>.

FIGURE 8-7 Spreadsheet to find
concentrations of species in aqueous NH_3
using Excel Goal Seek. Cell B3 contains
$pK_b = -\log K_b$. If $pK_b = 4.755$, $K_b = 10^{-4.755}$.

	A	B	C	D	E
1	Using Goal Seek for Ammonia Equilibrium				
2					
3	$pK_b =$	4.755	$K_b =$	1.76E-05	= 10^-B3
4	$pK_w =$	14.00	$K_w =$	1.00E-14	= 10^-B4
5	F =	0.01			
6					
7	pH =	9	Initial value is estimate		
8	$[H^+] =$	1.00E-09	= 10^-B7		
9	$[NH_4^+] = K_w/[H^+] - [H^+] =$	1.00E-05	= D4/B8-B8		
10	$[OH^-] = K_w/[H^+] =$	1.00E-05	= D4/B8		
11	$[NH_3] = F - K_w/[H^+] + [H^+] =$	9.99E-03	= B5-D4/B8+B8		
12	$Q = [NH_4^+][OH^-]/[NH_3] =$	1.00E-08	= B9*B10/B11		
13	$K_b - [NH_4^+][OH^-]/[NH_3] =$	1.76E-05	= D3-B12		

Goal Seek ? X

Set cell: B13

To value: 0

By changing cell: B7

OK   Cancel

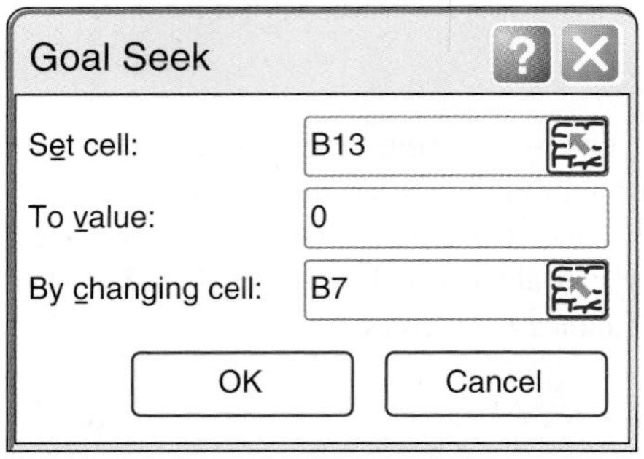

(a)

	A	B
7	pH =	10.61339622
8	$[H^+] =$	2.44E-11
9	$[NH_4^+] = K_w/[H^+] - [H^+] =$	4.11E-04
10	$[OH^-] = K_w/[H^+] =$	4.11E-04
11	$[NH_3] = F - K_w/[H^+] + [H^+] =$	9.59E-03
12	$Q = [NH_4^+][OH^-]/[NH_3] =$	1.76E-05
13	$K_b - [NH_4^+][OH^-]/[NH_3] =$	5.14E-18

(b)

Click OK and Excel varies the value in B7 until the value in B13 is close to 0. The final pH appears in cell B7 in Figure 8-8b. Concentrations of all species appear in cells B8:B11. That was pretty easy once we set up the spreadsheet!

The concentrations of $[NH_4^+]$ and $[NH_3]$ in cells B9 and B11 after executing Goal Seek in Figure 8-8b confirm that ammonia is a weak base. The fraction that has reacted with water is just 4.1%:

$$\text{Fraction reacted} = \frac{[NH_4^+]}{[NH_3] + [NH_4^+]} = \frac{[4.11 \times 10^{-4} \text{ M}]}{[9.59 \times 10^{-3} \text{ M}] + [4.11 \times 10^{-4} \text{ M}]} = 4.1\%$$

You should now appreciate that applying the systematic treatment of equilibrium to even the simplest of problems is not simple. In most equilibrium problems, we make simplifying approximations to reach a good answer with a reasonable effort. After solving a problem, we must always verify that our approximations are valid.

Here is an approximation to simplify the ammonia problem. Ammonia is a base, so we expect that $[OH^-] \gg [H^+]$. For example, suppose the pH comes out to 9. Then $[H^+] = 10^{-9}$ M and $[OH^-](= K_w/[H^+]) = 10^{-14}/10^{-9} = 10^{-5}$ M. That is, $[OH^-] \gg [H^+]$. In the first term of the numerator of Equation 8-21, we can neglect $[H^+]$ in comparison with $K_w/[H^+]$. In the denominator, we can also neglect $[H^+]$ in comparison with $K_w/[H^+]$. With these approximations, Equation 8-21 becomes

The solution to quadratic equation 8-22 is $[OH^-] = 4.11 \times 10^{-4}$ M, giving $[H^+] = K_w/[OH^-] = 2.44 \times 10^{-11}$ M, confirming the approximation that $[OH^-] \gg [H^+]$.

$$K_b = \frac{\left(\dfrac{K_w}{[H^+]} - \cancel{[H^+]}\right)\left(\dfrac{K_w}{[H^+]}\right)}{\left(F - \dfrac{K_w}{[H^+]} + \cancel{[H^+]}\right)} = \frac{\left(\dfrac{K_w}{[H^+]}\right)\left(\dfrac{K_w}{[H^+]}\right)}{\left(F - \dfrac{K_w}{[H^+]}\right)} = \frac{[OH^-]^2}{F - [OH^-]} \qquad (8\text{-}22)$$

Equation 8-22 is quadratic with one variable, $[OH^-]$. We can use algebra to solve for $[OH^-]$. We will deal with equations of this type extensively in the next chapter on acids and bases.

Now we introduce a more general spreadsheet approach that does not require us to reduce an equilibrium problem to one equation in one unknown.

CHAPTER 8 Activity and the Systematic Treatment of Equilibrium

Solubility and Hydrolysis of Thallium Azide

Consider the dissolution of thallium(I) azide, followed by base hydrolysis of azide:

$$\underset{\text{Thallium(I) azide}}{TlN_3(s)} \xrightleftharpoons{K_{sp}} Tl^+ + \underset{\text{Azide}}{N_3^-} \qquad K_{sp} = 10^{-3.66} \qquad (8\text{-}23)$$

$$N_3^- + H_2O \xrightleftharpoons{K_b} \underset{\text{Hydrazoic acid}}{HN_3} + OH^- \qquad K_b = 10^{-9.35} \qquad (8\text{-}24)$$

$$H_2O \xrightleftharpoons{K_w} H^+ + OH^- \qquad K_w = 10^{-14.00} \qquad (8\text{-}25)$$

Our goal is to find $[Tl^+]$, $[N_3^-]$, $[HN_3]$, $[H^+]$, and $[OH^-]$.

Step 1 Pertinent reactions. The three reactions are 8-23 to 8-25.

Step 2 Charge balance. The sum of positive charge equals the sum of negative charge:

$$[Tl^+] + [H^+] = [N_3^-] + [OH^-] \qquad (8\text{-}26)$$

Step 3 Mass balance. Dissolution of TlN_3 gives equal number of Tl^+ and N_3^-. Some azide becomes hydrazoic acid. The concentration of Tl^+ equals the sum of concentrations of N_3^- and HN_3.

$$[Tl^+] = [N_3^-] + [HN_3] \qquad (8\text{-}27)$$

Step 4 Equilibrium expressions

$$K_{sp} = [Tl^+]\gamma_{Tl^+}[N_3^-]\gamma_{N_3^-} = 10^{-3.66} \qquad (8\text{-}28)$$

$$K_b = \frac{[HN_3]\gamma_{HN_3}[OH^-]\gamma_{OH^-}}{[N_3^-]\gamma_{N_3^-}} = 10^{-9.35} \qquad (8\text{-}29)$$

$$K_w = [H^+]\gamma_{H^+}[OH^-]\gamma_{OH^-} = 10^{-14.00} \qquad (8\text{-}30)$$

Step 5 Count equations and unknowns. There are five equations, 8-26 to 8-30, and five unknowns ($[Tl^+]$, $[N_3^-]$, $[HN_3]$, $[H^+]$, and $[OH^-]$), so we have enough information to solve the problem.

Step 6 Solve. We now introduce a spreadsheet method with broad application to equilibrium problems.[13] We ignore activity coefficients for now, but will eventually deal with them.

With five unknowns and three equilibria, 8-28 to 8-30, the spreadsheet method begins with an *estimate* for two of the unknown concentrations:

Number of concentrations to estimate =
$$\text{(number of unknowns)} - \text{(number of equilibria)} = 5 - 3 = 2 \qquad (8\text{-}31)$$

We then write expressions for the remaining three concentrations in terms of the two estimated concentrations. It is helpful to estimate concentrations of species that appear in two or more equilibria. In Equations 8-28 to 8-30, the species N_3^- and OH^- each appear twice, so we will estimate their concentrations. From these estimates, we find the remaining concentrations from the equilibrium expressions:

$$[Tl^+][N_3^-] = K_{sp} \Rightarrow [Tl^+] = K_{sp}/[N_3^-] \qquad (8\text{-}32)$$

$$\frac{[HN_3][OH^-]}{[N_3^-]} = K_b \Rightarrow [HN_3] = \frac{K_b[N_3^-]}{[OH^-]} \qquad (8\text{-}33)$$

$$[H^+][OH^-] = K_w \Rightarrow [H^+] = K_b/[OH^-] \qquad (8\text{-}34)$$

Figure 8-9 sets out the work of finding five unknown concentrations in cells C6:C10. Type this spreadsheet, entering the required constants in cells B12:B14 and the formulas in cells C6:C10, F6:F8, and D12:D14.

Now we need to *estimate* concentrations for N_3^- and OH^-. Concentrations of different species in equilibrium could range over many orders of magnitude, which creates difficulties in computer arithmetic. Therefore, we express the concentrations of N_3^- and OH^- by their negative logarithms in cells B6 and B7. To estimate $[N_3^-] = 10^{-2}$ M, enter 2 in cell B6.

Just as $pH = -\log[H^+]$, we define pC as the negative logarithm of a concentration:

$$pC \equiv -\log C \qquad (8\text{-}35)$$

Azide is a linear ion with two equivalent N=N bonds:

$$\overset{..}{\underset{..}{N}} = \overset{+}{N} = \overset{..}{\underset{..}{N}}^-$$

Azide

$$\overset{..}{\underset{..}{N}}^- = \overset{+}{N} = \overset{..}{N} \diagdown_{\bullet \; H}$$

Hydrazoic acid

We continue to neglect activity coefficients until the next example.

The p function is the negative logarithm of a quantity:

$$pC \equiv -\log C \qquad pK = -\log K$$

If $K = 10^{-3.66}$, $pK = 3.66$.

FIGURE 8-9 Thallium azide solubility
spreadsheet without activity coefficients. Initial
estimates pN$_3^-$ = 2 and pOH$^-$ = 4 appear in
cells B6 and B7. From these two numbers, the
spreadsheet computes concentrations in cells
C6:C10. Solver then varies pN$_3^-$ and pOH$^-$ in
cells B6 and B7 until the charge and mass
balances in cell F8 are satisfied.

	A	B	C	D	E	F
1	Thallium azide equilibria					
2	1. *Estimate* values of pC = −log[C] for N$_3^-$ and OH$^-$ in cells B6 and B7					
3	2. Use Solver to adjust the values of pC to minimize the sum in cell F8					
4						
5	Species	pC	C (= 10^-pC)		Mass and charge balances	b$_i$
6	N$_3^-$	2	0.01	C6 = 10^-B6	b$_1$ = 0 = [Tl$^+$] − [N$_3^-$] − [HN$_3$] =	1.19E-02
7	OH$^-$	4	0.0001	C7 = 10^-B7	b$_2$ = 0 = [Tl$^+$] + [H$^+$] − [N$_3^-$] − [OH$^-$] =	1.18E-02
8	Tl$^+$		0.021877616	C8 = D12/C6	Σb$_i^2$ =	2.80E-04
9	HN$_3$		4.46684E-08	C9 = D13*C6/C7	F6 = C8-C6-C9	
10	H$^+$		1E-10	C10 = D14/C7	F7 = C8+C10-C6-C7	
11					F8 = F6^2+F7^2	
12	pK$_{sp}$ =	3.66	K$_{sp}$ =	0.000218776	= 10^-B12	
13	pK$_b$ =	9.35	K$_b$ =	4.46684E-10	= 10^-B13	
14	pK$_w$ =	14.00	K$_w$ =	1E-14	= 10^-B14	

where C is a concentration. Similarly, **pK** is the negative logarithm of an equilibrium constant. For example, $pK_w = -\log K_w$. If $K_w = 1.00 \times 10^{-14}$, $pK_w = 14.00$.

How do you estimate [N$_3^-$] and [OH$^-$]? The value of K_{sp} for TlN$_3$ is $10^{-3.66}$. If there were no reaction of N$_3^-$ with water, the concentrations would be [Tl$^+$] = [N$_3^-$] = $\sqrt{K_{sp}} = 10^{-3.66/2} = 10^{-1.83}$ M. We only need an estimate for [N$_3^-$], so pN$_3^-$ ≈ 2 is good enough in cell B6, giving [N$_3^-$] = 10^{-2} M in cell C6. But N$_3^-$ reacts with H$_2$O to make [OH$^-$] by Reaction 8-24. With no calculation whatsoever, we can estimate that [OH$^-$] might be ~10^{-4} M in a weakly basic solution. Therefore, enter pOH$^-$ = 4 in cell B7. From [N$_3^-$] and [OH$^-$] in cells C6 and C7, the spreadsheet computes [Tl$^+$], [HN$_3$], and [H$^+$] in cells C8:C10 from Equations 8-32 to 8-34. None of these values is correct. They are just starting estimates from which Excel can compute better values.

The charge balance (8-26) can be rearranged to the form

$$b_1 \equiv [Tl^+] + [H^+] - [N_3^-] - [OH^-] = 0 \qquad (8\text{-}26a)$$

and the mass balance (8-27) can be rearranged to the form

$$b_2 \equiv [Tl^+] - [N_3^-] - [HN_3] = 0 \qquad (8\text{-}27a)$$

The sums labeled b_1 and b_2 in Equations 8-26a and 8-27a would both be 0 if the concentrations are correct. These sums, calculated in cells F6 and F7, are not 0 because the concentrations [N$_3^-$] and [OH$^-$] are not correct.

To find correct concentrations, *we choose to minimize the sum* $\Sigma b_i^2 = b_1^2 + b_2^2$ *in cell F8.* Our criterion for correct concentrations is that they satisfy the mass and charge balances (as well as the equilibrium expressions). The sum $\Sigma b_i^2 = b_1^2 + b_2^2$ is always positive. We minimize the sum by using Excel Solver.

In the Excel 2010 Data ribbon on a PC, you might find Solver as an option. If not, click the File menu and select Excel Options and Add-Ins. Select Solver Add-In, click GO, and then OK to load the Solver. In Excel 2011 on a Mac, Solver is located in the Tools menu.

In the Data ribbon, select Analysis and click on Solver. In the Solver Parameters window (Figure 8-10a), click on Options. In Solver Options, we will only be concerned with the tab for All Methods. Set Constraint Precision = 1e-15, Use Automatic Scaling, Max time = 100 s, and Iterations = 200. Leave Ignore Integer Constraints unchecked and leave Integer Optimality at its default value. Your window should look like Figure 8-10b. Click OK.

Click on Solver. In the Solver Parameters window in Figure 8-10a, Set Objective F8 Equal To Min By Changing Cells B6:B7 (Figure 8-10a). Select a Solving Method should be at its default value of GRG Nonlinear. You just instructed Solver to vary pN$_3^-$ and pOH$^-$ in cells B6:B7 until the combined charge and mass balance in cell F8 is as close to zero as possible. Click Solve and Solver returns pN$_3^-$ = 1.830 038 and pOH$^-$ = 5.589 69 in cells B6:B7, which give $\Sigma b_i^2 \approx 10^{-30}$ in cell F8 in Figure 8-11. Concentrations that satisfy the equilibrium conditions and the charge and mass balances appear in cells C6:C10. Create the spreadsheet in Figure 8-9 and try this procedure. You will really like it.

We had estimated that [N$_3^-$] = 10^{-2} M and [OH$^-$] = 10^{-4} M. Solver tells us that [N$_3^-$] = 0.014 8 M and [OH$^-$] = 2.57×10^{-6} M. It is a good idea when using Solver to try

An important check: In an empty cell in Figure 8-11, compute the quotient K_b = [HN$_3$][OH$^-$]/[N$_3^-$] = C9*C7/C6. No matter what values are in cells C9, C7, and C6, the quotient must be K_b = 4.467E-10. If it is not, check the formulas for each concentration.

CHAPTER 8 Activity and the Systematic Treatment of Equilibrium

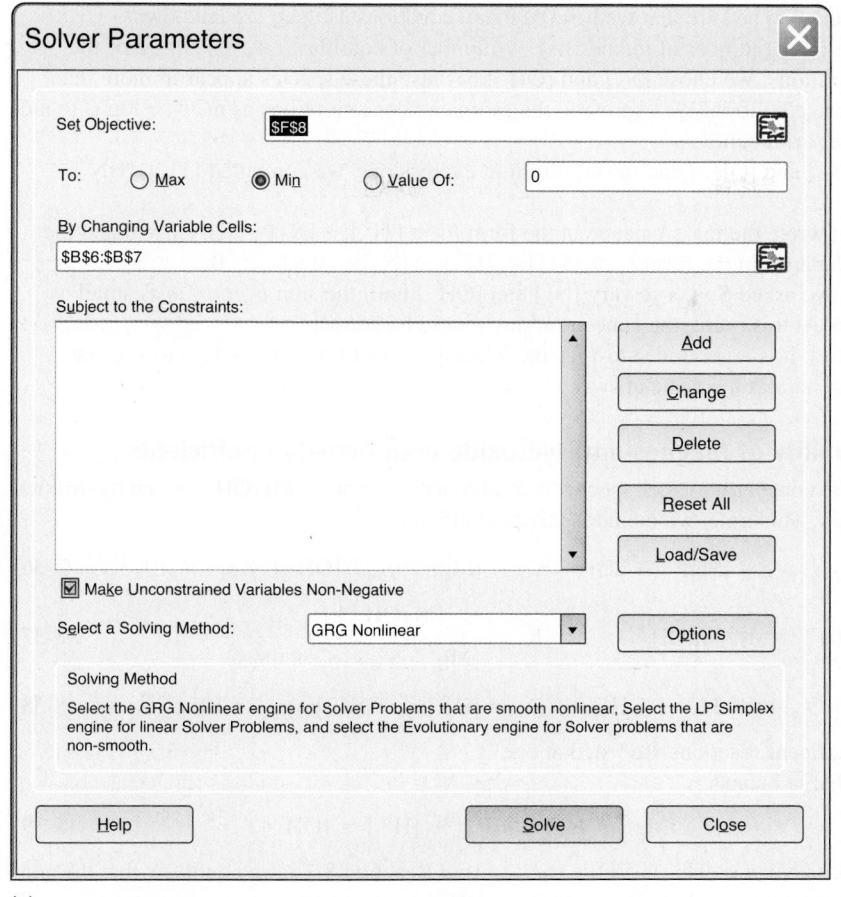

Solver Parameters

Set Objective: $F$8

To: ○ Max ● Min ○ Value Of: 0

By Changing Variable Cells:
$B$6:$B$7

Subject to the Constraints:

Add
Change
Delete
Reset All
Load/Save

☑ Make Unconstrained Variables Non-Negative

Select a Solving Method: GRG Nonlinear Options

Solving Method

Select the GRG Nonlinear engine for Solver Problems that are smooth nonlinear, Select the LP Simplex engine for linear Solver Problems, and select the Evolutionary engine for Solver problems that are non-smooth.

Help Solve Close

(a)

Options

All Methods | GRG Nonlinear | Evolutionary

Constraint Precision 1e-15

☑ Use Automatic Scaling

☐ Show Iteration Results

Solving with Integer Constraints

☐ Ignore Integer Constraints

Integer Optimality (%): 0.1

Solving Limits

Max Time (Seconds): 100

Iterations: 200

Evolutionary and Integer Constraints:

Max Subproblems:

Max Feasible Solutions:

OK Cancel

(b)

FIGURE 8-10 Solver Parameters and Solver Options windows.

	A	B	C	D	E	F
1	Thallium azide equilibria					
2	1. *Estimate* values of pC = –log[C] for N$_3^-$ and OH$^-$ in cells B6 and B7					
3	2. Use Solver to adjust the values of pC to minimize the sum in cell F8					
4						
5	Species	pC	C (= 10^-pC)		Mass and charge balances	b$_i$
6	N$_3^-$	1.830038	0.0147898	C6 = 10^-B6	b$_1$ = 0 = [Tl$^+$] – [N$_3^-$] – [HN$_3$] =	–1.42E-15
7	OH$^-$	5.58969	2.57223E-06	C7 = 10^-B7	b$_2$ = 0 = [Tl$^+$] + [H$^+$] – [N$_3^-$] – [OH$^-$] =	–1.29E-15
8	Tl$^+$		0.014792368	C8 = D12/C6	Σb$_i^2$ =	3.67E-30
9	HN$_3$		2.56834E-06	C9 = D13*C6/C7	F6 = C8–C6–C9	
10	H$^+$		3.88768E-09	C10 = D14/C7	F7 = C8+C10-C6-C7	
11					F8 = F6^2+F7^2	
12	pK$_{sp}$ =	3.66	K$_{sp}$ =	0.000218776	= 10^-B12	
13	pK$_b$ =	9.35	K$_b$ =	4.46684E-10	= 10^-B13	
14	pK$_w$ =	14.00	K$_w$ =	1E-14	= 10^-B14	

FIGURE 8-11 Thallium azide solubility spreadsheet after Solver has finished its work.

a few different starting values for pN_3^- and pOH^- to see if Solver produces the same answer each time. The closer your initial guesses are to the correct solution, the more likely Solver will find the correct solution. Figure 8-11 also tells us that $[HN_3] = 2.57 \times 10^{-6}$ M, which is nearly equal to $[OH^-]$ and is not an accident. Hydrolysis 8-24 produces one HN_3 for each OH^-. The fraction of azide that hydrolyzes is $[HN_3]/([N_3^-] + [HN_3]) = 0.017\%$. This fraction is sensible because N_3^- is a weak base with $K_b = 10^{-9.35}$.

Let's review the complex process that we just exercised:

1. We listed three chemical reactions that we think would occur when $TlN_3(s)$ dissolves in water and wrote their equilibrium constant expressions.
2. We wrote the charge and mass balances.

3. We checked to be sure that we have as many equations as there are unknowns.
4. We estimated (number of unknowns) − (number of equilibria) = 5 − 3 = 2 of the concentrations. We chose $[N_3^-]$ and $[OH^-]$ because these species appear in more than one of the equilibria. We expressed the estimated concentrations as $pC = -\log C$ to aid computer arithmetic.
5. From $[N_3^-]$ and $[OH^-]$ and the equilibrium expressions, we computed $[Tl^+]$, $[HN_3]$, and $[H^+]$.
6. We then wrote the mass balance in the form $b_1 \equiv [Tl^+] - [N_3^-] - [HN_3] = 0$ and the charge balance in the form $b_2 \equiv [Tl^+] + [H^+] - [N_3^-] - [OH^-] = 0$.
7. Finally, we asked Solver to vary $[N_3^-]$ and $[OH^-]$ until the sum $b_1^2 + b_2^2$ is as small as possible. At this point, the concentrations should be correct.
8. As a check it is a good idea to vary the initial values of $[N_3^-]$ and $[OH^-]$ to see that Solver generates the same answer.

Solubility of Magnesium Hydroxide with Activity Coefficients

Let's find the concentrations of species in a saturated solution of $Mg(OH)_2$, given the following chemistry. This time, we include activity coefficients.[14]

$$Mg(OH)_2(s) \xrightleftharpoons{K_{sp}} Mg^{2+} + 2OH^- \quad K_{sp} = [Mg^{2+}]\gamma_{Mg^{2+}}[OH^-]^2\gamma_{OH^-}^2 = 10^{-11.15} \quad \text{(8-36)}$$

$$Mg^{2+} + OH^- \xrightleftharpoons{K_1} MgOH^+ \qquad K_1 = \frac{[MgOH^+]\gamma_{MgOH^+}}{[Mg^{2+}]\gamma_{Mg^{2+}}[OH^-]\gamma_{OH^-}} = 10^{2.6} \quad \text{(8-37)}$$

$$H_2O \xrightleftharpoons{K_w} H^+ + OH^- \qquad K_w = [H^+]\gamma_{H^+}[OH^-]\gamma_{OH^-} = 10^{-14.00} \quad \text{(8-38)}$$

Step 1 Pertinent reactions are listed above.
Step 2 Charge balance:

$$2[Mg^{2+}] + [MgOH^+] + [H^+] = [OH^-] \quad \text{(8-39)}$$

Step 3 Mass balance. This is a little tricky. From Reaction 8-36, we could say that the concentrations of all species containing OH^- equal two times the concentrations of all magnesium species. However, Reaction 8-38 also creates 1 OH^- for each H^+. The mass balance accounts for both sources of OH^-:

$$\underbrace{[OH^-] + [MgOH^+]}_{\text{Species containing } OH^-} = 2\underbrace{\{[Mg^{2+}] + [MgOH^+]\}}_{\text{Species containing } Mg^{2+}} + [H^+] \quad \text{(8-40)}$$

After all this work, Equation 8-40 is equivalent to Equation 8-39.
Step 4 Equilibrium constant expressions are in Equations 8-36 through 8-38.
Step 5 Count equations and unknowns. We have four equations (8-36 to 8-39) and four unknowns: $[Mg^{2+}]$, $[MgOH^+]$, $[H^+]$, and $[OH^-]$.
Step 6 Solve. We will use the spreadsheet approach introduced in the TlN_3 problem, but including activity coefficients this time. With four unknowns and three equilibria, we will estimate one concentration:

Number of concentrations to estimate =
(number of unknowns) − (number of equilibria) = 4 − 3 = 1

The strategy is to estimate one concentration and then let Excel Solver optimize that concentration for us. The correct ionic strength is a by-product of the optimization.

Mg^{2+} appears in two equilibria and OH^- appears in three. It would be fine to estimate either concentration. I choose Mg^{2+}. For the sake of *estimating* $[Mg^{2+}]$, let's just consider the solubility equilibrium 8-36 and neglect activity coefficients. Reaction 8-36 creates 2 OH^- for each Mg^{2+}. If $x = [Mg^{2+}]$, then $[OH^-] = 2x$. The K_{sp} expression gives

$$K_{sp} \approx \underset{x}{[Mg^{2+}]}\underset{2x}{[OH^-]^2} = 7.1 \times 10^{-12}$$

$$(x)(2x)^2 = 4x^3 = 7.1 \times 10^{-12} \Rightarrow x = \left(\frac{7.1 \times 10^{-12}}{4}\right)^{1/3} = 1.2 \times 10^{-4}\,M$$

The solution is $x = [Mg^{2+}] = 1.2 \times 10^{-4}$ M or $pMg^{2+} = -\log(1.2 \times 10^{-4}) = 3.9$. I choose to *estimate* $pMg^{2+} = 4$ for the initial value.

Type the spreadsheet in Figure 8-12 with the value 0 in cell B5 and $pMg^{2+} = 4$ in cell B8. Cell C8 contains $[Mg^{2+}]$ computed from $[Mg^{2+}] = 10^{\wedge}$-B8. Formulas are listed at the

Note that K_1 is the same equilibrium constant as β_1 in Box 6-2 and Appendix I.

	A	B	C	D	E	F	G	H
1	Magnesium hydroxide equilibria							
2	1. *Estimate* pMg in cell B8							
3	2. Use Solver to adjust B8 to minimize sum in cell H15							
4	Ionic strength							
5	μ	3.793E-04				Extended	Activity	
6			Size			Debye-Hückel	Coefficient	
7	Species	pC	C (M)	α (pm)	Charge	log γ	γ	
8	Mg^{2+}	3.9115263	1.226E-04	800	2	−3.780E-02	9.166E-01	G8 = 10^F8
9	OH^-		2.567E-04	350	−1	−9.715E-03	9.779E-01	
10	$MgOH^+$		1.148E-05	500	1	−9.625E-03	9.781E-01	
11	H^+		4.071E-11	900	1	−9.392E-03	9.786E-01	
12								
13	pK_{sp} =	11.15	K_{sp} =	7.08E-12		Mass and charge balances:		b_i
14	pK_1 =	−2.60	K_1 =	3.98E+02	$b_1 = 0 = 2[Mg^{2+}] + [MgOH^+] + [H^+] − [OH^-]$ =			−4.79E-14
15	pK_w =	14.00	K_w =	1.00E-14			Σb_i^2 =	2.29E-27
16							H14 = 2*C8+C10+C11-C9	
17	Ion size estimate:						H15 = H14^2	
18	$MgOH^+$ size $\approx$ 500 pm						C8 = 10^-B8	
19							C9 = SQRT(D13/(C8*G8))/G9	
20	Initial value:						C10 = D14*C8*G8*C9*G9/G10	
21	pMg =	4					C11 = D15/(C9*G9*G11)	
22						F8 = −0.51*E8^2*SQRT($B$5)/(1+D8*SQRT($B$5)/305)		
23						B5 = 0.5*(E8^2*C8+E9^2*C9+E10^2*C10+E11^2*C11)		

FIGURE 8-12 Magnesium hydroxide solubility spreadsheet with activity coefficients after Solver has operated.

lower right of the spreadsheet. Cells C9:C11 compute $[OH^-]$, $[MgOH^+]$, and $[H^+]$ from the equilibrium expressions 8-36 to 8-38, including activity coefficients.

Now we wish to compute the ionic strength in cell B5 from the concentrations in cells C8:C11. However, the concentrations depend on the ionic strength. We say there is a *circular reference*, because concentrations depend on ionic strength and ionic strength depends on concentrations. You must enable Excel to handle the circular reference. In Excel 2010, select the File tab and select Options. In the Options window, select Formulas. In Calculation Options, check "Enable iterative calculation" and set Maximum Change to 1e-15. Click OK and your spreadsheet is ready to handle circular references. Now, change the value 0 in cell B5 to the formula "=0.5*(E8^2*C8+E9^2*C9+E10^2*C10+E11^2*C11)" shown at the bottom of the spreadsheet.

Ion sizes for Mg^{2+}, $[OH^-]$, and $[H^+]$ appear in Table 8-1. We do not know the size of $MgOH^+$. Mg^{2+} is probably best represented as $Mg(OH_2)_6^{2+}$ and $MgOH^+$ is $Mg(OH_2)_5(OH)^+$. $Mg(OH_2)_5(OH)^+$ should be similar in size to $Mg(OH_2)_6^{2+}$, except that $Mg(OH_2)_5(OH)^+$ has a charge of +1 and $Mg(OH_2)_6^{2+}$ has a charge of +2. Recall that the greater the charge of an ion, the more strongly it attracts solvent molecules and the larger is its hydrated radius. The size of $Mg^{2+} = Mg(OH_2)_6^{2+}$ in Table 8-1 is 800 pm. I *guess* that the size of $Mg(OH_2)_5(OH)^+$ is 500 pm. At the end of the equilibrium problem, you could change the size of $Mg(OH_2)_5(OH)^+$ to see if there is much effect on the answer (there is not).

Column F of Figure 8-12 computes log γ from the extended Debye-Hückel equation 8-6 with the formula in cell H22. Column G computes activity coefficients $\gamma = 10^{\wedge}(\log \gamma)$. These activity coefficients appear in the equilibrium expressions used to find concentrations in cells C9:C11 with formulas in cells H18:H20.

The charge balance in cell H14 is $b_1 = 2[Mg^{2+}] + [MgOH^+] + [H^+] − [OH^-]$. In this particular problem, the mass balance is identical to the charge balance, so we do not use the mass balance. The function we minimize in cell H15 is:

Function to minimize: $\qquad \Sigma b_i^2 = b_1^2 + b_2^2$

In this particular problem, there is no mass balance, so $\Sigma b_i^2 = b_1^2$.

So, finally, here we go. Compose the spreadsheet in Figure 8-12. Use an initial estimate $pMg^{2+} = 4$ in cell B8. In the Data tab, select Solver and then Options. Establish settings in the Option window as in Figure 8-10 and click OK. In the Solver Parameters window, Set Target Cell <u>H15</u> Equal To <u>Min</u> By Changing Cells <u>B8</u>. Click Solve. Solver varies pMg^{2+}

in B8 until the square of the charge balance in cell H15 is a minimum. Solver returns $\mu = 0.000\ 379$ M and $pMg^{2+} = 3.911\ 5$ in cells B5 and B8, giving $\Sigma b_i^2 \approx 10^{-27}$ in cell H15.

Final results shown in Figure 8-12 are

$$[Mg^{2+}] = 1.23 \times 10^{-4}\ M \qquad [OH^-] = 2.57 \times 10^{-4}\ M$$

$$[MgOH^+] = 1.15 \times 10^{-5}\ M \qquad [H^+] = 4.07 \times 10^{-11}\ M$$

$$\mu = 3.79 \times 10^{-4}\ M$$

Approximately 10% of the Mg^{2+} undergoes hydrolysis to $MgOH^+$. The pH of the solution is

$$pH = -\log[H^+]\gamma_{H^+} = -\log(4.07 \times 10^{-11})(0.979) = 10.40$$

The spreadsheet in Figure 8-12 provides a tool with which you can tackle a variety of modest equilibrium problems including activity coefficients.

Solver works best when your initial estimates are close to the actual values and when you do not ask Solver to find too many variables at once. Always try some different initial estimates to see if Solver reaches the same conclusion. Try executing Solver successively with values from one run input to the next run to see if the solution improves, as judged by the Σb_i^2 becoming smaller. If Solver is unable to find two or more variables in a complex problem, solve for one or two variables at a time while holding others fixed. A good solution is indicated by Σb_i^2 becoming very small and ceasing to change in successive cycles.

Solubility of Calcium Sulfate with Activity Coefficients

Let's find the concentrations of major species in a saturated solution of $CaSO_4$.

Step 1 Pertinent reactions. Even in such a simple system, there are quite a few reactions:

$$CaSO_4(s) \xrightleftharpoons{K_{sp}} Ca^{2+} + SO_4^{2-} \qquad pK_{sp} = 4.62 \qquad \text{(8-41)}$$

$$CaSO_4(s) \xrightleftharpoons{K_{ion\ pair}} CaSO_4(aq) \qquad pK_{ion\ pair} = 2.26* \qquad \text{(8-42)}$$

$$SO_4^{2-} + H_2O \xrightleftharpoons{K_{base}} HSO_4^- + OH^- \qquad pK_{base} = 12.01 \qquad \text{(8-43)}$$

$$Ca^{2+} + H_2O \xrightleftharpoons{K_{acid}} CaOH^+ + H^+ \qquad pK_{acid} = 12.70 \qquad \text{(8-44)}$$

$$H_2O \xrightleftharpoons{K_w} H^+ + OH^- \qquad pK_w = 14.00 \qquad \text{(8-45)}$$

There is no way you can be expected to come up with all of these reactions, so you will be given help with this step.

Step 2 Charge balance. Equating positive and negative charges gives

$$2[Ca^{2+}] + [CaOH^+] + [H^+] = 2[SO_4^{2-}] + [HSO_4^-] + [OH^-] \qquad \text{(8-46)}$$

Step 3 Mass balance. Reaction 8-41 produces 1 mole of sulfate for each mole of calcium. No matter what happens to these ions, the total concentration of all species with sulfate must equal the total concentration of all species with calcium:

$$[\text{total calcium}] = [\text{total sulfate}]$$

$$[Ca^{2+}] + [CaSO_4(aq)] + [CaOH^+] = [SO_4^{2-}] + [HSO_4^-] + [CaSO_4(aq)] \qquad \text{(8-47)}$$

Step 4 Equilibrium constant expressions. There is one for each chemical reaction.

$$K_{sp} = [Ca^{2+}]\gamma_{Ca^{2+}}[SO_4^{2-}]\gamma_{SO_4^{2-}} = 10^{-4.62} \qquad \text{(8-48)}$$

$$K_{ion\ pair} = [CaSO_4(aq)] = 10^{-2.26} \qquad \text{(8-49)}$$

$$K_{base} = \frac{[HSO_4^-]\gamma_{HSO_4^-}[OH^-]\gamma_{OH^-}}{[SO_4^{2-}]\gamma_{SO_4^{2-}}} = 10^{-12.01} \qquad \text{(8-50)}$$

$$K_{acid} = \frac{[CaOH^+]\gamma_{CaOH^+}[H^+]\gamma_{H^+}}{[Ca^{2+}]\gamma_{Ca^{2+}}} = 10^{-12.70} \qquad \text{(8-51)}$$

$$K_w = [H^+]\gamma_{H^+}[OH^-]\gamma_{OH^-} = 1.0 \times 10^{-14} \qquad \text{(8-52)}$$

Step 4 is the only one where activity coefficients come in.

Caveat emptor! (Buyer beware!) In all equilibrium problems, we are limited by how much of the system's chemistry is understood. Without all relevant equilibria, the calculated composition could be in error or species could be missing.

Multiply $[Ca^{2+}]$ and $[SO_4^{2-}]$ by 2 because 1 mol of each ion has 2 mol of charge.

The activity coefficient of neutral $CaSO_4(aq)$ is 1.

---

*The reaction $CaSO_4(s) \xrightleftharpoons{K_{ion\ pair}} CaSO_4(aq)$ is the sum of reactions $CaSO_4(s) \xrightleftharpoons{K_{sp}} Ca^{2+} + SO_4^{2-}$ from Appendix F and $Ca^{2+} + SO_4^{2-} \rightleftharpoons CaSO_4(aq)$ from Appendix I.

CHAPTER 8 Activity and the Systematic Treatment of Equilibrium

Step 5 Count equations and unknowns. There are seven equations (8-46 through 8-52) and seven unknowns: $[Ca^{2+}]$, $[SO_4^{2-}]$, $[CaSO_4(aq)]$, $[HSO_4^-]$, $[CaOH^+]$, $[H^+]$, and $[OH^-]$.

Step 6 Solve. We will create the same kind of spreadsheet used for $Mg(OH)_2$ in Figure 8-12. The number of concentrations that we need to estimate is (number of unknowns) − (number of equilibria) = 7 − 5 = 2.

Type the spreadsheet in Figure 8-13 with an initial ionic strength of 0 in cell B5. As described for Figure 8-12, enable circular references. Then change cell B5 to the formula "= 0.5*(E8^2*C8+E9^2*C9+E10^2*C10+E11^2*C11+E12^2*C12+E13^2*C13+E14^2*C14)" shown at the bottom of the spreadsheet. We need to estimate two starting concentrations for which I choose pCa^{2+} and pH in cells B8 and B9. We did not pick SO_4^{2-} for estimation because the value of $[Ca^{2+}]$ fixes the value of $[SO_4^{2-}]$ by the K_{sp} equilibrium. Cells C10:C14 contain the concentrations of other species computed from the equilibrium constant expressions, including activity coefficients. Ion sizes appear in column D, with 500 pm being a guess for the size of $CaOH^+$. Charges appear in column E. Activity coefficients are computed in columns F and G.

Now look at the values of the equilibrium constants. K_{base} and K_{acid} are 8 orders of magnitude smaller than K_{sp}. To a good approximation, the base and acid reactions 8-43 and 8-44 will have little effect on the solution composition. As a first guess, we estimate $[Ca^{2+}] \approx \sqrt{K_{sp}} = 10^{-2.3}$ M (or $pCa^{2+} = 2.3$). Since the acid and base reactions are negligible, the pH will be approximately 7. These are the initial estimates in cells B8 and B9.

The mass and charge balances in cells J17 and J18 will be close to zero when the concentrations are correct. Use Solver to minimize $\Sigma b_i^2 = b_1^2 + b_2^2$ in cell J19 by varying pCa^{2+} and pH in cells B8 and B9. Before using Solver, set Solver Options as shown in Figure 8-10b. My first execution of Solver returned $pCa^{2+} = 2.013\,35$ and pH $= 7.000\,05$. Your spreadsheet could give different results depending on settings you have checked. *These p values are good enough*, but you could try further optimization. Try varying just pCa^{2+} while holding pH fixed and then vary pH while holding pCa^{2+} fixed. Solver barely changes Σb_i^2 in cell J19, but we can see that varying pH has a slight effect. In this particular example, I found that changing pH *by hand* in the third decimal place affected Σb_i^2 in cell J19 and quickly found that the minimum in cell J19 was obtained with pH $= 6.995$. Values in cells B8, B9, and J19 in

	A	B	C	D	E	F	G	H	I	J
1	Calcium sulfate equilibria									
2	1. *Estimate* values in cells B8 and B9									
3	2. Use Solver to adjust B8 and B9 to minimize sum in cell J19									
4	Ionic strength					Extended				
5	μ =	0.038789				Debye-				
6				Size		Hückel	Activity			
7	Species	pC	C (M)	α (pm)	Charge	log γ	Coefficient, γ			
8	Ca^{2+}	2.01335	9.6973E-03	600	2	−2.896E-01	5.134E-01			
9	H^+	6.995	1.0116E-07	900	1	−6.353E-02	8.639E-01			
10	SO_4^{2-}		9.6972E-03	500	−2	−3.037E-01	4.969E-01			
11	$CaSO_4(aq)$		5.4954E-03		0	0.000E+00	1.0000E+00			
12	$CaOH^+$		1.3537E-08	500	1	−7.593E-02	8.396E-01			
13	HSO_4^-		4.9231E-08	450	−1	−7.783E-02	8.359E-01			
14	OH^-		1.3818E-07	350	−1	−8.193E-02	8.281E-01			
15										
16							Mass and charge balances:			b_i
17	$pK_{sp} =$	4.62	$K_{sp} =$	2.40E-05		$b_1 = 0 = [Ca^{2+}] + [CaOH^+] - [SO_4^{2-}] - [HSO_4^-] =$				7.27E-10
18	$pK_{ion\,pair} =$	2.26	$K_{ion\,pair} =$	5.50E-03		$b_2 = 0 = 2[Ca^{2+}] + [CaOH^+] + [H^+] - 2[SO_4^{2-}] - [HSO_4^-] - [OH^-] =$				1.22E-10
19	$pK_{base} =$	12.01	$K_{base} =$	9.77E-13			$\Sigma b_i^2 =$			5.43E-19
20	$pK_{acid} =$	12.70	$K_{acid} =$	2.00E-13			J17 = C8+C12-C10-C13			
21	$pK_w =$	14.00	$pK_w =$	1.00E-14			J18 = 2*C8+C12+C9-2*C10-C13-C14			
22	Ion size estimates:							J19 = J17^2 + J18^2		
23	$CaOH^+ \approx 500$ pm							C8 = 10^−B8		
24	HSO_4^- size ≈ HSO_3^- size = 450 pm							C9 = 10^−B9		
25	Initial values:							C10 = D17/(C8*G8*G10)		
26	pCa = 2.3	pH = 7						C11 = D18		
27	Optimize both pCa and pH together for a few cycles						C12 = D20*C8*G8/(G12*C9*G9)			
28	Then optimize just pCa and just pH alternately						C13 = D19*C10*G10/(G13*C14*G14)			
29	Continue optimization as long as Σb_i^2 keeps getting smaller					C14 = D21/(C9*G9*G14)				
30			B5 = 0.5*(E8^2*C8+E9^2*C9+E10^2*C10+E11^2*C11+E12^2*C12+E13^2*C13+E14^2*C14)							

FIGURE 8-13 Calcium sulfate solubility spreadsheet with activity coefficients after Solver operation. You can find this spreadsheet at www.whfreeman.com/qca9e.

Figure 8-13 were obtained by fixing pH = 6.995 and allowing Solver to optimize pCa^{2+}. It is unlikely that you will produce the same exact numbers in your spreadsheet. However, concentrations should agree with those in Figure 8-13 to two or more decimal places.

Results shown in Figure 8-13 are

$$[Ca^{2+}] = 9.70 \times 10^{-3}\ M \quad [SO_4^{2-}] = 9.70 \times 10^{-3}\ M$$
$$[CaOH^+] = 1.35 \times 10^{-8}\ M \quad [HSO_4^-] = 4.92 \times 10^{-8}\ M$$
$$[CaSO_4(aq)] = 5.50 \times 10^{-3}\ M \quad [OH^-] = 1.38 \times 10^{-7}\ M$$
$$\mu = 0.038\ 8\ M \quad [H^+] = 1.01 \times 10^{-7}\ M$$

Total concentration of dissolved sulfate
$$= [SO_4^{2-}] + [CaSO_4(aq)] + [HSO_4^-]$$
$$= 0.009\ 7 + 0.005\ 5 + 5 \times 10^{-8} = 0.015\ 2\ M$$

Congratulations, Dan! Your calculation agrees with the experimental result in Figure 6-1!

These concentrations confirm that the principal chemistry is the dissolution of $CaSO_4(s)$ to give $Ca^{2+}(aq)$, $SO_4^{2-}(aq)$, and the ion pair $CaSO_4(aq)$. Hydrolysis of Ca^{2+} and SO_4^{2-} to $CaOH^+$ and HSO_4^- is minor.

We Will Usually Omit Activity Coefficients

Although it is proper to write equilibrium constants in terms of activities, the complexity of manipulating activity coefficients is a nuisance. In most of this book, we will omit activity coefficients unless there is a particular point to be made. Occasional problems will remind you how to use activities.

Terms to Understand

activity	extended Debye-Hückel	ionic strength	pH
activity coefficient	equation	mass balance	pK
charge balance	ionic atmosphere		

Summary

The thermodynamic equilibrium constant for the reaction $aA + bB \rightleftharpoons cC + dD$ is $K = \mathcal{A}_C^c \mathcal{A}_D^d / (\mathcal{A}_A^a \mathcal{A}_B^b)$, where $\mathcal{A}_i$ is the activity of the ith species. The activity is the product of the concentration (c) and the activity coefficient (γ): $\mathcal{A}_i = c_i \gamma_i$. For nonionic compounds and gases, $\gamma_i \approx 1$. For ionic species, the activity coefficient depends on the ionic strength, defined as $\mu = \frac{1}{2}\Sigma c_i z_i^2$, where z_i is the charge of an ion. The activity coefficient decreases as ionic strength increases, at least for low ionic strengths ($\leq 0.1\ M$). Dissociation of ionic compounds increases with ionic strength because the ionic atmosphere of each ion diminishes the attraction of ions for one another. You should be able to estimate activity coefficients by interpolation in Table 8-1. pH is defined in terms of the activity of H^+: $pH = -\log \mathcal{A}_{H^+} = -\log[H^+]\gamma_{H^+}$. By analogy, pK is the negative logarithm of an equilibrium constant.

In the systematic treatment of equilibrium, write pertinent equilibrium expressions, as well as the charge and mass balances. The charge balance states that the sum of all positive charges in solution equals the sum of all negative charges. The mass balance states that the moles of all forms of an element in solution must equal the moles of that element delivered to the solution. Be sure that you have as many equations as unknowns and then solve for the concentrations by using algebra with approximations or spreadsheets with the Solver routine. For Solver, we *estimate* (number of unknowns) − (number of equilibria) initial pC values and then let Solver find the pC values (and ionic strength) that minimize the sum of squares of the charge and mass balances. The ionic strength value is a by-product of the optimization.

Exercises

8-A. Assuming complete dissociation of the salts, calculate the ionic strength of **(a)** 0.2 mM KNO_3; **(b)** 0.2 mM Cs_2CrO_4; **(c)** 0.2 mM $MgCl_2$ plus 0.3 mM $AlCl_3$.

8-B. Find the activity (not the activity coefficient) of the $(C_3H_7)_4N^+$ (tetrapropylammonium) ion in a solution containing 0.005 0 M $(C_3H_7)_4N^+Br^-$ plus 0.005 0 M $(CH_3)_4N^+Cl^-$.

8-C. Using activities, find $[Ag^+]$ in 0.060 M KSCN saturated with $AgSCN(s)$.

8-D. Using activities, calculate the pH and concentration of H^+ in 0.050 M LiBr at 25°C.

8-E. A 40.0-mL solution of 0.040 0 M $Hg_2(NO_3)_2$ was titrated with 60.0 mL of 0.100 M KI to precipitate Hg_2I_2 ($K_{sp} = 4.6 \times 10^{-29}$).

(a) Show that 32.0 mL of KI are needed to reach the equivalence point.

(b) When 60.0 mL of KI have been added, virtually all Hg_2^{2+} has precipitated, along with 3.20 mmol of I^-. Considering all ions remaining in the solution, calculate the ionic strength when 60.0 mL of KI have been added.

(c) Using activities, calculate pHg_2^{2+} ($= -\log \mathcal{A}_{Hg_2^{2+}}$) for part **(b)**.

8-F. (a) Write the mass balance for $CaCl_2$ in water if the species are Ca^{2+} and Cl^-.

(b) Write the mass balance if the species are Ca^{2+}, Cl^-, $CaCl^+$, and $CaOH^+$.

(c) Write the charge balance for part **(b)**.

8-G. Write the charge and mass balances for dissolving CaF_2 in water if the reactions are

$$CaF_2(s) \rightleftharpoons Ca^{2+} + 2F^-$$
$$Ca^{2+} + H_2O \rightleftharpoons CaOH^+ + H^+$$
$$Ca^{2+} + F^- \rightleftharpoons CaF^+$$
$$CaF_2(s) \rightleftharpoons CaF_2(aq)$$
$$F^- + H^+ \rightleftharpoons HF(aq)$$
$$HF(aq) + F^- \rightleftharpoons HF_2^-$$

8-H. Write charge and mass balances for aqueous $Ca_3(PO_4)_2$ if the species are Ca^{2+}, $CaOH^+$, $CaPO_4^-$, PO_4^{3-}, HPO_4^{2-}, $H_2PO_4^-$, and H_3PO_4.

8-I. (a) *Using activities*, find the concentrations of species in 0.10 M $NaClO_4$ saturated with $Mn(OH)_2$. Modify the spreadsheet in Figure 8-12 by fixing $\mu = 0.1$ in cell B5 and changing equilibrium constants and ion sizes. Suppose that the size of $MnOH^+$ is 400 pm and the chemistry is

$$Mn(OH)_2(s) \xrightleftharpoons{K_{sp}} Mn^{2+} + 2OH^- \qquad K_{sp} = 10^{-12.8}$$
$$Mn^{2+} + OH^- \xrightleftharpoons{K_1} MnOH^+ \qquad K_1 = 10^{3.4}$$
$$H_2O \xrightleftharpoons{K_w} H^+ + OH^- \qquad K_w = 10^{-14.00}$$

(b) Solve the same problem if there is no $NaClO_4$ in the solution.

(c) Why is the solubility of $Mn(OH)_2$ greater when $NaClO_4$ is present? Find the quotient ($[Mn^{2+}]$ in 0.1 M $NaClO_4$/($[Mn^{2+}]$ without $NaClO_4$).

Problems

Activity Coefficients

8-1. Explain why the solubility of an ionic compound increases as the ionic strength of the solution increases (at least up to ~0.5 M).

8-2. Which statements are true? In the ionic strength range 0–0.1 M, activity coefficients decrease with **(a)** increasing ionic strength, **(b)** increasing ionic charge, **(c)** decreasing hydrated radius.

8-3. Color Plate 4 shows how the color of the acid-base indicator bromocresol green (H_2BG) changes as NaCl is added to an aqueous solution of $(H^+)(HBG^-)$. Explain why the color changes from pale green to pale blue as NaCl is added.

HBG⁻ (yellow)

BG²⁻ (blue)

H⁺ counterion is not shown
A *counterion* is the other ion in a solution needed for charge neutrality

8-4. Calculate the ionic strength of **(a)** 0.008 7 M KOH and **(b)** 0.000 2 M $La(IO_3)_3$ (assuming complete dissociation at this low concentration and no hydrolysis reaction to make $LaOH^{2+}$).

8-5. Find the activity coefficient of each ion at the indicated ionic strength:

(a) SO_4^{2-} ($\mu = 0.01$ M)
(b) Sc^{3+} ($\mu = 0.005$ M)
(c) Eu^{3+} ($\mu = 0.1$ M)
(d) $(CH_3CH_2)_3NH^+$ ($\mu = 0.05$ M)

8-6. Interpolate in Table 8-1 to find the activity coefficient of H^+ when $\mu = 0.030$ M.

8-7. Calculate the activity coefficient of Zn^{2+} when $\mu = 0.083$ M by using **(a)** Equation 8-6; **(b)** linear interpolation in Table 8-1.

8-8. Calculate the activity coefficient of Al^{3+} when $\mu = 0.083$ M by linear interpolation in Table 8-1.

8-9. The equilibrium constant for dissolution of a nonionic compound, such as diethyl ether ($CH_3CH_2OCH_2CH_3$), in water can be written

$$\text{ether}(l) \rightleftharpoons \text{ether}(aq) \qquad K = [\text{ether}(aq)]\gamma_{\text{ether}}$$

At low ionic strength, $\gamma \approx 1$ for neutral compounds. At high ionic strength, most neutral molecules can be *salted out* of aqueous solution. That is, when a high concentration (typically >1 M) of a salt such as NaCl is added to an aqueous solution, neutral molecules usually become *less* soluble. Does the activity coefficient, γ_{ether}, increase or decrease at high ionic strength?

8-10. Including activity coefficients, find $[Hg_2^{2+}]$ in saturated Hg_2Br_2 in 0.001 00 M KBr.

8-11. Including activity coefficients, find the concentration of Ba^{2+} in a 0.100 M $(CH_3)_4NIO_3$ solution saturated with $Ba(IO_3)_2$.

8-12. Find the activity coefficient of H^+ in a solution containing 0.010 M HCl plus 0.040 M $KClO_4$. What is the pH of the solution?

8-13. Using activities, calculate the pH of a solution containing 0.010 M NaOH plus 0.012 0 M $LiNO_3$. What would be the pH if you neglected activities?

8-14. The temperature-dependent form of the extended Debye-Hückel equation 8-6 is

$$\log \gamma = \frac{(-1.825 \times 10^6)(\varepsilon T)^{-3/2} z^2 \sqrt{\mu}}{1 + \alpha \sqrt{\mu}/(2.00\sqrt{\varepsilon T})}$$

where ε is the (dimensionless) dielectric constant* of water, T is temperature (K), z is the charge of the ion, μ is ionic strength (mol/L), and α is the ion size parameter (pm). The dependence of ε on temperature is

$$\varepsilon = 79.755e^{(-4.6 \times 10^{-3})(T - 293.15)}$$

Calculate the activity coefficient of SO_4^{2-} at 50.00°C when $\mu = 0.100$ M. Compare your value with the one in Table 8-1.

---

*The dimensionless *dielectric constant*, ε, measures how well a solvent can separate oppositely charged ions. The force of attraction (newtons) between ions of charge q_1 and q_2 (coulombs) separated by distance r (meters) is

$$\text{Force} = -(8.988 \times 10^9)\frac{q_1 q_2}{\varepsilon r^2}$$

The larger the value of ε, the smaller the attraction between ions. Water, with $\varepsilon \approx 80$, separates ions very well. Here are some values of ε: methanol, 33; ethanol, 24; benzene, 2; vacuum and air, 1. Ionic compounds dissolved in solvents less polar than water may exist predominantly as ion pairs, not separate ions.

8-15. *Activity coefficient of a neutral molecule.* We use the approximation that the activity coefficient (γ) of neutral molecules is 1.00. A more accurate relation is $\log \gamma = k\mu$, where μ is ionic strength and $k \approx 0.11$ for NH_3 and CO_2 and $k \approx 0.2$ for organic molecules. With activity coefficients for HA, A^-, and H^+, predict the value of the quotient below for benzoic acid (HA $\equiv C_6H_5CO_2H$). The observed quotient is 0.63 ± 0.03.[15]

$$\text{Concentration quotient} = \frac{\dfrac{[H^+][A^-]}{[HA]} \ (\text{at } \mu = 0)}{\dfrac{[H^+][A^-]}{[HA]} \ (\text{at } \mu = 0.1 \text{ M})}$$

Systematic Treatment of Equilibrium

8-16. State the meaning of the charge and mass balance equations.

8-17. Why do activity coefficients not appear in the charge and mass balance equations?

8-18. Write the charge balance for a solution containing H^+, OH^-, Ca^{2+}, HCO_3^-, CO_3^{2-}, $Ca(HCO_3)^+$, $Ca(OH)^+$, K^+, and ClO_4^-.

8-19. Write the charge balance for a solution of H_2SO_4 in water if the H_2SO_4 ionizes to HSO_4^- and SO_4^{2-}.

8-20. Write the charge balance for an aqueous solution of arsenic acid, H_3AsO_4, in which the acid can dissociate to $H_2AsO_4^-$, $HAsO_4^{2-}$, and AsO_4^{3-}. Look up the structure of arsenic acid in Appendix G and write the structure of $HAsO_4^{2-}$.

8-21. (a) Write the charge and mass balances for a solution made by dissolving $MgBr_2$ to give Mg^{2+}, Br^-, $MgBr^+$, and $MgOH^+$.
(b) Modify the mass balance if the solution was made by dissolving 0.2 mol $MgBr_2$ in 1 L.

8-22. What would happen if charge balance did not exist in a solution? The force between two charges was given in the footnote to Problem 8-14. Find the force between two beakers separated by 1.5 m of air if one contains 250 mL with 1.0×10^{-6} M excess negative charge and the other has 250 mL with 1.0×10^{-6} M excess positive charge. There are 9.648×10^4 coulombs per mole of charge. Convert force from N into pounds (0.224 8 pounds/N). Could two elephants hold the beakers apart?

8-23. For a 0.1 M aqueous solution of sodium acetate, $Na^+CH_3CO_2^-$, one mass balance is simply $[Na^+] = 0.1$ M. Write a mass balance involving acetate.

8-24. Consider the dissolution of the compound X_2Y_3, which gives $X_2Y_2^{2+}$, X_2Y^{4+}, $X_2Y_3(aq)$, and Y^{2-}. Use the mass balance to find an expression for $[Y^{2-}]$ in terms of the other concentrations. Simplify your answer as much as possible.

8-25. Write a mass balance for a solution of $Fe_2(SO_4)_3$ if the species are Fe^{3+}, $Fe(OH)^{2+}$, $Fe(OH)_2^+$, $Fe_2(OH)_2^{4+}$, $FeSO_4^+$, SO_4^{2-}, and HSO_4^-.

8-26. ▦ *Ammonia equilibrium solved with Goal Seek.* Modify Figure 8-7 to find the concentrations of species in 0.05 M NH_3. The only change required is the value of F. How do the pH and fraction of ammonia hydrolysis ($= [NH_4^+]/([NH_4^+] + [NH_3])$) change when the formal concentration of NH_3 increases from 0.01 to 0.05 M?

8-27. ▦ *Ammonia equilibrium treated by Solver.* We now use the Solver spreadsheet introduced in Figure 8-9 for TlN_3 solubility to find the concentrations of species in 0.01 M ammonia solution, neglecting activity coefficients. In the systematic treatment of equilibrium for NH_3 hydrolysis, we have four unknowns ($[NH_3]$, $[NH_4^+]$, $[H^+]$, and $[OH^-]$) and two equilibria (8-13 and 8-14). Therefore, we will estimate the concentrations of (4 unknowns) $-$ (2 equilibria) $=$ 2 species, for which I choose NH_4^+ and OH^-. Set up the spreadsheet shown below, in which the estimates $pNH_4^+ = 3$ and $pOH^- = 3$ appear in cells B6 and B7. (Estimates come from the K_b equilibrium 8-17 with $[NH_4^+] = [OH^-] = \sqrt{K_b[NH_3]} \approx \sqrt{10^{-4.755}[0.01]} \Rightarrow pNH_4^+ = pOH^- \approx 3$. Estimates do not have to be very good for Solver to work.) The formula in cell C8 is $[NH_3] = [NH_4^+] \cdot [OH^-]/K_b$ and the formula in cell C9 is $[H^+] = K_w/[OH^-]$. The mass balance b_1 appears in cell F6 and the charge balance b_2 appears in cell F7. Cell F8 has the sum $b_1^2 + b_2^2$. As described for TlN_3 on page 176, open the Solver window and set the Solver Options. Then use Solver to Set Target Cell F8 Equal To Min By Changing Cells B6:B7. What are the concentrations of the species? What fraction of ammonia ($= [NH_4^+]/([NH_4^+] + [NH_3])$) is hydrolyzed? Your answers should agree with those from Goal Seek in Figure 8-8.

	A	B	C	D	E	F
1	Ammonia equilibrium					F =
2	1. *Estimate* values of pC = –log[C] for NH_4^+ and OH^- in cells B6 and B7					0.01
3	2. Use Solver to adjust the values of pC to minimize the sum in cell F8					
4						
5	Species	pC			Mass and charge balances	b_i
6	NH_4^+	3.00000000	1.00E-03	C6 = 10^-B6	$b_1 = 0 = F - [NH_4^+] - [NH_3] =$	–4.79E-02
7	OH^-	3.00000000	0.001	C7 = 10^-B7	$b_2 = 0 = [NH_4^+] + [H^+] - [OH^-] =$	1.00E-11
8	NH_3		0.056885293	C8 = C6*C7/D12	$\Sigma b_i^2 =$	2.29E-03
9	H^+		1E-11	C9 = D13/C7	F6 = F2-C6-C8	
10					F7 = C6+C9-C7	
11					F8 = F6^2+F7^2	
12	$pK_b =$	4.755	$K_b =$	1.76E-05	= 10^-B12	
13	$pK_w =$	14.00	$K_w =$	1.00E-14	= 10^-B13	

Spreadsheet for Problem 8-27.

CHAPTER 8 Activity and the Systematic Treatment of Equilibrium

8-28. 🖩 *Sodium acetate hydrolysis treated by Solver with activity coefficients.*

(a) Following the NH_3 example in Section 8-5, write the equilibria and charge and mass balances needed to find the composition of 0.01 M sodium acetate (Na^+A^-). *Include activity coefficients* where appropriate. The two reactions are hydrolysis ($pK_b = 9.244$) and ionization of H_2O.

(b) *Including activity coefficients*, set up a spreadsheet analogous to Figure 8-12 to find the concentrations of all species. Assign an initial value of ionic strength = 0.01. After the rest of the spreadsheet is set up, change the ionic strength from the numerical value 0.01 to the correct formula for ionic strength. This two-step process of beginning with a numerical value and then going to a formula is necessary because of circular references between ionic strength and concentrations that depend on ionic strength. There are four unknowns and two equilibria, so use Solver to find $4 - 2 = 2$ concentrations (pC values). Solver does not find both pC values at the same time well in this problem. Execute one pass to find both pC values by varying pA and pOH to minimize Σb_i^2. Then vary only pA to minimize Σb_i^2. Then vary only pOH to minimize Σb_i^2. Continue alternating to solve for one value at a time as long as Σb_i^2 continues to decrease. Find $[A^-]$, $[OH^-]$, $[HA]$, and $[H^+]$. Find the ionic strength, pH = $-\log([H^+]\gamma_{H^+})$ and the fraction of hydrolysis = $[HA]/F$.

8-29. (a) Following the example of $Mg(OH)_2$ in Section 8-5, write the equations needed to find the solubility of $Ca(OH)_2$. *Include activity coefficients* where appropriate. Look up the equilibrium constants in Appendixes F and I.

(b) 🖩 Suppose that the size of $CaOH^+ = Ca(H_2O)_5(OH)^+$ is 500 pm. Including activity coefficients, compute the concentrations of all species, the fraction of hydrolysis (= $[CaOH^+]/\{[Ca^{2+}] + [CaOH^+]\}$), and the solubility of $Ca(OH)_2$ in g/L. The *Handbook of Chemistry and Physics* lists the solubility of $Ca(OH)_2$ as 1.85 g/L at 0°C and 0.77 g/L at 100°C.

8-30. 🖩 *Systematic treatment of equilibrium for ion pairing.* Let's derive the fraction of ion pairing for the salts in Box 8-1, which are 0.025 F NaCl, Na_2SO_4, $MgCl_2$, and $MgSO_4$. Each case is somewhat different. All of the solutions will be near neutral pH because hydrolysis reactions of Mg^{2+}, SO_4^{2-}, Na^+, and Cl^- have small equilibrium constants. Therefore, we assume that $[H^+] = [OH^-]$ and omit these species from the calculations. We work $MgCl_2$ as an example and then you are asked to work each of the others. The ion-pair equilibrium constant, K_{ip}, comes from Appendix J.

Pertinent reaction:

$$Mg^{2+} + Cl^- \rightleftharpoons MgCl^+(aq)$$

$$K_{ip} = \frac{[MgCl^+(aq)]\gamma_{MgCl^+}}{[Mg^{2+}]\gamma_{Mg^{2+}}[Cl^-]\gamma_{Cl^-}} \quad \log K_{ip} = 0.6 \quad pK_{ip} = -0.6 \quad \text{(A)}$$

Charge balance (omitting H^+ and OH^- whose concentrations are both small in comparison with $[Mg^{2+}]$, $[MgCl^+]$, and $[Cl^-]$):

$$2[Mg^{2+}] + [MgCl^+] = [Cl^-] \quad \text{(B)}$$

Mass balances:

$$[Mg^{2+}] + [MgCl^+] = F = 0.025 \text{ M} \quad \text{(C)}$$

$$[Cl^-] + [MgCl^+] = 2F = 0.050 \text{ M} \quad \text{(D)}$$

Only two of the three Equations (B), (C), and (D) are independent. If you double (C) and subtract (D), you will produce (B). We choose (C) and (D) as independent equations.

Equilibrium constant expression: Equation (A)

Count: 3 equations (A, C, D) and 3 unknowns ($[Mg^{2+}]$, $[MgCl^+]$, $[Cl^-]$)

Solve: We will use Solver to find (number of unknowns) − (number of equilibria) = $3 - 1 = 2$ unknown concentrations.

	A	B	C	D	E	F	G	H
1	Magnesium chloride ion pairing with activities							
2	1. *Estimate* pMg^{2+} and PCl$^-$ in cells B8 and B9				Formal conc = F =		0.025	M
3	2. Use Solver to adjust B8 and B9 to minimize sum in cell H16				[Mg^{2+}] + [MgCl$^-$(aq)] = F			
4	Ionic strength				[Cl$^-$] + [MgCl$^+$(aq)] = 2F			
5	μ	7.093E-02				Extended	Activity	
6				Size		Debye-Hückel	Coefficient	
7	Species	pC	C (M)	α (pm)	Charge	log γ	γ	
8	Mg^{2+}	1.638971597	2.296E-02	800	2	−3.199E-01	0.479	G8 = 10^F8
9	Cl$^-$	1.319093767	4.796E-02	300	−1	−1.076E-01	0.780	
10	MgCl$^+$(aq)		2.037E-03	500	1	−9.455E-02	0.804	
11								
12								
13	pK$_{ip}$ =	−0.60	K$_{ip}$ =	3.98E+00			Mass balances:	b$_i$
14						b$_1$ = 0 = F − [Mg^{2+}] − [MgCl$^+$(aq)] =		3.37E-13
15	Ion pair fraction = [MgCl$^+$(aq)]/F =		0.0815			b$_2$ = 0 = 2F − [Cl$^-$] − [MgCl$^+$(aq)] =		2.11E-13
16				D15 = C10/G2			Σb$_i^2$ =	1.58E-25
17	Ion size estimate:						H14 = G2-C8-C10	
18	MgCl$^+$(aq) =	500					H15 = 2*G2-C9-C10	
19	Initial value:						H16 = H14^2 + H15^2	
20	pMg^{2+} =	1.7					C8 =10^-B8	
21	pCl$^-$ =	1.4					C9 =10^-B9	
22							C10 = D13*C8*G8*C9*G9/G10	
23	Check:				F8 = −0.51*E8^2*SQRT($B$5)/(1+D8*SQRT($B$5)/305)			
24	Total Mg =	0.02500	= C8+C10		B5 = 0.5*(E8^2*C8+E9^2*C9+E10^2*C10)			
25	Total Cl =	0.05000	= C9+C10					
26	K$_{ip}$ =	3.981E+00	=C10*G10/(C8*G8*C9*G9)					

Spreadsheet for Problem 8-30. You can find this spreadsheet at www.whfreeman.com/qca9e.

The spreadsheet shows the work. Formal concentration F = 0.025 M appears in cell G2. We estimate pMg^{2+} and pCl^- in cells B8 and B9. The ionic strength in cell B5 is given by the formula in cell H24. Excel must be set to allow for circular definitions as described on page 179. The sizes of Mg^{2+} and Cl^- are from Table 8-1 and the size of $MgCl^+$ is a guess. Activity coefficients are computed in columns E and F. Mass balances $b_1 = F - [Mg^{2+}] - [MgCl^+]$ and $b_2 = 2F - [Cl^-] - [MgCl^+]$ appear in cells H14 and H15, and the sum of squares $b_1^2 + b_2^2$ appears in cell H16. The charge balance is not used because it is not independent of the two mass balances. Solver is invoked to minimize $b_1^2 + b_2^2$ in cell H16 by varying pMg^{2+} and pCl^- in cells B8 and B9. From the optimized concentrations, the ion-pair fraction $= [MgCl^+]/F = 0.0815$ is computed in cell D15. This fraction is shown in the table in Box 8-1. As a check, total Mg and total Cl are correctly found to be 0.025 and 0.05 M in cells B24 and B25 and K_{ip} is correct in cell B26.

The Problem: Create a spreadsheet like the one for $MgCl_2$ to find the concentrations, ionic strength, and ion-pair fraction in 0.025 F NaCl. The ion-pair formation constant from Appendix J is log $K_{ip} = 10^{-0.5}$ for the reaction $Na^+ + Cl^- \rightleftharpoons NaCl(aq)$. The two mass balances are $[Na^+] + [NaCl(aq)] = F$ and $[Na^+] = [Cl^-]$ (which is also the charge balance). Estimate pNa^+ and pCl^- for input and then minimize the sum of squares of the two mass balances.

8-31. 🖩 *Ion pairing.* As in Problem 8-30, find the concentrations, ionic strength, and ion-pair fraction in 0.025 F Na_2SO_4. Assume that the size of $NaSO_4^-$ is 500 pm.

8-32. (a) 🖩 *Ion pairing.* As in Problem 8-30, find the concentrations, ionic strength, and ion-pair fraction in 0.025 F $MgSO_4$.

(b) Two possibly important reactions that we did not consider are acid hydrolysis of Mg^{2+} ($Mg^{2+} + H_2O \rightleftharpoons MgOH^+ + H^+$) and base hydrolysis of SO_4^{2-}. Write these two reactions and find their equilibrium constants in Appendices I and G. With the assumed pH near 7.0, and neglecting activity coefficients, show that both reactions are negligible.

8-33. 🖩 *Solubility with activity.* Find the concentrations of the major species in a saturated aqueous solution of LiF. Consider these reactions:

$$LiF(s) \rightleftharpoons Li^+ + F^- \qquad K_{sp} = [Li^+]\gamma_{Li^+}[F^-]\gamma_{F^-}$$
$$LiF(s) \rightleftharpoons LiF(aq) \qquad K_{ion\ pair} = [LiF(aq)]\gamma_{LiF(aq)}$$
$$F^- + H_2O \rightleftharpoons HF + OH^- \qquad K_b = K_w/K_a (\text{for HF})$$
$$H_2O \overset{K_w}{\rightleftharpoons} H^+ + OH^- \qquad K_w = [H^+]\gamma_{H^+}[OH^-]\gamma_{OH^-}$$

(a) Look up the equilibrium constants in the appendixes and write their pK values. The ion pair reaction is the sum of $LiF(s) \rightleftharpoons Li^+ + F^-$ from Appendix F and $Li^+ + F^- \rightleftharpoons LiF(aq)$ from Appendix J. Write the equilibrium constant expressions and the charge and mass balances.

(b) Create a spreadsheet that uses activities to find the concentrations of all species and the ionic strength. Use pF and pOH as independent variables to estimate. It does not work to choose pF and pLi because either concentration fixes that of the other through the relation $K_{sp} = [Li^+]\gamma_{Li^+}[F^-]\gamma_{F^-}$.

8-34. *Heterogeneous equilibria and calcite solubility.* If river water in Box 8-2 is saturated with calcite ($CaCO_3$), $[Ca^{2+}]$ is governed by the following equilibria:

$$CaCO_3(s) \rightleftharpoons Ca^{2+} + CO_3^{2-} \qquad K_{sp} = 4.5 \times 10^{-9}$$
$$CO_2(g) \rightleftharpoons CO_2(aq) \qquad K_{CO_2} = 0.032$$
$$CO_2(aq) + H_2O \rightleftharpoons HCO_3^- + H^+ \qquad K_1 = 4.46 \times 10^{-7}$$
$$HCO_3^- \rightleftharpoons CO_3^{2-} + H^+ \qquad K_2 = 4.69 \times 10^{-11}$$

(a) Combine these reactions to find the equilibrium constant for the reaction

$$CaCO_3(s) + CO_2(aq) + H_2O \rightleftharpoons Ca^{2+} + 2HCO_3^- \quad K = ? \quad (A)$$

(b) The mass balance for Reaction A is $[HCO_3^-] = 2[Ca^{2+}]$. Find $[Ca^{2+}]$ (in mol/L and in mg/L) in equilibrium with atmospheric CO_2 if $P_{CO_2} = 4.0 \times 10^{-4}$ bar $= 10^{-3.4}$ bar. Locate this point on the line in Box 8-2.

(c) The concentration of Ca^{2+} in the Don River is 80 mg/L. What effective P_{CO_2} is in equilibrium with this much Ca^{2+}? How can the river have this much CO_2?

9 | Monoprotic Acid-Base Equilibria

MEASURING pH INSIDE CELLULAR COMPARTMENTS

(*a*) Macrophage beginning to engulf a round cancer cell as phagocytosis begins. [© Microworks/ Phototake.] (*b*) Macrophages with ingested 1.6-μm-diameter fluorescent beads.
(*c*) Fluorescence image of panel *b*. [*b* and *c* from K. P. McNamara, T. Nguyen, G. Dumitrascu, J. Ji, N. Rosenzweig, and Z. Rosenzweig, "Synthesis, Characterization, and Application of Fluorescence Sensing Lipobeads for Intracellular pH Measurements," *Anal. Chem.* **2001,** *73,* 3240. Reprinted with permission © 2001 American Chemical Society.]

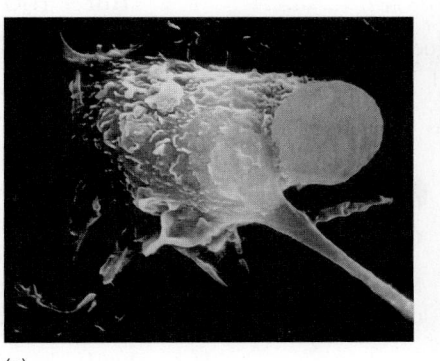 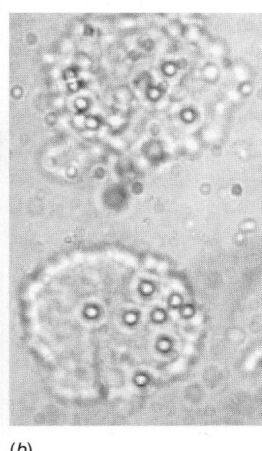 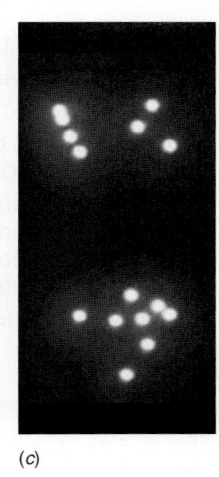

(a)　　　　(b)　　　　(c)

Macrophages (panel *a*) are white blood cells that fight infection by ingesting and dissolving foreign cells—a process called *phagocytosis*. The compartment containing the ingested foreign cell merges with compartments called *lysosomes*, which contain digestive enzymes that are most active in acid. Low enzyme activity above pH 7 protects the cell from enzymes that leak into the cell.

One way to measure pH inside the compartment containing the ingested particle and digestive enzymes is to present macrophages with polystyrene beads (panels *b* and *c*) coated with a lipid membrane to which fluorescent (light-emitting) dyes are covalently bound. Panel *d* shows that fluorescence intensity from the dye fluorescein depends on pH, but fluorescence from tetramethylrhodamine does not. The ratio of emission from the dyes is a measure of pH. Panel *e* shows the fluorescence intensity ratio changing in 3 s as the bead is ingested and the pH around the bead drops from 7.3 to 5.7. The drop in pH allows digestion to commence.

(*d*) Fluorescence spectra of lipobeads in solutions at pH 5–8. (*e*) pH change during phagocytosis of a single bead by a macrophage. [Data from McNamara et al., ibid.]

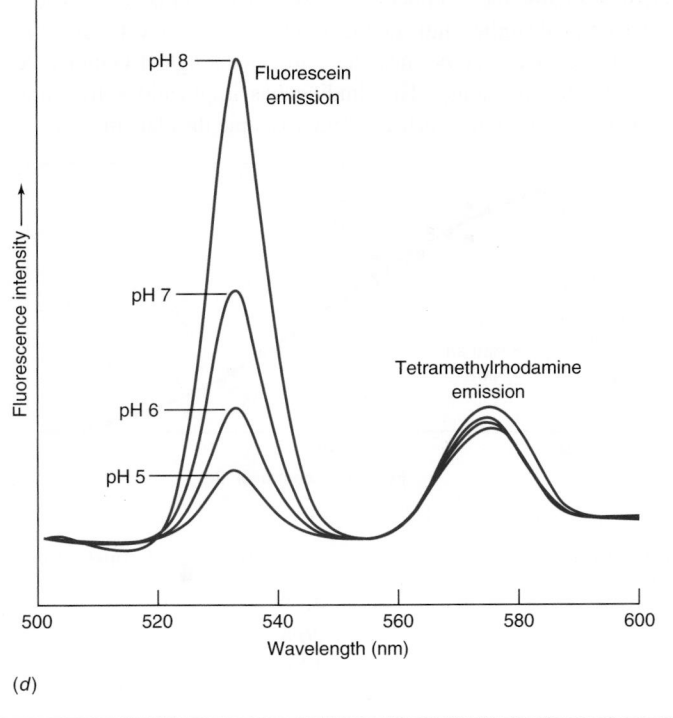 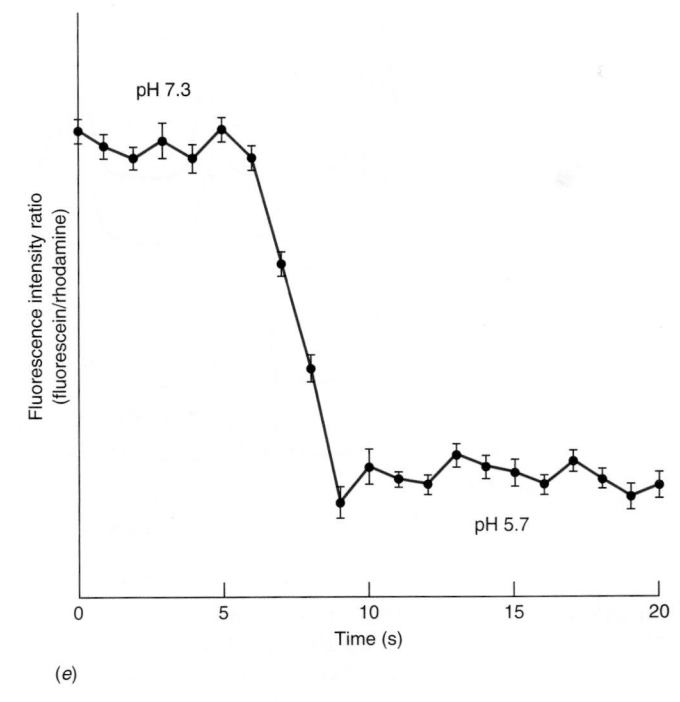

(d)　　　　(e)

cids and bases are essential to virtually every application of chemistry and for the intelligent use of analytical procedures such as chromatography and electrophoresis. It would be difficult to have a meaningful discussion of, say, protein purification or the weathering of rocks without understanding acids and bases. This chapter covers acid-base equilibria and buffers. Chapter 10 treats polyprotic systems involving two or more acidic protons. Nearly every biological macromolecule is polyprotic. Chapter 11 describes acid-base titrations. Now is the time to review fundamentals of acids and bases in Sections 6-5 through 6-7.

9-1 Strong Acids and Bases

Table 6-2 gave a list of strong acids and bases that you must memorize.

Equilibrium constants for the reaction[1]
$$HX(aq) + H_2O \rightleftharpoons H_3O^+ + X^-$$
HCl	$K_a = 10^{3.9}$
HBr	$K_a = 10^{5.8}$
HI	$K_a = 10^{10.4}$
HNO_3	$K_a = 10^{1.4}$

HNO_3 is discussed in Box 9-1.

What could be easier than calculating the pH of 0.10 M HBr? HBr is a **strong acid,** so the reaction

$$HBr + H_2O \rightarrow H_3O^+ + Br^-$$

goes to completion, and the concentration of H_3O^+ is 0.10 M. It will be our custom to write H^+ instead of H_3O^+, and we can say

$$pH = -\log[H^+] = -\log(0.10) = 1.00$$

EXAMPLE Activity Coefficient in a Strong-Acid Calculation

Calculate the pH of 0.10 M HBr, using activity coefficients.

Solution The ionic strength of 0.10 M HBr is $\mu = 0.10$ M, at which the activity coefficient of H^+ is 0.83 (Table 8-1). Remember that pH is $-\log \mathcal{A}_{H^+}$, not $-\log[H^+]$:

$$pH = -\log[H^+]\gamma_{H^+} = -\log(0.10)(0.83) = 1.08$$

TEST YOURSELF Calculate the pH of 0.010 M HBr in 0.090 M KBr. (*Answer:* 2.08)

BOX 9-1 Concentrated HNO_3 Is Only Slightly Dissociated[2]

Strong acids in dilute solution are essentially completely dissociated. As concentration increases, the degree of dissociation decreases. The figure shows a *Raman spectrum* of solutions of nitric acid of increasing concentration. The spectrum measures scattering of light whose energy corresponds to vibrational energies of molecules. The sharp signal at 1 049 cm^{-1} in the spectrum of 5.1 M $NaNO_3$ is characteristic of free NO_3^- anion.

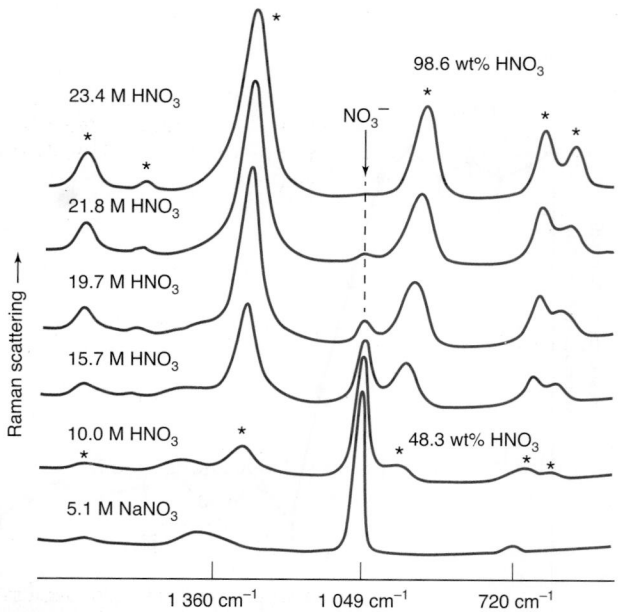

Raman spectrum of aqueous HNO_3 at 25°C. Signals at 1 360, 1 049, and 720 cm^{-1} arise from NO_3^- anion. Signals denoted by asterisks are from undissociated HNO_3. The wavenumber unit, cm^{-1}, is 1/wavelength.

A 10.0 M HNO_3 solution has a strong NO_3^- signal at 1 049 cm^{-1} from dissociated acid. Bands denoted by asterisks arise from *undissociated* HNO_3. As concentration increases, the 1 049 cm^{-1} signal disappears and signals attributed to undissociated HNO_3 increase. The graph shows the fraction of dissociation deduced from spectroscopic measurements. It is instructive to realize that, in 20 M HNO_3, there are fewer H_2O molecules than there are molecules of HNO_3. Dissociation decreases because there is not enough solvent to stabilize the free ions.

Theoretical studies indicate that dilute HNO_3 at a water-air *interface* is also a weak acid because there are not enough H_2O molecules to solvate the free ions.[3] This finding has implications for atmospheric chemistry at the surface of microscopic droplets in clouds.

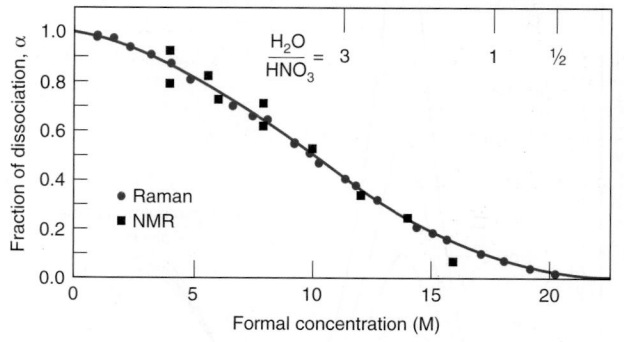

Temperature (°C)	Acid dissociation constant (K_a)
0	46.8
25	26.8
50	14.9

Now that we have reminded you of activity coefficients, you can breathe a sigh of relief because we will neglect activity coefficients unless there is a specific point to be made.

How do we calculate the pH of 0.10 M KOH? KOH is a **strong base** (completely dissociated), so $[OH^-] = 0.10$ M. Using $K_w = [H^+][OH^-]$, we write

$$[H^+] = \frac{K_w}{[OH^-]} = \frac{1.0 \times 10^{-14}}{0.10} = 1.0 \times 10^{-13} \text{ M}$$

$$pH = -\log[H^+] = 13.00$$

From $[OH^-]$, you can always find $[H^+]$:

$$[H^+] = \frac{K_w}{[OH^-]}$$

Finding the pH of other concentrations of KOH is pretty trivial:

$[OH^-]$ (M)	$[H^+]$ (M)	pH
$10^{-3.00}$	$10^{-11.00}$	11.00
$10^{-4.00}$	$10^{-10.00}$	10.00
$10^{-5.00}$	$10^{-9.00}$	9.00

A generally useful relation is

Relation between pH and pOH:

$$pH + pOH = -\log K_w = 14.00 \text{ at } 25°C \qquad (9\text{-}1)$$

The temperature dependence of K_w was given in Table 6-1.

The Dilemma

Well, life seems simple enough so far. Now we ask, "What is the pH of 1.0×10^{-8} M KOH?" Applying our usual reasoning, we calculate

$$[H^+] = K_w/(1.0 \times 10^{-8}) = 1.0 \times 10^{-6} \text{ M} \Rightarrow pH = 6.00$$

How can the base KOH produce an acidic solution (pH < 7)? It's impossible.

Adding base to water cannot *lower* the pH. (Lower pH is more *acidic*.) There must be something wrong.

The Cure

Clearly, there is something wrong with our calculation. In particular, we have not considered the contribution of OH^- from the ionization of water. In pure water, $[OH^-] = 1.0 \times 10^{-7}$ M, which is greater than the amount of KOH added to the solution. To handle this problem, we resort to the systematic treatment of equilibrium.

Step 1 *Pertinent reactions.* The only one is $H_2O \xrightleftharpoons{K_w} H^+ + OH^-$
Step 2 *Charge balance.* The species in solution are K^+, OH^-, and H^+. So,

$$[K^+] + [H^+] = [OH^-] \qquad (9\text{-}2)$$

Step 3 *Mass balance.* All K^+ comes from the KOH, so $[K^+] = 1.0 \times 10^{-8}$ M
Step 4 *Equilibrium constant.* $K_w = [H^+][OH^-] = 1.0 \times 10^{-14}$
Step 5 *Count.* There are three equations and three unknowns ($[H^+]$, $[OH^-]$, $[K^+]$), so we have enough information to solve the problem.
Step 6 *Solve.* We seek the pH, so let's set $[H^+] = x$. Writing $[K^+] = 1.0 \times 10^{-8}$ M in Equation 9-2, we get

$$[OH^-] = [K^+] + [H^+] = 1.0 \times 10^{-8} + x$$

If we had been using activities, step 4 is the only point at which activity coefficients would have entered.

Putting $[OH^-] = 1.0 \times 10^{-8} + x$ into the K_w equilibrium enables us to solve the problem:

$$[H^+][OH^-] = K_w$$

$$(x)(1.0 \times 10^{-8} + x) = 1.0 \times 10^{-14}$$

$$x^2 + (1.0 \times 10^{-8})x - (1.0 \times 10^{-14}) = 0$$

$$x = \frac{-1.0 \times 10^{-8} \pm \sqrt{(1.0 \times 10^{-8})^2 - 4(1)(-1.0 \times 10^{-14})}}{2(1)}$$

$$= 9.6 \times 10^{-8} \text{ M} \quad \text{or} \quad -1.1 \times 10^{-7} \text{ M}$$

Solution of a quadratic equation:

$$ax^2 + bx + c = 0$$

$$x = \frac{-b \pm \sqrt{b^2 - 4ac}}{2a}$$

Retain all digits in your calculator because b^2 is sometimes nearly equal to $4ac$. If you round off before computing $b^2 - 4ac$, your answer may be garbage.

Rejecting the negative concentration, we conclude that

$$[H^+] = 9.6 \times 10^{-8} \text{ M} \Rightarrow pH = -\log[H^+] = 7.02$$

This pH is eminently reasonable, because 10^{-8} M KOH should be slightly basic.

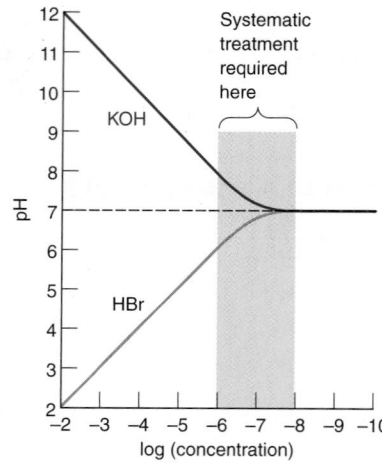

FIGURE 9-1 Calculated pH as a function of concentration of strong acid or strong base in water.

Any acid or base suppresses water ionization, as predicted by Le Châtelier's principle.

Question What concentrations of H^+ and OH^- are produced by H_2O dissociation in 0.01 M NaOH?

Of course, you know that K_a should really be expressed in terms of activities, not concentrations:

$$K_a = \mathcal{A}_{H^+} \mathcal{A}_{A^-} / \mathcal{A}_{HA}$$

Hydrolysis is reaction with water.

As K_a increases, pK_a decreases. Smaller pK_a means stronger acid.

HA and A^- are a **conjugate acid-base pair.** B and BH^+ also are conjugate.

Figure 9-1 shows the pH calculated for different concentrations of strong base or strong acid in water. There are three regions:

1. When the concentration is "high" ($\geq 10^{-6}$ M), pH is calculated by just considering the added H^+ or OH^-. That is, the pH of $10^{-5.00}$ M KOH *is* 9.00.
2. When the concentration is "low" ($\leq 10^{-8}$ M), the pH is 7.00. We have not added enough acid or base to change the pH of the water itself.
3. At intermediate concentrations of 10^{-6} to 10^{-8} M, the effects of water ionization and the added acid or base are comparable. Only in this region is a systematic equilibrium calculation necessary.

Region 1 is the only practical case. Unless you were to protect 10^{-7} M KOH from the air, dissolved CO_2, not KOH, would govern the pH.

Water Almost Never Produces 10^{-7} M H^+ and 10^{-7} M OH^-

The misconception that dissociation of water always produces 10^{-7} M H^+ and 10^{-7} M OH^- is true *only* in pure water with no added acid or base. In 10^{-4} M HBr, for example, the pH is 4. The concentration of OH^- is $[OH^-] = K_w/[H^+] = 10^{-10}$ M. But the only source of $[OH^-]$ is dissociation of water. If water produces 10^{-10} M OH^-, it must also produce 10^{-10} M H^+ because it makes one H^+ for every OH^-. In 10^{-4} M HBr solution, water dissociation produces only 10^{-10} M OH^- and 10^{-10} M H^+.

9-2 Weak Acids and Bases

Let's review the meaning of the **acid dissociation constant,** K_a, for the acid HA:

Weak-acid equilibrium:
$$HA \xrightleftharpoons{K_a} H^+ + A^- \qquad K_a = \frac{[H^+][A^-]}{[HA]} \qquad (9\text{-}3)$$

A **weak acid** is one that is not completely dissociated. That is, Reaction 9-3 does not go to completion. For a base, B, the **base hydrolysis constant,** K_b, is defined by the reaction

Weak-base equilibrium:
$$B + H_2O \xrightleftharpoons{K_b} BH^+ + OH^- \qquad K_b = \frac{[BH^+][OH^-]}{[B]} \qquad (9\text{-}4)$$

A **weak base** is one for which Reaction 9-4 does not go to completion.

   pK is the negative logarithm of an equilibrium constant:

$$pK_w = -\log K_w$$
$$pK_a = -\log K_a$$
$$pK_b = -\log K_b$$

As K increases, pK decreases, and vice versa. Comparing formic and benzoic acids, we see that formic acid is stronger, with a larger K_a and smaller pK_a, than benzoic acid.

$$\underset{\text{Formic acid}}{\overset{\displaystyle O \atop \displaystyle \|}{HCOH}} \rightleftharpoons H^+ + \underset{\text{Formate}}{\overset{\displaystyle O \atop \displaystyle \|}{HCO^-}} \qquad \begin{matrix} K_a = 1.80 \times 10^{-4} \\ pK_a = 3.744 \end{matrix}$$

$$\underset{\text{Benzoic acid}}{\bigcirc\!\!-\!\!\overset{\displaystyle O \atop \displaystyle \|}{C}OH} \rightleftharpoons \underset{\text{Benzoate}}{\bigcirc\!\!-\!\!\overset{\displaystyle O \atop \displaystyle \|}{C}O^-} + H^+ \qquad \begin{matrix} K_a = 6.28 \times 10^{-5} \\ pK_a = 4.202 \end{matrix}$$

The acid HA and its corresponding base, A^-, are said to be a **conjugate acid-base pair,** because they are related by the gain or loss of a proton. Similarly, B and BH^+ are a conjugate pair. The important relation between K_a and K_b for a conjugate acid-base pair is

Relation between K_a and K_b for conjugate pair:
$$K_a \cdot K_b = K_w \qquad (9\text{-}5)$$

Weak Is Conjugate to Weak

The conjugate base of a weak acid is a weak base. The conjugate acid of a weak base is a weak acid. Consider a weak acid, HA, with $K_a = 10^{-4}$. The conjugate base, A^-, has $K_b = K_w/K_a = 10^{-10}$. That is, if HA is a weak acid, A^- is a weak base. If K_a were 10^{-5}, then K_b would be 10^{-9}. As HA becomes a weaker acid, A^- becomes a stronger base (but never a strong base). Conversely, the greater the acid strength of HA, the less the base strength of A^-. However, if either A^- or HA is weak, so is its conjugate. If HA is strong (such as HCl), its conjugate base (Cl^-) is *so* weak that it is not a base at all in water.

The conjugate base of a weak acid is a weak base. The conjugate acid of a weak base is a weak acid. *Weak is conjugate to weak.*

Using Appendix G

Appendix G lists acid dissociation constants. Each compound is shown in its *fully protonated form*. Diethylamine, for example, is shown as $(CH_3CH_2)_2NH_2^+$, which is really the diethylammonium ion. The value of K_a (1.0×10^{-11}) given for diethylamine is actually K_a for the diethylammonium ion. To find K_b for diethylamine, we write $K_b = K_w/K_a = 1.0 \times 10^{-14}/1.0 \times 10^{-11} = 1.0 \times 10^{-3}$.

For polyprotic acids and bases, several K_a values are given. Pyridoxal phosphate is given in its fully protonated form as follows:[4]

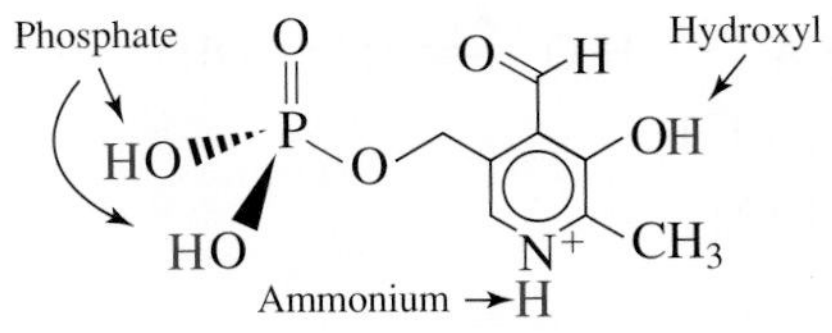

pKₐ	Kₐ
1.4 (POH)	0.04
3.51 (OH)	3.1×10^{-4}
6.04 (POH)	9.1×10^{-7}
8.25 (NH)	5.6×10^{-9}

Pyridoxal phosphate
(a derivative of vitamin B₆)

pK_1 (1.4) is for dissociation of one of the phosphate protons, and pK_2 (3.51) is for the hydroxyl proton. The third most acidic proton is the other phosphate proton, for which $pK_3 = 6.04$, and the NH^+ group is the least acidic ($pK_4 = 8.25$).

Species drawn in Appendix G are fully protonated. If a structure in Appendix G has a charge other than 0, it is not the structure that belongs with the name in the appendix. *Names refer to neutral molecules.* The neutral molecule pyridoxal phosphate is not the species drawn above, which has a +1 charge. The neutral molecule pyridoxal phosphate is

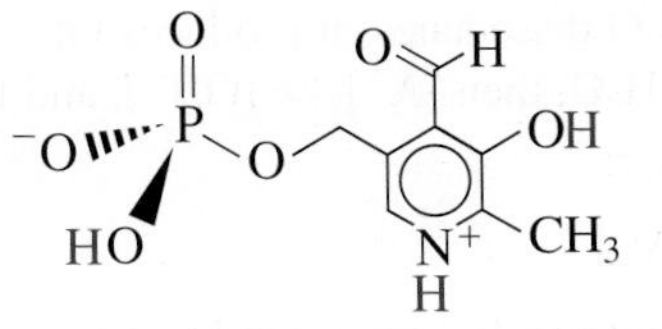

We took away a POH proton, not the NH^+ proton, because POH is the most acidic group in the molecule ($pK_a = 1.4$).

As another example, consider the molecule piperazine:

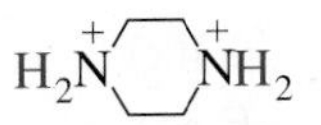

Structure shown for piperazine in Appendix G

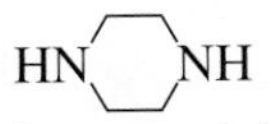

Actual structure of piperazine, which *must be neutral*

Appendix G gives pK_a for ionic strengths of 0 and 0.1 M, when available. We will use pK_a for $\mu = 0$ unless there is no value listed or we need $\mu = 0.1$ M for a specific purpose. For pyridoxal phosphate, we used values for $\mu = 0.1$ M because none are listed for $\mu = 0$.

K_a at $\mu = 0$ is the *thermodynamic acid dissociation constant* that applies at any ionic strength when activity coefficients are used for that ionic strength:

$$K_a = \frac{\mathcal{A}_{H^+}\mathcal{A}_{A^-}}{\mathcal{A}_{HA}} = \frac{[H^+]\gamma_{H^+}[A^-]\gamma_{A^-}}{[HA]\gamma_{HA}}$$

K_a at $\mu = 0.1$ M is the quotient of concentrations when the ionic strength is 0.1 M:

$$K_a (\mu = 0.1 \text{ M}) = \frac{[H^+][A^-]}{[HA]}$$

9-3 Weak-Acid Equilibria

Let's compare the ionization of *ortho*- and *para*-hydroxybenzoic acids:

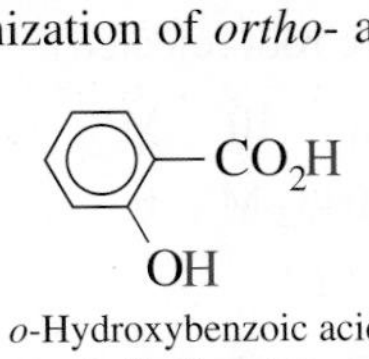

o-Hydroxybenzoic acid
(salicylic acid)
$pK_a = 2.97$

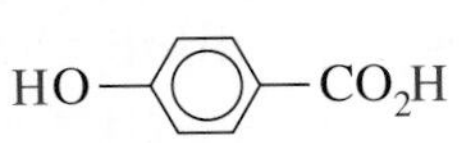

p-Hydroxybenzoic acid
$pK_a = 4.54$

Why is the *ortho* isomer 30 times more acidic than the *para* isomer? Any effect that stabilizes the product of a reaction drives the reaction forward. In the *ortho* isomer, the product of the acid dissociation reaction can form a strong, internal hydrogen bond.

The *para* isomer cannot form such a bond because the —OH and —CO$_2^-$ groups are too far apart. By stabilizing the product, the internal hydrogen bond is thought to make *o*-hydroxybenzoic acid more acidic than *p*-hydroxybenzoic acid.

A Typical Weak-Acid Problem

Formal concentration is the total number of moles of a compound dissolved in a liter. The formal concentration of a weak acid is the total amount of HA placed in the solution, even though some has changed into A$^-$.

The problem is to find the pH of a solution of the weak acid HA, given the formal concentration of HA and the value of K_a.[5] Let's call the formal concentration F and use the systematic treatment of equilibrium:

Reactions:
$$HA \xrightarrow{K_a} H^+ + A^- \qquad H_2O \xrightarrow{K_w} H^+ + OH^-$$

Charge balance:
$$[H^+] = [A^-] + [OH^-] \qquad (9\text{-}6)$$

Mass balance:
$$F = [A^-] + [HA] \qquad (9\text{-}7)$$

Equilibrium expressions:
$$K_a = \frac{[H^+][A^-]}{[HA]} \qquad (9\text{-}8)$$

$$K_w = [H^+][OH^-]$$

There are four equations and four unknowns ($[A^-]$, $[HA]$, $[H^+]$, $[OH^-]$), so the problem is solved if we can just do the algebra.

But it's not so easy to solve these simultaneous equations. If you combine them, you will discover that a cubic equation results. Just at this point, The Good Chemist comes galloping down from the mountain on her white stallion to rescue us and cries, "Wait! There is no reason to solve a cubic equation. We can make an excellent, simplifying approximation. (Besides, I have trouble solving cubic equations.)"

For any respectable weak acid, $[H^+]$ from HA will be much greater than $[H^+]$ from H$_2$O. When HA dissociates, it produces A$^-$. When H$_2$O dissociates, it produces OH$^-$. If dissociation of HA is much greater than dissociation of H$_2$O, then $[A^-] \gg [OH^-]$, and Equation 9-6 reduces to

$$[H^+] \approx [A^-] \qquad (9\text{-}9)$$

$x = [H^+]$ in weak-acid problems.

To solve the problem, first set $[H^+] = x$. Equation 9-9 says that $[A^-]$ also is equal to x. Equation 9-7 says that $[HA] = F - [A^-] = F - x$. Putting these expressions into Equation 9-8 gives

$$K_a = \frac{[H^+][A^-]}{[HA]} = \frac{(x)(x)}{F - x}$$

Setting F = 0.050 0 M and $K_a = 1.0_7 \times 10^{-3}$ for *o*-hydroxybenzoic acid, we can solve the equation, because it is just a quadratic equation.

$$\frac{x^2}{0.050\,0 - x} = 1.0_7 \times 10^{-3}$$

$$x^2 = (1.0_7 \times 10^{-3})(0.050\,0 - x)$$

$$x^2 + (1.07 \times 10^{-3})x - 5.35 \times 10^{-5} = 0$$

$$x = 6.8_0 \times 10^{-3}\text{ M (negative root rejected)}$$

$$[H^+] = [A^-] = x = 6.8_0 \times 10^{-3}\text{ M}$$

$$[HA] = F - x = 0.043_2\text{ M}$$

$$pH = -\log x = 2.17$$

For uniformity, we express pH to the 0.01 decimal place, regardless of the number of places justified by significant figures. Ordinary pH measurements are not more accurate than ±0.02 pH units.

Was the approximation $[H^+] \approx [A^-]$ justified? The calculated pH is 2.17, which means that $[OH^-] = K_w/[H^+] = 1.5 \times 10^{-12}$ M.

$$[A^-](\text{from HA dissociation}) = 6.8 \times 10^{-3} \text{ M}$$
$$\Rightarrow [H^+] \text{ from HA dissociation} = 6.8 \times 10^{-3} \text{ M}$$
$$[OH^-](\text{from H}_2\text{O dissociation}) = 1.5 \times 10^{-12} \text{ M}$$
$$\Rightarrow [H^+] \text{ from H}_2\text{O dissociation} = 1.5 \times 10^{-12} \text{ M}$$

The assumption that H^+ is derived mainly from HA is excellent.

In a solution of a weak acid, H^+ is derived almost entirely from HA, not from H_2O.

Fraction of Dissociation

The **fraction of dissociation, α,** is defined as the fraction of the acid in the form A^-:

Fraction of dissociation of an acid:
$$\alpha = \frac{[A^-]}{[A^-] + [HA]} = \frac{x}{x + (F - x)} = \frac{x}{F} \quad (9\text{-}10)$$

α is the fraction of HA that has dissociated:
$$\alpha = \frac{[A^-]}{[A^-] + [HA]}$$

For 0.050 0 M *o*-hydroxybenzoic acid, we find

$$\alpha = \frac{6.8 \times 10^{-3} \text{ M}}{0.050 \text{ 0 M}} = 0.14$$

That is, the acid is 14% dissociated at a formal concentration of 0.050 0 M.

The variation of α with formal concentration is shown in Figure 9-2. **Weak electrolytes** (compounds that are only partially dissociated) dissociate more as they are diluted. *o*-Hydroxybenzoic acid is more dissociated than *p*-hydroxybenzoic acid at the same formal concentration because the *ortho* isomer is a stronger acid. Box 9-2 provides an everyday application of the fraction of dissociation of a weak acid.

The Essence of a Weak-Acid Problem

When faced with finding the pH of a weak acid, you should immediately realize that $[H^+] = [A^-] = x$ and proceed to set up and solve the equation

Equation for weak acids:
$$\frac{[H^+][A^-]}{[HA]} = \frac{x^2}{F - x} = K_a \quad (9\text{-}11)$$

where F is the formal concentration of HA. The approximation $[H^+] = [A^-]$ would be poor only if the acid were too dilute or too weak, neither of which constitutes a practical problem.

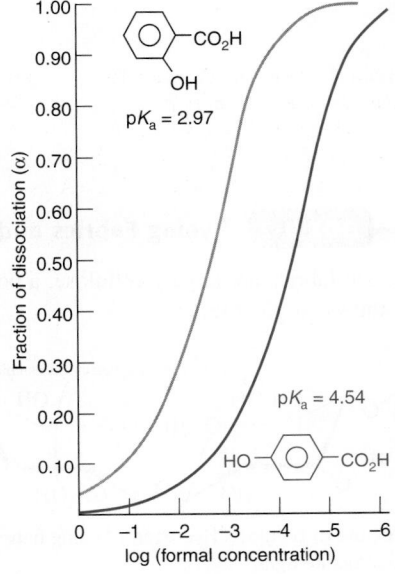

FIGURE 9-2 Fraction of dissociation of a weak electrolyte increases as electrolyte is diluted. The stronger acid is more dissociated than the weaker acid at all concentrations.

EXAMPLE A Weak-Acid Problem

Find the pH of 0.050 M trimethylammonium chloride.

Trimethylammonium chloride

Solution We assume that ammonium halide salts are completely dissociated to give $(CH_3)_3NH^+$ and Cl^-.* We then recognize that trimethylammonium ion is a weak acid, being the conjugate acid of trimethylamine, $(CH_3)_3N$, a weak base. Cl^- has no basic or

Cl^- has no acidic or basic properties because it is the conjugate base of the strong acid, HCl. If Cl^- had appreciable basicity, HCl would not be completely dissociated.

---

*$R_4N^+X^-$ salts are not completely dissociated, because there are some *ion pairs*, $R_4N^+X^-(aq)$ (Box 8-1). Equilibrium constants for $R_4N^+ + X^- \rightleftharpoons R_4N^+X^-(aq)$ are given below. For 0.050 F solutions, the fraction of ion pairing, calculated with activity coefficients, is 4% in $(CH_3)_4N^+Br^-$, 7% in $(CH_3CH_2)_4N^+Br^-$, and 9% in $(CH_3CH_2CH_2)_4N^+Br^-$.

R_4N^+	X^-	$K_{\text{ion pair}}$ ($\mu = 0$)	R_4N^+	X^-	$K_{\text{ion pair}}$ ($\mu = 0$)
Me_4N^+	Cl^-	1.1	Me_4N^+	I^-	2.0
Bu_4N^+	Cl^-	2.5	Et_4N^+	I^-	2.9
Me_4N^+	Br^-	1.4	Pr_4N^+	I^-	4.6
Et_4N^+	Br^-	2.4	Bu_4N^+	I^-	6.0
Pr_4N^+	Br^-	3.1			

$Me = CH_3-$, $Et = CH_3CH_2-$, $Pr = CH_3CH_2CH_2-$, $Bu = CH_3CH_2CH_2CH_2-$

acidic properties and should be ignored. In Appendix G, we find trimethylammonium ion listed as trimethylamine but drawn as trimethylammonium ion, with $pK_a = 9.799$ at an ionic strength of $\mu = 0$. So,

$$K_a = 10^{-pK_a} = 10^{-9.799} = 1.59 \times 10^{-10}$$

From here, everything is downhill.

$$(CH_3)_3NH^+ \xrightleftharpoons{K_a} (CH_3)_3N + H^+$$

$$\; F - x \qquad\qquad x \qquad x$$

$$\frac{x^2}{0.050 - x} = 1.59 \times 10^{-10} \tag{9-12}$$

$$x = 2.8 \times 10^{-6} \, M \Rightarrow pH = 5.55$$

TEST YOURSELF Find the pH of 0.050 M triethylammonium bromide. (**Answer:** 6.01)

BOX 9-2   Dyeing Fabrics and the Fraction of Dissociation[6]

Cotton fabrics are largely cellulose, a polymer with repeating units of the sugar glucose:

These oxygen atoms can bind to dye

Structure of cellulose. Hydrogen bonding between glucose units helps make the structure rigid.

Dyes are colored molecules that can form covalent bonds to fabric. For example, Procion Brilliant Blue M-R is a dye with a blue *chromophore* (the colored part) attached to a reactive dichlorotriazine ring:

Chlorine atoms can be replaced by oxygen atoms of cellulose

Blue chromophore   Procion Brilliant Blue M-R fabric dye

Oxygen atoms of the —CH_2OH groups on cellulose can replace Cl atoms of the dye to form covalent bonds that fix the dye permanently to the fabric:

Chemically reactive form of cellulose is deprotonated anion

After the fabric has been dyed in cold water, excess dye is removed with a hot wash. During the hot wash, the second Cl group of the dye is replaced by a second cellulose or by water (giving dye—OH).

The chemically reactive form of cellulose is the conjugate base:

$$\text{cellulose}—CH_2OH \xrightleftharpoons{K_a \approx 10^{-15}} \text{cellulose}—CH_2O^- + H^+$$

$$ROH RO^-$$

$$(\text{conjugate base})$$

To promote dissociation of the cellulose—CH_2OH proton, dyeing is carried out in sodium carbonate solution with a pH around 10.6. The fraction of reactive cellulose is given by the fraction of dissociation of the weak acid at pH 10.6:

$$\text{Fraction of dissociation} = \frac{[RO^-]}{[ROH] + [RO^-]} \approx \frac{[RO^-]}{[ROH]}$$

Because the fraction of dissociation of the very weak acid is so small, $[ROH] \gg [RO^-]$ in the denominator, which is therefore approximately just $[ROH]$. The quotient $[RO^-]/[ROH]$ can be calculated from K_a and the pH:

$$K_a = \frac{[RO^-][H^+]}{[ROH]} \Rightarrow \frac{[RO^-]}{[ROH]} = \frac{K_a}{[H^+]} \approx \frac{10^{-15}}{10^{-10.6}}$$

$$= 10^{-4.4} \approx \text{fraction of dissociation}$$

Only about one cellulose—CH_2OH group in 10^4 is in the reactive form at pH 10.6.

A handy tip: Equation 9-11 can always be solved with the quadratic formula. However, an easier method worth trying *first* is to neglect x in the denominator. If x comes out $\leq 1\%$ of F, then your approximation was good and you need not use the quadratic formula. For Equation 9-12, the approximation works like this:

$$\frac{x^2}{0.050 - x} \approx \frac{x^2}{0.050} = 1.59 \times 10^{-10} \Rightarrow x = \sqrt{(0.050)(1.59 \times 10^{-10})} = 2.8 \times 10^{-6} \text{ M}$$

The approximate solution ($x \approx 2.8 \times 10^{-6}$) is $\leq 1\%$ of 0.050 in the denominator of Equation 9-12. Therefore, the approximate solution is fine.

> **Approximation** Neglect x in denominator. If x comes out to be less than 1% of F, accept the approximation.

9-4 Weak-Base Equilibria

The treatment of weak bases is almost the same as that of weak acids.

$$B + H_2O \underset{}{\overset{K_b}{\rightleftharpoons}} BH^+ + OH^- \qquad K_b = \frac{[BH^+][OH^-]}{[B]}$$

> As K_b increases, pK_b decreases and the base becomes stronger.

We suppose that nearly all OH^- comes from the reaction of $B + H_2O$, and little comes from dissociation of H_2O. Setting $[OH^-] = x$, we must also set $[BH^+] = x$, because one BH^+ is produced for each OH^-. Calling the formal concentration of base F ($= [B] + [BH^+]$), we write

$$[B] = F - [BH^+] = F - x$$

Plugging these values into the K_b equilibrium expression, we get

Equation for weak base:
$$\frac{[BH^+][OH^-]}{[B]} = \frac{x^2}{F - x} = K_b \qquad (9\text{-}13)$$

which looks a lot like a weak-acid problem, except that now $x = [OH^-]$.

> A weak-base problem has the same algebra as a weak-acid problem, except $K = K_b$ and $x = [OH^-]$.

A Typical Weak-Base Problem

Consider the commonly occurring weak base, cocaine.

If the formal concentration is 0.037 2 M, the problem is formulated as follows:

$$B + H_2O \rightleftharpoons BH^+ + OH^-$$
$$0.037\ 2 - x \qquad\qquad x \qquad x$$

$$\frac{x^2}{0.037\ 2 - x} = 2.6 \times 10^{-6} \Rightarrow x = 3.1 \times 10^{-4} \text{ M}$$

Because $x = [OH^-]$, we can write

$$[H^+] = K_w/[OH^-] = 1.0 \times 10^{-14}/3.1 \times 10^{-4} = 3.2 \times 10^{-11} \text{ M}$$

$$pH = -\log[H^+] = 10.49$$

> **Question** What concentration of OH^- is produced by H_2O dissociation in this solution? Was it justified to neglect water dissociation as a source of OH^-?

This is a reasonable pH for a weak base.

What fraction of cocaine has reacted with water? We can formulate α for a base, called the **fraction of association:**

Fraction of association of a base:
$$\alpha = \frac{[BH^+]}{[BH^+] + [B]} = \frac{x}{F} = 0.008\ 3 \qquad (9\text{-}14)$$

> For a base, α is the fraction that has reacted with water.

Only 0.83% of the base has reacted.

Conjugate Acids and Bases—Revisited

HA and A⁻ are a conjugate acid-base pair. So are BH⁺ and B.

Earlier, we noted that **the conjugate base of a weak acid is a weak base**, and **the conjugate acid of a weak base is a weak acid.** We also derived an exceedingly important relation between the equilibrium constants for a conjugate acid-base pair: $K_a \cdot K_b = K_w$.

In Section 9-3, we considered o- and p-hydroxybenzoic acids, designated HA. Now consider their conjugate bases. For example, the salt sodium o-hydroxybenzoate dissolves to give Na^+ (which has no acid-base chemistry) and o-hydroxybenzoate, which is a weak base.

The acid-base chemistry is the reaction of o-hydroxybenzoate with water:

In aqueous solution,

$\bigcirc$—CO₂Na
        OH

gives

$\bigcirc$—CO₂⁻ + Na⁺
        OH

o–Hydroxybenzoate

$$\bigcirc\text{—CO}_2^- + H_2O \rightleftharpoons \bigcirc\text{—CO}_2H + OH^-$$

$\quad$ OH $\qquad\qquad\qquad\qquad$ OH

A^- (o-hydroxybenzoate) $\qquad\qquad$ HA

$\quad$ F − x $\qquad\qquad\qquad\qquad$ x $\qquad\quad$ x

(9-15)

$$\frac{x^2}{F - x} = K_b$$

From K_a for each isomer, we calculate K_b for the conjugate base.

Isomer of hydroxybenzoic acid	K_a	$K_b = K_w/K_a$
ortho	$1.0_7 \times 10^{-3}$	9.3×10^{-12}
para	2.9×10^{-5}	3.5×10^{-10}

Using each value of K_b and letting F = 0.050 0 M, we find

pH of 0.050 0 M o-hydroxybenzoate = 7.83

pH of 0.050 0 M p-hydroxybenzoate = 8.62

These are reasonable pH values for solutions of weak bases. Furthermore, as expected, the conjugate base of the stronger acid is the weaker base.

EXAMPLE **A Weak-Base Problem**

Find the pH of 0.10 M ammonia.

Solution When ammonia is dissolved in water, its reaction is

$$NH_3 + H_2O \overset{K_b}{\rightleftharpoons} NH_4^+ + OH^-$$

Ammonia $\qquad\qquad\qquad$ Ammonium ion

F − x $\qquad\qquad\qquad\qquad$ x $\qquad$ x

In Appendix G, we find ammonium ion, NH_4^+, listed next to ammonia. pK_a for ammonium ion is 9.245. Therefore, K_b for NH_3 is

$$K_b = \frac{K_w}{K_a} = \frac{10^{-14.00}}{10^{-9.245}} = 1.76 \times 10^{-5}$$

To find the pH of 0.10 M NH_3, we set up and solve the equation

$$\frac{[NH_4^+][OH^-]}{[NH_3]} = \frac{x^2}{0.10 - x} = K_b = 1.76 \times 10^{-5}$$

$$x = [OH^-] = 1.3_2 \times 10^{-3} \text{ M}$$

$$[H^+] = \frac{K_w}{[OH^-]} = 7.6 \times 10^{-12} \text{ M} \Rightarrow pH = -\log[H^+] = 11.12$$

TEST YOURSELF Find the pH of 0.10 M methylamine. (*Answer:* 11.80)

9-5 Buffers

A buffered solution resists changes in pH when acids or bases are added or when dilution occurs. The **buffer** is a mixture of an acid and its conjugate base. There must be comparable amounts of the conjugate acid and base (within a factor of ~10) to exert significant buffering.

The importance of buffers in all areas of science is immense. At the opening of this chapter, we saw that digestive enzymes in lysosomes operate best in acid, a constraint that allows a cell to protect itself from its own enzymes. If enzymes leak from the acidic lysosome into the buffered, neutral cytoplasm, they have low reactivity and do less damage to the cell than they would at their optimum pH. Conversely, Figure 9-3 shows the pH dependence of an enzyme-catalyzed reaction that is fastest near pH 8.0 and which would be slow if the enzyme were in an acidic lysosome. For an organism to survive, it must control the pH of each subcellular compartment so that each reaction proceeds at the proper rate.

Mixing a Weak Acid and Its Conjugate Base

If you mix A moles of a weak acid with B moles of its conjugate base, the moles of acid remain close to A and the moles of base remain close to B. Little reaction occurs to change either concentration.

To understand why this should be so, look at the K_a and K_b reactions in terms of Le Châtelier's principle. Consider an acid with $pK_a = 4.00$ and its conjugate base with $pK_b = 10.00$. Let's calculate the fraction of acid that dissociates in a 0.10 M solution of HA.

$$HA \rightleftharpoons H^+ + A^- \qquad pK_a = 4.00$$
$$ 0.10 - x \quad\quad x \quad\quad x$$

$$\frac{x^2}{F - x} = K_a \Rightarrow x = 3.1 \times 10^{-3} \text{ M}$$

$$\text{Fraction of dissociation} = \alpha = \frac{x}{F} = 0.031$$

The acid is only 3.1% dissociated under these conditions.

In a solution containing 0.10 mol of A^- dissolved in 1.00 L, the extent of reaction of A^- with water is even smaller.

$$A^- + H_2O \rightleftharpoons HA + OH^- \qquad pK_b = 10.00$$
$$0.10 - x x \quad\quad x$$

$$\frac{x^2}{F - x} = K_b \Rightarrow x = 3.2 \times 10^{-6}$$

$$\text{Fraction of association} = \alpha = \frac{x}{F} = 3.2 \times 10^{-5}$$

HA dissociates very little, and adding extra A^- to the solution makes HA dissociate even less. Similarly, A^- does not react much with water, and adding extra HA makes A^- react even less. If 0.050 mol of A^- plus 0.036 mol of HA are added to water, there will be close to 0.050 mol of A^- and close to 0.036 mol of HA in the solution at equilibrium.

Henderson-Hasselbalch Equation

The central equation for buffers is the **Henderson-Hasselbalch equation,** which is merely a rearranged form of the K_a equilibrium expression.

$$K_a = \frac{[H^+][A^-]}{[HA]}$$

$$\log K_a = \log \frac{[H^+][A^-]}{[HA]} = \log[H^+] + \log \frac{[A^-]}{[HA]}$$

$$\underbrace{-\log[H^+]}_{pH} = \underbrace{-\log K_a}_{pK_a} + \log \frac{[A^-]}{[HA]}$$

Henderson-Hasselbalch equation for an acid:
$$HA \overset{K_a}{\rightleftharpoons} H^+ + A^-$$

$$pH = pK_a + \log \frac{[A^-]}{[HA]} \tag{9-16}$$

The Henderson-Hasselbalch equation tells us the pH of a solution, provided we know the ratio of the concentrations of conjugate acid and base, as well as pK_a for the acid. If a solution is prepared from the weak base B and its conjugate acid, the analogous equation is

Henderson-Hasselbalch equation for a base:
$$BH^+ \overset{K_a = K_w/K_b}{\rightleftharpoons} B + H^+$$

$$pH = pK_a + \log \frac{[B]}{[BH^+]} \tag{9-17}$$

pK_a applies to *this* acid

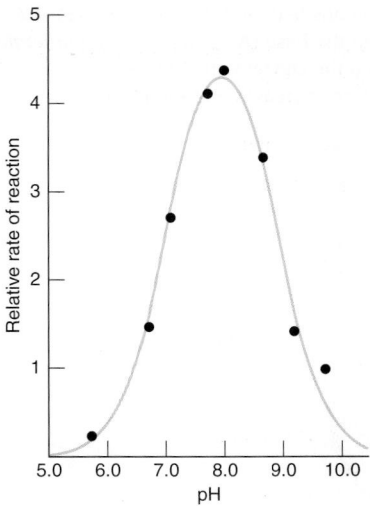

FIGURE 9-3 pH dependence of the rate of cleavage of an amide bond by the enzyme chymotrypsin, which helps digest proteins in your intestine. [Data from M. L. Bender, G. E. Clement, F. J. Kézdy, and H. A. Heck, "The Correlation of the pH (pD) Dependence and the Stepwise Mechanism of α-Chymotrypsin-Catalyzed Reactions," *J. Am. Chem. Soc.* **1964,** *86,* 3680.]

$$\overset{O}{\overset{\|}{RC}} - NHR'$$
$\uparrow$
Amide bond

When you mix a weak acid with its conjugate base, you get what you mix!

The approximation that the concentrations of HA and A^- remain unchanged breaks down for dilute solutions or at extremes of pH. We test the validity of the approximation on page 203.

$\log xy = \log x + \log y$

L. J. Henderson was a physician who wrote $[H^+] = K_a[\text{acid}]/[\text{salt}]$ in a physiology article in 1908, a year before the word "buffer" and the concept of pH were invented by the biochemist Sørensen. Henderson's contribution was the approximation of setting [acid] equal to the concentration of HA placed in solution and [salt] equal to the concentration of A^- placed in solution. In 1916, K. A. Hasselbalch wrote what we call the Henderson-Hasselbalch equation in a biochemical journal.[7]

Equations 9-16 and 9-17 are only sensible when the base (A⁻ or B) is in the *numerator*. When the concentration of base increases, the log term increases and the pH increases.

where pK_a is the acid dissociation constant of the weak acid BH^+. The important features of Equations 9-16 and 9-17 are that the base (A^- or B) appears in the numerator of both equations, and the equilibrium constant is K_a of the acid in the denominator.

Challenge   Show that, when activities are included, the Henderson-Hasselbalch equation is

$$pH = pK_a + \log \frac{[A^-]\gamma_{A^-}}{[HA]\gamma_{HA}} \qquad (9\text{-}18)$$

The Henderson-Hasselbalch equation is not an approximation. It is just a rearranged form of the equilibrium expression. The approximations we make are the values of $[A^-]$ and $[HA]$. In most cases, it is valid to assume that what we mix is what we get in solution. At the end of this chapter, we treat the case in which what we mix is not what we get because the solution is too dilute or the acid is too strong.

Properties of the Henderson-Hasselbalch Equation

In Equation 9-16, we see that, if $[A^-] = [HA]$, then $pH = pK_a$.

When $[A^-] = [HA]$, $pH = pK_a$.

Of course, you recall that $\log 1 = 0$.

$$pH = pK_a + \log \frac{[A^-]}{[HA]} = pK_a + \log 1 = pK_a$$

Regardless of how complex a solution may be, whenever $pH = pK_a$, for a particular acid, $[A^-]$ must equal $[HA]$ for that acid.

All equilibria must be satisfied simultaneously in any solution at equilibrium. If there are 10 different acids and bases in the solution, the 10 forms of Equation 9-16 will have 10 different quotients $[A^-]/[HA]$, but all 10 equations must give the same pH, because **there can be only one concentration of H^+ in a solution.**

Another feature of the Henderson-Hasselbalch equation is that, for every power-of-10 change in the ratio $[A^-]/[HA]$, the pH changes by one unit (Table 9-1). As the base (A^-) increases, the pH goes up. As the acid (HA) increases, the pH goes down. For any conjugate acid-base pair, you can say, for example, that if $pH = pK_a - 1$, there must be 10 times as much HA as A^-. Ten-elevenths is in the form HA and one-eleventh is in the form A^-.

TABLE 9-1	**Effect of [A⁻]/[HA] on pH**
[A⁻]/[HA]	**pH**
100 : 1	$pK_a + 2$
10 : 1	$pK_a + 1$
1 : 1	pK_a
1 : 10	$pK_a - 1$
1 : 100	$pK_a - 2$

$10^{\log z} = z$

EXAMPLE   **Using the Henderson-Hasselbalch Equation**

Sodium hypochlorite (NaOCl, the active ingredient of almost all bleaches) was dissolved in a solution buffered to pH 6.20. Find the ratio $[OCl^-]/[HOCl]$ in this solution.

Solution   In Appendix G, we find that $pK_a = 7.53$ for hypochlorous acid, HOCl. The pH is known, so the ratio $[OCl^-]/[HOCl]$ can be calculated from the Henderson-Hasselbalch equation.

$$HOCl \rightleftharpoons H^+ + OCl^-$$

$$pH = pK_a + \log \frac{[OCl^-]}{[HOCl]}$$

$$6.20 = 7.53 + \log \frac{[OCl^-]}{[HOCl]}$$

$$-1.33 = \log \frac{[OCl^-]}{[HOCl]}$$

$$10^{-1.33} = 10^{\log([OCl^-]/[HOCl])} = \frac{[OCl^-]}{[HOCl]}$$

$$0.047 = \frac{[OCl^-]}{[HOCl]}$$

The ratio $[OCl^-]/[HOCl]$ is set by pH and pK_a. We do not need to know how much NaOCl was added, or the volume.

TEST YOURSELF   Find $[OCl^-]/[HOCl]$ if pH is raised by one unit to 7.20. (*Answer:* 0.47)

Buffer in Action

For illustration, we choose a widely used buffer called "tris," which is short for tris(hydroxymethyl)-aminomethane.

$$
\begin{array}{cc}
\overset{+}{N}H_3 & NH_2 \\
| & | \\
C & C \\
HOCH_2\text{''''}\blacktriangle CH_2OH \rightleftharpoons HOCH_2\text{''''}\blacktriangle CH_2OH + H^+ \\
HOCH_2 & HOCH_2 \\
BH^+ & B \\
pK_a = 8.072 & \text{This form is "tris"}
\end{array}
$$

In Appendix G, we find $pK_a = 8.072$ for the conjugate acid of tris. An example of a salt containing the BH^+ cation is tris hydrochloride, which is BH^+Cl^-. When BH^+Cl^- is dissolved in water, it dissociates to BH^+ and Cl^-.

EXAMPLE A Buffer Solution

Find the pH of a 1.00-L aqueous solution prepared with 12.43 g of tris (FM 121.135) plus 4.67 g of tris hydrochloride (FM 157.596).

Solution The concentrations of B and BH^+ added to the solution are

$$
[B] = \frac{12.43 \text{ g/L}}{121.135 \text{ g/mol}} = 0.102\ 6 \text{ M} \qquad [BH^+] = \frac{4.67 \text{ g/L}}{157.596 \text{ g/mol}} = 0.029\ 6 \text{ M}
$$

Assuming that what we mixed stays in the same form, we plug these concentrations into the Henderson-Hasselbalch equation to find the pH:

$$
pH = pK_a + \log \frac{[B]}{[BH^+]} = 8.072 + \log \frac{0.102\ 6}{0.029\ 6} = 8.61
$$

TEST YOURSELF Find the pH if we add another 1.00 g of tris hydrochloride. (*Answer:* 8.53)

Notice that *the volume of solution is irrelevant*, because volume cancels in the numerator and denominator of the log term:

The pH of a buffer is nearly independent of volume.

$$
pH = pK_a + \log \frac{\text{moles of B/L of solution}}{\text{moles of } BH^+/\text{L of solution}}
$$

$$
= pK_a + \log \frac{\text{moles of B}}{\text{moles of } BH^+}
$$

BOX 9-3 Strong Plus Weak Reacts Completely

Strong acid reacts with a weak base essentially "completely" because the equilibrium constant is large.

$$
\begin{array}{ccccc}
B & + & H^+ & \rightleftharpoons & BH^+ \\
\text{Weak} & & \text{Strong} & & \\
\text{base} & & \text{acid} & &
\end{array} \qquad K = \frac{1}{K_a \text{ (for } BH^+\text{)}}
$$

If B is tris(hydroxymethyl)aminomethane, the equilibrium constant for reaction with HCl is

$$
K = \frac{1}{K_a} = \frac{1}{10^{-8.072}} = 1.2 \times 10^8
$$

Strong base reacts "completely" with a weak acid because the equilibrium constant is, again, very large.

$$
\begin{array}{ccccc}
OH^- & + & HA & \rightleftharpoons & A^- + H_2O \\
\text{Strong} & & \text{Weak} & & \\
\text{base} & & \text{acid} & &
\end{array} \qquad K = \frac{1}{K_b \text{ (for } A^-\text{)}}
$$

If HA is acetic acid, then the equilibrium constant for reaction with NaOH is

$$
K = \frac{1}{K_b} = \frac{K_a \text{ (for HA)}}{K_w} = 1.7 \times 10^9
$$

The reaction of a strong acid with a strong base is even more complete than a strong plus weak reaction:

$$
\begin{array}{ccccc}
H^+ & + & OH^- & \rightleftharpoons & H_2O \\
\text{Strong} & & \text{Strong} & & \\
\text{acid} & & \text{base} & &
\end{array} \qquad K = \frac{1}{K_w} = 10^{14}
$$

If you mix a strong acid, a strong base, a weak acid, and a weak base, the strong acid and strong base will neutralize each other until one is used up. The remaining strong acid or strong base will then react with the weak base or weak acid.

EXAMPLE Effect of Adding Acid to a Buffer

If we add 12.0 mL of 1.00 M HCl to the solution in the previous example, what will be the new pH?

Solution The key is to realize that, *when a strong acid is added to a weak base, they react completely to give BH$^+$* (Box 9-3). We are adding 12.0 mL of 1.00 M HCl, which contains $(0.012\ 0\ L)(1.00\ mol/L) = 0.012\ 0$ mol of H$^+$. This much H$^+$ consumes 0.012 0 mol of B to create 0.012 0 mol of BH$^+$:

	B	+	H$^+$	→	BH$^+$
	Tris		From HCl		
Initial moles	0.102 6		0.012 0		0.029 6
Final moles	0.090 6		—		0.041 6
	(0.102 6 − 0.012 0)				(0.029 6 + 0.012 0)

Information in the table allows us to calculate the pH:

$$pH = pK_a + \log \frac{\text{moles of B}}{\text{moles of BH}^+}$$

$$= 8.072 + \log \frac{0.090\ 6}{0.041\ 6} = 8.41$$

The volume of the solution is irrelevant.

TEST YOURSELF Find the pH if only 6.0 instead of 12.0 mL HCl were added. (*Answer:* 8.51)

> **Question** Does the pH change in the right direction when HCl is added?

A buffer resists changes in pH . . .

. . . because the buffer consumes the added acid or base.

We see that *the pH of a buffer does not change very much when a limited amount of strong acid or base is added.* Addition of 12.0 mL of 1.00 M HCl changed the pH from 8.61 to 8.41. Addition of 12.0 mL of 1.00 M HCl to 1.00 L of unbuffered solution would have lowered the pH to 1.93.

But *why* does a buffer resist changes in pH? It does so because the strong acid or base is consumed by B or BH$^+$. If you add HCl to tris, B is converted into BH$^+$. If you add NaOH, BH$^+$ is converted into B. As long as you don't use up B or BH$^+$ by adding too much HCl or NaOH, the log term of the Henderson-Hasselbalch equation does not change much and pH does not change much. Demonstration 9-1 illustrates what happens when buffer does get used up. The buffer has its maximum capacity to resist changes of pH when pH = pK$_a$. We return to this point later.

EXAMPLE Calculating How to Prepare a Buffer Solution

How many milliliters of 0.500 M NaOH should be added to 10.0 g of tris hydrochloride to give a pH of 7.60 in a final volume of 250 mL?

> **Question** Why do we add base (NaOH) to tris hydrochloride to make a buffer?
>
> *Answer:* Tris hydrochloride is a weak acid, BH$^+$. We need to convert some BH$^+$ into B to make a buffer, which is a mixture of BH$^+$ and B.

> **Question** HEPES is a common buffer in biochemistry with pK$_a$ = 7.56 in Table 9-2. HEPES is a neutral compound. Draw its structure. Would you add NaOH or HCl to HEPES to make a buffer?
>
> *Answer:* The form drawn in Table 9-2 is the neutral acid HA. We need to add NaOH to remove H$^+$ to make a mixture of HA and A$^-$.

Solution The moles of tris hydrochloride are $(10.0\ g)/(157.596\ g/mol) = 0.063\ 5$ mol. We can make a table to help solve the problem:

Reaction with OH$^-$:	BH$^+$	+	OH$^-$	→	B
Initial moles	0.063 5		x		—
Final moles	0.063 5 − x		—		x

The Henderson-Hasselbalch equation allows us to find x, because we know pH and pK$_a$.

$$pH = pK_a + \log \frac{\text{mol B}}{\text{mol BH}^+}$$

$$7.60 = 8.072 + \log \frac{x}{0.063\ 5 - x}$$

$$-0.472 = \log \frac{x}{0.063\ 5 - x}$$

$$10^{-0.472} = \frac{x}{0.063\ 5 - x} \Rightarrow x = 0.016\ 0\ mol$$

This many moles of NaOH is contained in

$$\frac{0.016\,0 \text{ mol}}{0.500 \text{ mol/L}} = 0.032\,0 \text{ L} = 32.0 \text{ mL}$$

TEST YOURSELF How many mL of 0.500 M NaOH should be added to 10.0 g of tris hydrochloride to give a pH of 7.40 in a final volume of 500 mL? (**Answer:** 22.3 mL)

A buffer resists changes in pH because added acid or base is consumed by buffer. As buffer is used up, it becomes less resistant to changes in pH.

In this demonstration,[8] a mixture containing approximately a $10:1$ mole ratio of $HSO_3^-:SO_3^{2-}$ is prepared. Because pK_a for HSO_3^- is 7.2, the pH should be approximately

$$pH = pK_a + \log \frac{[SO_3^{2-}]}{[HSO_3^-]} = 7.2 + \log \frac{1}{10} = 6.2$$

When formaldehyde is added, the net reaction is the consumption of HSO_3^-, but not of SO_3^{2-}:

$$H_2C{=}O + HSO_3^- \rightarrow H_2C\overset{O^-}{\underset{SO_3H}{\diagdown}} \rightarrow H_2C\overset{OH}{\underset{SO_3^-}{\diagdown}} \quad (A)$$

Formaldehyde   Bisulfite

$$H_2C{=}O + SO_3^{2-} \rightarrow H_2C\overset{O^-}{\underset{SO_3^-}{\diagdown}}$$

Sulfite

(B)

$$H_2C\overset{O^-}{\underset{SO_3^-}{\diagdown}} + HSO_3^- \rightarrow H_2C\overset{OH}{\underset{SO_3^-}{\diagdown}} + SO_3^{2-}$$

(In sequence A, bisulfite is consumed directly. In sequence B, the net reaction is destruction of HSO_3^-, with no change in the SO_3^{2-} concentration.)

We can prepare a table showing how the pH should change as the HSO_3^- reacts.

Percentage of reaction completed	$[SO_3^{2-}]:[HSO_3^-]$	Calculated pH
0	1:10	6.2
90	1:1	7.2
99	1:0.1	8.2
99.9	1:0.01	9.2
99.99	1:0.001	10.2

Through 90% completion, the pH should rise by just 1 unit. In the next 9% of the reaction, the pH will rise by another unit. At the end of the reaction, the change in pH is very abrupt.

In the formaldehyde *clock reaction*,[9] formaldehyde is added to a solution containing HSO_3^-, SO_3^{2-}, and phenolphthalein indicator. Phenolphthalein is colorless below pH 8 and red above this pH. The

solution remains colorless for more than a minute. Suddenly pH shoots up and the liquid turns pink. Monitoring pH with a glass electrode gave results in the graph.

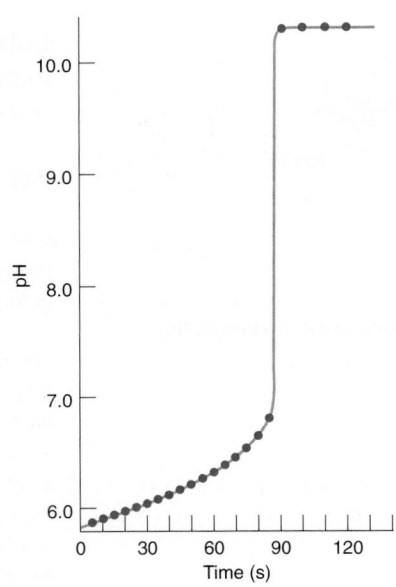

Graph of pH versus time in the formaldehyde clock reaction.

Procedure: Solutions must be fresh. Prepare a solution of formaldehyde by diluting 9 mL of 37 wt% formaldehyde (density = 1.08 g/mL) to 100 mL (to give 1.20 M CH_2O). Dissolve 1.4 g of $Na_2S_2O_5$ (sodium metabisulfite, 7.4 mmol)[10] and 0.18 g of Na_2SO_3 (1.4 mmol) in 400 mL of water, and add $\sim$1 mL of phenolphthalein solution (Table 11-3). Add 23 mL (28 mmol) of formaldehyde solution to the well-stirred solution to initiate the reaction. Reaction time can be adjusted by changing temperature, concentrations, or volume.

A less toxic variant of this demonstration uses glyoxal

$$\overset{O}{\overset{\|}{(HC}}-\overset{O}{\overset{\|}{CH)}}$$

(HC—CH) in place of formaldehyde.[11] A day before the demonstration, dilute 2.9 g 40 wt% glyoxal (20.0 mmol) to 25 mL. Dissolve 0.90 g $Na_2S_2O_5$ (4.7 mmol), 0.15 g Na_2SO_3 (1.2 mmol), and 0.18 g $Na_2EDTA \cdot 2H_2O$ (0.48 mmol, to protect sulfite from metal-catalyzed air oxidation) in 50 mL. A mole of $Na_2S_2O_5$ makes 2 moles of HSO_3^- by reaction with H_2O. For the demonstration, add 0.5 mL phenol red indicator (Table 11-3) to 400 mL H_2O plus 5.0 mL sulfite solution. Add 2.5 mL glyoxal solution (2.0 mmol) to the well-stirred sulfite solution to start the clock reaction.

Reasons why a calculation could be wrong:

1. Activity coefficients were not considered.

2. The temperature might not be 25°C, for which pK_a is listed.

3. The approximations $[HA] = F_{HA}$ and $[A^-] = F_{A^-}$ could be in error.

4. pK_a listed for tris in your favorite table is probably not exactly what you would measure.

5. Other ions besides the acid and base species affect pH by ion-pairing reactions with acid and base species.

Choose a buffer whose pK_a is close to the desired pH.

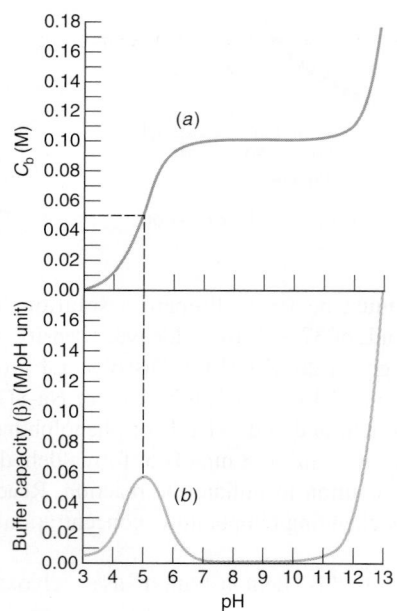

FIGURE 9-4 (a) C_b versus pH for a solution containing 0.100 F HA with $pK_a = 5.00$. (b) Buffer capacity versus pH for the same system reaches a maximum when pH = pK_a. The lower curve is the derivative of the upper curve.

Preparing a Buffer in Real Life!

If you really wanted to prepare a tris buffer of pH 7.60, you would *not* do it by calculating what to mix. Suppose that you wish to prepare 1.00 L of buffer containing 0.100 M tris at a pH of 7.60. You have available solid tris hydrochloride and approximately 1 M NaOH. Here's how I would do it:

1. Weigh out 0.100 mol of tris hydrochloride and dissolve it in a beaker containing about 800 mL of water.
2. Place a calibrated pH electrode in the solution and monitor the pH.
3. Add NaOH solution until the pH is exactly 7.60.
4. Transfer the solution to a volumetric flask and wash the beaker a few times. Add the washings to the volumetric flask.
5. Dilute to the mark and mix.

You do not simply add the calculated quantity of NaOH because it would not give exactly the desired pH. The reason for using 800 mL of water in the first step is so that the volume will be reasonably close to the final volume during pH adjustment. Otherwise, the pH will change slightly when the sample is diluted to its final volume and the ionic strength changes.

Buffer Capacity[12]

Buffer capacity, β, is a measure of how well a solution resists changes in pH when strong acid or base is added. Buffer capacity is defined as

Buffer capacity:
$$\beta = \frac{dC_b}{d\text{pH}} = -\frac{dC_a}{d\text{pH}} \qquad (9\text{-}19)$$

where C_a and C_b are the number of moles of strong acid and strong base per liter needed to produce a unit change in pH. The greater the buffer capacity, the more resistant the solution is to pH change.

Figure 9-4a shows C_b versus pH for a solution containing 0.100 F HA with $pK_a = 5.00$. The ordinate (C_b) is the formal concentration of strong base needed to be mixed with 0.100 F HA to give the indicated pH. For example, a solution containing 0.050 F OH^- plus 0.100 F HA would have a pH of 5.00 (neglecting activities).

Figure 9-4b, which is the derivative of the upper curve, shows buffer capacity for the same system. Buffer capacity reaches a maximum when pH = pK_a. That is, *a buffer is most effective in resisting changes in pH when pH = pK_a* (that is, when $[HA] = [A^-]$).

In choosing a buffer, *seek one whose pK_a is as close as possible to the desired pH. The useful pH range of a buffer is usually considered to be $pK_a \pm 1$ pH unit.* Outside this range, there is not enough of either the weak acid or the weak base to react with added base or acid. Buffer capacity can be increased by increasing the concentration of the buffer.

Buffer capacity in Figure 9-4b continues upward at high pH (and at low pH, which is not shown) simply because there is a high concentration of OH^- at high pH (or H^+ at low pH). Addition of a small amount of acid or base to a large amount of OH^- (or H^+) will not have a large effect on pH. A solution of high pH is buffered by the H_2O/OH^- conjugate acid-conjugate base pair. A solution of low pH is buffered by the H_3O^+/H_2O conjugate acid-conjugate base pair.

How Much Buffer to Use?

You do not need a fancy calculation to decide how much buffer to use. If there are sufficiently more moles of buffer present than there are moles of acid or base that could be introduced into the solution by a chemical reaction or by addition of other reagents, then the pH is not going to change very much.

EXAMPLE How Much Will the pH Change?

Suppose that a buffered solution contains 50 mmol of the buffer species HA and 50 mmol of the buffer species A^-. The pH will be equal to pK_a for the buffer. How much will the pH change if 20 mmol of some other acid is created by a chemical reaction?

Solution In the worst case, the acid generated by the reaction is strong enough to convert an equivalent quantity of A^- into HA. The moles of HA would then be $50 + 20 = 70$ mmol. The moles of A^- would then be $50 - 20 = 30$ mmol. The pH would be

$$\text{pH} = pK_a + \log \frac{\text{mol } A^-}{\text{mol HA}} = pK_a + \log \frac{30 \text{ mmol}}{70 \text{ mmol}} = pK_a - 0.37$$

The pH would drop by 0.37 units. If this change is acceptable, you have enough buffer.

TEST YOURSELF If, instead, 15 mmol of strong base were generated, by how much would the pH of the original buffer rise? (*Answer:* 0.27 pH units)

Buffer pH Depends on Ionic Strength and Temperature

The correct Henderson-Hasselbalch equation, 9-18, includes activity coefficients. The most important reason why the calculated pH of a buffer is not equal to the observed pH is because the ionic strength is not 0, so activity coefficients are not 1. Table 9-2 lists pK_a values for common buffers that are widely used in biochemistry. Values are listed for ionic strengths of 0 and 0.1 M. If a buffer solution has an ionic strength closer to 0.1 M than to 0, it is sensible to use pK_a for $\mu = 0.1$ to obtain a more realistic calculation of pH.

If we mix 0.200 mol boric acid with 0.100 mol NaOH in 1.00 L, we generate a 1:1 mixture of boric acid and its conjugate base with an ionic strength of 0.10 M:

$$\underset{\text{Boric acid (HA)}}{B(OH)_3} + OH^- \rightarrow \underset{\text{borate (A}^-)}{(HO)_2BO^-} + H_2O$$

For boric acid in Table 9-2 we find $pK_a = 9.24$ at $\mu = 0$ and $pK_a = 8.98$ at $\mu = 0.1$ M. We predict that the pH of a 1:1 mixture of boric acid and borate will have pH near $pK_a = 9.24$ at low ionic strength and near 8.98 at $\mu = 0.1$ M. As another example of ionic strength effects, when a 0.5 M stock solution of phosphate buffer at pH 6.6 is diluted to 0.05 M, the pH rises to 6.9—a rather significant effect.

For almost all problems in this book we use pK_a for $\mu = 0$. As a practical matter, when there is no value of pK_a listed for $\mu = 0$, use pK_a for $\mu = 0.1$ M.

Buffer pK_a depends on temperature, as indicated in the last column of Table 9-2. Tris has an exceptionally large dependence, -0.028 pK_a units per degree, near room temperature. A solution of tris with pH 8.07 at 25°C will have pH $\approx$ 8.7 at 4°C and pH $\approx$ 7.7 at 37°C.

Changing ionic strength changes pH.

Changing temperature changes pH.

When What You Mix Is Not What You Get

In dilute solution or at extremes of pH, the concentrations of HA and A$^-$ are not equal to their formal concentrations. Suppose we mix F_{HA} moles of HA and F_{A^-} moles of the salt Na^+A^-. The mass and charge balances are

What you mix is not what you get in dilute solutions or at extremes of pH.

Mass balance: $$F_{HA} + F_{A^-} = [HA] + [A^-]$$

Charge balance: $$[Na^+] + [H^+] = [OH^-] + [A^-]$$

The substitution $[Na^+] = F_{A^-}$ and a little algebra leads to the equations

$$[HA] = F_{HA} - [H^+] + [OH^-] \tag{9-20}$$

$$[A^-] = F_{A^-} + [H^+] - [OH^-] \tag{9-21}$$

So far we have assumed that $[HA] \approx F_{HA}$ and $[A^-] \approx F_{A^-}$, and we used these values in the Henderson-Hasselbalch equation. A more rigorous procedure is to use Equations 9-20 and 9-21. If F_{HA} or F_{A^-} is small, or if $[H^+]$ or $[OH^-]$ is large, then the approximations $[HA] \approx F_{HA}$ and $[A^-] \approx F_{A^-}$ are not good. In acidic solutions, $[H^+] \gg [OH^-]$, so $[OH^-]$ can be ignored in Equations 9-20 and 9-21. In basic solutions, $[H^+]$ can be neglected.

EXAMPLE **A Dilute Buffer Prepared from a Moderately Strong Acid**

What will be the pH if 0.010 0 mol of HA (with $pK_a = 2.00$) and 0.010 0 mol of A$^-$ are dissolved in water to make 1.00 L of solution?

Solution The solution is acidic (pH $\approx pK_a = 2.00$), so we neglect $[OH^-]$ in Equations 9-20 and 9-21. Setting $[H^+] = x$ in Equations 9-20 and 9-21, we use the K_a equation to find $[H^+]$:

$$\underset{0.010\ 0 - x}{HA} \rightleftharpoons \underset{x}{H^+} + \underset{0.010\ 0 + x}{A^-}$$

$$K_a = \frac{[H^+][A^-]}{[HA]} = \frac{(x)(0.010\ 0 + x)}{(0.010\ 0 - x)} = 10^{-2.00} \tag{9-22}$$

$$\Rightarrow x = 0.004\ 14\ M \quad \Rightarrow \quad pH = -\log[H^+] = 2.38$$

TABLE 9-2 **Structures and pK$_a$ values for common buffers**[a,b,c,d]

Name	Structure	pK$_a$[e] $\mu = 0$	pK$_a$[e] $\mu = 0.1$ M	Formula mass	$\Delta(pK_a)/\Delta T$ (K^{-1})
N-2-Acetamidoiminodiacetic acid (ADA)	O=C(NH$_2$)CH$_2$$\overset{+}{N}$H with CH$_2CO_2$H and CH$_2CO_2$H	—(CO$_2$H)	1.59	190.15	—
N-Tris(hydroxymethyl)methylglycine (TRICINE)	(HOCH$_2$)$_3$C$\overset{+}{N}$H$_2$CH$_2$CO$_2$H	2.02 (CO$_2$H)	—	179.17	−0.003
Phosphoric acid	H$_3$PO$_4$	2.15 (pK$_1$)	1.92	98.00	0.005
N,N-Bis(2-hydroxyethyl)glycine (BICINE)	(HOCH$_2$CH$_2$)$_2$$\overset{+}{N}HCH_2CO_2$H	2.23 (CO$_2$H)	—	163.17	—
ADA	(see above)	2.48 (CO$_2$H)	2.31	190.15	—
Piperazine-N,N'-bis(2-ethanesulfonic acid) (PIPES)	$^-$O$_3$SCH$_2$CH$_2$$\overset{+}{N}$H HN$\overset{+}{}CH_2CH_2SO_3^-$	— (pK$_1$)	2.67	302.37	
Citric acid	HO$_2$CCH$_2$$\overset{\text{OH}}{\underset{\text{CO}_2\text{H}}{\text{C}}}CH_2CO_2$H	3.13 (pK$_1$)	2.90	192.12	−0.002
Glycylglycine	H$_3$$\overset{+}{N}CH_2$C(=O)NHCH$_2CO_2$H	3.14 (CO$_2$H)	3.11	132.12	0.000
Piperazine-N,N'-bis(3-propanesulfonic acid) (PIPPS)	$^-$O$_3$S(CH$_2$)$_3$$\overset{+}{N}$H H$\overset{+}{N}$(CH$_2$)$_3SO_3^-$	— (pK$_1$)	3.79	330.42	—
Piperazine-N,N'-bis(4-butanesulfonic acid) (PIPBS)	$^-$O$_3$S(CH$_2$)$_4$$\overset{+}{N}$H H$\overset{+}{N}$(CH$_2$)$_4SO_3^-$	— (pK$_1$)	4.29	358.47	—
N,N'-Diethylpiperazine dihydrochloride (DEPP-2HCl)	CH$_3$CH$_2$$\overset{+}{N}$H H$\overset{+}{N}CH_2CH_3$·2Cl$^-$	— (pK$_1$)	4.48	215.16	—
Citric acid	(see above)	4.76 (pK$_2$)	4.35	192.12	−0.001
Acetic acid	CH$_3$CO$_2$H	4.76	4.62	60.05	0.000
N,N'-Diethylethylenediamine- N,N'-bis(3-propanesulfonic acid) (DESPEN)	$^-$O$_3$S(CH$_2$)$_3$$\overset{+}{N}HCH_2CH_2H\overset{+}{N}$(CH$_2$)$_3SO_3^-$ with CH$_3$CH$_2$ and CH$_2$CH$_3$	— (pK$_1$)	5.62	360.49	—
2-(N-Morpholino)ethanesulfonic acid (MES)	O—$\overset{+}{N}$HCH$_2$CH$_2$SO$_3^-$	6.27	6.06	195.24	−0.009
Citric acid	(see above)	6.40 (pK$_3$)	5.70	192.12	0.002
N,N,N',N'-Tetraethylethylenediamine dihydrochloride (TEEN · 2HCl)	Et$_2$$\overset{+}{N}HCH_2CH_2H\overset{+}{N}Et_2$·2Cl$^-$	— (pK1)	6.58	245.23	—
l,3-Bis[tris(hydroxymethyl)methyl-amino] propane hydrochloride (BIS-TRIS propane-2HCl)	(HOCH$_2$)$_3$C$\overset{+}{N}$H$_2$(CH$_2$)$_3$$\overset{+}{N}H_2$·2Cl$^-$ with (HOCH$_2$)$_3$C	6.65 (pK$_1$)	—	355.26	—
ADA	(see above)	6.84 (NH)	6.67	190.15	−0.007

a. The protonated form of each molecule is shown. Acidic hydrogen atoms are shown in **bold** type. pK$_a$ is for 25°C.

b. Many buffers in this table are widely used in biomedical research because of their weak metal binding and physiologic inertness (C. L. Bering, J. Chem. Ed. **1987**, 64, 803). In one study, where MES and MOPS had no discernible affinity for Cu^{2+}, a minor impurity in HEPES and HEPPS had a strong affinity for Cu^{2+} and MOPSO bound Cu^{2+} stoichiometrically (H. E. Marsh, Y.-P. Chin, L. Sigg, R. Hari, and H. Xu, Anal. Chem. **2003**, 75, 671). ADA, BICINE, ACES, and TES have some metal-binding ability (R. Nakon and C. R. Krishnamoorthy, Science **1983**, 221, 749). Lutidine buffers for the pH range 3 to 8 with limited metal-binding power have been described by U. Bips, H. Elias, M. Hauröder, G. Kleinhans, S. Pfeifer, and K. J. Wannowius, Inorg. Chem. **1983**, 22, 3862.

c. Some data from R. N. Goldberg, N. Kishore, and R. M. Lennen, J. Phys. Chem. Ref. Data **2002**, 31, 231. This paper gives the temperature dependence of pK$_a$.

d. Temperature and ionic strength dependence of pK$_a$ for buffers: HEPES—D. Feng, W. F. Koch, and Y. C. Wu, Anal. Chem. **1989**, 61, 1400; MOPSO—Y. C. Wu, P. A. Berezansky, D. Feng, and W. F. Koch, Anal. Chem. **1993**, 65, 1084; ACES and CHES—R. N. Roy, J. Bice, J. Greer, J. A. Carlsten, J. Smithson, W. S. Good, C. P. Moore, L. N. Roy, and K. M. Kuhler, J. Chem. Eng. Data **1997**, 42, 41; TEMN, TEEN, DEPP, DESPEN, PIPES, PIPPS, PIPBS, MES, MOPS, and MOBS—A. Kandegedara and D. B. Rorabacher, Anal. Chem. **1999**, 71, 3140. This last set of buffers was specifically developed for low metal-binding ability (Q. Yu, A. Kandegedara, Y. Xu, and D. B. Rorabacher, Anal. Biochem. **1997**, 253, 50).

e. See marginal note on page 191 for the distinction between pK$_a$ at $\mu = 0$ and at $\mu = 0.1$ M.

Name	Structure	pK_a[e]		Formula mass	$\Delta(pK_a)/\Delta T$ (K^{-1})
		$\mu = 0$	$\mu = 0.1$ M		
N-2-Acetamido-2-aminoethane-sulfonic acid (ACES)	$H_2NCCH_2\overset{+}{N}H_2CH_2CH_2SO_3^-$ (with O double bond on C)	6.85	6.75	182.20	−0.018
3-(N-Morpholino)-2-hydroxy-propanesulfonic acid (MOPSO)	$O\bigcirc\overset{+}{N}HCH_2\overset{OH}{C}HCH_2SO_3^-$	6.90	—	225.26	−0.015
Imidazole hydrochloride	(imidazolium structure) $\cdot Cl^-$	6.99	7.00	104.54	−0.022
PIPES	(see above)	7.14 (pK_2)	6.93	302.37	−0.007
3-(N-Morpholino)propanesulfonic acid (MOPS)	$O\bigcirc\overset{+}{N}HCH_2CH_2CH_2SO_3^-$	7.18	7.08	209.26	−0.012
Phosphoric acid	H_3PO_4	7.20 (pK_2)	6.71	98.00	−0.002
4-(N-Morpholino)butanesulfonic acid (MOBS)	$O\bigcirc\overset{+}{N}HCH_2CH_2CH_2CH_2SO_3^-$	—	7.48	223.29	
N-Tris(hydroxymethyl)methyl-2-aminoethanesulfonic acid (TES)	$(HOCH_2)_3C\overset{+}{N}H_2CH_2CH_2SO_3^-$	7.55	7.60	229.25	−0.019
N-2-Hydroxyethylpiperazine-N'-2-ethanesulfonic acid (HEPES)	$HOCH_2CH_2N\bigcirc\overset{+}{N}HCH_2CH_2SO_3^-$	7.56	7.49	238.30	−0.012
PIPPS	(see above)	— (pK_2)	7.97	330.42	—
N-2-Hydroxyethylpiperazine-N'-3-propanesulfonic acid (HEPPS)	$HOCH_2CH_2N\bigcirc\overset{+}{N}H(CH_2)_3SO_3^-$	7.96	7.87	252.33	−0.013
Glycine amide hydrochloride	$H_3\overset{+}{N}CH_2CNH_2\cdot Cl^-$ (with O double bond on C)	—	8.04	110.54	—
Tris(hydroxymethyl)aminomethane hydrochloride (TRIS$\cdot$HCl)	$(HOCH_2)_3C\overset{+}{N}H_3\cdot Cl^-$	8.07	8.10	157.60	−0.028
TRICINE	(see above)	8.14 (NH)	—	179.17	−0.018
Glycylglycine	(see above)	8.26 (NH)	8.09	132.12	−0.026
BICINE	(see above)	8.33 (NH)	8.22	163.17	−0.015
PIPBS	(see above)	— (pK_2)	8.55	358.47	—
DEPP$\cdot$2HCl	(see above)	— (pK_2)	8.58	207.10	—
DESPEN	(see above)	— (pK_2)	9.06	360.49	—
BIS-TRIS propane$\cdot$2HCl	(see above)	9.10 (pK_2)	—	355.26	—
Ammonia	NH_4^+	9.24	—	17.03	−0.031
Boric acid	$B(OH)_3$	9.24 (pK_1)	8.98	61.83	−0.008
Cyclohexylaminoethanesulfonic acid (CHES)	$\bigcirc-\overset{+}{N}H_2CH_2CH_2SO_3^-$	9.39	—	207.29	−0.023
TEEN$\cdot$2HCl	(see above)	— (pK_2)	9.88	245.23	—
3-(Cyclohexylamino)propanesulfonic acid (CAPS)	$\bigcirc-\overset{+}{N}H_2CH_2CH_2CH_2SO_3^-$	10.50	10.39	221.32	−0.028
N,N,N',N'−Tetraethylmethylene-diamine$\cdot$2HCl (TEMN$\cdot$2HCl)	$Et_2\overset{+}{N}HCH_2\overset{+}{H}NEt_2\cdot 2Cl^-$	— (pK_2)	11.01	231.21	—
Phosphoric acid	H_3PO_4	12.38 (pK_3)	11.52	98.00	−0.009
Boric acid	$B(OH)_3$	12.74 (pK_2)	—	61.83	—

HA in this solution is more than 40% dissociated. The acid is too strong for the approximation $[HA] \approx F_{HA}$.

The pH is 2.38 instead of 2.00. $[HA]$ and $[A^-]$ are not what we mixed:

$$[HA] = F_{HA} - [H^+] = 0.005\ 86\ M$$
$$[A^-] = F_{A^-} + [H^+] = 0.014\ 1\ M$$

In this example, HA is too strong and the concentrations are too low for HA and A^- to be equal to their formal concentrations.

TEST YOURSELF Find pH if $pK_a = 3.00$ instead of 2.00. Does the answer make sense? (**Answer:** 3.07)

The Henderson-Hasselbalch equation (with activity coefficients) is *always* true, because it is just a rearrangement of the K_a equilibrium expression. Approximations that are not always true are the statements $[HA] \approx F_{HA}$ and $[A^-] \approx F_{A^-}$.

In summary, a buffer consists of a mixture of a weak acid and its conjugate base. The buffer is most useful when $pH \approx pK_a$. Over a reasonable range of concentration, the pH of a buffer is nearly independent of concentration. A buffer resists changes in pH because it reacts with added acids or bases. If too much acid or base is added, the buffer will be consumed and will no longer resist changes in pH.

EXAMPLE   Excel's Goal Seek Tool and Naming of Cells

Goal Seek provides numerical solution to equations of almost any complexity. In setting up Equation 9-22, we made the (superb) approximation $[H^+] \gg [OH^-]$ and neglected $[OH^-]$. With Goal Seek, it is easy to use Equations 9-20 and 9-21 without approximations:

$$K_a = \frac{[H^+][A^-]}{[HA]} = \frac{[H^+](F_{A^-} + [H^+] - [OH^-])}{F_{HA} - [H^+] + [OH^-]} \qquad (9\text{-}23)$$

The spreadsheet illustrates Goal Seek and the naming of cells to make formulas more meaningful. In column A, enter labels for K_a, K_w, F_{HA}, F_A ($= F_{A^-}$), H ($= [H^+]$), and OH ($= [OH^-]$). Write numerical values for K_a, K_w, F_{HA}, and F_{A^-} in B1:B4. In cell B5, enter a *guess* for $[H^+]$.

	A	B	C	D	E
1	Ka =	0.01		Reaction quotient	
2	Kw =	1.00E-1.4		for Ka =	
3	FHA =	0.01		[H+][A–]/[HA] =	
4	FA =	0.01		0.001222222	
5	H =	1.000E-03		<–Vary H with Goal Seek until D4 = Ka	
6	OH = Kw/H =	1E-11		D4 = H*(FA + H – OH)/(FHA – H + OH)	
7	pH = –log(H) =	3.00			

Now we want to name cells B1 through B6. In Excel 2010 or 2007, select cell B1, go to the Formulas ribbon, and click on Define Name. The dialog box will ask if you want to use the name "Ka" that appears in cell A1. If you like this name, click OK. Now when you select cell B1, the name box at the upper left of the spreadsheet displays Ka instead of B1. Name the other cells in column B "Kw", "FHA", "FA", "H", and "OH". Now when you write a formula referring to cell B2, you can write Kw instead of B2. Kw is an *absolute reference* to cell $B$2.

In cell B6 enter the formula "=Kw/H" and Excel returns the value 1E-11 for $[OH^-]$. The beauty of naming cells is that "=Kw/H" is easier to understand than "=$B$2/$B$5". In cell B7 enter the formula "= –log(H)" for pH.

In cell D4, write "=H*(FA+H-OH)/(FHA-H+OH)", which is the quotient in Equation 9-23. Excel returns the value 0.001 222 based on the guess $[H^+] = 0.001$ in cell B5.

Now use Goal Seek to vary $[H^+]$ in cell B5 until the reaction quotient in cell D4 equals 0.01, which is the value of K_a. Before using Goal Seek in Excel 2010, click the File menu and select Options. In the Options window, select Formulas. Set Maximum Change to 1e-15 to find an answer with high precision. To execute Goal Seek, go to the Data ribbon, click on

What-If Analysis and then on Goal Seek. In the dialog box, Set cell <u>D4</u> To value <u>0.01</u> By changing cell <u>B5</u>. Click OK and Excel varies cell B5 until the value $[H^+] = 4.142 \times 10^{-3}$ gives a reaction quotient of 0.01 in cell D4. Different guesses for H might give negative solutions or might not reach a solution. Only one positive value of H satisfies Equation 9-23.

TEST YOURSELF   Find H if $K_a = 0.001$. (*Answer:* H $= 8.44 \times 10^{-4}$, pH $= 3.07$)

Terms to Understand

acid dissociation constant, K_a	fraction of association, α	pK	weak base
base hydrolysis constant, K_b	(of a base)	strong acid	weak electrolyte
buffer	fraction of dissociation, α	strong base	
buffer capacity	(of an acid)	weak acid	
conjugate acid-base pair	Henderson-Hasselbalch equation		

Summary

Strong acids or bases. For practical concentrations ($\gtrsim 10^{-6}$ M), pH or pOH is obtained directly from the formal concentration of acid or base. When the concentration is near 10^{-7} M, we use the systematic treatment of equilibrium to calculate pH. At still lower concentrations, the pH is 7.00, set by autoprotolysis of the solvent.

Weak acids. For the reaction HA $\rightleftharpoons$ H$^+$ + A$^-$, we set up and solve the equation $K_a = x^2/(F - x)$, where $[H^+] = [A^-] = x$, and $[HA] = F - x$. The fraction of dissociation is given by $\alpha = [A^-]/([HA] + [A^-]) = x/F$. The term p$K_a$ is defined as p$K_a = -\log K_a$.

Weak bases. For the reaction B + H$_2$O $\rightleftharpoons$ BH$^+$ + OH$^-$, we set up and solve the equation $K_b = x^2/(F - x)$, where $[OH^-] = [BH^+] = x$, and $[B] = F - x$. The conjugate acid of a weak base is a weak acid, and the conjugate base of a weak acid is a weak base. For a conjugate acid-base pair, $K_a \cdot K_b = K_w$.

Buffers. A buffer is a mixture of a weak acid and its conjugate base. It resists changes in pH because it reacts with added acid or base. The pH is given by the Henderson-Hasselbalch equation:

$$pH = pK_a + \log \frac{[A^-]}{[HA]}$$

where pK_a applies to the species in the denominator. The concentrations of HA and A$^-$ are essentially unchanged from those used to prepare the solution. The pH of a buffer is nearly independent of dilution, but buffer capacity increases as the concentration of buffer increases. The maximum buffer capacity is at pH $= pK_a$, and the useful range is pH $= pK_a \pm 1$.

The conjugate base of a weak acid is a weak base. The weaker the acid, the stronger the base. However, if one member of a conjugate pair is weak, so is its conjugate. The relation between K_a for an acid and K_b for its conjugate base in aqueous solution is $K_a \cdot K_b = K_w$. When a strong acid (or base) is added to a weak base (or acid), they react nearly completely.

Exercises

9-A. Using activity coefficients correctly, find the pH of 1.0×10^{-2} M NaOH.

9-B. (Without activities), calculate the pH of

(a) 1.0×10^{-8} M HBr

(b) 1.0×10^{-8} M H$_2$SO$_4$ (H$_2$SO$_4$ dissociates completely to 2H$^+$ plus SO$_4^{2-}$ at this low concentration).

9-C. What is the pH of a solution prepared by dissolving 1.23 g of 2-nitrophenol (FM 139.11) in 0.250 L?

9-D. The pH of 0.010 M *o*-cresol is 6.16. Find pK_a for this weak acid.

o-Cresol

9-E. Calculate the limiting value of the fraction of dissociation (α) of a weak acid (p$K_a = 5.00$) as the concentration of HA approaches 0. Repeat the same calculation for p$K_a = 9.00$.

9-F. Find the pH of 0.050 M sodium butanoate (the sodium salt of butanoic acid, also called butyric acid).

9-G. The pH of 0.10 M ethylamine is 11.82.

(a) Without referring to Appendix G, find K_b for ethylamine.

(b) Using results from part **(a)**, calculate the pH of 0.10 M ethylammonium chloride.

9-H. Which of the following bases would be most suitable for preparing a buffer of pH 9.00? (i) NH$_3$ (ammonia, $K_b = 1.76 \times 10^{-5}$); (ii) C$_6H_5NH_2$ (aniline, $K_b = 3.99 \times 10^{-10}$); (iii) H$_2NNH_2$ (hydrazine, $K_b = 1.05 \times 10^{-6}$); (iv) C$_5H_5$N (pyridine, $K_b = 1.58 \times 10^{-9}$).

9-I. A solution contains 63 different conjugate acid-base pairs. Among them is acrylic acid and acrylate ion, with the equilibrium ratio [acrylate]/[acrylic acid] = 0.75. What is the pH of the solution?

$$H_2C{=}CHCO_2H \qquad pK_a = 4.25$$
Acrylic acid

9-J. (a) Find the pH of a solution prepared by dissolving 1.00 g of glycine amide hydrochloride (Table 9-2) plus 1.00 g of glycine amide in 0.100 L.

$$H_2N-\overset{\displaystyle O}{\overset{\|}{C}}-NH_2$$

Glycine amide
$C_2H_6N_2O$
FM 74.08

(b) How many grams of glycine amide should be added to 1.00 g of glycine amide hydrochloride to give 100 mL of solution with pH 8.00?

(c) What would be the pH if the solution in part **(a)** were mixed with 5.00 mL of 0.100 M HCl?

(d) What would be the pH if the solution in part **(c)** were mixed with 10.00 mL of 0.100 M NaOH?

(e) What would be the pH if the solution in part **(a)** were mixed with 90.46 mL of 0.100 M NaOH? (This is exactly the quantity of NaOH required to neutralize the glycine amide hydrochloride.)

9-K. Select a compound from Table 9-2 that you could use to make 250 mL of 0.2 M buffer with a pH of 6.0. Explain how you would make the buffer.

9-L. A solution with an ionic strength of 0.10 M containing 0.010 0 M phenylhydrazine has a pH of 8.13. Using activity coefficients correctly, find pK_a for the phenylhydrazinium ion found in phenylhydrazine hydrochloride. Assume that $\gamma_{BH^+} = 0.80$.

$$\bigcirc\!\!-NHNH_2 \qquad \bigcirc\!\!-NH\overset{+}{N}H_3Cl^-$$

Phenylhydrazine   Phenylhydrazine hydrochloride
B $BH^+ Cl^-$

9-M. ▦ Use the Goal Seek spreadsheet at the end of the chapter to find the pH of 1.00 L of solution containing 0.030 mol HA ($pK_a = 2.50$) and 0.015 mol NaA. What would the pH be with the approximations [HA] = 0.030 and $[A^-] = 0.015$?

Problems

Strong Acids and Bases

9-1. Why doesn't water produce 10^{-7} M H^+ and 10^{-7} M OH^- when HBr is added?

9-2. Neglecting activity coefficients, calculate the pH of **(a)** 1.0×10^{-3} M HBr; **(b)** 1.0×10^{-2} M KOH.

9-3. Neglecting activity coefficients, calculate the pH of 5.0×10^{-8} M $HClO_4$. What fraction of H^+ is derived from dissociation of water?

9-4. (a) The measured pH of 0.100 M HCl at 25°C is 1.092. From this information, calculate the activity coefficient of H^+ and compare your answer with that in Table 8-1.

(b) The measured pH of 0.010 0 M HCl + 0.090 0 M KCl at 25°C is 2.102. From this information, calculate the activity coefficient of H^+ in this solution.

(c) The ionic strengths of the solutions in parts **(a)** and **(b)** are the same. What can you conclude about the dependence of activity coefficients on the particular ions in a solution?

Weak-Acid Equilibria

9-5. Write the chemical reaction whose equilibrium constant is

(a) K_a for benzoic acid, $C_6H_5CO_2H$

(b) K_b for benzoate ion, $C_6H_5CO_2^-$

(c) K_b for aniline, $C_6H_5NH_2$

(d) K_a for anilinium ion, $C_6H_5NH_3^+$

9-6. Find the pH and fraction of dissociation (α) of a 0.100 M solution of the weak acid HA with $K_a = 1.00 \times 10^{-5}$.

9-7. $BH^+ClO_4^-$ is a salt formed from the base B ($K_b = 1.00 \times 10^{-4}$) and perchloric acid. It dissociates into BH^+, a weak acid, and ClO_4^-, which is neither an acid nor a base. Find the pH of 0.100 M $BH^+ClO_4^-$.

9-8. Find the pH and concentrations of $(CH_3)_3N$ and $(CH_3)_3NH^+$ in 0.060 M trimethylammonium chloride.

9-9. Use the reaction quotient, Q, to explain why the fraction of dissociation of weak acid, HA, increases when the solution is diluted by a factor of 2.

9-10. *When is a weak acid weak and when is a weak acid strong?* Show that the weak acid HA will be 92% dissociated when dissolved in water if the formal concentration is one-tenth of K_a (F = $K_a/10$). Show that the fraction of dissociation is 27% when F = $10K_a$. At what formal concentration will the acid be 99% dissociated? Compare your answer with the left-hand curve in Figure 9-2.

9-11. A 0.045 0 M solution of benzoic acid has a pH of 2.78. Calculate pK_a for this acid.

9-12. A 0.045 0 M solution of HA is 0.60% dissociated. Calculate pK_a for this acid.

9-13. Barbituric acid dissociates as follows:

$$\underset{\substack{\text{Barbituric acid}\\\text{HA}}}{\overset{\substack{O\qquad\quad O}}{\underset{O}{\underset{}{HN\quad NH}}}} \overset{K_a = 9.8 \times 10^{-5}}{\rightleftharpoons} \underset{A^-}{\overset{\substack{O\qquad\quad O}}{\underset{O}{HN\quad N:^-}}} + H^+$$

(a) Calculate the pH and fraction of dissociation of $10^{-2.00}$ M barbituric acid.

(b) Calculate the pH and fraction of dissociation of $10^{-10.00}$ M barbituric acid.

9-14. Using activity coefficients, find the pH and fraction of dissociation of 50.0 mM hydroxybenzene (phenol) in 0.050 M LiBr. Take the size of $C_6H_5O^-$ to be 600 pm.

9-15. Cr^{3+} is acidic by virtue of the hydrolysis reaction

$$Cr^{3+} + H_2O \overset{K_{a1}}{\rightleftharpoons} Cr(OH)^{2+} + H^+$$

[Further reactions produce $Cr(OH)_2^+$, $Cr(OH)_3$, and $Cr(OH)_4^-$.] Find the value of K_{a1} in Figure 6-9. Considering only the K_{a1} reaction, find the pH of 0.010 M $Cr(ClO_4)_3$. What fraction of chromium is in the form $Cr(OH)^{2+}$?

9-16. From the dissociation constant of HNO_3 at 25°C in Box 9-1, find the percent dissociated in 0.100 M HNO_3 and in 1.00 M HNO_3.

9-17. 🖩 *Excel Goal Seek.* Solve the equation $x^2/(F - x) = K$ by using Goal Seek. Guess a value of x in cell A4 and evaluate $x^2/(F - x)$ in cell B4. Use Goal Seek to vary the value of x until $x^2/(F - x)$ is equal to K. Use your spreadsheet to check your answer to Problem 9-6.

	A	B
1	Using Excel GOAL SEEK	
2		
3	x =	x²/(F-x) =
4	0.01	1.1111E-03
5	F =	
6	0.1	

Weak-Base Equilibria

9-18. Covalent compounds generally have higher vapor pressure than ionic compounds. The "fishy" smell of fish arises from amines in the fish. Explain why squeezing lemon (which is acidic) onto fish reduces the fishy smell (and taste).

9-19. Find the pH and fraction of association (α) of a 0.100 M solution of the weak base B with $K_b = 1.00 \times 10^{-5}$.

9-20. Find the pH and concentrations of $(CH_3)_3N$ and $(CH_3)_3NH^+$ in a 0.060 M solution of trimethylamine.

9-21. Find the pH of 0.050 M NaCN.

9-22. Calculate the fraction of association (α) for 1.00×10^{-1}, 1.00×10^{-2}, and 1.00×10^{-12} M sodium acetate. Does α increase or decrease with dilution?

9-23. A 0.10 M solution of a base has pH = 9.28. Find K_b.

9-24. A 0.10 M solution of a base is 2.0% hydrolyzed ($\alpha = 0.020$). Find K_b.

9-25. Show that the limiting fraction of association of a base in water, as the concentration of base approaches 0, is $\alpha = 10^7 K_b/(1 + 10^7 K_b)$. Find the limiting value of α for $K_b = 10^{-4}$ and for $K_b = 10^{-10}$.

Buffers

9-26. Describe how to prepare 100 mL of 0.200 M acetate buffer, pH 5.00, starting with pure liquid acetic acid and solutions containing ~3 M HCl and ~3 M NaOH.

9-27. Describe how to prepare 250 mL of 1.00 M ammonia buffer, pH 9.00, starting with 28 wt% NH_3 ("concentrated ammonium hydroxide" listed on the back inside cover of the book) and "concentrated" HCl (37.2 wt%) or "concentrated" NaOH (50.5 wt%).

9-28. Consider a reaction mixture containing 100.0 mL of 0.100 M borate buffer at pH = pK_a = 9.24. At pH = pK_a, we know that $[H_3BO_3] = [H_2BO_3^-] = 0.050\,0$ M. Suppose that a chemical reaction whose pH we wish to control will be generating acid. To avoid changing the pH very much we do not want to generate more acid than would use up half of the $[H_2BO_3^-]$. How many moles of acid could be generated without using up more than half of the $[H_2BO_3^-]$? What would be the pH?

9-29. Why is the pH of a buffer nearly independent of concentration?

9-30. Why does buffer capacity increase as the concentration of buffer increases?

9-31. Why does buffer capacity increase as a solution becomes very acidic (pH ≈ 1) or very basic (pH ≈ 13)?

9-32. Why does the buffer capacity reach a maximum when pH = pK_a?

9-33. Explain the following statement: The Henderson-Hasselbalch equation (with activity coefficients) is *always* true; what may not be correct are the values of $[A^-]$ and $[HA]$ that we choose to use in the equation.

9-34. Which of the following acids would be most suitable for preparing a buffer of pH 3.10? (i) hydrogen peroxide; (ii) propanoic acid; (iii) cyanoacetic acid; (iv) 4-aminobenzenesulfonic acid.

9-35. A buffer was prepared by dissolving 0.100 mol of the weak acid HA ($K_a = 1.00 \times 10^{-5}$) plus 0.050 mol of its conjugate base Na^+A^- in 1.00 L. Find the pH.

9-36. Write the Henderson-Hasselbalch equation for a solution of formic acid. Calculate the quotient $[HCO_2^-]/[HCO_2H]$ at **(a)** pH 3.000; **(b)** pH 3.744; **(c)** pH 4.000.

9-37. Calculate the quotient $[HCO_2^-]/[HCO_2H]$ at pH 3.744 if the ionic strength is 0.1 M by using the effective equilibrium constant listed for $\mu = 0.1$ in Appendix G.

9-38. Given that pK_b for nitrite ion (NO_2^-) is 10.85, find the quotient $[HNO_2]/[NO_2^-]$ in a solution of sodium nitrite at **(a)** pH 2.00; **(b)** pH 10.00.

9-39. **(a)** Would you need NaOH or HCl to bring the pH of 0.050 0 M HEPES (Table 9-2) to 7.45?

(b) Describe how to prepare 0.250 L of 0.050 0 M HEPES, pH 7.45.

9-40. How many milliliters of 0.246 M HNO_3 should be added to 213 mL of 0.006 66 M 2,2′-bipyridine to give a pH of 4.19?

9-41. **(a)** Write the chemical reactions whose equilibrium constants are K_b and K_a for imidazole and imidazole hydrochloride, respectively.

(b) Calculate the pH of a solution prepared by mixing 1.00 g of imidazole with 1.00 g of imidazole hydrochloride and diluting to 100.0 mL.

(c) Calculate the pH of the solution if 2.30 mL of 1.07 M $HClO_4$ are added.

(d) How many milliliters of 1.07 M $HClO_4$ should be added to 1.00 g of imidazole to give a pH of 6.993?

9-42. Calculate the pH of a solution prepared by mixing 0.080 0 mol of chloroacetic acid plus 0.040 0 mol of sodium chloroacetate in 1.00 L of water.

(a) First do the calculation by assuming that the concentrations of HA and A^- equal their formal concentrations.

(b) Then do the calculation, using the real values of [HA] and $[A^-]$ in the solution.

(c) Using first your head, and then the Henderson-Hasselbalch equation, find the pH of a solution prepared by dissolving all the following compounds in one beaker containing a total volume of 1.00 L: 0.180 mol $ClCH_2CO_2H$, 0.020 mol $ClCH_2CO_2Na$, 0.080 mol HNO_3, and 0.080 mol $Ca(OH)_2$. Assume that $Ca(OH)_2$ dissociates completely.

9-43. Calculate how many milliliters of 0.626 M KOH should be added to 5.00 g of MOBS (Table 9-2) to give a pH of 7.40.

9-44. (a) Use Equations 9-20 and 9-21 to find the pH and concentrations of HA and A^- in a solution prepared by mixing 0.002 00 mol of acetic acid plus 0.004 00 mol of sodium acetate in 1.00 L of water.

(b) After working part **(a)** by hand, use Excel Goal Seek to find the same answers.

9-45. (a) Calculate the pH of a solution prepared by mixing 0.010 0 mol of the base B ($K_b = 10^{-2.00}$) with 0.020 0 mol of BH^+Br^- and diluting to 1.00 L. First calculate the pH by assuming [B] = 0.010 0 and [BH^+] = 0.020 0 M. Compare this answer with the pH calculated without making such an assumption.

(b) After working part **(a)** by hand, use Excel Goal Seek to find the same answers.

9-46. *Effect of ionic strength on pK_a.* K_a for the $H_2PO_4^-/HPO_4^{2-}$ buffer is

$$K_a = \frac{[HPO_4^{2-}][H^+]\gamma_{HPO_4^{2-}}\gamma_{H^+}}{[H_2PO_4^-]\gamma_{H_2PO_4^-}} = 10^{-7.20}$$

If you mix a 1:1 mole ratio of $H_2PO_4^-$ and HPO_4^{2-} at 0 ionic strength, the pH is 7.20. Using activity coefficients from Table 8-1, calculate the pH of a 1:1 mixture of $H_2PO_4^-$ and HPO_4^{2-} at an ionic strength of 0.10. Remember that pH $= -\log\mathscr{A}_{H^+} = -\log[H^+]\gamma_{H^+}$.

9-47. *Interpreting spectral data.* The graph shows the ^1H-nuclear magnetic resonance chemical shift of the H_4 proton on pyridine as a function of pH. Chemical shift is related to the environment of a proton in a molecule. If the environment changes, the chemical shift changes. Suggest an explanation for why the chemical shift changes

between low pH and high pH. Estimate pK_a for pyridinium ion ($C_5H_5NH^+$).

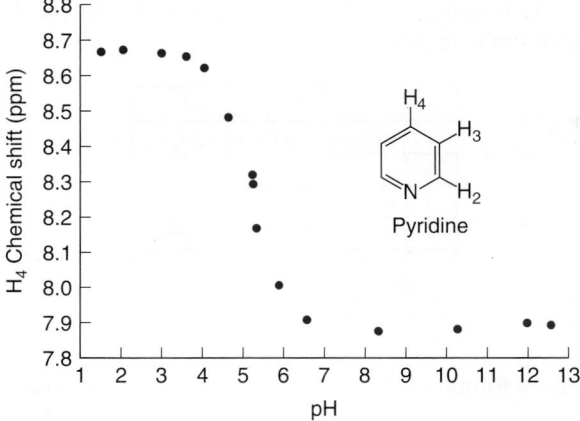

NMR chemical shift of H_4 of pyridine as a function of pH. [Data from A. D. Gift, S. M. Stewart, and P. K. Bokashanga, "Experimental Determination of pK_a Values by Use of NMR Chemical Shifts," *J. Chem. Ed.* **2012,** *89,* 1458.]

9-48. (a) *Systematic treatment of equilibrium.* The acidity of Al^{3+} is determined by the following reactions. Write the equations needed to find the pH of $Al(ClO_4)_3$ at a formal concentration F.

$$Al^{3+} + H_2O \underset{}{\overset{\beta_1}{\rightleftharpoons}} AlOH^{2+} + H^+$$

$$Al^{3+} + 2H_2O \underset{}{\overset{\beta_2}{\rightleftharpoons}} Al(OH)_2^+ + 2H^+$$

$$2Al^{3+} + 2H_2O \underset{}{\overset{K_{22}}{\rightleftharpoons}} Al_2(OH)_2^{4+} + 2H^+$$

$$Al^{3+} + 3H_2O \underset{}{\overset{\beta_3}{\rightleftharpoons}} Al(OH)_3(aq) + 3H^+$$

$$Al^{3+} + 4H_2O \underset{}{\overset{\beta_4}{\rightleftharpoons}} Al(OH)_4^- + 4H^+$$

$$3Al^{3+} + 4H_2O \underset{}{\overset{K_{43}}{\rightleftharpoons}} Al_3(OH)_4^{5+} + 4H^+$$

(b) Explain how you would use a spreadsheet such as Figure 8-9 to find the pH and concentrations of all species if you know the values of the equilibrium constants and the formal concentration F.

10 Polyprotic Acid-Base Equilibria

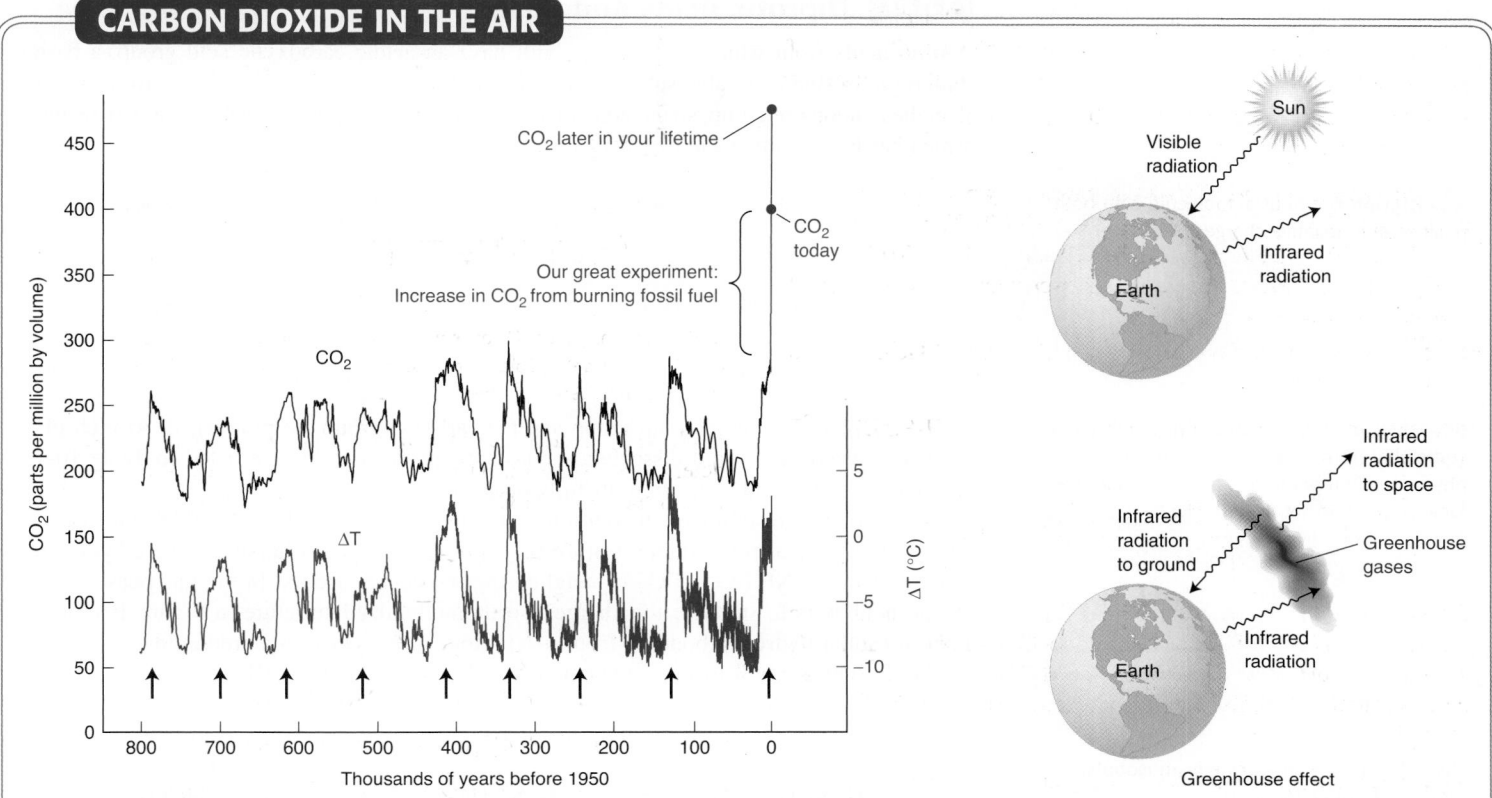

Upper curve: Atmospheric CO_2 deduced from air trapped in Antarctic ice and from direct atmospheric measurements. *Lower curve:* Atmospheric temperature at the level where precipitation forms is deduced from isotopic composition of ice. [Data from J. M. Barnola, D. Raynaud, C. Lorius, and N. I. Barkov, http://cdiac.esd.ornl.gov/ftp/trends/co2/vostok.icecore.co2.]

Perhaps the largest chemical and physical experiment ever conducted is our injection of carbon dioxide into the atmosphere in sufficient quantity to alter the cycle of CO_2 concentration that has persisted for at least 800 000 years. CO_2 is produced by combustion of fossil fuels (coal, oil, wood, natural gas), which are our predominant source of energy. In 2011, humans released 3.16×10^{13} kg of CO_2 from fossil fuel.[1] Mean atmospheric CO_2 increased by 2.1 ppm (μL/L) from 390.6 to 392.7 ppm. If all CO_2 produced in 2011 remained in the atmosphere, CO_2 would have risen by 4.0 ppm.[2] Instead, about half of the CO_2 dissolved in the ocean or was incorporated into plants.

CO_2 acts as a *greenhouse gas* to affect Earth's surface temperature. Earth absorbs sunlight and then emits infrared radiation. The balance between sunlight absorbed and radiation sent back to space determines the surface temperature. A **greenhouse gas** absorbs infrared radiation and reradiates some of it back to the ground. By intercepting some of Earth's radiation, CO_2 keeps our planet warmer than it would otherwise be.

The graph shows peaks in atmospheric temperature and CO_2 marked by arrows roughly every 100 000 years. Temperature change is attributed principally to cyclic changes in Earth's orbit and tilt. Small increases in temperature drive dissolved CO_2 from the ocean into the atmosphere. Increased atmospheric CO_2 further increases warming by the greenhouse effect. Cooling brought on by orbital changes redissolves CO_2 in the ocean, thereby causing further cooling. Temperature and CO_2 followed each other until 200 years ago. We are beginning to experience effects of adding CO_2 to the atmosphere. Each of the last three decades has been warmer than the one before it.[3] Climatic effects include sea-level rise, longer growing seasons, changes in river flows, increases in heavy downpours, earlier snowmelt, and extended ice-free seasons in oceans, lakes, and rivers.

olyprotic acids and bases are those that can donate or accept more than one proton. After we have studied **diprotic** systems (with two acidic or basic sites), the extension to three or more acidic sites is straightforward. Then we step back and take a qualitative look at the big picture and think about which species are dominant at any given pH.

10-1 Diprotic Acids and Bases

Amino acids from which proteins are built have an acidic carboxylic acid group, a basic amino group, and a variable substituent designated R. The carboxyl group is a stronger acid than the ammonium group, so the nonionized form rearranges spontaneously to the **zwitterion,** which has both positive and negative sites:

Amino group ⟶ H_2N $\overset{+}{H_3N}$ ⟵ Ammonium group

        CH—R ⟶   CH—R

Carboxylic acid ⟶ $HO-C$ ^-O-C ⟵ Carboxyl group

        O O

                Zwitterion

At low pH, both the ammonium group and the carboxyl group are protonated. At high pH, neither is protonated. Acid dissociation constants of amino acids are listed in Table 10-1, where each compound is drawn in its fully protonated form.

pKₐ values of amino acids in living cells differ somewhat from those in Table 10-1 because physiologic temperature is not 25°C and the ionic strength is not 0.

Zwitterions are stabilized in solution by interactions of $-NH_3^+$ and $-CO_2^-$ with water. The zwitterion is also the stable form of the amino acid in the solid state, where hydrogen bonding from $-NH_3^+$ to $-CO_2^-$ of neighboring molecules occurs. In the gas phase, there are no neighbors to stabilize the charges, so the nonionized structure in Figure 10-1 with intramolecular hydrogen bonding from $-NH_2$ to a carboxyl oxygen predominates.

Our discussion will focus on the amino acid leucine, designated HL.

The substituent R in leucine is an isobutyl group: $(CH_3)_2CHCH_2-$.

$$H_3\overset{+}{N}CHCO_2H \underset{pK_{a1} = 2.328}{\rightleftharpoons} H_3\overset{+}{N}CHCO_2^- \underset{pK_{a2} = 9.744}{\rightleftharpoons} H_2NCHCO_2^-$$

                          Leucine

     H_2L^+ HL L^-

The equilibrium constants refer to the following reactions:

Diprotic acid: $H_2L^+ \rightleftharpoons HL + H^+$ $K_{a1} \equiv K_1$ **(10-1)**

                      $HL \rightleftharpoons L^- + H^+$ $K_{a2} \equiv K_2$ **(10-2)**

Diprotic base: $L^- + H_2O \rightleftharpoons HL + OH^-$ K_{b1} **(10-3)**

                $HL + H_2O \rightleftharpoons H_2L^+ + OH^-$ K_{b2} **(10-4)**

We customarily omit the subscript "a" in K_{a1} and K_{a2}. We will always write the subscript "b" in K_{b1} and K_{b2}.

Recall that the relations between the acid and base equilibrium constants are

Relations between $K_{a1} \cdot K_{b2} = K_w$ **(10-5)**

K_a and K_b:

                      $K_{a2} \cdot K_{b1} = K_w$ **(10-6)**

We now set out to calculate the pH and composition of individual solutions of 0.050 0 M H_2L^+, 0.050 0 M HL, and 0.050 0 M L^-. Our methods are general. They do not depend on the charge type of the acids and bases. That is, *we would use the same procedure to find the pH of the diprotic H_2A, where A is anything, or H_2L^+, where HL is leucine.*

The Acidic Form, H_2L^+

Leucine hydrochloride contains the protonated species, H_2L^+, which can dissociate twice (Reactions 10-1 and 10-2). Because $K_1 = 4.70 \times 10^{-3}$, H_2L^+ is a weak acid. HL is an even weaker acid with $K_2 = 1.80 \times 10^{-10}$. It appears that H_2L^+ will dissociate only partly, and the resulting HL will hardly dissociate at all. For this reason, we make the (superb) approximation that a solution of H_2L^+ behaves as a monoprotic acid, with $K_a = K_1$.

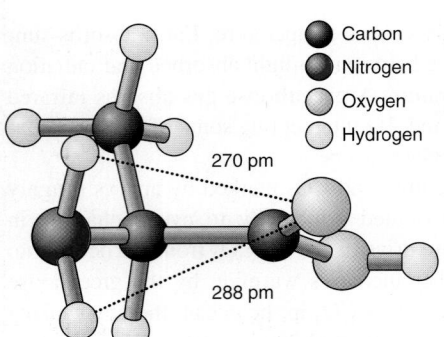

● Carbon
● Nitrogen
○ Oxygen
○ Hydrogen

270 pm

288 pm

FIGURE 10-1 Gas phase structure of alanine, determined by microwave spectroscopy.
[Information from S. Blanco, A. Lesarri, J. C. López, and J. L. Alonso, "The Gas-Phase Structure of Alanine," *J. Am. Chem. Soc.* **2004,** *126,* 11675.]

With this approximation, finding the pH of 0.050 0 M H_2L^+ is easy.

$$H_3\overset{+}{N}CHCO_2H \underset{}{\overset{K_a = K_{a1} = K_1}{\rightleftharpoons}} H_3\overset{+}{N}CHCO_2^- + H^+$$

H_2L^+	HL	H^+
$0.050\ 0 - x$	x	x

H_2L^+ can be treated as monoprotic with $K_a = K_{a1}$.

$$K_a = K_1 = 4.70 \times 10^{-3}$$

$$\frac{x^2}{F - x} = K_a \Rightarrow x = 1.32 \times 10^{-2}\ M$$

Solve for x with the quadratic equation.

$$[HL] = x = 1.32 \times 10^{-2}\ M$$
$$[H^+] = x = 1.32 \times 10^{-2}\ M \Rightarrow pH = -\log[H^+] = 1.88$$
$$[H_2L^+] = F - x = 3.68 \times 10^{-2}\ M$$

TABLE 10-1 **Acid dissociation constants of amino acids**

Amino acid[a]	Substituent[a]	Carboxylic acid[b] pK_a	Ammonium[b] pK_a	Substituent[b] pK_a	Formula mass
Alanine (A)	$-CH_3$	2.344	9.868		89.09
Arginine (R)	$-CH_2CH_2CH_2NHC\overset{\overset{+}{N}H_2}{\underset{NH_2}{\big\|}}$	1.823	8.991	(12.1^c)	174.20
Asparagine (N)	$-CH_2\overset{\overset{O}{\|}}{C}NH_2$	2.16^c	8.73^c		132.12
Aspartic acid (D)	$-CH_2CO_2H$	1.990	10.002	3.900	133.10
Cysteine (C)	$-CH_2SH$	(1.7)	10.74	8.36	121.16
Glutamic acid (E)	$-CH_2CH_2CO_2H$	2.16	9.96	4.30	147.13
Glutamine (Q)	$-CH_2CH_2\overset{\overset{O}{\|}}{C}NH_2$	2.19^c	9.00^c		146.15
Glycine (G)	$-H$	2.350	9.778		75.07
Histidine (H)	$-CH_2-$ (imidazole)	(1.6)	9.28	5.97	155.16
Isoleucine (I)	$-CH(CH_3)(CH_2CH_3)$	2.318	9.758		131.17
Leucine (L)	$-CH_2CH(CH_3)_2$	2.328	9.744		131.17
Lysine (K)	$-CH_2CH_2CH_2CH_2NH_3^+$	(1.77)	9.07	10.82	146.19
Methionine (M)	$-CH_2CH_2SCH_3$	2.18^c	9.08^c		149.21
Phenylalanine (F)	$-CH_2-$ (phenyl)	2.20	9.31		165.19
Proline (P)	(Structure of entire amino acid)	1.952	10.640		115.13
Serine (S)	$-CH_2OH$	2.187	9.209		105.09
Threonine (T)	$-CH(CH_3)(OH)$	2.088	9.100		119.12
Tryptophan (W)	$-CH_2-$ (indole)	2.37^c	9.33^c		204.23
Tyrosine (Y)	$-CH_2-\bigcirc-OH$	2.41^c	8.67^c	11.01^c	181.19
Valine (V)	$-CH(CH_3)_2$	2.286	9.719		117.15

a. The acidic protons are shown in **bold** type. Each amino acid is written in its fully protonated form. Standard abbreviations are shown in parentheses.
b. pK_a values refer to 25°C and zero ionic strength unless marked by c. Values considered to be uncertain are enclosed in parentheses. Appendix G gives pK_a for $\mu = 0.1$ M.
c. For these entries, the ionic strength is 0.1 M, and the constant refers to a product of concentrations instead of activities.

SOURCE: A. E. Martell and R. J. Motekaitis, NIST Database 46 (Gaithersburg, MD: National Institute of Standards and Technology, 2001).

What is the concentration of L^- in the solution? We have already assumed that it is very small, but it cannot be 0. We can calculate $[L^-]$ from the K_2 equation, with the concentrations of HL and H^+ that we just computed.

$$K_2 = \frac{[H^+][L^-]}{[HL]} \Rightarrow [L^-] = \frac{K_2[HL]}{[H^+]} \tag{10-7}$$

$$[L^-] = \frac{(1.80 \times 10^{-10})(1.32 \times 10^{-2})}{(1.32 \times 10^{-2})} = 1.80 \times 10^{-10}\ M\ (= K_2)$$

The approximation $[H^+] \approx [HL]$ reduces Equation 10-7 to $[L^-] = K_2$.

Our approximation that the diprotic acid can be treated as monoprotic is confirmed by this last result. The concentration of L^- is about eight orders of magnitude smaller than that of HL. The dissociation of HL is indeed negligible relative to the dissociation of H_2L^+. For most diprotic acids, K_1 is sufficiently larger than K_2 for this approximation to be valid. Even if K_2 were just 10 times less than K_1, $[H^+]$ calculated by ignoring the second ionization would be in error by only 4%. The error in pH would be only 0.01 pH unit. In summary, *a solution of a diprotic acid behaves like a solution of a monoprotic acid, with $K_a = K_1$.*

Dissolved carbon dioxide is one of the most important diprotic acids in Earth's ecosystem. Box 10-1 describes imminent danger to the entire ocean food chain as a result of increasing atmospheric CO_2 dissolving in the oceans. Reaction A in Box 10-1 lowers the concentration of CO_3^{2-} in the oceans. As a result, $CaCO_3$ shells and skeletons of creatures at the bottom of the food chain will dissolve by Reaction B.

BOX 10-1   Carbon Dioxide in the Ocean

The opening of this chapter shows that atmospheric CO_2 has oscillated between about 180 and 280 ppm by volume ($\mu L/L$) for 800 000 years. Burning of fossil fuel and destruction of Earth's forests since 1800 caused an exponential increase in CO_2 that is altering Earth's climate in your lifetime.

Increasing atmospheric CO_2 increases the concentration of CO_2 in the ocean, which consumes dissolved carbonate and lowers the pH:[4]

$$CO_2(aq) + H_2O + CO_3^{2-} \rightleftharpoons 2HCO_3^- \tag{A}$$

$$\underset{\text{Carbonate}}{\phantom{CO_3^{2-}}} \qquad \underset{\text{Bicarbonate}}{}$$

The pH of the ocean in panel a has already decreased from its preindustrial value of 8.16 to 8.04 today.[5] Without changes in our activities, the pH could be 7.7 by 2100.[6]

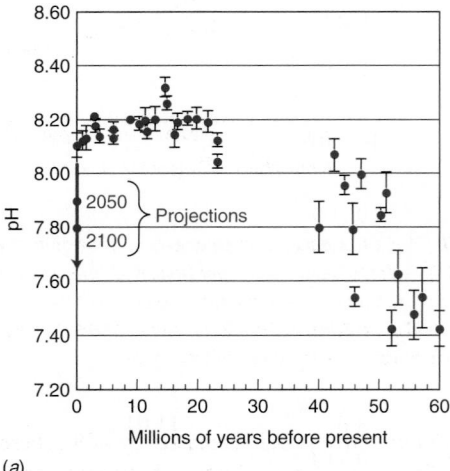

(a)

Equatorial Pacific Ocean surface pH deduced from $^{11}B/^{10}B$ ratio in fossil shells.
[Data from P. N. Pearson and M. R. Palmer, "Atmospheric Carbon Dioxide Concentrations Over the Past 60 Million Years," *Nature* **2000**, *406*, 695.]

Low carbonate concentration promotes dissolution of solid calcium carbonate:

$$\underset{\text{Calcium carbonate}}{CaCO_3(s)} \rightleftharpoons Ca^{2+} + CO_3^{2-} \tag{B}$$

Le Châtelier's principle tells us that decreasing $[CO_3^{2-}]$ draws the reaction to the right

If $[CO_3^{2-}]$ in the ocean decreases enough, organisms such as plankton and coral with $CaCO_3$ shells or skeletons will not survive.[7] Calcium carbonate has two crystalline forms called calcite and aragonite. Aragonite is more soluble than calcite. Aquatic organisms have either calcite or aragonite in their shells and skeletons.

Pteropods are a type of zooplankton that are also known as winged snails (panel b). When pteropods collected from the subarctic Pacific Ocean are kept in water that is less than saturated with aragonite, their shells begin to dissolve within 48 h. Animals such as the pteropod lie at the base of the food chain. Their destruction would reverberate through the entire ocean.

Today, ocean surface waters contain more than enough CO_3^{2-} to sustain aragonite and calcite. As atmospheric CO_2 inexorably increases during the twenty-first century, ocean surface waters will become undersaturated with respect to aragonite—killing off organisms that depend on this mineral for their structure. Polar regions will suffer this fate first because CO_2 is more soluble in cold water than in warm water and K_{a1} and K_{a2} at low temperature favor HCO_3^- and $CO_2(aq)$ relative to CO_3^{2-} (Problem 10-11).

Panel c shows the predicted concentration of CO_3^{2-} in polar ocean surface water as a function of atmospheric CO_2.[8] The upper horizontal line is the concentration of CO_3^{2-} below which aragonite dissolves. Atmospheric CO_2 is presently near 400 ppm and $[CO_3^{2-}]$ is near 100 $\mu mol/kg$ of seawater—more than enough to precipitate aragonite or calcite. When atmospheric CO_2 reaches 600 ppm, which is likely in the present century, $[CO_3^{2-}]$ will decrease to

CHAPTER 10 Polyprotic Acid-Base Equilibria

The Basic Form, L⁻

The species L^-, found in a salt such as sodium leucinate, can be prepared by treating leucine (HL) with an equimolar quantity of NaOH. Dissolving sodium leucinate in water gives a solution of L^-, the fully basic species. K_b values for this dibasic anion are

$$L^- + H_2O \rightleftharpoons HL + OH^- \qquad K_{b1} = K_w/K_{a2} = 5.55 \times 10^{-5}$$

$$HL + H_2O \rightleftharpoons H_2L^+ + OH^- \qquad K_{b2} = K_w/K_{a1} = 2.13 \times 10^{-12}$$

K_{b1} tells us that L^- will not *hydrolyze* very much to give HL. K_{b2} tells us that the resulting HL is such a weak base that hardly any further reaction to make H_2L^+ will occur.

We therefore treat L^- as monobasic, with $K_b = K_{b1}$. The results of this (fantastic) approximation are outlined as follows:

> **Hydrolysis** is the reaction of anything with water. Specifically, the reaction $L^- + H_2O \rightleftharpoons HL + OH^-$ is called hydrolysis.

$$\underset{\substack{L^- \\ 0.050\,0 - x}}{H_2NCHCO_2^-} + H_2O \underset{K_b = K_{b1}}{\rightleftharpoons} \underset{\substack{HL \\ x}}{\overset{+}{H_3}NCHCO_2^-} + \underset{\substack{OH^- \\ x}}{OH^-}$$

> L^- can be treated as monobasic, with $K_b = K_{b1}$.

$$K_b = K_{b1} = \frac{K_w}{K_{a2}} = 5.55 \times 10^{-5}$$

$$\frac{x^2}{F - x} = 5.55 \times 10^{-5} \Rightarrow x = 1.64 \times 10^{-3}\ M$$

$$[HL] = x = 1.64 \times 10^{-3}\ M$$

$$[H^+] = K_w/[OH^-] = K_w/x = 6.11 \times 10^{-12}\ M \Rightarrow pH = -\log[H^+] = 11.21$$

$$[L^-] = F - x = 4.84 \times 10^{-2}\ M$$

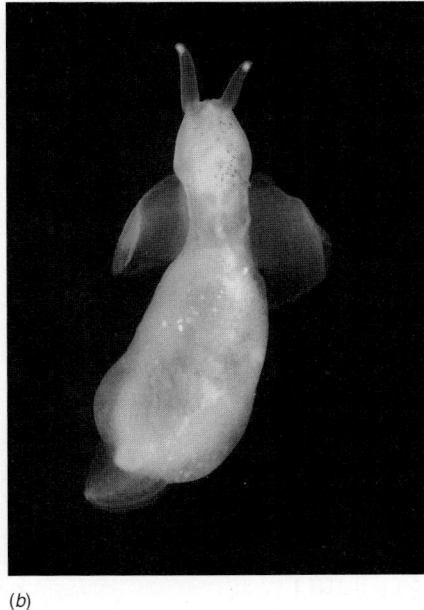

(b)

Pteropod. The shell of a live pteropod begins to dissolve after 48 h in water that is undersaturated with aragonite. [© David Shale/Nature Picture Library.]

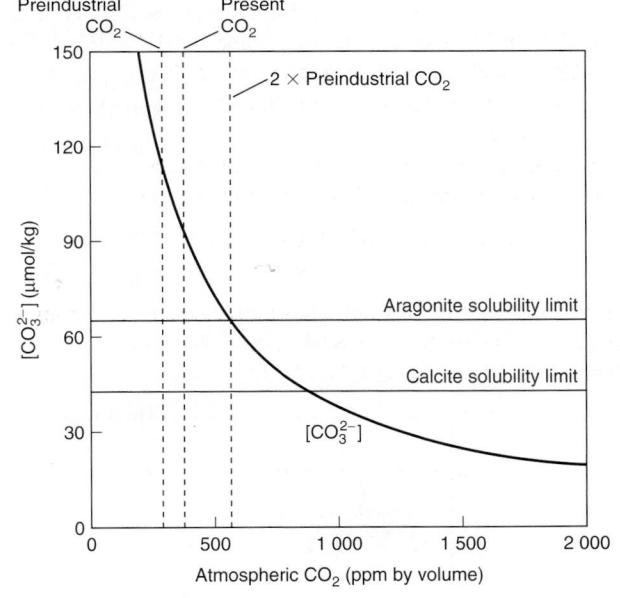

(c)

Calculated $[CO_3^{2-}]$ in polar ocean surface waters as a function of atmospheric CO_2. When $[CO_3^{2-}]$ drops below the upper horizontal line, argonite dissolves. [Information from J. C. Orr et al., "Anthropogenic Ocean Acidification over the Twenty-first Century and Its Impact on Calcifying Organisms," *Nature* **2005**, *437*, 681.] Reference 4 gives equations to calculate the curve in this figure.

60 μmol/kg and creatures with aragonite structures will begin to disappear from polar waters. At still higher atmospheric CO_2 concentration, extinctions will move to lower latitudes and will overtake organisms with calcite structures as well as aragonite structures.

Nature counteracts some changes. For example, phytoplankton called coccolithophores are marine organisms with a $CaCO_3$ skeleton several micrometers in diameter. These organisms produce about one-third of all $CaCO_3$ in the ocean. In the past 220 years, as atmospheric CO_2 increased, the average mass of the coccolithophore species *Emiliania huxleyi* has increased by 40%, thereby removing some CO_2 from the ocean.[9] Coccolithophores can mitigate CO_2 increases up to a point. It is not likely that any *calcifying* ($CaCO_3$-producing) marine organism can survive if CO_2 increases to the level where $CaCO_3$ is no longer thermodynamically stable.

The concentration of H_2L^+ can be found from the K_{b2} (or K_{a1}) equilibrium.

$$K_{b2} = \frac{[H_2L^+][OH^-]}{[HL]} = \frac{[H_2L^+]x}{x} = [H_2L^+]$$

We find that $[H_2L^+] = K_{b2} = 2.13 \times 10^{-12}$ M, and the approximation that $[H_2L^+]$ is insignificant relative to [HL] is well justified. In summary, if there is any reasonable separation between K_{a1} and K_{a2} (and, therefore, between K_{b1} and K_{b2}), *the fully basic form of a diprotic acid can be treated as monobasic, with $K_b = K_{b1}$.*

A tougher problem.

The Intermediate Form, HL

A solution prepared from leucine, HL, is more complicated than one prepared from either H_2L^+ or L^-, because HL is both an acid and a base.

HL is both an acid and a base.

$$HL \rightleftharpoons H^+ + L^- \qquad K_a = K_{a2} = 1.80 \times 10^{-10} \qquad \text{(10-8)}$$

$$HL + H_2O \rightleftharpoons H_2L^+ + OH^- \qquad K_b = K_{b2} = 2.13 \times 10^{-12} \qquad \text{(10-9)}$$

A molecule that can both donate and accept a proton is said to be **amphiprotic.** The acid dissociation reaction (10-8) has a larger equilibrium constant than the base hydrolysis reaction (10-9), so we expect that a solution of leucine will be acidic.

However, we cannot simply ignore Reaction 10-9, even if K_a and K_b differ by several orders of magnitude. Both reactions proceed to nearly equal extent, because H^+ produced in Reaction 10-8 reacts with OH^- from Reaction 10-9, thereby driving Reaction 10-9 to the right.

To treat this case, we resort to the systematic treatment of equilibrium. We derive results for leucine, whose intermediate form (HL) has no net charge. However, conclusions apply to the intermediate form of *any* diprotic acid, regardless of its charge.

For Reactions 10-8 and 10-9, the charge balance is

$$[H^+] + [H_2L^+] = [L^-] + [OH^-] \quad \text{or} \quad [H_2L^+] - [L^-] + [H^+] - [OH^-] = 0$$

From the acid dissociation equilibria, we replace $[H_2L^+]$ with $[HL][H^+]/K_1$, and $[L^-]$ with $[HL]K_2/[H^+]$. Also, we can always write $[OH^-] = K_w/[H^+]$. Putting these expressions into the charge balance gives

$$\frac{[HL][H^+]}{K_1} - \frac{[HL]K_2}{[H^+]} + [H^+] - \frac{K_w}{[H^+]} = 0$$

which can be solved for $[H^+]$. First, multiply all terms by $[H^+]$:

$$\frac{[HL][H^+]^2}{K_1} - [HL]K_2 + [H^+]^2 - K_w = 0$$

Then rearrange and factor out $[H^+]^2$:

$$[H^+]^2\left(\frac{[HL]}{K_1} + 1\right) = K_2[HL] + K_w$$

$$[H^+]^2 = \frac{K_2[HL] + K_w}{\dfrac{[HL]}{K_1} + 1}$$

Now multiply the numerator and denominator by K_1 and take the square root:

$$[H^+] = \sqrt{\frac{K_1K_2[HL] + K_1K_w}{K_1 + [HL]}} \qquad \text{(10-10)}$$

Up to this point, we have made no approximations, except to neglect activity coefficients. We solved for $[H^+]$ in terms of known constants plus the single unknown, [HL]. Where do we proceed from here?

The missing insight!

Just as we are feeling desperate, the Good Chemist again gallops down from the mountains to provide the missing insight: "The major species is HL," she cries, "because it is both a weak acid and a weak base. Neither Reaction 10-8 nor Reaction 10-9 goes very far. For [HL] in Equation 10-10, you can simply substitute the formal concentration, 0.050 0 M."

Taking the Good Chemist's advice, we write Equation 10-10 in its most useful form.

Intermediate form of diprotic acid:
$$[H^+] \approx \sqrt{\frac{K_1 K_2 F + K_1 K_w}{K_1 + F}} \qquad (10\text{-}11)$$

K_1 and K_2 in this equation are both *acid* dissociation constants (K_{a1} and K_{a2}).

where F is the formal concentration of HL (= 0.050 0 M in the present case).

At long last, we can calculate the pH of 0.050 0 M leucine:

$$[H^+] = \sqrt{\frac{(4.70 \times 10^{-3})(1.80 \times 10^{-10})(0.050\ 0) + (4.70 \times 10^{-3})(1.0 \times 10^{-14})}{4.70 \times 10^{-3} + 0.050\ 0}}$$

$$= 8.80 \times 10^{-7}\ M \Rightarrow pH = -\log[H^+] = 6.06$$

The concentrations of H_2L^+ and L^- can be found from the K_1 and K_2 equilibria, using $[H^+] = 8.80 \times 10^{-7}$ M and $[HL] = 0.050\ 0$ M.

$$[H_2L^+] = \frac{[H^+][HL]}{K_1} = \frac{(8.80 \times 10^{-7})(0.050\ 0)}{4.70 \times 10^{-3}} = 9.36 \times 10^{-6}\ M$$

$$[L^-] = \frac{K_2[HL]}{[H^+]} = \frac{(1.80 \times 10^{-10})(0.050\ 0)}{8.80 \times 10^{-7}} = 1.02 \times 10^{-5}\ M$$

Was the approximation $[HL] \approx 0.050\ 0$ M a good one? It certainly was, because $[H_2L^+]$ (= 9.36×10^{-6} M) and $[L^-]$ (= 1.02×10^{-5} M) are small in comparison with $[HL]$ ($\approx 0.050\ 0$ M). Nearly all leucine remains in the form HL. Note also that $[H_2L^+]$ is nearly equal to $[L^-]$. This result confirms that Reactions 10-8 and 10-9 proceed equally, even though K_a is 84 times bigger than K_b for leucine.

If $[H_2L^+] + [L^-]$ is not much less than $[HL]$ and if you wish to refine your values of $[H_2L^+]$ and $[L^-]$, try the method in Box 10-2.

BOX 10-2   **Successive Approximations**

The method of *successive approximations* is a good way to deal with difficult equations that do not have simple solutions. For example, Equation 10-11 is not a good approximation when the concentration of the intermediate species of a diprotic acid is not close to F, the formal concentration. This situation arises when K_1 and K_2 are nearly equal and F is small. Consider a solution of 1.00×10^{-3} M HM^-, the intermediate form of malic acid.

Malic acid
H_2M HM^-

$K_1 = 3.5 \times 10^{-4}$
$pK_1 = 3.46$

$K_2 = 7.9 \times 10^{-6}$
$pK_2 = 5.10$

M^{2-}

For a first approximation, assume that $[HM^-] \approx 1.00 \times 10^{-3}$ M. Plugging this value into Equation 10-10, we calculate first approximations for $[H^+]$, $[H_2M]$, and $[M^{2-}]$.

$$[H^+]_1 = \sqrt{\frac{K_1 K_2(0.001\ 00) + K_1 K_w}{K_1 + (0.001\ 00)}} = 4.53 \times 10^{-5}\ M$$

$$\Rightarrow [H_2M]_1 = \frac{[H^+][HM^-]}{K_1} = \frac{(4.53 \times 10^{-5})(1.00 \times 10^{-3})}{3.5 \times 10^{-4}}$$

$$= 1.29 \times 10^{-4}\ M$$

$$[M^{2-}]_1 = \frac{K_2[HM^-]}{[H^+]} = \frac{(7.9 \times 10^{-6})(1.00 \times 10^{-3})}{4.53 \times 10^{-5}}$$

$$= 1.75 \times 10^{-4}\ M$$

Clearly, $[H_2M]$ and $[M^{2-}]$ are not negligible relative to F = 1.00×10^{-3} M, so we need to revise our estimate of $[HM^-]$. The mass balance gives us a second approximation:

$$[HM^-]_2 = F - [H_2M]_1 - [M^{2-}]_1$$
$$= 0.001\ 00 - 0.000\ 129 - 0.000\ 175 = 0.000\ 696\ M$$

Inserting $[HM^-]_2 = 0.000\ 696$ into Equation 10-10 gives

$$[H^+]_2 = \sqrt{\frac{K_1 K_2(0.000\ 696) + K_1 K_w}{K_1 + (0.000\ 696)}} = 4.29 \times 10^{-5}\ M$$

$$\Rightarrow [H_2M]_2 = 8.53 \times 10^{-5}\ M$$
$$[M^{2-}]_2 = 1.28 \times 10^{-4}\ M$$

$[H_2M]_2$ and $[M^{2-}]_2$ can be used to calculate a third approximation for $[HM^-]$:

$$[HM^-]_3 = F - [H_2M]_2 - [M^{2-}]_2 = 0.000\ 786\ M$$

Plugging $[HM^-]_3$ into Equation 10-10 gives

$$[H^+]_3 = 4.37 \times 10^{-5}\ M$$

and the procedure can be repeated to get

$$[H^+]_4 = 4.35 \times 10^{-5}\ M$$

We are homing in on an estimate of $[H^+]$ in which the precision is already better than 1%. The fourth approximation gives pH = 4.36, compared with pH = 4.34 from the first approximation and pH = 4.28 from the formula pH $\approx \frac{1}{2}(pK_1 + pK_2)$. Considering the uncertainty in pH measurements, all this calculation was hardly worth the effort. However, the concentration $[HM^-]_5$ is 0.000 768 M, which is 23% less than the original estimate of 0.001 00 M. Successive approximations can be carried out by hand, but the process is more reliably performed with a spreadsheet. Problem 10-9 shows how to execute all of the iterations automatically in one step in Excel.

A summary of results for leucine is given here. Notice the relative concentrations of H_2L^+, HL, and L^- in each solution and notice the pH of each solution.

Solution	pH	$[H^+]$ (M)	$[H_2L^+]$ (M)	$[HL]$ (M)	$[L^-]$ (M)
0.050 0 M H_2A	1.88	1.32×10^{-2}	3.68×10^{-2}	1.32×10^{-2}	1.80×10^{-10}
0.050 0 M HA^-	6.06	8.80×10^{-7}	9.36×10^{-6}	5.00×10^{-2}	1.02×10^{-5}
0.050 0 M HA^{2-}	11.21	6.11×10^{-12}	2.13×10^{-12}	1.64×10^{-3}	4.84×10^{-2}

Simplified Calculation for the Intermediate Form

Usually Equation 10-11 is a fair-to-excellent approximation. An even simpler form results from two conditions that usually exist. First, if $K_2F \gg K_w$, the second term in the numerator of Equation 10-11 can be dropped.

$$[H^+] \approx \sqrt{\frac{K_1K_2F + K_1K_w}{K_1 + F}}$$

Then, if $K_1 \ll F$, the first term in the denominator also can be neglected.

$$[H^+] \approx \sqrt{\frac{K_1K_2F}{K_1 + F}}$$

Canceling F in the numerator and denominator gives

$$[H^+] \approx \sqrt{K_1K_2}$$

or

$$\log[H^+] \approx \frac{1}{2}(\log K_1 + \log K_2)$$

$$-\log[H^+] \approx -\frac{1}{2}(\log K_1 + \log K_2)$$

Recall that
$\log(x^{1/2}) = \frac{1}{2}\log x$
$\log(xy) = \log x + \log y$
$\log(x/y) = \log x - \log y$

Intermediate form of diprotic acid:

$$\boxed{pH \approx \frac{1}{2}(pK_1 + pK_2)} \qquad (10\text{-}12)$$

The pH of the intermediate form of a diprotic acid is close to midway between the two pK_a values and is almost independent of concentration.

Equation 10-12 is really good to keep in your head. It gives a pH of 6.04 for leucine, compared with pH = 6.06 from Equation 10-11. Equation 10-12 says that *the pH of the intermediate form of a diprotic acid is close to midway between pK_1 and pK_2, regardless of the formal concentration.*

EXAMPLE pH of the Intermediate Form of a Diprotic Acid

Potassium hydrogen phthalate, KHP, is a salt of the intermediate form of phthalic acid. Calculate the pH of 0.10 M and 0.010 M KHP.

Phthalic acid
H_2P

$pK_1 = 2.950$

Monohydrogen phthalte
HP^-

$pK_2 = 5.408$

Phthalate
P^{2-}

(Potassium hydrogen phthalate = K^+HP^-)

Solution With Equation 10-12, the pH of potassium hydrogen phthalate is estimated as $\frac{1}{2}(pK_1 + pK_2) = 4.18$, regardless of concentration. With Equation 10-11, we calculate pH = 4.18 for 0.10 M K^+HP^- and pH = 4.20 for 0.010 M K^+HP^-.

TEST YOURSELF Find the pH of 0.002 M K^+HP^- with Equation 10-11. (*Answer:* 4.28)

Advice When faced with the intermediate form of a diprotic acid, use Equation 10-11 to calculate pH. The answer should be close to $\frac{1}{2}(pK_1 + pK_2)$.

Summary of Diprotic Acid Calculations

Here is how we calculate the pH and composition of solutions prepared from different forms of a diprotic acid (H_2A, HA^-, or A^{2-}).

Solution of H_2A

1. Treat H_2A as a monoprotic acid with $K_a = K_1$ to find $[H^+]$, $[HA^-]$, and $[H_2A]$.

$$\underset{F-x}{H_2A} \xrightleftharpoons{K_1} \underset{x}{H^+} + \underset{x}{HA^-} \qquad \frac{x^2}{F-x} = K_1$$

2. Use the K_2 equilibrium to solve for $[A^{2-}]$.

$$[A^{2-}] = \frac{K_2[\cancel{HA^-}]}{\cancel{[H^+]}} = K_2$$

Solution of HA^-

1. Use the approximation $[HA^-] \approx F$ and find the pH with Equation 10-11.

$$[H^+] = \sqrt{\frac{K_1 K_2 F + K_1 K_w}{K_1 + F}}$$

The pH should be close to $\frac{1}{2}(pK_1 + pK_2)$.

2. With $[H^+]$ from step 1 and $[HA^-] \approx F$, solve for $[H_2A]$ and $[A^{2-}]$, using the K_1 and K_2 equilibria.

$$[H_2A] = \frac{[HA^-][H^+]}{K_1} \qquad [A^{2-}] = \frac{K_2[HA^-]}{[H^+]}$$

Solution of A^{2-}

1. Treat A^{2-} as monobasic, with $K_b = K_{b1} = K_w/K_{a2}$ to find $[A^{2-}]$, $[HA^-]$, and $[H^+]$.

$$\underset{F-x}{A^{2-}} + H_2O \xrightleftharpoons{K_{b1}} \underset{x}{HA^-} + \underset{x}{OH^-} \qquad \frac{x^2}{F-x} = K_{b1} = \frac{K_w}{K_{a2}}$$

$$[H^+] = \frac{K_w}{[OH^-]} = \frac{K_w}{x}$$

2. Use the K_1 equilibrium to solve for $[H_2A]$.

$$[H_2A] = \frac{[HA^-][H^+]}{K_{a1}} = \frac{[\cancel{HA^-}](K_w/[\cancel{OH^-}])}{K_{a1}} = K_{b2}$$

The calculations we have been doing are really important to understand and use. However, we should not get too cocky about our great powers, because there could be equilibria we have not considered. For example, Na^+ or K^+ in solutions of HA^- or A^{2-} form weak ion pairs that we have neglected:[10]

$$K^+ + A^{2-} \rightleftharpoons \{K^+A^{2-}\}$$
$$K^+ + HA^- \rightleftharpoons \{K^+HA^-\}$$

10-2 Diprotic Buffers

A buffer made from a diprotic (or polyprotic) acid is treated in the same way as a buffer made from a monoprotic acid. For the acid H_2A, we can write *two* Henderson-Hasselbalch equations, both of which are *always* true. If we happen to know $[H_2A]$ and $[HA^-]$, then we will use the pK_1 equation. If we know $[HA^-]$ and $[A^{2-}]$, we will use the pK_2 equation.

$$pH = pK_1 + \log \frac{[HA^-]}{[H_2A]} \qquad pH = pK_2 + \log \frac{[A^{2-}]}{[HA^-]}$$

All Henderson-Hasselbalch equations (with activity coefficients) are always true for a solution at equilibrium.

EXAMPLE A Diprotic Buffer System

Find the pH of a solution prepared by dissolving 1.00 g of potassium hydrogen phthalate and 1.20 g of disodium phthalate in 50.0 mL of water.

Solution Monohydrogen phthalate and phthalate were shown in the preceding example. The formula masses are $KHP = C_8H_5O_4K = 204.221$ and $Na_2P = C_8H_4O_4Na_2 = 210.094$. We know $[HP^-]$ and $[P^{2-}]$, so we use the pK_2 Henderson-Hasselbalch equation to find the pH:

$$pH = pK_2 + \log \frac{[P^{2-}]}{[HP^-]} = 5.408 + \log \frac{(1.20 \text{ g})/(210.094 \text{ g/mol})}{(1.00 \text{ g})/(204.221 \text{ g/mol})} = 5.47$$

K_2 is the acid dissociation constant of HP^-, which appears in the denominator of the log term. Notice that the volume of solution was not used to answer the question.

TEST YOURSELF Find the pH with 1.50 g Na_2P instead of 1.20 g. (*Answer:* 5.57)

EXAMPLE **Preparing a Buffer in a Diprotic System**

How many milliliters of 0.800 M KOH should be added to 3.38 g of oxalic acid to give a pH of 4.40 when diluted to 500 mL?

$$\underset{\substack{\text{Oxalic acid} \\ (\text{H}_2\text{Ox}) \\ \text{Formula mass} = 90.035}}{\text{HOCCOH}} \quad \substack{pK_1 = 1.250 \\ pK_2 = 4.266}$$

Solution The desired pH is above pK_2. We know that a 1:1 mole ratio of $\text{HOx}^-:\text{Ox}^{2-}$ would have pH = pK_2 = 4.266. If the pH is to be 4.40, there must be more Ox^{2-} than HOx^- present. We must add enough base to convert all H_2Ox into HOx^-, plus enough additional base to convert the right amount of HOx^- into Ox^{2-}.

$$\text{H}_2\text{Ox} + \text{OH}^- \rightarrow \text{HOx}^- + \text{H}_2\text{O}$$
$$\uparrow$$
$$\text{pH} \approx \tfrac{1}{2}(pK_1 + pK_2) = 2.76$$

$$\text{HOx}^- + \text{OH}^- \rightarrow \text{Ox}^{2-} + \text{H}_2\text{O}$$

A 1:1 mixture would have
pH = pK_2 = 4.266

In 3.38 g of H_2Ox, there are $0.037\,5_4$ mol. The volume of 0.800 M KOH needed to react with this much H_2Ox to make HOx^- is $(0.037\,5_4\text{ mol})/(0.800\text{ M}) = 46.9_3$ mL.

To produce a pH of 4.40 requires an additional x mol of OH^-:

	HOx^-	+	OH^-	$\rightarrow$	Ox^{2-}
Initial moles	$0.037\,5_4$		x		—
Final moles	$0.037\,5_4 - x$		—		x

$$\text{pH} = pK_2 + \log \frac{[\text{Ox}^{2-}]}{[\text{HOx}^-]}$$

$$4.40 = 4.266 + \log \frac{x}{0.037\,5_4 - x} \Rightarrow x = 0.021\,6_6 \text{ mol}$$

The volume of KOH needed to deliver $0.021\,6_6$ mole is $(0.021\,6_4\text{ mol})/(0.800\text{ M}) = 27.0_5$ mL. The total volume of KOH needed to bring the pH to 4.40 is $46.9_3 + 27.0_5 = 73.9_8$ mL.

TEST YOURSELF What volume of KOH would bring the pH to 4.50? (*Answer:* 76.5_6 mL)

10-3 Polyprotic Acids and Bases

The treatment of diprotic acids and bases can be extended to polyprotic systems. By way of review, let's write the pertinent equilibria for a triprotic system.

$$\text{H}_3\text{A} \rightleftharpoons \text{H}_2\text{A}^- + \text{H}^+ \qquad K_{a1} = K_1$$
$$\text{H}_2\text{A}^- \rightleftharpoons \text{HA}^{2-} + \text{H}^+ \qquad K_{a2} = K_2$$
$$\text{HA}^{2-} \rightleftharpoons \text{A}^{3-} + \text{H}^+ \qquad K_{a3} = K_3$$
$$\text{A}^{3-} + \text{H}_2\text{O} \rightleftharpoons \text{HA}^{2-} + \text{OH}^- \qquad K_{b1} = K_w/K_{a3}$$
$$\text{HA}^{2-} + \text{H}_2\text{O} \rightleftharpoons \text{H}_2\text{A}^- + \text{OH}^- \qquad K_{b2} = K_w/K_{a2}$$
$$\text{H}_2\text{A}^- + \text{H}_2\text{O} \rightleftharpoons \text{H}_3\text{A} + \text{OH}^- \qquad K_{b3} = K_w/K_{a1}$$

We deal with triprotic systems as follows:

1. H_3A is treated as a monoprotic weak acid, with $K_a = K_1$.
2. H_2A^- is treated as the intermediate form of a diprotic acid.

The K values in Equations 10-13 and 10-14 are K_a values for the triprotic acid.

$$[\text{H}^+] \approx \sqrt{\frac{K_1 K_2 F + K_1 K_w}{K_1 + F}} \qquad (10\text{-}13)$$

3. HA^{2-} is also treated as the intermediate form of a diprotic acid. However, HA^{2-} is "surrounded" by H_2A^- and A^{3-}, so the equilibrium constants to use are K_2 and K_3, instead of K_1 and K_2.

$$[H^+] \approx \sqrt{\frac{K_2K_3F + K_2K_w}{K_2 + F}} \qquad \text{(10-14)}$$

4. A^{3-} is treated as monobasic, with $K_b = K_{b1} = K_w/K_{a3}$.

EXAMPLE **A Triprotic System**

Find the pH of 0.10 M H_3His^{2+}, 0.10 M H_2His^+, 0.10 M HHis, and 0.10 M His^-, where His stands for the amino acid histidine.

Solution *0.10 M H_3His^{2+}:* Treating H_3His^{2+} as a monoprotic acid, we write

$$H_3His^{2+} \rightleftharpoons H_2His^+ + H^+$$
$$\quad F - x \qquad\qquad x \qquad\quad x$$

$$\frac{x^2}{F - x} = K_1 = 10^{-1.6} \Rightarrow x = 3._9 \times 10^{-2} \text{ M} \Rightarrow pH = 1.41$$

0.10 M H_2His^+: Using Equation 10-13, we find

$$[H^+] = \sqrt{\frac{(10^{-1.6})(10^{-5.97})(0.10) + (10^{-1.6})(1.0 \times 10^{-14})}{10^{-1.6} + 0.10}}$$

$$= 1.4_7 \times 10^{-4} \text{ M} \Rightarrow pH = 3.83$$

which is close to $\frac{1}{2}(pK_1 + pK_2) = 3.78$.

0.10 M HHis: Equation 10-14 gives

$$[H^+] = \sqrt{\frac{(10^{-5.97})(10^{-9.28})(0.10) + (10^{-5.97})(1.0 \times 10^{-14})}{10^{-5.97} + 0.10}}$$

$$= 2.3_7 \times 10^{-8} \text{ M} \Rightarrow pH = 7.62$$

which is the same as $\frac{1}{2}(pK_2 + pK_3) = 7.62$.

0.10 M His^-: Treating His^- as monobasic, we can write

$$His^- + H_2O \rightleftharpoons HHis + OH^-$$
$$\quad F - x \qquad\qquad x \qquad\quad x$$

$$\frac{x^2}{F - x} = K_{b1} = \frac{K_w}{K_{a3}} = 1.9 \times 10^{-5} \Rightarrow x = 1.3_7 \times 10^{-3} \text{ M}$$

$$pH = -\log\left(\frac{K_w}{x}\right) = 11.14$$

TEST YOURSELF Compute the pH of 0.010 M HHis. (*Answer:* 7.62)

We have reduced acid-base problems to just three types. When you encounter an acid or base, decide whether you are dealing with an *acidic*, *basic*, or *intermediate* form. Then do the appropriate arithmetic to answer the question at hand.

Three forms of acids and bases:
• acidic
• basic
• intermediate (amphiprotic)

10-4 Which Is the Principal Species?

We sometimes must identify which species of acid, base, or intermediate predominates under given conditions. For example, "What is the principal form of benzoic acid at pH 8?"

$$\langle\!\!\!\bigcirc\!\!\!\rangle\!\!-\!\!CO_2H \quad pK_a = 4.20$$

Benzoic acid

pK_a for benzoic acid is 4.20. So, at pH 4.20, there is a 1:1 mixture of benzoic acid (HA) and benzoate ion (A$^-$). At pH = pK_a + 1 (= 5.20), the quotient [A$^-$]/[HA] is 10:1. At pH = pK_a + 2 (= 6.20), the quotient [A$^-$]/[HA] is 100:1. As pH increases, the quotient [A$^-$]/[HA] increases still further.

For a monoprotic system, the basic species A$^-$ is the predominant form when pH > pK_a. The acidic species, HA, is the predominant form when pH < pK_a. The predominant form of benzoic acid at pH 8 is the benzoate anion, $C_6H_5CO_2^-$.

EXAMPLE Principal Species—Which One and How Much?

What is the predominant form of ammonia in a solution at pH 7.0? Approximately what fraction is in this form?

Solution In Appendix G, we find pK_a = 9.24 for the ammonium ion (NH$_4^+$, the conjugate acid of ammonia, NH$_3$). At pH = 9.24, [NH$_4^+$] = [NH$_3$]. Below pH 9.24, NH$_4^+$ will be the predominant form. Because pH = 7.0 is about 2 pH units below pK_a, the quotient [NH$_4^+$]/[NH$_3$] will be about 100:1. More than 99% is in the form NH$_4^+$.

TEST YOURSELF Approximately what fraction of ammonia is in the form NH$_3$ at pH 11? (**Answer:** somewhat less than 99% because pH is almost 2 units above pK_a)

For polyprotic systems, our reasoning is similar, but there are several values of pK_a. Consider oxalic acid, H$_2$Ox, with pK_1 = 1.25 and pK_2 = 4.27. At pH = pK_1, [H$_2$Ox] = [HOx$^-$]. At pH = pK_2, [HOx$^-$] = [Ox^{2-}]. The chart in the margin shows the major species in each pH region.

EXAMPLE Principal Species in a Polyprotic System

The amino acid arginine has the following forms:

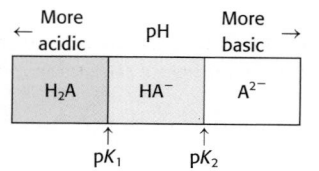

Appendix G tells us that the α-ammonium group (at the left) is more acidic than the substituent (at the right). What is the principal form of arginine at pH 10.0? Approximately what fraction is in this form? What is the second most abundant form at this pH?

An example of when you need to know principal species is when you design a chromatographic or electrophoretic separation. You would use different strategies for separating cations, anions, and neutral compounds.

$$pH = pK_a + \log\frac{[A^-]}{[HA]}$$

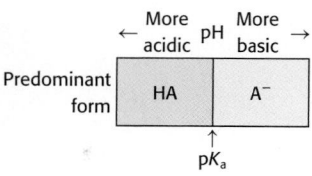

pH	Major species
< pK_a	HA
> pK_a	A$^-$

pH	Major species
pH < pK_1	H$_2$A
pK_1 < pH < pK_2	HA$^-$
pH > pK_2	A^{2-}

Solution We know that at pH = pK_2 = 8.99, $[H_2Arg^+]$ = [HArg]. At pH = pK_3 = 12.1, [HArg] = $[Arg^-]$. At pH = 10.0, the major species is HArg. Because pH 10.0 is about one pH unit higher than pK_2, we can say that $[HArg]/[H_2Arg^+]$ ≈ 10:1. About 90% of arginine is in the form HArg. The second most important species is H_2Arg^+, which makes up about 10% of the arginine.

TEST YOURSELF What is the predominant form of arginine at pH 11? What is the second major species? (*Answer:* HArg, Arg^-)

EXAMPLE **More on Polyprotic Systems**

In the pH range 1.82 to 8.99, H_2Arg^+ is the principal form of arginine. Which is the second most prominent species at pH 6.0? At pH 5.0?

Solution We know that the pH of the pure intermediate (amphiprotic) species, H_2Arg^+, is

$$\text{pH of } H_2Arg^+ \approx \frac{1}{2}(pK_1 + pK_2) = 5.40$$

Above pH 5.40 (and below pH = pK_2), HArg, the conjugate base of H_2Arg^+, will be the second most important species. Below pH 5.40 (and above pH = pK_1), H_3Arg^{2+} will be the second most important species.

TEST YOURSELF At what pH does $[H_2Arg^+]$ = $[Arg^-]$? (*Answer:* 10.54)

Figure 10-2 summarizes how we think of a triprotic system. We determine the principal species by comparing the pH of the solution with the pK_a values.

Go back and read the Example, "Preparing a Buffer in a Diprotic System," on page 220. See if it makes more sense now.

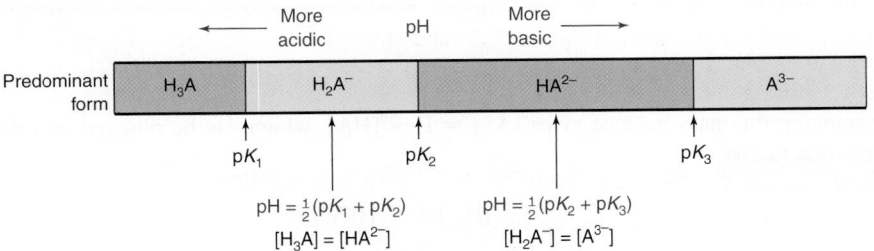

FIGURE 10-2 The predominant molecular form of a triprotic system (H_3A) in the various pH intervals.

Speciation describes the distribution of analyte among possible species. For an acid or base, speciation describes how much of each protonated form is present. Box 10-3 points out that partially protonated polyprotic acids and bases have multiple possible species with H^+ residing at different sites. When you ingest inorganic arsenic ($AsO(OH)_3$ and $As(OH)_3$) from drinking water, it is methylated to species such as $(CH_3)AsO(OH)_2$, $(CH_3)As(OH)_2$, $(CH_3)_2AsO(OH)$, $(CH_3)_2As(OH)$, $(CH_3)_3AsO$, and $(CH_3)_3As$. Speciation describes the forms and quantities that are present.

10-5 Fractional Composition Equations

We now derive equations that give the fraction of each species of acid or base at a given pH. These equations will be useful for acid-base and EDTA titrations, as well as for electrochemical equilibria. They are of key value in Chapter 13.

Monoprotic Systems

Our goal is to find an expression for the fraction of an acid in each form (HA and A^-) as a function of pH. We can do this by combining the equilibrium constant with the mass balance. Consider an acid with formal concentration F:

$$HA \overset{K_a}{\rightleftharpoons} H^+ + A^- \qquad K_a = \frac{[H^+][A^-]}{[HA]}$$

$$\text{Mass balance:} \quad F = [HA] + [A^-]$$

BOX 10-3   **Microequilibrium Constants**

Any polyprotic acid with distinguishable acidic sites has an equilibrium constant for acid dissociation *at each site*. Consider 9-methyladenine in the graph. Adenine is one of the building blocks of DNA and RNA (Appendix L). Adenine is attached to the DNA or RNA backbone through N9, which is bound to a methyl group in this example.

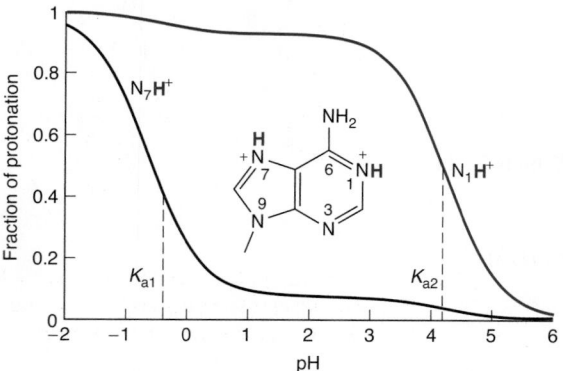

Fraction of protonation at N7 and N1 in 9-methyladenine based on microequilibrium constants. [Information from H. Sigel, "Acid-Base Properties of Purine Residues and the Effect of Metal Ions: Quantification of Rare Nucleobase Tautomers," *Pure Appl. Chem.* **2004,** *76,* 1869.]

Calling adenine A, the species drawn in the graph is H_2A^{2+}. As for any diprotic acid, 9-methyladenine has two sequential acid dissociation constants:

$$H_2A^{2+} \xrightleftharpoons{\;K_{a1}=10^{0.4}\;} HA^+ \xrightleftharpoons{\;K_{a2}=10^{-4.20}\;} A$$

N7 is more acidic (less basic) than N1, so we customarily associate K_{a1} with H^+ dissociation from N7 and K_{a2} with H^+ dissociation from N1. In fact, there is a *microequilibrium constant, k,* for loss of H^+ from each position (N7 and N1). "HA^+" is a mixture of two forms with H^+ residing at either N7 or N1:

$$
\begin{array}{ccc}
& N_7\!-\!N_1H^+ & \\
k_7=10^{0.61}\nearrow & & \searrow k_{71}=10^{-4.07} \\
{}^+HN_7\!-\!N_1H^+ & & N_7\!-\!N_1 \\
k_1=10^{-0.5}\searrow & & \nearrow k_{17}=10^{-2.96} \\
& {}^+HN_7\!-\!N_1 &
\end{array}
$$

The microequilibrium constant for loss of H^+ from N7 is k_7. k_1 is the microequilibrium constant for loss of H^+ from N1. The microequilibrium constant k_{71} is the acid dissociation constant for loss of H^+ from N1 after H^+ has already been lost from N7.

N7 is the more acidic site ($k_7 > k_1$), but there is an equilibrium with some H^+ at each site in HA^+. At pH = 1.9, which is midway between pK_{a1} and pK_{a2}, the graph shows that N7 is 93% deprotonated and N1 is 8% deprotonated. Box 11-2 discusses the meaning of negative pH values that appear in the graph. You can read more about microequilibrium constants in the supplementary topics for this book at www.whfreeman.com/qca9e.

Rearranging the mass balance gives $[A^-] = F - [HA]$, which can be plugged into the K_a expression to give

$$K_a = \frac{[H^+](F - [HA])}{[HA]}$$

or, with a little algebra,

$$[HA] = \frac{[H^+]F}{[H^+] + K_a} \tag{10-15}$$

α_{HA} = fraction of species in the form HA
α_{A^-} = fraction of species in the form A^-

$$\boxed{\alpha_{HA} + \alpha_{A^-} = 1}$$

The *fraction* of molecules in the form HA is called α_{HA}.

$$\alpha_{HA} = \frac{[HA]}{[HA] + [A^-]} = \frac{[HA]}{F} \tag{10-16}$$

Dividing Equation 10-15 by F gives

Fraction in the form HA: $\alpha_{HA} = \dfrac{[HA]}{F} = \dfrac{[H^+]}{[H^+] + K_a}$ (10-17)

The fraction denoted here as α_{A^-} is the same thing we called the **fraction of dissociation** (α) previously.

In a similar manner, the fraction in the form A^-, designated α_{A^-}, can be obtained:

Fraction in the form A^-: $\alpha_{A^-} = \dfrac{[A^-]}{F} = \dfrac{K_a}{[H^+] + K_a}$ (10-18)

Figure 10-3 shows α_{HA} and α_{A^-} for a system with $pK_a = 5.00$. At low pH, almost all of the acid is in the form HA. At high pH, almost everything is in the form A^-.

Diprotic Systems

The derivation of fractional composition equations for a diprotic system follows the same pattern used for the monoprotic system.

$$H_2A \overset{K_1}{\rightleftharpoons} H^+ + HA^-$$
$$HA^- \overset{K_2}{\rightleftharpoons} H^+ + A^{2-}$$

$$K_1 = \frac{[H^+][HA^-]}{[H_2A]} \Rightarrow [HA^-] = [H_2A]\frac{K_1}{[H^+]}$$

$$K_2 = \frac{[H^+][A^{2-}]}{[HA^-]} \Rightarrow [A^{2-}] = [HA^-]\frac{K_2}{[H^+]} = [H_2A]\frac{K_1K_2}{[H^+]^2}$$

Mass balance: $\quad F = [H_2A] + [HA^-] + [A^{2-}]$

$$F = [H_2A] + \frac{K_1}{[H^+]}[H_2A] + \frac{K_1K_2}{[H^+]^2}[H_2A]$$

$$F = [H_2A]\left(1 + \frac{K_1}{[H^+]} + \frac{K_1K_2}{[H^+]^2}\right) = [H_2A]\left(\frac{[H^+]^2 + [H^+]K_1 + K_1K_2}{[H^+]^2}\right)$$

For a diprotic system, we designate the fraction in the form H_2A as α_{H_2A}, the fraction in the form HA^- as α_{HA^-}, and the fraction in the form A^{2-} as $\alpha_{A^{2-}}$. From the definition of α_{H_2A}, we can write

Fraction in the form H_2A: $\quad \alpha_{H_2A} = \dfrac{[H_2A]}{F} = \dfrac{[H^+]^2}{[H^+]^2 + [H^+]K_1 + K_1K_2}$ $\qquad$ **(10-19)**

In a similar manner, we can derive the following equations:

Fraction in the form HA^-: $\quad \alpha_{HA^-} = \dfrac{[HA^-]}{F} = \dfrac{K_1[H^+]}{[H^+]^2 + [H^+]K_1 + K_1K_2}$ $\qquad$ **(10-20)**

Fraction in the form A^{2-}: $\quad \alpha_{A^{2-}} = \dfrac{[A^{2-}]}{F} = \dfrac{K_1K_2}{[H^+]^2 + [H^+]K_1 + K_1K_2}$ $\qquad$ **(10-21)**

Figure 10-4 shows the fractions α_{H_2A}, α_{HA^-}, and $\alpha_{A^{2-}}$ for fumaric acid, whose two pK_a values are only 1.46 units apart. α_{HA^-} rises only to 0.73 because the two pK values are so close together. There are substantial quantities of H_2A and A^{2-} in the region $pK_1 < pH < pK_2$.

Equations 10-19 through 10-21 apply equally well to B, BH^+, and BH_2^{2+} obtained from the base B. The fraction α_{H_2A} applies to the acidic form BH_2^{2+}. α_{HA^-} applies to BH^+ and $\alpha_{A^{2-}}$ applies to B. The constants K_1 and K_2 are the *acid* dissociation constants of BH_2^{2+} ($K_1 = K_w/K_{b2}$ and $K_2 = K_w/K_{b1}$).

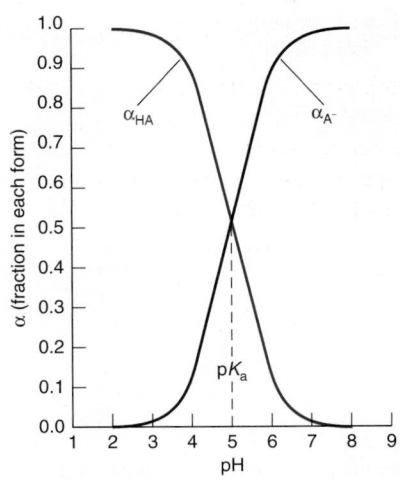

FIGURE 10-3 Fractional composition diagram of a monoprotic system with $pK_a = 5.00$. Below pH 5, HA is the dominant form, whereas, above pH 5, A^- dominates.

α_{H_2A} = fraction of species in the form H_2A
α_{HA^-} = fraction of species in the form HA^-
$\alpha_{A^{2-}}$ = fraction of species in the form A^{2-}

$$\alpha_{H_2A} + \alpha_{HA^-} + \alpha_{A^{2-}} = 1$$

The general form of α for the polyprotic acid H_nA is

$$\alpha_{H_nA} = \frac{[H^+]^n}{D}$$

$$\alpha_{H_{n-1}A} = \frac{K_1[H^+]^{n-1}}{D}$$

$$\alpha_{H_{n-j}A} = \frac{K_1K_2\cdots K_j[H^+]^{n-j}}{D}$$

where $D = [H^+]^n + K_1[H^+]^{n-1} + K_1K_2[H^+]^{n-2} + \cdots + K_1K_2K_3\cdots K_n$.

How to apply fractional composition equations to bases

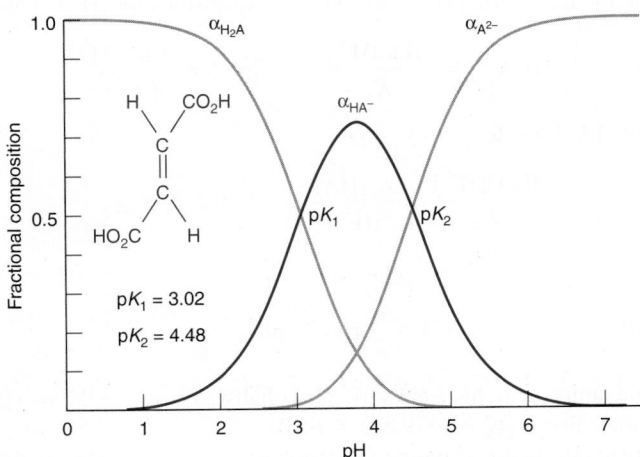

FIGURE 10-4 Fractional composition diagram for fumaric acid (*trans*-butenedioic acid). At low pH, H_2A is dominant. At intermediate pH, HA^- is dominant; at high pH, A^{2-} dominates. Because pK_1 and pK_2 are not separated very much, the fraction of HA^- never gets very close to unity.

10-6 Isoelectric and Isoionic pH

Biochemists speak of the isoelectric or isoionic pH of polyprotic molecules, such as proteins. These terms can be understood in terms of a diprotic system, such as the amino acid alanine.

$$\underset{\substack{\text{Alanine cation}\\ H_2A^+}}{\overset{\overset{\displaystyle CH_3}{\displaystyle |}}{H_3\overset{+}{N}CHCO_2H}} \rightleftharpoons \underset{\substack{\text{Neutral zwitterion}\\ HA}}{\overset{\overset{\displaystyle CH_3}{\displaystyle |}}{H_3\overset{+}{N}CHCO_2^-}} + H^+ \qquad pK_1 = 2.34$$

$$\underset{}{\overset{\overset{\displaystyle CH_3}{\displaystyle |}}{H_3\overset{+}{N}CHCO_2^-}} \rightleftharpoons \underset{\substack{\text{Alanine anion}\\ A^-}}{\overset{\overset{\displaystyle CH_3}{\displaystyle |}}{H_2NCHCO_2^-}} + H^+ \qquad pK_2 = 9.87$$

Isoionic pH is the pH of the pure, neutral, polyprotic acid.

Isoelectric pH is the pH at which average charge of the polyprotic acid is 0.

Alanine is the intermediate form of a diprotic acid, so we use Equation 10-11 to find the pH.

The **isoionic point** (or isoionic pH) is the pH obtained when the pure, neutral polyprotic acid HA (the neutral zwitterion) is dissolved in water. The only ions are H_2A^+, A^-, H^+, and OH^-. Most alanine is in the form HA, and the concentrations of H_2A^+ and A^- are *not* equal to each other.

The **isoelectric point** (or isoelectric pH) is the pH at which the *average* charge of the polyprotic acid is 0. Most of the molecules are in the uncharged form HA, and the concentrations of H_2A^+ and A^- *are* equal to each other. There is always some H_2A^+ and some A^- in equilibrium with HA.

When alanine is dissolved in water, the pH of the solution, by definition, is the *isoionic* pH. Because alanine (HA) is the intermediate form of the diprotic acid, H_2A^+, $[H^+]$ is given by

Isoionic point:

$$[H^+] = \sqrt{\frac{K_1K_2F + K_1K_w}{K_1 + F}} \qquad (10\text{-}22)$$

where F is the formal concentration of alanine. For 0.10 M alanine, the isoionic pH is found from

$$[H^+] = \sqrt{\frac{K_1K_2(0.10) + K_1K_w}{K_1 + (0.10)}} = 7.7 \times 10^{-7}\,M \Rightarrow pH = 6.11$$

From $[H^+]$, K_1, and K_2, you could calculate $[H_2A^+] = 1.68 \times 10^{-5}$ M and $[A^-] = 1.76 \times 10^{-5}$ M for pure alanine in water (the *isoionic* solution). There is a slight excess of A^- because HA is a slightly stronger acid than it is a base. It dissociates to make A^- a little more than it reacts with water to make H_2A^+.

The *isoelectric* point is the pH at which $[H_2A^+] = [A^-]$, and, therefore, the average charge of alanine is 0. To go from the *isoionic* solution (pure HA in water) to the isoelectric solution, we could add just enough strong acid to decrease $[A^-]$ and increase $[H_2A^+]$ until they are equal. Adding acid necessarily lowers the pH. For alanine, the isoelectric pH must be lower than the isoionic pH.

We calculate the isoelectric pH by first writing expressions for $[H_2A^+]$ and $[A^-]$:

$$[H_2A^+] = \frac{[HA][H^+]}{K_1} \qquad [A^-] = \frac{K_2[HA]}{[H^+]}$$

Setting $[H_2A^+] = [A^-]$, we find

$$\frac{[HA][H^+]}{K_1} = \frac{K_2[HA]}{[H^+]} \Rightarrow [H^+] = \sqrt{K_1K_2}$$

which gives

The isoelectric point is midway between the two pK_a values "surrounding" the neutral, intermediate species.

Isoelectric point:

$$pH = \frac{1}{2}(pK_1 + pK_2) \qquad (10\text{-}23)$$

For a diprotic amino acid, the isoelectric pH is halfway between the two pK_a values. The isoelectric pH of alanine is $\frac{1}{2}(2.34 + 9.87) = 6.10$.

The isoelectric and isoionic points for a polyprotic acid are almost the same. At the isoelectric pH, the average charge of the molecule is 0; thus $[H_2A^+] = [A^-]$ and pH = $\frac{1}{2}(pK_1 + pK_2)$. At the isoionic point, the pH is given by Equation 10-22, and $[H_2A^+]$ is not exactly equal to $[A^-]$.

Proteins Are Polyprotic Acids and Bases

Proteins perform biological functions such as structural support, catalysis of chemical reactions, immune response to foreign substances, transport of molecules across membranes, and control of genetic expression. The three-dimensional structure and function of a protein is determined by the sequence of *amino acids* from which the protein is made. The diagram below shows how amino acids are connected to make a *polypeptide*.

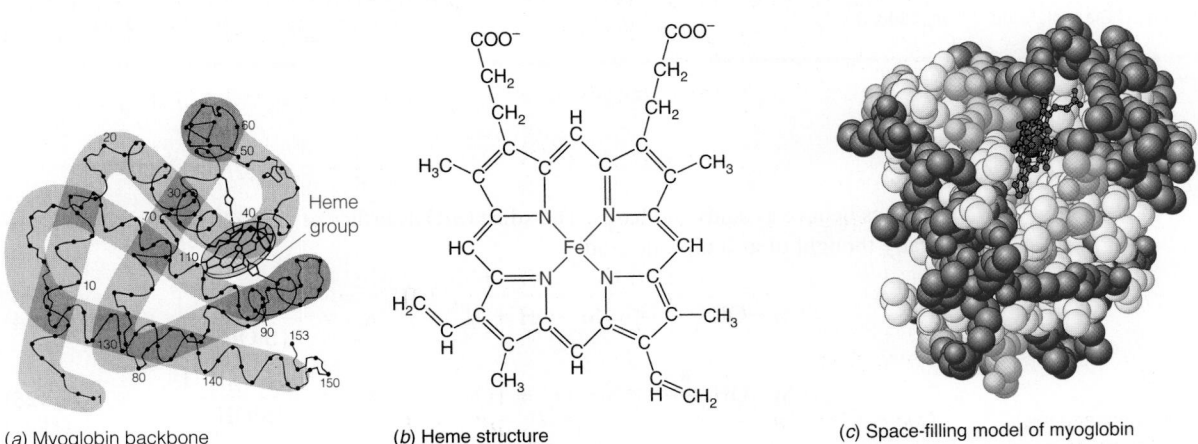

Of the 20 common amino acids in Table 10-1, three have basic substituents and four have acidic substituents.

The protein myoglobin in Figure 10-5 folds into several helical (spiral) regions that control access of oxygen and other small molecules to the heme group, whose function is to store O_2 in muscle cells. Of the 153 amino acids in sperm-whale myoglobin, 35 have basic substituents and 23 are acidic.

For a protein, the *isoionic* pH is the pH of a solution of pure protein with no other ions except H^+ and OH^-. Proteins are usually isolated in a charged form together with counterions such as Na^+, NH_4^+, or Cl^-. When the protein is subjected to intensive *dialysis* (see Demonstration 27-1) against pure water, the pH in the protein compartment approaches the isoionic point if the counterions are free to pass through the semipermeable dialysis membrane that retains the protein. The *isoelectric* point is the pH at which the protein has no net charge. Box 10-4 describes separation of proteins on the basis of their different isoelectric points.

Related properties—of interest to geology, environmental science, and ceramics—are the surface acidity of a solid[12] and the pH of zero charge.[13] Mineral, clay, and even organic

(a) Myoglobin backbone   (b) Heme structure   (c) Space-filling model of myoglobin

FIGURE 10-5 (*a*) Amino acid backbone of the protein myoglobin, which stores oxygen in muscle. Substituents (R groups from Table 10-1) are omitted for clarity. The flat *heme* group at the right side of the protein contains an iron atom that can bind O_2, CO, and other small molecules. [Information from M. F. Perutz, "The Hemoglobin Molecule." Copyright © 1964 by Scientific American, Inc.] (*b*) Structure of heme. (*c*) Space-filling model of myoglobin with *charged* acidic and basic amino acids in dark color. Lightly colored amino acids are *hydrophilic* (polar, water loving), but not charged. *Hydrophobic* (nonpolar, water repelling) amino acids in white. The surface of this water-soluble protein is dominated by charged and hydrophilic groups. [Information from J. M. Berg, J. L. Tymoczko, and L. Stryer, *Biochemistry*, 5th ed. (New York: Freeman, 2002).]

At its *isoelectric point*, the average charge of all forms of a protein is 0. It will therefore not migrate in an electric field at its isoelectric pH. This effect is the basis of a technique of protein separation called **isoelectric focusing.** A protein mixture is subjected to a strong electric field in a medium specifically designed to have a pH gradient. Positively charged molecules move toward the negative pole and negatively charged molecules move toward the positive pole. Each protein migrates until it reaches the point where the pH is the same as its isoelectric pH. At this point, the protein has no net charge and no longer moves. Each protein in the mixture is focused in one region at its isoelectric pH.[11]

Isoelectric focusing in a 6-mm-long × 100 μm-wide × 25-μm-deep capillary etched into silica glass is shown at the left below. Panel (i) shows fluorescent markers with known isoelectric pH (called pI) run as standards. Panels (ii) and (iii) show separations of fluorescence-labeled proteins. Proteins migrate until they reach their isoelectric pH and then stop migrating. If a molecule diffuses out of its isoelectric region, it becomes charged and immediately migrates back to its isoelectric zone. The graph shows measured pH versus distance in the capillary. Separations or reactions conducted in capillaries in glass or polymer chips are examples of *lab-on-a-chip* operations (Section 26-8).

The figure at the lower right shows separation of whole yeast cells at three different stages of growth (called early log, mid log, and stationary phase) by isoelectric focusing in a silica capillary tube. Acid-base properties (and, therefore, pI) of cell surfaces change during growth of the colony.

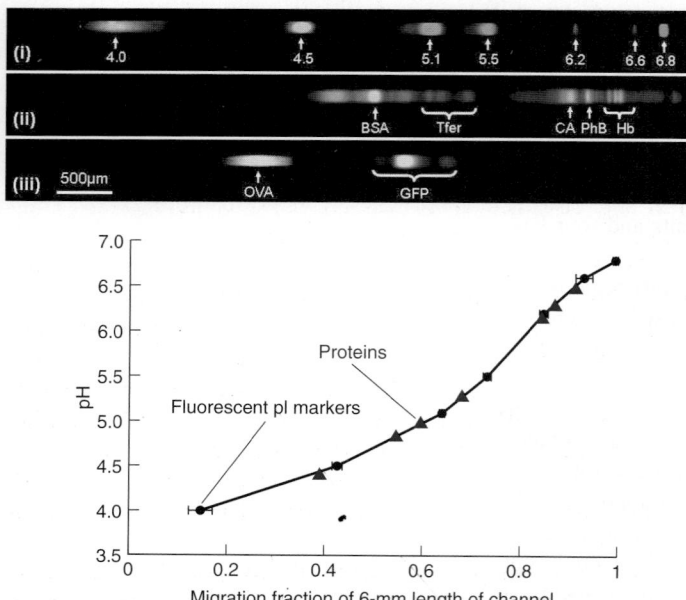

Lab-on-a-chip isoelectric focusing. (i) Fluorescent pI markers. (ii) and (iii) Separation of fluorescence-labeled proteins: (OVA) ovalbumin; (GFP) green fluorescent protein; (BSA) bovine serum albumin; (Tfer) transferrin; (CA) carbonic anhydrase; (PhB) phosphorylase B; and (Hb) hemoglobin. [Information from G. J. Sommer, A. K. Singh, and A. V. Hatch, "On-Chip Isoelectric Focusing Using Photopolymerized Immobilized pH Gradients," *Anal. Chem.* **2008**, *80*, 3327.]

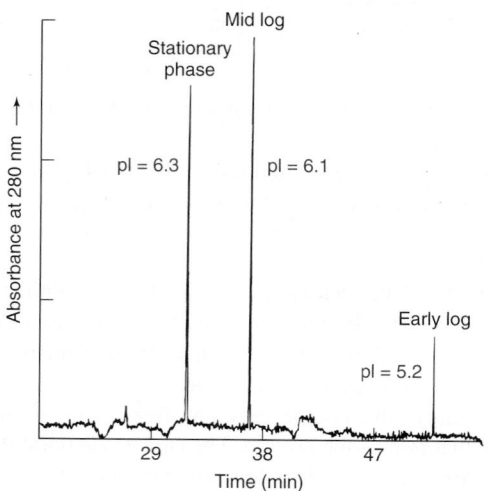

Capillary isoelectric focusing of whole yeast cells taken from three growth stages. After cells have been focused at their isoelectric pH, the inlet end of the capillary was elevated and liquid drained out of the capillary past an ultraviolet detector, creating three peaks observed here. The abscissa is the time required for the bands to reach the detector. [Information from R. Shen, S. J. Berger, and R. D. Smith, "Capillary Isoelectric Focusing of Yeast Cells," *Anal. Chem.* **2000**, *72*, 4603.]

surfaces behave as acids and bases. The silica (SiO_2) surface of sand or glass can be simplistically thought of as a diprotic acid:

$$\equiv Si-OH_2^+ \xrightleftharpoons{K_{a1}} \equiv Si-OH + H^+ \qquad K_{a1} = \frac{\{SiOH\}[H^+]}{\{SiOH_2^+\}} \qquad (10\text{-}24)$$

$$\equiv Si-OH \xrightleftharpoons{K_{a2}} \equiv Si-O^- + H^+ \qquad K_{a2} = \frac{\{SiO^-\}[H^+]}{\{SiOH\}} \qquad (10\text{-}25)$$

The notation $\equiv Si$ represents a surface silicon atom. Silanol groups ($\equiv Si-OH$) can donate or accept a proton to give the surface a negative or positive charge. In the equilibrium constants, the concentrations of the surface species $\{SiOH_2^+\}$, $\{SiOH\}$, and $\{SiO^-\}$ are measured in moles per gram of solid.

The *pH of zero charge* is the pH at which $\{SiOH_2^+\} = \{SiO^-\}$ and, therefore, the surface has no net charge. Like the isoelectric point of a diprotic acid, the pH of zero charge is $\frac{1}{2}(pK_{a1} + pK_{a2})$. *Colloidal particles* (those with diameters in the range 1–100 nm) tend to remain dispersed when they are charged, but they *flocculate* (come together and precipitate) near the pH of zero charge. In capillary electrophoresis (Chapter 26), the surface charge of the silica capillary governs the rate at which solvent moves through the capillary.

Terms to Understand

amino acid	greenhouse gas	isoelectric point	speciation
amphiprotic	hydrolysis	isoionic point	zwitterion
diprotic acids and bases	isoelectric focusing	polyprotic acids and bases	

Summary

Diprotic acids and bases fall into three categories:

1. The fully acidic form, H_2A, behaves as a monoprotic acid, $H_2A \rightleftharpoons H^+ + HA^-$, for which we solve the equation $K_{a1} = x^2/(F - x)$, where $[H^+] = [HA^-] = x$, and $[H_2A] = F - x$. From $[HA^-]$ and $[H^+]$, $[A^{2-}]$ can be found from the K_{a2} equilibrium.

2. The fully basic form, A^{2-}, behaves as a base, $A^{2-} + H_2O \rightleftharpoons HA^- + OH^-$, for which we solve the equation $K_{b1} = x^2/(F - x)$, where $[OH^-] = [HA^-] = x$, and $[A^{2-}] = F - x$. From these concentrations, $[H_2A]$ can be found from the K_{a1} or K_{b2} equilibrium.

3. The intermediate (amphiprotic) form, HA^-, is both an acid and a base. Its pH is given by

$$[H^+] = \sqrt{\frac{K_1 K_2 F + K_1 K_w}{K_1 + F}}$$

where K_1 and K_2 are acid dissociation constants for H_2A, and F is the formal concentration of the intermediate. In most cases, this equation reduces to the form pH $\approx \frac{1}{2}(pK_1 + pK_2)$, in which pH is independent of concentration.

In triprotic systems, there are two intermediate forms. The pH of each is found with an equation analogous to that for the intermediate form of a diprotic system. Triprotic systems also have one fully acidic and one fully basic form; these species can be treated as monoprotic for the purpose of calculating pH. For polyprotic buffers, we write the Henderson-Hasselbalch equation connecting the two principal species in the system. pK_a in this equation is the one that applies to the acid in the denominator of the log term. If acidic sites of a molecule are chemically distinct, there is a microequilibrium constant for H^+ dissociation from each site. Intermediate species are really a mixture of species with H^+ in equilibrium between the acidic sites.

The principal species of a monoprotic or polyprotic system is found by comparing the pH with the pK_a values. For pH $< pK_1$, the fully protonated species, H_nA, is the predominant form. For $pK_1 < pH < pK_2$, the form $H_{n-1}A^-$ is favored; and, at each successive pK value, the next deprotonated species becomes principal. Finally, at pH values higher than the highest pK, the fully basic form (A^{n-}) is dominant. The fractional composition of a solution is expressed by α, given in Equations 10-17 and 10-18 for a monoprotic system and Equations 10-19 through 10-21 for a diprotic system.

The isoelectric pH of a polyprotic compound is that pH at which the average charge of all species is 0. For a diprotic amino acid whose amphiprotic form is neutral, the isoelectric pH is given by pH $= \frac{1}{2}(pK_1 + pK_2)$. The isoionic pH of a polyprotic species is the pH that would exist in a solution containing only ions derived from the neutral polyprotic species and from H_2O. For a diprotic amino acid whose amphiprotic form is neutral, the isoionic pH is found from $[H^+] = \sqrt{(K_1 K_2 F + K_1 K_w)/(K_1 + F)}$, where F is the formal concentration of the amino acid.

Exercises

10-A. Find the pH and the concentrations of H_2SO_3, HSO_3^-, and SO_3^{2-} in each solution: **(a)** 0.050 M H_2SO_3; **(b)** 0.050 M $NaHSO_3$; **(c)** 0.050 M Na_2SO_3.

10-B. (a) How many grams of $NaHCO_3$ (FM 84.01) must be added to 4.00 g of K_2CO_3 (FM 138.21) to give a pH of 10.80 in 500 mL of water?

(b) What will be the pH if 100 mL of 0.100 M HCl are added to the solution in part **(a)**?

(c) How many milliliters of 0.320 M HNO_3 should be added to 4.00 g of K_2CO_3 to give a pH of 10.00 in 250 mL?

10-C. How many milliliters of 0.800 M KOH should be added to 5.02 g of 1,5-pentanedioic acid ($C_5H_8O_4$, FM 132.11) to give a pH of 4.40 when diluted to 250 mL?

10-D. Calculate the pH of a 0.010 M solution of each amino acid in the form drawn here.

(a)
$$NH_2$$
$$|$$
$$C=O$$
$$|$$
$$CH_2$$
$$|$$
$$CH_2$$
$$|$$
$$H_3\overset{+}{N}CHCO_2^-$$
Glutamine

(b)
$$S^-$$
$$|$$
$$CH_2$$
$$|$$
$$H_3\overset{+}{N}CHCO_2^-$$
Cysteine

(c)
$$H_2\overset{+}{N} \diagdown \diagup NH_2$$
$$C$$
$$|$$
$$NH$$
$$|$$
$$CH_2$$
$$|$$
$$CH_2$$
$$|$$
$$CH_2$$
$$|$$
$$H_2NCHCO_2^-$$
Arginine

10-E. (a) Draw the structure of the predominant form (principal species) of 1,3-dihydroxybenzene at pH 9.00 and at pH 11.00.

(b) What is the second most prominent species at each pH?

(c) Calculate the percentage in the major form at each pH.

10-F. Draw the structures of the predominant forms of glutamic acid and tyrosine at pH 9.0 and pH 10.0. What is the second most abundant species at each pH?

10-G. Calculate the isoionic pH of 0.010 M lysine.

10-H. Neutral lysine can be written HL. The other forms of lysine are H_3L^{2+}, H_2L^+, and L^-. The isoelectric point is the pH at which the *average* charge of lysine is 0. Therefore, at the isoelectric point, $2[H_3L^{2+}] + [H_2L^+] = [L^-]$. Use this condition to calculate the isoelectric pH of lysine.

Problems

Diprotic Acids and Bases

10-1. Consider HA^-, the intermediate form of a diprotic acid. K_a for this species is 10^{-4} and K_b is 10^{-8}. Nonetheless, the K_a and K_b reactions proceed to nearly the same extent when NaHA is dissolved in water. Explain.

10-2. Write the general structure of an amino acid. Why do some amino acids in Table 10-1 have two pK values and others three?

10-3. Write the chemical reactions whose equilibrium constants are K_{b1} and K_{b2} for the amino acid proline. Find the values of K_{b1} and K_{b2}.

10-4. Consider the diprotic acid H_2A with $K_1 = 1.00 \times 10^{-4}$ and $K_2 = 1.00 \times 10^{-8}$. Find the pH and concentrations of H_2A, HA^-, and A^{2-} in **(a)** 0.100 M H_2A; **(b)** 0.100 M NaHA; **(c)** 0.100 M Na_2A.

10-5. We will abbreviate malonic acid, $CH_2(CO_2H)_2$, as H_2M. Find the pH and concentrations of H_2M, HM^-, and M^{2-} in **(a)** 0.100 M H_2M; **(b)** 0.100 M NaHM; **(c)** 0.100 M Na_2M.

10-6. Calculate the pH of 0.300 M piperazine. Calculate the concentration of each form of piperazine in this solution.

10-7. Use three cycles of approximation by the method of Box 10-2 to calculate $[H^+]$, $[H_2A]$, $[HA^-]$, and $[A^{2-}]$ in 0.001 00 M monosodium oxalate, NaHA.

10-8. *Activity.* In this problem, we calculate the pH of the intermediate form of a diprotic acid, taking activities into account.

(a) Including activity coefficients, derive Equation 10-11 for potassium hydrogen phthalate (K^+HP^- in the example following Equation 10-12).

(b) Calculate the pH of 0.050 M KHP, using the results in part **(a)**. Assume that the sizes of both HP^- and P^{2-} are 600 pm. For comparison, Equation 10-11 gives pH = 4.18.

10-9. ▦ *Intermediate form of diprotic acid by iteration with Excel.* Enable a new spreadsheet to handle circular references. In Excel 2010, select the File tab and select Options. In the Options window, select Formulas. In Calculation Options, check "Enable iterative calculation" and set Maximum Change to 1e-12. Click OK. Create the spreadsheet below with all formulas shown, except write "=B8" for $[HA^-]$ in cell B11. At this point, values for all concentrations should be the same as the first approximation for malic acid in Box 10-2. Make sure that they are.

Now change the formula for $[HA^-]$ in cell B11 to "=B8−B10−B12". Excel iterates until circularly defined variables are within the allowed maximum change. Answers in the spreadsheet below will be displayed. Make sure that they are.

(a) Copy column B of your spreadsheet and paste it into column G. Change K_1 to 10^{-4} in cell G5 and change K_2 to 10^{-8} in cell G6. Change F to 0.01 in cell G8. Column G now contains concentrations for the amphiprotic salt Na^+HA^- with $K_1 = 10^{-4}$, $K_2 = 10^{-8}$, and F = 0.01 M. Check the answers by hand beginning with pH ≈ $\frac{1}{2}(pK_1 + pK_2)$. With $[HA^-] ≈ F$, calculate $[H_2A]$ and $[A^{2-}]$. Then find $[HA^-] ≈ F - [H_2A] - [A^{2-}]$.

(b) Copy column G of your spreadsheet and paste it into column H. Change K_2 to 10^{-5} in cell G6. Column G now contains the concentrations for the intermediate form of a diprotic acid with $K_1 = 10^{-4}$, $K_2 = 10^{-5}$, and F = 0.01 M. You should observe $[HA^-] = 6.13 \times 10^{-3}$ M and pH = 4.50.

	A	B	C	D	E
1	Successive approximations by circular reference				
2	for intermediate form of diprotic acid				
3					
4	H_2A malic acid				
5	$K_1 =$	3.50E-04			
6	$K_2 =$	7.90E-06			
7	$K_w =$	1.00E-14			
8	F =	1.000E-03			
9	$[H^+] =$	4.356E-05	$= SQRT((K_1{}^*K_2{}^*[HA^-]+K_1{}^*K_w)/(K_1+[HA^-]))$		
10	$[H_2A] =$	9.532E-05	$= [H^+][HA^-]/K_1$		
11	$[HA^-] =$	7.658E-04	$= F-[H_2A]-[A^{2-}]$		
12	$[A^{2-}] =$	1.389E-04	$= K_2[HA^-]/[H^+]$		
13	pH =	4.360887	$= -log[H^+]$		
14					
15	1. Set Excel Options Formulas to Enable Iterative Calculation				
16	with Maximum Change = 1E-12				
17	2. Begin by setting $[HA^-] = F$				
18	3. Write correct formulas for other concentrations				
19	4. Then change $[HA^-]$ to = $F-[H_2A]-[A^{2-}]$				
20	5. Correct answers now appear in all cells				

10-10. *Heterogeneous equilibrium.* CO_2 dissolves in water to give "carbonic acid" (which is mostly dissolved CO_2, as described in Box 6-4).

$$CO_2(g) \rightleftharpoons CO_2(aq) \qquad K = 10^{-1.5}$$

(The equilibrium constant is called the *Henry's law constant* for carbon dioxide, because Henry's law states that the solubility of a gas in a liquid is proportional to the pressure of the gas.) The acid dissociation constants listed for "carbonic acid" in Appendix G apply to $CO_2(aq)$. Given that P_{CO_2} in the atmosphere is $10^{-3.4}$ atm, find the pH of water in equilibrium with the atmosphere.

10-11. *Effect of temperature on carbonic acid acidity and the solubility of $CaCO_3$.*[14] Box 10-1 states that marine life with $CaCO_3$ shells and skeletons will be threatened with extinction in cold polar waters before that will happen in warm tropical waters. The following equilibrium constants apply to seawater at 0° and 30°C, when concentrations are measured in moles per kilogram of seawater and pressure is in bars:

$$CO_2(g) \rightleftharpoons CO_2(aq) \tag{A}$$

$$K_H = \frac{[CO_2(aq)]}{P_{CO_2}} = 10^{-1.2073} \text{ mol kg}^{-1} \text{ bar}^{-1} \text{ at } 0°C$$

$$= 10^{-1.6048} \text{ mol kg}^{-1} \text{ bar}^{-1} \text{ at } 30°C$$

$$CO_2(aq) + H_2O \rightleftharpoons HCO_3^- + H^+ \tag{B}$$

$$K_{a1} = \frac{[HCO_3^-][H^+]}{[CO_2(aq)]} = 10^{-6.1004} \text{ mol kg}^{-1} \text{ at } 0°C$$

$$= 10^{-5.8008} \text{ mol kg}^{-1} \text{ at } 30°C$$

$$HCO_3^- \rightleftharpoons CO_3^{2-} + H^+ \tag{C}$$

$$K_{a2} = \frac{[CO_3^{2-}][H^+]}{[HCO_3^-]} = 10^{-9.3762} \text{ mol kg}^{-1} \text{ at } 0°C$$

$$= 10^{-8.8324} \text{ mol kg}^{-1} \text{ at } 30°C$$

$$CaCO_3(s, aragonite) \rightleftharpoons Ca^{2+} + CO_3^{2-} \tag{D}$$

$$K_{sp}^{arg} = [Ca^{2+}][CO_3^{2-}] = 10^{-6.1113} \text{ mol}^2 \text{ kg}^{-2} \text{ at } 0°C$$

$$= 10^{-6.1391} \text{ mol}^2 \text{ kg}^{-2} \text{ at } 30°C$$

$$CaCO_3(s, calcite) \rightleftharpoons Ca^{2+} + CO_3^{2-} \tag{E}$$

$$K_{sp}^{cal} = [Ca^{2+}][CO_3^{2-}] = 10^{-6.3652} \text{ mol}^2 \text{ kg}^{-2} \text{ at } 0°C$$

$$= 10^{-6.3713} \text{ mol}^2 \text{ kg}^{-2} \text{ at } 30°C$$

The first equilibrium constant is called K_H for Henry's law (Problem 10-10). Units are given to remind you what units you must use.

(a) Combine the expressions for K_H, K_{a1}, and K_{a2} to find an expression for $[CO_3^{2-}]$ in terms of P_{CO_2} and $[H^+]$.

(b) From the result of (a), calculate $[CO_3^{2-}]$ (mol kg^{-1}) at $P_{CO_2} = 800$ μbar and pH = 7.8 at temperatures of 0° (polar ocean) and 30°C (tropical ocean). These are conditions that could possibly be reached around the year 2100.

(c) The concentration of Ca^{2+} in the ocean is 0.010 M. Predict whether aragonite and calcite will dissolve under the conditions in (b).

Diprotic Buffers

10-12. How many grams of Na_2CO_3 (FM 105.99) should be mixed with 5.00 g of $NaHCO_3$ (FM 84.01) to produce 100 mL of buffer with pH 10.00?

10-13. How many milliliters of 0.202 M NaOH should be added to 25.0 mL of 0.023 3 M salicylic acid (2-hydroxybenzoic acid) to adjust the pH to 3.50?

10-14. Describe how you would prepare exactly 100 mL of 0.100 M picolinate buffer, pH 5.50. Possible starting materials are pure picolinic acid (pyridine-2-carboxylic acid, FM 123.11), 1.0 M HCl, and 1.0 M NaOH. Approximately how many milliliters of the HCl or NaOH will be required?

10-15. How many grams of Na_2SO_4 (FM 142.04) should be added to how many grams of sulfuric acid (FM 98.08) to give 1.00 L of buffer with pH 2.80 and a total sulfur (= SO_4^{2-} + HSO_4^- + H_2SO_4) concentration of 0.200 M?

Polyprotic Acids and Bases

10-16. Phosphate at 0.01 M is one of the main buffers in blood plasma, whose pH is 7.45. Would phosphate be as useful if the plasma pH were 8.5?

10-17. Starting with the fully protonated species, write the stepwise acid dissociation reactions of the amino acids glutamic acid and tyrosine. Be sure to remove the protons in the correct order. Which species are the neutral molecules that we call glutamic acid and tyrosine?

10-18. (a) Calculate the quotient $[H_3PO_4]/[H_2PO_4^-]$ in 0.050 0 M KH_2PO_4.

(b) Find the same quotient for 0.050 0 M K_2HPO_4.

10-19. (a) Which two of the following compounds would you mix to make a buffer of pH 7.45: H_3PO_4 (FM 98.00), NaH_2PO_4 (FM 119.98), Na_2HPO_4 (FM 141.96), and Na_3PO_4 (FM 163.94)?

(b) If you wanted to prepare 1.00 L of buffer with a total phosphate concentration of 0.050 0 M, how many grams of each of the two selected compounds would you mix together?

(c) If you did what you calculated in part (b), you would not get a pH of exactly 7.45. Explain how you would really prepare this buffer in the lab.

10-20. Find the pH and the concentration of each species of lysine in a solution of 0.010 0 M lysine·HCl, lysine monohydrochloride. The notation "lysine·HCl" refers to a neutral lysine molecule that has taken on one extra proton by addition of one mole of HCl. A more meaningful notation shows the salt (lysineH$^+$)(Cl$^-$) formed in the reaction.

10-21. How many milliliters of 1.00 M KOH should be added to 100 mL of solution containing 10.0 g of histidine hydrochloride [His · HCl = (HisH$^+$)(Cl$^-$), FM 191.62] to get a pH of 9.30?

10-22. (a) Using activity coefficients, calculate the pH of a solution containing a 2.00 : 1.00 mole ratio of $HC^{2-} : C^{3-}$ (H_3C = citric acid). The ionic strength is 0.010 M.

(b) What will be the pH if the ionic strength is raised to 0.10 M and the mole ratio $HC^{2-} : C^{3-}$ is kept constant?

Which Is the Principal Species?

10-23. The acid HA has pK_a = 7.00.

(a) Which is the principal species, HA or A$^-$, at pH 6.00?

(b) Which is the principal species at pH 8.00?

(c) What is the quotient $[A^-]/[HA]$ at pH 7.00? At pH 6.00?

10-24. The diprotic acid H_2A has pK_1 = 4.00 and pK_2 = 8.00.

(a) At what pH is $[H_2A] = [HA^-]$?

(b) At what pH is $[HA^-] = [A^{2-}]$?

(c) Which is the principal species at pH 2.00: H_2A, HA^-, or A^{2-}?

(d) Which is the principal species at pH 6.00?

(e) Which is the principal species at pH 10.00?

10-25. The base B has pK_b = 5.00.

(a) What is the value of pK_a for the acid BH$^+$?

(b) At what pH is $[BH^+] = [B]$?

(c) Which is the principal species at pH 7.00: B or BH$^+$?

(d) What is the quotient $[B]/[BH^+]$ at pH 12.00?

10-26. Draw the structure of the predominant form of pyridoxal-5-phosphate at pH 7.00.

Fractional Composition Equations

10-27. The acid HA has $pK_a = 4.00$. Use Equations 10-17 and 10-18 to find the fraction in the form HA and the fraction in the form A^- at pH = 5.00. Does your answer agree with what you expect for the quotient $[A^-]/[HA]$ at pH 5.00?

10-28. A dibasic compound, B, has $pK_{b1} = 4.00$ and $pK_{b2} = 6.00$. Find the fraction in the form BH_2^{2+} at pH 7.00, using Equation 10-19. Note that K_1 and K_2 in Equation 10-19 are acid dissociation constants for BH_2^{2+} ($K_1 = K_w/K_{b2}$ and $K_2 = K_w/K_{b1}$).

10-29. What fraction of ethane-1,2-dithiol is in each form (H_2A, HA^-, A^{2-}) at pH 8.00? at pH 10.00?

10-30. Calculate α_{H_2A}, α_{HA^-}, and $\alpha_{A^{2-}}$ for *cis*-butenedioic acid at pH 1.00, 1.92, 6.00, 6.27, and 10.00.

10-31. **(a)** Derive equations for α_{H_3A}, $\alpha_{H_2A^-}$, $\alpha_{HA^{2-}}$, and $\alpha_{A^{3-}}$ for a triprotic system.
(b) Calculate the values of these fractions for phosphoric acid at pH 7.00.

10-32. A solution containing acetic acid, oxalic acid, ammonia, and pyridine has a pH of 9.00. What fraction of ammonia is not protonated?

10-33. A solution was prepared from 10.0 mL of 0.100 M cacodylic acid and 10.0 mL of 0.080 0 M NaOH. To this mixture was added 1.00 mL of 1.27×10^{-6} M morphine. Calling morphine B, calculate the fraction of morphine present in the form BH^+.

$(CH_3)_2AsOH$
Cacodylic acid
$K_a = 6.4 \times 10^{-7}$

Morphine
$K_b = 1.6 \times 10^{-6}$

10-34. *Fractional composition in a diprotic system.* Create a spreadsheet with Equations 10-19 through 10-21 to compute the three curves in Figure 10-4. Plot the three curves in a beautifully labeled figure.

10-35. *Fractional composition in a triprotic system.* For a triprotic system, the fractional composition equations are

$$\alpha_{H_3A} = \frac{[H^+]^3}{D} \qquad \alpha_{HA^{2-}} = \frac{K_1K_2[H^+]}{D}$$

$$\alpha_{H_2A^-} = \frac{K_1[H^+]^2}{D} \qquad \alpha_{A^{3-}} = \frac{K_1K_2K_3}{D}$$

where $D = [H^+]^3 + K_1[H^+]^2 + K_1K_2[H^+] + K_1K_2K_3$. Use these equations to create a fractional composition diagram analogous to Figure 10-4 for the amino acid tyrosine. What is the fraction of each species at pH 10.00?

10-36. *Fractional composition in a tetraprotic system.* Prepare a fractional composition diagram analogous to Figure 10-4 for the tetraprotic system derived from hydrolysis of Cr^{3+}:

$$Cr^{3+} + H_2O \rightleftharpoons Cr(OH)^{2+} + H^+ \qquad K_{a1} = 10^{-3.80}$$
$$Cr(OH)^{2+} + H_2O \rightleftharpoons Cr(OH)_2^+ + H^+ \qquad K_{a2} = 10^{-6.40}$$
$$Cr(OH)_2^+ + H_2O \rightleftharpoons Cr(OH)_3(aq) + H^+ \qquad K_{a3} = 10^{-6.40}$$
$$Cr(OH)_3(aq) + H_2O \rightleftharpoons Cr(OH)_4^- + H^+ \qquad K_{a4} = 10^{-11.40}$$

(Yes, the values of K_{a2} and K_{a3} are equal.)

(a) Use these equilibrium constants to prepare a fractional composition diagram for this tetraprotic system.

(b) You should do this part with your head and your calculator, not your spreadsheet. The solubility of $Cr(OH)_3$ is given by

$$Cr(OH)_3(s) \rightleftharpoons Cr(OH)_3(aq) \qquad K = 10^{-6.84}$$

What concentration of $Cr(OH)_3(aq)$ is in equilibrium with solid $Cr(OH)_3(s)$?

(c) What concentration of $Cr(OH)^{2+}$ is in equilibrium with $Cr(OH)_3(s)$ if the solution pH is adjusted to 4.00?

Isoelectric and Isoionic pH and Proteins

10-37. Which four amino acids in Table 10-1 have acidic substituents (that donate a proton) and which three have basic substituents (that accept a proton)?

10-38. What is the difference between the isoelectric pH and the isoionic pH of a protein with many different acidic and basic substituents?

10-39. Explain what is wrong with the following statement: At its isoelectric point, the charge on all molecules of a particular protein is 0.

10-40. Calculate the isoelectric and isoionic pH of 0.010 M threonine.

10-41. Explain how isoelectric focusing works.

11 Acid-Base Titrations

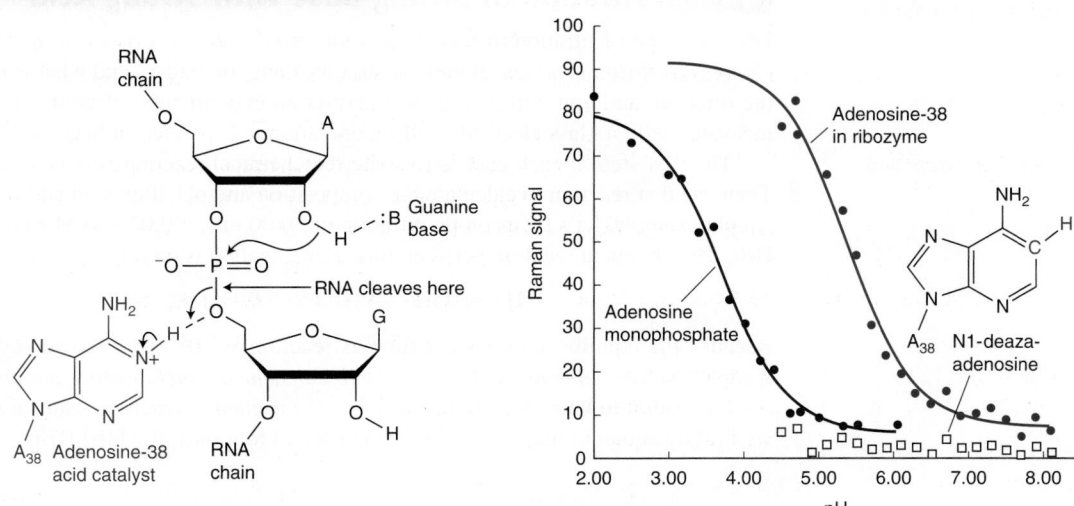

(*Left*) Proposed mechanism of cleavage of RNA chain with adenosine-38 serving as an acid catalyst to protonate a phosphate oxygen. (*Right*) Titrations of adenosine-38 or N1-deaza-adenosine-38 in RNA, and titration of free adenosine monophosphate in solution. [Data from M. Guo, R. C. Spitale, R. Volpini, J. Krucinska, G. Cristalli, P. R. Carey, and J. E. Wedekind, "Direct Raman Measurement of an Elevated Base pK_a in the Active Site of a Small Ribozyme in a Precatalytic Conformation," *J. Am. Chem. Soc.* **2009,** *131,* 12908.]

In addition to its role in translating genetic information from DNA (Appendix L) into the structure of proteins, ribonucleic acid (RNA) functions as a catalyst for chemical reactions. RNA that acts as a catalyst is called a *ribozyme*, by analogy with the word *enzyme*, which is a catalytic protein.

The chemical structure shows the proposed mechanism by which a "hairpin ribozyme" cleaves catenated RNA strands into smaller functional strands.[1] A key step involves transfer of a proton from adenosine-38 of the ribozyme to a phosphate to help break the phosphate—ribose sugar bond. At the same time, a guanine base (B) accepts a proton from the adjacent ribose, allowing that ribose oxygen to attack the phosphate. The ribozyme acts near neutral pH in a cell. However, pK_a for H^+ of adenosine in solution is near 4.0. Near pH 7, there would be too little protonated adenosine for the ribozyme to function. The local environments in enzymes and ribozymes can alter acid-base properties of amino acids and nucleotides. It was proposed that pK_a of adenosine-38 is raised so that its NH^+ is available for catalysis near neutral pH.

The graph shows titrations of adenosine monophosphate in solution (black circles), adenosine-38 of the ribozyme (colored circles), and a synthetic nucleoside in which NH^+ is replaced by an inert C—H bond at the adenosine-38 position. In each case, the Raman vibrational spectrum was measured at a series of different pH values established with buffers. When adenosine becomes protonated, some molecular vibrational frequencies change. The graph shows the intensity of the spectrum of the protonated form as a function of pH. At low pH, adenosine in solution or in RNA is fully protonated. At pH = pK_a, adensosine is half protonated. At high pH, adenosine is unprotonated. From the shape of the titration curve, we deduce that pK_a is 3.68 for free adenosine monophosphate and pK_a is 5.46 for adenosine-38 in the ribozyme. With pK_a elevated by almost 2 units, there is sufficient NH^+ near neutral pH for the ribozyme to function. To confirm that adenosine-38 is being observed in the titration, synthetic RNA with NH^+ replaced by C—H in adenosine-38 was also titrated. As shown by the squares in the graph, the inert C—H bond shows no response to changing the pH. This experiment confirms that the signal in the graph is from A-38 and not from several other adenosine units in the ribozyme. Acid-base titrations have broad applications in scientific research.

Lipophilicity is a measure of solubility in nonpolar solvents. It is determined from the equilibrium distribution of a drug between water and octanol.

Drug(*aq*) $\rightleftharpoons$ drug (*in octanol*)

Lipophilicity = $\log\left(\dfrac{[\text{drug (}in\ octanol\text{)}]}{[\text{drug(}aq\text{)}]}\right)$

Generally, the more lipophilic a drug, the easier it is to pass through a cell membrane.

First, write the reaction between *titrant* and *analyte*.

The titration reaction.

$\text{mL} \times \dfrac{\text{mol}}{\text{L}} = \text{mL} \times \dfrac{\text{mmol}}{\text{mL}} = \text{mmol}$

Before the equivalence point, there is excess OH^-.

In your sleep,

$\dfrac{\text{mmol}}{\text{mL}} = \dfrac{\text{mol}}{\text{L}} = M$

From an acid-base titration curve, we can deduce the quantities and pK_a values of acidic and basic substances in a mixture. In medicinal chemistry, the pK_a and *lipophilicity* of a candidate drug predict how easily it will cross cell membranes. From pK_a and pH, we can compute the charge of a polyprotic acid. Usually, the more highly charged a drug, the harder it is for that drug to cross a cell membrane. In this chapter, we learn to predict the shapes of titration curves and to find end points with electrodes and indicators.

11-1 Titration of Strong Base with Strong Acid

For each type of titration in this chapter, *our goal is to construct a graph showing how pH changes as titrant is added*. If you can do this, then you understand what is happening during the titration, and you will be able to interpret an experimental titration curve. pH is usually measured with a glass electrode, whose operation is explained in Section 15-5.

The first step in each case is to write the chemical reaction between titrant and analyte. Then use that reaction to calculate the composition and pH after each addition of titrant. As a simple example, let's focus on the titration of 50.00 mL of 0.020 00 M KOH with 0.100 0 M HBr. The chemical reaction between titrant and analyte is merely

$$H^+ + OH^- \rightarrow H_2O \qquad K = 1/K_w = 10^{14}$$

Because the equilibrium constant for this reaction is 10^{14}, it is fair to say that it "goes to completion." *Any amount of H^+ added will consume a stoichiometric amount of OH^-.*

It is useful to know the volume of HBr (V_e) needed to reach the equivalence point, which we find by equating moles of KOH being titrated to moles of added HBr:

$$\underbrace{(V_e\ \cancel{(L)})\left(0.100\ 0\ \frac{\text{mol}}{\cancel{L}}\right)}_{\substack{\text{mmol of HBr} \\ \text{at equivalence point}}} = \underbrace{(0.050\ 00\ \cancel{L})\left(0.020\ 00\ \frac{\text{mol}}{\cancel{L}}\right)}_{\substack{\text{mmol of OH}^- \\ \text{being titrated}}} \Rightarrow V_e = 0.010\ 00\ \text{L}$$

Instead of multiplying $\cancel{L} \times (\text{mol}/\cancel{L})$ to get mol, we frequently multiply mL $\times$ (mol/L), which is the same as $\cancel{mL} \times (\text{mmol}/\cancel{mL})$:

$$\underbrace{(V_e(\text{mL}))(0.100\ 0\ \text{M})}_{\substack{\text{mol of HBr} \\ \text{at equivalence point}}} = \underbrace{(50.00\ \text{mL})(0.020\ 00\ \text{M})}_{\substack{\text{mol of OH}^- \\ \text{being titrated}}} \Rightarrow V_e = 10.00\ \text{mL}$$

When 10.00 mL of HBr have been added, the titration is complete. Prior to this point, there is excess, unreacted OH^- present. After V_e, there is excess H^+ in the solution.

In the titration of any strong base with any strong acid, there are three regions of the titration curve that require different kinds of calculations:

1. Before the equivalence point, pH is determined by excess OH^- in the solution.
2. At the equivalence point, H^+ is just sufficient to react with all OH^- to make H_2O. The pH is determined by dissociation of water.
3. After the equivalence point, pH is determined by excess H^+ in the solution.

We will do one sample calculation for each region. Complete results are shown in Table 11-1 and Figure 11-1. As a reminder, the *equivalence point* occurs when added titrant is exactly enough for stoichiometric reaction with the analyte. The equivalence point is the ideal result we seek in a titration. What we actually measure is the *end point*, which is marked by a sudden physical change, such as indicator color or an electrode potential.

Region 1: Before the Equivalence Point

We will do this calculation once by the method you would have learned in general chemistry, and then we will streamline the process. When 3.00 mL of HBr have been added, the total volume is 53.00 mL. HBr is consumed by NaOH, leaving excess NaOH. The moles of added HBr are $(0.100\ 0\ \text{M})(0.003\ 00\ \text{L}) = 0.300 \times 10^{-3}$ mol HBr = 0.300 mmol HBr. The initial moles of NaOH were $(0.020\ 00\ \text{M})(0.050\ 00\ \text{L}) = 1.000 \times 10^{-3}$ mol NaOH = 1.000 mmol NaOH. Unreacted OH^- is the difference 1.000 mmol − 0.300 mmol = 0.700 mmol. The concentration of unreacted OH^- is (0.700 mmol)/(53.00 mL) = 0.013 2 M. So, $[H^+] = K_w/[OH^-] = 7.57 \times 10^{-13}$ M and pH = $-\log[H^+]$ = 12.12.

Now here is the streamlined calculation: When 3.00 mL of HBr have been added, the reaction is three-tenths complete because $V_e = 10.00$ mL. The fraction of OH^- left unreacted

	mL HBr added (V_a)	Concentration of unreacted OH⁻ (M)	Concentration of excess H⁺ (M)	pH
	0.00	0.020 0		12.30
	1.00	0.017 6		12.24
	2.00	0.015 4		12.18
	3.00	0.013 2		12.12
	4.00	0.011 1		12.04
Region 1	5.00	0.009 09		11.95
(excess OH⁻)	6.00	0.007 14		11.85
	7.00	0.005 26		11.72
	8.00	0.003 45		11.53
	9.00	0.001 69		11.22
	9.50	0.000 840		10.92
	9.90	0.000 167		10.22
	9.99	0.000 016 6		9.22
Region 2	10.00	—	—	7.00
	10.01		0.000 016 7	4.78
	10.10		0.000 166	3.78
	10.50		0.000 826	3.08
	11.00		0.001 64	2.79
Region 3	12.00		0.003 23	2.49
(excess H⁺)	13.00		0.004 76	2.32
	14.00		0.006 25	2.20
	15.00		0.007 69	2.11
	16.00		0.009 09	2.04

TABLE 11-1 Calculation of the titration curve for 50.00 mL of 0.020 00 M KOH treated with 0.100 0 M HBr

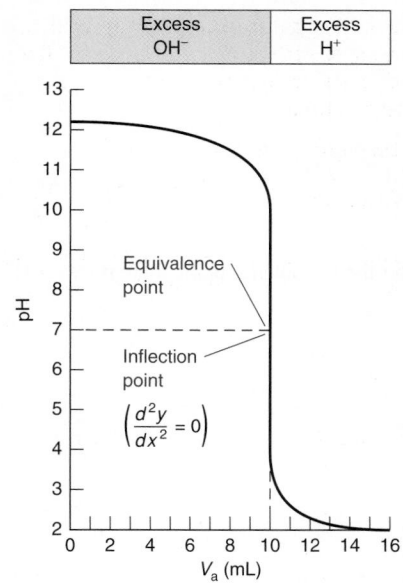

Excess OH⁻	Excess H⁺

FIGURE 11-1 Calculated titration curve, showing how pH changes as 0.100 0 M HBr is added to 50.00 mL of 0.020 00 M KOH. The equivalence point is an inflection point at which the second derivative is zero.

is seven-tenths. The concentration of remaining OH⁻ is the product of the fraction remaining, the initial concentration, and a dilution factor:

$$[\text{OH}^-] = \underbrace{\left(\frac{10.00 \text{ mL} - 3.00 \text{ mL}}{10.00 \text{ mL}}\right)}_{\substack{\text{Fraction} \\ \text{of OH}^- \\ \text{remaining}}} \underbrace{(0.020\ 00 \text{ M})}_{\substack{\text{Initial} \\ \text{concentration} \\ \text{of OH}^-}} \underbrace{\left(\frac{50.00 \text{ mL}}{50.00 \text{ mL} + 3.00 \text{ mL}}\right)}_{\substack{\text{Dilution} \\ \text{factor}}} = 0.013\ 2 \text{ M} \quad \textbf{(11-1)}$$

Initial volume of OH⁻

Total volume of solution

$$[\text{H}^+] = \frac{K_\text{w}}{[\text{OH}^-]} = \frac{1.0 \times 10^{-14}}{0.013\ 2} = 7.5_7 \times 10^{-13} \text{ M} \Rightarrow \text{pH} = 12.12$$

Equation 11-1 tells us that [OH⁻] is equal to a certain fraction of the initial concentration, with a correction for dilution. The dilution factor equals the initial volume of analyte divided by the total volume of solution.

In Table 11-1, the volume of acid added is designated V_a. pH is expressed to the 0.01 decimal place, regardless of what is justified by significant figures. We do this for the sake of consistency and also because 0.01 is near the limit of accuracy in pH measurements.

Region 2: At the Equivalence Point

Region 2 is the equivalence point, where just enough H⁺ has been added to consume OH⁻. We could prepare the same solution by dissolving KBr in water. pH is determined by dissociation of water:

$$\text{H}_2\text{O} \rightleftharpoons \underset{x}{\text{H}^+} + \underset{x}{\text{OH}^-}$$

$$K_\text{w} = x^2 \Rightarrow x = 1.00 \times 10^{-7} \text{ M} \Rightarrow \text{pH} = 7.00$$

The pH at the equivalence point in the titration of any strong base (or acid) with strong acid (or base) will be 7.00 at 25°C.

Challenge Using a setup similar to Equation 11-1, calculate [OH⁻] when 6.00 mL of HBr have been added. Check your pH against the value in Table 11-1.

At the equivalence point, pH = 7.00, but *only* in a strong-acid–strong-base reaction.

We will soon discover that *the pH is **not** 7.00 at the equivalence point in the titration of weak acids or bases*. The pH is 7.00 only if the titrant and analyte are both strong.

Region 3: After the Equivalence Point

Beyond the equivalence point, we are adding excess HBr to the solution. The concentration of excess H^+ at, say, 10.50 mL is given by

After the equivalence point, there is excess H^+.

$$[H^+] = \underbrace{(0.100\ 0\ M)}_{\substack{\text{Initial} \\ \text{concentration} \\ \text{of } H^+}}\underbrace{\left(\frac{\overbrace{0.50\ mL}^{\substack{\text{Volume of} \\ \text{excess } H^+}}}{\underbrace{50.00\ mL\ +\ 10.50\ mL}_{\substack{\text{Total volume} \\ \text{of solution}}}}\right)}_{\substack{\text{Dilution} \\ \text{factor}}} = 8.26 \times 10^{-4}\ M$$

$$pH = -\log[H^+] = 3.08$$

At $V_a = 10.50$ mL, there is an excess of just $V_a - V_e = 10.50 - 10.00 = 0.50$ mL of HBr. That is the reason why 0.50 appears in the dilution factor.

The Titration Curve

The titration curve in Figure 11-1 exhibits a rapid change in pH near the equivalence point. The equivalence point is where the slope (dpH/dV_a) is greatest (and the second derivative is 0, which makes it an *inflection point*). To repeat an important statement, the pH at the equivalence point is 7.00 *only* in a strong-acid–strong-base titration. If one or both of the reactants is weak, the equivalence point pH is *not* 7.00.

11-2 Titration of Weak Acid with Strong Base

The titration of a weak acid with a strong base allows us to put all our knowledge of acid-base chemistry to work. The example we examine is the titration of 50.00 mL of 0.020 00 M MES with 0.100 0 M NaOH. MES is an abbreviation for 2-(*N*-morpholino)ethanesulfonic acid, which is a weak acid with $pK_a = 6.27$.

The *titration reaction* is

Always start by writing the titration reaction.

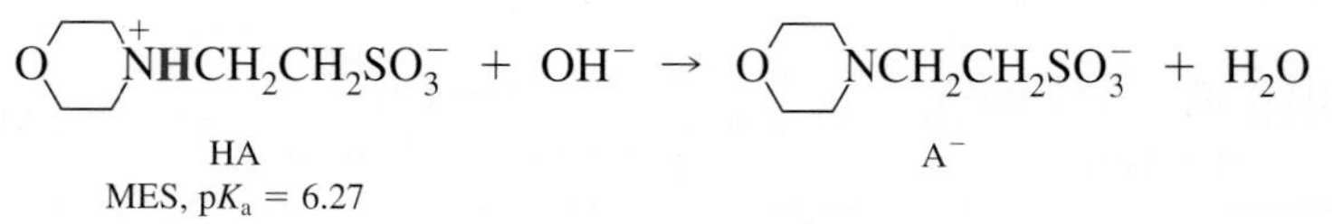

$$O\overset{\overset{+}{}}{\bigcirc}NHCH_2CH_2SO_3^- + OH^- \rightarrow O\bigcirc NCH_2CH_2SO_3^- + H_2O \qquad (11\text{-}2)$$

$$\underset{\substack{HA \\ \text{MES, } pK_a = 6.27}}{} \qquad\qquad\qquad \underset{A^-}{}$$

Reaction 11-2 is the reverse of the K_b reaction for the base A^-. Therefore, the equilibrium constant for Reaction 11-2 is $K = 1/K_b = 1/(K_w/K_a\ (\text{for HA})) = 5.4 \times 10^7$. The equilibrium constant is so large that we can say that the reaction goes "to completion" after each addition of OH^-. As we saw in Box 9-3, *strong plus weak react completely*.

Strong + weak → complete reaction

Let's first calculate the volume of base, V_b, needed to reach the equivalence point:

$$\underbrace{(V_b(mL))(0.100\ 0\ M)}_{\text{mmol of base}} = \underbrace{(50.00\ mL)(0.020\ 00\ M)}_{\text{mmol of HA}} \Rightarrow V_b = 10.00\ mL$$

The titration calculations for this problem are of four types:

1. Before any base is added, the solution contains just HA in water. This is a weak acid whose pH is determined by the equilibrium

$$HA \xrightleftharpoons{K_a} H^+ + A^-$$

2. From the first addition of NaOH until immediately before the equivalence point, there is a mixture of unreacted HA plus the A^- produced by Reaction 11-2. *Aha! A buffer!* We can use the Henderson-Hasselbalch equation to find the pH.

3. At the equivalence point, "all" HA has been converted into A^-. The same solution could have been made by dissolving A^- in water. We have a weak base whose pH is determined by the reaction

$$A^- + H_2O \xrightleftharpoons{K_b} HA + OH^-$$

4. Beyond the equivalence point, excess NaOH is being added to a solution of A^-. To a good approximation, pH is determined by the strong base. We calculate the pH as if we had simply added excess NaOH to water. We neglect the tiny effect of A^-.

Region 1: Before Base Is Added

Before adding any base, we have a solution of 0.020 00 M HA with $pK_a = 6.27$. This is simply a weak-acid problem.

$$HA \rightleftharpoons H^+ + A^- \qquad K_a = 10^{-6.27}$$
$$F-x \qquad x \qquad x$$

The initial solution contains just the *weak acid* HA.

$$\frac{x^2}{0.020\ 00 - x} = K_a \Rightarrow x = 1.03 \times 10^{-4} \Rightarrow pH = 3.99$$

Region 2: Before the Equivalence Point

Adding OH^- creates a mixture of HA and A^-. This mixture is a buffer whose pH can be calculated with the Henderson-Hasselbalch equation (9-16) from the quotient $[A^-]/[HA]$.

Suppose we wish to calculate $[A^-]/[HA]$ when 3.00 mL of OH^- have been added. Because $V_e = 10.00$ mL, we have added enough base to react with three-tenths of the HA. We can make a table showing the relative concentrations before and after the reaction:

Before the equivalence point, there is a mixture of HA and A^-, which is a *buffer*.

We only need *relative* concentrations because pH depends on the quotient $[A^-]/[HA]$.

Titration reaction:	HA	$+ OH^-$	$\rightarrow A^-$	$+ H_2O$
Relative initial quantities (HA $\equiv$ 1)	1	$\frac{3}{10}$	—	—
Relative final quantities	$\frac{7}{10}$	—	$\frac{3}{10}$	—

Once we know the *quotient* $[A^-]/[HA]$ in any solution, we know its pH:

$$pH = pK_a + \log\left(\frac{[A^-]}{[HA]}\right) = 6.27 + \log\left(\frac{3/10}{7/10}\right) = 5.90$$

The point at which the volume of titrant is $\frac{1}{2}V_e$ is a special one in any titration.

Titration reaction:	HA	$+ OH^-$	$\rightarrow A^-$	$+ H_2O$
Relative initial quantities	1	$\frac{1}{2}$	—	—
Relative final quantities	$\frac{1}{2}$	—	$\frac{1}{2}$	—

$$pH = pK_a + \log\left(\frac{1/2}{1/2}\right) = pK_a$$

When $V_b = \frac{1}{2}V_e$, pH = pK_a. This is a landmark in any titration.

When the volume of titrant is $\frac{1}{2}V_e$, pH = pK_a for the acid HA (neglecting activity coefficients). From an experimental titration curve, you can find the approximate value of pK_a by reading the pH when $V_b = \frac{1}{2}V_e$, where V_b is the volume of added base. (To find the true value of pK_a requires activity coefficients.)

Advice As soon as you recognize a mixture of HA and A^- in any solution, *you have a buffer!* You can calculate the pH from the quotient $[A^-]/[HA]$.

$$pH = pK_a + \log\left(\frac{[A^-]}{[HA]}\right)$$

Learn to recognize buffers! They lurk in every corner of acid-base chemistry.

Region 3: At the Equivalence Point

At the equivalence point, there is exactly enough NaOH to consume HA.

At the equivalence point, HA has been converted into A^-, a *weak base*.

Titration reaction:	HA	$+ OH^-$	$\rightarrow A^-$	$+ H_2O$
Relative initial quantities	1	1	—	—
Relative initial quantities	—	—	1	—

The solution contains "just" A^-. We could have prepared the same solution by dissolving the salt Na^+A^- in distilled water. *A solution of Na^+A^- is merely a solution of a weak base.*

To compute the pH of a weak base, we write the reaction of the weak base with water:

$$A^- + H_2O \rightleftharpoons HA + OH^- \qquad K_b = \frac{K_w}{K_a}$$
$$\underset{F-x}{} \qquad\qquad\qquad \underset{x}{} \quad \underset{x}{}$$

The only tricky point is that the formal concentration of A^- is no longer 0.020 00 M, which was the initial concentration of HA. A^- has been diluted by NaOH from the buret:

$$F' = \underbrace{(0.020\ 00\ M)}_{\substack{\text{Initial} \\ \text{concentration} \\ \text{of HA}}} \underbrace{\left(\frac{50.00\ mL}{50.00\ mL + 10.00\ mL}\right)}_{\text{Dilution factor}} = 0.016\ 7\ M$$

Initial volume of HA (arrow pointing to 50.00 mL)

Total volume of solution (arrow pointing to denominator)

With this value of F', we can solve the problem:

$$\frac{x^2}{F'-x} = \frac{x^2}{0.016\ 7-x} = K_b = \frac{K_w}{K_a} = 1.86 \times 10^{-8} \Rightarrow x = 1.76 \times 10^{-5}\ M$$

$$pH = -\log[H^+] = -\log\left(\frac{K_w}{x}\right) = 9.25$$

The pH at the equivalence point in this titration is 9.25. **It is not 7.00.** The equivalence-point pH will *always* be above 7 for the titration of a weak acid, because the acid is converted into its conjugate base at the equivalence point.

> The pH is always higher than 7 at the equivalence point in the titration of a weak acid with a strong base.

Region 4: After the Equivalence Point

> Here we assume that the pH is governed by the excess OH^-.

Now we are adding NaOH to a solution of A^-. The base NaOH is so much stronger than the base A^- that it is fair to say that the pH is determined by the excess OH^-.

Let's calculate the pH when $V_b = 10.10$ mL, which is just 0.10 mL past V_e. The concentration of excess OH^- is

> **Challenge** Compare the concentration of OH^- from excess titrant at $V_b = 10.10$ mL to the concentration of OH^- from hydrolysis of A^-. Satisfy yourself that it is fair to neglect the contribution of A^- to the pH after the equivalence point.

$$[OH^-] = \underbrace{(0.100\ 0\ M)}_{\substack{\text{Initial } OH^- \\ \text{concentration}}} \underbrace{\left(\frac{0.10\ mL}{50.00\ mL + 10.10\ mL}\right)}_{\text{Dilution factor}} = 1.66 \times 10^{-4}\ M$$

Volume of excess OH^- (arrow pointing to 0.10 mL)

Total volume of solution (arrow pointing to denominator)

$$pH = -\log\left(\frac{K_w}{[OH^-]}\right) = 10.22$$

The Titration Curve

> **Landmarks in a titration:**
> At $V_b = V_e$, the curve is steepest.
> At $V_b = \frac{1}{2}V_e$, pH = pK_a and the slope is minimal.

> The **buffer capacity** measures the ability of the solution to resist changes in pH.

Calculations for the titration of MES with NaOH are shown in Table 11-2. The calculated titration curve in Figure 11-2 has two easily identified points. One is the equivalence point, which is the steepest part of the curve. The other landmark is the point where $V_b = \frac{1}{2}V_e$ and pH = pK_a. This latter point is also an inflection point, having the minimum slope.

If you look back at Figure 9-4b, you will note that the maximum *buffer capacity* occurs when pH = pK_a. That is, the solution is most resistant to pH changes when pH = pK_a (and $V_b = \frac{1}{2}V_e$); the slope (dpH/dV_b) is therefore at its minimum.

Figure 11-3 shows how the titration curve depends on the acid dissociation constant of HA and on the concentrations of reactants. As HA becomes a weaker acid (with higher pK_a), or as the concentrations of analyte and titrant decrease, the inflection near the equivalence point decreases, until the equivalence point becomes too shallow to detect. *It is not practical to titrate an acid or base when its strength is too weak or its concentration is too dilute.*

11-3 Titration of Weak Base with Strong Acid

The titration of a weak base with a strong acid is the reverse of the titration of a weak acid with a strong base. The *titration reaction* is

$$B + H^+ \rightarrow BH^+$$

TABLE 11-2	Calculation of the titration curve for 50.00 mL of 0.020 00 M MES treated with 0.100 0 M NaOH	
	mL base added (V_b)	pH
Region 1 (weak acid)	0.00	3.99
	0.50	4.99
	1.00	5.32
	2.00	5.67
	3.00	5.90
	4.00	6.09
Region 2 (buffer)	5.00	6.27
	6.00	6.45
	7.00	6.64
	8.00	6.87
	9.00	7.22
	9.50	7.55
	9.90	8.27
Region 3 (weak base)	10.00	9.25
	10.10	10.22
	10.50	10.91
	11.00	11.21
Region 4 (excess OH⁻)	12.00	11.50
	13.00	11.67
	14.00	11.79
	15.00	11.88
	16.00	11.95

FIGURE 11-2 Calculated titration curve for the reaction of 50.00 mL of 0.020 00 M MES with 0.100 0 M NaOH. Landmarks occur at half of the equivalence volume (pH = pK_a) and at the equivalence point, which is the steepest part of the curve.

Because the reactants are a weak base and a strong acid, the reaction goes essentially to completion after each addition of acid. There are four distinct regions of the titration curve:

1. Before acid is added, the solution contains just the weak base, B, in water. The pH is determined by the K_b reaction.

$$\underset{F-x}{B} + H_2O \xrightarrow{K_b} \underset{x}{BH^+} + \underset{x}{OH^-}$$

When V_a (= volume of added acid) = 0, we have a *weak-base* problem.

2. Between the initial point and the equivalence point, there is a mixture of B and BH⁺— *Aha! A buffer!* The pH is computed by using

$$pH = pK_a \text{ (for BH}^+) + \log\left(\frac{[B]}{[BH^+]}\right)$$

When $0 < V_a < V_e$, we have a *buffer*.

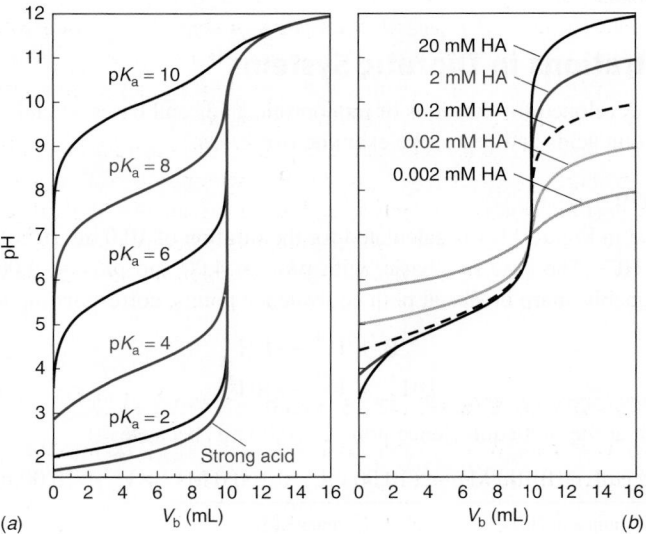

(a)

(b)

FIGURE 11-3 (*a*) Calculated curves showing the titration of 50.0 mL of 0.020 0 M HA with 0.100 M NaOH. (*b*) Calculated curves showing the titration of 50.0 mL of HA (pK_a = 5) with NaOH whose concentration is five times greater than that of HA. As the acid becomes weaker or more dilute, the end point becomes less distinct.

In adding acid (increasing V_a), we reach the special point where $V_a = \frac{1}{2}V_e$ and $\text{pH} = \text{p}K_a$ (for BH^+). As before, $\text{p}K_a$ can be determined easily from the titration curve.

3. At the equivalence point, B has been converted into BH^+, a weak acid. The pH is calculated by considering the acid dissociation reaction of BH^+.

When $V_a = V_e$, the solution contains the *weak acid* BH^+.

$$\underset{F' - x}{\text{BH}^+} \rightleftharpoons \underset{x}{B} + \underset{x}{\text{H}^+} \qquad K_a = \frac{K_w}{K_b}$$

The formal concentration of BH^+, F', is not the original formal concentration of B because some dilution has occurred. The solution contains BH^+ at the equivalence point, so it is acidic. *The pH at the equivalence point must be below 7.*

For $V_a > V_e$, there is excess *strong acid*.

4. After the equivalence point, the excess strong acid determines the pH. We neglect the contribution of weak acid, BH^+.

EXAMPLE Titration of Pyridine with HCl

Consider the titration of 25.00 mL of 0.083 64 M pyridine with 0.106 7 M HCl.

$$K_b = 1.59 \times 10^{-9} \Rightarrow K_a = \frac{K_w}{K_b} = 6.31 \times 10^{-6} \qquad \text{p}K_a = 5.20$$

Pyridine

The titration reaction is

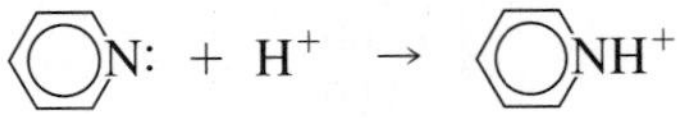

$$\text{N:} + \text{H}^+ \rightarrow \text{NH}^+$$

and the equivalence point occurs at 19.60 mL:

$$\underbrace{(V_e(\text{mL}))(0.106\ 7\ \text{M})}_{\text{mmol of HCl}} = \underbrace{(25.00\ \text{mL})(0.083\ 64\ \text{M})}_{\text{mmol of pyridine}} \Rightarrow V_e = 19.60\ \text{mL}$$

Find the pH when $V_a = 4.63$ mL.

Solution Part of the pyridine has been neutralized, so there is a mixture of pyridine and pyridinium ion—*Aha! A buffer!* The fraction of pyridine that has been titrated is $4.63/19.60 = 0.236$, because it takes 19.60 mL to titrate the whole sample. The fraction of pyridine remaining is $1 - 0.236 = 0.764$. The pH is

$$\text{pH} = \text{p}K_a + \log\left(\frac{[B]}{[\text{BH}^+]}\right)$$

$$= 5.20 + \log\frac{0.764}{0.236} = 5.71$$

TEST YOURSELF Find the pH when $V_a = 14.63$ mL. (*Answer:* 4.73)

11-4 Titrations in Diprotic Systems

The principles developed for titrations of monoprotic acids and bases extend directly to titrations of polyprotic acids and bases. We examine two cases.

A Typical Case

The upper curve in Figure 11-4 is calculated for the titration of 10.0 mL of 0.100 M base (B) with 0.100 M HCl. The base is dibasic, with $\text{p}K_{b1} = 4.00$ and $\text{p}K_{b2} = 9.00$. The titration curve has reasonably sharp breaks at both equivalence points, corresponding to the reactions

$$B + \text{H}^+ \rightarrow \text{BH}^+$$
$$\text{BH}^+ + \text{H}^+ \rightarrow \text{BH}_2^{2+}$$

The volume at the first equivalence point is 10.00 mL because

$$\underbrace{(V_e(\text{mL}))(0.100\ \text{M})}_{\text{mmol of HCl}} = \underbrace{(10.00\ \text{mL})(0.100\ 0\ \text{M})}_{\text{mmol of B}} \Rightarrow V_e = 10.00\ \text{mL}$$

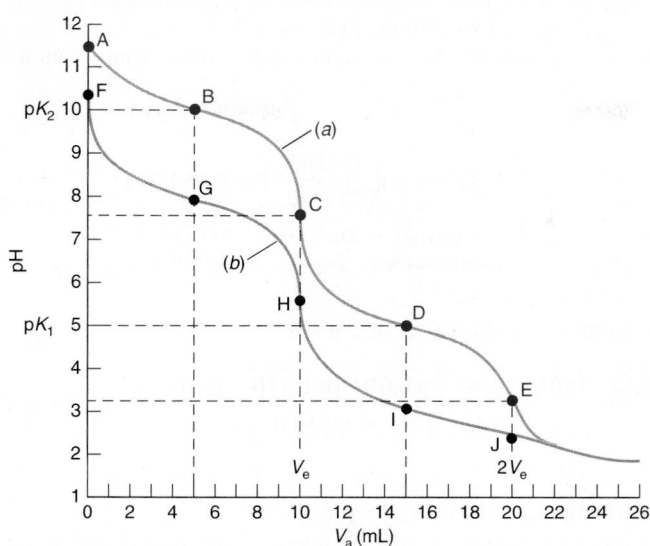

FIGURE 11-4 (*a*) Titration of 10.0 mL of 0.100 M base (pK_{b1} = 4.00, pK_{b2} = 9.00) with 0.100 M HCl. The two equivalence points are C and E. Points B and D are the half-neutralization points, whose pH values equal pK_{a2} and pK_{a1}, respectively. (*b*) Titration of 10.0 mL of 0.100 M nicotine (pK_{b1} = 6.15, pK_{b2} = 10.85) with 0.100 M HCl. There is no sharp break at the second equivalence point, J, because the pH is too low.

The volume at the second equivalence point must be $2V_e$, because the second reaction requires the same number of moles of HCl as the first reaction.

The pH calculations are similar to those for corresponding points in the titration of a monobasic compound. Let's examine points A through E in Figure 11-4.

$V_{e2} = 2V_{e1}$, always.

Point A

Before acid is added, the solution contains just weak base, B, whose pH is governed by the reaction

$$\underset{0.100\,-\,x}{B} + H_2O \overset{K_{b1}}{\rightleftharpoons} \underset{x}{BH^+} + \underset{x}{OH^-}$$

$$\frac{x^2}{0.100 - x} = 1.00 \times 10^{-4} \Rightarrow x = 3.11 \times 10^{-3}$$

$$[H^+] = \frac{K_w}{x} \Rightarrow pH = 11.49$$

The fully basic form of a dibasic compound can be treated as if it were monobasic. (The K_{b2} reaction can be neglected.)

Point B

At any point between A (the initial point) and C (the first equivalence point), we have a buffer containing B and BH^+. Point B is halfway to the equivalence point, so $[B] = [BH^+]$. The pH is calculated from the Henderson-Hasselbalch equation *for the weak acid*, BH^+, whose acid dissociation constant is K_{a2} (for BH_2^{2+}) = K_w/K_{b1} = $10^{-10.00}$.

$$pH = pK_{a2} + \log\frac{[B]}{[BH^+]} = 10.00 + \log 1 = 10.00$$

Of course, you remember that

$$K_{a2} = \frac{K_w}{K_{b1}} \qquad K_{a1} = \frac{K_w}{K_{b2}}$$

So the pH at point B is just pK_{a2}.

To calculate the quotient $[B]/[BH^+]$ at any point in the buffer region, just find what fraction of the way from point A to point C the titration has progressed. For example, if V_a = 1.5 mL, then

$$\frac{[B]}{[BH^+]} = \frac{8.5\ \text{mL}}{1.5\ \text{mL}}$$

because 10.0 mL are required to reach the equivalence point and we have added just 1.5 mL. The pH at V_a = 1.5 mL is

$$pH = 10.00 + \log\frac{8.5}{1.5} = 10.75$$

Point C

At the first equivalence point, *B has been converted into BH^+, the intermediate form of the diprotic acid, BH_2^{2+}. BH^+ is both an acid and a base.* From Equation 10-11, we know that

BH^+ is the *intermediate form* of a diprotic acid.

$$\text{pH} \approx \frac{1}{2}(\text{p}K_1 + \text{p}K_2)$$

$$[\text{H}^+] \approx \sqrt{\frac{K_1 K_2 \text{F} + K_1 K_\text{w}}{K_1 + \text{F}}} \qquad (11\text{-}3)$$

where K_1 and K_2 are the acid dissociation constants of BH_2^{2+}.

The formal concentration of BH^+ is calculated by considering dilution of the original solution of B.

$$\text{F} = \underbrace{(0.100\ \text{M})}_{\substack{\text{Original} \\ \text{concentration} \\ \text{of B}}} \underbrace{\left(\frac{10.0\ \text{mL}}{20.0\ \text{mL}} \right)}_{\substack{\text{Dilution} \\ \text{factor}}} = 0.050\ 0\ \text{M}$$

Initial volume of B

Total volume of solution

Plugging all the numbers into Equation 11-3 gives

$$[\text{H}^+] = \sqrt{\frac{(10^{-5})(10^{-10})(0.050\ 0) + (10^{-5})(10^{-14})}{10^{-5} + 0.050\ 0}} = 3.16 \times 10^{-8}$$

$$\text{pH} = 7.50$$

Note that, in this example, $\text{pH} = \frac{1}{2}(\text{p}K_{a1} + \text{p}K_{a2})$.

Point C in Figure 11-4 shows where the intermediate form of a diprotic acid lies on a titration curve. This is the *least-buffered* point on the whole curve, because the pH changes most rapidly if small amounts of acid or base are added. There is a misconception that the intermediate form of a diprotic acid behaves as a buffer when, in fact, it is the *worst choice* for a buffer.

The intermediate form of a polyprotic acid is the worst possible choice for a buffer.

Point D

At any point between C and E, there is a buffer containing BH^+ (the base) and BH_2^{2+} (the acid). When $V_\text{a} = 15.0\ \text{mL}$, $[\text{BH}^+] = [\text{BH}_2^{2+}]$ and

$$\text{pH} = \text{p}K_{a1} + \log \frac{[\text{BH}^+]}{[\text{BH}_2^{2+}]} = 5.00 + \log 1 = 5.00$$

Challenge Show that, if V_a were 17.2 mL, the ratio in the log term would be

$$\frac{[\text{BH}^+]}{[\text{BH}_2^{2+}]} = \frac{20.0\ \text{mL} - 17.2\ \text{mL}}{17.2\ \text{mL} - 10.0\ \text{mL}} = \frac{2.8}{7.2}$$

Point E

Point E is the second equivalence point, at which the solution is formally the same as one prepared by dissolving BH_2Cl_2 in water. The formal concentration of BH_2^{2+} is

$$\text{F} = (0.100\ \text{M})\left(\frac{10.0\ \text{mL}}{30.0\ \text{mL}} \right) = 0.033\ 3\ \text{M}$$

Original volume of B

Total volume of solution

The pH is determined by the acid dissociation reaction of BH_2^{2+}.

At the second equivalence point, we have made BH_2^{2+}, which can be treated as a monoprotic weak acid.

$$\underset{\text{F} - x}{\text{BH}_2^{2+}} \rightleftharpoons \underset{x}{\text{BH}^+} + \underset{x}{\text{H}^+} \qquad K_{a1} = \frac{K_\text{w}}{K_{b2}}$$

$$\frac{x^2}{0.033\ 3 - x} = 1.0 \times 10^{-5} \Rightarrow x = 5.72 \times 10^{-4} \Rightarrow \text{pH} = 3.24$$

Beyond the second equivalence point ($V_\text{a} > 20.0\ \text{mL}$), the pH of the solution can be calculated from the volume of strong acid added to the solution. For example, at $V_\text{a} = 25.00\ \text{mL}$, there is an excess of 5.00 mL of 0.100 M HCl in a total volume of $10.00 + 25.00 = 35.00\ \text{mL}$. The pH is found by writing

$$[\text{H}^+] = (0.100\ \text{M})\left(\frac{5.00\ \text{mL}}{35.00\ \text{mL}} \right) = 1.43 \times 10^{-2}\ \text{M} \Rightarrow \text{pH} = 1.85$$

Blurred End Points

When the equivalence-point pH is too low or too high or when pK_a values are too close together, end points are obscured.

Titrations of many diprotic acids or bases show two clear end points, as in curve *a* in Figure 11-4. Some titrations do not show both end points, as illustrated by curve *b*, which is calculated for the titration of 10.0 mL of 0.100 M nicotine (pK_{b1} = 6.15, pK_{b2} = 10.85) with 0.100 M HCl. The two reactions are

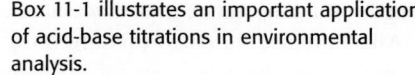

Nicotine (B) BH^+ BH_2^{2+}

There is no perceptible break at the second equivalence point (J), because BH_2^{2+} is too strong an acid (or, equivalently, BH^+ is too weak a base). As the titration approaches low pH($\lesssim$ 3), the approximation that HCl reacts completely with BH^+ to give BH_2^{2+} breaks down. To calculate pH between points I and J requires the systematic treatment of equilibrium. Later in this chapter, we describe how to calculate the whole curve with a spreadsheet.

11-5 Finding the End Point with a pH Electrode

Titrations are commonly performed to find out how much analyte is present or to measure equilibrium constants. We can obtain the information necessary for both purposes by monitoring pH during the titration. Figure 2-10 showed an *autotitrator*, which performs the entire operation automatically.[4] The instrument waits for pH to stabilize after each addition of titrant, before adding the next increment. The end point is computed automatically by finding the maximum slope in the titration curve.

Figure 11-5a shows experimental results for the manual titration of a hexaprotic weak acid, H_6A, with NaOH. Because the compound is difficult to purify, only a tiny amount was available for titration. Just 1.430 mg was dissolved in 1.000 mL of aqueous solution and titrated with microliter quantities of 0.065 92 M NaOH, delivered with a Hamilton syringe.

Box 11-1 illustrates an important application of acid-base titrations in environmental analysis.

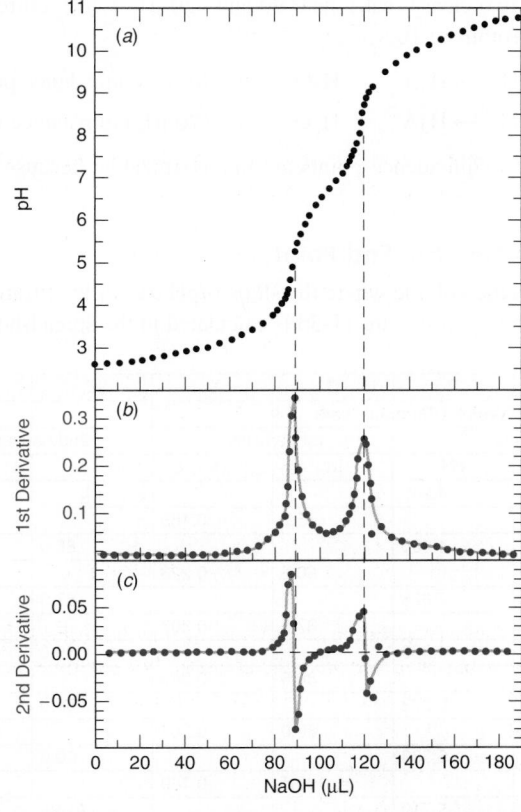

FIGURE 11-5 (*a*) Experimental points in the titration of 1.430 mg of xylenol orange, a hexaprotic acid, dissolved in 1.000 mL of aqueous 0.10 M $NaNO_3$. The titrant was 0.065 92 M NaOH. (*b*) The first derivative, $\Delta pH/\Delta V$, of the titration curve. (*c*) The second derivative, $\Delta(\Delta pH/\Delta V)/\Delta V$, which is the derivative of the curve in panel *b*. Derivatives for the first end point are calculated in Figure 11-6. End points are taken as maxima in the derivative curve and zero crossings of the second derivative.

BOX 11-1   **Alkalinity and Acidity**

The *alkalinity* of a sample of natural water is defined as the moles of HCl equivalent to the excess moles of basic species in the sample from weak acids with $pK_a > 4.5$ at 25°C and zero ionic strength.[2] Alkalinity is approximately equal to the moles of HCl needed to bring 1 kg of water to pH 4.5, which is the second equivalence point in the titration of CO_3^{2-}. To a good approximation,

$$\text{Alkalinity} \approx [OH^-] + 2[CO_3^{2-}] + [HCO_3^-]$$

When water is titrated with HCl to pH 4.5, OH^-, CO_3^{2-}, and HCO_3^- will have reacted. Minor species that could contribute to alkalinity in natural waters include phosphate, borate, silicate, fluoride, ammonia, sulfide, and organic compounds. In oceanography, alkalinity is used to estimate anthropogenic (man-made) CO_2 penetration into the ocean and in measuring the marine $CaCO_3$ budget (sources and sinks for $CaCO_3$).[3] Oceanographers must account for salinity (ionic strength) and temperature in the measurement of alkalinity.[2]

Alkalinity and *hardness* (dissolved Ca^{2+} and Mg^{2+}, Box 12-2) are important characteristics of irrigation water. Alkalinity in excess of the $Ca^{2+} + Mg^{2+}$ content is called "residual sodium carbonate." Water with a residual sodium carbonate content equivalent to ≥ 2.5 mmol H^+/kg is not suitable for irrigation. Residual sodium carbonate between 1.25 and 2.5 mmol H^+/kg is marginal, whereas ≤ 1.25 mmol H^+/kg is suitable for irrigation.

Acidity of natural waters refers to the total acid content that can be titrated to pH 8.3 with NaOH. This pH is the second equivalence

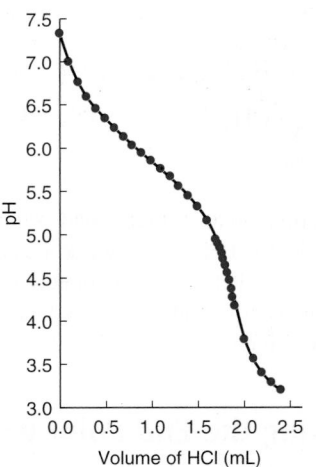

Alkalinity titration of 165.4 mL of saltwater at 20.05°C with 0.209 5 M HCl in a closed cell to prevent escape of CO_2. HCl contains NaCl so ionic strength remains constant. [Data from A. G. Dickson, Oak Ridge National Laboratory.]

point for titration of carbonic acid (H_2CO_3) with OH^-. Almost all weak acids in the water also will be titrated in this procedure. Acidity is expressed as millimoles of OH^- needed to bring 1 kg of water to pH 8.3.

Figure 11-5a shows two clear breaks, near 90 and 120 μL, which correspond to titration of the *third* and *fourth* protons of H_6A.

$$H_4A^{2-} + OH^- \rightarrow H_3A^{3-} + H_2O \qquad (\sim 90 \text{ μL equivalence point})$$
$$H_3A^{3-} + OH^- \rightarrow H_2A^{4-} + H_2O \qquad (\sim 120 \text{ μL equivalence point})$$

The first two and last two equivalence points are unrecognizable, because they occur at too low or too high a pH.

Using Derivatives to Find the End Point

The end point is taken as the volume where the slope (dpH/dV) of the titration curve is greatest. The slope (first derivative) in Figure 11-5b is calculated in the spreadsheet in Figure 11-6.

End point:

- maximum slope
- second derivative = 0

	A	B	C	D	E	F
1	Derivatives of a Titration Curve					
2	Data		1st derivative		2nd derivative	
3	μL NaOH	pH	μL	ΔpH/ΔμL		Δ(ΔpH/ΔμL)
4	85.0	4.245			μL	ΔμL
5			85.5	0.155		
6	86.0	4.400			86.0	0.0710
7			86.5	0.226		
8	87.0	4.626			87.0	0.0810
9			87.5	0.307		
10	88.0	4.933			88.0	0.0330
11			88.5	0.340		
12	89.0	5.273			89.0	−0.0830
13			89.0	0.257		
14	90.0	5.530			90.0	−0.0680
15			90.5	0.189		
16	91.0	5.719			91.25	−0.0390
17			92.0	0.131		
18	93.0	5.980				
19	Representative formulas:					
20	C5 = (A6+A4)/2			E6 = (C7+C5)/2		
21	D5 = (B6−B4)/(A6−A4)			F6 = (D7−D5)/(C7−C5)		

FIGURE 11-6 Spreadsheet for computing first and second derivatives near 90 μL in Figure 11-5.

The first two columns contain experimental volume and pH. (The pH meter was precise to three digits, even though accuracy ends in the second decimal place.) To compute the first derivative, each pair of volumes is averaged and the quantity $\Delta pH/\Delta V$ is calculated. The change in pH between consecutive readings is ΔpH and the change in volume between consecutive additions is ΔV. Figure 11-5c and the last two columns of the spreadsheet give the second derivative, computed in an analogous manner. The end point is the volume at which the second derivative is 0. Figure 11-7 allows us to make good estimates of the end points.

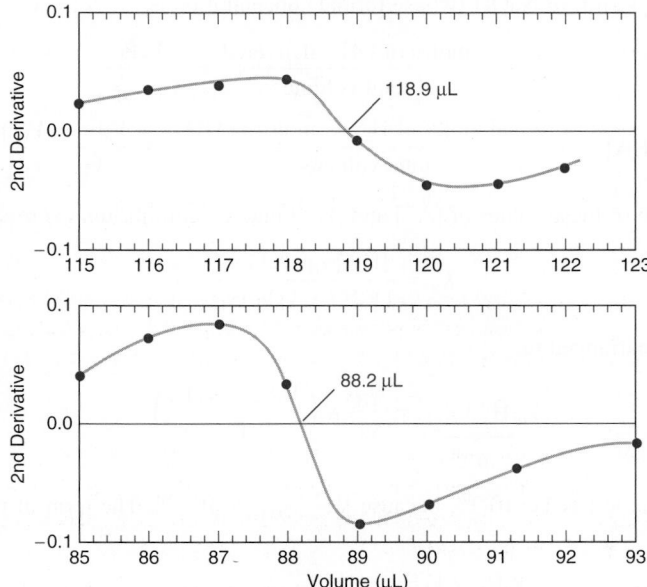

FIGURE 11-7 Enlargement of the end-point regions in the second derivative curve shown in Figure 11-5c.

EXAMPLE **Computing Derivatives of a Titration Curve**

Let's see how the first and second derivatives in Figure 11-6 are calculated.

Solution The first number in the third column, 85.5, is the average of the first two volumes (85.0 and 86.0) in the first column. The derivative $\Delta pH/\Delta V$ is calculated from the first two pH values and the first two volumes:

$$\frac{\Delta pH}{\Delta V} = \frac{4.400 - 4.245}{86.0 - 85.0} = 0.155$$

The coordinates ($x = 85.5$, $y = 0.155$) are one point in the graph of the first derivative in Figure 11-5b.

The second derivative is computed from the first derivative. The first entry in the fifth column of Figure 11-6 is 86.0, which is the average of 85.5 and 86.5. The second derivative is

$$\frac{\Delta(\Delta pH/\Delta V)}{\Delta V} = \frac{0.226 - 0.155}{86.5 - 85.5} = 0.071$$

The coordinates ($x = 86.0$, $y = 0.071$) are plotted in the second derivative graph in Figure 11-5c.

TEST YOURSELF Verify the derivative in cell D7 of Figure 11-6.

Using a Gran Plot to Find the End Point[5,6]

A problem with using derivatives to find the end point is that titration data are least accurate right near the end point, because buffering is minimal and electrode response is sluggish. A **Gran plot** uses data from before the end point (typically from 0.8 V_e or 0.9 V_e up to V_e) to locate the end point.

A related method uses data from the middle of the titration (not near the equivalence point) to deduce V_e and K_a.[7]

Consider the titration of a weak acid, HA:

$$HA \rightleftharpoons H^+ + A^- \qquad K_a = \frac{[H^+]\gamma_{H^+}[A^-]\gamma_{A^-}}{[HA]\gamma_{HA}}$$

It will be necessary to include activity coefficients in this discussion because a pH electrode responds to hydrogen ion *activity*, not concentration.

Prior to the equivalence point, it is a good approximation to say that each mole of NaOH converts 1 mol of HA into 1 mol of A^-. If we titrate V_a mL of HA (whose formal concentration is F_a) with V_b mL of NaOH (whose formal concentration is F_b), we can write

$$[A^-] = \frac{\text{moles of } OH^- \text{ delivered}}{\text{total volume}} = \frac{V_b F_b}{V_b + V_a}$$

$$[HA] = \frac{\text{original moles of } HA - \text{moles of } OH^-}{\text{total volume}} = \frac{V_a F_a - V_b F_b}{V_a + V_b}$$

Substitution of these values of $[A^-]$ and $[HA]$ into the equilibrium expression gives

$$K_a = \frac{[H^+]\gamma_{H^+} V_b F_b \gamma_{A^-}}{(V_a F_a - V_b F_b)\gamma_{HA}}$$

which can be rearranged to

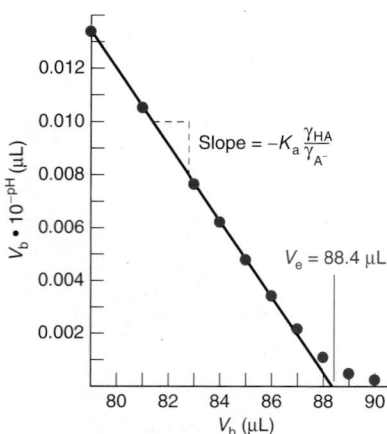

$$\underbrace{V_b[H^+]\gamma_{H^+}}_{10^{-pH}} = \frac{\gamma_{HA}}{\gamma_{A^-}}K_a\left(\frac{V_a F_a - V_b F_b}{F_b}\right) \qquad (11\text{-}4)$$

The term on the left is $V_b \cdot 10^{-pH}$, because $[H^+]\gamma_{H^+} = 10^{-pH}$. The term in parentheses on the right is

$$\frac{V_a F_a - V_b F_b}{F_b} = \frac{V_a F_a}{F_b} - V_b = V_e - V_b$$

Equation 11-4 can, therefore, be written in the form

Gran plot equation: $\qquad V_b \cdot 10^{-pH} = \frac{\gamma_{HA}}{\gamma_{A^-}}K_a(V_e - V_b) \qquad (11\text{-}5)$

A graph of $V_b \cdot 10^{-pH}$ versus V_b is called a *Gran plot*. If γ_{HA}/γ_{A^-} is constant, the graph is a straight line with a slope of $-K_a\gamma_{HA}/\gamma_{A^-}$ and an x-intercept of V_e. Figure 11-8 shows a Gran plot for the titration in Figure 11-5. Any units can be used for V_b, but the same units must be used on both axes. In Figure 11-8, V_b is expressed in microliters on both axes.

The beauty of a Gran plot is that it enables us to use data taken *before* the end point to find the end point. The slope of the Gran plot enables us to find K_a. Although we derived the Gran function for a monoprotic acid, the same plot ($V_b \cdot 10^{-pH}$ versus V_b) applies to polyprotic acids (such as H_6A in Figure 11-5).

The Gran function, $V_b \cdot 10^{-pH}$, does not actually go to 0, because 10^{-pH} is never 0. The curve must be extrapolated to find V_e. The reason the function does not reach 0 is that we have used the approximation that every mole of OH^- generates 1 mol of A^-, which is not true as V_b approaches V_e. Only the linear portion of the Gran plot is used.

Another source of curvature in the Gran plot is changing ionic strength, which causes γ_{HA}/γ_{A^-} to vary. In Figure 11-8, this variation was avoided by maintaining nearly constant ionic strength with $NaNO_3$. Even without added salt, the last 10–20% of data before V_e gives a fairly straight line because the quotient γ_{HA}/γ_{A^-} does not change very much.

Challenge Show that, when weak base, B, is titrated with a strong acid, the Gran function is

$$V_a \cdot 10^{+pH} = \left(\frac{1}{K_a} \cdot \frac{\gamma_B}{\gamma_{BH^+}}\right)(V_e - V_a) \qquad (11\text{-}6)$$

where V_a is the volume of strong acid and K_a is the acid dissociation constant of BH^+. A graph of $V_a \cdot 10^{+pH}$ versus V_a should be a straight line with a slope of $-\gamma_B/(\gamma_{BH^+}K_a)$ and an x-intercept of V_e.

Strong plus weak react completely.

$\mathcal{A}_{H^+} = [H^+]\gamma_{H^+} = 10^{-pH}$

$V_a F_a = V_e F_b \Rightarrow V_e = \frac{V_a F_a}{F_b}$

Gran plot:
- Plot $V_b \cdot 10^{-pH}$ versus V_b
- x-intercept $= V_e$
- Slope $= -K_a\gamma_{HA}/\gamma_{A^-}$

FIGURE 11-8 Gran plot for the first equivalence point of Figure 11-5. This plot gives an estimate of V_e that differs from that in Figure 11-7 by 0.2 μL (88.4 versus 88.2 μL). The last 10–20% of volume prior to V_e is normally used for a Gran plot.

11-6 Finding the End Point with Indicators

An acid-base **indicator** is itself an acid or base whose various protonated species have different colors. An example is thymol blue.

An **indicator** is an acid or a base whose various protonated forms have different colors.

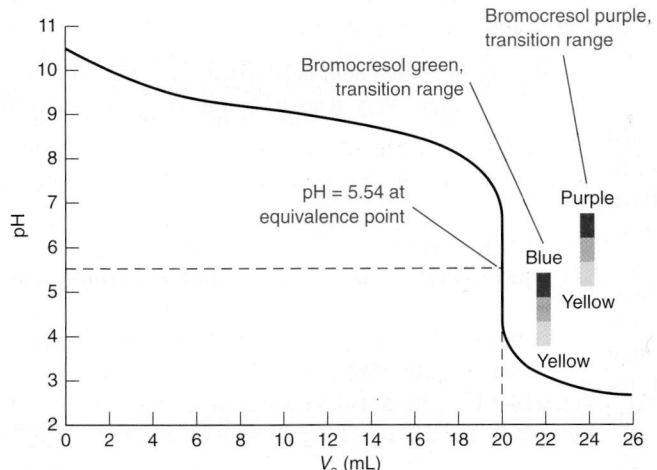

Red (R)
Thymol blue

Yellow (Y⁻)

Blue (B²⁻)

$$pK_1 = 1.7 \qquad pK_2 = 8.9$$

Below pH 1.7, the predominant species is red; between pH 1.7 and pH 8.9, the predominant species is yellow; above pH 8.9, the predominant species is blue (Color Plate 5). For simplicity, we designate the three species R, Y⁻, and B²⁻.

The equilibrium between R and Y⁻ can be written

$$R \overset{K_1}{\rightleftharpoons} Y^- + H^+ \qquad pH = pK_1 + \log \frac{[Y^-]}{[R]} \tag{11-7}$$

pH	[Y⁻]:[R]	Color
0.7	1:10	red
1.7	1:1	orange
2.7	10:1	yellow

At pH 1.7 ($= pK_1$), there will be a 1:1 mixture of the yellow and red species, which appears orange. As a crude rule of thumb, we can say that the solution will appear red when $[Y^-]/[R]$ $\lesssim 1/10$ and yellow when $[Y^-]/[R] \gtrsim 10/1$. From Equation 11-7, we see that the solution will be red when pH $\approx pK_1 - 1$ and yellow when pH $\approx pK_1 + 1$. In tables of indicator colors, thymol blue is listed as red below pH 1.2 and yellow above pH 2.8. The pH values predicted by our rule of thumb are 0.7 and 2.7. Between pH 1.2 and 2.8, the indicator exhibits various shades of orange. The pH range (1.2 to 2.8) over which the color changes is called the **transition range.** Whereas most indicators have a single color change, thymol blue has another transition, from yellow to blue, between pH 8.0 and pH 9.6. In this range, various shades of green are seen.

Acid-base indicator color changes are featured in Demonstration 11-1. Box 11-2 shows how optical absorption by indicators allows us to measure pH.

Choosing an Indicator

A titration curve for which pH = 5.54 at the equivalence point is shown in Figure 11-9. An indicator with a color change near this pH would be useful for an end point. In Figure 11-9,

One of the most common indicators is phenolphthalein, usually used for its colorless-to-pink transition at pH 8.0–9.6.

Colorless phenolphthalein
pH < 8.0

$2H^+ \ \| \ 2OH^-$

$+ \ 2H_2O$

Pink phenolphthalein
pH > 9.6

In strong acid, the colorless form of phenolphthalein turns orange-red. In strong base, the red species loses its color.[8]

Orange-red
(in 65–98% H_2SO_4)

Colorless
pH > 11

pH = 5.54 at equivalence point

Bromocresol green, transition range

Bromocresol purple, transition range

Purple

Blue

Yellow

Yellow

FIGURE 11-9 Calculated titration curve for the reaction of 100 mL of 0.010 0 M base ($pK_b = 5.00$) with 0.050 0 M HCl.

BOX 11-2   **What Does a Negative pH Mean?**

In the 1930s, Louis Hammett and his students measured the strengths of very weak acids and bases, using a weak reference base (B), such as p-nitroaniline ($pK_a = 0.99$), whose base strength could be measured in aqueous solution.

$$p\text{-Nitroanilinium ion} \quad \xrightarrow{\;pK_a = 0.99\;} \quad p\text{-Nitroaniline} + H^+$$

p-Nitroanilinium ion
BH$^+$

p-Nitroaniline
B

Suppose that some p-nitroaniline and a second base, C, are dissolved in a strong acid, such as 2 M HCl. The pK_a of CH$^+$ can be measured relative to that of BH$^+$ by first writing a Henderson-Hasselbalch equation for each acid:

$$pH = pK_a\,(\text{for BH}^+) + \log\frac{[B]\gamma_B}{[BH^+]\gamma_{BH^+}}$$

$$pH = pK_a\,(\text{for CH}^+) + \log\frac{[C]\gamma_C}{[CH^+]\gamma_{CH^+}}$$

Setting the two equations equal (because there is only one pH) gives

$$\underbrace{pK_a(\text{for CH}^+) - pK_a(\text{for BH}^+)}_{\Delta pK_a} = \log\frac{[B][CH^+]}{[C][BH^+]} + \log\frac{\gamma_B\gamma_{CH^+}}{\gamma_C\gamma_{BH^+}}$$

The ratio of activity coefficients is close to unity, so the second term on the right is close to 0. Neglecting this last term gives an operationally useful result:

$$\Delta pK_a \approx \log\frac{[B][CH^+]}{[C][BH^+]}$$

That is, if you have a way to find the concentrations of B, BH$^+$, C, and CH$^+$ and if you know pK_a for BH$^+$, then you can find pK_a for CH$^+$.

Concentrations can be measured with a spectrophotometer[10] or by nuclear magnetic resonance,[11] so pK_a for CH$^+$ can be determined. Then, with CH$^+$ as the reference, the pK_a for another compound, DH$^+$, can be measured. This procedure can be extended to measure the strengths of successively weaker bases (such as nitrobenzene, $pK_a = -11.38$), far too weak to be protonated in water.

The acidity of a strongly acidic solvent that protonates the weak base, B, is defined as the **Hammett acidity function:**

Hammett acidity function:
$$H_0 = pK_a\,(\text{for BH}^+) + \log\frac{[B]}{[BH^+]}$$

For dilute aqueous solutions, H_0 approaches pH. For concentrated acids, H_0 is a measure of the acid strength. The weaker a base, B, the stronger the acidity of the solvent must be to protonate the base. Acidity of strongly acidic solvents is now measured more conveniently by electrochemical methods.[12]

When we refer to *negative* pH, we usually mean H_0 values. For example, as measured by its ability to protonate very weak bases, 8 M HClO$_4$ has a "pH" close to -4. The figure shows that HClO$_4$ is a stronger acid than other mineral acids. Values of H_0 for several powerfully acidic solvents are shown in the table. The

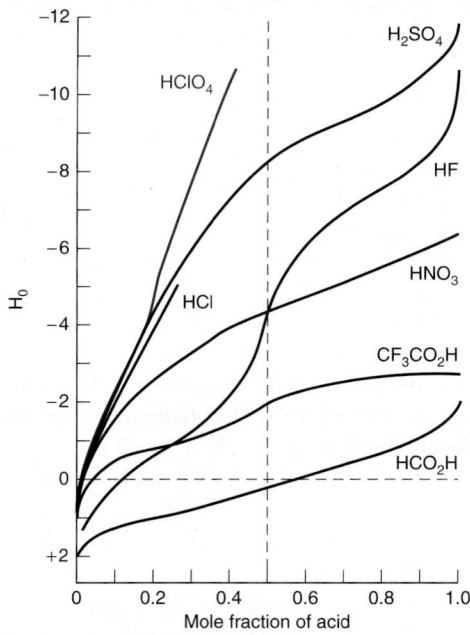

Hammett acidity function, H_0, for aqueous solutions of acids. [Data from R. A. Cox and K. Yates, "Acidity Functions," *Can. J. Chem.* **1983,** *61*, 2225.]

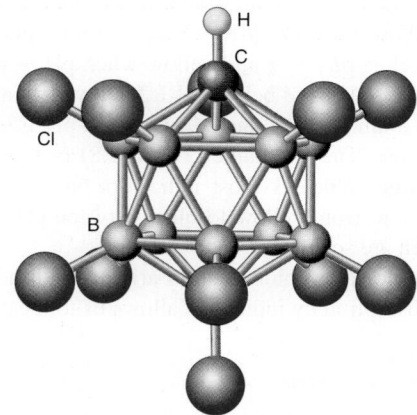

Icosahedral structure of carborane anion of [CHB$_{11}$Cl$_{11}$]$^-$H$^+$, the strongest known acid.[13] Icosahedral H$_2$[B$_{12}$Cl$_{12}$] is the strongest known diprotic acid.

strongest known acid is [CHB$_{11}$Cl$_{11}$]$^-$H$^+$, with an icosahedral carborane cage. Very weakly bound H$^+$ is attached to Cl in the gas and solid phase.[13]

Acid	Name	H_0
H$_2$SO$_4$ (100%)	sulfuric acid	-11.93
H$_2$SO$_4 \cdot$ SO$_3$	fuming sulfuric acid (oleum)	-14.14
HSO$_3$F	fluorosulfuric acid	-15.07
HSO$_3$F + 10% SbF$_5$	"super acid"	-18.94
HSO$_3$F + 7% SbF$_5 \cdot$ 3SO$_3$	—	-19.35

This one is just plain fun.[9] Place 900 mL of water and a magnetic stirring bar in each of two 1-L graduated cylinders. Add 10 mL of 1 M NH_3 to each. Then put 2 mL of phenolphthalein indicator solution in one and 2 mL of bromothymol blue indicator solution in the other. Both indicators will have the color of their basic species.

Drop a few chunks of Dry Ice (solid CO_2) into each cylinder. As the CO_2 bubbles through each cylinder, the solutions become more acidic. First the pink phenolphthalein color disappears. After some time, the pH drops low enough for bromothymol blue to change from blue to green, but not all the way to yellow.

Add 20 mL of 6 M HCl to *the bottom* of each cylinder, using a length of Tygon tubing attached to a funnel. Then stir each solution for a few seconds on a magnetic stirrer. Explain what happens. The sequence of events is shown in Color Plate 6.

the pH drops steeply (from 7 to 4) over a small volume interval. Therefore, any indicator with a color change in this pH interval would provide a fair approximation to the equivalence point. The closer the point of color change is to pH 5.54, the more accurate will be the end point. The difference between the observed end point (color change) and the true equivalence point is called the **indicator error.**

If you dump half a bottle of indicator into your reaction, you will introduce a different indicator error. Because indicators are acids or bases, they react with analyte or titrant. The moles of indicator must be negligible relative to the moles of analyte. Never use more than a few drops of dilute indicator solution.

Several of the indicators in Table 11-3 would be useful for the titration in Figure 11-9. If bromocresol purple were used, we would use the purple-to-yellow color change as the end point. The last trace of purple should disappear near pH 5.2, which is quite close to the true equivalence point in Figure 11-9. If bromocresol green were used as the indicator, a color change from blue to green (= yellow + blue) would mark the end point.

TABLE 11-3 Common indicators

Indicator	Transition range (pH)	Acid color	Base color	Preparation
Methyl violet	0.0–1.6	Yellow	Violet	0.05 wt% in H_2O
Cresol red	0.2–1.8	Red	Yellow	0.1 g in 26.2 mL 0.01 M NaOH. Then add ~225 mL H_2O.
Thymol blue	1.2–2.8	Red	Yellow	0.1 g in 21.5 mL 0.01 M NaOH. Then add ~225 mL H_2O.
Cresol purple	1.2–2.8	Red	Yellow	0.1 g in 26.2 mL 0.01 M NaOH. Then add ~225 mL H_2O.
Erythrosine, disodium	2.2–3.6	Orange	Red	0.1 wt% in H_2O
Methyl orange	3.1–4.4	Red	Yellow	0.01 wt% in H_2O
Congo red	3.0–5.0	Violet	Red	0.1 wt% in H_2O
Bromophenol blue	3.0–4.6	Yellow	Blue	0.1 g in 14.9 mL of 0.01 M NaOH. Then add ~225 mL H_2O.
Ethyl orange	3.4–4.8	Red	Yellow	0.1 wt% in H_2O
Bromocresol green	3.8–5.4	Yellow	Blue	0.1 g in 14.3 mL 0.01 M NaOH. Then add ~225 mL H_2O.
Methyl red	4.8–6.0	Red	Yellow	0.02 g in 60 mL ethanol. Then add 40 mL H_2O.
Chlorophenol red	4.8–6.4	Yellow	Red	0.1 g in 23.6 mL 0.01 M NaOH. Then add ~225 mL H_2O.
Bromocresol purple	5.2–6.8	Yellow	Purple	0.1 g in 18.5 mL 0.01 M NaOH. Then add ~225 mL H_2O.
p-Nitrophenol	5.6–7.6	Colorless	Yellow	0.1 wt% in H_2O
Litmus	5.0–8.0	Red	Blue	0.1 wt% in H_2O
Bromothymol blue	6.0–7.6	Yellow	Blue	0.1 g in 16.0 mL 0.01 M NaOH. Then add ~225 mL H_2O.
Phenol red	6.4–8.0	Yellow	Red	0.1 g in 28.2 mL 0.01 M NaOH. Then add ~225 mL H_2O.
Neutral red	6.8–8.0	Red	Yellow	0.01 g in 50 mL ethanol. Then add 50 mL H_2O.
Cresol red	7.2–8.8	Yellow	Red	See above.
α-Naphtholphthalein	7.3–8.7	Pink	Green	0.1 g in 50 mL ethanol. Then add 50 mL H_2O.
Cresol purple	7.6–9.2	Yellow	Purple	See above.
Thymol blue	8.0–9.6	Yellow	Blue	See above.
Phenolphthalein	8.0–9.6	Colorless	Pink	0.05 g in 50 mL ethanol. Then add 50 mL H_2O.
Thymolphthalein	8.3–10.5	Colorless	Blue	0.04 g in 50 mL ethanol. Then add 50 mL H_2O.
Alizarin yellow	10.1–12.0	Yellow	Orange-red	0.01 wt% in H_2O
Nitramine	10.8–13.0	Colorless	Orange-brown	0.1 g in 70 mL ethanol. Then add 30 mL H_2O.
Tropaeolin O	11.1–12.7	Yellow	Orange	0.1 wt% in H_2O

TABLE 11-4 Primary standards

Compound	Density (g/mL) for buoyancy corrections	Notes
ACIDS		
Potassium hydrogen phthalate FM 204.221	1.64	The pure commercial material is dried at 105°C and used to standardize base. A phenolphthalein end point is satisfactory.
HCl Hydrochloric acid FM 36.461	—	HCl and water distill as an *azeotrope* (a mixture), whose composition (~6 M) depends on pressure and is tabulated as a function of the pressure during distillation. See Problem 11-56 for more information.
$KH(IO_3)_2$ Potassium hydrogen iodate FM 389.912	—	This is a strong acid, so any indicator with an end point between ~5 and ~9 is adequate.
Benzoic acid FM 122.121	1.27	Primary standard for nonaqueous titrations in solvents such as ethanol. A glass electrode is used to find the end point.
Sulfosalicylic acid double salt FM 550.639	—	1 mol of commercial grade sulfosalicylic acid is combined with 0.75 mol of reagent-grade $KHCO_3$, recrystallized several times from water, and dried at 110°C to produce the double salt with 3 K^+ ions and one titratable H^+.[14] Phenolphthalein is used as the indicator for titration with NaOH.
$H_3\overset{+}{N}SO_3^-$ Sulfamic acid FM 97.094	2.15	Sulfamic acid is a strong acid with one acidic proton, so any indicator with an end point between ~5 and ~9 is suitable.
BASES		
$H_2NC(CH_2OH)_3$ Tris(hydroxymethyl)aminomethane (also called tris or tham) FM 121.135	1.33	The pure commercial material is dried at 100°–103°C and titrated with strong acid. The end point is in the range pH 4.5–5. $$H_2NC(CH_2OH)_3 + H^+ \rightarrow H_3\overset{+}{N}C(CH_2OH)_3$$
HgO Mercuric oxide FM 216.59	11.1	Pure HgO is dissolved in a large excess of I^- or Br^-, whereupon 2 OH^- are liberated: $$HgO + 4I^- + H_2O \rightarrow HgI_4^{2-} + 2OH^-$$ The base is titrated, using an indicator end point.
Na_2CO_3 Sodium carbonate FM 105.988	2.53	Primary-standard-grade Na_2CO_3 is commercially available. Alternatively, recrystallized $NaHCO_3$ can be heated for 1 h at 260°–270°C to produce pure Na_2CO_3. Sodium carbonate is titrated with acid to an end point of pH 4–5. Just before the end point, the solution is boiled to expel CO_2.
$Na_2B_4O_7 \cdot 10H_2O$ Borax FM 381.372	1.73	The recrystallized material is dried in a chamber containing an aqueous solution saturated with NaCl and sucrose. This procedure gives the decahydrate in pure form.[15] The standard is titrated with acid to a methyl red end point. $$\text{"}B_4O_7 \cdot 10H_2O^{2-}\text{"} + 2H^+ \rightarrow 4B(OH)_3 + 5H_2O$$

In general, *we seek an indicator whose transition range overlaps the steepest part of the titration curve as closely as possible.* The steepness of the titration curve near the equivalence point in Figure 11-9 ensures that the indicator error caused by the noncoincidence of the end point and equivalence point will be small. If the indicator end point were at pH 6.4 (instead of 5.54), the error in V_e would be only 0.25% in this particular case. You can estimate the indicator error by calculating the volume of titrant required to attain pH 6.4 instead of pH 5.54.

Choose an indicator whose transition range overlaps the steepest part of the titration curve.

11-7 Practical Notes

Acids and bases in Table 11-4 can be obtained pure enough to be *primary standards*.[16] NaOH and KOH are not primary standards because they contain carbonate (from reaction with atmospheric CO_2) and adsorbed water. Solutions of NaOH and KOH must be standardized against a primary standard such as potassium hydrogen phthalate. Solutions of NaOH for titrations are prepared by diluting a stock solution of 50 wt% aqueous NaOH. Sodium carbonate is insoluble in this stock solution and settles to the bottom.

Procedures for preparing standard acid and base are given at the end of this chapter.

Alkaline solutions (for example, 0.1 M NaOH) must be protected from the atmosphere; otherwise, they absorb CO_2:

$$OH^- + CO_2 \rightarrow HCO_3^-$$

CO_2 changes the concentration of strong base over a period of time and decreases the extent of reaction near the end point in the titration of weak acids. If solutions are kept in tightly capped polyethylene bottles, they can be used for about a week with little change.

Standard solutions are commonly stored in high-density polyethylene bottles with screw caps. Evaporation from the bottle slowly changes the reagent concentration. The chemical supplier Sigma-Aldrich reports that an aqueous solution stored in a tightly capped bottle became 0.2% more concentrated in 2 years at 23°C and 0.5% more concentrated in 2 years at 30°C. Enclosing the bottle in a sealed, aluminized bag reduced evaporation by a factor of 10. The lesson is that a standard solution has a finite shelf life.

Strongly basic solutions attack glass and are best stored in plastic containers. Such solutions should not be kept in a buret longer than necessary. Boiling 0.01 M NaOH in a flask for 1 h decreases the molarity by 10%, because OH^- reacts with glass.[17]

11-8 Kjeldahl Nitrogen Analysis

Developed in 1883, the **Kjeldahl nitrogen analysis** is one of the most widely used methods for determining nitrogen in organic substances.[18] Protein is the main nitrogen-containing constituent of food. Most proteins contain close to 16 wt% nitrogen, so measuring nitrogen is a surrogate for measuring protein (Box 11-3). The other common way to measure nitrogen in food is by combustion analysis (Section 27-4).

In the Kjeldahl method, a sample is first *digested* (decomposed and dissolved) in boiling sulfuric acid, which converts amine and amide nitrogen into ammonium ion, NH_4^+, and oxidizes other elements present:[24]

Kjeldahl digestion: organic C, H, N $\xrightarrow[H_2SO_4]{\text{boiling}}$ $NH_4^+ + CO_2 + H_2O$ **(11-8)**

Each atom of nitrogen in starting material is converted into one NH_4^+ ion.

Mercury, copper, and selenium compounds catalyze the digestion. To speed the reaction, the boiling point of concentrated (98 wt%) sulfuric acid (338°C) is raised by adding K_2SO_4. Digestion is carried out in a long-neck *Kjeldahl flask* (Figure 11-10) that prevents loss of sample from spattering. Alternative digestion procedures employ H_2SO_4 plus H_2O_2 or $K_2S_2O_8$ plus NaOH[25] in a microwave bomb (a pressurized vessel shown in Figure 28-7).

After digestion is complete, the solution containing NH_4^+ is made basic, and the liberated NH_3 is steam distilled (with a large excess of steam) into a receiver containing a known amount of HCl (Figure 11-11) (or $B(OH)_3$ in a different procedure in Problem 11-59).[26] Excess, unreacted HCl is then titrated with standard NaOH to determine how much HCl was consumed by NH_3.

BOX 11-3   Kjeldahl Nitrogen Analysis Behind the Headlines

In 2007, dogs and cats in North America suddenly began to die from kidney failure. The illness was traced to animal food containing ingredients from China. Melamine, used to make plastics, had been deliberately added to food ingredients "in a bid to meet the contractual demand for the amount of protein in the products."[19] Cyanuric acid, used to disinfect swimming pools, was also found in the food. Melamine alone does not cause kidney failure, but the combination of melamine and cyanuric acid makes a crystalline product that leads to kidney failure. Microbes residing in the gut can transform melamine into cyanuric acid.[20]

Melamine
(66.6 wt% nitrogen)

Cyanuric acid
(32.6 wt% nitrogen)

What do these compounds have to do with protein? Nothing—except that they are high in nitrogen. Protein, which contains ~16 wt% nitrogen, is the main source of nitrogen in food. Kjeldahl nitrogen analysis is a surrogate measurement for protein in food. For example, if food contains 10 wt% protein, it will contain ~16% of 10% = 1.6 wt% N. If you measure 1.6 wt% N in food, you could conclude that the food contains ~10 wt% protein. Melamine contains 66.6 wt% N, which is four times more than protein. Adding 1 wt% melamine to food makes it appear that the food contains an additional 4 wt% protein.

Protein source	Weight % nitrogen
meat	16.0
blood plasma	15.3
milk	15.6
flour	17.5
egg	14.9

SOURCE: D. J. Holme and H. Peck, *Analytical Biochemistry*, 3rd ed. (New York: Addison Wesley Longman, 1998), p. 388.

Incredibly, in the summer of 2008, approximately 300 000 Chinese babies became sick and at least six died of kidney failure.[21] Chinese companies had diluted milk with water and added melamine to make the protein content appear normal. Poisoned milk products were sold in the domestic and export markets. Two people were executed in 2009 for their role in producing tainted milk. In 2010, Chinese authorities found 40 tons of powdered milk containing melamine. In response to the appearance of melamine in food, at least one company developed a colorimetric assay that distinguishes protein nitrogen from nonprotein nitrogen[22] and many methods for the measurement of melamine in food have appeared.[23]

Another means to measure nitrogen in food is the *Dumas method*. Organic material mixed with CuO is heated in CO_2 at 650°–700°C to produce CO_2, H_2O, N_2, and nitrogen oxides. Products are carried by a stream of CO_2 through hot Cu to convert nitrogen oxides to N_2. The gases are bubbled through concentrated $KOH(aq)$ to capture CO_2. The volume of N_2 is measured in a gas buret. This method does not distinguish protein from melamine.

Neutralization of NH_4^+:

$$NH_4^+ + OH^- \rightarrow NH_3(g) + H_2O \qquad (11\text{-}9)$$

Distillation of NH_3 into standard HCl:

$$NH_3 + H^+ \rightarrow NH_4^+ \qquad (11\text{-}10)$$

Titration of unreacted HCl with NaOH:

$$H^+ + OH^- \rightarrow H_2O \qquad (11\text{-}11)$$

An alternative to a titration is to neutralize the acid and raise the pH with a buffer, followed by addition of reagents that form a colored product with NH_3.[27] The absorbance of the colored product gives the concentration of NH_3 from the digestion.

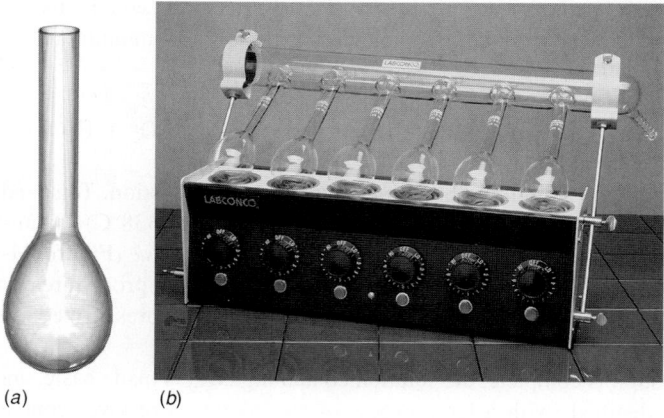

(a)   (b)

FIGURE 11-10 (*a*) Kjeldahl digestion flask with long neck to minimize loss from spattering. (*b*) Six-port manifold for multiple samples provides for exhaust of fumes. [© Courtesy Labconco Corporation.]

EXAMPLE Kjeldahl Analysis

A typical protein contains 16.2 wt% nitrogen. A 0.500-mL aliquot of protein solution was digested, and the liberated NH_3 was distilled into 10.00 mL of 0.021 40 M HCl. Unreacted HCl required 3.26 mL of 0.019 8 M NaOH for complete titration. Find the concentration of protein (mg protein/mL) in the original sample.

Solution The initial quantity of HCl in the receiver was (10.00 mL)(0.021 40 mmol/mL) = 0.214 0 mmol. The NaOH required for titration of unreacted HCl in Reaction 11-11 was (3.26 mL)(0.019 8 mmol/mL) = 0.064 5 mmol. The difference, 0.214 0 − 0.064 5 = 0.149 5 mmol, must be the quantity of NH_3 produced in Reaction 11-9 and distilled into the HCl.

Because 1 mol of N in the protein produces 1 mol of NH_3, there must have been 0.149 5 mmol of N in the protein, corresponding to

$$(0.149\ 5\ \text{mmol})\left(14.007\ \frac{\text{mg N}}{\text{mmol}}\right) = 2.094\ \text{mg N}$$

If the protein contains 16.2 wt% N, there must be

$$\frac{2.094\ \text{mg N}}{0.162\ \text{mg N/mg protein}} = 12.9\ \text{mg protein} \Rightarrow \frac{12.9\ \text{mg protein}}{0.500\ \text{mL}} = 25.8\ \frac{\text{mg protein}}{\text{mL}}$$

TEST YOURSELF Find mg protein/mL if 3.00 mL of NaOH were required. (*Answer: 26.7 mg/mL*)

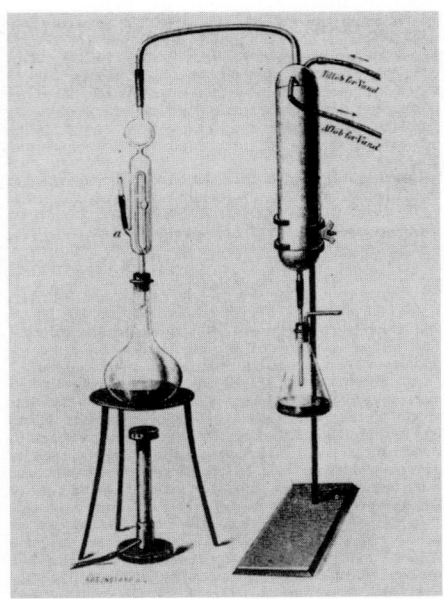

FIGURE 11-11 Original apparatus used by Dutch chemist J. Kjeldahl (1849–1900). [See D. T. Burns, "Kjeldahl, the Man, the Method and the Carlsberg Laboratory," *Anal. Proc.* (Royal Society of Chemistry) **1984,** *21,* 210. Johan Frederik Rosenstand, Danish artist (1820–1887).]

11-9 The Leveling Effect

The strongest acid that can exist in water is H_3O^+ and the strongest base is OH^-. If an acid stronger than H_3O^+ is dissolved in water, it protonates H_2O to make H_3O^+. If a base stronger than OH^- is dissolved in water, it deprotonates H_2O to make OH^-. Because of this **leveling effect,** $HClO_4$ and HCl behave as if they had the same acid strength; both are *leveled* to H_3O^+:

$$HClO_4 + H_2O \rightarrow H_3O^+ + ClO_4^-$$
$$HCl + H_2O \rightarrow H_3O^+ + Cl^-$$

The equilibrium constants for both reactions above are large. The $HClO_4$ and HCl are both converted into H_3O^+.

In acetic acid solvent, which is less basic than H_2O, $HClO_4$ and HCl are not leveled to the same strength:

$$HClO_4 + CH_3CO_2H \rightleftharpoons CH_3CO_2H_2^+ + ClO_4^- \qquad K = 1.3 \times 10^{-5}$$
$$\text{Acetic acid solvent}$$

$$HCl + CH_3CO_2H \rightleftharpoons CH_3CO_2H_2^+ + Cl^- \qquad K = 2.8 \times 10^{-9}$$

In acetic acid solution, $HClO_4$ is a stronger acid than HCl; but, in aqueous solution, both are leveled to the strength of H_3O^+.

The equilibrium constants for both reactions are small. However, the equilibrium constant for $HClO_4$ reacting with CH_3CO_2H is 10^4 times greater than the equilibrium constant for HCl reacting with CH_3CO_2H. These equilibrium indicate that $HClO_4$ is a stronger acid than HCl in acetic acid solvent.

Figure 11-12 shows a titration curve for a mixture of five acids titrated with 0.2 M tetrabutylammonium hydroxide in methyl isobutyl ketone solvent. This solvent is not protonated to a great extent by any of the acids. We see that perchloric acid is a stronger acid than HCl in this solvent as well.

Now consider a base such as urea, $(H_2N)_2C{=}O$ ($K_b = 1.3 \times 10^{-14}$), that is too weak to give a distinct end point when titrated with a strong acid in water.

Titration with $HClO_4$ in H_2O: $\qquad B + H_3O^+ \rightleftharpoons BH^+ + H_2O$

The end point cannot be recognized because the equilibrium constant for the titration reaction is not large enough. If an acid stronger than H_3O^+ were available, the titration reaction might have an equilibrium constant large enough to give a distinct end point. If the same

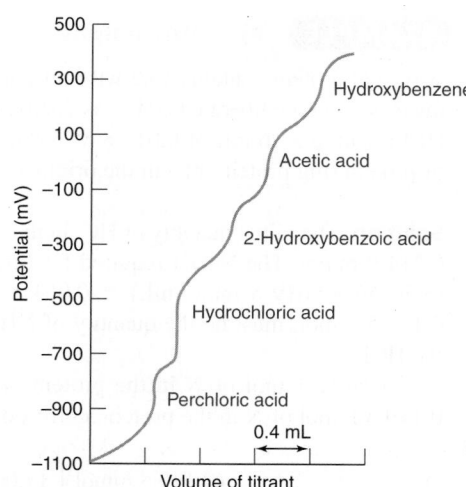

FIGURE 11-12 Titration of a mixture of acids with tetrabutylammonium hydroxide in methyl isobutyl ketone solvent shows that the order of acid strength is $HClO_4$ > HCl > 2-hydroxybenzoic acid > acetic acid > hydroxybenzene. Measurements were made with a glass electrode and a platinum reference electrode. The ordinate is proportional to pH, with increasing pH as the potential becomes more positive. [Data from D. B. Bruss and G. E. A. Wyld, "Methyl Isobutyl Ketone as a Wide-Range Solvent for Titration of Acid Mixtures and Nitrogen Bases," *Anal. Chem.* **1957,** *29,* 232.]

Question Where do you think the end point for the acid $H_3O^+ClO_4^-$ would come in Figure 11-12?

Answer: $H_3O^+ClO_4^-$ would come probably between HCl and 2-hydroxybenzoic acid because the acid is H_3O^+.

A base too weak to be titrated by H_3O^+ in water might be titrated by $HClO_4$ in acetic acid solvent.

Dielectric constant is discussed in Problem 8-14.

Compounds that can be protonated in acetonitrile by perchloric acid plus acetic acid, $CH_3C(OH)_2^+ClO_4^-$:

Thioacetamide

4-Nitrobenzamide

Experiment 10, "Fitting a Titration Curve with Excel Solver," at www.whfreeman.com/qca applies equations developed in this section.

base were dissolved in acetic acid and titrated with $HClO_4$ in acetic acid, a clear end point might be observed. The reaction

Titration with $HClO_4$ in CH_3CO_2H:
$$B + HClO_4 \rightleftharpoons \underbrace{BH^+ClO_4^-}_{\text{An ion pair}}$$

might have a large equilibrium constant, because $HClO_4$ is a much stronger acid than H_3O^+. (The product in this reaction is written as an ion pair because the *dielectric constant* of acetic acid is too low to allow ions to separate extensively.) Titrations that are not feasible in water might be feasible in other solvents.[28]

In *electrophoresis* (Chapter 26), ions are separated by their different mobilities in an electric field. Compounds shown in the margin are such weak bases that they cannot be protonated in aqueous solution, and, therefore, cannot be made into charged species for aqueous electrophoresis. However, in anhydrous acetonitrile solvent, they are protonated by $HClO_4$ in anhydrous acetic acid and can be separated as cations.[29]

11-10   Calculating Titration Curves with Spreadsheets

This chapter has been critical for developing your understanding of the chemistry that occurs during titrations. However, the approximations we used are of limited value when concentrations are too dilute or equilibrium constants are not of the right magnitude or K_a values are too closely spaced, like those in a protein. This section develops equations to deal with titrations in a general manner, using spreadsheets.[30]

Titrating a Weak Acid with a Strong Base

Consider the titration of a volume V_a of acid HA (initial concentration C_a) with a volume V_b of NaOH of concentration C_b. The charge balance for this solution is

Charge balance:
$$[H^+] + [Na^+] = [A^-] + [OH^-]$$

and the concentration of Na^+ is just

$$[Na^+] = \frac{C_bV_b}{V_a + V_b}$$

because we have diluted C_bV_b moles of NaOH to a total volume of $V_a + V_b$. Similarly, the formal concentration of the weak acid is

$$F_{HA} = [HA] + [A^-] = \frac{C_aV_a}{V_a + V_b}$$

because we have diluted C_aV_a moles of HA to a total volume of $V_a + V_b$.

Now we use the fractional composition equations from Section 10-5. The concentration of A^- can be written in terms of α_{A^-} defined in Equation 10-18:

α_{A^-} = fraction of acid in the form A^-:
$$\alpha_{A^-} = \frac{[A^-]}{F_{HA}}$$

$$[A^-] = \alpha_{A^-} \cdot F_{HA} = \frac{\alpha_{A^-} \cdot C_aV_a}{V_a + V_b} \qquad \textbf{(11-12)}$$

where $\alpha_{A^-} = K_a/([H^+] + K_a)$ and K_a is the acid dissociation constant of HA. Substituting for $[Na^+]$ and $[A^-]$ in the charge balance gives

$$[H^+] + \frac{C_b V_b}{V_a + V_b} = \frac{\alpha_{A^-} \cdot C_a V_a}{V_a + V_b} + [OH^-]$$

which you can rearrange to the form

Fraction of titration for weak acid by strong base:
$$\phi \equiv \frac{C_b V_b}{C_a V_a} = \frac{\alpha_{A^-} - \dfrac{[H^+] - [OH^-]}{C_a}}{1 + \dfrac{[H^+] - [OH^-]}{C_b}} \qquad (11\text{-}13)$$

$\phi = C_b V_b/C_a V_a$ is the fraction of the way to the equivalence point:

ϕ	Volume of base
0.5	$V_b = \frac{1}{2}V_e$
1	$V_b = V_e$
2	$V_b = 2V_e$

At last! Equation 11-13 is really useful. It relates the volume of titrant (V_b) to the pH and a bunch of constants. The quantity ϕ, which is the quotient $C_b V_b/C_a V_a$, is the fraction of the way to the equivalence point, V_e. When $\phi = 1$, the volume of base added, V_b, is equal to V_e. Equation 11-13 works backward from the way you are accustomed to thinking, because you need to put in pH on the right to get out volume on the left. Let me say that again: We *put in a concentration of H^+* and *get out the volume of titrant* that produces that concentration.

Let's use Equation 11-13 to calculate the titration curve for 50.00 mL of the weak acid 0.020 00 M MES with 0.100 0 M NaOH, which was shown in Figure 11-2 and Table 11-2. The equivalence volume is $V_e = 10.00$ mL. Quantities in Equation 11-13 are:

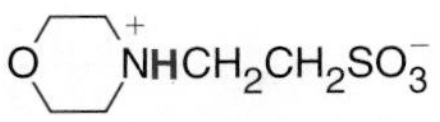

$$C_b = 0.1 \text{ M} \qquad\qquad [H^+] = 10^{-pH}$$
$$C_a = 0.02 \text{ M} \qquad\qquad [OH^-] = K_w/[H^+]$$
$$V_a = 50 \text{ mL}$$
$$K_a = 5.3_7 \times 10^{-7} \qquad \alpha_{A^-} = \frac{K_a}{[H^+] + K_a}$$
$$K_w = 10^{-14}$$

pH is the input $\qquad\qquad V_b = \dfrac{\phi C_a V_a}{C_b}$ is the output

2-(*N*-Morpholino)ethanesulfonic acid
MES, pK_a = 6.27

The input to the spreadsheet in Figure 11-13 is pH in column B and the output is V_b in column G. From the pH, the values of $[H^+]$, $[OH^-]$, and α_{A^-} are computed in columns C, D, and E. Equation 11-13 is used in column F to find the fraction of titration, ϕ. From this value, we calculate the volume of titrant, V_b, in column G.

	A	B	C	D	E	F	G
1	Titration of weak acid with strong base						
2							
3	C_b =	pH	$[H^+]$	$[OH^-]$	$\alpha(A^-)$	ϕ	V_b (mL)
4	0.1	3.90	1.26E-04	7.94E-11	0.004	-0.002	-0.020
5	C_a =	3.99	1.02E-04	9.77E-11	0.005	0.000	0.001
6	0.02	5.00	1.00E-05	1.00E-09	0.051	0.050	0.505
7	V_a =	6.00	1.00E-06	1.00E-08	0.349	0.349	3.493
8	50	6.27	5.37E-07	1.86E-08	0.500	0.500	5.000
9	K_a =	7.00	1.00E-07	1.00E-07	0.843	0.843	8.430
10	5.37E-07	8.00	1.00E-08	1.00E-06	0.982	0.982	9.818
11	K_w =	9.00	1.00E-09	1.00E-05	0.998	0.999	9.987
12	1.E-14	9.25	5.62E-10	1.78E-05	0.999	1.000	10.000
13		10.00	1.00E-10	1.00E-04	1.000	1.006	10.058
14		11.00	1.00E-11	1.00E-03	1.000	1.061	10.606
15		12.00	1.00E-12	1.00E-02	1.000	1.667	16.667
16							
17	C4 = 10^-B4			F4 = (E4-(C4-D4)/$A$6)/(1+(C4-D4)/$A$4)			
18	D4 = $A$12/C4			G4 = F4*$A$6*$A$8/$A$4			
19	E4 = $A$10/(C4+$A$10)						

FIGURE 11-13 Spreadsheet uses Equation 11-13 to calculate the titration curve for 50 mL of the weak acid, 0.02 M MES (pK_a = 6.27) treated with 0.1 M NaOH. We provide pH as input in column B, and the spreadsheet tells us what volume of base is required to generate that pH.

TABLE 11-5 Titration equations for spreadsheets

CALCULATION OF ϕ

Titrating strong acid with strong base

$$\phi = \frac{C_b V_b}{C_a V_a} = \frac{1 - \dfrac{[H^+] - [OH^-]}{C_a}}{1 + \dfrac{[H^+] - [OH^-]}{C_b}}$$

Titrating strong base with strong acid

$$\phi = \frac{C_a V_a}{C_b V_b} = \frac{1 + \dfrac{[H^+] - [OH^-]}{C_b}}{1 - \dfrac{[H^+] - [OH^-]}{C_a}}$$

Titrating weak acid (HA) with weak base (B)

$$\phi = \frac{C_b V_b}{C_a V_a} = \frac{\alpha_{A^-} - \dfrac{[H^+] - [OH^-]}{C_a}}{\alpha_{BH^+} + \dfrac{[H^+] - [OH^-]}{C_b}}$$

Titrating H_2A with strong base $(\rightarrow \rightarrow A^{2-})$

$$\phi = \frac{C_b V_b}{C_a V_a} = \frac{\alpha_{HA^-} + 2\alpha_{A^{2-}} - \dfrac{[H^+] - [OH^-]}{C_a}}{1 + \dfrac{[H^+] - [OH^-]}{C_b}}$$

Titrating dibasic B with strong acid $(\rightarrow \rightarrow BH_2^{2+})$

$$\phi = \frac{C_a V_a}{C_b V_b} = \frac{\alpha_{BH^+} + 2\alpha_{BH_2^{2+}} + \dfrac{[H^+] - [OH^-]}{C_b}}{1 - \dfrac{[H^+] - [OH^-]}{C_a}}$$

Titrating weak acid (HA) with strong base

$$\phi = \frac{C_b V_b}{C_a V_a} = \frac{\alpha_{A^-} - \dfrac{[H^+] - [OH^-]}{C_a}}{1 + \dfrac{[H^+] - [OH^-]}{C_b}}$$

Titrating weak base (B) with strong acid

$$\phi = \frac{C_a V_a}{C_b V_b} = \frac{\alpha_{BH^+} + \dfrac{[H^+] - [OH^-]}{C_b}}{1 - \dfrac{[H^+] - [OH^-]}{C_a}}$$

Titrating weak base (B) with weak acid (HA)

$$\phi = \frac{C_a V_a}{C_b V_b} = \frac{\alpha_{BH^+} + \dfrac{[H^+] - [OH^-]}{C_b}}{\alpha_{A^-} - \dfrac{[H^+] - [OH^-]}{C_a}}$$

Titrating H_3A with strong base $(\rightarrow \rightarrow \rightarrow A^{3-})$

$$\phi = \frac{C_b V_b}{C_a V_a} = \frac{\alpha_{H_2A^-} + 2\alpha_{HA^{2-}} + 3\alpha_{A^{3-}} - \dfrac{[H^+] - [OH^-]}{C_a}}{1 + \dfrac{[H^+] - [OH^-]}{C_b}}$$

Titrating tribasic B with strong acid $(\rightarrow \rightarrow \rightarrow BH_3^{3+})$

$$\phi = \frac{C_a V_a}{C_b V_b} = \frac{\alpha_{BH^+} + 2\alpha_{BH_2^{2+}} + 3\alpha_{BH_3^{3+}} + \dfrac{[H^+] - [OH^-]}{C_b}}{1 - \dfrac{[H^+] - [OH^-]}{C_a}}$$

SYMBOLS

ϕ = fraction of the way to the first equivalence point

C_a = initial concentration of acid

C_b = initial concentration of base

α = fraction of dissociation of acid or fraction of association of base

V_a = volume of acid

V_b = volume of base

CALCULATION OF α

Monoprotic systems

$$\alpha_{HA} = \frac{[H^+]}{[H^+] + K_a}$$

$$\alpha_{BH^+} = \frac{[H^+]}{[H^+] + K_{BH^+}}$$

$$\alpha_{A^-} = \frac{K_a}{[H^+] + K_a}$$

$$\alpha_B = \frac{K_{BH^+}}{[H^+] + K_{BH^+}}$$

SYMBOLS

K_a = acid dissociation constant of HA

K_{BH^+} = acid dissociation constant of BH^+ $(= K_w/K_b)$

Diprotic systems

$$\alpha_{H_2A} = \alpha_{BH_2^{2+}} = \frac{[H^+]^2}{[H^+]^2 + [H^+]K_1 + K_1K_2}$$

$$\alpha_{HA^-} = \alpha_{BH^+} = \frac{[H^+]K_1}{[H^+]^2 + [H^+]K_1 + K_1K_2}$$

$$\alpha_{A^{2-}} = \alpha_B = \frac{K_1K_2}{[H^+]^2 + [H^+]K_1 + K_1K_2}$$

SYMBOLS

K_1 and K_2 for the acid are the acid dissociation constants of H_2A and HA^-, respectively.

K_1 and K_2 for the base are the acid dissociation constants of BH_2^{2+} and BH^+, respectively: $K_1 = K_w/K_{b2}$; $K_2 = K_w/K_{b1}$.

TRIPROTIC SYSTEMS

$$\alpha_{H_3A} = \frac{[H^+]^3}{[H^+]^3 + [H^+]^2K_1 + [H^+]K_1K_2 + K_1K_2K_3}$$

$$\alpha_{H_2A^-} = \frac{[H^+]^2K_1}{[H^+]^3 + [H^+]^2K_1 + [H^+]K_1K_2 + K_1K_2K_3}$$

$$\alpha_{HA^{2-}} = \frac{[H^+]K_1K_2}{[H^+]^3 + [H^+]^2K_1 + [H^+]K_1K_2 + K_1K_2K_3}$$

$$\alpha_{A^{3-}} = \frac{K_1K_2K_3}{[H^+]^3 + [H^+]^2K_1 + [H^+]K_1K_2 + K_1K_2K_3}$$

How do we know what pH values to put in? Trial-and-error allows us to find the starting pH, by putting in a pH and seeing whether V_b is positive or negative. In a few tries, it is easy to home in on the pH at which $V_b = 0$. In Figure 11-13, a pH of 3.90 is too low, because ϕ and V are both negative. Input values of pH are spaced as closely as you like, so that you can generate a smooth titration curve. To save space, we show only a few points in Figure 11-13, including the midpoint (pH 6.27 $\Rightarrow V_b = 5.00$ mL) and the end point (pH 9.25 $\Rightarrow V_b = 10.00$ mL). This spreadsheet reproduces Table 11-2 without approximations other than neglecting activity coefficients. It gives correct results even when the approximations used in Table 11-2 fail.

In the margin:

In Figure 11-13, you could use Excel Goal Seek (page 173) to vary pH in cell B5 until V_b in cell G5 is 0.

Titrating a Weak Acid with a Weak Base

Now consider the titration of V_a mL of acid HA (initial concentration C_a) with V_b mL of base B whose concentration is C_b. Let the acid dissociation constant of HA be K_a and the acid dissociation constant of BH^+ be K_{BH^+}. The charge balance is

Charge balance: $\qquad [H^+] + [BH^+] = [A^-] + [OH^-]$

As before, we can say that $[A^-] = \alpha_{A^-} \cdot F_{HA}$, where $\alpha_{A^-} = K_a/([H^+] + K_a)$ and $F_{HA} = C_a V_a/(V_a + V_b)$.

We can write an analogous expression for $[BH^+]$, which is a monoprotic weak acid. If the acid were HA, we would use Equation 10-17 to say

$$[HA] = \alpha_{HA} \cdot F_{HA} \qquad \alpha_{HA} = \frac{[H^+]}{[H^+] + K_a}$$

where K_a applies to the acid HA. For the weak acid BH^+, we write

$$[BH^+] = \alpha_{BH^+} \cdot F_B \qquad \alpha_{BH^+} = \frac{[H^+]}{[H^+] + K_{BH^+}}$$

where the formal concentration of base is $F_B = C_b V_b/(V_a + V_b)$.

Substituting for $[BH^+]$ and $[A^-]$ in the charge balance gives

$$[H^+] + \frac{\alpha_{BH^+} \cdot C_b V_b}{V_a + V_b} = \frac{\alpha_{A^-} \cdot C_a V_a}{V_a + V_b} + [OH^-]$$

which can be rearranged to the useful result

Fraction of titration for weak acid by weak base:

$$\phi = \frac{C_b V_b}{C_a V_a} = \frac{\alpha_{A^-} - \dfrac{[H^+] - [OH^-]}{C_a}}{\alpha_{BH^+} + \dfrac{[H^+] - [OH^-]}{C_b}} \qquad (11\text{-}14)$$

In the margin:

α_{HA} = fraction of acid in the form HA

$$\alpha_{HA} = \frac{[HA]}{F_{HA}}$$

α_{BH^+} = fraction of base in the form BH^+

$$\alpha_{BH^+} = \frac{[BH^+]}{F_B}$$

Equation 11-14 for a weak base looks just like Equation 11-13 for a strong base, except that α_{BH^+} replaces 1 in the denominator.

Table 11-5 gives useful equations derived by writing a charge balance and substituting fractional compositions for various concentrations. For titration of the diprotic acid, H_2A, ϕ is the fraction of the way to the first equivalence point. When $\phi = 2$, we are at the second equivalence point. It should not surprise you that, when $\phi = 0.5$, pH $\approx pK_1$ and when $\phi = 1.5$, pH $\approx pK_2$. When $\phi = 1$, we have the intermediate HA^- and pH $\approx \frac{1}{2}(pK_1 + pK_2)$.

Terms to Understand

Gran plot	indicator	Kjeldahl nitrogen analysis	transition range
Hammett acidity function	indicator error	leveling effect	

Summary

Key equations used to calculate titration curves:

Strong acid/strong base titration

$$H^+ + OH^- \rightarrow H_2O$$

pH is determined by the concentration of excess unreacted H^+ or OH^-

Weak acid titrated with OH^-

$$HA + OH^- \rightarrow A^- + H_2O \quad (V_e = \text{equivalence volume})$$
$$(V_b = \text{volume of added base})$$

$V_b = 0$: pH determined by K_a $(HA \overset{K_a}{\rightleftharpoons} H^+ + A^-)$

$0 < V_b < V_e$: $pH = pK_a + \log([A^-]/[HA])$

$pH = pK_a$ when $V_b = \frac{1}{2}V_e$ (neglecting activity)

At V_e: pH governed by K_b $(A^- + H_2O \overset{K_b}{\rightleftharpoons} HA + OH^-)$

After V_e: pH is determined by excess OH^-

Weak base titrated with H^+

$$B + H^+ \rightarrow BH^+ \quad (V_e = \text{equivalence volume})$$
$$(V_a = \text{volume of added acid})$$

$V_a = 0$: pH determined by K_b $(B + H_2O \overset{K_b}{\rightleftharpoons} BH^+ + OH^-)$

$0 < V_a < V_e$: $pH = pK_{BH^+} + \log([B]/[BH^+])$

$pH = pK_{BH^+}$ when $V_a = \frac{1}{2}V_e$

At V_e: pH governed by K_{BH^+} $(BH^+ \overset{K_{BH^+}}{\rightleftharpoons} B + H^+)$

After V_e: pH is determined by excess H^+

H_2A titrated with OH^-

$$H_2A \overset{OH^-}{\rightarrow} HA^- \overset{OH^-}{\rightarrow} A^{2-}$$

Equivalence volumes: $V_{e2} = 2V_{e1}$

$V_b = 0$: pH determined by $K_1 (H_2A \overset{K_1}{\rightleftharpoons} H^+ + HA^-)$

$0 < V_b < V_{e1}$: $pH = pK_1 + \log([HA^-]/[H_2A])$

$pH = pK_1$ when $V_b = \frac{1}{2}V_{e1}$

At V_{e1}: $[H^+] = \sqrt{\dfrac{K_1 K_2 F' + K_1 K_w}{K_1 + F'}}$

$\Rightarrow pH \approx \frac{1}{2}(pK_1 + pK_2)$

$F' = $ formal concentration of HA^-

$V_{e1} < V_b < V_{e2}$: $pH = pK_2 + \log([A^{2-}]/[HA^-])$

$pH = pK_2$ when $V_b = \frac{3}{2}V_{e1}$

At V_{e2}: pH governed by K_{b1}

$$(A^{2-} + H_2O \overset{K_{b1}}{\rightleftharpoons} HA^- + OH^-)$$

After V_{e2}: pH is determined by excess OH^-

Behavior of derivatives at the equivalence point

First derivative: $\Delta pH/\Delta V$ has greatest magnitude

Second derivative: $\Delta(\Delta pH/\Delta V)/\Delta V = 0$

Gran plot

Plot $V_b \cdot 10^{-pH}$ versus V_b

$x = $ intercept $= V_e$; slope $= -K_a \gamma_{HA}/\gamma_{A^-}$

$K_a = $ acid dissociation constant

$\gamma = $ activity coefficient

Choosing an indicator: Color transition range should match pH at V_e. Preferably the color change should occur entirely within the steep portion of the titration curve.

Kjeldahl nitrogen analysis: A nitrogen-containing organic compound is digested with a catalyst in boiling H_2SO_4. Nitrogen is converted into NH_4^+, which is converted into NH_3 with base and distilled into standard HCl. Excess, unreacted HCl is titrated with standard NaOH to tell us how much nitrogen was present in the original analyte.

Exercises

11-A. Calculate the pH at each of the following points in the titration of 50.00 mL of 0.010 0 M NaOH with 0.100 M HCl. Volume of acid added: 0.00, 1.00, 2.00, 3.00, 4.00, 4.50, 4.90, 4.99, 5.00, 5.01, 5.10, 5.50, 6.00, 8.00, and 10.00 mL. Make a graph of pH versus volume of HCl added.

11-B. Calculate the pH at each point listed for the titration of 50.0 mL of 0.050 0 M formic acid with 0.050 0 M KOH. The points to calculate are $V_b = 0.0$, 10.0, 20.0, 25.0, 30.0, 40.0, 45.0, 48.0, 49.0, 49.5, 50.0, 50.5, 51.0, 52.0, 55.0, and 60.0 mL. Draw a graph of pH versus V_b.

11-C. Calculate the pH at each point listed for the titration of 100.0 mL of 0.100 M cocaine (Section 9-4, $K_b = 2.6 \times 10^{-6}$) with 0.200 M HNO_3. The points to calculate are $V_a = 0.0$, 10.0, 20.0, 25.0, 30.0, 40.0, 49.0, 49.9, 50.0, 50.1, 51.0, and 60.0 mL. Draw a graph of pH versus V_a.

11-D. Consider the titration of 50.0 mL of 0.050 0 M malonic acid with 0.100 M NaOH. Calculate the pH at each point listed and sketch the titration curve: $V_b = 0.0$, 8.0, 12.5, 19.3, 25.0, 37.5, 50.0, and 56.3 mL.

11-E. Write the chemical reactions (including structures of reactants and products) that occur when the amino acid histidine is titrated with perchloric acid. (Histidine is a molecule with no net charge.) A solution containing 25.0 mL of 0.050 0 M histidine was titrated with 0.050 0 M $HClO_4$. Calculate the pH at the following values of V_a: 0, 4.0, 12.5, 25.0, 26.0, and 50.0 mL.

11-F. Select indicators from Table 11-3 that would be useful for the titrations in Figures 11-1 and 11-2 and the $pK_a = 8$ curve in Figure 11-3. Select a different indicator for each titration and state what color change you would use as the end point.

11-G. When 100.0 mL of a weak acid were titrated with 0.093 81 M NaOH, 27.63 mL were required to reach the equivalence point. The pH at the equivalence point was 10.99. What was the pH when only 19.47 mL of NaOH had been added?

11-H. A 0.100 M solution of the weak acid HA was titrated with 0.100 M NaOH. The pH measured when $V_b = \frac{1}{2}V_e$ was 4.62. Using activity coefficients, calculate pK_a. The size of the A^- anion is 450 pm.

11-I. 🖩 *Finding the end point from pH measurements.* Here are data points around the second apparent end point in Figure 11-5:

V_b (μL)	pH	V_b (μL)	pH
107.0	6.921	117.0	7.878
110.0	7.117	118.0	8.090
113.0	7.359	119.0	8.343
114.0	7.457	120.0	8.591
115.0	7.569	121.0	8.794
116.0	7.705	122.0	8.952

(a) Prepare a spreadsheet or table analogous to Figure 11-6, showing the first and second derivatives. Plot both derivatives versus V_b and locate the end point in each plot.

(b) Prepare a Gran plot analogous to Figure 11-8. Use the least-squares procedure to find the best straight line and find the end point. You will have to use your judgment as to which points lie on the "straight" line.

11-J. *Indicator error.* Consider the titration in Figure 11-2 in which the equivalence-point pH in Table 11-2 is 9.25 at a volume of 10.00 mL.

(a) Suppose you used the yellow-to-blue transition of thymol blue indicator to find the end point. According to Table 11-3, the last trace of green disappears near pH 9.6. What volume of base is required to reach pH 9.6? The difference between this volume and 10 mL is the indicator error.

(b) If you used cresol red, with a color change at pH 8.8, what would be the indicator error?

11-K. *Spectrophotometry with indicators.** Acid-base indicators are themselves acids or bases. Consider an indicator, HIn, which dissociates according to the equation

$$\text{HIn} \xrightleftharpoons{K_a} \text{H}^+ + \text{In}^-$$

The molar absorptivity, ε, is 2 080 $M^{-1}\,cm^{-1}$ for HIn and 14 200 $M^{-1}\,cm^{-1}$ for In$^-$, at a wavelength of 440 nm.

(a) Write an expression for the absorbance of a solution containing HIn at a concentration [HIn] and In$^-$ at a concentration In$^-$ in a cell of pathlength 1.00 cm. The total absorbance is the sum of absorbances of each component.

(b) A solution containing indicator at a formal concentration of 1.84×10^{-4} M is adjusted to pH 6.23 and found to exhibit an absorbance of 0.868 at 440 nm. Calculate pK_a for this indicator.

*This problem is based on Beer's law in Section 18-2.

Problems

Titration of Strong Acid with Strong Base

11-1. Distinguish the terms *end point* and *equivalence point*.

11-2. Consider the titration of 100.0 mL of 0.100 M NaOH with 1.00 M HBr. Find the pH at the following volumes of acid added and make a graph of pH versus V_a: V_a = 0, 1, 5, 9, 9.9, 10, 10.1, and 12 mL.

11-3. Why does an acid-base titration curve (pH versus volume of titrant) have an abrupt change at the equivalence point?

Titration of Weak Acid with Strong Base

11-4. Sketch the general appearance of the curve for the titration of a weak acid with a strong base. What chemistry governs the pH in each of the four distinct regions of the curve?

11-5. Why is it not practical to titrate an acid or base that is too weak or too dilute?

11-6. A weak acid HA (pK_a = 5.00) was titrated with 1.00 M KOH. The acid solution had a volume of 100.0 mL and a molarity of 0.100 M. Find the pH at the following volumes of base added and make a graph of pH versus V_b: V_b = 0, 1, 5, 9, 9.9, 10, 10.1, and 12 mL.

11-7. Consider the titration of the weak acid HA with NaOH. At what fraction of V_e does pH = pK_a − 1? At what fraction of V_e does pH = pK_a + 1? Use these two points, plus V_b = 0, $\frac{1}{2}V_e$, V_e, and 1.2V_e to sketch the titration curve for the reaction of 100 mL of 0.100 M anilinium bromide ("aminobenzene · HBr") with 0.100 M NaOH.

11-8. What is the pH at the equivalence point when 0.100 M hydroxyacetic acid is titrated with 0.050 0 M KOH?

11-9. Find the equilibrium constant for the reaction of MES (Table 9-2) with NaOH.

11-10. When 22.63 mL of aqueous NaOH were added to 1.214 g of cyclohexylaminoethanesulfonic acid (FM 207.29, structure in Table 9-2) dissolved in 41.37 mL of water, the pH was 9.24. Calculate the molarity of the NaOH.

11-11. *Use activity coefficients* to calculate the pH after 10.0 mL of 0.100 M trimethylammonium bromide were titrated with 4.0 mL of 0.100 M NaOH.

Titration of Weak Base with Strong Acid

11-12. Sketch the general appearance of the curve for the titration of a weak base with a strong acid. What chemistry governs the pH in each of the four distinct regions of the curve?

11-13. Why is the equivalence-point pH necessarily below 7 when a weak base is titrated with strong acid?

11-14. A 100.0-mL aliquot of 0.100 M weak base B (pK_b = 5.00) was titrated with 1.00 M HClO$_4$. Find the pH at the following volumes of acid added and make a graph of pH versus V_a: V_a = 0, 1, 5, 9, 9.9, 10, 10.1, and 12 mL.

11-15. At what point in the titration of a weak base with a strong acid is the maximum buffer capacity reached? This is the point at which a given small addition of acid causes the least pH change.

11-16. What is the equilibrium constant for the reaction between benzylamine and HCl?

11-17. A 50.0-mL solution of 0.031 9 M benzylamine was titrated with 0.050 0 M HCl. Calculate the pH at the following volumes of added acid: V_a = 0, 12.0, $\frac{1}{2}V_e$, 30.0, V_e, and 35.0 mL.

11-18. Calculate the pH of a solution made by mixing 50.00 mL of 0.100 M NaCN with

(a) 4.20 mL of 0.438 M HClO$_4$

(b) 11.82 mL of 0.438 M HClO$_4$

(c) What is the pH at the equivalence point with 0.438 M HClO$_4$?

Titrations in Diprotic Systems

11-19. Sketch the general appearance of the curve for the titration of a weak diprotic acid with NaOH. What chemistry governs the pH in each distinct region of the curve?

11-20. The graph shows the titration curve for a protein containing 124 amino acids with 16 basic and 20 acidic substituents. The curve is smooth without clear breaks because 29 groups are titrated in the pH interval shown. The 29 end points are so close together that a nearly uniform rise results. The isoionic point is the pH of the pure protein with no ions present except H$^+$ and

OH^-. The isoelectric point is the pH at which the average charge on the protein is zero. Is the average charge of the protein positive, negative, or neutral at its isoionic point? How do you know?

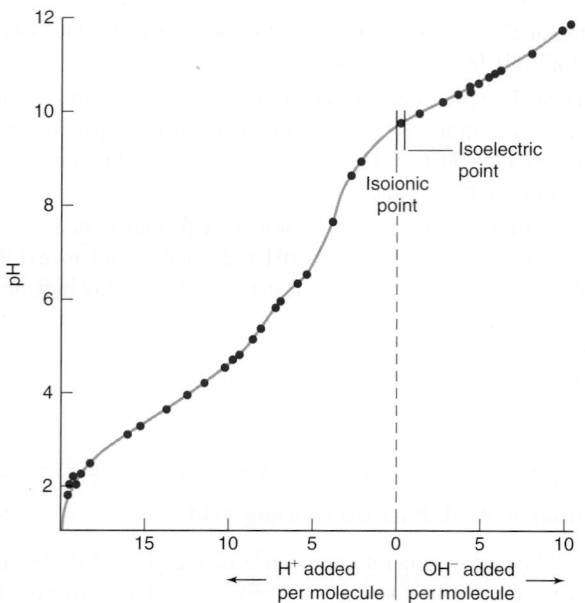

Acid-base titration of the protein ribonuclease. [Data from C. T. Tanford and J. D. Hauenstein, "Hydrogen Ion Equilibria of Ribonuclease," *J. Am. Chem. Soc.* **1956**, *78*, 5287.]

11-21. The base Na^+A^-, whose anion is dibasic, was titrated with HCl to give curve *b* in Figure 11-4. Is the first equivalence point H the isoelectric point or the isoionic point?

11-22. The figure compares the titration of a monoprotic weak acid with a monoprotic weak base and the titration of a diprotic acid with strong base.

(a) Write the reaction between the weak acid and the weak base and show that the equilibrium constant is $10^{7.78}$. This large value means that the reaction goes "to completion" after each addition of reagent.

(b) Why does pK_2 intersect the upper curve at $\frac{3}{2}V_e$ and the lower curve at $2V_e$? On the lower curve, "pK_2" is pK_a for the acid, BH^+.

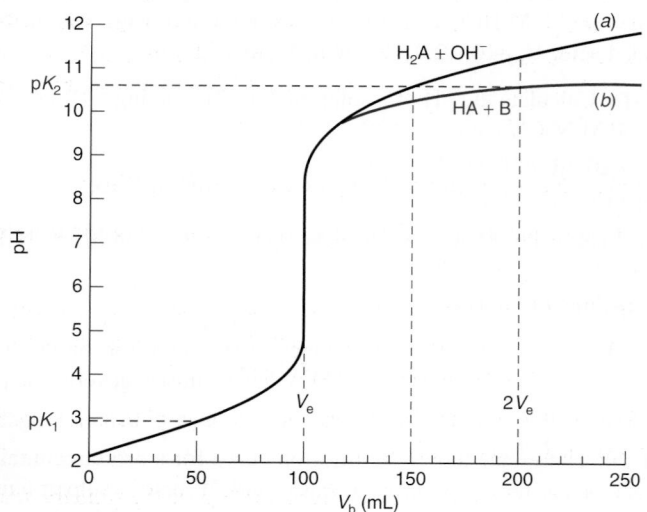

(*a*) Titration of 100 mL of 0.050 M H_2A (pK_1 = 2.86, pK_2 = 10.64) with 0.050 M NaOH. (*b*) Titration of 100 mL of the weak acid HA (0.050 M, pK_a = 2.86) with the weak base B (0.050 M, pK_b = 3.36).

11-23. The dibasic compound B (pK_{b1} = 4.00, pK_{b2} = 8.00) was titrated with 1.00 M HCl. The initial solution of B was 0.100 M and had a volume of 100.0 mL. Find the pH at the following volumes of acid added and make a graph of pH versus V_a: V_a = 0, 1, 5, 9, 10, 11, 15, 19, 20, and 22 mL.

11-24. A 100.0-mL aliquot of 0.100 M diprotic acid H_2A (pK_1 = 4.00, pK_2 = 8.00) was titrated with 1.00 M NaOH. Find the pH at the following volumes of base added and make a graph of pH versus V_b: V_b = 0, 1, 5, 9, 10, 11, 15, 19, 20, and 22 mL.

11-25. Calculate the pH at 10.0-mL intervals (from 0 to 100 mL) in the titration of 40.0 mL of 0.100 M piperazine with 0.100 M HCl. Make a graph of pH versus V_a.

11-26. Calculate the pH when 25.0 mL of 0.020 0 M 2-aminophenol have been titrated with 10.9 mL of 0.015 0 M $HClO_4$.

11-27. Consider the titration of 50.0 mL of 0.100 M sodium glycinate, $H_2NCH_2CO_2Na$, with 0.100 M HCl.

(a) Calculate the pH at the second equivalence point.

(b) Show that our approximate method of calculations gives incorrect (physically unreasonable) values of pH at V_a = 90.0 and V_a = 101.0 mL.

11-28. A solution containing 0.100 M glutamic acid (the molecule with no net charge) was titrated to its first equivalence point with 0.025 0 M RbOH.

(a) Draw the structures of reactants and products.

(b) Calculate the pH at the first equivalence point.

11-29. Find the pH of the solution when 0.010 0 M tyrosine is titrated to the equivalence point with 0.004 00 M $HClO_4$.

11-30. This problem deals with the amino acid cysteine, which we will abbreviate H_2C.

(a) A 0.030 0 M solution was prepared by dissolving dipotassium cysteine, K_2C, in water. Then 40.0 mL of this solution were titrated with 0.060 0 M $HClO_4$. Calculate the pH at the first equivalence point.

(b) Calculate the quotient $[C^{2-}]/[HC^-]$ in a solution of 0.050 0 M cysteinium bromide (the salt $H_3C^+Br^-$).

11-31. How many grams of dipotassium oxalate (FM 166.22) should be added to 20.0 mL of 0.800 M $HClO_4$ to give a pH of 4.40 when the solution is diluted to 500 mL?

11-32. When 5.00 mL of 0.103 2 M NaOH were added to 0.112 3 g of alanine (FM 89.093) in 100.0 mL of 0.10 M KNO_3, the measured pH was 9.57. *Use activity coefficients* to find pK_2 for alanine. Consider the ionic strength of the solution to be 0.10 M and consider each ionic form of alanine to have an activity coefficient of 0.77.

Finding the End Point with a pH Electrode

11-33. What is a Gran plot used for?

11-34. Data for the titration of 100.00 mL of a weak acid by NaOH are given below. Find the end point by preparing a Gran plot, using the last 10% of the volume prior to V_e.

mL NaOH	pH	mL NaOH	pH	mL NaOH	pH
0.00	4.14	20.75	6.09	22.70	6.70
1.31	4.30	21.01	6.14	22.76	6.74
2.34	4.44	21.10	6.15	22.80	6.78
3.91	4.61	21.13	6.16	22.85	6.82
5.93	4.79	21.20	6.17	22.91	6.86
7.90	4.95	21.30	6.19	22.97	6.92
11.35	5.19	21.41	6.22	23.01	6.98
13.46	5.35	21.51	6.25	23.11	7.11
15.50	5.50	21.61	6.27	23.17	7.20
16.92	5.63	21.77	6.32	23.21	7.30
18.00	5.71	21.93	6.37	23.30	7.49
18.35	5.77	22.10	6.42	23.32	7.74
18.95	5.82	22.27	6.48	23.40	8.30
19.43	5.89	22.37	6.53	23.46	9.21
19.93	5.95	22.48	6.58	23.55	9.86
20.48	6.04	22.57	6.63		

11-35. Prepare a second derivative graph to find the end point from the following titration data.

mL NaOH	pH	mL NaOH	pH	mL NaOH	pH
10.679	7.643	10.725	6.222	10.750	4.444
10.696	7.447	10.729	5.402	10.765	4.227
10.713	7.091	10.733	4.993		
10.721	6.700	10.738	4.761		

Finding the End Point with Indicators

11-36. Explain the origin of the rule of thumb that indicator color changes occur at $pK_{HIn} \pm 1$.

11-37. Why does a properly chosen indicator change color near the equivalence point in a titration?

11-38. The pH of microscopic vesicles (compartments) in living cells can be estimated by infusing an indicator (HIn) into the compartment and measuring the quotient $[In^-]/[HIn]$ from the spectrum of the indicator inside the vesicle. Explain how this tells us the pH.

11-39. Write the chemical formula of a compound with a negative pK_a.

11-40. Consider the titration in Figure 11-2, for which the pH at the equivalence point is calculated to be 9.25. If thymol blue is used as an indicator, what color will be observed through most of the titration prior to the equivalence point? At the equivalence point? After the equivalence point?

11-41. What color do you expect to observe for cresol purple indicator (Table 11-3) at the following pH values?
(a) 1.0; (b) 2.0; (c) 3.0.

11-42. Cresol red has *two* transition ranges listed in Table 11-3. What color would you expect it to be at the following pH values?
(a) 0; (b) 1; (c) 6; (d) 9.

11-43. Would the indicator bromocresol green, with a transition range of pH 3.8–5.4, ever be useful in the titration of a weak acid with a strong base?

11-44. (a) What is the pH at the equivalence point when 0.030 0 M NaF is titrated with 0.060 0 M $HClO_4$?

(b) Why would an indicator end point probably not be useful in this titration?

11-45. A titration curve for Na_2CO_3 titrated with HCl is shown here. Suppose that *both* phenolphthalein and bromocresol green are present in the titration solution. State what colors you expect to observe at the following volumes of added HCl:
(a) 2 mL; (b) 10 mL; (c) 19 mL.

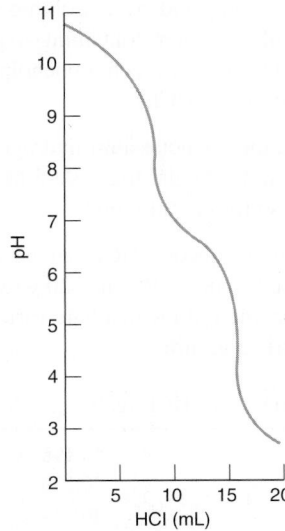

11-46. In the Kjeldahl nitrogen determination, the final product is a solution of NH_4^+ in HCl solution. It is necessary to titrate the HCl without titrating NH_4^+.
(a) Calculate the pH of pure 0.010 M NH_4Cl.
(b) Select an indicator that would allow you to titrate HCl but not NH_4^+.

11-47. A 10.231-g sample of window cleaner containing ammonia was diluted with 39.466 g of water. Then 4.373 g of solution were titrated with 14.22 mL of 0.106 3 M HCl to reach a bromocresol green end point. Find the weight percent of NH_3 (FM 17.031) in the cleaner.

11-48. A procedure to measure the alkalinity (Box 11-1) of home swimming pool water is to titrate a fixed volume of pool water by counting the number of drops of standard H_2SO_4 to reach the bromocresol green end point.[31] Explain what is measured in this titration and why bromocresol green was chosen.

Practical Notes, Kjeldahl Analysis, and Leveling Effect

11-49. Give the name and formula of a primary standard used to standardize (a) HCl and (b) NaOH.

11-50. Why is it more accurate to use a primary standard with a high equivalent mass (the mass required to provide or consume 1 mol of H^+) than one with a low equivalent mass?

11-51. Explain how to use potassium hydrogen phthalate to standardize a solution of NaOH.

11-52. A solution was prepared from 1.023 g of the primary standard tris (Table 11-4) plus 99.367 g of water; 4.963 g of the solution were titrated with 5.262 g of aqueous HNO_3 to reach the methyl red end point. Calculate the concentration of the HNO_3 (expressed as mol HNO_3/kg solution).

11-53. The balance says that you have weighed out 1.023 g of tris to standardize a solution of HCl. Use the buoyancy correction in Section 2-3 and the density in Table 11-4 to determine how many grams you have really weighed out. The volume of HCl required to react with the tris was 28.37 mL. Does the buoyancy correction introduce a random or a systematic error into the calculated molarity of HCl? What is the magnitude of the error expressed as a percentage? Is the calculated molarity of HCl higher or lower than the true molarity?

11-54. A solution was prepared by dissolving 0.194 7 g of HgO (Table 11-4) in 20 mL of water containing 4 g of KBr. Titration with HCl required 17.98 mL to reach a phenolphthalein end point. Calculate the molarity of the HCl.

11-55. How many grams of potassium hydrogen phthalate should be weighed into a flask to standardize ~0.05 M NaOH if you wish to use ~30 mL of base for the titration?

11-56. Constant-boiling aqueous HCl can be used as a primary standard for acid-base titrations. When ~20 wt% HCl (FM 36.461) is distilled, the composition of the distillate varies in a regular manner with the barometric pressure:

P (Torr)	HCl[a] (g/100 g solution)
770	20.196
760	20.220
750	20.244
740	20.268
730	20.292

a. The composition of distillate is from C. W. Foulk and M. Hollingsworth, J. Am. Chem. Soc. **1923**, 45, 1223, with numbers corrected for the current values of atomic masses.

(a) Make a graph of the data in the table to find the weight percent of HCl collected at 746 Torr.

(b) What mass of distillate (weighed in air, using weights whose density is 8.0 g/mL) should be dissolved in 1.000 0 L to give 0.100 00 M HCl? The density of distillate over the whole range in the table is close to 1.096 g/mL. You will need this density to change the mass measured in vacuum to mass measured in air. See Section 2-3 for buoyancy corrections.

11-57. (a) *Uncertainty in formula mass.* In an extremely-high-precision gravimetric titration, the uncertainty in formula mass of the primary standard could contribute to the uncertainty of the result. Read about uncertainty in molecular mass in Appendix B. Express the formula mass of potassium hydrogen phthalate, $C_8H_5O_4K$, with its 95% confidence interval based on a rectangular distribution of uncertainty of atomic mass with a coverage factor of $k = 2$.

(b) *Systematic uncertainty in reagent purity.* The manufacturer of potassium hydrogen phthalate states that the purity is $1.000\ 00 \pm 0.000\ 05$. In the absence of further information, we assume that the distribution of uncertainty is rectangular. What standard uncertainty would you use for the purity of this reagent?

11-58. The Kjeldahl procedure was used to analyze 256 µL of a solution containing 37.9 mg protein/mL. The liberated NH_3 was collected in 5.00 mL of 0.033 6 M HCl, and the remaining acid required 6.34 mL of 0.010 M NaOH for complete titration. What is the weight percent of nitrogen in the protein?

11-59. In the Kjeldahl method, an alternative to trapping NH_3 in HCl is to trap it in ~4 wt% aqueous boric acid, $B(OH)_3$, to make an equilibrium mixture of NH_4^+, NH_3, $B(OH)_3$, $BO(OH)_2^-$ (borate), and polyborates including triborate ($B_3O_3(OH)_4^-$), tetraborate ($B_4O_5(OH)_4^{2-}$), and

pentaborate ($B_5O_6(OH)_4^-$).[26] This mixture is then titrated with standard HCl to an end point pH of ~3.8 measured with a pH electrode. An advantage of using boric acid is than only one standard reagent (HCl) is required instead of two (HCl and NaOH). The boric acid procedure takes less time and expense than the method in the text.

Distillation of NH_3 into $B(OH)_3$:

$$NH_3 + B(OH)_3 + H_2O \rightleftharpoons NH_4^+ + B(OH)_4^- \qquad \textbf{(A)}$$

$$NH_3 + B(OH)_3 + H_2O \rightleftharpoons NH_4^+ + \text{polyborates} \qquad \textbf{(B)}$$

Titration with standard HCl:

$$H^+ + B(OH)_4^- + \text{polyborates} \rightarrow B(OH)_3 + H_2O \qquad \textbf{(C)}$$

(a) The Kjeldahl procedure was used to analyze 256 µL of a solution containing 37.9 mg protein/mL. Liberated NH_3 was collected in ~5 mL of ~4 wt% $B(OH)_3$, and the resulting solution required 8.28 mL of 0.050 M HCl for titration. Reaction C requires $1H^+$ for each NH_3 in reactions A and B. What is the wt% of nitrogen in the protein?

(b) Suppose that 5.00 mL of 4.00 wt% $B(OH)_3$ (density 1.00 g/mL, FM 61.84) were treated with 0.414 mmol NH_3 with no change in volume. If reaction A went to completion to create 0.414 mmol $B(OH)_4^-$ plus the remaining unreacted $B(OH)_3$, what would be the pH? Disregard reaction B for this question.

(c) At the pH in **(b)** what fraction of ammonia would not be protonated? NH_3 can evaporate from the trapping solution because some is not protonated.

(d) Find the equilibrium constant for reaction A.

11-60. What is meant by the leveling effect?

11-61. Considering the following pK_a values,[32] explain why dilute sodium methoxide ($NaOCH_3$) and sodium ethoxide ($NaOCH_2CH_3$) are leveled to the same base strength in aqueous solution. Write the chemical reactions that occur when these bases are added to water.

CH_3OH	$pK_a = 15.54$
CH_3CH_2OH	$pK_a = 16.0$
HOH	$pK_a = 15.74$ (for $K_a = [H^+][OH^-]/[H_2O]$)

11-62. The base B is too weak to titrate in aqueous solution.

(a) Which solvent, pyridine or acetic acid, would be more suitable for the titration of B with $HClO_4$? Why?

(b) Which solvent would be more suitable for the titration of a very weak acid with tetrabutylammonium hydroxide? Why?

11-63. Explain why sodium amide ($NaNH_2$) and phenyl lithium (C_6H_5Li) are leveled to the same base strength in aqueous solution. Write the chemical reactions that occur when these reagents are added to water.

11-64. Pyridine is half protonated in aqueous phosphate buffer at pH 5.2. If you mix 45 mL of phosphate buffer with 55 mL of methanol, the buffer must have a pH of 3.2 to half protonate pyridine. Suggest a reason why.

Calculating Titration Curves with Spreadsheets

11-65. Derive the following equation for the titration of potassium hydrogen phthalate (K^+HP^-) with NaOH:

$$\phi = \frac{C_bV_b}{C_aV_a} = \frac{\alpha_{HP^-} + 2\alpha_{P^{2-}} - 1 - \dfrac{[H^+] - [OH^-]}{C_a}}{1 + \dfrac{[H^+] - [OH^-]}{C_b}}$$

11-66. 📖 *Effect of pK_a in the titration of weak acid with strong base.* Use Equation 11-13 to compute and plot the family of curves at the left side of Figure 11-3. For a strong acid, choose a large K_a, such as $K_a = 10^2$ or $pK_a = -2$.

11-67. 📖 *Effect of concentration in the titration of weak acid with strong base.* Use your spreadsheet from Problem 11-66 to prepare a family of titration curves for $pK_a = 6$, with the following combinations of concentrations: **(a)** $C_a = 20$ mM, $C_b = 100$ mM; **(b)** $C_a = 2$ mM, $C_b = 10$ mM; **(c)** $C_a = 0.2$ mM, $C_b = 1$ mM.

11-68. 📖 *Effect of pK_b in the titration of weak base with strong acid.* Using the appropriate equation in Table 11-5, compute and plot a family of curves analogous to the left part of Figure 11-3 for the titration of 50.0 mL of 0.020 0 M B ($pK_b = -2.00$, 2.00, 4.00, 6.00, 8.00, and 10.00) with 0.100 M HCl. (The value $pK_b = -2.00$ represents a strong base.) In the expression for α_{BH^+}, $K_{BH^+} = K_w/K_b$.

11-69. 📖 *Titrating weak acid with weak base.*
(a) Prepare a family of graphs for the titration of 50.0 mL of 0.020 0 M HA ($pK_a = 4.00$) with 0.100 M B ($pK_b = 3.00$, 6.00, and 9.00).
(b) Write the acid-base reaction that occurs when acetic acid and sodium benzoate (the salt of benzoic acid) are mixed, and find the equilibrium constant for the reaction. Find the pH of a solution prepared by mixing 212 mL of 0.200 M acetic acid with 325 mL of 0.050 0 M sodium benzoate.

11-70. 📖 *Titrating diprotic acid with strong base.* Prepare a family of graphs for the titration of 50.0 mL of 0.020 0 M H_2A with 0.100 M NaOH. Consider the following cases: **(a)** $pK_1 = 4.00$, $pK_2 = 8.00$; **(b)** $pK_1 = 4.00$, $pK_2 = 6.00$; **(c)** $pK_1 = 4.00$, $pK_2 = 5.00$.

11-71. 📖 *Titrating nicotine with strong acid.* Prepare a spreadsheet to reproduce the lower curve in Figure 11-4.

11-72. 📖 *Titrating triprotic acid with strong base.* Prepare a spreadsheet to graph the titration of 50.0 mL of 0.020 0 M histidine · 2HCl with 0.100 M NaOH. Treat histidine · 2HCl with the triprotic acid equation in Table 11-5.

11-73. 📖 *A tetraprotic system.* Write an equation for the titration of tetrabasic base with strong acid ($B + H^+ \rightarrow \rightarrow \rightarrow \rightarrow BH_4^{4+}$). You can do this by inspection of Table 11-5 or you can derive it from the charge balance for the titration reaction. Graph the titration of 50.0 mL of 0.020 0 M sodium pyrophosphate ($Na_4P_2O_7$) with 0.100 M $HClO_4$. Pyrophosphate is the anion of pyrophosphoric acid.

Using Beer's Law with Indicators*

11-74. Spectrophotometric properties of a particular indicator are given below:

$$HIn \underset{\xleftarrow{\hspace{0.8cm}}}{\overset{pK_a = 7.95}{\rightleftharpoons}} In^- + H^+$$

$\lambda_{max} = 395$ nm $\qquad\qquad \lambda_{max} = 604$ nm

$\varepsilon_{395} = 1.80 \times 10^4 \text{ M}^{-1} \text{ cm}^{-1} \qquad \varepsilon_{604} = 4.97 \times 10^4 \text{ M}^{-1} \text{ cm}^{-1}$

$\varepsilon_{604} = 0$

A solution with a volume of 20.0 mL containing 1.40×10^{-5} M indicator plus 0.050 0 M benzene-1,2,3-tricarboxylic acid was treated with 20.0 mL of aqueous KOH. The resulting solution had an absorbance at 604 nm of 0.118 in a 1.00-cm cell. Calculate the molarity of the KOH solution.

11-75. A certain acid-base indicator exists in three colored forms:

$$H_2In \overset{pK_1 = 1.00}{\rightleftharpoons} HIn^- \overset{pK_2 = 7.95}{\rightleftharpoons} In^{2-}$$

H_2In	HIn^-	In^{2-}
$\lambda_{max} = 520$ nm	$\lambda_{max} = 435$ nm	$\lambda_{max} = 572$ nm
$\varepsilon_{520} = 5.00 \times 10^4$	$\varepsilon_{435} = 1.80 \times 10^4$	$\varepsilon_{572} = 4.97 \times 10^4$
Red	Yellow	Red
$\varepsilon_{435} = 1.67 \times 10^4$	$\varepsilon_{520} = 2.13 \times 10^3$	$\varepsilon_{520} = 2.50 \times 10^4$
$\varepsilon_{572} = 2.30 \times 10^4$	$\varepsilon_{572} = 2.00 \times 10^2$	$\varepsilon_{435} = 1.15 \times 10^4$

The units of molar absorptivity, ε, are $\text{M}^{-1} \text{ cm}^{-1}$. A solution containing 10.0 mL of 5.00×10^{-4} M indicator was mixed with 90.0 mL of 0.1 M phosphate buffer (pH 7.50). Calculate the absorbance of this solution at 435 nm in a 1.00-cm cell.

---

*These problems are based on Beer's law in Section 18-2.

Reference Procedure: Preparing Standard Acid and Base

Standard 0.1 M NaOH

1. Prepare 50 wt% aqueous NaOH solution in advance and allow the Na_2CO_3 precipitate to settle overnight. (Na_2CO_3 is insoluble in this solution.) Store the solution in a tightly sealed polyethylene bottle and avoid disturbing the precipitate when supernate is taken. The density is close to 1.50 g of solution per milliliter.

2. Dry primary-standard-grade potassium hydrogen phthalate for 1 h at 110°C and store it in a desiccator.

Potassium hydrogen
phthalate FM 204.221

3. Boil 1 L of water for 5 min to expel CO_2. Pour the water into a polyethylene bottle, which should be tightly capped whenever possible. Calculate the volume of 50 wt% NaOH needed (~5.3 mL) to produce 1 L of ~0.1 M NaOH. Use a graduated cylinder to transfer this much NaOH to the bottle of water. Mix well and allow the solution to cool to room temperature (preferably overnight).

4. Weigh out four ~0.51-g portions of potassium hydrogen phthalate and dissolve each in ~25 mL of distilled water in a 125-mL flask. Each sample should require ~25 mL of 0.1 M NaOH. Add 3 drops of phenolphthalein indicator (Table 11-3) to each, and titrate one of them rapidly to find the approximate end point. The buret should have a loosely fitted cap to minimize entry of CO_2.

5. Calculate the volume of NaOH required for each of the other three samples and titrate them carefully. During each titration, periodically tilt and rotate the flask to wash liquid from the walls into the solution. When very near the end, deliver less than 1 drop of titrant at a time. To do this, carefully suspend a fraction of a drop from the buret tip, touch it to the inside wall of the flask, wash it into the bulk solution by careful tilting, and swirl the solution. The end point is the first appearance of pink color that persists for 15 s. The color will slowly fade as CO_2 from air dissolves in the solution.

6. Calculate the average molarity ($\bar{x}$), the standard deviation (s), and the relative standard deviation ($s/\bar{x}$). If you have used some care, the relative standard deviation should be <0.2%.

Standard 0.1 M HCl

1. The inside cover of this book tells us that 8.2 mL of ~37 wt% HCl should be added to 1 L of water to produce ~0.1 M HCl. Prepare this solution in a capped polyethylene bottle, using a graduated cylinder to deliver the HCl.

2. Dry primary-standard-grade Na_2CO_3 for 1 h at 110°C and cool it in a desiccator.

3. Weigh four samples containing enough Na_2CO_3 to react with ~25 mL of 0.1 M HCl from a buret and place each in a 125-mL flask. When ready to titrate each one, dissolve it in ~25 mL of distilled water.

$$2HCl + Na_2CO_3 \rightarrow CO_2 + 2NaCl + H_2O$$
$$FM\ 105.988$$

Add 3 drops of bromocresol green indicator (Table 11-3) and titrate one sample rapidly to a green color to find the approximate end point.

4. Carefully titrate each of the other samples until it just turns from blue to green. Then boil the solution to expel CO_2. The solution should return to a blue color. Carefully add HCl from the buret until the solution turns green again.

5. Titrate one blank prepared from 3 drops of indicator plus 50 mL of 0.05 M NaCl. Subtract the blank volume of HCl from that required to titrate Na_2CO_3.

6. Calculate the mean HCl molarity, standard deviation, and relative standard deviation.

12 EDTA Titrations

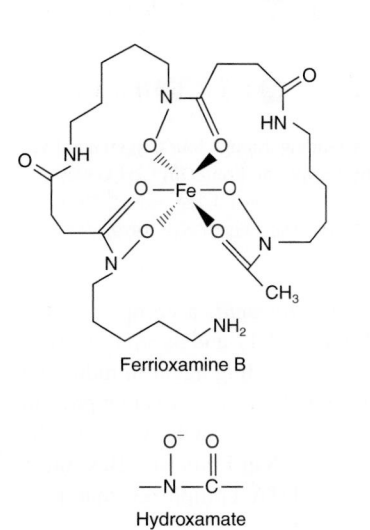

Ferrioxamine B

Hydroxamate group

Key
= Fe
= O
= N
= C

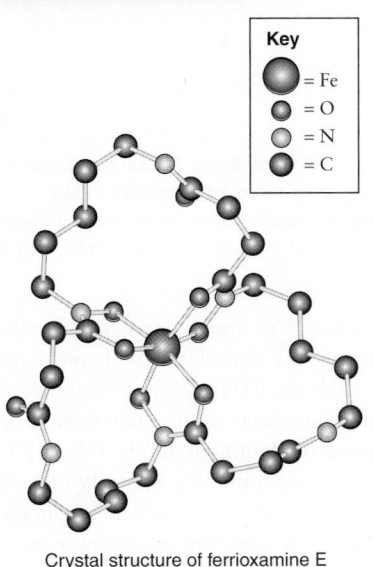

Crystal structure of ferrioxamine E

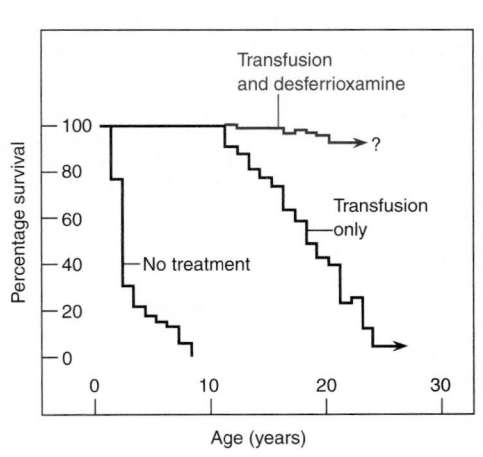

Structures show the iron complex ferrioxamine B and the related compound, ferrioxamine E, in which the chelate has a cyclic structure. Graph shows success of transfusions and transfusions plus chelation therapy. [Crystal structure data provided by M. Neu, Los Alamos National Laboratory, based on information from D. Van der Helm and M. Poling, *J. Am. Chem. Soc.* **1976,** *98,* 82. Graph data from P. S. Dobbin and R. C. Hider, "Iron Chelation Therapy," *Chem. Br.* **1990,** *26,* 565.]

3 ... Deferiprone ... + Fe³⁺ ⇌

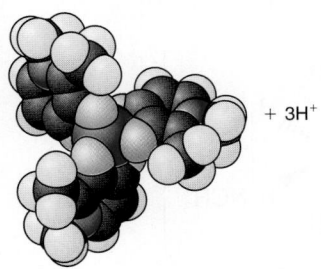

+ 3H⁺

Oxygen (O_2) in the human circulatory system is bound to iron in the protein hemoglobin, which consists of two pairs of subunits, designated α and β. β-Thalassemia major is a genetic disease in which β subunits are not synthesized in adequate quantities. A child afflicted with this disease survives only with frequent transfusions of normal red blood cells. However, the child accumulates 4–8 g of iron per year from hemoglobin in the transfused cells. Our bodies have no mechanism for excreting large quantities of iron, and most patients who received only transfusions died by age 20 from toxic effects of iron overload.

A ligand that binds to a metal ion through multiple ligand atoms is called a *chelate,* pronounced KEE-late. Chelation therapy enhances iron excretion from thalassemia patients. The most successful drug is *desferrioxamine B,* produced by the microbe *Streptomyces pilosus.*[1] Its Fe^{3+} complex, ferrioxamine B, has a formation constant of $10^{30.6}$. Used in conjunction with ascorbic acid (vitamin C), which reduces Fe^{3+} to soluble Fe^{2+}, desferrioxamine clears grams of iron per year from an overloaded patient. The Fe^{3+}-complex is excreted in the urine. In those patients for whom desferrioxamine is effective, there is a 91% rate of cardiac disease-free survival after 15 years of therapy.[2] Too high a dose of this expensive drug stunts a child's growth.

Desferrioxamine is expensive and must be taken by overnight subcutaneous infusion five to seven nights per week. It is not absorbed through the intestine. Many potent iron chelators have been tested to find an effective one that can be taken orally, but few have entered clinical practice.[3] The orally administered chelator deferiprone, introduced in 1987, is used with positive effect in more than 50 countries (but is not licensed in the U.S. and Canada). Combined use of desferrioxamine and deferiprone increases survival and reduces the incidence of cardiac disease, in comparison with desferrioxamine treatment only. An orally administered chelator called deferasirox was approved for use in the U.S. in 2005. The continued search for new chelators indicates that no current treatment is fully effective. In the long term, bone marrow transplants or gene therapy might cure the disease.

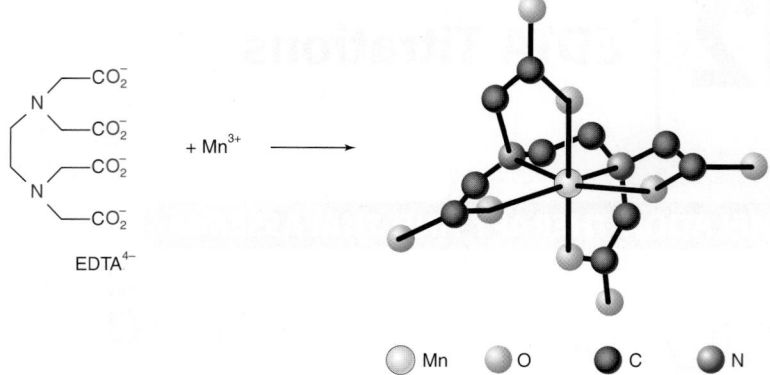

FIGURE 12-1 EDTA forms strong 1:1 complexes with most metal ions, binding through four oxygen and two nitrogen atoms. The six-coordinate geometry of Mn³⁺-EDTA found in the compound KMnEDTA · 2H₂O was deduced from X-ray crystallography. [Information from J. Stein, J. P. Fackler, Jr., G. J. McClune, J. A. Fee, and L. T. Chan, "Reactions of Mn-EDTA and MnCyDTA Complexes with O₂⁻: X-Ray Structure of KMnEDTA · 2H₂O," *Inorg. Chem.* **1979**, *18*, 3511.]

$\mathbf{E}$DTA is a merciful abbreviation for *ethylenediaminetetraacetic acid*, a compound that forms strong 1:1 complexes with most metal ions (Figure 12-1) and is employed in quantitative analysis. EDTA plays a larger role as a strong metal-binding agent in industrial processes and in products such as detergents, cleaning agents, and food additives that prevent metal-catalyzed oxidation of food. Metal-EDTA complexes then find their way into the environment. For example, the majority of nickel discharged into San Francisco Bay and a significant fraction of the iron, lead, copper, and zinc are EDTA complexes that pass unscathed through wastewater treatment plants.

12-1 Metal-Chelate Complexes

Metal ions are **Lewis acids,** accepting electron pairs from electron-donating ligands that are **Lewis bases.** Cyanide (CN^-) is called a **monodentate** ligand because it binds to a metal ion through only one atom (the carbon atom). Most transition metal ions bind six ligand atoms. A ligand that attaches to a metal ion through more than one ligand atom is said to be **multidentate** ("many toothed"), or a **chelating ligand.**[4]

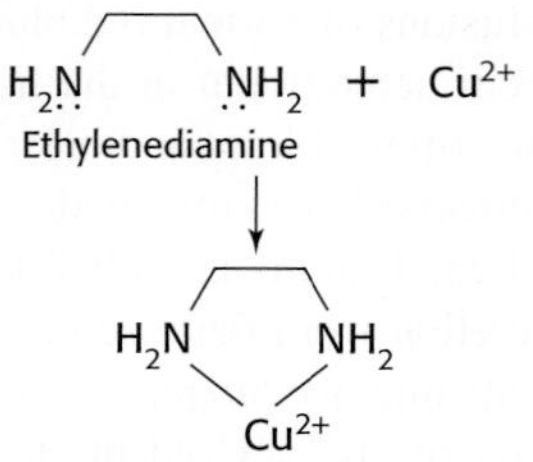

A simple chelating ligand is 1,2-diaminoethane ($H_2NCH_2CH_2NH_2$, also called ethylenediamine), whose binding to a metal ion is shown in the margin. We say that ethylenediamine is *bidentate* because it binds to the metal through two ligand atoms.

The **chelate effect** is the ability of multidentate ligands to form more stable metal complexes than those formed by similar monodentate ligands.[5,6] For example, the reaction of $Cd(H_2O)_6^{2+}$ with two molecules of ethylenediamine is more favorable than its reaction with four molecules of methylamine:

$$Cd(H_2O)_6^{2+} + 2H_2N\!\!-\!\!NH_2 \rightleftharpoons \left[\begin{array}{c} \text{complex} \end{array} \right]^{2+} + 4H_2O$$

Ethylenediamine

$$K \equiv \beta_2 = 8 \times 10^9 \qquad (12\text{-}1)$$

$$Cd(H_2O)_6^{2+} + 4CH_3\ddot{N}H_2 \rightleftharpoons \left[\begin{array}{c} \text{complex} \end{array} \right]^{2+} + 4H_2O$$

Methylamine

$$K \equiv \beta_4 = 4 \times 10^6 \qquad (12\text{-}2)$$

Margin notes (left column):

$$Ag^+ + 2\,:\!\!\overset{..}{C}\!\!\equiv\!\!N\!: \rightleftharpoons$$

Lewis acid (electron pair acceptor) Lewis base (electron pair donor)

$$(:\!N\!\!\equiv\!\!C\!\!-\!\!Ag\!\!-\!\!C\!\!\equiv\!\!N\!:)^-$$

Complex ion

$H_2N\quad NH_2 + Cu^{2+}$

Ethylenediamine

$H_2N\quad NH_2$
Cu^{2+}

Bidentate coordination

Nomenclature for formation constants (K and β) was discussed in Box 6-2.

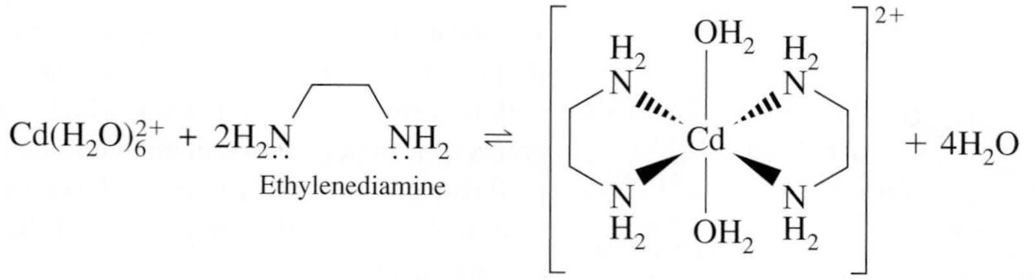

We have drawn *trans* isomers of the octahedral complexes (with H_2O ligands at opposite poles), but *cis* isomers are also possible (with H_2O ligands adjacent to each other).

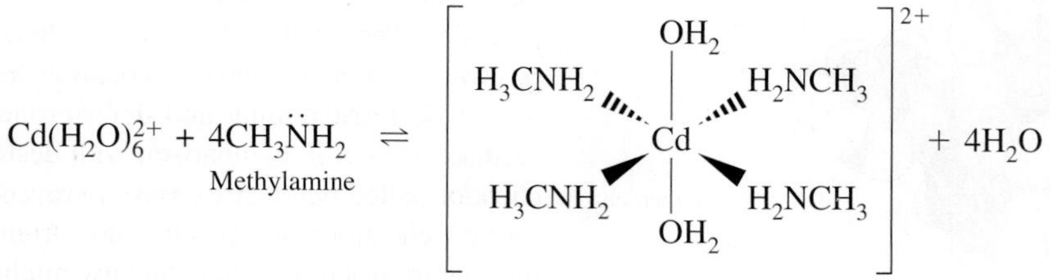

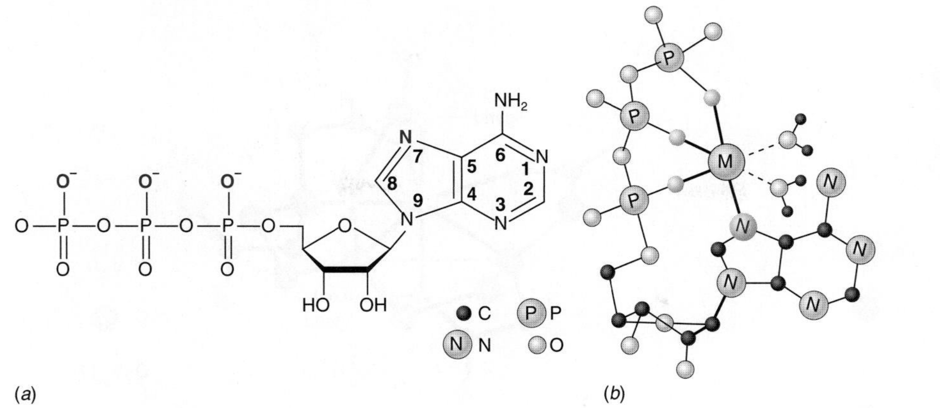

FIGURE 12-2 (*a*) Structure of adenosine triphosphate (ATP), with ligand atoms shown in color. (*b*) Possible structure of a metal-ATP complex; the metal, M, has four bonds to ATP and two bonds to H_2O ligands.

At pH 12 in the presence of 2 M ethylenediamine and 4 M methylamine, the quotient $[Cd(ethylenediamine)_2^{2+}]/[Cd(methylamine)_4^{2+}]$ is 30.

An important *tetradentate* ligand is adenosine triphosphate (ATP), which binds to divalent metal ions (such as Mg^{2+}, Mn^{2+}, Co^{2+}, and Ni^{2+}) through four of their six coordination positions (Figure 12-2). The fifth and sixth positions are occupied by water molecules. The biologically active form of ATP is generally the Mg^{2+} complex.

Metal-chelate complexes are ubiquitous in biology. Bacteria such as *Escherichia coli* and *Salmonella enterica* in your gut excrete a powerful iron chelator called enterobactin (Figure 12-3) to scavenge iron that is essential for bacterial growth. The iron-enterobactin complex is recognized at specific sites on the bacterial cell surface and taken into the cell. Iron is then released inside the bacterium by enzymatic disassembly of the chelator. To fight bacterial infection, your immune system produces a protein called siderocalin to sequester and inactivate enterobactin.[7] The opening of the chapter described an important medical application of chelates.

Aminocarboxylic acids in Figure 12-4 are synthetic chelating agents. Amine N atoms and carboxylate O atoms are potential ligand atoms in these molecules (Figures 12-5 and 12-6). When these molecules bind to a metal ion, the ligand atoms lose their protons. A medical application of the ligand DTPA in Figure 12-4 is illustrated by the tightly bound complex Gd^{3+}-DTPA, which is injected into humans at a concentration of ~0.5 mM to provide contrast in magnetic resonance imaging.[8] Enough gadolinium contrast agents are used in medical diagnosis for the gadolinium complexes to be observed intact in rivers and plant life downstream of sewage plants.[9]

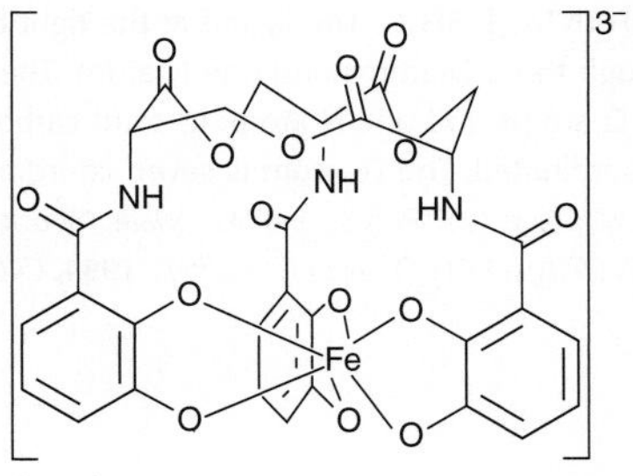

FIGURE 12-3 Iron(III)-enterobactin complex. Certain bacteria secrete enterobactin to capture iron and bring it into the cell. Enterobactin is one of several known chelates—designated *siderophores*—released by microbes to capture iron. [Information from R. J. Abergel, J. A. Warner, D. K. Shuh, and K. N. Raymond, "Enterobactin Protonation and Iron Release," *J. Am. Chem. Soc.* **2006,** *128,* 8920.]

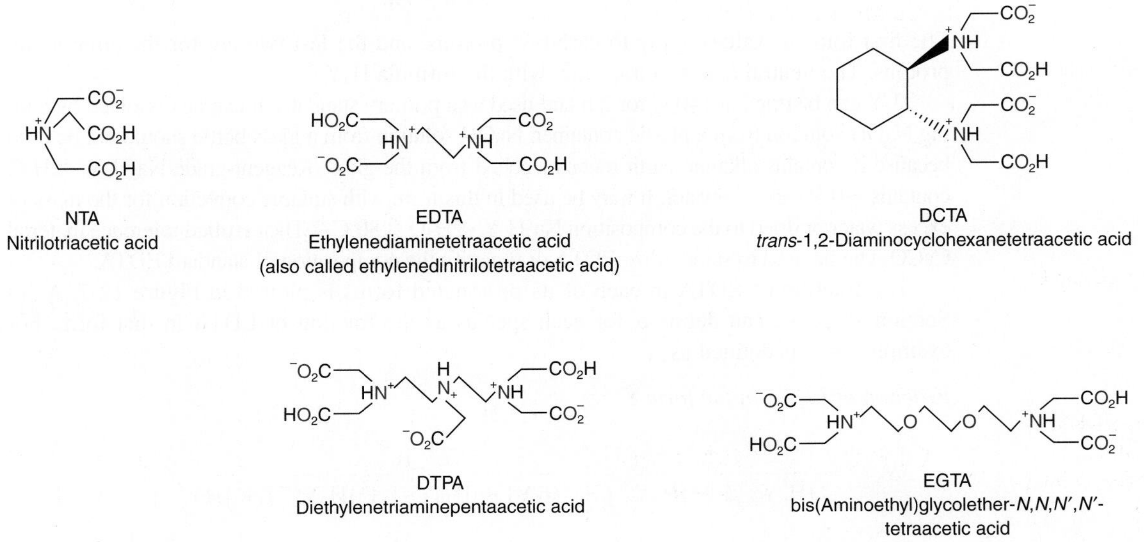

NTA	EDTA	DCTA
Nitrilotriacetic acid	Ethylenediaminetetraacetic acid (also called ethylenedinitrilotetraacetic acid)	*trans*-1,2-Diaminocyclohexanetetraacetic acid

DTPA	EGTA
Diethylenetriaminepentaacetic acid	bis(Aminoethyl)glycolether-*N,N,N',N'*-tetraacetic acid

FIGURE 12-4 Structures of analytically useful chelating agents. Nitrilotriacetic acid (NTA) tends to form 2:1 (ligand:metal) complexes with metal ions, whereas the others form 1:1 complexes.

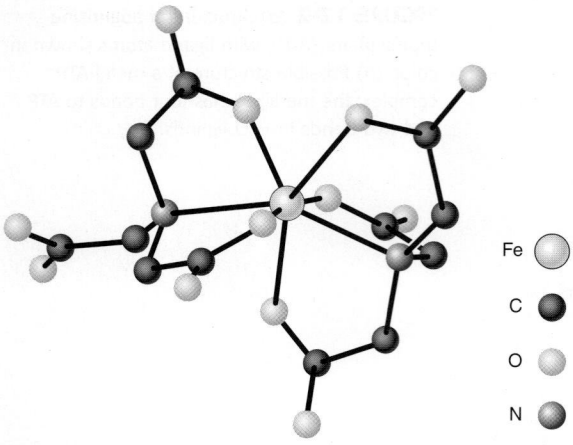

FIGURE 12-5 Structure of $Fe(NTA)_2^{3-}$ in the salt $Na_3[Fe(NTA)_2] \cdot 5H_2O$. The ligand at the right binds to Fe through three O atoms and one N atom. The other ligand uses two O atoms and one N atom. Its third carboxylate group is uncoordinated. The Fe atom is seven coordinate. [Information from W. Clegg, A. K. Powell, and M. J. Ware, "Structure of $Na_3[Fe(NTA)_2] \cdot 5H_2O$," *Acta Crystallogr.* **1984**, *C40*, 1822.]

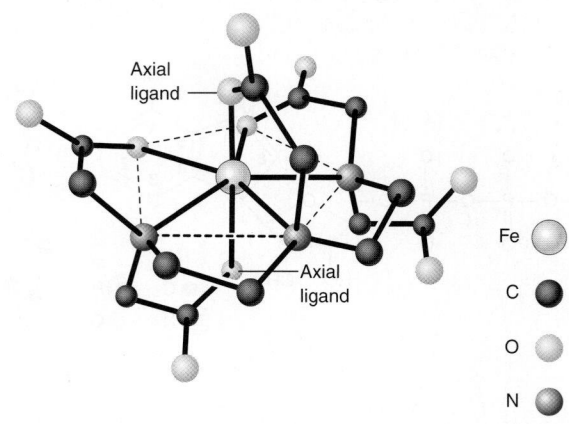

FIGURE 12-6 Structure of $Fe(DTPA)^{2-}$ found in the salt $Na_2[Fe(DTPA)] \cdot 2H_2O$. The seven-coordinate pentagonal bipyramidal coordination environment of the iron atom features 3 N and 2 O ligands in the equatorial plane (dashed lines) and two axial O ligands. The axial Fe—O bond lengths are 11 to 19 pm shorter than those of the more crowded equatorial Fe—O bonds. One carboxyl group of the ligand is uncoordinated. [Information from D. C. Finnen, A. A. Pinkerton, W. R. Dunham, R. H. Sands, and M. O. Funk, Jr., "Structures and Spectroscopic Characterization of Fe(III)-DTPA Complexes," *Inorg. Chem.* **1991**, *30*, 3960.]

12-2 EDTA

One mole of EDTA reacts with one mole of metal ion.

A titration based on complex formation is called a **complexometric titration.** Ligands other than NTA in Figure 12-4 form strong 1:1 complexes with all metal ions except univalent ions such as Li^+, Na^+, and K^+. *The stoichiometry is 1:1 regardless of the charge on the ion.* EDTA is, by far, the most widely used chelator in analytical chemistry. By direct titration or through an indirect sequence of reactions, virtually every element of the periodic table can be measured with EDTA.

Acid-Base Properties

EDTA is a hexaprotic system, designated H_6Y^{2+}. The highlighted, acidic hydrogen atoms are the ones that are lost upon metal-complex formation.

$$
\begin{array}{l}
HO_2CCH_2 \diagdown \qquad\qquad \diagup CH_2CO_2H \\
\qquad\quad \overset{+}{H}NCH_2CH_2\overset{+}{N}H \\
HO_2CCH_2 \diagup \qquad\qquad \diagdown CH_2CO_2H \\
\qquad\qquad\quad H_6Y^{2+}
\end{array}
$$

$pK_1 = 0.0\ (CO_2H)$	$pK_4 = 2.69\ (CO_2H)$
$pK_2 = 1.5\ (CO_2H)$	$pK_5 = 6.13\ (NH^+)$
$pK_3 = 2.00\ (CO_2H)$	$pK_6 = 10.37\ (NH^+)$

pK applies at 25°C and $\mu = 0.1$ M, except pK_1 applies at $\mu = 1$ M

The first four pK values apply to carboxyl protons, and the last two are for the ammonium protons. The neutral acid is tetraprotic, with the formula H_4Y.

H_4Y can be dried at 140°C for 2 h and used as a primary standard. It can be dissolved by adding NaOH solution from a plastic container. NaOH solution from a glass bottle should not be used because it contains alkaline earth metals leached from the glass. Reagent-grade $Na_2H_2Y \cdot 2H_2O$ contains ~0.3% excess water. It may be used in this form with suitable correction for the mass of excess water or dried to the composition $Na_2H_2Y \cdot 2H_2O$ at 80°C.[10] The certified reference material $CaCO_3$ can be used to standardize EDTA or to verify the composition of standard EDTA.

The fraction of EDTA in each of its protonated forms is plotted in Figure 12-7. As in Section 10-5, we can define α for each species as the fraction of EDTA in that form. For example, $\alpha_{Y^{4-}}$ is defined as

Fraction of EDTA in the form Y^{4-}:

$$
\alpha_{Y^{4-}} = \frac{[Y^{4-}]}{[H_6Y^{2+}] + [H_5Y^+] + [H_4Y] + [H_3Y^-] + [H_2Y^{2-}] + [HY^{3-}] + [Y^{4-}]}
$$

$$
\alpha_{Y^{4-}} = \frac{[Y^{4-}]}{[EDTA]} \tag{12-3}
$$

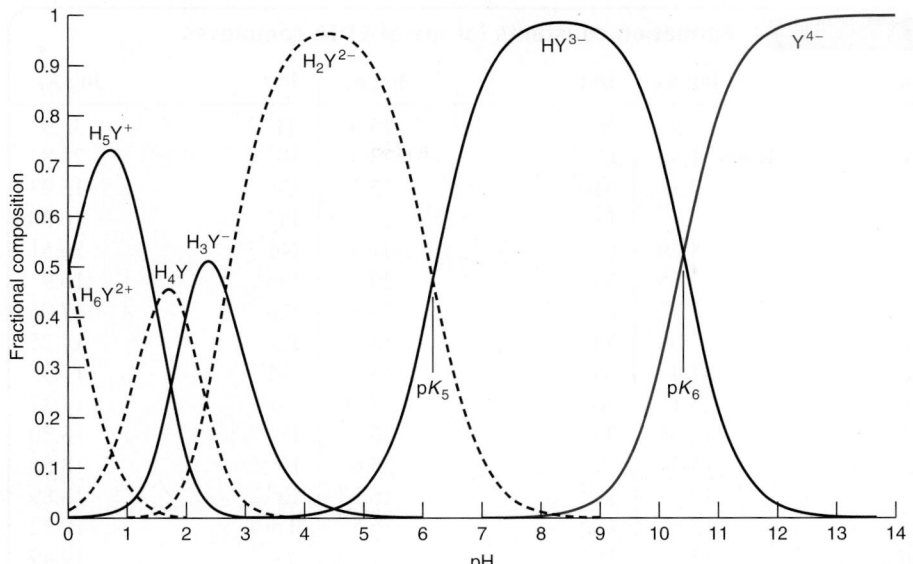

FIGURE 12-7 Fractional composition diagram for EDTA.

where [EDTA] is the total concentration of all *free* EDTA species in the solution. By "free," we mean EDTA not complexed to metal ions. Following the derivation in Section 10-5, we can show that $\alpha_{Y^{4-}}$ is given by

$$\alpha_{Y^{4-}} = \frac{K_1 K_2 K_3 K_4 K_5 K_6}{D} \qquad (12\text{-}4)$$

where $D = [H^+]^6 + [H^+]^5 K_1 + [H^+]^4 K_1 K_2 + [H^+]^3 K_1 K_2 K_3 + [H^+]^2 K_1 K_2 K_3 K_4 + [H^+] K_1 K_2 K_3 K_4 K_5 + K_1 K_2 K_3 K_4 K_5 K_6$. Table 12-1 gives values for $\alpha_{Y^{4-}}$ as a function of pH.

EXAMPLE   What Does $\alpha_{Y^{4-}}$ Mean?

The fraction of all free EDTA in the form Y^{4-} is called $\alpha_{Y^{4-}}$. At pH 6.00 and a formal concentration of 0.10 M, the composition of an EDTA solution is

$[H_6Y^{2+}] = 8.9 \times 10^{-20}$ M $[H_5Y^+] = 8.9 \times 10^{-14}$ M $[H_4Y] = 2.8 \times 10^{-7}$ M

$[H_3Y^-] = 2.8 \times 10^{-5}$ M $[H_2Y^{2-}] = 0.057$ M $[HY^{3-}] = 0.043$ M

$$[Y^{4-}] = 1.8 \times 10^{-6} \text{ M}$$

Find $\alpha_{Y^{4-}}$.

Solution $\alpha_{Y^{4-}}$ is the fraction in the form Y^{4-}:

$$\alpha_{Y^{4-}} = \frac{[Y^{4-}]}{[H_6Y^{2+}] + [H_5Y^+] + [H_4Y] + [H_3Y^-] + [H_2Y^{2-}] + [HY^{3-}] + [Y^{4-}]}$$

$$= \frac{1.8 \times 10^{-6}}{(8.9 \times 10^{-20}) + (8.9 \times 10^{-14}) + (2.8 \times 10^{-7}) + (2.8 \times 10^{-5}) + (0.057) + (0.043) + (1.8 \times 10^{-6})}$$

$$= 1.8 \times 10^{-5}$$

TEST YOURSELF   At what pH does $\alpha_{Y^{4-}} = 0.50$? (***Answer:*** pH = pK_6 = 10.37)

TABLE 12-1	Values of $\alpha_{Y^{4-}}$ for EDTA at 25°C and $\mu = 0.10$ M
pH	$\alpha_{Y^{4-}}$
0	1.3×10^{-23}
1	1.4×10^{-18}
2	2.6×10^{-14}
3	2.1×10^{-11}
4	3.0×10^{-9}
5	2.9×10^{-7}
6	1.8×10^{-5}
7	3.8×10^{-4}
8	4.2×10^{-3}
9	0.041
10	0.30
11	0.81
12	0.98
13	1.00
14	1.00

Question From Figure 12-7, which species has the greatest concentration at pH 6? At pH 7? At pH 11?

EDTA Complexes

The equilibrium constant for the reaction of a metal with a ligand is called the **formation constant**, K_f, or the **stability constant**:

Formation constant: $M^{n+} + Y^{4-} \rightleftharpoons MY^{n-4}$ $K_f = \dfrac{[MY^{n-4}]}{[M^{n+}][Y^{4-}]}$ (12-5)

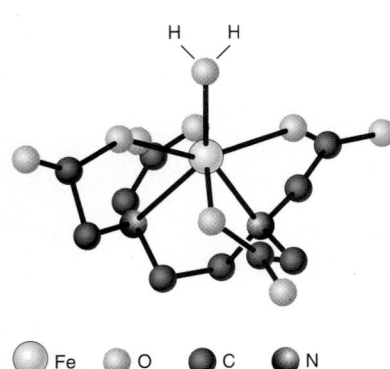

H H

Fe O C N

FIGURE 12-8 Seven-coordinate geometry of Fe(EDTA)(H$_2$O)$^-$. Other metal ions that form seven-coordinate EDTA complexes include Fe^{2+}, Mg^{2+}, Cd^{2+}, Co^{2+}, Mn^{2+}, Ru^{3+}, Cr^{3+}, Co^{3+}, V^{3+}, Ti^{3+}, In^{3+}, Sn^{4+}, Os^{4+}, and Ti^{4+}. Some of these same ions also form six-coordinate EDTA complexes. Eight-coordinate complexes are formed by Ca^{2+}, Er^{3+}, Yb^{3+}, and Zr^{4+}.

[Information from T. Mizuta, J. Wang, and K. Miyoshi, "A 7-Coordinate Structure of Fe(III)-EDTA," *Bull. Chem. Soc. Japan* **1993,** *66,* 2547.]

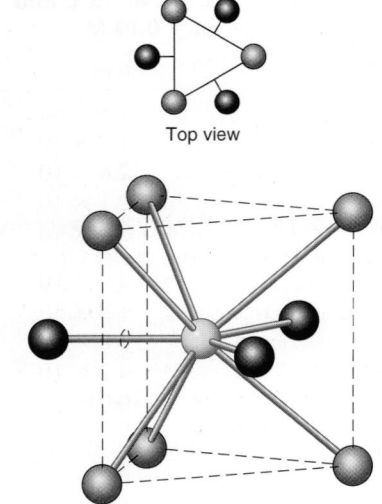

Top view

FIGURE 12-9 Tricapped trigonal prismatic structure of many Ln(III) and An(III) complexes, where Ln is a lanthanide element and An is an actinide element. In M(H$_2$O)$_9^{3+}$, bonds from the metal to the 6 O atoms at the corners of the prism are shorter than bonds from the metal to the 3 O atoms projecting out from the rectangular faces.

Only some of the free EDTA is in the form Y^{4-}.

TABLE 12-2		Formation constants for metal-EDTA complexes			
Ion	**log K_f**	**Ion**	**log K_f**	**Ion**	**log K_f**
Li$^+$	2.95	V^{3+}	25.9^a	Tl^{3+}	35.3
Na$^+$	1.86	Cr^{3+}	23.4^a	Bi^{3+}	27.8^a
K$^+$	0.8	Mn^{3+}	25.2	Ce^{3+}	15.93
Be^{2+}	9.7	Fe^{3+}	25.1	Pr^{3+}	16.30
Mg^{2+}	8.79	Co^{3+}	41.4	Nd^{3+}	16.51
Ca^{2+}	10.65	Zr^{4+}	29.3	Pm^{3+}	16.9
Sr^{2+}	8.72	Hf^{4+}	29.5	Sm^{3+}	17.06
Ba^{2+}	7.88	VO^{2+}	18.7	Eu^{3+}	17.25
Ra^{2+}	7.4	VO$_2^+$	15.5	Gd^{3+}	17.35
Sc^{3+}	23.1^a	Ag$^+$	7.20	Tb^{3+}	17.87
Y^{3+}	18.08	Tl$^+$	6.41	Dy^{3+}	18.30
La^{3+}	15.36	Pd^{2+}	25.6^a	Ho^{3+}	18.56
V^{2+}	12.7^a	Zn^{2+}	16.5	Er^{3+}	18.89
Cr^{2+}	13.6^a	Cd^{2+}	16.5	Tm^{3+}	19.32
Mn^{2+}	13.89	Hg^{2+}	21.5	Yb^{3+}	19.49
Fe^{2+}	14.30	Sn^{2+}	18.3^b	Lu^{3+}	19.74
Co^{2+}	16.45	Pb^{2+}	18.0	Th^{4+}	23.2
Ni^{2+}	18.4	Al^{3+}	16.4	U^{4+}	25.7
Cu^{2+}	18.78	Ga^{3+}	21.7		
Ti^{3+}	21.3	In^{3+}	24.9		

NOTE: *The stability constant is the equilibrium constant for the reaction* M^{n+} + Y^{4-} ⇌ MY^{n-4}. *Values in table apply at 25°C and ionic strength 0.1 M unless otherwise indicated.*

a. 20°C, ionic strength = 0.1 M.　　　　*b. 20°C, ionic strength = 1 M.*

SOURCE: *A. E. Martell, R. M. Smith, and R. J. Motekaitis, NIST Critically Selected Stability Constants of Metal Complexes, NIST Standard Reference Database 46, Gaithersburg, MD, 2001.*

Note that K_f for EDTA is defined in terms of the species Y^{4-} reacting with the metal ion. The equilibrium constant could have been defined for any of the other six forms of EDTA in the solution. Equation 12-5 should not be interpreted to mean that only Y^{4-} reacts with metal ions. Table 12-2 shows that formation constants for most EDTA complexes are large and tend to be larger for more positively charged cations.

In many transition metal complexes, EDTA engulfs the metal ion, forming the six-coordinate species in Figure 12-1. If you try to build a space-filling model of a six-coordinate metal-EDTA complex, you will find strain in the chelate rings. This strain is relieved when the O ligands are drawn back toward the N atoms. Such distortion opens up a seventh coordination position, which can be occupied by H$_2$O, as in Figure 12-8. In some complexes, such as Ca(EDTA)(H$_2$O)$_2^{2-}$, the metal ion is so large that it accommodates eight ligand atoms.[11] Larger metal ions require more ligand atoms. Even if H$_2$O is attached to the metal ion, the formation constant is still given by Equation 12-5. The relation remains true because solvent (H$_2$O) is omitted from the reaction quotient.

Lanthanide and actinide elements typically have a coordination number of 9, with the shape of a tricapped trigonal prism (Figure 12-9).[12] Eu(III) forms mixed complexes of the type Eu(EDTA)(NTA) in which EDTA provides six ligand atoms and NTA provides three ligand atoms (Figure 12-4).[13]

Conditional Formation Constant

The formation constant $K_f = [MY^{n-4}]/[M^{n+}][Y^{4-}]$ describes the reaction between Y^{4-} and a metal ion. As you see in Figure 12-7, most EDTA is not Y^{4-} below pH 10.37. The species HY^{3-}, H$_2$Y^{2-}, and so on, predominate at lower pH. From the definition $\alpha_{Y^{4-}} = [Y^{4-}]/[EDTA]$, we can express the concentration of Y^{4-} as

$$[Y^{4-}] = \alpha_{Y^{4-}}[EDTA]$$

where [EDTA] is the total concentration of all EDTA species not bound to metal ion.

The formation constant can now be rewritten as

$$K_f = \frac{[MY^{n-4}]}{[M^{n+}][Y^{4-}]} = \frac{[MY^{n-4}]}{[M^{n+}]\alpha_{Y^{4-}}[EDTA]}$$

If the pH is fixed by a buffer, then $\alpha_{Y^{4-}}$ is a constant that can be combined with K_f:

Conditional formation constant: $\qquad K_f' = \alpha_{Y^{4-}}K_f = \dfrac{[MY^{n-4}]}{[M^{n+}][EDTA]}$ $\qquad$ (12-6)

The number $K_f' = \alpha_{Y^{4-}}K_f$ is called the **conditional formation constant,** or the *effective formation constant.* It describes the formation of MY^{n-4} at any particular pH.

The conditional formation constant allows us to look at EDTA complex formation as if the uncomplexed EDTA were all in one form:

$$M^{n+} + EDTA \rightleftharpoons MY^{n-4} \qquad K_f' = \alpha_{Y^{4-}}K_f$$

At any given pH, we can find $\alpha_{Y^{4-}}$ and evaluate K_f'.

> With the conditional formation constant, we can treat EDTA complex formation as if all free EDTA were in one form.

EXAMPLE **Using the Conditional Formation Constant**

The formation constant in Table 12-2 for CaY^{2-} is $10^{10.65}$. Calculate the concentration of free Ca^{2+} in a solution of 0.10 M CaY^{2-} at pH 10.00 and at pH 6.00.

Solution The complex formation reaction is

$$Ca^{2+} + EDTA \rightleftharpoons CaY^{2-} \qquad K_f' = \alpha_{Y^{4-}}K_f$$

where EDTA on the left side of the equation refers to all forms of unbound EDTA (Y^{4-}, HY^{3-}, H_2Y^{2-}, H_3Y^-, and so on). Using $\alpha_{Y^{4-}}$ from Table 12-1, we find

At pH 10.00: $\qquad K_f' = (0.30)(10^{10.65}) = 1.3_4 \times 10^{10}$

At pH 6.00: $\qquad K_f' = (1.8 \times 10^{-5})(10^{10.65}) = 8.0 \times 10^5$

Dissociation of CaY^{2-} must produce equal quantities of Ca^{2+} and EDTA, so we can write

	Ca^{2+}	+	EDTA	$\rightleftharpoons$	CaY^{2-}
Initial concentration (M)	0		0		0.10
Final concentration (M)	x		x		$0.10 - x$

$$\frac{[CaY^{2-}]}{[Ca^{2+}][EDTA]} = \frac{0.10 - x}{x^2} = K_f' = 1.3_4 \times 10^{10} \quad \text{at pH 10.00}$$

$$= 8.0 \times 10^5 \quad \text{at pH 6.00}$$

Solving for x ($= [Ca^{2+}] = [EDTA]$), we find $[Ca^{2+}] = 2.7 \times 10^{-6}$ M at pH 10.00 and 3.5×10^{-4} M at pH 6.00. *Using the conditional formation constant at a fixed pH, we treat the dissociated EDTA as if it were a single species.*

TEST YOURSELF Find $[Ca^{2+}]$ in 0.10 M CaY^{2-} at pH 8.00 (***Answer:*** 2.3×10^{-5} M)

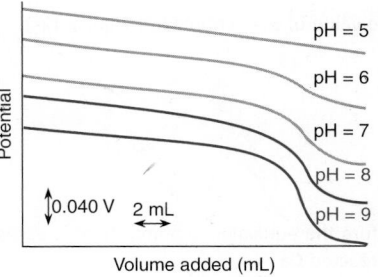

FIGURE 12-10 Titration of Ca^{2+} with EDTA as a function of pH. As the pH is lowered, the end point becomes less distinct. The potential was measured with mercury and calomel electrodes, as described in Exercise 15-B. [Data from C. N. Reilley and R. W. Schmid, "Chelometric Titration with Potentiometric End Point Detection. Mercury as a pM Indicator Electrode," *Anal. Chem.* **1958,** *30,* 947.]

You can see from the example that a metal-EDTA complex becomes less stable at lower pH. For a titration reaction to be effective, it must go "to completion" (say, 99.9%) which means that the equilibrium constant is large—the analyte and titrant are essentially completely reacted at the equivalence point. Figure 12-10 shows how pH affects the titration of Ca^{2+} with EDTA. Below pH ≈ 8, the end point is not sharp enough to allow accurate determination. The conditional formation constant for CaY^{2-} is just too small for "complete" reaction at low pH.

> pH can be used to select which metals will be titrated by EDTA and which will not. Metals with higher formation constants can be titrated at lower pH. If a solution containing both Fe^{3+} and Ca^{2+} is titrated at pH 4, Fe^{3+} is titrated without interference from Ca^{2+}.

12-3 **EDTA Titration Curves**

Now we calculate the concentration of free M^{n+} during its titration with EDTA.[14] The titration reaction is

$$M^{n+} + EDTA \rightleftharpoons MY^{n-4} \qquad K_f' = \alpha_{Y^{4-}}K_f \qquad (12\text{-}7)$$

If K_f' is large, we can consider the reaction to be complete at each point in the titration.

> K_f' is the effective formation constant at the fixed pH of the solution.

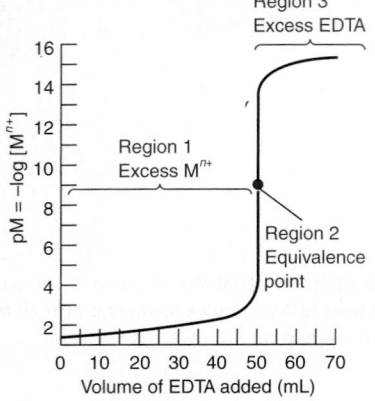

FIGURE 12-11 Three regions in an EDTA titration illustrated for reaction of 50.0 mL of 0.050 0 M M^{n+} with 0.050 0 M EDTA, assuming $K_f' = 1.15 \times 10^{16}$. The concentration of free M^{n+} decreases as the titration proceeds.

The titration curve is a graph of pM ($= -\log[M^{n+}]$) versus the volume of added EDTA. The curve is analogous to plotting pH versus volume of titrant in an acid-base titration. There are three natural regions of the titration curve in Figure 12-11.

Region 1: Before the Equivalence Point

In this region, there is excess M^{n+} left in solution after the EDTA has been consumed. The concentration of free metal ion is equal to the concentration of excess, unreacted M^{n+}. The dissociation of MY^{n-4} is negligible.

Region 2: At the Equivalence Point

There is exactly as much EDTA as metal in the solution. We can treat the solution as if it had been made by dissolving pure MY^{n-4}. Some free M^{n+} is generated by the slight dissociation of MY^{n-4}:

$$MY^{n-4} \rightleftharpoons M^{n+} + EDTA$$

In this reaction, EDTA represents free EDTA in all its forms. At the equivalence point, $[M^{n+}] = [EDTA]$.

Region 3: After the Equivalence Point

Now there is excess EDTA, and virtually all the metal ion is in the form MY^{n-4}. The concentration of free EDTA can be equated to the concentration of excess EDTA added after the equivalence point.

Titration Calculations

The value of $\alpha_{Y^{4-}}$ comes from Table 12-1.

Let's calculate the shape of the titration curve for the reaction of 50.0 mL of 0.040 0 M Ca^{2+} (buffered to pH 10.00) with 0.080 0 M EDTA:

$$Ca^{2+} + EDTA \rightarrow CaY^{2-}$$
$$K_f' = \alpha_{Y^{4-}} K_f = (0.30)(10^{10.65}) = 1.3_4 \times 10^{10}$$

Because K_f' is large, it is reasonable to say that the reaction goes to completion with each addition of titrant. We want to make a graph in which pCa^{2+} ($= -\log[Ca^{2+}]$) is plotted versus milliliters of added EDTA. The equivalence volume is 25.0 mL.

Region 1: Before the Equivalence Point

Before the equivalence point, there is excess unreacted Ca^{2+}.

Consider the addition of 5.0 mL of EDTA. Because the equivalence point requires 25.0 mL of EDTA, one-fifth of the Ca^{2+} will be consumed and four-fifths remains.

$$[Ca^{2+}] = \underbrace{\left(\frac{25.0\ mL - 5.0\ mL}{25.0\ mL}\right)}_{\substack{\text{Fraction of } Ca^{2+} \\ \text{remaining} \\ (= 4/5)}}\underbrace{(0.0400\ M)}_{\substack{\text{Original} \\ \text{Concentration} \\ \text{of } Ca^{2+}}}\underbrace{\left(\frac{50.0\ mL}{55.0\ mL}\right)}_{\substack{\text{Dilution} \\ \text{factor}}}$$

Initial volume of Ca^{2+}

Total volume of solution

$$= 0.029\ 1\ M \Rightarrow pCa^{2+} = -\log[Ca^{2+}] = 1.54$$

In a similar manner, we could calculate pCa^{2+} for any volume of EDTA less than 25.0 mL.

Region 2: At the Equivalence Point

At the equivalence point, the major species is CaY^{2-}, in equilibrium with small, equal amounts of free Ca^{2+} and EDTA.

Virtually all the metal is in the form CaY^{2-}. With negligible dissociation, the concentration of CaY^{2-} is equal to the original concentration of Ca^{2+}, with a correction for dilution.

Initial volume of Ca^{2+}

$$[CaY^{2-}] = \underbrace{(0.040\ 0\ M)}_{\substack{\text{Original} \\ \text{concentration} \\ \text{of } Ca^{2+}}}\underbrace{\left(\frac{50.0\ mL}{75.0\ mL}\right)}_{\substack{\text{Dilution} \\ \text{factor}}} = 0.026\ 7\ M$$

Total volume of solution

The concentration of free Ca^{2+} is small and unknown. We can write

	Ca^{2+}	+	EDTA	$\rightleftharpoons$	CaY^{2-}
Initial concentration (M)	—		—		0.026 7
Final concentration (M)	x		x		0.026 7 − x

$$\frac{[\text{CaY}^{2-}]}{[\text{Ca}^{2+}][\text{EDTA}]} = K_f' = 1.3_4 \times 10^{10}$$

$$\frac{0.026\,7 - x}{x^2} = 1.3_4 \times 10^{10} \Rightarrow x = 1.4 \times 10^{-6}\,\text{M}$$

$$\text{pCa}^{2+} = -\log[\text{Ca}^{2+}] = -\log x = 5.85$$

Region 3: After the Equivalence Point

In this region, virtually all the metal is in the form CaY^{2-}, and there is excess, unreacted EDTA. The concentrations of CaY^{2-} and excess EDTA are known. For example, at 26.0 mL, there is 1.0 mL of excess EDTA.

[EDTA] refers to the total concentration of all forms of EDTA not bound to metal.

$$[\text{EDTA}] = (0.080\,0\,\text{M})\underbrace{\left(\frac{1.0\,\text{mL}}{76.0\,\text{mL}}\right)}_{} = 1.05 \times 10^{-3}\,\text{M}$$

Volume of excess EDTA

Original concentration of EDTA — Dilution factor — Total volume of solution

After the equivalence point, virtually all the metal is in the form CaY^{2-}. There is a known excess of EDTA. A small amount of free Ca^{2+} exists in equilibrium with CaY^{2-} and EDTA.

$$[\text{CaY}^{2-}] = (0.040\,0\,\text{M})\underbrace{\left(\frac{50.0\,\text{mL}}{76.0\,\text{mL}}\right)}_{} = 2.63 \times 10^{-2}\,\text{M}$$

Original volume of Ca^{2+}

Original concentration of Ca^{2+} — Dilution factor — Total volume of solution

The concentration of Ca^{2+} is governed by

$$\frac{[\text{CaY}^{2-}]}{[\text{Ca}^{2+}][\text{EDTA}]} = K_f' = 1.3_4 \times 10^{10}$$

$$\frac{[2.63 \times 10^{-2}]}{[\text{Ca}^{2+}](1.05 \times 10^{-3})} = 1.3_4 \times 10^{10}$$

$$[\text{Ca}^{2+}] = 1.9 \times 10^{-9}\,\text{M} \Rightarrow \text{pCa}^{2+} = 8.73$$

The same sort of calculation can be used for any volume past the equivalence point.

The Titration Curve

Calculated titration curves for Ca^{2+} and Sr^{2+} in Figure 12-12 show a distinct break at the equivalence point, where the slope is greatest. The Ca^{2+} end point is more distinct than the Sr^{2+} end point because the conditional formation constant, $\alpha_{Y^{4-}}K_f$, for CaY^{2-} is greater than that of SrY^{2-}. If the pH is lowered, the conditional formation constant decreases (because $\alpha_{Y^{4-}}$ decreases), and the end point becomes less distinct, as we saw in Figure 12-10. The pH cannot be raised arbitrarily high, because metal hydroxide might precipitate.

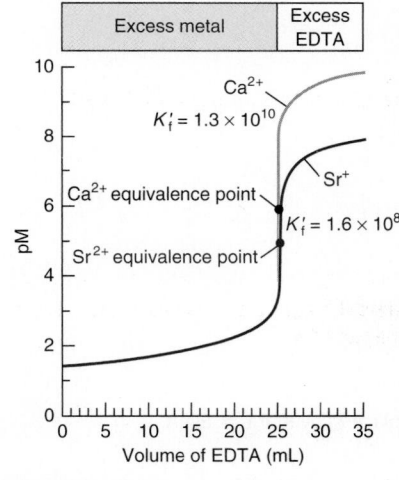

FIGURE 12-12 Theoretical titration curves for the reaction of 50.0 mL of 0.040 0 M metal ion with 0.080 0 M EDTA at pH 10.00.

12-4 ▦ Do It with a Spreadsheet

Let's see how to reproduce the EDTA titration curves in Figure 12-12 by using one equation that applies to the entire titration. Because the reactions are carried out at fixed pH, the equilibria and mass balances are sufficient to solve for all unknowns.

Consider the titration of metal ion M (initial concentration $= C_M$, volume $= V_M$) with a solution of ligand L (concentration $= C_L$, volume added $= V_L$) to form a 1:1 complex:

$$M + L \rightleftharpoons ML \qquad K_f = \frac{[\text{ML}]}{[\text{M}][\text{L}]} \Rightarrow [\text{ML}] = K_f[\text{M}][\text{L}] \qquad \textbf{(12-8)}$$

The mass balances for metal and ligand are

Mass balance for M: $\qquad [\text{M}] + [\text{ML}] = \dfrac{C_M V_M}{V_M + V_L}$

Mass balance for L: $\qquad [\text{L}] + [\text{ML}] = \dfrac{C_L V_L}{V_M + V_L}$

Total metal concentration
$$= \frac{\text{initial moles of metal}}{\text{total volume}}$$
$$= \frac{C_M V_M}{V_M + V_L}$$

Total ligand concentration
$$= \frac{\text{moles of ligand added}}{\text{total volume}}$$
$$= \frac{C_L V_L}{V_M + V_L}$$

FIGURE 12-13 Spreadsheet for the titration of 50.0 mL of 0.040 0 M Ca^{2+} with 0.080 0 M EDTA at pH 10.00. This spreadsheet reproduces calculations of Section 12-3. pM was varied by trial and error to find volumes of 5.00, 25.00, and 26.00 mL used in the preceding section. Better yet, use Goal Seek (page 173) to vary pM in cell B9 until the volume in cell E9 is 25.000 mL.

	A	B	C	D	E
1	Titration of 50 mL of 0.04 M Ca^{2+} with 0.08 M EDTA				
2					
3	C_M =	pM	M	Phi	V(ligand)
4	0.04	1.398	4.00E-02	0.000	0.002
5	V_M =	1.537	2.90E-02	0.201	5.026
6	50	2.00	1.00E-02	0.667	16.667
7	C(ligand) =	3.00	1.00E-03	0.963	24.074
8	0.08	4.00	1.00E-04	0.996	24.906
9	K_f' =	5.85	1.41E-06	1.000	25.0000
10	1.34E+10	7.00	1.00E-07	1.001	25.019
11		8.00	1.00E-08	1.007	25.187
12		8.73	1.86E-09	1.040	26.002
13	C4 = 10^-B4				
14	Equation 12-11:				
15	D4 = (1+$A$10*C4-(C4+C4*C4*$A$10)/$A$4)/				
16	(C4*$A$10+(C4+C4*C4*$A$10)/$A$8)				
17	E4 = D4*$A$4*$A$6/$A$8				

Substituting $K_f[M][L]$ (from Equation 12-8) for [ML] in the mass balances gives

$$[M](1 + K_f[L]) = \frac{C_M V_M}{V_M + V_L} \tag{12-9}$$

$$[L](1 + K_f[M]) = \frac{C_L V_L}{V_M + V_L} \Rightarrow [L] = \frac{\dfrac{C_L V_L}{V_M + V_L}}{1 + K_f[M]} \tag{12-10}$$

Now substitute the expression for [L] from Equation 12-10 back into Equation 12-9

$$[M]\left(1 + K_f \frac{\dfrac{C_L V_L}{V_M + V_L}}{1 + K_f[M]}\right) = \frac{C_M V_M}{V_M + V_L}$$

and do about five lines of algebra to solve for the fraction of titration, ϕ:

Spreadsheet equation for titration of M with L:

$$\phi = \frac{C_L V_L}{C_M V_M} = \frac{1 + K_f[M] - \dfrac{[M] + K_f[M]^2}{C_M}}{K_f[M] + \dfrac{[M] + K_f[M]^2}{C_L}} \tag{12-11}$$

As in acid-base titrations in Table 11-5, ϕ is the fraction of the way to the equivalence point. When $\phi = 1$, $V_L = V_e$. When $\phi = \frac{1}{2}$, $V_L = \frac{1}{2}V_e$. And so on.

For a titration with EDTA, you can follow the derivation through and find that the formation constant, K_f, should be replaced in Equation 12-11 by the conditional formation constant, K_f', which applies at the fixed pH of the titration. Figure 12-13 shows a spreadsheet in which Equation 12-11 is used to calculate the Ca^{2+} titration curve in Figure 12-12. As in acid-base titrations, your input in column B is pM = $-\log[Ca^{2+}]$ and the output in column E is volume of titrant. To find the initial point, vary pM until V_1 is close to 0.

If you reverse the process and titrate ligand with metal ion, the fraction of the way to the equivalence point is the inverse of the fraction in Equation 12-11:

Replace K_f by K_f' if L = EDTA.

Spreadsheet equation for titration of L with M:

$$\phi = \frac{C_M V_M}{C_L V_L} = \frac{K_f[M] + \dfrac{[M] + K_f[M]^2}{C_L}}{1 + K_f[M] - \dfrac{[M] + K_f[M]^2}{C_M}} \tag{12-12}$$

12-5 Auxiliary Complexing Agents

EDTA titration conditions in this chapter were selected to prevent metal hydroxide precipitation at the chosen pH. To permit many metals to be titrated in alkaline solutions with EDTA, we use an **auxiliary complexing agent.** This reagent is a ligand, such as ammonia, tartrate,

Replace K_f by K_f' if L = EDTA.

citrate, or triethanolamine, that binds the metal strongly enough to prevent metal hydroxide from precipitating, but weakly enough to give up the metal when EDTA is added. Zn^{2+} is usually titrated in ammonia buffer, which fixes the pH and complexes the metal ion to keep it in solution. Let's see how this works.

Metal-Ligand Equilibria[15]

Consider a metal ion that forms two complexes with the auxiliary complexing ligand L:

$$M + L \rightleftharpoons ML \qquad \beta_1 = \frac{[ML]}{[M][L]} \qquad (12\text{-}13)$$

$$M + 2L \rightleftharpoons ML_2 \qquad \beta_2 = \frac{[ML_2]}{[M][L]^2} \qquad (12\text{-}14)$$

The equilibrium constants, β_i, are called *overall* or **cumulative formation constants.** The fraction of metal ion in the uncomplexed state, M, can be expressed as

$$\alpha_M = \frac{[M]}{M_{tot}} \qquad (12\text{-}15)$$

where M_{tot} is the total concentration of all forms of M (= M, ML, and ML_2 in this case).

Now let's find a useful expression for α_M. The mass balance for metal is

$$M_{tot} = [M] + [ML] + [ML_2]$$

Equations 12-13 and 12-14 allow us to say $[ML] = \beta_1[M][L]$ and $[ML_2] = \beta_2[M][L]^2$. Therefore,

$$M_{tot} = [M] + \beta_1[M][L] + \beta_2[M][L]^2$$
$$= [M]\{1 + \beta_1[L] + \beta_2[L]^2\}$$

Substituting this last result into Equation 12-15 gives the desired result.

Fraction of free metal ion:
$$\alpha_M = \frac{[M]}{[M]\{1 + \beta_1[L] + \beta_2[L]^2\}} = \frac{1}{1 + \beta_1[L] + \beta_2[L]^2} \qquad (12\text{-}16)$$

> If the metal forms more than two complexes, Equation 12-16 takes the form
> $$\alpha_M = \frac{1}{1 + \beta_1[L] + \beta_2[L]^2 + \cdots + \beta_n[L]^n}$$

EXAMPLE **Ammonia Complexes of Zinc**

Zn^{2+} and NH_3 form the complexes $Zn(NH_3)^{2+}$, $Zn(NH_3)_2^{2+}$, $Zn(NH_3)_3^{2+}$, and $Zn(NH_3)_4^{2+}$. If the concentration of free, *unprotonated* NH_3 is 0.10 M, find the fraction of zinc in the form Zn^{2+}. (At any pH, there will also be some NH_4^+ in equilibrium with NH_3.)

Solution Appendix I gives formation constants for the complexes $Zn(NH_3)^{2+}$ ($\beta_1 = 10^{2.18}$), $Zn(NH_3)_2^{2+}$ ($\beta_2 = 10^{4.43}$), $Zn(NH_3)_3^{2+}$ ($\beta_3 = 10^{6.74}$), and $Zn(NH_3)_4^{2+}$ ($\beta_4 = 10^{8.70}$). The appropriate form of Equation 12-16 is

$$\alpha_{Zn^{2+}} = \frac{1}{1 + \beta_1[L] + \beta_2[L]^2 + \beta_3[L]^3 + \beta_4[L]^4} \qquad (12\text{-}17)$$

Equation 12-17 gives the fraction of zinc in the form Zn^{2+}. Putting in [L] = 0.10 M and the four values of β_i gives $\alpha_{Zn^{2+}} = 1.8 \times 10^{-5}$, which means there is very little free Zn^{2+} in the presence of 0.10 M NH_3.

TEST YOURSELF Find $\alpha_{Zn^{2+}}$ if free, unprotonated $[NH_3] = 0.02$ M. (*Answer:* 0.007 2)

EDTA Titration with an Auxiliary Complexing Agent

Now consider a titration of Zn^{2+} by EDTA in the presence of NH_3. The extension of Equation 12-6 requires a new conditional formation constant to account for the fact that only some of the EDTA is in the form Y^{4-} and only some of the zinc not bound to EDTA is in the form Zn^{2+}:

$$K_f'' = \alpha_{Zn^{2+}}\alpha_{Y^{4-}}K_f \qquad (12\text{-}18)$$

In this expression, $\alpha_{Zn^{2+}}$ is given by Equation 12-17 and $\alpha_{Y^{4-}}$ is given by Equation 12-4. For particular values of pH and $[NH_3]$, we compute K_f'' and proceed with titration calculations as

> K_f'' is the effective formation constant at a fixed pH and fixed concentration of auxiliary complexing agent. Box 12-1 describes the influence of metal ion hydrolysis on the effective formation constant.

HO   OH

HO_2C CO_2H
Tartaric acid

HO CO_2H

HO_2C CO_2H
Citric acid

$N(CH_2CH_2OH)_3$
Triethanolamine

Equation 12-18 states that the effective (conditional) formation constant for an EDTA complex is the product of the formation constant, K_f, times the fraction of metal in the form M^{m+} times the fraction of EDTA in the form Y^{4-}: $K_f'' = \alpha_{M^{m+}}\alpha_{Y^{4-}}K_f$. Table 12-1 tells us that $\alpha_{Y^{4-}}$ increases with pH until it levels off at 1 near pH 11.

In Section 12-3, we had no auxiliary complexing ligand and we implicitly assumed that $\alpha_{M^{m+}} = 1$. In fact, metal ions react with water to form $M(OH)_n$ species. Combinations of pH and metal ion in Section 12-3 were selected so that hydrolysis to $M(OH)_n$ is negligible. We can find such conditions for most M^{2+} ions, but not for M^{3+} or M^{4+}. Even in acidic solution, Fe^{3+} hydrolyzes to $Fe(OH)^{2+}$ and $Fe(OH)_2^{+}$.[16] (Appendix I gives formation constants for hydroxide complexes.) The graph shows that $\alpha_{Fe^{3+}}$ is close to 1 between pH 1 and 2 ($\log \alpha_{Fe^{3+}} \approx 0$), but then drops as hydrolysis occurs. At pH 5, the fraction of Fe(III) in the form Fe^{3+} is $\sim 10^{-5}$.

The effective formation constant for FeY^{-} in the graph has three contributions:

$$K_f''' = \frac{\alpha_{Fe^{3+}}\,\alpha_{Y^{4-}}}{\alpha_{FeY^{-}}}K_f$$

As pH increases, $\alpha_{Y^{4-}}$ increases, so K_f''' increases. As pH increases, metal hydrolysis occurs, so $\alpha_{Fe^{3+}}$ decreases. The increase in $\alpha_{Y^{4-}}$ is canceled by the decrease in $\alpha_{Fe^{3+}}$, so K_f''' is nearly constant above pH 3. The third contribution to K_f''' is $\alpha_{FeY^{-}}$, which is the fraction of the EDTA complex in the form FeY^{-}. At low pH, some of the complex gains a proton to form FeHY, which decreases $\alpha_{FeY^{-}}$ near

pH 1. In the pH range 2 to 5, $\alpha_{FeY^{-}}$ is nearly constant at 1. In neutral and basic solution, complexes such as $Fe(OH)Y^{2-}$ and $[Fe(OH)Y]_2^{4-}$ are formed and $\alpha_{FeY^{-}}$ decreases.

Take-home message: In this book, we restrict ourselves to cases in which there is no hydrolysis and $\alpha_{M^{m+}}$ is controlled by a deliberately added auxiliary ligand. In reality, hydrolysis of M^{m+} and MY influences most EDTA titrations and makes the theoretical analysis more complicated than we pretend in this chapter.

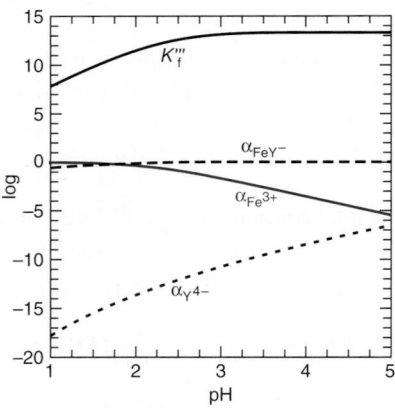

Contributions of $\alpha_{Y^{4-}}$, $\alpha_{Fe^{3+}}$, and $\alpha_{FeY^{-}}$ to the effective formation constant, K_f''', for FeY^{-}. Curves were calculated by considering the species H_6Y^{2+}, H_5Y^{+}, H_4Y, H_3Y^{-}, H_2Y^{2-}, HY^{3-}, Y^{4-}, Fe^{3+}, $Fe(OH)^{2+}$, $Fe(OH)_2^{+}$, FeY^{-}, and FeHY.

---

in Section 12-3, substituting K_f'' for K_f'. An assumption in this process is that EDTA is a much stronger complexing agent than ammonia, so essentially all EDTA binds Zn^{2+} until the metal ion is consumed.

EXAMPLE   **EDTA Titration in the Presence of Ammonia**

Consider the titration of 50.0 mL of 1.00×10^{-3} M Zn^{2+} with 1.00×10^{-3} M EDTA at pH 10.00 in the presence of 0.10 M NH_3. (This is the concentration of NH_3. There is also NH_4^+ in the solution.) The equivalence point is at 50.0 mL. Find pZn^{2+} after addition of 20.0, 50.0, and 60.0 mL of EDTA.

Solution With Equation 12-17, we found $\alpha_{Zn^{2+}} = 1.8 \times 10^{-5}$. Table 12-1 tells us that $\alpha_{Y^{4-}} = 0.30$. With K_f from Table 12-2, the conditional formation constant is

$$K_f'' = \alpha_{Zn^{2+}}\alpha_{Y^{4-}}K_f = (1.8 \times 10^{-5})(0.30)(10^{16.5}) = 1.7 \times 10^{11}$$

(a) *Before the equivalence point—20.0 mL:* Because the equivalence point is 50.0 mL, the fraction of Zn^{2+} remaining is 30.0/50.0. The dilution factor is 50.0/70.0. Therefore, the concentration of zinc not bound to EDTA is

$$C_{Zn^{2+}} = \underbrace{\left(\frac{30.0\ mL}{50.0\ mL}\right)}_{\substack{\text{Fraction of} \\ Zn^{2+}\ \text{remaining}}}\underbrace{(1.00 \times 10^{-3}\ M)}_{\substack{\text{Original}\ Zn^{2+} \\ \text{concentration}}}\underbrace{\left(\frac{50.0\ mL}{70.0\ mL}\right)}_{\substack{\text{Dilution} \\ \text{factor}}} = 4.3 \times 10^{-4}\ M$$

However, nearly all zinc not bound to EDTA is bound to NH_3. The concentration of free Zn^{2+} is

The relation $[Zn^{2+}] = \alpha_{Zn^{2+}}C_{Zn^{2+}}$ follows from Equation 12-15.

$$[Zn^{2+}] = \alpha_{Zn^{2+}}C_{Zn^{2+}} = (1.8 \times 10^{-5})(4.3 \times 10^{-4}\ M) = 7.7 \times 10^{-9}\ M$$

$$\Rightarrow pZn^{2+} = -\log[Zn^{2+}] = 8.11$$

Let's try a reality check: The product $[Zn^{2+}][OH^{-}]^2$ is $[10^{-8.11}][10^{-4.00}]^2 = 10^{-16.11}$, which does not exceed the solubility product of $Zn(OH)_2$ ($K_{sp} = 10^{-15.52}$) in Appendix F.

(b) *At the equivalence point—50.0 mL:* At the equivalence point, the dilution factor is $(50.0 \text{ mL}/100.0 \text{ mL})$, so $[\text{ZnY}^{2-}] = (50.0/100.0)(1.00 \times 10^{-3} \text{ M}) = 5.00 \times 10^{-4} \text{ M}$. We then create a table of concentrations:

	$C_{\text{Zn}^{2+}}$	$+$ EDTA	$\rightleftharpoons$	ZnY^{2-}
Initial concentration (M)	0	0		5.00×10^{-4}
Final concentration (M)	x	x		$5.00 \times 10^{-4} - x$

$$K_f'' = 1.7 \times 10^{11} = \frac{[\text{ZnY}^{2-}]}{[C_{\text{Zn}^{2+}}][\text{EDTA}]} = \frac{5.00 \times 10^{-4} - x}{x^2}$$

$$\Rightarrow x = C_{\text{Zn}^{2+}} = 5.4 \times 10^{-8} \text{ M}$$

$$[\text{Zn}^{2+}] = \alpha_{\text{Zn}^{2+}} C_{\text{Zn}^{2+}} = (1.8 \times 10^{-5})(5.4 \times 10^{-8} \text{ M}) = 9.7 \times 10^{-13} \text{ M}$$

$$\Rightarrow p\text{Zn}^{2+} = -\log[\text{Zn}^{2+}] = 12.01$$

(c) *After the equivalence point—60.0 mL:* Almost all zinc is in the form ZnY^{2-}. With a dilution factor of $(50.0 \text{ mL}/110.0 \text{ mL})$ for zinc, we find

$$[\text{ZnY}^{2-}] = \left(\frac{50.0}{110.0}\right)(1.00 \times 10^{-3} \text{ M}) = 4.5 \times 10^{-4} \text{ M}$$

We know the concentration of excess EDTA, whose dilution factor is $10.0 \text{ mL}/110.0 \text{ mL}$:

$$[\text{EDTA}] = \left(\frac{10.0}{110.0}\right)(1.00 \times 10^{-3} \text{ M}) = 9.1 \times 10^{-5} \text{ M}$$

Once we know $[\text{ZnY}^{2-}]$ and $[\text{EDTA}]$, we can use the equilibrium constant to find $[\text{Zn}^{2+}]$:

$$\frac{[\text{ZnY}^{2-}]}{[\text{Zn}^{2+}][\text{EDTA}]} = \alpha_{\text{Y}^{4-}} K_f = K_f' = (0.30)(10^{16.5}) = 9.5 \times 10^{15}$$

$$\frac{[4.5 \times 10^{-4}]}{[\text{Zn}^{2+}][9.1 \times 10^{-5}]} = 9.5 \times 10^{15} \Rightarrow [\text{Zn}^{2+}] = 5.3 \times 10^{-16} \text{ M}$$

$$\Rightarrow p\text{Zn}^{2+} = 15.28$$

After the equivalence point, the problem does not depend on the presence of NH_3, because we know both $[\text{ZnY}^{2-}]$ and $[\text{EDTA}]$.

TEST YOURSELF Find $p\text{Zn}^{2+}$ after adding 30.0 and 51.0 mL of EDTA. (*Answer:* 8.35, 14.28)

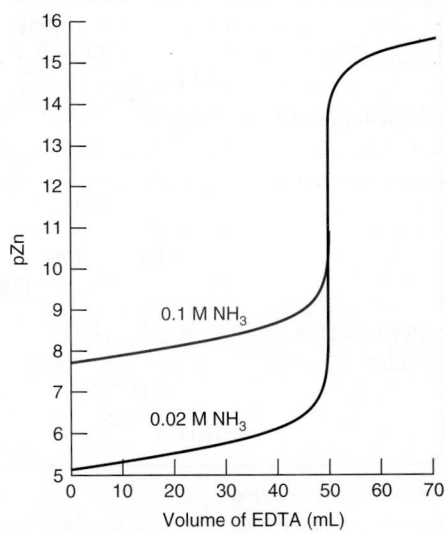

FIGURE 12-14 Calculated titration curves for the reaction of 50.0 mL of $1.00 \times 10^{-3} \text{ M Zn}^{2+}$ with $1.00 \times 10^{-3} \text{ M}$ EDTA at pH 10.00 in the presence of two different concentrations of NH_3.

Figure 12-14 compares the calculated titration curves for Zn^{2+} in the presence of different concentrations of auxiliary complexing agent. The greater the concentration of NH_3, the smaller the change of $p\text{Zn}^{2+}$ near the equivalence point. The auxiliary ligand must be kept below the level that would obliterate the end point of the titration. Color Plate 7 shows the appearance of a Cu^{2+}-ammonia solution during an EDTA titration.

12-6 Metal Ion Indicators

The most common way to detect the end point in EDTA titrations is with a metal ion indicator. Alternatives include a mercury electrode (Figure 12-10 and Exercise 15-B) and an ion-selective electrode (Section 15-6). A pH electrode follows the course of the titration in unbuffered solution, because $H_2\text{Y}^{2-}$ releases 2H^+ when it forms a metal complex.

Metal ion indicators (Table 12-3) are compounds that change color when they bind to a metal ion. *Useful indicators must bind metal less strongly than EDTA does.*

A typical titration is illustrated by the reaction of Mg^{2+} with EDTA at pH 10 with Calmagite indicator.

$$\underset{\text{Red}}{\text{MgIn}} + \underset{\text{Colorless}}{\text{EDTA}} \rightarrow \underset{\text{Colorless}}{\text{MgEDTA}} + \underset{\text{Blue}}{\text{In}} \qquad (12\text{-}19)$$

End-point detection methods:

1. Metal ion indicators

2. Mercury electrode

3. Ion-selective electrode

4. Glass (pH) electrode

The indicator must release its metal to EDTA.

TABLE 12-3 Common metal ion indicators

Name	Structure	pK_a	Color of free indicator	Color of metal ion complex
Calmagite	(H_2In^-)	$pK_2 = 8.1$ $pK_3 = 12.4$	H_2In^- red HIn^{2-} blue In^{3-} orange	Wine red
Eriochrome black T	(H_2In^-)	$pK_2 = 6.3$ $pK_3 = 11.6$	H_2In^- red HIn^{2-} blue In^{3-} orange	Wine red
Murexide	(H_4In^-)	$pK_2 = 9.2$ $pK_3 = 10.9$	H_4In^- red-violet H_3In^{2-} violet H_2In^{3-} blue	Yellow (with Co^{2+}, Ni^{2+}, Cu^{2+}); red with Ca^{2+}
Xylenol orange	(H_3In^{3-})	$pK_2 = 2.32$ $pK_3 = 2.85$ $pK_4 = 6.70$ $pK_5 = 10.47$ $pK_6 = 12.23$	H_5In^- yellow H_4In^{2-} yellow H_3In^{3-} yellow H_2In^{4-} violet HIn^{5-} violet In^{6-} violet	Red
Pyrocatechol violet	(H_3In^-)	$pK_1 = 0.2$ $pK_2 = 7.8$ $pK_3 = 9.8$ $pK_4 = 11.7$	H_4In red H_3In^- yellow H_2In^{2-} violet HIn^{3-} red-purple	Blue

PREPARATION AND STABILITY:

Calmagite: 0.05 g/100 mL H_2O; solution is stable for a year in the dark.

Eriochrome black T: Dissolve 0.1 g of the solid in 7.5 mL of triethanolamine plus 2.5 mL of absolute ethanol; solution is stable for months; best used for titrations above pH 6.5.

Murexide: Grind 10 mg of murexide with 5 g of reagent NaCl in a clean mortar; use 0.2–0.4 g of the mixture for each titration.

Xylenol orange: 0.5 g/100 mL H_2O; solution is stable indefinitely.

Pyrocatechol violet: 0.1 g/100 mL; solution is stable for several weeks.

At the start of the experiment, a small amount of indicator (In) is added to the colorless solution of Mg^{2+} to form a red complex. As EDTA is added, it reacts first with free, colorless Mg^{2+}. When free Mg^{2+} is used up, the last EDTA added before the equivalence point displaces indicator from the red MgIn complex. The change from the red MgIn to blue unbound In signals the end point of the titration (Demonstration 12-1).

Most metal ion indicators are also acid-base indicators, with pK_a values listed in Table 12-3. Because the color of free indicator is pH dependent, most indicators can be used only in certain pH ranges. For example, xylenol orange (pronounced ZY-leen-ol) changes from yellow to red when it binds to a metal ion at pH 5.5. This color change is easy to observe. At pH 7.5, the change is from violet to red and difficult to see. A spectrophotometer can measure the color change, but it is more convenient if we can see it. Figure 12-15 shows pH ranges in which many metals can be titrated and indicators that are useful in different ranges.

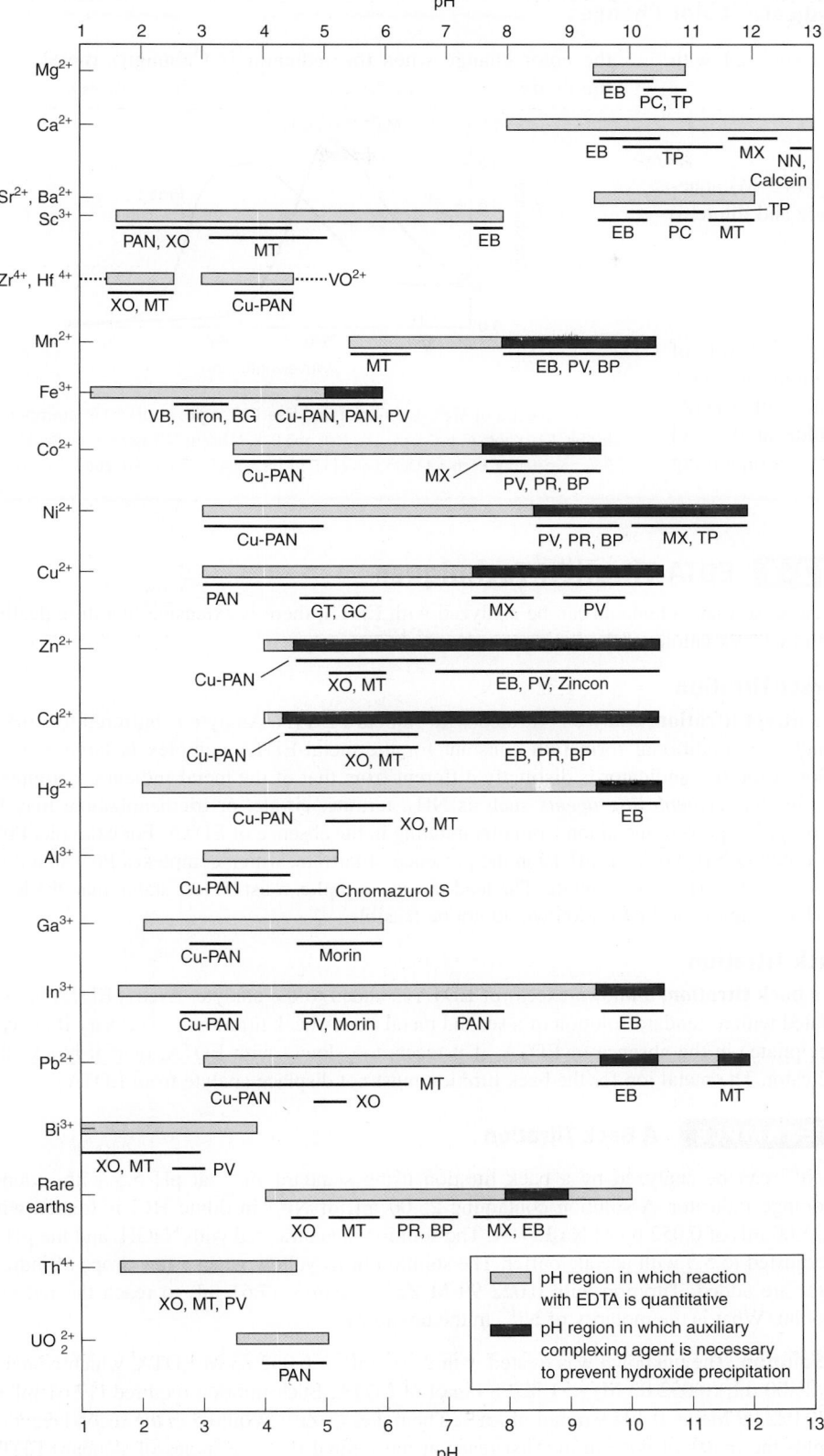

FIGURE 12-15 Guide to EDTA titrations of common metals. Light color shows pH region in which reaction with EDTA is quantitative. Dark color shows pH region in which auxiliary complexing agent is required to prevent metal from precipitating. Calmagite is more stable than Eriochrome black T (EB) and can be substituted for EB. [Data from K. Ueno, "Guide for Selecting Conditions of EDTA Titrations," *J. Chem. Ed.* **1965,** *42,* 432.]

Abbreviations for indicators:
BG, Bindschedler's green leuco base
BP, Bromopyrogallol red
EB, Eriochrome black T
GC, Glycinecresol red
GT, Glycinethymol blue
MT, Methylthymol blue
MX, Murexide
NN, Patton & Reeder's dye
PAN, Pyridylazonaphthol
Cu-PAN, PAN plus Cu-EDTA
PC, *o*-Cresolphthalein complexone
PR, Pyrogallol red
PV, Pyrocatechol violet
TP, Thymolphthalein complexone
VB, Variamine blue B base
XO, Xylenol orange

$pK_4(OH) = 12.5$
$pK_3(OH) = 7.6$

1,2-dihydroxybenzene-
3,5-disulfonic acid,
disodium salt

Tiron is an indicator for EDTA titration of Fe(III) at pH 2–3 at 40°C. Color change is from blue to pale yellow.

The indicator must give up its metal ion to the EDTA. If a metal does not freely dissociate from an indicator, the metal is said to **block** the indicator. Eriochrome black T is blocked by Cu^{2+}, Ni^{2+}, Co^{2+}, Cr^{3+}, Fe^{3+}, and Al^{3+}. It cannot be used for the *direct titration* of any of these metals. It can be used for a *back titration,* however. For example, excess standard EDTA can be added to Cu^{2+}. Then indicator is added and the excess EDTA is back-titrated with Mg^{2+}.

Question What color change occurs when the back titration is performed at pH 10? (*Answer:* blue → wine red)

This demonstration illustrates the color change associated with Reaction 12-19.

STOCK SOLUTIONS

Buffer (pH 10.0): Add 142 mL of concentrated (14.5 M) aqueous ammonia to 17.5 g of ammonium chloride and dilute to 250 mL with water.

$MgCl_2$: 0.05 M

EDTA: 0.05 M $Na_2H_2EDTA \cdot 2H_2O$

Prepare a solution containing 25 mL of $MgCl_2$, 5 mL of buffer, and 300 mL of water. Add six drops of Eriochrome black T or Calmagite indicator (Table 12-3) and titrate with EDTA. Note the color change from wine red to pale blue at the end point (Color Plate 8). The spectroscopic change accompanying

the color change when the indicator is Calmagite is shown in the figure.

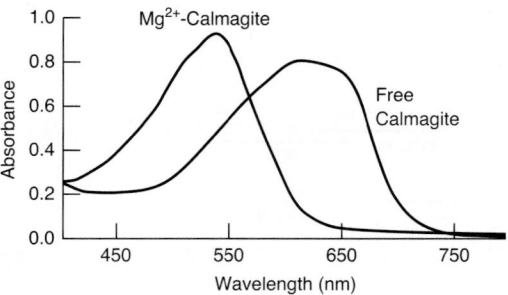

Visible spectra of Mg^{2+}-Calmagite and free Calmagite at pH 10 in ammonia buffer. [Data from C. E. Dahm, J. W. Hall, and B. E. Mattioni, "A Laser Pointer-Based Spectrometer for Endpoint Detection of EDTA Titrations," *J. Chem. Ed.* **2004**, *81*, 1787.]

12-7 EDTA Titration Techniques

Because so many elements can be analyzed with EDTA, there is extensive literature dealing with many variations of the basic procedure.[14, 17]

Direct Titration

In a **direct titration,** analyte is titrated with standard EDTA. Analyte is buffered to a pH at which the conditional formation constant for the metal-EDTA complex is large and the color of the free indicator is distinctly different from that of the metal-indicator complex.

Auxiliary complexing agents such as NH_3, tartrate, citrate, or triethanolamine may be employed to prevent metal ion from precipitating in the absence of EDTA. For example, Pb^{2+} is titrated in NH_3 buffer at pH 10 in the presence of tartrate, which complexes Pb^{2+} and does not allow $Pb(OH)_2$ to precipitate. The lead-tartrate complex must be less stable than the lead-EDTA complex, or the titration would not be feasible.

Back Titration

In a **back titration,** a known excess of EDTA is added to the analyte. Excess EDTA is then titrated with a standard solution of a second metal ion. A back titration is necessary if analyte precipitates in the absence of EDTA, if it reacts too slowly with EDTA, or if it blocks the indicator. The metal ion for the back titration must not displace analyte from EDTA.

Phytoremediation.[18, 19] An approach to removing toxic metals from contaminated soil is to grow plants that accumulate 1–15 g metal/g dry mass of plant. The plant is harvested to recover metals such as Pb, Cd, and Ni. Phytoremediation is enhanced by adding EDTA to mobilize insoluble metals. Unfortunately, rain spreads soluble metal-EDTA complexes through the soil, so phytoremediation is limited to locations where the connection with groundwater is blocked or where leaching is not important. The natural chelate, EDDS, mobilizes metal and biodegrades before it can spread very far.

S,S-Ethylenediaminedisuccinic acid (EDDS)

EXAMPLE A Back Titration

Ni^{2+} can be analyzed by a back titration using standard Zn^{2+} at pH 5.5 with xylenol orange indicator. A solution containing 25.00 mL of Ni^{2+} in dilute HCl is treated with 25.00 mL of 0.052 83 M Na_2EDTA. The solution is neutralized with NaOH, and the pH is adjusted to 5.5 with acetate buffer. The solution turns yellow when a few drops of indicator are added. Titration with 0.022 99 M Zn^{2+} requires 17.61 mL to reach the red end point. What is the molarity of Ni^{2+} in the unknown?

Solution The unknown was treated with 25.00 mL of 0.052 83 M EDTA, which contains (25.00 mL)(0.052 83 M) = 1.320 8 mmol of EDTA. Back titration required (17.61 mL) · (0.022 99 M) = 0.404 9 mmol of Zn^{2+}. The moles of Zn^{2+} required in the second reaction plus the moles of Ni^{2+} in the first reaction must equal the total moles of standard EDTA delivered to the solution:

$$0.404\ 9 \text{ mmol } Zn^{2+} + x \text{ mmol } Ni^{2+} = 1.320\ 8 \text{ mmol EDTA}$$

$$x = 0.915\ 9 \text{ mmol } Ni^{2+}$$

The concentration of Ni^{2+} is 0.915 9 mmol/25.00 mL = 0.036 64 M.

TEST YOURSELF If back titration required 13.00 mL Zn^{2+}, what was the original concentration of Ni^{2+}? (*Answer:* 0.040 88 M)

Back titration prevents precipitation of analyte. For example, $Al(OH)_3$ precipitates at pH 7 in the absence of EDTA. An acidic solution of Al^{3+} can be treated with excess EDTA, adjusted to pH 7–8 with sodium acetate, and boiled to ensure complete formation of stable, soluble $Al(EDTA)^-$. The solution is then cooled, Calmagite indicator is added, and back titrated with standard Zn^{2+}.

Displacement Titration

Hg^{2+} does not have a satisfactory indicator, but a **displacement titration** is feasible. Hg^{2+} is treated with excess $Mg(EDTA)^{2-}$ to displace Mg^{2+}, which is titrated with standard EDTA.

$$Hg^{2+} + MgY^{2-} \rightarrow HgY^{2-} + Mg^{2+} \qquad \text{(12-20)}$$

The conditional formation constant for $Hg(EDTA)^{2-}$ must be greater than K'_f for $Mg(EDTA)^{2-}$, or else Mg^{2+} will not be displaced from $Mg(EDTA)^{2-}$.

There is no suitable indicator for Ag^+. However, Ag^+ will displace Ni^{2+} from tetracyanonickelate(II) ion:

$$2Ag^+ + Ni(CN)_4^{2-} \rightarrow 2Ag(CN)_2^- + Ni^{2+}$$

The liberated Ni^{2+} can then be titrated with EDTA to find out how much Ag^+ was added.

Indirect Titration

Anions that precipitate with certain metal ions can be analyzed with EDTA by **indirect titration.** For example, sulfate can be analyzed by precipitation with excess Ba^{2+} at pH 1. The $BaSO_4(s)$ is washed and then boiled with excess standard EDTA at pH 10 to bring Ba^{2+} back into solution as $Ba(EDTA)^{2-}$. Excess EDTA is back-titrated with Mg^{2+}.

Alternatively, an anion can be precipitated with excess standard metal ion. The precipitate is filtered and washed, and excess metal in the filtrate is titrated with EDTA. Anions such as CO_3^{2-}, CrO_4^{2-}, S^{2-}, and SO_4^{2-} can be determined by indirect titration with EDTA.[20]

Masking

A **masking agent** is a reagent that protects some component of the analyte from reaction with EDTA. For example, Al^{3+} in a mixture of Mg^{2+} and Al^{3+} can be measured by first masking the Al^{3+} with F^-, thereby leaving only the Mg^{2+} to react with EDTA.

Cyanide masks Cd^{2+}, Zn^{2+}, Hg^{2+}, Co^{2+}, Cu^+, Ag^+, Ni^{2+}, Pd^{2+}, Pt^{2+}, Fe^{2+}, and Fe^{3+}, but not Mg^{2+}, Ca^{2+}, Mn^{2+}, or Pb^{2+}. When cyanide is added to a solution containing Cd^{2+} and Pb^{2+}, only Pb^{2+} reacts with EDTA. (**CAUTION:** Cyanide forms toxic gaseous HCN below pH 11. Cyanide solutions should be strongly basic and only handled in a hood.) Fluoride masks Al^{3+}, Fe^{3+}, Ti^{4+}, and Be^{2+}. (**CAUTION:** HF formed by F^- in acidic solution is extremely

> Masking prevents one species from interfering in the analysis of another. Masking is not restricted to EDTA titrations. Box 12-2 gives an important application of masking.

BOX 12-2 Water Hardness

Hardness is the total concentration of alkaline earth (Group 2) ions, which are mainly Ca^{2+} and Mg^{2+}, in water. Hardness is commonly expressed as the equivalent number of milligrams of $CaCO_3$ per liter. Thus, if $[Ca^{2+}] + [Mg^{2+}] = 1$ mM, we would say that the hardness is 100 mg $CaCO_3$ per liter because 100 mg $CaCO_3 = 1$ mmol $CaCO_3$. Water of hardness less than 60 mg $CaCO_3$ per liter is considered to be "soft." If the hardness is above 270 mg/L, the water is considered to be "hard."

Hard water reacts with soap to form insoluble curds:

$$Ca^{2+} + 2RCO_2^- \rightarrow Ca(RCO_2)_2(s) \qquad \text{(A)}$$
$$\underset{\text{Soap}}{} \qquad \underset{\text{Precipitate}}{}$$

R is a long-chain hydrocarbon such as $C_{17}H_{35}$—

Enough soap to consume Ca^{2+} and Mg^{2+} must be used before soap is available for cleaning. Hard water leaves solid deposits called *scale* on pipes when it evaporates, but hard water is not known to be unhealthy. Hardness is beneficial in irrigation water because alkaline earth ions tend to *flocculate* (cause to aggregate) colloidal particles in soil and thereby increase the permeability of the soil to water. Soft water etches concrete, plaster, and grout.

To measure hardness, the sample is treated with ascorbic acid (or hydroxylamine) to reduce Fe^{3+} to Fe^{2+} and Cu^{2+} to Cu^+ and then with cyanide to mask Fe^{2+}, Cu^+, and several other minor metal ions. Titration with EDTA at pH 10 in NH_3 buffer then gives the total concentrations of Ca^{2+} and Mg^{2+}. Ca^{2+} can be determined separately if the titration is carried out at pH 13 without ammonia. At this pH, $Mg(OH)_2$ precipitates and is inaccessible to EDTA. Interference by many metal ions can be reduced by the right choice of indicators.[21]

Insoluble carbonates are converted into soluble bicarbonates by excess carbon dioxide:

$$CaCO_3(s) + CO_2 + H_2O \rightarrow Ca(HCO_3)_2(aq) \qquad \text{(B)}$$

Heat converts bicarbonate into carbonate (driving off CO_2) and precipitates $CaCO_3$ scale that clogs boiler pipes. The fraction of hardness due to $Ca(HCO_3)_2(aq)$ is called *temporary hardness* because this calcium is lost (by precipitation of $CaCO_3$) upon heating. Hardness arising from other salts (mainly dissolved $CaSO_4$) is called *permanent hardness* because it is not removed by heating.

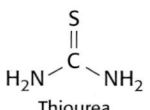

SH
|
HOCH₂CHCH₂SH
2,3-Dimercapto-1-propanol

hazardous and should not contact skin and eyes. It may not be immediately painful, but the affected area should be flooded with water and then treated with calcium gluconate gel that you have on hand *before* the accident. First-aid providers must wear rubber gloves to protect themselves.) Triethanolamine (page 275) masks Al^{3+}, Fe^{3+}, and Mn^{2+}; 2,3-dimercapto-1-propanol masks Bi^{3+}, Cd^{2+}, Cu^{2+}, Hg^{2+}, and Pb^{2+}.

Demasking releases metal ion from a masking agent. Cyanide complexes can be demasked with formaldehyde:

$$M(CN)_m^{n-m} + mH_2CO + mH^+ \rightarrow mH_2C \overset{OH}{\underset{CN}{<}} + M^{n+}$$

Formaldehyde

Thiourea masks Cu^{2+} by reducing it to Cu^+ and complexing the Cu^+. Copper can be liberated from thiourea by oxidation with H_2O_2. Selectivity afforded by masking, demasking, and pH control allows individual components of complex mixtures of metal ions to be analyzed by EDTA titration.

S
‖
H₂N—C—NH₂
Thiourea

Terms to Understand

auxiliary complexing agent	complexometric titration	displacement titration	masking agent
back titration	conditional formation constant	formation constant	metal ion indicator
blocking	cumulative formation constant	indirect titration	monodentate
chelate effect	demasking	Lewis acid	multidentate
chelating ligand	direct titration	Lewis base	stability constant

Summary

In a complexometric titration, analyte and titrant form a complex ion, and the equilibrium constant is called the formation constant, K_f. Chelating (multidentate) ligands form more stable complexes than do monodentate ligands. Synthetic aminocarboxylic acids such as EDTA have large metal-binding constants and form 1:1 complexes with most metals whose charge is ≥ 2.

Formation constants for EDTA are expressed in terms of $[Y^{4-}]$, even though there are six protonated forms of EDTA. Because the fraction ($\alpha_{Y^{4-}}$) of free EDTA in the form Y^{4-} depends on pH, we define a conditional (or effective) formation constant as $K_f' = \alpha_{Y^{4-}} K_f = [MY^{n-4}]/[M^{n+}][EDTA]$. This constant describes the hypothetical reaction $M^{n+} + EDTA \rightleftharpoons MY^{n-4}$, where EDTA refers to all forms of EDTA not bound to metal ion. Titration calculations fall into three categories. When excess unreacted M^{n+} is present, pM is calculated directly from $pM = -\log[M^{n+}]$. When excess EDTA is present, we know both $[MY^{n-4}]$ and [EDTA], so $[M^{n+}]$ can be calculated from the conditional formation constant. At the equivalence point, the condition $[M^{n+}] = [EDTA]$ allows us to solve for $[M^{n+}]$. A single spreadsheet equation applies in all three regions of the titration curve.

The greater the effective formation constant, the sharper is the EDTA titration curve. Addition of auxiliary complexing agents, which compete with EDTA for the metal ion and thereby limit the sharpness of the titration curve, is often necessary to keep the metal in solution. Calculations for a solution containing EDTA and an auxiliary complexing agent utilize the conditional formation constant $K_f'' = \alpha_M \alpha_{Y^{4-}} K_f$, where α_M is the fraction of free metal ion not complexed by the auxiliary ligand.

For end-point detection, we commonly use metal ion indicators or an ion-selective electrode. When a direct titration is not suitable, because the analyte is unstable, reacts slowly with EDTA, or has no suitable indicator, a back titration of excess EDTA or a displacement titration of $Mg(EDTA)^{2-}$ may be feasible. Masking prevents interference by unwanted species. Indirect EDTA titrations are available for many anions and other species that do not react directly with the reagent.

Exercises

12-A. Potassium ion in a 250.0 (± 0.1) mL water sample was precipitated with sodium tetraphenylborate:

$$K^+ + (C_6H_5)_4B^- \rightarrow KB(C_6H_5)_4(s)$$

The precipitate was filtered, washed, dissolved in an organic solvent, and treated with excess $Hg(EDTA)^{2-}$:

$$4HgY^{2-} + (C_6H_5)_4B^- + 4H_2O \rightarrow$$
$$H_3BO_3 + 4C_6H_5Hg^+ + 4HY^{3-} + OH^-$$

The liberated EDTA was titrated with 28.73 (± 0.03) mL of 0.043 7 (± 0.000 1) M Zn^{2+}. Find $[K^+]$ (and its absolute uncertainty) in the original sample.

12-B. A 25.00-mL sample containing Fe^{3+} and Cu^{2+} required 16.06 mL of 0.050 83 M EDTA for complete titration. A 50.00-mL sample of the unknown was treated with NH_4F to protect the Fe^{3+}. Then Cu^{2+} was reduced and masked by thiourea. Addition of 25.00 mL of 0.050 83 M EDTA liberated Fe^{3+} from its fluoride complex to form an EDTA complex. The excess EDTA required 19.77 mL of 0.018 83 M Pb^{2+} to reach a xylenol orange end point. Find $[Cu^{2+}]$ in the unknown.

12-C. Calculate pCu^{2+} (to the 0.01 decimal place) at each of the following points in the titration of 50.0 mL of 0.040 0 M EDTA with 0.080 0 M $Cu(NO_3)_2$ at pH 5.00: 0.1, 5.0, 10.0, 15.0, 20.0, 24.0, 25.0, 26.0, and 30.0 mL. Make a graph of pCu^{2+} versus volume of titrant.

12-D. Calculate the concentration of H_2Y^{2-} at the equivalence point in Exercise 12-C.

12-E. Suppose that 0.010 0 M Mn^{2+} is titrated with 0.005 00 M EDTA at pH 7.00.

(a) What is the concentration of free Mn^{2+} at the equivalence point?

(b) What is the quotient $[H_3Y^-]/[H_2Y^{2-}]$ in the solution when the titration is just 63.7% of the way to the equivalence point?

12-F. A solution containing 20.0 mL of 1.00×10^{-3} M Co^{2+} in the presence of 0.10 M $C_2O_4^{2-}$ at pH 9.00 was titrated with 1.00×10^{-2} M EDTA. Using formation constants from Appendix I for $Co(C_2O_4)$ and $Co(C_2O_4)_2^{2-}$, calculate pCo^{2+} for the following volumes of added EDTA: 0, 1.00, 2.00, and 3.00 mL. Consider the concentration of $C_2O_4^{2-}$ to be fixed at 0.10 M. Sketch a graph of pCo^{2+} versus milliliters of added EDTA.

12-G. Iminodiacetic acid forms 2:1 complexes with many metal ions:

$$H_2N^+ \overset{CH_2CO_3H}{\underset{CH_2CO_3H}{\diagdown}} \equiv H_3X^+$$

$$\alpha_{X^{2-}} = \frac{[X^{2-}]}{[H_3X^+] + [H_2X] + [HX^-] + [X^{2-}]}$$

$$Cu^{2+} + 2X^{2-} \rightleftharpoons CuX_2^{2-} \qquad K = \beta_2 = 3.5 \times 10^{16}$$

A 25.0 mL solution containing 0.120 M iminodiacetic acid buffered to pH 7.00 was titrated with 25.0 mL of 0.050 0 M Cu^{2+}. Given that $\alpha_{X^{2-}} = 4.6 \times 10^{-3}$ at pH 7.00, calculate $[Cu^{2+}]$ in the resulting solution.

Problems

EDTA

12-1. What is the chelate effect?

12-2. State (in words) what $\alpha_{Y^{4-}}$ means. Calculate $\alpha_{Y^{4-}}$ for EDTA at **(a)** pH 3.50 and **(b)** pH 10.50.

12-3. (a) Find the conditional formation constant for $Mg(EDTA)^{2-}$ at pH 9.00.

(b) Find the concentration of free Mg^{2+} in 0.050 M $Na_2[Mg(EDTA)]$ at pH 9.00.

12-4. *Metal ion buffers.* By analogy to a hydrogen ion buffer, a metal ion buffer tends to maintain a particular metal ion concentration in solution. A mixture of the acid HA and its conjugate base A^- maintains $[H^+]$ defined by the equation $K_a = [A^-][H^+]/[HA]$. A mixture of CaY^{2-} and Y^{4-} serves as a Ca^{2+} buffer governed by the equation $1/K_f' = [EDTA][Ca^{2+}]/[CaY^{2-}]$. How many grams of $Na_2EDTA \cdot 2H_2O$ (FM 372.23) should be mixed with 1.95 g of $Ca(NO_3)_2 \cdot 2H_2O$ (FM 200.12) in a 500-mL volumetric flask to give a buffer with $pCa^{2+} = 9.00$ at pH 9.00?

12-5. *Purification by reprecipitation and predominant species of polyprotic acids.* To measure oxygen isotopes in SO_4^{2-} for geologic studies, SO_4^{2-} was precipitated with excess Ba^{2+}.[22] In the presence of HNO_3, $BaSO_4$ precipitate is contaminated by NO_3^-. The solid can be purified by washing, redissolving in the absence of HNO_3, and reprecipitating. For purification, 30 mg of $BaSO_4$ were dissolved in 15 mL of 0.05 M DTPA (Figure 12-4) in 1 M NaOH with vigorous shaking at 70°C. $BaSO_4$ was reprecipitated by adding 10 M HCl dropwise to obtain pH 3–4 and allowing the mixture to stand for 1 h. The solid was washed twice by centrifuging, removing the liquid, and resuspending in deionized water. The molar ratio NO_3^-/SO_4^{2-} was reduced from 0.25 in the original precipitate to 0.001 after two cycles of dissolution and reprecipitation. What are the predominant species of sulfate and DTPA at pH 14 and pH 3? Explain why $BaSO_4$ dissolves in DTPA in 1 M NaOH and then reprecipitates when the pH is lowered to 3–4.

EDTA Titration Curves

12-6. A 100.0 mL solution of 0.050 0 M M^{n+} buffered to pH 9.00 was titrated with 0.050 0 M EDTA.

(a) What is the equivalence volume, V_e, in milliliters?

(b) Calculate the concentration of M^{n+} at $V = \frac{1}{2}V_e$.

(c) What fraction ($\alpha_{Y^{4-}}$) of free EDTA is in the form Y^{4-} at pH 9.00?

(d) The formation constant (K_f) is $10^{12.00}$. Calculate the value of the conditional formation constant $K_f' (= \alpha_{Y^{4-}}K_f)$.

(e) Calculate the concentration of M^{n+} at $V = V_e$.

(f) What is the concentration of M^{n+} at $V = 1.100 V_e$?

12-7. Calculate pCo^{2+} at each of the following points in the titration of 25.00 mL of 0.020 26 M Co^{2+} by 0.038 55 M EDTA at pH 6.00: **(a)** 12.00 mL; **(b)** V_e; **(c)** 14.00 mL.

12-8. Consider the titration of 25.0 mL of 0.020 0 M $MnSO_4$ with 0.010 0 M EDTA in a solution buffered to pH 8.00. Calculate pMn^{2+} at the following volumes of added EDTA and sketch the titration curve:

(a) 0 mL	**(d)** 49.0 mL	**(g)** 50.1 mL
(b) 20.0 mL	**(e)** 49.9 mL	**(h)** 55.0 mL
(c) 40.0 mL	**(f)** 50.0 mL	**(i)** 60.0 mL

12-9. For the same volumes used in Problem 12-8, calculate pCa^{2+} for the titration of 25.00 mL of 0.020 00 M EDTA with 0.010 00 M $CaSO_4$ at pH 10.00. Sketch the titration curve.

12-10. Calculate $[HY^{3-}]$ in a solution prepared by mixing 10.00 mL of 0.010 0 M $VOSO_4$, 9.90 mL of 0.010 0 M EDTA, and 10.0 mL of buffer with a pH of 4.00.

12-11. 🖩 *Titration of metal ion with EDTA.* Use Equation 12-11 to compute curves (pM versus mL of EDTA added) for the titration of 10.00 mL of 1.00 mM M^{2+} (= Cd^{2+} or Cu^{2+}) with 10.0 mM EDTA at pH 5.00. Plot both curves on one graph.

12-12. 🖩 *Effect of pH on the EDTA titration.* Use Equation 12-11 to compute curves (pCa^{2+} versus mL of EDTA added) for the titration of 10.00 mL of 1.00 mM Ca^{2+} with 1.00 mM EDTA at pH 5.00, 6.00, 7.00, 8.00, and 9.00. Plot all curves on one graph and compare your results with Figure 12-10.

12-13. 🖩 *Titration of EDTA with metal ion.* Use Equation 12-12 to reproduce the results of Exercise 12-C.

Auxiliary Complexing Agents

12-14. State the purpose of an auxiliary complexing agent and give an example of its use.

12-15. According to Appendix I, Cu^{2+} forms two complexes with acetate:

$$Cu^{2+} + CH_3CO_2^- \rightleftharpoons Cu(CH_3CO_2)^+ \qquad \beta_1 (= K_1)$$

$$Cu^{2+} + 2CH_3CO_2^- \rightleftharpoons Cu(CH_3CO_2)_2(aq) \qquad \beta_2$$

(a) Referring to Box 6-2, find K_2 for the reaction

$$Cu(CH_3CO_2)^+ + CH_3CO_2^- \rightleftharpoons Cu(CH_3CO_2)_2(aq) \qquad K_2$$

(b) Consider 1.00 L of solution prepared by mixing 1.00×10^{-4} mol $Cu(ClO_4)_2$ and 0.100 mol CH_3CO_2Na. Use Equation 12-16 to find the fraction of copper in the form Cu^{2+}.

12-16. Calculate pCu^{2+} at each of the following points in the titration of 50.00 mL of 0.001 00 M Cu^{2+} with 0.001 00 M EDTA at pH 11.00 in a solution with $[NH_3]$ *fixed* at 1.00 M:

(a) 0 mL   **(c)** 45.00 mL   **(e)** 55.00 mL

(b) 1.00 mL   **(d)** 50.00 mL

12-17. Consider the derivation of the fraction α_M in Equation 12-16.
(a) Derive the following expressions for the fractions α_{ML} and α_{ML_2}:

$$\alpha_{ML} = \frac{\beta_1[L]}{1 + \beta_1[L] + \beta_2[L]^2} \qquad \alpha_{ML_2} = \frac{\beta_2[L]^2}{1 + \beta_1[L] + \beta_2[L]^2}$$

(b) Calculate the values of α_{ML} and α_{ML_2} for the conditions in Problem 12-15.

12-18. *Microequilibrium constants for binding of metal to a protein.* The iron-transport protein, transferrin, has two distinguishable metal-binding sites, designated a and b. The *microequilibrium* formation constants for each site are defined as follows:

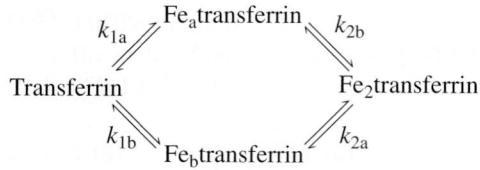

For example, the formation constant k_{1a} refers to the reaction Fe^{3+} + transferrin $\rightleftharpoons$ Fe_atransferrin, in which the metal ion binds to site a:

$$K_{1a} = \frac{[Fe_a\,transferrin]}{[Fe^{3+}][transferrin]}$$

(a) Write the chemical reactions corresponding to the conventional macroscopic formation constants, K_1 and K_2.
(b) Show that $K_1 = k_{1a} + k_{1b}$ and $K_2^{-1} = k_{2a}^{-1} + k_{2b}^{-1}$.
(c) Show that $k_{1a}k_{2b} = k_{1b}k_{2a}$. This expression means that, if you know any three of the microequilibrium constants, you automatically know the fourth one.
(d) *A challenge to your sanity:* From the equilibrium constants below, *find the equilibrium fraction of each of the four species (shown in the diagram) in circulating blood that is 40% saturated with iron* (that is, Fe/transferrin = 0.80, because each protein can bind 2 Fe).

Effective formation constants for blood plasma at pH 7.4

$k_{1a} = 6.0 \times 10^{22}$	$k_{2a} = 2.4 \times 10^{22}$
$k_{1b} = 1.0 \times 10^{22}$	$k_{2b} = 4.2 \times 10^{21}$
$K_1 = 7.0 \times 10^{22}$	$K_2 = 3.6 \times 10^{21}$

The binding constants are so large that you may assume that there is negligible free Fe^{3+}. To get started, let's use the abbreviations [T] = [transferrin], [FeT] = [Fe$_a$T] + [Fe$_b$T], and [Fe$_2$T] = [Fe$_2$transferrin]. Now we can write

Mass balance for protein: $[T] + [FeT] + [Fe_2T] = 1$ **(A)**

Mass balance for iron:

$$\frac{[FeT] + 2[Fe_2T]}{[T] + [FeT] + [Fe_2T]} = [FeT] + 2[Fe_2T] = 0.8 \qquad \textbf{(B)}$$

Combined equilibria: $\dfrac{K_1}{K_2} = 19._{44} = \dfrac{[FeT]^2}{[T][Fe_2T]}$ **(C)**

Now you have three equations with three unknowns and should be able to tackle this problem.

12-19. *Spreadsheet equation for auxiliary complexing agent.* Consider the titration of metal M (initial concentration = C_M, initial volume = V_M) with EDTA (concentration = C_{EDTA}, volume added = V_{EDTA}) in the presence of an auxiliary complexing ligand (such as ammonia). Follow the derivation in Section 12-4 to show that the master equation for the titration is

$$\phi = \frac{C_{EDTA}V_{EDTA}}{C_M V_M} = \frac{1 + K_f''[M]_{free} - \dfrac{[M]_{free} + K_f''[M]_{free}^2}{C_M}}{K_f''[M]_{free} + \dfrac{[M]_{free} + K_f''[M]_{free}^2}{C_{EDTA}}}$$

where K_f'' is the conditional formation constant in the presence of auxiliary complexing agent at the fixed pH of the titration (Equation 12-18) and $[M]_{free}$ is the total concentration of metal not bound to EDTA. $[M]_{free}$ is the same as $[M]$ in Equation 12-15. The result is equivalent to Equation 12-11, with [M] replaced by $[M]_{free}$ and K_f replaced by K_f''.

12-20. Auxiliary complexing agent. Use the equation derived in Problem 12-19.

(a) Prepare a spreadsheet to reproduce the 20-, 50-, and 60-mL points in the EDTA titration of Zn^{2+} in the presence of NH_3 in the example in Section 12-5.

(b) Use your spreadsheet to plot the curve for the titration of 50.00 mL of 5.00 mM Ni^{2+} by 10.0 mM EDTA at pH 11.00 with a constant oxalate concentration of 0.100 M.

12-21. *Spreadsheet equation for formation of the complexes ML and ML_2.* Consider the titration of metal M (initial concentration = C_M, initial volume = V_M) with ligand L (concentration = C_L, volume added = V_L), which can form 1:1 and 2:1 complexes:

$$M + L \rightleftharpoons ML \qquad \beta_1 = \frac{[ML]}{[M][L]}$$

$$M + 2L \rightleftharpoons ML_2 \qquad \beta_2 = \frac{[ML_2]}{[M][L]^2}$$

Let α_M be the fraction of metal in the form M, α_{ML} be the fraction in the form ML, and α_{ML_2} be the fraction in the form ML_2. Following the derivation in Section 12-5, you could show that these fractions are given by

$$\alpha_M = \frac{1}{1 + \beta_1[L] + \beta_2[L]^2} \qquad \alpha_{ML} = \frac{\beta_1[L]}{1 + \beta_1[L] + \beta_2[L]^2}$$

$$\alpha_{ML_2} = \frac{\beta_2[L]^2}{1 + \beta_1[L] + \beta_2[L]^2}$$

The concentrations of ML and ML_2 are

$$[ML] = \alpha_{ML}\frac{C_M V_M}{V_M + V_L} \qquad [ML_2] = \alpha_{ML_2}\frac{C_M V_M}{V_M + V_L}$$

because $C_M V_M/(V_M + V_L)$ is the total concentration of all metal in the solution. The mass balance for ligand is

$$[L] + [ML] + 2[ML_2] = \frac{C_L V_L}{V_M + V_L}$$

By substituting expressions for [ML] and [ML$_2$] into the mass balance, show that the master equation for a titration of metal by ligand is

$$\phi = \frac{C_L V_L}{C_M V_M} = \frac{\alpha_{ML} + 2\alpha_{ML_2} + ([L]/C_M)}{1 - ([L]/C_L)}$$

12-22. [icon] *Titration of M with L to form ML and ML$_2$.* Use the equation from Problem 12-21, where M is Cu^{2+} and L is acetate. Consider adding 0.500 M acetate to 10.00 mL of 0.050 0 M Cu^{2+} at pH 7.00 (so that all ligand is present as $CH_3CO_2^-$, not CH_3CO_2H). Formation constants for $Cu(CH_3CO_2)^+$ and $Cu(CH_3CO_2)_2$ are given in Appendix I. Construct a spreadsheet in which the input is pL and the output is [L], V_L, [M], [ML], and [ML$_2$]. Prepare a graph showing concentrations of L, M, ML, and ML$_2$ as V_L ranges from 0 to 3 mL.

Metal Ion Indicators

12-23. Explain why the change from red to blue in Reaction 12-19 occurs suddenly at the equivalence point instead of gradually throughout the entire titration.

12-24. List four methods for detecting the end point of an EDTA titration.

12-25. Calcium ion was titrated with EDTA at pH 11, using Calmagite as indicator (Table 12-3). Which is the principal species of Calmagite at pH 11? What color was observed before the equivalence point? After the equivalence point?

12-26. Pyrocatechol violet (Table 12-3) is to be used as a metal ion indicator in an EDTA titration. The procedure is as follows:

1. Add a known excess of EDTA to the unknown metal ion.

2. Adjust the pH with a suitable buffer.

3. Back-titrate the excess chelate with standard Al^{3+}.

From the following available buffers, select the best buffer, and then state what color change will be observed at the end point. Explain your answer.

(i) pH 6–7   (ii) pH 7–8   (iii) pH 8–9   (iv) pH 9–10

EDTA Titration Techniques

12-27. Give three circumstances in which an EDTA back titration might be necessary.

12-28. Describe what is done in a displacement titration and give an example.

12-29. Give an example of the use of a masking agent.

12-30. What is meant by water hardness? Explain the difference between temporary and permanent hardness.

12-31. How many milliliters of 0.050 0 M EDTA are required to react with 50.0 mL of 0.010 0 M Ca^{2+}? With 50.0 mL of 0.010 0 M Al^{3+}?

12-32. A 50.0-mL sample containing Ni^{2+} was treated with 25.0 mL of 0.050 0 M EDTA to complex all the Ni^{2+} and leave excess EDTA in solution. The excess EDTA was then back-titrated, requiring 5.00 mL of 0.050 0 M Zn^{2+}. What was the concentration of Ni^{2+} in the original solution?

12-33. A 50.0-mL aliquot of solution containing 0.450 g of $MgSO_4$ (FM 120.37) in 0.500 L required 37.6 mL of EDTA solution for titration. How many milligrams of $CaCO_3$ (FM 100.09) will react with 1.00 mL of this EDTA solution?

12-34. Cyanide solution (12.73 mL) was treated with 25.00 mL of Ni^{2+} solution (containing excess Ni^{2+}) to convert the cyanide into tetracyanonickelate(II):

$$4CN^- + Ni^{2+} \rightarrow Ni(CN)_4^{2-}$$

Excess Ni^{2+} was then titrated with 10.15 mL of 0.013 07 M EDTA. $Ni(CN)_4^{2-}$ does not react with EDTA. If 39.35 mL of EDTA were required to react with 30.10 mL of the original Ni^{2+} solution, calculate the molarity of CN^- in the 12.73-mL sample.

12-35. A 1.000-mL sample of unknown containing Co^{2+} and Ni^{2+} was treated with 25.00 mL of 0.038 72 M EDTA. Back titration with 0.021 27 M Zn^{2+} at pH 5 required 23.54 mL to reach the xylenol orange end point. A 2.000-mL sample of unknown was passed through an ion-exchange column that retards Co^{2+} more than Ni^{2+}. The Ni^{2+} that passed through the column was treated with 25.00 mL of 0.038 72 M EDTA and required 25.63 mL of 0.021 27 M Zn^{2+} for back titration. The Co^{2+} emerged from the column later. It, too, was treated with 25.00 mL of 0.038 72 M EDTA. How many milliliters of 0.021 27 M Zn^{2+} will be required for back titration?

12-36. A 50.0-mL solution containing Ni^{2+} and Zn^{2+} was treated with 25.0 mL of 0.045 2 M EDTA to bind all the metal. The excess unreacted EDTA required 12.4 mL of 0.012 3 M Mg^{2+} for complete reaction. An excess of the reagent 2,3-dimercapto-1-propanol was then added to displace the EDTA from zinc. Another 29.2 mL of Mg^{2+} were required for reaction with the liberated EDTA. Calculate the molarity of Ni^{2+} and Zn^{2+} in the original solution.

12-37. Sulfide ion was determined by indirect titration with EDTA. To a solution containing 25.00 mL of 0.043 32 M $Cu(ClO_4)_2$ plus 15 mL of 1 M acetate buffer (pH 4.5) were added 25.00 mL of unknown sulfide solution with vigorous stirring. The CuS precipitate was filtered and washed with hot water. Ammonia was added to the filtrate (which contained excess Cu^{2+}) until the blue color of $Cu(NH_3)_4^{2+}$ was observed. Titration of the filtrate with 0.039 27 M EDTA required 12.11 mL to reach the murexide end point. Calculate the molarity of sulfide in the unknown.

12-38. *Indirect EDTA determination of cesium.* Cesium ion does not form a strong EDTA complex, but it can be analyzed by adding a known excess of $NaBiI_4$ in cold concentrated acetic acid containing excess NaI. Solid $Cs_3Bi_2I_9$ is precipitated, filtered, and removed. The excess yellow BiI_4^- is then titrated with EDTA. The end point occurs when the yellow color disappears. (Sodium thiosulfate is used in the reaction to prevent the liberated I^- from being oxidized to yellow aqueous I_2 by O_2 from the air.) The precipitation is fairly selective for Cs^+. The ions Li^+, Na^+, K^+, and low concentrations of Rb^+ do not interfere, although Tl^+ does. Suppose that 25.00 mL of unknown containing Cs^+ were treated with 25.00 mL of 0.086 40 M $NaBiI_4$ and the unreacted BiI_4^- required 14.24 mL of 0.043 7 M EDTA for complete titration. Find the concentration of Cs^+ in the unknown.

12-39. The sulfur content of insoluble sulfides that do not readily dissolve in acid can be measured by oxidation with Br_2 to SO_4^{2-}.[23]

Metal ions are then replaced with H^+ by an ion-exchange column, and sulfate is precipitated as $BaSO_4$ with a known excess of $BaCl_2$. The excess Ba^{2+} is then titrated with EDTA to determine how much was present. (To make the indicator end point clearer, a small, known quantity of Zn^{2+} also is added. The EDTA titrates both the Ba^{2+} and the Zn^{2+}.) Knowing the excess Ba^{2+}, we can calculate how much sulfur was in the original material. To analyze the mineral sphalerite (ZnS, FM 97.46), 5.89 mg of powdered solid were suspended in a mixture of CCl_4 and H_2O containing 1.5 mmol Br_2.

After 1 h at 20°C and 2 h at 50°C, the powder dissolved and the solvent and excess Br_2 were removed by heating. The residue was dissolved in 3 mL of water and passed through an ion-exchange column to replace Zn^{2+} with H^+. Then 5.000 mL of 0.014 63 M $BaCl_2$ were added to precipitate all sulfate as $BaSO_4$. After the addition of 1.000 mL of 0.010 00 M $ZnCl_2$ and 3 mL of ammonia buffer, pH 10, the excess Ba^{2+} and Zn^{2+} required 2.39 mL of 0.009 63 M EDTA to reach the Calmagite end point. Find the weight percent of sulfur in the sphalerite. What is the theoretical value?

13 Advanced Topics in Equilibrium

St. Paul's Cathedral, London. [© Kamira/Shutterstock.]

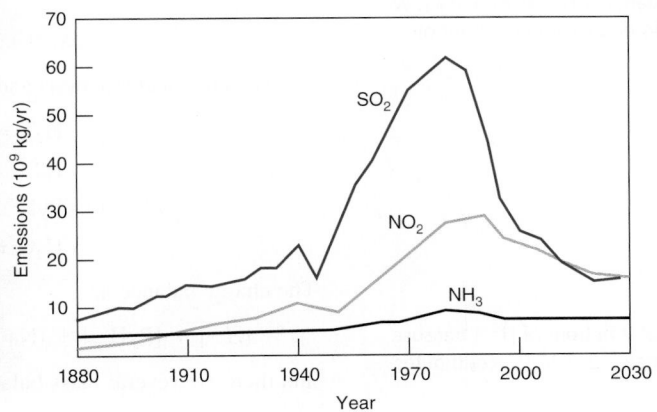

Estimated emissions over Europe. [Data from R. F. Wright, T. Larssen, L. Camarero, B. J. Crosby, R. C. Ferrier, R. Helliwell, M. Forsius, A. Jenkins, J. Kopáček, V. Majer, F. Moldan, M. Posch, M. Rogora, and W. Schöpp, "Recovery of Acidified European Surface Waters," *Environ. Sci. Technol.* **2005,** *39,* 64A.]

Limestone and marble are building materials whose main constituent is calcite, the common crystalline form of calcium carbonate. This mineral is not very soluble in neutral or basic solution ($K_{sp} = 4.5 \times 10^{-9}$), but it dissolves in acid by virtue of two **coupled equilibria,** in which the reactions have a species in common—carbonate, in this case:

$$CaCO_3(s) \rightleftharpoons Ca^{2+} + CO_3^{2-}$$
Calcite   Carbonate

$$CO_3^{2-} + H^+ \rightleftharpoons HCO_3^-$$
Bicarbonate

Carbonate produced in the first reaction is protonated to form bicarbonate in the second reaction. Le Châtelier's principle tells us that, if we remove a product of the first reaction, we will draw the reaction to the right, making calcite more soluble. This chapter deals with coupled equilibria in chemical systems.

Between 1980 and 1990, half a millimeter of the thickness of the external stone walls of St. Paul's Cathedral in London was dissolved by acidic rainfall. A corner of the building facing a power station dissolved at 10 times the rate of the rest of the building until the station was closed. The power station and other industries that burn coal emit SO_2, which is a major source of acid rain (Box 15-1). Loss of heavy industry and laws limiting emissions decreased atmospheric SO_2 from as high as 100 ppb in the 1970s to 10 ppb in 2000. Only a quarter of a millimeter of St. Paul's external stone disappeared between 1990 and 2000.[1]

T his optional chapter provides tools to compute the concentrations of species in systems with many simultaneous equilibria.[2] The most important tool is the systematic treatment of equilibrium from Chapter 8, with numerical solutions using Excel. Later chapters in this book do not depend on Chapter 13.

13-1 General Approach to Acid-Base Systems

Part of the approach to equilibrium problems in this chapter is adapted from Julian Roberts, University of Redlands.

We first illustrate a general approach to find the concentrations of species in mixtures of acids and bases. Consider a solution made by dissolving 20.0 mmol sodium tartrate (Na^+HT^-), 15.0 mmol pyridinium chloride (PyH^+Cl^-), and 10.0 mmol KOH in a volume of 1.00 L. The problem is to find the pH and concentrations of all species in the solution.

D-Tartaric acid
H_2T
$pK_1 = 3.036$, $pK_2 = 4.366$

Pyridinium chloride
PyH^+Cl^-
$pK_a = 5.20$

For this example, we designate the two acid dissociation constants of H_2T as K_1 and K_2. We designate the acid dissociation constant of PyH^+ as K_a.

The chemical reactions and equilibrium constants at 0 ionic strength are

$$H_2T \rightleftharpoons HT^- + H^+ \qquad K_1 = 10^{-3.036} \qquad \text{(13-1)}$$

$$HT^- \rightleftharpoons T^{2-} + H^+ \qquad K_2 = 10^{-4.366} \qquad \text{(13-2)}$$

$$PyH^+ \rightleftharpoons Py + H^+ \qquad K_a = 10^{-5.20} \qquad \text{(13-3)}$$

$$H_2O \rightleftharpoons H^+ + OH^- \qquad K_w = 10^{-14.00} \qquad \text{(13-4)}$$

The charge balance is

There is a factor of 2 in front of $[T^{2-}]$ because the ion has a charge −2. 1 M T^{2-} contributes a charge of 2 M.

$$[H^+] + [PyH^+] + [Na^+] + [K^+] = [OH^-] + [HT^-] + 2[T^{2-}] + [Cl^-] \qquad \text{(13-5)}$$

and there are several mass balances:

$$[Na^+] = 0.020\,0\,M \qquad [K^+] = 0.010\,0\,M \qquad [Cl^-] = 0.015\,0\,M$$

$$[H_2T] + [HT^-] + [T^{2-}] = 0.020\,0\,M \qquad [PyH^+] + [Py] = 0.015\,0\,M$$

There are 10 independent equations and 10 species, so we have enough information to solve for all the concentrations.

There is a systematic way to handle this problem without algebraic gymnastics.

"Independent" equations cannot be derived from one another. As a trivial example, the equations $a = b + c$ and $2a = 2b + 2c$ are not independent. The three equilibrium expressions for K_a, K_b, and K_w for a weak acid and its conjugate base provide only two independent equations because we can derive K_b from K_a and K_w: $K_b = K_w/K_a$.

Step 1 Write a *fractional composition equation* from Section 10-5 for each acid or base that appears in the charge balance.

Step 2 Substitute the fractional composition expressions into the charge balance and enter known values for $[Na^+]$, $[K^+]$, and $[Cl^-]$. Also, write $[OH^-] = K_w/[H^+]$. At this point, you will have a complicated equation in which the only variable is $[H^+]$.

Step 3 Use your trusty spreadsheet to solve for $[H^+]$.

Here is a recap of the fractional composition equations from Section 10-5 for *any* monoprotic acid HA and *any* diprotic acid H_2A.

$F_{HA} = [HA] + [A^-]$

Monoprotic system:
$$[HA] = \alpha_{HA}F_{HA} = \frac{[H^+]F_{HA}}{[H^+] + K_a} \qquad \text{(13-6a)}$$

$$[A^-] = \alpha_{A^-}F_{HA} = \frac{K_aF_{HA}}{[H^+] + K_a} \qquad \text{(13-6b)}$$

$F_{H_2A} = [H_2A] + [HA^-] + [A^{2-}]$

Diprotic system:
$$[H_2A] = \alpha_{H_2A}F_{H_2A} = \frac{[H^+]^2F_{H_2A}}{[H^+]^2 + [H^+]K_1 + K_1K_2} \qquad \text{(13-7a)}$$

$$[HA^-] = \alpha_{HA^-}F_{H_2A} = \frac{K_1[H^+]F_{H_2A}}{[H^+]^2 + [H^+]K_1 + K_1K_2} \qquad \text{(13-7b)}$$

Table 11-5 gave fractional composition equations for H_3A.

$$[A^{2-}] = \alpha_{A^{2-}}F_{H_2A} = \frac{K_1K_2F_{H_2A}}{[H^+]^2 + [H^+]K_1 + K_1K_2} \qquad \text{(13-7c)}$$

In each equation, α_i is the fraction in each form. For example, $\alpha_{A^{2-}}$ is the fraction of diprotic acid in the form A^{2-}. When we multiply $\alpha_{A^{2-}}$ times F_{H_2A} (the total or formal concentration of H_2A), the product is the concentration of A^{2-}.

Applying the General Procedure

Now let's apply the general procedure to the mixture of 0.020 0 M sodium tartrate (Na^+HT^-), 0.015 0 M pyridinium chloride (PyH^+Cl^-), and 0.010 0 M KOH. We designate the formal concentrations as $F_{H_2T} = 0.020\ 0$ M and $F_{PyH^+} = 0.015\ 0$ M.

Step 1 Write a *fractional composition equation* for each acid or base that appears in the charge balance.

$$[PyH^+] = \alpha_{PyH^+}F_{PyH^+} = \frac{[H^+]F_{PyH^+}}{[H^+] + K_a} \quad (13\text{-}8)$$

$$[HT^-] = \alpha_{HT^-}F_{H_2T} = \frac{K_1[H^+]F_{H_2T}}{[H^+]^2 + [H^+]K_1 + K_1K_2} \quad (13\text{-}9)$$

$$[T^{2-}] = \alpha_{T^{2-}}F_{H_2T} = \frac{K_1K_2F_{H_2T}}{[H^+]^2 + [H^+]K_1 + K_1K_2} \quad (13\text{-}10)$$

All quantities on the right side of these expressions are known, except for $[H^+]$.

Step 2 Substitute the fractional composition expressions into the charge balance 13-5. Enter values for $[Na^+]$, $[K^+]$, and $[Cl^-]$, and write $[OH^-] = K_w/[H^+]$.

$$[H^+] + [PyH^+] + [Na^+] + [K^+] = [OH^-] + [HT^-] + 2[T^{2-}] + [Cl^-] \quad (13\text{-}5)$$

$$[H^+] + \alpha_{PyH^+}F_{PyH^+} + [0.020\ 0] + [0.010\ 0]$$
$$= \frac{K_w}{[H^+]} + \alpha_{HT^-}F_{H_2T} + 2\alpha_{T^{2-}}F_{H_2T} + [0.015\ 0] \quad (13\text{-}11)$$

K_a, K_1, K_2, and $[H^+]$ are contained in the α expressions. The only variable in Equation 13-11 is $[H^+]$.

Step 3 The spreadsheet in Figure 13-1 solves Equation 13-11 for $[H^+]$.

In Figure 13-1, shaded boxes contain input. Everything else is computed by the spreadsheet. Values for F_{H_2T}, pK_1, pK_2, F_{PyH^+}, pK_a, and $[K^+]$ were given in the problem. The initial value of pH in cell H13 is a *guess*. We will use Excel Solver to vary pH until the sum of charges in cell E15 is 0. Species in the charge balance are in cells B10:B13. $[H^+]$ in cell B10 is computed from the pH we guessed in cell H13. $[PyH^+]$ in cell B11 is computed with Equation 13-8. Known values are entered for $[Na^+]$, $[K^+]$, and $[Cl^-]$.

Key step: *Guess* a value for $[H^+]$ and use Excel Solver to vary $[H^+]$ until it satisfies the charge balance.

	A	B	C	D	E	F	G	H	I
1	Mixture of 0.020 M Na⁺HT⁻, 0.015 M PyH⁺Cl⁻, and 0.010 M KOH								
2									
3	F_{H2T} =	0.020		F_{PyH+} =	0.015		$[K^+]$ =	0.010	
4	pK_1 =	3.036		pK_a =	5.20		K_w =	1.00E-14	
5	pK_2 =	4.366		K_a =	6.31E-06				
6	K_1 =	9.20E-04							
7	K_2 =	4.31E-05							
8									
9	Species in charge balance:						Other concentrations:		
10	$[H^+]$ =	1.00E-06		$[OH^-]$ =	1.00E-08		$[H_2T]$ =	4.93E-07	
11	$[PyH^+]$ =	2.05E-03		$[HT^-]$ =	4.54E-04		$[Py]$ =	1.29E-02	
12	$[Na^+]$ =	0.020		$[T^{2-}]$ =	1.95E-02				
13	$[K^+]$ =	0.010		$[Cl^-]$ =	0.015		pH =	6.000	←initial value
14									is a guess
15	Positive charge minus negative charge =				-2.25E-02	←vary pH in H13 with Solver to make this 0			
16					E15 = B10+B11+B12+B13-E10-E11-2*E12-E13				
17	Check: $[PyH^+]$ + $[Py]$ =			0.01500	(= B11+H11)				
18	Check: $[H_2T]$ + $[HT^-]$ + $[T^{2-}]$ =			0.02000	(= H10+E11+E12)				
19									
20	Formulas:								
21	B6 = 10^-B4		B7 = 10^-B5		E5 = 10^-E4		E10 = H4/B10		
22	B10 = 10^-H13		B12 = B3		B13 = H3		E13 = E3		
23	E11 = B6*B10*B3/(B10^2+B10*B6+B6*B7)						B11 = B10*E3/(B10+E5)		
24	E12 = B6*B7*B3/(B10^2+B10*B6+B6*B7)						H11 = E5*E3/(B10+E5)		
25	H10 = B10^2*B3/(B10^2+B10*B6+B6*B7)								

FIGURE 13-1 Spreadsheet for mixture of acids and bases uses Solver to find the value of pH in cell H13 that satisfies the charge balance in cell E15. The sums $[PyH^+]$ + $[Py]$ in cell D17 and $[H_2T]$ + $[HT^-]$ + $[T^{2-}]$ in cell D18 are computed to verify that the formulas for each of the species do not have mistakes. These sums are independent of pH.

In equilibrium problems in Section 8-5, we minimized both the charge and mass balances. In Section 13-1, we already used the mass balance to derive the fractional composition equations 13-8 to 13-10. Therefore, we only minimize the charge balance with the spreadsheet.

$[OH^-]$ is computed from $K_w/[H^+]$. $[HT^-]$ and $[T^{2-}]$ in cells E11 and E12 are computed with Equations 13-9 and 13-10.

The sum of charges, $[H^+] + [PyH^+] + [Na^+] + [K^+] - [OH^-] - [HT^-] - 2[T^{2-}] - [Cl^-]$, is computed in cell E15. If we had guessed the correct pH in cell H13, the sum of the charges would be 0. Instead, the sum is -2.25×10^{-2} M. We use Excel Solver to vary pH in cell H13 until the sum of charges in cell E15 is 0.

Using Excel Solver

For Excel 2010 on a PC, in the Data tab of your spreadsheet, select Analysis and click on Solver. (Excel is found in the Tools menu on a Mac.) In the Solver Parameters window (Figure 8-10a), click on Options. In Solver Options (Figure 8-10b), select the tab for All Methods. Set Constraint Precision = <u>1E-15</u>, Use Automatic Scaling, Max time = <u>100</u> s, and Iterations = <u>200</u>. Click OK.

In the Solver Parameters window (Figure 8-10a), Set Objective <u>E15</u> Equal To Value of <u>0</u> By Changing Cells <u>H13</u>. Then click Solve. Solver will vary the pH in cell H13 to make the net charge in cell E15 equal to 0. The concentrations after executing Solver are

	A	B	C	D	E	F	G	H
9	Species in charge balance:						Other concentrations:	
10	$[H^+] =$	5.04E-05		$[OH^-] =$	1.99E-10		$[H_2T] =$	5.73E-04
11	$[PyH^+] =$	1.33E-02		$[HT^-] =$	1.05E-02		$[Py] =$	1.67E-03
12	$[Na^+] =$	0.020		$[T^{2-}] =$	8.95E-03			
13	$[K^+] =$	0.010		$[Cl^-] =$	0.015		pH =	4.298
14								
15	Positive charge minus negative charge =				0.00E+000			

Ignorance Is Bliss: A Complication of Ion Pairing

We should not get too cocky with our newfound power to handle complex problems, because we have oversimplified the actual situation. For one thing, we have not included activity coefficients, which ordinarily affect the answer by a few tenths of a pH unit. In Section 13-2, we will incorporate activity coefficients.

Even with activity coefficients, we are always limited by chemistry that we do not know. In the mixture of sodium hydrogen tartrate (Na^+HT^-), pyridinium chloride (PyH^+Cl^-), and KOH, several possible ion-pair equilibria are

Equilibrium constants from A. E. Martell, R. M. Smith, and R. J. Motekaitis, *NIST Standard Reference Database 46*, Version 6.0, 2001.

$$Na^+ + T^{2-} \rightleftharpoons NaT^- \qquad K_{NaT^-} = \frac{[NaT^-]}{[Na^+][T^{2-}]} = 8 \qquad \text{(13-12)}$$

$$Na^+ + HT^- \rightleftharpoons NaHT \qquad K_{NaHT} = \frac{[NaHT]}{[Na^+][HT^-]} = 1.6 \qquad \text{(13-13)}$$

$$Na^+ + Py \rightleftharpoons PyNa^+ \qquad K_{PyNa^+} = 1.0 \qquad \text{(13-14)}$$

$$K^+ + T^{2-} \rightleftharpoons KT^- \qquad K_{KT^-} = 3$$

$$K^+ + HT^- \rightleftharpoons KHT \qquad K_{KHT} = ?$$

$$PyH^+ + Cl^- \rightleftharpoons PyH^+Cl^- \qquad K_{PyHCl} = ?$$

$$PyH^+ + T^{2-} \rightleftharpoons PyHT^- \qquad K_{PyHT^-} = ?$$

Some equilibrium constants at 0 ionic strength are listed above. Values for the other reactions are not available, but there is no reason to believe that the reactions do not occur.

How could we add ion pairing to our spreadsheet? For simplicity, we outline how to add just Reactions 13-12 and 13-13. With these reactions, the mass balance for sodium is

$$[Na^+] + [NaT^-] + [NaHT] = F_{Na} = F_{H_2T} = 0.020\ 0\ M \qquad \text{(13-15)}$$

From the ion-pair equilibria, we write $[NaT^-] = K_{NaT^-}[Na^+][T^{2-}]$ and $[NaHT] = K_{NaHT}[Na^+][HT^-]$. With these substitutions for $[NaT^-]$ and $[NaHT]$ in the mass balance for sodium, we find an expression for $[Na^+]$:

$$[Na^+] = \frac{F_{H_2T}}{1 + K_{NaT^-}[T^{2-}] + K_{NaHT}[HT^-]} \qquad \text{(13-16)}$$

A bigger nuisance when considering ion pairs is that the fractional composition equations for $[H_2T]$, $[HT^-]$, and $[T^{2-}]$ also are changed because the mass balance for H_2T now has five species in it instead of three:

$$F_{H_2T} = [H_2T] + [HT^-] + [T^{2-}] + [NaT^-] + [NaHT] \qquad \textbf{(13-17)}$$

We must derive new equations analogous to 13-9 and 13-10 from the mass balance 13-17.

The new fractional composition equations are messy, so we reserve this case for Problem 13-17. The end result is that ion-pair equilibria 13-12 and 13-13 change the calculated pH from 4.30 to 4.26. This change is not large, so neglecting ion pairs with small equilibrium constants does not lead to serious error. We find that 7% of sodium is tied up in ion pairs. *Our ability to compute the distribution of species in a solution is limited by our knowledge of relevant equilibria.*

13-2 Activity Coefficients

Even if we know all reactions and equilibrium constants for a given system, we cannot compute concentrations accurately without activity coefficients. Chapter 8 gave the extended Debye-Hückel equation 8-6 for activity coefficients with size parameters in Table 8-1. Many ions of interest are not in Table 8-1, and we do not know their size. In Section 8-5 we just guessed the sizes of ions that were not in the table. Now, we introduce the *Davies equation*, which has no size parameter (and is not more accurate than guessing the size):

$$\text{Davies equation:} \quad \log \gamma = -0.51z^2\left(\frac{\sqrt{\mu}}{1 + \sqrt{\mu}} - 0.3\mu\right) \qquad \text{(at 25°C)} \qquad \textbf{(13-18)}$$

where γ is the activity coefficient for an ion of charge z at ionic strength μ. Equation 13-18 can be used up to $\mu \approx 0.5$ M (Figure 13-2). For best accuracy, Pitzer's equations should be used (Chapter 8, reference 8).

Now consider a primary standard buffer containing 0.025 0 m KH_2PO_4 and 0.025 0 m Na_2HPO_4. Its pH at 25°C is 6.865 ± 0.006.[3] The concentration unit, m, is *molality*, which means moles of solute per kilogram of solvent. For precise chemical measurements, concentrations are expressed in molality, rather than molarity, because molality is independent of temperature. Uncertainties in equilibrium constants are usually sufficiently great to render the ~0.3% difference between molality and molarity of dilute solutions unimportant.

Acid-base equilibrium constants for H_3PO_4 at $\mu = 0$ at 25°C are

$$H_3PO_4 \rightleftharpoons H_2PO_4^- + H^+ \qquad K_1 = \frac{[H_2PO_4^-]\gamma_{H_2PO_4^-}[H^+]\gamma_{H^+}}{[H_3PO_4]\gamma_{H_3PO_4}} = 10^{-2.148} \qquad \textbf{(13-19)}$$

$$H_2PO_4^- \rightleftharpoons HPO_4^{2-} + H^+ \qquad K_2 = \frac{[HPO_4^{2-}]\gamma_{HPO_4^{2-}}[H^+]\gamma_{H^+}}{[H_2PO_4^-]\gamma_{H_2PO_4^-}} = 10^{-7.198} \qquad \textbf{(13-20)}$$

$$HPO_4^{2-} \rightleftharpoons PO_4^{3-} + H^+ \qquad K_3 = \frac{[PO_4^{3-}]\gamma_{PO_4^{3-}}[H^+]\gamma_{H^+}}{[HPO_4^{2-}]\gamma_{HPO_4^{2-}}} = 10^{-12.375} \qquad \textbf{(13-21)}$$

Equilibrium constants can be determined by measuring concentration quotients at several low ionic strengths and extrapolating to 0 ionic strength.

For $\mu \neq 0$, we rearrange the equilibrium constant expression to incorporate activity coefficients into an effective equilibrium constant, K' at the given ionic strength.

$$K_1' = K_1\left(\frac{\gamma_{H_3PO_4}}{\gamma_{H_2PO_4^-}\gamma_{H^+}}\right) = \frac{[H_2PO_4^-][H^+]}{[H_3PO_4]} \qquad \textbf{(13-22)}$$

$$K_2' = K_2\left(\frac{\gamma_{H_2PO_4^-}}{\gamma_{HPO_4^{2-}}\gamma_{H^+}}\right) = \frac{[HPO_4^{2-}][H^+]}{[H_2PO_4^-]} \qquad \textbf{(13-23)}$$

$$K_3' = K_3\left(\frac{\gamma_{HPO_4^{2-}}}{\gamma_{PO_4^{3-}}\gamma_{H^+}}\right) = \frac{[PO_4^{3-}][H^+]}{[HPO_4^{2-}]} \qquad \textbf{(13-24)}$$

For ionic species, we compute activity coefficients with the Davies equation 13-18. For the neutral species H_3PO_4, we assume that $\gamma \approx 1.00$.

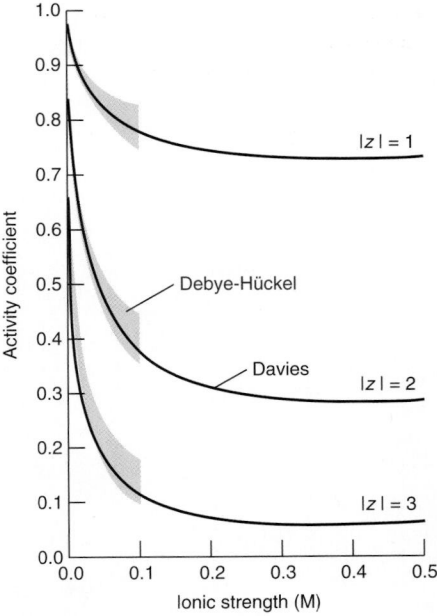

FIGURE 13-2 Activity coefficients from extended Debye-Hückel and Davies equations. Shaded areas give Debye-Hückel activity coefficients for the range of ion sizes in Table 8-1.

A 0.5 wt% solution of K_2HPO_4 has a *molarity* of 0.028 13 mol/L and a *molality* of 0.028 20 mol/kg. The difference is 0.25%.

K_2 gives the *concentration quotient*

$$\frac{[HPO_4^{2-}][H^+]}{[H_2PO_4^-]}$$

at a specified ionic strength.

Values of K_w are found in Table 6-1.

Now we remind ourselves of the equations for water ionization

$$H_2O \rightleftharpoons H^+ + OH^- \qquad K_w = [H^+]\gamma_{H^+}[OH^-]\gamma_{OH^-} = 10^{-13.995}$$

$$K_w' = \frac{K_w}{\gamma_{H^+}\gamma_{OH^-}} = [H^+][OH^-] \Rightarrow [OH^-] = K_w'/[H^+] \qquad \textbf{(13-25)}$$

$$pH = -\log([H^+]\gamma_{H^+}) \qquad \textbf{(13-26)}$$

Let's find the pH of 0.025 0 m KH_2PO_4 plus 0.025 0 m Na_2HPO_4 including activity coefficients. The chemical reactions are 13-19 through 13-21 plus water ionization. Mass balances are $[K^+] = 0.025\ 0\ m$, $[Na^+] = 0.050\ 0\ m$, and total phosphate $\equiv F_{H_3P} = 0.050\ 0\ m$. The charge balance is

$$[Na^+] + [K^+] + [H^+] = [H_2PO_4^-] + 2[HPO_4^{2-}] + 3[PO_4^{3-}] + [OH^-] \qquad \textbf{(13-27)}$$

Our strategy is to substitute expressions into the charge balance to obtain an equation in which the only variable is $[H^+]$. For this purpose, we use the fractional composition equations for the triprotic acid, H_3PO_4, which we abbreviate as H_3P:

$$[P^{3-}] = \alpha_{P^{3-}}F_{H_3P} = \frac{K_1'K_2'K_3'F_{H_3P}}{[H^+]^3 + [H^+]^2K_1' + [H^+]K_1'K_2' + K_1'K_2'K_3'} \qquad \textbf{(13-28)}$$

$$[HP^{2-}] = \alpha_{HP^{2-}}F_{H_3P} = \frac{[H^+]K_1'K_2'F_{H_3P}}{[H^+]^3 + [H^+]^2K_1' + [H^+]K_1'K_2' + K_1'K_2'K_3'} \qquad \textbf{(13-29)}$$

$$[H_2P^-] = \alpha_{H_2P^-}F_{H_3P} = \frac{[H^+]^2K_1'F_{H_3P}}{[H^+]^3 + [H^+]^2K_1' + [H^+]K_1'K_2' + K_1'K_2'K_3'} \qquad \textbf{(13-30)}$$

$$[H_3P] = \alpha_{H_3P}F_{H_3P} = \frac{[H^+]^3F_{H_3P}}{[H^+]^3 + [H^+]^2K_1' + [H^+]K_1'K_2' + K_1'K_2'K_3'} \qquad \textbf{(13-31)}$$

Tricks necessary to deal with ionic strength and activity coefficients in Excel

Figure 13-3 puts everything together in a spreadsheet. We need the ionic strength to calculate concentrations and we need the concentrations to calculate ionic strength. We say that there is a *circular reference* because concentration depends on ionic strength and ionic strength depends on concentrations. To handle the circular reference in Excel 2010, select the File tab and select Options. In the Options window, select Formulas. In Calculation Options, check "Enable iterative calculation" and set Maximum Change to 1e-15. Click OK and your spreadsheet is ready to handle circular references.

In spreadsheets with activity coefficients, you might observe the error message #NUM! in many places when you first set up the spreadsheet. If so, enter a *number* (such as 0) instead of a *formula* for ionic strength. After all formulas have been entered, go back to ionic strength and replace the number 0 with the formula for ionic strength.

Input values for $F_{KH_2PO_4}$, $F_{Na_2HPO_4}$, pK_1, pK_2, pK_3, and pK_w are in the shaded cells in Figure 13-3. We *guess* a value for pH in cell H15. The ionic strength is computed in cell E19. The initial pH is only a guess, so the concentrations and ionic strength are not correct yet. Cells A9:H10 compute activities with the Davies equation. Cells A13:H16 compute concentrations. $[H^+]$ in cell B13 is $(10^{-pH})/\gamma_{H^+} = (10\wedge\text{-H15})/\text{B9}$. Cell E18 computes the sum of charges.

An initial guess of pH = 7 in cell H15 gives a net charge of $-0.003\ 7\ m$ in cell E18 and an ionic strength of 0.105 5 m in cell E19. These values are not shown in Figure 13-3. Now use Excel Solver to vary pH in cell H15 to produce a net charge near 0 in cell E18. Solver Options should be set as described under "Using Excel Solver" (page 290) in Section 13-1. Figure 13-3 shows that executing Solver gives pH = 6.876 in cell H15 and a net charge of $-1 \times 10^{-17}\ m$ in cell E18. The calculated ionic strength in cell E19 is 0.100 m. The work is done!

The calculated pH of 6.876 differs from the certified pH of 6.865 by 0.011. This difference is about as close an agreement between a measured and computed pH as you will encounter. If we had used extended Debye-Hückel activity coefficients for $\mu = 0.1\ m$ in Table 8-1, the computed pH would be 6.859, differing from the stated pH by just 0.006.

Sometimes Solver cannot find a solution if Precision is set too small in the Options window. You can make Precision larger (such as 1E-10) and see if Solver can find a solution. You can also try a different initial guess for pH.

	A	B	C	D	E	F	G	H
1	Mixture of KH_2PO_4 and Na_2HPO_4 including activity coefficients from Davies equation							
2								
3	$F_{KH_2PO_4}$ =	0.0250		pK_1 =	2.148		K_1'=	1.17E-02
4	$F_{Na_2PO_4}$ =	0.0250		pK_2 =	7.198		K_2'=	1.70E-07
5	F_{H_3P} =	0.0500	(=B3+B4)	pK_3 =	12.375		K_3'=	1.86E-12
6				pK_w =	13.995		K_w'=	1.66E-14
7								
8	Activity coefficients:							
9	H^+ =	0.78		H_3P =	1.00	(fixed at 1)	HP^{2-} =	0.37
10	OH^- =	0.78		H_2P^- =	0.78		P^{3-} =	0.11
11								
12	Species in charge balance:						Other concentrations:	
13	$[H^+]$ =	1.70E-07		$[OH^-]$ =	9.74E-08		$[H_3P]$ =	3.65E-07
14	$[Na^+]$ =	0.050000		$[H_2P^-]$ =	2.50E-02			
15	$[K^+]$ =	0.025000		$[HP^{2-}]$ =	2.50E-02		pH =	6.876
16				$[P^{3-}]$ =	2.73E-07		↑ initial value is a guess	
17								
18	Positive charge minus negative charge =				-1.15E-17			
19			Ionic strength =		0.1000	=0.5*(B13+B14+B15+E13+E14		
20						+4*E15+9*E16)		
21	Formulas:							
22	H3 = 10^-E3*E9/(E10*B9)				H4 = 10^-E4*E10/(H9*B9)			
23	H5 = 10^-E5*H9/(H10*B9)				H6 = 10^-E6/(B9*B10)			
24	B9 = B10 = E10 = 10^(-0.51*1^2*(SQRT($E$19)/(1+SQRT($E$19))-0.3*$E$19))							
25	H9 = 10^(-0.51*2^2*(SQRT($E$19)/(1+SQRT($E$19))-0.3*$E$19))							
26	H10 = 10^(-0.51*3^2*(SQRT($E$19)/(1+SQRT($E$19))-0.3*$E$19))							
27	B13 = (10^-H15)/B9	B14 = 2*B4		B15 = B3			E13 = H6/(B13)	
28	E14 = B13^2*H3*B5/(B13^3+B13^2*H3+B13*H3*H4+H3*H4*H5)							
29	E15 = B13*H3*H4*B5/(B13^3+B13^2*H3+B13*H3*H4+H3*H4*H5)							
30	E16 = H3*H4*H5*B5/(B13^3+B13^2*H3+B13*H3*H4+H3*H4*H5)							
31	H13 = B13^3*B5/(B13^3+B13^2*H3+B13*H3*H4+H3*H4*H5)							
32	E18 = B13+B14+B15-E13-E14-2*E15-3*E16							

FIGURE 13-3 Spreadsheet solved for the system 0.025 0 m KH_2PO_4 plus 0.025 0 m Na_2HPO_4. The spreadsheet was set to handle the circular references between activity coefficients and ionic strength. Solver was employed to vary the pH in cell H15 until the charge balance in cell E18 was satisfied.

Back to Basics

A spreadsheet operating on the charge balance to reduce the net charge to zero is an excellent, general method for solving complex equilibrium problems. However, we did learn how to find the pH of a mixture of KH_2PO_4 plus Na_2HPO_4 back in Chapter 9 by a simple, less rigorous method. Recall that, when we mix a weak acid ($H_2PO_4^-$) and its conjugate base (HPO_4^{2-}), *what we mix is what we get*. The pH is estimated from the Henderson-Hasselbalch equation 9-18 with activity coefficients:

$$pH = pK_a + \log\frac{[A^-]\gamma_{A^-}}{[HA]\gamma_{HA}} = pK_2 + \log\frac{[HPO_4^{2-}]\gamma_{HPO_4^{2-}}}{[H_2PO_4^-]\gamma_{H_2PO_4^-}} \qquad (9\text{-}18)$$

For 0.025 m KH_2PO_4 plus 0.025 m Na_2HPO_4, the ionic strength is

$$\mu = \frac{1}{2}\sum_i c_i z_i^2 = \frac{1}{2}([K^+]\cdot(+1)^2 + [H_2PO_4^-]\cdot(-1)^2 + [Na^+]\cdot(+1)^2 + [HPO_4^{2-}]\cdot(-2)^2)$$

$$= \frac{1}{2}([0.025]\cdot 1 + [0.025]\cdot 1 + [0.050]\cdot 1 + [0.025]\cdot 4) = 0.100\ m$$

In Table 8-1, the activity coefficients at $\mu = 0.1\ m$ are 0.775 for $H_2PO_4^-$ and 0.355 for HPO_4^{2-}. Inserting these values into Equation 9-18 gives

$$pH = 7.198 + \log\frac{[0.025]0.355}{[0.025]0.775} = 6.859$$

$pK_a = pK_2$ in Equation 9-18 applies at $\mu = 0$.

The answer is the same that we generate with the spreadsheet because the approximation that what we mix is what we get is excellent in this case.

So you already knew how to find the pH of this buffer by a simple calculation. The value of the general method with the charge balance in the spreadsheet is that it applies even when what you mix is not what you get if the concentrations are low or K_2 is not so small or there are additional equilibria.

Ignorance Is Still Bliss

Even in such a simple solution as KH_2PO_4 plus Na_2HPO_4, for which we were justifiably proud of computing an accurate pH, we have overlooked numerous ion-pair equilibria:

$$PO_4^{3-} + Na^+ \rightleftharpoons NaPO_4^{2-} \qquad K = 27 \qquad HPO_4^{2-} + Na^+ \rightleftharpoons NaHPO_4^- \qquad K = 12$$

$$H_2PO_4^- + Na^+ \rightleftharpoons NaH_2PO_4 \qquad K = 2 \qquad NaPO_4^{2-} + Na^+ \rightleftharpoons Na_2PO_4^- \qquad K = 14$$

$$Na_2PO_4^- + H^+ \rightleftharpoons Na_2HPO_4 \qquad K = 5.4 \times 10^{10}$$

The reliability of the calculated concentrations depends on knowing all relevant equilibria and having the fortitude to include them all in the computation, which is by no means trivial.

The effective pK_2 for H_3PO_4 listed in the *NIST Critically Selected Stability Constants Database 46* (2001) for an ionic strength of 0.1 M has the following values: 6.71 for Na^+ background ions, 6.75 for K^+ background ions, and 6.92 for unspecified tetraalkylammonium background ions. The dependence of the effective pK on the nature of the background ions strongly suggests that ion-pairing reactions play an observable role in the solution chemistry.

13-3 Dependence of Solubility on pH

An important example of the effect of pH on solubility is tooth decay. Tooth enamel contains the mineral hydroxyapatite, which is insoluble near neutral pH, but dissolves in acid because both phosphate and hydroxide in the hydroxyapatite react with H^+:

$$\underset{\text{Calcium hydroxyapatite}}{Ca_{10}(PO_4)_6(OH)_2(s)} + 14H^+ \rightleftharpoons 10Ca^{2+} + 6H_2PO_4^- + 2H_2O$$

Bacteria on our teeth metabolize sugars to produce lactic acid, which lowers the pH enough to slowly dissolve tooth enamel. Fluoride inhibits tooth decay because it forms fluorapatite, $Ca_{10}(PO_4)_6F_2$, which is more acid resistant than hydroxyapatite.

Solubility of CaF₂

The mineral *fluorite* (CaF_2), also called *fluorspar*, has a cubic crystal structure shown in Figure 13-4 and often cleaves to form nearly perfect octahedra (eight-sided solids with equilateral triangular faces). Depending on impurities, the mineral takes on a variety of colors and may fluoresce when irradiated with an ultraviolet lamp. Fluorspar is converted to hydrofluoric acid (HF) for the synthesis of refrigerants and fluoropolymers. A large fraction of the world's supply of fluorspar is found in China.

There is an analogous set of reactions for K^+, whose equilibrium constants are similar to those for Na^+.

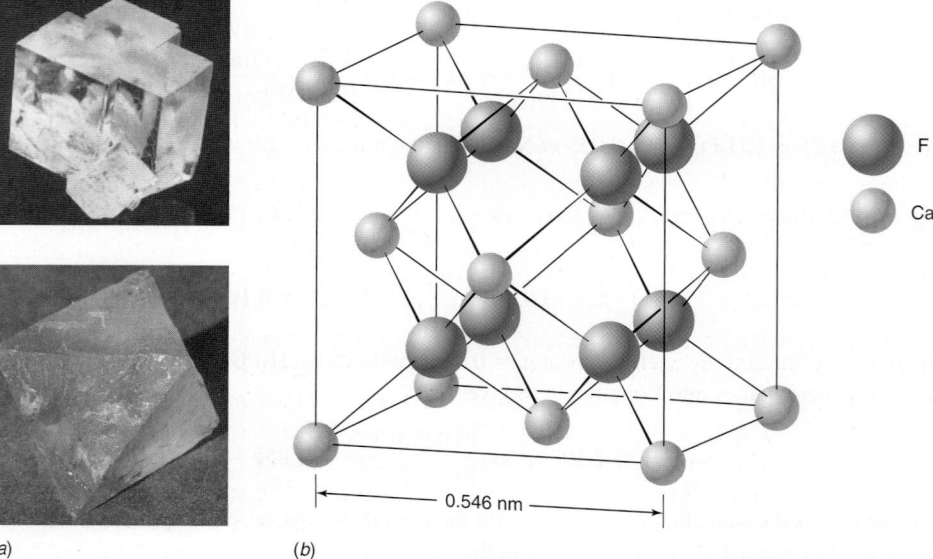
L-Lactic acid

FIGURE 13-4 Crystals of the mineral fluorite, CaF_2. Each Ca^{2+} ion is surrounded by eight F^- ions at the corners of a cube. Each F^- ion is surrounded by four Ca^{2+} ions at the corners of a tetrahedron. If you imagine the next unit cell above this one, you should see that the Ca^{2+} ion at the center of the top face of the cell shown here is adjacent to four F^- ions in this cell and four F^- ions in the next cell above it. [Top photo, © Mark A. Schneider/Science Source; bottom photo, © Joyce Photographics/Science Source.]

F

Ca

0.546 nm

(a)   (b)

The solubility of CaF_2 is governed by K_{sp} for the salt, hydrolysis of F^- and of Ca^{2+}, and by ion pairing between Ca^{2+} and F^-:

$$CaF_2(s) \rightleftharpoons Ca^{2+} + 2F^- \qquad K_{sp} = [Ca^{2+}]\gamma_{Ca^{2+}}[F^-]^2\gamma_{F^-}^2 = 10^{-10.50} \qquad (13\text{-}32)$$

$$HF \rightleftharpoons H^+ + F^- \qquad K_{HF} = \frac{[H^+]\gamma_{H^+}[F^-]\gamma_{F^-}}{[HF]\gamma_{HF}} = 10^{-3.17} \qquad (13\text{-}33)$$

$$Ca^{2+} + H_2O \rightleftharpoons CaOH^+ + H^+ \qquad K_a = \frac{[CaOH^+]\gamma_{CaOH^+}[H^+]\gamma_{H^+}}{[Ca^{2+}]\gamma_{Ca^{2+}}} = 10^{-12.70} \qquad (13\text{-}34)$$

$$Ca^{2+} + F^- \rightleftharpoons CaF^+ \qquad K_{ip} = \frac{[CaF^+]\gamma_{CaF^+}}{[Ca^{2+}]\gamma_{Ca^{2+}}[F^-]\gamma_{F^-}} = 10^{0.63} \qquad (13\text{-}35)$$

$$H_2O \rightleftharpoons H^+ + OH^- \qquad K_w = [H^+]\gamma_{H^+}[OH^-]\gamma_{OH^-} = 10^{-14.00} \qquad (13\text{-}36)$$

The charge balance is

Charge balance: $\quad [H^+] + 2[Ca^{2+}] + [CaOH^+] + [CaF^+] = [OH^-] + [F^-] \qquad (13\text{-}37)$

To find the mass balance, we need to realize that all calcium and fluoride species come from CaF_2. Therefore, the total fluoride equals two times the total calcium:

$$2[\text{total calcium species}] = [\text{total fluoride species}]$$

$$2\{[Ca^{2+}] + [CaOH^+] + [CaF^+]\} = [F^-] + [HF] + [CaF^+]$$

Mass balance: $\quad 2[Ca^{2+}] + 2[CaOH^+] + [CaF^+] = [F^-] + [HF] \qquad (13\text{-}38)$

There are seven independent equations and seven unknowns, so we have enough information. And…you already know how to solve this problem because you studied Chapter 8 well!

The dissolution of CaF_2 is very much like the dissolution of $CaSO_4$, which we studied in Section 8-5. There are seven unknowns and five equilibria, so we will let Solver find (number of unknowns – number of equilibria) = two concentrations for us, as we did in Figure 8-13. The spreadsheet for CaF_2 is shown in Figure 13-5. We begin with the estimates $pCa^{2+} = 4$ and $pH = 7$ in cells B8 and B9. Cells D17:D21 compute *effective equilibrium constants K'*, which incorporate the activity coefficients in Equations 13-39 through 13-43. Cells C10:C14 compute concentrations from the equilibrium constant expressions:

$$[F^-] = \sqrt{\frac{K'_{sp}}{[Ca^{2+}]}} \qquad K'_{sp} = \frac{K_{sp}}{\gamma_{Ca^{2+}}\gamma_{F^-}^2} \qquad (13\text{-}39)$$

$$[CaF^+] = K'_{ip}[Ca^{2+}][F^-] \qquad K'_{ip} = \frac{K_{ip}\gamma_{Ca^{2+}}\gamma_{F^-}}{\gamma_{CaF^+}} \qquad (13\text{-}40)$$

$$[CaOH^+] = \frac{K'_a[Ca^{2+}]}{[H^+]} \qquad K'_a = \frac{K_a\gamma_{Ca^{2+}}}{\gamma_{H^+}\gamma_{CaOH^+}} \qquad (13\text{-}41)$$

$$[HF] = \frac{[H^+][F^-]}{K'_{HF}} \qquad K'_{HF} = \frac{K_{HF}\gamma_{HF}}{\gamma_{H^+}\gamma_{F^-}} \qquad (13\text{-}42)$$

$$[OH^-] = \frac{K'_w}{[H^+]} \qquad K'_w = \frac{K_w}{\gamma_{H^+}\gamma_{OH^-}} \qquad (13\text{-}43)$$

Equations 13-39 through 13-43 are rearrangements of the equilibrium constant expressions 13-32 through 13-36. The rearranged forms incorporate activity coefficients into the effective equilibrium constant K'. Expressions for concentrations, such as $[HF] = [H^+][F^-]/K'_{HF}$, are then unencumbered by activity coefficients, which are implicit in K'.

Activity coefficients are computed in cells E8:F14 from the Davies equation 13-18 using the ionic strength computed in cell B5 from the concentrations. The spreadsheet is set to handle circular definitions (following the instructions on page 292) in which ionic strength depends on concentrations and concentrations depend on ionic strength.

The mass balance b_1 appears in cell J17 and the charge balance b_2 is in cell J18. Use Excel Solver to vary pCa^{2+} and pH in cells B8 and B9 to minimize $b_1^2 + b_2^2$ in cell J19. Results shown in Figure 13-5 are $pCa^{2+} = 3.676$ and $pH = 7.087$ in cells B8 and B9. Solver is not good at finding both unknowns together in this particular problem. I first solved for both variables together. Then I alternately held pCa^{2+} constant while solving for pH and then held pH constant while solving for pCa^{2+}. This cycle was repeated several times as long as Σb_i^2 in cell J19 continued to decrease. When Solver had gone as far as it could, I reduced Σb_i^2

Solubility products are in Appendix F. The acid dissociation constant of HF is from Appendix G. The hydrolysis constant for Ca^{2+} is derived from the formation constant of $CaOH^+$ in Appendix I plus the K_w equation. The ion-pair formation constant for CaF^+ is listed in Appendix J.

	A	B	C	D	E	F	G	H	I	J
1	Calcium fluoride equilibria with Davies equation for activity coefficients									
2	1. *Estimate* values in cells B8 and B9									
3	2. Use Solver to adjust B8 and B9 to minimize sum in cell J19									
4	Ionic strength									
5	μ =	0.000633	= 0.5*(D8^2*C8+D9^2*C9+D10^2*C10+D11^2*C11+D12^2*C12+D13^2*C13+D14^2*C14)							
6					Davies	Activity				
7	Species	pC	C(M)	Charge	log γ	coefficient, γ				
8	Ca^{2+}	3.6760486	2.1084E-04	2	-4.968E-02	8.919E-01	C8 = 10^-B8			F8 = 10^E8
9	H^+	7.0874	8.1771E-08	1	-1.242E-02	9.718E-01	C9 = 10^-B9			
10	F^-		4.2197E-04	-1	-1.242E-02	9.718E-01	C10 = SQRT(D17/C8)			
11	CaF^+		3.38498E-07	1	-1.242E-02	9.718E-01	C11 = D20*C8*C10			
12	$CaOH^+$		4.85863E-10	1	-1.242E-02	9.718E-01	C12 = D19*C8/C9			
13	HF		4.81996E-08	0	0.000E+00	1.000E+00	C13 = C9*C10/D18			
14	OH^-		1.29491E-07	-1	-1.242E-02	9.718E-01	C14 = D21/C9			
15										
16			K' (with activity coefficients)				Mass and charge balances:			b_i
17	pK_{sp} =	10.50	K_{sp}' =	3.75E-11	$b_1 = 0 = 2[Ca^{2+}] + 2[CaOH^+] + [CaF^+] - [F^-] - [HF]$ =					-1.60E-11
18	pK_{HF} =	3.17	K_{HF}' =	7.16E-04	$b_2 = 0 = [H^+] + 2[Ca^{2+}] + [CaOH^+] + [CaF^+] - [OH^-] - [F^-]$ =					-2.17E-11
19	pK_a =	12.70	K_a' =	1.88E-13					Σb_i^2 =	7.24E-22
20	pK_{ip} =	-0.63	K_{ip}' =	3.80E+00			J17 = 2*C8+2*C12+C11-C10-C13			
21	pK_w =	14.00	K_w' =	1.06E-14			J18 = C9+2*C8+C12+C11-C14-C10			
22							J19 = J17^2 + J18^2			
23	Initial values:						D17 = (10^-B17)/(F8*F10^2)			
24	pCa = 4	pH = 7					D18 = (10^-B18)*F13/(F9*F10)			
25							D19 = (10^-B19)*F8/(F9*F12)			
26	Optimize both pCa and pH together for a few cycles						D20 = (10^-B20)*F8*F10/F11			
27	Then optimize just pCa or just pH while holding the other constant						D21 = (10^-B21)/(F9*F14)			
28	Continue optimization as long as Σb_i^2 keeps getting smaller						E8 = -0.51*D8^2*(SQRT($B$5)/(1+SQRT($B$5)) - 0.3*$B$5)			

FIGURE 13-5 Spreadsheet using Solver and activities for saturated CaF_2 in water.

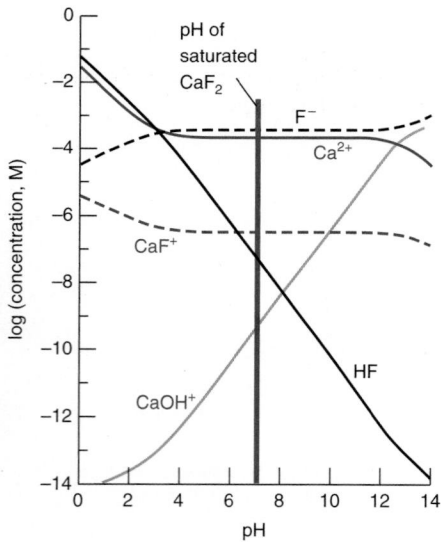

FIGURE 13-6 pH dependence of species in a saturated solution of CaF_2. As pH is lowered, H^+ reacts with F^- to make HF, and $[Ca^{2+}]$ increases. Note the logarithmic ordinate.

further by varying pH in the third and fourth decimal places by hand to obtain the results in Figure 13-5.

Figure 13-6 shows how the concentrations vary with pH. At low pH, H^+ reacts with F^- to produce HF and increases the solubility of CaF_2. The species CaF^+ and $CaOH^+$ are minor at most pH values, but $CaOH^+$ becomes the major form of calcium above pH 12.7, which is pK_a for Reaction 13-34. A reaction that we did not consider was precipitation of $Ca(OH)_2(s)$. Comparison of the product $[Ca^{2+}][OH^-]^2$ with K_{sp} for $Ca(OH)_2$ indicates that $Ca(OH)_2$ should precipitate at a pH between 13 and 14.

You could create Figure 13-6 from the spreadsheet in Figure 13-5 by fixing $\mu = 0$ in cell B5 and fixing pH at successive values between 0 to 14 in cell B9. For each fixed pH, execute Solver to find the value of pCa^{2+} in cell B8 that reduces the mass balance in cell J17 to near zero. The spreadsheet then gives the concentrations of all species at one fixed pH. The procedure has to be repeated for each pH value to obtain the data in Figure 13-6. Note that when we fix the pH, we have added new species, such as a buffer, that change the charge balance and ionic strength, but not the mass balance relating calcium to fluoride. Therefore, we use the mass balance, but not the charge balance, for fixed pH. We also neglected activity coefficients in generating Figure 13-6 because the ionic strength depends on unspecified reagent that was added to obtain each pH.

Acid Rain Dissolves Minerals and Creates Environmental Hazards

In general, salts of basic ions such as F^-, OH^-, S^{2-}, CO_3^{2-}, $C_2O_4^{2-}$, and PO_4^{3-} have increased solubility at low pH because the anions react with H^+. Figure 13-7 shows that marble, which is largely $CaCO_3$, dissolves more readily as the acidity of rain increases. Much of the acid in rain comes from SO_2 emissions from combustion of fuels containing sulfur and from nitrogen oxides produced by all types of combustion. SO_2, for example, reacts in the air to make sulfuric acid $(SO_2 + H_2O \rightarrow H_2SO_3 \xrightarrow{\text{oxidation}} H_2SO_4)$, which finds its way back to the ground in rainfall.

Aluminum is the third most abundant element on Earth (after oxygen and silicon), but it is tightly locked into insoluble minerals such as kaolinite ($Al_2(OH)_4Si_2O_5$) and bauxite (AlOOH). Acid rain from human activities is a recent change for our planet, and it is introducing soluble forms of aluminum (and lead and mercury) into the environment.[4] Figure 13-8 shows that, below pH 5, aluminum is mobilized from minerals and its concentration in lake water rises rapidly. At a concentration of 130 μg/L, aluminum kills fish. In humans, high concentrations of aluminum are associated with dementia, softening of bones, and anemia. Aluminum is suspected as a possible cause of Alzheimer's disease. Although metallic elements from minerals are liberated by acid, the concentration and availability of metal ions in the environment tend to be regulated by organic matter that binds metal ions.

Solubility of Barium Oxalate

Now consider the dissolution of $Ba(C_2O_4)$, whose anion is *dibasic* and whose cation is a weak acid.[5] We neglect activity coefficients for this example. The chemistry is

$$Ba(C_2O_4)(s) \rightleftharpoons Ba^{2+} + \underset{\text{Oxalate}}{C_2O_4^{2-}} \qquad K_{sp} = [Ba^{2+}][C_2O_4^{2-}] = 10^{-6.85} \qquad \textbf{(13-44)}$$

$$H_2C_2O_4 \rightleftharpoons HC_2O_4^- + H^+ \qquad K_1 = \frac{[H^+][HC_2O_4^-]}{[H_2C_2O_4]} = 10^{-1.25} \qquad \textbf{(13-45)}$$

$$HC_2O_4^- \rightleftharpoons C_2O_4^{2-} + H^+ \qquad K_2 = \frac{[H^+][C_2O_4^{2-}]}{[HC_2O_4^-]} = 10^{-4.27} \qquad \textbf{(13-46)}$$

$$Ba^{2+} + H_2O \rightleftharpoons BaOH^+ + H^+ \qquad K_a = \frac{[H^+][BaOH^+]}{[Ba^{2+}]} = 10^{-13.36} \qquad \textbf{(13-47)}$$

$$Ba^{2+} + C_2O_4^{2-} \rightleftharpoons Ba(C_2O_4)(aq) \qquad K_{ip} = \frac{[Ba(C_2O_4)(aq)]}{[Ba^{2+}][C_2O_4^{2-}]} = 10^{2.31} \qquad \textbf{(13-48)}$$

The value of K_{sp} is estimated for $\mu = 0$ and 20°C. K_{ip} is for $\mu = 0$ and 18°C. K_1, K_2, and K_a apply at $\mu = 0$ and 25°C.

The charge balance is

Charge balance: $[H^+] + 2[Ba^{2+}] + [BaOH^+] = [OH^-] + [HC_2O_4^-] + 2[C_2O_4^{2-}]$ **(13-49)**

The mass balance states that the total moles of barium equal the total moles of oxalate:

$$[\text{Total barium}] = [\text{total oxalate}]$$

$$[Ba^{2+}] + [BaOH^+] + \cancel{[Ba(C_2O_4)(aq)]} = [H_2C_2O_4] + [HC_2O_4^-] + [C_2O_4^{2-}] + \cancel{[Ba(C_2O_4)(aq)]}$$

Mass balance: $\underbrace{[Ba^{2+}] + [BaOH^+]}_{F_{Ba}} = \underbrace{[H_2C_2O_4] + [HC_2O_4^-] + [C_2O_4^{2-}]}_{F_{H_2Ox}}$ **(13-50)**

We are defining F_{Ba} and F_{H_2Ox} to *exclude* the ion pair $Ba(C_2O_4)(aq)$.

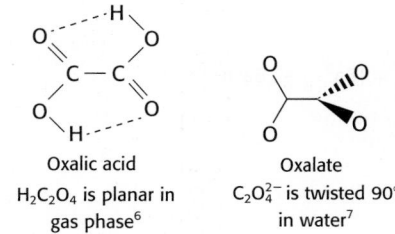

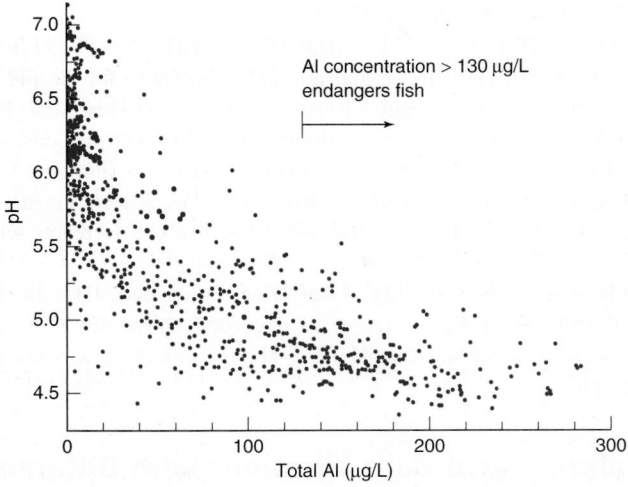

FIGURE 13-8 Relation of total aluminum (including dissolved and suspended species) in 1 000 Norwegian lakes as a function of the pH of the lake water. The more acidic the water, the greater the aluminum concentration. [Data from G. Howells, *Acid Rain and Acid Waters*, 2nd ed. (Hertfordshire: Ellis Horwood, 1995).]

FIGURE 13-7 Measured calcium in acid rain runoff from marble stone (which is largely $CaCO_3$) roughly increases as [H^+] in the rain increases. [Data from P. A. Baedecker and M. M. Reddy, "The Erosion of Carbonate Stone by Acid Rain," *J. Chem. Ed.* **1993**, *70*, 104.]

Oxalic acid
$H_2C_2O_4$ is planar in gas phase[6]

Oxalate
$C_2O_4^{2-}$ is twisted 90° in water[7]

There are eight unknowns and eight independent equations (including $[OH^-] = K_w/[H^+]$), so we have enough information to find the concentrations of all species.

We deal with ion pairing by adding Reactions 13-44 and 13-48 to find

$$Ba(C_2O_4)(s) \rightleftharpoons Ba(C_2O_4)(aq) \quad K = [Ba(C_2O_4)(aq)] = K_{sp}K_{ip} = 10^{-4.54} \quad (13\text{-}51)$$

Therefore, $[Ba(C_2O_4)(aq)] = 10^{-4.54}$ M as long as undissolved $Ba(C_2O_4)(s)$ is present.

> The ion pair $Ba(C_2O_4)(aq)$ has a constant concentration in this system.

Now for our old friends, the fractional composition equations. Abbreviating oxalic acid as H_2Ox, we can write

> $F_{H_2Ox} = [H_2C_2O_4] + [HC_2O_4^-] + [C_2O_4^{2-}]$

$$[H_2Ox] = \alpha_{H_2Ox}F_{H_2Ox} = \frac{[H^+]^2 F_{H_2Ox}}{[H^+]^2 + [H^+]K_1 + K_1K_2} \quad (13\text{-}52)$$

$$[HOx^-] = \alpha_{HOx^-}F_{H_2Ox} = \frac{K_1[H^+] F_{H_2Ox}}{[H^+]^2 + [H^+]K_1 + K_1K_2} \quad (13\text{-}53)$$

$$[Ox^{2-}] = \alpha_{Ox^{2-}}F_{H_2Ox} = \frac{K_1K_2 F_{H_2Ox}}{[H^+]^2 + [H^+]K_1 + K_1K_2} \quad (13\text{-}54)$$

Also, Ba^{2+} and $BaOH^+$ are a conjugate acid-base pair. Ba^{2+} behaves as a monoprotic acid HA, and $BaOH^+$ is the conjugate base A^-.

> $F_{Ba} = [Ba^{2+}] + [BaOH^+]$

$$[Ba^{2+}] = \alpha_{Ba^{2+}}F_{Ba} = \frac{[H^+] F_{Ba}}{[H^+] + K_a} \quad (13\text{-}55)$$

$$[BaOH^+] = \alpha_{BaOH^+}F_{Ba} = \frac{K_a F_{Ba}}{[H^+] + K_a} \quad (13\text{-}56)$$

Suppose that the pH is fixed by adding a buffer (and therefore the charge balance 13-49 is no longer valid). From K_{sp}, we can write

$$K_{sp} = [Ba^{2+}][C_2O_4^{2-}] = \alpha_{Ba^{2+}}F_{Ba}\alpha_{Ox^{2-}}F_{H_2Ox}$$

But the mass balance 13-50 told us that $F_{Ba} = F_{H_2Ox}$. Therefore,

$$K_{sp} = \alpha_{Ba^{2+}}F_{Ba}\alpha_{Ox^{2-}}F_{H_2Ox} = \alpha_{Ba^{2+}}F_{Ba}\alpha_{Ox^{2-}}F_{Ba}$$

$$\Rightarrow F_{Ba} = \sqrt{\frac{K_{sp}}{\alpha_{Ba^{2+}}\alpha_{Ox^{2-}}}} \quad (13\text{-}57)$$

In the spreadsheet in Figure 13-9, pH is specified in column A. From this pH, plus K_1 and K_2, the fractions α_{H_2Ox}, α_{HOx^-}, and $\alpha_{Ox^{2-}}$ are computed with Equations 13-52 through 13-54 in columns C, D, and E. From the pH and K_a, the fractions $\alpha_{Ba^{2+}}$ and α_{BaOH^+} are computed with Equations 13-55 and 13-56 in columns F and G. The total concentrations of barium and oxalate, F_{Ba} and F_{H_2Ox} are equal and calculated in column H from Equation 13-57. In a real spreadsheet, we would have continued to the right with column I. To fit on this page, the spreadsheet was continued in row 18. In this lower section, the concentrations of $[Ba^{2+}]$ and $[BaOH^+]$ are computed with Equations 13-55 and 13-56. $[H_2C_2O_4]$, $[HC_2O_4^-]$, and $[C_2O_4^{2-}]$ are found with Equations 13-52 through 13-54.

The net charge $(= [H^+] + 2[Ba^{2+}] + [BaOH^+] - [OH^-] - [HC_2O_4^-] - 2[C_2O_4^{2-}])$ is computed beginning in cell H19. If we had not added buffer to fix the pH, the net charge would be 0. Net charge goes from positive to negative between pH 6 and 8. By using Solver, we find the pH in cell A11 that makes the net charge 0 in cell H23 (with Constraint Precision = 1e-15 in Solver Options). This pH, 7.45, is the pH of unbuffered solution.

Figure 13-10 shows that the solubility of barium oxalate is steady near $10^{-3.4}$ M in the middle pH range. Solubility increases below pH 5 because $C_2O_4^{2-}$ reacts with H^+ to make $HC_2O_4^-$.

A final point is to see if the solubility of $Ba(OH)_2(s)$ is exceeded. Evaluation of the product $[Ba^{2+}][OH^-]^2$ shows that $K_{sp} = 10^{-6.85}$ is exceeded above pH 12.3. We predict that $Ba(OH)_2(s)$ will begin to precipitate at pH 12.3. This precipitation was not included in our calculations or graph.

> Niels Bjerrum (1879–1958) was a Danish physical chemist who made fundamental contributions to inorganic coordination chemistry and is responsible for much of our understanding of acids and bases and titration curves.[9]

13-4 Analyzing Acid-Base Titrations with Difference Plots[8]

A *difference plot*, also called a *Bjerrum plot*, is an excellent means to extract metal-ligand formation constants or acid dissociation constants from titration data obtained with electrodes. We will apply the difference plot to an acid-base titration curve.

	A	B	C	D	E	F	G	H
1	Finding the concentrations of species in saturated barium oxalate solution							
2								
3	K_{sp} =	1.41E-07		K_1 =	5.62E-02		K_{ip} =	2.04E+02
4	K_a =	4.37E-14		K_2 =	5.37E-05		K_w =	1.00E-14
5								F_{Ba}
6	pH	$[H^+]$	$\alpha(H_2Ox)$	$\alpha(HOx^-)$	$\alpha(Ox^{2-})$	$\alpha(Ba^{2+})$	$\alpha(BaOH^+)$	= F_{H2Ox}
7	0	1.E+00	9.5E-01	5.3E-02	2.9E-06	1.0E+00	4.4E-14	2.2E-01
8	2	1.E-02	1.5E-01	8.5E-01	4.5E-03	1.0E+00	4.4E-12	5.6E-03
9	4	1.E-04	1.2E-03	6.5E-01	3.5E-01	1.0E+00	4.4E-10	6.4E-04
10	6	1.E-06	3.3E-07	1.8E-02	9.8E-01	1.0E+00	4.4E-08	3.8E-04
11	7.451	4.E-08	4.1E-10	6.6E-04	1.0E+00	1.0E+00	1.2E-06	3.8E-04
12	8	1.E-08	3.3E-11	1.9E-04	1.0E+00	1.0E+00	4.4E-06	3.8E-04
13	10	1.E-10	3.3E-15	1.9E-06	1.0E+00	1.0E+00	4.4E-04	3.8E-04
14	12	1.E-12	3.3E-19	1.9E-08	1.0E+00	9.6E-01	4.2E-02	3.8E-04
15	14	1.E-14	3.3E-23	1.9E-10	1.0E+00	1.9E-01	8.1E-01	8.7E-04
16								
17								net
18	pH	$[Ba^{2+}]$	$[BaOH^+]$	$[H_2Ox]$	$[HOx^-]$	$[Ox^{2-}]$	$[OH^-]$	charge
19	0	2.2E-01	9.7E-15	2.1E-01	1.2E-02	6.4E-07	1.0E-14	1.4E+00
20	2	5.6E-03	2.4E-14	8.4E-04	4.7E-03	2.5E-05	1.0E-12	1.6E-02
21	4	6.4E-04	2.8E-13	7.4E-07	4.1E-04	2.2E-04	1.0E-10	5.1E-04
22	6	3.8E-04	1.7E-11	1.2E-10	6.9E-06	3.7E-04	1.0E-08	7.9E-06
23	7.451	3.8E-04	4.6E-10	1.6E-13	2.5E-07	3.8E-04	2.8E-07	0.0E+00
24	8	3.8E-04	1.6E-09	1.2E-14	7.0E-08	3.8E-04	1.0E-06	-9.2E-07
25	10	3.8E-04	1.6E-07	1.2E-18	7.0E-10	3.8E-04	1.0E-04	-1.0E-04
26	12	3.7E-04	1.6E-05	1.3E-22	7.1E-12	3.8E-04	1.0E-02	-1.0E-02
27	14	1.6E-04	7.1E-04	2.9E-26	1.6E-13	8.7E-04	1.0E+00	-1.0E+00
28								
29	B7 = 10^-A7						B19 = F7*H7	
30	C7 = B7^2/(B7^2+B7*$E$3+$E$3*$E$4)						C19 = G7*H7	
31	D7 = B7*$E$3/(B7^2+B7*$E$3+$E$3*$E$4)						D19 = C7*H7	
32	E7 = $E$3*$E$4/(B7^2+B7*$E$3+$E$3*$E$4)						E19 = D7*H7	
33	F7 = B7/(B7+$B$4)						F19 = E7*H7	
34	G7 = $B$4/(B7+$B$4)						G19 = $H$4/B7	
35	H7 = SQRT($B$3/(E7*F7))			H19 = B7+2*B19+C19-G19-E19-2*F19				

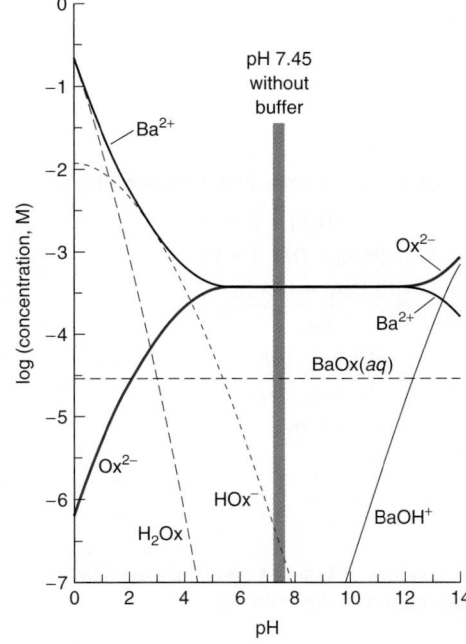

FIGURE 13-9 Spreadsheet for saturated BaC_2O_4. Solver was used to find the pH in cell A11 necessary to make the net charge 0 in cell H23.

FIGURE 13-10 pH dependence of the concentrations of species in saturated BaC_2O_4. As pH is lowered, H^+ reacts with $C_2O_4^{2-}$ to make $HC_2O_4^-$ and $H_2C_2O_4$, and the concentration of Ba^{2+} increases.

We derive the key equation for a diprotic acid, H_2A, and extend it to a general acid, H_nA. The mean fraction of protons bound to H_2A ranges from 0 to 2 and is defined as

$$\bar{n}_H = \frac{\text{moles of bound } H^+}{\text{total moles of weak acid}} = \frac{2[H_2A] + [HA^-]}{[H_2A] + [HA^-] + [A^{2-}]} \qquad (13\text{-}58)$$

We can measure $\bar{n}_H$ by a titration beginning with a mixture of A mmol of H_2A and C mmol of HCl in V_0 mL. We add HCl to increase the degree of protonation of H_2A, which is partially dissociated in the absence of HCl. We titrate the solution with standard NaOH at a concentration of C_b mol/L. After adding v mL of NaOH, the number of mmol of Na^+ in the solution is $C_b v$.

To maintain a nearly constant ionic strength, the solution of H_2A plus HCl contains 0.10 M KCl and the concentrations of H_2A and HCl are much less than 0.10 M. NaOH is sufficiently concentrated so that the added volume is small relative to V_0.

The charge balance for the titration solution is

$$[H^+] + [Na^+] + [K^+] = [OH^-] + [Cl^-]_{HCl} + [Cl^-]_{KCl} + [HA^-] + 2[A^{2-}]$$

where $[Cl^-]_{HCl}$ is from HCl and $[Cl^-]_{KCl}$ is from KCl. But $[K^+] = [Cl^-]_{KCl}$, so we cancel these terms. The net charge balance is

$$[H^+] + [Na^+] = [OH^-] + [Cl^-]_{HCl} + [HA^-] + 2[A^{2-}] \qquad (13\text{-}59)$$

The denominator of Equation 13-58 is $F_{H_2A} = [H_2A] + [HA^-] + [A^{2-}]$. The numerator can be written as $2F_{H_2A} - [HA^-] - 2[A^{2-}]$. Therefore

$$\bar{n}_H = \frac{2F_{H_2A} - [HA^-] - 2[A^{2-}]}{F_{H_2A}} \tag{13-60}$$

From Equation 13-59, we can write: $-[HA^-] - 2[A^{2-}] = [OH^-] + [Cl^-]_{HCl} - [H^+] - [Na^+]$. Substituting this expression into the numerator of Equation 13-60 gives

$$\bar{n}_H = \frac{2F_{H_2A} + [OH^-] + [Cl^-]_{HCl} - [H^+] - [Na^+]}{F_{H_2A}} = 2 + \frac{[OH^-] + [Cl^-]_{HCl} - [H^+] - [Na^+]}{F_{H_2A}}$$

For the general polyprotic acid H_nA, the mean fraction of bound protons turns out to be

$$\bar{n}_H = n + \frac{[OH^-] + [Cl^-]_{HCl} - [H^+] - [Na^+]}{F_{H_nA}} \tag{13-61}$$

Each term on the right side of Equation 13-61 is known during the titration. From the reagents that were mixed, we can say

$$F_{H_2A} = \frac{\text{mmol } H_2A}{\text{total volume}} = \frac{A}{V_0 + v} \qquad [Cl^-]_{HCl} = \frac{\text{mmol HCl}}{\text{total volume}} = \frac{C}{V_0 + v}$$

$$[Na^+] = \frac{\text{mmol NaOH}}{\text{total volume}} = \frac{C_b v}{V_0 + v}$$

$[H^+]$ and $[OH^-]$ are measured with a pH electrode and calculated as follows: Let the effective value of K_w that applies at $\mu = 0.10$ M be $K'_w = K_w/(\gamma_{H^+}\gamma_{OH^-}) = [H^+][OH^-]$ (Equation 13-25). Remembering that pH $= -\log([H^+]\gamma_{H^+})$, we write

$$[H^+] = \frac{10^{-pH}}{\gamma_{H^+}} \qquad [OH^-] = \frac{K'_w}{[H^+]} = 10^{(pH - pK'_w)} \cdot \gamma_{H^+}$$

Substituting into Equation 13-61 gives the measured fraction of bound protons:

Experimental fraction of protons bound to polyprotic acid

$$\bar{n}_H(\text{measured}) = n + \frac{10^{(pH - pK'_w)} \cdot \gamma_{H^+} + C/(V_0 + v) - (10^{-pH})/\gamma_{H^+} - C_b v/(V_0 + v)}{A/(V_0 + v)} \tag{13-62}$$

In acid-base titrations, a **difference plot,** or *Bjerrum plot,* is a graph of the mean fraction of protons bound to an acid versus pH. The mean fraction is $\bar{n}_H$ calculated with Equation 13-62. For complex formation, the difference plot gives the mean number of ligands bound to a metal versus pL ($= -\log[\text{ligand}]$).

Equation 13-62 gives the measured value of $\bar{n}_H$. What is the theoretical value? For a diprotic acid, the theoretical mean fraction of bound protons is

Equation 13-63 comes from Equation 13-58:

$$\bar{n}_H = \frac{2[H_2A] + [HA^-]}{[H_2A] + [HA^-] + [A^{2-}]}$$

$$= \frac{2[H_2A] + [HA^-]}{F_{H_2A}}$$

$$= \frac{2[H_2A]}{F_{H_2A}} + \frac{[HA^-]}{F_{H_2A}}$$

$$= 2\alpha_{H_2A} + \alpha_{HA^-}$$

$$\bar{n}_H(\text{theoretical}) = 2\alpha_{H_2A} + \alpha_{HA^-} \tag{13-63}$$

where α_{H_2A} is the fraction of acid in the form H_2A and α_{HA^-} is the fraction in the form HA^-. You should be able to write expressions for α_{H_2A} and α_{HA^-} in your sleep by now.

$$\alpha_{H_2A} = \frac{[H^+]^2}{[H^+]^2 + [H^+]K_1 + K_1K_2} \qquad \alpha_{HA^-} = \frac{[H^+]K_1}{[H^+]^2 + [H^+]K_1 + K_1K_2} \tag{13-64}$$

We extract K_1 and K_2 from an experimental titration by constructing a difference plot with Equation 13-62. This plot is a graph of $\bar{n}_H(\text{measured})$ versus pH. We then fit the theoretical curve (Equation 13-63) to the experimental curve by the method of least squares to find the values of K_1 and K_2 that minimize the sum of the squares of the residuals:

Best values of K_1 and K_2 minimize the sum of the squares of the residuals.

$$\Sigma(\text{residuals})^2 = \Sigma[\bar{n}_H(\text{measured}) - \bar{n}_H(\text{theoretical})]^2 \tag{13-65}$$

$H_3NCH_2CO_2H$
   Glycine

$pK_1 = 2.35$ at $\mu = 0$
$pK_2 = 9.78$ at $\mu = 0$

Experimental data for a titration of the amino acid glycine are given in Figure 13-11. The initial 40.0-mL solution contained 0.190 mmol of glycine plus 0.232 mmol of HCl to increase the fraction of fully protonated $^+H_3NCH_2CO_2H$. Aliquots of 0.490 5 M NaOH were added and the pH was measured after each addition. Volumes and pH are listed in columns A and B beginning in row 16. pH was precise to the 0.001 decimal place, but the accuracy of pH measurement is, at best, ± 0.02.

	A	B	C	D	E	F	G	H	I
1	Difference plot for glycine								
2				C16 = 10^-B16/$B$8					
3	Titrant NaOH =	0.4905	C_b (M)	D16 = 10^-$B$9/C16					
4	Initial volume =	40	V_0 (mL)	E16 = $B$7+($B$6-$B$3*A16-(C16-D16)*($B$4+A16))/$B$5					
5	Glycine =	0.190	L (mmol)	F16 = $C16^2/($C16^2+$C16*$E$10+$E$10*$E$11)					
6	HCl added =	0.232	A (mmol)	G16 = $C16*$E$10/($C16^2+$C16*$E$10+$E$10*$E$11)					
7	Number of H⁺ =	2	n	H16 = 2*F16+G16					
8	Activity coeff =	0.78	γ_H	I16 = (E16-H16)^2					
9	pK_w' =	13.807							
10	pK_1 =	2.312		K_1 =	0.0048713	= 10^-B10			
11	pK_2 =	9.625		K_2 =	2.371E-10	= 10^-B11			
12	$\Sigma(resid)^2$ =	0.0048	= sum of column I						
13									
14	v	pH	[H⁺] =	[OH⁻] =	Measured			Theoretical	(residuals)² =
15	mL NaOH		$(10^{-pH})/\gamma_H$	$(10^{-pKw})/[H^+]$	n_H	α_{H2A}	α_{HA^-}	n_H	$(n_{meas} - n_{theor})^2$
16	0.00	2.234	7.48E-03	2.08E-12	1.646	0.606	0.394	1.606	0.001656
17	0.02	2.244	7.31E-03	2.13E-12	1.630	0.600	0.400	1.600	0.000879
18	0.04	2.254	7.14E-03	2.18E-12	1.612	0.595	0.405	1.595	0.000319
19	0.06	2.266	6.95E-03	2.24E-12	1.601	0.588	0.412	1.588	0.000174
20	0.08	2.278	6.76E-03	2.30E-12	1.589	0.581	0.419	1.581	0.000056
21	0.10	2.291	6.56E-03	2.38E-12	1.578	0.574	0.426	1.574	0.000020
22	:								
23	0.50	2.675	2.71E-03	5.75E-12	1.353	0.357	0.643	1.357	0.000022
24	:								
25	1.56	11.492	4.13E-12	3.77E-03	0.016	0.000	0.017	0.017	0.000000
26	1.58	11.519	3.88E-12	4.01E-03	0.018	0.000	0.016	0.016	0.000004
27	1.60	11.541	3.69E-12	4.22E-03	0.015	0.000	0.015	0.015	0.000000

FIGURE 13-11 Spreadsheet for difference plot of the titration of 0.190 mmol glycine plus 0.232 mmol HCl in 40.0 mL with 0.490 5 M NaOH. Cells A16:B27 give only a fraction of the experimental data. [Complete data listed in Problem 13-15 were provided by A. Kraft, Heriot-Watt University.]

Input values of concentration, volume, and moles are in cells B3:B6 in Figure 13-11. Cell B7 has the value 2 to indicate that glycine is a diprotic acid. Cell B8 has the activity coefficient of H⁺ computed with the Davies equation, 13-18. Cell B9 begins with the effective value of $pK_w' = 13.797$ in 0.1 M KCl.[10] We allowed the spreadsheet to vary pK_w' for best fit of the experimental data, giving 13.807 in cell B9. Cells B10 and B11 began with estimates of pK_1 and pK_2 for glycine. We used the values 2.35 and 9.78 from Table 10-1, which apply at $\mu = 0$. As explained in the next section, we use Solver to vary pK_1, pK_2, and pK_w' for best fit of the experimental data, giving 2.312 and 9.625 in cells B10 and B11.

The spreadsheet in Figure 13-11 computes [H⁺] and [OH⁻] in columns C and D beginning in row 16. The mean fraction of protonation, $\bar{n}_H$(measured) from Equation 13-62, is in column E. The Bjerrum difference plot in Figure 13-12 shows $\bar{n}_H$(measured) versus pH. Values of α_{H_2A} and α_{HA^-} from Equations 13-64 are computed in columns F and G and $\bar{n}_H$(theoretical) was computed with Equation 13-63 in column H. Column I contains the

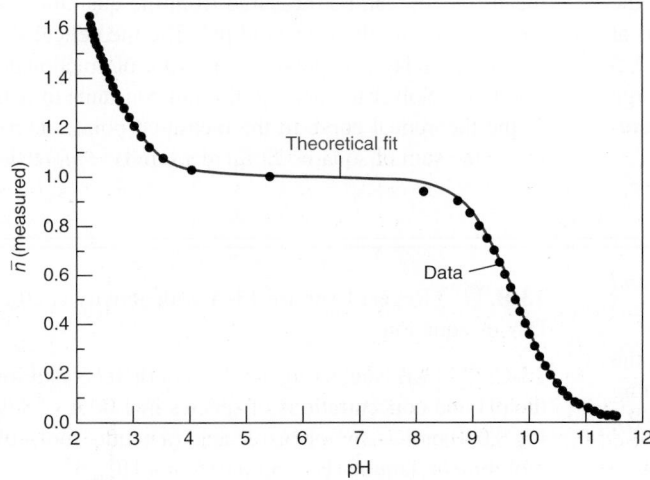

FIGURE 13-12 Bjerrum difference plot for the titration of glycine. Many experimental points are omitted from the figure for clarity.

squares of the residuals, $[\bar{n}_H(\text{measured}) - \bar{n}_H(\text{theoretical})]^2$. The sum of the squares of the residuals is in cell B12.

Using Excel Solver to Optimize More Than One Parameter

We want values of pK'_w, pK_1, and pK_2 that minimize the sum of squares of residuals in cell B12. Select Solver. In the Solver window, Set Target Cell <u>B12</u> Equal To <u>Min</u> By Changing Cells <u>B9, B10, B11</u>. Then click Solve, and Solver finds the best values in cells B9, B10, and B11 to minimize the sum of squares of residuals in cell B12. Starting with 13.797, 2.35, and 9.78 in cells B9, B10, and B11 gives a sum of squares of residuals equal to 0.110 in cell B12. After Solver is executed, cells B9, B10, and B11 become 13.807, 2.312, and 9.625. The sum in cell B12 is reduced to 0.004 8. When you use Solver to optimize several parameters at once, it is a good idea to try different starting values to see if the same solution is reached. Sometimes a local minimum can be reached that is not as low as another that might be reached elsewhere.

The theoretical curve $\bar{n}_H(\text{theoretical}) = 2\alpha_{H_2A} + \alpha_{HA^-}$ is plotted from values in column H in Figure 13-11 and shown by the solid curve in Figure 13-12. The curve fits the experimental data well, a result suggesting that we have found reliable values of pK_1 and pK_2.

It may seem inappropriate to allow pK'_w to vary, because we claimed to know pK'_w at the outset. The change of pK'_w from 13.797 to 13.807 significantly improved the fit. The value 13.797 gave values of $\bar{n}_H(\text{measured})$ that became level near 0.04 at the end of the titration in Figure 13-12. This behavior is qualitatively incorrect, because $\bar{n}_H$ must approach 0 at high pH. A small change in pK'_w markedly improves the fit as $\bar{n}_H$ approaches 0.

Strictly speaking, pK_1 and pK_2 should have primes to indicate that they apply in 0.10 M KCl. We left off the primes to avoid complicating the symbols. We did distinguish K_w, which applies at $\mu = 0$, from K'_w, which applies at $\mu = 0.10$ M.

Terms to Understand

coupled equilibria   difference plot

Summary

Coupled equilibria are reversible reactions that have a species in common. Therefore each reaction has an effect on the other.

The general treatment of acid-base systems begins with charge and mass balances and equilibrium expressions. There should be as many independent equations as chemical species. Substitute a fractional composition equation for each acid or base into the charge balance. After you enter known concentrations of species such as Na^+ and Cl^-, and substitute $K_w/[H^+]$ for $[OH^-]$, the remaining variable should be $[H^+]$. Use Excel Solver to find $[H^+]$ and then solve for all other concentrations from $[H^+]$. If there are equilibria in addition to acid-base reactions, such as ion pairing, then you need the full systematic treatment of equilibrium. Make maximum use of fractional composition equations to simplify the problem.

To use activity coefficients, calculate the effective equilibrium constant K' for each chemical reaction with activities from the Davies equation. K' is the equilibrium quotient of concentrations at a particular ionic strength. Spreadsheets in Figures 13-3 and 13-5 find the concentrations that minimize the charge balance or charge and mass balances and find the ionic strength in an automatic, iterative manner with circular definitions.

We considered solubility problems in which the cation and anion could each undergo one or more acid-base reactions and in which ion pairing could occur. Substitute fractional composition expressions for all acid-base species into the mass balance. In some systems, such as barium oxalate, the resulting equation contains the formal concentrations of anion and cation and $[H^+]$. The solubility product provides a relation between the formal concentrations of anion and cation, so you can eliminate one of them from the mass balance. By assuming a value for $[H^+]$, you can solve for the remaining formal concentration and, therefore, for all concentrations. By this means, you can find the composition as a function of pH. The pH of unbuffered solution is the pH at which the charge balance is satisfied.

To extract acid dissociation constants from an acid-base titration curve, we can construct a difference plot, or Bjerrum plot, which is a graph of the mean fraction of bound protons, $\bar{n}_H$, versus pH. This mean fraction can be measured from the quantities of reagents that were mixed and the measured pH. The theoretical shape of the difference plot is an expression in terms of fractional compositions. Use Excel Solver to vary equilibrium constants to obtain the best fit of the theoretical curve to the measured points. This process minimizes the sum of squares $\Sigma[\bar{n}_H(\text{measured}) - \bar{n}_H(\text{theoretical})]^2$.

Exercises

Instructors: Most of these exercises are quite long. Please be kind when you assign them.

13-A. ▦ Neglecting activity coefficients and ion pairing, find the pH and concentrations of species in 1.00 L of solution containing 0.010 mol hydroxybenzene (HA), 0.030 mol dimethylamine (B), and 0.015 mol HCl.

13-B. ▦ Repeat Exercise 13-A with activity coefficients from the Davies equation.

13-C. ▦ (a) Neglecting activity coefficients and ion pairing, find the pH and concentrations of species in 1.00 L of solution containing 0.040 mol 2-aminobenzoic acid (a neutral molecule, HA), 0.020 mol dimethylamine (B), and 0.015 mol HCl.

(b) What fraction of HA is in each of its three forms? What fraction of B is in each of its two forms? Compare your answers with what you would find if HCl reacted with B and then excess B reacted with HA. What pH do you predict from this simple estimate?

13-D. 🔲 Include activity coefficients from the Davies equation to find the pH and concentrations of species in the mixture of sodium tartrate, pyridinium chloride, and KOH in Section 13-1. Consider only Reactions 13-1 through 13-4.

13-E. 🔲 Consider a saturated solution of AgCN with the following chemistry without activity coefficients:

$$AgCN(s) \rightleftharpoons Ag^+ + CN^- \qquad pK_{sp} = 15.66$$

$$HCN(aq) \rightleftharpoons CN^- + H^+ \qquad pK_{HCN} = 9.21$$

$$Ag^+ + H_2O \rightleftharpoons AgOH(aq) + H^+ \qquad pK_{Ag} = 12.0$$

$$H_2O \rightleftharpoons H^+ + OH^- \qquad pK_w = 14.00$$

If pH is fixed by addition of unspecified reagent (such as a buffer), we cannot write a charge balance for the solution. The mass balance is

moles of dissolved silver = moles of dissolved cyanide

Express the concentration of each species in terms of $[Ag^+]$ and $[H^+]$. Put the expression for each species into the mass balance, which now contains $[Ag^+]$ and $[H^+]$ as the only concentrations. Solve the equation for $[Ag^+]$. Prepare a spreadsheet with the following columns and compute all of the concentrations:

pH	$[H^+]$	$[Ag^+]$	$[CN^-]$	$[HCN]$	$[AgOH]$	$[OH^-]$	Net charge
0	1						
1	0.1						
·							
·							
·							
14	10^{-14}						
No buffer	?						

For unbuffered solution, the pH is that value which gives 0 net charge. Use Solver to find the pH giving 0 net charge. Prepare a graph like Figure 13-6 showing log[concentration] versus pH for each species. Considering the equilibrium $Ag_2O(s) + H_2O \rightleftharpoons 2\,Ag^+ + 2OH^-$ with $pK_{Ag_2O} = 15.42$, will $Ag_2O(s)$ precipitate from the unbuffered solution?

13-F. 🔲 *Difference plot.* A solution containing 3.96 mmol acetic acid plus 0.484 mmol HCl in 200 mL of 0.10 M KCl was titrated with 0.490 5 M NaOH to measure K_a for acetic acid.

(a) Write expressions for the experimental mean fraction of protonation, $\bar{n}_H(measured)$, and the theoretical mean fraction of protonation, $\bar{n}_H(theoretical)$.

(b) From the following data, prepare a graph of $\bar{n}_H(measured)$ versus pH. Find the best values of pK_a and pK_w' by minimizing the sum of the squares of the residuals, $\Sigma[\bar{n}_H(measured) - \bar{n}_H(theoretical)]^2$.

v (mL)	pH	v (mL)	pH	v (mL)	pH	v (mL)	pH
0.00	2.79	2.70	4.25	5.40	4.92	8.10	5.76
0.30	2.89	3.00	4.35	5.70	4.98	8.40	5.97
0.60	3.06	3.30	4.42	6.00	5.05	8.70	6.28
0.90	3.26	3.60	4.50	6.30	5.12	9.00	7.23
1.20	3.48	3.90	4.58	6.60	5.21	9.30	10.14
1.50	3.72	4.20	4.67	6.90	5.29	9.60	10.85
1.80	3.87	4.50	4.72	7.20	5.38	9.90	11.20
2.10	4.01	4.80	4.78	7.50	5.49	10.20	11.39
2.40	4.15	5.10	4.85	7.80	5.61	10.50	11.54

Data from A. Kraft, J. Chem. Ed. **2003**, *80, 554.*

Problems

Instructors: Most of these problems are quite long. Please be kind when you assign them.

13-1. Why does the solubility of a salt of a basic anion increase with decreasing pH? Write chemical reactions for the minerals galena (PbS) and cerussite ($PbCO_3$) to explain how acid rain mobilizes traces of metal from relatively inert forms into the environment, where the metals can be taken up by plants and animals.

13-2. 🔲 **(a)** Considering just acid-base chemistry, not ion pairing and not activity coefficients, use the systematic treatment of equilibrium to find the pH of 1.00 L of solution containing 0.010 0 mol hydroxybenzene (HA) and 0.005 0 mol KOH.

(b) What pH would you have predicted from your knowledge of Chapter 11?

(c) Find the pH if [HA] and [KOH] were both reduced by a factor of 100.

13-3. 🔲 Repeat part **(a)** of Problem 13-2 with Davies activity coefficients. Remember that $pH = -\log([H^+]\gamma_{H^+})$.

13-4. From pK_1 and pK_2 for glycine at $\mu = 0$ in Table 10-1, compute pK_1' and pK_2' that apply at $\mu = 0.1$ M. Use the Davies equation for activity coefficients. Compare your answer with experimental values in cells B10 and B11 of Figure 13-11.

13-5. 🔲 Considering just acid-base chemistry, not ion pairing and not activity coefficients, use the systematic treatment of equilibrium to find the pH and concentrations of species in 1.00 L of solution containing 0.100 mol ethylenediamine and 0.035 mol HBr. Compare the pH to that found by the methods of Chapter 11.

13-6. 🔲 Considering just acid-base chemistry, not ion pairing and not activity coefficients, find the pH and concentrations of species in 1.00 L of solution containing 0.040 mol benzene-1,2,3-tricarboxylic acid (H_3A), 0.030 mol imidazole (a neutral molecule, HB), and 0.035 mol NaOH.

13-7. 🔲 Considering just acid-base chemistry, not ion pairing and not activity coefficients, find the pH and concentrations of species in 1.00 L of solution containing 0.020 mol arginine, 0.030 mol glutamic acid, and 0.005 mol KOH.

13-8. Solve Problem 13-7 by using Davies activity coefficients.

13-9. A solution containing 0.008 695 m KH$_2$PO$_4$ and 0.030 43 m Na$_2$HPO$_4$ is a primary standard buffer with a stated pH of 7.413 at 25°C. Calculate the pH of this solution by using the systematic treatment of equilibrium with activity coefficients from (a) the Davies equation and (b) the extended Debye-Hückel equation.

13-10. Considering just acid-base chemistry, not ion pairing and not activity coefficients, find the pH and composition of 1.00 L of solution containing 0.040 mol H$_4$EDTA (EDTA = ethylenedinitrilotetraacetic acid $\equiv$ H$_4$A), 0.030 mol lysine (neutral molecule $\equiv$ HL), and 0.050 mol NaOH.

13-11. The solution with no added KNO$_3$ for Figure 8-1 contains 5.0 mM Fe(NO$_3$)$_3$, 5.0 μM NaSCN, and 15 mM HNO$_3$. Use Davies activity coefficients to find the concentrations of all species in the following reactions:

$$Fe^{3+} + SCN^- \rightleftharpoons Fe(SCN)^{2+} \quad \log \beta_1 = 3.03 \ (\mu = 0)$$
$$Fe^{3+} + 2SCN^- \rightleftharpoons Fe(SCN)_2^+ \quad \log \beta_2 = 4.6 \ (\mu = 0)$$
$$Fe^{3+} + H_2O \rightleftharpoons FeOH^{2+} + H^+ \quad pK_a = 2.195 \ (\mu = 0)$$

(a) Write the four equilibrium expressions (including K_w). Express the effective equilibrium constants in terms of equilibrium constants and activity coefficients. For example, $K_w' = K_w/\gamma_{H^+}\gamma_{OH^-}$. Write expressions for [Fe(SCN)$^{2+}$], [Fe(SCN)$_2^+$], [FeOH^{2+}], and [OH$^-$] in terms of [Fe^{3+}], [SCN$^-$], and [H$^+$].

(b) Write the charge balance.

(c) Write mass balances for iron, thiocyanate, Na$^+$, and NO$_3^-$.

(d) With 7 unknowns ([Fe^{3+}], [SCN$^-$], [H$^+$], [Fe(SCN)$^{2+}$], [Fe(SCN)$_2^+$], [FeOH^{2+}], and [OH$^-$]) and 4 equilibrium expressions, we would like to use Excel Solver to find $7 - 4 = 3$ unknowns. It is logical to select [Fe^{3+}], [SCN$^-$], and [H$^+$] as the unknowns and to use the equations from (a) in the spreadsheet. Unfortunately, sometimes Solver is not good at finding three unknowns. This problem is one of those cases. We can reduce the problem to finding two unknowns if we can express one of the selected unknowns in terms of the others. Substitute expressions from part (a) into the mass balance for thiocyanate to derive the equation below, which gives [Fe^{3+}] in terms of [SCN$^-$].

$$[Fe^{3+}] = \frac{F_{SCN} - [SCN^-]}{\beta_1'[SCN^-] + 2\beta_2'[SCN^-]^2} \quad \text{(A)}$$

(e) Create a spreadsheet like Figure 13-5 to find all concentrations in the solution composed of 5.0 mM Fe(NO$_3$)$_3$, 5.0 μM NaSCN, and 15 mM HNO$_3$. Let pSCN and pH be the two unknowns that will be found by Solver. Use Davies activity coefficients. As an important check, show that the sum of species containing Fe is 5.000 mM and the sum of species containing thiocyanate (SCN$^-$) is 5.000 μM.

(f) The solution contains 0.015 8 M H$^+$. Nitric acid provides 0.015 0 M H$^+$. Where does the remaining 0.000 8 M H$^+$ come from?

(g) Find the quotient [Fe(SCN)$^{2+}$]/(\{[Fe^{3+}] + [FeOH^{2+}]\}[SCN$^-$]). This is the point for [KNO$_3$] = 0 in Figure 8-1. Compare your answer with Figure 8-1. The ordinate of Figure 8-1 is labeled [Fe(SCN)$^{2+}$]/([Fe^{3+}][SCN$^-$]), but [Fe^{3+}] really refers to the total concentration of iron not bound to thiocyanate.

(h) Find the quotient in part (g) when the solution also contains 0.20 M KNO$_3$. Compare your answer with Figure 8-1.

13-12. (a) Follow the steps of Problem 13-11 to solve this one. From the following equilibria, find the concentrations of species and the pH of 1.0 mM La$_2$(SO$_4$)$_3$. Use Davies activity coefficients.

$$La^{3+} + SO_4^{2-} \rightleftharpoons La(SO_4)^+ \quad \beta_1 = 10^{3.64} \ (\mu = 0)$$
$$La^{3+} + 2SO_4^{2-} \rightleftharpoons La(SO_4)_2^- \quad \beta_2 = 10^{5.3} \ (\mu = 0)$$
$$La^{3+} + H_2O \rightleftharpoons LaOH^{2+} + H^+ \quad K_a = 10^{-8.5} \ (\mu = 0)$$

Choose pSO$_4^{2-}$ and pH as the independent variables and find an expression for [La^{3+}] in terms of [SO$_4^{2-}$] and [H$^+$]. When I called Solver to minimize the sum of squares of the charge and mass balances with Constraint Precision = 1E-15, Solver could not converge to a solution. So I set Constraint Precision to 1E-10 and Solver found a solution. Then I reset Constraint Precision to 1E-15 and continued to alternately refine the values of [SO$_4^{2-}$] and [H$^+$].

(b) If La$_2$(SO$_4$)$_3$ were a strong electrolyte, what would be the ionic strength of 1.0 mM La$_2$(SO$_4$)$_3$? What is the actual ionic strength of this solution?

(c) What fraction of lanthanum is La^{3+}?

(d) Why did we not consider hydrolysis of SO$_4^{2-}$ to give HSO$_4^-$?

(e) Will La(OH)$_3$(s) precipitate in this solution?

13-13. Find the composition of a saturated solution of AgCN containing 0.10 M KCN adjusted to pH 12.00 with NaOH and state which species is the principal form of silver. Consider the following equilibria and use Davies activity coefficients.

$$AgCN(s) \rightleftharpoons Ag^+ + CN^- \quad pK_{sp} = 15.66$$
$$HCN(aq) \rightleftharpoons CN^- + H^+ \quad pK_{HCN} = 9.21$$
$$Ag^+ + H_2O \rightleftharpoons AgOH(aq) + H^+ \quad pK_a = 12.0$$
$$Ag^+ + CN^- + OH^- \rightleftharpoons Ag(OH)(CN)^- \quad pK_{Ag} = -13.22$$
$$Ag^+ + 2CN^- \rightleftharpoons Ag(CN)_2^- \quad p\beta_2 = -20.48$$
$$Ag^+ + 3CN^- \rightleftharpoons Ag(CN)_3^{2-} \quad p\beta_3 = -21.7$$

Suggested procedure: We know that [K$^+$] = 0.1 M and [H$^+$] = $(10^{-pH})/\gamma_{H^+}$. Let pCN and pNa be the master variables to adjust with Solver to minimize the mass and charge balances. We can find [Ag$^+$] = K_{sp}'/[CN$^-$]. Use equilibrium expressions to find [OH$^-$], [HCN], [AgOH], [Ag(OH)(CN)$^-$], [Ag(CN)$_2^-$], and [Ag(CN)$_3^{2-}$]. The mass balance is \{total silver\} + [K$^+$] = \{total cyanide\}.

13-14. Consider the reactions of Fe^{2+} with the amino acid glycine:

$$Fe^{2+} + G^- \rightleftharpoons FeG^+ \quad p\beta_1 = -4.31$$
$$Fe^{2+} + 2G^- \rightleftharpoons FeG_2(aq) \quad p\beta_2 = -7.65$$
$$Fe^{2+} + 3G^- \rightleftharpoons FeG_3^- \quad p\beta_3 = -8.87$$
$$Fe^{2+} + H_2O \rightleftharpoons FeOH^+ + H^+ \quad pK_a = 9.4$$
$$^+H_3NCH_2CO_2H \text{ glycine, } H_2G^+ \quad pK_1 = 2.350, \ pK_2 = 9.778$$

Suppose that 0.050 mol of FeG$_2$ is dissolved in 1.00 L and enough HCl is added to adjust the pH to 8.50. Use Davies activity coefficients to find the composition of the solution. What fraction of iron is in each of its forms, and what fraction of glycine is in each of its forms? From the distribution of species, explain the principal chemistry that requires addition of HCl to obtain a pH of 8.50.

Suggested procedure: There are six equilibrium constants listed above, plus the K_w equilibrium, for a total of seven equilibria. There are 11 unknown concentrations including [Cl$^-$] from HCl added to fix the pH. You can write mass balances for iron and for glycine and you can write a charge balance. With 11 unknowns and seven equilibria, you would need to solve for four unknowns. However, [H$^+$]

is known from the fixed pH. So you will need to solve for three unknown concentrations with Solver. Use equilibrium constants to express all concentrations in terms of $[Fe^{2+}]$, $[G^-]$, and $[H^+]$, of which $[H^+]$ is known. Let $[Fe^{2+}]$, $[G^-]$, and $[Cl^-]$ be the independent variables. Use Solver to find the values of pFe, pG, and pCl that satisfy the combined charge and mass balances.

13-15. Data for the glycine difference plot in Figure 13-12 are given below.

v (mL)	pH	v (mL)	pH	v (mL)	pH	v (mL)	pH
0.00	2.234	0.40	2.550	0.80	3.528	1.20	10.383
0.02	2.244	0.42	2.572	0.82	3.713	1.22	10.488
0.04	2.254	0.44	2.596	0.84	4.026	1.24	10.595
0.06	2.266	0.46	2.620	0.86	5.408	1.26	10.697
0.08	2.278	0.48	2.646	0.88	8.149	1.28	10.795
0.10	2.291	0.50	2.675	0.90	8.727	1.30	10.884
0.12	2.304	0.52	2.702	0.92	8.955	1.32	10.966
0.14	2.318	0.54	2.736	0.94	9.117	1.34	11.037
0.16	2.333	0.56	2.768	0.96	9.250	1.36	11.101
0.18	2.348	0.58	2.802	0.98	9.365	1.38	11.158
0.20	2.363	0.60	2.838	1.00	9.467	1.40	11.209
0.22	2.380	0.62	2.877	1.02	9.565	1.42	11.255
0.24	2.397	0.64	2.920	1.04	9.660	1.44	11.296
0.26	2.413	0.66	2.966	1.06	9.745	1.46	11.335
0.28	2.429	0.68	3.017	1.08	9.830	1.48	11.371
0.30	2.448	0.70	3.073	1.10	9.913	1.50	11.405
0.32	2.467	0.72	3.136	1.12	10.000	1.52	11.436
0.34	2.487	0.74	3.207	1.14	10.090	1.54	11.466
0.36	2.506	0.76	3.291	1.16	10.183	1.56	11.492
0.38	2.528	0.78	3.396	1.18	10.280	1.58	11.519
						1.60	11.541

Data from A. Kraft, J. Chem. Ed. **2003**, 80, 554.

(a) Reproduce the spreadsheet in Figure 13-11 and show that you get the same values of pK_1 and pK_2 in cells B10 and B11 after executing Solver. Start with different values of pK_1 and pK_2 and see if Solver finds the same solutions.

(b) Use Solver to find the best values of pK_1 and pK_2 while fixing pK'_w at its expected value of 13.797. Describe how $\bar{n}_H$(measured) behaves when pK'_w is fixed.

13-16. *Difference plot.* A solution containing 0.139 mmol of the triprotic acid tris(2-aminoethyl)amine · 3HCl plus 0.115 mmol HCl in 40 mL of 0.10 M KCl was titrated with 0.490 5 M NaOH to measure acid dissociation constants.

$$N(CH_2CH_2NH_3^+)_3 \cdot 3Cl^-$$
Tris(2-aminoethyl)amine · 3HCl
$$H_3A^{3+} \cdot 3Cl^-$$

(a) Write expressions for the experimental mean fraction of protonation, $\bar{n}_H$(measured), and the theoretical mean fraction of protonation, $\bar{n}_H$(theoretical).

(b) From the following data, prepare a graph of $\bar{n}_H$(measured) versus pH. Find the best values of pK_1, pK_2, pK_3, and pK'_w by minimizing the sum of the squares of the residuals, $\Sigma[\bar{n}_H$(measured) $- \bar{n}_H$(theoretical)$]^2$.

v (mL)	pH	v (mL)	pH	v (mL)	pH	v (mL)	pH
0.00	2.709	0.36	8.283	0.72	9.687	1.08	10.826
0.02	2.743	0.38	8.393	0.74	9.748	1.10	10.892
0.04	2.781	0.40	8.497	0.76	9.806	1.12	10.955
0.06	2.826	0.42	8.592	0.78	9.864	1.14	11.019
0.08	2.877	0.44	8.681	0.80	9.926	1.16	11.075
0.10	2.937	0.46	8.768	0.82	9.984	1.18	11.128
0.12	3.007	0.48	8.851	0.84	10.042	1.20	11.179
0.14	3.097	0.50	8.932	0.86	10.106	1.22	11.224
0.16	3.211	0.52	9.011	0.88	10.167	1.24	11.268
0.18	3.366	0.54	9.087	0.90	10.230	1.26	11.306
0.20	3.608	0.56	9.158	0.92	10.293	1.28	11.344
0.22	4.146	0.58	9.231	0.94	10.358	1.30	11.378
0.24	5.807	0.60	9.299	0.96	10.414	1.32	11.410
0.26	6.953	0.62	9.367	0.98	10.476	1.34	11.439
0.28	7.523	0.64	9.436	1.00	10.545	1.36	11.468
0.30	7.809	0.66	9.502	1.02	10.615	1.38	11.496
0.32	8.003	0.68	9.564	1.04	10.686	1.40	11.521
0.34	8.158	0.70	9.626	1.06	10.756		

Data from A. Kraft, J. Chem. Ed. **2003**, 80, 554.

(c) Create a fractional composition graph showing the fractions of H_3A^{3+}, H_2A^{2+}, HA^+, and A as a function of pH.

13-17. *Ion-pairing in acid-base systems.* This problem incorporates ion-pair equilibria 13-12 and 13-13 into the acid-base chemistry of Section 13-1.

(a) From mass balance 13-15, derive Equation 13-16.

(b) Substitute equilibrium expressions into mass balance 13-17 to derive an expression for $[T^{2-}]$ in terms of $[H^+]$, $[Na^+]$, and various equilibrium constants.

(c) With the same approach used in part **(b)**, derive expressions for $[HT^-]$ and $[H_2T]$.

(d) Add the species $[NaT^-]$ and $[NaHT]$ to the spreadsheet in Figure 13-1 and compute the composition and pH of the solution. Compute $[Na^+]$ with Equation 13-16. Compute $[H_2T]$, $[HT^-]$, and $[T^{2-}]$ from the expressions derived in parts **(b)** and **(c)**. Excel will indicate a *circular reference* problem because, for example, the formula for $[Na^+]$ depends on $[T^{2-}]$ and the formula for $[T^{2-}]$ depends on $[Na^+]$. Enable Excel to handle circular references (page 292). Then use Solver to find the pH in cell H13 that reduces the net charge in cell E15 to (near) zero.

14 Fundamentals of Electrochemistry

LITHIUM-ION BATTERY

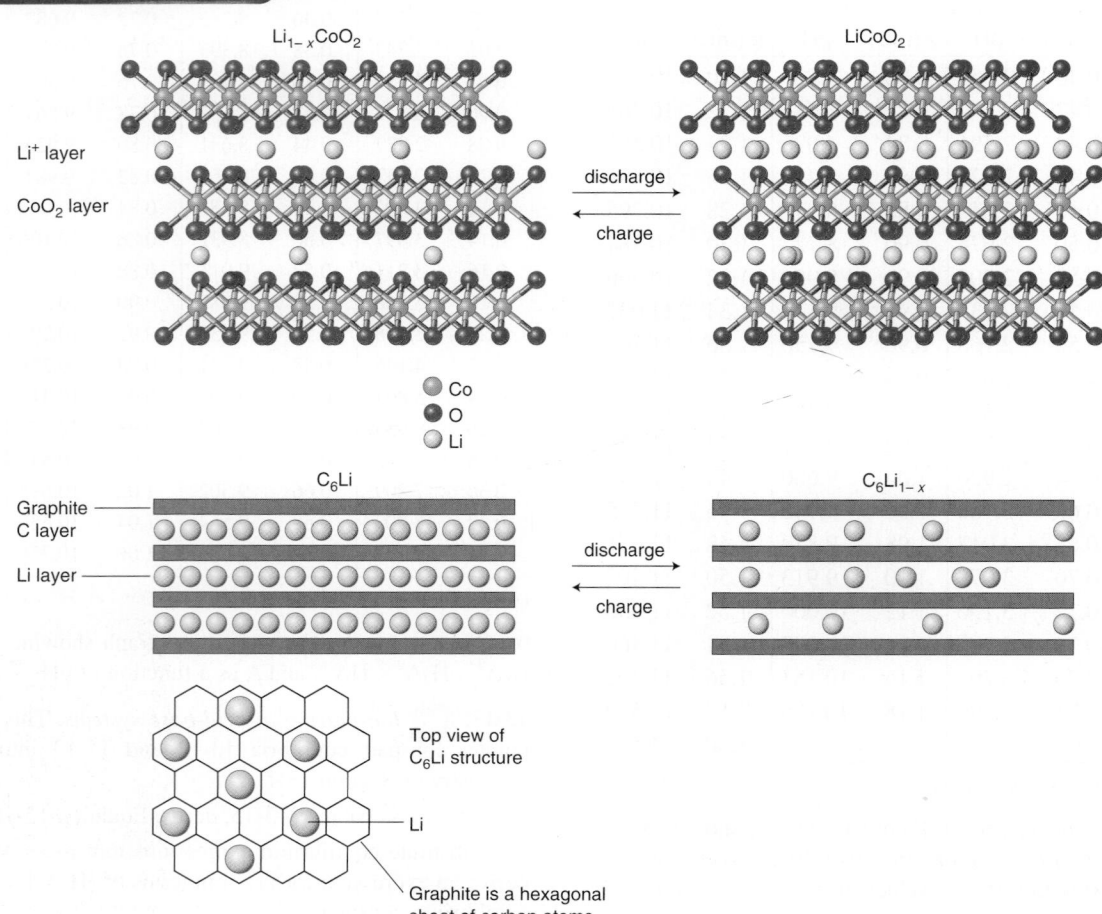

$Li_{1-x}CoO_2$

$LiCoO_2$

Li⁺ layer

CoO_2 layer

discharge →
← charge

Co
O
Li

C_6Li

C_6Li_{1-x}

Graphite
C layer

Li layer

discharge →
← charge

Top view of C_6Li structure

Li

Graphite is a hexagonal sheet of carbon atoms

High-capacity, rechargeable lithium-ion batteries, such as those in cell phones and laptop computers, are a shining example of the fruits of materials chemistry research. The approximate chemistry is

$$C_6Li + 2Li_{0.5}CoO_2 \underset{\text{Charge}}{\overset{\text{Discharge}}{\rightleftharpoons}} C_6 + 2LiCoO_2$$

In C_6Li, lithium atoms reside between layers of carbon in graphite. Atoms or molecules located between layers of a structure are said to be *intercalated*. During operation of the battery, lithium ions migrate from graphite to cobalt oxide. Lithium atoms leave electrons behind in the graphite, and the resulting Li^+ ions become intercalated between CoO_2 layers. To go from graphite to cobalt oxide, Li^+ passes through an electrolyte consisting of a lithium salt dissolved in a high-boiling organic solvent. A porous polymer separator between graphite and cobalt oxide is an electrical insulator that permits Li^+ ions to pass. Electrons travel from graphite through the external circuit to reach the cobalt oxide and maintain electroneutrality. During recharging, Li^+ goes from $LiCoO_2$ to graphite under the influence of an externally applied electric field.

Batteries are *galvanic cells*, which are the subject of this chapter. A galvanic cell uses a favorable chemical reaction to produce electricity. A single-cell lithium-ion battery produces ~3.7 volts. These batteries store twice as much energy per unit mass as the nickel-metal hydride batteries they replaced. Ongoing research is aimed at improved materials and high-area microstructures for the electrodes and the separator layer. Goals include higher energy density, longer life, and safer operation. In 2013, two fires from lithium batteries grounded the fleet of Boeing 787 jet aircraft for four months. Though these were rare incidents, safe operation on an airplane is imperative.

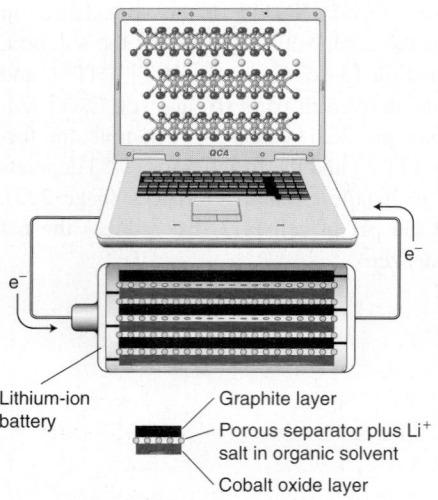

e⁻

e⁻

Lithium-ion battery

Graphite layer

Porous separator plus Li⁺ salt in organic solvent

Cobalt oxide layer

Direction of electron flow from lithium battery to laptop computer.

Electrochemistry is a major branch of analytical chemistry that uses electrical measurements of chemical systems for analytical purposes.[1] For example, the opening of Chapter 1 showed an electrode being used to measure neurotransmitter molecules released by a nerve cell. Electrochemistry also refers to the use of electricity to drive a chemical reaction or to the use of a chemical reaction to produce electricity.

14-1 Basic Concepts

A **redox reaction** involves transfer of electrons from one species to another. A species is said to be **oxidized** when it *loses electrons*. It is **reduced** when it *gains electrons*. An **oxidizing agent,** also called an **oxidant,** takes electrons from another substance and becomes reduced. A **reducing agent,** also called a **reductant,** gives electrons to another substance and is oxidized in the process. In the reaction

$$\underset{\substack{\text{Oxidizing} \\ \text{agent}}}{Fe^{3+}} + \underset{\substack{\text{Reducing} \\ \text{agent}}}{V^{2+}} \rightarrow Fe^{2+} + V^{3+} \tag{14-1}$$

Fe^{3+} is the oxidizing agent because it takes an electron from V^{2+}. V^{2+} is the reducing agent because it gives an electron to Fe^{3+}. Fe^{3+} is reduced, and V^{2+} is oxidized as the reaction proceeds from left to right. Appendix D reviews oxidation numbers and balancing of redox equations.

Oxidation: loss of electrons
Reduction: gain of electrons

Oxidizing agent: takes electrons
Reducing agent: gives electrons

$Fe^{3+} + e^- \rightarrow Fe^{2+}$
$V^{2+} \rightarrow V^{3+} + e^-$

Chemistry and Electricity

When electrons from a redox reaction flow through an electric circuit, we can learn something about the reaction by measuring current and voltage. Electric current is proportional to the rate of reaction, and the cell voltage is proportional to the free energy change for the electrochemical reaction.

Electric Charge

Electric charge (q) is measured in **coulombs** (C). The magnitude of the charge of a single electron or proton is 1.602×10^{-19} C, so a mole of electrons or protons has a charge of $(1.602 \times 10^{-19} \text{ C})(6.022 \times 10^{23} \text{ mol}^{-1}) = 9.649 \times 10^4$ C, which is called the **Faraday constant,** F. For N moles of a species with n charges per molecule, the moles of charge are nN. For example, for Fe^{3+}, $n = 3$ because each ion carries three units of charge. The electric charge in coulombs is

Relation between charge and moles:

$$\underset{\text{Coulombs}}{q} = \underset{\substack{\text{Unit charges} \\ \text{per molecule}}}{n} \cdot \underset{\text{Moles}}{N} \cdot \underset{\frac{\text{Coulombs}}{\text{mole } e^-}}{F} \tag{14-2}$$

The units work because the number of unit charges per molecule, n, is dimensionless. The charge on one mole of Fe^{3+} is $q = nNF = (3)(1 \text{ mol})(9.649 \times 10^4 \text{ C/mol}) = 2.89 \times 10^5$ C.

Michael Faraday (1791–1867) was a self-educated English "natural philosopher" (the old term for "scientist") who discovered that the extent of an electrochemical reaction is proportional to the electric charge passing through a cell. Faraday discovered many fundamental laws of electromagnetism. He gave us the electric motor, electric generator, and electric transformer, as well as the terms *ion, cation, anion, electrode, cathode, anode,* and *electrolyte*. His gift for lecturing is best remembered from his Christmas lecture demonstrations for children at the Royal Institution. Faraday "took great delight in talking to [children], and easily won their confidence.... They felt as if he belonged to them; and indeed he sometimes, in his joyous enthusiasm, appeared like an inspired child."[2]

[Science Source.]

> **EXAMPLE** Relating Coulombs to Quantity of Reaction
>
> If 5.585 g of Fe^{3+} were reduced in Reaction 14-1, how many coulombs of charge must have been transferred from V^{2+} to Fe^{3+}?
>
> **Solution** The moles of iron reduced are $(5.585 \text{ g})/(55.845 \text{ g/mol}) = 0.100\ 0$ mol Fe^{3+}. Each Fe^{3+} ion requires $n = 1$ electron in Reaction 14-1. Using the Faraday constant, we find that $0.100\ 0$ mol of electrons corresponds to
>
> $$q = nNF = (1)(0.100\ 0 \text{ mol } e^-)\left(9.649 \times 10^4 \frac{C}{\text{mol } e^-}\right) = 9.649 \times 10^3 \text{ C}$$
>
> **TEST YOURSELF** How many moles of Sn^{4+} are reduced to Sn^{2+} by 1.00 C of electric charge? (*Answer:* 5.18 μmol)

Electric Current

Electric **current** is the quantity of charge flowing each second through a circuit. The unit of current is the **ampere,** abbreviated A. A current of 1 ampere represents a charge of 1 coulomb per second flowing past a point in a circuit.

$1 \text{ A} = 1 \text{ C/s}$

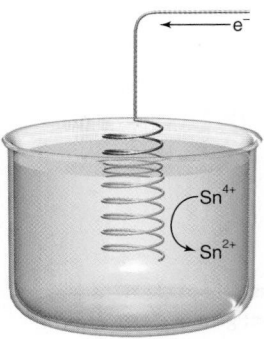

FIGURE 14-1 Electrons flowing into a coil of Pt wire at which Sn^{4+} ions in solution are reduced to Sn^{2+}. This process could not happen by itself, because there is no complete circuit. If Sn^{4+} is to be reduced at this Pt electrode, some other species must be oxidized at some other place.

It takes work to move like charges toward one another. Work can be done when opposite charges move toward one another.

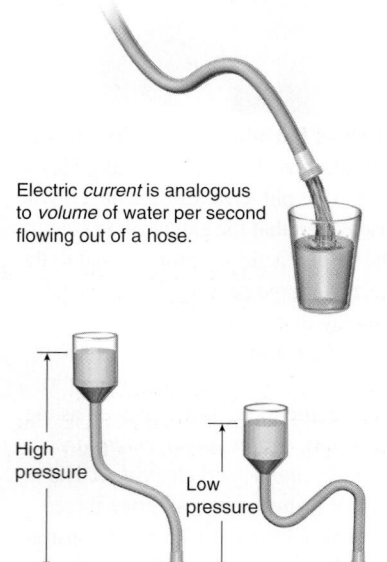

Electric *current* is analogous to *volume* of water per second flowing out of a hose.

Electric *potential difference* is analogous to the hydrostatic *pressure* pushing water through a hose. High pressure gives high flow.

FIGURE 14-2 Analogy between the flow of water through a hose and the flow of electricity through a wire.

$1 V = 1 J/C$

Relating Current to Rate of Reaction

Suppose that electrons are forced into a platinum wire immersed in a solution containing Sn^{4+} (Figure 14-1), which is reduced to Sn^{2+} at a constant rate of 4.24 mmol/h. How much current passes through the solution?

Solution *Two* electrons are required to reduce *one* Sn^{4+} ion:

$$Sn^{4+} + 2e^- \rightarrow Sn^{2+}$$

Electrons flow at a rate of $(2 \text{ mmol } e^-/\text{mmol } Sn^{4+})(4.24 \text{ mmol } Sn^{4+}/h) = 8.48 \text{ mmol } e^-/h$, which corresponds to

$$\frac{8.48 \text{ mmol } e^-/h}{3\,600 \text{ s/h}} = 2.356 \times 10^{-3} \frac{\text{mmol } e^-}{s} = 2.356 \times 10^{-6} \frac{\text{mol } e^-}{s}$$

To find the current, we convert moles of electrons per second into coulombs per second:

$$\text{Current} = \frac{\text{charge}}{\text{time}} = \frac{\text{coulombs}}{\text{second}} = \frac{\text{moles } e^-}{\text{second}} \cdot \frac{\text{coulombs}}{\text{mole}}$$

$$= \left(2.356 \times 10^{-6} \frac{\text{mol}}{s}\right)\left(9.649 \times 10^4 \frac{C}{\text{mol}}\right)$$

$$= 0.227 \text{ C/s} = 0.227 \text{ A} = 227 \text{ mA}$$

TEST YOURSELF What current reduces Sn^{4+} at a rate of 1.00 mmol/h? (*Answer:* 53.6 mA)

In Figure 14-1, we encountered a Pt **electrode,** which conducts electrons into or out of a chemical species in the redox reaction. Platinum is a common *inert* electrode. It does not participate in the redox chemistry except as a conductor of electrons.

Voltage, Work, and Free Energy

Positive and negative charges attract each other. Positive charges repel other positive charges; negative charges repel other negative charges. The presence of electric charge creates an **electric potential** that attracts or repels charged particles. The electric *potential difference*, E, between two points is the work per unit charge that is needed or can be done when charge moves from one point to the other. *Potential difference* is measured in **volts** (V). The greater the potential difference between two points, the more work is required or can be done when a charged particle travels between those points.

A good analogy for understanding current and potential is to think of water flowing through a garden hose (Figure 14-2). Current is the electric charge flowing past a point in a wire each second. Current is analogous to the volume of water flowing past a point in the hose each second. The potential difference is analogous to the pressure on the water in the hose. The greater the pressure, the faster the water flows.

When a charge, q, moves through a potential difference, E, the work done is

Relation between work and voltage:

$$\text{Work} = E \cdot q \qquad (14\text{-}3)$$

Joules   Volts   Coulombs

Work has the dimensions of energy, whose units are **joules** (J). One *joule* of energy is gained or lost when 1 *coulomb* of charge moves between points whose potentials differ by 1 *volt*. Equation 14-3 tells us that the dimensions of volts are joules per coulomb.

Electrical Work

How much work can be done if 2.4 mmol of electrons fall through a potential difference of 0.27 V?

Solution To use Equation 14-3, we must convert moles of electrons into coulombs of charge. Each electron has one unit of charge ($n = 1$), so

$$q = nNF = (1)(2.4 \times 10^{-3} \text{ mol})(9.649 \times 10^4 \text{ C/mol}) = 2.3 \times 10^2 \text{ C}$$

The work that could be done is

$$\text{Work} = E \cdot q = (0.27\ \text{V})(2.3 \times 10^2\ \text{C}) = 62\ \text{J}$$

TEST YOURSELF What must be the potential drop (V) for 1.00 μmol e⁻ to do 1.00 J of work? (*Answer:* 10.4 V)

In the garden hose analogy, suppose that one end of a hose is raised 1 m above the other end and a volume of 1 L of water flows through the hose. The flowing water goes through a mechanical device to do a certain amount of work. If one end of the hose is raised 2 m above the other, the amount of work that can be done by the falling water is twice as great. The elevation difference between the ends of the hose is analogous to electric potential difference and the volume of water is analogous to electric charge. The greater the electric potential difference between two points in a circuit, the more work can be done by the charge flowing between those two points.

The free energy change, ΔG, for a chemical reaction conducted reversibly at constant temperature and pressure equals the maximum possible electrical work that can be done by the reaction on its surroundings:

Section 6-2 gave a brief discussion of ΔG.

$$\text{Work done on surroundings} = -\Delta G \qquad (14\text{-}4)$$

The negative sign in Equation 14-4 indicates that the free energy of a system decreases when the work is done on the surroundings.

Combining Equations 14-2, 14-3, and 14-4 produces a relation of utmost importance:

$$\Delta G = -\text{work} = -E \cdot q$$

Relation between free energy difference and electric potential difference:

$$\underset{\substack{\text{Joules} \\ \text{(J)}}}{\Delta G} \ = \ \underset{\substack{\text{Unit charges} \\ \text{per molecule}}}{-n} \ \cdot \ \underset{\text{Moles}}{N} \ \cdot \ \underset{\text{C/mol}}{F} \ \cdot \ \underset{\substack{\text{Volts} \\ \text{(V)}}}{E} \qquad (14\text{-}5)$$

$q = nNF$

Equation 14-5 relates the free energy change of a chemical reaction to the electric potential difference (that is, the voltage) that can be generated by the reaction. Recall that n is the dimensionless number of charges per molecule and N is the number of moles, so nN is the moles of charge transferred in the reaction.

Ohm's Law

Ohm's law states that current, I, is directly proportional to the potential difference (voltage) across a circuit and inversely proportional to the **resistance, R,** of the circuit.

Ohm's law:

$$I = \frac{E}{R} \qquad (14\text{-}6)$$

The greater the voltage, the more current will flow. The greater the resistance, the less current will flow.

Units of resistance are **ohms,** assigned the Greek symbol Ω (omega). A current of 1 ampere flows through a circuit with a potential difference of 1 volt if the resistance is 1 ohm. From Equation 14-6, the unit ampere (A) is equivalent to V/Ω.

Box 14-1 shows measurements of the resistance of single molecules by measuring current and voltage and applying Ohm's law.

Power

Power, P, is the work done per unit time. The SI unit of power is J/s, better known as the **watt** (W).

$$P = \frac{\text{work}}{\text{s}} = \frac{E \cdot q}{\text{s}} = E \cdot \frac{q}{\text{s}} \qquad (14\text{-}7)$$

Because q/s is the current, I, we can write

$$P = E \cdot I \qquad (14\text{-}8)$$

power (watts) = work per second
$P = E \cdot I = (IR) \cdot I = I^2 R$

A cell capable of delivering 1 ampere at a potential difference of 1 volt has a power output of 1 watt.

BOX 14-1 **Ohm's Law, Conductance, and Molecular Wire³**

The electrical conductance of a single molecule suspended between two gold electrodes is known from measurement of voltage and current by applying Ohm's law. Conductance is 1/resistance, so it has the units 1/ohm ≡ siemens (S).

To make molecular junctions, the sharp gold tip of a scanning tunneling microscope was moved in and out of contact with a flat gold substrate in the presence of a solution containing a test molecule terminated by thiol (—SH) groups. Thiols spontaneously bind to gold, forming bridges such as that shown here. Nanoampere currents were observed with a potential difference of 0.1 V between the gold surfaces.

The graph shows four observations of conductance as the scanning tunneling microscope tip was pulled away from the Au substrate. Plateaus are observed at multiples of 19 nS. An interpretation is that a single molecule connecting two Au surfaces has a conductance of 19 nS (or a resistance of 50 MΩ). If two molecules form parallel bridges, conductance increases to 38 nS. Three molecules give a conductance of 57 nS. If there are three bridges and the electrodes are pulled apart, one of the bridges breaks and conductance drops to 38 nS. When the second bridge breaks, conductance drops to 19 nS. The exact conductance varies because the environment of each molecule on the Au surface is not identical. A histogram of >500 observations in the inset shows peaks at 19, 38, and 57 nS.

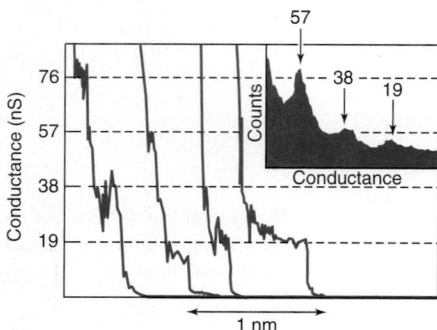

Change in conductance as Au scanning tunneling microscope tip immersed in dithiol solution is withdrawn from Au substrate. [Data from X. Xiao, B. Xu, and N. Tao, "Conductance Titration of Single-Peptide Molecules," *J. Am. Chem. Soc.* **2004,** *126,* 5370.]

Alkane hydrocarbons can be thought of as prototypical electrical insulators. The conductance of alkane dithiols decreases exponentially as chain length increases:⁴

$HS(CH_2)_8SH$ conductance = 16.1 nS

$HS(CH_2)_{10}SH$ conductance = 1.37 nS

$HS(CH_2)_{12}SH$ conductance = 0.35 nS

The conductance of conjugated aromatic bipyridines shown below is orders of magnitude greater than that of saturated hydrocarbons of a similar length. In fact, the conductance of 2.9 nS for the compound with six repeating units and length of 11 nm is almost three orders of magnitude higher than that reported for aromatic molecular wires of comparable length containing only carbon atoms.

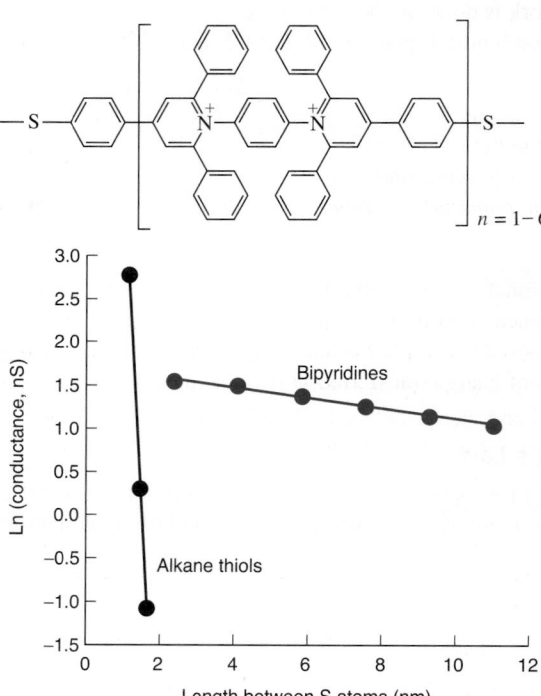

Dependence of conductivity on chain length from sulfur to sulfur. [Data for bipyridines from V. Kolivoška et al., "Single-Molecule Conductance in a Series of Extended Viologen Molecules," *J. Phys. Chem. Lett.,* **2013,** *4,* 589.]

EXAMPLE **Using Ohm's Law**

In the circuit in Figure 14-3, the battery generates a potential difference of 3.0 V, and the resistor has a resistance of 100 Ω. The wire connecting the battery and the resistor has negligible resistance. Find the current and power delivered by the battery.

Solution The current is

$$I = \frac{E}{R} = \frac{3.0 \text{ V}}{100 \text{ }\Omega} = 0.030 \text{ A} = 30 \text{ mA}$$

The power produced by the battery is

$$P = E \cdot I = (3.0 \text{ V})(0.030 \text{ A}) = 90 \text{ mW}$$

TEST YOURSELF What voltage is required to produce 180 mW of power? (**Answer:** 4.24 V)

What happens to the power generated by the circuit? *The energy appears as heat in the resistor.* The power (90 mW) equals the rate at which heat is produced in the resistor.

Here is a summary of symbols, units, and relations from the last few pages:

Relation between charge and moles:

$$q = n \cdot N \cdot F$$

Charge   Unit charges   Moles   C/mol
(coulombs, C)   per molecule   (Faraday constant)

Relation between work and voltage:

$$\text{Work} = E \cdot q \qquad \text{(Units: J/C = V)}$$

Joules   Volts   Coulombs
(J)   (V)   (C)

Relation between free energy difference and electric potential difference:

$$\Delta G = -n \cdot N \cdot F \cdot E$$

Joules   Unit charges   Moles   C/mol   Volts
(J)   per molecule   (V)

Ohm's law:

$$I = E / R$$

Current   Volts   Resistance
(A)   (V)   (ohms, Ω)

Electric power:

$$P = \frac{\text{work}}{\text{s}} = E \cdot I$$

Power   J/s   Volts   Amperes
(watts, W)

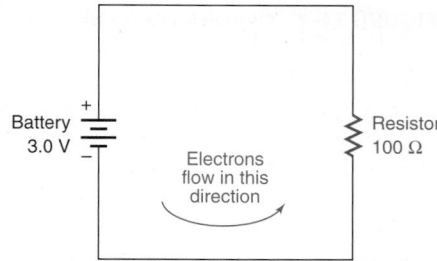

FIGURE 14-3 A circuit with a battery and a resistor. Benjamin Franklin investigated static electricity in the 1740s.[5] He thought electricity was a fluid that flows from a silk cloth to a glass rod when the rod is rubbed with the cloth. We now know that electrons flow from glass to silk. However, Franklin's convention for the direction of electric current has been retained, so we say that current flows from positive to negative—in the opposite direction of electron flow.

14-2 Galvanic Cells

A **galvanic cell** (also called a *voltaic cell*) uses a *spontaneous* chemical reaction to generate electricity. To accomplish this, one reagent must be oxidized and another must be reduced. The two cannot be in contact, or electrons would flow directly from the reducing agent to the oxidizing agent. Instead, oxidizing and reducing agents are physically separated, and electrons are forced to flow through a wire to go from one reactant to the other.

A Cell in Action

Figure 14-4 shows a galvanic cell with two electrodes suspended in a solution of $CdCl_2$. One electrode is cadmium; the other is metallic silver coated with solid AgCl. The reactions are

Reduction: $2AgCl(s) + 2e^- \rightleftharpoons 2Ag(s) + 2Cl^-(aq)$

Oxidation: $Cd(s) \rightleftharpoons Cd^{2+}(aq) + 2e^-$

Net reaction: $Cd(s) + 2AgCl(s) \rightleftharpoons Cd^{2+}(aq) + 2Ag(s) + 2Cl^-(aq)$ **(14-9)**

The net reaction is composed of a reduction and an oxidation, each of which is called a **half-reaction.** The two half-reactions are written with equal numbers of electrons so that their sum includes no free electrons.

The **potentiometer** in the circuit measures the difference in electric potential (voltage) between the two metal electrodes. The measured voltage is the difference $E_{measured} = E_+ - E_-$, where E_+ is the potential of the electrode attached to the positive terminal of the potentiometer and E_- is the potential of the electrode attached to the negative terminal. If electrons flow into the negative terminal, as in this illustration, the voltage is positive. The potentiometer has high electrical resistance so that little current flows through the meter. Ideally, no current would flow through the meter and we would say that the measured potential difference is the *open-circuit potential*, which is the hypothetical potential difference that would be observed if the electrodes were not connected to each other.

Oxidation of Cd metal to $Cd^{2+}(aq)$ provides electrons that flow through the circuit to the Ag electrode in Figure 14-4. At the Ag surface, Ag^+ (from AgCl) is reduced to $Ag(s)$.

The battery invented by Alessandro Volta (1745–1827) in 1799 consisted of layers of Zn and Ag separated by cardboard soaked in brine. The "voltaic pile" on display at the Royal Institution in London was given by Volta to Humphry Davy and Michael Faraday when they visited Italy in 1814. Using electrolysis, Davy was the first to isolate Na, K, Mg, Ca, Sr, and Ba. Faraday used piles to discover laws of electricity and magnetism. [Courtesy Daniel Harris.]

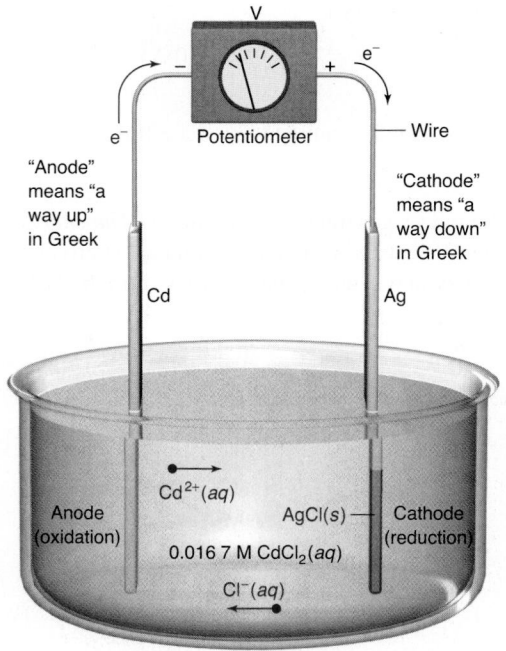

FIGURE 14-4 A simple galvanic cell.

Recall that ΔG is *negative* for a spontaneous reaction.

Chloride from AgCl goes into solution. The free energy change for the net reaction, −150 kJ per mole of Cd, provides the driving force that pushes electrons through the circuit.

EXAMPLE **Voltage Produced by a Chemical Reaction**

Calculate the voltage that would be measured by the potentiometer in Figure 14-4.

Solution Because $\Delta G = -150$ kJ per mol of Cd, we can use Equation 14-5 (where n is moles of electrons transferred in the balanced net reaction) to write

$$E = -\frac{\Delta G}{nNF} = -\frac{-150 \times 10^3 \text{ J}}{\left(2\frac{\text{electrons}}{\text{atom}}\right)(1 \text{ mol Cd})\left(9.649 \times 10^4\frac{\text{C}}{\text{mol}}\right)}$$

$$= +0.777 \text{ J/C} = +0.777 \text{ V}$$

A spontaneous chemical reaction (negative ΔG) produces a *positive voltage*.

Reminder: 1 J/C = 1 volt

$n = $ e⁻/atom is dimensionless

TEST YOURSELF Find E if $\Delta G = +150$ kJ and $n = 1$ e⁻/atom. (*Answer:* −1.55 V)

Cathode: where reduction occurs
Anode: where oxidation occurs

Michael Faraday wanted to describe his discoveries with terms that would "advance the general cause of science" and not "retard its progress." He sought the aid of William Whewell in Cambridge, who coined words such as "anode" and "cathode," meaning "a way up" and "a way down" (Figure 14-4).

Chemists define the electrode at which *reduction* occurs as the **cathode.** The **anode** is the electrode at which *oxidation* occurs. In Figure 14-4, Ag is the cathode because reduction takes place at its surface ($2AgCl + 2e^- \rightarrow 2Ag + 2Cl^-$). Cd is the anode because it is oxidized ($Cd \rightarrow Cd^{2+} + 2e^-$).

Electrons Move Toward More Positive Electric Potential

Being negatively charged, *electrons move toward more positive electric potential.* In Figure 14-4, the Ag electrode is positive with respect to the Cd electrode. Therefore, electrons move from Cd to Ag through the circuit. When we study the Nernst equation, you will learn how to find the electrode potentials and, therefore, to predict the direction of electron flow.

Electrons are not readily conducted through a solution. They must move through the wire. Ions are not conducted through a wire. They must move through the solution. Electroneutrality is maintained by a balance between the flow of electrons and the flow of ions so there is no significant buildup of charge in any region.

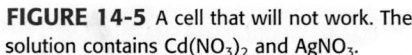

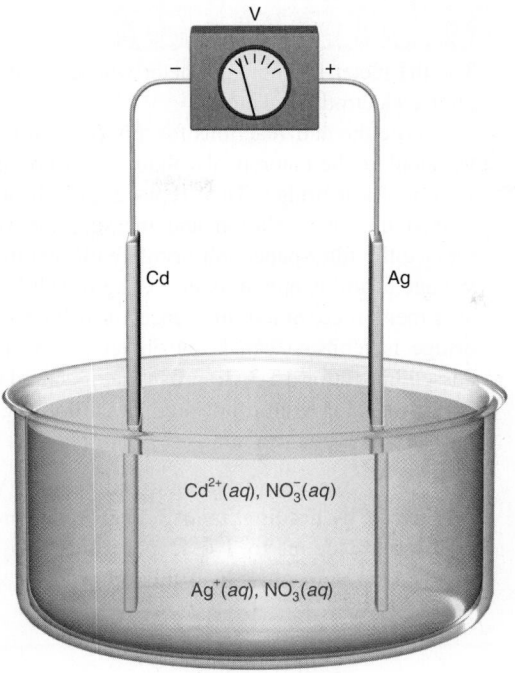

Salt Bridge

Consider the cell in Figure 14-5, in which the reactions are intended to be

Cathode:	$2Ag^+(aq) + 2e^- \rightleftharpoons 2Ag(s)$	
Anode:	$Cd(s) \rightleftharpoons Cd^{2+}(aq) + 2e^-$	

Net reaction: $\quad Cd(s) + 2Ag^+(aq) \rightleftharpoons Cd^{2+}(aq) + 2Ag(s)$ **(14-10)**

The net reaction is spontaneous, but little current flows through the circuit because Ag^+ is not forced to be reduced at the Ag electrode. Aqueous Ag^+ can react directly at the $Cd(s)$ surface, giving the same net reaction with no flow of electrons through the external circuit.

We can separate the reactants into two *half-cells*[8] if we connect the two halves with a **salt bridge,** as shown in Figure 14-6. The salt bridge is a U-shaped tube filled with a gel containing a high concentration of KNO_3 (or other electrolyte that does not affect the cell reaction). The ends of the bridge are porous glass disks that allow ions to diffuse but minimize mixing of solutions inside and outside the bridge. When the galvanic cell is operating, K^+ from the bridge migrates into the cathode compartment and a small amount of NO_3^- migrates from the cathode into the bridge. Ion migration offsets the charge buildup that would otherwise occur as electrons flow into the silver electrode. In the absence of a salt bridge, negligible reaction can occur because of charge buildup. The migration of ions out of the bridge is greater than the migration of ions into the bridge because the salt concentration in the bridge

The cell in Figure 14-5 is *short-circuited*.

A salt bridge maintains electroneutrality (no charge buildup) throughout the cell. See Demonstration 14-1.

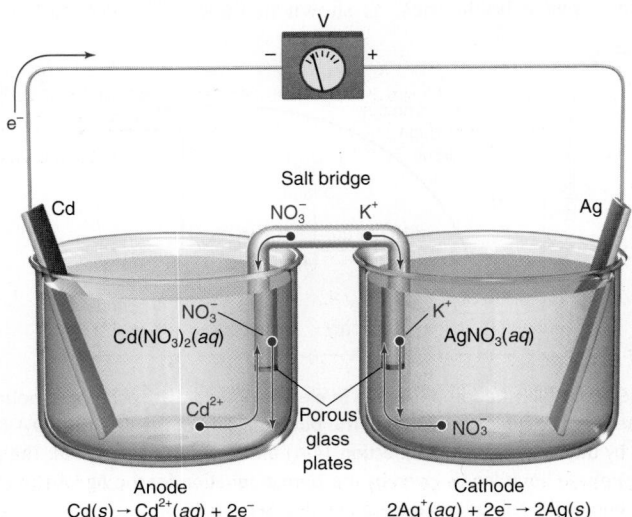

Anode	Cathode
$Cd(s) \rightarrow Cd^{2+}(aq) + 2e^-$	$2Ag^+(aq) + 2e^- \rightarrow 2Ag(s)$

FIGURE 14-6 A cell that works—thanks to the salt bridge!

A salt bridge is an ionic medium with a *semipermeable* barrier on each end. Small molecules and ions can cross a semipermeable barrier, but large molecules cannot. Demonstrate a "proper" salt bridge by filling a U-tube with agar and KCl as described in the text and construct the cell shown here.

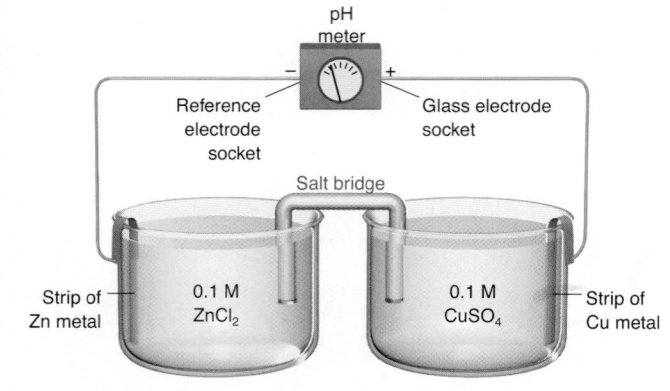

The pH meter is a potentiometer whose negative terminal is the reference electrode socket.

Write the half-reactions for this cell and use the Nernst equation to calculate the theoretical voltage. Measure the voltage with a conventional salt bridge. Then replace the salt bridge with filter paper soaked in NaCl solution and measure the voltage again. Finally, replace the filter-paper salt bridge with two fingers and measure the voltage again. A human is just a bag of salt housed in a semipermeable membrane. Small differences in voltage observed when the salt bridge is replaced can be attributed to the junction potential discussed in Section 15-3. To prove that it is hard to distinguish a chemistry instructor from a hot dog, use a hot dog as a salt bridge[6] and measure the voltage again.

Challenge One hundred eighty students at Virginia Tech made a salt bridge by holding hands.[7] Their resistance was lowered from 10^6 Ω per student to 10^4 Ω per student by wetting everyone's hands. Can your class beat this record?

---

is much higher than the salt concentration in the half-cells. At the left side of the salt bridge, NO_3^- migrates into the anode compartment and a little Cd^{2+} migrates into the bridge to prevent buildup of positive charge.

For reactions that do not involve Ag^+ or other species that react with Cl^-, the salt bridge usually contains KCl electrolyte. A typical salt bridge is prepared by heating 3 g of agar with 30 g of KCl in 100 mL of water until a clear solution is obtained. The solution is poured into the U-tube and allowed to gel. The bridge is stored in saturated aqueous KCl.

Line Notation

The salt bridge symbol ‖ represents the two phase boundaries on either side of the bridge.

Electrochemical cells are described by a notation employing just two symbols:

$$| \text{ phase boundary} \qquad \| \text{ salt bridge}$$

The cell in Figure 14-4 is represented by the *line diagram*

$$Cd(s)|CdCl_2(aq)|AgCl(s)|Ag(s)$$

Each phase boundary is indicated by a vertical line. The electrodes are shown at the extreme left- and right-hand sides of the line diagram. The cell in Figure 14-6 is

$$Cd(s)|Cd(NO_3)_2(aq)\|AgNO_3(aq)|Ag(s)$$

There is a change in electric potential at most phase boundaries in an electrochemical cell. The potential increase from Cd to Ag occurs principally at the $Cd(s) | Cd(NO_3)_2(aq)$ and $AgNO_3(aq) | Ag(s)$ phase boundaries, as shown in Figure 14-7 for the case of negligible

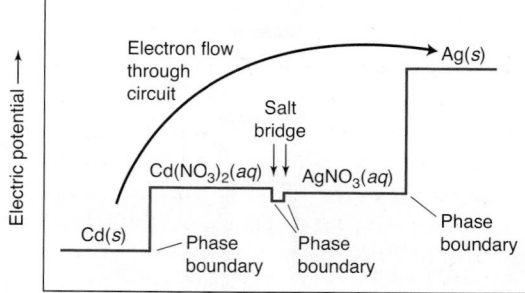

FIGURE 14-7 Schematic illustration of change in electric potential at each phase boundary of the cell in Figure 14-6 if there is negligible current flow. The potential step at the Cd | $Cd(NO_3)_2(aq)$ phase boundary is given by the *Nernst equation* (Section 14-4) for the Cd | Cd^{2+} half-cell. The potential step at the Ag | $AgNO_3(aq)$ phase boundary is given by the Nernst equation for the Ag | Ag^+ half-cell. *Junction potentials* at each end of the salt bridge are explained in Section 15-3.

BOX 14-2 Hydrogen-Oxygen Fuel Cell

1.5 kW *Apollo* fuel cell. Apollo used two of these units. [© DaffodilPhotography/Alamy.]

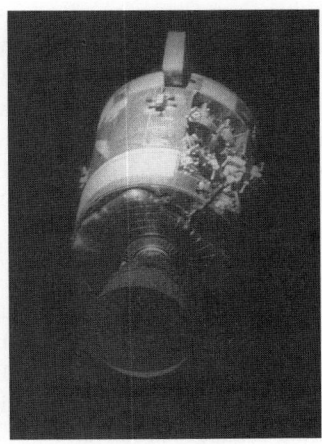

Ruptured *Apollo 13* service module observed by the crew after separation from the command module before re-entry into earth's atmosphere. [NASA.]

"Houston, we've had a problem." With these words, Commander Jim Lovell informed Mission Control that *Apollo 13* was in trouble on the second day of its trip to the moon in 1970. A tank holding liquid oxygen for the spacecraft's fuel cells had exploded. These fuel cells were developed in the 1960s as the most efficient method to provide electricity in space. As a result of the explosion, three astronauts were forced to use their lunar module as a "lifeboat" with almost no power or water for nearly four days as they flew to the moon and back to splash down in the Pacific Ocean. One of many technical highlights of the journey was adapting a LiOH canister from the command module to remove CO_2 from the atmosphere of the lunar module so the astronauts could survive ($2LiOH(s) + CO_2(g) \rightarrow Li_2CO_3(s) + H_2O(g)$). A billion people on

Earth, transfixed on the fate of three people, burst into cheers when the crew was safely recovered.

A modern H_2–O_2 polymer electrolyte fuel cell derives its energy from the net combination of H_2 and O_2 to make H_2O. *Fuel*, $H_2(g)$, flows into the cell at the left through a 10-μm-thick electrically conductive porous carbon sheet to the anode, which contains 2-nm catalytic particles of Pt (<0.5 mg/cm^2). H_2 dissociates to give Pt-bound H atoms, which go on to make H^+ and electrons. Electrons are conducted through the porous carbon into a circuit where they can do useful work. H^+ is conducted through the Nafion® polymer electrolyte membrane. Hydrated sulfonic acid groups of the polymer transport H^+ from one sulfonic acid group to another.

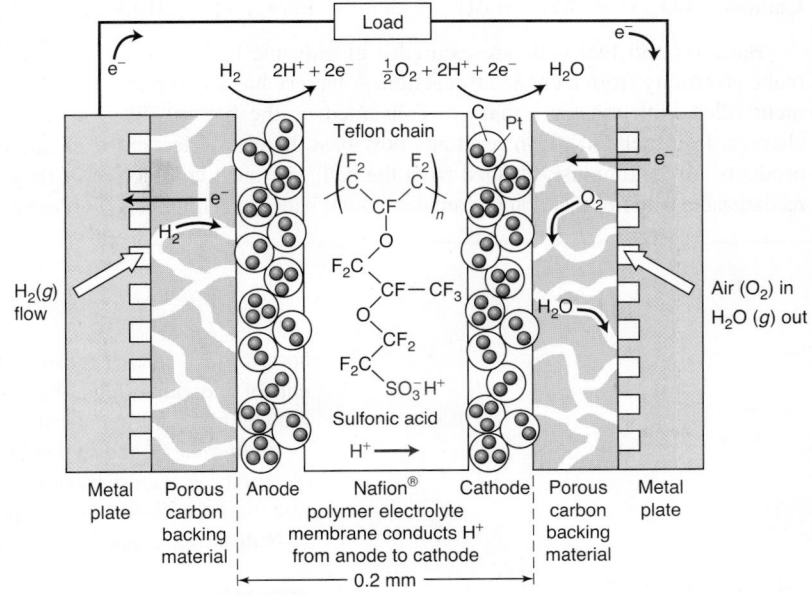

Schematic cross section of polymer electrolyte membrane hydrogen-oxygen fuel cell. [Information from S. Thomas and M. Zalbowitz, *Fuel Cells: Green Power* (Los Alamos National Laboratory, New Mexico, 1999), http://www.lanl.gov/orgs/mpa/mpa11/Green%20Power.pdf.]

When H^+ reaches the cathode, it combines on Pt catalyst particles with O_2 and electrons to make H_2O. O_2 is provided by air pumped in at the right. $H_2O(g)$ produced at the cathode exits the cell in the airstream. A key feature enabling the fuel cell to produce electricity is that the polymer electrolyte membrane conducts H^+ but not electrons.

An ideal cell would produce 1.16 V at 80°C if no current were flowing. The operating voltage is typically ~0.7 V when current flows and the cell does useful work. The cell is 60% (= 0.7 V/1.16 V) efficient at converting chemical energy into electrical energy. The other 40% of energy is converted to heat that is removed by air flowing through the cathode to maintain a temperature of 80°C. The cell produces an impressive ~0.5 A per square centimeter of its area. High voltage can be produced by stacking cells in series.

Other H_2–O_2 fuel cells with different electrolytes operate at higher temperature. Some cells are capable of generating megawatts of power with 85% efficiency. For comparison, automobiles with internal combustion engines convert ~20% of the energy content of gasoline into motion of the car. Some fuel cells extract hydrogen from natural gas (methane) and use catalytic ceramic oxides in place of expensive noble metals.

BOX 14-3 Lead-Acid Battery

A 12-V lead-acid battery consists of six cells that each deliver 2 V.[10] Invented in 1859 by the French physicist Gaston Planté at the age of 25, this was the first rechargeable battery. Its electrodes are metallic lead grids with a large surface area. Solid PbO_2 is pressed onto the cathode. The cell is filled with aqueous H_2SO_4, which is ~35 wt% $H_2SO_4 \approx 5.5\ m$ (molal) (~4.4 M) when fully charged. During discharge (when the battery is producing electricity), Pb is oxidized to $PbSO_4(s)$ at the anode. At the cathode, PbO_2 is reduced to $PbSO_4(s)$. As the cell discharges, both electrodes become coated with $PbSO_4(s)$. Both reactions consume H_2SO_4, whose concentration can fall to ~22 wt% $\approx 2.9\ m$ during discharge.

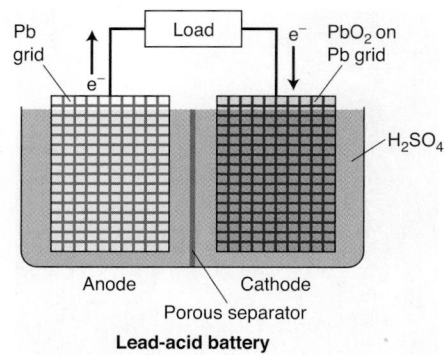

Lead-acid battery

Anode: $\quad\quad\quad Pb(s) + SO_4^{2-} \rightleftharpoons PbSO_4(s) + 2e^-$

Cathode: $PbO_2(s) + SO_4^{2-} + 4H^+ + 2e^- \rightleftharpoons PbSO_4(s) + 2H_2O$

Batteries and fuel cells are examples of galvanic cells, which make electricity from a chemical reaction. A battery has a compartment filled with reactants that are consumed as the battery discharges. In a fuel cell, fresh reactants flow past the electrodes and products are continuously flushed from the cell. The most common rechargeable batteries that you use are the lithium battery in computers

and mobile telephones and the lead-acid battery in a car. One reason why we use lead-acid batteries is that they can deliver several hundred amperes of current for a short time to start the engine. Alkaline batteries used in flashlights and toys are not rechargeable. Alkaline and lithium batteries should be discarded as hazardous waste when they no longer work. Lead-acid batteries are recycled by dealers who sell them.

current flow. There is also a small change in potential, called the *junction potential*, at each end of the salt bridge. Saturated KCl or saturated KNO_3 is often used in a salt bridge to reduce the junction potential to a few millivolts or less, as discussed in Section 15-3.

Batteries[9, 10] and fuel cells[11-13] are galvanic cells that consume their reactants to generate electricity. A *battery* has a static compartment filled with reactants. In a *fuel cell*, fresh reactants flow past the electrodes and products are continuously flushed from the cell. Boxes 14-2 and 14-3 describe important fuel cells and batteries.

14-3 Standard Potentials

The voltage measured in Figure 14-6 is the difference in electric potential between the Ag electrode on the right and the Cd electrode on the left. Voltage tells us how much work can be done by electrons flowing from one side to the other (Equation 14-3). The potentiometer (voltmeter) indicates a positive voltage when electrons flow into the negative terminal, as in Figure 14-6. If electrons flow the other way, the voltage is negative.

Sometimes the negative terminal of a voltmeter is labeled "common." It may be colored black and the positive terminal red. When a pH meter with a BNC socket is used as a potentiometer, the center wire is the positive input and the outer connection is the negative input. In older pH meters, the negative terminal is the narrow receptacle to which the reference electrode is connected.

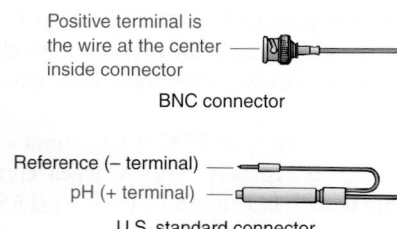

Positive terminal is the wire at the center inside connector

BNC connector

Reference (– terminal)

pH (+ terminal)

U.S. standard connector

In General, We Will Write All Half-Reactions as Reductions

From now on, *we will generally write all half-reactions as reductions*. The Nernst equation in the next section enables us to find the electrode potentials for each half-reaction and, therefore, to predict the direction of electron flow. Electrons flow through the wire from the more negative electrode to the more positive electrode.

Measuring the Standard Reduction Potential

Each half-reaction is assigned a **standard reduction potential,** $E°$, measured by an experiment shown in an idealized form in Figure 14-8. The half-reaction of interest is

$$Ag^+ + e^- \rightleftharpoons Ag(s) \quad\quad\quad (14-11)$$

which occurs in the half-cell at the right connected to the *positive* terminal of the potentiometer. *Standard* means that activities of all species are unity. For Reaction 14-11 under standard conditions, $\mathscr{A}_{Ag^+} = 1$ and, by definition, the activity of Ag(s) also is unity.

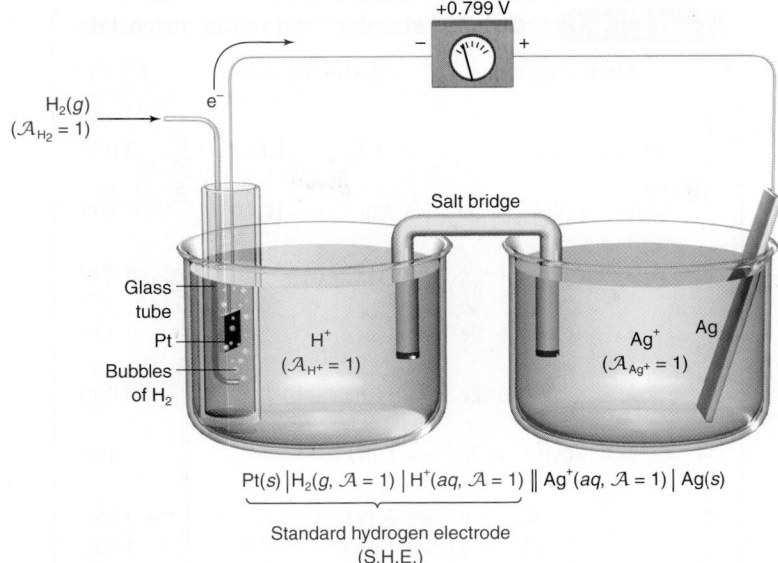

+0.799 V

H₂(g)
($\mathcal{A}_{H_2}$ = 1)

e⁻

Salt bridge

Glass
tube

Pt

Bubbles
of H₂

H⁺
($\mathcal{A}_{H^+}$ = 1)

Ag⁺   Ag
($\mathcal{A}_{Ag^+}$ = 1)

$Pt(s)\,|\,H_2(g, \mathcal{A} = 1)\,|\,H^+(aq, \mathcal{A} = 1)\,\|\,Ag^+(aq, \mathcal{A} = 1)\,|\,Ag(s)$

Standard hydrogen electrode
(S.H.E.)

FIGURE 14-8   Cell for measuring standard reduction potential of $Ag^+ + e^- \rightleftharpoons Ag(s)$. This cell is hypothetical because it is usually not possible to adjust the activity of a species to 1.

The potentiometer measures the difference in potential of the electrode attached to the positive terminal of the meter minus the potential of the electrode attached to the negative terminal of the meter:

$$E = E_+ - E_-$$

The left half-cell, connected to the *negative* terminal of the potentiometer, is called the **standard hydrogen electrode** (S.H.E.). It consists of a catalytic Pt surface[14] in contact with an acidic solution in which $\mathcal{A}_{H^+} = 1$. A stream of $H_2(g)$ bubbled through the electrode saturates the solution with $H_2(aq)$. The activity of $H_2(g)$ is unity if the pressure of $H_2(g)$ is 1 bar. We describe the reaction that comes to equilibrium at the surface of the Pt electrode as

Question What is the pH of the standard hydrogen electrode?

S.H.E. half-reaction: $H^+(aq, \mathcal{A} = 1) + e^- \rightleftharpoons \dfrac{1}{2}H_2(g, \mathcal{A} = 1)$ **(14-12)**

$H_2(g)$ dissolves to give $H_2(aq)$, which is in equilibrium with $H^+(aq)$ on the Pt surface.

We *arbitrarily* assign a potential of 0 to the standard hydrogen electrode at 25°C. The voltage measured by the meter in Figure 14-8 can therefore be *assigned* to Reaction 14-11, which occurs in the right half-cell. The measured value $E° = +0.799$ V is the standard reduction potential for Reaction 14-11. The positive sign tells us that electrons flow from Pt to Ag through the meter.

We arbitrarily *assign* a potential to Reaction 14-12 and then use the reaction as a reference point to measure other half-cell potentials. An analogy is the assignment of 0°C to the freezing point of water and 100°C to the boiling point of water at 1 atmosphere pressure. On this scale, hexane boils at 69° and benzene boils at 80°. The difference between boiling points is $80° - 69° = 11°$. If we were to assign the freezing point of water to be 200°C and the boiling point to be 300°C, hexane would boil at 269° and benzene would boil at 280°. The difference is still 11°. Regardless of where we set zero on the scale, differences between benzene and hexane remains constant.

By convention, $E° = 0$ for S.H.E. Walther Nernst appears to have been the first to assign the potential of the hydrogen electrode as 0 in 1897.[15]

The line notation for the cell in Figure 14-8 is

$$Pt(s)\,|\,H_2(g, \mathcal{A} = 1)|H^+(aq, \mathcal{A} = 1)\,\|\,Ag^+(aq, \mathcal{A} = 1)|Ag(s)$$

\underbrace{\hspace{5cm}}_{S.H.E.}

The standard reduction potential is the *difference* between the potential of the reaction of interest on the right and the potential of S.H.E on the left, which we have arbitrarily set to 0.

To measure the standard reduction potential of the half-reaction

$$Cd^{2+} + 2e^- \rightleftharpoons Cd(s) \qquad \text{(14-13)}$$

we construct the cell

$$S.H.E. \,\|\, Cd^{2+}(aq, \mathcal{A} = 1)|Cd(s)$$

with the cadmium half-cell connected to the positive terminal of the potentiometer. In this case, we observe a *negative* voltage of -0.402 V. The negative sign means that electrons flow from Cd to Pt, a direction opposite that of the cell in Figure 14-8.

Appendix H contains standard reduction potentials arranged alphabetically by element. If the half-reactions were arranged according to descending value of $E°$ (as in Table 14-1), we would find the strongest oxidizing agents at the upper left and the strongest reducing agents

TABLE 14-1	**Ordered standard reduction potentials**		
	Oxidizing agent	**Reducing agent**	$E°(V)$
	$F_2(g) + 2e^- \rightleftharpoons 2F^-$		2.890
	$O_3(g) + 2H^+ + 2e^- \rightleftharpoons O_2(g) + H_2O$		2.075
	$MnO_4^- + 8H^+ + 5e^- \rightleftharpoons Mn^{2+} + 4H_2O$		1.507
	$Ag^+ + e^- \rightleftharpoons Ag(s)$		0.799
	$Cu^{2+} + 2e^- \rightleftharpoons Cu(s)$		0.339
	$2H^+ + 2e^- \rightleftharpoons H_2(g)$		0.000
	$Cd^{2+} + 2e^- \rightleftharpoons Cd(s)$		-0.402
	$K^+ + e^- \rightleftharpoons K(s)$		-2.936
	$Li^+ + e^- \rightleftharpoons Li(s)$		-3.040

(Oxidizing power increases ↑ — Reducing power increases ↓)

at the lower right. If we connected the two half-cells represented by Reactions 14-11 and
14-13, Ag^+ would be reduced to $Ag(s)$ as $Cd(s)$ is oxidized to Cd^{2+}.

14-4 Nernst Equation

Le Châtelier's principle tells us that increasing reactant concentrations drives a reaction to the
right and increasing the product concentrations drives a reaction to the left. The net driving
force for a redox reaction is expressed by the **Nernst equation,** whose two terms include the
driving force under standard conditions ($E°$, which applies when all activities are unity) and
a term showing the dependence on reagent concentrations.

Nernst Equation for a Half-Reaction

For the half-reaction

$$a\text{A} + ne^- \rightleftharpoons b\text{B}$$

the Nernst equation giving the half-cell potential, E, is

Nernst equation:
$$E = E° - \frac{RT}{nF} \ln \frac{\mathcal{A}_B^b}{\mathcal{A}_A^a} \tag{14-14}$$

where $E°$ = standard reduction potential ($\mathcal{A}_A = \mathcal{A}_B = 1$)

$\quad R$ = gas constant (8.314 J/(K · mol) = 8.314 (V · C)/(K · mol))
$\quad T$ = temperature (K)
$\quad n$ = number of electrons in the half-reaction
$\quad F$ = Faraday constant (9.649×10^4 C/mol)
$\quad \mathcal{A}_i$ = activity of species i

The logarithmic term in the Nernst equation is the **reaction quotient, Q.**

$$Q = \mathcal{A}_B^b / \mathcal{A}_A^a \tag{14-15}$$

Q has the same form as the equilibrium constant, but the activities need not have their equilibrium values. Pure solids, pure liquids, and solvents are omitted from Q because their
activities are unity (or close to unity). Concentrations of solutes are expressed as moles per
liter and concentrations of gases are expressed as pressures in bars. When all activities are
unity, $Q = 1$ and $\ln Q = 0$, thus giving $E = E°$.

Converting the natural logarithm in Equation 14-14 into the base 10 logarithm and inserting $T = 298.15$ K ($25.00°$C) gives the most useful form of the Nernst equation:

Nernst equation at 25°C:
$$E = E° - \frac{0.059\,16 \text{ V}}{n} \log \frac{\mathcal{A}_B^b}{\mathcal{A}_A^a} \tag{14-16}$$

The potential changes by 59.16/n mV for each factor-of-10 change in Q at 25°C.

EXAMPLE Writing the Nernst Equation for a Half-Reaction

Let's write the Nernst equation for the reduction of white phosphorus to phosphine gas:

$$\frac{1}{4}P_4(s, \text{white}) + 3H^+ + 3e^- \rightleftharpoons \underset{\text{Phosphine}}{PH_3(g)} \qquad E° = -0.046 \text{ V}$$

Solution We omit solids from the reaction quotient, and the concentration of a gas is expressed as the pressure of the gas. Therefore, the Nernst equation is

$$E = -0.046 - \frac{0.059\ 16}{3}\log\frac{P_{PH_3}}{[H^+]^3}$$

TEST YOURSELF With $E°$ from Appendix H, write the Nernst equation for $ZnS(s) + 2e^- \rightleftharpoons$ $Zn(s) + S^{2-}$. (*Answer:* $E = -1.405 - \dfrac{0.059\ 16}{2}\log[S^{2-}]$)

Phosphine is a highly poisonous gas with the odor of decaying fish. Solid white phosphorus glows faintly in the dark from spontaneous oxidation by O_2 in air.

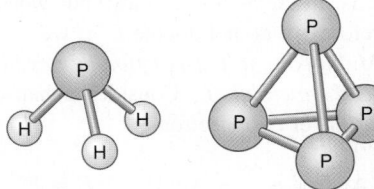

Phosphine   White phosphorus consists of tetrahedral P_4 molecules

EXAMPLE Multiplying a Half-Reaction Does Not Change $E°$

If we multiply a half-reaction by any factor, $E°$ does not change. However, the factor n before the log term and the form of the reaction quotient, Q, do change. Let's write the Nernst equation for the reaction in the preceding example, multiplied by 2:

$$\frac{1}{2}P_4(s, \text{white}) + 6H^+ + 6e^- \rightleftharpoons 2PH_3(g) \qquad E° = -0.046 \text{ V}$$

Solution

$$E = -0.046 - \frac{0.059\ 16}{6}\log\frac{P_{PH_3}^2}{[H^+]^6}$$

Even though this Nernst equation does not look like the one in the preceding example, Box 14-4 shows that the numerical value of E is unchanged. The squared reaction quotient cancels the doubled value of n in front of the log term.

TEST YOURSELF Write the Nernst equation for $P_4 + 12H^+ + 12e^- \rightleftharpoons 4PH_3$. From Box 14-4, show that E is the same as if the reaction were written with $\frac{1}{2}P_4$ or $\frac{1}{4}P_4$.

Nernst Equation for a Complete Reaction

In Figure 14-6, the measured voltage is the difference between the potentials of the two electrodes:

Nernst equation for a complete cell: $\qquad E = E_+ - E_- \qquad$ (14-17)

where E_+ is the potential of the electrode attached to the positive terminal of the potentiometer and E_- is the potential of the electrode attached to the negative terminal. The potential of each half-reaction (*written as a reduction*) is governed by a Nernst equation, and the voltage for the complete reaction is the difference between the two half-cell potentials.

Here is a procedure for writing a net cell reaction and finding its voltage:

Step 1   Write *reduction* half-reactions for both half-cells and find $E°$ for each in Appendix H. Multiply the half-reactions as necessary so that they both contain the same number of electrons. When you multiply a reaction, you *do not* multiply $E°$.

Step 2   Write a Nernst equation for the right half-cell, which is attached to the positive terminal of the potentiometer. This is E_+.

Step 3   Write a Nernst equation for the left half-cell, which is attached to the negative terminal of the potentiometer. This is E_-.

Step 4   Find the net cell voltage by subtraction: $E = E_+ - E_-$.

Step 5   Write the net cell reaction by subtracting the left half-reaction from the right half-reaction. (Subtraction is equivalent to reversing the left half-reaction and adding.)

Electrons flow through the circuit from the more negative to the more positive electrode.

Electrons flow spontaneously through the circuit from the more negative electrode to the more positive electrode. If the net cell voltage that you calculated, E ($= E_+ - E_-$), is positive, then electrons flow through the wire from the left-hand electrode to the right-hand electrode. If the net cell voltage is negative, then electrons flow in the opposite direction.

EXAMPLE Nernst Equation for a Complete Reaction

Find the voltage of the cell in Figure 14-6 if the right half-cell contains 0.50 M AgNO$_3$(aq) and the left half-cell contains 0.010 M Cd(NO$_3$)$_2$(aq). Write the net cell reaction and state in which direction electrons flow.

Solution

Step 1 Right electrode: $\quad 2Ag^+ + 2e^- \rightleftharpoons 2Ag(s) \qquad E°_+ = 0.799\ V$

Left electrode: $\quad Cd^{2+} + 2e^- \rightleftharpoons Cd(s) \qquad E°_- = -0.402\ V$

Step 2 Nernst equation for right electrode:

Pure solids, pure liquids, and solvents are omitted from Q.

$$E_+ = E°_+ - \frac{0.059\ 16}{2}\log\frac{1}{[Ag^+]^2} = 0.799 - \frac{0.059\ 16}{2}\log\frac{1}{[0.50]^2} = 0.781\ V$$

Step 3 Nernst equation for left electrode:

$$E_- = E°_- - \frac{0.059\ 16}{2}\log\frac{1}{[Cd^{2+}]} = -0.402 - \frac{0.059\ 16}{2}\log\frac{1}{[0.010]} = -0.461\ V$$

The silver electrode potential is most positive, so *electrons flow from Cd to Ag through the circuit* (Figure 14-9).

Step 4 Cell voltage: $\quad E = E_+ - E_- = 0.781 - (-0.461) = +1.242\ V$

Step 5 Net cell reaction:

Subtracting a reaction is the same as reversing the reaction and adding.

$$
\begin{array}{r}
2Ag^+ + 2e^- \rightleftharpoons 2Ag(s) \\
-\ \underline{Cd^{2+} + 2e^- \rightleftharpoons Cd(s)} \\
Cd(s) + 2Ag^+ \rightleftharpoons Cd^{2+} + 2Ag(s)
\end{array}
$$

TEST YOURSELF Does the reaction go in the same direction if the cells contain 5.0 μM AgNO$_3$ and 1.0 M Cd(NO$_3$)$_2$? (***Answer:*** Yes: $E_+ = 0.485\ V$, $E_- = -0.402\ V$, $E = +0.887\ V$)

What if you had written the Nernst equation for the right half-cell with just one electron instead of two?

$$Ag^+ + e^- \rightleftharpoons Ag(s)$$

Would the net cell voltage be different from what we calculated? It better not be, because the chemistry is still the same. Box 14-4 shows that *neither $E°$ nor E depend on how we write the*

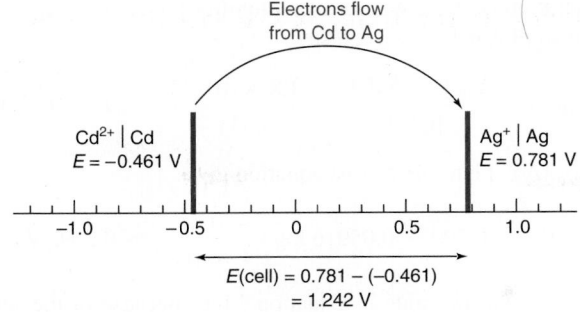

FIGURE 14-9 Electrons always flow from the more negative to the more positive electrode. That is, they always flow to the right in this diagram.[16]

reaction. Box 14-5 shows how to derive standard reduction potentials for half-reactions that are the sum of other half-reactions.

Different Descriptions of the Same Reaction

In Figure 14-4, the right half-reaction can be written

$$AgCl(s) + e^- \rightleftharpoons Ag(s) + Cl^- \qquad E_+^\circ = 0.222 \text{ V} \qquad \textbf{(14-18)}$$

$$E_+ = E_+^\circ - 0.059\ 16 \log[Cl^-] = 0.222 - 0.059\ 16 \log(0.033\ 4) = 0.309_3 \text{ V} \quad \textbf{(14-19)}$$

The Cl^- in the silver half-reaction was derived from 0.016 7 M $CdCl_2(aq)$.

Suppose that a different, less handsome, author had written this book and had chosen to describe the half-reaction differently:

$$Ag^+ + e^- \rightleftharpoons Ag(s) \qquad E_+^\circ = 0.799 \text{ V} \qquad \textbf{(14-20)}$$

This description is just as valid as the previous one. In both cases, Ag(I) is reduced to Ag(0).

If the two descriptions are equally valid, then they should predict the same voltage. The Nernst equation for Reaction 14-20 is

$$E_+ = 0.799 - 0.059\ 16 \log\frac{1}{[Ag^+]}$$

BOX 14-5 **Latimer Diagrams: How to Find $E°$ for a New Half-Reaction**

A **Latimer diagram** displays standard reduction potentials, $E°$, connecting various oxidation states of an element.[17] For example, in acid solution, the following standard reduction potentials are observed:

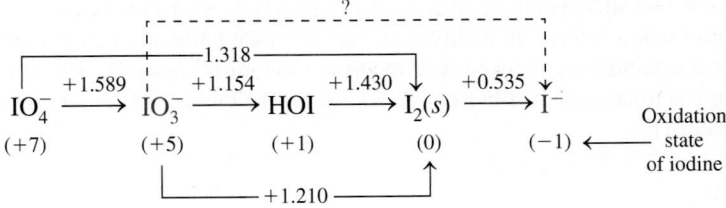

The notation $IO_3^- \xrightarrow{+1.154} HOI$ stands for the balanced equation

$$IO_3^- + 5H^+ + 4e^- \rightleftharpoons HOI + 2H_2O \qquad E° = +1.154 \text{ V}$$

We can derive reduction potentials for arrows that are not shown in the diagram by using $\Delta G°$. For example, the reaction shown by the dashed line in the Latimer diagram is

$$IO_3^- + 6H^+ + 6e^- \rightleftharpoons I^- + 3H_2O$$

To find $E°$ for this reaction, express the reaction as a sum of reactions whose potentials are known.

The standard free energy change, $\Delta G°$, for a reaction is given by Equation 14-5:

$$\Delta G° = -nNFE°$$

In this equation, $n = 1$ unit charge per electron (a dimensionless number) and N is the number of moles of electrons in the half-reaction. The product nN is numerically the same as the number of electrons in the half-reaction and has the dimensions of mol.

When two reactions are added, $\Delta G°$ *is the sum of* $\Delta G°$ *values for each reaction.* To apply free energy to our problem, we write reactions whose sum is the desired reaction:

$$IO_3^- + 6H^+ + 5e^- \xrightarrow{E_1°=1.210} \tfrac{1}{2}I_2(s) + 3H_2O \quad \Delta G_1° = -5F(1.210)$$

$$\tfrac{1}{2}I_2(s) + e^- \xrightarrow{E_2°=0.535} I^- \qquad \Delta G_2° = -1F(0.535)$$

$$\overline{IO_3^- + 6H^+ + 6e^- \xrightarrow{E_3°=?} I^- + 3H_2O \qquad \Delta G_3° = -6FE_3°}$$

But, because $\Delta G_1° + \Delta G_2° = \Delta G_3°$, we can solve for $E_3°$:

$$\Delta G_3° = \Delta G_1° + \Delta G_2°$$

$$-6FE_3° = -5F(1.210) - 1F(0.535)$$

$$E_3° = \frac{5(1.210) + 1(0.535)}{6} = 1.098 \text{ V}$$

To find the concentration of Ag^+, we use the solubility product for AgCl. The cell contains 0.033 4 M Cl^- and solid AgCl, so we can say

$$K_{sp} = [Ag^+][Cl^-]$$

$$[Ag^+] = \frac{K_{sp} \text{ (for AgCl)}}{[Cl^-]} = \frac{1.8 \times 10^{-10}}{0.033\ 4} = 5.4 \times 10^{-9} \text{ M}$$

Putting this value of $[Ag^+]$ into the Nernst equation gives

$$E_+ = 0.799 - 0.05916 \log\frac{1}{5.4 \times 10^{-9}} = 0.309_9 \text{ V}$$

which differs slightly from the value in Equation 14-19 because of the accuracy of K_{sp} and the neglect of activity coefficients. Reactions 14-18 and 14-20 give the same voltage in the Nernst equation because they describe the same cell.

The cell voltage cannot depend on how we write the reaction!

Advice for Finding Relevant Half-Reactions

How to figure out the half-cell reactions

When faced with a cell drawing or a line diagram, first write reduction reactions for each half-cell. To do this, *look in the cell for an element in two oxidation states.* For the cell

$$Pb(s)|PbF_2(s)|F^-(aq)\|Cu^{2+}(aq)|Cu(s)$$

we see Pb in two oxidation states, as $Pb(s)$ and $PbF_2(s)$, and Cu in two oxidation states, as Cu^{2+} and $Cu(s)$. Thus, the half-reactions are

Right half-cell:	$Cu^{2+} + 2e^- \rightleftharpoons Cu(s)$	
Left half-cell:	$PbF_2(s) + 2e^- \rightleftharpoons Pb(s) + 2F^-$	**(14-21)**

You might have chosen to write the Pb half-reaction as

Left half-cell:	$Pb^{2+} + 2e^- \rightleftharpoons Pb(s)$	**(14-22)**

because you know that if $PbF_2(s)$ is present, there must be some Pb^{2+} in the solution. Reactions 14-21 and 14-22 are equally valid and must predict the same cell voltage. Your choice of reactions depends on whether the F^- or Pb^{2+} concentration is easier to figure out.

Don't invent species not shown in the cell. Use what is shown in the line diagram to select the half-reactions.

We described the left half-cell in terms of a redox reaction involving Pb because Pb is the element that appears in two oxidation states. We would not write a reaction such as $F_2(g) + 2e^- \rightleftharpoons 2F^-$ because $F_2(g)$ is not shown in the line diagram of the cell.

The Nernst Equation Is Used in Measuring Standard Reduction Potentials

Problem 14-22 gives an example of the use of the Nernst equation to find $E°$.

The standard reduction potential would be observed if the half-cell of interest (with unit activities) were connected to a standard hydrogen electrode, as it is in Figure 14-8. It is nearly impossible to construct such a cell, because we have no way to adjust concentrations and ionic strength to give unit activities. In reality, activities less than unity are used in each half-cell, and the Nernst equation is employed to extract the value of $E°$ from the cell voltage.[18] In the hydrogen electrode, standard buffers with known pH (Table 15-3) are used to obtain known activities of H^+.

14-5 $E°$ and the Equilibrium Constant

A galvanic cell produces electricity because the cell reaction is not at equilibrium. The potentiometer allows negligible current to flow through it (Box 14-6), so concentrations in the cell remain unchanged. If we replaced the potentiometer with a wire, there would be much more current and concentrations would change until the cell reached equilibrium. At that point, nothing would drive the reaction, and E would be 0. When a battery (which is a galvanic cell) runs down to 0 V, the chemicals inside have reached equilibrium and the battery is "dead."

At equilibrium, E (not $E°$) = 0.

Now let's relate E for a whole cell to the reaction quotient, Q, for the net cell reaction. For the two half-reactions

Net cell reaction:
$aA + bB \rightleftharpoons cC + dD$

Right electrode:	$aA + ne^- \rightleftharpoons cC$	$E_+°$
Left electrode:	$dD + ne^- \rightleftharpoons bB$	$E_-°$
Net reaction:	$aA + bB \rightleftharpoons cC + dD$	$E°$

Why doesn't operation of a cell change the concentrations in the cell? Cell voltage is measured under conditions of *negligible current flow*. The resistance of a high-quality pH meter is 10^{13} Ω. If a cell produces 1 V, the current through the circuit is

$$I = \frac{E}{R} = \frac{1 \text{ V}}{10^{13} \, \Omega} = 10^{-13} \text{ A}$$

The electron flow is

$$\frac{10^{-13} \text{ C/s}}{9.649 \times 10^4 \text{ C/mol}} = 10^{-18} \text{ mol e}^-/\text{s}$$

which produces negligible oxidation and reduction of reagents in the cell. *The meter measures the voltage of the cell without affecting concentrations in the cell.*

If a salt bridge were left in a real cell for very long, concentrations and ionic strength would change because of diffusion between each compartment and the salt bridge. Cells should be set up for such a short time that mixing is insignificant.

the Nernst equation looks like this:

$$E = E_+ - E_- = E_+^\circ - \frac{0.059 \ 16}{n} \log \frac{\mathcal{A}_C^c}{\mathcal{A}_A^a} - \left(E_-^\circ - \frac{0.059 \ 16}{n} \log \frac{\mathcal{A}_B^b}{\mathcal{A}_D^d} \right)$$

$$E = \underbrace{(E_+^\circ - E_-^\circ)}_{E^\circ} - \frac{0.059 \ 16}{n} \log \underbrace{\frac{\mathcal{A}_C^c \mathcal{A}_D^d}{\mathcal{A}_A^a \mathcal{A}_B^b}}_{Q} = E^\circ - \frac{0.059 \ 16}{n} \log Q \qquad \textbf{(14-23)}$$

$\log a + \log b = \log ab$

Equation 14-23 is true at any time. In the special case when the cell is at equilibrium, $E = 0$ and $Q = K$, the equilibrium constant. Therefore, Equation 14-23 is transformed into these most important forms at equilibrium:

Finding E° from K: $E^\circ = \dfrac{0.059 \ 16}{n} \log K$ (at 25°C) $\textbf{(14-24)}$

Finding K from E°: $K = 10^{nE^\circ / 0.059 \ 16}$ (at 25°C) $\textbf{(14-25)}$

To go from Equation 14-24 to 14-25:

$$\frac{0.059 \ 16}{n} \log K = E^\circ$$

$$\log K = \frac{nE^\circ}{0.059 \ 16}$$

$$10^{\log K} = 10^{nE^\circ / 0.059 \ 16}$$

$$K = 10^{nE^\circ / 0.059 \ 16}$$

Equation 14-25 allows us to deduce the equilibrium constant from E°. Alternatively, we can find E° from K with Equation 14-24.

EXAMPLE   Using E° to Find the Equilibrium Constant

Find the equilibrium constant for the reaction

$$\text{Cu}(s) + 2\text{Fe}^{3+} \rightleftharpoons 2\text{Fe}^{2+} + \text{Cu}^{2+}$$

Solution Divide the reaction into two half-reactions found in Appendix H:

$$\begin{array}{ll} 2\text{Fe}^{3+} + 2e^- \rightleftharpoons 2\text{Fe}^{2+} & E_+^\circ = 0.771 \text{ V} \\ - \quad \underline{\text{Cu}^{2+} + 2e^- \rightleftharpoons \text{Cu}(s)} & E_-^\circ = 0.339 \text{ V} \\ \text{Cu}(s) + 2\text{Fe}^{3+} \rightleftharpoons 2\text{Fe}^{2+} + \text{Cu}^{2+} & \end{array}$$

We associate E_-° with the half-reaction that must be *reversed* to get the desired net reaction.

Then find E° for the net reaction

$$E^\circ = E_+^\circ - E_-^\circ = 0.771 - 0.339 = 0.432 \text{ V}$$

and compute the equilibrium constant with Equation 14-25:

$$K = 10^{(2)(0.432)/(0.059 \ 16)} = 4 \times 10^{14}$$

A modest value of E° produces a large equilibrium constant. The value of K is correctly expressed with one significant figure because E° has three digits. Two are used for the exponent (14), and one is left for the multiplier (4).

Significant figures for logs and exponents were discussed in Section 3-2.

TEST YOURSELF   Find K for the reaction $\text{Cu}(s) + 2\text{Ag}^+ \rightleftharpoons 2\text{Ag}(s) + \text{Cu}^{2+}$. (**Answer:** $E^\circ = 0.460$ V, $K = 4 \times 10^{15}$)

Finding K for Net Reactions That Are Not Redox Reactions

Consider the following half-reactions whose difference is the solubility reaction for iron(II) carbonate (which is not a redox reaction):

$$\begin{array}{ll} FeCO_3(s) + 2e^- \rightleftharpoons Fe(s) + CO_3^{2-} & E_+^\circ = -0.756 \text{ V} \\ \underline{- \quad Fe^{2+} + 2e^- \rightleftharpoons Fe(s)} & \underline{E_-^\circ = -0.44 \text{ V}} \\ FeCO_3(s) \rightleftharpoons Fe^{2+} + CO_3^{2-} & E^\circ = -0.756 - (-0.44) = -0.31_6 \text{ V} \end{array}$$

Iron(II)
carbonate

$$K = K_{sp} = 10^{(2)(-0.31_6)/(0.059\ 16)} = 10^{-11}$$

From E° for the net reaction, we can compute K_{sp} for iron(II) carbonate. Potentiometric measurements allow us to find equilibrium constants that are too small or too large to measure by determining concentrations of reactants and products directly.

"Wait!" you protest. "How can there be a redox potential for a reaction that is not a redox reaction?" Box 14-5 shows that the redox potential is just another way of expressing the free energy change of the reaction. The more energetically favorable the reaction (the more negative ΔG°), the more positive is E°.

The general form of a problem involving the relation between E° values for half-reactions and K for a net reaction is

$$\begin{array}{ll} \text{Half-reaction} & E_+^\circ \\ \underline{- \quad \text{Half-reaction}} & \underline{E_-^\circ} \\ \text{Net reaction} & E^\circ = E_+^\circ - E_-^\circ \qquad K = 10^{nE^\circ/0.059\ 16} \end{array}$$

If you know E_-° and E_+°, you can find E° and K for the net cell reaction. Alternatively, if you know E° and either E_-° or E_+°, you can find the missing standard potential. If you know K, you can calculate E° and use it to find either E_-° or E_+°, provided you know one of them.

EXAMPLE Relating E° and K

From the formation constant of Ni(glycine)$_2$ plus E° for the Ni^{2+} | Ni(s) couple,

$$\begin{array}{ll} Ni^{2+} + 2 \text{ glycine}^- \rightleftharpoons Ni(glycine)_2 & K \equiv \beta_2 = 1.2 \times 10^{11} \\ Ni^{2+} + 2e^- \rightleftharpoons Ni(s) & E^\circ = -0.236 \text{ V} \end{array}$$

deduce the value of E° for the reaction

$$Ni(glycine)_2 + 2e^- \rightleftharpoons Ni(s) + 2 \text{ glycine}^- \qquad (14\text{-}26)$$

Solution We need to see the relations among the three reactions:

$$\begin{array}{ll} \quad Ni^{2+} + 2e^- \rightleftharpoons Ni(s) & E_+^\circ = -0.236 \text{ V} \\ \underline{- \quad Ni(glycine)_2 + 2e^- \rightleftharpoons Ni(s) + 2 \text{ glycine}^-} & \underline{E_-^\circ = ?} \\ Ni^{2+} + 2 \text{ glycine}^- \rightleftharpoons Ni(glycine)_2 & E^\circ = ? \quad K = 1.2 \times 10^{11} \end{array}$$

We know that $E_+^\circ - E_-^\circ$ must equal E°, so we can deduce the value of E_-° if we can find E°. But E° can be determined from the equilibrium constant for the net reaction:

$$E^\circ = \frac{0.059\ 16}{n} \log K = \frac{0.059\ 16}{2} \log(1.2 \times 10^{11}) = 0.328 \text{ V}$$

Hence, the standard reduction potential for half-reaction 14-26 is

$$E_-^\circ = E_+^\circ - E^\circ = -0.236 - 0.328 = -0.564 \text{ V}$$

TEST YOURSELF Select half-reactions in Appendix H to find the formation constant β_2 for Cu$^+$ + 2 ethylenediamine $\rightleftharpoons$ Cu(ethylenediamine)$_2^+$. (*Answer:* $E^\circ = 0.637$ V, $\beta_2 = 6 \times 10^{10}$)

14-6 Cells as Chemical Probes[19]

It is essential to distinguish two classes of equilibria associated with galvanic cells:

1. Equilibrium *between* the two half-cells
2. Equilibrium *within* each half-cell

Margin notes:

E° for dissolution of iron(II) carbonate is negative, which means that the reaction is "not spontaneous." "Not spontaneous" simply means $K < 1$.

Possible structure of Ni(glycine)$_2$

The more negative potential of -0.564 V for reducing Ni(glycine)$_2$ compared to -0.236 V for reducing Ni^{2+} tells us that it is harder to reduce Ni(glycine)$_2$ than Ni^{2+}. Complexation with glycine makes it harder to reduce Ni^{2+}.

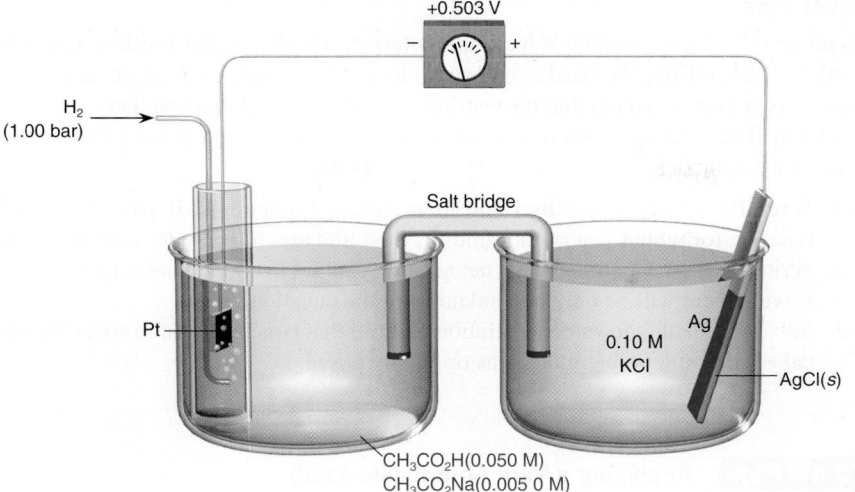

+0.503 V

H₂
(1.00 bar)

Salt bridge

Pt

Ag

0.10 M
KCl

AgCl(s)

CH₃CO₂H(0.050 M)
CH₃CO₂Na(0.005 0 M)

$Pt(s) | H_2(1.00 \text{ bar}) | CH_3CO_2H(0.050 \text{ M}), \; CH_3CO_2Na(0.005 \text{ 0 M}) \| Cl^-(0.10 \text{ M}) | AgCl(s) | Ag(s)$

Aha! A buffer!

If a galvanic cell has a nonzero voltage, then the net cell reaction is not at equilibrium. We say that equilibrium *between* the two half-cells has not been established.

*We allow half-cells to stand long enough to come to chemical equilibrium **within** each half-cell.* For example, in the right-hand half-cell in Figure 14-10, the reaction

$$AgCl(s) \rightleftharpoons Ag^+(aq) + Cl^-(aq)$$

is at equilibrium. It is not part of the net cell reaction. It simply occurs when AgCl(s) is in contact with water. In the left half-cell, the reaction

$$CH_3CO_2H \rightleftharpoons CH_3CO_2^- + H^+$$

has also come to equilibrium. Neither reaction is part of the net cell redox reaction.

The redox reaction for the right half-cell of Figure 14-10 is

$$AgCl(s) + e^- \rightleftharpoons Ag(s) + Cl^-(aq, 0.10 \text{ M}) \qquad E_+^\circ = 0.222 \text{ V}$$

But what is the reaction in the left half-cell? The only element we find in two oxidation states is hydrogen. We see that $H_2(g)$ bubbles into the cell, and we also realize that every aqueous solution contains H^+. Therefore, hydrogen is present in two oxidation states, and the half-reaction that takes place on the catalytic Pt surface can be written as

$$2H^+(aq, ? \text{ M}) + 2e^- \rightleftharpoons H_2(g, 1.00 \text{ bar}) \qquad E_-^\circ = 0$$

The net cell reaction is not at equilibrium, because the measured voltage is 0.503 V, not 0 V.

The Nernst equation for the net cell reaction is

$$E = E_+ - E_- = (0.222 - 0.059\,16 \log[Cl^-]) - \left(0 - \frac{0.059\,16}{2} \log \frac{P_{H_2}}{[H^+]^2} \right)$$

After inserting the known quantities, we discover that the only unknown is $[H^+]$. *The measured voltage therefore allows us to find $[H^+]$ in the left half-cell:*

$$0.503 = (0.222 - 0.059\,16 \log[0.10]) - \left(0 - \frac{0.059\,16}{2} \log \frac{1.00}{[H^+]^2} \right)$$

$$\Rightarrow [H^+] = 1.8 \times 10^{-4} \text{ M}$$

This, in turn, allows us to evaluate the equilibrium constant for the acid-base reaction that has come to equilibrium in the left half-cell:

$$K_a = \frac{[CH_3CO_2^-][H^+]}{[CH_3CO_2H]} = \frac{(0.005\,0)(1.8 \times 10^{-4})}{0.050} = 1.8 \times 10^{-5}$$

The cell in Figure 14-10 acts as a *probe* to measure $[H^+]$ in the left half-cell. Using this type of cell, we could determine the equilibrium constant for acid dissociation or base hydrolysis in the left half-cell.

A chemical reaction that occurs *within one half-cell* will reach equilibrium and is assumed to remain at equilibrium. Such a reaction is not the net cell reaction.

Question Why can we assume that the concentrations of acetic acid and acetate ion are equal to their initial (formal) concentrations?

Survival Tips

Problems in this chapter include some brainbusters designed to bring together your knowledge of electrochemistry, chemical equilibrium, solubility, complex formation, and acid-base chemistry. They require you to find the equilibrium constant for a reaction that occurs in only one half-cell. The reaction of interest is not the net cell reaction and is not a redox reaction. Here is a good approach:

Half-reactions that you write *must* involve species that appear in two oxidation states in the cell.

Step 1 Write the two half-reactions and their standard potentials. If you choose a half-reaction for which you cannot find $E°$, then find another way to write the reaction.

Step 2 Write a Nernst equation for the net reaction and put in all the known quantities. If all is well, there will be only one unknown in the equation.

Step 3 Solve for the unknown concentration and use that concentration to solve the chemical equilibrium problem that was originally posed.

EXAMPLE   Analyzing a Very Complicated Cell

The cell in Figure 14-11 measures the formation constant (K_f) of $Hg(EDTA)^{2-}$. The right-hand compartment contains 0.500 mmol of Hg^{2+} and 2.00 mmol of EDTA in 0.100 L buffered to pH 6.00. The voltage is +0.342 V. Find the value of K_f for $Hg(EDTA)^{2-}$.

Solution

Step 1 The left half-cell is a standard hydrogen electrode for which we can say $E_- = 0$. In the right half-cell, mercury is in two oxidation states. So let's write the half-reaction

$$Hg^{2+} + 2e^- \rightleftharpoons Hg(l) \qquad E_+° = 0.852 \text{ V}$$

$$E_+ = 0.852 - \frac{0.059\ 16}{2}\log\left(\frac{1}{[Hg^{2+}]}\right)$$

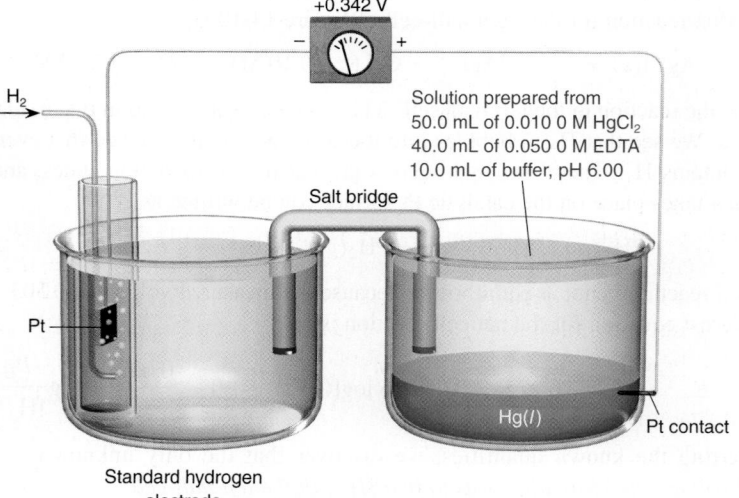

FIGURE 14-11 A galvanic cell that can be used to measure the formation constant for $Hg(EDTA)^{2-}$.

+0.342 V

H₂

Solution prepared from
50.0 mL of 0.010 0 M HgCl₂
40.0 mL of 0.050 0 M EDTA
10.0 mL of buffer, pH 6.00

Salt bridge

Pt

Hg(*l*)

Pt contact

Standard hydrogen
electrode

S.H.E. ‖ Hg(EDTA)²⁻(*aq*, 0.005 00 M), EDTA(*aq*, 0.015 0 M) | Hg(*l*)

In the right half-cell, the reaction between Hg^{2+} and EDTA is

$$Hg^{2+} + Y^{4-} \overset{K_f}{\rightleftharpoons} HgY^{2-}$$

Because we expect K_f to be large, we assume that virtually all the Hg^{2+} has reacted to make HgY^{2-}. Therefore, the concentration of HgY^{2-} is 0.500 mmol/100 mL = 0.005 00 M. The remaining EDTA has a total concentration of (2.00 − 0.50) mmol/100 mL = 0.015 0 M. The right-hand compartment therefore contains 0.005 00 M HgY^{2-}, 0.015 0 M EDTA, and a small, unknown concentration of Hg^{2+}.

The formation constant for HgY^{2-} can be written

$$K_f = \frac{[HgY^{2-}]}{[Hg^{2+}][Y^{4-}]} = \frac{[HgY^{2-}]}{[Hg^{2+}]\alpha_{Y^{4-}}[EDTA]}$$

Recall that $[Y^{4-}] = \alpha_{Y^{4-}}[EDTA]$.

where [EDTA] is the formal concentration of EDTA not bound to metal. In this cell, [EDTA] = 0.015 0 M. The fraction of EDTA in the form Y^{4-} is $\alpha_{Y^{4-}}$ (Section 12-2). Because we know that $[HgY^{2-}] = 0.005\ 00$ M, all we need to find is $[Hg^{2+}]$ in order to evaluate K_f.

Step 2 The Nernst equation for the net cell reaction is

$$E = 0.342 = E_+ - E_- = \left[0.852 - \frac{0.059\ 16}{2} \log\left(\frac{1}{[Hg^{2+}]}\right) \right] - (0)$$

in which the only unknown is $[Hg^{2+}]$.

Step 3 Now we solve the Nernst equation to find $[Hg^{2+}] = 5._7 \times 10^{-18}$ M, and this value of $[Hg^{2+}]$ allows us to evaluate the formation constant for HgY^{2-}:

$$K_f = \frac{[HgY^{2-}]}{[Hg^{2+}]\alpha_{Y^{4-}}[EDTA]} = \frac{(0.005\ 00)}{(5._7 \times 10^{-18})(1.8 \times 10^{-5})(0.015\ 0)}$$

$$= 3 \times 10^{21}$$

$\alpha_{Y^{4-}}$ comes from Table 12-1.

The mixture of EDTA plus $Hg(EDTA)^{2-}$ in the cathode serves as a mercuric ion "buffer" that fixes the concentration of Hg^{2+}. This concentration, in turn, determines the cell voltage.

TEST YOURSELF Find K_f if the cell voltage had been 0.300 V. (**Answer:** 8×10^{22})

14-7 Biochemists Use $E^{\circ\prime}$

In respiration, O_2 oxidizes molecules from food to yield energy or metabolic intermediates. The standard reduction potentials that we have been using so far apply to systems in which all activities of reactants and products are unity. If H^+ is involved in the reaction, E° applies when pH = 0 ($\mathcal{A}_{H^+} = 1$). *Whenever H^+ appears in a redox reaction, or whenever reactants or products are acids or bases, reduction potentials are pH dependent.*

Because the pH inside a plant or animal cell is about 7, reduction potentials that apply at pH 0 are not particularly appropriate. For example, at pH 0, ascorbic acid (vitamin C) is a more powerful reducing agent than succinic acid. However, at pH 7, this order is reversed. It is the reducing strength at pH 7, not at pH 0, that is relevant to a living cell.

The *standard potential* for a redox reaction is defined for a galvanic cell in which all activities are unity. The **formal potential** is the reduction potential that applies under a *specified* set of conditions (including pH, ionic strength, and concentration of complexing agents). Biochemists call the formal potential at pH 7 $E^{\circ\prime}$ (read "E zero prime"). Table 14-2 lists $E^{\circ\prime}$ values for some biological redox couples.

The formal potential at pH = 7 is called $E^{\circ\prime}$.

Relation Between E° and $E^{\circ\prime}$

Consider the half-reaction

$$aA + ne^- \rightleftharpoons bB + mH^+ \qquad E^\circ$$

in which A is an oxidized species and B is a reduced species. Both A and B might be acids or bases, as well. The Nernst equation for this half-reaction is

$$E = E^\circ - \frac{0.059\ 16}{n} \log \frac{[B]^b[H^+]^m}{[A]^a}$$

To find $E^{\circ\prime}$, we rearrange the Nernst equation to a form in which the log term contains only the *formal concentrations* of A and B raised to the powers a and b, respectively.

Recipe for $E^{\circ\prime}$: $E = \underbrace{E^\circ + \text{other terms}}_{\substack{\text{All of this is called } E^{\circ\prime} \\ \text{when pH} = 7}} - \frac{0.059\ 16}{n} \log \frac{F_B^b}{F_A^a}$ (14-27)

If we had included activity coefficients, they would appear in $E^{\circ\prime}$ also.

The entire collection of terms over the brace, evaluated at pH = 7, is called $E^{\circ\prime}$.

TABLE 14-2 Reduction potentials of biological interest

Reaction	$E°$ (V)	$E°'$ (V)
$O_2 + 4H^+ + 4e^- \rightleftharpoons 2H_2O$	+1.229	+0.815
$Fe^{3+} + e^- \rightleftharpoons Fe^{2+}$	+0.771	+0.771
$I_2 + 2e^- = 2I^-$	+0.535	+0.535
Cytochrome a $(Fe^{3+}) + e^- \rightleftharpoons$ cytochrome a (Fe^{2+})	+0.290	+0.290
$O_2(g) + 2H^+ + 2e^- \rightleftharpoons H_2O_2$	+0.695	+0.281
Cytochrome c $(Fe^{3+}) + e^- \rightleftharpoons$ cytochrome c (Fe^{2+})	—	+0.254
2, 6-Dichlorophenolindophenol $+ 2H^+ + 2e^- \rightleftharpoons$ reduced 2,6-dichlorophenolindophenol	—	+0.22
Dehydroascorbate $+ 2H^+ + 2e^- \rightleftharpoons$ ascorbate $+ H_2O$	+0.390	+0.058
Fumarate $+ 2H^+ + 2e^- \rightleftharpoons$ succinate	+0.433	+0.031
Methylene blue $+ 2H^+ + 2e^- \rightleftharpoons$ reduced product	+0.532	+0.011
Glyoxylate $+ 2H^+ + 2e^- \rightleftharpoons$ glycolate	—	−0.090
Oxaloacetate $+ 2H^+ + 2e^- \rightleftharpoons$ malate	+0.330	−0.102
Pyruvate $+ 2H^+ + 2e^- \rightleftharpoons$ lactate	+0.224	−0.190
Riboflavin $+ 2H^+ + 2e^- \rightleftharpoons$ reduced riboflavin	—	−0.208
$FAD + 2H^+ + 2e^- \rightleftharpoons FADH_2$	—	−0.219
$(Glutathione-S)_2 + 2H^+ + 2e^- \rightleftharpoons$ 2 glutathione-SH	—	−0.23
Safranine T $+ 2e^- \rightleftharpoons$ leucosafranine T	−0.235	−0.289
$(C_6H_5S)_2 + 2H^+ + 2e^- \rightleftharpoons 2C_6H_5SH$	—	−0.30
$NAD^+ + H^+ + 2e^- \rightleftharpoons NADH$	−0.105	−0.320
$NADP^+ + H^+ + 2e^- \rightleftharpoons NADPH$	—	−0.324
Cystine $+ 2H^+ + 2e^- \rightleftharpoons$ 2 cysteine	—	−0.340
Acetoacetate $+ 2H^+ + 2e^- \rightleftharpoons$ L-β-hydroxybutyrate	—	−0.346
Xanthine $+ 2H^+ + 2e^- \rightleftharpoons$ hypoxanthine $+ H_2O$	—	−0.371
$2H^+ + 2e^- \rightleftharpoons H_2$	0.000	−0.414
Gluconate $+ 2H^+ + 2e^- \rightleftharpoons$ glucose $+ H_2O$	—	−0.44
$SO_4^{2-} + 2e^- + 2H^+ \rightleftharpoons SO_3^{2-} + H_2O$	—	−0.454
$2SO_3^{2-} + 2e^- + 4H^+ \rightleftharpoons S_2O_4^{2-} + 2H_2O$	—	−0.527

To convert [A] or [B] into F_A or F_B, we use fractional composition equations (Section 10-5), which relate the formal (that is, total) concentration of *all* forms of an acid or a base to its concentration in a *particular* form:

For a monoprotic acid:

$F = [HA] + [A^-]$

For a diprotic acid:

$F = [H_2A] + [HA^-] + [A^{2-}]$

Monoprotic system:

$$[HA] = \alpha_{HA}F = \frac{[H^+]F}{[H^+] + K_a} \qquad (14\text{-}28)$$

$$[A^-] = \alpha_{A^-}F = \frac{K_aF}{[H^+] + K_a} \qquad (14\text{-}29)$$

Diprotic system:

$$[H_2A] = \alpha_{H_2A}F = \frac{[H^+]^2F}{[H^+]^2 + [H^+]K_1 + K_1K_2} \qquad (14\text{-}30)$$

$$[HA^-] = \alpha_{HA^-}F = \frac{K_1[H^+]F}{[H^+]^2 + [H^+]K_1 + K_1K_2} \qquad (14\text{-}31)$$

$$[A^{2-}] = \alpha_{A^{2-}}F = \frac{K_1K_2F}{[H^+]^2 + [H^+]K_1 + K_1K_2} \qquad (14\text{-}32)$$

where F is the formal concentration of HA or H_2A, K_a is the acid dissociation constant for HA, and K_1 and K_2 are the acid dissociation constants for H_2A.

One way to *measure* $E°'$ is to establish a half-cell in which the formal concentrations of the oxidized and reduced species are equal and the pH is adjusted to 7. Then the log term in Equation 14-27 is zero and the measured potential (versus S.H.E.) is $E°'$.

EXAMPLE **Calculating the Formal Potential**

Find $E^{\circ\prime}$ for the reaction

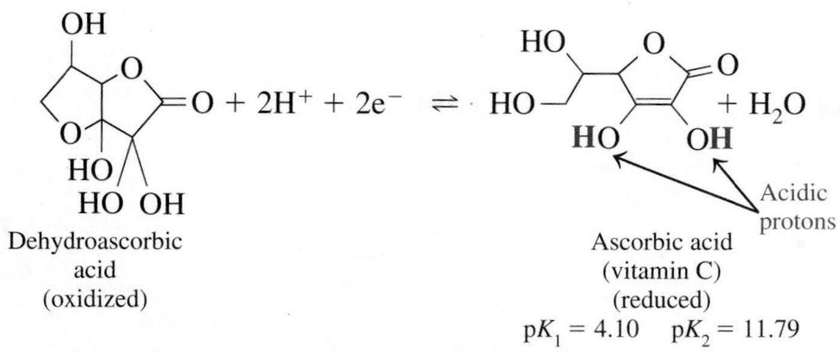

$$=O + 2H^+ + 2e^- \rightleftharpoons HO- \quad\quad + H_2O \quad E^\circ = 0.390 \text{ V} \quad \textbf{(14-33)}$$

Acidic protons

Dehydroascorbic
acid
(oxidized)

Ascorbic acid
(vitamin C)
(reduced)

$pK_1 = 4.10 \quad pK_2 = 11.79$

Solution Abbreviating dehydroascorbic acid[20] as D, and ascorbic acid as H_2A, we rewrite the reduction as

$$D + 2H^+ + 2e^- \rightleftharpoons H_2A + H_2O$$

for which the Nernst equation is

$$E = E^\circ - \frac{0.059\ 16}{2}\log\frac{[H_2A]}{[D][H^+]^2} \quad\quad \textbf{(14-34)}$$

D is not an acid or a base, so its formal concentration equals its molar concentration: $F_D = [D]$. For the diprotic acid H_2A, we use Equation 14-30 to express $[H_2A]$ in terms of F_{H_2A}:

$$[H_2A] = \frac{[H^+]^2 F_{H_2A}}{[H^+]^2 + [H^+]K_1 + K_1K_2}$$

Putting these values into Equation 14-34 gives

$$E = E^\circ - \frac{0.059\ 16}{2}\log\left(\frac{\dfrac{[H^+]^2\ F_{H_2A}}{[H^+]^2 + [H^+]K_1 + K_1K_2}}{F_D[H^+]^2}\right)$$

which can be rearranged to the form

$$E = E^\circ - \underbrace{\frac{0.059\ 16}{2}\log\left(\frac{1}{[H^+]^2 + [H^+]K_1 + K_1K_2}\right)}_{\substack{\text{Formal potential} (= E^{\circ\prime} \text{ if pH} = 7) \\ = +0.062 \text{ V}}} - \frac{0.059\ 16}{2}\log\frac{F_{H_2A}}{F_D} \quad \textbf{(14-35)}$$

Putting the values of E°, K_1, and K_2 into Equation 14-35 and setting $[H^+] = 10^{-7.00}$, we find $E^{\circ\prime} = +0.062$ V.

TEST YOURSELF Compute $E^{\circ\prime}$ for the reaction $O_2 + 4H^+ + 4e^- \rightleftharpoons 2H_2O$. (*Answer:* 0.815 V)

Curve *a* in Figure 14-12 shows how the calculated formal potential for Reaction 14-33 depends on pH. The potential decreases as the pH increases, until pH $\approx pK_2 = 11.79$. Above pK_2, A^{2-} is the dominant form of ascorbic acid, and no protons are involved in the net redox reaction. Therefore, the potential becomes independent of pH.

A biological example of $E^{\circ\prime}$ is the reduction of Fe(III) in the protein transferrin. This protein has two Fe(III)-binding sites, one in each half of the molecule designated C and N for the carboxyl and amino terminals of the peptide chain. Transferrin carries Fe(III) through the blood to cells that require iron. Membranes of these cells have a receptor that binds Fe(III)-transferrin and takes it into a compartment called an endosome into which H^+ is pumped to lower the pH to ~5.8. Iron is released from transferrin in the endosome and continues into the cell as Fe(II) attached to an intracellular metal-transport protein. The entire cycle of transferrin uptake, metal removal, and transferrin release back to the bloodstream takes 1−2 min. The time required for Fe(III) to dissociate from transferrin at pH 5.8 is ~6 min, which is too long to account for release in the endosome. The reduction potential of Fe(III)-transferrin at pH 5.8 is $E^{\circ\prime} = -0.52$ V, which is too low for physiologic reductants.

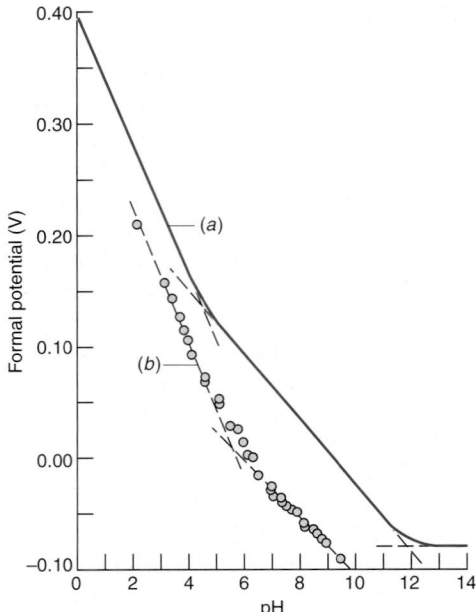

FIGURE 14-12 Reduction potential of ascorbic acid, showing its dependence on pH. (*a*) Graph of the function labeled formal potential in Equation 14-35. (*b*) Experimental polarographic half-wave reduction potential of ascorbic acid in a medium of ionic strength = 0.2 M. The half-wave potential (Chapter 17) is nearly the same as the formal potential. At high pH (>12), the half-wave potential does not level off to a slope of 0, as Equation 14-35 predicts. Instead, hydrolysis of ascorbic acid occurs and the chemistry is more complex than Reaction 14-33. [Data from J. J. Ruiz, A. Aldaz, and M. Dominguez, *Can. J. Chem.* **1977,** *55,* 2799; ibid., **1978,** *56,* 1533.]

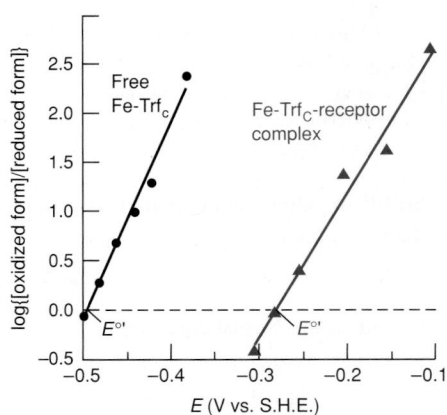

FIGURE 14-13 Spectroscopic measurement of log{[Fe(III)Trf$_C$]/[Fe(II)Trf$_C$]} versus potential at pH 5.8. [Data from S. Dhungana, C. H. Taboy, O. Zak, M. Larvie, A. L. Crumbliss, and P. Aisen, "Redox Properties of Human Transferrin Bound to Its Receptor," *Biochemistry* **2004,** *43,* 205.]

The mystery of how Fe(III) is released from transferrin in the endosome was solved by measuring $E^{\circ'}$ for the Fe(III)-transferrin-receptor complex at pH 5.8. To simplify the chemistry, transferrin was cleaved and only the C-terminal half of the protein (designated Trf$_C$) was used. Figure 14-13 shows measurements of log{[Fe(III)Trf$_C$]/[Fe(II)Trf$_C$]} for free protein and for the protein-receptor complex. In Equation 14-27, $E = E^{\circ'}$ when the log term is zero (that is, when [Fe(III)Trf$_C$] = [Fe(II)Trf$_C$]). Figure 14-13 shows that $E^{\circ'}$ for Fe(III)Trf$_C$ is near -0.50 V, but $E^{\circ'}$ for the Fe(III)Trf$_C$-receptor complex is -0.29 V. The reducing agents NADH and NADPH in Table 14-2 are strong enough to reduce Fe(III)Trf$_C$ bound to its receptor at pH 5.8, but not strong enough to reduce free Fe(III)-transferrin.

Terms to Understand

ampere	Faraday constant	oxidant	reduction
anode	formal potential	oxidation	resistance
cathode	galvanic cell	oxidizing agent	salt bridge
coulomb	half-reaction	potentiometer	standard hydrogen electrode
current	joule	power	standard reduction potential
$E^{\circ'}$	Latimer diagram	reaction quotient	volt
electric potential	Nernst equation	redox reaction	watt
electrochemistry	ohm	reducing agent	
electrode	Ohm's law	reductant	

Summary

Electric current is the number of coulombs of charge per second passing a point. The electric potential difference, E (volts), between two points is the work (joules) per unit charge that is needed or can be done when charge moves from one point to the other. For N moles of a species with n charges per molecule, the electric charge in coulombs is $q = nNF$, where F is the Faraday constant (C/mol). Work done when a charge of q coulombs passes through a potential difference of E volts is work = $E \cdot q$. The maximum work that can be done on the surroundings by a spontaneous chemical reaction is related to the free energy change for the reaction: work = $-\Delta G$. If a chemical change produces a potential difference, E, the relation between free energy and potential difference is $\Delta G = -nNFE$, where nN = moles of charge taken through the potential difference. Ohm's law ($I = E/R$) describes the relation between current, voltage, and resistance in an electric circuit. It can be combined with the definitions of work and power (P = work per second) to give $P = E \cdot I = I^2R$.

A galvanic cell uses a spontaneous redox reaction to produce electricity. The electrode at which oxidation occurs is the anode, and the electrode at which reduction occurs is the cathode. Two half-cells are usually separated by a salt bridge that allows ions to migrate from one side to the other to maintain charge neutrality, but prevents reactants in the two half-cells from mixing. The standard reduction potential of a half-reaction is the difference in potential measured when that reaction is connected to a standard hydrogen electrode. The term "standard" means that activities of reactants and products are unity. If several half-reactions are added to give another half-reaction, the standard potential of the net half-reaction can be found by equating the free energy of the net half-reaction to the sum of free energies of the component half-reactions.

In a galvanic cell, electrons flow through a wire from the less positive electrode to the more positive electrode. The voltage of

the cell is the difference between the potentials of the two half-reactions: $E = E_+ - E_-$, where E_+ is the potential of the half-cell connected to the positive terminal of the potentiometer and E_- is the potential of the half-cell connected to the negative terminal. The potential of each half-reaction is given by the Nernst equation: $E = E° - (0.059\ 16/n) \log Q$ (at 25°C), where each reaction is written as a reduction and Q is the reaction quotient. The reaction quotient has the same form as the equilibrium constant, but it is evaluated with concentrations existing at the time of interest.

Complex equilibria can be studied by making them part of an electrochemical cell. If we measure the voltage and know the concentrations (activities) of all but one of the reactants and products, the Nernst equation allows us to compute the concentration of the unknown species. The electrochemical cell serves as a probe for that species.

Biochemists use the formal potential of a half-reaction at pH 7 ($E°'$) instead of the standard potential ($E°$), which applies at pH 0. $E°'$ is found by writing the Nernst equation for the half-reaction and grouping together all terms except the logarithm containing the formal concentrations of reactant and product. The combination of terms, evaluated at pH 7, is $E°'$.

Exercises

14-A. In olden days, mercury batteries with the following chemistry were used to power heart pacemakers:

$$Zn(s) + HgO(s) \rightarrow ZnO(s) + Hg(l) \qquad E° = 1.35 \text{ V}$$

(Mercury cells were phased out because they create environmental hazards when disposed.) What is the cell voltage? If the power required to operate the pacemaker is 0.010 0 W, how many kilograms of HgO (FM 216.59) will be consumed in 365 days? How many pounds of HgO is this? (1 pound = 453.6 g)

14-B. Calculate $E°$ and K for each of the following reactions.
(a) $I_2(s) + 5Br_2(aq) + 6H_2O \rightleftharpoons 2IO_3^- + 10Br^- + 12H^+$
(b) $Cr^{2+} + Fe(s) \rightleftharpoons Fe^{2+} + Cr(s)$
(c) $Mg(s) + Cl_2(g) \rightleftharpoons Mg^{2+} + 2Cl^-$
(d) $5MnO_2(s) + 4H^+ \rightleftharpoons 3Mn^{2+} + 2MnO_4^- + 2H_2O$
(e) $Ag^+ + 2S_2O_3^{2-} \rightleftharpoons Ag(S_2O_3)_2^{3-}$
(f) $CuI(s) \rightleftharpoons Cu^+ + I^-$

14-C. Calculate the voltage of each of the following cells. State the direction of electron flow.

(a) $Fe(s) \mid FeBr_2(0.010 \text{ M}) \parallel NaBr(0.050 \text{ M}) \mid Br_2(l) \mid Pt(s)$
(b) $Cu(s) \mid Cu(NO_3)_2(0.020 \text{ M}) \parallel Fe(NO_3)_2(0.050 \text{ M}) \mid Fe(s)$
(c) $Hg(l) \mid Hg_2Cl_2(s) \mid KCl(0.060\text{ M}) \parallel KCl(0.040 \text{ M}) \mid Cl_2(g,\ 0.50$ bar$) \mid Pt(s)$

14-D. The left half-reaction of the cell drawn here can be written in *either* of two ways:

$$AgI(s) + e^- \rightleftharpoons Ag(s) + I^- \qquad (1)$$
$$Ag^+ + e^- \rightleftharpoons Ag(s) \qquad (2)$$

The right half-cell reaction is

$$H^+ + e^- \rightleftharpoons \frac{1}{2}H_2(g) \qquad (3)$$

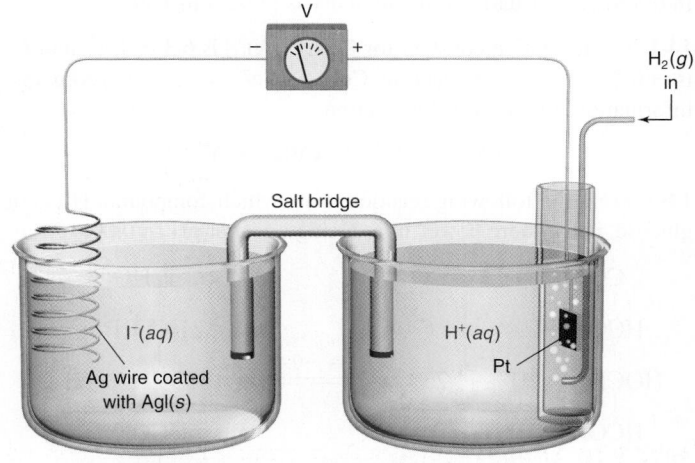

$Ag(s) \mid AgI(s) \mid NaI\ (0.10\ \text{M}) \parallel HCl\ (0.10\ \text{M}) \mid H_2(g,\ 0.20\ \text{bar}) \mid Pt(s)$

(a) Using Reactions 2 and 3, calculate $E°$ and write the Nernst equation for the cell.
(b) Use the value of K_{sp} for AgI to compute $[Ag^+]$ and find the cell voltage. In which direction do electrons flow?
(c) Suppose, instead, that you wish to describe the cell with Reactions 1 and 3. We know that the cell voltage (E, not $E°$) must be the same, no matter which description we use. Write the Nernst equation for Reactions 1 and 3 and use it to find $E°$ in Reaction 1. Compare your answer with the value in Appendix H.

14-E. Calculate the voltage of the cell

$$Cu(s)|Cu^{2+}(0.030 \text{ M})\|K^+Ag(CN)_2^-(0.010 \text{ M}),$$
$$HCN(0.10 \text{ F}), \text{ buffer to pH } 8.21|Ag(s)$$

by considering the following reactions:

$$Ag(CN)_2^- + e^- \rightleftharpoons Ag(s) + 2CN^- \qquad E° = -0.310 \text{ V}$$
$$HCN \rightleftharpoons H^+ + CN^- \qquad pK_a = 9.21$$

In which direction do electrons flow?

14-F. (a) Write a balanced equation for the reaction $PuO_2^+ \rightarrow Pu^{4+}$ and calculate $E°$ for the reaction.

$$PuO_2^{2+} \xrightarrow{+0.966} PuO_2^+ \xrightarrow{?} Pu^{4+} \xrightarrow{+1.006} Pu^{3+}$$
$$\underset{1.021}{\underbrace{\qquad\qquad\qquad\qquad}}$$

(b) Predict whether an equimolar mixture of PuO_2^{2+} and PuO_2^+ will oxidize H_2O to O_2 at a pH of 2.00 and $P_{O_2} = 0.20$ bar. Will O_2 be liberated at pH 7.00?

14-G. Calculate the voltage of the following cell, in which KHP is potassium hydrogen phthalate, the monopotassium salt of phthalic acid. In which direction do electrons flow?

$$Hg(l)|Hg_2Cl_2(s)|KCl(0.10 \text{ M})\|KHP(0.050 \text{ M})|$$
$$H_2(g, 1.00 \text{ bar})|Pt(s)$$

14-H. The following cell has a voltage of 0.083 V:

$$Hg(l)|Hg(NO_3)_2(0.001\ 0 \text{ M}), KI(0.500 \text{ M})\|S.H.E.$$

From this voltage, calculate the equilibrium constant for the reaction

$$Hg^{2+} + 4I^- \rightleftharpoons HgI_4^{2-}$$

In 0.5 M KI, virtually all the mercury is present as HgI_4^{2-}.

14-I. The formation constant for $Cu(EDTA)^{2-}$ is 6.3×10^{18}, and $E°$ is $+0.339$ V for the reaction $Cu^{2+} + 2e^- \rightleftharpoons Cu(s)$. From this information, find $E°$ for the reaction

$$CuY^{2-} + 2e^- \rightleftharpoons Cu(s) + Y^{4-}$$

14-J. From the following reaction, state which compound, $H_2(g)$ or glucose, is the more powerful reducing agent at pH = 0.00.

$$
\begin{array}{c}
CO_2H \\
| \\
HCOH \\
| \\
HOCH + 2H^+ + 2e^- \\
| \\
HCOH \\
| \\
HCOH \\
| \\
CH_2OH \\
\text{Gluconic acid} \\
pK_a = 3.56
\end{array}
\xrightleftharpoons{E°' = -0.45 \text{ V}}
\begin{array}{c}
CHO \\
| \\
HCOH \\
| \\
HOCH + H_2O \\
| \\
HCOH \\
| \\
HCOH \\
| \\
CH_2OH \\
\text{Glucose} \\
\text{(no acidic protons)}
\end{array}
$$

14-K. Living cells convert energy derived from sunlight or combustion of food into energy-rich ATP (adenosine triphosphate)

molecules. For ATP synthesis, $\Delta G = +34.5$ kJ/mol. This energy is then made available to the cell when ATP is hydrolyzed to ADP (adenosine diphosphate). In animals, ATP is synthesized when protons pass through a complex enzyme in the mitochondrial membrane.[21] Two factors account for the movement of protons through this enzyme into the mitochondrion (see the figure): (1) $[H^+]$ is higher outside the mitochondrion than inside because protons are *pumped* out of the mitochondrion by enzymes that catalyze the oxidation of food. (2) The inside of the mitochondrion is negatively charged with respect to the outside.

(a) The synthesis of one ATP molecule requires $2H^+$ to pass through the phosphorylation enzyme. The difference in free energy when a molecule travels from a region of high activity to a region of low activity is

$$\Delta G = -RT \ln \frac{\mathcal{A}_{\text{high}}}{\mathcal{A}_{\text{low}}}$$

How big must the pH difference be (at 298 K) if the passage of two protons is to provide enough energy to synthesize one ATP molecule?

(b) pH differences this large have not been observed in mitochondria. How great an electric potential difference between inside and outside is necessary for the movement of two protons to provide energy to synthesize ATP? In answering this question, neglect any contribution from the pH difference.

(c) The energy for ATP synthesis is thought to be provided by *both* the pH difference and the electric potential. If the pH difference is 1.00 pH unit, what is the magnitude of the potential difference?

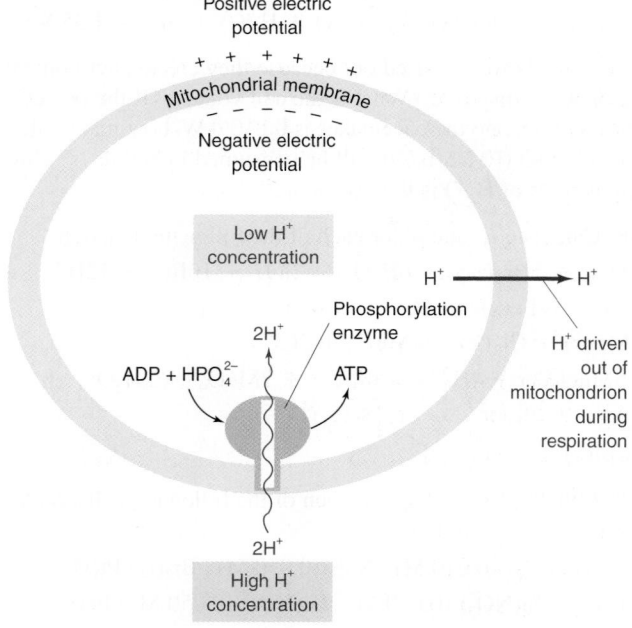

Problems

Basic Concepts

14-1. Explain the difference between electric charge (q, coulombs), electric current (I, amperes), and electric potential (E, volts).

14-2. (a) How many electrons are in one coulomb?

(b) How many coulombs are in one mole of charge?

14-3. The basal rate of consumption of O_2 by a 70-kg human is about 16 mol of O_2 per day. This O_2 oxidizes food and is reduced to H_2O, providing energy for the organism:

$$O_2 + 4H^+ + 4e^- \rightleftharpoons 2H_2O$$

(a) To what current (in amperes = C/s) does this respiration rate correspond? (Current is defined by the flow of electrons from food to O_2.)

(b) Compare your answer in part **(a)** with the current drawn by a refrigerator using 5.00×10^2 W at 115 V. Remember that power (in watts) = work/s = $E \cdot I$.

(c) If the electrons flow from nicotinamide adenine dinucleotide (NADH) to O_2, they experience a potential drop of 1.1 V. What is the power output (in watts) of our human friend?

14-4. A 6.00-V battery is connected across a 2.00-kΩ resistor.

(a) How much current (A) and how many e^-/s flow through the circuit?

(b) How many joules of heat are produced for each electron?

(c) If the circuit operates for 30.0 min, how many moles of electrons will have flowed through the resistor?

(d) What voltage would the battery need to deliver for the power to be 1.00×10^2 W?

14-5. Consider the redox reaction

$$I_2 + 2S_2O_3^{2-} \rightleftharpoons 2I^- + S_4O_6^{2-}$$
$$\text{Thiosulfate} \qquad \qquad \text{Tetrathionate}$$

(a) Identify the oxidizing agent on the left side of the reaction and write its balanced half-reaction.

(b) Identify the reducing agent of the left side of the reaction and write its balanced half-reaction.

(c) How many coulombs of charge are passed from reductant to oxidant when 1.00 g of thiosulfate reacts?

(d) If the rate of reaction is 1.00 g of thiosulfate consumed per minute, what current (in amperes) flows from reductant to oxidant?

14-6. The space shuttle's expendable booster engines derive their power from solid reactants:

$$6NH_4^+ClO_4^-(s) + 10Al(s) \rightarrow 3N_2(g) + 9H_2O(g)$$
FM 117.49
$$\qquad \qquad \qquad \qquad + 5Al_2O_3(s) + 6HCl(g)$$

(a) Find the oxidation numbers of the elements N, Cl, and Al in reactants and products. Which reactants act as reducing agents and which act as oxidants?

(b) The heat of reaction is $-9\,334$ kJ for every 10 mol of Al consumed. Express this as heat released per gram of total reactants.

Galvanic Cells

14-7. Explain how a galvanic cell uses a spontaneous chemical reaction to generate electricity.

14-8. Write a line notation and two reduction half-reactions for each cell pictured.

14-9. Draw a picture of the following cell and write reduction half-reactions for each electrode:

$$Pt(s)|Fe^{3+}(aq), Fe^{2+}(aq)\|Cr_2O_7^{2-}(aq), Cr^{3+}(aq),$$
$$HA(aq)|Pt(s)$$

14-10. Consider the rechargeable battery:

$$Zn(s)|ZnCl_2(aq)\|Cl^-(aq)|Cl_2(l)|C(s)$$

(a) Write reduction half-reactions for each electrode. From which electrode will electrons flow from the battery into a circuit if the electrode potentials are not too different from $E°$ values?

(b) If the battery delivers a constant current of 1.00×10^3 A for 1.00 h, how many kilograms of Cl_2 will be consumed?

14-11. *Lithium-ion battery.*

(a) Write a half-reaction for each electrode of the lithium-ion battery at the opening of this chapter. Which is the anode and which is the cathode? What is the total formula mass of reactants $(2Li_{0.5}CoO_2 + LiC_6)$?

(b) The charge capacity of a battery can be expressed as A·h/kg, which is the number of amperes delivered by 1 kg of reactants for 1 hour. How many coulombs and how many moles of electrons are in 1 A·h?

(c) Show that the theoretical capacity of the battery is 100 A·h/kg of reactants.

(d) The specific energy content of a battery per unit mass of reactants can be expressed as W·h/kg. A Li^+-ion battery delivers 100 A·h/kg of reactants at 3.7 V. Express the specific energy as W·h/g $LiCoO_2$.

Standard Potentials

14-12. Which will be the strongest oxidizing agent under standard conditions (that is, all activities = 1): HNO_2, Se, UO_2^{2+}, Cl_2, H_2SO_3, or MnO_2?

14-13. (a) Cyanide ion causes $E°$ for Fe(III) to decrease:

$$Fe^{3+} + e^- \rightleftharpoons Fe^{2+} \qquad \qquad E° = 0.771 \text{ V}$$
$$\text{Ferric} \qquad \quad \text{Ferrous}$$

$$Fe(CN)_6^{3-} + e^- \rightleftharpoons Fe(CN)_6^{4-} \qquad E° = 0.356 \text{ V}$$
$$\text{Ferricyanide} \qquad \text{Ferrocyanide}$$

Which ion, Fe(III) or Fe(II), is stabilized more by complexing with CN^-?

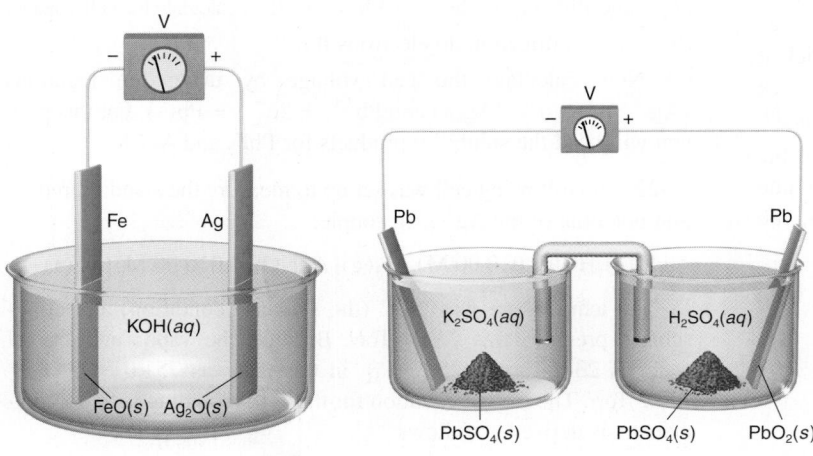

(b) Using Appendix H, answer the same question when the ligand is phenanthroline instead of cyanide.

Phenanthroline

Nernst Equation

14-14. What is the difference between E and $E°$ for a redox reaction? Which one runs down to 0 when the complete cell comes to equilibrium?

14-15. (a) Write the Nernst equations for the half-reactions in Demonstration 14-1. In which direction do electrons move through the circuit?

(b) If you use your fingers as a salt bridge in Demonstration 14-1, will your body take in Cu^{2+} or Zn^{2+}?

14-16. Write the Nernst equation for the following half-reaction and find E when pH = 3.00 and P_{AsH_3} = 1.0 mbar.

$$As(s) + 3H^+ + 3e^- \rightleftharpoons AsH_3(g) \qquad E° = -0.238 \text{ V}$$
Arsine

14-17. (a) Write the line notation for the following cell.

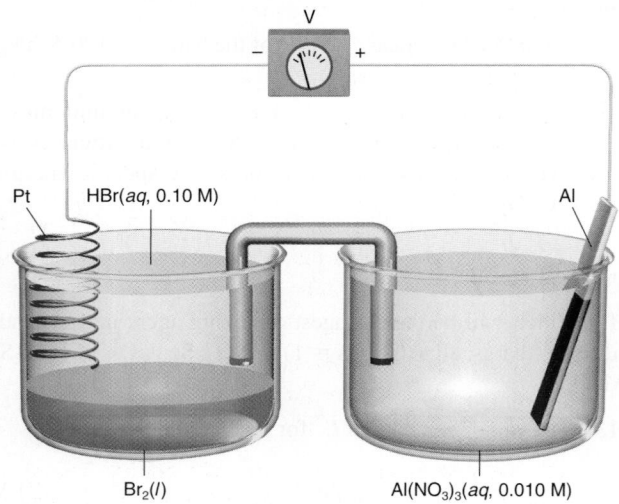

(b) Calculate the potential of each half-cell and the cell voltage, E. In which direction will electrons flow through the circuit? Write the spontaneous net cell reaction.

(c) The left half-cell was loaded with 14.3 mL of $Br_2(l)$ (density = 3.12 g/mL). The aluminum electrode contains 12.0 g of Al. Which element, Br_2 or Al, is the limiting reagent? (That is, which reagent will be used up first?)

(d) If the cell is somehow operated under conditions in which it produces a constant voltage of 1.50 V, how much electrical work will have been done when 0.231 mL of $Br_2(l)$ has been consumed?

(e) If the potentiometer is replaced by a 1.20-kΩ resistor and if the heat dissipated by the resistor is 1.00×10^{-4} J/s, at what rate (grams per second) is $Al(s)$ dissolving? (In this question, the voltage is not 1.50 V.)

14-18. A nickel-metal hydride rechargeable battery formerly used in laptop computers is based on the following chemistry:

Cathode:

$$NiOOH(s) + H_2O + e^- \underset{charge}{\overset{discharge}{\rightleftharpoons}} Ni(OH)_2(s) + OH^-$$

Anode:

$$MH(s) + OH^- \underset{charge}{\overset{discharge}{\rightleftharpoons}} M(s) + H_2O + e^-$$

The anode material, MH, is a transition metal hydride or rare earth alloy hydride. Explain why the voltage remains nearly constant during the entire discharge cycle.

14-19. (a) Write the half-reactions for a H_2–O_2 fuel cell. Find the theoretical cell voltage at 25°C if P_{H_2} = 1.0 bar, P_{O_2} = 0.2 bar, and $[H^+]$ = 0.5 M at both electrodes. (Real fuel cells operate at temperatures of 60° – 1000°C and produce ~0.7 V.)

(b) If the cell is 70% efficient at converting chemical energy into electrical energy, and a stack of cells produces 20 kW at 220 V, how many grams of H_2 are consumed in an hour?

(c) In the U.S., engines are frequently rated in "horsepower." Use Table 1-4 to convert 20 kW to horsepower.

14-20. *Alkaline battery.* Use the web to look up the construction and chemistry of an alkaline battery, such as a commonly used AA battery.

(a) Draw a diagram of the battery and label its components.

(b) Write the anode and cathode half-reactions and the net cell reaction.

(c) From your reading, suggest what is the white chemical that sometimes leaks from an old battery.

(d) *Extra credit:* The zinc half-reaction is listed in Appendix H. The manganese oxide half-reaction can be derived from four half-reactions in the appendix:

$$2MnO_2(s) + 8H^+ + 4e^- \rightleftharpoons 2Mn^{2+} + 4H_2O \qquad \textbf{(1)}$$
$$Mn_2O_3(s) + 6H^+ + 2e^- \rightleftharpoons 2Mn^{2+} + 3H_2O \qquad \textbf{(2)}$$
$$2H_2O + 2e^- \rightleftharpoons H_2(g) + 2OH^- \qquad \textbf{(3)}$$
$$2H^+ + 2e^- \rightleftharpoons H_2(g) \qquad \textbf{(4)}$$

Use the method of Box 14-5 to find the standard potential for the manganese oxide half-reaction. Find the standard potential for the net cell reaction.

The manganese oxide half-reaction is the sum (1) − (2) + (3) − (4). For each half-reaction, $\Delta G° = -nNFE°$, where nN is the number of moles of electrons in that half-reaction. $\Delta G°$ for the net reaction is $\Delta G_1° - \Delta G_2° + \Delta G_3° - \Delta G_4°$. When you multiply a half-reaction in Appendix H, you *do not* multiply $E°$ because the work done *per electron* is independent of how many electrons are transferred.

14-21. Suppose that the concentrations of NaF and KCl were each 0.10 M in the cell $Pb(s) \mid PbF_2(s) \mid F^-(aq) \parallel Cl^-(aq) \mid AgCl(s) \mid Ag(s)$.
(a) Using the half-reactions $2AgCl(s) + 2e^- \rightleftharpoons 2Ag(s) + 2Cl^-$ and $PbF_2(s) + 2e^- \rightleftharpoons Pb(s) + 2F^-$, calculate the cell voltage.

(b) In which direction do electrons flow?

(c) Now calculate the cell voltage by using the reactions $2Ag^+ + 2e^- \rightleftharpoons 2Ag(s)$ and $Pb^{2+} + 2e^- \rightleftharpoons Pb(s)$. For this part, you will need the solubility products for PbF_2 and AgCl.

14-22. The following cell was set up to measure the standard reduction potential of the $Ag^+ \mid Ag$ couple:

$$Pt(s) \mid HCl(0.010\ 00 \text{ M}), H_2(g) \parallel AgNO_3(0.010\ 00 \text{ M}) \mid Ag(s)$$

The temperature was 25°C (the standard condition) and atmospheric pressure was 751.0 Torr. Because the vapor pressure of water is 23.8 Torr at 25°C, P_{H_2} in the cell was 751.0 − 23.8 = 727.2 Torr. The Nernst equation for the cell, including activity coefficients, is derived as follows:

Right electrode: $\quad Ag^+ + e^- \rightleftharpoons Ag(s) \qquad E_+^\circ = E_{Ag^+|Ag}^\circ$

Left electrode: $\quad H^+ + e^- \rightleftharpoons \frac{1}{2}H_2(g) \qquad E_-^\circ = 0\ V$

$$E_+ = E_{Ag^+|Ag}^\circ - 0.059\,16 \log\left(\frac{1}{[Ag^+]\gamma_{Ag^+}}\right)$$

$$E_- = 0 - 0.059\,16 \log\left(\frac{P_{H_2}^{1/2}}{[H^+]\gamma_{H^+}}\right)$$

$$E = E_+ - E_- = E_{Ag^+|Ag}^\circ - 0.059\,16 \log\left(\frac{[H^+]\gamma_{H^+}}{P_{H_2}^{1/2}[Ag^+]\gamma_{Ag^+}}\right)$$

Given a measured cell voltage of +0.798 3 V and using activity coefficients from Table 8-1, find $E_{Ag^+|Ag}^\circ$. Be sure to express P_{H_2} in bar in the reaction quotient.

14-23. Write a balanced chemical equation (in acidic solution) for the reaction represented by the question mark.[22] As in Box 14-5, calculate E° for the reaction.

$$
\begin{array}{c}
\overset{1.441}{\overbrace{}} \\[2pt]
BrO_3^- \overset{1.491}{\longrightarrow} HOBr \overset{1.584}{\longrightarrow} Br_2(aq) \overset{1.098}{\longrightarrow} Br^- \\[2pt]
\underset{?}{\underbrace{}}
\end{array}
$$

14-24. What must be the relation between E_1° and E_2° if the species X^+ is to disproportionate spontaneously under standard conditions to X^{3+} and $X(s)$? Write a balanced equation for the disproportionation.

$$X^{3+} \overset{E_1^\circ}{\rightarrow} X^+ \overset{E_2^\circ}{\rightarrow} X(s)$$

14-25. *Including activities*, calculate the voltage of the cell Ni(s) | NiSO$_4$(0.002 0 M) ‖ CuCl$_2$(0.003 0 M) | Cu(s). Assume that the salts are completely dissociated (that is, neglect ion-pair formation). By the reasoning in Figure 14-9, in which direction do electrons flow?

14-26. *Lead-acid battery with activities.*[10] Refer to Box 14-3.

(a) Write reduction half-reactions from Appendix H for the anode and cathode during battery discharge. Write the net reaction and find E° for the net reaction.

(b) Write a line diagram for the battery including PbSO$_4$ on each electrode.

(c) Write the reactions that occur when the battery is being recharged when the car engine is running. Write one reduction and one oxidation.

(d) Write the Nernst equation for each half-reaction for a fully charged lead-acid battery, *including activity coefficients*. Express concentrations of solutes in the Nernst equations in molality, m. In a fully charged battery, the electrolyte concentration is 5.5 m (molal) H$_2$SO$_4$ (35 wt% H$_2$SO$_4$). Activities of solids are unity. However, the activity of H$_2$O in 35 wt% H$_2$SO$_4$ is not unity because the acid is so concentrated. Write $\mathcal{A}_{H_2O} = m_{H_2O}\gamma_{H_2O}$ for water, $\mathcal{A}_{SO_4^{2-}} = m_{SO_4^{2-}}\gamma_{SO_4^{2-}}$ for sulfate, and $\mathcal{A}_{H^+} = m_{H^+}\gamma_{H^+}$ for H$^+$. Combine the cathode and anode equations into a single Nernst equation with a single term.

(e) The activity of H$_2$O in 35 wt% H$_2$SO$_4$, measured by lowering of the vapor pressure of H$_2$O, is $\mathcal{A}_{H_2O} = m_{H_2O}\gamma_{H_2O} = 0.66$ at 25°C.[23] Activities of SO$_4^{2-}$ and H$^+$ cannot be measured separately, but the *mean activity* can be measured. For a salt C$_m$A$_n$ with cation C^{n+} and anion A^{m-}, the *mean activity coefficient* is defined as $\gamma_\pm = (\gamma_+^m \gamma_-^n)^{1/(m+n)}$, where γ_+ and γ_- are the individual activity coefficients. The mean activity coefficient is a thermodynamically defined, measureable quantity. For 5.5 m H$_2$SO$_4$, $\gamma_\pm = (\gamma_{H^+}^2 \gamma_{SO_4^{2-}})^{1/3} = 0.22$ at 25°C[23]

(as measured from galvanic cells containing H$_2$SO$_4$). Use $\mathcal{A}_{H_2O}$ and $\gamma_\pm$ in the Nernst equation to calculate the voltage of the lead-acid battery.

14-27. Cabin air in the *Apollo* spacecraft was circulated through a canister of solid LiOH to remove CO$_2$ to prevent poisoning the astronauts with excess CO$_2$. Write the chemical reaction between LiOH(s) and CO$_2$(g). Why was LiOH used instead of the less expensive NaOH or KOH?

Relation of E° and the Equilibrium Constant

14-28. For the reaction CO + $\frac{1}{2}$O$_2$ $\rightleftharpoons$ CO$_2$, $\Delta G^\circ = -257$ kJ per mole of CO at 298 K. Find E° and the equilibrium constant for the reaction.

14-29. Calculate E°, ΔG°, and K for the following reactions.

(a) 4Co^{3+} + 2H$_2$O $\rightleftharpoons$ 4Co^{2+} + O$_2$(g) + 4H$^+$

(b) Ag(S$_2$O$_3$)$_2^{3-}$ + Fe(CN)$_6^{4-}$ $\rightleftharpoons$ Ag(s) + 2S$_2$O$_3^{2-}$ + Fe(CN)$_6^{3-}$

14-30. A solution contains 0.100 M Ce^{3+}, 1.00 × 10^{-4} M Ce^{4+}, 1.00 × 10^{-4} M Mn^{2+}, 0.100 M MnO$_4^-$, and 1.00 M HClO$_4$.

(a) Write a balanced net reaction that can occur between species in this solution.

(b) Calculate ΔG° and K for the reaction.

(c) Calculate E for the conditions given.

(d) Calculate ΔG for the conditions given.

(e) At what pH would the given concentrations of Ce^{4+}, Ce^{3+}, Mn^{2+}, and MnO$_4^-$ be in equilibrium at 298 K?

14-31. For the cell Pt(s)|VO^{2+}(0.116 M), V^{3+}(0.116 M), H$^+$(1.57 M) ‖ Sn^{2+}(0.031 8 M), Sn^{4+}(0.031 8 M) | Pt(s), E (not E°) = −0.289 V. Write the net cell reaction and calculate its equilibrium constant. Do not use E° values from Appendix H to answer this question.

14-32. Calculate E° for the half-reaction Pd(OH)$_2$(s) + 2e$^-$ $\rightleftharpoons$ Pd(s) + 2OH$^-$ given that K_{sp} for Pd(OH)$_2$ is 3 × 10^{-28} and E° = 0.915 V for the reaction Pd^{2+} + 2e$^-$ $\rightleftharpoons$ Pd(s).

14-33. From the standard potentials for reduction of Br$_2$(aq) and Br$_2$(l) in Appendix H, calculate the solubility of Br$_2$ in water at 25°C. Express your answer as g/L.

14-34. **(a)** Henry's law states that the concentration of a dissolved gas is proportional to the partial pressure of that gas above the solution. For dissolution of Cl$_2$, Henry's law is [Cl$_2$(aq)] = $K_h P_{Cl_2}$. Using half-reactions in Appendix H, find the concentration of Cl$_2$(aq) in equilibrium with Cl$_2$(g, 1 bar) at 298.15 K.

(b) For modest temperature excursions away from 298.15 K (25°C), the standard reduction potential for a half-reaction can be written in the form

$$E^\circ(T) = E^\circ + \left(\frac{dE^\circ}{dT}\right)\Delta T$$

where E° is the standard reduction potential at 298.15 K, $E^\circ(T)$ is the standard reduction potential at temperature T (K), and ΔT is $(T - 298.15)$. Using dE°/dT in Appendix H, find K_h for Cl$_2$ at 323.15 K. Does the solubility of Cl$_2$(g) increase or decrease when the temperature is raised from 298.15?

14-35. Given the following information, calculate the standard potential for the reaction FeY$^-$ + e$^-$ $\rightleftharpoons$ FeY^{2-}, where Y is EDTA.

$$FeY^- + e^- \rightleftharpoons Fe^{2+} + Y^{4-} \qquad E° = -0.730 \text{ V}$$

FeY^{2-}: $\qquad K_f = 2.1 \times 10^{14}$

FeY$^-$: $\qquad K_f = 1.3 \times 10^{25}$

14-36. Find $E°$ for the half-reaction $Al^{3+} + 3e^- \rightleftharpoons Al(s)$ at 50°C. See Problem 14-34(b) for background information.

14-37. This problem is slightly tricky. Calculate $E°$, $\Delta G°$, and K for the reaction

$$2Cu^{2+} + 2I^- + HO\!-\!\!\!\langle\ \rangle\!\!\!-\!OH \rightleftharpoons$$

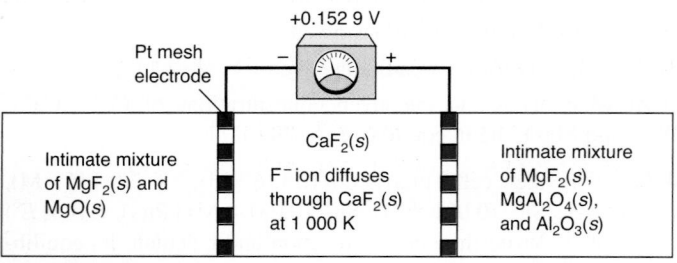
Hydroquinone

$$2CuI(s) + O\!=\!\!\langle\ \rangle\!\!=\!O + 2H^+$$
Quinone

which is the sum of *three* half-reactions listed in Appendix H. Use $\Delta G° \ (= -nFE°)$ for each of the half-reactions to find $\Delta G°$ for the net reaction. Note that, if you reverse the direction of a reaction, you reverse the sign of $\Delta G°$.

14-38. *Thermodynamics of a solid-state reaction.* The following electrochemical cell is reversible at 1 000 K in an atmosphere of flowing $O_2(g)$:[24]

Left half-cell: $MgF_2(s) + \frac{1}{2}O_2(g) + 2e^- \rightleftharpoons MgO(s) + 2F^-$

Right half-cell: $MgF_2(s) + Al_2O_3(s) + \frac{1}{2}O_2(g) + 2e^- \rightleftharpoons$
$$MgAl_2O_4(s) + 2F^-$$

(a) Write a Nernst equation for each half-cell. Write the net reaction and its Nernst equation. The activity of $O_2(g)$ is the same on both sides. The activity of F^- is the same on both sides, governed by F^- ions diffusing through $CaF_2(s)$. Show that the observed voltage is $E°$ for the net reaction.

(b) From the relation $\Delta G° = -nFE°$, find $\Delta G°$ for the net reaction. Note that 1 V = 1 J/C.

(c) The cell voltage in the temperature range T = 900 to 1 250 K is $E(V) = 0.122\ 3 + 3.06 \times 10^{-5}\ T$. Assuming that $\Delta H°$ and $\Delta S°$ are constant, find $\Delta H°$ and $\Delta S°$ from the relation $\Delta G° = \Delta H° - T\Delta S°$.

Using Cells as Chemical Probes

14-39. With Figure 14-11 as an example, explain what we mean when we say that there is equilibrium *within* each half-cell but not necessarily *between* the two half-cells.

14-40. The cell $Pt(s) \mid H_2(g,\ 1.00\ \text{bar}) \mid H^+(aq,\ \text{pH} = 3.60) \parallel Cl^-$ $(aq,\ x\ \text{M}) \mid AgCl(s) \mid Ag(s)$ can be used as a probe to find the concentration of Cl^- in the right compartment.

(a) Write reactions for each half-cell, a balanced net cell reaction, and the Nernst equation for the net cell reaction.

(b) Given a measured cell voltage of 0.485 V, find $[Cl^-]$ in the right compartment.

14-41. The quinhydrone electrode was introduced in 1921 as a means of measuring pH.[25]

$Pt(s) \mid 1{:}1$ mole ratio of quinone(aq) and hydroquinone(aq),
unknown pH $\parallel Cl^-(aq,\ 0.50\ \text{M}) \mid Hg_2Cl_2(s) \mid Hg(l) \mid Pt(s)$

The solution whose pH is to be measured is placed in the left half-cell, which also contains a 1:1 mole ratio of quinone and hydroquinone. The half-cell reaction is

$$O\!=\!\!\langle\ \rangle\!\!=\!O + 2H^+ + 2e^- \rightleftharpoons HO\!-\!\!\langle\ \rangle\!\!-\!OH$$
Quinone $\qquad\qquad$ Hydroquinone

(a) Write half-reactions and Nernst equations for each half-cell.

(b) Ignoring activities, rearrange the Nernst equation for the net reaction to the form $E(\text{cell}) = A + (B \cdot \text{pH})$, where A and B are constants. Calculate A and B at 25°C.

(c) If the pH were 4.50, in which direction would electrons flow through the potentiometer?

14-42. The voltage for the following cell is 0.490 V. Find K_b for the organic base RNH_2.

$Pt(s) \mid H_2(1.00\ \text{bar}) \mid$
$\qquad RNH_2(aq,\ 0.10\ \text{M}),\ RNH_3^+Cl^-(aq,\ 0.050\ \text{M}) \parallel \text{S.H.E.}$

14-43. The voltage of the cell shown here is −0.246 V. The right half-cell contains the metal ion, M^{2+}, whose standard reduction potential is −0.266 V.

$$M^{2+} + 2e^- \rightleftharpoons M(s) \qquad E° = -0.266 \text{ V}$$

Calculate K_f for the metal-EDTA complex.

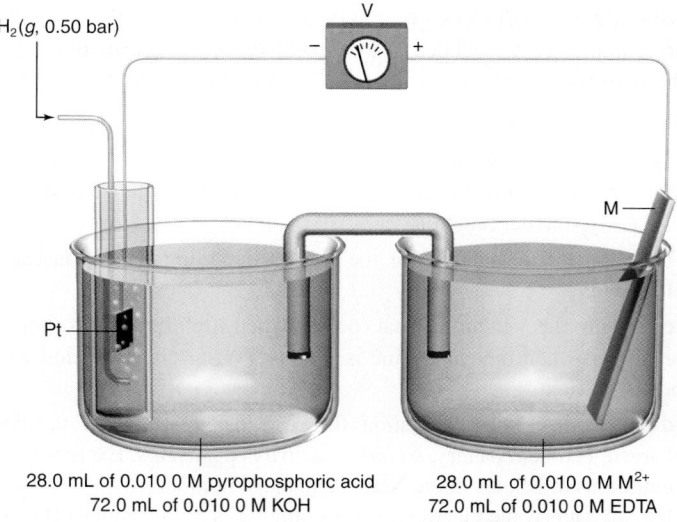

28.0 mL of 0.010 0 M pyrophosphoric acid
72.0 mL of 0.010 0 M KOH

28.0 mL of 0.010 0 M M^{2+}
72.0 mL of 0.010 0 M EDTA
buffered to pH 8.00

14-44. The following cell was constructed to find the difference in K_{sp} between two naturally occurring forms of $CaCO_3(s)$, called *calcite* and *aragonite*.[26]

$Pb(s) \mid PbCO_3 \mid CaCO_3(s,\ \text{calcite}) \mid \text{buffer(pH 7.00)} \parallel$
$\qquad \text{buffer(pH 7.00)} \mid CaCO_3(s,\ \text{aragonite}) \mid PbCO_3(s) \mid Pb(s)$

Each compartment of the cell contains a mixture of solid $PbCO_3$ ($K_{sp} = 7.4 \times 10^{-14}$) and either calcite or aragonite, both of which have $K_{sp} \approx 5 \times 10^{-9}$. Each solution was buffered to pH 7.00 with an inert buffer, and the cell was completely isolated from atmospheric CO_2. The measured cell voltage was −1.8 mV. Find the ratio of solubility products, K_{sp} (for calcite)/K_{sp} (for aragonite)

14-45. *Do not ignore activity coefficients in this problem.* If the voltage for the following cell is 0.512 V, find K_{sp} for $Cu(IO_3)_2$. Neglect any ion pairing.

$$Ni(s)\,|\,NiSO_4(0.002\ 5\ M)\,\|\,KIO_3(0.10\ M)\,|\,Cu(IO_3)_2(s)\,|\,Cu(s)$$

Biochemists Use $E^{\circ\prime}$

14-46. Explain what $E^{\circ\prime}$ is and why it is preferred over E° in biochemistry.

14-47. We are going to find $E^{\circ\prime}$ for the reaction

$$C_2H_2(g) + 2H^+ + 2e^- \rightleftharpoons C_2H_4(g)$$

(a) Write the Nernst equation for the half-reaction, using E° from Appendix H.

(b) Rearrange the Nernst equation to the form

$$E = E^{\circ} + \text{other terms} - \frac{0.059\ 16}{2}\log\left(\frac{P_{C_2H_4}}{P_{C_2H_2}}\right)$$

(c) The quantity (E° + other terms) is $E^{\circ\prime}$. Evaluate $E^{\circ\prime}$ for pH = 7.00.

14-48. Evaluate $E^{\circ\prime}$ for the half-reaction

$$\underset{\text{Cyanogen}}{(CN)_2(g)} + 2H^+ + 2e^- \rightleftharpoons \underset{\text{Hydrogen cyanide}}{2HCN(aq)}$$

14-49. Calculate $E^{\circ\prime}$ for the reaction

$$\underset{\text{Oxalic acid}}{H_2C_2O_4} + 2H^+ + 2e^- \rightleftharpoons \underset{\text{Formic acid}}{2HCO_2H} \qquad E^{\circ} = 0.204\ V$$

14-50. HOx is a monoprotic acid with $K_a = 1.4 \times 10^{-5}$ and H_2Red^- is a diprotic acid with $K_1 = 3.6 \times 10^{-4}$ and $K_2 = 8.1 \times 10^{-8}$. Find E° for the reaction

$$HOx + e^- \rightleftharpoons H_2Red^- \qquad E^{\circ\prime} = 0.062\ V$$

14-51. Given the following information, find K_a for nitrous acid, HNO_2.

$$NO_3^- + 3H^+ + 2e^- \rightleftharpoons HNO_2 + H_2O \qquad \begin{array}{l} E^{\circ} = 0.940\ V \\ E^{\circ\prime} = 0.433\ V \end{array}$$

14-52. Using the reaction

$$HPO_4^{2-} + 2H^+ + 2e^- \rightleftharpoons HPO_3^{2-} + H_2O \qquad E^{\circ} = -0.234\ V$$

and acid dissociation constants from Appendix G, calculate E° for the reaction

$$H_2PO_4^- + H^+ + 2e^- \rightleftharpoons HPO_3^{2-} + H_2O$$

14-53. This problem requires Beer's law from Chapter 18. The oxidized form (Ox) of a flavoprotein that functions as a one-electron reducing agent has a molar absorptivity (ε) of $1.12 \times 10^4\ M^{-1}cm^{-1}$ at 457 nm at pH 7.00. For the reduced form (Red), $\varepsilon = 3.82 \times 10^3$ $M^{-1}cm^{-1}$ at 457 nm at pH 7.00.

$$Ox + e^- \rightleftharpoons Red \qquad E^{\circ\prime} = -0.128\ V$$

The substrate (S) is the molecule reduced by the protein.

$$Red + S \rightleftharpoons Ox + S^-$$

Both S and S^- are colorless. A solution at pH 7.00 was prepared by mixing enough protein plus substrate (Red + S) to produce initial concentrations [Red] = [S] = 5.70×10^{-5} M. The absorbance at 457 mm was 0.500 in a 1.00-cm cell.

(a) Calculate the concentrations of Ox and Red from the absorbance data.

(b) Calculate the concentrations of S and S^-.

(c) Calculate the value of $E^{\circ\prime}$ for the reaction $S + e^- \rightleftharpoons S^-$.

15 Electrodes and Potentiometry

DNA SEQUENCING BY COUNTING PROTONS

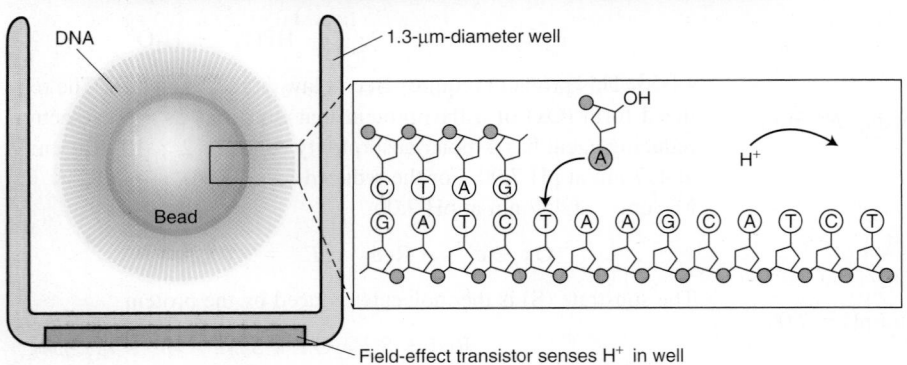

DNA — 1.3-µm-diameter well

Bead

Field-effect transistor senses H⁺ in well

Ⓐ Nucleotide base   ⬠ Deoxyribose sugar   ● Phosphate

Ion Torrent® sequencer measures H^+ released each time one base (A, T, C, or G) is added to a growing chain of DNA attached to a bead contained in a micron-size well. See Appendix L for the chemistry of DNA chain replication. [Information from J. M. Rothberg et al., "An Integrated Semiconductor Device Enabling Non-Optical Genome Sequencing," *Nature* **2011**, *475*, 348.]

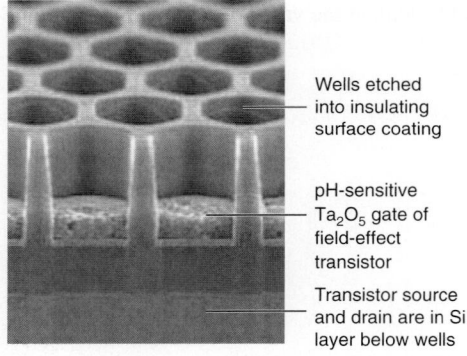

Wells etched into insulating surface coating

pH-sensitive Ta_2O_5 gate of field-effect transistor

Transistor source and drain are in Si layer below wells

Each 1.3-µm-diameter hexagonal well in this chip has a pH-sensitive layer of tantalum oxide at the bottom. [Reprinted by permission from Macmillan Publishers Ltd: From J. M. Rothberg et al., "An Integrated Semiconductor Device Enabling Non-Optical Genome Sequencing," *Nature* **2011**, *475*, p. 348, Figure S7.]

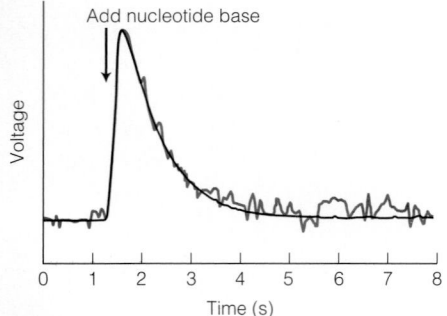

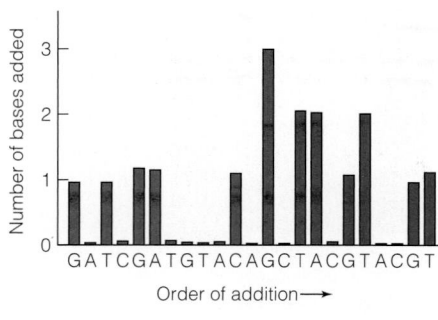

Left: Signal observed from one well when a nucleotide base is added to DNA. Smooth line is physical model fit to the data. *Right:* Integrated signal from successive addition of bases to one well. Base sequence read from the chart is GTGACGGGTTAAGTTGT. [Data from J. M. Rothberg et al., "An Integrated Semiconductor Device Enabling Non-Optical Genome Sequencing," *Nature* **2011**, *475*, 348.]

The Human Genome Project in the 1990s invested $3 billion over a decade to decipher the sequence of nucleotide bases (A, T, C, and G) in human DNA for the first time. Commercial instruments have an eventual goal of sequencing DNA from one person in one day at a cost of $1000. One instrument uses the field-effect transistor described in Section 15-8 to measure H^+ released when each base is added to a strand of DNA.

The heart of the instrument is a 2×2 cm chip with $\sim 10^9$ etched wells. DNA to be sequenced is sheared randomly into fragments with a length of ~ 150 base pairs. Micron-size polyacrylamide beads are coated with 10^5–10^6 copies of one strand of one fragment. Different beads are coated with different fragments. The enzyme DNA polymerase and DNA primers needed to replicate DNA are added to the fragments on the beads. One bead is placed into each well. Different wells contain different DNA fragments.

For sequencing, a solution containing one base is passed over the chip and the base diffuses into each well. If the base is the one required for the next position of the DNA in that well, the base is added to the DNA, H^+ is released, and the pH in that well decreases by ~ 0.02. If the base was not the next one required, no reaction takes place in that well. The field-effect transistor registers a voltage change for a few seconds before H^+ diffuses out of the well. Then reagents are washed away and fresh solution with a different base is added. The sequencer records the response of each well to each addition of nucleotide base to find the sequence of the DNA in that well. A computer constructs a map of $\sim 3 \times 10^9$ bases in human DNA from the sequences of short overlapping fragments.

C lever chemists have designed electrodes that respond selectively to specific analytes in solution or in the gas phase. Typical ion-selective electrodes are about the size of your pen. Ion-sensing field effect transistors that are just 1 μm in size and used for sequencing DNA are shown on the facing page. The use of electrodes to measure voltages that provide chemical information is called **potentiometry.**

In the simplest case, analyte is an *electroactive species* that is part of a galvanic cell. An **electroactive species** is one that can donate or accept electrons at an electrode. We turn the unknown solution into a half-cell by inserting an electrode, such as a Pt wire, that can transfer electrons to or from the analyte. Because this electrode responds to analyte, it is called the **indicator electrode** or, equivalently, the **working electrode.** We connect this half-cell to a second half-cell by a salt bridge. The second half-cell has a fixed composition, so it has a constant potential. Because of its constant potential, the second half-cell is called a **reference electrode.** The cell voltage is the difference between the variable potential of the analyte half-cell and the constant potential of the reference electrode.

Indicator (working) electrode: responds to analyte activity

Reference electrode: maintains a fixed (reference) potential

15-1 Reference Electrodes[1]

Suppose you want to measure the relative amounts of Fe^{2+} and Fe^{3+} in a solution. You can make this solution part of a galvanic cell by inserting a Pt wire and connecting the cell to a constant-potential half-cell by a salt bridge, as shown in Figure 15-1. We will designate the potential of the half-cell connected to the positive terminal of the potentiometer (voltmeter) E_+ in Figure 15-1. The potential of the half-cell connected to the negative terminal of the potentiometer will be written E_-. These labels do not tell us that a half-cell potential is positive or negative. The labels simply tell us how the cells are connected to the potentiometer.

The two half-reactions (written as *reductions*) are

Right electrode: $\quad\quad\quad\quad\quad\quad Fe^{3+} + e^- \rightleftharpoons Fe^{2+}\quad\quad\quad\quad\quad E_+^\circ = 0.771\ V$

Left electrode: $\quad\quad\quad\quad\quad\quad AgCl(s) + e^- \rightleftharpoons Ag(s) + Cl^-\quad\quad E_-^\circ = 0.222\ V$

where E° for each half-cell is the *standard potential* when the activities of reactants and products are unity. The electrode potentials are

$$E_+ = 0.771 - 0.059\ 16\ \log\left(\frac{[Fe^{2+}]}{[Fe^{3+}]}\right)$$

$$E_- = 0.222 - 0.059\ 16\ \log[Cl^-]$$

E_+ is the potential of the electrode attached to the positive input of the potentiometer. E_- is the potential of the electrode attached to the negative input of the potentiometer.

and the cell voltage is the difference $E_+ - E_-$:

$$E = \left\{0.771 - 0.059\ 16\ \log\left(\frac{[Fe^{2+}]}{[Fe^{3+}]}\right)\right\} - \{0.222 - 0.059\ 16\ \log[Cl^-]\}$$

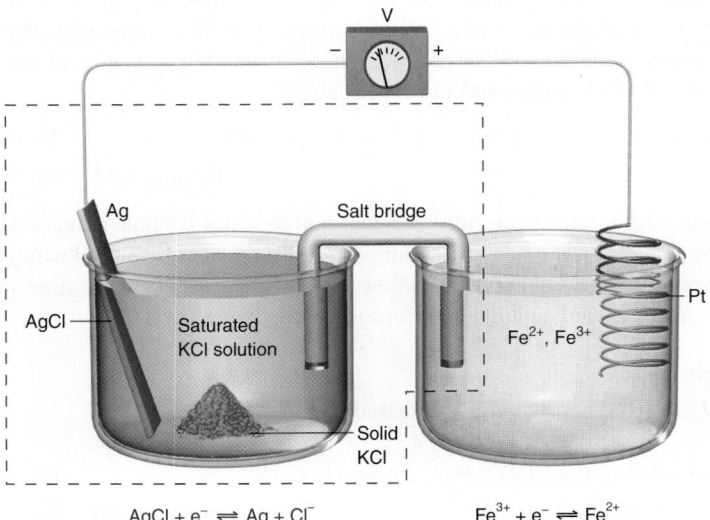

$$Ag(s)\,|\,AgCl(s)\,|\,Cl^-(aq)\,\|\,Fe^{2+}(aq),\,Fe^{3+}(aq)\,|\,Pt(s)$$

FIGURE 15-1 A galvanic cell that can be used to measure the quotient $[Fe^{2+}]/[Fe^{3+}]$ in the right half-cell. The Pt wire is the *indicator electrode,* and the entire left half-cell plus salt bridge (enclosed by the dashed line) can be considered a *reference electrode.*

The voltage really tells us the quotient of activities, $\mathcal{A}_{Fe^{2+}}/\mathcal{A}_{Fe^{3+}}$. We will generally neglect activity coefficients and write the Nernst equation with concentrations instead of activities.

But [Cl⁻] in the left half-cell is constant, fixed by the solubility of KCl, with which the solution is saturated. Therefore, the cell voltage changes only when the quotient [Fe²⁺]/[Fe³⁺] changes.

The half-cell on the left in Figure 15-1 acts as a *reference electrode*. We can picture the cell and salt bridge enclosed by the dashed line as a single unit dipped into the analyte solution, as shown in Figure 15-2. The Pt wire is the indicator electrode, whose potential responds to the quotient [Fe²⁺]/[Fe³⁺]. The reference electrode completes the redox reaction and provides a *constant potential* to the left side of the potentiometer. Changes in the cell voltage result from changes in the quotient [Fe²⁺]/[Fe³⁺].

Silver-Silver Chloride Reference Electrode[2]

The half-cell enclosed by the dashed line in Figure 15-1 is called a **silver-silver chloride electrode.** Figure 15-3 shows how the electrode is reconstructed as a thin tube that can be dipped into analyte solution. Figure 15-4 shows a *double-junction electrode* that minimizes contact between analyte solution and KCl from the electrode. Silver-silver chloride and calomel reference electrodes (described next) are used because they are convenient. A standard hydrogen electrode (S.H.E.) is difficult to use because it requires H_2 gas and a freshly prepared catalytic Pt surface that is easily poisoned in many solutions.

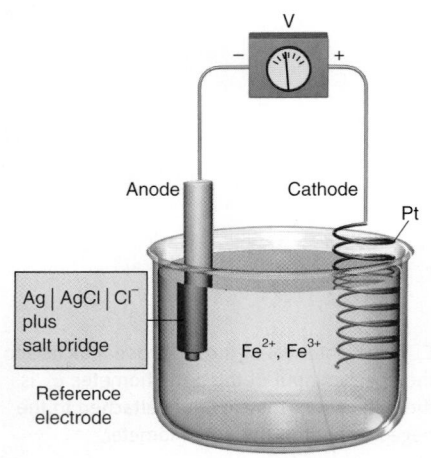

FIGURE 15-2 Another view of Figure 15-1. The contents of the dashed box in Figure 15-1 are now considered to be a reference electrode dipped into the analyte solution.

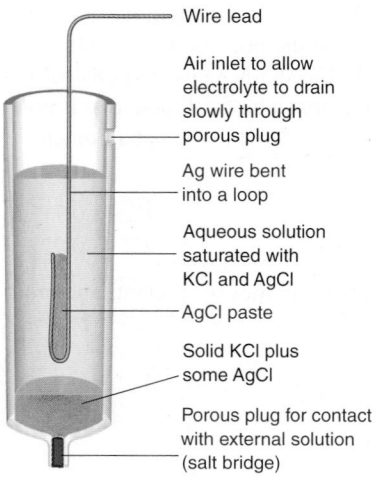

FIGURE 15-3 Silver-silver chloride reference electrode.

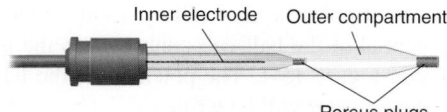

FIGURE 15-4 Double-junction reference electrode. The inner electrode is the same as the one in Figure 15-3. The solution in the outer compartment is compatible with analyte solution. For example, if you do not want Cl⁻ to contact the analyte, the outer electrode can be filled with KNO_3 solution. The inner and outer solutions slowly mix, so the outer compartment must be refilled periodically with fresh KNO_3 solution.

The standard reduction potential for the AgCl|Ag couple is $+0.222$ V at 25°C. This would be the potential of a silver-silver chloride electrode if $\mathcal{A}_{Cl^-}$ were unity. But the activity of Cl⁻ in a saturated solution of KCl at 25°C is not unity, and the potential of the electrode in Figure 15-3 is $+0.197$ V with respect to S.H.E. at 25°C.

Ag | AgCl electrode: $\qquad AgCl(s) + e^- \rightleftharpoons Ag(s) + Cl^- \qquad\qquad E° = +0.222$ V

$$E(\text{saturated KCl*}) = +0.197 \text{ V}$$

A problem with reference electrodes is that porous plugs become clogged, thus causing sluggish, unstable electrical response. Some designs incorporate a free-flowing capillary in place of the porous plug. Other designs allow you to force fresh solution from the electrode through the electrode-analyte junction prior to a measurement.

Calomel Electrode

The **calomel electrode** in Figure 15-5 is based on the reaction

Calomel electrode: $\quad \dfrac{1}{2}Hg_2Cl_2(s) + e^- \rightleftharpoons Hg(l) + Cl^- \qquad E° = +0.268$ V

Mercury(I) chloride $\qquad\qquad\qquad\qquad E(\text{saturated KCl*}) = +0.241$ V
(Calomel)

The standard potential for this reaction is $+0.268$ V. If the cell is saturated with KCl at 25°C, the potential is $+0.241$ V. A calomel electrode saturated with KCl is called a **saturated**

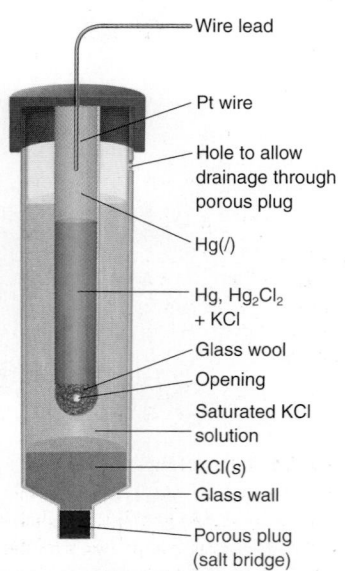

FIGURE 15-5 A saturated calomel electrode (S.C.E.).

---

*Molarity of saturated aqueous KCl at 25°C ≈ 4.2 M (~26.5 wt%).

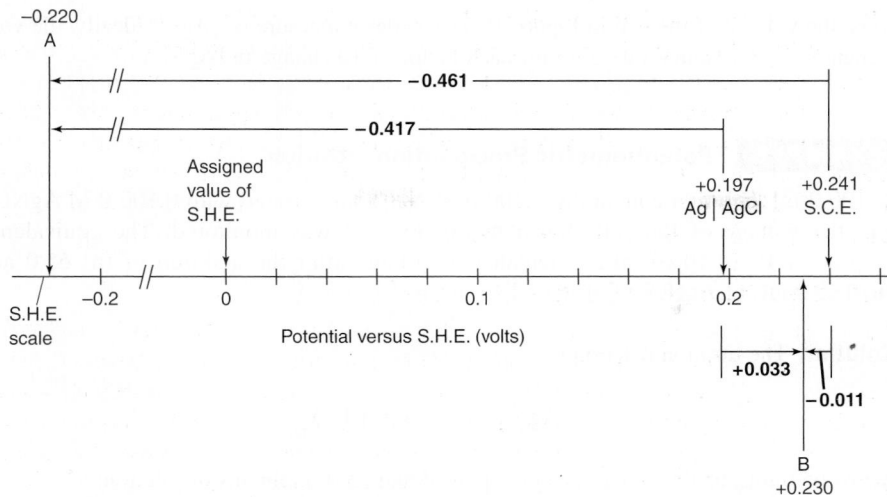

FIGURE 15-6 A diagram that helps us convert electrode potential between different reference scales.

calomel electrode, abbreviated **S.C.E.** The advantage in using saturated KCl is that [Cl⁻] does not change if some liquid evaporates.

Voltage Conversions Between Different Reference Scales

If an electrode has a potential of -0.461 V with respect to a calomel electrode, what is the potential with respect to a silver-silver chloride electrode? What would be the potential with respect to the standard hydrogen electrode?

To answer these questions, draw a diagram like Figure 15-6 showing the positions of the calomel and silver-silver chloride electrodes with respect to the standard hydrogen electrode. We see that point A, which is -0.461 V from S.C.E., is -0.417 V from the silver-silver chloride electrode and -0.220 V with respect to S.H.E. Point B, whose potential is $+0.033$ V from silver-silver chloride, is -0.011 V from S.C.E. and $+0.230$ V from S.H.E.

15-2 Indicator Electrodes

We will study two broad classes of indicator electrodes. *Metal electrodes* described in this section develop an electric potential in response to a redox reaction at the metal surface. *Ion-selective electrodes* described later are not based on redox processes. Instead, selective binding of one type of ion to a membrane generates an electric potential.

The most common metal indicator electrode is platinum, which is relatively *inert*—it does not participate in many chemical reactions. Its purpose is simply to transmit electrons to or from species in solution. Gold electrodes are even more inert than Pt. Various types of carbon are used as indicator electrodes because the rates of many redox reactions on the carbon surface are fast. A metal electrode works best when its surface is large and clean. To clean a Pt electrode, dip it in hot 8 M HNO_3 in a fume hood and rinse with distilled water.

Figure 15-7 shows how a silver electrode can be used with a reference electrode to measure Ag^+ concentration.[3] The reaction at the Ag indicator electrode is

$$Ag^+ + e^- \rightleftharpoons Ag(s) \qquad E_+^\circ = 0.799 \text{ V}$$

The calomel reference half-cell reaction is

$$Hg_2Cl_2(s) + 2e^- \rightleftharpoons 2Hg(l) + 2Cl^- \qquad E_- = 0.241 \text{ V}$$

The reference potential (E_-, not E_-°) is fixed at 0.241 V because the reference cell is saturated with KCl. The Nernst equation for the entire cell is therefore

$$E = E_+ - E_- = \underbrace{\left\{ 0.799 - 0.059\ 16 \log\left(\frac{1}{[Ag^+]}\right) \right\}}_{\substack{\text{Potential of } Ag \mid Ag^+ \\ \text{indicator electrode}}} - \underbrace{\{0.241\}}_{\substack{\text{Potential of} \\ \text{S.C.E. reference electrode}}}$$

$$E = 0.558 + 0.059\ 16 \log[Ag^+] \qquad \text{(15-1)}$$

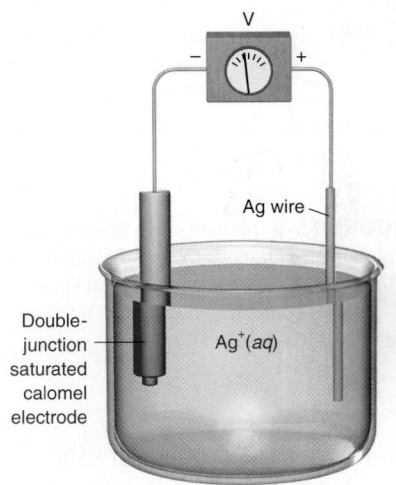

FIGURE 15-7 Use of Ag and calomel electrodes to measure [Ag⁺]. The calomel electrode has a double junction, like that in Figure 15-4. The outer compartment of the electrode is filled with KNO₃, so there is no direct contact between Cl⁻ in the inner compartment and Ag⁺ in the beaker.

That is, the voltage of the cell in Figure 15-7 provides a measure of $[Ag^+]$. Ideally, the voltage changes by 59.16 mV (at 25°C) for each factor-of-10 change in $[Ag^+]$.

EXAMPLE **Potentiometric Precipitation Titration**

A 100.0-mL solution containing 0.100 0 M NaCl was titrated with 0.100 0 M AgNO₃, and the voltage of the cell shown in Figure 15-7 was monitored. The equivalence volume is $V_e = 100.0$ mL. Calculate the voltage after the addition of (a) 65.0 and (b) 135.0 mL of AgNO₃.

Solution The titration reaction is

$$Ag^+ + Cl^- \rightarrow AgCl(s)$$

(a) At 65.0 mL, 65.0% of Cl^- has precipitated and 35.0% remains in solution:

$$[Cl^-] = \underbrace{(0.350)}_{\substack{\text{Fraction} \\ \text{remaining}}} \underbrace{(0.100\ 0\ M)}_{\substack{\text{Original} \\ \text{concentration} \\ \text{of } Cl^-}} \underbrace{\left(\frac{100.0}{165.0}\right)}_{\substack{\text{Dilution} \\ \text{factor}}} = 0.021\ 2\ M$$

Initial volume of Cl^-

Total volume of solution

To find the cell voltage in Equation 15-1, we need to know $[Ag^+]$:

$$[Ag^+][Cl^-] = K_{sp} \Rightarrow [Ag^+] = \frac{K_{sp}}{[Cl^-]} = \frac{1.8 \times 10^{-10}}{0.021\ 2\ M} = 8.5 \times 10^{-9}\ M$$

The cell voltage is therefore

$$E = 0.558 + 0.059\ 16 \log(8.5 \times 10^{-9}) = 0.081\ V$$

(b) At 135.0 mL, there is an excess of 35.0 mL of AgNO₃ = 3.50 mmol Ag^+ in a total volume of 235.0 mL. Therefore, $[Ag^+] = (3.50\ \text{mmol})/(235.0\ \text{mL}) = 0.014\ 9\ M$. The cell voltage is

$$E = 0.558 + 0.059\ 16 \log(0.014\ 9) = 0.450\ V$$

TEST YOURSELF Find the voltage after addition of 99.0 mL of AgNO₃. (*Answer:* 0.177 V)

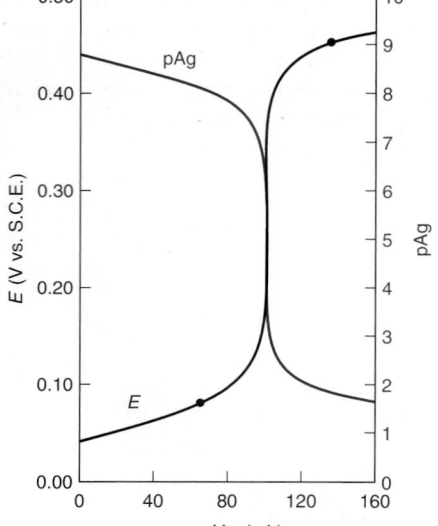

FIGURE 15-8 Titration curve computed for the addition of 0.100 0 M Ag^+ to 100.0 mL of 0.100 0 M Cl^- with the electrodes in Figure 15-7. Points computed at 65.0 and 135.0 mL are shown. Colored line is pAg = $-\log[Ag^+]$.

The cell responds to a change in $[Cl^-]$, which necessarily changes $[Ag^+]$ because $[Ag^+][Cl^-] = K_{sp}$.

Demonstration 15-1 is a great example of indicator and reference electrodes.

Figure 15-8 shows the titration curve for the preceding example. There is a strong analogy to acid-base titrations, with Ag^+ replacing H^+ and Cl^- acting as a base being titrated. As the acid-base titration proceeds, $[H^+]$ increases and pH decreases. As the Ag^+/Cl^- titration proceeds, $[Ag^+]$ increases and pAg($\equiv -\log[Ag^+]$) decreases. The silver electrode measures pAg, which you can see by substituting pAg = $-\log[Ag^+]$ into Equation 15-1:

$$E = 0.558 - 0.059\ 16\ \text{pAg} \qquad (15\text{-}2)$$

A silver electrode is also a halide electrode, if solid silver halide is present.[4] If the solution contains $AgCl(s)$, we substitute $[Ag^+] = K_{sp}/[Cl^-]$ into Equation 15-1 to find an expression relating the cell voltage to $[Cl^-]$:

$$E = 0.558 + 0.059\ 16 \log\left(\frac{K_{sp}}{[Cl^-]}\right) \qquad (15\text{-}3)$$

Metals including Ag, Cu, Zn, Cd, and Hg can be used as indicator electrodes for their aqueous ions. Most metals, however, are unsuitable for this purpose because the equilibrium $M^{n+} + ne^- \rightleftharpoons M$ is not readily established at the metal surface.

The Belousov-Zhabotinskii reaction is a cerium-catalyzed oxidation of malonic acid by bromate, in which the quotient $[Ce^{3+}]/[Ce^{4+}]$ oscillates by a factor of 10 to 100.[6]

$$3CH_2(CO_2H)_2 + 2BrO_3^- + 2H^+ \rightarrow$$

Malonic acid   Bromate

$$2BrCH(CO_2H)_2 + 3CO_2 + 4H_2O$$
Bromomalonic acid

When the Ce^{4+} concentration is high, the solution is yellow. When Ce^{3+} predominates, the solution is colorless. With redox indicators (Section 15-2), this reaction oscillates through a sequence of colors.[7]

Oscillation between yellow and colorless is set up in a 300-mL beaker with the following solutions:

160 mL of 1.5 M H_2SO_4
40 mL of 2 M malonic acid
30 mL of 0.5 M $NaBrO_3$ (or saturated $KBrO_3$)
4 mL of saturated ceric ammonium sulfate,
    $(Ce(SO_4)_2 \cdot 2(NH_4)_2SO_4 \cdot 2H_2O)$

After an induction period of 5 to 10 min with magnetic stirring, oscillations can be initiated by adding 1 mL of ceric ammonium sulfate solution. The reaction may need more Ce^{4+} over a 5-min period to initiate oscillations.

A galvanic cell is built around the reaction as shown in the figure. The quotient $[Ce^{3+}]/[Ce^{4+}]$ is monitored by Pt and calomel electrodes. You should be able to write the cell reactions and a Nernst equation for this experiment.

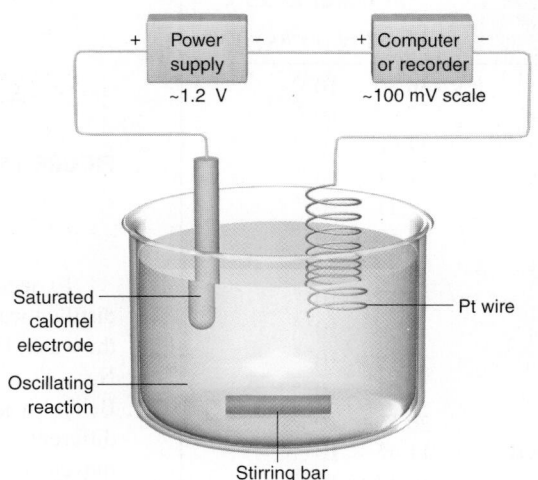

Apparatus used to monitor the quotient $[Ce^{3+}]/[Ce^{4+}]$ for an oscillating reaction. [Information from George Rossman, California Institute of Technology.]

In place of a potentiometer (a pH meter), use a computer or recorder to show the oscillations. Because the potential oscillates over a range of ~100 mV but is centered near ~1.2 V, the cell voltage is offset by ~1.2 V with any available power supply.[8] Trace *a* shows what is usually observed. The potential changes rapidly during the abrupt colorless-to-yellow change and gradually during the gentle yellow-to-colorless change. Trace *b* shows two different cycles superimposed in the same solution. This unusual event occurred in a reaction that had been oscillating normally for about 30 min.[9]

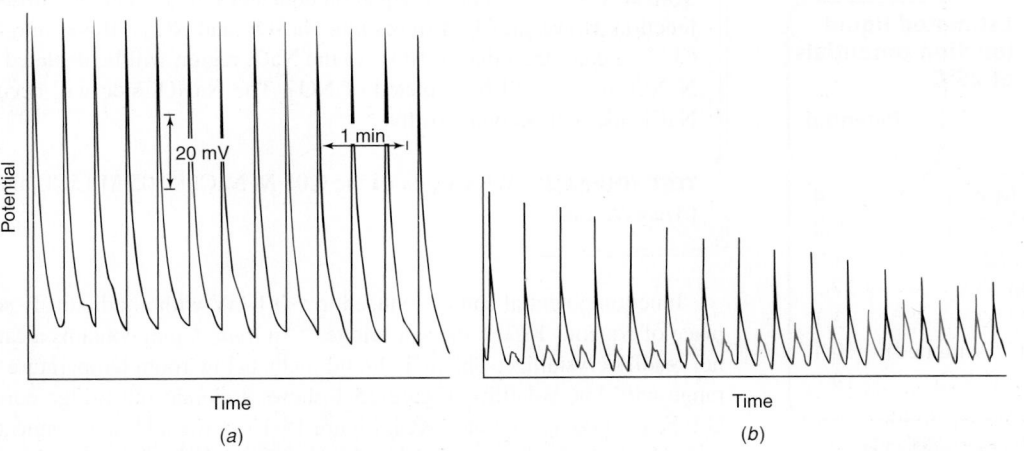

15-3   What Is a Junction Potential?

Whenever dissimilar electrolyte solutions are in contact, a voltage difference called the **junction potential** develops at their interface. This usually small voltage (usually a few millivolts) is found at each end of a salt bridge connecting two half-cells. *The junction potential puts a fundamental limitation on the accuracy of direct potentiometric measurements,* because we usually do not know the contribution of the junction to the measured voltage.

$$E_{observed} = E_{cell} + E_{junction}$$

Junction potential is usually unknown, so E_{cell} is uncertain.

TABLE 15-1	Mobilities of ions in water at 25°C
Ion	Mobility $[m^2/(s \cdot V)]^a$
H^+	36.30×10^{-8}
Rb^+	7.92×10^{-8}
K^+	7.62×10^{-8}
NH_4^+	7.61×10^{-8}
La^{3+}	7.21×10^{-8}
Ba^{2+}	6.59×10^{-8}
Ag^+	6.42×10^{-8}
Ca^{2+}	6.12×10^{-8}
Cu^{2+}	5.56×10^{-8}
Na^+	5.19×10^{-8}
Li^+	4.01×10^{-8}
OH^-	20.50×10^{-8}
$Fe(CN)_6^{4-}$	11.45×10^{-8}
$Fe(CN)_6^{3-}$	10.47×10^{-8}
SO_4^{2-}	8.27×10^{-8}
Br^-	8.13×10^{-8}
I^-	7.96×10^{-8}
Cl^-	7.91×10^{-8}
NO_3^-	7.40×10^{-8}
ClO_4^-	7.05×10^{-8}
F^-	5.70×10^{-8}
HCO_3^-	4.61×10^{-8}
$CH_3CO_2^-$	4.24×10^{-8}

a. The mobility of an ion is the terminal velocity that the particle achieves in an electric field of 1 V/m. Mobility = velocity/field. The units of mobility are therefore $(m/s)/(V/m) = m^2/(s \cdot V)$.

TABLE 15-2	Estimated liquid junction potentials at 25°C
Junction	Potential (mV)
0.1 M NaCl | 0.1 M KCl	−6.4
0.1 M NaCl | 3.5 M KCl	−0.2
1 M NaCl | 3.5 M KCl	−1.9
0.1 M HCl | 0.1 M KCl	+27
0.1 M HCl | 3.5 M KCl	+3.1
0.1 M NaOH | KCl (saturated)	−0.4
0.1 M NaOH | 0.1 M KCl	−19

NOTE: A positive sign means that the right side of the junction becomes positive with respect to the left side.

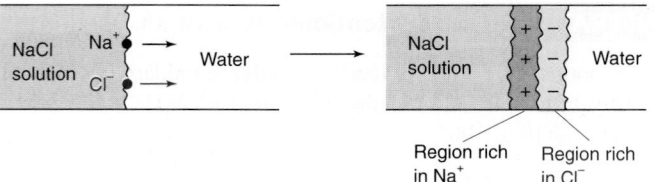

FIGURE 15-9 Development of the junction potential caused by unequal mobilities of Na^+ and Cl^-.

To see why the junction potential occurs, consider a solution of NaCl in contact with distilled water (Figure 15-9). Na^+ and Cl^- ions begin to diffuse from the NaCl solution into the water. However, Cl^- ion has a greater **mobility** than Na^+. That is, Cl^- diffuses faster than Na^+. As a result, a region rich in Cl^-, with excess negative charge, develops at the front. Behind it is a positively charged region depleted of Cl^-. The result is an electric potential difference at the junction of the NaCl and H_2O phases. The junction potential opposes the movement of Cl^- and accelerates the movement of Na^+. The steady-state junction potential represents a balance between the unequal mobilities that create a charge imbalance and the tendency of the resulting charge imbalance to retard the movement of Cl^-.

Table 15-1 shows mobilities of several ions and Table 15-2 lists some liquid junction potentials. Saturated KCl is used in a salt bridge because K^+ and Cl^- have similar mobilities. Junction potentials at the two interfaces of a KCl salt bridge are small.

Nonetheless, the junction potential of 0.1 M HCl | 3.5 M KCl is 3.1 mV. A pH electrode has a response of 59 mV per pH unit. A pH electrode dipped into 0.1 M HCl will have a junction potential of ~3 mV, or an error of 0.05 pH units (12% error in $[H^+]$).

EXAMPLE Junction Potential

A 0.1 M NaCl solution was placed in contact with a 0.1 M $NaNO_3$ solution. Which side of the junction is positive?

Solution Because $[Na^+]$ is equal on both sides, there is no net diffusion of Na^+ across the junction. However, Cl^- diffuses into $NaNO_3$, and NO_3^- diffuses into NaCl. The mobility of Cl^- is greater than that of NO_3^-, so the NaCl region will be depleted of Cl^- faster than the $NaNO_3$ region will be depleted of NO_3^-. The $NaNO_3$ side will become negative, and the NaCl side will become positive.

TEST YOURSELF Which side of the 0.05 M NaCl | 0.05 M LiCl junction will be positive? (*Answer:* LiCl)

Junction potential can be reduced to ~0.1 mV with a judiciously selected ionic liquid in place of aqueous KCl in the salt bridge.[10] An *ionic liquid* contains a cation and anion that do not readily crystallize. The ionic liquid melts below room temperature and has a wide liquid range with low volatility. Figure 15-1 shows a classic salt bridge consisting of an inverted U-tube containing saturated KCl. Figure 15-10 shows a U-tube connecting two electrochemical cells from the *bottom*. The U-tube contains only ionic liquid whose solubility in water is <1 mM. The mobilities of the cation and anion are matched within 3% and are one-third as great as those of K^+ and Cl^-.

Errors of ~10–100 mV can occur if a reference electrode has a *nanoporous* glass plug instead of a *microporous* glass plug to separate the inner electrode solution from the sample solution.[11] An example of a glass plug is drawn at the bottom of the electrode in Figure 15-3. Nanoporous plugs have trade names such as Vycor® and CoralPor®. Microporous glass has passages on the order of ~0.1–3 μm in diameter through which liquid flows slowly and diffuses. Channels in nanoporous glass are ~4–20 nm in diameter. Silanol groups (Si—OH) on glass are deprotonated above pH 3 to give a negatively charged surface with Si—O⁻. In the close quarters of a nanopore, anions do not freely pass by the negative surface, but cations pass through readily (Figure 15-11). This electrostatic screening can produce solution-dependent electrode potentials that are not observed if an electrode has a microporous plug.

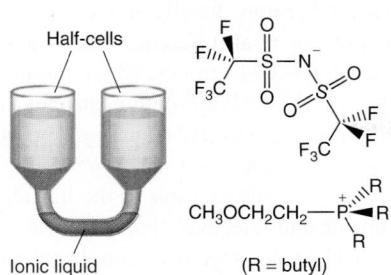

Ionic liquid is tributyl(2-methoxyethyl)phosphonium
bis(pentafluorethanesulfonyl)amidate

FIGURE 15-10 Salt bridge filled with an
ionic liquid can reduce the junction potential
to ~0.1 mV.

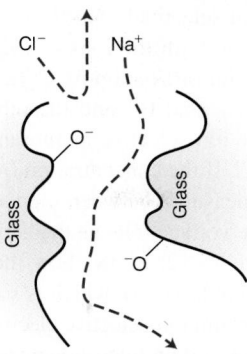

FIGURE 15-11 Negative charge on the glass
wall of a narrow channel blocks passage of
anions, but permits cations to flow. The
separation of charge creates potential
differences across a nanoporous glass plug.

15-4 How Ion-Selective Electrodes Work[12]

Ion-selective electrodes discussed in the remainder of this chapter respond selectively to one
ion. These electrodes are fundamentally different from metal electrodes in that ion-selective
electrodes do not involve redox processes. The key feature of an ideal ion-selective electrode
is a thin membrane capable of binding only the intended ion.

Consider the *liquid-based ion-selective electrode* in Figure 15-12a. The electrode is said
to be "liquid based" because the ion-selective membrane is a *hydrophobic* organic polymer
impregnated with a viscous organic solution containing an ion exchanger and, sometimes, a

Hydrophobic: "water hating" (does not mix
with water)

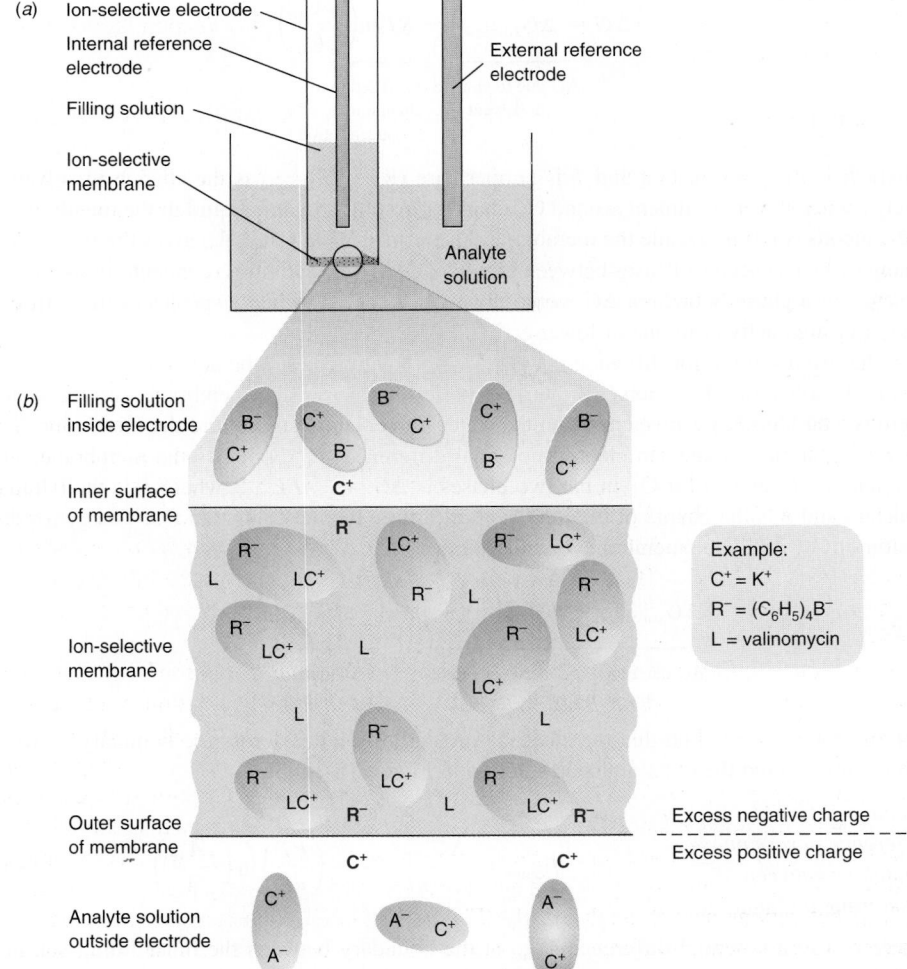

Example:
$C^+ = K^+$
$R^- = (C_6H_5)_4B^-$
L = valinomycin

FIGURE 15-12 (*a*) Ion-selective electrode
immersed in aqueous solution containing
analyte cation, C^+. Typically, the membrane is
made of poly(vinyl chloride) impregnated with
the *plasticizer* dioctyl sebacate, a nonpolar
liquid that softens the membrane and dissolves
the ion-selective ionophore (L), the complex
(LC^+), and a hydrophobic anion (R^-). (*b*) Close-
up of membrane. Ellipses encircling pairs of ions
are a guide for the eye to count the charge in
each phase. Bold colored ions represent excess
charge in each phase. The electric potential
difference across each surface of the membrane
depends on the activity of analyte ion in the
aqueous solution contacting the membrane.

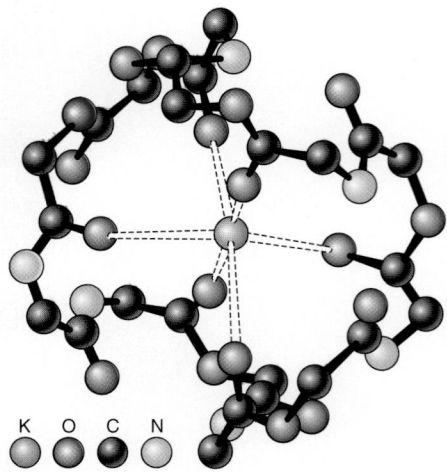

FIGURE 15-13 Valinomycin-K$^+$ complex has six carbonyl oxygen atoms in octahedral coordination around K$^+$. [Information from L. Stryer, *Biochemistry*, 4th ed. (New York: W. H. Freeman and Company, 1995), p. 273.]

K   O   C   N

L has some ability to bind other ions besides C$^+$, so those other ions interfere to some extent with the measurement of C$^+$. An ion-selective electrode uses a ligand with a strong preference to bind the desired ion.

Example of hydrophobic anion, R$^-$:

Tetraphenylborate, $(C_6H_5)_4B^-$

ligand that selectively binds the analyte cation, C$^+$. The inside of the electrode contains filling solution with the ions C$^+$(aq) and B$^-$(aq). The outside of the electrode is immersed in analyte solution containing C$^+$(aq), A$^-$(aq) and, perhaps, other ions. Ideally, it does not matter what A$^-$ and B$^-$ and the other ions are. The electric potential difference (the voltage) across the ion-selective membrane is measured by two reference electrodes, which might be Ag | AgCl. If the concentration (really, the activity) of C$^+$ in the analyte solution changes, the voltage measured between the two reference electrodes also changes. By using a calibration curve, the voltage tells us the activity of C$^+$ in the analyte solution.

Figure 15-12b shows how the electrode works. The key in this example is the ligand, L (called an *ionophore*), which is soluble inside the membrane and selectively binds analyte ion. In a potassium ion-selective electrode, for example, L could be valinomycin, a natural antibiotic secreted by certain microorganisms to carry K$^+$ ion across cell membranes (Figure 15-13). *The ligand, L, is chosen to have a high affinity for analyte cation, C$^+$, and low affinity for other ions.* In an ideal electrode, L binds only C$^+$. Real electrodes always have some affinity for other cations, so these cations interfere to some degree with the measurement of C$^+$. For charge neutrality, the membrane also contains a hydrophobic anion, R$^-$, such as tetraphenylborate, $(C_6H_5)_4B^-$, that is soluble in the membrane and poorly soluble in water.

Almost all analyte ion inside the membrane in Figure 15-12b is bound in the complex LC$^+$, which is in equilibrium with a small amount of free C$^+$ in the membrane. The membrane also contains excess free L. C$^+$ can diffuse across the interface. In an ideal electrode, R$^-$ cannot leave the membrane, because it is not soluble in water, and the aqueous anion A$^-$ cannot enter the membrane, because it is not soluble in the organic phase. When a few C$^+$ ions diffuse from the membrane into the aqueous phase, there is excess positive charge in the aqueous phase. This imbalance creates an electric potential difference that opposes diffusion of more C$^+$ into the aqueous phase. The region of charge imbalance extends just a few nanometers into the membrane and into the neighboring solution.

When C$^+$ diffuses from a region of activity $\mathcal{A}_m$ in the membrane to a region of activity $\mathcal{A}_o$ in the outer solution, the free energy change is

$$\Delta G = \underbrace{\Delta G_{\text{solvation}}}_{\substack{\Delta G \text{ due to change} \\ \text{in solvent}}} - \underbrace{RT \ln\left(\frac{\mathcal{A}_m}{\mathcal{A}_o}\right)}_{\substack{\Delta G \text{ due to} \\ \text{change in activity} \\ \text{(concentration)}}}$$

where R is the gas constant and T is temperature (K). $\Delta G_{\text{solvation}}$ is the change in solvation energy when the environment around C$^+$ changes from the organic liquid in the membrane to the aqueous solution outside the membrane. The term $-RT \ln (\mathcal{A}_m/\mathcal{A}_o)$ gives the free energy change when a species diffuses between regions of different activities (concentrations). In the absence of a phase boundary, ΔG would always be negative when a species diffuses from a region of high activity to one of lower activity.

The driving force for diffusion of C$^+$ from the membrane to the aqueous solution is the favorable solvation of the ion by water. As C$^+$ diffuses from the membrane into the water, there is a buildup of positive charge in the water immediately adjacent to the membrane. The charge separation creates an electric potential difference (E_{outer}) across the membrane. The free energy difference for C$^+$ in the two phases is $\Delta G = -nFE_{\text{outer}}$, where F is the Faraday constant and n is the charge of the ion. At equilibrium, the net change in free energy for diffusion of C$^+$ across the membrane boundary must be 0:

$$\underbrace{\Delta G_{\text{solvation}} - RT \ln\left(\frac{\mathcal{A}_m}{\mathcal{A}_o}\right)}_{\substack{\Delta G \text{ due to transfer between phases} \\ \text{and activity difference}}} + \underbrace{(-nFE_{\text{outer}})}_{\substack{\Delta G \text{ due to} \\ \text{charge imbalance}}} = 0$$

Solving for E_{outer}, we find that the electric potential difference across the boundary between the membrane and the outer aqueous solution in Figure 15-12b is

Electric potential difference across phase boundary between membrane and analyte:

$$E_{\text{outer}} = \frac{\Delta G_{\text{solvation}}}{nF} - \left(\frac{RT}{nF}\right) \ln\left(\frac{\mathcal{A}_m}{\mathcal{A}_o}\right) \qquad \textbf{(15-4)}$$

There is also a potential difference E_{inner} at the boundary between the inner filling solution and the membrane, with terms analogous to those in Equation 15-4.

The potential difference between the outer analyte solution and the inner filling solution is the difference $E = E_{outer} - E_{inner}$. In Equation 15-4, E_{outer} depends on the activities of C^+ in the analyte solution and in the membrane near its outer surface. E_{inner} is constant because the activity of C^+ in the filling solution is constant.

But the activity of C^+ in the membrane ($\mathcal{A}_m$) is very nearly constant for the following reason: The high concentration of LC^+ in the membrane is in equilibrium with free L and a small concentration of free C^+ in the membrane. The hydrophobic anion R^- is poorly soluble in water and therefore cannot leave the membrane. *Very little* C^+ can diffuse out of the membrane because each C^+ that enters the aqueous phase leaves behind one R^- in the membrane. (This separation of charge is the source of the potential difference at the phase boundary.) As soon as a tiny fraction of C^+ diffuses from the membrane into solution, further diffusion is prevented by excess positive charge in the solution near the membrane.

So the potential difference between the outer and the inner solutions is

$$E = E_{outer} - E_{inner} = \frac{\Delta G_{solvation}}{nF} - \left(\frac{RT}{nF}\right)\ln\left(\frac{\mathcal{A}_m}{\mathcal{A}_o}\right) - E_{inner}$$

$$E = \underbrace{\frac{\Delta G_{solvation}}{nF}}_{\text{Constant}} + \left(\frac{RT}{nF}\right)\ln \mathcal{A}_o - \underbrace{\left(\frac{RT}{nF}\right)\ln \mathcal{A}_m}_{\text{Constant}} - \underbrace{E_{inner}}_{\text{Constant}}$$

$\ln\dfrac{x}{y} = \ln x - \ln y$

Combining the constant terms, we find that the potential difference across the membrane depends only on the activity of analyte in the outer solution:

$$E = \text{constant} + \left(\frac{RT}{nF}\right)\ln \mathcal{A}_o$$

Converting ln into log and inserting values of R, T, and F gives a useful expression for the potential difference across the membrane:

From Appendix A, $\ln x = (\ln 10)(\log x) = 2.303 \log x$

Electric potential difference for ion-selective electrode:

$$E = \text{constant} + \frac{0.059\ 16}{n}\log \mathcal{A}_o \text{ (volts at 25°C)} \qquad \textbf{(15-5)}$$

The number 0.059 16 V is $\dfrac{RT \ln 10}{F}$ at 25°C

where n is the charge of the analyte ion and $\mathcal{A}_o$ is its activity in the outer (unknown) solution. Equation 15-5 applies to any ion-selective electrode, including a glass pH electrode. If the analyte is an anion, the sign of n is negative. Later, we will modify the equation to account for interfering ions.

A difference of 59.16 mV (at 25°C) builds up across a glass pH electrode for every factor-of-10 change in activity of H^+ in the analyte solution. Because a factor-of-10 difference in activity of H^+ is 1 pH unit, a difference of, say, 4.00 pH units would lead to a potential difference of $4.00 \times 59.16 = 237$ mV. The charge of a calcium ion is $n = 2$, so a potential difference of $59.16/2 = 29.58$ mV is expected for every factor-of-10 change in activity of Ca^{2+} in the analyte measured with a calcium ion-selective electrode.

15-5 pH Measurement with a Glass Electrode

The **glass electrode** used to measure pH is the most common *ion-selective electrode*. A typical pH **combination electrode,** incorporating both glass and reference electrodes in one body, is shown in Figure 15-14. A line diagram of this cell can be written as follows:

M. Cremer at the Institute of Physiology at Munich discovered in 1906 that a potential difference of 0.2 V developed across a glass membrane with acid on one side and neutral saline solution on the other. The student Klemensiewicz, working with F. Haber in Karlsruhe in 1908, improved the glass electrode and carried out the first acid-base titration to be monitored with a glass electrode.[13]

Glass membrane
selectively binds H^+

$$\underbrace{Ag(s)|AgCl(s)|Cl^-(aq)}_{\substack{\text{Outer reference} \\ \text{electrode}}}\|\underbrace{H^+(aq, \text{outside})}_{\substack{H^+ \text{ outside} \\ \text{glass electrode} \\ \text{(analyte solution)}}}\underbrace{H^+(aq, \text{inside}),}_{\substack{H^+ \text{ inside} \\ \text{glass} \\ \text{electrode}}} \underbrace{Cl^-(aq)|AgCl(s)|Ag(s)}_{\substack{\text{Inner reference} \\ \text{electrode}}}$$

The pH-sensitive part of the electrode is the thin glass bulb or cone at the bottom of the electrodes in Figures 15-14 and 15-15. The reference electrode at the left in the line diagram is the coiled Ag | AgCl electrode in the combination electrode in Figure 15-14. The reference electrode at the right side of the line diagram is the straight Ag | AgCl electrode at the center of the electrode in Figure 15-14. The two reference electrodes measure the electric potential

FIGURE 15-14 Diagram of a glass combination electrode with a silver-silver chloride reference electrode. The glass electrode is immersed in a solution of unknown pH so that the porous plug on the lower right is below the surface of the liquid. The two Ag | AgCl electrodes measure the voltage across the glass membrane.

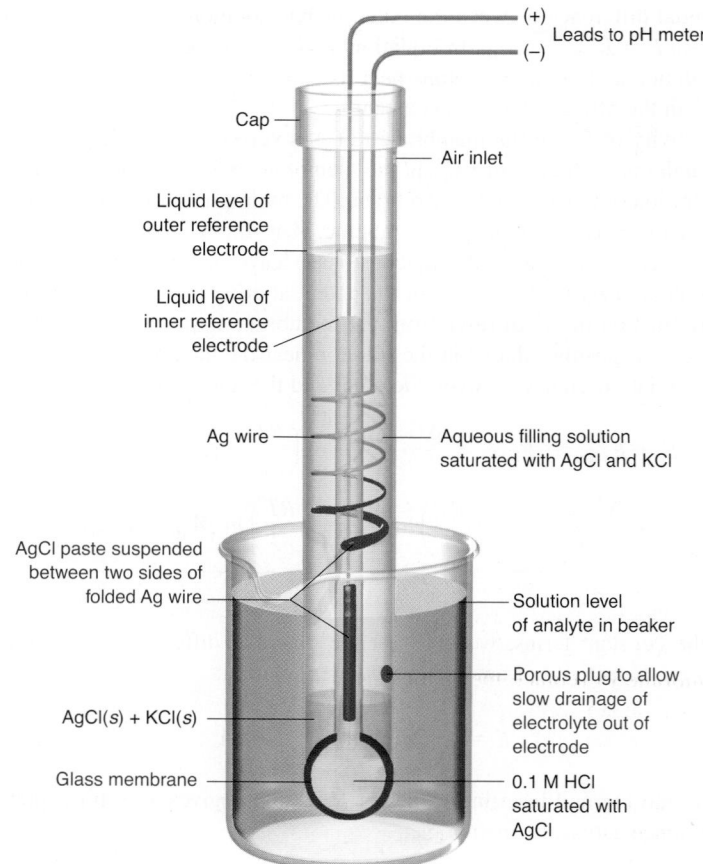

Leads to pH meter

(+)
(−)

Cap

Air inlet

Liquid level of outer reference electrode

Liquid level of inner reference electrode

Ag wire

Aqueous filling solution saturated with AgCl and KCl

AgCl paste suspended between two sides of folded Ag wire

Solution level of analyte in beaker

Porous plug to allow slow drainage of electrolyte out of electrode

AgCl(s) + KCl(s)

Glass membrane

0.1 M HCl saturated with AgCl

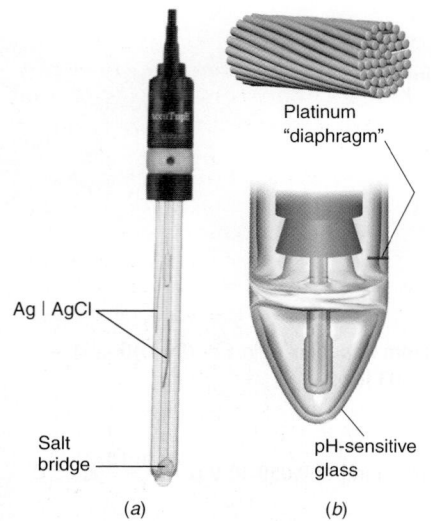

Platinum "diaphragm"

Ag | AgCl

pH-sensitive glass

Salt bridge

(a)   (b)

FIGURE 15-15 (a) Glass-body combination electrode with pH-sensitive glass bulb at the bottom. The porous ceramic plug (the salt bridge) connects analyte solution to the reference electrode. Two silver wires coated with AgCl are visible inside the electrode. [Courtesy Thermo Fisher Scientific, Inc., Pittsburgh PA.] (b) A pH electrode with a platinum diaphragm (a bundle of Pt wires), which is said to be less prone to clogging than a ceramic plug. [Information from W. Knappek, *Am. Lab. News Ed.*, July 2003, p. 14]

Crystalline: repeating structure
Amorphous: irregular structure without long range order

difference across the glass membrane. The salt bridge in the line diagram is the porous plug at the bottom right side of the combination electrode in Figure 15-14.

Figure 15-16a depicts the irregular structure of silicate glass, which is in the bulb of a glass pH electrode. Negatively charged oxygen atoms in glass can bind cations of suitable size. Monovalent cations, particularly Na^+, can move sluggishly through the glass. Figure 15-16b is an atomic resolution micrograph showing crystalline and amorphous regions of pure silica glass (pure SiO_2). Figure 15-16c is a micrograph of an amorphous region of silica

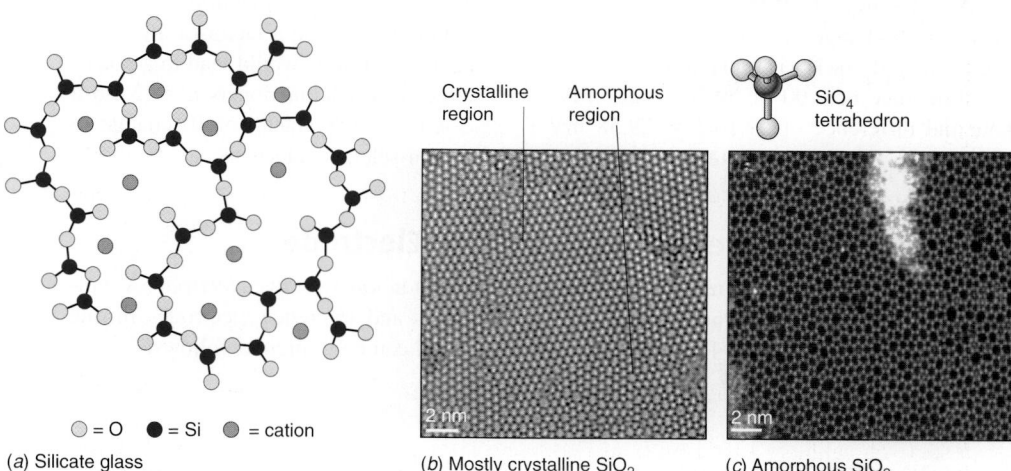

○ = O   ● = Si   ◐ = cation

(a) Silicate glass

Crystalline region   Amorphous region

SiO_4 tetrahedron

2 nm

(b) Mostly crystalline SiO_2

2 nm

(c) Amorphous SiO_2

FIGURE 15-16 (a) Schematic structure of silicate glass, which consists of an irregular network of SiO_4 tetrahedra connected through oxygen atoms. Cations such as Li^+, Na^+, K^+, and Ca^{2+} are coordinated to oxygen atoms. The silicate network is not planar. The diagram is a projection of each tetrahedron onto the plane of the page. [Information from G. A. Perley, "Glasses for Measurement of pH," *Anal. Chem.* **1949**, *21*, 394.] (b) and (c) Transmission electron micrographs of thin layer of SiO_2 deposited on graphene. Each vertex of each polygon is looking down the axis of an SiO_4 tetrahedron connected to adjacent SiO_4 tetrahedra through oxygen atoms. [P. Y. Huang et al., "Direct Imaging of a Two-Dimensional Silica Glass on Graphene," *Nano Lett.* **2012**, *12*, 1081. See also P. Y. Huang et al., "Imaging Atomic Rearrangements in Two-Dimensional Silica Glass: Watching Silica's Dance," *Science* **2013**, *342*, 224. Reprinted with permission © 2012 American Chemical Society.]

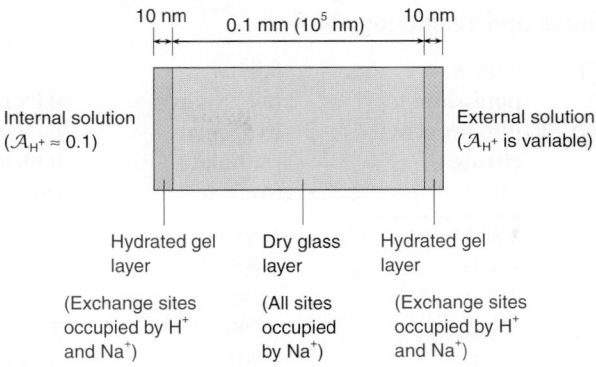

FIGURE 15-17 Schematic cross section of the glass membrane of a pH electrode.

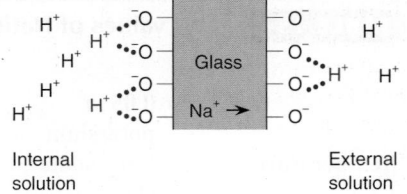

FIGURE 15-18 Ion-exchange equilibria on surfaces of a glass membrane: H^+ replaces metal ions bound to the negatively charged oxygen atoms. The pH of the internal solution is fixed. As the pH of the external solution (the sample) changes, the electric potential difference across the glass membrane changes.

glass. Crystalline regions have a repeating array of rings with six SiO_4 units. Amorphous regions have a mixture of ring sizes with irregular orientations. Amorphous silica glass is an approximation of the structure of silicate glass.

A schematic cross section of the glass membrane of a pH electrode is shown in Figure 15-17. The two surfaces swell as they absorb water. Metal ions in these *hydrated gel* regions of the membrane diffuse out of the glass and into solution. H^+ can diffuse into the membrane to replace metal ions. The reaction in which H^+ replaces cations in the glass is an **ion-exchange equilibrium** (Figure 15-18). A pH electrode responds selectively to H^+ because H^+ is the main ion that binds significantly to the hydrated gel layer.

To perform an electrical measurement, at least some tiny current must flow through the entire circuit—even across the glass pH electrode membrane. Studies with tritium (radioactive ^3H) show that H^+ does not cross the glass membrane. However, Na^+ sluggishly crosses the membrane. The H^+-sensitive membrane may be thought of as two surfaces electrically connected by Na^+ transport. The membrane's resistance is typically $10^8\ \Omega$, so little current actually flows across it.

The potential difference between the inner and outer silver-silver chloride electrodes in Figure 15-14 depends on the chloride concentration in each electrode compartment and on the potential difference across the glass membrane. Because $[Cl^-]$ is fixed in each compartment and because $[H^+]$ is fixed on the inside of the glass membrane, the only variable is the pH of analyte solution outside the glass membrane. Equation 15-5 states that *the voltage of the ideal pH electrode changes by 59.16 mV for every pH-unit change of analyte activity at 25°C.*

The response of real glass electrodes is described by the Nernst-like equation

Response of glass $E = \text{constant} + \beta(0.059\ 16)\log\mathcal{A}_{H^+}(\text{outside})$
electrode: $E = \text{constant} - \beta(0.059\ 16)\,\text{pH(outside)}$ (at 25°C)   **(15-6)**

The value of β, the *electromotive efficiency*, is close to 1.00 (typically >0.98). We measure the constant and β when we calibrate the electrode in *at least two solutions* of known pH.

Calibrating a Glass Electrode

A pH electrode should be calibrated with two (or more) standard buffers such that the pH of the unknown lies within the range of the standards. Standards in Table 15-3 are accurate to ±0.01 pH unit.[15] Problem 15-30 shows how the pH of standard buffers is measured.

When you calibrate an electrode with standard buffers, you measure a voltage with the electrode in each buffer (Figure 15-19). The pH of buffer S1 is pH_{S1} and the measured electrode potential in this buffer is E_{S1}. The pH of buffer S2 is pH_{S2} and the measured electrode potential is E_{S2}. The equation of the line through the two standard points is

$$\frac{E - E_{S1}}{pH - pH_{S1}} = \frac{E_{S2} - E_{S1}}{pH_{S2} - pH_{S1}} \qquad \textbf{(15-7)}$$

The slope of the line is $\Delta E/\Delta pH = (E_{S2} - E_{S1})/(pH_{S2} - pH_{S1})$, which is 59.16 mV/pH unit at 25°C for an ideal electrode and $\beta(59.16)$ mV/pH unit for a real electrode, where β is the correction factor in Equation 15-6.

So little current flows across a glass electrode that it was not practical when discovered in 1906. One of the first people to use a vacuum tube amplifier to measure pH with a glass electrode was an undergraduate, W. H. Wright, at the University of Illinois in 1928, who knew about electronics from amateur radio. In 1935, Arnold Beckman at Caltech invented a portable, rugged, vacuum tube pH meter, a device that revolutionized chemical instrumentation.[14]

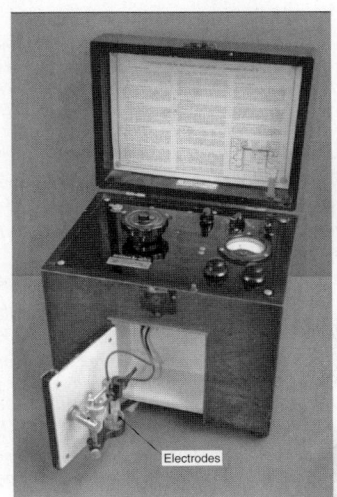

Beckman's pH meter. [Canada Science and Technology Museum; 1978.0006.]

The pH electrode measures H^+ activity, not H^+ concentration.

A pH electrode must be calibrated at the same temperature as the unknown before it can be used. It should be calibrated at least every 2 h during sustained use.

TABLE 15-3 **pH values of National Institute of Standards and Technology buffers**

Temperature (°C)	0.05 m potassium tetroxalate (1)	Saturated (25°C) potassium hydrogen tartrate (2)	0.05 m potassium dihydrogen citrate (3)	0.05 m potassium hydrogen phthalate (4)	0.08 m MOPSO 0.08 m NaMOPSO 0.08 m NaCl (5)
0	1.667	—	3.863	4.003	7.268
5	1.666	—	3.840	3.999	7.182
10	1.665	—	3.820	3.998	7.098
15	1.669	—	3.802	3.999	7.018
20	1.672	—	3.788	4.002	6.940
25	1.677	3.557	3.776	4.008	6.865
30	1.681	3.552	3.766	4.015	6.792
35	1.688	3.549	3.759	4.024	6.722
37	—	3.548	3.756	4.028	6.695
40	1.694	3.547	3.753	4.035	6.654
45	1.699	3.547	3.750	4.047	6.588
50	1.706	3.549	3.749	4.060	6.524
55	1.713	3.554	—	4.075	—
60	1.722	3.560	—	4.091	—
70	—	3.580	—	4.126	—
80	—	3.609	—	4.164	—
90	—	3.650	—	4.205	—
95	—	3.674	—	4.227	—

NOTE: *The designation m stands for molality. Masses in the buffer recipes below are apparent masses measured in air.*

In the buffer solution preparations, it is essential to use high-purity materials and freshly distilled or deionized water of resistivity greater than 2 000 ohm · m. Solutions having pH 6 or above should be stored in plastic containers, preferably ones with an NaOH trap to prevent ingress of atmospheric carbon dioxide. They can normally be kept for 2–3 weeks, or slightly longer in a refrigerator. Buffer materials in this table are available as Standard Reference Materials from the National Institute of Standards and Technology. http://www.nist.gov/srm/. pH standards for D₂O and aqueous-organic solutions can be found in P. R. Mussini, T Mussini, and S. Rondinini, Pure Appl. Chem. ***1997,** 69, 1007.*

1. 0.05 m potassium tetroxalate, $KHC_2O_4 \cdot H_2C_2O_4$. Dissolve 12.71 g of potassium tetroxalate dehydrate (Standard Reference Material used without drying) in 1 kg of water. pH values from P. M. Juusola, J. I. Partanen, K. P. Vahteristo, P. O. Minkkinen, and A. K. Covington, J. Chem. Eng. Data **2007,** 52, 973.

2. Saturated (25°C) potassium hydrogen tartrate, $KHC_4H_4O_6$. An excess of the salt is shaken with water, and it can be stored in this way. Before use, it should be filtered or decanted at a temperature between 22° and 28°C.

3. 0.05 m potassium dihydrogen citrate, $KH_2C_6H_5O_7$. Dissolve 11.41 g of the salt in 1 L of solution at 25°C.

4. 0.05 m potassium hydrogen phthalate. Although this is not usually essential, the crystals may be dried at 100°C for 1 h, then cooled in a desiccator. At 25°C, 10.12 g $C_6H_4(CO_2H)(CO_2K)$ are dissolved in water, and the solution made up to 1 L.

5. 0.08 m MOPSO ((3-N-morpholino)-2-hydroxypropanesulfonic acid, Table 9-2), 0.08 m sodium salt of MOPSO, 0.08 m NaCl. Buffers 5 and 7 are recommended for 2-point standardization of electrodes for pH measurements of physiologic fluids. MOPSO is crystallized twice from 70 wt% ethanol and dried in vacuum at 50°C for 24 h. NaCl is dried at 110°C for 4 h. Na^+MOPSO^- may be prepared by neutralization of MOPSO with standard NaOH. The sodium salt is also available as a Standard Reference Material. Dissolve 18.00 g MOPSO, 19.76 g Na^+MOPSO^-, and 4.674 g NaCl in 1.000 kg H_2O.

6. 0.025 m disodium hydrogen phosphate, 0.025 m potassium dihydrogen phosphate. The anhydrous salts are best; each should be dried for 2 h at 120°C and cooled in a desiccator, because they are slightly hygroscopic. Higher drying temperatures should be avoided to prevent formation of condensed phosphates. Dissolve 3.53 g Na_2HPO_4 and 3.39 g KH_2PO_4 in water to give 1 L of solution at 25°C.

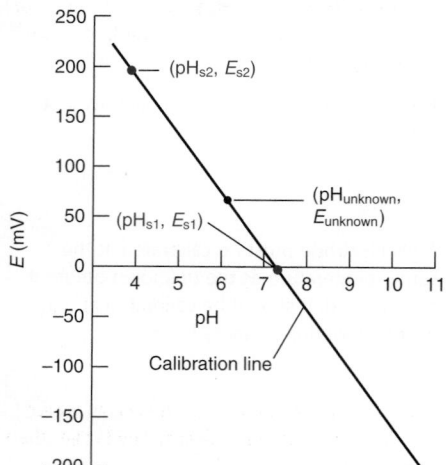

FIGURE 15-19 Two-point calibration of a pH electrode.

To measure the pH of an unknown, measure the potential of the unknown with the calibrated electrode and find pH by substitution in Equation 15-7:

$$\frac{E_{\text{unknown}} - E_{S1}}{\text{pH}_{\text{unknown}} - \text{pH}_{S1}} = \frac{E_{S2} - E_{S1}}{\text{pH}_{S2} - \text{pH}_{S1}} \qquad (15\text{-}8)$$

Alas, modern pH meters are "black boxes" that do these calculations for us by applying Equations 15-7 and 15-8 and automatically displaying pH. When measuring pH with a glass electrode, the junction potential at the salt bridge opening at the side of the electrode (Figure 15-15) could change in different solutions. Such a change introduces an error in the derived pH.

Equation 15-8 is an "operational" definition prescribing how to measure pH. The definition pH = $-\log \mathcal{A}_{H+}$ is not rigorous because the activity of any single ion, such as H^+, is not thermodynamically defined.[16] Only the *mean activity* of pairs of ions (such as H^+Cl^-) is thermodynamically defined and measurable. The procedure behind Equation 15-8, using standard buffers in Table 15-3, is intended to come as close as possible to complying with the idealization pH = $-\log \mathcal{A}_{H+}$. The valid limits for Equation 15-8 are approximately $2 \leq \text{pH} \leq 12$ and ionic strength ≤ 0.1 mol/kg. (Physical chemists usually express concentrations in

0.025 m potassium dihydrogen phosphate 0.025 m disodium hydrogen phosphate (6)	0.08 m HEPES 0.08 m NaHEPES 0.08 m NaCl (7)	0.008 695 m potassium dihydrogen phosphate 0.030 43 m disodium hydrogen phosphate (8)	0.01 m borax (9)	0.025 m sodium bicarbonate 0.025 m sodium carbonate (10)	Saturated (at 25°C) Ca(OH)$_2$ (11)
6.984	7.853	7.534	9.464	10.317	13.42
6.951	7.782	7.500	9.395	10.245	13.21
6.923	7.713	7.472	9.332	10.179	13.00
6.900	7.645	7.448	9.276	10.118	12.81
6.881	7.580	7.429	9.225	10.062	12.63
6.865	7.516	7.413	9.180	10.012	12.45
6.853	7.454	7.400	9.139	9.966	12.29
6.844	7.393	7.389	9.102	9.925	12.07
6.840	7.370	7.385	9.088	9.910	11.98
6.838	7.335	7.380	9.068	9.889	11.71
6.834	7.278	7.373	9.038	9.856	—
6.833	7.223	7.367	9.011	9.828	—
6.834	—	—	8.985	—	—
6.836	—	—	8.962	—	—
6.845	—	—	8.921	—	—
6.859	—	—	8.885	—	—
6.877	—	—	8.850	—	—
6.886	—	—	8.833	—	—

7. *0.08 m HEPES (N-2-hydroxyethylpiperazine-N′-2-ethanesulfonic acid, Table 9-2, 0.08 m sodium salt of HEPES, 0.08 m NaCl. Buffers 5 and 7 are recommended for 2-point standardization of electrodes for pH measurements of physiologic fluids. HEPES is crystallized twice from 80 wt% ethanol and dried in vacuum at 50°C for 24 h. NaCl is dried at 110°C for 4 h. Na$^+$HEPES$^-$ may be prepared by neutralization of HEPES with standard NaOH. The sodium salt is also available as a Standard Reference Material. Dissolve 19.04 g HEPES, 20.80 g Na$^+$HEPES$^-$, and 4.674 g NaCl in 1.000 kg H$_2$O.*

8. *0.008 695 m potassium dihydrogen phosphate, 0.030 43 m disodium hydrogen phosphate. Prepare like Buffer 6; dissolve 1.179 g KH$_2$PO$_4$ and 4.30 g Na$_2$HPO$_4$ in water to give 1 L of solution at 25°C.*

9. *0.01 m sodium tetraborate decahydrate. Dissolve 3.80 g Na$_2$B$_4$O$_7$ ·10H$_2$O in water to give 1 L of solution. This borax solution is particularly susceptible to pH change from carbon dioxide absorption, and it should be correspondingly protected.*

10. *0.025 m sodium bicarbonate, 0.025 m sodium carbonate. Primary standard grade Na$_2$CO$_3$ is dried at 250°C for 90 min and stored over CaCl$_2$ and Drierite. Reagent-grade NaHCO$_3$ is dried over molecular sieves and Drierite for 2 days at room temperature. Do not heat NaHCO$_3$, or it may decompose to Na$_2$CO$_3$. Dissolve 2.092 g of NaHCO$_3$ and 2.640 g of Na$_2$CO$_3$ in 1 L of solution at 25°C.*

11. *Ca(OH)$_2$ is a secondary standard whose pH is not as accurate as that of primary standards. Wash low-alkali grade CaCO$_3$ well with H$_2$O to remove alkali metals. Heat the powder to 1 000°C in a Pt dish for 45 min and cool in a desiccator. Add the resulting CaO slowly with stirring to H$_2$O. Boil the suspension, cool it, and filter through a medium porosity sintered glass funnel. Dry the solid at 110°C and crush to a fine granular state. If [OH$^-$] of the saturated solution at 25°C exceeds 0.020 6 M (measured by titration with strong acid), there were probably soluble alkali metals in the CaCO$_3$. From R. G. Bates, V. E. Bower, and E. R. Smith, J. Res. Natl. Bur. Std., 1956, 56, 305; http://www.nist.gov/nvl/jrespastpapers.cfm.*

SOURCES: *R. P. Buck, S. Rondinini, A. K. Covington, F. G. K. Baucke, C. M. A. Brett, M. F. Camoes, M. J. T. Milton, T. Mussini, R. Naumann, K. W. Pratt, P. Spitzer, and G. S. Wilson, "Measurement of pH. Definitions, Standards, and Procedures," Pure Appl. Chem. 2002, 74, 2169. R. G. Bates, J. Res. Natl. Bureau Stds. 1962, 66A, 179; B. R. Staples and R. G. Bates, J. Res. Natl. Bureau Stds. 1969, 73A, 37. Data on HEPES and MOPSO are from Y. C Wu, P. A. Berezansky, D. Feng, and W. F. Koch, Anal. Chem. 1993, 65, 1084, and D. Feng, W. F Koch, and Y. C Wu, Anal. Chem. 1989, 61, 1400. Instructions for preparing some of these solutions are from G. Mattock in C. N. Reilley, ed., Advances in Analytical Chemistry and Instrumentation (New York: Wiley, 1963), Vol. 2, p. 45. See also R. G. Bates, Determination of pH: Theory and Practice, 2nd ed. (New York: Wiley, 1973), Chap. 4.*

molality, which is a temperature-independent quantity. Molarity changes with temperature because solutions usually expand when heated.)

Before using a pH electrode, be sure that the air inlet near the upper end of the electrode in Figure 15-14 is not capped. (This hole is capped during storage to prevent evaporation of the electrode filling solution.) Wash the electrode with distilled water and *blot* it dry with a tissue. Do not *wipe* it, because this action might produce a static charge on the glass.

To calibrate the electrode, dip it in a standard buffer whose pH is near 7 and allow the electrode to equilibrate with stirring for at least a minute. Following the manufacturer's instructions, press a key that might say "Calibrate" or "Read" on a microprocessor-controlled meter or adjust the reading of an analog meter to indicate the pH of the standard buffer. Then wash the electrode with water, blot it dry, and immerse it in a second standard whose pH is further from 7 than the pH of the first standard. Enter the pH of the second buffer on the meter. Finally, dip the electrode in the unknown, stir the liquid, allow the reading to stabilize, and read the pH.

Store a glass electrode in aqueous solution to prevent dehydration of the glass. Ideally, the solution should be similar to that inside the reference compartment of the electrode. If the electrode has dried, recondition it in dilute acid for several hours. If the electrode is to be used above pH 9, soak it in a high-pH buffer. (The field effect transistor pH electrode in Section 15-8 is stored dry. Prior to use, scrub it gently with a soft brush and soak it in pH 7 buffer for 10 min.)

pH measurements outside the range 2–12 or above an ionic strength of 0.1 mol/kg could have large errors.

Do not leave a glass electrode out of water (or in a nonaqueous solvent) any longer than necessary.

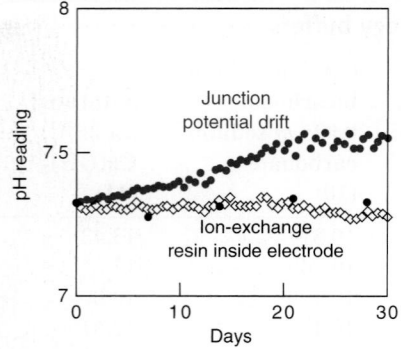

FIGURE 15-20 Solid colored circles show the drift in apparent pH of a low-conductivity industrial water supply measured continuously by a single electrode. Individual measurements with a freshly calibrated electrode (black circles) demonstrate that the pH is not drifting. Drift is attributed to slow clogging of the electrode's porous plug with AgCl(s). When a cation-exchange resin was placed inside the reference electrode near the porous plug, Ag(I) was bound by the resin and did not precipitate. This electrode gave the drift-free, continuous reading shown by open diamonds. [Data from S. Ito, H. Hachiya, K. Baba, Y. Asano, and H. Wada, "Improvement of the Ag | AgCl Reference Electrode and Its Application to pH Measurement," *Talanta* **1995**, *42*, 1685.]

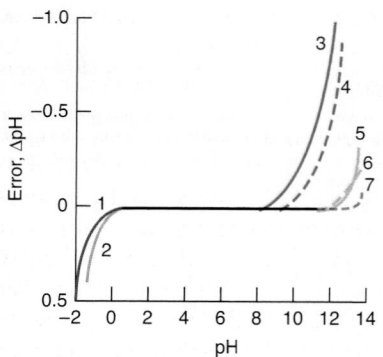

FIGURE 15-21 Acid and alkaline errors of some glass electrodes. 1: Corning 015, H_2SO_4. 2: Corning 015, HCl. 3: Corning 015, 1 M Na^+. 4: Beckman-GP, 1 M Na^+. 5: L & N Black Dot, 1 M Na^+. 6: Beckman Type E, 1 M Na^+. 7: Ross electrode.[22] [Data from R. G. Bates, *Determination of pH: Theory and Practice*, 2nd ed. (New York: Wiley, 1973). Ross electrode data are from Orion, *Ross pH Electrode Instruction Manual*.]

Challenge Use Equation 15-6 to show that the potential of the glass electrode changes by 1.3 mV when $\mathcal{A}_{H^+}$ changes by 5.0%. Show that 1.3 mV = 0.02 pH unit.

Moral: A small uncertainty in voltage (1.3 mV) or pH (0.02 unit) corresponds to a large uncertainty (5%) in analyte concentration. Similar uncertainties arise in other potentiometric measurements.

If a glass electrode responds sluggishly, or if it cannot be calibrated properly, try soaking it in 6 M HCl, followed by water. As a last resort, soak the electrode in 20 wt% aqueous ammonium bifluoride, NH_4HF_2, for 1 min in a plastic beaker. This reagent dissolves glass and exposes fresh surface. Wash the electrode with water and try calibrating it again. *Avoid contact with ammonium bifluoride, which produces a dangerous HF burn.*

Errors in pH Measurement

1. *Standards.* A pH measurement cannot be more accurate than our standards, which are typically ±0.01 pH unit.

2. *Junction potential.* A *junction potential* exists at the porous plug near the bottom of the electrode in Figure 15-14. If the ionic composition of the analyte solution is different from that of the standard buffer, the junction potential will change *even if the pH of the two solutions is the same* (Box 15-1). This effect gives an uncertainty of at least ~0.01 pH unit.

3. *Junction potential drift.* Most combination electrodes have a Ag | AgCl reference electrode containing saturated KCl solution. More than 350 mg Ag(I) per liter dissolve in the KCl, mainly as $AgCl_4^{3-}$ and $AgCl_3^{2-}$. In the porous plug, KCl is diluted and AgCl can precipitate. If analyte solution contains a reducing agent, Ag(s) also can precipitate in the plug. Both effects change the junction potential, causing a slow drift of the pH reading (solid colored circles in Figure 15-20). You can compensate for this error by recalibrating the electrode every 2 h.

4. *Sodium error.* When $[H^+]$ is very low and $[Na^+]$ is high, the electrode responds to Na^+ and the apparent pH is lower than the true pH. This is called the *sodium error* or *alkaline error* (Figure 15-21).

5. *Acid error.* In strong acid, the measured pH is higher than the actual pH (Figure 15-21), perhaps because the glass is saturated with H^+ and cannot be further protonated.

6. *Equilibration time.* It takes time for an electrode to equilibrate with a solution. A well-buffered solution requires ~30 s with adequate stirring. A poorly buffered solution (such as one near the equivalence point of a titration) needs many minutes.

7. *Hydration of glass.* A dry electrode requires several hours of soaking before it responds to H^+ correctly.

8. *Temperature.* A pH meter should be calibrated at the same temperature at which the measurement will be made.

9. *Cleaning.* If an electrode has been exposed to a hydrophobic liquid, such as oil, it should be cleaned with a solvent that will dissolve the liquid and then conditioned in aqueous solution. The reading of an improperly cleaned electrode can drift for hours while the electrode re-equilibrates with aqueous solution.

Errors 1 and 2 limit the accuracy of pH measurement with the glass electrode to ±0.02 pH unit, at best. Measurement of pH *differences* between solutions can be accurate to about ±0.002 pH unit, but knowledge of the true pH will still be at least an order of magnitude more uncertain. An uncertainty of ±0.02 pH unit corresponds to an uncertainty of ±5% in $\mathcal{A}_{H^+}$.

Not All pH Electrodes Are Glass

Glass electrodes are the most common means, but not the only means, for measuring pH. Solid-state pH electrodes based on the field effect transistor are described at the end of this chapter. The DNA sequencer at the opening of this chapter uses a field effect transistor with a Ta_2O_5 layer[23] to sense H^+ released when a nucleotide base is incorporated into DNA. Liquid-based ion-selective electrodes for H^+ are described in Section 15-6.

An anhydrous IrO_2 layer formed by oxidation of iridium wire responds to pH by a half-reaction that might be[24]

$$IrO_2(s) + H^+ + e^- \rightleftharpoons IrOOH(s)$$

$$E = E° - 0.059\ 16 \log\left(\frac{1}{[H^+]}\right) = E° - 0.059\ 16\ pH$$

Other metal oxide electrodes have been used under extreme conditions. For example, a ZrO_2 electrode can measure pH up to 300°C.[25]

The *Phoenix Mars Lander* described later in Box 15-3 had two liquid-based ion-selective electrodes in each Wet Chemistry Lab to measure the pH of Martian soil suspended in water. It was not certain that these electrodes would survive the temperatures and pressures encountered during the mission, so a rugged IrO_2 pH electrode was added. The IrO_2 electrode remains accurate at pH > 9, a condition at which ion-selective electrodes were nonresponsive.

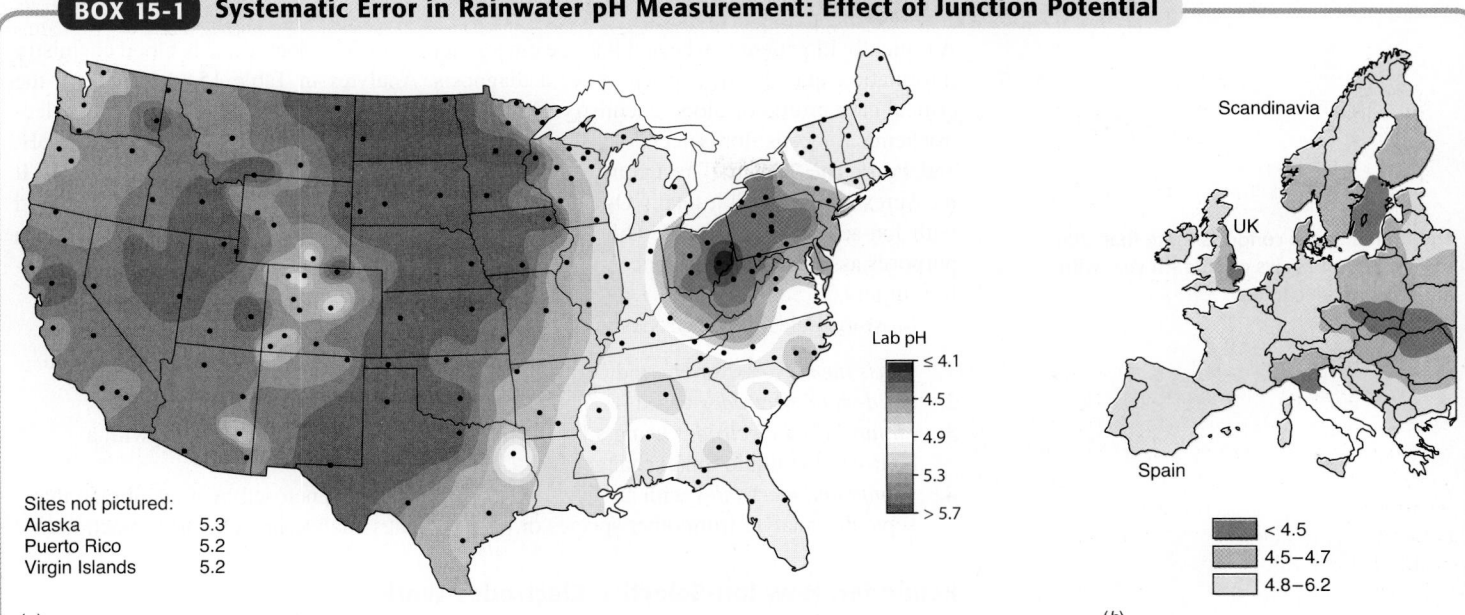

Lab pH

≤ 4.1
4.5
4.9
5.3
> 5.7

Sites not pictured:
Alaska   5.3
Puerto Rico   5.2
Virgin Islands 5.2

(a)

Scandinavia

UK

Spain

< 4.5
4.5−4.7
4.8−6.2

(b)

(a) pH of precipitation in the U.S. in 2011. The lower the pH, the more acidic the water. [Data from National Atmospheric Deposition Program (NRSP-3). (2007). NADP Program Office, Illinois State Water Survey, 2204 Griffith Dr., Champaign, IL 61820, http://nadp.sws.uiuc.edu.]

(b) pH of rain in Europe. No values are reported for Italy and Greece. [Data from H. Rodhe, F. Dentener, and M. Schulz, *Environ. Sci. Technol.* **2002,** *36,* 4382.]

Combustion products from automobiles and factories include nitrogen oxides and sulfur dioxide, which react with water in the atmosphere to produce acids.[17]

$$SO_2 + H_2O \rightarrow H_2SO_3 \xrightarrow{\text{oxidation}} H_2SO_4$$

Sulfurous acid   Sulfuric acid

Acid rain in North America is most severe in the East, downwind of many coal-fired power plants. In the three-year period 1995–1997, after SO_2 emissions were limited by a new law, concentrations of SO_4^{2-} and H^+ in precipitation decreased by 10–25% in the eastern United States.[18]

Acid rain threatens lakes and forests throughout the world. Monitoring the pH of rainwater is a critical component of programs to measure and reduce the production of acid rain.

To identify systematic errors in the measurement of pH of rainwater, a careful study was conducted with 17 laboratories.[19] Eight samples were provided to each laboratory, along with instructions on how to conduct the measurements. Each laboratory used two buffers to standardize pH meters. Sixteen laboratories successfully measured the pH of Unknown A

(within ±0.02 pH unit), which was 4.008 at 25°C. One lab whose measurement was 0.04 pH unit low had a faulty commercial standard buffer.

Panel (c) shows typical results for the pH of rainwater. The average of the 17 measurements is given by the horizontal line at pH 4.14, and the letters s, t, u, v, w, x, y, and z identify types of pH electrodes. Types s and w had relatively large systematic errors. The type s electrode was a combination electrode (Figure 15-14) containing a reference electrode liquid junction with an exceptionally large area. Electrode type w had a reference electrode filled with a gel.

A hypothesis was that variations in the liquid junction potential (Section 15-3) led to variations among the pH measurements. Standard buffers have ionic strengths of 0.05 to 0.1 M, whereas rainwater samples have ionic strengths two or more orders of magnitude lower. To see whether junction potential caused systematic errors, 2×10^{-4} M HCl was used as a pH standard in place of high ionic strength buffers. Panel (d) shows good results from all but the first lab. The standard deviation of 17 measurements was reduced from 0.077 pH unit (with standard buffer) to 0.029 pH unit (with HCl standard). It was concluded that junction potential caused most of the variability between labs and that a low ionic strength standard is appropriate for rainwater pH measurements.[20, 21]

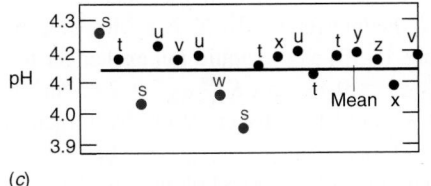

(c)

(c) pH of rainwater from identical samples measured at 17 different labs using standard buffers for calibration. Letters designate different types of pH electrodes.

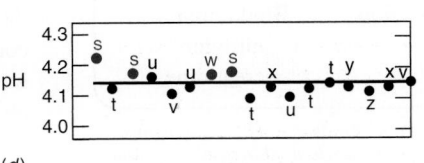

(d)

(d) Rainwater pH measured by using low ionic strength HCl for calibration.

15-6 Ion-Selective Electrodes[26,27]

A critically ill patient is wheeled into the emergency room. The doctor needs blood chemistry information quickly to help her make a diagnosis. Analytes in Table 15-4 are part of the critical care profile of blood chemistry. Every analyte in the table can be measured by electrochemical means. Ion-selective electrodes are the method of choice for Na^+, K^+, Cl^-, pH, and P_{CO_2}. The "Chem 7" test constitutes up to 70% of tests performed in the hospital lab. It measures Na^+, K^+, Cl^-, total CO_2, glucose, urea, and creatinine, four of which are analyzed with ion-selective electrodes. Other ion-selective electrodes are being developed for such purposes as monitoring the concentration of the anticoagulant drug heparin administered during surgery.[28]

Most ion-selective electrodes fall into one of the following classes:

1. *Glass membranes* for H^+ and certain monovalent cations
2. *Solid-state electrodes* based on inorganic crystals or conductive polymers
3. *Liquid-based electrodes* using a hydrophobic polymer membrane saturated with a hydrophobic liquid ion exchanger
4. *Compound electrodes* with an analyte-selective electrode enclosed by a membrane that separates analyte from other species or that generates analyte in a chemical reaction.

Reminder: How Ion-Selective Electrodes Work

In Figure 15-12, analyte ions equilibrate with ion-exchange ligand L in the ion-selective membrane. Diffusion of some analyte ions out of the membrane creates a slight charge imbalance (an electric potential difference) across the interface between the membrane and the analyte solution. Changes in analyte ion concentration in the solution change the potential difference across the outer boundary of the ion-selective membrane. A calibration curve relates the potential difference to analyte concentration.

An ion-selective electrode responds to the activity of *free analyte*, not complexed analyte. For example, when Pb^{2+} in tap water at pH 8 was measured with an ion-selective electrode, the result was $[Pb^{2+}] = 2 \times 10^{-10}$ M.[29] When lead in the same tap water was measured by inductively coupled plasma–mass spectrometry (Section 21-7), the result was more than 10 times greater: 3×10^{-9} M. The discrepancy arose because the inductively coupled plasma measures *all* lead and the ion-selective electrode measures *free Pb^{2+}*. In tap water at pH 8, much of the lead is complexed by CO_3^{2-}, OH^-, and other anions. When the pH of tap water was adjusted to 4, Pb^{2+} dissociated from its complexes and the concentration indicated by the ion-selective electrode was 3×10^{-9} M—the same value found by inductively coupled plasma.

Selectivity Coefficient

No electrode responds exclusively to one kind of ion, but the glass pH electrode is among the most selective. Sodium ion is the principal interfering species. Its effect on the pH reading is only significant when $[H^+] \geq 10^{-12}$ M and $[Na^+] \geq 10^{-2}$ M (Figure 15-21).

An electrode intended to measure ion A also responds to interfering ion X. The **selectivity coefficient** gives the relative response to different species with the same charge:

$$\text{Selectivity coefficient:} \qquad K_{A,X}^{Pot} = \frac{\text{response to X}}{\text{response to A}} \qquad (15\text{-}9)$$

The superscipt "Pot" for "potentiometric" is customary in the chemical literature. The smaller the selectivity coefficient, the less the interference by X. A K^+ ion-selective electrode that uses the chelator valinomycin (Figure 15-13) as the liquid ion exchanger has selectivity coefficients $K_{K^+,Na^+}^{Pot} = 1 \times 10^{-5}$, $K_{K^+,Cs^+}^{Pot} = 0.44$, and $K_{K^+,Rb^+}^{Pot} = 2.8$. Na^+ hardly interferes with the measurement of K^+, but Cs^+ and Rb^+ interfere strongly. In fact, the electrode responds more to Rb^+ than to K^+.

If the response to each ion is Nernstian, then the response of an ion-selective electrode to its primary ion (A) and to interfering ions of the *same charge* (X) is[12, 30]

$$\text{Response of ion-selective electrode:} \qquad E = \text{constant} \pm \frac{0.059\,16}{z_A} \log\left[\mathscr{A}_A + \sum_X K_{A,X}^{Pot}\, \mathscr{A}_X \right] \qquad (15\text{-}10)$$

where z_A is the magnitude of the charge of A, $\mathscr{A}_A$ and $\mathscr{A}_X$ are activities, and $K_{A,X}^{Pot}$ is the selectivity coefficient for each interfering ion. If the ion-selective electrode is connected to the positive terminal of the potentiometer, the sign of the log term is positive if A is a cation

---

The United States conducts more than 200 million clinical assays of K^+ each year with ion-selective electrodes.

Analyte ions establish an ion-exchange equilibrium at the surface of the ion-selective membrane. Other ions that bind to the same sites interfere.

The ion-selective electrode responds to Pb^{2+} with little response to $Pb(OH)^+$ or $Pb(CO_3)(aq)$.

TABLE 15-4	Critical care profile
Function	**Analyte**
Conduction	K^+, Ca^{2+}
Contraction	Ca^{2+}, Mg^{2+}
Energy level	Glucose, P_{O_2}, lactate, hematocrit
Ventilation	P_{O_2}, P_{CO_2}
Perfusion	Lactate, $SO_2\%$, hematocrit
Acid-base	pH, P_{CO_2}, HCO_3^-
Osmolality	Na^+, glucose
Electrolyte balance	Na^+, K^+, Ca^{2+}, Mg^{2+}
Renal function	Blood urea nitrogen, creatinine

SOURCE: C. C. Young, "Evolution of Blood Chemistry Analyzers Based on Ion Selective Electrodes," J. Chem. Ed. *1997*, 74, 177.

When measuring selectivity coefficients, you must *demonstrate* that the response of the electrode to each interfering ion follows the Nernst equation.[31-33] This is not as simple as it sounds. An ion-selective membrane that has equilibrated with its primary ion might respond slowly to weakly bound interfering ions.

The graph shows the *separate solution method* for measuring selectivity coefficients. In this method, a calibration curve is constructed

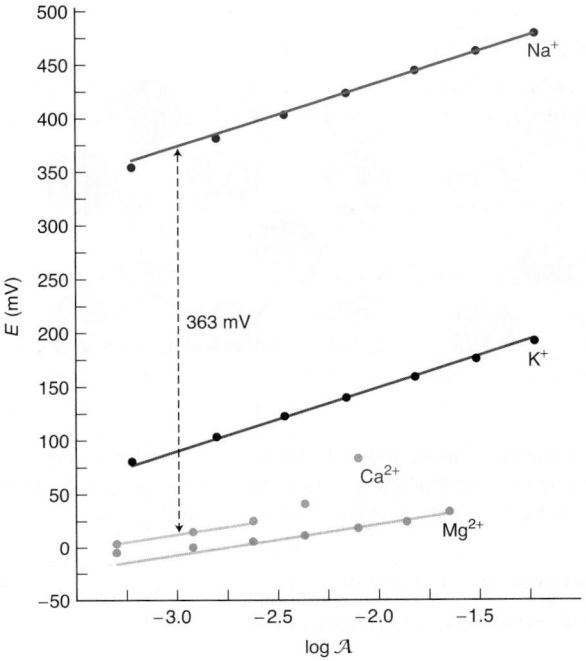

Measurement of selectivity coefficients of Na^+ ion-selective electrode. Activities on the abscissa were calculated from concentrations and activity coefficients. [Data from E. Bakker, "Determination of Unbiased Selectivity Coefficients of Neutral Carrier-Based Cation-Selective Electrodes," *Anal. Chem.* **1997**, *69*, 1061.]

for each ion by itself. Other common procedures are the *fixed interference method* and the *matched potential method.*[31]

The graph shows the response of a sodium ion-selective electrode to the interfering ions K^+, Ca^{2+}, and Mg^{2+}. To obtain a Nernstian response to interfering ions, the electrode was prepared in the absence of Na^+. The electrode was filled with 0.01 M KCl and soaked in 0.01 M KCl overnight to condition the ion-selective membrane prior to measurements. After measuring K^+, Ca^{2+}, and Mg^{2+}, Na^+ was measured. For subsequent use to measure Na^+, the inner filling solution was replaced by 0.01 M NaCl.

The data demonstrate a near-Nernstian response for each ion. At the laboratory temperature of 21.5°C, Nernstian response would be $(RT \ln 10)/zF = 58.5/z$ mV per decade (factor of 10) of ion activity, where z is the charge of the ion. Measured slopes are 61.3 ± 1.5 mV for Na^+, 56.3 ± 0.6 mV for K^+, 26.0 ± 1.0 mV for Mg^{2+}, and 31.2 ± 0.7 mV for Ca^{2+}. The deviation of Ca^{2+} from a straight line above an activity of $10^{-2.5}$ is attributed to Na^+ impurity in high-purity $CaCl_2$. Electrode response to Na^+ is much greater than to Ca^{2+}, so a little Na^+ has a large effect.

To find the selectivity coefficient, we measure the difference between the line for an interfering ion and the Na^+ calibration line at any chosen activity and apply the formula

$$\log K_{A,X}^{Pot} = \frac{z_A F(E_X - E_A)}{RT \ln 10} + \log\left(\frac{\mathcal{A}_A}{(\mathcal{A}_X)^{z_A/z_X}}\right) \quad (15\text{-}11)$$

where A = Na^+ with charge $z_A = 1$ and X is an interfering ion of charge z_X. At an activity of 10^{-3}, the dashed line shows a difference of $E_{Ca^{2+}} - E_{Na^+} = -363$ mV. The selectivity coefficient is

$$\log K_{Na^+, Ca^{2+}}^{Pot} = \frac{(+1)F(-0.363\ \mathrm{V})}{RT \ln 10} + \log\left(\frac{10^{-3}}{(10^{-3})^{1/2}}\right) = -7.0$$

We could choose a different activity to measure $E_{Ca^{2+}} - E_{Na^+}$ and should find nearly the same $K_{Na^+, Ca^{2+}}^{Pot}$. The other lines in the graph give $\log K_{Na^+, Mg^{2+}}^{Pot} = -8.0$ and $\log K_{Na^+, K^+}^{Pot} = -4.9$.

and negative if A is an anion. Box 15-2 describes how selectivity coefficients are measured. Problem 15-46 provides a formula for estimating the error in measuring primary ion A caused by interfering ion X, which need not have the same charge as A.

EXAMPLE Using the Selectivity Coefficient

A fluoride ion-selective electrode has a selectivity coefficient $K_{F^-, OH^-}^{Pot} = 0.1$. What will be the change in electrode potential when 1.0×10^{-4} M F^- at pH 5.5 is raised to pH 10.5?

Solution From Equation 15-10, we find the potential with negligible OH^- at pH 5.5:

$$E = \text{constant} - 0.059\ 16 \log[1.0 \times 10^{-4}] = \text{constant} + 236.6\ \text{mV}$$

At pH 10.50, $[OH^-] = 3.2 \times 10^{-4}$ M, so the electrode potential is

$$E = \text{constant} - 0.059\ 16 \log[1.0 \times 10^{-4} + (0.1)(3.2 \times 10^{-4})]$$
$$= \text{constant} + 229.5\ \text{mV}$$

The change is $229.5 - 236.6 = -7.1$ mV, which is quite significant. If you didn't know about the pH change, you would think that the concentration of F^- had increased by 32%.

TEST YOURSELF Find the change in potential when 1.0×10^{-4} M F^- at pH 5.5 is raised to pH 9.5. (*Answer:* -0.8 mV)

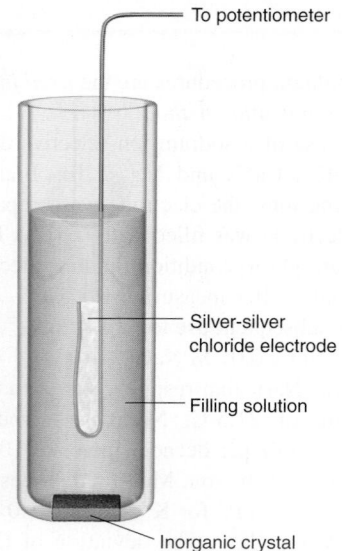

FIGURE 15-22 Diagram of an ion-selective electrode employing an inorganic salt crystal as the ion-selective membrane.

To potentiometer

Silver-silver chloride electrode

Filling solution

Inorganic crystal

Solid-State Electrodes

A **solid-state ion-selective electrode** based on an inorganic crystal is shown in Figure 15-22. A common electrode of this type is the fluoride electrode, employing a crystal of LaF_3 doped with Eu^{2+}. *Doping* means adding a small amount of Eu^{2+} in place of La^{3+}. The filling solution contains 0.1 M NaF and 0.1 M NaCl. Fluoride electrodes are used to monitor the fluoridation of municipal water supplies.

To conduct a tiny electric current, F^- migrates across the LaF_3 crystal, as shown in Figure 15-23. Anion vacancies are created within the crystal when we dope LaF_3 with EuF_2. An adjacent fluoride ion can jump into the vacancy, leaving a new vacancy behind. In this manner, F^- diffuses from one side to the other.

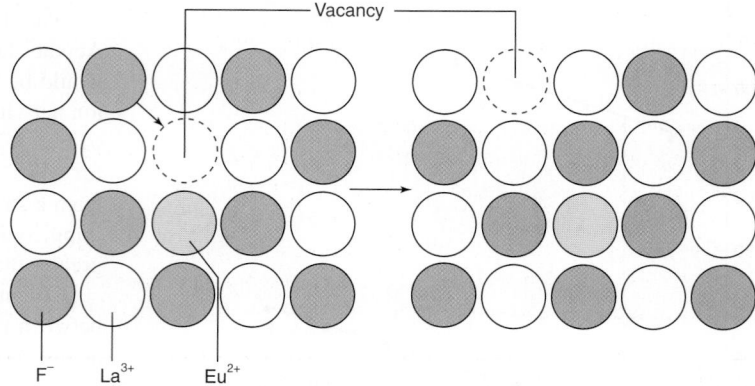

Vacancy

F^- La^{3+} Eu^{2+}

FIGURE 15-23 Migration of F^- through LaF_3 doped with EuF_2. Because Eu^{2+} has less charge than La^{3+}, an anion vacancy occurs for every Eu^{2+}. A neighboring F^- can jump into the vacancy, thereby moving the vacancy to another site. Repetition of this process moves F^- through the lattice.

By analogy with the pH electrode, the response of the F^- electrode is

Response of F^- electrode: $E = \text{constant} - \beta(0.059\ 16) \log \mathcal{A}_{F^-(\text{outside})}$ **(15-12)**

where β is close to 1.00. Equation 15-12 has a negative sign before the log term because fluoride is an anion. The F^- electrode has a nearly Nernstian response over a F^- concentration range from about 10^{-6} M to 1 M (Figure 15-24). The electrode is more responsive to F^- than to most other ions by $>1\ 000$. The only interfering species is OH^-, for which the selectivity coefficient is $K^{\text{Pot}}_{F^-,\ OH^-} = 0.1$. At low pH, F^- is converted to HF ($pK_a = 3.17$), to which the electrode is insensitive.

A routine procedure for measuring F^- is to dilute the unknown in a high ionic strength buffer containing acetic acid, sodium citrate, NaCl, and NaOH to adjust the pH to 5.5. The buffer keeps all standards and unknowns at a constant ionic strength, so the fluoride activity coefficient is constant in all solutions (and can therefore be ignored).

$$E = \text{constant} - \beta(0.059\ 16) \log[F^-]\gamma_{F^-}$$
$$= \underbrace{\text{constant} - \beta(0.059\ 16) \log \gamma_{F^-}}_{\substack{\text{This expression is constant because}\\ \gamma_{F^-} \text{ is constant at constant ionic strength}}} - \beta(0.059\ 16) \log[F^-]$$

At pH 5.5, there is no interference by OH^- and little conversion of F^- to HF. Citrate complexes Fe^{3+} and Al^{3+}, which would otherwise bind F^- and interfere with the analysis.

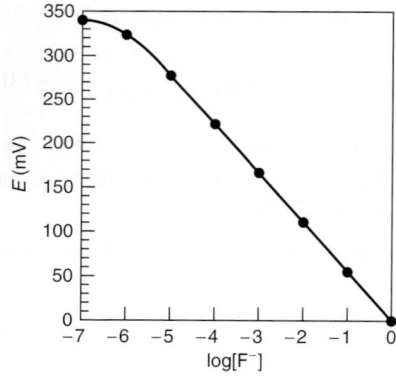

FIGURE 15-24 Calibration curve for fluoride ion-selective electrode. [Data from M. S. Frant and J. W. Ross, Jr., "Electrode for Sensing Fluoride Ion Activity in Solution," *Science* **1966**, *154*, 1553.]

EXAMPLE Response of an Ion-Selective Electrode

When a fluoride electrode was immersed in standard solutions (maintained at a constant ionic strength of 0.1 M with $NaNO_3$), the following potentials (versus S.C.E.) were observed:

$[F^-]$ (M)	E (mV)
1.00×10^{-5}	100.0
1.00×10^{-4}	41.5
1.00×10^{-3}	-17.0

Because the ionic strength is constant, the response should depend on the logarithm of the F^- *concentration*. Find $[F^-]$ in an unknown that gave a potential of 0.0 mV.

Solution We fit the calibration data with Equation 15-12:

$$E = m \underbrace{\log [F^-]}_{x} + b$$
$$\,_{y}$$

Plotting E versus $\log [F^-]$ gives a straight line with a slope $m = -58.5$ mV and a y-intercept $b = -192.5$ mV. Setting $E = 0.0$ mV, solve for $[F^-]$:

$$0.0 \text{ mV} = (-58.5 \text{ mV})\log [F^-] - 192.5 \text{ mV} \Rightarrow [F^-] = 5.1 \times 10^{-4} \text{ M}$$

TEST YOURSELF Find $[F^-]$ if $E = 81.2$ mV. Is the calibration curve valid for $E = 110.7$ mV? (*Answer:* 2.1×10^{-4} M, no because calibration points do not go above 100 mV)

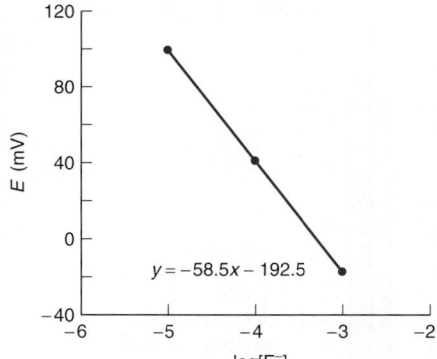

Another common inorganic crystal electrode uses Ag_2S for the membrane. This electrode responds to Ag^+ and to S^{2-}. If we dope the electrode with CuS, CdS, or PbS, we can prepare electrodes sensitive to Cu^{2+}, Cd^{2+}, or Pb^{2+}, respectively (Table 15-5).

Figure 15-25 illustrates the mechanism by which a CdS crystal responds selectively to certain ions. The crystal can be cleaved to expose planes of Cd atoms or S atoms. The Cd plane in Figure 15-25a selectively adsorbs HS^- ions, whereas an S plane does not interact strongly with HS^-. Figure 15-25b shows strong response of the exposed Cd face to HS^-, but

TABLE 15-5	Properties of solid-state ion-selective electrodes			
Ion	Concentration range (M)	Membrane material	pH range	Interfering species
F^-	10^{-6}–1	LaF_3	5–8	OH^- (0.1 M)
Cl^-	10^{-4}–1	AgCl	2–11	$CN^-, S^{2-}, I^-, S_2O_3^{2-}, Br^-$
Br^-	10^{-5}–1	AgBr	2–12	CN^-, S^{2-}, I^-
I^-	10^{-6}–1	AgI	3–12	S^{2-}
SCN^-	10^{-5}–1	AgSCN	2–12	$S^{2-}, I^-, CN^-, Br^-, S_2O_3^{2-}$
CN^-	10^{-6}–10^{-2}	AgI	11–13	S^{2-}, I^-
S^{2-}	10^{-5}–1	Ag_2S	13–14	

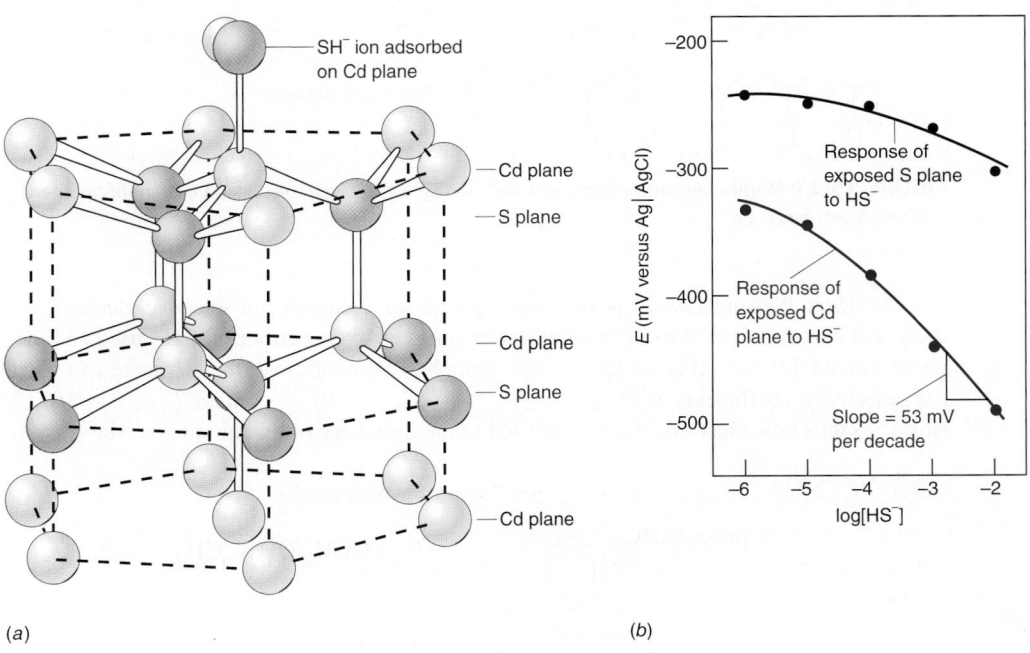

(a) (b)

FIGURE 15-25 (*a*) Crystal structure of hexagonal CdS, showing alternating planes of Cd and S along the vertical axis (the crystal *c*-axis). HS^- is shown adsorbed to the uppermost Cd plane. (*b*) Potentiometric response of exposed crystal faces to HS^-. [Data from K. Uosaki, Y. Shigematsu, H. Kita, Y. Umezawa, and R. Souda, "Crystal-Face-Specific Response of a Single-Crystal Cadmium Sulfide Based Ion-Selective Electrode," *Anal. Chem.* **1989,** *61,* 1980.]

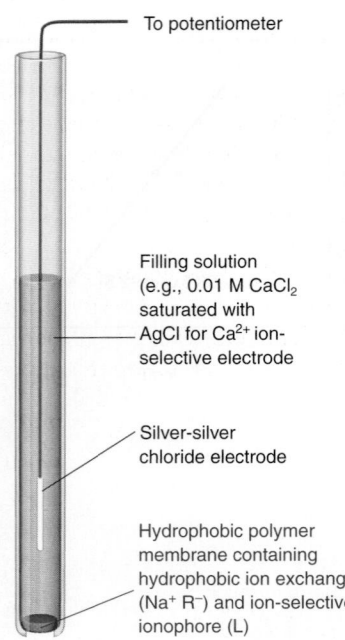

FIGURE 15-26 Calcium ion-selective electrode with a liquid ion exchanger.

To potentiometer

Filling solution
(e.g., 0.01 M CaCl₂
saturated with
AgCl for Ca²⁺ ion-
selective electrode

Silver-silver
chloride electrode

Hydrophobic polymer
membrane containing
hydrophobic ion exchanger
(Na⁺ R⁻) and ion-selective
ionophore (L)

only weak response when the S face is exposed. The opposite behavior is observed in response to Cd^{2+} ions. The partial response of the S face to HS^- in the upper curve is attributed to about 10% of the exposed atoms actually being Cd instead of S.

Liquid-Based Ion-Selective Electrodes

A **liquid-based ion-selective electrode** is similar to the solid-state electrode in Figure 15-22, except that the liquid-based electrode has a hydrophobic membrane impregnated with a hydrophobic ion exchanger (called an *ionophore*) that is selective for analyte ion (Figure 15-26). The response of a Ca^{2+} ion-selective electrode is given by

Response of Ca^{2+} electrode: $\quad E = \text{constant} + \beta\left(\dfrac{0.059\ 16}{2}\right)\log \mathcal{A}_{Ca^{2+}}(\text{outside})$ **(15-13)**

where β is close to 1.00. Equations 15-13 and 15-12 have different signs before the log term because one involves an anion and the other a cation. Note also that the charge of Ca^{2+} requires a factor of 2 in the denominator before the logarithm.

The membrane at the bottom of the electrode in Figure 15-26 is made from poly(vinyl chloride) impregnated with ion exchanger. One particular Ca^{2+} liquid ion exchanger consists of the neutral hydrophobic ligand (L), called an *ionophore*, and a salt of the hydrophobic anion (Na^+R^-) dissolved in a hydrophobic liquid (Figure 15-27) in the poly(vinyl chloride) membrane. The most serious interference comes from Sr^{2+}. The selectivity coefficient in Equation 15-9 is $K_{Ca^{2+},\ Sr^{2+}}^{Pot} = 0.13$, which means that the response to Sr^{2+} is 13% as great as the response to an equal concentration of Ca^{2+}. For most cations, $K_{Ca^{2+},X}^{Pot} < 10^{-3}$.

Hydrophobic anion (R⁻)
Tetrakis[3,5-bis(trifluoromethyl)phenyl]borate

Hydrophobic liquid solvent
2-Nitrophenyl octyl ether

Hydrophobic Ca²⁺-binding ligand (L)
N, N-Dicyclohexyl-N′,N′-dioctadecyl-3-oxapentanediamide

Poly(vinyl chloride)

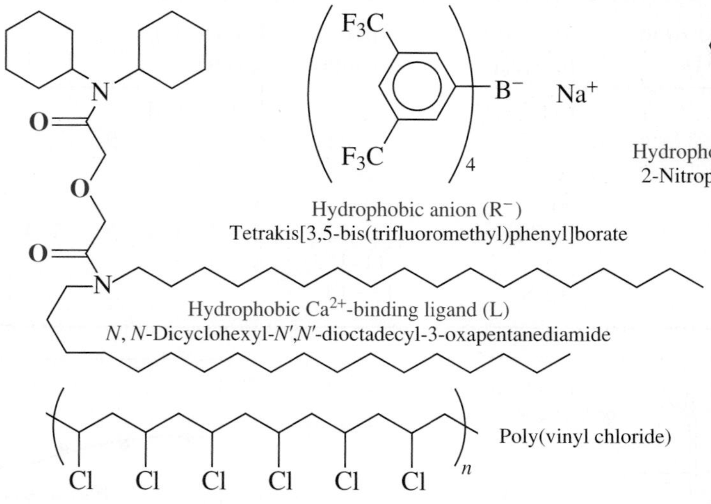

FIGURE 15-27 Membrane components of a Ca^{2+} ion-selective electrode. Ligand L is an ionophore that selectively binds Ca^{2+}.

The *Mars Phoenix Lander* at the opening Chapter 7 carried ion-selective electrodes to study soil chemistry on Mars. The liquid-based ion-selective electrodes for H^+ used an ionophore called ETH 2418 (Figure 15-28). This ionophore responds over the pH range 1 to 9 and has selectivity coefficients $K_{H^+,Na^+}^{Pot} = 10^{-8.6}$, $K_{H^+,K^+}^{Pot} = 10^{-9.7}$, and $K_{H^+,Ca^{2+}}^{Pot} = 10^{-7.8}$. Box 15-3 tells how electrode interference led to the discovery of perchlorate on Mars.

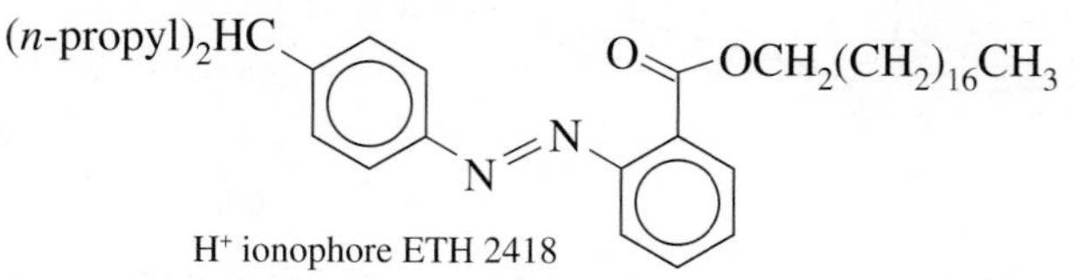

$(n\text{-propyl})_2HC$ $\qquad O\!=\!OCH_2(CH_2)_{16}CH_3$

H^+ ionophore ETH 2418

FIGURE 15-28 Ionophore ETH 2418 for liquid-based H^+ ion-selective electrodes. ETH stands for the Swiss Federal Institute of Technology (Eidgenössische Technische Hochschule Zürich) where many ionophores were synthesized.

BOX 15-3 How Was Perchlorate Discovered on Mars?[34]

Nobody expected perchlorate (ClO_4^-) to be abundant on Mars, so the *Phoenix Mars Lander* Wet Chemistry Laboratory was not designed to look for ClO_4^-. However, the nitrate ion-selective electrode sent to Mars was 1 000 times *more sensitive* to ClO_4^- than to NO_3^-. That is, $K_{NO_3^-, ClO_4^-}^{Pot} = 10^3$. Aqueous solution used to leach ions from Martian soil contained 1 mM NO_3^-. Nitrate would only be detected if its concentrations were >1 mM.

Imagine the surprise in the eyes of Professor Sam Kounaves and his students at Tufts University when the ion-selective electrodes that they helped design and build had an unexpected, huge response. When salts were leached from 1 g of Martian soil by water in the Wet Chemistry Laboratory, the NO_3^- electrode potential changed by 200 mV, as if the NO_3^- concentration were >1 M, which would have

required more NO_3^- than the mass of soil that was analyzed. However, 4–6 mg of ClO_4^- would have produced the observed response. Perchlorate occurs at similar levels on Earth in arid regions including the Atacama Desert and Antarctic dry valleys.[35] On Earth, ClO_4^- is thought to arise from photochemical reactions of ozone (O_3) with chlorine species in the atmosphere. On Mars, ClO_4^- could be made by ultraviolet photooxidation of solid chlorides with mineral catalysis.[36, 37] Results from the 2012 *Curiosity Rover* are consistent with the presence of hydrated calcium perchlorate on Mars.[37] Independent evidence for ClO_4^- comes from the observation that Martian soil liberates a species with a mass of 32 upon thermal decomposition around 450°C. Perchlorate would liberate O_2 (mass 32) at this temperature.

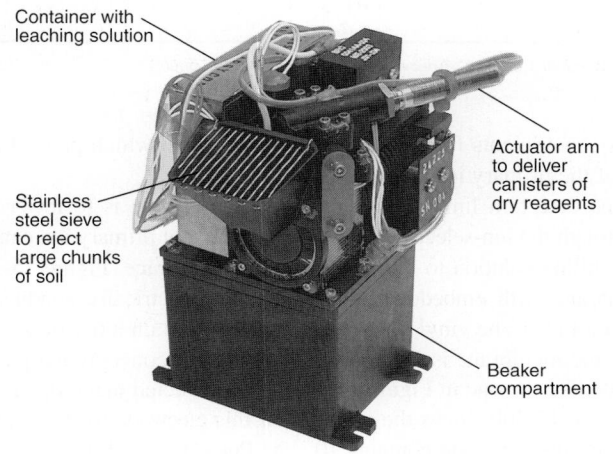

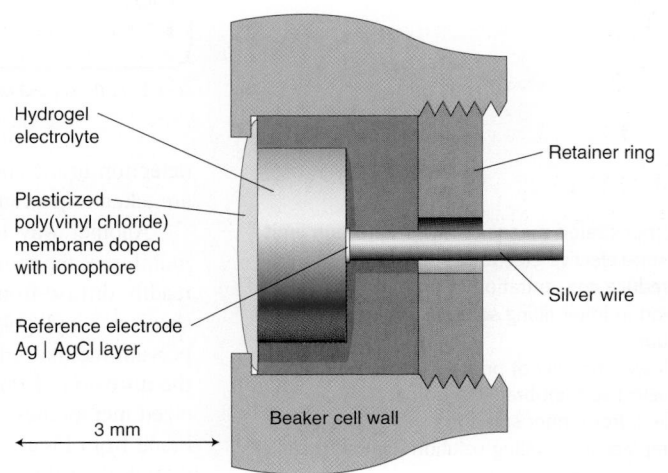

Left: One of four Wet Chemistry Laboratory cells for soil analysis on 2008 *Phoenix Mars Lander*, whose robotic arm is shown at the opening of Chapter 7. Sensors are embedded in walls of the 40-mL epoxy plastic "beaker." Fifteen ion-selective electrodes measured Ca^{2+}, Mg^{2+}, K^+, NO_3^-, NH_4^+, SO_4^{2-}, Cl^-, Br^-, I^-, and H^+ leached from soil into aqueous solution. Other electrodes measured conductivity, reduction potential, redox couples, and reducible metals including Cu^{2+}, Cd^{2+}, Pb^{2+}, Fe^{2+}, Fe^{3+}, and Hg^{2+}. [NASA/JPL-Caltech/University of Arizona/Max Planck Institute.] *Right: Phoenix* liquid-based ion-selective electrode. Hydrogel electrolyte retains aqueous 1 mM M^+Cl^-, where M^+ is analyte cation. [Republished with permission from John Wiley & Sons Inc. From S. P. Kounaves et al., "The MECA Wet Chemistry Laboratory on the 2007 Phoenix Mars Scout Lander," *J. Geophys. Res.* **2009**, *114*, E00A19, Figure 11. Permission conveyed through Copyright Clearance Center, Inc.]

Lowering the Detection Limit for Ion-Selective Electrodes[29]

The U.S. Environmental Protection Agency requires water suppliers to take action to eliminate lead if more than 10% of tap water samples contain more than 15 ppb (7×10^{-8} M) lead. The black curve in Figure 15-29 shows that typical liquid-based ion-selective electrodes cannot measure lead below 10^{-6} M. The internal filling solution in the electrode contains 0.5 mM $PbCl_2$.

The colored curve in Figure 15-29 was obtained with the same electrode, but the internal filling solution was replaced by a *metal ion buffer* (Section 15-7) that fixes $[Pb^{2+}]$ at 10^{-12} M. Now the electrode responds to changes in $[Pb^{2+}]$ down to $\sim 10^{-11}$ M and would be useful for measuring lead in drinking water.

Detection limits for liquid-based ion-selective electrodes had been limited by leakage of the primary ion (Pb^{2+}, in this case) from the internal filling solution through the ion-exchange membrane. Leakage provides a substantial concentration of primary ion at the external surface of the membrane. If analyte concentration is below 10^{-6} M, leakage from the electrode maintains an effective concentration near 10^{-6} M at the outer surface of the electrode. By lowering $[Pb^{2+}]$ inside the electrode, the concentration leaking outside the membrane is reduced and the detection limit is reduced.

The response of the electrode with 10^{-12} M Pb^{2+} in the filling solution is limited by interference from Na^+ in the internal solution, which contains 0.05 M Na_2EDTA from the metal ion buffer. However, not only is the detection limit for Pb^{2+} improved by 10^5, but the selectivity for Pb^{2+} over other cations increases by several orders of magnitude. Table 15-6 gives

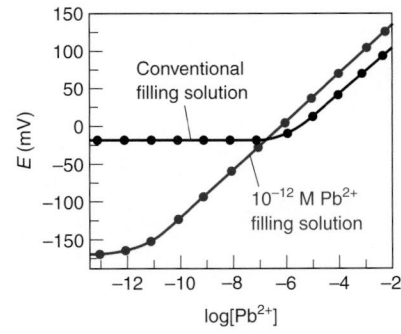

FIGURE 15-29 Response of Pb^{2+} liquid-based ion-selective electrode with (*black curve*) conventional filling solution containing 0.5 mM Pb^{2+} or (*colored curve*) metal ion buffer filling solution in which $[Pb^{2+}] = 10^{-12}$ M. [Data from T. Sokalski, A. Ceresa, T. Zwickl, and E. Pretsch, "Large Improvement of the Lower Detection Limit of Ion-Selective Polymer Membrane Electrodes," *J. Am. Chem. Soc.* **1997**, *119*, 11347.]

TABLE 15-6	Detection limits and selectivity coefficients for liquid-based ion-selective electrodes operating without leakage of primary ion	
Primary ion ion (A)	**Detection limit for A (μM)**	**Selectivity coefficient for some interfering ions (X) log $K_{A,X}^{Pot}$ (Equation 15-9)**
Na^+	30	H^+, −4.8; K^+, −2.7; Ca^{2+}, −6.0
K^+	5	Na^+, −4.2; Mg^{2+}, −7.6; Ca^{2+}, −6.9
NH_3	20	
Cs^+	8	Na^+, −4.7; Mg^{2+}, −8.7; Ca^{2+}, −8.5
Ca^{2+}	0.1	H^+, −4.9; Na^+, −4.8; Mg^{2+}, −5.3
Ag^+	0.03	H^+, −10.2; Na^+, −10.3; Ca^{2+}, −11.3
Pb^{2+}	0.06	H^+, −5.6; Na^+, −5.6; Mg^{2+}, −13.8
Cd^{2+}	0.1	H^+, −6.7; Na^+, −8.4; Mg^{2+}, −13.4
Cu^{2+}	2	H^+, −0.7; Na^+, $<$−5.7; Mg^{2+}, $<$−6.9
ClO_4^-	20	OH^-, −5.0; Cl^-, −4.9; NO_3^-, −3.1
I^-	2	OH^-, −1.7

SOURCE: *E. Bakker and E. Pretsch, "Modern Potentiometry," Angew. Chem. Int. Ed.* **2007,** *46, 5660.*

Demonstrated means to lower detection limit of ion-selective electrode
- reduce concentration of primary (analyte) ion in inner filling solution with metal ion buffer
- lower mobility of primary ion in the ion-selective membrane so primary ion cannot leak from inner solution
- replace inner filling solution with electrically conductive polymer

detection limits and selectivity coefficients for ion-selective electrodes in which precautions are taken to prevent leakage of the primary ion.

Another way to reduce the detection limit of an ion-selective electrode is to lower the mobility of the primary ion through the ion-selective membrane so that that primary ion cannot readily diffuse from the inner filling solution to the outside of the membrane. Figure 15-30a shows a vinyl polymer membrane with embedded nanoparticles of electrically conductive polyaniline that selectively bind Pb^{2+}. The vinyl membrane does not contain a plasticizer, so the diffusion of Pb^{2+} through the membrane is 10^6 times slower than in conventional plasticized membranes. An electrode like the one in Figure 15-26 was constructed using the membrane from Figure 15-30a. Figure 15-30b shows the response of this electrode to Pb^{2+}. Even though the filling solution inside the electrode contains 10^{-5} M $Pb(NO_3)_2$, so little Pb^{2+} diffuses through the membrane that the detection limit of the ion-selective electrode is 2×10^{-11} M. An added bonus of this electrode design is that the ion-selective electrode functions with little degradation for at least 6 months. Membranes of electrodes whose response is shown in Figure 15-29 have a lifetime of ~1 week.

Another way to reduce the flux of primary ion out of an ion-selective electrode is to eliminate the internal filling solution. Box 15-4 describes an ion-selective electrode in which an electrically conductive polymer replaces the inner filling solution. Methods used to lower the limit of detection for liquid-based ion-selective electrodes do not work for

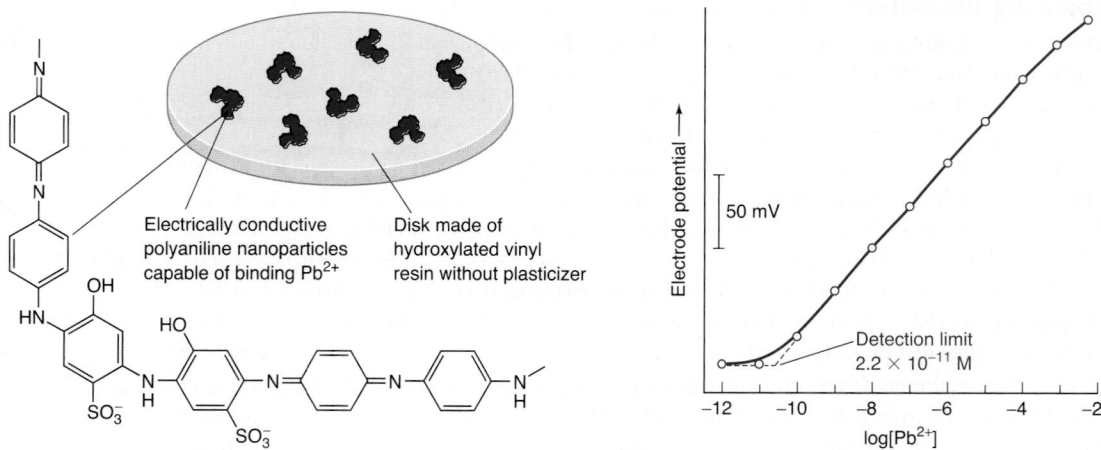

FIGURE 15-30 (*a*) Vinyl polymer membrane containing sodium tetraphenylborate and embedded nanoparticles of electrically conductive polyaniline that selectively binds Pb^{2+}. Partial structure of the polymer is shown. (*b*) Response of ion-selective electrode made with the vinyl polymer membrane. [Data from X.-G. Li, H. Feng, M.-R. Huang, G.-L. Gu, and M. G. Moloney, "Ultrasensitive Pb(II) Potentiometric Sensor Based on Copolyaniline Nanoparticles in a Plasticizer-Free Membrane with Long Lifetime," *Anal. Chem.* **2012,** *84,* 134.]

CHAPTER 15 Electrodes and Potentiometry

It is possible to reduce interference by ions from the filling solution of a liquid-based ion-selective electrode by replacing the filling solution in Figure 15-26 with an electrically conductive polymer. The filling solution or the conductive polymer communicates the electric potential difference at the ion-exchange membrane to the inner electrode.

The electrode shown here has a gold wire coated with a thin layer of the electrically conductive poly(3-octylthiophene). When the polymer is oxidized, electrons can move along the conjugated backbone of the molecule. (*Conjugated* means that the molecule contains alternating single and double bonds.) The conductivity of the oxidized molecule can be up to ~0.1% that of copper metal. The coated wire fills a plastic 10-μL pipet tip whose opening is overcoated with an ion-exchange membrane containing a ligand (L in Figure 15-12) that is selective for Ag^+. When conditioned in 1 nM $AgNO_3$ solution, this electrode exhibits linear response down to 10 nM Ag^+ and a detection limit of ~2 nM.

The sensitive silver analysis can be transformed into a sensitive protein analysis by a *sandwich immunoassay* using antibodies. An **antibody** is a protein produced by the immune system of an animal in response to a foreign molecule, which is called an **antigen**. An antibody specifically recognizes and binds to the antigen that stimulated its synthesis.

In the sandwich immunoassay, the antigen is the analyte protein. An antibody for this protein is bound to a gold surface. When analyte is introduced, it binds to the antibody. Then a second antibody that binds to another site on the analyte is introduced. The second antibody contains covalently attached gold particles with a diameter of ~13 nm containing ~10^5 gold atoms. After washing away unbound antibody, Ag metal is catalytically deposited on the surface of the gold nanoparticles. Approximately 100 atoms of Ag are deposited for every atom of Au in the original particle. Therefore, there are ~10^7 Ag atoms per antibody molecule.

To complete the analysis, Ag metal is oxidized to Ag^+ with hydrogen peroxide (H_2O_2) and the liberated Ag^+ is measured by the ion-selective electrode. For each molecule of protein analyte, roughly 10^7 ions of Ag^+ are produced. We say that the assay *amplifies* the analyte signal by a factor of 10^7. The assay detects ~12 pmol (12×10^{-12} mol) of protein analyte in a 50 μL sample. An analogous assay of ribonucleic acid (RNA) with the Ag^+ ion-selective electrode detects 0.2 amol (0.2×10^{-18} mol, 120 000 molecules) in a 4 μL sample.

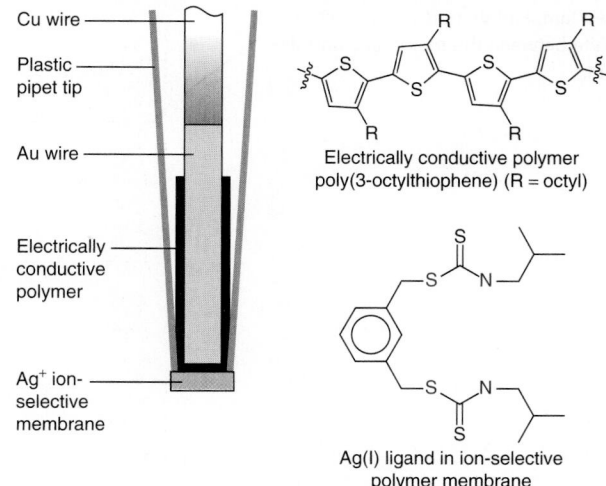

Electrically conductive polymer
poly(3-octylthiophene) (R = octyl)

Ag(I) ligand in ion-selective
polymer membrane

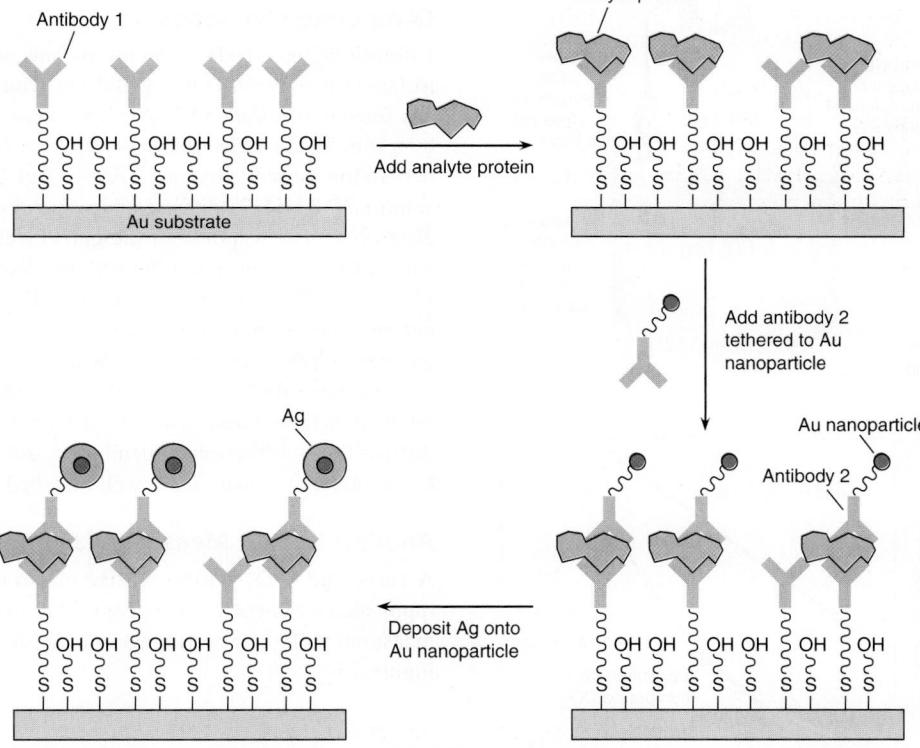

Sandwich immunoassay with deposition of silver metal on Au nanoparticles. [Information from K. Y. Chumbimuni-Torres, Z. Dai, N. Rubinova, Y. Xiang, E. Prëtsch, J. Wang, and E. Bakker, "Potentiometric Biosensing of Proteins with Ultrasenstive Ion-Selective Microelectrodes and Nanoparticle Labels," *J. Am. Chem. Soc.* **2006,** *128,* 13676. See also *Anal. Chem.* **2006,** *78,* 1318 and *Sensors and Actuators B* **2007,** *121,* 135.]

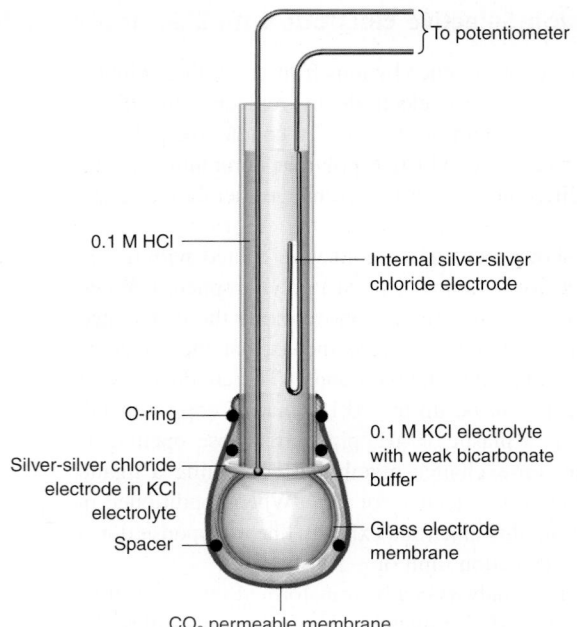

FIGURE 15-31 CO_2 gas-sensing Severinghaus electrode.[38] The membrane is stretched taut, and there is a thin layer of electrolyte between the membrane and the glass bulb.

To potentiometer

0.1 M HCl

Internal silver-silver chloride electrode

O-ring

Silver-silver chloride electrode in KCl electrolyte

Spacer

0.1 M KCl electrolyte with weak bicarbonate buffer

Glass electrode membrane

CO_2 permeable membrane

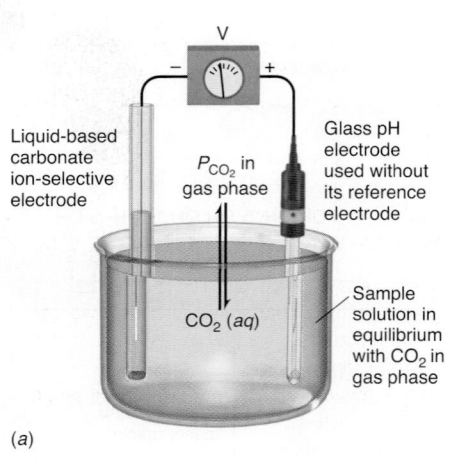

V

Liquid-based carbonate ion-selective electrode

P_{CO_2} in gas phase

$CO_2 (aq)$

Glass pH electrode used without its reference electrode

Sample solution in equilibrium with CO_2 in gas phase

(a)

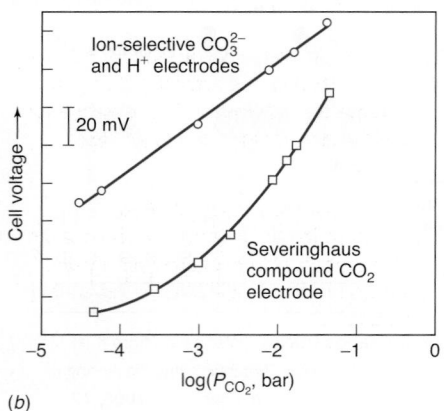

Ion-selective CO_3^{2-} and H^+ electrodes

20 mV

Cell voltage

Severinghaus compound CO_2 electrode

-5   -4   -3   -2   -1   0

$\log(P_{CO_2}, \text{bar})$

(b)

FIGURE 15-32 (a) CO_3^{2-} and H^+ ion-selective electrodes used together to measure CO_2. The H^+ electrode could be a glass electrode used without its reference electrode. (b) Response of the ion-selective electrode pair and of a compound electrode (Figure 15-31) to CO_2.
[Data from X. Xie and E. Bakker, "Non-Severinghaus Potentiometric Dissolved CO_2 Sensor with Improved Characteristics," *Anal. Chem.* **2013**, *85*, 1332.]

solid-state ion-selective electrodes because analyte concentration adjacent to the electrode is governed by the solubility of the inorganic salt crystal in the ion-sensitive membrane.

Compound Electrodes

Compound electrodes contain a conventional electrode surrounded by a membrane that isolates (or generates) the analyte to which the electrode responds. The Severinghaus CO_2 gas-sensing electrode in Figure 15-31 consists of an ordinary glass pH electrode surrounded by a thin layer of electrolyte solution enclosed in a semipermeable membrane made of rubber, Teflon, or polyethylene.[38] A Ag | AgCl reference electrode is immersed in the electrolyte solution. When CO_2 diffuses through the semipermeable membrane, it lowers the pH in the electrolyte. The response of the glass electrode to the change in pH is a measure of the dissolved CO_2 concentration outside the electrode. Other acidic or basic gases, including NH_3, SO_2, H_2S, NO_x (nitrogen oxides), and HN_3 (hydrazoic acid) can be detected in the same manner and can interfere with the CO_2 measurement. These electrodes can be used to measure gases in solution or *in the gas phase*.

The Severinghaus compound electrode for CO_2 in Figure 15-31 is extremely useful in medical patient monitoring. However, it suffers from slow response because CO_2 must diffuse through the outer membrane and it is not sensitive to low concentrations of CO_2. Fortunately, its sensitivity is well matched to physiologic CO_2 levels.

Another Way to Measure Dissolved CO_2

A carbonate (CO_3^{2-}) ion-selective electrode[39] plus a pH electrode used together provide a rapid measurement of dissolved CO_2 over a wider range of concentrations than can be measured with the Severinghaus electrode. The voltage measured by this pair of electrodes in Figure 15-32a is

$$E_+ = c_1 + S \log \mathcal{A}_{H^+}$$

$$E_- = c_2 - \left(\frac{S}{2}\right) \log \mathcal{A}_{CO_3^{2-}}$$

$$E_{cell} = E_+ - E_- = c_1 - c_2 + S \log \mathcal{A}_{H^+} + \left(\frac{S}{2}\right) \log \mathcal{A}_{CO_3^{2-}}$$

$$E_{cell} = (c_1 - c_2) + \left(\frac{S}{2}\right) \log \mathcal{A}_{CO_3^{2-}} \mathcal{A}_{H^+}^2 \qquad \text{(15-14)}$$

where S is the temperature dependent slope (ideally 0.059 16 V at 25°C) and c_1 and c_2 are constants.

Problem 10-11 (page 230) showed equilibria for the sequence $CO_2(g) \overset{K_H}{\rightleftharpoons} CO_2(aq) \overset{K_{a1}}{\rightleftharpoons} HCO_3^- \overset{K_{a2}}{\rightleftharpoons} CO_3^{2-}$, where K_H is the Henry's law constant for the solubility of

CHAPTER 15 Electrodes and Potentiometry

$CO_2(g)$ in aqueous solution and K_{a1} and K_{a2} are the acid dissociation constants of "carbonic acid" (which is mainly $CO_2(aq)$). Combining the equilibrium expressions gives

$$P_{CO_2}K_H = \mathcal{A}_{CO_2(aq)} = \left(\frac{\mathcal{A}_{CO_3^{2-}}\,\mathcal{A}_{H^+}^2}{K_{a1}\,K_{a2}}\right) \tag{15-15}$$

where P_{CO_2} is the pressure of $CO_2(g)$ in equilibrium with $\mathcal{A}_{CO_2(aq)}$.

The electrodes in Figure 15-32 measure the product $\mathcal{A}_{CO_3^{2-}}\,\mathcal{A}_{H^+}^2$ (Equation 15-14), from which P_{CO_2} and $\mathcal{A}_{CO_2(aq)}$ can be found with Equation 15-15. The electrodes can be calibrated in solutions equilibrated with known P_{CO_2}, as shown in Figure 15-32b.

We see that the pair of ion-selective electrodes gives a linear response to P_{CO_2} over more than three orders of magnitude, going down to $\sim 10^{-4.5}$ bar or less. By contrast, the compound electrode in Figure 15-31 has a linear response only in the approximate range 10^{-1} to 10^{-2} bar and the sensitivity decreases at lower levels of P_{CO_2}.

15-7 Using Ion-Selective Electrodes

Ion-selective electrodes respond linearly to the logarithm of analyte activity over four to six orders of magnitude. Electrodes do not consume unknowns, and they introduce negligible contamination. Response time is seconds or minutes, so electrodes are used to monitor flow streams in industrial applications. Color and turbidity do not hinder electrodes. Microelectrodes can be used inside living cells.

Precision is rarely better than 1% and is usually worse. Proteins or other organic solutes can foul electrodes, which leads to sluggish, drifting response. Certain ions interfere with or poison particular electrodes. Some electrodes are fragile and have limited shelf life.

Electrodes respond to the *activity* of *uncomplexed* analyte ion. Therefore, ligands must be absent or masked. Because we usually wish to know concentrations, not activities, an inert salt is often used to bring all standards and samples to a high, constant ionic strength. If activity coefficients are constant, the electrode potential gives concentrations directly.

Human blood plasma has eight major calcium-containing species that can be separated by capillary electrophoresis and measured by inductively coupled plasma atomic emission spectrometry (Figure 15-33). You will learn about these techniques later in this book. Of the eight species, one with a concentration of 1.05 mM was identified as free Ca^{2+}. In the other seven species, with a total concentration of 1.21 mM, Ca^{2+} is bound to proteins or other ligands. When Ca^{2+} in blood was measured with an ion-selective electrode, only free Ca^{2+} was observed.[40] Calcium bound to ligands is invisible to an ion-selective electrode.

Standard Addition with Ion-Selective Electrodes

When ion-selective electrodes are used, it is important that the composition of the standard solution closely approximates the composition of the unknown. The medium in which the analyte exists is called the **matrix.** For complex or unknown matrixes, the **standard addition** method (Section 5-3) can be used. In this technique, the electrode is immersed in unknown and the potential is recorded. Then a small volume of standard solution is added, so as not to perturb the ionic strength of the unknown. The change in potential tells how the electrode responds to analyte and, therefore, how much analyte was in the unknown. It is best to add several successive aliquots and use a graphical procedure to extrapolate back to the concentration of unknown. Standard addition is best if the additions increase analyte to 1.5 to 3 times its original concentration.

A graphical standard addition procedure is based on the equation for the response of the ion-selective electrode, which we will write in the form

$$E = k + \beta\left(\frac{RT\ln 10}{nF}\right)\log[X] \tag{15-16}$$

where E is the meter reading and $[X]$ is the concentration of analyte. This reading is the difference in potential between the ion-selective electrode and the reference electrode. The constants k and β depend on the particular ion-selective electrode. The factor $(RT/F)\ln 10$ is 0.059 16 V at 298.15 K. If $\beta \approx 1$, we say that the response is Nernstian. Let's abbreviate the term $(\beta RT/nF)\ln 10$ as S for slope.

Let the initial volume of unknown be V_0 and the initial concentration of analyte be c_X. Let the volume of added standard be V_S and the concentration of standard be c_S. Then the total

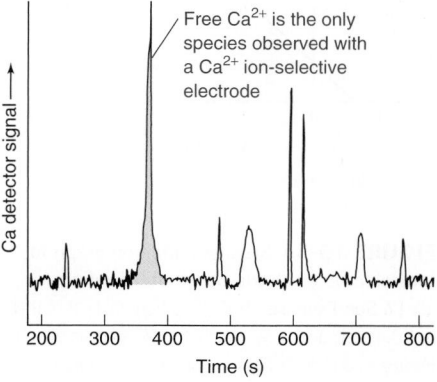

Henry's law:

$$\mathcal{A}_{CO_2(aq)} = K_H P_{CO_2}$$

FIGURE 15-33 Separation of calcium-containing species in human blood plasma. The largest peak is free Ca^{2+}. Other peaks are proteins or small molecules to which Ca^{2+} is bound. The detector measures calcium. [Data from B. Deng, P. Zhu, Y. Wang, J. Feng, X. Li, X. Xu, H. Lu, and Q. Xu, "Determination of Free Calcium and Calcium-Containing Species in Human Plasma by Capillary Electrophoresis-Inductively Coupled Plasma Optical Emission Spectrometry," *Anal. Chem.* **2008,** *80,* 5721.]

Advantages of ion-selective electrodes:
- less expensive than competing techniques such as atomic spectroscopy and ion chromatography
- linear response to log $\mathcal{A}$ over a wide range
- nondestructive
- noncontaminating
- short response time
- unaffected by color or turbidity

A 1-mV error in potential corresponds to a 4% error in monovalent ion activity. A 5-mV error corresponds to a 22% error. The relative error *doubles* for divalent ions and *triples* for trivalent ions.

Electrodes respond to the *activity* of *uncomplexed* ion. If ionic strength is constant, concentration is proportional to activity and the electrode can be calibrated in terms of concentration.

R = gas constant
T = temperature (K)
n = charge of the ion being detected
F = Faraday constant

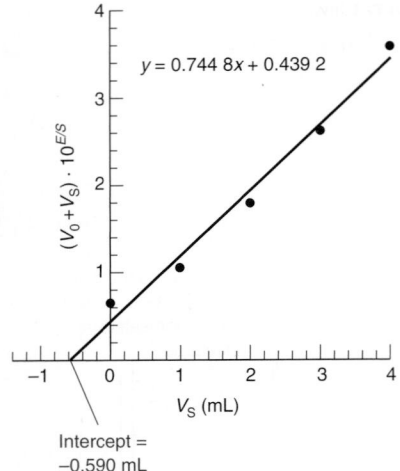

$$y = 0.744\,8x + 0.439\,2$$

Intercept = −0.590 mL

FIGURE 15-34 Standard addition graph for ion-selective electrode based on Equation 15-17. See Exercise 15-F. [Data from G. Li, B. J. Polk, L. A. Meazell, and D. W. Hatchett, "ISE Analysis of Hydrogen Sulfide in Cigarette Smoke," *J. Chem. Ed.* **2000**, *77*, 1049.]

Plastic bottles are better than glass for very dilute solutions of metal salts because ions are adsorbed by glass.

[EDTA] = total concentration of all forms of EDTA not bound to metal ion

$\alpha_{Y^{4-}}$ = fraction of unbound EDTA in the form Y^{4-}

concentration of analyte after standard is added is $(V_0c_X + V_Sc_S)/(V_0 + V_S)$. Substituting this expression for [X] in Equation 15-16 and some rearrangement gives

Standard addition plot for ion-selective electrode:

$$\underbrace{(V_0 + V_S)\,10^{E/S}}_{y} = \underbrace{10^{k/S}\,V_0c_X}_{b} + \underbrace{10^{k/S}\,c_S}_{m}\underset{\substack{\uparrow \\ x}}{V_S} \qquad (15\text{-}17)$$

A graph of $(V_0 + V_S)10^{E/S}$ on the y-axis versus V_S on the x-axis has a slope $m = 10^{k/S}c_S$ and a y-intercept of $10^{k/S}V_0c_X$ (Figure 15-34). The x-intercept is found by setting $y = 0$:

$$x\text{-intercept} = -\frac{b}{m} = -\frac{10^{k/S}V_0c_X}{10^{k/S}c_S} = -\frac{V_0c_X}{c_S} \qquad (15\text{-}18)$$

Equation 15-18 gives us the concentration of unknown, c_X, from V_0, c_S, and the x-intercept.

A weakness of standard addition with ion-selective electrodes is that we cannot measure β in Equation 15-16 in the unknown matrix. We could measure β in a series of standard solutions (not containing unknown) and use this value to compute S in the function $(V_0 + V_S)10^{E/S}$ in Equation 15-17. A better procedure is to add concentrated, known matrix to the unknown and to all standards so that the matrix is essentially the same in all solutions.

Metal Ion Buffers

It is pointless to dilute $CaCl_2$ to 10^{-6} M for standardizing an ion-selective electrode. At this low concentration, Ca^{2+} might be lost by adsorption on glass or reaction with impurities.

A **metal ion buffer** made from metal ion plus an excess of a suitable ligand (such as $CaCl_2$ plus excess EDTA) can regulate the metal ion concentration at a chosen level. Consider the reaction of Ca^{2+} with EDTA at pH 6.00 at which the fraction of EDTA in the form Y^{4-} is $\alpha_{Y^{4-}} = 1.8 \times 10^{-5}$ (Table 12-1)

$$Ca^{2+} + Y^{4-} \rightleftharpoons CaY^{2-}$$

$$K_f = 10^{10.65} = \frac{[CaY^{2-}]}{[Ca^{2+}]\alpha_{Y^{4-}}[EDTA]} \qquad (15\text{-}19)$$

If equal concentrations of CaY^{2-} and EDTA are present in a solution,

$$[Ca^{2+}] = \frac{[CaY^{2-}]}{K_f\alpha_{Y^{4-}}[EDTA]} = \frac{\cancel{[CaY^{2-}]}}{(10^{10.65})(1.8 \times 10^{-5})\cancel{[EDTA]}} = 1.2 \times 10^{-6}\ \text{M}$$

More accurate calculations would use activity coefficients.

EXAMPLE **Preparing a Metal Ion Buffer**

What concentration of EDTA should be added to 0.010 M CaY^{2-} at pH 6.00 to give $[Ca^{2+}] = 1.0 \times 10^{-6}$ M?

Solution From Equation 15-19, we write

$$[EDTA] = \frac{[CaY^{2-}]}{K_f\alpha_{Y^{4-}}[Ca^{2+}]} = \frac{0.010}{(10^{10.65})(1.8 \times 10^{-5})(1.00 \times 10^{-6})} = 0.012_4\ \text{M}$$

These are practical concentrations of CaY^{2-} and of EDTA.

TEST YOURSELF What concentration of EDTA should be added to 0.010 M CaY^{2-} at pH 6.00 to give $[Ca^{2+}] = 1.0 \times 10^{-7}$ M? (*Answer:* 0.12_4 M)

A metal ion buffer is the only way to obtain $[Pb^{2+}] \approx 10^{-12}$ M for the electrode filling solution in Figure 15-29.

15-8 Solid-State Chemical Sensors

Solid-state chemical sensors are fabricated by the same technology used for microelectronic chips. The field effect transistor (FET) is the heart of many sensors such as the pH electrode in Figure 15-35. The opening of this chapter described a chip containing 10^9 pH-sensitive field effect transistors in a 2×2 cm area for sequencing DNA by measuring H^+ released each time a nucleotide base is added to the DNA.

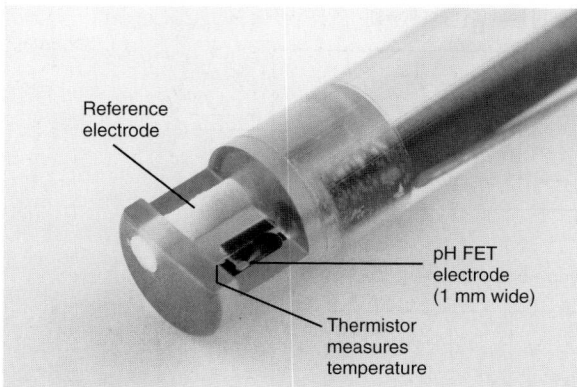

FIGURE 15-35 Combination pH electrode based on field effect transistor. The thermistor senses temperature and is used for automatic temperature compensation. [Courtesy SENTRON, Europe BV.]

Reference electrode

pH FET electrode (1 mm wide)

Thermistor measures temperature

Semiconductors and Diodes

Semiconductors such as Si (Figure 15-36), Ge, and GaAs are materials whose electric *resistivity*[41] lies between those of conductors and insulators. The four valence electrons of the pure materials are all involved in bonds between atoms (Figure 15-37a). A phosphorus impurity, with five valence electrons, provides one extra **conduction electron** that is free to move through the crystal (Figure 15-37b). An aluminum impurity has one less bonding electron than necessary, creating a vacancy called a **hole,** which behaves as a positive charge carrier. When a neighboring electron fills the hole, a new hole appears at an adjacent position (Figure 15-37c). A semiconductor with excess conduction electrons is called *n-type*, and one with excess holes is called *p-type*.

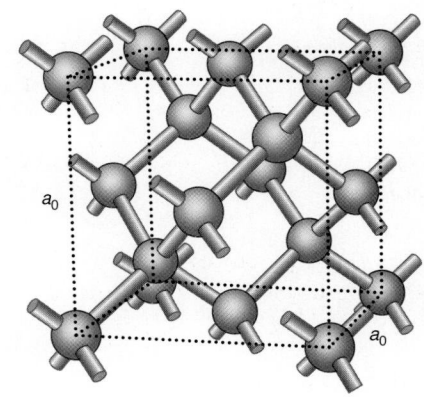

$a_0 = 0.357$ nm in diamond
$a_0 = 0.543$ nm in silicon

FIGURE 15-36 Face centered cubic crystal structure of diamond and silicon. Each atom is tetrahedrally bonded to four neighbors, with C—C bond lengths of 154 pm in diamond and Si—Si bond lengths of 235 pm in silicon.

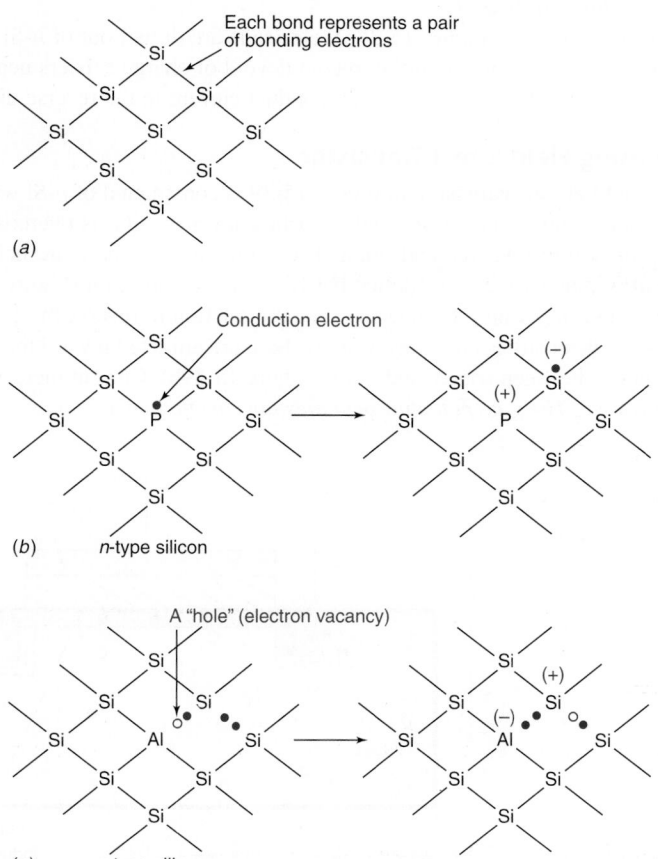

Each bond represents a pair of bonding electrons

(a)

Conduction electron

(b) *n*-type silicon

A "hole" (electron vacancy)

(c) *p*-type silicon

FIGURE 15-37 (a) The electrons of pure silicon are all involved in the sigma-bonding framework. (b) An impurity atom such as phosphorus adds one extra electron (●), which is relatively free to move through the crystal. (c) An aluminum impurity atom lacks one electron needed for the sigma-bonding framework. The hole (○) introduced by the Al atom can be occupied by an electron from a neighboring bond, effectively moving the hole to the neighboring bond.

An *activation energy* is needed to get the charge carriers to move across the diode. For Si, ~0.6 V of forward bias is required before current will flow. For Ge, the requirement is ~0.2 V.

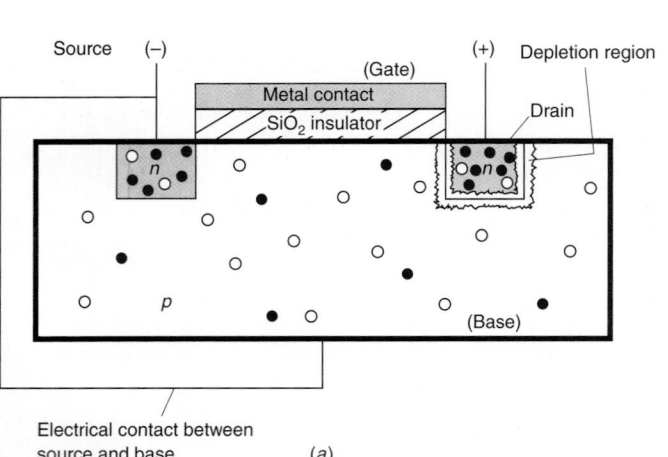

For moderate reverse bias voltages, no current flows. If the voltage is sufficiently negative, *breakdown* occurs and current flows in the reverse direction.

(a) Forward bias (current flows)

(+)   p-Si   n-Si   (−)

(b) Reverse bias (no current flows)

(−)   p-Si   Depletion region (~10^{-6} m)   n-Si   (+)

⊖ Negative electron

⊕ Positive hole

FIGURE 15-38 Behavior of a *pn* junction, showing that current (*a*) can flow under forward bias conditions, but (*b*) is prevented from flowing under reverse bias.

A **diode** is a *pn* junction (Figure 15-38a). If *n*-Si is made negative with respect to *p*-Si, electrons flow from the external circuit into the *n*-Si. At the *pn* junction, electrons and holes combine. As electrons move from the *p*-Si into the circuit, a fresh supply of holes is created in the *p*-Si. The net result is that current flows when *n*-Si is negative with respect to *p*-Si. The diode is said to be *forward biased*.

If the polarity is reversed (Figure 15-38b), electrons are drawn out of *n*-Si and holes are drawn out of *p*-Si, leaving a thin *depletion region* devoid of charge carriers near the *pn* junction. The diode is *reverse biased* and does not conduct current in the reverse direction.

Chemical-Sensing Field Effect Transistor

The *base* of the **field effect transistor** in Figure 15-39 is constructed of *p*-Si with two *n*-type regions called *source* and *drain*. An insulating surface layer of SiO_2 is overcoated by a conductive metal *gate* between source and drain. The source and the base are held at the same electric potential. When a voltage is applied between source and drain (Figure 15-39a), little current flows because the drain-base interface is a *pn* junction in reverse bias.

The more positive the gate, the more current can flow between source and drain.

If the gate is made positive, electrons from the base are attracted toward the gate and form a conductive channel between source and drain (Figure 15-39b). Current increases as the gate is made more positive. *The gate potential regulates the current between source and drain.*

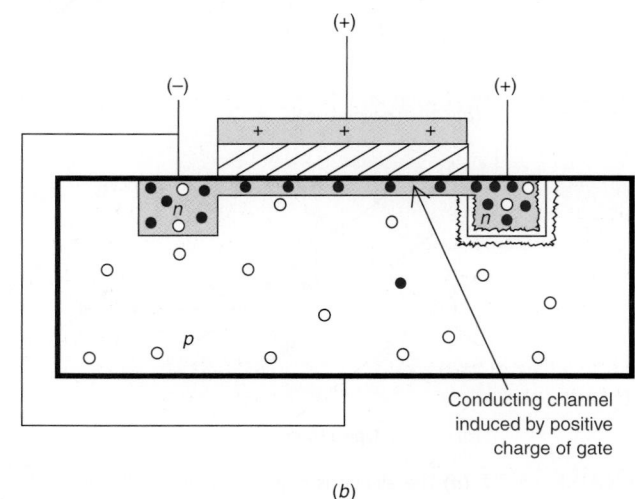

FIGURE 15-39 Operation of a field effect transistor. (*a*) Nearly random distribution of holes and electrons in the base in the absence of gate potential. (*b*) Positive gate potential attracts electrons that form a conductive channel beneath the gate. Current can flow through this channel between source and drain.

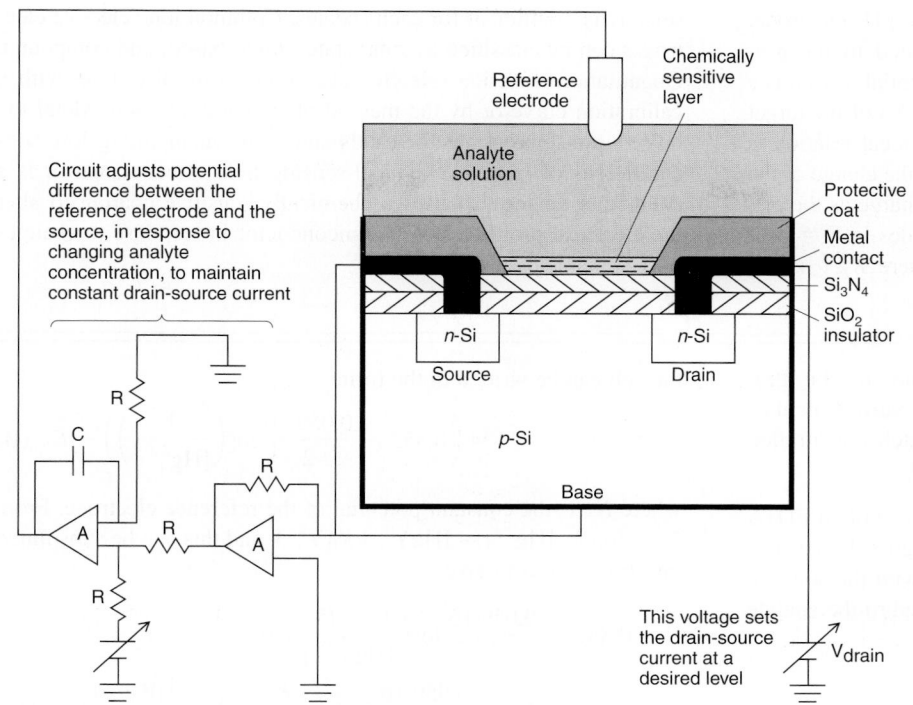

Circuit adjusts potential difference between the reference electrode and the source, in response to changing analyte concentration, to maintain constant drain-source current

This voltage sets the drain-source current at a desired level

FIGURE 15-40 Operation of a chemical-sensing field effect transistor. The transistor is coated with a thin, insulating SiO_2 layer and a second thin layer of Si_3N_4 (silicon nitride), which is impervious to ions and improves electrical stability. The circuit at the lower left adjusts the potential difference between the reference electrode and the source in response to changes in the analyte solution, such that a constant drain-source current is maintained.

The essential feature of the chemical-sensing field effect transistor in Figure 15-40 is the chemically sensitive layer over the gate. An example is a layer of AgBr. When exposed to silver nitrate solution, Ag^+ is adsorbed on the AgBr (Figure 27-3), thereby giving it a positive charge and increasing current between source and drain. *The voltage that must be applied by an external circuit to bring the current back to its initial value is the response to Ag^+.* Figure 15-41 shows that Ag^+ makes the gate more positive and Br^- makes the gate more negative. The response is close to 59 mV for a 10-fold concentration change. The transistor is smaller (Figure 15-35) and more rugged than ion-selective electrodes. The sensing surface is typically only 1 mm^2.

The chemically sensitive surface over the gate of a field effect transistor can be made of a variety of chemically modified materials including silicon,[42] graphene (single atomic layers of graphite),[43,44] carbon nanotubes,[45] semiconducting nanowires,[46–48] and gold.[49] The chemically sensitive surface could be remote from the transistor and connected to the gate by a wire.[48,49] Chemical-sensing field effect transistors have been designed to measure small molecules and ions, as well as proteins,[44,48,49] DNA,[47] drugs such as the anticoagulant heparin,[42] whole bacteria,[45] and the insides of living cells.[46] Low detection limits such as 2 fM (2×10^{-15} M) for the protein lectin have been reported.[48]

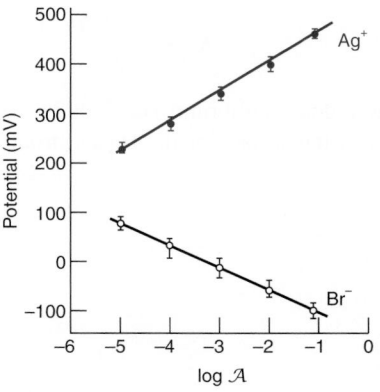

FIGURE 15-41 Response of a silver bromide-coated field effect transistor. Error bars are 95% confidence intervals for data obtained from 195 sensors prepared from different chips. [Data from R. P. Buck and D. E. Hackleman, "Field Effect Potentiometric Sensors," *Anal. Chem.* **1977,** *49,* 2315.]

Terms to Understand

antibody	glass electrode	metal ion buffer	silver-silver chloride electrode
antigen	hole	mobility	solid-state ion-selective
calomel electrode	indicator electrode	potentiometry	electrode
combination electrode	ion-exchange equilibrium	reference electrode	standard addition
compound electrode	ion-selective electrode	saturated calomel electrode	working electrode
conduction electron	junction potential	(S.C.E.)	
diode	liquid-based ion-selective	selectivity coefficient	
electroactive species	electrode	semiconductor	
field effect transistor	matrix		

Summary

In potentiometric measurements, the indicator electrode responds to changes in the activity of analyte, and the reference electrode is a self-contained half-cell with a constant potential. The most common reference electrodes are calomel and silver-silver chloride. Common indicator electrodes include (1) the inert Pt electrode, (2) a silver electrode responsive to Ag^+, halides, and other ions that react with Ag^+, and (3) ion-selective electrodes. Unknown junction potentials at liquid-liquid interfaces limit the accuracy of most potentiometric measurements.

Ion-selective electrodes, including the glass pH electrode, respond mainly to one ion that is selectively bound to the ion-exchange membrane of the electrode. The potential difference across the membrane, E, depends on the activity ($\mathcal{A}_o$) of the target ion in the external analyte solution. At 25°C, the ideal relation is $E(V) = \text{constant} + (0.059\ 16/n) \log \mathcal{A}_o$, where n is the charge of the target ion. For interfering ions (X) with the same charge as the primary ion (A), the response of ion-selective electrodes is $E = \text{constant} \pm (0.059\ 16/n) \log[\mathcal{A}_A + \Sigma K_{A,X}^{Pot} \mathcal{A}_X]$, where $K_{A,X}^{Pot}$ is the selectivity coefficient for each species. Common ion-selective electrodes can be classified as solid state, liquid based, and compound. Quantitation with ion-selective electrodes is usually done with a calibration curve or by the method of standard addition. Metal ion buffers are appropriate for establishing and maintaining low concentrations of ions. A chemical-sensing field effect transistor is a solid-state device that uses a chemically sensitive coating to alter the electrical properties of a semiconductor in response to changes in the chemical environment.

Exercises

15-A. The apparatus in Figure 15-7 was used to monitor the titration of 50.0 mL of 0.100 M $AgNO_3$ with 0.200 M NaBr. Calculate the cell voltage at each volume of NaBr, and sketch the titration curve: 1.0, 12.5, 24.0, 24.9, 25.1, 26.0, and 35.0 mL.

15-B. The apparatus in the figure can follow the course of an EDTA titration and was used to generate the curves in Figure 12-10. The heart of the cell is a pool of liquid Hg in contact with the solution and with a Pt wire. A small amount of HgY^{2-} added to the analyte equilibrates with a very tiny amount of Hg^{2+}:

$$Hg^{2+} + Y^{4-} \rightleftharpoons HgY^{2-}$$

$$K_f = \frac{[HgY^{2-}]}{[Hg^{2+}][Y^{4-}]} = 10^{21.5} \quad \textbf{(A)}$$

The redox equilibrium $Hg^{2+} + 2e^- \rightleftharpoons Hg(l)$ is established rapidly at the surface of the Hg electrode, so the Nernst equation for the cell can be written in the form

$$E = E_+ - E_- = \left(0.852 - \frac{0.059\ 16}{2} \log\left(\frac{1}{[Hg^{2+}]}\right)\right) - E_- \quad \textbf{(B)}$$

where E_- is the constant potential of the reference electrode. From Equation A, $[Hg^{2+}] = [HgY^{2-}]/K_f[Y^{4-}]$, and this can be substituted into Equation B to give

$$E = 0.852 - \frac{0.059\ 16}{2} \log\left(\frac{[Y^{4-}]K_f}{[HgY^{2-}]}\right) - E_-$$

$$= 0.852 - E_- - \frac{0.059\ 16}{2} \log\left(\frac{K_f}{[HgY^{2-}]}\right) - \frac{0.059\ 16}{2} \log[Y^{4-}]$$

$$\textbf{(C)}$$

where K_f is the formation constant for HgY^{2-}. This apparatus thus responds to the changing EDTA concentration during an EDTA titration.

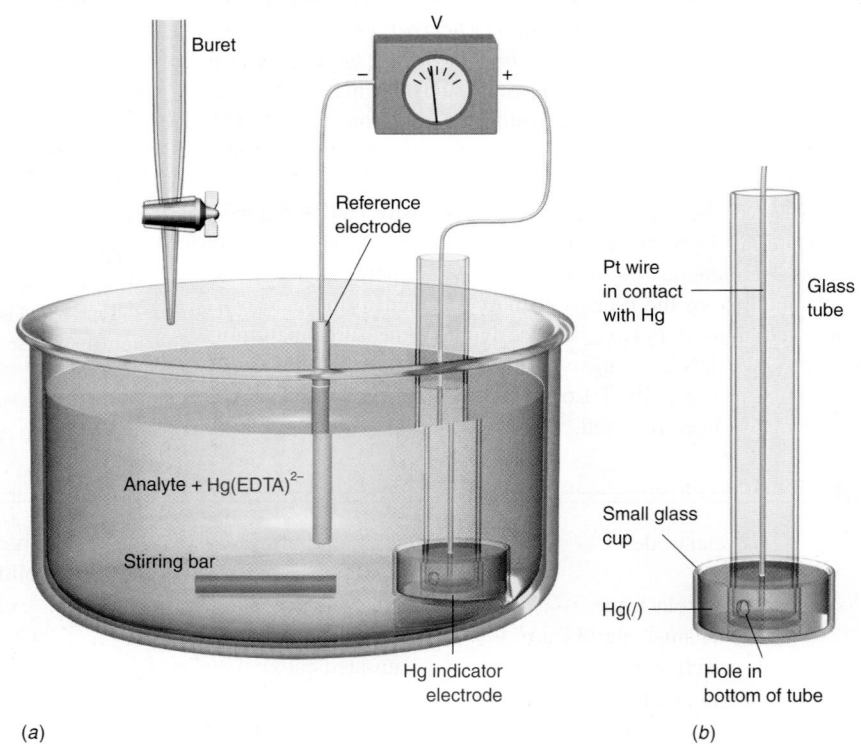

(a) Apparatus for Exercise 15-B. (b) Enlarged view of mercury electrode.

Suppose that you titrate 50.0 mL of 0.010 0 M $MgSO_4$ with 0.020 0 M EDTA at pH 10.0, using the apparatus in the figure with an S.C.E. reference electrode. Analyte contains 1.0×10^{-4} M $Hg(EDTA)^{2-}$ added at the beginning of the titration. Calculate the cell voltage at the following volumes of added EDTA, and draw a graph of millivolts versus milliliters: 0, 10.0, 20.0, 24.9, 25.0, and 26.0 mL.

15-C. A solid-state fluoride ion-selective electrode responds to F^- but not to HF. It also responds to hydroxide ion at high concentration when $[OH^-] \approx [F^-]/10$. Suppose that such an electrode gave a potential of +100 mV (versus S.C.E.) in 10^{-5} M NaF and +41 mV in 10^{-4} M NaF. Sketch qualitatively how the potential would vary if the electrode were immersed in 10^{-5} M NaF and the pH ranged from 1 to 13.

CHAPTER 15 Electrodes and Potentiometry

15-D. One glass-membrane sodium ion-selective electrode has a selectivity coefficient $K_{Na^+, H^+}^{Pot} = 36$. When this electrode was immersed in 1.00 mM NaCl at pH 8.00, a potential of -38 mV (versus S.C.E.) was recorded.

(a) Neglecting activity coefficients, calculate the potential with Equation 15-10 if the electrode were immersed in 5.00 mM NaCl at pH 8.00.

(b) What would the potential be for 1.00 mM NaCl at pH 3.87? You can see that pH is a critical variable for this sodium electrode.

15-E. An ammonia gas-sensing electrode gave the following calibration points when all solutions contained 1 M NaOH.

NH_3 (M)	E (V)	NH_3 (M)	E (V)
1.00×10^{-5}	268.0	5.00×10^{-4}	368.0
5.00×10^{-5}	310.0	1.00×10^{-3}	386.4
1.00×10^{-4}	326.8	5.00×10^{-3}	427.6

A dry food sample weighing 312.4 mg was digested by the Kjeldahl procedure (Section 11-8) to convert all nitrogen into NH_4^+. The digestion solution was diluted to 1.00 L, and 20.0 mL were transferred to a 100-mL volumetric flask. The 20.0-mL aliquot was treated with 10.0 mL of 10.0 M NaOH plus enough NaI to complex the Hg catalyst from the digestion and diluted to 100.0 mL. When measured with the ammonia electrode, this solution gave a reading of 339.3 mV. Calculate the wt% nitrogen in the food sample.

15-F. H_2S from cigarette smoke was collected by bubbling smoke through aqueous NaOH and measured with a sulfide ion-selective electrode. Standard additions of volume V_S containing Na_2S at concentration $c_S = 1.78$ mM were then made to $V_0 = 25.0$ mL of unknown and the electrode response, E, was measured.

V_S (mL)	E (V)	V_S (mL)	E (V)
0	0.046 5	3.00	0.030 0
1.00	0.040 7	4.00	0.026 5
2.00	0.034 4		

From a separate calibration curve, it was found that $\beta = 0.985$ in Equation 15-16. Using $T = 298.15$ K and $n = -2$ (the charge of S^{2-}), prepare a standard addition graph with Equation 15-17 and find the concentration of sulfide in the unknown.

Problems

Reference Electrodes

15-1. (a) Write the half-reactions for the silver-silver chloride and calomel reference electrodes.

(b) Predict the voltage for the following cell.

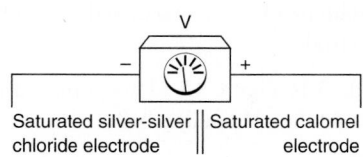

Saturated silver-silver chloride electrode ‖ Saturated calomel electrode

15-2. Draw a diagram like Figure 15-6 to convert the following potentials. The Ag | AgCl and calomel reference electrodes are saturated with KCl.

(a) 0.523 V versus S.H.E. = ? versus Ag | AgCl

(b) -0.111 V versus Ag | AgCl = ? versus S.H.E.

(c) -0.222 V versus S.C.E. = ? versus S.H.E.

(d) 0.023 V versus Ag | AgCl = ? versus S.C.E.

(e) -0.023 V versus S.C.E. = ? versus Ag | AgCl

15-3. Suppose a saturated calomel electrode replaces the silver-silver chloride electrode in Figure 15-2. Calculate the cell voltage if $[Fe^{2+}]/[Fe^{3+}] = 2.5 \times 10^{-3}$.

15-4. From the following potentials, calculate the *activity* of Cl^- in 1 M KCl.

$$E°(\text{calomel electrode}) = 0.268 \text{ V}$$
$$E(\text{calomel electrode, 1 M KCl}) = 0.280 \text{ V}$$

15-5. For a silver-silver chloride electrode, the following potentials are observed:

$$E° = 0.222 \text{ V} \qquad E(\text{saturated KCl}) = 0.197 \text{ V}$$

From these potentials, find the activity of Cl^- in saturated KCl. Calculate E for a calomel electrode saturated with KCl, given that $E°$ for the calomel electrode is 0.268 V. (Your answer will not be exactly the value 0.241 used in this book.)

Indicator Electrodes

15-6. A cell was prepared by dipping a Cu wire and a saturated calomel electrode into 0.10 M $CuSO_4$ solution. The Cu wire was attached to the positive terminal of a potentiometer and the calomel electrode was attached to the negative terminal.

(a) Write a half-reaction for the Cu electrode.

(b) Write the Nernst equation for the Cu electrode.

(c) Calculate the cell voltage.

15-7. Explain why a silver electrode can be an indicator electrode for Ag^+ and for halides.

15-8. A 10.0-mL solution of 0.050 0 M $AgNO_3$ was titrated with 0.025 0 M NaBr in the cell

$$\text{S.C.E.} \parallel \text{titration solution} | Ag(s)$$

Find the cell voltage for 0.1 and 30.0 mL of titrant.

15-9. A solution containing 50.0 mL of 0.100 M EDTA buffered to pH 10.00 was titrated with 50.0 mL of 0.020 0 M $Hg(ClO_4)_2$ in the cell shown in Exercise 15-B:

$$\text{S.C.E.} \parallel \text{titration solution} | Hg(l)$$

From the cell voltage $E = -0.027$ V, find the formation constant of $Hg(EDTA)^{2-}$.

15-10. Consider the cell S.C.E. ‖ cell solution | Pt(s), whose voltage is -0.126 V. The cell solution contains 2.00 mmol of $Fe(NH_4)_2(SO_4)_2$, 1.00 mmol of $FeCl_3$, 4.00 mmol of Na_2EDTA, and lots of buffer, pH 6.78, in a volume of 1.00 L.

(a) Write a reaction for the right half-cell.

(b) Find the quotient $[Fe^{2+}]/[Fe^{3+}]$ in the cell solution. (This expression gives the ratio of *uncomplexed* ions.)

(c) Find the quotient of formation constants: (K_f for $FeEDTA^-$)/ (K_f for $FeEDTA^{2-}$).

15-11. *An equilibrium challenge:* Here's a cell you'll really like:

$$Ag(s) \,|\, AgCl(s) \,|\, KCl(aq, \text{ saturated}) \,\|\, \text{cell solution} \,|\, Cu(s)$$

The cell solution was made by mixing

    25.0 mL of 4.00 mM KCN
    25.0 mL of 4.00 mM $KCu(CN)_2$
    25.0 mL of 0.400 M acid, HA, with $pK_a = 9.50$
    25.0 mL of KOH solution

The measured voltage was -0.440 V. *Calculate the molarity of the KOH solution.* Assume that essentially all copper(I) is $Cu(CN)_2^-$. A little HCN comes from the reaction of KCN with HA. Neglect the small amount of HA consumed by reaction with HCN. For the right half-cell, the reaction is $Cu(CN)_2^- + e^- \rightleftharpoons Cu(s) + 2CN^-$. *Suggested procedure:* From E, find $[CN^-]$. From $[CN^-]$, find pH. From pH, figure out how much OH^- had been added.

Junction Potential

15-12. What causes a junction potential? How does this potential limit the accuracy of potentiometric analyses? Identify a cell in the illustrations in Section 14-2 that has no junction potential.

15-13. Why is the 0.1 M HCl | 0.1 M KCl junction potential of opposite sign and greater magnitude than the 0.1 M NaCl | 0.1 M KCl potential in Table 15-2?

15-14. Which side of the liquid junction 0.1 M KNO_3 | 0.1 M NaCl will be negative?

15-15. Why do the liquid junctions potentials 0.1 M HCl | 0.1 M KCl and 0.1 M NaOH | 0.1 M KCl have opposite signs in Table 15-2? Why is the junction potential for 0.1 M NaOH | 0.1 M KCl so much more negative than 0.1 M NaOH | KCl (saturated)?

15-16. In problem 15-1, we neglected to include the junction potentials at each side of the salt bridge connecting the saturated Ag | AgCl and saturated calomel electrodes. What would be the most sensible salt bridge to use for the cell so that the junction potentials will be negligible? Why?

15-17. Refer to the footnote in Table 15-1. How many seconds will it take for **(a)** H^+ and **(b)** NO_3^- to migrate a distance of 12.0 cm in a field of 7.80×10^3 V/m?

15-18. Suppose that an ideal hypothetical cell such as that in Figure 14-8 were set up to measure $E°$ for the half-reaction $Ag^+ + e^- \rightleftharpoons Ag(s)$.

(a) Calculate the equilibrium constant for the net cell reaction.

(b) If there were a junction potential of $+2$ mV (increasing E from 0.799 to 0.801 V), by what percentage would the calculated equilibrium constant increase?

(c) Answer parts **(a)** and **(b)** using 0.100 V instead of 0.799 V for the value of $E°$ for the silver reaction.

15-19. Explain how the cell $Ag(s) \,|\, AgCl(s) \,|\, 0.1$ M HCl | 0.1 M KCl | AgCl(s) | Ag(s) can be used to measure the 0.1 M HCl | 0.1 M KCl junction potential.

15-20. 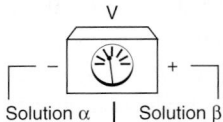 *Henderson equation.* The junction potential, E_j, between solutions α and β can be estimated with the Henderson equation:

$$E_j \approx \frac{\sum_i \frac{|z_i| u_i}{z_i} [C_i(\beta) - C_i(\alpha)]}{\sum_i |z_i| u_i [C_i(\beta) - C_i(\alpha)]} \frac{RT}{F} \ln \frac{\sum_i |z_i| u_i C_i(\alpha)}{\sum_i |z_i| u_i C_i(\beta)}$$

where z_i is the charge of species i, u_i is the *mobility* of species i (Table 15-1), $C_i(\alpha)$ is the concentration of species i in phase α, and $C_i(\beta)$ is the concentration in phase β. (Activity coefficients are neglected in this equation.)

(a) Using your calculator, show that the junction potential of 0.1 M HCl | 0.1 M KCl is 26.9 mV at 25°C. (Note that $(RT/F) \ln x = 0.059\,16 \log x$.)

(b) Set up a spreadsheet to reproduce the result in part **(a)**. Then use your spreadsheet to compute and plot the junction potential for 0.1 M HCl | x M KCl, where x varies from 1 mM to 4 M.

(c) Use your spreadsheet to explore the behavior of the junction potential for y M HCl | x M KCl, where $y = 10^{-4}$, 10^{-3}, 10^{-2}, and 10^{-1} M and $x = 1$ mM or 4 M.

pH Measurement with a Glass Electrode

15-21. Describe how you would calibrate a pH electrode and measure the pH of blood ($\sim$7.5) *at 37°C.* Use standard buffers in Table 15-3.

15-22. List the sources of error associated with pH measurement with the glass electrode.

15-23. If Electrode 3 in Figure 15-21 were placed in a solution of pH 11.0, what would the pH reading be?

15-24. Which National Institute of Standards and Technology buffer(s) would you use to calibrate an electrode for pH measurements in the range 3–4?

15-25. Why do glass pH electrodes tend to indicate a pH lower than the actual pH in strongly basic solution?

15-26. Suppose that the Ag | AgCl outer electrode in Figure 15-14 is filled with 0.1 M NaCl instead of saturated KCl. Suppose that the electrode is calibrated in a dilute buffer containing 0.1 M KCl at pH 6.54 at 25°C. The electrode is then dipped in a second buffer *at the same pH* and same temperature, but containing 3.5 M KCl. Use Table 15-2 to estimate how much the indicated pH will change.

15-27. (a) When the difference in pH across the membrane of a glass electrode at 25°C is 4.63 pH units, how much voltage is generated by the pH gradient?

(b) What would the voltage be for the same pH difference at 37°C?

15-28. When calibrating a glass electrode, 0.025 *m* potassium dihydrogen phosphate/0.025 *m* disodium hydrogen phosphate buffer (Table 15-3) gave a reading of -18.3 mV at 20°C and 0.05 *m* potassium hydrogen phthalate buffer gave a reading of $+146.3$ mV. What is the pH of an unknown giving a reading of $+50.0$ mV? What is the slope of the calibration curve (mV/pH unit) and what is the theoretical slope at 20°C? Find the value of β in Equation 15-6.

15-29. *Activity problem.* The 0.0250 m KH$_2$PO$_4$/0.0250 m Na$_2$HPO$_4$ buffer (6) in Table 15-3 has a pH of 6.865 at 25°C.

(a) Show that the ionic strength of the buffer is $\mu = 0.100\ m$.

(b) From the pH and K_2 for phosphoric acid, find the quotient of activity coefficients, $\gamma_{HPO_4^{2-}}/\gamma_{H_2PO_4^-}$, at $\mu = 0.100\ m$.

(c) You have the urgent need to prepare a pH 7.000 buffer to be used as a calibration standard.[50] You can use the activity coefficient ratio from part **(b)** to accurately prepare such a buffer if the ionic strength is kept at 0.100 m. What molalities of KH$_2$PO$_4$ and Na$_2$HPO$_4$ should be mixed to obtain a pH of 7.000 and $\mu = 0.100\ m$?

15-30. *How pH of primary standards is measured.*[15, 51] Two reproducible electrochemical cells without a liquid junction are used for measurements. In these equations, m stands for molality and γ is the activity coefficient on a molality scale. Molality and molarity in dilute solutions are almost identical.

Cell 1:

Pt(s) | H$_2$(g) | primary standard buffer(aq), NaCl(aq) | AgCl(s) | Ag(s)

Ag electrode: AgCl(s) + e$^-$ $\rightleftharpoons$ Ag(s) + Cl$^-$(aq)

$$E_{Ag} = E^\circ_{Ag|AgCl} - \frac{RT\ln 10}{F}\log m_{Cl}\gamma_{Cl}$$

Pt electrode: H$^+$(aq) + e$^-$ $\rightleftharpoons$ $\frac{1}{2}$H$_2$(g)

$$E_{Pt} = E^\circ_{H_2|H^+} - \frac{RT\ln 10}{F}\log\frac{P_{H_2}^{1/2}}{m_H\gamma_H}$$

$$E_{cell\ 1} = (E^\circ_{Ag|AgCl} - E^\circ_{H_2|H^+}) - \frac{RT\ln 10}{F}\log\frac{m_H\gamma_H m_{Cl}\gamma_{Cl}}{P_{H_2}^{1/2}}$$

Cell 2:

Pt(s) | H$_2$(g) | HCl(aq, 0.01 m) | AgCl(s) | Ag(s)

$$E_{cell\ 2} = (E^\circ_{Ag|AgCl} - E^\circ_{H_2|H^+}) - \frac{RT\ln 10}{F}\log\frac{m_H m_{Cl}\gamma_H\gamma_{Cl}}{P_{H_2}^{1/2}}$$

$$E_{cell\ 2} = (E^\circ_{Ag|AgCl} - E^\circ_{H_2|H^+}) - \frac{RT\ln 10}{F}\log\frac{(0.01)^2\gamma_\pm^2}{P_{H_2}^{1/2}}$$

$$E_{cell\ 1} - E_{cell\ 2} = -\frac{RT\ln 10}{F}(\log\underbrace{m_H\gamma_H m_{Cl}\gamma_{Cl}}_{\text{activities in cell 1}} - \log\underbrace{[(0.01)^2\gamma_\pm^2]}_{\text{activities in cell 2}})\quad\text{(A)}$$

Both cells have Pt|H$_2$|H$^+$ and Ag|AgCl electrodes and both use the same pressure P_{H_2} at the same temperature T. Cell 1 contains the buffer whose pH ($\equiv -\log\mathcal{A}_H = -\log m_H\gamma_H$) is to be measured and it also contains NaCl at molality m_{Cl}. Cell 2 contains 0.01 m HCl instead of buffer. The difference in voltage between cells 1 and 2 in Equation A depends only on the activities of H$^+$ and Cl$^-$ in the two cells.

The mean activity coefficient of 0.01 m HCl, $\gamma_\pm \equiv (\gamma_H\gamma_{Cl})^{1/2} = 0.905$, is a well defined quantity measured with a galvanic cell.[52] Molality m_{Cl} is the known concentration of NaCl in cell 1. Everything in Equation A is known except the product $m_H\gamma_H\gamma_{Cl} = \mathcal{A}_H\gamma_{Cl}$ in cell 1. The activity $\mathcal{A}_H$ is what we are trying to measure. Rearrange Equation A to isolate unknowns on the left:

$$-\log\underset{\substack{\text{H}^+\text{ and Cl}^-\\ \text{in cell 1}}}{\mathcal{A}_H\gamma_{Cl}} = \frac{E_{cell\ 1} - E_{cell\ 2}}{(RT\ln 10)/F} - \frac{RT\ln 10}{F}\times\underset{\substack{0.01\ m\ \text{HCl}\\ \text{in cell 2}}}{\log[(0.01)^2(0.905)^2]} + \underset{\substack{\text{NaCl}\\ \text{in cell 1}}}{\log m_{Cl}}\quad\text{(B)}$$

The left side can also be written $-\log\mathcal{A}_H\gamma_{Cl} \equiv p(\mathcal{A}_H\gamma_{Cl})$.

Consider buffer 6 in Table 15-3 made of 0.025 m KH$_2$PO$_4$ plus 0.025 m Na$_2$HPO$_4$, whose ionic strength is 0.1 m. Our goal is to measure its pH ($= -\log\mathcal{A}_H$). Three solutions are made by adding 0.005, 0.01, and 0.02 m NaCl to the buffer. The quantity p($\mathcal{A}_H\gamma_{Cl}$) $\equiv -\log\mathcal{A}_H\gamma_{Cl}$ measured from Equation B is shown in the graph and extrapolated to p($\mathcal{A}_H\gamma_{Cl}$)$^\circ$ = 6.972 for zero added NaCl. At zero added NaCl, γ_{Cl} is the activity coefficient of Cl$^-$ if it were present in the buffer.

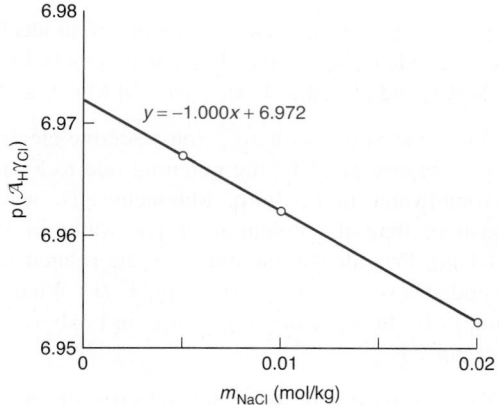

(a) Draw a picture of cells 1 and 2.

(b) There is no rigorously correct way to find the pH of the buffer because the activity coefficient for a single ion (γ_{Cl}) cannot be measured because ions only come in pairs with counterions. However, the extended Debye-Hückel equation 8-6 allows us to *estimate* the single-ion activity coefficients.

$$\log\gamma_{Cl} = \frac{-Az^2\sqrt{\mu}}{1 + B\alpha\sqrt{\mu}}\quad\text{(8-6)}$$

where μ is the ionic strength ($= 0.1$ mol/kg) of the buffer, A and B are temperature-dependent constants from the Debye-Hückel theory, z is the charge of the ion (-1 for Cl$^-$), and α is the ion size parameter in Table 8-1. The value of A at 25°C is 0.511. The *Bates-Guggenheim convention* chosen to estimate γ_{Cl} for measuring the pH of standard buffers is $B\alpha = 1.5$ (mol/kg)$^{-1/2}$. This value of $B\alpha$ is equivalent to choosing a size parameter of 458 pm for Cl$^-$. (In Table 8-1, the size of Cl$^-$ is listed as 300 pm.) Using $B\alpha = 1.5$ (mol/kg)$^{-1/2}$, calculate the pH of the standard buffer at 25°C and compare your answer to the value in Table 15-3.

Ion-Selective Electrodes

15-31. **(a)** Explain the principle of operation of an ion-selective electrode.

(b) How does a compound electrode differ from an ion-selective electrode?

15-32. What does the selectivity coefficient tell us? Is it better to have a large or a small selectivity coefficient?

15-33. What makes a liquid-based ion-selective electrode specific for one analyte?

15-34. Why is it preferable to use a metal ion buffer to achieve pM = 8 rather than just dissolving enough M to give a 10^{-8} M solution?

15-35. To determine the *concentration* of a dilute analyte with an ion-selective electrode, why do we use standards with a constant, high concentration of an inert salt?

15-36. A cyanide ion-selective electrode obeys the equation

$$E = \text{constant} - 0.059\,16 \log[CN^-]$$

The potential was -0.230 V when the electrode was immersed in 1.00 mM NaCN.

(a) Evaluate the constant in the equation.

(b) Using result from **(a)**, find $[CN^-]$ if $E = -0.300$ V.

(c) Without using the constant from **(a)**, find $[CN^-]$ if $E = -0.300$ V.

15-37. By how many volts will the potential of an ideal Mg^{2+} ion-selective electrode change if the electrode is removed from 1.00×10^{-4} M $MgCl_2$ and placed in 1.00×10^{-3} M $MgCl_2$ at 25°C?

15-38. When measured with a F^- ion-selective electrode with a Nernstian response at 25°C, the potential due to F^- in unfluoridated groundwater in Foxboro, Massachusetts, was 40.0 mV more positive than the potential of tap water in Providence, Rhode Island. Providence maintains its fluoridated water at the recommended level of 1.00 ± 0.05 mg F^-/L. What is the concentration of F^- in mg/L in groundwater in Foxboro? (Disregard the uncertainty.)

15-39. The selectivities of a Li^+ ion-selective electrode are indicated on the following diagram. Which alkali metal (Group 1) ion causes the most interference? Which alkaline earth (Group 2) ion causes the most interference? How much greater must be $[K^+]$ than $[Li^+]$ for the two ions to give equal response?

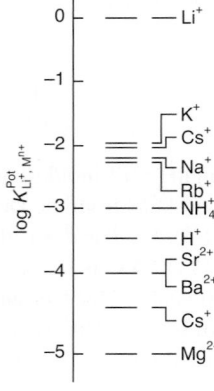

15-40. A metal ion buffer was prepared from 0.030 M ML and 0.020 M L, where ML is a metal-ligand complex and L is free ligand.

$$M + L \rightleftharpoons ML \qquad K_f = 4.0 \times 10^8$$

Calculate the concentration of free metal ion, M, in this buffer.

15-41. 🖳 *Calibration curve and propagation of uncertainty for exponents.* The following data were obtained when a Ca^{2+} ion-selective electrode was immersed in standard solutions whose ionic strength was constant at 2.0 M.

Ca^{2+} (M)	E (mV)
3.38×10^{-5}	-74.8
3.38×10^{-4}	-46.4
3.38×10^{-3}	-18.7
3.38×10^{-2}	$+10.0$
3.38×10^{-1}	$+37.7$

(a) Prepare a calibration curve and find the least-squares slope and intercept and their standard deviations.

(b) Calculate the value of β in Equation 15-13.

(c) For a measured potential, the calibration curve gives us $\log[Ca^{2+}]$. We can compute $[Ca^{2+}] = 10^{\log[Ca^{2+}]}$. Using rules for propagation of uncertainty in Table 3-1, calculate $[Ca^{2+}]$ and its associated uncertainty for a sample that gave a reading of -22.5 (± 0.3) mV in four replicate measurements.

15-42. The selectivity coefficient, $K^{Pot}_{Li^+, H^+}$, for a Li^+ ion-selective electrode is 4×10^{-4}. When this electrode is placed in 3.44×10^{-4} M Li^+ solution at pH 7.2, the potential is -0.333 V versus S.C.E. What would be the potential if the pH were lowered to 1.1 and the ionic strength were kept constant?

15-43. *Standard addition.* A particular CO_2 compound electrode like the one in Figure 15-31 obeys the equation $E = \text{constant} - [\beta RT \,(\ln 10)/2F]\log[CO_2]$, where R is the gas constant, T is temperature (303.15 K), F is the Faraday constant, and $\beta = 0.933$ (measured from a calibration curve). $[CO_2]$ represents all forms of dissolved carbon dioxide at the pH of the experiment, which was 5.0. Standard additions of volume V_S containing a standard concentration $c_S = 0.020\,0$ M $NaHCO_3$ were made to an unknown solution whose initial volume was $V_0 = 55.0$ mL.

V_S (mL)	E (V)	V_S (mL)	E (V)
0	0.079 0	0.300	0.058 8
0.100	0.072 4	0.800	0.050 9
0.200	0.065 3		

Prepare a graph with Equation 15-17 and find $[CO_2]$ in the unknown.

15-44. *Standard addition with confidence interval.* Ammonia in seawater was measured with an ammonia-selective electrode. A 100.0 mL aliquot of seawater was treated with 1.00 mL of 10 M NaOH to convert NH_4^+ to NH_3. Therefore, $V_0 = 101.0$ mL. A reading was then taken with the electrode. Then 10.00-mL aliquots of standard $NH_4^+Cl^-$ were added and results are shown in the table.

V_S (mL)	E (V)	V_S (mL)	E (V)
0	$-0.084\,4$	30.00	$-0.039\,4$
10.00	$-0.058\,1$	40.00	$-0.034\,7$
20.00	$-0.046\,9$		

Data derived from H. Van Ryswyk, E. W. Hall, S. J. Petesch, and A. E. Wiedeman, "Extending the Marine Microcosm Laboratory," J. Chem. Ed. **2007**, *84, 306.*

The standard contains 100.0 ppm (mg/L) of nitrogen in the form of $NH_4^+Cl^-$. A separate experiment determined that the electrode slope $\beta RT\,(\ln 10)/F$ is 0.056 6 V.

(a) Prepare a standard addition graph. Find the concentration and 95% confidence interval for ammonia nitrogen (ppm) in the 100.0 mL of seawater.

(b) Standard addition is best if the additions increase analyte to 1.5 to 3 times its original concentration. Does this experiment fall in that range? A criticism of this experiment is that too much added standard creates error because the standards contribute too much to the computed result and the reading from the initial solution is not weighted heavily enough.

15-45. Data below come from the graph in Box 15-2. The separate solutions method was used to measure selectivity coefficients for a sodium ion-selective electrode at 21.5°C. Use Equation 15-11 in Box 15-2 to calculate log K^{Pot} for each line below.

$(E_{Mg^{2+}} - E_{Na^+})$ at $\mathcal{A} = 10^{-3} = -0.385$ V $\Rightarrow$ log $K^{Pot}_{Na^+,Mg^{2+}} = ?$

$(E_{Mg^{2+}} - E_{Na^+})$ at $\mathcal{A} = 10^{-2} = -0.418$ V $\Rightarrow$ log $K^{Pot}_{Na^+,Mg^{2+}} = ?$

$(E_{K^+} - E_{Na^+})$ at $\mathcal{A} = 10^{-3} = -0.285$ V $\Rightarrow$ log $K^{Pot}_{Na^+,K^+} = ?$

$(E_{K^+} - E_{Na^+})$ at $\mathcal{A} = 10^{-1.5} = -0.285$ V $\Rightarrow$ log $K^{Pot}_{Na^+,K^+} = ?$

15-46. The H^+ ion-selective electrode on the *Mars Phoenix Lander* has selectivity coefficients $K^{Pot}_{H^+,Na^+} = 10^{-8.6}$ and $K^{Pot}_{H^+,Ca^{2+}} = 10^{-7.8}$. Let A be the primary ion sensed by the electrode and let its charge be z_A. Let X be an interfering ion with charge z_X. The relative error in primary ion activity due to an interfering ion is[53]

$$\% \text{ error in } \mathcal{A}_A = \frac{(K^{Pot}_{A,X})^{z_X/z_A} \mathcal{A}_X}{\mathcal{A}_A^{z_X/z_A}} \times 100$$

This expression applies for errors less than ~10%. If the pH is 8.0 ($\mathcal{A}_{H^+} = 10^{-8.0}$) and $\mathcal{A}_{Na^+} = 10^{-2.0}$, what is the relative error in measuring $\mathcal{A}_{H^+}$? If the pH is 8.0 and $\mathcal{A}_{Ca^{2+}} = 10^{-2.0}$, what is the relative error in measuring $\mathcal{A}_{H^+}$?

15-47. A Ca^{2+} ion-selective electrode was calibrated in metal ion buffers with ionic strength fixed at 0.50 M. Using the following electrode readings, write an equation for the response of the electrode to Ca^{2+} and Mg^{2+}.

$[Ca^{2+}]$ (M)	$[Mg^{2+}]$ (M)	mV
1.00×10^{-6}	0	-52.6
2.43×10^{-4}	0	$+16.1$
1.00×10^{-6}	3.68×10^{-3}	-38.0

15-48. The Pb^{2+} ion buffer used inside the electrode for the colored curve in Figure 15-29 was prepared by mixing 1.0 mL of 0.10 M $Pb(NO_3)_2$ with 100.0 mL of 0.050 M Na_2EDTA. At the measured pH of 4.34, $\alpha_{Y^{4-}} = 1.46 \times 10^{-8}$ (Equation 12-4). Show that $[Pb^{2+}] = 1.4 \times 10^{-12}$ M.

15-49. Solutions with a wide range of Hg^{2+} concentrations were prepared to calibrate an experimental Hg^{2+} ion-selective electrode. For the range $10^{-5} < [Hg^{2+}] < 10^{-1}$ M, $Hg(NO_3)_2$ was used directly. The range $10^{-11} < [Hg^{2+}] < 10^{-6}$ M could be covered by the buffer system $HgCl_2(s) + KCl(aq)$ (based on pK_{sp} for $HgCl_2 = 13.16$). The range $10^{-15} < [Hg^{2+}] < 10^{-11}$ M was obtained with $HgBr_2(s) + KBr(aq)$ (based on pK_{sp} for $HgBr_2 = 17.43$). The resulting calibration curve is shown in the figure. Calibration points

for the $HgCl_2/KCl$ buffer are not in line with the other data. Suggest a possible explanation.

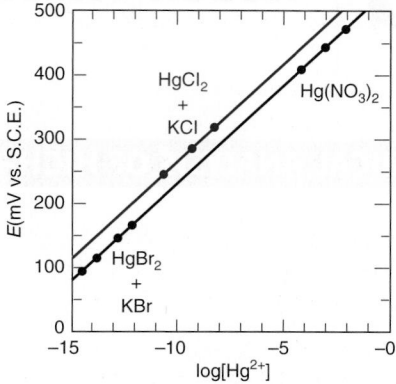

Hg^{2+} ion-selective electrode calibration curves, presumably all at the same ionic strength. [Data from J. A. Shatkin, H. S. Brown, and S. Licht, "Composite Graphite Ion Selective Electrode Array Potentiometry for the Detection of Mercury and Other Relevant Ions in Aquatic Systems," *Anal. Chem.* **1995,** *67,* 1147.]

15-50. *Activity problem.* Citric acid is a triprotic acid (H_3A) whose anion (A^{3-}) forms stable complexes with many metal ions.

$$Ca^{2+} + A^{3-} \xrightleftharpoons{K_f} CaA^-$$

When a Ca^{2+} ion-selective electrode with a slope of 29.58 mV was immersed in a solution having $\mathcal{A}_{Ca^{2+}} = 1.00 \times 10^{-3}$, the reading was $+2.06$ mV. Calcium citrate solution was prepared by mixing equal volumes of solutions 1 and 2.

Solution 1:
$$[Ca^{2+}] = 1.00 \times 10^{-3} \text{ M, pH} = 8.00, \mu = 0.10 \text{ M}$$

Solution 2:
$$[Citrate]_{total} = 1.00 \times 10^{-3} \text{ M, pH} = 8.00, \mu = 0.10 \text{ M}$$

The calcium citrate solution gave an electrode reading of -25.90 mV.

(a) Refer to the discussion with Figure A-2 in Appendix A. Calculate the activity of Ca^{2+} in the calcium citrate solution.

(b) Calculate the formation constant, K_f, for CaA^-. Assume that the size of CaA^- is 500 pm. At pH 8.00 and $\mu = 0.10$ M, the fraction of free citrate in the form A^{3-} is 0.998.

Solid-State Chemical Sensors

15-51. What does analyte do to a chemical-sensing field effect transistor to produce a signal related to the activity of analyte?

15-52. Explain how the device at the opening of this chapter measures the sequence of nucleotide bases in DNA. What is the purpose of the chemically sensitive field effect transistor in the device?

16 | Redox Titrations

Permanent magnet levitates above superconducting disk cooled in a pool of liquid nitrogen. Redox titrations are crucial in measuring the chemical composition of a superconductor. [U.S. government photo courtesy D. Cornelius with materials from T. Vanderah.]

Superconductors are materials that lose all electric resistance when cooled below a critical temperature. Prior to 1987, all known superconductors required cooling to temperatures near that of liquid helium (4 K), a process that is costly and impractical for all but a few applications. In 1987, a giant step was taken when "high-temperature" superconductors that retain their superconductivity above the boiling point of liquid nitrogen (77 K) were discovered. The most startling characteristic of a superconductor is magnetic levitation, shown above. When a magnetic field is applied to a superconductor, current flows in the outer skin of the material such that the applied magnetic field is exactly canceled by the induced magnetic field, and the net field inside the specimen is zero. Current flow in the skin of the superconductor repels the magnet and causes it to float above the superconductor. Expulsion of a magnetic field from a superconductor is called the *Meissner effect*.

A prototypical high-temperature superconductor is yttrium barium copper oxide, $YBa_2Cu_3O_7$, in which two-thirds of the copper is in the $+2$ oxidation state and one-third is in the unusual $+3$ state. Another example is $Bi_2Sr_2(Ca_{0.8}Y_{0.2})Cu_2O_{8.295}$, in which the average oxidation state of copper is $+2.105$ and the average oxidation state of bismuth is $+3.090$ (which is formally a mixture of Bi^{3+} and Bi^{5+}). The most reliable means to unravel these complex formulas is through "wet" oxidation-reduction titrations described in this chapter.

Iron and its compounds are environmentally acceptable redox agents that are finding increased use in remediating toxic waste in groundwaters:[1]

$$3H_2S + 8HFeO_4^- + 6H_2O \rightarrow$$

$$\underset{\text{Pollutant}}{3H_2S} + \underset{\substack{\text{Ferrate(VI)} \\ \text{(oxidant)}}}{8HFeO_4^-} + 6H_2O \rightarrow$$

$$\underbrace{8Fe(OH)_3(s) + 3SO_4^{2-}}_{\text{(safe products)}} + 2OH^-$$

A redox titration is based on an oxidation-reduction reaction between analyte and titrant. In addition to the many common analytes in chemistry, biology, and environmental and materials science that can be measured by redox titrations, exotic oxidation states of elements in uncommon materials such as superconductors and laser materials are measured by redox titrations. For example, chromium added to laser crystals to increase their efficiency is found in the common oxidation states $+3$ and $+6$, and the unusual $+4$ state. A redox titration is a good way to unravel the nature of this complex mixture of chromium ions.[2]

This chapter introduces the theory of redox titrations and discusses some common reagents. A few of the oxidants and reductants in Table 16-1 can be used as titrants.[3] Most reductants used as titrants react with O_2 and, therefore, require protection from air.

TABLE 16-1 — Oxidizing and reducing agents

Oxidants		Reductants	
BiO_3^-	Bismuthate	(structure)	Ascorbic acid (vitamin C)
BrO_3^-	Bromate		
Br_2	Bromine		
Ce^{4+}	Ceric		
CH_3—⬡—$SO_2NCl^-Na^+$	Chloramine T	BH_4^-	Borohydride
Cl_2	Chlorine	Cr^{2+}	Chromous
ClO_2	Chlorine dioxide	$S_2O_4^{2-}$	Dithionite
$Cr_2O_7^{2-}$	Dichromate	Fe^{2+}	Ferrous
FeO_4^{2-}	Ferrate(VI)	N_2H_4	Hydrazine
H_2O_2	Hydrogen peroxide	HO—⬡—OH	Hydroquinone
$Fe^{2+} + H_2O_2$	Fenton reagent[4]		
OCl^-	Hypochlorite	NH_2OH	Hydroxylamine
IO_3^-	Iodate	H_3PO_2	Hypophosphorous acid
I_2	Iodine	(structure)	Retinol (vitamin A)
$Pb(acetate)_4$	Lead(IV) acetate		
HNO_3	Nitric acid		
O	Atomic oxygen		
O_2	Dioxygen (oxygen)	Sn^{2+}	Stannous
O_3	Ozone	SO_3^{2-}	Sulfite
$HClO_4$	Perchloric acid	SO_2	Sulfur dioxide
IO_4^-	Periodate	$S_2O_3^{2-}$	Thiosulfate
MnO_4^-	Permanganate	(structure)	
$S_2O_8^{2-}$	Peroxydisulfate		α-Tocopherol (vitamin E)[5]

16-1 The Shape of a Redox Titration Curve

Consider the titration of iron(II) with standard cerium(IV), monitored potentiometrically with Pt and calomel electrodes as shown in Figure 16-1. The titration reaction is

Titration reaction:

$$Ce^{4+} + Fe^{2+} \rightarrow Ce^{3+} + Fe^{3+} \tag{16-1}$$

Ceric titrant   Ferrous analyte   Cerous   Ferric

> The titration reaction goes to completion after each addition of titrant. The equilibrium constant is $K = 10^{nE°/0.059}$ [16] at 25°C. Strong acid prevents hydrolysis reactions such as $Fe^{3+} + H_2O \rightleftharpoons Fe(OH)^{2+} + H^+$.

for which $K \approx 10^{16}$ in 1 M $HClO_4$. Each mole of ceric ion oxidizes 1 mol of ferrous ion rapidly and quantitatively. The titration reaction creates a mixture of Ce^{4+}, Ce^{3+}, Fe^{2+}, and Fe^{3+} in the beaker in Figure 16-1. Box 16-1 shows the likely mechanism of Reaction 16-1.

At the Pt indicator electrode, *two* reactions come to equilibrium:

Indicator half-reaction: $Fe^{3+} + e^- \rightleftharpoons Fe^{2+}$ $E° = 0.767$ V   **(16-2)**

Indicator half-reaction: $Ce^{4+} + e^- \rightleftharpoons Ce^{3+}$ $E° = 1.70$ V   **(16-3)**

> Equilibria 16-2 and 16-3 are both established at the Pt electrode.

The potentials are the formal potentials that apply in 1 M $HClO_4$. The Pt indicator electrode responds to the relative concentrations (really, activities) of Ce^{4+} and Ce^{3+} or Fe^{3+} and Fe^{2+}.

We now set out to calculate how the cell voltage changes as Fe^{2+} is titrated with Ce^{4+}. The titration curve has three regions.

Region 1: Before the Equivalence Point

As each aliquot of Ce^{4+} is added, titration reaction 16-1 consumes Ce^{4+} and creates an equal number of moles of Ce^{3+} and Fe^{3+}. Prior to the equivalence point, excess unreacted Fe^{2+} remains in solution. Therefore, we can find the concentrations of Fe^{2+} and Fe^{3+} without difficulty. On the other hand, we cannot find the concentration of Ce^{4+} without solving a fancy

> We can use either Reaction 16-2 or Reaction 16-3 to describe the cell voltage at any time. However, because we know $[Fe^{2+}]$ and $[Fe^{3+}]$, it is more convenient for now to use Reaction 16-2.

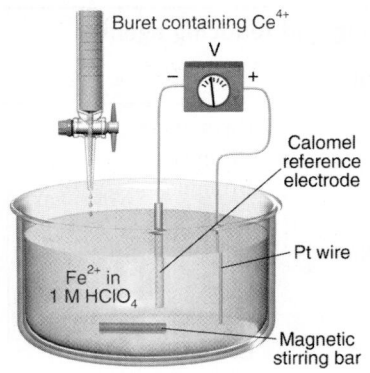

FIGURE 16-1 Apparatus for potentiometric titration of Fe^{2+} with Ce^{4+}.

Buret containing Ce^{4+}

Calomel reference electrode

Pt wire

Fe^{2+} in 1 M $HClO_4$

Magnetic stirring bar

little equilibrium problem. Because the amounts of Fe^{2+} and Fe^{3+} are both known, it is *convenient* to calculate the cell voltage by using Reaction 16-2 instead of Reaction 16-3.

E_+ is the potential of the Pt electrode connected to the positive terminal of the potentiometer in Figure 16-1. E_- is the potential of the calomel electrode connected to the negative terminal.

$$E = E_+ - E_- \tag{16-4}$$

$$E = \left[\underset{\substack{\uparrow \\ \text{Formal potential for } Fe^{3+} \\ \text{reduction in} \\ \text{1 M } HClO_4}}{0.767} - 0.059\ 16 \log\left(\frac{[Fe^{2+}]}{[Fe^{3+}]}\right) \right] - \underset{\substack{\uparrow \\ \text{Potential of} \\ \text{saturated calomel} \\ \text{electrode}}}{0.241} \tag{16-5}$$

$$E = 0.526 - 0.059\ 16 \log\left(\frac{[Fe^{2+}]}{[Fe^{3+}]}\right) \tag{16-6}$$

For Reaction 16-2, $E_+ = E°(Fe^{3+}|Fe^{2+})$ when $V = \frac{1}{2}V_e$.

One special point is reached before the equivalence point. When the volume of titrant is one-half of the amount required to reach the equivalence point ($V = \frac{1}{2}V_e$), $[Fe^{3+}] = [Fe^{2+}]$. In this case, the log term is 0, and $E_+ = E°$ for the $Fe^{3+}|Fe^{2+}$ couple. *The point at which $V = \frac{1}{2}V_e$ is analogous to the point at which $pH = pK_a$ when $V = \frac{1}{2}V_e$ in an acid-base titration.*

The voltage at zero titrant volume cannot be calculated, because we do not know how much Fe^{3+} is present. If $[Fe^{3+}] = 0$, the voltage calculated with Equation 16-6 would be $-\infty$. In fact, there must be some Fe^{3+} in each reagent, either as an impurity or from oxidation of Fe^{2+} by atmospheric oxygen. In any case, the voltage could never be lower than that needed to reduce the solvent ($H_2O + e^- \rightarrow \frac{1}{2}H_2 + OH^-$).

Region 2: At the Equivalence Point

Exactly enough Ce^{4+} has been added to react with all the Fe^{2+}. Virtually all cerium is in the form Ce^{3+}, and virtually all iron is in the form Fe^{3+}. Tiny amounts of Ce^{4+}

BOX 16-1 **Many Redox Reactions Are Atom-Transfer Reactions**

Reaction 16-1 appears as if an electron moves from Fe^{2+} to Ce^{4+} to give Fe^{3+} and Ce^{3+}. In fact, this reaction and many others are thought to proceed through atom transfer, not electron transfer.[6] In this case, a hydrogen atom (a proton plus an electron) could be transferred from aqueous Fe^{2+} to aqueous Ce^{4+}:

"Ce^{4+}"

$Ce(H_2O)_7(OH)^{3+}$

"Fe^{2+}"

$Fe(H_2O)_6^{2+}$

"Ce^{3+}"

$Ce(H_2O)_8^{3+}$

"Fe^{3+}"

$Fe(H_2O)_5(OH)^{2+}$

Other common redox reactions between metallic species could proceed via transfer of oxygen atoms or halogen atoms to effect net electron transfer from one metal to another.

and Fe^{2+} are present at equilibrium. From the stoichiometry of Reaction 16-1, we can say that

$$[Ce^{3+}] = [Fe^{3+}] \qquad (16\text{-}7)$$

$$[Ce^{4+}] = [Fe^{2+}] \qquad (16\text{-}8)$$

To understand why Equations 16-7 and 16-8 are true, imagine that *all* the cerium and the iron have been converted into Ce^{3+} and Fe^{3+}. Because we are at the equivalence point, $[Ce^{3+}] = [Fe^{3+}]$. Now let Reaction 16-1 come to equilibrium:

$$Fe^{3+} + Ce^{3+} \rightleftharpoons Fe^{2+} + Ce^{4+} \qquad \text{(reverse of Reaction 16-1)}$$

If a little bit of Fe^{3+} goes back to Fe^{2+}, an equal number of moles of Ce^{4+} must be made. So $[Ce^{4+}] = [Fe^{2+}]$.

At any time, Reactions 16-2 and 16-3 are *both* in equilibrium at the Pt electrode. At the equivalence point, it is *convenient* to use both reactions to describe the cell voltage. The Nernst equations for these reactions are

At the equivalence point, we use both Reactions 16-2 and 16-3 to calculate the cell voltage. This is strictly a matter of algebraic convenience.

$$E_+ = 0.767 - 0.059\,16 \log\left(\frac{[Fe^{2+}]}{[Fe^{3+}]}\right) \qquad (16\text{-}9)$$

$$E_+ = 1.70 - 0.059\,16 \log\left(\frac{[Ce^{3+}]}{[Ce^{4+}]}\right) \qquad (16\text{-}10)$$

Here is where we stand: Equations 16-9 and 16-10 are each a statement of algebraic truth. But neither one alone allows us to find E_+, because we do not know exactly what tiny concentrations of Fe^{2+} and Ce^{4+} are present. It is possible to solve the four simultaneous equations 16-7 through 16-10 by first *adding* Equations 16-9 and 16-10:

$$2E_+ = 0.767 + 1.70 - 0.059\,16 \log\left(\frac{[Fe^{2+}]}{[Fe^{3+}]}\right) - 0.059\,16 \log\left(\frac{[Ce^{3+}]}{[Ce^{4+}]}\right)$$

$$\log a + \log b = \log ab$$

$$-\log a - \log b = -\log ab$$

$$2E_+ = 2.46_7 - 0.059\,16 \log\left(\frac{[Fe^{2+}][Ce^{3+}]}{[Fe^{3+}][Ce^{4+}]}\right)$$

But, because $[Ce^{3+}] = [Fe^{3+}]$ and $[Ce^{4+}] = [Fe^{2+}]$ at the equivalence point, the ratio of concentrations in the log term is unity. Therefore, the logarithm is 0 and

$$2E_+ = 2.46_7 \text{ V} \Rightarrow E_+ = 1.23 \text{ V}$$

The cell voltage is

$$E = E_+ - E(\text{calomel}) = 1.23 - 0.241 = 0.99 \text{ V} \qquad (16\text{-}11)$$

In this particular titration, the equivalence-point voltage is independent of the concentrations and volumes of the reactants.

Region 3: After the Equivalence Point

Now virtually all iron is Fe^{3+}. The moles of Ce^{3+} equal the moles of Fe^{3+}, and there is a known excess of unreacted Ce^{4+}. Because we know both $[Ce^{3+}]$ and $[Ce^{4+}]$, it is *convenient* to use Reaction 16-3 to describe the chemistry at the Pt electrode:

After the equivalence point, we use Reaction 16-3 because we can easily calculate $[Ce^{3+}]$ and $[Ce^{4+}]$. It is not convenient to use Reaction 16-2, because we do not know the concentration of Fe^{2+}, which has been "used up."

$$E = E_+ - E(\text{calomel}) = \left[1.70 - 0.059\,16 \log\left(\frac{[Ce^{3+}]}{[Ce^{4+}]}\right) \right] - 0.241 \qquad (16\text{-}12)$$

At the special point when $V = 2V_e$, $[Ce^{3+}] = [Ce^{4+}]$ and $E_+ = E°(Ce^{4+} \mid Ce^{3+}) = 1.70$ V.

Before V_e, the indicator electrode potential is fairly steady near $E°(Fe^{3+} \mid Fe^{2+}) = 0.77$ V.[7] After V_e, the indicator electrode potential levels off near $E°(Ce^{4+} \mid Ce^{3+}) = 1.70$ V. At V_e, there is a rapid rise in voltage.

EXAMPLE Potentiometric Redox Titration

Suppose that we titrate 100.0 mL of 0.050 0 M Fe^{2+} with 0.100 M Ce^{4+}, using the cell in Figure 16-1. The equivalence point occurs when $V_{Ce^{4+}} = 50.0$ mL. Calculate the cell voltage at 36.0, 50.0, and 63.0 mL.

Anyone with a serious need to calculate redox titration curves should use a spreadsheet with a more general set of equations than we use in this section.[8] The supplement at www.whfreeman.com/qca explains how to use spreadsheets to compute redox titration curves.

Solution *At 36.0 mL:* This is 36.0/50.0 of the way to the equivalence point. Therefore, 36.0/50.0 of the iron is in the form Fe^{3+} and 14.0/50.0 is in the form Fe^{2+}. Putting $[Fe^{2+}]/[Fe^{3+}] = 14.0/36.0$ into Equation 16-6 gives $E = 0.550$ V.

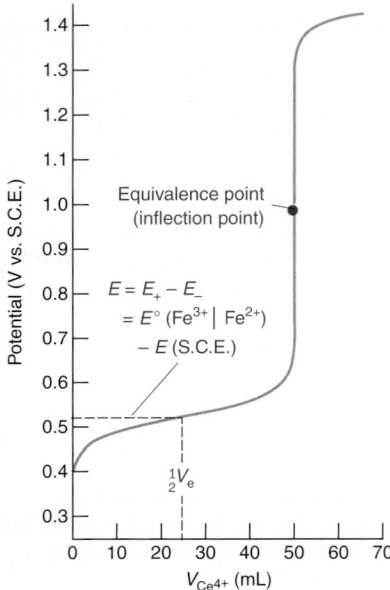

FIGURE 16-2 Theoretical curve for titration of 100.0 mL of 0.050 0 M Fe^{2+} with 0.100 M Ce^{4+} in 1 M $HClO_4$. You cannot calculate the potential for zero titrant, but you can start at a small volume such as $V_{Ce^{4+}} = 0.1$ mL.

At 50.0 mL: Equation 16-11 tells us that the cell voltage at the equivalence point is 0.99 V, regardless of the concentrations of reagents for this particular titration.

At 63.0 mL: The first 50.0 mL of cerium were converted into Ce^{3+}. There is an excess of 13.0 mL of Ce^{4+}, so $[Ce^{3+}]/[Ce^{4+}] = 50.0/13.0$ in Equation 16-12, and $E = 1.424$ V.

TEST YOURSELF Find E at $V_{Ce^{4+}} = 20.0$ and 51.0 mL (*Answer:* 0.516, 1.358 V)

Shapes of Redox Titration Curves

The calculations above allow us to plot the titration curve in Figure 16-2, which shows potential as a function of the volume of added titrant. The equivalence point is marked by a steep rise in voltage. The value of E_+ at $\frac{1}{2}V_e$ is the formal potential of the $Fe^{3+}|Fe^{2+}$ couple, because the quotient $[Fe^{2+}]/[Fe^{3+}]$ is unity at this point. The voltage at any point in this titration depends only on the *ratio* of reactants; their *concentrations* do not figure in any calculations in this example. We expect, therefore, that the curve in Figure 16-2 will be independent of dilution. We should observe the same curve if both reactants were diluted by a factor of 10.

For Reaction 16-1, the titration curve in Figure 16-2 is symmetric near the equivalence point because the reaction stoichiometry is 1:1. Figure 16-3 shows the curve calculated for the titration of Tl^+ by IO_3^- in 1.00 M HCl.

$$IO_3^- + 2Tl^+ + 2Cl^- + 6H^+ \rightarrow ICl_2^- + 2Tl^{3+} + 3H_2O \qquad (16\text{-}13)$$

The curve is *not symmetric about the equivalence point* because the stoichiometry of reactants is 2:1, not 1:1. Still, the curve is so steep near the equivalence point that negligible error is introduced if the center of the steep part is taken as the end point. Demonstration 16-1 provides an example of an asymmetric titration curve whose shape also depends on the pH of the reaction medium.

The voltage change near the equivalence point increases as the difference between $E°$ of the two redox couples in the titration increases. The larger the difference in $E°$, the greater the equilibrium constant for the titration reaction. For Figure 16-2, half-reactions 16-2 and 16-3 differ by 0.93 V, and there is a large break at the equivalence point in the titration curve. For Figure 16-3, the half-reactions differ by 0.47 V, so there is a smaller break at the equivalence point.

$$IO_3^- + 2Cl^- + 6H^+ + 4e^- \rightleftharpoons ICl_2^- + 3H_2O \qquad E° = 1.24 \text{ V}$$
$$Tl^{3+} + 2e^- \rightleftharpoons Tl^+ \qquad E° = 0.77 \text{ V}$$

Clearest results are achieved with the strongest oxidizing and reducing agents. The same rule applies to acid-base titrations where strong-acid or strong-base titrants give the sharpest break at the equivalence point.

16-2 Finding the End Point

As in acid-base titrations, indicators and electrodes are commonly used to find the end point of a redox titration.

Redox Indicators

A **redox indicator** is a compound that changes color when it goes from its oxidized to its reduced state. The indicator ferroin changes from pale blue (almost colorless) to red.

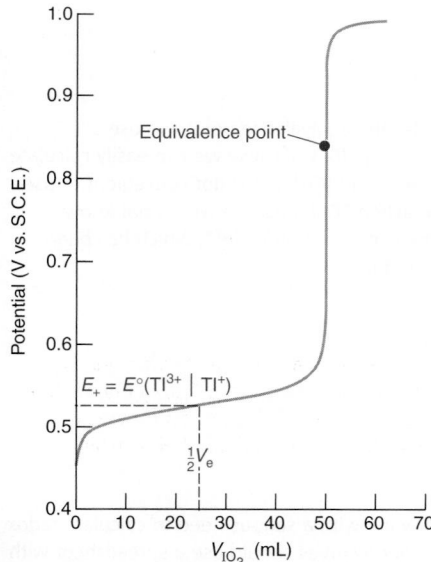

FIGURE 16-3 Theoretical curve for titration of 100.0 mL of 0.010 0 M Tl^+ with 0.010 0 M IO_3^- in 1.00 M HCl. The equivalence point at 0.842 V is not at the center of the steep part of the curve because the stoichiometry of the reaction is not 1:1.

Oxidized ferroin
(pale blue)
In(oxidized)

Reduced ferroin
(red)
In(reduced)

To predict the potential range over which the indicator color will change, we first write a Nernst equation for the indicator.

$$In(\text{oxidized}) + ne^- \rightleftharpoons In(\text{reduced})$$

$$E = E° - \frac{0.059\ 16}{n} \log\left(\frac{[In(\text{reduced})]}{[In(\text{oxidized})]}\right) \qquad (16\text{-}14)$$

This reaction illustrates many principles of potentiometric titrations:

$$MnO_4^- + 5Fe^{2+} + 8H^+ \rightarrow Mn^{2+} + 5Fe^{3+} + 4H_2O \quad \text{(A)}$$
$$\underset{\text{Titrant}}{} \quad \underset{\text{Analyte}}{}$$

Dissolve 0.60 g of $Fe(NH_4)_2(SO_4)_2 \cdot 6H_2O$ (FM 392.13; 1.5 mmol) in 400 mL of 1 M H_2SO_4. Titrate the well-stirred solution with 0.02 M $KMnO_4$ ($V_e \approx 15$ mL), using Pt and calomel electrodes with a pH meter as a potentiometer. Before use, zero the meter by connecting the two inputs directly to each other and setting the millivolt scale to 0.

Calculate some points on the theoretical titration curve before performing the experiment. Then compare the theoretical and experimental results. Also note the coincidence of the potentiometric and visual end points.

Question Potassium permanganate is purple, and all the other species in this titration are colorless (or very faintly colored). What color change is expected at the equivalence point?

To calculate the theoretical curve, we use the following half-reactions:

$$Fe^{3+} + e^- \rightleftharpoons Fe^{2+} \qquad E° = 0.68 \text{ V in 1 M } H_2SO_4 \quad \text{(B)}$$

$$MnO_4^- + 8H^+ + 5e^- \rightarrow Mn^{2+} + 4H_2O \qquad E° = 1.507 \text{ V} \quad \text{(C)}$$

Prior to V_e, calculations are similar to those in Section 16-1 for the titration of Fe^{2+} by Ce^{4+}, but with $E° = 0.68$ V. After V_e, you can find the potential by using Reaction C. For example, suppose that you titrate 0.400 L of 3.75 mM Fe^{2+} with 0.020 0 M $KMnO_4$. From the stoichiometry of Reaction A, $V_e = 15.0$ mL. When you have added 17.0 mL of $KMnO_4$, the concentrations of species in Reaction C are $[Mn^{2+}] = 0.719$ mM, $[MnO_4^-] = 0.095$ 9 mM, and $[H^+] = 0.959$ M (neglecting the small quantity of H^+ consumed in the titration). The voltage should be

$$E = E_+ - E(\text{calomel})$$

$$= \left[1.507 - \frac{0.059\ 16}{5} \log\left(\frac{[Mn^{2+}]}{[MnO_4^-][H^+]^8} \right) \right] - 0.241$$

$$= \left[1.507 - \frac{0.059\ 16}{5} \log\left(\frac{7.19 \times 10^{-4}}{(9.59 \times 10^{-5})(0.959)^8} \right) \right] - 0.241$$

$$= 1.254 \text{ V}$$

To calculate the voltage at V_e, we add the Nernst equations for Reactions B and C, as we did for the cerium and iron reactions in Section 16-1. Before doing so, however, multiply the permanganate equation by 5 so that we can add the log terms:

$$E_+ = 0.68 - 0.059\ 16 \log\left(\frac{[Fe^{2+}]}{[Fe^{3+}]} \right)$$

$$5E_+ = 5\left[1.507 - \frac{0.059\ 16}{5} \log\left(\frac{[Mn^{2+}]}{[MnO_4^-][H^+]^8} \right) \right]$$

Now add the two equations to get

$$6E_+ = 8.215 - 0.059\ 16 \log\left(\frac{[Mn^{2+}][Fe^{2+}]}{[MnO_4^-][Fe^{3+}][H^+]^8} \right) \quad \text{(D)}$$

The stoichiometry of the titration reaction A tells us that, at V_e, $[Fe^{3+}] = 5[Mn^{2+}]$ and $[Fe^{2+}] = 5[MnO_4^-]$. Substituting these values into Equation D gives

$$6E_+ = 8.215 - 0.059\ 16 \log\left(\frac{[Mn^{2+}](5[MnO_4^-])}{[MnO_4^-](5[Mn^{2+}])[H^+]^8} \right)$$

$$6E_+ = 8.215 - 0.059\ 16 \log\left(\frac{1}{[H^+]^8} \right) \quad \text{(E)}$$

Inserting the concentration of $[H^+]$, which is (400 mL/415 mL) $\times$ (1.00 M) = 0.964 M, we find

$$6E_+ = 8.215 - 0.059\ 16 \log\left(\frac{1}{(0.964)^8} \right) \Rightarrow E_+ = 1.368 \text{ V}$$

The predicted cell voltage at V_e is $E = E_+ - E(\text{calomel}) = 1.368 - 0.241 = 1.127$ V.

As with acid-base indicators, the color of In(reduced) will be observed when

$$\frac{[\text{In(reduced)}]}{[\text{In(oxidized)}]} \gtrsim \frac{10}{1}$$

and the color of In(oxidized) will be observed when

$$\frac{[\text{In(reduced)}]}{[\text{In(oxidized)}]} \lesssim \frac{1}{10}$$

Putting these quotients into Equation 16-14 tells us that the color change will occur over the range

$$E = \left(E° \pm \frac{0.059\ 16}{n} \right) \text{volts}$$

For ferroin, with $E° = 1.147$ V (Table 16-2), we expect the color change to occur in the approximate range 1.088 V to 1.206 V with respect to the standard hydrogen electrode. With a saturated calomel reference electrode, the indicator transition range will be

$$\begin{pmatrix} \text{Indicator transition} \\ \text{range versus calomel} \\ \text{electrode (S.C.E.)} \end{pmatrix} = \begin{pmatrix} \text{Transition range} \\ \text{versus standard hydrogen} \\ \text{electrode (S.H.E.)} \end{pmatrix} - E(\text{calomel}) \quad \text{(16-15)}$$

$$= (1.088 \text{ to } 1.206) - (0.241)$$

$$= 0.847 \text{ to } 0.965 \text{ V (versus S.C.E.)}$$

A redox indicator changes color over an approximate range of $\pm(59/n)$ mV, centered at $E°$ for the indicator. n is the number of electrons in the indicator half-reaction.

Figure 15-6 should help you understand Equation 16-15.

The indicator transition range should overlap the steep part of the titration curve.

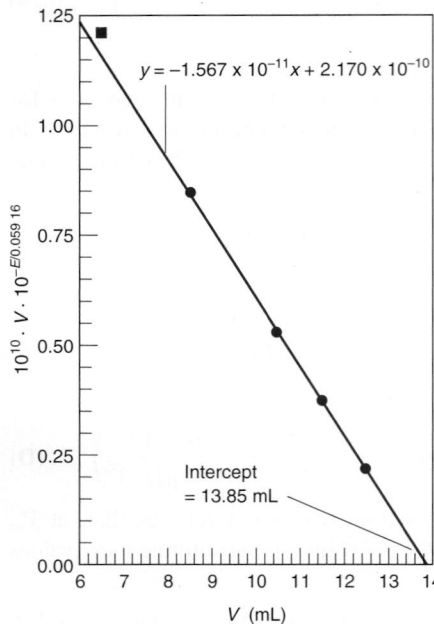

$y = -1.567 \times 10^{-11}x + 2.170 \times 10^{-10}$

Intercept = 13.85 mL

FIGURE 16-4 Gran plot for titration of Fe^{2+} by Ce^{4+} in Exercise 16-D.[9] The line was fit to the four points shown by circles. In the function on the ordinate, the value of n is 1. Numerical values were multiplied by 10^{10} for ease of display. Multiplication does not change the x-intercept.

The constant 0.059 16 V is $(RT \ln 10)/nF$, where R is the gas constant, T is 298.15 K, F is the Faraday constant, and n is the number of electrons in the $Fe^{3+} \mid Fe^{2+}$ redox half-reaction ($n = 1$). For $T \neq 298.15$ K or $n \neq 1$, the number 0.059 16 will change.

TABLE 16-2 Redox indicators

Indicator	Color		$E°$
	Oxidized	Reduced	
Phenosafranine	Red	Colorless	0.28
Indigo tetrasulfonate	Blue	Colorless	0.36
Methylene blue	Blue	Colorless	0.53
Diphenylamine	Violet	Colorless	0.75
4'-Ethoxy-2,4-diaminoazobenzene	Yellow	Red	0.76
Diphenylamine sulfonic acid	Red-violet	Colorless	0.85
Diphenylbenzidine sulfonic acid	Violet	Colorless	0.87
Tris(2,2'-bipyridine)iron	Pale blue	Red	1.120
Tris(1,10-phenanthroline)iron (ferroin)	Pale blue	Red	1.147
Tris(5-nitro-1,10-phenanthroline)iron	Pale blue	Red-violet	1.25
Tris(2,2'-bipyridine)ruthenium	Pale blue	Yellow	1.29

Ferroin would therefore be a useful indicator for the titrations in Figures 16-2 and 16-3.

The larger the difference in standard potential between titrant and analyte, the greater the break in the titration curve at the equivalence point. A redox titration is usually feasible if the difference between analyte and titrant is $\gtrsim 0.2$ V. However, the end point of such a titration is not very sharp and is best detected potentiometrically. If the difference in formal potentials is $\gtrsim 0.4$ V, then a redox indicator usually gives a satisfactory visual end point.

Gran Plot

With the apparatus in Figure 16-1, we measure electrode potential, E, versus volume of titrant, V, during a redox titration. The end point is the maximum of the first derivative, $\Delta E/\Delta V$, or the zero crossing of the second derivative, $\Delta(\Delta E/\Delta V)/\Delta V$ (Figure 11-5).

A more accurate way to use potentiometric data is to prepare a Gran plot[9,10] as we did for acid-base titrations in Section 11-5. The Gran plot uses data from well before the equivalence point (V_e) to locate V_e. Potentiometric data taken close to V_e are the least accurate because electrodes are slow to equilibrate with species in solution when one member of a redox couple is nearly used up.

For the oxidation of Fe^{2+} to Fe^{3+}, the potential prior to V_e is

$$E = \left[E°' - 0.059\ 16 \log\left(\frac{[Fe^{2+}]}{[Fe^{3+}]}\right) \right] - E_{ref} \qquad \text{(16-16)}$$

where $E°'$ is the formal potential for $Fe^{3+} \mid Fe^{2+}$ and E_{ref} is the potential of the reference electrode (which we have also been calling E_-). If the volume of titrant is V and if the reaction goes "to completion" with each addition of titrant, then $[Fe^{2+}]/[Fe^{3+}] = (V_e - V)/V$. Substituting this expression into Equation 16-16 and rearranging, we eventually stumble onto a linear equation of the form $y = mx + b$:

$$\underbrace{V \cdot 10^{-nE/0.059\ 16}}_{y} = \underbrace{V_e \cdot 10^{-n(E_{ref} - E°')/0.059\ 16}}_{b} - \underbrace{V}_{x} \underbrace{10^{-n(E_{ref} - E°')/0.059\ 16}}_{m} \qquad \text{(16-17)}$$

where n is the number of electrons in the half-reaction at the indicator electrode.

A graph of $V \cdot 10^{-nE/0.059\ 16}$ versus V should be a straight line with x-intercept $= V_e$ (Figure 16-4). If the ionic strength of the reaction is constant, the activity coefficients are constant, and Equation 16-17 gives a straight line over a wide volume range. If ionic strength varies as titrant is added, we just use the last 10–20% of the data prior to V_e.

The Starch-Iodine Complex

Numerous analytical procedures are based on redox titrations involving iodine. Starch[11] is the indicator of choice for these procedures because it forms an intense blue complex with iodine. Starch is not a redox indicator; it responds specifically to the presence of I_2, not to a change in redox potential.

The active fraction of starch is amylose, a polymer of the sugar α-D-glucose, with the repeating unit shown in Figure 16-5. Small molecules can fit into the center of the coiled,

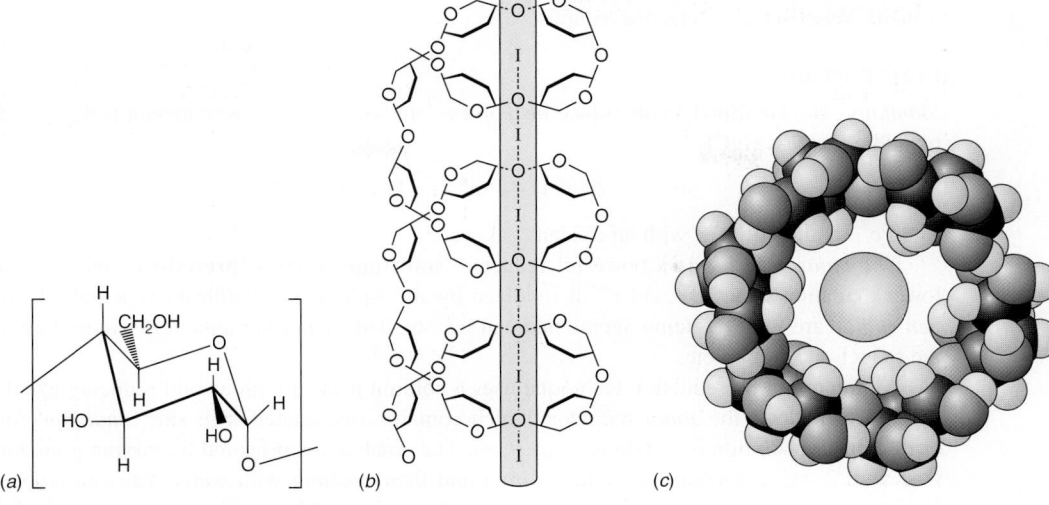

(a) (b) (c)

FIGURE 16-5 (*a*) Structure of the repeating unit of amylose. (*b*) Schematic structure of the starch-iodine complex. The amylose chain forms a helix around I_6 units. [Information from A. T. Calabrese and A. Khan, "Amylose-Iodine Complex Formation with KI: Evidence for Absence of Iodide Ions within the Complex," *J. Polymer Sci.* **1999**, *A37*, 2711.] (*c*) View down the starch helix, showing iodine inside the helix.[11] [Data from R. D. Hancock, Power Engineering, Salt Lake City.]

helical polymer. In the presence of starch, iodine forms I_6 chains inside the amylose helix and the color turns dark blue.

$$I \cdot I \cdot I \cdot I \cdot I \cdot I$$

Starch is readily biodegraded, so it should be freshly dissolved or the solution should contain a preservative, such as HgI_2 (~1 mg/100 mL) or thymol. A hydrolysis product of starch is glucose, which is a reducing agent. Therefore, partially hydrolyzed starch solution can be a source of error in a redox titration.

16-3 Adjustment of Analyte Oxidation State

Sometimes we need to adjust the oxidation state of analyte before it can be titrated. For example, Mn^{2+} can be **preoxidized** to MnO_4^- and then titrated with standard Fe^{2+}. Preadjustment must be quantitative, and you must eliminate excess preadjustment reagent so that it will not interfere in the subsequent titration.

Preoxidation

Several powerful oxidants can be easily removed after preoxidation. *Peroxydisulfate* ($S_2O_8^{2-}$, also called *persulfate*) is a strong oxidant that requires Ag^+ as a catalyst.

$$S_2O_8^{2-} + Ag^+ \rightarrow 2SO_4^{2-} + \underset{\substack{\text{Powerful} \\ \text{oxidant}}}{Ag^{3+}}$$

Excess reagent is destroyed by boiling the solution after oxidation of analyte is complete.

$$2S_2O_8^{2-} + 2H_2O \xrightarrow{\text{boiling}} 4SO_4^{2-} + O_2 + 4H^+$$

The $S_2O_8^{2-}$ and Ag^+ mixture oxidizes Mn^{2+} to MnO_4^-, Ce^{3+} to Ce^{4+}, Cr^{3+} to $Cr_2O_7^{2-}$, and VO^{2+} to VO_2^+.

Silver(I, III) oxide ($Ag^IAg^{III}O_2$, usually written as AgO)[12] dissolved in concentrated mineral acid has an oxidizing power similar to $S_2O_8^{2-}$ plus Ag^+. Excess Ag^{3+} can be removed by boiling:

$$Ag^{3+} + H_2O \xrightarrow{\text{boiling}} Ag^+ + \tfrac{1}{2}O_2 + 2H^+$$

Solid *sodium bismuthate* ($NaBiO_3$) has an oxidizing strength similar to that of Ag^+ plus $S_2O_8^{2-}$. Excess solid oxidant is removed by filtration.

Hydrogen peroxide is a good oxidant in basic solution. It can transform Co^{2+} into Co^{3+}, Fe^{2+} into Fe^{3+}, and Mn^{2+} into MnO_2. In acidic solution it can *reduce* $Cr_2O_7^{2-}$ to Cr^{3+} and MnO_4^- to Mn^{2+}. Excess H_2O_2 spontaneously **disproportionates** in boiling water.

$$2H_2O_2 \xrightarrow{\text{boiling}} O_2 + 2H_2O$$

In **disproportionation**, a reactant is converted into products in higher and lower oxidation states. The compound oxidizes and reduces *itself.*

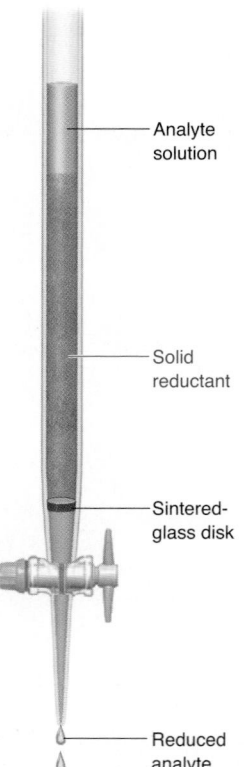

Analyte
solution

Solid
reductant

Sintered-
glass disk

Reduced
analyte

FIGURE 16-6 A column filled with a solid reagent used for prereduction of analyte is called a *reductor*.

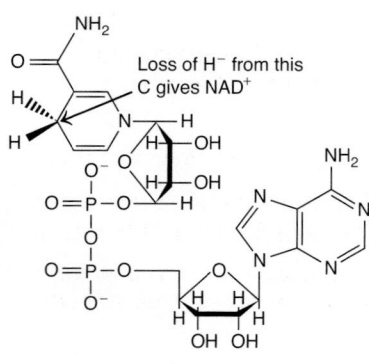

NADH
β-Nicotinamide adenine dinucleotide

NH_2

Loss of H⁻ from this C gives NAD^+

Slow disproportionation of H_2O_2 at room temperature limits the shelf life of household H_2O_2 used as an antiseptic to cleanse wounds.

Prereduction

Stannous chloride ($SnCl_2$) will reduce Fe^{3+} to Fe^{2+} in hot HCl. Excess reductant is destroyed by adding excess $HgCl_2$:

$$Sn^{2+} + 2HgCl_2 \rightarrow Sn^{4+} + Hg_2Cl_2 + 2Cl^-$$

The Fe^{2+} is then titrated with an oxidant.

Chromous chloride is a powerful reductant sometimes used to **prereduce** analyte to a lower oxidation state. Excess Cr^{2+} is oxidized by atmospheric O_2. *Sulfur dioxide* and *hydrogen sulfide* are mild reducing agents that can be expelled by boiling an acidic solution after the reduction is complete.

An important prereduction technique uses a column packed with a solid reducing agent. Figure 16-6 shows the *Jones reductor*, which contains zinc coated with zinc *amalgam*. An **amalgam** is a solution of anything in mercury. The amalgam is prepared by mixing granular zinc with 2 wt% aqueous $HgCl_2$ for 10 min and then washing with water. You can reduce Fe^{3+} to Fe^{2+} by passage through a Jones reductor, using 1 M H_2SO_4 as solvent. Wash the column well with water and titrate the combined washings with standard MnO_4^-, Ce^{4+}, or $Cr_2O_7^{2-}$. Pass a solution containing only the matrix through the reductor in the same manner as the unknown for a blank determination.

Most reduced analytes are reoxidized by atmospheric oxygen. To prevent oxidation, the reduced analyte can be collected in an acidic solution of Fe^{3+}. Ferric ion is reduced to Fe^{2+}, which is stable in acid. The Fe^{2+} is then titrated with an oxidant. By this means, elements such as Cr, Ti, V, and Mo can be indirectly analyzed.

Zinc is a powerful reducing agent, with $E° = -0.764$ for the reaction $Zn^{2+} + 2e^- \rightleftharpoons Zn(s)$, so the Jones reductor is not very selective. The *Walden reductor*, filled with solid Ag and 1 M HCl, is more selective. The reduction potential for Ag | AgCl (0.222 V) is high enough that species such as Cr^{3+} and TiO^{2+} are not reduced and therefore do not interfere in the analysis of a metal such as Fe^{3+}.

Mercury in the Jones reductor or any other form creates a toxic waste hazard, so its use should be minimized or substitute methods should be found. The next paragraphs show how a substitute method for reduction with toxic cadmium was found.

U.S. environmental regulations set a maximum of 10 ppm NO_3^- nitrogen in drinking water. Metallic Cd has been the most common reducing agent used in the measurement of NO_3^- in water. Passing nitrate through a Cd-filled column reduces NO_3^- to NO_2^-, for which a spectrophotometric analysis is available. However, Cd is toxic and creates a hazardous waste threat to the environment.

Therefore, a commercial field test for NO_3^- was developed with the biological reducing agent β-nicotinamide adenine dinucleotide (NADH) instead of Cd. The enzyme nitrate reductase from genetically engineered yeast catalyzes the reduction:

$$NO_3^- + NADH + H^+ \xrightarrow[\text{pH 7}]{\text{nitrate reductase}} NO_2^- + NAD^+ + H_2O$$

Nitrate   Nitrite

Excess NADH is then oxidized to NAD^+ to eliminate interference with color development when NO_2^- is measured by a chemical reaction that forms a colored product. The color intensity is measured in the field in the range 0.05–10 ppm nitrate nitrogen with a battery-operated spectrophotometer. A classroom adaptation of the field test measures NO_3^- in an aquarium.[13]

16-4 Oxidation with Potassium Permanganate

Potassium permanganate ($KMnO_4$) is a strong oxidant with an intense violet color. In strongly acidic solutions (pH ≲ 1), it is reduced to colorless Mn^{2+}.

$$MnO_4^- + 8H^+ + 5e^- \rightleftharpoons Mn^{2+} + 4H_2O \qquad E° = 1.507 \text{ V}$$

Permanganate   Manganous

In neutral or alkaline solution, the product is the brown solid, MnO_2.

$$MnO_4^- + 4H^+ + 3e^- \rightleftharpoons MnO_2(s) + 2H_2O \qquad E° = 1.692 \text{ V}$$

Manganese
dioxide

In strongly alkaline solution (2 M NaOH), green manganate ion is produced.

$$MnO_4^- + e^- \rightleftharpoons MnO_4^{2-} \qquad E° = 0.56 \text{ V}$$
Manganate

Representative permanganate titrations are listed in Table 16-3. For titrations in strongly acidic solution, $KMnO_4$ serves as its own indicator because the product, Mn^{2+}, is colorless (Color Plate 9). The end point is taken as the first persistent appearance of pale pink MnO_4^-. If the titrant is too dilute to be seen, an indicator such as ferroin can be used.

Preparation and Standardization

Potassium permanganate is not a primary standard because traces of MnO_2 are invariably present. In addition, distilled water usually contains enough organic impurities to reduce some freshly dissolved MnO_4^- to MnO_2. To prepare a 0.02 M stock solution, dissolve $KMnO_4$ in distilled water, boil it for an hour to hasten the reaction between MnO_4^- and organic impurities, and filter the resulting mixture through a clean, sintered-glass filter to remove precipitated MnO_2. Do not use filter paper (organic matter!) for the

> $KMnO_4$ serves as its own indicator in acidic solution.

> $KMnO_4$ is not a primary standard.

TABLE 16-3   Analytical applications of permanganate titrations

Species analyzed	Oxidation reaction	Notes
Fe^{2+}	$Fe^{2+} \rightleftharpoons Fe^{3+} + e^-$	Fe^{3+} is reduced to Fe^{2+} with Sn^{2+} or a Jones reductor. Titration is carried out in 1 M H_2SO_4 or 1 M HCl containing Mn^{2+}, H_3PO_4, and H_2SO_4. Mn^{2+} inhibits oxidation of Cl^- by MnO_4. H_3PO_4 complexes Fe^{3+} to prevent formation of yellow Fe^{3+}-chloride complexes.
$H_2C_2O_4$	$H_2C_2O_4 \rightleftharpoons 2CO_2 + 2H^+ + 2e^-$	Add 95% of titrant at 25°C, then complete titration at 55°–60°C.
Br^-	$Br^- \rightleftharpoons \frac{1}{2}Br_2(g) + e^-$	Titrate in boiling 2 M H_2SO_4 to remove $Br_2(g)$.
H_2O_2	$H_2O_2 \rightleftharpoons O_2(g) + 2H^+ + 2e^-$	Titrate in 1 M H_2SO_4.
HNO_2	$HNO_2 + H_2O \rightleftharpoons NO_3^- + 3H^+ + 2e^-$	Add excess standard $KMnO_4$ and back-titrate after 15 min at 40°C with Fe^{2+}.
As^{3+}	$H_3AsO_3 + H_2O \rightleftharpoons H_3AsO_4 + 2H^+ + 2e^-$	Titrate in 1 M HCl with KI or ICl catalyst.
Sb^{3+}	$H_3SbO_3 + H_2O \rightleftharpoons H_3SbO_4 + 2H^+ + 2e^-$	Titrate in 2 M HCl.
Mo^{3+}	$Mo^{3+} + 2H_2O \rightleftharpoons MoO_2^{2+} + 4H^+ + 3e^-$	Reduce Mo in a Jones reductor, and run the Mo^{3+} into excess Fe^{3+} in 1 M H_2SO_4. Titrate the Fe^{2+} formed.
W^{3+}	$W^{3+} + 2H_2O \rightleftharpoons WO_2^{2+} + 4H^+ + 3e^-$	Reduce W with Pb(Hg) at 50°C and titrate in 1 M HCl.
U^{4+}	$U^{4+} + 2H_2O \rightleftharpoons UO_2^{2+} + 4H^+ + 2e^-$	Reduce U to U^{3+} with a Jones reductor. Expose to air to produce U^{4+}, which is titrated in 1 M H_2SO_4.
Ti^{3+}	$Ti^{3+} + H_2O \rightleftharpoons TiO^{2+} + 2H^+ + e^-$	Reduce Ti to Ti^{3+} with a Jones reductor, and run the Ti^{3+}into excess Fe^{3+} in 1 M H_2SO_4. Titrate the Fe^{2+} that is formed.
Mg^{2+}, Ca^{2+}, Sr^{2+}, Ba^{2+}, Zn^{2+}, Co^{2+}, La^{3+}, Th^{4+}, Pb^{2+}, Ce^{3+}, BiO^+, Ag^+	$H_2C_2O_4 \rightleftharpoons 2CO_2 + 2H^+ + 2e^-$	Precipitate the metal oxalate. Dissolve in acid and titrate the $H_2C_2O_4$.
$S_2O_8^{2-}$	$S_2O_8^{2-} + 2Fe^{2+} + 2H^+ \rightleftharpoons 2Fe^{3+} + 2HSO_4^-$	Peroxydisulfate is added to excess standard Fe^{2+} containing H_3PO_4. Unreacted Fe^{2+} is titrated with MnO_4^-.
PO_4^{3-}	$Mo^{3+} + 2H_2O \rightleftharpoons MoO_2^{2+} + 4H^+ + 3e^-$	$(NH_4)_3PO_4 \cdot 12MoO_3$ is precipitated and dissolved in H_2SO_4. The Mo(VI) is reduced (as above) and titrated.

---

filtration. Store the reagent in a dark glass bottle. Aqueous $KMnO_4$ is unstable by virtue of the reaction

$$4MnO_4^- + 2H_2O \rightarrow 4MnO_2(s) + 3O_2 + 4OH^-$$

which is slow in the absence of MnO_2, Mn^{2+}, heat, light, acids, and bases. Permanganate should be standardized often for the most accurate work. Prepare and standardize fresh dilute solutions from 0.02 M stock solution, using water distilled from alkaline $KMnO_4$.

Potassium permanganate can be standardized by titration of sodium oxalate ($Na_2C_2O_4$) by Reaction 7-1 or pure electrolytic iron wire. Dissolve dry (105°C, 2 h) sodium oxalate (available as a 99.9–99.95% pure primary standard) in 1 M H_2SO_4 and treat it with 90–95% of the required $KMnO_4$ solution at room temperature. Then warm the solution to 55–60°C and complete the titration by slow addition of $KMnO_4$. Subtract a blank value to account for the quantity of titrant (usually one drop) needed to impart a pink color to the solution.

If pure Fe wire is used as a standard, dissolve it in warm 1.5 M H_2SO_4 under N_2. The product is Fe^{2+}, and the cooled solution can be used to standardize $KMnO_4$ (or other oxidants) with no special precautions. Adding 5 mL of 86 wt% H_3PO_4 per 100 mL of solution masks the yellow color of Fe^{3+} and makes the end point easier to see. Ferrous ammonium sulfate, $Fe(NH_4)_2(SO_4)_2 \cdot 6H_2O$, and ferrous ethylenediammonium sulfate, $Fe(H_3NCH_2CH_2NH_3)(SO_4)_2 \cdot 2H_2O$, are sufficiently pure to be standards for most purposes.

16-5 Oxidation with Ce^{4+}

Reduction of Ce^{4+} to Ce^{3+} proceeds cleanly in acidic solutions. The aquo ion, $Ce(H_2O)_n^{4+}$, probably does not exist in any of these solutions, because Ce(IV) binds anions (ClO_4^-, SO_4^{2-}, NO_3^-, Cl^-). Variation of the $Ce^{4+} \mid Ce^{3+}$ formal potential with the medium is indicative of these interactions:

Different formal potentials imply that different cerium species are present in each solution.

$$Ce^{4+} + e^- \rightleftharpoons Ce^{3+} \qquad \text{Formal potential} \begin{cases} 1.70 \text{ V in 1 F } HClO_4 \\ 1.61 \text{ V in 1 F } HNO_3 \\ 1.47 \text{ V in 1 F } HCl \\ 1.44 \text{ V in 1 F } H_2SO_4 \end{cases}$$

Ce^{4+} is yellow and Ce^{3+} is colorless, but the color change is not distinct enough for cerium to be its own indicator. Ferroin and other substituted phenanthroline redox indicators (Table 16-2) are well suited to titrations with Ce^{4+}.

Ce^{4+} can be used in place of $KMnO_4$ in most procedures. In the oscillating reaction in Demonstration 15-1, Ce^{4+} oxidizes malonic acid to CO_2 and formic acid:

$$\underset{\text{Malonic acid}}{CH_2(CO_2H)_2} + 2H_2O + 6Ce^{4+} \rightarrow 2CO_2 + \underset{\text{Formic acid}}{HCO_2H} + 6Ce^{3+} + 6H^+$$

This reaction can be used for quantitative analysis of malonic acid by heating a sample in 4 M $HClO_4$ with excess standard Ce^{4+} and back-titrating unreacted Ce^{4+} with Fe^{2+}. Analogous procedures are available for many alcohols, aldehydes, ketones, and carboxylic acids.

Preparation and Standardization

Primary-standard-grade ammonium hexanitratocerate(IV), $(NH_4)_2Ce(NO_3)_6$, can be dissolved in 1 M H_2SO_4 and used directly. Although the oxidizing strength of Ce^{4+} is greater in $HClO_4$ or HNO_3, these solutions undergo slow photochemical decomposition with concomitant oxidation of water. Ce^{4+} in H_2SO_4 is stable indefinitely, even though the reduction potential of 1.44 V is great enough to oxidize H_2O to O_2. Reaction with water is slow, even though it is thermodynamically favorable. Solutions in HCl are unstable because Cl^- is oxidized to Cl_2—rapidly, when the solution is hot. Sulfuric acid solutions of Ce^{4+} can be used for titrations of unknowns in HCl because the reaction with analyte is fast, whereas reaction with Cl^- is slow. Less expensive salts, including $Ce(HSO_4)_4$, $(NH_4)_4Ce(SO_4)_4 \cdot 2H_2O$, and $CeO_2 \cdot xH_2O$ (also called $Ce(OH)_4$), are adequate for preparing titrants that are subsequently standardized with $Na_2C_2O_4$ or Fe as described for MnO_4^-.

16-6 Oxidation with Potassium Dichromate

In acidic solution, orange dichromate ion is a powerful oxidant that is reduced to chromic ion:

$$Cr_2O_7^{2-} + 14H^+ + 6e^- \rightleftharpoons 2Cr^{3+} + 7H_2O \qquad E° = 1.36 \text{ V}$$

Dichromate ⟶ Chromic

Cr(VI) waste is carcinogenic and should not be poured down the drain. See page 25 for disposal method.

In 1 M HCl, the formal potential is just 1.00 V and, in 2 M H_2SO_4, it is 1.11 V; so dichromate is a less powerful oxidizing agent than MnO_4^- or Ce^{4+}. In basic solution, $Cr_2O_7^{2-}$ is converted into yellow chromate ion (CrO_4^{2-}), whose oxidizing power is nil:

$$CrO_4^{2-} + 4H_2O + 3e^- \rightleftharpoons Cr(OH)_3(s, \text{ hydrated}) + 5OH^- \qquad E° = -0.12 \text{ V}$$

Potassium dichromate, $K_2Cr_2O_7$, is a primary standard. Its solutions are stable, and it is cheap. Dichromate is orange and Cr^{3+} complexes range from green to violet, so indicators with distinctive color changes, such as diphenylamine sulfonic acid or diphenylbenzidine sulfonic acid, are used to find a dichromate end point. Alternatively, reactions can be monitored with Pt and calomel electrodes.

$K_2Cr_2O_7$ is not as strong an oxidant as $KMnO_4$ or Ce^{4+}. It is employed chiefly for the determination of Fe^{2+} and, indirectly, for species that will oxidize Fe^{2+} to Fe^{3+}. For indirect analyses, the unknown is treated with a measured excess of Fe^{2+}. Then unreacted Fe^{2+} is titrated with $K_2Cr_2O_7$. For example, ClO_3^-, NO_3^-, MnO_4^-, and organic peroxides can be analyzed this way. Box 16-2 describes the use of dichromate in water pollution analysis.

Diphenylbenzidine sulfonate (reduced, colorless)

Diphenylbenzidine sulfonate (oxidized, violet)
+ 2H⁺ + 2e⁻

16-7 Methods Involving Iodine

When a reducing analyte is titrated with iodine (to give I^-), the method is called *iodimetry*. In *iodometry*, an oxidizing analyte is added to excess I^- to produce iodine, which is then titrated with standard thiosulfate solution.

Iodimetry: titration *with* iodine

Iodometry: titration *of* iodine produced by a chemical reaction

Molecular iodine is only slightly soluble in water (1.3×10^{-3} M at 20°C), but its solubility is enhanced by complexation with iodide.

$$I_2(aq) + I^- \rightleftharpoons I_3^- \qquad K = 7 \times 10^2$$

Iodine   Iodide   Triiodide

A typical 0.05 M solution of I_3^- for titrations is prepared by dissolving 0.12 mol of KI plus 0.05 mol of I_2 in 1 L of water. When we speak of using iodine as a titrant, we almost always mean that we are using a solution of I_2 plus excess I^-.

A solution made from 1.5 mM I_2 + 1.5 mM KI in water contains[23]

0.9 mM I_2	5 μM I_5^-
0.9 mM I^-	40 nM I_6^{2-}
0.6 mM I_3^-	0.3 μM HOI

Use of Starch Indicator

Starch (Figure 16-5) is an indicator for iodine. In a solution with no other colored species, it is possible to see the color of ~5 μM I_3^-. With starch, the limit of detection is ten times lower.

In iodimetry (titration *with* I_3^-), starch can be added at the beginning of the titration. The first drop of excess I_3^- after the equivalence point causes the solution to turn dark blue.

In iodometry (titration *of* I_3^-), I_3^- is present throughout the reaction up to the equivalence point. *Starch should not be added until immediately before the equivalence point* (as detected visually, by fading of the I_3^-; Color Plate 11). Otherwise some iodine tends to remain bound to starch particles after the equivalence point is reached.

Starch-iodine complexation is temperature dependent. At 50°C, the color is only one-tenth as intense as at 25°C. If maximum sensitivity is required, cooling in ice water is recommended.[25] Organic solvents decrease the affinity of iodine for starch and markedly reduce the utility of the indicator.

An alternative to using starch is to add a few milliliters of *p*-xylene to the vigorously stirred titration vessel. After each addition of reagent near the end point, stop stirring long enough to examine the color of the organic phase. I_2 is 400 times more soluble in *p*-xylene than in water, and its color is readily detected in the organic phase.[24]

Preparation and Standardization of I_3^- Solutions

Triiodide (I_3^-) is prepared by dissolving solid I_2 in excess KI. Sublimed I_2 is pure enough to be a primary standard, but it is seldom used as a standard because it evaporates while it is being weighed. Instead, the approximate amount is rapidly weighed, and the solution of I_3^- is standardized with a pure sample of analyte or $Na_2S_2O_3$.

Acidic solutions of I_3^- are unstable because the excess I^- is slowly oxidized by air:

$$6I^- + O_2 + 4H^+ \rightarrow 2I_3^- + 2H_2O$$

There is a significant vapor pressure of toxic I_2 above solid I_2 and aqueous I_3^-. Vessels containing I_2 or I_3^- should be capped and kept in a fume hood. Waste solutions of I_3^- should not be dumped into a sink in the open lab.

In neutral solutions, oxidation is insignificant in the absence of heat, light, and metal ions. At pH ≳ 11, triiodide disproportionates to hypoiodous acid, iodate, and iodide.

HOI   hypoiodous acid
IO_3^- iodate

BOX 16-2 **Environmental Carbon Analysis and Oxygen Demand**

Drinking water and industrial waste streams are partially characterized and regulated on the basis of their carbon content and oxygen demand.[14] *Inorganic carbon* (IC) is the $CO_2(g)$ liberated when water is acidified to pH < 2 with H_3PO_4 and purged with Ar or N_2. IC corresponds to CO_3^{2-} and HCO_3^{-} in the sample. After inorganic carbon is removed, *total organic carbon* (TOC) is the CO_2 produced by oxidizing the remaining organic matter:

TOC analysis: Organic carbon $\xrightarrow[\text{metal catalyst}]{O_2/\sim700°C}$ CO_2

Total carbon (TC) is defined as the sum TC = TOC + IC.

Different oxidation techniques produce different values for TOC, because not all organic matter is oxidized by each technique. The state of the art is such that TOC is defined by the result obtained with a particular instrument.

Commercial instruments that measure TOC by thermal oxidation have detection limits of 4–50 ppb (4–50 µg C/L). A typical 20-µL water sample is analyzed in 3 min using infrared absorption to measure CO_2. Other instruments oxidize organic matter by irradiating a suspension of solid TiO_2 catalyst (0.2 g/L) in water at pH 3.5 with ultraviolet light.[15] Light creates electron-hole pairs (Section 15-8) in the TiO_2. Holes oxidize H_2O to hydroxyl radical (HO·), a powerful oxidant that converts organic carbon into CO_2, which is measured by the electrical conductivity of carbonic acid. (Pure TiO_2 hardly absorbs visible light, so it cannot use sunlight efficiently. By doping TiO_2 with ~1 wt% carbon, the efficiency of using visible light is markedly increased.[16]) Color Plate 10 shows an instrument in which $K_2S_2O_8$ (potassium persulfate) in acid is exposed to ultraviolet radiation to generate sulfate radical (SO_4^{-}), which oxidizes organic matter to CO_2. Other instruments generate sulfate radical simply by heating the aqueous $K_2S_2O_8$ to 100°C.

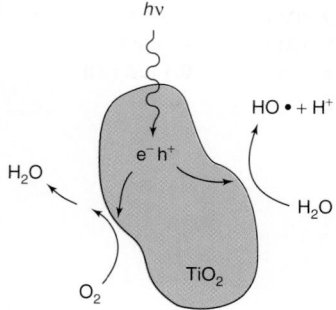

Ultraviolet radiation absorbed by TiO_2 creates an electron-hole pair. The hole oxidizes H_2O to the powerful oxidant HO·. The electron reduces O_2 to H_2O by a chain of reactions. TiO_2 is a catalyst and O_2 is consumed in the net reaction: organic C + O_2 → CO_2.

TOC is widely used to determine compliance with discharge laws. Municipal and industrial wastewater typically has TOC > 1 mg C/mL. Tap water TOC is 50–500 ng C/mL. High-purity water for the electronics industry has TOC < 1 ng C/mL.

Total oxygen demand (TOD) tells us how much O_2 is required for complete combustion of pollutants in a waste stream.[18] A volume of N_2 containing a known quantity of O_2 is mixed with the sample and complete combustion is carried out by passage through a catalyst at 900°C. The remaining O_2 is measured by an electronic sensor. Different species in the waste stream consume different amounts of O_2. For example, urea consumes five times as much O_2 as formic acid does. Species such as NH_3 and H_2S also contribute to TOD.

Pollutants can be oxidized by refluxing with dichromate ($Cr_2O_7^{2-}$). *Chemical oxygen demand* (COD) is defined as the O_2 that is chemically equivalent to the $Cr_2O_7^{2-}$ consumed in this process. Each $Cr_2O_7^{2-}$ consumes $6e^-$ (to make $2Cr^{3+}$) and each O_2 consumes $4e^-$ (to make H_2O). Therefore, 1 mol of $Cr_2O_7^{2-}$ is chemically equivalent to 1.5 mol of O_2 for this computation. COD analysis is carried out by

refluxing polluted water for 2 h with excess standard $Cr_2O_7^{2-}$ in H_2SO_4 solution containing Ag^+ catalyst. Unreacted $Cr_2O_7^{2-}$ is measured by titration with standard Fe^{2+} or by spectrophotometry. Permits for industry may include COD limits for waste streams. "Oxidizability," which is used in Europe, is analogous to COD. Oxidizability is measured by refluxing with permanganate in acid solution at 100°C for 10 min. Each MnO_4^{-} consumes five electrons and is chemically equivalent to 1.25 O_2. Electrochemical methods based on photooxidation with TiO_2 could replace cumbersome refluxing with $Cr_2O_7^{2-}$ or MnO_4^{-}. Problem 17-24 describes one proven method.

Biochemical oxygen demand (BOD) is the O_2 required for degradation of organic matter by aquatic microorganisms.[19,20] A healthy body of natural water could have a BOD of 2–3 mg O_2/L. Polluted water could require 10–30 mg O_2/L and sewage could have a BOD > 1 000 mg O_2/L. The procedure calls for incubating a sealed container of wastewater with no extra air space for 5 days at 20°C in the dark while microbes metabolize organic compounds in the waste. Dissolved O_2 is measured with a Clark electrode (Box 17-2) before and after incubation. The difference is BOD. BOD also measures species such as HS^- and Fe^{2+} that may be in the water. Inhibitors are added to prevent oxidation of nitrogen species such as NH_3. There is interest in developing a rapid analysis to provide information equivalent to BOD. This goal could be achieved by substituting ferricyanide ($Fe(CN)_6^{3-}$) for O_2 as the electron sink for bacterial degradation of organic matter. Ferricyanide requires just 3 h, and results are similar to that of the 5-day standard procedure.[21]

Bound nitrogen includes all nitrogen-containing compounds, except N_2, dissolved in water. Kjeldahl nitrogen analysis described in Section 11-8 is excellent for amines and amides but fails to respond to many other forms of nitrogen. Combustion can convert most forms of nitrogen in aqueous samples into NO, which can be measured by chemiluminescence after reaction with ozone:[22]

Bound nitrogen analysis: Nitrogen compounds $\xrightarrow[\text{catalyst}]{O_2/\sim1\,000°C}$ NO

$$\underset{\text{Nitrous oxide}}{NO} + \underset{\text{Ozone}}{O_3} \rightarrow \underset{\substack{\text{Nitric oxide}\\\text{Excited electronic state}}}{NO_2^*}$$

$$NO_2^* \rightarrow NO_2 + \underset{\substack{\text{Characteristic}\\\text{light emission}}}{h\nu}$$

Azide (N_3^-) and hydrazines ($RNHNH_2$) are not quantitatively converted into NO by combustion. Bound nitrogen measurements are required for compliance with wastewater discharge regulations.

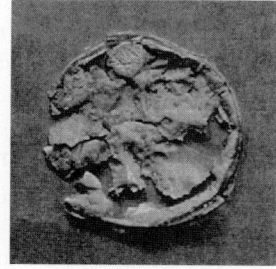

TiO_2-blended PVC before irradiation   After irradiation for 20 days

Here is a "green" idea: TiO_2 can be blended into poly(vinyl chloride) (PVC) plastic so that the plastic is degraded by sunlight.[17] Ordinary PVC lasts many years in municipal landfills after it is discarded. TiO_2-blended PVC would decompose in a short time. [Courtesy H. Hidaka and S. Horikoshi, Meisei University, Tokyo.]

An excellent way to prepare standard I_3^- is to add a weighed quantity of the primary standard potassium iodate (KIO_3) to a small excess of KI.[26] Then add excess strong acid (giving $pH \approx 1$) to produce I_3^- by quantitative reverse disproportionation:

$$IO_3^- + 8I^- + 6H^+ \rightleftharpoons 3I_3^- + 3H_2O \qquad \text{(16-18)}$$

Freshly acidified iodate plus iodide can be used to standardize thiosulfate. I_3^- must be used immediately or it will be oxidized by air. The disadvantage of KIO_3 is its low molecular mass relative to the number of electrons it accepts, giving a larger-than-desirable relative weighing error in preparing solutions.

Use of Sodium Thiosulfate

Sodium thiosulfate is the almost universal titrant for triiodide. In neutral or acidic solution, triiodide oxidizes thiosulfate to tetrathionate:

$$I_3^- + 2S_2O_3^{2-} \rightleftharpoons 3I^- + O=\overset{\displaystyle O}{\underset{\displaystyle O^-}{\overset{\|}{\underset{|}{S}}}}-S-S-\overset{\displaystyle O}{\underset{\displaystyle O^-}{\overset{\|}{\underset{|}{S}}}}=O \qquad \text{(16-19)}$$

Thiosulfate   Tetrathionate

One mole of I_3^- in Reaction 16-19 is equivalent to one mole of I_2. I_2 and I_3^- are interchangeable through the equilibrium $I_2 + I^- \rightleftharpoons I_3^-$. In basic solution, I_3^- disproportionates to I^- and HOI, which can oxidize $S_2O_3^{2-}$ to SO_4^{2-}. Reaction 16-19 should be carried out below pH 9. Acidic solutions of I_3^- can be titrated, but should be protected from air oxidation by blanketing with N_2. The common form of thiosulfate, $Na_2S_2O_3 \cdot 5H_2O$, is not pure enough to be a primary standard. Instead, thiosulfate is usually standardized by reaction with a fresh solution of I_3^- prepared from KIO_3 plus KI.

Anhydrous, primary standard $Na_2S_2O_3$ can be prepared from the pentahydrate.[27]

A stable solution of $Na_2S_2O_3$ can be prepared by dissolving the reagent in high-quality, freshly boiled distilled water. Dissolved CO_2 makes the solution acidic and promotes disproportionation of $S_2O_3^{2-}$:

$$S_2O_3^{2-} + H^+ \rightleftharpoons HSO_3^- + S(s) \qquad \text{(16-20)}$$

Bisulfite   Sulfur

and metal ions catalyze atmospheric oxidation of thiosulfate:

$$2Cu^{2+} + 2S_2O_3^{2-} \rightarrow 2Cu^+ + S_4O_6^{2-}$$

$$2Cu^+ + \frac{1}{2}O_2 + 2H^+ \rightarrow 2Cu^{2+} + H_2O$$

Thiosulfate solutions should be stored in the dark. Addition of 0.1 g of sodium carbonate per liter maintains the pH in an optimum range for stability of the solution. Three drops of chloroform added to each bottle of thiosulfate solution helps prevent bacterial growth. An acidic solution of thiosulfate is unstable, but the reagent can be used to titrate I_3^- in acidic solution because the reaction with triiodide is faster than Reaction 16-20.

Analytical Applications of Iodine

Reducing agents can be titrated directly with standard I_3^- in the presence of starch, until reaching the intense blue starch-iodine end point (Table 16-4). An example is the iodimetric determination of vitamin C:

Reducing agent + $I_3^- \rightarrow 3I^-$

Ascorbic acid   Dehydroascorbic
(vitamin C)   acid[28]

Oxidizing agents can be treated with excess I^- to produce I_3^- (Table 16-5, Box 16-3). The iodometric analysis is completed by titrating the liberated I_3^- with standard thiosulfate. Starch is not added until just before the end point.

Oxidizing agent + $3I^- \rightarrow I_3^-$

Species analyzed	Oxidation reaction	Notes
As^{3+}	$H_3AsO_3 + H_2O \rightleftharpoons H_3AsO_4 + 2H^+ + 2e^-$	Titrate directly in $NaHCO_3$ solution with I_3^-.
Sn^{2+}	$SnCl_4^{2-} + 2Cl^- \rightleftharpoons SnCl_6^{2-} + 2e^-$	Sn(IV) is reduced to Sn(II) with granular Pb or Ni in 1 M HCl and titrated in the absence of oxygen.
N_2H_4	$N_2H_4 \rightleftharpoons N_2 + 4H^+ + 4e^-$	Titrate in $NaHCO_3$ solution.
SO_2	$SO_2 + H_2O \rightleftharpoons H_2SO_3$ $H_2SO_3 + H_2O \rightleftharpoons SO_4^{2-} + 4H^+ + 2e^-$	Add SO_2 (or H_2SO_3 or HSO_3^- or SO_3^{2-}) to excess standard I_3^- in dilute acid and back-titrate unreacted I_3^- with standard thiosulfate.
H_2S	$H_2S \rightleftharpoons S(s) + 2H^+ + 2e^-$	Add H_2S to excess I_3^- in 1 M HCl and back-titrate with thiosulfate.
$Zn^{2+}, Cd^{2+}, Hg^{2+}, Pb^{2+}$	$M^{2+} + H_2S \rightarrow MS(s) + 2H^+$ $MS(s) \rightleftharpoons M^{2+} + S + 2e^-$	Precipitate and wash metal sulfide. Dissolve in 3 M HCl with excess standard I_3^- and back-titrate with thiosulfate.
Cysteine, glutathione, thioglycolic acid, mercaptoethanol	$2RSH \rightleftharpoons RSSR + 2H^+ + 2e^-$	Titrate the sulfhydryl compound at pH 4–5 with I_3^-.
HCN	$I_2 + HCN \rightleftharpoons ICN + I^- + H^+$	Titrate in carbonate-bicarbonate buffer, using *p*-xylene as an extraction indicator.
$H_2C{=}O$	$H_2CO + 3OH^- \rightleftharpoons HCO_2^- + 2H_2O + 2e^-$	Add excess I_3^- plus NaOH to the unknown. After 5 min, add HCl and back-titrate with thiosulfate.
Glucose (and other reducing sugars)	$\overset{\displaystyle O}{\overset{\displaystyle \|}{R}}CH + 3OH^- \rightleftharpoons RCO_2^- + 2H_2O + 2e^-$	Add excess I_3^- plus NaOH to the sample. After 5 min, add HCl and back-titrate with thiosulfate.
Ascorbic acid (vitamin C)	$Ascorbate + H_2O \rightleftharpoons dehydroascorbate + 2H^+ + 2e^-$	Titrate directly with I_3^-.
H_3PO_3	$H_3PO_3 + H_2O \rightleftharpoons H_3PO_4 + 2H^+ + 2e^-$	Titrate in $NaHCO_3$ solution.

Species analyzed	Reaction	Notes
Cl_2	$Cl_2 + 3I^- \rightleftharpoons 2Cl^- + I_3^-$	Reaction in dilute acid.
HOCl	$HOCl + H^+ + 3I^- \rightleftharpoons Cl^- + I_3^- + H_2O$	Reaction in 0.5 M H_2SO_4.
Br_2	$Br_2 + 3I^- \rightleftharpoons 2Br^- + I_3^-$	Reaction in dilute acid.
BrO_3^-	$BrO_3^- + 6H^+ + 9I^- \rightleftharpoons Br^- + 3I_3^- + 3H_2O$	Reaction in 0.5 M H_2SO_4.
IO_3^-	$2IO_3^- + 16I^- + 12H^+ \rightleftharpoons 6I_3^- + 6H_2O$	Reaction in 0.5 M HCl.
IO_4^-	$2IO_4^- + 22I^- + 16H^+ \rightleftharpoons 8I_3^- + 8H_2O$	Reaction in 0.5 M HCl.
O_2	$O_2 + 4Mn(OH)_2 + 2H_2O \rightleftharpoons 4Mn(OH)_3$ $2Mn(OH)_3 + 6H^+ + 3I^- \rightleftharpoons 2Mn^{2+} + I_3^- + 6H_2O$	The sample is treated with Mn^{2+}, NaOH, and KI. After 1 min, it is acidified with H_2SO_4, and the I_3^- is titrated.
H_2O_2	$H_2O_2 + 3I^- + 2H^+ \rightleftharpoons I_3^- + 2H_2O$	Reaction in 1 M H_2SO_4.
O_3^a	$O_3 + 3I^- + 2H^+ \rightleftharpoons O_2 + I_3^- + H_2O$	O_3 is passed through neutral 2 wt% KI solution. Add H_2SO_4 and titrate.
NO_2^-	$2HNO_2 + 2H^+ + 3I^- \rightleftharpoons 2NO + I_3^- + 2H_2O$	The nitric oxide is removed (by bubbling CO_2 generated in situ) prior to titration of I_3^-.
As^{5+}	$H_3AsO_4 + 2H^+ + 3I^- \rightleftharpoons H_3AsO_3 + I_3^- + H_2O$	Reaction in 5 M HCl.
$S_2O_8^{2-}$	$S_2O_8^{2-} + 3I^- \rightleftharpoons 2SO_4^{2-} + I_3^-$	Reaction in neutral solution. Then acidify and titrate.
Cu^{2+}	$2Cu^{2+} + 5I^- \rightleftharpoons 2CuI(s) + I_3^-$	NH_4HF_2 is used as a buffer.
$Fe(CN)_6^{3-}$	$2Fe(CN)_6^{3-} + 3I^- \rightleftharpoons 2Fe(CN)_6^{4-} + I_3^-$	Reaction in 1 M HCl.
MnO_4^-	$2MnO_4^- + 16H^+ + 15I^- \rightleftharpoons 2Mn^{2+} + 5I_3^- + 8H_2O$	Reaction in 0.1 M HCl.
MnO_2	$MnO_2(s) + 4H^+ + 3I^- \rightleftharpoons Mn^{2+} + I_3^- + 2H_2O$	Reaction in 0.5 M H_3PO_4 or HCl.
$Cr_2O_7^{2-}$	$Cr_2O_7^{2-} + 14H^+ + 9I^- \rightleftharpoons 2Cr^{3+} + 3I_3^- + 7H_2O$	Reaction in 0.4 M HCl requires 5 min for completion and is particularly sensitive to air oxidation.
Ce^{4+}	$2Ce^{4+} + 3I^- \rightleftharpoons 2Ce^{3+} + I_3^-$	Reaction in 1 M H_2SO_4.

*a. The pH must be $\gtrsim$ 7 when O_3 is added to I^-. In acidic solution, each O_3 produces 1.25 I_3^-, not 1 I_3^-. [N. V. Klassen, D. Marchington, and H. C. E. McGowan, Anal. Chem. **1994**, 66, 2921.]*

BOX 16-3 **Iodometric Analysis of High-Temperature Superconductors**

An important application of superconductors is in powerful electromagnets needed for medical magnetic resonance imaging. Ordinary conductors in such magnets require a huge amount of electric power. Because electricity moves through a superconductor with no resistance, the voltage can be removed from the electromagnetic coil once the current has started. Current continues to flow with no power consumption.

A breakthrough in superconductor technology came with the discovery[29] of yttrium barium copper oxide, $YBa_2Cu_3O_7$, whose crystal structure is shown here. When heated, the material readily loses oxygen atoms from the Cu—O chains, and any composition between $YBa_2Cu_3O_7$ and $YBa_2Cu_3O_6$ is observable.

When high-temperature superconductors were discovered, the oxygen content in the formula $YBa_2Cu_3O_x$ was unknown. $YBa_2Cu_3O_7$ represents an unusual set of oxidation states. Common states of yttrium and barium are Y^{3+} and Ba^{2+}, and the common states of copper are Cu^{2+} and Cu^+. If all the copper were Cu^{2+}, the formula of the superconductor would be $(Y^{3+})(Ba^{2+})_2(Cu^{2+})_3(O^{2-})_{6.5}$, with a cation charge of +13 and an anion charge of −13. The composition $YBa_2Cu_3O_7$ requires Cu^{3+}, which is rather rare. Formally, $YBa_2Cu_3O_7$ can be thought of as $(Y^{3+})(Ba^{2+})_2(Cu^{2+})_2(Cu^{3+})(O^{2-})_7$ with a cation charge of +14 and an anion charge of −14.

Redox titrations proved to be the most reliable way to measure the oxidation state of copper and thereby deduce the oxygen content of $YBa_2Cu_3O_x$.[30] An iodometric method includes two experiments.

In *Experiment A*, $YBa_2Cu_3O_x$ is dissolved in dilute acid, in which Cu^{3+} is converted into Cu^{2+}. For simplicity, we write reactions for $YBa_2Cu_3O_7$, but you could balance these equations for $x \neq 7$.[31]

$$YBa_2Cu_3O_7 + 13H^+ \rightarrow$$

$$Y^{3+} + 2Ba^{2+} + 3Cu^{2+} + \frac{13}{2}H_2O + \frac{1}{4}O_2 \quad (1)$$

The total copper content is measured by treatment with iodide

$$Cu^{2+} + \frac{5}{2}I^- \rightarrow CuI(s) + \frac{1}{2}I_3^- \quad (2)$$

and titration of the liberated triiodide with standard thiosulfate by Reaction 16-19. Each mole of Cu in $YBa_2Cu_3O_7$ is equivalent to 1 mol of $S_2O_3^{2-}$ in Experiment A.

In *Experiment B*, $YBa_2Cu_3O_x$ is dissolved in dilute acid containing I^-. Each mole of Cu^{2+} produces 0.5 mol of I_3^- by Reaction (2) and each mole of Cu^{3+} produces 1 mol of I_3^-:

$$Cu^{3+} + 4I^- \rightarrow CuI(s) + I_3^- \quad (3)$$

The moles of $S_2O_3^{2-}$ required in Experiment A equal the total moles of Cu in the superconductor. The difference in $S_2O_3^{2-}$ required between Experiments B and A gives the Cu^{3+} content. From this difference, you can find x in the formula $YBa_2Cu_3O_x$.[32]

Although we can balance cation and anion charges in the formula $YBa_2Cu_3O_7$ by including Cu^{3+} in the formula, there is no evidence for discrete Cu^{3+} ions in the crystal. There is also no evidence that some of the oxygen is in the form of peroxide, O_2^{2-}, which also would balance the cation and anion charges. The best description of the valence state in the solid crystal involves electrons and holes delocalized in the Cu—O planes and chains. Nonetheless, the formal designation of Cu^{3+} and Reactions 1 through 3 accurately describe the redox chemistry of $YBa_2Cu_3O_7$. Problem 16-37 describes titrations that separately measure oxidation numbers of Cu and Bi in the superconductor $Bi_2Sr_2(Ca_{0.8}Y_{0.2})Cu_2O_{8.295}$.

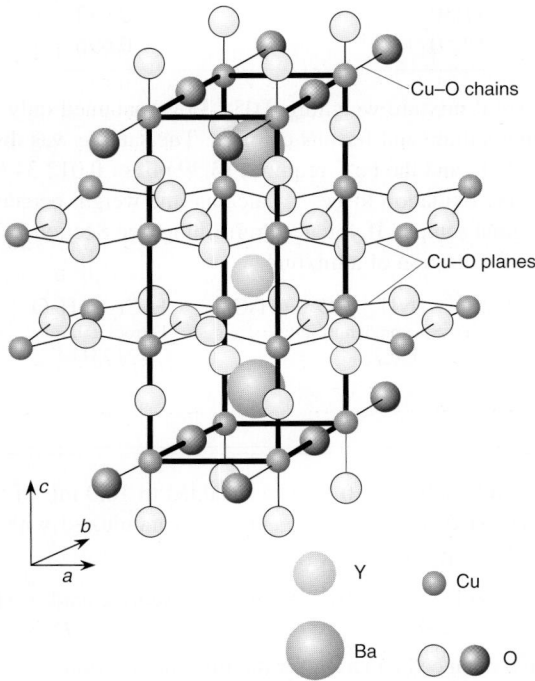

Cu—O chains

Cu—O planes

Y

Cu

Ba

O

c

b

a

Structure of $YBa_2Cu_3O_7$. One-dimensional Cu—O chains (shown in color) run along the crystallographic b-axis and two-dimensional Cu—O sheets lie in the a–b plane. Loss of colored oxygen atoms from the chains at elevated temperature results in $YBa_2Cu_3O_6$. [Information from G. F. Holland and A. M. Stacy, "Physical Properties of the Quaternary Oxide Superconductor $YBa_2Cu_3O_x$," *Acc. Chem. Res.* **1988**, *21*, 8.]

Terms to Understand

amalgam   preoxidation   redox indicator
disproportionation   prereduction   redox titration

Summary

Redox titrations are based on an oxidation-reduction reaction between analyte and titrant. Sometimes a quantitative chemical preoxidation (with reagents such as $S_2O_8^{2-}$, $Ag^IAg^{III}O_2$, $NaBiO_3$, or H_2O_2) or prereduction (with reagents such as $SnCl_2$, $CrCl_2$, SO_2, H_2S, or a metallic reductor column) is necessary to adjust the oxidation state of the analyte prior to analysis. The end point of a redox titration is commonly detected by potentiometry or with a redox indicator. A useful indicator must have a transition range

$(= E°(\text{indicator}) \pm 0.059\ 16/n$ V) that overlaps the abrupt change in potential of the titration curve.

The greater the difference in reduction potential between analyte and titrant, the sharper will be the end point. Plateaus before and after the equivalence point are centered near $E°(\text{analyte})$ and $E°(\text{titrant})$. Prior to the equivalence point, the half-reaction involving analyte is used to find the voltage because the concentrations of both the oxidized and the reduced forms of analyte are known. After the equivalence point, the half-reaction involving titrant is employed. At the equivalence point, both half-reactions are used to find the voltage.

Common oxidizing titrants include $KMnO_4$, Ce^{4+}, and $K_2Cr_2O_7$. Many procedures are based on oxidation with I_3^- or titration of I_3^- liberated in a chemical reaction.

Exercises

16-A. A 20.0-mL solution of 0.005 00 M Sn^{2+} in 1 M HCl was titrated with 0.020 0 M Ce^{4+} to give Sn^{4+} and Ce^{3+}. Calculate the potential (versus S.C.E.) at the following volumes of Ce^{4+}: 0.100, 1.00, 5.00, 9.50, 10.00, 10.10, and 12.00 mL. Sketch the titration curve.

16-B. Would indigo tetrasulfonate be a suitable redox indicator for the titration of $Fe(CN)_6^{4-}$ with Tl^{3+} in 1 M HCl? (*Hint:* The potential at the equivalence point must be between the potentials for each redox couple.)

16-C. Compute the titration curve for Demonstration 16-1, in which 400.0 mL of 3.75 mM Fe^{2+} are titrated with 20.0 mM MnO_4^- at a *fixed pH* of 0.00 in 1 M H_2SO_4. Calculate the potential versus S.C.E. at titrant volumes of 1.0, 7.5, 14.0, 15.0, 16.0, and 30.0 mL and sketch the titration curve.

16-D. A titration of 50.0 mL of unknown Fe^{2+} with 0.100 M Ce^{4+} at 25°C, monitored with Pt and calomel electrodes, gave data in the table.[9] Prepare a Gran plot and decide which data lie on a straight line. Find the *x*-intercept of this line, which is the equivalence volume. Calculate the molarity of Fe^{2+} in the unknown.

Titrant volume, V (mL)	E (volts)
6.50	0.635
8.50	0.651
10.50	0.669
11.50	0.680
12.50	0.696

16-E. A solid mixture weighing 0.054 85 g contained only ferrous ammonium sulfate and ferrous chloride. The sample was dissolved in 1 M H_2SO_4, and the Fe^{2+} required 13.39 mL of 0.012 34 M Ce^{4+} for complete oxidation to Fe^{3+}. Calculate the weight percent of Cl in the original sample. If you need refreshing, see Section 7-2 for an example of a titration of a mixture.

$$FeSO_4 \cdot (NH_4)_2SO_4 \cdot 6H_2O \qquad FeCl_2 \cdot 6H_2O$$

Ferrous ammonium sulfate	Ferrous chloride
FM 392.13	FM 234.84

Problems

Shape of a Redox Titration Curve

16-1. Consider the titration in Figure 16-2.

(a) Write a balanced titration reaction.

(b) Write two different half-reactions for the indicator electrode.

(c) Write two different Nernst equations for the cell voltage.

(d) Calculate E at the following volumes of Ce^{4+}: 10.0, 25.0, 49.0, 50.0, 51.0, 60.0, and 100.0 mL. Compare your results with Figure 16-2.

16-2. Consider the titration of 100.0 mL of 0.010 0 M Ce^{4+} in 1 M $HClO_4$ by 0.040 0 M Cu^+ to give Ce^{3+} and Cu^{2+}, using Pt and saturated Ag | AgCl electrodes to find the end point.

(a) Write a balanced titration reaction.

(b) Write two different half-reactions for the indicator electrode.

(c) Write two different Nernst equations for the cell voltage.

(d) Calculate E at the following volumes of Cu^+: 1.00, 12.5, 24.5, 25.0, 25.5, 30.0, and 50.0 mL. Sketch the titration curve.

16-3. Consider the titration of 25.0 mL of 0.010 0 M Sn^{2+} by 0.050 0 M Tl^{3+} in 1 M HCl, using Pt and saturated calomel electrodes to find the end point.

(a) Write a balanced titration reaction.

(b) Write two different half-reactions for the indicator electrode.

(c) Write two different Nernst equations for the cell voltage.

(d) Calculate E at the following volumes of Tl^{3+}: 1.00, 2.50, 4.90, 5.00, 5.10, and 10.0 mL. Sketch the titration curve.

16-4. Ascorbic acid (0.010 0 M) was added to 10.0 mL of 0.020 0 M Fe^{3+} at pH 0.30, and the potential was monitored with Pt and saturated Ag | AgCl electrodes.

$$\text{Dehydroascorbic acid} + 2H^+ + 2e^- \rightleftharpoons \text{ascorbic acid} + H_2O$$
$$E° = 0.390\ V$$

(a) Write a balanced equation for the titration reaction.

(b) Using $E° = 0.767$ V for the $Fe^{3+} | Fe^{2+}$ couple, calculate the cell voltage when 5.0, 10.0, and 15.0 mL of ascorbic acid have been added. (*Hint:* Refer to the calculations in Demonstration 16-1.)

16-5. Consider the titration of 25.0 mL of 0.050 0 M Sn^{2+} with 0.100 M Fe^{3+} in 1 M HCl to give Fe^{2+} and Sn^{4+}, using Pt and calomel electrodes.

(a) Write a balanced titration reaction.

(b) Write two half-reactions for the indicator electrode.

(c) Write two Nernst equations for the cell voltage.

(d) Calculate E at the following volumes of Fe^{3+}: 1.0, 12.5, 24.0, 25.0, 26.0, and 30.0 mL. Sketch the titration curve.

Finding the End Point

16-6. Select indicators from Table 16-2 that would be suitable for finding the end point in Figure 16-3. What color changes would be observed?

16-7. Would tris(2,2'-bipyridine)iron be a useful indicator for the titration of Sn^{2+} with $Mn(EDTA)^-$? (*Hint:* The potential at the equivalence point must be between the potentials for each redox couple.)

Adjustment of Analyte Oxidation State

16-8. Explain what we mean by *preoxidation* and *prereduction*. Why is it important to be able to destroy the reagents used for these purposes?

16-9. Write balanced reactions for the destruction of $S_2O_8^{2-}$, Ag^{3+}, and H_2O_2 by boiling.

16-10. What is a Jones reductor and what is it used for?

16-11. Why don't Cr^{3+} and TiO^{2+} interfere in the analysis of Fe^{3+} when a Walden reductor, instead of a Jones reductor, is used for prereduction?

Redox Reactions of KMnO₄, Ce(IV), and K₂Cr₂O₇

16-12. From information in Table 16-3, explain how you would use $KMnO_4$ to find the content of $(NH_4)_2S_2O_8$ in a solid mixture with $(NH_4)_2SO_4$. What is the purpose of phosphoric acid in the procedure?

16-13. Write balanced half-reactions in which MnO_4^- acts as an oxidant at

(a) pH = 0;   **(b)** pH = 10;   **(c)** pH = 15.

16-14. When 25.00 mL of unknown were passed through a Jones reductor, molybdate ion (MoO_4^{2-}) was converted into Mo^{3+}. The filtrate required 16.43 mL of 0.010 33 M $KMnO_4$ to reach the purple end point.

$$MnO_4^- + Mo^{3+} \rightarrow Mn^{2+} + MoO_2^{2+}$$

A blank required 0.04 mL. Balance the reaction and find the molarity of molybdate in the unknown.

16-15. A 25.00-mL volume of commercial hydrogen peroxide solution was diluted to 250.0 mL in a volumetric flask. Then 25.00 mL of the diluted solution were mixed with 200 mL of water and 20 mL of 3 M H_2SO_4 and titrated with 0.021 23 M $KMnO_4$. The first pink color was observed with 27.66 mL of titrant. A blank prepared from water in place of H_2O_2 required 0.04 mL to give visible pink color. Using the H_2O_2 reaction in Table 16-3, find the molarity of the commercial H_2O_2.

16-16. Two possible reactions of MnO_4^- with H_2O_2 to produce O_2 and Mn^{2+} are

Scheme 1:	$MnO_4^- \rightarrow Mn^{2+}$
	$H_2O_2 \rightarrow O_2$
Scheme 2:	$MnO_4^- \rightarrow O_2 + Mn^{2+}$
	$H_2O_2 \rightarrow H_2O$

(a) Complete the half-reactions for both schemes by adding e^-, H_2O, and H^+ and write a balanced net equation for each scheme.

(b) Sodium peroxyborate tetrahydrate, $NaBO_3 \cdot 4H_2O$ (FM 153.86), produces H_2O_2 when dissolved in acid: $BO_3^- + 2H_2O \rightarrow H_2O_2 + H_2BO_3^-$. To decide whether Scheme 1 or Scheme 2 occurs, students at the U.S. Naval Academy[33] weighed 1.023 g $NaBO_3 \cdot 4H_2O$ into a 100-mL volumetric flask, added 20 mL of 1 M H_2SO_4, and diluted to the mark with H_2O. Then they titrated 10.00 mL of this solution with 0.010 46 M $KMnO_4$ until the first pale pink color persisted. How many mL of $KMnO_4$ are required in Scheme 1 and in Scheme 2? (The Scheme 1 stoichiometry was observed.)

16-17. A 50.00-mL sample containing La^{3+} was treated with sodium oxalate to precipitate $La_2(C_2O_4)_3$, which was washed, dissolved in acid, and titrated with 18.04 mL of 0.006 363 M $KMnO_4$. Write the titration reaction and find $[La^{3+}]$ in the unknown.

16-18. Aqueous glycerol solution weighing 100.0 mg was treated with 50.0 mL of 0.083 7 M Ce^{4+} in 4 M $HClO_4$ at 60°C for 15 min to oxidize glycerol to formic acid.

$$
\begin{array}{cc}
\mathrm{CH_2-CH-CH_2} & \\
|\quad\ |\quad\ | & \mathrm{HCO_2H} \\
\mathrm{OH\ \ \ OH\ \ \ OH} & \\
\text{Glycerol} & \text{Formic acid} \\
\text{FM 92.095} &
\end{array}
$$

The excess Ce^{4+} required 12.11 mL of 0.044 8 M Fe^{2+} to reach a ferroin end point. Find wt% glycerol in the unknown.

16-19. Nitrite (NO_2^-) can be determined by oxidation with excess Ce^{4+}, followed by back titration of unreacted Ce^{4+}. A 4.030-g sample of solid containing only $NaNO_2$ (FM 68.995) and $NaNO_3$ was dissolved in 500.0 mL. A 25.00-mL sample of this solution was treated with 50.00 mL of 0.118 6 M Ce^{4+} in strong acid for 5 min, and excess Ce^{4+} was back-titrated with 31.13 mL of 0.042 89 M ferrous ammonium sulfate.

$$2Ce^{4+} + NO_2^- + H_2O \rightarrow 2Ce^{3+} + NO_3^- + 2H^+$$
$$Ce^{4+} + Fe^{2+} \rightarrow Ce^{3+} + Fe^{3+}$$

What is the formula for ferrous ammonium sulfate? Calculate wt% $NaNO_2$ in the solid.

16-20. Calcium fluorapatite ($Ca_{10}(PO_4)_6F_2$, FM 1 008.6) laser crystals were doped with chromium to improve their efficiency. It was suspected that the chromium could be in the +4 oxidation state.[2]

1. To measure the total oxidizing power of chromium in the material, a crystal was dissolved in 2.9 M $HClO_4$ at 100°C, cooled to 20°C, and titrated with standard Fe^{2+}, using Pt and Ag | AgCl electrodes to find the end point. Chromium above the +3 state should oxidize an equivalent amount of Fe^{2+} in this step. That is, Cr^{4+} would consume one Fe^{2+}, and Cr^{6+} in $Cr_2O_7^{2-}$ would consume three Fe^{2+}:

$$Cr^{4+} + Fe^{2+} \rightarrow Cr^{3+} + Fe^{3+}$$
$$\tfrac{1}{2}Cr_2O_7^{2-} + 3Fe^{2+} \rightarrow Cr^{3+} + 3Fe^{3+}$$

2. In a second step, the total chromium content was measured by dissolving a crystal in 2.9 M $HClO_4$ at 100°C and cooling to 20°C. Excess $S_2O_8^{2-}$ and Ag^+ were then added to oxidize all chromium to $Cr_2O_7^{2-}$. Unreacted $S_2O_8^{2-}$ was destroyed by boiling, and the remaining solution was titrated with standard Fe^{2+}. In this step, each Cr in the original unknown reacts with three Fe^{2+}.

$$Cr^{x+} \xrightarrow{\ S_2O_8^{2-}\ } Cr_2O_7^{2-}$$
$$\tfrac{1}{2}Cr_2O_7^{2-} + 3Fe^{2+} \rightarrow Cr^{3+} + 3Fe^{3+}$$

In step 1, 0.437 5 g of laser crystal required 0.498 mL of 2.786 mM Fe^{2+} (prepared by dissolving $Fe(NH_4)_2(SO_4)_2 \cdot 6H_2O$ in 2 M $HClO_4$). In step 2, 0.156 6 g of crystal required 0.703 mL of the same Fe^{2+} solution. Find the average oxidation number of Cr in the crystal and find the total micrograms of Cr per gram of crystal.

16-21. Primary-standard-grade arsenic(III) oxide (As_4O_6) is a useful (but carcinogenic) reagent for standardizing oxidants including MnO_4^- and I_3^-. To standardize MnO_4^-, As_4O_6 is dissolved in base and then titrated with MnO_4^- in acid. A small amount of iodide (I^-) or iodate (IO_3^-) catalyzes the reaction between H_3AsO_3 and MnO_4^-.

$$As_4O_6 + 8OH^- \rightleftharpoons 4HAsO_3^{2-} + 2H_2O$$

$$HAsO_3^{2-} + 2H^+ \rightleftharpoons H_3AsO_3$$

$$5H_3AsO_3 + 2MnO_4^- + 6H^+ \rightarrow 5H_3AsO_4 + 2Mn^{2+} + 3H_2O$$

(a) A 3.214-g aliquot of $KMnO_4$ (FM 158.034) was dissolved in 1.000 L of water, heated to cause any reactions with impurities to occur, cooled, and filtered. What is the theoretical molarity of this solution if $KMnO_4$ were pure and if none was consumed by impurities?

(b) What mass of As_4O_6 (FM 395.68) would be just sufficient to react with 25.00 mL of $KMnO_4$ solution in part **(a)**?

(c) It was found that 0.146 8 g of As_4O_6 required 29.98 mL of $KMnO_4$ solution for the faint color of unreacted MnO_4^- to appear. In a blank titration, 0.03 mL of MnO_4^- was required to produce enough color to be seen. Calculate the molarity of the permanganate solution.

Methods Involving Iodine

16-22. Why is iodine almost always used in a solution containing excess I^-?

16-23. State two ways to make standard triiodide solution.

16-24. In which technique, iodimetry or iodometry, is starch indicator not added until just before the end point? Why?

16-25. The pathogenic bacterium *Salmonella enterica* uses tetrathionate found in the human gut as an oxidant— just as we use O_2 to metabolize our food.[34] Write the half-reaction in which tetrathionate serves as an oxidant. Is tetrathionate as powerful an oxidant as O_2?

16-26. (a) Potassium iodate solution was prepared by dissolving 1.022 g of KIO_3 (FM 214.00) in a 500-mL volumetric flask. Then 50.00 mL of the solution were pipetted into a flask and treated with excess KI (2 g) and acid (10 mL of 0.5 M H_2SO_4). How many moles of I_3^- are created by the reaction?

(b) The triiodide from part **(a)** reacted with 37.66 mL of $Na_2S_2O_3$ solution. What is the concentration of the $Na_2S_2O_3$ solution?

(c) A 1.223-g sample of solid containing ascorbic acid and inert ingredients was dissolved in dilute H_2SO_4 and treated with 2 g of KI and 50.00 mL of KIO_3 solution from part **(a)**. Excess triiodide required 14.22 mL of $Na_2S_2O_3$ solution from part **(b)**. Find the weight percent of ascorbic acid (FM 176.13) in the unknown.

(d) Does it matter whether starch indicator is added at the beginning or near the end point in the titration in part **(c)**?

16-27. A 3.026-g portion of a copper(II) salt was dissolved in a 250-mL volumetric flask. A 50.0-mL aliquot was analyzed by adding 1 g of KI and titrating the liberated iodine with 23.33 mL of 0.046 68 M $Na_2S_2O_3$. Find the weight percent of Cu in the salt. Should starch indicator be added to this titration at the beginning or just before the end point?

16-28. *Winkler titration for dissolved O_2.* Dissolved O_2 is a prime indicator of the ability of a body of natural water to support aquatic life. If excessive nutrients run into a lake from fertilizer or sewage, algae and phytoplankton thrive. When algae die and sink to the bottom of the lake, their organic matter is decomposed by bacteria that consume O_2 from the water. Eventually, the water can be sufficiently depleted of O_2 so that fish cannot live. The process by which a body of water becomes enriched in nutrients, some forms of life thrive, and the water eventually becomes depleted of O_2 is called *eutrophication*. One way to measure dissolved O_2 is by the Winkler method that involves an iodometric titration:[35]

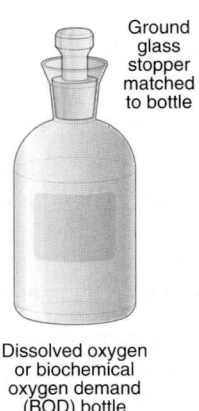

Ground glass stopper matched to bottle

Dissolved oxygen or biochemical oxygen demand (BOD) bottle

1. Collect water in a ~300 mL bottle with a tightly fitting, individually matched ground glass stopper. The manufacturer indicates the volume of the bottle (± 0.1 mL) with the stopper inserted on the bottle. Submerge the stoppered bottle at the desired depth in the water to be sampled. Remove the stopper and fill the bottle with water. Dislodge any air bubbles before inserting the stopper while the bottle is still submerged.

2. Immediately pipet 2.0 mL of 2.15 M $MnSO_4$ and 2.0 mL of alkali solution containing 500 g NaOH/L, 135 g NaI/L, and 10 g NaN_3/L (sodium azide). The pipet should be below the liquid surface during addition to avoid introducing air bubbles. The dense solutions sink and displace close to 4.0 mL of natural water from the bottle.

3. Stopper the bottle tightly, remove displaced liquid from the cup around the stopper, and mix by inversion. O_2 is consumed and $Mn(OH)_3$ precipitates:

$$4Mn^{2+} + O_2 + 8OH^- + 2H_2O \rightarrow 4Mn(OH)_3(s)$$

Azide consumes any nitrite (NO_2^-) in the water so that nitrite cannot subsequently interfere in the iodometric titration:

$$2NO_2^- + 6N_3^- + 4H_2O \rightarrow 10N_2 + 8OH^-$$

4. Back at the lab, slowly add 2.0 mL of 18 M H_2SO_4 below the liquid surface, stopper the bottle tightly, remove the displaced liquid from the cup, and mix by inversion. Acid dissolves $Mn(OH)_3$, which reacts quantitatively with I^-:

$$2Mn(OH)_3(s) + 3H_2SO_4 + 3I^- \rightarrow 2Mn^{2+} + I_3^- + 3SO_4^{2-} + 6H_2O$$

5. Measure 200.0 mL of the liquid into an Erlenmeyer flask and titrate with standard thiosulfate. Add 3 mL of starch solution just before the end point and complete the titration.

A bottle of 297.6 mL of water from a creek at 0°C in the winter was collected and required 14.05 mL of 10.22 mM thiosulfate.

(a) What fraction of the 297.6 mL sample remains after treatment with $MnSO_4$ and alkali solution?

(b) What fraction remains after treatment with H_2SO_4? Assume that H_2SO_4 sinks into the bottle and displaces 2.0 mL of solution prior to mixing.

(c) How many mL of the original sample are contained in the 200.0 mL that are titrated?

(d) How many moles of I_3^- are produced by each mole of O_2 in the water?

(e) Express the dissolved O_2 content in mg O_2/L.

(f) Pure water that is saturated with O_2 contains 14.6 mg O_2/L at 0°C. What is the fraction of saturation of the creek water with O_2?

(g) Write a reaction of NO_2^- with I^- that would interfere with the titration if N_3^- were not introduced. See Table 16-5.

16-29. H_2S was measured by slowly adding 25.00 mL of aqueous H_2S to 25.00 mL of acidified standard 0.010 44 M I_3^- to precipitate elemental sulfur. (If $[H_2S] > 0.01$ M, then precipitated sulfur traps some I_3^- solution, which is not subsequently titrated.) The remaining I_3^- was titrated with 14.44 mL of 0.009 336 M $Na_2S_2O_3$. Find the molarity of the H_2S solution. Should starch indicator be added to this titration at the beginning or just before the end point?

16-30. From the following reduction potentials

$$I_2(s) + 2e^- \rightleftharpoons 2I^- \qquad E° = 0.535 \text{ V}$$
$$I_2(aq) + 2e^- \rightleftharpoons 2I^- \qquad E° = 0.620 \text{ V}$$
$$I_3^- + 2e^- \rightleftharpoons 3I^- \qquad E° = 0.535 \text{ V}$$

(a) Calculate the equilibrium constant for $I_2(aq) + I^- \rightleftharpoons I_3^-$.

(b) Calculate the equilibrium constant for $I_2(s) + I^- \rightleftharpoons I_3^-$.

(c) Calculate the solubility (g/L) of $I_2(s)$ in water.

16-31. The Kjeldahl analysis in Section 11-8 is used to measure the nitrogen content of organic compounds, which are digested in boiling sulfuric acid to decompose to ammonia, which, in turn, is distilled into standard acid. The remaining acid is then back-titrated with base. Kjeldahl himself had difficulty in 1880 discerning by lamplight the methyl red indicator end point in the back titration. He could have refrained from working at night, but instead he chose to complete the analysis differently. After distilling the ammonia into standard sulfuric acid, he added a mixture of KIO_3 and KI to the acid. The liberated iodine was then titrated with thiosulfate, using starch for easy end-point detection—even by lamplight.[36] Explain how the thiosulfate titration is related to the nitrogen content of the unknown. Derive a relationship between moles of NH_3 liberated in the digestion and moles of thiosulfate required for titration of iodine.

16-32. Some people have an allergic reaction to the food preservative sulfite (SO_3^{2-}), which can be measured by instrumental methods[37] or by a redox titration: To 50.0 mL of wine were added 5.00 mL of solution containing (0.804 3 g KIO_3 + 6.0 g KI)/100 mL. Acidification with 1.0 mL of 6.0 M H_2SO_4 quantitatively converted IO_3^- into I_3^-. The I_3^- reacted with SO_3^{2-} to generate SO_4^{2-}, leaving excess I_3^- in solution. The excess I_3^- required 12.86 mL of 0.048 18 M $Na_2S_2O_3$ to reach a starch end point.

(a) Write the reaction that occurs when H_2SO_4 is added to KIO_3 + KI and explain why 6.0 g KI were added to the stock solution. Is it necessary to measure out 6.0 g accurately? Is it necessary to measure 1.0 mL of H_2SO_4 accurately?

(b) Write a balanced reaction between I_3^- and sulfite.

(c) Find the concentration of sulfite in the wine. Express your answer in mol/L and in mg SO_3^{2-} per liter.

(d) *t test.* Another wine was found to contain 277.7 mg SO_3^{2-}/L with a standard deviation of ±2.2 mg/L for three determinations by the iodimetric method. A spectrophotometric method gave 273.2 ± 2.1 mg/L

in three determinations. Are these results significantly different at the 95% confidence level?

16-33. Potassium bromate, $KBrO_3$, is a primary standard for the generation of Br_2 in acidic solution:

$$BrO_3^- + 5Br^- + 6H^+ \rightleftharpoons 3Br_2(aq) + 3H_2O$$

The Br_2 can be used to analyze many unsaturated organic compounds. Al^{3+} was analyzed as follows: An unknown was treated with 8-hydroxyquinoline (oxine) at pH 5 to precipitate aluminum oxinate, $Al(C_9H_7ON)_3$. The precipitate was washed, dissolved in warm HCl containing excess KBr, and treated with 25.00 mL of 0.020 00 M $KBrO_3$.

The excess Br_2 was reduced with KI, which was converted into I_3^-. The I_3^- required 8.83 mL of 0.051 13 M $Na_2S_2O_3$ to reach a starch end point. How many milligrams of Al were in the unknown?

16-34. *Iodometric analysis of high-temperature superconductor.* The procedure in Box 16-3 was carried out to find the effective copper oxidation state, and therefore the number of oxygen atoms, in the formula $YBa_2Cu_3O_{7-z}$, where $0 \leq z \leq 0.5$.

(a) In Experiment A of Box 16-3, 1.00 g of superconductor required 4.55 mmol of $S_2O_3^{2-}$. In Experiment B, 1.00 g of superconductor required 5.68 mmol of $S_2O_3^{2-}$. Calculate the value of z in the formula $YBa_2Cu_3O_{7-z}$ (FM 666.246 − 15.999 4z).

(b) *Propagation of uncertainty.* In several replications of Experiment A, the thiosulfate required was 4.55 (±0.10) mmol of $S_2O_3^{2-}$ per gram of $YBa_2Cu_3O_{7-z}$. In Experiment B, the thiosulfate required was 5.68 (±0.05) mmol of $S_2O_3^{2-}$ per gram. Calculate the uncertainty of x in the formula $YBa_2Cu_3O_x$.

16-35. Here is a description of an analytical procedure for superconductors containing unknown quantities of Cu(I), Cu(II), Cu(III), and peroxide (O_2^{2-}):[38] "The possible trivalent copper and/or peroxide-type oxygen are reduced by Cu(I) when dissolving the sample (*ca.* 50 mg) in deoxygenated HCl solution (1 M) containing a known excess of monovalent copper ions (*ca.* 25 mg CuCl). On the other hand, if the sample itself contained monovalent copper, the amount of Cu(I) in the solution would increase upon dissolving the sample. The excess Cu(I) was then determined by coulometric back-titration...in an argon atmosphere." The abbreviation "*ca.*" means "approximately." *Coulometry* is an electrochemical method in which the electrons liberated in the reaction $Cu^+ \rightarrow Cu^{2+} + e^-$ are measured from the charge flowing through an electrode. Explain with your own words and equations how this analysis works.

16-36. $Li_{1+y}CoO_2$ is an anode for lithium batteries. Cobalt is present as a mixture of Co(III) and Co(II). Most preparations also contain inert lithium salts and moisture. To find the stoichiometry, Co was measured by atomic absorption and its average oxidation state was measured by a potentiometric titration.[39] For the titration, 25.00 mg of solid were dissolved under N_2 in 5.000 mL containing

0.100 0 M Fe^{2+} in 6 M H_2SO_4 plus 6 M H_3PO_4 to give a clear pink solution:

$$Co^{3+} + Fe^{2+} \rightarrow Co^{2+} + Fe^{3+}$$

Unreacted Fe^{2+} required 3.228 mL of 0.015 93 M $K_2Cr_2O_7$ for complete titration.

(a) How many mmol of Co^{3+} are contained in 25.00 mg of the material?

(b) Atomic absorption found 56.4 wt% Co in the solid. What is the average oxidation state of Co?

(c) Find y in the formula $Li_{1+y}CoO_2$.

(d) What is the theoretical quotient wt% Li/wt% Co in the solid? The observed quotient, after washing away inert lithium salts, was $0.138\ 8 \pm 0.000\ 6$. Is the observed quotient consistent with the average cobalt oxidation state?

16-37. *Warning! The Surgeon General has determined that this problem is hazardous to your health.* The oxidation numbers of Cu and Bi in high-temperature superconductors of the type $Bi_2Sr_2(Ca_{0.8}Y_{0.2})Cu_2O_x$ (which could contain Cu^{2+}, Cu^{3+}, Bi^{3+}, and Bi^{5+}) can be measured by the following procedure.[40] In Experiment A, the superconductor is dissolved in 1 M HCl containing excess 2 mM CuCl. Bi^{5+} (written as BiO_3^-) and Cu^{3+} consume Cu^+ to make Cu^{2+}:

$$BiO_3^- + 2Cu^+ + 4H^+ \rightarrow BiO^+ + 2Cu^{2+} + 2H_2O \quad (1)$$
$$Cu^{3+} + Cu^+ \rightarrow 2Cu^{2+} \quad (2)$$

The excess, unreacted Cu^+ is then titrated by *coulometry* (described in Chapter 17). In Experiment B, the superconductor is dissolved in 1 M HCl containing excess 1 mM $FeCl_2 \cdot 4H_2O$. Bi^{5+} reacts with the Fe^{2+} but Cu^{3+} does not react with Fe^{2+}.[41]

$$BiO_3^- + 2Fe^{2+} + 4H^+ \rightarrow BiO^+ + 2Fe^{3+} + 2H_2O \quad (3)$$
$$Cu^{3+} + \frac{1}{2}H_2O \rightarrow Cu^{2+} + \frac{1}{4}O_2 + H^+ \quad (4)$$

The excess, unreacted Fe^{2+} is then titrated by coulometry. The total oxidation number of Cu + Bi is measured in Experiment A, and the oxidation number of Bi is determined in Experiment B. The difference gives the oxidation number of Cu.

(a) In Experiment A, a sample of $Bi_2Sr_2CaCu_2O_x$ (FM 760.37 + 15.999 4x) (containing no yttrium) weighing 102.3 mg was dissolved in 100.0 mL of 1 M HCl containing 2.000 mM CuCl. After reaction with the superconductor, coulometry detected 0.108 5 mmol of unreacted Cu^+ in the solution. In Experiment B, 94.6 mg of superconductor were dissolved in 100.0 mL of 1 M HCl containing 1.000 mM $FeCl_2 \cdot 4H_2O$. After reaction with the superconductor, coulometry detected 0.057 7 mmol of unreacted Fe^{2+}. Find the average oxidation numbers of Bi and Cu in the superconductor and the oxygen stoichiometry coefficient, x.

(b) Find the uncertainties in the oxidation numbers and x if the quantities in Experiment A are 102.3 (±0.2) mg and 0.108 5 ($\pm0.000\ 7$) mmol and the quantities in Experiment B are 94.6 (±0.2) mg and 0.057 7 ($\pm0.000\ 7$) mmol. Assume negligible uncertainty in other quantities.

HOW SWEET IT IS!

(a) Electrochemical detector measures sugars emerging from a chromatography column. Sugars are oxidized at the Cu electrode, whose potential is regulated with respect to the Ag|AgCl reference electrode. Water is reduced ($H_2O + e^- \rightarrow \frac{1}{2}H_2 + OH^-$) at the stainless steel exit arm, and electric current is measured between Cu and steel. [Information from Bioanalytical Systems, West Lafayette, IN.]

(b) Anion-exchange separation of sugars in 0.1 M NaOH with CarboPac PA1 column. Upper chromatogram shows a standard mixture of (1) fucose, (2) methylglucose, (3) arabinose, (4) glucose, (5) fructose, (6) lactose, (7) sucrose, and (8) cellobiose. Lower chromatogram was obtained with Bud Dry beer diluted by a factor of 100 with water and filtered through a 0.45-μm membrane to remove particles. [Data from P. Luo, M. Z. Luo, and R. P. Baldwin, "Determination of Sugars in Food Products," *J. Chem. Ed.* **1993,** *70,* 679.]

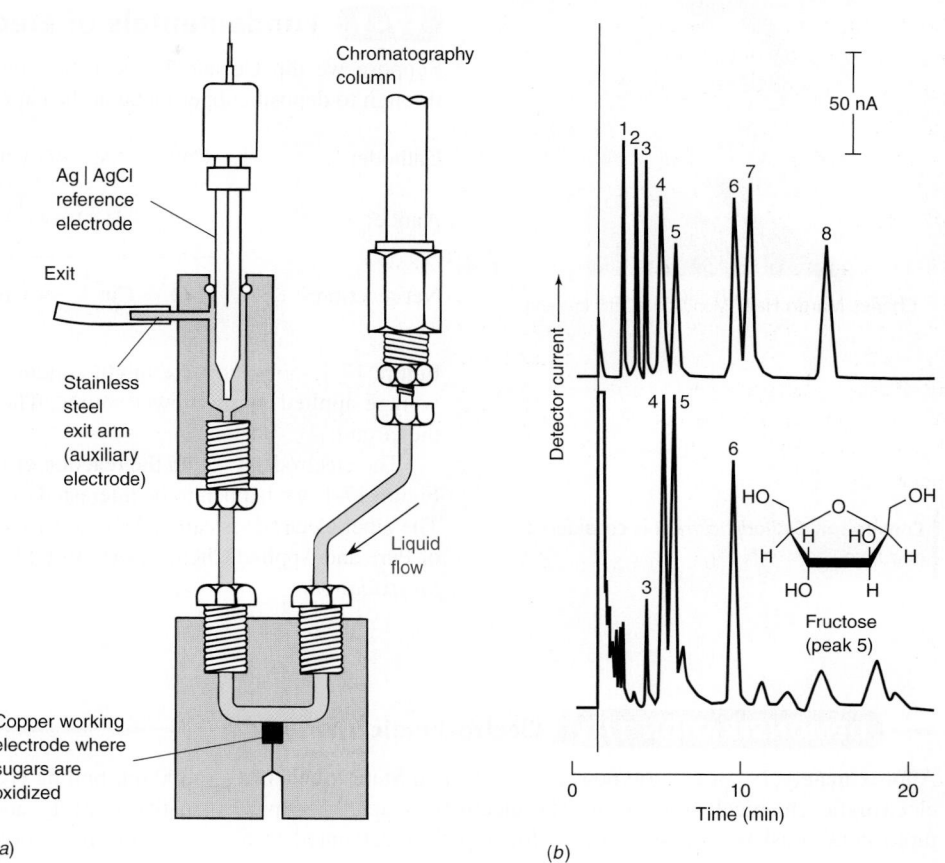

(a)

(b)

You can measure sugars in your favorite beverage by separating the sugars by anion-exchange chromatography (Chapter 26) in strongly basic solution and detecting them with an electrode as they emerge from the column.[1] The —OH groups of sugars such as fructose, whose structure is shown in the chromatogram, partially dissociate to —O⁻ anions in 0.1 M NaOH. Anions are separated from one another when they pass through a column that has fixed positive charges. As sugars emerge from the column, they are detected by oxidation at a Cu electrode poised at a potential of +0.55 V versus Ag|AgCl. The chromatogram is a graph of detector current versus time. Each sugar gives a peak whose area is proportional to the moles exiting the column.

Brand	Sugar concentration (g/L)			
	Glucose	**Fructose**	**Lactose**	**Maltose**
Budweiser	0.54	0.26	0.84	2.05
Bud Dry	0.14	0.29	0.46	—
Coca Cola	45.1	68.4	—	1.04
Pepsi	44.0	42.9	—	1.06
Diet Pepsi	0.03	0.01	—	—

Electrolytic production of aluminum by the Hall-Héroult process consumes ~3.5% of global electricity.[2] Al^{3+} in a molten solution of Al_2O_3 and cryolite (Na_3AlF_6) is reduced to Al at the cathode of a cell that typically draws 250 kA. Charles Hall invented this process in 1886 when he was 22 years old, just after graduating from Oberlin College.[3]

Charles Martin Hall. [Mondadori/Getty Images.]

Convention *Cathodic current* is considered *negative*.

$\mathbf{P}$revious chapters dealt with *potentiometry*—in which voltage was measured in the absence of significant current. Now we consider electroanalytical methods in which current is essential.[4] Techniques in this chapter are all examples of **electrolysis**—the process in which a chemical reaction is forced to occur at an electrode by an imposed voltage (Demonstration 17-1). The home glucose monitor described in this chapter, with sales of $2.5 billion in the United States in 2009 and projected worldwide sales of $14 billion in 2014, is the single largest electroanalytical application.

17-1 Fundamentals of Electrolysis

Suppose we dip Cu and Pt electrodes into a solution of Cu^{2+} and force electric current through to deposit copper metal at the cathode and to liberate O_2 at the anode.

Cathode: $$Cu^{2+} + 2e^- \rightleftharpoons Cu(s)$$

Anode: $$H_2O \rightleftharpoons \frac{1}{2}O_2(g) + 2H^+ + 2e^-$$

Net reaction: $$H_2O + Cu^{2+} \rightleftharpoons Cu(s) + \frac{1}{2}O_2(g) + 2H^+ \qquad (17\text{-}1)$$

Figure 17-1 shows how we might conduct the experiment. The potentiometer measures the voltage applied by the power source. The ammeter measures the current flowing through the circuit.

The electrode at which the reaction of interest occurs is called the **working electrode.** In Figure 17-1, we happen to be interested in reduction of Cu^{2+}, so Cu is the working electrode. The other electrode is called the *counter electrode*. The convention of the International Union of Pure and Applied Chemistry (IUPAC) is that *current is negative for reduction and positive for oxidation.*

DEMONSTRATION 17-1 **Electrochemical Writing[5]**

Approximately 7% of electric power in the United States goes into electrolytic chemical production. The electrolysis apparatus pictured here consists of a sheet of Al foil taped or cemented to a wood surface. Any size will work, but an area of 15 cm on a side is convenient for a classroom demonstration. Tape to the metal foil (at one edge only) a sandwich consisting of filter paper, printer paper, and another sheet of filter paper. Make a stylus from Cu wire (18 gauge or thicker) looped at one end and passed through a length of glass tubing.

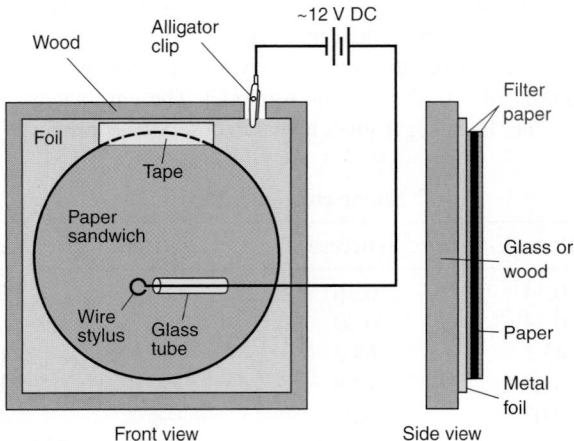

Front view   Side view

Prepare 1 wt% starch by making a paste from 5 g soluble starch and 5 mg HgI_2 (a preservative[6]) in 50 mL H_2O. Pour the paste into

500 mL boiling H_2O and boil until the solution is clear. (As an alternative to HgI_2, add a few drops of chloroform to the clear starch solution after cooling. Another alternative is to make fresh starch solution without preservative.) Prepare a fresh solution from 1.6 g KI, 20 mL water, 5 mL starch solution, and 5 mL phenolphthalein indicator solution. (If the solution darkens after several days, decolorize it with a few drops of dilute $Na_2S_2O_3$.) Soak the three layers of paper with the KI-starch-phenolphthalein solution. Connect the stylus and foil to a 12-V DC power source, and write on the paper with the stylus.

When the stylus is the cathode, water is reduced to H_2 plus OH^- and pink color appears from the reaction of OH^- with phenolphthalein.

Cathode: $$H_2O + e^- \rightarrow \frac{1}{2}H_2(g) + OH^-$$

When the polarity is reversed and the stylus is the anode, I^- is oxidized to I_2; a black (very dark blue) color appears from the reaction of I_2 with starch.

Anode: $$I^- \rightarrow \frac{1}{2}I_2 + e^-$$

Pick up the top sheet of filter paper and the printer paper, and you will discover that the writing appears in the opposite color on the bottom sheet of filter paper (Color Plate 12).

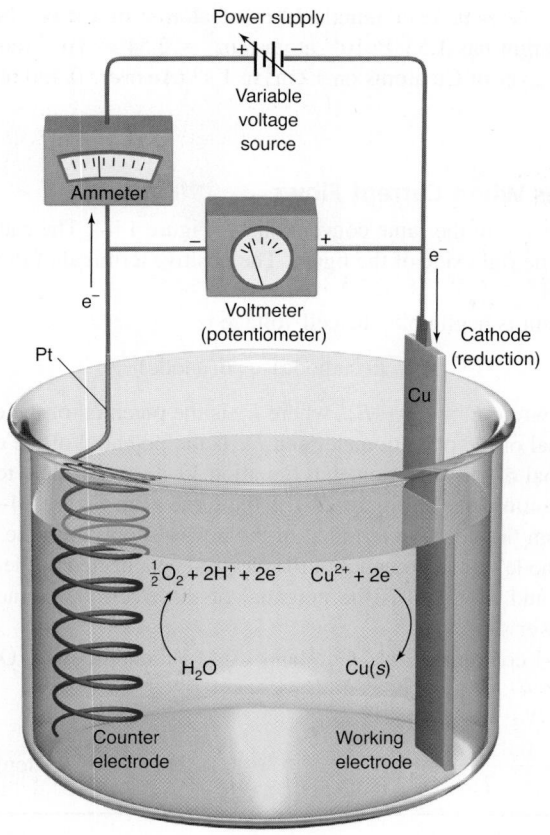

Power supply

+ ┤|┤|├ −

Variable
voltage
source

Ammeter

Voltmeter
(potentiometer)

− +

e^-

Cathode
(reduction)

Pt

Cu

$\frac{1}{2}O_2 + 2H^+ + 2e^-$ $Cu^{2+} + 2e^-$

H_2O $Cu(s)$

Counter
electrode

Working
electrode

e^-

FIGURE 17-1 Electrolysis experiment. The
power supply ┤|┤|├ is a variable voltage
source. The potentiometer measures voltage
and the ammeter measures current.

Current Measures the Rate of Reaction

If a current I flows for a time t, the charge q passing any point in the circuit is

*Relation of charge
to current and time:*

$$q = I \cdot t \qquad (17\text{-}2)$$

Coulombs   Amperes · Seconds

The number of moles of electrons is

$$\text{Moles of } e^- = \frac{\text{coulombs}}{\text{coulombs/mole}} = \frac{I \cdot t}{F}$$

If a reaction requires n electrons per molecule, the quantity reacting in time t is

*Relation of moles to
current and time:*

$$\text{Moles reacted} = \frac{I \cdot t}{nF} \qquad (17\text{-}3)$$

An **ampere** is an electric current of 1 coulomb
per second.
A **coulomb** contains $6.241\,5 \times 10^{18}$ electrons.
Faraday constant:

$$F = 9.648\,5 \times 10^4 \text{ C/mol}$$

moles of electrons $= \dfrac{I \cdot t}{F}$.

EXAMPLE   **Relating Current, Time, and Amount of Reaction**

If a current of 0.17 A flows for 16 min through the cell in Figure 17-1, how many grams of
$Cu(s)$ will be deposited?

Solution We first calculate the moles of e^- flowing through the cell:

$$\text{Moles of } e^- = \frac{I \cdot t}{F} = \frac{\left(0.17 \dfrac{C}{s}\right)(16 \text{ min})\left(60 \dfrac{s}{\text{min}}\right)}{96\,485\left(\dfrac{C}{\text{mol}}\right)} = 1.6_9 \times 10^{-3} \text{ mol}$$

The cathode half-reaction requires $2e^-$ for each Cu deposited. Therefore,

$$\text{Moles of } Cu(s) = \frac{1}{2}(\text{moles of } e^-) = 8.4_5 \times 10^{-4} \text{ mol}$$

The mass of $Cu(s)$ deposited is $(8.4_5 \times 10^{-4} \text{ mol})(63.546 \text{ g/mol}) = 0.054$ g.

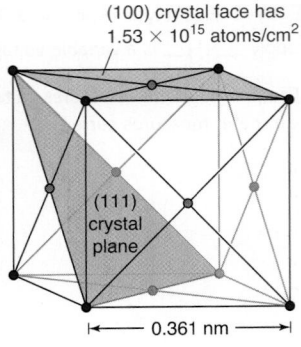

(100) crystal face has
1.53×10^{15} atoms/cm^2

(111) crystal plane

|← 0.361 nm →|

Face-centered cubic crystal

To use $E = E(\text{cathode}) - E(\text{anode})$, you must write both reactions as *reductions*. $E(\text{cathode}) - E(\text{anode})$ is the *open-circuit* voltage measured with negligible current flow between cathode and anode.

The Nernst equation should really be written with activities, not concentrations.

Free energy change for Reaction 17-1:

$$\begin{array}{ccccccc} \Delta G & = & -n & \cdot & N & \cdot & F & \cdot & E \text{ (14-5)} \\ \text{Joules} & & \text{Unit} & & \text{Moles} & & \text{C/mol} & & \text{Volts} \\ \text{(J)} & & \text{charges per} & & & & & & \text{(V)} \\ & & \text{molecule} & & & & & & \end{array}$$

$$\Delta G/N = -(2)\left(96\,485\frac{C}{mol}\right)(-0.911\ V)$$

$$= +1.76 \times 10^5\ C \cdot V/mol$$

$$= +1.76 \times 10^5\ J/mol = 176\ kJ/mol$$

Note that C × V = J

TEST YOURSELF A *monolayer* (single layer of atoms) of Cu on the (100) crystal face shown in the margin has 1.53×10^{15} atoms/cm^2 = 2.54×10^{-9} mol/cm^2. What current can deposit one layer of Cu atoms on 1 cm^2 in 1 s? (***Answer:*** 0.490 mA)

Voltage Changes When Current Flows

Figure 17-1 is drawn with the same conventions as Figure 14-4. The cathode—where reduction occurs—is at the right side of the figure. The positive terminal of the potentiometer is on the right-hand side.

If electric current is negligible, the cell voltage is

$$E = E(\text{cathode}) - E(\text{anode}) \tag{17-4}$$

In Chapter 14, we wrote $E = E_+ - E_-$, where E_+ is the potential of the electrode attached to the positive terminal of the potentiometer and E_- is the potential of the electrode attached to the negative terminal of the potentiometer. Equation 17-4 is equivalent to $E = E_+ - E_-$. The polarity of the potentiometer in Figure 17-1 is the same as in Figure 14-4. In an electrolysis, electrons come from the negative terminal of the power supply into the cathode of the electrolysis cell. $E(\text{cathode})$ is the potential of the electrode connected to the negative terminal of the power supply and $E(\text{anode})$ is the potential of the electrode connected to the positive terminal of the power supply.

If Reaction 17-1 contains 0.20 M Cu^{2+} and 1.0 M H$^+$ and liberates O$_2$ at a pressure of 1.0 bar, we find

$$E = \underbrace{\left\{0.339 - \frac{0.059\,16}{2}\log\left(\frac{1}{[\text{Cu}^{2+}]}\right)\right\}}_{E(\text{cathode})} - \underbrace{\left\{1.229 - \frac{0.059\,16}{2}\log\left(\frac{1}{P_{\text{O}_2}^{1/2}[\text{H}^+]^2}\right)\right\}}_{E(\text{anode})}$$

$$= \left\{0.339 - \frac{0.059\,16}{2}\log\left(\frac{1}{[0.20]}\right)\right\} - \left\{1.229 - \frac{0.059\,16}{2}\log\left(\frac{1}{(1.0)^{1/2}[1.0]^2}\right)\right\}$$

$$= 0.318 - 1.229 = -0.911\ V$$

This voltage would be read on the potentiometer in Figure 17-1 if there were negligible current. The voltage is negative because the positive terminal of the potentiometer is connected to the negative side of the power supply. The free-energy change computed in the margin is positive because the reaction is not spontaneous. We need the power supply to force the reaction to occur. If current is not negligible, *overpotential*, *ohmic potential*, and *concentration polarization* can change the voltage required to drive the reaction.

Overpotential is the voltage required to overcome the *activation energy* for a reaction at an electrode (Figure 17-2).[7] The faster you wish to drive the reaction, the greater the overpotential that must be applied. Electric current measures the rate of electron transfer. Applying a greater overpotential will sustain a higher *current density* (current per unit area of electrode surface, A/m^2). Table 17-1 shows that the overpotential for liberation of H$_2$ at a Cu surface must be increased from 0.479 to 1.254 V to increase the current density from 10 A/m^2 to 10 000 A/m^2. Activation energy depends on the nature of the surface. H$_2$ is evolved at a Pt surface with little overpotential, whereas a Hg surface requires ~1 V to drive the reaction.

FIGURE 17-2 (*a*) Schematic energy profile for electron transfer from a metal to H$_3$O$^+$, leading to liberation of H$_2$. (*b*) Applying a potential to the metal raises the energy of the electron in the metal and decreases the activation energy for electron transfer.

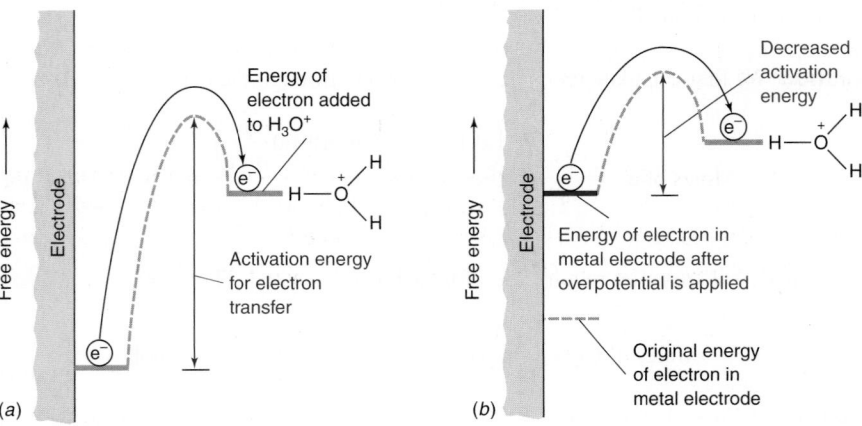

TABLE 17-1	Overpotential (V) for gas evolution at various current densities (A/m²) at 25°C							
	10 A/m^2		100 A/m^2		$1\,000 \text{ A/m}^2$		$10\,000 \text{ A/m}^2$	
Electrode	H_2	O_2	H_2	O_2	H_2	O_2	H_2	O_2
Platinized Pt	0.015 4	0.398	0.030 0	0.521	0.040 5	0.638	0.048 3	0.766
Smooth Pt	0.024	0.721	0.068	0.85	0.288	1.28	0.676	1.49
Cu	0.479	0.422	0.584	0.580	0.801	0.660	1.254	0.793
Ag	0.475 1	0.580	0.761 8	0.729	0.874 9	0.984	1.089 0	1.131
Au	0.241	0.673	0.390	0.963	0.588	1.244	0.798	1.63
Graphite	0.599 5		0.778 8		0.977 4		1.220 0	
Pb	0.52		1.090		1.179		1.262	
Zn	0.716		0.746		1.064		1.229	
Hg	0.9		1.0		1.1		1.1	

SOURCE: International Critical Tables, *1929*, 6, 339. *This reference also gives overpotentials for* Cl_2, Br_2, *and* I_2.

Ohmic potential is the voltage needed to overcome electric resistance, R, of the solution in the electrochemical cell when a current, I, is flowing:

Ohmic potential: $$E_{ohmic} = IR \qquad (17\text{-}5)$$

If the cell has a resistance of 2 ohms and a current of 20 mA is flowing, the voltage required to overcome the resistance is $E = (2\ \Omega)(20\text{ mA}) = 0.040$ V.

Concentration polarization occurs when the concentrations of reactants or products at the surface of an electrode are not the same as they are in bulk solution. For Reaction 17-1, the Nernst equation should be written

$$E(\text{cathode}) = 0.339 - \frac{0.059\ 16}{2}\log\left(\frac{1}{[Cu^{2+}]_s}\right)$$

where $[Cu^{2+}]_s$ is the concentration in solution *at the surface of the electrode*. If reduction of Cu^{2+} occurs rapidly, $[Cu^{2+}]_s$ could be very small because Cu^{2+} cannot diffuse to the electrode as fast as it is consumed. As $[Cu^{2+}]_s$ decreases, $E(\text{cathode})$ becomes more negative.

Overpotential, ohmic potential, and concentration polarization make electrolysis more difficult. They make the cell voltage more negative, requiring more voltage from the power supply in Figure 17-1 to drive the reaction forward.

$$E = \underbrace{E(\text{cathode}) - E(\text{anode})}_{} - IR - \text{overpotentials} \qquad (17\text{-}6)$$

These terms include the effects of concentration polarization

There can be concentration polarization and overpotential at both the cathode and the anode.

Resistance is measured in ohms, whose symbol is capital Greek omega, Ω.

Electrodes respond to concentrations (activities) of reactants and products adjacent to the electrode, not to concentrations in the bulk solution.

If $[Cu^{2+}]_s$ were reduced from 0.2 M to 2 μM, $E(\text{cathode})$ would change from 0.318 to 0.170 V.

EXAMPLE Effects of Ohmic Potential, Overpotential, and Concentration Polarization

Suppose we wish to electrolyze I^- to I_3^- in a 0.10 M KI solution containing 3.0×10^{-5} M I_3^- at pH 10.00 with $P_{H_2} = 1.00$ bar.

$$3I^- + 2H_2O \rightarrow I_3^- + H_2(g) + 2OH^-$$

(a) Find the cell voltage if no current is flowing. **(b)** Then suppose that electrolysis increases $[I_3^-]_s$ to 3.0×10^{-4} M, but other concentrations are unaffected. Suppose that the cell resistance is 2.0 Ω, the current is 63 mA, the cathode overpotential is 0.382 V, and the anode overpotential is 0.025 V. What voltage is needed to drive the reaction?

Solution (a) The open-circuit voltage is $E(\text{cathode}) - E(\text{anode})$:

Cathode: $2H_2O + 2e^- \rightarrow H_2(g) + 2OH^-$ $E^\circ = -0.828$ V

Anode: $I_3^- + 2e^- \rightarrow 3I^-$ $E^\circ = 0.535$ V

$$E(\text{cathode}) = -0.828 - \frac{0.059\ 16}{2}\log(P_{H_2}[OH^-]^2)$$

$$= -0.828 - \frac{0.059\ 16}{2}\log[(1.00)(1.0 \times 10^{-4})^2] = -0.591\ \text{V}$$

$$E(\text{anode}) = 0.535 - \frac{0.059\ 16}{2}\log\left(\frac{[I^-]^3}{[I_3^-]}\right)$$

$$= 0.535 - \frac{0.059\ 16}{2}\log\left(\frac{[0.10]^3}{[3.0 \times 10^{-5}]}\right) = 0.490\ \text{V}$$

$$E = E(\text{cathode}) - E(\text{anode}) = -1.081\ \text{V}$$

We would have to apply -1.081 V to force the reaction to occur.

(b) Now $E(\text{cathode})$ is unchanged but $E(\text{anode})$ changes because $[I_3^-]_s$ is different from $[I_3^-]$ in bulk solution.

$$E(\text{anode}) = 0.535 - \frac{0.059\ 16}{2}\log\left(\frac{[0.10]^3}{[3.0 \times 10^{-4}]}\right) = 0.520\ \text{V}$$

$$E = E(\text{cathode}) - E(\text{anode}) - IR - \text{overpotentials}$$

$$= -0.591\ \text{V} - 0.520\ \text{V} - (2.0\ \Omega)(0.063\ \text{A}) - 0.382\ \text{V} - 0.025\ \text{V}$$

$$= -1.644\ \text{V}$$

Instead of -1.081 V, we need to apply -1.644 V to drive the reaction. Notice that both overpotentials contribute in a manner that increases the magnitude of the required voltage.

TEST YOURSELF   Find the voltage in part **(b)** if $[I^-]_s = 0.01$ M. (*Answer:* -1.732 V)

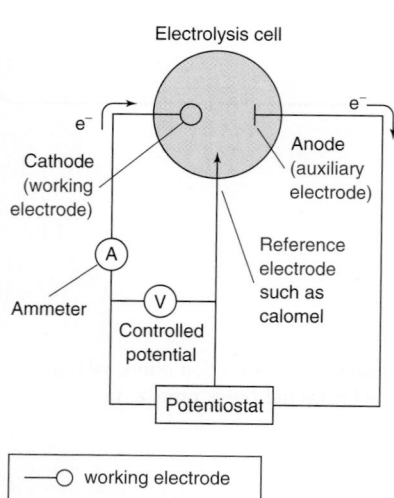

Electrolysis cell

Cathode (working electrode)

Anode (auxiliary electrode)

Reference electrode such as calomel

Ammeter

Controlled potential

Potentiostat

— ◯ working electrode
— ⊣ auxiliary electrode
— → reference electrode

FIGURE 17-3 Circuit used for controlled-potential electrolysis with a three-electrode cell.

Working electrode: where the analytical reaction occurs

Auxiliary electrode: the other electrode needed for current flow

Reference electrode: used to measure the potential of the working electrode

The chromatographic detector at the opening of this chapter has a Cu working electrode, a stainless steel auxiliary electrode, and a Ag|AgCl reference electrode.

Controlled-Potential Electrolysis with a Three-Electrode Cell

An **electroactive species** is one that can be oxidized or reduced at an electrode. We regulate the potential of the working electrode to control which species react and which do not. Metal electrodes are said to be **polarizable,** which means that their potentials are easily changed when small currents flow. A reference electrode such as calomel or Ag|AgCl is said to be **nonpolarizable,** because its potential does not vary much unless a significant current is flowing. Ideally, we want to measure the potential of a polarizable working electrode with respect to a nonpolarizable reference electrode. How can we do this if there is to be significant current at the working electrode and negligible current at the reference electrode?

The answer is to introduce a third electrode (Figure 17-3). The **working electrode** is the one at which the reaction of interest occurs. A calomel or other **reference electrode** is used to measure the potential of the working electrode. The **auxiliary electrode** (also called the *counter electrode*) is the current-supporting partner of the working electrode. Current flows between the working and the auxiliary electrodes. Current going through the working electrode is measured by the ammeter (A) located next to the working electrode. Negligible current flows through the reference electrode, so its potential is unaffected by ohmic potential, concentration polarization, and overpotential. Therefore, the reference electrode potential is constant. In **controlled-potential electrolysis,** the voltage difference between the working and reference electrodes in a three-electrode cell is measured by the potentiometer (V) and regulated by an electronic device called a **potentiostat.**

Figure 17-4 illustrates the measurement of overpotential and ohmic potential in a student experiment. A potential difference of 1.000 V was imposed by a potentiostat connected to two copper electrodes immersed in 1.0 M $CuSO_4(aq)$. The anode reaction is $Cu(s) \rightleftharpoons Cu^{2+} + 2e^-$ and the cathode reaction is $Cu^{2+} + 2e^- \rightleftharpoons Cu(s)$.

A potentiometer measures the voltage between the Cu electrode at the left in Figure 17-4 and a Ag|AgCl reference electrode immersed in 1.0 M $CuSO_4$ and connected to the bulk solution via a *Luggin capillary*. There is negligible current between the tip of the capillary and the reference electrode because there is negligible current at the reference electrode. There is ohmic loss over distance d between the working electrode and the tip of the capillary. Even with no current at the reference electrode, there is some unknown junction potential at the porous plug at the bottom of the reference electrode where the internal electrode solution (such as KCl) meets 1.0 M $CuSO_4$.

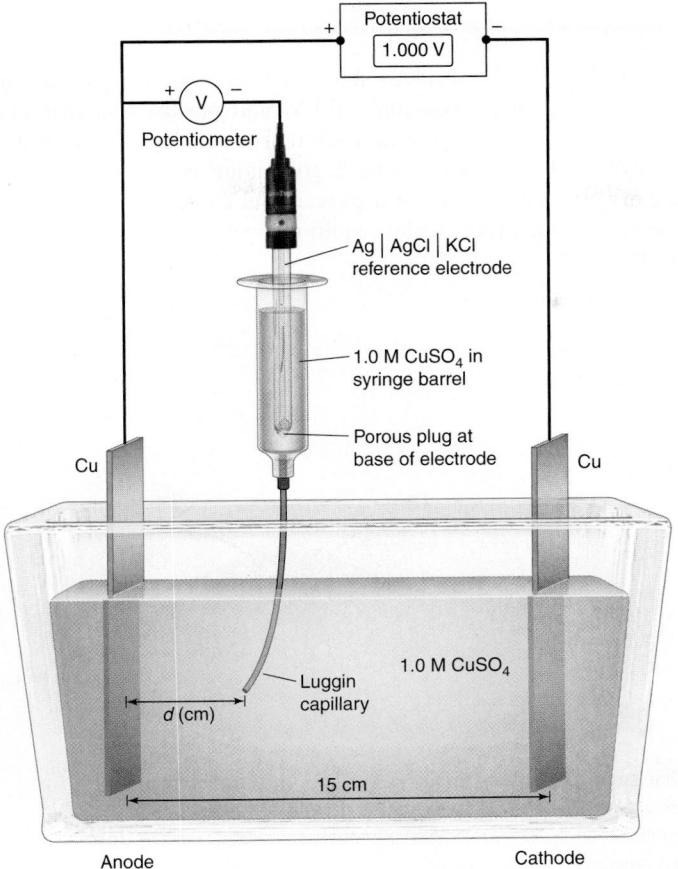

FIGURE 17-4 Measurement of copper electrode potential versus Ag | AgCl reference electrode using a Luggin capillary placed at distance *d* from the copper electrode. The reference electrode is dipped in a syringe barrel sealed to the electrode by Parafilm® to prevent liquid from draining out of the syringe. Parafilm is not shown in the diagram. Plastic capillary tubing is attached to the outlet of the syringe. [Information from F. J. Vidal-Iglesias, J. Solla-Gullón, A. Rodes, E. Herrero, and A. Aldaz, "Understanding the Nernst Equation and Other Electrochemical Concepts," *J. Chem. Ed.* **2012,** *89,* 936.]

Ohmic loss is the potential difference (the voltage) between two points in a solution due to electrical resistance between the points. If the resistance is R and the current is I, the ohmic loss is IR. Resistance is proportional to the distance between the points. The voltage between the anode and the outlet of the Luggin capillary in Figure 17-4 changes by 64.7 mV/cm in Figure 17-5. The change between the cathode and reference electrode is −65.0 mV/cm. The difference between the magnitude of the slopes is ascribed to experimental error.

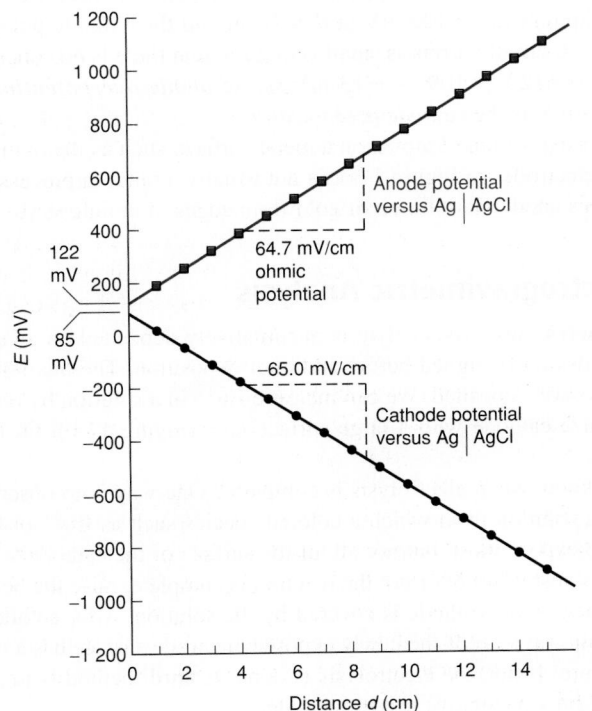

FIGURE 17-5 Student data from the experiment in Figure 17-4 for cathode and anode areas of 16.5 cm² immersed in solution. [Data from F. J. Vidal-Iglesias, J. Solla-Gullón, A. Rodes, E. Herrero, and A. Aldaz,, "Understanding the Nernst Equation and Other Electrochemical Concepts," *J. Chem. Ed.* **2012,** *89,* 936.]

BOX 17-1   Metal Reactions at Atomic Steps

An atom on the surface of a crystal is bound to atoms around it on the surface and to atoms below it. The most reactive site for deposition or dissolution of atoms from a metal surface is often at an atomic step, where each atom makes fewer chemical bonds to its nearest neighbors. The micrograph shows terraces on the surface of a gold crystal. Each terrace intersecting the line SS' is just one atom tall (0.25 nm). A crystal surface of such high quality is typically described as "atomically flat."

The diagram shows the mechanism of anodic dissolution of gold at a potential <0.1 V more positive than that required to initiate dissolution in a solution containing 0.1 M $HClO_4$ plus 3 mM HCl. Chloride attacks gold atoms at the edge of a terrace to form $AuCl_4^-$ (aq). Terraces recede at a rate of 400–600 atoms/min under the experimental conditions.

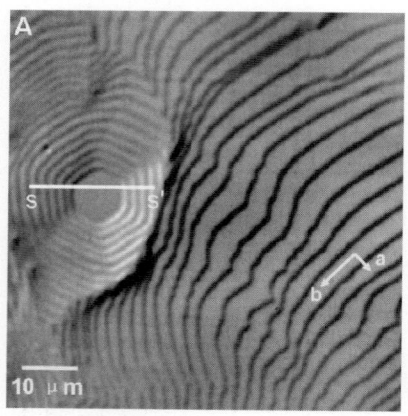

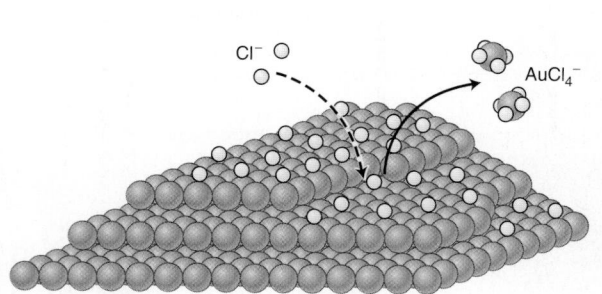

Terraces on (111) crystal surface of gold and mechanism of Cl^--assisted anodic dissolution. The location of a (111) crystal plane is shown on page 398. [R. Wen, A. Lahiri, M. Azhagurajan, S. Kobayashi, and K. Itaya, "A New in Situ Optical Microscope with Single Atomic Layer Resolution for Observation of Electrochemical Dissolution of Au(111)," *J. Am. Chem. Soc.* **2010,** *132,* 13657, Figure 2. Reprinted with permission © 2010, American Chemical Society.]

With no applied voltage between the Cu electrodes, the measured potential of each Cu electrode was +109 mV when the Luggin capillary was placed against the Cu surface. This voltage is the equilibrium (zero current) potential of Cu immersed in 1.0 M $CuSO_4$ measured versus Ag | AgCl | 3 M KCl. When 1.000 V is imposed between the Cu anode and Cu cathode, current flows through the solution. With current flowing, the anode potential in Figure 17-5 extrapolates to +122 mV at $d = 0$ cm and the cathode potential extrapolates to +85 mV at $d = 0$ cm. If current is small enough so that there is no concentration polarization, the difference $122 - 109 = +13$ mV is the *anode overpotential.* The difference $85 - 109 = -22$ mV is the *cathode overpotential.*

Dissolution or deposition of atoms on a metal surface, such as the oxidation or reduction of copper at the electrodes in Figure 17-4, is not usually a random process on a flat surface. Box 17-1 describes anodic dissolution of gold from edges of atomic steps.

17-2   Electrogravimetric Analysis

In **electrogravimetric analysis,** analyte is quantitatively deposited on an electrode by electrolysis. The electrode is weighed before and after deposition. The increase in mass tells us how much analyte was deposited. We can measure Cu^{2+} in a solution by reducing it to Cu(s) on a clean Pt gauze cathode with a large surface area (Figure 17-6). O_2 is liberated at the counter electrode.

How do you know when electrolysis is complete? One way is to observe the disappearance of color in a solution from which a colored species such as Cu^{2+} or Co^{2+} is removed. Another way is to expose most, but not all, of the surface of the cathode to the solution during electrolysis. To test whether or not the reaction is complete, raise the beaker or add water so that fresh surface of the cathode is covered by the solution. After an additional period of electrolysis (15 min, say), see if the newly exposed electrode surface has a deposit. If it does, repeat the procedure. If not, the electrolysis is done. A third method is to remove a drop of solution and perform a qualitative test for analyte.

Tests for completion of the deposition:
- disappearance of color
- no deposition on freshly exposed electrode surface
- qualitative test for analyte in solution

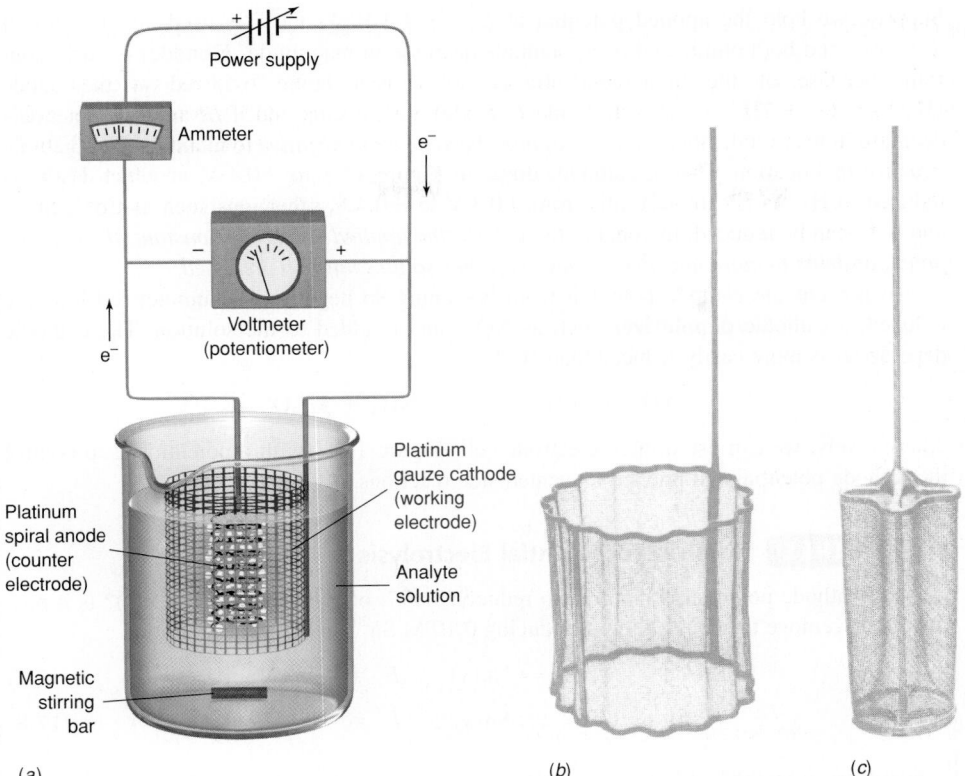

FIGURE 17-6 (*a*) Electrogravimetric analysis. Analyte is deposited on the large Pt gauze electrode. If analyte is to be oxidized, rather than reduced, the polarity of the power supply is reversed so that deposition still occurs on the large electrode. (*b*) Outer Pt gauze electrode. (*c*) Optional inner Pt gauze electrode designed to be spun by a motor in place of magnetic stirring.

On page 398, we calculated that -0.911 V needs to be applied between the electrodes to deposit Cu(*s*) on the cathode. The actual behavior of the electrolysis in Figure 17-7 shows that nothing special happens at -0.911 V. Near -2 V, the reaction begins in earnest. At low voltage, a small *residual current* is observed from reduction at the cathode and an equal amount of oxidation at the anode. Reduction might involve traces of dissolved O_2, impurities such as Fe^{3+}, or surface oxide on the electrode.

Table 17-1 shows that an overpotential of ~1 V is required for O_2 formation at the smooth Pt anode. Overpotential is the main reason why not much happens in Figure 17-7 until -2 V is applied. Beyond -2 V, the rate of reaction (the current) increases steadily. Around -4.6 V, the current increases more rapidly with the onset of reduction of H_3O^+ to H_2. Gas bubbles at the working electrode interfere with deposition of solids.

The voltage between the two electrodes is

$$E = E(\text{cathode}) - E(\text{anode}) - IR - \text{overpotentials} \qquad (17\text{-}6)$$

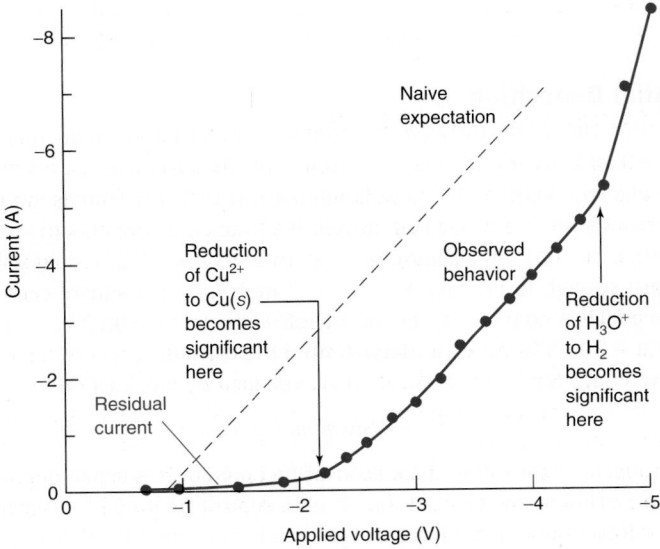

FIGURE 17-7 Observed current-voltage relation for electrolysis of 0.2 M $CuSO_4$ in 1 M $HClO_4$ under N_2, using the apparatus in Figure 17-6. By convention, current is negative for reduction at the working electrode.

Suppose we hold the applied potential at $E = -2.0$ V. As Cu^{2+} is used up, the current decreases and both ohmic and overpotentials decrease in magnitude. E(anode) is fairly constant because of the high concentration of solvent being oxidized at the anode $(H_2O \rightarrow \frac{1}{2}O_2 + 2H^+ + 2e^-)$. If E and E(anode) are constant and if IR and overpotentials decrease in magnitude, then E(cathode) *must become more negative* to maintain the algebraic equality in Equation 17-6. E(cathode) drops in Figure 17-8 to -0.4 V, at which H_3O^+ is reduced to H_2. As E(cathode) falls from $+0.3$ V to -0.4 V, other ions such as Co^{2+}, Sn^{2+}, and Ni^{2+} can be reduced. In general, then, *when the applied voltage is constant, the cathode potential drifts to more negative values and other solutes might be reduced.*

To prevent the cathode potential from becoming so negative that unintended ions are reduced, a cathodic **depolarizer** such as NO_3^- can be added to the solution. The cathodic depolarizer is more easily reduced than H_3O^+:

$$NO_3^- + 10H^+ + 8e^- \rightarrow NH_4^+ + 3H_2O$$

Alternatively, we can use a three-electrode cell (Figure 17-3) with a potentiostat to control the cathode potential and prevent unwanted side reactions.

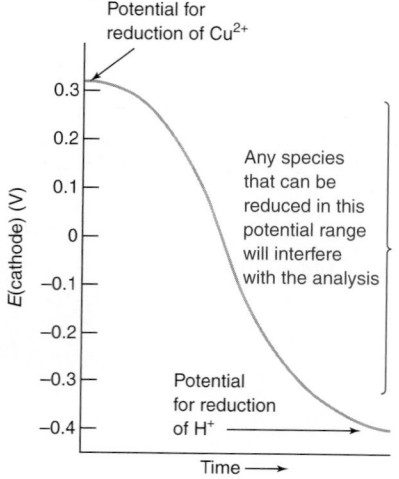

A **cathodic depolarizer** is reduced in preference to solvent. For *oxidation* reactions, **anodic depolarizers** include N_2H_4 (hydrazine) and NH_2OH (hydroxylamine).

FIGURE 17-8 E(cathode) becomes more negative with time when electrolysis is conducted in a two-electrode cell with a constant voltage between the electrodes.

Figure labels: Potential for reduction of Cu^{2+}; Any species that can be reduced in this potential range will interfere with the analysis; Potential for reduction of H^+; E(cathode) (V); Time →

EXAMPLE Controlled-Potential Electrolysis

What cathode potential is required to reduce 99.99% of 0.10 M Cu^{2+} to $Cu(s)$? Is it possible to remove this Cu^{2+} without reducing 0.10 M Sn^{2+} in the same solution?

$$Cu^{2+} + 2e^- \rightleftharpoons Cu(s) \qquad E° = 0.339 \text{ V} \qquad \text{(17-7)}$$

$$Sn^{2+} + 2e^- \rightleftharpoons Sn(s) \qquad E° = -0.141 \text{ V} \qquad \text{(17-8)}$$

Solution If 99.99% of Cu^{2+} were reduced, the concentration of remaining Cu^{2+} would be 1.0×10^{-5} M, and the required cathode potential would be

$$E(\text{cathode}) = 0.339 - \frac{0.059\,16}{2}\log\left(\frac{1}{\underset{[Cu^{2+}]}{1.0 \times 10^{-5}}}\right) = 0.19 \text{ V}$$

The cathode potential required to reduce Sn^{2+} is

$$E(\text{cathode, for reduction of } Sn^{2+}) = -0.141 - \frac{0.059\,16}{2}\log\left(\frac{1}{\underset{[Sn^{2+}]}{0.10}}\right) = -0.17 \text{ V}$$

We do not expect reduction of Sn^{2+} at a cathode potential more positive than -0.17 V. The reduction of 99.99% of Cu^{2+} without reducing Sn^{2+} appears feasible.

TEST YOURSELF Will E(cathode) $= 0.19$ V reduce 0.10 M SbO^+ at pH 2 by the reaction $SbO^+ + 2H^+ + 3e^- \rightleftharpoons Sb(s) + H_2O$, $E° = 0.208$ V? (*Answer:* E(cathode) for $SbO^+ = 0.11$ V, so reduction should not occur at 0.19 V)

Underpotential Deposition

When Sn^{2+} in 1 M HCl is electrolyzed at a gold working electrode, reduction is observed at E(cathode) $= -0.18$ V by the technique of cyclic voltammetry, which we study later in this chapter. From what you know so far, you should not expect that potentials more positive than -0.18 V will reduce Sn^{2+}. Yet, a small current is observed at E(cathode) $= +0.12$ V. The coulombs required at -0.18 V increase in proportion to $[Sn^{2+}]$. The coulombs required at $+0.12$ V are just enough to produce 8.7×10^{-10} mol $Sn(s)$ per square centimeter of gold electrode surface.[8] Then no more current flows at E(cathode) $= +0.12$ V.

Reduction at $+0.12$ V is called **underpotential deposition.** It occurs at a potential that is not predicted to reduce Sn^{2+} to bulk $Sn(s)$. It is explained by the reaction

$$Sn^{2+} + 2e^- \rightleftharpoons Sn(\textit{monolayer on Au}) \qquad \text{(17-9)}$$

in which the product is a one-atom-thick layer of tin on gold. It is apparently more favorable to deposit a layer of tin atoms on gold than it is to deposit tin on bulk tin metal. Therefore, the potential for Reaction 17-9 is more positive than the potential for Reaction 17-8.

17-3 Coulometry

Coulometry is a chemical analysis based on counting the electrons used in a reaction. For example, cyclohexene can be titrated with Br_2 generated by electrolytic oxidation of Br^-:

$$2Br^- \rightarrow Br_2 + 2e^- \qquad \text{(17-10)}$$

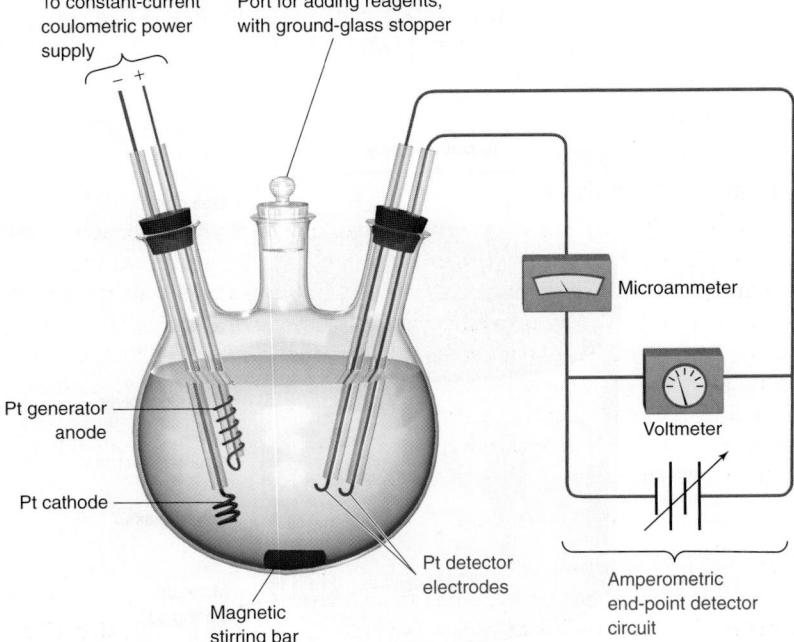

$$\text{Cyclohexene} \qquad trans\text{-1,2-Dibromocyclohexane}$$

(17-11)

The initial solution contains an unknown quantity of cyclohexene and a large amount of Br^-. When Reaction 17-10 has generated just enough Br_2 to react with all the cyclohexene, the moles of electrons liberated in Reaction 17-10 are equal to twice the moles of Br_2 and therefore twice the moles of cyclohexene.

The reaction is carried out *at a constant current* with the apparatus in Figure 17-9. Br_2 generated at the Pt anode at the left reacts with cyclohexene. When cyclohexene is consumed, the concentration of Br_2 suddenly rises, signaling the end of the reaction.

The rise in Br_2 concentration is detected by measuring the current between the two detector electrodes at the right in Figure 17-9. A voltage of 0.25 V applied between these two electrodes is not enough to electrolyze any solute, so only a tiny current of <1 μA flows through the microammeter. At the equivalence point, cyclohexene is consumed, $[Br_2]$ suddenly increases, and detector current flows by virtue of the reactions:

$$
\begin{aligned}
\textit{Detector anode:} \qquad & 2Br^- \rightarrow Br_2 + 2e^- \\
\textit{Detector cathode:} \qquad & Br_2 + 2e^- \rightarrow 2Br^-
\end{aligned}
$$

In practice, enough Br_2 is first generated in the absence of cyclohexene to give a detector current of 20.0 μA. When cyclohexene is added, the current decreases to a tiny value because Br_2 is consumed. Br_2 is then generated by the coulometric circuit, and the end point is taken when the detector again reaches 20.0 μA. Because the reaction is begun with Br_2 present, impurities that can react with Br_2 before analyte is added are eliminated.

Electrolysis current (not to be confused with the detector current) for the Br_2-generating electrodes could be controlled by a hand-operated switch. As the detector current approaches 20.0 μA, you could close the switch for shorter and shorter intervals. This practice is analogous to adding titrant dropwise from a buret near the end of a titration. The switch in the coulometer circuit serves as a "stopcock" for addition of Br_2 to the reaction. Modern instruments automate the entire procedure.

Coulometric methods are based on measuring the number of electrons that participate in a chemical reaction.

Both Br_2 and Br^- must be present for the detector half-reactions to occur. Prior to the equivalence point, there is Br^-, but virtually no Br_2.

FIGURE 17-9 Apparatus for coulometric titration of cyclohexene with Br_2. The solution contains cyclohexene, 0.15 M KBr, and 3 mM mercuric acetate in a mixed solvent of acetic acid, methanol, and water. Mercuric acetate catalyzes the addition of Br_2 to the olefin. [Information from D. H. Evans, "Coulometric Titration of Cyclohexene with Bromine," *J. Chem. Ed.* **1968**, *45*, 88.]

EXAMPLE Coulometric Titration

A 2.000-mL volume containing 0.611 3 mg of cyclohexene/mL is to be titrated in Figure 17-9. How much time is required for titration at a constant current of 4.825 mA?

Solution The moles of cyclohexene are

$$\frac{(2.000 \text{ mL})(0.611 \text{ 3 mg/mL})}{(82.146 \text{ mg/mmol})} = 0.014 \text{ 88 mmol}$$

In Reactions 17-10 and 17-11, 1 molecule of cyclohexene requires 1 molecule of Br_2, which requires $n = 2$ electrons. Equation 17-3 relates reaction time to moles of reaction:

$$\text{Moles of cyclohexene} = \frac{I \cdot t}{nF} \Rightarrow t = \frac{(\text{mol cyclohexene})nF}{I}$$

$$t = \frac{(0.014 \text{ 88} \times 10^{-3} \text{ mol})(2)(96 \text{ 485 C/mol})}{(4.825 \times 10^{-3} \text{ C/s})} = 595.1 \text{ s}$$

The reaction requires just under 10 min for completion.

n = 2 is the *dimensionless* number of electrons required per molecule of cyclohexene.

TEST YOURSELF How much time is required to titrate 1.000 mg of cyclohexene at 4.000 mA? (*Answer:* 587.3 s)

Commercial coulometers deliver electrons with an accuracy of ~0.1%. With extreme care, the Faraday constant has been measured to within several parts per million by coulometry.[9] Coulometers can generate species such as H^+, OH^-, Ag^+, I_2, and Ce^{4+} to titrate CO_2, sulfites in food, sulfide in wastewater, and Fe^{2+} in dietary supplements.[10] Unstable reagents such as Ag^{2+}, Cu^+, Mn^{3+}, and Ti^{3+} can be generated and used in situ.

Advantages of coulometry:
• precision
• sensitivity
• generation of unstable reagents in situ

The Latin *in situ* means "in place." Reagent is used right where it is generated.

In Figure 17-9, the reactive species (Br_2) is generated at the anode. The cathode products (H_2 from solvent and Hg from the catalyst) do not interfere with the reaction of Br_2 and cyclohexene. In some cases, however, H_2 or Hg could react with analyte. Then it is desirable to separate the counter electrode from the analyte, using the cell in Figure 17-10. H_2 bubbles innocuously out of the cathode chamber without mixing with the bulk solution.

Types of Coulometry

Coulometry employs either a *constant current* or a *controlled potential*. Constant-current methods, like the preceding Br_2/cyclohexene example, are called **coulometric titrations.** If we know the current and the time of reaction, we know how many coulombs have been delivered from Equation 17-2: $q = I \cdot t$.

Controlled-potential coulometry in a three-electrode cell is more selective than constant-current coulometry. Because the working electrode potential is constant, current decreases exponentially as analyte is consumed. Charge is measured by integrating current over the time of the reaction:

$$q = \int_0^t I \, dt \tag{17-12}$$

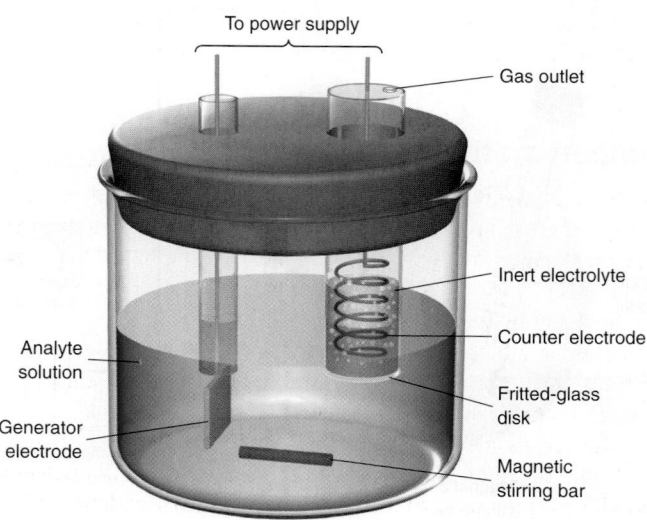

FIGURE 17-10 Isolating the counter electrode from analyte. Ions can flow through the porous fritted-glass disk. The liquid level in the counter electrode compartment should be higher than the liquid in the reactor so that analyte solution does not flow into the compartment.

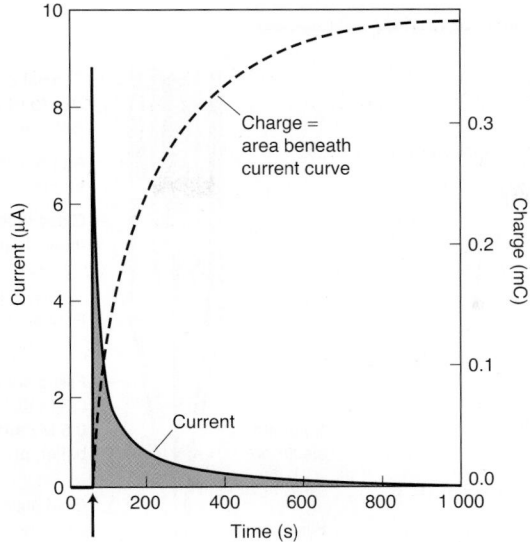

FIGURE 17-11 Controlled-potential coulometric oxidation of the sugar D-fructose in a three-electrode cell by the enzyme D-fructose dehydrogenase adsorbed on carbon particles on a carbon electrode at +500 mV versus Ag | AgCl. No current is observed in the absence of enzyme. A Pt auxiliary electrode in another compartment connected by a salt bridge completes the circuit and reduces a component of the buffer solution. [Data from S. Tsujimura, A. Nishina, Y. Kamitaka, and K. Kano, "Coulometric D-Fructose Biosensor Based on Direct Electron Transfer Using D-Fructose Dehydrogenase," *Anal. Chem.* **2009,** *81,* 9383.]

In Figure 17-11, the sugar D-fructose is oxidized at an electrode by controlled-potential coulometry catalyzed by the enzyme D-fructose dehydrogenase adsorbed on a porous carbon electrode held at +500 mV versus Ag | AgCl. Electrons from D-fructose go through the enzyme and into the electrode. Addition of D-fructose at the time indicated by the arrow gives a spike of current followed by exponential decay. The total charge transferred from D-fructose into the electrode is the area under the current-versus-time curve, given by the integral in Equation 17-12.

D-Fructose   5-Keto-D-fructose

With exponentially decreasing current, you need to decide when to stop integrating the area under the current-versus-time curve. One choice is to let the current decay to a set value. For example, the current (*above* the residual current) will ideally be 0.1% of its initial value when 99.9% of the analyte has been consumed. Alternatively, the current-versus-time curve can be fit to a theoretical mathematical shape that permits the integration.

17-4   Amperometry

In **amperometry,** we measure the electric current between a pair of electrodes that are driving an electrolysis reaction. One reactant is the intended analyte, and the measured current is proportional to the concentration of analyte. The measurement of dissolved O_2 with the **Clark electrode** in Box 17-2 is based on amperometry. A different kind of sensor based on conductivity—the "electronic nose"—is described in Box 17-3.

Biosensors[17] use biological components such as *enzymes, antibodies,* or DNA for a highly selective response to one analyte. Biosensors that generate electric or optical signals are most common. Examples of amperometric biosensors are those that measure lactate in perspiration or injured tissue,[18] H_2O_2 release from damaged liver cells,[19] biomarkers for breast cancer,[20] and specific proteins, nucleic acids, and small molecules.[21,22] We now describe blood glucose monitors, which are, by far, the most widely used biosensors.

Amperometry: Electric current is proportional to the concentration of analyte.

Coulometry: Total number of electrons used for a reaction tells us how much analyte is present.

Enzyme: A protein that catalyzes a biochemical reaction. The enzyme increases the rate of reaction by many orders of magnitude.

Antibody: A protein that binds to a specific target molecule called an **antigen.** Antibodies bind to foreign cells that infect your body to initiate their destruction or identify them for attack by immune system cells.

BOX 17-2 Clark Oxygen Electrode

The Clark oxygen electrode[11] is widely used in medicine and biology to measure dissolved oxygen by amperometry. Leland Clark, who invented the electrode, also invented the glucose monitor and the heart-lung machine.

The glass body of the microelectrode in the figure is drawn to a fine point with a 5-μm opening at the base. Inside the opening is a 10- to 40-μm-long plug of silicone rubber, which is permeable to O_2. Oxygen diffuses into the electrode through the rubber and is reduced at the Au tip on the Pt wire, which is held at -0.75 V with respect to the Ag | AgCl reference electrode:

Pt | Au cathode: $O_2 + 4H^+ + 4e^- \rightarrow 2H_2O$

Ag | AgCl anode: $4Ag + 4Cl^- \rightarrow 4AgCl + 4e^-$

A Clark electrode is calibrated by placing it in solutions of known O_2 concentration, and a graph of current versus $[O_2]$ is constructed. The electrode also contains a silver *guard electrode* extending most of the way to the bottom. The guard electrode is kept at a negative potential so that any O_2 diffusing in from the top of the electrode is reduced and does not interfere with measurement of O_2 diffusing in through the silicone membrane at the bottom. Similar electrodes have been designed for detection of NO and CO,[12] and NO_2^-.[13]

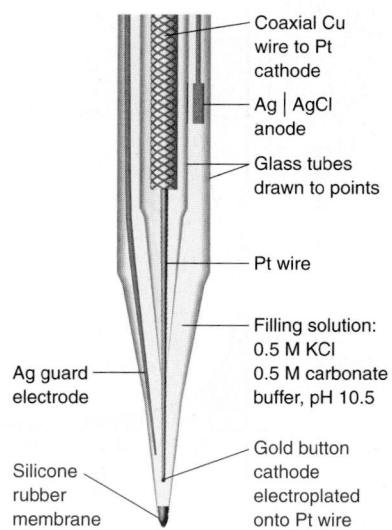

Clark oxygen microelectrode used to measure dissolved O_2 in small volumes. The tip of the cathode is plated with Au, which is less prone than Pt to fouling by adsorption of species from the test solution. [Information from N. P. Revsbech, "An Oxygen Microsensor with a Guard Column," *Limnol. Oceanogr.* **1989**, *34*, 474.]

BOX 17-3 What Is an "Electronic Nose"?

In the "old days," chemists prided themselves on their ability to identify compounds by odor. Smelling unknown chemicals is a bad idea because some vapors are toxic. Chemists are developing "electronic noses" that recognize odors to assess the freshness of meat, to find out if fruit is internally bruised, and to detect adulteration of food products.[14]

One approach to recognizing vapors is to coat interdigitated electrodes with an electronically conductive polymer, such as a derivative of polypyrrole.

Polypyrrole

When the polymer absorbs gaseous odor molecules, the electrical conductivity of the polymer changes. Different gases affect conductivity in different ways. Other sensor coatings are polymers containing conductive particles of silver or graphite. When the polymer absorbs small molecules, it swells and the conductivity decreases.

Pt/Au interdigitated electrodes

Interdigitated electrodes coated with conductive polymer to create an electronic nose. The conductivity of the polymer changes when it absorbs odor molecules. The spacing between "fingers" is ~0.25 mm.

One commercial "nose" has 32 sets of electrodes, each coated with a different polymer. The sensor yields 32 different responses when exposed to a vapor. The 32 changes are a "fingerprint" of the vapor. The sensor is "trained" by pattern recognition algorithms to recognize an odor by its fingerprint. Another conductivity sensor detects a cancer marker protein in blood.[15] Other electronic noses measure electric potential at the gates of field effect transistors (Section 15-8) or changes in optical absorption of polymers at the tips of optical fibers. Color changes in an array of 36 pigments detect toxic industrial chemicals at part per billion concentrations.[16]

Blood Glucose Monitor

Diabetes afflicts approximately 9% of the U.S. population. Many people with diabetes must monitor their blood sugar (glucose) several times a day to control their disease through diet and insulin injections. Figure 17-12 shows a home glucose monitor featuring a disposable test strip with two carbon working electrodes and a Ag | AgCl reference electrode. As little as 4 μL of blood applied in the circular opening at the right of the figure is wicked over all three electrodes by a thin *hydrophilic* ("water loving") mesh. A 20-s measurement begins when the liquid reaches the reference electrode.

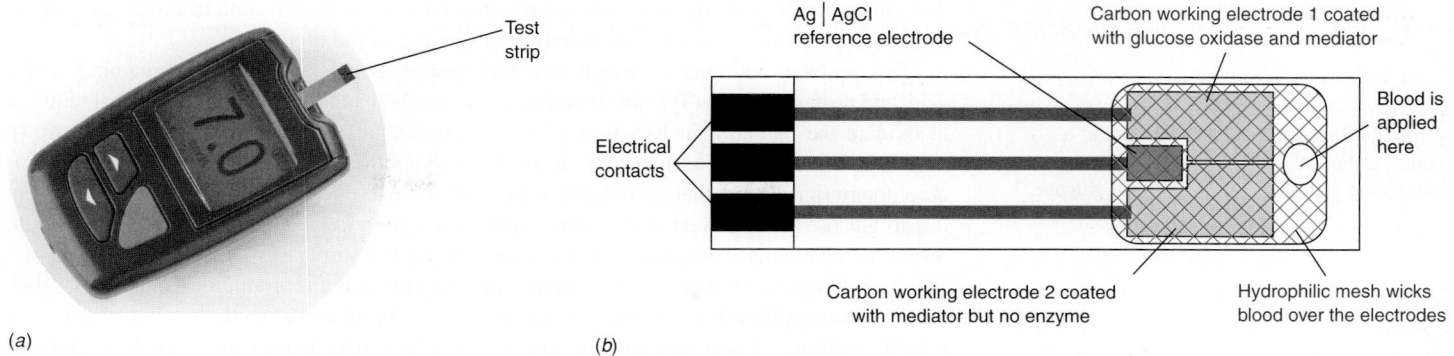

(a)   (b)

FIGURE 17-12 (*a*) Personal glucose monitor used by diabetics to measure blood sugar level. [© Roger Ashford/Alamy.] (*b*) Details of disposable test strip to which a drop of blood is applied.

Working electrode 1 is coated with the enzyme *glucose oxidase*[23] and a *mediator*, which we describe soon. The enzyme catalyzes the reaction of glucose with O_2:

Reaction in coating above working electrode 1:

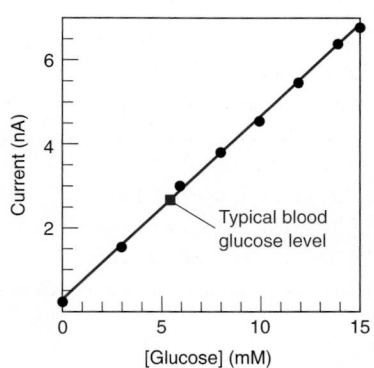

FIGURE 17-13 Response of an amperometric glucose electrode with dissolved O_2 concentration corresponding to an oxygen pressure of 0.027 bar, which is 20% lower than the typical concentration in subcutaneous tissue. [Data from S.-K. Jung and G. W. Wilson, "Polymeric Mercaptosilane-Modified Platinum Electrodes for Elimination of Interferants in Glucose Biosensors," *Anal. Chem.* **1996,** *68,* 591.]

In the absence of enzyme, the oxidation of glucose is negligible.

Early glucose monitors measured H_2O_2 from Reaction 17-13 by oxidation at a single working electrode, which was held at +0.6 V versus Ag | AgCl:

*Reaction at work-
ing electrode 1:*
$$H_2O_2 \rightarrow O_2 + 2H^+ + 2e^- \qquad (17\text{-}14)$$

The current is proportional to the concentration of H_2O_2, which, in turn, is proportional to the glucose concentration in blood (Figure 17-13).

One problem with early glucose monitors is that their response depended on the concentration of O_2 in the enzyme layer, because O_2 participates in Reaction 17-13. If the O_2 concentration is low, the monitor responds as though the glucose concentration were low.

A good way to reduce O_2 dependence is to incorporate into the enzyme layer a species that substitutes for O_2 in Reaction 17-13. A substance that transports electrons between the analyte (glucose, in this case) and the electrode is called a **mediator.**

A **mediator** transports electrons between the analyte and the working electrode. The mediator undergoes no net reaction itself.

Reaction in coating above working electrode 1:

$$\text{Glucose} + 2 \begin{bmatrix} \text{Fe} \\ \end{bmatrix}^+ \xrightarrow{\text{glucose oxidase}} \text{gluconolactone} + 2 \text{Fe} + 2H^+ \quad (17\text{-}15)$$

1,1'-Dimethylferricinium cation
Mediator

1,1'-Dimethylferrocene

Ferrocene contains flat, five-membered rings, similar to benzene. Each ring formally carries one negative charge, so the oxidation state of Fe, which sits between the rings, is +2. This molecule is called a *sandwich complex.*

The mediator consumed in Reaction 17-15 is then regenerated at the working electrode:

Reaction at working electrode 1:

$$\text{Fe} \xrightarrow[-e^-]{\text{working electrode}} \begin{bmatrix} \text{Fe} \end{bmatrix}^+ \qquad (17\text{-}16)$$

The mediator lowers the required working electrode potential from 0.6 V to 0.2 V versus Ag|AgCl, thereby improving the stability of the sensor and eliminating some interference by other species in the blood.

The current at the working electrode is proportional to the concentration of ferrocene, which, in turn, is proportional to the concentration of glucose in the blood.

One problem with glucose monitors is that species such as ascorbic acid (vitamin C), uric acid, and acetaminophen (Tylenol) found in blood can be oxidized at the same potential required to oxidize the mediator in Reaction 17-16. To correct for this interference, the test strip in Figure 17-12 has a second indicator electrode coated with mediator, *but not with glucose oxidase*. Interfering species that are reduced at electrode 1 are also reduced at electrode 2. The current due to glucose is the current at electrode 1 *minus* the current at electrode 2 (both measured with respect to the reference electrode). Now you see why the test strip has two working electrodes.

A challenge is to manufacture glucose monitors in such a reproducible manner that they do not require calibration. A user expects to add a drop of blood to the test strip and get a reliable reading without first constructing a calibration curve from known concentrations of glucose in blood. Each lot of test strips is highly reproducible and calibrated at the factory.

An ingenious method uses a glucose monitor to measure a range of analytes, including drugs, small biological molecules, disease marker proteins, and heavy metal ions.[25] Target analyte liberates a bound form of the enzyme invertase that converts table sugar (sucrose) into glucose, which is measured by the glucose monitor. Through this approach, glucose monitors might be broadly applied in medical diagnostics and environmental monitoring.

"Electrical Wiring" of Enzymes and Mediators for Blood Glucose Monitor

Demand for glucose monitors provides an economic stimulus for research that continues to improve the performance of home glucose monitors.[26] Noteworthy advances include (1) monitoring the reaction by coulometry instead of amperometry, (2) using a different enzyme to catalyze glucose oxidation, and (3) "electrical wiring" to increase the rate of reaction and to prevent reactants from diffusing away from the working electrode.

In *amperometry*, current flowing during glucose oxidation is measured. In *coulometry*, the number of electrons required to oxidize the glucose in a blood sample is counted. Amperometry measures the *rate of oxidation*. Coulometry measures the *number of molecules that have been oxidized*. The rate of reaction, and therefore the current, depends on temperature, but total charge transferred during oxidation is independent of temperature. Therefore, the coulometric measurement is independent of temperature. Total charge transferred is also insensitive to the activity of the enzyme (how quickly it works) and the mobility of the mediator, both of which affect current. Current is also affected by depletion of glucose during the measurement. In coulometry, the goal is to use up all the glucose.

Replacing the enzyme glucose oxidase with *glucose dehydrogenase* eliminates O_2 as a reactant. A *cofactor* called PQQ, which is bound to glucose dehydrogenase, receives $2H^+ + 2e^-$ during the oxidation.

Glucose   +   Pyrroloquinoline quinone (PQQ) cofactor in glucose dehydrogenase $\xrightarrow{\text{glucose dehydrogenase}}$

Gluconolactone   +   PQQH$_2$ (17-17)

Unlike Reaction 17-13, O_2 is not involved in Reaction 17-17. Therefore, there is no dependence of the response to dissolved O_2.

In an "electrically wired" polymer gel on the surface of a carbon electrode (Figure 17-14), the enzyme and an osmium mediator are tethered to a polymer backbone. The $PQQH_2$ product of Reaction 17-17 is oxidized back to $PQQ + 2H^+$ by a nearby Os^{3+}. Os^{3+} is reduced to Os^{2+}

A modified sensor measures glucose at a concentration of 2 fM in a 30-μL volume containing just 36 000 molecules of glucose.[24]

A **cofactor** is a small, nonprotein molecule that is bound to an enzyme and is necessary for the activity of the enzyme.

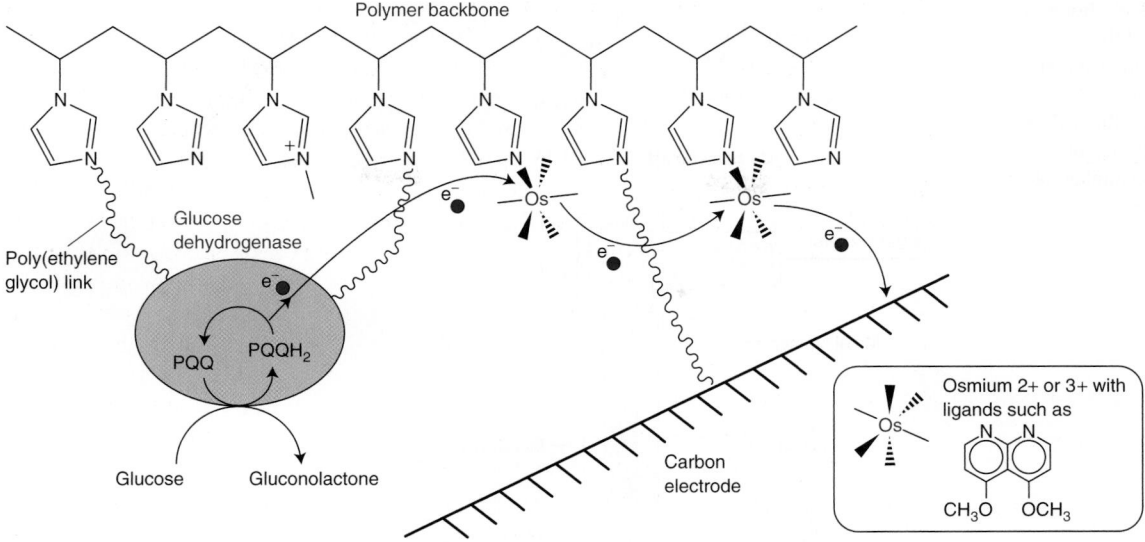

Polymer backbone

Poly(ethylene glycol) link

Glucose dehydrogenase

Glucose → Gluconolactone

PQQ PQQH₂

Carbon electrode

Osmium 2+ or 3+ with ligands such as

CH_3O OCH_3

FIGURE 17-14 "Electrically wired" glucose dehydrogenase. The enzyme catalyzes the oxidation of glucose, reducing PQQ to $PQQH_2$. $PQQH_2$ is oxidized back to PQQ + $2H^+$ by Os^{3+}. Electrons move through successive osmium atoms until they reach the carbon anode. All members of the redox chain are bound to a polymer backbone.

in the process. Os^{2+} can exchange an electron with another Os^{3+}. Electrons are transported rapidly from Os to Os until they reach a carbon anode. Electrons then flow through a circuit to the Ag | AgCl counter electrode where AgCl is reduced to Ag + Cl^-.

"Electrical wiring" of the enzyme and the osmium mediators increases the current by a factor of 10 to 100 compared with an enzyme/mediator layer deposited onto an electrode. Covalent attachment of osmium to the polymer prevents the mediator from diffusing to the counter electrode, where it would react and create a large background current. Ligands for osmium are chosen so that the mildest possible potential (+0.1 V versus Ag | AgCl) can be applied to the electrode to oxidize glucose. At this potential, common oxidizable interferents produce acceptably small errors in glucose measurement.

The newest glucose monitor test strips require only 0.3 μL of blood for a measurement, significantly reducing the pain experienced by people who need to measure their glucose several times each day. Glucose in the entire volume is oxidized in about 1 minute and current is measured as a function of time. Integrating current versus time (Equation 17-12) gives the total charge required to reduce glucose.

Rotating Disk Electrode

A molecule has three ways to reach the surface of an electrode: (1) *diffusion* through a concentration gradient; (2) *convection*, which is the movement of bulk fluid by physical means such a stirring or boiling; and (3) *migration*, which is the attraction or repulsion of an ion by a charged surface. A common working electrode for amperometry is the **rotating disk electrode,** for which convection and diffusion control the flux of analyte to the electrode.[27,28]

When the electrode in Figure 17-15a is spun at ~1 000 revolutions per minute, a vortex is established that brings analyte near the electrode rapidly by convection. If the potential is great enough and the reaction is fast enough, the concentration of analyte at the electrode surface is nearly 0. The concentration gradient is shown schematically in Figure 17-15b. Analyte must traverse the final, short distance (~10–100 μm) by diffusion alone.

The rate at which analyte diffuses from bulk solution to the surface of the electrode is proportional to the concentration difference between the two regions:

$$\text{Current} \propto \text{rate of diffusion} \propto [C]_0 - [C]_s \qquad \textbf{(17-18)}$$

where $[C]_0$ is the concentration in bulk solution and $[C]_s$ is the concentration at the surface of the electrode. At sufficiently great potential, the rate of reaction at the electrode is so fast that $[C]_s \ll [C]_0$ and Equation 17-18 reduces to

$$\text{Limiting current} = \text{diffusion current} \propto [C]_0 \qquad \textbf{(17-19)}$$

Three ways for analyte to reach an electrode:
• diffusion
• convection
• migration

The symbol ∝ means "is proportional to."

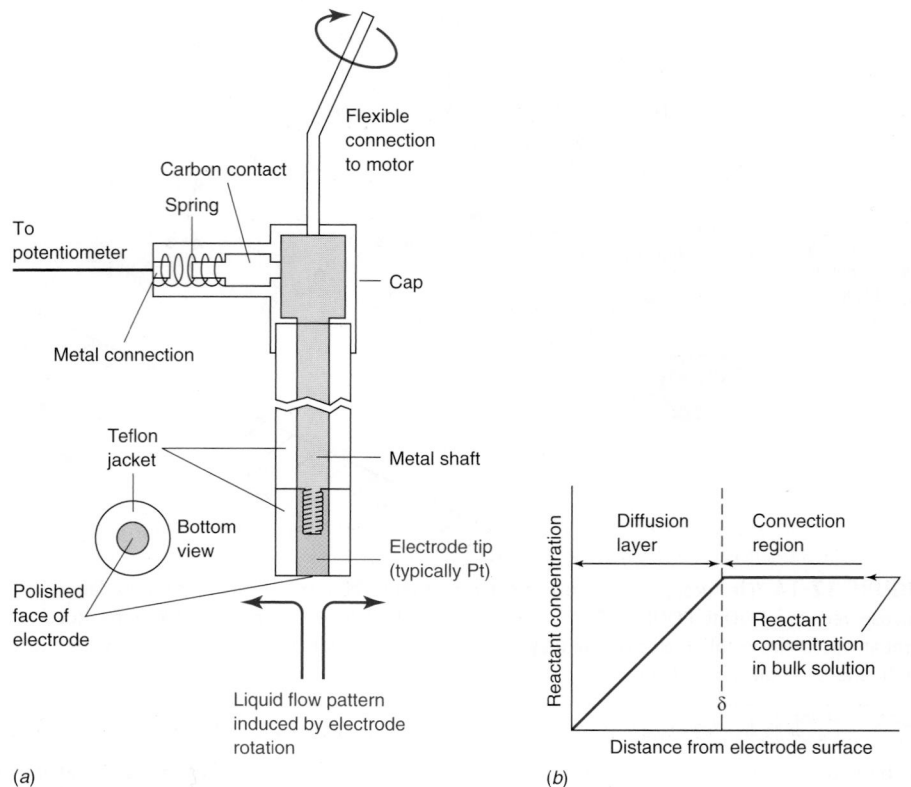

(a)

(b)

The limiting current is called the **diffusion current** because it is governed by the rate at which analyte diffuses to the electrode. The proportionality of diffusion current to bulk-solute concentration is the basis for quantitative analysis by amperometry and, in the next section, voltammetry.

The faster a rotating disk electrode spins, the thinner is the diffusion layer in Figure 17-15b and the greater is the diffusion current. A rapidly rotating Pt electrode can measure 20 nM H_2O_2 in rainwater.[29] H_2O_2 is oxidized to O_2 at $+0.4$ V (versus S.C.E.) at the Pt surface, and the current is proportional to $[H_2O_2]$ in the rainwater.

17-5 Voltammetry

Voltammetry is a collection of techniques in which the relation between current and voltage is observed during electrochemical processes.[30] The **voltammogram** in Figure 17-16a is a graph of current versus working electrode potential for a mixture of ferricyanide and ferrocyanide being oxidized or reduced at a rotating disk electrode. By convention, current is positive when analyte is oxidized at the working electrode. The limiting (diffusion) current for oxidation of $Fe(CN)_6^{4-}$ is observed at potentials above $+0.5$ V (versus S.C.E.).

$$Fe(CN)_6^{4-} \rightarrow Fe(CN)_6^{3-} + e^-$$

Ferrocyanide   Ferricyanide
Fe(II)   Fe(III)

In this region, current is governed by the rate at which $Fe(CN)_6^{4-}$ diffuses to the electrode. Diffusion current is proportional to the bulk concentration of $Fe(CN)_6^{4-}$ (Figure 17-16b) just as it is for the rotating disk electrode in Equation 17-19. Below 0 V, there is another plateau corresponding to the diffusion current for reduction of $Fe(CN)_6^{3-}$, whose concentration is constant in all the solutions.

The relatively inert metal electrodes, platinum and gold, are mainly used for oxidations. Their useful potential ranges are shown in Table 17-2. At negative potentials, water is reduced to H_2 on Pt and Au, which limits their utility. Glassy carbon electrodes are suitable for oxidation and reduction. Boron-doped chemical-vapor-deposited diamond (Figure 17-17) is an exceptionally inert, commercially available carbon electrode with a wide potential window, low background current,[31] resistance to surface oxidation,[32] high overpotential for O_2 evolution,[32] and visible and infrared transparency. Single atomic layers of graphite are called

Diamond is an electrical insulator. Replacing one of every 500 C atoms with B makes diamond a semiconductor with a conductivity 10^{-5} to 10^{-4} that of metals.

Question Is B-doped diamond p-type or n-type? Consult Figure 15-37.

Answer: p-type

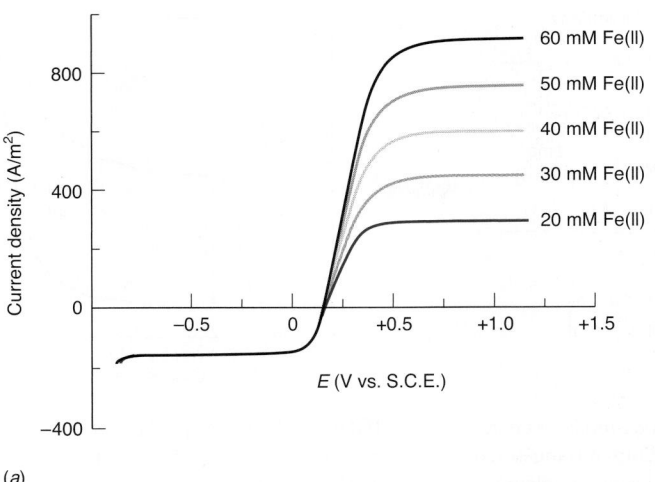

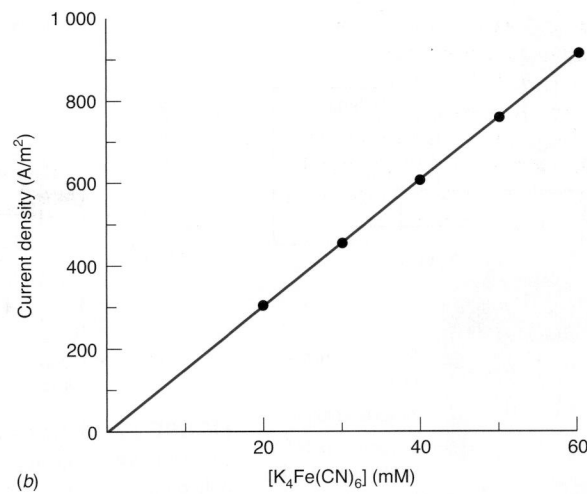

(a)

(b)

FIGURE 17-16 (*a*) Voltammograms for a mixture of 10 mM $K_3Fe(CN)_6$ and 20–60 mM $K_4Fe(CN)_6$ in 0.1 M Na_2SO_4 at a glassy carbon rotating electrode. Rotation speed = 2 000 revolutions/min and voltage sweep rate = 5 mV/s. (*b*) Dependence of limiting current on $K_4Fe(CN)_6$ concentration. [Data from J. Nikolic, E. Expósito, J. Iniesta, J. González-Garcia, and V. Montiel, "Theoretical Concepts and Applications of a Rotating Disk Electrode," *J. Chem. Ed.* **2000,** *77,* 1191.]

TABLE 17-2 Approximate working electrode potential range in 1 M H_2SO_4

Electrode	Potential range (V vs. S.C.E.)
Pt	−0.2 to +0.9 V
Au	−0.3 to +1.4 V
Hg	−1.3 to +0.1 V
Glassy carbon	−0.8 to +1.1 V
B-doped diamond[a]	−1.5 to +1.7 V
Fluorinated B-doped diamond[b]	−2.5 to +2.5 V

a. A. E. Fischer, Y. Show, and G. M. Swain, "Electrochemical Performance of Diamond Thin-Film Electrodes from Different Commercial Sources," Anal. Chem. **2004,** 76, 2553; Y. Dai, G. M. Swain, M. D. Porter, and J Zak, "Optically Transparent Carbon Electrodes," Anal. Chem. **2008,** 80, 14; J. Stotter, Y. Show, S. Wang, and G. Swain, "Comparison of Electrical, Optical, and Electrochemical Properties of Diamond and Indium Tin Oxide Thin-Film Electrodes," Chem. Mater. **2005,** 17, 4880.

b. S. Ferro and A. De Battisti, "The 5-V Window of Polarizability of Fluorinated Diamond Electrodes in Aqueous Solution," Anal. Chem. **2003,** 75, 7040.

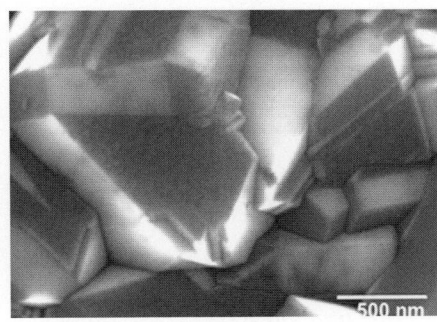

FIGURE 17-17 Scanning electron micrograph of boron-doped diamond electrode polycrystalline surface. [Y. Song and G. M. Swain, "Development of a Method for Total Inorganic Arsenic Analysis Using Anodic Stripping Voltammetry and a Au-coated, Diamond Thin-Film Electrode," *Anal. Chem.* **2007,** *79,* 2412. Reprinted with permission © 2007, American Chemical Society.]

graphene. Electrochemical oxidation (called "anodization") of graphene at 2.0 V versus Ag | AgCl in phosphate buffer produces an electrode with an electrochemical potential range comparable to that of B-doped diamond.[33]

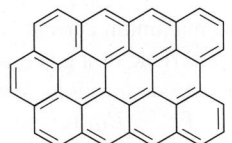

Graphene is a single plane of hexagonal carbon. The Nobel Prize in Physics in 2010 was awarded to A. Geim and K. Novoselov for their isolation of graphene.

Polarography

Voltammetry conducted with a **dropping-mercury electrode** is called **polarography.** The dispenser in Figure 17-18 suspends one drop of mercury from the bottom of the capillary. After current and voltage are measured, the drop is mechanically dislodged. Then a fresh drop is suspended and the next measurement is made. Freshly exposed Hg yields reproducible current-potential behavior. The current for other electrodes, such as Pt, depends on surface condition and is not as reproducible as that of freshly dispensed Hg.

Most reactions studied with the Hg electrode are reductions. At a Pt surface, reduction of H^+ competes with reduction of many analytes:

$$2H^+ + 2e^- \rightarrow H_2(g) \qquad E^\circ = 0$$

Polarography has been largely replaced by voltammetry with electrodes that do not present the toxicity hazard of Hg. Mercury is still the electrode of choice for stripping analysis, which is the most sensitive voltammetric technique. For cleaning up Hg spills, see note 34.

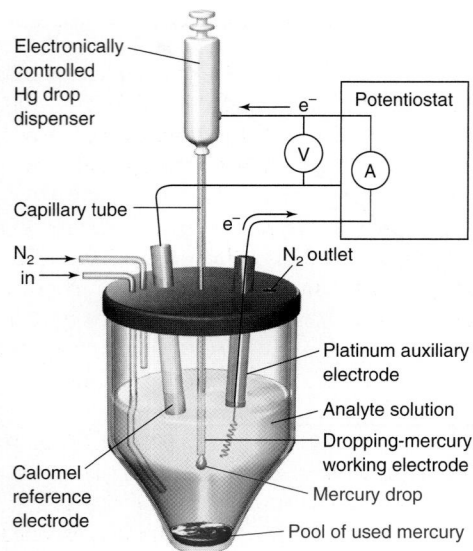

FIGURE 17-18 Polarography apparatus featuring a dropping-mercury working electrode. J. Heyrovský invented polarography in 1922, for which he received the Nobel Prize in 1959.

An **amalgam** is anything dissolved in Hg.

Polarograms in the older literature have large oscillations superimposed on the wave in Figure 17-20a. For the first 50 years of polarography, current was measured continuously as Hg flowed from a capillary tube. Each drop grew until it fell off and was replaced by a new drop. Current oscillated from a low value when the drop was small to a high value when the drop was big.

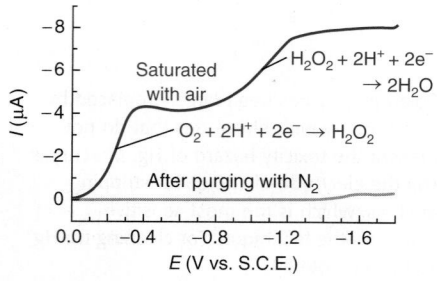

FIGURE 17-21 Sampled current polarogram of 0.1 M KCl saturated with air or purged (bubbled) with N_2 to remove O_2.

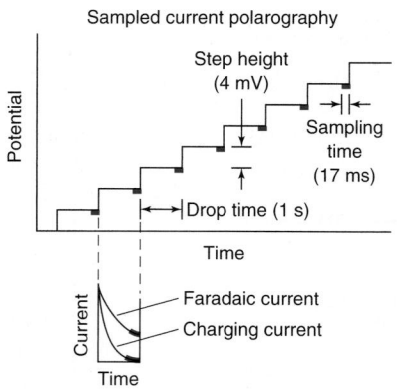

FIGURE 17-19 *Staircase voltage profile* used in *sampled current polarography*. Current is measured only during the intervals shown by heavy, colored lines. Potential is scanned toward more negative values as the experiment progresses. Lower graph shows that charging current decays more rapidly than faradaic current after each voltage step.

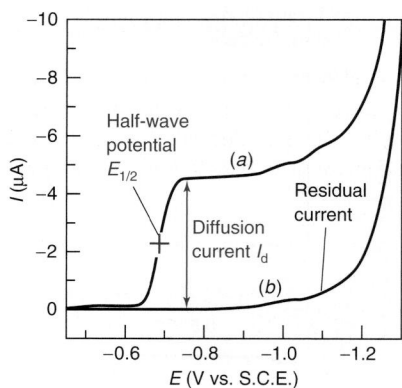

FIGURE 17-20 *Sampled current polarograms* of (a) 5 mM Cd^{2+} in 1 M HCl and (b) 1 M HCl alone.

Table 17-1 shows a large *overpotential* for reduction of H^+ at the Hg surface. Reactions that are thermodynamically less favorable than reduction of H^+ can be carried out without competitive reduction of H^+. In neutral or basic solutions, even alkali metal (Group 1) cations are reduced more easily than H^+. Furthermore, reduction of a metal into a mercury *amalgam* is more favorable than reduction to the solid state:

$$K^+ + e^- \rightarrow K(s) \qquad E° = -2.936 \text{ V}$$
$$K^+ + e^- + Hg \rightarrow K(in\ Hg) \qquad E° = -1.975 \text{ V}$$

Mercury is not useful for studying oxidations because Hg is oxidized in noncomplexing media near $+0.25$ V (versus S.C.E.). If the concentration of Cl^- is 1 M, Hg is oxidized near 0 V because Hg(II) is stabilized by Cl^-:

$$Hg(l) + 4Cl^- \rightleftharpoons HgCl_4^{2-} + 2e^-$$

One way to conduct a measurement is by **sampled current polarography** with the *staircase voltage ramp* in Figure 17-19. After each drop of Hg is dispensed, the potential is made more negative by 4 mV. After waiting almost 1 s, current is measured during the last 17 ms of the life of each Hg drop. The **polarographic wave** in Figure 17-20a results from reduction of Cd^{2+} analyte to form an amalgam:

$$Cd^{2+} + 2e^- \rightarrow Cd(in\ Hg)$$

The potential at which half the maximum current is reached in Figure 17-20a, called the **half-wave potential** ($E_{1/2}$), is characteristic of a given analyte in a given medium and can be used for qualitative analysis. For electrode reactions in which reactants and products are both in solution, such as $Fe^{3+} + e^- \rightleftharpoons Fe^{2+}$, $E_{1/2}$ (expressed with respect to S.H.E.) is nearly equal to $E°$ for the half-reaction.

For quantitative analysis, the *diffusion current* in the plateau region is proportional to the concentration of analyte. Diffusion current is measured from the baseline recorded without analyte in Figure 17-20b. **Residual current** in the absence of analyte is due to reduction of impurities in solution and on the surface of the electrodes. Near -1.2 V in Figure 17-20, current increases rapidly as reduction of H^+ to H_2 commences.

For quantitative analysis, the limiting current should be controlled by the rate at which analyte can diffuse to the electrode. We minimize convection by using an unstirred solution. We minimize migration (electrostatic attraction of analyte) by using a high concentration of *supporting electrolyte*, such as 1 M HCl in Figure 17-20.

Oxygen must be absent because O_2 gives two polarographic waves when it is reduced to H_2O_2 and then to H_2O (Figure 17-21). Typically, N_2 is bubbled through analyte solution for 10 min to remove O_2.[35] Then N_2 flow in the gas phase is continued to keep O_2 out. The liquid should not be purged with N_2 during a measurement, because we do not want convection of analyte to the electrode.

When a power supply pumps electrons into or out of an electrode, the charged surface of the electrode attracts ions of opposite charge. The charged electrode and the oppositely charged ions next to it constitute the **electric double layer.**

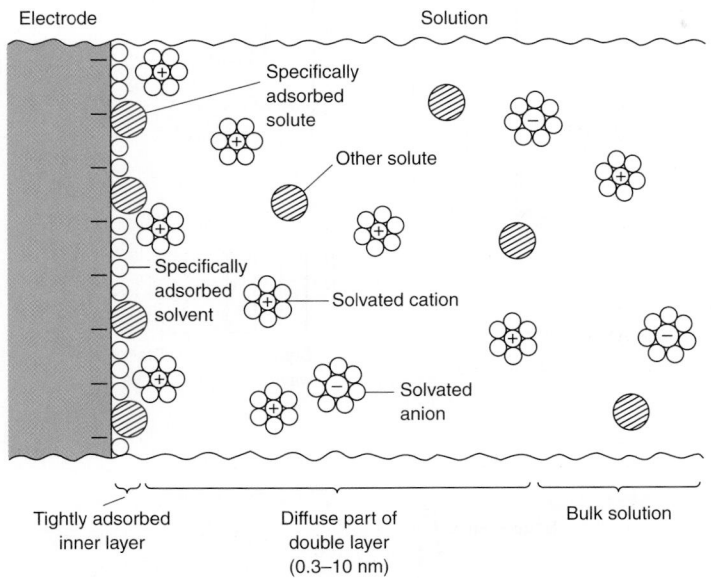

Electrode-solution interface. The tightly adsorbed inner layer (also called the *compact, Helmholtz,* or *Stern* layer) may include solvent and solute molecules. Cations in the inner layer do not completely balance the charge of the electrode. Therefore, excess cations are required in the *diffuse part of the double layer* for charge balance.

A given solution has a *potential of zero charge* at which there is no excess charge on the electrode. This potential is -0.58 V (versus a calomel electrode containing 1 M KCl) for a mercury electrode immersed in 0.1 M KBr. It shifts to -0.72 V for the same electrode in 0.1 M KI.

The first layer of molecules at the surface of the electrode is *specifically adsorbed* by van der Waals and electrostatic forces. The adsorbed solute could be neutral molecules, anions, or cations. Iodide is more strongly adsorbed than bromide, so the potential of zero charge for KI is more negative than for KBr. A more negative potential is required to expel adsorbed iodide from the electrode surface.

The next layer beyond the specifically adsorbed layer is rich in cations attracted by the negative electrode. The excess of cations decreases with increasing distance from the electrode. This region, whose composition is different from that of bulk solution, is called the *diffuse part of the double layer* and is typically 0.3–10 nm thick. The thickness is controlled by the balance between attraction toward the electrode and randomization by thermal motion.

When a species is created or destroyed by an electrochemical reaction, its concentration near the electrode is different from its concentration in bulk solution (Figure 17-15b and Color Plate 13). The region containing excess product or decreased reactant is called the *diffusion layer* (not to be confused with the diffuse part of the double layer).

Faradaic and Charging Currents

The current we seek to measure in voltammetry is **faradaic current** due to reduction or oxidation of analyte at the working electrode. In Figure 17-20a, faradaic current comes from reduction of Cd^{2+} at the Hg electrode. Another current, called **charging current** (or *capacitor current* or *condenser current*) interferes with every measurement. We step the working electrode to a more negative potential by forcing electrons into the electrode from the potentiostat. In response, cations in solution flow toward the electrode, and anions flow away from the electrode (Box 17-4). This flow of ions and electrons, called *charging current*, is not from redox reactions. We try to minimize charging current, which obscures faradaic current. Charging current usually controls the detection limit in voltammetry.

The bottom of Figure 17-19 shows the behavior of faradaic and charging currents after each potential step. Faradaic current decays because analyte cannot diffuse to the electrode fast enough to sustain the high reaction rate. Charging current decays even faster because ions near the electrode redistribute themselves rapidly. A second after each potential step, faradaic current is still significant and charging current is small.

Square Wave Voltammetry

The most efficient voltage profile for voltammetry, called **square wave voltammetry,** uses the waveform in Figure 17-22, which consists of a square wave superimposed on a staircase.[36] During each cathodic pulse, analyte is reduced at the electrode surface. During the anodic pulse, analyte that was just reduced is reoxidized. The square wave polarogram in Figure 17-23 is the *difference* in current between intervals 1 and 2 in Figure 17-22. Electrons flow from the electrode to analyte at point 1 and in the reverse direction at point 2. Because the two currents have opposite signs, their difference is larger than either current alone. When

Faradaic current: due to redox reaction at the electrode

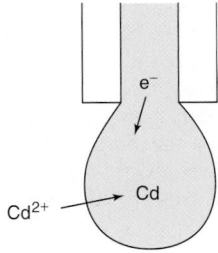

Charging current: due to electrostatic attraction or repulsion of ions in solution and electrons in the electrode

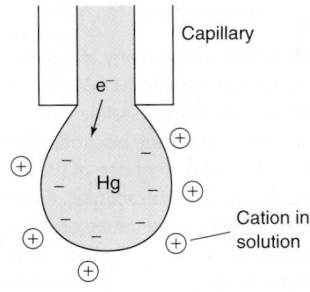

Faradaic current is the signal of interest. Charging current obscures the signal of interest, so we seek to minimize charging current.

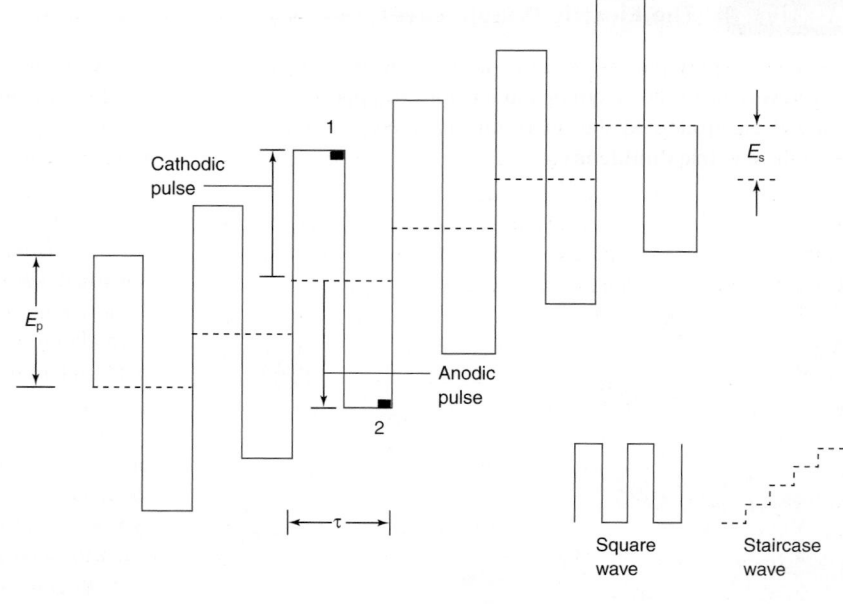

FIGURE 17-22 Waveform for square wave voltammetry. Typical parameters are pulse height (E_p) = 25 mV, step height (E_s) = 10 mV, and pulse period (τ) = 5 ms. Current is measured in regions 1 and 2. Optimum values are E_p = 50/n mV and E_s = 10/n mV, where n is the number of electrons in the half-reaction.

Advantages of square wave voltammetry:
- increased signal
- derivative (peak) shape permits better resolution of neighboring signals
- faster measurement

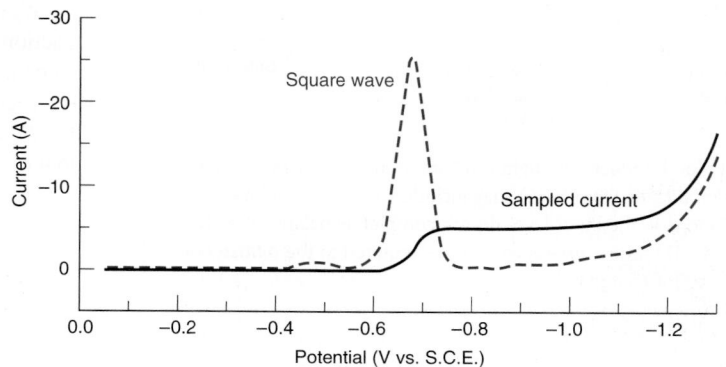

FIGURE 17-23 Comparison of polarograms of 5 mM Cd^{2+} in 1 M HCl. Waveforms are shown in Figures 17-19 and 17-22. Sampled current: drop time = 1 s, step height = 4 mV, sampling time = 17 ms. Square wave: drop time = 1 s, step height = 4 mV, pulse period = 67 ms, pulse height = 25 mV, sampling time = 17 ms.

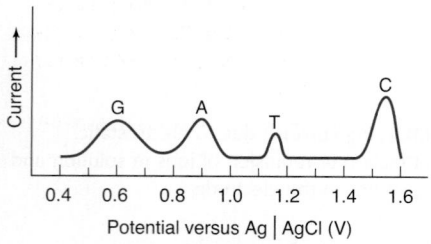

FIGURE 17-24 Voltammetry of double stranded DNA (30 mg/L in 10 mM KCl plus 10 mM phosphate-buffered saline solution, pH 7) with an anodized graphene electrode. Each peak corresponds to oxidation of one of the four nucleotide bases: guanine, adenine, thymine, or cytosine. Four peaks could not be observed with graphene that was not anodized nor with boron-doped diamond or glassy carbon electrodes.
[Information from C. X. Lim, H. Y. Hoh, P. K. Ang, and K. P. Loh, "Direct Voltammetric Detection of DNA and pH Sensing on Epitaxial Graphene," *Anal. Chem.* **2010,** *82,* 7387.]

the difference is plotted, the shape of the square wave polarogram in Figure 17-23 is essentially the derivative of the sampled current polarogram.

The signal in square wave voltammetry is increased relative to a sampled current voltammogram and the wave becomes peak shaped. Signal is increased because reduced product from each cathodic pulse is right at the surface of the electrode waiting to be oxidized by the next anodic pulse. Each anodic pulse provides a high concentration of reactant at the surface of the electrode for the next cathodic pulse. The detection limit is reduced from $\sim 10^{-5}$ M for sampled current voltammetry to $\sim 10^{-7}$ M in square wave voltammetry. Because it is easier to resolve neighboring peaks than neighboring waves, square wave voltammetry can resolve species whose half-wave potentials differ by ~ 0.05 V, whereas potentials must differ by ~ 0.2 V to be resolved in sampled current voltammetry. Figure 17-24 shows the complete resolution of the four nucleotide bases in double stranded DNA by voltammetry on an anodized graphene electrode.

Square wave voltammetry is faster than other voltammetric techniques. The square wave polarogram in Figure 17-23 was recorded in one-fifteenth of the time required for the sampled current polarogram. In principle, the shorter the pulse period, τ, in Figure 17-22, the greater the current that will be observed. With τ = 5 ms (a practical lower limit) and E_s = 10 mV, an entire square wave polarogram with a 1-V width is obtained in 0.5 s. Such rapid sweeps allow voltammograms to be recorded on individual components as they emerge from a chromatography column. Box 17-5 describes a powerful clinical analysis that uses square wave voltammetry for detection.

CHAPTER 17 Electroanalytical Techniques

BOX 17-5 Aptamer Biosensor for Clinical Use

Aptamers are ~15–40-base molecules of DNA (deoxyribonucleic acid, Appendix L) or RNA (ribonucleic acid) that *strongly* and *selectively* bind to a specific molecule[37] or the surface of one type of living cell. An aptamer for a desired target molecule is chosen from a pool of ~10^{15} random DNA or RNA sequences by successive cycles of binding to the target, removing unbound material, and replicating the bound nucleic acid. Once the sequence of nucleic acids in an aptamer for a specific target is known, that aptamer can be synthesized in large quantities. The aptamer behaves as a custom-made, synthetic "antibody." It can bind to a section of a macromolecule, such as a protein, or it can engulf a small molecule, as shown in panel *a*. Unlike antibodies, which are fragile proteins that must be refrigerated for storage, aptamers have a long shelf life at room temperature and are therefore more practical candidates for highly specific chemical sensors.[38]

Panel *b* shows the principle of operation of a biosensor that can measure the concentration of the cancer chemotherapeutic agent doxorubicin in blood continuously in real time. The heart of the sensor is a DNA aptamer bound to a gold electrode through a sulfur atom on one end. A redox-active molecule of methylene blue is attached to the other end. In the absence of doxorubicin, the aptamer is elongated and methylene blue is far from the electrode. When doxorubicin binds to the aptamer, the aptamer conformation changes and brings methylene blue close to the electrode. The rate of electron transfer from the electrode to methylene blue is monitored by square wave voltammetry. Current increases in the presence of doxorubicin over the therapeutically relevant range of 0.2 to 8 μM. Different analytes could be measured by the same technique by changing the aptamer.

Two key technical feats were required for a practical sensor: (1) Fouling of the electrode by macromolecular components of blood was eliminated by slow flow of a 125-μm-thick layer of buffer between the electrode surface and the blood. The small doxorubicin molecule diffuses across the buffer layer, but macromolecules do not. (2) Slow drift of the electrical response was reduced by a factor of 15 and the signal-to-noise ratio was increased by a factor of 3 by manipulating voltammetry parameters to emphasize the different response to the two conformations of the aptamer.

Drugs are metabolized at different rates in different patients. A goal is to measure drugs in the blood of a patient during treatment to customize the dose for that person.

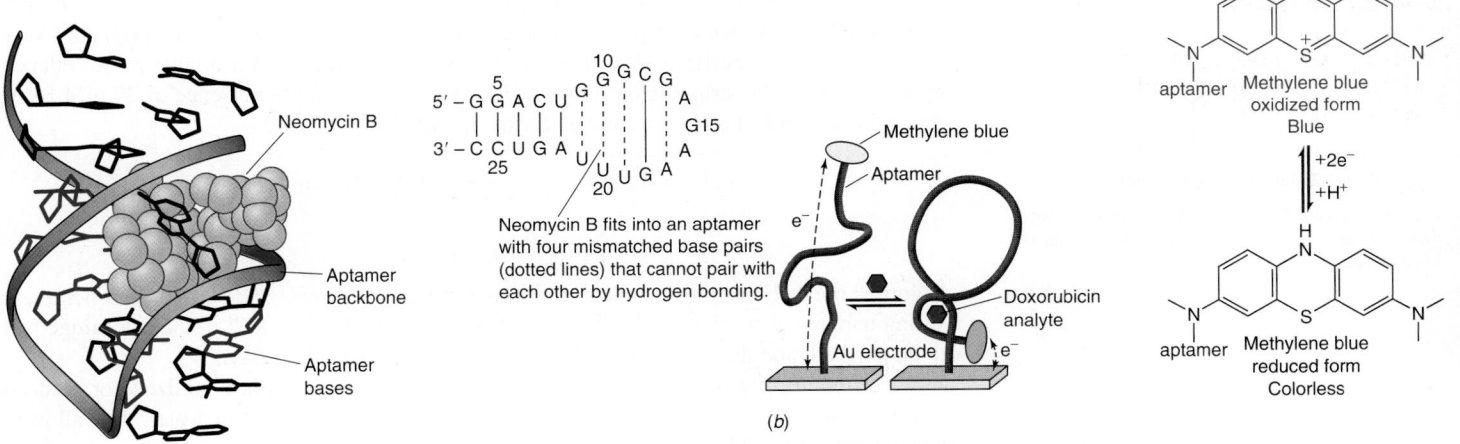

(a)

(a) RNA aptamer with 26 bases binds antibiotic neomycin B. Nucleotide bases U, G, A, and T are shown in Appendix L. [Information from L. Jiang, A. Majumdar, W. Hu, T. J. Jaishree, W. Xu, and D. J. Patel, "Saccharide-RNA Recognition in a Complex Formed Between Neomycin B and an RNA Aptamer," *Structure* **1999**, *7*, 817.]

(b) Principle of operation of biosensor. When doxorubicin binds to aptamer, redox-active methylene blue is brought close to electrode and current increases. [Information from B. S. Ferguson, D. A. Hoggarth, D. Maliniak, K. Ploense, R. J. White, N. Woodward, K. Hsieh, A. J. Bonham, M. Eisenstein, T. E. Kippin, K. W. Plaxco, and H. T. Soh, "Real-Time, Aptamer-Based Tracking of Circulating Therapeutic Agents in Living Animals," *Sci. Transl. Med.* **2013**, *5*, 213ra165.]

Stripping Analysis

In **stripping analysis,** analyte from a dilute solution is concentrated into a thin film of Hg or other electrode material, usually by electroreduction. The electroactive species is then *stripped* from the electrode by reversing the direction of the voltage sweep. The potential becomes more *positive*, *oxidizing* the species back into solution. Peak current measured during oxidation is proportional to the quantity of analyte that was deposited. Figure 17-25 shows an anodic stripping voltammogram of Cd, Pb, and Cu from honey.

Stripping is the most sensitive voltammetric technique (Table 17-3), because analyte is concentrated from a dilute solution. The longer the period of concentration, the more sensitive is the analysis. Only a fraction of analyte from the solution is deposited, so deposition must be done for a reproducible time (such as 5 min) with reproducible stirring.

There is incentive to replace Hg as an electrode because Hg is toxic and expensive to discard. Films of bismuth[39,40] or tin[41] partially mimic desirable properties of Hg in providing a fairly reproducible surface and reaching strongly reducing potentials before reducing H_2O to H_2.

Anodic stripping analysis:

1. Concentrate analyte onto electrode by reduction
2. Reoxidize analyte by scanning to more positive potential
3. Peak current during oxidation is proportional to analyte concentration

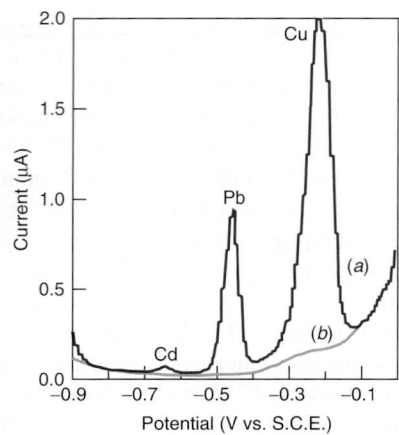

FIGURE 17-25 (a) Anodic stripping voltammogram of honey dissolved in water and acidified to pH 1.2 with HCl. Cd, Pb, and Cu were reduced from solution into a thin film of Hg for 5 min at -1.4 V (versus S.C.E.) prior to recording the voltammogram. (b) Voltammogram obtained without 5-min reduction step. The concentrations of Cd and Pb in the honey were 7 and 27 ng/g (ppb), respectively. Precision was 2–4%.

[Data from Y. Li, F. Wahdat, and R. Neeb, "Digestion-Free Determination of Heavy Metals in Honey," *Fresenius J. Anal. Chem.* **1995**, *351*, 678.]

2,3-Dihydroxynaphthalene

Presumably, the polarizable pi electron cloud of naphthalene is bound to the polarizable Hg by van der Waals forces.

TABLE 17-3 Detection limits for stripping analysis

Analyte	Stripping mode	Detection limit
Ag^+	Anodic	2×10^{-12} M[a]
Testosterone	Anodic	2×10^{-10} M[b]
As(III)	Anodic	1×10^{-11} M[c]
I^-	Cathodic	1×10^{-10} M[d]
DNA or RNA	Cathodic	2–5 pg/mL[e]
Fe^{3+}	Cathodic	5×10^{-12} M[f]

a. S. Dong and Y. Wang, Anal. Chim. Acta **1988**, 212, 341.

b. J. Wang, "Adsorptive Stripping Voltammetry." EG&G Princeton Applied Research Application Note A-7 (1985).

c. C. Gao, X.-Y. Yu, S.-Q. Xiong, J.-H. Liu, and X.-J. Huang, Anal. Chem. **2013**, 85, 2673.

d. G. W. Luther III, C. Branson Swartz, and W. J. Ullman, Anal. Chem. **1988**, 60, 1721. I^- is deposited onto the mercury drop by anodic oxidation: $Hg(l) + I^- \rightleftharpoons \frac{1}{2} Hg_2I_2(adsorbed\ on\ Hg) + e^-$.

e. S. Reher, Y. Lepka, and G. Schwedt, Fresenius J. Anal. Chem. **2000**, 368, 720; J. Wang, Anal. Chim. Acta **2003**, 500, 247.

f. L. M. Laglera, J. Santos-Echeandia, S. Caprara, and D. Monticelli, Anal. Chem. **2013**, 85, 2486.

The detection limit for Fe(III) in seawater can be reduced to 5×10^{-12} M by *catalytic stripping*.[42] Seawater is acidified to pH 2.0 with HCl in the presence of 30 μM 2,3-dihydroxynaphthalene (L) and left to equilibrate for 24 h. A sample is then adjusted to pH 8.7 in the presence of 20 mM bromate (BrO_3^-) and purged with N_2 to remove O_2. Dihydroxynaphthalene forms a complex, $L_nFe(III)$, which adsorbs onto a mercury drop poised at 0 V versus Ag | AgCl during 60 s of vigorous stirring. After stirring is stopped and the solution becomes stationary, the potential is scanned from -0.1 to -1.15 V. Near -0.6 V, Fe(III) is reduced to Fe(II), which begins to diffuse away from the electrode. Before Fe(II) goes very far, BrO_3^- oxidizes Fe(II) back to Fe(III), which is readsorbed and available to be reduced again. The cathodic stripping current is ~300 times greater in the presence of 20 mM BrO_3^- than without BrO_3^-. Fe(II) is a catalyst for the net reduction of BrO_3^-.

$$L_nFe(III)_{adsorbed} + e^- \xrightarrow[stripping]{Cathodic} L_nFe(II)_{in\ diffusion\ layer}$$
$$L_nFe(III)_{in\ diffusion\ layer} \qquad BrO_3^-$$

Rigorous precautions must be taken to remove iron from reagents and equipment when measuring nanomolar-to-picomolar concentrations. For example, 3 M KCl in the salt bridge had to be purified, and the bridge itself was made of Teflon instead of glass.

A different form of stripping in which the target analyte is neither oxidized nor reduced is exemplified by the measurement of perchlorate (ClO_4^-) in drinking water. The allowed upper limit for perchlorate in drinking water in California was set at 6 μg/L (6 ppb) in 2007 after it was determined that ClO_4^- interferes with the uptake of I^- by the thyroid gland and can decrease production of thyroid hormones.

Figure 17-26 shows the electrode for measuring part per billion concentrations of ClO_4^- by cathodic stripping. A gold electrode is coated with a film of poly(3-octylthiophene). When this polymer is oxidized, some sulfur atoms become positive and the polymer becomes conductive. The polymer is overcoated with a ~0.7-μm-thick layer of poly(vinyl chloride) (PVC), whose structure was shown at the bottom of Figure 15-27. The coated electrode is immersed in a sample of drinking water containing 1 mM Li_2SO_4 as background electrolyte and spun at 4 000 rpm to transport liquid by convection toward the electrode (Figure 17-15).

The measurement of ClO_4^- begins with oxidation of poly(3-octylthiophene) on the spinning electrode at 0.83 V versus Ag | AgCl for 10 min. Of common inorganic anions in drinking water, only ClO_4^- is soluble in the hydrophobic PVC membrane. During the oxidation step, ClO_4^- migrates into the PCV membrane to neutralize the charge of oxidized poly(3-octylthiophene). The right side of Figure 17-26 shows the electrode after ClO_4^- has been concentrated in the PVC membrane. To measure ClO_4^- by cathodic stripping, the electrode potential is then swept from 0.83 to 0.3 V at 0.1 V/s to reduce poly(3-octylthiophene) back to its neutral form and expel ClO_4^- from the PVC membrane. The peak current (above a substantial background current) observed during stripping is a linear function of ClO_4^- concentration in the drinking water. The detection limit of 0.2 nM (0.2 ppb) is suitable for water quality monitoring.

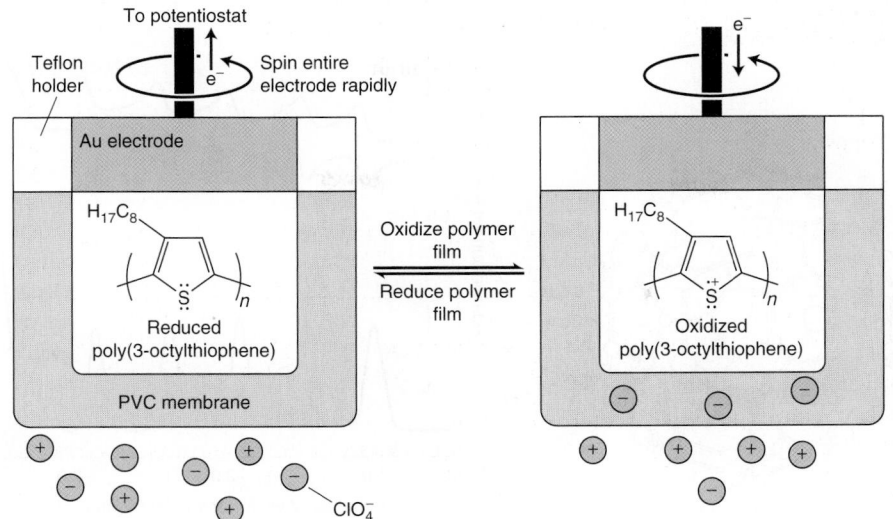

FIGURE 17-26 Principle of cathodic stripping of perchlorate on a rotating gold electrode with two thin polymer layers. When polymer in the electrode is oxidized, it becomes positive. Perchlorate anions are attracted into the PVC membrane by the positive oxidized polymer.
[Information from Y. Kim and S. Amemiya, "Stripping Analysis of Nanomolar Perchlorate in Drinking Water with a Voltammetric Ion-Selective Electrode Based on Thin-Layer Liquid Membranes," *Anal. Chem.* **2008**, *80*, 6056; A. Izadyar, U. Kim, M. M. Ward, and S. Amemiya, "Double-Polymer-Modified Pencil Lead for Stripping Voltammetry of Perchlorate in Drinking Water," *J. Chem. Ed.* **2012**, *89*, 1323.]

Cyclic Voltammetry

In **cyclic voltammetry,** we apply the triangular waveform in Figure 17-27 to the working electrode. After the application of a linear voltage ramp between times t_0 and t_1 (typically a few seconds), the ramp is reversed to bring the potential back to its initial value at time t_2. The cycle might be repeated many times.

The upper portion of the cyclic voltammogram in Figure 17-28a, from time t_0 to time t_1, exhibits a *cathodic wave.* Instead of leveling off at the top of the wave, current decreases at more negative potential because analyte becomes depleted near the electrode. Diffusion is too slow to replenish analyte near the electrode. At the time of peak voltage (t_1), cathodic current has decayed below the peak value. After t_1, the applied potential is reversed and, eventually, reduced product near the electrode is oxidized. This reaction produces an *anodic wave* between times t_1 and t_2. As reduced product is depleted, current approaches its initial value at t_2.

Figure 17-28a illustrates a *reversible* reaction that is fast enough to maintain equilibrium concentrations of reactant and product *at the electrode surface.* Peak anodic

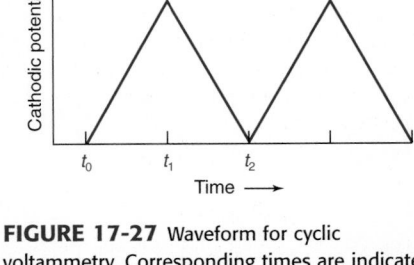

FIGURE 17-27 Waveform for cyclic voltammetry. Corresponding times are indicated in Figure 17-28.

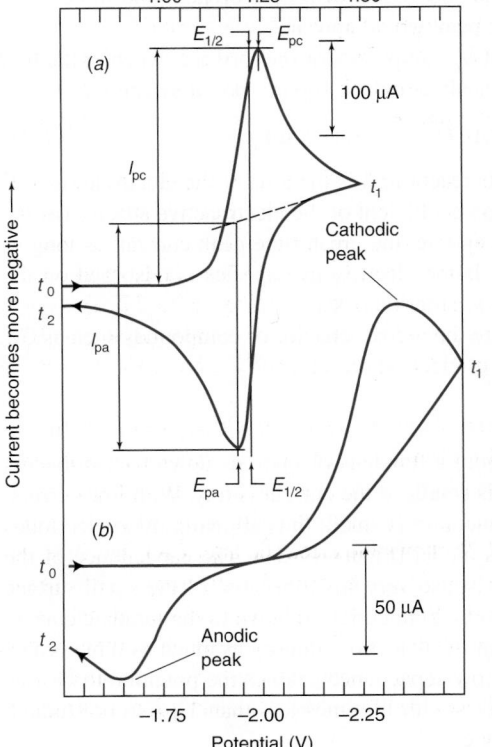

FIGURE 17-28 Cyclic voltammograms of (*a*) 1 mM O_2 in acetonitrile with 0.10 M $(C_2H_5)_4N^+ClO_4^-$ electrolyte and (*b*) 0.060 mM 2-nitropropane in acetonitrile with 0.10 M $(n\text{-}C_7H_{15})_4N^+ClO_4^-$ electrolyte. The reaction in curve *a* is

$$O_2 + e^- \rightleftharpoons O_2^-$$
Superoxide

Working electrode, Hg; reference electrode, Ag|0.001 M $AgNO_3(aq)$|0.10 M $(C_2H_5)_4N^+ClO_4^-$ in acetonitrile; scan rate = 100 V/s. I_{pa} is the peak anodic current and I_{pc} is the peak cathodic current. E_{pa} and E_{pc} are the potentials at which these currents are observed. Curves (*a*) and (*b*) are displaced vertically. The current is near zero at time t_0 for each curve.
[Information from D. H. Evans, K. M. O'Connell, R. A. Petersen, and M. J. Kelly, "Cyclic Voltammetry," *J. Chem. Ed.* **1983**, *60*, 290.]

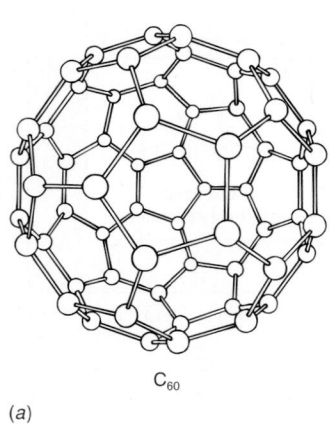

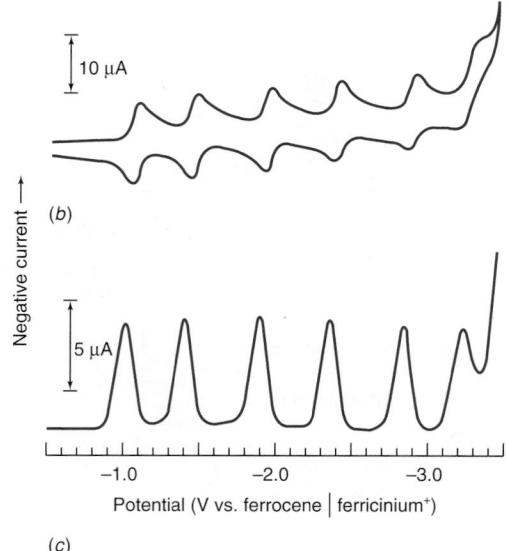

C_{60}

(a)

(c)

FIGURE 17-29 *(a)* Structure of C_{60} (buckminsterfullerene). *(b)* Cyclic voltammetry and *(c)* polarogram of 0.8 mM C_{60}, showing six waves for reduction to C_{60}^-, C_{60}^{2-}, . . . , C_{60}^{6-}. The acetonitrile/toluene solution was at $-10°C$ with $(n\text{-}C_4H_9)_4N^+PF_6^-$ supporting electrolyte. The reference electrode contains the ferrocene|ferricinium$^+$ redox couple. Ferrocene is $(C_5H_5)_2Fe$ and ferricinium cation is $(C_5H_5)_2Fe^+$. The structure of ferrocene was shown in Reaction 17-15. [Data from Q. Xie, E. Pérez-Cordero, and L. Echegoyen, "Electrochemical Detection of C_{60}^{6-} and C_{70}^{6-}," *J. Am. Chem. Soc.* **1992,** *114,* 3978.]

and peak cathodic currents have equal magnitudes in a reversible process, and are separated by

$$E_{pa} - E_{pc} = \frac{2.22RT}{nF} = \frac{57.0}{n}(\text{mV}) \quad (\text{at } 25°C) \qquad \textbf{(17-20)}$$

where E_{pa} and E_{pc} are the potentials at which the *peak anodic* and *peak cathodic* currents are observed and n is the number of electrons in the half-reaction. The half-wave potential, $E_{1/2}$, lies midway between the two peak potentials. Figure 17-28b is the cyclic voltammogram of an *irreversible* reaction, which is too slow to maintain equilibrium concentrations of reactant and product at the electrode surface. Cathodic and anodic peaks are broader and more separated. If oxidation were very slow, no anodic peak would appear.

For a reversible reaction, the peak current (I_{pc}, amperes) for the forward sweep of the first cycle is proportional to the concentration of analyte and the square root of sweep rate:

$$I_{pc} = (2.69 \times 10^8)n^{3/2}ACD^{1/2}v^{1/2} \quad (\text{at } 25°C) \qquad \textbf{(17-21)}$$

where n is the number of electrons in the half-reaction, A is the area of the electrode (m^2), C is the concentration (mol/L), D is the diffusion coefficient of the electroactive species (m^2/s), and v is sweep rate (V/s). The faster the sweep rate, the greater the peak current, as long as the reaction is fast enough to be reversible. If the electroactive species is adsorbed on the electrode, the peak current is proportional to v rather than $\sqrt{v}$.

Cyclic voltammetry is used to characterize the redox behavior of compounds such as C_{60} in Figure 17-29 and to elucidate the kinetics of electrode reactions.[43]

Microelectrodes

Microelectrodes have working dimensions from a few tens of microns down to nanometers (Figure 17-30).[44] The electrode surface area is small, so the current is tiny. With low current, the ohmic drop ($= IR$) in a highly resistive medium is small, thus allowing microelectrodes to be used in poorly conducting nonaqueous media (Figure 17-31). The capacitance of the double layer (Box 17-4) of a microelectrode is also very small because of the small surface area. Low capacitance gives low background charging current relative to the faradaic current of a redox reaction. The result is a lowering of the detection limit by as much as three orders of magnitude over conventional electrodes. Low capacitance enables the potential to be varied at rates up to 10^6 V/s, thus allowing species with lifetimes less than 1 µs to be studied. Sufficiently small electrodes fit inside a living cell.[46]

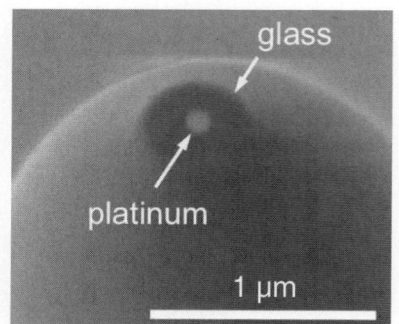

FIGURE 17-30 Platinum electrode with diameter of ~110 nm sealed in a glass capillary. [N. Nioradze, R. Chen, J. Kim, M. Shen, P. Santhosh, and S. Amemiya, "Origins of Nanoscale Damage to Glass-Sealed Platinum Electrodes with Submicrometer and Nanometer Size," *Anal. Chem.* **2013,** *85,* 6198, Figure 3a. Reprinted with permission © 2013, American Chemical Society.]

Advantages of microelectrodes:

- fit into small places
- useful in resistive, nonaqueous media (because of small ohmic losses)
- rapid voltage scans (possible because of small double-layer capacitance) allow short-lived species to be studied
- detection limits increased by orders of magnitude because of low charging current

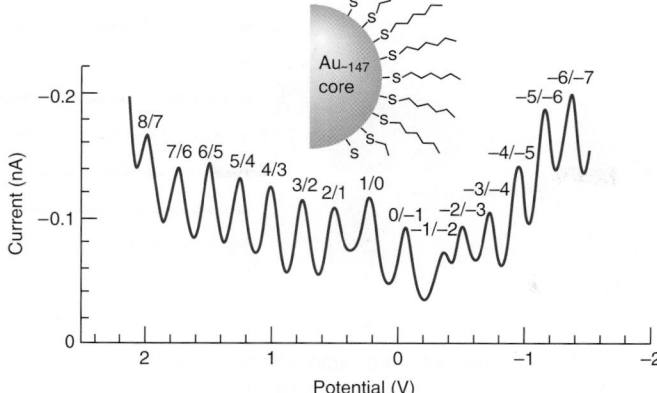

FIGURE 17-31 Voltammogram of gold nanoparticles[45] $Au_{~147}$ capped with ~50 hexanethiol molecules) in 1,2-dichloroethane solution recorded with 25-μm-diameter Pt working electrode. The nanoparticle exhibits oxidation states from −7 to +8 over the potential range of this scan. Supporting electrolyte: 10 mM $[(C_6H_5)_3P=N=P(C_6H_5)_3]^+[(C_6F_5)_4B]^-$. Potential measured versus "quasi-reference" electrode—a silver wire whose potential is ~0.1 V versus Ag|AgCl. [Data from B. M. Quinn, P. Liljeroth, V. Ruiz, T. Laaksonen, and K. Kontturi, "Electrochemical Resolution of 15 Oxidation States for Monolayer Protected Gold Nanoparticles," *J. Am. Chem. Soc.* **2003,** *125,* 6644.]

It is common in cyclic voltammetry with microelectrodes to employ conditions in which the peak in Figure 17-28a is not observed. Instead the *sigmoidal* curve in Figure 17-32a is seen in which current reaches a flat plateau. Current in Figure 17-28a decreases after the peak because diffusion of analyte to the electrode cannot keep up with the rate of electrolysis. At a low scan rate, current would reach a flat plateau because the rate of electrolysis would not exceed the rate of diffusion. For microelectrodes, diffusion is fed from a hemispheric volume with increasing surface area at increasing distance from the electrode (Figure 17-32b). It is common with microelectrodes to see the shape in Figure 17-32a because the rate of electrolysis does not exceed the rate of diffusion to the electrode.

Figure 17-33 shows a carbon fiber coated with a cation-exchange membrane called Nafion, which has fixed negative charges. Cations diffuse through the membrane, but anions are excluded. The electrode can measure the cationic neurotransmitter, dopamine, in a rat brain.[47] Negatively charged ascorbate, which ordinarily interferes with dopamine analysis, is excluded by Nafion. The response to dopamine is 1 000 times higher than the response to ascorbate.

Figure 17-34 shows an application of a microelectrode array in biology. Cells called PC 12 from a tumor of the adrenal gland release dopamine when stimulated by K^+. Neurotransmitters are contained in small intracellular compartments called *vesicles*. To release their content outside the cell, vesicles fuse with the cell membrane and open up. This process is called *exocytosis*. The microelectrode array consists of 5-μm-diameter carbon fibers in a seven-barrel glass capillary pulled to a fine point. Each carbon fiber in Figure 17-34a

$$-[(CF_2CF_2)_n-CFCF_2]_x-$$

Nafion membrane permits cations to pass, but not anions

$$\begin{array}{c} | \\ O \\ | \\ CF_2 \\ | \\ CFCF_3 \\ | \\ O \\ | \\ CF_2CF_2SO_3^-Na^+ \end{array}$$

Dopamine

$$+ 2H^+ + 2e^-$$

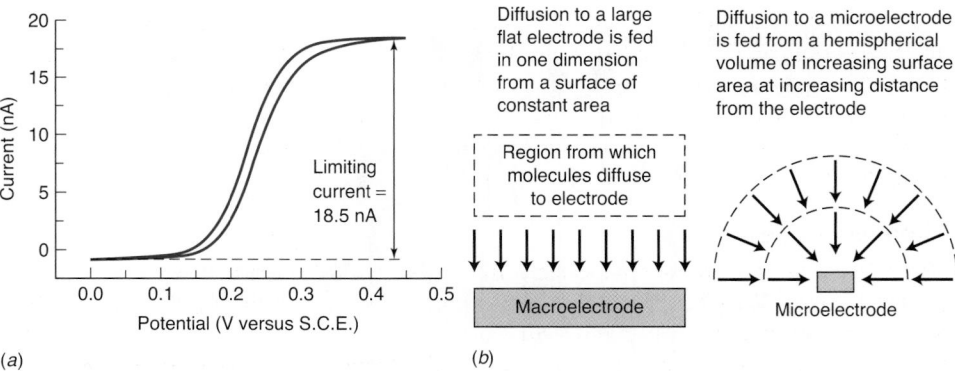

(a)

Diffusion to a large flat electrode is fed in one dimension from a surface of constant area

Region from which molecules diffuse to electrode

Macroelectrode

Diffusion to a microelectrode is fed from a hemispherical volume of increasing surface area at increasing distance from the electrode

Microelectrode

(b)

FIGURE 17-32 (*a*) Sigmoidal shape of cyclic voltammogram commonly observed with microelectrodes. Voltammogram of 10 mM ferrocyanide in aqueous 1 M NaF at a scan rate of 5 mV/s on a circular disk Pt microelectrode. [Data from U. K. Sur, A. Dhason, and V. Lakshminarayanan, "A Simple and Low-Cost Ultramicroelectrode Fabrication and Characterization Method for Undergraduate Students," *J. Chem. Ed.* **2012,** *89,* 168.] (*b*) Diffusion of analyte to a large, planar electrode is from a constant, flat area. Diffusion of analyte to a microelectrode is from a hemispheric region of solution whose area increases with increasing distance from the electrode.

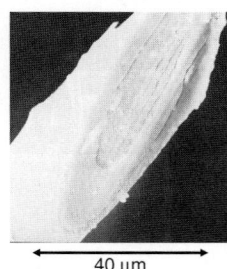

40 μm

FIGURE 17-33 Electron micrograph of the tip of a Nafion-coated carbon-fiber electrode. The carbon inside the electrode has a diameter of 10 μm. Nafion permits cations to pass but excludes anions. [R. M. Wightman, L. J. May, and A. C. Michael, "Detection of Dopamine Dynamics in the Brain," *Anal. Chem.* **1988,** *60,* 769A. Reprinted with permission © 1988, American Chemical Society.]

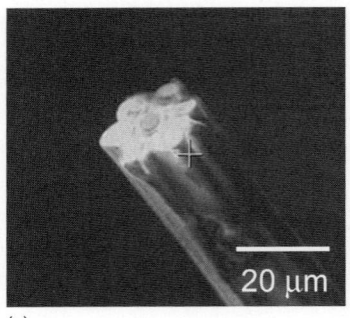
(a)

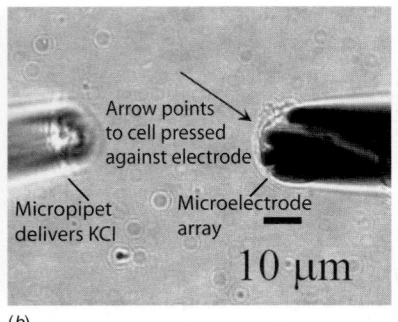
Arrow points to cell pressed against electrode

Micropipet delivers KCl

Microelectrode array

10 μm

(b)

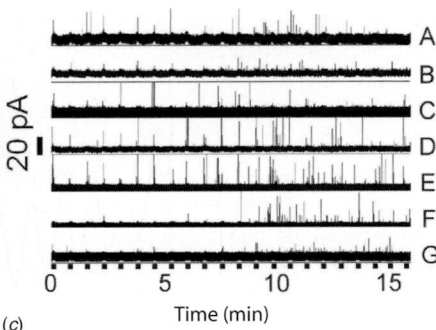
20 pA

0   5   10   15
Time (min)

(c)

FIGURE 17-34 (*a*) Carbon-fiber microelectrode array with seven electrodes. (*b*) Microelectrode array (right) pressed against PC 12 cell distorted into a crescent shape by the electrode. Micropipet delivers 0.1 M KCl to stimulate dopamine release from the cell. A Ag|AgCl reference/auxiliary electrode is not visible. (*c*) Amperometric traces from seven electrodes during exocytotic release of neurotransmitter.

[B. Zhang, K. L. Adams, S. J. Luber, D. J. Eves, M. L. Heien, and A. G. Ewing, "Spatially and Temporally Resolved Single-Cell Exocytosis Utilizing Individually Addressable Carbon Microelectrode Arrays," *Anal. Chem.* **2008,** *80,* 1394, Figure 2A. Reprinted with permission © 2008, American Chemical Society.]

is an independent working electrode that samples a region ~5 μm in diameter. Figure 17-34b shows the electrode pressed against a single cell, distorting the cell into a crescent shape. When 0.1 M KCl from the micropipet in Figure 17-34b is injected into the medium, the cell releases dopamine in discrete exocytotic events. Figure 17-34c shows traces from electrodes A−G over 16 min during which KCl is injected every 45 s. All traces are different, indicating that different patches of cell surface respond differently. For example, patches near electrodes F and G are inactive for the first 8 min. Something changes after 8 min and these two patches become active.

17-6 Karl Fischer Titration of H_2O

The **Karl Fischer titration,**[48] which measures traces of water in transformer oil, solvents, foods, polymers, and other substances, might be performed half a million times each day.[49] The titration is usually performed by delivering titrant from an automated buret or by coulometric generation of titrant. The volumetric procedure tends to be appropriate for larger amounts of water (but can go as low as ~1 mg H_2O), and the coulometric procedure tends to be appropriate for smaller amounts of water.

We illustrate the coulometric procedure in Figure 17-35, in which the main compartment contains anode solution plus unknown. The smaller compartment at the left has an internal Pt electrode immersed in cathode solution and an external Pt electrode immersed in the anode

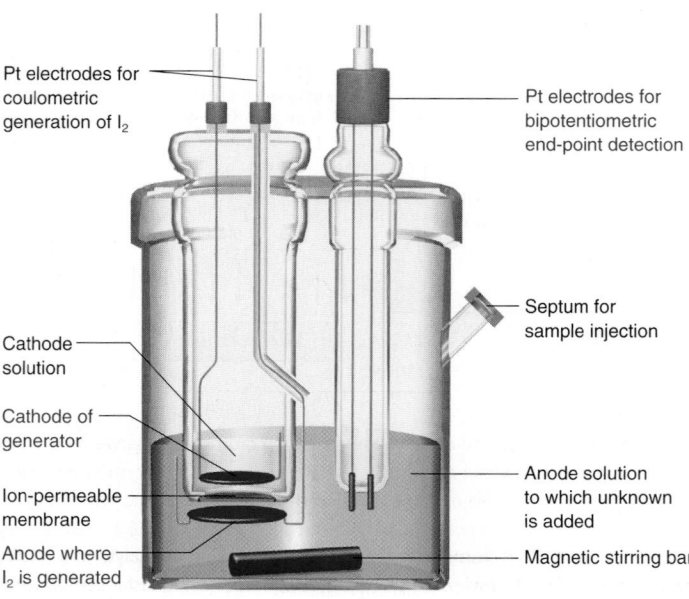

Pt electrodes for coulometric generation of I_2

Pt electrodes for bipotentiometric end-point detection

Septum for sample injection

Cathode solution

Cathode of generator

Ion-permeable membrane

Anode where I_2 is generated

Anode solution to which unknown is added

Magnetic stirring bar

FIGURE 17-35 Apparatus for coulometric Karl Fischer titration.

solution of the main compartment. The two compartments are separated by an ion-permeable membrane. Two Pt electrodes are used for end-point detection.

Anode solution contains an alcohol, a base, SO_2, I^-, and possibly other organic solvent. Methanol and diethylene glycol monomethyl ether ($CH_3OCH_2CH_2OCH_2CH_2OH$) were typical alcohols used in the past. Typical bases are imidazole and diethanolamine. The solvent often contained chloroform, dichloromethane, carbon tetrachloride, or formamide. It is desirable to avoid chlorinated solvents to reduce environmental hazards. In a commercial, proprietary solvent with less environmental impact, methanol was replaced by ethanol, which is less toxic but has a low dielectric constant that leads to low electrical conductivity and slow reactions. Proprietary additives increase the conductivity and reaction rates. The new solvent eliminates the need for chlorinated solvents in some applications. Formamide, chloroform, or xylene can be added to the commercial solvent to increase the solubility of some samples or temperature can be increased to 50°C to enhance solubility. Nonpolar substances such as transformer oil require sufficient solvent, such as chloroform, to make the reaction homogeneous. Otherwise, moisture trapped in oily emulsions is inaccessible. (An *emulsion* is a fine suspension of liquid-phase droplets in another liquid.)

The anode at the lower left in Figure 17-35 generates I_2 by oxidation of I^-. In the presence of H_2O, reactions occur between the alcohol (ROH), base (B), SO_2, and I_2.

$$ROH + SO_2 + B \rightarrow BH^+ + ROSO_2^- \qquad \text{(17-22)}$$
$$H_2O + I_2 + ROSO_2^- + 2B \rightarrow ROSO_3^- + 2BH^+I^- \qquad \text{(17-23)}$$

The net reaction is oxidation of SO_2 by I_2, with formation of $ROSO_3^-$. One mole of I_2 is consumed for each mole of H_2O when the solvent is methanol. In other solvents, the stoichiometry can be more complex.[49]

In a typical procedure, the main compartment in Figure 17-35 is filled with anode solution and the coulometric generator is filled with cathode solution that may contain reagents designed to be reduced at the cathode. Current is run until moisture in the main compartment is consumed, as indicated by the end-point detection system described after the Example. An unknown is injected through the septum, and the coulometer is run again until moisture has been consumed. Two moles of electrons correspond to 1 mol of H_2O if the $I_2:H_2O$ stoichiometry is 1:1.

Imidazole   Diethanolamine

Maintain pH in the range 4 to 7. Above pH 8, nonstoichiometric side reactions occur. Below pH 3, the reaction is slow.

EXAMPLE **Standardization and Blank Correction in Karl Fischer Titration**

It is routine to standardize Karl Fischer reagents, or even a coulometer, with a standard such as lincomycin hydrochloride monohydrate, which contains 3.91 wt% H_2O. The coulometer is run until the end point is reached, indicating that the Karl Fischer reagent is dry. A port is opened briefly to add solid lincomycin, which is then titrated to the same end point. Then an unknown is added and titrated in the same manner. Find the wt% H_2O in the unknown.

Milligrams lincomycin	μg H_2O observed	μg H_2O theoretical	Difference (μg) = blank correction
3.89	172.4	152.1	172.4 − 152.1 = 20.3
13.64	556.3	533.3	556.3 − 533.3 = 23.0
19.25	771.4	752.7	771.4 − 752.7 = 18.7
			Average correction = 20.7

Milligrams unknown	μg H_2O observed	μg H_2O corrected (= observed − 20.7)	wt% H_2O in unknown
24.17	540.8	520.1	520.1 μg/24.17 mg = 2.15%
17.08	387.6	366.9	366.9 μg/17.08 mg = 2.15%

SOURCE: *Data from W. C. Schinzer, Pfizer Co., Michigan Pharmaceutical Sciences, Portage, MI.*

Lincomycin hydrochloride monohydrate
$C_{18}H_{37}N_2O_7SCl$, FM 461.01
3.91 wt% H_2O

Solution For lincomycin, we observe ~20.7 μg more H_2O than expected, *independent of the sample size*. Excess H_2O comes from the atmosphere when the port is opened to add solid. To determine moisture in unknowns, subtract this blank from the total moisture titrated. This procedure can generate very reproducible data.

TEST YOURSELF The observed H_2O in 20.33 mg of unknown was 888.8 μg. Apply the correction and find the wt% H_2O in the unknown. (*Answer:* 4.27%)

A **bipotentiometric** measurement is the most common way to detect the end point of a Karl Fischer titration. The detector circuit maintains a *constant current* (usually 5 or 10 μA) between the two detector electrodes at the right in Figure 17-35 while measuring the voltage needed to sustain the current. Prior to the equivalence point, the solution contains I^-, but little I_2 (which is consumed in Reaction 17-23 as fast as it is generated). To maintain a current of 10 μA, the cathode potential must be negative enough to reduce some component of the solvent system. At the equivalence point, excess I_2 appears and current can be carried at low voltage by Reactions 17-24 and 17-25. The abrupt voltage drop marks the end point.

Cathode:	$I_3^- + 2e^- \rightarrow 3I^-$	(17-24)
Anode:	$3I^- \rightarrow + I_3^- + 2e^-$	(17-25)

A trend in Karl Fischer coulometric instrumentation is to eliminate the separate cathode compartment in Figure 17-35 to reduce conditioning time required before samples can be analyzed and to eliminate clogging of the membrane.[50] The challenge is to minimize interference by products of the cathodic reaction.

End points in Karl Fischer titrations tend to drift because of slow reactions and water leaking into the cell from the air. Some instruments measure the rate at which I_2 must be generated to maintain the end point and then compare this rate with that measured before sample was added. Other instruments allow you to set a "persistence of end point" time, typically 5 to 60 s, during which the detector voltage must be stable to define the end point.

A round robin study of accuracy and precision of the coulometric procedure identified sources of systematic error.[51] In some labs, either the instruments were inaccurate or workers did not measure the quantity of standards accurately. In other cases, solvent was not appropriate. Commercial reagents recommended by instrument manufacturers should be used with each instrument.

Alternatives to the Karl Fischer titration have desirable attributes for some applications. Water in solvents can be measured by a gas chromatographic method using an ionic liquid stationary phase (as on page 637) with thermal conductivity detection.[52] The method works for samples with high or low levels of H_2O and has a lower limit of detection of 2 ng H_2O. The method is insensitive to common interferences and side reactions of the Karl Fischer titration. Another method uses cathodic stripping voltammetry of gold oxide formed on a gold electrode in the presence of H_2O. The method requires no addition of reagents to measure H_2O in ionic liquids.[53]

Terms to Understand

amalgam	coulomb	faradaic current	reference electrode
ampere	coulometric titration	half-wave potential	residual current
amperometry	coulometry	Karl Fischer titration	rotating disk electrode
aptamer	cyclic voltammetry	mediator	sampled current polarography
auxiliary electrode	depolarizer	nonpolarizable electrode	square wave voltammetry
biosensor	diffusion current	ohmic potential	stripping analysis
bipotentiometric titration	dropping-mercury electrode	overpotential	underpotential deposition
charging current	electric double layer	polarizable electrode	voltammetry
Clark electrode	electroactive species	polarographic wave	voltammogram
concentration polarization	electrogravimetric analysis	polarography	working electrode
controlled-potential electrolysis	electrolysis	potentiostat	

Summary

In electrolysis, a chemical reaction is forced to occur by the flow of electricity through a cell. The moles of electrons flowing through the cell are It/nF, where I is current, t is time, n is the number of electrons per molecule, and F is the Faraday constant. The magnitude of the voltage that must be applied to an electrolysis cell is $E = E(\text{cathode}) - E(\text{anode}) - IR - \text{overpotentials}$.

1. Overpotential is the voltage required to overcome the activation energy of an electrode reaction. A greater overpotential is required to drive a reaction at a faster rate.

2. Ohmic potential ($= IR$) is that voltage needed to overcome internal resistance of the cell. A Luggin capillary allows us to measure an electrode potential with minimal ohmic loss.

3. Concentration polarization occurs when the concentration of electroactive species near an electrode is not the same as its concentration in bulk solution. Concentration polarization is embedded in the terms $E(\text{cathode})$ and $E(\text{anode})$.

Overpotential, ohmic potential, and concentration polarization always oppose the desired reaction and require a greater voltage to be applied for electrolysis.

Controlled-potential electrolysis is conducted in a three-electrode cell in which the potential of the working electrode is measured with respect to a reference electrode to which negligible current flows. Current flows between the working and auxiliary electrodes.

In electrogravimetric analysis, analyte is deposited on an electrode, whose increase in mass is then measured. With a constant voltage in a two-electrode cell, electrolysis is not very selective, because the working electrode potential changes as the reaction proceeds.

In coulometry, the moles of electrons needed for a chemical reaction are measured. In a coulometric (constant current) titration, the time needed for complete reaction measures the number of electrons consumed. Controlled-potential coulometry is more selective than constant-current coulometry, but slower. Electrons consumed in the reaction are measured by integrating the current-versus-time curve.

In amperometry, current at the working electrode is proportional to analyte concentration. The Clark electrode measures dissolved O_2 by amperometry. A glucose biosensor generates H_2O_2 by enzymatic oxidation of glucose, and the H_2O_2 is measured by amperometric oxidation at an electrode. A mediator is employed to rapidly shuttle electrons between electrode and analyte. A coulometric glucose monitor counts electrons released by oxidation of all glucose in a small blood sample. "Electrical wiring" of the enzyme and mediator in a glucose monitor increases signal from the desired reaction and decreases background current from mediator diffusing to the auxiliary electrode. Aptamers are small molecules of DNA or RNA that strongly and selectively bind a target molecule and can be used to make sensors for such molecules.

Electroactive species can reach the surface of an electrode by diffusion, convection, and migration (electrostatic attraction). A rotating disk electrode promotes convective transfer of species toward the electrode. Molecules must cross the final tens of microns to the electrode by diffusion. The limiting current governed by the rate of diffusion is called the diffusion current.

Voltammetry is a collection of methods in which the dependence of current on the applied potential of the working electrode is observed. Polarography is voltammetry with a dropping-mercury working electrode. This electrode gives reproducible results because fresh surface is always exposed. Hg is useful for reductions because the high overpotential for H^+ reduction on Hg prevents interference by H^+ reduction. Hg is too easily oxidized to be useful for oxidizing many analytes. Oxidations can be done with Pt or Au electrodes. Carbon electrodes (including diamond and graphene) are useful for a range of oxidations and reductions. For quantitative analysis by voltammetry, the diffusion current is proportional to analyte concentration if there is a sufficient concentration of supporting electrolyte. The half-wave potential is characteristic of a particular analyte in a particular medium.

Sampled current voltammetry uses a staircase voltage profile to reduce the contribution of charging current relative to the desired measurement of faradaic current. One second after each voltage step, charging current is nearly 0, but there is still substantial faradaic current from the redox reaction.

Square wave polarography achieves increased sensitivity and a derivative peak shape by applying a square wave superimposed on a staircase voltage ramp. With each cathodic pulse, there is a rush of analyte to be reduced at the electrode surface. During the anodic pulse, reduced analyte is reoxidized. The polarogram is the difference between the cathodic and anodic currents. Square wave polarography permits fast, real-time measurements not possible with other electrochemical methods.

Stripping is the most sensitive form of voltammetry. In anodic stripping voltammetry, analyte is concentrated into a single drop or thin film of mercury or other electrodes by reduction at a fixed voltage for a fixed time. The potential is then made more positive, and current is measured as analyte is reoxidized. In the ClO_4^- electrode, analyte is concentrated into a hydrophobic membrane to neutralize the charge of a conducting polymer during oxidation. Reduction of the polymer liberates ClO_4^- from the membrane.

In cyclic voltammetry, a triangular waveform is applied, and cathodic and anodic processes are observed in succession. Reduction or oxidation currents go through a peak and then decrease if diffusion is too slow to replenish analyte at the electrode surface. A reversible reaction is fast enough to maintain equilibrium concentrations of reactant and product at the electrode surface. Increased separation and broadening of cathodic and anodic peaks occur in irreversible reactions that are not fast enough to maintain equilibrium concentrations at the electrode.

Microelectrodes fit into small places and their low current allows them to be used in resistive, nonaqueous media. Their low capacitance increases sensitivity by reducing charging current and permits rapid voltage scanning, which allows very short-lived species to be studied. Voltammograms from microelectrodes at modest voltage scan rates reach a steady diffusion current without going through a peak current because radial diffusion of analyte from a hemispheric volume keeps up with the rate of the electrode reaction.

The Karl Fischer titration of water uses a buret to deliver reagent or coulometry to generate reagent. In bipotentiometric end-point detection, the voltage needed to maintain a constant current between two Pt electrodes is measured. Voltage changes abruptly at the equivalence point, when one member of a redox couple is either created or destroyed.

Exercises

17-A. A dilute Na_2SO_4 solution is to be electrolyzed with a pair of smooth Pt electrodes at a current density of 100 A/m² and a current of 0.100 A. The products are $H_2(g)$ and $O_2(g)$ at 1.00 bar. Calculate the required voltage if the cell resistance is 2.00 Ω and there is no concentration polarization. What voltage would be required if the Pt electrodes were replaced by Au electrodes?

17-B. (a) At what cathode potential will Sb(s) deposition commence from 0.010 M SbO^+ solution at pH 0.00? Express this potential versus S.H.E. and versus $Ag|AgCl$. (Disregard overpotential, about which you have no information.)

$$SbO^+ + 2H^+ + 3e^- \rightleftharpoons Sb(s) + H_2O \qquad E° = 0.208 \text{ V}$$

(b) What percentage of 0.10 M Cu^{2+} could be reduced electrolytically to Cu(s) before 0.010 M SbO^+ in the same solution begins to be reduced at pH 0.00?

17-C. Calculate the cathode potential (versus S.C.E.) needed to reduce cobalt(II) to 1.0 μM in each of the following solutions. In each case, Co(s) is the product of the reaction. (Disregard any overpotential.)

(a) 0.10 M $HClO_4$

(b) 0.10 M $C_2O_4^{2-}$ (Find the potential at which $[Co(C_2O_4)_2^{2-}] = 1.0$ μM.)

$$Co(C_2O_4)_2^{2-} + 2e^- \rightleftharpoons Co(s) + 2C_2O_4^{2-} \qquad E° = -0.474 \text{ V}$$

(c) 0.10 M EDTA at pH 7.00 (Find the potential at which $[Co(EDTA)^{2-}] = 1.0$ μM.)

17-D. Ions that react with Ag^+ can be determined electrogravimetrically by deposition on a silver working anode:

$$Ag(s) + X^- \rightarrow AgX(s) + e^-$$

(a) What will be the final mass of a silver anode used to electrolyze 75.00 mL of 0.023 80 M KSCN if the initial mass of the anode is 12.463 8 g?

(b) At what electrolysis voltage (versus S.C.E.) will AgBr(s) be deposited from 0.10 M Br^-? (Consider negligible current flow, so that there is no ohmic potential, concentration polarization, or overpotential.)

(c) Is it theoretically possible to separate 99.99% of 0.10 M KI from 0.10 M KBr by controlled-potential electrolysis?

17-E. Chlorine has been used for decades to disinfect drinking water. An undesirable side effect of this treatment is reaction with organic impurities to create organochlorine compounds, some of which could be toxic. Monitoring total organic halide (designated TOX) is required for many water providers. A standard procedure for TOX is to pass water through activated charcoal, which adsorbs organic compounds. Then the charcoal is combusted to liberate hydrogen halides:

$$\text{Organic halide (RX)} \xrightarrow{O_2/800°C} CO_2 + H_2O + HX$$

HX is absorbed into aqueous solution and measured by coulometric titration with a silver anode:

$$X^-(aq) + Ag(s) \rightarrow AgX(s) + e^-$$

When 1.00 L of drinking water was analyzed, a current of 4.23 mA was required for 387 s. A blank prepared by oxidizing charcoal required 6 s at 4.23 mA. Express the TOX of the drinking water as μmol halogen/L. If all halogen is chlorine, express the TOX as μg Cl/L.

17-F. Cd^{2+} was used as an internal standard in the analysis of Pb^{2+} by square wave polarography. Cd^{2+} gives a reduction wave at -0.60 V and Pb^{2+} gives a reduction wave at -0.40 V. It was first verified that the ratio of peak heights is proportional to the ratio of concentrations over the whole range employed in the experiment. Here are results for known and unknown mixtures:

Analyte	Concentration (M)	Current (μA)
Known		
Cd^{2+}	$3.23 (\pm0.01) \times 10^{-5}$	$1.64 (\pm0.03)$
Pb^{2+}	$4.18 (\pm0.01) \times 10^{-5}$	$1.58 (\pm0.03)$
Unknown + Internal Standard		
Cd^{2+}	?	$2.00 (\pm0.03)$
Pb^{2+}	?	$3.00 (\pm0.03)$

The unknown mixture was prepared by mixing 25.00 (±0.05) mL of unknown (containing only Pb^{2+}) plus 10.00 (±0.05) mL of 3.23 (±0.01) × 10^{-4} M Cd^{2+} and diluting to 50.00 (±0.05) mL.

(a) Disregarding uncertainties, find $[Pb^{2+}]$ in the undiluted unknown.

(b) Find the absolute uncertainty for the answer to part (a).

17-G. Consider the cyclic voltammogram of the Co^{3+} compound $Co(B_9C_2H_{11})_2^-$. Suggest a chemical reaction to account for each wave. Are the reactions reversible? How many electrons are involved in each step? Sketch the sampled current and square wave polarograms expected for this compound.

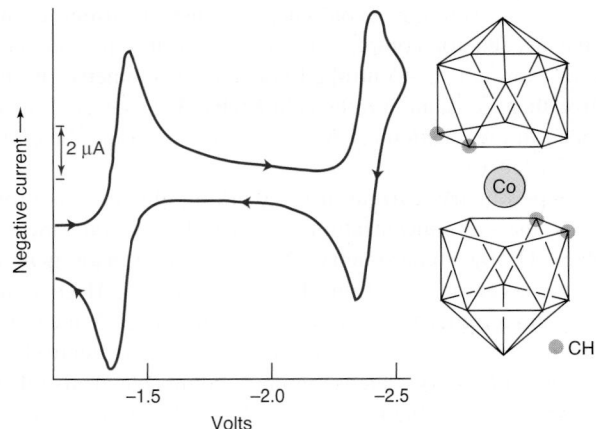

Cyclic voltammogram of $Co(B_9C_2H_{11})_2^-$. [Data from W. E. Geiger, Jr., W. L. Bowden, and N. El Murr, "An Electrochemical Study of the Protonation Site of the Cobaltocene Anion and of Cyclopentadienylcobalt(I) Dicarbollides," *Inorg. Chem.* **1979,** *18,* 2358.]

$E_{1/2}$ (V vs S.C.E)	I_{pa}/I_{pc}	$E_{pa} - E_{pc}$ (mV)
-1.38	1.01	60
-2.38	1.00	60

17-H. In a coulometric Karl Fischer water analysis, 25.00 mL of pure "dry" methanol required 4.23 C to generate enough I_2 to react with residual H_2O in the methanol. A suspension of 0.847 6 g of finely ground polymeric material in 25.00 mL of the same "dry" methanol required 63.16 C. Find the wt% H_2O in the polymer.

Problems

Fundamentals of Electrolysis

17-1. The figure shows the behavior of Pt and Ag cathodes at which reduction of H_3O^+ to $H_2(g)$ occurs. Explain why the two curves are not superimposed.

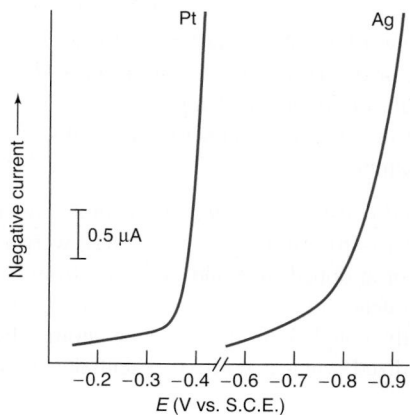

Current versus voltage for Pt and Ag electrodes in O_2-free, aqueous H_2SO_4 adjusted to pH 3.2. [Data from D. Marín, F. Mendicuti, and C. Teijeiro, "An Electrochemistry Experiment: Hydrogen Evolution Reaction on Different Electrodes," *J. Chem. Ed.* **1994,** *71,* A277.]

17-2. How many hours are required for 0.100 mol of electrons to flow through a circuit if the current is 1.00 A?

17-3. The standard free energy change for the formation of $H_2(g) + \frac{1}{2}O_2(g)$ from $H_2O(l)$ is $\Delta G° = +237.13$ kJ. The reactions are

Cathode: $2H_2O + 2e^- \rightleftharpoons H_2(g) + 2OH^-$

Anode: $H_2O \rightleftharpoons \frac{1}{2}O_2(g) + 2H^+ + 2e^-$

Calculate the standard voltage ($E°$) needed to decompose water into its elements by electrolysis. What does the word *standard* mean in this question?

17-4. Consider the following electrolysis reactions.

Cathode: $H_2O(l) + e^- \rightleftharpoons \frac{1}{2}H_2(g, 1.0\ bar) + OH^-(aq, 0.10\ M)$

Anode: $Br^-(aq, 0.10\ M) \rightleftharpoons \frac{1}{2}Br_2(l) + e^-$

(a) Calculate the voltage needed to drive the net reaction if current is negligible.

(b) Suppose that the cell has a resistance of 2.0 Ω and a current of 100 mA. How much voltage is needed to overcome the cell resistance? This is the ohmic potential.

(c) Suppose that the anode reaction has an overpotential of 0.20 V and that the cathode overpotential is 0.40 V. What voltage is needed to overcome these effects combined with those of parts (a) and (b)?

(d) Suppose that concentration polarization occurs. $[OH^-]_s$ at the cathode surface increases to 1.0 M and $[Br^-]_s$ at the anode surface decreases to 0.010 M. What voltage is needed to overcome these effects combined with those of (b) and (c)?

17-5. (a) Which voltage, V_1 or V_2, in the diagram is constant in controlled-potential electrolysis? Which are the working, auxiliary, and reference electrodes in the diagram?

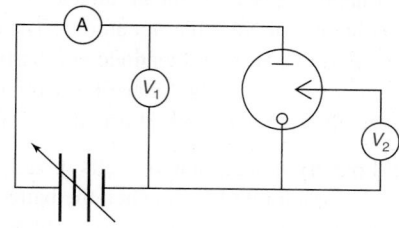

(b) Explain how the Luggin capillary in Figure 17-4 measures the electric potential at the opening of the capillary.

17-6. (a) The cell in Figure 17-4 is:

$$Cu(s) \mid 1.0\ M\ CuSO_4(aq) \mid KCl\ (aq,\ 3\ M) \mid AgCl(s) \mid Ag(s)$$

Write half-reactions for this cell. Neglecting activity coefficients and the junction potential between $CuSO_4(aq)$ and KCl (aq), predict the equilibrium (zero-current) voltage expected when the Luggin capillary contacts the Cu electrode. For this purpose, suppose that the reference electrode potential is 0.197 V vs. S.H.E. Why is the observed equilibrium potential $+109$ mV, not the value you calculated?

(b) How would the overpotentials change if >1.000 V were imposed by the potentiostat?

17-7. The Weston cell is a stable voltage standard formerly used in potentiometers. (The potentiometer compares an unknown voltage with that of the standard. In contrast with the conditions of this problem, very little current may be drawn from the cell if it is to be a voltage standard.)

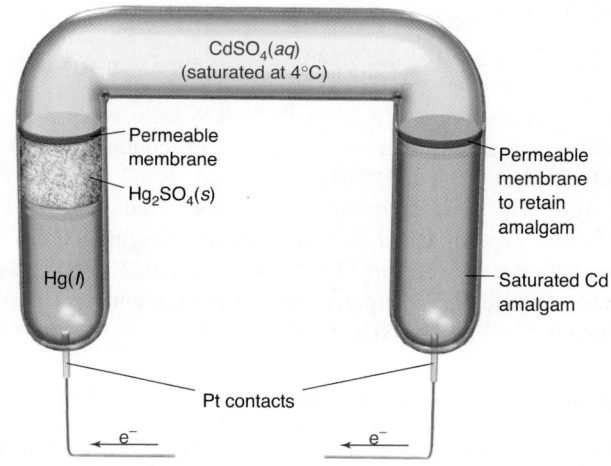

$$Hg_2SO_4 + Cd(in\ Hg) \rightleftharpoons 2Hg + CdSO_4$$

(a) How much work (J) can be done by the Weston cell if the voltage is 1.02 V and 1.00 mL of Hg (density = 13.53 g/mL) is deposited?

(b) If the cell passes current through a 100-Ω resistor that dissipates heat at a rate of 0.209 J/min, how many grams of Cd are oxidized each hour? (This question is not meant to be consistent with part (a). The voltage is no longer 1.02 volts.)

17-8. The chlor-alkali process,[54] in which seawater is electrolyzed to make Cl_2 and NaOH, is the second most important commercial electrolysis, behind Al production.

Anode: $Cl^- \rightarrow \frac{1}{2}Cl_2 + e^-$

Hg cathode: $Na^+ + H_2O + e^- \rightarrow NaOH + \frac{1}{2}H_2$

The Nafion membrane (page 421) used to separate the anode and cathode compartments resists chemical attack. Its anionic side chains permit conduction of Na^+, but not anions. The cathode compartment contains pure water, and the anode compartment contains seawater from which Ca^{2+} and Mg^{2+} have been removed. Explain how the membrane allows NaOH to be formed free of NaCl.

17-9. A lead-acid battery in a car has six cells in series, each delivering close to 2.0 V for a total of 12 V when the battery is discharging. Recharging requires ~2.4 V per cell, or ~14 V for the entire battery.[55] Explain these observations in terms of Equation 17-6.

Electrogravimetric Analysis

17-10. A 0.326 8-g unknown containing $Pb(CH_3CHOHCO_2)_2$ (lead lactate, FM 385.3) plus inert material was electrolyzed to produce 0.111 1 g of PbO_2 (FM 239.2). Was the PbO_2 deposited at the anode or at the cathode? Find the weight percent of lead lactate in the unknown.

17-11. A solution of Sn^{2+} is to be electrolyzed to reduce the Sn^{2+} to $Sn(s)$. Calculate the cathode potential (versus S.H.E.) needed to reduce $[Sn^{2+}]$ to 1.0×10^{-8} M if no concentration polarization occurs. What would be the potential versus S.C.E. instead of S.H.E? Would the potential be more positive or more negative if concentration polarization occurred?

17-12. What cathode potential (versus S.H.E.) is required to reduce 99.99% of Cd(II) from a solution containing 0.10 M Cd(II) in 1.0 M ammonia if there is negligible current? Consider the following reactions and assume that nearly all Cd(II) is in the form $Cd(NH_3)_4^{2+}$.

$$Cd^{2+} + 4NH_3 \rightleftharpoons Cd(NH_3)_4^{2+} \qquad \beta_4 = 3.6 \times 10^6$$

$$Cd^{2+} + 2e^- \rightleftharpoons Cd(s) \qquad E° = -0.402 \text{ V}$$

17-13. *Electroplating efficiency.*[56] Nickel was electrolytically plated onto a carbon electrode from a bath containing 290 g/L $NiSO_4 \cdot 6H_2O$, 30 g/L $B(OH)_3$, and 8 g/L NaCl at −1.2 V vs Ag | AgCl. The most important side reaction is reduction of H^+ to H_2. In one experiment, a carbon electrode weighing 0.477 5 g before deposition weighed 0.479 8 g after 8.082 C had passed through the circuit. What percentage of the current went into the reaction $Ni^{2+} + 2e^- \rightarrow Ni(s)$?

Coulometry

17-14. Explain how amperometric end-point detection in Figure 17-9 operates.

17-15. What does a mediator do?

17-16. The sensitivity of a coulometer is governed by the delivery of its minimum current for its minimum time. Suppose that 5 mA can be delivered for 0.1 s.

(a) How many moles of electrons are delivered by 5 mA for 0.1 s?

(b) How many milliliters of a 0.01 M solution of a two-electron reducing agent are required to deliver the same number of electrons?

17-17. The experiment in Figure 17-9 required 5.32 mA for 964 s for complete reaction of a 5.00-mL aliquot of unknown cyclohexene solution.

(a) How many moles of electrons passed through the cell?

(b) How many moles of cyclohexene reacted?

(c) What was the molarity of cyclohexene in the unknown?

17-18. $H_2S(aq)$ can be analyzed by titration with coulometrically generated I_2.

$$H_2S + I_2 \rightarrow S(s) + 2H^+ + 2I^-$$

To 50.00 mL of sample were added 4 g of KI. Electrolysis required 812 s at 52.6 mA. Calculate the concentration of H_2S (μg/mL) in the sample.

17-19. In Figure 17-11, 2.00 nmol of fructose were introduced at the time of the arrow. How many electrons are lost in the oxidation of one molecule of fructose? Compare the theoretical number of coulombs with the observed number of coulombs for complete oxidation of the sample.

17-20. Diamond consists of carbon atoms in a face centered cubic crystal (Figure 15-36). Amino groups ($—NH_2$) were added onto the surface of a boron-doped diamond electrode from a plasma made by a radio frequency discharge in $NH_3(g)$. Ferrocene groups were then chemically coupled to the nitrogen atoms. Ferrocene was reversibly oxidized when a positive potential was applied to the electrode.[57]

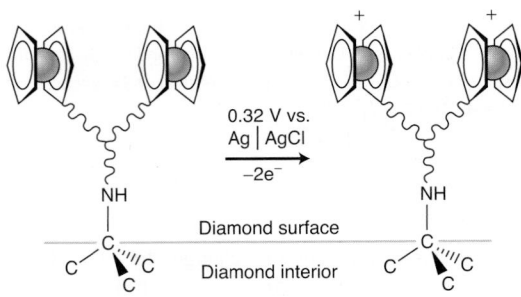

(a) Referring to Section 15-8, describe how boron doping transforms diamond from an insulator to a semiconductor. Is the semiconductor *p*-type or *n*-type? Explain your reasoning.

(b) The surface density of carbon atoms on diamond is approximately 1.7×10^{15} atoms/cm^2. (Different crystal planes have somewhat different atom densities. The electrode in Figure 17-17 has a variety of exposed crystal planes.) Oxidation of bound ferrocene groups on the 0.38 cm^2 surface of a diamond electrode required 23 μC of charge. Find the ferrocene surface density (molecules/cm^2) on diamond. If there are two ferrocene groups attached to each N atom, how many N atoms are there in each square centimeter? Approximately what fraction of C surface atoms are substituted by N atoms?

17-21. Ti^{3+} is to be generated in 0.10 M $HClO_4$ for coulometric reduction of azobenzene.

$$TiO^{2+} + 2H^+ + e^- \rightleftharpoons Ti^{3+} + H_2O \qquad E° = 0.100 \text{ V}$$

$$4Ti^{3+} + C_6H_5N{=}NC_6H_5 + 4H_2O \rightarrow 2C_6H_5NH_2 + 4TiO^{2+} + 4H^+$$

Azobenzene   Aniline

At the counter electrode, water is oxidized, and O_2 is liberated at a pressure of 0.20 bar. Both electrodes are made of smooth Pt, and each has a total surface area of 1.00 cm^2. The rate of reduction of the azobenzene is 25.9 nmol/s, and the resistance of the solution between the generator electrodes is 52.4 Ω.

(a) Calculate the current density (A/m^2) at the electrode surface. Use Table 17-1 to estimate the overpotential for O_2 liberation.

(b) Calculate the cathode potential (versus S.H.E.) assuming that $[TiO^{2+}]_{surface} = [TiO^{2+}]_{bulk} = 0.050$ M and $[Ti^{3+}]_{surface} = 0.10$ M.

(c) Calculate the anode potential (versus S.H.E.).

(d) What should the applied voltage be?

17-22. *Propagation of uncertainty.* In an extremely careful measurement of the Faraday constant, a pure silver anode was oxidized to Ag^+ with a constant current of 0.203 639 0 ($\pm$0.000 000 4) A for 18 000.075 ($\pm$0.010) s to give a mass loss of 4.097 900 ($\pm$0.000 003) g from the anode. Given that atomic mass of Ag = 107.868 2 ($\pm$0.000 2), find the value of the Faraday constant and its uncertainty.

17-23. *Coulometric titration of sulfite in wine.*[58] Sulfur dioxide is added to many foods as a preservative. In aqueous solution, the following species are in equilibrium:

$$SO_2 \rightleftharpoons H_2SO_3 \rightleftharpoons HSO_3^- \rightleftharpoons SO_3^{2-} \qquad \text{(A)}$$

Sulfur dioxide · Sulfurous acid · Bisulfite · Sulfite

Bisulfite reacts with aldehydes in food near neutral pH:

$$
\begin{array}{c}
\text{H} \\
\underset{\text{R}}{\overset{}{>}} \text{C=O} + HSO_3^- \rightleftharpoons \underset{\text{HO}\;\;\;\text{R}}{\overset{\text{H}}{>}}\!\!-SO_3^-
\end{array}
\qquad \text{(B)}
$$

Aldehyde · Adduct

Sulfite is released from the adduct in 2 M NaOH and can be analyzed by its reaction with I_3^- to give I^- and sulfate. Excess I_3^- must be present for quantitative reaction.

Here is a coulometric procedure for analysis of total sulfite in white wine. Total sulfite means all species in Reaction A and the adduct in Reaction B. We use white wine so that we can see the color of a starch-iodine end point.

1. Mix 9.00 mL of wine plus 0.8 g NaOH and dilute to 10.00 mL. The NaOH releases sulfite from its organic adducts.
2. Generate I_3^- at the working electrode (the anode) by passing a known current for a known time through the cell in Figure 17-10. The cell contains 30 mL of 1 M acetate buffer (pH 3.7) plus 0.1 M KI. In the cathode compartment, H_2O is reduced to $H_2 + OH^-$. The frit retards diffusion of OH^- into the main compartment, where it would react with I_3^- to give IO^-.
3. Generate I_3^- at the anode with a current of 10.0 mA for 4.00 min.
4. Inject 2.000 mL of the wine/NaOH solution into the cell, where the sulfite reacts with I_3^-, leaving excess I_3^-.
5. Add 0.500 mL of 0.050 7 M thiosulfate to consume I_3^- by Reaction 16-19 and leave excess thiosulfate.
6. Add starch indicator to the cell and generate fresh I_3^- with a constant current of 10.0 mA. A time of 131 s was required to consume excess thiosulfate and reach the starch end point.

(a) In what pH range is each form of sulfurous acid predominant?

(b) Write balanced half-reactions for the anode and cathode.

(c) At pH 3.7, the dominant form of sulfurous acid is HSO_3^- and the dominant form of sulfuric acid is HSO_4^{2-}. Write balanced reactions between I_3^- and HSO_3^- and between I_3^- and thiosulfate.

(d) Find the concentration of total sulfite in undiluted wine.

17-24. *Chemical oxygen demand by coulometry.* An electrochemical device incorporating photooxidation on a TiO_2 surface could replace refluxing with $Cr_2O_7^{2-}$ to measure chemical oxygen demand (Box 16-2). The diagram shows a working electrode held at +0.30 V versus Ag | AgCl and coated with nanoparticles of TiO_2. Ultraviolet irradiation generates electrons and holes in TiO_2. Holes oxidize

organic matter at the surface. Electrons reduce H_2O at the auxiliary electrode in a compartment connected to the working compartment by a salt bridge. The sample compartment is only 0.18 mm thick with a volume of 13.5 μL. It requires ~1 min for all organic matter to diffuse to the TiO_2 surface and be exhaustively oxidized.

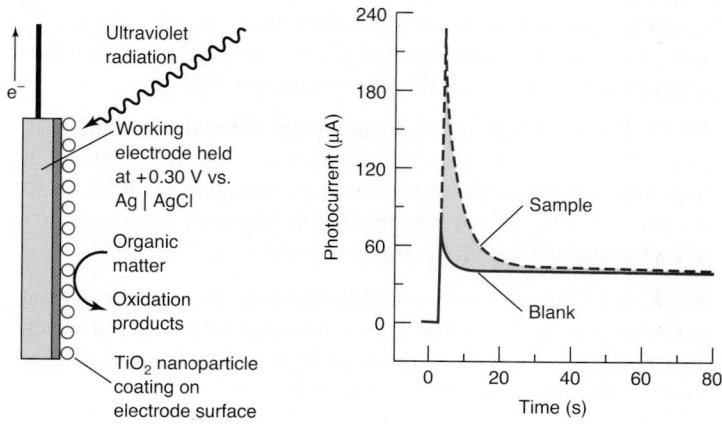

Left: Working electrode. Right: Photocurrent response for sample and blank. Both solutions contain 2 M $NaNO_3$. [Data from H. Zhao, D. Jiang, S. Zhang, K. Catterall, and R. John, "Development of a Direct Photoelectrochemical Method for Determination of Chemical Oxygen Demand," *Anal. Chem.* **2004**, *76*, 155.]

The blank curve in the graph shows the response when the sample compartment contains just electrolyte. Before irradiation, no current is observed. Ultraviolet radiation causes a spike in the current, followed by a decrease to a steady level near 40 μA. This current arises from oxidation of water at the TiO_2 surface under ultraviolet exposure. The upper curve shows the same experiment, but with wastewater in the sample compartment. The increased current arises from oxidation of organic matter. When the organic matter is consumed, the current decreases to the blank level. The area between the two curves tells us how many electrons flow from oxidation of organic matter in the sample.

(a) Balance the oxidation half-reaction that occurs in this cell:

$$C_cH_hO_oN_nX_x + AH_2O \rightarrow BCO_2 + CX^- + DNH_3 + EH^+ + Fe^-$$

where X is any halogen. Express the stoichiometry coefficients A, B, C, D, E, and F in terms of c, h, o, n, and x.

(b) How many molecules of O_2 are required to balance the half-reaction in part **(a)** by reduction of oxygen ($O_2 + 4H^+ + 4e^- \rightarrow 2H_2O$)?

(c) The area between the two curves in the graph is $\int_0^\infty (I_{sample} - I_{blank})dt = 9.43$ mC. This is the number of electrons liberated by complete oxidation of the sample. How many moles of O_2 would be required for the same oxidation?

(d) Chemical oxygen demand (COD) is expressed as mg of O_2 required to oxidize 1 L of sample. Find the COD for this sample.

(e) If the only oxidizable substance in the sample were $C_9H_6NO_2ClBr_2$, what is its concentration in mol/L?

Amperometry

17-25. What is a Clark electrode, and how does it work?

17-26. (a) How does the amperometric glucose monitor in Figure 17-12 work?

(b) Why is a mediator advantageous in the glucose monitor?

(c) How does the coulometric glucose monitor in Figure 17-14 work?

(d) Why does the signal in the amperometric measurement depend on the temperature of the blood sample, whereas the signal in coulometry is independent of temperature? Do you expect the signal to increase or decrease with increasing temperature in amperometry?

(e) Glucose ($C_6H_{12}O_6$, FM 180.16) is present in normal human blood at a concentration near 1 g/L. How many microcoulombs are required for complete oxidation of glucose in 0.300 μL of blood in a home glucose monitor if the concentration is 1.00 g/L?

17-27. Explain why each voltammogram from the rotating disk electrode in Figure 17-16 reaches a plateau at low potential and at high potential. What chemistry occurs in each plateau? Why do all the curves overlap at low potential? How might the current density in each plateau change if the rotation speed were decreased?

17-28. For a rotating disk electrode operating at sufficiently great potential, the redox reaction rate is governed by the rate at which analyte diffuses through the diffusion layer to the electrode (Figure 17-15b). The thickness of the diffusion layer is

$$\delta = 1.61 D^{1/3} \nu^{1/6} \omega^{-1/2}$$

where D is the diffusion coefficient of reactant (m^2/s), ν is the kinematic viscosity of the liquid (= viscosity/density = m^2/s), and ω is the rotation rate (radians/s) of the electrode. There are 2π radians in a circle. The current density (A/m^2) is

$$\text{Current density} = 0.62 n F D^{2/3} \nu^{-1/6} \omega^{1/2} C_0$$

where n is the number of electrons in the half-reaction, F is the Faraday constant, and C_0 is the concentration of the electroactive species in bulk solution (mol/m^3, not mol/L). Consider the oxidation of $Fe(CN)_6^{4-}$ in a solution of 10.0 mM $K_3Fe(CN)_6$ + 50.0 mM $K_4Fe(CN)_6$ at +0.90 V (versus S.C.E.) at a rotation speed of 2.00×10^3 revolutions per minute.[27] The diffusion coefficient of $Fe(CN)_6^{4-}$ is 2.5×10^{-9} m^2/s, and the kinematic viscosity is 1.1×10^{-6} m^2/s. Calculate the thickness of the diffusion layer and the current density. If you are careful, the current density should look like the value in Figure 17-16b.

Voltammetry

17-29. In 1 M NH_3/1 M NH_4Cl solution, Cu^{2+} is reduced to Cu^+ near −0.3 V (versus S.C.E.), and Cu^+ is reduced to Cu(in Hg) near −0.6 V.

(a) Sketch a qualitative sampled current polarogram for a solution of Cu^+.

(b) Sketch a polarogram for a solution of Cu^{2+}.

(c) Suppose that Pt, instead of Hg, were used as the working electrode. Which, if any, reduction potential would you expect to change?

17-30. (a) Explain the difference between charging current and faradaic current.

(b) What is the purpose of waiting 1 s after a voltage pulse before measuring current in sampled current voltammetry?

(c) Why is square wave voltammetry more sensitive than sampled current voltammetry?

17-31. Suppose that the diffusion current in a polarogram for reduction of Cd^{2+} at a mercury electrode is 14 μA. If the solution contains 25 mL of 0.50 mM Cd^{2+}, what percentage of Cd^{2+} is reduced in the 3.4 min required to scan from −0.6 to −1.2 V?

17-32. The drug Librium gives a polarographic wave with $E_{1/2}$ = −0.265 V (versus S.C.E.) in 0.05 M H_2SO_4. A 50.0-mL sample containing Librium gave a wave height of 0.37 μA. When 2.00 mL of 3.00 mM Librium in 0.05 M H_2SO_4 were added to the sample, the wave height increased to 0.80 μA. Find the molarity of Librium in the unknown.

17-33. Explain what is done in anodic stripping voltammetry. Why is stripping the most sensitive voltammetric technique?

17-34. The figure shows a series of standard additions of Cu^{2+} to acidified tap water measured by anodic stripping voltammetry at an iridium electrode. The unknown and all standard additions were made up to the same final volume.

(a) What chemical reaction occurs during the concentration stage of the analysis?

(b) What chemical reaction occurs during the stripping stage of the analysis?

(c) Find the concentration of Cu^{2+} in the tap water.

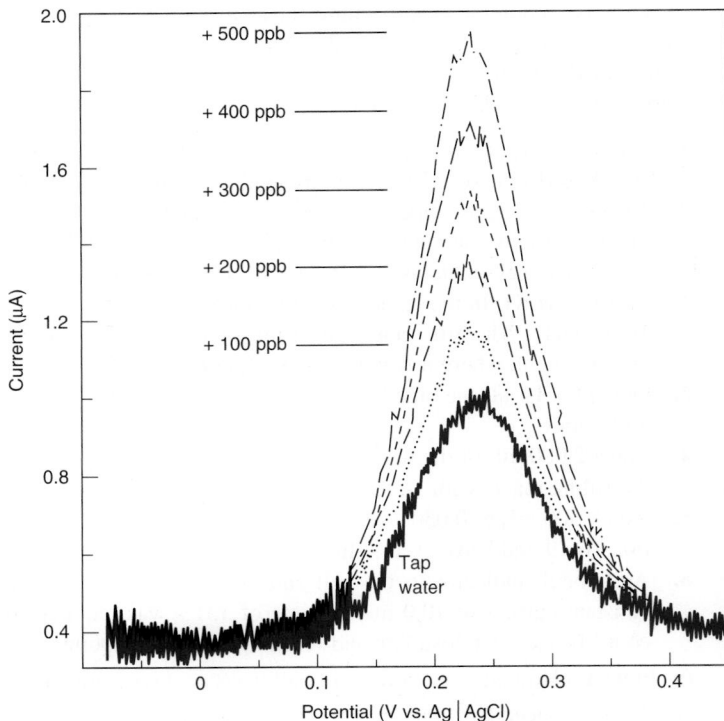

Anodic stripping voltammograms of tap water and five standard additions of 100 ppb Cu^{2+}. [Data from M. A. Nolan and S. P. Kounaves, "Microfabricated Array of Ir Microdisks for Determination of Cu^{2+} or Hg^{2+} Using Square Wave Stripping Voltammetry," *Anal. Chem.* **1999,** *71,* 3567.]

17-35. From the two standard additions of 50 pm Fe(III) in the figure, find the concentration of Fe(III) in the seawater. Estimate where the baseline should be drawn for each trace and measure the peak height from the baseline. Consider the volume to be constant for all three solutions.

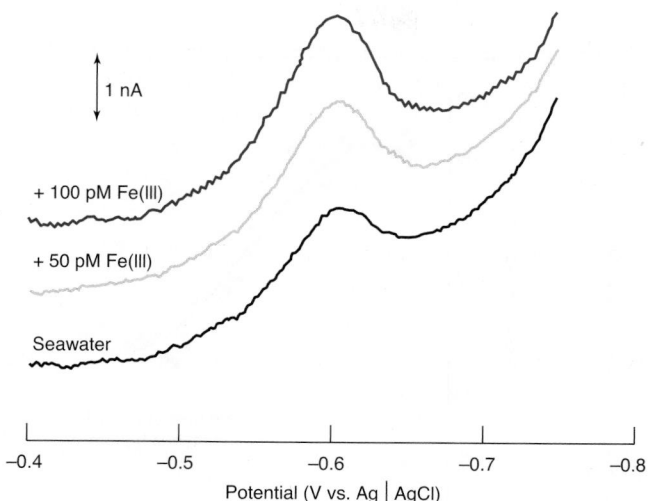

Sampled current cathodic stripping voltammogram of Fe(III) in seawater plus two standard additions of 50 pM Fe(III). [Data from H. Obata and C. M. G. van den Berg, "Determination of Picomolar Levels of Iron in Seawater Using Catalytic Cathodic Stripping Voltammetry," *Anal. Chem.* **2001,** *73,* 2522. See also C. M. G. van den Berg, "Chemical Speciation of Iron in Seawater by Cathodic Stripping Voltammetry with Dihydroxynaphthalene," *Anal. Chem.* **2006,** *78,* 156.]

17-36. Cathodic stripping of ClO_4^- in Figure 17-26 does not involve oxidation or reduction of ClO_4^-. Explain how this measurement works.

17-37. The cyclic voltammogram of the antibiotic chloramphenicol (abbreviated RNO_2) is shown here. The first cathodic scan goes from 0 to -1.0 V. The first cathodic wave, A, is from the reaction $RNO_2 + 4e^- + 4H^+ \rightarrow RNHOH + H_2O$. Peak B in the reverse anodic scan could be assigned to $RNHOH \rightarrow RNO + 2H^+ + 2e^-$. In the second cathodic scan from $+0.9$ to -0.4 V, the new peak C appears. Write a reaction for peak C and explain why peak C was not seen in the initial scan.

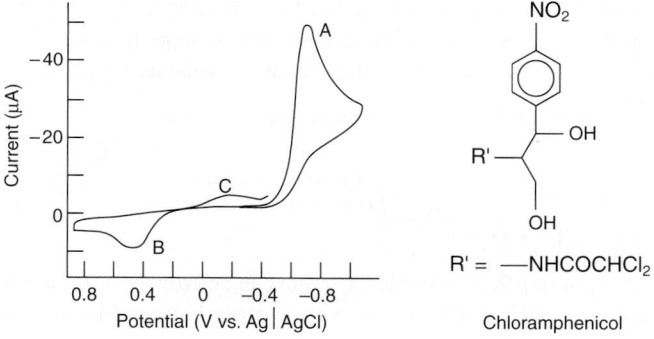

Cyclic voltammogram of 3.7×10^{-4} chloramphenicol in 0.1 M acetate buffer, pH 4.62. The voltage of the carbon paste working electrode was scanned at a rate of 350 mV/s. [Data from P. T. Kissinger and W. R. Heineman, "Cyclic Voltammetry," *J. Chem. Ed.* **1983,** *60,* 702.]

17-38. Peak current (I_p) and scan rate (v) are listed for cyclic voltammetry of the reversible reaction Fe(II) → Fe(III) of a water-soluble ferrocene derivative in 0.1 M NaCl.[59]

Scan rate (V/s)	Peak anodic current (μA)
0.019 2	2.18
0.048 9	3.46
0.075 1	4.17
0.125	5.66
0.175	6.54
0.251	7.55

If a graph of I_p versus $\sqrt{v}$ gives a straight line, then the reaction is diffusion controlled. Prepare such a graph and use it to find the diffusion coefficient of the reactant from Equation 17-21 for this one-electron oxidation. The area of the working electrode is 0.020 1 cm^2, and the concentration of reactant is 1.00 mM.

17-39. What are the advantages of using a microelectrode for voltammetric measurements?

17-40. What is the purpose of the Nafion membrane in Figure 17-33?

17-41. *Measuring the size of a microelectrode by cyclic voltammetry.*

(a) Redox chemistry for ferrocyanide in Figure 17-32 was given at the beginning of Section 17-5. Write the analyte half-reaction that occurs at the upper plateau near 0.4 V and at the lower plateau near 0 V (versus S.C.E.).

(b) The limiting current I_{limit}, which is the difference between the upper and lower plateaus, is related to the radius of the disk-shaped electrode (r) and the diffusion coefficient (D) and bulk concentration (C) of analyte:

$$I_{limit} = 4nFDCr$$

where n is the number of electrons in the half-reaction and F is the Faraday constant. In this equation, the units of concentration should be mol/m^3 to be consistent with the other quantities in SI units. The diffusion coefficient for ferrocyanide cited in the reference for Figure 17-32 is 9.2×10^{-10} m^2/s in water at 25°C. Calculate the radius of the microelectrode.

Karl Fischer Titration

17-42. Write the chemical reactions that show that 1 mol of I_2 is required for 1 mol of H_2O in a Karl Fischer titration.

17-43. Explain how the end point is detected in a Karl Fischer titration in Figure 17-35.

18 | Fundamentals of Spectrophotometry

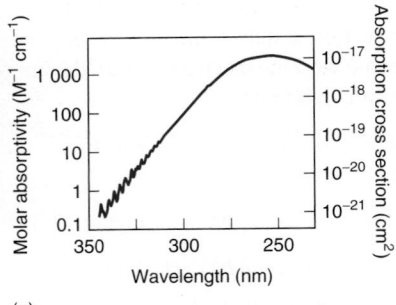

(a)

Spectrum of ozone, showing maximum absorption of ultraviolet radiation at a wavelength near 260 nm. At this wavelength, a layer of ozone is more opaque than a layer of gold of the same mass. [Data from R. P. Wayne, *Chemistry of Atmospheres* (Oxford: Clarendon Press, 1991).]

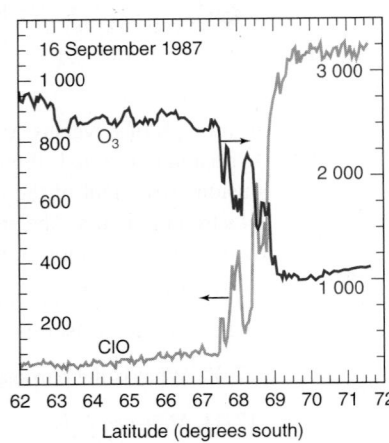

(b)

Spectroscopically measured concentrations of O_3 and ClO (measured in ppb = nL/L) in the stratosphere near the South Pole in 1987. The loss of O_3 at latitudes where ClO concentration increases is consistent with catalytic destruction of O_3 in steps 2 through 4 below. [Data from J. G. Anderson, W. H. Brune, and M. H. Proffitt, *J. Geophys. Res.* **1989**, *94D*, 11465.]

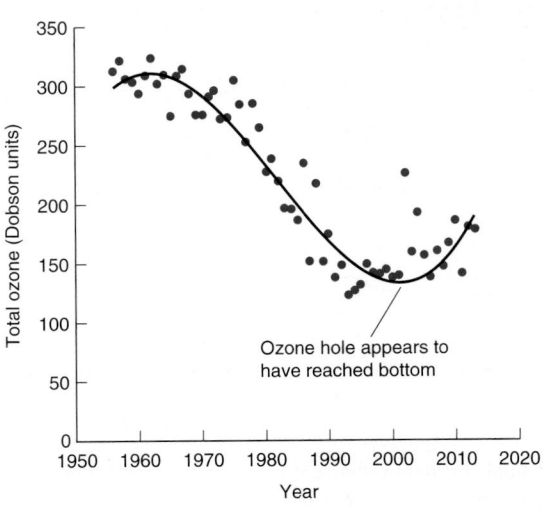

(c)

Mean atmospheric O_3 at Halley in Antarctica in October. *Dobson units* are defined in Problem 18-16. [Data from J. D. Shanklin, British Antarctic Survey, http://www.antarctica.ac.uk/met/jds/ozone/.]

Ozone, formed at altitudes of 20 to 40 km by the action of solar ultraviolet radiation ($h\nu$) on O_2, absorbs ultraviolet radiation that causes sunburn and skin cancer.

$$O_2 \xrightarrow{h\nu} 2O \qquad O + O_2 \rightarrow O_3$$
$$\text{Ozone}$$

In 1985, the British Antarctic Survey reported that ozone in the Antarctic stratosphere had decreased by 50% in early spring (October), relative to levels observed in the preceding 20 years. Ground, airborne, and satellite observations have since shown that this "ozone hole" occurs only in early spring (Figure 1-1) and continued to deepen until the year 2000.

An explanation begins with chlorofluorocarbons such as Freon-12 (CCl_2F_2), formerly used as a refrigerant and propellant in spray cans. These long-lived compounds, which are not found in nature,[2] diffuse to the stratosphere, where they catalyze ozone decomposition.

(1) $CCl_2F_2 \xrightarrow{h\nu} {}^{\bullet}CClF_2 + Cl^{\bullet}$ Photochemical Cl$^{\bullet}$ formation

(2) $Cl^{\bullet} + O_3 \rightarrow {}^{\bullet}ClO + O_2$

(3) $O_3 \xrightarrow{h\nu} O + O_2$ Net reaction of (2)–(4): Catalytic O_3 destruction $2O_3 \rightarrow 3O_2$

(4) $O + {}^{\bullet}ClO \rightarrow Cl^{\bullet} + O_2$

Cl produced in step 4 reacts again in step 2, so a single Cl$^{\bullet}$ atom can destroy $>10^5$ molecules of O_3. The chain is terminated when Cl$^{\bullet}$ or $^{\bullet}$ClO reacts with hydrocarbons or NO_2 to form HCl or $ClONO_2$.

Stratospheric clouds[3] formed during the Antarctic winter catalyze the reaction of HCl with $ClONO_2$ to form Cl_2, which is split by sunlight into Cl$^{\bullet}$ atoms to initiate O_3 destruction:

$$HCl + ClONO_2 \xrightarrow[\text{polar clouds}]{\text{surface of}} Cl_2 + HNO_3 \qquad Cl_2 \xrightarrow{h\nu} 2Cl^{\bullet}$$

Polar stratospheric clouds require winter cold to form. Only when the sun is rising in September and October, and clouds are still present, are conditions right for O_3 destruction.

To protect life from ultraviolet radiation, the Montreal Protocol treaty, negotiated in 1987, phased out chlorofluorocarbons. Safer substitutes are now in use.

Spectrophotometry is any technique that uses light to measure chemical concentrations. A procedure based on absorption of visible light is called *colorimetry*. Chapter 18 is intended to give a stand-alone overview of spectrophotometry sufficient for introductory purposes. Chapter 19 goes further into applications and Chapter 20 discusses instrumentation.

[C. Calvin/UCAR/NOAA]

Following the discovery of the Antarctic ozone "hole" in 1985, atmospheric chemist Susan Solomon led the first expedition in 1986 specifically intended to make chemical measurements of the Antarctic atmosphere by using balloons and ground-based spectroscopy. The expedition discovered that ozone depletion occurred after polar sunrise and that the concentration of chemically active chlorine in the stratosphere was ~100 times greater than had been predicted from gas-phase chemistry. Solomon's group identified chlorine as the culprit in ozone destruction and polar stratospheric clouds as the catalytic surface for the release of so much chlorine.

18-1 Properties of Light

It is convenient to describe light in terms of both particles and waves. Light waves consist of perpendicular, oscillating electric and magnetic fields. For simplicity, a *plane-polarized* wave is shown in Figure 18-1. In this figure, the electric field is in the *xy* plane, and the magnetic field is in the *xz* plane. **Wavelength,** λ, is the crest-to-crest distance between waves. **Frequency,** ν, is the number of complete oscillations that the wave makes each second. The unit of frequency is 1/second. One oscillation per second is called one **hertz** (Hz). A frequency of 10^6 s^{-1} is therefore said to be 10^6 Hz, or 1 *megahertz* (MHz).

The relation between frequency and wavelength is

Relation between frequency and wavelength: $\quad\quad \nu\lambda = c \quad\quad\quad\quad\quad$ **(18-1)**

where c is the speed of light (2.998×10^8 m/s in vacuum). In a medium other than vacuum, the speed of light is c/n, where n is the **refractive index** of that medium. For visible wavelengths in most substances, $n > 1$, so visible light travels more slowly through matter than through vacuum. When light moves between media with different refractive indexes, the frequency remains constant but the wavelength changes.

With regard to energy, it is more convenient to think of light as particles called **photons.** Each photon carries energy E given by

Relation between energy and frequency: $\quad\quad E = h\nu \quad\quad\quad\quad\quad$ **(18-2)**

where h is *Planck's constant* ($= 6.626 \times 10^{-34}$ J·s).

Equation 18-2 states that energy is proportional to frequency. Combining Equations 18-1 and 18-2, we can write

$$E = \frac{hc}{\lambda} = hc\tilde{\nu} \quad\quad\quad\quad\quad \textbf{(18-3)}$$

where $\tilde{\nu}$ ($= 1/\lambda$) is called **wavenumber.** Energy is inversely proportional to wavelength and directly proportional to wavenumber. Red light, with a longer wavelength than blue light, is less energetic than blue light. The most common unit of wavenumber is cm^{-1}, read "reciprocal centimeters" or "wavenumbers."

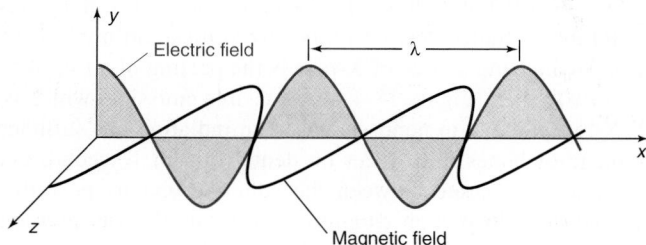

FIGURE 18-1 Plane-polarized electromagnetic radiation of wavelength λ, propagating along the *x*-axis. The electric field of plane-polarized light is confined to a single plane. Ordinary, unpolarized light has electric field components in all planes parallel to the direction of travel.

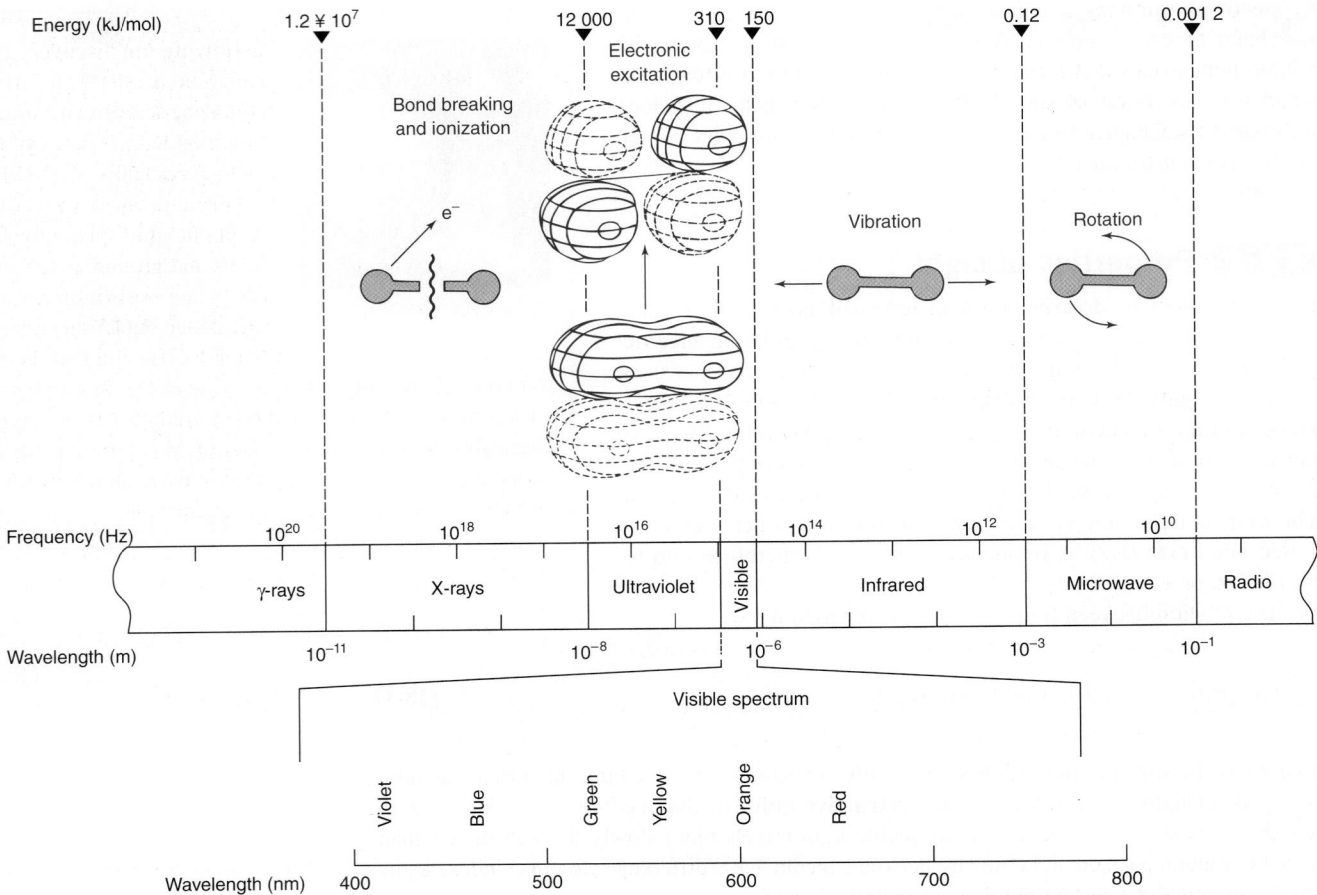

FIGURE 18-2 Electromagnetic spectrum, showing representative molecular processes that occur when light in each region is absorbed. The visible spectrum spans the wavelength range 380–780 nanometers (1 nm = 10^{-9} m).

The names of regions of the **electromagnetic spectrum** in Figure 18-2 are historical. There are no abrupt changes in characteristics as we go from one region to the next, such as visible to infrared. Visible light—the kind of electromagnetic radiation we see—represents only a small fraction of the electromagnetic spectrum.

18-2 Absorption of Light

When a molecule absorbs a photon, the energy of the molecule increases. We say that the molecule is promoted to an **excited state** (Figure 18-3). If a molecule emits a photon, the energy of the molecule is lowered. The lowest energy state of a molecule is called the **ground state.** Figure 18-2 indicates that microwave radiation stimulates rotation of molecules when it is absorbed. Infrared radiation stimulates vibrations. Visible and ultraviolet radiation promote electrons to higher energy orbitals.

X-rays and short-wavelength ultraviolet radiation are harmful because they break chemical bonds and ionize molecules, which is why you should minimize your exposure to medical X-rays. An amazing source of X-rays is the peeling of a roll of Scotch® tape in a vacuum of 10^{-5} to 10^{-6} bar.[4] Figure 18-4 shows visible emission, which is accompanied by bursts of 10^5 X-ray photons in nanoseconds. The radiation has sufficient intensity to make an X-ray image of bones in a finger on dental film in 1 second. Peeling the tape causes electric charge to separate between the adhesive and its polyethylene backing. Periodically, the electric field is high enough to cause an electric discharge in the low-pressure gas between the adhesive and the backing. Highly accelerated electrons liberate burst of X-rays (called bremsstrahlung radiation) when they strike the positively charged adhesive and are suddenly decelerated. X-rays are not emitted when tape is unpeeled in air at atmospheric pressure.

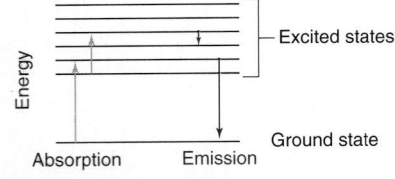

FIGURE 18-3 Absorption of light increases the energy of a molecule. Emission of light decreases its energy.

EXAMPLE Photon Energies

By how many kilojoules per mole is the energy of O_2 increased when it absorbs ultraviolet radiation with a wavelength of 147 nm? How much is the energy of CO_2 increased when it absorbs infrared radiation with a wavenumber of 2 300 cm^{-1}?

Solution For the ultraviolet radiation, the energy increase is

$$\Delta E = h\nu = h\frac{c}{\lambda}$$

$$= (6.626 \times 10^{-34} \text{ J} \cdot \text{s})\left[\frac{(2.998 \times 10^8 \text{ m/s})}{(147 \text{ nm})(10^{-9} \text{ m/nm})}\right] = 1.35 \times 10^{-18} \text{ J/molecule}$$

$$(1.35 \times 10^{-18} \text{ J/molecule})(6.022 \times 10^{23} \text{ molecules/mol}) = 814 \text{ kJ/mol}$$

This is enough energy to break the O=O bond in oxygen. For CO_2, the energy increase is

$$\Delta E = h\nu = h\frac{c}{\lambda} = hc\tilde{\nu} \qquad \left(\text{recall that } \tilde{\nu} = \frac{1}{\lambda}\right)$$

$$= (6.626 \times 10^{-34} \text{ J} \cdot \text{s})(2.998 \times 10^8 \text{ m/s})(2\ 300 \text{ cm}^{-1})(100 \text{ cm/m})$$

$$= 4.6 \times 10^{-20} \text{ J/molecule} = 28 \text{ kJ/mol}$$

Infrared absorption increases the amplitude of the vibrations of the CO_2 bonds.

TEST YOURSELF What is the wavelength, wavenumber, and name of radiation with an energy of 100 kJ/mol? (*Answer:* 1.20 μm, 8.36×10^3 cm^{-1}, infrared)

FIGURE 18-4 Visible (blue) emission from peeling a roll of Scotch® tape in 1.3 μbar of air. Visible light is accompanied by bursts of X-rays.
[Courtesy C. Camara, Tribogenics and S. Putterman, University of California, Los Angeles.]

When light is absorbed by a sample, the *irradiance* of the beam of light is decreased. **Irradiance,** P, is the energy per second per unit area of the light beam. A rudimentary spectrophotometric experiment is illustrated in Figure 18-5. Light passes through a **monochromator** (a prism, a grating, or a filter) to select one wavelength (Color Plate 14). Light with a narrow range of wavelengths is said to be **monochromatic** ("one color"). The monochromatic light, with irradiance P_0, strikes a sample of length b. The irradiance of the beam emerging from the other side of the sample is P. Some of the light may be absorbed by the sample, so $P \leq P_0$.

Irradiance is the energy per unit time per unit area in the light beam (watts per square meter, W/m²). The terms *intensity* or *radiant power* have been used for this same quantity.

Monochromatic light consists of "one color" (one wavelength). The better the monochromator, the narrower is the range of wavelengths in the emerging beam.

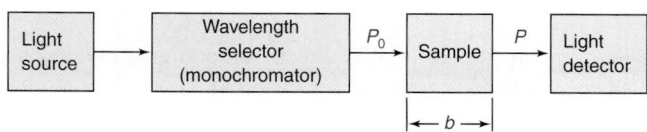

FIGURE 18-5 Schematic diagram of a single-beam spectrophotometer. P_0, irradiance of beam entering sample; P, irradiance of beam emerging from sample; b, length of path through sample.

Transmittance, T, is defined as the fraction of the original light that passes through the sample.

Transmittance:
$$T = \frac{P}{P_0} \qquad \qquad (18\text{-}4)$$

Therefore, T has the range 0 to 1. The *percent transmittance* ($= 100T$) ranges between 0 and 100%. **Absorbance** is defined as

Absorbance:
$$A = \log\left(\frac{P_0}{P}\right) = -\log T \qquad \qquad (18\text{-}5)$$

When no light is absorbed, $P = P_0$ and $A = 0$. If 90% of the light is absorbed, 10% is transmitted and $P = P_0/10$. This ratio gives $A = 1$. If only 1% of the light is transmitted, $A = 2$. Absorbance is sometimes called *optical density*.

Relation between transmittance and absorbance:

P/P_0	% T	A
1	100	0
0.1	10	1
0.01	1	2

Beer's law, Equation 18-6, states that *absorbance* is proportional to the concentration of the absorbing species. The fraction of light passing through a sample (the *transmittance*) is related logarithmically, not linearly, to the sample concentration. Why should this be?

Imagine light of irradiance P passing through an *infinitesimally thin* layer of solution whose thickness is dx. A physical model of the absorption process suggests that, within the infinitesimally thin layer, decrease in power (dP) ought to be proportional to the incident power (P), to the concentration of absorbing species (c), and to the thickness of the section (dx):

$$dP = -\beta P c \, dx \qquad \textbf{(A)}$$

where β is a constant of proportionality and the minus sign indicates a decrease in P as x increases. The rationale for saying that the decrease in power is proportional to the incident power may be understood from a numerical example. If 1 photon out of 1 000 incident photons is absorbed in a thin layer, we would expect that 2 out of 2 000 incident photons would be absorbed. The decrease in photons (power) is proportional to the incident flux of photons (power).

Equation A can be rearranged and integrated to find an expression for P:

$$-\frac{dP}{P} = \beta c \, dx \Rightarrow -\int_{P_0}^{P} \frac{dP}{P} = \beta c \int_{0}^{b} dx$$

The limits of integration are $P = P_0$ at $x = 0$ and $P = P$ at $x = b$. The integral on the left side is $\int dP/P = \ln P$, so

$$-\ln P - (-\ln P_0) = \beta c b \Rightarrow \ln\left(\frac{P_0}{P}\right) = \beta c b$$

Finally, converting $\ln$ to $\log$, using the relation $\ln z = (\ln 10)(\log z)$, gives Beer's law:

$$\underbrace{\log\left(\frac{P_0}{P}\right)}_{\text{Absorbance}} = \underbrace{\left(\frac{\beta}{\ln 10}\right)}_{\text{Constant} \equiv \varepsilon} cb \Rightarrow A = \varepsilon cb$$

The logarithmic relation of P_0/P to concentration arises because, in each infinitesimal portion of the total volume, *the decrease in power is proportional to the power incident upon that section.* As light travels through a sample, power loss in each succeeding layer decreases because less incident power reaches that layer. Molar absorptivity ranges from 0 (if the probability for photon absorption is 0) to approximately $10^5 \text{ M}^{-1} \text{ cm}^{-1}$ (when the probability for photon absorption approaches unity).

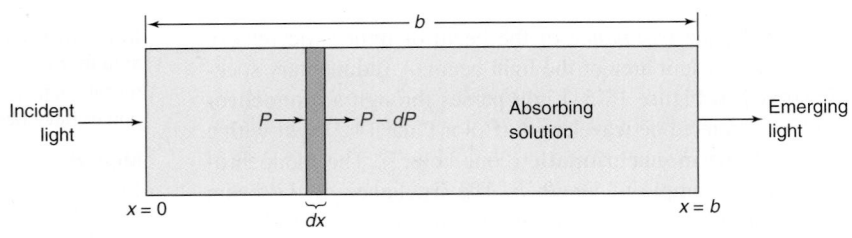

Absorbance is so important because it is directly proportional to the concentration, c, of the light-absorbing species in the sample (Color Plate 15).

Beer's law: $\qquad\qquad\qquad A = \varepsilon bc \qquad\qquad\qquad$ **(18-6)**

Equation 18-6, which is the heart of spectrophotometry as applied to analytical chemistry, is called the *Beer-Lambert law*,[6] or simply **Beer's law.** Absorbance is dimensionless, but some people write "absorbance units" after absorbance. The concentration of the sample, c, is usually given in units of moles per liter (M). The pathlength, b, is commonly expressed in centimeters. The quantity ε (epsilon) is called the **molar absorptivity** (or *extinction coefficient* in the older literature) and has the units $\text{M}^{-1} \text{ cm}^{-1}$ to make the product εbc dimensionless. Molar absorptivity tells how much light is absorbed at a particular wavelength by a particular substance.

Box 18-1 explains why absorbance, not transmittance, is directly proportional to concentration.

EXAMPLE   Absorbance, Transmittance, and Beer's Law

Find the absorbance and transmittance of a 0.002 40 M solution of a substance with a molar absorptivity of $313 \text{ M}^{-1} \text{ cm}^{-1}$ in a cell with a 2.00-cm pathlength.

Solution Equation 18-6 gives us the absorbance.

$$A = \varepsilon bc = (313 \text{ M}^{-1} \text{ cm}^{-1})(2.00 \text{ cm})(0.002 \, 40 \text{ M}) = 1.50$$

Transmittance is obtained from Equation 18-5 by raising 10 to the power equal to the expression on each side of the equation:

$$\log T = -A$$
$$T = 10^{\log T} = 10^{-A} = 10^{-1.50} = 0.031\ 6$$

Just 3.16% of the incident light emerges from this solution.

TEST YOURSELF   The transmittance of a 0.010 M solution of a compound in a 0.100-cm-pathlength cell is $T = 8.23\%$. Find the absorbance (A) and the molar absorptivity (ε). (*Answer:* 1.08, 1.08×10^3 M^{-1} cm^{-1})

Equation 18-6 could be written

$$A_\lambda = \varepsilon_\lambda bc$$

because A and ε depend on the wavelength of light. The quantity ε is a coefficient of proportionality between absorbance and the product bc. The greater the molar absorptivity, the greater the absorbance. An **absorption spectrum** (Demonstration 18-1) is a graph showing how A (or ε) varies with wavelength.

The plural of "spectrum" is "spectra."

The part of a molecule responsible for light absorption is called a **chromophore.** Any substance that absorbs visible light appears colored when white light is transmitted through it or reflected from it. (White light contains all the colors in the visible spectrum.) The substance absorbs certain wavelengths of the white light, and our eyes detect the wavelengths that are not absorbed. Table 18-1 gives a rough guide to colors.[10] The observed color is called the *complement* of the absorbed color. For example, bromophenol blue has maximum absorbance at 591 nm and its observed color is blue. Color Plates 16 and 17 display absorption spectra and observed colors.

The color of a solution is the **complement** of the color of the light that it absorbs. The color we perceive depends not only on the wavelength of light, but on its intensity.

When Beer's Law Fails

Beer's law states that absorbance is proportional to the concentration of the absorbing species. It applies to *monochromatic* radiation, and it works very well for dilute solutions ($\lesssim 0.01$ M) of most substances. "Monochromatic" means that the bandwidth of the light must be substantially smaller than the width of the absorption band of the chromophore.[11]

In concentrated solutions, solute molecules influence one another as a result of their proximity. At very high concentration, the solute *becomes* the solvent. Properties of a molecule are not exactly the same in different environments. Nonabsorbing solutes in a solution can also interact with the absorbing species and alter the absorptivity.

If the absorbing molecule participates in a concentration-dependent chemical equilibrium, the absorptivity changes with concentration. For example, in concentrated solution, a weak acid, HA, may be mostly undissociated. As the solution is diluted, dissociation increases. If the absorptivity of A$^-$ is not the same as that of HA, the solution will appear not to obey Beer's law as it is diluted.

Beer's law works for monochromatic radiation passing through a dilute solution in which the absorbing species is not participating in a concentration-dependent equilibrium.

TABLE 18-1	Colors of visible light	
Wavelength of maximum absorption (nm)	Color absorbed	Color observed
380–420	Violet	Green-yellow
420–440	Violet-blue	Yellow
440–470	Blue	Orange
470–500	Blue-green	Red
500–520	Green	Purple-red
520–550	Yellow-green	Violet
550–580	Yellow	Violet-blue
580–620	Orange	Blue
620–680	Red	Blue-green
680–780	Red	Green

The spectrum of visible light can be projected on a screen in a darkened room in the following manner: Mount four layers of plastic diffraction grating* on a cardboard frame having a square hole large enough to cover the lens of an overhead projector. Tape the assembly over the projector lens facing the screen. Place an opaque cardboard surface with two 1×3 cm slits on the working surface of the projector.

When the lamp is turned on, the white image of each slit is projected on the center of the screen. A visible spectrum appears on either side of each image. By placing a beaker of colored solution over one slit, you can see its color projected on the screen where the white image previously appeared. The spectrum beside the colored image loses intensity at wavelengths absorbed by the colored species.

Color Plate 16a shows the spectrum of white light and the spectra of three different colored solutions. Potassium dichromate, which appears orange or yellow, absorbs blue wavelengths. Bromophenol blue absorbs orange and appears blue to our eyes. Phenolphthalein absorbs the center of the visible spectrum. For comparison, spectra of these three solutions recorded with a spectrophotometer are shown in Color Plate 16b.

This same setup can be used to demonstrate fluorescence and the properties of colors.[7] Other demonstrations of absorption and emission spectra[8] and decomposition of spectra into chromaticity coordinates have been described.[9]

*Edmund Scientific Co., www.edmundoptics.com, catalog no. NT40-267.

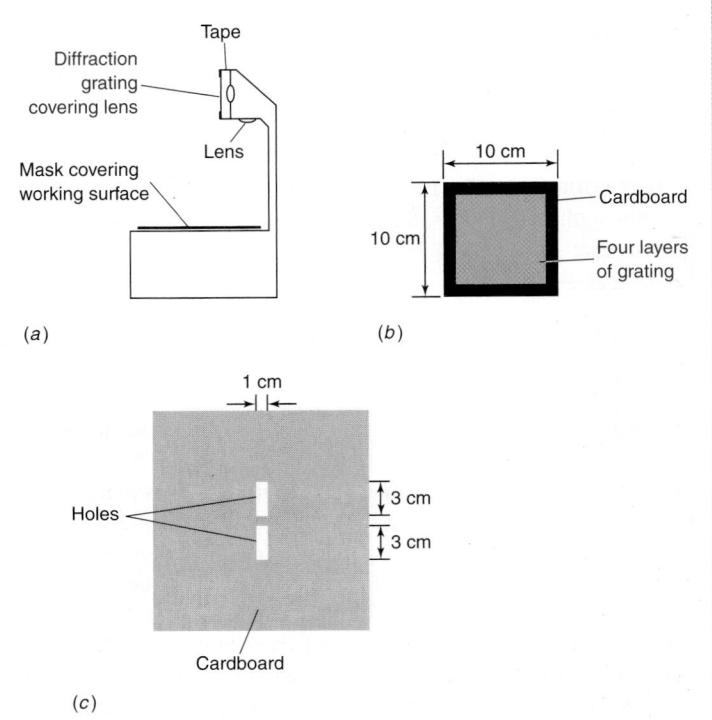

(a) Overhead projector. (b) Diffraction grating mounted on cardboard. (c) Mask for working surface.

18-3 Measuring Absorbance

The minimum requirements for a spectrophotometer (a device to measure absorbance of light) were shown in Figure 18-5. Light from a continuous source is passed through a monochromator, which selects a narrow band of wavelengths from the incident beam. This "monochromatic" light travels through a sample of pathlength b, and the irradiance of the emergent light is measured.

For visible and ultraviolet spectroscopy, a liquid sample is usually contained in a cell called a **cuvet** that has flat, fused-silica (SiO_2) faces (Figure 18-6). Glass is suitable for visible, but not ultraviolet spectroscopy because it absorbs ultraviolet radiation. The most common cuvets have a 1.000-cm pathlength and are sold in matched sets for sample and reference.

For infrared measurements, cells are commonly constructed of NaCl or KBr. For the 400 to 50 cm^{-1} far-infrared region, polyethylene is a transparent window. Solid samples are commonly ground to a fine powder, which can be added to mineral oil (a viscous hydrocarbon also called Nujol) to give a dispersion that is called a *mull* and is pressed between two KBr plates. The analyte spectrum is obscured in a few regions in which the mineral oil absorbs infrared radiation. Alternatively, a 1 wt% mixture of solid sample with KBr can be ground to a fine powder and pressed into a translucent pellet at a pressure of ~60 MPa (600 bar). Solids and powders can also be examined by *diffuse reflectance*, in which reflected infrared radiation, instead of transmitted infrared radiation, is observed. Wavelengths absorbed by the sample are not reflected as well. This technique is sensitive only to the surface of the sample.

Gases are more dilute than liquids and require cells with longer pathlengths, typically ranging from 10 cm to many meters. A pathlength of many meters is obtained by reflecting light so that it traverses the sample many times before reaching the detector.

Figure 18-5 outlines a *single-beam* instrument, which has only one beam of light. We do not measure the incident irradiance, P_0, directly. Rather, the irradiance of light passing through a reference cuvet containing pure solvent (or a reagent blank) is *defined* as P_0. This cuvet is then removed and replaced by an identical one containing sample. The irradiance of light striking the detector after passing through the sample is the quantity P. Knowing both P and P_0 allows T and A to be determined. The reference cuvet compensates for reflection, scattering, and absorption by the cuvet and solvent. The reference cuvet containing pure solvent

Approximate low-energy cutoff for common infrared windows:

sapphire (Al_2O_3)	1 500 cm^{-1}
NaCl	650 cm^{-1}
KBr	350 cm^{-1}
AgCl	350 cm^{-1}
CsBr	250 cm^{-1}
CsI	200 cm^{-1}

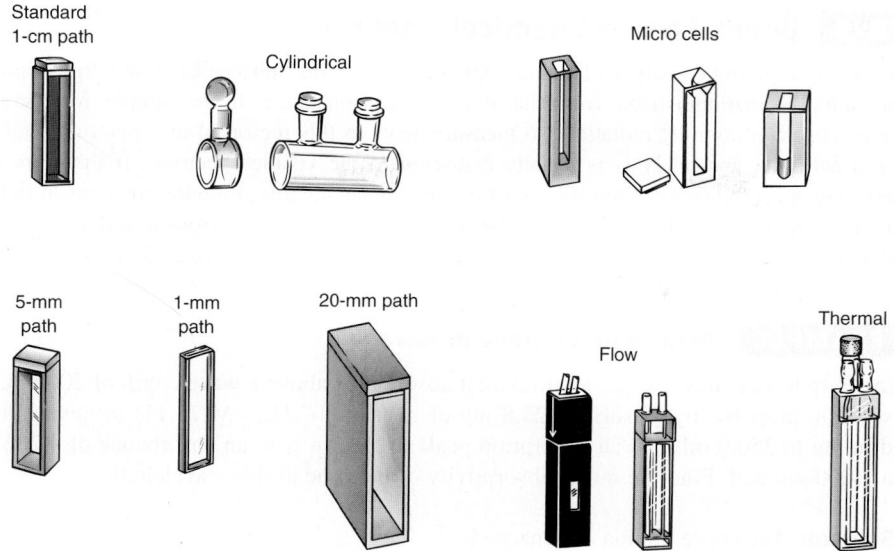

FIGURE 18-6 Common cuvets for visible and ultraviolet spectroscopy. Flow cells permit continuous flow of solution through the cell. In the thermal cell, liquid from a constant-temperature bath flows through the cell jacket to maintain a desired temperature. [Information from A. H. Thomas Co., Philadelphia, PA.]

5-mm path   1-mm path   20-mm path   Flow   Thermal

acts as a blank in the measurement of transmittance. A *double-beam* instrument, which splits the light to pass alternately between sample and reference cuvets, is described in Chapter 20.

In recording an absorbance spectrum, first record a *baseline spectrum* with reference solutions (pure solvent or a reagent blank) in both the sample and reference cuvets. Cuvets are sold in matched pairs that are as identical as possible to each other. If the cuvets and the instrument were perfect, the baseline would be 0 everywhere. In our imperfect world, the baseline usually exhibits small positive and negative absorbance. Subtract the baseline absorbance from the sample absorbance to obtain the true absorbance at each wavelength.

For spectrophotometric analysis, we normally choose the wavelength of maximum absorbance for two reasons: (1) The sensitivity of the analysis is greatest at maximum absorbance; that is, we get the maximum response for a given concentration of analyte. (2) The curve is relatively flat at the maximum, so there is little variation in absorbance if the monochromator drifts a little or if the width of the transmitted band changes slightly. Beer's law is obeyed when absorbance is constant across the selected waveband.

Modern spectrophotometers are most precise (reproducible) at intermediate levels of absorbance ($A \approx 0.3$ to 2). If too little light gets through the sample (high absorbance), intensity is hard to measure. If too much light gets through (low absorbance), it is hard to distinguish transmittance of the sample from that of the reference. It is desirable to adjust sample concentration so that absorbance falls in this intermediate range. Compartments must be tightly closed to avoid stray light, which leads to false readings.

Figure 18-7 shows the relative standard deviation of replicate measurements made at 350 nm with a diode array spectrometer. Filled circles are from replicate measurements in which the sample was not removed from the cuvet holder between measurements. Open circles are from measurements in which the sample was removed and then replaced in the cuvet holder between measurements. Relative standard deviation is below 0.1% in both cases for absorbance in the range 0.3 to 2. Data points represented by open squares were obtained when a 10-year-old cuvet holder was used and the sample was removed and replaced in the holder between measurements. Variability in the position of the cuvet more than doubles the relative standard deviation. The conclusion is that modern spectrometers with modern cell holders provide excellent reproducibility. Precision was degraded when an old cell holder was used and the sample was removed and inserted between measurements.

Always keep containers covered to exclude dust, which scatters light and appears to the spectrophotometer to be absorbance. Filtering the final solution through a very fine filter (for example, 0.5 μm) may be necessary in critical work. Handle cuvets with a tissue to avoid putting fingerprints on the faces, and keep cuvets scrupulously clean.

Slight mismatch between sample and reference cuvets leads to systematic errors in spectrophotometry. Place cuvets in the spectrophotometer as reproducibly as possible. Variation in apparent absorbance arises from slightly misplacing the cuvet in its holder, or turning a flat cuvet around by 180°, or rotating a circular cuvet.

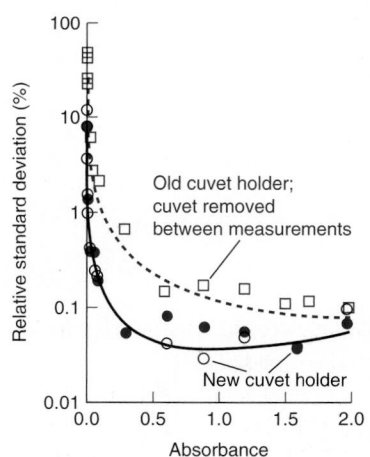

FIGURE 18-7 Precision of replicate absorbance measurements with a diode array spectrometer at 350 nm with a dichromate solution. Filled circles are from replicate measurements in which the sample was not removed from the cuvet holder between measurements. Open circles are from measurements in which the sample was removed and then replaced in the cuvet holder between measurements. Best reproducibility is observed at intermediate absorbance ($A \approx 0.3$ to 2). Note logarithmic ordinate. Lines are least-squares fit of data to theoretical equations.
[Data from J. Galbán, S. de Marcos, I. Sanz, C. Ubide, and J. Zuriarrain, "Uncertainty in Modern Spectrometers," *Anal. Chem.* **2007,** *79,* 4763.]

18-4 Beer's Law in Chemical Analysis

For a compound to be analyzed by spectrophotometry, it must absorb light, and this absorption should be distinguishable from that due to other substances in the sample. Most compounds absorb ultraviolet radiation, so measurements in this region of the spectrum tend to be inconclusive, and analysis is usually restricted to the visible spectrum. If there are no interfering species, however, ultraviolet absorbance is satisfactory. Proteins are assayed in the ultraviolet region at 280 nm because the aromatic amino acids tyrosine and tryptophan (Table 10-1) present in most proteins absorb near 280 nm (see Problem 18-21).

EXAMPLE Measuring Benzene in Hexane

This example illustrates the measurement of molar absorptivity from a single solution. It is better to measure several concentrations to obtain a more reliable absorptivity and to demonstrate that Beer's law is obeyed.

(a) Pure hexane has negligible ultraviolet absorbance above a wavelength of 200 nm. A solution prepared by dissolving 25.8 mg of benzene (C_6H_6, FM 78.11) in hexane and diluting to 250.0 mL had an absorption peak at 256 nm with an absorbance of 0.266 in a 1.000-cm cell. Find the molar absorptivity of benzene at this wavelength.

Solution The concentration of benzene is

$$[C_6H_6] = \frac{(0.025\ 8\ \text{g})/(78.11\ \text{g/mol})}{0.250\ 0\ \text{L}} = 1.32_1 \times 10^{-3}\ \text{M}$$

We find the molar absorptivity from Beer's law:

$$\text{Molar absorptivity} = \varepsilon = \frac{A}{bc} = \frac{(0.266)}{(1.000\ \text{cm})(1.32_1 \times 10^{-3}\ \text{M})} = 201._3\ \text{M}^{-1}\ \text{cm}^{-1}$$

(b) A sample of hexane contaminated with benzene had an absorbance of 0.070 at 256 nm in a cuvet with a 5.000-cm pathlength. Find the concentration of benzene in mg/L.

Solution Using Beer's law with the molar absorptivity from part (a), we find:

$$[C_6H_6] = \frac{A}{\varepsilon b} = \frac{0.070}{(201._3\ \text{M}^{-1}\ \text{cm}^{-1})(5.000\ \text{cm})} = 6.9_5 \times 10^{-5}\ \text{M}$$

$$[C_6H_6] = \left(6.9_5 \times 10^{-5}\ \frac{\text{mol}}{\text{L}}\right)\left(78.11 \times 10^3\ \frac{\text{mg}}{\text{mol}}\right) = 5.4\ \frac{\text{mg}}{\text{L}}$$

TEST YOURSELF 0.10 mM $KMnO_4$ has an absorbance maximum of 0.26 at 525 nm in a 1.000-cm cell. Find the molar absorptivity and the concentration of a solution whose absorbance is 0.52 at 525 nm in the same cell. (*Answer:* 2 600 M^{-1} cm^{-1}, 0.20 mM)

Serum Iron Determination

Iron for biosynthesis is transported through the bloodstream by the protein *transferrin*, whose Fe^{3+}-binding site is shown in Figure 18-8. The following procedure measures the Fe content of transferrin in blood.[12] This analysis requires about 1 μg of Fe for an accuracy of 2–5%. Human blood usually contains about 45 vol% cells and 55 vol% *plasma* (liquid). If blood is collected without an anticoagulant, the blood clots; the liquid that remains is called *serum*. Serum normally contains about 1 μg of Fe/mL attached to transferrin.

The serum iron determination has three steps:

Step 1 Reduce Fe^{3+} in transferrin to Fe^{2+}, which is released from the protein. Suitable reducing agents include hydroxylamine hydrochloride ($NH_3OH^+Cl^-$), thioglycolic acid, and ascorbic acid.

$$2Fe^{3+} + 2HSCH_2CO_2H \rightarrow 2Fe^{2+} + HO_2CCH_2S\!-\!SCH_2CO_2H + 2H^+$$
Thioglycolic acid

Step 2 Add trichloroacetic acid (Cl_3CCO_2H) to precipitate proteins, leaving Fe^{2+} in solution. Centrifuge the mixture to remove the precipitate. If protein were left in the solution, it would partially precipitate in the final solution. Light scattered by the precipitate would be mistaken for absorbance.

$$\text{Protein}(aq) \xrightarrow{\ \text{CCl}_3\text{CO}_2\text{H}\ } \text{protein}(s) \downarrow$$

OH
|
H⟋⁺N⟍⟍H
H
Cl⁻

Hydroxylamine hydrochloride

OH
HO⟍⟋O⟍O
HO OH

Ascorbic acid
(vitamin C)

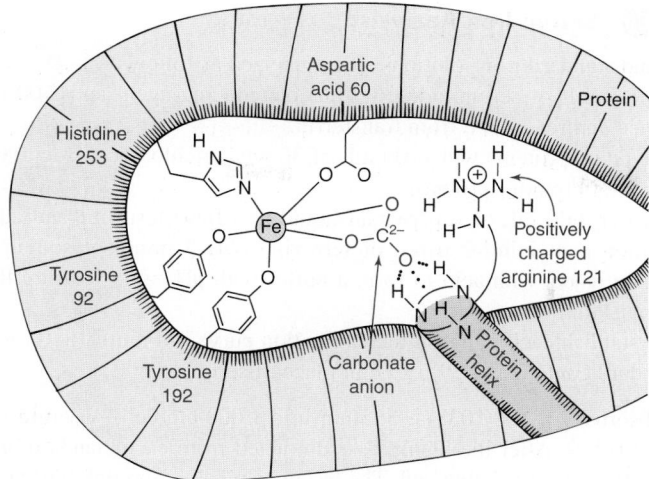

FIGURE 18-8 Each of the two Fe-binding sites of transferrin is located in a cleft in the protein. Each site has one nitrogen ligand from the amino acid histidine and three oxygen ligands from tyrosine and aspartic acid. Two oxygen ligands come from a carbonate anion (CO_3^{2-}) anchored by electrostatic attraction to positively charged arginine and by hydrogen bonding to the protein helix. When a cell takes up transferrin, the transferrin is brought into a compartment whose pH is lowered to 5.5. H^+ reacts with carbonate to make HCO_3^-, thereby releasing Fe^{3+} from the protein. [Information from E. N. Baker, B. F. Anderson, H. M. Baker, M. Haridas, G. E. Norris, S. V. Rumball, and C. A. Smith, "Metal and Anion Binding Sites in Lactoferrin and Related Proteins," *Pure Appl. Chem.* **1990,** *62,* 1067.]

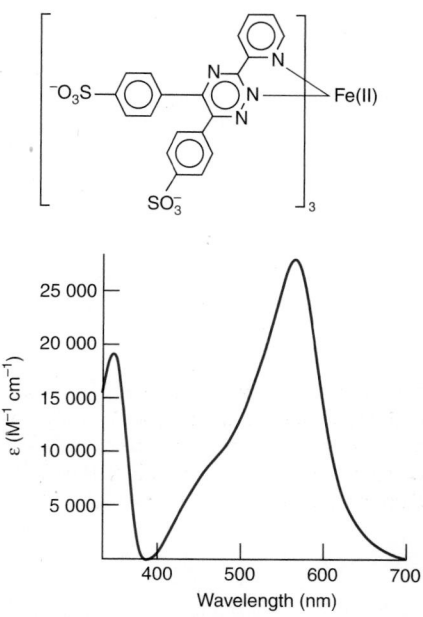

FIGURE 18-9 Visible absorption spectrum of the complex (ferrozine)$_3$Fe(II) used in the colorimetric analysis of iron.

Step 3 Transfer a measured volume of supernatant liquid from step 2 to a fresh vessel. Add buffer plus excess ferrozine to form a purple complex. Measure the absorbance at the 562-nm peak (Figure 18-9). Buffer sets a pH at which complex formation is complete.

$$Fe^{2+} + 3\ \text{ferrozine}^{2-} \rightarrow (\text{ferrozine})_3 Fe^{4-}$$

Purple complex
$\lambda_{max} = 562$ nm

Supernate is the liquid layer above the solid that collects at the bottom of a tube during centrifugation.

In most spectrophotometric analyses, it is important to prepare a **reagent blank** containing all reagents, but with analyte replaced by distilled water. Any absorbance of the blank is due to the color of uncomplexed ferrozine plus the color caused by the iron impurities in the reagents and glassware. *Subtract the blank absorbance from the absorbance of samples and standards before doing any calculations.*

Prepare a series of iron standards for a *calibration curve* (Figure 18-10) to show that Beer's law is obeyed. Standards are prepared by the same procedure as unknowns. The absorbance of the unknown should fall within the region covered by the standards. Pure iron wire with a shiny, rust-free surface is dissolved in acid to prepare standards (Appendix K). Ferrous ammonium sulfate ($Fe(NH_4)_2(SO_4)_2 \cdot 6H_2O$) and ferrous ethylenediammonium sulfate ($Fe(H_3NCH_2CH_2NH_3)(SO_4)_2 \cdot 4H_2O$) are suitable standards for less accurate work.

If unknowns and standards are prepared with identical volumes, then the quantity of iron in the unknown can be calculated from the least-squares equation for the calibration line. For example, in Figure 18-10, if the unknown has an absorbance of 0.357 (after subtracting the absorbance of the blank), the sample contains 3.59 μg of iron.

In the procedure we just described, results would be about 10% high because serum copper also forms a colored complex with ferrozine. Interference is eliminated if neocuproine or thiourea is added. These reagents **mask** Cu^+ by forming strong complexes that prevent Cu^+ from reacting with ferrozine.

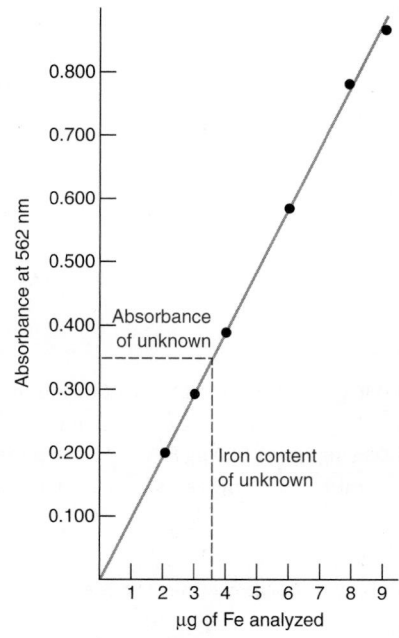

FIGURE 18-10 Calibration curve showing the validity of Beer's law for the (ferrozine)$_3$Fe(II) complex used in serum iron determination. Each sample was diluted to a final volume of 5.00 mL. Therefore, 1.00 μg of iron is equivalent to a concentration of 3.58×10^{-6} M.

Neocuproine
A **masking agent** for copper

Tightly bound complex with low absorbance at 562 nm

To find the uncertainty in μg Fe, use Equation 4-27.

EXAMPLE Serum Iron Analysis

Serum iron and standard iron solutions were analyzed as follows:

Step 1 To 1.00 mL of serum, add 2.00 mL of reducing agent and 2.00 mL of acid to reduce and release Fe from transferrin.

Step 2 Precipitate proteins with 1.00 mL of 30 wt% trichloroacetic acid. Centrifuge the mixture to remove protein.

Step 3 Transfer 4.00 mL of supernatant liquid to a fresh test tube and add 1.00 mL of solution containing 2 μmol of ferrozine plus 3 mmol of sodium acetate. The acetate reacts with acid to form a buffer with pH ≈ 5. Measure the absorbance after 10 min.

Step 4 To establish each point on the calibration curve in Figure 18-10, use 1.00 mL of standard containing 2–9 μg Fe in place of serum.

The blank absorbance was 0.038 at 562 nm in a 1.000-cm cell. A serum sample had an absorbance of 0.129. After the blank was subtracted from each standard absorbance, the points in Figure 18-10 were obtained. The least-squares line through the standard points is

$$\text{Absorbance} = 0.067_0 \times (\mu\text{g Fe in initial sample}) + 0.001_5$$

According to Beer's law, the intercept should be 0, not 0.001_5. We will use the small, observed intercept for our analysis. Find the concentration of iron in the serum.

Solution Rearranging the least-squares equation of the calibration line and inserting the corrected absorbance (observed − blank = 0.129 − 0.038 = 0.091) of unknown, we find

$$\mu\text{g Fe in unknown} = \frac{\text{absorbance} - 0.001_5}{0.067_0} = \frac{0.091 - 0.001_5}{0.067_0} = 1.33_6\,\mu\text{g}$$

The concentration of Fe in the serum is

$$[\text{Fe}] = \text{moles of Fe/liters of serum}$$

$$= \left(\frac{1.33_6 \times 10^{-6}\,\text{g Fe}}{55.845\,\text{g Fe/mol Fe}}\right)/(1.00 \times 10^{-3}\,\text{L}) = 2.39 \times 10^{-5}\,\text{M}$$

TEST YOURSELF If the observed absorbance is 0.200 and the blank absorbance is 0.049, what is the concentration of Fe (μg/mL) in the serum? (***Answer:*** 2.23 μg/mL)

EXAMPLE Preparing Standard Solutions

Points for the calibration curve in Figure 18-10 were obtained by using 1.00 mL of standard containing ~2, 3, 4, 6, 8, and 9 μg Fe in place of serum. A stock solution was made by dissolving 1.086 g of pure, clean (Appendix K) iron wire in 40 mL of 12 M HCl and diluting to 1.000 L with H_2O. This solution contains 1.086 g Fe/L in 0.48 M HCl. How can we make Fe standards in ~0.1 M HCl for the calibration curve from this stock solution? (HCl prevents precipitation of $Fe(OH)_3$.)

$$\left(\frac{1.086\,\text{g Fe}}{\text{L}}\right)\left(\frac{1\,000\,\text{mg/g}}{1\,000\,\text{mL/L}}\right) = \left(\frac{1.086\,\text{mg Fe}}{\text{mL}}\right)$$

$$\left(\frac{1.086\,\text{mg Fe}}{\text{mL}}\right)\left(\frac{1\,000\,\mu\text{g Fe}}{\text{mg Fe}}\right) = \left(\frac{1\,086\,\mu\text{g Fe}}{\text{mL}}\right)$$

Solution Stock solution contains 1.086 g Fe/L = 1.086 mg Fe/mL = 1 086 μg Fe/mL. We can make a ~2 μg Fe/mL standard by a 2:1000 dilution to give $\frac{2\,\text{mL}}{1\,000\,\text{mL}}$ (1 086 μg Fe/mL) = 2.172 μg Fe/mL. Transfer 2.000 mL of stock solution with a Class A transfer pipet into a 1 L volumetric flask and dilute to volume with 0.1 M HCl. In a similar manner use Class A transfer pipets to dilute the following volumes to 1 L with 0.1 M HCl:

Volume of stock solution (mL)	Class A pipet	Fe delivered (μg)
3.000	3 mL	3 × 1.086 = 3.258
4.000	4 mL	4.344
6.000	2 × 3 mL	6.516
8.000	2 × 4 mL	8.688
9.000	(5 mL + 4 mL) or (3 × 3 mL)	9.774

Glassware (including pipets) for this procedure must be acid washed (page 32) to remove traces of metal from glass surfaces.

TEST YOURSELF Dilution by mass is more accurate than dilution by volume. A stock solution contains 1.044 g Fe/kg solution in 0.48 M HCl. How many µg Fe/g solution are in a solution made by mixing 2.145 g stock solution with 243.27 g 0.1 M HCl? (*Answer:* 9.125 µg Fe/g)

Serial dilution—a sequence of consecutive dilutions—is an important means to reduce the generation of large volumes of chemicals waste by using smaller volumes for dilution.

Serial dilution is a sequence of consecutive dilutions.

EXAMPLE Serial Dilution with Limited Glassware

Starting from a stock solution containing 1 000 µg Fe/mL in 0.5 M HCl, how could you make a 3 µg Fe/mL standard in 0.1 M HCl Fe if you have only 100-mL volumetric flasks and 1-, 5-, and 10-mL transfer pipets?

Solution One way to solve this problem is by these two steps:

1. Pipet three 10.00-mL portions of stock solution into a 100-mL flask and dilute to the mark with 0.1 M HCl. $[Fe] = \left(\dfrac{30}{100}\right)(1\ 000\ \mu g\ Fe/mL) = 300\ \mu g\ Fe/mL.$

2. Transfer 1.00 mL of solution from step 1 to a 100-mL flask and dilute to the mark with 0.1 M HCl. $[Fe] = \left(\dfrac{1}{100}\right)(300\ \mu g\ Fe/mL) = 3\ \mu g\ Fe/mL.$

TEST YOURSELF Describe a different set of dilutions to produce 3 µg Fe/mL from 1 000 µg Fe/mL (*Answer:* Dilute 10 mL of stock to 100 mL to get [Fe] = 100 µg Fe/mL. Dilute 3 × 1.00 mL of the new solution up to 100 mL to get 3 µg Fe/mL.)

18-5 Spectrophotometric Titrations

In a **spectrophotometric titration,** we monitor changes in absorbance during a titration to tell when the equivalence point has been reached. A solution of the iron-transport protein, transferrin (Figure 18-8), can be titrated with iron to measure the transferrin content. Transferrin without iron, called apotransferrin, is colorless. Each molecule, with a molecular mass near 80 000, binds two Fe^{3+} ions. When Fe^{3+} binds to the protein, a red color with an absorbance maximum at a wavelength of 465 nm develops. The absorbance is proportional to the concentration of iron bound to the protein. Therefore, absorbance may be used to monitor a titration of an unknown amount of apotransferrin with a standard solution of Fe^{3+}.

$$\text{Apotransferrin} + 2Fe^{3+} \rightarrow (Fe^{3+})_2\text{transferrin}$$
$$\text{(colorless)} \qquad\qquad\qquad \text{(red)}$$

This titration works well for a solution of purified transferrin, but it is not very good for serum because of the background color in serum.

Figure 18-11 shows the titration of 2.000 mL of apotransferrin with 1.79 mM ferric nitrilotriacetate. As iron is added to the protein, red color develops and absorbance increases. When protein is saturated with iron, no more iron can bind and the curve levels off. The extrapolated intersection of the two straight portions of the titration curve at 203 µL in Figure 18-11 is taken as the end point. Absorbance rises slowly after the equivalence point because ferric nitrilotriacetate has some absorbance at 465 nm.

The quantity of Fe^{3+} required for complete reaction in Figure 18-11 is $(203 \times 10^{-6}\ L) \times (1.79 \times 10^{-3}\ mol/L) = 0.363\ \mu mol$. Each protein molecule binds 2 Fe^{3+} ions, so the moles of protein in the sample must be $\frac{1}{2}(0.363\ \mu mol) = 0.182\ \mu mol$.

To construct the graph in Figure 18-11, dilution must be considered because the volume is different at each point. Each point on the graph represents the absorbance that would be observed *if the solution had not been diluted from its original volume of 2.000 mL.*

$$\text{Corrected absorbance} = \left(\frac{\text{total volume}}{\text{initial volume}}\right)(\text{observed absorbance}) \qquad \text{(18-7)}$$

FIGURE 18-11 Spectrophotometric titration of transferrin with ferric nitrilotriacetate. Absorbance is corrected as if no dilution had taken place. The initial absorbance of the solution, before iron is added, is due to a colored impurity.

Ferric nitrilotriacetate is soluble at neutral pH. In the absence of nitrilotriacetate, Fe^{3+} precipitates as $Fe(OH)_3$ in neutral solution. Nitrilotriacetate binds Fe^{3+} through four atoms shown in **bold** type:

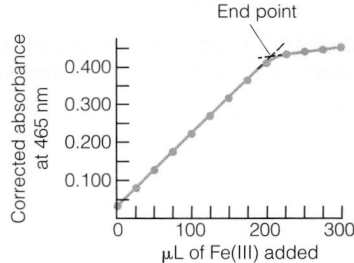

Nitrilotriacetate anion

EXAMPLE Correcting Absorbance for the Effect of Dilution

The absorbance measured after adding 125 μL (= 0.125 mL) of ferric nitrilotriacetate to 2.000 mL of apotransferrin was 0.260. Calculate the corrected absorbance that should be plotted in Figure 18-11.

Solution The total volume was 2.000 + 0.125 = 2.125 mL. If the volume had been 2.000 mL, the absorbance would have been greater than 0.260 by a factor of 2.125/2.000.

$$\text{Corrected absorbance} = \left(\frac{2.125\ \text{mL}}{2.000\ \text{mL}}\right)(0.260) = 0.276$$

The absorbance plotted in Figure 18-11 is 0.276.

TEST YOURSELF In a different titration, the absorbance after adding 75 μL of ferric nitrilotriacetate to 1.500 mL of apotransferrin was 0.222. Find the corrected absorbance. (*Answer:* 0.233)

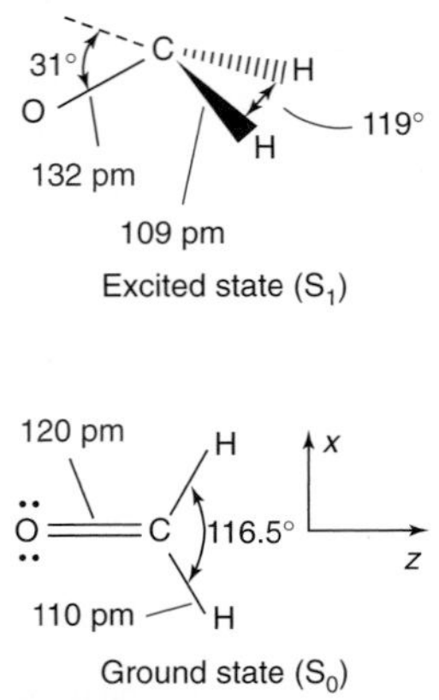

FIGURE 18-12 Geometry of formaldehyde in its ground state S_0 and lowest excited singlet state S_1.

A triplet state splits into three slightly different energy levels in a magnetic field, but a singlet state is not split by a magnetic field.

The shorter the wavelength of electromagnetic radiation, the greater the energy.

18-6 What Happens When a Molecule Absorbs Light?

When a molecule absorbs a photon, the molecule is promoted to a more energetic *excited state* (Figure 18-3). Conversely, when a molecule emits a photon, the energy of the molecule falls by an amount equal to the energy of the photon.

For example, consider formaldehyde in Figure 18-12. In its ground state, the molecule is planar, with a double bond between carbon and oxygen. From the electron dot description of formaldehyde, we expect two pairs of nonbonding electrons to be localized on the oxygen atom. The double bond consists of a sigma bond between carbon and oxygen and a pi bond made from the $2p_y$ (out-of-plane) atomic orbitals of carbon and oxygen.

Electronic States of Formaldehyde

Molecular orbitals describe the distribution of electrons in a molecule, just as *atomic orbitals* describe the distribution of electrons in an atom. In the molecular orbital diagram for formaldehyde in Figure 18-13, one of the nonbonding orbitals of oxygen is mixed with the three sigma bonding orbitals. These four orbitals, labeled σ_1 through σ_4, are each occupied by a pair of electrons with opposite spin (spin quantum numbers $= +\frac{1}{2}$ and $-\frac{1}{2}$). At higher energy is an occupied pi bonding orbital (π), made of the p_y atomic orbitals of carbon and oxygen. The highest energy occupied orbital is the nonbonding orbital (n), composed principally of the oxygen $2p_x$ atomic orbital. The lowest energy unoccupied orbital is the pi antibonding orbital (π^*). Electrons in this orbital produce repulsion, rather than attraction, between the carbon and oxygen atoms.

In an **electronic transition,** an electron from one molecular orbital moves to another orbital, with a concomitant increase or decrease in the energy of the molecule. The lowest energy electronic transition of formaldehyde promotes a nonbonding (n) electron to the antibonding pi orbital (π^*).[13] There are two possible transitions, depending on the spin quantum numbers in the excited state (Figure 18-14). The state in which the spins are opposed is called a **singlet state.** If the spins are parallel, we have a **triplet state.**

The lowest energy excited singlet and triplet states are called S_1 and T_1, respectively. In general, T_1 has lower energy than S_1. In formaldehyde, the transition $n \rightarrow \pi^*(T_1)$ requires absorption of visible light with a wavelength of 397 nm. The $n \rightarrow \pi^*(S_1)$ transition occurs when ultraviolet radiation with a wavelength of 355 nm is absorbed.

With an electronic transition near 397 nm, you might expect from Table 18-1 that formaldehyde appears green-yellow. In fact, formaldehyde is colorless because the probability of undergoing a transition between singlet and triplet states, such as $n(S_0) \rightarrow \pi^*(T_1)$, is exceedingly small. Formaldehyde absorbs so little light at 397 nm that our eyes do not detect any absorbance at all. Singlet-to-singlet transitions such as $n(S_0) \rightarrow \pi^*(S_1)$ are much more probable, so the ultraviolet absorption is more intense.

Although formaldehyde is planar in its ground state, it has a pyramidal structure in both the S_1 (Figure 18-12) and T_1 excited states. Promotion of a nonbonding electron to an antibonding C—O orbital lengthens the C—O bond and changes the molecular geometry.

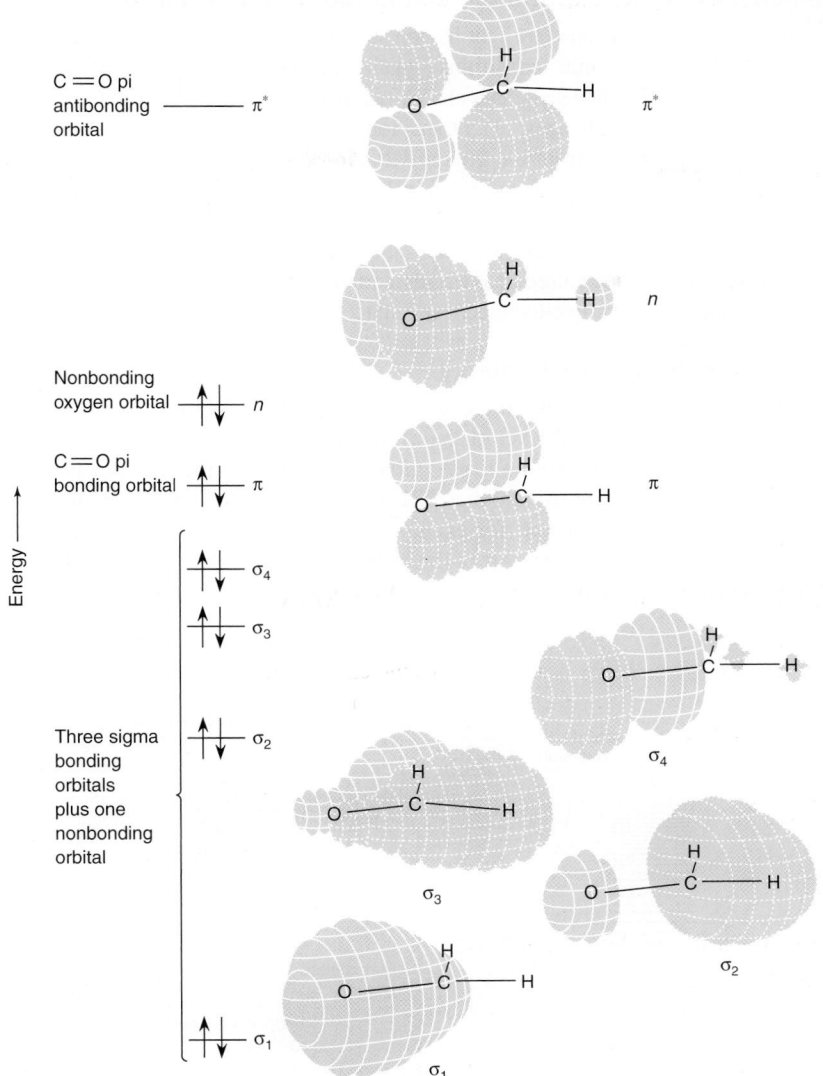

C=O pi
antibonding ——— π^*
orbital

Nonbonding
oxygen orbital ⬆⬇ n

C=O pi
bonding orbital ⬆⬇ π

Energy ⬆

⬆⬇ σ_4

⬆⬇ σ_3

Three sigma
bonding
orbitals
plus one
nonbonding
orbital

⬆⬇ σ_2

⬆⬇ σ_1

π^*

n

π

σ_4

σ_3

σ_2

σ_1

FIGURE 18-13 Molecular orbital diagram of formaldehyde, showing energy levels and orbital shapes. The coordinate system of the molecule was shown in **Figure 18-12**. [Information from W. L. Jorgensen and L. Salem, *The Organic Chemist's Book of Orbitals* (New York: Academic Press, 1973).]

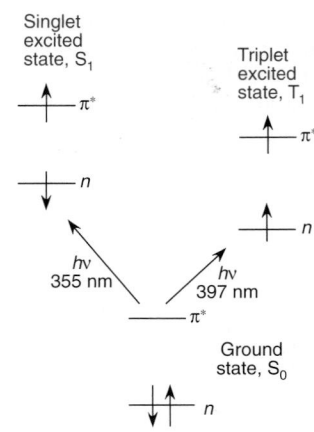

Singlet
excited
state, S_1

⬆ π^*

⬆ n

$h\nu$
355 nm

Triplet
excited
state, T_1

⬆ π^*

⬆ n

$h\nu$
397 nm

π^*

Ground
state, S_0

⬆⬇ n

FIGURE 18-14 Two possible electronic states arise from an $n \rightarrow \pi^*$ transition. The terms *singlet* and *triplet* are used because a triplet state splits into three slightly different energy levels in a magnetic field, but a single state is not split.

Vibrational and Rotational States of Formaldehyde

Absorption of visible or ultraviolet radiation promotes electrons to higher energy orbitals in formaldehyde. Infrared and microwave radiation are not energetic enough to induce electronic transitions, but they can change the vibrational or rotational motion of the molecule.

Each of the four atoms of formaldehyde can move along three axes in space, so the entire molecule can move in $4 \times 3 = 12$ different ways. Three of these motions correspond to translation of the entire molecule in the x, y, and z directions. Another three motions correspond to rotation about x-, y-, and z-axes placed at the center of mass of the molecule. The remaining six motions are vibrations shown in Figure 18-15.

When formaldehyde absorbs an infrared photon with a wavenumber of 1 251 cm^{-1} (= 14.97 kJ/mol), the asymmetric bending vibration in Figure 18-15 is stimulated: Oscillations of the atoms increase in amplitude, and the energy of the molecule increases.

Spacings between rotational energy levels of a molecule are smaller than vibrational energy spacings. A molecule in the rotational ground state could absorb microwave photons with energies of 0.029 07 or 0.087 16 kJ/mol (wavelengths of 4.115 or 1.372 mm) to be promoted to the two lowest excited states. Absorption of microwave radiation causes the molecule to rotate faster than it does in its ground state.

A nonlinear molecule with n atoms has $3n - 6$ vibrational modes and three rotations. A linear molecule can rotate about only two axes; it therefore has $3n - 5$ vibrational modes and two rotations.

The C—O stretching vibration of formaldehyde is reduced from 1 746 cm^{-1} in the S_0 state to 1 183 cm^{-1} in the S_1 state because the strength of the C—O bond decreases when the antibonding π^* orbital is populated.

Your microwave oven heats food by transferring rotational energy to water molecules in the food.

Vibrational transitions usually involve simultaneous rotational transitions. Electronic transitions usually involve simultaneous vibrational and rotational transitions.

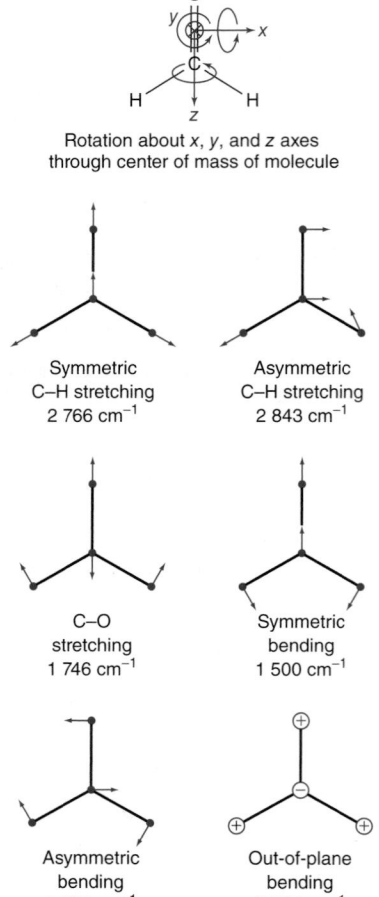

Rotation about x, y, and z axes through center of mass of molecule

Symmetric
C–H stretching
2 766 cm^{-1}

Asymmetric
C–H stretching
2 843 cm^{-1}

C–O
stretching
1 746 cm^{-1}

Symmetric
bending
1 500 cm^{-1}

Asymmetric
bending
1 251 cm^{-1}

Out-of-plane
bending
1 167 cm^{-1}

FIGURE 18-15 The six modes of vibration of formaldehyde. The wavenumber of infrared radiation needed to stimulate each kind of motion is given in units of reciprocal centimeters, cm^{-1}. The molecule rotates about x-, y-, and z-axes located at the center of mass near the middle of the C=O bond.

Internal conversion is a nonradiative transition between states with the same spin (e.g., $S_1 \rightarrow S_0$).

Intersystem crossing is a nonradiative transition between states with different spin (e.g., $T_1 \rightarrow S_0$).

Fluorescence: emission of a photon during a transition between states with the same spin (e.g., $S_1 \rightarrow S_0$).

Phosphorescence: emission of a photon during a transition between states with different spin (e.g., $T_1 \rightarrow S_0$).

Combined Electronic, Vibrational, and Rotational Transitions

In general, when a molecule absorbs light having sufficient energy to cause an electronic transition, **vibrational** and **rotational transitions**—that is, changes in the vibrational and rotational states—occur as well. Formaldehyde can absorb one photon with just the right energy to cause the following simultaneous changes: (1) a transition from the S_0 to the S_1 electronic state; (2) a change in vibrational energy from the ground vibrational state of S_0 to an excited vibrational state of S_1; and (3) a transition from one rotational state of S_0 to a different rotational state of S_1.

Electronic absorption bands are usually broad (Figure 18-9) because many different vibrational and rotational levels are available at slightly different energies. A molecule can absorb photons with a wide range of energies and still be promoted from the ground electronic state to one particular excited electronic state.

What Happens to Absorbed Energy?

Suppose that absorption promotes a molecule from the ground electronic state, S_0, to a vibrationally and rotationally excited level of the excited electronic state S_1 (Figure 18-16). Usually, the first process after absorption is *vibrational relaxation* to the lowest vibrational level of S_1. In this *nonradiative* transition, labeled R_1 in Figure 18-16, vibrational energy is transferred to other molecules (solvent, for example) through collisions, not by emission of a photon. The net effect is to convert part of the energy of the absorbed photon into heat spread throughout the entire medium.

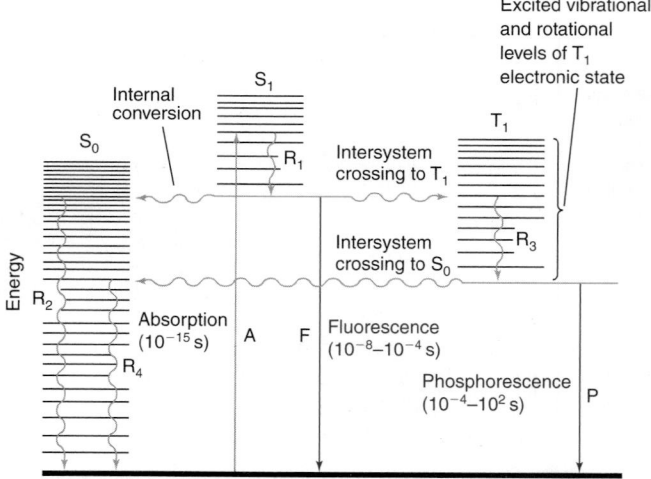

FIGURE 18-16 Physical processes that can occur after a molecule absorbs an ultraviolet or visible photon. S_0 is the ground electronic state. S_1 and T_1 are the lowest excited singlet and triplet electronic states. Straight arrows represent processes involving photons, and wavy arrows are nonradiative transitions. R denotes vibrational relaxation. Absorption could terminate in any of the vibrational levels of S_1, not just the one shown. Fluorescence and phosphorescence can terminate in any of the vibrational levels of S_0.

At the S_1 level, several events can happen. The molecule could enter a highly excited vibrational level of S_0 having the same energy as S_1. This is called *internal conversion* (IC). From this excited state, the molecule can relax back to the ground vibrational state and transfer its energy to neighboring molecules through collisions. This nonradiative process is labeled R_2. If a molecule follows the path A–R_1–IC–R_2 in Figure 18-16, the entire energy of the photon will have been converted into heat.

Alternatively, the molecule could cross from S_1 into an excited vibrational level of T_1. Such an event is known as *intersystem crossing* (ISC). After the nonradiative vibrational relaxation R_3, the molecule finds itself at the lowest vibrational level of T_1. From here, the molecule might undergo a second intersystem crossing to S_0, followed by the nonradiative relaxation R_4. All processes mentioned so far simply convert light into heat.

A molecule could also relax from S_1 or T_1 to S_0 by emitting a photon. The radiational transition $S_1 \rightarrow S_0$ is called **fluorescence** (Box 18-2), and the radiational transition $T_1 \rightarrow S_0$ is called **phosphorescence.** The relative rates of internal conversion, intersystem crossing, fluorescence, and phosphorescence depend on the molecule, the solvent, and conditions such as temperature and pressure. The energy of phosphorescence is less than the energy of fluorescence, so phosphorescence comes at longer wavelengths than fluorescence (Figure 18-17).

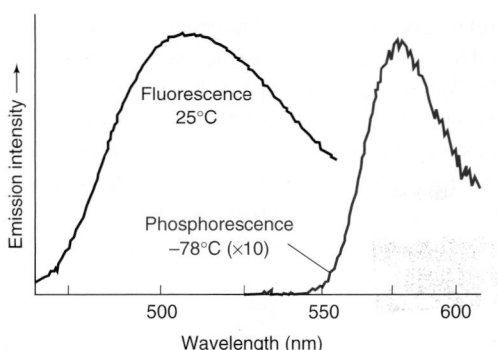

FIGURE 18-17 Emission spectra showing that phosphorescence comes at lower energy than fluorescence from the same molecule. The phosphorescence is ~10 times weaker than the fluorescence and is only observed when the sample is cooled. [Data from J. C. Fister, III, J. M. Harris, D. Rank, and W. Wacholtz, "Molecular Photophysics of Acridine Yellow Studied by Phosphorescence and Delayed Fluorescence," *J. Chem. Ed.* **1997**, *74*, 1208.]

BOX 18-2   Fluorescence All Around Us

Most white fabrics fluoresce. Just for fun, turn on an ultraviolet lamp in a darkened room containing several people (*but do not look directly at the lamp*). You will discover emission from white fabrics (shirts, pants, shoelaces, unmentionables) containing fluorescent compounds to enhance whiteness. You might see fluorescence from teeth and from recently bruised areas of skin that show no surface damage. Many demonstrations of fluorescence and phosphorescence have been described.[14]

A fluorescent whitener from laundry detergent

A fluorescent lamp is a glass tube filled with Hg vapor. The inner walls are coated with a blend of red and green *phosphors* (luminescent substances). The red phosphor is Eu^{3+} doped into Y_2O_3. The green phosphor is Tb^{3+} doped into $CeMgAl_{11}O_{19}$. Hg vapor in the lamp is excited by an electric current and emits ultraviolet radiation at 185 and 254 nm, and a series of visible lines in panel *a*. Hg emission by itself looks blue to our eyes. When the ultraviolet radiation is absorbed by the phosphors, Eu^{3+} emits red light at 612 nm and Tb^{3+} emits green light at 542 nm. Combined blue, red, and green emission appears white to us.

Fluorescent lamps are more efficient than incandescent lamps in converting electricity into light. Replacing a 75-W incandescent bulb with an 18-W compact fluorescent bulb saves 57 W. Over the 10 000-h lifetime of the fluorescent bulb, you will reduce CO_2 emission by ~600 kg and will put 10 kg less SO_2 into the atmosphere (Problem 18-30).

Alas, fluorescent bulbs contain toxic Hg and should be recycled at a collection center where Hg is captured. Fluorescent bulbs should not be discarded as ordinary waste. Burning a metric ton (1 000 kg) of coal to make electricity releases ~0.18 g Hg into the atmosphere. A study in 2008 estimated that the U.S. would reduce annual Hg emission by 25 000 metric tons by switching all incandescent lights to fluorescent lights, even if only 25% of the fluorescent bulbs were recycled.[15] Reduction in Hg emission results from decreased use of electricity.

Light-emitting diodes are even more efficient than fluorescent lights, have longer lifetimes, and do not contain Hg. White light diodes, such as those found in flashlights, contain a blue light emitting diode made of semiconductors such as gallium nitride and indium gallium nitride. The transparent cap of the diode contains a phosphor whose yellow emission combines with blue light from the diode to produce white light.

Lamp	Approximate efficiency (lumens per watt)[a, b]
Flashlight (incandescent)	<6
Incandescent light bulb	15
Long fluorescent tube	80
Compact fluorescent lamp[c]	60
White light-emitting diode (LED)[d]	100–150
High-pressure sodium street lamp	130

a. C. J. Humphreys, "Solid State Lighting," Mater. Res. Bull., **2008**, 33, 459.

b. Lumen (lm) is a measure of luminous flux. 1 lm = radiant energy emitted in a solid angle of 1 steradian (sr) from a source that radiates 1/683 W/sr uniformly in all directions at a frequency of 540 THz (near the middle of the visible spectrum).

c. Lifetime of compact fluorescent lamp is reduced by a factor of 10 if it is frequently turned on and then off after just a few minutes.

d. LED lamp lifetime ≈ 10^5 h; fluorescent lamp lifetime ≈ 10^4 h; incandescent lamp lifetime ≈ 10^3 h.

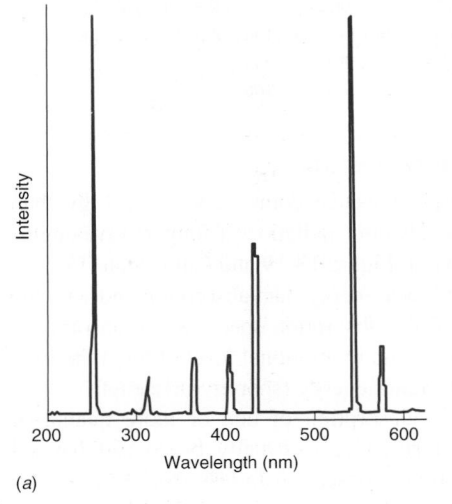

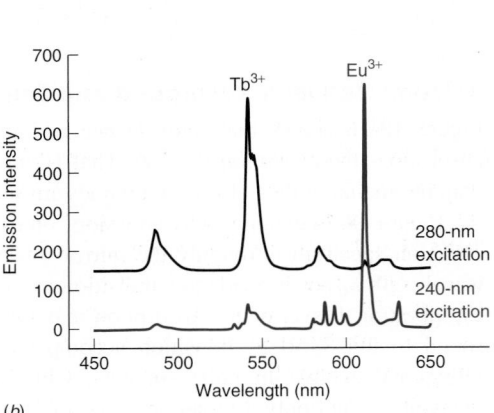

(*a*) Atomic emission spectrum of Hg vapor. [Data from S. R. Goode and L. A. Metz, "Emission Spectroscopy in the Undergraduate Laboratory," *J. Chem. Ed.* **2003**, *80*, 1455. (*b*) Emission spectra of phosphors scraped from the inside of a compact fluorescent lamp. Tb^{3+} is selectively excited at 280 nm and Eu^{3+} is selectively excited at 240 nm. [Data from C. Degli Esposti and L. Bizzochi, "The Radiative Decay of Green and Red Photoluminescent Phosphors," *J. Chem. Ed.* **2008**, *85*, 839.]

The **lifetime** of a state is the time needed for the population of that state to decay to 1/e of its initial value, where e is the base of natural logarithms.

Fluorescence and phosphorescence are relatively rare. Molecules generally decay from the excited state by nonradiative transitions. The *lifetime* of fluorescence is 10^{-8} to 10^{-4} s. The lifetime of phosphorescence is longer (10^{-4} to 10^2 s) because phosphorescence involves a change in spin quantum numbers (from two unpaired electrons to no unpaired electrons), which is improbable. A few materials, such as strontium aluminate doped with europium and dysprosium ($SrAl_2O_4$:Eu:Dy) phosphoresce for *hours* after exposure to light.[16] One application for this material is in signs leading to emergency exits when power is lost.

18-7 Luminescence

Fluorescence and phosphorescence are examples of **luminescence,** which is emission of light from an excited state of a molecule. Luminescence is inherently more sensitive than absorption. Imagine yourself in a stadium at night; the lights are off, but each of the 50 000 raving fans is holding a lighted candle. If 500 people blow out their candles, you will hardly notice the difference. Now imagine that the stadium is completely dark; then 500 people suddenly light their candles. In this case, the effect is dramatic. The first example is analogous to changing transmittance from 100% to 99%, which is equivalent to an absorbance of $-\log 0.99 = 0.004\ 4$. It is hard to measure such a small absorbance because the background is so bright. The second example is analogous to observing fluorescence from 1% of the molecules in a sample. Against a black background, fluorescence is striking.

Luminescence is sensitive enough to observe *single molecules.*[17] Figure 18-18 shows *observed* tracks of two molecules of the highly fluorescent Rhodamine 6G at 0.78-s intervals in a thin layer of silica gel. These observations confirm the "random walk" of diffusing molecules postulated by Albert Einstein in 1905.

Rhodamine 6G

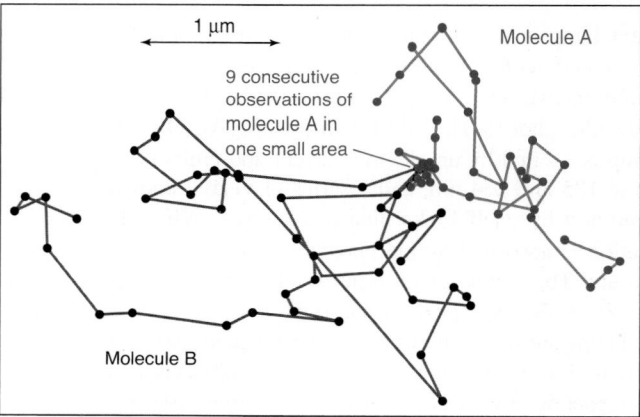

FIGURE 18-18 Tracks of two molecules of 20 pM Rhodamine 6G in silica gel observed by fluorescence integrated over 0.20-s periods at 0.78-s intervals. Some points are not connected, because the molecule disappeared above or below the focal plane in the 0.45-μm thick film and was not observed in a particular observation interval. In the nine periods when molecule A was in one location, it might have been adsorbed to a particle of silica. An individual molecule emits thousands of photons in 0.2 s as the molecule cycles between the ground and excited state. Only a fraction of these photons reach the detector, which generates a burst of ~10–50 electrons. [Data from K. S. McCain, D. C. Hanley, and J. M. Harris, "Single-Molecule Fluorescence Trajectories for Investigating Molecular Transport in Thin Silica Sol-Gel Films," *Anal. Chem.* **2003,** *75,* 4351.]

Relation Between Absorption and Emission Spectra

Figure 18-16 shows that fluorescence and phosphorescence come at lower energy than absorption (the *excitation* energy). That is, molecules emit radiation at longer wavelengths than the radiation they absorb. Examples are shown in Figure 18-19 and Color Plate 18.

Figure 18-20 explains why emission comes at lower energy than absorption and why the emission spectrum is roughly the mirror image of the absorption spectrum. In absorption, wavelength λ_0 corresponds to a transition from the ground vibrational level of S_0 to the lowest vibrational level of S_1. Absorption maxima at higher energy (shorter wavelength) correspond to the S_0 to S_1 transition accompanied by absorption of one or more quanta of vibrational energy. In polar solvents, vibrational structure is usually broadened beyond recognition, and only a broad envelope of absorption is observed. In less polar or nonpolar solvents, vibrational structure is observed.

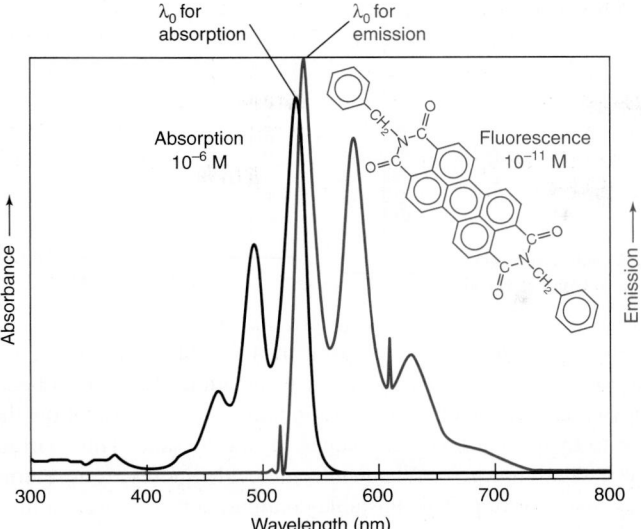

FIGURE 18-19 Absorption (black line) and emission (colored line) of bis(benzylimido) perylene in dichloromethane solution, illustrating the approximate mirror image relation between absorption and emission. The 10^{-11} M solution used for emission had an average of just 10 analyte molecules in the volume probed by the 514-nm excitation laser. [Data from P. J. G. Goulet, N. P. W. Pieczonka, and R. F. Aroca, "Overtones and Combinations in Single-Molecule Surface-Enhanced Resonance Raman Scattering Spectra," *Anal. Chem.* **2003,** *75,* 1918.]

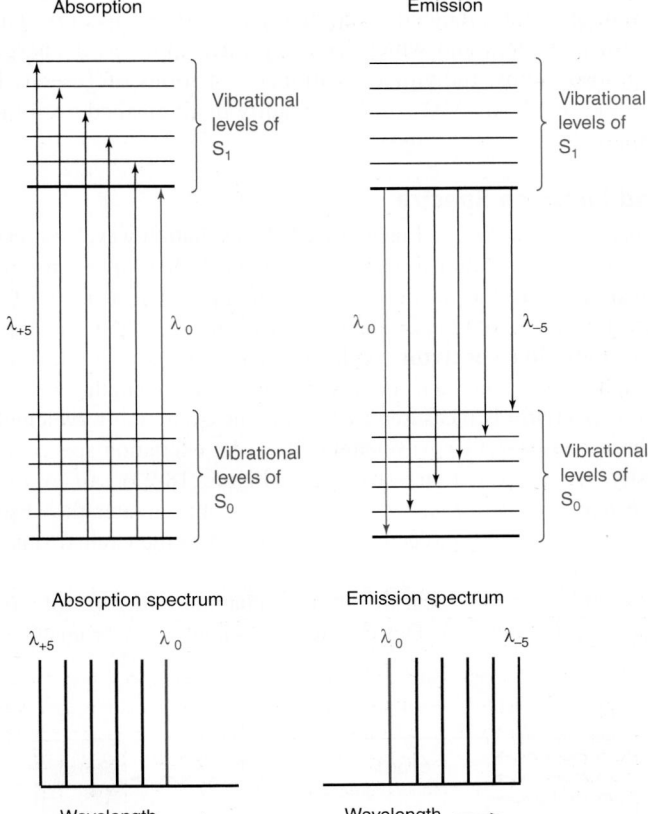

FIGURE 18-20 Energy-level diagram showing why structure is seen in the absorption and emission spectra and why the spectra are roughly mirror images of each other. In absorption, wavelength λ_0 comes at lowest energy, and λ_{+5} is at highest energy. In emission, wavelength λ_0 comes at highest energy, and λ_{-5} is at lowest energy.

After absorption, the vibrationally excited S_1 molecule relaxes back to the lowest vibrational level of S_1 prior to emitting any radiation. Emission from S_1 in Figure 18-20 can go to any vibrational level of S_0. The highest energy transition comes at wavelength λ_0, with a series of peaks following at longer wavelength. Absorption and emission spectra will have an approximate mirror image relation if the spacings between vibrational levels are roughly equal and if the transition probabilities are similar.

The λ_0 transitions in Figure 18-19 (and later in Figure 18-23) do not exactly overlap. In the emission spectrum, λ_0 comes at slightly lower energy than in the absorption spectrum. The reason is explained in Figure 18-21. A molecule absorbing radiation is initially in its electronic ground state, S_0. This molecule possesses a certain geometry and solvation. The

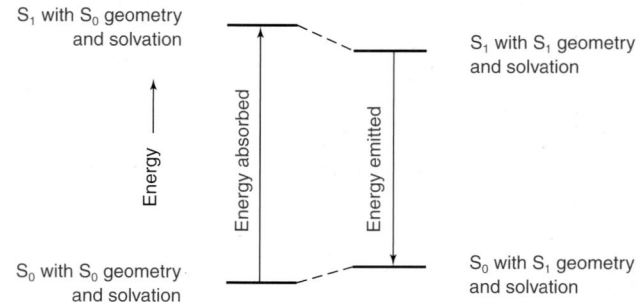

FIGURE 18-21 Diagram showing why the λ_0 transitions do not exactly overlap in Figures 18-19 and 18-23.

S$_1$ with S$_0$ geometry and solvation

S$_1$ with S$_1$ geometry and solvation

Energy →

Energy absorbed

Energy emitted

S$_0$ with S$_0$ geometry and solvation

S$_0$ with S$_1$ geometry and solvation

Electronic transitions are fast, relative to nuclear motion, so each atom has nearly the same position and momentum before and after a transition. This is called the *Franck-Condon principle.*

electronic transition to excited state S_1 is faster than the vibrational motion of atoms or the translational motion of solvent molecules. When radiation is first absorbed, the excited S_1 molecule still possesses S_0 geometry and solvation. Shortly after excitation, the geometry and solvation change to their most favorable values for the S_1 state. This rearrangement lowers the energy of the excited molecule. When an S_1 molecule fluoresces, it returns to the S_0 state with S_1 geometry and solvation. This unstable configuration must have a higher energy than that of an S_0 molecule with S_0 geometry and solvation. The net effect in Figure 18-21 is that the λ_0 emission energy is less than the λ_0 excitation energy.

Solution-phase spectra are broadened because absorbing molecules are surrounded by solvent molecules with a variety of orientations that create slightly different energy levels for different absorbing molecules. Simple gas-phase molecules, which are not in close contact with neighbors and which have a limited number of energy levels, have extremely sharp absorptions. Individual rotational transitions of gaseous $H_2{}^{16}O$, $H_2{}^{17}O$, and $H_2{}^{18}O$ with line widths of 0.02 cm^{-1} are readily distinguished even though they are just 0.2 cm^{-1} apart.

Excitation and Emission Spectra

An emission experiment is shown in Figure 18-22. An excitation wavelength (λ_{ex}) is selected by one monochromator, and luminescence is observed through a second monochromator, usually positioned at 90° to the incident light to minimize the intensity of scattered light reaching the detector. If we hold the excitation wavelength fixed and scan through the emitted radiation, an **emission spectrum** such as Figure 18-19 is produced. An emission spectrum is a graph of emission intensity versus emission wavelength.

Emission spectrum: constant λ_{ex} and variable λ_{em}

Excitation spectrum: variable λ_{ex} and constant λ_{em}

An **excitation spectrum** is measured by varying the excitation wavelength and measuring emitted light at one particular wavelength (λ_{em}). An excitation spectrum is a graph of emission intensity versus excitation wavelength (Figure 18-23). *An excitation spectrum looks very much like an absorption spectrum* because, the greater the absorbance at the excitation wavelength, the more molecules are promoted to the excited state and the more emission will be observed.

In emission spectroscopy, we measure emitted radiation, rather than the fraction of incident radiation striking the detector. Detector response and monochromator efficiency vary

FIGURE 18-22 Essentials of a luminescence experiment. The sample is irradiated at one wavelength and emission is observed over a range of wavelengths. The excitation monochromator selects the excitation wavelength (λ_{ex}) and the emission monochromator selects one wavelength at a time (λ_{em}) to observe.

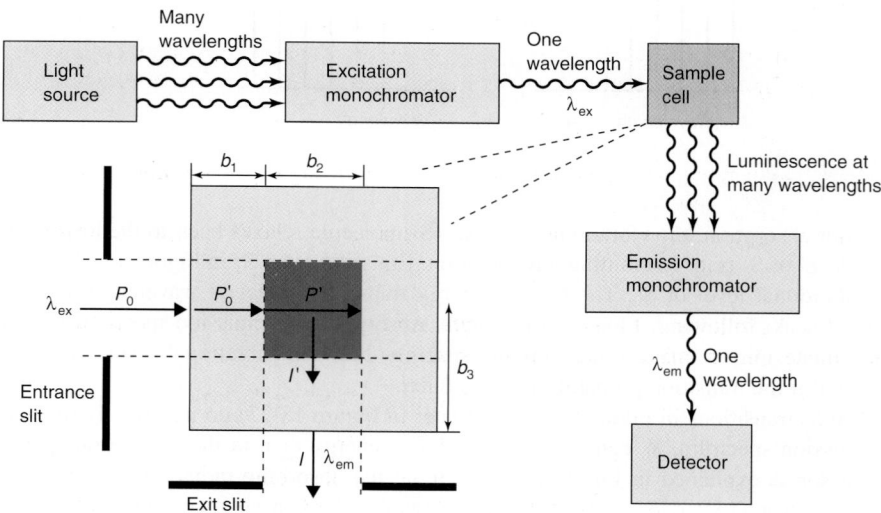

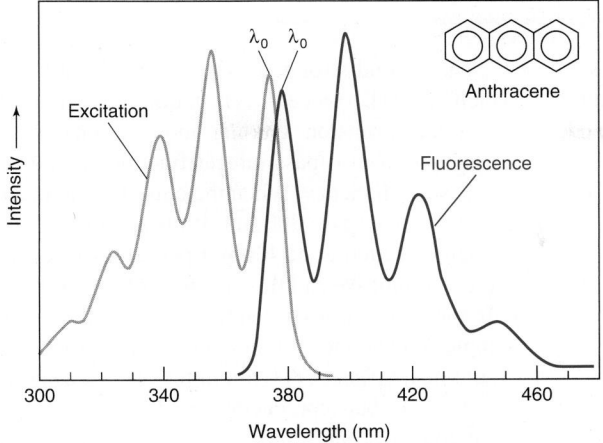

FIGURE 18-23 Excitation and emission spectra of anthracene have the same mirror image relationship as the absorption and emission spectra in Figure 18-19. An excitation spectrum is nearly the same as an absorption spectrum. [Data from C. M. Byron and T. C. Werner, "Experiments in Synchronous Fluorescence Spectroscopy for the Undergraduate Instrumental Chemistry Course," *J. Chem. Ed.* **1991,** *68,* 433.]

with wavelength, so the recorded emission spectrum is not a true profile of emitted irradiance versus emission wavelength. For quantitative analysis employing a single emission wavelength, this effect is inconsequential. If a true emission profile is required, it is necessary to calibrate the response at each emission wavelength for a particular combination of detector and emission monochromator. When recording an excitation spectrum, the true profile is not observed because the source lamp intensity and source monochromator efficiency vary with wavelength. To obtain a true excitation profile, the instrument must be calibrated for a particular combination of source lamp and source monochromator. Box 18-3 describes common forms of light scattering that can be confused with emission when you are interpreting an experimental spectrum.

Luminescence Intensity

A simplified view of processes occurring during a luminescence measurement is shown in the enlarged sample cell at the lower left of Figure 18-22. We expect emission to be proportional to the irradiance absorbed by the sample. In Figure 18-22, the irradiance (W/m^2) incident on the sample cell is P_0. Some is absorbed over the pathlength b_1, so the irradiance striking the central region of the cell is

$$\text{Irradiance striking central region} = P_0' = P_0 \cdot 10^{-\varepsilon_{ex}b_1c} \tag{18-8}$$

where ε_{ex} is the molar absorptivity at the wavelength λ_{ex} and c is the concentration of analyte. The irradiance of the beam when it has traveled the additional distance b_2 is

$$P' = P_0' \cdot 10^{-\varepsilon_{ex}b_2c} \tag{18-9}$$

Emission intensity I is proportional to irradiance absorbed in the central region of the cell:

$$\text{Emission intensity} = I' = k'(P_0' - P') \tag{18-10}$$

where k' is a constant of proportionality. Not all radiation emitted from the center of the cell in the direction of the exit slit is observed. Some is absorbed between the center and the edge of the cell. Emission intensity I emerging from the cell is given by Beer's law:

$$I = I' \cdot 10^{-\varepsilon_{em}b_3c} \tag{18-11}$$

where ε_{em} is the molar absorptivity at the emission wavelength and b_3 is the distance from the center to the edge of the cell.

Combining Equations 18-10 and 18-11 gives an expression for emission intensity:

$$I = k'(P_0' - P')10^{-\varepsilon_{em}b_3c}$$

Substituting expressions for P_0' and P' from Equations 18-8 and 18-9, we obtain a relation between incident irradiance and emission intensity:

$$I = k'(P_0 \cdot 10^{-\varepsilon_{ex}b_1c} - P_0 \cdot 10^{-\varepsilon_{ex}b_1c} \cdot 10^{-\varepsilon_{ex}b_2c})10^{-\varepsilon_{em}b_3c}$$

$$= k'P_0 \cdot \underbrace{10^{-\varepsilon_{ex}b_1c}}_{\substack{\text{Loss of intensity} \\ \text{in region 1}}}\underbrace{(1 - 10^{-\varepsilon_{ex}b_2c})}_{\substack{\text{Absorption} \\ \text{in region 2}}}\underbrace{10^{-\varepsilon_{em}b_3c}}_{\substack{\text{Loss of intensity} \\ \text{in region 3}}} \tag{18-12}$$

Equation 18-8 comes from Equations 18-5 and 18-6. If species other than analyte absorbed at the wavelengths of interest, we would have to include them.

BOX 18-3   **Rayleigh and Raman Scattering**

Emission spectra can exhibit confusing features in addition to fluorescence and phosphorescence. The chart shows an emission spectrum of aqueous dichlorofluorescein (colored line) and, for reference, an emission spectrum of pure water (black line). The excitation wavelength is 400 nm. The only difference between the two traces is fluorescence from dichlorofluorescein with a peak at 522 nm.

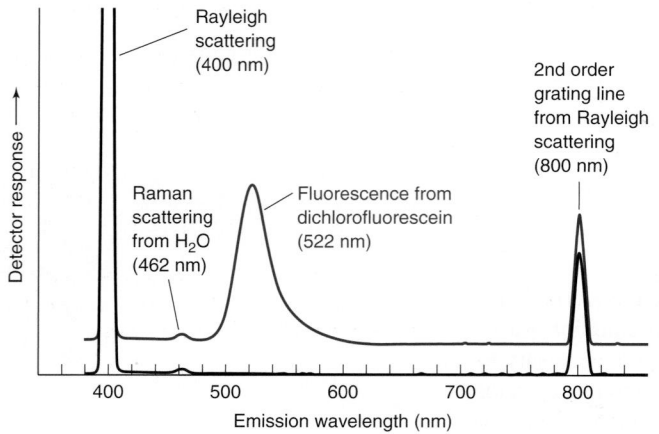

Upper trace: Emission spectrum of aqueous dichlorofluorescein. *Lower trace:* Spectrum observed from pure water. [Information from Kris Varazo, Francis Marion University. See R. J. Clark and A. Oprysa, "Fluorescence and Light Scattering," *J. Chem. Ed.* **2004,** *81,* 705.]

What are the other peaks? The strongest peak, which is off-scale, is observed at the excitation wavelength of 400 nm. It is called **Rayleigh scattering** after the same Lord Rayleigh (J. W. Strutt) who discovered argon (page 75). The oscillating electromagnetic field of the excitation light source causes electrons in water molecules to oscillate at the same frequency as the incident radiation. Oscillating electrons emit this same frequency of radiation in all directions. The time required for scattering is essentially the period of one oscillation of the incoming electromagnetic wave, which is ~10^{-15} s for 400-nm light. By comparison, the time for fluorescence is ~10^{-8} to 10^{-4} s. Rayleigh scattering is always present and is usually filtered out so that it is not displayed in the emission spectrum.

The second strongest peak in this example comes at 800 nm, which is exactly twice the excitation wavelength. It is an artifact of the monochromator. Grating monochromators designed to pass wavelength λ also pass integer fractions λ/2, λ/3, and so on, with decreasing efficiency. When the emission monochromator in Figure 18-22 is set to pass 800 nm, it also passes some light at 400 nm. Rayleigh scattering at 400 nm passes through the monochromator set to 800 nm. We call this *second-order diffraction* from the monochromator. If we used a filter to block 400-nm light between the sample cell and the emission monochromator in Figure 18-22, there would be no peak at 800 nm in the emission spectrum.

A weak, but reproducible, peak is observed in both water and dichlorofluorescein solution at 462 nm. The difference in energy between incident light at 400 nm and the peak at 462 nm corresponds to a vibrational energy of H_2O. The peak at 462 nm is called **Raman scattering** after the Indian physicist C. V. Raman, who discovered this phenomenon in 1928 and was awarded the Nobel Prize in 1930. In this type of scattering, which also occurs in a time frame of ~10^{-15} s, a small fraction of incident photons gives up one quantum of molecular vibrational energy to H_2O. Scattered radiation emerges with less energy than the excitation energy. Vibrational energy is customarily expressed as the wavenumber (cm^{-1}) of a photon with that energy. Liquid H_2O has a broad range of vibrational energies centered near 3 404 cm^{-1}. The wavenumber of the exciting radiation is 1/wavelength = 1/400 nm = 25 000 cm^{-1}. In Raman scattering, an incident photon with energy of 25 000 cm^{-1} gives up 3 404 cm^{-1} and emerges at (25 000 − 3 404) = 21 596 cm^{-1}. The wavelength is 1/(21 596 cm^{-1}) ≈ 463 nm. The observed peak is at 462 nm.

What are the lessons from this example? First, compare the spectrum of pure solvent to the spectrum of a sample under study so that you can disregard peaks due to solvent. Second, fluorescence comes at a fixed location, such as 522 nm for dichlorofluorescein. The wavelength of scattered radiation varies with incident wavelength. If we used an excitation wavelength of 410 nm instead of 400, the second-order grating line would be seen at 820 nm and the water Raman line would be at an energy that is 3 404 cm^{-1} less than the exciting light, or 477 nm. Scattered radiation shifts with the incident wavelength, but fluorescence and phosphorescence do not.

Consider the limit of low concentration, which means that the exponents $\varepsilon_{ex}b_1c$, $\varepsilon_{ex}b_2c$, and $\varepsilon_{em}b_3c$ are all very small. The terms $10^{-\varepsilon_{ex}b_1c}$, $10^{-\varepsilon_{ex}b_2c}$, and $10^{-\varepsilon_{em}b_3c}$ are all close to unity. We can replace $10^{-\varepsilon_{ex}b_1c}$ and $10^{-\varepsilon_{em}b_3c}$ with 1 in Equation 18-12. We cannot replace $10^{-\varepsilon_{ex}b_2c}$ with 1 because we are subtracting this term from 1 and would be left with 0. Instead, we expand $10^{-\varepsilon_{ex}b_2c}$ in a power series:

The series 18-13 follows from the relation $10^{-A} = (e^{\ln 10})^{-A} = e^{-A \ln 10}$ and the expansion of e^x:

$$10^{-\varepsilon_{ex}b_2c} = 1 - \varepsilon_{ex}b_2c \ln 10 + \frac{(\varepsilon_{ex}b_2c \ln 10)^2}{2!} - \frac{(\varepsilon_{ex}b_2c \ln 10)^3}{3!} + \cdots \quad \textbf{(18–13)}$$

$$e^x = 1 + \frac{x^1}{1!} + \frac{x^2}{2!} + \frac{x^3}{3!} + \cdots$$

Each term of 18-13 becomes smaller and smaller, so we just keep the first two terms. The central factor in Equation 18-12 becomes $(1 - 10^{-\varepsilon_{ex}b_2c}) = (1 - [1 - \varepsilon_{ex}b_2c \ln 10]) = \varepsilon_{ex}b_2c \ln 10$, so the entire equation can be written

Emission intensity at low concentration: $k'P_0(\varepsilon_{ex}b_2c \ln 10) = \boxed{I = kP_0c}$ **(18-14)**

where $k = k'\varepsilon_{ex}b_2 \ln 10$ is a constant.

Equation 18-14 says that, *at low concentration, emission intensity is proportional to analyte concentration.* Data for anthracene in Figure 18-24 are linear below 10^{-6} M. Blank samples invariably scatter light and must be run in every analysis. Equation 18-14 tells us that doubling the incident irradiance (P_0) will double the emission intensity (up to a point). In contrast, doubling P_0 has no effect on absorbance, which is a *ratio* of two intensities. The sensitivity of a luminescence measurement can be increased by more than a factor of 3 by placing mirror coatings on the two walls of the sample cell opposite the slits in Figure 18-22.[18]

For higher concentrations, we need all the terms in Equation 18-12, or we need an even more accurate equation.[19] As concentration increases, maximum emission is reached. Then emission decreases because absorption increases more rapidly than the emission. We say the emission is **quenched** (decreased) by **self-absorption,** which is the absorption of excitation or emission energy by analyte molecules in the solution. Quenching by self-absorption is also called the *inner filter effect*. At high concentration, even the *shape* of the emission spectrum can change because absorption and emission both depend on wavelength.

Example: Fluorimetric Assay of Selenium in Brazil Nuts

Selenium is a trace element essential to life. For example, the selenium-containing enzyme glutathione peroxidase catalyzes the destruction of peroxides (ROOH) that are harmful to cells. Conversely, at high concentration, selenium can be toxic.

To measure selenium in Brazil nuts, 0.1 g of nut is digested with 2.5 mL of 70 wt% HNO_3 in a Teflon *bomb* (Figure 28-7) in a microwave oven. Hydrogen selenate (H_2SeO_4) in the digest is reduced to hydrogen selenite (H_2SeO_3) with hydroxylamine (NH_2OH). Selenite is then **derivatized** to form a fluorescent product that is extracted into cyclohexane.

$$\text{2,3-Diaminonaphthalene} + H_2SeO_3 \xrightarrow[50°C]{pH = 2} \text{Fluorescent product} + 3H_2O \qquad (18\text{–}15)$$

Maximum response of the fluorescent product was observed with an excitation wavelength of 378 nm and an emission wavelength of 518 nm. In Figure 18-25, emission is proportional to concentration only up to ~0.1 μg Se/mL. Beyond 0.1 μg Se/mL, the response becomes curved, eventually reaches a maximum, and finally *decreases* with increasing concentration as self-absorption dominates. Equation 18-12 predicts this behavior.

Luminescence in Analytical Chemistry[20]

Some analytes, such as riboflavin (vitamin B_2)[21] and polycyclic aromatic compounds (an important class of carcinogens), are naturally fluorescent and can be analyzed directly. Most compounds are not luminescent. However, coupling to a fluorescent moiety such as *fluorescein* provides a route to sensitive analysis. Ca^{2+} can be measured from the fluorescence of a complex it forms with a fluorescein derivative called *calcein*. Molecules whose luminescence responds selectively to part per billion levels of Hg(II) or Pb(II)[22] or adenosine triphosphate (ATP)[23] are available. Box 18-4 describes the thought behind the design of a fluorescent sensor molecule for CN^-.

Fluorescein

Chelating aminodiacetate group

Calcein

Molecular biologists use *DNA microarrays* ("gene chips") to monitor gene expression and mutations and to detect and identify pathological microorganisms.[25] A single chip can contain thousands of known single-strand DNA sequences in known locations. The chip is incubated with unknown single-strand DNA that has been tagged with fluorescent labels. After the unknown DNA has bound to its complementary strands on the chip, the amount bound to each spot on the chip is measured by fluorescence intensity.

Light from a firefly or light stick[26] is an example of **chemiluminescence**—emission of light from a chemical reaction.[27] Chemiluminescence detectors for sulfur and nitrogen in organic compounds are employed in gas chromatography (Section 24-3). Nitric oxide (NO),[28] which transmits signals between living cells, and bacterial quorum sensing molecules[29] can be measured at part per billion levels by their chemiluminescent reactions with the compound luminol. Organosulfur compounds derived from living organisms in natural waters can be measured at part per trillion levels by chemiluminescence.[30] Light emitted from a redox reaction at an electrode is called *electrochemiluminescence*.[31]

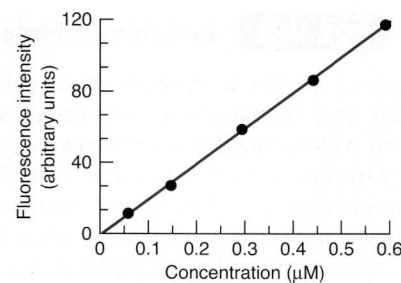

FIGURE 18-24 Linear calibration curve for fluorescence of anthracene measured at the wavelength of maximum fluorescence in Figure 18-23. [Data from C. M. Byron and T. C. Werner, "Experiments in Synchronous Fluorescence Spectroscopy for the Undergraduate Instrumental Chemistry Course," *J. Chem. Ed.* **1991,** *68,* 433.]

Derivatization is the chemical alteration of analyte so that it can be detected easily or separated easily from other species.

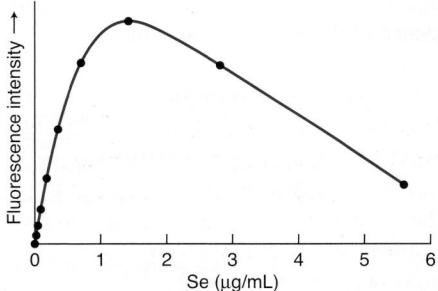

FIGURE 18-25 Fluorescence calibration curve for the selenium-containing product in Reaction 18-15. The curvature and maximum are due to self-absorption. [Data from M.-C. Sheffield and T. M. Nahir, "Analysis of Selenium in Brazil Nuts by Microwave Digestion and Fluorescence Detection," *J. Chem. Ed.* **2002,** *79,* 1345.]

Luminol
(5-amino-2,3-dihydro-1,4-phthalazinedione)

oxidant (such as NO or H_2O_2)

OH^-, metal catalyst

+ N_2 + blue light

BOX 18-4 | **Designing a Molecule for Fluorescence Detection**

Here we present an example of a deliberately designed, weakly fluorescent molecule that becomes strongly fluorescent when it reacts with the nonfluorescent analyte cyanide.

Structurally rigid, conjugated aromatic systems such as anthracene in Figure 18-23 are strongly fluorescent. *Trans*-stilbene shown below is not very fluorescent, but when the structure is made rigid by encasing the central double bond in a system of fused rings, the molecule becomes very fluorescent.[24]

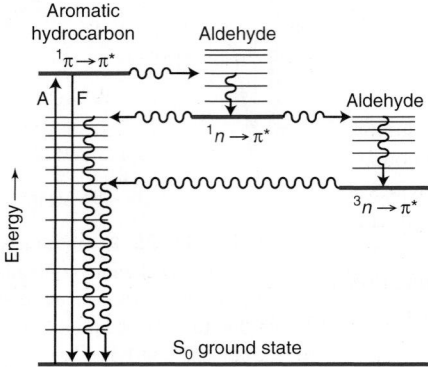

Trans-stilbene
Weakly fluorescent

Rigid derivative of *trans*-stilbene
Strongly fluorescent

Figures 18-13 and 18-14 showed that the lowest energy electronic transition of formaldehyde is the type $n \rightarrow \pi^*$ (nonbonding $\rightarrow$ pi antibonding). Molecules whose lowest excited singlet state is $n \rightarrow \pi^*$ are typically nonemissive.[24] If an aldehyde group is located adjacent to fused aromatic rings that are normally fluorescent, and if the aldehyde $n \rightarrow \pi^*$ energy lies below the $\pi \rightarrow \pi^*$ energy of the aromatic system, energy is transferred from $\pi \rightarrow \pi^*$ to $n \rightarrow \pi^*$ and lost by nonradiative processes. The aldehyde *quenches* the fluorescence of the aromatic molecule.

Fluorescence (F) from an aromatic molecule is reduced by transfer of excitation energy to a proximate aldehyde group followed by nonradiative processes shown by wiggly lines.

The molecule labeled **I** was deliberately designed to be a sensor for cyanide. The fluorescence of **I** increases sevenfold when CN^- reacts with the aldehyde to form a cyanohydrin. The C=O double bond is converted to a single bond that does not quench fluorescence. The aromatic part of the molecule is held rigid by hydrogen bonding before and after reaction of the aldehyde to enhance fluorescence of the product. If the diphenylacetylene unit were free to rotate about the central C≡C bond, fluorescence would be weaker. The part of a molecule that fluoresces is called the *fluorophore*, just as the part of a molecule that absorbs light is called the chromophore.

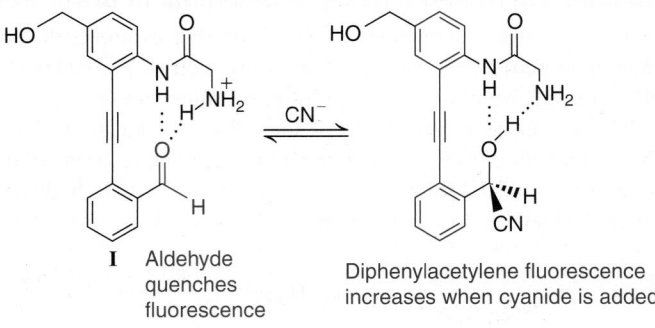

I Aldehyde quenches fluorescence

Diphenylacetylene fluorescence increases when cyanide is added

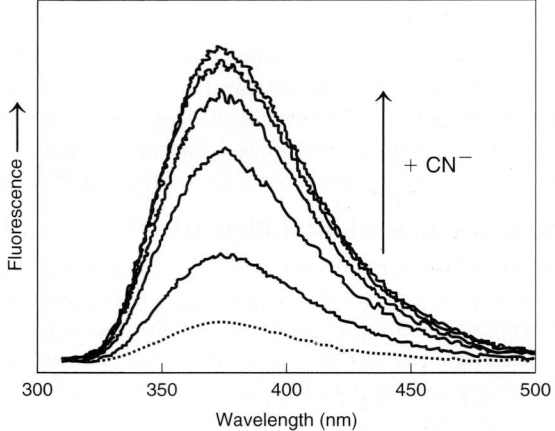

Increase in fluorescence of 5.0 μM **I** in 10.0 mM aqueous HEPES buffer (Table 9-2) at pH 7.0 with addition of 0, 0.2, 0.5, 0.7, 0.9, 1.2, and 1.5 mM NaCN and $\lambda_{ex} = 270$ nm. Fluorescence of **I** was not affected by 15 other anions at a concentration of 1.5 mM. [Information from J. Jo and D. Lee, "Turn-On Fluorescence Detection of Cyanide in Water: Activation of Latent Fluorophores Through Remote Hydrogen Bonds that Mimic Peptide β-Turn Motif," *J. Am. Chem. Soc.* **2009**, *131*, 16283.]

Terms to Understand

absorbance	excitation spectrum	molecular orbital	rotational transition
absorption spectrum	excited state	monochromatic light	self-absorption
Beer's law	fluorescence	monochromator	singlet state
chemiluminescence	frequency	phosphorescence	spectrophotometric titration
chromophore	ground state	photon	spectrophotometry
cuvet	hertz	quenching	transmittance
derivatization	irradiance	Raman scattering	triplet state
electromagnetic spectrum	luminescence	Rayleigh scattering	vibrational transition
electronic transition	masking	reagent blank	wavelength
emission spectrum	molar absorptivity	refractive index	wavenumber

Summary

Light can be thought of as waves whose wavelength (λ) and frequency (ν) have the important relation $\lambda\nu = c$, where c is the speed of light. Alternatively, light may be viewed as consisting of photons whose energy (E) is given by $E = h\nu = hc/\lambda = hc\tilde{\nu}$, where h is Planck's constant and $\tilde{\nu}(= 1/\lambda)$ is the wavenumber. Absorption of light is commonly measured by absorbance (A) or transmittance (T), defined as $A = \log(P_0/P)$ and $T = P/P_0$, where P_0 is the incident irradiance and P is the exiting irradiance. Absorption spectroscopy is useful in quantitative analysis because absorbance is proportional to the concentration of the absorbing species in dilute solution (Beer's law): $A = \varepsilon bc$. In this equation, b is pathlength, c is concentration, and ε is the molar absorptivity (a constant of proportionality). You should be able to prepare a series of standard solutions for calibration by serial dilution of a concentrated standard.

Basic components of a spectrophotometer include a radiation source, a monochromator, a sample cell, and a detector. To minimize errors in spectrophotometry, samples should be free of particles, cuvets must be clean, and they should be positioned reproducibly in the sample holder. Measurements should be made at a wavelength of maximum absorbance. Instrument errors tend to be minimized if the absorbance falls in the range $A \approx 0.3$–2.

In a spectrophotometric titration, absorbance is monitored as titrant is added. For many reactions, there is an abrupt change in slope when the equivalence point is reached.

When a molecule absorbs light, it is promoted to an excited state from which it returns to the ground state by nonradiative processes or by fluorescence (singlet → singlet emission) or phosphorescence (triplet → singlet emission). Emission intensity is proportional to concentration at low concentration. At sufficiently high concentration, emission decreases because of self-absorption by the analyte. An excitation spectrum (a graph of emission intensity versus excitation wavelength) is similar to an absorption spectrum (a graph of absorbance versus wavelength). An emission spectrum (a graph of emission intensity versus emission wavelength) is observed at lower energy than the absorption spectrum and tends to be the mirror image of the absorption spectrum. A molecule that is not fluorescent can be analyzed by attaching a fluorophore to it or by its ability to enhance or diminish the fluorescence of a sensor compound. Light emitted by a chemical reaction—chemiluminescence—is also used for quantitative analysis.

Exercises

18-A. (a) What value of absorbance corresponds to 45.0% T?

(b) If a 0.010 0 M solution exhibits 45.0% T at some wavelength, what will be the percent transmittance for a 0.020 0 M solution of the same substance?

18-B. (a) A 3.96×10^{-4} M solution of compound A exhibited an absorbance of 0.624 at 238 nm in a 1.000-cm cuvet; a blank solution containing only solvent had an absorbance of 0.029 at the same wavelength. Find the molar absorptivity of compound A.

(b) The absorbance of an unknown solution of compound A in the same solvent and cuvet was 0.375 at 238 nm. Find the concentration of A in the unknown.

(c) A concentrated solution of compound A in the same solvent was diluted from an initial volume of 2.00 mL to a final volume of 25.00 mL and then had an absorbance of 0.733. What is the concentration of A in the concentrated solution?

18-C. Ammonia can be determined spectrophotometrically by reaction with phenol in the presence of hypochlorite (OCl^-):

$$2 \bigcirc\!\!-OH + NH_3 \xrightarrow{OCl^-} O=\bigcirc=N-\bigcirc-O^-$$

Phenol (colorless)   Ammonia (colorless)   Blue product, $\lambda_{max} = 625$ nm

A 4.37-mg sample of protein was chemically digested to convert its nitrogen into ammonia and then diluted to 100.0 mL. Then 10.0 mL of the solution were placed in a 50-mL volumetric flask and treated with 5 mL of phenol solution plus 2 mL of sodium hypochlorite solution. The sample was diluted to 50.0 mL, and the absorbance at 625 nm was measured in a 1.00-cm cuvet after 30 min. For reference, a standard solution was prepared from 0.010 0 g of NH_4Cl (FM 53.49) dissolved in 1.00 L of water. Then 10.0 mL of this standard were placed in a 50-mL volumetric flask and analyzed in the same manner as the unknown. A reagent blank was prepared by using distilled water in place of unknown.

Sample	Absorbance at 625 nm
Blank	0.140
Reference	0.308
Unknown	0.592

(a) Calculate the molar absorptivity of the blue product.

(b) Calculate the weight percent of nitrogen in the protein.

18-D. Cu^+ reacts with neocuproine to form the colored complex $(neocuproine)_2Cu^+$, with an absorption maximum at 454 nm. Neocuproine reacts with few other metals. The copper complex is soluble in 3-methyl-1-butanol (isoamyl alcohol), an organic solvent that does not dissolve appreciably in water. In other words, when isoamyl alcohol is added to water, a two-layered mixture results, with the denser water layer at the bottom. If $(neocuproine)_2Cu^+$ is present, virtually all of it goes into the organic phase. For the purpose of this problem, assume that the isoamyl alcohol does not dissolve in the water at all and that all of the colored complex will be in the organic phase. Suppose that the following procedure is carried out:

1. A rock containing copper is pulverized, and all metals are extracted from it with strong acid. The acidic solution is neutralized with base and made up to 250.0 mL in flask A.

2. Next, 10.00 mL of the solution are transferred to flask B and treated with 10.00 mL of reducing agent to convert Cu^{2+} to Cu^+. Then 10.00 mL of buffer are added so that the pH is suitable for complex formation with neocuproine.

3. 15.00 mL of this solution are withdrawn and placed in flask C. To the flask are added 10.00 mL of an aqueous solution containing neocuproine and 20.00 mL of isoamyl alcohol. After the mixture

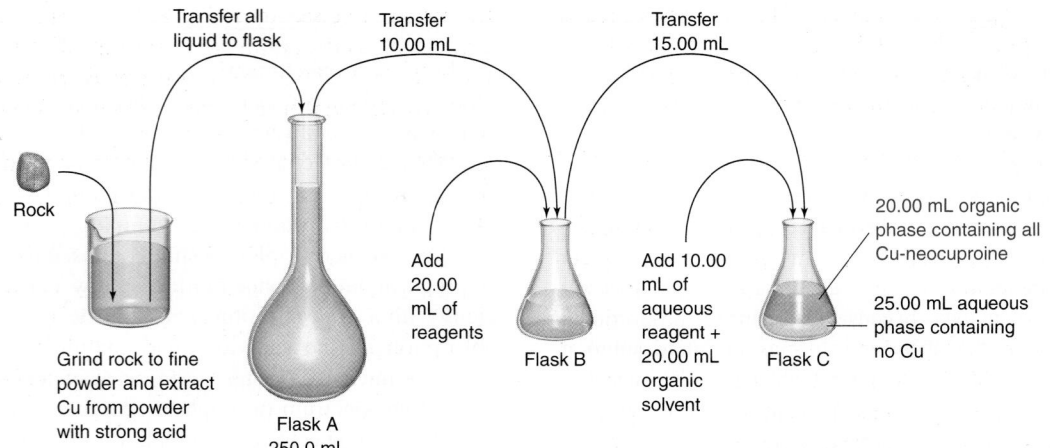

Grind rock to fine powder and extract Cu from powder with strong acid

Flask A
250.0 mL

Rock

Transfer all liquid to flask

Add 20.00 mL of reagents

Transfer 10.00 mL

Flask B

Add 10.00 mL of aqueous reagent + 20.00 mL organic solvent

Transfer 15.00 mL

Flask C

20.00 mL organic phase containing all Cu-neocuproine

25.00 mL aqueous phase containing no Cu

For Exercise 18-D.

has been shaken well and the phases allowed to separate, all $(\text{neocuproine})_2\text{Cu}^+$ is in the organic phase.

4. A few milliliters of the upper layer are withdrawn, and the absorbance at 454 nm is measured in a 1.00-cm cell. A blank carried through the same procedure gives an absorbance of 0.056.

(a) Suppose that the rock contained 1.00 mg of Cu. What will be the concentration of Cu (mol/L) in the isoamyl alcohol phase?

(b) If the molar absorptivity of $(\text{neocuproine})_2\text{Cu}^+$ is $7.90 \times 10^3 \text{ M}^{-1}$ cm^{-1}, what will be the observed absorbance? Remember that a blank carried through the same procedure gave an absorbance of 0.056.

(c) A rock is analyzed and found to give a final absorbance of 0.874 (uncorrected for the blank). How many milligrams of Cu are in the rock?

18-E. Semi-xylenol orange is a yellow compound at pH 5.9 but turns red when it reacts with Pb^{2+}. A 2.025-mL sample of semi-xylenol orange at pH 5.9 was titrated with 7.515×10^{-4} M $\text{Pb(NO}_3)_2$, with the following results:

Total µL Pb^{2+} added	Absorbance at 490 nm wavelength	Total µL Pb^{2+} added	Absorbance at 490 nm wavelength
0.0	0.227	42.0	0.425
6.0	0.256	48.0	0.445
12.0	0.286	54.0	0.448
18.0	0.316	60.0	0.449
24.0	0.345	70.0	0.450
30.0	0.370	80.0	0.447
36.0	0.399		

Make a graph of absorbance versus microliters of Pb^{2+} added. Be sure to correct the absorbances for dilution. Corrected absorbance is what would be observed if the volume were not changed from its initial value of 2.025 mL. Assuming that the reaction of semi-xylenol orange with Pb^{2+} has a 1:1 stoichiometry, find the molarity of semi-xylenol orange in the original solution.

Problems

Properties of Light

18-1. Fill in the blanks.

(a) If you double the frequency of electromagnetic radiation, you _____ the energy.

(b) If you double the wavelength, you _____ the energy.

(c) If you double the wavenumber, you _____ the energy.

18-2. (a) How much energy (in kilojoules) is carried by one mole of photons of red light with $\lambda = 650$ nm?

(b) How many kilojoules are carried by one mole of photons of violet light with $\lambda = 400$ nm?

18-3. Calculate the frequency (Hz), wavenumber (cm^{-1}), and energy (J/photon and J/[mol of photons]) of visible light with a wavelength of 562 nm.

18-4. Which molecular processes correspond to the energies of microwave, infrared, visible, and ultraviolet photons?

18-5. Characteristic orange light produced by sodium in a flame is due to an intense emission called the sodium D line, which is actually a doublet, with wavelengths (measured in vacuum) of 589.157 88 and

589.755 37 nm. The index of refraction of air at a wavelength near 589 nm is 1.000 292 6. Calculate the frequency, wavelength, and wavenumber of each component of the D line, measured in air.

Absorption of Light and Measuring Absorbance

18-6. Box 3-1 (page 50) showed a single-beam spectrophotometric ozone monitor. The emission spectrum of the mercury vapor lamp is shown in Box 18-2.

(a) What is the light source?

(b) What is the detector?

(c) What is the monochromator?

(d) What are the components of the optical pathlength, b?

(e) How does this instrument measure O_3?

18-7. How do transmittance, absorbance, and molar absorptivity differ? Which one is proportional to concentration?

18-8. What is an absorption spectrum?

18-9. Why does a compound whose visible absorption maximum is at 480 nm (blue-green) appear to be red?

CHAPTER 18 Fundamentals of Spectrophotometry

18-10. *Color and absorption spectra.* Color Plate 17 shows colored solutions and their spectra. From Table 18-1, predict the color of each solution from the wavelength of maximum absorption. Do observed colors agree with predicted colors?

18-11. Why is it most accurate to measure absorbances in the range $A = 0.3–2$?

18-12. The absorbance of a 2.31×10^{-5} M solution of a compound is 0.822 at a wavelength of 266 nm in a 1.00-cm cell. Calculate the molar absorptivity at 266 nm.

18-13. What color would you expect to observe for a solution of Fe(ferrozine)$_3^{4-}$, which has a visible absorbance maximum at 562 nm?

18-14. When I was a boy, Uncle Wilbur let me watch as he analyzed the iron content of runoff from his banana ranch. A 25.0-mL sample was acidified with nitric acid and treated with excess KSCN to form a red complex. (KSCN itself is colorless.) The solution was then diluted to 100.0 mL and put in a variable-pathlength cell. For comparison, a 10.0-mL reference sample of 6.80×10^{-4} M Fe^{3+} was treated with HNO$_3$ and KSCN and diluted to 50.0 mL. The reference was placed in a cell with a 1.00-cm light path. The runoff sample exhibited the same absorbance as the reference when the pathlength of the runoff cell was 2.48 cm. What was the concentration of iron in Uncle Wilbur's runoff?

18-15. Vapor at a pressure of 30.3 μbar from the solid compound pyrazine had a transmittance of 24.4% at a wavelength of 266 nm in a 3.00-cm cell at 298 K.

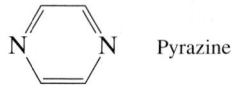 Pyrazine

(a) Convert transmittance to absorbance.

(b) Convert pressure to concentration (mol/L) with the ideal gas law (Problem 1-18).

(c) Find the molar absorptivity of gaseous pyrazine at 266 nm.

18-16. The *absorption cross section* on the ordinate of the ozone absorption spectrum at the beginning of this chapter is defined by the relation

$$\text{Transmittance } (T) = e^{-n\sigma b}$$

where n is the number of absorbing molecules per cubic centimeter, σ is the absorption cross section (cm^2), and b is the pathlength (cm). The total ozone in the atmosphere is approximately 8×10^{18} molecules above each square centimeter of Earth's surface (from the surface up to the top of the atmosphere). If this were compressed into a 1-cm-thick layer, the concentration would be 8×10^{18} molecules/cm^3.

(a) Using the ozone spectrum at the beginning of the chapter, estimate the transmittance and absorbance of this 1-cm^3 sample at 325 and 300 nm.

(b) Sunburns are caused by radiation in the 295- to 310-nm region. At the center of this region, the transmittance of atmospheric ozone is 0.14. Calculate the absorption cross section for $T = 0.14$, $n = 8 \times 10^{18}$ molecules/cm^3, and $b = 1$ cm. By what percentage does the transmittance increase if the ozone concentration decreases by 1% to 7.92×10^{18} molecules/cm^3?

(c) Atmospheric O$_3$ is measured in *Dobson units* (1 unit = 2.69×10^{16} molecules O$_3$ above each cm^2 of Earth's surface). (Dobson unit $\equiv$ thickness [in hundredths of a millimeter] that the O$_3$ column would occupy if it were compressed to 1 atm at 0°C.) The graph shows variations in O$_3$ concentration as a function of latitude and season. Using an

absorption cross section of 2.5×10^{-19} cm^2, calculate the transmittance in the winter and in the summer at 30°–50° N latitude, at which O$_3$ varies from 290 to 350 Dobson units. By what percentage is the ultraviolet transmittance greater in winter than in summer?

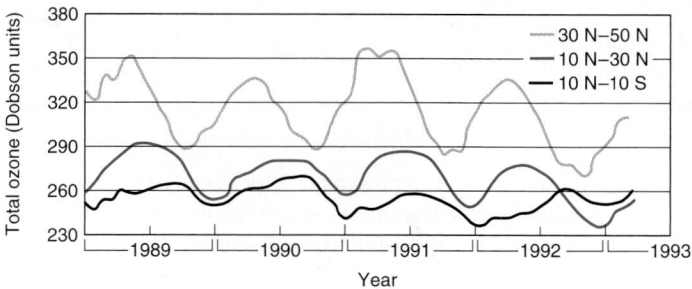

Variation in atmospheric ozone at different latitudes. [Data from P. S. Zurer, *Chem. Eng. News*, 24 May 1993, p. 8.]

Beer's Law in Chemical Analysis

18-17. What is the purpose of neocuproine in the serum iron analysis?

18-18. A compound with molecular mass 292.16 g/mol was dissolved in a 5-mL volumetric flask. A 1.00-mL aliquot was withdrawn, placed in a 10-mL volumetric flask, and diluted to the mark. The absorbance at 340 nm was 0.427 in a 1.000-cm cuvet. The molar absorptivity at 340 nm is $\varepsilon_{340} = 6\,130$ M^{-1} cm^{-1}.

(a) Calculate the concentration of compound in the cuvet.

(b) What was the concentration of compound in the 5-mL flask?

(c) How many milligrams of compound were used to make the 5-mL solution?

18-19. If a sample for spectrophotometric analysis is placed in a 10-cm cell, the absorbance will be 10 times greater than the absorbance in a 1-cm cell. Will the absorbance of the reagent-blank solution also be increased by a factor of 10?

18-20. You have been sent to India to investigate the occurrence of goiter disease attributed to iodine deficiency. As part of your investigation, you must make field measurements of traces of iodide (I$^-$) in groundwater. The procedure is to oxidize I$^-$ to I$_2$ and convert the I$_2$ into an intensely colored complex with the dye brilliant green in the organic solvent toluene.

(a) A 3.15×10^{-6} M solution of the colored complex exhibited an absorbance of 0.267 at 635 nm in a 1.000-cm cuvet. A blank solution made from distilled water in place of groundwater had an absorbance of 0.019. Find the molar absorptivity of the colored complex.

(b) The absorbance of an unknown solution prepared from groundwater was 0.175. Find the concentration of the unknown.

18-21. *Protein concentration.* The molar absorptivity of a protein in water at 280 nm can be estimated within ~5–10% from its content of the amino acids tyrosine and tryptophan (Table 10-1) and from the number of disulfide linkages (R—S—S—R) between cysteine residues.[32]

$$\varepsilon_{280\text{ nm}}(\text{M}^{-1}\text{ cm}^{-1}) \approx 5\,500\,n_{\text{Trp}} + 1\,490\,n_{\text{Tyr}} + 125\,n_{\text{S-S}}$$

where n_{Trp} is the number of tryptophans, n_{Tyr} is the number of tyrosines, and $n_{\text{S-S}}$ is the number of disulfides. This information is obtained from the amino acid sequence and from additional studies such as X-ray crystallography and nuclear magnetic resonance to locate disulfide linkages.

The protein human serum transferrin has 678 amino acids including 8 tryptophans, 26 tyrosines, 19 disulfide linkages, and two carbohydrate chains of variable length.[33] The molecular mass with the dominant carbohydrates is 79 550.

(a) Predict the molar absorptivity of transferrin.

(b) Predict the absorbance of a 1 wt% transferrin solution in a 1.000-cm-pathlength cuvet. (The value tabulated by Pace[32] is 11.10.)

(c) From your answer to **(a)**, estimate wt% and mg/mL of a transferrin solution with an absorbance of 1.50 at 280 nm.

18-22. Nitrite ion, NO_2^-, is a preservative for bacon and other foods, but it is potentially carcinogenic. A spectrophotometric determination of NO_2^- makes use of the following reactions:

$$HO_3S\!-\!\langle\bigcirc\rangle\!-\!NH_2 + NO_2^- + 2H^+ \longrightarrow$$

Sulfanilic acid

$$HO_3S\!-\!\langle\bigcirc\rangle\!-\!\overset{+}{N}\!\equiv\!N + 2H_2O$$

$$HO_3S\!-\!\langle\bigcirc\rangle\!-\!\overset{+}{N}\!\equiv\!N + \langle\bigcirc\bigcirc\rangle\!-\!NH_2 \longrightarrow$$

1-Aminonaphthalene

$$HO_3S\!-\!\langle\bigcirc\rangle\!-\!N\!=\!N\!-\!\langle\bigcirc\bigcirc\rangle\!-\!NH_2 + H^+$$

Colored product, λ_{max} = 520 nm

Here is an abbreviated procedure for the determination:

1. To 50.0 mL of unknown solution containing nitrite is added 1.00 mL of sulfanilic acid solution.
2. After 10 min, 2.00 mL of 1-aminonaphthalene solution and 1.00 mL of buffer are added.
3. After 15 min, the absorbance is read at 520 nm in a 5.00-cm cell.

The following solutions were analyzed:

A. 50.0 mL of food extract known to contain no nitrite (that is, a negligible amount); final absorbance = 0.153.
B. 50.0 mL of food extract suspected of containing nitrite; final absorbance = 0.622.
C. Same as B, but with 10.0 μL of 7.50×10^{-3} M $NaNO_2$ added to the 50.0-mL sample; final absorbance = 0.967.

(a) Calculate the molar absorptivity, ε, of the colored product. Remember that a 5.00-cm cell was used.

(b) How many micrograms of NO_2^- were present in 50.0 mL of food extract?

18-23. *Preparing standards for a calibration curve.*

(a) How much ferrous ethylenediammonium sulfate $(Fe(H_3NCH_2CH_2NH_3)(SO_4)_2 \cdot 4H_2O$, FM 382.15) should be dissolved in a 500-mL volumetric flask with 1 M H_2SO_4 to obtain a stock solution with ~500 μg Fe/mL?

(b) When making stock solution **(a)**, you actually weighed out 1.627 g of reagent. What is the Fe concentration in μg Fe/mL?

(c) How would you prepare 500 mL of standard containing ~1, 2, 3, 4, and 5 μg Fe/mL in 0.1 M H_2SO_4 from stock solution **(b)** using any Class A pipets from Table 2-4 with only 500-mL volumetric flasks?

(d) To reduce the generation of chemical waste, describe how you could prepare 50 mL of standard containing ~1, 2, 3, 4, and 5 μg Fe/mL in 0.1 M H_2SO_4 from stock solution **(b)** by serial dilution using any Class A pipets from Table 2-3 with only 50-mL volumetric flasks?

18-24. *Preparing standards for a calibration curve.*

(a) How much ferrous ammonium sulfate $(Fe(NH_4)_2(SO_4)_2 \cdot 6H_2O$, FM 392.15) should be dissolved in a 500-mL volumetric flask with 1 M H_2SO_4 to obtain a stock solution with 1 000 μg Fe/mL?

(b) When making stock solution **(a)**, you weighed out 3.627 g of reagent. What is the Fe concentration in μg Fe/mL?

(c) How would you prepare 250 mL of standard containing ~1, 2, 3, 4, 5, 7, 8, and 10 (±20%) μg Fe/mL in 0.1 M H_2SO_4 from stock solution **(b)** using only 5- and 10-mL Class A pipets, only 250-mL volumetric flasks, and only two consecutive dilutions of the stock solution? For example, to prepare a solution with ~4 μg Fe/mL, you could first dilute 15 mL (= 10 + 5 mL) of stock solution up to 250 mL to get $\sim(\frac{15}{250})(1\,000$ μg Fe/mL$) = \sim 60$ μg Fe/mL. Then dilute 15 mL of the new solution up to 250 mL again to get $\sim(\frac{15}{250})(60$ μg Fe/mL$) = \sim 3.6$ μg Fe/mL.

18-25. *Uncertainty in preparing standard solution.* Using uncertainties of Class A volumetric flasks and pipets in Tables 2-3 and 2-4, estimate the absolute and relative uncertainty (%) in concentration of a solution made by diluting two 3-mL portions of stock solution containing 1.086 ± 0.002 g Fe/L in 0.48 M HCl to 1 L with 0.1 M HCl.

Spectrophotometric Titrations

18-26. A 2.00-mL solution of apotransferrin was titrated as illustrated in Figure 18-11. It required 163 μL of 1.43 mM ferric nitrilotriacetate to reach the end point.

(a) Why does the slope of the absorbance versus volume graph change abruptly at the equivalence point?

(b) How many moles of Fe(III) (= ferric nitrilotriacetate) were required to reach the end point?

(c) Each apotransferrin molecule binds two ferric ions. Find the molar concentration of apotransferrin in the 2.00-mL solution.

18-27. The iron-binding site of transferrin in Figure 18-8 can accommodate certain other metal ions besides Fe^{3+} and certain other anions besides CO_3^{2-}. Data are given in the table for the titration of transferrin (3.57 mg in 2.00 mL) with 6.64 mM Ga^{3+} solution in the presence of the anion oxalate, $C_2O_4^{2-}$, and in the absence of a suitable anion. Prepare a graph similar to Figure 18-11, showing both sets of data. Indicate the theoretical equivalence point for the binding of one and two Ga^{3+} ions per molecule of protein and the observed end point. How many Ga^{3+} ions are bound to transferrin in the presence and in the absence of oxalate?

Titration in presence of $C_2O_4^{2-}$		Titration in absence of anion	
Total μL Ga^{3+} added	Absorbance at 241 nm	Total μL Ga^{3+} added	Absorbance at 241 nm
0.0	0.044	0.0	0.000
2.0	0.143	2.0	0.007
4.0	0.222	6.0	0.012
6.0	0.306	10.0	0.019
8.0	0.381	14.0	0.024
10.0	0.452	18.0	0.030
12.0	0.508	22.0	0.035
14.0	0.541	26.0	0.037
16.0	0.558		
18.0	0.562		
21.0	0.569		
24.0	0.576		

18-28. Gold nanoparticles (Figure 17-31) can be titrated with the oxidizing agent TCNQ in the presence of excess of Br^- to oxidize $Au(0)$ to $AuBr_2^-$ in deaerated toluene. Gold atoms in the interior of the particle are $Au(0)$. Gold atoms bound to $C_{12}H_{25}S$—(dodecanethiol) ligands on the surface of the particle are $Au(I)$ and are not titrated.

$$Au(0) + \text{[TCNQ]} + 2Br^- \rightarrow AuBr_2^- + TCNQ^{-}$$

TCNQ
Tetracyanoquinodimethane

Reduced radical anion
$\lambda_{max} = 856$ nm

Reduced $TCNQ^{-}$ has a low-energy electronic absorption peak at 856 nm.

The table gives the absorbance at 856 nm when 0.700 mL of 1.00×10^{-4} M TCNQ + 0.05 M $(C_8H_{17})_4N^+Br^-$ in toluene is titrated with gold nanoparticles (1.43 g/L in toluene) from a microsyringe with a Teflon-coated needle. Absorbance in the table has already been corrected for dilution.

Total μL nanoparticles	Absorbance at 856 nm	Total μL nanoparticles	Absorbance at 856 nm
4.9	0.208	19.1	0.706
8.0	0.301	22.0	0.770
11.0	0.405	25.0	0.784
13.7	0.502	30.0	0.785
16.2	0.610	35.9	0.784

Data from G. Zotti, B. Vercelli, and A. Berlin, "Reaction of Gold Nanoparticles with Tetracyanoquinoidal Molecules," Anal. Chem. 2008, 80, 815.

(a) Make a graph of absorbance versus volume of titrant and estimate the equivalence point. Calculate the mmol of $Au(0)$ in 1.00 g of nanoparticles.

(b) From other analyses of similarly prepared nanoparticles, it is estimated that 25 wt% of the mass of the particle is dodecanethiol ligand ($C_{12}H_{25}S$—, FM 201.40). Calculate mmol of $C_{12}H_{25}S$ in 1.00 g of nanoparticles.

(c) The $Au(I)$ content of 1.00 g of nanoparticles should be $1.00 -$ mass of $Au(0) -$ mass of $C_{12}H_{25}S$. Calculate the micromoles of $Au(I)$ in 1.00 g of nanoparticles and the mole ratio $Au(I):C_{12}H_{25}S$. In principle, this ratio should be 1:1. The difference is most likely because $C_{12}H_{25}S$ was not measured for this specific nanoparticle preparation.

18-29. *Biotin-streptavidin fluorescence titration. Biotin* is a *cofactor* in enzymatic carboxylation reactions. Biotin activates CO_2 for biosynthetic reactions.

CO_2 binds to N

Biotin

Streptavidin is a protein isolated from the bacterium *Streptomyces avidinii* that binds biotin with a formation constant of $\sim 10^{14}$ M^{-1}. The biotin-streptavidin complex is widely used in biotechnology because the noncovalent complex is stable in the presence of detergents, protein denaturants, and organic solvents, and at extremes of pH and temperature.

The stoichiometry of the biotin-streptavidin complex was measured by a fluorescence titration. Fluorescein (page 453) covalently attached to biotin via the biotin carboxyl group fluoresces at 520 nm when irradiated at 493 nm. When biotin-fluorescein (BF) binds to streptavidin (SA), fluorescence decreases. The table gives emission intensity for addition of BF to SA and also for addition of SA to BF. Data are already corrected for dilution.

Addition of BF to SA[a]		Addition of SA to BF[b]	
Mole ratio [BF]:[SA]	Fluorescence (10^5 counts/s)	Mole ratio [SA]:[BF]	Fluorescence (10^6 counts/s)
0.00	0.000	0.000	2.958
0.51	0.202	0.061	2.394
1.02	0.333	0.122	1.613
1.53	0.476	0.183	0.884
2.04	0.595	0.244	0.212
2.55	0.678	0.306	0.144
3.06	0.749	0.367	0.144
3.57	0.892	0.428	0.144
4.08	1.927	0.489	0.144
4.59	3.770	0.550	0.144
5.10	5.970		
5.61	8.230		
6.12	10.609		

a. Titration of ~0.11 μM SA with BF at pH 7.3 in 10 mM triethanolamine buffer.
b. Titration of ~0.72 μM BF with SA at pH 7.3 in 10 mM triethanolamine buffer.
Data from D. P. Mascotti and M. J. Waner "Complementary Spectroscopic Assays for Investigating Protein-Ligand Binding Activity," J. Chem. Ed. 2010, 87, 735.

(a) Make a graph of fluorescence versus mole ratio for each titration and state the stoichiometry of binding of biotin to streptavidin.

(b) Explain the shape of each titration curve.

18-30. *Greenhouse gas reduction.* An 18-W compact fluorescent bulb produces approximately the same amount of light as a 75-W incandescent bulb that screws into the same socket. The fluorescent bulb lasts ~10 000 h and the incandescent bulb lasts ~750 h. Over the lifetime of the fluorescent bulb, the electricity savings is $(75 - 18$ W$) \cdot (10^4$ h$) = 570$ kW $\cdot$ h. One kilogram of coal produces ~2 kW $\cdot$ h of electricity. If coal contains 60 wt% carbon, how many more kilograms of CO_2 are produced by running the incandescent bulb rather than the fluorescent bulb? If the coal contains 2 wt% sulfur, how many more kilograms of SO_2 are produced?

Luminescence

18-31. In formaldehyde, the transition $n \rightarrow \pi^*(T_1)$ occurs at 397 nm, and the $n \rightarrow \pi^*(S_1)$ transition comes at 355 nm. What is the difference in energy (kJ/mol) between the S_1 and T_1 states? This difference is due to the different electron spins in the two states.

18-32. What is the difference between fluorescence and phosphorescence?

18-33. What is the difference between luminescence and chemiluminescence?

18-34. Explain what happens in Rayleigh scattering and Raman scattering. How much faster is scattering of visible light than fluorescence?

18-35. Consider a molecule that can fluoresce from the S_1 state and phosphoresce from the T_1 state. Which is emitted at longer wavelength, fluorescence or phosphorescence? Make a sketch showing absorption, fluorescence, and phosphorescence on a single spectrum.

18-36. What is the difference between a fluorescence excitation spectrum and a fluorescence emission spectrum? Which one resembles an absorption spectrum?

18-37. ▦ The wavelengths of maximum absorption and emission of anthracene in Figure 18-23 are approximately 357 and 402 nm. Molar absorptivities at these wavelengths are $\varepsilon_{ex} = 9.0 \times 10^3\ M^{-1}\ cm^{-1}$ and $\varepsilon_{em} = 5 \times 10^1\ M^{-1}\ cm^{-1}$. Consider a fluorescence experiment in Figure 18-22 with cell dimensions $b_1 = 0.30$ cm, $b_2 = 0.40$ cm, and $b_3 = 0.5$ cm. Calculate the relative fluorescence intensity with Equation 18-12 as a function of concentration over the range 10^{-8} to 10^{-3} M. Explain the shape of the curve. Up to approximately what concentration is fluorescence proportional to concentration (within 5%)? Is the calibration range in Figure 18-24 sensible?

18-38. ▦ *Standard addition.* Selenium from 0.108 g of Brazil nuts was converted into the fluorescent product in Reaction 18-15, and extracted into 10.0 mL of cyclohexane. Then 2.00 mL of the cyclohexane solution were placed in a cuvet for fluorescence measurement. Standard additions of fluorescent product containing 1.40 µg Se/mL are given in the table. Construct a standard addition graph to find the concentration of Se in the 2.00-mL unknown solution. Find the wt% of Se in the nuts and its uncertainty and 95% confidence interval.

Volume of standard added (µL)	Fluorescence intensity (arbitrary units)
0	41.4
10.0	49.2
20.0	56.4
30.0	63.8
40.0	70.3

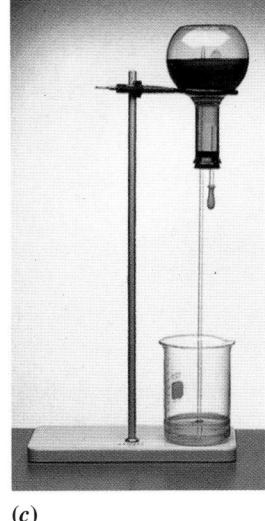

(a)   (b)   (c)

COLOR PLATE 1 HCl Fountain (Demonstration 6-2) (a) Basic indicator solution in beaker. (b) Indicator is drawn into flask and changes to acidic color. (c) Solution levels at end of experiment. [© Macmillan, Photo by Ken Karp.]

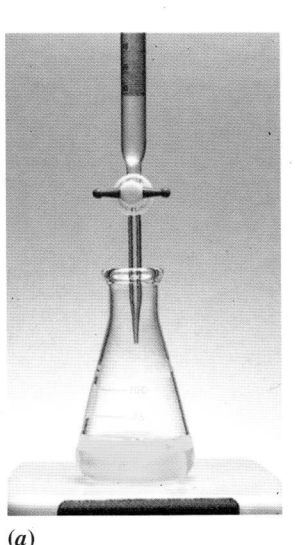

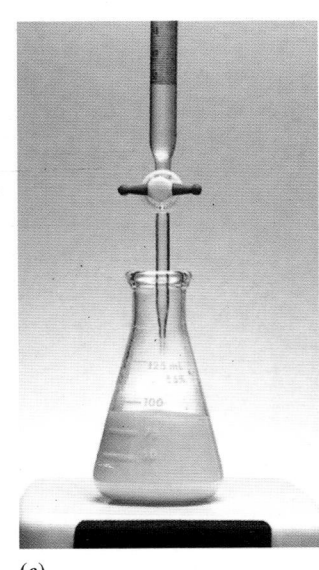

(a)   (b)   (c)

COLOR PLATE 2 Fajans Titration of Cl$^-$ with AgNO$_3$, Using Dichlorofluorescein (Demonstration 7-1) (a) Indicator before beginning titration. (b) AgCl precipitate before end point. (c) Indicator adsorbed on precipitate after end point. [© Macmillan, Photo by Ken Karp.]

(a)   (b)

COLOR PLATE 3 Effect of Ionic Strength on Ionic Dissociation (Demonstration 8-1) (a) Two beakers containing identical solutions with Fe(SCN)$^{2+}$, Fe^{3+}, and SCN$^-$. (b) The red color of Fe(SCN)$^{2+}$ fades when KNO$_3$ is added to the right-hand beaker because the equilibrium Fe^{3+} + SCN$^-$ $\rightleftharpoons$ Fe(SCN)$^{2+}$ shifts to the left. [© Macmillan, Photo by Ken Karp.]

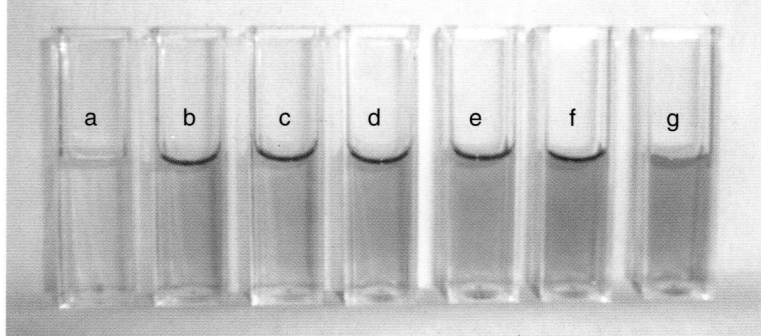

COLOR PLATE 4 Effect of Ionic Strength on Color of Bromocresol Green (Problem 8-3) Color of ~16 μM bromocresol green (H_2BG) acid-base indicator in water changes from green to blue as NaCl is added. (*a*) Yellow acidic form HBG^- obtained in 0.5 mM HCl solution with no added NaCl. (*g*) Blue form BG^{2-} obtained in 0.5 mM NaOH solution with no added NaCl. (*b–f*) BG solutions with no added acid or base, but increasing concentrations of NaCl (0, 0.003 9, 0.010 6, 0.077 4, 0.203 2 M). [From H. B. Rodriguez and M. Mirenda, "A Simplified Undergraduate Laboratory Experiment To Evaluate the Effect of the Ionic Strength on the Equilibrium Concentration Quotient of the Bromcresol Green Dye," *J. Chem. Ed.* **2012,** *89,* 1201. Courtesy M. Mirenda, Universidad de Buenos Aires. Reprinted with permission © 2012, American Chemical Society.]

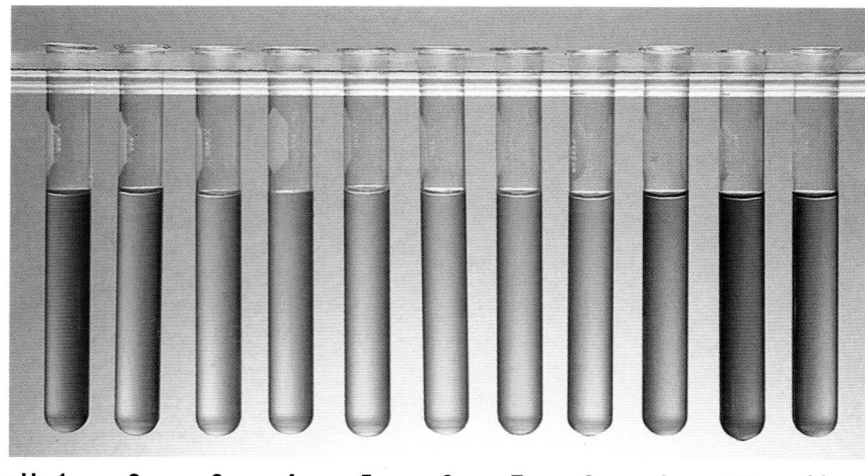

pH: 1   2   3   4   5   6   7   8   9   10   11

COLOR PLATE 5 Thymol Blue (Section 11-6) Acid-base indicator thymol blue between pH 1 and 11. The p*K* values are 1.7 and 8.9. [© Macmillan, Photo by Ken Karp.]

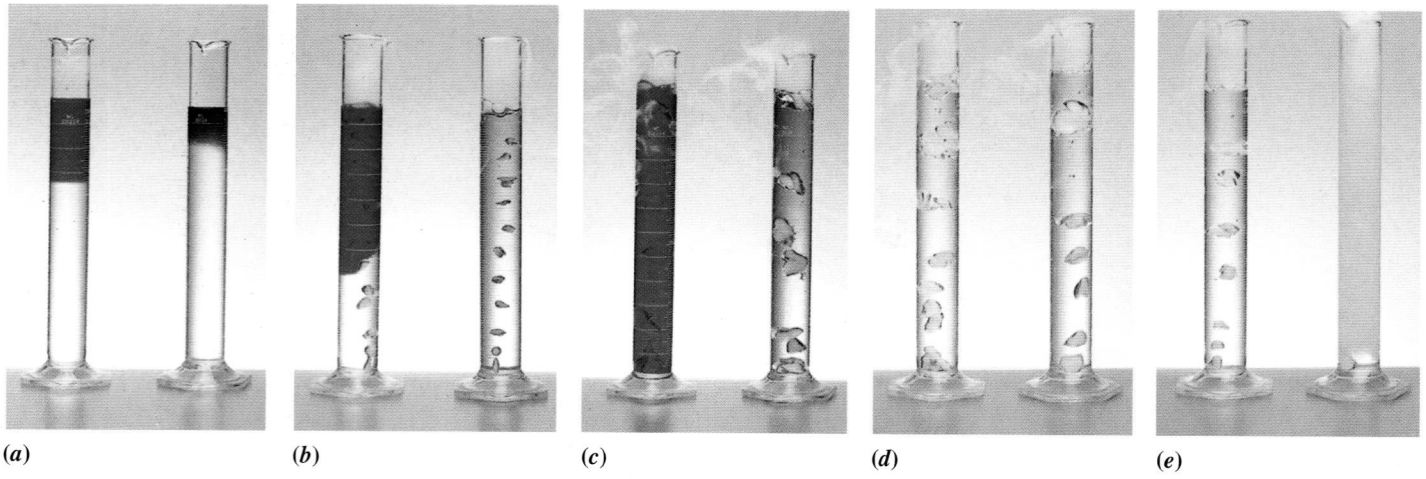

(*a*)   (*b*)   (*c*)   (*d*)   (*e*)

COLOR PLATE 6 Indicators and Acidity of CO_2 (Demonstration 11-1) (*a*) Cylinders before adding Dry Ice. Ethanol indicator solutions of phenolphthalein (left) and bromothymol blue (right) have not yet mixed with entire cylinder. (*b*) Adding Dry Ice causes bubbling and mixing. (*c*) Further mixing. (*d*) Phenolphthalein changes to its colorless acidic form. Color of bromothymol blue is due to mixture of acidic and basic forms. (*e*) After addition of HCl and stirring of right-hand cylinder, bubbles of CO_2 can be seen leaving solution, and indicator changes completely to its acidic color. [© Macmillan, Photo by Ken Karp.]

COLOR PLATE 7 Titration of Cu(II) with EDTA, Using Auxiliary Complexing Agent (Section 12-5) 0.02 M $CuSO_4$ before titration (left). Color of Cu(II)-ammonia complex after adding ammonia buffer, pH 10 (center). End-point color when EDTA has displaced all ammonia ligands (right).
[© Macmillan, Photo by Ken Karp.]

COLOR PLATE 8 Titration of Mg^{2+} by EDTA, Using Eriochrome Black T Indicator (Demonstration 12-1) Before (left), near (center), and after (right) equivalence point.
[© Macmillan, Photo by Ken Karp.]

COLOR PLATE 9 Titration of VO^{2+} with Potassium Permanganate (Section 16-4) Blue VO^{2+} solution prior to titration (left). Mixture of blue VO^{2+} and yellow VO_2^+ observed during titration (center). Dark color of MnO_4^- at end point (right). [© Macmillan, Photo by Ken Karp.]

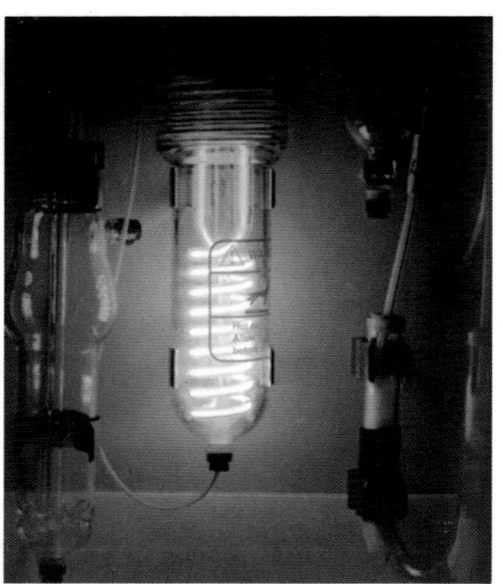

COLOR PLATE 10 Photolytic Environmental Carbon Analyzer (Box 16-2) A measured water sample is injected into the chamber at the left, where it is acidified with H_3PO_4 and sparged (bubbled with Ar or N_2) to remove CO_2 derived from HCO_3^- and CO_3^{2-}. The CO_2 is measured by its infrared absorbance. The sample is then forced into the digestion chamber, where $S_2O_8^{2-}$ is added and the sample is exposed to ultraviolet radiation from an immersion lamp (the coil at the center of the photo). Sulfate radicals (SO_4^-) formed by irradiation oxidize most organic compounds to CO_2, which is measured by infrared absorbance. The U-tube at the right contains Sn and Cu granules to scavenge volatile acids, such as HCl and HBr, liberated in the digestion. [Photo from U.S. Environmental Protection Agency, Cincinnati, OH.]

COLOR PLATE 11 Iodometric Titration (Section 16-7)
I_3^- solution (left). I_3^- solution before end point in titration with $S_2O_3^{2-}$ (left center). I_3^- solution immediately before end point with starch indicator present (right center). At the end point (right).
[© Macmillan, Photo by Ken Karp.]

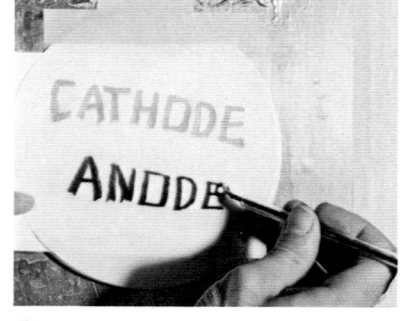

(a)   *(b)*   *(c)*

COLOR PLATE 12 Electrochemical Writing (Demonstration 17-1) (*a*) Stylus used as cathode. (*b*) Stylus used as anode. (*c*) The polarity of the foil backing is opposite that of the stylus and produces reverse color on the bottom sheet of paper. [Dan Harris.]

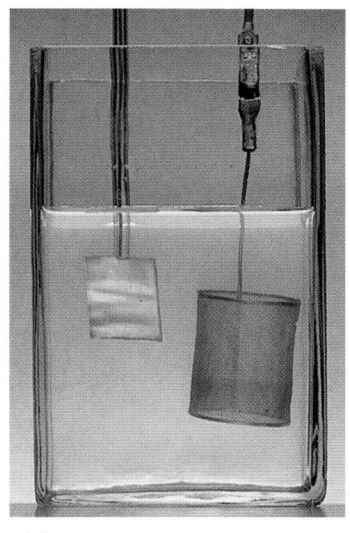
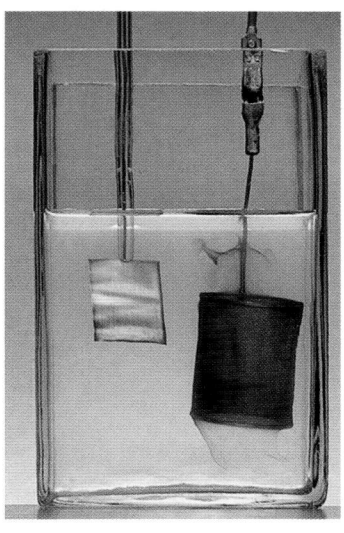

(a) (b)

COLOR PLATE 13 Formation of Diffusion Layer During Electrolysis (Box 17-4) (*a*) Cu electrode (flat plate, left) and Pt electrode (mesh basket, right) immersed in solution containing KI and starch, with no electric current. (*b*) Starch-iodine complex forms at surface of Pt anode when current flows. [© Macmillan, Photo by Ken Karp.]

COLOR PLATE 14 Grating Dispersion (Section 18-2) Visible spectrum produced by a grating inside a spectrophotometer. [© Macmillan, Photo by Ken Karp.]

COLOR PLATE 15 Beer's Law (Section 18-2) $Fe(phenanthroline)_3^{2+}$ standards for spectrophotometric analysis. Volumetric flasks contain $Fe(phenanthroline)_3^{2+}$ with Fe concentrations ranging from 1 mg/L (left) to 10 mg/L (right). The absorbance, as evidenced by the intensity of the color, is proportional to the iron concentration. [© Macmillan, Photo by Ken Karp.]

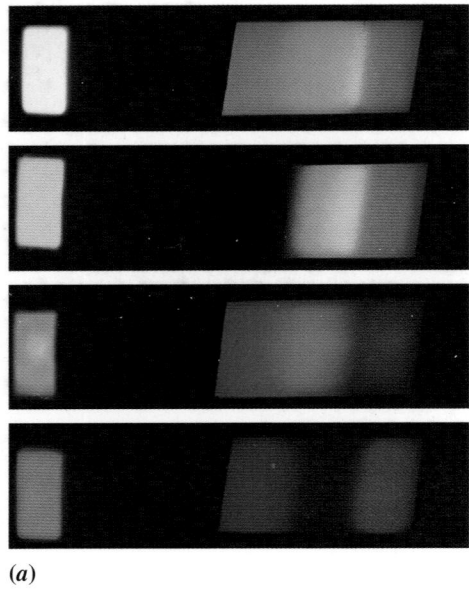

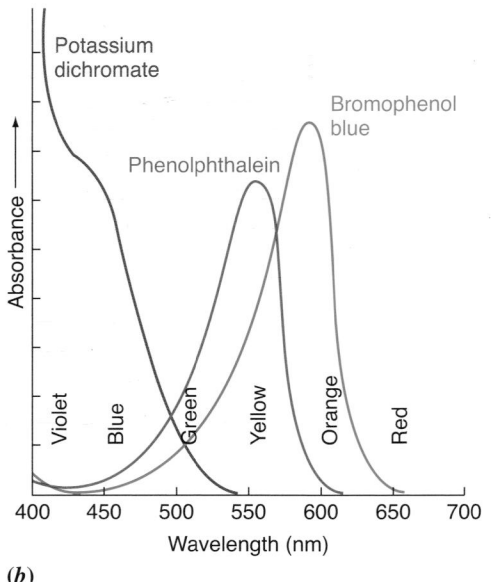

Potassium
dichromate

Bromophenol
blue

Phenolphthalein

Absorbance

Violet | Blue | Green | Yellow | Orange | Red

400   450   500   550   600   650   700
Wavelength (nm)

(a)   (b)

COLOR PLATE 16 Absorption Spectra (Demonstration 18-1)
(a) Projected visible spectra of (from top to bottom) white light, potassium dichromate, bromophenol blue, and phenolphthalein.
(b) Visible absorption spectra of the same compounds recorded with a spectrophotometer. [Dan Harris.]

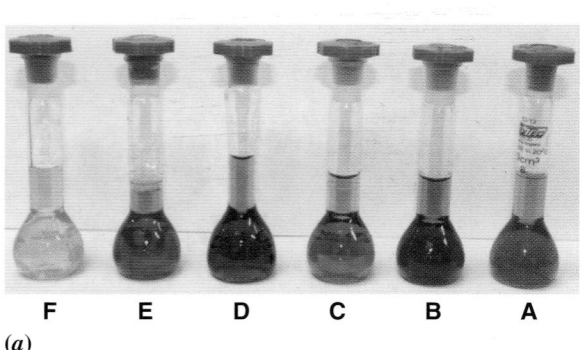

F   E   D   C   B   A

(a)

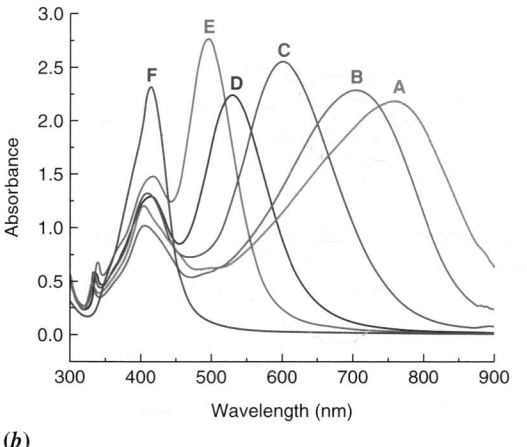

3.0
2.5
2.0
1.5
1.0
0.5
0.0

Absorbance

E
F   D   C   B   A

300   400   500   600   700   800   900
Wavelength (nm)

(b)

COLOR PLATE 17 Absorption Spectra and Color (Section 18-2 and Problem 18-10) Flasks contain suspensions of silver nanoparticles whose color depends on the size and shape of the particles, which are approximately triangular plates with edge lengths of ~50–100 nm. The visible absorption spectrum of each suspension is shown in the graph. Stable suspensions of nanoparticles are called *colloids* (Demonstration 27-1). [Republished with permission of Royal Society of Chemistry, from D. M. Ledwith, A. M. Whelan, and J. M. Kelly, "A Rapid, Straight-Forward Method for Controlling the Morphology of Stable Silver Nano Particles," *J. Mater. Chem.* **2007,** *17,* 2459, Figure 1. Permission conveyed through Copyright Clearance Center, Inc. Courtesy J. M. Kelly and D. Ledwith, Trinity College, University of Dublin.]

COLOR PLATE 18 Luminescence (Section 18-7) (a) Green crystal of yttrium aluminum garnet containing a small amount of Cr^{3+}. (b) When irradiated with high-intensity blue light from a laser at the right side, Cr^{3+} absorbs blue light and emits lower energy red light. When the laser is removed, the crystal appears green again. [Courtesy M. Seltzer, Michelson Laboratory, China Lake, CA.]

(a)   (b)

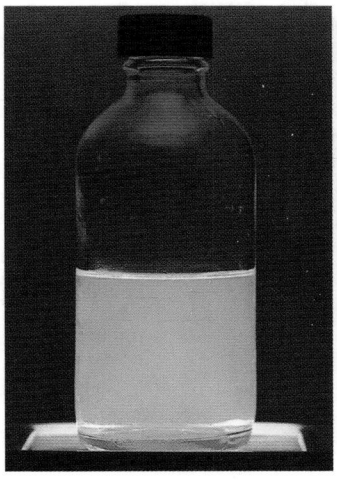

COLOR PLATE 19 Quenching of Ru(II) Luminescence by O_2 (Section 19-6) Left: Orange-red luminescence from ~5 μM $(bipyridyl)_3RuCl_2$ in methanol after most of the air was removed by bubbling with Dry Ice. Right: Luminescence is quenched (decreased) after bubbling O_2 through the solution for 30 s. [Dan Harris.]

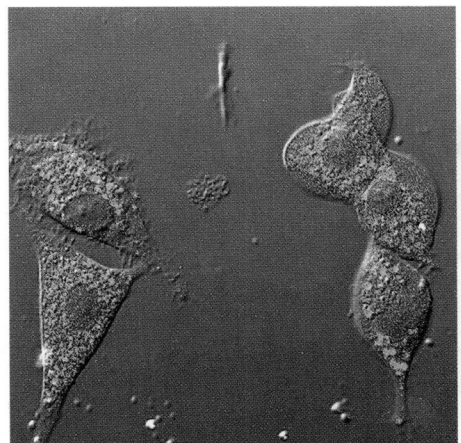

Oregon green reference dye—insensitive to O_2

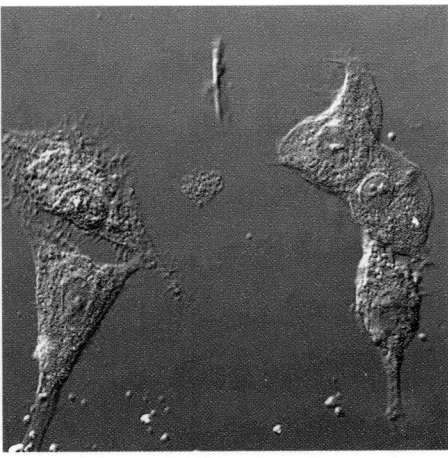

Ru(II) dye—sensitive to O_2

COLOR PLATE 20 Fluorescence from O_2-Indicator Beads Inside Living Cells (Section 18-6) Green light emitted from the dye Oregon Green is independent of local O_2 concentration. Red light emitted from tris(4,7-diphenyl-1,10-phenanthroline)ruthenium(II) chloride decreases in the presence of O_2. The ratio of intensities at the red and green wavelengths tells us the concentration of O_2 inside the cell. [Courtesy R. Kopelman, University of Michigan. From H. Xu, J. W. Aylott, R. Kopelman, T. J. Miller, and M. A. Philbert, "A Real-Time Ratiometric Method for the Determination of Molecular Oxygen Inside Living Cells Using Sol—Gel-Based Spherical Optical Nanosensors with Applications to Rat C6 Glioma," *Anal. Chem.* **2001**, *73*, 4124. Reprinted with permission © 2001, American Chemical Society.]

COLOR PLATE 21 Upconversion (Box 19-2) Low-energy green light from a 5-mW laser is converted to high-energy blue fluorescence. Does this violate the conservation of energy? Box 19-2 provides an answer. [Courtesy F. N. Castellano and T. N. Singh-Rachford, Bowling Green State University. See R. R. Islangulov, D. V. Kozlov, and F. N. Castellano, *Chem. Commun.* **2005**, 3776.]

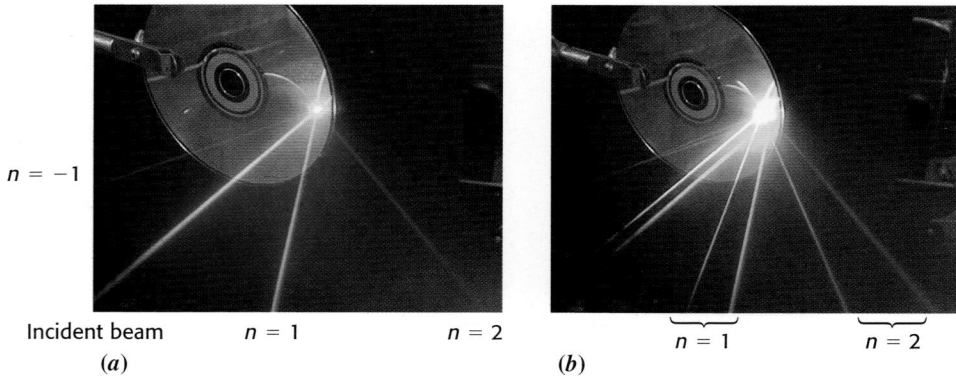

$n = -1$

Incident beam $n = 1$ $n = 2$

(a) $n = 1$ $n = 2$

(b)

COLOR PLATE 22 Laser Diffraction from a Compact Disc (Section 20-2) Grooves in an audio compact disc or a computer compact disc have a spacing of 1.6 μm. *(a)* When a red laser strikes a disk at normal incidence ($\theta = 0$ in Figure 20-7 and Equation 20-2), three diffracted beams with orders $n = +1$, $+2$, and -1 are observed. *(b)* Red and green lasers strike the disc at normal incidence. Green light has a shorter wavelength than red light, and so, according to Equation 20-2, green light is diffracted at a smaller angle (ϕ). Beams have been made visible by "fog" from liquid nitrogen. [Courtesy Joel Tellinghuisen, Vanderbilt University. See J. Tellinghuisen, *J. Chem. Ed.* **2002,** *79,* 703 and F. Wakabayashi and K. Hamada, *J. Chem. Ed.* **2006,** *83,* 56.]

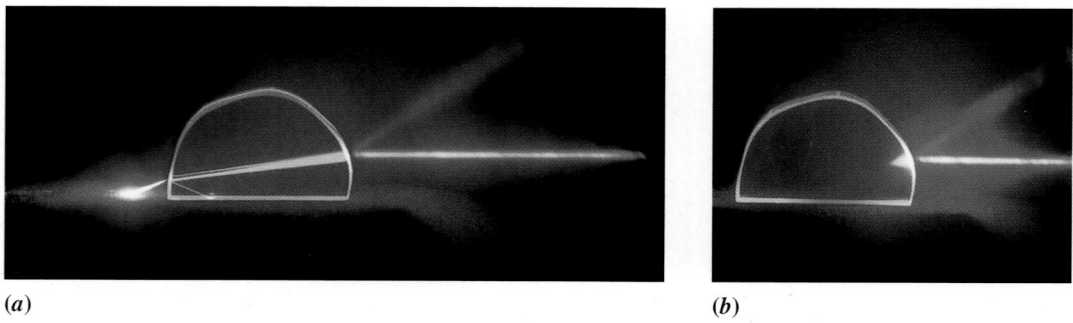

(a)   *(b)*

COLOR PLATE 23 Transmission, Reflection, Refraction, Absorption, and Luminescence (Section 20-4) *(a)* Blue-green laser is directed into a crystal of yttrium aluminum garnet doped with Er^{3+}, which emits yellow light. Light entering the crystal from the right is refracted (bent) and partially reflected at the right-hand surface of the crystal. The laser beam appears yellow inside the crystal because of luminescence from Er^{3+}. As it exits the crystal at the left side, the laser beam is refracted again, and partially reflected back into the crystal. *(b)* Same experiment, but with blue light instead of blue-green light. Blue light is absorbed by Er^{3+} and does not penetrate very far into the crystal. [Courtesy M. Seltzer, Michelson Laboratory, China Lake, CA.]

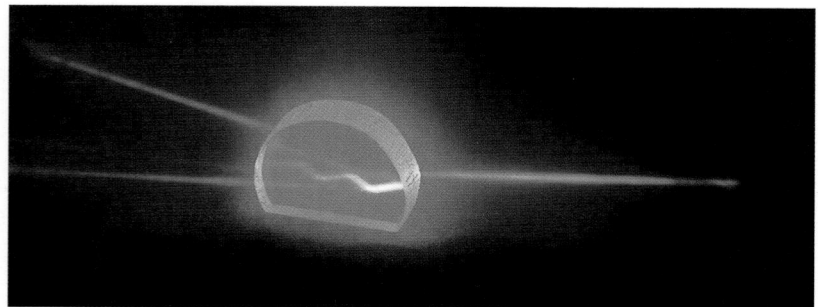

COLOR PLATE 24 Multiple Internal Reflections in a Bulk Crystal (Section 20-4) Multiple internal reflections observed when blue laser light enters a crystal of yttrium aluminum garnet doped with Ho^{3+}. The beam entering from the right is mostly reflected back into the crystal at each face, creating a zigzag pattern inside the crystal. Part of the light is transmitted out of the crystal at each face. In an optical fiber, the angle of incidence is such that the beam is totally reflected within the fiber. [Courtesy M. Seltzer, Michelson Laboratory, China Lake, CA.]

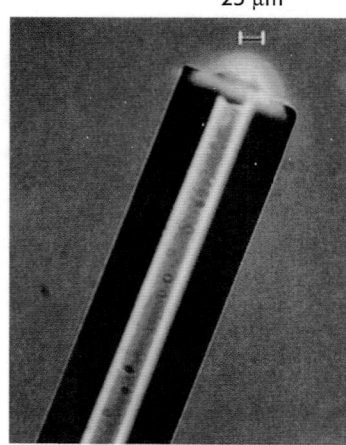
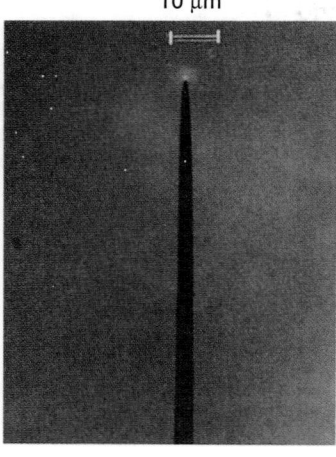

25 μm   10 μm

COLOR PLATE 25 Oxygen Optode (Section 20-4) (*a*) Sensor prepared from a 100-μm-diameter optical fiber. The active layer at the end contains tris(1,10-phenanthroline)Ru(II) chloride dissolved in polyacrylamide that is covalently bound to the fiber. Light from the fiber excites the Ru compound, which emits characteristic orange-red light that is collected with a microscope. When immersed in a sample containing O_2, emission is decreased. The decrease is a measure of O_2 concentration. (*b*) An optode with a submicrometer tip pulled from a larger fiber. This fiber can detect 10 amol of O_2. [From Z. Rosenzweig and R. Kopelman, "Development of a Submicro-meter Optical Fiber Oxygen Sensor," *Anal. Chem.* **1995,** *67,* 2650. Reprinted with permission © 1995, American Chemical Society.]

COLOR PLATE 26 Polychromator for Inductively Coupled Plasma Atomic Emission Spectrometer with One Detector for Each Element (Section 21-4) Light emitted by a sample in the plasma enters the polychromator at the right and is dispersed into its component wavelengths by a grating at the bottom of the diagram. Each different emission wavelength (shown schematically by colored lines) is diffracted at a different angle and directed to a different photomultiplier detector on the focal curve. Each detector sees only one preselected element, and all elements are measured simultaneously. [Courtesy Thermo Fisher Scientific Inc.]

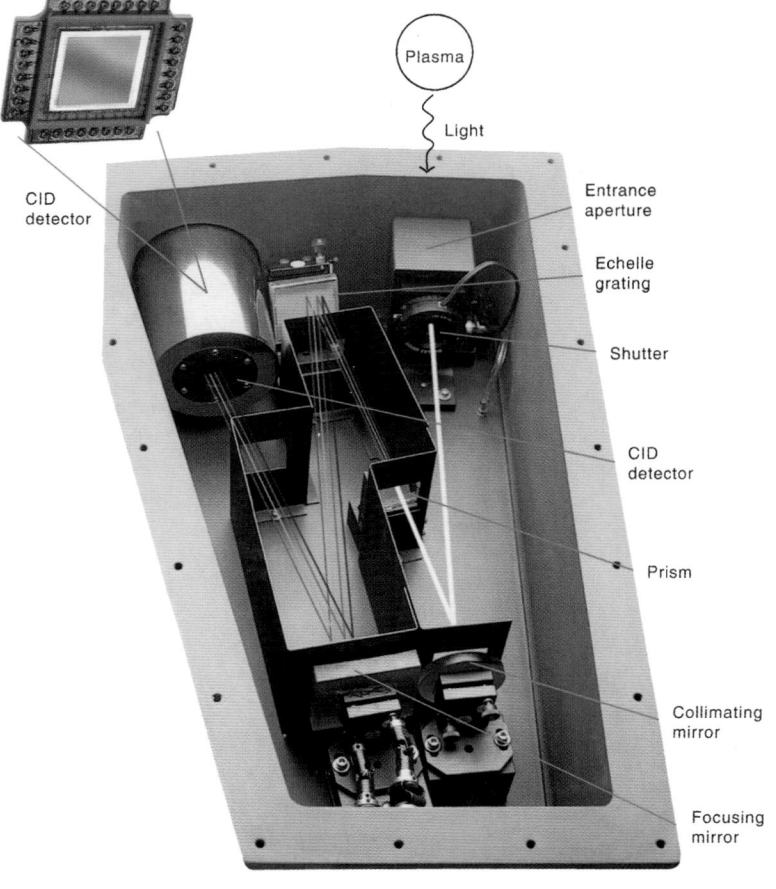

CID detector

Plasma

Light

Entrance aperture

Echelle grating

Shutter

CID detector

Prism

Collimating mirror

Focusing mirror

COLOR PLATE 27 Polychromator for Inductively Coupled Plasma Atomic Emission Spectrometer with One Detector for All Elements (Section 21-4) Light emitted by a sample in the plasma enters the polychromator at the upper right and is dispersed vertically by a prism and then horizontally by a grating. The resulting two-dimensional pattern of wavelengths from 165 to 1 000 nm is detected by a charge injection device detector with 262 000 pixels. All elements are detected simultaneously. [Courtesy Thermo Fisher Scientific Inc.]

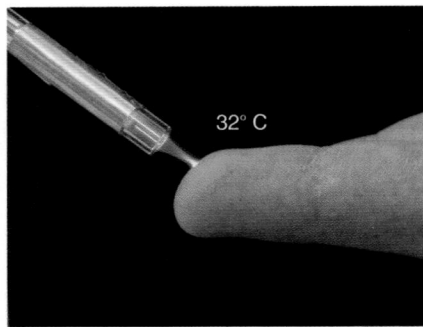

32° C

COLOR PLATE 28 Low-Temperature Plasma Ionizes Substances from Surfaces for Mass Spectral Analysis (Section 22-6) Room-temperature plasma is made by passing He, Ar, N_2, or ambient air through a glass tube with a coaxial grounded wire. The tube is wrapped on the outside with a copper cladding to which is applied a 3-kV alternating current at a frequency of 2.5 kHz and power of 1 W. Excited-state species in the plasma ionize and dislodge molecules from a surface such as skin. The surface would be placed adjacent to the inlet of a mass spectrometer to record a spectrum of the ions. There is no electrical shock to the skin. [Courtesy R. G. Cooks, Purdue University. From J. D. Harper, N. A. Charipar, C. C. Mulligan, X. Zhang, R. G. Cooks, and Z. Ouyang, "Low-Temperature Plasma Probe for Ambient Desorption Ionization," *Anal. Chem.* **2008,** *80,* 9097. Reprinted with permission © 2008, American Chemical Society.]

After addition of phase transfer agent

COLOR PLATE 29 Addition of Phase Transfer Agent Extracts Colored Anions from Water into Ether (Box 23-1) Lower panel: Vials with aqueous lower phase and diethyl ether upper phase. Colored anions are present in the aqueous phase. Upper panel: After adding trioctylmethylammonium chloride to each vial and mixing well, trioctylmethylammonium cation extracts colored anion into ether phase. [From A. J. Pezhathinal, K. Rocke, L. Susanto, D. Handke, R. Chan-Yu-King, and P. Gordon, "Colorful Chemical Demonstrations on the Extraction of Anionic Species from Water into Ether Mediated by Tricaprylylmethylammonium Chloride (Aliquat 336), a Liquid-Liquid Phase-Transfer Agent," *J. Chem. Ed.* **2006,** *83,* 1161. Reprinted with permission © 2006, American Chemical Society.]

Before addition of phase transfer agent

| on: CrO_4^{2-} | Food color | Gatorade drink | Orange II | MnO_4^- | $Cr_2O_7^{2-}$ | Orange red |

COLOR PLATE 30 Mechanism of Chromatography: Metal Ion Extraction with Ionic Liquid (Section 23-2) Left: Nd(III) and Fe(III) in 9 M HCl aqueous phase of a two-phase mixture. Right: Upon mixing, Fe(III) is extracted into the upper ionic liquid (IL) phase and Nd(III) remains in the aqueous phase. This separation was developed for the recovery of the rare earth element (Nd) from neodymium iron boride ($Nd_2Fe_{14}B$) permanent magnets, such as those found in computer hard disk drives. [Republished with permission of Royal Society of Chemistry, from T. Vander Hoogerstraete, S. Wellens, and K. Verachtert, "Removal of Transition Metals from Rare Earths by Solvent Extraction with an Undiluted Phosphonium Ionic Liquid: Separations Relevant to Rare-Earth Magnet Recycling," *Green Chem.* **2013,** *15,* 919, Figure 2. Permission conveyed through Copyright Clearance Center, Inc.]

Separation of metals occurs because Fe(III) is in the anionic form $FeCl_4^-$:

$$FeCl_4^- (aq) + (n\text{-}C_6H_{13})_3 \overset{+}{P}(n\text{-}C_{14}H_{29}) Cl^- (org) \underset{D=7\,000}{\rightleftharpoons} Cl^- (aq) + (n\text{-}C_6H_{13})_3 \overset{+}{P}(n\text{-}C_{14}H_{29}) FeCl_4^- (org)$$

Ionic liquid　　　　　　　　　　　　　　　Ion pair in ionic liquid

COLOR PLATE 31 Thin-Layer Chromatography (Section 25-1) The mixture to be separated is placed in tiny spots near the base of a plastic or glass plate coated with an adsorptive stationary phase. When the plate is placed in a shallow pool of solvent in a closed chamber, liquid migrates *up* the plate by capillary action. Different components of the mixture are carried along by the solvent to different extents, depending on how strongly they are adsorbed on the stationary phase. The stronger the adsorption, the slower a component travels. (*a*) Solvent ascends past a mixture of dyes near the bottom of the plate. (*b*) Separation achieved after solvent has ascended most of the way up the plate. [© Macmillan, Photo by Ken Karp.]

(*a*) (*b*) (*c*)

COLOR PLATE 32 Supercritical Carbon Dioxide (Box 25-3) (*a*) Liquid CO_2 in a 60-mL steel chamber at 30°C and 6.9 MPa. The red color is from a little I_2 added to the liquid to make it visible. (*b*) Beginning of the supercritical phase transition as the temperature is raised. (*c*) Single-phase supercritical carbon dioxide. [From H. Black, "Supercritical Carbon Dioxide: The 'Greener' Solvent," *Environ. Sci. Technol.* **1996,** *30,* 124A. Photos courtesy D. Pesiri and W. Tumas, Los Alamos National Laboratory. Reprinted with permission © 1996, American Chemical Society.]

COLOR PLATE 33 Velocity Profiles for Hydrodynamic and Electroosmotic Flow (Section 26-6) A fluorescent dye was imaged inside a capillary tube at times 0, 66, and 165 ms after initiating flow. The highest concentration of dye is represented by blue and the lowest concentration is red in these images in which different colors are assigned to different fluorescence intensities. [From P. H. Paul, M. G. Garguilo, and D. J. Rakestraw, "Imagining of Pressure and Electrokinetically Driven Flows through Open Capillaries," *Anal. Chem.* **1998,** *70,* 2459. Reprinted with permission © 1998, American Chemical Society. See also D. Ross, T. J. Johnson, and L. E. Locascio, *Anal. Chem.* **2001,** *73,* 2509.]

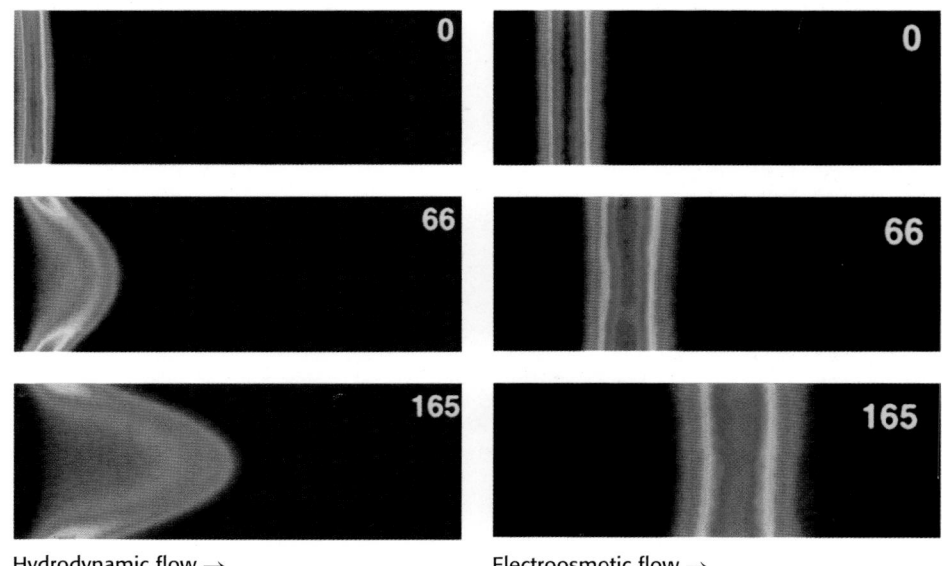

Hydrodynamic flow →
100-μm-diameter capillary

Electroosmotic flow →
75-μm-diameter capillary

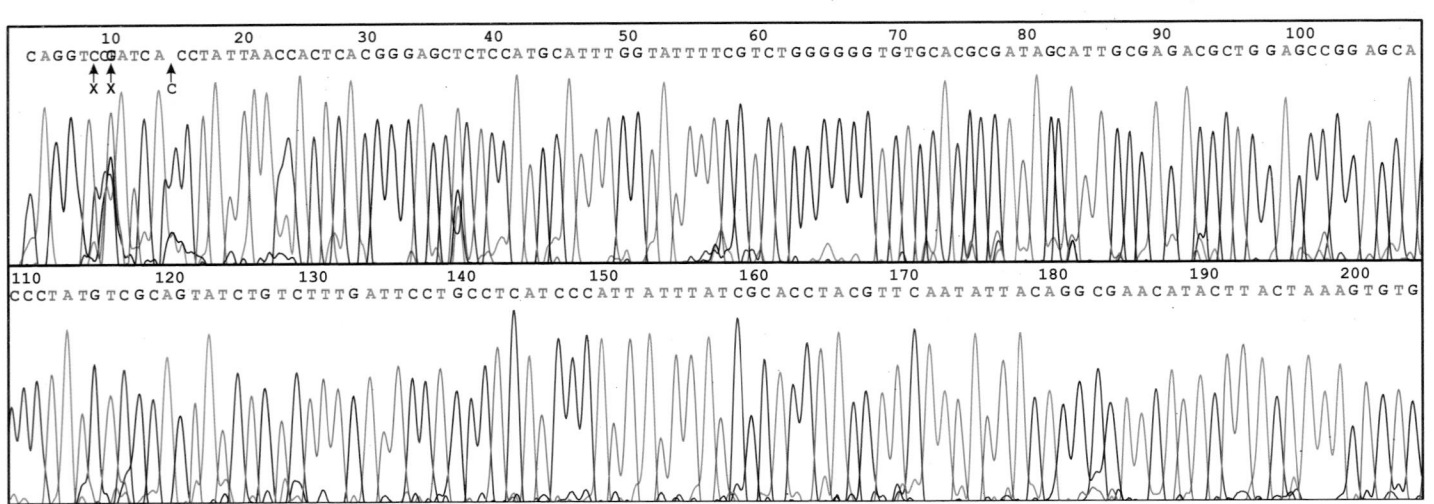

COLOR PLATE 34 **DNA Sequencing by Capillary Gel Electrophoresis with Fluorescence Detection (Section 26-7)** Portion of DNA nucleotide base sequence made with a microfluidic device ("lab-on-a-chip") capable of reading a length of 365 bases with 99% accuracy. Successive peaks correspond to lengths of DNA with one more base. DNA strands terminated in each of the four different bases, A, T, C, and G, are tagged with a different fluorescent label that identifies the terminal base when it passes through a fluorescence detector. Different lengths of DNA are separated by sieving through an 18-cm-long electrophoresis channel containing polyacrylamide gel with 6 M urea to stabilize single strands. Injected sample volume is 30 nL, which contains 100 amol (60 million molecules) of DNA. [From R. G. Blazej, P. Kumaresan, S. A. Cronier, and R. A. Mathies, "Inline Injection Microdevice for Attomole-Scale Sanger DNA Sequencing," *Anal. Chem.* **2007,** *79,* 4499. Reprinted with permission © 2007, American Chemical Society.]

(a)   (b)   (c)

COLOR PLATE 35 **Colloids and Dialysis (Demonstration 27-1)** (*a*) Colloidal Fe(III) (left) and ordinary aqueous Fe(III) (right). (*b*) Dialysis bags containing colloidal Fe(III) (left) and a solution of Cu(II) (right) immediately after placement in flasks of water. (*c*) After 24 h of dialysis, the Cu(II) has diffused out and is dispersed uniformly between the bag and the flask, but the colloidal Fe(III) remains inside the bag. [Dan Harris.]

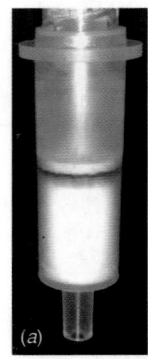

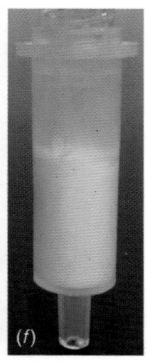

(a) Load column with 1 mL aqueous sample

(b) Begin eluting with 4 vol% 2-propanol

(c) After 3 mL of 4 vol% 2-propanol

(d) After additional 3 mL of 4 vol% 2-propanol

(e) After 3 mL of 5 vol% 2-propanol

(f) After 3 mL of CH₃OH

COLOR PLATE 36 **Solid-Phase Extraction (Section 28-3)** 1 mL of water containing FD&C Blue No. 1 dye and FD&C Red No. 40 dye was passed through a solid-phase extraction cartridge containing 500 mg of C_{18}-silica. Prior to applying dye solution, the cartridge had been washed with 5 mL of 3 mM HCl in CH_3OH, followed by 5 mL of deionized H_2O. Three aqueous 2-propanol washes eluted all of the red dye and one methanol wash eluted all of the blue dye. [From Karyn M. Usher, Metropolitan State University, Saint Paul, Minnesota. For a student experiment, see H. F. Rossi, III, J. Rizzo, D. C. Zimmerman, and K. M. Usher, *J. Chem. Ed.* **2012,** *89,* 1551.]

19 Applications of Spectrophotometry

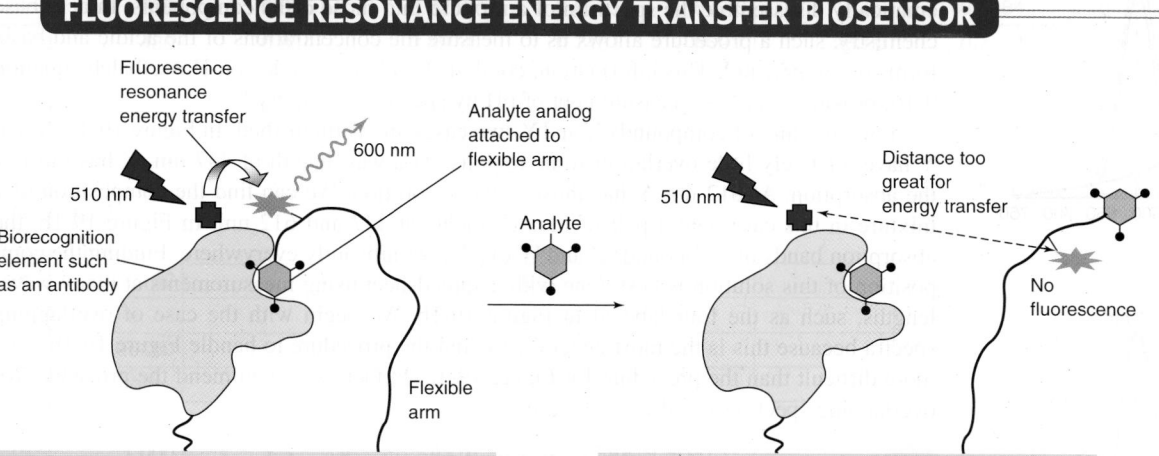

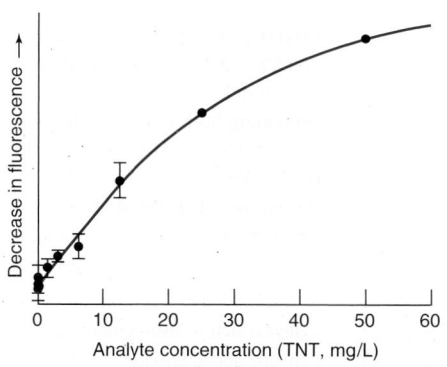

Response of biosensor to TNT is a *decrease* in fluorescence with increasing concentration of analyte. [Data from I. L. Medintz, E. R. Goldman, M. E. Lassman, A. Hayhurst, A. W. Kusterbeck, and J. R. Deschamps, "Self-Assembled TNT Biosensor Based on Modular Multifunctional Surface-Tethered Components," *Anal. Chem.* **2005**, *77*, 365.]

A *biosensor* is a device that uses biological components such as enzymes, antibodies, or DNA, in combination with electric, optical, or other signals, to achieve a selective response to one analyte.[1] The biosensor in the diagram consists of two surface-tethered components. The biorecognition element can be an antibody, DNA, RNA, or carbohydrate with specific affinity for an analyte. A structural analog of the analyte is bound to a flexible arm adjacent to the recognition element. In the absence of analyte, the tethered analog binds to the recognition element.

A chromophore ▦ that efficiently absorbs radiant energy is attached to the recognition element adjacent to the recognition site. The fluorescent chromophore ✷ is attached to the flexible arm adjacent to the analyte structural analog. At the left side of the figure, absorbing and emitting chromophores are close to each other. Light with a wavelength of 510 nm is absorbed by ▦ and efficiently transferred through space to ✷, which fluoresces strongly at 600 nm. This *fluorescence resonance energy transfer* decreases with the sixth power of the distance between the donor and acceptor.[2,3]

When analyte is added, it displaces the tethered analog from the biorecognition element. The higher the concentration of analyte, the more displacement occurs. When the analog is displaced from the binding site, ✷ and ▦ are no longer close enough for energy transfer, and fluorescence is decreased. In the graph, analyte is trinitrotoluene (the explosive, TNT) and the detection limit is 0.1 mg/L (0.1 ppm). After a measurement, the sensor is washed to remove analyte and then reused for more analyses.

This chapter describes applications of absorption and emission of electromagnetic radiation in chemical analysis. *Flow injection analysis* is introduced as an important analytical method that commonly uses absorption or emission to measure analytical response. We use Excel Solver and spreadsheet matrix operations as powerful tools for numerical analysis.

19-1 Analysis of a Mixture

Consider two chemical species X and Y whose visible absorption spectra are shown in Figure 19-1a. The absorbance at any wavelength of a solution containing X and Y is the sum of absorbances of X and Y at that wavelength, as shown by the dashed line in Figure 19-1a. For

mixture at any wavelength is the sum of absorbances of each component at that wavelength. For wavelengths λ' and λ'',

At wavelength $\lambda' = 447$ nm: $\qquad A' = \varepsilon_X' b[X] + \varepsilon_Y' b[Y]$

At wavelength $\lambda'' = 612$ nm: $\qquad A'' = \varepsilon_X'' b[X] + \varepsilon_Y'' b[Y]$ (19-5)

where the ε values apply to each species at each wavelength. The absorptivities of X and Y at wavelengths λ' and λ'' must be measured in separate experiments.

We can solve Equations 19-5 for the two unknowns [X] and [Y]. The result is

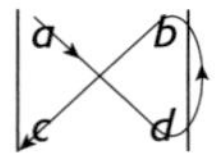

Analysis of a mixture when spectra are resolved:
$$[X] = \frac{\begin{vmatrix} A' & \varepsilon_Y' b \\ A'' & \varepsilon_Y'' b \end{vmatrix}}{\begin{vmatrix} \varepsilon_X' b & \varepsilon_Y' b \\ \varepsilon_X'' b & \varepsilon_Y'' b \end{vmatrix}} \qquad [Y] = \frac{\begin{vmatrix} \varepsilon_X' b & A' \\ \varepsilon_X'' b & A'' \end{vmatrix}}{\begin{vmatrix} \varepsilon_X' b & \varepsilon_Y' b \\ \varepsilon_X'' b & \varepsilon_Y'' b \end{vmatrix}}$$ (19-6)

In Equations 19-6, each symbol $\begin{vmatrix} a & b \\ c & d \end{vmatrix}$ is a *determinant*. It is a shorthand way of writing the product $(a \times d) - (b \times c)$. Thus, $\begin{vmatrix} 1 & 2 \\ 3 & 4 \end{vmatrix}$ means $(1 \times 4) - (2 \times 3) = -2$.

$\begin{vmatrix} a & b \\ c & d \end{vmatrix} = (a \times d) - (b \times c)$

EXAMPLE   Analysis of a Mixture, Using Equations 19-6

The molar absorptivities of compounds X and Y (not the ones in Figure 19-1a) were measured with pure samples of each:

λ (nm)	ε (M^{-1} cm^{-1})	
	X	Y
272	16 400	3 870
327	3 990	6 420

A mixture of compounds X and Y in a 1.000-cm cell had an absorbance of 0.957 at 272 nm and 0.559 at 327 nm. Find the concentrations of X and Y in the mixture.

Solution Using Equations 19-6 and setting $b = 1.000$, we find

$$[X] = \frac{\begin{vmatrix} 0.957 & 3\,870 \\ 0.559 & 6\,420 \end{vmatrix}}{\begin{vmatrix} 16\,400 & 3\,870 \\ 3\,990 & 6\,420 \end{vmatrix}} = \frac{(0.957)(6\,420) - (3\,870)(0.559)}{(16\,400)(6\,420) - (3\,870)(3\,990)} = 4.43 \times 10^{-5} \text{ M}$$

$$[Y] = \frac{\begin{vmatrix} 16\,400 & 0.957 \\ 3\,990 & 0.559 \end{vmatrix}}{\begin{vmatrix} 16\,400 & 3\,870 \\ 3\,990 & 6\,420 \end{vmatrix}} = 5.95 \times 10^{-5} \text{ M}$$

TEST YOURSELF Find the concentration of [X] if the absorbances are 0.700 at 272 nm and 0.550 at 327 nm. (**Answer:** 2.63×10^{-5} M)

To analyze a mixture of two compounds, it is necessary to measure absorbance at two wavelengths and to know ε at each wavelength for each compound. Similarly, a mixture of n components can be analyzed by making n absorbance measurements at n wavelengths.

Solving Simultaneous Linear Equations with Excel

Excel solves systems of linear equations with a single statement. If the following matrix mathematics below is unfamiliar to you, disregard it. The important result is the template in Figure 19-4 for solving simultaneous equations. You can use this template by following the instructions in the last paragraph of this section, even if the math is not familiar.

FIGURE 19-4 Solving simultaneous linear equations with Excel.

	A	B	C	D	E	F	G
1	Solving Simultaneous Linear Equations with Excel Matrix Operations						
2							
3	Wavelength	Coefficient Matrix		Absorbance		Concentrations	
4	(nm)	(M^{-1}cm^{-1})		of unknown		in mixture (M)	
5	272	16440	3870	0.957		4.41782E-05	← [X]
6	327	3990	6420	0.559		5.96151E-05	← [Y]
7		K		A		C	
8							
9	1. Enter matrix of coefficients εb in cells B5:C6						
10	2. Enter absorbance of unknown at each wavelength (cells D5:D6)						
11	3. Highlight block of blank cells required for solution (F5 and F6)						
12	4. Type the formula "= MMULT(MINVERSE(B5:C6),D5:D6)"						
13	5. Press CONTROL+SHIFT+ENTER on a PC or COMMAND+RETURN on a Mac						
14	6. Behold! The answer appears in cells F5 and F6						

The simultaneous equations of the preceding example are

$$A' = \varepsilon'_X b[X] + \varepsilon'_Y b[Y]$$
$$A'' = \varepsilon''_X b[X] + \varepsilon''_Y b[Y]$$
$$\Rightarrow$$
$$0.957 = 16\,440[X] + 3\,870[Y]$$
$$0.559 = 3\,990[X] + 6\,420[Y]$$

which can be written in matrix notation in the form

$$\begin{bmatrix} 0.957 \\ 0.559 \end{bmatrix} = \begin{bmatrix} 16\,400 & 3\,870 \\ 3\,990 & 6\,420 \end{bmatrix} \begin{bmatrix} [X] \\ [Y] \end{bmatrix} \qquad (19\text{-}7)$$
$$\mathbf{A} \quad = \quad\quad \mathbf{K} \quad\quad\quad \mathbf{C}$$

K is the *matrix* of molar absorptivity times pathlength, εb. **A** is the matrix of absorbance of the unknown. A matrix such as **A**, with only one column or one row, is called a *vector*. **C** is the vector of unknown concentrations.

A matrix **K**$^{-1}$, called the *inverse* of **K**, is such that the products **K K**$^{-1}$ or **K**$^{-1}$ **K** are equal to a unit matrix with ones on the diagonal and zero elsewhere.[5] We can solve Equation 19-7 for the concentration vector, **C**, by multiplying both sides of the equation by **K**$^{-1}$:

$$\mathbf{KC} = \mathbf{A}$$
$$\underbrace{\mathbf{K}^{-1}\mathbf{KC}}_{= \mathbf{C}} = \mathbf{K}^{-1}\mathbf{A}$$

To solve simultaneous equations, find the inverse matrix **K**$^{-1}$ *and multiply it times* **A**. *The product is* **C**, *the concentrations in the unknown mixture.*

In Figure 19-4, we enter the wavelengths in column A just to keep track of information. We will not use these wavelengths for computation. Enter the products εb for pure X in column B and εb for pure Y in column C. The array in cells B5:C6 is the matrix **K**. The Excel function MINVERSE(B5:C6) gives the inverse matrix, **K**$^{-1}$. The function MMULT(matrix 1, matrix 2) gives the product of two matrices (or a matrix and a vector). The concentration vector, **C**, is equal to **K**$^{-1}$**A**, which we get with the single statement

$$= \text{MMULT}(\underbrace{\text{MINVERSE(B5:C6)}}_{\mathbf{K}^{-1}}, \underbrace{\text{D5:D6}}_{\mathbf{A}})$$

To use the template in Figure 19-4, enter the coefficients, εb, measured from the pure compounds, in cells B5:C6. Enter the absorbance of the unknown in cells D5:D6. Highlight cells F5:F6 and type the formula "=MMULT(MINVERSE(B5:C6),D5:D6)". Press CONTROL+SHIFT+ENTER on a PC or COMMAND(⌘)+RETURN on a Mac. The concentrations [X] and [Y] in the mixture now appear in cells F5:F6.

The product **K**$^{-1}$**KC** is just **C**:
$$\underbrace{\begin{bmatrix} 1 & 0 \\ 0 & 1 \end{bmatrix}}_{\mathbf{K}^{-1}\mathbf{K}} \underbrace{\begin{bmatrix} [X] \\ [Y] \end{bmatrix}}_{\mathbf{C}} = \underbrace{\begin{bmatrix} [X] \\ [Y] \end{bmatrix}}_{\mathbf{C}}$$

Procedure for solving simultaneous equations with Excel

Isosbestic Points

Often one absorbing species, X, is converted into another absorbing species, Y, in the course of a chemical reaction. This transformation leads to an obvious behavior shown in Figure 19-5. If the spectra of pure X and pure Y cross each other at any wavelength, then every spectrum

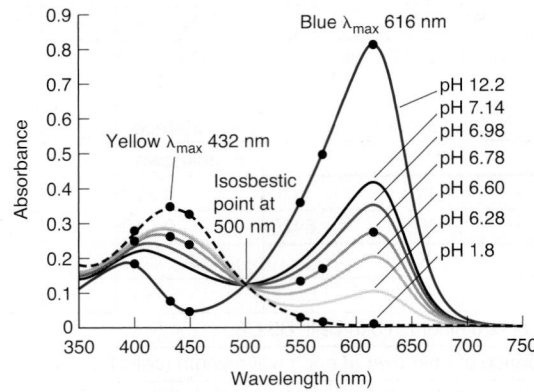

FIGURE 19-5 Absorption spectrum of 10 μM bromothymol blue at different pH values. Absorbance at points shown by dots is used in Problem 19-8. [Data from E. Klotz, R. Doyle, E. Gross, and B. Mattson, "The Equilibrium Constant for Bromothymol Blue," *J. Chem. Ed.* **2011,** *88,* 637.]

recorded during this chemical reaction will cross at that same point, called an **isosbestic point.** *An isosbestic point is good evidence that only two principal species are present.*[6]

Consider the indicator bromothymol blue, which changes between yellow (HIn) and blue (In⁻) near pH 7.1:

Bromothymol blue (HIn)
yellow

In⁻
blue

Because the spectra of HIn and In⁻ (at the same concentration) happen to cross at 500 nm, all spectra in Figure 19-5 cross at this point. If the spectra of HIn and In⁻ crossed at several points, there would be several isosbestic points.

To see why there is an isosbestic point, we write an equation for the absorbance of the solution at 500 nm:

$$A^{500} = \varepsilon^{500}_{HIn}b[HIn] + \varepsilon^{500}_{In}b[In^-] \tag{19-8}$$

But because the spectra of pure HIn and pure In⁻ (at the same concentration) cross at 500 nm, ε^{500}_{HIn} must be equal to ε^{500}_{In}. Setting $\varepsilon^{500}_{HIn} = \varepsilon^{500}_{In} = \varepsilon^{500}$, we can factor Equation 19-8:

$$A^{500} = \varepsilon^{500}b([HIn] + [In^-]) \tag{19-9}$$

An isosbestic point occurs when $\varepsilon_X = \varepsilon_Y$ and $[X] + [Y]$ is constant.

In Figure 19-5, all solutions contain the same total concentration $[HIn] + [In^-]$. Only the pH varies. The sum of concentrations in Equation 19-9 is constant, so A^{500} is constant.

19-2 Measuring an Equilibrium Constant

To measure an equilibrium constant, we measure concentrations (actually activities) of species at equilibrium. Spectrophotometry can be used for this purpose. We examine a graphical procedure called the Scatchard plot that is widely used in biochemistry and then give a more powerful procedure using Excel Solver.

Absorbance is proportional to *concentration* (not activity), so concentrations must be converted into activities to get true equilibrium constants.

Scatchard Plot

Suppose that species P (such as a protein) and ligand X react to form PX.

$$P + X \rightleftharpoons PX \tag{19-10}$$

Neglecting activity coefficients, we can write

$$K = \frac{[PX]}{[P][X]} \tag{19-11}$$

Consider a series of solutions in which increments of X are added to a constant amount of P. Letting P_0 be the total concentration of P (in the forms P and PX), we can write

Clearing the cobwebs from your brain, you realize that Equation 19-12 is a mass balance.

$$[P] = P_0 - [PX] \tag{19-12}$$

Now the equilibrium expression, Equation 19-11, can be rearranged as follows:

$$\frac{[PX]}{[X]} = K[P] = K(P_0 - [PX])\qquad(19\text{-}13)$$

A graph of $[PX]/[X]$ versus $[PX]$ with slope $-K$ is called a **Scatchard plot**[7] (Figure 19-6).

If we know $[PX]$, we can find $[X]$ with the mass balance

$$X_0 = [\text{total X}] = [PX] + [X]$$

To measure $[PX]$, we might use spectrophotometric absorbance. Suppose that P and PX each have some absorbance at wavelength λ, but X has no absorbance at this wavelength. For simplicity, let all measurements be made in a cell of pathlength 1.000 cm so that we can omit b ($=1.000$ cm) when writing Beer's law.

The absorbance at some wavelength is the sum of absorbances of PX and P:

$$A = \varepsilon_{PX}[PX] + \varepsilon_P[P]$$

Substituting $[P] = P_0 - [PX]$, we can write

$$A = \varepsilon_{PX}[PX] + \underbrace{\varepsilon_P P_0}_{A_0} - \varepsilon_P[PX]\qquad(19\text{-}14)$$

But $\varepsilon_P P_0$ is A_0, the initial absorbance before any X is added. Therefore,

$$A = [PX](\varepsilon_{PX} - \varepsilon_P) + A_0 \Rightarrow [PX] = \frac{\Delta A}{\Delta\varepsilon}\qquad(19\text{-}15)$$

where $\Delta\varepsilon = \varepsilon_{PX} - \varepsilon_P$ and ΔA ($= A - A_0$) is the observed absorbance after each addition of X minus the initial absorbance.

Substituting $[PX]$ from Equation 19-15 into Equation 19-13 gives

Scatchard equation:
$$\frac{\Delta A}{[X]} = K\Delta\varepsilon P_0 - K\Delta A\qquad(19\text{-}16)$$

A graph of $\Delta A/[X]$ versus ΔA should be a straight line with a slope of $-K$. Absorbance measured while P is titrated with X can be used to find K for the reaction of X with P.

Two cases commonly arise in the application of Equation 19-16. If K is small, then large concentrations of X are needed to produce PX. Therefore, $X_0 \gg P_0$, and $[X] \approx X_0$. Alternatively, if K is not small, then $[X] \neq X_0$, and $[X]$ must be measured, either at another wavelength or by measuring a different physical property.

Errors in a Scatchard plot could be substantial. We define the fraction of saturation as

$$\text{Fraction of saturation} = S = \frac{[PX]}{P_0}\qquad(19\text{-}17)$$

The most accurate data are obtained for $0.2 \leq S \leq 0.8$.[8] A range representing ~75% of the total saturation curve should be measured before concluding that equilibrium (19-10) is obeyed. People have made mistakes by exploring too little of the binding curve.

A More General Procedure with Excel Solver

Riboflavin, also called vitamin B2, is essential for energy production and biosynthesis in many types of cells. Riboflavin-binding protein from egg white is thought to provide vitamin B2 to the embryo and to deny vitamin B2 to bacteria in the egg. Neutral red is a dye with some structural similarity to riboflavin. Figure 19-7 shows spectrophotometric data for binding of neutral red to riboflavin-binding protein. Our goal is to use these data to find the equilibrium constant K for formation of the protein-neutral red complex.

$$\underset{\text{Protein}}{P} + \underset{\text{Neutral red}}{X} \rightleftharpoons \underset{\text{Complex}}{PX} \qquad K = \frac{[PX]}{[P][X]}\qquad(19\text{-}18)$$

As increments of protein (P) are added to a fixed quantity of ligand (X = neutral red), absorption of X at 450 nm decreases while product (PX) absorption at 545 nm increases. Both absorptions are from the neutral red chromophore free in solution or bound to the protein. Protein P without X has no visible absorption. Cells B16:B26 in Figure 19-8 give the total concentration of ligand ($X_0 = [X] + [PX]$) and cells C16:C26 give the total concentration of

---

A **Scatchard plot** is a graph of $[PX]/[X]$ versus $[PX]$. The slope is $-K$.

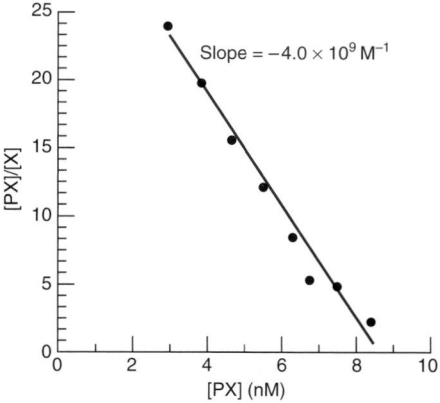

FIGURE 19-6 Scatchard plot for binding of antigen (X) to antibody (P). The antibody binds the explosive, trinitrotoluene (TNT). The antigen is a fluorescent analog of TNT. From the slope, the binding constant for the reaction $P + X \rightleftharpoons PX$ is $K = 4.0 \times 10^9\ M^{-1}$. [Data from Figure 4 of A. Bromberg and R. A. Mathies, "Homogeneous Immunoassay for Detection of TNT on a Capillary Electrophoresis Chip," *Anal. Chem.* **2003,** *75,* 1188.]

Equation 19-13 can be recast as $S/[X] = K(1 - S)$. Plot $S/[X]$ versus S.

Riboflavin

Neutral red
(an analog of riboflavin)

FIGURE 19-7 Spectrophotometric changes observed when ten successive 10 μL aliquots of 0.310 mM riboflavin binding protein are added to 1.00 mL of 8.4 μM neutral red in 0.05 M tris-HCl buffer (Table 9-2) at pH 9.0. [Data from P. Chenprakhon, J. Sucharitakul, B. Panijpan, and P. Chaiyen, "Measuring Binding Affinity of Protein-Ligand Interactions Using Spectrophotometry," *J. Chem. Ed.* **2010**, *87*, 829.]

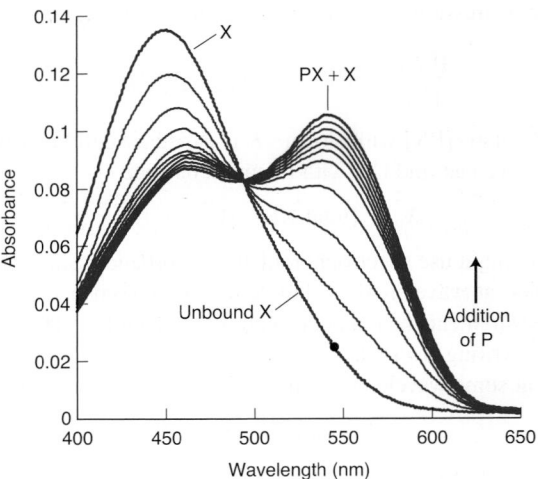

protein ($P_0 = [P] + [PX]$). Observed absorbance (A_{obs}) at the PX peak wavelength of 545 nm appears in cells D16:D26.

The final curve in Figure 19-7 represents a mixture of PX, P, and unreacted X. We can calculate the molar absorptivity (ε_X) for X at 545 nm from the absorbance of pure X shown by the black dot in Figure 19-7. With a pathlength of $b = 1.000$ cm, $\varepsilon_X = A_{545}/bX_0 = (0.026\ 4)/(1.000\ \text{cm} \times 8.4\ \mu\text{M}) = 3.143 \times 10^3\ \text{M}^{-1}\ \text{cm}^{-1}$. For subsequent calculations, we retain extra, insignificant digits. We do not know the molar absorptivity (ε_{PX}) of product PX because it is found in an unknown mixture with X.

	A	B	C	D	E	F	G	H	I	J
1	Finding K for Binding of Neutral Red (X) to Riboflavin-Binding Protein (P)									
2	10 μL aliquots of 310 μM protein were added to 1.00 mL of 8.4 μM neutral red									
3	X_o = total concentration of neutral red									
4	P_o = total concentration of riboflavin-binding protein									
5	K = [PX]/([P][X])									
6	$[PX] = [K(P_o+X_o)+1 \pm \text{sqrt}\{[K(P_o+X_o)+1]^2 - 4K^2X_oP_o\}]/(2K)$					← (Eq. A)				
7	Absorbance at 545 nm = A = $\varepsilon_X[X]+\varepsilon_{PX}$ [PX]					← (Eq. B)				
8	ε_X = (initial A_{545})/(initial X_o) = 0.0264/8.4 μM =				3143	$\text{M}^{-1}\text{cm}^{-1}$				
9	Parameters to find with Solver:									
10	K =	1.91E+05		initial estimate =	4.3E+05					
11	ε_{PX} =	1.63E+04		initial estimate =	1.5E+04					
12										
13	Protein				[PX] (M)					
14	added	X_o	P_o	A_{obs}	Eq. A using	[X] =	[P] =	A_{calc}	$(A_{obs}-A_{calc})^2$	$[PX]/X_o$
15	(μL)	(M)	(M)	at 545 nm	minus sign	X_o-[PX]	P_o-[PX]	Eq. B		fraction of
16	0	8.400E-06	0	0.0264	0	8.400E-06	0	0.0264		X reacted
17	10	8.317E-06	3.069E-06	0.0452	1.712E-06	6.605E-06	1.357E-06	0.0487	1.233E-05	0.000
18	20	8.235E-06	6.324E-06	0.0645	3.124E-06	5.111E-06	3.200E-06	0.0671	6.605E-06	0.206
19	30	8.155E-06	9.029E-06	0.0808	3.997E-06	4.158E-06	5.032E-06	0.0783	6.118E-06	0.379
20	40	8.077E-06	1.192E-05	0.0897	4.686E-06	3.391E-06	7.237E-06	0.0872	6.377E-06	0.490
21	50	8.000E-06	1.476E-05	0.0940	5.174E-06	2.826E-06	9.588E-06	0.0934	4.013E-07	0.580
22	60	7.925E-06	1.755E-05	0.0970	5.521E-06	2.404E-06	1.203E-05	0.0977	4.96E-07	0.647
23	70	7.850E-06	2.028E-05	0.0995	5.769E-06	2.081E-06	1.451E-05	0.1007	1.516E-06	0.697
24	80	7.778E-06	2.296E-05	0.1018	5.948E-06	1.830E-06	1.702E-05	0.1029	1.139E-06	0.735
25	90	7.706E-06	2.560E-05	0.1041	6.076E-06	1.630E-06	1.952E-05	0.1043	5.463E-08	0.788
26	100	7.636E-06	2.818E-05	0.1066	6.169E-06	1.467E-06	2.201E-05	0.1053	1.602E-06	0.808
27								sum =	3.664E-05	

FIGURE 19-8 Spreadsheet using Solver to find the equilibrium constant in Equation 19-18 from the spectrophotometric data in Figure 19-7.

Our procedure will be to *estimate* K and ε_{PX} and to calculate the resulting absorbance from Beer's law for a mixture of PX and X:

$$A_{calc} = \varepsilon_{PX}[X] + \varepsilon_{PX}[PX] \qquad (19\text{-}19)$$

where the pathlength $b = 1.000$ cm is omitted for simplicity. We then use Excel Solver to adjust the values of K and ε_{PX} until the calculated absorbance at 545 nm is as close as possible to the observed absorbance for all ten additions of P to X in Figure 19-7. The least-squares criterion is to minimize the sum of squares of the differences between observed and calculated absorbance $\Sigma(A_{obs} - A_{calc})^2$ in cell I27 of Figure 19-8.

At any point in the titration, the mass balances say that $[X] = X_0 - [PX]$ and $[P] = P_0 - [PX]$. Substitute these identities into the equilibrium expression to find an equation for $[PX]$:

Mass balances for P + X $\rightleftharpoons$ PX
$X_0 = [X] + [PX]$
$P_0 = [P] + [PX]$

$$K = \frac{[PX]}{[P][X]} = \frac{[PX]}{(P_0 - [PX])(X_0 - [PX])} \qquad (19\text{-}20)$$

Equation 19-20 is a quadratic equation for $[PX]$ whose two solutions are

$$[PX] = \frac{[K(P_0 + X_0) + 1] \pm \sqrt{[K(P_0 + X_0) + 1]^2 - 4K^2 X_0 P_0}}{2K} \qquad (19\text{-}21)$$

Equation 19-21 is labeled "Eq. A" in row 6 of the spreadsheet in Figure 19-8. The solution with the negative sign in the numerator is the one we want. The solution with the positive sign gives $[PX] > P_0$, which is not possible. $[PX]$ is computed in cells E16:E26 from K and ε_{PX} in cells B10:B11.

How can we estimate K and ε_{PX} to begin the calculations? We usually do not need very accurate estimates for Solver to refine. Just *suppose* that the final solution made with 100 μL of P has 90% of X bound to P. In row 26 of the spreadsheet, the known formal concentrations are $X_0 = 7.64$ μM and $P_0 = 28.18$ μM. If 90% of X is in the form PX, then $[PX] = 6.89$ μM and $[X] = 0.75$ μM, leaving $[P] = P_0 - [PX] = 21.29$ μM. Our estimate for $K = [PX]/([P][X])$ is $[6.89\ \mu M]/([21.29\ \mu M][0.75\ \mu M]) = 4.3 \times 10^5$. From the estimates $[X] = 0.75$ μM and $[P] = 21.29$ μM, we estimate ε_{PX} from Beer's law:

$$A_{obs} = \varepsilon_{PX}b[X] + \varepsilon_{PX}b[PX])$$
$$0.106\ 6 = (3.143 \times 10^3\ M^{-1}\ cm^{-1})(1\ cm)[0.75\ \mu M] + \varepsilon_{PX}(1\ cm)[6.89\ \mu M])$$
$$\Rightarrow \varepsilon_{PX} = 1.5 \times 10^4\ M^{-1}\ cm^{-1}$$

In Figure 19-8, enter the estimated values $K = 4.3 \times 10^5$ and $\varepsilon_{PX} = 1.5 \times 10^4\ M^{-1}\ cm^{-1}$ in cells B10:B11. With these values, the spreadsheet calculates all quantitites in cells E16:J26 and the least squares sum $\Sigma(A_{obs} - A_{calc})^2$ in cell I27. In Excel 2010, go to the Data ribbon and select Solver in the Analysis section. Set Objective I27 To Min By Changing Variable Cells B10:B11. The solving method is the default GRG Nonlinear. In Options, set Constraint Precision to 1E-14 and check Use Automatic Scaling. Click OK in Options and then click Solve. In a flash, Solver finds the values $K = 1.91 \times 10^5$ and $\varepsilon_{PX} = 1.63 \times 10^4\ M^{-1}\ cm^{-1}$ in cells B10:B11. Figure 19-9 shows that these values of K and ε_{PX} provide a good fit to the observed

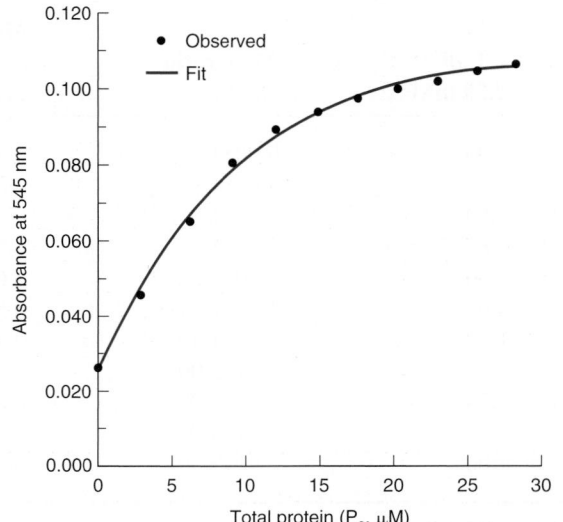

FIGURE 19-9 Observed and calculated changes in absorbance at 454 nm from using Solver to find K and ε_{PX} in Figure 19-8.

absorbance at 545 nm. Cell J26 tells us that the fraction of X that reacted with 100 μL of P was 0.808. We had guessed that this fraction was 0.9 to make initial estimates of K and ε_{PX}.

Solver allowed us to find the molar absorptivity of the product PX and the equilibrium constant. We could not have used the Scatchard plot without knowing ε_{PX}. Solver provides a versatile method for finding an equilibrium constant from experimental data. Column J of the spreadsheet tells us that the experimental data span the fraction of reaction from 0.2 to 0.8, which is highly desirable.

19-3 The Method of Continuous Variation

Suppose that several complexes can form between species P and X:

$$P + X \rightleftharpoons PX$$
$$P + 2X \rightleftharpoons PX_2$$
$$P + 3X \rightleftharpoons PX_3$$

If one complex (say, PX_2) predominates, the **method of continuous variation** (also called *Job's method*[9]) allows us to identify the stoichiometry of the predominant complex.

The classic procedure is to mix P and X and dilute to constant volume so that the total concentration $[P] + [X]$ is constant. For example, 2.50 mM solutions of P and X could be mixed as shown in Table 19-1 to give various X:P ratios, but constant total concentration. The absorbance of each solution is measured, typically at λ_{max} for the complex, and a graph is made showing *corrected* absorbance (defined in Equation 19-22) versus mole fraction of X. *Maximum absorbance is reached at the composition corresponding to the stoichiometry of the predominant complex.*

Corrected absorbance is the measured absorbance minus the absorbance that would be produced by free P and free X alone:

$$\text{Corrected absorbance} = \text{measured absorbance} - \varepsilon_P b P_T - \varepsilon_X b X_T \qquad \text{(19-22)}$$

where ε_P and ε_X are the molar absorptivities of pure P and pure X, b is the pathlength, and P_T and X_T are the total concentrations of P and X in the solution. For the first solution in Table 19-1, $P_T = (1.00/25.0)(2.50 \text{ mM}) = 0.100 \text{ mM}$ and $X_T = (9.00/25.0)(2.50 \text{ mM}) = 0.900 \text{ mM}$. If P and X do not absorb at the wavelength of interest, no absorbance correction is needed.

Maximum absorbance occurs at the mole fraction of X corresponding to the stoichiometry of the complex (Figure 19-10). If the predominant complex is PX_2, the maximum occurs at:

$$\text{Mole fraction of X in } P_aX_b = \frac{b}{b + a}(= 0.667 \text{ when } b = 2 \text{ and } a = 1)$$

If P_3X were predominant, the maximum would occur at (mole fraction of X) = $1/(1 + 3)$ = 0.250.

For the reaction $P + nX \rightleftharpoons PX_n$, you could show that $[PX_n]$ reaches a maximum when the initial concentrations have the ratio $[X]_0 = n[P]_0$. To do this, write

$$K = \frac{[PX_n]}{([P]_0 - [PX_n])([X]_0 - n[PX_n])}$$

and set the partial derivatives $\partial[PX_n]/\partial[P]_0$ and $\partial[PX_n]/\partial[X]_0$ equal to 0.

Method of continuous variation:

$$P + nX \rightleftharpoons PX_n$$

Maximum absorbance occurs when (mole fraction of X) = $n/(n + 1)$.

TABLE 19-1	Solutions for the method of continuous variation		
mL of 2.50 mM P	mL of 2.50 mM X	Mole ratio (X:P)	Mole fraction of X $\left(\dfrac{\text{mol X}}{\text{mol X + mol P}}\right)$
1.00	9.00	9.00:1	0.900
2.00	8.00	4.00:1	0.800
2.50	7.50	3.00:1	0.750
3.33	6.67	2.00:1	0.667
4.00	6.00	1.50:1	0.600
5.00	5.00	1.00:1	0.500
6.00	4.00	1:1.50	0.400
6.67	3.33	1:2.00	0.333
7.50	2.50	1:3.00	0.250
8.00	2.00	1:4.00	0.200
9.00	1.00	1:9.00	0.100

NOTE: *All solutions are diluted to a total volume of 25.0 mL with a buffer.*

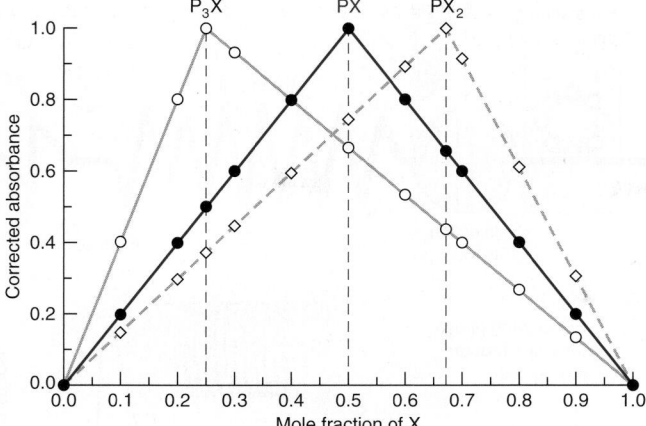

FIGURE 19-10 Ideal behavior of Job plots for formation of the complexes P_3X, PX, and PX_2.

Here are some precautions for applying the method of continuous variation:

1. Verify that the complex follows Beer's law.
2. Use constant ionic strength and pH, if applicable.
3. Take readings at more than one wavelength; the maximum should occur at the same mole fraction for each wavelength.
4. Do experiments at different total concentrations of P + X. If a second set of solutions were prepared in the proportions given in Table 19-1, but from stock concentrations of 5.00 mM, the maximum should still occur at the same mole fraction.

The method of continuous variation can be carried out with many separate solutions, as in Table 19-1. However, a titration is more sensible. Figure 19-11a shows a titration of EDTA with Cu^{2+}. In Figure 19-11b, the abscissa has been transformed into mole fraction of Cu^{2+} (=[moles of Cu^{2+}]/[moles of Cu^{2+} + moles of EDTA]) instead of volume of Cu^{2+}. The sharp maximum at a mole fraction of 0.5 indicates formation of a 1:1 complex. If the equilibrium constant is not large, the maximum is more curved than in Figure 19-11b. The curvature can be used to estimate the equilibrium constant.[10]

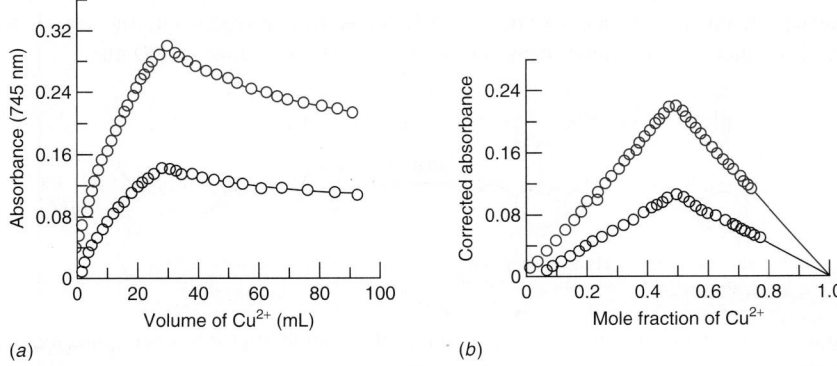

(a)   (b)

FIGURE 19-11 (*a*) Spectrophotometric titration of 30.0 mL of EDTA in acetate buffer with $CuSO_4$ in the same buffer. *Upper curve:* [EDTA] = [Cu^{2+}] = 5.00 mM. *Lower curve:* [EDTA] = [Cu^{2+}] = 2.50 mM. The absorbance has not been "corrected" in any way. (*b*) Transformation of data into mole fraction format. The absorbance of free $CuSO_4$ at the same formal concentration has been subtracted from each point in panel *a*. EDTA is transparent at this wavelength. [Data from Z. D. Hill and P. MacCarthy, "Novel Approach to Job's Method," *J. Chem. Ed.* **1986,** *63,* 162.]

19-4 Flow Injection Analysis and Sequential Injection

In **flow injection analysis,** a liquid sample is injected into a *continuously flowing* liquid carrier containing a reagent that reacts with the sample.[11–15] Additional reagents might be added further downstream. As sample flows from the injector to the detector, the sample zone broadens and reacts with reagent to form a product to which the detector responds. Advantages of flow injection over "batch processes" in which individual samples are separately analyzed include speed, automation of solution handling, and low cost. It is normal for flow

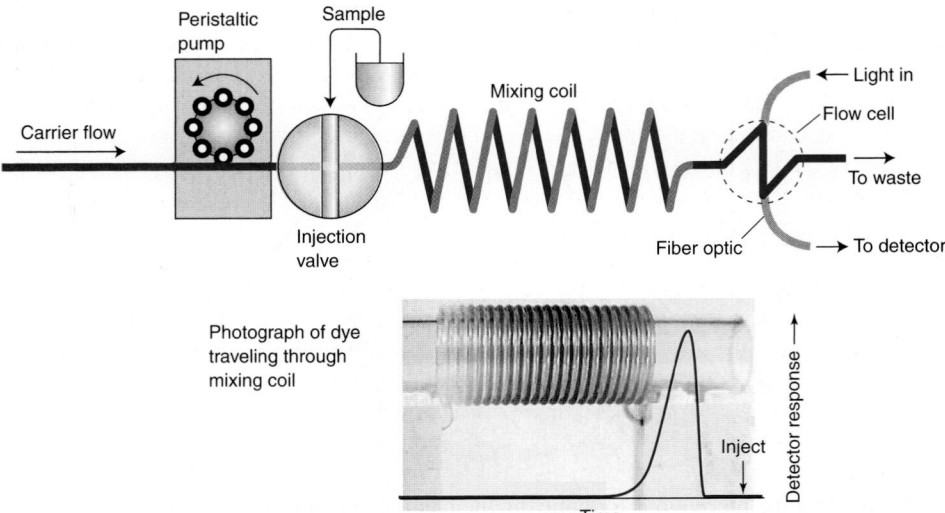

FIGURE 19-12 Schematic diagram of flow injection analysis in which sample is injected into carrier flow containing reagents that form a color with the analyte. The peristaltic pump pushes liquid through flexible tubing by the action of eight rollers along the tubing. Photograph and graph show dispersion of a dye injected into the carrier. [Information from tutorial by J. Ruzicka, *Flow Injection Analysis,* 4th ed., 2009, available free from www.flowinjection.com/freecd.aspx.]

injection to handle 100 samples per hour. Autosamplers holding hundreds of samples are essential for automated analysis. Flow injection is a workhorse in many soil and water analysis labs, which routinely measure large numbers of samples.

Figure 19-12 shows a representative analysis for traces of the herbicide acetochlor in food.[16] The sample is prepared by homogenizing a food such as grain or flour, extracting herbicide from the food with an organic solvent, and hydrolyzing the extract. Aqueous hydrolysis product is injected into the carrier stream for flow injection analysis in Figure 19-12.

Acetochlor herbicide   0.2 M HCl / 100°C   Hydrolysis product

The carrier stream contains a diazonium salt (the *reagent*) that reacts with the sample to give a colored product, which can be measured by its visible absorbance at 400 nm.

Diazonium reagent in carrier stream + Hydrolysis product in sample → analysis → Colored product (an azo dye)

Figure 19-13 shows dispersion and reaction of the sample after it has been injected into the carrier stream. If there were no reagent, the sample would just begin to spread out (disperse) as it traveled downstream. When liquid flows relatively slowly through a cylindrical tube, the flow is *laminar*. Friction with the walls of the tube slows the flow to zero at the walls. The flow at the center of the channel is twice as fast as the average flow. Between the center and walls of the tube, there is a parabolic velocity profile. The top panel in Figure 19-13 shows curved leading

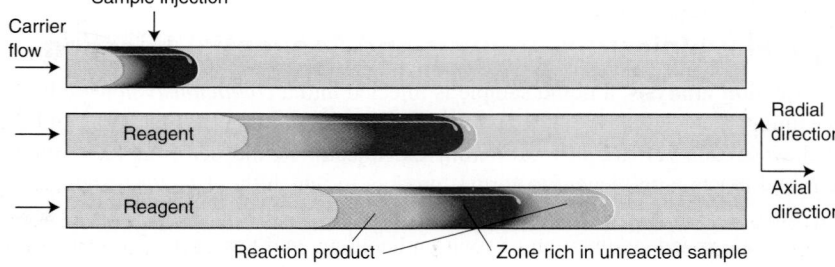

FIGURE 19-13 Dispersion and reaction of sample as it travels downstream after injection into the carrier. [Information from tutorial by J. Ruzicka, *Flow Injection Analysis,* 4th ed., 2009, available free from www.flowinjection.com.]

CHAPTER 19 Applications of Spectrophotometry

and trailing edges of the sample zone. Liquid near the walls mixes with liquid in the bulk by radial diffusion. The narrower the tube, the faster is radial mixing. With a typical tube diameter of 0.5 to 0.75 mm, diffusion of liquid away from the walls is significant in a few seconds. In Figure 19-12, the bulk of the path between injection and detection is a helical mixing coil. Curvature and sharp bends in the flow path create turbulence, which promotes mixing.

As the sample plug is transported through the mixing coil, reaction with reagent in the carrier stream occurs from the leading and trailing edges of the sample zone. Turbulence is required for good mixing of the reagent with the sample because the length of the sample zone is much greater than the diameter of the tubing. Formation of product depends on the rate of the chemical reaction, as well as the rate of mixing of the zones. Typical times between injection and detection are just tens of seconds. In contrast to most methods of chemical analysis, the mixing of analyte with reagents is incomplete, and chemical equilibrium is not reached in flow injection analysis.

The key to analytical precision is the repeatability of flow injection. The concentration profile of product passing through the detector depends on many conditions including volume of sample, flow rate, reaction rate, and temperature. Reproducible conditions provide a reproducible response. For replicate injections of the herbicide acetochlor, the standard deviation was 1.6% when measuring 1 ppm of the herbicide in food.

The detector flow cell in Figure 19-12 has a Z-shaped liquid path. Monochromatic light is brought by way of a fiber optic. Light that has passed through the flow cell travels to the detector via another fiber optic. Common flow cells have a pathlength of 10 mm with a volume of 60 μL. Typical volumes of injected samples are tens of microliters. Peak height, rather than peak area, is commonly taken as the analytical signal in flow injection.

Figure 19-14 shows a flow injection setup designed for continuous shipboard analysis of nanomolar concentrations of ammonia in seawater. The peristaltic pump feeds three liquids into the reaction coil, each at a rate of 160 μL/min. One liquid is seawater collected 1 m below the ocean surface. Reagent 1 (25 mM o-phthaldialdehyde) and reagent 2 (10 mM sodium sulfite plus 5 mM formaldehyde) mix with the seawater before the stream enters the 1.0-mm-diameter × 2-m-long reaction coil, which is kept at 65°C to speed the reaction to produce a fluorescent product. In the detector, the solution is irradiated with 365-nm ultraviolet light brought through a fiber optic. Fluorescence at 423 nm is collected by a second fiber optic at a right angle to the excitation beam. Calibration in Figure 19-15 is carried out by periodically replacing seawater with 0.958-mL volumes of NH_4Cl standards through the injection valve. The standard deviation of replicate injections is 2.2% for 200 nM NH_4Cl and 6.7% for 1.0 nM NH_4Cl.

Analysis time in Figure 19-15 is unusually long. Large-volume standards run slowly to boost sensitivity and reduce reagent consumption were measured at a rate of 8 per hour. By contrast, continuous measurement of seawater produced 3 600 readings per hour. Flow injection typically measures 100 samples per hour.

Sequential Injection[11, 17]

Sequential injection is distinguished from flow injection by *flow programming* and *flow reversal* under computer control. Flow is *not continuous*. Smaller volumes of reagents are

Flow injection is a dynamic process in which equilibrium is not reached. Reproducibility is obtained by repeating the same conditions in every run.

The terms "mixing coil" and "reaction coil" are used interchangeably.

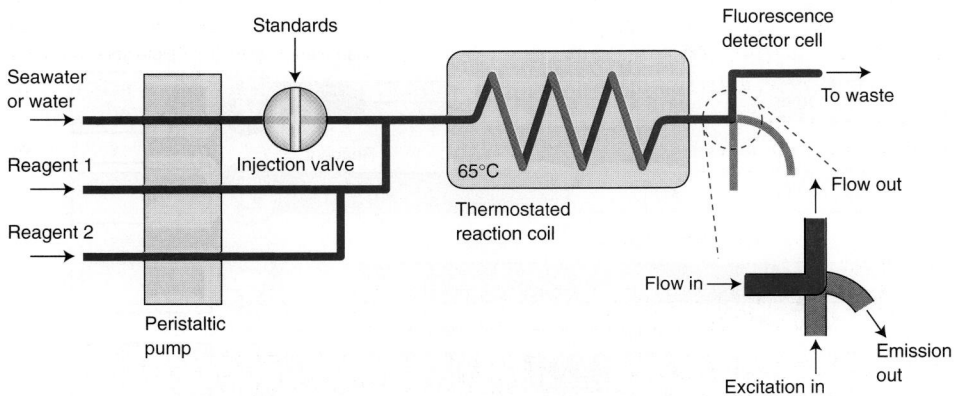

FIGURE 19-14 Flow injection analysis with two reagent channels for continuous analysis of nanomolar concentrations of NH_3 in seawater. [Information from N. Amornthammarong and J.-Z. Zhang, "Shipboard Fluorometric Flow Analyzer for High-Resolution Underway Measurement of Ammonium in Seawater," *Anal. Chem.* **2008**, *80*, 1019.]

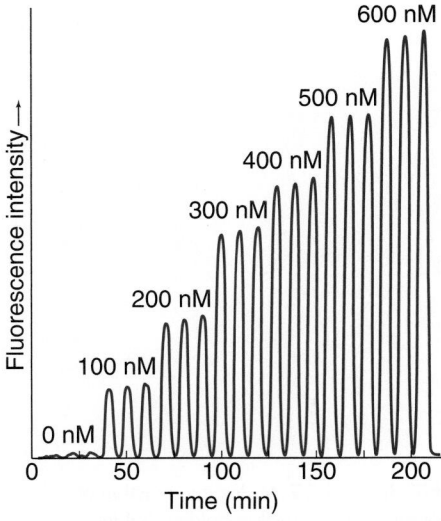

FIGURE 19-15 Response to replicate injections of NH_4Cl standards. [Data from N. Amornthammarong and J.-Z. Zhang, "Shipboard Fluorometric Flow Analyzer for High-Resolution Underway Measurement of Ammonium in Seawater," *Anal. Chem.* **2008**, *80*, 1019.]

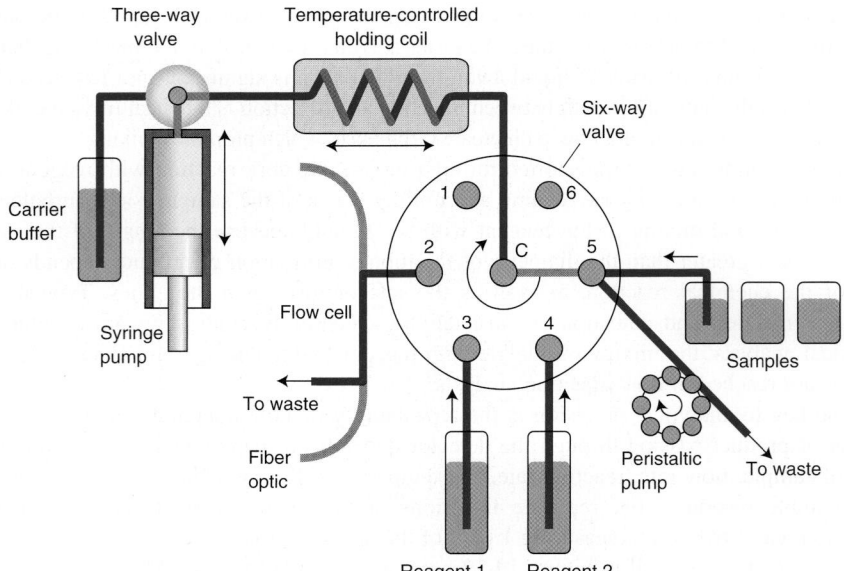

FIGURE 19-16 Schematic diagram of sequential injection equipment. Rotation of the valve can connect central port C to any of the ports 1 through 6. [Information from tutorial by J. Ruzicka, *Flow Injection Analysis,* 4th ed., 2009, www.flowinjection.com.]

Three-way valve

Temperature-controlled holding coil

Six-way valve

Carrier buffer

Syringe pump

Flow cell

To waste

Fiber optic

Samples

Peristaltic pump

To waste

Reagent 1   Reagent 2

required and less waste is generated than in flow injection. Sequential injection has been miniaturized to the point where it is also called a "lab-on-a-valve." Miniaturization and discontinuous flow reduce the consumption of expensive reagents such as enzymes and antibodies for biochemical assays. Sequential injection is used for continuous monitoring of environmental and industrial processes, even at remote locations.

The central feature of the sequential injection apparatus in Figure 19-16 is a six-way valve. In this example, ports 2, 3, 4, and 5 are used. The figure shows a connection between port 5 and central port C. Liquid samples are brought to port 5 by an auxiliary peristaltic pump. Rotation of the valve by computer control can connect any one of the other ports to C. Near the upper left is a stepper-motor-controlled syringe pump, which moves precise volumes of liquid forward or backward under computer control. The valve at the top of the syringe pump can connect the syringe to a reservoir of carrier buffer or to the holding coil.

Figure 19-17 shows how a reaction of a sample with one reagent could be conducted. Sample from port 5 of Figure 19-16 is first taken into the holding coil. Then the six-way valve is rotated to take reagent 1 from port 3 into the holding coil. Flow is then stopped to allow sample and reagent to mix and react in the holding coil. After a selected time, flow is reversed and the liquid is sent out via port 2 through the flow cell in which absorbance is monitored at a selected wavelength. Volumes of each reagent and time in the mixing coil are under computer control. In Figure 19-16, two different reagents could be mixed with the sample. A variant of the procedure is to stop the flow when the product zone reaches the detector and to measure the change of absorbance versus time as more product forms in the detector. Figure 19-18 shows sequential injection equipment, which can be smaller than a laptop computer. Figure 20-24 shows another example of sequential injection.

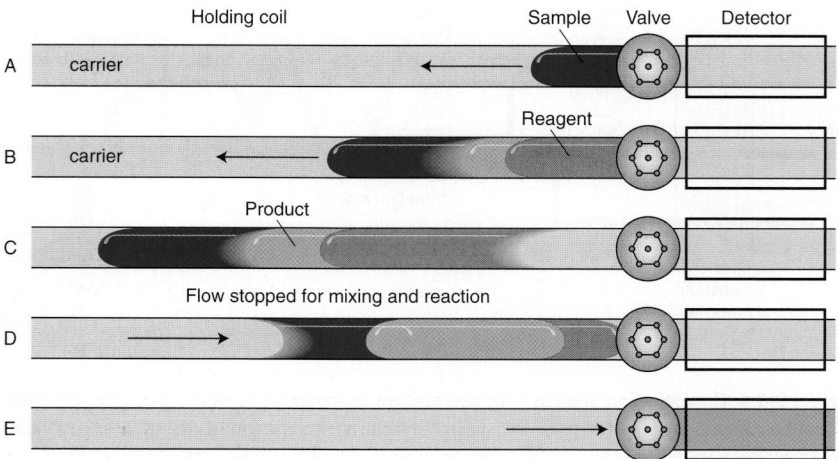

FIGURE 19-17 Mixing and reaction of sample with reagent in holding coil in sequential injection. After taking in sample (A) and reagent (B), flow is stopped (C) to allow product to form. Flow is then reversed (D) to pass product through the detector (E). [Information from tutorial by J. Ruzicka, *Flow Injection Analysis,* 4th ed., 2009, www.flowinjection.com.]

Holding coil   Sample   Valve   Detector

A   carrier

B   carrier   Reagent

C   Product

         Flow stopped for mixing and reaction

D

E

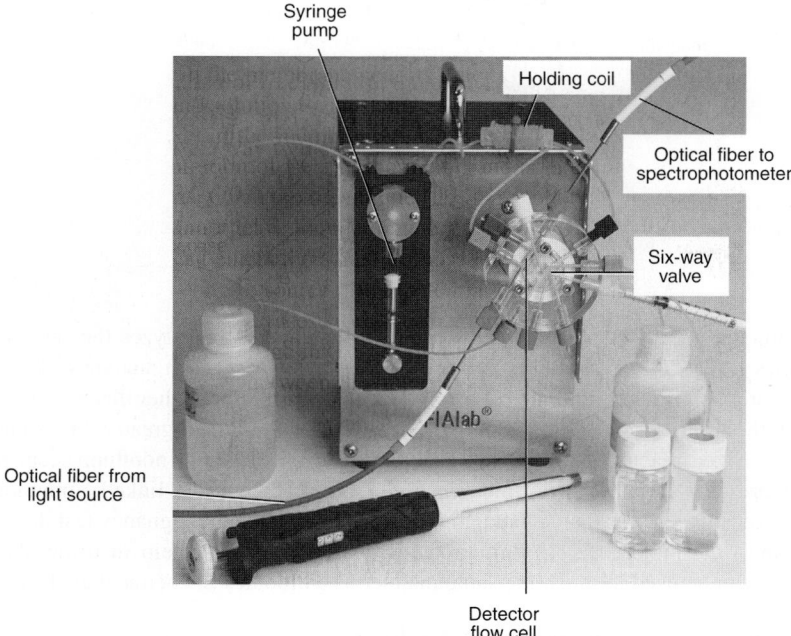

Syringe
pump

Holding coil

Optical fiber to
spectrophotometer

Six-way
valve

Optical fiber from
light source

Detector
flow cell

FIGURE 19-18 Sequential injection apparatus with spectrophotometric detection. The central feature of the equipment is the six-way valve, so sequential injection is sometimes called "lab-on-a-valve." [Information from tutorial by J. Ruzicka, *Flow Injection Analysis,* 4th ed., 2009, www.flowinjection.com]

19-5 Immunoassays

An important application of light absorption and emission is in **immunoassays,** which employ antibodies to detect analyte. An *antibody* is a protein produced by the immune system of an animal in response to a foreign molecule called an *antigen*. An antibody recognizes a specific antigen. The formation constant for the antibody-antigen complex is large, whereas the binding of the antibody to other molecules is weak.

Figure 19-19 illustrates the principle of an *enzyme-linked immunosorbent assay,* abbreviated ELISA in biochemical literature. Antibody 1, which is specific for the analyte of interest (the antigen), is bound to a polymeric support. In steps 1 and 2, analyte is incubated with the polymer-bound antibody to form a complex. The fraction of antibody sites that bind analyte increases with increasing concentration of analyte in the unknown. The surface is then washed to remove unbound substances. In steps 3 and 4, the antibody-antigen complex is treated with antibody 2, which recognizes a different region of the analyte. Antibody 2 was specially prepared for the immunoassay by covalent attachment of an enzyme that will be used later in the process. Again, excess unbound substances are washed away.

The enzyme attached to antibody 2 is critical for quantitative analysis. Figure 19-20 shows two ways in which the enzyme can be used. The enzyme can transform a colorless

Rosalyn Yalow received the Nobel Prize in Medicine in 1977 for developing immunoassay techniques in the 1950s, using proteins labeled with radioactive ^{131}I to enable their detection.[18] Yalow, a physicist, worked with Solomon Berson, a medical doctor, in this pioneering effort.

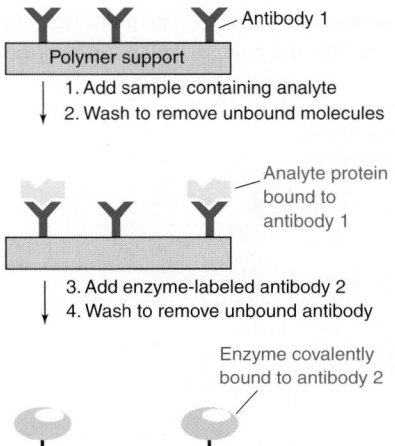

Antibody 1

Polymer support

1. Add sample containing analyte
2. Wash to remove unbound molecules

Analyte protein
bound to
antibody 1

3. Add enzyme-labeled antibody 2
4. Wash to remove unbound antibody

Enzyme covalently
bound to antibody 2

Antibody 2

FIGURE 19-19 Enzyme-linked immunosorbent assay. Antibody 1, which is specific for the analyte of interest, is bound to a polymer support and treated with unknown. After excess, unbound molecules have been washed away, the analyte remains bound to antibody 1. The bound analyte is then treated with antibody 2, which recognizes a different site on the analyte and to which an enzyme is covalently attached. After unbound material has been washed away, each molecule of analyte is coupled to an enzyme that will be used in Figure 19-20.

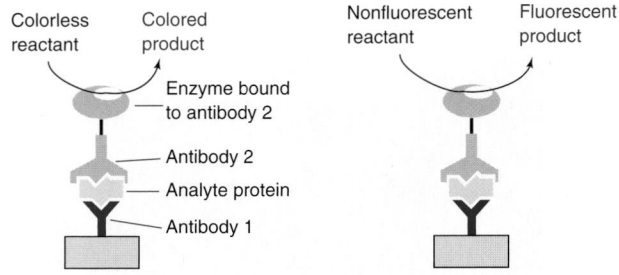

FIGURE 19-20 Enzyme bound to antibody 2 can catalyze reactions that produce colored or fluorescent products. Each molecule of analyte bound in the immunoassay leads to many molecules of colored or fluorescent product that are easily measured.

reactant into a colored product. Because one enzyme molecule catalyzes the same reaction many times, many molecules of colored product are created for each analyte molecule. The enzyme thereby *amplifies* the signal in the chemical analysis. The higher the concentration of analyte in the original unknown, the more enzyme is bound and the greater the extent of the enzyme-catalyzed reaction. Alternatively, the enzyme can convert a nonfluorescent reactant into a fluorescent product. Colorimetric and fluorometric enzyme-linked immunosorbent assays are sensitive to less than a nanogram of analyte. The home pregnancy test described at the opening of Chapter 0 is an immunoassay of a placental protein in urine. Aptamers (Box 17-5) can be used in much the same manner as antibodies in chemical analysis.[19]

Immunoassays in Environmental Analysis

Commercial immunoassay kits are available for screening and analysis of pesticides, industrial chemicals, explosives, and microbial toxins at the parts per trillion to parts per million levels in groundwater, soil, and food.[20] An advantage of screening in the field is that uncontaminated regions that require no further attention are readily identified. An immunoassay can be 20–40 times less expensive than a chromatographic analysis and can be completed in 0.3–3 h in the field, using 1-mL samples. Chromatography generally must be done in a lab and might require several days, because analyte must first be extracted or concentrated from liter-quantity samples to obtain a sufficient concentration.

Figure 19-21 shows an immunoassay to measure trace levels of Hg^{2+} in environmental waters. Prior to the process in Figure 19-21, Hg^{2+} in the sample is converted to $HgCl_4^{2-}$ by adding HCl to a level of 0.1 M. A 5-mL sample is then passed through a solid-phase extraction column (page 785) containing 0.1 mL of anion-exchange resin (Chapter 26) that binds $HgCl_4^{2-}$ and separates it from transition metal ions and the potentially interfering species Cd^{2+} and Mn^{2+}. $HgCl_4^{2-}$ is converted to Hg^{2+}-EDTA complex (Chapter 12) and washed from the column by EDTA solution at pH 7.5. The sample is then treated with a mouse antibody that recognizes Hg^{2+}-EDTA. The solution now contains antibody-bound Hg^{2+}-EDTA and some excess antibody, shown in the dashed oval at the upper left of Figure 19-21. The more Hg^{2+} in the water sample, the more antibody is bound to Hg^{2+}-EDTA and the less free antibody is present.

This solution is next passed over a 1-mm^2 well containing 100-μm diameter poly(methylmethacrylate) polymer beads to which Hg^{2+}-EDTA is covalently attached. Some free antibody in the sample binds to Hg^{2+}-EDTA on the surface of the beads. The more Hg^{2+} in the water sample, the less free antibody is available to bind to the beads. In the final step in

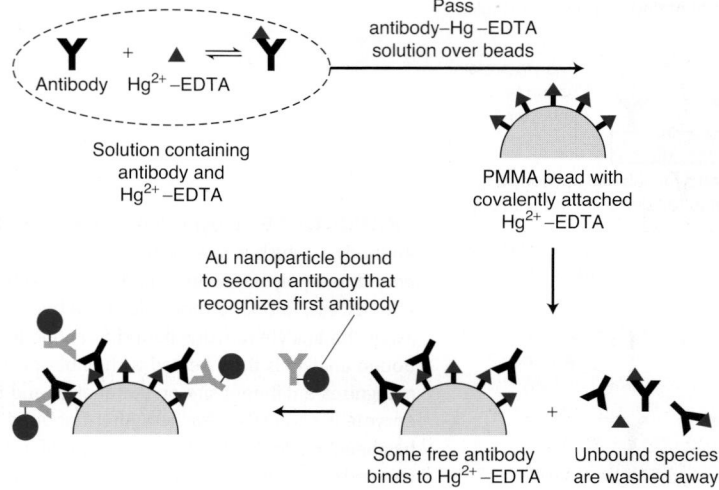

FIGURE 19-21 Immunoassay for Hg^{2+} in natural water. [Information from Y. Date, A. Aota, S. Terakado, K. Sasaki, N. Matsumoto, Y. Watanabe, T. Matsue, and N. Ohmura, "Trace-Level Mercury Ion (Hg^{2+}) in Aqueous Sample Based on Solid-Phase Extraction Followed by Microfluidic Immunoassay," *Anal. Chem.* **2013**, *85*, 434.]

CHAPTER 19 Applications of Spectrophotometry

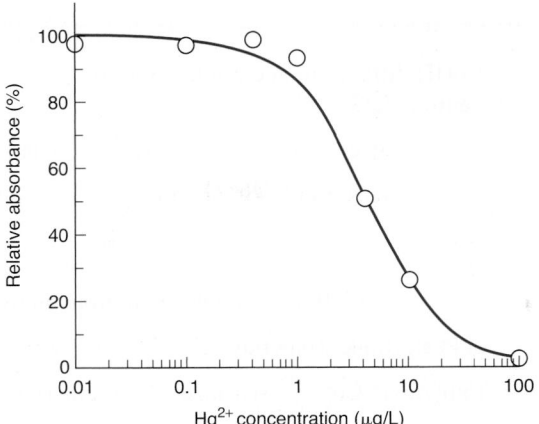

FIGURE 19-22 Standard curve for immunoassay for Hg^{2+} in natural water. [Data from Y. Date, A. Aota, S. Terakado, K. Sasaki, N. Matsumoto, Y. Watanabe, T. Matsue, and N. Ohmura, "Trace-Level Mercury Ion (Hg^{2+}) in Aqueous Sample Based on Solid-Phase Extraction Followed by Microfluidic Immunoassay," *Anal. Chem.* **2013,** *85,* 434.]

Figure 19-21, a second antibody is passed over the beads. This second antibody recognizes the first mouse antibody. A 40-nm-diameter gold nanoparticle is attached to each molecule of the second antibody. The nanoparticle has an intense red color that is measured by a 520-nm light-emitting diode with a photodiode detector. The more Hg^{2+} that was present in the natural water, the less antibody is bound to the beads and the fewer gold nanoparticles end up bound to the beads. The red color decreases with increasing Hg^{2+} in the natural water, as shown by the standard curve in Figure 19-22. The lower limit of detection is 0.8 μg Hg/L (0.8 ppb), which satisfies guidelines for drinking water analytical capability. The entire assay is done with a portable microfluidic device—a "lab on a chip" (Section 26-8)—that can be used in the field.

Time-Resolved Fluorescence Immunoassays[21]

The sensitivity of fluorescence immunoassays can be enhanced by a factor of 100 (to detect 10^{-13} M analyte) with time-resolved measurements of luminescence from the lanthanide ion Eu^{3+}. Organic chromophores such as fluorescein are plagued by background fluorescence at 350–600 nm from solvent, solutes, and particles. This background decays to a negligible level within 100 μs after excitation. Sharp luminescence at 615 nm from Eu^{3+}, however, has a much longer lifetime, falling to 1/e (= 37%) of its initial intensity in approximately 700 μs. In a time-resolved fluorescence measurement (Figure 19-23), luminescence is measured between 200 and 600 μs after a brief laser pulse at 340 nm. The next pulse is flashed at 1 000 μs, and the cycle is repeated approximately 1 000 times per second. Rejecting emission within 200 μs of the excitation eliminates most of the background fluorescence.

Figure 19-24 shows how Eu^{3+} can be incorporated into an immunoassay. A chelating group that binds lanthanide ions is attached to antibody 2 in Figure 19-19. While bound to the antibody, Eu^{3+} has weak luminescence. After all steps in Figure 19-19 have been completed, the pH of the solution is lowered in the presence of a chelator that extracts Eu^{3+} into solution. Strong luminescence from the soluble metal ion is then easily detected by a time-resolved measurement.

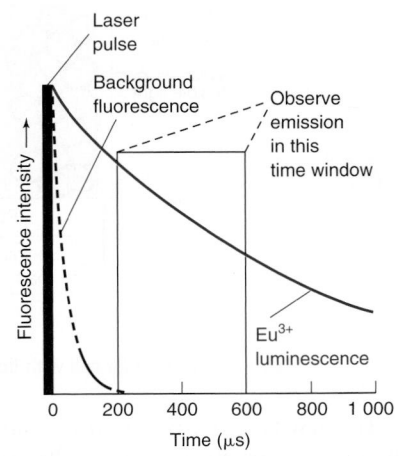

FIGURE 19-23 Emission intensity in a time-resolved fluorescence experiment.

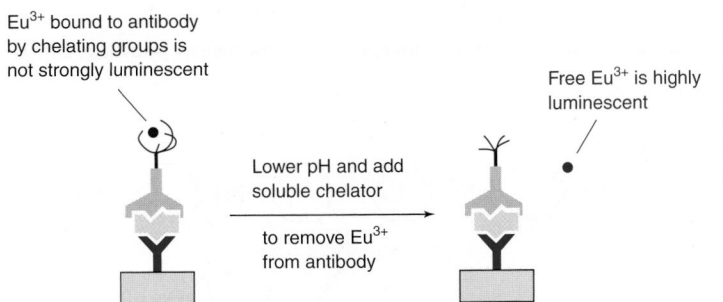

FIGURE 19-24 Antibody 2 in the immunosorbent assay of Figure 19-19 can be labeled with Eu^{3+}, which is not strongly luminescent when bound to the antibody. To complete the analysis, the pH of the solution is lowered to liberate strongly luminescent Eu^{3+} ion.

Eu³⁺ bound to antibody by chelating groups is not strongly luminescent

Lower pH and add soluble chelator

to remove Eu³⁺ from antibody

Free Eu³⁺ is highly luminescent

19-6 Sensors Based on Luminescence Quenching

When a molecule absorbs a photon, it is promoted to an excited state from which it can lose energy as heat or it can emit a lower energy photon (Figure 18-16). Box 19-1 describes how absorbed light can be converted into electricity. In this section, we discuss how excited-state molecules can serve as chemical sensors (Figure 19-25).

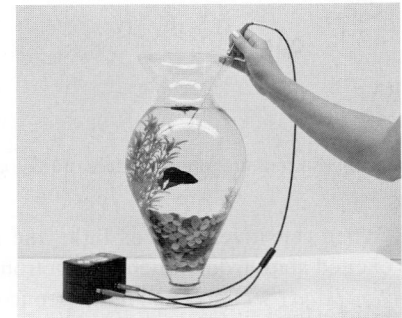

FIGURE 19-25 Fiber-optic sensor measures O_2 by its ability to quench the luminescence of Ru(II) at the tip of the fiber. A blue-light-emitting diode provides excitation energy. [© Courtesy Ocean Optics, Dunedin, FL/Ocean Optics, Inc.]

BOX 19-1 **Converting Light into Electricity**

Earth's deserts receive 250–300 W/m² of solar irradiance. If solar energy could be used with 10% efficiency, 5% of the sunlight falling on the deserts would provide all the energy used in the world in 2010. Single-crystal silicon solar cells have a solar-to-electric power conversion efficiency of ~25%, but high cost limits their use. Less expensive polycrystalline silicon photocells used on rooftops are ~15% efficient. Commercially available dye-sensitized solar cells have efficiencies of 15% or more.[22] Dye-sensitized cells might provide an affordable means to use ambient indoor light to replace batteries in some electronics.

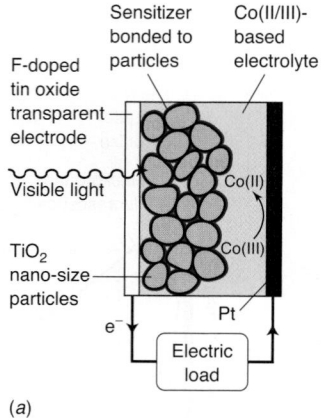

(a)

Dye-sensitized solar cell with liquid electrolyte.

The new type of *dye-sensitized solar cell*[23,24] in panel *a* has a power conversion efficiency of ~12%. Sunlight enters the cell from the left through a transparent, electrically conductive layer of fluorine-doped tin oxide. The electrode is coated with a ~5- to 10-μm-thick layer of nanometer-size TiO₂ particles coated with a *sensitizer* in panel *b* that absorbs visible light. Nanoparticles have so much surface area that the monolayer of sensitizer absorbs most of the visible light striking the cell.

(b)

Zinc-porphyrin sensitizer on the surface of TiO₂ nanoparticles absorbs sunlight.

When sensitizer absorbs light, the molecule is promoted to an excited state from which an electron is injected into the *conduction band* (Section 15-8) of the semiconductor TiO₂ in <100 ps. In this process, the sensitizer is oxidized to a radical cation. Electrons flow through an external circuit (where they can do useful work) and arrive at the Pt electrode where Co(III) in solution is reduced to Co(II). The cycle is completed when

Co(II) diffuses to and reduces the oxidized sensitizer back to its neutral state:

$$\text{sensitizer} + h\nu \rightarrow \text{sensitizer*} \quad (\text{*excited state})$$

$$\text{sensitizer*} \rightarrow \text{sensitizer}^{+\bullet} + e^- \text{ injected into TiO}_2$$

Radical cation

e⁻ flows through circuit from tin oxide to Pt

at Pt electrode: $(\text{bipyridyl})_3\text{Co}^{3+} + e^- \rightarrow (\text{bipyridyl})_3\text{Co}^{2+}$

$(\text{bipyridyl})_3\text{Co}^{2+} + \text{sensitizer}^{+\bullet} \rightarrow (\text{bipyridyl})_3\text{Co}^{3+} + \text{sensitizer}$

The acetonitrile liquid phase has several components in addition to (bipyridyl)₃Co(II,III). Evaporation of acetonitrile is one of the life-limiting factors for the cell.

A prospect for less expensive and longer-lasting solar cells with efficiency comparable to that of dye-sensitized cells is an all-solid-state cell with a perovskite photoactive layer and an organic hole-transport matrix to enable charge separation.[25] Recall from Section 15-8 that a *hole* is the absence of an electron. Perovskites are materials with the formula ABX₃, where A is a large cation, B is a small cation, and X is an anion, with a crystal structure in panel *c*. One useful perovskite is $(\text{CH}_3\text{NH}_3)\text{PbI}_3$, where A = CH_3NH_3^+, B = Pb²⁺, and X = I⁻. A demonstrated hole-transport layer is the organic material spiro-OMeTAD in panel *d*.

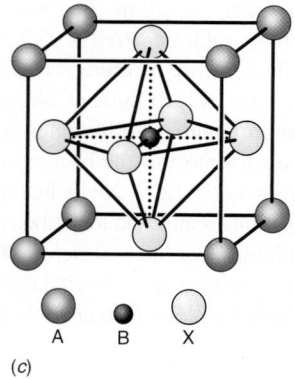

(c)

ABX₃ perovskite cubic crystal structure named for the mineral perovskite, CaTiO₃.

(d)

2,2′,7,7′-tetrakis-(N,N-di-p-methoxyphenylamine)-9,9′spirofluorene
Spiro-OMeTAD hole-transport material

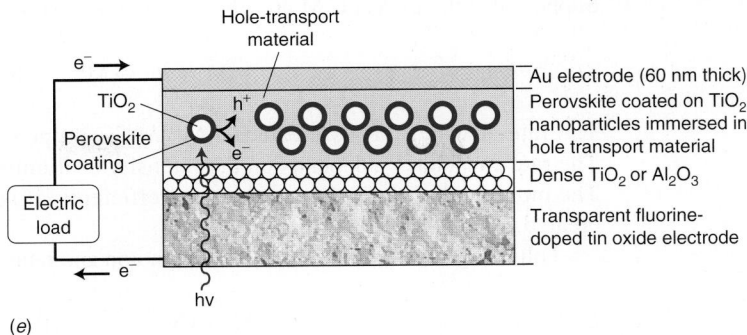

(e)

Layout of all-solid-state, <2-μm-thick, dye-sensitized solar cell. [Information from M. Jacoby, "Tapping Solar Power with Perovskites," *Chem. Eng. News,* 24 February 2014, p. 10.]

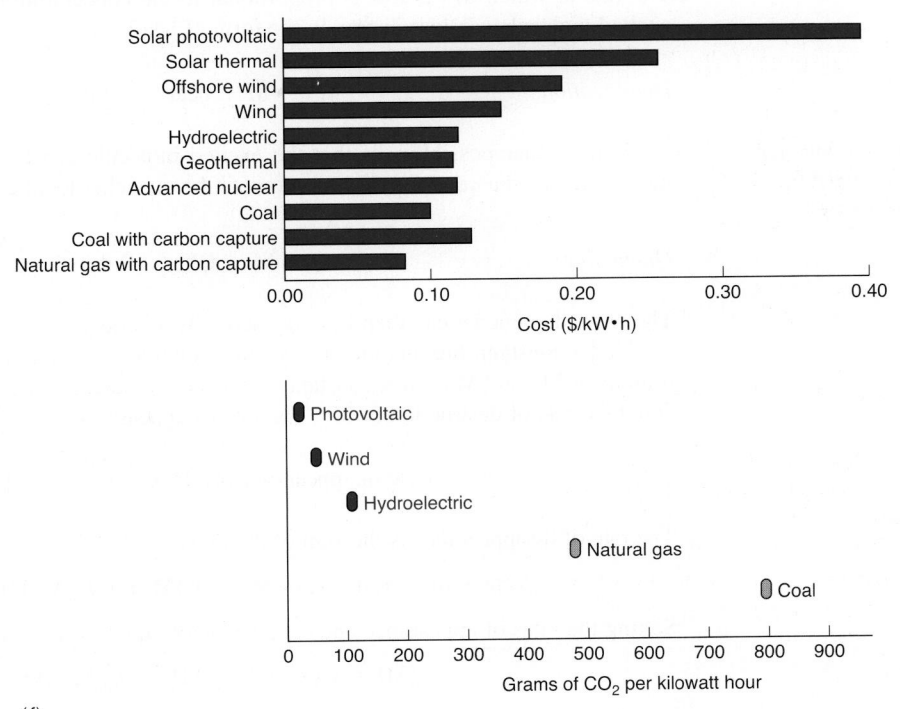

(f)

Estimated electricity costs and CO$_2$ emissions. Solar thermal electric power refers to sunlight concentrated by mirrors to produce high-temperature steam for generating electricity. [Cost data is from L. Glickman, "Energy Efficiency in the Built Environment," *Physics Today,* July 2008, p. 35. CO$_2$ emissions from S. Pacca and A. Horvath, "Greenhouse Gas Emissions from Building and Operating Electric Power Plants in the Upper Colorado River Basin," *Environ. Sci. Technol.* **2002,** *36,* 3194. Reprinted with permission © 2002, American Chemical Society.]

In panel *e,* light passes through the fluorine-doped tin oxide electrode to reach the active layer containing perovskite sensitizer on the surface of TiO$_2$ nanoparticles. These particles are immersed in a matrix of hole-transport material. When perovskite absorbs a photon, an electron-hole pair is created. The hole is transported into the spiro-OMeTAD matrix in contact with the rear surface gold electrode. The electron passes to the front surface tin oxide electrode. Electrons flow through a wire from tin oxide to gold, where they are injected into the hole-transport medium to reduce it back to neutrality.

Converting sunlight into electricity is good, but converting sunlight into a fuel such as H$_2$ or CH$_3$OH would be even better.[26] Fuel can be used when and where it is needed. Green leaves of plants use sunlight to reduce CO$_2$ to carbohydrates, a fuel that is oxidized back to CO$_2$ by animals and plants to provide energy.

We are rapidly exhausting Earth's supply of fossil fuels (coal, oil, wood, and gas), which are also raw materials for plastics, fabrics, and many essential items. Moreover, burning fossil fuels increases atmospheric CO$_2$, which acidifies the ocean and warms the Earth.

Why burn irreplaceable raw materials? The answer is that electricity from fossil fuel is relatively inexpensive. In 2008, 85% of global energy production was from oil, coal, and natural gas. Panel *f* shows estimated costs of generating electricity and estimated CO$_2$ emissions from each source. Fossil fuels produce the most CO$_2$. We will probably consume fossil fuel until this increasingly scarce resource becomes more expensive than renewable energy—which is likely to happen in your lifetime.

Nuclear energy is cost competitive with fossil fuel,[27] has negligible greenhouse gas emission, and creates far less air pollution. However, catastrophic consequences of accidents and intractable issues of waste containment have almost halted construction of nuclear power plants and led to shutdown of many existing plants since three nuclear reactors melted down as a result of the tsunami at Fukushima, Japan, in 2011. Radioactive material continues to leak from that site and cleanup will take decades at a huge cost.

Luminescence Quenching

Suppose that the molecule M absorbs light and is promoted to the excited state M*:

Absorption: $\qquad\qquad$ M $+ h\nu \rightarrow$ M* $\qquad$ Rate $= \dfrac{d[\text{M*}]}{dt} = k_a[\text{M}]$

The rate at which M* is created, $d[\text{M*}]/dt$, is proportional to the concentration of M. The rate constant, k_a, depends on the intensity of illumination and the absorptivity of M. The more intense the light and the more efficiently it is absorbed, the faster M* will be created.

Following absorption, M* can emit a photon and return to the ground state:

Emission: $\qquad\qquad$ M* $\rightarrow$ M $+ h\nu$ $\qquad$ Rate $= -\dfrac{d[\text{M*}]}{dt} = k_e[\text{M*}]$

The rate at which M* is lost is proportional to the concentration of M*. Alternatively, the excited molecule can lose energy in the form of heat:

Deactivation: $\qquad\qquad$ M* $\rightarrow$ M $+$ heat $\qquad$ Rate $= -\dfrac{d[\text{M*}]}{dt} = k_d[\text{M*}]$

Still another possibility is that the excited molecule can transfer energy to a different molecule, called a *quencher* (Q), to promote the quencher to an excited state (Q*):

Quenching: $\qquad\qquad$ M* $+$ Q $\rightarrow$ M $+$ Q* $\qquad$ Rate $= -\dfrac{d[\text{M*}]}{dt} = k_q[\text{M*}][\text{Q}]$

The excited quencher can then lose its energy by a variety of processes.

Under constant illumination, the system soon reaches a steady state in which the concentrations of M* and M remain constant. In the steady state, the rate of appearance of M* must equal the rate of destruction of M*. The rate of appearance is

$$\text{Rate of appearance of M*} = \frac{d[\text{M*}]}{dt} = k_a[\text{M}]$$

The rate of disappearance is the sum of the rates of emission, deactivation, and quenching

$$\text{Rate of disappearance of M*} = k_e[\text{M*}] + k_d[\text{M*}] + k_q[\text{M*}][\text{Q}]$$

Setting the rates of appearance and disappearance equal to each other gives

$$k_a[\text{M}] = k_e[\text{M*}] + k_d[\text{M*}] + k_q[\text{M*}][\text{Q}] \qquad (19\text{-}23)$$

The **quantum yield** for a photochemical process is the fraction of absorbed photons that produce a desired result. If the result occurs every time a photon is absorbed, then the quantum yield is unity. The quantum yield is a number between 0 and 1.

The quantum yield for emission from M* is the rate of emission divided by the rate of absorption. In the absence of quencher, we designate this quantum yield Φ_0:

$$\Phi_0 = \frac{\text{photons emitted per second}}{\text{photons absorbed per second}} = \frac{\text{emission rate}}{\text{absorption rate}} = \frac{k_e[\text{M*}]}{k_a[\text{M}]}$$

Substituting in the value of $k_a[\text{M}]$ from Equation 19-23 and setting [Q] = 0 gives an expression for the quantum yield of emission in the steady state:

$$\Phi_0 = \frac{k_e[\text{M*}]}{k_e[\text{M*}] + k_d[\text{M*}] + k_q[\text{M*}][0]} = \frac{k_e}{k_e + k_d} \qquad (19\text{-}24)$$

If [Q] $\neq$ 0, then the quantum yield for emission (Φ_Q) is

$$\Phi_Q = \frac{k_e[\text{M*}]}{k_e[\text{M*}] + k_d[\text{M*}] + k_q[\text{M*}][\text{Q}]} = \frac{k_e}{k_e + k_d + k_q[\text{Q}]} \qquad (19\text{-}25)$$

In luminescence-quenching experiments, we measure emission in the absence and presence of a quencher. Equations 19-24 and 19-25 tell us that the relative yields are

Stern-Volmer equation: $\qquad$ $\dfrac{\Phi_0}{\Phi_Q} = \dfrac{k_e + k_d + k_q[\text{Q}]}{k_e + k_d} = 1 + \left(\dfrac{k_q}{k_e + k_d}\right)[\text{Q}]$ $\qquad$ **(19-26)**

The *Stern-Volmer equation* says that, if we measure relative emission (Φ_0/Φ_Q) as a function of quencher concentration and plot this quantity versus [Q], we should observe a straight line. The quantity Φ_0/Φ_Q in Equation 19-26 is equivalent to I_0/I_Q, where I_0 is the emission intensity in the absence of quencher and I_Q is the intensity in the presence of quencher.

A Luminescent Intracellular O_2 Sensor

We illustrate our discussion with Ru(II) complexes, which strongly absorb visible light, efficiently emit light at significantly longer wavelengths than they absorb, are stable for long periods, and have a long-lived excited state whose emission is quenched by O_2 (Color Plate 19).[28] A widely used luminescent Ru(II) complex is $Ru(dpp)_3^{2+} \cdot 2Cl^-$.

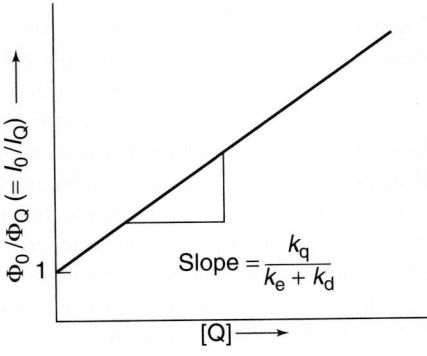

Graph of Stern-Volmer equation:

$$\text{Slope} = \frac{k_q}{k_e + k_d}$$

(dpp)₃Ru(II)

dpp = 4,7-diphenyl-1,10-phenanthroline

Oxygen is a good quencher because its ground state has two unpaired electrons—it is a *triplet* state designated $^3O_2$. O_2 has a low-lying *singlet* state with no unpaired electrons. Figure 18-16 showed that the lowest excited state of many molecules is a triplet. This triplet excited state $^3M^*$ can transfer energy to $^3O_2$ to produce a ground-state singlet molecule and excited $^1O_2^*$.

$$^3M^* \quad + \quad ^3O_2 \quad \rightarrow \quad ^1M \quad + \quad ^1O_2^*$$

Excited state + Ground state → Ground state + Excited state

There are two electron spins up and two down in both the reactants and products. This energy transfer therefore conserves the net spin and is more rapid than processes that change the spin.

Color Plate 20 shows tiny dots of light from fluorescent silica beads shot into living cells by a "gene-gun" that is usually used to infect cells with DNA-coated particles. Two dyes are bound inside each porous bead, whose size is in the range 100 to 600 nm.

When illuminated with blue light, one dye emits green light near 525 nm and the other emits red light near 610 nm (Figure 19-26). Oxygen does not affect the green dye, but it does affect the red dye. The ratio of red-to-green emission intensity tells us the concentration of O_2

The ground state of Ru(II) is a singlet, and the lowest excited state is a triplet. When Ru(II) absorbs visible light, the excited singlet relaxes to the luminescent triplet state. O_2 quenches the luminescence by providing a radiationless pathway by which the triplet is converted into the ground-state singlet. Luminescence quenching of C_{70} is even more sensitive than that of Ru(II) and responds to part per billion levels of O_2.[29]

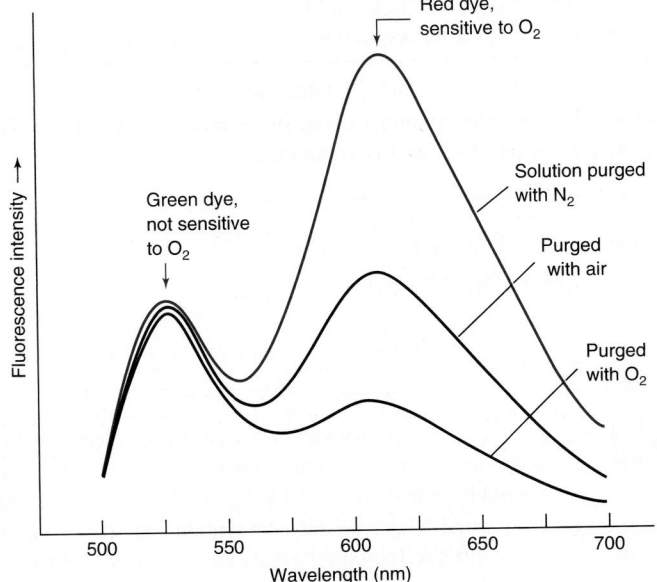

Red dye, sensitive to O_2

Green dye, not sensitive to O_2

Solution purged with N_2

Purged with air

Purged with O_2

Wavelength (nm)

Fluorescence intensity

FIGURE 19-26 Fluorescence from O_2-indicator beads showing constant intensity near 525 nm and variable intensity near 610 nm. The ratio of emission intensity at these two wavelengths is related to O_2 concentration. [Data from H. Xu, J. W. Aylott, R. Kopelman, T. J. Miller, and M. A. Philbert, "Real-Time Method for Determination of O_2 Inside Living Cells Using Optical Nanosensors," *Anal. Chem.* **2001**, *73*, 4124.]

BOX 19-2 **Upconversion**

Fluorescence and phosphorescence always come at lower energy than the excitation energy, as in Figure 18-19, because part of the excitation energy is converted to heat by vibrational relaxation such as R1 and R3 in Figure 18-16. Color Plate 21 shows *green* laser light shining into a solution that is emitting *blue* fluorescence.[30] Blue photons carry *more* energy than green photons. This *upconversion*, which creates high-energy photons from low-energy photons, does not violate conservation of energy because it requires *two* green photons to create *one* blue photon.

The solution contains a ruthenium(II) complex plus 9,10-diphenylanthracene in a deaerated organic solvent.

L = 4,4' dimethyl-2,2'-bipyridine

9,10-diphenyl-anthracene

Absorption of green laser light promotes the ground state (S_0) Ru complex to its excited singlet state (S_1), which decays to the triplet state (T_1).

$$L_3Ru(S_0) \xrightarrow{\text{green photon}} L_3Ru^*(S_1) \xrightarrow{-\text{heat}} L_3Ru^*(T_1)$$

We denote an excited state with an asterisk. The excited triplet $L_3Ru^*(T_1)$ is relatively long lived in the absence of O_2. It can transfer its excitation energy to ground-state anthracene to make an excited triplet state of anthracene.

$$L_3Ru^*(T_1) + A(S_0) \longrightarrow L_3Ru(S_0) + A^*(T_1)$$

The reaction of $L_3Ru^*(T_1)$ with $A(S_0)$ conserves electron spin angular momentum, with two spins up and two spins down in reactants and products. In general, reactions that conserve spin angular momentum are faster than reactions in which spin changes. All reactions below conserve spin angular momentum.

Triplet anthracene survives long enough to allow two triplets to collide. One is promoted to the excited singlet state S_1 and the other is demoted to the ground state S_0.

$$A^*(T_1) + A^*(T_1) \longrightarrow A^*(S_1) + A(S_0)$$

Finally, excited singlet anthracene can emit a blue photon to return to the ground state.

$$A^*(S_1) \xrightarrow{\text{blue fluorescence}} A(S_0)$$

The net result is the conversion of two green photons absorbed by $L_3Ru(II)$ into one blue photon emitted by anthracene from its excited singlet state. This rare combination of cleverly selected reactions converts green light into blue light.

The measured *quantum yield* for this process is 3.3%. That is, for every 100 green photons absorbed, 3.3 blue photons are emitted.

Upconverting phosphors are used in commercial medical diagnostic kits[31] and have been applied to the measurement of nitric oxide in live cells and tissues.[32] The advantage of upconversion for biomedical probes is that low-energy, near-infrared (800 to 1 000 nm) incident radiation stimulates little background emission from the complex biological matrix that can be highly fluorescent under visible radiation.

---

in the vicinity of the beads. More than three-fourths of available intracellular O_2 is consumed by cells within 2 min at 21°C after removing O_2 from the external liquid medium.

Conditions	Intracellular O_2 (ppm)
Air-saturated buffer solution	8.8 ± 0.8
Cells in air-saturated buffer	7.9 ± 2.1
Cells in buffer 25 s after removing O_2	6.5 ± 1.7
Cells in buffer 120 s after removing O_2	≤ 1.5

We end this chapter with Color Plate 21, which shows blue fluorescence from a solution irradiated with green light. Blue photons carry more energy than green photons, so how could this happen? Read Box 19-2 for an explanation.

Terms to Understand

flow injection analysis
immunoassay
isosbestic point

method of continuous variation
quantum yield
quenching

Scatchard plot
sequential injection

Summary

The absorbance of a mixture is the sum of absorbances of the individual components. At a minimum, you should be able to find the concentrations of two species in a mixture by writing and solving two simultaneous equations for absorbance at two wavelengths. This procedure is most accurate if the two absorption spectra have regions where they do not overlap very much. With a spreadsheet, you should be able to use

matrix operations to solve n simultaneous Beer's law equations for n components in a solution, with measurements at n wavelengths. You should be able to use Excel Solver to decompose a spectrum into a sum of spectra of the components by minimizing the function $(A_{calc} - A_m)^2$.

Isosbestic (crossing) points are observed when a solution contains variable proportions of two components with a constant total concentration. A Scatchard plot is used to measure an equilibrium constant. You should be able to apply Excel Solver to fit nonlinear experimental measurements such as absorbance or fluorescence to the equilibrium expression to find an equilibrium constant. The method of continuous variation (Job's plot) allows us to determine the stoichiometry of a complex.

In flow injection analysis, sample injected into a flowing carrier stream is mixed with a color-forming reagent and passed into a flow-through detector. Analyte disperses and reacts with reagent without reaching equilibrium. Precision depends on carrying out the process reproducibly. In sequential injection, sample and reagent are separately aspirated into the carrier by a computer-controlled syringe pump. After being stationary in the reaction coil, analyte, product, and reagent are pushed through the detector.

Immunoassays use antibodies to detect the analyte of interest. In an enzyme-linked immunosorbent assay, signal is amplified by coupling analyte to an enzyme that catalyzes many cycles of a reaction producing a colored or fluorescent product. Time-resolved fluorescence measurements provide sensitivity by separating analyte fluorescence in time and wavelength from background fluorescence.

Luminescence intensity is proportional to the concentration of the emitting species if the concentration is low enough. We can measure some analytes, such as O_2, by their ability to quench (decrease) the luminescence of another compound.

Exercises

19-A. [icon] This problem can be worked with Equations 19-6 on a calculator or with the spreadsheet in Figure 19-4. Transferrin is the iron-transport protein found in blood. It has a molecular mass of 81 000 and carries two Fe^{3+} ions. Desferrioxamine B is a chelator used to treat patients with iron overload (see the opening of Chapter 12). It has a molecular mass of about 650 and can bind one Fe^{3+}. Desferrioxamine can take iron from many sites within the body and is excreted (with its iron) through the kidneys. Molar absorptivities of these compounds (saturated with iron) at two wavelengths are given in the table. Both compounds are colorless (no visible absorption) in the absence of iron.

	ε (M^{-1} cm^{-1})	
λ (nm)	Transferrin	Desferrioxamine
428	3 540	2 730
470	4 170	2 290

(a) A solution of transferrin exhibits an absorbance of 0.463 at 470 nm in a 1.000-cm cell. Calculate the concentration of transferrin in milligrams per milliliter and the concentration of bound iron in micrograms per milliliter.

(b) After adding desferrioxamine (which dilutes the sample), the absorbance at 470 nm was 0.424, and the absorbance at 428 nm was 0.401. Calculate the fraction of iron in transferrin and the fraction in desferrioxamine. Remember that transferrin binds two iron atoms and desferrioxamine binds only one.

19-B. [icon] The spreadsheet lists molar absorptivities of three dyes and the absorbance of a mixture of the dyes in a 1.000-cm cell. Use the least-squares procedure in Figure 19-3 to find the concentration of each dye in the mixture.

	A	B	C	D	E
1	Mixture of Dyes				
2					Absorbance
3	Wavelength	———	Molar Absorptivity	———	of mixture
4	(nm)	Tartrazine	Sunset Yellow	Ponceau 4R	A_m
5	350	6.229E+03	2.019E+03	4.172E+03	0.557
6	375	1.324E+04	4.474E+03	2.313E+03	0.853
7	400	2.144E+04	7.403E+03	3.310E+03	1.332
8	425	2.514E+04	8.551E+03	4.534E+03	1.603
9	450	2.200E+04	1.275E+04	6.575E+03	1.792
10	475	1.055E+04	1.940E+04	1.229E+04	2.006
11	500	1.403E+03	1.869E+04	1.673E+04	1.821
12	525	0.000E+00	7.641E+03	1.528E+04	1.155
13	550	0.000E+00	3.959E+02	9.522E+03	0.445
14	575	0.000E+00	0.000E+00	1.814E+03	0.084

SOURCE: J. J. B. Nevado, J. R. Flores, and M. J. V. Llerena, "Simultaneous Spectrophotometric Determination of Tartrazine, Sunset Yellow, and Ponceau 4R in Commercial Products," Fresenius J. Anal. Chem. **1998**, 361, 465.

19-C. The protein bovine serum albumin can bind several molecules of the dye methyl orange. To measure the binding constant K for one dye molecule, solutions were prepared with a fixed concentration (X_0) of dye and a larger, variable concentration of protein (P). The equilibrium is Reaction 19-18, with X = methyl orange.

$$(H_3C)_2N-\!\!\!\!\!\!\bigcirc\!\!\!-\!N\!\!=\!\!N\!-\!\!\!\bigcirc\!\!\!-SO_3^- Na^+$$

Methyl orange

Experimental data are shown in cells A16-D20 in the spreadsheet on the next page. The authors report the increase in absorbance (ΔA) at 490 nm as P is added to X. X and PX absorb visible light, but P does not. Equilibrium expression 19-20 applies and [PX] is given by Equation 19-21. Before P is added, the absorbance is $\varepsilon_X X_0$. The increase in absorbance when P is added is

$$\Delta A = \varepsilon_X[X] + \varepsilon_{PX}[PX] - \varepsilon_X X_0 = \varepsilon_X([X] - X_0) + \varepsilon_{PX}[PX] = \Delta\varepsilon[PX]$$

$$= -\varepsilon_X[PX] \text{ from mass balance } [X] = X_0 - [PX] \qquad \Delta\varepsilon = \varepsilon_{PX} - \varepsilon_X$$

The spreadsheet uses Solver to vary K and $\Delta\varepsilon$ in cells B10:B11 to minimize the sum of squares of differences between observed and calculated ΔA in solutions with different amounts of P. Cell E16 computes [PX] from Equation 19-21, which is Equation A on line 6 of the spreadsheet. Cells F16 and G16 find [X] and [P] from mass balances. Cell H16 computes $\Delta A_{calc} = \Delta\varepsilon[PX]$, which is Equation B on line 7.

To estimate a value of K in cell B10, *suppose* that 50% of X has reacted in row 20 of the spreadsheet. The total concentration of X is $X_0 = 5.7$ μM. If half is reacted, then $[X] = [PX] = 2.85$ μM and $[P] = P_0 - [PX] = 40.4 - 2.85 = 37.55$ μM. The binding constant is $K = [PX]/([P][X]) = [2.85 \text{ μM}]/([37.55 \text{ μM}][2.85 \text{ μM}]) = 2.7 \times 10^4$, which we enter as our guess for K in cell B10. We estimate $\Delta\varepsilon$ in cell B11 by *supposing* that 50% of X has reacted in row 20. In Equation B on line 7, $\Delta A = \Delta\varepsilon[PX]$. The measured value of ΔA in row 20 is 0.0291 and we just estimated that [PX] = 2.85 μM. Therefore, our guess for $\Delta\varepsilon$ in cell B11 is $\Delta\varepsilon = \Delta A/[PX] = (0.0291)/(2.85 \text{ μM}) = 1.0 \times 10^4$.

Your assignment is to write formulas in columns E through J of the spreadsheet to reproduce what is shown and to find values in cells E17:J20. Then use Solver to find K and $\Delta\varepsilon$ in cells B10:B11 to minimize $\Sigma(A_{obs} - A_{calc})^2$ in cell I21.

	A	B	C	D	E	F	G	H	I	J
1	Finding K for Binding of Methyl Orange (X) to Bovine Serum Albumin (P)									
2	Aliquots of protein were added to 1.00 mL of 5.7 µM methyl orange									
3	X_o = total concentration of methyl orange									
4	P_o = total concentration of protein									
5	K = [PX]/([P][X])									
6	$[PX] = [K(P_o+X_o)+1 \pm \sqrt{[K(P_o+X_o)+1]^2 - 4K^2X_oP_o}]/(2K)$					← (Eq. A)				
7	Change in absorbance at 490 nm = $\Delta A = [PX]\Delta\varepsilon$ for 1 cm pathlength					← (Eq. B)				
8	$\Delta\varepsilon = \varepsilon_{PX} - \varepsilon_X$									
9	Parameters to find with Solver:									
10	K =	2.70E+04		initial estimate =	2.7E+04					
11	$\Delta\varepsilon$ =	1.00E+04		initial estimate =	1.0E+04	$M^{-1}cm^{-1}$				
12										
13					[PX] (M)					[PX]/X_o
14	Protein	X_o	P_o	ΔA_{obs}	Eq. A using	[X] =	[P] =	ΔA_{calc}	$(\Delta A_{obs}-A_{calc})^2$	fraction of
15	addition	(M)	(M)	at 490 nm	minus sign	X_o-[PX]	P_o-[PX]	Eq. B		X reacted
16	1	5.7E-06	8.0E-06	0.0118	9.153E-07	4.785E-06	7.085E-06	0.0092	7.0086E-06	0.161
17	2	5.7E-06	1.14E-05	0.0148						
18	3	5.7E-06	1.63E-05	0.0187						
19	4	5.7E-06	3.28E-05	0.0268						
20	5	5.7E-06	4.04E-05	0.0291						
21								sum =	7.0086E-06	

Spreadsheet for Exercise 19-C.

SOURCE: A. Örstan and J. F. Wojcik, "Spectroscopic Determination of Protein-Ligand Binding Constants," J. Chem. Ed. **1987**, 64, 814.

19-D. Complex formation by 3-aminopyridine and picric acid in chloroform solution gives a yellow product with an absorbance maximum at 400 nm. Neither starting material absorbs significantly at this wavelength. Stock solutions containing 1.00×10^{-4} M of each compound were mixed as indicated, and the absorbances of the mixtures were recorded. Prepare a graph of absorbance versus mole fraction of 3-aminopyridine and find the stoichiometry of the complex.

Picric acid (mL)	3-Aminopyridine (mL)	Absorbance at 400 nm
2.70	0.30	0.106
2.40	0.60	0.214
2.10	0.90	0.311
1.80	1.20	0.402
1.50	1.50	0.442
1.20	1.80	0.404
0.90	2.10	0.318
0.60	2.40	0.222
0.30	2.70	0.110

SOURCE: E. Bruneau, D. Lavabre, G. Levy, and J. C. Micheau, "Quantitative Analysis of Continuous-Variation Plots with a Comparison of Several Methods," J. Chem. Ed. **1992**, 69, 833.

Picric acid

3-Aminopyridine

Problems

Analysis of a Mixture

19-1. This problem can be worked by calculator or with the spreadsheet in Figure 19-4. Consider compounds X and Y in the example labeled "Analysis of a Mixture, Using Equations 19-6" on page 464. Find [X] and [Y] in a solution whose absorbance is 0.233 at 272 nm and 0.200 at 327 nm in a 0.100-cm cell.

19-2. The figure shows spectra of 1.00×10^{-4} M MnO_4^-, 1.00×10^{-4} M $Cr_2O_7^{2-}$, and an unknown mixture of both, all in 1.000-cm-pathlength cells. Absorbances are given in the table. Use the least-squares procedure in Figure 19-3 to find the concentration of each species in the mixture.

Wavelength (nm)	MnO_4^- standard	$Cr_2O_7^{2-}$ standard	Mixture
266	0.042	0.410	0.766
288	0.082	0.283	0.571
320	0.168	0.158	0.422
350	0.125	0.318	0.672
360	0.056	0.181	0.366

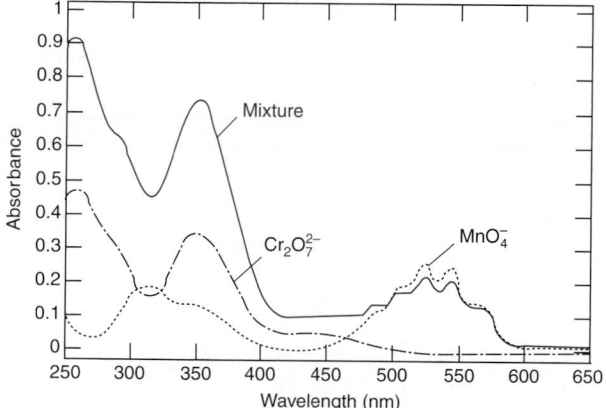

Visible spectrum of MnO_4^-, $Cr_2O_7^{2-}$, and an unknown mixture containing both ions. [Data from M. Blanco, H. Iturriaga, S. Maspoch, and P. Tarín, "A Simple Method for Spectrophotometric Determination of Two-Components with Overlapped Spectra," J. Chem. Ed. **1989**, 66, 178.]

19-3. When are isosbestic points observed and why?

19-4. The metal ion indicator xylenol orange (Table 12-3) is yellow at pH 6 ($\lambda_{max} = 439$ nm). The spectral changes that occur as VO^{2+} is added to the indicator at pH 6 are shown here. The mole ratio VO^{2+}/xylenol orange at each point is

Trace	Mole ratio	Trace	Mole ratio	Trace	Mole ratio
0	0	6	0.60	12	1.3
1	0.10	7	0.70	13	1.5
2	0.20	8	0.80	14	2.0
3	0.30	9	0.90	15	3.1
4	0.40	10	1.0	16	4.1
5	0.50	11	1.1		

Suggest a sequence of chemical reactions to explain the spectral changes, especially the isosbestic points at 457 and 528 nm.

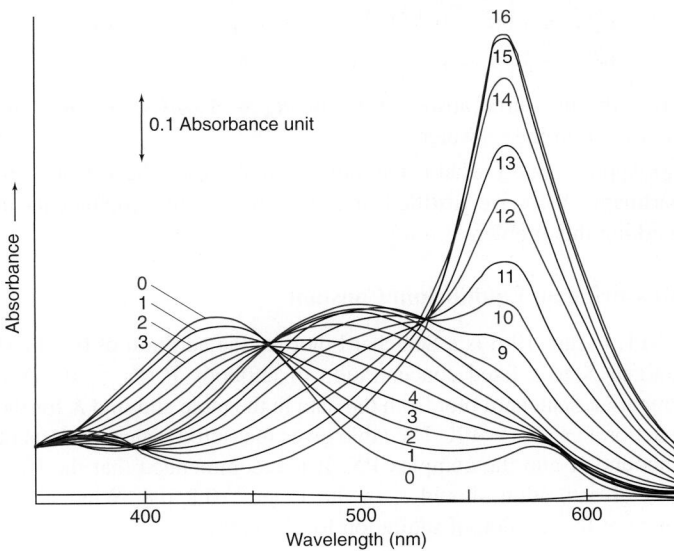

Absorption spectra for the reaction of xylenol orange with VO^{2+} at pH 6.0.
[Data from D. C. Harris and M. H. Gelb, "Binding of Xylenol Orange to Transferrin: Demonstration of Metal-Anion Linkage," *Biochim. Biophys. Acta* **1980**, *623, 1.*]

19-5. Infrared spectra are customarily recorded on a transmittance scale so that weak and strong bands can be displayed on the same scale. The region near 2 000 cm^{-1} in the infrared spectra of compounds A and B is shown in the figure. Note that absorption corresponds to a downward peak on this scale. The spectra were recorded from a 0.010 0 M solution of each, in cells with 0.005 00-cm pathlengths. A mixture of A and B in a 0.005 00-cm cell gave a transmittance of 34.0% at 2 022 cm^{-1} and 38.3% at 1 993 cm^{-1}. Find the concentrations of A and B.

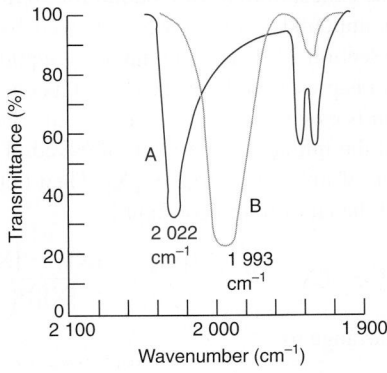

Wavenumber	Pure A	Pure B
2 022 cm^{-1}	31.0% *T*	97.4% *T*
1 993 cm^{-1}	79.7% *T*	20.0% *T*

19-6. Spectroscopic data for the indicators thymol blue (TB), semithymol blue (STB), and methylthymol blue (MTB) are shown in the table. A solution of TB, STB, and MTB in a 1.000-cm cuvet had absorbances of 0.412 at 455 nm, 0.350 at 485 nm, and 0.632 at 545 nm. Modify the spreadsheet in Figure 19-4 to handle three simultaneous equations and find [TB], [STB], and [MTB] in the mixture.

λ (nm)	ε(M^{-1} cm^{-1})		
	TB	STB	MTB
455	4 800	11 100	18 900
485	7 350	11 200	11 800
545	36 400	13 900	4 450

*SOURCE: S. Kiciak, H. Gontarz, and E. Krzyżanowska, "Monitoring the Synthesis of Semimethylthymol Blue and Methylthymol Blue," Talanta **1995**, 42, 1245.*

19-7. The spreadsheet gives the product εb for four pure compounds and a mixture at infrared wavelengths. Modify Figure 19-4 to solve four equations and find the concentration of each compound. You can treat the coefficient matrix as if it were molar absorptivity because the pathlength was constant (but unknown) for all measurements.

Wavelength (μm)	Coefficient matrix (εb)				Absorbance of unknown
	p-xylene	*m*-xylene	*o*-xylene	ethylbenzene	
12.5	1.5020	0.0514	0	0.0408	0.1013
13.0	0.0261	1.1516	0	0.0820	0.09943
13.4	0.0342	0.0355	2.532	0.2933	0.2194
14.3	0.0340	0.0684	0	0.3470	0.03396

*SOURCE: Z. Zdravkovski, "Mathcad in Chemistry Calculations. II. Arrays," J. Chem. Ed. **1992**, 69, 242A.*

19-8. *Two ways to analyze a mixture.* Figure 19-5 shows the spectrum of the indicator bromothymol blue adjusted to several pH values. The spectrum at pH 12.2 is that of the pure blue form and the spectrum at pH 1.8 is that of the pure yellow form. At other pH values, there is a mixture of the two forms. The total concentration is 10 μM and the pathlength is 1.0 cm in all spectra. For the purpose of calculation, assume that there are more than two significant digits in concentration and pathlength. Absorbance at the dots on three of the curves in Figure 19-5 is given in the table.

Wavelength (nm)	Absorbance of standards		Absorbance of mixture pH 6.78
	Blue (pH 12.2) In$^-$	Yellow (pH 1.8) HIn	
400	0.182	0.274	0.248
432	0.076	0.344	0.265
450	0.046	0.323	0.237
550	0.356	0.027	0.131
570	0.570	0.011	0.193
616	0.811	0.005	0.272

*SOURCE: E. Klotz, R. Doyle, E. Gross, and B. Mattson, "The Equilibrium Constant for Bromothymol Blue," J. Chem. Ed. **2011**, 88, 637.*

(a) Prepare a spreadsheet like Figure 19-3 to use absorption at all six wavelengths to find [In$^-$] and [HIn] in the mixture. Comment on the sum [In$^-$] + [HIn].

(b) From [In$^-$] and [HIn] in the mixture, and from pK_a = 7.10 for HIn, calculate the pH of the mixture. (This calculation is the source of pH labels in the figure.)

(c) Use Equations 19-6 at the peak wavelengths of 432 and 616 nm to find [In$^-$] and [HIn] in the mixture. Compare your answers to those in **(a)**. Which answers, **(a)** or **(c)**, are probably more accurate? Why?

19-9. *Challenging your acid-base prowess.* A solution was prepared by mixing 25.00 mL of 0.080 0 M aniline, 25.00 mL of 0.060 0 M sulfanilic acid, and 1.00 mL of 1.23×10^{-4} M HIn and then diluting to 100.0 mL. (HIn stands for protonated indicator.)

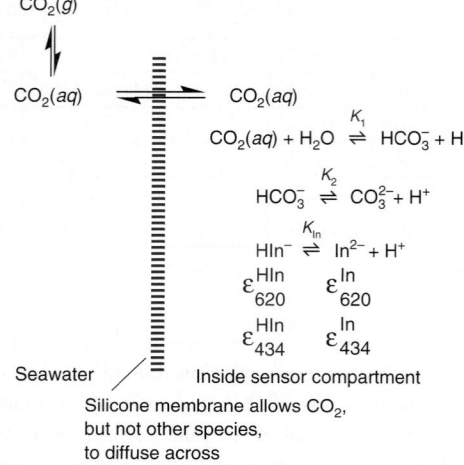

Anilinium ion
$pK_a = 4.601$

Sulfanilic acid
$pK_a = 3.232$

$$HIn \rightleftharpoons H^+ + In^-$$

$\varepsilon_{325} = 2.45 \times 10^4 \ M^{-1} \ cm^{-1}$ $\varepsilon_{325} = 4.39 \times 10^3 \ M^{-1} \ cm^{-1}$
$\varepsilon_{550} = 2.26 \times 10^4 \ M^{-1} \ cm^{-1}$ $\varepsilon_{550} = 1.53 \times 10^4 \ M^{-1} \ cm^{-1}$

The absorbance measured at 550 nm in a *5.00-cm cell* was 0.110. Find the concentrations of HIn and In⁻ and pK_a for HIn.

19-10. *Chemical equilibrium and analysis of a mixture.* (*Warning! This is a long problem.*) A remote optical sensor for CO_2 in the ocean was designed to operate without the need for calibration.[33]

CO₂(g)

CO₂(aq)   CO₂(aq)

$$CO_2(aq) + H_2O \overset{K_1}{\rightleftharpoons} HCO_3^- + H^+$$

$$HCO_3^- \overset{K_2}{\rightleftharpoons} CO_3^{2-} + H^+$$

$$HIn^- \overset{K_{In}}{\rightleftharpoons} In^{2-} + H^+$$

ε_{620}^{HIn} ε_{620}^{In}

ε_{434}^{HIn} ε_{434}^{In}

Seawater   Inside sensor compartment

Silicone membrane allows CO_2, but not other species, to diffuse across

The sensor compartment is separated from seawater by a silicone membrane through which CO_2, but not dissolved ions, can diffuse. Inside the sensor, CO_2 equilibrates with HCO_3^- and CO_3^{2-}. For each measurement, the sensor is flushed with fresh solution containing 50.0 μM bromothymol blue indicator (NaHIn) and 42.0 μM NaOH. All indicator is in the form HIn^- or In^{2-} near neutral pH, so we can write two mass balances: (1) $[HIn^-] + [In^{2-}] = F_{In} = 50.0$ μM and (2) $[Na^+] = F_{Na} = 50.0$ μM $+ 42.0$ μM $= 92.0$ μM. HIn^- has an absorbance maximum at 434 nm and In^{2-} has a maximum at 620 nm. The sensor measures the absorbance *ratio* $R_A = A_{620}/A_{434}$ reproducibly without need for calibration. From this ratio, we can find $[CO_2(aq)]$ in the seawater as outlined here:

(a) From Beer's law for the mixture, write equations for $[HIn^-]$ and $[In^{2-}]$ in terms of the absorbances at 620 and 434 nm (A_{620} and A_{434}). Then show that

$$\frac{[In^{2-}]}{[HIn^-]} = \frac{R_A \varepsilon_{434}^{HIn} - \varepsilon_{620}^{HIn}}{\varepsilon_{620}^{In^{2-}} - R_A \varepsilon_{434}^{In^{2-}}} \equiv R_{In} \tag{A}$$

(b) From the mass balance (1) and the acid dissociation constant K_{In}, show that

$$[HIn^-] = \frac{F_{In}}{R_{In} + 1} \tag{B}$$

$$[In^{2-}] = \frac{K_{In} F_{In}}{[H^+](R_{In} + 1)} \tag{C}$$

(c) Show that $[H^+] = K_{In}/R_{In}$.   **(D)**

(d) From the carbonic acid dissociation equilibria, show that

$$[HCO_3^-] = \frac{K_1[CO_2(aq)]}{[H^+]} \tag{E}$$

$$[CO_3^{2-}] = \frac{K_1 K_2[CO_2(aq)]}{[H^+]^2} \tag{F}$$

(e) Write the charge balance for the solution in the sensor compartment. Substitute in expressions B, C, E, and F for $[HIn^-]$, $[In^{2-}]$, $[HCO_3^-]$, and $[CO_3^{2-}]$.

(f) Suppose that the various constants have the following values:

$\varepsilon_{434}^{HIn} = 8.00 \times 10^3 \ M^{-1} \ cm^{-1}$ $K_1 = 3.0 \times 10^{-7}$
$\varepsilon_{620}^{HIn} = 0$ $K_2 = 3.3 \times 10^{-11}$
$\varepsilon_{434}^{In^{2-}} = 1.90 \times 10^3 \ M^{-1} \ cm^{-1}$ $K_{In} = 2.0 \times 10^{-7}$
$\varepsilon_{620}^{In^{2-}} = 1.70 \times 10^4 \ M^{-1} \ cm^{-1}$ $K_w = 6.7 \times 10^{-15}$

From the measured absorbance ratio $R_A = A_{620}/A_{434} = 2.84$, find $[CO_2(aq)]$ in the seawater.

(g) Approximately what is the ionic strength inside the sensor compartment? Were we justified in neglecting activity coefficients in working this problem?

Measuring an Equilibrium Constant

19-11. Figure 19-6 is a Scatchard plot for the addition of 0–20 nM antigen X to a fixed concentration of antibody P ($P_0 = 10$ nM). Prepare a Scatchard plot from the data in the table and find K for the reaction $P + X \rightleftharpoons PX$. The table gives measured concentrations of unbound X and the complex PX. It is recommended that the fraction of saturation should span the range ~0.2–0.8. What is the range of the fraction of saturation for the data?

[X] (nM)	[PX] (nM)	[X] (nM)	[PX] (nM)
0.12_0	2.8_7	0.73_1	6.2_9
0.19_2	3.8_0	1.2_2	6.7_7
0.29_6	4.6_6	1.5_0	7.5_2
0.45_0	5.5_4	3.6_1	8.4_5

19-12. *Scatchard plot for binding of estradiol to albumin.* Data in the table come from a student experiment to measure the binding constant of the radioactively labeled hormone estradiol (X) to the protein, bovine serum albumin (P). Estradiol (7.5 nM) was equilibrated with various concentrations of albumin for 30 min at 37°C. A small fraction of unbound estradiol was removed by solid-phase microextraction (Section 24-4) and measured by liquid scintillation counting. Albumin is present in large excess, so its concentration in any given solution is essentially equal to its initial concentration in that solution. Call the initial concentration of estradiol $[X]_0$ and the final concentration of unbound estradiol $[X]$. Then bound estradiol is $[X]_0 - [X]$ and the equilibrium constant is

$$X + P \rightleftharpoons PX \quad K = \frac{[PX]}{[X][P]} = \frac{[X]_0 - [X]}{[X][P]}$$

which you can rearrange to

$$\frac{[X]_0}{[X]} = K[P] + 1$$

A graph of $[X]_0/[X]$ versus $[P]$ should be a straight line with a slope of K. The quotient $[X]_0/[X]$ is equal to the counts of radioactive estradiol extracted from a solution without albumin divided by the counts of estradiol extracted from a solution with estradiol.

(a) Prepare a graph of $[X]_0/[X]$ versus $[P]$ and find K. From the standard deviation of the slope, find the 95% confidence interval for K.

(b) What fraction of estradiol is bound to albumin at the first and last points?

$[X]_0/[X]$	$[P]$ (μM)	$[X]_0/[X]$	$[P]$ (μM)
1.26	6.3	3.33	50.0
1.62	10.0	4.19	60.0
2.16	20.0	4.13	70.0
2.51	30.0	4.36	80.0
3.34	40.0		

SOURCE: *P. Liang, B. Adhyaru, W. L. Pearson, and K. R. Williams, "The Binding Constant of Estradiol to Bovine Serum Albumin," J. Chem. Ed.* **2006**, *83, 294.*

19-13. Iodine reacts with mesitylene to form a complex with an absorption maximum at 332 nm in CCl_4 solution:

$$K = \frac{[\text{complex}]}{[I_2][\text{mesitylene}]}$$

Iodine Mesitylene Complex
$\varepsilon_{332} \approx 0$ $\varepsilon_{332} \approx 0$ $\lambda_{max} = 332$ nm

(a) Given that the product absorbs at 332 nm, but neither reactant has significant absorbance at this wavelength, use the equilibrium constant, K, and Beer's law to show that

$$\frac{A}{[\text{mesitylene}][I_2]_{tot}} = K\varepsilon - \frac{KA}{[I_2]_{tot}}$$

where A is the absorbance at 332 nm, ε is the molar absorptivity of the complex at 332 nm, [mesitylene] is the concentration of free mesitylene, and $[I_2]_{tot}$ is the total concentration of iodine in the solution $(= [I_2] + [\text{complex}])$. The cell pathlength is 1.000 cm.

(b) Spectrophotometric data for this reaction are shown in the table. Because $[\text{mesitylene}]_{tot} \gg [I_2]$, we can say that $[\text{mesitylene}] \approx [\text{mesitylene}]_{tot}$. Prepare a graph of $A/([\text{mesitylene}][I_2]_{tot})$ versus $A/[I_2]_{tot}$ and find the equilibrium constant and molar absorptivity of the complex.

$[\text{Mesitylene}]_{tot}$ (M)	$[I_2]_{tot}$ (M)	Absorbance at 332 nm
1.690	7.817×10^{-5}	0.369
0.921 8	2.558×10^{-4}	0.822
0.633 8	3.224×10^{-4}	0.787
0.482 9	3.573×10^{-4}	0.703
0.390 0	3.788×10^{-4}	0.624
0.327 1	3.934×10^{-4}	0.556

SOURCE: *P. J. Ogren and J. R. Norton, "Applying a Simple Linear Least-Squares Algorithm to Data with Uncertainties in Both Variables," J. Chem. Ed.* **1992**, *69, A130.*

19-14. 🖳 Now we use Solver to find K for the previous problem. The only absorbing species at 332 nm is the complex, so, from Beer's law, $[\text{complex}] = A/\varepsilon$ (because pathlength $= 1.000$ cm). I_2 is either free or bound in the complex, so $[I_2] = [I_2]_{tot} - [\text{complex}]$. There is a huge excess of mesitylene, so $[\text{mesitylene}] \approx [\text{mesitylene}]_{tot}$.

$$K = \frac{[\text{complex}]}{[I_2][\text{mesitylene}]} = \frac{A/\varepsilon}{([I_2]_{tot} - A/\varepsilon)[\text{mesitylene}]_{tot}}$$

The spreadsheet shows some of the data. You will need to use all the data. Column A contains [mesitylene] and column B contains

$[I_2]_{tot}$. Column C lists the measured absorbance. *Guess* a value of the molar absorptivity of the complex, ε, in cell A7. Then compute the concentration of the complex $(= A/\varepsilon)$ in column D. The equilibrium constant in column E is given by E2 = [complex]/([I2][mesitylene]) = (D2)/((B2−D2)*A2).

	A	B	C	D	E
1	[Mesitylene]	[I2]tot	A	[Complex] = A/ε	Keq
2	1.6900	7.817E-05	0.369	7.380E-05	9.99282
3	0.9218	2.558E-04	0.822	1.644E-04	1.95128
4	0.6338	3.224E-04	0.787	1.574E-04	1.50511
5				Average =	4.48307
6	Guess for ε:			Standard Dev =	4.77680
7	5.000E+03			Stdev/Average =	1.06552

What should we minimize with Solver? We want to vary ε in cell A7 until the values of K in column E are as constant as possible. We would like to minimize a function like $\Sigma(K_i - K_{average})^2$, where K_i is the value in each line of the table and $K_{average}$ is the average of all computed values. The problem with $\Sigma(K_i - K_{average})^2$ is that we can minimize this function simply by making K_i very small, but not necessarily constant. What we really want is for all the K_i to be clustered around the mean value. A good way to do this is to minimize the *relative standard deviation* of the K_i, which is (standard deviation)/average. In cell E5, we compute the average value of K and in cell E6 the standard deviation. Cell E7 contains the relative standard deviation. Use Solver to minimize cell E7 by varying cell A7. Compare your answer with that of Problem 19-13.

Method of Continuous Variation

19-15. *Method of continuous variation.* Make a graph of absorbance versus mole fraction of thiocyanate from the data in the table.

mL Fe^{3+} solution[a]	mL SCN^- solution[b]	Absorbance at 455 nm
30.00	0	0.001
27.00	3.00	0.122
24.00	6.00	0.226
21.00	9.00	0.293
18.00	12.00	0.331
15.00	15.00	0.346
12.00	18.00	0.327
9.00	21.00	0.286
6.00	24.00	0.214
3.00	27.00	0.109
0	30.00	0.002

a. Fe^{3+} solution: 1.00 mM $Fe(NO_3)_3$ + 10.0 mM HNO_3.

b. SCN^- solution: 1.00 mM KSCN + 15.0 mM HCl.

SOURCE: *Z. D. Hill and P. MacCarthy, "Novel Approach to Job's Method," J. Chem. Ed.* **1986**, *63, 162.*

(a) What is the stoichiometry of the predominant $Fe(SCN)_n^{3-n}$ species?

(b) Why is the peak not as sharp as those in Figure 19-10?

(c) Why does one solution contain 10.0 mM acid and the other 15.0 mM acid?

19-16. The indicator xylenol orange (Table 12-3) forms a complex with Zr(IV) in HCl solution. Prepare a Job plot from the data in the table and suggest the stoichiometry of the complex (xylenol orange)$_x$Zr$_z$.

Mole fraction xylenol orange	Relative absorbance at $\lambda_{max} = 550$ nm	Mole fraction xylenol orange	Relative absorbance at $\lambda_{max} = 550$ nm
0.100	0.110	0.600	0.236
0.200	0.235	0.700	0.145
0.300	0.352	0.800	0.088
0.400	0.412	0.900	0.045
0.500	0.348		

SOURCE: L. Perring and B. Bourqui, "An Alternative Method for Fluoride Determination by Ion Chromatography Using Post-Column Reaction," Am. Lab. News Ed., March 2002, p. 28.

19-17. *Simulating a Job's plot.* Consider the reaction A + 2B $\rightleftharpoons$ AB$_2$, for which $K = [AB_2]/[A][B]^2$. Suppose that the following mixtures of A and B at a fixed total concentration of 10^{-4} M are prepared:

$[A]_{total}$ (M)	$[B]_{total}$ (M)	$[A]_{total}$ (M)	$[B]_{total}$ (M)
1.00×10^{-5}	9.00×10^{-5}	5.00×10^{-5}	5.00×10^{-5}
2.00×10^{-5}	8.00×10^{-5}	6.00×10^{-5}	4.00×10^{-5}
2.50×10^{-5}	7.50×10^{-5}	7.00×10^{-5}	3.00×10^{-5}
3.00×10^{-5}	7.00×10^{-5}	8.00×10^{-5}	2.00×10^{-5}
3.33×10^{-5}	6.67×10^{-5}	9.00×10^{-5}	1.00×10^{-5}
4.00×10^{-5}	6.00×10^{-5}		

(a) Prepare a spreadsheet to find the concentration of AB$_2$ for each mixture, for equilibrium constants of $K = 10^6$, 10^7, and 10^8. One way to do this is to enter the values of $[A]_{total}$ and $[B]_{total}$ in columns A and B, respectively. Then put a trial (guessed) value of $[AB_2]$ in column C. From the mass balances $[A]_{total} = [A] + [AB_2]$ and $[B]_{total} = [B] + 2[AB_2]$, we can write $K = [AB_2]/[A][B]^2 = [AB_2]/\{([A]_{total} - [AB_2])([B]_{total} - 2[AB_2])^2\}$. In column D, enter the reaction quotient $[AB_2]/[A][B]^2$. For example, cell D2 has the formula "=C2/((A2-C2)(B2-2*C2)^2)". Then vary the value of $[AB_2]$ in cell C2 with Solver until the reaction quotient in cell D2 is equal to the desired equilibrium constant (such as 10^8).

(b) Prepare a graph by the method of continuous variation in which you plot $[AB_2]$ versus mole fraction of A for each equilibrium constant. Explain the shapes of the curves.

19-18. A study was conducted with derivatives of the DNA nucleotide bases adenine and thymine bound inside micelles (Box 26-1) in aqueous solution.

O
‖
CH$_3$CNH

N
N
N
N

Adenine derivative

O
HN
O
N

N(CH$_3$)$_3^+$Br$^-$

Thymine derivative

O
‖
CH$_3$(CH$_2$)$_{11}$OSO$^-$Na$^+$
‖
O

⫴

$\sim\sim\sim\sim$ $\ominus\oplus$

Sodium dodecyl sulfate

Sodium dodecyl sulfate forms micelles with the hydrocarbon tails pointed inward and ionic headgroups exposed to water. It was hypothesized that the bases would form a 1:1 hydrogen-bonded complex inside the micelle as they do in DNA:

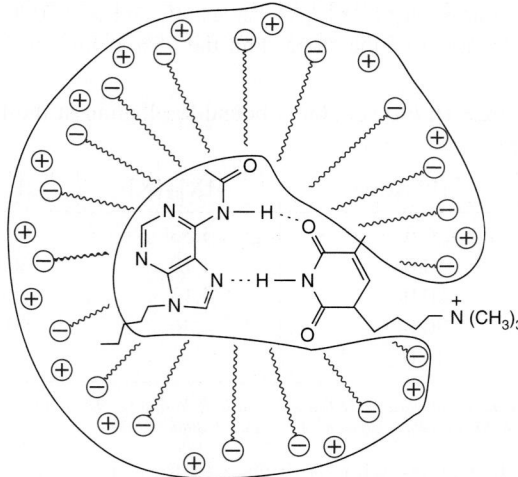

Hydrogen-bonded base pair inside micelle, with hydrocarbon tails anchoring the bases to the micelle.

To test the hypothesis, aliquots of 5.0 mM adenine derivative were mixed with aliquots of 5.0 mM thymine derivative in proportions shown in the table. Each solution also contained 20 mM sodium dodecyl sulfate. The concentration of product measured by nuclear magnetic resonance also is shown in the table. Are the results consistent with formation of a 1:1 complex? Explain your answer.

Adenine volume (mL)	Thymine volume (mL)	Concentration of product (mM)
0.450	0.050	0.118 ± 0.009
0.400	0.100	0.202 ± 0.038
0.350	0.150	0.265 ± 0.021
0.300	0.200	0.307 ± 0.032
0.250	0.250	0.312 ± 0.060
0.200	0.300	0.296 ± 0.073
0.150	0.350	0.260 ± 0.122
0.100	0.400	0.187 ± 0.110
0.050	0.450	0.103 ± 0.104

Flow Injection, Luminescence, and Immunoassay

19-19. Explain what is done in flow injection analysis and sequential injection. What is the principal difference between the two techniques? Which one is called "lab-on-a-valve"?

19-20. Explain how signal amplification is achieved in enzyme-linked immunosorbent assays.

19-21. What is the advantage of a time-resolved emission measurement with Eu^{3+} versus measurement of fluorescence from organic chromophores?

19-22. The end of Box 19-2 states that "the advantage of upconversion for biomedical probes is that low energy near-infrared (800 to 1000 nm) incident radiation stimulates little background emission from the complex biological matrix that can be highly fluorescent under visible radiation." Suggest why near-infrared radiation stimulates less emission than visible radiation and why this behavior is useful.

19-23. Here is an *immunoassay* to measure explosives such as trinitrotoluene (TNT) in organic solvent extracts of soil. The assay employs a *flow cytometer*, which counts small particles (such as living cells) flowing through a narrow tube past a detector. The cytometer in this experiment irradiates the particles with a green laser and measures fluorescence from each particle as it flows past the detector.

1. Antibodies that bind TNT are chemically attached to 5-μm-diameter latex beads.
2. The beads are incubated with a fluorescent derivative of TNT to saturate the antibodies, and excess TNT derivative is removed. The beads are resuspended in aqueous detergent.

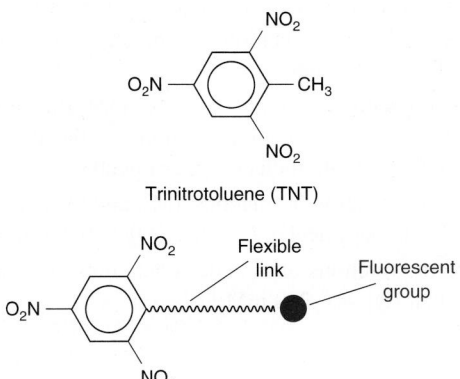

Trinitrotoluene (TNT)

Fluorescently labeled TNT

3. 5 μL of the suspension are added to 100 μL of sample or standard. TNT in the sample or standard displaces some derivatized TNT from bound antibodies. The higher the concentration of TNT, the more derivatized TNT is displaced.
4. An aliquot is injected into the flow cytometer, which measures fluorescence of individual beads as they pass the detector. The figure shows median fluorescence intensity ± standard deviation. TNT can be quantified in the ppb to ppm range.

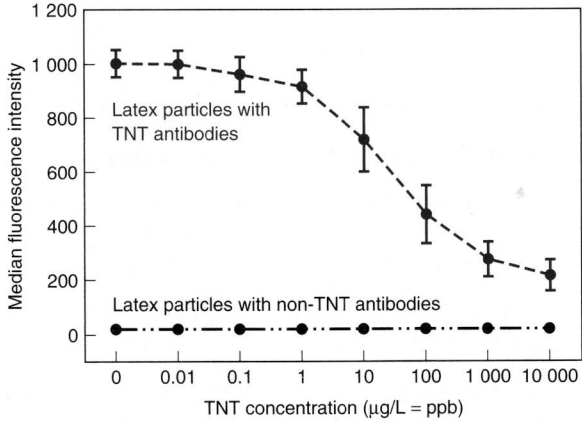

Fluorescence of TNT-antibody-beads versus TNT concentration. [Data from G. P. Anderson, S. C. Moreira, P. T. Charles, I. L. Medintz, E. R. Goldman, M. Zeinali, and C. R. Taitt, "TNT Detection Using Multiplexed Liquid Array Displacement Immunoassays," *Anal. Chem.* **2006,** *78,* 2279.]

Draw pictures showing the state of the beads in steps 1, 2, and 3 and explain how this method works.

19-24. The graph shows the effect of pH on quenching of luminescence of tris(2,2'-bipyridine)Ru(II) by 2,6-dimethylphenol. The ordinate, K_{SV}, is the collection of constants, $k_q/(k_e + k_d)$, in the Stern-Volmer equation 19-26. The greater K_{SV}, the greater the quenching. Suggest an explanation for the shape of the graph and estimate pK_a for 2,6-dimethylphenol.

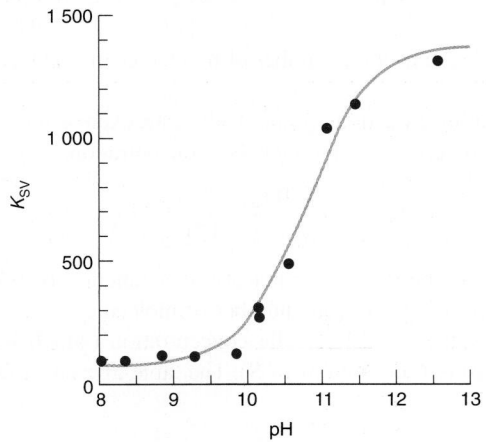

Tris(2,2'-bipyridine)ruthenium(II)

2,6-Dimethylphenol

Quenching of tris(2,2'-bipyridine)Ru(II) by 2,6-dimethylphenol as a function of pH. [Data from H. Gsponer, G. A. Argüello, and G. A. Argüello, "Determinations of pK_a from Luminescence Quenching Data," *J. Chem. Ed.* **1997,** *74,* 968.]

19-25. *Fluorescence quenching in micelles.* Consider an aqueous solution with a high concentration of micelles (Box 26-1) and relatively low concentrations of the fluorescent molecule pyrene and a quencher (cetylpyridinium chloride, designated Q), both of which dissolve in the micelles.

Pyrene
(fluorescent species)

$CH_2(CH_2)_{14}CH_3$

Cetylpyridinium chloride
(quencher, Q)

Quenching occurs if pyrene and Q are in the same micelle. Let the total concentration of quencher be [Q] and the concentration of micelles be [M]. The average number of quenchers per micelle is $\overline{Q} = [Q]/[M]$. If Q is randomly distributed among the micelles, then the probability that a particular micelle has *n* molecules of Q is given by the *Poisson distribution:*[34]

$$\text{Probability of } n \text{ molecules of Q in micelle} \equiv P_n = \frac{\overline{Q}^n}{n!}e^{-\overline{Q}} \quad (1)$$

where *n*! is *n* factorial ($= n[n-1][n-2]\ldots[1]$). The probability that there are no molecules of Q in a micelle is

$$\text{Probability of 0 molecules of Q in micelle} = \frac{\overline{Q}^0}{0!}e^{-\overline{Q}} = e^{-\overline{Q}} \quad (2)$$

because $0! \equiv 1$.

Let I_0 be the fluorescence intensity of pyrene in the absence of Q and let I_Q be the intensity in the presence of Q (both measured at the same concentration of micelles). The quotient I_Q/I_0 must be $e^{-\overline{Q}}$, which is the probability that a micelle does not possess a quencher molecule. Substituting $\overline{Q} = [Q]/[M]$ gives

$$I_Q/I_0 = e^{-\overline{Q}} = e^{-[Q]/[M]} \quad (3)$$

Micelles are made of the surfactant molecule, sodium dodecyl sulfate (shown in Problem 19-18). When surfactant is added to a solution, no micelles form until a minimum concentration called the *critical micelle concentration* (CMC) is attained. When the total concentration of surfactant, [S], exceeds the critical concentration, then the surfactant found in micelles is [S] − [CMC]. The molar concentration of micelles is

$$[M] = \frac{[S] - [CMC]}{N_{av}} \qquad (4)$$

where N_{av} is the average number of molecules of surfactant in each micelle.

Combining Equations 3 and 4 gives an expression for fluorescence as a function of total quencher concentration, [Q]:

$$\ln\frac{I_0}{I_Q} = \frac{[Q]N_{av}}{[S] - [CMC]} \qquad (5)$$

By measuring fluorescence intensity as a function of [Q] at fixed [S], we can find the average number of molecules of S per micelle if we know the critical micelle concentration (which is independently measured in solutions of S). The table gives data for 3.8 μM pyrene in a micellar solution with a total concentration of sodium dodecyl sulfate [S] = 20.8 mM.

Q (μM)	I_0/I_Q	Q (μM)	I_0/I_Q	Q (μM)	I_0/I_Q
0	1	158	2.03	316	4.04
53	1.28	210	2.60	366	5.02
105	1.61	262	3.30	418	6.32

SOURCE: M. F. R. Prieto, M. C. R. Rodríguez, M. M. González, A. M. R. Rodríguez, and J. C. M. Fernández, "Fluorescene Quenching in Microheterogeneous Media," J. Chem. Ed. **1995**, 72, 662.

(a) If micelles were not present, quenching would be expected to follow the Stern-Volmer equation 19-26. Show that the graph of I_0/I_Q versus [Q] is not linear.

(b) The critical micelle concentration is 8.1 mM. Prepare a graph of $\ln(I_0/I_Q)$ versus [Q]. Use Equation 5 to find N_{av}, the average number of sodium dodecyl sulfate molecules per micelle.

(c) Find the concentration of micelles, [M], and the average number of molecules of Q per micelle, $\overline{Q}$, when [Q] = 0.200 mM.

(d) Compute the fractions of micelles containing 0, 1, and 2 molecules of Q when [Q] = 0.200 mM.

20 | Spectrophotometers

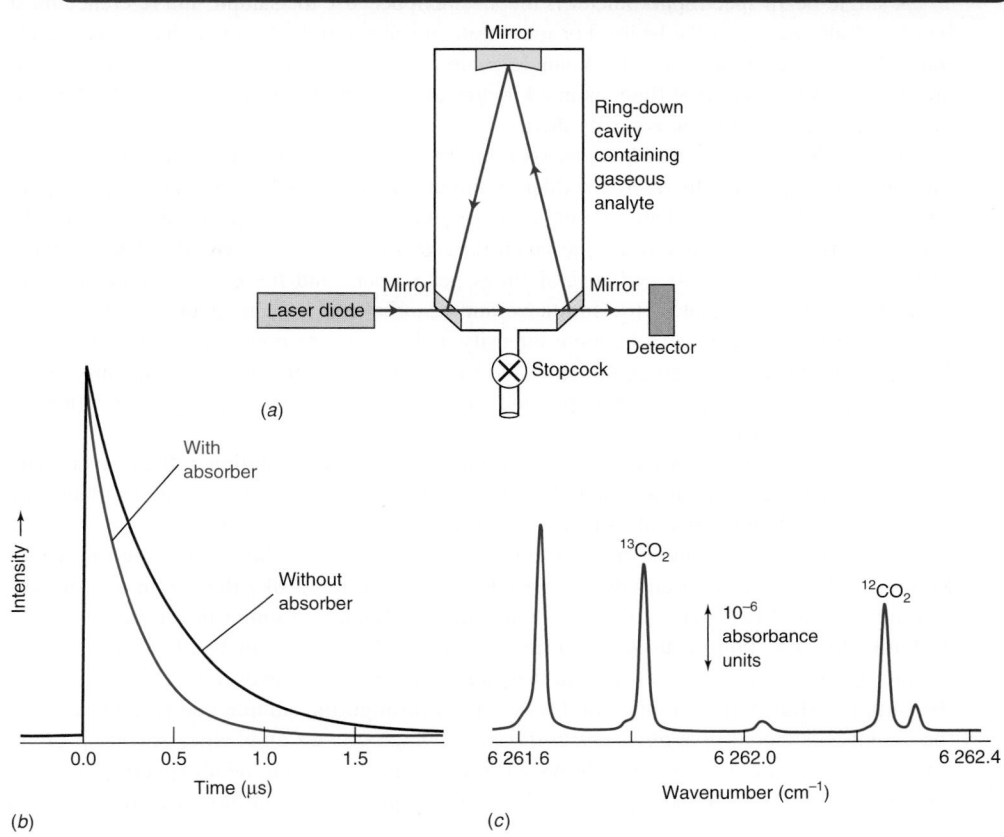

(a)

(b)

Cavity ring-down spectrum of ~3 mbar of CO_2, which is similar to the concentration in human breath. [Data from E. R. Crosson, K. N. Ricci, B. A. Richman, F. C. Chilese, T. G. Owano, R. A. Provencal, M. W. Todd, J. Glasser, A. A. Kachanov, B. A. Paldus, T. G. Spence, and R. N. Zare, "Stable Isotope Ratios Using Cavity Ring-Down Spectroscopy: Determination of $^{13}C/^{12}C$ for Carbon Dioxide in Human Breath," *Anal. Chem.* **2002,** *74,* 2003.]

(c)

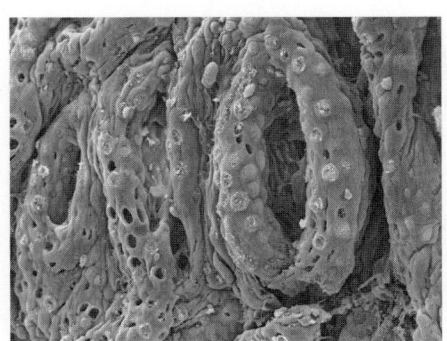

An ulcer: Mucosal surface of the bowel of a patient suffering from ulcerative colitis. [SPL/Science Source.]

Urea

Cavity ring-down can measure absorbance as low as ~10^{-6} and has the potential to provide sensitive detectors for chromatography.[1] In panel *a*, a laser is directed into a cavity with a path defined by three highly reflective mirrors. If the mirror reflectivity is 99.999%, then 0.001% of the power penetrates each mirror in each pass through the cavity. After less than 1 ms, the cavity is filled with circulating light and the source is abruptly shut off. Each time a circulating light ray reaches the mirror at the lower right, ~0.001% of the power leaks out to the detector. Graph *b* shows decay of the detector signal from a cavity containing a nonabsorbing gas or liquid. If an absorbing species is present, decay is faster because signal is lost by absorption during each pass between the mirrors. The difference in signal decay time with and without absorber provides a measure of absorbance. In one commercial instrument with a cavity length of 25 cm, the effective pathlength is ~20 km as the light circulates ~10^5 times while a little leaks out on each pass. This instrument measures isotopes of gases such as CO_2, CH_4, NH_3, H_2S, HF, H_2CO, and C_2H_4.[2]

Spectrum *c* shows absorbance measured for $CO_2(g)$ with the natural mixture of 98.9% ^{12}C and 1.1% ^{13}C. Peaks arise from transitions between rotational levels of two vibrational states. The spectral region was chosen to include a strong absorption of $^{13}CO_2$ and a weak absorption of $^{12}CO_2$, so the intensities of the isotopic peaks are similar. Each point in the spectrum was obtained by varying the laser frequency.

The areas of the $^{13}CO_2$ and $^{12}CO_2$ peaks from human breath were used to determine whether a patient was infected with *Helicobacter pylori*, a bacterium that causes ulcers. After a patient ingests ^{13}C-urea, *H. pylori* converts ^{13}C-urea into $^{13}CO_2$, which appears in the patient's breath. The ratio $^{13}C/^{12}C$ in the breath of an infected person increases by 1–5%, whereas the ratio $^{13}C/^{12}C$ from an uninfected person is constant to within 0.1%.

Figure 18-5 showed the essential features of a *single-beam spectrophotometer*. Light from a *source* is separated into narrow bands of wavelength by a *monochromator*, passed through a sample, and measured by a *detector*. We measure the *irradiance* (P_0, watts/m²) striking the detector with a *reference* cell (a solvent blank or reagent blank) in the sample compartment. When the reference is replaced by the sample of interest, some radiation is usually absorbed and the irradiance striking the detector, P, is smaller than P_0. The quotient P/P_0, which is a number between 0 and 1, is the *transmittance, T. Absorbance*, which is proportional to concentration, is $A = \log P_0/P = -\log T$.

A single-beam spectrophotometer is inconvenient because the sample and reference must be placed alternately in the beam. For measurements at multiple wavelengths, the reference must be run at each wavelength. A single-beam instrument is poorly suited to measuring absorbance as a function of time, as in a kinetics experiment, because both the source intensity and the detector response slowly drift.

Figure 20-1 shows a *double-beam spectrophotometer*, in which light alternately passes through the sample and the reference (blank), directed by a rotating mirror (the *chopper*) into and out of the light path. When light passes through the sample, the detector measures irradiance P. When the chopper diverts the beam through the reference cuvet, the detector measures P_0. The beam is chopped several times per second, and the circuitry automatically compares P and P_0 to obtain transmittance and absorbance. This procedure provides automatic correction for changes of source intensity and detector response with time and wavelength because the power emerging from the two samples is compared so frequently. Most research-quality spectrophotometers provide automatic wavelength scanning and continuous recording of absorbance versus wavelength.

It is routine to first record a *baseline* spectrum with reference solution in both cuvets. The baseline absorbance at each wavelength is then subtracted from the measured absorbance of the sample to obtain the true absorbance of the sample at each wavelength.

Components of a double-beam ultraviolet-visible spectrophotometer are shown in Figure 20-2. Visible light comes from a quartz-halogen lamp like that in an automobile headlight. The ultraviolet source is a deuterium arc lamp that emits in the range 200 to 400 nm. Only one lamp is used at a time. Grating 1 selects a narrow band of wavelengths to enter the monochromator, which in turn selects an even narrower band to pass through the sample. After being chopped and transmitted through the sample and reference cells, the signal is detected by a *photomultiplier tube*, which creates an electric current proportional to irradiance. Figure 20-3 shows a research-quality double-beam spectrophotometer. Now we describe the components of the spectrophotometer in more detail.

20-1 Lamps and Lasers: Sources of Light

A *tungsten lamp* is an excellent source of continuous visible and near-infrared radiation. A typical tungsten filament operates at a temperature near 3 000 K and produces useful radiation in the range 320 to 2 500 nm (Figure 20-4). This range covers the entire visible region and parts of the infrared and ultraviolet regions as well. Ultraviolet spectroscopy normally employs a *deuterium arc lamp* in which a controlled electric discharge causes D_2 to dissociate and emit ultraviolet radiation from 110 to 400 nm (Figure 20-4). In a typical ultraviolet-visible spectrophotometer, a switch is made between deuterium and tungsten lamps when

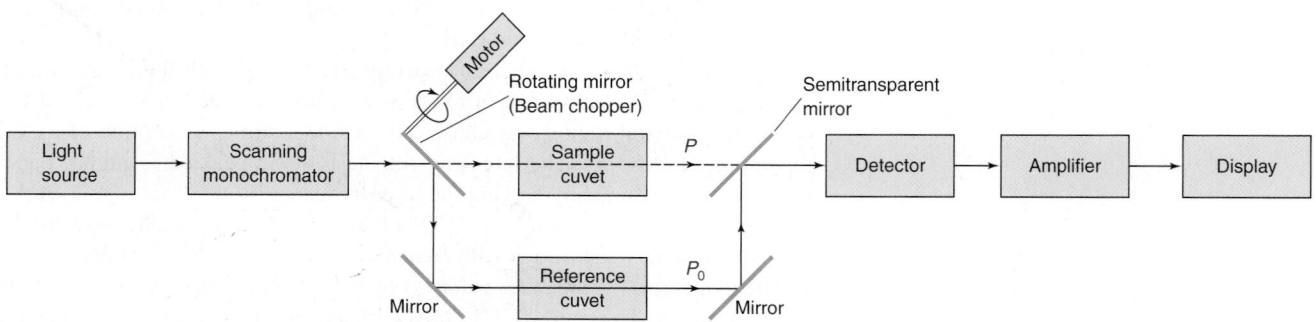

FIGURE 20-1 Schematic diagram of a double-beam scanning spectrophotometer. The rotating beam chopper passes the incident beam alternately through the sample and reference cuvets.

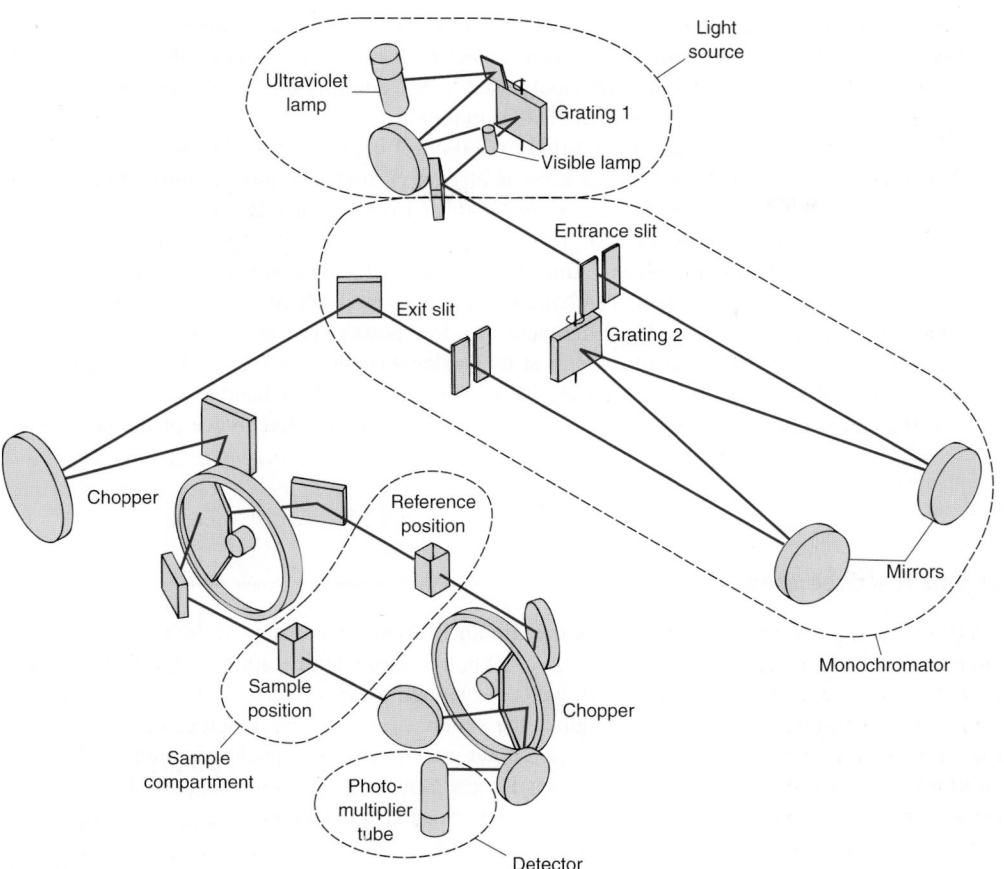

FIGURE 20-2 Optical train of Varian Cary 3E ultraviolet-visible double-beam spectrophotometer.
[Information from Varian Australia Pty. Ltd., Victoria, Australia.]

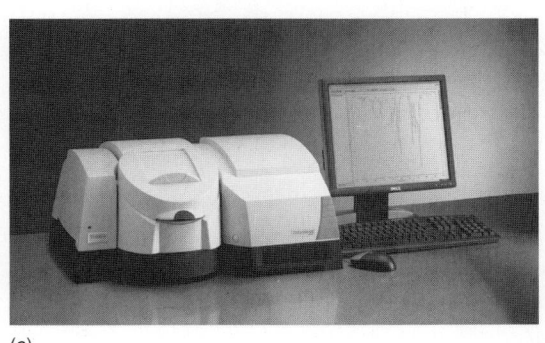

(a)

FIGURE 20-3 (a) Thermo Scientific Evolution 600 ultraviolet-visible double-beam spectrophotometer. (b) Optical train of Evolution 600, showing layout of components. [© Courtesy Thermo Fisher Scientific, Madison, WI.]

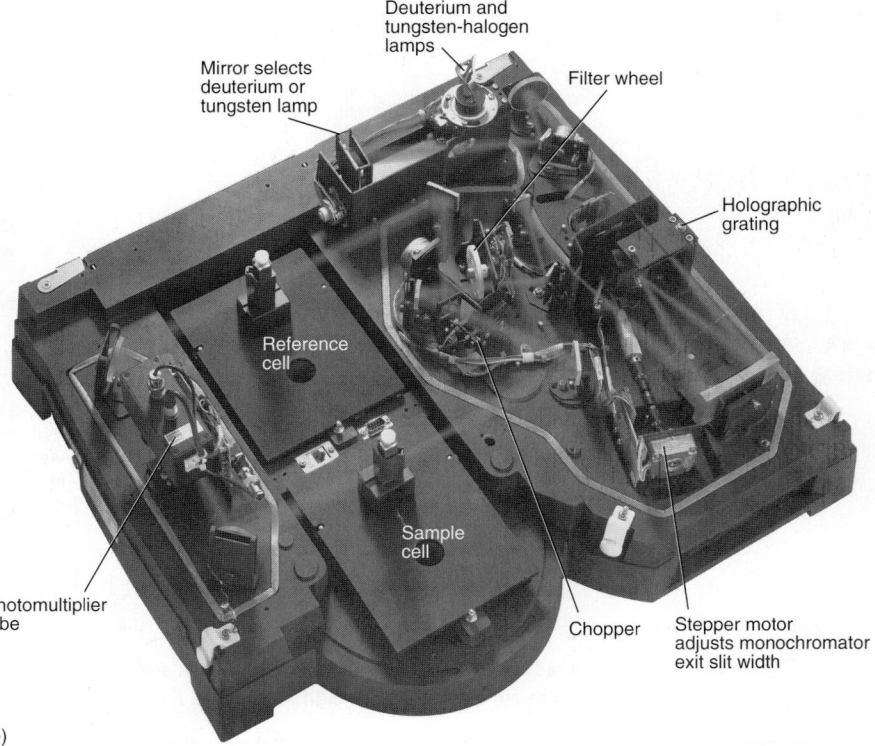

(b)

20-1 Lamps and Lasers: Sources of Light

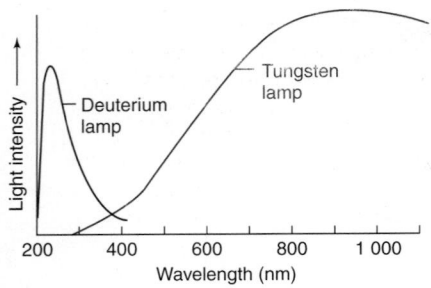

FIGURE 20-4 Intensities of a tungsten filament at 3 200 K and a deuterium lamp.

passing through 360 nm, so that the source with the highest intensity is always employed. For selected visible and ultraviolet frequencies, electric discharge arc lamps filled with mercury vapor (Box 18-2) or xenon gas are widely used. Light-emitting diodes provide narrow bands of visible and near-infrared (close to visible) radiation.[3]

Infrared radiation in the range 4 000 to 200 cm^{-1} is commonly obtained from a silicon carbide *globar*, heated to near 1 500 K by an electric current. The globar emits radiation with approximately the same spectrum as a *blackbody* at 1 000 K (Box 20-1).

Lasers provide isolated lines of a single wavelength for many applications. A laser with a wavelength of 3 μm might have a **bandwidth** (range of wavelengths, also called *linewidth*) of 3×10^{-14} to 3×10^{-8} μm. The bandwidth is measured where the radiant power falls to half of its maximum value. The brightness of a low-power laser at its output wavelength is 10^{13} times greater than that of the sun at its brightest (yellow) wavelength. (Of course, the sun emits all wavelengths, whereas the laser emits only a narrow band. The total brightness of the sun is much greater than that of the laser.) The angular divergence of the laser beam

---

BOX 20-1 Blackbody Radiation and the Greenhouse Effect

When an object is heated, it emits radiation—it glows. Even at room temperature, objects emit infrared radiation. Imagine a hollow sphere whose inside surface is perfectly black. That is, the surface absorbs all radiation striking it. If the sphere is at constant temperature, it must emit as much radiation as it absorbs, or else its temperature would rise. If a small hole were made in the wall, we would observe that the escaping radiation has a continuous spectral distribution. The object is called a *blackbody*, and the radiation is called **blackbody radiation.** Emission from real objects such as the tungsten filament of a light bulb resembles that from an ideal blackbody.

The power per unit area radiating from the surface of an object is called the *exitance* (or *emittance*), M. For a blackbody, exitance is given by

Exitance from blackbody: $M = \sigma T^4$

where σ is the Stefan-Boltzmann constant ($5.669\ 8 \times 10^{-8}$ W / (m^2 · K^4)). A blackbody whose temperature is 1 000 K radiates 5.67×10^4 watts per square meter of surface area. If the temperature is doubled, the exitance increases by a factor of $2^4 = 16$.

The graph shows that maximum blackbody emission for an object near 300 K occurs at infrared wavelengths (~10 μm). The surface of the sun behaves like a blackbody with a temperature near 5 800 K, emitting mainly *visible* light (~0.5 μm = 500 nm).

The mean solar energy flux at Earth's orbit is ~1 365 W/m^2. Averaged over the spherical surface of the planet, the radiant flux reaching Earth's upper atmosphere is 340 W/m^2. The atmosphere absorbs 23% of this radiant energy and reflects 22% back into space. Earth absorbs 47% of the solar flux and reflects 7%. Radiation reaching Earth should be just enough to keep the surface temperature at 254 K, which would not support life as we know it.

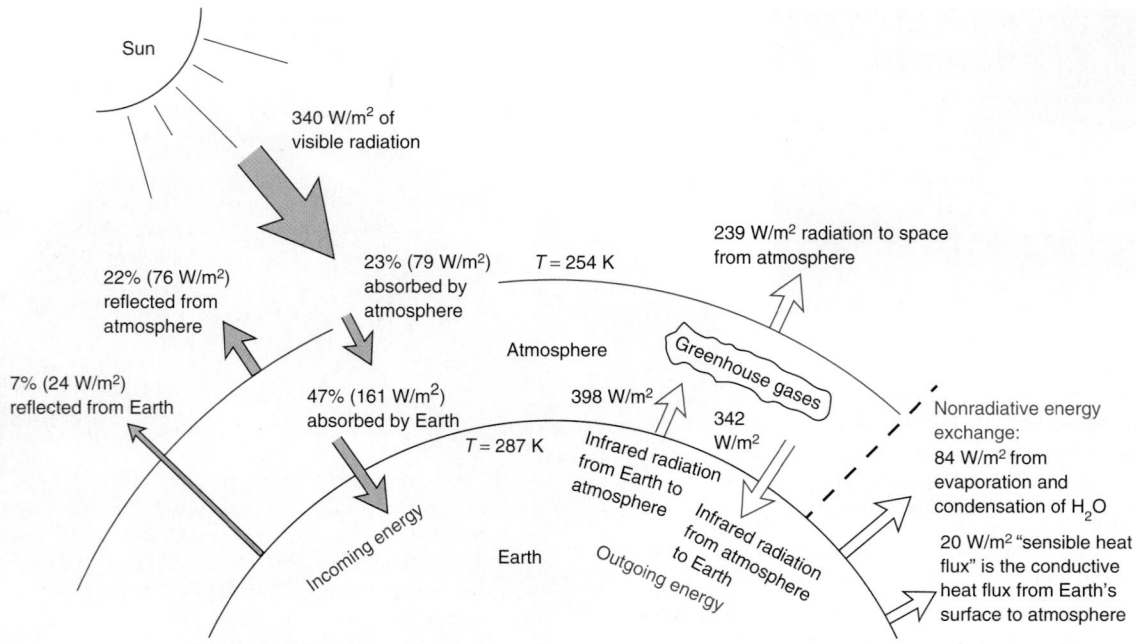

Balance between radiant energy reaching Earth from the sun and energy reradiated to space. Exchange of infrared radiation between Earth and its atmosphere keeps Earth's surface 33 K warmer than the upper atmosphere.

from its direction of travel is typically less than 0.05°, a property that allows us to illuminate a small target. Laser light is typically *plane polarized*, with the electric field oscillating in one plane perpendicular to the direction of travel (Figure 18-1). Laser light is *coherent*, which means that all waves emerging from the laser oscillate in phase with one another.

A necessary condition for lasing is *population inversion*, in which a higher energy state has a greater population, *n*, than a lower energy state in the lasing medium. In Figure 20-5a, this condition occurs when the population of state E_2 exceeds that of E_1. Molecules in ground state E_0 of the lasing medium are *pumped* to excited state E_3 by broadband radiation from a powerful lamp or by an electric discharge. Molecules in state E_3 rapidly relax to E_2, which has a relatively long lifetime. After a molecule in E_2 decays to E_1, it rapidly relaxes to the ground state, E_0 (thereby keeping the population of E_2 greater than the population of E_1).

A photon with an energy that exactly spans two states can be absorbed to raise a molecule to an excited state. Alternatively, that same photon can stimulate the excited molecule to emit a photon and return to the lower state. This is called *stimulated emission*. When a photon emitted by a

Properties of laser light:

monochromatic:	one wavelength
extremely bright:	high power at one wavelength
collimated:	parallel rays
polarized:	electric field oscillates in one plane
coherent:	all waves in phase

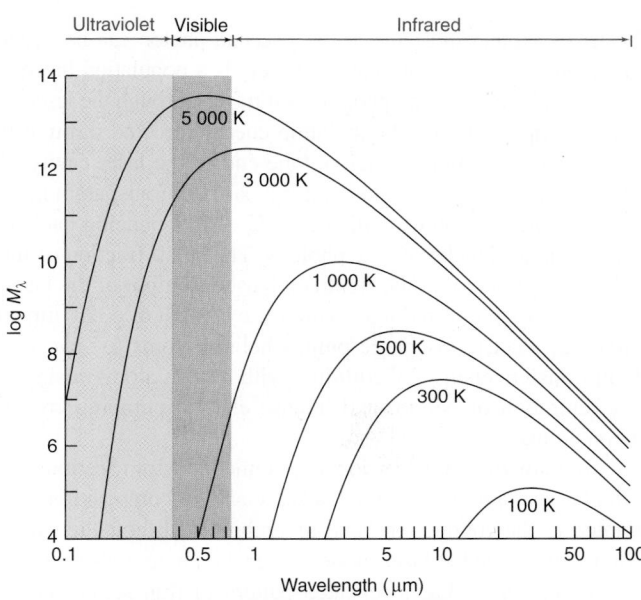

Spectral distribution of blackbody radiation. Both axes are logarithmic. The family of curves is called the *Planck distribution* after the German physicist, Max Planck (1858–1947). His derivation of the law governing blackbody radiation in 1900 relied on the hypothesis that electromagnetic energy could only be emitted in discrete quanta. Planck received the Nobel Prize in 1918.

Why does the average temperature of Earth's surface stay at a comfortable 287 K?

The blackbody curves tell us that Earth radiates mainly *infrared* radiation, rather than visible light. Although the atmosphere is transparent to incoming visible light, it absorbs outgoing infrared radiation. The main absorbers, called **greenhouse gases,** are water[4] and CO_2 and, to lesser extents, O_3, CH_4, chlorofluorocarbons, and N_2O. Radiation emitted from Earth is absorbed by the atmosphere and part of it is reradiated back to Earth. The atmosphere behaves like an insulating blanket, maintaining Earth's surface temperature 33 K warmer than the temperature of the upper atmosphere.[5]

Human activities since the dawn of the Industrial Revolution have increased atmospheric CO_2 from ~280 ppm in the year 1750

to ~400 ppm today through the burning of fossil fuel. The additional CO_2 greenhouse gas contributes an additional ~2 W/m² of radiative heating of Earth's surface. CO_2 is the largest anthropogenic source of radiative heating. The graph shows that Earth's surface is ~0.8°C warmer now than it was in 1850–1899. "It is certain that Global Mean Surface Temperature has increased since the late 19th century. Each of the past three decades has been successively warmer at the Earth's surface than all the previous decades in the instrumental record, and the first decade of the 21st century has been the warmest."[6]

One response to global warming is a rise in sea level from thermal expansion of water, which will flood low-lying coastal areas. Will there be disastrous climatic changes? Will there be compensating responses? We cannot accurately answer these questions, but prudence suggests that we should avoid making such large changes in our atmosphere.

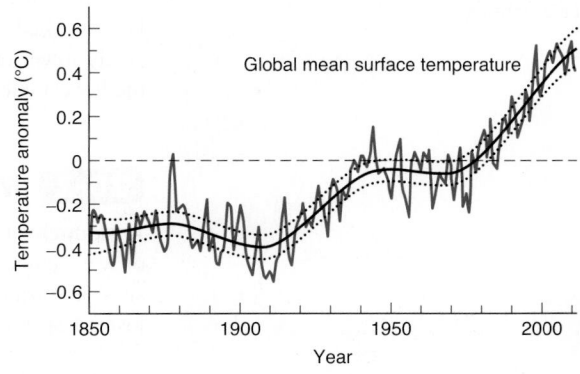

Global mean surface temperature based on current measurements and proxy data such as tree rings and isotope ratios in sediments and ice cores. All ten of the warmest years have occurred since 1997. The dotted lines are the 90% confidence interval for the smooth solid trend line. [Data from *Fifth Assessment Report of the Intergovernmental Panel on Climate Change, 2013*.[6] See also M. E. Mann and P. D. Jones, "Global Surface Temperatures Over the Past Two Millennia," *Geophys. Res. Lett.* **2003,** *30*, 1820.]

FIGURE 20-5 (*a*) Energy-level diagram illustrating the principle of operation of a laser. (*b*) Basic components of a laser. The population inversion is created in the lasing medium. Pump energy might be derived from either intense lamps or an electric discharge.

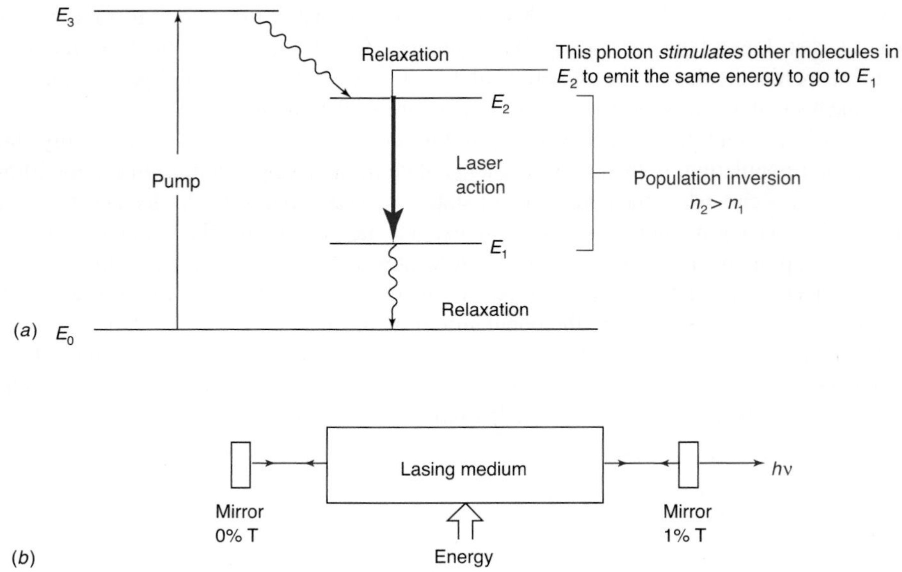

molecule falling from E_2 to E_1 strikes another molecule in E_2, a second photon can be emitted with the same phase and polarization as the incident photon. If there is a population inversion ($n_2 > n_1$), one photon stimulates the emission of many photons as it travels through the laser.

Figure 20-5b shows essential components of a laser. Pump energy directed through the side of the lasing medium creates the population inversion. One end of the laser cavity is a mirror that reflects all light (0% transmittance). The other end is a partially transparent mirror that reflects most light (1% transmittance). Photons with energy $E_2 - E_1$ bouncing back and forth between the mirrors stimulate an avalanche of new photons. The small fraction of light passing through the partially transparent mirror at the right is the useful output of the laser.

A helium-neon laser is a common source of red light with a wavelength of 632.8 nm and an output power of 0.1–25 mW. An electric discharge pumps helium atoms to state E_3 in Figure 20-5. The excited helium transfers energy by colliding with a neon atom, raising the neon to state E_2. The high concentration of helium and intense electric pumping create a population inversion among neon atoms.

In a *laser diode*, population inversion of charge carriers in a semiconductor (Section 15-8) is achieved by a high electric field across a *pn* junction in GaAs,[7] GaN,[8] or compositions such as $Al_xGa_{1-x}N$. Electrons promoted to the conduction band recombine with holes left in the valence band to emit light. Commonly available laser diodes cover the range 360–1 550 nm. An infrared-emitting *quantum cascade laser* has a repeated pattern of thin semiconductor layers called a *superlattice* that creates a series of equally spaced energy gaps at progressively lower energy in the material. A single electron transiting in steps from the highest to the lowest energy level emits multiple photons of the same energy.

20-2 Monochromators

A **monochromator** disperses light into its component wavelengths and selects a narrow band of wavelengths to pass on to the sample or detector. The monochromator in Figure 20-2 consists of entrance and exit slits, mirrors, and a *grating* to disperse the light. *Prisms* were used instead of gratings in older instruments.

Gratings[9]

A **grating** is a reflective or transmissive optical component with a series of closely ruled lines. When light is reflected from or transmitted through the grating, each line behaves as a separate source of radiation. Different wavelengths of light are reflected or transmitted at different angles from the grating (Color Plate 14). The bending of light rays by a grating is called **diffraction.** (Bending of light rays by a prism or lens, called **refraction,** is discussed in Section 20-4.)

In the grating monochromator in Figure 20-6, *polychromatic* radiation from the entrance slit is *collimated* (made into a beam of parallel rays) by a concave mirror. These rays fall on a reflection grating, whereupon different wavelengths are diffracted at different angles. The light strikes a second concave mirror, which focuses each wavelength at a different point on

The 2014 Nobel Prize in Physics went to I. Akasaki, H. Amano, and S. Nakamura for the invention of efficient blue light-emitting (LED) diodes. These diodes, which are not lasers, are the foundation of energy-saving white LED lighting.

Grating: optical element with closely spaced lines

Diffraction: bending of light by a grating

Refraction: bending of light by a lens or prism

Polychromatic: many wavelengths

Monochromatic: one wavelength

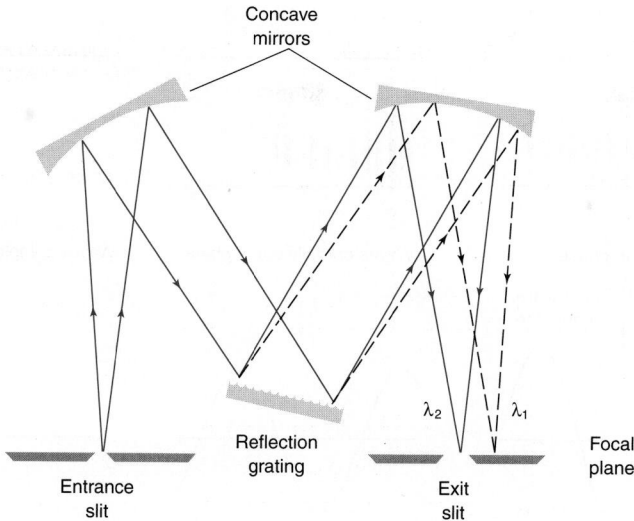

FIGURE 20-6 Czerny-Turner grating monochromator.

Concave
mirrors

λ_2 λ_1

Reflection
grating

Focal
plane

Entrance
slit

Exit
slit

the focal plane. The orientation of the reflection grating directs only one narrow band of wavelengths to the exit slit of the monochromator. Rotation of the grating allows different wavelengths to pass through the exit slit.

The reflection grating in Figure 20-7 is ruled with a series of closely spaced, parallel grooves with a repeat distance d. The grating is coated with aluminum to make it reflective. A thin layer of silica (SiO_2) on top of the aluminum protects the metal from oxidation, which would reduce its reflectivity. When light is reflected from the grating, each groove behaves as a source of radiation. When adjacent light rays are in phase, they reinforce one another. When they are not in phase, they partially or completely cancel one another (Figure 20-8).

Consider the incident and emerging light rays shown in Figure 20-7. Fully constructive interference occurs when the difference in length of the two paths is an integral multiple of the wavelength of light. The difference in pathlength is $a - b$ in Figure 20-7. Constructive interference occurs if

$$n\lambda = a - b \qquad (20\text{-}1)$$

where the diffraction order $n = \pm 1, \pm 2, \pm 3, \pm 4, \ldots$. The interference maximum for which $n = \pm 1$ is called *first-order diffraction*. When $n = \pm 2$, we have *second-order diffraction*, and so on.

In Figure 20-7, the incident angle θ is defined to be positive. The diffraction angle ϕ goes in the opposite direction, so, by convention, ϕ is negative. It is possible for ϕ to be on the same side of the facet normal as θ, in which case ϕ would be positive. In Figure 20-7, $a = d \sin \theta$ and $b = -d \sin \phi$ (because ϕ is negative, $\sin \phi$ is negative). Substituting into Equation 20-1 gives the condition for constructive interference:

Grating equation: $\qquad n\lambda = d(\sin \theta + \sin \phi) \qquad (20\text{-}2)$

where d is the distance between adjacent grooves. For each incident angle, θ, there is a series of reflection angles, ϕ, at which a given wavelength will produce maximum constructive interference (Color Plate 22).

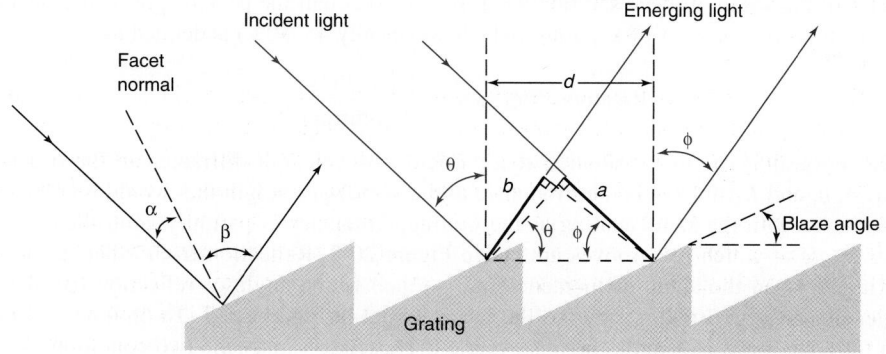

FIGURE 20-7 Principle of a reflection grating.

Incident light

Emerging light

Facet
normal

d

θ

b a

ϕ

α

β

θ ϕ

Blaze angle

Grating

FIGURE 20-8 Interference of adjacent waves that are (a) 0°, (b) 90°, and (c) 180° out of phase.

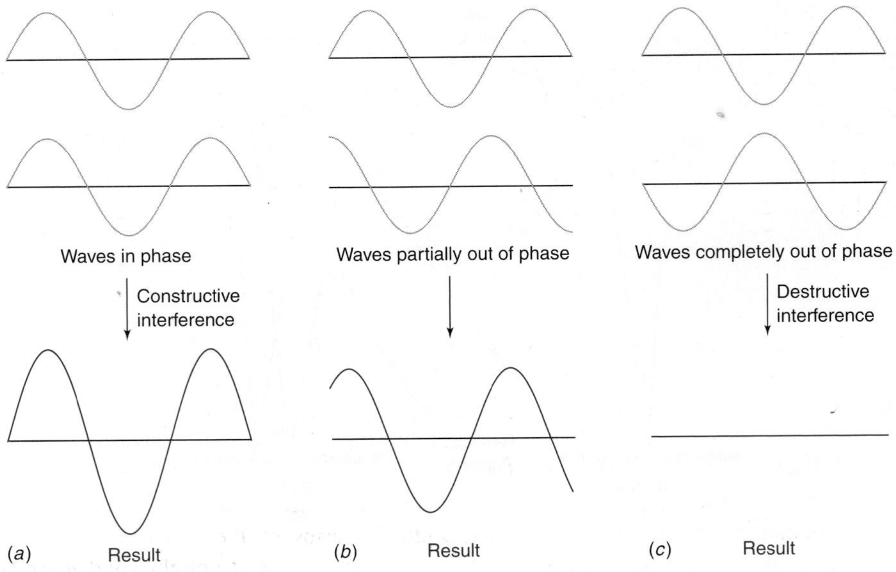

Resolution, Dispersion, and Efficiency of a Grating

Resolution measures the ability to separate two closely spaced peaks. The greater the resolution, the smaller is the difference ($\Delta\lambda$) between two wavelengths that can be distinguished from each other. The precise definition (which is beyond the scope of this discussion) means that the valley between the two peaks is about three-fourths of the height of the peaks when they are just barely resolved. The resolution of a grating is given by

Resolution of grating: $$\frac{\lambda}{\Delta\lambda} = nN \qquad (20\text{-}3)$$

where λ is wavelength, n is the diffraction order in Equation 20-1, and N is the number of grooves of the grating that are illuminated. The more grooves in a grating, the better the resolution between closely spaced wavelengths. If we need to resolve lines that are 0.05 nm apart at a wavelength of 500 nm, the required resolution is $\lambda/\Delta\lambda = 500$ nm/0.05 nm $= 10^4$. Equation 20-3 tells us that, if we desire a first-order resolution of 10^4, there must be 10^4 grooves in the grating. If the grating has a ruled length of 10 cm, we require 10^3 grooves/cm.

Dispersion measures the ability to separate wavelengths differing by $\Delta\lambda$ through the difference in angle, $\Delta\phi$ (radians). For the grating in Figure 20-7, the dispersion is

Dispersion of grating: $$\frac{\Delta\phi}{\Delta\lambda} = \frac{n}{d\cos\phi} \qquad (20\text{-}4)$$

where n is the diffraction order. Dispersion and resolution both increase with decreasing groove spacing. Equation 20-4 tells us that a grating with 10^3 grooves/cm provides a resolution of 0.102 radians (5.8°) per micrometer of wavelength if $n = 1$ and $\phi = 10°$. Wavelengths differing by 1 μm would be separated by an angle of 5.8°.

To select a narrower band of wavelengths from the monochromator, we decrease the exit slit width in Figure 20-6. Decreasing the exit slit width decreases the energy reaching the detector. Thus, *resolution of closely spaced absorption bands is achieved at the expense of decreased signal-to-noise ratio*. For quantitative analysis, a monochromator bandwidth that is ≲1/5 of the width of the absorption band (measured at half the peak height) is reasonable.

The relative *efficiency* of a grating (which is typically 45–80%) is defined as

$$\text{Relative efficiency} = \frac{E_\lambda^n(\text{grating})}{E_\lambda(\text{mirror})} \qquad (20\text{-}5)$$

where E_λ^n(grating) is the irradiance at a particular wavelength diffracted in the order of interest, n, and E_λ(mirror) is the irradiance at the same wavelength that would be reflected by a mirror with the same coating as the grating. Efficiency is partially controlled by the *blaze angle* at which the grooves are cut in Figure 20-7. Reflection from a flat surface is maximum when the angle of incidence (α) is equal to the angle of reflection (β). These angles are measured with respect to the facet normal in Figure 20-7. To optimize diffraction for a certain wavelength, the blaze angle is chosen to satisfy the two conditions $n\lambda =$

Resolution: ability to distinguish two closely spaced peaks

Dispersion: ability to produce angular separation of adjacent wavelengths

Trade-off between resolution and signal: The narrower the exit slit, the greater the resolution and the noisier the signal.

Specular reflection: For a flat reflective surface, such as a mirror, the angle of incidence (α in Figure 20-7) is equal to the angle of reflection (β).

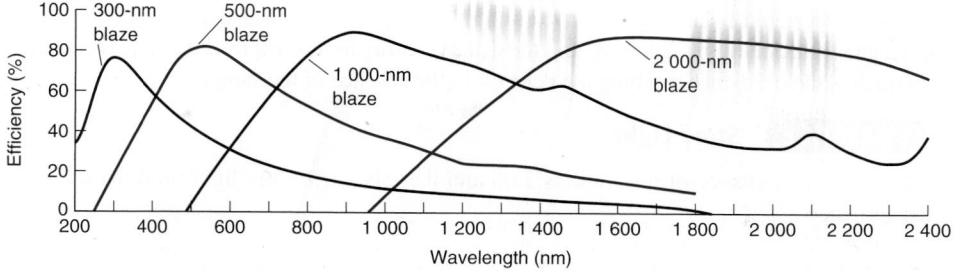

FIGURE 20-9 Efficiencies of diffraction gratings with 3 000 grooves/cm and blaze angles optimized for different wavelengths.
[Data from Princeton Instruments, Trenton, NJ.]

$d(\sin \theta + \sin \phi)$ and $\alpha = \beta$. Each grating is optimized for a limited range of wavelengths (Figure 20-9), so a spectrophotometer might require several different gratings to scan through its entire spectral range.

Choosing the Monochromator Bandwidth

The wider the exit slit in Figure 20-6, the wider the band of wavelengths selected by the monochromator. We usually measure slit width in terms of the bandwidth of radiation selected by the slit. Instead of saying that a slit is 0.3 mm wide, we might say that the *bandwidth* getting through the slit is 1.0 nm.

A wide slit increases the energy reaching the detector and gives a high signal-to-noise ratio, leading to good precision in measuring absorbance. However, Figure 20-10 shows that, if the bandwidth is large relative to the width of the peak being measured, peak shape is distorted. We choose a bandwidth as wide as the spectrum permits to allow the most possible light to reach the detector. A monochromator bandwidth that is one-fifth as wide as the absorption peak generally gives acceptably small distortion of the peak shape.[10]

> Monochromator bandwidth should be as large as possible, but small compared with the width of the peak being measured.

Stray Light

In every instrument, some **stray light** (wavelengths outside the bandwidth expected from the monochromator) reaches the detector. Stray light coming through the monochromator from the light source arises from diffraction into unwanted orders and angles and unintended scatter from optical components and walls. Stray light can also come from outside the instrument if the sample compartment is not perfectly sealed. Entry holes for tubing or electrical wires required at the sample for some experiments should be sealed to reduce stray light. Error from stray light is

> High-quality spectrometers could have two monochromators in series (called a **double monochromator**) to reduce stray light. Unwanted radiation that passes through the first monochromator is rejected by the second monochromator.

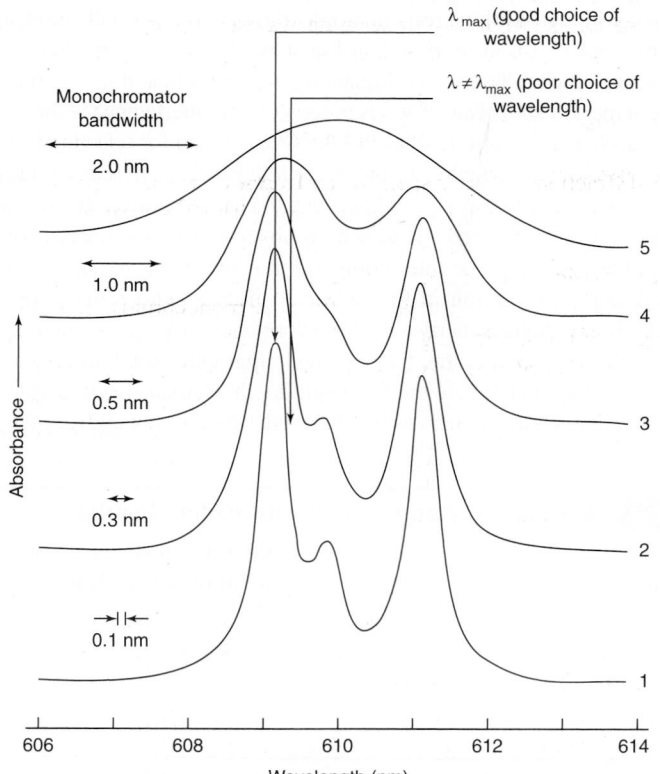

FIGURE 20-10 Increasing monochromator bandwidth broadens the bands and decreases the apparent absorbance of Pr^{3+} in a crystal of yttrium aluminum garnet (a laser material).
[Data from M. D. Seltzer, Michelson Laboratory, China Lake, CA.]

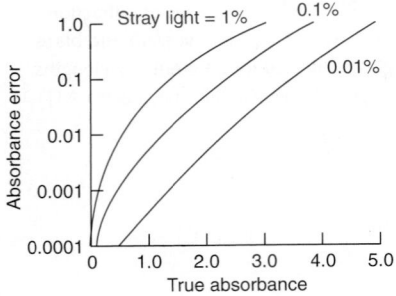

FIGURE 20-11 Absorbance error introduced by different levels of stray light. Stray light is expressed as a percentage of the irradiance incident on the sample. [Data from M. R. Sharp, "Stray Light in UV-VIS Spectrophotometers," *Anal. Chem.* **1984,** *53,* 339A.]

Filters permit certain bands of wavelength to pass through.

most serious when the sample absorbance is high (Figure 20-11) because the stray light constitutes a large fraction of the light reaching the detector. Stray light is expressed as a percentage of P_0, which is the irradiance reaching the detector in the absence of the sample.

EXAMPLE Stray Light

If the true absorbance of a sample is 2.00 and there is 1.0% stray light, find the apparent absorbance.

Solution A true absorbance of 2.00 means that the true transmittance is $T = 10^{-A} = 10^{-2.00} = 0.010 = 1.0\%$. Transmittance is the irradiance passing through the sample, P, divided by the irradiance passing through the reference, P_0: $T = P/P_0$. If stray light with irradiance S passes through both the sample and the reference, the apparent transmittance is

$$\text{Apparent transmittance} = \frac{P + S}{P_0 + S} \qquad (20\text{-}6)$$

If $P/P_0 = 0.010$ and there is 1.0% stray light, then $S = 0.010$ and the apparent transmittance is

$$\text{Apparent transmittance} = \frac{P + S}{P_0 + S} = \frac{0.010 + 0.010}{1 + 0.010} = 0.019_8$$

The apparent absorbance is $-\log T = -\log(0.019_8) = 1.70$, instead of 2.00.

TEST YOURSELF What level of stray light gives an absorbance error of 0.01 at an absorbance of 2? That is, what value of S gives an apparent absorbance of 1.99? (***Answer:*** $S = 0.000\ 24 = 0.024\%$)

Stray light in research-quality instruments can be 0.01% to 0.000 1%, or even less.

Table 20-1 gives the absorbance of a solution that you can prepare to test the accuracy of absorbance measurements on your spectrophotometer. Absorbance accuracy is affected by all components of the spectrophotometer, as well as stray light. Two wavelength accuracy standards are a solution of holmium oxide[11] for absorption and a mercury-argon discharge lamp for emission.[12]

Filters

It is frequently necessary to *filter* (remove) bands of radiation from a signal. For example, the grating monochromator in Figure 20-6 directs first-order diffraction of a small wavelength band to the exit slit. (By "first order," we mean $n = 1$ in Equation 20-2.) Let λ_1 be the wavelength whose first-order diffraction reaches the exit slit. Equation 20-2 shows that, if $n = 2$, the wavelength $\frac{1}{2}\lambda_1$ also reaches the same exit slit because $\frac{1}{2}\lambda_1$ gives constructive interference at the same angle as λ_1. For $n = 3$, the wavelength $\frac{1}{3}\lambda_1$ also reaches the slit. One solution for selecting just λ_1 is to place a filter in the beam, so that wavelengths $\frac{1}{2}\lambda_1$ and $\frac{1}{3}\lambda_1$ will be blocked. To scan a wide range of wavelengths, it may be necessary to use several filters and to change them as the wavelength changes.

The simplest filter is colored glass, which absorbs a broad portion of the spectrum and transmits other portions. For finer control, *interference filters* and *holographic filters* are constructed to pass radiation in the region of interest and reflect other wavelengths (Figure 20-12). These devices derive their performance from constructive or destructive interference of light waves within the filter. Some holographic notch filters have such a sharp cutoff that it is possible to attenuate the Rayleigh line in Raman spectroscopy (Box 18-3) by a factor of 10^6 while allowing signals to be observed 100 cm^{-1} away from the Rayleigh line.

TABLE 20-1 Calibration standards for ultraviolet absorbance	
Wavelength (nm)	$K_2Cr_2O_7$ absorbance 60.06 mg/L in 5.0 mM H_2SO_4 in 1-cm cell
235	0.748 ± 0.010
257	0.865 ± 0.010
313	0.292 ± 0.010
350	0.640 ± 0.010

SOURCE: S. Ebel, "Validation of Analysis Methods," Fresenius J. Anal. Chem. *1992,* 342, 769.

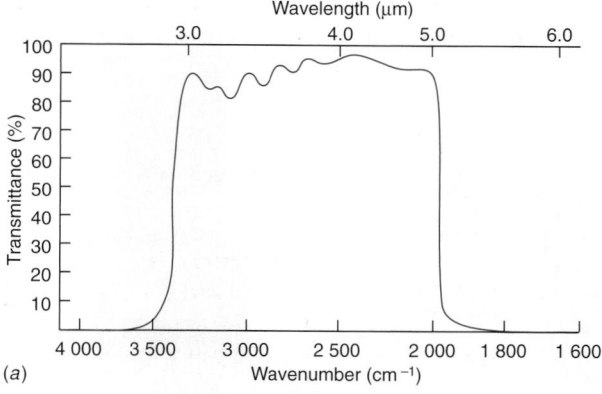

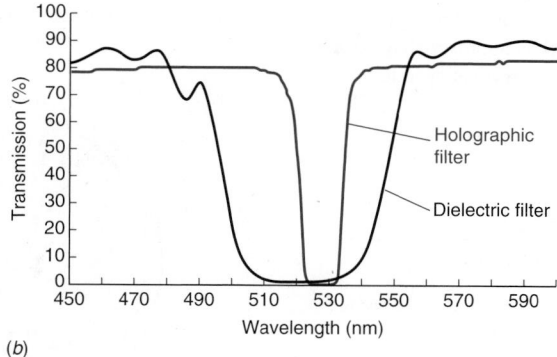

FIGURE 20-12 Transmission of filters. (*a*) Wide band-pass dielectric interference filter has ~90% transmission in the 3- to 5-μm wavelength range but <0.01% transmittance outside this range. [Information from Barr Associates, Westford, MA.] (*b*) Holographic interference filter provides greater attenuation and narrower band-pass than dielectric filter. [Information from H. Owen, "The Impact of Volume Phase Holographic Filters and Gratings on the Development of Raman Instrumentation," *J. Chem. Ed.* **2007**, *84*, 61.]

20-3 Detectors

A detector produces an electric signal when it is struck by photons. For example, a **phototube** emits electrons from a photosensitive, negatively charged surface (the cathode) when struck by visible light or ultraviolet radiation. Electrons flow through a vacuum to a positively charged collector whose current is proportional to the radiation intensity.

Figure 20-13 shows that detector response depends on the wavelength of the incident photons. For example, for a given radiant power (W/m²) of 420-nm light, the S-20 photomultiplier produces a current about four times greater than the current produced for the same radiant power of 300-nm radiation. The response below 280 nm and above 800 nm is essentially 0. In a single-beam spectrophotometer, the 100% transmittance control must be readjusted each time the wavelength is changed. This calibration adjusts the spectrophotometer to the maximum detector output that can be obtained at each wavelength. Subsequent readings are scaled to the 100% reading.

Detector response is a function of wavelength of incident light.

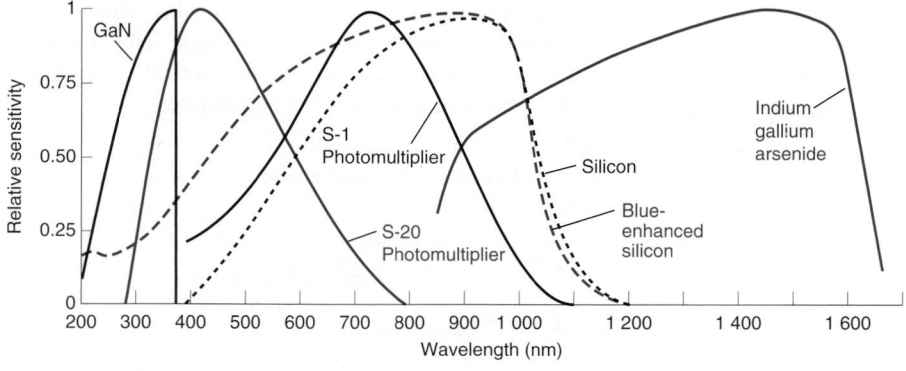

FIGURE 20-13 Detector response. The greater the sensitivity, the greater the current or voltage produced by the detector for a given incident irradiance (W/m²) of photons. Each curve is normalized to a maximum value of 1. $In_xGa_{1-x}As$ response can be shifted to longer or shorter wavelength by varying the composition. [Information from Barr Associates, Westford, MA. GaN data from APA Optics, Blaine, MN.]

Photomultiplier Tube

A **photomultiplier tube** (Figure 20-14) is a very sensitive device in which electrons emitted from the photosensitive surface strike a second surface, called a *dynode*, which is positive with respect to the photosensitive emitter. Electrons are accelerated and strike the dynode with more than their original kinetic energy. Each energetic electron knocks more than one electron from the dynode. These new electrons are accelerated toward a second dynode, which is more positive than the first dynode. Upon striking the second dynode, even more electrons are knocked off and accelerated toward a third dynode. This process is repeated several times, so more than 10^6 electrons are finally collected for each photon striking the first surface. Extremely low light intensities are translated into measurable electric signals. Your eye is even more sensitive than a photomultiplier (Box 20-2).

FIGURE 20-14 Photomultiplier tube with nine dynodes. Amplification of the signal occurs at each dynode, which is approximately 90 volts more positive than the previous dynode.
[© David J. Green/Alamy.]

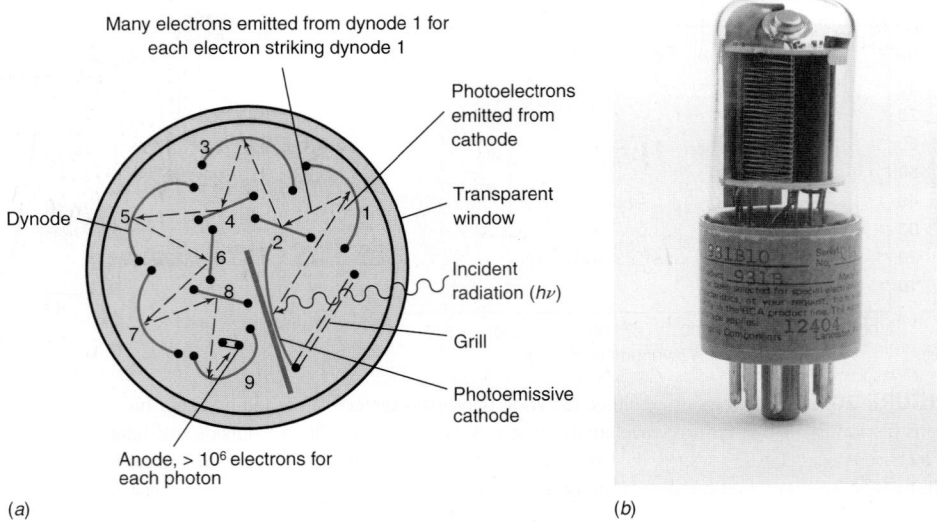

Many electrons emitted from dynode 1 for each electron striking dynode 1

Dynode

Photoelectrons emitted from cathode

Transparent window

Incident radiation ($h\nu$)

Grill

Photoemissive cathode

Anode, > 10^6 electrons for each photon

(a)

(b)

BOX 20-2   **The Most Important Photoreceptor**

The retina at the back of your eye contains photosensitive cells, called *rods* and *cones*, that are sensitive to levels of light varying over seven orders of magnitude. Light impinging on these cells is translated into nerve impulses that are transmitted by the optic nerve to the brain. Rod cells detect the dimmest light but cannot distinguish colors. Cone cells operate in bright light and give us color vision.

A stack of about 1 000 *disks* in each rod cell contains the light-sensing protein *rhodopsin*, in which the chromophore 11-*cis*-retinal (from vitamin A) is attached to the protein *opsin*. When a photon is absorbed by rhodopsin, the *cis* double bond isomerizes within pico-seconds to a *trans* geometry, moving attached protein atoms about 0.5 nm. The resulting conformational change of the protein triggers a series of changes that affect the transport of ions across the cell membrane and generates a nerve signal to the brain. A rod cell can respond to a single photon and the brain perceives light when fewer than 10 rod cells have been triggered. Absorption of a photon leads to the release of all *trans*-retinal from rhodopsin. The pigment is now *bleached* (loses all color) and cannot respond to more light until retinal isomerizes back to the 11-*cis* form and recombines with the protein.

In the dark, there is a continuous flow of 10^9 Na$^+$ ions per second out of the rod cell's inner segment, through the adjoining medium, and into the cell's outer segment. An energy-dependent process using adenosine triphosphate (ATP) and oxygen pumps Na$^+$ out of the cell. Another process involving a molecule called cyclic GMP keeps the gates of the outer segment open for ions to flow back into the cell. When light is absorbed and rhodopsin is bleached, a series of reactions leads to destruction of cyclic GMP and shutdown of the channels through which Na$^+$ flows into the cell. A single photon reduces the ion current by 3%—corresponding to a decreased current of 3×10^7 ions per second. This *amplification* is greater than that of a photomultiplier tube, which is one of the most sensitive man-made photodetectors. The ion current returns to its dark value as the protein and retinal recombine and cyclic GMP is restored to its initial concentration.

Na$^+$

Disks

Plasma membrane

Cytoplasm

Na$^+$

Nucleus

Synapse

Outer segment

Inner segment

Rod cell

11-*cis* double bond isomerizes to *trans* when photon is absorbed

Protein atoms move 0.5 nm when 11-*cis* double bond isomerizes

11-*cis*-Retinal bound to rhodospin

Chromophore

Absorption of light

trans double bond

CHAPTER 20 Spectrophotometers

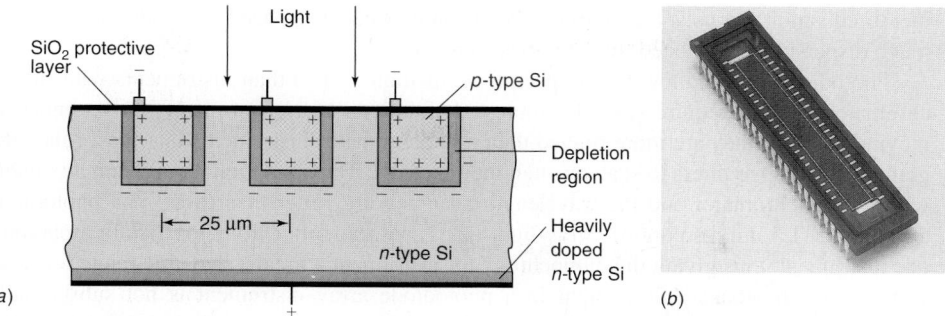

FIGURE 20-15 (*a*) Schematic cross-sectional view of photodiode array. (*b*) Array with 46 rectangular elements, each with an active area of 0.9 x 4.4 mm. The central black rectangle is photosensitive silicon that responds to wavelengths from 190 to 1 100 nm. The entire chip is 5 cm in length. [Courtesy Hamamatsu Photonics.]

All photodetectors produce a small response in the absence of light. This *dark current* could arise from spontaneous emission of electrons from the cathode of a photomultiplier tube or spontaneous generation of electrons and holes in a semiconductor device. Atomic vibrations can provide enough energy to an electron for it to escape from the cathode. The higher the temperature of the cathode, the greater the dark current.

Photodiode Array

Spectrophotometers with a photomultiplier tube detector scan slowly through a spectrum one wavelength at a time. A photodiode array spectrophotometer records the entire spectrum at once in a fraction of a second. One application of rapid scanning is chromatography, in which the full spectrum of a compound is recorded as it emerges from the column.

At the heart of rapid spectroscopy is the **photodiode array** shown in Figure 20-15 (or the charge coupled device described later). Rows of *p*-type silicon on a substrate (the underlying body) of *n*-type silicon create a series of *pn* junction diodes. A reverse bias is applied to each diode, drawing electrons and holes away from the junction. In the depletion region at each junction, there are few electrons and holes. The junction acts as a capacitor, with charge stored on either side of the depletion region. At the beginning of the measurement cycle, each diode is fully charged.

When radiation strikes the semiconductor, free electrons and holes are created and migrate to regions of opposite charge, partially discharging the capacitor. The more radiation that strikes each diode, the less charge remains at the end of the measurement. The longer the array is irradiated between readings, the more each capacitor is discharged. The state of each capacitor is determined at the end of the cycle by measuring the current needed to recharge the capacitor.

In a *dispersive spectrometer* with a monochromator (Figure 20-1), only one narrow band of wavelengths reaches the detector at any time. In a *photodiode array spectrophotometer* (Figure 20-16), different wavelengths reach different parts of the detector array. All wavelengths are recorded simultaneously, allowing more rapid acquisition of the spectrum or higher signal-to-noise ratio, or some combination of both. In the photodiode array spectrophotometer, *white light* (with all wavelengths) passes through the sample. The light then enters a **polychromator,** which disperses the light into its component wavelengths and directs the light at the diode array. Each diode receives a *different wavelength*, and all wavelengths are

For a refresher on semiconductors, see Section 15-8.

A photodiode array spectrophotometer measures all wavelengths at once, giving faster acquisition and higher signal-to-noise ratio.

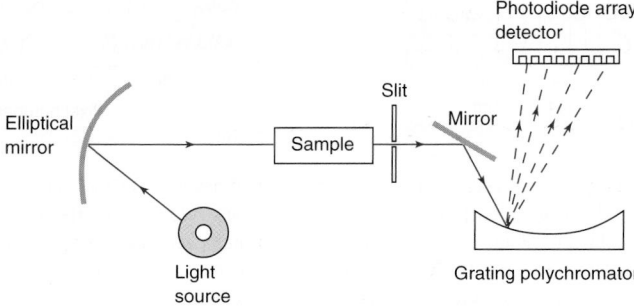

FIGURE 20-16 Design of photodiode array spectrophotometer.

measured simultaneously. Resolution depends on how closely spaced the diodes are and how much dispersion is produced by the polychromator.

Photodiode arrays allow faster spectral acquisition (<1 s) than instruments with monochromators (which require several minutes). Photodiode array instruments have almost no moving parts, so they are more rugged than monochromator instruments that must rotate the grating and change filters to scan through the spectrum. The resolution of ~0.1 nm attainable with a monochromator and the wavelength accuracy are better than those of a photodiode array (~0.5–1.5 nm resolution). Stray light is less with a monochromator than in a photodiode instrument, thus giving the monochromator instrument a greater dynamic range for measuring high absorbance. Stray light in a photodiode array instrument is not substantially increased when the sample compartment is open. With a monochromator instrument, the compartment must be tightly closed.

Charge Coupled Device[13]

A **charge coupled device** is an extremely sensitive detector that stores photo-generated charge in a two-dimensional array. A charge coupled device can produce a higher signal-to-noise ratio than can be obtained with a photomultiplier tube. The device in Figure 20-17a is constructed of p-doped Si on an n-doped substrate. The structure is capped with an insulating layer of SiO_2, on top of which is placed a pattern of conducting Si electrodes. When light is absorbed in the p-doped region, an electron is introduced into the conduction band and a hole is left in the valence band. The electron is attracted to the region beneath the positive electrode, where it is stored. The hole migrates to the n-doped substrate, where it combines with an electron. Each electrode can store ~10^5 electrons before electrons spill out into adjacent elements.

The charge coupled device is a two-dimensional array, as shown in Figure 20-17b. After the desired observation time, electrons stored in each *pixel* (picture element) of the top row are moved into the serial register at the top and then moved, one pixel at a time, to the top right position, where the charge is read out. Then the next row is moved up and read out, and the sequence is repeated until the entire array has been read. The transfer of stored charges is carried out by an array of electrodes considerably more complex than we have indicated in Figure 20-17a. Charge transfer from one pixel to the next is extremely efficient, with a loss of approximately five of every million electrons. Operation of a charge coupled device is somewhat analogous to a series of buckets (pixels) being filled by raindrops (electrons). For readout, the content of each bucket is dumped and measured.

Digital cameras use charge coupled devices to record the image. W. S. Boyle and G. E. Smith of Bell Laboratories shared a Nobel Prize in 2009 for inventing the charge coupled device in 1969.

Electrons from adjacent pixels can be combined to create a single, larger picture element. This process, called *binning*, increases the sensitivity of the charge coupled device at the expense of resolution.

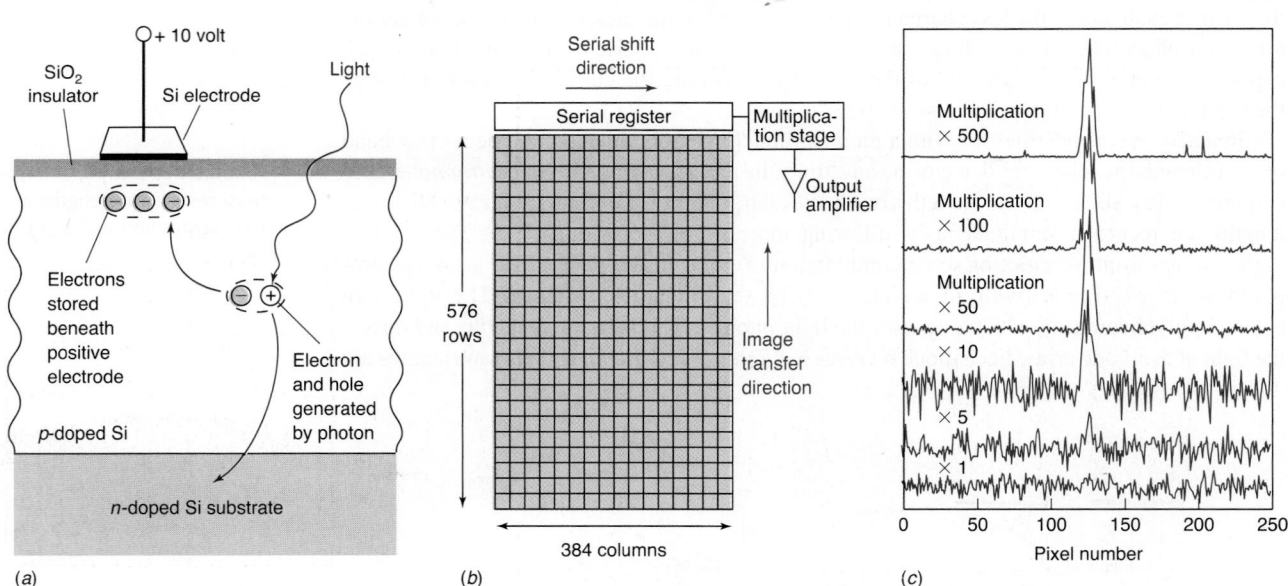

(a)   (b)   (c)

FIGURE 20-17 Schematic representation of a charge coupled device. (a) Cross-sectional view, indicating charge generation and storage in each pixel. (b) Top view showing two-dimensional nature of an array. An actual array is about the size of a postage stamp. (c) Effect of multiplier stage on signal-to-noise ratio of a signal that is too weak to be seen without multiplication. [Data from C. G. Coates, Andor Technology, Belfast, Ireland "A Sensitive Detector of Ultralow-Light Imaging," *Am. Lab.*, August 2004, p. 13.]

TABLE 20-2 Minimum detectable signal (photons/s/detector element) of ultraviolet/visible detectors

Signal acquisition time (s)	Photodiode array		Photomultiplier tube		Charge coupled device	
	Ultraviolet	Visible	Ultraviolet	Visible	Ultraviolet	Visible
1	6 000	3 300	30	122	31	17
10	671	363	6.3	26	3.1	1.7
100	112	62	1.8	7.3	0.3	0.2

SOURCE: *R. B. Bilhorn, J. V. Sweedler, P. M. Epperson, and M. B. Denton, "Charge Transfer Device Detectors for Analytical Optical Spectroscopy,"* Appl. Spectros. *1987, 41, 1114.*

The minimum detectable signal for visible light in Table 20-2 is 17 photons/s. The sensitivity of the charge coupled device is derived from its high *quantum efficiency* (electrons generated per incident photon), low background electrical noise (thermally generated free electrons), and low noise associated with readout.

The most sensitive charge coupled devices have a "multiplication stage," which multiplies the signal by $\sim 10^2$ to 10^3 between the serial register and the output amplifier. Noise that occurs during signal collection is also multiplied, but noise associated with readout is not. For the weakest signals in which readout is the dominant source of noise, multiplication increases the signal-to-noise ratio (Figure 20-17c). For cases in which the dominant noise occurs during signal collection, multiplication cannot improve the signal-to-noise ratio, but multiplication decreases the time needed to collect the signal.

The *image-intensified charge coupled device* in Figure 20-18 amplifies incoming photons by $\sim 10^6$ prior to detection. An image intensifier is used when the available light is very dim (such as weak fluorescence) or to convert radiation outside the visible range (such as near infrared radiation with a wavelength near 1 000 nm) into visible light for detection. A photon enters the evacuated device through a window at the left and strikes a *photocathode* that is optimized for ultraviolet, visible, or infrared wavelengths. When the photocathode is irradiated by light, it emits electrons. The quantum efficiency for generating one electron from one photon can be ~ 10 to 50%.

Electrons from the photocathode strike the *microchannel plate*, which is a two-dimensional array of 20-μm-diameter glass tubes. The tubes have electrically resistive walls coated with ~ 5 to 10 nm of a secondary electron emitter. The faces of the plate have a conductive coating to allow a potential difference of $\sim 1\ 000$ V to be applied between the faces. The potential becomes more positive from left to right in each channel in the enlargement at the right side of Figure 20-18. A single electron striking a channel causes several secondary electrons to be

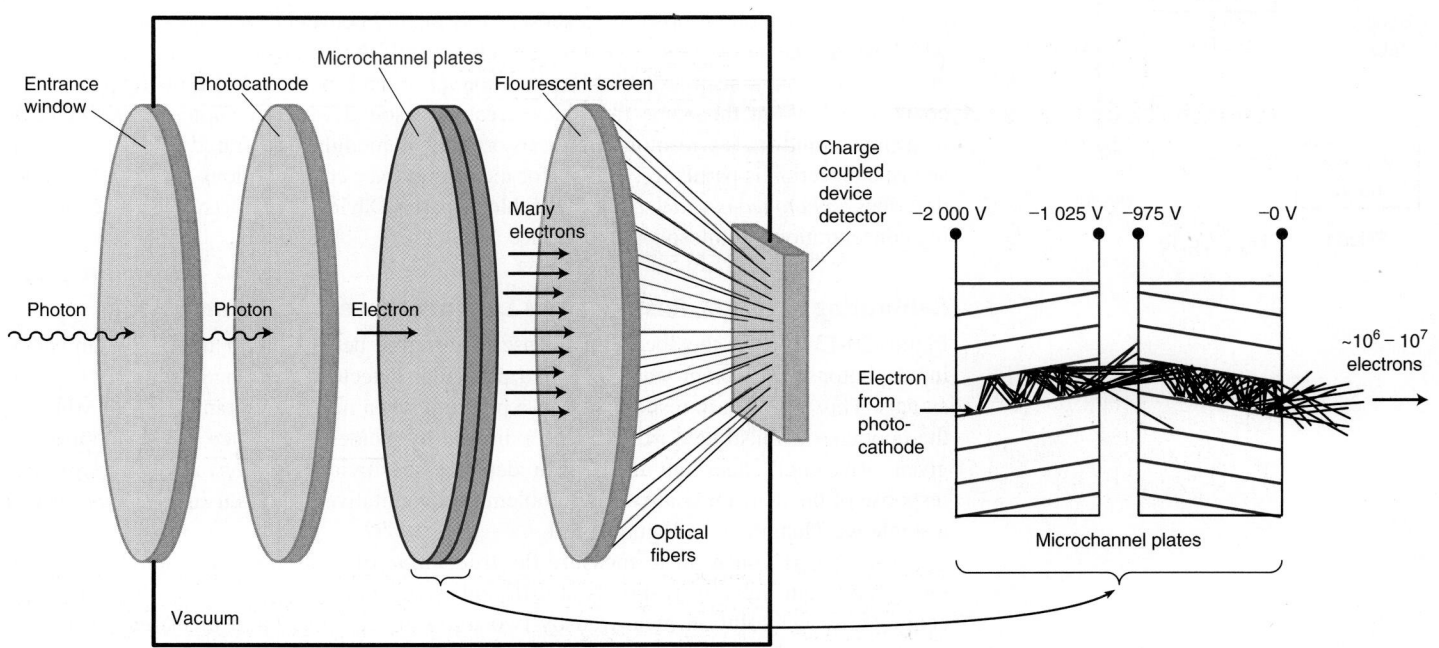

FIGURE 20-18 Image-intensified charge coupled device uses microchannel plates to amplify incoming signal.

emitted. Emitted electrons are accelerated by the more positive potential deeper into the channel. Each secondary electron strikes the wall further into the channel and emits several more electrons. A microchannel behaves like a photomultiplier tube with continuous walls instead of discrete plates. Each of the two plates provides a gain of ~10^3 to 10^4, giving ~10^6 to 10^7 outgoing electrons at the right side of the second plate for every electron entering the first plate. Electrons emerging from the microchannel plates strike a fluorescent screen that emits photons that are carried through optical fibers onto the pixels of the charge coupled device at the right.

Infrared Detectors

Detectors for visible and ultraviolet radiation rely on incoming photons to eject electrons from a photosensitive surface or to promote electrons from the valence band of silicon to the conduction band. Infrared photons do not have sufficient energy to generate a signal in either kind of detector. Therefore, other kinds of devices are used for infrared detection.

A **thermocouple** is a junction between two different electrical conductors. Electrons have lower free energy in one conductor than in the other, so they flow from one to the other until the resulting voltage difference prevents further flow. The junction potential is temperature dependent because electrons flow back to the high-energy conductor at higher temperature. If a thermocouple is blackened to absorb radiation, its temperature (and hence voltage) becomes sensitive to radiation. A typical sensitivity is 6 V per watt of radiation absorbed.

A **ferroelectric material,** such as deuterated triglycine sulfate, has a permanent electric polarization because of alignment of the molecules in the crystal. One face of the crystal is positively charged and the opposite face is negative. The polarization is temperature dependent, and its variation with temperature is called the *pyroelectric effect*. When the crystal absorbs infrared radiation, its temperature and polarization change. The voltage change is the signal in a pyroelectric detector. Deuterated triglycine sulfate is a common detector in Fourier transform spectrometers described later in this chapter.

A **photoconductive detector** is a semiconductor whose conductivity increases when infrared radiation excites electrons from the valence band to the conduction band. **Photovoltaic detectors** contain *pn* junctions, across which an electric field exists. Absorption of infrared radiation creates electrons and holes, which are attracted to opposite sides of the junction and which change the voltage across the junction. Mercury cadmium telluride ($Hg_{1-x}Cd_xTe$, $0 < x < 1$) is a detector material whose sensitivity to different wavelengths is affected by the stoichiometry coefficient, x. Photoconductive and photovoltaic devices can be cooled to 77 K (liquid nitrogen temperature) to reduce thermal electrical noise by more than an order of magnitude.

Infrared radiation absorbed by a substance is rapidly converted to heat (atomic motions), which warms the air or gas surrounding the substance. If the infrared radiation is modulated at an audio frequency, such as 1 000 Hz, then the air around the sample warms (expands) and cools (contracts) at this same frequency, creating a sound wave. A *photoacoustic detector* measures sound waves resulting from absorption of modulated infrared radiation. Photoacoustic detection is particularly useful for measuring trace concentrations of gases. Box 20-3 describes a *photoacoustic detector* that enabled Charles David Keeling to measure the increasing concentration of atmospheric CO_2 since 1958.

Calibrating Detector Response for Luminescence Measurements

Figure 20-13 showed the spectral response of different detectors. For the same number of input photons at different wavelengths, a particular detector will generate different output signals. This variation in response is not a problem when measuring transmittance, which is the quotient of transmitted irradiance (P) divided by incident irradiance (P_0). The ratio at a given wavelength does not depend on detector sensitivity at that wavelength. Spectral response of the detector is also not a problem for quantitative analysis using luminescence at a single wavelength, as in Figure 18-24.

However, if you want to measure the true *shape* of a luminescence band, you must know how your detector responds at different wavelengths. Figure 20-19 shows fluorescence from one solution measured by two fluorometers. The band measured by instrument A lies at shorter wavelength than the same fluorescence measured by instrument B.

In a **ferroelectric material**, dipole moments of molecules remain aligned in the absence of an external field. This alignment gives the material a permanent electric polarization.

Infrared radiation promotes electrons from the valence band of silicon to the conduction band. Semiconductors that are used as infrared detectors have smaller band gaps than silicon.

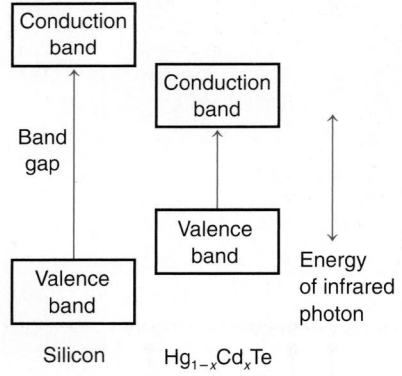

The blue line at the upper right of the opening graph in Chapter 10 (page 211) shows measurements of atmospheric CO_2 on the Mauna Loa volcano in Hawaii beginning in 1958.[14] When Charles David Keeling proposed continuous monitoring of atmospheric CO_2 in 1956, he chose a *nondispersive, photoacoustic* infrared analyzer that is very different from today's instruments.[15] "Nondispersive" means that there was no prism or grating to spread the radiation into its component wavelengths.

The infrared source is a resistively heated wire at ~525°C. Radiation is split into two beams and "chopped" (interrupted) at 20 Hz by a rotating wheel. Chopping is key for photoacoustic detection. The sample cell contains dry air pumped in from outside the observatory. The reference cell contains dry, CO_2-free air. CO_2 in the sample cell absorbs some infrared radiation, but gas in the reference cell does not. Detector cells contain CO_2 in argon. CO_2 in the detector absorbs infrared radiation, so the detector gas alternately warms (expands) and cools (contracts) at a frequency of 20 Hz. Warming and cooling create a pressure oscillation at 20 Hz that is detected by a microphone in each detector cell. Use of a microphone to observe the oscillating pressure wave caused by infrared absorption is called *photoacoustic detection*. The recorder displays the difference in the responses of the two microphones. The more CO_2 in the sample, the less

radiation reaches the detector through the sample cell, and the larger the difference in response.

The observatory on Mauna Loa at an altitude of 3.4 km is intended to measure pristine air over the Pacific Ocean. Four air intakes located 90° apart are each 7 m above the ground and 175 m from the observatory. The two upwind intakes are selected. The spectrometer monitors air from one intake for 10 min, then monitors the other for 10 min, and then measures a reference gas for 10 min.

A strip chart recorder shows the average difference between air and reference gas from four air measurements each hour. Sometimes readings are steady and other times they vary when CO_2 is emitted from volcanic vents on Mauna Loa. Data representing pristine air were obtained by rejecting readings for any hour when the variation in CO_2 was more than 0.5 ppm. A reading for a given day required that there be at least 6 consecutive hours of steady data from which to compose an average. If readings varied too much, no value was reported for that day.

The key to accuracy was knowing the CO_2 concentration in the reference gas, which was measured by precision manometry in Keeling's lab in California.[14] Experimental uncertainty for CO_2 in air was estimated as ±0.2 ppm for levels of 300–400 ppm. Annually increasing CO_2 since 1958 has awakened us to our influence on Earth's atmosphere.

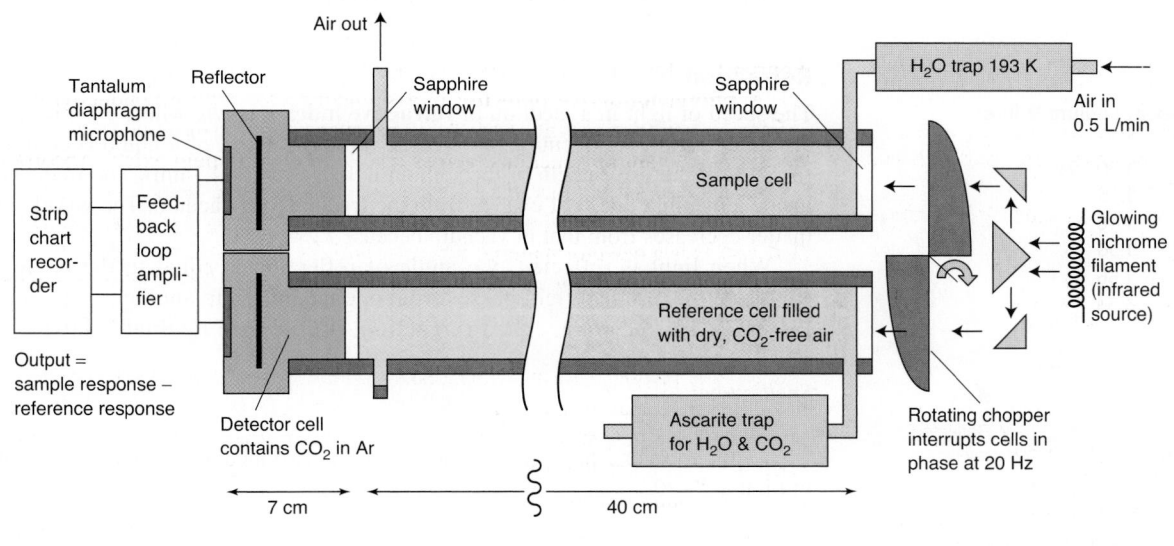

The detector in instrument B is more sensitive at long wavelength than is the detector in instrument A.

To obtain the true shape of a luminescence spectrum, we must measure the relative response of the detector at each wavelength. Calibration can be done with certified luminescence standards whose fluorescence has been measured with calibrated detectors.[16] Apparent fluorescence measured by your instrument is compared with the known fluorescence to obtain a calibration factor at each wavelength. When a measured spectrum is multiplied by the calibration factor at each wavelength, a true spectral shape is obtained. Corrected spectra in Figure 20-19 coincide—as they must—because they come from the same solution.

FIGURE 20-19 Uncorrected fluorescence spectra of the same solution recorded with two different spectrophotometers differ from each other. After correcting for response of each detector, the spectra are superimposed. [Data from U. Resch-Genger and P. Nording, Sigma-Aldrich Certified Luminescence Standards application note.]

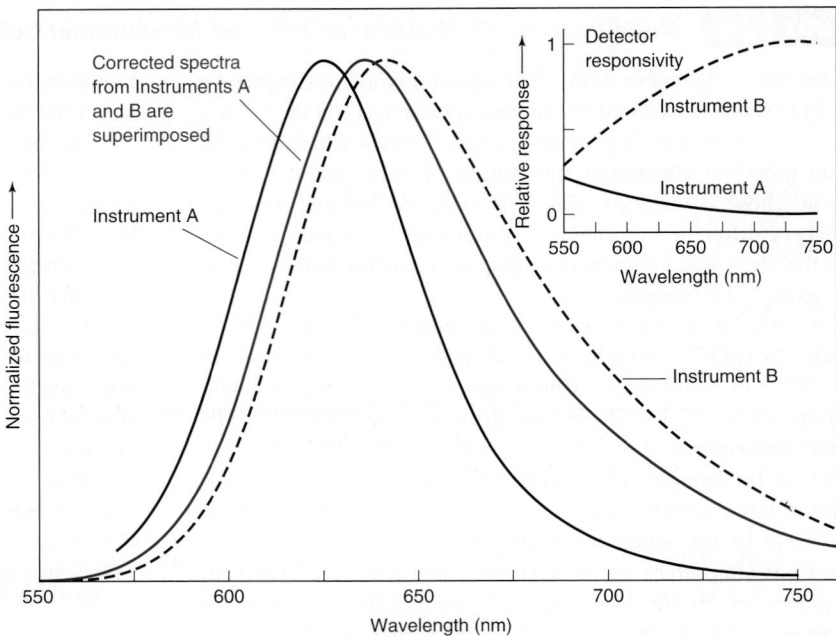

20-4 Optical Sensors

An *optode* is a chemical sensor based on an *optical fiber*. To understand how optodes work, we first need to know a little about refraction of light.

Refraction

Refractive index at sodium D line:

vacuum	1
air (0°, 1 bar)	1.000 29
water	1.33
fused silica	1.46
benzene	1.50
bromine	1.66
iodine	3.34

The speed of light in a medium of **refractive index** n is c/n, where c is the speed of light in vacuum. That is, for vacuum, $n = 1$. The refractive index of a liquid is commonly reported for 20°C at the wavelength of the sodium D line ($\lambda = 589.3$ nm). The frequency of light, ν, inside a medium does not change from the frequency in vacuum. The wavelength of light in matter decreases from that in vacuum because $\lambda\nu = c/n$.

When light is reflected, the angle of reflection is equal to the angle of incidence (Figure 20-20). When light passes from one medium into another, its path is bent (Color Plate 23). This bending, called **refraction,** is described by **Snell's law:**

Snell's law:
$$n_1 \sin \theta_1 = n_2 \sin \theta_2 \qquad (20\text{-}7)$$

where n_1 and n_2 are the refractive indexes of the two media and θ_1 and θ_2 are angles defined in Figure 20-20.

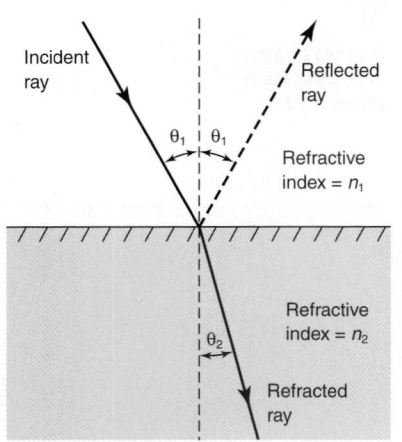

FIGURE 20-20 Snell's law: $n_1 \sin \theta_1 = n_2 \sin \theta_2$. When light passes from air into any medium, the greater the refractive index of the medium, the smaller is θ_2.

How to deal with the inverse sine function and radians.

EXAMPLE Refraction of Light by Water

Visible light travels from air (medium 1) into water (medium 2) at a 45° angle (θ_1 in Figure 20-20). At what angle, θ_2, does the light ray pass through the water?

Solution The refractive index is close to 1 for air and 1.33 for water. From Snell's law,

$$(1.00)(\sin 45°) = (1.33)(\sin \theta_2) \Rightarrow \theta_2 = 32°$$

If your radians and inverse trigonometric functions are rusty, here is how to solve for θ_2: $\sin \theta_2 = (\sin 45°)/1.33 = 0.531\ 7$, so $\theta_2 = \sin^{-1}(0.531\ 7) \equiv \arcsin(0.513\ 7)$. In Excel, the inverse sine function is ASIN and angles are expressed in radians. The Excel function ASIN(0.531 7) returns a value of 0.560 6 radians. There are π radians in 180 degrees. So,

$$\text{Degrees} = 180 \times \frac{\text{radians}}{\pi} = 180 \times \frac{0.560\ 6}{\pi} = 32°$$

What is θ_2 if the incident ray is perpendicular to the surface (that is, $\theta_1 = 0°$)?

$$(1.00)(\sin 0°) = (1.33)(\sin \theta_2) \Rightarrow \theta_2 = 0°$$

A perpendicular ray is not refracted.

TEST YOURSELF Visible light strikes the surface of benzene at a 45° angle. At what angle does the light ray pass through the benzene? (***Answer:*** 28°)

Optical Fibers

Optical fibers carry light by *total internal reflection*. Optical fibers are replacing electrical wires for communication because fibers are immune to electrical noise, transmit data at a higher rate, and can handle more signals. Optical fibers can bring an optical signal from inside a chemical reactor out to a spectrophotometer for process monitoring.

A flexible optical fiber has a high-refractive-index, transparent core enclosed in a lower-refractive-index, transparent cladding (Figure 20-21a). The cladding is enclosed in a protective plastic jacket. The core and coating can be made from glass or polymer.

Consider the light ray striking the wall of the core in Figure 20-21b at the angle of incidence θ_i. Part of the ray is reflected inside the core, and part might be transmitted into the cladding at the angle of refraction, θ_r (Color Plate 24). If the index of refraction of the core is n_1 and the index of the cladding is n_2, Snell's law (Equation 20-7) tells us that

$$n_1 \sin \theta_i = n_2 \sin \theta_r \Rightarrow \sin \theta_r = \frac{n_1}{n_2} \sin \theta_i \qquad \textbf{(20-8)}$$

If $(n_1/n_2) \sin \theta_i$ is greater than 1, no light is transmitted into the cladding because $\sin \theta_r$ cannot be greater than 1. In such case, θ_i exceeds the *critical angle* for total internal reflection. *If $n_1/n_2 > 1$, there is a range of angles θ_i in which essentially all light is reflected at the walls of the core, and a negligible amount enters the cladding.* All rays entering one end of the fiber within a certain cone of acceptance emerge from the other end of the fiber with little loss.

> Light traveling from a region of high refractive index (n_1) to a region of low refractive index (n_2) is totally reflected if the angle of incidence exceeds the *critical angle* given by $\sin \theta_{critical} = n_2/n_1$.

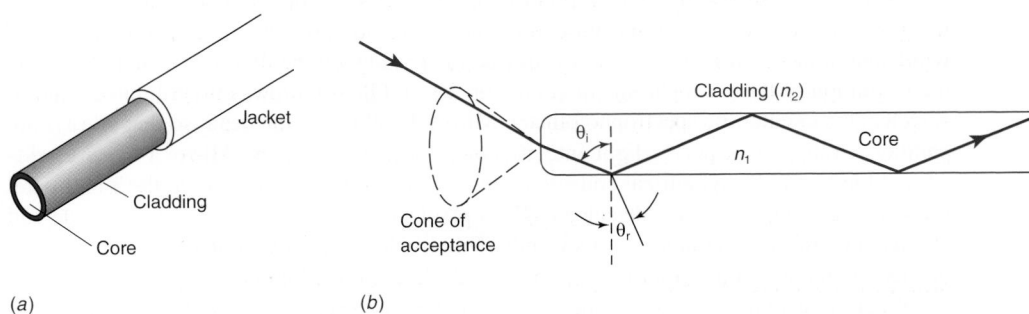

(a) (b)

FIGURE 20-21 (a) Optical fiber construction and (b) principle of operation. Any light ray entering within the cone of acceptance will be totally reflected at the wall of the fiber.

Optodes

We can create optical sensors for specific analytes by placing a chemically sensitive layer at the end of the fiber.[17] An optical fiber sensor is called an **optode** (or *optrode*), derived from the words "optical" and "electrode." Optodes have been designed to respond to analytes such as dissolved O_2 and pH, Li^+ in blood, NO in cells, sulfites in food, and atmospheric NO_2 and O_3.[18]

The end of the O_2 optode in Color Plate 25 is coated with a Ru(II) complex in a layer of polymer. Luminescence from Ru(II) is *quenched* (decreased) by O_2, as discussed in Section 19-6. The optode is inserted into a liquid sample as small as 100 fL (100×10^{-15} L) on the stage of a microscope. The degree of quenching tells us the concentration of O_2. The detection limit is 10 amol of O_2. By incorporating the enzyme glucose oxidase (Reaction 17-13), the optode becomes a glucose sensor with a detection limit of 1 fmol of glucose.[19] An optode for measuring biochemical oxygen demand (Box 16-2) employs yeast cells immobilized in a membrane to consume O_2 and uses Ru(II) luminescence to measure O_2.[20]

> **Question** How many molecules are in 10 amol? (***Answer:*** 6 million)

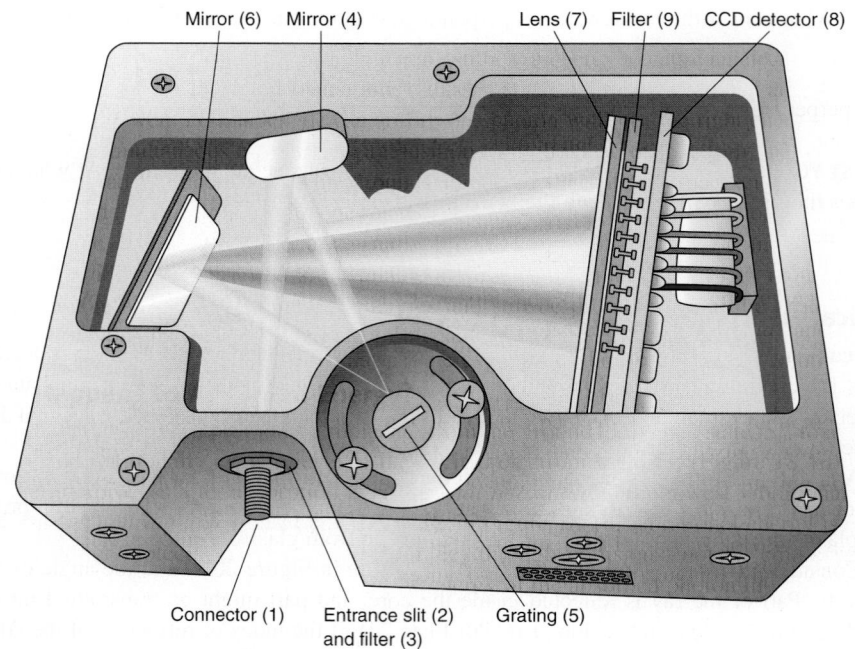

FIGURE 20-22 Optical train of Ocean Optics USB 4000 fiber-optic spectrophotometer used in flow injection and sequential injection analysis. This miniature spectrophotometer fits in your hand and weighs 190 g. [Information from Ocean Optics, Dunedin, FL.]

Mirror (6)   Mirror (4)   Lens (7)   Filter (9)   CCD detector (8)

Connector (1)   Entrance slit (2) and filter (3)   Grating (5)

Fiber-Optic Spectrophotometer

You are now ready to appreciate the fiber-optic spectrophotometer used in flow injection and sequential injection analysis. Figure 19-12 showed the main features of flow injection, including a flow cell to measure absorbance. Sample flows through a cylindrical cell that is irradiated at one end with visible light delivered by a fiber optic from a tungsten-halogen lamp. On the other end of the flow cell, a fiber optic delivers transmitted light to a spectrophotometer. Figure 19-18 shows a sequential analyzer with a flow cell. Figure 19-14 employs a fiber-optic spectrophotometer to measure fluorescence. Optical fibers and a compact spectrophotometer are enabling technologies for these analytical capabilities.

Figure 20-22 shows the optical train of the compact spectrophotometer. An optical fiber from the flow cell delivers light through connector 1 and entrance slit 2. The entrance slit width determines how many lines of the diffraction grating will be illuminated and, therefore, the resolution of the monochromator (Equation 20-3). Filter 3 allows only a limited band of wavelengths to enter the spectrophotometer. Mirror 4 collimates the beam so that all rays are parallel. Grating 5 disperses light into its component wavelengths. Mirror 6 focuses diffracted light onto the cylindrical collection lens 7 which directs the light onto the 3 648-pixel linear charge coupled device detector 8. Each pixel, whose physical size is 8 μm wide and 200 μm tall, receives a range of wavelengths. Filter 9, located between lens 7 and detector 8, blocks second- and third-order ($n = 2$ and $n = 3$) diffracted radiation.

When a customer orders the spectrophotometer, she specifies the wavelength range of interest for her application. The manufacturer installs the grating and filters at the factory for the desired wavelength range. The spectrometer is capable of operating in parts of the 200 to 1 100 nm range with different gratings and filters. Visible stray light is ~0.05 to 0.1%. The instrument requires just 4 ms to measure a spectrum. Signal can be integrated for as long as 10 s to improve the signal-to-noise ratio (Section 20-6). There are no moving parts in this rugged and relatively inexpensive instrument.

Attenuated Total Reflectance

Figure 20-21 showed total internal reflection of a light ray traveling through an optical fiber. The same behavior is observed in a flat layer of material whose refractive index, n_1, is greater than the refractive index of the surroundings, n_2. A planar layer in which light is totally reflected is called a **waveguide.** A chemical sensor can be fabricated by placing a chemically sensitive layer on a waveguide.[21]

When light in Figure 20-21 strikes the wall, the ray is totally reflected if θ_i exceeds the critical angle given by $\sin \theta_{\text{critical}} = n_2/n_1$. Even though light is totally reflected, the electric field of the light penetrates the cladding to some extent. Figure 20-23 shows that the field dies out exponentially inside the cladding. The part of the light that penetrates the wall of an optical fiber or waveguide is called an *evanescent wave*.

Evanescent means "vanishing" or "fleeting." Light "escapes" from the waveguide, but it vanishes over a short distance.

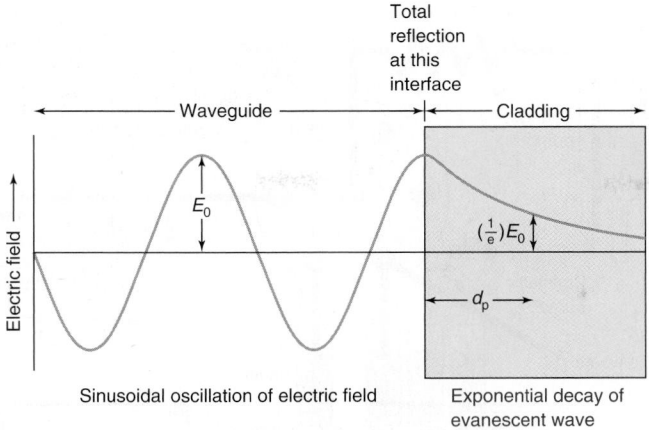

For nonabsorbing media, the electric field, E, of the evanescent wave in Figure 20-23 decays as

$$\frac{E}{E_0} = e^{-x/d_p} \qquad \left(d_p = \frac{\lambda/n_1}{2\pi\sqrt{\sin^2\theta_i - (n_2/n_1)^2}} \right) \qquad (20\text{-}9)$$

where E_0 is the magnitude of the field at the reflective interface, x is the distance into the cladding, and λ is the wavelength of light in a vacuum. The *penetration depth*, d_p, is the distance at which the evanescent field dies down to 1/e of its value at the interface. Consider a waveguide with $n_1 = 1.70$ and $n_2 = 1.45$, giving a critical angle of 58.5°. If light with a wavelength of 590 nm has a 70° angle of incidence, then the penetration depth is 140 nm. This depth is great enough to allow the light to interact with many layers of large molecules, such as proteins, whose dimensions are on the order of 10 nm.

Figure 20-24a shows an **attenuated total reflectance** infrared sensor for measuring caffeine in soft drinks. "Attenuated" means "decreased." The diamond crystal at the right side of the diagram acts as a waveguide; it has a circular upper surface whose diameter is 3 mm. When light passes through the crystal, it is totally reflected three times. The upper surface is in contact with a liquid channel carved into a block of poly(tetrafluoroethylene) (Teflon). At the center of the channel are 5 mg of hydrophobic sorbent beads made from polystyrene-divinylbenzene (Figure 26-1) held in place by cotton wool. When soft drink flows through the channel, caffeine is adsorbed by the beads and becomes concentrated in the beads. Sugar and caramel colorant present in the soft drink are not retained by the beads.

The evanescent wave from infrared radiation traveling through the diamond waveguide extends into the sorbent beads. When caffeine is present, wavelengths absorbed by caffeine are attenuated. The resulting spectrum is shown in Figure 20-24b. Each peak in the spectrum arises from a vibrational mode of caffeine. The integrated area beneath the entire spectrum is taken as the signal in this experiment.

Sequential injection apparatus (Section 19-4) in Figure 20-24a carries out the following operations. First, 2 mL of water are pumped through the detector cell and a background spectrum is recorded. Then 3 mL of soft drink or standard are pumped through the cell. The beads retain caffeine while sugar and caramel colorant in the soft drink are not retained. After washing the beads with 2 mL of water to remove sugar and colorant, a second spectrum is recorded. The background spectrum is subtracted from the second spectrum to obtain a spectrum whose integrated intensity is proportional to the concentration of caffeine in the soft drink. Finally, the beads are washed with 0.25 mL of acetonitrile, which quantitatively removes caffeine. The procedure can be repeated indefinitely to measure more unknowns and standards.

Surface Plasmon Resonance[22]

Conduction electrons in a metal are nearly free to move within the metal in response to an applied electric field. A *surface plasma wave*, also called a *surface plasmon*, is an electromagnetic wave that propagates along the boundary between a metal and a *dielectric* (an electrical insulator). The electromagnetic field decreases exponentially into both layers but is concentrated in the dielectric layer.

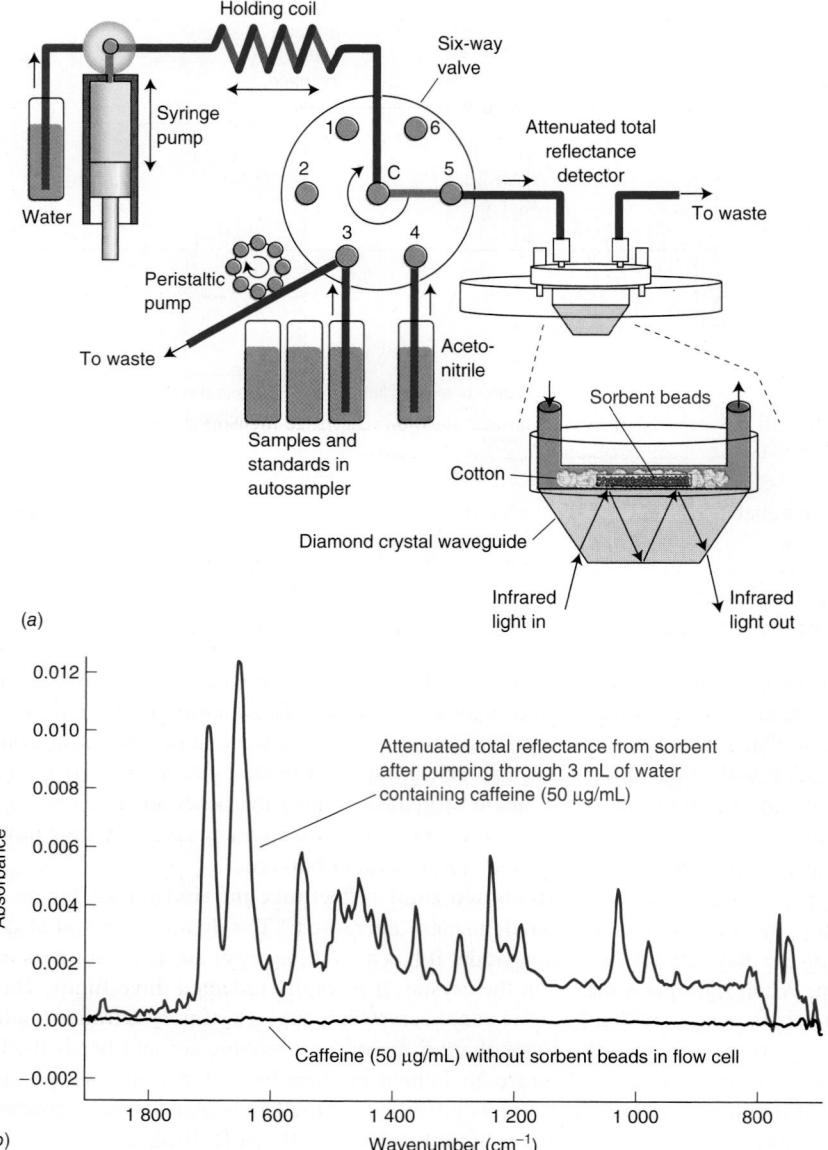

FIGURE 20-24 (*a*) Sequential injection apparatus for measuring caffeine in soft drinks by attenuated total reflectance from analyte retained by sorbent beads. (*b*) Attenuated total reflectance spectrum of caffeine adsorbed from 3 mL of standard containing 50 μg caffeine/mL. [Information from M. C. Alcudia-León, R. Lucena, S. Cárdenas, and M. Valcárcel, "Characterization of an Attenuated Total Reflection-Based Sensor for Integrated Solid-Phase Extraction and Infrared Detection," *Anal. Chem.* **2008**, *80*, 1146.]

Figure 20-25a shows essentials of one common **surface plasmon resonance** measurement. Monochromatic light whose electric field oscillates in the plane of the page is directed into a prism whose bottom face is coated with ~50 nm of gold. The bottom surface of the gold is coated with a chemical layer (~2–20 nm) that selectively binds an analyte of interest. At low angles of incidence, θ, much, but not all of the light is reflected by the gold. When θ increases to the critical angle for total internal reflection, the reflectivity is ideally 100%. As θ increases further, a surface plasmon (oscillating electron cloud) is set up, absorbing energy from the incident light. Because some of the energy is absorbed in the gold layer, the reflectivity decreases from 100%. There is a small range of angles at which the plasmon is in resonance with the incident light, a condition creating the sharp dip in the curve in Figure 20-25b. As θ increases beyond the resonance condition, less energy is absorbed and reflectivity increases.

The angle at which reflectivity is minimum depends on the refractive indexes of all the layers in Figure 20-25a. The angle for minimum reflectivity in Figure 20-25b changes by ~0.1° when a thin layer of polymer is coated onto the gold. When analyte binds to a chemically sensitive layer on the gold, the refractive index of that layer changes slightly and the angle for minimum reflectivity changes slightly. Commercial instruments can measure changes in the surface plasmon resonance angle with a precision of ~10^{-4} to 10^{-5} degrees. For biosensors, the chemically sensitive layer might contain an antibody or antigen, DNA or RNA, a protein, or a carbohydrate that has a selective interaction with some analyte.

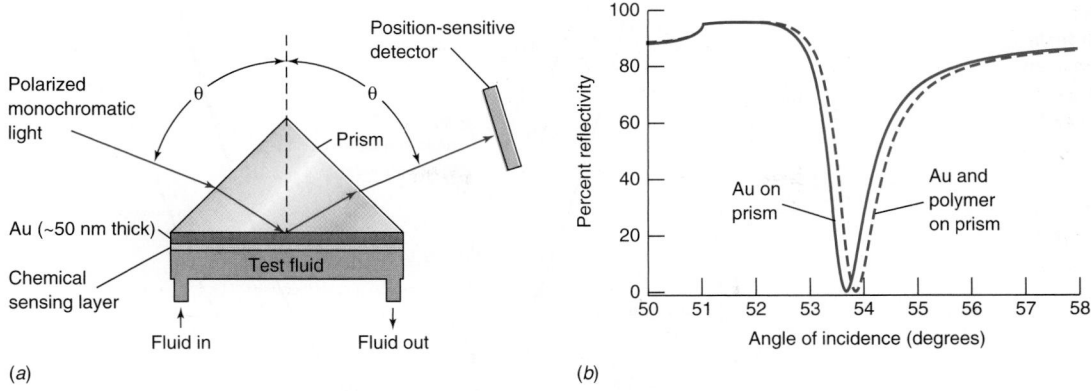

FIGURE 20-25 (*a*) Essentials of a surface plasmon resonance measurement. (*b*) Reflectivity versus angle, θ. [Information from J. M. Brockman, B. P. Nelson, and R. M. Corn, "Surface Plasmon Resonance Imaging Measurements of Ultrathin Organic Films," *Annu. Rev. Phys. Chem.* **2000,** *51,* 41.]

Figures 20-26 shows a scheme by which surface plasmon resonance can detect a specific sequence of DNA. A gold coating on a prism is treated with a synthetic peptide nucleic acid and mercaptohexanol. *Peptide nucleic acid* (abbreviated PNA) has the same bases as DNA (abbreviated A, T, C, and G), but the deoxyribose sugar-phosphate backbone of DNA is replaced by a peptide backbone. A sulfur atom at one end of the PNA chain binds to the gold surface. Mercaptohexanol occupies binding sites on gold not occupied by PNA. The complex (called a *hybrid*) between PNA and a complementary strand of DNA is stronger than the complex between two strands of complementary DNA. The sensor is designed to detect a specific, short sequence of DNA that is complementary to the synthetic PNA. When a mixture of DNA strands is incubated with PNA, only complementary DNA binds tightly. Noncomplementary DNA can be washed away. Hybrids of DNA to PNA are more stable to variations in temperature and ionic strength than are hybrids of DNA to DNA.

Prior to exposing unknown DNA to the Au-PNA surface, *biotin* is attached to one end of each DNA strand. Biotin is an essential B vitamin. The protein *streptavidin* is a 53 000-dalton tetrameric protein isolated from the bacterium *Streptomyces avidinii*. It has a strong, specific affinity for biotin, with a formation constant of 10^{15}. When biotin-DNA is exposed to streptavidin, the protein binds tightly to biotin (Figure 20-26).

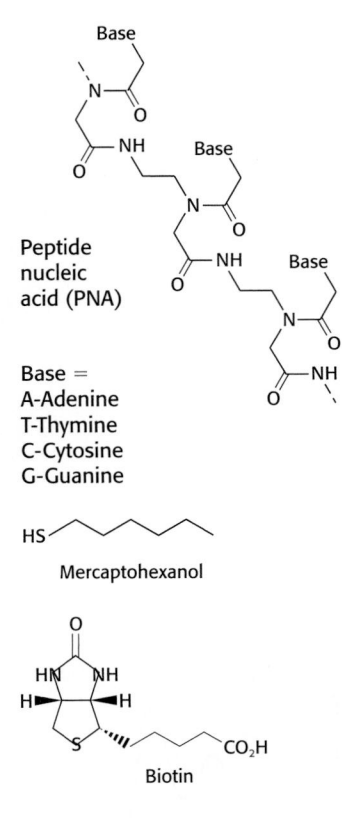

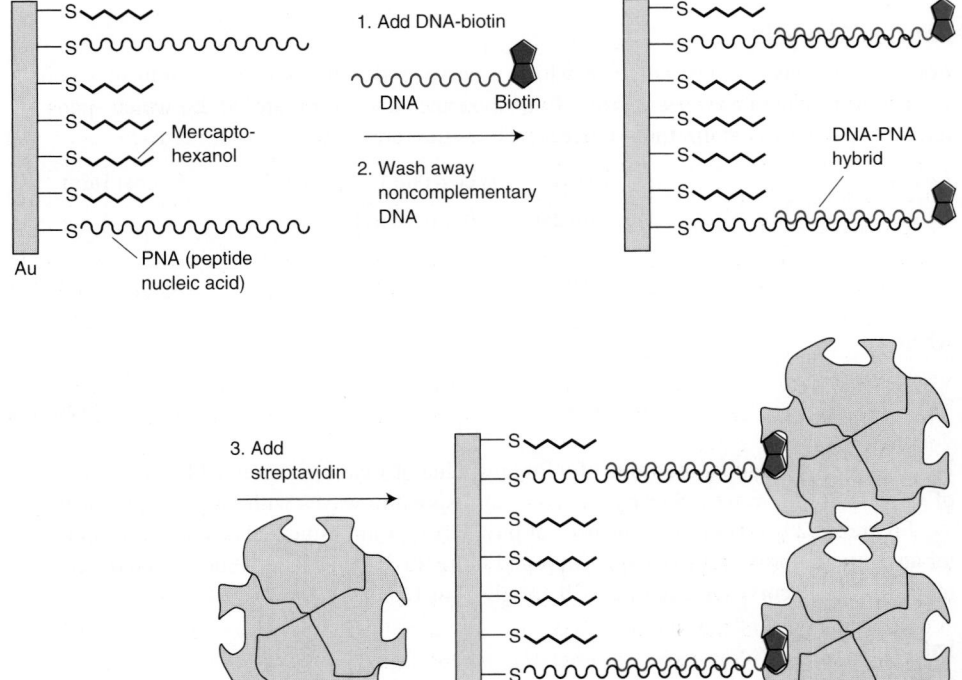

FIGURE 20-26 Scheme for detection of specific DNA sequence by surface plasmon resonance.

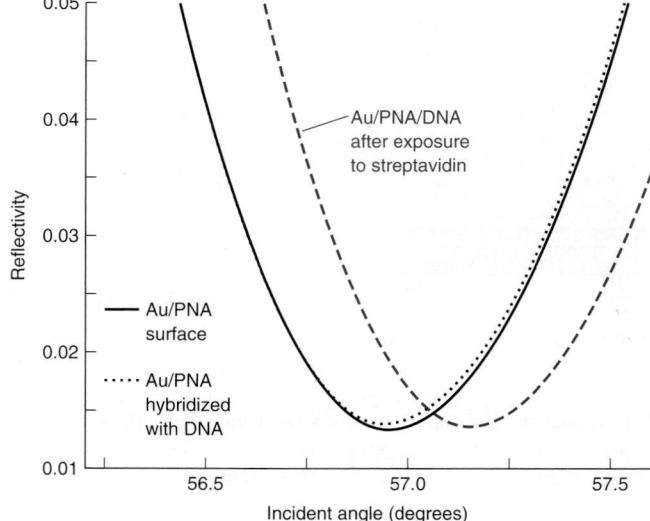

FIGURE 20-27 A small shift in surface plasmon resonance angle occurs when a small strand of DNA binds to complementary PNA on a gold surface. A large shift occurs when the protein streptavidin binds to biotin at the end of the DNA strand. [Data from J. Liu, S. Tian, L. Tiefenauer, P. E. Nielsen, and W. Knoll, "Simultaneously Amplified Electrochemical and Surface Plasmon Detection of DNA Hybridization Based on Ferrocene-Streptavidin Conjugates," *Anal. Chem.* **2005,** *77,* 2756.]

The solid line in Figure 20-27 is the surface plasmon resonance of the gold film with bound PNA and mercaptohexanol. When complementary DNA labeled with biotin binds to the PNA, the resonance shown by the dotted line is almost imperceptibly shifted because the amount of DNA that is bound is very small and the refractive index of the thin surface layer is barely changed. When streptavidin is added, the large protein binds to the biotin on the end of every DNA. The massive protein changes the refractive index of the organic film enough to shift the resonance position by 0.3° (dashed line in Figure 20-27). The signal in Figure 20-27 arises from $\sim 1 \times 10^{12}$ PNA molecules per square centimeter of gold surface. DNA with 12 nucleotide bases and one mismatch with the PNA does not bind to the PNA.

20-5 Fourier Transform Infrared Spectroscopy[23]

A photodiode array or charge coupled device can measure an entire spectrum at once. The spectrum is spread into its component wavelengths, and each wavelength is directed onto one detector element. For the infrared region, the most important method for observing the entire spectrum at once is *Fourier transform spectroscopy.*

Fourier Analysis

Fourier analysis is a procedure in which a curve is decomposed into a sum of sine and cosine terms, called a *Fourier series.* To analyze the curve in Figure 20-28, which spans the interval $x_1 = 0$ to $x_2 = 10$, the Fourier series has the form

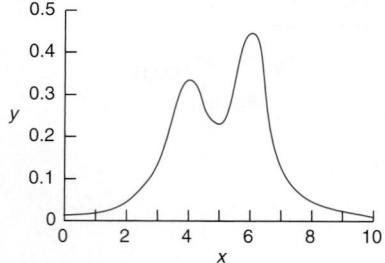

FIGURE 20-28 A curve to be decomposed into a sum of sine and cosine terms by Fourier analysis.

Fourier series:
$$y = a_0 \sin(0\omega x) + b_0 \cos(0\omega x) + a_1 \sin(1\omega x) + b_1 \cos(1\omega x)$$
$$+ a_2 \sin(2\omega x) + b_2 \cos(2\omega x) + \dots$$
$$= \sum_{n=0}^{\infty} [a_n \sin(n\omega x) + b_n \cos(n\omega x)] \tag{20-10}$$

where

$$\omega = \frac{2\pi}{x_2 - x_1} = \frac{2\pi}{10 - 0} = \frac{\pi}{5} \tag{20-11}$$

Equation 20-10 says that the value of y for any value of x can be expressed by an infinite sum of sine and cosine waves. Successive terms correspond to waves with increasing frequency.

Figure 20-29 shows how sequences of three, five, or nine sine and cosine waves give better and better approximations to the curve in Figure 20-28. The coefficients a_n and b_n required to construct the curves in Figure 20-29 are given in Table 20-3.

Interferometry

The heart of a Fourier transform infrared spectrophotometer is the **interferometer** in Figure 20-30. Radiation from the source at the left strikes a *beamsplitter,* which transmits some

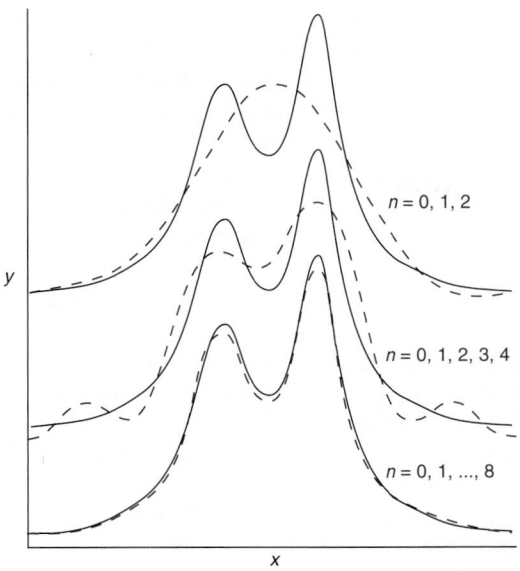

FIGURE 20-29 Fourier series reconstruction of the curve in Figure 20-28. Solid line is the original curve and dashed lines are made from a series of $n = 0$ to $n = 2$, 4, or 8 in Equation 20-10. Coefficients a_n and b_n are given in Table 20-3.

TABLE 20-3	Fourier coefficients for Figure 20-29	
n	a_n	b_n
0	0	0.136 912
1	−0.006 906	−0.160 994
2	0.015 185	0.037 705
3	−0.014 397	0.024 718
4	0.007 860	−0.043 718
5	0.000 089	0.034 864
6	−0.004 813	−0.018 858
7	0.006 059	0.004 580
8	−0.004 399	0.003 019

light and reflects some light. For ease of explanation, consider a beam of monochromatic radiation. (In fact, the Fourier transform spectrophotometer uses a continuum source of infrared radiation, not a monochromatic source.) Suppose that the beamsplitter reflects half of the light and transmits half. When light strikes the beamsplitter at point O, some is reflected to a stationary mirror at a distance OS and some is transmitted to a movable mirror at a distance OM. Rays reflected by the mirrors travel back to the beamsplitter, where half of each ray is transmitted and half is reflected. One recombined ray travels in the direction of the detector, and another heads back to the source.

In general, the paths OM and OS are not equal, so the two waves reaching the detector are not in phase. If the two waves are in phase, they interfere constructively to give a wave with twice the amplitude, as shown in Figure 20-8. If the waves are one-half wavelength (180°) out of phase, they interfere destructively and cancel. For any intermediate phase difference, there is partial cancellation.

Albert Michelson developed the interferometer around 1880 and conducted the Michelson-Morley experiment in 1887, in which it was found that the speed of light is independent of the motion of the source and the observer. This crucial experiment led Einstein to the theory of relativity. Michelson also used the interferometer to create the predecessor of today's length standard based on the wavelength of light. He received the Nobel Prize in 1907 "for precision optical instruments and the spectroscopic and metrological investigations carried out with their aid."

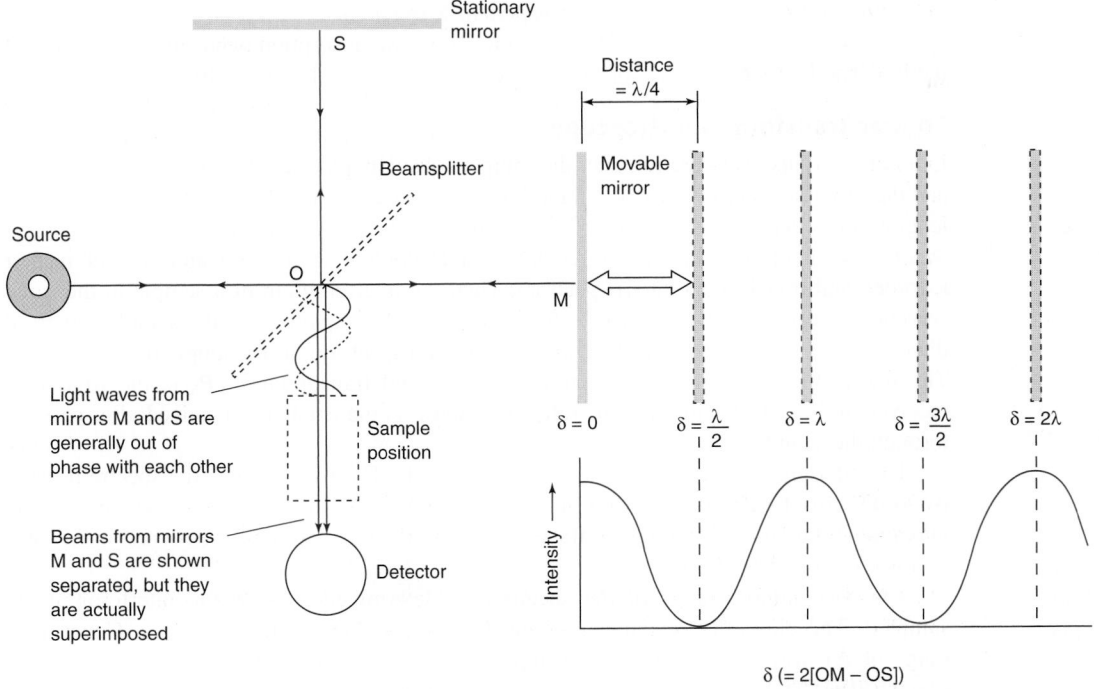

FIGURE 20-30 Schematic diagram of Michelson interferometer. Detector response as a function of retardation (= 2[OM − OS]) is shown for monochromatic incident radiation of wavelength λ.

The difference in pathlength followed by the two waves in Figure 20-30 is 2(OM − OS). This difference is called the *retardation*, δ. Constructive interference occurs whenever δ is an integer multiple of the wavelength, λ. A minimum appears when δ is a half-integer multiple of λ. If mirror M moves away from the beamsplitter at a constant speed, light reaching the detector goes through a sequence of maxima and minima as the interference alternates between constructive and destructive phases.

A graph of output light intensity versus retardation, δ, is called an **interferogram.** If the light from the source is monochromatic, the interferogram is a cosine wave:

$$I(\delta) = B(\tilde{\nu}) \cos\left(\frac{2\pi\delta}{\lambda}\right) = B(\tilde{\nu}) \cos(2\pi\tilde{\nu}\delta) \qquad (20\text{-}12)$$

where $I(\delta)$ is the intensity of light reaching the detector and $\tilde{\nu}$ is the wavenumber (= $1/\lambda$) of the light. Clearly, I is a function of the retardation, δ. $B(\tilde{\nu})$ is a constant that accounts for the intensity of the light source, efficiency of the beamsplitter (which never gives exactly 50% reflection and 50% transmission), and response of the detector. All these factors depend on $\tilde{\nu}$. For monochromatic light, there is only one value of $\tilde{\nu}$.

Figure 20-31a shows the interferogram produced by monochromatic radiation of wavenumber $\tilde{\nu}_0 = 2 \text{ cm}^{-1}$. The wavelength (repeat distance) of the interferogram can be seen in the figure to be $\lambda = 0.5$ cm, which is equal to $1/\tilde{\nu}_0 = 1/(2 \text{ cm}^{-1})$. Figure 20-31b shows the interferogram from a source with two monochromatic waves ($\tilde{\nu}_0 = 2$ and $\tilde{\nu}_0 = 8 \text{ cm}^{-1}$) with relative intensities 1 : 1. A short wave oscillation ($\lambda = \frac{1}{8}$ cm) is superimposed on a long wave oscillation ($\lambda = \frac{1}{2}$ cm). The interferogram is a sum of two terms:

$$I(\delta) = B_1 \cos(2\pi\tilde{\nu}_1\delta) + B_2 \cos(2\pi\tilde{\nu}_2\delta) \qquad (20\text{-}13)$$

where $B_1 = 1$, $\tilde{\nu}_1 = 2 \text{ cm}^{-1}$, $B_2 = 1$, and $\tilde{\nu}_2 = 8 \text{ cm}^{-1}$.

Fourier analysis of the interferogram gives back the spectrum from which the interferogram is made. *The spectrum is the Fourier transform of the interferogram.*

Fourier analysis decomposes a curve into its component wavelengths. Fourier analysis of the interferogram in Figure 20-31a gives the (trivial) result that the interferogram is made from a single wavelength function, with $\lambda = \frac{1}{2}$ cm. Fourier analysis of the interferogram in Figure 20-31b gives the slightly more interesting result that the interferogram is composed of two wavelengths ($\lambda = \frac{1}{2}$ and $\lambda = \frac{1}{8}$ cm) with relative contributions 1 : 1. We say that the spectrum is the *Fourier transform* of the interferogram.

The interferogram in Figure 20-31c is derived from a spectrum with an absorption band centered at $\tilde{\nu}_0 = 4 \text{ cm}^{-1}$. The interferogram is the sum of contributions from all source wavelengths. The Fourier transform of the interferogram in Figure 20-31c is indeed the third spectrum in Figure 20-31c. That is, decomposition of the interferogram into its component wavelengths gives back the band centered at $\tilde{\nu}_0 = 4 \text{ cm}^{-1}$. *Fourier analysis of the interferogram gives back the intensities of its component wavelengths.*

The interferogram in Figure 20-31d comes from two absorption bands in the spectrum at the left. The Fourier transform of this interferogram gives back the spectrum to its left.

Fourier Transform Spectroscopy

In a Fourier transform spectrometer, the sample is usually placed between the interferometer and the detector, as in Figures 20-30 and 20-32. Because the sample absorbs at certain wavelengths, *the interferogram contains the spectrum of the source minus the spectrum of the sample.* An interferogram of a reference sample containing the cell and solvent is first recorded and transformed into a spectrum. Then the interferogram of a sample in the same solvent is recorded and transformed into a spectrum. The quotient of the sample spectrum divided by the reference spectrum is the transmission spectrum of the sample (Figure 20-33). The quotient is the same as computing P/P_0 to find transmittance. P_0 is the irradiance received at the detector through the reference, and P is the irradiance received after passage through the sample.

The interferogram loses intensity at those wavelengths absorbed by the sample.

The interferogram is recorded at discrete intervals. The *resolution* of the spectrum (ability to discern closely spaced peaks) is approximately equal to $(1/\Delta) \text{ cm}^{-1}$, where Δ is the maximum retardation. If the mirror travel is ±2 cm, the retardation is ±4 cm and the resolution is $1/(4 \text{ cm}) = 0.25 \text{ cm}^{-1}$.

resolution ≈ $(1/\Delta) \text{ cm}^{-1}$
Δ = maximum retardation (cm)

For a spectral range of $\Delta\tilde{\nu} \text{ cm}^{-1}$, points must be taken at retardation intervals of $1/(2\Delta\tilde{\nu})$.

Capital delta has two different purposes in this section:

$\Delta\tilde{\nu}$ = spectral range (cm^{-1})
Δ = maximum retardation (cm)

The wavenumber range of the spectrum is determined by how the interferogram is sampled. The closer the spacing between data points, the greater the range. Covering a range of $\Delta\tilde{\nu}$ wavenumbers requires sampling the interferogram at retardation intervals of $\delta = 1/(2\Delta\tilde{\nu})$. If $\Delta\tilde{\nu}$ is $8\,000 \text{ cm}^{-1}$, sampling must occur at intervals of $\delta = 1/(2 \cdot 8\,000 \text{ cm}^{-1}) = 0.625 \times 10^{-4} \text{ cm} = 0.625 \text{ μm}$. This sampling interval corresponds to a mirror motion of

Spectrum

Interferogram

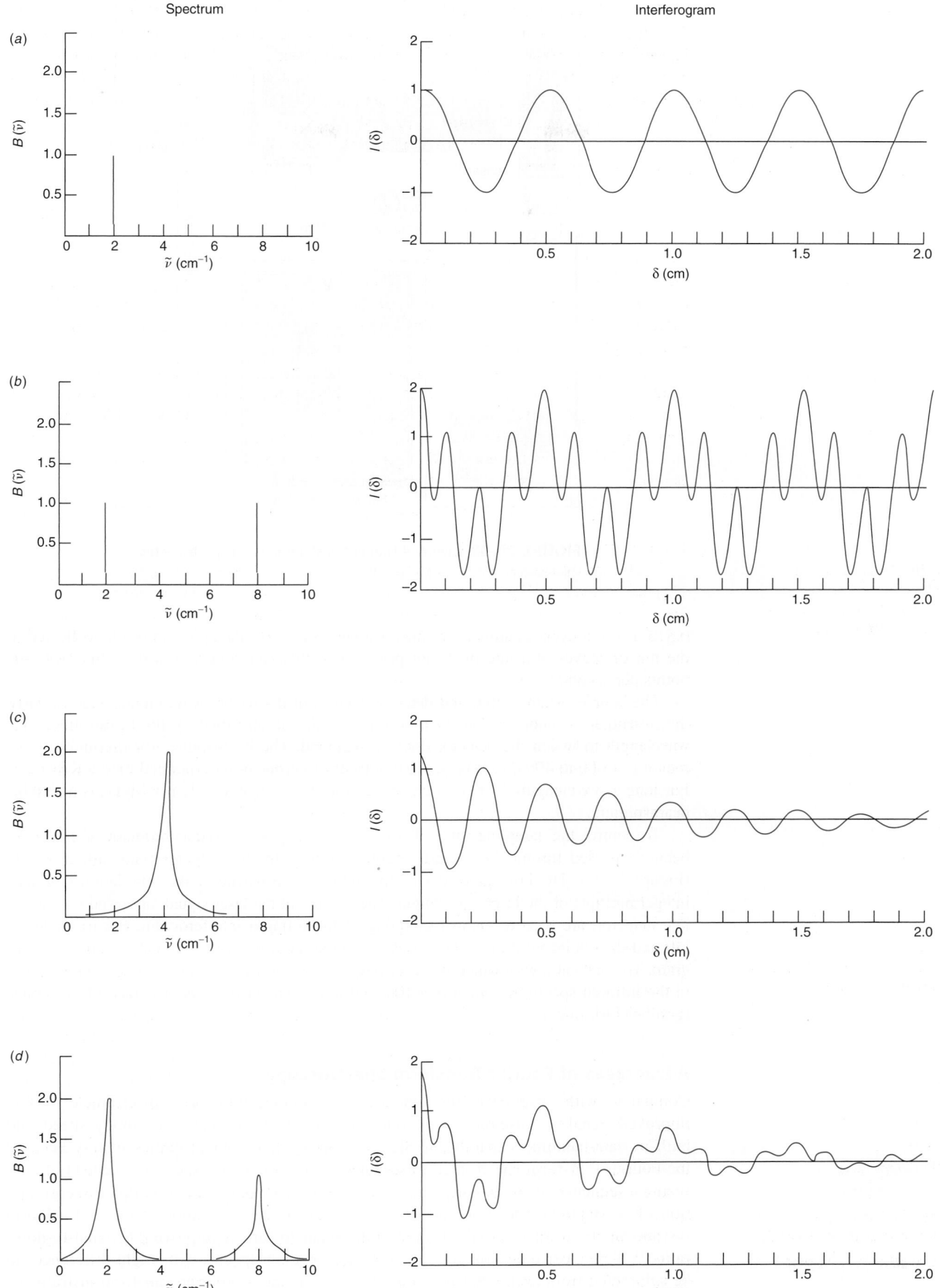

FIGURE 20-31 Interferograms produced by different spectra.

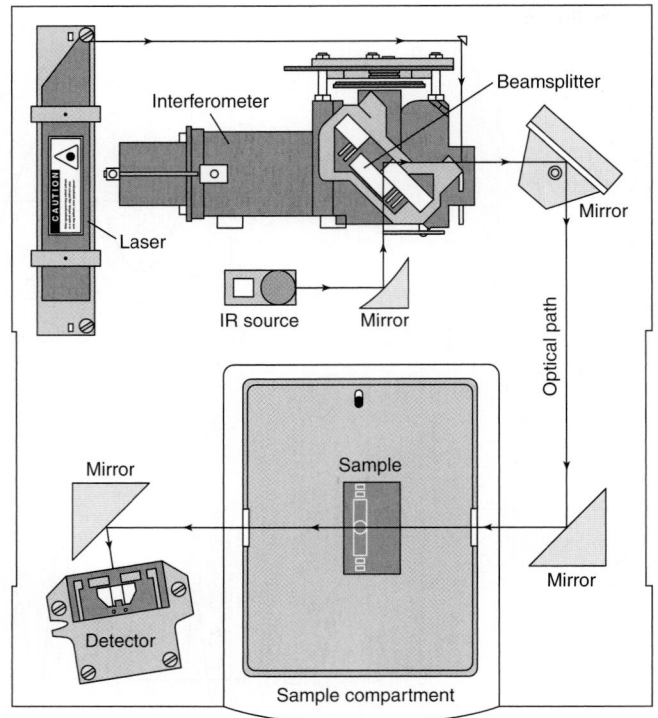

FIGURE 20-32 Layout of Fourier transform infrared spectrometer.
[Information from Nicolet, Madison, WI.]

0.312 µm. For every centimeter of mirror travel, 3.2×10^4 data points must be collected. If the mirror moves at a rate of 2 mm per second, the data collection rate would be 6 400 points per second.

The source, beamsplitter, and detector each limit the usable wavelength range. Clearly, the instrument cannot respond to a wavelength that is absorbed by the beamsplitter or a wavelength to which the detector does not respond. The beamsplitter for the mid-infrared region (~4 000 to 400 cm^{-1}) is typically a layer of germanium evaporated onto a KBr plate. For longer wavelengths ($\tilde{\nu} < 400$ cm^{-1}), a film of the organic polymer Mylar is a suitable beamsplitter.

To control the sampling interval for the interferogram, a monochromatic visible laser beam is passed through the interferometer along with the polychromatic infrared light (Figure 20-32). The laser gives destructive interference whenever the retardation is a half-integer multiple of the laser wavelength. These zeros in the laser signal, observed with a visible detector, are used to control sampling of the infrared interferogram. For example, an infrared data point might be taken at every second zero point of the visible-light interferogram. The precision with which the laser frequency is known gives an accuracy of 0.01 cm^{-1} in the infrared spectrum, which is a 100-fold improvement over the accuracy of dispersive (grating) instruments.

Advantages of Fourier Transform Spectroscopy

Compared with dispersive instruments, the Fourier transform spectrometer offers improved signal-to-noise ratio at a given resolution, better frequency accuracy, speed, and built-in data-handling capabilities. Signal-to-noise improvement comes mainly because the Fourier transform spectrometer uses energy from the entire spectrum, instead of analyzing a sequence of small wavebands available from a monochromator. Precise reproduction of wavenumber position from one spectrum to the next allows Fourier transform instruments to average signals from multiple scans to further improve the signal-to-noise ratio. Wavenumber precision and low noise levels allow spectra with slight differences to be subtracted from each other to expose those differences. Fourier transform instruments are not as accurate as some dispersive spectrometers for measuring transmittance. Advantages of Fourier transform infrared spectrometers are so great that it is nearly impossible to purchase a dispersive infrared spectrometer.

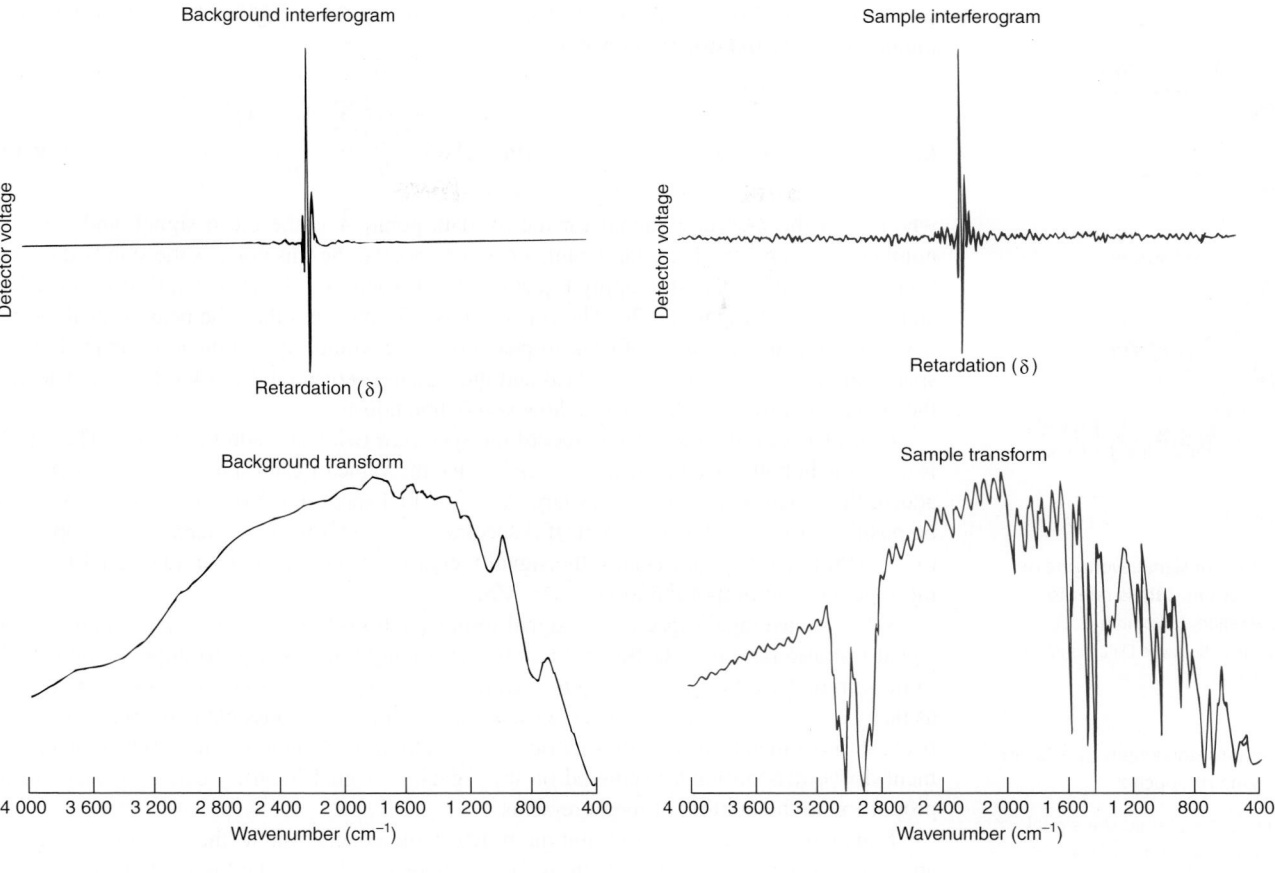

Background interferogram

Sample interferogram

Detector voltage

Detector voltage

Retardation (δ)

Retardation (δ)

Background transform

Sample transform

Wavenumber (cm^{-1})

Wavenumber (cm^{-1})

Transmission spectrum

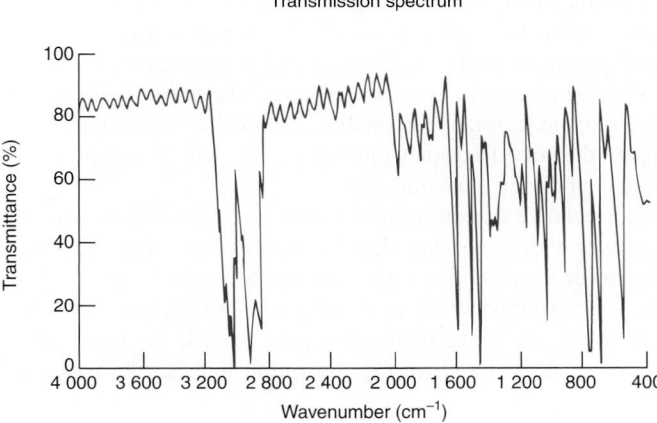

Transmittance (%)

Wavenumber (cm^{-1})

FIGURE 20-33 Fourier transform infrared spectrum of polystyrene film. The Fourier transform of the background interferogram gives a spectrum determined by the source intensity, beamsplitter efficiency, detector response, and absorption by traces of H_2O and CO_2 in the atmosphere. The sample compartment is purged with dry N_2 to reduce the levels of H_2O and CO_2. The transform of the sample interferogram is a measure of all the instrumental factors, plus absorption by the sample. The transmission spectrum is obtained by dividing the sample transform by the background transform. Each interferogram is an average of 32 scans and contains 4 096 data points, giving a nominal resolution of 8 cm^{-1}. The mirror velocity was 0.693 cm/s. [Data from M. P. Nadler, Michelson Laboratory, China Lake, CA.]

20-6 Dealing with Noise[24]

An advantage of Fourier transform spectroscopy is that the entire interferogram is recorded in a few seconds and stored in a computer. The signal-to-noise ratio can be improved by collecting tens or hundreds of interferograms and averaging them.

Signal Averaging

Signal averaging can improve the quality of data, as illustrated in Figure 20-34.[25] The lowest trace contains a great deal of noise. A simple way to estimate the noise level is to measure the maximum amplitude of the noise in a region free of signal. The signal is measured from the middle of the baseline noise to the middle of the noisy peak. By this criterion, the lowest trace of Figure 20-34 has a signal-to-noise ratio of 14/9 = 1.6.

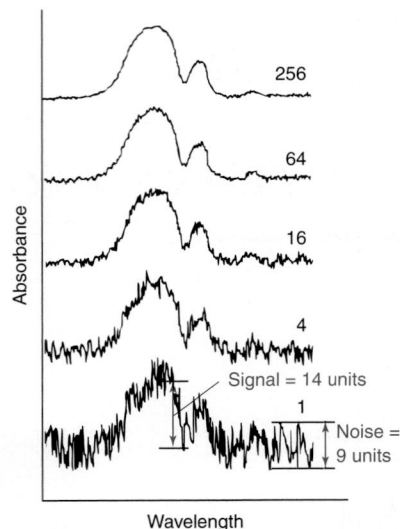

FIGURE 20-34 Effect of signal averaging on a simulated noisy spectrum. Labels refer to number of scans averaged. [Data from R. Q. Thompson, "Experiments in Software Data Handling," *J. Chem. Ed.* **1985,** *62,* 866.]

To improve the signal-to-noise ratio by a factor of n requires averaging n^2 spectra.

Question By what factor should the signal-to-noise ratio be improved when 16 spectra are averaged? Measure the noise level in Figure 20-34 to test your prediction.

$$\text{Detection limit} \equiv \frac{3s}{m}$$

s = standard deviation of noise
m = slope of linear calibration curve

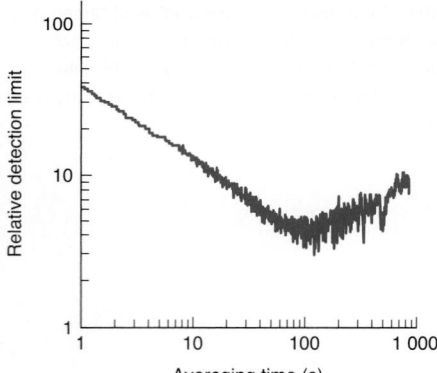

FIGURE 20-35 Reduction in relative detection limit with increase in signal averaging time for photoacoustic measurement of O_2 in air. Ten measurements were averaged every second. [Reprinted with permission from D. K. Havey, P. A. Bueno, K. A. Gillis, J. T. Hodges, G. W. Mulholland, R. D. van Zee, and M. R. Zachariah, "Photoacoustic Spectrometer with a Calculable Cell Constant for Measurements of Gases and Aerosols," *Anal. Chem.* **2010,** *82,* 7935. © 2010 American Chemical Society.]

A more common measurement of noise, which requires a digitized signal, is the **root-mean-square (rms) noise,** defined as

Root-mean-square noise:
$$\text{rms noise} = \sqrt{\frac{\sum_i (A_i - \overline{A})^2}{n}} \qquad (20\text{-}14)$$

where A_i is the measured signal for the ith data point, $\overline{A}$ is the mean signal, and n is the number of data points. For a large number of data points, the rms noise is the standard deviation of the noise. It is best to apply Equation 20-14 where the signal is flat, as it is at the left and right sides of Figure 20-34. The rms noise is ~5 times less than the peak-to-peak noise. If we used rms noise instead of peak-to-peak noise, we would say that the noise in the bottom spectrum of Figure 20-34 is 9/5 = 1.8 and the signal-to-noise ratio is 14/1.8 = 7.8. Clearly, the signal-to-noise ratio depends on how you define noise.

Consider what happens if you record the spectrum twice and add the results. The signal is the same in both spectra and adds to give twice the value of each spectrum. If n spectra are added, the signal will be n times as large as in the first spectrum. Noise is random, so it may be positive or negative at any point. If n spectra are added, the noise increases in proportion to $\sqrt{n}$ (Problem 20-34). Because the signal increases in proportion to n, the signal-to-noise ratio increases in proportion to $n/\sqrt{n} = \sqrt{n}$.

When we average n spectra, the signal-to-noise ratio is improved by $\sqrt{n}$. To improve the signal-to-noise ratio by a factor of 2 requires averaging four spectra. To improve the signal-to-noise ratio by a factor of 10 requires averaging 100 spectra. Spectroscopists might record as many as 10^4 to 10^5 scans to observe weak signals. It is rarely possible to do better than this because instrumental instabilities cause drift in addition to random noise. When an instrument drifts, it is no longer centered on the signal, so signal intensity ceases to increase in proportion to the number of measurements.

Equation 5-5 stated that the minimum detectable concentration (the detection limit) for an analytical method is often taken as $3s/m$, where s is the standard deviation for multiple measurements of a blank (or a region where there is no signal expected) and m is the slope of the linear calibration curve. The slope is the change in signal per unit change in analyte concentration, such as mV of signal per (mole per liter) change in analyte concentration. We can obtain a lower detection limit by decreasing the standard deviation (the *noise*) in the signal, if the slope m of the response remains constant for smaller and smaller concentrations of analyte. When averaging a signal, the noise frequency must be greater than the sampling rate so that noise will tend to average out during a measurement.

Figure 20-35 shows how the relative detection limit for O_2 in air from photoacoustic spectroscopy (page 506) decreases with increasing signal averaging time. The more time spent averaging out noise, the lower the detection limit, up to an averaging time of ~80 s. Thereafter, instabilities such as frequency drift of the diode laser source cause the detection limit to get worse (larger) with increasing signal averaging time. If the laser source drifts, it is no longer in resonance with the sharp O_2 absorption line, so the measured absorption decreases.

Types of Noise

Figure 20-36 shows three common types of noise in electrical instruments.[26] The graphs show the amplitude of the noise versus the frequency of the noise. The upper trace is *white noise*, also called *Gaussian noise*. The noise amplitude is independent of frequency. One source of white noise, called *Johnson noise*, is random fluctuations of electrons in an electronic device. Lowering the operating temperature is one way to reduce Johnson noise. *Shot noise* is another form of white noise attributable to the quantized nature of charge carriers and photons. At low signal levels, noise arises from random variation in the small number of photons reaching a detector or the small number of electrons and holes generated in a semiconductor.

The second trace in Figure 20-36 shows *1/f noise*, also called *drift*, which is greatest at zero frequency and decreases in proportion to 1/frequency. An example of low-frequency noise in laboratory instruments is flickering or drifting of a light source in a spectrophotometer or a flame in atomic spectroscopy. Drift arises from causes such as slow changes in instrument components with temperature and with age and variation of power-line voltage to an instrument. The classical way to detect and account for drift is to periodically measure standards and correct the instrument reading for any observed change.[27] In

CHAPTER 20 Spectrophotometers

Figure 20-35, signal averaging ceases to improve the detection limit after ~80 s of averaging time because the wavelength of the diode laser used to measure O_2 drifts out of resonance with the O_2 absorption.

The bottom trace of Figure 20-36 shows *line noise* (also called *interference* or *whistle noise*) at discrete frequencies such as the 60-Hz transmission-line frequency or the 0.2-Hz vibrational frequency when elephants walk through the basement of your building. Electrical shielding and grounding the shielding and instrument to the same ground point help reduce line noise.

Beam Chopping

Spectrophotometers in Figures 20-1 through 20-3 have a rotating mirror called a *chopper* that alternately sends light through sample and reference cells. Chopping allows both cells to be sampled almost continuously, and it provides a means of noise reduction. **Beam chopping** moves the analytical signal from zero frequency to the frequency of the chopper. The chopping frequency can be selected so that $1/f$ noise and line noise are minimal. High-frequency detector circuits are required to take advantage of beam chopping.

A Low-Noise Spectrophotometer

Spectrophotometer noise is broadly attributable to (i) sources that are independent of the light level, (ii) sources that are proportional to photo-generated current, and (iii) variation of the intensity of the light source.[28] For several decades, precision has been governed by variation of intensity of the light source, which produces a root-mean-square noise equivalent to an apparent absorbance of ~0.000 03 at visible wavelengths. That is, absorbance below ~0.000 03 cannot be measured because it is hidden in the noise.

Figure 20-37 shows a system designed to cancel the noise generated by variation of the light source. Visible light from a tungsten-halogen lamp goes through a monochromator and is then split into two beams that pass through sample and reference cells. Light from the sample cell goes to a photodetector that generates current I_{sample}. Light from the reference cell goes to a photodetector that generates current I_{ref}. Electronics then convert the currents to voltages V_{sample} and V_{ref} and the *difference voltage* $V_{difference} = V_{sample} - V_{ref}$. Noise from variation of the light source intensity affects V_{sample} and V_{ref} equally. That is, if the intensity momentarily increases by 0.1%, the voltages recorded by each detector increase by 0.1%. The difference $V_{difference}$ should be *zero* in the absence of sample absorption. The traces at the right of Figure 20-37 are simulations showing a weak absorption superimposed on strong noise. The signal is too weak to see in the sample spectrum (V_{sample}), but is clear in the difference spectrum ($V_{difference}$).

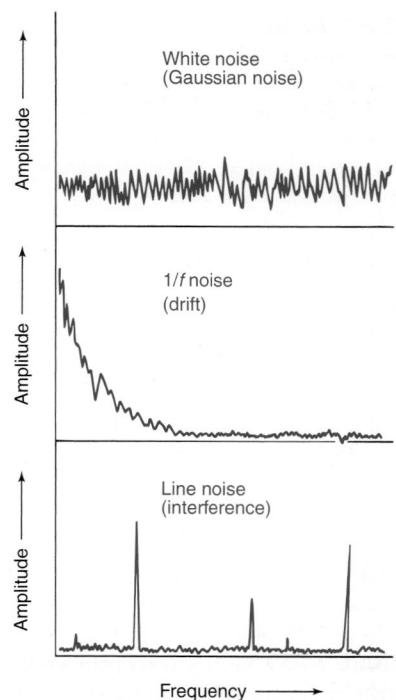

FIGURE 20-36 Three types of noise in electrical instruments. White noise is always present. A beam chopping frequency can be selected to reduce $1/f$ noise and line noise to insignificant values.

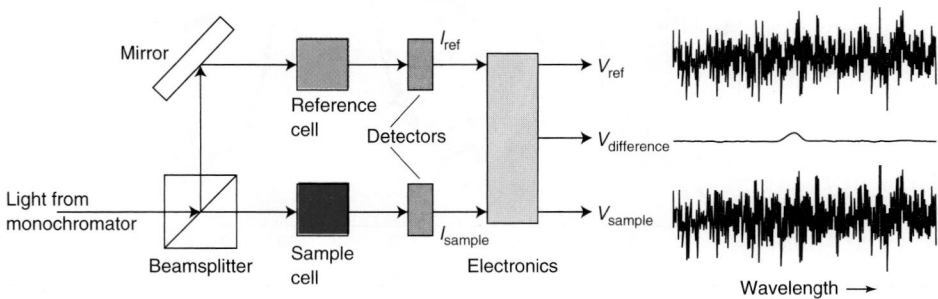

FIGURE 20-37 Principle of low-noise spectrophotometer. [Information from Z. Xu and D. W. Larsen, "Development of Ultra-Low-Noise Spectrophotometry for Analytical Applications," *Anal. Chem.* **2005,** *77,* 6463.]

Figure 20-38 is an experimental implementation of the scheme to cancel source intensity noise. The upper trace is the absorption spectrum of a 9-nM dye solution with a peak absorbance of 0.000 2. The lower spectrum was obtained from the same instrument modified to display absorbance computed from the difference voltage $V_{difference}$.[29] The signal-to-noise ratio is improved by a factor of ten. Hardware modifications could enable an additional factor-of-ten improvement for a total signal-to-noise ratio reduction of 100.

Savitzky-Golay Polynomial Smoothing of Noisy Data

When possible, signal averaging is an excellent first approach to noise reduction. If further noise reduction is needed after signal averaging reaches a practical limit, algorithms are available for numerical manipulation of data to reduce noise. Figure 20-39 shows

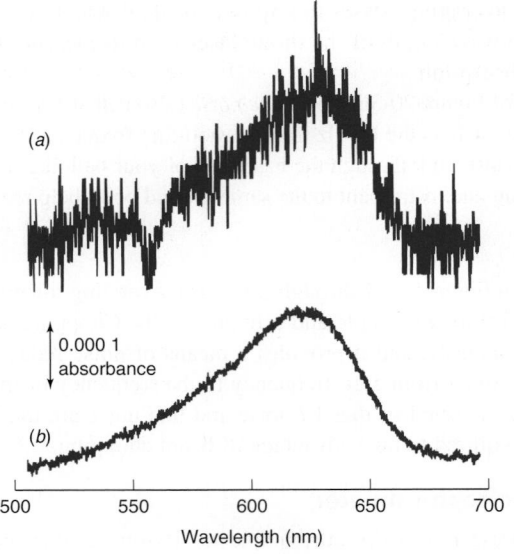

FIGURE 20-38 Spectrum of 2 nM Nile Blue in methanol recorded with (*a*) commercial spectrophotometer and (*b*) same instrument modified to cancel light source variation. [Data from Z. Xu and D. W. Larsen, "Development of Ultra-Low-Noise Spectrophotometry for Analytical Applications," *Anal. Chem.* **2005,** *77,* 6463.]

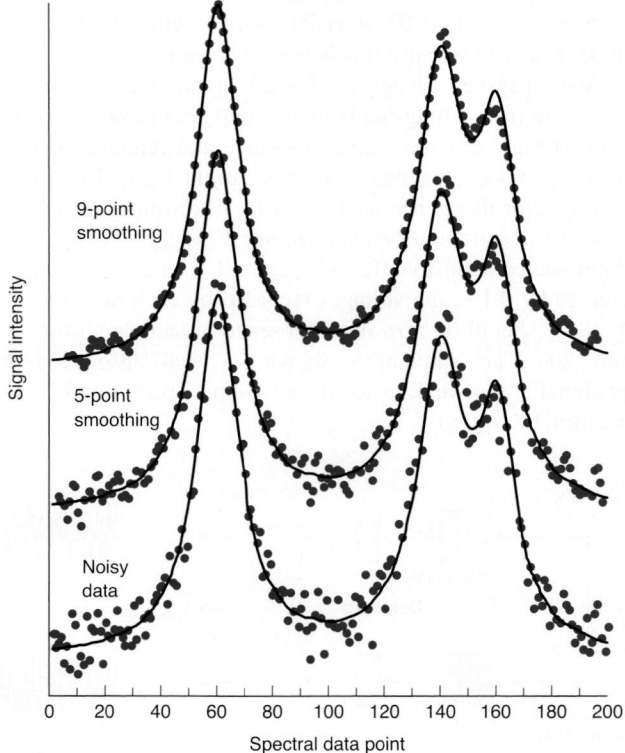

FIGURE 20-39 Savitzky-Golay polynomial smoothing of noisy data. Solid lines show ideal signal. Points in the lowest spectrum illustrate data with noise. Middle spectrum shows 5-point cubic polynomial smoothing of the data. Upper spectrum illustrates 9-point cubic polynomial smoothing.

smoothing by polynomial fitting of a cubic curve to sets of five or nine data points.[30,31] The solid line in the lower trace shows an ideal signal with no noise. Dots in the lower trace simulate noisy experimental data. A caveat for polynomial fitting is that the spacing of data points on the abscissa (*x*-axis) must be uniform, which is typically the way most instruments record data.

There are 200 data points in the spectral region shown Figure 20-39. Points 8 to 12 of the original noisy data are shown as solid colored circles in Figure 20-40. To adjust point 10 for noise reduction, a cubic curve is fit by the method of least squares to the five points 8 to 12. Point 10 is then moved up or down to the *y*-position of the cubic curve at $x = 10$. The open circle is the smoothed position of point 10. To smooth point 11, a cubic curve is fit to the original noisy data points 9, 10, 11, 12, and 13. This procedure is continued until every point from 3 to 198 is adjusted. The result is the 5-point-smoothed data set in the middle of Figure 20-39. We lose points 1, 2, 199, and 200 because there are not five points available on both sides of these points.

Points in the upper trace of Figure 20-39 come from fitting a cubic curve to sets of nine points. To adjust point 10, a curve is fit to points 6, 7, 8, 9, 10, 11, 12, 13, and 14 of the original noisy data at the bottom of the figure. Point 10 is then moved to the y-position of the cubic curve at $x = 10$. In a similar manner, every point from 5 to 195 is adjusted. We observe that 9-point smoothing gives less noise than 5-point smoothing. That is, points in the upper trace lie closer to the ideal (solid) curve than do points in the middle trace. However, increasing the number of points used for smoothing can produce spectral distortion. For example, the amplitude (height) of the peak at $x = 160$ is reduced by 9-point smoothing relative the amplitude for 5-point smoothing.

The smoothing algorithm is simple to execute in a spreadsheet. The y-coordinate y_0(adjusted) of the central point for five-point cubic curve fitting is

$$y_0(\text{adjusted}) = \frac{-3y_{-2} + 12y_{-1} + 17y_0 + 12y_1 - 3y_2}{35} \qquad (20\text{-}15)$$

where y_i are the five measured values of y around the point being adjusted. For example, the adjusted value of y_{10} is

$$y_{10}(\text{adjusted}) = \frac{-3y_8 + 12y_9 + 17y_{10} + 12y_{11} - 3y_{12}}{35} \qquad (20\text{-}16)$$

The points in the middle spectrum of Figure 20-39 were obtained by applying Equation (20-15) to each point from $x = 3$ to $x = 198$ of the original, noisy data set.

For 9-point smoothing, the corresponding formula is

$$y_0(\text{adjusted}) = \frac{-21y_{-4} + 14y_{-3} + 39y_{-2} + 54y_{-1} + 59y_0 + 54y_1 + 39y_2 + 14y_3 - 21y_4}{231} \qquad (20\text{-}17)$$

Equation 20-17 was applied to points 5 through 196 of the original data to obtain the smoothed data at the top of Figure 20-39.

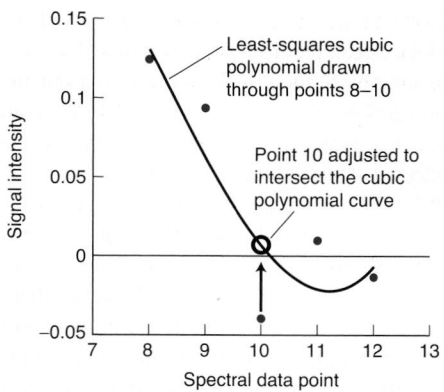

FIGURE 20-40 How the Savitzky-Golay procedure adjusts point 10 by 5-point cubic polynomial smoothing. Colored circles show original noisy data. Open circle shows smoothed position of point 10.

Terms to Understand

attenuated total reflectance	grating	photodiode array	signal averaging
bandwidth	greenhouse gas	photomultiplier tube	Snell's law
beam chopping	interferogram	phototube	stray light
blackbody radiation	interferometer	photovoltaic detector	surface plasmon resonance
charge coupled device	laser	polychromator	thermocouple
diffraction	monochromator	refraction	waveguide
dispersion	optical fiber	refractive index	
ferroelectric material	optode	resolution	
Fourier analysis	photoconductive detector	root-mean-square (rms) noise	

Summary

Spectrophotometer components include the source, sample cell, monochromator, and detector. Tungsten and deuterium lamps provide visible and ultraviolet radiation; a silicon carbide globar is a good infrared source. Tungsten and silicon carbide lamps behave approximately as blackbodies, which are objects that absorb all light striking them and reradiate a characteristic spectrum of radiation. Emission of radiant energy from the surface of the blackbody is proportional to the fourth power of temperature and shifts to shorter wavelengths as temperature increases. Lasers provide high-intensity, coherent, monochromatic radiation by stimulated emission from a medium in which an excited state has been pumped to a higher population than that of a lower state. Sample cells must be transparent to the radiation of interest. A reference sample compensates for reflection and scattering by the cell and solvent. A grating monochromator disperses light into its component wavelengths. The finer a grating is ruled, the higher the resolution and the greater the dispersion of wavelengths over angles. Narrow slits improve resolution but increase noise, because less light reaches the detector. A bandwidth that is one-fifth of the width of the spectral peak is a good compromise between maximizing signal-to-noise ratio and minimizing peak distortion. Stray light introduces absorbance errors that are most serious when the transmittance of a sample is very low. Filters pass wide bands of wavelength and reject other bands.

A photomultiplier tube is a sensitive detector of visible and ultraviolet radiation; photons cause electrons to be ejected from a metallic cathode. The signal is amplified at each successive dynode on which the photoelectrons impinge. Photodiode arrays and charge coupled devices are solid-state detectors in which photons create electrons and holes in semiconductor materials. Coupled to a polychromator, these devices can record all wavelengths of a spectrum simultaneously, with resolution limited by the number and spacing of detector elements. An image intensified charge coupled device converts incoming photons to electrons, amplifies the electrons with a microchannel plate, converts electrons back to photons with a fluorescent screen, and then detects the photons. Common infrared detectors include thermocouples, ferroelectric materials, and photoconductive and photovoltaic devices. Photoacoustic spectroscopy detects infrared absorption by converting some absorbed energy to sound waves that are detected. Cavity ringdown spectroscopy measures very low absorption by gases or liquids by circulating a laser beam through the sample thousands of times to multiply the pathlength. The decay time of the circulating laser power is a measure of absorption by the sample.

When light passes from a region of refractive index n_1 to a region of refractive index n_2, the angle of refraction (θ_2) is related to the angle of incidence (θ_1) by Snell's law: $n_1 \sin \theta_1 = n_2 \sin \theta_2$. Optical fibers and flat waveguides transmit light by a series of total internal reflections. Optodes are sensors based on optical fibers. Some optodes have a layer of material whose absorbance or fluorescence changes in the presence of analyte. Light can be carried to and from the tip by the optical fiber. When light is transmitted through a fiber optic or a waveguide by total internal reflectance, some light, called the evanescent wave, penetrates through the reflective interface during each reflection. In attenuated total reflectance devices, the waveguide is coated with a substance that absorbs light in the presence of analyte. In a surface plasmon resonance sensor, we measure the change in the angle of minimum reflectance from a gold film coated with a chemically sensitive layer on the back face of a prism. A chemical process that changes the refractive index of the layer on the gold film changes the angle of minimum reflectance.

Fourier analysis decomposes a signal into its component wavelengths. An interferometer contains a beamsplitter, a stationary mirror, and a movable mirror. Reflection of light from the two mirrors creates an interferogram. Fourier analysis of the interferogram tells us what frequencies went into the interferogram. In a Fourier transform spectrophotometer, the interferogram of the source is first measured without a sample. Then the sample is placed in the beam and a second interferogram is recorded. The transforms of the interferograms tell what amount of light at each frequency reaches the detector with and without the sample. The quotient of the two transforms is the transmission spectrum. The resolution of a Fourier transform spectrum is approximately $1/\Delta$, where Δ is the maximum retardation. To cover a wavenumber range $\Delta\tilde{\nu}$ requires sampling the interferogram at intervals of $\delta = 1/(2\Delta\tilde{\nu})$.

If you average n scans, the signal-to-noise ratio should increase by $\sqrt{n}$. White noise, which is independent of frequency, arises from sources such as random fluctuations of electrons in components (Johnson noise) and the discrete nature of charge carriers and photons (shot noise). $1/f$ noise decreases with increasing frequency. Drift and flicker of the intensity of a light source or brightness of a flame in atomic spectroscopy are sources of $1/f$ noise. Line noise occurs at discrete frequencies such as the 60-Hz line frequency of a power supply. Beam chopping in a dual-beam spectrophotometer reduces $1/f$ and line noise. A spectrometer modified to record the difference between the sample and reference signals can reduce noise from lamp flicker by at least an additional factor of 10. Polynomial smoothing is one of many digital procedures to reduce noise in data.

Exercises

20-A. (a) If a diffraction grating has a resolution of 10^4, is it possible to distinguish two spectral lines with wavelengths of 10.00 and 10.01 μm?

(b) With a resolution of 10^4, how close in wavenumbers (cm^{-1}) is the closest line to 1 000 cm^{-1} that can barely be resolved?

(c) Calculate the resolution of a 5.0-cm-long grating ruled at 250 lines/mm for first-order ($n = 1$) diffraction and tenth-order ($n = 10$) diffraction.

(d) Find the angular dispersion ($\Delta\phi$, in radians and degrees) between light rays with wavenumbers of 1 000 and 1 001 cm^{-1} for second-order diffraction ($n = 2$) from a grating with 250 lines/mm and $\phi = 30°$.

20-B. The true absorbance of a sample is 1.000, but the monochromator passes 1.0% stray light. Add the light coming through the sample to the stray light to find the apparent transmittance of the sample. Convert this answer back into absorbance and find the relative error in the calculated concentration of the sample.

20-C. Refer to the Fourier transform infrared spectrum in Figure 20-33.

(a) The interferogram was sampled at retardation intervals of 1.266 0 $\times$ 10^{-4} cm. What is the theoretical wavenumber range (0 to ?) of the spectrum?

(b) A total of 4 096 data points were collected from $\delta = -\Delta$ to $\delta = +\Delta$. Compute the value of Δ, the maximum retardation.

(c) Calculate the approximate resolution of the spectrum.

(d) The interferometer mirror velocity is given in the figure caption. How many microseconds elapse between each datum?

(e) How many seconds were required to record each interferogram once?

(f) What kind of beamsplitter is typically used for the region 400 to 4 000 cm^{-1}? Why is the region below 400 cm^{-1} not observed?

20-D. The table shows signal-to-noise ratios recorded in a nuclear magnetic resonance experiment. Construct graphs of **(a)** signal-to-noise ratio versus n and **(b)** signal-to-noise ratio versus $\sqrt{n}$, where n is the number of scans. Draw error bars corresponding to the standard deviation at each point. Is the signal-to-noise ratio proportional to $\sqrt{n}$? Find the 95% confidence interval for each row of the table.

Signal-to-noise ratio at the aromatic protons
of 1% ethylbenzene in CCl_4

Number of experiments	Number of accumulations	Signal-to-noise ratio	Standard deviation
8	1	18.9	1.9
6	4	36.4	3.7
6	9	47.3	4.9
8	16	66.7	7.0
6	25	84.6	8.6
6	36	107.2	10.7
6	49	130.3	13.3
4	64	143.2	15.1
4	81	146.2	15.0
4	100	159.4	17.1

*Data from M. Henner, P. Levior, and B. Ancian, "An NMR Spectrometer-Computer Interface Experiment," J. Chem. Ed. **1979**, 56, 685.*

Problems

The Spectrophotometer

20-1. Describe the role of each component of the spectrophotometer in Figure 20-1.

20-2. Explain how a laser generates light. List important properties of laser light.

20-3. Would you use a tungsten or a deuterium lamp as a source of 300-nm radiation? What kind of lamp provides radiation at 4-μm wavelength?

20-4. Which variables increase the resolution of a grating? Which variables increase the dispersion of the grating? How is the blaze angle chosen to optimize a grating for a particular wavelength?

20-5. What is the role of a filter in a grating monochromator?

20-6. What are the advantages and disadvantages of decreasing monochromator slit width?

20-7. Explain how the following detectors of visible radiation work: (a) photomultiplier tube, (b) photodiode array, (c) charge coupled device, and (d) image intensified charge coupled device.

20-8. Deuterated triglycine sulfate (abbreviated DTGS) is a common ferroelectric infrared detector material. Explain how it works.

20-9. Consider a reflection grating operating with an incident angle of 40° in Figure 20-7.

(a) How many lines per centimeter should be etched in the grating if the first-order diffraction angle for 600 nm (visible) light is to be −30°?

(b) Answer the same question for 1 000 cm^{-1} (infrared) light.

20-10. Show that a grating with 10^3 grooves/cm provides a dispersion of 5.8° per μm of wavelength if $n = 1$ and $\phi = 10°$ in Equation 20-4.

20-11. (a) What resolution is required for a diffraction grating to resolve wavelengths of 512.23 and 512.26 nm?

(b) With a resolution of 10^4, how close in nm is the closest line to 512.23 nm that can barely be resolved?

(c) Calculate the fourth-order resolution of a grating that is 8.00 cm long and is ruled at 185 lines/mm.

(d) Find the angular dispersion ($\Delta\phi$) between light rays with wavelengths of 512.23 and 512.26 nm for first-order diffraction ($n = 1$) and thirtieth-order diffraction from a grating with 250 lines/mm and $\phi = 3.0°$.

20-12. (a) The true absorbance of a sample is 1.500, but 0.50% stray light reaches the detector. Find the apparent transmittance and apparent absorbance of the sample.

(b) How much stray light can be tolerated if the absorbance error is not to exceed 0.001 at a true absorbance of 2?

(c) A research-quality spectrophotometer has a stray light level of <0.000 05% at 340 nm. What will be the maximum absorbance error for a sample with a true absorbance of 2? Of 3?

20-13. The pathlength of a cell for infrared spectroscopy can be measured by counting *interference fringes* (ripples in the transmission spectrum). The following spectrum shows 30 interference maxima between 1 906 and 698 cm^{-1} obtained by placing an empty KBr cell in a spectrophotometer.

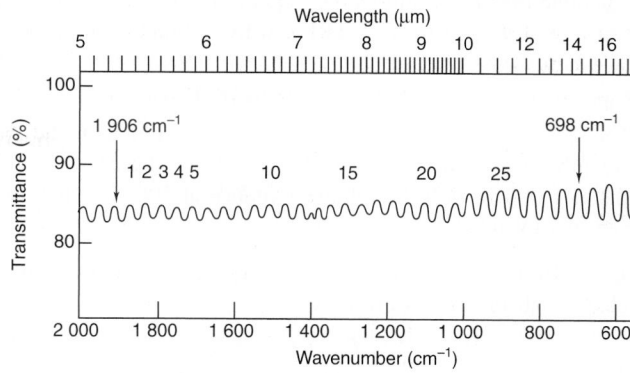

The fringes arise because light reflected from the cell compartment interferes constructively or destructively with the unreflected beam.

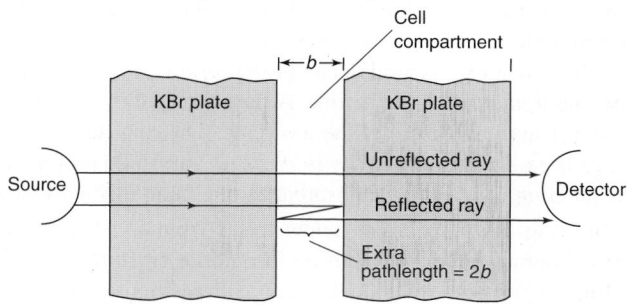

If the reflected beam travels an extra distance λ, it will interfere constructively with the unreflected beam. If the reflection pathlength is λ/2, destructive interference occurs. Peaks therefore arise when $m\lambda = 2b$ and troughs occur when $m\lambda/2 = 2b$, where m is an integer. If the medium between the KBr plates has refractive index n, the wavelength in the medium is λ/n, so the equations become $m\lambda/n = 2b$ and $m\lambda/2n = 2b$. The cell pathlength can be shown to be given by

$$b = \frac{N}{2n} \cdot \frac{\lambda_1\lambda_2}{\lambda_2 - \lambda_1} = \frac{N}{2n} \cdot \frac{1}{\tilde{\nu}_1 - \tilde{\nu}_2}$$

where N maxima occur between wavelengths λ_1 and λ_2. Calculate the pathlength of the cell that gave the interference fringes shown earlier.

20-14. Calculate the power per unit area (the exitance, W/m^2) radiating from a blackbody at 77 K (liquid nitrogen temperature) and at 298 K (room temperature).

20-15. The exitance (power per unit area per unit wavelength) from a blackbody (Box 20-1) is given by the *Planck distribution*:

$$M_\lambda = \frac{2\pi hc^2}{\lambda^5}\left(\frac{1}{e^{hc/\lambda kT} - 1}\right)$$

where λ is wavelength, T is temperature (K), h is Planck's constant, c is the speed of light, and k is Boltzmann's constant. The area under each curve between two wavelengths in the blackbody graph in Box 20-1 is equal to the power per unit area (W/m^2) emitted between those two wavelengths. We find the area by integrating the Planck function between wavelengths λ_1 and λ_2:

$$\text{Power emitted} = \int_{\lambda_1}^{\lambda_2} M_\lambda d\lambda$$

For a narrow wavelength range, Δλ, the value of M_λ is nearly constant and the power emitted is simply the product $M_\lambda\Delta\lambda$.

(a) Evaluate M_λ at $\lambda = 2.00\ \mu m$ and at $\lambda = 10.00\ \mu m$ at $T = 1\,000$ K.

(b) Calculate the power emitted per square meter at 1 000 K in the interval $\lambda = 1.99\ \mu m$ to $\lambda = 2.01\ \mu m$ by evaluating the product $M_\lambda \Delta\lambda$, where $\Delta\lambda = 0.02\ \mu m$.

(c) Repeat part **(b)** for the interval 9.99 to 10.01 μm.

(d) The quantity $M_{2\ \mu m}/M_{10\ \mu m}$ is the relative exitance at the two wavelengths. Compare the relative exitance at these two wavelengths at 1 000 K with the relative exitance at 100 K. What does your answer mean?

20-16. (a) In the cavity ring-down measurement at the opening of this chapter, absorbance is given by

$$A = \frac{L}{c \ln 10}\left(\frac{1}{\tau} - \frac{1}{\tau_0}\right)$$

where L is the length of the triangular path in the cavity, c is the speed of light, τ is the ring-down lifetime with sample in the cavity, and τ_0 is the ring-down lifetime with no sample in the cavity. Ring-down lifetime is obtained by fitting the observed ring-down signal intensity I to an exponential decay of the form $I = I_0 e^{-t/\tau}$, where I_0 is the initial intensity and t is time. A measurement of CO_2 is made at a wavelength absorbed by the molecule. The ring-down lifetime for a 21.0-cm-long empty cavity is 18.52 μs and 16.06 μs for a cavity containing CO_2. Find the absorbance of CO_2 at this wavelength.

(b) The ring-down spectrum below arises from $^{13}CH_4$ and $^{12}CH_4$ from 1.9 ppm (vol/vol) of methane in outdoor air at 0.13 bar. The spectrum arises from individual rotational transitions of the ground vibrational state to a second excited C—H vibrational state of the molecule. **(i)** Explain what quantity is plotted on the ordinate (y-axis). **(ii)** The peak for $^{12}CH_4$ is at 6 046.954 6 cm^{-1}. What is the wavelength of this peak in nm? What is the name of the spectral region where this peak is found?

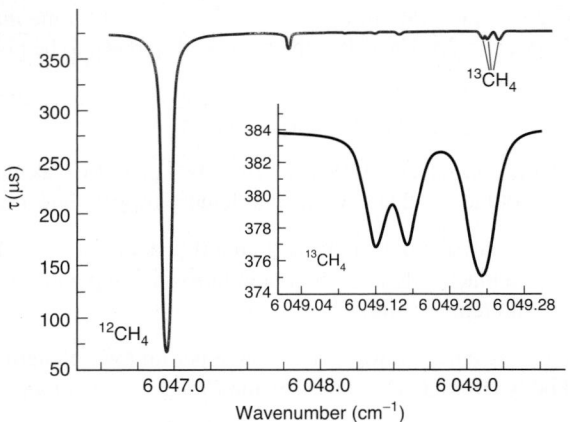

Cavity ring-down spectrum of methane in air. [Reprinted with permission from Y. Chen, K. K. Lehmann, J. Kessler, B. S. Lollar, G. L. Couloume, and T. C. Onstott, "Measurement of the $^{13}C/^{12}C$ of Atmospheric CH_4 using Near-Infrared Cavity Ring-Down Spectroscopy," *Anal. Chem.* **2013,** *85,* 11250. Copyright © 2013 American Chemical Society.]

Optical Sensors

20-17. Light passes from benzene (medium 1) to water (medium 2) in Figure 20-20 at **(a)** $\theta_1 = 30°$ or **(b)** $\theta_1 = 0°$. Find the angle θ_2 in each case.

20-18. Explain how an optical fiber works. Why does the fiber still work when it is bent?

20-19. The photograph of upconversion in Color Plate 21 shows total internal reflection of the blue ray inside the cuvet. The angle

of incidence of the blue ray on the wall of the cuvet is ~55°. We estimate that the refractive index of the organic solvent is 1.50 and the refractive index of the fused-silica cuvet is 1.46. Calculate the critical angle for total internal reflection at the solvent/silica interface and at the silica/air interface. From your calculation, which interface is responsible for total internal reflection in the photo?

20-20. Explain how the attenuated total reflection sensor in Figure 20-24 works.

20-21. Consider a planar waveguide used for attenuated total reflection measurement of a film coated on one surface of the waveguide. For a given angle of incidence, the sensitivity of attenuated total reflectance increases as the thickness of the waveguide decreases. Explain why. (The waveguide can be less than 1 μm thick.)

20-22. (a) Find the critical value of θ_i in Figure 20-21 beyond which there is total internal reflection in a ZrF_4-based infrared optical fiber whose core refractive index is 1.52 and whose cladding refractive index is 1.50.

(b) The loss of radiant power (due to absorption and scatter) in an optical fiber of length ℓ is expressed in decibels per meter (dB/m), defined as

$$\frac{\text{Power out}}{\text{Power in}} = 10^{-\ell(\text{dB/m})/10}$$

Calculate the quotient power out/power in for a 20.0-m-long fiber with a loss of 0.010 0 dB/m.

20-23. Find the minimum angle θ_i for total reflection in the optical fiber in Figure 20-21 if the index of refraction of the cladding is 1.400 and the index of refraction of the core is **(a)** 1.600 or **(b)** 1.800.

20-24. The prism shown here is used to totally reflect light at a 90° angle. No surface of this prism is silvered. Use Snell's law to explain why total reflection occurs. What is the minimum refractive index of the prism for total reflection?

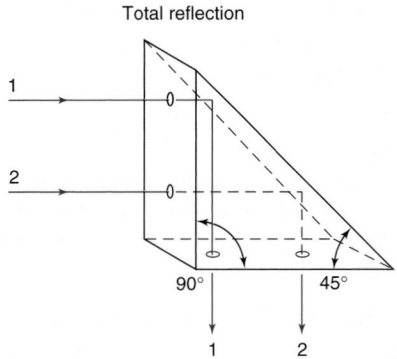

Total reflection

20-25. Here is an extremely sensitive method for measuring nitrite (NO_2^-) down to 1 nM in natural waters. The water sample is treated with sulfanilamide and N-(1-naphthylethylenediamine) in acid solution to produce a colored product with a molar absorptivity of $4.5 \times 10^4\ M^{-1}\ cm^{-1}$ at 540 nm. The colored solution is pumped into a 4.5-meter-long, coiled Teflon tube whose fluorocarbon wall has a refractive index of 1.29. The aqueous solution inside the tube has a refractive index near 1.33. The colored solution is pumped through the coiled tube. An optical fiber delivers white light into one end of the tube, and an optical fiber at the other end leads to a polychromator and detector.

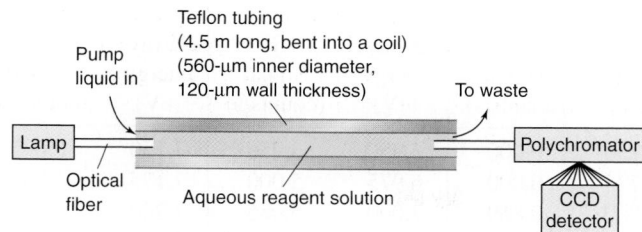

Long-pathlength spectrometer. [Information from W. Yao, R. H. Byrne, and R. D. Waterbury, "Determination of Nanomolar Concentrations of Nitrite and Nitrate Using Long Path Length Absorbance Spectroscopy," *Environ. Sci. Technol.* **1998**, *32*, 2646.]

(a) Explain the purpose of the coiled Teflon tube and explain how it functions.

(b) What is the critical angle of incidence for total internal reflection at the Teflon/water interface?

(c) What is the predicted absorbance of a 1.0 nM solution of colored reagent?

20-26. (a) A particular silica glass waveguide is reported to have a loss coefficient of 0.050 dB/cm (power out/power in, defined in Problem 20-22) for 514-nm-wavelength light. The thickness of the waveguide is 0.60 μm and the length is 3.0 cm. The angle of incidence (θ_i in the figure) is 70°. What fraction of the incident radiant intensity is transmitted through the waveguide?

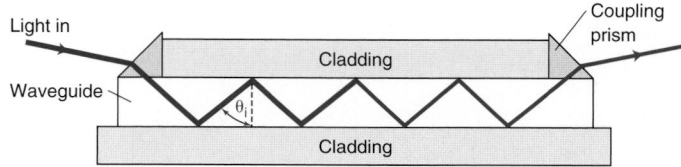

(b) If the index of refraction of the waveguide material is 1.5, what is the wavelength of light inside the waveguide? What is the frequency?

20-27. ▦ The variation of refractive index, *n*, with wavelength for fused silica is given by

$$n^2 - 1 = \frac{(0.696\ 166\ 3)\lambda^2}{\lambda^2 - (0.068\ 404\ 3)^2} + \frac{(0.407\ 942\ 6)\lambda^2}{\lambda^2 - (0.116\ 241\ 4)^2} + \frac{(0.897\ 479\ 4)\lambda^2}{\lambda^2 - (9.896\ 161)^2}$$

where λ is expressed in μm.

(a) Make a graph of *n* versus λ with points at the following wavelengths: 0.2, 0.4, 0.6, 0.8, 1, 2, 3, 4, 5, and 6 μm.

(b) The ability of a prism to spread apart (disperse) neighboring wavelengths increases as the slope $dn/d\lambda$ increases. Is the dispersion of fused silica greater for blue light or red light?

Fourier Transform Spectroscopy

20-28. The interferometer mirror of a Fourier transform infrared spectrophotometer travels ±1 cm.

(a) How many centimeters is the maximum retardation, Δ?

(b) State what is meant by resolution.

(c) What is the approximate resolution (cm^{-1}) of the instrument?

(d) At what retardation interval, δ, must the interferogram be sampled (converted into digital form) to cover a spectral range of 0 to 2 000 cm^{-1}?

20-29. Explain why the transmission spectrum in Figure 20-33 is calculated from the quotient (sample transform)/(background transform) instead of the difference (sample transform) − (background transform).

Dealing with Noise

20-30. Describe three general types of noise that have a different dependence on frequency. Give an example of the source of each kind of noise.

20-31. Explain how beam chopping reduces line noise and noise from source drift.

20-32. Explain how the difference voltage in Figure 20-37 reduces noise from source flicker.

20-33. A spectrum has a signal-to-noise ratio of 8/1. How many spectra must be averaged to increase the signal-to-noise ratio to 20/1?

20-34. A measurement with a signal-to-noise ratio of 100/1 can be thought of as a signal, *S*, with 1% uncertainty, *e*. That is, the measurement is $S \pm e = 100 \pm 1$.

(a) Use the rules for propagation of uncertainty to show that, if you add two such signals, the result is total signal $= 200 \pm \sqrt{2}$, giving a signal-to-noise ratio of $200/\sqrt{2} = 141/1$.

(b) Show that if you add four such measurements, the signal-to-noise ratio increases to 200/1.

(c) Show that averaging *n* measurements increases the signal-to-noise ratio by a factor of $\sqrt{n}$ compared with the value for one measurement.

20-35. In Figure 20-35, the relative detection limit is 37 after 1 s of signal averaging, 12.5 after 10 s, and 5.6 after 40 s of signal averaging. Based on the value of 37 at 1 s, what are the expected detection limits at 10 s and 40 s in an ideal experiment?

20-36. Results of an electrochemical experiment are shown in the figure. In each case, a voltage is applied between two electrodes at time $= 20$ s and the absorbance of a solution decreases until the voltage is stepped back to its initial value at time $= 60$ s. The upper traces show the average results for 100, 300, and 1 000 repetitions of the experiment. The measured signal-to-rms noise ratio in the upper trace is 60.0. Predict the signal-to-noise ratios expected for 300 cycles, 100 cycles, and 1 cycle and compare your answers with the observed values in the figure.

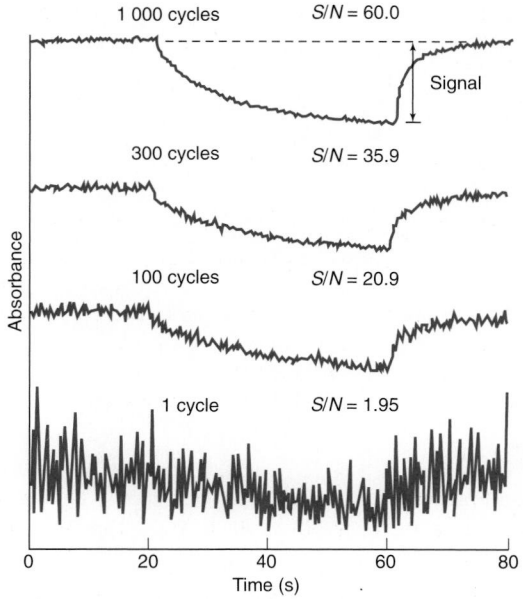

Signal averaging in an experiment in which absorbance is measured after the electric potential is changed at 20 s. [Data from A. F. Slaterbeck, T. H. Ridgeway, C. J. Seliskar, and W. R. Heineman, "Spectroelectrochemical Sensing Based on Multimode Selectivity Simultaneously Achievable in a Single Device," *Anal. Chem.* **1999**, *71*, 1196.]

20-37. *Savitzky-Golay polynomial smoothing.* Experimental data from a noisy X-ray fluorescence spectrum are given in the table.

(a) Compute the smoothed value of the ordinate (y) for an abscissa of $x = 7.025$ eV by using Equation 20-15 for 5-point cubic polynomial smoothing and Equation 20-17 for 9-point smoothing.

(b) ▦ Compute 5-point cubic polynomial smoothed values for the entire data. Overlay the original data and the smoothed data on a graph.

x X-ray energy (eV)	y Signal (counts/s)	x X-ray energy (eV)	y Signal (counts/s)	x X-ray energy (eV)	y Signal (counts/s)
6.750	0.000	6.950	6.956	7.150	8.695
6.775	0.000	6.975	5.000	7.175	6.086
6.800	0.869	7.000	6.086	7.200	1.739
6.825	1.086	7.025	14.347	7.225	2.391
6.850	0.000	7.050	14.130	7.250	2.391
6.875	1.086	7.075	13.913	7.275	1.086
6.900	0.869	7.100	11.739	7.300	0.652
6.925	1.956	7.125	9.782		

21 | Atomic Spectroscopy

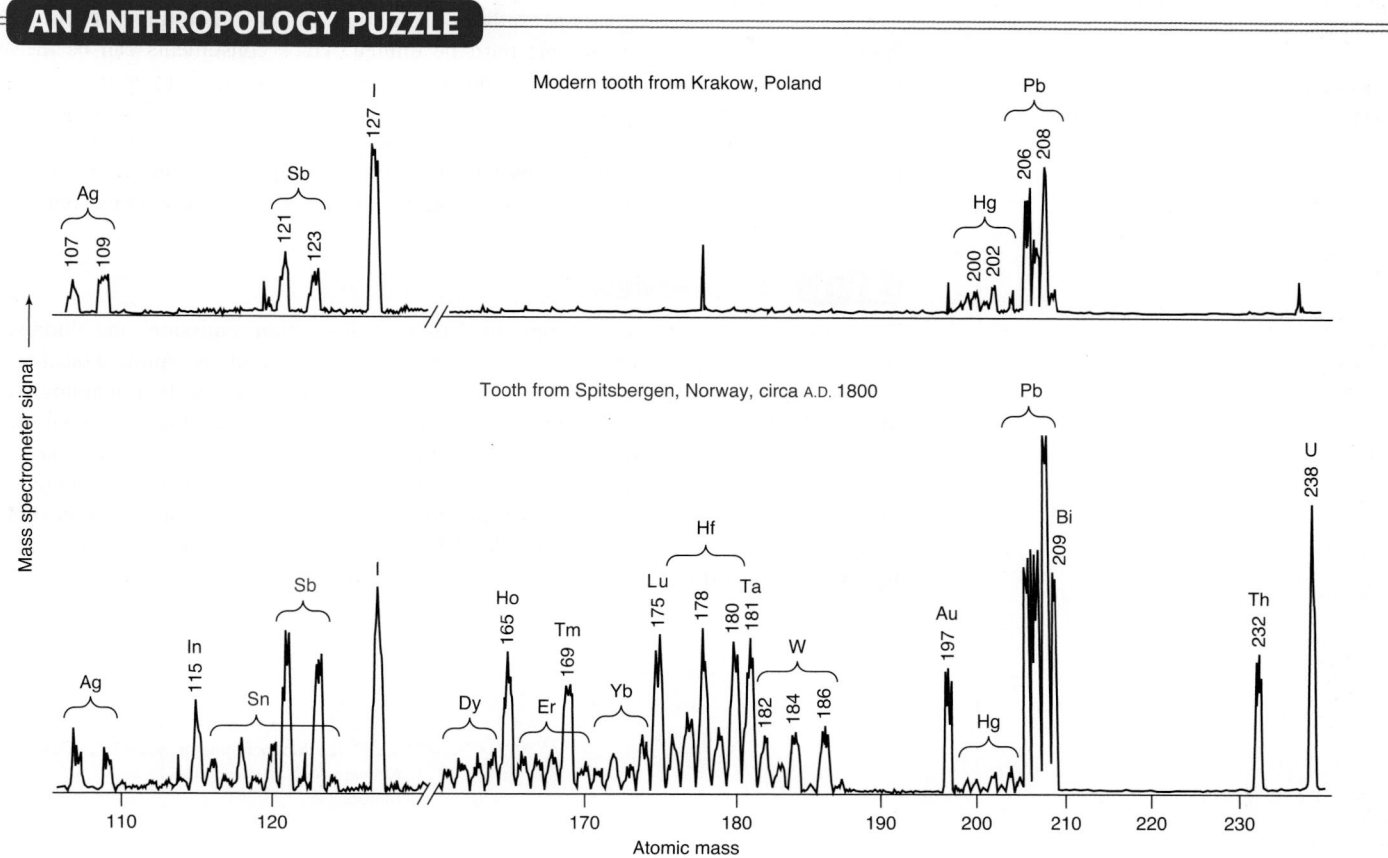

Trace element profile of a tooth from a modern man and from a person who lived in Scandinavia 200 years ago. [Data from A. Cox, F. Keenan, M. Cooke, and J. Appleton, "Trace Element Profiling of Dental Tissues Using Laser Ablation Inductively Coupled Plasma–Mass Spectrometry," *Fresenius J. Anal. Chem.* **1996,** *354,* 254.]

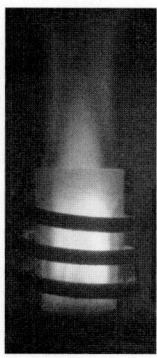

Inductively coupled argon plasma atomizes substances at 6 000 K. [© The Natural History Museum/Alamy.]

In atomic spectroscopy, a substance is decomposed into atoms in a flame, furnace, or *plasma*. A **plasma** is a gas that is hot enough to contain ions and free electrons. Each element is measured by absorption or emission of ultraviolet or visible radiation by the gaseous atoms. To measure trace elements in a tooth, tiny portions of the tooth are vaporized (*ablated*) by a laser pulse and swept into a plasma. The plasma ionizes some of the atoms, which pass into a mass spectrometer that separates ions by their mass and measures their quantity.

Elements are incorporated into teeth from the diet or by inhalation. The figure shows trace element profiles measured by laser ablation–plasma ionization–mass spectrometry of the dentine of teeth from a modern person and one who lived in Scandinavia about A.D. 1800. The contrast is striking. The old tooth contains significant amounts of tin and bismuth, which are nearly absent in the modern tooth. The old tooth contains more lead and antimony than the modern tooth. Tin and lead are constituents of pewter, which was used for cooking vessels and utensils. Bismuth and antimony also might come from pewter.

Even more striking in the old tooth is the abundance of rare earths (dysprosium, holmium, erbium, thulium, ytterbium, and lutetium) and the elements tantalum, tungsten, gold, thorium, and uranium. Rare earth minerals are found in Scandinavia (in fact, many rare earth elements were discovered there), but what were they used for? Did people prepare food with them? Did they somehow get into the food chain?

In *atomic spectroscopy*, samples vaporized at 2 000–8 000 K decompose into atoms and ions whose concentrations are measured by emission or absorption of characteristic wavelengths of radiation. Because of its high sensitivity, its ability to distinguish one element from another in a complex sample, its ability to perform simultaneous multielement analyses, and the ease with which many samples can be automatically analyzed, atomic spectroscopy is a principal tool of analytical chemistry.[1,2] Ions in the vapor can also be measured by a mass spectrometer. Equipment for atomic spectroscopy is expensive, but widely available.

Analyte is measured at parts per million ($\mu g/g$) to parts per trillion (pg/g) levels. To analyze major constituents, the sample must be diluted. Trace constituents can be measured directly without *preconcentration*. The precision of atomic spectroscopy is typically a few percent (depending on the type of sample and matrix), which is not as good as that of some wet chemical methods. When carried out in the most careful manner, inductively coupled plasma–atomic emission spectroscopy has an accuracy and precision on the order of 0.1% and can be used to certify DNA reference materials on the basis of phosphorus content.[3]

Preconcentration: concentrating a dilute
analyte to a level high enough to be analyzed

21-1 An Overview

Three forms of atomic spectroscopy are based on absorption, emission, and fluorescence (Figure 21-1).[4] In **atomic absorption** in Figure 21-2, liquid sample is aspirated (sucked) into a flame whose temperature is 2 000–3 000 K. Liquid evaporates and the remaining solid is **atomized** (broken into atoms) in the flame, which replaces the cuvet of conventional spectrophotometry. The pathlength of the flame is typically 10 cm. The *hollow-cathode lamp* at the left in Figure 21-2 has an iron cathode. When the cathode is bombarded with energetic Ne^+ or Ar^+ ions, excited Fe atoms vaporize and emit light with the same frequencies absorbed by analyte Fe in the flame. At the right side of Figure 21-2, a detector measures the amount of light that passes through the flame.

Types of atomic spectroscopy:
- emission from a thermally populated excited state
- absorption of sharp lines from hollow-cathode lamp
- fluorescence following absorption of laser radiation

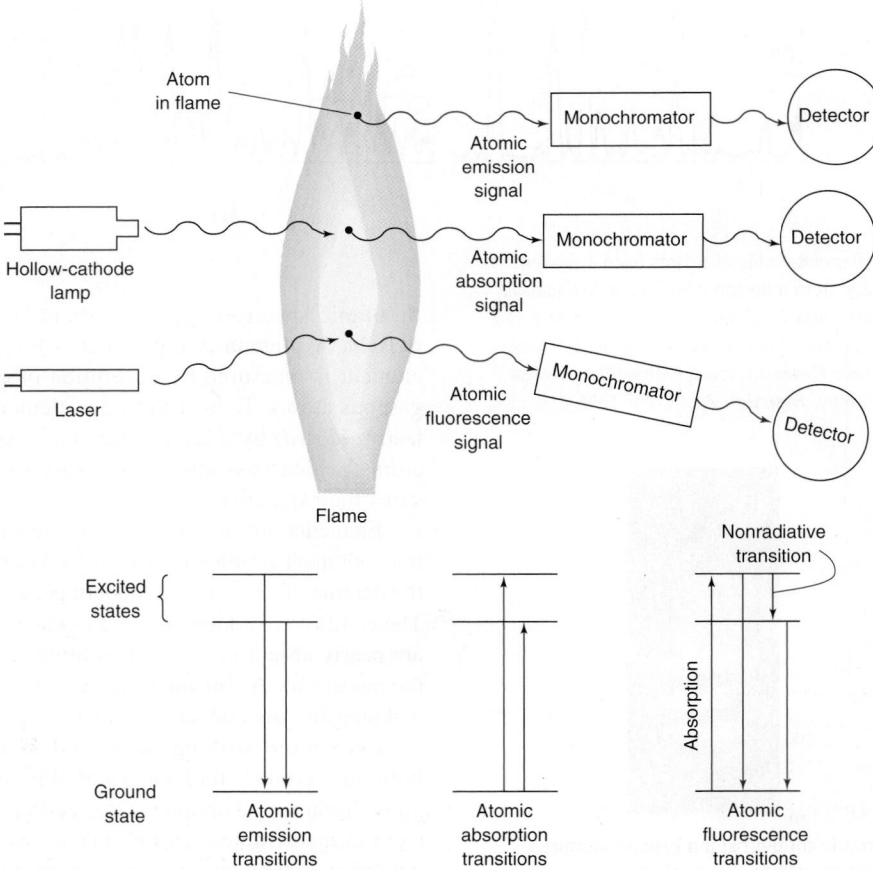

FIGURE 21-1 Emission, absorption, and fluorescence by atoms in a flame. In atomic absorption, atoms absorb part of the light from the source and the remainder of the light reaches the detector. Atomic emission comes from atoms that are in an excited state because of the high thermal energy of the flame. To observe atomic fluorescence, atoms are excited by an external lamp or laser. An excited atom can fall to a lower state and emit radiation.

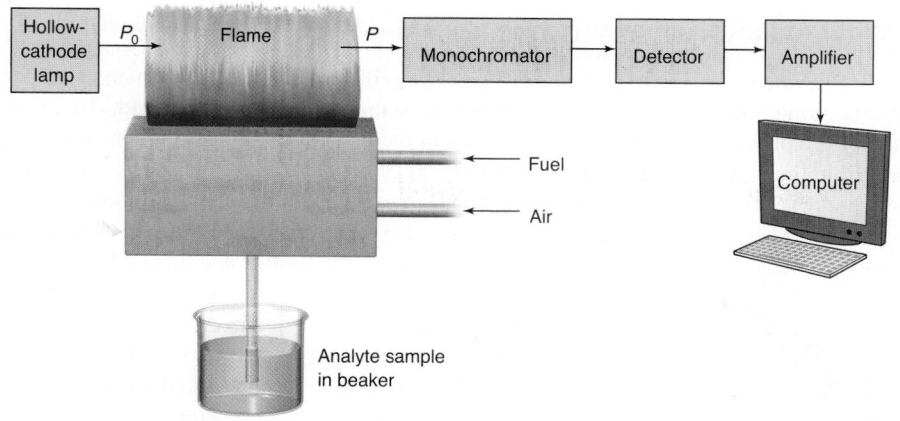

FIGURE 21-2 Atomic absorption experiment. As in Figure 18-5, transmittance = $T = P/P_0$ and absorbance = $A = -\log T$. In practice, P_0 is the irradiance reaching the detector when no sample is going into the flame, and P is measured while sample is present.

An important difference between atomic and molecular spectroscopy is the width of absorption or emission bands. Absorption and emission spectra of liquids and solids typically have bandwidths of ~10 to 100 nm, as in Figures 18-9 and 18-19. In contrast, spectra of gaseous atoms consist of sharp lines with widths of ~0.001 nm (Figure 21-3). Lines are so sharp that there is usually little overlap between lines from different elements in the same sample. Therefore, some instruments can measure more than 70 elements simultaneously. We will see later that sharp analyte absorption lines require that the light source also have sharp lines.

Figure 21-1 also illustrates an **atomic fluorescence** experiment. A laser irradiates atoms in the flame to promote the atoms to an excited electronic state from which they can fluoresce to return to the ground state. Figure 21-4 shows atomic fluorescence from 2 ppb of lead in tap water. Atomic fluorescence is potentially a thousand times more sensitive than atomic absorption, but equipment for atomic fluorescence is not common. An important example of atomic fluorescence is in the analysis of mercury (Box 21-1).

Fluorescence is more sensitive than absorption because we can observe a weak fluorescence signal above a dark background. In absorption, we are looking for small differences between large amounts of light reaching the detector.

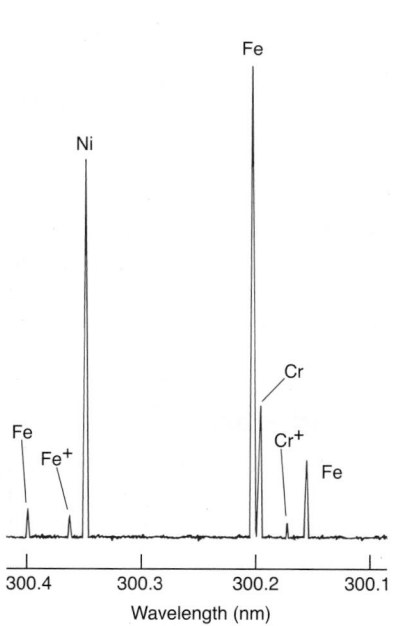

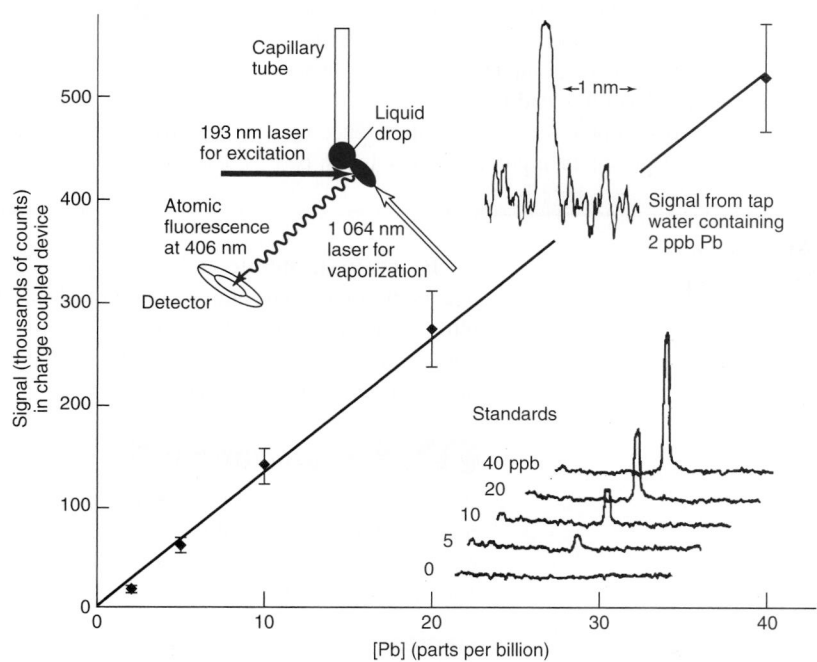

FIGURE 21-3 A portion of the emission spectrum of a steel hollow-cathode lamp, showing lines from gaseous Fe, Ni, and Cr atoms and weak lines from Cr^+ and Fe^+ ions. The monochromator resolution is 0.001 nm, which is comparable to the true linewidths. [Data from A. P. Thorne, "Fourier Transform Spectrometry in the Ultraviolet," *Anal. Chem.* **1991**, *63*, 57A.]

FIGURE 21-4 Atomic fluorescence from Pb at 405.8 nm. Water containing parts per billion of colloidal $PbCO_3$ was ejected from a capillary tube and exposed to a 6-ns pulse of 1 064-nm laser radiation focused on the drop. This pulse created a plume of vapor moving toward the laser. After 2.5 μs, the plume was exposed to a 193-nm laser pulse, creating excited Pb atoms whose fluorescence was measured for 0.1 μs with an optical system whose resolution was 0.2 nm. The figure shows a calibration curve constructed from colloidal $PbCO_3$ standards and the signal from tap water containing 2 ppb Pb. [Data from S. K. Ho and N. H. Cheung, "Sub-Part-per-Billion Analysis of Aqueous Lead Colloids by ArF Laser Induced Atomic Fluorescence," *Anal. Chem.* **2005**, *77*, 193.]

BOX 21-1   Mercury Analysis by Cold Vapor Atomic Fluorescence

Mercury is a volatile toxic pollutant. The map shows Hg(0) concentrations in the air near Earth's surface. Mercury is also found as Hg(II)(*aq*) in clouds and on particles in the atmosphere. Approximately two-thirds of atmospheric mercury comes from human activities, including coal burning, waste incineration, and Cl_2 production by the chlor-alkali process (Problem 17-8).

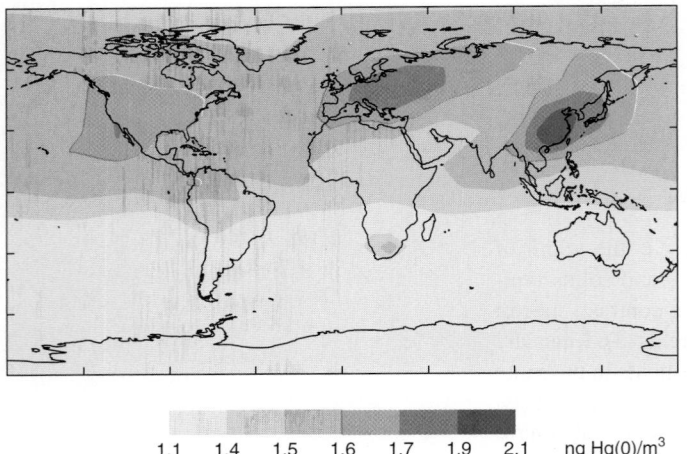

1.1   1.4   1.5   1.6   1.7   1.9   2.1   ng Hg(0)/m³

Global annual average surface concentration of Hg(0). [Data from C. Seigneur, K. Vijayaraghavan, P. Karamchandani, and C. Scott, "Global Source Attribution of Mercury Deposition in the United States," *Environ. Sci. Technol.* **2004**, *38*, 555. Reprinted with permission © 2004 American Chemical Society.]

A sensitive method to measure mercury in matrices such as water, soil, and fish generates Hg(*g*), which is measured by atomic absorption or fluorescence.[5] For most environmental samples, automated digestion/analysis equipment is available.[6] For water analysis by one standard method, all mercury is first oxidized to Hg(II) with BrCl in the purge flask at the lower left in the diagram. Halogens are reduced to halides with hydroxylamine (NH_2OH), and Hg(II) is then reduced to Hg(0) with $SnCl_2$. This two-step process prevents interference by I^-, which tightly binds Hg(I). Hg(*g*) is then purged from solution by

bubbling purified Ar or N_2. Hg(0) is collected at room temperature in the sample trap, which contains gold-coated silica sand.[7] Hg binds to Au, while other gases in the purge stream pass through. The sample trap is then heated to 450°C to release Hg(*g*), which is caught in the analytical trap at room temperature. Two traps are used so that all other gaseous impurities are removed prior to analysis. Hg(*g*) is then released from the analytical trap by heating and flows into the fluorescence cell. Fluorescence intensity strongly depends on gaseous impurities that can quench the emission from Hg.

The lower limit of quantitation is ~0.5 ng/L (parts per trillion). To measure such small quantities requires extraordinary care at every stage of analysis to prevent contamination.[7,8] Mercury amalgam fillings in a worker's teeth can contaminate samples exposed to exhaled breath.

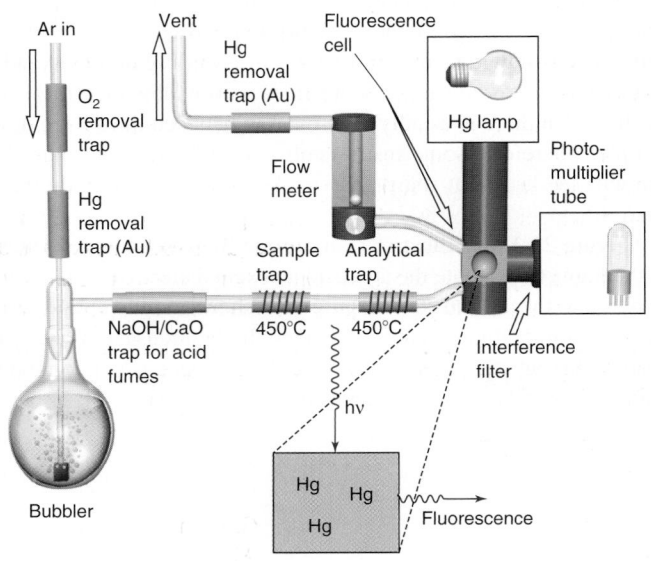

Information from Mercury analysis by U.S. Environmental Protection Agency Method 1631.

By contrast, **atomic emission** (Figure 21-1) is widely used.[9] Collisions in a flame or *plasma* promote some atoms to excited electronic states from which they can emit photons to return to lower energy states. No lamp is required. Emission intensity is proportional to the concentration of the element in the sample. Emission from atoms in a plasma is now the dominant form of atomic spectroscopy.

21-2   Atomization: Flames, Furnaces, and Plasmas

In atomic spectroscopy, analyte is *atomized* in a flame, an electrically heated furnace, or a plasma. Flames were used for decades, but they have been replaced by the inductively coupled plasma and the graphite furnace. We begin our discussion with flames because they are still common in teaching labs.

Flames

Most flame spectrometers use a *premix burner*, such as that in Figure 21-5, in which fuel, oxidant, and sample are mixed before introduction into the flame. Sample solution is drawn into the *pneumatic nebulizer* by the rapid flow of oxidant (usually air) past the tip of the sample capillary. Liquid breaks into a fine mist as it leaves the capillary. The spray is directed against a glass bead, upon which the droplets break into smaller particles. The formation of small droplets is termed **nebulization.** A fine suspension of liquid (or solid) particles in a gas is called an **aerosol.** The nebulizer creates an aerosol from the liquid sample. The mist, oxidant, and fuel flow past baffles that promote further mixing and block large droplets of

Organic solvents with surface tensions lower than that of water are excellent for atomic spectroscopy because they form smaller droplets, thus leading to more efficient atomization.

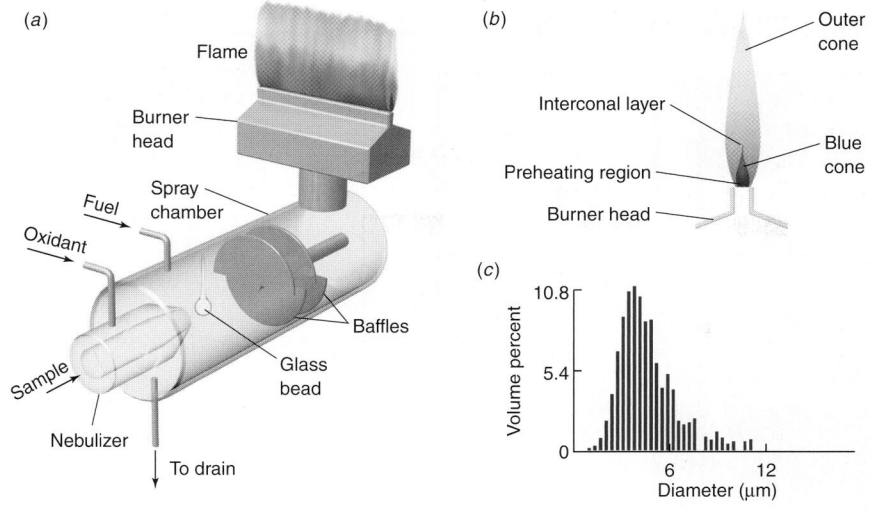

(a)

Flame

Burner head

Spray chamber

Fuel

Oxidant

Sample

Nebulizer

To drain

Baffles

Glass bead

(b)

Outer cone

Interconal layer

Preheating region

Blue cone

Burner head

(c)

10.8

Volume percent

5.4

0

6   12

Diameter (μm)

FIGURE 21-5 (*a*) Premix burner. (*b*) End view of flame. The slot in the burner head is about 0.5 mm wide. (*c*) Distribution of droplet sizes produced by a particular nebulizer. [Data from R. H. Clifford, I. Ishii, A. Montaser, and G. A. Meyer, "Droplet-Size and Velocity Distributions of Aerosols from Commonly Used Nebulizers," *Anal. Chem.* **1990**, *62*, 390.]

liquid. Excess liquid collects at the bottom of the spray chamber and flows out to a drain. Aerosol reaching the flame contains only about 5% of the initial sample.

The common fuel-oxidizer combination is acetylene and air, which produces a flame temperature of 2 400–2 700 K (Table 21-1). If a hotter flame is required to atomize high-boiling elements (called *refractory* elements), acetylene and nitrous oxide can be used. In the flame profile in Figure 21-5b, gas entering the preheating region is heated by conduction and radiation from the primary reaction zone (the small blue cone at the base of the flame). Combustion is completed in the outer cone, where surrounding air is drawn into the flame. Flames emit light that must be subtracted from the total signal to obtain the analyte signal.

TABLE 21-1	Maximum flame temperatures	
Fuel	**Oxidant**	**Temperature (K)**
Acetylene, HC≡CH	Air	2 400–2 700
Acetylene	Nitrous oxide, N_2O	2 900–3 100
Acetylene	Oxygen	3 300–3 400
Hydrogen	Air	2 300–2 400
Hydrogen	Oxygen	2 800–3 000
Cyanogen, N≡C—C≡N	Oxygen	4 800

Droplets entering the flame evaporate; then the remaining solid vaporizes and decomposes into atoms. Many elements form oxides and hydroxides in the outer cone. Molecules do not have the same spectra as atoms, so the atomic signal is lowered. Molecules also emit broad radiation that must be subtracted from the sharp atomic signals. If the flame is relatively rich in fuel (a "rich" flame), excess carbon tends to reduce metal oxides and hydroxides and thereby increases sensitivity. A "lean" flame, with excess oxidant, is hotter. Different elements require either rich or lean flames for best analysis. The height in the flame at which maximum atomic absorption or emission is observed depends on the element being measured and the flow rates of sample, fuel, and oxidizer.[10] Box 21-2 describes how atomic emission from a Bunsen burner flame can be used to measure Na^+ in a beverage.

Furnaces[11]

An electrically heated **graphite furnace** (Figure 21-6) is more sensitive than a flame and requires less sample. From 1 to 100 μL of sample are injected into the furnace tube through the hole at the center. Light from a hollow-cathode lamp travels through windows at each end of the tube. To prevent oxidation of the graphite, Ar gas is passed over the furnace and the maximum recommended temperature is 2 550°C for not more than 7 s.

In flame spectroscopy, the *residence time* of analyte in the optical path is <1 s as it rises through the flame. A graphite furnace confines the atomized sample in the optical path for several seconds, thereby affording higher sensitivity. Whereas 1–2 mL is the minimum volume of solution necessary for flame analysis, as little as 1 μL is adequate for a furnace. Precision is rarely better than 5–10% with manual sample injection, but automated injection improves reproducibility to ~1%.

FIGURE 21-6 Pyrocoated graphite tubes family. Electrically heated graphite furnace tubes for atomic spectroscopy, typically 38-mm long. Graphite is coated with a thin layer of pyrolytic carbon to reduce porosity of the surface. [© PerkinElmer, Inc. All rights reserved. Printed with permission.]

Furnaces offer increased sensitivity and require less sample than a flame.

BOX 21-2 **Measuring Sodium with a Bunsen Burner Photometer**

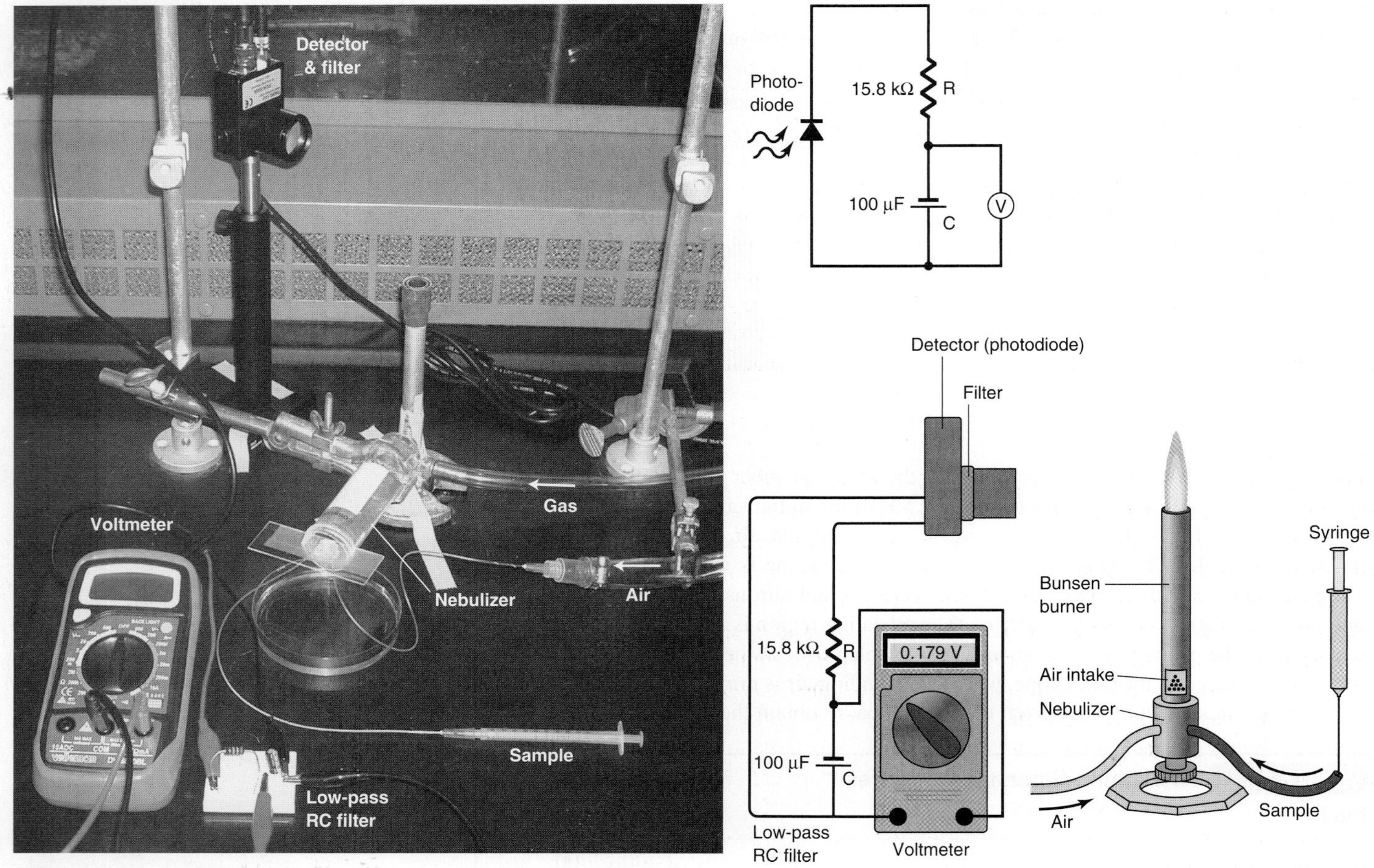

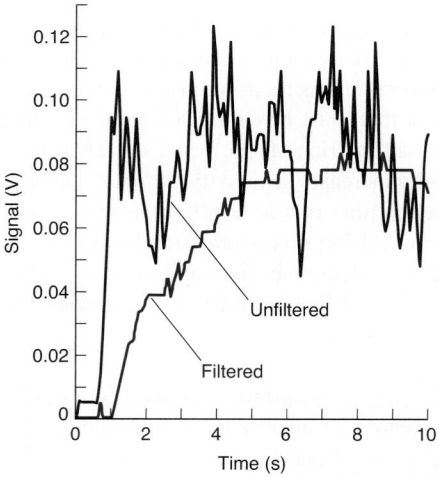

Student flame photometer uses atomic emission from a Bunsen burner to measure Na$^+$ in drinks. [Information from C. N. LaFratta, S. Jain, I. Pelse, O. Simoska, and K. Elvy, "Using a Homemade Flame Photometer to Measure Sodium Concentration in a Sports Drink," *J. Chem. Ed.* **2013**, *90*, 372. Photo courtesy C. N. LaFratta, Bard College, Annandale on Hudson, New York.]

Metal ion solutions emit characteristic colors when they are introduced into a flame on the tip of a Pt wire. Such flame tests were a staple of inorganic *qualitative* analysis in the past. Sodium emits intense yellow light at 589 nm when the atom falls from the first electronic excited state $(1s)^2(2s)^2(2p)^6(3p)^1$ to the ground state $(1s)^2(2s)^2(2p)^6(3s)^1$. The experiment in the photograph uses a Bunsen burner flame to measure sodium in drinks such as Gatorade.

A 20-fold dilution of Gatorade with water is injected into an air stream in the homemade nebulizer to send a mist of droplets into the Bunsen burner flame. Yellow atomic emission is measured by the photodiode detector fitted with an optical filter that only passes radiation in the range 590 ± 5 nm. Flicker of the flame creates a rapidly varying amount of light. To overcome this source of noise, output from the detector is passed through a *low-pass RC filter* made from a resistor (R) and a capacitor (C). Voltage is measured across the capacitor. This circuit rejects signals with a frequency greater than ~0.3 Hz. Slowly varying signal is accepted, but rapidly varying signal from flicker of the flame is rejected. The graph shows a noisy unfiltered signal and a smoother filtered signal. A calibration curve is constructed from standard solutions of NaCl. Emission

from the unknown corresponds to a sodium concentration within 6% of the 460 mg/L stated on the Gatorade label.

Raw signal has noise from flicker of the flame. In a separate measurement, a low-pass filter reduces noise. [Data from C. N. LaFratta et al., ibid.]

When you inject sample, the droplet should contact the floor of the furnace and remain in a small area (Figure 21-7a). If you inject the droplet too high (Figure 21-7b), it splashes and spreads, leading to poor precision. In the worst case, the drop adheres to the tip of the pipet and is finally deposited around the injector hole when the pipet is withdrawn.

Compared with flames, furnaces require more operator skill to find proper conditions for each type of sample. The furnace is heated in three or more steps to properly atomize the sample. To measure Fe in the iron-storage protein ferritin, a 10-μL sample containing ~0.1 ppm Fe is injected into the furnace at ~90°C. The furnace is programmed to *dry* the sample at 125°C for 20 s. Drying is followed by 60 s of *charring* at 1 400°C to destroy organic matter. Charring is also called *pyrolysis*, which means decomposing with heat. Charring creates smoke that would interfere with the Fe determination. After charring, the sample is *atomized* at 2 100°C for 10 s. Absorbance reaches a maximum and then decreases as Fe evaporates from the oven. The analytical signal is the time-integrated absorbance (the peak area) during atomization. After atomization, the furnace is heated to 2 500°C for 3 s to clean out remaining residue.

The furnace is purged with Ar or N_2 during each step except atomization to remove volatile material. Gas flow is halted during atomization to avoid blowing analyte out of the furnace. When developing a method for a new kind of sample, it is important to record the signal as a function of time, because signals are also observed from smoke during charring and from the glow of the red-hot furnace in the last part of atomization. A skilled operator interprets which signal is due to analyte so that the right peak is integrated.

The furnace in Figure 21-8a performs better than a simple graphite tube. Sample is injected onto a platform that is heated by radiation from the furnace wall, so its temperature lags behind that of the wall. Analyte does not vaporize until the wall reaches constant temperature (Figure 21-8b). At constant furnace temperature, the area under the absorbance peak in Figure 21-8b is a reliable measure of the analyte. A heating rate of 2 000 K/s rapidly dissociates molecules and increases the concentration of free atoms in the furnace.

The furnace in Figure 21-8a is heated *transversely* (from side to side) to provide nearly uniform temperature over the whole furnace. In furnaces with *longitudinal* (end-to-end) heating, the center of the furnace is hotter than the ends. Atoms from the central region condense at the ends, where they can vaporize during the next sample run. Interference from previous runs, called a *memory effect*, is reduced in a transversely heated furnace. To further reduce memory effects, ordinary graphite is coated with a dense layer of *pyrolytic carbon* formed by thermal decomposition of an organic vapor. The coating seals the relatively porous graphite to reduce absorption of foreign atoms.

A sample can be *preconcentrated* by injecting and evaporating multiple aliquots in the graphite furnace prior to analysis.[12] To measure traces of As in drinking water, a 30-μL aliquot of water plus matrix modifier was injected and evaporated. The procedure was repeated five more times so that the total sample volume was 180 μL. The detection limit for As was 0.3 μg/L (parts per billion). Without preconcentration, the detection limit would have been 1.8 μg/L. This increased capability is critical because As is considered to be a health hazard at concentrations of just a few parts per billion.

You must determine reasonable time and temperature for each stage of the analysis. Once a program is established, it can be applied to a large number of similar samples.

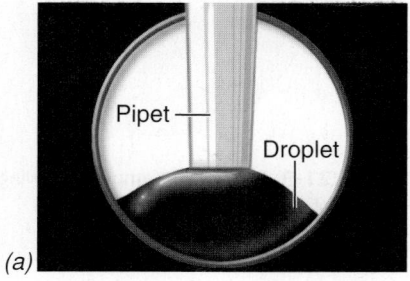

(a)

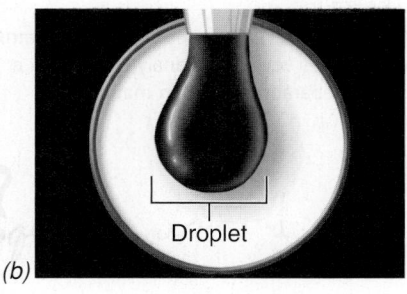

(b)

FIGURE 21-7 (*a*) Correct position for injecting sample into a graphite furnace deposits the droplet in a small volume on the floor of the furnace. (*b*) If injection is too high, sample splatters and precision is poor. [Information from P. K. Booth, "Improvements in Method Development for Graphite Furnace Atomic Absorption Spectrometry," *Am. Lab.*, February 1995, p. 48X.]

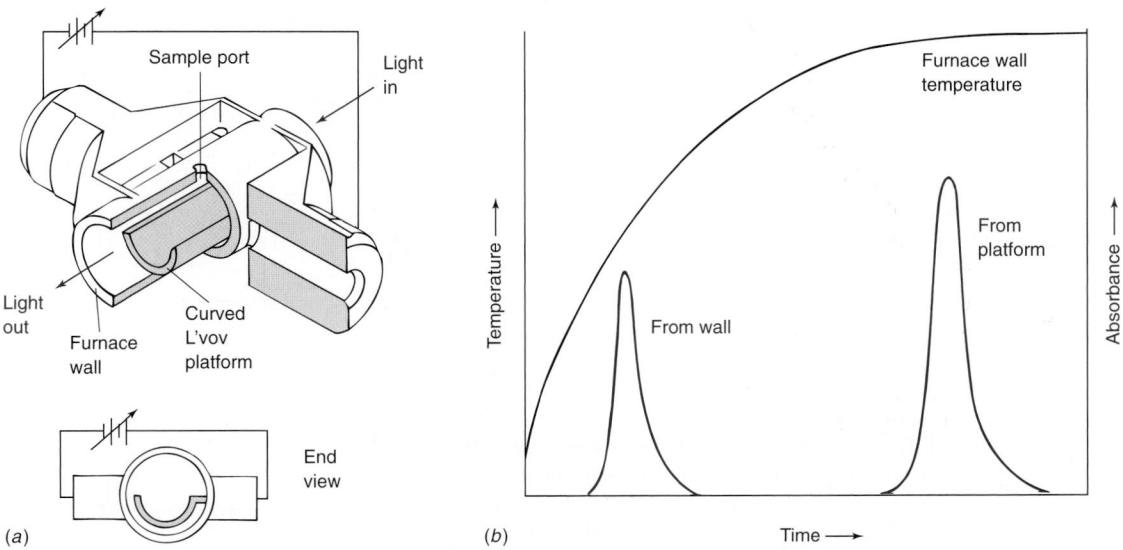

(a)

(b)

FIGURE 21-8 (*a*) Transversely heated graphite furnace maintains nearly constant temperature over its whole length, thereby reducing memory effect from previous runs. The *L'vov platform* is uniformly heated by radiation from the outer wall, not by conduction. The platform is attached to the wall by one small connection that is hidden from view. [Information from Perkin-Elmer Corp., Norwalk, CT.] (*b*) Heating profiles comparing analyte evaporation from wall and from platform. [Data from W. Slavin, "Atomic Absorption Spectroscopy," *Anal. Chem.* **1982,** *54,* 685A.]

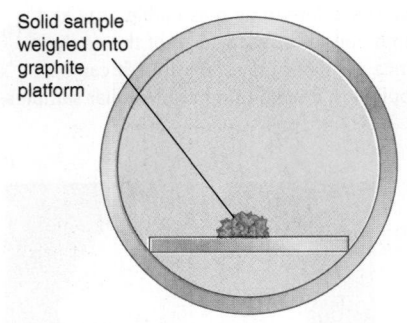

FIGURE 21-9 Direct solid sampling showing end view of furnace.

Matrix modifier: increases volatility of matrix or decreases volatility of analyte to obtain a cleaner separation between matrix and analyte

Liquid samples are ordinarily used in furnaces. In *direct solid sampling*, a solid is analyzed without sample preparation (Figure 21-9). For example, trace impurities in tungsten powder used to make components for industry can be analyzed by weighing 0.1 to 100 mg of powder onto a graphite platform.[13] The platform is transferred into the furnace and heated to 2 600°C to atomize impurities in the tungsten, but not the tungsten itself, which melts at 3 410°C. After several runs, residual tungsten is scraped off the platform and the platform could be reused 400 times. Because so much more sample is analyzed when solid is injected than when liquid is injected, detection limits can be 100 times lower than those obtained for liquid injection. For example, Zn could be detected at a level of 10 pg/g (10 parts per trillion) in 100 mg of tungsten. Calibration curves are obtained by injecting standard solutions of the trace elements and analyzing them as conventional liquids. Results obtained from direct solid sampling are in good agreement with results obtained by laboriously dissolving the solid. Other solids that have been analyzed by direct solid sampling include graphite, silicon carbide, cement, river sediments, hair, and vegetable matter.[14]

Matrix Modifiers for Furnaces

Everything in a sample other than analyte is called the **matrix.** Ideally, matrix decomposes and vaporizes during the charring step. A **matrix modifier** is a substance added to the sample to reduce the loss of analyte during charring by making the matrix more volatile or making analyte less volatile.

The matrix modifier ammonium nitrate can be added to seawater to increase the volatility of the matrix NaCl. Figure 21-10a shows a graphite furnace heating profile used to analyze Mn in seawater. When 0.5 M NaCl solution is subjected to this profile, signals are observed at the analytical wavelength of Mn (Figure 21-10b). Much of the apparent absorbance is probably due to optical scatter by smoke created by heating NaCl. Absorption at the start of the atomization step interferes with the measurement of Mn. Adding NH_4NO_3 to the sample in Figure 21-10c reduces matrix absorption peaks. NH_4NO_3 plus NaCl gives NH_4Cl and $NaNO_3$, which cleanly evaporate instead of making smoke.

The matrix modifier $Pd(NO_3)_2$ is added to seawater to decrease the volatility of the analyte Sb. In the absence of modifier, 90% of Sb is lost during charring at 1 250°C. With the modifier, seawater matrix can be largely evaporated at 1 400°C without loss of Sb.[15]

FIGURE 21-10 Reduction of interference by using a matrix modifier. (*a*) Graphite furnace temperature profile for analysis of Mn in seawater. (*b*) Absorbance profile from 10 μL of 0.5 M reagent-grade NaCl subjected to temperature profile in panel *a*. Absorbance is monitored at the Mn wavelength of 279.5 nm with a monochromator bandwidth of 0.5 nm. (*c*) Reduced absorbance from 10 μL of 0.5 M NaCl plus 10 μL of 50 wt% NH_4NO_3 matrix modifier. [Data from M. N. Quigley and F. Vernon, "Matrix Modification Experiment for Electrothermal Atomic Absorption Spectrophotometry," *J. Chem. Ed.* **1996,** *73,* 980.]

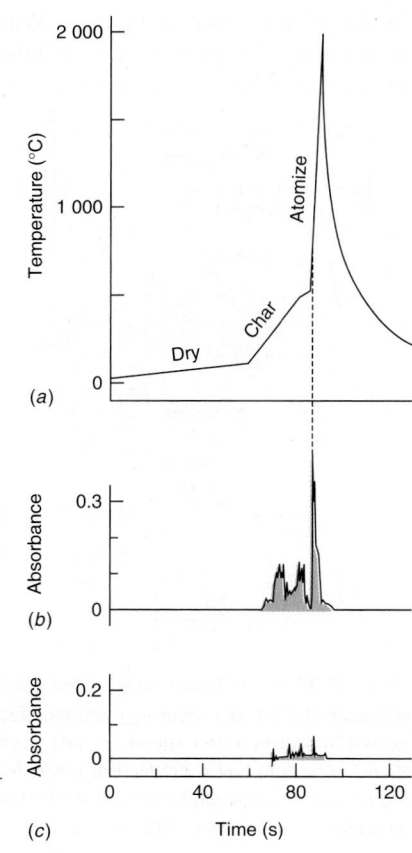

CHAPTER 21 Atomic Spectroscopy

The matrix modifier $Mg(NO_3)_2$ raises the temperature for atomization of Al analyte.[16] At high temperature, $Mg(NO_3)_2$ decomposes to $MgO(g)$ and Al is converted into Al_2O_3. At sufficiently high temperature, Al_2O_3 decomposes to Al and O, and Al evaporates. Evaporation of Al is retarded by $MgO(g)$ by virtue of the reaction

$$3MgO(g) + 2Al(s) \rightleftharpoons 3Mg(g) + Al_2O_3(s) \qquad (21\text{-}1)$$

When MgO has evaporated, Reaction 21-1 no longer occurs and Al_2O_3 decomposes and evaporates. A matrix modifier that raises the boiling temperature of analyte allows a higher charring temperature to be used to remove matrix without losing analyte.

It is important to monitor the absorption signal from a graphite furnace as a function of time, as in Figure 21-10b. Peak shapes help you decide how to adjust time and temperature for each step to obtain a clean signal from analyte. Also, a graphite furnace has a finite lifetime. Degradation of peak shape or a change in the slope of the calibration curve tells you that it is time to change the furnace.

Inductively Coupled Plasma

The **inductively coupled plasma**[17] shown at the beginning of the chapter is twice as hot as a combustion flame (Figure 21-11). The high temperature, stability, and relatively inert Ar environment eliminate much of the interference encountered with flames. Simultaneous multi-element analysis described in Section 21-4 is routine for inductively coupled plasma–atomic emission spectroscopy, which has replaced flame atomic absorption. The plasma instrument costs more to purchase and operate than a flame instrument.

The cross-sectional view of an inductively coupled plasma burner in Figure 21-12 shows two turns of a 27- or 41-MHz induction coil wrapped around the upper opening of the quartz apparatus. High-purity Ar gas is fed through the plasma gas inlet. After a spark from a Tesla coil ionizes Ar, free electrons are accelerated by the radio-frequency field. Electrons collide with atoms and transfer energy to the entire gas, maintaining a temperature of 6 000 to 10 000 K. The quartz torch is protected from overheating by Ar coolant gas. Figure 21-13 shows one way to introduce

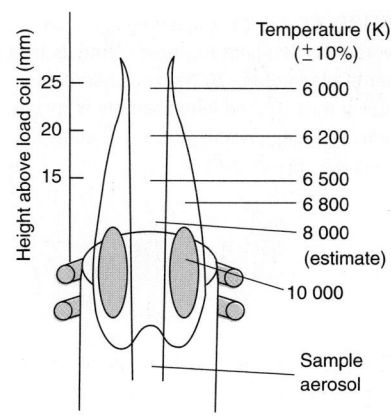

FIGURE 21-11 Temperature profile of inductively coupled plasma. [Data from V. A. Fassel, "Simultaneous or Sequential Determination of the Elements at All Concentration Levels," *Anal. Chem.* **1979,** *51,* 1290A.]

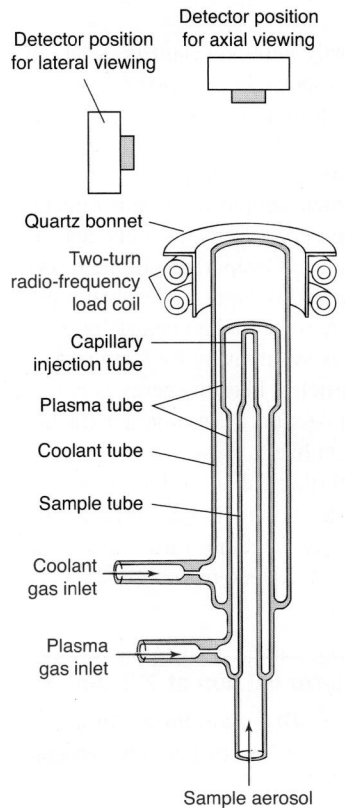

FIGURE 21-12 Inductively coupled plasma burner and locations for lateral and axial viewing. [Data from R. N. Savage and G. M. Hieftje, "Miniature Inductively Coupled Plasma Source for Atomic Emission Spectrometry," *Anal. Chem.* **1979,** *51,* 408.]

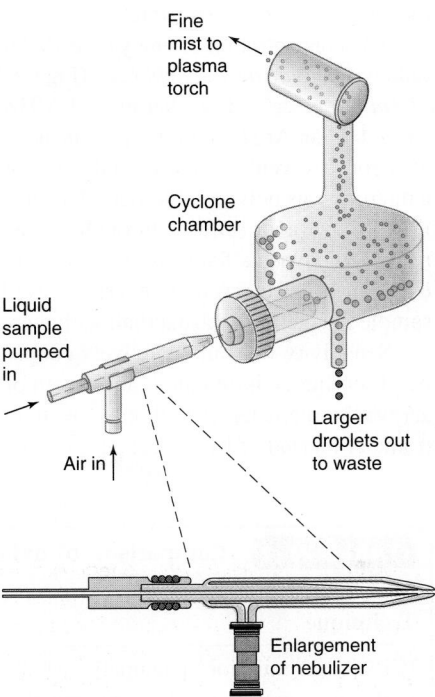

FIGURE 21-13 Cyclonic nebulizer for inductively coupled plasma torch. [Information from P. Liddell, Glass Expansion, West Melbourne, Australia.]

FIGURE 21-14 (*a*) Ultrasonic nebulizer lowers detection limit for most elements by an order of magnitude. (*b*) Aerosol chamber showing mist created when sample is sprayed against vibrating crystal. [Courtesy Teledyne CETAC Technologies, Omaha, NE.]

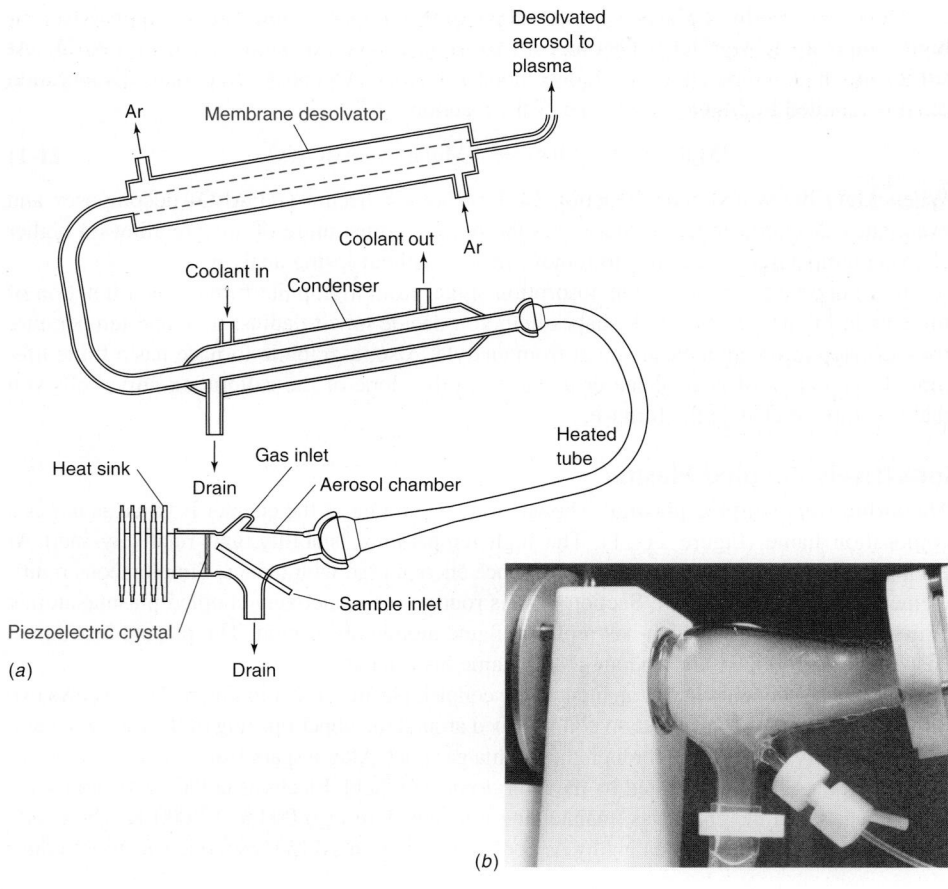

(*a*)

(*b*)

A **piezoelectric crystal** is one whose dimensions change in an applied electric field. A sinusoidal voltage applied between two faces of the crystal causes it to oscillate. Quartz is the most common piezoelectric material.

a fine mist into the torch. Liquid sample is pumped slowly into the nebulizer at the lower left. Liquid emerging from the capillary tip of the nebulizer is sprayed by a coaxial flow of Ar into a cyclone chamber in which coarse particles fall to the bottom. Fine mist exiting the chamber at the top goes to the plasma torch.

The concentration of analyte needed for adequate signal is reduced by an order of magnitude with an *ultrasonic nebulizer* (Figure 21-14), in which sample solution is directed onto a *piezoelectric* crystal oscillating at 1 MHz. The vibrating crystal creates a fine aerosol that is carried by an Ar stream through a heated tube where solvent evaporates. In the next refrigerated zone, solvent condenses and is removed. Then the stream enters a desolvator containing a microporous polytetrafluoroethylene membrane in a chamber maintained at 160°C. Remaining solvent vapor diffuses through the membrane and is swept away by flowing Ar. Analyte reaches the plasma flame as an aerosol of dry, solid particles. Plasma energy is not needed to evaporate solvent, so more energy is available for atomization. Also, a larger fraction of the sample reaches the plasma than with a conventional nebulizer.

Sensitivity with an inductively coupled plasma is further enhanced by a factor of 3 to 10 by observing emission along the length of the plasma (axial view in Figure 21-12) instead of across the diameter. Additional sensitivity is obtained by detecting ions with a mass spectrometer instead of by optical emission (Table 21-2), as described in Section 21-7.

TABLE 21-2	Comparison of detection limits for Ni^+ ion at 231 nm
Technique[a]	**Detection limits for different instruments (ng/mL)**
ICP/atomic emission (pneumatic nebulizer)	3–50
ICP/atomic emission (ultrasonic nebulizer)	0.3–4
Graphite furnace/atomic absorption	0.02–0.06
ICP/mass spectrometry	0.001–0.2

a. ICP = *inductively coupled plasma.*

SOURCE: *J. M. Mermet and E. Poussel, "ICP Emission Spectrometers: Analytical Figures of Merit," Appl. Spectros.* **1995,** *49, 12A.*

There are efforts to make miniature plasmas that might be practical for analysis in the field. One experimental device creates a ~0.5-mm-diameter plasma in a 1-cm-long channel in a Teflon and quartz chip for atomic emission.[18] Another experimental device creates a plasma from a liquid in a narrow channel on a microfluidic chip.[19]

21-3 How Temperature Affects Atomic Spectroscopy

Temperature determines the degree to which a sample breaks down into atoms and the extent to which a given atom is found in its ground, excited, or ionized states. Each of these effects influences the strength of the signal we observe.

The Boltzmann Distribution

Consider an atom with energy levels E_0 and E^* separated by ΔE (Figure 21-15). An atom (or molecule) may have more than one state at a given energy. Figure 21-15 shows three states at E^* and two at E_0. The number of states at each energy is called the *degeneracy*. We will call the degeneracies g_0 and g^*.

The **Boltzmann distribution** describes the relative populations of different states at thermal equilibrium. If equilibrium exists (which is not true in the blue cone of a flame but is probably true above the blue cone), the relative population (N^*/N_0) of any two states is

Boltzmann distribution:
$$\frac{N^*}{N_0} = \frac{g^*}{g_0}e^{-\Delta E/kT} \tag{21-2}$$

where T is temperature (K) and k is Boltzmann's constant (1.381×10^{-23} J/K).

Effect of Temperature on Excited-State Population

The lowest excited state of a sodium atom lies 3.371×10^{-19} J/atom above the ground state. The degeneracy of the excited state is 2, whereas that of the ground state is 1. The fraction of Na in the excited state in an acetylene-air flame at 2 600 K is, from Equation 21-2,

$$\frac{N^*}{N_0} = \left(\frac{2}{1}\right)e^{-(3.371\times 10^{-19}\text{ J})/[(1.381\times 10^{-23}\text{ J/K})(2\,600\text{ K})]} = 1.67 \times 10^{-4}$$

That is, less than 0.02% of the atoms are in the excited state.

If the temperature were 2 610 K, the fraction of atoms in the excited state would be

$$\frac{N^*}{N_0} = \left(\frac{2}{1}\right)e^{-(3.371\times 10^{-19}\text{ J})/[(1.381\times 10^{-23}\text{ J/K})(2\,610\text{ K})]} = 1.74 \times 10^{-4}$$

The fraction of atoms in the excited state is still less than 0.02%, but that fraction has increased by $100(1.74 - 1.67)/1.67 = 4\%$.

The Effect of Temperature on Absorption and Emission

We see that more than 99.98% of the sodium atoms are in their ground state at 2 600 K. *Varying the temperature by 10 K hardly affects the ground-state population and would not noticeably affect the signal in atomic absorption.*

How would emission intensity be affected by a 10 K rise in temperature? In Figure 21-15, absorption arises from ground-state atoms, but emission arises from excited-state atoms. Emission intensity is proportional to the population of the excited state. *Because the excited-state population changes by 4% when the temperature rises 10 K, emission intensity rises by 4%.* It is critical in atomic *emission* spectroscopy that the flame be very stable, or emission intensity will vary significantly. In atomic *absorption* spectroscopy, temperature variation is important but not as critical.

Almost all atomic emission is carried out with an inductively coupled plasma, whose temperature is more stable than that of a flame. Plasma is normally used for emission, not absorption, because it is so hot that there is a substantial population of excited-state atoms and ions. Table 21-3 compares excited-state populations for a flame at 2 500 K and a plasma at 6 000 K. Although the fraction of excited atoms is small, each atom emits many photons per second because it is rapidly promoted back to the excited state by collisions.

Energy levels of halogen atoms (F, Cl, Br, I) are so high that they emit ultraviolet radiation below 200 nm. This spectral region is called *vacuum ultraviolet* because O_2 absorbs radiation below 200 nm, so spectrometers for the far-ultraviolet were customarily evacuated. Some plasma emission spectrometers are purged with N_2 to exclude air so that the region 130 to 200 nm is accessible and Cl, Br, I, P, and S can be analyzed.[20] In another application, nitrogen in

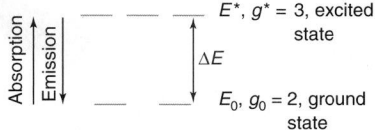

FIGURE 21-15 Two energy levels with different degeneracies. Ground-state atoms that absorb light are promoted to the excited state. Excited-state atoms can emit light to return to the ground state.

The Boltzmann distribution applies to a system at thermal equilibrium.

A 10-K temperature rise changes the excited-state population by 4% in this example.

Atomic absorption is not as sensitive to temperature as atomic emission, which is exponentially sensitive to temperature.

TABLE 21-3	Effect of energy difference and temperature on population of excited states			
Wavelength difference of states (nm)	Energy difference of states (J/atom)	Excited-state fraction $(N^*/N_0)^a$		
		2 500 K	6 000 K	
250	7.95×10^{-19}	1.0×10^{-10}	6.8×10^{-5}	
500	3.97×10^{-19}	1.0×10^{-5}	8.3×10^{-3}	
750	2.65×10^{-19}	4.6×10^{-4}	4.1×10^{-2}	

a. Based on the equation $N^*/N_0 = (g^*/g_0)e^{-\Delta E/kT}$ in which $g^* = g_0 = 1$.

fertilizers is measured along with other major elements in the fertilizer. The torch is designed to be purged with Ar to exclude N_2 from air. Unknowns are purged with He to remove dissolved air. Emission from N is observed near 174 nm.

21-4 Instrumentation

Fundamental requirements for an atomic absorption experiment are shown in Figure 21-2. Principal differences between atomic and ordinary molecular spectroscopy lie in the light source (or lack of a light source in atomic emission), the sample container (the flame, furnace, or plasma), and the need to subtract background emission.

Atomic Linewidths[21]

The linewidth of the source must be narrower than the linewidth of the atomic vapor for Beer's law to be obeyed. "Linewidth" and "bandwidth" are used interchangeably, but "lines" are narrower than "bands."

Beer's law requires that the linewidth of the radiation source should be substantially narrower than the linewidth of the absorbing sample. Otherwise, the measured absorbance will not be proportional to the sample concentration. Atomic absorption lines are very sharp, with an intrinsic width of only ~10^{-4} nm.

Linewidth is governed by the **Heisenberg uncertainty principle,** which says that the shorter the lifetime of the excited state, the more uncertain is its energy:

ΔE in Equation 21-2 is the difference in energy between ground and excited states. δE in Equation 21-3 is the uncertainty in ΔE. δE is a small fraction of ΔE.

Heisenberg uncertainty principle: $$\delta E \delta t \gtrsim \frac{h}{4\pi} \qquad (21\text{-}3)$$

where δE is the uncertainty in the energy difference between ground and excited states, δt is the lifetime of the excited state before it decays to the ground state, and h is Planck's constant. Equation 21-3 says that the uncertainty in the energy difference between two states multiplied by the lifetime of the excited state is at least $h/4\pi$. If δt decreases, then δE increases. The lifetime of an excited state of an isolated gaseous atom is ~10^{-9} s. Therefore, the uncertainty in its energy is

$$\delta E \gtrsim \frac{h}{4\pi\delta t} = \frac{6.6 \times 10^{-34}\,\text{J}\cdot\text{s}}{4\pi(10^{-9}\,\text{s})} \approx 10^{-25}\,\text{J}$$

Suppose that the energy difference (ΔE) between the ground and the excited state of an atom corresponds to visible light with a wavelength of $\lambda = 500$ nm. This energy difference is $\Delta E = hc/\lambda = 4.0 \times 10^{-19}$ J (Equation 18-3). In this equation, c is the speed of light. The relative uncertainty in the energy difference is $\delta E/\Delta E \lesssim (10^{-25}\,\text{J})/(4.0 \times 10^{-19}\,\text{J}) \approx 2 \times 10^{-7}$. The relative uncertainty in wavelength ($\delta\lambda/\lambda$) is the same as the relative uncertainty in energy:

$\delta\lambda$ is the width of an absorption or emission line measured at half the height of the peak.

$$\frac{\delta\lambda}{\lambda} = \frac{\delta E}{\Delta E} \gtrsim 2 \times 10^{-7} \Rightarrow \delta\lambda \gtrsim (2 \times 10^{-7})(500\,\text{nm}) = 10^{-4}\,\text{nm} \qquad (21\text{-}4)$$

The inherent linewidth of an atomic absorption or emission signal is ~10^{-4} nm because of the short lifetime of the excited state.

Two mechanisms broaden the lines to 10^{-3} to 10^{-2} nm in atomic spectroscopy. One is the **Doppler effect.** An atom moving toward the radiation source experiences more oscillations of the electromagnetic wave in a given time period than one moving away from the source (Figure 21-16). That is, an atom moving toward the source "sees" higher frequency light than that encountered by one moving away. In the laboratory frame of reference, the atom moving toward the source absorbs lower frequency light than that absorbed by the one moving away. The linewidth, $\delta\lambda$, due to the Doppler effect, is

Doppler linewidth: $$\delta\lambda \approx \lambda(7 \times 10^{-7})\sqrt{\frac{T}{M}} \qquad (21\text{-}5)$$

where T is temperature (K) and M is the mass of the atom (daltons). For an emission line near $\lambda = 300$ nm from Fe ($M = 56$ Da) at 2 500 K in Figure 21-3, the Doppler linewidth is $(300 \text{ nm})(7 \times 10^{-7})\sqrt{2\,500/56} = 0.001\,4$ nm, which is an order of magnitude greater than the natural linewidth.

Linewidth is also affected by **pressure broadening** from collisions between atoms. Collisions shorten the lifetime of the excited state. Uncertainty in the frequency of atomic absorption and emission lines is roughly numerically equal to the collision frequency between atoms and is proportional to pressure. The Doppler effect and pressure broadening are similar in magnitude and yield linewidths of 10^{-3} to 10^{-2} nm in atomic spectroscopy.

Hollow-Cathode Lamp

Monochromators generally cannot isolate lines narrower than 10^{-3} to 10^{-2} nm. To produce narrow lines of the correct frequency, we use a **hollow-cathode lamp** containing a vapor of the same element as that being analyzed.

The hollow-cathode lamp in Figure 21-17 is filled with Ne or Ar at a pressure of ~130–700 Pa (1–5 Torr). The cathode is made of the element whose emission lines we want. When ~500 V is applied between the anode and the cathode, gas is ionized and positive ions are accelerated toward the cathode. After ionization occurs, the lamp is maintained at a constant current of 2–30 mA by a lower voltage. Cations strike the cathode with enough energy to "sputter" metal atoms from the cathode into the gas phase. Gaseous atoms excited by collisions with high-energy electrons emit photons. This atomic radiation has the same frequency absorbed by analyte in the flame or furnace. Atoms in the lamp are cooler than atoms in a flame, so lamp emission is sufficiently narrower than the width of the absorption line of atoms in the flame to be nearly "monochromatic" (Figure 21-18). The purpose of a monochromator in atomic spectroscopy is to select one line from the hollow-cathode lamp and to reject as much emission from the flame or furnace as possible. A different lamp is usually required for each element, although some lamps are made with more than one element in the cathode.

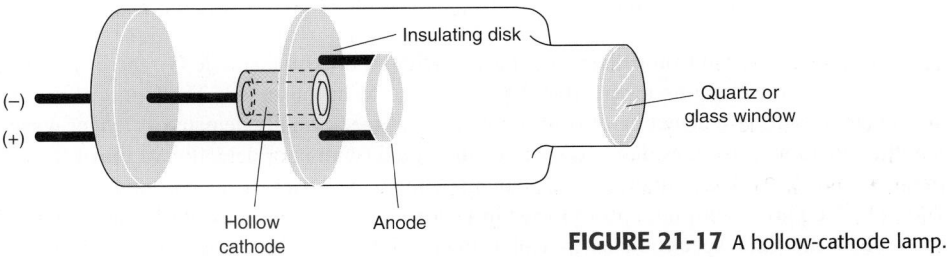

FIGURE 21-17 A hollow-cathode lamp.

Insulating disk

Quartz or glass window

(−)
(+)

Hollow cathode   Anode

Multielement Detection with the Inductively Coupled Plasma

An inductively coupled plasma emission spectrometer does not require any lamps and can measure up to 70 elements simultaneously. Color Plates 26 and 27 illustrate two designs for multielement analysis. In Color Plate 26, atomic emission enters the polychromator and is dispersed into its component wavelengths by the grating at the bottom. One photomultiplier detector (Figure 20-14) is required at the correct position for each element.

In Color Plate 27, atomic emission entering from the top right is reflected by a collimating mirror (which makes light rays parallel), dispersed in the vertical plane by a prism, and then dispersed in the horizontal plane by a grating. Dispersed radiation lands on a *charge injection device* (CID) detector, which is related to the charge coupled device (CCD) in Figure 20-17. Different wavelengths are spread over the 262 000 pixels of the CID shown at the upper left of Color Plate 27. In a CCD detector, each pixel must be read one at a time in row-by-row order. Each pixel of a CID detector can be read individually at any time. Selectively reading relevant pixels eliminates time spent reading pixels that are of no interest. A given pixel can be monitored and read before it becomes full. The charge in the pixel is then neutralized to reset the pixel to zero. The pixel can then accumulate more charge and be read several times while other pixels are filling up at a slower pace. This process increases the dynamic range of the detector by allowing large signals to be measured in some pixels, while small signals are measured in other pixels. Another advantage of the CID detector over the CCD detector is that strong signals in one pixel are less prone to spill over into neighboring pixels (a process called *blooming* in CCD detectors). CID detectors can therefore measure weak emission signals adjacent to strong signals. Figure 21-19 shows a spectrum seen by a CID detector.

Doppler and pressure effects broaden the atomic lines by one to two orders of magnitude relative to their inherent linewidths.

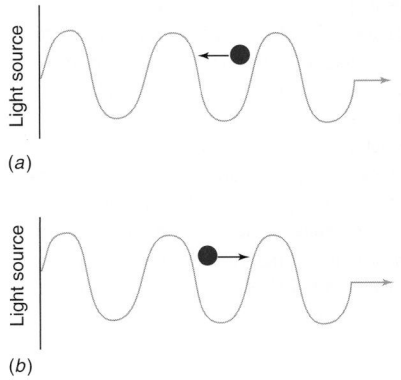

(a)

(b)

FIGURE 21-16 The Doppler effect. A molecule moving (a) toward the radiation source "feels" the electromagnetic field oscillate more often than one moving (b) away from the source.

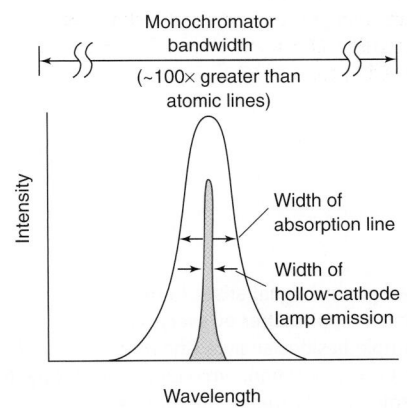

Monochromator bandwidth (~100× greater than atomic lines)

Width of absorption line

Width of hollow-cathode lamp emission

Wavelength

FIGURE 21-18 Relative widths of hollow-cathode emission line, atomic absorption line, and monochromator bandwidth. Linewidths are measured at half the signal height. The linewidth from the hollow cathode is relatively narrow because gas temperature in the lamp is lower than flame temperature (so there is less Doppler broadening) and pressure in the lamp is lower than pressure in a flame (so there is less pressure broadening).

Capabilities of CID detector:

- pixels are individually addressed
- rapidly filling pixel can be read, re-zeroed, and read again
- filled pixel does not bloom into neighbors

FIGURE 21-19 "Constellation image" of inductively coupled plasma emission from 200 μg Fe/mL seen by charge injection detector. Almost all peaks are from iron. Horizontally blurred "galaxies" near the top are Ar plasma emission. A prism spreads wavelengths of 200–400 nm over most of the detector. Wavelengths >400 nm are bunched together at the top. A grating provides high resolution in the horizontal direction. Selected peaks are labeled with wavelength (in nanometers) and diffraction order (*n* in Equation 20-1) in parentheses. Two Fe peaks labeled in color at the lower left and lower right are both the same wavelength (238.204 nm) diffracted into different orders by the grating. [Courtesy M. D. Seltzer, Michelson Laboratory, China Lake, CA.]

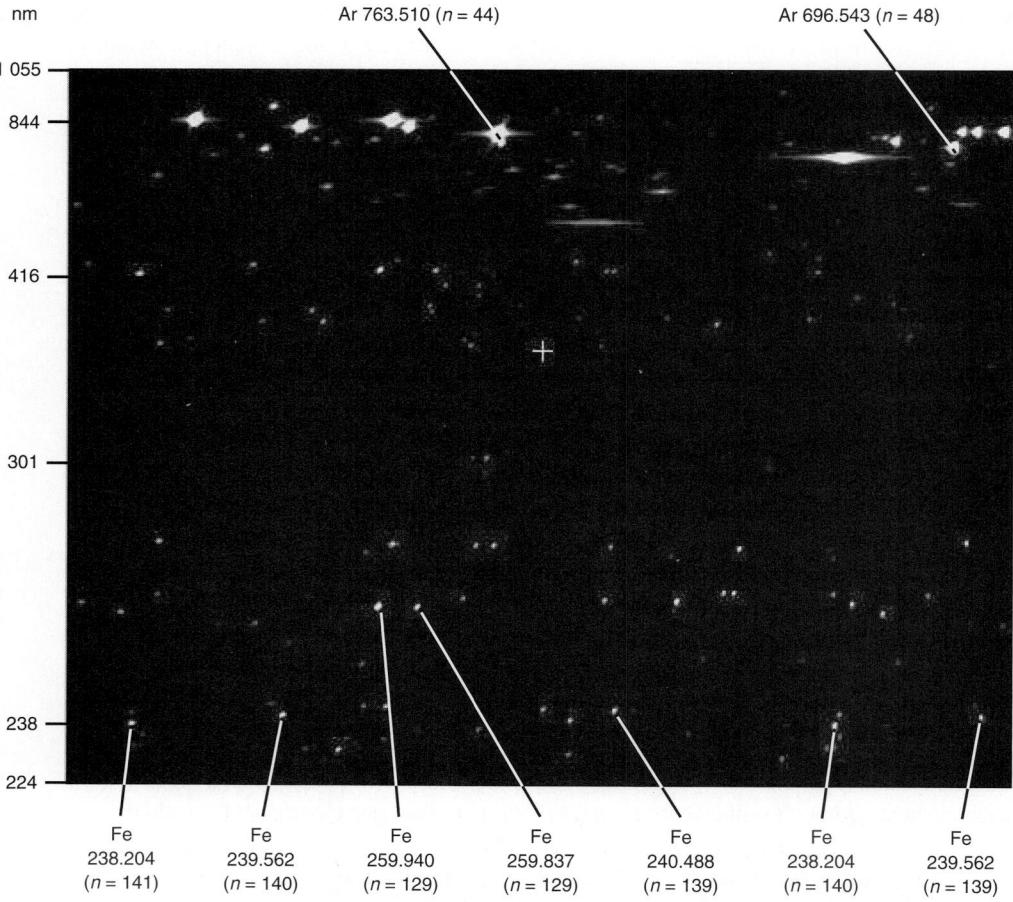

The spectrometer in Color Plate 27 is purged with N_2 or Ar to exclude O_2, thereby allowing ultraviolet wavelengths in the 100–200-nm range to be observed. This spectral region permits more sensitive detection of some elements that are normally detected at longer wavelengths and allows halogens, P, S, and N to be measured (with poor detection limits of tens of parts per million). These nonmetallic elements cannot be observed at wavelengths above 200 nm. The photomultiplier spectrometer in Color Plate 26 is more expensive and complicated than the CID spectrometer in Color Plate 27 but provides lower detection limits because a photomultiplier tube is more sensitive than a CID detector.

Background Correction

Background signal arises from absorption, emission, or scatter by everything in the sample besides analyte (the *matrix*), as well as from absorption, emission, or scatter by the flame, the plasma, or the furnace.

Atomic spectroscopy must provide **background correction** to distinguish analyte signal from absorption, emission, and optical scattering of the sample matrix, the flame, plasma, or red-hot graphite furnace. Figure 21-20 shows the spectrum of a sample analyzed in a graphite furnace. Sharp atomic signals with a maximum absorbance near 1.0 are superimposed on a broad background with an absorbance of 0.3. If we did not subtract the background absorbance, significant errors would result. Background correction is critical for graphite furnaces, which tend to contain residual smoke from charring. Optical scatter from smoke must be distinguished from optical absorption by analyte.

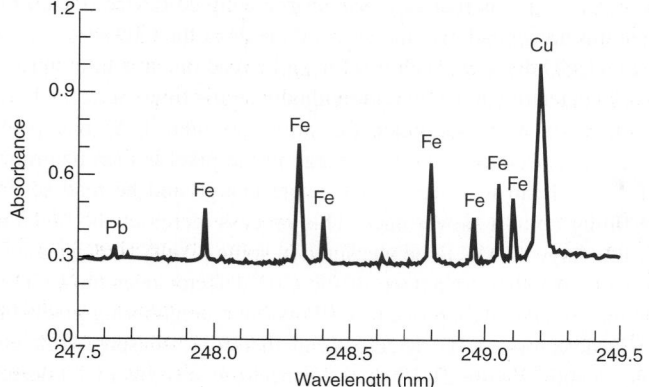

FIGURE 21-20 Graphite furnace absorption spectrum of bronze dissolved in HNO_3. [Data from B. T. Jones, B. W. Smith, and J. D. Winefordner, "Continuum Source Atomic Absorption Spectrometry in a Graphite Furnace with Photodiode Array Detection," *Anal. Chem.* **1989,** *61,* 1670.]

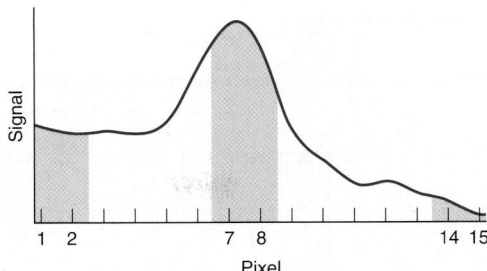

FIGURE 21-21 Data from charge injection detector illustrating baseline correction in plasma emission spectrometry. The mean value of pixels on either side of a peak is subtracted from the mean value of pixels under the peak. [Data from M. D. Seltzer, Michelson Laboratory, China Lake, CA.]

Figure 21-21 shows how background is subtracted in an emission spectrum collected with a charge injection device detector. The figure shows 15 pixels from one row of the detector centered on an analytical peak. (The spectrum was manipulated by a computer algorithm to make it look smooth, even though the original data consisted of a single reading for each pixel and would look like a bar graph.) Pixels 7 and 8 were selected to represent the peak. Pixels 1 and 2 represent the baseline at the left and pixels 14 and 15 represent the baseline at the right. The mean baseline is the average of pixels 1, 2, 14, and 15. The mean peak amplitude is the average of pixels 7 and 8. The corrected peak height is the mean peak amplitude minus the mean baseline amplitude.

For atomic absorption, *beam chopping* or electrical *modulation* of the hollow-cathode lamp (pulsing it on and off) can distinguish the signal of the flame from the atomic line at the same wavelength. Figure 21-22 shows light from the lamp being periodically blocked by a rotating chopper. Signal reaching the detector while the beam is blocked must be from flame emission. Signal reaching the detector when the beam is not blocked is from the lamp and the flame. The difference between the two signals is the desired analytical signal.

Beam chopping corrects for flame emission but not for scattering. Most spectrometers provide an additional means to correct for scattering and broad background absorption. Deuterium lamps and Zeeman correction systems are most common.

For *deuterium lamp background correction*, broad emission from a D_2 lamp (Figure 20-4) is passed through the flame in alternation with that from the hollow cathode. The monochromator bandwidth is so wide that analyte atomic transitions absorb a negligible fraction of D_2 radiation. Light from the hollow-cathode lamp is absorbed by analyte and absorbed and scattered by background. Light from the D_2 lamp is absorbed and scattered only by background. The difference between absorbance measured with the hollow-cathode lamp and absorbance measured with the D_2 lamp is the absorbance of analyte.

An excellent, but expensive, background correction technique for a graphite furnace for many elements relies on the *Zeeman effect* (pronounced ZAY-mon). When a magnetic field is applied parallel to the light path through a furnace, the absorption (or emission) line of analyte atoms is split into three components. Two are shifted to slightly lower and higher wavelengths (Figure 21-23), and one component is unshifted. The unshifted component does not have the correct electromagnetic polarization to absorb light traveling parallel to the magnetic field and is therefore "invisible."

To use the Zeeman effect for background correction, a strong magnetic field is pulsed on and off. Sample and background are observed when the field is off. Background alone is observed when the field is on. The difference is the corrected signal.

The advantage of Zeeman background correction is that it operates at the analytical wavelength. In contrast, D_2 background correction is made over a broad band. A structured or sloping background is averaged by this process, potentially misrepresenting the true background signal at the analytical wavelength. Background correction in Figure 21-21 is similar to D_2 background correction, but the wavelength range in Figure 21-21 is restricted to the immediate vicinity of the analytical peak.

Detection Limits

One common definition of **detection limit** is the concentration of an element that gives a signal equal to three times the standard deviation of signal from a blank (Equation 5-5). Equivalently, the detection limit is the concentration of an element that gives a signal equal to three times the root-mean-square noise (Equation 20-14) in the baseline adjacent to the analyte signal.

Figure 21-24 compares detection limits for flame, furnace, and inductively coupled plasma analyses on instruments from one manufacturer. The detection limit for furnaces is

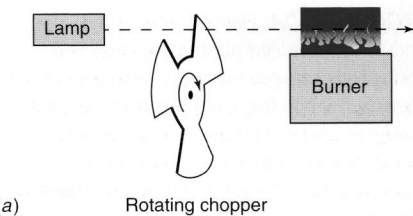

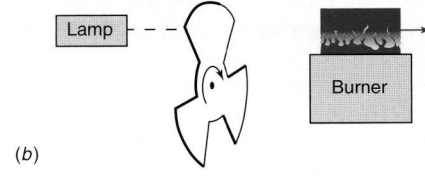

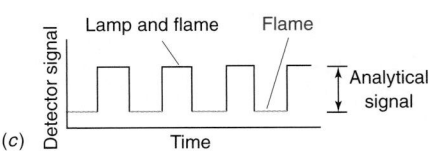

FIGURE 21-22 Beam chopper for flame background correction. (*a*) Lamp and flame emission reach detector. (*b*) Only flame emission reaches detector. (*c*) Resulting square wave signal.

Background correction methods:
- subtract adjacent pixels of CID display
- beam chopping
- D_2 lamp
- Zeeman

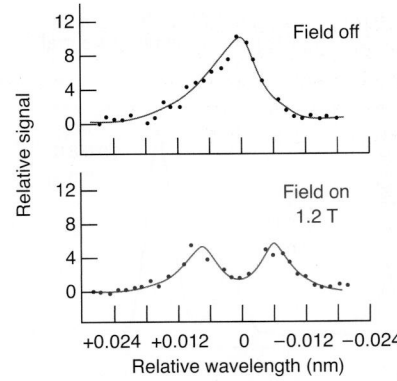

FIGURE 21-23 Zeeman effect on Co fluorescence in a graphite furnace with excitation at 301 nm and detection at 341 nm. The magnetic field strength for the lower spectrum is 1.2 tesla. [Data from J. P. Dougherty, F. R. Preli, Jr., J. T. McCaffrey, M. D. Seltzer, and R. G. Michel, "Instrumentation for Zeeman Electrothermal Atomizer Laser Excited Atomic Fluorescence Spectrometry," *Anal. Chem.* **1987**, *59*, 1112.]

FIGURE 21-24 Flame, furnace, and inductively coupled plasma emission and inductively coupled plasma–mass spectrometry detection limits (ng/g = ppb) with instruments from GBC Scientific Equipment, Australia. Accurate quantitative analysis requires concentrations 10–100 times greater than the detection limit. [Data for flame, furnace, ICP from R. J. Gill, *Am. Lab.,* November 1993, p. 24F. ICP–MS from T. T. Nham, *Am. Lab.,* August 1998, p. 17A. Data for Cl, Br, and I are from reference 20.]

Detection limits (ng/g), values listed per element in order:
ICP emission / Flame atomic absorption / Graphite furnace atomic absorption / ICP–mass spectrometry

Element	ICP emission	Flame AA	Graphite furnace AA	ICP–MS
Li	0.7	2	0.1	0.0002
Be	0.07	1	0.02	0.0009
Fe (legend)	0.7	5	0.02	0.008
B	1	500	15	0.0008
C	10	—	—	
Na	3	0.2	0.005	0.0002
Mg	0.08	0.3	0.004	0.0003
Al	2	30	0.01	0.0002
Si	5	100	0.1	<0.0001
P	7	40 000	30	<0.0001
S	3	—	—	0.0001
Cl	60	—	—	
K	20	3	0.1	0.0002
Ca	0.07	0.5	0.01	0.007
Sc	0.3	40	—	0.0002
Ti	0.4	70	0.5	0.004
V	0.7	50	0.2	0.0003
Cr	2	3	0.01	0.0003
Mn	0.2	2	0.01	0.0002
Fe	0.7	5	0.02	0.008
Co	1	4	0.02	0.0002
Ni	3	90	0.1	0.001
Cu	0.9	1	0.02	0.0005
Zn	0.6	0.5	0.001	0.003
Ga	10	60	0.5	0.006
Ge	20	200	—	0.002
As	7	200	0.2	0.003
Se	10	250	0.5	0.05
Br	150	—	—	0.02
Rb	1	7	0.05	0.0003
Sr	0.2	2	0.1	0.0003
Y	0.6	200	—	0.0003
Zr	2	1000	0.6	0.0006
Nb	5	2000	—	0.0008
Mo	3	—	0.02	0.002
Ru	10	60	1	0.001
Rh	20	4	—	0.0003
Pd	4	10	0.3	0.001
Ag	0.8	2	0.005	0.0007
Cd	0.5	0.4	0.003	0.0008
In	20	40	1	0.0003
Sn	9	30	0.2	0.0009
Sb	9	40	0.1	0.001
Te	4	30	0.1	0.02
I	40	—	—	0.002
Cs	40 000	4	0.2	0.0003
Ba	0.6	10	0.04	0.0003
La	1	2000	—	0.0003
Hf	4	2000	—	0.0008
Ta	10	2000	—	0.0005
W	10	1000	—	0.002
Re	3	600	—	0.0007
Os	0.2	100	—	0.002
Ir	7	400	—	0.0004
Pt	7	100	0.2	0.001
Au	2	10	0.1	0.0009
Hg	7	150	2	0.0009
Tl	10	20	0.1	0.0004
Pb	10	10	0.05	0.0006
Bi	7	40	0.1	0.0005
Ce	2	—	—	0.0003
Pr	9	6000	—	0.0002
Nd	10	1000	0.5	0.001
Sm	10	1000	—	0.001
Eu	0.9	20	0.5	0.0004
Gd	5	2000	—	0.001
Tb	6	500	0.1	0.0002
Dy	2	30	1	0.0009
Ho	2	40	—	0.0002
Er	2	30	2	0.0007
Tm	2	900	—	0.0002
Yb	0.3	4	—	0.001
Lu	0.3	300	—	0.0002
Th	7	—	—	0.0003
U	60	40 000	—	0.0005

Other elements shown without data: N, O, F, Ne, Ar, Tc, Xe, Po, At, Rn, Pm, Pa, Np, Pu, Am, Cm, Bk, Cf, Es, Fm, Md, No, Lr.

☐ Requires N_2O/C_2H_2 flame and is therefore better analyzed by inductively coupled plasma

☐ Best analyzed by emission

typically two orders of magnitude lower than that observed with a flame because the sample is confined in the small volume of the furnace for a relatively long time. Detection limits for the inductively coupled plasma are intermediate between the flame and the furnace. With ultrasonic nebulization (Figure 21-14) and axial plasma viewing (Figure 21-12), the sensitivity of the inductively coupled plasma is close to that of the graphite furnace.

Commercial standard solutions for flame atomic absorption are not necessarily suitable for plasma and furnace analyses. The latter methods require purer grades of water and acids for standard solutions and, especially, for dilutions. For the most sensitive analyses, solutions are prepared in a dust-free clean room or hood with a filtered air supply to reduce background contamination that *will be* detected by your instruments.

Any standard solution has a limited shelf life. High-density polyethylene plastic bottles containing standards for atomic spectroscopy are packaged in sealed, aluminized bags to retard evaporation. In one study, the concentration of standard not stored in an aluminized bag increased by 0.11% per year at 23°C and 0.26% per year at 30°C because water evaporated from the sealed bottle.[22] Evaporation from bottles sealed in aluminized bags was 0.01% per year at 23°C and 0.08% per year at 30°C.

Standards evaporate—even from sealed bottles

21-5 Interference

Interference is any effect that changes the signal while analyte concentration remains unchanged. Interference can be corrected by removing the source of interference or by preparing standards that exhibit the same interference.

Types of Interference

Types of interference:
- spectral: unwanted signals overlapping analyte signal
- chemical: chemical reactions decreasing the concentration of analyte atoms
- ionization: ionization of analyte atoms decreasing the concentration of neutral atoms

Spectral interference refers to the overlap of analyte signal with signals due to other elements or molecules in the sample or with signals due to the flame or furnace. Interference from the flame can be subtracted by using D_2 or Zeeman background correction. The best means of dealing with overlap between lines of different elements in the sample is to choose another wavelength for analysis. High-resolution spectrometers eliminate interference from other elements by resolving closely spaced lines (Figure 21-25).

A single element such as Nb gives rise to over 1 000 discrete emission lines, so some lines from different elements overlap—even at high resolution. Software in one commercial instrument measures each element from several different spectral lines.[23] It converts each intensity to a concentration, using a different calibration curve for every line. If there were no spectral interference, all lines for one element would give the same concentration. Results that deviate from most results are attributed to spectral interference and are discarded.

Elements that form very stable diatomic oxides are incompletely atomized at the temperature of a flame or furnace. The spectrum of a molecule is much broader and more complex than that of an atom, because vibrational and rotational transitions are combined with electronic transitions (Section 18-6). The broad spectrum leads to spectral interference at

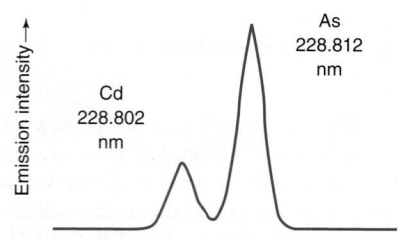

FIGURE 21-25 A Cd line at 228.802 nm causes spectral interference with the As line at 228.812 nm in most spectrometers. With sufficiently high resolution, peaks are separated and there is no interference. The instrument used for this spectrum has a 1-m Czerny-Turner monochromator (Figure 20-6) with a resolution of 0.005 nm from 160 to 320 nm and 0.010 nm from 320 to 800 nm. [Data from Jobin Yvon Horiba Group, Longjumeau Cedex, France.]

many wavelengths. Figure 21-26 shows a plasma containing Y and Ba atoms as well as YO molecules. Note how broad the molecular emission is relative to the atomic emission.

When trace impurities in tungsten powder are analyzed by graphite furnace atomic absorption using direct solid sampling (Figure 21-9), WO_3 from the surface of the powder sublimes and fills the furnace with vapor, creating spectral interference throughout the visible and ultraviolet regions. Heating the powder under H_2 at $1\ 000-1\ 200°C$ in the furnace prior to atomization reduces WO_3 to metallic tungsten and eliminates the interference.

Chemical interference is caused by any component of the sample that decreases the extent of atomization of analyte. For example, SO_4^{2-} and PO_4^{3-} hinder the atomization of Ca^{2+}, perhaps by forming nonvolatile salts. **Releasing agents** are chemicals added to a sample to decrease chemical interference. EDTA and 8-hydroxyquinoline protect Ca^{2+} from interference by SO_4^{2-} and PO_4^{3-}. La^{3+} is a releasing agent, apparently because it preferentially reacts with PO_4^{3-} and frees the Ca^{2+}. A fuel-rich flame reduces certain oxidized analyte species that would otherwise hinder atomization. Higher flame temperatures eliminate many kinds of chemical interference.

Ionization interference can be a problem in the analysis of alkali metals at relatively low temperature and in the analyses of other elements at higher temperature. For any element, we can write a gas-phase ionization reaction:

$$M(g) \rightleftharpoons M^+(g) + e^-(g) \qquad K = \frac{[M^+][e^-]}{[M]} \qquad (21\text{-}6)$$

Because alkali metals have low ionization potentials, they are most extensively ionized. At 2 450 K and a pressure of 0.1 Pa, sodium is 5% ionized. With its lower ionization potential, potassium is 33% ionized. Ions have energy levels different from those of neutral atoms, so the desired signal is decreased. If there is a strong signal from the ion, you could use the ion signal rather than the atomic signal.

An **ionization suppressor** decreases the extent of ionization of analyte. In the analysis of potassium, it is recommended that solutions contain 1 000 ppm of CsCl, because cesium is more easily ionized than potassium. By producing a high concentration of electrons in the flame, ionization of Cs suppresses ionization of K. Ionization suppression is desirable in a low-temperature flame in which we want to observe neutral atoms.

The *method of standard addition* (Section 5-3) compensates for many types of interference by adding known quantities of analyte to the unknown in its complex matrix. For example, Figure 21-27 shows the analysis of strontium in aquarium water. The slope of the standard addition curve is 0.018 8 absorbance units/ppm. If, instead, Sr is added to distilled water, the slope is 0.030 8 absorbance units/ppm. That is, in distilled water, the absorbance increases 0.030 8/0.018 8 = 1.64 times more than it does in aquarium water for each addition of standard Sr. We attribute the lower response in aquarium water to interference by other species. The absolute value of the *x*-intercept of the standard addition curve, 7.41 ppm, is a reliable measure of Sr in the aquarium.

The most accurate work demands exact matrix matching between samples and standards. For example, the Be mass fraction of a beryllium oxide Standard Reference Material was certified by inductively coupled plasma–atomic emission.[24] There was a 2.3% systematic error for Be and Mn (an internal standard) when the matrix was changed from 2.1 vol% H_2SO_4:H_2O to (2.0 vol% H_2SO_4 plus 0.07 vol% HCl). When four laboratories used exact matrix matching and the most care, agreement among results from the different labs had a standard deviation of 0.074%.

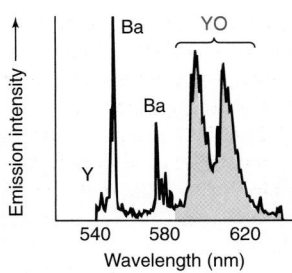

FIGURE 21-26 Emission from a plasma produced by laser irradiation of the high-temperature superconductor $YBa_2Cu_3O_7$. Solid is vaporized by the laser, and excited atoms and molecules such as YO emit light at characteristic wavelengths. [Data from W. A. Weimer, "Plasma Emission from Laser Ablation of $YBa_2Cu_3O_7$," *Appl. Phys. Lett.* **1988,** *52,* 2171.]

Le Châtelier's principle tells us that adding electrons to the right side of Reaction 21-6 drives the reaction back to the left.

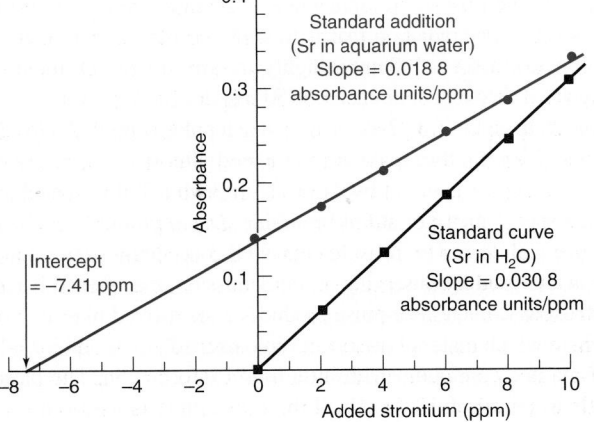

FIGURE 21-27 Atomic absorption calibration curve for Sr added to distilled water and standard addition of Sr to aquarium water. All solutions are made up to a constant volume, so the ordinate is the concentration of added Sr. [Data from L. D. Gilles de Pelichy, C. Adams, and E. T. Smith, "Analysis of Sr in Marine Aquariums by Atomic Absorption Spectroscopy," *J. Chem. Ed.* **1997,** *74,* 1192.]

Virtues of the Inductively Coupled Plasma

An inductively coupled argon plasma eliminates many common interferences.[25] The plasma is twice as hot as a conventional flame, and the residence time of analyte in the plasma is about twice as long. Therefore, atomization is more complete and signal is enhanced. Formation of analyte oxides and hydroxides is negligible. The plasma is remarkably free of background radiation where sample emission is observed 15–35 mm above the load coil.

Concentrations of excited atoms in the cooler, outer part of the flame are lower than in the warmer, central part of the flame. Emission from the central region is absorbed in the outer region. This **self-absorption** increases with increasing concentration of analyte and gives non-linear calibration curves. In a plasma, the temperature is more uniform, and self-absorption is not nearly so important. Plasma emission calibration curves are linear over five orders of magnitude. In flames and furnaces, the linear range is two orders of magnitude. For inductively coupled plasma–mass spectrometry, the linear range is eight orders of magnitude (Table 21-4).

TABLE 21-4 Comparison of atomic analysis methods				
	Flame absorption	**Furnace absorption**	**Plasma emission**	**Plasma–mass spectrometry**
Detection limits (ng/g)	10–1 000	0.01–1	0.1–10	0.000 01–0.000 1
Linear range	10^2	10^2	10^5	10^8
Precision				
short term (5–10 min)	0.1–1%	0.5–5%	0.1–2%	0.5–2%
long term (hours)	1–10%	1–10%	1–5%	<5%
Interferences				
spectral	very few	very few	many	few
chemical	many	very many	very few	some
mass	—	—	—	many
Sample throughput	10–15 s/element	3–4 min/element	6–60 elements/min	all elements in 2–5 min
Dissolved solid	0.5–5%	>20% slurries and solids	1–20%	0.1–0.4%
Sample volume	large	very small	medium	medium
Purchase cost	1	2	4–9	10–15

SOURCE: *Adapted from TJA Solutions, Franklin, MA.*

21-6 Sampling by Laser Ablation

At the opening of this chapter, we saw an example of *laser ablation–inductively coupled plasma–mass spectrometry*[26] for the analysis of teeth. In laser **ablation,** a pulsed laser beam is focused onto a microscopic spot on a solid sample, creating an explosion of particles, atoms, electrons, and ions into the gas phase. A Nd:YAG (neodymium-doped yttrium aluminum garnet) laser with a wavelength of 1.064 μm is often used. Multiplication of the laser frequency to give wavelengths of 532, 266, or 213 nm provides radiation that is less intense but more strongly absorbed by many samples. A deep ultraviolet laser at 193 nm is highly absorbed by most samples. A 10-ns pulse with 10 mJ of energy focused onto a spot diameter of 50 μm delivers a power of 50 GW/cm². Material is typically removed to a depth of 0.02 to 5 μm by each pulse (Figure 21-28). Only nanograms of material are ablated by each pulse, making the method almost nondestructive. Ablation product generated in a sealed chamber is swept by Ar or He through a Teflon-coated tube into the plasma for analysis by mass spectrometry or atomic emission. *Depth profiling* can be done by successive pulses probing deeper and deeper to measure elemental concentrations as a function of depth.

If the plasma is analyzed by observing atomic emission, the method is called **laser-induced breakdown spectroscopy.** Each laser pulse produces a short-lived plasma with a temperature of 10 000 to 20 000 K in which material dissociates into excited atoms and ions. Initial optical emission from the plasma is a continuum with little useful information. The plasma expands supersonically and cools to temperatures at which discrete atomic emission lines are observed after

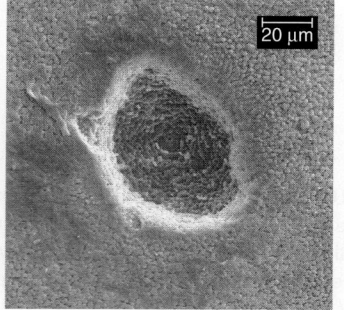

FIGURE 21-28 Microscopic crater ablated into a mussel shell by 10 pulses from a 266-nm laser with a beam energy of 4.5 mJ per 10-ns pulse and a repetition rate of 10 Hz. [From V. R. Bellotto and N. Miekely, "Improvements in Calibration Procedures for the Quantitative Determination of Trace Elements in Carbonate Material (Mussel Shells) by Laser Ablation ICP–MS," *Fresenius J. Anal. Chem.* **2000,** *367,* 635. © Springer-Verlag 2000. With kind permission of Springer Science+Business Media.]

~1−10 μs. The detector is *gated* to record atomic emission several microseconds after the laser pulse. One commercial system observes plasma emission through seven optical fibers directed to seven polychromators, each containing a linear 2 048-pixel charge coupled device to view a different spectral region. Spectral resolution is 0.1 nm over the range 200–980 nm. Box 21-3 describes laser-induced breakdown spectroscopy used on the surface of Mars to measure soil and rock composition.

Quantitative analysis by laser induced breakdown spectroscopy requires careful attention to detail[27] and could be inaccurate by ±10% or more. Elements can be selectively ablated, selectively transported to the plasma, or selectively atomized in the plasma. Therefore, the relative numbers of ions detected are not necessarily equal to relative quantities in the solid sample. The most reliable—but usually unattainable—calibration is a standard sample containing elements of interest in the same matrix as the unknown. Mussel shell in Figure 21-28 is composed principally of $CaCO_3$. A calibration standard was made by dissolving known quantities of metals with a large excess of Ca^{2+} in acid and precipitating everything with CO_3^{2-}. The carbonate precipitate was washed, dried, and pressed into a dense pellet whose ablation behavior is similar to that of mussel shell. In the absence of *matrix-matched standards*, results from laser ablation can be compared with results obtained by completely digesting a material and analyzing the homogeneous solution.

21-7 Inductively Coupled Plasma–Mass Spectrometry

The 15.8 electron volt (eV) ionization energy of Ar is higher than those of all elements except He, Ne, and F. In an Ar plasma, analyte elements can be ionized by collisions with Ar^+, excited Ar atoms, or energetic electrons. The plasma can be directed into a mass spectrometer (Chapter 22), which separates and measures ions according to their mass-to-charge ratio.[28] For the most accurate measurements of isotope ratios, the mass spectrometer has one detector for each desired isotope.[29]

The trace element profile of teeth at the opening of this chapter was obtained by inductively coupled plasma–mass spectrometry. Figure 21-29 shows an example in which Hawaiian and Cuban coffee beans were extracted with trace-metal-grade nitric acid and the aqueous extract was analyzed by inductively coupled plasma–mass spectrometry. Coffee brewed from either bean contains ~15 ng Pb/mL. However, the Cuban beans also contain Hg at a concentration similar to that of Pb.

The difficulty in sampling anything with a mass spectrometer is that the spectrometer requires high vacuum to avoid collisions between ions and background gas molecules that

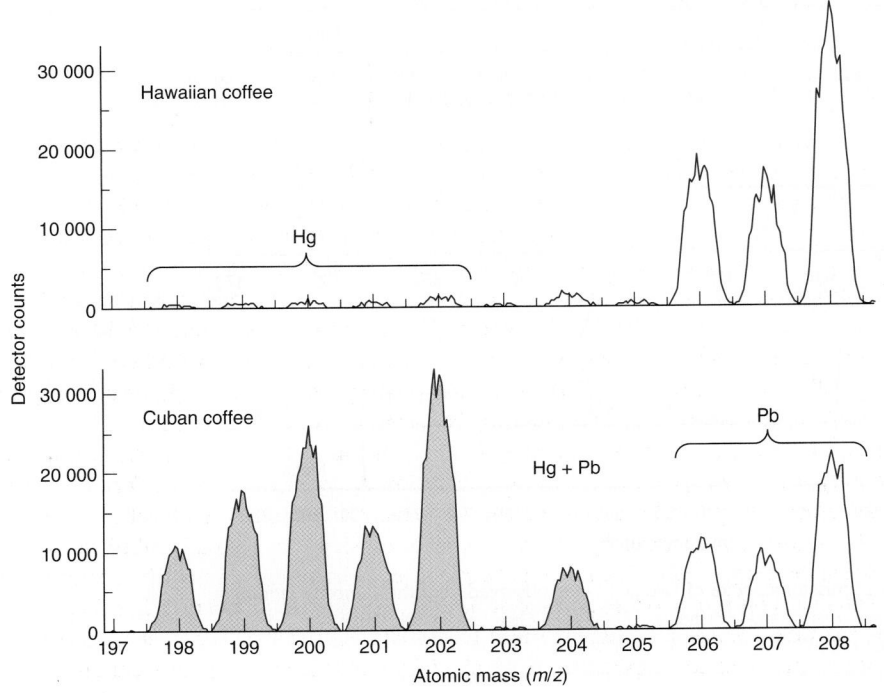

FIGURE 21-29 Partial elemental profile of coffee beans by inductively coupled plasma–mass spectrometry. The Cuban beans have a much higher Hg content than the Hawaiian beans. A blank has not been subtracted from either spectrum. [Data from G. S. Ostrom and M. D. Seltzer, Michelson Laboratory, China Lake, CA.]

BOX 21-3 **Atomic Emission Spectroscopy on Mars**

(a)

[NASA/JPL-CalTech.]

Since 2012, the *Mars Science Laboratory* rover, *Curiosity*, has been exploring Mars with a suite of 10 scientific instruments. The Chem-Cam laser-induced breakdown spectroscopy unit measures chemical composition of rocks and soil up to seven meters away. Scientists select a target with the rover's high-resolution telescope. A 1.067-μm-wavelength laser then fires a series of pulses through the telescope to vaporize a ~0.4-mm-diameter area. Each pulse with a power density $>1 \text{ GW/cm}^2$ creates a luminous plasma of atoms. Optical emission from the plasma is collected through the telescope

and directed to three spectrographs to generate a profile of atomic composition of the irradiated area.

Spectra shown here come from two types of soil. Fine and coarse grain mafic soil contains silicate that is rich in magnesium and iron. Coarse grain felsic soil is rich in silicon, aluminum, calcium, and sodium. The inset shows emission from hydrogen and carbon. Hydrogen is thought to arise from adsorbed water or water of hydration within amorphous particles. Water represents ~1–3 wt% of the Martian surface and ~5–9 wt% of certain amorphous components of the surface.

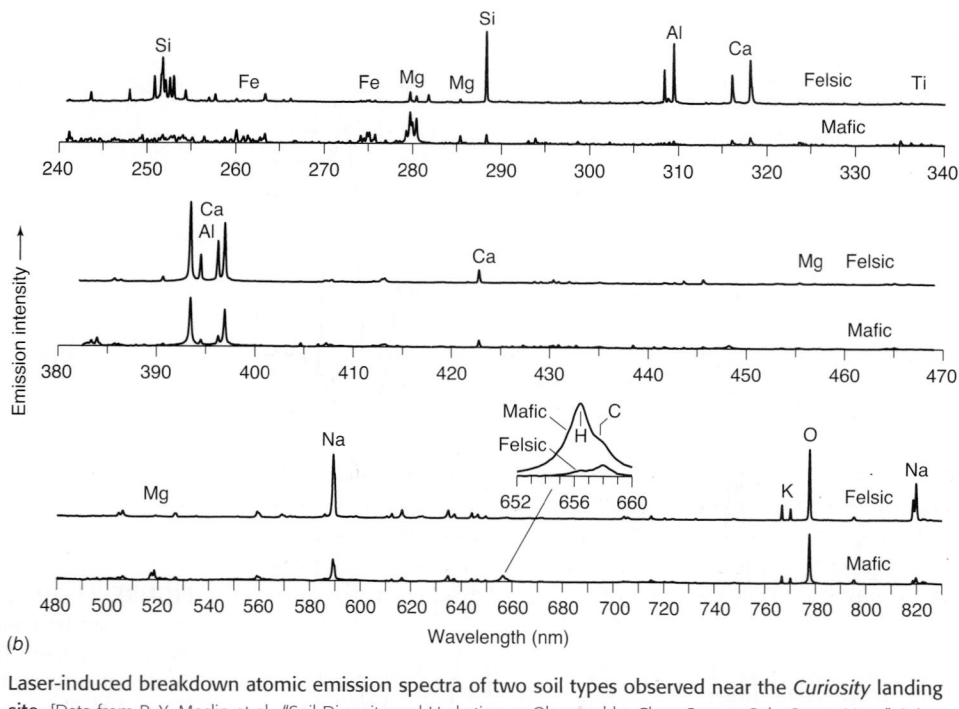

(b)

Laser-induced breakdown atomic emission spectra of two soil types observed near the *Curiosity* **landing site.** [Data from P.-Y. Meslin et al., "Soil Diversity and Hydration as Observed by ChemCam at Gale Crater, Mars," *Science* **2013**, *341*, DOI: 10.10.1126/science.1238670.]

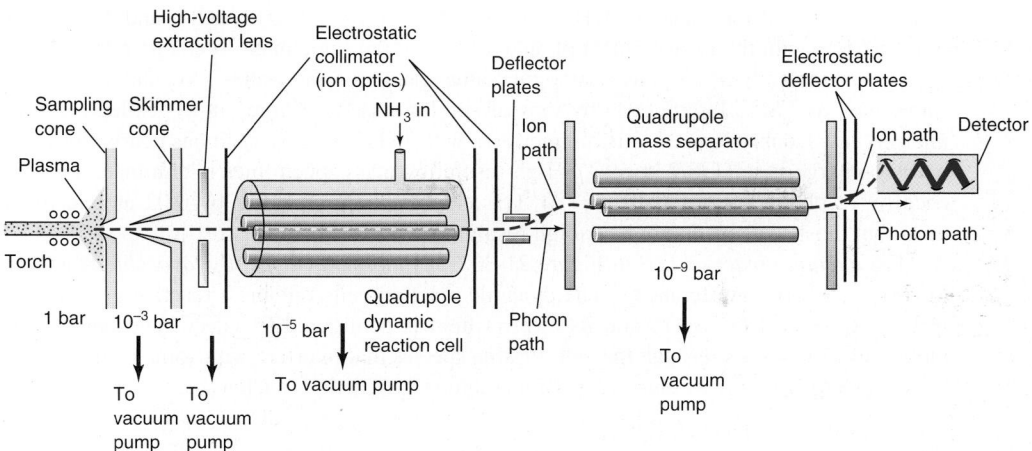

FIGURE 21-30 Interface between inductively coupled plasma and mass spectrometer is shown with a dynamic reaction cell to reduce *isobaric* interference by plasma species with the same mass as some analytes. Chapter 22 discusses mass spectrometry.

divert the ions from their trajectory in a magnetic field. Figure 21-30 shows an interface between a horizontal Ar plasma and a mass spectrometer. The plasma at the left is directed onto a water-cooled Ni sampling cone with a 1-mm-diameter orifice through which a fraction of the plasma can pass. Behind the sampling cone is a water-cooled skimmer cone with an even smaller orifice. The extraction lens behind the skimmer cone has a high negative potential to attract positive ions from the plasma. Pressure is reduced in each successive section of the instrument. From the skimmer cone, ions enter a *dynamic reaction cell* that might contain NH_3, CH_4, or other reactive molecules. The dynamic reaction cell removes *isobaric* interfering species (such as Ar^+ and ArO^+) from the plasma. **Isobaric interference** occurs when the interfering species has the same mass-to-charge ratio as analyte ion. Chemistry in the dynamic reaction cell will be described shortly. Following the dynamic reaction cell, ions are separated by a mass spectrometer. Ions of selected mass-to-charge ratio are deflected into the detector (at the right of the diagram), where they are counted. Photons from the plasma do not hit the detector, or they would generate a signal.

Sub part-per-trillion detection limits for many elements with the dynamic reaction cell are so low as to tax the cleanliness of reagents, glassware, and procedures.[30] Solutions must be made from extremely pure water and trace-metal-grade HNO_3 in Teflon or polyethylene vessels protected from dust. HCl, H_2SO_4, and H_3PO_4 are normally avoided because they create isobaric interferences. However, a dynamic reaction cell permits the analysis of H_2SO_4, and H_3PO_4 solutions by removing isobaric interferences.[31] The plasma–mass spectrometer interface cannot tolerate high concentrations of dissolved solids that clog the orifice of the sampling cone. The plasma reduces organic matter to carbon that can clog the orifice. Organic material can be analyzed if some O_2 is fed into the plasma to oxidize the carbon.

Matrix effects on the yield of ions in the plasma are important, so calibration standards should be in the same matrix as the unknown. Standard addition is highly appropriate to correct for matrix effects. When trace elements in blood were measured by inductively coupled plasma–mass spectrometry, matrix interference produced values for $^{75}As^+$ and $^{78}Se^+$ that were almost a factor of 2 above certified values for Standard Reference Materials.[32] Calibration by standard addition reduced the errors to 0-4%.

An internal standard for inductively coupled plasma–mass spectrometry should be chosen to have nearly the same ionization energy as analyte. For example, Tm can be used as an internal standard for U. The ionization energies of these two elements are 5.81 and 6.08 eV, respectively, so they should ionize to nearly the same extent in different matrices. If possible, internal standards with just one major isotope should be selected for maximum response.

Isobaric Interference and the Dynamic Reaction Cell

Ar is an "inert" gas with virtually no chemistry. However, Ar^+ has the same electronic configuration as Cl, and their chemistry is analogous. The inductively coupled plasma is a rich

Ar^+ is similar to Cl in its chemical reactivity.

source of Ar^+ and ions such as ArH^+, ArC^+, ArN^+, $ArNH^+$, ArO^+, $ArCl^+$, and Ar_2^+. These ions interfere with the measurement of analyte ions of the same mass-to-charge ratio.

For example, $^{40}Ar^{16}O^+$ has nearly the same mass as $^{56}Fe^+$, and $^{40}Ar_2^+$ has nearly the same mass as $^{80}Se^+$. Interference by ions of similar mass-to-charge ratio is called *isobaric interference*. Doubly ionized $^{138}Ba^{2+}$ interferes with $^{69}Ga^+$ because each has nearly the same mass-to-charge ratio ($138/2 = 69/1$). High-resolution mass spectrometers eliminate interference by resolving species such as $^{40}Ar^{16}O^+$ and $^{56}Fe^+$, which differ by 0.02 atomic mass units, but most systems do not have high resolution.

The *dynamic reaction cell* in Figure 21-30, uses *thermodynamically favorable* reactions to reduce isobaric interference.[33] The dynamic reaction cell contains a reactive gas such as NH_3, CH_4, N_2O, CO, or O_2, and its electric field is configured to select lower and upper masses of ions to pass through the cell. Plasma species that interfere with some elements can be reduced by as many as nine orders of magnitude by reactions such as

Electron transfer from NH_3:

$$^{40}Ar^+ + NH_3 \rightarrow NH_3^+ + Ar \qquad\qquad ^{40}Ar^{14}N^+ + NH_3 \rightarrow NH_3^+ + Ar + N$$

$$^{40}Ar^{12}C^+ + NH_3 \rightarrow NH_3^+ + Ar + C \qquad\qquad ^{40}Ar^{16}O^+ + NH_3 \rightarrow NH_3^+ + Ar + O$$

Proton transfer to NH_3:

$$^{40}ArH^+ + NH_3 \rightarrow NH_4^+ + Ar \qquad\qquad ^{35,37}Cl^{16}OH^+ + NH_3 \rightarrow NH_4^+ + ClO$$

$$^{40}Ar^{14}NH^+ + NH_3 \rightarrow NH_4^+ + Ar + N$$

For example, $^{40}Ar^{12}C^+$ interferes with $^{52}Cr^+$ and $^{35}Cl^{17}OH^+$ interferes with $^{53}Cr^+$. A dynamic reaction cell with NH_3 permits the measurement of Cr by removing ArC^+ and $ClOH^+$. $^{40}Ar^{16}O^+$ interferes with $^{56}Fe^+$. Either ArO^+ can be removed by reaction with NH_3 or Fe^+ can be shifted to a different mass by reaction with N_2O:

$$Fe^+ + N_2O \rightarrow FeO^+ + N_2$$

In natural waters with traces of iron and high concentrations of calcium, $^{40}Ca^{16}O^+$ interferes with $^{56}Fe^+$. CaO^+ can be removed by reaction with CO:

$$CaO^+ + CO \rightarrow Ca^+ + CO_2$$

21-8 X-ray Fluorescence[34,35]

Dmitri Mendeleev created a periodic table in 1869 in which the known elements were arranged according to their chemical properties and (with some exceptions) by increasing atomic mass. The atomic number Z assigned to each element was just its order in the periodic table, beginning with hydrogen = 1. No physical meaning was attached to Z until 1913 when Harry Moseley—a 26-year-old graduate student working with Ernest Rutherford in Manchester, England—found evidence from X-ray fluorescence that Z is the number of positive charges in the atomic nucleus. According to Frederick Soddy, who received the Nobel Prize in 1921 for the study of radioactive decay and the theory of isotopes, "Moseley . . . called the roll of the elements, so that for the first time we could say definitely the number of possible elements between the beginning and the end, and the number that still remained to be found."[36] Alas, when the Great War broke out in 1914, Moseley volunteered for military service and was killed at Gallipoli in 1915. Had he lived, Moseley would likely have received the Nobel Prize.

X-rays are high-energy photons generated when high-energy (10 to 50 kV [kilovolt]) electrons are accelerated into an anode made of material such as W, Mo, Ag, or Rh. Moseley directed X-rays at different elements and observed X-ray fluorescence at a few discrete wavelengths. He measured wavelength from the angle of diffraction of the X-rays by a crystal of potassium ferrocyanide onto a photographic plate. **X-ray fluorescence** is the emission of X-rays following the absorption of X-rays by a material.

Figure 21-31 shows an X-ray fluorescence spectrum of soil. Each element that is present emits two characteristic X-ray peaks (narrow bands). The abscissa (x-axis, if you have forgotten) is X-ray energy in keV (thousand electron volts), rather than wavelength. The energy (E) of a photon is

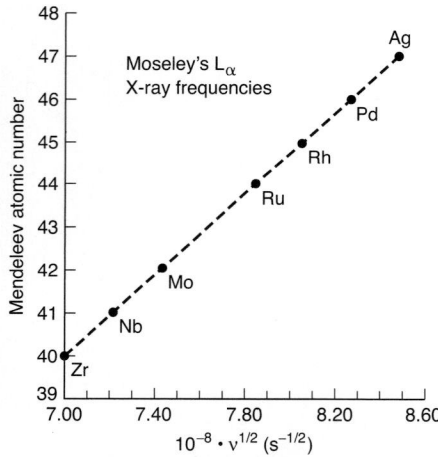

Moseley showed that X-ray fluorescence frequencies vary in a smooth manner with the atomic number from Mendeleev's periodic table [*Philosophical Magazine* **1914**, *27*, 703]. The graph shows that an element is missing between Mo and Ru and tells us what the X-ray frequency would be for that element when it is found.

Example: Convert 6.4 keV X-ray energy to wavelength:

$E = (6\,400 \text{ eV})(1.602 \times 10^{-19} \text{ J/eV}) = 1.025 \times 10^{-15} \text{ J}$
(conversion from Table 1-4)

$\lambda = hc/E$

$= (6.626 \times 10^{-34} \text{ J} \cdot \text{s})(2.998 \times 10^8 \text{ m/s})/(1.025 \times 10^{-15} \text{ J})$

$= 1.94 \times 10^{-10} \text{ m} = 0.194 \text{ nm}$

Wavelength = 0.194 nm (3 000 times smaller than visible wavelengths)

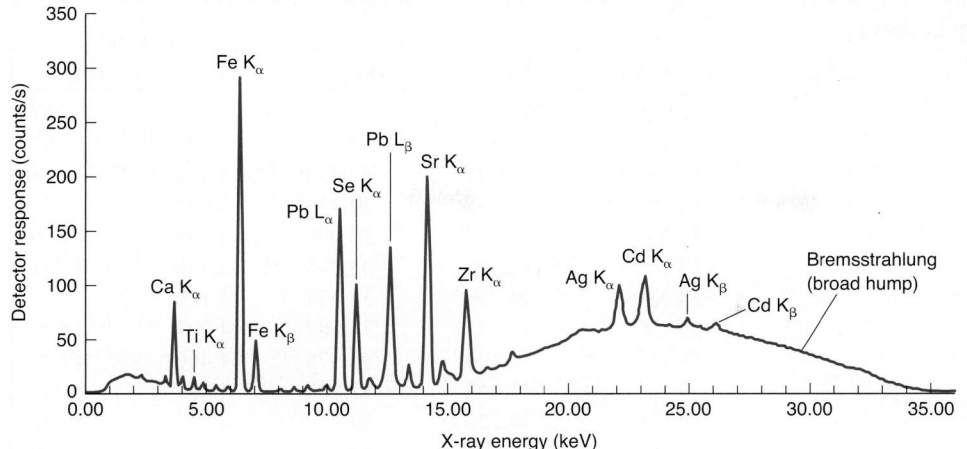

FIGURE 21-31 X-ray fluorescence spectrum of a soil sample. [Data from P. T. Palmer, San Francisco State University.]

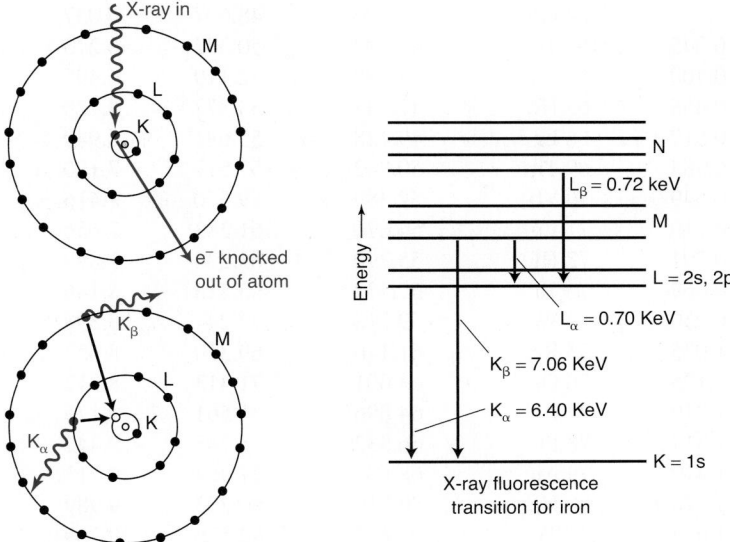

FIGURE 21-32 Upper left: Incoming X-ray photon knocks a K shell (1s) electron out of an atom. Lower left: Electron from L or M shell falls into the K shell vacancy, emitting K_α or K_β X-ray fluorescence. Right: X-ray fluorescence transitions of iron.

inversely proportional to its wavelength (λ): $E = hc/\lambda$, where h is Planck's constant and c is the speed of light. Elements are identified by their peak energies and quantified by the number of photons in each peak.

The origin of X-ray fluorescence is depicted in Figure 21-32, which shows *shells* of energy levels of electrons in an atom. Shells were historically labeled K, L, M, . . . before their structure was explained by quantum mechanics. We now call K the 1s orbital and L encompasses the 2s and 2p orbitals. Only the innermost shells are well separated in energy.

The process in Figure 21-32 begins at the upper left with an incoming X-ray knocking an electron from the K or L shell out of the atom. At the lower left, higher-energy electrons fall into the vacancy, emitting X-rays in the process. An electron falling from L to K emits an X-ray called K_α. Emission from M to K is called K_β. The vacancy left in the L or M shells is filled from higher-energy electrons accompanied by more X-rays. Emission from M to L is called L_α and emission from N to L is called L_β. Table 21-5 lists X-ray fluorescence energies for the elements, which are independent of the chemical form (±0.002 keV) because the electrons are deep inside the atom. Fe(0) in steel and Fe(VI) in Na_2FeO_4 have the same K and L energies.

TABLE 21-5 **X-ray fluorescence energies (keV)**

Element	$K_{\alpha 1}$	$K_{\beta 1}$	$L_{\alpha 1}$	$L_{\beta 1}$	Element	$K_{\alpha 1}$	$K_{\beta 1}$	$L_{\alpha 1}$	$L_{\beta 1}$
3 Li	0.054				48 Cd	23.174	26.096	3.134	3.317
4 Be	0.109				49 In	24.210	27.276	3.287	3.487
5 B	0.183				50 Sn	25.271	28.486	3.444	3.663
6 C	0.277				51 Sb	26.359	29.726	3.605	3.844
7 N	0.392				52 Te	27.472	30.996	3.769	4.030
8 O	0.525				53 I	28.612	32.295	3.938	4.221
9 F	0.677				54 Xe	29.779	33.624	4.110	
10 Ne	0.849				55 Cs	30.973	34.987	4.287	4.620
11 Na	1.041	1.071			56 Ba	32.194	36.378	4.466	4.828
12 Mg	1.254	1.302			57 La	33.442	37.801	4.651	5.042
13 Al	1.487	1.557			58 Ce	34.720	39.257	4.840	5.262
14 Si	1.740	1.836			59 Pr	36.026	40.748	5.034	5.489
15 P	2.014	2.139			60 Nd	37.361	42.271	5.230	5.722
16 S	2.308	2.464			61 Pm	38.725	43.826	5.433	5.961
17 Cl	2.622	2.816			62 Sm	40.118	45.413	5.636	6.205
18 Ar	2.958	3.191			63 Eu	41.542	47.038	5.846	6.456
19 K	3.314	3.590			64 Gd	42.996	48.697	6.057	6.713
20 Ca	3.692	4.013	0.341	0.345	65 Tb	44.482	50.382	6.273	6.978
21 Sc	4.091	4.461	0.395	0.400	66 Dy	45.998	52.119	6.495	7.248
22 Ti	4.511	4.932	0.452	0.458	67 Ho	47.547	53.877	6.720	7.525
23 V	4.952	5.427	0.511	0.519	68 Er	49.128	55.681	6.949	7.811
24 Cr	5.415	5.947	0.573	0.583	69 Tm	50.742	57.517	7.180	8.101
25 Mn	5.899	6.490	0.637	0.649	70 Yb	52.389	59.370	7.416	8.402
26 Fe	6.404	7.058	0.705	0.719	71 Lu	54.070	61.283	7.656	8.709
27 Co	6.930	7.649	0.776	0.791	72 Hf	55.790	63.234	7.899	9.023
28 Ni	7.478	8.265	0.852	0.869	73 Ta	57.532	65.223	8.146	9.343
29 Cu	8.048	8.905	0.930	0.950	74 W	59.318	67.244	8.398	9.672
30 Zn	8.639	9.572	1.012	1.035	75 Re	61.140	69.310	8.653	10.010
31 Ga	9.252	10.264	1.098	1.125	76 Os	63.001	71.413	8.912	10.355
32 Ge	9.886	10.982	1.188	1.219	77 Ir	64.896	73.561	9.175	10.708
33 As	10.544	11.726	1.282	1.317	78 Pt	66.832	75.748	9.442	11.071
34 Se	11.222	12.496	1.379	1.419	79 Au	68.804	77.984	9.713	11.442
35 Br	11.924	13.291	1.480	1.526	80 Hg	70.819	80.253	9.989	11.823
36 Kr	12.649	14.112	1.586	1.637	81 Tl	72.872	82.576	10.269	12.213
37 Rb	13.395	14.961	1.694	1.752	82 Pb	74.969	84.936	10.552	12.614
38 Sr	14.165	15.836	1.807	1.872	83 Bi	77.108	87.343	10.839	13.024
39 Y	14.958	16.738	1.923	1.996	84 Po	79.290	89.800	11.131	13.447
40 Zr	15.775	17.668	2.042	2.124	85 At	81.520	92.300	11.427	13.876
41 Nb	16.615	18.623	2.166	2.257	86 Rn	83.780	94.870	11.727	14.316
42 Mo	17.479	19.608	2.293	2.395	87 Fr	86.100	97.470	12.031	14.770
43 Tc	18.367	20.619	2.424	2.538	88 Ra	88.470	100.130	12.340	15.236
44 Ru	19.279	21.657	2.559	2.683	89 Ac	90.884	102.850	12.652	15.713
45 Rh	20.216	22.724	2.697	2.834	90 Th	93.350	105.609	12.969	16.202
46 Pd	21.177	23.819	2.839	2.990	91 Pa	95.868	108.427	13.291	16.702
47 Ag	22.163	24.942	2.984	3.151	92 U	98.439	111.300	13.615	17.220

*Tabulation from J. B. Kortright and A. C. Thompson, X-Ray Data Booklet, http://xdb.lbl.gov/Section1/Sec_1-2.html; Original source: J. A. Bearden, "X-Ray Wavelengths," Rev. Mod. Phys. **1967**, 39, 78.*

K_{α}/K_{β} intensity ratio $\approx$ 5:1. For elements with Z > 50, L_{α}/L_{β} intensity ratio $\approx$ 1:1. Subscript 1 (as in $K_{\alpha 1}$) is written because there can be several closely spaced transitions of each type. You will only see one peak of each type with the resolution of a handheld X-ray fluorescence analyzer.

Figure 21-33 shows a handheld X-ray fluorescence analyzer designed for rapid qualitative and quantitative analysis in the field. Instead of using a monochromator, this spectrometer uses a *silicon drift detector* to measure both energy and the number of photons at that energy (expressed as counts per second). A sample is placed as close as possible to the X-ray tube and detector. Handheld instruments are commonly used to screen for lead in paint and toys, metals in ores, and toxic elements in soil and food. Museums use X-ray fluorescence for nondestructive characterization of artifacts and art to help determine authenticity and provenance.

Handheld X-ray fluorescence analyzers can detect levels as low as 1 ppm (1 μg/g) for heavy elements such as As, Pb, and Hg (Table 21-6). Detection limits are near 1 wt% for the light elements Mg, Al, and Si, which are not efficiently excited or detected with typical analyzers and whose fluorescence is absorbed by air. Lighter elements such as H, C, N, and O are not detected.

Although the X-ray intensity from handheld analyzers is far lower than those used in dental and medical X-ray equipment, *safety should be a prime consideration when using these devices*. In the laboratory, a handheld analyzer should be mounted in a lead-lined test stand to prevent user exposure to X-rays. In the field, you should remain behind the analyzer, which should not be pointed at a person.

For qualitative and semiquantitative analysis within a factor of ~2, no sample preparation is necessary and the analysis requires about a minute to perform. For analysis with a precision of 10% or better, samples should be homogenized (by grinding, for example) and calibration standards should be made in the same matrix as the sample. *Standard addition* of analyte to a finely ground sample provides the most accurate quantitative results.

Manufacturers' software can give false positives, false negatives, and incorrect element concentrations if you blindly accept what the instrument reports. *Always verify that both K peaks or both L peaks in Table 21-5 (±0.05 keV) are observed at the proper intensity ratio for an element purported to be present.* In Figure 21-31, you cannot conclude that iron is present unless you see both K_α and K_β peaks with relative intensities of ~5:1 according to the note in Table 21-5. You cannot conclude that lead is present unless you see both L_α and L_β peaks with relative intensities of ~1:1. For calcium, K_α is labeled at 3.69 keV. K_β is expected at 4.01 keV from Table 21-5 and a small peak at 4.01 keV is visible, but not labeled in the spectrum. In general, the region of the spectrum below 3 keV is not meaningful for a handheld analyzer, probably because the source and detector provide poor sensitivity at these energies.

A number of artifacts may be observed in a raw spectrum. A broad peak labeled *bremsstrahlung* between 15 and 35 keV in Figure 21-31 is due to backscattering of X-rays from the sample to the detector. Peaks from the X-ray source and secondary filters can appear in the sample spectrum at their normal energies (Rayleigh scattering), at slightly lower energies (called Compton scattering), or at both energies. Two photons from intense analyte fluorescence can arrive simultaneously at the detector to give weak "sum" peaks, which for Fe occur at $2K_\alpha$, $2K_\beta$, and $K_\alpha + K_\beta$. Silicon in the detector can absorb Si K_α energy (1.74 eV) to give weak "escape" peaks shifted by 1.74 eV to lower energy from strong analyte peaks. Peaks can appear from elements present in the sample matrix, sample holder, and materials used in the X-ray tube source, filters, and optical components. These artifacts complicate the interpretation of spectra. Manufacturers' user guides and a tutorial in the Analytical Sciences Digital Library provide practical guidance for spectral interpretation and methods of quantitative analysis.

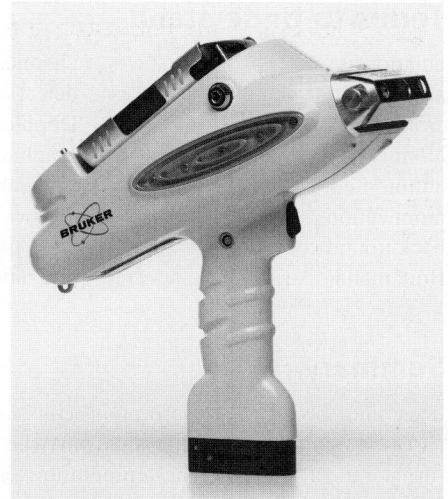

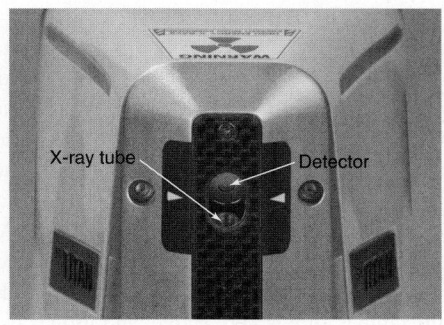

FIGURE 21-33 Handheld X-ray fluorescence analyzer for field analysis. Close-up shows proximity of X-ray tube and fluorescence detector. [Courtesy Bruker Elemental.]

TABLE 21-6	Approximate detection limits for a handheld XRF analyzer	
Elements	Range of atomic number (Z)	Detection limit (wt%)
Mg to Si	12–14	<1% (requires vacuum)
P	15	<0.5%
S to Ar	16–18	<200 ppm
K to Sc	19–21	<50 ppm
Ti to Cu	22–29	<10 ppm
Zn to Ru	30–44	<5 ppm
Rh to In	45–49	<10 ppm
Sn to Lu	50–71	<20 ppm
Hf to Au	72–79	<10 ppm
Hg to Pu	80–94	<5 ppm

Data from Olympus Innov-X estimated for 1–2 min test time for low-density samples, such as soil, powder, and liquid.

Terms to Understand

ablation	chemical interference	ionization suppressor	plasma
aerosol	detection limit	isobaric interference	pressure broadening
atomic absorption	Doppler effect	laser-induced breakdown	releasing agent
atomic emission	graphite furnace	spectroscopy	self-absorption
atomic fluorescence	Heisenberg uncertainty principle	matrix	spectral interference
atomization	hollow-cathode lamp	matrix modifier	X-ray fluorescence
background correction	inductively coupled plasma	nebulization	
Boltzmann distribution	ionization interference	piezoelectric crystal	

Summary

In atomic spectroscopy, absorption, emission, or fluorescence from gaseous atoms is measured. Liquids may be atomized by a plasma, a furnace, or a flame. Flame temperatures are usually in the range 2 300–3 400 K. The choice of fuel and oxidant determines the temperature of the flame and affects the extent of spectral, chemical, or ionization interference that will be encountered. Temperature instability affects atomization in atomic absorption and has an even larger effect on atomic emission, because the excited-state population is exponentially sensitive to temperature. An electrically heated graphite furnace requires less sample than a flame and has a lower detection limit. In an inductively coupled plasma, a radio-frequency induction coil heats Ar^+ ions to 6 000–10 000 K. At this high temperature, emission is observed from electronically excited atoms and ions. There is little chemical interference in an inductively coupled plasma, the temperature is very stable, and little self-absorption is observed.

Plasma emission spectroscopy does not require a light source and is capable of measuring ~70 elements simultaneously with a charge injection device detector. Background correction for a given emission peak is based on subtracting the intensity of neighboring pixels in the detector. The lowest detection limits are obtained by directing the plasma into a mass spectrometer that separates and measures ions from the plasma. In flame and furnace atomic absorption spectroscopy, a hollow-cathode lamp made of the analyte element provides spectral lines sharper than those of the atomic vapor. The inherent linewidth of atomic lines is limited by the Heisenberg uncertainty principle. Lines in a flame, furnace, or plasma are broadened by a factor of 10–100 by the Doppler effect and by atomic collisions. Correction for background emission from the flame is possible by electrically pulsing the lamp on and off or mechanically chopping the beam. Light scattering and spectral background can be subtracted by measuring absorption with a deuterium lamp or by Zeeman background correction, in which the atomic energy levels are alternately shifted in and out of resonance with the lamp frequency by a magnetic field. Chemical interference can be reduced by addition of releasing agents, which prevent the analyte from reacting with interfering species. Ionization interference in flames is suppressed by adding easily ionized elements such as Cs.

In inductively coupled plasma–mass spectrometry, isobaric interference occurs between species with the same mass and charge. Interference can be eliminated if mass spectral resolution is sufficiently great, but this is not often the case. A dynamic reaction cell reduces isobaric interference by using an exothermic reaction of a gas such as NH_3, N_2O, or CO to remove interfering molecular ions such as ArO^+ or to transform analyte into a molecular ion that can be measured without interference.

For qualitative and semiquantitative analysis, solids and liquids can be sampled by laser ablation. Ablated material is swept through an inductively coupled plasma to detect multiple elements by atomic emission or mass spectrometry. When emission is measured, the technique is called laser-induced breakdown spectroscopy. Matrix-matched standards are necessary for semiquantitative analysis.

X-ray fluorescence is the emission of X-rays by a material after it has absorbed X-rays. Handheld spectrometers enable qualitative and semiquantitative measurements of elemental composition in the field. Each element emits characteristic, narrow X-ray lines, whose energies identify the element and whose integrated intensities are proportional to the concentration of the element. Many elements can be measured at levels below 0.01 wt% with ~10% accuracy when suitable care is exercised. X-rays from the spectrometer source ionize electrons from the inner (K and L) shells of atoms. When outer electrons fall into the vacancies, characteristic X-rays called K_α, K_β, L_α, and L_β are emitted. You must verify that both expected lines for each element are observed in the spectrum before you conclude that the element is present.

Exercises

21-A. Li was determined by atomic emission with the method of standard addition. Prepare a standard addition graph (Section 5-3) to find the concentration of Li and its uncertainty in pure unknown. The Li standard contained 1.62 μg Li/mL.

Unknown (mL)	Standard (mL)	Final volume (mL)	Emission intensity (arbitrary units)
10.00	0.00	100.0	309
10.00	5.00	100.0	452
10.00	10.00	100.0	600
10.00	15.00	100.0	765
10.00	20.00	100.0	906

21-B. Mn was used as an internal standard for measuring Fe by atomic absorption. A standard mixture containing 2.00 μg Mn/mL and 2.50 μg Fe/mL gave a quotient (Fe signal/Mn signal) = 1.05/1.00. A mixture with a volume of 6.00 mL was prepared by mixing 5.00 mL of unknown Fe solution with 1.00 mL containing 13.5 μg Mn/mL. The absorbance of this mixture at the Mn wavelength was 0.128, and the absorbance at the Fe wavelength was 0.185. Find the molarity of the unknown Fe solution.

21-C. (a) The atomic absorption signal shown here was obtained with 0.048 5 μg Fe/mL in a graphite furnace. The root-mean-square noise in the baseline, measured by the instrument's computer, is $s = 0.30$ vertical units, where each horizontal line on the chart is 1 vertical unit. Estimate where the baseline is beneath the tall signal and measure the height of the signal. Estimate the detection limit for Fe, defined as the concentration of Fe that gives a signal height of $3s$.

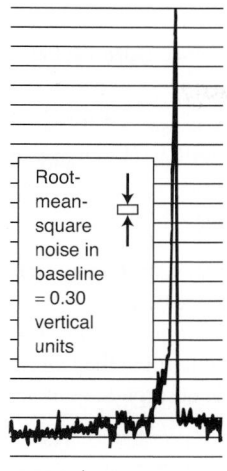

Root-mean-square noise in baseline = 0.30 vertical units

(b) Seven replicate measurements of a standard containing 1.00 ng Hg/L gave readings of 0.88, 1.48, 0.94, 1.12, 1.03, 1.40, and 1.14 ng/L in cold vapor atomic absorption (Box 21-1). From Equations 5-5 and 5-6, estimate the detection and quantitation limits. (Note that in Equations 5-5 and 5-6, the quotient s/m is the standard deviation in concentration.)

21-D. The measurement of Li in brine (salt water) is used by geochemists to help determine the origin of this fluid in oil fields. Flame atomic emission and absorption of Li are subject to interference by scattering, ionization, and overlapping spectral emission from other elements. Atomic absorption analysis of replicate samples of a marine sediment gave results in the table.

Sample and treatment	Li found (µg/g)	Analytical method	Flame type
1. None	25.1	standard curve	air/C_2H_2
2. Dilute to 1/10 with H_2O	64.8	standard curve	air/C_2H_2
3. Dilute to 1/10 with H_2O	82.5	standard addition	air/C_2H_2
4. None	77.3	standard curve	N_2O/C_2H_2
5. Dilute to 1/10 with H_2O	79.6	standard curve	N_2O/C_2H_2
6. Dilute to 1/10 with H_2O	80.4	standard addition	N_2O/C_2H_2

SOURCE: B. Baraj, L. F. H. Niencheski, R. D. Trapaga, R. G. França, V. Cocoli, and D. Robinson, "Interference in the Flame Atomic Absorption Determination of Li," Fresenius J. Anal. Chem. **1999**, 364, 678.

(a) Suggest a reason for the increasing apparent concentration of Li in samples 1 through 3.

(b) Why do samples 4 through 6 give an almost constant result?

(c) What value would you recommend for reporting the real concentration of Li in the sample?

21-E. Assign as many of the peaks as you can in the X-ray fluorescence spectrum shown here. For each K_α peak, state where the K_β peak should appear and state whether there is a plausible peak at the K_β position. There must be a K_β for every K_α, but the K_β intensity should be ~1/5 of the K_α intensity.

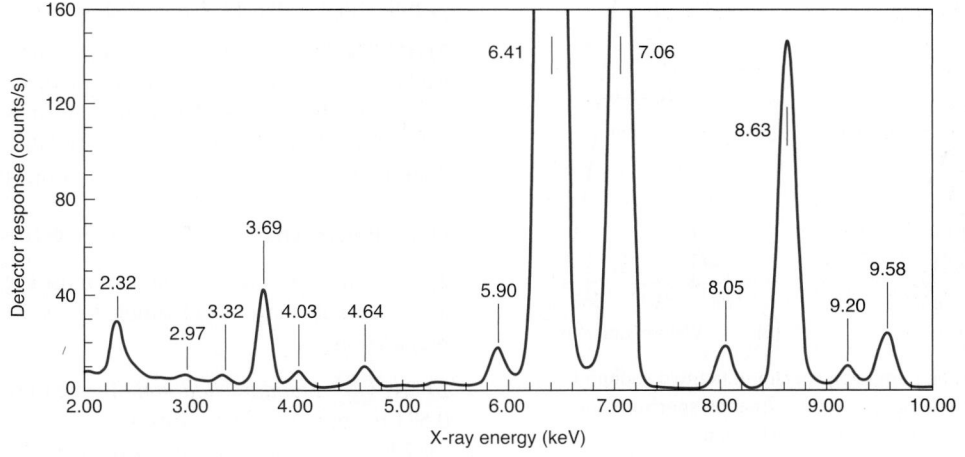

Partial X-ray fluorescence spectrum of a soil sample. [Data from P. T. Palmer, San Francisco State University.]

Problems

Techniques of Atomic Spectroscopy

21-1. In which technique, atomic absorption or atomic emission, is flame temperature stability more critical? Why?

21-2. State the advantages and disadvantages of a furnace compared with a flame in atomic absorption spectroscopy.

21-3. Figure 21-10 shows a temperature profile for a furnace atomic absorption experiment. Explain the purpose of each different part of the heating profile.

21-4. State the advantages and disadvantages of the inductively coupled plasma compared with a flame in atomic spectroscopy.

21-5. Explain what is meant by the Doppler effect. Rationalize why Doppler broadening increases with increasing temperature and decreasing mass in Equation 21-5.

21-6. Explain how the following background correction techniques work: **(a)** beam chopping; **(b)** deuterium lamp; **(c)** Zeeman.

21-7. Explain what is meant by spectral, chemical, ionization, and isobaric interference.

21-8. Bone consists of the protein collagen and the mineral hydroxyapatite, $Ca_{10}(PO_4)_6(OH)_2$. The Pb content of archaeological human skeletons measured by graphite furnace atomic absorption sheds light on customs and economic status of individuals in historical times.[37] Explain why La^{3+} is added to bone samples to suppress matrix interference in Pb analysis.

21-9. (a) Explain the purpose of the dynamic reaction cell in Figure 21-30.

(b) In geologic strontium isotopic analysis, there is isobaric interference between $^{87}Rb^+$ and $^{87}Sr^+$. A dynamic reaction cell with CH_3F converts Sr^+ to SrF^+ but does not convert Rb^+ to RbF^+. How does this reaction eliminate interference?

21-10. What is the purpose of a matrix-matched standard in laser-induced breakdown spectroscopy?

21-11. Laser atomic fluorescence excitation and emission spectra of sodium in an air-acetylene flame are shown here. In the *excitation* spectrum, the laser (linewidth = 0.03 nm) was scanned through various wavelengths, while the detector monochromator (bandwidth = 1.6 nm) was held fixed near 589 nm. In the *emission* spectrum, the laser was fixed at 589.0 nm, and the detector monochromator wavelength was varied. Explain why the emission spectrum gives one broad band, whereas the excitation spectrum gives two sharp lines. How can the excitation linewidths be much narrower than the detector monochromator bandwidth?

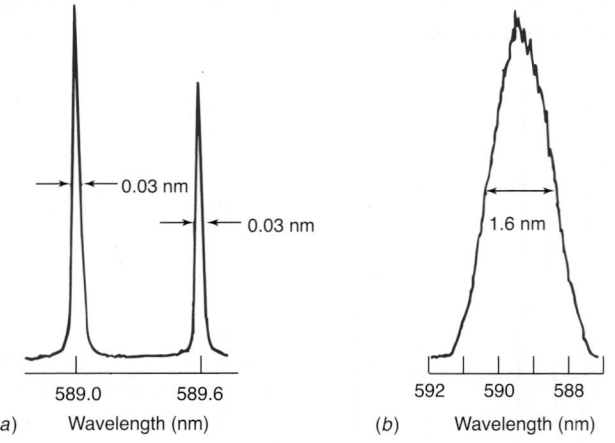

(a) Wavelength (nm)   (b) Wavelength (nm)

Fluorescence excitation and emission spectra of the two sodium D lines in an air-acetylene flame. (*a*) In the excitation spectrum, the laser was scanned. (*b*) In the emission spectrum, the monochromator was scanned. The monochromator slit width was the same for both spectra. [Data from S. J. Weeks, H. Haraguchi, and J. D. Winefordner, "Improvement of Detection Limits in Laser-Excited Atomic Fluorescence Flame Spectrometry," *Anal. Chem.* **1978,** *50,* 360.]

21-12. Concentrations (pg per g of snow) of metals by atomic fluorescence in the Agassiz Ice Cap in Greenland for the period 1988–1992 are[38] Pb, $1.0_4 (\pm 0.1_7) \times 10^2$; Tl, $0.43 \pm 0.08_7$; Cd, $3.5 \pm 0.8_7$; Zn, $1.7_4 (\pm 0.2_6) \times 10^2$; and Al, $6._1 (\pm 1._7) \times 10^3$. The mean annual snowfall was 11.5 g/cm². Calculate the mean annual flux of each metal in units of ng/cm². *Flux* means how much metal falls on each cm².

21-13. Calculate the emission wavelength (nm) of excited atoms that lie 3.371×10^{-19} J per molecule above the ground state.

21-14. Derive the entries for 500 nm in Table 21-3. Find N^*/N_0 at 6 000 K if $g^* = 3$ and $g_0 = 1$?

21-15. Calculate the Doppler linewidth for the 589-nm line of Na and for the 254-nm line of Hg, both at 2 000 K.

21-16. The first excited state of Ca is reached by absorption of 422.7-nm light.

(a) Find the energy difference (kJ/mol) between ground and excited states.

(b) The degeneracies are $g^*/g_0 = 3$ for Ca. Find N^*/N_0 at 2 500 K.

(c) By what percentage will N^*/N_0 change with a 15-K rise in temperature?

(d) Find N^*/N_0 at 6 000 K.

21-17. An *electron volt* (eV) is the energy change of an electron moved through a potential difference of 1 volt: $eV = (1.602 \times 10^{-19} \text{ C})(1 \text{ V}) = 1.602 \times 10^{-19}$ J per electron = 96.49 kJ per mole of electrons. Use the Boltzmann distribution to fill in the table and explain why Br is not readily observed in atomic absorption or atomic emission.

	Na	Cu	Br
Excited state energy (eV)	2.10	3.78	8.04
Wavelength (nm)			
Degeneracy ratio (g^*/g_0)	3	3	2/3
N^*/N_0 at 2 600 K in flame			
N^*/N_0 at 6 000 K in plasma			

21-18. MgO prevents premature evaporation of Al in a furnace by maintaining the aluminum as Al_2O_3. Another type of matrix modifier prevents loss of signal from the atom X that readily forms the molecular carbide XC in a graphite furnace (a source of carbon). For example, adding yttrium to a sample containing barium increases the Ba signal by 30%. The bond dissociation energy of YC is greater than that of BaC. Explain what is happening to increase the Ba signal.

21-19. The 20-μm-radius laser ablation pit in Figure 21-28 was created by a laser pulse with a duration of 10 ns and an energy of 2.4 mJ. Express the laser power density in units of W/cm². Recall that 1 W = 1 J/s. If the depth of the pit is 1 μm and the density of the material is 4 g/mL, how much mass is removed by one pulse?

Quantitative Analysis by Atomic Spectroscopy

21-20. Why is an internal standard most appropriate for quantitative analysis when unavoidable sample losses are expected during sample preparation?

21-21. *Standard addition.* To measure Ca in breakfast cereal, 0.521 6 g of crushed Cheerios was ashed in a crucible at 600°C in air for 2 h.[39] The residue was dissolved in 6 M HCl, quantitatively transferred to a volumetric flask, and diluted to 100.0 mL. Then 5.00-mL aliquots were transferred to 50-mL volumetric flasks. Each was treated with standard Ca^{2+} (containing 20.0 μg/mL), diluted to volume with H_2O, and analyzed by flame atomic absorption. Construct a standard addition graph and use the method of least squares to find the *x*-intercept and its uncertainty. Find wt% Ca in Cheerios and its uncertainty.

Ca^{2+} standard (mL)	Absorbance	Ca^{2+} standard (mL)	Absorbance
0	0.151	8.00	0.388
1.00	0.185	10.00	0.445
3.00	0.247	15.00	0.572
5.00	0.300	20.00	0.723

21-22. *Internal standard.* A solution was prepared by mixing 10.00 mL of unknown (X) with 5.00 mL of standard (S) containing 8.24 μg S/mL and diluting the mixture to 50.0 mL. The measured signal quotient was (signal due to X/signal due to S) = 1.690/1.000.

(a) In a separate experiment, in which the concentrations of X and S were equal, the quotient was (signal due to X/signal due to S) = 0.930/1.000. What is the concentration of X in the unknown?

(b) Answer the same question if, in a separate experiment in which the concentration of X was 3.42 times the concentration of S, the quotient was (signal due to X/signal due to S) = 0.930/1.000.

21-23. [image] Potassium standards gave the following emission intensities at 404.3 nm. Emission from the unknown was 417. Find $[K^+]$ and its uncertainty in the unknown.

Sample (μg K/mL):	0	5.00	10.00	20.00	30.00
Relative emission:	0	124	243	486	712

21-24. *Quality assurance.* Tin is leached (dissolved) into canned foods from the tin-plated steel can.[40] For analysis by inductively coupled plasma–atomic emission, food is digested by microwave heating in a Teflon bomb (Figure 28-7) in three steps with HNO_3, H_2O_2, and HCl.

(a) CsCl is added to the final solution at a concentration of 1 g/L. What is the purpose of the CsCl?

(b) [image] Calibration data are shown in the table. Find the slope and intercept and their standard deviations and R^2, which is a measure of the goodness of fit of the data to a line. Draw the calibration curve.

Sn (μg/L)	Emission at 189.927 nm	Sn (μg/L)	Emission at 189.927 nm
0	4.0	40.0	31.1
10.0	8.5	60.0	41.7
20.0	19.6	100.0	78.8
30.0	23.6	200.0	159.1

(c) Interference by high concentrations of other elements was assessed at different emission lines of Sn. Foods containing little tin were digested and spiked with Sn at 100.0 μg/L. Then other elements were deliberately added. The table shows selected results. Which elements interfere at each of the two wavelengths? Which wavelength is preferred for the analysis?

Element added at 50 mg/L	Sn found (μg/L) with 189.927-nm emission line	Sn found (μg/L) with 235.485-nm emission line
None	100.0	100.0
Ca	96.4	104.2
Mg	98.9	92.6
P	106.7	104.6
Si	105.7	102.9
Cu	100.9	116.2
Fe	103.3	intense emission
Mn	99.5	126.3
Zn	105.3	112.8
Cr	102.8	76.4

(d) *Limits of detection and quantitation.* The slope of the calibration curve in part **(b)** is 0.782 units per (μg/L) of Sn. Food containing little Sn gave a mean signal of 5.1 units for seven replicates. Food spiked with 30.0 μg Sn/L gave a mean signal of 29.3 units with a standard deviation of 2.4 units for seven replicates. Use Equations 5-5 and 5-6 to estimate the limits of detection and quantitation.

(e) A 2.0-g food sample was digested and eventually diluted to 50 mL for analysis. Express the limit of quantitation from part **(d)** in terms of mg Sn/kg food.

21-25. Titanocene dichloride, $(\pi\text{-}C_5H_5)_2TiCl_2$, is a potential antitumor drug thought to be carried to cancer cells by the protein transferrin (Figure 18-8). To measure the Ti(IV) binding capacity of transferrin, the protein was treated with excess titanocene dichloride. After allowing time for Ti(IV) binding to the protein, excess small molecules were removed by dialysis (Demonstration 27-1). The protein was then digested with 2 M NH_3 and used to prepare a series of solutions with standard additions for chemical analysis. All solutions were made to the same total volume. Titanium and sulfur in each solution were measured by inductively coupled plasma-atomic emission spectrometry, with results in the table. Each transferrin molecule contains 39 sulfur atoms. Find the molar ratio Ti/transferrin in the protein.

Added Ti (mg/L)	ICP-AES signal	Added S (mg/L)	ICP-AES signal
0	0.86	0	0.017 4
3.00	1.10	37.0	0.022 1
6.00	1.34	74.0	0.026 8
12.0	1.82	148.0	0.036 2

Data derived from A. Cardona and E. Meléndez, "Determination of the Titanium Content of Human Transferrin by Inductively-Coupled Plasma-Atomic-Emission Spectroscopy," Anal. Bioanal. Chem. 2006, 386, 1689.

X-ray Fluorescence

21-26. Explain why X-ray fluorescence is observed when matter absorbs X-rays of sufficient energy. Why does each element have a unique X-ray signature?

21-27. Where would K_β emission peaks for Ti, Se, and Zr be found in Figure 21-31? Why are they not labeled?

21-28. Why are L_α and L_β peaks, but not K_α and K_β peaks for lead identified in Figure 21-31? Why are L_α and L_β peaks not identified for iron in Figure 21-31?

21-29. How much energy in kJ/mol is released when nitrogen emits K_α radiation at 0.392 keV? Compare the K_α energy to 945 kJ/mol, which is the energy required to break the triple bond in N_2 (one of the strongest chemical bonds).

21-30. Identify as many of peaks as you can in the X-ray fluorescence spectrum (next page) of soil from the location where water from the roof of a house drips onto the ground. For each K_α peak, state where the K_β peak should appear and state whether there is a plausible peak at the K_β position. There must be a K_β for every K_α, but the K_β intensity should only be ~1/5 of the K_α intensity.

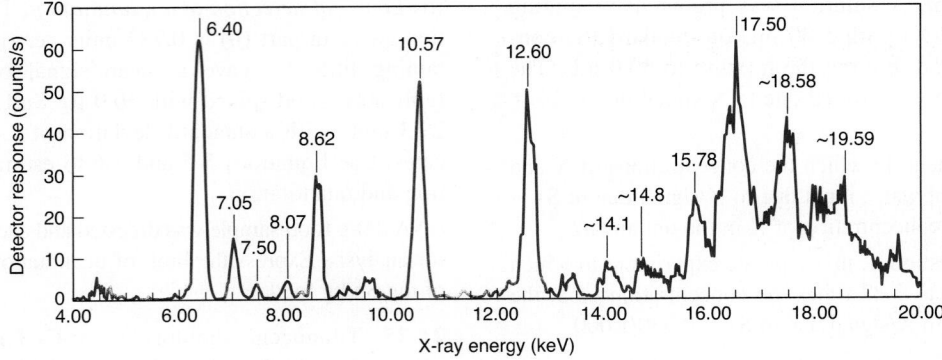

Spectrum for Problem 21-30.

X-ray fluorescence spectrum of soil taken from the location where water drips from the roof of a house.
[Data from S. J. Bachofer, "Sampling the Soils Around a Residence Containing Lead-Based Paints: An X-ray Fluorescence Experiment," *J. Chem. Ed.* **2008,** *85,* 980 was smoothed by Savitzky-Golay 5-point polynomial Equation 20-15.]

21-31. Ayurveda is a system of medicine practiced in India. In one study, 20% of U.S.- and Indian-manufactured Ayurvedic medicines purchased through the Internet in 2005 contained detectable levels of several toxic elements.[41] Identify as many peaks as you can in the X-ray fluorescence spectrum of the Ayurvedic medicine shown here.

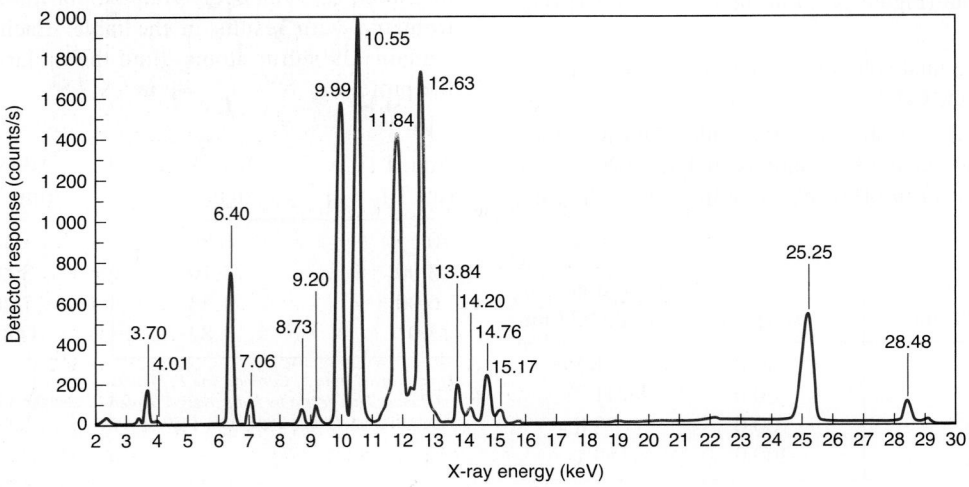

Spectrum for Problem 21-31.

X-ray fluorescence spectrum of Ayurvedic medicine. [Data from P. T. Palmer, San Francisco State University.]

22 Mass Spectrometry

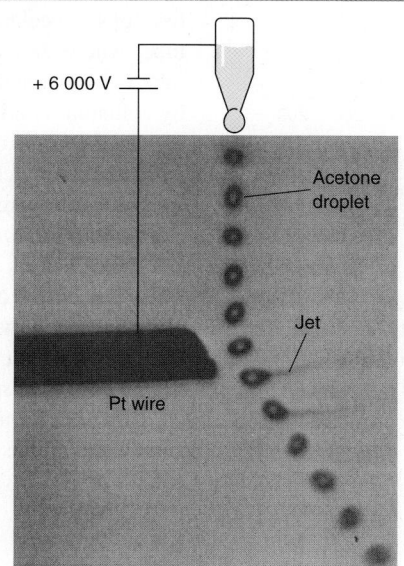

Electrospray is one method for expelling charged protein molecules into the gas phase for mass spectrometry. In the experiment shown here, droplets of acetone with a diameter of 16 μm fall past a Pt wire held at +6 000 V with respect to the nozzle from which the droplets came. High voltage creates a glowing electric corona discharge (a plasma containing electrons and positive ions) around the wire, but the discharge is not visible in this photograph. Droplets falling through the discharge become positively charged and are repelled by the wire, thus deflecting their path to the right. When positively charged droplets come close to the wire, we see a mist of liquid jetting away from the positively charged wire. Microscopic droplets in the mist rapidly evaporate. If the liquid had been an aqueous protein solution, the evaporated droplets would leave charged protein molecules in the gas phase.

Francis W. Aston (1877–1945) built a "mass spectrograph" in 1919 that could separate ions differing in mass by 1% and focus them onto a photographic plate. He found that neon consists of two isotopes (^{20}Ne and ^{22}Ne) and went on to discover 212 of the 281 naturally occurring isotopes. He received the Nobel Prize for Chemistry in 1922.

Mass spectrometry has long been used to measure isotopes and decipher organic structures. Hydrogen and oxygen isotopes in Antarctic ice cores provide a record of Earth's temperature, as shown at the opening of Chapter 10. The back cover of this book shows that isotopic composition of carbonate in the teeth of some Jurassic dinosaurs suggests, surprisingly, that they were not cold blooded like modern reptiles. Mass spectrometry can elucidate the amino acid sequence in a protein,[1] the sequence of nucleic acids in DNA, the structure of complex carbohydrates, and the types of lipids in a single organism. Mass spectrometry can measure masses of individual cells[2] and viruses.[3] Mass spectrometry is the most powerful detector for chromatography, offering both qualitative and quantitative information, providing high sensitivity, and distinguishing different substances with the same retention time. Chromatography–mass spectrometry is a key tool for guarding the safety of our food supply by monitoring for toxic substances such as pesticide residues.

22-1 What Is Mass Spectrometry?

Mass spectrometry is a technique for studying the masses of atoms or molecules or fragments of molecules.[1,4] A **mass spectrum** displays the number of ions detected at each value of **mass-to-charge ratio,** *m/z.* To obtain a mass spectrum, gaseous species desorbed from condensed phases are ionized, the ions are accelerated by an electric field, and then ions are separated according to their mass-to-charge ratio. If all charges are +1, then *m/z* is numerically equal to

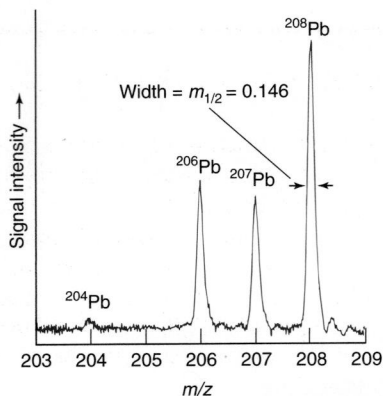

FIGURE 22-1 Mass spectrum showing natural isotopes of Pb. [Data from Y. Su, Y. Duan, and Z. Jin, "Development and Evaluation of a Glow Discharge Microwave-Induced Plasma Tandem Source for Time-of-Flight Mass Spectrometry," *Anal. Chem.* **2000,** *72,* 5600.] **Variability of isotopic abundances in Pb from natural sources creates a large uncertainty in the atomic mass (207.2 ± 0.1) in the periodic table.**

the mass. If an ion has a charge of +2, for example, then m/z is half of the mass. The mass spectrum in Figure 22-1 displays detector response versus m/z, showing four natural isotopes of Pb^+ ions. The area of each peak is proportional to the abundance of each isotope. Box 22-1 defines *nominal mass*, which is the mass we usually speak of in this chapter.

The three essential components of any mass spectrometer are (1) a source of ions, (2) a mass separator, and (3) a detector. Figure 22-2 shows a **magnetic sector mass spectrometer,** which uses a magnetic field to allow ions of a selected m/z to pass from the ion source to the detector.[5] Gaseous molecules entering at the upper left are converted into ions (usually positive ions), accelerated by an electric field in the ion source, and expelled into the analyzer tube, where they encounter a magnetic field perpendicular to their direction of travel. The tube is maintained under high vacuum ($\sim 10^{-5}$ Pa $= 10^{-10}$ bar) so that ions are not deflected by collision with background gas molecules. The magnet deflects ions toward the detector at the far end of the tube (see Box 22-2). Heavy ions are not deflected enough and light ones are deflected too much to reach the detector. The spectrum of masses is obtained by varying the magnetic field strength.

At the *electron multiplier* detector[6] in Figure 22-2, each arriving ion starts a cascade of electrons, just as a photon starts a cascade of electrons in a photomultiplier tube (Figure 20-14). A series of dynodes multiplies the number of electrons by $\sim 10^6$ to 10^8 before they reach the anode where current is measured. The mass spectrum shows detector response as a function of m/z selected by the magnetic field. Some detectors have a digital mode in which each count corresponds to one ion reaching the detector. Most detectors display an analog signal proportional to the rate at which ions reach the detector. Mass spectrometers can measure negative ions by reversing voltages where the ions are formed and detected.

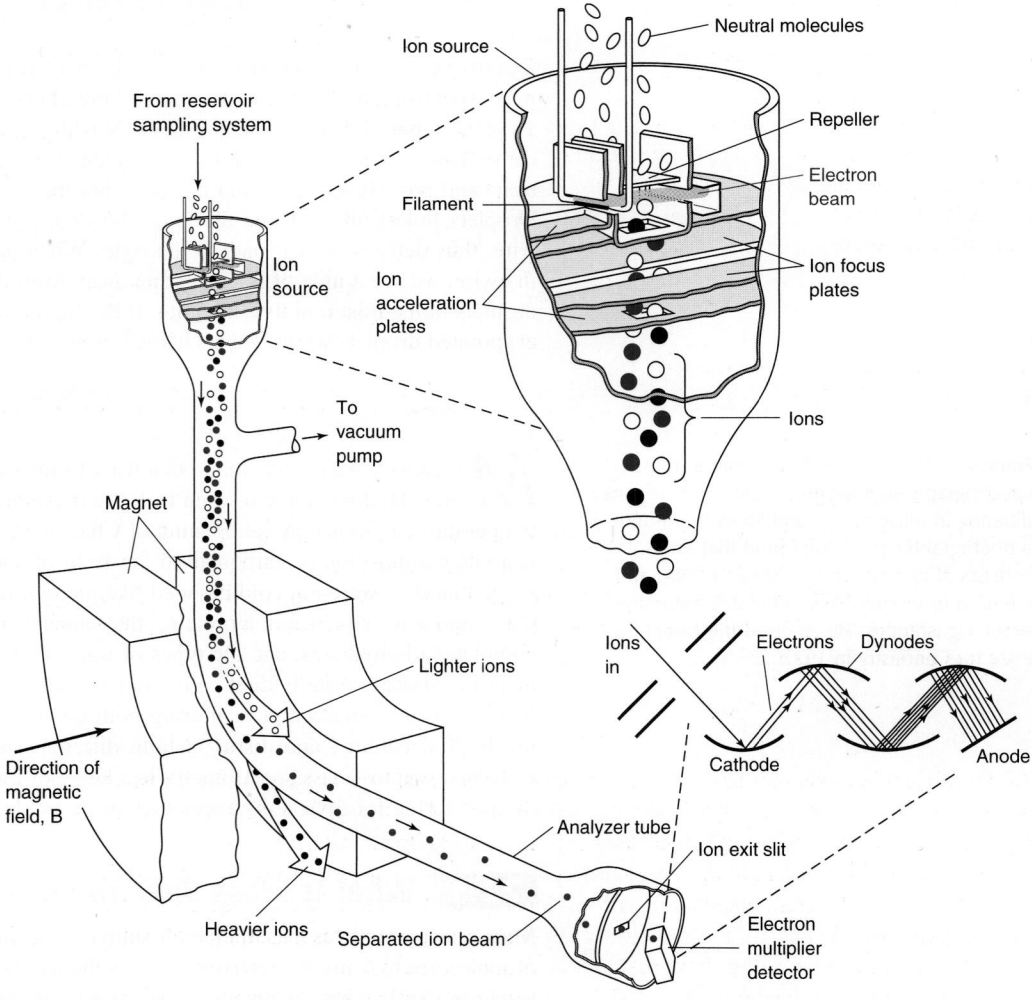

FIGURE 22-2 Magnetic sector mass spectrometer. [Information from F. W. McLafferty, *Interpretation of Mass Spectra* (New York: Benjamin, 1966).]

Atomic mass (also called *atomic weight*) is the weighted average of the masses of the isotopes of an element. The unit of atomic mass, called the *unified atomic mass unit* (u) or the *dalton* (Da), is defined as 1/12 of the mass of an atom of ^{12}C. Bromine consists of 50.69% ^{79}Br with a mass of 78.918 34 Da and 49.31% ^{81}Br with a mass of 80.916 29 Da. Therefore, its atomic mass is (0.506 9)(78.918 34 Da) + (0.493 1)(80.916 29 Da) = 79.904 Da. There is no atom of Br with a mass of 79.904 Da, which is only a weighted average. The units u and Da are equivalent to the unit g/mol.

The **molecular mass** of a molecule or an ion is the sum of atomic masses listed in the periodic table. For bromoethane, C_2H_5Br,

the molecular mass is (2 × 12.010 6 Da) + (5 × 1.007 98 Da) + (1 × 79.904 Da) = 108.965 Da.

The **nominal mass** of a molecule or ion is the *integer* mass of the species with the most abundant isotope of each of the constituent atoms. For carbon, hydrogen, and bromine the most abundant isotopes are ^{12}C, ^{1}H, and ^{79}Br. Therefore, the nominal mass of C_2H_5Br is (2 × 12) + (5 × 1) + (1 × 79) = 108. The *monoisotopic mass* is the exact mass of the species with the most abundant isotope of each of the constituent atoms. For C_2H_5Br, the monoisotopic mass is (2 × 12 Da) + (5 × 1.007 825 Da) + (1 × 78.918 34 Da) = 107.957 46 Da.

Electron ionization in the ion source of the mass spectrometer in Figure 22-2 creates positive ions, M^{z+}, with different masses. Let the mass of an ion be m and let its charge be $+ze$ (where e is the magnitude of the charge of an electron). When the ion is accelerated through a potential difference V by the ion acceleration plates in the ion source, it acquires a kinetic energy equal to the electric potential difference:

$$\underbrace{\frac{1}{2}mv^2}_{\substack{\text{Kinetic energy} \\ (v = \text{velocity})}} = \underbrace{zeV}_{\text{Potential energy}} \Rightarrow v = \sqrt{\frac{2zeV}{m}} \qquad \text{(A)}$$

An ion with charge ze and velocity v traveling perpendicular to a magnetic field B experiences a force $zevB$ that is perpendicular to both the velocity vector and the magnetic field vector. This force deflects the ion through a circular path of radius r. The centripetal force (mv^2/r) required to deflect the particle is provided by the magnetic field.

$$\underbrace{\frac{mv^2}{r}}_{\text{Centripetal force}} = \underbrace{zevB}_{\text{Magnetic force}} \Rightarrow v = \frac{zeBr}{m} \qquad \text{(B)}$$

Equating velocities from Equations A and B gives

$$\frac{zeBr}{m} = \sqrt{\frac{2zeV}{m}} \Rightarrow \boxed{\frac{m}{z} = \frac{eB^2r^2}{2V}} \qquad \text{(C)}$$

Equation C tells us the radius of curvature of the path traveled by an ion with mass m and charge z. The radius of curvature is fixed by the geometry of the hardware. Ions can be selected to reach the detector by adjusting the magnetic field, B, or the accelerating voltage, V. Normally, B is varied to select ions and V is fixed near 3 000 V. Transmission of ions and detector response both decrease when V is decreased.

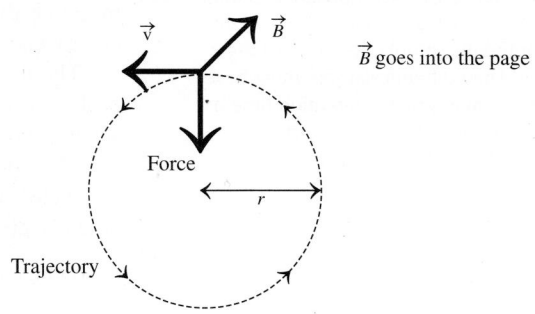

$\vec{B}$ goes into the page

Electron Ionization

Molecules entering the ion source in Figure 22-2 are converted into ions by **electron ionization.** Electrons emitted from a hot filament (like the one in an incandescent light bulb) are accelerated through 70 V before interacting with incoming molecules. Some (~0.01%) molecules (M) absorb as much as 12–15 electron volts (1 eV = 96.5 kJ/mol), which is enough for ionization:

$$\text{M} + \underset{70 \text{ eV}}{\text{e}^-} \rightarrow \underset{\text{molecular ion}}{\text{M}^{+\bullet}} + \underset{\sim 55 \text{ eV}}{\text{e}^-} + \underset{0.1 \text{ eV}}{\text{e}^-}$$

Almost all stable molecules have an even number of electrons. When one electron is lost, the resulting cation with one unpaired electron is designated $M^{+\bullet}$, the **molecular ion.** After ionization, $M^{+\bullet}$ usually has enough internal energy (~1 eV) to break into fragments.

A small positive potential on the repeller plate of the ion source pushes ions toward the analyzer tube, and a small potential on the ion focus plates creates a focused beam. A potential difference of ~1 000–10 000 V between the ion acceleration plates imparts a high velocity to ions as they are expelled from the bottom of the ion source.

The electron kinetic energy of 70 eV is much greater than the ionization energy of molecules. Consider formaldehyde in Figure 22-3, whose molecular orbitals were shown in Figure 18-13. The most easily lost electron comes from a nonbonding ("lone pair") orbital

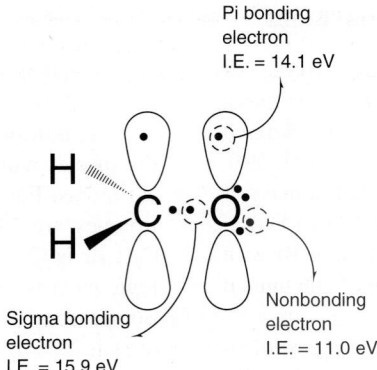

FIGURE 22-3 Ionization energies (I.E.) of valence electrons in formaldehyde. [Data from C. R. Brundle, M. B. Robin, N. A. Kuebler, and H. Basch, "Perfluoro Effects in Photoelectron Spectroscopy," *J. Am. Chem. Soc.* **1972,** *94,* 1451.] Here are other first ionization energies:

$CH_3CH_2CH_2CH_3$ 10.6 eV (sigma)

$CH_2\!=\!CHCH_2CH_3$ 9.6 eV (pi)

$(CH_3CH_2)_2\ddot{O}$ 9.6 eV (nonbonding)

8.6 eV (nonbonding)

9.2 eV (pi)

centered on oxygen, with an ionization energy of 11.0 eV. To remove a pi bonding electron from neutral formaldehyde requires 14.1 eV, and to remove the highest energy sigma bonding electron from the neutral molecule requires 15.9 eV.

Interaction with a 70-eV electron most likely removes the electron with lowest ionization energy. The resulting molecular ion, $M^{+\bullet}$, can have so much extra energy that it breaks into fragments. There might be so little $M^{+\bullet}$ that its peak is small or absent in the mass spectrum. The electron ionization mass spectrum at the left side of Figure 22-4 does not exhibit an $M^{+\bullet}$ peak that would be at *m/z* 226. Instead, there are peaks at *m/z* 197, 156, 141, 112, 98, 69, and 55, arising from fragmentation of $M^{+\bullet}$. These peaks provide clues about the structure of the molecule. A computer search is commonly used to match the spectrum of an unknown to similar spectra in a library.[7]

If you lower the kinetic energy of electrons in the ionization source to, say, 20 eV, there will be a lower yield of ions and less fragmentation. You would likely observe a greater abundance of molecular ions. We customarily use 70 eV because it gives reproducible fragmentation patterns that can be compared with library spectra. If, prior to ionization, a gas is cooled by going through a supersonic expansion nozzle, the cold molecules have little vibrational energy and the molecular ion can be observed.[8]

The most intense peak in a mass spectrum is called the **base peak.** Intensities of other peaks are expressed as a percentage of the base peak intensity. In the electron ionization spectrum in Figure 22-4, the base peak is at *m/z* 141.

Electrons with energy near 70 eV almost exclusively create cationic molecular products. If the electron energy is lower, it is possible to form negative ions from molecules with a sufficiently great electron affinity:

Resonance capture:

$$M \quad + \quad e^- \quad \rightarrow \quad M^{-\bullet}$$
$$_{<0.1 \text{ eV}}$$

Dissociative capture:

$$XY \quad + \quad e^- \quad \rightarrow \quad X \quad + \quad Y^{-\bullet}$$
$$_{0.1 - 10 \text{ eV}}$$

> A reasonable match of an experimental spectrum to one in the computer library is *not* proof of molecular structure—it is just a clue. You must be able to explain all major peaks (and even minor peaks at high *m/z*) in the spectrum in terms of the proposed structure, and you should obtain a matching spectrum from an authentic sample before reaching a conclusion. The authentic sample must have the same chromatographic retention time as the proposed unknown. Many isomers produce nearly identical mass spectra.

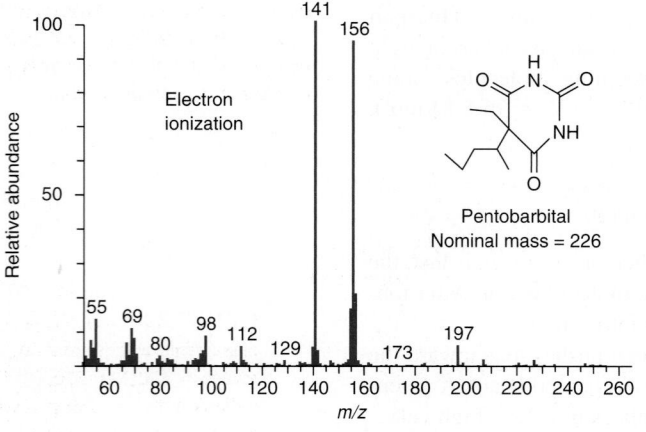

Electron ionization

Pentobarbital
Nominal mass = 226

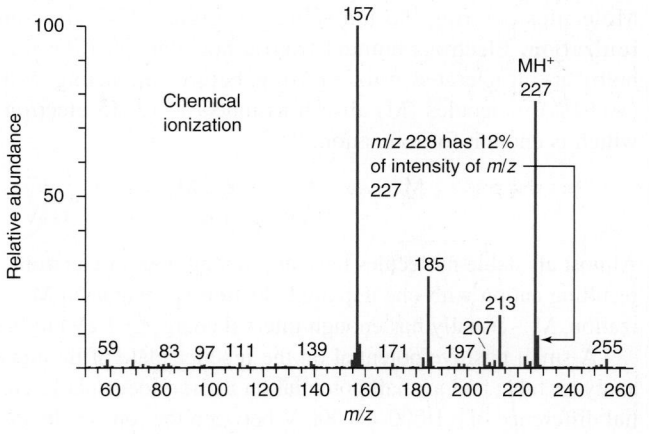

Chemical ionization

m/z 228 has 12% of intensity of *m/z* 227

MH^+ 227

FIGURE 22-4 Mass spectra of the sedative pentobarbital from electron ionization (left) or chemical ionization (right). The molecular ion ($M^{+\bullet}$, *m/z* 226) is not evident with electron ionization. The dominant ion from chemical ionization is MH^+. The peak at *m/z* 255 in the chemical ionization spectrum is from $M(C_2H_5)^+$. [Data from Varian Associates, Sunnyvale, CA.]

CHAPTER 22 Mass Spectrometry

Chemical ionization produces less fragmentation than electron ionization. For chemical ionization, the ionization source is filled with a *reagent gas* such as methane, isobutane, or ammonia, at a pressure of ~1 mbar. Energetic electrons (100–200 eV) convert CH_4 into a variety of reactive products:

$$CH_4 + e^- \rightarrow CH_4^{+\cdot} + 2e^-$$
$$CH_4^{+\cdot} + CH_4 \rightarrow CH_5^+ + {}^\cdot CH_3$$
$$CH_4^{+\cdot} \rightarrow CH_3^+ + H^\cdot$$
$$CH_3^+ + CH_4 \rightarrow C_2H_5^+ + H_2$$

CH_5^+ is a potent proton donor that reacts with analyte to give the *protonated molecule*, MH^+, which is usually the most abundant ion in the chemical ionization mass spectrum.

$$CH_5^+ + M \rightarrow CH_4 + MH^+$$

In the chemical ionization spectrum in Figure 22-4, MH^+ at m/z 227 is the second strongest peak and there are fewer fragments than in the electron ionization spectrum.

Ammonia or isobutane are used in place of CH_4 to reduce the fragmentation of MH^+. These reagents bind H^+ more strongly than CH_4 does and impart less energy to MH^+ when the proton is transferred to M. Another mild, versatile ionization reagent is NO^+ generated from NO by radioactive ^{210}Po.[10]

Methane and NH_3 also promote formation of the parent anion $M^{-\cdot}$, as well as some fragment anions. Electrons are slowed down when they pass through either gas, and analyte molecules can capture the less energetic electrons to make analyte anions.[11] Ammonia also forms NH_2^-, which can transfer an electron to analyte. Other negative chemical ionization reagents include O_2^-, F^-, and SF_6^-.[10] Anions such as Cl^- and OH^- in solution can produce negative ions in the mass spectrum by reactions such as

$$M + Cl^- \rightarrow (M + Cl)^-$$
$$M + OH^- \rightarrow (M - H)^- + H_2O$$

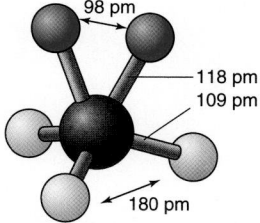

CH_5^+ is a CH_3 tripod with an added H_2 unit. **[H–C–H]** is held together by two electrons distributed over three atoms. Atoms of the H_2 unit rapidly exchange with atoms of the CH_3 unit.[9]

The **molecular ion, $M^{+\cdot}$**, can be formed by reactions such as

$$CH_4^{+\cdot} + M \rightarrow CH_4 + M^{+\cdot}$$

MH^+ is the **protonated molecule**, not the molecular ion.

$(M - H)^-$ is the anion formed by the loss of H^+ from M.

Resolving Power

Mass spectra in Figure 22-4 are computer-generated bar graphs. In contrast, Figure 22-1 shows actual detector signal. Each peak has a width that limits how closely two peaks could be spaced and still be resolved. If peaks are too close, they appear to be a single peak.

The higher the **resolving power** of a mass spectrometer, the better it is able to separate two peaks with similar mass.

$$\text{Resolving power} = \frac{m}{\Delta m} \quad \text{or} \quad \frac{m}{m_{1/2}} \qquad (22\text{-}1)$$

$$\underset{\text{Figure 22-5a}}{} \qquad \underset{\text{Figure 22-5b}}{}$$

where m is the smaller value of m/z. The denominator is defined in two different ways. In Figure 22-5a, the denominator is the separation of the two peaks (Δm) when the overlap at their base is 10% of the maximum peak height. The resolving power in Figure 22-5a is $m/\Delta m = 500/1.00 = 5.00 \times 10^2$. In Figure 22-5b, the denominator is taken as $m_{1/2}$, the width of the peak at half the maximum height, which is 0.481 m/z units. By this definition, the resolving power is $m/m_{1/2} = 500/0.481 = 1.04 \times 10^3$ for the same two peaks. Figure 22-5c

The expression $m/m_{1/2}$ gives a value about twice as great as $m/\Delta m$ for resolving power.

Resolution is the smallest difference in m/z values that can be detected as separate peaks and should be reported with the m/z value where it is measured. Examples: "$\Delta m = 0.1$ at $m/z = 1\,000$" or "$\Delta m = 0.1$ throughout the m/z scale." **Resolving power** is a big number; *resolution* is a small number.

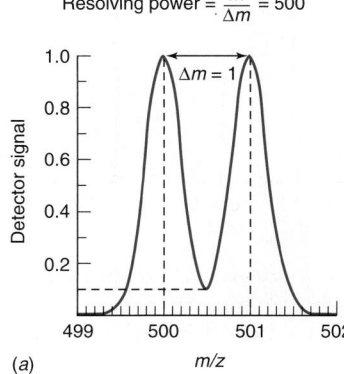

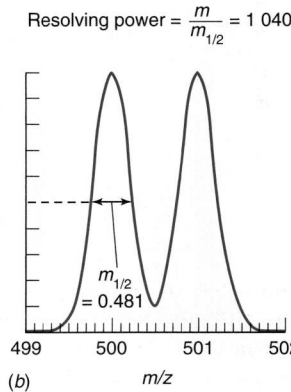

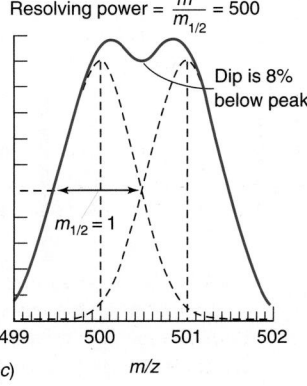

FIGURE 22-5 Resolving power. (*a*) By one definition, the resolving power is $m/\Delta m = 500/1 = 500$. (*b*) By a second definition, the resolving power for the same pair of peaks is $m/m_{1/2} = 500/0.481 = 1\,040$. (*c*) With the second definition, two peaks at m/z 500 and 501 are just barely discernible if the resolving power is 500.

shows that, when the expression $m/m_{1/2}$ gives a resolving power of 5.00×10^2, two peaks at m/z 500 and 501 are barely discernible. Specify which definition you use when expressing resolving power.

EXAMPLE Resolving Power

Using the ^{208}Pb peak in Figure 22-1, find the resolving power from the expression $m/m_{1/2}$.

Solution The width at half-height is 0.146 m/z units. Therefore,

$$\text{Resolving power} = \frac{m}{m_{1/2}} = \frac{208}{0.146} = 1.42 \times 10^3$$

An instrument with a resolving power of 1.42×10^3 separates peaks well near m/z 200 but gives a barely discernible separation at m/z 1 420.

TEST YOURSELF Measure the resolving power $m/m_{1/2}$ from the ^{31}P peak in Figure 22-10. (***Answer:*** $m/m_{1/2}$ = 30.974/0.003 3 = 9 400)

22-2 Oh, Mass Spectrum, Speak to Me!*

*An expression borrowed from O. David Sparkman, a master teacher.

Each mass spectrum has a story to tell. The molecular ion, $M^{+\bullet}$, tells us the molecular mass of an unknown. Unfortunately, with electron ionization, some compounds do not exhibit a molecular ion, because $M^{+\bullet}$ breaks apart so efficiently. However, fragments provide clues to the structure of an unknown. To find the molecular mass, we can obtain a chemical ionization mass spectrum, which usually has a strong peak for MH^+.

The **nitrogen rule** helps us propose compositions for molecular ions: If a compound has an odd number of nitrogen atoms—in addition to any number of C, H, halogens, O, S, Si, and P—then $M^{+\bullet}$ has an odd nominal mass. For a compound with an even number of nitrogen atoms (0, 2, 4,...), $M^{+\bullet}$ has an even nominal mass. A molecular ion at m/z 128 can have 0 or 2 N atoms, but it cannot have 1 N atom.

Molecular Ion and Isotope Patterns

Electron ionization of aromatic compounds (those containing benzene rings) usually gives significant intensity for $M^{+\bullet}$. $M^{+\bullet}$ is the base peak (most intense) in the spectra of benzene and biphenyl in Figure 22-6.

Although we write the molecular ion as $M^{+\bullet}$, we use the notation M+1 and M−29 for other peaks, without indicating positive charge. M+1 refers to an ion with a mass one unit greater than that of $M^{+\bullet}$.

The next higher mass peak, M+1, provides information on elemental composition. Table 22-1 lists the natural abundance of several isotopes. For carbon, 98.93% of atoms are ^{12}C and 1.07% are ^{13}C. The ratio of the two isotopes is ^{13}C/^{12}C = 1.07/98.93 = 0.010 8. Nearly all hydrogen is ^{1}H, with 0.012% ^{2}H. Applying the factors in Table 22-2 to C_nH_m, the intensity of the M+1 peak should be

Intensity of M + 1 relative to molecular ion for C_nH_m:

$$\text{Intensity} = \underbrace{n \times 1.08\%}_{\text{From }^{13}\text{C}} + \underbrace{m \times 0.012\%}_{\text{From }^2\text{H}} \qquad (22\text{-}2)$$

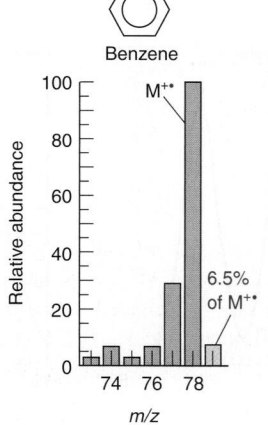

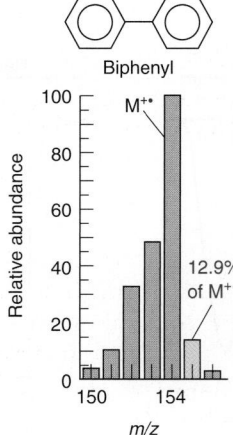

FIGURE 22-6 Electron ionization (70 eV) mass spectra of molecular ion region of benzene (C_6H_6) and biphenyl ($C_{12}H_{10}$). [NIST/EPA/NIH Mass Spectral Database.[7]]

TABLE 22-1 Isotopes of selected elements

Element	Mass number	Mass (Da)[a]	Abundance (atom%)[b]	Element	Mass number	Mass (Da)[a]	Abundance (atom%)[b]
Proton	1	1.007 276 467	—	Cl	35	34.968 85	75.78
Neutron	1	1.008 664 916	—		37	36.965 90	24.22
Electron	—	0.000 548 580	—	Ar	36	35.967 55	0.336
H	1	1.007 825	99.988		38	37.962 73	0.063
	2	2.014 10	0.012		40	39.962 38	99.600
B	10	10.012 94	19.9	Fe	54	53.939 61	5.845
	11	11.009 31	80.1		56	55.934 94	91.754
C	12	12(exact)	98.93		57	56.935 40	2.119
	13	13.003 35	1.07		58	57.933 28	0.282
N	14	14.003 07	99.632	Br	79	78.918 34	50.69
	15	15.000 11	0.368		81	80.916 29	49.31
O	16	15.994 91	99.757	I	127	126.904 47	100
	17	16.999 13	0.038	Hg	196	195.965 81	0.15
	18	17.999 16	0.205		198	197.966 75	9.97
F	19	18.998 40	100		199	198.968 26	16.87
Si	28	27.976 93	92.230		200	199.968 31	23.10
	29	28.976 49	4.683		201	200.970 29	13.18
	30	29.973 77	3.087		202	201.970 63	29.86
P	31	30.973 76	100		204	203.973 48	6.87
S	32	31.972 07	94.93	Pb	204	203.973 03	1.4
	33	32.971 46	0.76		206	205.974 45	24.1
	34	33.967 87	4.29		207	206.975 88	22.1
	36	35.967 08	0.02		208	207.976 64	52.4

a. 1 dalton (Da) ≡ 1/12 of the mass of ^{12}C = 1.660 538 86 (28) × 10^{-27} kg (from http://physics.nist.gov/constants). Nuclide masses from G. Audi, A. H. Wapsta, and C. Thibault, Nucl. Phys. **2003**, A729, 337, found at www.nndc.bnl.gov/masses. This source provides more significant figures for atomic mass than are cited in this table.

b. Abundance is representative of what is found in nature. Significant variations are observed. For example, ^{18}O in natural substances has been found in the range 0.188 to 0.222 atom%. The latest list of isotope abundances, which is slightly different from this table, is found in J. K. Böhlke et al., "Isotopic Compositions of the Elements, 2001," J. Phys. Chem. Ref. Data **2005**, 34, 57.

TABLE 22-2 Isotope abundance factors (%) for interpreting mass spectra

Element	X+1	X+2	X+3	X+4	X+5	X+6
H	0.012n					
C	1.08n	0.005 8n(n − 1)				
N	0.369n					
O	0.038n	0.205n				
F	0					
Si	5.08n	3.35n	0.170n(n − 1)	0.056n(n − 1)		
P	0					
S	0.801n	4.52n	0.036n(n − 1)	0.102n(n − 1)		
Cl	—	32.0n	—	5.11n(n − 1)	—	0.544n(n − 1)(n − 2)
Br	—	97.3n	—	47.3n(n − 1)	—	15.3n(n − 1)(n − 2)
I	0					

EXAMPLE: For a peak at m/z = X containing n carbon atoms, the intensity from carbon at X+1 is n × 1.08% of the intensity at X. The intensity at X + 2 is n(n − 1) × 0.005 8% of the intensity at X. Contributions from isotopes of other atoms in the ion are additive.

For benzene, C_6H_6, $M^{+\bullet}$ is observed at m/z 78. The predicted intensity at m/z 79 is 6 × 1.08% + 6 × 0.012% = 6.55% of the abundance of $M^{+\bullet}$. The observed intensity in Figure 22-6 is 6.5%. An intensity within ±10% of the expected value (5.9 to 7.2%, in this case) is within the precision of ordinary mass spectrometers. For biphenyl ($C_{12}H_{10}$), we predict that M+1 should have 12 × 1.08% + 10 × 0.012% = 13.1% of the intensity of $M^{+\bullet}$. The observed value is 12.9%.

EXAMPLE Listening to a Mass Spectrum

In the chemical ionization mass spectrum of pentobarbital in Figure 22-4, the peak with the most significant intensity at the high end of the mass spectrum at m/z 227 is suspected to be MH^+. If this is so, then the nominal mass of M is 226. The nitrogen rule tells us that a molecule with an even mass must have an even number of nitrogen atoms. If you know from elemental analysis that the compound contains only C, H, N, and O, how many atoms of carbon would you suspect are in the molecule?

Solution The m/z 228 peak has 12.0% of the height of the m/z 227 peak. Table 22-2 tells us that n carbon atoms will contribute $n \times 1.08\%$ intensity at m/z 228 from ^{13}C. Contributions from ^{2}H and ^{17}O are small. ^{15}N makes a larger contribution, but there are probably few atoms of nitrogen in the compound. Our first guess is

$$\text{Number of C atoms} = \frac{\text{observed peak intensity for M + 1}}{\text{contribution per carbon atom}} = \frac{12.0\%}{1.08\%} = 11.1 \approx 11$$

The actual composition of MH^+ at m/z 227 is $C_{11}H_{19}O_3N_2$. From the factors in Table 22-2, the theoretical intensity at m/z 228 is

$$\text{Intensity} = \underbrace{11 \times 1.08\%}_{^{13}C} + \underbrace{19 \times 0.012\%}_{^2H} + \underbrace{3 \times 0.038\%}_{^{17}O} + \underbrace{2 \times 0.369\%}_{^{15}N} = 13.0\%$$

The theoretical intensity is within the 10% uncertainty of the observed value of 12.0%.

TEST YOURSELF A compound with a molecular ion at m/z 117 has a peak at m/z 118 with a relative intensity of 9.3%. Is there an odd or even number of N atoms? How many carbons are in the formula? (*Answer:* odd, 8C. It cannot be 9C because C_9N would be at $m/z >122$.)

Box 22-3 describes other information available from isotope ratios, including measuring the body temperature of dinosaurs.

Ions containing Cl or Br have distinctive isotopic peaks shown in Figure 22-7.[15] The mass spectrum of 1-bromobutane in Figure 22-8, has two nearly equal peaks at m/z 136 and 138, strongly indicating that the molecular ion contains one Br atom. The fragment at m/z 107 has a nearly equal partner at m/z 109, suggesting that this fragment ion contains Br.

BOX 22-3 Isotope Ratio Mass Spectrometry and Dinosaur Body Temperature

The opening of Chapter 24 describes how *isotope ratio mass spectrometry*[12] is used to detect steroid abuse by athletes. A compound eluted from a gas chromatography column is passed through a combustion furnace loaded with a metal catalyst (such as CuO/Pt at 820°C) to oxidize organic compounds to CO_2. The H_2O combustion by-product is removed by passage through a Nafion fluorocarbon tube (page 421). Water diffuses through the membrane, but other combustion products are retained. CO_2 then enters a mass spectrometer that monitors m/z 44 ($^{12}CO_2$) and 45 ($^{13}CO_2$). One detector is dedicated to each ion.

Natural carbon is composed of 98.9% ^{12}C and 1.1% ^{13}C. The chart shows consistent, small variations in ^{13}C from natural sources. The standard used to measure carbon isotope ratios is calcium carbonate from the Pee Dee belemnite (fossil shell) formation in South Carolina (designated PDB) with a ratio $R_{PDB} = {}^{13}C/{}^{12}C = 0.011\ 23_{72}$. (The composition is accurate to four significant figures, but small differences are precise to six significant figures.) The δ^{13}C scale expresses small variations in isotopic compositions:

$$\delta^{13}C \text{ (parts per thousand, \%o)} = 1\ 000\left(\frac{R_{sample} - R_{PDB}}{R_{PDB}}\right)$$

δ^{13}C of natural materials provides information about their biological and geographic origins.[13,14] As one example, the chart shows that δ^{13}C of caffeine in a beverage can distinguish whether the caffeine is of natural or synthetic origin.

On the back cover of this book, the isotopic composition of carbonate in dinosaur teeth is used as a measure of dinosaur body temperature. The lower the temperature at which calcium carbonate crystallizes, the more it is enriched with both ^{13}C and ^{18}O. Tooth enamel contains the mineral apatite, $Ca_{10}(PO_4)_6(OH)_2$, in which some phosphate is replaced by carbonate. When tooth enamel is treated with anhydrous phosphoric acid, $CO_2(g)$ is liberated. The CO_2 is pumped off, condensed at low temperature, purified by distillation and gas chromatography, and subjected to isotope ratio mass spectrometry. The amount of $^{13}C^{18}O\ ^{16}O^+$ at m/z 47 is compared with the amount of $^{12}C^{16}O_2^+$ at m/z 44:

$$\text{Enrichment (\%o)} = 1\ 000\left(\frac{R_{obs} - R_{random}}{R_{random}}\right)$$

in which $R_{obs} = (m/z\ 47)/(m/z\ 44)$ from a specimen such as a dinosaur tooth and $R_{random} = (m/z\ 47)/(m/z\ 44)$ for a random distribution

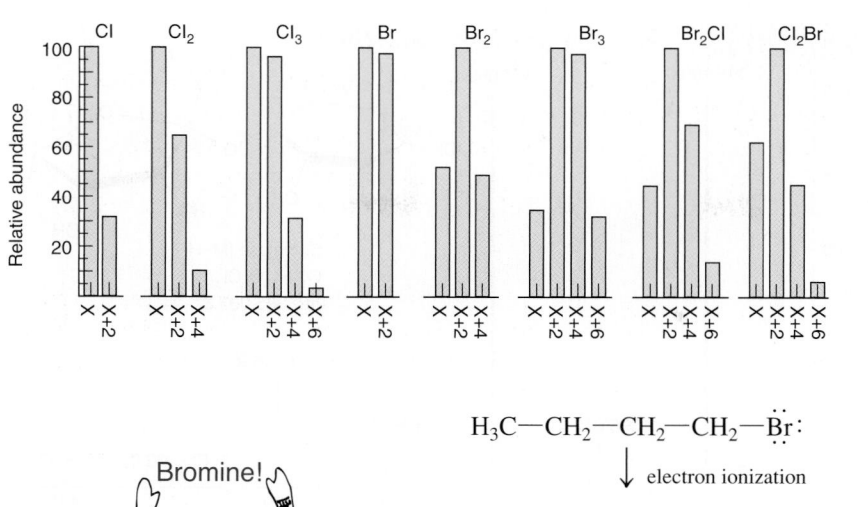

FIGURE 22-7 Calculated isotopic patterns for species containing Cl and Br.

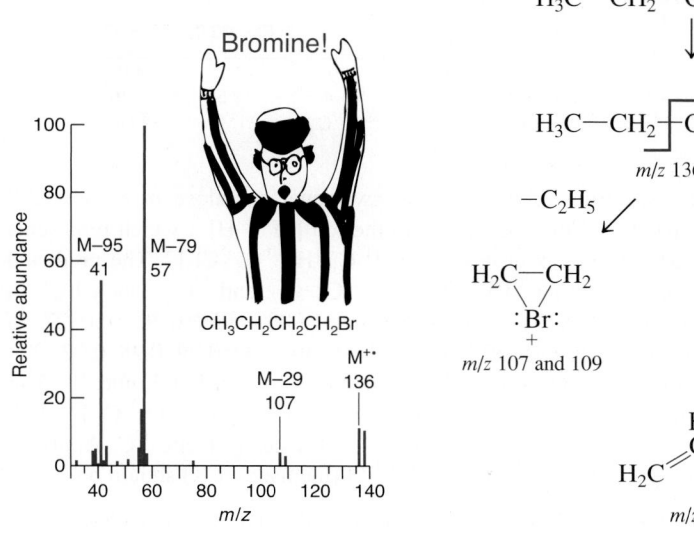

FIGURE 22-8 Electron ionization mass spectrum (70 eV) of 1-bromobutane. [Data from A. Illies, P. B. Shevlin, G. Childers, M. Peschke, and J. Tsai, "Mass Spectrometry for Large Undergraduate Laboratory Sections," *J. Chem. Ed.* **1995,** *72,* 717. Referee from Maddy Harris.]

of the naturally abundant isotopes. Enrichment is the ordinate of the graph on the back cover of the book. The calibration line shows points measured for CaCO₃ crystallized at known temperature in the lab and for coral growing at known temperature in the ocean.

Enrichment measured from teeth of modern animals agrees with the known body temperatures of those animals. Enrichment measured from dinosaur teeth probably indicates the body temperature of the dinosaurs.

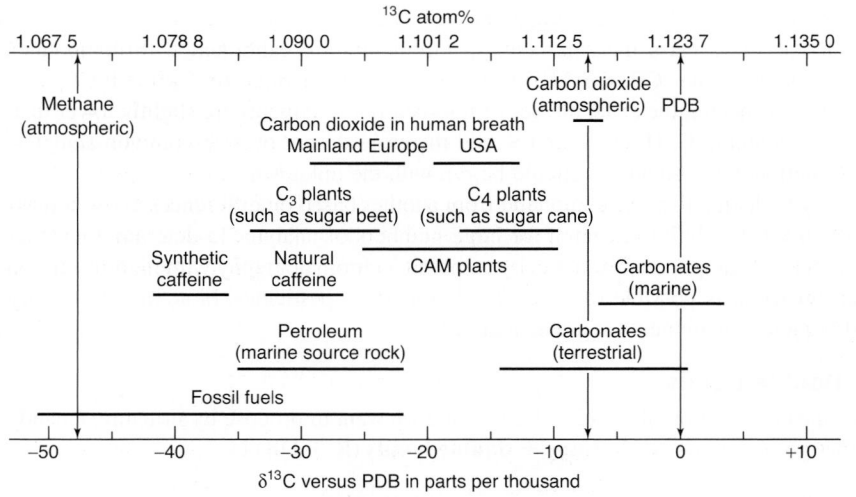

Variation in ^{13}C from natural sources. C_3, C_4, and CAM are types of plants with distinct metabolic pathways leading to different ^{13}C incorporation. [Information from W. Meier-Augenstein, *LCGC* **1997,** *15,* 244. Caffeine data from L. Zhang et al., *Anal. Chem.* **2012,** *84,* 2805.]

FIGURE 22-9 High-resolution time-of-flight mass spectrum of molecular ion region of sucralose anion [M − H]⁻. [Data from I. Ferrer, J. A. Zweigenbaum, and E. M. Thurman, "Analytical Methodologies for the Detection of Sucralose in Water," *Anal. Chem.* **2013**, *85*, 9581.]

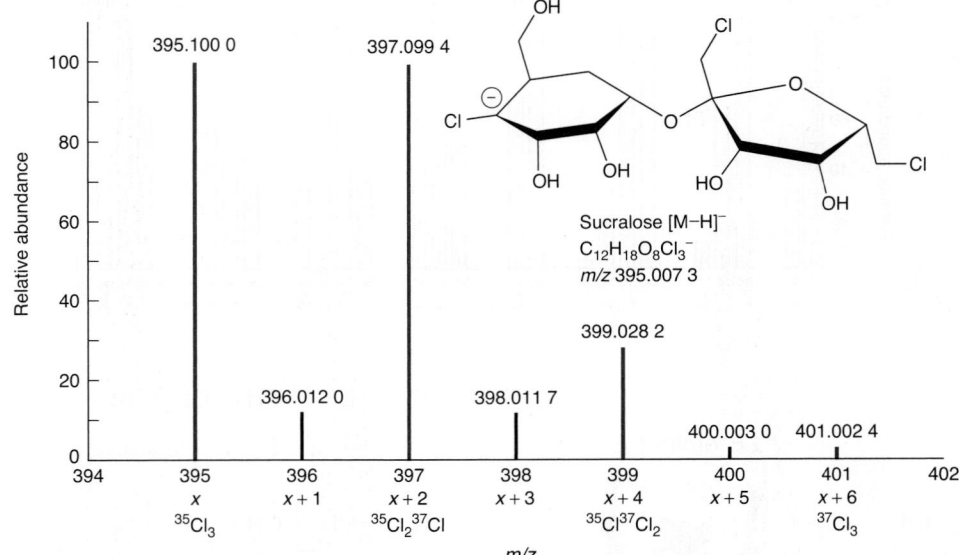

Exercise 22-C teaches you how to calculate the relative intensities of the peaks for a molecule with several Cl atoms.

The negative ion mass spectrum of the artificial sweetener sucralose in Figure 22-9 exhibits the characteristic pattern of three Cl atoms in the ion [M − H]⁻, which represents loss of one H atom from M⁻. The peak at *m/z* 395 is $[^{12}C_{12}{}^{1}H_{18}{}^{16}O_8{}^{35}Cl_3]^-$. The predicted isotope pattern for molecules containing $^{35}Cl_3$, $^{35}Cl_2{}^{37}Cl$, $^{35}Cl^{37}Cl_2$, and $^{37}Cl_3$, labeled Cl_3 in Figure 22-7, has relative intensities X:X+2:X+4:X+6 = 1.000:0.959:0.307:0.0327. In Figure 22-9, relative intensities at *m/z* 395, 397, 399, and 401 are 1.00:0.99:0.28:0.02.

Now we have to use our heads to predict the intensities at X+1, X+3, and X+5 in Figure 22-9. The composition of X+1 at *m/z* 396 is mainly $[^{13}C^{12}C_{11}{}^{1}H_{18}{}^{16}O_8{}^{35}Cl_3]^-$, with small contributions from molecules with one ^{2}H or one ^{17}O instead of one ^{13}C. Applying factors from Table 22-2, we predict that (intensity of X+1)/(intensity of X) should be

$$\text{Intensity at X} + 1 = \underbrace{12 \times 1.08\%}_{^{13}C} + \underbrace{18 \times 0.012\%}_{^{2}H} + \underbrace{8 \times 0.038\%}_{^{17}O} = 13.5\%$$

The observed intensity of X+1 in Figure 22-9 is 12.0% of the intensity of X at *m/z* 395.

The peak at X+2 is principally from the $^{35}Cl_2{}^{37}Cl$ molecule. X+3 is mainly the ^{13}C isotopic partner of X+2. By the exact same calculation that we just did, the intensity of X+3 should be 13.5% of the intensity of X+2. The observed intensity at X+3 is 11.8% of X+2.

High-Resolution Mass Spectrometry

An ion at *m/z* 84 could have a variety of elemental compositions, such as $C_5H_8O^+ = 84.056\ 96$ Da or $C_6H_{12}^+ = 84.093\ 35$ Da. We can determine which composition is correct if the spectrometer can distinguish sufficiently small differences in mass. With a *time-of-flight mass spectrometer* or an *orbitrap mass spectrometer* (Section 22-3), we can resolve differences of 0.001 or less at *m/z* 100. Figure 22-10 shows a high-resolution spectrum that distinguishes $^{31}P^+$, $^{15}N^{16}O^+$, and $^{14}N^{16}OH^+$—all with a nominal mass of 31 Da.

To ensure accurate mass measurement, spectrometers are calibrated with compounds such as perfluorokerosene ($CF_3(CF_2)_nCF_3$) or perfluorotributylamine (($CF_3CF_2CF_2CF_2)_3N$). In high-resolution spectra, the exact masses of fluorocarbon fragments are slightly lower than those of ions containing C, H, O, N, and S, and therefore do not overlap common analytes. For high-resolution work, standards should be run with the unknown.

The ability to distinguish one compound from another based on differences in exact mass has revolutionized the ability to search for large numbers of analytes to determine whether they are present in a sample. For example, by combining chromatography with high-resolution mass spectrometry, it is possible to screen foods for many pesticides, biotoxins, veterinary drugs, and organic contaminants at a practical rate.[17]

Predicted masses from Table 22-1 and masses observed by high-resolution mass spectrometry:

$C_5H_8O^{+ \cdot}$:

5C	5 × 12.000 00
8H	+8 × 1.007 825
1O	+1 × 15.994 91
−e⁻	−1 × 0.000 55
	84.056 96
observed:[16]	84.059 1

$C_6H_{12}^+$:

6C	6 × 12.000 00
12H	12 × 1.007 825
−e⁻	−1 × 0.000 55
	84.093 35
observed:[16]	84.093 9

Rings + Double Bonds

If we know the composition of a molecular ion, and we want to propose its structure, a handy equation that gives the number of **rings + double bonds** (R + DB) is

Rings + double bonds formula:

$$R + DB = c - h/2 + n/2 + 1 \tag{22-3}$$

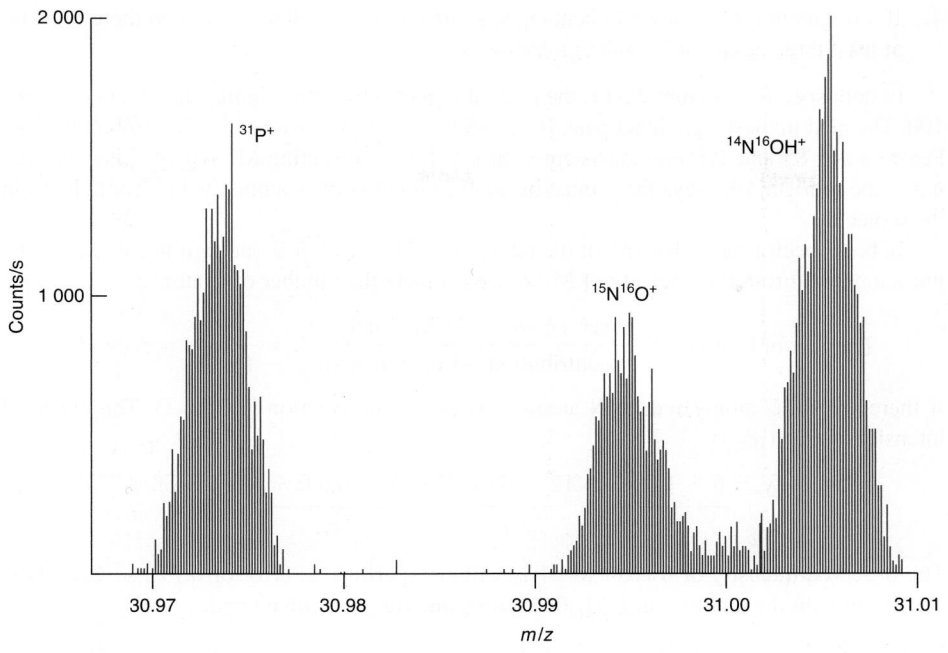

FIGURE 22-10 High-resolution double-focusing sector-field mass spectrum distinguishing $^{31}P^+$, $^{15}N^{16}O^+$, and $^{14}N^{16}OH^+$ with a resolution of $m/\Delta m = 4\ 000$. The spectrum is a bar graph of detector counts/s at each increment of m/z. [Data from J. S. Becker, S. F. Boulyga, C. Pickhardt, J. Becker, S. Buddrus, and M. Przybylski, "Determination of Phosphorus in Small Amounts of Protein Samples by ICP-MS," *Anal. Bioanal. Chem.* **2003,** *375*, 561.]

where c is the number of Group 14 atoms (C, Si, and so on, which all make four bonds), h is the number of (H + halogen) atoms (which make one bond), and n is the number of Group 15 atoms (N, P, As, and so on, which make three bonds). Group 16 atoms (O, S, and so on, which make two bonds) do not enter the formula. Here is an example:

$$R + DB = c - h/2 + n/2 + 1$$

$$R + DB = (14 + 1) - \frac{22 + 1 + 1}{2} + \frac{1 + 1}{2} + 1 = 5$$

The molecule has one ring + four double bonds.
Note that c includes C + Si, h includes H + Cl + Br, and n includes N + As.

$$\underbrace{C_{14}}_{c}\,Si\,\underbrace{H_{22}ClBr}_{h}\,\underbrace{NAs}_{n}\,O_3S$$

If P makes more than three bonds or if S makes more than two bonds, then Equation 22-3 does not include these extra bonds. Examples that violate Equation 22-3:

Identifying the Molecular Ion Peak

Figure 22-11 shows electron ionization mass spectra of isomers with the elemental composition $C_6H_{12}O$. $M^{+\bullet}$ is marked by the solid triangle at m/z 100.

If these spectra had been obtained from unknowns, how would we know that the peak at m/z 100 represents the molecular ion? Here are some guidelines:

1. $M^{+\bullet}$ will be at the highest m/z value of any of the "significant" peaks in the spectrum that cannot be attributed to isotopes or background. "Background" arises from sources such as pump oil in the spectrometer and stationary phase from a gas chromatography column. Experience is required to recognize these signals. With electron ionization, the molecular ion peak intensity is often no greater than 5–20% of the base peak and might not represent more than 1% of all ions.
2. Intensities of isotopic peaks at M+1, M+2, and so on, must be consistent with the proposed formula.
3. The peak for the heaviest fragment ion should not correspond to an improbable mass loss from $M^{+\bullet}$. It is rare to find a mass loss in the range 3–14 or 21–25 Da. Common losses include 15 (CH_3), 17 (OH or NH_3), 18 (H_2O), 29 (C_2H_5), 31 (OCH_3), 43 Da (CH_3CO or C_3H_7), and many others. If you think that m/z 150 represents $M^{+\bullet}$, but there is a significant peak at m/z 145, then $M^{+\bullet}$ is not assigned correctly, because a mass loss of 5 Da is improbable. Alternatively, the two peaks represent ions from different compounds or both peaks represent fragments from a compound whose mass is greater than 150 Da.

The small peak at m/z 86 is not a loss of 14 Da from $M^{+\cdot}$. It is the isotopic partner of m/z 85 containing ^{13}C.

We see no evidence at M+1, M+2, and M+3 for the elements Cl, Br, Si, or S.

Challenge What intensities would you see at M+2 if there were one Cl, one Br, one Si, or one S in the molecule?

(*Answer:* Cl, 0.32 M at M+2; Br 0.97 M at M+2; Si, 0.034 M at M+2; S, 0.045 M at M+2)

$R + DB = c - h/2 + n/2 + 1$
$= 6 - 12/2 + 0 + 1 = 1$

4. If a fragment ion is known to contain, say, three atoms of element X, then there must be at least three atoms of X in the molecular ion.

In both spectra in Figure 22-11, the peak at highest m/z with "significant" intensity is m/z 100. The next highest significant peak is m/z 85, representing a loss of 15 Da (probably CH_3). Peaks at m/z 85 and 100 are consistent with m/z 100 representing $M^{+\cdot}$. If $M^{+\cdot}$ has an even mass, the nitrogen rule says there must be an even number of N atoms (which could be 0) in the molecule.

In both spectra, M+1 has 6% of the intensity of $M^{+\cdot}$, with just one significant digit in the measurement. From the intensity of M+1, we estimate the number of C atoms:

$$\text{Number of C atoms} = \frac{\text{observed }(M+1)/M^{+\cdot}\text{ intensity}}{\text{contribution per carbon atom}} = \frac{6\%}{1.08\%} = 5_{.6} \approx 6$$

If there are six C atoms and no N atoms, a possible composition is $C_6H_{12}O$. The expected intensity of M+1 is

$$\text{Intensity} = \underbrace{6 \times 1.08\%}_{^{13}C} + \underbrace{12 \times 0.012\%}_{^2H} + \underbrace{1 \times 0.038\%}_{^{17}O} = 6.7\% \text{ of } M^{+\cdot}$$

The observed intensity of 6% for M+1 is within experimental error of 6.7%, so $C_6H_{12}O$ is consistent with the data so far. $C_6H_{12}O$ requires one ring or double bond.

Interpreting Fragmentation Patterns

Consider how the 2-hexanone molecular ion can break apart to give the many peaks in Figure 22-11. Reactions A and B in Figure 22-12 show $M^{+\cdot}$ derived from loss of a non-bonding electron from oxygen, which has the lowest ionization energy. In Reaction A, the C—C bond adjacent to C=O splits so that one electron goes to each C atom. The products are a neutral butyl radical ($\cdot C_4H_9$) and $CH_3CO^{+\cdot}$. Only the ion is detected by the mass spectrometer, giving the base peak at m/z 43. Cleavage of the C—C bond in Reaction B gives an ion with m/z 85, corresponding to loss of $\cdot CH_3$ from the molecular ion. Two other major peaks in the spectrum arise from C_4—C_5 bond cleavage to give $CH_3CH_2^+$ (m/z 29) and $^+CH_2CH_2COCH_3$ (m/z 71). The nitrogen rule told us that molecules containing only C, H, halogens, O, S, Si, P, and an even number of N atoms (such as 0) have an even mass. A fragment of a neutral molecule that is missing an H atom must have an odd mass.

We have yet to account for the second-tallest peak at m/z 58, which is special because it has an *even* mass. The molecular ion has an even mass (100 Da). Radical fragments, such as $CH_3CH_2^{\cdot}$, have an *odd* mass. All of the fragments discussed so far have an odd mass. The peak at m/z 58 results from the loss of a *neutral* molecule with an *even* mass of 42 Da.

Reaction D in Figure 22-12 shows a common rearrangement that leads to loss of a neutral molecule with even mass. In ketones with an H atom on the γ carbon atom (3 carbons away from the carbonyl carbon), the H atom can be transferred to O^+. Concomitantly, the C_α—C_β bond cleaves and a neutral molecule of propene ($CH_3CH=CH_2$, 42 Da) is lost. The resulting ion has a mass of 58 Da.

Mass spectra in Figure 22-11 allow us to distinguish one isomer of $C_6H_{12}O$ from the other. The main difference is a peak at m/z 71 for 2-hexanone that is absent in 4-methyl-2-pentanone. The peak at m/z 71 results from loss of ethyl radical, $CH_3CH_2^{\cdot}$, from $M^{+\cdot}$. Ethyl

FIGURE 22-11 Electron ionization (70 eV) mass spectra of isomeric ketones with the composition $C_6H_{12}O$. [NIST/EPA/NIH Mass Spectral Database.[7]]

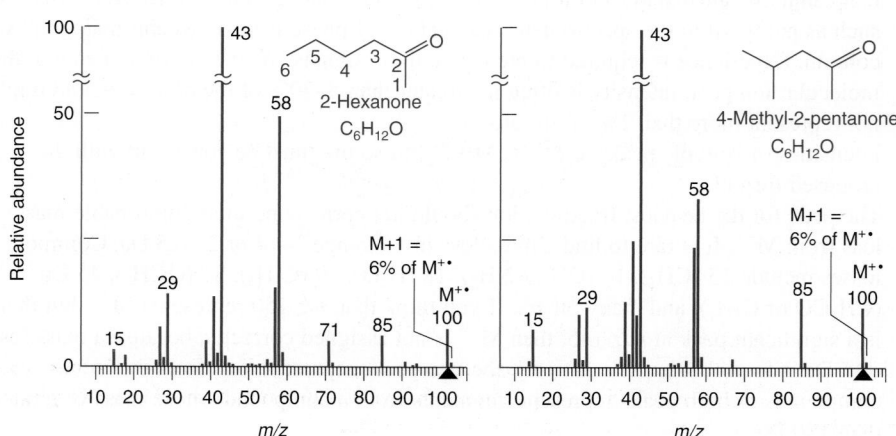

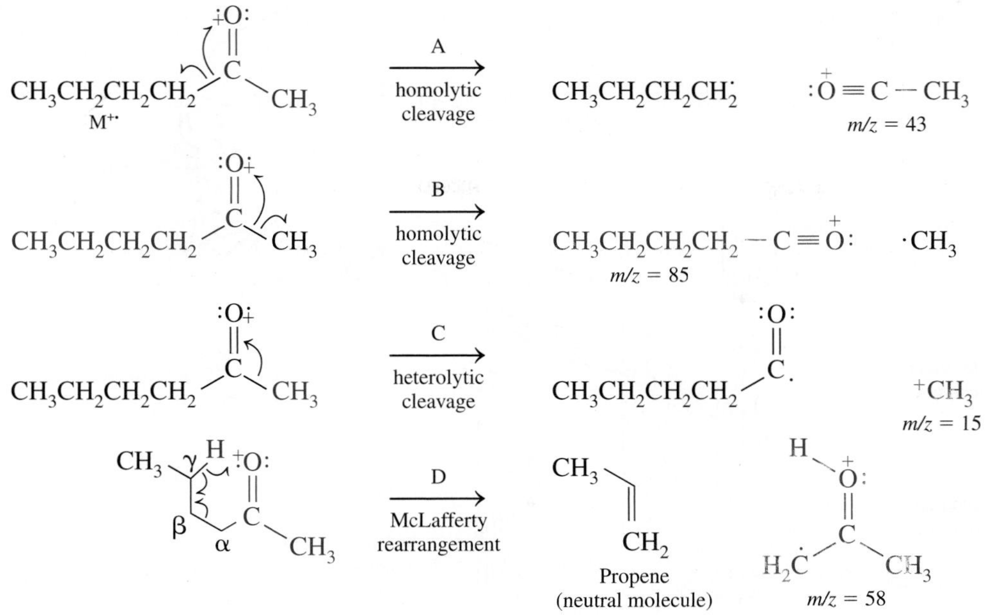

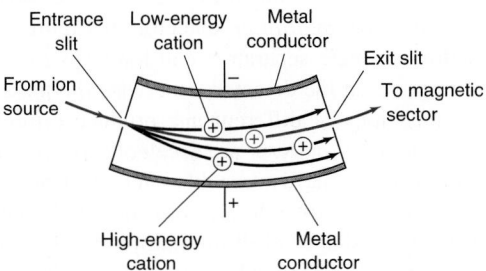

Reaction A (homolytic cleavage):

$CH_3CH_2CH_2CH_2\overset{+\cdot O:}{\underset{}{C}}CH_3 \;(M^{+\cdot}) \xrightarrow{\text{homolytic cleavage}} CH_3CH_2CH_2CH_2^\cdot \quad :\overset{+}{O}\equiv C-CH_3 \quad m/z = 43$

Reaction B (homolytic cleavage):

$CH_3CH_2CH_2CH_2\overset{:O+}{\underset{}{C}}CH_3 \xrightarrow{\text{homolytic cleavage}} CH_3CH_2CH_2CH_2-C\equiv\overset{+}{O}: \;(m/z = 85) \quad \cdot CH_3$

Reaction C (heterolytic cleavage):

$CH_3CH_2CH_2CH_2\overset{:O+}{\underset{}{C}}CH_3 \xrightarrow{\text{heterolytic cleavage}} CH_3CH_2CH_2CH_2\overset{:O:}{\underset{}{C\cdot}} \quad {}^+CH_3 \;(m/z = 15)$

Reaction D (McLafferty rearrangement):

$\underset{\beta}{CH_3}\underset{\alpha}{\gamma}\overset{H\;+O:}{\underset{}{C}}CH_3 \xrightarrow{\text{McLafferty rearrangement}} \overset{CH_3}{\underset{CH_2}{}} \;\text{Propene (neutral molecule)} \quad \overset{H}{\underset{}{\overset{\cdot}{O}:}}\,H_2C\!\!-\!\!C\!\!-\!\!CH_3 \;(m/z = 58)$

FIGURE 22-12 Four possible fragmentation pathways for the molecular ion of 2-hexanone.

⌒ is transfer of one electron
⌢ is transfer of two electrons

Types of bond breaking:

- **Homolytic cleavage:** One electron remains with each fragment
- **Heterolytic cleavage:** Both electrons stay with one fragment

In general, the most intense peaks correspond to the most stable fragments.

radical is derived from carbon atoms 5 and 6 of 2-hexanone. There is no simple way for an ethyl radical to cleave from 4-methyl-2-pentanone. The diagram in the margin shows how peaks at m/z 15, 85, 43, and 57 can arise from breaking bonds in 4-methyl-2-pentanone. The rearrangement at the bottom of Figure 22-12 accounts for m/z 58.

Other major peaks in Figure 22-11 might be $CH_2\!=\!C\!=\!O^{+\cdot}$ (m/z 42), $C_3H_5^+$ (m/z 41), $C_3H_3^+$ (m/z 39), $C_2H_5^+$ or $HC\equiv O^+$ (m/z 29), and $C_2H_3^+$ (m/z 27). Small fragments are common in many spectra and not very useful for structure determination.

Interpretation of mass spectra to elucidate molecular structure is an important field.[18] Fragmentation patterns can even unravel the structures of large biological molecules such as proteins and carbohydrates.

Bond cleavages of 4-methyl-2-pentanone leading to m/z 15, 85, 43, and 57:

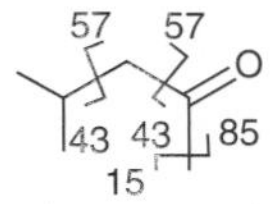

57 57
43 43 85
15

Challenge Draw a rearrangement like Reaction D in Figure 22-12 to show how m/z 58 arises from 4-methyl-2-pentanone.

22-3 Types of Mass Spectrometers

Figure 22-2 shows a *magnetic sector mass spectrometer* in which ions with different mass, but constant kinetic energy, are separated by their trajectories in a magnetic field. Kinetic energy is imparted to the ions by the voltage between the acceleration plates in the ion source. Ions have a small spread of kinetic energies because the acceleration that they experience depends on where in the ion source they were formed.

The resolving power of a mass spectrometer is limited by the ~0.1% variation in kinetic energy of ions emerging from the source. This variation limits the resolving power to ~1 000, corresponding to a resolution of 0.1 at m/z 100. In a **double-focusing mass spectrometer,** ions ejected from the source pass through an electric sector (Figure 22-13) before the magnetic sector. The electric sector only permits a narrow range of ion kinetic energies to pass to the magnetic sector. With both sectors in series, it is possible to achieve a resolving power of ~10^5, corresponding to a resolution of 0.001 at m/z 100.

FIGURE 22-13 Electric sector of a double-focusing mass spectrometer. Positive ions are attracted toward the negative plate. Trajectories of high-energy ions are changed less than trajectories of low-energy ions. Ions reaching the exit slit have a narrow range of kinetic energies.

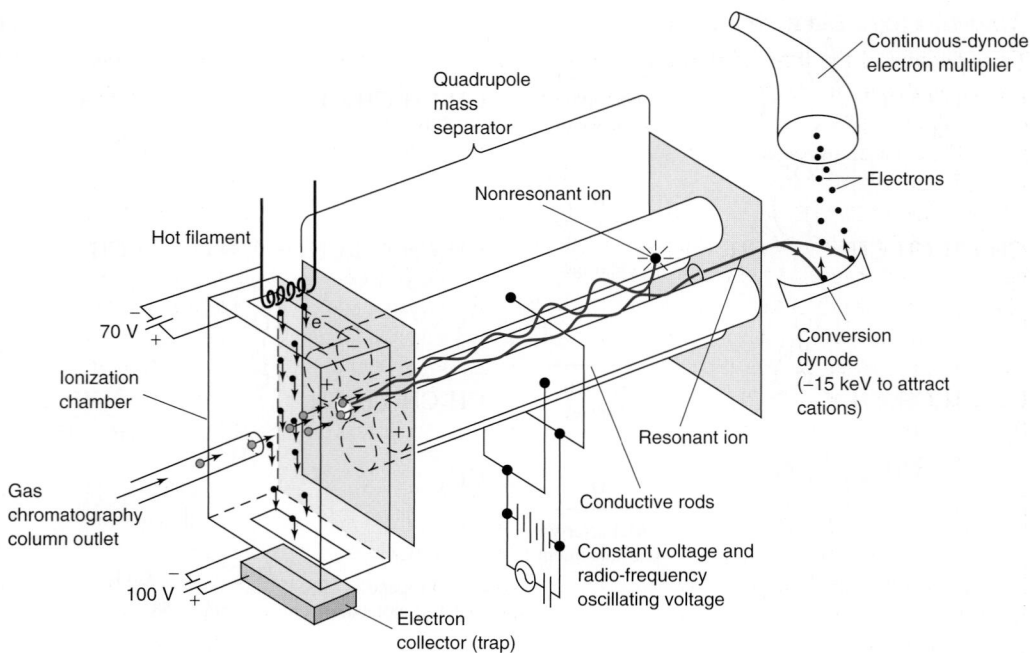

FIGURE 22-14 Quadrupole mass spectrometer. Ideally, the rods should have a hyperbolic cross section on the surfaces that face one another.

Transmission Quadrupole Mass Spectrometer

Magnetic sector and double-focusing mass spectrometers are not detectors of choice for chromatography. Figure 22-14 shows a **transmission quadrupole mass spectrometer**[19] connected to an open tubular gas chromatography column to record multiple spectra from each component as it is eluted. Species exiting the chromatography column pass through a heated connector into the electron ionization chamber, which is pumped to maintain a pressure of $\sim 10^{-9}$ bar, using a high-speed turbomolecular or oil diffusion pump. Ions are accelerated through a potential of 5–15 V before entering the quadrupole filter.

The quadrupole is a common mass separator because of its low cost. It consists of four parallel metal rods to which are applied both a constant voltage and a radio-frequency oscillating voltage. The electric field deflects ions in complex trajectories as they migrate from the ionization chamber toward the detector, allowing only ions with one particular mass-to-charge ratio to reach the detector. Other ions (nonresonant ions) collide with the rods and are lost before they reach the detector. Rapidly varying voltages select ions of different masses to reach the detector. Transmission quadrupoles can record 2–8 spectra per second, covering a range that is usually not more than 1 500 to 3 000 *m/z*. They typically achieve unit resolution (resolving $\Delta m/z = 1$) over the entire *m/z* range.

Transmission quadrupoles (and other quadrupole instruments described later) operate at *constant resolution*. That is, ions of *m/z* 100 and 101 are separated to the same extent as ions of *m/z* 500 and 501. Magnetic sector and double-focusing instruments, as well as time-of-flight instruments described next, operate at *constant resolving power*, which means that the separation between peaks decreases as *m/z* increases.

In place of the *discrete-dynode* detector in Figure 22-2, we show a *conversion dynode* and a *continuous-dynode electron multiplier* detector in Figure 22-14. The conversion dynode is used with quadrupole mass separators and ion traps to ensure that all ions produce a similar electrical response at the detector; otherwise the mass spectrum may exhibit *mass discrimination* in which ions of different *m/z* produce different responses. Cations emerging from the quadrupole mass analyzer are accelerated toward the conversion dynode held at −15 kV. A cation striking the dynode liberates electrons, which are accelerated toward the mouth of the continuous-dynode electron multiplier. The electrically resistive lead-doped glass wall of the horn-shaped electron multiplier has a potential of −2 kV at the mouth and ground potential at the narrow end (Figure 22-15). An electron striking the wall of the electron multiplier near the mouth liberates several electrons, which are accelerated toward more positive potential deeper into the horn. After many rebounds, each incident electron is multiplied to $\sim 10^6$ to 10^8 electrons at the narrow end of the horn.

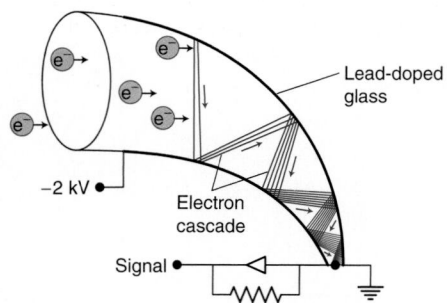

FIGURE 22-15 Continuous-dynode electron multiplier, also known as a Channeltron®. For each incident electron, $\sim 10^6$–10^8 electrons are detected at the far end of the horn. [Information from J. T. Watson and O. D. Sparkman, *Introduction to Mass Spectrometry,* 4th ed. (Chichester: Wiley, 2007).]

Time-of-Flight Mass Spectrometer

The principle of the **time-of-flight mass spectrometer** is shown in Figure 22-16.[20,21] Ions can be produced inside the source region at the upper left of the figure by electron ionization in the gas phase or by laser irradiation of a solid sample on the surface of the backplate. (The latter process is matrix-assisted laser desorption/ionization described later in Box 22-4.) Alternatively, ions produced externally are directed into the source region. About 3 000 to 20 000 times per second, ~20 000 V is applied to the backplate to accelerate ions toward the right and expel them from the ion source into the drift region, where there is no electric or magnetic field and no further acceleration. Ideally, all ions have the same kinetic energy, which is $\frac{1}{2}mv^2$, where m is the mass of the ion and v is its velocity. If ions have the same kinetic energy, but different masses, *lighter ions travel faster than heavier ions*. In its simplest incarnation, the time-of-flight mass spectrometer is just a long (~1 m), straight, evacuated tube with the source at one end and the detector at the other end. Ions expelled from the source drift to the detector in order of increasing mass, because the lighter ones travel faster.

The instrument in Figure 22-16 is designed for improved resolving power. The main limitation on resolving power is that all ions do not emerge from the source with the same kinetic energy. An ion formed close to the backplate is accelerated through a higher voltage difference than one formed near the grid, so the ion near the backplate gains more kinetic energy. Also, there is some distribution of kinetic energies among the ions, even in the absence of the accelerating voltage. Heavier ions with more-than-average kinetic energy reach the detector at the same time as lighter ions with less-than-average energy.

Consider two ions expelled from the source with different speeds at different times. The ion formed close to the grid has less kinetic energy but is expelled first. The ion formed close to the backplate has more kinetic energy but is expelled later. The faster ion catches up to the slower ion at the *space focus plane* at a distance $2s$ from the grid (where s is the distance between the backplate and the grid). It can be shown that all ions of a given mass reach the space focus plane at the same time. After passing the space focus plane, ions diverge again, with faster ions overtaking slower ones. In the absence of countermeasures, ions would be spread by the time they reach the detector and resolving power would be low.

To improve resolving power, ions are turned around by the "reflectron" at the right side of Figure 22-16. The reflectron is a series of hollow rings held at increasingly positive potential, terminated by a grid whose potential is more positive than the accelerating potential on the backplate of the source. Ions entering the reflectron are slowed down, stopped, and reflected back to the left. The more kinetic energy an ion had when it entered the reflectron, the further it penetrates before it is turned around. Reflected ions reach a new space focus plane at the grid in front of the detector. *All ions of the same mass reach this grid at the same time, regardless of their initial kinetic energies.*

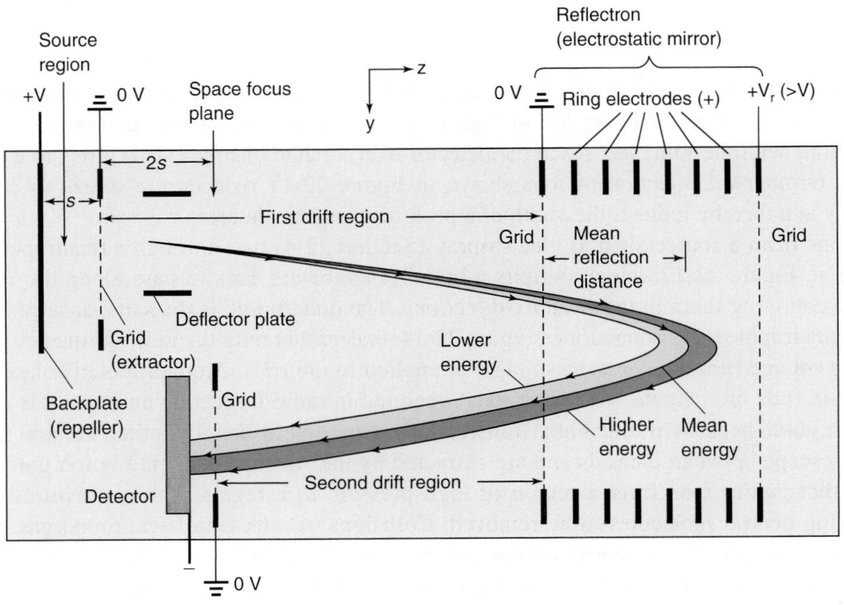

Ideally, the kinetic energy of an ion expelled from the source is zeV, where ze is the charge of the ion and V is the voltage on the backplate.

FIGURE 22-16 Time-of-flight mass spectrometer. Positive ions are accelerated out of the source by voltage +V periodically applied to the backplate. Light ions travel faster and reach the detector sooner than heavier ions.

Resolving power can be as high as ~50 000, and m/z accuracy is ~2 ppm. At m/z 500, 2 ppm accuracy is $(500)(2 \times 10^{-6}) = 0.001$ m/z. Other advantages of the time-of-flight spectrometer are its high acquisition rate (10^2 to 10^4 spectra/s) and its capability for measuring high masses ($m/z \approx 10^6$). The time-of-flight instrument requires a lower operating pressure than does the transmission quadrupole instrument (~10^{-10} bar versus ~10^{-9} bar) because the drift distance is longer and ions cannot be allowed to collide with background gas.

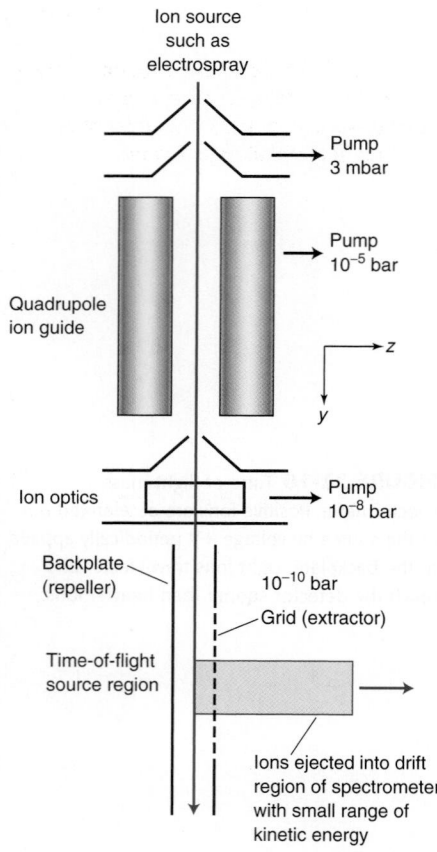

Ion source
such as
electrospray

Pump
3 mbar

Pump
10^{-5} bar

Quadrupole
ion guide

z

y

Ion optics

Pump
10^{-8} bar

Backplate
(repeller)

10^{-10} bar

Grid (extractor)

Time-of-flight
source region

Ions ejected into drift
region of spectrometer
with small range of
kinetic energy

FIGURE 22-17 Orthogonal injection of ions into a time-of-flight mass spectrometer. The ion optics, quadrupole ion guide, and ion source attach to the spectrometer at the upper left of Figure 22-16. [Information from I. V. Chernushevich, W. Ens, and K. G. Standing, "Orthogonal Injection TOFMS," *Anal. Chem.* **1999,** *71,* 452A.]

EXAMPLE **Time Resolution Required in a Time-of-Flight Spectrometer**

(a) How close to a peak at m/z 500 can another peak be resolved if the resolving power is $m/\Delta m = 50\,000$? **(b)** If ions are accelerated by 20 000 V from the source of the spectrometer and the distance from the source to the detector is 2.000 m, how long will it take for an ion to reach the detector? What is the difference in drift time for two ions that are just resolved?

Solution **(a)** The resolved mass difference is

$$\text{Resolving power} = \frac{m}{\Delta m} \Rightarrow \Delta m = \frac{m}{\text{resolving power}} = \frac{500}{50\,000} = 0.01$$

For $m/z = 500$, two peaks separated by $m/z = 0.01$ are resolved.

(b) Equation A in Box 22-2 tells us that the velocity (v) of an ion with mass m and charge z after acceleration through a voltage V is v $= \sqrt{2zeV/m}$, where e is the elementary charge. The footnote to Table 22-1 tells us 1 Da $= 1.661 \times 10^{-27}$ kg. The velocity of a 500 Da ion with a charge $z = 1$ is

$$\text{v} = \sqrt{\frac{2zeV}{m}} = \sqrt{\frac{2(1)(1.602 \times 10^{-19}\,\text{C})(20\,000\,\text{V})}{(500\,\text{Da})(1.661 \times 10^{-27}\,\text{kg/Da})}} = 87.839\,819 \text{ km/s}$$

The time to transit 2.000 m is $t = $ distance/velocity $= (2.000\text{ m})/(8.783\,981\,9 \times 10^4\text{ m/s})$ $= 22.768\,717$ μs. We retain extra insignificant digits because we will need to subtract two numbers to find a small difference in the next step.

The same calculation for $m/z = 500.01$, which is just resolved from m/z 500, gives a transit time of $22.768\,945$ μs. The difference in transit time is $0.000\,23$ μs $= 0.23$ ns. Two resolved ions reach the detector 0.23 ns apart. The instrument must collect multiple data points in 0.23 ns to see the shapes of the two peaks. Fast electronics are an enabling technology in mass spectrometry.

TEST YOURSELF Would the answers be different if $m/z = 500$ were an ion with a mass of 1 000 Da and $z = 2$? (**Answer:** no)

A limiting factor in resolution of a time-of-flight spectrometer is that ions ejected from the source region at the upper left of Figure 22-16 have a range of kinetic energies. Therefore, ions with the same m/z reach the detector over a range of times, giving rise to a broad peak. *Orthogonal injection* of ions shown in Figure 22-17 reduces the spread of kinetic energy and thereby reduces the width of a peak reaching the detector.

Ions from a source such as electrospray (Section 22-4) pass through a quadrupole *ion guide* in Figure 22-17, which permits a broad range of m/z ions to pass along the y-axis, while confining them in the z- and x-directions. The quadrupole is the same as a transmission quadrupole mass separator in Figure 22-14, except that only the radio-frequency oscillating voltage (not the constant voltage) is applied to the rods. A quadrupole (or hexapole with six rods or octapole with eight rods) operated in radio frequency mode only is called an ion guide because it efficiently transfers ions from one end to the other. Neutral molecules escape between the rods and are extracted by the vacuum system. The ion guide is a pipe that guides ions from a region of high pressure to a region of low pressure, while allowing neutral molecules to be removed. Collisions of ions with background gas in the ion guide reduce the spread of kinetic energy to near-thermal values. The combination of small orifices, quadrupole ion guide, and ion optics in Figure 22-17 produces a narrow beam of ions with a small spread of energy. Periodic voltage pulses on the backplate of the time-of-flight ion source expel ions with a small range of kinetic energy into the drift region of the spectrometer.

Three-Dimensional Quadrupole Ion-Trap Mass Spectrometer

The **three-dimensional quadrupole ion-trap** mass spectrometer[22] is a compact device that is well suited as a chromatography detector. In Figure 22-18, substances emerging from the chromatography column enter the cavity of the three-dimensional ion-trap mass analyzer from the lower left through a heated transfer line. The gate electrode periodically admits electrons from the filament at the top into the cavity through holes in the end cap. Molecules undergo electron ionization in the cavity formed by the two end caps and a ring electrode, all of which are electrically isolated from one another. Alternatively, chemical ionization is achieved by adding a reagent gas such as methane to the cavity, ionizing it with the electron beam to form reagent ions, and then allowing reagent ions to react with analyte molecules. Some gas chromatography–mass spectrometry systems and all liquid chromatography–mass spectrometry systems produce ions outside the ion trap and inject them into the trap.

A constant-frequency radio-frequency voltage applied to the central ring electrode causes ions to circulate in stable, three-dimensional orbits in the cavity, with the lowest m/z in the outermost orbits. Increasing the amplitude of the radio-frequency voltage destabilizes the orbits of ions of one m/z value at a time, sending them flying out of the two end caps. Ions expelled through the lower end cap in Figure 22-18 are detected by the electron multiplier with high sensitivity (~1–10 pg). Scans from m/z 10 to 650 can be conducted eight times per second. Resolving power is 1 000–4 000, m/z accuracy is 0.1, and the maximum m/z is ~20 000. In other mass analyzers, a small fraction of ions reaches the detector. With a three-dimensional ion trap, half of the ions reach the detector, giving this device 10 to 100 times more sensitivity than the transmission quadrupole.

The quadrupole ion trap contains He gas at 10^{-6} bar, which is higher than the allowed pressure in a transmission quadrupole (10^{-9} bar) or a time-of-flight instrument (10^{-10} bar). Gas in the ion trap cools the ions by collisions, which absorb excess vibrational and rotational energy from the ions. Cool ions have less random displacement from the ideal orbits established by the electric field of the quadrupole. Lower pressure is required in other mass filters to increase the mean free path of ions so that they are not diverted from the intended trajectory by collisions with background gas.

The capacity of the three-dimensional quadrupole to hold ions is limited by the small volume enclosed by the electrodes and by the total charge of the ions, which can alter the electric field inside the quadrupole. The field created by the ions is called the *space charge*. When the space charge is too great, performance of the ion trap is degraded. Mass spectrometers have feedback controls that limit the space charge to tolerable levels by controlling the number of ions allowed in the trap.

Linear Quadrupole Ion-Trap Mass Spectrometer

The **linear quadrupole ion trap** has higher trapping efficiency and higher storage capacity than the three-dimensional quadrupole ion trap. The quadrupole mass filter in Figure 22-14 used a direct current voltage and a radio-frequency voltage to select ions of a particular m/z to be transmitted through the filter. The linear ion trap in Figure 22-19 adds sections at each end of the quadrupole to create a potential well. If the ends are sufficiently positive with respect to the center section, cations become trapped in the center section. Ions are confined in the radial direction (the xy-plane) by a radio-frequency field applied to the central section. By manipulating the voltages, ions of a specific m/z value can be expelled through slits in the x-direction or out the ends in the z-direction to one or more detectors.

Three-dimensional quadrupole ion traps have low detection limits, but they are limited by the number of ions they can hold without exceeding the allowed space charge. Also, they only trap ~5% of injected ions. The linear ion trap has even lower detection limits because it can hold ~30 times more ions and can be more than 10 times as efficient at trapping injected ions.

Linear and three-dimensional quadrupole ion traps can trade spectral acquisition rate for resolution. They are commonly operated at unit resolution with scan rates on the order of 11 000 m/z units per second. At the expense of slowing the scan rate to 27 m/z units per second, a resolution of 0.05 m/z units can be attained.

At the University of Bonn, W. Paul showed in the 1950s that ions could be manipulated by quadrupole electric fields. He received the Nobel Prize for Physics in 1989.

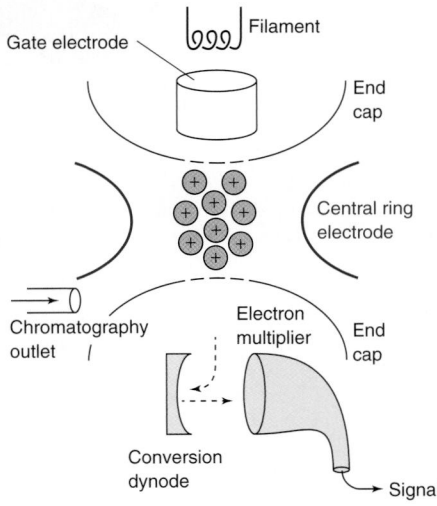

FIGURE 22-18 Schematic diagram of three-dimensional quadrupole ion-trap mass spectrometer. Mass analyzer consists of two end caps and central ring electrode.

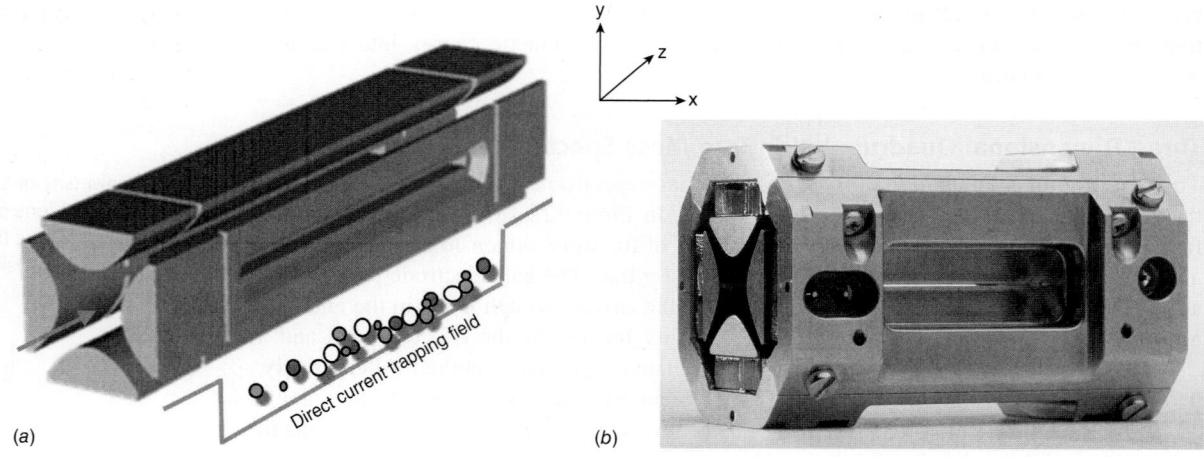

(a)

(b)

FIGURE 22-19 (*a*) **Linear ion trap.** [© Drawing from Z. Ouyang, G. Wu, Y. Song, H. Li, W. R. Plass, and R. G. Cooks, "Rectilinear Ion Trap: Concepts, Calculations, and Analytical Performance of a New Mass Analyzer," *Anal. Chem.* **2004,** *76,* 4595. Reprinted with permission © 2004, American Chemical Society.] (*b*) **Photograph of LTQ XL linear ion trap.** [Courtesy Thermo Fisher Scientific, San Jose, CA.]

Orbitrap Mass Spectrometer

The orbitrap is a high-resolution mass analyzer that does not require a magnetic field or a radio-frequency field (Figure 22-20). The **orbitrap mass spectrometer** provides a resolving power of ~140 000–480 000 at m/z 100 (Figure 22-21), m/z accuracy of 1–5 ppm with external calibration, upper m/z limit of ~6 000, and a dynamic range of several thousand. With internal calibration standards, sub-part-per-million m/z accuracy is attainable. The cutaway view in Figure 22-20a shows precisely machined central and outer electrodes. The central electrode is held at −5 kV, while the two outer electrodes (Figure 22-20b) are close to ground potential and electrically isolated from each other. Figure 22-20b shows electric field vectors perpendicular to the axis of symmetry (*z*) near the center of the orbitrap, but increasingly angled toward the center of the orbitrap at increasing distance from the center.

A packet of ions is introduced perpendicular to the plane of Figure 22-20b at the indicated point. The electric field pushes ions into an orbit toward the center of the orbitrap in Figure 22-20c. Initial momentum carries ions from the right side of the orbitrap to the left side until the electric field is strong enough to push the ions back to the right. When ions have

FIGURE 22-20 (*a*) Cutaway drawing of orbitrap. (*b*) Electric field in one longitudinal plane of orbitrap. (*c*) Initial path of ion entering the orbitrap and stable path for successive orbits. [Information from A. Makarov, "Electrostatic Axially Harmonic Orbital Trapping: A High-Performance Technique of Mass Analysis," *Anal. Chem.* **2000,** *72,* 1156. Reprinted with permission © 2000, American Chemical Society.] (*d*) Photograph of standard orbitrap with half of outer sheath removed. The corresponding length of the exposed section of a compact orbitrap is 21 mm. [Courtesy Thermo Fisher Scientific, San Jose, CA.]

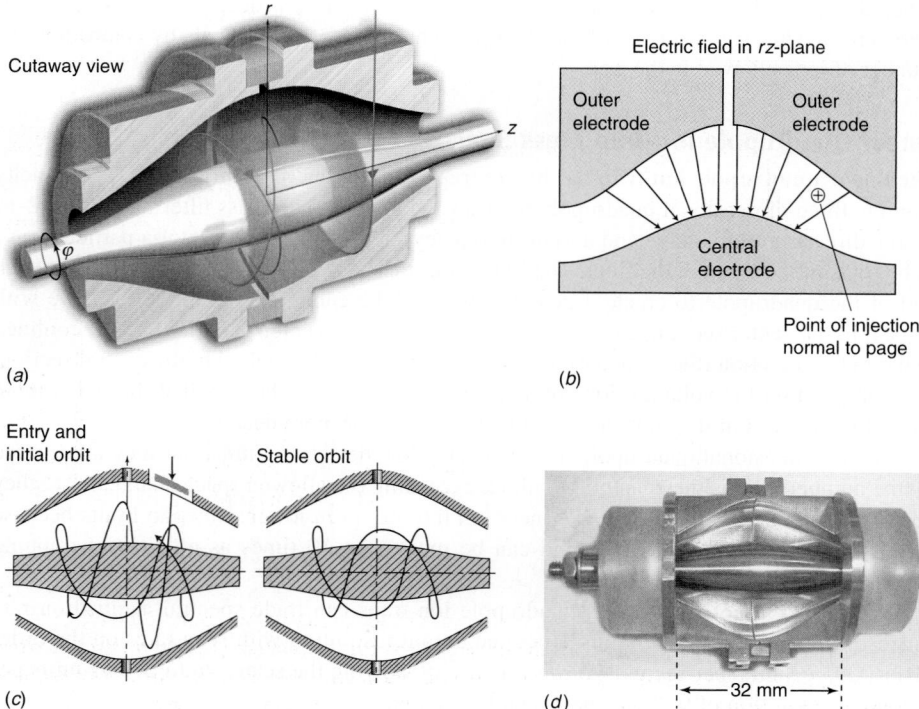

Cutaway view

(a)

Electric field in *rz*-plane

Outer electrode

Outer electrode

Central electrode

Point of injection normal to page

(b)

Entry and initial orbit

Stable orbit

(c)

32 mm

(d)

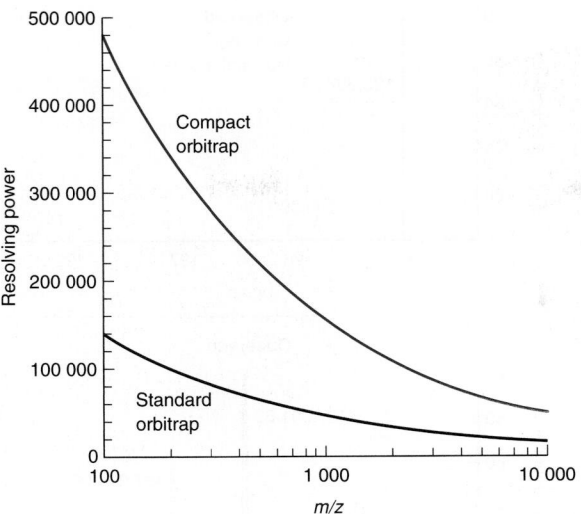

FIGURE 22-21 Orbitrap resolving power ($m/m_{1/2}$) for 0.76 s scan time. Resolving power decreases in proportion to $\sqrt{m/z}$ and increases in proportion to the scan time. Longer scans can achieve higher resolution, but the practical limit for scan time is ~3 s. [Data from R. A. Zubarev and A. Makarov, "Orbitrap Mass Spectrometry," *Anal. Chem.* **2013,** *85,* 5288.]

traveled far enough to the right, they are pushed back to the left. They circle back and forth around the central electrode as long as they do not strike another molecule. Maintaining the unperturbed orbit long enough requires the best vacuum of any mass spectrometer at $\sim 10^{-13}$ bar—providing a mean free path of 100 km. The oscillation frequency of an ion between the right and left halves of the orbitrap is proportional to $1/\sqrt{m/z}$.

Ions oscillating between the two halves of the orbitrap induce an opposite charge called the *image charge* in the outer electrode. A packet of cations in the right half of the orbitrap attracts electrons in the outer right electrode. A packet of cations in the left half of the orbitrap attracts electrons in the outer left electrode. An amplifier connected to the two halves of the split outer electrode measures the image current oscillating in synchrony with the ions. The orbitrap contains ions with different m/z values, each creating a component of current with a different frequency. The observed signal is the sum of currents from all m/z values. After recording the current for a predetermined time (~0.1–3 s), a computer decomposes the current into its component frequencies—and hence component m/z values—through a Fourier transform (Section 20-5).

Ions must be injected into the orbitrap in one small packet. Figure 22-22 shows one way this can be done. Ions from an electrospray source (Section 22-4) are accumulated in a linear

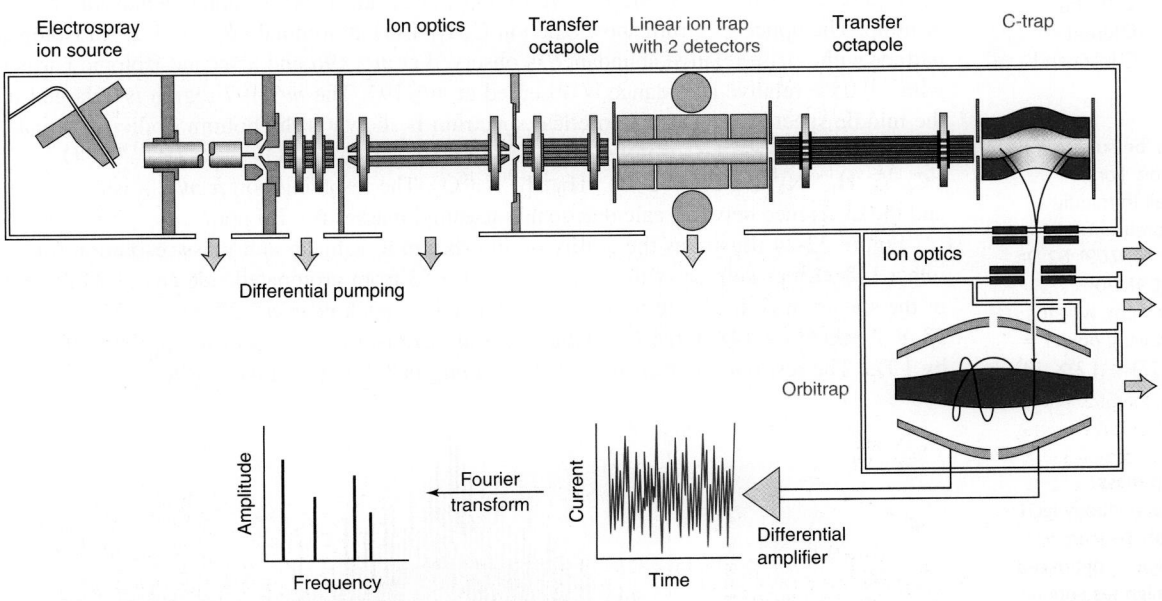

FIGURE 22-22 Arrangement of linear ion trap and C-trap for collecting ions, compressing them into a small packet, and feeding the packet into the orbitrap. [Reprinted with permission from A. Makarov, E. Denisov, A. Kholomeev, W. Balschun, O. Lange, K. Strupat, and S. Horning, "Performance Evaluation of a Hybrid Linear Ion Trap/ Orbitrap Mass Spectrometer," *Anal. Chem.* **2006,** *78,* 2113. © 2006, American Chemical Society.]

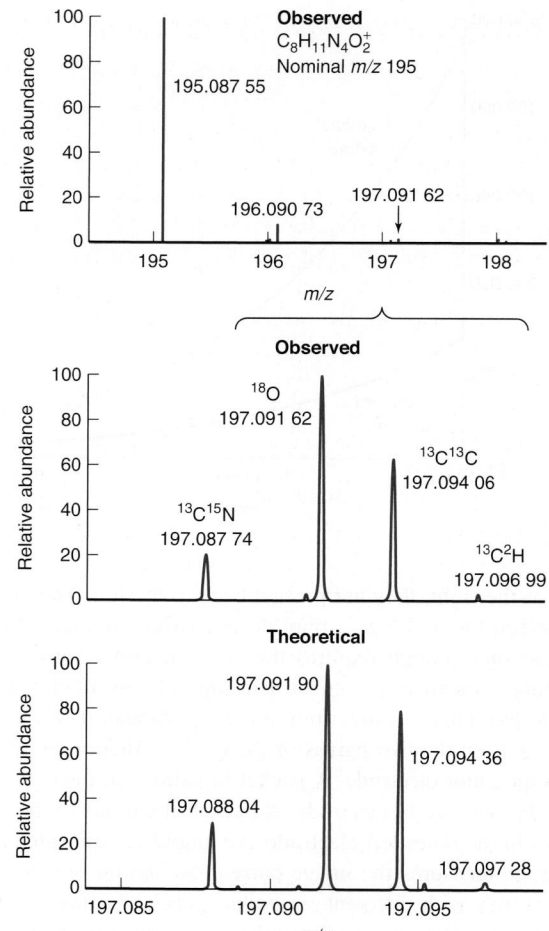

FIGURE 22-23 Example of high resolving power ($m/m_{1/2} = 1.3 \times 10^6$) and mass accuracy (1.5 ppm) achievable by orbitrap mass spectrometer with 3-s detection time and external calibration. Upper spectrum shows m/z 195–198 region. Middle spectrum enlargement of the m/z 197 region shows individual isotopologues. Lower spectrum is theoretical pattern for m/z 197. [E. Denisov, E. Damoc, O. Lang, and A. Makarov, "Orbitrap Mass Spectrometry with Resolving Powers Above 1,000,000," *Intl. J. Mass. Spectrom.* **2012,** *325,* 80, Figure 2. © Thermo Fisher Scientific (Bremen) GmbH; Republished with permission from John Wiley & Sons Inc. Permission conveyed through Copyright Clearance Center, Inc.]

Isotopologues: molecules or ions with different numbers of isotopes of one or more elements: $^{12}C_2H_4O$ and $^{12}C^{13}CH_4O$

Isotopomers: molecules or ions with the same isotopic composition, but different locations of the isotopes: $^{13}CH_3{}^{12}CH{=}O$ and $^{12}CH_3{}^{13}CH{=}O$

For theoretical mass of a cation, be sure to subtract the mass of one electron. For $^{12}C_8{}^1H_{11}{}^{14}N_4{}^{16}O_1{}^{18}O_1^+$ (tallest peak in middle spectrum of Figure 22-23), theoretical mass = 197.091 90 and observed mass = 197.091 62 Da. Difference in ppm = 10^6 (197.091 90 − 197.091 62)/(197.091 90) = 1.4 ppm. $w_{1/2}$ for the peak is 0.000 17 Da, so resolving power = $m/m_{1/2} = (197.091\ 62)/(0.000\ 17) = 1.2 \times 10^6$.

ion trap and then transferred in one batch to a C-trap, which electrodynamically squeezes them into a small packet. The packet is ejected through ion optics into the orbitrap. During the ~0.1-ms injection period, the electric field in the orbitrap is ramped up in such a manner that the ions begin to orbit the central electrode. After the voltage on the central electrode reaches its constant level for stable orbits, detection begins.

Figure 22-23 illustrates the high resolving power and mass accuracy achievable by an orbitrap. The upper spectrum shows the ion $C_8H_{11}N_4O_2^+$ at nominal m/z 195 Da. An isotopic partner with ~10% relative abundance is observed at m/z 196 and a second isotopic partner with ~0.05% relative abundance is observed at m/z 197. The m/z 197 region is enlarged in the middle spectrum and the theoretical spectrum is shown at the bottom. Individual peaks are observed for the *isotopologues* $^{12}C_7{}^{13}C_1{}^1H_{11}{}^{14}N_3{}^{15}N_1{}^{16}O_2^+$, $^{12}C_8{}^1H_{11}{}^{14}N_4{}^{16}O_1{}^{18}O_1^+$, $^{12}C_6{}^{13}C_2{}^1H_{11}{}^{14}N_4{}^{16}O_2^+$, and $^{12}C_7{}^{13}C_1{}^1H_{10}{}^2H_1{}^{14}N_4{}^{16}O_2^+$. The resolving power $m/m_{1/2}$ is 1.3×10^6 and the difference between calculated and measured masses is ~1.5 ppm.

Figure 22-24 illustrates the ability of an orbitrap to achieve unit mass resolution for an intact 148-kDa protein carrying a net charge of +53 from protonated side chains. Only part of the spectrum is displayed in the figure. The tallest peak near m/z 2 800.69 has a mass of $53 \times 2\ 800.69 = 148\ 436.6$ Da. Adjacent peaks are isotopologues of the protein differing by 1 Da. The resolving power measured by $m/m_{1/2}$ is 3.3×10^5 at m/z 2 800.

FIGURE 22-24 Partial orbitrap mass spectrum of 148 kDa monoclonal antibody IgG1 with charge of +53 obtained with 3-s acquisition time and many conditions optimized to observe the intact protein at high resolution. [J. Shaw and J. S. Brodbelt, "Extending the Isotopically Resolved Mass Range of Orbitrap Mass Spectrometers," *Anal. Chem.* **2013,** *85,* 8313, Figure 4C. Reprinted with permission © 2013, American Chemical Society.]

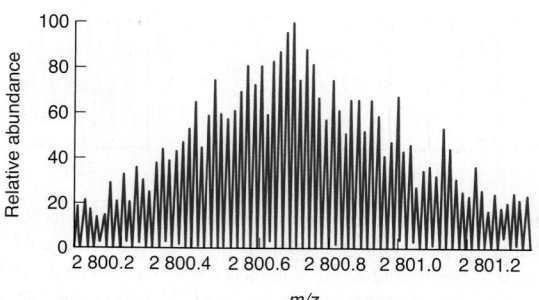

22-4 Chromatography–Mass Spectrometry Interfaces

Mass spectrometry is widely used as the detector in chromatography to provide both qualitative and quantitative information. The spectrometer can be highly selective for the analyte of interest. This selectivity eases the requirements for sample preparation or complete chromatographic separation of components in a mixture, and it increases the signal-to-noise ratio.

Mass spectrometry requires high vacuum to prevent molecular collisions during ion separation. Chromatography is inherently a high-pressure technique. The problem in marrying the two techniques is to remove the huge excess of matter between the chromatograph and the spectrometer. Gas chromatography felicitously evolved to employ narrow capillary columns whose eluate does not overwhelm the pumping capabilities of the mass spectrometer vacuum system. The capillary column is connected directly to the inlet of the mass spectrometer through a heated transfer line, as in Figures 22-14 and 22-18.[23]

Liquid chromatography creates a huge volume of gas when solvent vaporizes at the interface between the column and the mass spectrometer.[24] Almost all of this gas must be removed prior to ion separation. Nonvolatile mobile-phase additives (such as phosphate buffer), which are common in chromatography, must be avoided for mass spectrometry. *Electrospray ionization* and *atmospheric pressure chemical ionization* are dominant methods for introducing eluate from liquid chromatography into a mass spectrometer.

Volatile buffer components and additives for liquid chromatography that are compatible with mass spectrometry include NH_3, HCO_2H, CH_3CO_2H, CCl_3CO_2H, $(CH_3)_3N$, and $(C_2H_5)_3N$. Avoid additive concentrations >20 mM and detergent concentrations >10 μM.

Electrospray Ionization

Electrospray ionization,[26] is illustrated in Figure 22-25a. Liquid from the chromatography column enters the steel nebulizer capillary at the upper left, along with a coaxial flow of N_2 gas. For positive ion mass spectrometry, the nebulizer is held at 0 V and the spray chamber is held at −3 500 V. For negative ion mass spectrometry, all voltages are reversed. The strong electric field at the nebulizer outlet, combined with the coaxial flow of N_2 gas, creates a fine aerosol of charged particles.

Typically, but not always, *ions that vaporize from aerosol droplets were already in solution in the chromatography column.* For example, protonated bases (BH^+) and ionized acids (A^-) can be observed. Other gas-phase ions arise from complexation between analyte, M (which could be neutral or charged), and stable ions from the solution. Examples include

J. B. Fenn received part of the Nobel Prize in Chemistry in 2002[25] for electrospray ionization. Fenn stated that in ejecting proteins into the gas phase, "We learned to make elephants fly." K. Tanaka received part of the same prize for laser desorption/ionization, described in Box 22-4.

You, the chemist, need to adjust the pH of the chromatography solvent to favor BH^+ or A^- for mass spectrometric detection.

MH^+ (mass = M + 1) $M(K^+)$ (mass = M + 39)

$M(NH_4^+)$ (mass = M + 18) $M(HCO_2^-)$ (mass = M + 45)

$M(Na^+)$ (mass = M + 23) $M(CH_3CO_2^-)$ (mass = M + 59)

For protein electrospray, it is common to find multiply charged ions such as $[M + nH]^{n+}$ and, sometimes, $[M + nNa]^{n+}$ or $[M + nNH_4]^{n+}$. There is little fragmentation in electrospray.

Positive ions from the aerosol are attracted toward the glass capillary leading into the mass spectrometer by an even more negative potential of −4 500 V. Gas flowing from atmospheric pressure in the spray chamber transports ions to the right through the capillary to the exit, where the pressure is reduced to ~3 mbar by a vacuum pump.

Figure 22-25b provides more detail on ionization. Voltage imposed between the steel nebulizer capillary and the inlet to the mass spectrometer creates excess charge in the liquid by redox reactions. If the nebulizer is positively biased, oxidation enriches the liquid in positive ions by reactions such as

Electrospray requires low buffer concentration so that buffer ions do not overwhelm analyte ions in the mass spectrum. Low-surface-tension organic solvent is better than water. In reversed-phase chromatography (Section 25-1), it is good to use a stationary phase that strongly retains analyte so that a high fraction of organic solvent can be used. A flow rate of 0.05 to 0.4 mL/min is best for electrospray.

$$Fe(s) \rightarrow Fe^{2+} + 2e^-$$

$$H_2O(l) \rightarrow \frac{1}{2}O_2 + 2H^+ + 2e^-$$

Electrons from the oxidation flow through the external circuit and eventually neutralize gaseous positive ions at the inlet to the mass spectrometer. It is possible for analyte to be chemically altered by species such as $HO^\bullet$ generated during electrospray.[27]

Charged liquid exiting the capillary forms a cone and then a fine filament and finally breaks into a spray of fine droplets (see Figure 22-25c and the opening of this chapter). It is thought that a droplet shrinks to ~1 μm by solvent evaporation until the repulsive force of the excess charge equals the cohesive force of surface tension. At that point, the droplet breaks up by ejecting tiny droplets with diameters of ~10 nm. They evaporate, leaving their cargo of ions in the gas phase. Aerodynamic forces might also contribute to droplet breakup.[28]

In electrospray, little fragmentation of analyte occurs and mass spectra are simple. Fragmentation can be intentionally increased by **collisionally activated dissociation** (also

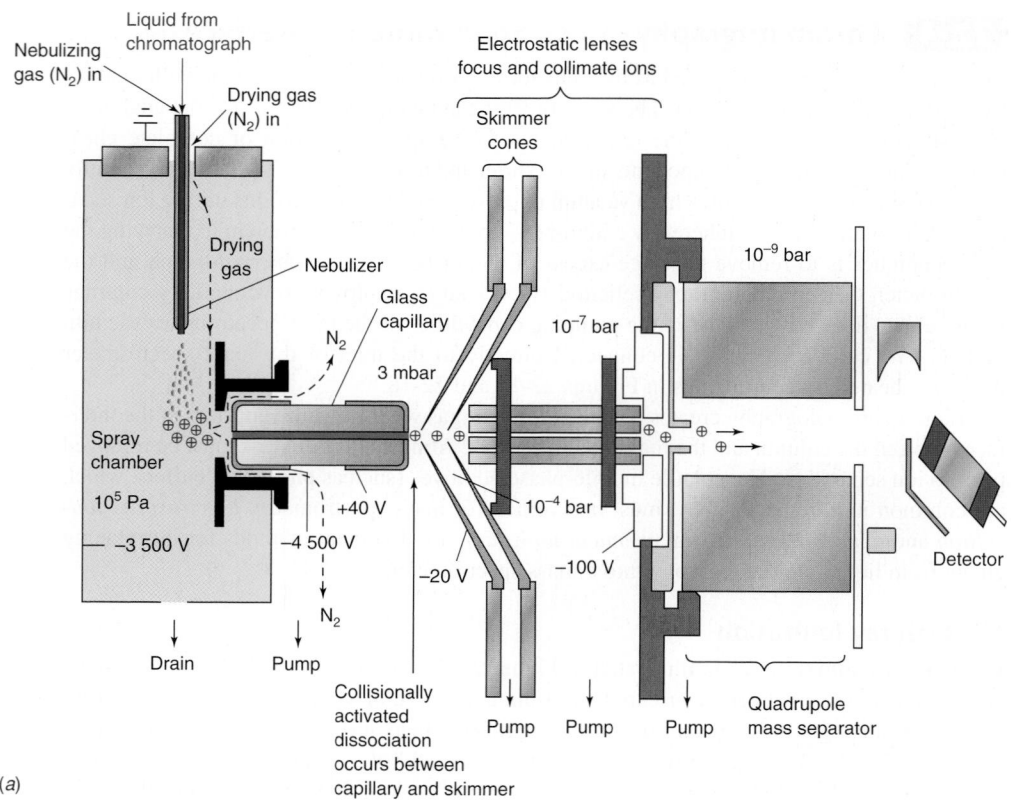

(a)

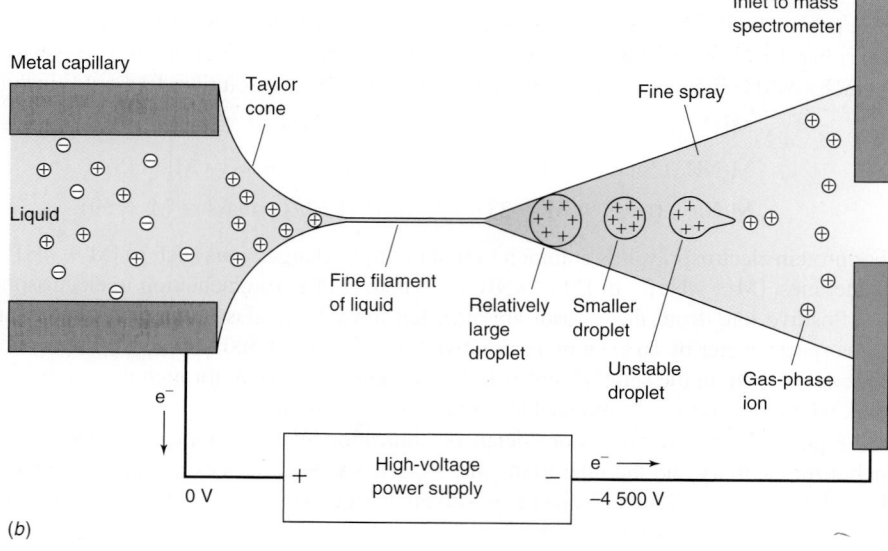

(b)

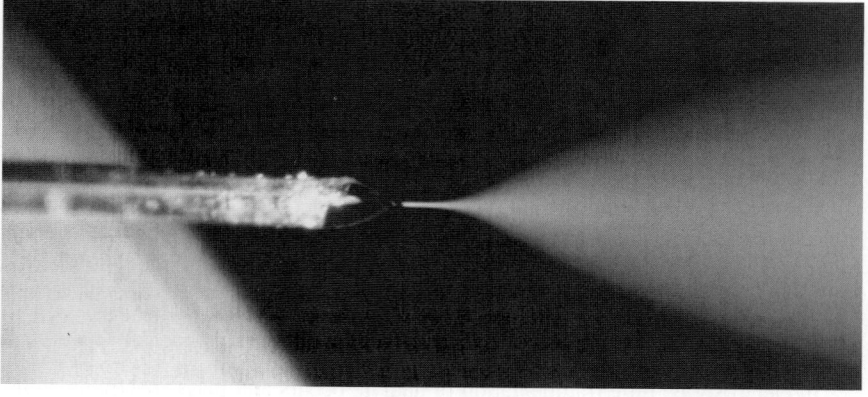

(c)

FIGURE 22-25 (*a*) Pneumatically assisted electrospray interface for mass spectrometry. (*b*) Gas-phase ion formation. [Information from E. C. Huang, T. Wachs, J. J. Conboy, and J. D. Henion, "Atmospheric Pressure Ionization Mass Spectrometry," *Anal. Chem.* **1990,** *62,* 713A and P. Kebarle and L. Tang, "From Ions in Solution to Ions in the Gas Phase," *Anal. Chem.* **1993,** *65,* 972A.] (*c*) Electrospray from a silica capillary. [Courtesy R. D. Smith, Pacific Northwest Laboratory, Richland, WA.]

referred to as *collision-induced dissociation*) with the 3 mbar of background N_2 gas in the region between the glass capillary and the skimmer cone in Figure 22-25a. In Figure 22-25a, the outlet of the glass capillary is coated with a metal layer that is held at +40 V. The potential difference between the metal skimmer cone and the capillary is $-20 - (40) = -60$ V. Positive ions accelerated through 60 V collide with N_2 molecules with enough energy to break into fragments. Adjusting the skimmer cone voltage controls the degree of fragmentation. A small voltage difference favors molecular ions, whereas a large voltage difference creates fragments that aid in identification of analyte. Collisionally activated dissociation also tends to break apart complexes such as $M(Na^+)$.

For example, with a cone voltage difference of –20 V, the positive ion spectrum of the drug acetaminophen exhibits a base peak at m/z 152 for the protonated molecule, $[M+H]^+$ (colored species in the margin). A smaller peak at m/z 110 probably corresponds to the fragment shown in the margin. When the cone voltage difference is −50 V, collisionally activated dissociation decreases the peak at m/z 152 and increases the fragment peak at m/z 110. The negative ion spectrum has a large peak at m/z 150 for $[M-H]^-$. As the cone voltage difference is raised from +20 V to +50 V, this peak decreases and a fragment at m/z 107 increases.

Figure 22-26 shows an electrospray interface for capillary electrophoresis. The silica capillary is contained in a stainless steel capillary held at the required outlet potential for electrophoresis. The steel makes electrical contact with the liquid inside the silica by a liquid sheath flowing between the capillaries. The sheath liquid, which is typically a mixed organic-aqueous solvent, constitutes ~90% of the aerosol.

Atmospheric Pressure Chemical Ionization

In **atmospheric pressure chemical ionization,** heat and a coaxial flow of N_2 convert eluate into a fine mist from which solvent and analyte evaporate (Figures 22-27 and 22-28). Like chemical ionization in the ion source of a mass spectrometer, atmospheric pressure chemical ionization creates new ions from gas-phase reactions between ions and molecules. The distinguishing feature of this technique is that a high voltage is applied to a metal needle in the path of the aerosol. An electric *corona* (a plasma containing charged particles) forms around the needle, injecting electrons into the aerosol and creating ions. For example, protonated analyte, MH^+, can be formed in the following manner:

$$N_2 + e^- \rightarrow N_2^{+\cdot} + 2e^-$$
$$N_2^{+\cdot} + 2N_2 \rightarrow N_4^{+\cdot} + N_2$$
$$N_4^{+\cdot} + H_2O \rightarrow H_2O^{+\cdot} + 2N_2$$
$$H_2O^{+\cdot} + H_2O \rightarrow H_3O^+ + {}^{\cdot}OH$$
$$H_3O^+ + nH_2O \rightarrow H_3O^+(H_2O)_n$$
$$H_3O^+(H_2O)_n + M \rightarrow MH^+ + (n+1)H_2O$$

Analyte M might also form a negative ion by electron capture:

$$M + e^- \rightarrow M^{-\cdot}$$

A molecule, X—Y, in the eluate might create a negative ion by the reaction

$$X{-}Y + e^- \rightarrow X^{\cdot} + Y^-$$

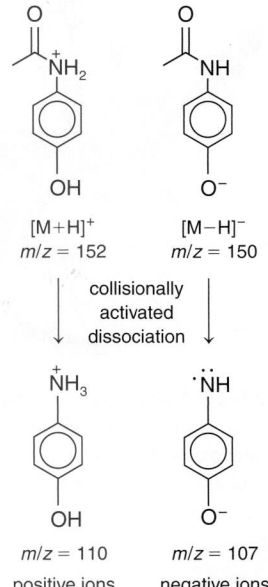

Collisionally activated dissociation of acetaminophen:

$[M+H]^+$ $m/z = 152$

$[M-H]^-$ $m/z = 150$

collisionally activated dissociation

$m/z = 110$ positive ions

$m/z = 107$ negative ions

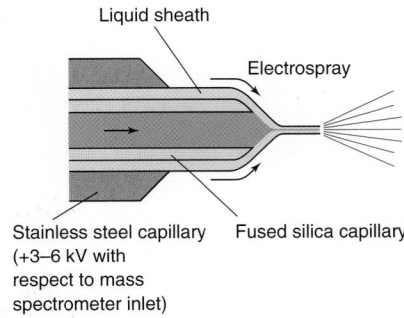

FIGURE 22-26 Electrospray interface for capillary electrophoresis–mass spectrometry.

Unlike electrospray, atmospheric pressure chemical ionization *creates gaseous ions from neutral analyte molecules*. Analyte must have some volatility. For nonvolatile molecules such as sugars and proteins, electrospray can be used.

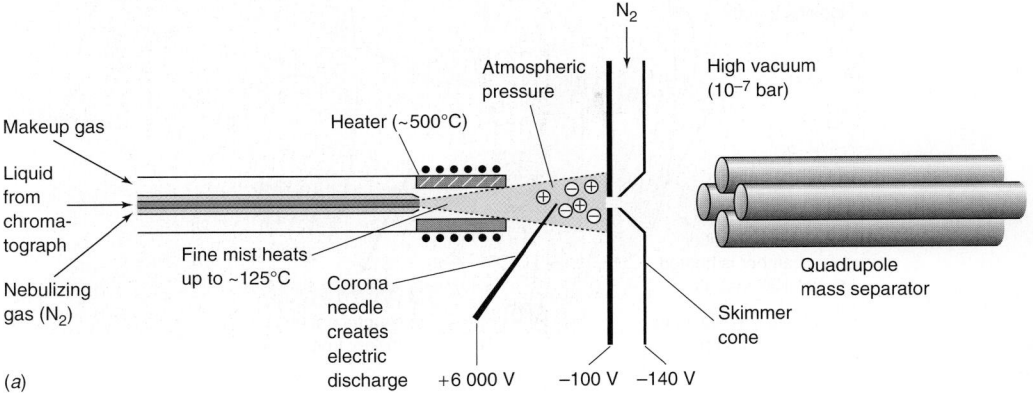

FIGURE 22-27 Atmospheric pressure chemical ionization interface between a liquid chromatography column and a mass spectrometer. A fine aerosol is produced by the nebulizing gas flow and the heater. The electric discharge from the corona needle creates gaseous ions from the analyte. [Information from E. C. Huang, T. Wachs, J. J. Conboy, and J. D. Henion, "Atmospheric Pressure Ionization Mass Spectrometry," *Anal. Chem.* **1990**, *62*, 713A.]

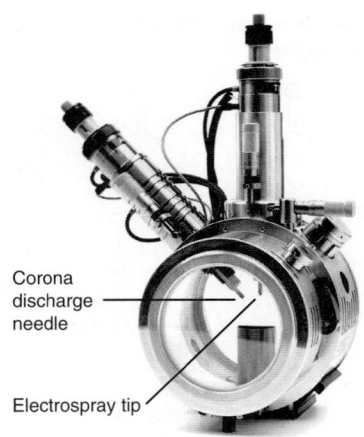

FIGURE 22-28 Dual-purpose interface for liquid chromatography–mass spectrometry performs electrospray through the electrospray tip or atmospheric pressure chemical ionization by activating the corona discharge needle. [Courtesy AP Sciex.]

The species Y^- could abstract a proton from a weakly acidic analyte, AH:

$$AH + Y^- \rightarrow A^- + HY$$

Atmospheric pressure chemical ionization handles a variety of analytes and accepts chromatography flow rates up to 2 mL/min. Generally, to be observed, analyte M must be capable of forming the protonated ion, MH^+. Atmospheric pressure chemical ionization tends to produce single-charge ions and is unsuitable for macromolecules such as proteins. There is little fragmentation, but the skimmer cone voltage difference can be adjusted to favor *collisionally activated dissociation* into a small number of fragments.

Direct Electron Ionization[29]

The direct ionization interface for capillary liquid chromatography in Figure 22-29 is analogous to the gas chromatography interface in Figure 22-14. Electron ionization cannot tolerate much background solvent or the process becomes chemical ionization. Therefore, direct electron ionization is restricted to use with capillary liquid chromatography columns with an inner diameter of 75 μm and solvent flow rates of ~100–500 nL/min so that the vacuum pump can remove solvent as rapidly as it evaporates. A fused silica capillary connected to the chromatography column is drawn to a 5-μm-diameter tip where it enters the ionization chamber through the rear of the cylindrical wall in Figure 22-29. The chamber is heated to 150–350°C, but most of the capillary is thermally isolated so liquid does not evaporate inside the capillary. Liquid exiting the capillary creates aerosol droplets from which solvent evaporates during their 8 mm transit across the evacuated ionization chamber. Evaporation of remaining solvent is completed when droplets strike the hot stainless steel conical surface of the repeller at the back of the chamber. Molecular (M^+) and fragment (A^+, B^+) ions from electron ionization are ejected into the mass spectrometer.

The resulting mass spectrum is similar to those in the library of 70 eV electron ionization spectra[7] and can be identified with library search software. Direct electron ionization is well suited for chromatography with aqueous acetonitrile and methanol solvents, tolerates non-volatile buffers such as 10 mM phosphate, and is insensitive to matrix effects that commonly enhance or suppress the signal in other ionization methods.[30] Only about 1 molecule in 10^4 becomes ionized. The detection limit is ~1–10 ng for a full mass spectrum and 100 times lower for selected ion monitoring.

Photoionization

The chromatography–mass spectrometry interface in Figure 22-30 uses intense vacuum ultraviolet radiation from a source such as a 118-nm (10.5 eV) laser, a 126-nm (9.8 eV) lamp,[31] or a krypton lamp (10.0 and 10.6 eV) to remove an electron from molecules whose ionization energy is less than the photon energy. The species $M^{+\bullet}$ or MH^+ are formed with little fragmentation. Ion formation can be enhanced by addition of dopants such as toluene or acetone at higher concentration than analyte. Ionized dopant can then ionize analyte

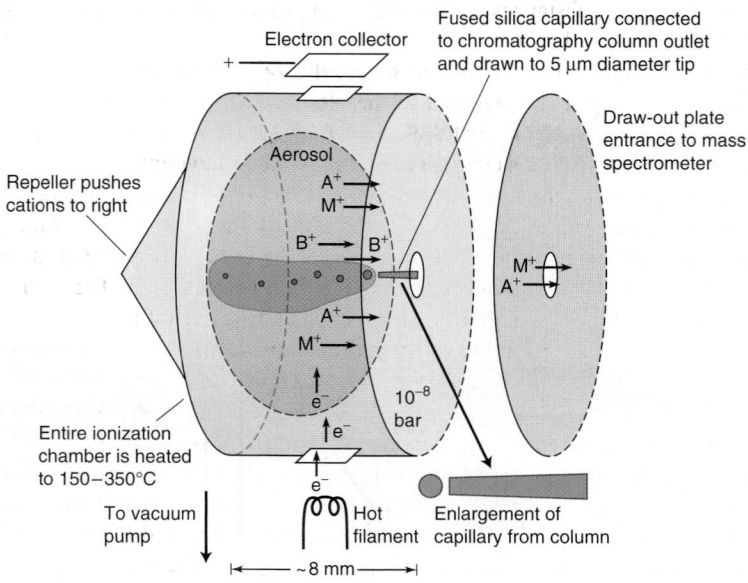

FIGURE 22-29 Direct electron ionization interface for capillary liquid chromatography. The fused silica capillary inlet goes through the cylindrical metal wall of the ionization chamber.

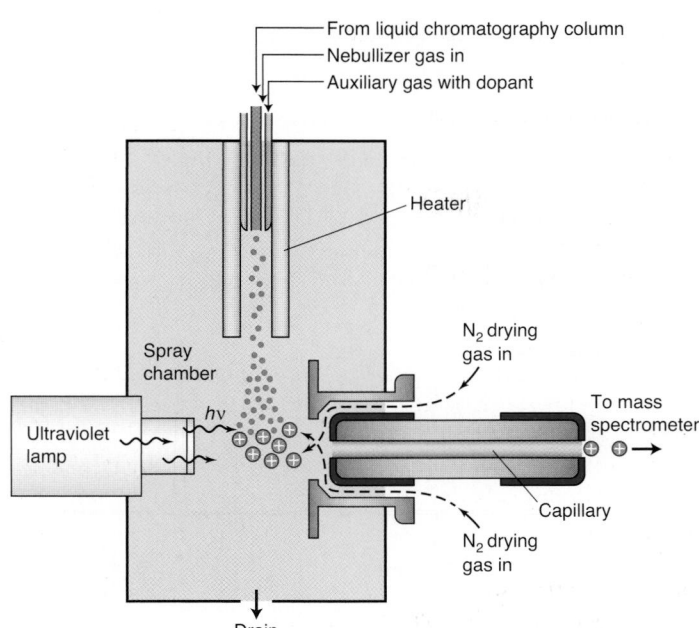

From liquid chromatography column
Nebullizer gas in
Auxiliary gas with dopant

Heater

Spray chamber

N_2 drying gas in

Ultraviolet lamp

hv

To mass spectrometer

Capillary

N_2 drying gas in

Drain

Possible chemistry for generating cations:

$$D + hv \ (10 \ eV) \rightarrow D^+ + e^-$$
$$D^+ + M \rightarrow D + M^+$$
$$D^+ + S \rightarrow [D - H] + SH^+$$
$$SH^+ + M \rightarrow S + MH^+$$

D = dopant (toluene, acetone, anisole)
S = solvent (methanol, acetonitrile, water)
M = analyte

Possible chemistry for generating anions:

$$D + hv \ (10 \ eV) \rightarrow D^+ + e^-$$
$$M + e^- \rightarrow M^-$$

FIGURE 22-30 Photoionization source for liquid chromatography–mass spectrometry. For an early report on this subject, see D. B. Robb, T. R. Covey, and A. P. Bruins, "Atmospheric Pressure Photoionization: An Ionization Method for Liquid Chromatography–Mass Spectrometry," *Anal. Chem.* **2000,** *72,* 3653.

molecules. Most organic molecules can be observed by photoionization. Ionization energies of the common chromatography solvents H_2O, CH_3OH, and CH_3CN are too high for them to be ionized, so they do not interfere with the observation of solute analytes.

22-5 Chromatography–Mass Spectrometry Techniques

The power of mass spectrometry as a chromatographic detector is that it can be used to selectively monitor one or a few analytes in a complex mixture, even if the separation of the components is not perfect. We strive to attain good separations, but mass spectrometry can ease the burden on the chromatography.

Selected Ion Monitoring and Extracted Ion Monitoring

Selected ion monitoring and *extracted ion monitoring* increase the selectivity of mass spectrometry for individual analytes and decrease the response to everything else (decreasing background noise). Figure 22-31a shows a liquid chromatogram with ultraviolet absorbance detection of a mixture of herbicides (designated 1–6) spiked at a level near 1 ppb into river water. (*Spiked* means deliberately added.) The broad hump underlying the analyte peaks arises from many natural substances in river water. The simplest way to use a mass spectrometer as a chromatographic detector is to add up the total current of all ions of all masses detected above a selected value. This **reconstructed total ion chromatogram** is shown in Figure 22-31b, which is just as congested as the ultraviolet chromatogram because all substances emerging at any moment contribute to the signal. This chromatogram is "reconstructed" from individual mass spectra recorded during chromatography.

To be more selective, we use **selected ion monitoring,** in which the mass spectrometer is set to monitor just a few values of *m/z* (never more than four or five in any time interval). Figure 22-31c shows the **selected ion chromatogram** for which just *m/z* 312 is monitored. The signal corresponds to MH^+ from herbicide 6, which is imazaquin. The signal-to-noise ratio in selected ion monitoring is greater than the signal-to-noise ratio in chromatograms *a* or *b* because (1) most of the spectral acquisition time is spent collecting data in a small mass range and (2) little but the intended analyte gives a signal at *m/z* 312.

An **extracted ion chromatogram** is like a selected ion chromatogram, but the extracted ion chromatogram does not benefit from taking all available time to measure just one or a few mass spectral peaks. To create an extracted ion chromatogram, the entire mass spectrum is recorded repeatedly during chromatography. Then one value of *m/z* is taken from each spectrum for display. For example, the chromatogram might display the intensity of *m/z* 312 observed as a function of time. In selected ion monitoring, only the *m/z* 312 signal would be

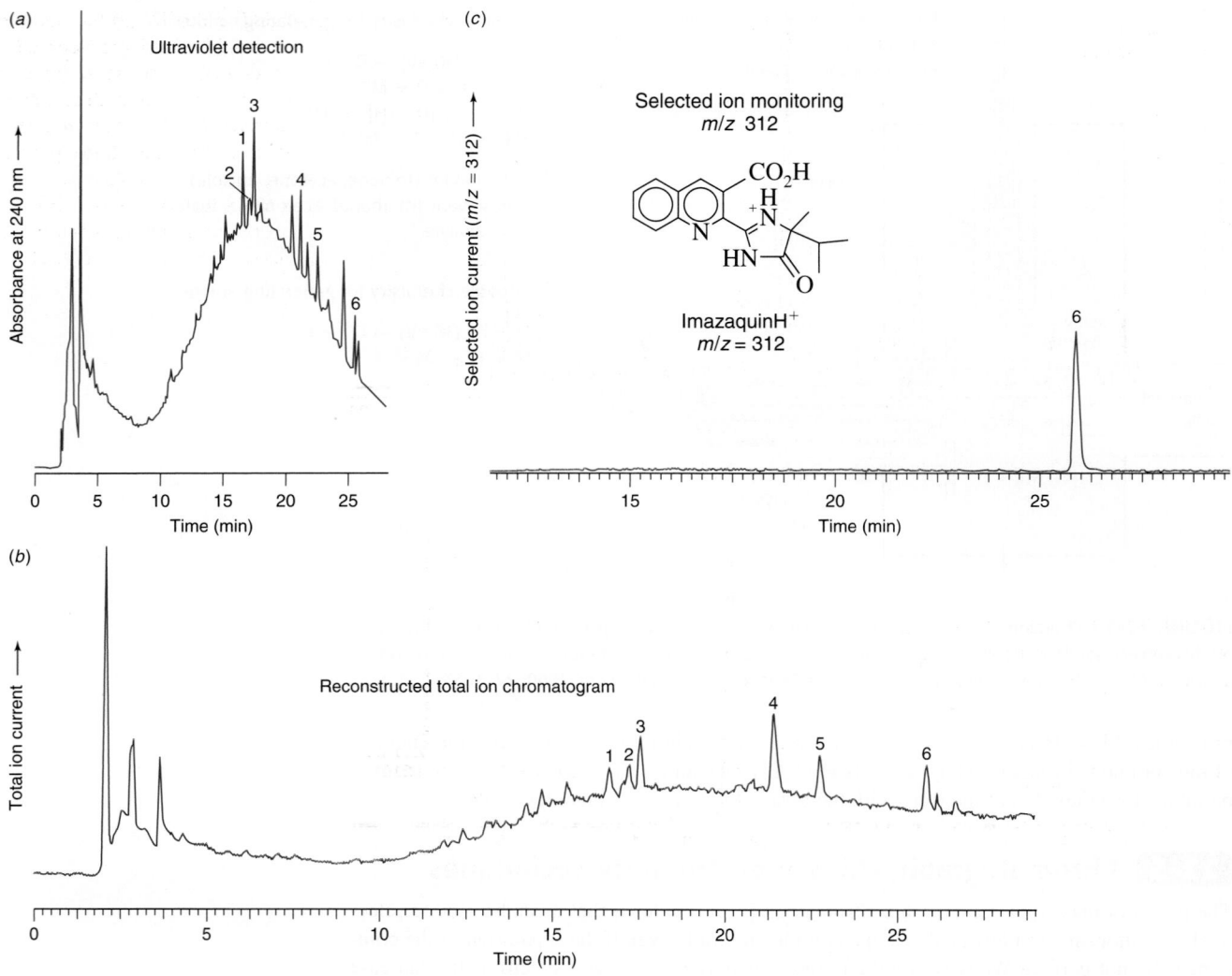

(a) Ultraviolet detection

Absorbance at 240 nm

3
1
2
4
5
6

Time (min)
0 5 10 15 20 25

(b) Reconstructed total ion chromatogram

Total ion current

4
3
1 2
5
6

Time (min)
0 5 10 15 20 25

(c) Selected ion monitoring
m/z 312

Selected ion current (m/z = 312)

ImazaquinH⁺
m/z = 312

6

Time (min)
15 20 25

FIGURE 22-31 Chromatograms of herbicides (designated 1–6) spiked into river water at a level near 1 ppb demonstrate increased signal-to-noise ratio in selected ion monitoring. (*a*) Ultraviolet detection at 240 nm. (*b*) Electrospray reconstructed total ion chromatogram. (*c*) Electrospray selected ion monitoring at *m/z* 312. [A. Laganà, G. Fago, and A. Marino, "Simultaneous Determination of Imidazolinone Herbicides from Soil and Natural Waters," *Anal. Chem.* **1998,** *70,* 121. Copyright © 1998, American Chemical Society.]

collected. For the extracted ion chromatogram, all values of *m/z* are measured, but only *m/z* 312 intensity is displayed. You should collect the entire mass spectrum when you do not know what you are looking for and you want to observe everything.

Figure 22-32 is an important example of an extracted ion chromatogram. By using liquid chromatography–time-of-flight mass spectrometry, it is possible to search for 100 pesticides at once in food extracts. The time-of-flight spectrometer provides nearly unique identification of small molecules, such as pesticides, by exact mass measurement with *m/z* uncertainty of ~1–2 ppm. Components of fruit-based soft drinks isolated by solid-phase extraction (Section 28-3) were separated by liquid chromatography to obtain the complex total ion chromatogram in the upper part of Figure 22-32. When the extracted ion chromatogram window was set for *m/z* 202.043 ± 0.01 to find the pesticide thiabendazole, one major peak at 9.31 min was observed. Combined levels of several pesticides found in the majority of tested soft drinks from several countries exceeded European allowed maximum residue levels for drinking water by factors of 10 to 35.

Selected Reaction Monitoring

Selectivity and signal-to-noise ratio in a chromatogram are increased markedly by **selected reaction monitoring,** illustrated in Figure 22-33 with a *triple quadrupole mass spectrometer*. A mixture of ions enters quadrupole Q1, which passes just one selected **precursor ion** to the second stage, Q2. The second stage is a radio frequency ion guide (as in Figure 22-17)

Selected reaction monitoring is one of several techniques carried out by consecutive mass filters. These techniques are collectively called **tandem mass spectrometry, mass spectrometry/mass spectrometry, MS/MS,** or **MS².**

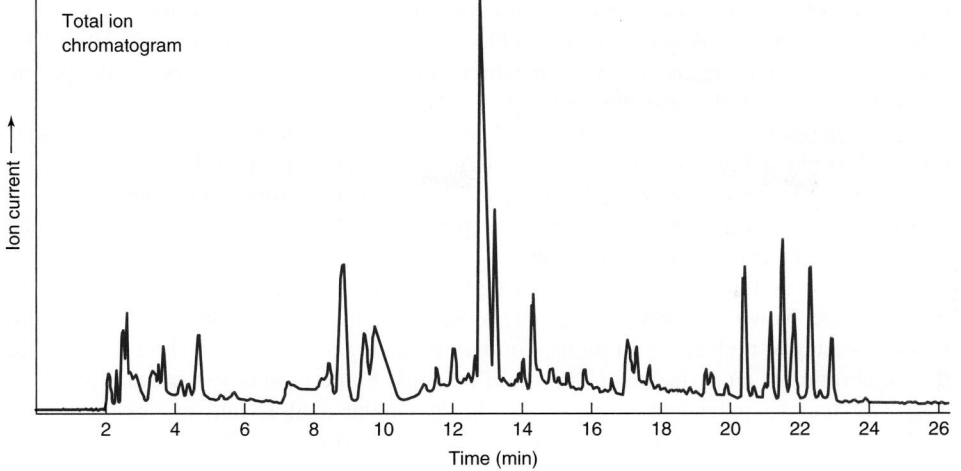

Total ion chromatogram

(a)

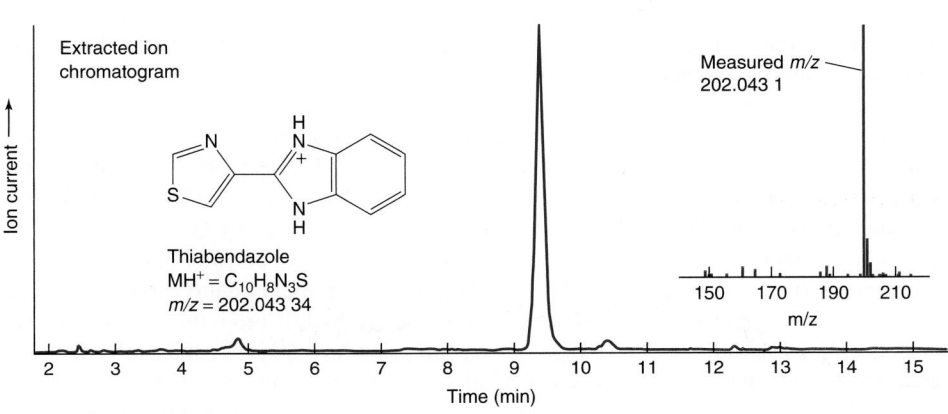

Extracted ion chromatogram

Thiabendazole
$MH^+ = C_{10}H_8N_3S$
$m/z = 202.043\ 34$

Measured m/z
202.043 1

(b)

FIGURE 22-32 (*a*) Total ion chromatogram of pesticides from fruit-based soft drinks from London's Gatwick Airport. (*b*) Extracted ion chromatogram set for thiabendazole with *m/z* window 202.043 ± 0.01. Inset shows mass spectrum of thiabendazole peak at 9.31 min. [Data from J. F. García-Reyes, B. Gilbert-López, A. Molina-Díaz, and A. R. Fernández-Alba, "Determination of Pesticide Residues in Fruit-Based Soft Drinks," *Anal. Chem.* **2008,** *80,* 8966.]

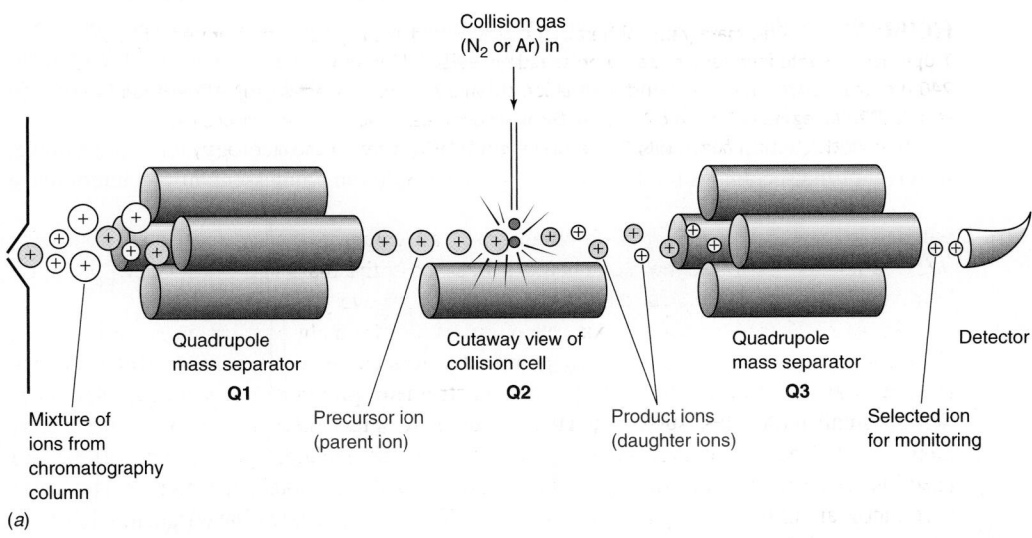

Collision gas
(N₂ or Ar) in

Quadrupole mass separator **Q1**

Cutaway view of collision cell **Q2**

Quadrupole mass separator **Q3**

Detector

Mixture of ions from chromatography column

Precursor ion (parent ion)

Product ions (daughter ions)

Selected ion for monitoring

The process in Q2 is called **collisionally activated dissociation** or **collision-induced dissociation.**

(a)

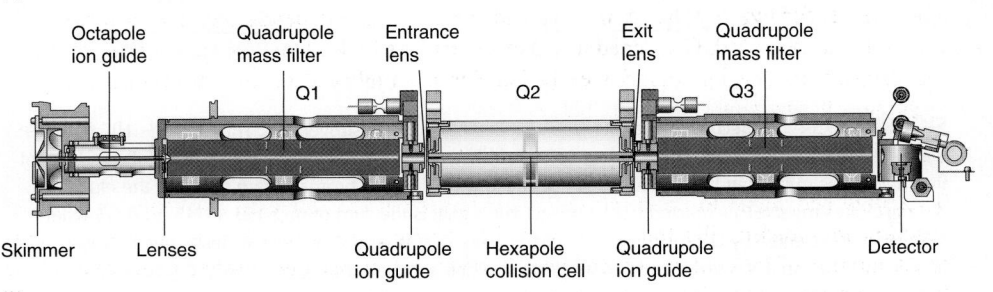

Octapole ion guide

Quadrupole mass filter
Q1

Entrance lens

Q2

Exit lens

Quadrupole mass filter
Q3

Skimmer

Lenses

Quadrupole ion guide

Hexapole collision cell

Quadrupole ion guide

Detector

(b)

FIGURE 22-33 (*a*) Principle of selected reaction monitoring, also called tandem mass spectrometry, mass spectrometry/mass spectrometry, MS/MS, or MS². C. Enke and coworkers invented this process in 1979.[32] (*b*) Cutaway view of triple quadrupole mass spectrometer. [Information from Agilent Technologies, Santa Clara, CA.]

that passes all ions of all *m/z* on to the third stage, Q3. However, while inside Q2, which is called a *collision cell*, the precursor ion collides with N_2 or Ar at a pressure of $\sim 10^{-8}$–10^{-6} bar and breaks into fragments called **product ions.** Quadrupole Q3 allows only specific product ions to reach the detector.

Selected reaction monitoring is highly selective for the analyte of interest. An example is provided by the analysis of human estrogens in sewage at parts per trillion levels (ng/L). Estrogens are hormones in the ovarian cycle. The synthetic estrogen, 17α-ethinylestradiol (designated EE_2) is a contraceptive. Even at parts per trillion levels, some estrogens can provoke reproductive disturbances in fish.

A project in Italy measured the estrogens entering the aquatic environment from human waste. Now think about this problem. Sewage contains thousands of organic compounds—many at high concentrations. To measure nanograms of one analyte was beyond the capabilities of analytical chemistry until recently. Some *sample preparation* was necessary to remove polar compounds from the less polar analyte and to *preconcentrate* analyte. Raw sewage (150 mL) was filtered to remove particles >1.5 μm and then passed through a *solid-phase extraction* cartridge (Section 28-3) containing carbon adsorbent that retained analyte. The cartridge was washed with polar solvents to remove polar materials. Estrogens were washed from the cartridge by a mixture of dichloromethane and methanol. Solvent was evaporated and the residue was dissolved in 200 μL of aqueous solution containing another estrogen as internal standard. A volume of 50 μL was injected for chromatography.

Figure 22-34a shows the collisionally activated dissociation mass spectrum of the deprotonated molecule of estrogen EE_2. The *precursor ion* $[M - H]^-$ (*m/z* 295) obtained by electrospray was isolated by mass separator Q1 in Figure 22-33 and sent into Q2 for *collisionally activated dissociation*. Then all fragments $>$*m/z* 140 were analyzed by Q3 to give the mass spectrum in Figure 22-34a. For subsequent *selected reaction monitoring*, only the *product ions* at *m/z* 159 and 145 were selected by Q3. The chromatogram in Figure 22-34b shows the signal from these product ions when *m/z* 295 was selected by Q1. From the area of the peak for EE_2, its concentration in the sewage was calculated to be 3.6 ng/L. Amazingly, there are other compounds that contribute mass spectral signals for this same set of masses (295 → 159 + 145) at elution times of 15–18 min. EE_2 was identified by its retention time and its complete mass spectrum.

The presence of pharmaceuticals and illegal drugs in municipal wastewater is now widely recognized.[33] A study in Spain in 2011 found the following doses of drugs per 1 000 inhabitants per day in urban sewage: amphetamine, 2.5; cocaine, 4.6; and cannabis, 68.[34] These substances find their way from wastewater into drinking water and the environment. Alarmingly, pharmaceuticals have now been found at the threshold for toxicological concern in vegetables irrigated with treated wastewater.[35]

Selected reaction monitoring is a powerful tool in forensic toxicology, able, for example, to screen urine specimens for 100 illicit drugs in a single run with essentially no interference

In *preconcentration*, analyte isolated from 150 mL was eventually dissolved in 200 μL. The concentration increased by a factor of

$$\frac{150 \times 10^{-3}\,\text{L}}{200 \times 10^{-6}\,\text{L}} = 750$$

3.6 ng/L in sewage became 750×3.6 ng/L = 2.7 μg/L for chromatography.

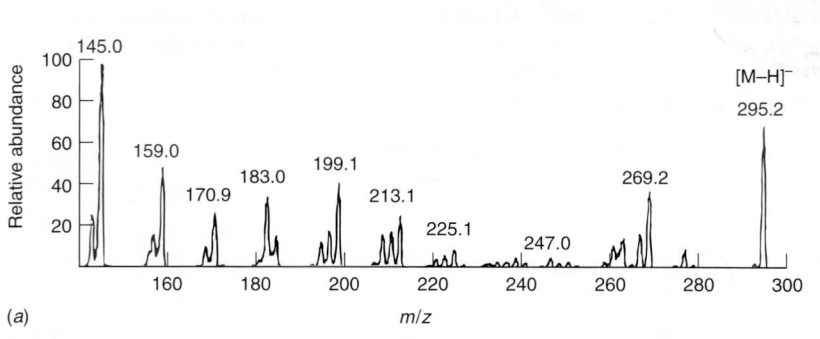
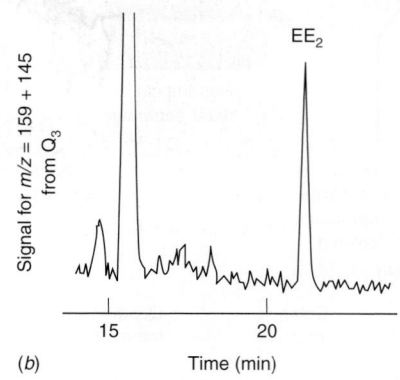

FIGURE 22-34 (*a*) Electrospray tandem mass spectrum of pure estrogen EE_2. The ion $[M - H]^-$ (*m/z* 295) was selected by quadrupole Q1 in Figure 22-33 and dissociated in the collision cell Q2. The full spectrum of fragments was measured by Q3. (*b*) Selected reaction monitoring chromatogram showing the elution of 3.6 ng/L of estrogen EE_2 extracted from sewage. The signal is the sum of *m/z* 159 + 145 from Q3 when *m/z* 295 was selected by Q1. [Data from C. Baronti, R. Curini, G. D'Ascenzo, A. di Corcia, A. Gentili, and R. Samperi, "Monitoring Estrogens at Activated Sludge Sewage Treatment Plants and in a Receiving River Water," *Environ. Sci. Tech.* **2000,** *34,* 5059.]

among analytes.[36] Selected reaction monitoring is so effective at eliminating interference that it can be used with flow injection *without chromatography to separate the analytes*. The time to measure six pesticide residues in food extracts was reduced from 15−30 min per run for liquid chromatography−mass spectrometry to 65 s per run with flow injection−mass spectrometry.[37] In Figure 22-33, the second stage of mass separation (quadrupole Q3) has just unit mass resolution. Selected reaction monitoring becomes even more immune to interference when the second stage of mass separation is done at high resolution with a time-of-flight or orbitrap spectrometer.

The three-dimensional quadrupole ion trap in Figure 22-18 and the linear ion trap in Figure 22-19[38] can each perform selected reaction monitoring without a second mass separator. Ions of different *m/z* are injected into the ion trap and then all but one *m/z* are intentionally ejected. The remaining ions with a single *m/z* are then accelerated to induce collisional dissociation with background He in the ion trap. After a period of dissociation, product ions are expelled to the detector to record a mass spectrum.

Figure 22-35 shows a miniature mass spectrometer under development for medical analysis. The 25-kg instrument uses a linear ion trap chosen for its ability to operate at relatively high pressure (10^{-5} bar) and to conduct selected reaction monitoring for adequate signal-to-noise ratio from samples in a complex matrix, such as therapeutic drugs in whole blood. A 2.5-μL blood sample containing an internal standard for quantitation is spotted onto a triangular, silica-coated paper and allowed to dry. The paper is mounted in the sample cassette and 40 μL of CH_3OH are applied by the solvent pump to dissolve small compounds from the blood. A 4.3 kV potential applied to the paper then creates a fine electrospray of ions from one vertex of the triangle into the mass spectrometer.

Selected reaction monitoring can be repeated by selecting a product ion for further dissociation. The repeated process is called MS^n, which denotes multiple repetitions of selected reaction monitoring. The beauty of MS^n with an ion trap is that the entire process takes place in one piece of hardware under software control.

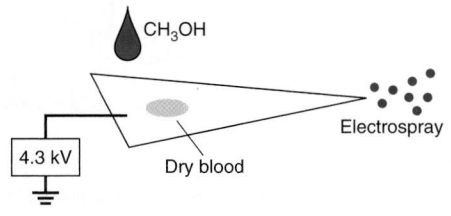

Paper spray sample introduction

MS^n: successive cycles of selected reaction monitoring. The product ion of one cycle becomes the precursor ion for the next cycle.

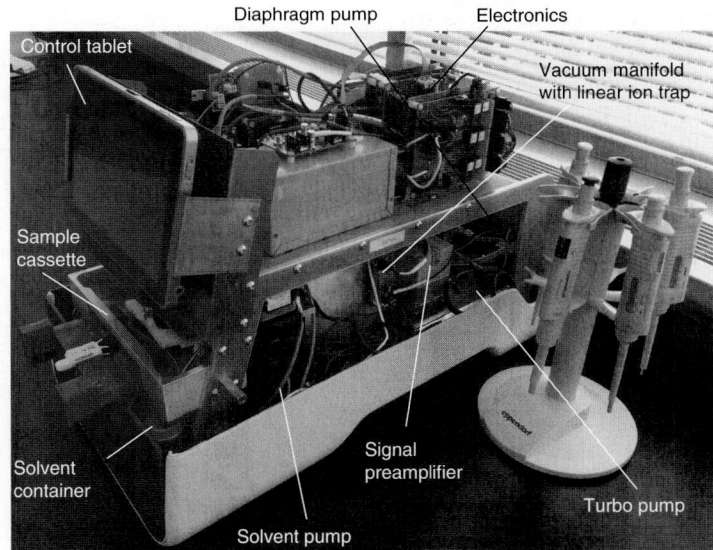

FIGURE 22-35 Developmental miniature mass spectrometer with linear ion trap for multiple reaction monitoring of complex samples such as drugs in whole blood. [Courtesy Zheng Ouyang, Purdue University.]

Electrospray of Proteins

Electrospray[39–41] and matrix-assisted laser desorption/ionization (MALDI, Box 22-4) are well suited to study charged macromolecules such as proteins. Proteins have carboxylic acid and amine side chains (Table 10-1) that can be positively or negatively charged, depending on pH. Electrospray (Figure 22-25) ejects protein ions from solution into the gas phase by mechanisms described in Box 22-5. One application of electrospray is the identification of hemoglobin disorders (such as sickle cell anemia) from dried blood spot samples.[44]

BOX 22-4   Matrix-Assisted Laser Desorption/Ionization

Major methods for introducing proteins and other macromolecules into mass spectrometers are electrospray and **matrix-assisted laser desorption/ionization (MALDI).**[42] Most often, MALDI is used with a time-of-flight mass spectrometer (Figure 22-16) to measure m/z up to $\sim10^6$. Typically, 1 μL of a 10-μM solution of analyte is mixed with 1 μL of a 1- to 100-mM solution of an ultraviolet-absorbing compound such as

2,5-dihydroxybenzoic acid (the *matrix*) directly on a probe that fits into the source of the spectrometer. Evaporation of the liquid leaves an intimate mixture of fine crystals of matrix plus analyte.

To introduce ions into the gas phase, a brief (~10 ns) ultraviolet or infrared laser pulse is directed onto the sample. The matrix vaporizes and expands into the gas phase, carrying analyte along with it. The high

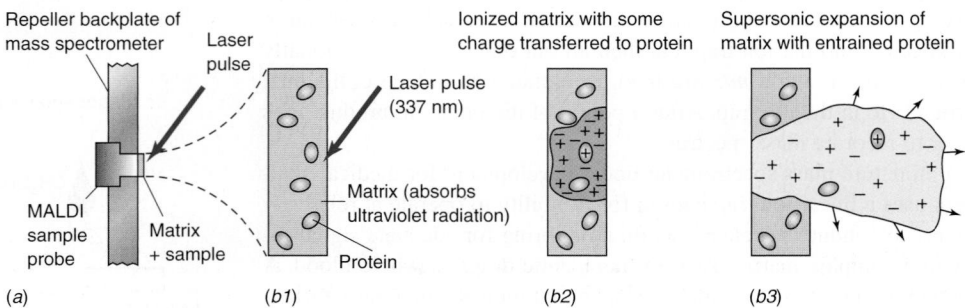

Sequence of events in matrix-assisted laser desorption/ionization. (*a*) Dried mixture of analyte and matrix on sample probe inserted into backplate of ion source. (*b1*) Enlarged view of laser pulse striking sample. (*b2*) Matrix is ionized and vaporized by laser and transfers some charge to analyte. (*b3*) Vapor expands in a supersonic plume.

BOX 22-5   Making Elephants Fly (Mechanisms of Protein Electrospray)

There are thought to be two mechanisms for releasing multiply charged protein cations into the gas phase from aerosol liquid droplets created by electrospray. Most native proteins have a compact globular conformation, such as that of myoglobin in Figure 10-5, with charged and hydrophilic amino acid residues on the outside and nonpolar residues inside. A globular protein is likely to be on the inside of an aqueous aerosol droplet so the polar residues are solvated by water. If acid is added, a preponderance of positively

charged residues can cause the protein to unfold. In the unfolded state, hydrophobic residues spontaneously move to the surface of the droplet, as oil goes to the surface of water.

Electrospray at a positive potential creates excess H^+ in the droplets by electrolysis: $H_2O \rightarrow \frac{1}{2}O_2 + 2H^+ + 2e^-$. The simulation in panel *a* shows the *chain ejection model* for a globular protein with six protonated side chains placed near the middle of an aerosol droplet containing 1 000 H_2O molecules plus four NH_4^+ ions. The protonated

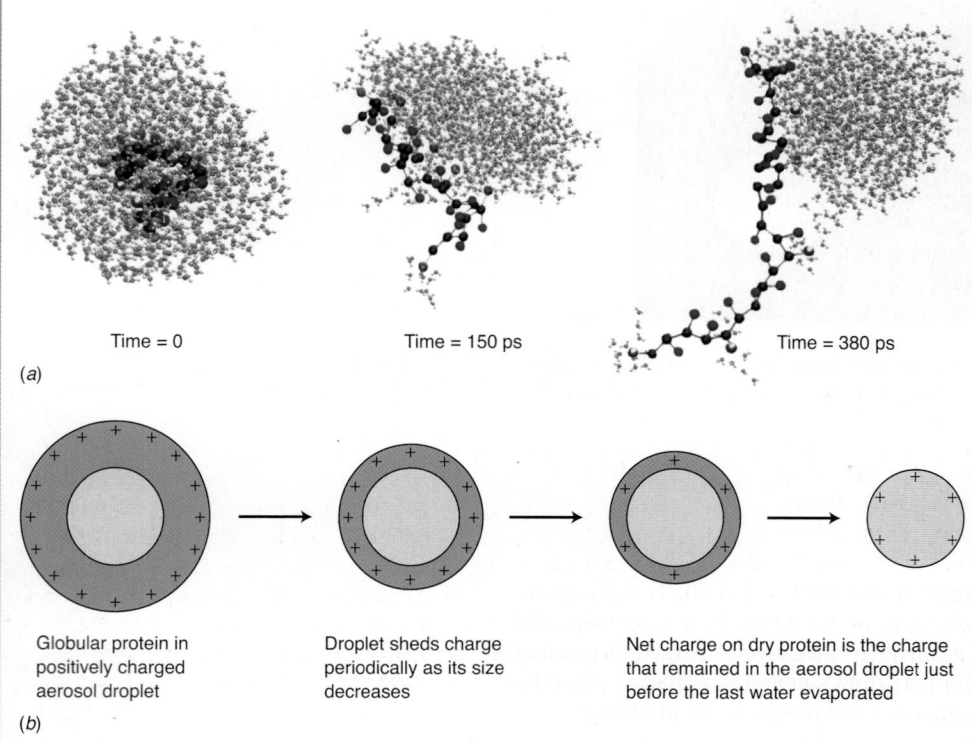

Time = 0   Time = 150 ps   Time = 380 ps

(*a*)

(*a*) *Chain ejection model.* Molecular dynamics simulation of unfolded protein with positively charged side chains being ejected from a positively charged aerosol droplet. [L. Konermann, E. Ahadi, A. D. Rodriguez, and S. Vahidi, "Unraveling the Mechanism of Electrospray Ionization," *Anal. Chem.* **2013**, *85*, 2, Figure 6. American Chemical Society publication. Reprinted with permission © 2013, American Chemical Society.]

(*b*)

Globular protein in positively charged aerosol droplet

Droplet sheds charge periodically as its size decreases

Net charge on dry protein is the charge that remained in the aerosol droplet just before the last water evaporated

(*b*) *Charged residue model.* As a droplet containing a globular protein gets smaller, it periodically ejects water and excess charge so that electrostatic repulsion is not greater than the cohesive surface tension that holds the drop together. Eventually, the drop is barely bigger than the globular protein it contains, and the excess charge is dictated by the size of the protein plus its last remaining water. When the last water evaporates, the excess charge remains with the protein. [Information from L. Konermann, E. Ahadi, A. D. Rodriguez, and S. Vahidi, "Unraveling the Mechanism of Electrospray Ionization," *Anal. Chem.* **2013**, *85*, 2.]

matrix-to-sample ratio inhibits association between analyte molecules and provides protonated or ionic species that transfer charge to analyte, much of which carries a single charge. Shortly after ions expand into the source, a voltage pulse applied to the backplate expels ions into the spectrometer. The spectrum below shows proteins from milk that has not undergone any sample preparation except for mixing with the matrix. In contrast to multiply charged protein ion formation by electrospray, MALDI typically creates singly charged proteins. It is possible to map differences in chemical composition in different regions by directing a laser at different parts of fixed cells, such as neurons.[43]

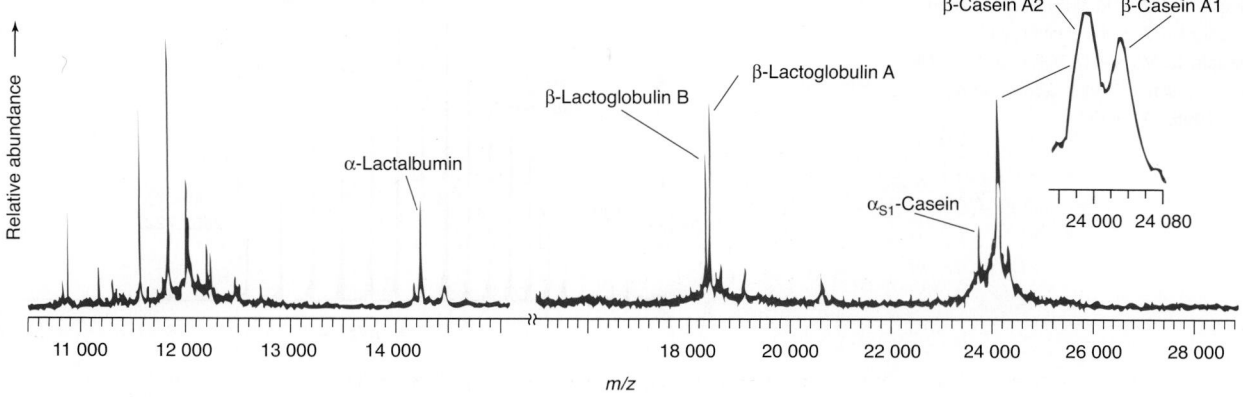

Partial mass spectrum of cow's milk (containing 2% milk fat) observed by MALDI/time-of-flight mass spectrometry. [Data from R. M. Whittal and L. Li, "Time-Lag Focusing MALDI-TOF Mass Spectrometry," *Am. Lab.* December 1997, p. 30.]

protein chain uncoils and pushes its way to the outside of the droplet in ~100 ps (10^{-10} s). Positively charged side chains are repelled by excess positive charge on the surface of the droplet. As the protein chain is expelled from the droplet, some excess H^+ from the droplet departs with the protein bound to protein side chains. Some H_2O remains attached to charged protein sites in the gas phase. In about 1 ns, the protein is fully ejected from the droplet. Water of solvation is lost from the protein by collisions with gas molecules.

If a protein remains globular because, for example, the original solution was buffered near neutral pH, it remains near the center of the aerosol droplet as water evaporates. In the *charged residue model* in panel *b*, the droplet shrinks until the repulsive force of its excess charge equals the cohesive force of surface tension. At that point, the droplet expels smaller droplets that carry away some charge. This process continues until the droplet is barely bigger than the protein it contains. When the last water evaporates, the excess charge remains with the protein. Globular proteins released into the gas phase have a net charge that depends mainly on their diameter and not on the equilibrium degree of protonation at the pH of the original solution. The process in panel *b* takes ~1 μs, which is 10^3 times longer than the process in panel *a*.

Panel *c* compares electrospray mass spectra of the protein myoglobin from solutions at pH 7 and pH 2. The globular protein at pH 7 is released by the charged residue mechanism. The two observe charge states are +9 and +8. The predicted charge of an aerosol droplet equal to the diameter of globular myoglobin is +9.5, in agreement with the observed charge in the mass spectrum. By contrast, electrospray of unfolded myoglobin from a solution at pH 2 gives a higher yield of protein cations and a range of charge states from +8 to +25. In the chain ejection mechanism, protein is ejected from droplets of many different sizes as the droplets shrink. The larger the droplet at the time of ejection, the more highly charged is the droplet and the more H^+ is associated with the ejected protein. The broad range of observed charge in the mass spectrum represents the broad range of droplet sizes at the time of protein ejection.

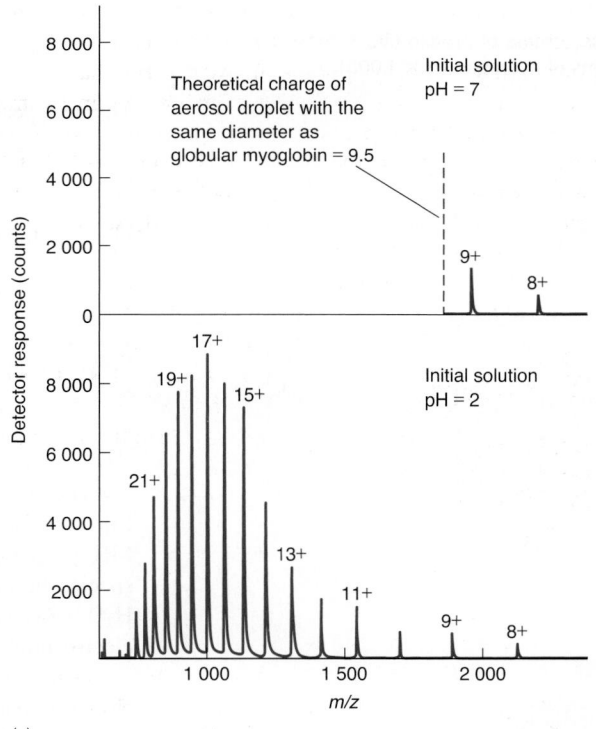

(c)

(c) Electrospray ionization mass spectra of myoglobin from solutions of initial pH = 7 and pH = 2. At pH 7, the globular protein is released by the charged residue mechanism with the charge of a droplet that is the size of the protein. At pH 2, unfolded protein is released by the chain ejection mechanism with a range of charges from a continuum of droplet sizes. [L. Konermann, E. Ahadi, A. D. Rodriguez, and S. Vahidi, "Unraveling the Mechanism of Electrospray Ionization," *Anal. Chem.* **2013**, *85*, 2, Figure 7. Reprinted with permission © 2013, American Chemical Society.]

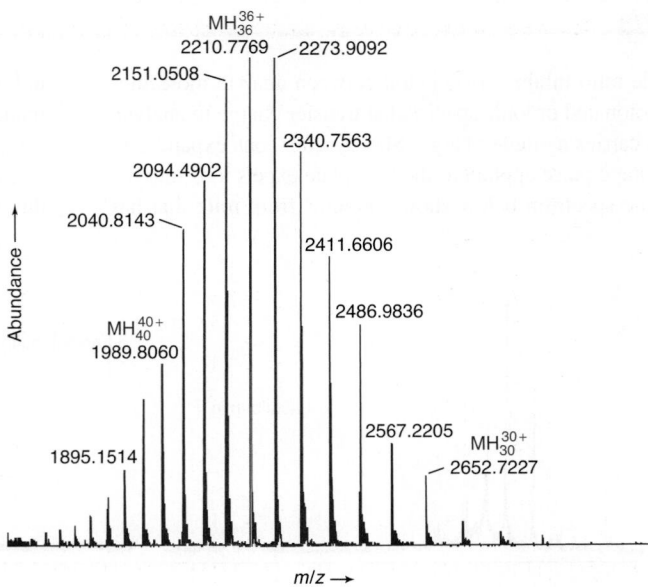

FIGURE 22-36 Electrospray time-of-flight mass spectrum of one anion-exchange chromatographic peak containing the protein transferrin with one particular set of carbohydrate substituents. Peaks arise from species with different numbers of protons, MH_n^{n+}. [Data from M. E. Del Castillo Busto, M. Montes-Bayón, E. Blanco-González, J. Meija, and A. Sanz-Medel, "Strategies to Study Human Serum Transferrin Isoforms Using Integrated Liquid Chromatography ICPMS, MALDI-TOF, and ESI-Q-TOF Detection: Application to Chronic Alcohol Abuse," *Anal. Chem.* **2005,** *77,* 5615.]

Mass = mass of protein (M) + mass of n atoms of hydrogen ($n \times 1.008$).

Each peak in the mass spectrum of the protein transferrin in Figure 22-36 arises from molecules with different numbers of protons, MH_n^{n+}.[45] Although we have already labeled charges on several peaks, we do not know what these charges are before analyzing the spectrum. If we can find the charge for each species, we can find the molecular mass, M, of the neutral protein.

To find the charge, consider a peak with $m/z = m_n$ derived from the neutral molecule plus n protons:

$$m_n = \frac{mass}{charge} = \frac{M + n(1.008)}{n} = \frac{M}{n} + 1.008 \Rightarrow \boxed{m_n - 1.008 = \frac{M}{n}} \quad (22\text{-}4)$$

The next peak at lower m/z should have $n + 1$ protons and a charge of $n + 1$. For this peak,

$$m_{n+1} = \frac{M + (n + 1)(1.008)}{n + 1} = \frac{M}{n + 1} + 1.008 \Rightarrow \boxed{m_{n+1} - 1.008 = \frac{M}{n + 1}} \quad (22\text{-}5)$$

Forming the quotient of the expressions in the boxes, we get

$$\frac{m_n - 1.008}{m_{n+1} - 1.008} = \frac{M/n}{M/(n + 1)} = \frac{n + 1}{n} \quad (22\text{-}6)$$

Solving Equation 22-6 for n gives the charge on peak m_n:

$$n = \frac{m_{n+1} - 1.008}{m_n - m_{n+1}} \quad (22\text{-}7)$$

The fourth column in Table 22-3 shows the charge of each peak, n, calculated with Equation 22-7. The charge of the peak at m/z 2 652.722 7 is $n = 30$. We assign this peak as MH_{30}^{30+}. The next peak at m/z 2 567.220 5 is MH_{31}^{31+}, and so on. We find highly protonated species because the chromatography solvent was acidic (95 vol% acetonitrile + 5 vol% H_2O + 0.2 vol% formic acid). Even without adding formic acid, electrolysis in positive-ion electrospray produces H^+ by oxidation of H_2O.

From any peak, we can find the mass of the neutral molecule by rearranging the right side of Equation 22-4:

$$M = n \times (m_n - 1.008) \quad (22\text{-}8)$$

Reproducible masses computed with Equation 22-8 appear in the last column of Table 22-3. The accuracy of molecular mass determination is limited by the accuracy of the m/z scale. For this work, m/z was calibrated with an external standard of poly(propylene glycol).

Electron-Transfer Dissociation for Protein Sequencing

We saw on page 227 that proteins are chains of amino acids held together by amide (also called peptide) bonds. Proteins are synthesized on *ribosomes*, which are assemblies of RNA and protein whose job is to translate the sequence of messenger RNA into a corresponding

TABLE 22-3 Analysis of electrospray mass spectrum of tetrasialo-transferrin in Figure 22-36

Observed $m/z \equiv m_n$	$m_{n+1} - 1.008$	$m_n - m_{n+1}$	Charge = n = $\dfrac{m_{n+1} - 1.008}{m_n - m_{n+1}}$	Molecular mass = $n \times (m_n - 1.008)$
2 652.722 7	2 566.212 5	85.502 2	30.013 ≈ 30	79 551.44
2 567.220 5	2 485.975 6	80.236 9	30.983 ≈ 31	79 552.59
2 486.983 6	2 410.652 6	75.323 0	32.004 ≈ 32	79 551.22
2 411.660 6	2 339.748 3	70.904 3	32.999 ≈ 33	79 551.54
2 340.756 3	2 272.901 2	66.847 1	34.001 ≈ 34	79 551.44
2 273.909 2	2 209.768 9	63.132 3	35.002 ≈ 35	79 551.54
2 210.776 9	2 150.042 8	59.726 1	35.998 ≈ 36	79 551.68
2 151.050 8	2 093.482 2	56.560 6	37.013 ≈ 37	79 551.58
2 094.490 2	2 039.806 3	53.675 9	38.002 ≈ 38	79 552.32
2 040.814 3	1 988.798 0	51.008 3	38.990 ≈ 39	79 552.45
1 989.806 0	1 894.143 4		40	79 551.92
				mean = 79 551.78 ± 0.48

SOURCE: *M. E. Del Castillo Busto, M. Montes-Bayón, E. Blanco-González, J. Meija, and A. Sanz-Medel, "Strategies to Study Human Serum Transferrin Isoforms Using Integrated Liquid Chromatography ICPMS, MALDI-TOF, and ESI-Q-TOF Detection: Application to Chronic Alcohol Abuse,"* Anal. Chem. **2005,** 77, 5615.

sequence of amino acids. After synthesis, some proteins are specifically modified by enzymes to add groups such as acetate, phosphate, carbohydrates, and lipids to specific amino acid side chains. A branch of biochemistry called *proteomics* seeks to characterize the structure and function of the full complement of proteins in an organism.

Mass spectrometry is a principal tool for deducing the sequence of amino acids in a protein. Protein is digested into shorter chains by enzymatic cleavage. Individual chains are cleaved to make fragments of all possible lengths. High-resolution mass spectrometry allows us to deduce which amino acids are in each fragment. A computer can use this information to reconstruct the sequence of amino acids.

Electron-transfer dissociation is a selective way to cleave polypeptides into fragments in a mass spectrometer.[46] This process involves the exothermic transfer of an electron from a gas-phase anion to a gas-phase polypeptide with concomitant bond breaking. In the chain of amino acids below,

cleavage occurs at each indicated position to form ten possible charged fragments designated c_1 to c_5 and z_1 to z_5. In the c fragments, charge resides on the fragment to the left of the broken bond. In the z fragments, charge resides on the fragment to the right of the broken bond. Unlike collisionally activated dissociation, electron-transfer dissociation does not break other bonds in the peptide chain or the side groups, including groups such as phosphate or carbohydrates bound to side chains. Electron-transfer dissociation provides structural information that is complementary to that produced by collisionally active dissociation.[47]

The mass spectrometer in Figure 22-22, when equipped with two electrospray sources, can carry out electron-transfer dissociation for polypeptide sequencing.[48] Polypeptide cations produced by electrospray for 0.2 s are collected in the linear ion trap and stored in the downstream section by application of appropriate voltages. The polypeptide source is then turned off and, after 0.4 s, 9-anthracenecarboxylic acid solution is electrosprayed from a second source for 0.2 s. Anthracenecarboxylate anions are collected in the upstream section of the linear ion trap. Collisionally activated decarboxylation in the ion trap produces anion A^-:

9-Anthracenecarboxylate A^-

Potentials are then adjusted to allow anions and cations that had been at opposite ends of the ion trap to mix with one another. A⁻ transfers an electron to a polypeptide P^{n+}, thereby inducing electron-transfer dissociation of one peptide bond:

$$A^- + P^{n+} \rightarrow A + P^{(n-1)+} \rightarrow \text{cleavage of one peptide bond}$$

Different bonds are cleaved in individual molecules to make all possible c and z fragments. The reaction is quenched (halted) by ejecting anions from the ion trap. Finally, peptide cations are expelled and exact m/z values are measured with the orbitrap in Figure 22-22. With m/z accuracy of 1 part per million, most electron-transfer dissociation product ions with $m/z < 1\ 000$ can be uniquely assigned and identified as N- or C-terminal peptides.[49]

22-6 Open-Air Sampling for Mass Spectrometry

Since 2004, new mass spectral sampling techniques have been introduced to vaporize and ionize analytes directly from the surface of an object in the open air with little harm to the object. This sampling capability opens up new vistas for mass spectrometry in qualitative analysis.[50]

Direct Analysis in Real Time (DART)

A **direct analysis in real time** (DART) source produces electronically excited He or vibrationally excited N_2, which are directed at the surface of an object to be sampled in open air. In Figure 22-37, heated gas flows through a needle held at +1 to +5 kV with respect to a grounded perforated disk upstream in the source. The glow discharge plasma contains electrons, ions, and excited neutral species. Electrodes 1 and 2 are held at positive potentials for positive ion mass spectrometry and at negative potentials for negative ion mass spectrometry. At positive potential, electrodes 1 and 2 prevent cations from exiting the DART source. At negative potential, anions and electrons are retained.

The DART gun is directed at an object to be sampled. With a helium source, excited He atoms with an energy of 19.8 eV (designated 2^3S and shown as He* in the margin) react with atmospheric water vapor to create protonated water clusters. These clusters can react with analyte M on the surface of an object to create MH^+. Other chemistry can produce $(M - H)^-$, M^-, or adducts such as $(M + NH_4)^+$ or $(M + Cl)^-$.

If the sample is a poppy seed from a bagel, two major species observed by a high-resolution time-of-flight mass spectrometer are the protonated molecules, (morphine)H^+ ($C_{17}H_{19}NO_3H^+$ at m/z 286.144 3) and (codeine)H^+ ($C_{18}H_{21}NO_3H^+$ at m/z 300.161 1).[51] Separate quantitative analysis shows that poppy seeds contain ~33 and ~14 μg/g (ppm) of morphine and codeine, respectively.

$$He^* + H_2O \rightarrow H_2O^+ + He + e^-$$
$$H_2O^+ + H_2O \rightarrow H_3O^+ + OH$$
$$H_3O^+ + (n-1)H_2O \rightarrow (H_2O)_nH^+$$

The process in which an excited He atom ionizes a molecule such as H_2O is called *Penning ionization*.

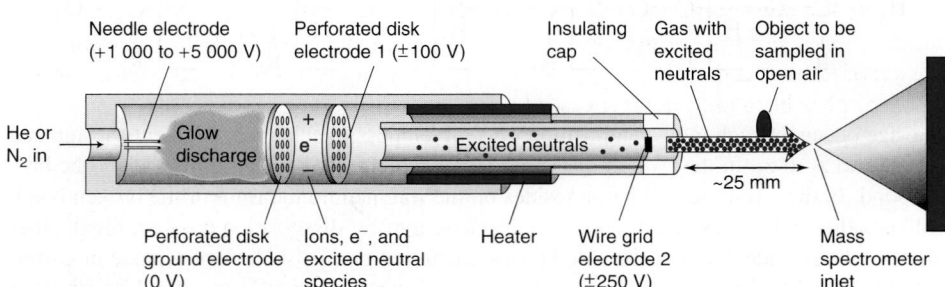

FIGURE 22-37 Direct analysis in real time (DART) source. [Information from R. B. Cody, J A. Laramée, and H. D. Durst, "Versatile New Ion Source for the Analysis of Materials in Open Air under Ambient Conditions," *Anal. Chem.* **2005,** *77,* 2297 and JEOL USA, Peabody, MA.]

Low-Temperature Plasma

In a technique related to DART, a low-temperature plasma is created by passing He, Ar, N_2, or ambient air through a glass tube with a grounded wire at the center (Figure 22-38 and Color Plate 28). The tube is wrapped on the outside with a copper sheath to which is applied a 3-kV alternating current. Excited-state species in the plasma ionize and dislodge molecules from a surface such as human skin into the source of a mass spectrometer. There is no electrical shock to the surface under study. A battery-operated, handheld plasma has been described.[52]

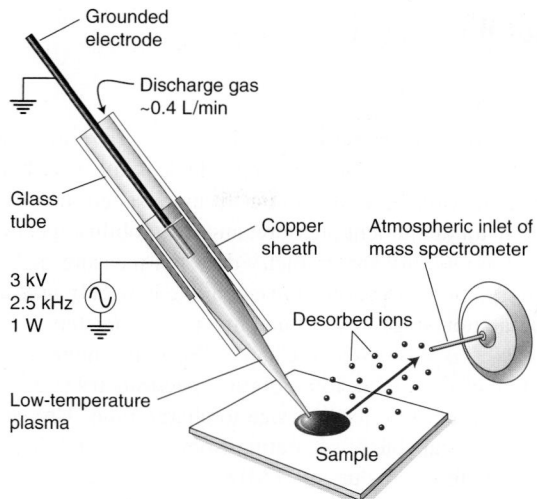

Grounded electrode

Discharge gas ~0.4 L/min

Glass tube

Copper sheath

Atmospheric inlet of mass spectrometer

3 kV
2.5 kHz
1 W

Desorbed ions

Low-temperature plasma

Sample

FIGURE 22-38 Low-temperature plasma for sampling a surface in the open air. [Information from J. D. Harper, N. A. Charipar, C. C. Mulligan, X. Zhang, R. G. Cooks, and Z. Ouyang, "Low-Temperature Plasma Probe for Ambient Desorption Ionization," *Anal. Chem.* **2008,** *80,* 9097.]

Desorption Electrospray Ionization (DESI)

In **desorption electrospray ionization** (DESI), micron-sized, charged droplets created by electrospray of analyte-free solvent (Figure 22-25) are directed onto the surface of an object under study.[53] Analyte on the surface dissolves in the droplets. Further bombardment knocks droplets into the air toward a mass spectrometer inlet. As in conventional electrospray, it is common to observe multiply charged ions and alkali metal adducts in the mass spectrum.

Figure 22-39 shows DESI mapping of inks on a page. A mixture of methanol and water is electrosprayed toward the paper, just 2 mm away from the spray tip. The mass spectrum (*m/z* 150–600) of rebounded droplets is recorded every 0.67 s. The paper is translated in small steps to make a two-dimensional map. The mass spectral signal observed at a blank area of the paper is subtracted from the signal observed from an inked area to obtain the spectrum of the ink.

In this example, the date "1432" was written with a blue pen. Then a second blue pen was used to change 4 to 9 and 3 to 8, so that the date would be "1982." The second pen had a blue ink different from that in the first pen, but the difference is subtle. The objective of the demonstration was to show that the original date had been altered.

The first ink was Basic Violet 3, with a moleclar ion M[+] at *m/z* 372.243. The second ink was Solvent Blue 2, with a molecular ion at *m/z* 484.275. Map A in Figure 22-39 shows extracted-ion monitoring of *m/z* 372.4 with unit mass resolution. White shows where *m/z* 372.4 is maximal and black shows where *m/z* 372.4 is minimal. Therefore, map A shows where Basic Violet 3 ink is located. Map B is made from extracted-ion monitoring of *m/z* 484.5 with unit resolution to show where Solvent Blue 2 ink is located. We see that the second ink was written over the first ink to change "1432" to "1982." Map C is the sum of maps A and B. Frame D is an optical image of the writing. Mass spectral imaging of inks has been used in legal cases to test the authenticity of documents.

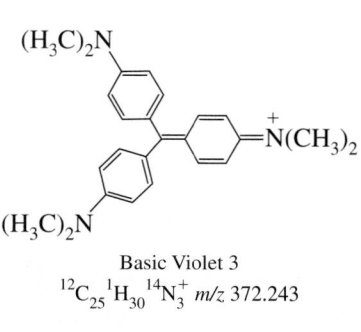

Basic Violet 3
$^{12}C_{25}{}^{1}H_{30}{}^{14}N_3^+$ *m/z* 372.243

Solvent Blue 2
$^{12}C_{34}{}^{1}H_{34}{}^{14}N_3^+$ *m/z* 484.275

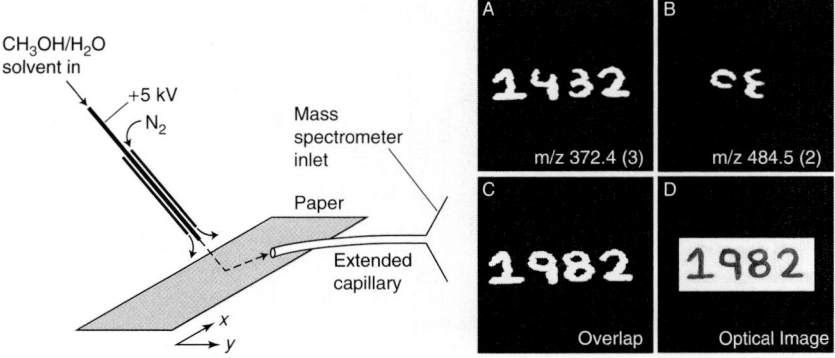

FIGURE 22-39 Desorption electrospray ionization (DESI) mapping of inks. Images A, B, and C at the right are maps made by scanning the page in the *x*- and *y*-directions beneath the electrospray source. D is an optical image of the surface. [Republished with permission of Royal Society of Chemistry, from D. R. Ifa, L. M. Gumaelius, L. S. Eberlin, N. E. Manicke, and R. G. Cooks, "Forensic Analysis of Inks by Imaging Desorption Electrospray Ionization (DESI) Mass Spectrometry," *Analyst* **2007,** *132,* 461, Figure 1, Figure 5. Permission conveyed through Copyright Clearance Center, Inc.]

22-7 Ion Mobility Spectrometry

Ion Mobility Spectrometer[54]

More than 10^4 **ion mobility spectrometers** are deployed at airport security checkpoints to detect explosives, and perhaps 10^5 hand-portable devices are used by military and civil defense personnel. Although functionally similar to mass spectrometers, ion mobility spectrometers are operated in air at ambient pressure, and ion mobility spectrometry is *not* a form of mass spectrometry. Ion mobility spectrometry does not measure molecular mass and provides no structural information. However, this technique is widely used and is now incorporated as an integral separation stage inside some mass spectrometers.

Electrophoresis, which is discussed in Chapter 26, is the migration of ions in solution under the influence of an electric field. Ion mobility spectrometry is *gas-phase electrophoresis*, which separates ions according to their size-to-charge ratio. Unlike mass spectrometry, ion mobility spectrometry is capable of separating isomers. Ion mobility can be used to study large protein assemblies with masses up to 50 MDa.[55]

At first sight, the ion mobility spectrometer in Figure 22-40a reminds us of a time-of-flight spectrometer. In portable units, the drift tube is 5–10 cm long with a flow of air (drift

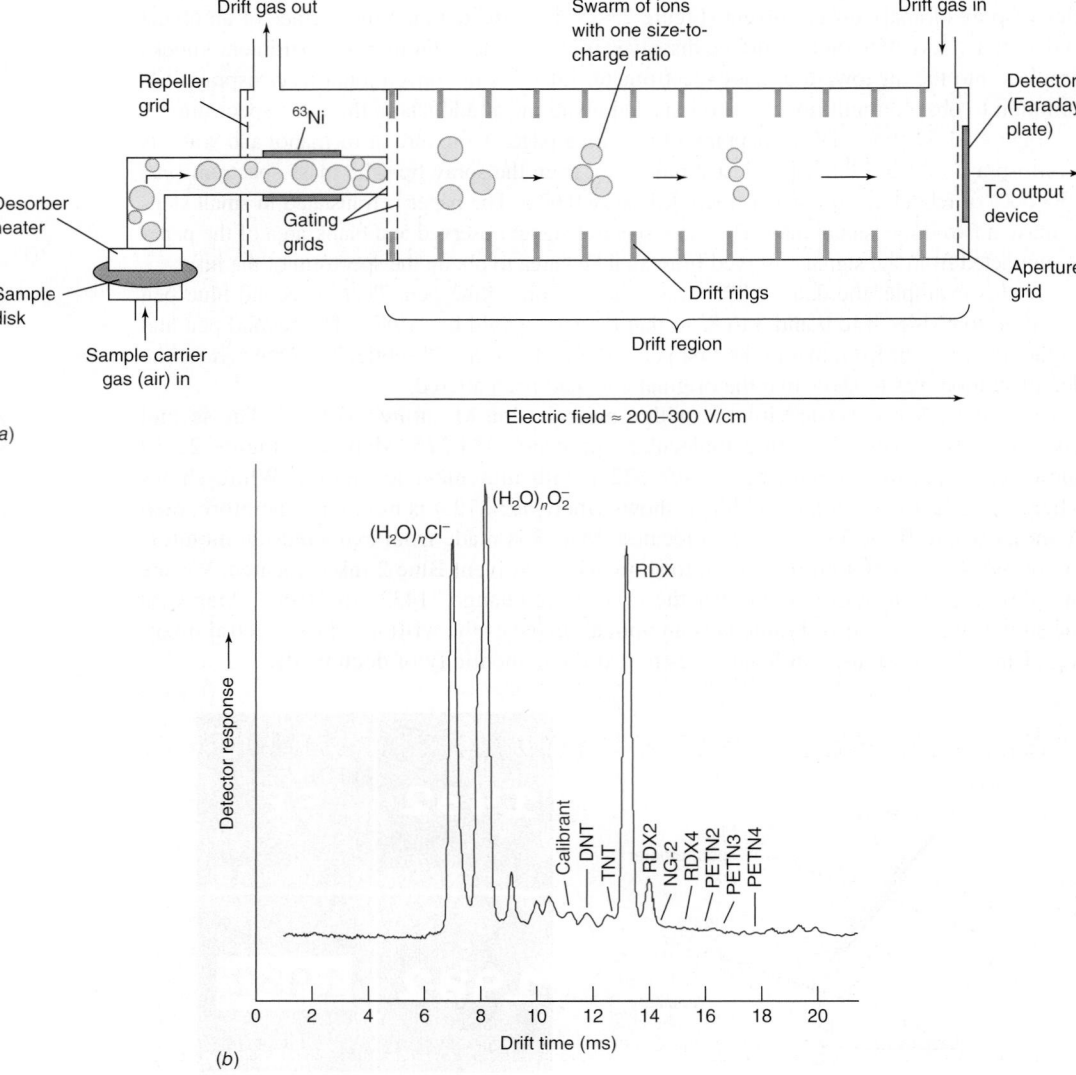

(a)

(b)

FIGURE 22-40 (*a*) Schematic illustration of ion mobility spectrometer. The aperture grid prevents serious line broadening. Ions approaching a bare detector plate induce current, which appears as signal before the arrival of the ions. The aperture grid protects the detector from induced current until ions are between the grid and the detector. (*b*) Negative ion mobility spectrum of explosives designated RDX, TNT, PETN, and so on. The calibrant 4-nitrobenzonitrile is an internal mobility standard. Cl_2 is the reagent gas for generating anions by chemical ionization. [Data from W. R. Stott, Smiths Detection, Toronto.]

gas) from right to left in the figure. Typically, sample adsorbed on a cotton swab is placed in the heated anvil at the left to desorb analyte vapor. Dry air doped with a chemical ionization reagent (such as Cl_2 for anions and acetone or NH_3 for cations) sweeps the vapor through a tube containing 10 millicuries of radioactive ^{63}Ni. Reagent gas ionized by β-emission from ^{63}Ni reacts with analyte to generate analyte ions.

A spectral scan is initiated when a ~250-μs voltage pulse on a gating grid admits a packet of ions into the drift tube. An electric field of 200–300 V/cm in the drift region is established by potential differences on the drift rings. The field causes either cations or anions to drift from left to right at ~1–2 m/s. Ions are retarded by collisions with drift gas (air) molecules at atmospheric pressure. Each ion travels at its own speed equal to KE, where E is the electric field and K is the *mobility*. Small ions have greater mobility than large ions of the same charge because large ions experience more drag.

The ion mobility "spectrum" in Figure 22-40b shows detector response versus drift time for several explosives. Peak area is proportional to the number of ions. Peaks are identified by their mobility, which is reproducibly measured relative to that of an internal standard. Runs are repeated ~20 times per second. The displayed spectrum averages many runs and requires 2–5 s to obtain.

Detection limits are 0.1–1 pg for compounds with favorable ionization chemistry. Mobility spectrometers have limited resolution, but combining mobility determinations with selective ionization can minimize false positives. Resolution can be improved by operating in a closed chamber with pressures of 2-4 bar.[56]

Figure 22-41 shows a variant of ion mobility separation called *field asymmetric waveform spectrometry* (FAIMS). Separation of ions is based on differences in their mobility in strong and weak electric fields. The mechanisms of separation in Figure 22-40 and 22-41 are different, and provide complementary means of resolving molecules. In Figure 22-41, analyte is swept by a carrier gas (He + N_2) flowing from left to right, which is opposite the direction of drift gas flow in Figure 22-40. An oscillating electric field is applied between two metal plates by impressing a radio frequency (RF) *dispersion voltage* of several kilovolts on one of the plates. A small direct current (DC) compensation voltage of just a few volts is applied to the other plate. The separation of the plates is ~2 mm and the length from left to right is ~50 mm. The residence time for an ion in the spectrometer is ~0.1–0.2 s.

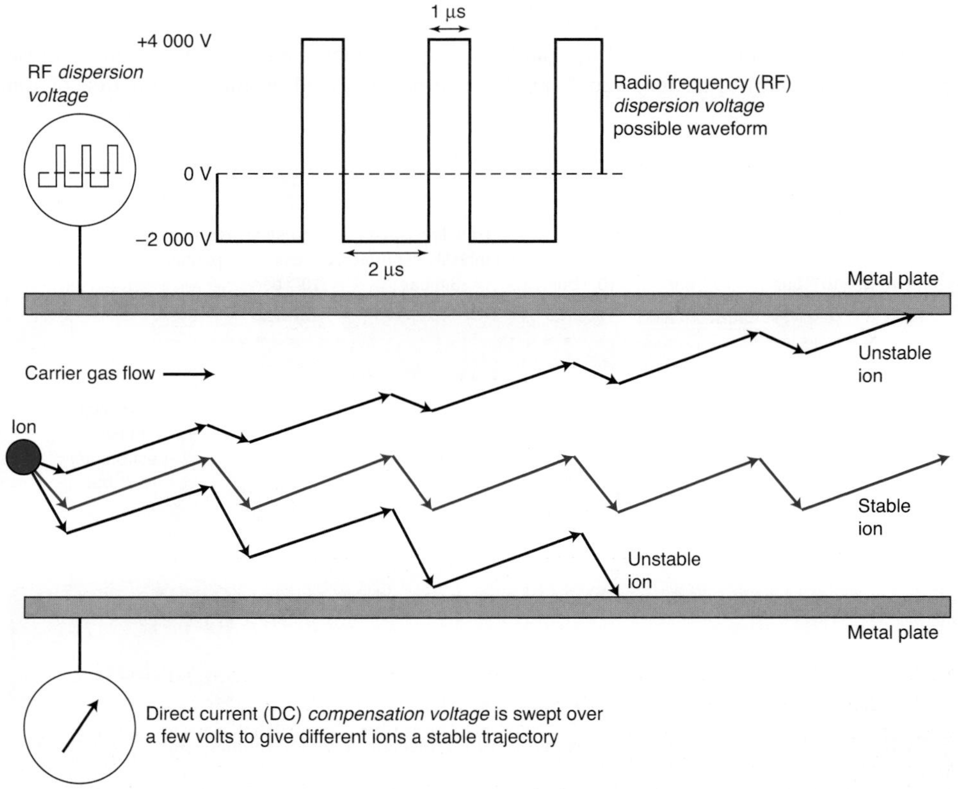

Small ions drift faster than large ions. For a description of ion mobility spectrometry in electrophoretic terms, see Problem 26-51.

FIGURE 22-41 Principle of operation of field asymmetric waveform ion mobility spectrometer (FAIMS).

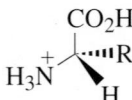

L-leucineH$^+$ (R = —CH$_2$CH(CH$_3$)$_2$
L-isoleucineH$^+$ (R = —CH(CH$_3$)(CH$_2$CH$_3$)

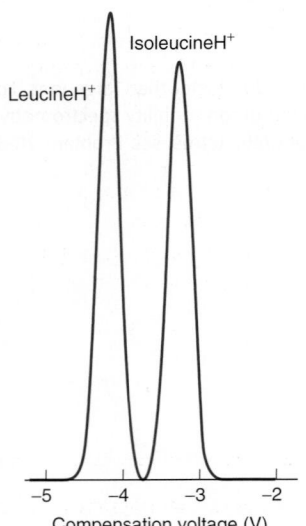

FIGURE 22-42 Ion mobility separation of protonated leucine and isoleucine cations with a field asymmetric waveform (FAIMS) spectrometer. Gap between electrodes = 1.88 mm. Peak dispersion voltage = 5.4 kV at 750 kHz frequency. Carrier gas is 50 vol% He in N$_2$. Ion residence time = 150 − 200 ms. [A. A. Schvartsburg, D. C. Prior, K. Tang, and R. D. Smith, "High-Resolution Differential Ion Mobility Separations Using Planar Analyzers at Elevated Dispersion Fields," *Anal. Chem.* **2010,** *82,* 7649, Figure 3. Reprinted with permission © 2010, American Chemical Society.]

At low electric field strength, the mobility of an ion is proportional to the electric field. If this were true in Figure 22-41, the average vertical deflection of an ion as it is swept from left to right would be zero because the positive dispersion voltage (+4 000 V) is applied for half of the time of the negative dispersion voltage (−2 000 V). In fact, in high electric fields, the mobility of an ion becomes a nonlinear function of the field strength. For different ions, the high field mobility could be more or less than the low field mobility. The result, in general, is that a given ion has a net motion up or down as it is swept from left to right. When an ion collides with either metal plate, it is neutralized and its ride is over. Therefore, a small direct current *compensation voltage* is applied to the lower plate to counteract the upward or downward drift of particular ions.

Figure 22-42 shows separation of the protonated amino acids leucine and isoleucine by sweeping the compensation voltage from −2 to −5 V. At −3.2 V, isoleucine traverses the spectrometer without colliding with the metal plates. At −4.1 V, leucine can pass through. These two ions have exactly the same mass and are not readily distinguished by a mass spectrometer. However, if an ion mobility separation precedes the mass spectrometer, the ions can be distinguished.

Some commercial mass spectrometers incorporate an ion mobility separation stage, as in Figure 22-43.[57] Ions produced by electrospray at the left are desolvated and focused by an ion guide and then pass through a quadrupole mass separator. Ions selected by the quadrupole can be stored in the trap cell, which consists of a stack of metal rings. A radio frequency electric field confines ions along the axis of the trap. Direct current voltage applied to selected rings can either confine ions inside the trap or expel ions toward the ion mobility stage.

The *traveling wave ion mobility separator* is also a stack of metal rings with a radio frequency voltage to confine ions along the axis. A direct current pulse applied to successive rings creates a "traveling wave" of electric potential that pushes ions from left to right through the separator. Ions with different sizes and shapes experience different amounts of drag from collisions with helium in the separator. Following this traveling wave ion mobility separation, the transfer cell focuses ions and passes them to the high-field pusher, which injects them orthogonally into a time-of-flight mass spectrometer.

Both the trap and transfer cells can be operated at high enough energy to carry out collisionally activated dissociation of ions. The instrument in Figure 22-43 is therefore capable of selected reaction monitoring with ion mobility separation of reactant or product ions. The traveling wave ion mobility separator has a lower resolving power than some drift tubes, but its higher transmission achieves lower detection limits.

The combination of ion mobility and mass spectrometry allows isobaric ions (those with the same *m/z*) to be resolved. Ions that can be separated by the combination of ion

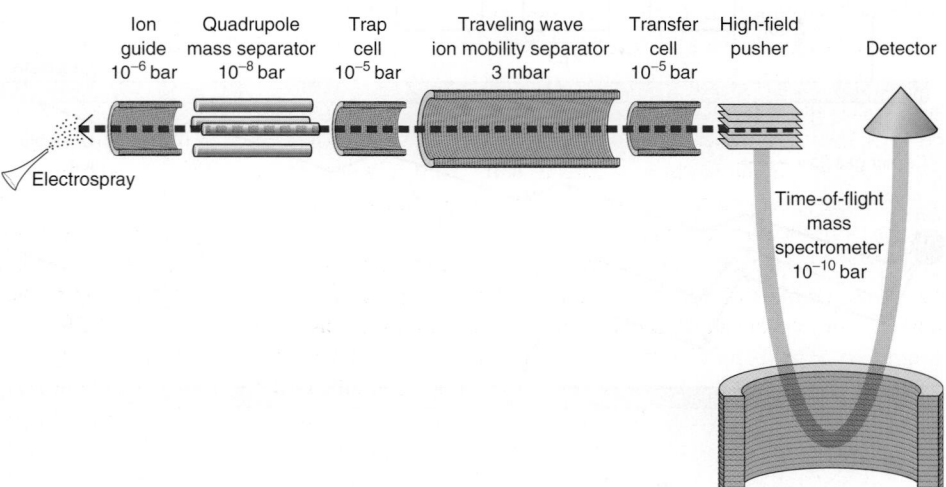

FIGURE 22-43 Layout of a mass spectrometer with a traveling wave ion mobility separator between the quadrupole and time-of-flight mass separators. [Courtesy Prof. Matthew F. Bush, Department of Chemistry, University of Washington.]

mobility and mass spectrometry include diastereomers and different conformations of peptides encountered in protein sequencing. An ion mobility stage can simplify a mass spectrum by dividing a mixture into a series of simpler subsets of ions. The ion mobility stage can increase the signal-to-noise ratio by isolating ions of interest from a background of other ions.[58]

Terms to Understand

atmospheric pressure chemical ionization
atomic mass
base peak
chemical ionization
collisionally activated dissociation
desorption electrospray ionization (DESI)
direct analysis in real time (DART)
double-focusing mass spectrometer
electron ionization

electrospray ionization
extracted ion chromatogram
ion mobility spectrometer
linear quadrupole ion trap
magnetic sector mass spectrometer
mass spectrometry
mass spectrum
mass-to-charge ratio, *m/z*
matrix-assisted laser desorption/ ionization (MALDI)

molecular ion
molecular mass
nitrogen rule
nominal mass
orbitrap mass spectrometer
precursor ion
product ion
reconstructed total ion chromatogram
resolving power
rings + double bonds formula

selected ion chromatogram
selected ion monitoring
selected reaction monitoring
three-dimensional quadrupole ion trap
time-of-flight mass spectrometer
transmission quadrupole mass spectrometer

Summary

Ions are created or desorbed in the ion source of a mass spectrometer. Neutral molecules are converted into ions by electron ionization (which produces a molecular ion, $M^{+\cdot}$, and many fragments) or by chemical ionization (which tends to create MH^+ and few fragments). A magnetic sector mass spectrometer separates gaseous ions by accelerating them in an electric field and deflecting ions of different mass-to-charge (m/z) ratio through different arcs in a magnetic field. Ions are detected by an electron multiplier, which works like a photomultiplier tube. A conversion dynode may be necessary prior to the electron multiplier to ensure that all ions produce a similar electrical response at the detector. The mass spectrum is a graph of detector response versus m/z value.

A double-focusing mass spectrometer attains high resolution by employing an electric sector with the magnetic sector to select ions with a narrow range of kinetic energy. Other mass separators include the transmission quadrupole mass spectrometer, the time-of-flight mass spectrometer, the three-dimensional quadrupole ion trap, the linear ion trap, and the orbitrap. The time-of-flight instrument is capable of high acquisition rates and has a nearly unlimited upper mass range. Resolving power is defined as $m/\Delta m$ or $m/m_{1/2}$, where m is the mass being measured, Δm is the difference in mass between two peaks that are separated with a 10% valley between them, and $m_{1/2}$ is the width of a peak at half-height. Time-of-flight and orbitrap spectrometers can provide high-resolution spectra with part per million mass accuracy.

In a mass spectrum, the molecular ion is found from the highest m/z value of any "significant" peak that cannot be attributed to isotopes or background signals. For a given composition, you should be able to predict the relative intensities of the isotopic peaks at $M+1$, $M+2$, and so on. Among common elements, Cl and Br have particularly diagnostic isotope patterns. From a molecular composition, the rings + double bonds equation helps us propose structures. An organic compound with an odd number of nitrogen atoms will have an odd mass. Fragment ions arising from bond cleavage and rearrangements provide clues to molecular structure.

Gas emerging from a capillary gas chromatography column can go directly into the ion source of a well-pumped mass spectrometer to provide qualitative and quantitative information about the components of a mixture. For capillary liquid chromatography with a flow rate of ~500 nL/min, liquid can be admitted directly to a well-pumped electron ionization source. For liquid chromatography with wider columns and higher flow rates, electrospray ionization employs high voltage at the exit of the column, combined with coaxial N_2 gas flow, to create a fine aerosol of charged droplets containing ions that were already present in the liquid phase. Analyte is often associated with other ions to give species such as $[MNa]^+$ or $[M(CH_3CO_2)]^-$. Control of pH helps ensure that selected analytes are in anionic or cationic form. Another interface between liquid chromatography and mass spectrometry is atmospheric pressure chemical ionization, which utilizes a corona discharge to create a variety of gaseous ions from aerosol droplets. Both atmospheric pressure chemical ionization and electrospray tend to produce unfragmented ions. Collisionally activated dissociation to produce fragment ions is controlled by the cone voltage at the mass spectrometer inlet. Electrospray of proteins typically creates an array of highly charged ions such as MH_n^{n+}. Matrix-assisted laser desorption/ionization is a gentle way to produce predominantly singly charged, intact protein ions for mass spectrometry.

A reconstructed total ion chromatogram shows the signal from all ions above a chosen m/z emerging from chromatography as a function of time. An extracted ion chromatogram shows the signal for one ion taken from the complete mass spectrum. Selected ion monitoring of one or a few values of m/z improves signal-to-noise because all of the time is spent measuring just that one or a few ions. In selected reaction monitoring, a precursor ion isolated by one mass filter passes into a collision cell in which it breaks into products. One (or more) product ion is then selected by a second mass filter for passage to the detector. This process is extremely selective for just one analyte and vastly increases the signal-to-noise ratio for this analyte. Electron-transfer dissociation is employed in protein sequencing to break amide bonds in a polypeptide without breaking other bonds.

Several methods can ionize molecules from the surface of an object in ambient atmosphere. Direct analysis in real time (DART) and a low-temperature plasma use electronically excited helium or nitrogen to ionize analytes. Desorption electrospray ionization (DESI) directs electrosprayed solvent onto a surface to dislodge ions.

An ion mobility spectrometer separates gas-phase ions by their different mobilities in an electric field at atmospheric pressure. Field asymmetric waveform spectrometry separates ions based on differences of mobility in a high electric field at which mobility is not a linear function of electric field. Ion mobility can add another dimension to mass spectrometry by separating isobaric ions according to differences in mobility.

Exercises

22-A. Measure the width at half-height of the peak at m/z 53 and calculate the resolving power of the spectrometer from the expression $m/m_{1/2}$. Would you expect to be able to resolve two peaks at 100 and 101 Da?

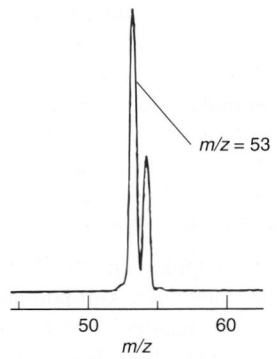

Mass spectrum. [Data from V. J. Angelico, S. A. Mitchell, and V. H. Wysocki, "Low-Energy Ion–Surface Reactions of Pyrazine with Two Classes of Self-Assembled Monolayers," *Anal. Chem.* **2000**, *72*, 2603.]

22-B. What resolving power is required to distinguish $CH_3CH_2^+$ from $HC{\equiv}O^+$?

22-C. *Isotope patterns.* Consider an element with two isotopes whose natural abundances are a and b ($a + b = 1$). If there are n atoms of the element in a compound, the probability of finding each combination of isotopes is derived from the expansion of the binomial $(a + b)^n$. For carbon, the abundances are $a = 0.989\ 3$ for ^{12}C and $b = 0.010\ 7$ for ^{13}C. The probability of finding 2 ^{12}C atoms in acetylene, $HC{\equiv}CH$, is given by the first term of the expansion of $(a + b)^2 = a^2 + 2ab + b^2$. The value of a^2 is $(0.989\ 3)^2 = 0.978\ 7$, so the probability of finding 2 ^{12}C atoms in acetylene is 0.978 7. The probability of finding 1 ^{12}C + 1 ^{13}C is $2ab = 2(0.989\ 3)(0.010\ 7) = 0.021\ 2$. The probability of finding 2 ^{13}C is $b^2 = (0.010\ 7)^2 = 0.000\ 114$. The molecular ion, by definition, contains 2 ^{12}C atoms. The M+1 peak contains 1 ^{12}C + 1 ^{13}C. The intensity of M+1 relative to $M^{+\bullet}$ will be $(0.021\ 2)/(0.978\ 7) = 0.021\ 7$. (We are ignoring ^2H because its natural abundance is small.) Predict the relative amounts of $C_6H_4{}^{35}Cl_2$, $C_6H_4{}^{35}Cl^{37}Cl$, and $C_6H_4{}^{37}Cl_2$ in 1,2-dichlorobenzene. Draw a stick diagram of the distribution, like Figure 22-7.

22-D. **(a)** Find the number of rings plus double bonds in a molecule with the composition $C_{14}H_{12}$ and draw one plausible structure.

(b) For an ion or radical, the rings + double bonds formula gives noninteger answers because the formula is based on valences in neutral molecules with all electrons paired. How many rings plus double bonds are predicted for $C_4H_{10}NO^+$? Draw one structure for $C_4H_{10}NO^+$.

22-E. **(a)** Spectra A and B belong to isomers of $C_6H_{12}O$. Explain how you can tell which isomer goes with each spectrum.

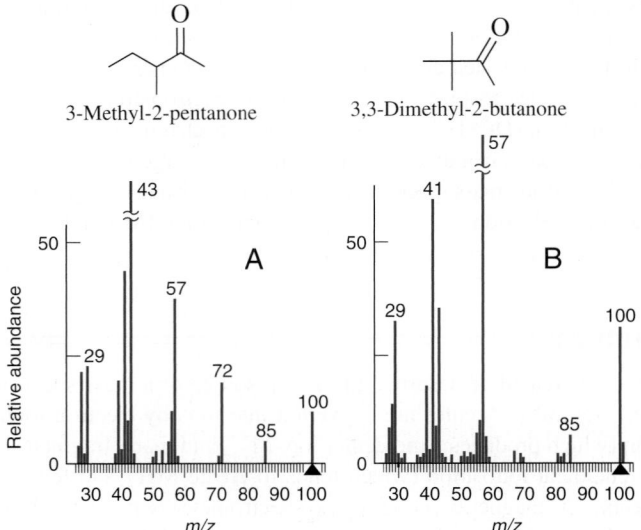

Mass spectra of isomeric ketones with the composition $C_6H_{12}O$. [NIST/EPA/NIH Mass Spectral Database.[7]]

(b) The intensity of the M+1 peak at m/z 101 must be incorrect in both spectra. It is entirely missing in spectrum A and too intense (15.6% of intensity of $M^{+\bullet}$) in spectrum B. What should be the intensity of M+1 relative to $M^{+\bullet}$ for the composition $C_6H_{12}O$?

22-F. (This is a long exercise suitable for group work.) Relative intensities for the molecular ion region of several compounds are listed in parts **(a)–(d)** and shown in the figure. Suggest a composition for each molecule and calculate the expected isotopic peak intensities.

(a) m/z (intensity): 94 (999), 95 (68), 96 (3)

(b) m/z (intensity): 156 (566), 157 (46), 158 (520), 159 (35)

(c) m/z (intensity): 224 (791), 225 (63), 226 (754), 227 (60), 228 (264), 229 (19), 230 (29)

(d) m/z (intensity): 154 (122), 155 (9), 156 (12) (*Hint:* Contains sulfur.)

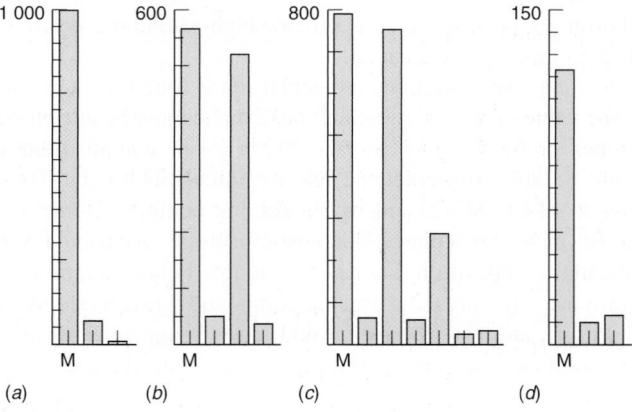

Mass spectra. [NIST/EPA/NIH Mass Spectral Database.[7]]

22-G. *Protein molecular mass from electrospray.* The enzyme lysozyme[59] exhibits MH_n^{n+} peaks at $m/z = 1\,789.1$, $1\,590.4$, $1\,431.5$, $1\,301.5$, and $1\,193.1$. Follow the procedure of Table 22-3 to find the mean molecular mass and its standard deviation.

22-H. *Quantitative analysis by selected ion monitoring.* Caffeine in beverages and urine can be measured by adding caffeine-D_3 as an internal standard and using selected ion monitoring to measure each compound by gas chromatography. The figure shows mass chromatograms of caffeine (m/z 194) and caffeine-D_3 (m/z 197), which have nearly the same retention time.

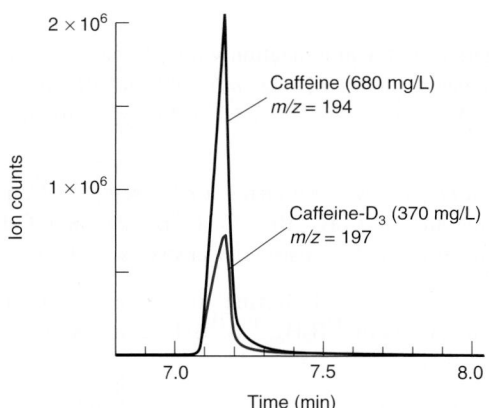

Selected ion monitoring mass chromatogram showing caffeine and caffeine-D_3 eluted from a capillary gas chromatography column. [Data from D. W. Hill, B. T. McSharry, and L. S. Trzupek, "Quantitative Analysis by Isotopic Dilution Using Mass Spectrometry," *J. Chem. Ed.* **1988**, *65*, 907.]

Caffeine (M = 194 Da)   Caffeine-D_3 (M = 197 Da)

Suppose that the following data were obtained for standard mixtures:

Caffeine (mg/L)	Caffeine-D_3 (mg/L)	Caffeine peak area	Caffeine-D_3 peak area
13.60×10^2	3.70×10^2	11 438	2 992
6.80×10^2	3.70×10^2	6 068	3 237
3.40×10^2	3.70×10^2	2 755	2 819

NOTE: *Injected volume was different in all three runs.*

(a) Compute the mean response factor in the equation

$$\frac{\text{Area of analyte signal}}{\text{Area of standard signal}} = F\left(\frac{\text{concentration of analyte}}{\text{concentration of standard}}\right)$$

(b) For analysis of a cola beverage, 1.000 mL of beverage was treated with 50.0 μL of standard solution containing 1.11 g/L caffeine-D_3 in methanol. The combined solution was passed through a solid-phase extraction cartridge that retains caffeine. Polar solutes were washed off with water. Then the caffeine was washed off the cartridge with an organic solvent and the solvent was evaporated to dryness. The residue was dissolved in 50 μL of methanol for gas chromatography. Peak areas were 1 144 for m/z 197 and 1 733 for m/z 194. Find the concentration of caffeine (mg/L) in the beverage.

Problems

What Is Mass Spectrometry?

22-1. Briefly describe how a magnetic sector mass spectrometer works.

22-2. How are ions created for each of the mass spectra in Figure 22-4? Why are the two spectra so different?

22-3. Define the unit *dalton*. From this definition, compute the mass of 1 Da (= 1 unified atomic mass unit = 1 u) in grams. The mean of 60 measurements of the mass of individual *E. coli* cells vaporized by MALDI and measured with a quadrupole ion trap was $5.03\ (\pm 0.14) \times 10^{10}$ Da.[2] Express this mass in femtograms.

22-4. Nickel has two major and three minor isotopes. For the purpose of this problem, suppose that the *only* isotopes are ^{58}Ni and ^{60}Ni. The atomic mass of ^{58}Ni is 57.935 3 Da and the mass of ^{60}Ni is 59.933 2 Da. From the amplitude of the peaks in the following spectrum, calculate the atomic mass of Ni and compare your answer with the value in the periodic table.

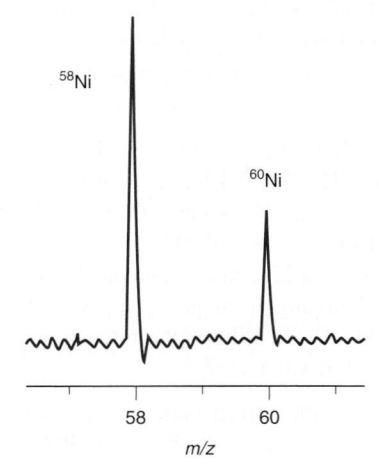

Mass spectrum. [Data from Y. Su, Y. Duan, and Z. Jin, "Helium Plasma Source Time-of-Flight Mass Spectrometry: Off-Cone Sampling for Elemental Analysis," *Anal. Chem.* **2000**, *72*, 2455.]

22-5. Measure the width at half-height of the tallest peak in the spectrum below and calculate the resolving power of the spectrometer from the expression $m/m_{1/2}$. Would you expect to be able to distinguish two peaks at 10 000 and 10 001 Da?

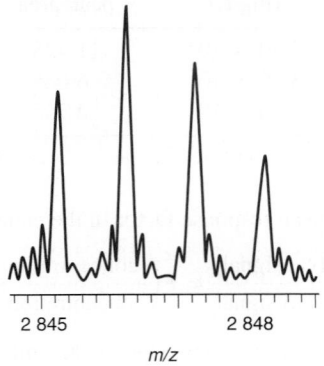

MALDI mass spectrum of the peptide melittin. [Data from P. B. O'Connor and C. E. Costello, "Application of Multishot Acquisition in Fourier Transform Mass Spectrometry," *Anal. Chem.* **2000,** *72,* 5125.]

22-6. The two peaks near m/z 31.00 in Figure 22-10 differ in mass by 0.010 Da. *Estimate* the resolving power of the spectrometer from the expression $m/\Delta m$ without making any measurements in the figure.

22-7. The highest resolution[60] and best mass accuracy[61] spectra are obtained by Fourier transform ion cyclotron resonance mass spectrometry. Molecular ions of two peptides (chains of seven amino acids) differing in mass by 0.000 45 Da were separated with a 10% valley between them.[62] The ions each have a mass of 906.49 Da and a width at half-height of 0.000 27 Da. Compute the resolving power by the 10% valley formula and by the half-width formula, both in Equation 22-1. Compare the difference in mass of these two compounds with the mass of an electron.

The Mass Spectrum

22-8. The mass of a fragment ion in a high-resolution spectrum is 83.086 5 Da. Which composition, $C_5H_7O^+$ or $C_6H_{11}^+$, better matches the observed mass?

22-9. Calculate the theoretical masses of the species in Figure 22-10 and compare your answers with the values observed in the figure.

22-10. An ion with the formula $C_7H_{10}NO_2^+$ appears at m/z 140.

(a) Write the formulas of the isotopologues contributing to m/z 141 and predict the intensity at m/z 141 relative to the intensity at m/z 140.

(b) Write the formulas of the three most abundant isotopologues contributing to m/z 142.

22-11. In Figure 22-9, the sucralose species with a nominal mass X = 395 is $[{}^{12}C_{12}{}^{1}H_{18}{}^{16}O_8{}^{35}Cl_3]^-$. X+1 arises from isotopologues containing one ${}^{13}C$, one ${}^{2}H$, or one ${}^{17}O$. The predicted intensity at X + 1 is 12 × 1.08% + 18 × 0.012% + 8 × 0.038% = 13.5% of X. Use factors in Table 22-2 to write the analogous equation for the intensity at X+2 arising from just $[{}^{13}C_2{}^{12}C_{10}{}^{1}H_{18}{}^{16}O_8{}^{35}Cl_3]^-$ and $[{}^{12}C_{12}{}^{1}H_{18}{}^{18}O_1{}^{16}O_7{}^{35}Cl_3]^-$. Why is the intensity at X + 2 so much greater than what you calculate?

22-12. Calculate the theoretical masses of the sucralose species in Figure 22-9 at nominal masses of 395, 397, 399, and 401. Find the difference in ppm between the observed and calculated m/z.

$$\text{ppm difference} = 10^6 \times \frac{\text{observed } m/z - \text{calculated } m/z}{\text{calculated } m/z}$$

22-13. Dioctyl phthalate is a ubiquitous background substance in most labs because it is a plasticizer used in Tygon tubing.

Dioctyl phthalate
$C_{24}H_{38}O_4$
Nominal mass 390

(a) Suggest a structure for the base peak cation at m/z 149 and the second most abundant peak at m/z 279 observed by electron ionization.[11]

(b) In negative chemical ionization with NH_3 reagent gas, there is a prominent parent anion $M^{-\bullet}$ at m/z 390 and dominant fragment anions are observed at m/z 277 and 221. Suggest structures for these anions.

22-14. *Isotope patterns.* Referring to Exercise 22-C, predict the relative amounts of $C_2H_2{}^{79}Br_2$, $C_2H_2{}^{79}Br^{81}Br$, and $C_2H_2{}^{81}Br_2$ in 1,2-dibromoethylene. Compare your answer with Figure 22-7.

22-15. *Isotope patterns.* Referring to Exercise 22-C, predict the relative abundances of ${}^{10}B_2H_6$, ${}^{10}B^{11}BH_6$, and ${}^{11}B_2H_6$ for diborane (B_2H_6).

22-16. *Isotope patterns.* From the natural abundance of ${}^{79}Br$ and ${}^{81}Br$, predict the relative amounts of $CH^{79}Br_3$, $CH^{79}Br_2{}^{81}Br$, $CH^{79}Br^{81}Br_2$, and $CH^{81}Br_3$. As in Exercise 22-C, the fraction of each isotopic molecule comes from the expansion of $(a + b)^3$, where a is the abundance of ${}^{79}Br$ and b is the abundance of ${}^{81}Br$.

$$(a + b)^n = a^n + \frac{n}{1!}a^{n-1}b^1 + \frac{(n)(n-1)}{2!}a^{n-2}b^2 + \frac{(n)(n-1)(n-2)}{3!}a^{n-3}b^3 + \cdots$$

Compare your answer with Figure 22-7.

22-17. Alkanes do not ordinarily exhibit a molecular ion $M^{+\bullet}$ in the electron ionization mass spectrum because $M^{+\bullet}$ breaks into fragments. If dilute alkane vapor in He is allowed to expand through a narrow orifice into a vacuum, expansion of the gas cools the molecules and reduces their vibrational energy. The molecular ion of the alkane $C_{72}H_{146}$ is the base peak when the molecule is sufficiently cold.[8]

(a) Find the nominal mass of the molecular ion $C_{72}H_{146}{}^{+\bullet}$.

(b) Find the monoisotopic mass of $C_{72}H_{146}{}^{+\bullet}$.

(c) ${}^{13}C$ has a natural abundance of 1.07%, whereas ${}^{2}H$ has a natural abundance of 0.012%. Therefore, the main isotopologues of the molecular ion are ${}^{12}C_{72}H_{146}$ and ${}^{12}C_{71}{}^{13}C_1H_{146}$. As in the previous problem, we can find the relative abundance of ${}^{12}C_{72}$ and ${}^{12}C_{71}{}^{13}C_1$ from the first two terms of the binomial expansion $(a + b)^{72}$, where $a = 0.989\ 3$ for ${}^{12}C$ and $b = 0.010\ 7$ for ${}^{13}C$. Predict the relative abundance of M and M+1 in the mass spectrum of $C_{72}H_{146}$.

22-18. Find the number of rings + double bonds in molecules with the following compositions and draw one plausible structure for each: **(a)** $C_{11}H_{18}N_2O_3$; **(b)** $C_{12}H_{15}BrNPOS$; **(c)** fragment in a mass spectrum with the composition $C_3H_5^+$.

22-19. (Each part of this problem is quite long and best worked by groups of students.) Peak intensities of the molecular ion region are listed in parts **(a)**–**(g)** and shown in the figure. Identify which peak represents the molecular ion, suggest a composition for it, and calculate the expected isotopic peak intensities. Restrict your attention to elements in Table 22-1.

(a) m/z (intensity): 112 (999), 113 (69), 114 (329), 115 (21)

(b) m/z (intensity): 146 (999), 147 (56), 148 (624), 149 (33), 150 (99), 151 (5)

(c) m/z (intensity): 90 (2), 91 (13), 92 (96), 93 (999), 94 (71), 95 (2)

(d) m/z (intensity): 226 (4), 227 (6), 228 (130), 229 (215), 230 (291), 231 (168), 232 (366), 233 (2), 234 (83) (Calculate expected intensities from isotopes of the major element present.)

(e) m/z (intensity): 172 (531), 173 (12), 174 (999), 175 (10), 176 (497)

(f) m/z (intensity): 177 (3), 178 (9), 179 (422), 180 (999), 181 (138), 182 (9)

(g) m/z (intensity): 182 (4), 183 (1), 184 (83), 185 (16), 186 (999), 187 (132), 188 (10)

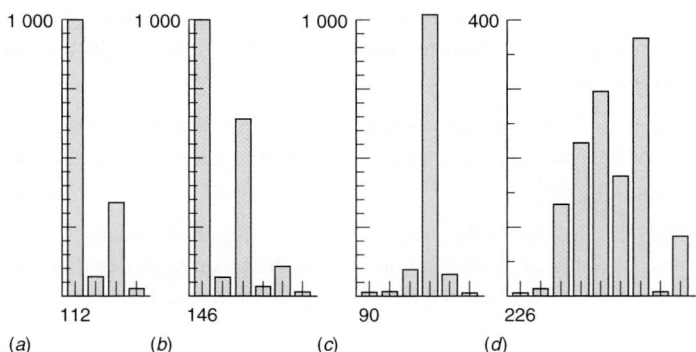

(a) (b) (c) (d)

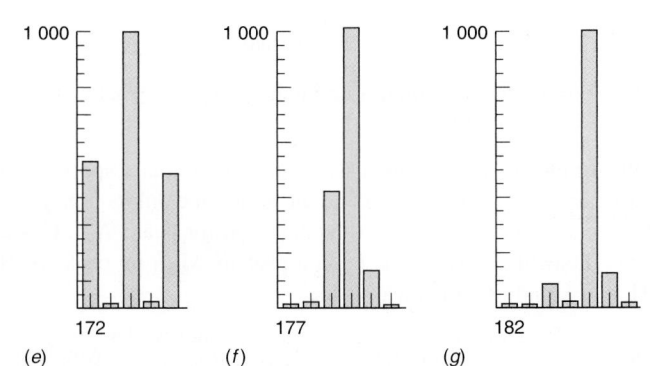

(e) (f) (g)

Mass spectra. [NIST/EPA/NIH Mass Spectral Database.[7]]

22-20. Suggest a composition for the halogen compound whose mass spectrum is shown in the figure. Assign each of the major peaks.

22-21. *Mass spectral interpretation.* The compound $C_9H_4N_2Cl_6$ is a byproduct found in chlorinated pesticides.

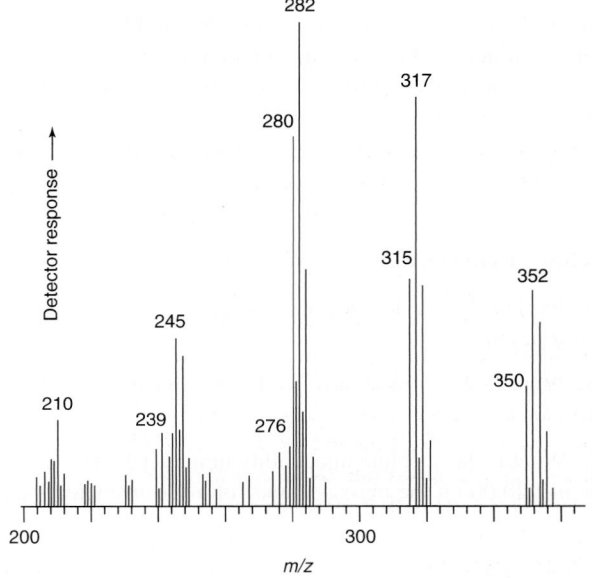

(a) Verify that the formula for rings + double bonds agrees with the structure.

(b) Find the nominal mass of $C_9H_4N_2Cl_6$.

(c) Suggest an assignment for m/z 350, 315, 280, 245, and 210 in the high-mass region of the electron ionization mass spectrum.

Mass spectrum of $C_9H_4N_2Cl_6$. [Data from W. Vetter and W. Jun, "Elucidation of a Polychlorinated Bipyrrole Structure Using Enantioselective GC," *Anal. Chem.* **2002,** 74, 4287.]

22-22. The chart in Box 22-3 shows that $\delta^{13}C$ for CO_2 in human breath in the United States is different from that of CO_2 in human breath in mainland Europe. Suggest an explanation.

22-23. (a) The mass of ^1H in Table 22-1 is 1.007 825 Da. Compare it with the sum of the masses of a proton and an electron given in the table.

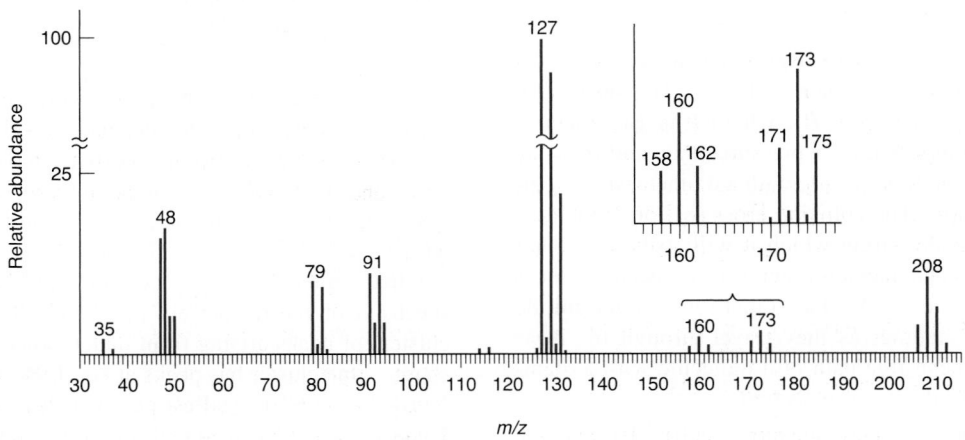

Mass spectrum for Problem 22-20. [NIST/EPA/NIH Mass Spectral Database.[7]]

(b) ^2H (deuterium) contains one proton, one neutron, and one electron. Compare the sum of the masses of these three particles with the mass of ^2H.

(c) The discrepancy in part **(b)** comes from the conversion of mass into binding energy that holds the nucleus together. The relation of mass, m, to energy, E is $E = mc^2$, where c is the speed of light. From the discrepancy in part **(b)**, calculate the binding energy of ^2H in joules and in kJ/mol. (1 Da = $1.660\ 5 \times 10^{-27}$ kg.)

(d) The binding energy (ionization energy) of the electron in a hydrogen or deuterium atom is 13.6 eV. Use Table 1-4 to convert this number into kJ/mol and compare it with the binding energy of the ^2H nucleus.

(e) A typical bond dissociation energy in a molecule is 400 kJ/mol. How many times larger is the nuclear binding energy of ^2H than a bond energy?

22-24. ▦ *Isotope patterns.* (Caution: This problem could lead to serious brain injury.) For an element with three isotopes with abundances a, b, and c, the distribution of isotopes in a molecule with n atoms is based on the expansion of $(a + b + c)^n$. By a process analogous to that in Exercise 22-C, predict what the mass spectrum of Si_2 will look like.

Mass Spectrometers

22-25. Explain how a double-focusing mass spectrometer achieves high resolution.

22-26. What is the purpose of the reflectron in a time-of-flight mass spectrometer?

22-27. What is the absolute uncertainty in mass (±? Da) at *m/z* 100 and at *m/z* 20 000 if the mass accuracy of a spectrometer is 2 ppm?

22-28. A limitation on how many spectra per second can be recorded by a time-of-flight mass spectrometer is the time it takes the slowest ions to go from the source to the detector. Suppose we want to scan up to *m/z* 500. Calculate the speed of this heaviest ion if it is accelerated through 5.00 kV in the source. How long would it take to drift 2.00 m through a spectrometer? At what frequency could you record spectra if a new extraction cycle were begun each time this heaviest ion reached the detector? What would be the frequency if you wanted to scan up to *m/z* 1 000?

22-29. If ions are accelerated through 20 kV in the source of a time-of-flight mass spectrometer with a 2.00 m drift distance, what are the transit times of ions with *m/z* = 100 and *m/z* = 1 000 000?

22-30. (a) The *mean free path* is the average distance a molecule travels before colliding with another molecule. The mean free path, λ, is given by $\lambda = kT/(\sqrt{2}\sigma P)$, where k is Boltzmann's constant, T is temperature (K), P is pressure (Pa), and σ is the collision cross section. For a molecule with a diameter d, the collision cross section is πd^2. The collision cross section is the area swept out by the molecule within which it will strike any other molecule it encounters. The magnetic sector mass spectrometer is maintained at a pressure of $\sim 10^{-5}$ Pa so that ions do not collide with (and deflect) one another as they travel through the mass analyzer. What is the mean free path of a molecule with a diameter of 1 nm at 300 K in the mass analyzer?

(b) The vacuum in an orbitrap mass separator is $\sim 10^{-8}$ Pa. Find the mean free path in the orbitrap for the same conditions as **(a)**.

Chromatography/Mass Spectrometry

22-31. Which liquid chromatography/mass spectrometry interface, atmospheric pressure chemical ionization or electrospray, requires analyte ions to be in solution prior to the interface? How does the other interface create gaseous ions from neutral species in solution?

22-32. Why is direct electron ionization in Figure 22-14 permissible for gas chromatography, but only for capillary columns in liquid chromatography in Figure 22-29?

22-33. Why does vacuum ultraviolet photoionization with $\sim$120-nm-wavelength radiation produce less molecular fragmentation than does 70 eV electron ionization?

22-34. What is collisionally activated dissociation? At what points in a mass spectrometer does it occur?

22-35. What are the differences between a reconstructed total ion chromatogram, an extracted ion chromatogram, and a selected ion chromatogram?

22-36. What is selected reaction monitoring? Why is it also called MS/MS? Why does it improve the signal-to-noise ratio for a particular analyte?

22-37. (a) To detect the drug ibuprofen by liquid chromatography–mass spectrometry, would you choose the positive or negative ion mode for the spectrometer? Would you choose acidic or neutral chromatography solvent? State your reasons.

Ibuprofen (FM 206)

(b) If the unfragmented ion has an intensity of 100, what should be the intensity of M+1?

22-38. An electrospray–transmission quadrupole mass spectrum of the α chain of hemoglobin from acidic solution exhibits nine peaks corresponding to MH_n^{n+}. Find the charge, n, for peaks A–I. Calculate the molecular mass of the neutral protein, M, from peaks A, B, G, H, and I, and find the mean value.

Peak	*m/z*	Amplitude	Peak	*m/z*	Amplitude
A	1 261.5	0.024	F	not stated	1.000
B	1 164.6	0.209	G	834.3	0.959
C	not stated	0.528	H	797.1	0.546
D	not stated	0.922	I	757.2	0.189
E	not stated	0.959			

22-39. The molecular ion region in the mass spectrum of a large molecule, such as a protein, consists of a cluster of peaks differing by 1 Da. This pattern occurs because a molecule with many atoms has a high probability of containing one or several atoms of ^{13}C, ^{15}N, ^{18}O, ^2H, and ^{34}S. In fact, the probability of finding a molecule with only ^{12}C, ^{14}N, ^{16}O, ^1H, and ^{32}S may be so small that the nominal molecular ion is not observed. The electrospray mass spectrum of the rat protein interleukin-8 consists of a series of clusters of peaks arising from intact molecular ions with different charge. One cluster has peaks at *m/z* 1 961.12, 1 961.35, 1 961.63, 1 961.88, 1 962.12 (tallest peak), 1 962.36, 1 962.60, 1 962.87, 1 963.10, 1 963.34, 1 963.59, 1 963.85, and 1 964.09. These peaks correspond to isotopic ions differing by 1 Da. From the observed peak separation, find the charge of the ions in this cluster. From m/z of the tallest peak, estimate the molecular mass of the protein.

22-40. Nanoparticles with a gold core and organothiol outer shell with well-defined stoichiometries $Au_x(SR)_y$ can be prepared and isolated.[63]

(a) The MALDI–time-of-flight mass spectrum of the nanoparticle with $R = CH_2CH_2C_6H_5$ exhibits a broad molecular ion at m/z 76.3 kDa. Electrospray ionization gives a series of sharp, multiply charged peaks at m/z 38 152.7 (charge $= +2$), 25 433.3 (+3), 19 075.2 (+4), and 15 260.4 (+5). Find the average molecular mass of $Au_x(SCH_2CH_2C_6H_5)_y$ from the four peaks.

(b) In another electrospray experiment, the (-3) ions for three different thiol shells were found at m/z 27 243 (R $= n$-$C_{12}H_{25}$), 25 440 (R $= CH_2CH_2C_6H_5$), and 24 876 (R $= n$-C_6H_{13}). Compute the mass of the molecular ion for the three $Au_x(SR)_y$ nanoparticles. Assuming that x is constant, find the value of y for each nanoparticle from the difference in molecular mass between the molecular ions. From the value of y, find the value of x. Note that Au is monoisotopic with a mass of 196.966 6.

22-41. Phytoplankton at the ocean surface maintain the fluidity of their cell membranes by altering their lipid (fat) composition when the temperature changes. When the ocean temperature is high, plankton synthesize relatively more 37:2 than 37:3.[64]

37:2 = $C_{37}H_{70}O$

37:3 = $C_{37}H_{68}O$

After they die, plankton sink to the ocean floor and end up buried in sediment. The deeper we sample a sediment, the further back into time we delve. By measuring the relative quantities of cell-membrane compounds at different depths in the sediment, we can infer the temperature of the ocean long ago.

The molecular ion regions of the chemical ionization mass spectra of 37:2 and 37:3 are listed in the table. Predict the expected intensities of M, M+1, and M+2 for each of the four species listed. Include contributions from C, H, O, and N, as appropriate. Compare your predictions with the observed values. Discrepant intensities in these data are typical unless care is taken to obtain high-quality data.

Compound	Species in mass spectrum	M	M+1	M+2
37:3	$[MNH_4]^+$ (m/z 546)^a	100	35.8	7.0
37:3	$[MH]^+$ (m/z 529)^b	100	23.0	8.0
37:2	$[MNH_4]^+$ (m/z 548)^a	100	40.8	3.7
37:2	$[MH]^+$ (m/z 531)^b	100	33.4	8.4

a. Chemical ionization with ammonia.
b. Chemical ionization with isobutane.

22-42. Chlorate (ClO_3^-), chlorite (ClO_2^-), bromate (BrO_3^-), and iodate (IO_3^-) can be measured in drinking water at the 1-ppb level with 1% precision by selected reaction monitoring.[65] Chlorate and chlorite arise from ClO_2 used as a disinfectant. Bromate and iodate can be formed from Br^- or I^- when water is disinfected with ozone (O_3). For the highly selective measurement of chlorate, the negative ion selected by Q1 in Figure 22-33 is m/z 83 and the negative ion selected by Q3 is m/z 67. Explain how this measurement works and how it distinguishes ClO_3^- from ClO_2^-, BrO_3^-, and IO_3^-.

22-43. *Quantitative analysis by isotope dilution.*[66] In *isotope dilution*, a known amount of an unusual isotope (called the *spike*) is added to an unknown as an internal standard for quantitative analysis. After the mixture has been homogenized, some of the element of interest must be isolated. The ratio of the isotopes is then measured. From this ratio, the quantity of the element in the original unknown can be calculated.

Natural vanadium has atom fractions $^{51}V = 0.997\ 5$ and $^{50}V = 0.002\ 5$. The atom fraction is defined as

$$\text{Atom fraction of } ^{51}V = \frac{\text{atoms of } ^{51}V}{\text{atoms of } ^{50}V + \text{atoms of } ^{51}V}$$

A spike enriched in ^{50}V has atom fractions $^{51}V = 0.639\ 1$ and $^{50}V = 0.360\ 9$.

(a) Let isotope A be ^{51}V and isotope B be ^{50}V. Let A_x be the atom fraction of isotope A ($=$ atoms of A/[atoms of A + atoms of B]) in an unknown. Let B_x be the atom fraction of B in an unknown. Let A_s and B_s be the corresponding atom fractions in a spike. Let C_x be the total concentration of all isotopes of vanadium (μmol/g) in the unknown and let C_s be the concentration in the spike. Let m_x be the mass of unknown and m_s be the mass of spike. After m_x grams of unknown are mixed with m_s grams of spike, the ratio of isotopes in the mixture is found to be R. Show that

$$R = \frac{\text{mol A}}{\text{mol B}} = \frac{A_x C_x m_x + A_s C_s m_s}{B_x C_x m_x + B_s C_s m_s} \quad \textbf{(A)}$$

(b) Solve Equation A for C_x to show that

$$C_x = \left(\frac{C_s m_s}{m_x}\right)\left(\frac{A_s - RB_s}{RB_x - A_x}\right) \quad \textbf{(B)}$$

(c) A 0.401 67-g sample of crude oil containing an unknown concentration of natural vanadium was mixed with 0.419 46 g of spike containing 2.243 5 μmol V/g enriched with ^{50}V (atom fractions: $^{51}V = 0.639\ 1$, $^{50}V = 0.360\ 9$).[67] After dissolution and equilibration of the oil and the spike, some of the vanadium was isolated by ion-exchange chromatography. The measured isotope ratio in the isolated vanadium was $R = {}^{51}V/{}^{50}V = 10.545$. Find the concentration of vanadium (μmol/g) in the crude oil.

(d) Examine the calculation in part **(c)** and express the answer with the correct number of significant figures.

23 Introduction to Analytical Separations

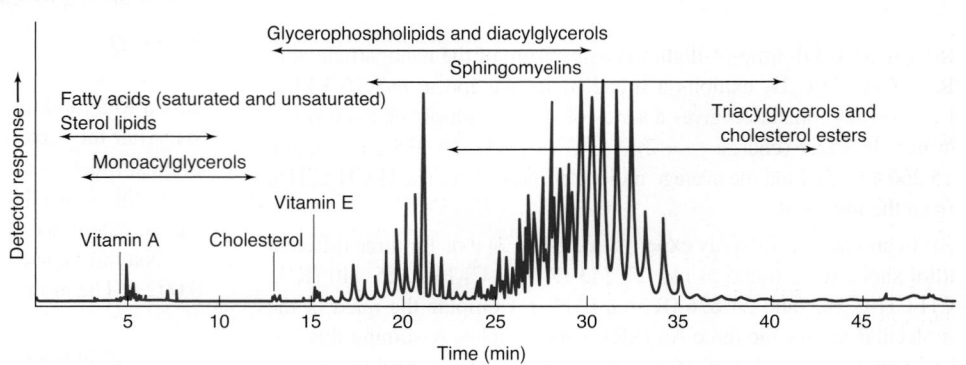

Reversed-phase liquid chromatography separation of breast milk extract with positive electrospray ionization mass spectrometric detection. [Data from A. Villaseñor, I. Garcia-Perez, A. Garcia, J. M. Posma, M. Fernández-López, A. J. Nicholas, N. Modi, E. Holmes, and C. Barbas, "Breast Milk Metabolome Characterization in a Single-Phase Extraction, Multiplatform Analytical Approach," *Anal. Chem.* **2014,** *86,* 8245.]

The World Health Organization recommends exclusive breastfeeding of infants for the first six months of life. The study described here seeks to identify biomarkers in breast milk associated with infant development and health. Breast milk is a complex mixture. It has butterfat droplets dispersed within water that contains carbohydrates, proteins, and many micronutrients. Fat droplets are largely triacylglycerols stabilized by surface-active molecules such as phospholipids. Breast milk composition changes during lactation and is strongly affected by the mother's diet. A method was developed to measure as many metabolites as possible using a single solvent extraction coupled with multiple chromatography–high resolution mass spectrometric techniques.

Individual or pooled 50-μL samples of breast milk were mixed with 350 μL of 1:1 (vol/vol) methanol:methyl-*tert*-butyl ether. Precipitated protein was removed by centrifugation, leaving a large fraction of metabolites dissolved in the single-phase liquid. After adding internal standard, many lipids could be identified and measured by liquid chromatography–positive electrospray ionization mass spectrometry (Section 22-4), as shown in the figure.

Carbohydrates were measured by liquid chromatography with negative electrospray ionization mass spectrometry. Other compounds including amino acids, organic acids, cholesterol, and saccharides were measured by derivatization to make them more volatile, followed by gas chromatography–mass spectrometry. The study identified changes in breast milk composition over time that might correspond to the changing needs of the infant.

In the vast majority of real analytical problems, we must separate, identify, and measure one or more components from a complex mixture. This chapter discusses fundamentals of analytical separations, and the next three chapters describe specific methods.

23-1 Solvent Extraction

Extraction is the physical transfer of a solute from one phase to another. Common reasons to carry out an extraction in analytical chemistry are to isolate or concentrate the desired analyte or to separate it from species that would interfere in the analysis. The most common case is the extraction of an aqueous solution with an organic solvent. Diethyl ether, toluene, and hexane are common solvents that are *immiscible* with and less dense than water. They form a separate phase that floats on top of the aqueous phase. Chloroform,

Two liquids are **miscible** if they form a single phase when they are mixed in any ratio. **Immiscible** liquids remain in separate phases. Organic solvents with low polarity are generally immiscible with water, which is highly polar.

dichloromethane, and carbon tetrachloride are common solvents that are denser than water.* In the two-phase mixture, one phase is predominantly water and the other phase is predominantly organic.

Suppose that solute S is partitioned between phases 1 and 2, as depicted in Figure 23-1. Solutes partition is based on "like dissolves like," which means that solute is more soluble in a phase whose polarity is similar to that of the solute. The **partition coefficient,** K, is the equilibrium constant for the reaction

$$S(\text{in phase 1}) \rightleftharpoons S(\text{in phase 2})$$

Partition coefficient:
$$K = \frac{\mathcal{A}_{S_2}}{\mathcal{A}_{S_1}} \approx \frac{[S]_2}{[S]_1} \tag{23-1}$$

where $\mathcal{A}_{S_1}$ refers to the activity of solute in phase 1. Lacking knowledge of the activity coefficients, we will write the partition coefficient in terms of concentrations.

Suppose that solute S in V_1 mL of solvent 1 (water) is extracted with V_2 mL of solvent 2 (toluene). Let m be the moles of S in the system and let q be the fraction of S remaining in phase 1 at equilibrium. The molarity in phase 1 is therefore qm/V_1. The fraction of total solute transferred to phase 2 is $(1 - q)$, and the molarity in phase 2 is $(1 - q)m/V_2$. Therefore,

$$K = \frac{[S]_2}{[S]_1} = \frac{(1 - q)m/V_2}{qm/V_1}$$

from which we can solve for q:

$$\begin{array}{c}\text{Fraction remaining in phase 1}\\\text{after 1 extraction}\end{array} = q = \frac{V_1}{V_1 + KV_2} \tag{23-2}$$

Equation 23-2 says that the fraction of solute remaining in the water (phase 1) depends on the partition coefficient and the volumes. If the phases are separated and fresh toluene (solvent 2) is added, the fraction of solute remaining in the water at equilibrium will be

$$\begin{array}{c}\text{Fraction remaining in phase 1}\\\text{after 2 extractions}\end{array} = q \cdot q = \left(\frac{V_1}{V_1 + KV_2}\right)^2$$

After n extractions, each with volume V_2, the fraction remaining in the water is

$$\begin{array}{c}\text{Fraction remaining in phase 1}\\\text{after } n \text{ extractions}\end{array} = q^n = \left(\frac{V_1}{V_1 + KV_2}\right)^n \tag{23-3}$$

EXAMPLE **Extraction Efficiency**

Solute A has a partition coefficient of 3 between toluene and water, with three times as much in the toluene phase. Suppose that 100 mL of a 0.010 M aqueous solution of A are extracted with toluene. What fraction of A remains in the aqueous phase **(a)** if one extraction with 500 mL is performed or **(b)** if five extractions with 100 mL are performed?

Solution **(a)** With water as phase 1 and toluene as phase 2, Equation 23-2 says that, after a 500-mL extraction, the fraction remaining in the aqueous phase is

$$q = \frac{100}{100 + (3)(500)} = 0.062 \approx 6\%$$

(b) With five 100-mL extractions, the fraction remaining is given by Equation 23-3:

$$\text{Fraction remaining} = \left(\frac{100}{100 + (3)(100)}\right)^5 = 0.000\,98 \approx 0.1\%$$

It is more efficient to do several small extractions than one big extraction.

TEST YOURSELF   If the partition coefficient is 10, what fraction remains in 100 mL of water after 1 and 5 extractions with 20 mL of toluene? (**Answer:** 33%, 0.41%)

FIGURE 23-1 Partitioning of a solute between two liquid phases requires physical movement of solute from one phase to the other. We say that there is **mass transfer** between the phases.

For simplicity, we assume that the two phases are not soluble in each other. A more realistic treatment considers that most liquids are partially soluble in each other.[1]

The larger the partition coefficient, the less solute remains in phase 1.

Example: If $q = \frac{1}{4}$, then $\frac{1}{4}$ of the solute remains in phase 1 after one extraction. A second extraction reduces the concentration to $(\frac{1}{4})(\frac{1}{4}) = \frac{1}{16}$ of initial concentration.

Many small extractions are more effective than a few large extractions.

The limit for extracting solute S from phase 1 (volume V_1) into phase 2 (volume V_2) is attained by dividing V_2 into an infinite number of infinitesimally small portions for extraction. With $K = [S_2]/[S_1]$, the limiting fraction of solute remaining in phase 1 is[2]

$$q_{\text{limit}} = e^{-(V_2/V_1)K}$$

*Whenever a choice exists between $CHCl_3$ and CCl_4, the less toxic $CHCl_3$ should be chosen. Hexane and toluene are greatly preferred over benzene, which is a carcinogen.

pH Effects

If a solute is an acid or base, its charge changes as the pH is changed. Usually, a neutral species is more soluble in an organic solvent and a charged species is more soluble in aqueous solution. Consider a basic amine whose neutral form, B, has partition coefficient K between aqueous phase 1 and organic phase 2. Suppose that the conjugate acid, BH^+, is soluble *only* in aqueous phase 1. Let's denote the acid dissociation constant of BH^+ as K_a. The **distribution coefficient,** D, is defined as

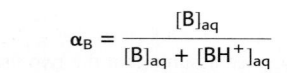

$$\alpha_B = \frac{[B]_{aq}}{[B]_{aq} + [BH^+]_{aq}}$$

α_B is the same as α_{A^-} in Equation 10-18.

$$\text{Distribution coefficient:} \qquad D = \frac{\text{total concentration in phase 2}}{\text{total concentration in phase 1}} \qquad (23\text{-}4)$$

which becomes

$$D = \frac{[B]_2}{[B]_1 + [BH^+]_1} \qquad (23\text{-}5)$$

Substituting $K = [B]_2/[B]_1$ and $K_a = [H^+][B]_1/[BH^+]_1$ into Equation 23-5 leads to

$$\text{Distribution of base} \atop \text{between two phases:} \qquad D = \frac{K \cdot K_a}{K_a + [H^+]} = K \cdot \alpha_B \qquad (23\text{-}6)$$

where α_B is the fraction of weak base in the neutral form, B, in the aqueous phase. *The distribution coefficient D is used in place of the partition coefficient K in Equation 23-2 when dealing with a species that has more than one chemical form, such as B and BH⁺.*

Charged species tend to be more soluble in water than in organic solvent. To extract a base into water, use a pH low enough to convert B into BH^+ (Figure 23-2). To extract the acid HA into water, use a pH high enough to convert HA into A^-.

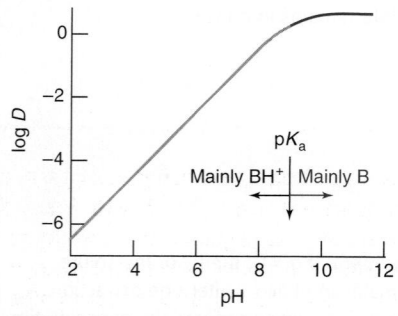

FIGURE 23-2 Effect of pH on the distribution coefficient for the extraction of a base into an organic solvent. In this example, $K = 3.0$ and pK_a for BH^+ is 9.00.

Challenge Suppose that acid HA (with dissociation constant K_a) is partitioned between aqueous phase 1 and organic phase 2 with partition coefficient K for HA. Assuming that A^- is not soluble in the organic phase, show that the distribution coefficient is given by

$$\text{Distribution of acid} \atop \text{between two phases:} \qquad D = \frac{K \cdot [H^+]}{[H^+] + K_a} = K \cdot \alpha_{HA} \qquad (23\text{-}7)$$

where α_{HA} is the fraction of weak acid in the form HA in the aqueous phase.

EXAMPLE Effect of pH on Extraction

Suppose that the partition coefficient for an amine, B, is $K = 3.0$ and the acid dissociation constant of BH^+ is $K_a = 1.0 \times 10^{-9}$. If 50 mL of 0.010 M aqueous amine are extracted with 100 mL of solvent, what will be the formal concentration remaining in the aqueous phase **(a)** at pH 10.00 and **(b)** at pH 8.00?

Solution **(a)** At pH 10.00, $D = KK_a/(K_a + [H^+]) = (3.0)(1.0 \times 10^{-9})/(1.0 \times 10^{-9} + 1.0 \times 10^{-10}) = 2.73$. *Using D in place of K,* Equation 23-2 says that the fraction remaining in the aqueous phase is

$$q = \frac{50}{50 + (2.73)(100)} = 0.15 \Rightarrow 15\% \text{ left in water}$$

The concentration of amine in the aqueous phase is 15% of 0.010 M = 0.001 5 M.
(b) At pH 8.00, $D = (3.0)(1.0 \times 10^{-9})/(1.0 \times 10^{-9} + 1.0 \times 10^{-8}) = 0.273$. Therefore,

$$q = \frac{50}{50 + (0.273)(100)} = 0.65 \Rightarrow 65\% \text{ left in water}$$

The concentration in the aqueous phase is 0.006 5 M. At pH 10, the base is predominantly in the form B and is extracted into the organic solvent. At pH 8, it is in the form BH^+ and remains in the water.

Extraction with a Metal Chelator

Most complexes that can be extracted into organic solvents are neutral. Charged complexes, such as Fe(EDTA)$^-$ or Fe(1,10-phenanthroline)$_3^{2+}$, are not very soluble in organic solvents. One scheme for separating metal ions from one another is to selectively complex one ion with an organic ligand and extract it into an organic solvent. Ligands such as dithizone (Demonstration 23-1), 8-hydroxyquinoline, and cupferron are commonly employed. Each is a weak acid, HL, which loses a proton when it binds to a metal ion through atoms shown in **bold** type.

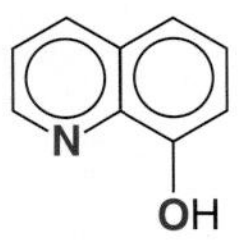

$$HL(aq) \rightleftharpoons H^+(aq) + L^-(aq) \qquad K_a = \frac{[H^+]_{aq}[L^-]_{aq}}{[HL]_{aq}} \qquad (23\text{-}8)$$

$$nL^-(aq) + M^{n+}(aq) \rightleftharpoons ML_n(aq) \qquad \beta = \frac{[ML_n]_{aq}}{[M^{n+}]_{aq}[L^-]_{aq}^n} \qquad (23\text{-}9)$$

8-Hydroxyquinoline
(oxine)

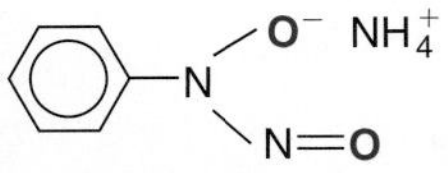

Cupferron

β is the overall formation constant defined in Box 6-2.

Each ligand can react with many different metal ions, but some selectivity is achieved by controlling the pH.

Let's derive an equation for the distribution coefficient of a metal between two phases when essentially all the metal in the aqueous phase (*aq*) is in the form M^{n+} and all the metal

---

DEMONSTRATION 23-1 Extraction with Dithizone

Dithizone (diphenylthiocarbazone) is a green-colored compound that is soluble in nonpolar organic solvents and insoluble in water below pH 7.[3] In alkaline aqueous solution, dithizone forms a soluble yellow ion. It forms red, hydrophobic complexes with di- and trivalent metals. Dithizone is used for analytical extractions and for colorimetric determinations of metal ions.

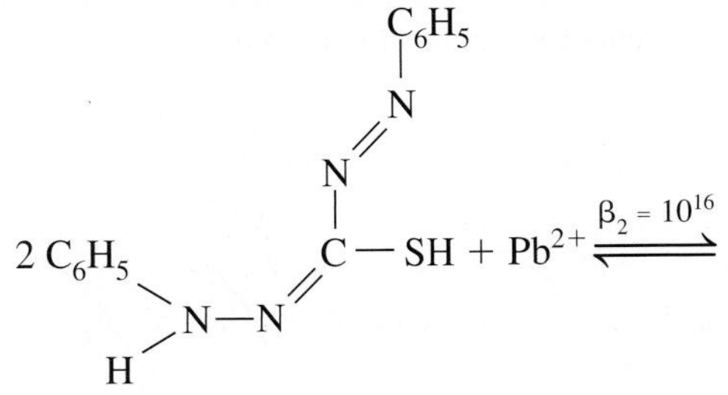

Dithizone
(green)

(colorless)

$pK_a \approx 4.5$

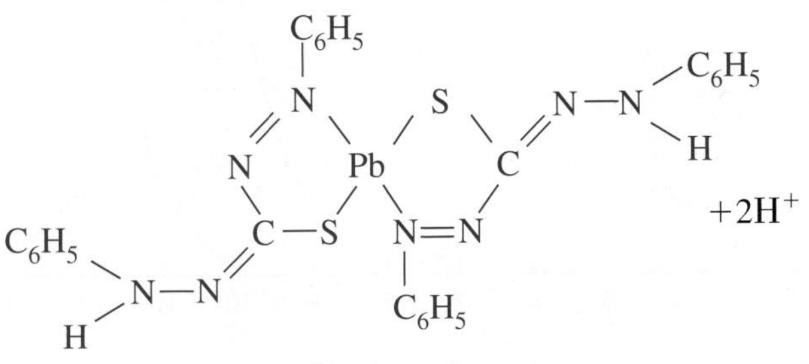

Metal complex
(red)

You can demonstrate the equilibrium between green ligand and red complex by using three large test tubes sealed with tightly fitting rubber stoppers. Place some hexane plus a few milliliters of dithizone solution (prepared by dissolving 1 mg of dithizone in 100 mL of CHCl$_3$) in each test tube. Add distilled water to tube A, tap water to tube B, and 2 mM Pb(NO$_3$)$_2$ to tube C. After shaking and settling, tubes B and C contain a red upper phase, whereas A remains green.

Proton equilibrium in the dithizone reaction is shown by adding a few drops of 1 M HCl to tube C. After shaking, the dithizone turns green again. Competition with a stronger ligand is shown by adding a few drops of 0.05 M EDTA solution to tube B. Again, shaking causes a reversion to the green color.

Practicing "Green" Chemistry

Chemical procedures that produce less waste or less hazardous waste are said to be "green" because they reduce harmful environmental effects. In chemical analyses with dithizone, you can substitute aqueous micelles (Box 26-1) for the organic phase (which has traditionally been chloroform, CHCl$_3$) to eliminate chlorinated solvent and the tedious extraction.[4] For example, a solution containing 5.0 wt% of the micelle-forming surfactant Triton X-100 dissolves 8.3×10^{-5} M dithizone at 25°C and pH <7. The concentration of dithizone inside the micelles, which constitute a small fraction of the volume of solution, is much greater than 8.3×10^{-5} M. Aqueous micellar solutions of dithizone can be used for the spectrophotometric analysis of metals such as Zn(II), Cd(II), Hg(II), Cu(II), and Pb(II) with results comparable to those obtained with an organic solvent.

in the organic phase (*org*) is in the form ML_n (Figure 23-3). We define the partition coefficients for ligand and complex as follows:

$$HL(aq) \rightleftharpoons HL(org) \qquad K_L = \frac{[HL]_{org}}{[HL]_{aq}} \qquad (23\text{-}10)$$

$$ML_n(aq) \rightleftharpoons ML_n(org) \qquad K_M = \frac{[ML_n]_{org}}{[ML_n]_{aq}} \qquad (23\text{-}11)$$

The distribution coefficient we seek is

$$D = \frac{[\text{total metal}]_{org}}{[\text{total metal}]_{aq}} \approx \frac{[ML_n]_{org}}{[M^{n+}]_{aq}} \qquad (23\text{-}12)$$

From Equations 23-11 and 23-9, we can write

$$[ML_n]_{org} = K_M[ML_n]_{aq} = K_M\beta[M^{n+}]_{aq}[L^-]^n_{aq}$$

Using $[L^-]_{aq}$ from Equation 23-8 gives

$$[ML_n]_{org} = \frac{K_M\beta[M^{n+}]_{aq}K_a^n[HL]^n_{aq}}{[H^+]^n_{aq}}$$

Putting this value of $[ML_n]_{org}$ into Equation 23-12 gives

$$D \approx \frac{K_M\beta K_a^n[HL]^n_{aq}}{[H^+]^n_{aq}}$$

Because most HL is in the organic phase, we substitute $[HL]_{aq} = [HL]_{org}/K_L$ to produce the most useful expression for the distribution coefficient:

Distribution of metal-chelate complex between phases:
$$D \approx \frac{K_M\beta K_a^n}{K_L^n} \frac{[HL]^n_{org}}{[H^+]^n_{aq}} \qquad (23\text{-}13)$$

We see that the distribution coefficient for metal ion extraction depends on pH and ligand concentration. It is sometimes possible to select a pH where D is large for one metal and small for another. For example, Figure 23-4 shows that Cu^{2+} could be separated from Pb^{2+} and Zn^{2+} by extraction with dithizone at pH 5. Demonstration 23-1 illustrates the pH dependence of an extraction. Box 23-1 describes *crown ethers* and *phase transfer agents* used to extract polar reagents into nonpolar solvents for chemical reactions.

M^{n+} is in the aqueous phase and ML_n is in the organic phase.

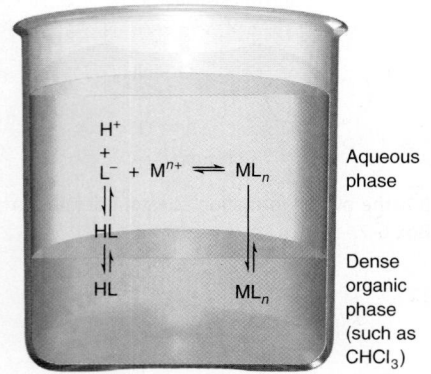

FIGURE 23-3 Extraction of a metal ion with a chelator. The predominant form of metal in the aqueous phase is M^{n+}, and the predominant form in the organic phase is ML_n.

You can select a pH to bring the metal into either phase.

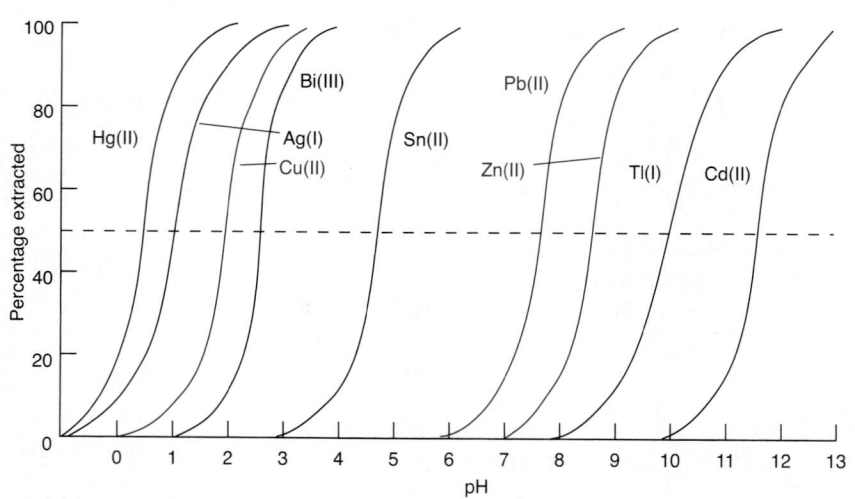

FIGURE 23-4 Extraction of metal ions by dithizone into CCl_4. At pH 5, Cu^{2+} is completely extracted into CCl_4, whereas Pb^{2+} and Zn^{2+} remain in the aqueous phase. [Information from G. H. Morrison and H. Freiser in C. L. Wilson and D. Wilson, eds., *Comprehensive Analytical Chemistry*, Vol. IA (New York: Elsevier, 1959).]

BOX 23-1 **Crown Ethers and Phase Transfer Agents**

Crown ethers are synthetic compounds that envelop metal ions (especially alkali metal cations) in a pocket of oxygen ligands. Crown ethers are used as *phase transfer catalysts* because they can extract water-soluble ionic reagents into nonpolar solvents, where reaction with hydrophobic compounds can occur. In the potassium complex of dibenzo-30-crown-10, K^+ is engulfed by 10 oxygen atoms, with K—O distances averaging 288 pm. Only the hydrophobic outside of the complex is exposed to solvent.

Hydrophobic cations and anions can function as phase transfer agents to bring ions of the opposite charge into organic solvents. For example, Color Plate 29 shows the extraction of colored anions from the lower aqueous phase into the upper diethyl ether layer

when trioctylmethylammonium chloride is added and the mixture is stirred. For $KMnO_4$ in the aqueous phase, the equilibrium below lies well to the right:

$$K^+(aq) + MnO_4^-(aq) + \underset{\text{Ion pair}}{(C_8H_{17})_3\overset{+}{N}CH_3\ Cl^-(org)} \rightleftharpoons$$

$$K^+(aq) + Cl^-(aq) + \underset{\text{Ion pair}}{(C_8H_{17})_3\overset{+}{N}CH_3\ MnO_4^-(org)}$$

An important application of this reaction is to bring permanganate into the organic phase to oxidize an organic compound. In geochemical studies to detect debris from supernova explosions, phase transfer is used in the analytical separation of hafnium from tungsten.[5]

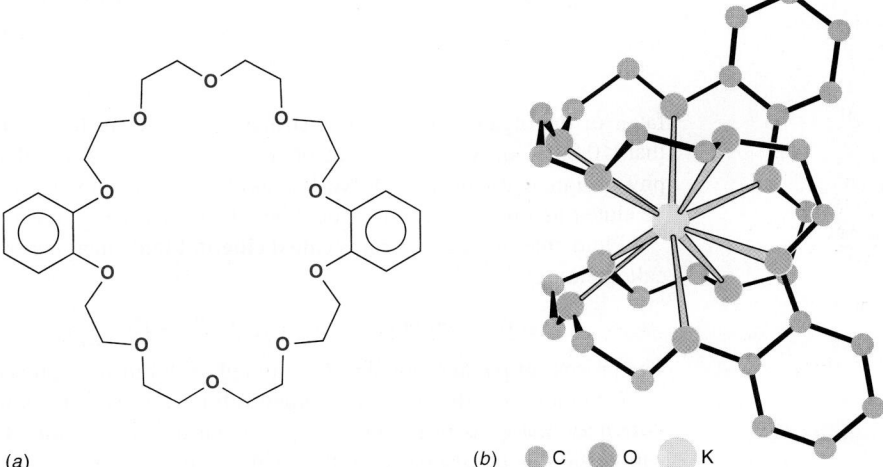

(*a*) Molecular structure of dibenzo-30-crown-10. (*b*) Three-dimensional structure of its K^+ complex. [Information from M. A. Bush and M. R. Truter, "Crystal Structures of Alkali-Metal Complexes with Cyclic Polyethers," *Chem. Commun.* **1970,** 1439.]

23-2 What Is Chromatography?

Chromatography operates on the same principle as extraction, but one phase is held in place while the other moves past it.[8,9] Figure 23-5 shows a solution containing solutes A and B placed on top of a column packed with solid particles and filled with solvent. When the outlet is opened, solutes A and B flow down into the column. Fresh solvent is then applied to the top of the column and the mixture is washed down the column by continuous solvent flow. If solute A is more strongly adsorbed than solute B on the solid particles, then solute A spends a smaller fraction of the time free in solution. Solute A moves down the column more slowly than solute B and emerges at the bottom after solute B. We have just separated a mixture into its components by *chromatography*.

The **mobile phase** (the solvent moving through the column) in chromatography is either a liquid or a gas. The **stationary phase** (the one that stays in place inside the column) is most commonly a viscous liquid chemically bonded to the inside of a capillary tube or onto the surface of solid particles packed in the column. Alternatively, as in Figure 23-5, the solid particles themselves may be the stationary phase. In any case, the equilibration of solutes between mobile and stationary phases gives rise to separation.

Color Plate 30 illustrates the mechanism of chromatography in which solutes are separated because of their unequal distribution coefficients between two phases that could represent the aqueous mobile phase and organic stationary phase. When 9 M HCl aqueous phase containing Nd(III) and Fe(III) is placed in contact with an *ionic liquid* organic phase and allowed to reach equilibrium, Fe(III) is extracted into the organic phase and Nd(III) remains in the aqueous phase. The distribution coefficient in Equation 23-4 for Fe(III) is 7 000 in

In 1903 in Warsaw, the botanist M. Tswett invented adsorption chromatography to separate plant pigments using a hydrocarbon solvent and inulin powder (a carbohydrate) as stationary phase.[6] The separation of colored bands led to the name **chromatography**, from the Greek *chromatos* ("color") and *graphein* ("to write")—"color writing."

Chromatography lay dormant until Tswett's methods were applied beginning in 1931 to biochemical separations by E. Lederer and R. Kuhn in Heidelberg, by P. Karrer in Zurich, and by L. Zechmeister in Hungary.[7] During the 1930s, adsorption chromatography became an established tool in biochemistry.

An **ionic liquid** contains a cation and anion that do not readily crystallize. It melts below room temperature and has a wide liquid range with low volatility. There is no added solvent in the ionic liquid phase.[10]

FIGURE 23-5 The idea behind chromatography: solute A, with a greater affinity than solute B for the stationary phase, remains on the column longer. Panel *f* is a reconstruction of the separation of pigments from red paprika skin from the work of L. Zechmeister in the 1930s. Bands marked by horizontal lines are different pigments. The lower stationary phase is $Ca(OH)_2$ and the upper stationary phase is $CaCO_3$. [Panel *f* information from L. S. Ettre, "The Rebirth of Chromatography 75 Years Ago," *LCGC North Am.* **2007**, *25*, 640.]

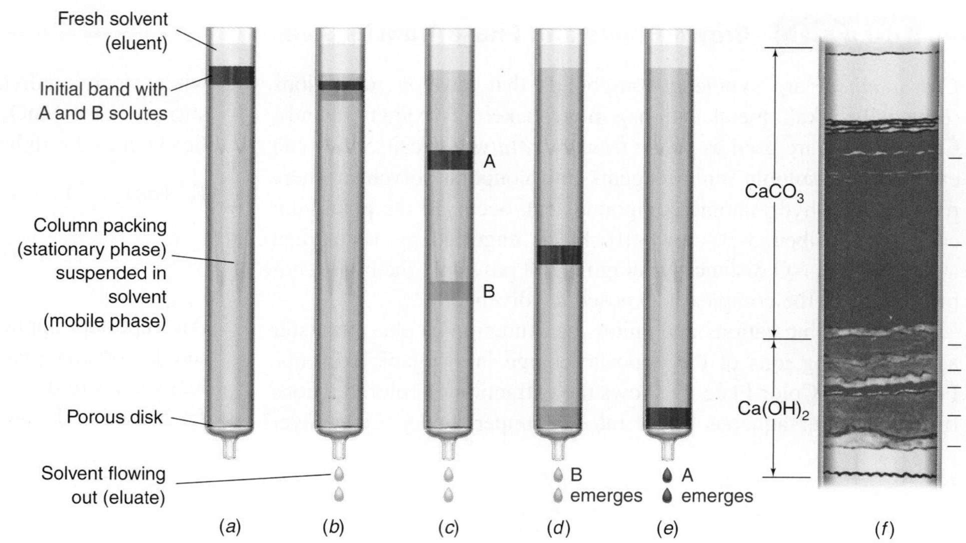

Figure labels: Fresh solvent (eluent); Initial band with A and B solutes; Column packing (stationary phase) suspended in solvent (mobile phase); Porous disk; Solvent flowing out (eluate); B emerges; A emerges; $CaCO_3$; $Ca(OH)_2$; (a) (b) (c) (d) (e) (f)

favor of the organic phase. The distribution coefficient for Nd(III) is $<10^{-5}$, which means that $<0.001\%$ of Nd goes into the organic phase. If these two phases were on a chromatography column and a mixture of Nd(III) and Fe(III) were applied to the column, Nd(III) would be eluted first because Fe(III) would be retained by the stationary phase.

Fluid entering the column is called **eluent.** Fluid emerging from the end of the column is called **eluate:**

$$\text{eluent in} \quad \rightarrow \quad \boxed{\text{COLUMN}} \quad \rightarrow \quad \text{eluate out}$$

The process of passing liquid or gas through a chromatography column is called **elution.**

Columns are either **packed** or **open tubular.** A packed column is filled with particles of stationary phase, as in Figure 23-5. An open tubular column is a narrow, hollow capillary with stationary phase coated on the inside walls.

Types of Chromatography

Chromatography is divided into categories on the basis of the mechanism of interaction of the solute with the stationary phase, as shown in Figure 23-6.

Adsorption chromatography. A solid stationary phase and a liquid or gaseous mobile phase are used. Solute is adsorbed on the surface of the solid particles. The more strongly a solute is adsorbed, the slower it travels through the column.

Partition chromatography. A liquid stationary phase is bonded to a solid surface, which is typically the inside of the fused silica (SiO_2) chromatography column in gas chromatography. Solute equilibrates between the stationary liquid and the mobile phase, which is a flowing gas in gas chromatography.

Ion-exchange chromatography. Anions such as $-SO_3^-$ or cations such as $-N(CH_3)_3^+$ are covalently attached to the stationary solid phase, usually a *resin.* Solute ions of the opposite charge are attracted to the stationary phase. The mobile phase is a liquid.

Molecular exclusion chromatography. Also called *size exclusion,* **gel filtration,** or **gel permeation chromatography.** This technique separates molecules by size, with the larger solutes passing through most quickly. In the ideal case of molecular exclusion, there is no attractive interaction between the stationary phase and the solute. Rather, the liquid or gaseous mobile phase passes through a porous gel. The pores are small enough to exclude large solute molecules but not small ones. Large molecules stream past without entering the pores. Small molecules take longer to pass through the column because they enter the gel and therefore must flow through a larger volume before leaving the column.

Affinity chromatography. This most selective kind of chromatography employs specific interactions between one kind of solute molecule and a second molecule that is covalently attached (immobilized) to the stationary phase. For example, the immobilized molecule might be an antibody to a particular protein. When a mixture containing a thousand proteins is passed through the column, only the one protein that interacts with the antibody binds to the column. After all other solutes have been washed from the column, the desired protein is dislodged by changing the pH or ionic strength.

---

elu*ent*—**in**
elu*ate*—**out**

Tswett invented adsorption chromatography in 1903.

For their pioneering work on liquid-liquid partition chromatography in 1941, A. J. P. Martin and R. L. M. Synge received the Nobel Prize in 1952.

B. A. Adams and E. L. Holmes developed the first synthetic ion-exchange resins in 1935. **Resins** are relatively hard, amorphous organic solids. **Gels** are relatively soft.

Large molecules pass through the column *faster* than small molecules.

FIGURE 23-6 Major types of chromatography.

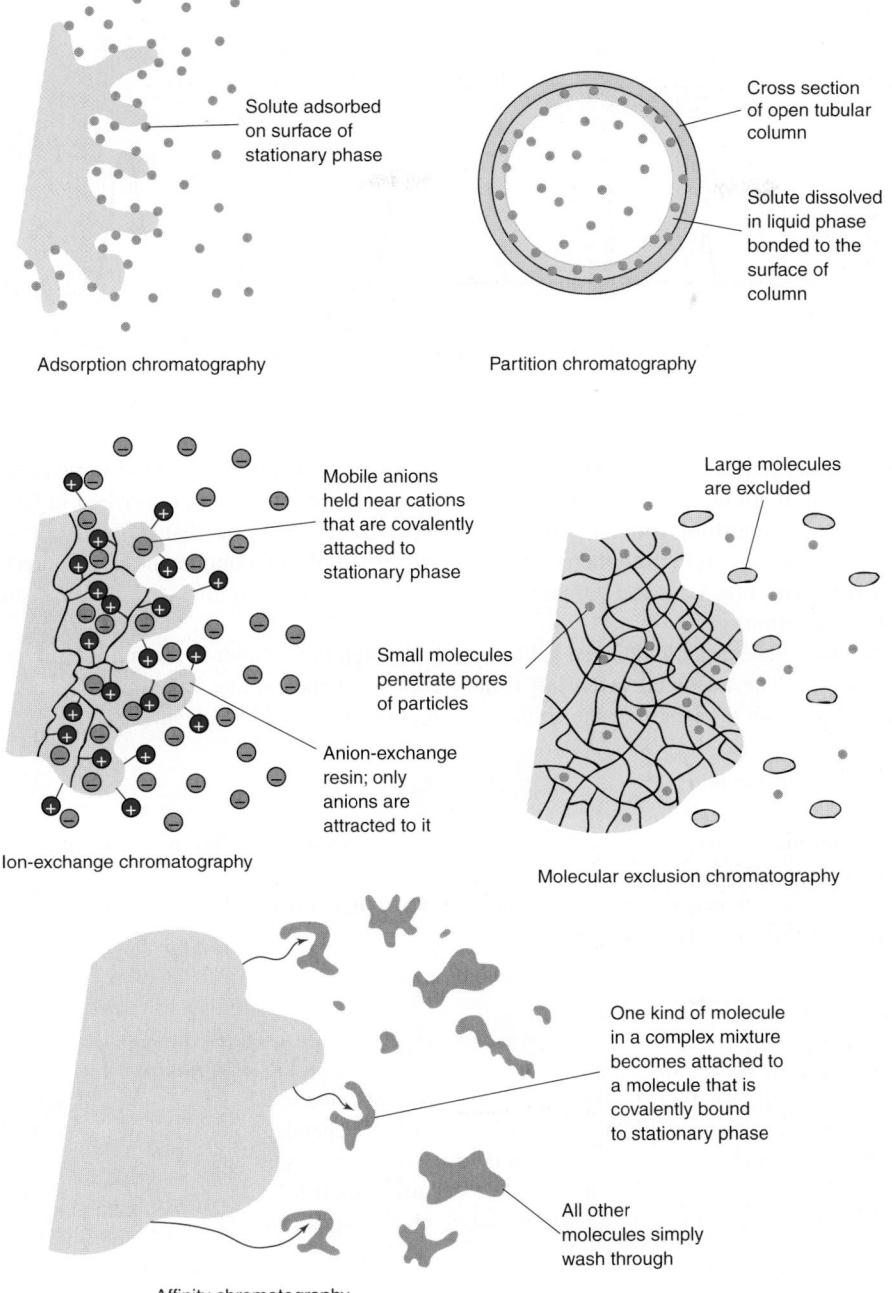

Solute adsorbed on surface of stationary phase

Adsorption chromatography

Cross section of open tubular column

Solute dissolved in liquid phase bonded to the surface of column

Partition chromatography

Mobile anions held near cations that are covalently attached to stationary phase

Small molecules penetrate pores of particles

Anion-exchange resin; only anions are attracted to it

Ion-exchange chromatography

Large molecules are excluded

Molecular exclusion chromatography

One kind of molecule in a complex mixture becomes attached to a molecule that is covalently bound to stationary phase

All other molecules simply wash through

Affinity chromatography

23-3 A Plumber's View of Chromatography

The speed of the mobile phase passing through a chromatography column is expressed either as a volume flow rate or as a linear velocity. Consider a liquid chromatography experiment in which the column has an inner diameter of 0.60 cm (radius $\equiv r = 0.30$ cm) and the mobile phase occupies 20% of the column volume. Each centimeter of column length has a volume of $\pi r^2 \times$ length $= \pi(0.30 \text{ cm})^2(1 \text{ cm}) = 0.283$ mL, of which 20% ($= 0.056\ 5$ mL) is mobile phase (solvent). The **volume flow rate,** such as 0.30 mL/min, tells how many milliliters of solvent per minute travel through the column. The **linear velocity** tells how many centimeters are traveled in 1 min by the solvent. Because 1 cm of column length contains 0.056 5 mL of mobile phase, 0.30 mL would occupy (0.30 mL)/(0.056 5 mL/cm) = 5.3 cm of column length. The linear velocity corresponding to 0.30 mL/min is 5.3 cm/min.

Volume flow rate = volume of solvent per unit time traveling through column

Linear velocity = distance per unit time traveled by solvent

The Chromatogram

Solutes eluted from a chromatography column are observed with detectors described in later chapters. A **chromatogram** is a graph showing the detector response as a function of

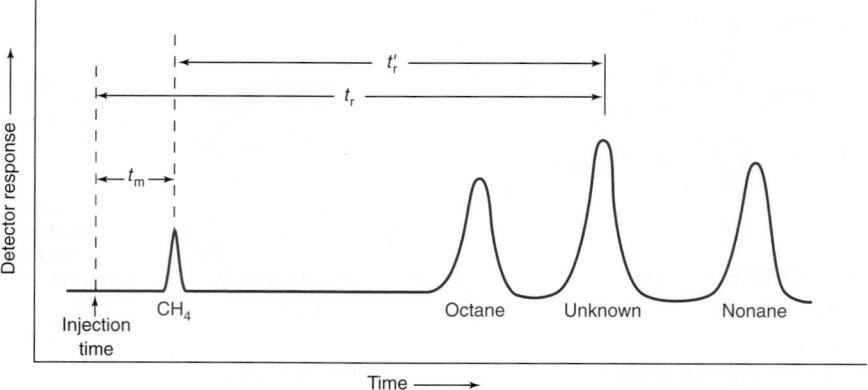

FIGURE 23-7 Schematic gas chromatogram showing measurement of retention times.

elution time. Figure 23-7 shows what might be observed when a mixture of octane, nonane, and an unknown are separated by gas chromatography, which is described in Chapter 24. The **retention time,** t_r, for each component is the time that elapses between injection of the mixture onto the column and the arrival of that component at the detector. **Retention volume,** V_r, is the volume of mobile phase required to elute a particular solute from the column.

Mobile phase or an unretained solute travels through the column in the minimum possible time, t_m. The **adjusted retention time,** t_r', for a retained solute is the additional time required to travel the length of the column, beyond that required by solvent:

Adjusted retention time: $$t_r' = t_r - t_m \qquad (23\text{-}14)$$

In gas chromatography, t_m is usually taken as the time needed for CH_4 to travel through the column (Figure 23-7).

For two components 1 and 2, the **relative retention,** α (also called *separation factor*), is the ratio of their adjusted retention times:

Relative retention: $$\alpha = \frac{t_{r2}'}{t_{r1}'} = \frac{k_2}{k_1} \qquad (23\text{-}15)$$

where $t_{r2}' > t_{r1}'$, so $\alpha > 1$. The greater the relative retention, the greater the separation between two components. Relative retention is fairly independent of flow rate and can therefore be used to help identify peaks when the flow rate changes.

For each peak in the chromatogram, the **retention factor,** k, is the time required to elute that peak minus the time t_m required for mobile phase to pass through the column, expressed in multiples of t_m.

Retention factor: $$k = \frac{t_r - t_m}{t_m} \qquad (23\text{-}16)$$

Retention factor used to be called *capacity factor, capacity ratio,* or *partition ratio* and was formerly written as k' instead of k.

The longer a component is retained by the column, the greater is the retention factor. It takes volume V_m to push solvent from the beginning of the column to the end of the column. If it takes an additional volume $3V_m$ to elute a solute, then the retention factor for that solute is 3.

EXAMPLE **Retention Parameters**

A mixture of benzene, toluene, and methane was injected into a gas chromatograph. Unretained methane gave a sharp spike in 42 s, whereas benzene required 251 s and toluene was eluted in 333 s. Find the adjusted retention time and retention factor for each solute, and the relative retention.

Solution The adjusted retention times are

Benzene: $t_r' = t_r - t_m = 251 - 42 = 209$ s   Toluene: $t_r' = 333 - 42 = 291$ s

The retention factors are

$$\text{Benzene: } k = \frac{t_r - t_m}{t_m} = \frac{251 - 42}{42} = 5.0 \qquad \text{Toluene: } k = \frac{333 - 42}{42} = 6.9$$

The relative retention is expressed as a number greater than unity.

$$\alpha = \frac{t_r'(\text{toluene})}{t_r'(\text{benzene})} = \frac{333 - 42}{251 - 42} = 1.39$$

TEST YOURSELF Ethylbenzene eluted at 353 s. Find its retention factor and the relative retention for ethylbenzene and toluene. (*Answer:* 7.4, 1.07)

Relation Between Retention Time and the Partition Coefficient

The retention factor in Equation 23-16 is equivalent to

$$k = \frac{\text{time solute spends in stationary phase}}{\text{time solute spends in mobile phase}} \qquad (23\text{-}17)$$

Let's see why this is true. If the solute spends all its time in the mobile phase and none in the stationary phase, it would be eluted in time t_m. Putting $t_r = t_m$ into Equation 23-16 gives $k = 0$, because solute spends no time in the stationary phase. Suppose that solute spends equal time in the stationary and mobile phases. The retention time would then be $t_r = 2t_m$ and $k = (2t_m - t_m)/t_m = 1$. If solute spends three times as much time in the stationary phase as in the mobile phase, $t_r = 4t_m$ and $k = (4t_m - t_m)/t_m = 3$.

If solute spends three times as much time in the stationary phase as in the mobile phase, there will be three times as many moles of solute in the stationary phase as in the mobile phase at any time. The quotient in Equation 23-17 is equivalent to

$$\frac{\text{Time solute spends in stationary phase}}{\text{Time solute spends in mobile phase}} = \frac{\text{moles of solute in stationary phase}}{\text{moles of solute in mobile phase}}$$

$$k = \frac{c_s V_s}{c_m V_m} \qquad (23\text{-}18)$$

where c_s is the concentration of solute in the stationary phase, V_s is the volume of the stationary phase, c_m is the concentration of solute in the mobile phase, and V_m is the volume of the mobile phase.

The quotient c_s/c_m is the ratio of concentrations of solute in the stationary and mobile phases. If the column is run slowly enough to be at equilibrium, the quotient c_s/c_m is the *partition coefficient*, K, introduced in connection with solvent extraction. Therefore, we cast Equation 23-18 in the form

Partition coefficient $= K = \dfrac{c_s}{c_m}$

Relation of retention time $\qquad k = K\dfrac{V_s}{V_m} \overset{\text{Eq. 23-16}}{=} \dfrac{t_r - t_m}{t_m} = \dfrac{t_r'}{t_m} \qquad (23\text{-}19)$
to partition coefficient:

which relates retention time to the partition coefficient and the volumes of stationary and mobile phases. Because $t_r' \propto k \propto K$, relative retention can also be expressed as

Relative retention: $\qquad \alpha = \dfrac{t_{r2}'}{t_{r1}'} = \dfrac{k_2}{k_1} = \dfrac{K_2}{K_1} \qquad (23\text{-}20)$

That is, the relative retention of two solutes is proportional to the ratio of their partition coefficients. This relation is the physical basis of chromatography.

> **Physical basis of chromatography**
> The greater the ratio of partition coefficients between mobile and stationary phases, the greater the separation between two components of a mixture.

EXAMPLE Retention Time and Partition Coefficient

In the preceding example, unretained methane gave a sharp spike in 42 s, whereas benzene required 251 s. The open tubular chromatography column has an inner diameter of 250 μm and is coated on the inside with a layer of stationary phase 1.0 μm thick. Estimate the partition coefficient ($K = c_s/c_m$) for benzene between stationary and mobile phases and state what fraction of the time benzene spends in the mobile phase.

Solution We need to calculate the relative volumes of the stationary and mobile phases. The column is an open tube with a thin coating of stationary phase on the inside wall.

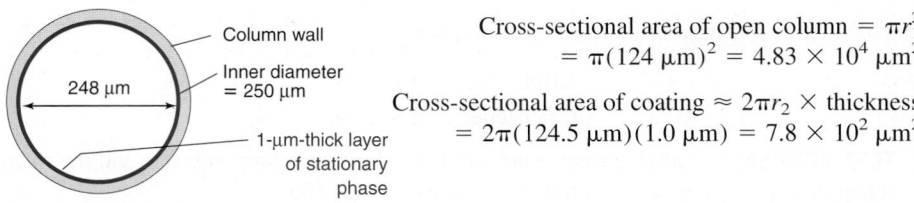

Column wall
Inner diameter = 250 μm
248 μm
1-μm-thick layer of stationary phase

$$\text{Cross-sectional area of open column} = \pi r_1^2$$
$$= \pi(124 \ \mu m)^2 = 4.83 \times 10^4 \ \mu m^2$$

$$\text{Cross-sectional area of coating} \approx 2\pi r_2 \times \text{thickness}$$
$$= 2\pi(124.5 \ \mu m)(1.0 \ \mu m) = 7.8 \times 10^2 \ \mu m^2$$

Radius of hollow cavity: $r_1 = 124 \ \mu m$
Radius to middle of stationary phase:
$r_2 = 124.5 \ \mu m$

The relative volumes of the phases are proportional to the relative cross-sectional areas of the phases. Therefore, $V_s/V_m = (7.8 \times 10^2 \ \mu m^2)/(4.83 \times 10^4 \ \mu m^2) = 0.016 \ 1$. In the preceding example, we found that the retention factor for benzene is $k = 5.0$. Substituting this value into Equation 23-19 gives the partition coefficient:

$$k = K\frac{V_s}{V_m} \Rightarrow 5.0 = K(0.0161) \Rightarrow K = 310$$

To find the fraction of time spent in the mobile phase, we use Equations 23-16 and 23-17:

$$k = \frac{\text{time in stationary phase}}{\text{time in mobile phase}} = \frac{t_r - t_m}{t_m} = \frac{t_s}{t_m} \Rightarrow t_s = kt_m$$

where t_s is the time in the stationary phase. The fraction of time in the mobile phase is

$$\text{Fraction of time in mobile phase} = \frac{t_m}{t_s + t_m} = \frac{t_m}{kt_m + t_m} = \frac{1}{k + 1} = \frac{1}{5.0 + 1} = 0.17$$

TEST YOURSELF Find the partition coefficient for toluene ($t_r = 333$ s) and state what fraction of the time it spends in the mobile phase. (**Answer:** 430, 0.13)

Retention volume, V_r, is the volume of mobile phase required to elute a particular solute from the column:

Retention volume: $$V_r = t_r \cdot u_v \qquad (23\text{-}21)$$

where u_v is the volume flow rate (volume per unit time) of the mobile phase. The retention volume of a particular solute is constant over a range of flow rates.

Scaling Up

We normally carry out chromatography for *analytical* purposes (to separate and identify or measure the components of a mixture) or for *preparative* purposes (to purify a significant quantity of a component of a mixture). Analytical chromatography is usually performed with narrow columns that provide good separation. For preparative chromatography, we use fatter columns that can handle more load (Figure 23-8).[11] Preparative chromatography is especially important in the pharmaceutical industry, which can afford the high cost of separating compounds such as *optical isomers* of drugs (Box 24-1).

If you have developed a procedure to separate 2 mg of a mixture on a column with a diameter of 1.0 cm, what size column should you use to separate 20 mg of the mixture? The most straightforward way to scale up is to maintain the same column length and to increase the cross-sectional area to maintain a constant ratio of sample mass to column volume. Cross-sectional area is πr^2, where r is the column radius, so the desired diameter is given by

Scaling equation: $$\frac{\text{Large mass}}{\text{Small mass}} = \left(\frac{\text{large column radius}}{\text{small column radius}}\right)^2 \qquad (23\text{-}22)$$

$$\frac{20 \text{ mg}}{2 \text{ mg}} = \left(\frac{\text{large column radius}}{0.50 \text{ cm}}\right)^2$$

Large column radius = 1.58 cm

A column with a diameter near 3 cm would be appropriate.

FIGURE 23-8 Industrial-scale preparative chromatography column. Some columns can purify up to a kilogram of material. [Maximilian Stock Ltd./Science Source.]

Volume is proportional to time, so any ratio of times can be written as the corresponding ratio of volumes. If V_m is the elution volume for unretained solute,

$$k = \frac{t_r - t_m}{t_m} = \frac{V_r - V_m}{V_m}$$

where V_r is the retention volume for solute.

Scaling rules:
- Keep column length constant
- Cross-sectional area of column $\propto$ mass of analyte:

$$\frac{\text{Mass}_2}{\text{Mass}_1} = \left(\frac{\text{radius}_2}{\text{radius}_1}\right)^2$$

(The symbol $\propto$ means "is proportional to.")
- Maintain constant linear velocity:

$$\frac{\text{Volume flow}_2}{\text{Volume flow}_1} = \left(\frac{\text{radius}_2}{\text{radius}_1}\right)^2$$

- Sample volume applied to column $\propto$ mass of analyte
- If you change column length, mass of sample can be increased in proportion to total length

To reproduce the conditions of the smaller column in the larger column, keep the *linear velocity* (not the volume flow rate) constant. Because the area (and hence volume) of the large column is 10 times greater than that of the small column in this example, the volume flow rate should be 10 times greater to maintain a constant linear velocity. If the small column had a volume flow rate of 0.3 mL/min, the large column should be run at 3 mL/min.

The mass of sample (g) that can be run in preparative chromatography on a reversed-phase (page 674) silica-based column is *roughly*

$$\text{Column capacity (g)} \approx (2.2 \times 10^{-7})Ld_c^2\sigma_g$$

where L is column length in mm, d_c is column diameter in mm, and σ_g is the surface area (m^2) per gram of stationary phase.[12] For $L = 250$ mm, $d_c = 50$ mm, and $\sigma_g = 200$ m^2/g, we estimate the column capacity as $(2.2 \times 10^{-7})(250)(50)^2(200) = 28$ g. This calculation presumes that the band will occupy the entire volume of the column, so it is surely an upper limit estimate. If you want the band to occupy just 20% of the column, so there is room for chromatography, the mass of sample would be $(0.2)(28 \text{ g}) = 5.6$ g.

23-4 Efficiency of Separation

Two factors contribute to how well compounds are separated by chromatography. One is the difference in elution times between peaks: the farther apart, the better their separation. The other factor is how broad the peaks are: the wider the peaks, the poorer their separation. This section discusses how we measure the efficiency of a separation.

Resolution

Solute moving through a chromatography column tends to spread into a Gaussian shape with standard deviation σ (Figure 23-9). The longer a solute resides in a column, the broader the band becomes. Common measures of breadth are (1) the width $w_{1/2}$ measured at a height equal to half of the peak height and (2) the width w at the *baseline* between tangents drawn to the steepest parts of the peak. From Equation 4-3 for a Gaussian peak, it is possible to show that $w_{1/2} = 2.35\sigma$ and $w = 4\sigma$.

In chromatography, the **resolution** of two peaks from each other is defined as

Resolution:
$$\text{Resolution} = \frac{\Delta t_r}{w_{av}} = \frac{\Delta V_r}{w_{av}} = \frac{0.589\Delta t_r}{w_{1/2av}} \qquad (23\text{-}23)$$

where Δt_r or ΔV_r is the separation between peaks (in units of time or volume) and w_{av} is the average width of the two peaks in corresponding units. (Peak width is measured at the base,

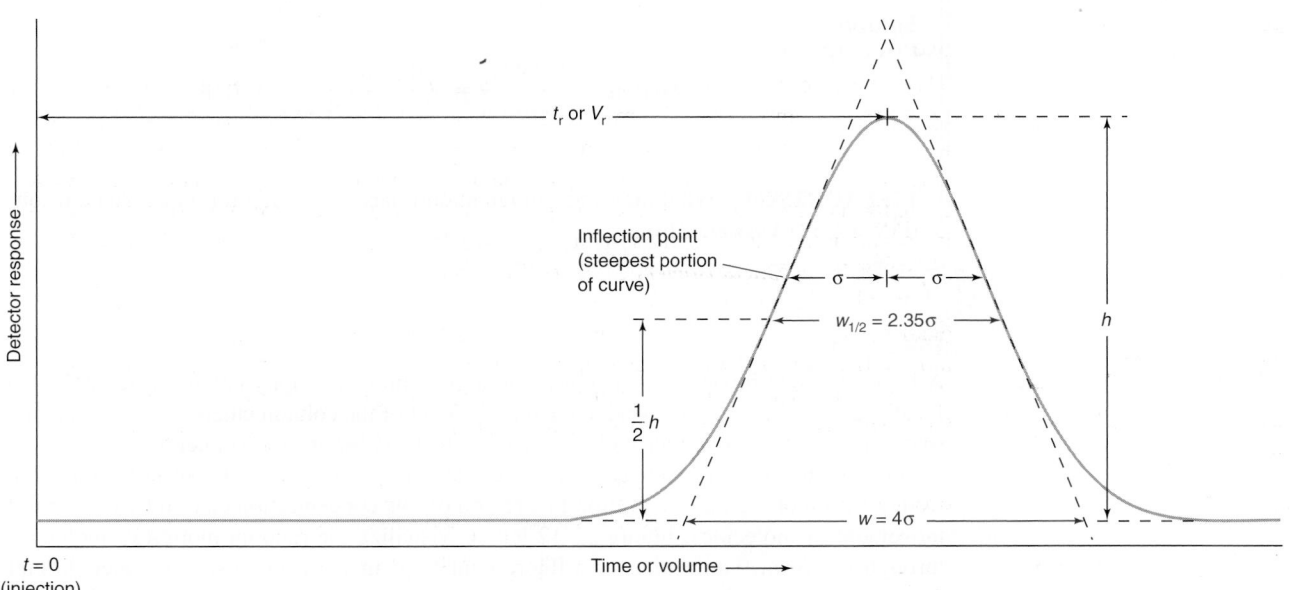

FIGURE 23-9 Idealized Gaussian chromatogram showing how w and $w_{1/2}$ are measured. The value of w is obtained by extrapolating the tangents to the inflection points down to the baseline.

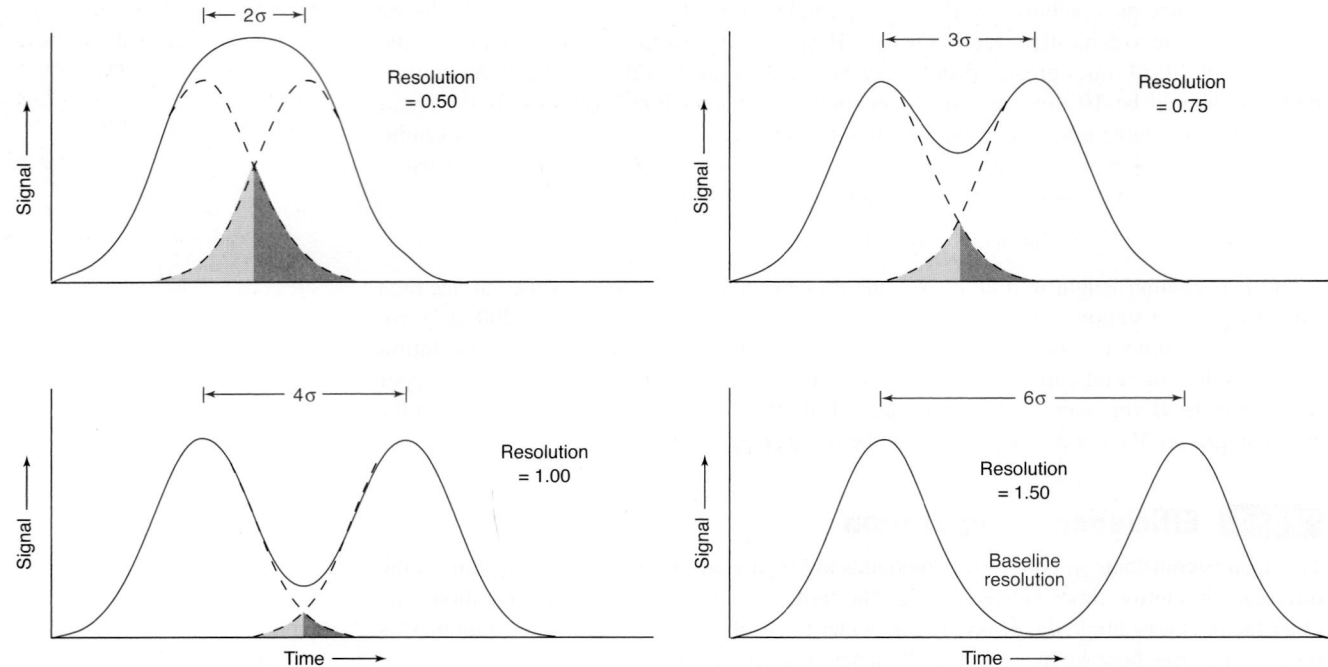

FIGURE 23-10 Resolution of Gaussian peaks of equal area and amplitude. Dashed lines show individual peaks, and solid lines are the sum of two peaks. Overlapping area is shaded.

Baseline resolution means that the signal returns to zero between peaks.

as shown in Figure 23-9.) Alternatively, the last expression in Equation 23-23 uses $w_{1/2av}$, the width at half-height of Gaussian peaks. The width at half-height is usually used because it is easiest to measure. Figure 23-10 shows the overlap of two peaks with different degrees of resolution. For quantitative analysis, a resolution >1.5 (*baseline resolution*) is highly desirable.

EXAMPLE **Measuring Resolution**

A peak with a retention time of 407 s has a width at half-height of 7.6 s. A neighboring peak is eluted 17 s later with a $w_{1/2} = 9.4$ s. Find the resolution for these two components.

Solution

$$\text{Resolution} = \frac{0.589\Delta t_r}{w_{1/2av}} = \frac{0.589(17 \text{ s})}{\frac{1}{2}(7.6 \text{ s} + 9.4 \text{ s})} = 1.18$$

TEST YOURSELF What difference in retention times is required for an adequate resolution of 1.5? (*Answer:* 21.6 s)

Diffusion

A band of solute broadens as it moves through a chromatography column (Figure 23-11). Ideally, an infinitely narrow band applied to the inlet of the column emerges with a Gaussian shape at the outlet. In less ideal circumstances, the band becomes asymmetric.

One main cause of band spreading is **diffusion,** which is the net transport of a solute from a region of high concentration to a region of low concentration caused by the random movement of molecules. Figure 23-12 lets us visualize the random motion of molecules through the *Brownian motion* of a fluorescent bead inside a microscopic water droplet. The bead is pushed by water molecules moving with random directions and speeds. Changes in the x and y coordinates of the bead in successive intervals follow a Gaussian distribution.

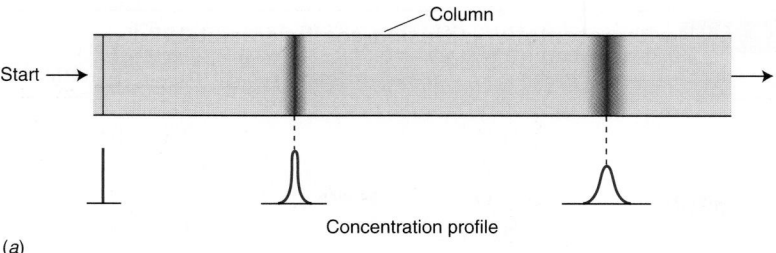

(a)

Concentration profile

FIGURE 23-11 (*a*) Schematic representation of broadening of an initially sharp band of solute as it moves through a chromatography column. (*b*) Experimentally observed diffusional broadening of a band after 2 and 26 min in a capillary electrophoresis column. (c) Expanded view of Gaussian bandshape at 26 min. [Data from M. U. Musheev, S. Javaherian, V. Okhonin, and S. N. Krylov, "Diffusion as a Tool of Measuring Temperature Inside a Capillary," *Anal. Chem.* **2008,** *80,* 6752.]

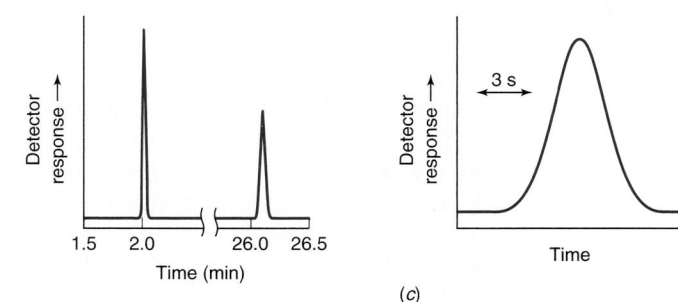

(b)

(c)

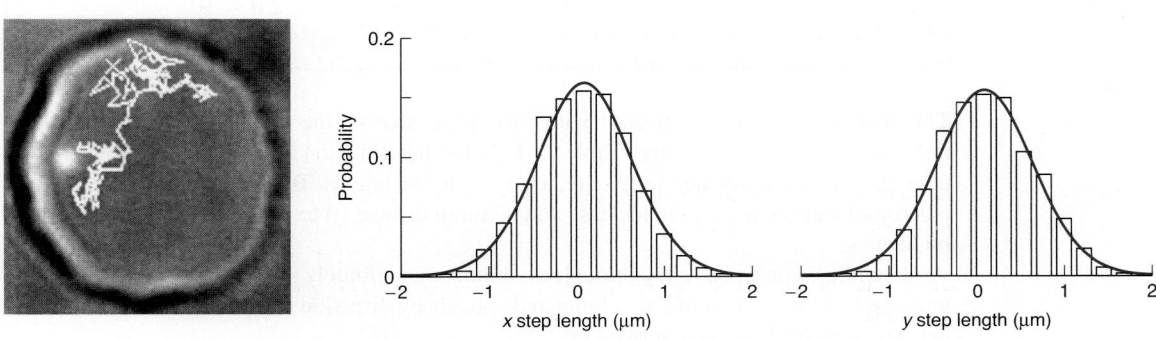

FIGURE 23-12 Brownian motion of a 290-nm-diameter fluorescent bead in a 20-μm-diameter water drop at 155-ms intervals. Histograms show Δ*x* and Δ*y* for each step in 800 photographs. Smooth curve is Gaussian fit. [Data from J. C. Gadd, C. L. Kuyper, B. S. Fujimoto, R. W. Allen, and D. T. Chiu, "Sizing Subcellular Organelles and Nanoparticles Confined within Aqueous Droplets," *Anal. Chem.* **2008,** *80,* 3450. Reprinted with permission © 2008, American Chemical Society.]

The **diffusion coefficient** measures the rate at which molecules move randomly from a region of high concentration to a region of low concentration. Figure 23-13 shows motion of solute across a plane with a concentration gradient *dc/dx*. The number of moles crossing each square meter per second, called the *flux, J,* is proportional to the concentration gradient (*dc/dx*):

Definition of diffusion coefficient:
(Fick's first law of diffusion)

$$\text{Flux}\left(\frac{\text{mol}}{\text{m}^2 \cdot \text{s}}\right) \equiv J = -D\frac{dc}{dx} \quad \textbf{(23-24)}$$

The constant of proportionality, *D*, is the diffusion coefficient, and the negative sign is necessary because net flux is from high concentration to low concentration. If concentration is expressed as mol/m^3, the units of *D* are m^2/s.

The diffusion coefficient is related to the thermal energy of a molecule and the friction it experiences while diffusing:

Stokes-Einstein equation:

$$D = \frac{kT}{f} \quad \textbf{(23-25)}$$

where *kT* is the *thermal energy* of the molecule (*k* = Boltzmann's constant and *T* = kelvins), and *f* is its *friction coefficient*. The greater the friction coefficient, the harder it is for a molecule to diffuse through a fluid. For a spherical particle of radius *r* moving through a fluid of *viscosity* η, the friction coefficient is

Stokes equation:

$$f = 6\pi\eta r \quad \textbf{(23-26)}$$

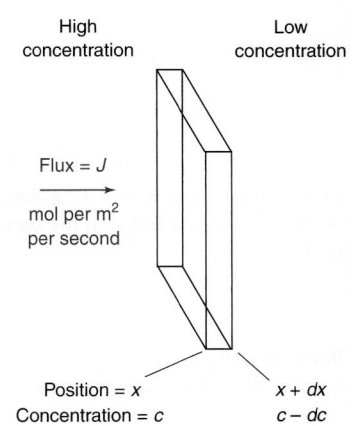

FIGURE 23-13 The flux of molecules diffusing across a plane of unit area is proportional to the concentration gradient and to the diffusion coefficient: $J = -D(dc/dx)$.

Viscosity measures resistance of a fluid to flow. The more viscous a liquid, the slower it flows at a given pressure.

TABLE 23-1	Representative diffusion coefficients at 298 K	
Solute	**Solvent**	**Diffusion coefficient (m^2/s)**
H_2O	H_2O	2.3×10^{-9}
Sucrose	H_2O	0.52×10^{-9}
Glycine	H_2O	1.1×10^{-9}
CH_3OH	H_2O	1.6×10^{-9}
Ribonuclease (FM 13 700)	H_2O (293 K)	0.12×10^{-9}
Serum albumin (FM 66 000)	H_2O (293 K)	0.059×10^{-9}
I_2	Hexane	4.0×10^{-9}
CCl_4	Heptane	3.2×10^{-9}
N_2	CCl_4	3.4×10^{-9}
$CS_2(g)$	Air (293 K)	1.0×10^{-5}
$O_2(g)$	Air (273 K)	1.8×10^{-5}
H^+	H_2O	9.3×10^{-9}
OH^-	H_2O	5.3×10^{-9}
Li^+	H_2O	1.0×10^{-9}
Na^+	H_2O	1.3×10^{-9}
K^+	H_2O	2.0×10^{-9}
Cl^-	H_2O	2.0×10^{-9}
I^-	H_2O	2.0×10^{-9}

The friction coefficient increases with increasing size of the diffusing particle and with increasing viscosity of the fluid. Table 23-1 shows that diffusion in liquids is 10^4 times slower than diffusion in gases due to the greater viscosity of liquids. Because of their large radius, macromolecules such as ribonuclease and albumin diffuse 10 to 100 times slower than small molecules.

If solute begins its journey through a column in an infinitely sharp layer with m moles per unit cross-sectional area of the column and spreads by diffusion as it travels, then the Gaussian profile of the band is described by

Broadening of chromatography band by diffusion:

$$c = \frac{m}{\sqrt{4\pi Dt}}e^{-x^2/(4Dt)} \tag{23-27}$$

where c is concentration (mol/m^3), t is time, and x is the distance along the column from the current center of the band. (The band center is always $x = 0$ in this equation.) Comparison of Equations 23-27 and 4-3 shows that the standard deviation of the band is

Standard deviation of band:

$$\sigma = \sqrt{2Dt} \tag{23-28}$$

Plate Height: A Measure of Column Efficiency

Equation 23-28 tells us that the standard deviation for diffusive band spreading is $\sqrt{2Dt}$. If solute has traveled a distance x at linear velocity u_x (m/s), then the time it has been on the column is $t = x/u_x$. Therefore

$$\sigma^2 = 2Dt = 2D\frac{x}{u_x} = \underbrace{\left(\frac{2D}{u_x}\right)}_{\text{Plate height} \equiv H}x = Hx$$

Plate height:

$$H = \sigma^2/x \tag{23-29}$$

Plate height, H, is the constant of proportionality between the variance, σ^2, of the band and the distance it has traveled, x. The name came from the theory of distillation in which separation could be performed in discrete stages called plates. Plate height is also called the *height equivalent to a theoretical plate (HETP)*. It is approximately the length of column required for one equilibration of solute between mobile and stationary phases. We explore this concept later in Box 23-2. *The smaller the plate height, the narrower the bandwidth.*

Bandwidth $\propto \sqrt{t}$. If elution time increases by a factor of 4, diffusion will broaden the band by a factor of 2.

u_x = **linear velocity** (distance/time)
u_v = **volume flow rate** (volume/time)

As a teenager, A. J. P. Martin, co-inventor of partition chromatography, built distillation columns in discrete sections from coffee cans. (We don't know what he was distilling!) When he formulated the theory of partition chromatography, he adopted terms from distillation theory.

The ability of a column to separate components of a mixture is improved by decreasing plate height. An efficient column has more theoretical plates than an inefficient column. Different solutes passing through the same column have different plate heights because they have different diffusion coefficients. Plate heights are ~0.1 to 1 mm in gas chromatography, ~10 μm in high-performance liquid chromatography, and <1 μm in capillary electrophoresis.

Plate height is the length σ^2/x, where σ is the standard deviation of the Gaussian band in Figure 23-9 and x is the distance traveled. For solute emerging from a column of length L, the number of plates, N, in the entire column is the length L divided by the plate height:

$$N = \frac{L}{H} = \frac{Lx}{\sigma^2} = \frac{L^2}{\sigma^2} = \frac{16L^2}{w^2}$$

because $x = L$ and $\sigma = w/4$. In this expression, w has units of length and the number of plates is dimensionless. If we express L and w (or σ) in units of time instead of length, N is still dimensionless. We obtain a useful expression for N by writing

Number of plates on column: $$N = \frac{16t_r^2}{w^2} = \frac{t_r^2}{\sigma^2} \qquad (23\text{-}30)$$

where t_r is the retention time of the peak and w is the width at the base in Figure 23-9 *in units of time*. If we use the width at half-height, also called the *half-width*, instead of the width at the base, we get

Number of plates on column: $$N = \frac{5.55t_r^2}{w_{1/2}^2} \qquad (23\text{-}31)$$

EXAMPLE Measuring Plates

A solute with a retention time of 407 s has a width at the base of 13.0 s on a column 12.2 m long. Find the number of plates and plate height.

Solution

$$N = \frac{16t_r^2}{w^2} = \frac{16 \cdot (407 \text{ s})^2}{(13.0 \text{ s})^2} = 1.57 \times 10^4$$

$$H = \frac{L}{N} = \frac{12.2 \text{ m}}{1.57 \times 10^4} = 0.78 \text{ mm}$$

TEST YOURSELF The half-width of the same peak is 7.6 s. Find the plate height. (*Answer:* 0.77 mm)

To estimate the number of theoretical plates for the asymmetric peak in Figure 23-14, draw a horizontal line across the peak at 1/10 of the maximum height. The quantities A and B can then be measured to determine the **asymmetry factor,** B/A. The Foley-Dorsey equation estimates the number of plates for a tailed peak as[13]

$$N \approx \frac{41.7(t_R/w_{0.1})^2}{(B/A) + 1.25} \qquad (23\text{-}32)$$

where $w_{0.1}$ ($= A + B$) is the width at 1/10 height. All quantities must be measured in the same units, such as time or length.

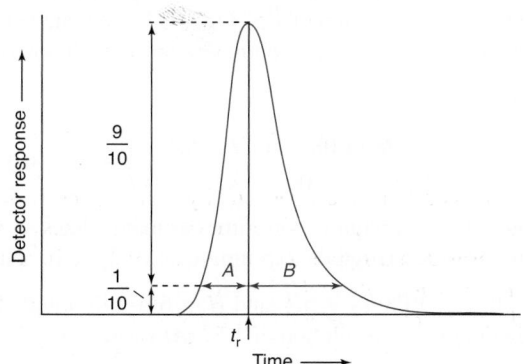

Small plate height ⇒ narrow peaks ⇒ better separations

Choose a peak with a retention factor $k > 5$ when you measure the number of plates on a column.

Challenge If N is constant, show that the width of a chromatographic peak increases in proportion to retention time. That is, successive peaks in a chromatogram should be increasingly broad.

U.S. Pharmacopeia defines *tailing factor* (T_f) for asymmetry measured at 1/20 of the peak height as $T_{f} = w_{0.05}/2A_{0.05}$. $T_f > 1$ is *tailing*. $T_f < 1$ is *fronting*. A tailing peak has a long tail and a fronting peak has a long front.

FIGURE 23-14 Asymmetric peak showing parameters used to estimate the number of theoretical plates.

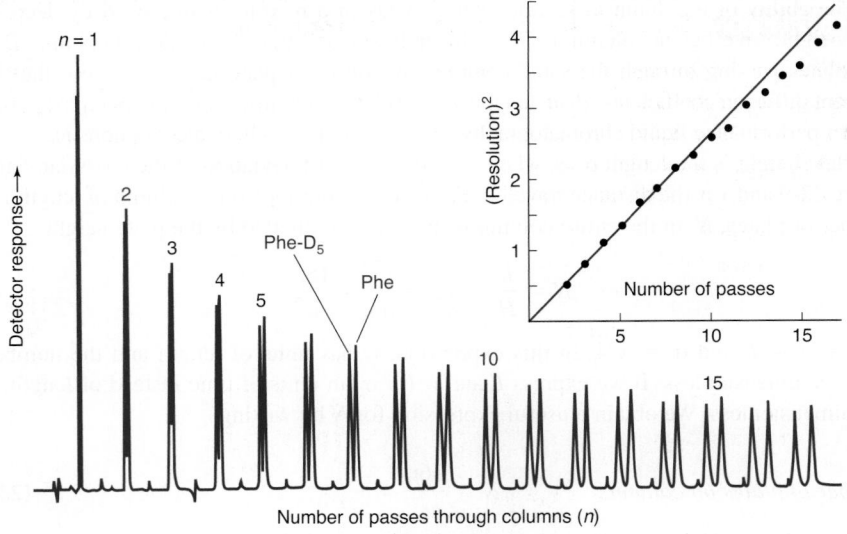

FIGURE 23-15 Separation of 0.5 mM L-phenylalanine and 0.5 mM L-phenylalanine-D$_5$ by repeated passes through a pair of Hypersil C$_8$ liquid chromatography columns (25 cm × 4.6 mm) eluted with 10:90 acetonitrile:water containing 25 mM Na$_2$SO$_4$ and 0.1% trifluoroacetic acid in the water. The relative retention on the first pass is α = 1.03. [Data from K. Lan and J. W. Jorgensen, "Pressure-Induced Retention Variations in Reversed-Phase Alternate-Pumping Recycle Chromatography," *Anal. Chem.* **1998,** *70,* 2773.]

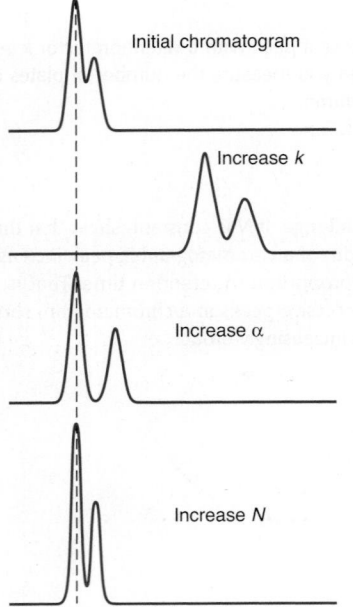

FIGURE 23-16 Effect of separation parameters on resolution. [Information from L. R. Snyder, J. J. Kirkland, and J. L. Glajch, *Practical HPLC Method Development,* 2nd ed. (New York: Wiley, 1997)].

Resolution $\propto \sqrt{N} \propto \sqrt{L}$

In gas chromatography, *k* is controlled by varying column temperature and α by varying stationary phase.

Factors Affecting Resolution

For two closely spaced, symmetric peaks, the relation governing resolution is[14]

Purnell equation: $$\text{Resolution} = \frac{\sqrt{N}}{4}\frac{(\alpha - 1)}{\alpha}\left(\frac{k_2}{1 + k_2}\right) \qquad (23\text{-}33)$$

where N is the number of theoretical plates, α is the relative retention of the two peaks (Equation 23-20), and k_2 is the retention factor for the more retained component (Equation 23-16). Equation 23-23 is used to measure resolution, while Equation 23-33 highlights key parameters that can be varied to achieve resolution.

One important feature of Equation 23-33 is that resolution is proportional to $\sqrt{N}$. Therefore, *doubling the column length increases resolution by* $\sqrt{2}$. Figure 23-15 shows the effect of column length on the separation of L-phenylalanine from L-phenylalanine-D$_5$, in which the phenyl ring bears five deuterium atoms. The mixture was cycled repeatedly through a pair of chromatography columns by an ingenious valving system. After two passes in Figure 23-15, the peaks are barely resolved (resolution = 0.7). By 10 passes, baseline separation (resolution >1.5) has been achieved. The inset shows that the square of resolution is proportional to the number of passes, as predicted by Equation 23-33.

However, increasing the column length increases separation time. If the plate height (*H*) for a column were made smaller, the number of theoretical plates (*N*) would increase with no change in separation time, so peaks would be narrower and better resolved (Figure 23-16). Factors controlling band spreading are discussed in the next section. Alternatively, if conditions were changed to reduce retention *k*, resolution would become worse. When *k* is made larger, resolution usually improves. If the relative retention α were increased, the peaks would move apart and so the resolution would increase. Important equations from chromatography are summarized in Table 23-2.

EXAMPLE **Plates Needed for Desired Resolution**

Two solutes have a relative retention of α = 1.08 and retention factors $k_1 = 5.0$ and $k_2 = 5.4$. The number of theoretical plates is nearly the same for both compounds. How many plates are required to give a resolution of 1.5? Of 3.0? If the plate height is $H = 0.5$ mm in gas chromatography, how long must the column be for a resolution of 1.5?

Solution We use Equation 23-33:

$$\text{Resolution} = 1.5 = \frac{\sqrt{N}}{4}\frac{(1.08 - 1)}{1.08}\left(\frac{5.4}{1 + 5.4}\right) \Rightarrow N = 9.2_2 \times 10^3 \text{ plates}$$

To double the resolution to 3.0 requires four times as many plates = $3.6_9 \times 10^4$ plates. For resolution = 1.5, the required length is (0.5 mm/plate)($9.2_2 \times 10^3$ plates) = 4.6 m.

TEST YOURSELF If α = 1.08, $k_2 = 5.4$, and $H = 6$ μm in liquid chromatography, what length of column in cm gives a resolution of 1.5? (***Answer:*** 5.5 cm)

TABLE 23-2 Summary of chromatography equations

Quantity	Equation	Parameters
Partition coefficient	$K = c_s/c_m$	c_s = concentration of solute in stationary phase c_m = concentration of solute in mobile phase
Adjusted retention time	$t_r' = t_r - t_m$	t_r = retention time of solute of interest t_m = retention time of unretained solute
Retention volume	$V_r = t_r \cdot u_v$	u_v = volume flow rate = volume per unit time
Retention factor	$k = t_r'/t_m = KV_s/V_m$ $k = \dfrac{t_s}{t_m} = \dfrac{t_r - t_m}{t_m}$	V_s = volume of stationary phase V_m = volume of mobile phase t_s = time solute spends in stationary phase = $t_r - t_m$ t_m = time solute spends in mobile phase
Relative retention (also called separation factor)	$\alpha = \dfrac{t_{r2}'}{t_{r1}'} = \dfrac{k_2}{k_1} = \dfrac{K_2}{K_1}$	Subscripts 1 and 2 refer to two solutes
Number of plates	$N = \dfrac{16t_r^2}{w^2} = \dfrac{5.55t_r^2}{w_{1/2}^2}$	w = width at base $w_{1/2}$ = width at half-height
Plate height	$H = \dfrac{\sigma^2}{x} = \dfrac{L}{N}$	σ = standard deviation of band x = distance traveled by center of band L = length of column N = number of plates on column
Resolution	$\text{Resolution} = \dfrac{\Delta t_r}{w_{av}} = \dfrac{\Delta V_r}{w_{av}}$	Δt_r = difference in retention times ΔV_r = difference in retention volumes w_{av} = average width measured at baseline in same units as numerator (time or volume)
	$\text{Resolution} = \dfrac{\sqrt{N}}{4}\dfrac{(\alpha - 1)}{\alpha}\left(\dfrac{k_2}{1 + k_2}\right)$	N = number of plates α = relative retention k_2 = retention factor for second peak

23-5 Why Bands Spread[15]

A band of solute invariably spreads as it travels through a chromatography column (Figure 23-11) and emerges at the detector with a standard deviation σ. Each individual mechanism contributing to broadening produces a standard deviation σ_i. The observed variance (σ_{obs}^2) is the sum of variances from all contributing mechanisms:

Variance is additive:
$$\sigma_{obs}^2 = \sigma_1^2 + \sigma_2^2 + \sigma_3^2 + \cdots = \sum \sigma_i^2 \qquad (23\text{-}34)$$

Variance is additive, but standard deviation is not.

Broadening Outside the Column[16]

Solute cannot be applied to the column in an infinitesimally small volume. If the sample injected has a volume ΔV (measured in units of volume), the contribution to the variance of the final bandwidth is

Variance due to injection or detection:
$$\sigma_{injection}^2 = \sigma_{detector}^2 = \frac{(\Delta V)^2}{12} \qquad (23\text{-}35)$$

The same relation holds for broadening in a detector whose volume is ΔV. Sometimes on-column detection is possible, thereby eliminating the problem of band spreading in a detector.

Components of a chromatograph (injector, column, and detector) are connected by capillary tubing. In **laminar flow** through the tubing, fluid velocity is greatest at the center and decreases to zero at the walls (Figure 19-13). There is no turbulent mixing of liquid near the walls with liquid near the center. The contribution of laminar flow to the bandwidth is

Variance due to connecting tubing:
$$\sigma_{tubing}^2 = \frac{\pi d_t^4 l_t u_v}{384D} \qquad (23\text{-}36)$$

where d_t and l_t are the diameter and length of the tubing, u_v is the volume flow rate, and D is the diffusion coefficient. Band spreading in the connecting tubing is minimized by keeping the tubing narrow and short.

Band Spreading Before and After the Column

A band from a column eluted at a rate of 1.35 mL/min has a width at half-height of 0.272 min. The sample was applied as a sharp plug with a volume of 0.30 mL, the detector volume is 0.20 mL, and the connecting tubing is 30 cm long with a 0.050-cm diameter. Find the variances introduced by injection, detection, and connecting tubing, assuming a solute diffusion coefficient of 1.0×10^{-9} m²/s. What would $w_{1/2}$ (in time units) be if broadening occurred only on the column?

Solution Figure 23-9 tells us that the width at half-height is $w_{1/2} = 2.35\sigma$. Therefore, the observed total variance in volume units is

$$\sigma_{obs}^2 = \left(\frac{w_{1/2}}{2.35}\right)^2 = \left(\frac{0.272 \text{ min} \cdot 1.35 \text{ mL/min}}{2.35}\right)^2 = 0.024\,42 \text{ mL}^2$$

Injection volume is $\Delta V_{injection} = 0.30$ mL. Therefore,

$$\sigma_{injection}^2 = \frac{\Delta V_{injection}^2}{12} = \frac{(0.30 \text{ mL})^2}{12} = 0.007\,50 \text{ mL}^2$$

The detector volume is $\Delta V_{detector} = 0.20$ mL, so $\sigma_{detector}^2 = \Delta V_{detector}^2 / 12 = 0.003\,33 \text{ mL}^2$. The diffusion coefficient is 1.0×10^{-9} m²/s or 6.0×10^{-4} cm²/min. Therefore,

$$\sigma_{tubing}^2 = \frac{\pi d_t^4 l_t u_v}{384D} = \frac{\pi (0.050 \text{ cm})^4 (30 \text{ cm})(1.35 \text{ cm}^3/\text{min})}{384(6 \times 10^{-4} \text{ cm}^2/\text{min})} = 0.003\,45 \text{ cm}^6 = 0.003\,45 \text{ mL}^2$$

The observed variance is

$$\sigma_{obs}^2 = \sigma_{column}^2 + \sigma_{injection}^2 + \sigma_{detector}^2 + \sigma_{tubing}^2$$

$$0.024\,42 \text{ mL}^2 = \sigma_{column}^2 + 0.007\,50 \text{ mL}^2 + 0.003\,33 \text{ mL}^2 + 0.003\,45 \text{ mL}^2$$

$$\Rightarrow \sigma_{column} = 0.101 \text{ mL (in volume)}$$

$$\text{or } \sigma_{column} = \frac{0.101 \text{ mL}}{1.35 \text{ mL/min}} = 0.074\,6 \text{ min (in time)}$$

The width due to column broadening alone is $w_{1/2} = 2.35\sigma_{column} = 0.175$ min, which is two-thirds of the observed width. Thus the extra-column components contributed one-third of the peak broadening.

TEST YOURSELF Predict the observed $w_{1/2}$ if the injected volume were decreased to 0.15 mL. (*Answer:* 0.239 min)

Modern chromatographic systems use small columns. Injection, detection, and tubing volumes must be correspondingly small.

Plate Height Equation

Plate height, H, is proportional to the variance of a chromatographic band (Equation 23-29): the smaller the plate height, the narrower the band. The **van Deemter equation** tells us how the column and flow rate affect plate height:

van Deemter equation
for plate height:

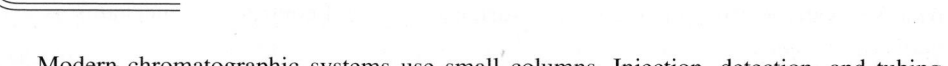

$$H \approx A + \frac{B}{u_x} + Cu_x \qquad (23\text{-}37)$$

| Multiple | Longitudinal | Equilibration time |
| paths | diffusion | (mass transfer) |

where u_x is the linear velocity and A, B, and C are constants for a given column and stationary phase. Changing the column and stationary phase changes A, B, and C. The van Deemter equation says there are band-broadening mechanisms that are proportional to linear velocity, inversely proportional to linear velocity, and independent of linear velocity (Figure 23-17). At the *optimum linear velocity*, the plate height H of the column is lowest, and so the number of plates N is greatest. Below the optimum linear velocity, longitudinal diffusion broadening, B, is most significant. Above the optimum, equilibration time

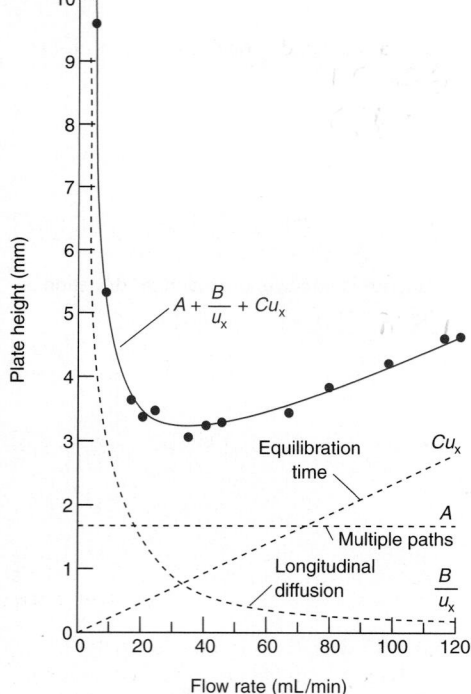

FIGURE 23-17 Application of van Deemter equation to gas chromatography: $A = 1.65$ mm, $B = 25.8$ mm · mL/min, and $C = 0.023\,6$ mm · min/mL. [Experimental points data from H. W. Moody, "The Evaluation of the Parameters in the van Deemter Equation," *J. Chem. Ed.* **1982,** *59,* 290.]

broadening, C, is dominant. At the optimum linear velocity, the B and C terms contribute equally to plate height.

In packed columns, all three terms contribute to band broadening. For open tubular columns, the multiple path term, A, is 0, so bandwidth decreases and resolution increases. In capillary electrophoresis (Chapter 26), both A and C go to 0, thereby reducing plate height to submicron values and providing extraordinary separation powers.

Longitudinal Diffusion

If you could apply a thin, disk-shaped band of solute to the center of a column, the band would slowly broaden as molecules diffuse from the high concentration within the band to regions of lower concentration ahead of and behind the band. Diffusional broadening of a band, called **longitudinal diffusion** because it takes place along the axis of the column, occurs while the band is transported along the column by solvent flow (Figure 23-18).

The term B/u_x in Equation 23-37 arises from longitudinal diffusion. The faster the linear flow, the less time is spent in the column and the less diffusional broadening occurs. Equation 23-28 told us that the variance from diffusion is

$$\sigma^2 = 2D_m t = \frac{2D_m L}{u_x}$$

Plate height due to longitudinal diffusion:
$$H_{\text{longitudinal diffusion}} = \frac{\sigma^2}{L} = \frac{2D_m}{u_x} \equiv \frac{B}{u_x} \qquad (23\text{-}38)$$

where D_m is the diffusion coefficient of solute in the mobile phase, t is time in the mobile phase, and $H_{\text{longitudinal diffusion}}$ is the plate height due to longitudinal diffusion. The time needed to travel the length of the column (L) is L/u_x, where u_x is the linear velocity.

Finite Equilibration Time Between Phases

The term Cu_x in Equation 23-37 comes from the finite time required for solute to equilibrate between mobile and stationary phases.[17] **Mass transfer** is the movement of solute from one phase to another. Solute must diffuse from the mobile phase to the surface of the stationary phase for this equilibration to occur (Figure 23-19). The time required depends on the distance that the solute must diffuse to get to the stationary phase, and inversely on how fast it diffuses. The faster the mobile phase velocity (u_x), the less time available for this transfer to occur.

Plate height from finite equilibration time, also called the *mass transfer term*, is

Plate height due to finite equilibration time:
$$H_{\text{mass transfer}} = Cu_x = (C_m + C_s)u_x \qquad (23\text{-}39)$$

where C_m describes mass transfer through the mobile phase and C_s describes mass transfer through the stationary phase. Specific equations for C_m and C_s depend on the type of chromatography and geometry of the column.

For gas chromatography in an open tubular column, the terms are

Mass transfer in mobile phase:
$$C_m = \frac{1 + 6k + 11k^2}{24(k+1)^2} \frac{r^2}{D_m} \qquad (23\text{-}40)$$

Mass transfer in stationary phase:
$$C_s = \frac{2k}{3(k+1)^2} \frac{d^2}{D_s} \qquad (23\text{-}41)$$

where k is the retention factor, r is the column radius, D_m is the diffusion coefficient of solute in the mobile phase, d is the thickness of stationary phase, and D_s is the diffusion coefficient of solute in the stationary phase. Figure 23-19 depicts C_m broadening. Decreasing column radius, r, reduces plate height by decreasing the distance through which solute must diffuse to reach the stationary phase. Slow diffusion of solute into and out of the stationary phase film causes C_s broadening (Eq. 23-41). Decreasing stationary phase thickness, d, reduces plate height and increases efficiency because solute can diffuse faster from the depths of the stationary phase into the mobile phase.

Packed columns: $A, B, C \neq 0$
Open tubular columns: $A = 0$
Capillary electrophoresis: $A = C = 0$

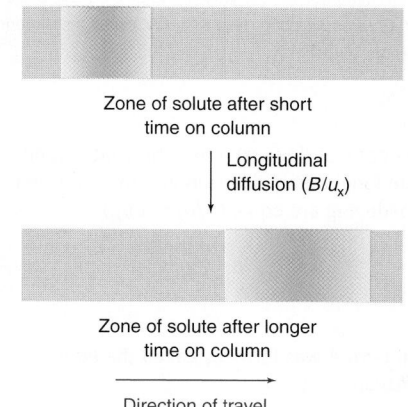

Zone of solute after short time on column

Longitudinal diffusion (B/u_x)

Zone of solute after longer time on column

Direction of travel

FIGURE 23-18 Longitudinal diffusion gives rise to B/u_x in the van Deemter equation. Solute continuously diffuses away from the concentrated center of its zone. The faster the flow, the less time is spent on the column and the less longitudinal diffusion occurs.

Longitudinal diffusion in a gas is much faster than diffusion in a liquid, so the optimum linear velocity in gas chromatography is higher than in liquid chromatography.

Laminar flow and slow diffusion contribute to C_m.

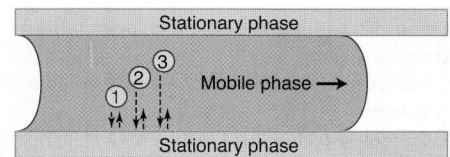

Fast mass transfer relative to mobile phase velocity

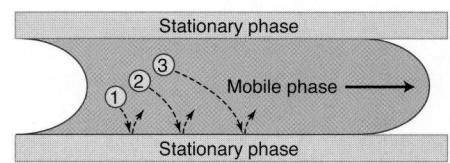

Slow mass transfer relative to mobile phase velocity

FIGURE 23-19 If mass transfer in the mobile phase (dashed arrows) occurs quickly, solute molecules do not experience additional broadening due to the C_m term. If mass transfer is slow relative to the mobile phase velocity, C_m broadening occurs. [Information from Fig. 13A.5 of J. F. Rubinson and K. A. Rubinson, *Contemporary Chemical Analysis* (Upper Saddle River, NJ: Prentice Hall, 1998).]

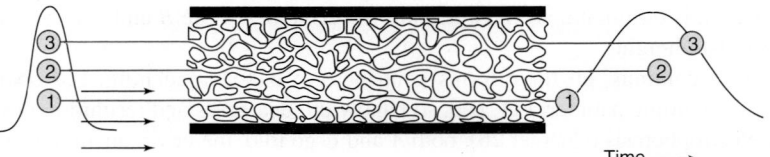

Time ⟶

H is optimal (lowest) when the contributions from longitudinal diffusion and mass transfer broadening are equal ($B/u_x = Cu_x$.)

In gas and liquid chromatography, mass transfer plate height is also decreased by increasing temperature, which increases the diffusion coefficient of solute in the stationary and mobile phases. Raising the temperature in liquid chromatography allows the linear velocity to be increased, while maintaining acceptable resolution.[18] Resolution is maintained because of the increased rate of mass transfer between phases at elevated temperature. Retention and relative retention are also affected by temperature.

Multiple Flow Paths

The term *A* was formerly called the *eddy diffusion* term.

The term A in the van Deemter equation (23-37) arises from multiple effects for which the theory is murky.[19] Figure 23-20 is a pictorial explanation of one effect. Because some flow paths are longer than others, molecules entering the column at the same time on the left are eluted at different times on the right. For simplicity, we approximate many different effects by the constant A in Equation 23-37.

Advantages of Open Tubular Columns

Compared with packed columns, open tubular columns can provide
• higher resolution
• shorter analysis time
• increased sensitivity
• lower sample capacity

In gas chromatography, we have a choice of using open tubular columns or packed columns. For similar analysis times, open tubular columns provide higher resolution and increased sensitivity to small quantities of analyte. Open tubular columns have small sample capacity, so they are not useful for preparative separations.

Particles in a packed column resist flow of the mobile phase, so the linear velocity cannot be very fast. For the same length of column and applied pressure, the linear velocity in an open tubular column is much higher than that of a packed column. Therefore, the open tubular column can be made 100 times longer than the packed column and still achieve a similar pressure drop and linear velocity. If plate height is the same, the longer column provides 100 times more theoretical plates, yielding $\sqrt{100} = 10$ times more resolution.

For a given pressure, linear velocity in an open tubular column is proportional to cross-sectional area of the column and inversely proportional to column length:

$$u_x \propto \frac{\text{area}}{\text{length}}$$

Plate height is reduced in an open tubular column because band spreading by multiple flow paths (Figure 23-20) cannot occur. In the van Deemter curve for the packed column in Figure 23-17, the A term accounts for half of the plate height at the most efficient flow rate (minimum H) near 30 mL/min. If A were deleted, the number of plates on the column would be doubled. To obtain high performance from an open tubular column, the radius of the column must be small and the stationary phase must be as thin as possible to ensure rapid mass transfer between mobile and stationary phases (small C_m and C_s terms).

Compared with packed columns, open tubular columns allow
• increased linear velocity or a longer column or both
• decreased plate height, which means higher resolution

Table 23-3 compares the performances of packed and open tubular gas chromatography columns with the same stationary phase. For similar analysis times, the open tubular column gives resolution seven times better (10.6 versus 1.5) than that of the packed column. Alternatively, speed could be traded for resolution. If the open tubular column were reduced to 5 m

TABLE 23-3	Comparison of packed and wall-coated open tubular gas chromatographic column performance[a]		
Property		**Packed**	**Open tubular**
Column length, L		2.4 m	100 m
Linear gas velocity		8 cm/s	16 cm/s
Plate height for methyl oleate		0.73 mm	0.34 mm
Retention factor, k, for methyl oleate		58.6	2.7
Theoretical plates, N		3 290	294 000
Resolution of methyl stearate and methyl oleate		1.5	10.6
Retention time of methyl oleate		29.8 min	38.5 min

a. Methyl stearate ($CH_3(CH_2)_{16}CO_2CH_3$) and methyl oleate (cis-$CH_3(CH_2)_7CH=CH(CH_2)_7CO_2CH_3$) were separated on columns with poly(diethylene glycol succinate) stationary phase at 180°C.

SOURCE: L. S. Ettre. *Introduction to Open Tubular Columns* (Norwalk, CT: Perkin-Elmer Corp., 1979), p. 26.

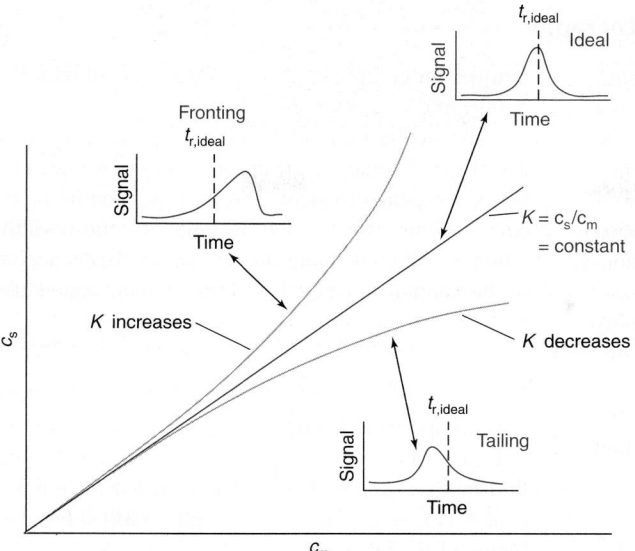

FIGURE 23-21 Common isotherms and their resulting chromatographic peak shapes. $t_{r,ideal}$ is the retention time for linear isotherm conditions.

in length, the same solutes could be separated with a resolution of 1.5, but the time would be reduced from 38.5 to 0.83 min.

A Touch of Reality: Asymmetric Peaks

A Gaussian peak shape results when the partition coefficient, K ($= c_s/c_m$), is independent of the concentration of solute on the column. In real columns, K changes as the concentration of solute increases, and chromatographic peaks are skewed.[20] A graph of c_s versus c_m at a given temperature is called an *isotherm*. Three common isotherms and their resulting peak shapes are shown in Figure 23-21. The ideal center isotherm gives a symmetric peak, with a retention time ($t_{r,ideal}$) that does not depend on solute concentration. This constancy allows use of retention time t_r to identify the solute.

The upper and lower isotherms in Figure 23-21 arise from an *overloaded* column in which too much solute has been applied to the column. The upper isotherm results in *fronting* peaks ($B/A < 1$ in Figure 23-14), where there is a gradual rise and an abrupt fall of the chromatographic peak. As the concentration of solute increases, the solute becomes more and more soluble in the stationary phase. There is so much solute in the stationary phase that the stationary phase begins to resemble solute. (As in extraction, the general rule is "like dissolves like.") The severity of fronting increases with the amount of excess solute injected.

The lower isotherm in Figure 23-21 results in *tailing* peaks ($B/A > 1$ in Figure 23-14), where the peak jumps up quickly and then gradually falls. If the column has limited retention capacity, injection of concentrated solute saturates a significant portion of sorption sites, leaving fewer sites available for retention, and resulting in a lower k. Such saturation causes the peak maximum to shift to a shorter retention time. The shift forward in retention time increases with increased overload.

Sites that bind solute strongly can also cause tailing. Silica surfaces of columns and stationary phase particles have silanol groups (—SiOH) that form hydrogen bonds with polar solutes, thereby leading to serious tailing. Column manufacturers use **silanization** to reduce tailing by blocking the silanol groups with nonpolar trimethylsilyl groups:

c_s = concentration of solute in stationary phase

c_m = concentration of solute in mobile phase

At low analyte concentration (infinite dilution) all three isotherms appear linear. So if the column capacity is high enough, Gaussian peaks are observed.

Overload in gas chromatography tends to cause fronting, while overload in liquid chromatography causes tailing.

Highly porous particles with high surface area are used in liquid chromatography to maximize surface area and thus minimize overloading.

Silanization is also called **end-capping**.

$$
\begin{array}{c}
\underset{\substack{\text{Solid phase}\\\text{with exposed}\\\text{—SiOH groups}}}{
\overset{\displaystyle \text{OH} \qquad \text{OH}}{\underset{\displaystyle \quad}{
\begin{array}{c} | \qquad\quad | \\ \text{—Si—O—Si—} \\ | \qquad\quad | \end{array}}}}
+ \underset{\text{Hexamethyldisilazine}}{(CH_3)_3SiNHSi(CH_3)_3}
\rightarrow
\underset{\text{Protected surface}}{
\overset{\displaystyle (CH_3)_3SiO \quad OSi(CH_3)_3}{
\begin{array}{c} | \qquad\qquad\quad | \\ \text{—Si—O—Si—} \\ | \qquad\qquad\quad | \end{array}}}
+ NH_3 \quad \textbf{(23-42)}
\end{array}
$$

Fused silica columns for gas chromatography can also be silanized to minimize interaction of solute with active sites on the walls.

Now that you have been exposed to many concepts, you might want to read about a microscopic model of chromatography in Box 23-2.

Stochastic theory provides a model for chromatography.[21] "Stochastic" implies the presence of a random variable. The model supposes that as a molecule travels through a column, it spends an average time τ_m in the mobile phase between adsorption events. The time between desorption and the next adsorption is random, but the *average* time is τ_m. The average time spent adsorbed to the stationary phase between one adsorption and one desorption is τ_s. While the molecule is adsorbed on the stationary phase, it does not move. When the molecule is in the mobile phase, it moves with the linear velocity u_x of the mobile phase. The probability that an adsorption or desorption occurs in a given time follows the Poisson distribution, which was described briefly in Problem 19-25.

We assume that all molecules spend total time t_m in the mobile phase. This is the retention time of an unretained solute. Important results of the stochastic model are

- A solute molecule is adsorbed and desorbed an average of n times as it flows through the column, where $n = t_m/\tau_m$ and t_m is the total time in the mobile phase.
- The adjusted retention time for a solute is

$$t'_r = n\tau_s \qquad \textbf{(A)}$$

This is the average time that the solute is bound to the stationary phase during its transit through the column.

- The width (standard deviation) of a peak due to effects of the stationary phase is

$$\sigma = \tau_s\sqrt{2n} \qquad \textbf{(B)}$$

Consider the idealized chromatogram in the illustration, with one unretained component and two retained substances, A and B. The chromatographic parameters are representative of a high-performance liquid chromatography separation on a 15-cm-long × 0.39-cm-diameter column packed with 5-μm-diameter spherical particles of C_{18}-silica (Section 25-1). With a volume flow rate of 1.0 mL/min, the linear velocity is $u_x = 2.4$ mm/s. From the measured width at half-height ($w_{1/2}$), the standard deviation (σ) of a Gaussian peak is computed from $w_{1/2} = 2.35\sigma$ (Figure 23-9). The plate

number for components A and B, computed with Equation 23-30, is $N = (t_r/\sigma)^2 = 1.00 \times 10^4$.

The stochastic model applies to processes involving the stationary phase. To analyze the chromatogram, we need to subtract contributions to peak broadening from dispersion in the mobile phase and extra-column effects such as finite injection width, finite detector volume, and connecting tubing. These effects account for the width of the unretained peak. To subtract the unwanted effects, we write

$$\sigma^2_{\text{observed}} = \sigma^2_{\text{stationary phase}} + \sigma^2_{\text{unretained peak}}$$

$$\sigma^2_{\text{stationary phase}} = \sigma^2_{\text{observed}} - \sigma^2_{\text{unretained peak}}$$

For component A, $\sigma^2_{\text{stationary phase}} = \sigma^2_{\text{observed}} - \sigma^2_{\text{unretained peak}} = (3.6\text{ s})^2 - (1.5\text{ s})^2 \Rightarrow \sigma_{\text{stationary phase}} = 3.27$ s. For component B, we find $\sigma_{\text{stationary phase}} = 5.81$ s. The adjusted retention time for component A is $t'_r = t_r - t_m = 360 - 60 = 300$ s. For component B, $t'_r = 600 - 60 = 540$ s.

Now we use t'_r and σ ($= \sigma_{\text{stationary phase}}$) for each component to find physically meaningful parameters. Combining Equations A and B above, we find

$$n = 2\left(\frac{t'_r}{\sigma}\right)^2 \qquad \tau_s = \frac{\sigma^2}{2t'_r}$$

and we already knew that $\tau_m = t_m/n$. From the parameters in the illustration, we compute the results in the table.

	Component A	Component B
n (number of times adsorbed)	16 800	17 300
τ_s (duration of one adsorption)	17.8 ms	31.2 ms
τ_m (time between adsorption events)	3.6 ms	3.5 ms
Distance between adsorptions ($= u_x\tau_m$)	8.6 μm	8.4 μm

Terms to Understand

adjusted retention time	eluent	miscible	resolution
adsorption chromatography	elution	mobile phase	retention factor
affinity chromatography	extraction	molecular exclusion	retention time
asymmetry factor	gel filtration chromatography	chromatography	retention volume
baseline resolution	gel permeation chromatography	open tubular column	silanization
chromatogram	ion-exchange chromatography	packed column	stationary phase
diffusion	laminar flow	partition chromatography	van Deemter equation
diffusion coefficient	linear velocity	partition coefficient	volume flow rate
distribution coefficient	longitudinal diffusion	plate height	
eluate	mass transfer	relative retention	

Summary

A solute can be extracted from one phase into another in which it is more soluble. The ratio of solute concentrations in each phase at equilibrium is called the partition coefficient. If more than one form of the solute exists, we use a distribution coefficient instead of a partition coefficient. We derived equations relating the fraction of

solute extracted to the partition or distribution coefficient, volumes, and pH. Many small extractions are more effective than a few large extractions. A metal chelator, soluble only in organic solvents, can extract metal ions from aqueous solutions, with selectivity achieved by adjusting pH. Crown ethers and salts containing a hydrophobic

Both components spend nearly the same time (~3.5 ms) in the mobile phase between adsorption events. Component A spends an average of 17.8 ms bound to the stationary phase each time it is adsorbed and component B spends an average of 31.2 ms. This difference in τ_s is the reason why A and B are separated from each other.

During its transit through the column, each substance becomes adsorbed $n \approx 17\,000$ times. The distance traveled between adsorptions is ~8.5 μm. The chromatogram was simulated for a column with $N = 10\,000$ theoretical plates. The plate height is 15 cm/(10 000 plates) = 15 μm. In Section 23-4, we stated that plate height is approximately the length of column required for one equilibration of solute between mobile and stationary phases. From the stochastic theory in this example, we find that there are approximately two equilibrations with the stationary phase in each length corresponding to the plate height.

The time required for a solute to flow past a stationary-phase particle whose diameter is $d = 5$ μm is $t = (5\ \mu m)/(2.4\ mm/s) = 2.1$ ms.

The stochastic theory predicts that the fraction of time that a molecule in the mobile phase will travel *less than* distance d is $1 - e^{-t/\tau_m} = 1 - e^{-(2.1\ ms)/(3.5\ ms)} = 0.55$. That is, approximately half of the time, a solute molecule does not travel as far as the next particle of stationary phase before becoming adsorbed again to the same particle from which it just desorbed. If we lined up spherical particles of stationary phase, it would take 30 000 particles to cover the 15-cm length of the column. Each solute molecule binds ~17 000 times as it transits the column, and half of those binding steps are to the same particle from which it just desorbed.

The model provides insight into microscopic events that occur during chromatography. The model omits some phenomena that occur in real columns. For example, in a porous stationary phase, the mobile phase could be stagnant inside the pores. When a molecule enters such a pore, it will adsorb and desorb many times from the same particle before escaping from the pore.

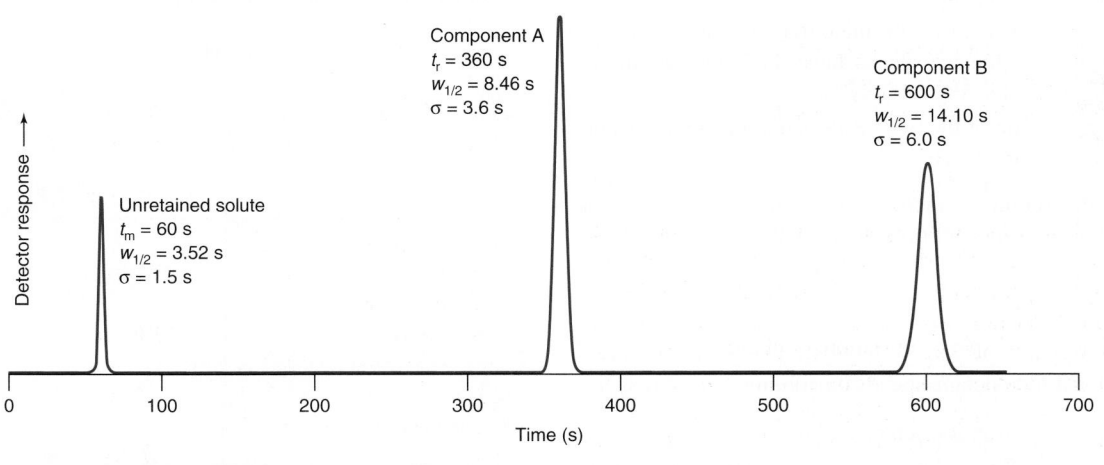

Component A
$t_r = 360$ s
$w_{1/2} = 8.46$ s
$\sigma = 3.6$ s

Component B
$t_r = 600$ s
$w_{1/2} = 14.10$ s
$\sigma = 6.0$ s

Unretained solute
$t_m = 60$ s
$w_{1/2} = 3.52$ s
$\sigma = 1.5$ s

Detector response →

Time (s)

Idealized liquid chromatographic separation of three components.

ion act as phase transfer agents to bring a hydrophilic ion from an aqueous phase into an organic phase.

In adsorption and partition chromatography, a continuous equilibration of solute between mobile and stationary phases occurs. Eluent goes into a column and eluate comes out. Columns may be packed with stationary phase, or may be open tubular with stationary phase bonded to the inner wall. In ion-exchange chromatography, the solute is attracted to the stationary phase by coulombic forces. In molecular exclusion chromatography, the fraction of stationary phase pore volume available to solute decreases as the size of the solute molecules increases. Affinity chromatography relies on specific, noncovalent interactions between the stationary phase and one solute in a complex mixture.

The relative retention of two components is the quotient of their adjusted retention times. The retention factor for a single component is the adjusted retention time divided by the elution time for solvent. Retention factor gives the ratio of time spent by solute in the stationary phase to time spent in the mobile phase. When a separation is scaled up from a small load to a large load, linear velocity is held constant and the cross-sectional area of the column should be increased in proportion to the loading.

Plate height ($H = \sigma^2/x$) is related to the breadth of a band emerging from the column. The smaller the plate height, the sharper the band. The number of plates for a Gaussian peak is $N = 5.55 t_r^2/w_{1/2}^2$. Plate height is approximately the length of column required for one equilibration of solute between mobile and stationary phases. Resolution of neighboring peaks is the difference in retention time divided by the average width (measured at the baseline, $w = 4\sigma$). Resolution is proportional to $\sqrt{N}$. Doubling the length of a column increases resolution by $\sqrt{2}$. Some retention ($k \approx 0.5$ to 20) is essential to achieve resolution, but greater retention increases separation time without substantial improvements in resolution. Relative retention (α) strongly influences chromatographic resolution.

The standard deviation of a diffusing band of solute is $\sigma = \sqrt{2Dt}$, where D is the diffusion coefficient and t is time. The van Deemter equation describes band broadening on the chromatographic column: $H \approx A + B/u_x + Cu_x$, where H is plate height, u_x is linear velocity, and A, B, and C are constants. The first term represents irregular flow paths, the second, longitudinal diffusion, and the third, the finite rate of mass transfer of solute between mobile and stationary phases. The optimum flow rate, which minimizes plate height, is faster for gas chromatography than for liquid chromatography. The number of plates and the optimal flow rate increase as the stationary phase particle size or open tubular column diameter decreases. In gas chromatography, open tubular columns can provide higher resolution or shorter analysis times than packed columns. Bands spread during injection, during detection, and in connecting tubing, as well as during passage through the separation column. The observed variance of the band is the sum of the variances for all mechanisms of spreading. Fronting and tailing can be corrected by injecting less sample and by masking strong adsorption sites on the stationary phase.

Exercises

23-A. A solute with a partition coefficient of 4.0 is extracted from 10 mL of phase 1 into phase 2.

(a) What volume of phase 2 is needed to extract 99% of the solute in one extraction?

(b) What is the total volume of solvent 2 needed to remove 99% of the solute in three equal extractions instead?

23-B. Consider a chromatography experiment in which two components with retention factors $k_1 = 4.00$ and $k_2 = 5.00$ are injected into a column with $N = 1.00 \times 10^3$ theoretical plates. The retention time for the less-retained component is $t_{r1} = 10.0$ min.

(a) Calculate t_m and t_{r2}. Find $w_{1/2}$ (width at half-height) and w (width at the base) for each peak.

(b) Using graph paper, sketch the chromatogram analogous to Figure 23-7, supposing that the two peaks have the same amplitude (height). Draw the half-widths accurately.

(c) Calculate the resolution of the two peaks and compare this value with those drawn in Figure 23-10.

23-C. (a) Find the retention factors for octane and nonane in Figure 23-7. When you measure distances, estimate them to the nearest 0.1 mm.

(b) Find the ratio

$$\frac{\text{Time octane spends in stationary phase}}{\text{Total time octane spends on column}}$$

(c) Find the relative retention for octane and nonane.

(d) Find the partition coefficient for octane by assuming that the volume of the stationary phase equals half the volume of the mobile phase.

23-D. A gas chromatogram of a mixture of toluene and ethyl acetate is shown here.

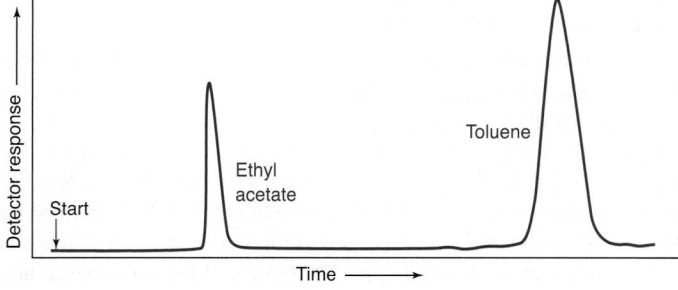

(a) Use the width of each peak (measured at the base) to calculate the number of theoretical plates in the column. Estimate all lengths to the nearest 0.1 mm.

(b) Using the width of the toluene peak at its base, calculate the width expected at half-height. Compare the measured and calculated values. When the thickness of the line is significant relative to the length being measured, it is important to take the pen line width into account. You can measure from the edge of one line to the corresponding edge of the other line, as shown here.

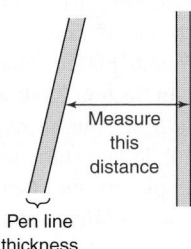

23-E. The three chromatograms shown here were obtained with 2.5, 1.0, and 0.4 µL of ethyl acetate injected on the same column under the same conditions. Explain why the peak becomes less symmetrical with increasing sample size.

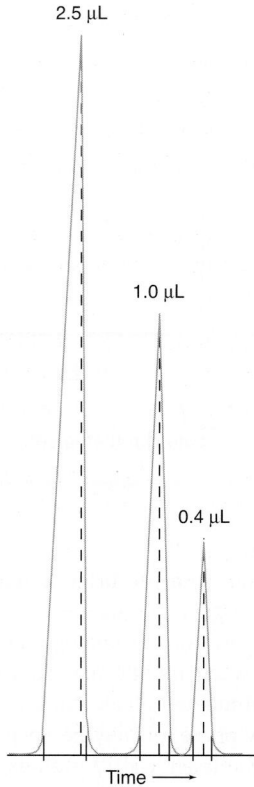

23-F. The relative retention for two compounds in gas chromatography is 1.068 on a column with a plate height of 0.520 mm. The retention factor for compound 1 is 5.16.

(a) Find the retention factor for compound 2.

(b) What length of column will separate the compounds with a resolution of 1.50?

(c) The retention time for air (t_m) is 2.00 min. If the number of plates is the same for both compounds, find t_r and $w_{1/2}$ for each peak.

(d) If the volume ratio of stationary phase to mobile phase is 0.30, find the partition coefficient for compound 1.

Problems

Solvent Extraction

23-1. If you are extracting a substance from water into ether, is it more effective to do one extraction with 300 mL of ether or three extractions with 100 mL?

23-2. When performing liquid-liquid extraction, it is common practice to shake the separatory funnel to disperse one phase into the other.

(a) Does shaking the separatory funnel change the partition coefficient?

(b) Explain the purpose of shaking the funnel.

23-3. If you wish to extract aqueous acetic acid into hexane, is it more effective to adjust the aqueous phase to pH 3 or pH 8?

23-4. (a) Why is it difficult to extract the EDTA complex of aluminum into an organic solvent but easy to extract the 8-hydroxyquinoline complex?

(b) If you need to bring the EDTA complex into the organic solvent, should you add a phase transfer agent with a hydrophobic cation or a hydrophobic anion?

23-5. Why is the extraction of a metal ion into an organic solvent with 8-hydroxyquinoline more complete at higher pH?

23-6. The distribution coefficient for extraction of a metal complex from aqueous to organic solvents is $D = [\text{total metal}]_{\text{org}}/[\text{total metal}]_{\text{aq}}$. Give physical reasons why β and K_a appear in the numerator of Equation 23-13, but K_L and $[H^+]_{\text{aq}}$ appear in the denominator.

23-7. Give a physical interpretation of Equations 23-6 and 23-7 in terms of the fractional composition equations for a monoprotic acid discussed in Section 10-5.

23-8. Solute S has a partition coefficient of 4.0 between water (phase 1) and chloroform (phase 2) in Equation 23-1.

(a) Calculate the concentration of S in chloroform if $[S]_{\text{aq}}$ is 0.020 M.

(b) If the volume of water is 80.0 mL and the volume of chloroform is 10.0 mL, find the quotient (mol S in chloroform)/(mol S in water).

23-9. The solute in Problem 23-8 is initially dissolved in 80.0 mL of water. It is then extracted six times with 10.0-mL portions of chloroform. Find the fraction of solute remaining in the aqueous phase.

23-10. The weak base B ($K_b = 1.0 \times 10^{-5}$) equilibrates between water (phase 1) and benzene (phase 2).

(a) Define the distribution coefficient, D, for this system.

(b) Explain the difference between D and K, the partition coefficient.

(c) Calculate D at pH 8.00 if $K = 50.0$.

(d) Will D be greater or less at pH 10 than at pH 8? Explain why.

23-11. Consider the extraction of M^{n+} from aqueous solution into organic solution by reaction with protonated ligand, HL:

$$M^{n+}(aq) + nHL(org) \rightleftharpoons ML_n(org) + nH^+(aq)$$

$$K_{\text{extraction}} = \frac{[ML_n]_{\text{org}}[H^+]_{\text{aq}}^n}{[M^{n+}]_{\text{aq}}[HL]_{\text{org}}^n}$$

Rewrite Equation 23-13 in terms of $K_{\text{extraction}}$ and express $K_{\text{extraction}}$ in terms of the constants in Equation 23-13. Give a physical reason why each constant increases or decreases $K_{\text{extraction}}$.

23-12. Butanoic acid has a partition coefficient of 3.0 (favoring benzene) when distributed between water and benzene. Find the formal concentration of butanoic acid in each phase when 100 mL of 0.10 M aqueous butanoic acid is extracted with 25 mL of benzene **(a)** at pH 4.00 and **(b)** at pH 10.00.

23-13. For a given value of $[HL]_{\text{org}}$ in Equation 23-13, over what pH range (how many pH units) will D change from 0.01 to 100 if $n = 2$?

23-14. For the extraction of Cu^{2+} by dithizone in CCl_4, $K_L = 1.1 \times 10^4$, $K_M = 7 \times 10^4$, $K_a = 3 \times 10^{-5}$, $\beta = 5 \times 10^{22}$, and $n = 2$.

(a) Calculate the distribution coefficient for extraction of 0.1 μM Cu^{2+} into CCl_4 by 0.1 mM dithizone at pH 1.0 and at pH 4.0.

(b) If 100 mL of 0.1 μM aqueous Cu^{2+} are extracted once with 10 mL of 0.1 mM dithizone at pH 1.0, what fraction of Cu^{2+} remains in the aqueous phase?

23-15. Consider the extraction of 100.0 mL of $M^{2+}(aq)$ by 2.0 mL of 1×10^{-5} M dithizone in $CHCl_3$, for which $K_L = 1.1 \times 10^4$, $K_M = 7 \times 10^4$, $K_a = 3 \times 10^{-5}$, $\beta = 5 \times 10^{18}$, and $n = 2$.

(a) Derive an expression for the fraction of metal ion extracted into the organic phase, in terms of the distribution coefficient and volumes of the two phases.

(b) Prepare a graph of the percentage of metal ion extracted over the pH range 0 to 5.

23-16. The theoretical limit for extracting solute S from phase 1 (volume V_1) into phase 2 (volume V_2) is attained by dividing V_2 into an infinite number of infinitesimally small portions and conducting an infinite number of extractions. With a partition coefficient $K = [S]_2/[S]_1$, the limiting fraction of solute remaining in phase 1 is $q_{\text{limit}} = e^{-(V_2/V_1)K}$. Let $V_1 = V_2 = 50$ mL and let $K = 2$. Let volume V_2 be divided into n equal portions to conduct n extractions. Find the fraction of S extracted into phase 2 for $n = 1, 2$, and 10 extractions. How many portions are required to attain 95% of the theoretical limit?

A Plumber's View of Chromatography

23-17. Match the terms in the first list with the characteristics in the second list.

1. adsorption chromatography

2. partition chromatography

3. ion-exchange chromatography

4. molecular exclusion chromatography

5. affinity chromatography

A. Ions in mobile phase are attracted to counterions covalently attached to stationary phase.

B. Solute in mobile phase is attracted to specific groups covalently attached to stationary phase.

C. Solute equilibrates between mobile phase and surface of stationary phase.

D. Solute equilibrates between mobile phase and film of liquid attached to stationary phase.

E. Different-sized solutes penetrate pores in stationary phase to different extents. Largest solutes are eluted first.

23-18. The partition coefficient for a solute in chromatography is $K = c_s/c_m$, where c_s is the concentration in the stationary phase and

c_m is the concentration in the mobile phase. The larger the partition coefficient, the longer it takes a solute to be eluted. Explain why.

23-19. (a) Write the meaning of the retention factor, k, in terms of time spent by solute in each phase.

(b) Write an expression in terms of k for the fraction of time spent by a solute molecule in the mobile phase.

(c) The *retention ratio* in chromatography is defined as

$$R = \frac{\text{time for solvent to pass through column}}{\text{time for solute to pass through column}} = \frac{t_m}{t_r}$$

Show that R is related to the retention factor by the equation $R = 1/(k + 1)$.

23-20. (a) A chromatography column with a length of 10.3 cm and inner diameter of 4.61 mm is packed with a stationary phase that occupies 61.0% of the volume. If the volume flow rate is 1.13 mL/min, find the linear velocity in cm/min.

(b) How long does it take for solvent (which is the same as unretained solute) to pass through the column?

(c) Find the retention time for a solute with a retention factor of 10.0.

23-21. An open tubular column is 30.1 m long and has an inner diameter of 0.530 mm. It is coated on the inside wall with a layer of stationary phase that is 3.1 μm thick. Unretained solute passes through in 2.16 min, whereas a particular solute has a retention time of 17.32 min.

(a) Find the linear velocity and volume flow rate.

(b) Find the retention factor for the solute and the fraction of time spent in the stationary phase.

(c) Find the partition coefficient, $K = c_s/c_m$, for this solute.

23-22. A chromatographic procedure separates 4.0 mg of unknown mixture on a column with a length of 40 cm and a diameter of 0.85 cm.

(a) What size column would you use to separate 100 mg of the same mixture?

(b) If the flow is 0.22 mL/min on the small column, what volume flow rate should be used on the large column?

(c) The mobile phase occupies 35% of the column volume. Calculate the linear velocity for the small column and the large column.

23-23. Solvent passes through a column in 3.0 min but solute requires 9.0 min.

(a) Calculate the retention factor, k.

(b) What fraction of time is the solute in the mobile phase in the column?

(c) The volume of stationary phase is 1/10 of the volume of the mobile phase in the column ($V_s = 0.10V_m$). Find the partition coefficient, K, for this system.

23-24. Solvent occupies 15% of the volume of a chromatography column whose inner diameter is 3.0 mm. If the volume flow rate is 0.2 mL/min, find the linear velocity.

23-25. Consider a chromatography column in which $V_s = V_m/5$. Find the retention factor if $K = 3$ and if $K = 30$.

23-26. The retention volume of a solute is 76.2 mL for a column with $V_m = 16.6$ mL and $V_s = 12.7$ mL. Calculate the retention factor and the partition coefficient for this solute.

23-27. An open tubular column has an inner diameter of 207 μm and the thickness of the stationary phase on the inner wall is 0.50 μm. Unretained solute passes through in 63 s and a particular solute emerges in 433 s. Find the partition coefficient for this solute and find the fraction of time spent in the stationary phase.

23-28. Which of the following columns will provide:

(a) highest number of plates?

(b) greatest retention?

(c) highest relative retention?

(d) best separation?

Column 1: $N = 1000$; $k_2 = 1.2$; $\alpha = 1.16$; resolution $= 0.6$
Column 2: $N = 5000$; $k_2 = 3.9$; $\alpha = 1.06$; resolution $= 0.8$
Column 3: $N = 500$; $k_2 = 4.7$; $\alpha = 1.31$; resolution $= 1.1$
Column 4: $N = 2000$; $k_2 = 2.4$; $\alpha = 1.24$; resolution $= 1.5$

23-29. In chromatography, resolution is governed by **(a)** the number of plates, **(b)** relative retention, and **(c)** retention factor. Construct a set of graphs to show the dependence of resolution on each of these three parameters. Comment on the nature of the dependence of resolution on each of these three parameters.

23-30. (a) Explain why the diffusion coefficient of CH_3OH is greater than that of sucrose in Table 23-1.

(b) Make an order-of-magnitude estimate of the diffusion coefficient of water vapor in air at 298 K.

23-31. Isotopic compounds (isotopologues) are separated in Figure 23-15 by repeated passage through a pair of columns. Each cycle in the figure represents one pass through length $L = 25$ cm containing N theoretical plates. The relative retention (α) is 1.03 and the retention factor for L-phenylalanine is $k_2 = 1.62$.

(a) The observed resolution after 10 cycles is 1.60. Calculate the number of theoretical plates, N, in column length L. The mixture has passed through length $10L$ in 10 cycles.

(b) Find the plate height in μm.

(c) Predict the resolution expected from two cycles. The observed value is 0.71.

Efficiency and Band Spreading

23-32. Chromatograms of compounds A and B were obtained at the same flow rate with two columns of equal length. The value of t_m is 1.3 min in both cases.

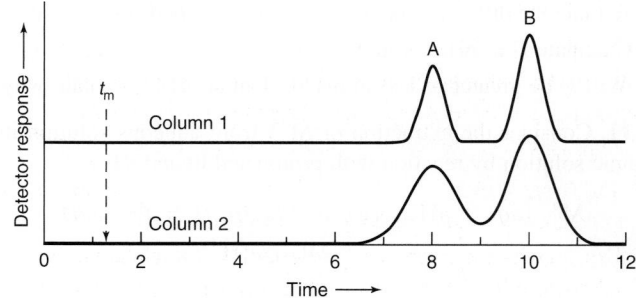

(a) Which column has more theoretical plates?

(b) Which column has a larger plate height?

(c) Which column gives higher resolution?

(d) Which column gives a greater relative retention?

(e) Which compound has a higher retention factor?

(f) Which compound has a greater partition coefficient?

(g) What is the numerical value of the retention factor of peak A?

(h) What is the numerical value of the retention factor of peak B?

(i) What is the numerical value of the relative retention?

23-33. For the separation of A and B by column 2 in Problem 23-32:

(a) If broadening is mainly due to longitudinal diffusion, how should the flow rate be changed to improve the resolution?

(b) If broadening is mainly due to the finite equilibrium time, how should the flow rate be changed to improve the resolution?

(c) If broadening is mainly due to multiple flow paths, what effect will flow rate have on the resolution?

23-34. Why does plate height depend on linear velocity, not volume flow rate?

23-35. Which column is more efficient: plate height = 0.1 mm or plate height =1 mm?

23-36. Why is longitudinal diffusion a more serious problem in gas chromatography than in liquid chromatography?

23-37. In chromatography, why is the optimal flow rate greater if the stationary phase particle size is smaller?

23-38. What is the optimal flow rate in Figure 23-17 for best separation of solutes?

23-39. Match statements 1–5 with the band broadening terms in the second list.

1. Depends on radius of open tubular column.

2. Not present in an open tubular column.

3. Depends on length and radius of connecting tubing.

4. Increases with diffusion coefficient of solute.

5. Increases with thickness of stationary phase film.

A Multiple paths

B Longitudinal diffusion

C_m Equilibration time in mobile phase

C_s Equilibration time in stationary phase

EC Extra column band broadening

23-40. Explain how silanization reduces tailing of chromatographic peaks.

23-41. Describe how nonlinear partition isotherms lead to non-Gaussian bandshapes. Draw the chromatographic peak shapes produced by an overloaded gas chromatography column and an overloaded liquid chromatography column.

23-42. A separation of 2.5 mg of an unknown mixture has been optimized on a column of length L and diameter d.

(a) Explain why you might not achieve the same resolution if the 2.5 mg of unknown mixture were injected in twice the injection volume.

(b) Explain why you might not achieve the same resolution if 5.0 mg of unknown mixture were injected in the original injection volume.

23-43. An infinitely sharp zone of solute is placed at the center of a column at time $t = 0$. After diffusion for time t_1, the standard deviation of the Gaussian band is 1.0 mm. After 20 min more, at time t_2, the standard deviation is 2.0 mm. What will be the width after another 20 min, at time t_3?

23-44. A chromatogram with ideal Gaussian bands has $t_r = 9.0$ min and $w_{1/2} = 2.0$ min.

(a) How many theoretical plates are present?

(b) Find the plate height if the column is 10 cm long.

23-45. (a) For the asymmetric chromatogram in Figure 23-14, calculate the asymmetry factor, B/A.

(b) The asymmetric chromatogram in Figure 23-14 has a retention time equal to 15.0 min and a $w_{0.1}$ of 44 s. Find the number of theoretical plates.

(c) The width of a Gaussian peak at a height equal to 1/10 of the peak height is 4.297σ. Suppose that the peak in (b) is symmetric with $A = B = 22$ s. Use Equations 23-30 and 23-32 to find the plate number.

23-46. (a) Two chromatographic peaks with widths, w, of 6 min are eluted at 24 and 29 min. Which diagram in Figure 23-10 will most closely resemble the chromatogram?

(b) Two chromatographic peaks are separated such that the signal returns to baseline between the two peaks. Based on Figure 23-10, what is the numerical value of the resolution between the peaks?

23-47. A chromatographic peak has a width, w, of 4.0 mL and a retention volume of 49 mL. What width is expected for a band with a retention volume of 127 mL? Assume that the only band spreading occurs on the column itself.

23-48. A peak for sucrose eluted with water from a column at 0.66 mL/min had a width at half-height, $w_{1/2}$, of 39.6 s. The sample was applied as a sharp plug with a volume of 0.40 mL, the detector volume was 0.25 mL, and there were 20 cm of 0.050-cm-diameter connecting tubing. Find the variances introduced by injection, detection, and connecting tubing. What would $w_{1/2}$ be if the only broadening occurred on the column?

23-49. Two compounds with partition coefficients of 15 and 18 are to be separated on a column with $V_m/V_s = 3.0$ and $t_m = 1.0$ min. Calculate the number of theoretical plates needed to produce a resolution of 1.5.

23-50. Calculate the number of theoretical plates needed to achieve a resolution of 2.0 if:

(a) $\alpha = 1.05$ and $k_2 = 5.00$.

(b) $\alpha = 1.10$ and $k_2 = 5.00$.

(c) $\alpha = 1.05$ and $k_2 = 10.00$.

(d) How can you increase N, α, and k_2 in a chromatography method? In this problem, which has a larger effect on resolution, α or k_2?

23-51. Consider the peaks for pentafluorobenzene and benzene in the gas chromatogram shown here. The elution time for unretained solute is 1.06 min. The open tubular column is 30.0 m in length and 0.530 mm in diameter, with a layer of stationary phase 3.0 μm thick on the inner wall.

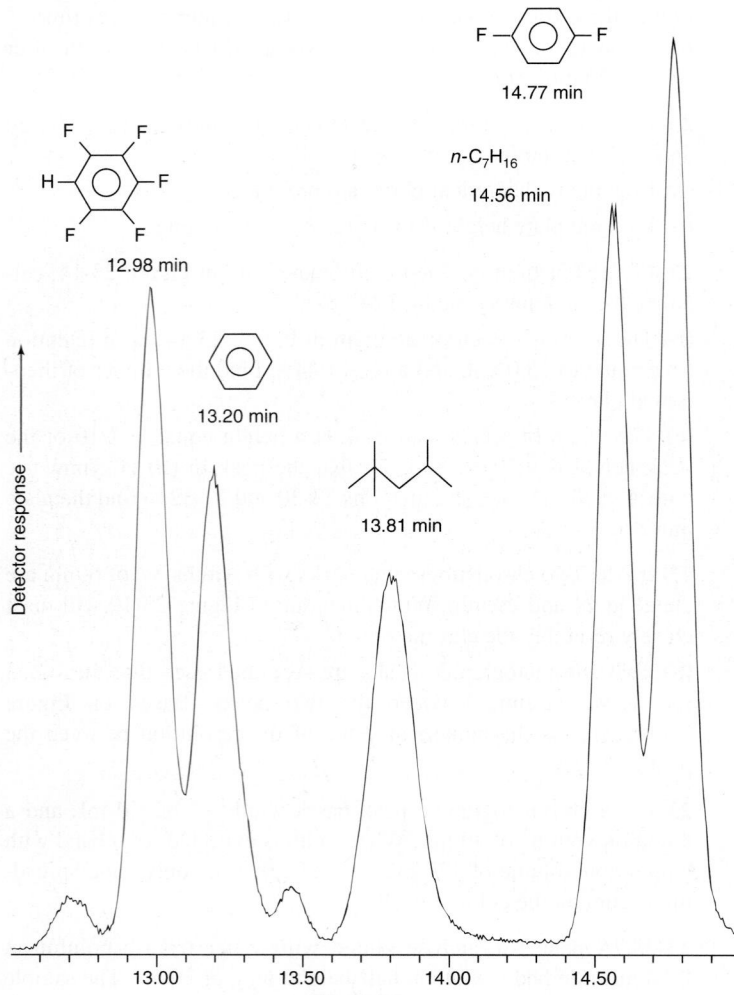

(a) Find the adjusted retention times and retention factors for both compounds.

(b) Find the relative retention, α.

(c) Measuring $w_{1/2}$ on the chromatogram, find the number of plates, N_1 and N_2, and the plate height for these two compounds.

(d) Measuring the width, w, at the baseline on the chromatogram, find the number of plates for these two compounds.

(e) Use your answer to **(d)** to find the resolution between the two peaks.

(f) Using the number of plates $N = \sqrt{N_1 N_2}$, with values from **(d)**, calculate what the resolution should be and compare your answer with the measured resolution in **(e)**.

23-52. A layer with negligible thickness containing 10.0 nmol of methanol ($D = 1.6 \times 10^{-9}$ m²/s) was placed in a tube of water 5.00 cm in diameter and allowed to spread by diffusion. Using Equation 23-27, prepare a graph showing the Gaussian concentration profile of the methanol zone after 1.00, 10.0, and 100 min. Prepare a second graph showing the same experiment with the enzyme ribonuclease ($D = 0.12 \times 10^{-9}$ m²/s).

23-53. A 0.25-mm-diameter open tubular gas chromatography column is coated with stationary phase that is 0.25 μm thick. The diffusion coefficient for a compound with a retention factor $k = 10$ is $D_m = 1.0 \times 10^{-5}$ m²/s in the gas phase and $D_s = 1.0 \times 10^{-9}$ m²/s in the stationary phase. Consider longitudinal diffusion and finite equilibration time in the mobile and stationary phases as sources of broadening. Prepare a graph showing the plate height from each of these three sources and the total plate height as a function of linear velocity (from 2 cm/s to 1 m/s). Then change the stationary phase thickness to 2 μm and repeat the calculations. Explain the difference in the two results.

23-54. Consider two Gaussian peaks with relative areas of 4:1. Construct a set of graphs to show the overlapping peaks if the resolution is 0.5, 1, or 2.

DOPING IN SPORTS

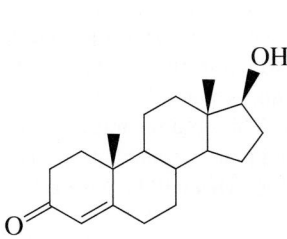

Testosterone

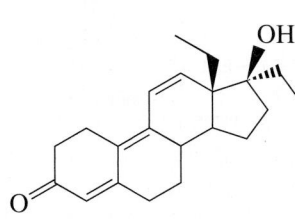

Tetrahydrogestrinone (THG)

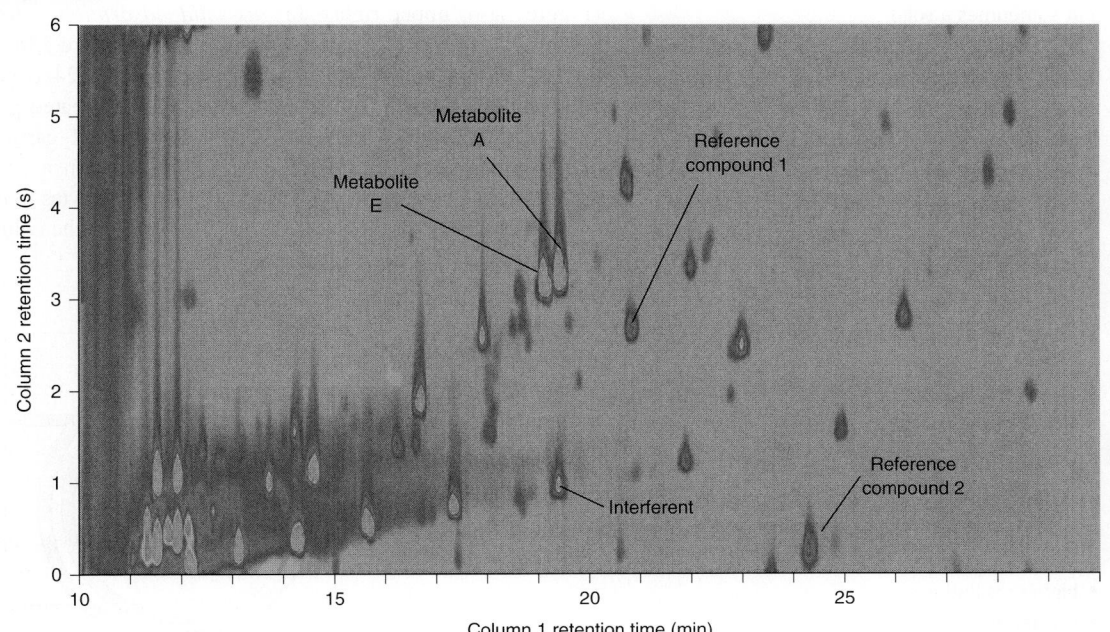

Two-dimensional gas chromatography–combustion isotope ratio mass spectrometry separation of athlete's urine showing signal for $^{12}CO_2$ (m/z 44) as a function of retention time on each of two columns. Highlighted peaks are testosterone metabolites and natural reference compounds. Column 1: 30 m × 0.25 mm with 1-μm film of 100% dimethylpolysiloxane. Column 2: 1.5 m × 0.1 mm with 0.1-μm film of (phenylpolysilphenylenesiloxane)$_{0.5}$(dimethylpolysiloxane)$_{0.5}$. [Information from H. J. Tobias, Y. Zhang, R. J. Auchus, and J. T. Brenna, "Detection of Synthetic Testosterone Use by Novel Comprehensive Two-Dimensional Gas Chromatography Combustion-Isotope Ratio Mass Spectrometry," *Anal. Chem.* **2011,** *83,* 7158, Figure 4A. Reprinted with permission © 2011, American Chemical Society.]

Doping is the unethical use of drugs to enhance performance in sports.[1] Steroids are the most common doping agents. Natural testosterone and synthetic anabolic steroids promote muscle development and repair. Steroid use adversely affects numerous organs. The World Anti-Doping Agency prohibits all anabolic steroids. Tetrahydrogestrinone is an example of a synthetic *designer steroid*, which is an untested, potent drug with unknown side effects.[2]

Urine provides a convenient, but complex, sample containing hundreds of natural substances. Performance enhancing drugs are present at very low levels. Measurement of steroids in urine by gas chromatography requires extensive, costly sample preparation to reduce the complexity of the sample.

Rigorous sample preparation can be relaxed with *two-dimensional gas chromatography*. In the example shown here, components are separated by 30-m-long column 1. Eluate is collected for 6-second intervals and then applied to 1.5-m-long column 2. The stationary phase in column 1 contains methyl groups and the stationary phase in column 2 is rich in benzene rings. Components eluted together in one 6-s interval from column 1, such as metabolite A and an interferent at 19.3 minutes, are separated on column 2 because the molecular retention forces are different. The two-dimensional chromatogram shows retention time for each column on one of the axes.

Products eluted from the second column are passed through a combustion chamber to convert every carbon atom to $^{12}CO_2$ or $^{13}CO_2$. An isotope ratio mass spectrometer measures the relative abundance of the two isotopes (Box 22-3). Synthetic testosterone contains less ^{13}C than natural testosterone due to differences in their biosynthetic pathways. A ^{13}C deficit of just 4.6‰ (parts per thousand) is beyond experimental uncertainty for confirming that synthetic steroids are present.

Chapter 23 gave a foundation for understanding chromatographic separations. Chapters 24 through 26 discuss specific methods and instrumentation. The goal is for you to understand how chromatographic methods work and what parameters you can control for best results.[3]

24-1 The Separation Process in Gas Chromatography

In **gas chromatography,**[4–6] gaseous analyte is transported through the column by a gaseous mobile phase, called the **carrier gas.** In *gas-liquid partition chromatography*, the stationary phase is a nonvolatile liquid bonded to the inside of the column or to a fine solid support (Figure 23-6, upper right). In *gas-solid adsorption chromatography*, analyte is adsorbed directly on solid particles of stationary phase (Figure 23-6, upper left).

In the schematic *gas chromatograph* in Figure 24-1, volatile liquid or gaseous sample is injected through a **septum** (a rubber disk) into a heated port, in which it rapidly evaporates. Vapor is swept through the column by He, N_2, or H_2 carrier gas, and separated analytes flow through a detector whose response is displayed on a computer. The column must be hot enough to provide sufficient vapor pressure for analytes to be eluted in a reasonable time. The detector is maintained at a higher temperature than the column so analytes will be gaseous.

Gas chromatography:

mobile phase: gas

stationary phase: usually a nonvolatile liquid, but sometimes a solid

analyte: gas or volatile liquid

The choice of carrier gas depends on the detector and the desired separation efficiency and speed.

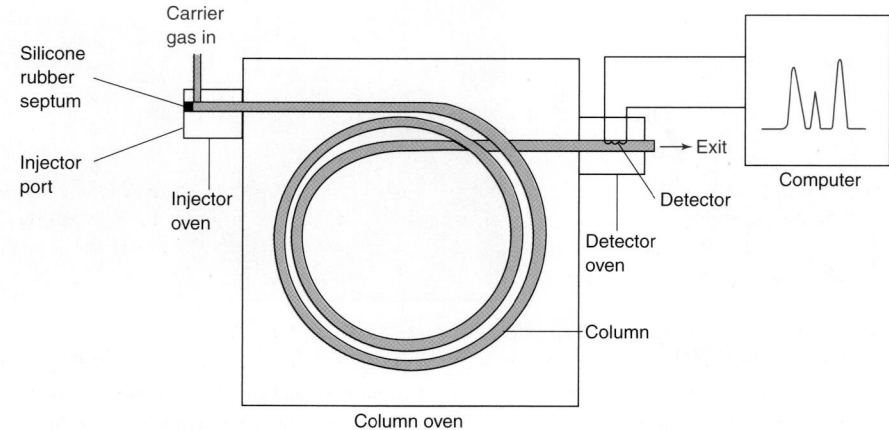

FIGURE 24-1 Schematic diagram of a gas chromatograph.[7]

Open Tubular Columns

The vast majority of analyses use long, narrow **open tubular columns** (Figure 24-2) made of fused silica (SiO_2) and coated with polyimide (a plastic capable of withstanding 350°C) for support and protection from atmospheric moisture.[8] As discussed in Section 23-5, open tubular columns offer higher resolution, shorter analysis time, and greater sensitivity than packed columns, but they have less sample capacity.

Compared with packed columns, open tubular columns offer
• higher resolution
• shorter analysis time
• greater sensitivity
• lower sample capacity

Wall-coated open tubular column: liquid stationary phase on inside wall of column

Porous-layer open tubular column: solid stationary phase particles on inside wall of column

Packed column: column filled with solid stationary phase particles

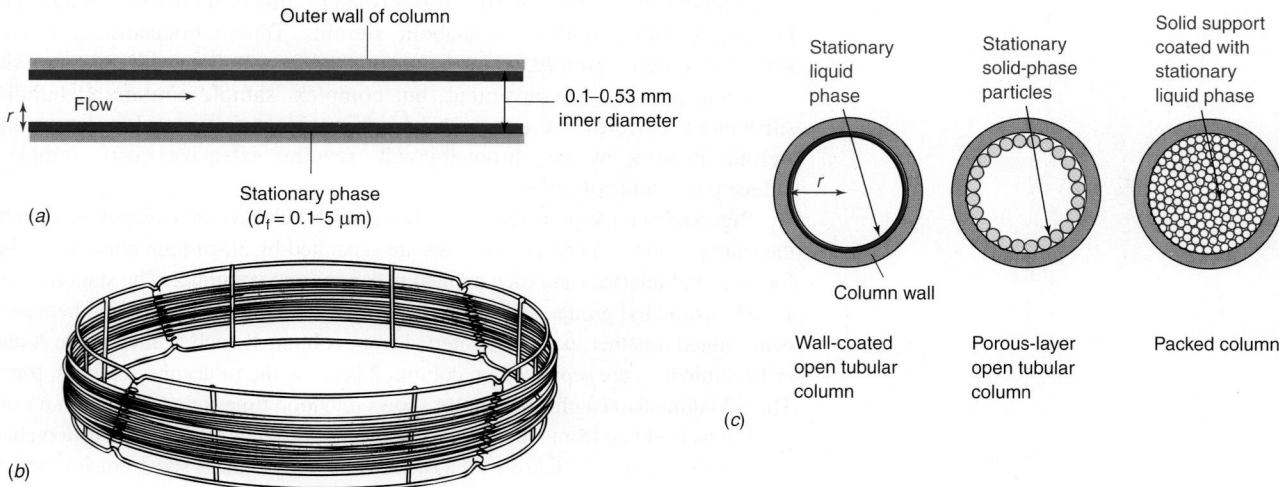

FIGURE 24-2 (*a*) Typical dimensions of open tubular gas chromatography column. (*b*) Fused-silica column with a cage diameter of 0.2 m and column length of 15–100 m. (c) Cross-sectional views of wall-coated, porous-layer, and packed columns.

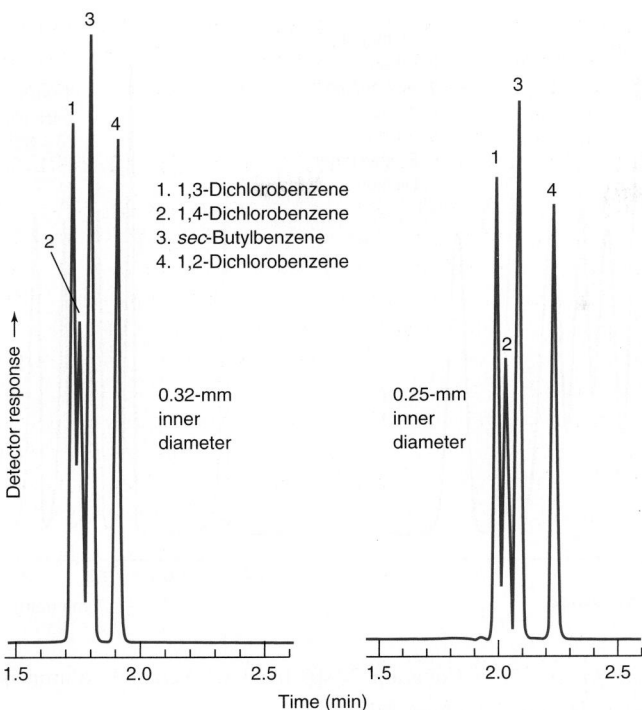

FIGURE 24-3 Effect of open tubular column inner diameter on resolution. Narrower columns provide higher resolution. Notice the increased resolution of peaks 1 and 2 in the narrow column. Conditions: DB-1 stationary phase (0.25 μm thick) in 15-m wall-coated column operated at 95°C with He linear velocity of 34 cm/s. [Data from J&W Scientific, Folsom, CA.]

The most common type of gas chromatography column is the *wall-coated* column in Figure 24-2c, with a 0.1- to 5-μm-thick film of high-molecular-weight stationary liquid phase coated on the inner wall. The less common porous-layer column has solid particles of stationary phase on the inner wall.

Column inner diameters are typically 0.10–0.53 mm and lengths are 15–100 m, with 30 m being common. Columns are coiled (Figure 24-2b) to fit within a compact temperature-controlled column oven. Narrow columns provide higher resolution than wider columns (Figure 24-3 and Equation 23-40), but require higher operating pressure and have less sample capacity. Diameters ≥0.32 mm tend to overload the vacuum system of a mass spectrometer, so the gas stream must be split and only a fraction sent to the spectrometer. The number of theoretical plates, N, on a column is proportional to length. In Equation 23-33, resolution is proportional to $\sqrt{N}$ and, therefore, to the square root of column length (Figure 24-4).

Analytes in gas chromatography may be below their boiling point, and so only partly in the vapor phase.

Equation 23-33:
$$\text{Resolution} = \frac{\sqrt{N}}{4} \frac{(\alpha - 1)}{\alpha} \left(\frac{k_2}{1 + k_2} \right)$$

N = plate number
α = relative retention
k_2 = retention factor of second peak in pair

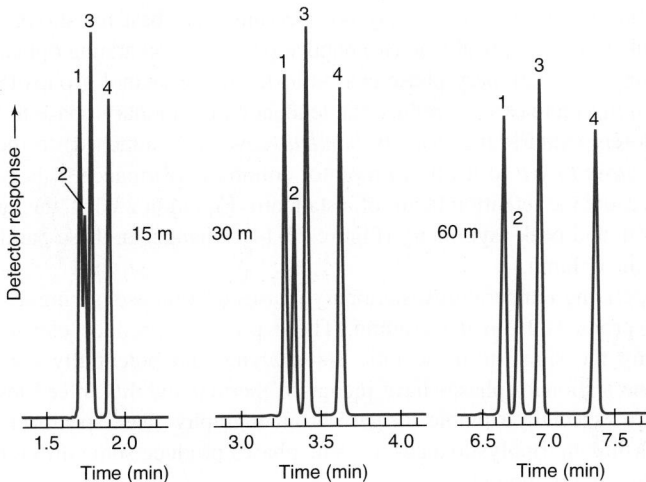

FIGURE 24-4 Resolution increases in proportion to the square root of column length. Notice the increased resolution of peaks 1 and 2 as length is increased. Conditions: DB-1 stationary phase (0.25 μm thick) in 0.32-mm-diameter wall-coated column operated at 95°C with He linear velocity of 34 cm/s. Compounds 1–4 are the same as in Figure 24-3. [Data from J&W Scientific, Folsom, CA.]

FIGURE 24-5 Effect of stationary phase thickness on open tubular column performance. Increasing thickness increases retention time and increases resolution of early-eluting peaks. Conditions: DB-1 stationary phase in wall-coated column (15 m × 0.32 mm) operated at 40°C with He linear velocity of 38 cm/s. [Data from J&W Scientific, Folsom, CA.]

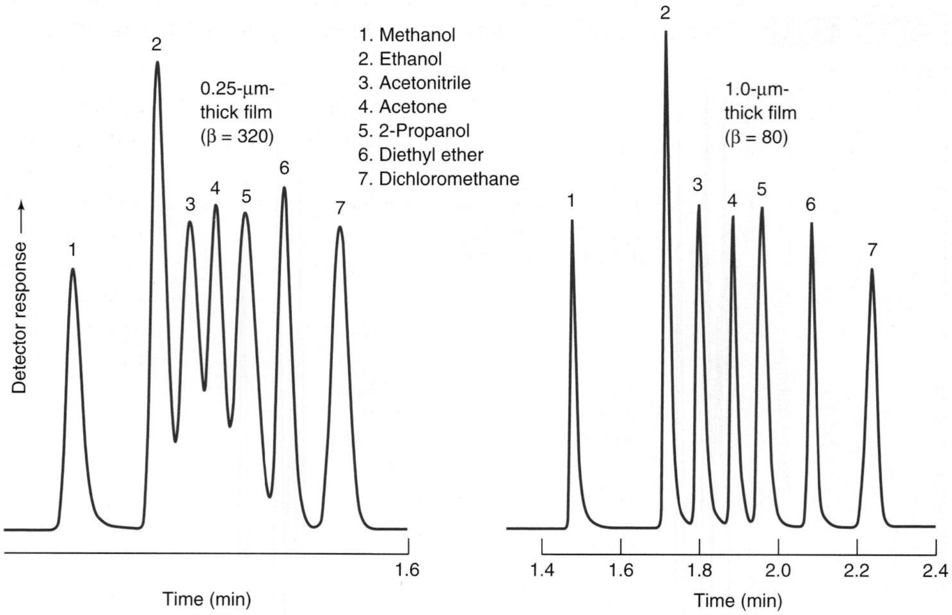

1. Methanol
2. Ethanol
3. Acetonitrile
4. Acetone
5. 2-Propanol
6. Diethyl ether
7. Dichloromethane

Retention factor: $k = \dfrac{t_r - t_m}{t_m}$

t_r = retention time of solute
t_m = transit time of mobile phase

Partition coefficient: $K = c_s/c_m$
c_s = solute concentration in stationary phase
c_m = solute concentration in mobile phase

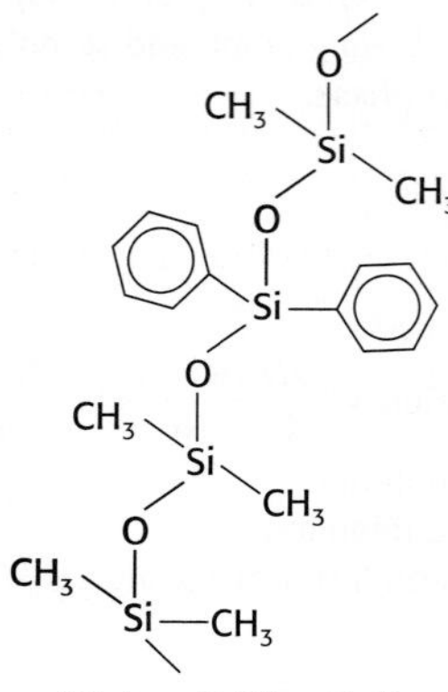

(Diphenyl)(dimethyl)-polysiloxane

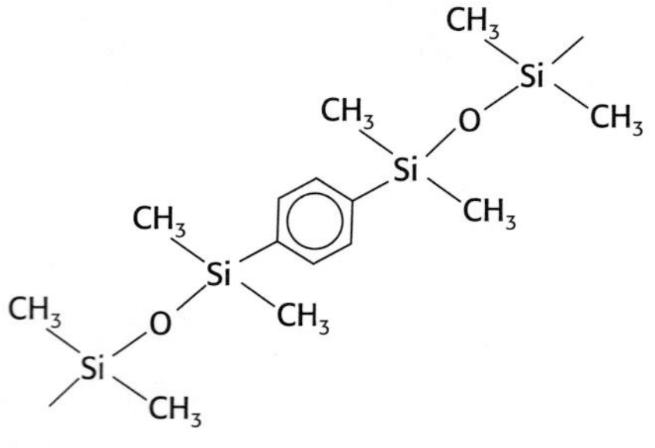

Arylene
polysiloxane

The *retention factor* (k) in Equation 23-16 for a wall-coated column is related to the *partition coefficient* (K) and the *phase ratio* (β):

$$k = \frac{K}{\beta} \tag{24-1}$$

The dimensionless **phase ratio** is the volume of the mobile phase divided by the volume of the stationary phase. For typical column dimensions, the phase ratio is

Phase ratio:
$$\beta = \frac{r}{2d_f} \tag{24-2}$$

where r is the column radius and d_f is the stationary phase film thickness in Figure 24-2a. Increasing the thickness of the stationary phase decreases β, which increases retention time and sample capacity in Figure 24-5 when the linear velocity of the mobile phase is constant. Increased retention factor increases resolution of early-eluting peaks ($k \lesssim 5$) (Equation 23-33). Thick films of stationary phase can shield analytes from the silica surface and reduce *tailing* (Figure 23-21), but they can also increase bleed (decomposition and evaporation) of the stationary phase at elevated temperature. A thickness of 0.25 μm is standard, but thicker films are used for volatile analytes.

The choice of liquid stationary phase (Table 24-1) is based on the rule "like dissolves like." Nonpolar columns are best for nonpolar solutes. Columns of intermediate polarity are best for intermediate polarity solutes, and strongly polar columns are best for strongly polar solutes. Box 24-1 describes *chiral* (optically active) bonded phases for separating optical isomers.

As a column ages, stationary phase can be lost, surface silanol groups (Si—O—H) are exposed, and tailing increases. To reduce the tendency of stationary phase to bleed from the column at high temperature, it is usually *bonded* (covalently attached) to the silica surface and covalently *cross-linked* to itself. To monitor column performance, it is good practice to periodically measure the retention factor of a standard (Equation 23-16), the number of plates (Equation 23-30), and peak asymmetry (Figure 23-14). Changes in these parameters indicate degradation of the column.

At upper operating temperatures, stationary phases decompose, giving a slow "bleed" of decomposition products from the column. These products produce elevated background signals, reducing the signal-to-noise ratio for analytes and potentially contaminating the detector. Arylene stationary phases have increased thermal stability, bleed less at high temperature, and are especially suitable for gas chromatography–mass spectrometry. Compared with (diphenyl)(dimethyl)polysiloxanes, arylene phases produce some differences in relative retention of different compounds.

To reduce interference from column bleeding, use the thinnest possible stationary phase and the narrowest and shortest column that provides adequate separation. Oxidation of the stationary phase by O_2 is also a major source of bleed. High-purity carrier gas should be used, and it should be passed through an O_2 scrubber before the column. Even 1 ppb of O_2 slowly

TABLE 24-1 Common stationary phases in capillary gas chromatography

Structure	Polarity[a]	Common applications	Temperature range[b]
(Diphenyl)$_x$(dimethyl)$_{1-x}$ polysiloxane	$x = 0$ Nonpolar van der Waals forces	Solvents, petroleum products, waxes, simulated distillation	$-60°–330°/350°C$
	$x = 0.05$ Nonpolar van der Waals forces	Flavors, environmental aromatic hydrocarbons	$-60°–330°/350°C$
	$x = 0.35$ Intermediate van der Waals forces	Pesticides, polychlorinated biphenyls, amines, nitrogen-containing herbicides	$40°–310°C$
	$x = 0.65$ Intermediate van der Waals forces	Triglycerides, phenols, free fatty acids	$50°–280°/300°C$
(Cyanopropylphenyl)$_{0.14}$ (dimethyl)$_{0.86}$ polysiloxane	Intermediate Dipole-dipole and van der Waals forces	Pesticides, polychlorinated biphenyls, alcohols, oxygenates	$-20°–270°/280°C$
—[CH$_2$CH$_2$—O]$_n$— Carbowax poly(ethylene glycol)	Strongly polar Dipole-dipole, hydrogen bonding, and van der Waals forces	Fatty acid methyl esters, flavors, amines, solvents, xylene isomers	$40°–250°/260°C$
(Biscyanopropyl)$_{0.9}$ (cyanopropylphenyl)$_{0.1}$ polysiloxane	Strongly polar Dipole-dipole and van der Waals forces	cis/trans-Fatty acid methyl esters, dioxin isomers	$0°–260°/275°C$

a. Dominant retention forces are shown beneath polarity. All molecules experience van der Waals forces, which are (i) attraction between an instantaneous dipole and a dipole induced in a neighbor and (ii) attraction between a permanent dipole and the dipole induced in a neighbor. Dipole-dipole force is attraction between permanent dipoles. Hydrogen bonding is attraction caused by partial sharing of electrons from an electronegative atom such as O with a positively polarized H-atom on a neighboring molecule.

b. Lower limit is the temperature at which the liquid stationary phase freezes. Two upper limits are given. The first is the isothermal temperature at which the column can routinely operate. The second is the maximum programmed temperature to which the column can only be exposed for short periods of time.

Data from Restek Chromatography Products Catalog, 2013–2014, Bellefonte, PA.

degrades the column. To a lesser extent, H_2O can break down the stationary phase by hydrolysis. To minimize bleed, manufacturers modify the silica surface of the capillary to eliminate exposed silanol groups (Si—OH), which can initiate breakdown of the stationary phase.

Ionic liquids are the newest type of stationary phase for gas chromatography.[14] They melt below room temperature and have a wide liquid range with low volatility at elevated temperature. Ionic liquids provide multiple types of solvation interactions. Thus they offer novel selectivities for polar analytes and increased operating temperature with low bleed. The upper phase in Color Plate 30 is an ionic liquid.

Ionic liquid

1,9-Di(3-vinylimidazolium)nonane bis(trifluoromethyl)sulfonylimidate
(Supelco SP-IL 100 stationary phase)

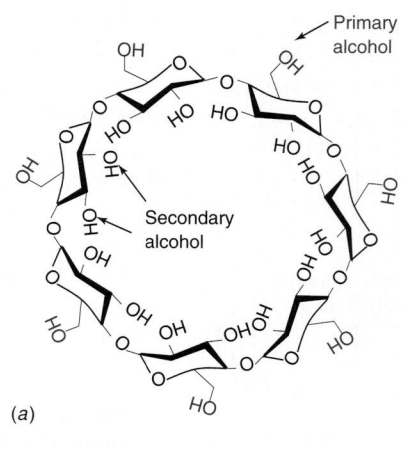
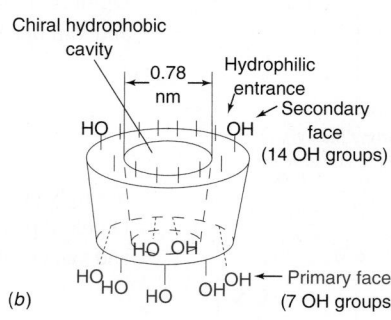
Wall-coated columns do not retain permanent gases (such as O_2, N_2, and CH_4) or low-boiling compounds (such as natural gas) sufficiently. *Porous-layer* columns in Figure 24-2c have high-surface-area porous solid particles adhering to the column wall. The highly retentive surface of these particles is the active stationary phase. **Molecular sieves** (Figure 24-6) are inorganic or organic materials with cavities into which small molecules enter and are partially retained.[15] Molecules such as He, Ar, O_2, N_2, CH_4, and CO can be separated from one another (Figure 24-7). Porous polymers, high-surface-area carbon, and *alumina* (Al_2O_3) can separate hydrocarbons in gas-solid adsorption chromatography.

Gases can be dried by passage through traps containing molecular sieves because water is strongly retained. Inorganic sieves can be regenerated (dried) by heating to 300°C in vacuum or under flowing N_2.

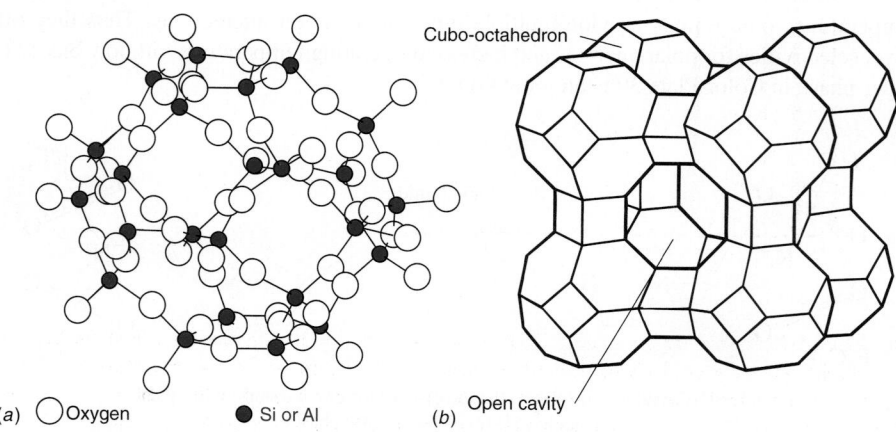

FIGURE 24-6 Structure of the molecular sieve, $Na_{12}(Al_{12}Si_{12}O_{48}) \cdot 27H_2O$.
(*a*) Aluminosilicate framework of one cubo-octahedron of a mineral class called *zeolites*.
(*b*) Interconnection of eight cubo-octahedra to produce a cavity into which small molecules can enter.

Enantiomers have different affinities for the cyclodextrin cavity, so they separate as they travel through the chromatography column. The chromatogram shows the separation of the two enantiomers of each of three polychlorinated biphenyls (PCB).

Production of polychlorinated biphenyls has been banned because they are probable human carcinogens, but they remain an environmental concern because they persist in the environment. When originally produced, the polychlorinated biphenyls were 1:1 (*racemic*) mixtures of the two enantiomers. The unequal mixture in the chromatogram is evidence that plants metabolize the two enantiomers at different rates.

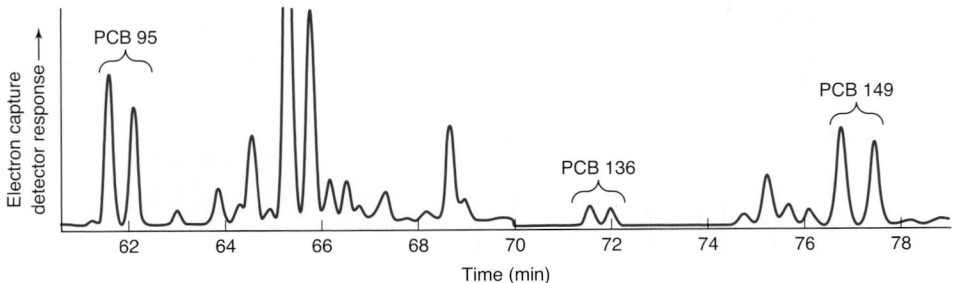

Chiral separation of polychlorinated biphenyl compounds from eucalyptus leaves using a 25 m × 0.25 mm open tubular column with a 0.25-μm-thick stationary phase containing methylated β-cyclodextrin chemically bonded to poly(dimethylsiloxane). [Data from S.-J. Chen, M. Tian, J. Zheng, Z.-C. Zhu, Y. Luo, X.-J. Luo, and B.-X. Mai, "Elevated Levels of Polychlorinated Biphenyls in Plants, Air, and Soils at an E–Waste Site in Southern China and Enantioselective Biotransformation of Chiral PCBs in Plants," *Env. Sci. Tech.* **2014**, *48*, 3847.]

Chiral polychlorinated biphenyl PCB 136. The two rings are perpendicular to each other. The mirror images are not superimposable because they cannot freely rotate about the C—C bond between the rings.

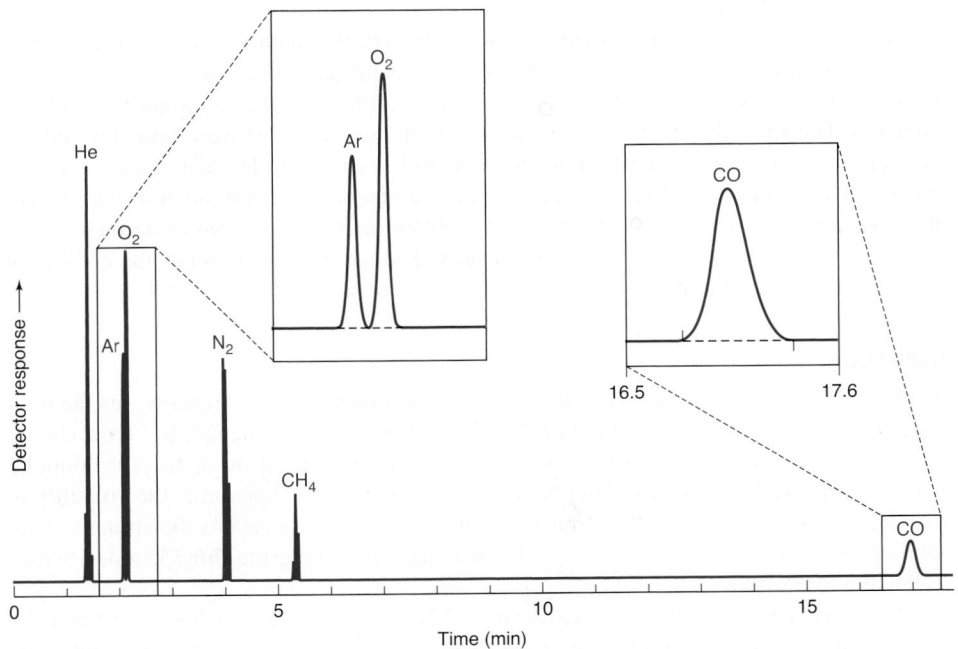

Gas-solid chromatography: Analyte is adsorbed directly on solid particles of stationary phase

Gas-liquid chromatography: Analyte partitions into a thin film of nonvolatile liquid on the stationary phase

FIGURE 24-7 Permanent gas analysis obtained with a 30 m × 0.53 mm column coated with a 50-μm-thick porous layer of MXT-MSieve 5A operated at 30°C; H_2 carrier gas with microthermal conductivity detection. [Data from J. de Zeeuw, "The Development and Applications of PLOT Columns in Gas-Solid Chromatography," *LCGC North Am.* **2010**, *28*, 848. Courtesy of Restek Corp., Middelburg, Netherlands.]

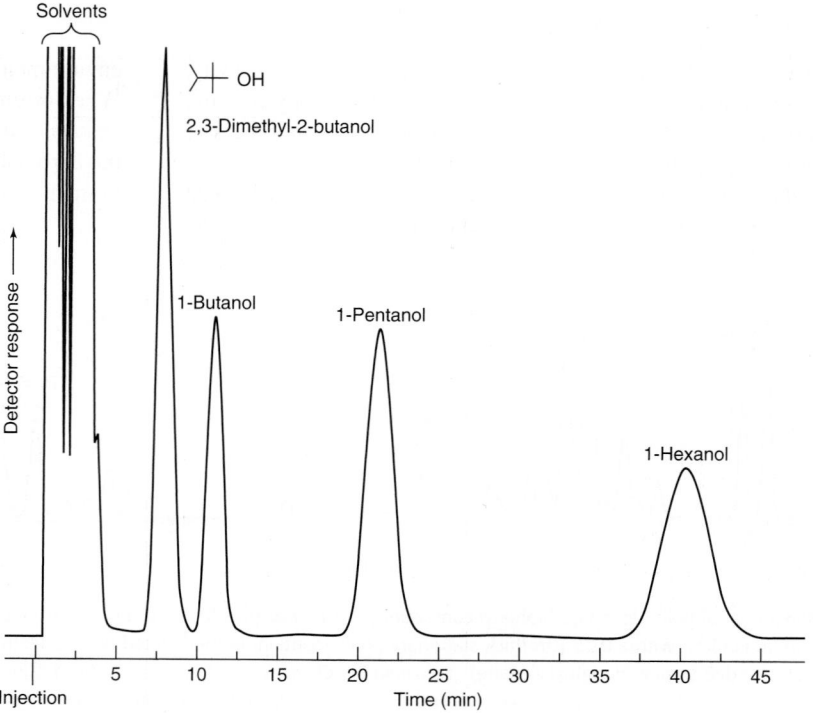

Teflon is a chemically inert polymer with the structure —CF_2—CF_2—CF_2—CF_2—.

Packed Columns

Packed columns (Figure 24-2c) are filled with fine particles of solid support coated with a nonvolatile liquid stationary phase, or the solid itself may be the stationary phase. Compared with open tubular columns, packed columns provide greater sample capacity but give broader peaks, longer retention times, and less resolution (Figure 24-8). Despite their inferior resolution, packed columns are used for preparative separations, which require a great deal of stationary phase, or to separate gases that are poorly retained. Packed columns are usually made of stainless steel or glass and are typically 1–5 m in length and 3–6 mm in diameter. The solid support is often silica that is *silanized* (Reaction 23-42) to reduce hydrogen bonding to polar solutes. For tenaciously binding solutes, Teflon is a useful support, but it is limited to <200°C.

In a packed column, uniform particle size decreases the multiple path term in the van Deemter equation (23-37), thereby reducing plate height and increasing resolution. Small particle size decreases the time required for solute equilibration, thereby improving column efficiency. However, the smaller the particle size, the less space between particles and the more pressure required to force mobile phase through the column. Particle size is expressed in micrometers or as a *mesh size*, which refers to the size of screens through which the particles are passed or retained (Table 28-2). A 100/200 mesh particle passes through a 100 mesh screen, but not through a 200 mesh screen. The mesh number equals the number of openings per linear inch of screen.

Retention

Figure 24-9 illustrates how the retention of polar and nonpolar solutes changes as the boiling point of the solute (T_{bp}) and polarity of the stationary phase change. In Figure 24-9a, 10 compounds are eluted nearly in order of increasing boiling point from a nonpolar stationary phase. The principal determinant of retention on this column is the volatility of the solutes. In Figure 24-9b, the strongly polar stationary phase retains the strongly polar solutes. The three alcohols are the last to be eluted, following the three ketones, which follow the four alkanes.

The superscript ° in ΔH°_{vap} means that reactants and products are in their standard states, which are pure liquid, gas at 1 bar, and solute with an activity of unity.

Retention is governed by thermodynamics.[16] The retention factor (k) depends on both the enthalpy of vaporization (ΔH°_{vap}) and the enthalpy of mixing (ΔH°_{mix}) of pure solute with the stationary phase liquid.[17]

$$\ln k = \frac{\Delta H^{\circ}_{vap}}{RT} + \frac{\Delta H^{\circ}_{mix}}{RT} + \text{constant} \qquad (24\text{-}3)$$

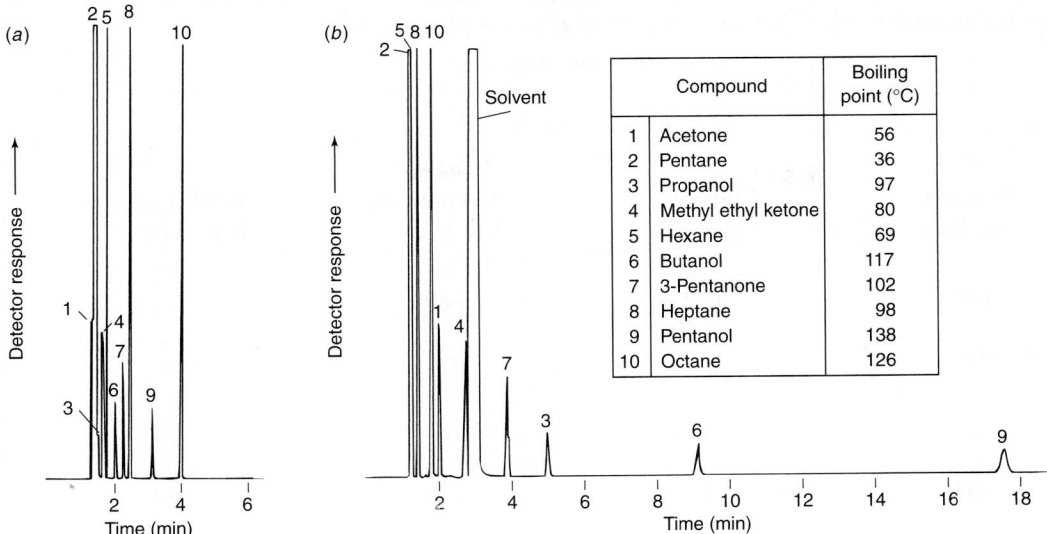

FIGURE 24-9 Separation of 10 compounds on (a) nonpolar poly(dimethylsiloxane) and (b) strongly polar poly(ethylene glycol) 1-μm-thick stationary phases (phase ratio β = 80) in 30 m × 0.32 mm open tubular columns at 70°C. [Data from Restek Co., Bellefonte, PA.]

For non-hydrogen-bonding solutes, Trouton's rule states that the heat of vaporization is proportional to the boiling temperature:[18]

$$\Delta H_{vap}^{\circ} \approx (88 \text{ J mol}^{-1} \text{ K}^{-1}) \cdot T_{bp} \qquad (24\text{-}4)$$

Combining Equations 24-3 and 24-4 yields the approximation

$$\ln k \approx \frac{(88 \text{ J mol}^{-1} \text{ K}^{-1})T_{bp}}{RT} + \frac{\Delta H_{mix}^{\circ}}{RT} + \text{constant} \qquad (24\text{-}5)$$

In Figure 24-9 retention increases with boiling point (T_{bp}) for compounds of the same chemical class (homologous series). Greater retention also results from "like dissolves like," which increases the enthalpy of mixing. Therefore, alcohols are more retained relative to alkanes on the polar poly(ethylene glycol) phase (Figure 24-9b). Relative retention on a polar phase depends on the nature of the polar interactions (Table 24-1). Hydrogen bonding to the polar stationary phase in Figure 24-9b is probably the strongest force leading to retention. Dipole interactions of the ketones are the second strongest force.

The Kovats **retention index**, I, for a linear alkane equals 100 times the number of carbon atoms. For octane, $I \equiv 800$; and for nonane, $I \equiv 900$. A compound eluted between octane and nonane (Figure 23-7) has a retention index between 800 and 900 computed by the formula

Retention index: $\qquad I = 100\left[n + (N - n)\dfrac{\log t_r' (\text{unknown}) - \log t_r' (n)}{\log t_r' (N) - \log t_r' (n)}\right] \qquad (24\text{-}6)$

where n is the number of carbon atoms in the *smaller* alkane; N is the number of carbon atoms in the *larger* alkane; $t_r'(n)$ is the adjusted retention time of the *smaller* alkane; and $t_r'(N)$ is the adjusted retention time of the *larger* alkane.

EXAMPLE Retention Index

If the retention times in Figure 23-7 are $t_r(CH_4) = 0.5$ min, $t_r(\text{octane}) = 14.3$ min, $t_r(\text{unknown}) = 15.7$ min, and $t_r(\text{nonane}) = 18.5$ min, find the retention index for the unknown.

Solution The index is computed with Equation 24-6:

$$I = 100\left[8 + (9 - 8)\frac{\log 15.2 - \log 13.8}{\log 18.0 - \log 13.8}\right] = 836$$

TEST YOURSELF Where would an unknown with a retention index of 936 be eluted in Figure 23-7? (**Answer:** after nonane)

Trouton discovered his rule while an undergraduate as a "result of an afternoon playing with numbers."

Retention index relates the retention time of a solute to the retention times of linear alkanes.

Adjusted retention time: $t_r' = t_r - t_m$
t_r = retention time for solute
t_m = time for unretained solute (CH_4) to pass through column

Retention index[a]

Phase	Benzene b.p. 80°C	Butanol b.p. 117°C	2-Pentanone b.p. 102°C	1-Nitropropane b.p. 132°C
Poly(dimethylsiloxane)	651	651	667	705
(Diphenyl)$_{0.05}$(dimethyl)$_{0.95}$–polysiloxane	667	667	689	743
(Diphenyl)$_{0.35}$ (dimethyl)$_{0.65}$–polysiloxane	746	733	773	867
(Cyanopropylphenyl)$_{0.14}$–(dimethyl)$_{0.86}$ polysiloxane	721	778	784	881
(Diphenyl)$_{0.65}$(dimethyl)$_{0.35}$–polysiloxane	794	779	825	938
Poly(ethylene glycol)	963	1 158	998	1 230

a. For reference, boiling points (b.p.) for various alkanes are: hexane, 69°C; heptane, 98°C; octane, 126°C; nonane, 151°C; decane, 174°C; undecane, 196°C; dodecane, 216°C. Retention indexes for the straight-chain alkanes are fixed values and do not vary with the stationary phase: hexane, 600; heptane, 700; octane, 800; nonane, 900; decane, 1 000; undecane, 1 100; dodecane, 1 200. Benzene accounts for dipole-induced dipole interactions; butanol for hydrogen bond donating and accepting; 2-pentanone and nitropropane for dipole-dipole interactions.

Data from Restek Chromatography Products Catalog, 2013–2014, Bellefonte, PA.

When identifying an eluted compound by comparing its mass spectrum with a mass spectral library, false matches are frequent. If retention index is used as a second characteristic, false matches are reduced.

The retention index of 651 for benzene on poly(dimethylsiloxane) in Table 24-2 means that benzene is eluted between hexane ($I \equiv 600$) and heptane ($I \equiv 700$) from this nonpolar stationary phase. Nitropropane is eluted just after heptane on the same column. As we go down the table, the stationary phases become more polar. For poly(ethylene glycol) at the bottom of the table, benzene is eluted after nonane, and nitropropane is eluted after dodecane. The average of the retention index of the probe compounds in Table 24-2 defines the overall polarity of the phases, but selectivity will depend on the types of polar interaction offered by each column. Both (diphenyl)$_{0.35}$(dimethyl)$_{0.65}$ polysiloxane and (cyanopropylphenyl)$_{0.14}$(dimethyl)$_{0.86}$ polysiloxane are intermediate polarity columns. Butanol shows greater relative retention on the cyanopropylphenyl phase due to the dipolar nature of this phase.

Temperature and Pressure Programming

A large fraction of all gas chromatography is run with **temperature programming,** in which the temperature of the column is raised *during* the separation to increase analyte vapor pressure and decrease retention times of late-eluting components. In gas chromatography, solutes equilibrate between the stationary and vapor phases. The vapor pressure of a solute depends on the temperature (T) and the heat of vaporization (ΔH_{vap}) of the solute:

Clausius-Clapeyron equation:

$$\log P_{solute} = -\frac{\Delta H_{vap}}{RT} + \text{constant} \tag{24-7}$$

Raising column temperature
- decreases retention time
- sharpens peaks

At a constant (*isothermal*) temperature of 150°C in Figure 24-10, the column is above the boiling temperatures of the small alkanes, and they are entirely in the vapor phase and so weakly retained that they emerge close together. Retention increases dramatically with the boiling point of compounds (Equation 24-5). Less volatile compounds may not even be eluted from the column. Figure 24-10a illustrates the **general elution problem,** which is the inability of a single isothermal separation to provide adequate separation within a reasonable run time for samples containing compounds of widely different boiling points.

If the column temperature is increased from 50°C to 250°C at a rate of 8°C/min (Figure 24-10b), all compounds are eluted and the separation of peaks is fairly uniform. The initial temperature (50°C) is far below the boiling temperature of most compounds, and so they are strongly retained (*trapped*) at the head of the column. For much of the temperature program, each compound's vapor pressure remains low and the compound remains at the head of the column. Once the column temperature is high enough for a compound to have appreciable vapor pressure, it starts moving down the column. As it moves, the temperature continues to increase and the compound becomes less and less retained, finally emerging as a sharp peak.

With **programmed temperature,** solutes are
- strongly retained (trapped) at column head for most of their time on column
- weakly retained when eluted due to the higher temperature, and so are sharp peaks

CHAPTER 24 Gas Chromatography

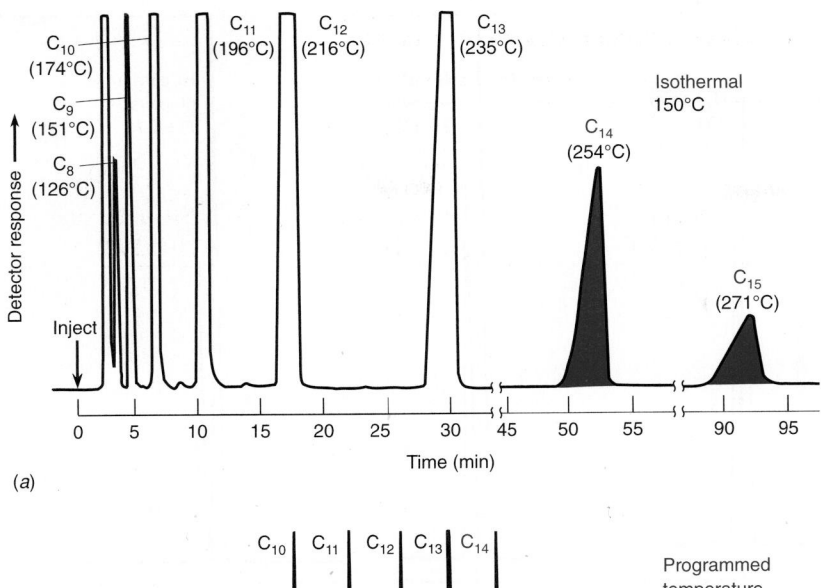

(a)

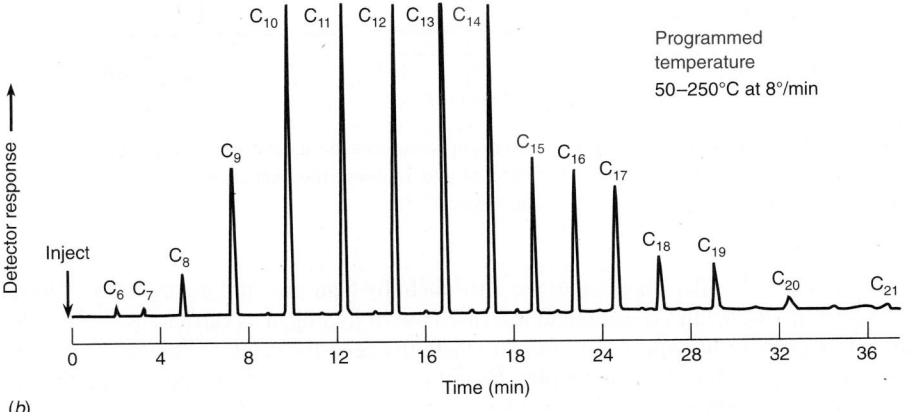

(b)

FIGURE 24-10 Comparison of (a) isothermal (constant temperature) and (b) programmed temperature chromatography. Each sample contains linear alkanes run on a 6 m × 1.6 mm column packed with 3% Apiezon L (nonpolar liquid phase) on 100/200 mesh VarAport 30 solid support with He flow rate of 10 mL/min. Boiling temperature is shown for each alkane in (a). Detector sensitivity is 16 times greater in (a) than in (b). [Data from H. M. McNair and E. J. Bonelli, *Basic Gas Chromatography* (Palo Alto, CA; Varian Instrument Division, 1968).]

Most gas chromatography columns come with a label showing their temperature limits (Table 24-1). The lower limit is the temperature at which the liquid phase solidifies. Operating the column below this minimum temperature will not do harm, but poor peak shape and other performance problems might occur. Two maximum temperature limits may be shown. The lower one is the isothermal temperature limit at which the column can be kept for a long time. The upper one is the programmed temperature limit to which the column should only be exposed for a few minutes at a time at the end of a programmed temperature run. High temperatures decompose the stationary phase and cause column "bleeding." An increase in baseline signal at low temperature is an indicator of column degradation. Other signs of column degradation are peak broadening, tailing, and changing retention times.

Many chromatographs are equipped with electronic pressure control of the carrier gas. Increasing the inlet pressure increases the flow of mobile phase and decreases retention time. In some cases, programmed pressure can be used instead of programmed temperature to reduce retention times of late-eluting components. At the end of a run, the pressure can be rapidly reduced to its initial value for the next run. Time is not wasted waiting for a hot column to cool before the next injection. Programmed pressure is useful for analytes that cannot tolerate high temperature.

Carrier Gas

Helium is the most common carrier gas and is compatible with most detectors. For a flame ionization detector, N_2 gives a lower detection limit than He. Figure 24-11 shows that H_2, He, and N_2 give essentially the same optimal plate height (0.3 mm) at significantly different velocities. Optimal velocity increases in the order $N_2 < He < H_2$. Fastest separations can be achieved with H_2 as carrier gas, and H_2 can be run much faster than its optimal velocity with little penalty in resolution.[19] Figure 24-12 shows the effect of carrier gas on the separation of two compounds on the same column with the same temperature program.

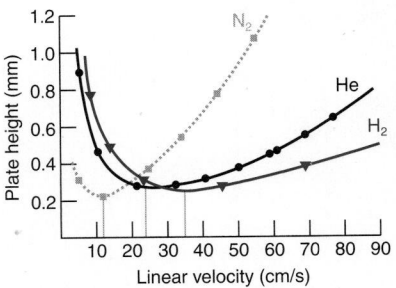

FIGURE 24-11 van Deemter curves for gas chromatography of n-$C_{17}H_{36}$ at 175°C, using N_2, He, or H_2 in a 25 m × 0.25 mm wall-coated column with nonpolar OV-101 stationary phase. [Data from R. R. Freeman, ed., *High Resolution Gas Chromatography* (Palo Alto, CA: Hewlett Packard Co., 1981).]

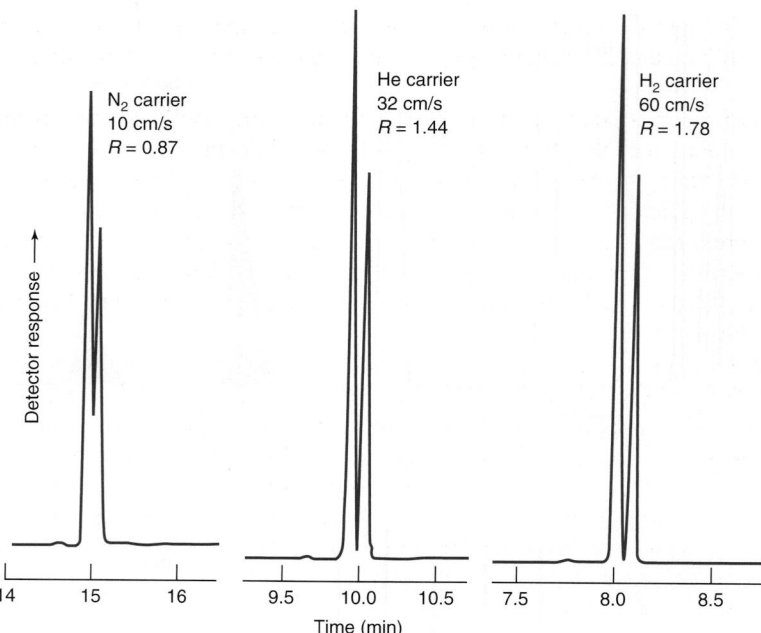

FIGURE 24-12 Separation of two polyaromatic hydrocarbons on a wall-coated open tubular column with different carrier gases. Resolution, *R*, increases and analysis time decreases as we change from N_2 to He to H_2 carrier gas. [Data from J&W Scientific, Folsom, CA.]

Gas cylinders are heavy and at high pressure, and must always be handled with care.[20]

van Deemter equation:

$$H \approx A + \frac{B}{u_x} + Cu_x$$

| Multiple paths | Longitudinal diffusion | Equilibration time |

Gases are most often supplied to the instrument by high-pressure gas cylinders. Decreasing availability of helium is leading to increased use of hydrogen as carrier gas.[21] Electrolytic generators produce high-purity H_2 and eliminate the need for tanks of compressed H_2. Low flow rates in capillary chromatography are unlikely to create a dangerous concentration of H_2. For gas chromatography–mass spectrometry, H_2 reduces the efficiency of a turbomolecular vacuum pump but has little effect on a diffusion pump. It is possible for H_2 to react with unsaturated compounds on metal surfaces.

H_2 and He give better resolution (smaller plate height) than N_2 at high flow rate because solutes diffuse more rapidly through H_2 and He than through N_2. The more rapidly a solute diffuses between phases, the smaller is the mass transfer (Cu_x) term in the van Deemter equation (23-37). Equations 23-40 and 23-41 describe the effects of the finite rate of mass transfer in an open tubular column. If the stationary phase is thin enough ($\lesssim 0.5\mu$m), mass transfer is dominated by slow diffusion through the *mobile phase* rather than through the *stationary phase*. That is, $C_s \ll C_m$ in Equations 23-40 and 23-41. For a column of a given radius, *r*, and a solute of a given retention factor, *k*, the only variable affecting the rate of mass transfer in the mobile phase (Equation 23-40) is the diffusion coefficient of solute through the mobile phase. Diffusion coefficients follow the order $H_2 > He > N_2$.

Most analyses are run at carrier gas velocities that are 1.5–2 times greater than the optimum velocity at the minimum of the van Deemter curve. The higher velocity is chosen to give maximum efficiency (most theoretical plates) per unit time. A decrease in resolution is tolerated in return for faster analyses.

Gas flow through a narrow column may be too low for best detector performance, so extra *makeup gas* is sometimes added between the column and the detector. Makeup gas that is optimum for detection can be a different gas from that used in the column.

Impurities in carrier gas degrade the stationary phase. High-quality gas should be used, and it should be passed through purifiers to remove traces of O_2, H_2O, and organic compounds. An oxygen indicator trap should be in line after the main purifier. Steel or copper tubing, rather than plastic or rubber, should be used for gas lines because metals are less permeable to air and do not release volatile contaminants into the gas stream. As with thermal degradation, symptoms of oxidative degradation of the stationary phase include increased baseline signal at low temperature, peak broadening and tailing, and altered retention times.

Guard Columns and Retention Gaps

In gas chromatography, a *guard column* and a *retention gap* are both typically a 3- to 10-m length of empty capillary in front of the capillary chromatography column. The capillary is silanized so that solutes are not retained by the bare silica wall. Physically, the guard column and the retention gap are identical, but they are employed for different purposes.

The purpose of a **guard column** is to accumulate nonvolatile substances that would otherwise contaminate the chromatography column and degrade its performance. Periodically, the beginning of the guard column should be cut off to eliminate nonvolatile residues. Trim the guard column when you observe irregular peak shapes from a column that had been producing symmetric peaks. It is a good idea to trim 10–20 cm off the guard column every time the injection inlet liner is changed. With electronic pneumatically controlled chromatographs, be sure to enter the new length of the guard column in the control software.

A **retention gap** is used to improve peak shapes under certain conditions. If you introduce a large volume of sample (>2 μL) by *splitless* or *on-column* injection (described in the next section), microdroplets of liquid solvent can persist inside the column for the first few meters. Solutes dissolved in the droplets are carried along with the droplets and give rise to a series of ragged bands. The retention gap allows solvent to evaporate prior to entering the chromatography column. Use at least 1 m of retention gap per microliter of solvent. Even small volumes of solvent with very different polarity from the stationary phase can cause irregular peak shapes. The retention gap helps separate solvent from solute to improve peak shapes.

We calculate plate number, N, with Equation 23-30 by using retention time and peak width. Plate height, H, is the length of the column, L, divided by N. Do not include the retention gap or guard column as part of L for calculating H.[22] For peaks with a retention factor $k < 5$, plate height might not be meaningful when a guard column or retention gap is used.

24-2 Sample Injection[23]

Liquid volumes of ~1 μL are injected through a rubber septum into a heated glass port. Gas volumes of ~10 μL up to 5 mL can be injected with a gas-tight syringe or introduced via a sample injection valve similar to that shown for liquid chromatography in Figure 25-20.

A 10-μL microsyringe is used to inject 0.1–2 μL of liquid.[24] To fill the microsyringe, first exclude all air by repeatedly drawing liquid into the syringe and rapidly expelling it back into the liquid. Then slowly draw up more liquid than will be injected. Hold the syringe with the needle up and depress the plunger until it reads the desired volume. Wipe the needle with a tissue and draw some air into the syringe, as shown in Figure 24-13. When the needle is inserted through the rubber septum into the heated injection port of the chromatograph, sample does not immediately evaporate because there is no sample in the needle. If there were sample in the needle, the most volatile components would begin to evaporate and would be depleted before the sample is injected.

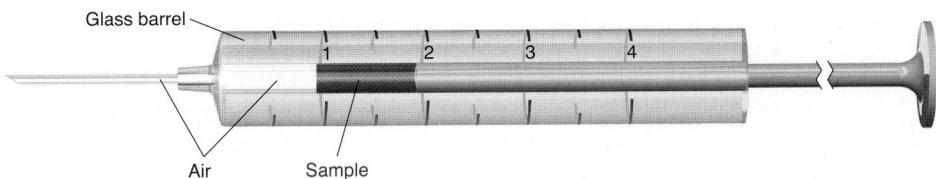

Glass barrel

Air   Sample

FIGURE 24-13 Microsyringe filled with 1 μL of liquid sample. Air is then drawn into the barrel to avoid premature evaporation of sample in the injection port.

An injection port with a silanized glass liner is shown in Figure 24-14. Carrier gas sweeps vaporized sample from the port into the chromatography column. Decomposed sample, nonvolatile components, and septum debris accumulate in the glass liner, which is periodically replaced. Glass wool is used near the bottom of some liners to trap particles and pyrolysis products to prevent them from reaching the column. The liner must seal properly or carrier gas will bypass the liner. The lifetime of the rubber septum could be as little as 20 manual injections or ~100 autosampler injections.

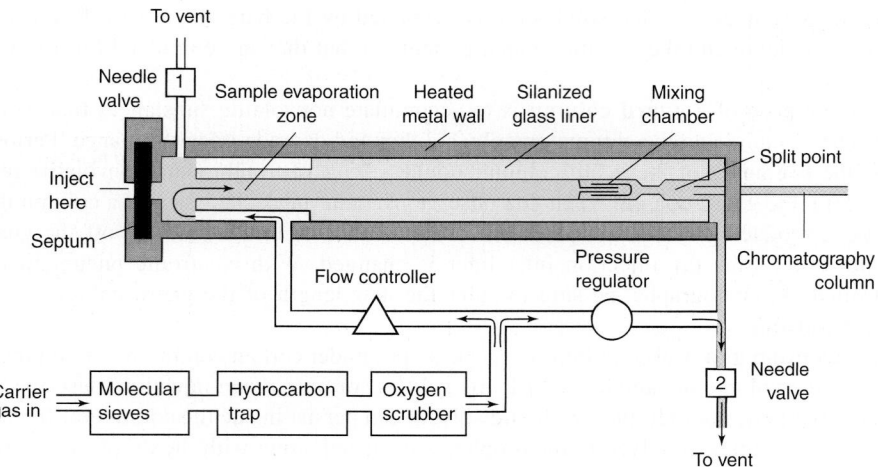

FIGURE 24-14 Injection port for split injection into an open tubular column. The glass liner is slowly contaminated by nonvolatile and decomposed samples and must be replaced periodically. For splitless injection, the glass liner is a straight tube with no mixing chamber. For dirty samples, split injection is used and a packing material can be placed inside the liner to adsorb undesirable components of the sample. Glass wool might be placed at the near end of the liner so that liquid on the outside of the syringe needle gets wiped off by the wool before the needle is withdrawn.

Injection into open tubular columns:

split: routine for introducing small sample volume into open tubular column

splitless: best for trace levels of high-boiling solutes in low-boiling solvents

on-column: best for thermally unstable solutes and high-boiling solvents; best for quantitative analysis

Split Injection

If analytes of interest constitute >0.1% of the sample, **split injection** is usually preferred. For high-resolution work, best results are obtained with the smallest amount of sample (≤1 μL) that can be adequately detected—preferably containing ≤1 ng of each component. A complete injection contains too much material for a 0.32-mm-diameter or smaller column. A split injection delivers only 0.2–2% of the sample to the column. In Figure 24-14, sample is injected rapidly (<1 s) through the septum into the evaporation zone. The injector temperature is kept high (for example, 350°C) to promote fast evaporation. A brisk flow of carrier gas sweeps the sample through the *mixing chamber*, where complete vaporization and good mixing occur. At the split point, a small fraction of vapor enters the chromatography column, but most passes through needle valve 2 to a waste vent. The pressure regulator leading to needle valve 2 controls the fraction of sample discarded. The proportion of sample that does not reach the column is called the *split ratio* and typically ranges from 50:1 to 500:1. The fatty acid analysis in Figure 24-15 used split injection with 99% of the sample diverted to waste. After sample has been flushed from the injection port (~30 s), needle valve 2 is closed and carrier gas flow is correspondingly reduced. Quantitative analysis with split injection can be inaccurate because the split ratio is not reproducible from run to run.

FIGURE 24-15 Resolution of 22 18-carbon fatty acid methyl esters with 1, 2, or 3 double bonds. Peaks 1, 2, 3, 4, 11, 12, 13, 15, 16, 17, 18, 19, 20, and 21 are all *trans* fats (see Figure 5-3). A Supelco IL100 ionic liquid (structure shown on page 637) wall-coated open tubular column (100 m × 0.25 mm with 0.20-μm-thick film, β = 310) was operated at 150°C with H₂ carrier gas at a linear flow rate of 25 cm/s and flame ionization detection. [Data from C. Ragonese, P. Q. Tranchida, P. Dugo, G. Dugo, L. M. Sidisky, M. V. Robillard, and L. Mondello, *Anal. Chem.* **2009,** *81,* 5561.]

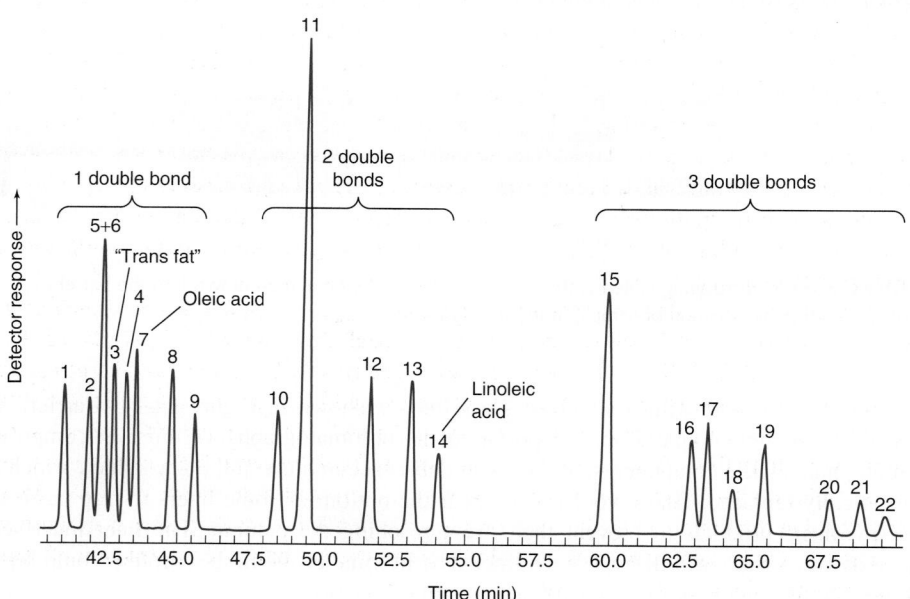

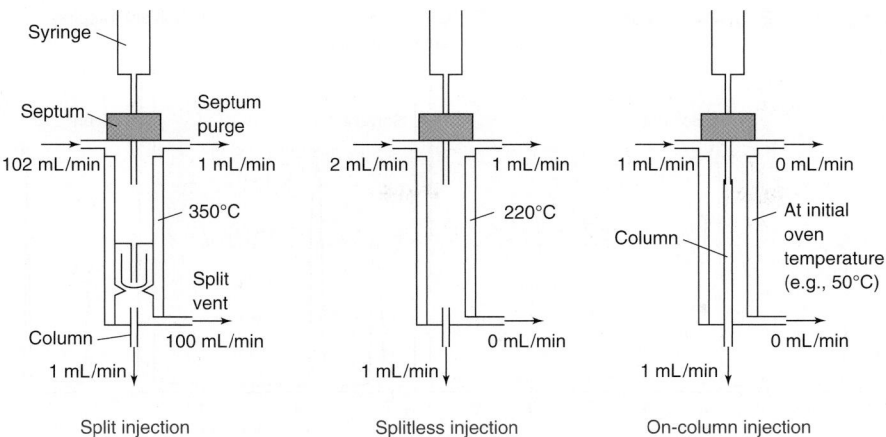

A 1-μL liquid injection creates roughly 0.5 mL of gas volume, which can fill the glass liner in Figure 24-14. Some vapor can escape backward toward the septum. Lower boiling components evaporate first and are more likely to escape than components with higher boiling points. The injection port temperature should be high enough to minimize this fractionation of the sample. However, if the injector temperature is too high, decomposition can occur. During injection and chromatography, *septum purge* gas flow through needle valve 1 in Figure 24-14 is run at ~1 mL/min to remove excess sample vapor and gas that bleeds from the hot rubber septum.

Splitless Injection

For trace analysis[25] of analytes that are less than 0.01% of the sample, **splitless injection** is appropriate. The same port shown for split injection in Figure 24-14 is used. However, the glass liner is a straight, empty tube with no mixing chamber, as shown in Figure 24-16. A large volume (~2 μL) of dilute solution in a low-boiling solvent is injected slowly (~2 s) into the liner, with the split vent closed. Slow flow through the septum purge is maintained during injection and chromatography to remove vapors that escape from the injection liner. Injector temperature for splitless injection is lower (~220°C) than that for split injection, because the sample spends more time in the port and we do not want it to decompose. The residence time of the sample in the glass liner is ~1 min, because carrier gas flows through the liner at the column flow rate, which is ~1 mL/min. In splitless injection, ~80% of the sample is applied to the column, and little fractionation occurs during injection.

The initial column temperature is set 40°C *below* the boiling point of the solvent, which therefore condenses at the beginning of the column. As solutes catch up with the condensed plug of solvent, they are trapped in the solvent in a narrow band at the beginning of the column. This **solvent trapping** leads to sharp chromatographic peaks. Without solvent trapping, the bands could not be sharper than the 1-min injection time. Chromatography is initiated by raising the column temperature to vaporize the solvent trapped at the head of the column. The boiling points of the first peaks of interest should be >30°C above that of the solvent.

For solvent trapping, sample should contain 10^4 times as much solvent as analyte and column temperature should be 40°C below the solvent's boiling point.

An alternative means of condensing solutes in a narrow band at the beginning of the column is called **cold trapping.** The initial column temperature is 150°C lower than the boiling points of the solutes of interest. Solvent and low-boiling components are eluted rapidly, but high-boiling solutes remain in a narrow band at the beginning of the column. The column is then rapidly warmed to initiate chromatography of the high-boiling solutes. For low-boiling solutes, *cryogenic focusing* is required, with an initial column temperature below room temperature.

For cold trapping, stationary phase film thickness must be ≥2 μm.

Figure 24-17 shows effects of operating parameters in split and splitless injections. Experiment A is a standard split injection with brisk flow through the split vent in Figure 24-16. The column was kept at 75°C. The injection liner was purged rapidly by carrier gas, and peaks are quite sharp. Experiment B shows the same sample injected in the same way, except the split vent was closed. Then the injection liner was purged slowly, and sample was applied to the column over a long time. Peaks are broad, and they tail badly because fresh carrier gas continuously mixes with vapor in the injector, making it more and more dilute but never completely flushing the sample from the injector. Peak areas in B are much greater than those in A because the entire sample reaches the column in B, whereas only a small fraction of sample reaches the column in A.

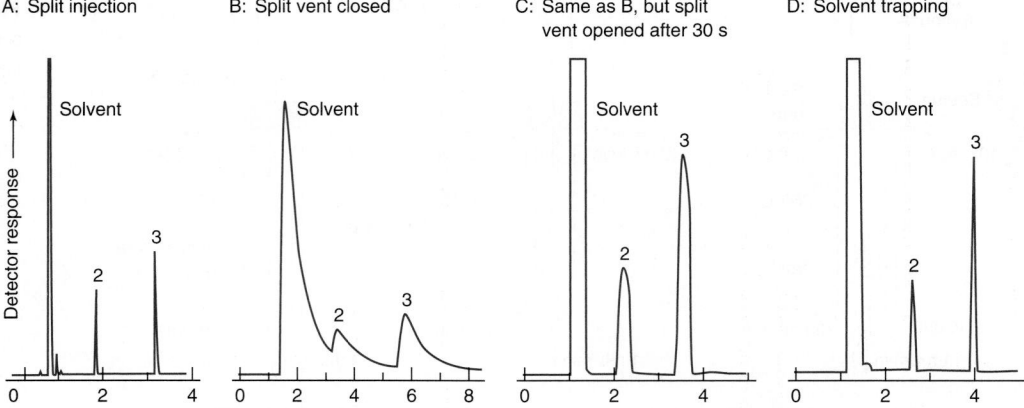

FIGURE 24-17 Split and splitless injections of a solution containing 1 vol% methyl isobutyl ketone (b.p. 118°C) and 1 vol% *p*-xylene (b.p. 138°C) in dichloromethane (b.p. 40°C) on a BP-10 moderately polar cyanopropylphenyl methyl silicone open tubular column (10 m × 0.22 mm with 0.25-μm-thick film, column temperature = 75°C). Vertical scale is the same for A–C. In D, signal heights should be multiplied by 2.33 to be on the same scale as A–C. [Data from P. J. Marriott and P. D. Carpenter, "Capillary Gas Chromatography Injection," *J. Chem. Ed.* **1996**, *73*, 96.]

Experiment C is the same as B, but the split vent was opened after 30 s to rapidly purge all vapors from the injection liner. The bands in chromatogram C would be similar to those in B, but the bands are truncated after 30 s. Experiment D was the same as C, except that the column was initially cooled to 25°C to trap solvent and solutes at the beginning of the column. This is the correct condition for splitless injection. Solute peaks are sharp because the solutes were applied to the column in a narrow band of trapped solvent. Detector response in D is different from A–C. Actual peak areas in D are greater than those in A because most of the sample is applied to the column in D, but only a small fraction is applied in A. To make experiment D a proper splitless injection, the sample would need to be much more dilute.

On-Column Injection

On-column injection is used for samples that decompose above their boiling points and is preferred for quantitative analysis. Solution is injected directly into the column, without going through a hot injector (Figure 24-16). The initial column temperature is low enough to condense solutes in a narrow zone. Warming the column initiates chromatography. Samples are subjected to the lowest possible temperature in this procedure, and little loss of any solute occurs. The needle of a standard microliter syringe fits inside a 0.53-mm-diameter column, but this column does not give the best resolution. For 0.20- to 0.32-mm-diameter columns, which give better resolution, special syringes with thin silica needles are required.

24-3 Detectors

For *qualitative analysis*, a mass spectrometer (Chapter 22) can identify a chromatographic peak by comparing its spectrum with a library of spectra. For mass spectral identification, sometimes two prominent mass spectral peaks are selected in the electron ionization spectrum. The *quantitation ion* is used for quantitative analysis. The *confirmation ion* is used for qualitative identification. For example, the confirmation ion might be expected to be 65% as abundant as the quantitation ion. If the observed abundance is not close to 65%, then we suspect that the compound is misidentified.

Another method to identify a peak is to compare its retention time with that of an authentic sample of the suspected compound. The most reliable way to compare retention times is by **spiking**, in which an authentic compound is added to the unknown. If the added compound is identical to a component of the unknown, then the relative area of that one peak will increase. Identification is tentative when carried out with one column but is firmer when carried out on several columns with different stationary phase polarity, such as in the confirmatory analysis of blood alcohol concentration in Figure 24-18. In a commercial test, gas from the *headspace* of a heated vial (60°C) of blood or urine is injected into a port leading to two parallel columns with different kinds of stationary phase. If a peak appears with the retention time of ethanol on both columns, it is very likely to be ethanol.

Headspace is the gas phase above a solid or liquid sample within a sealed vial.

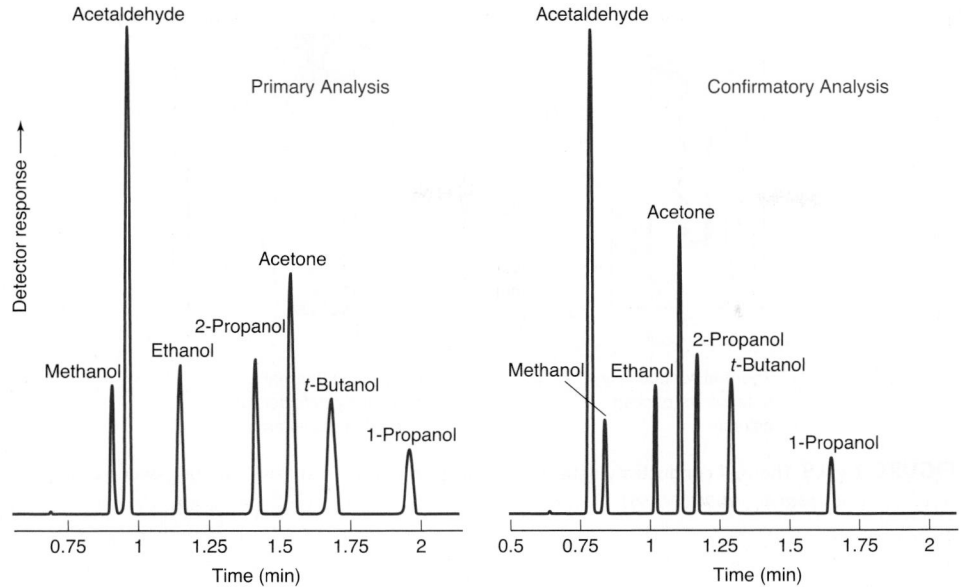

FIGURE 24-18 Confirmatory analysis of alcohol with a single sample measured on two columns with *orthogonal* stationary phases (having different retention mechanisms). Headspace equilibrated with a urine or blood sample is injected into a single port with 50:1 split injection. Carrier gas is directed in parallel to two different columns, each with its own flame ionization detector, for separation at 40°C. The columns give different orders of elution because of their different balance of hydrogen bonding and dipole forces. *t*-Butanol and 1-propanol are internal standards for measuring ethanol. The sample in this illustration was a laboratory standard, not a blood headspace sample. [Data from Restek Co., Bellefonte, PA.]

Quantitative analysis can be based on either the area or height of a chromatographic peak. Area is the preferred metric. Height is less affected by poor resolution or by noisy or sloped baselines, but it is affected by any change in band broadening. In the *linear response* concentration range, *the area of a peak reported by the instrument is proportional to the quantity of that component.* Judgment is required to draw baselines beneath peaks and to decide on the boundaries of the area to be measured.[26] The start and end of integration may be indicated by vertical ticks on the baseline, as for CO in Figure 24-7. If area must be measured by hand, and if the peak is Gaussian, then the area is

Linear response means that peak area is proportional to analyte concentration. For very narrow peaks, peak height is often substituted for peak area.

$$\text{Area of Gaussian peak} = 1.064 \times \text{peak height} \times w_{1/2} \qquad \textbf{(24-8)}$$

where $w_{1/2}$ is the width at half-height (Figure 23-9). Quantitative analysis is almost always performed by adding a known quantity of *internal standard* to the unknown (Section 5-4), as in the blood alcohol determination in Figure 24-18. In this example, *t*-butanol and 1-propanol are used as internal standards to measure ethanol. After measurement of the *response factor* with standard mixtures, the equation in the margin is used to measure the quantity of unknown.

Quantitative analysis with internal standard:

$$\frac{A_X}{[X]} = F \frac{A_S}{[S]}$$

A_X = area of analyte signal
A_S = area of internal standard
[X] = concentration of analyte
[S] = concentration of internal standard
F = response factor

Thermal Conductivity Detector

In the past, **thermal conductivity detectors** were common in gas chromatography because they are simple and *universal*—responding to all analytes. Thermal conductivity is useful for packed columns, but is less sensitive than other detectors for open tubular columns (Table 24-3). The thermal conductivity detector does not alter the sample, so this detector may be used in conjunction with other detectors.

TABLE 24-3	Detection limits and linear ranges of gas chromatography detectors	
Detector	**Approximate detection limit**	**Linear range**
Thermal conductivity	400 pg/mL (propane)	$>10^5$
Flame ionization	2 pg/s	$>10^7$
Electron capture	As low as 5 fg/s	10^4
Flame photometric	<1 pg/s (phosphorus)	$>10^4$
	<10 pg/s (sulfur)	$>10^3$
Nitrogen-phosphorus	100 fg/s	10^5
Sulfur chemiluminescence	100 fg/s (sulfur)	10^5
Photoionization	25 pg to 50 pg (aromatics)	$>10^5$
Vacuum ultraviolet absorbance	15–250 pg	10^3
Mass spectrometric	25 fg to 100 pg	10^5

*Most data are from D. G. Westmoreland and G. R. Rhodes, "Detectors for Gas Chromatography," Pure Appl. Chem. **1989**, 61, 1147.*

TABLE 24-4	Thermal conductivity at 273 K and 1 atm
Gas	Thermal conductivity J/(K · m · s)
H_2	0.170
He	0.141
NH_3	0.021 5
N_2	0.024 3
C_2H_4	0.017 0
O_2	0.024 6
CO	0.023 0
Ar	0.016 2
C_3H_8	0.015 1
CO_2	0.014 4
Cl_2	0.007 6

The energy per unit area per unit time flowing from a hot region to a cold region is given by

$$\text{Energy flux } (J/m^2 \cdot s) = -\kappa(dT/dx)$$

where κ is the thermal conductivity [units = J/(K · m · s)] and dT/dx is the temperature gradient (K/m). Thermal conductivity is to energy flux as the diffusion coefficient is to mass flux.

Thermal conductivity detector
- 10^5 linear response range
- H_2 and He give lowest detection limit
- sensitivity increases with
  increasing filament current
  decreasing flow rate
  lower detector block temperature
- nondestructive

Flame ionization detector
- N_2 carrier gas gives best detection limit
- signal proportional to number of susceptible carbon atoms
- 100-fold lower detection limit than thermal conductivity
- 10^7 linear response range

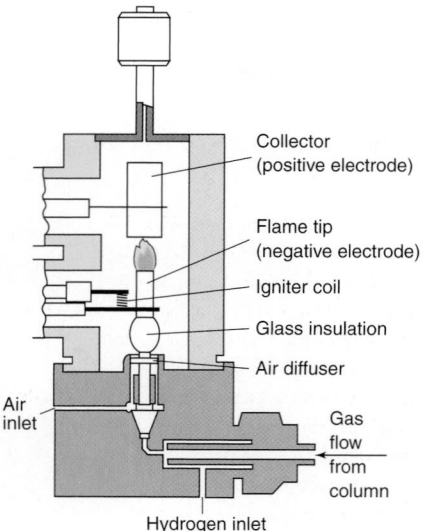

FIGURE 24-20 Flame ionization detector. [Information from Varian Associates, Palo Alto, CA.]

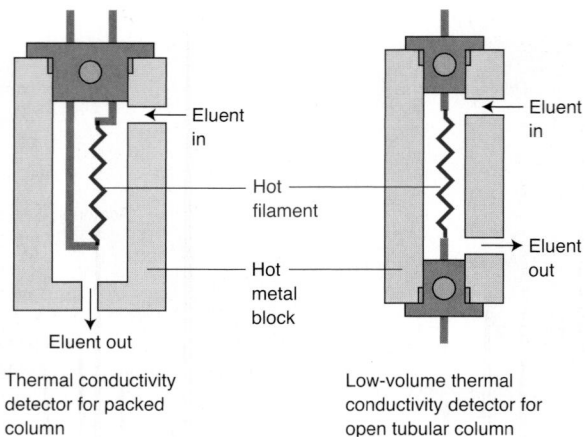

FIGURE 24-19 Thermal conductivity detectors. [Information from J. V. Hinshaw, "The Thermal Conductivity Detector," *LCGC North Am.* **2006**, *24*, 38.]

Thermal conductivity measures the ability of a substance to transport heat from a hot region to a cold region (Table 24-4). For thermal conductivity detection, carrier gas must have a very different thermal conductivity from the analytes. Helium is most commonly used. It has the second highest thermal conductivity (after H_2), so any analyte mixed with He lowers the conductivity of the gas stream. In Figure 24-19, eluate from the column flows over a hot tungsten-rhenium filament. When analyte emerges, the conductivity of the gas stream decreases, the filament gets hotter, its electrical resistance increases, and the voltage across the filament changes. The detector measures the change in voltage.

The thermal conductivity of eluate is measured relative to that of pure carrier gas. Each stream may be passed over a different filament or alternated over a single filament. Resistance of the sample filament is measured with respect to that of the reference filament. Sensitivity increases with the square of filament current, but the maximum recommended current should not be exceeded. The filament should be off when carrier gas is not flowing.

The sensitivity of a thermal conductivity detector (but *not* that of the flame ionization detector described next) is inversely proportional to flow rate: It is more sensitive at a lower flow rate. Sensitivity also increases with increasing temperature differences between the filament and the surrounding block in Figure 24-19. The block should therefore be maintained at the lowest temperature that allows all solutes to remain gaseous.

Flame Ionization Detector

In the **flame ionization detector** in Figure 24-20, eluate is burned in a mixture of H_2 and air.[27] Carbon atoms (except carbonyl and carboxyl carbons) produce CH radicals, which are thought to produce CHO^+ ions and electrons in the flame.

$$CH + O \rightarrow CHO^+ + e^-$$

Only about 1 in 10^5 carbon atoms produces an ion, but ion production is proportional to the number of susceptible carbon atoms entering the flame. In the absence of analyte, a current of $\sim 10^{-14}$ A flows between the flame tip and the collector, which is held at +200 to 300 V with respect to the flame tip. Analytes produce a current of $\sim 10^{-12}$ A, which is converted to voltage, amplified, filtered to remove high-frequency noise, and finally converted to a digital signal.

Response to organic compounds is proportional to solute mass over seven orders of magnitude. The detection limit is ~100 times smaller than that of thermal conductivity and is reduced by 50% when N_2 carrier gas is used instead of He. For open tubular columns, N_2 makeup gas is added to the H_2 or He eluate before it enters the detector. Flame ionization is sensitive enough for narrow-bore open tubular columns. It responds to most hydrocarbons with near constant response per carbon atom, but with lower response for C bonded to N or O.[28] Flame ionization is insensitive to nonhydrocarbons such as H_2, He, N_2, O_2, CO, CO_2, H_2O, NH_3, NO, H_2S, and SiF_4. Figure 24-21 shows a gas chromatogram of a tequila extract, where the effluent from the column was split and sent to

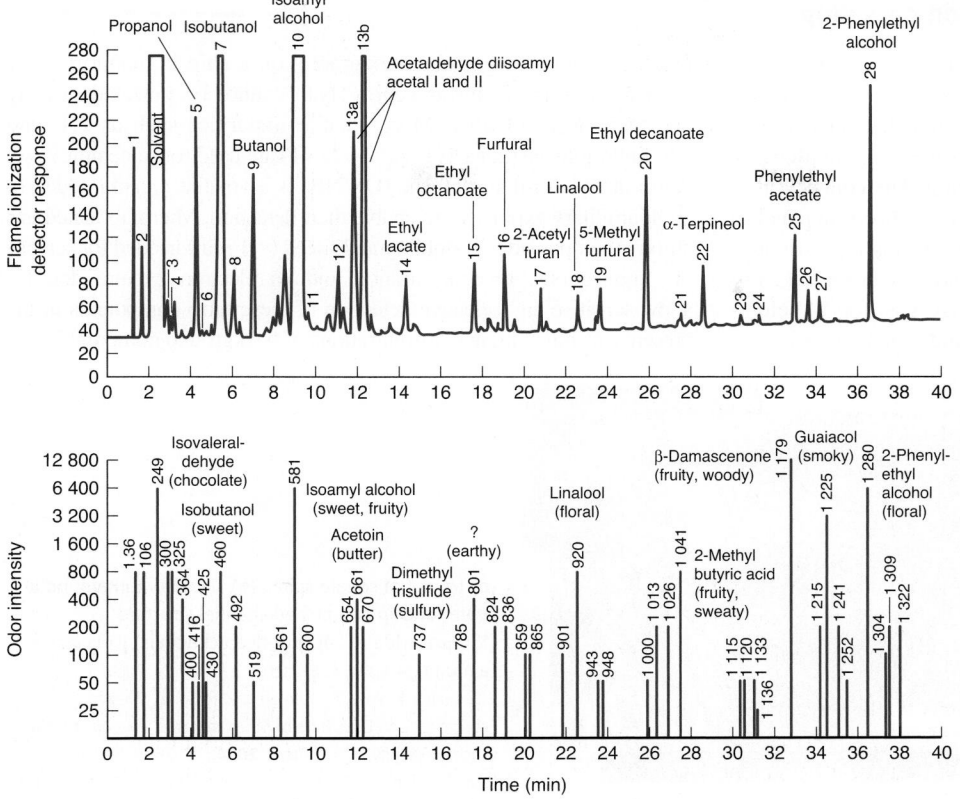

FIGURE 24-21 Gas chromatograms of extracts of tequila: (*a*) flame ionization detector response and (*b*) olfactometric (smell) response on a poly(ethylene glycol) open tubular column (30 m × 0.53 mm with 1-μm-thick film, temperature = 50°–230°C at 3°C/min). Numbered peaks are amongst 175 constituents identified by mass spectrometry and Kovats retention index. Many of the most potent odors of tequila, such as the smoky aroma of guaiacol at 34.5 min, are eluted at retention times where there is little or no response from the flame ionization detector. Retention indices based on straight-chain ethyl esters shown in chromatogram *b* allow correlation of the peaks to those run on other columns with other detectors. [Data from S. M. Benn and T. L. Peppard, "Characterization of Tequila Flavor by Instrumental and Sensory Analysis," *J. Agric. Food Chem.* **1996,** *44,* 557.]

two detectors. The upper chromatogram used a flame ionization detector. The lower chromatogram used *olfactometric detection*, in which a trained person smells the effluent and rates the intensity and character of the odor.[29]

Electron Capture Detector

Most detectors other than flame ionization and thermal conductivity respond to limited classes of analytes. The **electron capture detector** (Figure 24-22) is particularly sensitive to halogen-containing molecules (as in Box 24-1), conjugated carbonyls, nitriles, nitro compounds (Box 24-2), and organometallic compounds, but relatively insensitive to hydrocarbons, alcohols, and ketones. The detector is particularly useful for chlorinated pesticides and fluorocarbons in environmental samples. The carrier or makeup gas must be either N_2 or 5% methane in Ar. Moisture decreases the sensitivity. Gas entering the detector is ionized by high-energy electrons ("β-particles") emitted from a foil containing radioactive ^{63}Ni. Electrons in the plasma thus formed are attracted to an anode, producing a small current that is maintained at a constant level by variable frequency pulses applied between the cathode and anode. When analytes with a high electron affinity enter the detector, they capture some electrons and decrease the conductivity of the plasma. The detector responds by varying the frequency of voltage pulses to maintain a constant current. The frequency of the pulses is the detector signal. The electron capture detector is extremely sensitive (Table 24-3), with a detection limit comparable to that of mass spectrometric selected ion monitoring.

Other Detectors

The *nitrogen-phosphorus detector*, also called an *alkali flame detector*, is a modified flame ionization detector that is especially sensitive to compounds containing N and P.[30] Its response to N and P is 10^4–10^6 times greater than the response to carbon. It is particularly important for drug, pesticide, and herbicide analyses. The detector features a glass bead containing Rb_2SO_4 at the burner tip. Ions such as NO_2^-, CN^-, and PO_2^- produced by these elements when they contact the bead carry the current that is measured. N_2 from air is inert to this detector and does not interfere. The bead must be replaced periodically because Rb_2SO_4 is consumed.

A *flame photometric detector* measures optical emission from phosphorus, sulfur, lead, tin, or other selected elements. When eluate passes through a H_2-air flame, as in the flame ionization detector, excited atoms emit characteristic light. Phosphorus emission at 526 nm or sulfur emission at 394 nm can be isolated by a narrow-band interference filter and detected

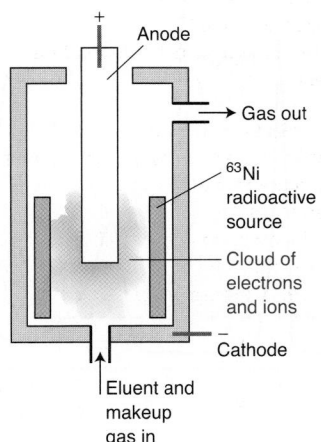

FIGURE 24-22 Electron capture detector.

Other gas chromatography detectors
electron capture: halogens, conjugated C=O, —C≡N, —NO₂
nitrogen-phosphorus: highlights P, N
flame photometer: individual selected elements, such as P, S, Sn, Pb
vacuum ultraviolet absorbance: most analytes
photoionization: aromatics, unsaturated compounds
sulfur chemiluminescence: S
nitrogen chemiluminescence: N
atomic emission: most elements (selected individually)
mass spectrometer: most analytes

BOX 24-2 **Chromatography Column on a Chip**

Microfabricated instruments are being developed for environmental monitoring, homeland security, medical diagnosis, and forensic science. The micrograph shows a 3-m-long square spiral channel etched into silicon whose edge dimension is 3.2 cm. In panel *a*, gas flows into the white spiral and out from the colored spiral. The connection between the two paths at the center of the structure is shown in panel *b*. A glass plate bonded to the top of the silicon creates a gas-tight channel, which is coated with a ~0.15-μm-thick layer of cross-linked poly(dimethylsiloxane) (Table 24-1). The carrier gas for a field instrument is air that is filtered to remove water and organic vapors.

One application of gas chromatography on a chip is monitoring for explosives. 2,4,6-Trinitrotoluene (TNT) cannot be detected directly due to its low volatility. More volatile impurities such as 2,4- and 2,6-dinitrotoluene (2,4-DNT and 2,6-DNT) are used for screening. Also, 2,3-dimethyl-2,3-dinitrobutane (DMNB) is a volatile *taggant* added to all nonmilitary explosives to enable their detection. Microfabricated gas chromatographs offer portability and speed (<1 min) needed for screening applications. Preconcentration and an electron capture detector allow sensitive and selective detection of these nitro compounds in the presence of much higher concentrations of background compounds.

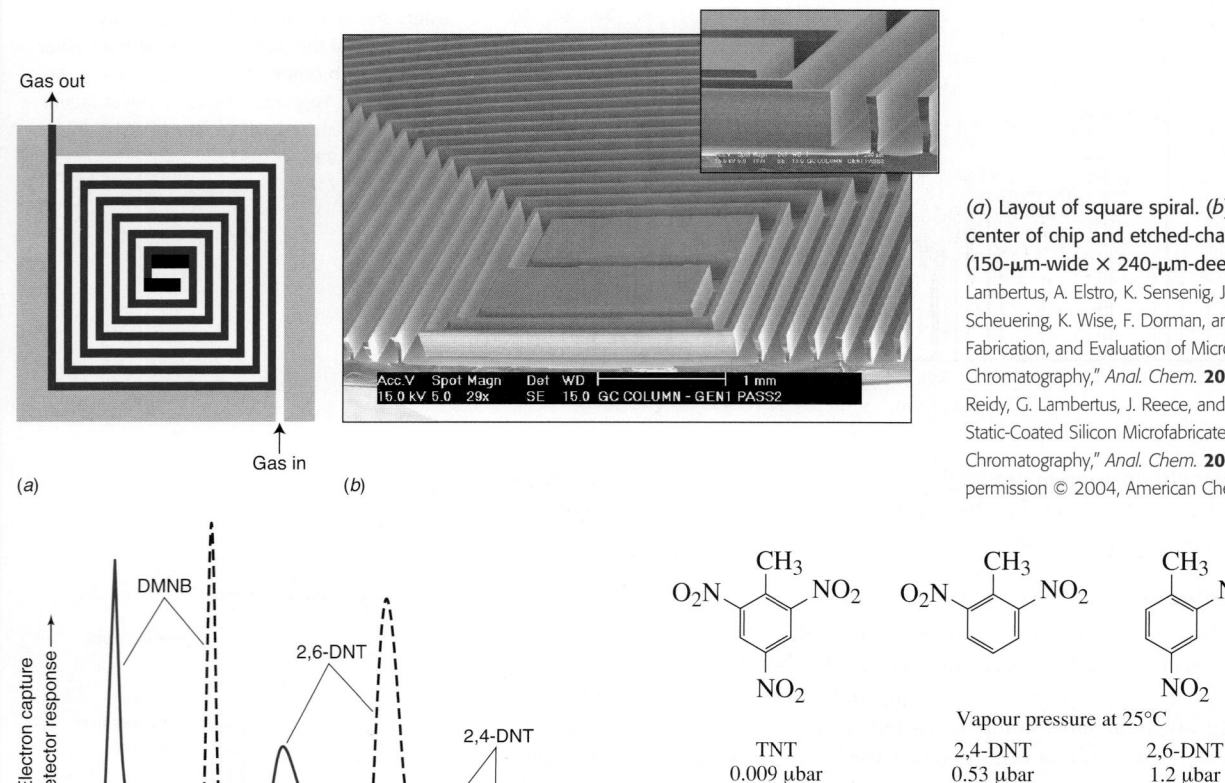

(*a*) Layout of square spiral. (*b*) Gas flow turnaround at center of chip and etched-channel structure (150-μm-wide × 240-μm-deep channels) [Data from G. Lambertus, A. Elstro, K. Sensenig, J. Potkay, M. Agah, S. Scheuering, K. Wise, F. Dorman, and R. Sacks, "Design, Fabrication, and Evaluation of Microfabricated Columns for Gas Chromatography," *Anal. Chem.* **2004,** *76,* 2629. See also S. Reidy, G. Lambertus, J. Reece, and R. Sacks, "High-Performance, Static-Coated Silicon Microfabricated Columns for Gas Chromatography," *Anal. Chem.* **2006,** *78,* 2623. Reprinted with permission © 2004, American Chemical Society.]

	Vapour pressure at 25°C		
TNT	2,4-DNT	2,6-DNT	DMNB
0.009 μbar	0.53 μbar	1.2 μbar	2.7 μbar

(*c*) Rapid detection of explosive markers using isothermal (120°C) column, 3 mL/min N$_2$ carrier gas, and electron capture detection. Dotted chromatogram shows improved peak shape due to cold trapping at 70°C for 15 s. [Data from W. R. Collin, G. Serrano, L. K. Wright, H. W. Chang, N. Nuñovero and E. T. Zellers, "Microfabricated Gas Chromatograph for Rapid, Trace-Level Determination of Gas-Phase Explosive Marker Compounds," *Anal. Chem.* **2014,** *86,* 655.]

with a photomultiplier tube. The response to sulfur or phosphorus compounds is 10^5 times greater than to hydrocarbons. The sensitive and selective response enables detection of trace sulfur compounds (shown in Figure 24-23) that contribute to the aroma of beer. A large hydrocarbon peak coeluted with a sulfur compound can quench the sulfur response.

A *vacuum ultraviolet detector* measures absorbance spectra in the range 115−185 nm.[31] Virtually all molecules absorb at these wavelengths. The spectrum provides qualitative information that complements mass spectrometry. Beer's law is obeyed over 3−4 orders of magnitude, with detection limits in the pg range.

A *photoionization detector* uses a vacuum ultraviolet source to ionize aromatic and unsaturated compounds; it has little response to saturated hydrocarbons or halocarbons. Electrons produced by the ionization are collected and measured.

A *sulfur chemiluminescence detector*[32] takes exhaust from a flame ionization detector, in which sulfur has been oxidized to SO, and mixes it with ozone (O$_3$) to form an excited state of SO$_2$ that emits blue and ultraviolet radiation. Emission intensity is proportional to the mass of sulfur eluted, regardless of the source, and the response to S is 10^7 times

Reactions thought to give *sulfur chemiluminescence:*

Sulfur compound $\xrightarrow{\text{H}_2\text{-O}_2 \text{ flame}}$ SO + products

SO + O$_3$ → SO$_2$* + O$_2$ (SO$_2$* = excited state)

SO$_2$* → SO$_2$ + $h\nu$

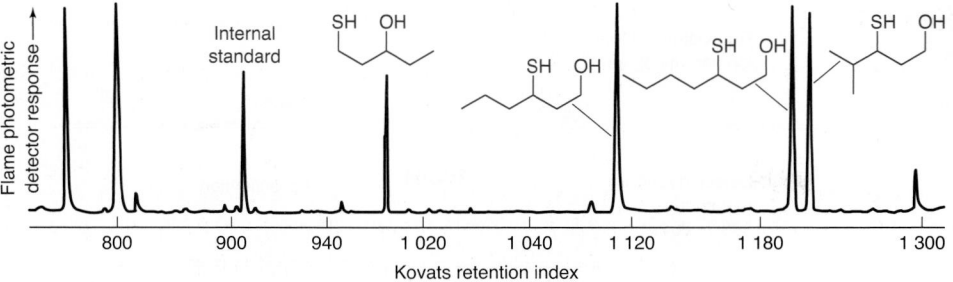

FIGURE 24-23 Gas chromatogram of an extract of hops using sulfur flame photometric detection. Open tubular column (50 m × 0.32 mm with 1.2-μm-thick film of poly(dimethylsiloxane), β = 67). Programmed temperature 36°–250°C. Kovats retention index based on *n*-alkanes was monitored separately with a flame ionization detector. [Data from J. Gros, S. Nizet, and S. Collin, "Occurrence of Odorant Polyfunctional Thiols in the Super Alpha Tomahawk Hop Cultivar. Comparison with the Thiol-Rich Nelson Sauvin Bitter Variety," *J. Agric. Food Chem.* **2011,** *59,* 8853].

greater than the response to C. A *nitrogen chemiluminescence detector*[32] is analogous. Combustion of eluate at 1 800°C converts nearly all nitrogen compounds (but not N_2) into NO, which reacts with O_3 to create a chemiluminescent product that emits near-infrared radiation at 1 200 nm. Response to N is 10^7 times greater than the response to C. The response only depends on the quantity of N, allowing measurement of all analytes (even unknown analytes) to within 10%.

Gas Chromatography–Mass Spectrometry

Mass spectrometry serves as a sensitive detector to provide both qualitative and quantitative information. With *selected ion monitoring* or *selected reaction monitoring* (Section 22-5), we can measure one component in a complex chromatogram of poorly separated compounds. Selected ion monitoring lowers the detection limit by a factor of 10^2–10^3 compared with *m/z* scanning, because more time is spent just collecting ions of interest in selected ion monitoring.

Figure 24-24 illustrates **selected ion monitoring.** The *reconstructed total ion chromatogram* in trace *a* was obtained from a portable gas chromatograph–mass spectrometer designed for identification of spills at accident sites. A total of 1 072 spectra of eluate were

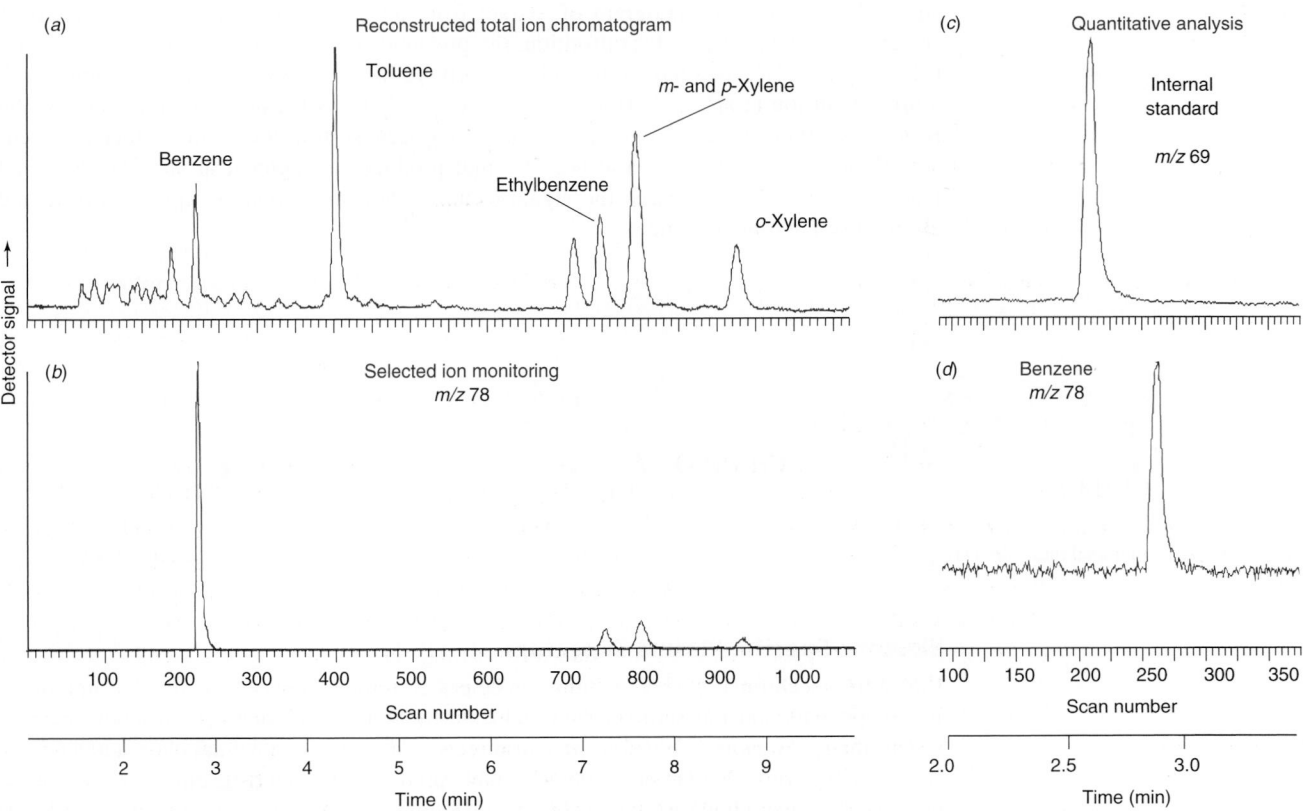

FIGURE 24-24 Selected ion monitoring in gas chromatography–mass spectrometry. (*a*) Reconstructed total ion chromatogram of automobile exhaust with electron ionization. (*b*) Selected ion monitoring at *m/z* 78. (*c*) and (*d*) Quantitative analysis of benzene after adding an internal standard with a prominent ion at *m/z* 69. [Data from Inficon, Syracuse, NY.]

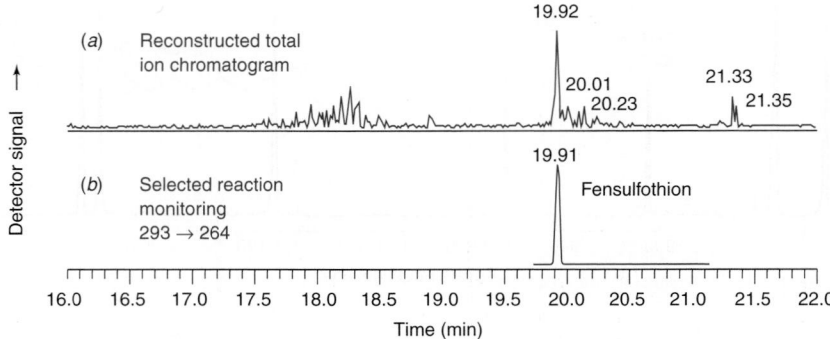

FIGURE 24-25 Selected reaction monitoring in gas chromatography–mass spectrometry. (*a*) Reconstructed total ion chromatogram of extract from orange peel with electron ionization. (*b*) Selected reaction monitoring with the precursor ion *m/z* 293 selected by mass filter Q1 in Figure 22-33 and product ion *m/z* 264 selected by mass filter Q3. The chromatogram in trace *b* is a graph of intensity at *m/z* 264 from Q3 versus time. [Data from Thermo Finnigan GC and GC/MS Division, San Jose, CA.]

recorded at equal time intervals between 1 and 10 min. The ordinate of the reconstructed total ion chromatogram is the sum of detector signal for all *m/z* above a selected cutoff. It measures everything eluted from the column. The selected ion chromatogram in trace *b* is obtained by parking the detector at *m/z* 78 and only measuring this one mass. By spending all the time measuring just one ion, the signal-to-noise ratio increases and the chromatogram is simplified. A peak is observed for benzene (nominal mass 78 Da) and minor peaks are observed for benzene derivatives at 7–9 min. For quantitative analysis, an internal standard with a signal at *m/z* 69 is added to the mixture. Even though this internal standard overlaps the congested part of the chromatogram near a retention time of 2 min, the selected ion chromatogram for *m/z* 69 has a single peak in trace *c*. To measure benzene, the area of the *m/z* 78 peak in trace *d* is compared with the area of *m/z* 69 in trace *c*.

Selected reaction monitoring is illustrated in Figure 24-25. Trace *a* is the reconstructed total ion chromatogram of extract from an orange peel. To make the analysis selective for the pesticide fensulfothion, the precursor ion *m/z* 293 selected by mass filter Q1 in Figure 22-33 was passed to collision cell Q2 in which it broke into fragments with a prominent ion at *m/z* 264. Trace *b* in Figure 24-25 shows the detector signal at *m/z* 264 from mass filter Q3. Only one peak is observed because few compounds other than fensulfothion give rise to an ion at *m/z* 293 that produces a fragment at *m/z* 264. Selected reaction monitoring increases the signal-to-noise ratio in chromatographic analysis and eliminates much interference.

Fensulfothion
(nominal mass 308 Da)

M − 15 (*m/z* 293)
Selected by Q1

M − 15 − 29 (*m/z* 264)
Selected by Q3

Element-Specific Plasma Detectors

Eluate from a chromatography column can be passed through a plasma to atomize and ionize its components and measure selected elements by atomic emission spectroscopy or mass spectrometry. An *atomic emission detector* directs eluate through a helium plasma in a microwave cavity. Every element of the periodic table produces characteristic emission that can be detected by a photodiode array polychromator (Figure 20-16). Sensitivity for sulfur can be 10 times better than the sensitivity of flame photometric detection.

The extremely sensitive inductively coupled plasma–mass spectrometer was described in Section 21-7.[33] Figure 24-26 shows 15 pesticides measured by gas chromatography–

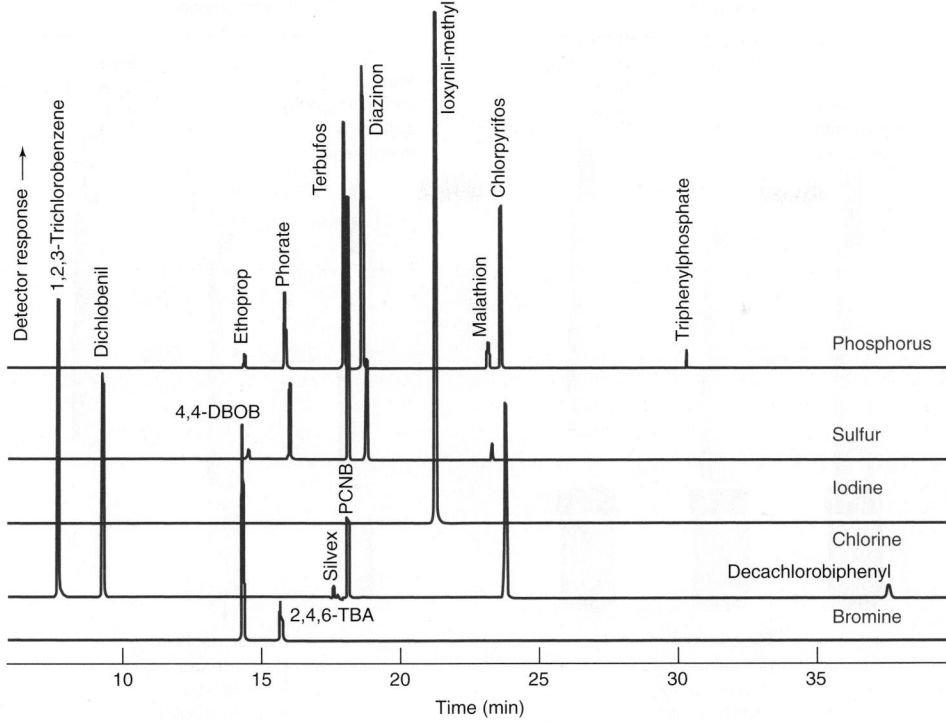

FIGURE 24-26 Gas chromatography–inductively coupled plasma–mass spectrometry of a mixture of 15 pesticides. Each trace responds to just one element. [Data from D. Profrock, P. Leonhard, S. Wilbur, and A. Prange, "Sensitive, Simultaneous Determination of P, S, Cl, Br, and I Containing Pesticides in Environmental Samples by GC Hyphenated with Collision-Cell ICP-MS," *J. Anal. Atom. Spectros.* **2004,** *19,* 1.]

inductively coupled plasma–mass spectrometry. Eluate was atomized and ionized in the plasma. Ions were measured by a mass spectrometer that could monitor any set of *m/z* values. The figure shows traces for P, S, I, Cl, and Br.

24-4 Sample Preparation

Sample preparation is the process of transforming a sample into a form that is suitable for analysis. This process might entail extracting analyte from a complex matrix, *preconcentrating* very dilute analytes to get a concentration high enough to measure, removing or masking interfering species, or chemically transforming (*derivatizing*) analyte into a more convenient or more easily detected form. Chapter 28 addresses sample preparation, so we now describe only techniques that are especially applicable to gas chromatography.

Solid-phase microextraction extracts compounds from liquids, air, or even sludge without using any solvent.[34] The key component is a fused-silica fiber coated with a 10- to 100-μm-thick film of stationary phase similar to those used in gas chromatography. Figure 24-27 shows the fiber attached to the base of a syringe with a fixed metal needle. The fiber can be extended from the needle or retracted inside the needle. Figure 24-28 demonstrates the procedure of exposing the fiber to a sample solution for a fixed time while stirring and, perhaps, heating. Only a fraction of the analyte in the sample is extracted into the fiber. It is best to determine by experiment how much time is required for the fiber to equilibrate with analyte and to allow this much time for extraction. If you use shorter times, the concentration of analyte in the fiber is likely to vary from sample to sample and more complex kinetic calibration is needed.[35]

After sample collection, retract the fiber and insert the syringe into a gas chromatograph equipped with a 0.7-mm-inner-diameter injection port liner. Extend the fiber into the hot injection liner, where analyte is thermally desorbed from the fiber in splitless mode for a fixed time. The narrow port maintains desorbed analyte in a narrow band. Collect desorbed analyte by *cold trapping* (page 647) at the head of the column prior to chromatography. If there will be a long time between sampling and injection, insert the needle into a septum to seal the fiber from the atmosphere.

Example of *derivatization:*

$$\underset{\substack{\text{Nonvolatile}\\\text{carboxylic acid}}}{\text{RCOH}} \longrightarrow \underset{\substack{\text{Volatile trimethylsilyl}\\\text{ester derivative}}}{\text{RCOSi(CH}_3)_3}$$

To maximize solid-phase microextraction recovery:

- "Like dissolves like" in selecting fiber
- Thick or porous film for volatiles (C_2–C_6)
- Agitate or stir sample
- Add 25% NaCl and adjust pH (reduce pH for acids and increase for bases)
- For headspace, heat sample to 40°–90°C

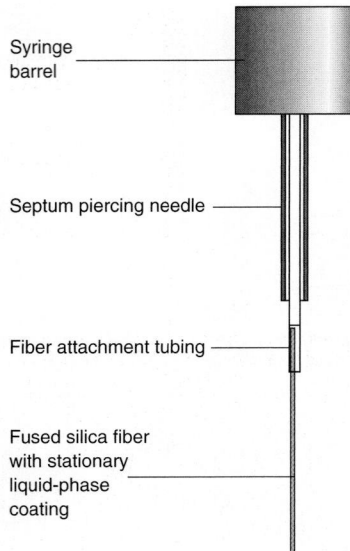

Syringe barrel

Septum piercing needle

Fiber attachment tubing

Fused silica fiber with stationary liquid-phase coating

FIGURE 24-27 Syringe for solid-phase microextraction. The fused-silica fiber is withdrawn inside the steel needle after sample collection and when the syringe is used to pierce a septum.

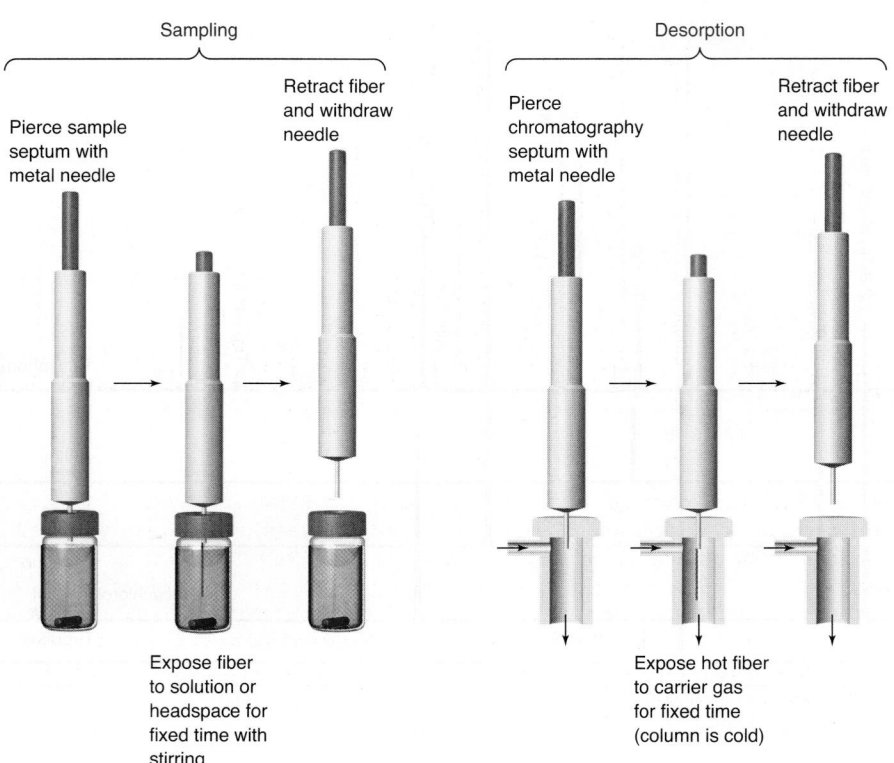

Sampling

Pierce sample septum with metal needle

Retract fiber and withdraw needle

Expose fiber to solution or headspace for fixed time with stirring

Desorption

Pierce chromatography septum with metal needle

Retract fiber and withdraw needle

Expose hot fiber to carrier gas for fixed time (column is cold)

FIGURE 24-28 Sampling by solid-phase microextraction and desorption of analyte from the coated fiber into a gas chromatograph. [Information from Supelco Chromatography Products catalog, Bellefonte, PA.]

Too much headspace volume reduces extraction efficiency. Fill vial ~2/3 with solid or liquid sample.

Stir-bar sorption

[Information from GERSTEL, Inc., Linthicum, MD.]

The stir bar can also be suspended in the headspace of the vial by a bent wire for *headspace extraction.*[37]

Analysis of complex samples can be simplified by analyzing the *headspace* gas above a solid or liquid within a sealed vial.[36] Volatile components partition into the gas phase, and so are separated from the nonvolatile matrix. Figure 24-29 shows a chromatogram of volatile compounds from roasted coffee beans determined by solid-phase microextraction of the headspace above the beans.

In solid-phase microextraction, the mass of analyte (m, µg) absorbed in the coated fiber is

Mass of analyte extracted: $$m = \frac{KV_f c_0 V_s}{KV_f + V_s} \qquad (24\text{-}9)$$

where V_f is the volume of film on the fiber, V_s is the volume of solution being extracted, and c_0 is the initial concentration (µg/mL) of analyte in the solution being extracted. K is the partition coefficient for solute between the film and the solution: $K = c_f/c_s$, where c_f is the concentration of analyte in the film and c_s is the concentration of analyte in the solution. If you extract a large volume of solution such that $V_s \gg KV_f$, then Equation 24-9 reduces to $m = KV_f c_0$. That is, the mass extracted is proportional to the concentration of analyte in solution. For quantitative analysis, you can construct a calibration curve by extracting known solutions. Alternatively, internal standards and standard additions are both useful for solid-phase microextraction.[35]

Stir-bar sorptive extraction is closely related to solid-phase microextraction, but is ~100 times more sensitive for trace analysis.[38] A magnetic stirring bar enclosed in a thin glass jacket is coated with a 0.5- to 1-mm-thick layer of sorbent such as poly(dimethylsiloxane)—the same as the stationary phase in nonpolar gas chromatography columns. The stirring bar is placed in an aqueous liquid sample such as fruit juice, wine, urine, or blood plasma and stirred for 0.5–4 h to absorb hydrophobic analytes. The mass of analyte extracted is given by Equation 24-9, but the volume of sorbent (V_f) is increased from ~0.5 µL in solid-phase microextraction to 25–125 µL in stir-bar sorptive extraction. Therefore, 50 to 250 times more analyte is extracted with the stir bar. After extraction, the stir bar is touched to a tissue to remove water droplets, possibly rinsed with a few milliliters of water, and then placed in a *thermal desorption tube*. Desorption is typically conducted by heating the tube to 250°C for 5 min in flowing carrier gas. Volatile analytes are collected by cold trapping and then separated by gas chromatography. Analytes at part per billion or lower concentration can be measured by

CHAPTER 24 Gas Chromatography

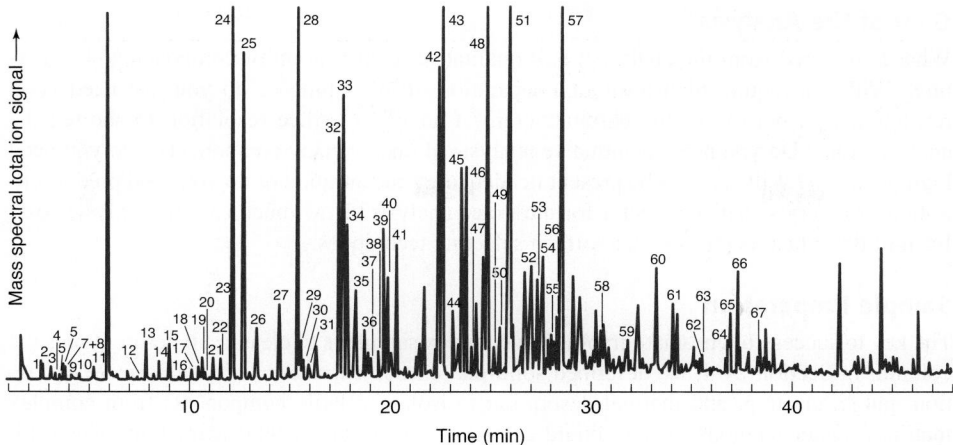

FIGURE 24-29 Gas chromatogram of volatile compounds from 40-min solid-phase microextraction of headspace above 2 g of roasted coffee beans at 60°C. To capture a range of compounds, the fiber had a 50-μm-thick coating of divinylbenzene and carbonaceous resin on 30-μm-thick poly(dimethylsiloxane). Analytes were desorbed from the fiber for 5 min at 260°C in splitless mode in the injection port. Column temperature was 40°C during desorption and then ramped at 4°C/min for chromatography. Column: 30 m × 0.25 mm with 0.25-μm-thick poly(ethylene glycol). Reconstructed total ion chromatogram sums all ions from m/z 40–400. Numbered peaks are identified components of coffee aroma. [Data from L. Mondello, University of Messina, Italy and Supleco, Bellefonte, PA.]

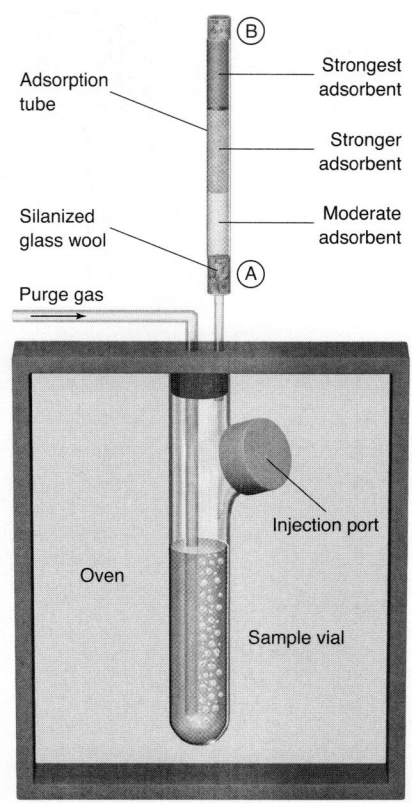

FIGURE 24-30 Purge and trap apparatus for extracting volatile substances from a liquid or solid by flowing gas.

Tenax

You need to establish the time and temperature required to purge 100% of the analyte from the sample in separate control experiments.

using standard addition, isotopic internal standards, or a calibration curve constructed with the same matrix.

Purge and trap is a method for removing volatile analytes from liquids or solids (such as groundwater or soil), concentrating the analytes, and introducing them into a gas chromatograph. In contrast with solid-phase microextraction, which removes only a portion of analyte from the sample, the goal in purge and trap is to remove 100% of analyte from the sample. Quantitative removal of polar analytes from polar matrices can be difficult.

Figure 24-30 shows apparatus for measuring volatile flavor components in carbonated beverages. Helium purge gas from a stainless steel needle is bubbled through the cola in the sample vial, which is heated to 50°C to aid evaporation of analytes. Purge gas exiting the sample vial passes through an adsorption tube containing three layers of adsorbent particles with increasing adsorbent strength. For example, the moderate adsorbent could be a nonpolar (diphenyl)(dimethyl)polysiloxane, the stronger adsorbent could be the polymer Tenax, and the strongest adsorbent could be carbon molecular sieves.[39]

During the purge and trap process, gas flows through the adsorbent tube from end A to end B in Figure 24-30. After purging all analyte from the sample into the adsorption tube, reverse the gas flow to go from B to A and purge the trap at 25°C to remove as much water or other solvent as possible from the adsorbents. Then connect outlet A of the adsorption tube to the injection port of a gas chromatograph operating in splitless mode and heat the trap to ~200°C. Desorbed analytes flow into the chromatography column where they are concentrated by cold trapping. After complete desorption from the trap, warm the chromatography column to initiate the separation.

Thermal desorption is a method for releasing volatile compounds from solid samples. A weighed sample is placed in a steel or glass tube and held in place with glass wool. The sample is purged with carrier gas to remove O_2, which is vented to the air, not into the chromatography column. The desorption tube is then connected to the chromatography column and heated to release volatile substances, which are collected by cold trapping at the beginning of the column. The column is then heated rapidly to initiate chromatography.

24-5 Method Development in Gas Chromatography

In general, there are several satisfactory choices of gas chromatography methods to solve a particular problem. As a broad guide, consider the following issues in this order: (1) goal of the analysis, (2) sample preparation, (3) detector, (4) column, and (5) injection.[40]

Order of decisions:

1. goal of analysis
2. sample preparation
3. detector
4. column
5. injection

Goal of the Analysis

What is required from the analysis? Is it qualitative identification of components in a mixture? Will you require high-resolution separation of everything or do you just need good resolution in a portion of the chromatogram? Can you sacrifice resolution to shorten the analysis time? Do you need quantitative analysis of one or many components? Do you need high precision? Will analytes be present in adequate concentration or do you need preconcentration or a very sensitive detector for ultratrace analysis? How much can the analysis cost? Each of these factors creates trade-offs in selecting techniques.

Sample Preparation

The key to successful chromatography of a complex sample is to clean it up before it sees the column. In Section 24-4, we described solid-phase microextraction, stir-bar sorptive extraction, purge and trap, and thermal desorption to isolate volatile components from complex matrices. Other methods include liquid extraction, supercritical fluid extraction, and solid-phase extraction, most of which are described in Chapter 28. These techniques isolate desired analytes from interfering substances, and they can concentrate dilute analytes up to detectable levels. If you do not clean up your samples, chromatograms could contain a broad "forest" of unresolved peaks, and nonvolatile substances will ruin the expensive chromatography column.

Garbage in—garbage out!

Choosing the Detector

The next step is to choose a detector. Do you need information about everything in the sample or do you want to detect a particular element or class of compounds?

The most general purpose detector for open tubular chromatography is a mass spectrometer. Flame ionization is probably the most popular detector, but it mainly responds to hydrocarbons and Table 24-3 shows that it is not as sensitive as electron capture, nitrogen-phosphorus, or chemiluminescence detectors. The flame ionization detector requires the sample to contain ≥ 10 ppm of each analyte for split injection. The thermal conductivity detector responds to all classes of compounds, but it is not very sensitive.

Sensitive detectors for ultratrace analysis each respond to a limited class of analytes. The electron capture detector is specific for halogen-containing molecules, nitriles, nitro compounds, and conjugated carbonyls. For split injection, the sample should contain ≥ 100 ppb of each analyte if an electron capture detector is to be used. A photoionization detector can be specific for aromatic and unsaturated compounds. The nitrogen-phosphorus detector has enhanced response to compounds containing either of these two elements, but it also responds to hydrocarbons. Sulfur and nitrogen chemiluminescence detectors each respond to just one element. Flame photometric detectors are specific for selected elements, such as S, P, Pb, or Sn. A selective detector might be chosen to simplify the chromatogram by not responding to everything that is eluted. Figure 24-23 shows only the sulfur compounds in a complex mixture. Mass spectrometry with selected reaction monitoring (Figure 24-25) is an excellent way to monitor one analyte of interest in a complex sample. The complete mass spectrum of an eluted component helps to identify it.

Selecting the Column

The basic choices are stationary phase, column diameter and length, and thickness of stationary phase. A nonpolar stationary phase from Table 24-1 is most useful. An intermediate polarity stationary phase will handle most separations that the nonpolar column cannot. For highly polar compounds, a strongly polar column might be necessary. Optical isomers and closely related geometric isomers require special stationary phases for separation.

Table 24-5 shows that there are only a few sensible combinations of column diameter and film thickness. Highest resolution is afforded by the narrowest columns. Thin-film, narrow-bore columns are especially good for separating mixtures of high-boiling-point compounds that are retained too strongly on thick-film columns. Short retention times provide high-speed analyses. However, thin-film, narrow-bore columns have very low sample capacity, require high-sensitivity detectors (flame ionization might not be adequate), do not retain low-boiling-point compounds well, and could suffer from exposure of active sites on the silica.

Thick-film, narrow-bore columns in Table 24-5 provide a good compromise between resolution and sample capacity. They can be used with most detectors (but usually not thermal conductivity) and with compounds of high volatility. Retention times are longer than those of thin-film columns. Thick-film, wide-bore columns are required for use with thermal

TABLE 24-5 Gas chromatography column comparisons

Description	Thin-film narrow-bore	Thick-film narrow-bore	Thick-film wide-bore
Inner diameter	0.10–0.32 mm	0.25–0.32 mm	0.53 mm
Film thickness	~0.2 μm	~1–2 μm	~2–5 μm
Phase ratio (β)	125–400	31–80	26–66

Advantages	High resolution
	Trace analysis
	Fast separations
	Low temperatures
	Elute high-b.p. compounds
		Good capacity
		Good resolution
		  (4 000 plates/m)
		Easy to use
		Retains volatile compounds
		Good for mass spectrometry
			High capacity (100 ng/solute)
			Good for thermal conductivity
			  detector
			Simple injection techniques

Disadvantages	Low capacity
	  (≤1 ng/solute)
	Requires high-sensitivity
	  detector (not mass spectrometry)
	Surface activity of exposed silica
		Moderate resolution
		Long retention time for
		  high-b.p. compounds
			Low resolution
			  (500–2 000 plates/m)
			Long retention time for
			  high-b.p. compounds

SOURCE: *S. Cram, "How to Develop, Validate, and Troubleshoot Capillary GC Methods," American Chemical Society Short Course, 1996.*

conductivity detectors. They have high sample capacity and can handle highly volatile compounds, but they give low resolution and have long retention times.

If a particular column meets most of your requirements but does not provide sufficient resolution, then a narrower column of the same type could be used (Equation 23-40). To obtain similar retention times for the same length of column, the phase ratio β should be held constant. That is, the thickness of the stationary phase should be decreased in proportion to the diameter. A column diameter of 0.15 mm is sensible for maximizing resolution without requiring a chromatograph specially designed for narrower columns.[41]

Doubling the length of a column doubles the number of plates and, according to Equation 23-33, increases resolution by $\sqrt{2}$. Doubling the column length is not the best way to increase resolution because it doubles retention time. A narrower column increases resolution with no penalty in retention time. Selecting another stationary phase changes the relative retention (α in Equation 23-33) and might resolve components of interest.

To increase the speed of analysis without decreasing resolution, a shorter, narrower column can be chosen. Another way to decrease retention time without sacrificing resolution is to switch carrier gas from He to H_2 and increase flow rate by a factor of 1.5 to 2 (Figure 24-11).[21]

If you have measured resolution of a few key components of a mixture under a small number of conditions, commercial software is available to optimize conditions (such as temperature and pressure programming) for the best separation.[43] For complex samples where numerous analytes must be quantified, two-dimensional separation (Box 24-3) may be required.

Figure 24-31 suggests approximate upper and lower analyte mass limits for different columns and detectors. The abscissa shows the mass of a given analyte that reaches the column or detector. A typical volume of injected liquid is 1 μL, with a mass of 1 mg. If analyte concentration is 1 ppm, the mass of analyte in 1 μL is 10^{-9} g = 1 ng. A vertical line at 10^{-9} g falls in the operating range of all columns and four of the detectors. The sample mass is too small for thermal conductivity, too great for electron capture, and near the upper limit for selected ion monitoring. For a 100:1 split injection, the mass of analyte introduced onto the column would be 100 times smaller, or 10^{-11} g. This mass is in range for all but the thermal

To *improve resolution*, use a
• narrower column
• longer column
• different stationary phase

BOX 24-3   Two-Dimensional Gas Chromatography

Two-dimensional chromatography (often called GC×GC) uses two columns with different retention mechanisms to resolve components of complex samples.[42] Panel *a* shows three components that are unresolved by nonpolar column 1. The flame ionization detector only shows the black summed signal that looks like one peak. Fractions 1–5 across the width of the peak were individually collected and injected, in turn, onto polar column 2 to give five chromatograms in panel *b*. Column 2 resolves the components that are coeluted from column 1.

The heart of two-dimensional chromatography is the capillary tube *modulator* positioned between the two columns in panel *c*. In the chapter opening, the modulator accumulates eluate from column 1 for 6 s and then rapidly admits the accumulated eluate onto column 2 for the second-dimension separation that must occur in <6 s. Panel *d* shows one mechanism by which a modulator temporarily retains eluate at one location by condensing it with a stream of cold liquid CO_2 or liquid N_2 for the desired number of seconds.

To obtain rapid separation in column 2, this column must be much shorter than column 1. For high resolution, column 2 is narrower than column 1 and has a thinner film of stationary phase, as described in the caption of the two-dimensional chromatogram at the opening of the chapter. All these factors permit high resolution in a short time.

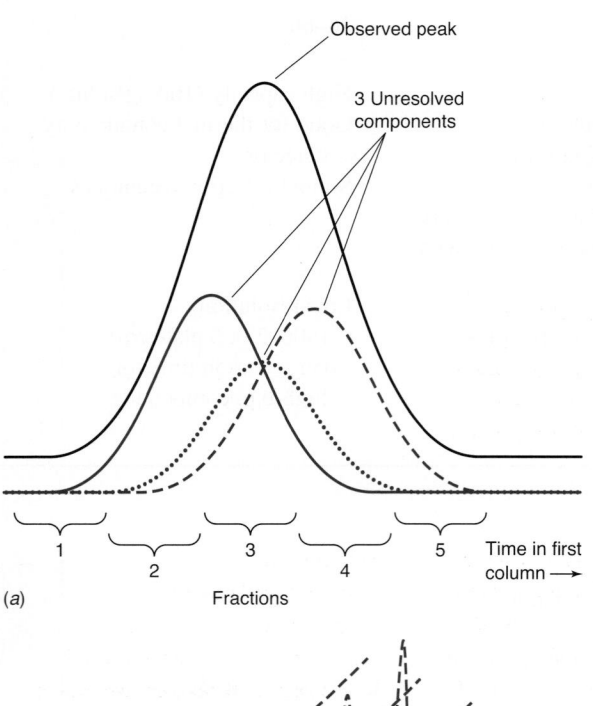

(a)   Fractions

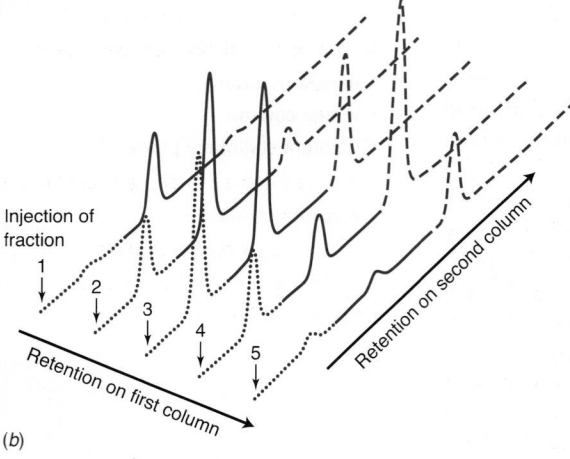

(b)

Unresolved components in panel *a* are resolved in panel *b* by injecting fractions 1–5 from the first column onto a second column of different polarity. [Data from J. Dallüge, J. Beens, and U. A. Th. Brinkman, "Comprehensive Two-Dimensional Gas Chromatography: A Powerful and Versatile Analytical Tool," *J. Chromatogr. A.* **2003**, *1000*, 69.]

A two-dimensional chromatogram, such as the one showing steroid metabolites at the chapter opening, plots retention time for column 1 on one axis and retention time for column 2 on the perpendicular axis. The intensity of the detector response is coded by color or brightness at each point.

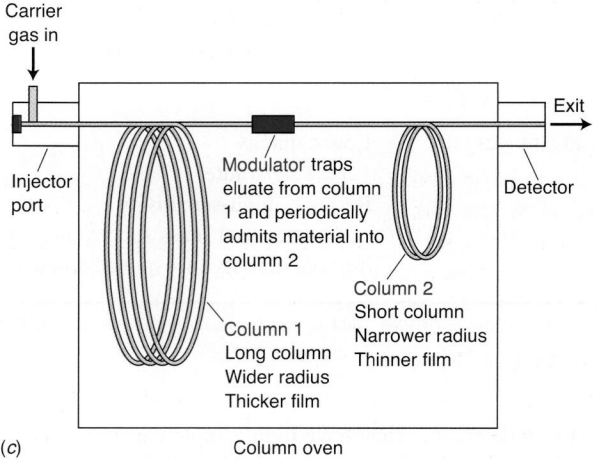

(c)   Column oven

(c) Layout of two-dimensional chromatography (GC×GC). The modulator retains eluate from column 1 and periodically admits it to column 2.

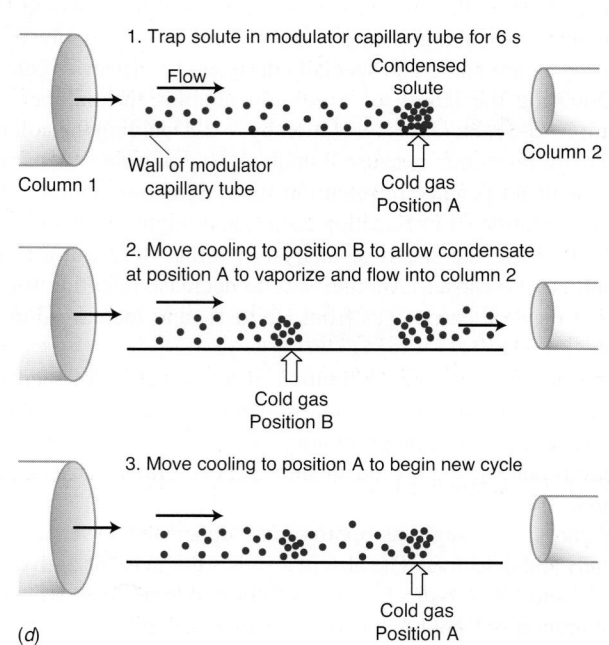

(d) One mechanism by which a modulator can work is to condense eluate with a cold stream of gas.

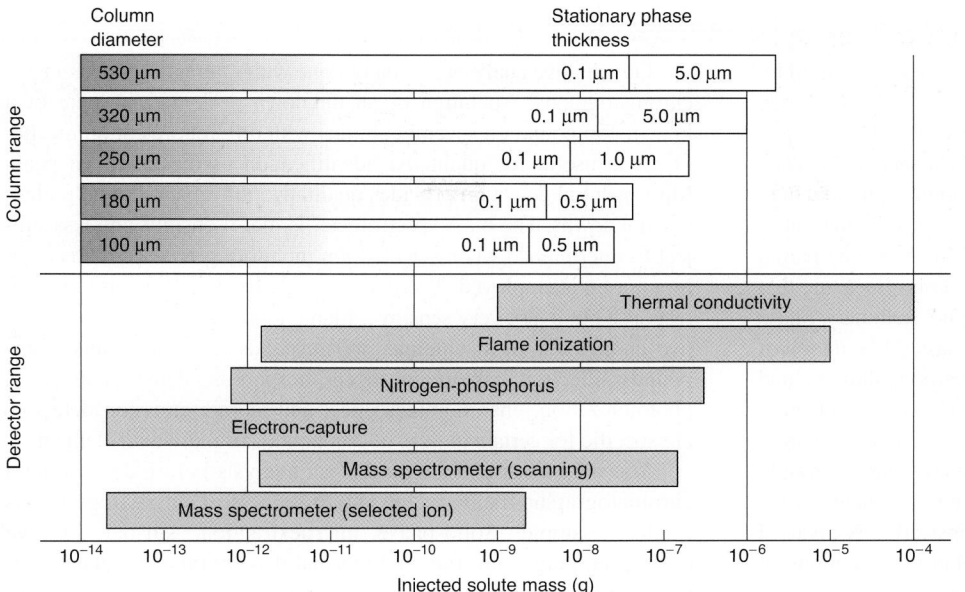

FIGURE 24-31 Approximate injected analyte mass limits for wall-coated open tubular gas chromatography columns and detectors. [Data from J. V. Hinshaw, "Setting Realistic Expectations for GC Optimization," *LCGC North Am.* **2006,** *24,* 1194.]

conductivity detector. For columns, the region to the left of 10^{-11} g is shaded because chromatography becomes progressively more problematic as the injected mass gets smaller. Analyte mass less than 10^{-11} g could be lost to adsorption or decomposition in the injector and column. Polar solutes might be lost to adsorption at even higher levels. At the upper end of the column scale, a dividing line recommends where to change from a thin-film stationary phase to a thick film to have adequate sample capacity.

Choosing the Injection Method

The last major decision is how to inject the sample. *Split injection* is best for high concentrations of analyte or gas analysis. Quantitative analysis is very poor. Less volatile components can be lost during injection. Split injection offers high resolution and can handle dirty samples if an adsorbent packing is added to the injection liner. Thermally unstable compounds can decompose during high-temperature injection.

Splitless injection is required for very dilute solutions. It offers high resolution but is poor for quantitative analysis because less volatile compounds can be lost during injection. (Less volatile compounds might not have transferred from the injection liner onto the column before the split valve is opened.) Splitless is better than split injection for compounds of moderate thermal stability because the injection temperature is lower. Splitless injection introduces sample onto the column slowly, so solvent trapping or cold trapping is required and temperature programming is necessary. Samples containing less than 100 ppm of each analyte can be analyzed with a column film thickness of <1 μm and splitless injection. Samples containing 100–1 000 ppm of each analyte require a column film thickness ≥1 μm.

On-column injection is best for quantitative analysis and for thermally sensitive compounds. It is strictly a low-resolution technique and cannot be used with columns whose inner diameter is less than 0.2 mm. It can handle dilute or concentrated solutions and relatively large or small volumes. Other column requirements are the same as in splitless injection.

Split injection:
• concentrated sample
• high resolution
• dirty samples (use packed liner)
• could cause thermal decomposition

Splitless injection:
• dilute sample
• high resolution
• requires solvent trapping or cold trapping

On-column injection:
• best for quantitative analysis
• thermally sensitive compounds
• low resolution

Terms to Understand

carrier gas	molecular sieve	sample preparation	splitless injection
cold trapping	on-column injection	selected ion monitoring	stir-bar sorptive extraction
electron capture detector	open tubular column	selected reaction monitoring	temperature programming
flame ionization detector	packed column	septum	thermal conductivity detector
gas chromatography	phase ratio	solid-phase microextraction	thermal desorption
general elution problem	purge and trap	solvent trapping	
guard column	retention gap	spiking	
headspace	retention index	split injection	

Summary

In gas chromatography, a volatile liquid or gaseous solute is carried by a gaseous mobile phase over a stationary phase on the inside of an open tubular column or on a solid support. Long, narrow, fused-silica open tubular columns have low capacity but give excellent separation. They can be wall-coated or porous layer. Packed columns provide high capacity but poor resolution. Solutes must be partially in the vapor phase to be eluted from a gas chromatography column. Solute partial pressure and retention are governed by thermodynamics, meaning that column temperature controls retention. Each liquid stationary phase most strongly retains solutes in its own polarity class ("like dissolves like"). Solid stationary phases, including porous carbon, alumina, and molecular sieves, are used to separate permanent gases. The retention index measures elution times in relation to those of linear alkanes. Temperature or pressure programming reduces elution times of strongly retained components. Without compromising separation efficiency, the linear velocity may be increased when H_2 or He, instead of N_2, is used as carrier gas. Split injection provides high-resolution separations of relatively concentrated samples. Splitless injection of very dilute samples requires solvent trapping or cold trapping to concentrate solutes at the start of the column to give sharp bands. On-column injection is best for quantitative analysis and for thermally unstable solutes.

Quantitative analysis is usually done with internal standards in gas chromatography. Coelution of an unknown peak with a spike of a known compound on several columns with different retention mechanisms is useful for qualitative identification of the unknown peak. Mass spectral detection provides qualitative information to help identify unknowns. The mass spectrometer is more sensitive and less subject to interference when selected ion monitoring or selected reaction monitoring is employed. Thermal conductivity detection has universal response but is not very sensitive. Flame ionization detection is sensitive enough for most columns and responds to most organic compounds. Electron capture, nitrogen-phosphorus, flame photometry, photoionization, chemiluminescence, and atomic emission detectors are specific for certain classes of compounds or individual elements.

You need to define the goal of an analysis before developing a chromatographic method. One key to successful chromatography is a clean sample. Solid-phase microextraction, stir-bar sorptive extraction, purge and trap, and thermal desorption can isolate volatile components from complex matrices. After sample preparation, the remaining decisions for method development are to select a detector, a column, and the injection method, in that order.

Exercises

24-A. (a) In Table 24-2, 2-pentanone has a retention index of 998 on a poly(ethylene glycol) column (also called Carbowax). Between which two straight-chain hydrocarbons is 2-pentanone eluted?

(b) An unretained solute is eluted from a certain column in 1.80 min. Decane ($C_{10}H_{22}$) is eluted in 15.63 min and undecane ($C_{11}H_{24}$) is eluted in 17.22 min. What is the retention time of a compound whose retention index is 1 050?

24-B. For isothermal elution of a homologous series of compounds (those with similar structures, but differing by the number of CH_2 groups in a chain), log t_r' is usually a linear function of the number of carbon atoms. A compound was known to be a member of the family

$$(CH_3)_2CH(CH_2)_nCH_2OSi(CH_3)_3$$

(a) From the gas chromatographic retention times given here, prepare a graph of log t_r' versus n and estimate the value of n in the chemical formula.

$n = 7$	4.0 min	CH_4	1.1 min
$n = 8$	6.5 min	unknown	42.5 min
$n = 14$	86.9 min		

(b) Calculate the retention factor for the unknown.

24-C. Resolution of two peaks (Equation 23-33) depends on the column plate number N, relative retention, α, and retention factor, k. Suppose you have two peaks with a resolution of 1.0 and you want to increase the resolution to 1.5 for baseline separation for quantitative analysis (Figure 23-10).

(a) You can increase the resolution to 1.5 by increasing column length. By what factor must the column length be increased? If flow rate is constant, how much more time will the separation require when the length is increased?

(b) You might change relative retention, α, by choosing a different stationary phase. If α were 1.016, to what value must it be increased to obtain a resolution of 1.5? If you were separating two alcohols with a (diphenyl)$_{0.05}$(dimethyl)$_{0.95}$ polysiloxane stationary phase (Table 24-1), what stationary phase would you choose to increase α? Will this change affect the time required for chromatography?

24-D. (a) *Review Section 5-4 on internal standards.* When a solution containing 234 mg of butanol (FM 74.12) and 312 mg of hexanol (FM 102.17) in 10.0 mL was separated by gas chromatography, the relative peak areas were butanol:hexanol = 1.00:1.45. Taking butanol as the internal standard, find the response factor for hexanol.

(b) Use Equation 24-8 to estimate the areas of the peaks for butanol and hexanol in Figure 24-8.

(c) The solution from which the chromatogram was generated contained 112 mg of butanol. What mass of hexanol was in the solution?

(d) What is the largest source in uncertainty in this problem? How great is this uncertainty?

24-E. *Review Section 5-4 on internal standards.* When 1.06 mmol of 1-pentanol and 1.53 mmol of 1-hexanol were separated by gas chromatography, they gave peak areas of 922 and 1 570 units, respectively. When 0.57 mmol of pentanol was added to an unknown containing hexanol, the peak areas were 843:816 (pentanol:hexanol). How much hexanol did the unknown contain?

Problems

24-1. (a) What are the relative advantages and disadvantages of packed and open tubular columns in gas chromatography?

(b) Explain the difference between wall-coated and porous-layer open tubular columns.

(c) What is the advantage of bonding (covalently attaching) the stationary phase to the column wall or cross-linking the stationary phase to itself?

24-2. Why do open tubular columns provide greater resolution than packed columns in gas chromatography?

24-3. (a) What are the advantages and disadvantages of using a narrower open tubular column?

(b) What are the advantages and disadvantages of using a longer open tubular column?

(c) What are the advantages and disadvantages of using a thicker film of stationary phase?

24-4. (a) What types of solutes are typically separated with a poly(dimethylsiloxane)-coated open tubular column?

(b) What types of solutes are typically separated with a poly(ethylene glycol)-coated open tubular column?

(c) What types of solutes are typically separated with a porous-layer open tubular column?

24-5. (a) What are the advantages and disadvantages of temperature programming in gas chromatography?

(b) What is the advantage of pressure programming?

24-6. (a) What are the characteristics of an ideal carrier gas?

(b) Why do H_2 and He allow more rapid linear velocities in gas chromatography than N_2 does, without loss of column efficiency (Figure 24-11)?

24-7. (a) When would you use split, splitless, or on-column injection in gas chromatography?

(b) Explain how solvent trapping and cold trapping work in splitless injection.

24-8. To which kinds of analytes do the following gas chromatography detectors respond?

(a) thermal conductivity   **(f)** photoionization

(b) flame ionization   **(g)** sulfur chemiluminescence

(c) electron capture   **(h)** atomic emission

(d) flame photometric   **(i)** mass spectrometer

(e) nitrogen-phosphorus   **(j)** vacuum ultraviolet absorbance

24-9. Why does a thermal conductivity detector respond to all analytes except the carrier gas? Why isn't the flame ionization detector universal?

24-10. Explain what is displayed in a reconstructed total ion chromatogram, in selected ion monitoring, and in selected reaction monitoring. Which technique is most selective and which is least selective and why?

24-11. What is the purpose of derivatization in chromatography? Give an example.

24-12. (a) Explain how solid-phase microextraction works. Why is cold trapping necessary during injection with this technique? Is all the analyte in an unknown extracted into the fiber in solid-phase microextraction?

(b) Explain the differences between stir-bar sorptive extraction and solid-phase microextraction. Which is more sensitive and why?

24-13. Why is splitless injection used with purge and trap sample preparation?

24-14. State the order of decisions in method development for gas chromatography.

24-15. This problem reviews concepts from Chapter 23. An unretained solute passes through a chromatography column in 3.7 min and analyte requires 8.4 min.

(a) Find the adjusted retention time and retention factor for the analyte.

(b) Find the phase ratio β for a 0.32-mm-diameter column with a 1.0-μm-thick film of stationary phase.

(c) Find the partition coefficient for the analyte.

(d) Determine the retention time on a similar length of 0.32-mm-diameter column with a 0.5-μm-thick film of the same stationary phase at the same temperature.

24-16. (a) If retention times in Figure 23-7 are 1.0 min for CH_4, 12.0 min for octane, 13.0 min for unknown, and 15.0 min for nonane, find the Kovats retention index for the unknown.

(b) What would be the Kovats index for the unknown if the phase ratio of the column were doubled?

(c) What would be the Kovats index for the unknown if the length of the column were halved?

24-17. Use Table 24-2 to predict the elution order of the following compounds from columns containing **(a)** poly(dimethylsiloxane), **(b)** (diphenyl)$_{0.35}$(dimethyl)$_{0.65}$polysiloxane, and **(c)** poly(ethylene glycol): hexane, heptane, octane, benzene, butanol, 2-pentanone.

24-18. Use Table 24-2 to predict the elution order of the following compounds from columns containing **(a)** poly(dimethylsiloxane), **(b)** (diphenyl)$_{0.35}$(dimethyl)$_{0.65}$polysiloxane, and **(c)** poly(ethylene glycol):

1. 1-pentanol (n-$C_5H_{11}OH$, b.p. 138°C)
2. 2-hexanone ($CH_3C(=O)C_4H_9$, b.p. 128°C)
3. heptane (n-C_7H_{16}, b.p. 98°C)
4. octane (n-C_8H_{18}, b.p. 126°C)
5. nonane (n-C_9H_{20}, b.p. 151°C)
6. decane (n-$C_{10}H_{22}$, b.p. 174°C)

24-19. (a) Use Trouton's rule, $\Delta H^{\circ}_{vap} \approx (88 \text{ J mol}^{-1} \text{ K}^{-1}) \cdot T_{bp}$, to estimate the enthalpy of vaporization of octane (b.p. 126°C).

(b) Use the form of the Clausius-Clapeyron equation below to estimate the vapor pressure of octane at the column temperature in Figure 24-9 (70°C).

$$\ln\left(\frac{P_1}{P_2}\right) = -\left(\frac{\Delta H_{vap}}{R}\right)\left(\frac{1}{T_1} - \frac{1}{T_2}\right)$$

(c) Calculate the vapor pressure for hexane (b.p. 69°C) at 70°C.

(d) What is the relationship between solute vapor pressure and retention?

(e) Why is the technique called "gas chromatography" if retained analytes are only partially vaporized?

24-20. Retention time depends on temperature, T, according to the equation $\log t'_r = (a/T) + b$, where a and b are constants for a specific compound on a specific column. A compound is eluted from a gas chromatography column at an adjusted retention time $t'_r = 15.0$ min when the column temperature is 373 K. At 363 K, $t'_r = 20.0$ min. Find the parameters a and b and predict t'_r at 353 K.

24-21. Describe how retention time of butanol on a poly(ethylene glycol) column will change with increasing temperature. Use the retention time for butanol in Figure 24-9b as the starting point.

24-22. This problem reviews concepts from Chapter 23 using Figure 24-7.

(a) Calculate the number of theoretical plates (N in Equation 23-30) and the plate height (H) for CO.

(b) Find the resolution (Equation 23-23) between argon and oxygen.

24-23. This problem reviews concepts from Chapter 23 using Figure 24-15.

(a) Calculate the retention factor for peak 11 given $t_m = 6.7$ min.

(b) Calculate the number of theoretical plates (N in Equation 23-31) and the plate height (H) for peak 11.

(c) Find the resolution (Equation 23-23) between peaks 16 and 17.

24-24. (a) Why is it illogical to use a thin stationary phase (0.2 μm) in a wide-bore (0.53-mm) open tubular column?

(b) Consider a narrow-bore (0.25 mm diameter), thin-film (0.10 μm) column with 5 000 plates per meter. Consider also a wide-bore (0.53 mm diameter), thick-film (5.0 μm) column with 1 500 plates per meter. The density of stationary phase is approximately 1.0 g/mL. What mass of stationary phase is in each column in a length equivalent to one theoretical plate? How many nanograms of analyte can be injected into each column if the mass of analyte is not to exceed 1.0% of the mass of stationary phase in one theoretical plate?

24-25. The graph shows van Deemter curves for *n*-nonane at 70°C in the 3.0-m-long microfabricated column in Box 24-2 with a 1- to 2-μm-thick stationary phase.

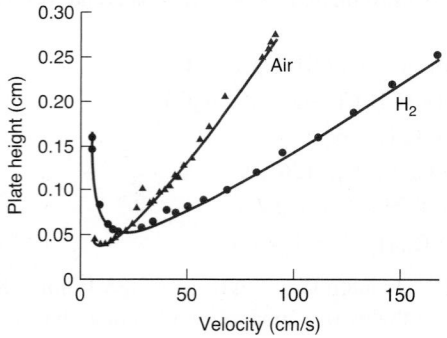

van Deemter curves. [Data from G. Lambertus, A. Elstro, K. Sensenig, J. Potkay, M. Agah, S. Scheuering, K. Wise, F. Dorman, and R. Sacks, "Design, Fabrication, and Evaluation of Microfabricated Columns for Gas Chromatography," *Anal. Chem.* **2004,** *76,* 2629.]

(a) Why would air be chosen as the carrier gas? What is the danger of using air as carrier gas?

(b) Measure the optimum velocity and plate height for air and for H_2 carrier gases.

(c) How many plates are there in the 3-m-long column for each carrier gas at optimum flow rate?

(d) How long does unretained gas take to travel through the column at optimum velocity for each carrier gas?

(e) If stationary phase is sufficiently thin with respect to column diameter, which of the two mass transfer terms (23-40 or 23-41) becomes negligible? Why?

(f) Why is the loss of column efficiency at high flow rates less severe for H_2 than for air carrier gas?

24-26. (a) How can you improve the resolution between two closely spaced peaks in gas chromatography?

(b) What approach from **(a)** would be most cost effective (not involve a purchase)?

24-27. (a) When a solution containing 234 mg of pentanol (FM 88.15) and 237 mg of 2,3-dimethyl-2-butanol (FM 102.17) in 10.0 mL was separated, relative peak areas were pentanol : 2,3-dimethyl-2-butanol = 0.913 : 1.00. Considering pentanol to be the internal standard, find the response factor for 2,3-dimethyl-2-butanol.

(b) Use Equation 24-8 to find the areas for pentanol and 2,3-dimethyl-2-butanol in Figure 24-8.

(c) The concentration of pentanol internal standard in the unknown solution was 93.7 mM. What was the concentration of 2,3-dimethyl-2-butanol?

24-28. A standard solution containing 6.3×10^{-8} M iodoacetone and 2.0×10^{-7} M *p*-dichlorobenzene (an internal standard) gave peak areas of 395 and 787, respectively, in a gas chromatogram. A 3.00-mL unknown solution of iodoacetone was treated with 0.100 mL of 1.6×10^{-5} M *p*-dichlorobenzene and the mixture was diluted to 10.00 mL. Gas chromatography gave peak areas of 633 and 520 for iodoacetone and *p*-dichlorobenzene, respectively. Find the concentration of iodoacetone in the 3.00 mL of original unknown.

24-29. Heptane, decane, and an unknown had adjusted retention times of 12.6 min (heptane), 22.9 min (decane), and 20.0 min (unknown). The retention indexes for heptane and decane are 700 and 1 000, respectively. Find the retention index for the unknown.

24-30. In the analysis of odorants in tequila in Figure 24-21, tequila was diluted with water and extracted four times with dichloromethane (CH_2Cl_2, b.p. 40°C). The 400 mL of CH_2Cl_2 was evaporated down to 1 mL and 1 μL of the extract was injected on-column onto a poly(ethylene glycol) open tubular column (30 m $\times$ 0.53 mm, film thickness = 1 μm) initially at 50°C and then ramped to 230°C.

(a) Why was the diluted tequila extracted four times with dichloromethane instead of once with a larger volume?

(b) Why was on-column injection used?

(c) Why was a poly(ethylene glycol) column chosen for this application?

(d) What was the phase ratio of the column?

(e) Why was a wide-bore 0.53-mm-diameter column chosen for this application?

24-31. The gasoline additive methyl *t*-butyl ether (MTBE) has been leaking into groundwater ever since its introduction in the 1990s. MTBE can be measured at parts per billion levels by solid-phase microextraction from groundwater to which 25% (wt/vol) NaCl has been added (*salting out*, Problem 8-9). After microextraction, analytes are thermally desorbed from the fiber in the port of a gas chromatograph. The figure on the next page shows a reconstructed total ion chromatogram and selected ion monitoring of substances desorbed from the extraction fiber.

Methyl *t*-butyl ether	Ethyl *t*-butyl ether	*t*-Amyl methyl ether
MTBE	ETBE	TAME
nominal mass: 88	102	102

(a) What is the purpose of adding NaCl prior to extraction?

(b) What nominal mass is being observed in selected ion monitoring? Why are only three peaks observed?

(c) Here is a list of major ions above *m/z* 50 in the mass spectra. The base (tallest) peak is marked by an asterisk. Given that MTBE and TAME have an intense peak at *m/z* 73 and there is no significant peak at m/z 73 for ETBE, suggest a structure for m/z 73. Suggest structures for all ions listed in the table.

MTBE	ETBE	TAME
73*	87	87
57	59*	73*
	57	71
		55

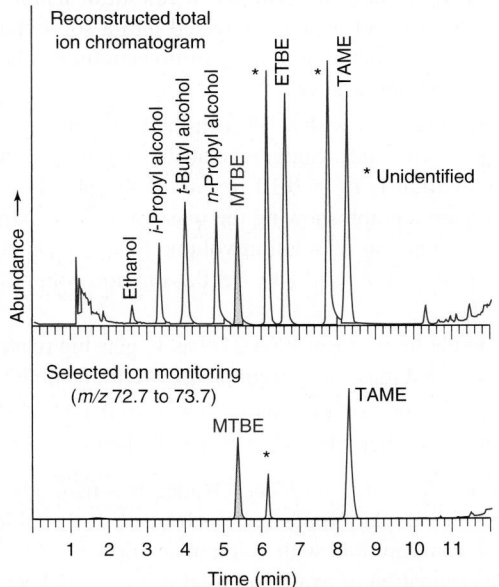

Reconstructed total ion chromatogram and selected ion monitoring of solid-phase microextract of groundwater. Chromatography conditions: 30 m × 0.32 mm column with 5-μm-thick film of poly(dimethylsiloxane). Temperature = 50°C for 4 min, then raised 20°C/min to 90°C, held for 3 min, and then raised 40°C/min to 200°C. [Data from D. A. Cassada, Y. Zhang, D. D. Snow, and R. F. Spalding, "Trace Analysis of Ethanol, MTBE, and Related Oxygenate Compounds in Water Using Solid-Phase Microextraction and Gas Chromatography/Mass Spectrometry," Anal. Chem. **2000**, 72, 4654.]

24-32. Here is a student procedure to measure nicotine in urine. A 1.00-mL sample of biological fluid was placed in a 12-mL vial containing 0.7 g Na_2CO_3 powder. After 5.00 μg of the internal standard 5-aminoquinoline were injected, the vial was capped with a Teflon-coated silicone rubber septum. The vial was heated to 80°C for 20 min and then a solid-phase microextraction needle was passed through the septum and left in the headspace for 5.00 min. The fiber was retracted and inserted into a gas chromatograph. Volatile substances were desorbed from the fiber at 250°C for 9.5 min in the injection port while the column was at 60°C. The column temperature was then raised to 260°C at 25°C/min and eluate was monitored by electron ionization mass spectrometry with selected ion monitoring at m/z 84 for nicotine and m/z 144 for internal standard. Calibration data from replicate standard mixtures taken through the same procedure are given in the table.

Nicotine in urine (μg/L)	Area ratio m/z 84/144
12	$0.05_6, 0.05_9$
51	$0.40_2, 0.39_1$
102	$0.68_4, 0.66_9$
157	$1.01_1, 1.06_3$
205	$1.27_8, 1.35_5$

Data from A. E. Wittner, D. M. Klinger, X. Fan, M. Lam, D. T. Mathers, and S. A. Mabury, "Quantitative Determination of Nicotine and Cotinine in Urine and Sputum Using a Combined SPME-GC/MS Method," J. Chem. Ed. 2002, 79, 1257.

(a) Why was the vial heated to 80°C before and during extraction?

(b) Why was the chromatography column kept at 60°C during thermal desorption of the extraction fiber?

(c) Suggest a structure for m/z 84 from nicotine. What is the m/z 144 ion from the internal standard, 5-aminoquinoline?

Nicotine
$C_{10}H_{14}N_2$

5-Aminoquinoline
$C_9H_8N_2$

(d) ⬛🔲 Urine from an adult female nonsmoker had an area ratio m/z 84/144 = 0.51 and 0.53 in replicate determinations. Urine from a nonsmoking girl whose parents are heavy smokers had an area ratio 1.18 and 1.32. Find the nicotine concentration (μg/L) and its standard uncertainty in the urine of each person.

24-33. Nitric oxide (NO) is a cell-signaling agent in physiologic processes including vasodilation, inhibition of clotting, and inflammation. A sensitive chromatography–mass spectrometry method was developed to measure two of its metabolites, nitrite (NO_2^-) and nitrate (NO_3^-), in biological fluids. Internal standards, $^{15}NO_2^-$ and $^{15}NO_3^-$, were added to the fluid at concentrations of 80.0 and 800.0 μM, respectively. Natural $^{14}NO_2^-$ and $^{14}NO_3^-$ plus the internal standards were then converted to volatile derivatives:

Pentafluorobenzyl bromide

m/z 62 for ^{14}N
m/z 63 for ^{15}N

m/z 46 for ^{14}N
m/z 47 for ^{15}N

Because biological fluids are so complex, the derivatives were first isolated by high-performance liquid chromatography. For quantitative analysis, liquid chromatography peaks corresponding to the two products were injected into a gas chromatograph, ionized by *negative ion* chemical ionization (giving major peaks for NO_2^- and NO_3^-) and the products measured by selected ion monitoring. Results are shown in the figure on the next page. If the ^{15}N internal standards undergo the same reactions and same separations at the same rate as the ^{14}N analytes, then the concentrations of analytes are simply

$$[^{14}NO_x^-] = [^{15}NO_x^-](R - R_{blank})$$

where R is the measured peak area ratio (m/z 46/47 for nitrite and m/z 62/63 for nitrate) and R_{blank} is the measured ratio of peak areas in a blank prepared from the same buffers and reagents with no added nitrite or nitrate. The ratios of peak areas are m/z 46/47 = 0.062 and m/z 62/63 = 0.538. The ratios for the blank were

m/z 46/47 = 0.040 and m/z 62/63 = 0.058. Find the concentrations of nitrite and nitrate in the urine.

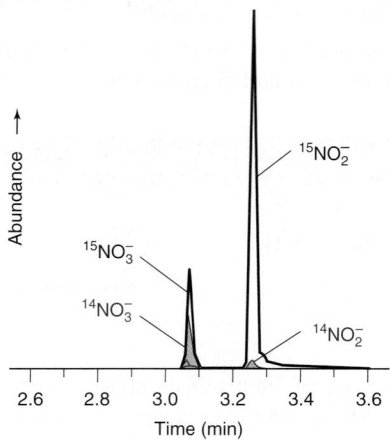

Selected ion chromatogram showing negative ions at m/z 46, 47, 62, and 63 obtained by derivatizing nitrite and nitrate plus internal standards ($^{15}NO_2^-$ and $^{15}NO_3^-$) in urine. [Data from D. Tsikas, "Derivatization and Quantification of Nitrite and Nitrate in Biological Fluids by Gas Chromatography/Mass Spectrometry," *Anal. Chem.* **2000,** *72,* 4064; *Anal. Chem.* **2010,** *82,* 2585.]

24-34. *van Deemter equation for open tubular column.* Equation 23-37 contains terms (A, B, and C) describing three band-broadening mechanisms.

(a) Which term is 0 for an open tubular column? Why?

(b) Express the value of B in terms of measurable physical properties.

(c) Express the value of C in terms of measurable physical quantities.

(d) The linear flow rate that produces minimum plate height is found by setting the derivative $dH/du_x = 0$. Find an expression of the minimum plate height in terms of the measurable physical quantities used to answer **(b)** and **(c)**.

24-35. *Theoretical performance in gas chromatography.* As the inside radius of an open tubular column is decreased, the maximum possible column efficiency increases and sample capacity decreases. For a thin stationary phase that equilibrates rapidly with analyte, the minimum theoretical plate height is given by

$$\frac{H_{min}}{r} = \sqrt{\frac{1 + 6k + 11k^2}{3(1 + k)^2}}$$

where r is the inside radius of the column and k is the retention factor.

(a) Find the limit of the square-root term as $k \to 0$ (unretained solute) and as $k \to \infty$ (infinitely retained solute).

(b) If the column radius is 0.10 mm, find H_{min} for the two cases in **(a)**.

(c) What is the maximum number of theoretical plates in a 50-m-long column with a 0.10-mm radius if $k = 5.0$?

(d) The phase ratio is defined as the volume of the mobile phase divided by the volume of the stationary phase ($\beta = V_m/V_s$). Derive the relationship between β and the thickness of the stationary phase in a wall-coated column (d_f) and the inside radius of the column (r).

(e) Find k if $K = 1\,000$, $d_f = 0.20\ \mu m$, and $r = 0.10$ mm.

24-36. Consider the chromatography of $n\text{-}C_{12}H_{26}$ on a 25 m $\times$ 0.53 mm open tubular column of 5% phenyl–95% methyl polysiloxane with a stationary phase thickness of 3.0 μm and He carrier gas at 125°C.

The observed retention factor for $n\text{-}C_{12}H_{26}$ is 8.0. Measurements were made of plate height, H, at various values of linear velocity, u_x (m/s). A least-squares curve through the data points is given by

$$H \text{ (m)} = (6.0 \times 10^{-5}\ m^2/s)/u_x + (2.09 \times 10^{-3}\ s)u_x$$

From the coefficients of the van Deemter equation, find the diffusion coefficient of $n\text{-}C_{12}H_{26}$ in the mobile and stationary phases. Why is one of these diffusion coefficients so much greater than the other?

24-37. 🖿 *Efficiency of solid-phase microextraction.* Equation 24-9 gives the mass of analyte extracted into a solid-phase microextraction fiber as a function of the partition coefficient between the fiber coating and the solution.

(a) A commercial fiber with a 100-μm-thick coating has a film volume of 6.9×10^{-4} mL. Suppose that the initial concentration of analyte in solution is $c_0 = 0.10\ \mu g/mL$ (100 ppb). Use a spreadsheet to prepare a graph showing the mass of analyte extracted into the fiber as a function of solution volume for partition coefficients of 10 000, 5 000, 1 000, and 100. Let the solution volume vary from 0 to 100 mL.

(b) Evaluate the limit of Equation 24-9 as V_s gets big relative to KV_f. Does the extracted mass in your graph approach this limit?

(c) What percentage of the analyte from 10.0 mL of solution is extracted into the fiber when $K = 100$ and when $K = 10\,000$?

24-38. *Literature search problem.* Oxalate is a naturally occurring substance found in plant foods such as fruits and vegetables. Within the body it can combine with calcium to form kidney or urinary stones. Determination of oxalate in food is important because of the potential for harmful effects on health. Search the literature for a headspace gas chromatography method for the determination of oxalate in food and answer the following questions.

(a) Give the citation (authors, title, journal name, year, volume, pages) for a research paper describing this analysis.

(b) What type of gas chromatography detector is used? Why is this detector appropriate?

(c) What are the precision, limit of quantification, and linear range of the method?

(d) How was the gas chromatographic determination validated?

(e) What alternative methods could be used for determining oxalate in food?

(f) Challenge question: What type of column was used?

24-39. *Literature search problem.* Genotoxic compounds damage DNA and cause mutations or cancer. Regulatory guidelines limit genotoxic impurities in pharmaceuticals. Mesityl oxide (4-methyl-3-penten-2-one) is an intermediate product in the synthesis of some drugs. Search the literature for a gas chromatography method for the determination of mesityl oxide in the pharmaceutical enalapril maleate and answer the following questions.

(a) Give the citation (authors, title, journal name, year, volume, pages) for a research paper describing this analysis.

(b) What is the instrumental method used?

(c) What type of column was used?

(d) What quantification method is used?

(e) What are the precision, limit of detection, limit of quantification, and linear range of the method?

(f) For how long could a standard solution be used?

25 High-Performance Liquid Chromatography

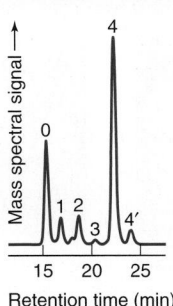

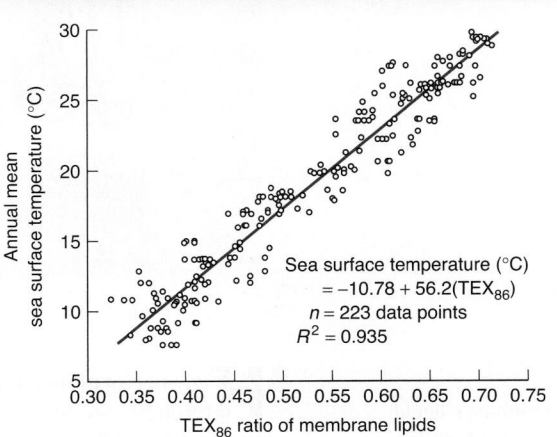

Correlation of today's sea surface temperature with TEX_{86} in 223 core-top sediments from the Atlantic, Pacific, and Indian Oceans. [Data from J.-H. Kim, S. Schouten, E. C. Hopmans, B. Donner, and J. S. Sinninghe Damsté, "Global Sediment Core-top Calibration of the TEX_{86} Paleothermometer in the Ocean," *Geochim. Cosmochim. Acta* **2008,** *72,* 1154.]

Liquid chromatogram of archaeal cell membrane lipids extracted from sea floor sediment. Compounds 4 and 4′ are isomers with an unknown relationship. Chromatography was run on a 150 × 2.1 mm column containing 3-μm Prevail Cyano phase. [Data from S. Schouten, C. Huguet, E. C. Hopmans, M. V. Kienhuis, and J. S. Sinninghe Damsté, "Analytical Methodology for TEX_{86} Paleothermometry by High-Performance Liquid Chromatography/Atmospheric Pressure Chemical Ionization-Mass Spectrometry," *Anal. Chem.* **2007,** *79,* 2940.]

Archaea are single-cell organisms that constitute a substantial fraction of life in the ocean. They manufacture cell membrane lipids, which can be measured by liquid chromatography. Lipids 0 and 4 are the major components. Of the minor components, type 1 is predominant in archaea living at low temperature (near 0°C). Types 2, 3, and 4′ are manufactured in increasing amounts at warmer ocean temperatures, presumably giving membranes proper fluidity. When archaea die, they fall to the ocean floor, where their lipids remain in the sediment. The surface of the sediment contains remains of the most recent archaea. The deeper we dig into the sediment, the older are the remains.

The graph shows the correlation between today's sea surface temperature and membrane lipid types found in the topmost sediment. The function in the graph is the sum of lipids 2, 3, and 4′ expressed as a fraction of lipids 1, 2, 3, and 4′:

$$TEX_{86} \equiv \frac{[2] + [3] + [4']}{[1] + [2] + [3] + [4']} \text{ (measured by chromatography)}$$

Empirical correlation: Sea surface temperature (°C) $= -10.78 + 56.2(TEX_{86})$

By measuring archaeal membrane lipids as a function of depth in ocean sediment and in sedimentary rock on land, it is possible to construct a history of ocean temperature and to compare the membrane lipid ratio to other indicators of climate.[1]

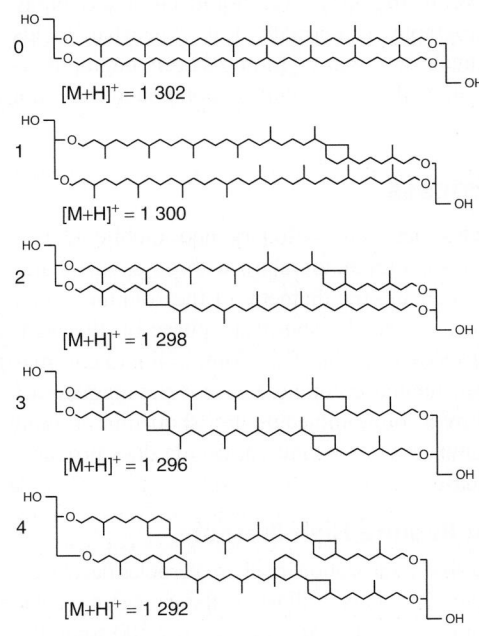

0   $[M+H]^+ = 1\ 302$

1   $[M+H]^+ = 1\ 300$

2   $[M+H]^+ = 1\ 298$

3   $[M+H]^+ = 1\ 296$

4   $[M+H]^+ = 1\ 292$

4′   Regio-isomer of 4
$[M+H]^+ = 1\ 292$

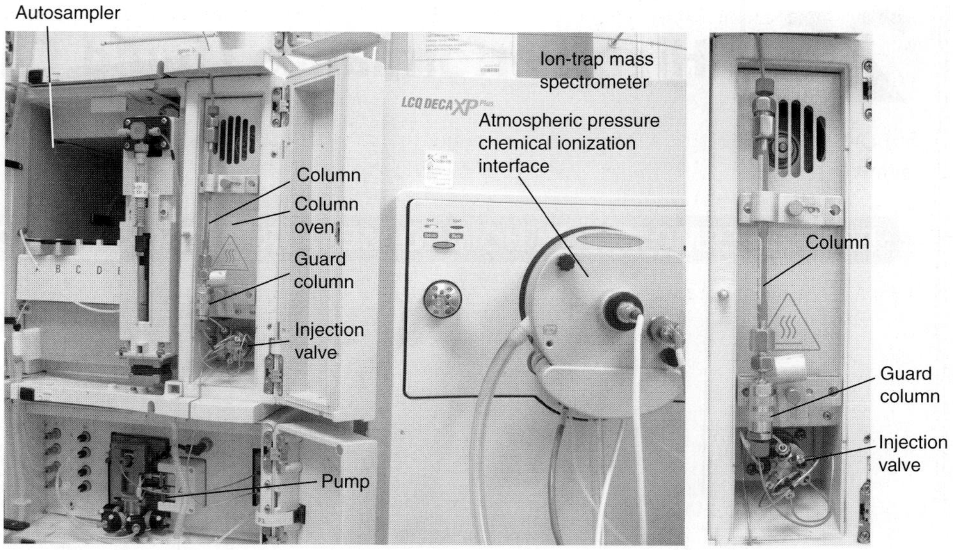

Autosampler

Ion-trap mass spectrometer

Atmospheric pressure chemical ionization interface

Column

Column oven

Guard column

Injection valve

Pump

Column

Guard column

Injection valve

The pioneer of high-performance liquid chromatography was C. Horváth at Yale University in 1965.

High-performance liquid chromatography (HPLC) uses high pressure to force solvent through closed columns containing fine particles that give high-resolution separations.[2-8] The HPLC system in Figure 25-1 consists of an autosampler, a solvent delivery system (a pump), a sample injection valve, a high-pressure chromatography column, and a mass spectrometer, which serves as a detector.[9] Not shown in the photograph are solvent reservoirs, a photodiode array absorbance detector, and a computer to control the hardware and display results. The column is housed in an oven whose door is normally closed to keep the column at constant temperature. In this chapter, we discuss liquid-liquid partition and liquid-solid adsorption chromatography. Chapter 26 deals with ion-exchange, molecular exclusion, affinity, and hydrophobic interaction chromatography.

Chromatographers generally choose gas chromatography over liquid chromatography when there is a choice, because gas chromatography is normally less expensive, yields greater separation efficiency, and generates much less waste. Liquid chromatography is important because many chemical, biochemical, and pharmaceutical compounds are not sufficiently volatile for gas chromatography.

25-1 The Chromatographic Process

Increasing the rate at which solute equilibrates between stationary and mobile phases increases the efficiency of chromatography. For gas chromatography with an open tubular column, rapid equilibration is accomplished by reducing the diameter of the column so that molecules can diffuse quickly between the channel and the stationary phase on the wall. Diffusion in liquids is 10^4 times slower than diffusion in gases. Therefore, in liquid chromatography, it is not generally feasible to use open tubular columns, because the diameter of the solvent channel is too great to be traversed by a solute molecule in a short time. Liquid chromatography is conducted with packed columns so that a solute molecule does not have to diffuse very far to encounter the stationary phase.

Increasing efficiency is equivalent to decreasing plate height, H, in the van Deemter equation (23-37):

$$H \approx A + \frac{B}{u_x} + Cu_x$$

u_x = linear velocity

Small Particles Give High Efficiency but Require High Pressure

The efficiency of a packed column increases as the size of the stationary phase particles decreases. Typical particle sizes in HPLC are 1.7–5 μm. Figure 25-2a and b illustrate the increased resolution afforded by decreasing particle size. Plate number increased from 2 000 to 7 500 when the particle size decreased, so the peaks are sharper with the smaller particle size. In Figure 25-2c, a stronger solvent was used to elute the peaks in less time. Decreasing particle size permits us to improve resolution or to maintain the same resolution while decreasing run time.

EXAMPLE Scaling Relations Between Columns[10]

Commonly, silica particles occupy ~40% of the column volume and solvent occupies ~60% of the column volume, regardless of particle size. The column used in Figure 25-2a has an inside diameter of 4.6 mm and was run at a volume flow rate (u_v) of 3.0 mL/min with a sample size of 20 μL. The column used in Figure 25-2b has a diameter of $d_c = 2.1$ mm. What flow rate should be used in trace *b* to achieve the same linear velocity (u_x) as in trace *a*? What sample volume should be injected?

Solution Column volume is proportional to the square of column diameter. Changing the diameter from 4.6 to 2.1 mm reduces volume by a factor of $(2.1/4.6)^2 = 0.208$. Therefore, u_v should be reduced by a factor of 0.208 to maintain the same linear velocity.

u_v(small column) $= 0.208 \times u_v$(large column) $= (0.208)(3.0$ mL/min$) = 0.62$ mL/min

To maintain the same ratio of injected sample to column volume,

Injection volume in small column $= 0.208 \times$ (injection volume in large column)

$= (0.208)(20$ μL$) = 4.2$ μL

TEST YOURSELF What should be the volume flow rate and injected volume for a 1.5-mm-diameter column? (*Answer:* 0.32 mL/min, 2.1 μL)

van Deemter plots of plate height versus linear flow rate in Figure 25-3 show that small particles reduce plate height and that plate height is not very sensitive to increased flow rate when the particles are small. For real samples on a column operating under typical conditions, the number of theoretical plates in a column of length L (cm) is *approximately*[11]

$$N \approx \frac{3\,000 \cdot L(\text{cm})}{d_p(\mu\text{m})} \qquad (25\text{-}1)$$

where d_p is the particle diameter in μm. The 5.0-cm-long column in Figure 25-2a with 4.0-μm-diameter particles is predicted to provide $\sim(3\,000)(5.0)/4.0 = 3\,800$ plates. The observed plate number for the second peak is 2 000. Perhaps the column was not run at optimum flow rate. When the stationary phase particle diameter is reduced to 1.7 μm, the optimum plate number is expected to be $\sim(3\,000)(5.0)/1.7 = 8\,800$. The observed value is 7 500.

One reason why small particles give better resolution is that they provide more uniform flow through the column, thereby reducing the multiple path term, A, in the van Deemter equation (23-37). A second reason is that the distance through which solute must diffuse in the mobile and stationary phases is on the order of the particle size. The smaller the particles, the less distance solute must diffuse. This effect decreases the C term in the van Deemter equation for finite equilibration time. Optimum flow rate for small particles is faster than for large particles because the distance through which the solutes must diffuse is smaller.

An added benefit of small particle size, coupled with a narrow column and higher flow, is that analyte is not diluted so much as it travels through the column. The limit of quantitation for conditions in Figure 25-2c (50 μg/L) is four times lower than the limit of quantitation in Figure 25-2a (200 μg/L).

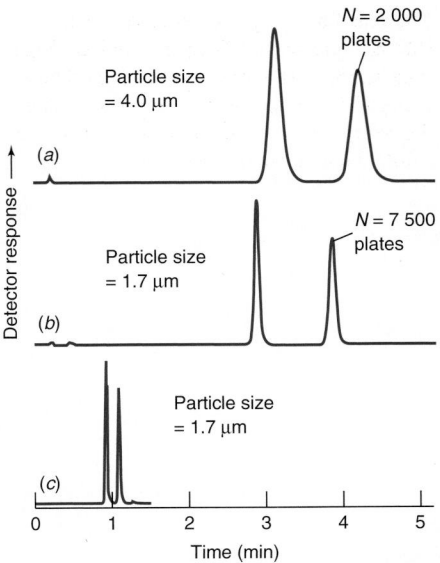

FIGURE 25-2 (*a* and *b*) Chromatograms of the same sample run at the same linear velocity on 5.0-cm-long columns packed with C_{18}-silica. (*c*) A stronger solvent was used to elute solutes more rapidly from the column in panel *b*. [Data from Y. Yang and C. C. Hodges, "Assay Transfer from HPLC to UPLC for Higher Analysis Throughput," *LCGC North Am. Supplement,* May 2005, p. S31.]

Optimal $H \approx 2 \times$ particle diameter $= 2d_p$

Smaller particle size leads to
• higher plate number
• higher pressure
• shorter optimum run time
• lower detection limit

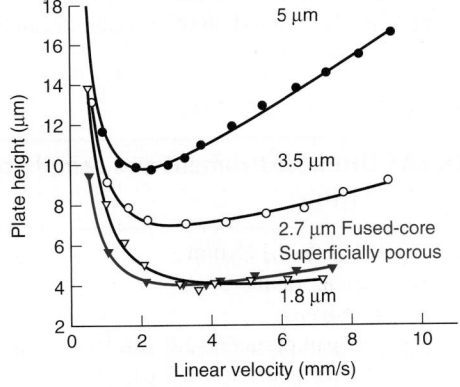

FIGURE 25-3 van Deemter curves: Plate height as a function of *linear velocity* (mm/s) for microporous (Figure 25-5) stationary phase particle diameters of 5.0, 3.5, and 1.8 μm, as well as *superficially porous* particles (Figure 25-9) with a diameter of 2.7 μm (0.5-μm porous layer thickness). Measurements for naphthalene eluted from C_{18}-silica (50 mm × 4.6 mm) with 60 vol% acetonitrile/40 vol% H_2O at 24°C. [Data from MAC-MOD Analytical, Chadds Ford, PA.]

TABLE 25-1	Performance as a function of particle diameter		
Particle size d_p (μm)	Retention time (min)	Plate number (N)	Required pressure (bar)
5.0	30	25 000	19
3.0	18	42 000	87
1.5	9	83 000	700
1.0	6	125 000	2 300

NOTE: *Theoretical performance of 25-cm-long × 33-μm-diameter capillary for minimum plate height for solute with retention factor k = 2 and diffusion coefficient = 6.7 × 10^{-10} m²/s in acetonitrile/water eluent.*

SOURCE: *J. E. MacNair, K. D. Patel, and J. W. Jorgenson, "Ultrahigh-Pressure Reversed-Phase Capillary Liquid Chromatography with 1.0-μm Particles," Anal. Chem. **1999**, 71, 700.*

The penalty for small particle size is resistance to solvent flow. The pressure required to drive solvent through a column is

Column pressure:
$$P = f\frac{u_v\eta L}{\pi r^2 d_p^2}$$
(25-2)

where u_v is volume flow rate, η is the viscosity of the solvent, L is the length of the column, r is column radius, and d_p is the particle diameter. The factor f depends on particle shape and particle packing. The physical significance of Equation 25-2 is that pressure in HPLC is proportional to flow rate and column length, and inversely proportional to the square of column radius (or diameter) and the square of particle size. The difference between traces a and b in Figure 25-2 is that particle size was decreased from 4.0 μm to 1.7 μm and column diameter was decreased from 4.6 mm to 2.1 mm. Therefore, the required pressure increases by a factor of (4.6 mm/2.1 mm)²(4.0 μm/1.7 μm)² = 27. That is, *27 times more pressure* is required to operate the column in Figure 25-2b.

Before 2004, HPLC operated at pressures of ~7–40 MPa (70–400 bar, 1 000–6 000 pounds/inch²) to attain flow rates of ~0.5–5 mL/min. Then commercial equipment became available to employ 1.5- to 2-μm-diameter particles at pressures up to 100 MPa (1 000 bar, 15 000 pounds/inch²). These instruments provide substantially increased resolution or decreased run time. Table 25-1 shows theoretical performance for different particle sizes; such performance was realized in research with ultrahigh-pressure equipment. Chromatography with 1.5- to 2-μm-diameter particles at high pressure is commonly called **UHPLC (ultra-high-performance liquid chromatography)**. Table 25-2 shows that smaller extra-column volumes and increased sample cleanliness (finer filtration) are required to maintain UHPLC performance. Peaks eluted from a UHPLC column are so narrow that fast detectors are required.

Another penalty of small particle size is increased frictional heating as solvent is forced through the particle bed.[12] The center of a column is warmer than the outer wall, and the outlet is warmer than the inlet. A 100-mm-long × 2.1-mm-diameter column containing 1.7-μm particles eluted with acetonitrile generates a temperature difference of ~10°C from the inlet to the outlet at a flow rate of 1.0 mL/min. The centerline of the column can be ~2°C warmer than the

Viscosity measures resistance of a fluid to flow. The more viscous a liquid, the greater the pressure for a given flow rate.

HPLC: High-Performance Liquid Chromatography

UHPLC: Ultra-High-Performance Liquid Chromatography

Particle diameter ≤2μm

Analogy of fluid flow with electric current:

Electric power (W) = current (A) × voltage (V)

Chromatographic heat generation:

Power (W) = volume flow rate × pressure drop
 m³/s Pa = kg/(m · s²)

TABLE 25-2	HPLC and UHPLC instrument and sample characteristics	
Property	HPLC	UHPLC
Connecting tubing	0.175–0.125 mm inner diameter	0.125–0.062 5 mm inner diameter
Extra column variance	≥40 μL²	≤10 μL²
Column frits	5-μm particles: 2.0-μm frits 3-μm particles: 0.5-μm frits	<2-μm particles: 0.2-μm frits
Sample filtration	5- or 3-μm particles: 0.5-μm filter or centrifuge	<2-μm particles: 0.2-μm filter

SOURCE: *J. W. Dolan, "UHPLC Tips and Techniques," LCGC North Am. **2010**, 28, 944; S. Fekete and J. Fekete, "The Impact of Extra-Column Band Broadening on the Chromatographic Efficiency of 5-cm-Long Narrow-Bore Very Efficient Columns," J. Chromatogr. A **2011**, 1218, 5286.*

wall. To avoid undue band broadening from temperature differences, column diameter should be ≤2.1 mm for sub-2-μm particles.

The Column

Columns are expensive and easily degraded by dust or particles in the sample or solvent and by irreversible adsorption of impurities from the sample or solvent. To avoid introducing particulate matter into the column, samples should be centrifuged and/or filtered through a ≤0.5-μm filter (Table 25-2) before they are loaded into vials for an autosampler or taken into a syringe for manual injection. An in-line 0.5-μm filter should be installed immediately downstream of the autosampler.

The HPLC equipment in Figure 25-1 uses steel or high-strength plastic columns that are 3–30 cm in length, with an inner diameter of 1–5 mm (Figure 25-4). The entrance to the main column is protected by a short **guard column** containing the same stationary phase as the main column. Fine particles and strongly adsorbed solutes are retained in the guard column, which is periodically replaced when column pressure increases or after a set number of injections or time in service. Although guard columns make sense with 10- to 30-cm-long chromatography columns, many people do not consider a guard column to be cost effective for ≤5-cm columns.

Heating a chromatography column[13] decreases the viscosity of the solvent, thereby reducing the required pressure or permitting faster flow. Increased temperature decreases retention times and affects relative retention. However, increased temperature can degrade the stationary phase and decrease column lifetime. Most HPLC stationary phases can withstand up to 60°C and some even higher. When column temperature is not controlled, retention fluctuates with the ambient temperature. Using a column heater set 10°C above room temperature improves the reproducibility of retention times and the precision of quantitative analysis. Some chromatographers routinely conduct all separations at 50° or 60°C. For a heated column, the mobile phase should be passed through a preheating metal coil between the injector and the column so that the solvent and column are at the same temperature. If their temperatures differ, peaks become distorted and retention times change.

In the recent past, the most common HPLC column diameter was 4.6 mm. Now, 2.1 mm is becoming most common. The narrow column is more compatible with mass spectrometry, which requires low solvent flow. The narrow column requires less sample and produces less waste. Instruments that used a 4.6-mm column can also operate with a 2.1-mm column. Columns narrower than 2.1 mm require specially designed instruments to reduce band broadening outside the column (Table 25-2). Capillary columns as narrow as 25 μm can be used.

The Stationary Phase

The most common support is highly pure, spherical, **microporous particles** of silica (Figure 25-5) that have a surface area of several hundred square meters per gram. Greater than 99% of this surface area is inside the pores. Pores must be wide enough for the solute and solvent to enter freely. Most silica cannot be used above pH 8, because it dissolves in base. Figure 25-6 shows the structure of ordinary silica and ethylene-bridged silica, which resists hydrolysis up to pH 12.[14] For separation of basic compounds at pH 8–12, ethylene-bridged silica or polymeric supports such as polystyrene (Figure 26-1) can be used.

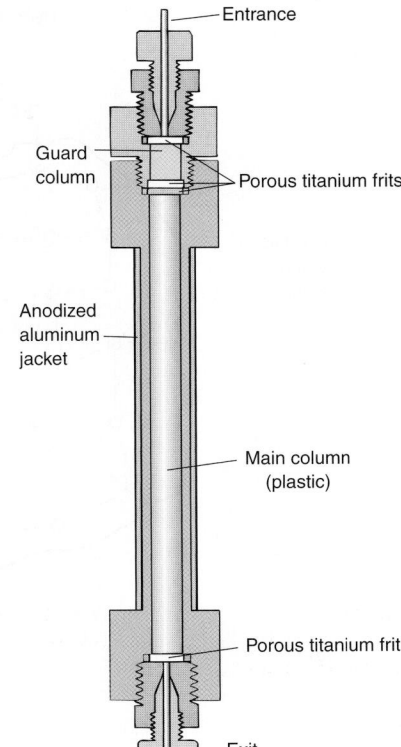

FIGURE 25-4 HPLC column with replaceable guard column to collect irreversibly adsorbed impurities. Titanium frits retain the stationary phase and distribute liquid evenly over the diameter of the column. [Information from Upchurch Scientific, Oak Harbor, WA.] **In Figure 25-1, flow direction is bottom to top, which is opposite the flow direction in this figure.**

Band broadening outside the column was discussed in Section 23-5.

Small molecules: 6- to 12-nm pores
Polypeptides and proteins: >30-nm pores

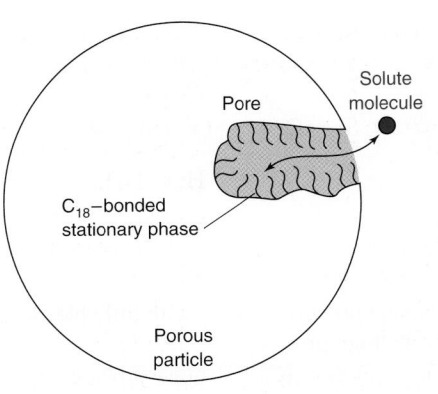

(a)

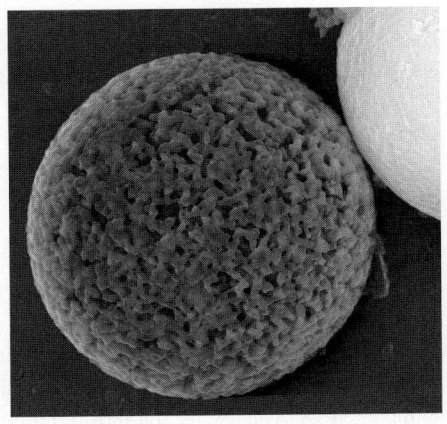

(b)

FIGURE 25-5 (a) Schematic structure of porous particle. The pore diameter is typically ~0.2% of the diameter of the particle. (b) Scanning electron micrograph of 4.4-μm-diameter microporous silica chromatography particle from an experimental batch made by K. Wyndham at Waters Corporation. [Courtesy J. Jorgensen, University of North Carolina.]

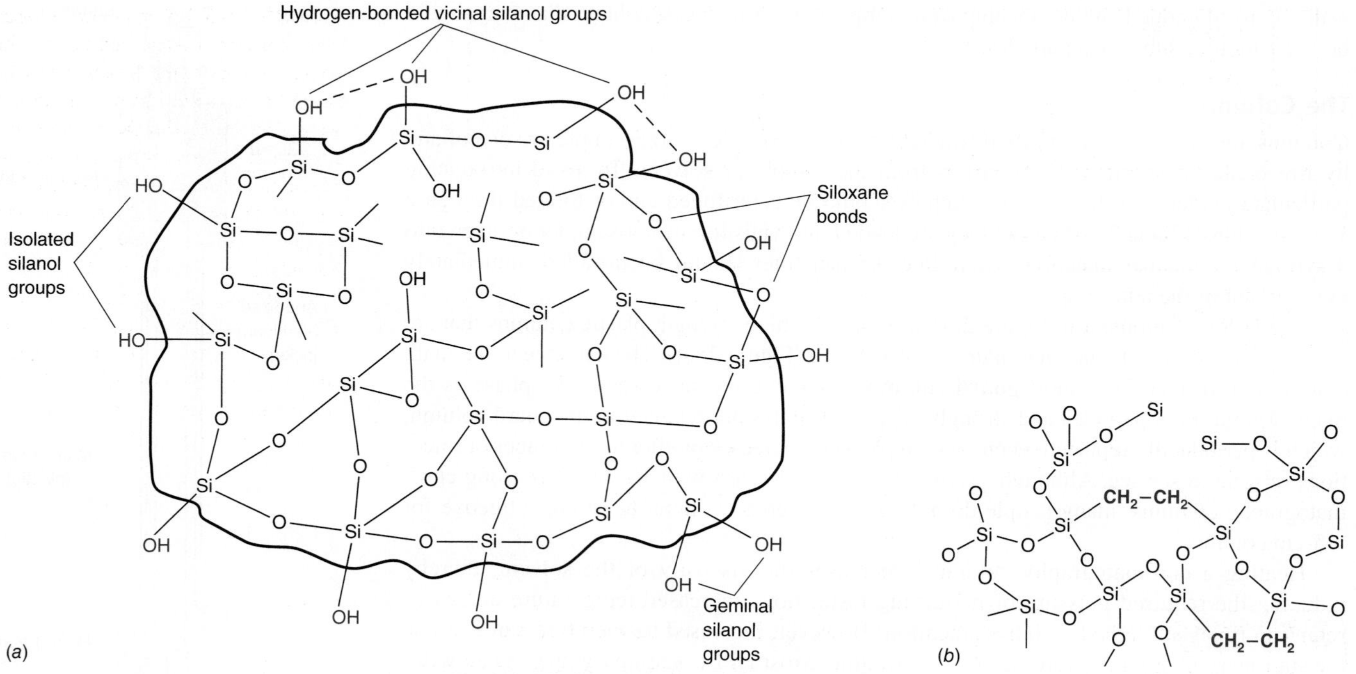

FIGURE 25-6 (*a*) Chemical nature of silica particle. [Data from R. E. Majors, *LCGC Supplement*, May 1997, p. S8.] (*b*) Base-hydrolysis-resistant silica incorporates ethylene bridges in place of oxide bridges between some silicon atoms. Ethylene-bridged structure is more rigid and well suited for particles that are <2 μm in diameter, which must withstand high pressure.

A silica surface (Figure 25-6) has up to 8 μmol of silanol groups (Si—OH) per square meter. Silanol groups are protonated at pH ~2–3. They dissociate to negative Si—O⁻ over a broad pH range above 3. In the old Type A silica, exposed Si—O⁻ groups strongly retained protonated bases (for example, RNH_3^+) and lead to tailing (Figure 25-7). Metallic impurities in Type A silica also cause tailing. Use of Type A silica is not recommended. Type B silica in Figure 25-7, which has fewer exposed silanol groups and fewer metallic impurities, is the most common form used today.

Bare silica can be used as the stationary phase for adsorption chromatography. Most commonly, liquid-liquid partition chromatography is conducted with a **bonded stationary phase** covalently attached to the silica surface by reactions such as

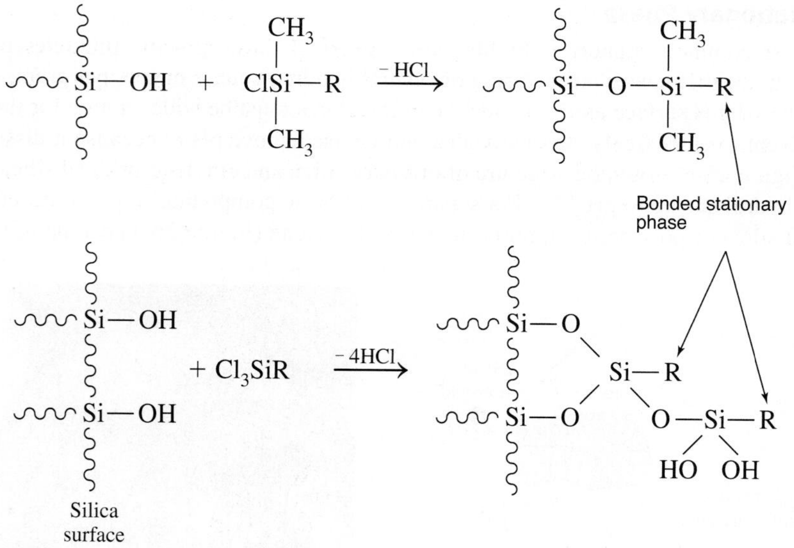

Residual silanol groups on the silica surface are capped with trimethylsilyl groups by reaction with $ClSi(CH_3)_3$ to eliminate polar adsorption sites that cause tailing.

There are ~4 μmol of R groups per square meter of support surface area, with little bleeding of the stationary phase from the column during chromatography.

Table 25-3 shows common bonded phases. The octadecyl (C_{18}) stationary phase (often abbreviated ODS) is by far the most common. Retention factors for a given solute on different nonpolar bonded phases (such as C_4, C_8, and C_{18}) are different. Retention factors for a given

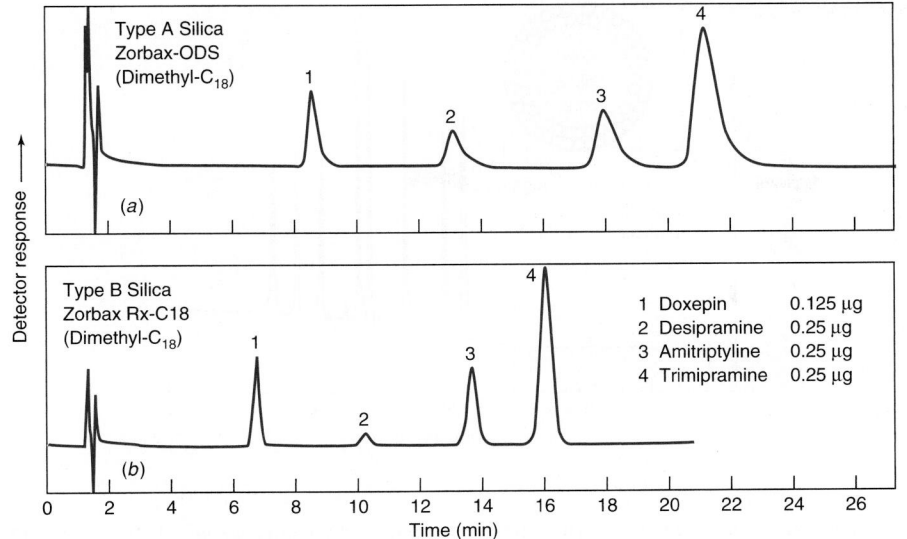

FIGURE 25-7 Tailing of amine bases on silica: (*a*) Type A silica support gives distorted peaks. (*b*) Less acidic Type B silica with fewer Si—OH groups and less metallic impurity gives symmetric peaks with shorter retention time. In both cases, chromatography was performed with a 15 × 0.46 cm column eluted at 1.0 mL/min at 40°C with 30 vol% acetonitrile/70 vol% sodium phosphate buffer (pH 2.5) containing 0.2 wt% triethylamine and 0.2 wt% trifluoroacetic acid. The detector measured ultraviolet absorbance at 254 nm. Additives such as triethylamine and trifluoroacetic acid are often used to mask strong adsorption sites and thereby reduce tailing. [Data from J. J. Kirkland, *Am. Lab.*, June 1994, p. 28K.]

TABLE 25-3 **Some common bonded phases for liquid chromatography**

Bonded polar phases		Bonded nonpolar phases	
R = spacer—(C=O)NH$_2$	amide	R = (CH$_2$)$_{17}$CH$_3$	octadecyl
R = (CH$_2$)$_3$NH$_2$	amino	R = (CH$_2$)$_7$CH$_3$	octyl
R = (CH$_2$)$_3$C≡N	cyano	R = (CH$_2$)$_3$C$_6$H$_5$	phenyl
R = (CH$_2$)$_2$OCH$_2$CH(OH)CH$_2$OH	diol	R = (CH$_2$)$_3$C$_6$F$_5$	pentafluorophenyl
R = spacer—CH$_2$$\overset{+}{N}$(CH$_3$)$_2$(CH$_2$)$_3SO_3^-$	ZIC-HILIC®	R = (CH$_2$)$_3$CH$_3$	butyl
		R = (CH$_2$)$_3$C≡N	cyano

Nonpolar phases are ordered from most nonpolar at the top of the table to least nonpolar at the bottom. Cyanopropyl exhibits polar and nonpolar retention.

---

solute on C$_{18}$ columns from different manufacturers can vary, in part due to differences in surface area.[15] Cyanopropyl phases retain solutes by either polar or nonpolar interactions. Weak, nonpolar retention by the cyanopropyl phase is useful for the separation of highly hydrophobic compounds such as cell membrane lipids at the opening of this chapter.

The siloxane (Si—O—SiR) bond hydrolyzes below pH 2, so HPLC with a bonded phase on a silica support is generally limited to the pH range 2–8. If bulky isobutyl groups are attached to the silicon atom of the bonded phase (Figure 25-8), the stationary phase is protected from attack by H$_3$O$^+$ and is stable for long periods at low pH, even at elevated temperature (for example, pH 0.9 at 90°C). Due to the steric bulk of the protecting groups, these phases have lower bonded ligand density (~2 μmol of R groups per meter square of surface), and so exhibit lower retention.

General retentivity of reversed phases:
Cyano < C$_4$ < phenyl < C$_8$ < C$_{18}$

Bidentate C$_{18}$ stationary phase provides increase stability above pH 8:

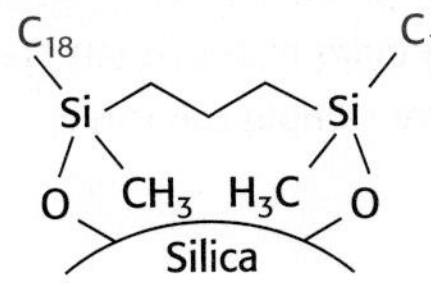

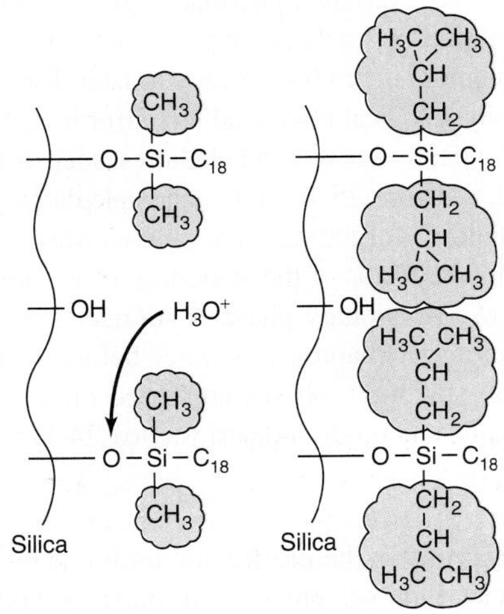

FIGURE 25-8 Bulky isobutyl groups protect siloxane bonds from hydrolysis at low pH.
[Information from J. J. Kirkland, *Am. Lab.* June 1994, p. 28K.]

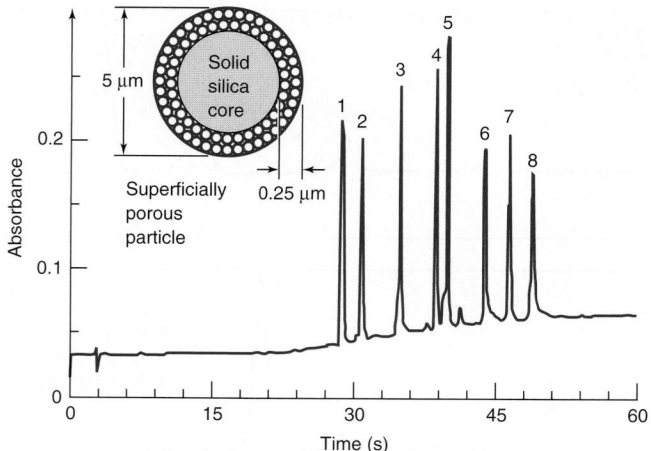

FIGURE 25-9 Rapid separation of proteins on superficially porous C₁₈-silica in 75 × 2.1 mm column containing Poroshell 300SB-C18. Mobile phase A: 0.1 wt% trifluoroacetic acid in H_2O. Mobile phase B: 0.07 wt% trifluoroacetic acid in acetonitrile. Solvent was changed continuously from 95 vol% A/5 vol% B to 100% B over 1 min. Flow = 3 mL/min at 70°C at 26 MPa (260 bar) with ultraviolet detection at 215 nm. Peaks: 1, angiotensin II; 2, neurotensin; 3, ribonuclease; 4, insulin; 5, lysozyme; 6, myoglobin; 7, carbonic anhydrase; 8, ovalbumin. [Data from R. E. Majors, *LCGC North Am. Column Technology Supplement,* June 2004, p. S8. From Agilent Technologies.]

Nonpolar bonded phase with embedded polar groups offers different selectivity from C₁₈, has improved peak shape for bases, and tolerates 100% aqueous eluent.

Superficially porous particles:

- rapid mass transfer of macromolecules into thin porous layer
- less eddy diffusion broadening for small molecules
- does not require high column pressure
- ~25% lower sample capacity

Another type of nonpolar stationary phase has a *polar embedded group*. The example in the margin consists of a long hydrocarbon chain with a polar amide group near its base. Embedded polar groups provide different selectivities from C₁₈ stationary phases, improved peak shapes for bases, and compatibility with 100% aqueous phase. Other nonpolar stationary phases should not be exposed to 100% aqueous phase because they become very difficult to re-equilibrate with mobile phase.

Figure 25-9 shows a rapid separation of proteins on **superficially porous particles** (also called *fused-core* or *core-shell* particles), which consist of a 0.25-μm-thick porous silica layer on a 4.5-μm-diameter nonporous silica core. A stationary phase such as C₁₈ is bonded throughout the thin, porous outer layer. The rationale for developing superficially porous particles was that mass transfer into the 0.25-μm-thick porous layer would be more rapid than mass transfer into fully porous particles with a diameter of 5 μm. This explanation is correct for macromolecules (5–500 kDa), which diffuse more slowly than small molecules. However, for small molecules (<500 Da), the improved efficiency of superficially porous particles is due mainly to decreased eddy diffusion broadening arising from more uniform column packing, not to an increased rate of mass transfer.[16] Figure 25-3 shows that the van Deemter curve for superficially porous particles with a total diameter of 2.7 μm and a porous layer thickness of 0.5 μm is similar to that of a totally porous particle with a diameter of 1.8 μm. The superficially porous particle enables separations similar to those achieved with 1.8-μm totally porous particles without requiring such high pressure.

Box 25-1 describes high-resolution, high-speed chromatography with *colloidal crystals* of silica in capillary columns. The phenomenon of *slip flow* described in the box decreases plate height and resistance to flow to enable million-plate resolution.

Porous graphitic carbon[18] is a stationary phase that exhibits increased retention of nonpolar compounds relative to retention by C₁₈. Graphite has high affinity for polar compounds and separates isomers that cannot be separated on C₁₈. The stationary phase is stable in 10 M acid and 10 M base.

Pharmaceutical companies often separate the two enantiomers (mirror image isomers) of a drug because each enantiomer has a different pharmacological effect. The drug thalidomide, prescribed in the 1960s to prevent morning sickness in pregnant women, produced gross birth defects in more than 10 000 children before it was banned. It was later discovered that one enantiomer of thalidomide has a desirable physiological effect and its mirror image caused birth defects.

To resolve enantiomers, optically active bonded phases, such as those shown in Figure 25-10 and Exercise 25-B, are used.[19] Figure 25-10 shows the calculated geometry of the chiral drug naproxen binding to one enantiomer of the stationary phase. Mirror image forms of the drug are designated *R* and *S*. Mirror image forms of the stationary phase are designated (*R,R*) and (*S,S*). Binding of (*S*)-naproxen to (*S,S*)-stationary phase is stronger than binding of (*R*)-naproxen to (*S,S*)-stationary phase . Therefore, (*R*)-naproxen is eluted before (*S*)-naproxen from (*S,S*)-stationary phase. Some other chiral stationary phases are based on substituted cellulose, on cyclic peptides with sugar substituents, and on cyclodextrins (Box 24-1).

The Elution Process

Reversed-phase chromatography is the most common mode of HPLC. Its stationary phase is nonpolar or weakly polar and the solvent is more polar. In *partition chromatography,* the

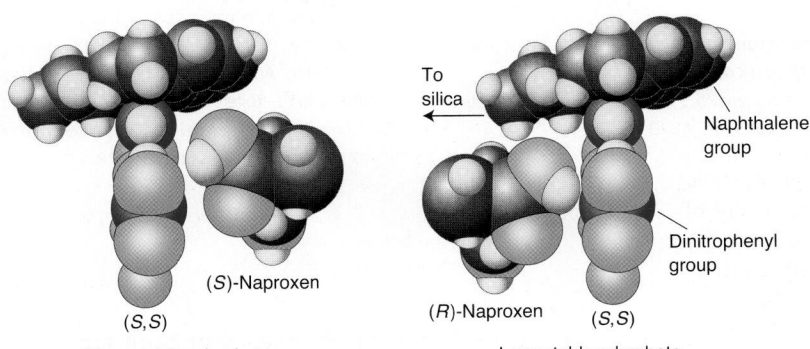

Chiral stationary phase Whelk-O 1

(R,R)

O₂N

(S,S)

NO₂

(S)-Naproxen

(R)-Naproxen

FIGURE 25-10 Interaction of enantiomers of the drug naproxen with (S,S) chiral stationary phase Whelk-O 1. (S)-Naproxen is adsorbed more strongly and is therefore retained longer by the column. [Data from S. Ahuja, "A Strategy for Developing HPLC Methods for Chiral Drugs," *LCGC North Am.* **2007,** *25,* 1112.]

Interaction of (R)- and (S)-naproxen with (S,S) stationary phase

(S)-Naproxen

(S,S)

More stable adsorbate

To silica

Naphthalene group

Dinitrophenyl group

(R)-Naproxen (S,S)

Less stable adsorbate

solute is either dissolved in mobile phase or dissolved in the bonded phase attached to the silica surface (Figure 25-11a). The **polarity index** P' in Table 25-4 ranks solvents based on their ability to dissolve polar solutes. The higher the index, the more polar is the solvent. A solvent's polarity arises from its dipole and its ability to donate hydrogen to a hydrogen bond (H-bond acidity in Table 25-4) or to accept a hydrogen to form a hydrogen bond (H-bond basicity in Table 25-4). Acetonitrile's polarity is due mainly to its strong dipole, whereas methanol forms hydrogen bonds. Retention is decreased by making the mobile phase more like the stationary phase. *In reversed-phase chromatography, a less polar solvent is a stronger mobile phase*, which elutes solutes more rapidly from the column. Box 25-2 describes the solvent-stationary phase interface in reversed-phase chromatography.

In *adsorption chromatography*, solvent molecules compete with solute molecules for discrete retention sites on the stationary phase (Figure 25-11b and Color Plate 31). Adsorption chromatography on bare silica is an example of **normal-phase chromatography,** in which we use a polar stationary phase and a less polar solvent. *A more polar solvent has a higher eluent strength*. The relative abilities of different solvents to elute a given solute from the adsorbent are nearly independent of the nature of the solute. Elution occurs when solvent displaces solute from the stationary phase.

The **eluent strength** in Table 25-4 is a measure of the solvent adsorption energy on bare silica, with the value for pentane defined as 0. The more polar the solvent, the greater is its eluent strength for adsorption chromatography with bare silica. The greater the eluent strength,

Reversed-phase chromatography:
- nonpolar stationary phase
- less polar solvent has higher mobile phase strength

Normal-phase chromatography:
- polar stationary phase
- more polar solvent has higher mobile phase strength

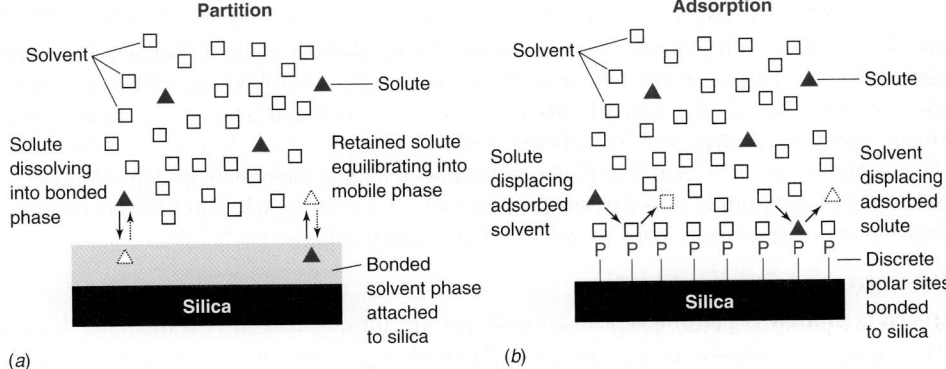

Partition

Solvent

Solute

Solute dissolving into bonded phase

Retained solute equilibrating into mobile phase

Bonded solvent phase attached to silica

Silica

(a)

Adsorption

Solvent

Solute

Solute displacing adsorbed solvent

Solvent displacing adsorbed solute

Discrete polar sites bonded to silica

Silica

(b)

FIGURE 25-11 (*a*) In *partition chromatography,* solute equilibrates between the mobile phase and the stationary bonded phase. The more time the solute spends in the mobile phase, the earlier it is eluted. (*b*) In *adsorption chromatography,* solvent molecules compete with solute molecules for binding sites on the stationary phase. The greater the eluent strength of the solvent, the more easily it displaces the solute.

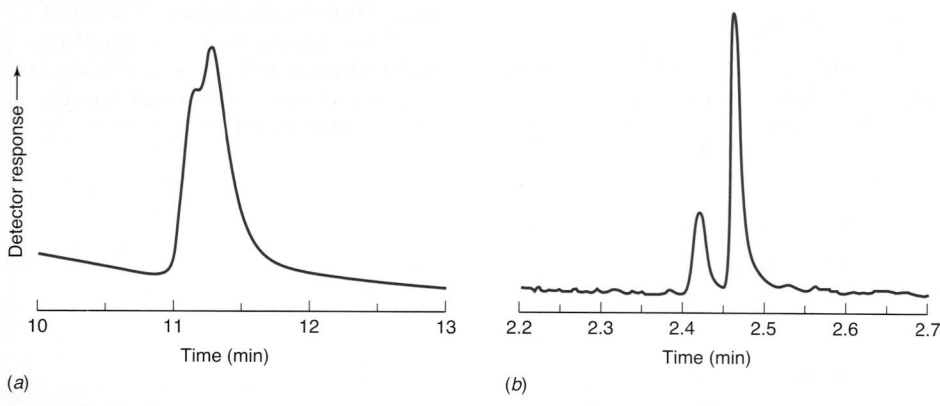

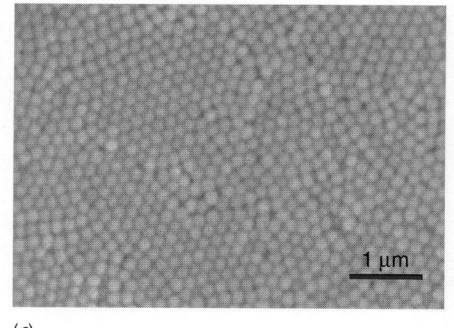

(c) Colloidal crystal of 125-nm spherical silica particles. [Information from B. J. Rogers and M. J. Wirth, "Slip Flow Through Colloid Crystals of Varying Particle Diameter," *ACS Nano* **2013**, *7*, 725. Courtesy Benjamin J. Rogers and Mary J. Wirth, Department of Chemistry, Purdue University. Reprinted with permission © 2013, American Chemical Society.]

Gradient separation of two forms of fluorescently labeled bovine serum albumin on (a) 50 × 2.1 mm UHPLC column of 1.7-μm C_4-silica particles and (b) 10 mm × 75 μm colloidal crystal column of 470-nm C_4-silica nanoparticles. [Data from B. J. Rogers, R. E. Birdsall, Z. Wu, and M. J. Wirth, "RPLC of Intact Proteins using Sub-0.5 μm Particles and Commercial Instrumentation," *Anal. Chem.* **2013**, *85*, 6820.]

Improvements in separation science are needed for applications in proteomics and pharmaceutical biotechnology and in discovery of biomarkers for diseases. For instance, two forms of fluorescently labeled bovine serum albumin are barely resolved by a 50-mm-long UHPLC column in panel *a*, but baseline resolution is achieved in panel *b* by a 10-mm-long capillary column packed with 470-nm C_4 (butyl)-silica nanoparticles.[17] Such *colloidal crystal* columns have achieved plate numbers over one million, with plate heights *H* that are 500 times lower than for conventional protein separations. *Colloids* are particles with diameters of ~1–500 nm. A *colloidal crystal* is a mostly regularly packed array of such particles (panel *c*), achieved by carefully loading the colloid into the capillary.

In colloidal crystal columns, the number of plates is 15 times higher than expected from the particle size and linear velocity. The pressure is five times less than predicted by Equation 25-2. Consider the flow between adjacent particles at the top of panel *d*. Frictional drag on a stationary particle surface typically causes fluid velocity to be zero at the surface, yielding the parabolic velocity profile of laminar flow (left side of panel *d*). If there is weak interaction between the fluid and the particle surface, *slip flow* occurs in which fluid at the surface continues to move. Slip flow requires less pressure to move liquid between particles and gives more uniform velocity between particles (right side of panel *d*) The contact angle of water on a butyl-silica surface is 83°, indicating weak interaction with the hydrophobic surface. (If water wets a surface well, it spreads out with a nearly zero contact angle.) The 100 nm *slip length* in panel *d* is small relative to HPLC particle size, but has significant effects with nanoparticles.

In slip flow with a colloidal crystal column, *C*-term mass transfer broadening in the van Deemter equation (23-37) is reduced by

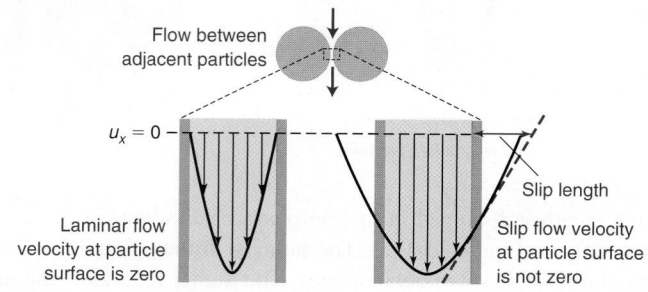

(d) Flow in channels between particles: In laminar flow, velocity at particle surface is zero and parabolic velocity profile is steep. In slip flow, velocity at particle surface is not zero because aqueous solvent does not wet the hydrophobic surface very well. Resulting parabolic velocity profile is not very steep. [Information from B. J. Rogers, B. C. Wei, and M. J. Wirth, "Ultrahigh-Efficiency Protein Separations with Submicrometer Silica using Slip Flow," *LCGC North Am.* **2012**, *30*, 890.]

the shorter diffusion distances afforded by the <500-nm-diameter particles. *A*-term multiple-path broadening is decreased by the uniform crystal-like arrangement of nanoparticles, which gives the column a colorful opal-like appearance.

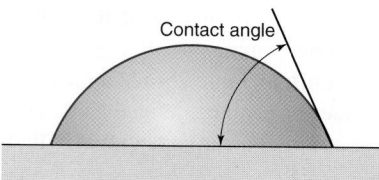

the more rapidly will solutes be eluted from a normal-phase column. Some differences between polarity index and eluent strength are evident due to differences in their retention mechanisms. Normal-phase chromatography is sensitive to small amounts of water in the eluent, but reversed-phase chromatography is not.

Given the variety of modes of liquid chromatography and their contrary dependences on polarity, the term **mobile phase strength** is used to describe the generic ability of the mobile phase to elute components from a liquid chromatography column.

Isocratic and Gradient Elution

Isocratic elution is performed with a single solvent (or constant solvent mixture). If one solvent does not provide sufficiently rapid elution of all components, then **gradient elution**

Isocratic elution: constant mobile phase composition

Gradient elution: continuous change of solvent composition to increase mobile phase strength. Gradient elution in HPLC is analogous to temperature programming in gas chromatography. Increased mobile phase strength is required to elute more strongly retained solutes.

TABLE 25-4

Properties of common HPLC solvents

| Solvent | Type of Polarity | | | Polarity Index (P') | Eluent strength (bare silica) | Ultraviolet cutoff (nm) |
	H-bond acidity	H-bond basicity	Dipolar			
Alkanes	0	0	0	0	≤0.01	≤200
Toluene	17%	83%	0%	2.4	0.22	284
Methyl *t*-butyl ether	0%	~60%	~40%	~2.4	0.48	210
Chloroform	43%	0%	57%	2.7	0.26	245
Diethyl ether	0%	64%	36%	2.8	0.43	215
Dichloromethane	27%	0%	73%	3.1	0.30	233
2-Propanol	36%	40%	24%	3.9	0.60	205
Tetrahydrofuran (THF)	0%	49%	51%	4.0	0.53	212
Ethyl acetate	0%	45%	55%	4.4	0.48	256
Methanol (CH_3OH)	43%	29%	28%	5.1	0.70	205
Acetone	6%	38%	56%	5.1	0.53	330
Acetonitrile (CH_3CN)	15%	25%	60%	5.8	0.52	190
Water	43%	18%	39%	10.2	—	190

Data from L. R. Snyder, J. J. Kirkland, and J. W. Dolan, in *Introduction to Modern Liquid Chromatography*, 3rd ed. (Hoboken, NJ: Wiley, 2010), revised using M. Vitha and P. W. Carr, "The Chemical Interpretation and Practice of Linear Solvation Energy Relationships in Chromatography," J. Chromatogr. A *2006*, 1126, 143.

Colored solvents are typically used for reversed-phase chromatography. Unshaded solvents are typically used for normal-phase chromatography. Gray shaded solvents may be used in either mode.

BOX 25-2   Structure of the Solvent-Bonded Phase Interface

Structured layers of solvent at the interface between the stationary and mobile phases influence retention in liquid chromatography. The diagrams show solvent layers at the surface of C_{18}-silica in the presence of methanol/water (40:60 vol:vol) and acetonitrile/water (30:70 vol:vol), as deduced from adsorption studies of phenol and caffeine. The 40:60 and 30:70 compositions were chosen because they have nearly the same eluent strength.

Methanol forms a *monolayer* (one molecule thick) of adsorbed solvent on the C_{18} surface with a thickness of 0.25 nm. Acetonitrile forms a pool of adsorbed solvent with a thickness of 1.3 nm and a high capacity for dissolved solute. Adsorbed phenol and caffeine molecules can reside inside the C_{18} layer or on the outer surface of the C_{18} layer, or they can be dissolved in the CH_3CN pool. Each type of adsorption site has a different binding energy and binding capacity and affects the shape of the eluted solute band differently at different solute concentrations. The pool of adsorbed CH_3CN is a stronger eluent than the bulk organic/aqueous solution and explains why acetonitrile is a stronger eluent than methanol.

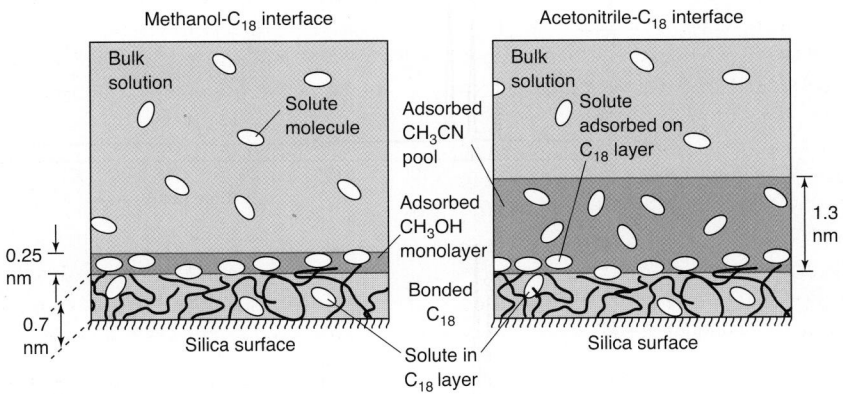

Structure of adsorbed solvent on C_{18}-bonded phase in reversed-phase chromatography. [Information from F. Gritti and G. Guiochon, "Adsorption Mechanism in RPLC. Effect of the Nature of the Organic Modifier," *Anal. Chem.* **2005,** *77,* 4257.]

---

can be used. In this case, increasing amounts of solvent B are added to solvent A during the separation to create a continuous gradient.

Figure 25-12 shows the effect of increasing mobile phase strength in the isocratic elution of eight compounds from a reversed-phase column. In a reversed-phase separation, mobile phase strength *decreases* as the solvent becomes *more* polar. Mobile phase polarity is

- Aqueous buffer for HPLC is prepared and the pH adjusted *prior* to mixing with organic solvent.[20]
- Ultrapure water for HPLC should be freshly prepared by a purification train or by distillation. Water slowly extracts impurities from polyethylene or glass.
- To prepare 70% B, for example, mix 70 mL of B with 30 mL of A. *The result is different* from placing 70 mL of B in a volumetric flask and diluting to 100 mL with A because there is a volume change when A and B are mixed.

General elution problem: For a complex mixture, isocratic conditions can often be found to produce adequate separation of early-eluting peaks or late-eluting peaks, but not both. This problem drives us to use gradient elution.

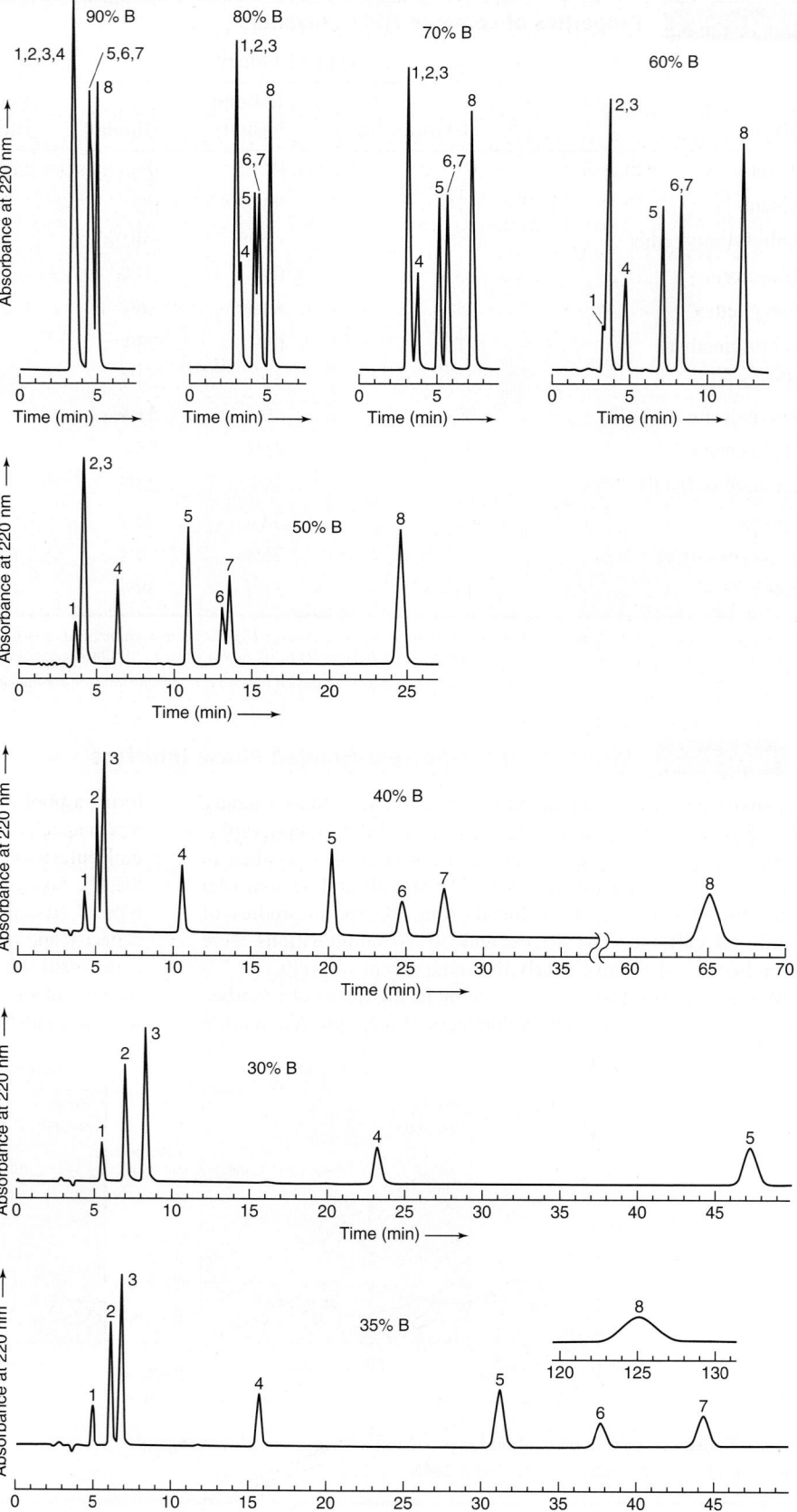

FIGURE 25-12 Isocratic HPLC separation of a mixture of aromatic compounds at 1.0 mL/min on a 25 × 0.46-cm Hypersil ODS column (C$_{18}$ on 5-μm silica) at ambient temperature (~22°C): (1) benzyl alcohol; (2) phenol; (3) 3′,4′-dimethoxyacetophenone; (4) benzoin; (5) ethyl benzoate; (6) toluene; (7) 2,6-dimethoxytoluene; (8) o-methoxybiphenyl. Eluent consisted of aqueous buffer (designated A) and acetonitrile (designated B). The notation "90% B" in the first chromatogram means 10 vol% A and 90 vol% B. The buffer contained 25 mM KH$_2$PO$_4$ plus 0.1 g/L sodium azide adjusted to pH 3.5 with HCl.

adjusted by mixing two (or more) miscible solvents. It is customary to call the aqueous solvent A and the organic solvent B. The polarity index of the mixture (P') is

$$P'_{AB} = \Phi_A P'_A + \Phi_B P'_B \qquad (25\text{-}3)$$

where Φ_A and Φ_B are the volume fraction of solvents A and B. The first chromatogram (upper left) was obtained with a solvent consisting of 90 vol% acetonitrile and 10 vol% aqueous buffer. Acetonitrile has a high mobile phase strength (lower polarity), and all compounds are eluted rapidly. Only three peaks are observed because of overlap. The first chromatogram was obtained with 90% B. When mobile phase polarity is increased by changing the solvent to 80% B, there is slightly more separation and five peaks are observed. At 60% B, we begin to see a sixth peak. At 40% B, there are eight clear peaks, but compounds 2 and 3 are not fully resolved. At 30% B, all peaks would be resolved, but the separation takes too long. Each 10% change in %B causes about a threefold change in retention factor. Backing up to 35% B (the bottom trace) separates all peaks in a little over 2 h (which is too long for many purposes).

Freeware to simulate reversed-phase chromatography is found at www.hplcsimulator.org.[7]

The empirical *linear-solvent-strength model* supposes a logarithmic relationship between retention factor k for a given solute and the mobile phase composition Φ:

$$\log k \approx \log k_W - S\Phi \qquad (25\text{-}4)$$

where $\log k_w$ is the extrapolated retention factor for 100% aqueous eluent, Φ is the fraction of organic solvent ($\Phi = 0.4$ for 40 vol% organic solvent/60 vol% H_2O), and S is a constant for each compound, with a typical value of ~4 for small molecules. Figure 25-13 shows the linear-solvent-strength model fit to the retention in Figure 25-12.

An *empirical* model is based on observations, not on theory. Equation 25-4 is empirical in that it approximately describes the relation between k and Φ. The relationship is not predicted by a theory based on first principles.

From the isocratic elutions in Figure 25-12, the gradient in Figure 25-14 was selected to resolve all peaks in 38 min. First, 30% B (B = acetonitrile) was run for 8 min to separate components 1, 2, and 3. Then mobile phase strength was increased over 5 min to 45% B and held there for 15 min to elute peaks 4 and 5. Finally, the solvent was changed to 80% B over 2 min and held there to elute the last peaks.

Box 25-3 describes gradient elution in **supercritical fluid chromatography.**

Hydrophilic Interaction Chromatography (HILIC)[24]

Hydrophilic substances are soluble in water or attract water to their surfaces. Polar organic and biochemical molecules have hydrophilic regions. **Hydrophilic interaction chromatography** (HILIC) is used to separate molecules that are too polar to be retained by reversed-phase

Hydrophilic: "water loving"—soluble in water, surface is wetted by water

Hydrophobic: "water hating"—insoluble in water, surface is not wetted by water

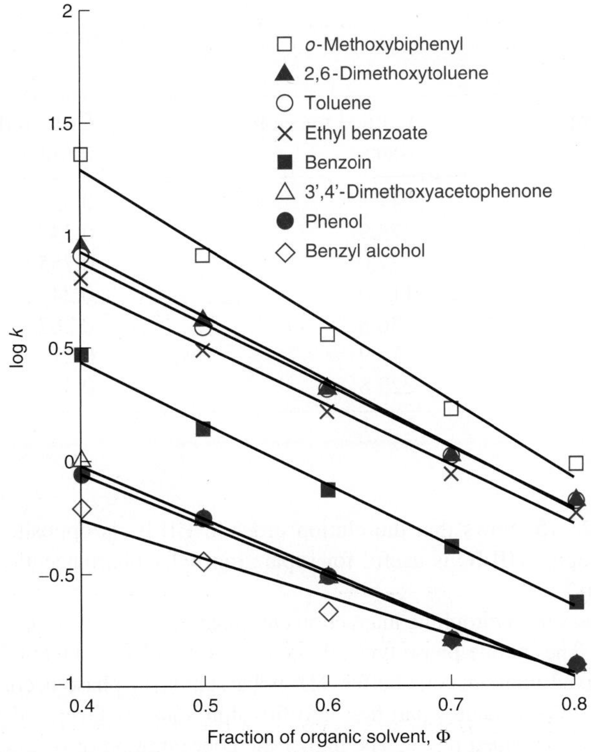

FIGURE 25-13 Linear-solvent-strength model. Graph of log k vs. Φ for eight aromatic compounds in Figure 25-12. The curvature evident in the *o*-methoxybiphenyl data shows that the linear-solvent-strength model is a reasonable approximation only over limited ranges of mobile phase composition.

BOX 25-3   "Green" Technology: Supercritical Fluid Chromatography

In the phase diagram for carbon dioxide, solid CO_2 (Dry Ice) is in equilibrium with gaseous CO_2 at a temperature of $-78.7°C$ and a pressure of 1.00 bar. The solid *sublimes* without turning into liquid. Above the *triple point* at $-56.6°C$ (at which solid, liquid, and gas are in equilibrium), liquid and vapor coexist as separate phases. For example, at 0°C, liquid is in equilibrium with gas at 34.9 bar. Moving up the liquid-gas boundary, we see that two phases always exist until the *critical point* is reached at 31.3°C and 73.9 bar. *Above this temperature, only one phase exists, no matter what the pressure.* We call this phase a **supercritical fluid** (Color Plate 32). Its density and viscosity are between those of the gas and liquid, as is its ability to act as a solvent.

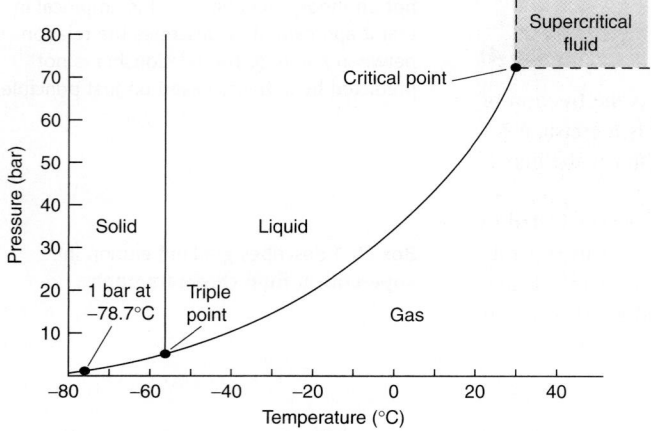

Phase diagram for CO_2.

An interesting supercritical fluid to demonstrate is SF_6.[21] The photographs show different changes as fluid is warmed and cooled through its supercritical temperature.

Supercritical fluid chromatography with a mixture of CO_2 and organic solvent is a "green" technology that reduces organic solvent use by up to 90% for the separation of kilograms of compounds and enantiomers in the pharmaceutical industry.[22] The low viscosity of the supercritical fluid also permits faster flow to increase productivity. Though CO_2 is a weak solvent by itself, when mixed with some organic solvent, it is capable of dissolving a variety of compounds.

Supercritical fluid chromatography provides increased speed and resolution, relative to liquid chromatography, because of higher diffusion coefficients of solutes in supercritical fluids. Unlike gases, supercritical fluids can dissolve nonvolatile solutes. When pressure is released, the solvent turns to gas, leaving solute in the gas phase for easy detection. Carbon dioxide is the supercritical fluid of choice for chromatography because it is compatible with flame ionization and ultraviolet detectors, it does not absorb ultraviolet radiation above 190 nm,[23] it has a low critical temperature, and it is nontoxic.

Equipment for supercritical fluid chromatography is similar to that for HPLC with packed columns or open tubular columns. Eluent strength is increased in HPLC by gradient elution and in gas chromatography by raising the temperature. In supercritical fluid chromatography, eluent strength is increased by making the solvent *denser* by increasing the pressure, or by gradient elution. The chromatogram shows gradient elution using simultaneous monitoring with multiple HPLC detectors.

Critical constants

Compound	Critical temperature (°C)	Critical pressure (bar)	Critical density (g/mL)
Argon	−122.5	47	0.53
Carbon dioxide	31.3	73.9	0.448
Sulfur hexafluoride	45.6	37.0	0.755
Ammonia	132.2	113.0	0.24
Diethyl ether	193.6	36.8	0.267
Methanol	240.5	79.9	0.272
Water	374.4	229.8	0.344

columns. Figure 25-15 shows that the elution order in HILIC is opposite that in reversed-phase chromatography. HILIC is useful for separating polar pharmaceuticals, peptides, and saccharides (sugars).

Stationary phases for hydrophilic interaction chromatography, such as those in Figure 25-16, are strongly polar. The mobile phase typically contains (60–97 vol%) CH_3CN or other aprotic organic solvent mixed with aqueous buffer. The polar stationary phase becomes coated with a thin layer of water. Polar solutes partition into this thin aqueous layer and may also interact directly with the polar bonded phase. The higher the concentration of organic solvent, the less soluble is the polar solute in the mobile phase.

Gradient elution goes from a weak mobile phase to a strong mobile phase (Figure 25-15). The solvent is usually an aqueous/organic mixture. In reversed-phase chromatography,

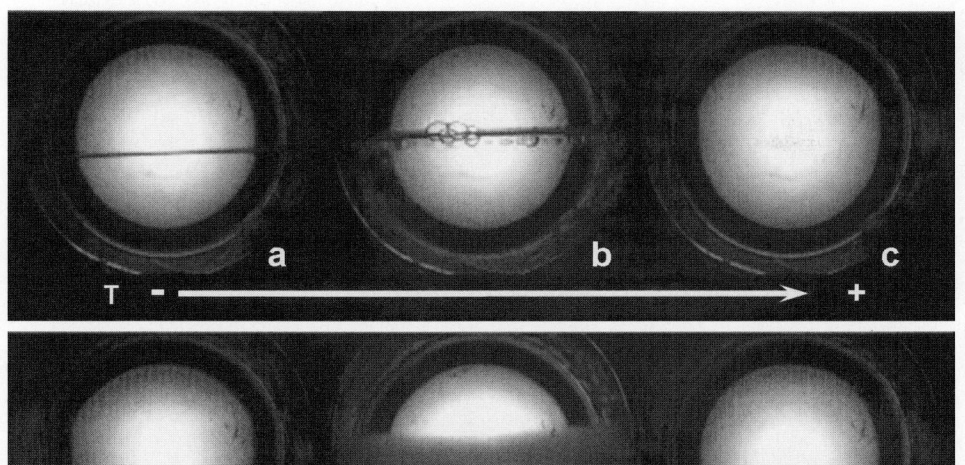

Warming and cooling SF_6 through its supercritical temperature. Upon warming (upper panel), liquid boils and the meniscus rises as the liquid expands. Upon cooling (lower panel), droplets of liquid form throughout the fluid. Gravity pulls droplets down, creating a "storm" before separation into two distinct phases. [Republished with permission of Royal Society of Chemistry, from P. Licence, D. Litchfield, M. P. Dellar, and M. Poliakoff, "'Supercriticality'; a Dramatic but Safe Demonstration of the Critical Point," *Green Chem.* **2004,** *6,* 352, Figures 3 and 4. Courtesy Peter Licence, University of Nottingham. Permission conveyed through Copyright Clearance Center, Inc.]

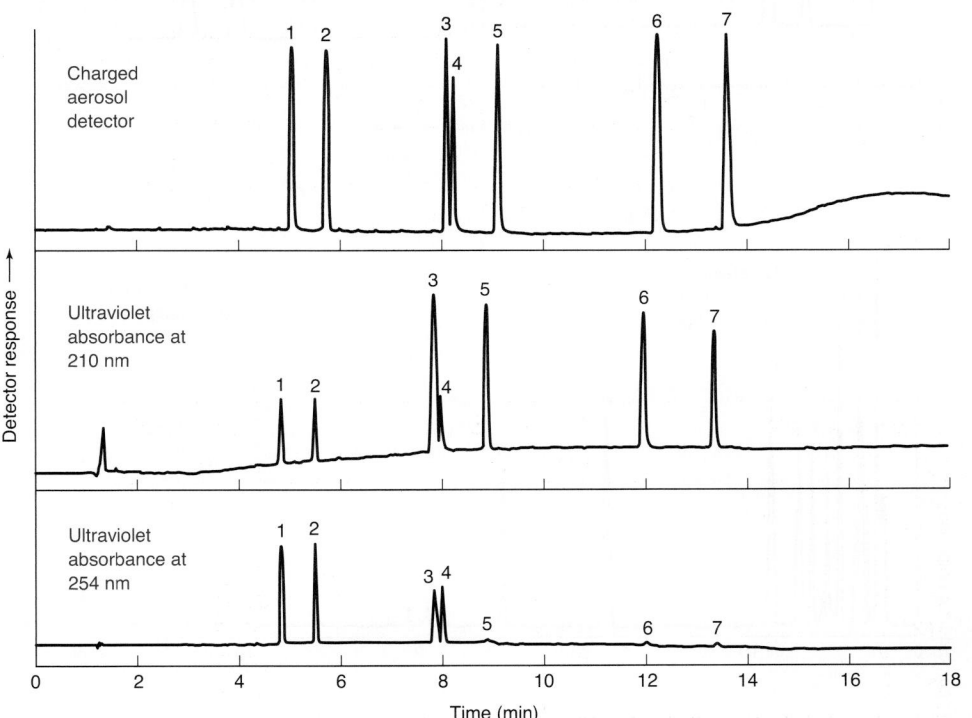

Packed-column supercritical chromatogram of seven steroids (100 mg/L) using 10–35% methanol gradient in CO_2 at 40°C and 10 MPa (100 bar) outlet pressure: (1) 7-methyltestosterone, (2) testosterone, (3) progesterone, (4) cortisone, (5) estrone, (6) estradiol, and (7) estriol. [Data from C. Brunelli, T. Górecki, Y. Zhao, and P. Sandra, "Corona-Charged Aerosol Detection in Supercritical Fluid Chromatography for Pharmaceutical Analysis," *Anal. Chem.* **2007,** *79,* 2472.]

mobile phase strength is increased by *decreasing* the fraction of water in the eluent to increase the solubility of solutes in the mobile phase. In HILIC, mobile phase strength is increased by *increasing* the fraction of water in the mobile phase. Gradient elution goes from low aqueous content to high aqueous content. In normal-phase chromatography, the mobile phase is nonaqueous, and gradients go from nonpolar to increasingly polar organic solvents (for example, from 5% CH_2Cl_2 in hexane to 30% CH_2Cl_2 in hexane).

Selecting the Separation Mode

There can be many ways to separate components of a given mixture. Figure 25-17 is a decision tree for choosing a starting point. If the molecular mass of analyte is below 2 000, use the upper part of the figure; if the molecular mass is greater than 2 000, use the lower part. In

There are no firm rules in the decision tree in Figure 25-17. Methods in either part of the diagram might work for molecules whose size belongs to the other part.

FIGURE 25-14 Gradient elution of the same mixture of aromatic compounds in Figure 25-12 with the same column, flow rate, and solvents. The upper trace is the *segmented gradient* profile, so named because it is divided into several different segments.

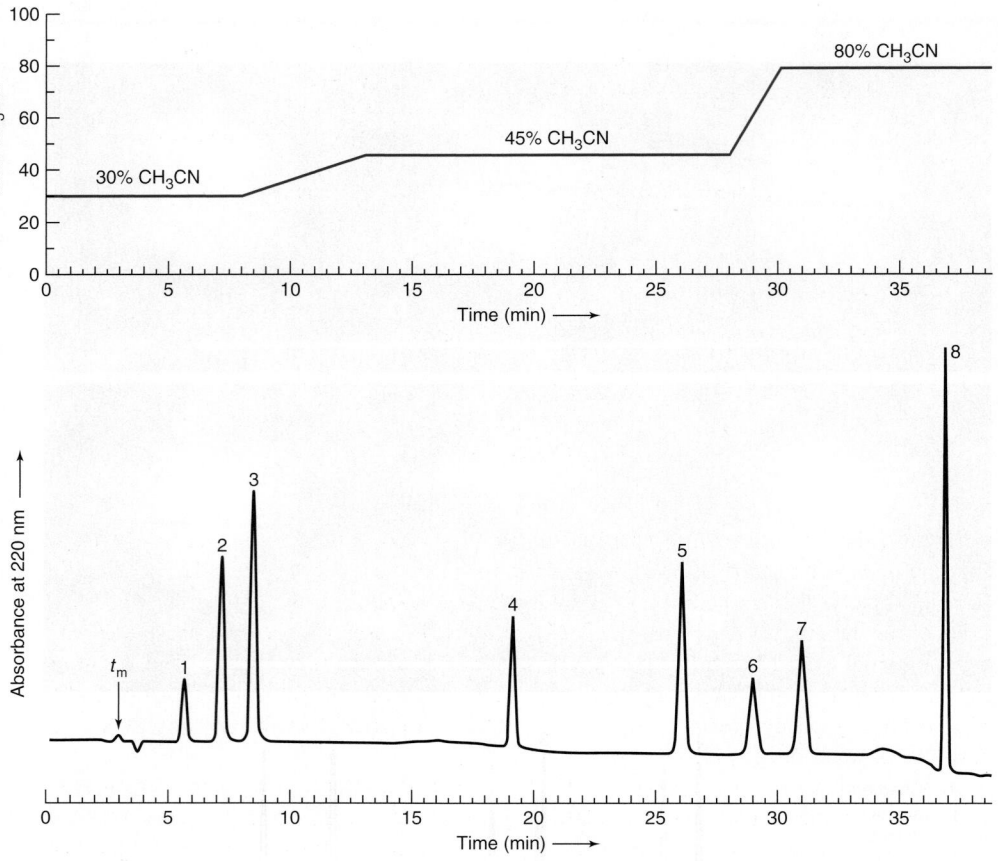

FIGURE 25-15 Gradient elution of derivatized amino acids by reversed-phase and hydrophilic interaction liquid chromatography. Polar amino acids such as glycine (G), serine (S), and glutamine (Q) that are weakly retained by reversed-phase are strongly retained in HILIC. The reversed-phase separation was on a 15 × 0.1-cm column of 3.5-μm Zorbax XDB C$_{18}$ using a gradient from 5% to 90% methanol over 64 min with constant 0.1% formic acid. The HILIC separation was on a 25 × 0.1 cm TSKgel Amide-80 column (amide 5-μm silica) at 55 μL/min using a gradient from 85% to 55% CH$_3$CN over 45 min with 15 mM ammonium acetate buffer (pH 5.5) in the aqueous phase. Derivatization with natural or isotopically labeled formaldehyde enabled detection of over 400 amine-containing compounds in urine. Amino acid abbreviations are in Table 10-1. [Data from K. Guo, C. J. Ji, and L. Li, "Stable-Isotope Dimethylation Labeling Combined with LC-ESI MS for Quantification of Amine-Containing Metabolites in Biological Samples," *Anal. Chem.* **2007,** *79,* 8631.]

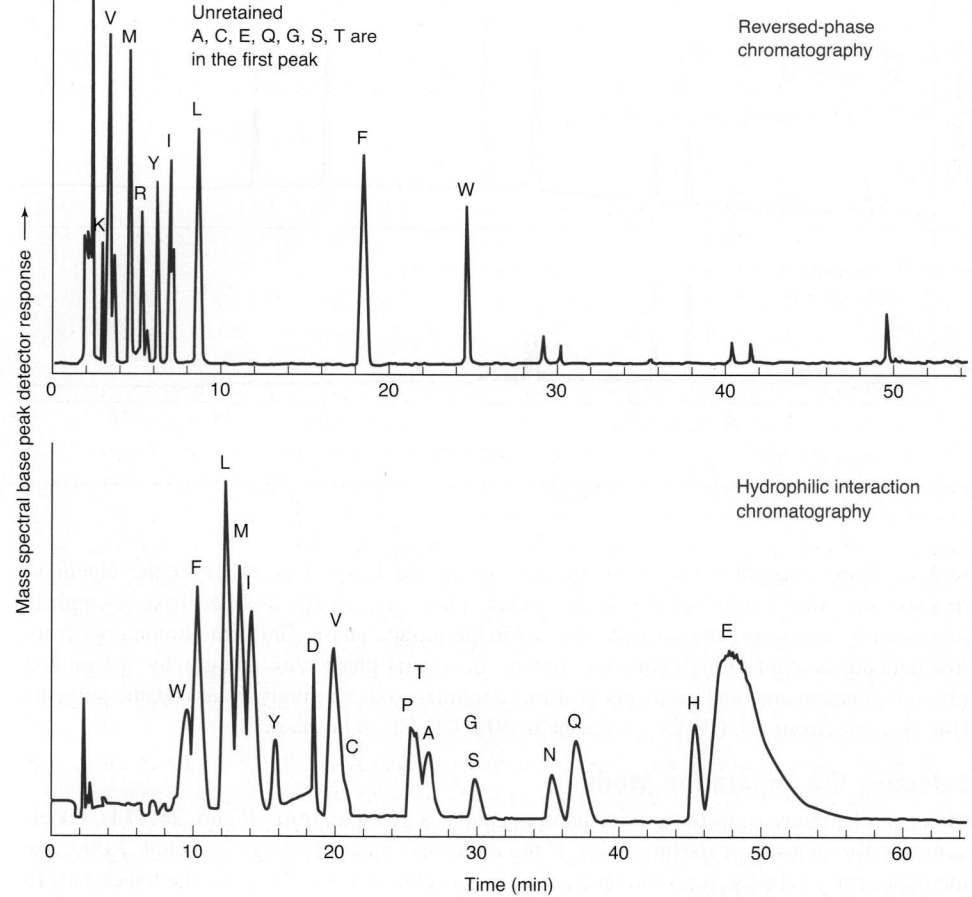

either part, the first question is whether solutes dissolve in water or in organic solvents. Suppose we have a mixture of small molecules (molecular mass <2 000) soluble in methanol. Table 25-4 ranks solvent polarity, with the most polar solvents at the bottom. The polarity index of methanol (5.1) places it among solvents typically used for reversed-phase chromatography. Figure 25-17 suggests that we try reversed-phase chromatography. The decision path is shown in color. Our column choices include bonded phases containing octadecyl (C_{18}), octyl (C_8), phenyl, pentafluorophenyl, and cyano groups. Mobile phases can be mixtures of methanol/water or acetonitrile/water.

If solutes dissolve only in water, then the charge of the solutes governs the decision. If the solutes are not charged, the decision tree suggests that we try reversed-phase chromatography or hydrophilic interaction chromatography. If the solutes include a weak acid or base, buffers or ion-pairing agents (Figure 26-11) are needed. Ion-exchange chromatography, described in Section 26-1, may also be used.

If molecular masses of solutes are >2 000, and if they are soluble in organic solvents, Figure 25-17 tells us to try molecular exclusion chromatography, described in Section 26-3. If molecular masses of solutes are >2 000, and they are soluble in water, the decision is based on which characteristic distinguishes the macromolecules within the sample—do they differ in size, in charge, or in hydrophobicity? The corresponding modes of liquid chromatography are described in Chapter 26.

Solvents

Pure HPLC-grade (expensive) solvents are required to prevent degradation of costly columns with impurities and to minimize detector background signals from contaminants. Before use, solvents are purged with He or evacuated to remove dissolved air.[25] Vacuum filtering with ≤0.5-μm-pore filters also removes particulate matter from the solvent. Air bubbles create difficulties for pumps, columns, and detectors. Dissolved O_2 may also affect the response of ultraviolet, fluorescence, and electrochemical detectors. Many instruments now have on-line *degassers* to remove dissolved gases. A filter is used on the intake tubing in the solvent reservoir to exclude >0.5 μm particulate matter for HPLC and >0.2 μm for UHPLC.

Sample and solvent may pass through a short, expendable *guard column* (Figures 25-1 and 25-4) that has the same stationary phase as the analytical column and retains strongly adsorbed species. After a series of runs, or prior to storage, the analytical column should be cleaned. For example, if the eluent was acetonitrile/aqueous buffer (40:60 vol/vol), first replace aqueous buffer with water[26] and wash the column with 5–10 mobile phase volumes (V_m) of acetonitrile/water (40:60 vol/vol). Then wash the column with 10–20 mobile phase volumes (V_m) of strong eluent such as acetonitrile/water (95:5 vol/vol) to remove strongly retained solutes. Store the column with this solvent to inhibit bacterial growth. This procedure is suitable for alkyl, aryl, cyano, embedded polar phase, and HILIC columns. A different procedure is recommended for normal-phase columns.[27]

Normal-phase separations are sensitive to water in the solvent. To speed the equilibration of stationary phase when changing eluents, organic solvents for normal-phase chromatography should be 50% saturated with water. Saturation can be achieved by adding a few milliliters of water to dry solvent and stirring. Then the wet solvent is separated from the excess water and mixed with an equal volume of dry solvent.

HPLC solvents are flammable and should be stored in a secure metal cabinet when not in use.

Purging with gas is called **sparging**.

Replace a guard column when plate number or pressure changes by more than 10%. If changing the guard column does not improve performance, wash the column.

Mobile phase volume = $V_m \approx L d_c^2/2$, where L and d_c are the column length and diameter in cm. V_m is the total volume of solvent in the column between stationary phase particles and in the pores.

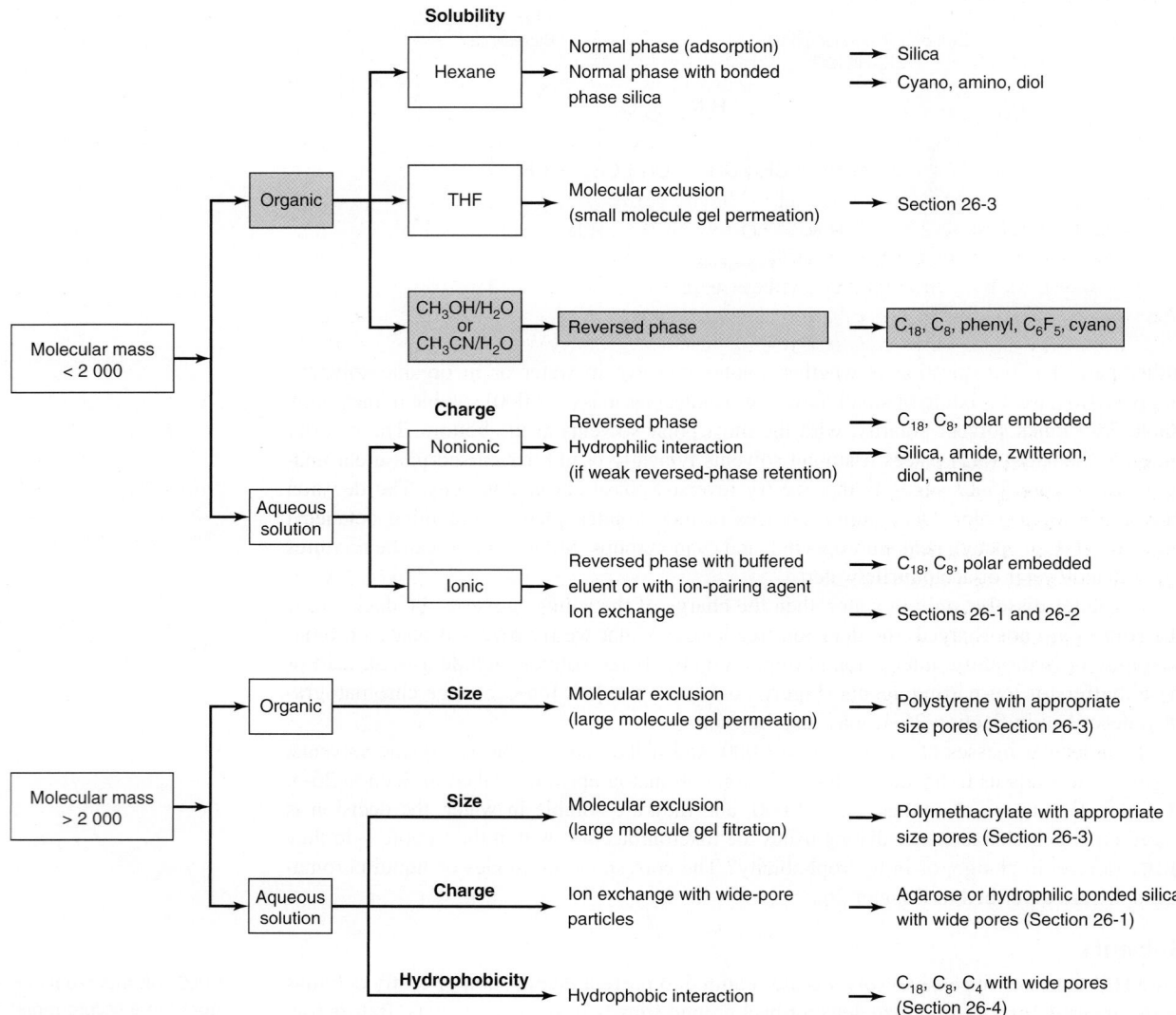

Analyte molecular mass	Analyte soluble in	Differentiating characteristic	Chromatography mode	Typical HPLC columns

FIGURE 25-17 Guide to HPLC mode selection. THF stands for tetrahydrofuran. [Information from *The LC Handbook,* publication no. 5990-7595EN (Agilent Technologies, 2011).]

Maintaining Symmetric Bandshape[28]

HPLC columns should provide narrow, symmetric peaks. If a new column does not reproduce the quality of the separation of a standard mixture whose chromatogram comes with the column, and if you have satisfied yourself that the problem is not in the rest of your system, return the column.

The asymmetry factor B/A in Figure 23-14 should rarely be outside the range 0.9–1.5. If the peak shape changed suddenly, check if the change coincides with a new batch of eluent or a new column. For weak acids and bases, the pH and buffer capacity of the mobile phase can strongly affect retention and peak shape. If a guard column is being used, try replacing it. If the peak shape gradually worsened, wash the column as described in the previous section. Older methods may include triethylamine in the mobile phase. The additive binds to sites on impure (Type A) silica that would otherwise strongly bind analyte and cause tailing (Figure 25-7). Improved grades of silica (Type B) have reduced tailing and, therefore, reduced the need for such additives.

If all peaks exhibit tailing or splitting, the frit at the beginning of the column may be clogged with particulate matter.[29] You can try to unclog the frit by disconnecting and reversing

Inject a standard mixture each day to evaluate the HPLC system. Changes in peak shapes or retention times alert you to a problem.

CHAPTER 25 High-Performance Liquid Chromatography

the column and flushing it with 20–30 mL. The column should not be connected to the detector during reverse flushing. If peaks remain distorted, replace the column.

Doubled peaks or altered retention times (Figure 25-18) sometimes occur if the solvent in which sample is dissolved has much greater mobile phase strength than the mobile phase. Try to dissolve the sample in a solvent of lower mobile phase strength or in the mobile phase itself.

Overloading causes the distorted shape shown in Figure 23-21.[30] To see if overloading is occurring, reduce the sample mass by a factor of 10 and see if retention times increase or peaks become narrower. If either change occurs, reduce the mass again until the injected mass does not affect retention time and peak shape. In general, reversed-phase columns can handle 1–10 μg of sample per gram of silica. A 4.6-mm-diameter column contains 1 g of silica in a length of 10 cm. To prevent peak broadening from too large an injection volume, the injected volume should be less than 15% of the peak volume measured at the baseline. For example, if a peak eluted at 1 mL/min has a width of 0.2 min, the peak volume is 0.2 mL. The injection volume should not exceed 15% of 0.2 mL, or 30 μL.

The volume of a chromatography system outside of the column from the point of injection to the point of detection is called the *extra-column volume*. Excessive extra-column volume allows bands to broaden by diffusion or mixing. Tailing of early eluting peaks is suggestive of broadening due to the extra-column volume.[31] Use short, narrow tubing of the diameters indicated in Table 25-2, and be sure that connections are made with matched fittings to minimize gaps and thereby minimize extra-column band spreading.

HPLC columns have a typical lifetime of 500–2 000 injections.[11] You can monitor the health of a column by keeping a record of pressure, resolution, and peak shape. The pressure required to maintain a given flow rate increases as a column ages. System wear becomes serious when pressure exceeds 17 MPa (2 500 pounds/inch^2). It is desirable to develop methods in which pressure does not exceed 14 MPa (2 000 pounds/inch^2). When the pressure reaches 17 MPa, the in-line 0.5-μm frit or the guard column should be replaced. To determine which component is the problem, replace the suspect part with one that you know is good.[32] If the problem is corrected, you are done. If the problem persists, replace the original part and replace the second suspect part. Continue this process until the problem is solved. If the problem is not solved, it is probably time for a new column. If you use a column for repetitive analyses, replace the column when the required resolution is lost or when tailing becomes significant. Resolution and tailing criteria should be established during method development.

25-2 Injection and Detection in HPLC

We now consider the hardware required to introduce sample and solvent onto the column and to detect compounds as they leave the column. Mass spectrometric detection, which is extremely powerful and important, was discussed in Sections 22-4 and 22-5.

Pumps and Injection Valves

Figure 25-19 shows the basic design of a pump that produces a programmable, constant flow rate up to 10 mL/min at pressures up to 40 MPa (400 bar) for HPLC and up to 100 MPa (1 000 bar) for UHPLC.[33] A sapphire piston draws solvent from a reservoir into the piston chamber and then delivers the solvent to the column. Two check valves govern the direction of solvent flow through

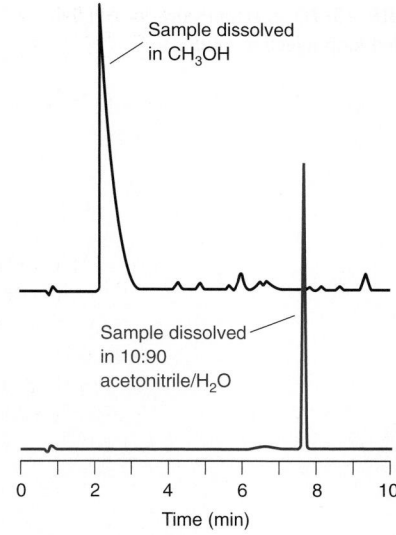

FIGURE 25-18 Effect of sample solvent on retention time and peak shape of *n*-butylaniline. Eluent (1 mL/min) is 10:90 (vol/vol) acetonitrile/H$_2$O with 0.1 wt% trifluoroacetic acid. Lower sample was dissolved in eluent. Upper sample was dissolved in methanol, which is a much stronger solvent than eluent. Column: 15 cm × 4.6 mm, 5-μm-diameter C$_{18}$-silica, 30°C. Injection: 10 μL containing 0.5 μg analyte. Ultraviolet detection at 254 nm. [Data from Supelco, Bellefonte, PA.]

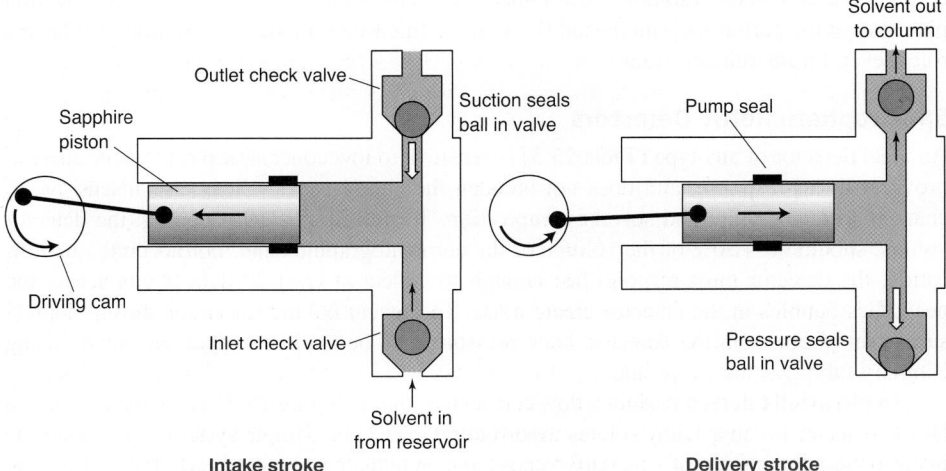

Intake stroke **Delivery stroke**

FIGURE 25-19 High-pressure reciprocating piston pump for HPLC. The sapphire piston is moved back and forth by a rotating cam. Check valves open or close depending on the direction of flow. When the piston moves to the left, solvent is drawn into the piston chamber through the inlet port. Suction causes the ball in the outlet valve to block the outlet port. When the piston moves to the right, solvent is forced out of the piston chamber to the column. Flow causes the ball in the inlet valve to block the inlet port. Flow lifts the ball in the outlet check valve, allowing liquid to be delivered to the column. Delivery rate is controlled by stroke frequency or volume. [Information from J. W. Dolan, "LC Pumps," *LCGC North Am.* **2008,** *26,* 1168.]

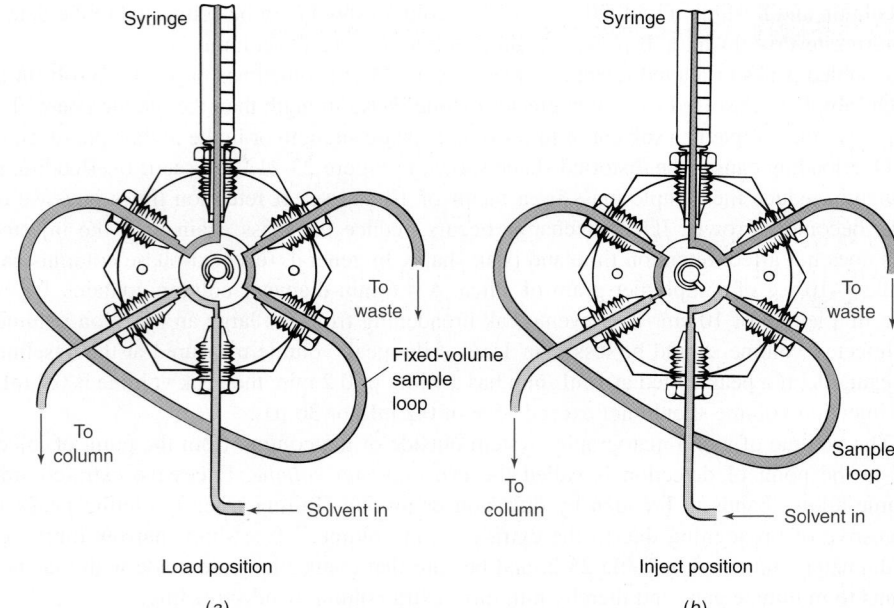

FIGURE 25-20 Injection valve for HPLC in the filled-loop injection mode.

Syringe

To waste

Fixed-volume sample loop

To column

Solvent in

Load position

(a)

Syringe

To waste

Sample loop

To column

Solvent in

Inject position

(b)

Solvents should be filtered to remove particulate matter that can cause check valves to leak. Pump seals should be replaced every 6–12 months.

High-pressure mixing pump:
• smaller dwell volume
• limited to binary gradients (two solvents)

Low-pressure mixing pump:
• less expensive
• gradients of up to four solvents

Injection tips:
• Pass samples through a 0.5-μm filter to remove particulate matter prior to injection.
• Sample solvent strength should be less than or equal to eluent solvent strength.
• HPLC syringe needle is *blunt*, not pointed, to prevent damage to the injection port.
• Reproducibly load injector loop with <0.5 or >3 loop volumes.

Linear range: analyte concentration range over which detector response is proportional to concentration

Dynamic range: range over which detector responds in any manner (not necessarily linearly) to changes in analyte concentration (see Figure 4-14)

Detection limit: concentration of analyte that gives 3 : 1 signal-to-noise ratio

the pump. The quality of a pump for HPLC is measured by how steady and reproducible a flow it can produce. A fluctuating flow can create detector noise that obscures weak peaks. Most HPLC pumps use multiple pistons to provide smoother flow.

Gradients can be delivered in two ways. In a *high-pressure mixing pump* each solvent is delivered by a different pump (solvent A from pump 1 and solvent B from pump 2). The relative flow rates from each pump will determine the mobile phase composition. Alternatively, in a *low-pressure mixing pump*, gradients made from up to four solvents are constructed by proportioning liquids through a four-way valve at low pressure before the piston and then pumping the mixture at high pressure into the column. Either way, the gradient is controlled electronically and programmable in 0.1 vol% increments.

The *injection valve* in Figure 25-20 has interchangeable sample loops, each of which holds a fixed volume. Loops of different sizes hold volumes that range from 2 to 1 000 μL. In the load position, a blunt-tipped syringe is used to rinse and load the loop with fresh sample at atmospheric pressure. High-pressure flow from the pump to the column passes through the lower left segment of the valve. When the valve is rotated 60° counterclockwise in Figure 25-20b, the pump is connected to the right side of the sample loop and flow from the pump transfers the content of the sample loop onto the column.

Figure 25-20 shows a *filled-loop injection* in which sample is rinsed through the loop until excess sample exits through the waste port. At least three loop volumes of sample should be flushed through the injector to ensure complete loop filling. To change the injection volume, the loop must be replaced with a tube of different volume. Alternatively, in *partial-loop injection*, a carefully measured volume of less than half the loop volume is introduced into the injector. When the injection valve is turned, the loop contents are flushed onto the column. Argentometric titrations determined that autosamplers can achieve 0.6% injection precision in the partial-loop mode and 0.1% in the filled-loop mode.[34] (Titrations still have a role—even for instrumental analysis.)

Spectrophotometric Detectors

An ideal detector of any type (Table 25-5) is sensitive to low concentrations of every analyte, provides linear response, and does not broaden the eluted peaks.[35] It is also insensitive to changes in temperature and solvent composition. To prevent peak broadening, the detector volume should be <20% of the volume of the chromatographic band. For accurate quantification, the detector must respond fast enough to collect at least 20 data points across the peak. Gas bubbles in the detector create noise. To prevent bubble formation during depressurization of eluate in the detector, back pressure may be applied to the detector by using capillary tubing as the waste line.

An **ultraviolet detector** using a flow cell such as that in Figure 25-21 is the most common HPLC detector because many solutes absorb ultraviolet light. Simple systems employing the intense 254-nm emission of a mercury vapor lamp were the backbone of early HPLC systems,

TABLE 25-5 Comparison of common HPLC detectors

Detector	Response*	Limit of detection	Linear range	Useful with gradient?
Ultraviolet	Selective ≥210 nm universal ≤210 nm	ng	10^5	Yes
Refractive index	Universal	μg	10^3	No
Evaporative light scattering	Universal	High ng	10^3	Yes
Charged aerosol	Universal	Low ng	10^4	Yes
Electrochemical	Very selective	fg–pg	10^5	No
Fluorescence	Very selective	pg	10^3–10^4	Yes
Nitrogen chemiluminescence	Selective	Sub ng		Yes
$N \xrightarrow{\text{combustion}} NO \xrightarrow{O_3} NO_2^* \rightarrow h\nu$				
Conductivity	Selective	High ng	10^4	No
Mass spectrometry	Selective or universal	ng		Yes

*Selective response means that the detector responds to a limited class of analytes. For example, the nitrogen chemiluminescence detector responds only to compounds containing nitrogen atoms. Universal response means that the detector responds to almost all compounds. For example, the charged aerosol detector responds to all nonvolatile compounds, regardless of their structure. Data from M. Swartz, "HPLC Detectors: A Brief Review," J. Liq. Chromatogr. **2010**, 33, 1130.

but are little used today. More versatile, variable-wavelength detectors have broadband deuterium, xenon, or tungsten lamps and a monochromator, so you can choose an optimum wavelength for your analytes. At wavelengths above 210 nm, detection is selective for compounds with an absorbing chromophore. Many compounds absorb wavelengths below 210 and so ultraviolet detection < 210 nm is nearly universal. The system in Figure 25-22 uses a *photodiode array* to record the spectrum of each solute as it is eluted. The spectrum can be matched with library spectra to help identify the peak. Spectra collected across a peak can be compared to evaluate peak purity and detect coeluting components.

The high efficiency of UHPLC systems requires small extra-column volumes (Table 25-2), whereas the sensitivity of absorbance detection requires a longer pathlength. Figure 25-22 shows a total internal reflection (Section 20-4) flow cell design that reduces the cell volume while increasing the pathlength to 6 cm.

High-quality detectors provide full-scale absorbance ranges from 0.000 5 to 3 absorbance units, with a noise level near 1% of full scale. The linear range extends over five orders of

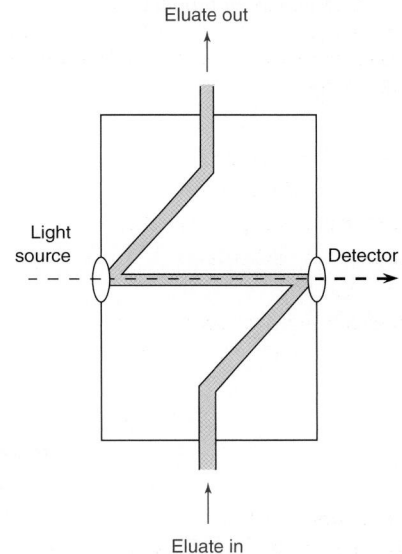

FIGURE 25-21 Light path in a spectrophotometric micro Z-path flow cell. One common cell with a pathlength of 1 cm has a volume of 8 μL. Another cell with a 0.5-cm pathlength contains only 2.5 μL. In the future, it should be possible to lower detection limits by 1–2 orders of magnitude by placing mirrors on both ends of the cavity. Light transits the cell multiple times, with ~1% of the power leaking through the mirror to the detector on each pass.[36]

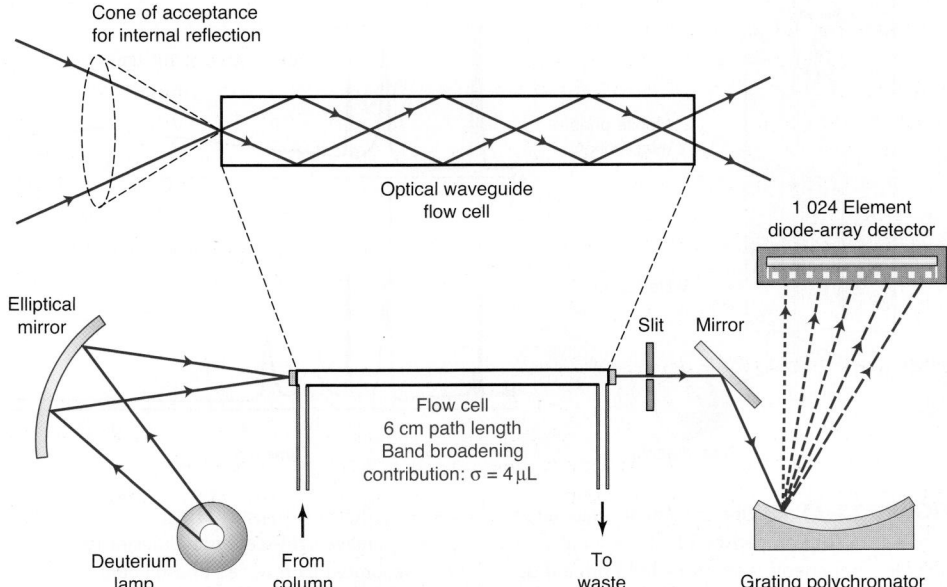

FIGURE 25-22 Diode array ultraviolet detector for UHPLC. Broadband emission from a deuterium lamp is directed into a light-guided flow cell detector. Transmitted light is diffracted by a grating and directed across a photodiode array (Section 20-3). The light-guided flow cell is equivalent to an optical fiber (Section 20-4). The 6-cm light-guided flow cell has a dispersion (σ) of 4 μL.

A photodiode array was used to record the ultraviolet spectrum of each peak in Figure 25-12 as it was eluted. By this means, it was possible to determine which compounds were in each peak.

Derivatization:

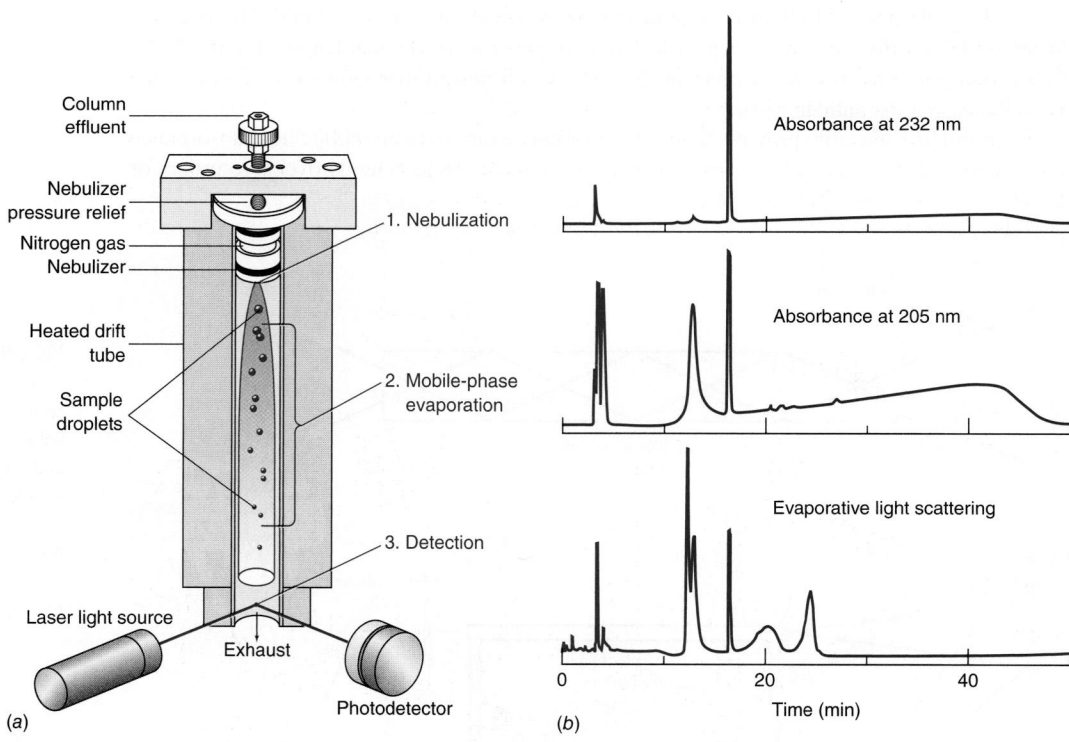

magnitude of solute concentration (which is another way of saying that Beer's law is obeyed over this range). Ultraviolet detectors are good for gradient elution with nonabsorbing solvents. Table 25-4 lists approximate cutoff wavelengths below which a solvent absorbs too strongly.

Fluorescence detectors excite the eluate with a laser and measure fluorescence at longer wavelengths (Figure 18-22). These detectors are up to 100 times more sensitive than an ultraviolet detector, but respond only to the few analytes that fluoresce. To increase the utility of fluorescence and electrochemical detectors (described later), fluorescent or electroactive groups can be covalently attached to the analyte. This process of **derivatization** can be performed on the mixture prior to chromatography, as in Box 25-1, or by addition of reagents to the eluate between the column and the detector (called *post-column derivatization*).[37] For example, the nonfluorescent *o*-phthaldialdehyde reacts under alkaline conditions with primary amine compounds to form a fluorescent product, which can be excited near 345 nm and emits strongly at 400–500 nm. UHPLC separation with post-column derivatization allows detection of histamine, cadaverine, spermidine, tyramine, and putrescine, all indicators of food spoilage, with detection limits in the low fmol range within 6 min.[38]

Evaporative Light-Scattering Detector

An **evaporative light-scattering detector** responds to any analyte that is significantly less volatile than the mobile phase.[39] In Figure 25-23a, eluate enters the detector at the top. In the nebulizer, eluate is mixed with nitrogen gas and forced through a small-bore needle to form a uniform dispersion of droplets. Solvent evaporates from the droplets in the heated drift tube, leaving a fine mist of solid particles to enter the detection zone at the bottom. Particles are detected by the light that they scatter from a diode laser to a photodiode. Detector response is related to the mass of analyte, not to the structure or molecular mass of the analyte.

Evaporative light scattering is compatible with gradient elution. Also, there are no peaks associated with the solvent front, so there is no interference with early eluting peaks. What do we mean by solvent front? In Figure 25-14, you can see small positive and negative signals at

FIGURE 25-23 (*a*) Operation of an evaporative light-scattering detector. [Information from Alltech Associates, Deerfield, IL.] (*b*) Comparison of ultraviolet absorbance and evaporative light-scattering detector response. Soluble components were extracted from a drug tablet and separated by reversed-phase liquid chromatography on a 150 × 4.6 mm column containing C₈-silica with a 30-nm pore size for separating polymers. Solvent A is 0.01 wt% trifluoroacetic acid in H_2O. Solvent B is 45 wt% propanol and 0.01 wt% trifluoroacetic acid in H_2O. A gradient was run from 10 vol% B up to 90 vol% B at 1 mL/min. Gradient time was not stated. [Data from L. A. Doshier, J. Hepp, and K. Benedek, "Method Development Tools for the Analysis of Complex Pharmaceutical Samples," *Am. Lab.*, December 2002, p. 18.]

3–4 min. These signals arise from changes in the refractive index of the mobile phase due to solvent in which the sample was dissolved. This change displaces the ultraviolet detector signal at the time t_m required for unretained mobile phase to pass through the column. If a peak is eluted close to the time t_m, it can be distorted by the solvent front peaks. An evaporative light-scattering detector has no solvent front peaks.

If you use a buffer in the eluent, it must be volatile or else it will evaporate down to solid particles that scatter light and obscure the analyte. Low-concentration buffers made from acetic, formic, or trifluoroacetic acid, ammonium acetate, diammonium phosphate, ammonia, or triethylamine are suitable. Buffers for evaporative light scattering are the same as those for mass spectrometric detection.

Evaporative light-scattering detectors are particularly attractive for determining nonvolatile compounds that do not absorb above 200 nm (for example, saturated hydrocarbons, steroids, surfactants, sugars). Figure 25-23b compares evaporative light scattering with ultraviolet absorbance for the detection of soluble components of a drug tablet containing polymers and small-molecule active ingredients. Two components have absorbance at 232 nm. Four or five components are evident with 205-nm detection. The solvent gradient produces a sloping baseline at 205 nm. All components are observed by evaporative light scattering, and there is no slope to the baseline. Some broad peaks arise from polymers with a distribution of molecular masses.

Charged Aerosol Detector

The **charged aerosol detector** is a sensitive, almost universal detector with nearly equal response to equal masses of nonvolatile analytes, as shown in the chromatogram in Box 25-3.[40] For example, it is useful for measuring the material balance in pharmaceutical preparations because the relative area of each peak in the chromatogram is nearly equal to the relative mass of that component in the preparation. Lipids and carbohydrates, which have little response from an ultraviolet detector, give the same response as other components of the mixture. A minor peak in the charged aerosol chromatogram must be a minor component of the mixture, whereas it could be a major component in the ultraviolet chromatogram.

At the top left of the charged aerosol nebulizer in Figure 25-24, eluate and N_2 gas enter a nebulizer similar to the premix burner in Figure 21-5. Fine mist from the nebulizer reaches the drying tube, while larger droplets fall to the drain. In the drying tube, solvent evaporates at ambient temperature, leaving an aerosol containing ~1% of the original analyte. Meanwhile, part of the N_2 stream passes over a Pt needle held at ~+10 kV with respect to the outer case of the corona charging chamber to form N_2^+. A chain of events, perhaps as written for atmospheric pressure chemical ionization on page 581, transfers positive charge to aerosol particles that flow from the charging chamber through a small-ion trap. Charged plates of the

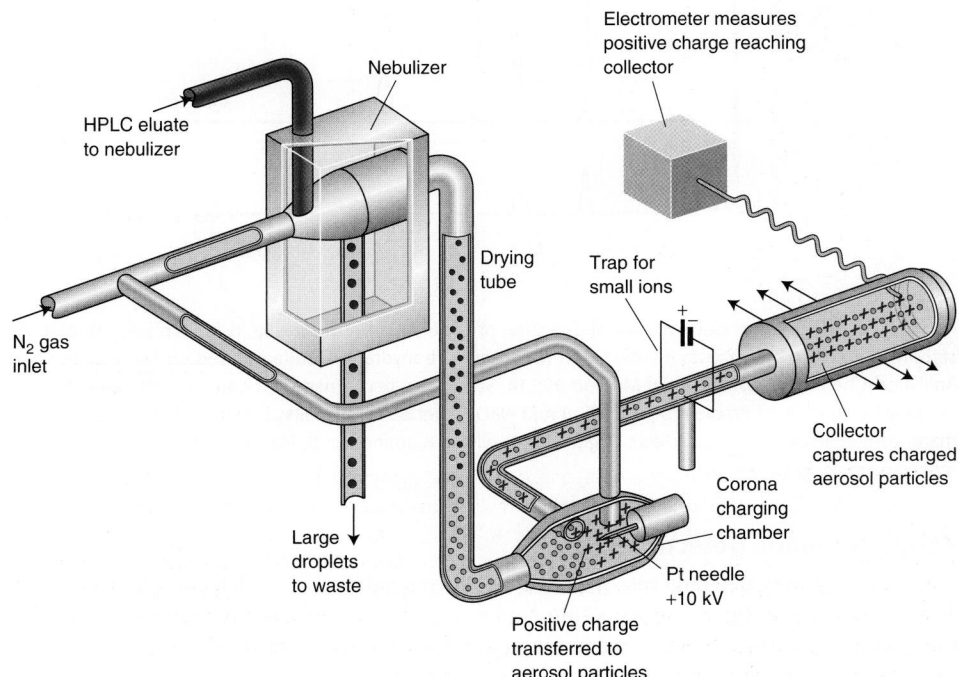

FIGURE 25-24 Operation of the charged aerosol detector. [Information from ESA, Inc., Chelmsford MA.]

trap attract and remove small mobile ions. Aerosol particles are too massive to be deflected and pass through the trap to the collector. Total charge reaching the collector is measured by an electrometer, which produces the detector signal for the chromatogram.

The dynamic range of the detector spans 4–5 orders of magnitude in concentration. Response is approximately proportional to $\sqrt{m}$. Equal masses of different analytes give equal response within ~15% in isocratic elution. Response depends on solvent composition, with higher response for higher percentage of volatile organic solvent and lower response for water. As the organic composition of a gradient increases, so does the response. However, the detector can still be used with gradient elution.[41]

Electrochemical Detector[42]

An **electrochemical detector** responds to analytes that can be oxidized or reduced, such as phenols, aromatic amines, peroxides, mercaptans, ketones, aldehydes, conjugated nitriles, aromatic halogen compounds, and aromatic nitro compounds. The opening of Chapter 17 shows a detector in which eluate is oxidized or reduced at the working electrode. The potential is maintained at a selected value with respect to the Ag|AgCl reference electrode, and current is measured between the working electrode and the stainless steel auxiliary electrode. Carbon, gold, silver, and platinum working electrodes are common. Current is proportional to solute concentration over six orders of magnitude. Aqueous or other polar solvents containing dissolved electrolytes are required, and they must be rigorously freed of oxygen. Metal ions extracted from tubing can be masked by adding EDTA to the mobile phase. The detector is sensitive to flow rate and temperature changes.

Pulsed electrochemical measurements at Au or Pt working electrodes expand the classes of detectable compounds to include alcohols, carbohydrates, and sulfur compounds.[43] Figure 25-25 shows the determination of carbohydrates in instant coffee by anion-exchange chromatography and pulsed electrochemical detection. The electrode is held at +0.8 V (versus a saturated calomel electrode) for 120 ms to oxidatively desorb organic compounds from the electrode surface and to oxidize the metal surface. Then the electrode is brought to −0.6 V for 200 ms to reduce the oxide to pristine metal. The electrode is then brought to a constant working potential (typically in the range +0.4 to −0.4 V), at which analyte is oxidized or reduced. After a delay of 400 ms to allow the charging current (Figure 17-19) to decay to 0, current is integrated for the next 200 ms to measure analyte. The sequence of pulses is then repeated to measure successive data points as eluate emerges from the column.

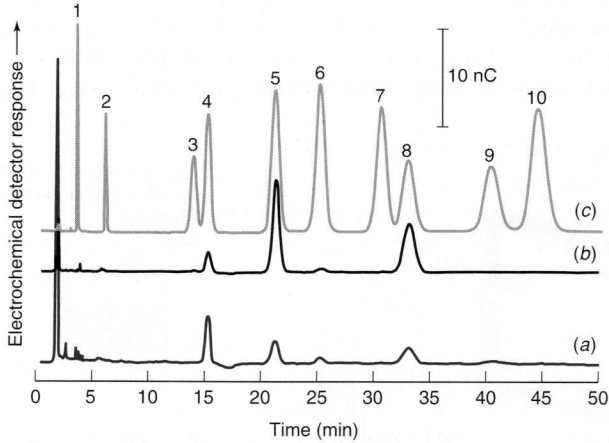

FIGURE 25-25 Pulsed electrochemical detection of (*a*) free carbohydrates in instant coffee, (*b*) total carbohydrate extract from instant coffee, and (*c*) mixed carbohydrates standard based on Association of Analytical Chemists (AOAC) Official Method 995.13. Separation performed on a CarboPac PA1 anion-exchange column with gradient of 10 to 300 mM NaOH over 50 min. Peaks: 1, mannitol; 2, fucose; 3, rhamnose; 4, arabinose; 5, galactose; 6, glucose; 7, xylose; 8, mannose; 9, fructose; 10, ribose. [Data from Thermo Scientific, Sunnyvale, CA.]

Refractive Index Detector

A **refractive index detector** responds to almost every solute, but its detection limit is about 1 000 times poorer than that of the ultraviolet detector. A deflection-type detector has two triangular 5- to 10-μL compartments through which pure solvent or eluate passes. Collimated (parallel) visible light, filtered to remove infrared radiation that would heat the sample, passes through the cell with pure solvent in both compartments and is directed to a photodiode array. When solute with a different refractive index enters the cell, the beam is deflected and different pixels of the array are irradiated.

A refractive index detector is useless in gradient elution because we cannot match the sample and reference while solvent composition is changing. Refractive index is sensitive to changes in pressure and temperature ($\sim$0.01°C).[44] With its low sensitivity, a refractive index detector is not useful for trace analysis. It also has a small linear range, spanning only a factor of 500 in solute concentration. The primary appeal of this detector is its universal response to all solutes, including those that have little ultraviolet absorption.

Comments on Mass Spectrometric Detection in Liquid Chromatography

Electrospray ionization (Figure 22-25) is the most common liquid chromatography–mass spectrometry interface. It requires that analyte be charged in solution. Atmospheric pressure chemical ionization (Figure 22-27) is used for compounds that do not ionize well by electrospray ionization. These are typically more stable, lower molecular mass compounds and some nonpolar compounds. Atmospheric pressure ionization vaporizes the sample first and then uses a corona discharge to add charge to the analyte in the gas phase. This process is more likely to cause sample degradation than electrospray ionization. As the two methods use different ionization techniques, the response and selectivity of the two interfaces may differ significantly. Electrospray ionization is more susceptible to ionization suppression by the matrix.

Volatile, low-concentration ($\leq$10 mM) buffers used with mass spectrometers are made from trifluoroacetic acid, formic acid, acetic acid, and ammonia. HCl should not be used because it corrodes the metal interface of the mass spectrometer. Blank injections should be run to observe background ions generated in the absence of sample. Suitable organic solvents include acetonitrile, methanol, ethanol, propanol, and acetone. Tetrahydrofuran is less desirable because, for some types of samples, it can lead to large background signals in a total ion chromatogram. Ion-pair agents (Figure 26-11) and surfactants should not be used because they create background signals and suppress electrospray signals. Triethylamine suppresses ionization and gives an intense MH^+ peak at m/z 102.

Mass spectral sensitivity is reduced by aqueous mobile phase. More organic phase can be added through a "T" connection between the column and the mass spectrometer. Alternatively, a different stationary phase can be chosen to require more organic phase. If compounds are weakly retained by reversed-phase columns, hydrophilic interaction liquid chromatography allows use of an organic-rich mobile phase. The pH of the mobile phase can be adjusted between the column and the detector so that acidic or basic analytes are ionized. Atmospheric pressure chemical ionization and electrospray with positive and negative ion modes can all be tried to obtain adequate sensitivity. Eluate emerging from the column prior to t_m can be diverted to waste to avoid introducing unnecessary salts into the mass spectrometer.

Run a *performance qualification* sample or an internal standard frequently to verify that the separation and the mass spectrometer are behaving normally. When a liquid chromatography-mass spectrometry method develops a problem, but the performance qualification check is normal, the problem is most likely in the chromatography of the unknown. If the performance qualification check fails, the problem is likely in the mass spectrometer. You can test the mass spectrometer alone by direct flow injection of a standard sample or calibration solution into the mass spectrometer.

25-3 Method Development for Isocratic Reversed-Phase Separations[2,3,45]

Many separations encountered in industrial and research laboratories can be handled by reversed-phase chromatography. We now describe a general procedure for developing an isocratic separation of an unknown mixture with a reversed-phase column. The next section deals with gradient separations. In method development, the goals are to obtain adequate separation in a reasonable time. Ideally, the procedure should be *rugged*, which means that the separation will not be seriously degraded by gradual deterioration of the column, by *small* variations in solvent composition, pH, and temperature, or by use of a different batch of the same stationary phase, perhaps from a different manufacturer. If your column does not have temperature control, you can at least insulate it to reduce temperature fluctuations.

Reversed-phase chromatography is usually adequate to separate mixtures of low-molecular-mass neutral or charged organic compounds. If isomers do not separate well, normal-phase

Desired attributes of a new chromatographic method:
- adequate resolution of desired analytes
- short run time
- rugged (not drastically affected by small variations in conditions)

chromatography or porous graphitic carbon is recommended because solutes have stronger, more specific interactions with the stationary phase. For enantiomers, chiral stationary phases (Figure 25-10) are required. Separations of inorganic ions, polymers, and biological macromolecules are described in Chapter 26.

As in gas chromatography (Section 24-5), the first steps in method development are to (1) determine the goal of the analysis, (2) select a method of sample preparation to ensure a "clean" sample, and (3) choose a detector that allows you to observe the desired analytes. The remainder of method development described in the following sections assumes that steps 1 through 3 have been carried out.

Criteria for an Adequate Separation

In liquid chromatography, the entire space within a column accessible to solvent and small solutes is called the **dead volume,** V_m. This volume includes the pores in the particles and the volume between particles. If the time required for solvent or unretained solute to travel through the column is t_m, the dead volume is $V_m = t_m F$, where F is the volume flow rate.

The *retention factor k* in Equation 23-16 measures adjusted retention time, $t_r - t_m$, in units of the time t_m required for mobile phase to pass through the column. Reasonable separations demand that k for all peaks be in the range 0.5–20. If k is too small, the first peak is distorted by the solvent front. If k is too great, the run takes too long. In the lowest trace in Figure 25-12, t_m is the time when the first baseline disturbance is observed near 3 min. If you do not observe a baseline disturbance, you can estimate

$$\textit{Solvent front:} \qquad V_m \approx \frac{L d_c^2}{2} \quad \text{or, equivalently,} \quad t_m \approx \frac{L d_c^2}{2F} \qquad (25\text{-}5)$$

where V_m is the volume in mL at which unretained solute is eluted (= volume at which solvent front appears), L is column length (cm), d_c is column diameter (cm), and F is flow rate (mL/min). In reversed-phase chromatography, t_m could be measured by running the unretained solutes uracil (detected at 260 nm) or $NaNO_3$ (detected at 210 nm) through the column. In hydrophilic interaction chromatography, t_m can be measured using toluene (detected at 215 nm).

For quantification, a minimum resolution (Figure 23-10) of 1.5 between the two closest peaks is desired. For ruggedness, a resolution of 2 is desirable so that resolution remains adequate despite small changes in conditions or slow deterioration of the column.

A chromatographic method must not exceed the upper allowable operating pressure for the hardware. Keeping pressure below ~15 MPa (150 bar) prolongs the life of the pump, valves, seals, and autosampler. Pressure can double during the life of a column because of progressive clogging. Establishing an operating pressure of ≤15 MPa during method development allows for column degradation.

All peaks (certainly all peaks that need to be measured) should be symmetric, with an asymmetry factor B/A in Figure 23-14 in the range 0.9–1.5. Asymmetric peak shapes should be improved as described at the end of Section 25-1 (page 684) before optimizing a separation.

Optimization of Isocratic Retention

Method development focuses on the parameters governing resolution: retention factor k, the number of plates N, and the relative retention α. Method development is finished as soon as the separation meets your criteria. There is a good chance of attaining adequate separation without going through all the steps.

Step 1   Choose column and organic solvent.
Step 2   Optimize the separation by adjusting the retention factor k by varying the organic solvent/water composition of the mobile phase.
Step 3   Check for low plate number (Equation 25-1).
Step 4   Adjust relative retention by adjusting: **(i)** solvent composition, **(ii)** column temperature, **(iii)** type of organic solvent, or **(iv)** column type.
Step 5   Optimize column dimensions to either increase plate number or decrease separation time.

Step 1 is choosing starting conditions, as listed in Table 25-6. Typical conditions for reversed-phase HPLC are a 10- to 15-cm column packed with 3- or 5-μm C_{18} on Type B silica. The first solvent mixture to try is acetonitrile and water. Acetonitrile has low viscosity, which allows relatively low operating pressure, and it permits ultraviolet detection down to 190 nm (Table 25-4). At 190 nm, many analytes have some absorbance. Methanol is the second choice for organic solvent because it has a higher viscosity and longer wavelength

TABLE 25-6 **Starting conditions for reversed-phase chromatography**

Stationary phase:	C_{18} or C_8 on 5- or 3-μm-diameter spherical Type B silica for operation in the pH 2–7.5 range. For operation in the pH 8–12 range, use ethylene-bridged silica (Figure 25-6). For operation above 50°C, use sterically protected silica (Figure 25-8). Alternatively, polymer or zirconia stationary phases are stable over wide pH and temperature ranges.

Column:	15×0.21 cm column for 5-μm particles
	10×0.21 cm column for 3-μm particles (shorter run, similar resolution)
	   When changing particle size, maintain constant (column length/particle diameter) for
	   similar resolution at same linear flow rate.
	15×0.46 cm column for 5-μm particles
	10×0.46 cm column for 3-μm particles (shorter run, similar resolution)

Flow rate:	1.0 mL/min for 0.46-cm-diameter column 0.2 mL/min for 0.21-cm-diameter column $= \left(\dfrac{0.21 \text{ cm}}{0.46 \text{ cm}}\right)^2 \left(1.0 \dfrac{\text{mL}}{\text{min}}\right)$
Mobile phase:	CH_3CN/H_2O for neutral analytes CH_3CN/aqueous buffer[a] for ionic organic analytes 5 vol% CH_3CN in H_2O to 95 vol% CH_3CN in H_2O for gradient elution
Temperature:	30°–40°C with temperature control Some people routinely prefer 50°–60°C
Sample size:	0.21-cm-diameter column: 2 μL containing 25–50 μg of each analyte 0.46-cm-diameter column: 5–10 μL containing 25–50 μg of each analyte (Older ultraviolet absorbance detectors might require more sample.)
For mass spectrometry:	5×0.21 cm column with 3-μm particles of C_{18}-silica run at 30°–60°C Mobile phase is acetonitrile or methanol plus volatile, aqueous buffer. Aqueous buffer is made from 2 mM ammonium formate plus 1–10 g formic acid per L (giving pH $\approx$ 2.7–1.7). Ammonium formate buffer can be used for pH $\approx$ 2.8–4.8. Ammonium acetate buffer can be used for pH $\approx$ 3.8–5.8. Use NH_3, formic acid, or acetic acid to adjust pH. Ammonium carbonate can be used for pH $\approx$ 8–11. Atmospheric pressure chemical ionization or electrospray interface. Electrospray requires analyte to be charged in solution.

a. Buffer is 25 mM phosphate/pH 2–3 made by treating H_3PO_4 with KOH. K^+ is more soluble than Na^+ in organic solvents and leads to less tailing. This concentration of phosphate is suitable for ultraviolet detection. Add 0.2 g sodium azide per liter as a preservative if buffer will not be consumed quickly.

Information from L. R. Synder, J. J Kirkland, and J. L. Glajch, *Practical HPLC Method Development* (New York: Wiley, 1997) and S. Needham, "HPLC Method Development for LC/MS," short course presented at Pittcon 2009.

ultraviolet cutoff. Tetrahydrofuran is the third choice because it has less usable ultraviolet range, it is slowly oxidized and can form explosive peroxides,[47] it is incompatible with poly ether ether ketone (PEEK) tubing, and it equilibrates more slowly with stationary phase.

Step 2 is adjusting the mobile phase composition to achieve retention of $0.5 \leq k \leq 20$ for all compounds. Figure 25-12 showed such a set of experiments. The initial experiment was done with a high concentration of CH_3CN (90% B) to ensure the elution of all components of the unknown. Then %B was successively lowered to increase retention of all components. Eluent containing 35% B was selected as it resolved all components, albeit the last component took 2 h to be eluted.

With 35% B, peak 1 is eluted at 4.9 min and peak 8 is eluted at 125.2 min. The solvent front appears at $t_m = 2.7$ min. Therefore, k for peak 1 is $(4.9 - 2.7)/2.7 = 0.8$ and k for peak 8 is $(125.2 - 2.7)/2.7 = 45$. *Because it is not possible to fit all compounds within $0.5 \leq k \leq 20$, gradient elution is suggested* (Section 25-4).

If we were not concerned about measuring peaks 1 to 3, mobile phases of 40% B, 50%, and 60% B all provide suitable retention ($0.5 \leq k \leq 20$) of peaks 4 to 8. Additional optimization of the resolution is still needed.

Step 3: At this point it is good practice to check that the column is providing the plate number (*N*) expected. The observed plate number in Figure 25-12 (12 000 plates) is close to the theoretical maximum of 15 000 predicted by Equation 25-1. Thus, the column is performing well and we can proceed with method development. If the plate number is poor, the column should be washed or replaced, and the extra-column volume of the instrument should be checked.

"Rule of three": Decreasing %B by 10% increases retention factor, k, by a factor of ~3.

If all components are not eluted in the range $0.5 \leq k \leq 20$, try gradient elution.

Step 4: Optimize Relative Retention

Adjustment of the relative retention may be done in a number of ways:

  i. Fine-tune solvent composition (%B).
 ii. Vary column temperature.
iii. Change the polarity of organic solvent.
 iv. Change the stationary phase

i. The easiest and first to try is fine-tuning %B.[48] If we were developing a method for only compounds 4–8 in Figure 25-12, 45% B resolves toluene (6) and 2,6-dimethoxytoluene (7) and yields a reasonable overall run time. No further optimization of relative retention would be required, and we could jump to Step 5.

Changing %B has only subtle effects on resolution in Figures 25-12 and 25-13. Some samples exhibit greater changes in relative retention—even changing the elution order—when adjusting %B.

ii. If adjusting %B does not provide adequate resolution, a second means to fine-tune the relative retention is to vary the column temperature. Most HPLC instruments control column temperature to ensure reproducible retention. Column temperature could affect the relative retention of compounds, and elevated temperature permits faster chromatography to be conducted.[49] Figure 25-26 suggests a systematic procedure for method development in which solvent composition and temperature are the two independent variables. Changing one variable at a time, either %B or T, allows us to map out the selectivity changes due to each variable. An optimum combination of solvent composition and column temperature can thus be found. For elevated temperature operation, pH should be below 6 to retard dissolution of silica. Alternatively, zirconia-based stationary phases work up to at least 200°C.

iii. If neither adjusting %B nor the temperature provides adequate resolution, switching the organic solvent from acetonitrile to an organic solvent with a different type of polarity (Table 25-4) will change resolution.[50] Acetonitrile's polarity is predominantly dipolar; methanol is a much stronger hydrogen bonding solvent; and tetrahydrofuran cannot act as a hydrogen-bond donor.

Figure 25-27 illustrates the effect of changing solvent on a separation. To obtain chromatogram A, the acetonitrile/buffer mobile phase was optimized as per Step 2 so that all compounds are eluted within $0.5 \leq k \leq 20$. At the best composition, 30 vol% acetonitrile/70 vol% buffer, peaks 4 and 5 are not resolved adequately for quantitative analysis. Chromatogram B in Figure 25-27 shows the effect of changing to a methanol/buffer mobile phase. It is not necessary to start all over again with 90% methanol and try a series of decreasing percent

Retention decreases 1–3% per degree increase in temperature.

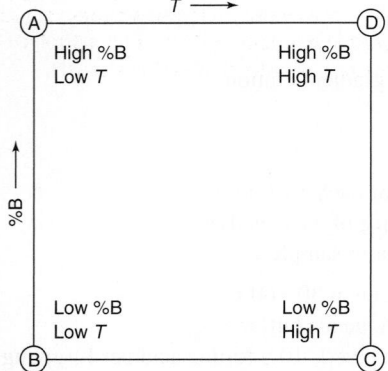

FIGURE 25-26 Isocratic method development for HPLC, using solvent composition (%B) and temperature (T) as independent variables. %B and T are each varied between selected low and high values. From the quality of the separation obtained with conditions A–D, we can select intermediate conditions to try to further improve the separation.

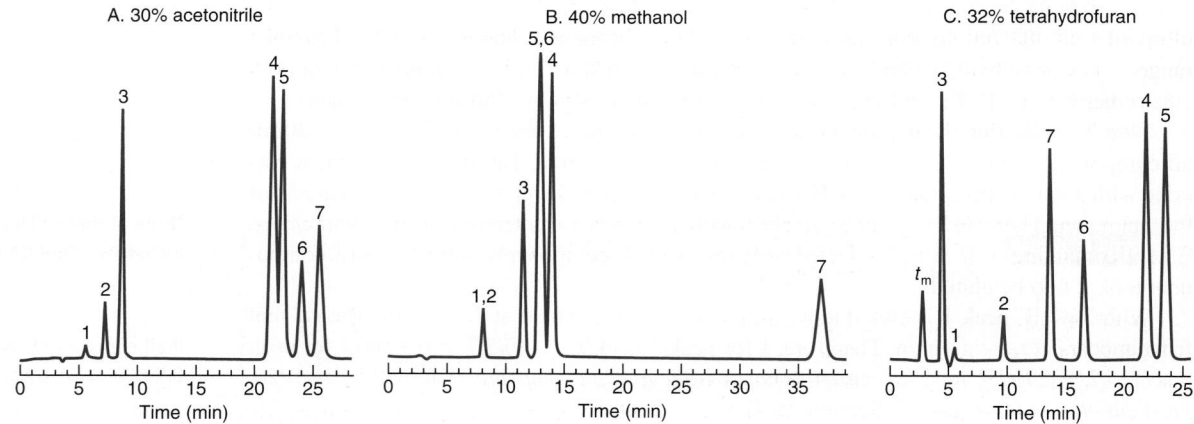

FIGURE 25-27 Effect of organic solvent on the separation of seven aromatic compounds by HPLC. Column: 25 × 0.46 cm Hypersil ODS (C_{18} on 5-μm silica) at ambient temperature (~22°C). Elution rate was 1.0 mL/min with the following solvents: (A) 30 vol% acetonitrile/70 vol% buffer; (B) 40% methanol/60% buffer; and (C) 32% tetrahydrofuran/68% buffer. The aqueous buffer contained 25 mM KH_2PO_4 plus 0.1 g/L NaN_3 adjusted to pH 3.5 with HCl. The negative dip in C between peaks 3 and 1 is associated with the solvent front. Peak identities were tracked with a photodiode array ultraviolet spectrophotometer: (1) benzyl alcohol, (2) phenol, (3) 3′,4′-dimethoxyacetophenone, (4), m-dinitrobenzene, (5) p-dinitrobenzene, (6) o-dinitrobenzene, and (7) benzoin.

CHAPTER 25 High-Performance Liquid Chromatography

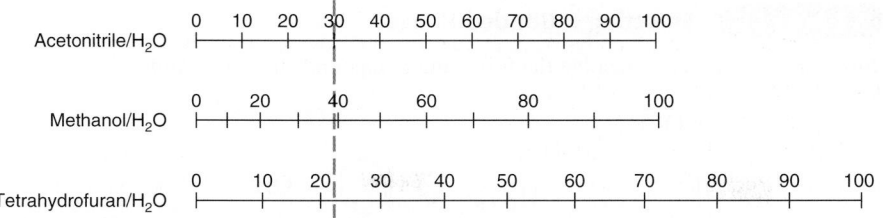

FIGURE 25-28 Nomograph showing volume percentage of solvents having the same eluent strength. A vertical line intersects each solvent line at the same eluent strength. For example, 30 vol% acetonitrile/70 vol% water has about the same eluent strength as 40 vol% methanol or 22 vol% tetrahydrofuran. [Data from L. R. Snyder, J. J. Kirkland, and J. L. Glajch, *Practical HPLC Method Development* (New York: Wiley, 1997).]

methanol. Figure 25-28 allows us to select a methanol/water mixture that has approximately the same mobile phase strength as a particular acetonitrile/water mixture. A vertical line drawn at 30% acetonitrile (the composition used in chromatogram A) intersects the methanol line near 40%. Therefore, 40% methanol has about the same mobile phase strength as 30% acetonitrile. The average retention time in chromatogram B in Figure 25-27 is indeed about the same as that in chromatogram A. With methanol, however, the seven components give only five peaks. Also when we changed from acetonitrile to methanol, the order of elution of some compounds changed.

Chromatogram C in Figure 25-27 shows the separation using tetrahydrofuran. Figure 25-28 tells us that 22% tetrahydrofuran has the same mobile phase strength as 30% acetonitrile. When 22% tetrahydrofuran was tried, elution times were too long. Trial and error demonstrated that 32% tetrahydrofuran was best. All seven compounds were cleanly separated in chromatogram C in an acceptable time. The order of elution with tetrahydrofuran differs from the order with acetonitrile. In general, changing solvent is a powerful way to change relative retention of different compounds. However, it is not possible to predict how the relative retention will change.

Mixing multiple organic solvents was a popular means of optimizing resolution in the past. For instance, solutes in Figure 25-27 are baseline resolved with $0.5 \leq k \leq 20$ using a mobile phase of 20% methanol/16% tetrahydrofuran/64% buffer. However a mobile phase containing three or more components is complex, and needs constant tweaking to maintain performance. Hence, multiple solvents should only be used after simpler strategies have failed.

iv. If other avenues of changing relative retention do not yield adequate resolution, we can change the column. C_{18}-silica is the most common stationary phase, and it separates a wide range of mixtures when mobile phase is chosen carefully, as in Figures 25-12 and 25-27. Retention and selectivity of C_{18} columns depends on their silica, pore size, bonded phase, and other characteristics. You cannot randomly replace one kind of C_{18} column with another type of C_{18} column and expect the same result.

Table 25-7 is a guide to selecting other bonded phases. Columns with Type A silica are not recommended as they yield tailed peaks for amines and less reproducible retention. Hydrophobicity of the phase depends on the surface area, bonded-phase density, and carbon length of the bonded ligand. The chemistry of the bonded phase and any polar embedded functionalities can introduce additional interactions that may alter relative retention. Databases of several hundred commercial HPLC columns are freely available to help you select columns with similar or different selectivity.[51]

The way we know which peak is which is either to inject more of one standard to see which peak grows or to use a photodiode array spectrometer to record the entire ultraviolet spectrum of each peak as it is eluted.

Simpler mobile phases are more robust.

Not all C_{18} columns are alike!

TABLE 25-7 **Selection of bonded stationary phases for reversed-phase liquid chromatography**

Column properties that affect relative retention
- Type of silica (Type A vs. Type B)
- Hydrophobicity of bonded phase (cyano < phenyl < C_8 < C_{18})
- Additional interactions (phenyl = pi-pi; cyano = dipole; pentafluorophenyl = fluorous)
- Polar embedded groups (hydrogen bonding, dipole)
- Type of end-capping (particularly affects amine separations)

Column properties that do not affect relative retention
- Particle size
- Column diameter
- Column length

Selecting a Bonded Phase

Suggest an approach to resolve the following compounds in a short time.

1. Anisole   **2.** 4-Butylbenzoic acid   **3.** Toluene

4. Mefenamic acid   **5.** Ethylbenzene   **6.** *trans*-Chalcone

Solution The compounds all have hydrophobic phenyl rings, making them amenable to reversed-phase separation with ultraviolet detection. C_{18} is the most common stationary phase used in reversed-phase chromatography, so it is tried first in Figure 25-29. The order of elution should be from most polar to least polar. However, the mixture of nonpolar and polar character of these compounds makes it difficult to predict the exact elution order. Compounds **4** and **6** coelute from the C_{18} column in the upper trace. Table 25-7 suggests that a phenyl stationary phase retains analytes by pi-pi interactions, which might be good for distinguishing **4** and **6**. Some improvement is achieved in the middle trace, but retention is also decreased. When trying to separate compounds, consider the *differences* in the coeluting compounds rather than their *similarities*. Compounds **4** and **6** differ substantially in their polar character. Table 25-7 indicates that polar embedded groups contribute polar interactions to retention. The polar embedded column resolves all of the compounds in the lower trace of Figure 25-9.

Chromatograms in Figure 25-29 are *simulated*, not experimental. Simulation is based on the linear-solvent-strength model (Equation 25-4) with measured retention behavior for different stationary phases. A simulation is never more than an approximate prediction. In the words of Izaak Kolthoff, one of the most prominent analytical chemists of the 20th century:

Theory guides, experiment decides

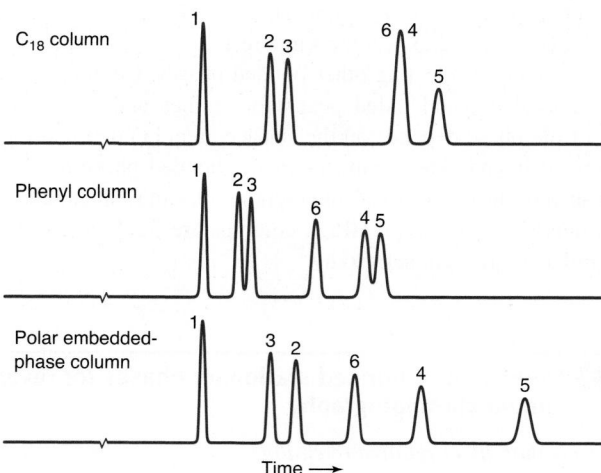

FIGURE 25-29 *Simulated* chromatograms of six compounds on C_{18}-, phenyl-, and polar embedded phase-silica columns, using 50:50 (vol/vol) acetonitrile/30 mM aqueous phosphate buffer (pH 2.8). Column: 150 × 4.6 mm, 5-μm particles; $N = 10\,000$. [Data from J. W. Dolan, "Selectivity in Reversed Phase LC Separations, Part III: Column-Type Selectivity," *LCGC North Am.* **2011,** *29,* 236.]

TEST YOURSELF Why does the retention order of peaks **2** and **3** change on the polar embedded column? (***Answer:*** Compound **2** has polar functionalities that could interact with the polar embedded groups, increasing retention. Compound **3** cannot interact with these groups.)

Step 5: Optimize Column Dimensions

Adjustment of retention (Step 2) and then relative retention (Step 4) will often yield an adequate separation. Further improvement in the resolution can be achieved by changing column conditions (length, particle size, or flow rate) to increase the plate number N.[45] For a given stationary phase, changing the column length or particle size will not change the retention factor (k) or relative retention (α), but may increase the separation time (longer column or slower flow rate).

Conversely, if the resolution is more than adequate (resolution $\gg$ 2), a lower number of plates might be sufficient and would yield a faster separation. For the separation of compounds 4–8 in Figure 25-12, 40% acetonitrile yields a resolution of 3.3 with the 25-cm column of 5-μm particles, but the run time is 68 min. A 10-cm column of the same particles would yield a resolution of 2.1 and reduce the run time to 27 min. Ultimately, the column conditions are a balance of the resolution, the run time, and the column pressure.[52,53]

EXAMPLE **Adjusting Column Conditions**

How can column conditions in Figure 25-12 be changed to yield the same resolution in less time?

Solution Equation 25-1 showed that the plate number of a column is related to its length-to-particle-size ratio (L/d_p). A 15-cm column packed with 3-μm particles would yield the same plate number as the 25-cm column of 5-μm particles used in Figure 25-12.

$$N \approx \frac{3\,000\,L\,(\text{cm})}{d_p\,(\mu\text{m})} = \frac{3\,000\,(25\text{ cm})}{(5\,\mu\text{m})} = 15\,000 = \frac{3\,000\,(15\text{ cm})}{(3\,\mu\text{m})}$$

The time required for a separation is proportional to the column length. The analysis time using a 15-cm column would be (15 cm/25 cm) or 60% of that in Figure 25-12.

TEST YOURSELF What length of column packed with 2-μm particles is needed to yield the same plate number as in Figure 25-12? How long would the separation take? (*Answer:* 10 cm, on which separation would take only 10/25 or 40% as much time as in Figure 25-12.)

How pH Affects Retention

If a solute is an acid or base, its charge depends on pH. The neutral form is retained by reversed-phase columns ("like dissolves like"), whereas the ionized form is hydrophilic and so weakly retained. The kinetics of acid dissociation equilibria are extremely fast compared to the time scale of HPLC separations. Therefore retention is governed by the "average" form of the acid or base. The ionized form is essentially unretained ($k_{A^-} \approx k_{BH^+} \approx 0$), so

Retention factor of weak acid: $\quad k_{obs} \approx \alpha_{HA} k_{HA}$ (25-6)

$$\alpha_{HA} = \frac{[HA]_{aq}}{[HA]_{aq} + [A^-]_{aq}}$$

where α_{HA} is the fraction of acid in the neutral form HA in the mobile phase and k_{HA} is the retention factor for the fully protonated form. Figure 25-30a shows that at low pH an acid such as benzoic acid ($pK_a = 4.20$) is near fully protonated ($\alpha_{HA} \approx 1$) and its retention is unaffected

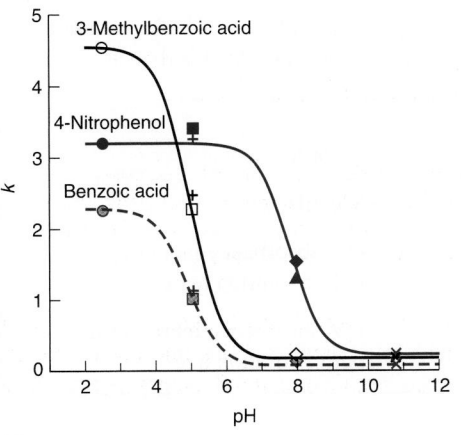
(a)

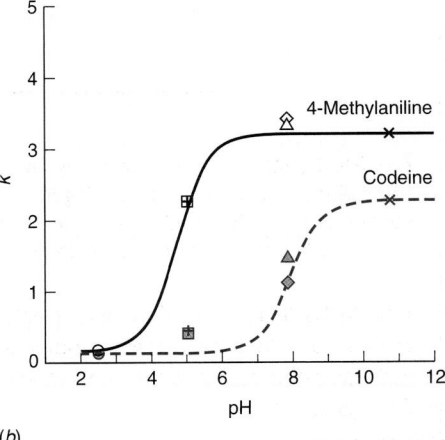
(b)

FIGURE 25-30 Effect of pH on reversed-phase retention of (*a*) weak acids and (*b*) weak bases. Different buffers are used to cover the pH range: ○ pH 2.48 phosphate; + 5.01 acetate; □ 5.01 piperazine; ◇ 7.90 phosphate; △ 7.91 tris; × 10.76 aminobutane. The effect of pH is much greater than the effect of buffer type. Separations were performed on a 15 × 0.46 cm pH-stable X-Terra MS-C$_{18}$ column at 25°C. Typical silica phases are limited to the pH range 2–8. [Data from L. G. Gagliardi, C. B. Castells, C. Ràfols, M. Rosés, and E. Bosch, "Modeling Retention and Selectivity as a Function of pH and Column Temperature in Liquid Chromatography," *Anal. Chem.* **2006,** *78,* 5858.]

by pH. As the pH of the mobile phase approaches pK_a, there is a dramatic decrease in the fraction of benzoic acid in the protonated form and thus in its retention. At pH well above the pK_a, ionized benzoate is the dominant form and is essentially unretained. Organic solvent in the mobile phase raises the pK_a of weak acids slightly, but aqueous phase pK_a effectively guides method development.

Similarly, for weak bases,

$$\text{Retention factor of weak base:} \qquad k_{obs} \approx \alpha_B k_B \qquad \qquad \text{(25-7)}$$

where α_B is the fraction of base in the neutral form B in the mobile phase and k_B is the retention factor for the neutral form of the base. Figure 25-30b shows that a base such as 4-methylaniline (pK_a = 5.08) is weakly retained at low pH and is more retained as the pH approaches its pK_a. Then retention is constant well above the pK_a when B is fully deprotonated. Organic solvents in the mobile phase shift pK_a for BH$^+$ to slightly lower values than their aqueous phase values.

When separating acids or bases, the aqueous component of the mobile phase must contain a buffer. Retention of neutral compounds is nearly unaffected by the presence of buffers. HPLC buffer must have adequate buffering capacity at the desired pH, be soluble in the mobile phase, and be compatible with the detector and instrument hardware. Table 25-8 shows common HPLC buffers. For separations with ultraviolet detection and pH ≤8, common buffers are phosphate, trifluoroacetate, formate, and acetate. Volatility of buffers is important with detectors such as electrospray mass spectrometry or evaporative light scattering. Buffers with ammonium ions are more soluble than buffers with potassium ions, which are more soluble than buffers with sodium ions. Buffers are most soluble in methanol, followed by acetonitrile, and least soluble in tetrahydrofuran. To ensure adequate buffer capacity, consider both the pK_a and buffer concentration (5–25 mM are typical).

Optimization of Isocratic Separations of Acids and Bases

The steps in optimizing the separation of acids and bases by reversed-phase chromatography are similar to those for neutral compounds, with a few important differences. We can use buffers of different pH to control the degree of protonation of the analytes. Retention changes

Aqueous buffer for HPLC is prepared and the pH adjusted *prior* to mixing with organic solvent. Quote pH and concentration of the aqueous solution before mixing.[20]

Test solubility of new buffers by mixing them with organic solvent in a beaker and watching for cloudiness. Buffer solubility:

$NH_4^+A^- > K^+A^- > Na^+A^-$

$CH_3CN > CH_3OH >$ tetrahydrofuran

Symptoms of poor buffering
- poor retention time reproducibility
- asymmetric "shark fin" peaks

TABLE 25-8	Common HPLC buffers			
Buffer	Aqueous pK_a	Buffering pH range	Ultraviolet (UV) cutoff (nm)[*]	Comments
Trifluoroacetic acid (TFA)	0.3		210 (0.1%)	Ion pairing, volatile
Phosphate pK_{a1}	2.1	1.1–3.1	<200 (10 mM)	Limited solubility
pK_{a2}	7.2	5.2–8.2	<200 (10 mM)	
Citrate pK_{a1}	3.1	2.1–4.1	230 (10 mM)	May corrode stainless steel
pK_{a2}	4.7	3.7–5.7	230 (10 mM)	
pK_{a3}	6.4	5.4–7.4	230 (10 mM)	
Formate	3.8	2.8–4.8	210 (10 mM)	Volatile
Acetate	4.8	3.8–5.8	210 (10 mM)	Volatile
BIS-TRIS propane · 2HCl pK_{a1}	6.8	5.8–7.8	215 (10 mM)	Air oxidation of buffer solution increases UV absorbance
pK_{a2}	9.0	8.0–10.0	215 (10 mM)	
Borate	9.1	8.1–10.1	200 (10 mM)	
Ammonia	9.2	8.2–10.2	200 (10 mM)	Volatile
Bicarbonate pK_{a2}	10.3	9.3–11.3	<200 (10 mM)	Volatile
1-Methylpiperidine · HCl	10.1	9.1–11.1	215 (10 mM)	UV absorbance increases as buffer solution ages
Triethylamine · HCl	11.0	10.0–12.0	<200 (10 mM)	UV absorbance increases as buffer solution ages

*Absorbance of aqueous solution <0.5 at wavelengths above cutoff.

Data from L. R. Snyder, J. J. Kirkland, and J. W. Dolan, *Introduction to Modern Liquid Chromatography, 3rd ed. (Hoboken, NJ: Wiley, 2010).*

dramatically when pH is near pK_a of carboxylic acids or amines (Figure 25-30). To ensure strong retention and robust methods, the pH should be well below the pK_a of carboxylic acids (4–5) or well above that of amines (5–10). The high pH (>8) needed to deprotonate most amines requires special columns to avoid dissolving the silica particles.

Method development commonly starts with a mobile phase of pH 2.5–3.0 using phosphate buffer for ultraviolet detection or ammonium formate for mass spectrometric detection. Start with a high %B to ensure that all compounds are eluted in a reasonable time, and adjust %B downward until there is reasonable retention of all compounds. If peaks are not within $0.5 \leq k \leq 20$, a gradient separation is needed. Relative retention can be powerfully adjusted with pH, but ideally the buffer pH should be >1 pH unit from the pK_a of critical components. If the optimized pH is close to a pK_a, best retention reproducibility is achieved by preparing buffers by weight or volume, rather than by adjustment of pH with a pH meter.

Steps for optimizing separations of ionizable compounds
1. Choose column, organic solvent, and *pH*.
2. Adjust %B to get $0.5 \leq k \leq 20$.
3. Adjust relative retention with *pH* (most important) > solvent ≈ column > %B > temperature ≫ *buffer concentration or type* (least important).

25-4 Gradient Separations[3]

Figure 25-12 showed an isocratic separation of eight compounds that required a run time of more than 2 h. When mobile phase strength was low enough to resolve early peaks (2 and 3), the elution of later peaks was very slow. To retain the desired resolution but decrease the analysis time, the *segmented gradient* (a gradient with several distinct parts) in Figure 25-14 was selected. Peaks 1–3 were separated with a low mobile phase strength (30% B). Between 8 and 13 min, B was increased linearly from 30 to 45% to elute the middle peaks. Between 28 and 30 min, B was increased linearly from 45 to 80% to elute the final peaks.

Begin Method Development with a Scouting Gradient

The quickest way to help decide whether to use isocratic or gradient elution is to run a broad gradient.[54] Figure 25-31a shows how the sample mixture in Figure 25-12 is separated by a linear gradient from 10 to 90% acetonitrile in 40 min. The *gradient time*, t_G, is the time over which the solvent composition is changed (40 min). Let Δt be the difference in the retention time between the first and the last peak in the chromatogram. In Figure 25-31a, $\Delta t = 35.5$ min $- 14.0$ min $= 21.5$ min. The criterion to choose whether to use a gradient is

$$\text{Use a gradient if } \Delta t/t_G > 0.40 \qquad \text{Use isocratic elution if } \Delta t/t_G < 0.25$$

If all peaks are eluted over a narrow solvent range, then isocratic elution is feasible. If a wide solvent range is required, then gradient elution is more practical. In Figure 25-31a, $\Delta t/t_G = 21.5$ min/40 min $= 0.54 > 0.40$. Therefore, gradient elution is recommended. Isocratic elution is possible, but the time required in Figure 25-12 is impractically long.

If isocratic elution is indicated because $\Delta t/t_G < 0.25$, then a good starting solvent is the composition at the point halfway through the interval Δt. If the first peak is eluted at 10 min and the last peak is eluted at 20 min, a reasonable isocratic solvent has the composition at 15 min in the gradient.

If $0.25 < \Delta t/t_G < 0.40$, then either isocratic or gradient elution may be appropriate. Other factors such as available equipment and complexity of the sample will influence which mode is selected.

The scouting run on a new mixture should be a gradient.
- If $\Delta t/t_G > 0.40$, use gradient elution.
- If $\Delta t/t_G < 0.25$, use isocratic elution.
- Isocratic solvent should have the composition applied to the column halfway through the period Δt.
- If $0.25 < \Delta t/t_G < 0.40$, either can be used. Decision depends on complexity of sample.

Developing a Gradient Separation

The first run should survey a broad range of mobile phase strength such as 10 to 90% B in 40 min in Figure 25-31a. Serendipitously, the first run in Figure 25-31 separated all eight peaks. We could stop at this point if we were willing to settle for a 36-min run time.

The next step in developing a gradient method is to spread the peaks out by choosing a shallower gradient. The gradient profile for Figure 25-31a is shown in Figure 25-32. Peak 1 was eluted at 14.0 min when the solvent was 28% B. Peak 8 was eluted near 35.5 min, when the solvent was 71% B. The portions of the gradient from 10 to 28% B and 71 to 90% B were not needed. Therefore, the second run could be made with a gradient from 28 to 71% B over the same t_G (40 min). Conditions chosen in Figure 25-31b were 30 to 82% B in 40 min. This gradient spread the peaks and reduced the run time to 32 min.

In Figure 25-31c, we want to see whether a steeper gradient could be used to reduce the run time. The gradient limits were the same as in trace *b*, but t_G was reduced to 20 min. Peaks 6 and 7 are not fully resolved with the shorter gradient time. Trace *b* represents reasonable conditions for the gradient separation.

Steps in gradient method development:
1. Run a wide gradient (e.g., 5 to 95% B) over 40–60 min. From this run, decide whether gradient or isocratic elution is best.
2. If gradient elution is chosen, eliminate portions of the gradient prior to the first peak and following the last peak. Use the same gradient time as in step 1.
3. If the separation in step 2 is acceptable, try reducing the gradient time to reduce the run time.

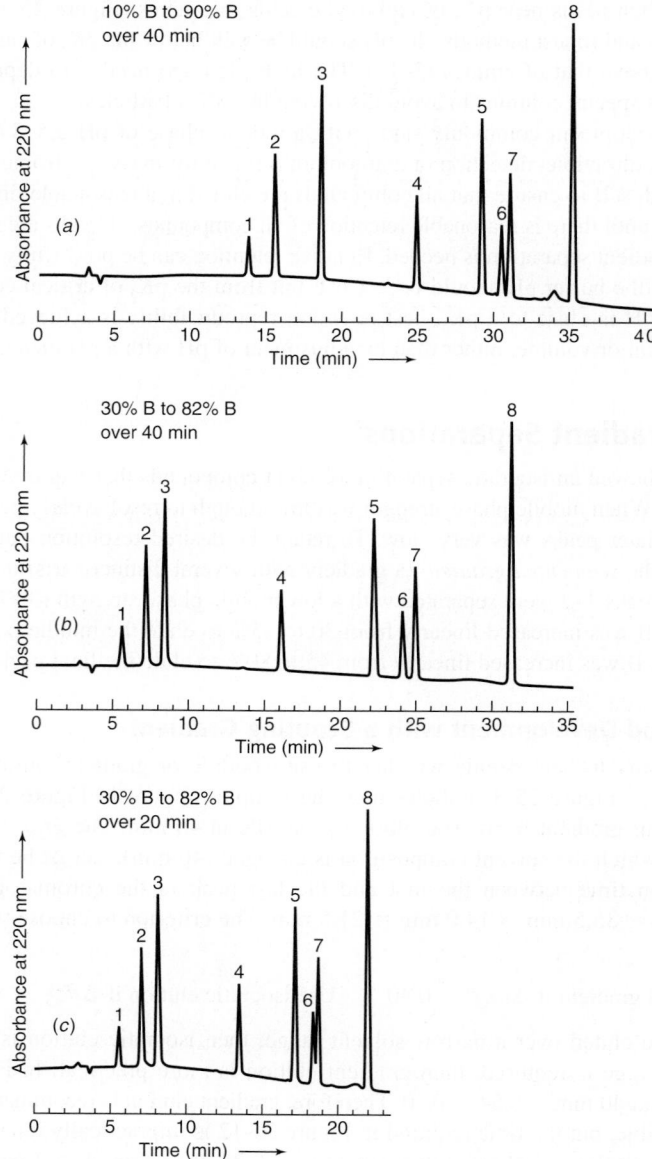

FIGURE 25-31 Linear gradient separations of the mixture from Figure 25-12 in the same column and solvent system [buffer (solvent A) with acetonitrile (solvent B)] at a flow rate of 1.0 mL/min. Dwell time = 5 min.

If the separation in Figure 25-31b were not acceptable, you could try to improve it by reducing the flow rate or going to a *segmented gradient*, as in Figure 25-14. The segmented gradient provides an appropriate solvent composition for each region of the chromatogram. It is easy to experiment with flow rate and gradient profiles. More difficult approaches to improve the separation are to change the solvent, use a longer column, use a smaller particle size, or change the stationary phase. Box 25-4 provides guidance on selecting gradient time and scaling gradients[55] from one size column to another.

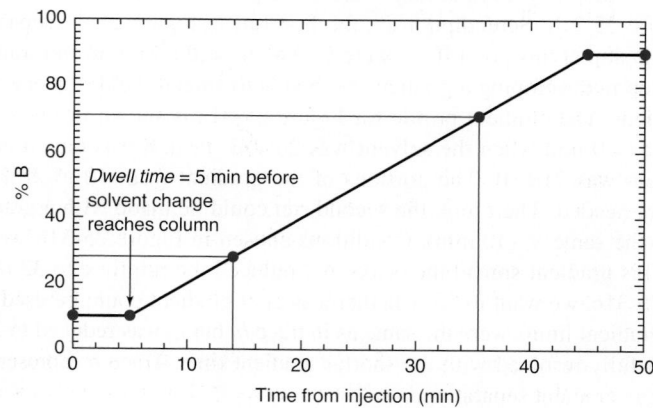

FIGURE 25-32 Solvent gradient for Figure 25-31a. The gradient was begun at the time of injection ($t = 0$), but the dwell time was 5 min. Therefore, the solvent was 10% B for the first 5 min. Then the composition increased linearly to 90% B over 40 min. After $t = 45$ min, the composition was constant at 90% B.

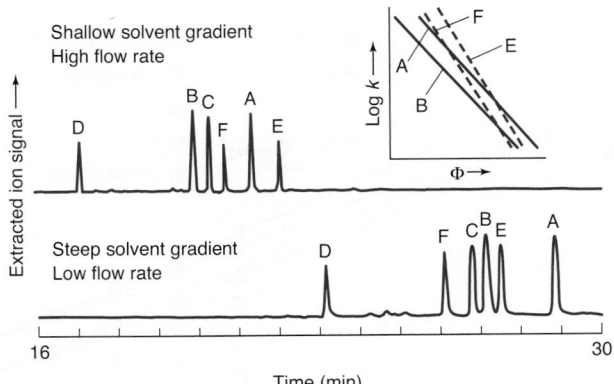

FIGURE 25-33 Effect of flow rate on gradient separation of peptides of different molecular weight. Order of elution of longer peptides E and F changes with respect to shorter peptides A and B when solvent gradient conditions are changed. [Data from M. Gilar, H. W. Xie, and A. Jaworski, "Utility of Retention Prediction Model for Investigation of Peptide Separation Selectivity in Reversed-Phase Liquid Chromatography: Impact of Concentration of Trifluoroacetic Acid, Column Temperature, Gradient Slope and Type of Stationary Phase," *Anal. Chem.* **2010,** *82,* 265.]

Elution order can change when the gradient conditions are changed. Plots of $\log k$ vs. Φ such as Figure 25-13 show whether elution order changes might occur. In the sample in Figure 25-13 all components respond similarly to changes in mobile phase, as evidenced by the nearly parallel lines. No elution order changes occur in Figure 25-31. In contrast, Figure 25-33 shows an example in which lines in the plot of $\log k$ vs. Φ are not parallel and cross each other, so elution order changes are observed when gradient conditions are changed.

In reversed-phase separations, 10–20 volumes (V_m) of initial solvent should be passed through the column after a run to equilibrate the stationary phase with solvent for the next run. Equilibration can take as long as the separation. Under some circumstances, the column does not need to reach equilibrium with the initial solvent to provide run-to-run repeatability.[56]

Tracking peak identities by their peak area or with a selective detector such as mass spectrometry aids development of gradient methods.

Dwell Volume and Dwell Time

The volume between the point at which solvents are mixed and the beginning of the column is called the **dwell volume.** The *dwell time*, t_D, is the time required for the gradient to reach the column. Dwell volumes range from 0.5 to 10 mL in different systems. In Figure 25-32, the dwell volume is 5 mL, and the flow rate is 1.0 mL/min. Therefore, the dwell time is 5 min. A solvent change initiated at 8 min does not reach the column until 13 min.

Differences in dwell volumes between different systems are an important reason why gradient separations on one chromatograph do not necessarily transfer to another.[53,57] Dwell volumes are smaller for high-pressure mixing pumps and larger for low-pressure mixing pumps. It is helpful to state the dwell volume for your system when you report a gradient separation. One way to compensate for dwell volume is to inject sample at the time t_D instead of at $t = 0$. For the gradient in Figure 25-32, it would have been better to inject the sample at $t = 5$ min, but this was not done.

You can measure dwell volume by first disconnecting the column and connecting the inlet tube directly to the outlet tube. Place water in reservoirs A and B of the solvent delivery system. Add 0.1 vol% acetone to reservoir B. Program the gradient to go from 0 to 100% B in 20 min and begin the gradient at $t = 0$. With the detector set to 260 nm, the response will ideally look like that in Figure 25-34. The delay between the start of the gradient and the first response at the detector is the dwell time, t_D.

$$t_D = \frac{\text{dwell volume (mL)}}{\text{flow rate (mL/min)}}$$

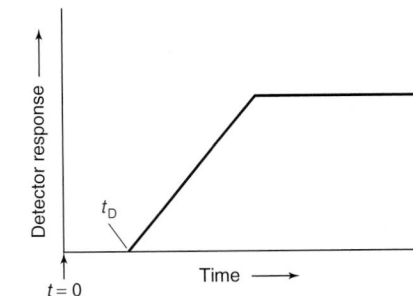

FIGURE 25-34 Measurement of dwell time using nonabsorbing solvent in reservoir A and a weak absorber in reservoir B. The gradient from 0 to 100% B is begun at time $t = 0$ but does not reach the detector until time t_D. The column is removed from the system for this measurement. Real response will be rounded instead of having the sharp intersections shown in this illustration.

25-5 Do It with a Computer

Method development is simplified by computer simulations using commercial software[3,58] or your own spreadsheet. With data from a few experiments, you can predict the effects of solvent composition and temperature in isocratic or gradient separations. You can estimate what will be optimum conditions in hours of work instead of days of work.

The basis for most simulations of reversed-phase separations is the empirical *linear-solvent-strength model* (Equation 25-4), which supposes a logarithmic relationship between retention factor k for a given solute and the mobile phase composition Φ. Figure 25-35 shows measurements for nine compounds in the range $\Phi = 0.65$ to 0.8 in methanol/water on a C_{18} column. The parameters S and $\log k_W$ in Equation 25-4 are the slopes and y-intercepts of the lines in Figure 25-35.

Equation 25-4 is not accurate over wide ranges of solvent composition Φ. In Figure 25-35, the authors only varied Φ from 0.65 to 0.8. When we extrapolate beyond the measured range of

Problem 25-46 introduces you to an instructive, freely available, online HPLC simulator.[7]

FIGURE 25-35 Linear-solvent-strength model. Graph of log k versus Φ for nine organic compounds eluted from a C_{18} column by methanol/water. [Data from R. A. Shalliker, S. Kayillo, and G. R. Dennis, "Optimizing Chromatographic Separation: An Experiment Using an HPLC Simulator," *J. Chem. Ed.* **2008,** *85,* 1265.]

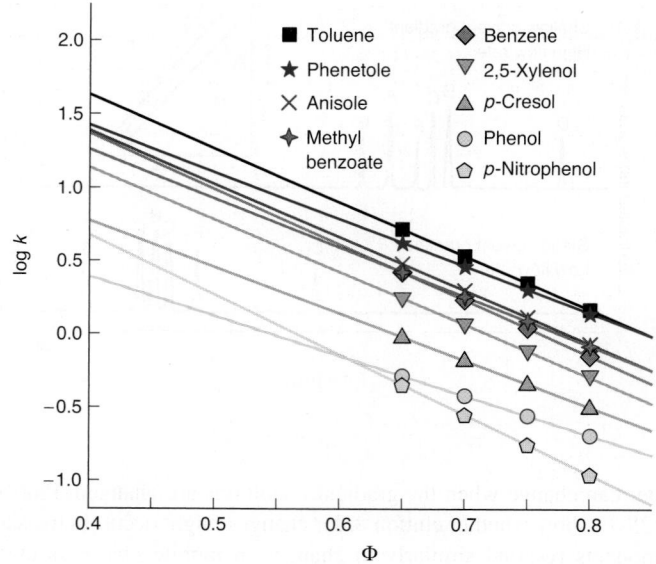

Φ, we are in uncharted territory. Data for Figure 25-35 were collected at high values of Φ to keep the run time short. In the measured range of Φ, *p*-nitrophenol is eluted first and phenol is eluted next. The extrapolated lines for these two components cross at $\Phi = 0.61$. For $\Phi < 0.61$, we predict that phenol will be eluted before *p*-nitrophenol.

To simulate a chromatogram for an isocratic solvent composition such as $\Phi = 0.6$ for 60 vol% CH_3OH/40 vol% H_2O on a particular HPLC column, we begin with solvent transit time $(t_m) = 1.85$ min and plate number $(N) = 7\ 000$. Then, for a chosen value of solvent strength Φ, the retention factor k for each component is computed from

$$k = 10^{(\log k_W - S\Phi)} \tag{25-8}$$

Retention time t_r is found by rearranging Equation 23-16 to the form:

$$t_r = t_m(k + 1) \tag{25-9}$$

Assuming Gaussian peak shape, the standard deviation of the band in Figure 23-9 is found by rearranging Equation 23-30:

$$\sigma = t_r/\sqrt{N} \tag{25-10}$$

The spreadsheet in Figure 25-36 simulates an isocratic separation. Highlighted cells require your input. The time step in cell C7 gives the interval between calculated points in the chromatogram. Relative areas in cells E14:E22 are arbitrary. You could set them all to 1 or you could try to vary them to match peak heights of an experimental chromatogram. The linear-solvent-strength parameters log k_W and S in cells C14:D22 are from experimental measurements in Figure 25-35. The spreadsheet computes k in cells F14:F22 with Equation 25-8. It computes t_r in cells G14:G22 with Equation 25-9 and the standard deviation of each Gaussian peak in cells H14:H22 with Equation 25-10.

The shape of each Gaussian chromatographic peak is given by

$$\text{Detector signal } (y) = \frac{\text{relative area}}{\sigma\sqrt{2\pi}} e^{-(t - t_r)^2/2\sigma^2} \tag{25-11}$$

where relative areas are the numbers you specified in cells E14:E22, t is time, t_r is the retention time in cells G14:G22, and σ is the standard deviation in cells H14:H22. The detector signal beginning in cell E30 is the sum of nine terms of the form of Equation 25-11—one term for each compound in the mixture. Each compound has its own σ, t_r, and relative area. The spreadsheet calculates detector signal for times beginning at $t = 0$ and proceeding past elution of the last peak.

Figure 25-37 shows simulations done with the spreadsheet. At a solvent strength $\Phi = 0.75$, all nine compounds are eluted within 6 min, but resolution of peaks 5, 6, and 7 is poor. At $\Phi = 0.60$, peaks 1 and 2 overlap and peaks 5 and 6 have changed their order of elution. At $\Phi = 0.56$, all peaks are resolved and the last peak is eluted in 22 min.

	A	B	C	D	E	F	G	H
1	Chromatogram simulator - Gaussian peaks							
2	Data for methanol:water with Waters C-18 column - Shalliker et al., J. Chem. Ed. **2008**, *85*, 1265							
3								
4			constants					
5			$t_m =$	1.85	min	(time for mobile phase to transit the column)		
6			$N =$	7000	plates	(plate number for column)		
7			time step =	0.01	min	(time between calculated points)		
8			sqrt(2*pi) =	2.50663				
9			$\Phi =$	0.56		(fraction of organic solvent)		
10								
11						k	Retention	σ (std dev
12		Compound			Relative	retention	time	peak width)
13	Number	Name	log k_w	S	area	factor	t_r (min)	(min)
14	1	p-nitrophenol	2.323	4.113	0.25	1.05	3.79	0.045
15	2	phenol	1.488	2.734	0.2	0.91	3.53	0.042
16	3	p-cresol	2.059	3.205	0.25	1.84	5.25	0.063
17	4	2,5-xylenol	2.591	3.619	0.4	3.67	8.63	0.103
18	5	benzene	2.895	3.806	0.8	5.80	12.59	0.150
19	6	methyl benzoate	2.617	3.392	0.8	5.22	11.50	0.137
20	7	anisole	2.840	3.646	0.4	6.28	13.48	0.161
21	8	phenetole	2.734	3.258	0.4	8.12	16.87	0.202
22	9	toluene	3.118	3.705	0.5	11.05	22.28	0.266

	C	D	E
27			y
28	time	time	detector
29	step	(min)	signal
30	0	0	0
31	1	0.01	0
32	2	0.02	0

F14 = 10^(C14-D14*$C$9)

G14 = $C$5*(F14+1)

H14 = G14/SQRT($C$6)

Detector signal:

E30 = ($E$14/($H$14*$C$8)*EXP(-((D30-$G$14)^2)/(2*$H$14^2)))

+ ($E$15/($H$15*$C$8)*EXP(-((D30-$G$15)^2)/(2*$H$15^2)))

+ an analogous term for each of the other compounds

FIGURE 25-36 A spreadsheet to simulate isocratic chromatographic separation.

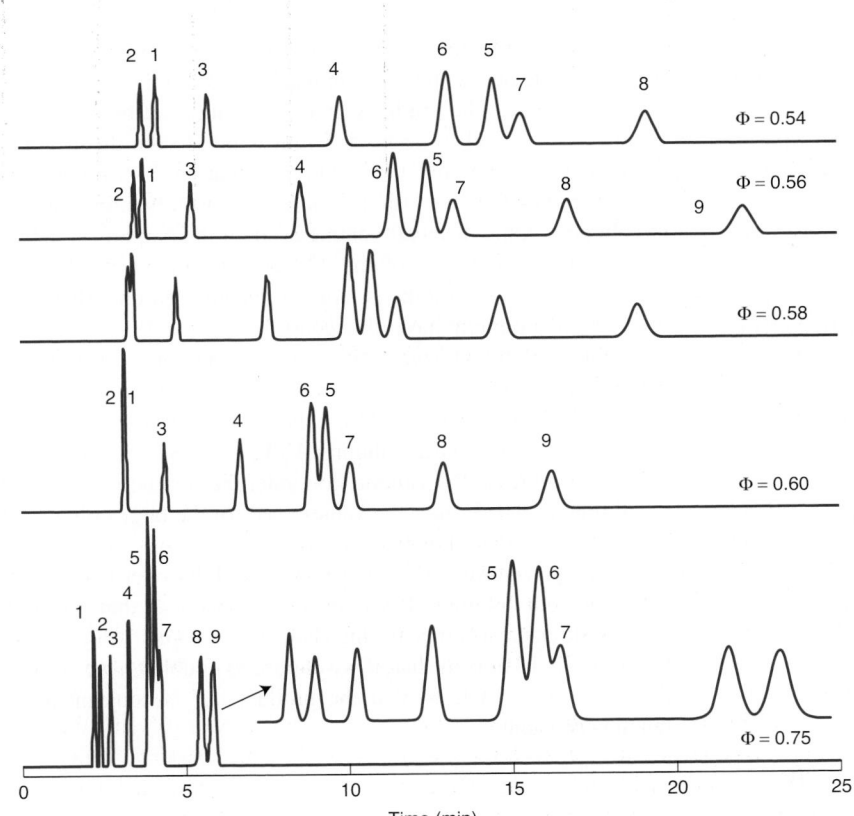

FIGURE 25-37 Chromatograms simulated with the spreadsheet in Figure 25-36. Peak numbers are *p*-nitrophenol (1), phenol (2), *p*-cresol (3), 2,5-xylenol (4), benzene (5), methyl benzoate (6), anisole (7), phenetole (8), and toluene (9).

Resolution $= \dfrac{\Delta t_r}{w_{av}}$ (Equation 23-23)

Δt_r = difference in retention time
w_{av} = average width at base of peak = 4σ

In supplementary topics at the website www.whfreeman.com/qca9e, you will find equations and a spreadsheet to simulate gradient and isocratic elution.

Solvent composition $\Phi = 0.56$ (56 vol% methanol/44 vol% aqueous buffer) is not good enough to make the chromatographic method robust. A *robust* separation is one that maintains adequate resolution despite *small* changes in conditions such as Φ, pH, and temperature. *Resolution* between closely spaced peaks is 1.5 for peaks 2 and 1, 1.8 for peaks 6 and 5, and 1.3 for peaks 5 and 7. For a robust separation, we desire a minimum resolution of 2.0, but this value cannot always be achieved. At solvent composition $\Phi = 0.54$, resolution of peaks 2 and 1 and peaks 6 and 5 is above 2.0, but the resolution of peaks 5 and 7 has decreased to 1.2. Peak 9, which is off the chart, has a retention time of 26 min. The composition $\Phi = 0.56$ appears to be about the best we can do with methanol and water on this column. To obtain better resolution, we could decrease flow rate, use smaller particles, increase column length, or change the temperature, solvent, or stationary phase.

From a few experiments to find S and log k, we can use the spreadsheet to *estimate* that $\Phi = 0.56$ is the optimum condition to try in the lab. Solvent composition $\Phi = 0.56$ is outside the measured range in Figure 25-35. The only way to know if the curves in Figure 25-36 remain linear down to $\Phi = 0.56$ is to try the experiment. With only a little more complexity, the linear-solvent-strength model enables us to simulate and optimize gradient separations.[3]

With the tools in this chapter, you can usually find a way to separate the components of a mixture if it does not contain too many compounds. If reversed-phase chromatography fails, hydrophilic interaction or normal-phase chromatography or one of the methods in Chapter 26 could be appropriate. Method development is part science, part art, and part luck.

BOX 25-4 Choosing Gradient Conditions and Scaling Gradients

We now discuss equations that allow you to select sensible linear gradient conditions and to scale gradients from one column to another. For gradient elution, the average retention factor k^* for each solute is the value of k when that solute is halfway through the column:

$$k^* = \dfrac{t_G F}{\Delta \Phi V_m S} \qquad (25\text{-}12)$$

where t_G is gradient time (min), F is flow rate (mL/min), $\Delta \Phi$ is the change in solvent composition during the gradient, V_m is the volume of mobile phase in the column (mL), and S is the slope in the linear-solvent-strength model (Equation 25-4). We take $S = 4$ for this discussion.

In isocratic elution, a retention factor $k \approx 5$ provides separation from the solvent front and does not require excessive time. For gradient elution, $k^* \approx 5$ is a reasonable starting condition. Let's calculate a sensible gradient time for the experiment in Figure 25-31a, in which we chose a gradient from 10 to 90% B ($\Delta \Phi = 0.8$) in a 25×0.46 cm column eluted at 1.0 mL/min. From Equation 25-5, $V_m \approx L d_c^2/2 = (25 \text{ cm})(0.46 \text{ cm})^2/2 = 2.6_5$ mL. We calculate the required gradient time by rearranging Equation 25-12:

$$t_G = \dfrac{k^* \Delta \Phi V_m S}{F} = \dfrac{(5)(0.8)(2.65 \text{ mL})(4)}{(1.0 \text{ mL/min})} = 42 \text{ min} \qquad (25\text{-}13)$$

A reasonable gradient time would be 42 min. In Figure 25-31a, t_G is 40 min, giving $k^* = 4.7$. In Figure 25-31b, we changed the gradient to $\Delta \Phi = 0.52$, giving better separation:

$$k^* = \dfrac{t_G F}{\Delta \Phi V_m S} = \dfrac{(40 \text{ min})(1.0 \text{ mL/min})}{(0.52)(2.65 \text{ mL})(4)} = 7.3$$

The separation is poorer in Figure 25-31c, for which $k^* = 3.6$.

If you have a successful gradient separation and want to transfer it from column 1 to column 2 whose dimensions are different, the scaling relations are

$$\dfrac{F_2}{F_1} = \dfrac{m_2}{m_1} = \dfrac{t_{D2}}{t_{D1}} = \dfrac{V_2}{V_1} \qquad (25\text{-}14)$$

where F is volume flow rate (mL/min), m is the mass of sample, t_D is the dwell time before the gradient reaches the column, and V is total column volume. The gradient time, t_G, should not be changed. In Figure 25-32, dwell time $t_D = 5$ min is due to the dwell volume between the mixer and the column. Equation 25-14 tells us to change volume flow rate, sample mass, and dwell time in proportion to column volume. If dwell volume is small in comparison with the volume of solvent on the column, V_m, the dwell time t_D can be inconsequential. If dwell volume is large, it becomes an important factor over which you might have little control.

Suppose that you have optimized a gradient on a 25×0.46 cm column and you want to transfer it to a 10×0.21 cm column. The quotient V_2/V_1 is $(\pi r^2 L)_2/(\pi r^2 L)_1$, where r is column radius and L is column length. For these columns, $V_2/V_1 = 0.083$. Equation 25-14 tells us to decrease the volume flow rate, the sample mass, and the dwell time to 0.083 times the values used for the large column. The gradient time should not be changed.

When you make these changes, you will discover that k^* is the same for both columns. If you change a condition that affects k^*, you should make a compensating change to restore k^*. For example, Equation 25-12 tells us that, if we choose to double t_G, we could cut the flow rate in half so that the product $t_G F$ is constant and k^* remains constant.

Terms to Understand

Summary

In high-performance liquid chromatography (HPLC), solvent is pumped at high pressure through a column containing stationary phase particles with diameters of 1.5–5 μm. The smaller the particle size, the more efficient the column, but the greater the resistance to flow. Microporous silica particles with a covalently bonded liquid phase such as octadecyl groups ($—C_{18}H_{37}$) are most common. Polarity index measures the ability of a solvent to dissolve polar molecules. Mobile phase strength measures the ability of a given solvent to elute solutes from the column. In normal-phase chromatography and hydrophilic interaction chromatography, the stationary phase is polar and a less polar solvent is used. Mobile phase strength increases as the polarity of the solvent increases. Reversed-phase chromatography employs a nonpolar stationary phase and polar solvent. Mobile phase strength increases as the polarity of the solvent decreases. Most separations of organic compounds can be done on reversed-phase columns. Polar compounds that are not retained on reversed-phase columns can be separated by hydrophilic interaction chromatography. Polar compounds that only dissolve in organic solvents can be separated by normal-phase chromatography. Normal-phase chromatography or porous graphitic carbon is good for separating isomers. Chiral phases are used for optical isomers.

If a solution of organic solvent and water is used in reversed-phase chromatography, mobile phase strength increases as the percentage of organic solvent increases. If the solvent has a fixed composition, the process is called isocratic elution. In gradient elution, mobile phase strength is increased during chromatography by increasing the percentage of strong (less polar) solvent.

A short guard column containing the same stationary phase as the analytical column is placed before the analytical column to protect it from contamination with particulate matter or irreversibly adsorbed solutes. A high-quality pump provides smooth solvent flow. The injection valve allows rapid, precise sample introduction. The column is best housed in an oven to maintain a reproducible temperature. Column efficiency increases with smaller particles and at elevated temperature because the rate of mass transfer between phases is increased. Mass spectrometric detection provides quantitative and qualitative information for each substance eluted from the column. Ultraviolet detection is most common, and it can provide qualitative information if a photodiode array is used to record a full spectrum of each analyte. Refractive index detection has universal response but is not very sensitive. Evaporative light scattering and the charged aerosol detectors respond to the mass of each nonvolatile solute. Electrochemical and fluorescence detectors are extremely sensitive, but selective. In supercritical fluid chromatography, nonvolatile solutes are separated by a process whose efficiency, speed, and detectors more closely resemble those of gas chromatography than those of liquid chromatography.

Steps in method development: (1) determine the goal of the analysis, (2) select a method of sample preparation, (3) choose a detector, and (4) use a systematic procedure to select the mobile phase for isocratic or gradient elution. Aqueous acetonitrile, methanol, and tetrahydrofuran are customary solvents for reversed-phase separations. Retention is optimized by varying the amount of organic solvent in the mobile phase. Relative retention is adjusted by fine-tuning the organic solvent concentration, altering the temperature, and by choice of a new organic solvent or column. Column length and particle size can be changed to improve efficiency or speed of analysis. Criteria for a successful separation are $0.5 \leq k \leq 20$, resolution ≥ 2.0, operating pressure ≤ 15 MPa (for conventional equipment), and asymmetry factor in the range 0.9–1.5. Retention of weak acids and bases is governed by their ionization state, which is controlled by the mobile phase pH. In selecting a mobile phase buffer, consider the desired pH, buffering capacity, solubility, and compatibility with detection.

A wide gradient is a good first choice to determine whether to use isocratic or gradient elution. If it is not possible to separate all components with $0.5 \leq k \leq 20$, gradient elution is needed. Retention factors measured at a few solvent compositions can be fit to the linear-solvent-strength model to enable computer optimization of a separation.

Exercises

25-A. A known mixture of compounds A and B gave the following HPLC results:

Compound	Concentration (mg/mL in mixture)	Peak area (arbitrary units)
A	1.03	10.86
B	1.16	4.37

A solution was prepared by mixing 12.49 mg of B plus 10.00 mL of unknown containing just A and diluting to 25.00 mL. Peak areas of 5.97 and 6.38 were observed for A and B, respectively. Find the concentration of A (mg/mL) in the unknown.

25-B. A bonded stationary phase for the separation of optical isomers has the structure

Optically active center

To resolve the enantiomers of amines, alcohols, or thiols, the compounds are first derivatized with a nitroaromatic group that increases their interaction with the bonded phase and makes them observable with a spectrophotometric detector.

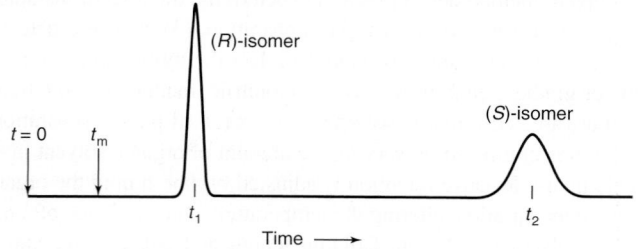

Mixture to be resolved (RNH$_2$)

Aromatic derivative (mixture of two enantiomers)

When the mixture is eluted with 20 vol% 2-propanol in hexane, the (R)-enantiomer is eluted before the (S)-enantiomer, with the following chromatographic parameters:

$$\text{Resolution} = \frac{\Delta t_r}{w_{av}} = 7.7 \qquad \text{Relative retention } (\alpha) = 4.53$$

$$k \text{ for } (R)\text{-isomer} = 1.35 \qquad t_m = 1.00 \text{ min}$$

where w_{av} is the average width of the two Gaussian peaks at their base.

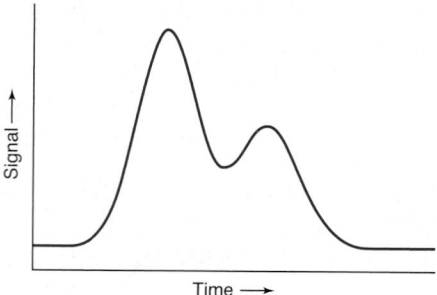

(a) Find t_1, t_2, and w_{av}, with units of minutes.

(b) The width of a peak at half-height is $w_{1/2}$ (Figure 23-9). If the plate number for each peak is the same, find $w_{1/2}$ for each peak.

(c) The area of a Gaussian peak is $1.064 \times$ peak height $\times$ $w_{1/2}$. Given that the areas under the two bands should be equal, find the relative peak heights (height$_R$/height$_S$).

25-C. Two peaks emerge from a reversed-phase chromatography column as sketched in the illustration.

According to Equation 23-33, resolution is given by

$$\text{Resolution} = \frac{\sqrt{N}}{4}\left(\frac{\alpha - 1}{\alpha}\right)\left(\frac{k_2}{1 + k_2}\right)$$

where N is plate number, α is relative retention (Equation 23-20), and k_2 is the retention factor for the more retained component (Equation 23-16).

(a) If you decrease the amount of organic solvent in the mobile phase, you will increase retention. Sketch the chromatogram if retention factors increase but N and α are constant.

(b) If you change the solvent type or the stationary phase, you will change the relative retention. Sketch the chromatogram if α increases but N and k_1 are constant.

(c) If you decrease particle size or increase column length, you can increase the plate number. Sketch the chromatogram if N increases by (i) decreasing particle size and (ii) increasing column length. Assume α and k_2 are constant.

25-D. After poisonous melamine and cyanuric acid appeared in milk in China (Box 11-3) in 2008, there was a flurry of activity to develop methods to measure these substances. An analytical method for milk is to treat 1 volume of milk with 9 volumes of H_2O/CH_3CN (20:80 vol/vol) to precipitate proteins. The mixture is centrifuged for 5 min to remove precipitate. The supernatant liquid is filtered through a 0.5-μm filter and injected into a HILIC liquid chromatography column (TSKgel Amide-80 stationary phase) and products are measured by mass spectrometry with selected reaction monitoring (Section 22-5). Melamine is measured in positive ion mode with the transition m/z 127 → 85. Cyanuric acid is measured in negative ion mode with the transition m/z 128 → 42.

(a) Write the formulas for the four ions and propose structures for all four ions.

(b) Even though milk is a complex substance, only one clean peak is observed for melamine and one for cyanuric acid spiked into milk. Explain why.

25-E. The graph shows retention data from a C$_8$-silica column with an acetonitrile/water mobile phase.

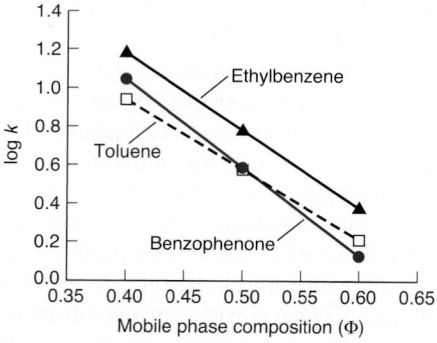

Linear-solvent-strength plot of retention on C$_8$-silica column (15 × 0.46 cm, 5-μm particles, 1.0 mL/min, pressure = 7–8 MPa). [Data from J. H. Zhao and P. W. Carr, "An Approach to the Concept of Resolution Optimization through Changes in the Effective Chromatographic Selectivity," *Anal. Chem.* **1999,** *71,* 2623.]

(a) What mobile phase composition provides greatest retention (k) for the components? Least retention? Coelution (equal k) of two components?

(b) Predict the retention time of each peak at 40% and 60% acetonitrile. Draw a chromatogram (a "stick diagram" representing each peak as a vertical line) of the separation at each mobile phase composition.

(c) Would 60% acetonitrile yield adequate resolution?

(d) Assuming Gaussian peaks, does the separation at 60% acetonitrile have the attributes of a good separation?

Problems

High-Performance Liquid Chromatography

25-1. (a) Why does mobile phase strength increase as solvent becomes *less* polar in reversed-phase chromatography, whereas mobile phase strength increases as solvent becomes *more* polar in normal-phase chromatography?

(b) What kind of gradient is used in supercritical fluid chromatography?

25-2. Why are the relative eluent strengths of solvents in adsorption chromatography fairly independent of solute?

25-3. In hydrophilic interaction chromatography (HILIC), why is eluent strength increased by *increasing* the fraction of water in the mobile phase?

25-4. (a) Why is high pressure needed in HPLC?

(b) For a given column length, why do smaller particles give a higher plate number?

(c) What is a bonded phase in liquid chromatography?

25-5. (a) Use Equation 25-1 to estimate the length of column required to achieve 1.0×10^4 plates if the stationary phase particle size is 10.0, 5.0, 3.0, or 1.5 μm.

(b) If the retention time was 20 min on the 10.0-μm-particle-size column, what is the retention time on the 5.0-, 3.0-, and 1.5-μm columns from part **(a)**? Assume that flow rate is constant for all columns.

(c) Use Equation 25-2 to estimate the pressure of the columns in **(a)** given that the pressure of the 10.0-μm column was 4.4 MPa.

(d) If the flow rate was 2.0 mL/min, what is the baseline width (in time and volume) for the peaks on 10.0-, 5.0-, 3.0-, or 1.5-μm columns from part **(a)**?

(e) Which of these column configurations would require a UHPLC instrument?

25-6. (a) Why are HPLC particles porous?

(b) Why are particles with 60- to 120-Å pores used for small molecules but wide-pore 300-Å stationary phases used to separate polypeptides and proteins?

25-7. If a 15-cm-long HPLC column has a plate height of 5.0 μm, what will be the half-width (in seconds) of a peak eluted at 10.0 min? If plate height = 25 μm, what will be $w_{1/2}$?

25-8. (a) UHPLC can provide exquisite resolution when run slowly on long columns or rapid separations with reasonable resolution if short columns are run fast. The drug acetaminophen run on a 50×2.1 mm C_{18} UHPLC column has a retention time of 0.63 min and a width at half-height of 2.3 s. Find the plate number and plate height. How many 1.7-μm-diameter particles placed side-by-side are equal to one theoretical plate?

(b) From Figure 25-3, we expect an optimum plate height of 4 μm. How many particles placed side-by-side are equal to one theoretical plate? Do you think the column in **(a)** is being run for maximum resolution or maximum speed?

25-9. Why are silica stationary phases generally limited to operating in the pH range 2–8? Why does the silica in Figure 25-8 have improved stability at low pH?

25-10. How do additives such as triethylamine reduce tailing of certain solutes?

25-11. HPLC peaks should generally not have an asymmetry factor, B/A in Figure 23-14, outside the range 0.9–1.5.

(a) Sketch the shape of a peak with an asymmetry of 1.8.

(b) What might you do to correct the asymmetry?

25-12. (a) Sketch a graph of the van Deemter equation (plate height versus linear flow rate). What would the curve look like if the multiple path term were 0? If the longitudinal diffusion term were 0? If the finite equilibration time term were 0?

(b) Explain why the van Deemter curve for 1.8-μm particles in Figure 25-3 is nearly flat at high flow rate. What can you say about each of the terms in the van Deemter equation for 1.8-μm particles?

(c) Explain why the 2.7-μm superficially porous particle enables separations similar to those achieved by 1.8-μm totally porous particles, but the superficially porous particle requires lower pressure.

25-13. The figure shows the separation of two enantiomers on a chiral stationary phase.

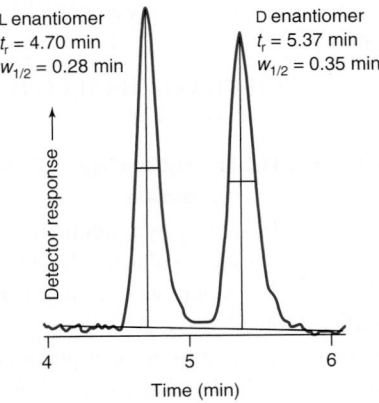

Separation of enantiomers of Ritalin by HPLC with a chiral stationary phase.
[Data from R. Bakhtiar, L. Ramos, and F. L. S. Tse, "Quantification of Methylphenidate in Plasma Using Chiral Liquid-Chromatography/Tandem Mass Spectrometry: Application to Toxicokinetic Studies," *Anal. Chim. Acta* **2002,** *469,* 261.]

(a) From t_r and $w_{1/2}$, find N for each peak.

(b) From t_r and $w_{1/2}$, find the resolution.

(c) Given $t_m = 1.62$ min, use Equation 23-33 with the average N to predict the resolution.

25-14.(a) According to Equation 25-2, if all conditions are constant, but particle size is reduced from 3 μm to 0.7 μm, by what factor must pressure be increased to maintain constant linear velocity?

(b) If all conditions except pressure are constant, by what factor will linear velocity increase if column pressure is increased by a factor of 10?

(c) With 0.7-μm particles in a 9 cm $\times$ 50 μm column, increasing pressure from 70 MPa to 700 MPa decreased analysis time by approximately a factor of 10 while increasing plate count from 12 000 to 45 000.[59] Explain why small particles permit 10-fold faster flow without losing efficiency or, in this case, with improved efficiency.

25-15. Use Figure 25-17 to suggest which type of liquid chromatography you could use to separate compounds in each of the following categories:

(a) Molecular mass <2 000, soluble in octane

(b) Molecular mass <2 000, soluble in methanol-water mixtures

(c) Molecular mass <2 000, weak acid

(d) Molecular mass <2 000, highly polar

(e) Molecular mass <2 000, ionic

(f) Molecular mass >2 000, soluble in water, nonionic, various sized solutes

(g) Molecular mass >2 000, soluble in water, variety of charges

(h) Molecular mass >2 000, soluble in tetrahydrofuran

25-16. Microporous silica particles with a density of 2.2 g/mL and a diameter of 10 μm have a measured surface area of 300 m²/g. Calculate the surface area of the spherical silica as if it were simply solid particles. What does this calculation tell you about the shape or porosity of the particles?

25-17. (a) Nonpolar aromatic compounds were separated by HPLC on an octadecyl (C_{18}) bonded phase. The eluent was 65 vol% methanol in water. How would the retention times be affected if 90% methanol were used instead?

(b) Octanoic acid and 1-aminooctane were passed through the same column described in **(a)**, using an eluent of 20% methanol/80% buffer (pH 3.0). State which compound is expected to be eluted first and why.

$$CH_3CH_2CH_2CH_2CH_2CH_2CH_2CO_2H$$
Octanoic acid

$$CH_3CH_2CH_2CH_2CH_2CH_2CH_2CH_2NH_2$$
1-Aminooctane

(c) Polar solutes were separated by hydrophilic interaction chromatography (HILIC) with a strongly polar bonded phase. How would retention times be affected if eluent were changed from 80 vol% to 90 vol% acetonitrile in water?

(d) Polar solutes were separated by normal-phase chromatography on bare silica using methyl *t*-butyl ether and 2-propanol solvent. How would retention times be affected if eluent were changed from 40 vol% to 60 vol% 2-propanol? (*Hint:* See Table 25-4.)

25-18. In *monolithic columns*[60] the stationary phase is a single porous piece of silica or polymer filling the entire column and synthesized within the column from liquid precursors. Monolithic columns offer similar plate height to HPLC particles, but with less resistance to flow. Therefore, faster flow or longer columns can be used. The figure shows separation of isotopic molecules on a long monolithic column. Packed columns have too much resistance to flow to be made so long.

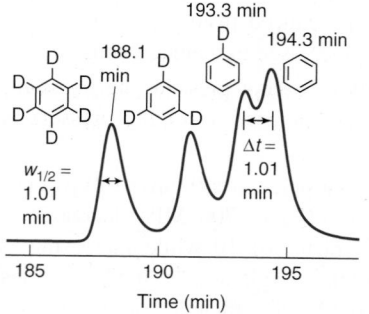

Separation of isotopic molecules on a 440-cm-long monolithic C_{18}-silica column eluted with CH_3CN/H_2O (30:70 vol/vol) at 30°C. [Data from K. Miyamoto, T. Hara, H. Kobayashi, H. Morisaka, D. Tokuda, K. Horie, K. Koduki, S. Makino, O. Nuñez, C. Yang, T. Kawabe, T. Ikegami, H. Takubo, Y. Ishihama, and N. Tanaka, "High-Efficiency Liquid Chromatographic Separation Utilizing Long Monolithic Silica Capillary Columns," *Anal. Chem.* **2008**, *80*, 8741.]

(a) Unretained thiourea is eluted in 41.7 min. Find the linear velocity u_x (mm/s).

(b) Find the retention factor k for C_6D_6.

(c) Find the plate number N and plate height for C_6D_6.

(d) Assuming that the peak widths for C_6H_5D and C_6H_6 are the same as that of C_6D_6, find the resolution of C_6H_5D and C_6H_6.

(e) Retention times for C_6H_5D and C_6H_6 are 193.3 and 194.3 min, respectively. Find the relative retention (α) between C_6H_5D and C_6H_6.

(f) If we just increased the column length to increase N, what value of N and what column length would be required for a resolution of 1.000?

(g) Without increasing the length of the column, and without changing the stationary phase, how might you improve the resolution?

(h) When the solvent was changed from CH_3CN/H_2O (30:70 vol/vol) to $CH_3CN/CH_3OH/H_2O$ (10:5:85), the relative retention for C_6H_5D and C_6H_6 increased to 1.008_8 and the retention factor for C_6H_6 changed to 17.0. If the plate number were unchanged, what would be the resolution?

25-19. The antitumor drug gimatecan is available as nearly pure (*S*)-enantiomer. Neither pure (*R*)-enantiomer nor a *racemic* (equal) mixture of the two enantiomers is available. To measure small quantities of (*R*)-enantiomer in nearly pure (*S*)-gimatecan, a preparation was subjected to normal-phase chromatography on each of the enantiomers of a commercial, chiral stationary phase designated (*S,S*)- and (*R,R*)-DACH-DNB. Chromatography on the (*R,R*)-stationary phase gave a slightly asymmetric peak at $t_r = 6.10$ min with retention factor $k = 1.22$. Chromatography on the (*S,S*)-stationary phase gave a slightly asymmetric peak at $t_r = 6.96$ min with $k = 1.50$. With the (*S,S*) stationary phase, a small peak with 0.03% of the area of the main peak was observed at 6.10 min.

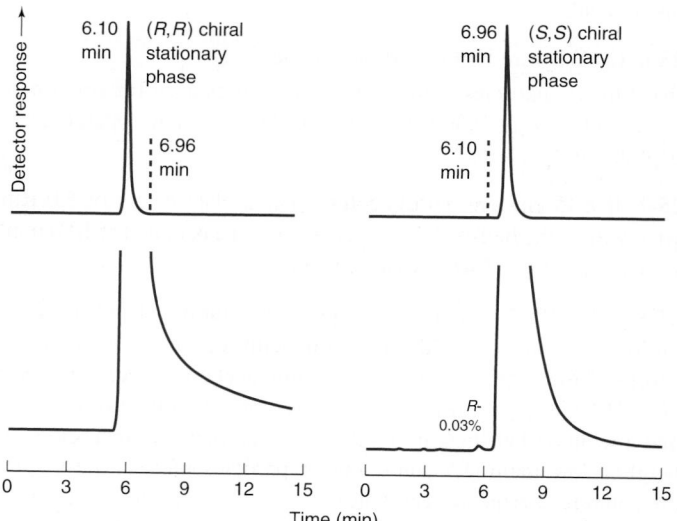

Chromatography of gimatecan on each enantiomer of a chiral stationary phase. Lower traces have enlarged vertical scale. [Data from E. Badaloni, W. Cabri, A. Ciogli, R. Deias, F. Gasparrini, F. Giorgi, A. Vigevani, and C. Villani, "Combination of HPLC 'Inverted Chirality Columns Approach' and MS/MS Detection for Extreme Enantiomeric Excess Determination Even in Absence of Reference Samples." *Anal. Chem.* **2007**, *79*, 6013.]

(a) Explain the appearance of the upper chromatograms. Dashed lines are position markers, not part of the chromatogram. What would the chromatogram of pure (R)-gimatecan look like on the same two stationary phases?

(b) Explain the appearance of the two lower chromatograms and why it can be concluded that the gimatecan contained 0.03% of the (R)-enantiomer. Why is the (R)-enantiomer not observed with the (R,R)-stationary phase?

(c) Find the relative retention (α) for the two enantiomers on the (S,S)-stationary phase.

(d) The column provides $N = 6\,800$ plates. What would be the resolution between the two equal peaks in a racemic (equal) mixture of (R)- and (S)-gimatecan? If the peaks were symmetric, does this resolution provide baseline separation in which signal returns to baseline before the next peak begins?

25-20. Suppose that an HPLC column produces Gaussian peaks. The detector measures absorbance at 254 nm. A sample containing equal moles of compounds A and B was injected into the column. Compound A ($\varepsilon_{254} = 2.26 \times 10^4 \text{ M}^{-1} \text{ cm}^{-1}$) has a height $h = 128$ mm and a half-width $w_{1/2} = 10.1$ mm. Compound B ($\varepsilon_{254} = 1.68 \times 10^4 \text{ M}^{-1} \text{ cm}^{-1}$) has $w_{1/2} = 7.6$ mm. What is the height of peak B in millimeters?

25-21. Morphine and morphine 3-β-D-glucuronide were separated on two different 50×4.6 mm columns with 3-μm particles.[61] Column A was C_{18}-silica run at 1.4 mL/min and column B was bare silica run at 2.0 mL/min.

Morphine   Morphine 3-β-D-glucuronide

(a) Estimate the volume, V_m, and time, t_m, at which unretained solute would emerge from each column. The observed times are 0.65 min for column A and 0.50 min for column B.

(b) Column A was eluted with 2 vol% acetonitrile in water containing 10 mM ammonium formate at pH 3. Morphine 3-β-D-glucuronide emerged at 1.5 min and morphine at 2.8 min. Explain the order of elution.

(c) Find the retention factor k for each solute on column A, using $t_m = 0.65$ min.

(d) Column B was eluted with a 5.0-min gradient beginning at 90 vol% acetonitrile in water and ending at 50 vol% acetonitrile in water. Both solvents contained 10 mM ammonium formate, pH 3. Morphine emerged at 1.3 min and morphine 3-β-D-glucuronide emerged at 2.7 min. Explain the order of elution. Why does the gradient go from high to low acetonitrile volume fraction?

(e) From Equation 25-12 in Box 25-4, estimate k^* on Column B assuming $S = 4$ and with $t_m = 0.50$ min.

25-22. The rate at which heat is generated inside a chromatography column from friction of flowing liquid is power (watts, W = J/s) = volume flow rate (m^3/s) × pressure drop (pascals, Pa = kg/[m·s²]).

(a) Explain the analogy between heat generated in a chromatography column and heat generated in an electric circuit (power = current × voltage).

(b) At what rate (watts = J/s) is heat generated for a flow of 1 mL/min with a pressure difference of 3 500 bar between the inlet and outlet? You will need to convert mL/min to m^3/s. Also 1 bar $\equiv 10^5$ Pa.

25-23. To which kinds of analytes do these liquid chromatography detectors respond?

(a) ultraviolet

(b) refractive index

(c) evaporative light scattering

(d) charged aerosol

(e) electrochemical

(f) fluorescence

(g) nitrogen chemiluminescence

(h) conductivity

25-24. The chromatogram in Box 25-3 shows the supercritical fluid chromatography separation of seven steroids monitored by three detectors.

(a) In the middle chromatogram, ultraviolet detection provides near universal response for the steroids, whereas in the lower chromatogram the ultraviolet detector provides a selective response for a few of the steroids. How can ultraviolet detection act as either a selective or universal detector?

(b) Why is a sloping baseline observed at 210 nm, but the baseline is flat at 254 nm?

(c) Use the baseline disturbance early in the 254 nm chromatogram to measure t_m. How does the measured value compare with that predicted using Equation 25-5 given that the column is 25×0.46 cm and the flow rate is 2.0 mL/min.

25-25. *Chromatography–mass spectrometry.* Cocaine metabolism in rats can be studied by injecting the drug and periodically withdrawing blood to measure levels of metabolites by HPLC–mass spectrometry. For quantitative analysis, isotopically labeled internal standards are mixed with the blood sample. Blood was analyzed by reversed-phase chromatography with an acidic eluent and atmospheric pressure chemical ionization mass spectrometry for detection. The mass spectrum of the collisionally activated dissociation products from the m/z 304 positive ion is shown in the figure on the next page. Selected reaction monitoring (m/z 304 from mass filter Q1 and m/z 182 from Q3 in Figure 22-33) gave a single chromatographic peak at 9.22 min for cocaine. The internal standard $^2\text{H}_5$-cocaine gave a single peak at 9.19 min for m/z 309 (Q1) → 182 (Q3).

(a) Draw the structure of the ion at m/z 304.

(b) Suggest a structure for the ion at m/z 182.

(c) The intense peaks at m/z 182 and 304 do not have ^{13}C isotopic partners at m/z 183 and 305. Explain why.

(d) Rat plasma is exceedingly complex. Why does the chromatogram show just one clean peak?

(e) Given that $^2\text{H}_5$-cocaine has only two major mass spectral peaks at m/z 309 and 182, which atoms are labeled with deuterium?

(f) Explain how you would use $^2\text{H}_5$-cocaine for measuring cocaine in blood.

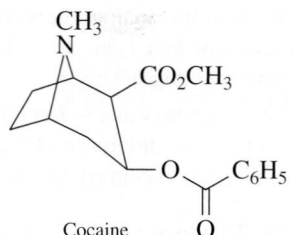

Cocaine

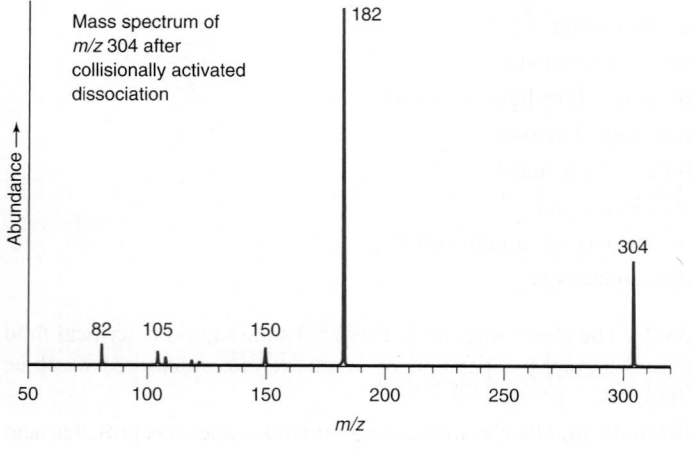

Mass spectrum of
m/z 304 after
collisionally activated
dissociation

Abundance →

50   100   150   200   250   300

m/z

182

304

82   105   150

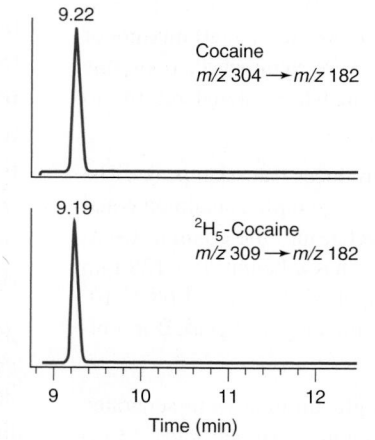

9.22

Cocaine
m/z 304 → *m/z* 182

9.19

$^2H_5$-Cocaine
m/z 309 → *m/z* 182

9   10   11   12

Time (min)

Spectrum for Problem 25-25.

Left: Mass spectrum of collisionally activated dissociation products from *m/z* 304 positive ion from atmospheric pressure chemical ionization mass spectrum of cocaine. Right: Chromatograms obtained by selected reaction monitoring. [Data from G. Singh, V. Arora, P. T. Fenn, B. Mets, and I. A. Blair, "Isotope Dilution Liquid Chromatography Tandem Mass Spectrometry Assay for Trace Analysis of Cocaine and Its Metabolites in Plasma," *Anal. Chem.* **1999,** *71,* 2021.]

25-26. *Chromatography–mass spectrometry.* HPLC separation of enantiomers of the drug Ritalin on a chiral stationary phase was shown in Problem 25-13.

Ritalin (methylphenidate)
$C_{14}H_{19}O_2N$
Nominal mass = 233

(a) Detection is by atmospheric pressure chemical ionization with selected reaction monitoring of the *m/z* 234 → 84 transition. Explain how this detection works and propose structures for *m/z* 234 and *m/z* 84.

(b) For quantitative analysis, the internal standard $^2H_3$-Ritalin with a deuterated methyl group was added. Deuterated enantiomers have the same retention times as unlabeled enantiomers. Which selected reaction monitoring transition should be monitored to produce a chromatogram of the internal standard in which unlabeled Ritalin will be invisible?

Method Development

25-27. (a) Explain how to measure *k* and resolution.

(b) State three methods for measuring t_m in reversed-phase chromatography.

(c) State three methods for measuring t_m in hydrophilic interaction liquid chromatography.

(d) Estimate t_m for a 15 × 0.46 cm column containing 5-μm particles operating at a flow rate of 1.5 mL/min. Estimate t_m if the particle size were 3.5 μm instead.

25-28. What is the difference between extra-column volume and dwell volume? How do each of these volumes affect a chromatogram?

25-29. What does it mean for a separation procedure to be "rugged" and why is it desirable?

25-30. What are criteria for an adequate isocratic chromatographic separation?

25-31. Explain how to use a gradient for the first run to decide whether isocratic or gradient elution would be more appropriate.

25-32. What are the general steps in developing an isocratic separation for reversed-phase chromatography?

25-33. (a) Use Figure 25-28 to select a tetrahydrofuran/water mobile phase strength of equivalent strength to 80% methanol.

(b) Describe how to prepare 1 L of this tetrahydrofuran mobile phase.

(c) What limitations would be imposed by the use of tetrahydrofuran?

25-34. What are the general steps in developing an isocratic separation for reversed-phase chromatography with one organic solvent and temperature as variables?

25-35. The "rule of three" states that the retention factor for a given solute decreases *approximately* threefold when the organic phase increases by 10%. In Figure 25-12, t_m = 2.7 min. Find *k* for peak 5 at 50% B. Predict the retention time for peak 5 at 40% B and compare the observed and predicted times.

25-36. (a) Make a graph showing retention times of peaks 6, 7, and 8 in Figure 25-12 as a function of %acetonitrile (%B). Predict the retention time of peak 8 at 45% B.

(b) *Linear-solvent-strength model:* In Figure 25-12, t_m = 2.7 min. Compute *k* for peaks 6, 7, and 8 as a function of %B. Make a graph of log *k* versus Φ, where Φ = %B/100. Find the equation of a straight line through a suitable linear range for peak 8. The slope is −*S* and the intercept is log k_w. From the line, predict t_r for peak 8 at 45% B and compare your answer with **(a)**.

(c) *Gradient elution:* A linear eluent gradient from 40 to 80% acetonitrile over 30 min is performed on the column in Figure 25-12. Assuming a dwell volume of 0 mL, use your data from **(b)** to plot the retention factor of peaks 6 and 8 during the gradient. What are the general characteristics of the plot?

(d) Why are the peaks in a gradient separation sharp?

25-37. A reversed-phase separation of a reaction mixture calls for isocratic elution with 48% methanol/52% water. If you want to change the procedure to use acetonitrile/water, what is a good starting percentage of acetonitrile to try?

25-38. (a) When you try separating an unknown mixture by reversed-phase chromatography with 50% acetonitrile/50% water, the peaks are too close together and are eluted in the range $k = 2$–6. Should you use a higher or lower concentration of acetonitrile in the next run?

(b) When you try separating an unknown mixture by normal-phase chromatography with 50% hexane/50% methyl t-butyl ether, the peaks are too close together and are eluted in the range $k = 2$–6. Should you use a higher or lower concentration of hexane in the next run?

25-39. The figure shows reversed-phase retention data for three compounds.

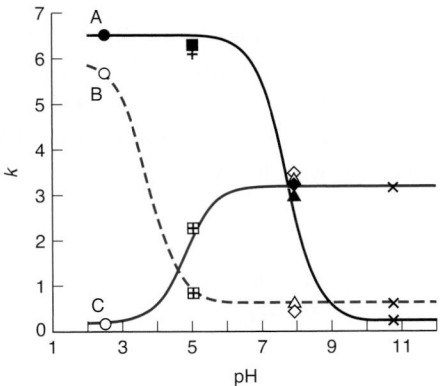

Effect of pH on retention of weak acids and weak bases on a reversed-phase column (15 × 0.46-cm pH-stable X-Terra MS-C$_{18}$ column at 25°C). [Data from L. G. Gagliardi, C. B. Castells, C. Ràfols, M. Rosés, and E. Bosch, "Modeling Retention and Selectivity as a Function of pH and Column Temperature in Liquid Chromatography," *Anal. Chem.* **2006**, *78*, 5858.]

(a) Identify whether compounds A, B, and C are weak acids or bases. For each compound, what is pK_a and the retention factor of the fully protonated form?

(b) Over what pH range would a method be least rugged with regard to retention of component C?

(c) Each different symbol in the plot indicates a different buffer (circle = pH 2.48 phosphate; plus = pH 5.01 acetate; and so on). Why are different buffers used for this experiment?

25-40. Use Figure 25-30 for the following questions:

(a) What pH would be best for the separation of benzoic acid, 4-nitrophenol, and 3-methylbenzoic acid?

(b) What pH would be best for the separation of benzoic acid, 3-methylbenzoic acid, and 4-methylaniline?

(c) What pH would be best for separation of 4-nitrophenol, 4-methylaniline, and codeine on a typical C$_{18}$-silica column?

25-41. Retention factors for three solutes separated on a C$_8$ nonpolar stationary phase are listed in the table. Eluent was a 70 : 30 (vol/vol) mixture of 50 mM citrate buffer (adjusted to pH with NH$_3$) plus methanol. Draw the dominant species of each compound at each pH in the table and explain the behavior of the retention factors.

	Retention factor		
Analyte	pH 3	pH 5	pH 7
Acetophenone	4.21	4.28	4.37
Salicylic acid	2.97	0.65	0.62
Nicotine	0.00	0.13	3.11

Acetophenone

Salicylic acid
pK_a = 2.97

Nicotine
pK_1 = 3.15
pK_2 = 7.85

25-42. A mixture of 14 compounds was subjected to a reversed-phase gradient separation going from 5% to 100% acetonitrile with a gradient time of 60 min. The sample was injected at t = dwell time. All peaks were eluted between 22 and 50 min.

(a) Is the mixture more suitable for isocratic or gradient elution?

(b) If the next run is a gradient, select the starting and ending %acetonitrile and the gradient time.

25-43. (a) List ways in which the resolution between two closely spaced peaks might be changed.

(b) After optimization of an isocratic elution with several solvents, the resolution of two peaks is 1.2. How might you improve the resolution without changing solvents or the kind of stationary phase?

25-44. (a) You wish to use a wide gradient from 5 vol% to 95 vol% B for the first separation of a mixture of small molecules to decide whether to use gradient or isocratic elution. What should be the gradient time, t_G, for a 15 × 0.46-cm column containing 3-μm particles with a flow of 1.0 mL/min?

(b) You optimized the gradient separation going from 20 vol% to 34 vol% B in 11.5 min at 1.0 mL/min. Find k^* for this optimized separation. To scale up to a 15 × 1.0 cm column, what should be the gradient time and the volume flow rate? If the sample load on the small column was 1 mg, what sample load can be applied to the large column? Verify that k^* is unchanged.

25-45. 🖩 *Simulating a separation with a spreadsheet.* Use the spreadsheet in Figure 25-36 to simulate the chromatograms for $\Phi = 0.75$ and $\Phi = 0.56$ in Figure 25-37.

25-46. *Online HPLC simulation:*[7] Go to www.hplcsimulator.org to download a web-based chromatography simulation. Open the simulator. If at any point in working this problem you encounter difficulties, closing the simulator and reopening will return you to the default conditions.

(a) To familiarize yourself with the simulator, find the following: solvent A and B, solvent B fraction, temperature, injection volume, flow rate, plate height (HETP), number of plates, pressure, first compound in table, its retention factor and retention time, and time required to elute all compounds.

(b) Click on the first compound name in the table. The corresponding peak in the chromatogram will be highlighted in red. Adjust solvent B fraction. What happens to retention of the compound? If you adjust solvent fraction by entering a number in the window, click on the graph to see the change.

(c) Adjust solvent B fraction to a value that gives a retention factor ~5. Vary the flow rate to create a van Deemter curve of plate height

(HETP) versus flow rate. Compare your results to Figure 25-3. What is the particle size used in this column?

(d) Close the simulator and reopen to return to default conditions. Adjust solvent B fraction to get $0.5 \leq k \leq 20$. Note the solvent B fraction of acetonitrile. Click on each of the compounds in the Compound table to highlight each component within the chromatogram. Click on the Solvent B tab and change the solvent to methanol. Adjust the solvent B fraction of methanol to get a comparable retention factor range. Note the solvent B fraction of methanol. Do your results agree with the %methanol predicted by the nomogram in Figure 25-28? Was the resolution improved by changing the organic solvent?

(e) Close the simulator and reopen to return to default conditions. Click on the Gradient elution mode radio button in the top left of the window. Click on the name of each compound in the lower table to evaluate the separation. Explore the effect of gradient time by double clicking on the lowest number under Time (min) under the Gradient elution mode radio button.

25-47. *Literature search problem:* The purity of cocaine bought on the street varies dramatically. Cutting agents include levamisole, a compound normally used to kill parasites. Search the literature for a reversed-phase liquid chromatography method with diode array detection for the determination of the purity of street cocaine.

(a) Give the citation (authors, title, journal name, year, volume, pages) for the research paper that fits the criteria of this analysis.

(b) What alternative methods could be used for analysis of adulterated street cocaine?

(c) What is the range of purity of street cocaine?

(d) What type of analytical column was used for the analysis?

(e) How was the method validated?

(f) What is "robustness of an analytical method?"

25-48. *Literature search problem:* Human serum albumin (HSA) is an important protein ingredient in cryopreservation media used in procedures such as *in vitro* fertilization. Search the literature for a high-performance liquid chromatography method for the determination of human serum albumin and the stabilizer *N*-acetyltryptophan in medical devices.

(a) Give the citation (authors, title, journal name, year, volume, pages) for the research paper that fits the criteria of this analysis.

(b) What alternative methods could be used for analysis of human serum albumin?

(c) What type of analytical column is used for the separation?

(d) How long was the gradient? How long were the additional wash and equilibration steps within the gradient method?

(e) What parameters were assessed in the method validation?

(f) Why were particles with 300 Å pores used?

26 | Chromatographic Methods and Capillary Electrophoresis

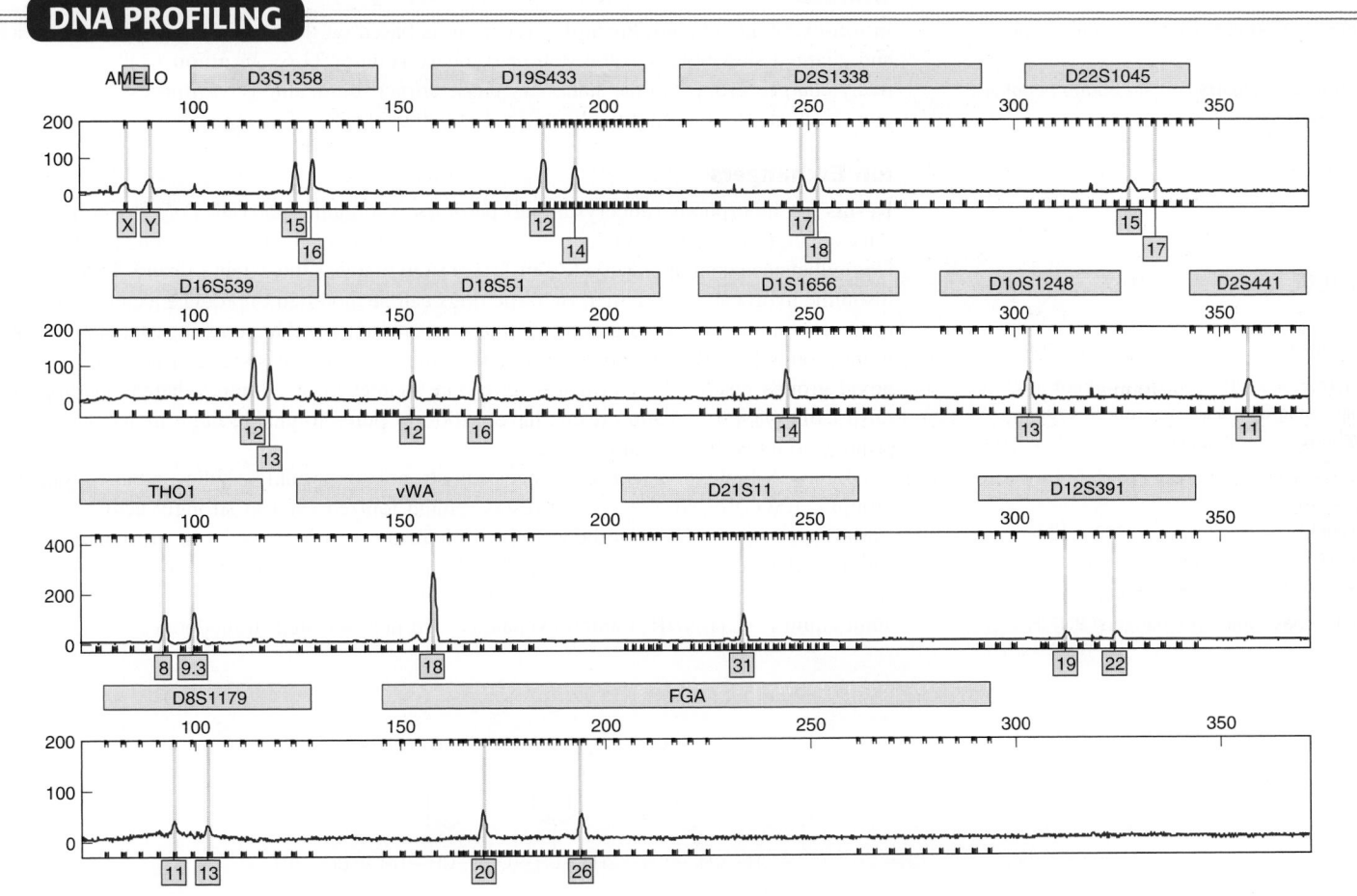

Forensic DNA profile. [Data from C. Hurth, University of Arizona.]

Forensic DNA analysis creates a genetic fingerprint from a sample such as saliva. The DNA profile links the sample to one person with a very high probability and excludes other people as the source of the sample.[1] The device described at the end of this chapter is intended to reduce profiling time to less than the time a suspect is in custody.

Each row of the profile is an *electropherogram* of DNA labeled with one of four fluorescent dyes. The number of nucleotide bases in DNA is shown above each row. Sixteen different DNA strands are labeled in gray boxes. In the top row, DNA D3S1358 could have 100, 104, 108, 112, . . . , 140, or 144 nucleotide bases, indicated by 12 tick marks. Each different length of DNA is called an *allele*. The twelve alleles of D3S1358 are numbered 9 through 20. The person who provided the saliva has alleles 15 and 16—one from each parent. If both parents had the same allele, there would only be one allele in the profile. DNA D1S1656 in the second row has only allele 14. The number of possible combinations of alleles of 16 kinds of DNA is so great that the profile can be confidently assigned to one person.

This chapter continues our discussion of pressure-driven chromatographic methods including ion-exchange, molecular exclusion, affinity, and hydrophobic interaction chromatography. We then introduce separations driven by an electric field. In electrophoresis, liquid moves through a capillary tube by *electroosmosis*. This process enables us to create miniature analytical chips in which fluids are driven through capillary channels etched into glass or plastic. Chemical reactions and separations are carried out on these chips. The last section of this chapter describes a *microfluidic* device that uses pressure to drive liquids through the steps of forensic DNA analysis, culminating in a separation of DNA by capillary electrophoresis.

26-1 Ion-Exchange Chromatography[2]

In **ion-exchange chromatography,** retention is based on the attraction between solute ions and charged sites bound to the stationary phase (Figure 23-6). In **anion exchangers,** positively charged groups on the stationary phase attract solute anions. **Cation exchangers** contain covalently bound, negatively charged sites that attract solute cations.

Ion Exchangers

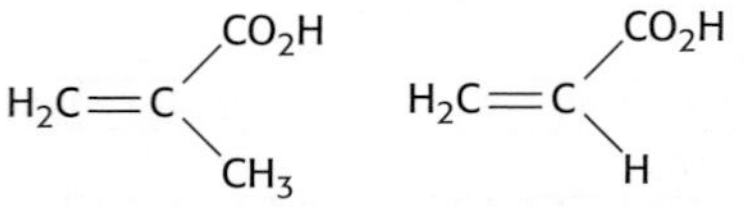

Resins are amorphous (noncrystalline) particles of organic material. *Polystyrene resins* for ion exchange are made by the copolymerization of styrene and divinylbenzene (Figure 26-1). Divinylbenzene content is varied from 1 to 16% to increase the extent of **cross-linking** of the insoluble hydrocarbon polymer. Benzene rings can be modified to produce a cation-exchange resin, containing sulfonate groups ($-SO_3^-$), or an anion-exchange resin, containing ammonium groups ($-NR_3^+$). If methacrylic acid is used in place of styrene, a polymer with carboxyl groups results. Typical resins have pore diameters of ~1 nm, whereas *macroporous* (also called *macroreticular*) resins have ~100 nm pores to enable large molecules such as proteins to move freely within the resin.

Table 26-1 classifies ion exchangers as strongly or weakly acidic or basic. Sulfonate groups ($-SO_3^-$) of strongly acidic resins remain ionized even in strongly acidic solutions. Carboxyl groups ($-CO_2^-$) of the weakly acidic resins are protonated near pH 4 and lose their cation-exchange capacity. "Strongly basic" quaternary ammonium groups ($-CH_2NR_3^+$) (which are not really basic at all) remain cationic at all values of pH. Weakly basic tertiary ammonium ($-CH_2NHR_2^+$) anion exchangers are deprotonated in moderately basic solution

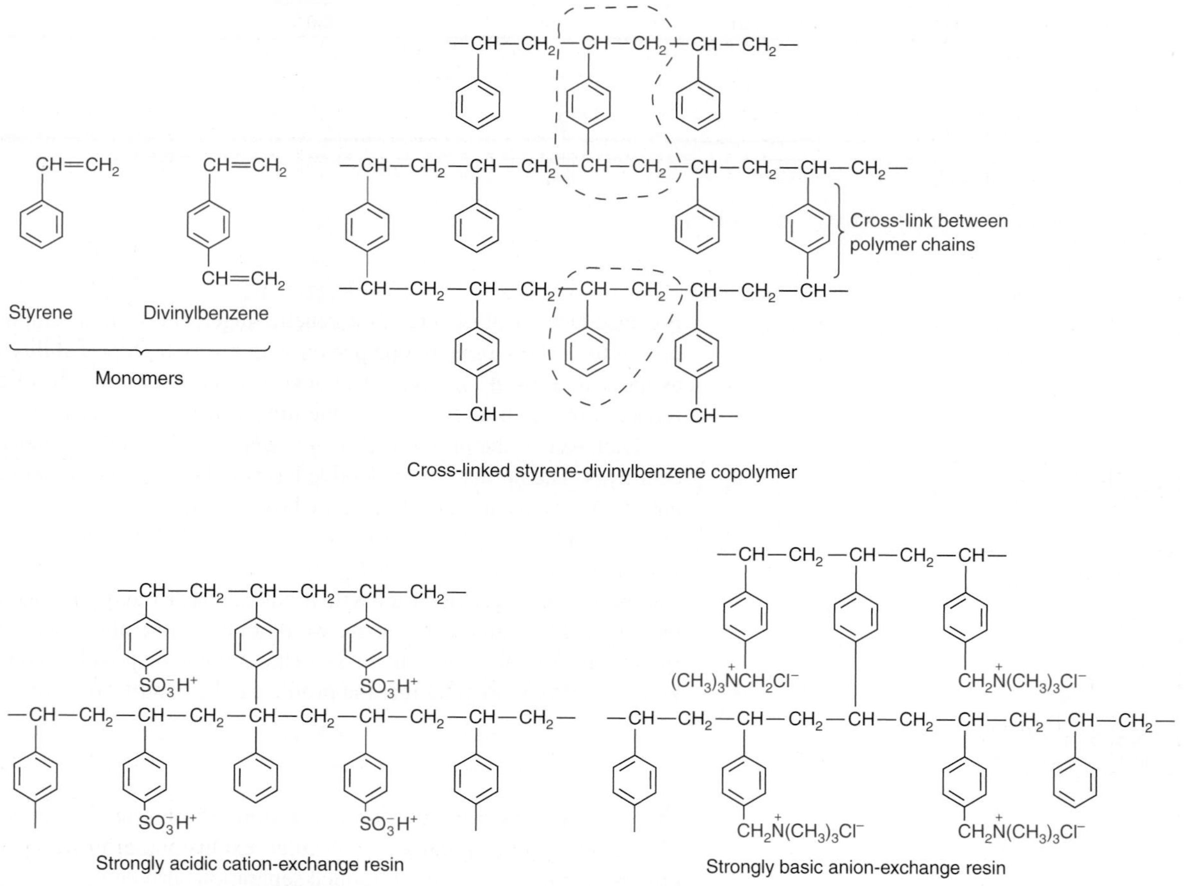

Anion exchangers contain bound *positive* groups.

Cation exchangers contain bound *negative* groups.

Poly(acrylic acid) in superabsorbent disposable diapers is a common cation exchanger that you can demonstrate in your classroom.[3]

Strongly acidic cation exchangers: RSO_3^-
Weakly acidic cation exchangers: RCO_2^-
"Strongly basic" anion exchangers: $RNR_3'^+$
Weakly basic anion exchangers: $RNR_2'H^+$

Cross-linked styrene-divinylbenzene copolymer

Strongly acidic cation-exchange resin

Strongly basic anion-exchange resin

FIGURE 26-1 Structures of styrene-divinylbenzene cross-linked ion-exchange resins.

TABLE 26-1	Ion-exchange resins			
Type	**Structure**		**Example resins**	**Capacity (meq/mL)**[a]
Cation exchangers				
Strong acid	Aryl—SO_3^- H^+		Dowex 50W-X4	1.1
			Dowex 50W-X8, Amberlite IR-120	1.7 – 1.9
			Dowex 50W-X12, Amberlite IR-122	2.1
Weak acid	R—CO_2^- Na^+		Duolite C-433, Bio-Rex 70	2.4 – 4.5
			Amberlite IRC-86	
Anion exchangers				
Strong base				
Type 1	Aryl—$CH_2\overset{+}{N}(CH_3)_3$ Cl^-		Dowex 1-X4	1.0
			Dowex 1-X8, Amberlite IRA-400	1.2 – 1.4
Type 2	Aryl—$CH_2\overset{+}{N}(CH_3)_2(CH_2)_2OH$ Cl^-		Dowex 2-X8, Amberlite IRA-410	1.2 – 1.4
Weak base	R—$\overset{+}{N}HR_2$ Cl^-		Dowex 4-X8, Amberlite IRA-68	1.6
Chelating				
Iminodiacetate	Aryl—$N(CH_2CO_2^-)_2$		Chelex 100®	0.4

a. 1 meq/mL means 1 milliequivalent of ion-exchange site per milliliter of wet resin. A milliequivalent is one millimole of charge on the resin.

SOURCE: J. A. Dean, ed., *Analytical Chemistry Handbook*, 2nd ed. (New York: McGraw-Hill, 2000).

and lose their ability to bind anions. Chelating resins such as iminodiacetate exchangers have a high preference for transition metal cations, and to a lesser extent, for alkaline earth cations. The resin loses its chelating ability below pH 4.

The extent of cross-linking may be indicated by the notation "-XN" after the name of the resin. For example, Dowex 1-X4 contains 4% divinylbenzene, and Bio-Rad AG 50W-X12 contains 12% divinylbenzene. The resin becomes more rigid and less porous as cross-linking increases. Lightly cross-linked resins permit rapid equilibration of solute between the inside and the outside of the particle. However, resins with little cross-linking swell in water. Hydration decreases both the density of ion-exchange sites and the selectivity of the resin for different ions. More heavily cross-linked resins exhibit less swelling and higher exchange capacity and selectivity, but they have longer equilibration times. The charge density of polystyrene ion exchangers is so great that highly charged macromolecules such as proteins may be irreversibly bound.

Cellulose, dextran, and agarose ion exchangers, which are polymers of sugar molecules, possess larger pore sizes and lower charge densities than those of polystyrene resins. They are well suited for ion exchange of macromolecules, such as proteins. Table 26-2 lists charged functional groups used to derivatize polysaccharide hydroxyl groups. DEAE-Sepharose, for example, is an anion exchanger containing diethylaminoethyl groups on cross-linked agarose. Other macroporous ion exchangers are based on polyacrylamide or polymethacrylate. For high-performance liquid chromatography, proprietary organic and inorganic ion exchangers

Analytical grade *"AG"* resins are extensively purified to remove organic and inorganic impurities.

Because they are much softer than polystyrene **resins,** dextran and its relatives are called **gels.**

Gel brand names can reflect their structure:
- **Sepha**cel and Hyper**Cel:** cross-linked **cel**lulose.
- **Sepha**dex: cross-linked **dex**tran.
- **Sepha**rose: cross-linked aga**rose.**

TABLE 26-2	Common active groups on ion-exchange gels		
Type	**Abbreviation**	**Name**	**Structure**
Cation exchangers			
Strong acid	SP	Sulfopropyl	—O—$CH_2CHOHCH_2OCH_2CH_2CH_2SO_3^-$
	S	Methyl sulfonate	—O—$CH_2CHOHCH_2OCH_2CHOHCH_2SO_3^-$
Weak acid	CM	Carboxymethyl	—O—$CH_2CO_2^-$
Anion exchangers			
Strong base	Q	Quaternary ammonium	—O—$CH_2\overset{+}{N}(CH_3)_3$
Weak base	DEAE	Diethylaminoethyl	—O—$CH_2CH_2\overset{+}{N}H(CH_2CH_3)_2$
	ANX	Diethylaminopropyl	—O—$CH_2CHOHCH_2\overset{+}{N}H(CH_2CH_3)_2$

SOURCE: *Ion Exchange and Chromatofocusing: Principles and Methods* (publication no. 11-0004-21 AB, GE Healthcare, 2010).

are designed to have appropriate pore size and charge density for protein interactions, combined with rigidity for operation at high pressure.

Ion-Exchange Selectivity and Equivalents

Consider the competition of Na^+ and H^+ for sites on the cation-exchange resin R^-:

Selectivity coefficient:
$$R^-H^+ + Na^+ \rightleftharpoons R^-Na^+ + H^+ \qquad K = \frac{[R^-Na^+][H^+]}{[R^-H^+][Na^+]} \qquad (26\text{-}1)$$

The equilibrium constant is called the **selectivity coefficient,** because it describes the relative selectivity of the resin for Na^+ and H^+. Selectivities of polystyrene resins in Table 26-3 tend to increase with the extent of cross-linking, because the pore size of the resin shrinks as cross-linking increases. Ions such as Li^+, with a large *hydrated radius* (Chapter 8 opener), do not have as much access to the resin as do smaller ions, such as Cs^+.

In general, ion exchangers favor the binding of ions of higher charge, decreased hydrated radius, and increased *polarizability*. A fairly general order of selectivity for cations is

$$Pu^{4+} \gg La^{3+} > Ce^{3+} > Pr^{3+} > Eu^{3+} > Y^{3+} > Sc^{3+} > Al^{3+} \gg$$
$$Tl^+ > Ba^{2+} > Ag^+ > Pb^{2+} > Sr^{2+} > Ca^{2+} > Ni^{2+} > Cd^{2+} >$$
$$Cu^{2+} > Co^{2+} > Zn^{2+} > Mg^{2+} > UO_2^{2+} \gg$$
$$Cs^+ > Rb^+ > K^+ > NH_4^+ > Na^+ > H^+ > Li^+$$

Reaction 26-1 can be driven in either direction, even though Na^+ is bound more tightly than H^+. Washing a column containing Na^+ with a substantial excess of H^+ will replace Na^+

Polarizability measures the ability of an ion's electron cloud to be deformed by nearby charges. Deformation of the cloud induces a dipole in the ion. The attraction between the induced dipole and the nearby charge increases the binding of the ion to the resin.

TABLE 26-3 Relative selectivity coefficients of ion-exchange resins

Sulfonic acid cation-exchange resin					Quaternary ammonium anion-exchange resin		
	Relative selectivity for divinylbenzene content of					Relative selectivity for resin of	
Cation	4%	8%	12%	16%	Anion	Type 1	Type 2
H^+ (reference)	1.0	1.0	1.0	1.0	OH^- (reference)	1.0	1.0
Li^+	0.90	0.85	0.81	0.74	F^-	1.6	0.3
Na^+	1.3	1.5	1.7	1.9	Propionate	2.6	0.3
NH_4^+	1.6	1.95	2.3	2.5	Acetate	3.2	0.5
K^+	1.75	2.5	3.05	3.35	Formate	4.6	0.5
Rb^+	1.9	2.6	3.1	3.4	IO_3^-	5.5	0.5
Cs^+	2.0	2.7	3.2	3.45	HCO_3^-	6.0	1.2
Ag^+	6.0	7.6	12.0	17.0	Cl^-	22	2.3
Mn^{2+}	2.2	2.35	2.5	2.7	NO_2^-	24	3
Mg^{2+}	2.4	2.5	2.6	2.8	BrO_3^-	27	3
Fe^{2+}	2.4	2.55	2.7	2.9	HSO_3^-	27	3
Zn^{2+}	2.6	2.7	2.8	3.0	CN^-	28	3
Co^{2+}	2.65	2.8	2.9	3.05	Br^-	50	6
Cu^{2+}	2.7	2.9	3.1	3.6	NO_3^-	65	8
Cd^{2+}	2.8	2.95	3.3	3.95	ClO_3^-	74	12
Ni^{2+}	2.85	3.0	3.1	3.25	HSO_4^-	85	15
Ca^{2+}	3.4	3.9	4.6	5.8	Phenolate	110	27
Sr^{2+}	3.85	4.95	6.25	8.1	I^-	175	17
Hg^{2+}	5.1	7.2	9.7	14.0	Salicylate	450	65
Pb^{2+}	5.4	7.5	10.1	14.5	Benzenesulfonate	>500	75
Ba^{2+}	6.15	8.7	11.6	16.5			

SOURCE: *F. de Dardel and T. V. Arden, "Ion Exchangers" in Ullmann's Encyclopedia of Industrial Chemistry, 7th ed. (New York: Wiley-VCH, 2010)*

with H^+ in a process called *regeneration*. Even though Fe^{2+} is bound more tightly than H^+, Fe^{2+} is removed from the resin by washing with excess acid. Washing a column in the H^+ form with Na^+ will convert the column into the Na^+ form.

Ion exchangers loaded with one kind of ion bind small amounts of a different ion nearly quantitatively. H^+-loaded resin will bind small amounts of Li^+ nearly quantitatively, even though the selectivity is greater for H^+. The Li^+ binding releases an *equivalent* charge from the resin in the form of H^+ ions. The same column binds large quantities of Ni^{2+} or Fe^{2+} because the resin has greater selectivity for these ions than for H^+. Binding of one divalent metal ion releases two H^+. The amount of charge exchanged is measured in **equivalents (eq),** which is the amount of cation that will exchange with one mole of a monovalent ion such as H^+

In the 1850s, J. Thomas Way discovered many of the fundamental properties of ion exchange while studying the "power of soil to absorb manure."[4]

$$\text{Equivalents:} \quad nR^-H^+ + M^{n+} \rightarrow R_n^-M^{n+} + nH^+ \qquad eq = n \times (\text{moles of } M^{n+}) \quad \text{(26-2)}$$

One mole of Ni^{2+} exchanges with two moles of H^+, so one mole of Ni^{2+} is two equivalents.

The **ion-exchange capacity** is the number of ionic sites on the resin that can participate in the exchange process. Typical styrene-divinylbenzene resins have capacities of 1–5 meq per mL of wet resin, whereas particles used for ion chromatography (Section 26-2) have lower capacity. For weak acid or base ion exchangers, the capacity can vary dramatically with pH.

EXAMPLE **Equivalents from a Cation-Exchange Column**

How much H^+ is released when 100 mL of 20 mM Na^+ are quantitatively retained on a cation-exchange column in the H^+ form? What if 40 mL of 5.0 mM Fe^{3+} are retained on the column?

Solution Each Na^+ that is retained releases one H^+. One mole of H^+ has 1 equivalent of charge. So

$$\begin{aligned} eq &= n \times (\text{moles of } M^{n+}) \\ &= 1 \text{ eq/mol} \times 0.100 \text{ L} \times 0.020 \text{ mol/L} \\ &= 0.002\ 0 \text{ eq} = 2.0 \text{ meq} \end{aligned}$$

If 40 mL of 5.0 mM Fe^{3+} are loaded on the column

$$\begin{aligned} eq &= n \times (\text{moles of } M^{n+}) \\ &= 3 \text{ eq/mol} \times 0.040 \text{ L} \times 0.005\ 0 \text{ mol/L} \\ &= 6.0 \times 10^{-4} \text{ eq} = 0.60 \text{ meq} \end{aligned}$$

TEST YOURSELF Find the milliequivalents of H^+ released if 100 mL of 26.3 mM Ni^{2+} were loaded on a cation exchange column in the H^+ form. (*Answer:* 5.26 meq)

Donnan Exclusion

With a strong acid (RSO_3^-) cation exchange resin, cations (M^{n+}) freely enter the pores of the resin to undergo ion exchange at the R^- sites. Neutral analytes may also freely enter the pores of an ion-exchange resin, even though they do not interact with the ion-exchange sites. Ions with the same charge as the resin, such as Cl^-, experience electrostatic repulsion from the R^- sites and are *excluded* from the pores of the resin, a process known as **Donnan exclusion.**

The high concentration of bound negative charges within the resin repels anions from the resin.

Donnan exclusion is the basis of *ion-exclusion chromatography*.[5] When a solution of NaCl and sugar is applied to a cation-exchange column with fixed R^- sites and mobile Na^+ counterions, NaCl emerges from the column *before* sugar because Cl^- cannot penetrate the pores, but sugar does. Partially ionized analytes, such weak acids, are separated based on their pK_a, which controls the fraction of deprotonated acid A^- at a given pH.

Conducting Ion-Exchange Chromatography

Ion-exchange *resins* are used for applications involving small molecules (FM $\lesssim$ 500), which can penetrate the small pores of the resin. A mesh size (Table 28-2) of 100/200 is suitable for most work. Higher mesh numbers (smaller particle size) lead to finer separations but slower column operation. For preparative separations, the sample may occupy 10 to 20% of the column volume. Ion-exchange *gels* are used for large molecules (such as proteins and nucleic

Three classes of ion exchangers:

1. resins

2. gels

3. inorganic exchangers

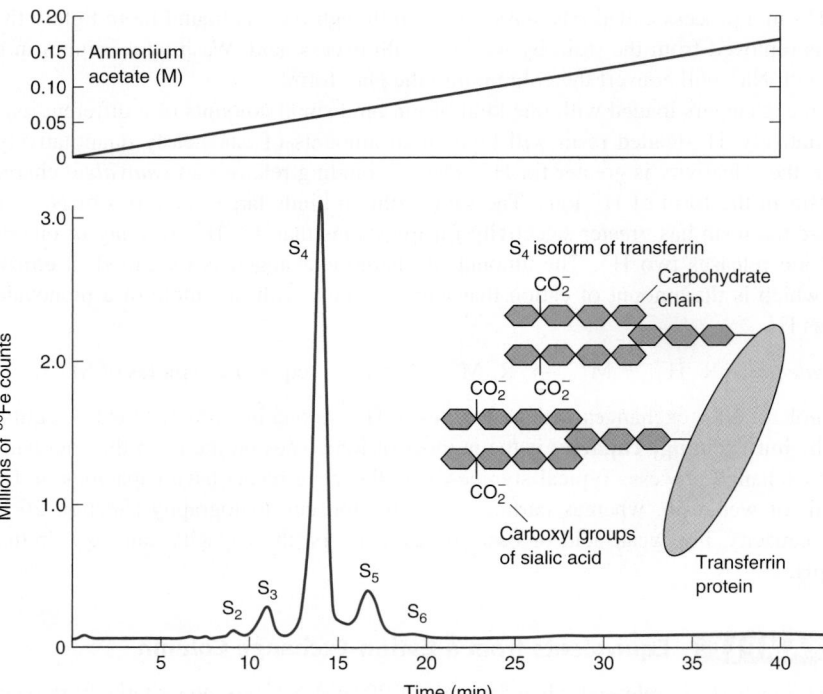

FIGURE 26-2 Ionic-strength-gradient anion-exchange separation of *isoforms* S_2 to S_6 of the protein transferrin bearing two to six sialic acid groups. The 50 × 5-mm column contained Pharmacia Mono-Q resin, which consists of 10-μm-diameter beads with quaternary ammonium anion-exchange sites. [Data from S. A. Rodríguez, E. B. González, G. A. Llamas, M. Montes-Bayón, and A. Sanz-Medel, "Detection of Transferrin Isoforms in Human Serum: Comparison of UV and ICP–MS Detection after CZE and HPLC Separations," *Anal. Bioanal. Chem.* **2005,** *383,* 390.]

An ionic strength gradient is analogous to a solvent or temperature gradient.

acids), which cannot penetrate the pores of resins. Separations involving harsh chemical conditions (high temperature, high radiation levels, strongly basic solution, or powerful oxidizing agents) employ *inorganic ion exchangers,* such as hydrous oxides of Zr, Ti, Sn, and W.

Gradient elution with increasing ionic strength or changing pH is common in ion-exchange chromatography. Consider a column to which anion A^- is bound more tightly than anion B^-. We separate A^- from B^- by elution with C^-, which is less tightly bound than either A^- or B^-. As the concentration of C^- is increased, B^- is eventually displaced and moves down the column. At a still higher concentration of C^-, the anion A^- also is eluted.

In Figure 26-2, a gradient of ammonium acetate was used for an anion-exchange separation of differently charged forms of the protein transferrin at pH 6.0. Transferrin (Figure 18-8) has a molecular mass near 81 000 and can bind two Fe^{3+} ions for transport through the blood. Attached to the protein are two carbohydrate chains, each terminated in one to four sialic acid sugars. The most common form, designated S_4 in Figure 26-2, contains four negatively charged sialic acid groups. Transferrin molecules containing different numbers of sialic acid sugars are called *isoforms* and are labeled S_2 through S_6. Relative amounts of different isoforms vary with certain pathological disorders. In Figure 26-2, transferrin was detected by directing eluate into an inductively coupled plasma–mass spectrometer to detect ^{56}Fe (Section 21-7). Serum species that do not contain iron are invisible to this detector.

Applications of Ion Exchange

Ion exchange can be used to convert one salt into another. For example, we can prepare tetrapropylammonium hydroxide from a tetrapropylammonium salt of some other anion:

$$(CH_3CH_2CH_2)_4N^+I^- \xrightarrow[\text{OH}^-\text{form}]{\text{anion exchanger}} (CH_3CH_2CH_2)_4N^+OH^-$$

$$\text{Tetrapropylammonium} \qquad\qquad\qquad \text{Tetrapropylammonium}$$
$$\text{iodide} \qquad\qquad\qquad\qquad \text{hydroxide}$$

Preconcentration: process of concentrating trace components of a sample prior to their analysis

Ion exchange is used for **preconcentration** of trace components of a solution to obtain enough for analysis. Ongoing research in oceanography is measuring Al, Mn, Fe, Cu, Zn, and Cd at part per trillion levels in the presence of orders-of-magnitude higher concentrations of Na^+, Ca^{2+}, and Mg^{2+} in the ocean. Trace elements are preconcentrated and separated from alkali (Group 1) and alkaline earth (Group 2) elements by passage through the Chelate-PA1 ion-exchange resin shown in Figure 26-3. Ethylenediaminetriacetic acid groups in the resin bind analyte elements quantitatively at pH 6 and above. Alkali and alkaline earth elements are weakly bound at pH 6. The procedure is to pass 125 g of seawater (adjusted to pH 6 with ammonium acetate buffer) through 0.5 g of resin and wash away loosely bound metals with 40 mL of pH 6 buffer. Strongly retained metal ions are then quantitatively eluted in the

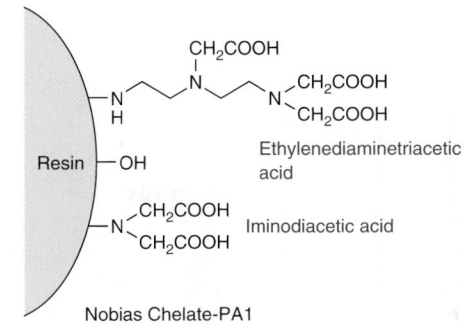

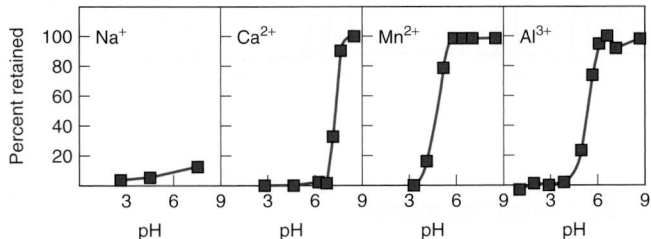

FIGURE 26-3 Nobias Chelate-PA1 resin used to preconcentrate trace metals from seawater contains ethylenediaminetriacetic acid and iminodiacetic acid groups covalently attached to hydrophilic methacrylate polymer resin. Graphs show fraction of metal ions retained by resin as a function of pH. Mn^{2+} and Al^{3+} are well retained at pH 6. Na^+ and Ca^{2+} are poorly retained at pH 6. [Data from Y. Sohrin, S. Urushihara, S. Nakatsuka, T. Kono, E. Higo, T. Minami, K. Norisuye, and S. Umetani, "Multielemental Determination of GEOTRACES Key Trace Metals in Seawater by ICPMS after Preconcentration Using an Ethylenediaminetriacetic Acid Chelating Resin," *Anal. Chem.* **2008,** *80,* 6267.]

reverse direction with 15 mL of 1 M HNO_3. Trace metals are concentrated by a factor of 8 from 125 g of seawater into ~15 g of eluate, and >99.9% of alkali and alkaline earth metals are removed. Eluate is analyzed by inductively coupled plasma–mass spectrometry.

Ion exchange is used to purify water. **Deionized water** is prepared by passing water through an anion-exchange resin in its OH^- form and a cation-exchange resin in its H^+ form. Suppose that $Cu(NO_3)_2$ is present in the water. The cation-exchange resin binds Cu^{2+} and replaces it with $2H^+$. The anion-exchange resin binds NO_3^- and replaces it with OH^-. The eluate is pure water:

$$\left.\begin{array}{l} Cu^{2+} \xrightarrow{\text{ } H^+ \text{ ion exchange } } 2H^+ \\[1em] 2NO_3^- \xrightarrow{\text{ } OH^- \text{ ion exchange } } 2OH^- \end{array}\right\} \longrightarrow \text{ pure } H_2O$$

In some laboratory buildings, tap water is purified first by passage through activated carbon to adsorb organic material, and then by *reverse osmosis*. In the latter process, water is forced by pressure through a membrane containing pores through which few molecules larger than H_2O can pass. Ions do not pass through the pores because their hydrated radii are too large. Reverse osmosis removes 95–99% of ions, organic molecules, bacteria, and particles from water.

"Water polishing" equipment is used in many laboratories after reverse osmosis. The water is passed through activated carbon and then through several ion-exchange cartridges that convert ions into H^+ and OH^-. The resulting high-purity water has a resistivity (Chapter 15, note 41) of 180 000 ohm · m (18 Mohm · cm), with concentrations of individual ions below 1 ng/mL (1 ppb).[6]

Cation-exchange chromatography has been used to separate enantiomers (mirror image isomers) of cationic metal complexes by elution with one enantiomer of tartrate anion.[7] Tartrate has a different ion-pair formation constant with each cation enantiomer and therefore elutes one cation enantiomer from the column before the other.

In pharmaceuticals, ion-exchange resins are used for drug stabilization and as aids for tablet disintegration. Ion exchangers are also used for taste masking, for sustained-release products, as topical products for application to skin, and for ophthalmic or nasal delivery.[8]

Simultaneous Separation of Anions and Cations on One Column

Section 25-1 described *hydrophilic interaction chromatography* (HILIC) with a polar stationary phase. The zwitterionic bonded stationary phase in Figure 25-16 has fixed positive and negative charges and is useful for the simultaneous separation of cations and anions. HILIC utilizes a mixed aqueous-organic eluent. The gradient in Figure 26-4 goes from 78 vol% organic to 60.5 vol% organic. Eluent strength *increases* as the fraction of acetonitrile *decreases*. The charged aerosol detector (Figure 25-24) responds to almost all analytes and is compatible with gradient elution.

Water softeners use ion exchange to remove Ca^{2+} and Mg^{2+} from "hard" water (Box 12-2).

In the Old Testament (Exodus 15:25), Moses used rotten wood to make brackish water drinkable. Carboxylated cellulose is an effective cation exchanger that would remove bitter cations from the water.

High-purity water should be used immediately. Contamination from storage vessels and equilibration with atmospheric CO_2 quickly degrade the resistivity to ~1 Mohm · cm.

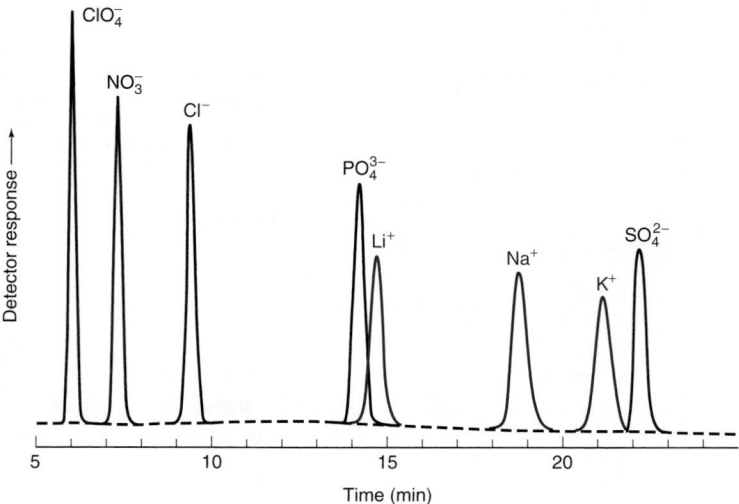

FIGURE 26-4 Simultaneous separation of anions and cations on a ZIC-HILIC zwitterionic stationary phase (Figure 25-16) with charged aerosol detection. The 150 × 4.6 mm column with 5-μm-diameter stationary phase was eluted with a gradient of 20–70% B over 26 min at a flow rate of 0.5 mL/min at 30°C. Solvent A: 15 vol% 100 mM ammonium acetate (pH 4.68) in water, 5% methanol, 20% 2-propanol, 60% acetonitrile. Solvent B: 50 vol% 30 mM ammonium acetate (pH 4.68) in water, 5% methanol, 20% 2-propanol, 25% acetonitrile. [Data from M. Swartz, "Ion Chromatography: An Overview and Recent Developments," *LCGC North Am.* **2010,** *28,* 530.]

What are the ions in pristine snow? Antarctic snow provides a measure of global atmospheric chemistry because there are no local sources of pollution. One study found the following species by ion chromatography:

Ion	Concentrations observed (μg/L = ppb)	
	Minimum	Maximum
F^-	0.10	6.20
Cl^-	25	40 100
Br^-	0.8	49.4
NO_3^-	8.6	354
SO_4^{2-}	10.6	4 020
$H_2PO_4^-$	1.8	49.0
HCO_2^-	1.1	45.7
$CH_3CO_2^-$	5.0	182
$CH_3SO_3^-$	1.1	281
NH_4^+	2.4	46.5
Na^+	15	17 050
K^+	3.1	740
Mg^{2+}	2.7	1 450
Ca^{2+}	12.6	1 010

SOURCE: *R. Udisti, S. Bellandi, and G. Piccardi, "Analysis of Snow from Antarctica,"* Fresenius J. Anal. Chem. **1994,** *349,* 289.

The **separator column** separates analytes, and the **suppressor** replaces ionic eluent with a nonionic species.

One fingerprint contains 80–520 ng of Cl^- plus[10]

750–2 800 ng lactate	30–150 ng phosphate
30–440 ng oxalate	70–220 ng nitrate
60–100 ng sulfate	50–60 ng nitrite

26-2 Ion Chromatography

Ion chromatography, a high-performance version of ion-exchange chromatography, is generally the method of choice for anion analysis.[9] In the semiconductor industry, it is used to monitor anions and cations at 0.1 ppb levels in deionized water.

Suppressed-Ion Anion and Cation Chromatography

In **suppressed-ion <u>anion</u> chromatography** (Figure 26-5a), a mixture of <u>anions</u> is separated by ion exchange and detected by electrical conductivity. The key feature of suppressed-ion chromatography is removal of unwanted electrolyte prior to conductivity measurement.

For the sake of illustration, consider a sample containing $NaNO_3$ and $CaSO_4$ injected into the *separator column*—an anion-exchange column in the hydroxide form—followed by elution with K^+OH^-. NO_3^- and SO_4^{2-} equilibrate with the resin and are slowly displaced by OH^- eluent. Na^+ and Ca^{2+} cations are not retained and simply wash through. Eventually, $K^+NO_3^-$ and $2K^+SO_4^{2-}$ are eluted from the separator column, as shown in the upper graph of Figure 26-5a. These species cannot be detected easily, however, because eluate contains a high concentration of K^+OH^-, whose high conductivity obscures that of analyte.

To remedy this problem, the solution next passes through a *suppressor*, in which cations are replaced by H^+. H^+ exchanges with K^+, in this example, through a cation-exchange membrane in the suppressor. H^+ diffuses from high concentration outside the membrane to low concentration inside the membrane. K^+ diffuses from high concentration inside to low concentration outside. K^+ is carried away outside the membrane, so its concentration is always low on the outside. The net result is that K^+OH^- eluent, which has high conductivity, is converted into H_2O, which has low conductivity. When analyte is present, $H^+NO_3^-$ or $(H^+)_2SO_4^{2-}$ with high conductivity is produced and detected.

Suppressed-ion <u>cation</u> chromatography is conducted in a similar manner, but the suppressor replaces Cl^- from eluent with OH^- through an anion-exchange membrane. Figure 26-5b illustrates the separation of $Na^+NO_3^-$ and $Ca^{2+}SO_4^{2-}$. With H^+Cl^- eluent, Na^+Cl^- and $Ca^{2+}2Cl^-$ emerge from the cation-exchange separator column, and Na^+OH^- and $Ca^{2+}(OH^-)_2$ emerge from the suppressor column. H^+Cl^- eluate is converted into H_2O in the suppressor.

Figure 26-6 illustrates analysis of trace anions and cations in South Pole ice cores. Eluent was OH^- for the anion separation and H^+ for the cation separation. After passing through the suppressor both eluents were converted to H_2O, which has low conductivity. The negative *water dip* in the chromatograms indicates the time t_m for unretained water to pass through each column. High-purity water and careful handling are necessary when performing low parts-per-billion analyses.

For a univalent eluent, such as H^+ and OH^-, retention in ion chromatography is governed by

$$\log k = \text{constant} - n \log[E^{+/-}] \quad (26\text{-}3)$$

where the constant is related to the column capacity and ion exchange selectivity, n is the magnitude of the charge on the analyte ion, and E is the eluent ion. Ion exchange selectivity is governed by the nature of the ion exchange site and the degree of cross-linking of the support.[11] Readily available software can be used to simulate and optimize ion chromatographic separations.[12]

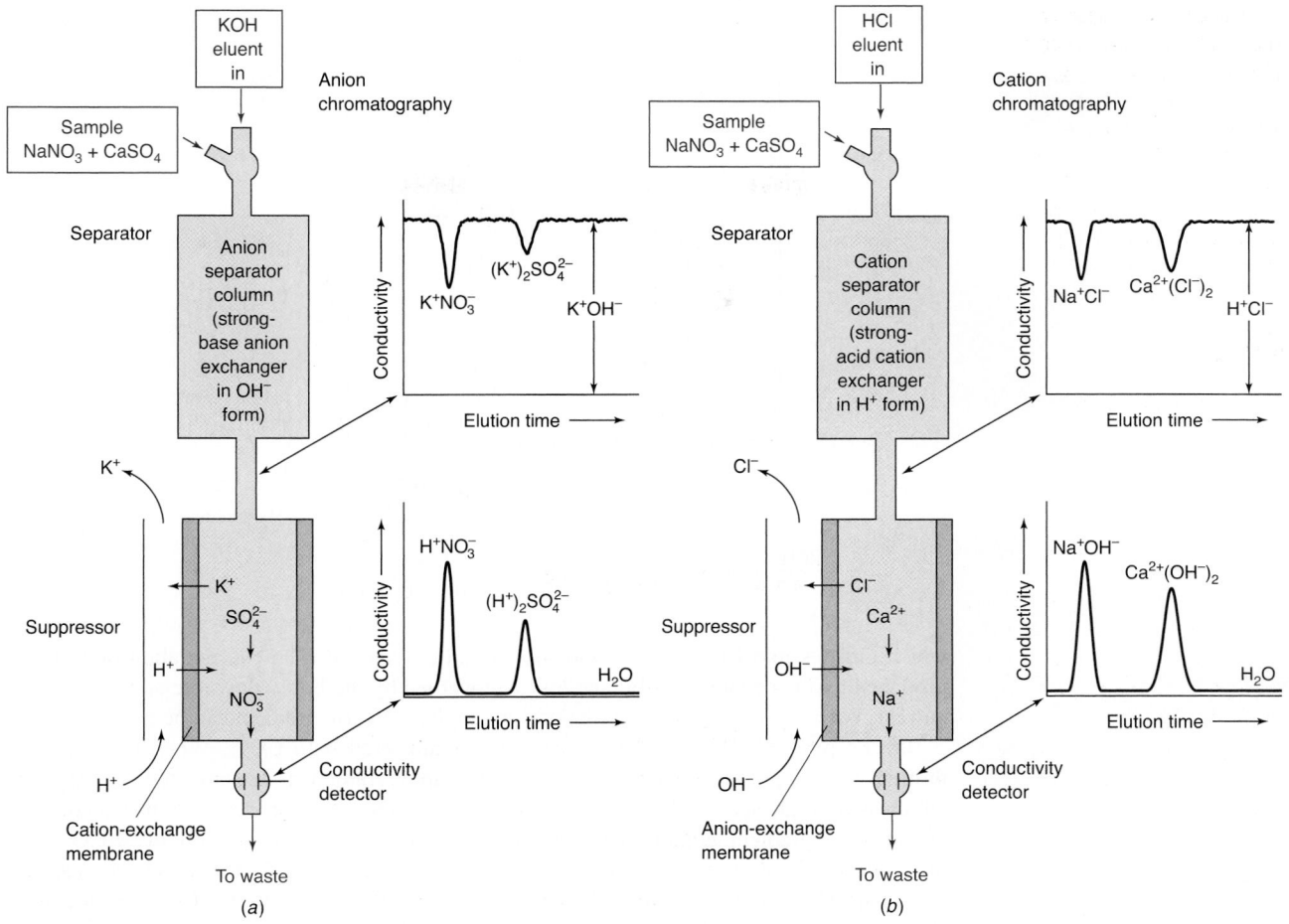

FIGURE 26-5 Schematic illustrations of (a) suppressed-ion anion chromatography and (b) suppressed-ion cation chromatography.

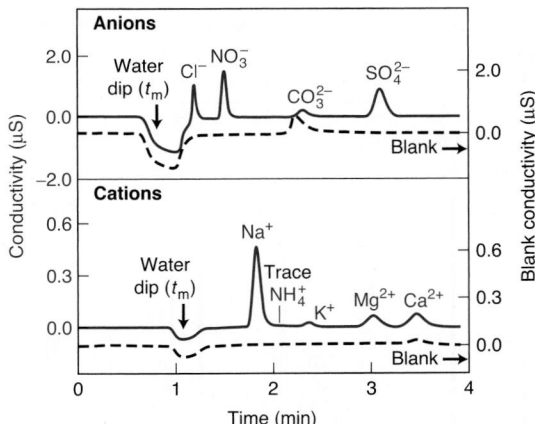

FIGURE 26-6 Ion chromatography of trace anions and cations in South Pole ice core samples. Transport of ice cores to the lab would contaminate samples and blanks. Connecting the ion chromatograph directly to the ice melter reduces handling of the cores, allowing detection of low-ppb ion concentrations. Anion analysis: IonPac AS11 column with 8.0 mM NaOH eluent and ion suppression. Cation analysis: IonPac CS12A column with 11 mM H_2SO_4 eluent and ion suppression. [Data from J. H. Cole-Dai, D. M. Budner, and D. G. Ferris, "High Speed, High Resolution, and Continuous Chemical Analysis of Ice Cores Using a Melter and Ion Chromatography," *Environ. Sci. Technol.* **2006,** *40,* 6764. Undergraduates S. Klein and C. Duval contributed to the laboratory testing of the apparatus.]

In automated systems, H^+ and OH^- eluents and suppressors are generated by electrolysis of H_2O. Figure 26-7 shows a system that generates K^+OH^- for >1 000 hours of isocratic or gradient elution before it is necessary to refresh the reagents. Water in the reservoir of aqueous K_2HPO_4 reacts at the metal anode to produce H^+ and $O_2(g)$. H^+ combines with HPO_4^{2-} to form $H_2PO_4^-$. For each H^+ ion that is generated, one K^+ ion migrates through the cation-exchange barrier membrane, which transports K^+, but not anions, and allows negligible liquid to pass. The barrier membrane must withstand the high pressure of liquid in the K^+OH^- generation chamber destined for the chromatography column. For each H^+ generated at the anode, one K^+ flows through the cation-exchange barrier and one OH^- is generated at the cathode. Liquid exiting the K^+OH^- generation chamber contains K^+OH^- and $H_2(g)$. The stream passes through an anion trap that removes traces of anions such as carbonate and degradation products from the ion-exchange resin. The trap is continuously replenished with electrolytically generated OH^-, which is not shown in the diagram. After the trap, liquid flows through a polymeric capillary that is permeable to $H_2(g)$,

Manually prepared HCO_3^-/CO_3^{2-} mixtures are a common eluent for anion chromatography. After suppression, neutral carbonic acid is formed:

$$HCO_3^- \text{ or } CO_3^{2-} \xrightarrow{H^+} H_2CO_3$$

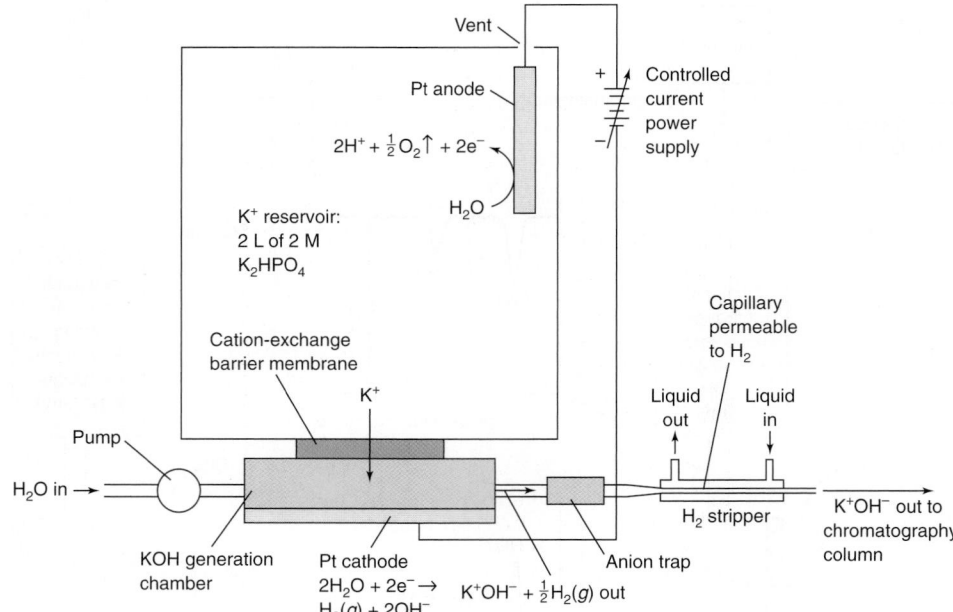

FIGURE 26-7 Electrolytic K⁺OH⁻ eluent generator for ion chromatography. [Information from Y. Liu, K. Srinivasan, C. Pohl, and N. Avdalovic, "Recent Developments in Electrolytic Devices for Ion Chromatography," *J. Biochem. Biophys. Methods* **2004,** *60,* 205.]

which diffuses into an external liquid stream and is removed. The concentration of K^+OH^- produced by the apparatus in Figure 26-7 is governed by the liquid flow rate and the electric current. With computer control of the power supply, a precise gradient can be generated.

In the past, K^+OH^- eluent was usually contaminated with CO_3^{2-}. When CO_3^{2-} passes through the suppressor after the ion chromatography column, it is converted into H_2CO_3, which is a weak acid with some electrical conductivity that interferes with detection of analytes. In gradient elution with increasing K^+OH^-, the concentration of H_2CO_3 also increases, and so does the background conductivity. The feedstock for the electrolytic generator is pure H_2O and the product is aqueous K^+OH^- containing very little CO_3^{2-}. Figure 26-8 shows a separation of 44 anions with a hydroxide gradient.

Suppressors in Figure 26-5 have been replaced by electrolytic units such as that in Figure 26-9, which generate H^+ or OH^- to neutralize eluate and require only H_2O as feedstock. With electrolytic eluent generation and electrolytic suppression, ion chromatography has been simplified and highly automated.

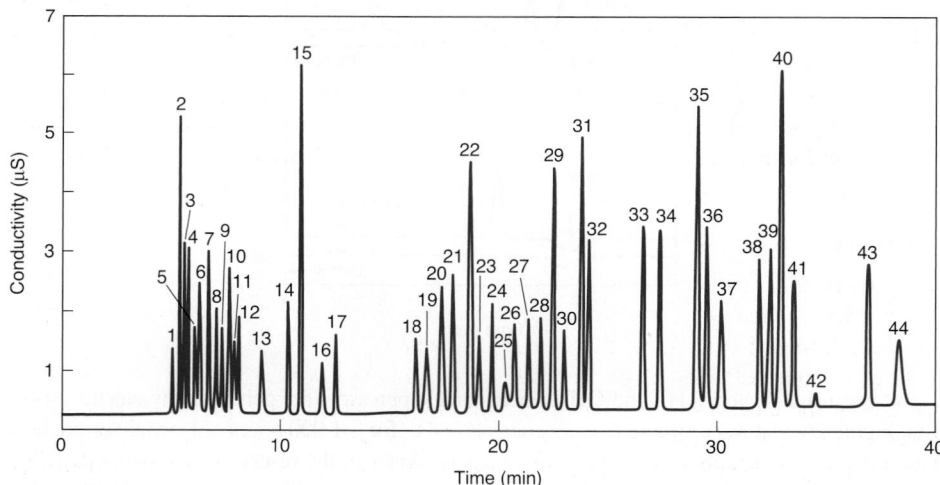

FIGURE 26-8 Anion separation by ion chromatography with a gradient of electrolytically generated K⁺OH⁻ and conductivity detection after suppression. Column: Thermo IonPac AS11-HC, 25 × 0.04 cm, 4-μm particles, flow = 15 μL/min. Eluent: 1 to 14 mM OH⁻ from 0 to 16 min, 14 to 55 mM OH⁻ from 16 to 40 min, injection = 0.4 μL of 0.6 to 4 ppm standards. Peaks: (1) quinate, (2) F⁻, (3) acetate, (4) lactate, (5) 2-hydroxybutyrate, (6) propionate, (7) formate, (8) butyrate, (9) 2-hydroxyvalerate, (10) pyruvate, (11) isovalerate, (12) ClO₂⁻, (13) valerate, (14) BrO₃⁻, (15) Cl⁻, (16) 2-oxovalerate, (17) NO₂⁻, (18) ethylphosphonate, (19) trifluoroacetate, (20) azide, (21) Br⁻, (22) NO₃⁻, (23) citramalate, (24) malate, (25) CO₃²⁻, (26) malonate, (27) citraconitate, (28) maleate, (29) SO₄²⁻, (30) α-ketoglutarate, (31) C₂O₄²⁻, (32) fumarate, (33) oxaloacetate, (34) WO₄²⁻, (35) MoO₄²⁻, (36) PO₄³⁻, (37) phthalate, (38) AsO₄³⁻, (39) citrate, (40) CrO₄²⁻, (41) iso-citrate, (42) *cis*-aconitate, (43) *trans*-aconitate, (44) I⁻. [Data from Thermo Scientific, Sunnyvale, CA.]

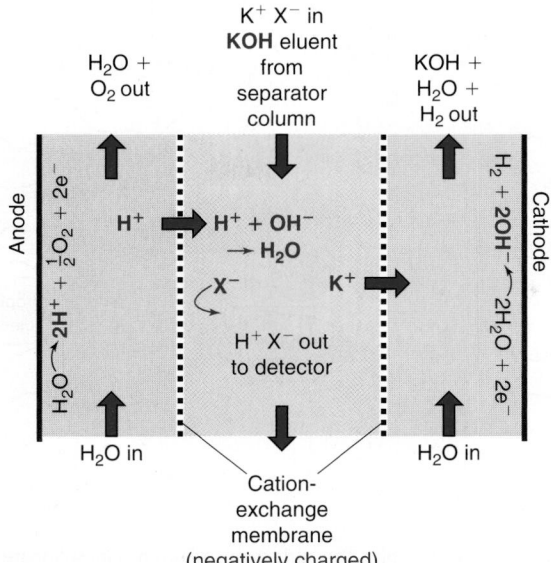

Diagram labels:

H₂O + O₂ out

K⁺ X⁻ in **KOH** eluent from separator column

KOH + H₂O + H₂ out

Anode

$H_2O \rightarrow 2H^+ + \frac{1}{2}O_2 + 2e^-$

H^+

$H^+ + OH^- \rightarrow H_2O$

X^-

K^+

$H^+ X^-$ out to detector

Cathode

$H_2 + 2OH^- \leftarrow 2H_2O + 2e^-$

H₂O in   H₂O in

Cation-exchange membrane (negatively charged)

FIGURE 26-9 Electrolytic suppression for anion chromatography replaces K⁺OH⁻ eluent with H₂O. H⁺ generated at the anode passes through the cation-exchange membrane and replaces K⁺ in the eluent. Anions such as X⁻ do not pass through the cation-exchange membrane due to Donnan exclusion.

In ion chromatography, limits of detection are in the low parts-per-billion range, but parts-per-trillion levels in ultra-pure water are monitored in the power and semiconductor industries using preconcentration with online ion chromatography.[13] Calibration curves are generally linear, although nonlinearity may be observed at low analyte concentrations or when HCO_3^-/CO_3^{2-} eluents are used.[14] Measurement uncertainties are typically greater than 1%, but can be as low as 0.2%.[15]

Ion Chromatography Without Suppression

If the ion-exchange capacity of the separator column is sufficiently low, and if dilute eluent is used, ion suppression is unnecessary. Also, anions of weak acids, such as borate, silicate, sulfide, and cyanide, cannot be determined with ion suppression, because these anions are converted into very weakly conductive products (such as H_2S).

For *nonsuppressed anion chromatography*, we use a resin with an exchange capacity near 5 μmol/g, with 10^{-4} M Na^+ or K^+ salts of benzoic, *p*-hydroxybenzoic, or phthalic acid as eluent. These eluents give a low background conductivity, and analyte anions are detected by a small *change* in conductivity as they emerge from the column. A judicious choice of pH produces an average eluent charge between 0 and -2 and provides control of eluent strength. Even dilute carboxylic acids, which are slightly ionized, are suitable eluents for some separations. *Nonsuppressed cation chromatography* is conducted with dilute HNO_3 eluent for monovalent ions and with ethylenediammonium salts ($^+H_2NCH_2CH_2NH_2^+$) for divalent ions.

Detectors[16]

Conductivity detectors respond to all ions. In suppressed-ion chromatography, it is easy to measure analyte because eluent conductivity is reduced to near 0 by suppression. Suppression also allows us to use eluent concentration gradients.

In nonsuppressed anion chromatography, the conductivity of the analyte anion is higher than that of the eluent, so conductivity increases when analyte emerges from the column. Detection limits are normally in the mid-ppb to low-ppm range but can be lowered by a factor of 10 by using carboxylic acid eluents instead of carboxylate salts. Most HPLC detectors discussed in Section 25-2 may also be used in ion chromatography. For example, some inorganic anions and carboxylic acids can be detected by their ultraviolet absorption.

For quantitative analysis with a conductivity detector, the response factor for each ion must be measured because different ions have different conductivity. A newly developed detector responding to the total *charge* in each eluted band, rather than to conductivity, permits quantification of both known and unknown ions.[17]

Benzoate or phthalate eluents enable sensitive (<1 ppm) **indirect detection** of anions. As each analyte emerges, nonabsorbing analyte anion replaces an equivalent amount of absorbing eluent anion (Figure 26-31). Absorbance therefore *decreases* when analyte appears, as in Figure 26-10. For cation chromatography, $CuSO_4$ is a suitable ultraviolet-absorbing eluent.

Benzoate

p-Hydroxybenzoate

Phthalate

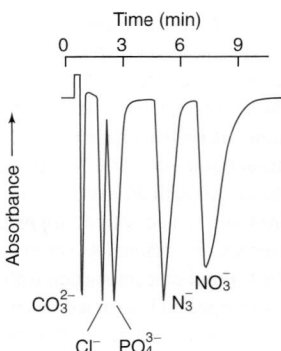

FIGURE 26-10 Indirect spectrophotometric detection of transparent ions. The column was eluted with 1 mM sodium phthalate plus 1 mM borate buffer, pH 10. The principle of indirect detection is illustrated in Figure 26-31. [Data from H. Small, "Indirect Photometric Chromatography," *Anal. Chem.* **1982**, *54*, 462.]

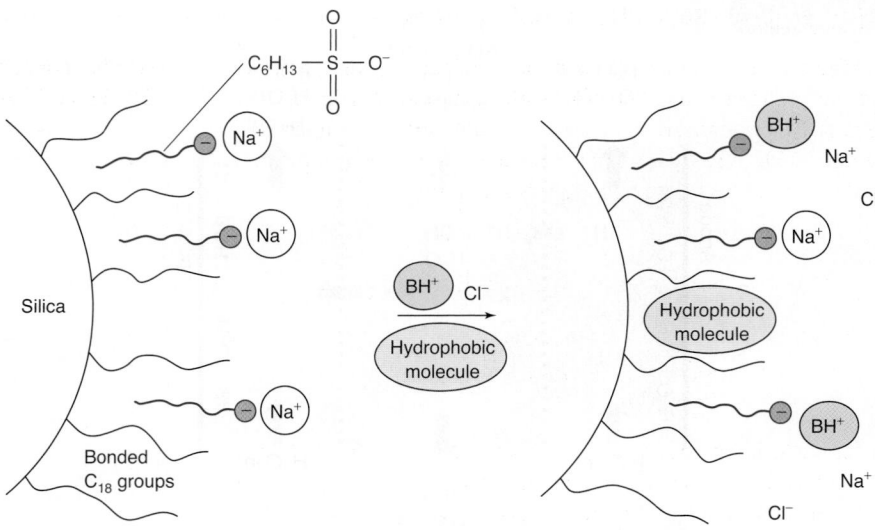

FIGURE 26-11 Ion-pair chromatography. The surfactant sodium hexanesulfonate added to the mobile phase binds to the nonpolar stationary phase. Negative sulfonate groups protruding from the stationary phase then act as ion-exchange sites for analyte cations, such as protonated organic bases, BH⁺. Reversed-phase retention of hydrophobic molecules may also occur.

Ion-Pair Chromatography

Ion-pair chromatography uses a reversed-phase HPLC column instead of an ion-exchange column to separate polar or ionic compounds.[18] To separate cations (such as protonated organic bases), an anionic *surfactant* (Box 26-1) such as $n\text{-}C_6H_{13}SO_3^-$ is added to the mobile phase. The surfactant lodges in the hydrophobic stationary phase, effectively transforming the stationary phase into an ion exchanger (Figure 26-11). Analyte cations are attracted to the surfactant anions.[19] The retention mechanism is a mixture of reversed-phase and ion-exchange interactions. In Figure 26-12, four neutral As species are retained by the C_{18} column. 4-Aminophenylarsonic acid is cationic and would be coeluted with weakly retained monomethylarsonic acid under reversed-phase conditions. However, ion pairing with $n\text{-}C_6H_{13}SO_3^-$ increases retention of the cation so that it is resolved. To separate anions, tetrabutylammonium salts can be added to the mobile phase as the ion-pair reagent.

The term **speciation** describes the distribution of an analyte among possible chemical forms, such as the five forms of arsenic in Figure 26-12. Various species of arsenic have widely different toxicity.

Speciation: distribution of an analyte among possible chemical forms.

FIGURE 26-12 Speciation of arsenic compounds in chicken livers by ion-pair chromatography using Zorbax SQ-Aq reversed-phase column. The column was equilibrated with an aqueous mobile phase containing 5 mM sodium hexane sulfonate ($n\text{-}C_6H_{13}SO_3^-$) and buffered to pH 2.0 with 20 mM citric acid. Chicken liver was spiked with 40 μg As L⁻¹ of each arsenic species. Neutral As compounds are retained by hydrophobic interaction with C_{18}. Cationic As compound is retained by interaction with adsorbed ion-pairing agent. Detection by As atomic fluorescence enables selective detection of arsenic species in complex matrices.
[Data from R. P. Monasterio, J. A. Londonio, S. S. Farias, P. Smichowski, and R. G. Wuilloud, "Organic Solvent-Free Reversed-Phase Ion-Pairing Liquid Chromatography Coupled to Atomic Fluorescence Spectrometry for Organoarsenic Species Determination in Several Matrices," *J. Agric. Food Chem.* **2011,** *59,* 3566.]

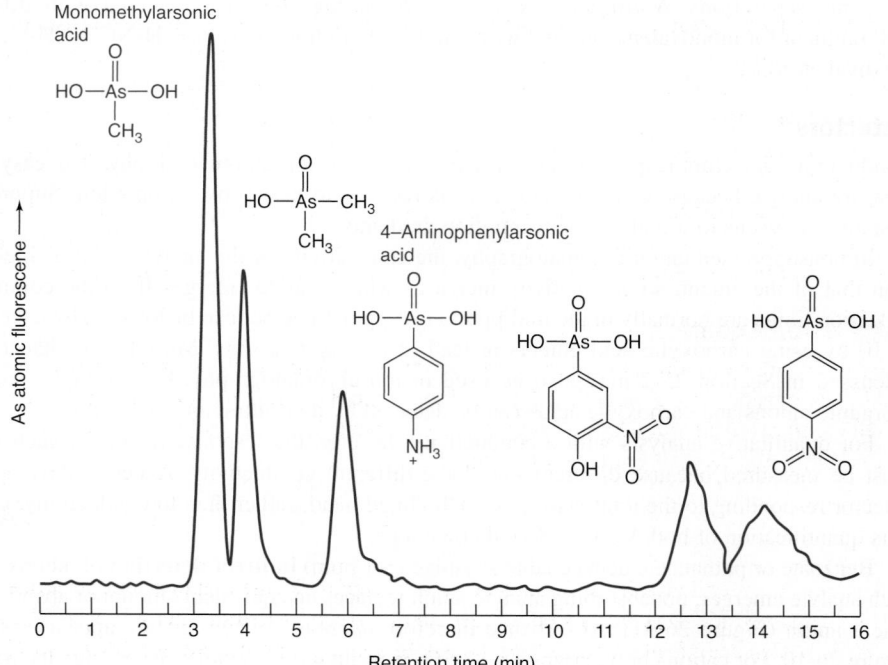

BOX 26-1 Surfactants and Micelles

A **surfactant** is a molecule that accumulates at the interface between two phases and modifies the surface tension. (*Surface tension* is the energy per unit area needed to form a surface or interface.) Molecules with long hydrophobic tails and ionic headgroups are a common class of surfactants for aqueous solution:

$$CH_3$$
$$N—CH_3 \quad Br^-$$
$$CH_3$$

Cetyltrimethylammonium bromide
$C_{16}H_{33}N(CH_3)_3^+ \, Br^-$

A **micelle** is an aggregate of surfactants. In water, the hydrophobic tails form clusters that are, in effect, little oil drops insulated from the aqueous phase by the ionic headgroups. At low concentration, surfactant molecules do not form micelles. When their concentration exceeds the *critical micelle concentration*, spontaneous aggregation into micelles occurs.[20] Isolated surfactant molecules exist in equilibrium with micelles. Nonpolar organic solutes are soluble inside micelles. Cetyltrimethylammonium bromide forms micelles containing ~61 molecules (mass ≈ 22 kDa) in water at 25°C at a critical micelle concentration of 0.9 mM.

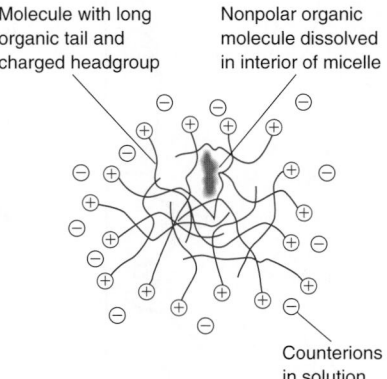

Molecule with long organic tail and charged headgroup

Nonpolar organic molecule dissolved in interior of micelle

Counterions in solution

Structure of a micelle formed when ionic molecules with long, nonpolar tails aggregate in aqueous solution. The interior of the micelle resembles a nonpolar organic solvent, whereas the exterior charged groups interact strongly with water. [Information from F. M. Menger, R. Zana, and B. Lindman, "Portraying the Structure of Micelles," *J. Chem. Ed.* **1998**, *75*, 115.]

Ion-pair chromatography is more complex than reversed-phase chromatography because equilibration of the surfactant with the stationary phase is slow, the separation is more sensitive to variations in temperature and pH, and the concentration of surfactant affects the separation. Methanol is the organic solvent of choice because ionic surfactants are more soluble in methanol/water mixtures than in acetonitrile/water mixtures. Strategies for method development analogous to the scheme in Figure 25-26 vary the pH and surfactant concentration with fixed methanol concentration and temperature.[21] Because of the slow equilibration of surfactant with the stationary phase, gradient elution is not recommended in ion-pair chromatography. Many ion-pair reagents have significant ultraviolet absorption, which makes detection of analytes in the low ultraviolet problematic. Trifluoroacetic acid is a weaker ion-pairing agent than traditional alkyl sulfonates. It is widely used in the separation of peptides and proteins, as it equilibrates quickly and can be used with gradient elution.

26-3 Molecular Exclusion Chromatography

In **molecular exclusion chromatography** (also called *size exclusion*, **gel filtration,** or *gel permeation chromatography*), molecules are separated according to size.[22] Small molecules penetrate the pores in the stationary phase, but large molecules do not (Figure 23-6). Because small molecules must pass through an effectively larger volume, *large molecules are eluted first* (Figure 26-13). This technique is widely used in biochemistry to purify macromolecules.

Salts of low molecular mass (or any small molecule) can be removed from solutions of large molecules by gel filtration because the large molecules are eluted first. This technique, called *desalting*, is useful for changing the buffer composition of a macromolecule solution.

> Large molecules pass through the column faster than small molecules do.
>
> **Gel filtration:** usually refers to a hydrophilic stationary phase and aqueous eluent
>
> **Gel permeation:** usually refers to hydrophobic stationary phase and organic eluent

The Elution Equation

The *total* volume of mobile phase in a chromatography column is V_m, which includes solvent inside and outside the gel particles. The volume of mobile phase *outside* the gel particles is called the **interstitial volume,** V_o. The volume of solvent *inside* the gel is therefore $V_m - V_o$. The quantity K_{av} (read "K average") is defined as

$$K_{av} = \frac{V_r - V_o}{V_m - V_o} \qquad (26\text{-}4)$$

where V_r is the retention volume for a solute. For a large molecule that does not penetrate the gel, $V_r = V_o$, and $K_{av} = 0$. For a small molecule that freely penetrates the gel, $V_r = V_m$, and $K_{av} = 1$. Molecules of intermediate size penetrate some gel pores, but not others, so K_{av} is

> In pure molecular exclusion, all molecules are eluted between $K_{av} = 0$ and $K_{av} = 1$.

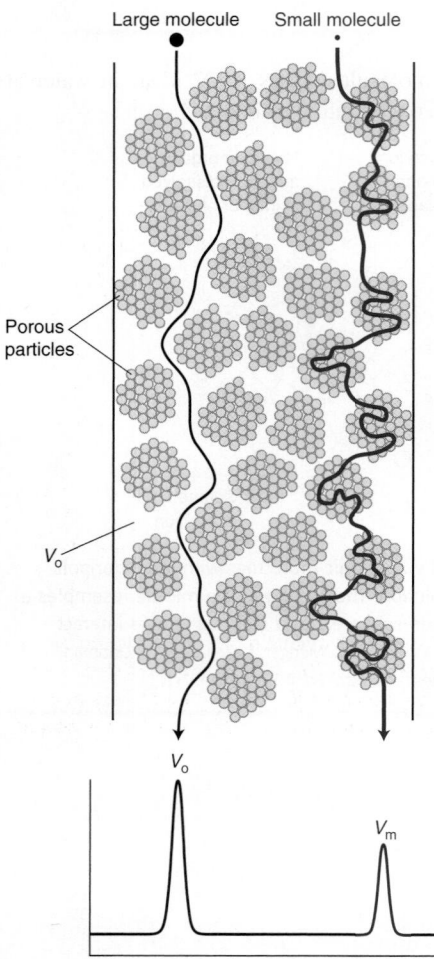

Large molecule   Small molecule

Porous
particles

V_o

V_o

V_m

Volume or time

FIGURE 26-13 Molecular exclusion
chromatography. Large molecules cannot
penetrate the pores of the stationary phase.
They are eluted at volume V_o (the volume
of mobile phase between particles). Small
molecules, which can penetrate the gel, require
a larger volume for elution. Molecules that
penetrate the smallest pores are eluted in
volume V_m (the total volume of solvent inside
and outside the gel particles).

HPLC columns for molecular exclusion are larger
than other HPLC columns, so as to increase the
separation volume between V_o and V_m.

between 0 and 1. Ideally, gel penetration is the only mechanism by which molecules are
retained in this type of chromatography. In fact, there may be some adsorption, so K_{av} can be
greater than 1.

Interstitial volume is measured by passing a large, inert molecule through the column. Its
elution volume is defined as V_o. Blue Dextran 2000, a blue dye of molecular mass 2×10^6 Da,
is commonly used for this purpose. The volume V_m can be calculated from the measured
column bed volume per gram of dry gel. For example, 1 g of dry Sephadex G-100 occupies 15
to 20 mL when swollen with aqueous solution. The solid phase of the swollen gel occupies
only ~1 mL, so V_m is 14 to 19 mL, or 93–95% of the total column volume. Equal masses of
different solid phases produce widely varying volumes when swollen with solvent.

Stationary Phase[23]

Gels for open-column, preparative-scale molecular exclusion include Sephadex (Table 26-4) and
Bio-Gel P, which is a polyacrylamide cross-linked by *N,N'*-methylenebisacrylamide. The smallest pore sizes in highly cross-linked gels exclude molecules with a molecular mass $\gtrsim 700$ Da,
whereas the largest pore sizes exclude molecules with molecular mass $\gtrsim 10^8$ Da. The finer the
particle size of the gel, the greater the resolution and the slower the flow rate of the column.

HPLC columns for molecular exclusion are typically 8 mm in diameter and 30 cm in
length. Hydrophilic HPLC packings are made of hydroxylated cross-linked polymethacrylate.
Silica (Table 26-4) with controlled pore size provides 10 000–20 000 plates for a column
packed with 10-μm particles. The silica is coated with a hydrophilic phase to minimize solute
adsorption. A hydroxylated polyether resin with a well-defined pore size can be used over the
pH range 2–12, whereas silica phases generally cannot be used above pH 8. Particles with different pore sizes can be mixed to give a wider molecular size separation range.

For HPLC of hydrophobic polymers, cross-linked polystyrene spheres are available with
pore sizes ranging from 5 nm up to hundreds of nanometers. Smaller particle sizes yield
better column efficiencies, with 3-μm diameter particles yielding ≥ 16 000 plates for a 15-cm
column length. Very high molecular weight polymers (>2 000 000 Da) are separated using
larger particles to avoid shear degradation of the polymer.

Molecular Mass Determination

Gel filtration is used mainly to separate molecules of significantly different molecular sizes
(Figure 26-14). The pore size chosen depends on the molecular mass of the molecules to be
separated. For each stationary phase, we construct a calibration curve, which is a graph of
log(molecular mass) versus elution volume (Figure 26-15). We estimate the molecular mass
of an unknown by comparing its elution volume with those of standards. We must exercise
caution in interpreting results, however, because molecules with the same molecular mass but
different shapes exhibit different elution characteristics. For proteins, it is important to use an
ionic strength high enough (>0.05 M) to eliminate electrostatic adsorption of solute by occasional charged sites on the gel.

Polymer samples must be prepared and separated using a solvent that dissolves the
polymer. The stationary phase must be compatible with the solvent and have a mass calibration range that encompasses the anticipated mass range of the sample. Synthetic polymers
contain a distribution of polymer chain lengths. The elution volume reflects the average
molecular mass of the polymer and the peak width reflects the distribution of chain lengths.

TABLE 26-4	Representative molecular exclusion media				
Gel filtration in open columns		**TSKgel SW silica for HPLC**			
Name	**Fractionation range for globular proteins (Da)**	**Name**	**Pore size (nm)**	**Fractionation range for globular proteins (Da)**	
Sephadex G-10	to 700	G2000SW	12.5	5 000–150 000	
Sephadex G-25	1 000–5 000	G3000SW	25	10 000–500 000	
Sephadex G-50	1 500–30 000	G4000SW	45	20 000–7 000 000	
Sephadex G-75	3 000–80 000				
Sephadex G-100	4 000–150 000				
Sephadex G-200	5 000–600 000				

NOTE: *Sephadex is manufactured by GE Amersham Biosciences. TSKgel SW silica is manufactured by Tosoh Bioscience.*

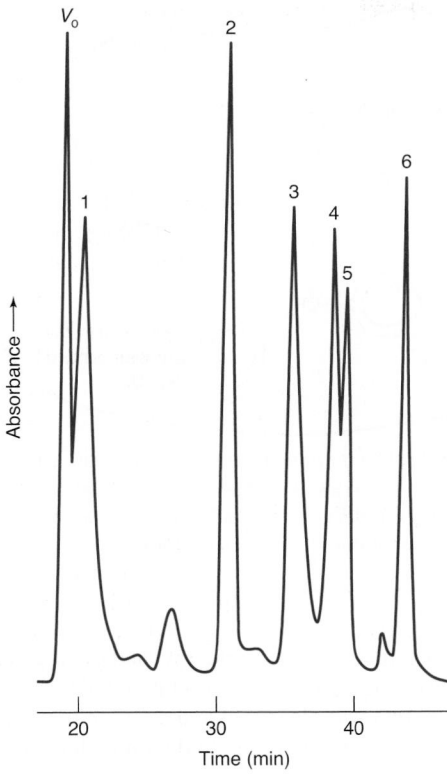

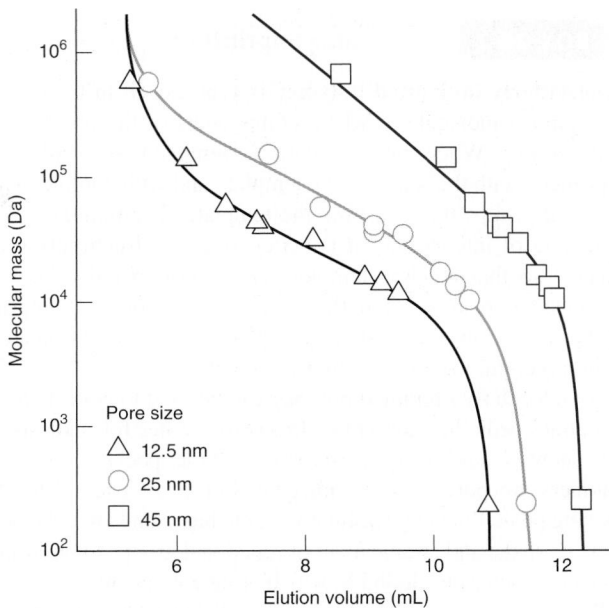

FIGURE 26-15 Molecular mass calibration graph for hydrophilic silica molecular exclusion columns (30 × 0.78 cm) of different pore size. Calibration based on biomolecules ranging from thyroglobin (660 000 Da) to glycine tetramer (246 Da). [Data from Tosoh Bioscience, Stuttgart, Germany.]

1. Thyroglobulin (660 000 Da)
2. Bovine serum albumin (67 000 Da)
3. β-Lactoglobulin (18 400 Da)
4. Myoglobin (16 900 Da)
5. Cytochrome c (12 400 Da)
6. Glycine tetramer (246 Da)

FIGURE 26-14 Separation of proteins by molecular exclusion chromatography on two 60 × 0.75 cm TSKgel G3000SW columns in series eluted at 1.0 mL/min with 0.3 M NaCl in 0.05 M phosphate buffer, pH 7. [Data from Tosoh Bioscience, Stuttgart, Germany.]

Large polymers scatter visible light, so a detector for scattered light is common in gel permeation chromatography.

26-4 Affinity Chromatography

Affinity chromatography is used to isolate a single compound or a class of compounds from a complex mixture. The technique is based on specific and reversible binding of that one compound to the stationary phase (Figure 23-6). When sample is passed through the column, only one solute is bound. After everything else has washed through, the one adhering solute is eluted by changing a condition such as pH or ionic strength to weaken its binding. Affinity chromatography is especially applicable in biochemistry and is based on specific interactions between enzymes and substrates, antibodies and antigens, or receptors and hormones.

Figure 26-16 shows the isolation of the protein immunoglobulin G (IgG) by affinity chromatography on a column containing covalently bound *protein A*. Protein A binds to one

Affinity media and protocols are commercially available for many biochemical applications.[24]

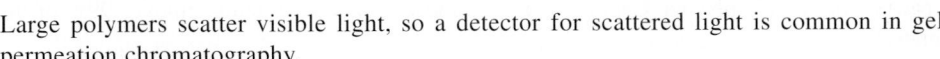

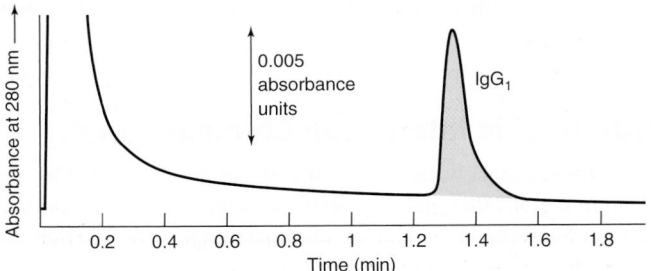

FIGURE 26-16 Purification of monoclonal antibody IgG by affinity chromatography on a 5 × 0.46 cm column containing protein A covalently attached to polymer support. Other proteins in the sample are eluted from 0 to 0.3 min at pH 7.6. When eluent pH is lowered to 2.6, IgG is freed from protein A and emerges from the column. [Data from B. J. Compton and L. Kreilgaard, "Chromatographic Analysis of Therapeutic Proteins," *Anal. Chem.* **1994,** *66,* 1175A.]

A **molecularly imprinted polymer** is synthesized in the presence of a template molecule to which components of the polymer have some affinity. When the template is removed, the polymer is "imprinted" with the shape of the template and with complementary functional groups that can bind the template. The template can be the analyte of interest, but it is better to use a structurally related molecule so that residual template in the polymer does not give false positive results when the polymer is employed. Imprinted polymer can be used as a stationary phase in affinity chromatography or a recognition element in a chemical sensor.

A molecularly imprinted polymer can be used to collect and preconcentrate penicillin antibiotics from river water for analysis. The figure shows a conceptual structure of a polymer pocket formed when monomers are polymerized with penicillin G as a template. After removing penicillin G by washing with methanol and HCl, the pocket retains its shape and arrangement of functional groups to bind similar molecules, such as penicillin V, which is shown in color.

When river water spiked with 30-ppb levels of eight different penicillin variants was passed through a column containing imprinted polymer, 90–99% of six of the penicillins was retained by the column. Two penicillin variants did not bind as well. Penicillins retained by the column were eluted by a small volume of 0.05 M

tetrabutylammonium hydrogen sulfate in methanol and analyzed by HPLC.

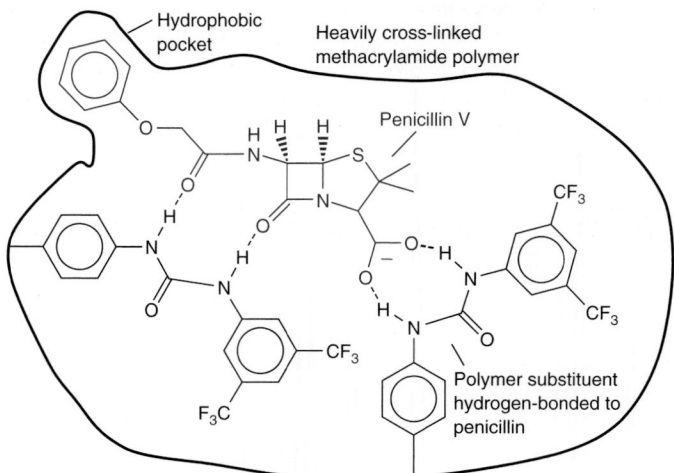

Notional structure of pocket in a polymer imprinted to bind penicillin derivatives. [Information from J. L. Urraca, M. C. Moreno-Bondi, A. J. Hall, and B. Sellergren, "Direct Extraction of Penicillin G and Derivatives from Aqueous Samples Using a Stoichiometrically Imprinted Polymer," *Anal. Chem.* **2007,** *79,* 695.]

specific region of IgG at pH $\gtrsim$ 7.2. When a crude mixture containing IgG and other proteins was passed through the column at pH 7.6, everything except IgG was eluted within 0.3 min. At 1 min, the eluent pH was lowered to 2.6 and IgG was cleanly eluted at 1.3 min as a purified and concentrated fraction. Purification can be several thousand fold with high recoveries of active material.

Affinity chromatography may also be used to selectively remove a specific contaminant or multiple matrix components. For instance, trace proteins in human serum or plasma that may be *biomarkers* of disease are difficult to discover in the presence of other proteins whose concentrations could be up to 10 orders of magnitude greater. Affinity columns with antibodies that capture and retain the 10 most abundant proteins remove 90% of the protein matrix from eluted plasma. Capture of the 22 most abundant proteins removes 99% of the matrix.[25]

Optical isomers of a drug can have completely different therapeutic effects. Affinity chromatography can be used to isolate individual optical isomers for drug evaluation.[26] The drug candidate in the margin has two chiral carbon atoms indicated by colored dots. With two possible geometries at each site, there are four stereoisomers. The mixture of isomers was covalently attached to protein and injected into mice to raise a mixture of antibodies to all stereoisomers. B cells in the spleen produce antibodies. One B cell produces only one kind of antibody. By isolating individual B cells, it is possible to isolate the gene for the antibody to each of the stereoisomers. The gene can be transplanted into *E. coli* cells for mass production of a single kind of antibody, called a *monoclonal antibody*. When the mixture of stereoisomers is passed through a column to which just one kind of antibody is attached, only one of the four stereoisomers is retained. By lowering the pH, the retained isomer is eluted in pure form.

Box 26-2 shows how *molecularly imprinted polymers* can be used as affinity media. *Aptamers* (Box 17-5) are another class of chromatographically useful compounds with high affinity for a selected target.[28]

Water does not *wet* a hydrophobic surface made of carbon nanotubes, so a drop remains almost spherical. The drop would flatten out on a hydrophilic surface, such as glass. [K. S. Lau, J. Bico, K. B. K. Teo, M. Chhowalla, G. A. J. Amaratunga, W. I. Milne, G. H. McKinley, and K. K. Gleason, "Superhydrophobic Carbon Nanotube Forests," *Nano Lett.* **2003,** *3,* 1701. Reprinted with permission © 2003, American Chemical Society.]

26-5 Hydrophobic Interaction Chromatography

Hydrophobic substances repel water, and their surfaces are not wetted by water.[29] A protein can have *hydrophilic* regions that make it soluble in water and *hydrophobic* regions capable of interacting with a hydrophobic stationary chromatography phase. High concentrations of ammonium sulfate cause proteins to *salt out* of solution. Ammonium, sodium, and potassium salts of phosphate and sulfate decrease the solubility of proteins in water. Thiocyanate, iodide, and perchlorate salts have the opposite effect of dissolving proteins.

Hydrophobic interaction chromatography is used principally for protein purification.[30] A common stationary phase shown in Figure 26-17 has hydrophobic phenyl or alkyl groups attached to agarose gel (a polysaccharide), whose pore size is large enough for proteins to enter. When protein solution with a high concentration (such as 1 M) of ammonium sulfate is applied to the column, the salt induces the protein to bind to the hydrophobic surface of the stationary phase. Then a gradient of *decreasing salt concentration* is applied to *increase* the solubility of proteins in water and elute them from the column.

26-6 Principles of Capillary Electrophoresis[31,32]

Late in 2007, more than 200 people receiving the anticoagulant *heparin* suffered acute, allergic reactions and died.[33] Heparin is a complex mixture of sulfate-substituted polysaccharides that have molecular masses of 2–50 kDa and are isolated from pig intestines. As soon as the problem was recognized in January 2008, U.S. distributors recalled heparin products and the U.S. Food and Drug Administration launched an investigation. Heparin is administered thousands of times every day to manage life-threatening conditions, so an immediate understanding and solution to the problem were required.

When exposed to the enzyme heparinase, heparin is cleaved into disaccharide units. Tainted heparin contained 20–50 wt% of macromolecular components that did not react with heparinase. *Capillary electrophoresis* was the first technique to observe two contaminants (Figure 26-18).[34] One was dermatan sulfate, which was not known to cause allergic reactions. The other was identified by nuclear magnetic resonance as oversulfated chondroitin sulfate. An animal study verified that oversulfated chondroitin sulfate caused the allergic reaction. By March 2008, deaths from contaminated heparin had ceased and emergency regulations were issued to incorporate capillary electrophoresis and nuclear magnetic resonance into required testing of heparin imported into the U.S. Contaminated heparin had been prepared in China. Oversulfated chondroitin sulfate might have been added because it has anticoagulant activity and costs less than heparin.

Electrophoresis is the migration of ions in solution under the influence of an electric field. The technique was pioneered in the 1930s by the Swedish biophysical chemist A. Tiselius, who received the Nobel Prize in 1948 for his work on electrophoresis and "discoveries concerning the complex nature of serum proteins."

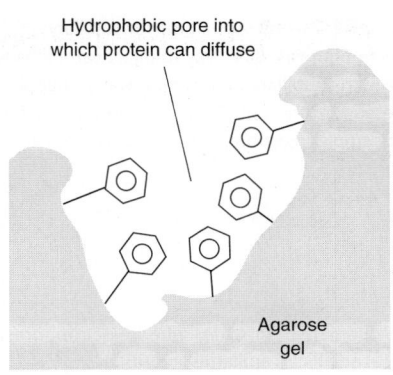

FIGURE 26-17 Stationary phase for hydrophobic interaction chromatography has ~10–20% as many bonded phenyl or alkyl groups per unit volume as a reversed-phase stationary phase.

Cations are attracted to the negative terminal (the cathode).

Anions are attracted to the positive terminal (the anode).

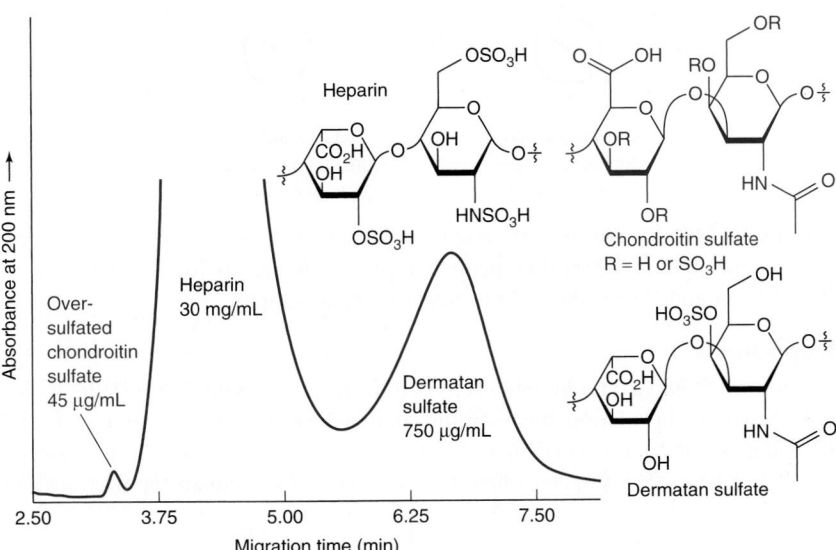

FIGURE 26-18 Electropherogram of heparin (30 mg/mL) spiked with oversulfated chondroitin sulfate and dermatan sulfate. Tainted heparin had ~200 times more oversulfated chondroitin sulfate than shown here. Conditions: −16 kV, 20°C, 30 cm × 25 μm capillary, detector at 21.5 cm. Background buffer was made by adding 0.60 M H_3PO_4 to 0.60 M Li_3PO_4 to reach pH 2.8. [Data from Robert Weinberger, CE Technologies, and Todd Wielgos, Baxter Healthcare. For details, see T. Wielgos, K. Havel, N. Ivanova, and R. Weinberger, "Determination of Impurities in Heparin by Capillary Electrophoresis Using High Molarity Phosphate Buffers," *J. Pharm. Biomed. Anal.* **2009,** *49*, 319.]

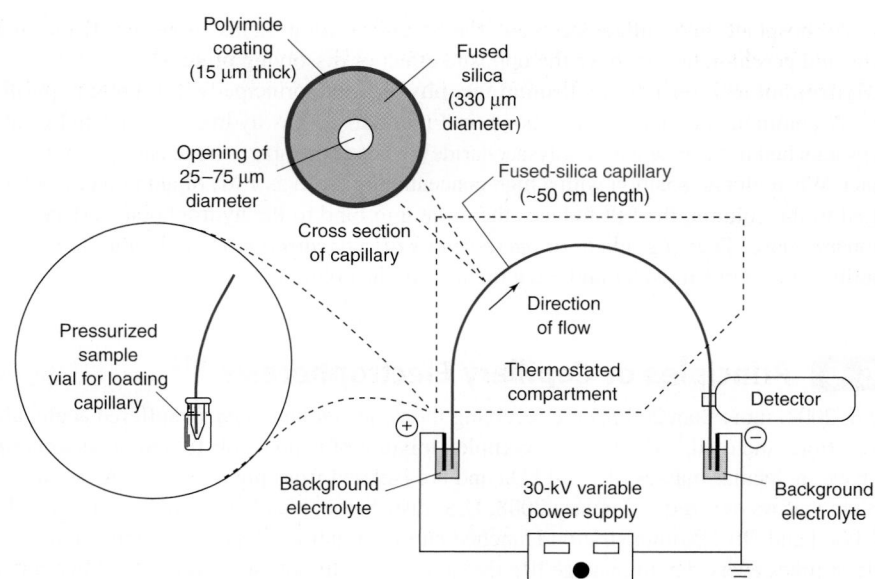

J. W. Jorgenson first described electrophoresis in fused silica capillaries in 1981.[35]

Electric potential difference = 30 kV

$$Electric\ field = \frac{30\ kV}{0.50\ m} = 60\ kV/m$$

In **capillary electrophoresis,** shown in Figure 26-19, components of a solution are separated by applying a voltage of ~30 kV from end to end of a fused-silica (SiO_2) capillary tube that is 50 cm long and has an inner diameter of 25–75 μm. Different ions have different *mobilities* and migrate through the capillary at different speeds.[36] Modifications of this procedure described later allow neutral molecules, as well as ions, to be separated. Rapid electrophoresis can monitor transient events such as neurotransmitter release.[37] The small diameter of a capillary enables the analysis of single cells, nuclei, vesicles, or mitochondria.[38] Near-physiological separation conditions allow probing of biomolecular interactions, even at the single enzyme-molecule level.[39]

Capillary electrophoresis provides high resolution. When we conduct chromatography in a packed column, peaks are broadened by three mechanisms in the van Deemter equation (23-37): multiple flow paths, longitudinal diffusion, and finite rate of mass transfer. An open tubular column eliminates multiple paths and thereby reduces plate height and improves resolution. Capillary electrophoresis reduces plate height further by knocking out the mass transfer term that comes from the finite time needed for solute to equilibrate between the mobile and stationary phases. In capillary electrophoresis, *there is no stationary phase*. The only fundamental source of broadening under ideal conditions is longitudinal diffusion:

$$H = \cancel{A} + \frac{B}{u_x} + \cancel{Cu_x} \qquad (26\text{-}5)$$

Multiple path term eliminated by open tubular column

Mass transfer term eliminated because there is no stationary phase

(Other sources of broadening in real systems are mentioned later.) Capillary electrophoresis routinely produces 50 000–500 000 theoretical plates (Figure 26-20), a performance that is an order-of-magnitude better than that of liquid chromatography.

Electrophoresis

When an ion with charge q (coulombs) is placed in an electric field E (V/m), the force on the ion is qE (newtons). In solution, the retarding frictional force is fu_{ep}, where u_{ep} is the velocity of the ion and f is the *friction coefficient*. The subscript "ep" stands for "electrophoresis." The ion quickly reaches a steady speed when the accelerating force equals the frictional force:

$$\begin{array}{cc} qE & = & fu_{ep} \\ \text{Accelerating} & & \text{Frictional} \\ \text{force} & & \text{force} \end{array}$$

Electrophoretic mobility:
$$u_{ep} = \frac{q}{f}E \equiv \mu_{ep}E \qquad (26\text{-}6)$$

Electrophoretic mobility

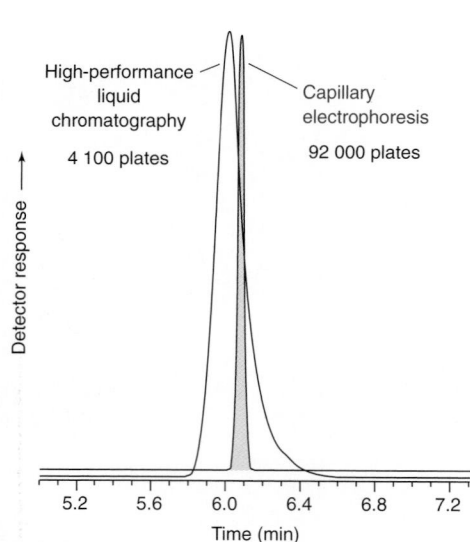

FIGURE 26-20 Comparison of peak widths for benzyl alcohol ($C_6H_5CH_2OH$) in capillary electrophoresis and HPLC. [Data from S. Fazio, R. Vivilecchia, L. Lesueur, and J. Sheridan, *Am. Biotech. Lab.,* January 1990, p. 10.]

Electrophoretic **mobility** (μ_{ep}) is the constant of proportionality between the speed of the ion and electric field strength. Mobility is proportional to the charge of the ion and inversely proportional to the friction coefficient. For molecules of similar size, mobility increases with charge:

We encountered **mobility** earlier in connection with junction potentials (Table 15-1).

$$\text{HO}_2\text{C} \overset{\overset{\displaystyle}{}}{\underset{\underset{\displaystyle \text{CO}_2^-}{}}{}} \text{CO}_2\text{H}$$

$$\mu_{ep} = -2.54 \times 10^{-8}\ \frac{\text{m}^2}{\text{V} \cdot \text{s}}$$

$$^-\text{O}_2\text{C} \overset{\overset{\displaystyle}{}}{\underset{\underset{\displaystyle \text{CO}_2\text{H}}{}}{}} \text{CO}_2^-$$

$$\mu_{ep} = -4.69 \times 10^{-8}\ \frac{\text{m}^2}{\text{V} \cdot \text{s}}$$
(solvent is H_2O at 25°C)

$$^-\text{O}_2\text{C} \overset{\overset{\displaystyle}{}}{\underset{\underset{\displaystyle \text{CO}_2^-}{}}{}} \text{CO}_2^-$$

$$\mu_{ep} = -5.95 \times 10^{-8}\ \frac{\text{m}^2}{\text{V} \cdot \text{s}}$$

For a spherical particle of radius r moving through a fluid of *viscosity* η, the friction coefficient, f, is

Stokes equation:
$$f = 6\pi\eta r \tag{26-7}$$

which was introduced on page 617 in the discussion of diffusion. Mobility is $\mu = q/f = q/6\pi\eta r$, so the greater the molecular radius, the lower the mobility. Most molecules are not spherical, but Equation 26-7 defines an effective *hydrodynamic radius* of a molecule, as if it were a sphere, based on its observed mobility.

Viscosity measures resistance to flow in a fluid. The units are kg m^{-1} s^{-1}. Relative to water, maple syrup is very viscous and hexane has low viscosity.

Electroosmosis

The inside wall of a fused silica capillary is covered with silanol (Si—OH) groups which have a negative charge (Si—O$^-$) above pH 3. Figure 26-21a shows the *electric double layer* (Box 17-4) at the capillary surface. The double layer consists of fixed negative charges on the wall and excess cations near the wall. A tightly adsorbed, immobile layer of cations partially neutralizes the negative charge. The remaining negative charge is neutralized by mobile cations in the *diffuse part of the double layer* in solution near the wall. The thickness of the diffuse part of the double layer ranges from ~10 nm when the ionic strength is 1 mM to ~0.3 nm when the ionic strength is 1 M.

In an electric field, cations are attracted to the cathode and anions are attracted to the anode (Figure 26-21b). Excess cations in the diffuse part of the double layer impart net momentum toward the cathode. This pumping action, called **electroosmosis** (or *electroendosmosis*), is driven by cations within ~10 nm of the walls and creates uniform pluglike *electroosmotic flow* of the entire solution toward the cathode (Figure 26-22a). This process is in sharp contrast with *hydrodynamic flow*, which is driven by a pressure difference. In hydrodynamic flow, the velocity profile through a cross section of the fluid is parabolic: It is fastest at the center and slows to 0 at the walls (Figure 26-22b and Color Plate 33).

Ions in the diffuse part of the double layer adjacent to the capillary wall are the "pump" that drives electroosmotic flow.

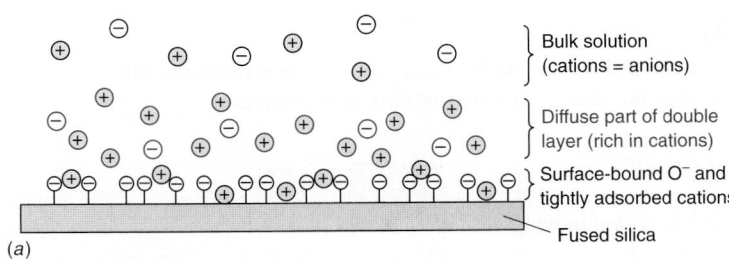

(a)

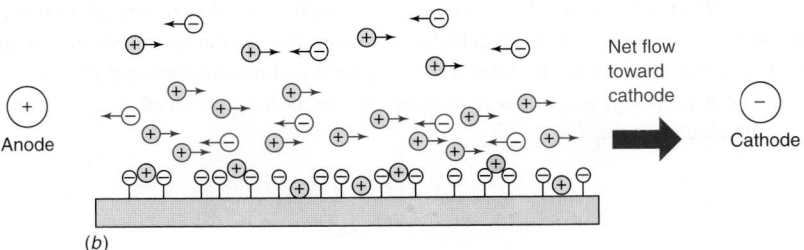

(b)

FIGURE 26-21 (*a*) Electric double layer created by negatively charged silica surface and nearby cations. (*b*) Predominance of cations in diffuse part of the double layer produces net electroosmotic flow toward the cathode when an external field is applied.

FIGURE 26-22 (*a*) Electroosmosis gives uniform flow over more than 99.9% of the cross section of the capillary. The speed decreases immediately adjacent to the capillary wall. (*b*) Parabolic velocity profile of hydrodynamic flow (also called *laminar flow*), with the highest velocity at the center of the tube and zero velocity at the walls. Experimentally observed velocity profiles are shown in Color Plate 33.

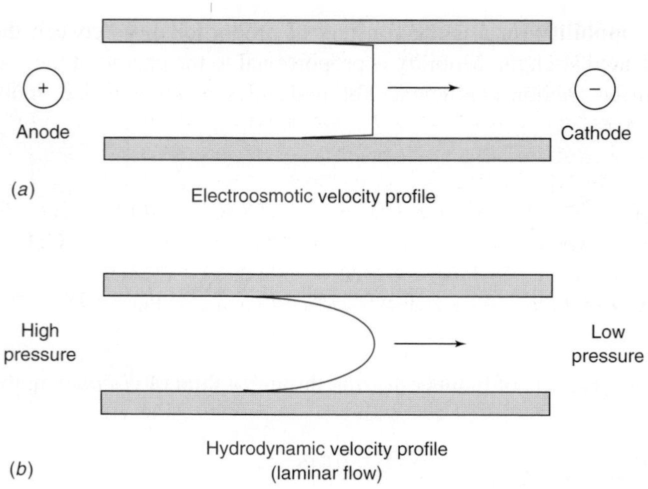

Electroosmotic velocity is measured by adding to the sample a *neutral* molecule to which the detector responds.

$$\text{Electroosmotic velocity} = \frac{\text{distance from injector to detector}}{\text{migration time of neutral molecule}}$$

The constant of proportionality between electroosmotic velocity, u_{eo}, and applied field is called *electroosmotic mobility*, μ_{eo}.

Electroosmotic mobility:
$$u_{eo} = \mu_{eo}E \qquad (26\text{-}8)$$

Electroosmotic mobility
(units = $m^2/[V \cdot s]$)

Electroosmotic mobility is proportional to the surface charge density on the silica and inversely proportional to the square root of ionic strength. Electroosmosis decreases at low pH (Si—O$^-$ → Si—OH decreases surface charge density) and high ionic strength. At pH 9 in 20 mM borate buffer, electroosmotic flow is ~2 mm/s. At pH 3, flow is reduced by an order of magnitude.

Uniform electroosmotic flow contributes to the high resolution of capillary electrophoresis. Any effect that decreases uniformity creates band broadening and decreases resolution. The flow of ions in the capillary generates heat (called *Joule heating*) at a rate of I^2R joules per second, where I is current (A) and R is the resistance of the solution (ohms) (Section 14-1). Most of the capillary in Figure 26-19 is in a thermostated compartment to control the temperature inside the capillary.[40] Typically, the centerline of the capillary channel is 0.02 to 0.3 K hotter than the edge of the channel. Lower viscosity in the warmer region disturbs the flat electroosmotic profile of the fluid. Joule heating is not a serious problem in a capillary tube with a diameter of 50 μm, but the temperature gradient would be prohibitive if the diameter were $\geq$100 μm.

The capillary must be thin enough to dissipate heat rapidly. Temperature gradients disturb the mobility and reduce resolution.

Mobility

The *apparent* (or observed) *mobility*, μ_{app}, of an ion is the sum of the electrophoretic mobility of the ion plus the electroosmotic mobility of the solution.

Apparent mobility:
$$\mu_{app} = \mu_{ep} + \mu_{eo} \qquad (26\text{-}9)$$

For an analyte *cation* moving in the same direction as the electroosmotic flow, μ_{ep} and μ_{eo} have the same sign, so μ_{app} is greater than μ_{ep}. Electrophoresis transports *anions* in the opposite direction from electroosmosis (Figure 26-21b), so for anions the two terms in Equation 26-9 have opposite signs. At neutral or high pH, brisk electroosmosis transports anions to the *cathode* because electroosmosis is usually faster than electrophoresis. At low pH, electroosmosis is weak and anions may never reach the detector. If you want to separate anions at low pH, you can reverse the polarity to make the sample end negative and the detector end positive.

The apparent mobility, μ_{app}, of a particular species is the net speed, u_{net}, of the species divided by the electric field, E:

Apparent mobility:
$$\mu_{app} = \frac{u_{net}}{E} = \frac{L_d/t}{V/L_t} \qquad (26\text{-}10)$$

$$\text{Speed} = \frac{\text{distance to detector}}{\text{migration time}} = \frac{L_d}{t}$$

$$\text{Electric field} = \frac{\text{applied voltage}}{\text{capillary length}} = \frac{V}{L_t}$$

where L_d is the length of column from injection to the detector, L_t is the total length of the column from end to end, V is the voltage applied between the two ends, and t is the time required for solute to migrate from the injection end to the detector. Electroosmotic flow is typically measured by adding an ultraviolet-absorbing neutral solute to the sample and measuring its *migration time*, $t_{neutral}$, to the detector.

For quantitative analysis by electrophoresis, *normalized peak areas* are required. The normalized peak area is the measured peak area divided by the migration time. In chromatography, each analyte passes through the detector at the same rate, so peak area is proportional to the quantity of analyte. In electrophoresis, analytes with different apparent mobilities pass through the detector at different rates. The higher the apparent mobility, the shorter the migration time and the less time the analyte spends in the detector. To correct for time spent in the detector, divide the peak area for each analyte by its migration time.

Electroosmotic mobility is the speed of the neutral species, $u_{neutral}$, divided by the electric field:

Electroosmotic mobility:
$$\mu_{eo} = \frac{u_{neutral}}{E} = \frac{L_d/t_{neutral}}{V/L_t} \qquad \textbf{(26-11)}$$

The *electrophoretic mobility* of an analyte is the difference $\mu_{app} - \mu_{eo}$.

For maximum precision, mobilities are measured relative to an internal standard. Absolute variation from run to run should not affect relative mobilities, unless there are time-dependent (nonequilibrium) interactions of the solute with the wall.

For molecules of similar size, the magnitude of the electrophoretic mobility increases with charge. If a molecule is an acid or base, its charge depends on pH.[41] The ionized form has a charge and thus an electrophoretic mobility, but its neutral form has no charge and so no electrophoretic mobility. The kinetics of acid dissociation equilibria are extremely fast compared with the time scale of capillary electrophoresis separations. Therefore the electrophoretic mobility is governed by the average charge of the acid or base. For a weak acid

Observed electrophoretic mobility:
$$\mu_{ep\ obs} = (\alpha_{A^-})(\mu_{ep\ A^-}) \qquad \textbf{(26-12)}$$

In Figure 26-23, benzoic acid ($pK_a = 4.20$) has an average charge of -0.5, and is separated from 2-methylbenzoic acid ($pK_a = 3.90$), whose average charge is -0.67. The migration time depends on the apparent mobility (Equation 26-9).

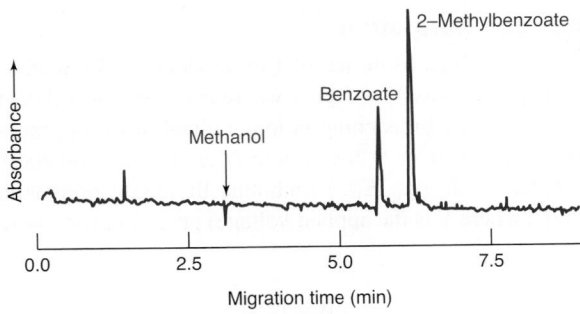

Migration time (min)

FIGURE 26-23 Capillary electrophoresis separation of substituted benzoic acids prepared by a student using the Grignard reaction. Electrophoresis carried out at pH 4.20 (pK_a of benzoic acid) using 5 mM acetate buffer at 2.50×10^4 V in a capillary with a length of 0.500 m and a distance to the detector of 0.400 m. The negative baseline deflection from methanol in the sample was used to measure the electroosmotic flow. [Data from N. S. Mills, J. D. Spence, and M. M. Bushey, "Capillary Electrophoresis of Substituted Benzoic Acids," *J. Chem. Ed.* **2005,** *82,* 1226.]

EXAMPLE **Mobility Calculations**

In Figure 26-23 the migration speed of substituted benzoic acids results from the combined effects of electroosmotic flow and electrophoretic mobility. The voltage applied to the 0.500-m-long capillary is 2.50×10^4 V. The neutral marker molecule carried by electroosmotic flow requires 188 s to travel 0.400 m from the inlet to the detector. Migration times of benzoate and 2-methylbenzoate are 340 s and 371 s, respectively. Find the electroosmotic velocity and electroosmotic mobility. Find the apparent and electrophoretic mobilities of benzoate and 2-methylbenzoate.

Side notes (right column):

For quantitative analysis, use

$$\text{normalized peak area} = \frac{\text{peak area}}{\text{migration time}}$$

$$\alpha_{A^-} = \frac{[A^-]}{[HA] + [A^-]}$$

Normal order of elution in capillary zone electrophoresis:

1. cations (highest mobility first)

2. all neutrals (unseparated)

3. anions (highest mobility last)

Solution Electroosmotic velocity, u_{eo}, is found from the migration time of the neutral marker:

$$\text{Electroosmotic velocity} = \frac{\text{distance to detector }(L_d)}{\text{migration time}} = \frac{0.400 \text{ m}}{188 \text{ s}} = 0.002\ 13 \text{ m/s}$$

Electric field is the voltage divided by the total length, L_t, of the column: $E = 2.50 \times 10^4$ V/ 0.500 m = 5.00×10^4 V/m. Mobility is the constant of proportionality between velocity and electric field:

$$u_{eo} = \mu_{eo}E \Rightarrow \mu_{eo} = \frac{u_{eo}}{E} = \frac{0.002\ 13 \text{ m/s}}{5.00 \times 10^4 \text{ V/m}} = 4.26 \times 10^{-8}\frac{m^2}{V \cdot s}$$

The mobility of the neutral marker, which we just calculated, is the electroosmotic mobility for the entire solution.

The apparent mobility of benzoate is obtained from its migration time:

$$\mu_{app} = \frac{u_{net}}{E} = \frac{0.400 \text{ m/340 s}}{5.00 \times 10^4 \text{ V/m}} = 2.35 \times 10^{-8}\frac{m^2}{V \cdot s}$$

Electrophoretic mobility describes the response of the ion to the electric field. Subtract electroosmotic mobility from apparent mobility to find electrophoretic mobility:

$$\mu_{app} = \mu_{ep} + \mu_{eo} \Rightarrow \mu_{ep} = \mu_{app} - \mu_{eo} = (2.35 - 4.26) \times 10^{-8}\frac{m^2}{V \cdot s} = -1.91 \times 10^{-8}\frac{m^2}{V \cdot s}$$

The electrophoretic mobility is negative because benzoate has a negative charge and migrates in the opposite direction from electroosmotic flow. Electroosmotic flow at pH 4.2 is faster than electromigration, so benzoate is carried to the detector. Similar calculations for 2-methylbenzoate give $\mu_{app} = 2.16 \times 10^{-8}$ m²/(V · s) and $\mu_{ep} = -2.10 \times 10^{-8}$ m²/(V · s). The electrophoretic mobility of 2-methylbenzoate is more negative than that of benzoate because it has a greater average negative charge.

TEST YOURSELF If electroosmotic mobility were 3.00×10^{-8} m²/(V · s), what would be the migration times of neutral marker and benzoate? (*Answer:* 267 s, 734 s)

Theoretical Plates and Resolution

Consider a capillary of length L_d from the inlet to the detector. In Section 23-4, we defined the number of theoretical plates as $N = L_d^2/\sigma^2$, where σ is the standard deviation of the band. If the only mechanism of zone broadening is longitudinal diffusion, the standard deviation was given by Equation 23-28: $\sigma = \sqrt{2Dt}$, where D is the diffusion coefficient and t is the migration time ($= L_d/u_{net} = L_d/[\mu_{app}E]$). Combining these equations with the definition of electric field ($E = V/L_t$, where V is the applied voltage) gives an expression for the number of plates:

Number of plates:
$$N = \frac{\mu_{app}V}{2D}\frac{L_d}{L_t} \tag{26-13}$$

How many theoretical plates might we hope to attain? Using a typical value of $\mu_{app} = 2 \times 10^{-8}$ m²/(V · s) (derived for a 10-min migration time in a capillary with $L_t = 60$ cm, $L_d = 50$ cm, and 25 kV) and using diffusion coefficients from Table 23-1, we find

For K^+:
$$N = \frac{[2 \times 10^{-8} \text{ m}^2/(V \cdot s)][25\ 000 \text{ V}]}{2(2 \times 10^{-9} \text{ m}^2/\text{s})}\frac{0.50 \text{ m}}{0.60 \text{ m}} = 1.0 \times 10^5 \text{ plates}$$

For serum albumin:
$$N = \frac{[2 \times 10^{-8} \text{ m}^2/(V \cdot s)][25\ 000 \text{ V}]}{2(0.059 \times 10^{-9} \text{ m}^2/\text{s})}\frac{0.50 \text{ m}}{0.60 \text{ m}} = 3.5 \times 10^6 \text{ plates}$$

For the small, rapidly diffusing K^+ ion, we expect 100 000 plates. For the slowly diffusing protein serum albumin (66 000 Da), we expect more than 3 million plates. High plate count means that bands are very narrow and resolution between adjacent bands is excellent.

In reality, additional sources of zone broadening include the finite width of the injected band (Equation 23-35), a parabolic mobility profile from heating inside the capillary, adsorption of solute on the capillary wall (which acts as a stationary phase), the finite length of the

---

Number of plates: $N = \dfrac{L_d^2}{\sigma^2}$

L_d = distance to detector
σ = standard deviation of Gaussian band
L_t = total length of column

Under special conditions in which a reverse hydrodynamic flow was imposed to slow the passage of analytes through the capillary, up to 17 million plates were observed in the separation of small molecules![42]

CHAPTER 26 Chromatographic Methods and Capillary Electrophoresis

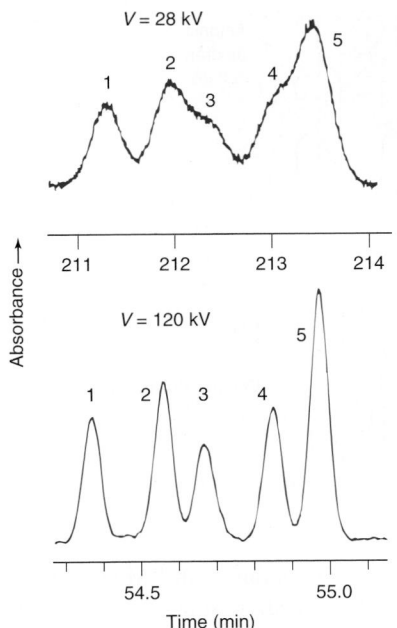

FIGURE 26-24 Small region from the electropherogram of a complex mixture shows that increasing voltage increases resolution. All conditions are the same in both runs except voltage, which is ordinarily limited to ~30 kV. Special precautions were required to prevent electric arcing and overheating at 120 kV. [Data from K. M. Hutterer and J. W. Jorgenson, "Ultrahigh-Voltage Capillary Zone Electrophoresis," *Anal. Chem.* **1999,** *71,* 1293.]

detection zone, and mobility mismatch of solute and buffer ions that leads to nonideal electrophoretic behavior. If these factors are properly controlled, ~10^5 plates are routinely achieved.

Equation 26-13 says that, for constant L_d/L_t, plate count is independent of capillary length. In contrast with chromatography, longer capillaries in electrophoresis do not give higher resolution.

Equation 26-13 also tells us that, the higher the voltage, the greater the number of plates (Figure 26-24). Voltage is ultimately limited by capillary heating, which produces a parabolic temperature profile that gives band broadening. The optimum voltage is found by making an *Ohm's law plot* of current versus voltage with *background electrolyte* (also called *run buffer*) in the capillary. In the absence of overheating, this curve should be a straight line. The maximum allowable voltage is the value at which the curve deviates from linearity (by, say, 5%). Buffer concentration and composition, thermostat temperature, and active cooling all play roles in how much voltage can be tolerated. Up to a point, higher voltage gives better resolution and faster separations.

Resolution between closely spaced peaks A and B in an electropherogram is governed by

Resolution:
$$\text{Resolution} = \frac{\sqrt{N}}{4} \frac{\Delta\mu_{app}}{\overline{\mu}_{app}} \qquad (26\text{-}14)$$

where N is the number of plates, $\Delta\mu_{app}$ is the difference in the apparent mobility of the ions, and $\overline{\mu}_{app}$ is their average apparent mobility. Increasing $\Delta\mu_{app}$ increases separation of peaks, and increasing N decreases their width. Decreasing the average apparent mobility allows more time for peak separation to occur.

> **Background electrolyte** (the solution in the capillary and the electrode reservoirs) controls pH and electrolyte composition in the capillary.

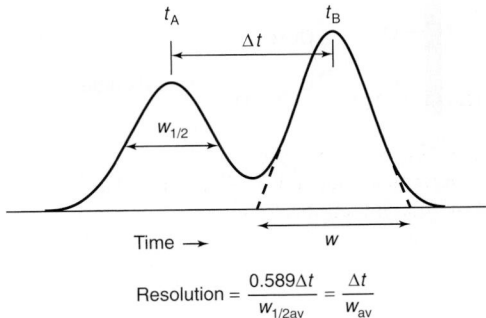

$$\text{Resolution} = \frac{0.589\Delta t}{w_{1/2av}} = \frac{\Delta t}{w_{av}}$$

26-7 Conducting Capillary Electrophoresis

Clever variations of electrophoresis allow us to separate neutral molecules as well as ions, to separate optical isomers, and to lower detection limits by up to 10^6.

Controlling the Environment Inside the Capillary

The inside capillary wall controls electroosmotic velocity and provides undesired adsorption sites for multiply charged molecules, such as proteins. A fused-silica capillary should be prepared for its *first* use by washing for 1 h with 1 M NaOH at a flow rate of ~4 column volumes/min, followed by 1 h with water, followed by 1 h with 6 M HCl, followed by 1 h with run buffer.[43] NaOH is thought to generate Si—OH groups on the silica surface and HCl removes metal ions from the surface. For subsequent use at high pH, wash for 2 min with 0.1 M NaOH, followed by 30 s with deionized water, and then for at least 5 min with run buffer.[44] If the

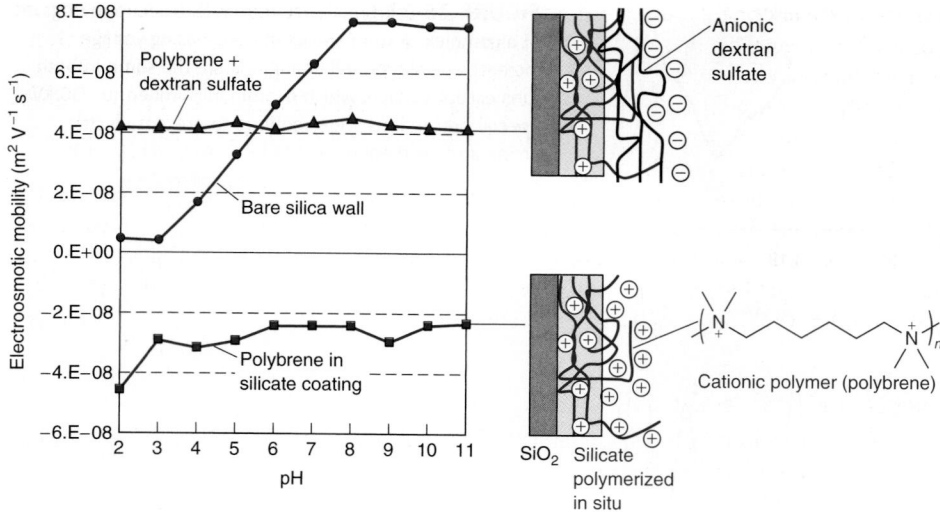

FIGURE 26-25 Effect of wall coating on electroosmotic mobility. Bare silica has little charge below pH 3 and high negative charge above pH 8. Polybrene cation embedded in silicate (lower structure) gives nearly constant positive charge to the wall. Anionic dextran sulfate adsorbed on polybrene (upper structure) gives constant negative charge to the wall.
[Data from M. R. N. Monton, M. Tomita, T. Soga, and Y. Ishihama, "Polymer Entrapment in Polymerized Silicate for Preparing Highly Stable Capillary Coatings for CE and CE–MS," *Anal. Chem.* **2007**, *79*, 7838.]

Foul-smelling 1,4-butanediamine is called putrescine.

Covalent coating helps prevent protein from sticking to a capillary and provides reproducible migration times:

capillary is being run with pH 2.5 phosphate buffer, wash between runs with 1 M phosphoric acid, deionized water, and run buffer.[45] When changing buffers, allow at least 5 min of flow for equilibration. For the pH range 4–6, at which equilibration of the wall with buffer is slow, the capillary needs frequent regeneration with 0.1 M NaOH if migration times become erratic. Buffer in both reservoirs should be replaced periodically because ions become depleted and because electrolysis raises the pH at the cathode and lowers the pH at the anode. The capillary inlet should be ~2 mm away from and below the electrode to minimize entry of electrolytically generated acid or base into the column.[46] Stored capillaries should be filled with distilled water.

Different separations require more or less electroosmotic flow. Small anions with high mobility and highly negatively charged proteins require brisk electroosmotic flow or they will not travel toward the cathode. At pH 3, there is little charge on the silanol groups and little electroosmotic flow. At pH 8, the wall is highly charged and electroosmotic flow is strong. Figure 26-25 shows that electroosmotic mobility in a bare silica capillary is small and positive below pH 3. Mobility increases and reaches a high, steady value above pH 8.

Proteins with many positively charged substituents can bind tightly to negatively charged silica. To control this, 1,4-butanediamine (which gives $^+H_3NCH_2CH_2CH_2CH_2NH_3^+$) may be added to the run buffer to neutralize charge on the wall. Wall charge can be reduced to near 0 by covalent attachment of silanes with neutral, hydrophilic substituents. However, many coatings are unstable under alkaline conditions.

You can reverse the direction of electroosmotic flow by adding a cationic surfactant, such as didodecyldimethylammonium bromide, to the run buffer.[47] This molecule has a positive charge at one end and two long hydrocarbon tails. The surfactant coats the negatively charged silica, with the tails pointing away from the surface (Figure 26-26). A second

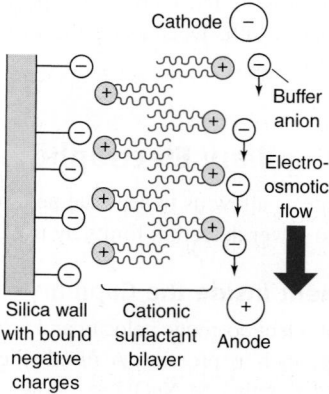

FIGURE 26-26 Charge reversal created by a cationic surfactant bilayer coated on the capillary wall. The diffuse part of the double layer contains excess anions, so electroosmotic flow is opposite that shown in Figure 26-21. The surfactant is didodecyldimethylammonium ion, $(n\text{-}C_{12}H_{25})N(CH_3)_2^+$, represented as ～～(+).

layer of surfactant orients itself in the opposite direction so that the tails form a nonpolar hydrocarbon layer. This *bilayer* adheres tightly to the wall of the capillary and effectively reverses the wall charge from negative to positive. Buffer anion movement creates electro-osmotic flow from cathode to anode when voltage is applied. Electroosmotic flow is in the direction opposite that shown in Figure 26-21. Best results are obtained when the capillary is freshly regenerated for each run.

Figure 26-25 shows a more stable cationic coating formed by embedding the cationic polymer polybrene in a silicate layer formed in situ on the capillary wall. The graph shows that electroosmotic flow is nearly constant in the pH range 3–11 and opposite that of bare silica. A stable, pH-independent negative surface can be made by adsorption of the anionic polymer dextran sulfate on the cationic polybrene surface.

Sample Injection and Composition

Hydrodynamic injection in Figure 26-19 uses pressure to force sample into the capillary. For hydrodynamic injection, the injected volume is

For quantitative analysis, it is critical to use an internal standard because the amount of sample injected into the capillary is not reproducible.

Hydrodynamic injection: $$\text{Volume} = \frac{\Delta P \pi d^4 t}{128 \eta L_t} \qquad \text{(26-15)}$$

where ΔP is the pressure difference between the ends of the capillary, d is the capillary inner diameter, t is injection time, η is sample viscosity, and L_t is the total length of the capillary.

EXAMPLE **Hydrodynamic Injection Time**

How much time is required to inject a sample equal to 2.0% of the length of a 50-cm capillary if the diameter is 50 μm and the pressure difference is 2.0×10^4 Pa (0.20 bar)? Assume that the viscosity is 0.001 0 kg/(m · s), which is close to the viscosity of water.

Solution The injection plug will be 1.0 cm long and occupy a volume of $\pi r^2 \times \text{length} = \pi (25 \times 10^{-6} \text{ m})^2 (1.0 \times 10^{-2} \text{ m}) = 1.9_6 \times 10^{-11} \text{ m}^3$. The required time is

$$t = \frac{128 \eta L_t (\text{volume})}{\Delta P \pi d^4} = \frac{128[0.001 \ 0 \text{ kg/(m · s)}](0.50 \text{ m})(1.9_6 \times 10^{-11} \text{m}^3)}{(2.0 \times 10^4 \text{ Pa}) \pi (50 \times 10^{-6} \text{m})^4} = 3.2 \text{ s}$$

The units work out when we realize that Pa = force/area = $(\text{kg · m/s}^2)/\text{m}^2 = \text{kg/(m · s}^2)$.

TEST YOURSELF How much time would be required to inject a 1.0-cm-long sample with twice the viscosity of water into a 40-cm-long column at the same ΔP? (*Answer:* 5.1 s)

Electrokinetic injection uses the electric field to drive sample into the capillary. The capillary is dipped into the sample and a voltage is applied between the ends of the capillary. The moles of each ion taken into the capillary in t seconds are

Electrokinetic injection: $$\text{Moles injected} = \mu_{app} \underbrace{\left(E \frac{\kappa_b}{\kappa_s} \right)}_{\text{Effective electric field} \equiv E_{eff}} t \pi r^2 C \qquad \text{(26-16)}$$

where μ_{app} is the apparent mobility of analyte ($= \mu_{ep} + \mu_{eo}$), E is the applied electric field (V/m), r is capillary radius, C is sample concentration (mol/m^3), and κ_b / κ_s is the ratio of conductivities of buffer and sample. Each analyte has a different mobility, so the injected sample does not have the same composition as the original sample. Electrokinetic injection is most useful for capillary gel electrophoresis (described later), in which liquid in the capillary is too viscous for hydrodynamic injection.

EXAMPLE **Electrokinetic Injection Time**

How much time is required to inject a sample equal to 2.0% of the length of a 50-cm capillary if the diameter is 50 μm and the injection electric field is 10 kV/m? Assume that the sample has 1/10 of the conductivity of background electrolyte and $\mu_{app} = 2.0 \times 10^{-8}$ m^2/(V · s).

Solution The factor κ_b/κ_s in Equation 26-16 is 10 in this case. The length of sample plug injected onto the column is (sample speed) × (time) = $\mu_{app}E_{eff}t$. The desired injection plug will be 10 mm long. The required time is

$$t = \frac{\text{plug length}}{\text{speed}} = \frac{\text{plug length}}{\mu_{app}\left(E\dfrac{\kappa_b}{\kappa_s}\right)} = \frac{0.010 \text{ m}}{[2.0 \times 10^{-8} \text{ m}^2/(\text{V} \cdot \text{s})](10\,000 \text{ V/m})(10)} = 5.0 \text{ s}$$

Equation 26-16 multiplies the plug length times its cross-sectional area to find its volume and then multiplies by concentration to find the moles in that volume.

TEST YOURSELF What is the effect on injection time if you decrease the applied voltage by a factor of 2? (***Answer:*** injection time is doubled)

Conductivity Effects: Stacking and Skewed Peaks

We choose conditions so that analyte is focused into narrow bands at the start of the capillary by a process called **stacking.** Without stacking, if you inject a zone with a length of 10 mm, no analyte band can be narrower than 10 mm when it reaches the detector.

Stacking depends on the relation between the electric field in the zone of injected sample and that in the background electrolyte on either side of the sample. Optimal buffer concentration in the sample solution is 1/10 of the background electrolyte concentration, and the sample concentration should be <1/500 of the background electrolyte concentration. If sample has a much lower ionic strength than the run buffer, the sample's conductivity is lower and its resistance is much greater. Electric field is inversely proportional to conductivity: the lower the conductivity, the greater the electric field. The electric field across the sample plug inside the capillary is higher than the electric field in the background electrolyte. Figure 26-27 shows ions in the sample plug migrating very fast, because the electric field is very high. When ions reach the zone boundary, they slow down because the field is lower outside the sample plug. This process of *stacking* continues until analyte cations are concentrated at one end of the sample plug and analyte anions are at the other end. The broad injection becomes concentrated into narrow bands of analyte cations or anions. Figure 26-28 shows signal enhancement by stacking. Various methods have been developed to concentrate analytes within the capillary.[48]

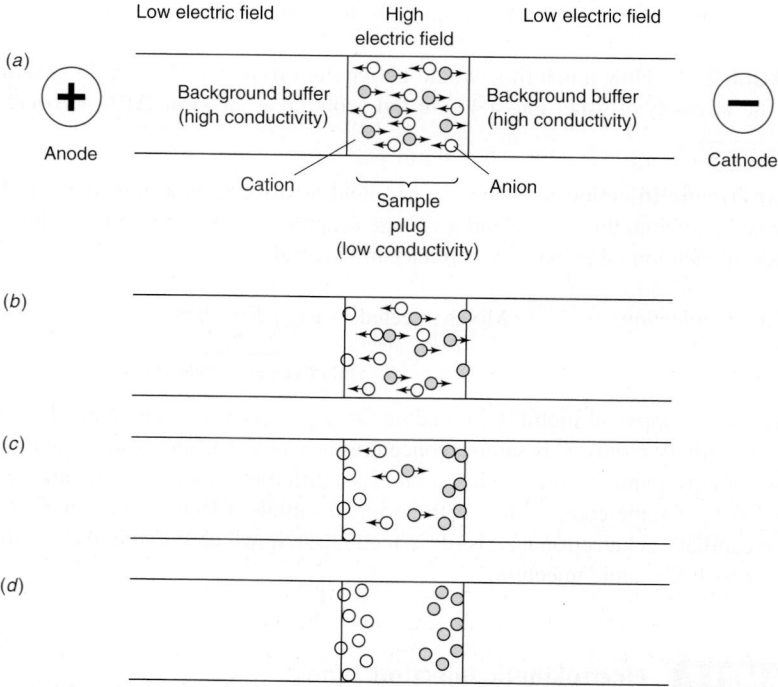

FIGURE 26-27 Stacking of anions and cations at opposite ends of low-conductivity sample plug occurs because the electric field in the sample plug is much higher than the field in background electrolyte. Time increases from panels *a* to *d*. Electroneutrality is maintained by migration of background electrolyte ions, which are not shown.

CHAPTER 26 Chromatographic Methods and Capillary Electrophoresis

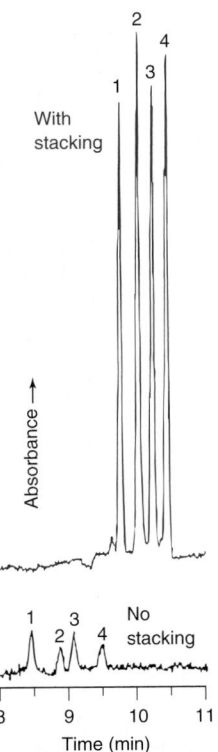

FIGURE 26-28 Lower trace: Sample injected electrokinetically for 2 s without stacking is limited in volume to prevent band broadening. Upper trace: With stacking, 15 times more sample could be injected (for 30 s), so the signal is 15 times stronger with no increase in bandwidth. [Data from Y. Zhao and C. E. Lunte, "pH-Mediated Field Amplification On-Column Preconcentration of Anions in Physiological Samples for Capillary Electrophoresis," *Anal. Chem.* **1999,** *71,* 3985.]

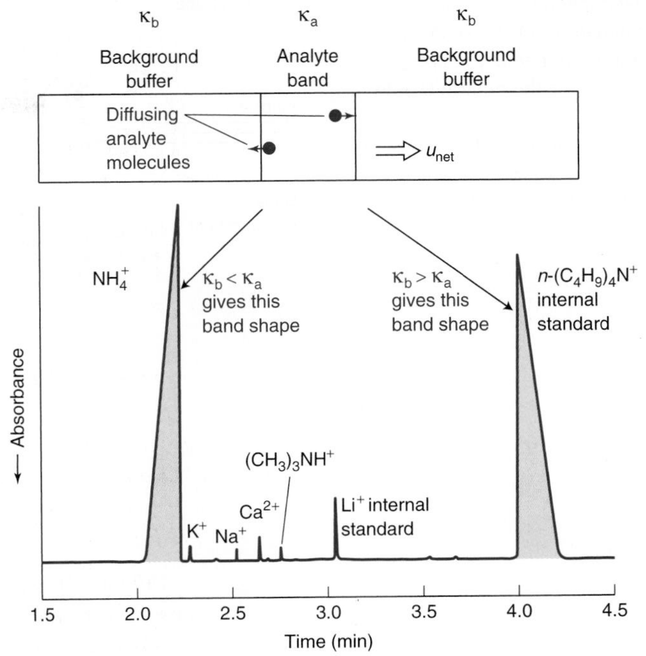

FIGURE 26-29 Irregular peak shapes arise when conductivity of analyte band, κ_a, is not the same as background conductivity, κ_b. Electropherogram shows cations extracted from the surface of a silicon semiconductor wafer. Background electrolyte contains imidazolium ion for *indirect* spectrophotometric detection, whose principle is shown in Figure 26-31. [Data from T. Ehmann, L. Fabry, L. Kotz, and S. Pahlke, "Monitoring of Ionic Contaminants on Silicon Wafer Surfaces Using Capillary Electrophoresis," *Am. Lab.,* June 2002, p. 18.]

If the conductivity of an analyte band differs significantly from the conductivity of background electrolyte, peak distortion occurs. Figure 26-29 shows a band containing one analyte (as opposed to the sample plug in Figure 26-27, which contains all analytes in the entire injection). If background conductivity is greater than analyte conductivity ($\kappa_b > \kappa_a$), the electric field is lower outside the analyte band than inside the band. The band migrates to the right in Figure 26-29. An analyte molecule that diffuses past the front on the right suddenly encounters a lower electric field and it slows down. Soon, the analyte zone catches up with the molecule and it is back in the zone. A molecule that diffuses out of the zone on the left encounters a lower electric field and it also slows down. The analyte zone is moving faster than the wayward molecule and pulls away. This condition leads to a sharp front and a broad tail, as shown in the lower right electropherogram in Figure 26-29. When $\kappa_b < \kappa_a$, we observe the opposite electropherogram.

To minimize band distortion, sample concentration must be much less than the background electrolyte concentration. Otherwise, it is necessary to choose a buffer *co-ion* that has the same mobility as the analyte ion.

Co-ion: buffer ion with same charge as analyte

Counterion: buffer ion with opposite charge from analyte

Detectors

Water is so transparent that *ultraviolet detectors* can operate at wavelengths as short as 185 nm, where most solutes have strong absorption. To take advantage of short-wavelength ultraviolet detection, background electrolyte must have very low absorption. Borate buffers are commonly used in electrophoresis because of their transparency.[49] Sensitivity is poor because the optical pathlength is only as wide as the capillary, which is 25–75 μm.

Fluorescence detection (Color Plate 34) is sensitive to naturally fluorescent analytes or to fluorescent derivatives,[50] with dynamic range up to 10^9.[51] *Contactless conductivity detection* uses electrodes placed outside of the capillary to measure the conductivity difference between the analyte zone and buffer.[52] *Amperometric detection* is sensitive to analytes that can be oxidized or reduced at an electrode (Figure 26-30).[52] *Electrospray mass spectrometry* (Figure 22-26) provides low detection limits and gives information about analyte structure.[53]

Approximate detection limits (M) for direct detection in electrophoresis:

Ultraviolet absorption	10^{-5}–10^{-7}
Laser induced fluorescence	10^{-9}–10^{-12}
Conductivity	10^{-6}–10^{-7}
Amperometry	10^{-10}–10^{-11}
Mass spectrometry	10^{-8}–10^{-9}

Data from H. H. Lauer and G. P. Rozing, *High Performance Capillary Electrophoresis,* 2nd ed. (Agilent Technologies, 2010).

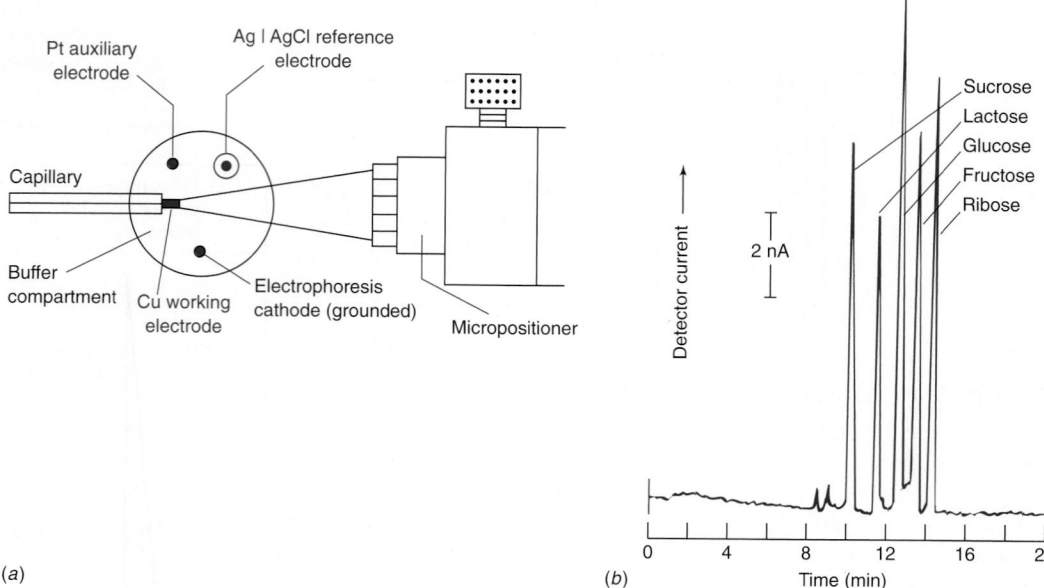

FIGURE 26-30 (a) Amperometric detection with macroscopic working electrode at the outlet of the capillary. (b) Electropherogram of sugars separated in 0.1 M NaOH, in which OH groups are partially ionized, thereby turning the molecules into anions. [Information from J. Ye and R. P. Baldwin, "Amperometric Detection in Capillary Electrophoresis with Normal Size Electrodes," *Anal. Chem.* **1993,** *65,* 3525.]

Detection limits for indirect detection are generally 10–100 times poorer than for direct detection.

Figure 26-31 shows the principle of *indirect detection*, which applies to fluorescence, absorbance, amperometry, conductivity, and other forms of detection. A substance with a steady background signal is added to the background electrolyte. In the analyte band, analyte molecules displace the chromophoric substance, so the detector signal *decreases* when analyte passes by. Figure 26-32 shows an undergraduate's impressive separation of Cl^- isotopes with indirect detection in the presence of the ultraviolet-absorbing anion chromate. Electroneutrality dictates that an analyte band containing Cl^- must have a lower concentration of CrO_4^{2-} than is found in the background electrolyte. With less CrO_4^{2-} to absorb ultraviolet radiation, a negative peak appears when Cl^- reaches the detector. Benzoate and phthalate are other anions useful for this purpose. Detection limits in capillary electrophoresis are generally about an order of magnitude higher than detection limits in ion chromatography, but one to two orders of magnitude lower than detection limits for ion-selective electrodes.

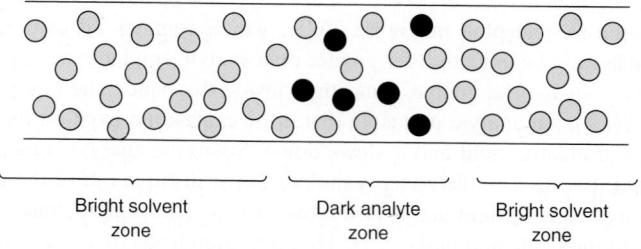

FIGURE 26-31 Principle of indirect detection. When analyte emerges from the capillary, the strong background signal decreases.

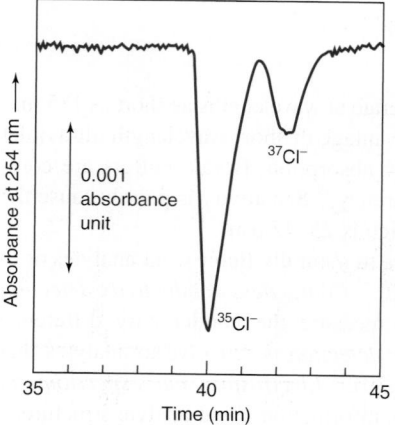

FIGURE 26-32 Separation of natural isotopes of 0.56 mM Cl^- by capillary electrophoresis with indirect spectrophotometric detection at 254 nm. Background electrolyte contains 5 mM CrO_4^{2-} to provide absorbance at 254 nm and 2 mM borate buffer, pH 9.2. The capillary had a diameter of 75 μm, a total length of 47 cm (length to detector = 40 cm) and an applied voltage of 20 kV. The difference in electrophoretic mobility of $^{35}Cl^-$ and $^{37}Cl^-$ is just 0.12%. Conditions were adjusted so that electroosmotic flow was nearly equal and opposite to electrophoretic flow. The resulting near-zero apparent mobilities gave the two isotopes maximum time to be separated by their slightly different mobilities. [Data from C. A. Lucy and T. L. McDonald, "Separation of Chloride Isotopes by Capillary Electrophoresis Based on the Isotope Effect on Ion Mobility," *Anal. Chem.* **1995,** *67,* 1074.]

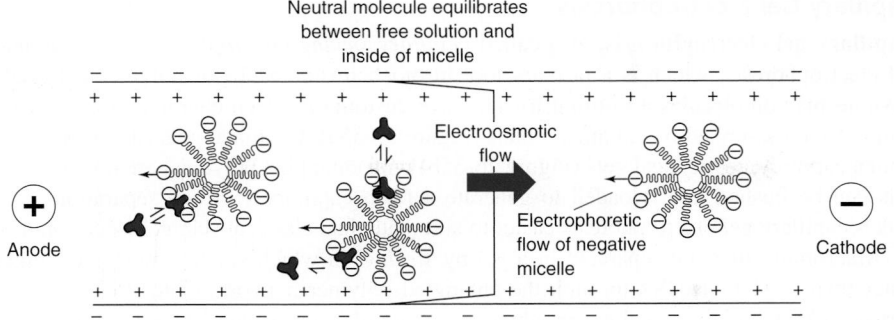

Neutral molecule equilibrates
between free solution and
inside of micelle

Electroosmotic
flow

Anode

Electrophoretic
flow of negative
micelle

Cathode

FIGURE 26-33 Negatively charged sodium dodecyl sulfate micelles migrate against electroosmotic flow. Neutral molecules (solid color) are in dynamic equilibrium between free solution and the inside of the micelle. The more time spent in the micelle, the more the neutral molecule lags behind electroosmotic flow.

Micellar Electrokinetic Chromatography[54]

The type of electrophoresis we have been discussing so far is called **capillary zone electrophoresis.** Separation is based on differences in electrophoretic mobility. If the capillary wall is negative, electroosmotic flow is toward the cathode (Figure 26-21) and the order of elution is cations before neutrals before anions. If the capillary wall charge is reversed by coating it with a cationic surfactant (Figure 26-26) and instrument polarity is reversed, then the order of elution is anions before neutrals before cations. Neither scheme separates neutral molecules from one another.

Micellar electrokinetic chromatography separates neutral molecules and ions. We illustrate a case in which the anionic surfactant sodium dodecyl sulfate is present above its *critical micelle concentration* (Box 26-1), so negatively charged micelles are formed.[56] In Figure 26-33, electroosmotic flow is to the right. Electrophoretic migration of negatively charged micelles is to the left, but net motion is to the right because electroosmotic flow dominates.

In the absence of micelles, all neutral molecules reach the detector in time t_0. Micelles injected with the sample reach the detector in time t_{mc}, which is longer than t_0 because micelles migrate upstream. If a neutral molecule equilibrates between free solution and the inside of the micelles, its migration time is increased, because it migrates at the slower rate of the micelle part of the time. The neutral molecule reaches the detector at a time between t_0 and t_{mc}. *The more time the neutral molecule spends inside the micelle, the longer is its migration time.* Migration times of cations and anions also are affected by micelles, because ions partition between solution and micelles and interact electrostatically with micelles.

Micellar electrokinetic chromatography is a form of chromatography because micelles behave as a *pseudostationary phase*. Micelles migrate in the electric field, but their concentration is constant because they are part of the run buffer. Separation of neutral molecules is based on partitioning between the solution and the pseudostationary phase. The mass transfer term Cu_x is no longer 0 in the van Deemter equation 26-5, but mass transfer into the micelles is fairly fast and band broadening is modest. Typical plate numbers, N, are a few hundred thousand.

Imagination runs wild with the variables in micellar electrokinetic chromatography. We can add anionic, cationic, zwitterionic, and neutral surfactants to change partition coefficients of analytes. (Cationic surfactants also change the charge on the wall and the direction of electroosmotic flow.) We can add solvents such as acetonitrile or *N*-methylformamide to increase the solubility of organic analytes and to change the partition coefficient between the solution and the micelles.[57] We can add cyclodextrins (Box 24-1) to separate optical isomers that spend different fractions of the time associated with the cyclodextrins.[58] In Figure 26-34, chiral micelles were used to separate enantiomers of chiral drugs.

Normal order of elution in capillary zone electrophoresis:

1. cations (highest mobility first)

2. all neutrals (*unseparated*)

3. anions (highest mobility last)

Sodium dodecyl sulfate (n-$C_{12}H_{25}OSO_3^-$ Na^+)

Micellar electrokinetic chromatography: The more time the neutral analyte spends inside the micelle, the longer is its migration time. S. Terabe introduced this technique in 1984.[55]

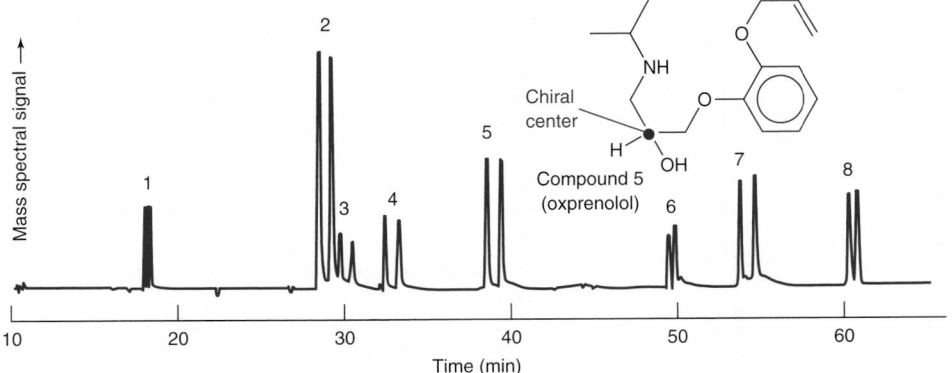

Chiral
center

Compound 5
(oxprenolol)

FIGURE 26-34 Separation of enantiomers of eight β-blocker drugs by micellar electrokinetic chromatography at pH 8.0 in a 120-cm capillary at 30 kV. The structure of one compound is shown. Micelles were formed by a polymer surfactant containing L-leucinate substituents for chiral recognition. [Data from C. Akbay, S. A. A. Rizvi, and S. A. Shamsi, "Simultaneous Enantioseparation and Tandem UV-MS Detection of Eight β-Blockers in Micellar Electrokinetic Chromatography Using a Chiral Molecular Micelle," *Anal. Chem.* **2005,** *77,* 1672.]

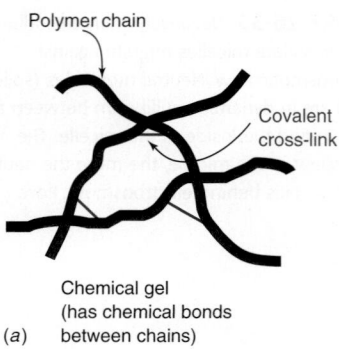

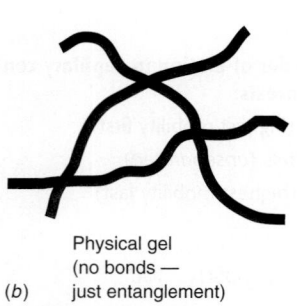

FIGURE 26-35 (a) A chemical gel contains covalent cross-links between different polymer chains. (b) A physical gel is not cross-linked but derives its properties from physical entanglement of the polymers.

Capillary Gel Electrophoresis

Capillary gel electrophoresis, also called *capillary sieving electrophoresis*, is a variant of gel electrophoresis, which is a primary tool in biochemistry. Slabs of polymer gel used to separate macromolecules according to size have customarily been chemical gels, in which chains are cross-linked by chemical bonds (Figure 26-35a). Chemical gels cannot be flushed from a capillary, so physical gels (Figure 26-35b) with entangled polymers are used. Physical gels can be flushed and reloaded to generate a fresh capillary for each separation, which makes capillary gel electrophoresis easier to automate than classical slab gel electrophoresis.

Macromolecules are separated in a gel by *sieving*, in which smaller molecules migrate faster than large molecules through the entangled polymer network. Color Plate 34 shows part of a DNA sequence analysis in which a mixture of fluorescence-labeled fragments with up to 400 nucleotides was separated in 32 minutes in a capillary containing 38 g/L of linear polyacrylamide sieving medium and 6 M urea to stabilize single strands of DNA. Each strand terminating in one of the four bases, A, T, C, or G, is tagged with one of four different fluorescent labels that identify the terminal bases as they pass through a fluorescence detector. Capillary electrophoresis was an enabling technology for determining the sequence of nucleic acids in the human genome.[59] Most DNA forensic analyses and sequencing, as in the opening of this chapter, are done on automated, multicapillary electrophoresis instruments known as *genetic analyzers*.[60]

Biochemists measure molecular mass of proteins by *sodium dodecyl sulfate (SDS)–gel electrophoresis*.[61] Proteins are first *denatured* (unfolded to random coils) by reducing their disulfide bonds (—S—S—) with excess 2-mercaptoethanol ($HSCH_2CH_2OH$) and adding sodium dodecyl sulfate ($C_{12}H_{25}OSO_3^-Na^+$). Dodecyl sulfate anion coats hydrophobic regions and gives the protein a large negative charge that is approximately proportional to the length of the protein. Denatured proteins are then separated by electrophoresis through a sieving gel. Large molecules are retarded more than small molecules, which is the opposite behavior from size exclusion chromatography. In Figure 26-36, the logarithm of molecular mass of the protein is proportional to 1/(migration time). Absolute migration times vary from run to run, so relative migration times are measured. The relative migration time is the migration time of a protein divided by the migration time of a fast-moving small dye molecule.

Method Development

Capillary electrophoresis is not used as much as liquid chromatography. Advantages of electrophoresis relative to chromatography include (1) higher resolution, (2) low waste production, and (3) generally simpler equipment. Drawbacks of electrophoresis include (1) higher limits of detection, (2) run-to-run irreproducibility of migration times, (3) insolubility of some analytes in common electrolyte solutions, and (4) inability to scale up to a preparative separation.

Liquid chromatography is two decades more mature than capillary electrophoresis, and so is often the first method analysts try. However, capillary electrophoresis is gradually becoming more widely used in the pharmaceutical industry, particularly for biopharmaceuticals.[62] For example, capillary electrophoresis is the pharmacopeia (that is, official) method to separate the isoforms erythropoietin, a hormone that increases red blood cell levels and so is

FIGURE 26-36 Calibration curve for protein molecular mass in sodium dodecyl sulfate–capillary gel electrophoresis. On the abscissa, t_{rel}, is the migration time of each protein divided by the migration time of a small dye molecule. Relative migration time (t_{rel}) is more reproducible than migration time. [Data from J. K. Grady, J. Zang, T. M. Laue, P. Arosio, and N. D. Chasteen, "Characterization of the H- and L-Subunit Ratios in Ferritins by Sodium Dodecyl Sulfate–Capillary Gel Electrophoresis," *Anal. Biochem.* **2002,** *302,* 263.]

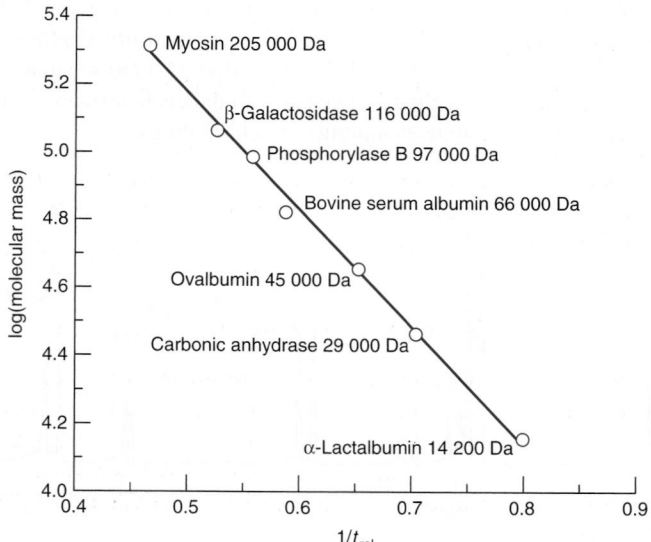

CHAPTER 26 Chromatographic Methods and Capillary Electrophoresis

sometimes used for doping in endurance sports. Electrophoresis has also displaced liquid chromatography as the preferred method for forensic analysis of alkaloids in opium and heroin.[63] The enabling technology for this application was dynamic coating of the capillary between runs to eliminate adsorption of analytes on the silica surface and decrease variations in migration times to less than 0.5%.

In developing a method, it is important to remember that electrophoresis is a fundamentally different separation technique than chromatography. Method development for capillary electrophoresis addresses the following points:[64]

1. Select a detection method that provides the required limit of detection. For ultraviolet absorption, select the optimum wavelength. If necessary, use indirect detection or derivatization.

2. If there is a choice, separate analytes as anions, which do not stick to the negatively charged wall. If separating polycations, such as proteins at low pH, select additives to coat the walls or to reverse the wall charge.

3. Dissolve the entire sample. If the sample is not soluble in dilute aqueous buffer, try adding 6 M urea or surfactants. Acetate buffer tends to dissolve more organic solutes than does phosphate buffer. (If it is necessary to dissolve the sample, nonaqueous solvents can be used.[65] However, the nonaqueous solvent must be compatible with plastic parts in the system. Acetonitrile and methanol are recommended, but electric current should be kept low to minimize outgassing and evaporation. Aqueous micelles or cyclodextrin solve some solubility problems that would otherwise require organic solvent.)

4. A capillary should be dedicated to a single type of analysis. The new capillary should be cut so that its ends are blunt, and conditioned with NaOH and buffer prior to first use.

5. Determine how many peaks are present. Assign each peak by using authentic samples of analytes and diode array ultraviolet detection or mass spectrometry.

6. For complex mixtures, use computerized experimental design to help optimize separation conditions.[66]

7. Use the short end of the capillary in Figure 26-19 for quick scouting runs to determine the direction of migration and whether broad peaks are present. Broad peaks indicate that wall effects are occurring and a coating could be required.

8. See if pH alone provides adequate separation. For acids, begin with 50 mM borate buffer, pH 9.3. For bases, try 50 mM phosphate, pH 2.5. If the separation is not adequate, try adjusting buffer pH close to the average pK_a of the solutes.

9. If pH does not provide adequate separation or if analytes are neutral, use surfactants for micellar electrokinetic capillary chromatography. For chiral solutes, try adding cyclodextrins.

10. Select a between-run capillary rinse procedure. If migration times are reproducible from run to run with no rinse, then no rinse is required. If migration times increase, rinse for 5–10 s with 0.1 M NaOH, followed by 5 min with buffer. If migration times still drift, try increasing or decreasing the NaOH rinse time by a few seconds. If migration times decrease, rinse with 0.1 M H_3PO_4. If proteins or other cations stick to the walls, try rinsing with 0.1 M sodium dodecyl sulfate or use a commercial dynamic coating.

11. If necessary, select a sample cleanup method. Cleanup could be required if resolution is poor, if the salt content is high, or if the capillary fouls. Cleanup might involve solid-phase extraction (Section 28-3), protein precipitation, or dialysis (Demonstration 27-1).

12. If the detection limit is not sufficient, select a stacking method to concentrate analyte in the capillary.

13. For quantitative analysis, determine the linear range required to measure the least concentrated and most concentrated analytes. If desired, select an internal standard. If the migration time or peak area of the standard changes, then you have an indication that some condition has drifted out of control.

26-8 Lab-on-a-Chip: DNA Profiling

The opening of this chapter showed a forensic DNA profile that can be assigned with high likelihood to one person and can be used to exclude other people as the source of a biological sample. A DNA profile can implicate or exonerate a person accused of a crime. Running a profile can take three days, during which a suspect in custody has a good chance of being released. It is highly desirable to automate the process so it can be done in a few hours, while a suspect is still in custody.

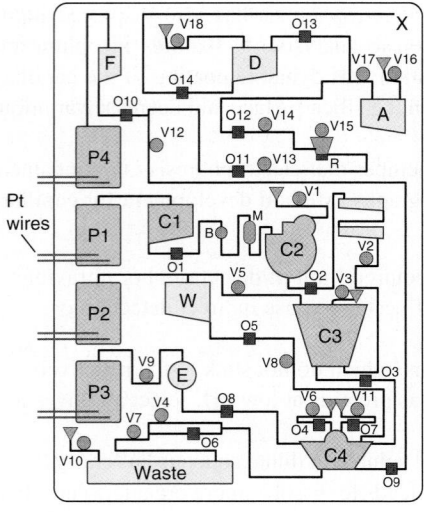

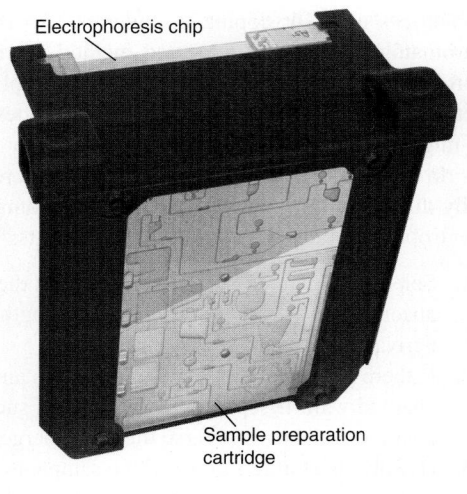

FIGURE 26-37 Microfluidic device for automated forensic DNA analysis developed by the University of Arizona and the British Forensic Science Service. Features labeled V and O in the diagram of the sample preparation cartridge are valves. Other items are mentioned in the text. [Information from F. Zenhausern, University of Arizona. From A. J. Hopwood, C. Hurth, J. Yang, Z. Cai, N. Moran, J. G. Lee-Edghill, A. Nordquist, R. Lenigk, M. D. Estes, J. P. Haley, C. R. McAlister, X. Chen, C. Brooks, S. Smith, K. Elliott, P. Koumi, F. Zenhausern, and G. Tully, "Integrated Microfluidic System for Rapid Forensic DNA Analysis: Sample Collection to DNA Profile." *Anal. Chem.* **2010,** *82,* 6991. Reprinted with permission © 2010 American Chemical Society.]

The European Space Agency ExoMars mission will search for signs of life on Mars with a lab-on-a-chip to measure amines, amino acids, and enantiomers of amino acids.[67] Sub-part-per-trillion detection limits are 1 000 times lower than that of the *Viking* gas chromatography–mass spectrometer, which did not detect organic compounds on Mars in 1976.

DNA profiles and other biological analyses are being handled increasingly by *microfluidic devices*. A device of this type is called "lab-on-a-chip."[68,69] Glass or plastic chips—often the size of a microscope slide—employ electroosmosis or pressure to move liquid with precise control through micrometer-size channels for a variety of applications.[70] Chemical reactions can be conducted by moving microliters or picoliters of fluid from different reservoirs, mixing and heating them, and analyzing products by electrophoresis in a narrow channel etched into glass or polymer. The microfluidic device in Figure 26-37 is the original version developed in 2010 to permit DNA profiling in 2 h. The 23 × 18 cm polycarbonate plastic sample preparation cartridge is designed for a single use to avoid contamination of a sample by previous samples.

In the first step of creating a DNA profile, cells swabbed from inside a person's cheek are *lysed* (broken open) with 1 mL of liquid and centrifuged, and 150 μL of liquid from the centrifuge tube are injected into chamber C1 in Figure 26-37. Pump P1 creates pressure by electrolyzing an aqueous NaCl solution with Pt electrodes to generate H_2 and O_2 gas that pushes liquid from the pump. Upon passage through chambers B, M, C2, and C3, the sample has been mixed with magnetic ion-exchange beads that bind negatively charged DNA. In chamber C4, the beads with their DNA are captured by a magnet. Liquid containing other cellular constituents is washed from the captured DNA into the waste compartment. Pump P2 then sends buffer from chamber W to raise the pH and change the charge of the beads from positive to neutral. DNA is released from the beads and 10 μL of DNA solution are sent to chamber R by pump P3.

In chamber R, the *polymerase chain reaction* (PCR) is conducted.[71] This ingenious reaction, which won a Nobel Prize for Kary Mullis, who invented the process in 1984, *amplifies* (makes many copies of) chosen sections of DNA. The microfluidic system conducts 27 cycles of amplification to make about 10^7 copies of each of 16 *short tandem repeat* sections of DNA found in the human genome. Each cycle of PCR amplification takes 4 min and requires changing the temperature of chamber R to 94°, 59°, and 72°C. Every repetition of the temperature sequence doubles the amount of DNA. Primers used to start each DNA replication are labeled with one of four different fluorescent dyes. Each of the 16 different kinds of DNA that is replicated is uniquely characterized by the combination of the color of its fluorescence and its number of base pairs.

Chamber F contains formamide solvent and a set of DNA standards used to calibrate the rate at which different lengths of DNA migrate in gel electrophoresis. These standards are labeled with a fluorescent dye that can be distinguished from the four dyes used in amplifying the unknown DNA. Pump P4 sends formamide solution from chamber F through chamber R to carry replicated DNA into chamber D, where it is denatured at 95°C. *Denaturation* means that hydrogen bonds are broken and the double helix unwinds into two separate chains. Denatured DNA is pumped from exit X through Teflon capillary tubing onto the glass electrophoresis chip at the top right of Figure 26-37.

The electrophoresis chip has a thin (25-μm deep × 50-μm wide) channel that is 13 cm long etched into glass. During gel electrophoresis on this chip, short strands of DNA migrate more rapidly than long strands. DNA migrating past a fluorescence detector located 11 cm from the injection point is identified by the wavelength of its fluorescence and its migration time. Comparison with the migration times of the internal standards tells us the number of nucleotides in each unknown DNA. The resulting DNA profile, such as that at the opening of this chapter, can be compared with the profile of authentic DNA from a suspect or used to interrogate an international database (such as CODIS) to identify unknown DNA.

Within four years of introducing the device in Figure 26-37, its functions were further automated and improved and the device was made smaller.[72,73] A buccal swab from inside a person's cheek is simply dropped into the analyzer, which does the rest of the job of extracting, amplifying, and analyzing DNA without human intervention. The device's footprint was made four times smaller by building it in six layers instead of one. A device of the size in Figure 26-37 can process four samples in parallel instead of one sample.

DNA evidence is good at excluding innocent people, but sloppy laboratory work could incriminate the wrong person. Forensic analysis requires meticulous execution without preconception of who is guilty. The automated microfluidic device not only reduces the time for DNA profiling, but also removes many sources of human error from the results.

Terms to Understand

affinity chromatography	electroosmosis	indirect detection	molecular exclusion
anion exchanger	electrophoresis	interstitial volume	chromatography
capillary electrophoresis	equivalent	ion chromatography	molecularly imprinted polymer
capillary gel electrophoresis	gel	ion-exchange capacity	preconcentration
capillary zone electrophoresis	gel filtration	ion-exchange chromatography	resin
cation exchanger	gradient elution	ion-pair chromatography	selectivity coefficient
cross-linking	hydrodynamic injection	micellar electrokinetic	speciation
deionized water	hydrophobic interaction	chromatography	stacking
Donnan exclusion	chromatography	micelle	suppressed-ion chromatography
electrokinetic injection	hydrophobic substance	mobility	surfactant

Summary

Ion-exchange chromatography employs resins and gels with covalently bound charged groups that attract solute counterions (and that exclude ions having the same charge as the resin). Ion-exchange binding increases with increased ion charge and polarizability, and it depends inversely on the radius of the hydrated ion. Ion binding results in release of an equivalent ionic charge from the resin. Polystyrene resins are useful for small ions. Greater cross-linking of the resin increases the capacity, selectivity, and time needed for equilibration. Ion-exchange gels based on cellulose and dextran have large pore sizes and low charge densities, suitable for the separation of macromolecules. Certain inorganic solids have ion-exchange properties and are useful at extremes of temperature or radiation. Ion exchangers operate by the principle of mass action, with a gradient of increasing ionic strength most commonly used to effect a separation. Zwitterionic hydrophilic interaction chromatography can separate anions and cations on the same column.

In suppressed-ion chromatography, a separator column separates ions of interest, and a suppressor membrane converts eluent into a nonionic form so that analytes can be detected by their electrical conductivity. Eluent and suppressor can be generated continuously by electrolysis. Alternatively, nonsuppressed ion chromatography uses an ion-exchange column and low-concentration eluent. Ion-pair chromatography employs an ionic surfactant in the eluent to make a reversed-phase column function as an ion-exchange column.

Molecular exclusion chromatography is used for separations based on size and for molecular mass determinations of macromolecules. Molecular exclusion is based on the inability of large molecules to enter pores in the stationary phase. Small molecules enter these pores and therefore exhibit longer elution times than large molecules. In affinity chromatography, the stationary phase retains one particular solute in a complex mixture. After all other components have been eluted, the desired species is liberated by a change in conditions. In hydrophobic interaction chromatography, high concentrations of ammonium sulfate induce proteins to adhere to a hydrophobic stationary phase. A gradient of decreasing salt concentration is applied to increase the solubility of proteins in water and to elute them from the column.

In capillary zone electrophoresis, ions are separated by differences in mobility in a strong electric field applied between the ends of a silica capillary tube. The greater the charge and the smaller the hydrodynamic radius, the greater the electrophoretic mobility. Normally, the capillary wall is negative, and solution is transported from anode to cathode by electroosmosis of cations in the electric double layer. Solute cations arrive first, followed by neutral species, followed by solute anions (if electroosmosis is stronger than electrophoresis, which typically is true). Apparent mobility is the sum of electrophoretic mobility and electroosmotic mobility (which is the same for all species). Zone dispersion (broadening) arises mainly from longitudinal diffusion and the finite length of the injected sample. Stacking of solute ions in the capillary occurs when the sample has a low conductivity. Electroosmotic flow is reduced at low pH because surface $Si—O^-$ groups are protonated. $Si—O^-$ groups can be masked by polyamine cations, and the wall charge can be reversed by a cationic surfactant that forms a bilayer along the wall. Covalent coatings reduce electroosmosis and wall adsorption. Hydrodynamic sample injection uses pressure; electrokinetic injection uses an electric field. Ultraviolet absorbance is commonly used for detection. Micellar electrophoretic chromatography uses micelles as a pseudostationary phase to separate neutral molecules and ions. Capillary gel electrophoresis separates macromolecules by sieving. In contrast with molecular exclusion chromatography, small molecules move fastest in gel electrophoresis. Microfluidic devices ("lab-on-a-chip") use electroosmotic or hydrodynamic flow in lithographically fabricated channels to conduct chemical reactions and chemical analysis.

Exercises

26-A. *Total salt experiment:* Inorganic cations can be quantified by passing a salt solution through a cation-exchange column in the H^+ form and titrating the released H^+ with strong base. The moles of OH^- titrant equal the equivalents of cation charge in solution.

(a) What volume of 0.023 1 M NaOH is needed to titrate the eluate when 10.00 mL of 0.045 8 M KNO_3 have been loaded on a cation exchange column in the H^+ form?

(b) A 0.269 2 g sample of unknown salt is dissolved in deionized water and loaded onto a cation exchange column in the H^+ form. The column is rinsed with deionized water and the combined loading and rinse solutions are titrated with 0.139 6 M KOH. A volume of 30.64 mL is required to reach the endpoint. How many equivalents of cation are in the sample? Find the milliequivalents of cation per gram of sample (meq/g).

(c) The mass of substance containing one equivalent is called the *equivalent mass*. If the cation has a +1 charge, the equivalent mass equals the molar mass. If the cation has a +2 charge, the equivalent mass is half the molar mass. What is the equivalent mass of the sample in **(b)**?

26-B. Vanadyl sulfate ($VOSO_4$, FM 163.00), as supplied commercially, is contaminated with H_2SO_4 (FM 98.08) and H_2O. A solution was prepared by dissolving 0.244 7 g of impure $VOSO_4$ in 50.0 mL of water. Spectrophotometric analysis indicated that the concentration of the blue VO^{2+} ion was 0.024 3 M. A 5.00-mL sample was passed through a cation-exchange column loaded with H^+. When

VO^{2+} from the 5.00-mL sample became bound to the column, the H^+ released required 13.03 mL of 0.022 74 M NaOH for titration. Find the weight percent of each component ($VOSO_4$, H_2SO_4, and H_2O) in the vanadyl sulfate.

26-C. Blue Dextran 2000 was eluted during gel filtration in a volume of 36.4 mL from a 40 × 2-cm (length × diameter) column of Sephadex G-50, which fractionates molecules in the molecular mass range 1 500–30 000.

(a) At what retention volume would hemoglobin (molecular mass 64 000) be expected?

(b) Suppose that radioactive $^{22}NaCl$, which is not adsorbed on the column, is eluted in a volume of 109.8 mL. What would be the retention volume of a molecule with $K_{av} = 0.65$?

26-D. Consider a capillary electrophoresis experiment conducted near pH 9, at which the electroosmotic flow is stronger than the electrophoretic flow.

(a) Draw a picture of the capillary, showing the placement of the anode, cathode, injector, and detector. Show the direction of electroosmotic flow and the direction of electrophoretic flow of a cation and an anion. Show the direction of net flow.

(b) Using Table 15-1, explain why Cl^- has a shorter migration time than I^-. Predict whether Br^- will have a shorter migration time than Cl^- or a greater migration time than I^-.

(c) Why is the mobility of I^- greater than that of Cl^-?

Problems

Ion-Exchange and Ion Chromatography

26-1. State the effects of increasing cross-linking on an ion-exchange column.

26-2. The exchange capacity of an ion-exchange resin can be defined as the number of moles of charged sites per gram of dry resin. Describe how you would measure the exchange capacity of an anion-exchange resin by using standard NaOH, standard HCl, or any other reagent you wish.

26-3. What does the designation 200/400 mesh mean on a bottle of chromatography stationary phase? What is the size range of such particles? (See Table 28-2.) Which particles are smaller, 100/200 or 200/400 mesh?

26-4. Consider the separation of inorganic and organic anions in Figure 26-8.

(a) What is the probable charge (X^{n-}) of pyruvate (peak 10), 2-oxovalerate (peak 16), and maleate (peak 28)? (Hint: Look at the ions around the peaks in question.)

(b) Iodide (peak 44) is a −1 ion. Explain its strong retention.

26-5. Consider a negatively charged protein adsorbed on an anion-exchange gel at pH 8.

(a) How will a gradient of eluent pH (from pH 8 to some lower pH) be useful for eluting the protein? Assume that the ionic strength of the eluent is kept constant.

(b) How would a gradient of ionic strength (at constant pH) be useful for eluting the protein?

26-6. Propose a scheme for separating trimethylamine, dimethylamine, methylamine, and ammonia from one another by ion-exchange chromatography.

26-7. What is deionized water? What kinds of impurities are not removed by deionization?

26-8. Low iron concentration (as low as 0.02 nM) in the open ocean limits phytoplankton growth. Preconcentration is required to determine such low concentrations. Trace Fe^{3+} from a large volume of seawater is concentrated onto a 1.2-mL chelating resin column. The column is then rinsed with 30 mL of high-purity water and eluted with 10 mL of 1.5 M high-purity HNO_3.

(a) For each sample, seawater is passed through the column for 17 hours at 10 mL/min. How much is the concentration of Fe^{3+} in the 10 mL of HNO_3 eluate increased by this preconcentration procedure?

(b) What is the concentration of Fe^{3+} in the seawater when 57 nM Fe^{3+} is found in the nitric acid eluate?

(c) Reagent-grade concentrated nitric acid is 15.7 M and contains ≤0.2 ppm iron. What would be the apparent concentration of Fe (nM) in a seawater blank if reagent-grade acid were used to prepare the 1.5 M HNO_3 eluent?

26-9. *Material balance.* If you intend to measure all the anions and cations in an unknown, one sanity check on your results is that the total positive charge should equal the total negative charge. The table lists concentrations of anions and cations measured in pond water in an undergraduate experiment and expressed in μg/mL.

Find the total concentration of negative and positive charge (mol/L) to assess the quality of the analysis. What do you conclude about this analysis?

Anion	μg/mL	Cation	μg/mL
F^-	0.26	Na^+	2.8
Cl^-	43.6	NH_4^+	0.2
NO_3^-	5.5	K^+	3.5
SO_4^{2-}	12.6	Mg^{2+}	7.3
		Ca^{2+}	24.0

*Data from K. Sinniah and K. Piers, "Ion Chromatography: Analysis of Ions in Pond Water," J. Chem. Ed. **2001**, 78, 358.*

26-10. In ion-exclusion chromatography, ions are separated from nonelectrolytes by an ion-exchange column. Nonelectrolytes penetrate the stationary phase, whereas ions of the same charge as the resin are repelled by the fixed charges. Because co-ions have access to less of the column volume, electrolytes are eluted before nonelectrolytes. The chromatogram shows the separation of trichloroacetic acid (TCA, $pK_a = -0.5$), dichloroacetic acid (DCA, $pK_a = 1.1$), and monochloroacetic acid (MCA, $pK_a = 2.86$) by passage through a cation-exchange resin eluted with 0.01 M HCl. Explain why the three acids are separated and why they emerge in the order shown.

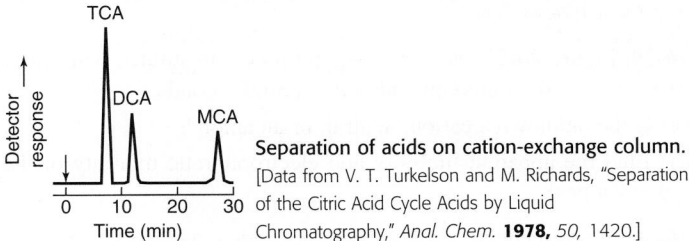

Separation of acids on cation-exchange column.
[Data from V. T. Turkelson and M. Richards, "Separation of the Citric Acid Cycle Acids by Liquid Chromatography," *Anal. Chem.* **1978**, 50, 1420.]

26-11. (a) Explain how anions and cations can be separated by hydrophilic interaction chromatography in Figure 26-4, using the zwitterionic stationary phase in Figure 25-16. Why does the gradient go from high CH_3CN to low CH_3CN content?

(b) Explain how the charged aerosol detector used in Figure 26-4 works.

26-12. State the purpose of the separator and suppressor in suppressed-ion chromatography. For cation chromatography, why is the suppressor an anion-exchange membrane?

26-13. (a) Suppose that the reservoir in Figure 26-7 contains 1.5 L of 2.0 M K_2PO_4. For how many hours can the reservoir provide 20 mM KOH at a flow rate of 1.0 mL/min if 75% consumption of K^+ in the reservoir is feasible?

(b) What starting and ending current would be required to produce a gradient from 5.0 mM KOH to 0.10 M KOH at 1.0 mL/min flow rate?

26-14. The system in Figure 26-7 can be adapted to produce the strong acid eluent methanesulfonic acid ($CH_3SO_3^-H^+$). For this purpose, the polarity of the electrodes is reversed and the reservoir contains $NH_4^+ CH_3SO_3^-$. The barrier membrane and the resin bed at the bottom of the figure must both be anion-exchangers loaded with $CH_3SO_3^-$. Draw this system and write the chemistry that occurs in each part.

26-15. (a) The suppressor in Figure 26-9 enables low parts-per-billion conductivity detection for anions such as Cl^- and Br^-, but very poor detection limits for anions such as CN^- and borate. Explain why.

(b) Mixtures of sodium carbonate and bicarbonate can be used as eluent in suppressed-ion anion chromatography. Detection limits are poorer than with hydroxide eluent due to a higher background conductivity. Explain why.

26-16. Decomposition of dithionite ($S_2O_4^{2-}$) was studied by chromatography on an anion-exchange column eluted with 20 mM trisodium 1,3,6-naphthalenetrisulfonate in 90 vol% H_2O/10 vol% CH_3CN with ultraviolet detection at 280 nm. A solution of sodium dithionite stored for 34 days in the absence of air gave five peaks identified as SO_3^{2-}, SO_4^{2-}, $S_2O_3^{2-}$, $S_2O_4^{2-}$, and $S_2O_5^{2-}$. All the peaks had a *negative* absorbance. Explain why.

26-17. Norepinephrine (NE) in human urine can be assayed by ion-pair chromatography by using an octadecylsilane stationary phase and sodium octyl sulfate as the mobile-phase additive. Electrochemical detection (oxidation at 0.65 V versus Ag|AgCl) is used, with 2,3-dihydroxybenzylamine (DHBA) as internal standard.

Norepinephrine cation (NE)   2,3-Dihydroxybenzylamine cation (DHBA)

$$CH_3CH_2CH_2CH_2CH_2CH_2CH_2CH_2OSO_3^- \, Na^+$$
Sodium octyl sulfate

(a) Explain the physical mechanism by which an ion-pair separation works.

(b) A urine sample containing an unknown amount of NE and a fixed, added concentration of DHBA gave a detector peak height ratio NE/DHBA = 0.298. Then small standard additions of NE were made, with the following results:

Added concentration of NE (ng/mL)	Peak height ratio NE/DHBA
12	0.414
24	0.554
36	0.664
48	0.792

Using the graphical treatment in Section 5-3, find the original concentration of NE in the specimen.

Molecular Exclusion, Affinity, and Hydrophobic Interaction Chromatography

26-18. Would molecular exclusion, affinity, or hydrophobic interaction chromatography be most appropriate for each of the following applications?

(a) Purifying and concentrating a crude mixture of an antibody.

(b) Desalting a solution containing a 30 kDa protein.

(c) Finding the molecular mass distribution of polystyrene with 15 kDa average mass.

(d) Separation of cytochrome *c* (12 400 Da) and ribonuclease A (12 600 Da). Cytochrome *c* has lower surface hydrophobicity than ribonuclease A.

26-19. (a) How can molecular exclusion chromatography be used to measure the molecular mass of a protein?

(b) Which pore size in Figure 26-15 is most suitable for chromatography of molecules with molecular mass near 100 000?

26-20. A gel-filtration column has a radius, r, of 0.80 cm and a length, l, of 20.0 cm.

(a) Calculate the volume, V_t, of the column, which is equal to $\pi r^2 l$.

(b) The interstitial volume, V_o, was 18.1 mL and the total volume of mobile phase was 35.8 mL. Find K_{av} for a solute eluted at 27.4 mL.

26-21. Ferritin (molecular mass 450 000), transferrin (molecular mass 80 000), and ferric citrate were separated by molecular exclusion chromatography on Bio-Gel P-300. The column had a length of 37 cm and a 1.5-cm diameter. Eluate fractions of 0.65 mL were collected. The maximum of each peak came at the following fractions: ferritin, 22; transferrin, 32; and ferric citrate, 84. (That is, the ferritin peak came at an elution volume of $22 \times 0.65 = 14.3$ mL.) Assuming that ferritin is eluted at the interstitial volume and that ferric citrate is eluted at V_m, find K_{av} for transferrin.

26-22. (a) The interstitial volume, V_o, in Figure 26-15 is the volume at which the curves rise vertically at the left. Find V_o for the 25-nm-pore-size column. To the nearest order of magnitude, what is the smallest molecular mass of molecules excluded from this column?

(b) What is the molecular mass of molecules eluted at 9.7 mL from the 12.5-nm column?

(c) V_m is the volume at which the curves drop vertically at the right. Find the largest molecular mass that can freely enter the 45-nm pores.

26-23. A polystyrene resin molecular exclusion HPLC column has a diameter of 7.8 mm and a length of 30 cm. The solid portions of the gel particles occupy 20% of the volume, the pores occupy 40%, and the volume between particles occupies 40%.

(a) At what volume would totally excluded molecules be expected to emerge?

(b) At what volume would the smallest molecules be expected?

(c) A mixture of polyethylene glycols of various molecular masses is eluted between 23 and 27 mL. What does this imply about the retention mechanism for these solutes on the column?

26-24. 🖳 The following substances were separated on a gel filtration column. Estimate the molecular mass of the unknown.

Compound	V_r (mL)	Molecular mass (Da)
Blue Dextran 2000	17.7	2×10^6
Aldolase	35.6	158 000
Catalase	32.3	210 000
Ferritin	28.6	440 000
Thyroglobulin	25.1	669 000
Unknown	30.3	?

26-25. In the separation of proteins by hydrophobic interaction chromatography, why does eluent strength increase with *decreasing* salt concentration in the aqueous eluent?

Capillary Electrophoresis

26-26. Electrophoretic mobility of the anionic form A^- of the weak acid phenol (HA = C_6H_5OH) and its derivatives are:

Analyte HA	pK_a	μ_{A^-} (m²/(V·s))
Phenol	9.98	-2.99×10^{-8}
4-Methylphenol	10.27	-2.59×10^{-8}
4-Ethylphenol	10.22	-2.39×10^{-8}
2,4,5-Trichlorophenol	6.83	-2.85×10^{-8}

Data from S. C. Smith and M. G. Khaledi, "Optimization of pH for the Separation of Organic Acids in Capillary Zone Electrophoresis," Anal. Chem. *1993*, 65, 193.

(a) Explain the trend in electrophoretic mobility from phenol to 4-methylphenol to 4-ethylphenol.

(b) Predict the electrophoretic mobility of the analytes at pH 10.0. Explain why the predicted mobility differs from μ_{A^-}.

(c) The electroosmotic mobility is toward the cathode and greater in magnitude than the analyte electrophoretic mobilities. In what order will the peaks appear in the electropherogram at pH 10.0?

26-27. What is electroosmosis?

26-28. Electroosmotic velocities of buffered solutions are shown for a bare silica capillary and one with aminopropyl groups (silica—Si—$CH_2CH_2CH_2NH_2$) covalently attached to the wall. A positive sign means that flow is toward the cathode. Explain the signs and relative magnitudes of the velocities.

Capillary wall	Electroosmotic velocity (mm/s) for $E = 4.0 \times 10^4$ V/m	
	pH 10	pH 2.5
Bare silica	+3.1	+0.2
Aminopropyl-modified silica	+1.8	-1.3

Data from K. Emoto, J. M. Harris, and M. Van Alstine, "Grafting Poly(ethylene glycol) Epoxide to Amino-Derivatized Quartz: Effect of Temperature and pH on Grafting Density," Anal. Chem. *1996*, 68, 3751.

26-29. Figure 26-23 shows the separation of substituted benzoates. There is a peak of unknown identity at 86.0 seconds.

(a) Is the unknown a cation, neutral, or an anion?

(b) Find the apparent mobility and electrophoretic mobility of the unknown peak.

26-30. Fluorescent derivatives of amino acids separated by capillary zone electrophoresis had migration times with the following order: arginine (fastest) < phenylalanine < asparagine < serine < glycine (slowest). Explain why arginine has the shortest migration time.

26-31. What is the principal source of zone broadening in ideal capillary electrophoresis?

26-32. Consider the electrophoresis of heparin in Figure 26-18.

(a) Electrophoresis was carried out at pH 2.8, at which sulfate groups are negative. Why was reverse polarity (detector end positive) used?

(b) The ionic strength of 30 mg/mL heparin samples is higher than that of typical samples for electrophoresis. What is the benefit of high buffer concentration (0.6 M phosphate)?

(c) A narrow capillary (25-μm diameter) was chosen to be compatible with the high-ionic-strength buffer. What is the advantage of the narrow capillary?

(d) Li^+ has lower mobility than Na^+. Explain why lithium phosphate can be used at higher electric field strength than sodium phosphate to generate the same current. What is the advantage of higher field strength for this separation?

26-33. State three different methods to reduce electroosmotic flow. Why does the direction of electroosmotic flow change when a silica capillary is washed with a cationic surfactant?

26-34. Explain how neutral molecules can be separated by micellar electrokinetic chromatography. Why is this a form of chromatography? Why are micelles called a pseudostationary phase?

26-35. (a) What pressure difference is required to inject a sample equal to 1.0% of the length of a 60.0-cm capillary in 4.0 s if the diameter is 50 μm? Assume that the viscosity of the solution is 0.001 0 kg/(m · s).

(b) Injection in some homebuilt capillary electrophoresis instruments is performed by raising the sample vial to create a siphon. The pressure exerted by a column of water of height h is $h\rho g$, where ρ is the density of water and g is the acceleration of gravity (9.8 m/s^2). To what height would you need to raise the sample vial to create the necessary pressure to load the sample in 4.0 s? Is it possible to raise the inlet of this column to this height? How could you obtain the desired pressure?

26-36. (a) How many moles of analyte are present in a 10.0-μM solution that occupies 1.0% of the length of a 60.0 cm × 25 μm capillary?

(b) What voltage is required to inject this many moles into a capillary in 4.0 s if the sample has 1/10 of the conductivity of background electrolyte, $\mu_{app} = 3.0 \times 10^{-8}$ m^2/(V · s), and the sample concentration is 10.0 μM?

26-37. Measure the number of plates for the electrophoretic peak in Figure 26-20. Use Equation 23-32 for asymmetric peaks to find the number of plates for the chromatographic peak.

26-38. (a) A long thin molecule has a greater friction coefficient than a short fat molecule. Predict whether fumarate or maleate will have greater electrophoretic mobility.

Fumarate   Maleate

(b) Electrophoresis is run with the injection end positive and the detection end negative. At pH 8.5, both anions have a charge of −2. The electroosmotic flow from the positive terminal to the negative terminal is greater than the electrophoretic flow, so these two anions have a net migration from the positive to the negative end of the capillary in electrophoresis. From your answer to part (a), predict the order of elution of these two species.

(c) At pH 4.0, both anions have a charge close to −1, and the electroosmotic flow is weak. Therefore electrophoresis is run with the injection end negative and the detection end positive. The anions migrate from the negative end of the capillary to the positive end. Predict the order of elution.

26-39. (a) A particular solution in a particular capillary has an electroosmotic mobility of 1.3×10^{-8} m^2/(V · s) at pH 2 and 8.1×10^{-8} m^2/(V · s) at pH 12. How long will it take a neutral solute to travel 52 cm from the injector to the detector if 27 kV is applied across the 62-cm-long capillary tube at pH 2? At pH 12?

(b) An analyte anion has an electrophoretic mobility of -1.6×10^{-8} m^2/(V · s). How long will it take to reach the detector at pH 2? At pH 12?

26-40. Figure 26-24 shows the effect on resolution of increasing voltage from 28 to 120 kV.

(a) What is the expected ratio of migration times ($t_{120\ kV}/t_{28\ kV}$) in the two experiments? Measure the migration times for peak 1 and find the observed ratio.

(b) What is the expected ratio of plates ($N_{120\ kV}/N_{28\ kV}$) in the two experiments?

(c) What is the expected ratio of bandwidths ($\sigma_{120\ kV}/\sigma_{28\ kV}$)?

(d) What is the physical reason why increasing voltage decreases bandwidth and increases resolution?

26-41. The observed behavior of benzyl alcohol ($C_6H_5CH_2OH$) in capillary electrophoresis is given here. Draw a graph showing the number of plates versus the electric field and explain what happens as the field increases.

Electric field (V/m)	Number of plates
6 400	38 000
12 700	78 000
19 000	96 000
25 500	124 000
31 700	124 000
38 000	96 000

26-42. Measure the width of the $^{35}Cl^-$ peak at half-height in Figure 26-32 and calculate the plate number. The capillary length to detector was 40.0 cm. Find the plate height.

26-43. The migration time for Cl^- in a capillary zone electrophoresis experiment is 17.12 min and the migration time for I^- is 17.78 min. From mobilities in Table 15-1, predict the migration time of Br^-. (The observed value is 19.6 min.)

26-44. ▦ *Molecular mass by sodium dodecylsulfate–gel electrophoresis.* Ferritin is a hollow iron-storage protein[74] consisting of 24 subunits that are a variable mixture of heavy (H) and light (L) chains, arranged in octahedral symmetry. The hollow center, with a diameter of 8 nm, can hold up to 4 500 iron atoms in the approximate form of the mineral ferrihydrite ($5Fe_2O_3 \cdot 9H_2O$). Iron(II) enters the protein through eight pores located on the threefold symmetry axes of the octahedron. Oxidation to Fe(III) occurs at catalytic sites on the H chains. Other sites on the inside of the L chains appear to nucleate the crystallization of ferrihydrite.

Migration times for protein standards and the ferritin subunits are given in the table. Prepare a graph of log(molecular mass) versus 1/(relative migration time), where relative migration time = (migration time)/(migration time of marker dye). Compute the molecular mass of the ferritin light and heavy chains. The masses of the chains, computed from amino acid sequences, are 19 766 and 21 099 Da.

Protein	Molecular mass (Da)	Migration time (min)
Orange G marker dye	small	13.17
α-Lactalbumin	14 200	16.46
Carbonic anhydrase	29 000	18.66
Ovalbumin	45 000	20.16
Bovine serum albumin	66 000	22.36
Phosphorylase B	97 000	23.56
β-Galactosidase	116 000	24.97
Myosin	205 000	28.25
Ferritin light chain		17.07
Ferritin heavy chain		17.97

SOURCE: J. K. Grady, J. Zang, T. M. Laue, P. Arosio, and N. D. Chasteen, "Characterization of the H- and L-Subunit Ratios in Ferritins by Sodium Dodecyl Sulfate–Capillary Gel Electrophoresis," Anal. Biochem. **2002**, 302, 263.

26-45. *Resolution.* Suppose that the electroosmotic mobility of a solution is $+1.61 \times 10^{-7}$ m^2/(V · s). How many plates are required to separate sulfate from bromide with a resolution of 2.0? Use Table 15-1 for mobilities and Equation 26-14 for resolution.

26-46. The water-soluble vitamins niacinamide (a neutral compound), riboflavin (a neutral compound), niacin (an anion), and thiamine (a cation) were separated by micellar electrokinetic chromatography in 15 mM borate buffer (pH 8.0) with 50 mM sodium dodecyl sulfate. The migration times were niacinamide (8.1 min), riboflavin (13.0 min), niacin (14.3 min), and thiamine (21.9 min). What would the order have been in the absence of sodium dodecyl sulfate? Which compound is most soluble in the micelles?

26-47. When the following three compounds are separated by micellar electrokinetic chromatography at pH 9.6, three peaks are observed. When 10 mM α-cyclodextrin is added to the run buffer, two of the three peaks split into two peaks, giving a total of five peaks. Explain this observation and predict which compound does not split.

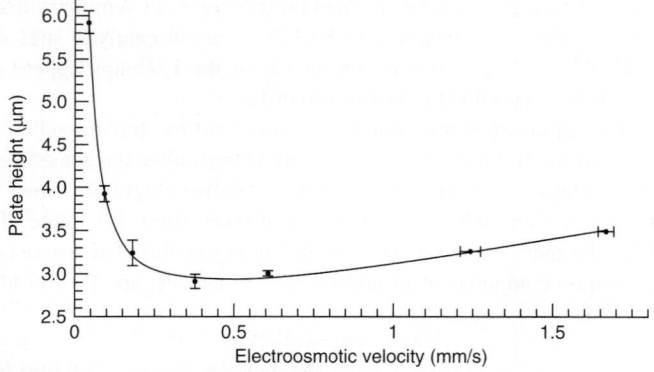

1	2	3
Cyclobarbital	Thiopental	Phenobarbital

26-48. Consider the van Deemter plot for the separation of neutral dyes by micellar electrokinetic chromatography:[75]

(a) Explain why plate height increases at low and high velocities.

(b) The irregular flow path term, A, in the van Deemter equation should really be 0 for the ideal case of micellar electrokinetic chromatography. The observed value of A is 2.32 μm, which accounts for two-thirds of the band broadening at the optimum velocity. Suggest some reasons why A is not 0.

26-49. To obtain the best separation of two weak acids in capillary electrophoresis, it makes sense to use the pH at which their charge difference is greatest. Prepare a spreadsheet to examine the charges of malonic and phthalic acid as a function of pH. At what pH is the difference greatest?

26-50. (a) *Ion mobility spectrometry* (Section 22-7) is *gas-phase electrophoresis*. Describe how ion mobility spectrometry works and state the analogies between this technique and capillary electrophoresis.

(b) As in electrophoresis, the velocity, u, of a gas-phase ion is $u = \mu E$, where μ is the mobility of the ion and E is the electric field ($E = V/L$, where V is the voltage difference across distance L). In ion mobility spectrometry, the time to go from the gate to the detector (Figure 22-40a) is called *drift time*, t_d. Drift time is related to voltage: $t_d = L/u = L/(\mu E) = L/(\mu(V/L)) = L^2/(\mu V)$. Plate number is $N = 5.55(t_d/w_{1/2})^2$, where $w_{1/2}$ is the width of the peak at half-height. Ideally, peak width depends only on the width of the gate pulse that admits ions to the drift tube and on diffusive broadening of ions while they migrate:[76]

$$w_{1/2}^2 = t_g^2 + \left(\frac{16kT\ln 2}{Vez}\right)t_d^2$$

Initial width   Diffusive
from gate pulse   broadening

where t_g is the time that the ion gate is open, k is Boltzmann's constant, T is temperature, V is the potential difference from the gate to the detector, e is the elementary charge, and z is the charge of the ion. Prepare a graph of N versus V ($0 \le V \le 20\,000$) for an ion with $\mu = 8 \times 10^{-5}$ m^2/(V · s), and $t_g = 0$, 0.05, or 0.2 ms at 300 K. Let the length of the drift region be $L = 0.2$ m. Explain the shapes of the curves. What is the disadvantage of using short t_g?

(c) Why does decreasing T increase N?

(d) In a well-optimized ion mobility spectrometer, protonated arginine ion ($z = 1$) had a drift time of 24.925 ms and $w_{1/2} = 0.154$ ms at 300 K. Find N. For $V = 12\,500$ V and $t_g = 0.05$ ms, what is the theoretical plate number?

(e) In a well-optimized ion mobility spectrometer with a 10-cm drift tube and 200 V/cm field strength, protonated leucine had $t_d = 22.5$ ms and protonated isoleucine had $t_d = 22.0$ ms. Both had $N \approx 80\,000$. What is the resolution of the two peaks?

27 Gravimetric and Combustion Analysis

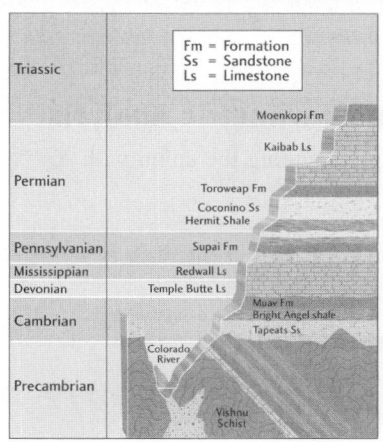

Layers of rock exposed in the Grand Canyon by the erosive action of the Colorado River provide a window on a billion years of Earth's history. [Left: Information from F. Press, R. Siever, J. Gratzinger, and T. H. Jordan. *Understanding Earth*, 4th ed. (New York: W. H. Freeman and Company, 2004). Right: Carol Polich.]

In the 1800s, geologists understood that new layers (*strata*) of rock are deposited on top of older layers. Characteristic fossils in each layer helped geologists to identify strata from the same geologic era all around the world. However, the actual age of each layer was unknown.

Ernest Rutherford, Frederick Soddy, Bertram Boltwood, and Robert Strutt showed in the early 1900s that uranium decays to lead plus eight atoms of helium with a half-life of several billion years. Rutherford estimated the age of a rock from its U and He content. Boltwood obtained more reliable ages of minerals by measuring the U and Pb content.

In 1910, Arthur Holmes, a 20-year-old student of Strutt at Imperial College in London, became the first person to assign ages to minerals formed in specific geological periods. Holmes conjectured that, when certain U-containing minerals crystallized from hot magma, the crystals would be relatively free of impurities such as Pb. Once the mineral solidified, Pb would begin to accumulate. The ratio Pb/U would tell how long ago the mineral crystallized. Holmes measured U by the rate of production of radioactive Rn gas. To measure Pb, he dissolved each mineral in molten borax, dissolved the fused mass in acid, and quantitatively precipitated milligrams of $PbSO_4$. The nearly constant ratio Pb/U = 0.045 g/g in 15 minerals from the Devonian period was consistent with the hypotheses that Pb is the end product of U decay and that little Pb had been present when the minerals crystallized. The calculated age of the Devonian minerals from the Devonian period was 370 million years—four times older than the most accepted age of the Earth at that time.

Geologic ages deduced by Holmes in 1911

Geologic period	Pb/U (g/g)	Millions of years	Today's accepted value
Carboniferous	0.041	340	330–362
Devonian	0.045	370	362–380
Silurian	0.053	430	418–443
Precambrian	0.125–0.20	1 025–1 640	900–2 500

SOURCE: C. Lewis, *The Dating Game* (Cambridge: Cambridge University Press, 2000); A. Holmes, "The Association of Lead with Uranium in Rock-Minerals, and Its Application to the Measurement of Geological Time," Proc. R. Soc. Lond. A **1911**, 85, 248.

In **gravimetric analysis,** the mass of a product is used to calculate the quantity of the original analyte (the species being analyzed). Careful gravimetric analysis by T. W. Richards and his colleagues early in the twentieth century determined the atomic masses of Ag, Cl, and N to six-figure accuracy.[1] This Nobel Prize–winning work allowed the accurate determination of atomic masses of many elements. In **combustion analysis,** a sample is burned in excess oxygen and products such as CO_2 and H_2O are measured. Combustion is typically used to measure C, H, N, S, and halogens in organic matter. To measure other elements, organic matter is burned in a closed system. Products and *ash* (solid residue) are then dissolved in acid or base and measured by inductively coupled plasma with atomic emission or mass spectrometry.

27-1 An Example of Gravimetric Analysis

Chloride can be measured by precipitating the anion with Ag^+ and finding the mass of AgCl.

$$Ag^+ + Cl^- \rightarrow AgCl(s) \qquad (27\text{-}1)$$

EXAMPLE A Gravimetric Calculation

A 10.00-mL solution containing Cl^- was treated with excess $AgNO_3$ to precipitate 0.436 8 g of AgCl. What was the molarity of Cl^- in the unknown?

Solution The formula mass of AgCl is 143.321. Precipitate weighing 0.436 8 g contains

$$\frac{0.436\ 8\ \text{g AgCl}}{143.321\ \text{g AgCl/mol AgCl}} = 3.048 \times 10^{-3}\ \text{mol AgCl}$$

Because 1 mol of AgCl contains 1 mol of Cl^-, there must have been 3.048×10^{-3} mol of Cl^- in the unknown.

$$[Cl^-] = \frac{3.048 \times 10^{-3}\ \text{mol}}{0.010\ 00\ \text{L}} = 0.304\ 8\ \text{M}$$

TEST YOURSELF How many grams of Br^- were in a sample that produced 1.000 g of AgBr precipitate (FM = 187.77)? (***Answer:*** 0.425 5 g)

EXAMPLE Marie Curie's Measurement of the Atomic Mass of Radium

In her Ph.D. research (*Radioactive Substances*, 1903), Marie Curie measured the atomic mass of radium, which she discovered.[2] From its chemical properties, she knew that radium is in the same family as barium, so the formula of radium chloride is $RaCl_2$. When 0.091 92 g of pure $RaCl_2$ was dissolved and treated with excess $AgNO_3$, 0.088 90 g of AgCl precipitated. How many moles of Cl^- were in the $RaCl_2$? From this measurement, find the atomic mass of Ra.

Solution AgCl precipitate weighing 0.088 90 g contains

$$\frac{0.088\ 90\ \text{g AgCl}}{143.321\ \text{g AgCl/mol AgCl}} = 6.202_9 \times 10^{-4}\ \text{mol AgCl}$$

Because 1 mol of AgCl contains 1 mol of Cl^-, there must have been $6.202_9 \times 10^{-4}$ mol of Cl^- in the $RaCl_2$. For 2 mol of Cl, there must be 1 mol of Ra, so

$$\text{mol radium} = \frac{6.202_9 \times 10^{-4}\ \text{mol Cl}}{2\ \text{mol Cl/mol Ra}} = 3.101_4 \times 10^{-4}\ \text{mol}$$

Let the formula mass of $RaCl_2$ be x. We found that 0.091 92 g $RaCl_2$ contains $3.101_4 \times 10^{-4}$ mol $RaCl_2$. Therefore

$$3.101_4 \times 10^{-4} \text{ mol } RaCl_2 = \frac{0.091\ 92 \text{ g } RaCl_2}{x \text{ g } RaCl_2/\text{mol } RaCl_2}$$

$$x = \frac{0.091\ 92 \text{ g } RaCl_2}{3.101_4 \times 10^{-4} \text{ mol } RaCl_2} = 296.3_8 \text{ g/mol}$$

The atomic mass of Cl is 35.452, so the formula mass of $RaCl_2$ is

Formula mass of $RaCl_2$ = atomic mass of Ra + 2(35.452 g/mol) = 296.3_8 g/mol
$$\Rightarrow \text{atomic mass of Ra} = 225.5 \text{ g/mol}$$

The inside cover of this book lists the atomic number (the integer mass) of the long-lived isotope of Ra, which is 226.

TEST YOURSELF How many grams of AgBr would have been formed from 0.100 g of $RaBr_2$? (*Answer:* 0.097 g)

Representative analytical precipitations are listed in Table 27-1. A few common organic **precipitants** (agents that cause precipitation) are listed in Table 27-2. Conditions must be controlled to selectively precipitate one species. Potentially interfering substances may need to be removed prior to analysis.

TABLE 27-1 Representative gravimetric analyses

Species analyzed	Precipitated form	Form weighed	Interfering species
K^+	$KB(C_6H_5)_4$*	$KB(C_6H_5)_4$	NH_4^+, Ag^+, Hg^{2+}, Tl^+, Rb^+, Cs^+
Mg^{2+}	$Mg(NH_4)PO_4 \cdot 6H_2O$	$Mg_2P_2O_7$	Many metals except Na^+ and K^+
Ca^{2+}	$CaC_2O_4 \cdot H_2O$	$CaCO_3$ or CaO	Many metals except Mg^{2+}, Na^+, K^+
Ba^{2+}	$BaSO_4$	$BaSO_4$	Na^+, K^+, Li^+, Ca^{2+}, Al^{3+}, Cr^{3+}, Fe^{3+}, Sr^{2+}, Pb^{2+}, NO_3^-
Ti^{4+}	TiO (5,7-dibromo-8-hydroxyquinoline)$_2$	Same	Fe^{3+}, Zr^{4+}, Cu^{2+}, $C_2O_4^{2-}$, citrate, HF
VO_4^{3-}	Hg_3VO_4	V_2O_5	Cl^-, Br^-, I^-, SO_4^{2-}, CrO_4^{2-}, AsO_4^{3-}, PO_4^{3-}
Cr^{3+}	$PbCrO_4$	$PbCrO_4$	Ag^+, NH_4^+
Mn^{2+}	$Mn(NH_4)PO_4 \cdot H_2O$	$Mn_2P_2O_7$	Many metals
Fe^{3+}	$Fe(HCO_2)_3$	Fe_2O_3	Many metals
Co^{2+}	Co(1-nitroso-2-naphthoate)$_2$	$CoSO_4$ (by reaction with H_2SO_4)	Fe^{3+}, Pd^{2+}, Zr^{4+}
Ni^{2+}	Ni(dimethylglyoximate)$_2$	Same	Pd^{2+}, Pt^{2+}, Bi^{3+}, Au^{3+}
Cu^{2+}	CuSCN (after reduction of Cu^{2+} to Cu^+ with HSO_3^-)	CuSCN	NH_4^+, Pb^{2+}, Hg^{2+}, Ag^+
Zn^{2+}	$Zn(NH_4)PO_4 \cdot H_2O$	$Zn_2P_2O_7$	Many metals
Ce^{4+}	$Ce(IO_3)_4$	CeO_2	Th^{4+}, Ti^{4+}, Zr^{4+}
Al^{3+}	Al(8-hydroxyquinolate)$_3$	Same	Many metals
Sn^{4+}	Sn(cupferron)$_4$	SnO_2	Cu^{2+}, Pb^{2+}, As(III)
Pb^{2+}	$PbSO_4$	$PbSO_4$	Ca^{2+}, Sr^{2+}, Ba^{2+}, Hg^{2+}, Ag^+, HCl, HNO_3
NH_4^+	$NH_4B(C_6H_5)_4$*	$NH_4B(C_6H_5)_4$	K^+, Rb^+, Cs^+
Cl^-	AgCl	AgCl	Br^-, I^-, SCN^-, S^{2-}, $S_2O_3^{2-}$, CN^-
Br^-	AgBr	AgBr	Cl^-, I^-, SCN^-, S^{2-}, $S_2O_3^{2-}$, CN^-
I^-	AgI	AgI	Cl^-, Br^-, SCN^-, S^{2-}, $S_2O_3^{2-}$, CN^-
SCN^-	CuSCN	CuSCN	NH_4^+, Pb^{2+}, Hg^{2+}, Ag^+
CN^-	AgCN	AgCN	Cl^-, Br^-, I^-, SCN^-, S^{2-}, $S_2O_3^{2-}$
F^-	$(C_6H_5)_3SnF$	$(C_6H_5)_3SnF$	Many metals (except alkali metals), SiO_4^{4-}, CO_3^{2-}
ClO_4^-	$KClO_4$	$KClO_4$	
SO_4^{2-}	$BaSO_4$	$BaSO_4$	Na^+, K^+, Li^+, Ca^{2+}, Al^{3+}, Cr^{3+}, Fe^{3+}, Sr^{2+}, Pb^{2+}, NO_3^-
PO_4^{3-}	$Mg(NH_4)PO_4 \cdot 6H_2O$	$Mg_2P_2O_7$	Many metals except Na^+, K^+
NO_3^-	Nitron nitrate	Nitron nitrate	ClO_4^-, I^-, SCN^-, CrO_4^{2-}, ClO_3^-, NO_2^-, Br^-, $C_2O_4^{2-}$
CO_3^{2-}	CO_2 (by acidification)	CO_2	(The liberated CO_2 is trapped with Ascarite and weighed.)

*The solubility of $KB(C_6H_5)_4$ in water is 1.8×10^{-4} M at 25°C and 1.3×10^{-4} M at 0°C. [V. P. Kozitskii, "Solubility of Potassium Tetraphenylborate in Mixtures of Acetone with Water at 0–50° and Its Solubility in Water at 0–97.5°," Bull. Acad. Sci. USSR Div. Chem. Sci., *1972*, 21, 6.] For a gravimetric procedure to measure K^+ in the presence of NH_4^+, see R. M. Engelbrecht and F. A. McCoy, "Determination of Potassium by a Tetraphenylborate Method," Anal. Chem. *1956*, 28, 1772.

TABLE 27-2 Common organic precipitating agents

Name	Structure	Ions precipitated
Dimethylglyoxime		Ni^{2+}, Pd^{2+}, Pt^{2+}
Cupferron		Fe^{3+}, VO_2^+, Ti^{4+}, Zr^{4+}, Ce^{4+}, Ga^{3+}, Sn^{4+}
8-Hydroxyquinoline (oxine)		Mg^{2+}, Zn^{2+}, Cu^{2+}, Cd^{2+}, Pb^{2+}, Al^{3+}, Fe^{3+}, Bi^{3+}, Ga^{3+}, Th^{4+}, Zr^{4+}, UO_2^{2+}, TiO^{2+}
Salicylaldoxime		Cu^{2+}, Pb^{2+}, Bi^{3+}, Zn^{2+}, Ni^{2+}, Pd^{2+}
1-Nitroso-2-naphthol		Co^{2+}, Fe^{3+}, Pd^{2+}, Zr^{4+}
Nitron		NO_3^-, ClO_4^-, BF_4^-, WO_4^{2-}
Sodium tetraphenylborate	$Na^+B(C_6H_5)_4^-$	K^+, Rb^+, Cs^+, NH_4^+, Ag^+, organic ammonium ions
Tetraphenylarsonium chloride	$(C_6H_5)_4As^+Cl^-$	$Cr_2O_7^{2-}$, MnO_4^-, ReO_4^-, MoO_4^{2-}, WO_4^{2-}, ClO_4^-, I_3^-

27-2 Precipitation

The ideal product of a gravimetric analysis should be pure, insoluble, and easily filterable, and should possess a known composition. Few substances meet these requirements, but appropriate techniques can help optimize properties of gravimetric precipitates.

Precipitated particles should not be so small that they clog or pass through the filter. Larger crystals have less surface area to which impurities can become attached. At the other extreme is a *colloidal suspension* of particles that have diameters in the approximate range 1–500 nm and pass through most filters (Figure 27-1 and Demonstration 27-1). Precipitation conditions determine the particle size.

FIGURE 27-1 (*a*) Particle size distribution of colloids formed when $FeSO_4$ was oxidized to Fe^{3+} in 10^{-4} M OH^- in the presence of phosphate (PO_4^{3-}), silicate (SiO_4^{4-}), or no added anions. [Data from M. L. Magnuson, D. A. Lytle, C. M. Frietch, and C. A. Kelty, *Anal. Chem.* **2001,** *73,* 4815.] (*b*) Electron micrograph of 7-nm-diameter colloidal particles of $[Fe(OH)_{\sim2.5}(NO_3)_{\sim0.5}]_{\sim1000}$ made by treating ferric nitrate with $2HCO_3^-$ per Fe^{3+}. [From T. G. Spiro, S. E. Allerton, J. Renner, A. Terzis, R. Bils, and P. Saltman "The Hydrolytic Polymerization of Iron (III)," *J. Am. Chem. Soc.* **1966,** *88,* 2721. Reprinted with permission © 1966, American Chemical Society.]

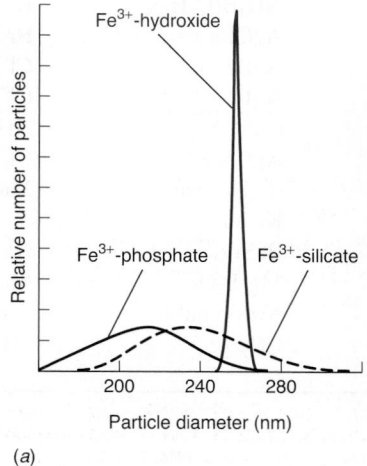

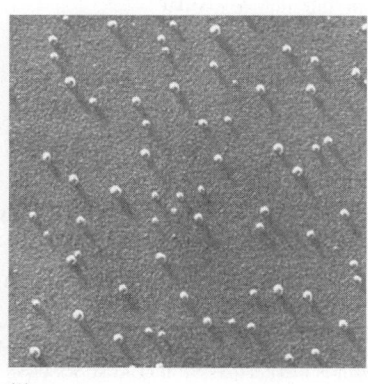

(*a*)　　　(*b*)

Colloids are particles with diameters of ~1–500 nm. They are larger than most molecules but too small to precipitate. They remain in solution, suspended by the *Brownian motion* (random movement) of solvent molecules.[3]

To prepare colloidal iron(III) hydroxide, heat 200 mL of distilled water in a beaker to 70–90°C and leave an identical beaker of water at room temperature. Add 1 mL of 1 M $FeCl_3$ to each beaker and stir. The warm solution turns brown-red in a few seconds, whereas the cold solution remains yellow (Color Plate 35). The yellow color is characteristic of low-molecular-mass Fe^{3+} compounds. The red color results from colloidal aggregates of Fe^{3+} ions held together by hydroxide, oxide, and some chloride ions. Each particle contains ~10^3 atoms of Fe with a molecular mass of 10^5 and a diameter of 10 nm.

You can demonstrate the size of colloidal particles by **dialysis,** a process in which two solutions are separated by a *semipermeable membrane* that has pores with diameters of 1–5 nm.[4] Small molecules diffuse through these pores, but large molecules (such as proteins or colloids) cannot.

Pour some brown-red colloid into a dialysis tube knotted at one end; then tie off the other end. Drop the tube into a flask of distilled water to show that the color remains inside the bag after several days (Color Plate 35). For comparison, leave an identical bag containing dark blue 1 M $CuSO_4 \cdot 5H_2O$ in another flask of water. Cu^{2+} diffuses out and the internal and external solution will be light blue in 24 h. The yellow food coloring, tartrazine, can be used in place of Cu^{2+}. If dialysis is conducted in hot water, it is completed during one class period.[5]

Dialysis is used to treat patients suffering from kidney failure. Blood is passed over a membrane through which metabolic waste products diffuse and are diluted into a large volume of liquid that is discarded. Protein molecules, which are a necessary part of the blood plasma, are too large to cross the membrane and are retained in the blood.

Microdialysis

A *microdialysis probe* is used in biology to sample small molecules in fluids without contamination by large molecules, such as proteins. For example, a probe made of a thin, rigid semipermeable tube can be inserted into the brain of an anesthetized rat to collect neurotransmitter molecules. Fluid pumped through the probe at a rate of 3 μL/min transports small molecules that diffused into the probe. Small molecules in the fluid exiting the probe (*dialysate*) are monitored by liquid chromatography (Chapter 25) or capillary electrophoresis (Chapter 26).

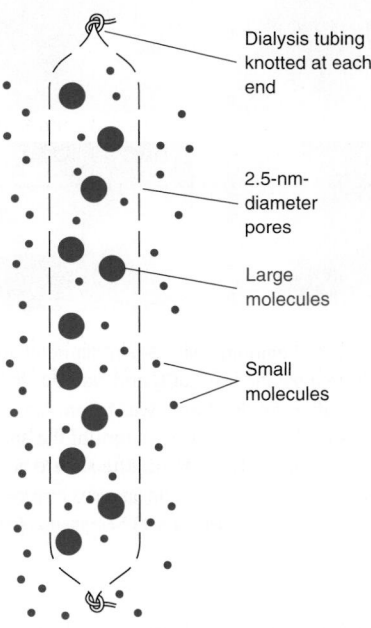

Large molecules remain trapped inside a dialysis bag, whereas small molecules diffuse through the membrane in both directions.

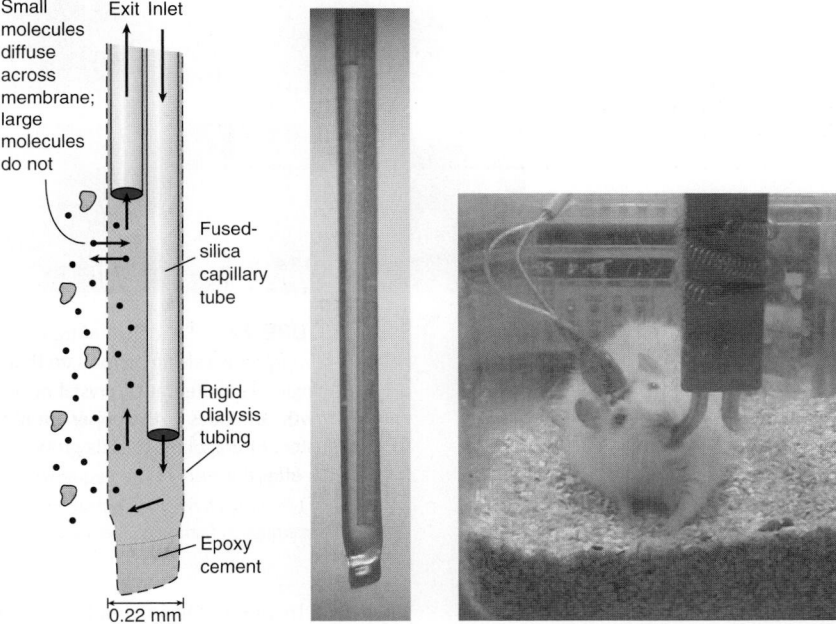

Microdialysis probe. Small molecules pass through the semipermeable membrane, but large molecules cannot. [Courtesy R. T. Kennedy and Z. D. Sandlin, University of Michigan.]

It is thermodynamically favorable for colloidal particles to grow into larger particles because the energy of atoms on the surface of a crystal is higher than the energy of atoms inside the crystal. As a particle grows, the fraction of atoms on the surface decreases. Particles with nanometer size are more soluble than particles with micrometer size. The process in which small particles dissolve and large particles grow is called *Ostwald ripening*.

Crystal Growth

For a short time after a precipitant is added to analyte, the solution contains more dissolved solute than should be present at equilibrium. Such a solution is said to be **supersaturated.** Crystallization then occurs in two stages: nucleation and particle growth. In **nucleation,** solutes form clusters of sufficient size, which then reorganize into an ordered structure capable of growing into larger particles.[6] Nucleation can occur on suspended impurity particles or scratches on a glass surface. When Fe(III) reacts with 0.1 M tetramethylammonium hydroxide at 25°C, nuclei of hydrated Fe(III) oxide (ferrihydrite, $\sim$5FeO(OH) $\cdot$ 2H$_2$O) form that are 4 nm in diameter and contain $\sim$50 Fe atoms.[7] In *particle growth*, molecules, ions, or other nuclei[8] condense onto the nucleus to form a larger crystal. Ferrihydrite transforms into goethite, FeO(OH), and grows into crystalline plates with lateral dimensions of $\sim$30 $\times$ 7 nm after 15 min at 60°C.

A study of calcium carbonate crystallization revealed simultaneous nucleation pathways in which crystals grow from amorphous (noncrystalline) particles and directly from solution or from nuclei that are too small to observe.[9] In *crystallization*, a solid with long-range order is created.

$$Ca^{2+}(aq) + CO_3^{2-}(aq) \rightleftharpoons \underset{\textit{Amorphous}}{(CaCO_3)_n(aq)} \rightleftharpoons \underset{\textit{Crystalline}}{(CaCO_3)_n(s)}$$

Amorphous particles have no crystalline order.

Aragonite: crystalline CaCO$_3$, which is not the most stable form at 25°C

Calcite: most stable crystalline form of CaCO$_3$ at 25°C

The sequence of micrographs in Figure 27-2 begins with a spherical, *amorphous* CaCO$_3$ particle that had grown to a diameter of 3 μm in 95 s after mixing solutions of CaCl$_2$ and NaHCO$_3$. Then *aragonite* CaCO$_3$ crystals begin to grow out of the surface and consume the amorphous ball in $\sim$20 s. Other micrographs from the same study show a more stable *calcite* CaCO$_3$ crystal nucleating on the surface of the less stable aragonite. Calcite grows, while aragonite dissolves.

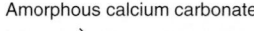
Amorphous calcium carbonate

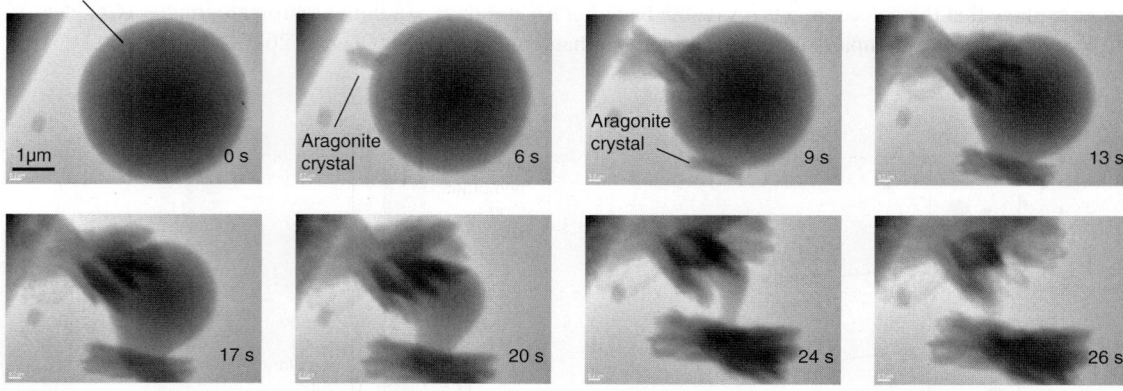

FIGURE 27-2 Sequence of transmission electron micrographs beginning with 3-μm-diameter sphere of amorphous calcium carbonate that had been growing from a solution of CaCl$_2$ and NaHCO$_3$ for $\sim$95 s. A single aragonite CaCO$_3$ crystal nucleates on the sphere 6 s after the first frame was taken. Aragonite growth continues at the upper left after 9 s and another crystal nucleates at the bottom of the sphere. Clusters of aragonite crystals grow at the expense of the sphere in both locations. The sphere is gone 26 s after the start of the sequence. [From M. H. Nielsen, S. Aloni, and J. J. De Yoreo, "In Situ TEM Imaging of CaCO$_3$ Nucleation Reveals Coexistence of Direct and Indirect Pathways," *Science* **2014**, *345*, 1158, Figure 2. Reprinted with permission from AAAS, permission conveyed through Copyright Clearance Center, Inc.]

In precipitation, nucleation procedes faster than particle growth in a highly supersaturated solution, producing tiny particles or, worse, a colloid. In a less supersaturated solution, nucleation is slower, so nuclei have can grow into larger, more tractable particles.

Techniques that promote particle growth include:

Supersaturation tends to decrease the particle size of a precipitate.

1. raising the temperature to increase solubility and thereby decrease supersaturation
2. adding precipitant slowly with vigorous mixing, to prevent a local, highly supersaturated condition where the stream of precipitant first enters the analyte
3. using a large volume of solution so that concentrations of analyte and precipitant are low

Homogeneous Precipitation

In our discussion so far, precipitation has been carried out by mixing a solution of precipitant with a solution of analyte. In **homogeneous precipitation,** precipitant is generated slowly from within an initially homogeneous solution by a chemical reaction (Table 27-3). This is beneficial because, when precipitation is slow, particle growth dominates over nucleation to give larger, purer particles that are easier to filter. When precipitation is rapid, nucleation tends to dominate over crystallization and the resulting particles are small and hard to filter.

CHAPTER 27 Gravimetric and Combustion Analysis

TABLE 27-3 **Common reagents used for homogeneous precipitation**

Precipitant	Reagent	Reaction	Some elements precipitated
OH^-	Urea	$(H_2N)_2CO + 3H_2O \rightarrow CO_2 + 2NH_4^+ + 2OH^-$	Al, Ga, Th, Bi, Fe, Sn
OH^-	Potassium cyanate	$HOCN + 2H_2O \rightarrow NH_4^+ + CO_2 + OH^-$ Hydrogen cyanate	Cr, Fe
S^{2-}	Thioacetamide[a]	$\overset{\overset{\displaystyle S}{\|\|}}{CH_3CNH_2} + H_2O \rightarrow \overset{\overset{\displaystyle O}{\|\|}}{CH_3CNH_2} + H_2S$	Sb, Mo, Cu, Cd
SO_4^{2-}	Sulfamic acid	$H_3\overset{+}{N}SO_3^- + H_2O \rightarrow NH_4^+ + SO_4^{2-} + H^+$	Ba, Ca, Sr, Pb
$C_2O_4^{2-}$	Dimethyl oxalate	$\overset{\overset{\displaystyle OO}{\|\| \|\|}}{CH_3OCCOCH_3} + 2H_2O \rightarrow 2CH_3OH + C_2O_4^{2-} + 2H^+$	Ca, Mg, Zn
PO_4^{3-}	Trimethyl phosphate	$(CH_3O)_3P{=}O + 3H_2O \rightarrow 3CH_3OH + PO_4^{3-} + 3H^+$	Zr, Hf
CrO_4^{2-}	Chromic ion plus bromate	$2Cr^{3+} + BrO_3^- + 5H_2O \rightarrow 2CrO_4^{2-} + Br^- + 10H^+$	Pb
8-Hydroxyquinoline	8-Acetoxyquinoline	(reaction structure) $+ H_2O \rightarrow$ (structure) $+ CH_3CO_2H$	Al, U, Mg, Zn

*a. Hydrogen sulfide is volatile and toxic; it should be handled only in a well-vented hood. Thioacetamide is a carcinogen that should be handled with gloves. If thioacetamide contacts your skin, wash yourself thoroughly immediately. Leftover reagent is destroyed by heating at 50°C with 5 mol of NaOCl per mole of thioacetamide. [H. Elo, "Is Thioacetamide a Serious Health Hazard in Inorganic Chemistry Laboratories?" J. Chem. Ed. **1987**, 64, A144.]*

An example of homogeneous precipitation is the slow formation of Fe(III) formate from a solution of Fe(III) plus formic acid. Precipitation is initiated by decomposing urea in boiling water to slowly produce OH^-:

$$H_2N\overset{\overset{\displaystyle O}{\|\|}}{C}NH_2 + 3H_2O \xrightarrow{\text{heat}} CO_2 + 2NH_4^+ + 2OH^- \qquad (27\text{-}2)$$
Urea

Gradual OH^- production enhances the particle size of Fe(III) formate precipitate:

$$H\overset{\overset{\displaystyle O}{\|\|}}{C}OH + OH^- \rightarrow HCO_2^- + H_2O \qquad (27\text{-}3)$$
Formic acid $\qquad$ Formate

$$3HCO_2^- + Fe^{3+} \rightarrow Fe(HCO_2)_3 \cdot nH_2O(s)\downarrow \qquad (27\text{-}4)$$
Fe(III) formate

The hydrated iron(III) formate product does not have a well-defined, constant composition. Instead, it is heated to 850°C for 1 h to decompose to Fe_2O_3, which can be weighed to find out how much Fe(III) was present.

Precipitation in the Presence of Electrolyte

Ionic compounds are usually precipitated in the presence of an electrolyte. To understand why, we must discuss how tiny colloidal crystallites *coagulate* (come together) into larger crystals. We illustrate the case of AgCl, which is commonly formed in 0.1 M HNO_3. Figure 27-3 shows a colloidal particle of AgCl growing in a solution containing excess Ag^+, H^+, and NO_3^-. The surface of the particle has excess positive charge due to the **adsorption** of extra silver ions on exposed chloride ions. (To be adsorbed means to be attached to the surface. In contrast,

An **electrolyte** is a compound that dissociates into ions when it dissolves.

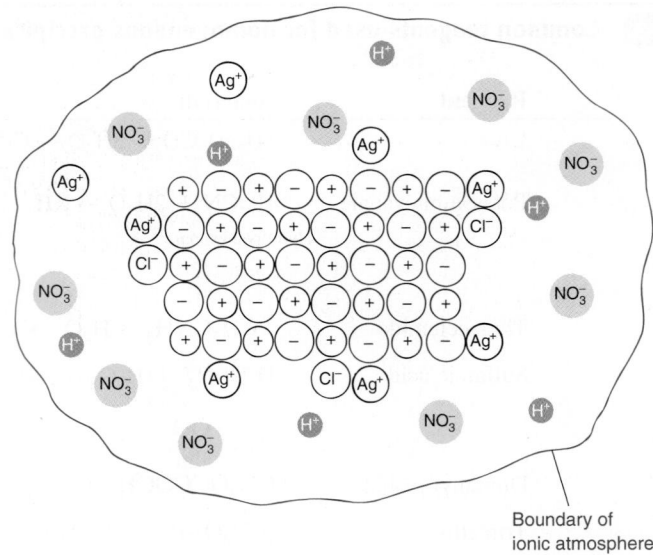

Boundary of
ionic atmosphere

FIGURE 27-3 Colloidal particle of AgCl growing in a solution containing excess Ag^+, H^+, and NO_3^-. The particle has a net positive charge because of adsorbed Ag^+ ions. The region of solution surrounding the particle is called the *ionic atmosphere*. It has a net negative charge because the particle attracts anions and repels cations.

Although it is common to find the excess common ion adsorbed on the crystal surface, other ions can be selectively adsorbed. Citrate is bound in preference to sulfate on the surface of $BaSO_4$.

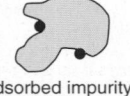

Adsorbed impurity   Absorbed impurity
(external)   (internal)

absorption means penetration beyond the surface, to the inside.) The positively charged surface attracts anions and repels cations from the *ionic atmosphere* surrounding the particle in Figure 27-3. The positively charged particle and the negatively charged ionic atmosphere together are called the **electric double layer.**

Colloidal particles must collide with one another to coalesce. However, the negatively charged ionic atmospheres of the particles repel one another. Particles must have enough kinetic energy to overcome electrostatic repulsion before they can coalesce.

Heating promotes coalescence by increasing kinetic energy. Increasing electrolyte concentration (HNO_3 for AgCl) decreases the thickness of the ionic atmosphere and allows particles to come closer together before electrostatic repulsion becomes significant. For this reason, most gravimetric precipitations are done in the presence of an electrolyte.

Figure 27-4 shows the measured force between a 10-μm-diameter spherical particle of crystalline Al_2O_3 and the surface of a flat Al_2O_3 crystal in 1 mM LiCl and in 1 M LiCl at pH 11 (LiOH). At this pH, both surfaces are negative, probably from adsorbed OH^-. The thickness of the ionic atmosphere adjacent to each surface in 1 mM LiCl is ~10 nm. As the sphere is moved toward the crystal, positively charged ionic atmospheres begin to repel each other at a separation of ~30 nm, reaching maximum repulsion near 7 nm. As separation decreases, attractive *van der Waals forces* dominate (Box 27-1). Maximum attraction occurs at a separation of ~0.5 nm. When pushed closer together, electron clouds begin to overlap and strong

---

BOX 27-1   van der Waals Attraction

van der Waals force exists between all types of molecules. Consider two electrically neutral molecules with no permanent dipole (no permanent charge separation with the molecules). At any instant, the movement of electrons in molecule A gives rise to an instantaneous separation of negative and positive charges within molecule A. The negative region repels electrons of neighboring molecule B, creating an instantaneous separation of charges in B. Electrostatic attraction between the negative region of A and the positive region of B is the van der Waals attraction. Attraction is only significant within a few nanometers and is constantly changing as electrons move within each molecule.

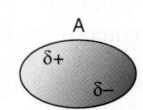

At any instant, movement of electrons creates negative and positive regions in neutral molecule A

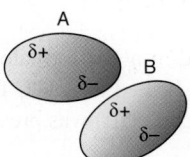

When neutral molecule B approaches A, the instantaneous charges in A induce instantaneous charge separation in B, resulting in electrostatic attraction between A and B

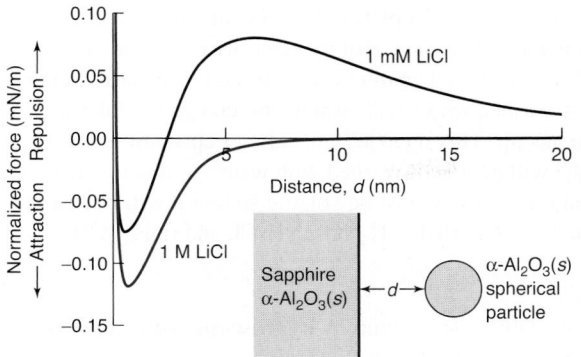

FIGURE 27-4 Measured force between a spherical 10-μm-diameter particle of crystalline α-alumina (Al_2O_3) and the *c* crystal plane surface of sapphire (Al_2O_3) in 1 mM LiCl or 1 M LiCl electrolyte at pH 11 (obtained with LiOH). Normalized force is force/$(2\pi r)$, where *r* is the radius of the sphere. [Data from H. Yilmaz, K. Sato, and K. Watari, "Ion-Specific Interaction of Alumina Surfaces," *J. Am. Ceram. Soc.* **2009**, *92*, 318.]

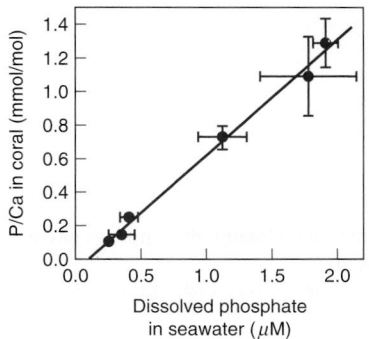

FIGURE 27-5 Coprecipitation of phosphate with calcium carbonate in coral skeleton. Coprecipitated phosphate is proportional to phosphate concentration in seawater. By measuring P/Ca in ancient coral, we can infer that phosphate concentration in the western Mediterranean Sea 11 200 years ago was twice as high as current values. [Data from P. Montagna, M. McCulloch, M. Taviani, C. Mazzoli, and B. Vendrell, "Phosphorus in Cold-Water Corals as a Proxy for Seawater Nutrient Chemistry," *Science* **2006**, *312*, 1788.]

repulsion occurs. Figure 27-4 also shows the same experiment conducted in 1 M LiCl. At this high electrolyte concentration, the thickness of the ionic atmospheres is ~0.3 nm, so long-distance repulsion is not observed. van der Waals attraction dominates, reaching a maximum at ~0.7 nm separation.

Digestion

Liquid from which a substance precipitates or crystallizes is called the **mother liquor.** After precipitation, most procedures require a period of standing in the presence of the hot mother liquor. This treatment, called **digestion,** promotes slow recrystallization of the precipitate. Particle size increases and impurities tend to be expelled from the crystal.

Purity

Adsorbed impurities are bound to the surface of a crystal. *Absorbed* impurities (within the crystal) are classified as *inclusions* or *occlusions*. Inclusions are impurity ions that randomly occupy sites in the crystal lattice normally occupied by ions that belong in the crystal. Inclusions are more likely when the impurity ion has a size and charge similar to those of one of the ions that belongs to the product. Occlusions are pockets of impurity that are literally trapped inside the growing crystal.

Adsorbed, occluded, and included impurities are said to be **coprecipitated.** That is, the impurity is precipitated along with the desired product, even though the solubility of the impurity has not been exceeded (Figure 27-5).

Coprecipitation tends to be worst in colloidal precipitates such as $BaSO_4$, $Al(OH)_3$, and $Fe(OH)_3$, which have a large surface area. Many procedures call for washing away the mother liquor, redissolving the precipitate, and *reprecipitating* the product. During the second precipitation, the concentration of impurities in solution is lower than during the first precipitation, and the degree of coprecipitation therefore tends to be lower.

A trace component of a solution can be isolated by coprecipitation with a major component or an added component.[10] Precipitate that collects the trace component is said to be a *gathering agent*, and the process is called **gathering.** Natural arsenic in drinking water in Bangladesh is a major health hazard. One way to remove arsenic is by coprecipitation with $Fe(OH)_3$.[11] Fe(II) or Fe(s) is added to the water and oxidized in air to precipitate $Fe(OH)_3$. After filtration through sand to remove solids, the water is drinkable.

Some impurities can be treated with a **masking agent** to prevent them from reacting with the precipitant. In the gravimetric analysis of Be^{2+}, Mg^{2+}, Ca^{2+}, or Ba^{2+} with the reagent *N-p*-chlorophenylcinnamohydroxamic acid, impurities such as Ag^+, Mn^{2+}, Zn^{2+}, Cd^{2+}, Hg^{2+}, Fe^{2+}, and Ga^{3+} are kept in solution by excess KCN. Pb^{2+}, Pd^{2+}, Sb^{3+}, Sn^{2+}, Bi^{3+}, Zr^{4+}, Ti^{4+}, V^{5+}, and Mo^{6+} are masked with a mixture of citrate and oxalate.

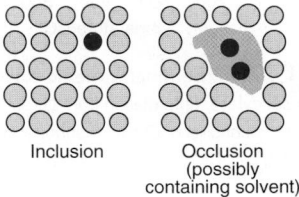

Inclusion   Occlusion (possibly containing solvent)

Removal of occluded NO_3^- from $BaSO_4$ by *reprecipitation*

	$[NO_3^-]/[SO_4^-]$ in precipitate
Initial precipitate	0.279
1st reprecipitation	0.028
2nd reprecipitation	0.001

[H. Bao, "Purifying Barite for Oxygen Isotope Measurement by Dissolution and Reprecipitation in a Chelating Solution," *Anal. Chem.* **2006**, *78*, 304.]

$$Ca^{2+} \ + \ 2RH \ \rightarrow \ CaR_2(s)\downarrow + 2H^+$$

Analyte   *N-p*-Chlorophenyl-cinnamohydroxamic acid   Precipitate

$$Mn^{2+} \ + \ 6CN^- \ \rightarrow \ Mn(CN)_6^{4-}$$

Impurity   Masking agent   Stays in solution

Even when a precipitate forms in a pure state, impurities might collect on the product while it is standing in the mother liquor. This process is called *postprecipitation* and usually involves a supersaturated impurity that does not readily crystallize. An example is the crystallization of MgC_2O_4 on CaC_2O_4.

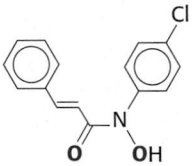

N-p-Chlorophenylcinnamo-hydroxamic acid (RH) (ligand atoms are **bold**)

Washing a precipitate on a filter helps remove droplets of liquid containing excess solute. Some precipitates can be washed with water, but many require electrolyte to maintain coherence. For these precipitates, the ionic atmosphere is required to neutralize the surface charge of the tiny particles. If electrolyte is washed away with water, the charged solid particles repel one another and the product breaks up. This breaking up, called **peptization**, results in loss of product through the filter. AgCl will peptize if washed with water, so it is washed with dilute HNO_3 instead. Electrolyte used for washing must be volatile so that it will be lost during drying. Volatile electrolytes include HNO_3, HCl, NH_4NO_3, NH_4Cl, and $(NH_4)_2CO_3$.

Ammonium chloride decomposes when heated:

$$NH_4Cl(s) \rightarrow NH_3(g) + HCl(g)$$

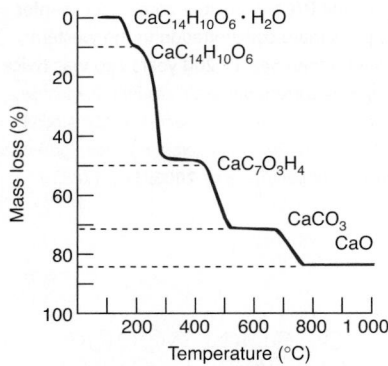

FIGURE 27-6 Thermogravimetric curve for calcium salicylate. [Information from G. Liptay, ed., *Atlas of Thermoanalytical Curves* (London: Heyden and Son, 1976).]

Product Composition

The final product must have a known, stable composition. A **hygroscopic substance** is one that picks up water from the air and is therefore difficult to weigh accurately. Many precipitates contain a variable quantity of water and must be dried under conditions that give a known (possibly zero) stoichiometry of H_2O.

Ignition (strong heating) is used to change the chemical form of some precipitates. For example, igniting $Fe(HCO_2)_3 \cdot nH_2O$ at 850°C for 1 h gives Fe_2O_3, and igniting $Mg(NH_4)PO_4 \cdot 6H_2O$ at 1 100°C gives $Mg_2P_2O_7$.

In **thermogravimetric analysis,** a substance is heated, and its mass is measured as a function of temperature. Figure 27-6 shows how the composition of calcium salicylate changes in four stages:

$$\text{(27-5)}$$

Calcium salicylate monohydrate   ~200°C   ~300°C

CaO   ~700°C $CaCO_3$ ~500°C

Calcium oxide   *Calcium carbonate*

The composition of the product depends on the temperature and duration of heating.

27-3 Examples of Gravimetric Calculations

We now examine some examples that illustrate how to relate the mass of a gravimetric precipitate to the quantity of the original analyte. The general approach is to relate the moles of product to the moles of reactant.

EXAMPLE Relating Mass of Product to Mass of Reactant

The piperazine content of an impure commercial material can be determined by precipitating and weighing the diacetate:[12]

$$\text{(27-6)}$$

Piperazine   Acetic acid   Piperazine diacetate
FM 86.136   FM 60.052   FM 206.240

In one experiment, 0.312 6 g of sample was dissolved in 25 mL of acetone, and 1 mL of acetic acid was added. After 5 min, the precipitate was filtered, washed with acetone, dried at 110°C, and found to weigh 0.712 1 g. Find the wt% of piperazine in the sample.

Solution For each mole of piperazine in the impure material, 1 mol of product is formed.

$$\text{Moles of product} = \frac{0.712\ 1\ \text{g}}{206.240\ \text{g/mol}} = 3.453 \times 10^{-3}\ \text{mol}$$

This many moles of piperazine corresponds to

$$\text{Grams of piperazine} = (3.453 \times 10^{-3}\ \text{mol})\left(86.136\ \frac{\text{g}}{\text{mol}}\right) = 0.297\ 4\ \text{g}$$

If you were performing this analysis, it would be important to determine that the impurities in the piperazine do not precipitate—otherwise the result will be high.

which gives

$$\text{Percentage of piperazine in analyte} = \frac{0.297\,4\text{ g}}{0.312\,6\text{ g}} \times 100 = 95.14\%$$

An alternative (but equivalent) way to work this problem is to realize that 206.240 g (1 mol) of product will be formed for every 86.136 g (1 mol) of piperazine analyzed. Because 0.712 1 g of product was formed, the amount of reactant is given by

$$\frac{x\text{ g piperazine}}{0.712\,1\text{ g product}} = \frac{86.136\text{ g piperazine}}{206.240\text{ g product}}$$

$$\Rightarrow x = \left(\frac{86.136\text{ g piperazine}}{206.240\text{ g product}}\right)0.712\,1\text{ g product} = 0.297\,4\text{ g piperazine}$$

The quantity 86.136/206.240 is the *gravimetric factor* relating the mass of starting material to the mass of product.

Gravimetric factor: relates mass of product to mass of analyte.

TEST YOURSELF A 0.385 4 g sample gave 0.800 0 g of product. Find the wt% of piperazine in the sample. (*Answer:* 86.69%)

For a reaction in which the stoichiometric relation between analyte and product is not 1 : 1, we must use the correct stoichiometry in formulating the gravimetric factor. For example, an unknown containing Mg^{2+} (atomic mass = 24.305 0) can be analyzed gravimetrically to produce magnesium pyrophosphate ($Mg_2P_2O_7$, FM 222.553). The gravimetric factor would be

$$\frac{\text{Grams of Mg in analyte}}{\text{Grams of }Mg_2P_2O_7\text{ formed}} = \frac{2 \times (24.305\,0)}{222.553}$$

because it takes two Mg^{2+} to make one $Mg_2P_2O_7$.

EXAMPLE **Calculating How Much Precipitant to Use**

(a) To measure the nickel content in steel, the alloy is dissolved in 12 M HCl and neutralized in the presence of citrate ion, which maintains iron in solution. The slightly basic solution is warmed, and dimethylglyoxime (DMG) is added to precipitate the red DMG-nickel complex quantitatively. The product is filtered, washed with cold water, and dried at 110°C.

$$Ni^{2+} + 2 \quad \longrightarrow \quad + 2H^+ \tag{27-7}$$

DMG
FM 58.69   FM 116.12

Bis(dimethylglyoximate)nickel(II)
FM 288.91

If the nickel content is known to be near 3 wt% and you wish to analyze 1.0 g of steel, what volume of 1.0 wt% alcoholic DMG solution should be used to give a 50% excess of DMG for the analysis? Assume that the density of the alcohol solution is 0.79 g/mL.

Solution Because the Ni content is about 3%, 1.0 g of steel will contain about 0.03 g of Ni, which corresponds to

$$\frac{0.03\text{ g Ni}}{58.69\text{ g Ni/mol Ni}} = 5.11 \times 10^{-4}\text{ mol Ni}$$

27-3 Examples of Gravimetric Calculations

This amount of metal requires

$$2(5.11 \times 10^{-4} \text{ mol Ni})(116.12 \text{ g DMG/mol Ni}) = 0.119 \text{ g DMG}$$

because 1 mol of Ni^{2+} requires 2 mol of DMG. A 50% excess of DMG would be $(1.5)(0.119 \text{ g}) = 0.178 \text{ g}$. This much DMG is contained in

$$\frac{0.178 \text{ g DMG}}{0.010 \text{ g DMG/g solution}} = 17.8 \text{ g solution}$$

which occupies a volume of

$$\frac{17.8 \text{ g solution}}{0.79 \text{ g solution/mL}} = 23 \text{ mL}$$

(b) If 1.163 4 g of steel gives 0.179 5 g of precipitate, what percentage of Ni is in the steel?

Solution For each mole of Ni in the steel, 1 mol of precipitate will be formed. Therefore, 0.179 5 g of precipitate corresponds to

$$\frac{0.179 \text{ 5 g Ni(DMG)}_2}{288.91 \text{ g Ni(DMG)}_2/\text{mol Ni(DMG)}_2} = 6.213 \times 10^{-4} \text{ mol Ni(DMG)}_2$$

The mass of Ni in the steel is $(6.213 \times 10^{-4} \text{ mol Ni})(58.69 \text{ g/mol Ni}) = 0.036 \text{ 46 g}$, and the wt% Ni in steel is

$$\frac{0.036 \text{ 46 g Ni}}{1.163 \text{ 4 g steel}} \times 100 = 3.134\%$$

A slightly simpler way to approach this problem comes from realizing that 58.69 g of Ni (1 mol) would give 288.91 g (1 mol) of product. Calling the mass of Ni in the sample x, we can write

$$\frac{\text{Grams of Ni analyzed}}{\text{Grams of product formed}} = \frac{x}{0.179 \text{ 5}} = \frac{58.69}{288.91} \Rightarrow \text{Ni} = 0.036 \text{ 46 g}$$

TEST YOURSELF An alloy contains ~2.0 wt% Ni. What volume of 0.83 wt% DMG should be used to provide a 50% excess of DMG for the analysis of 1.8 g of steel? What mass of $Ni(DMG)_2$ precipitate is expected? (**Answer:** 33 mL, 0.18 g)

EXAMPLE A Problem with Two Components

A mixture of the 8-hydroxyquinoline complexes of Al and Mg weighed 1.084 3 g. When ignited in a furnace open to the air, the mixture decomposed, leaving a residue of Al_2O_3 and MgO weighing 0.134 4 g. Find the weight percent of $Al(C_9H_6NO)_3$ in the original mixture.

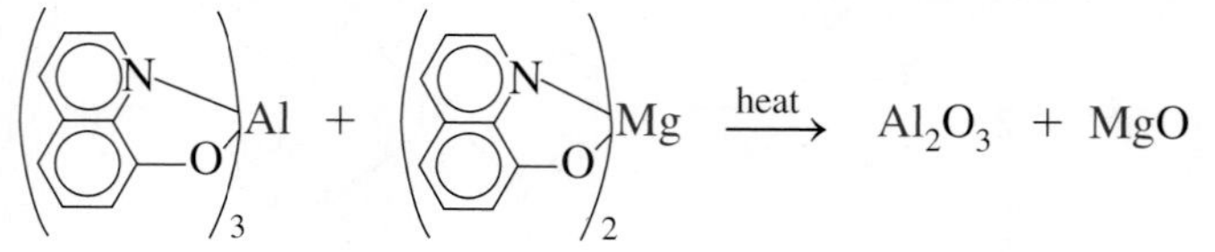

AlQ₃	MgQ₂		
FM 459.43	FM 312.61	FM 101.96	FM 40.304

Solution We will abbreviate the 8-hydroxyquinoline anion as Q. Letting the mass of AlQ_3 be x and the mass of MgQ_2 be y, we can write

$$x + y = 1.084 \text{ 3 g}$$

Mass of AlQ₃   Mass of MgQ₂

The moles of Al are $x/459.43$, and the moles of Mg are $y/312.61$. The moles of Al_2O_3 are one-half of the total moles of Al, because it takes 2 mol of Al to make 1 mol of Al_2O_3.

$$\text{Moles of Al}_2O_3 = \left(\frac{1}{2}\right)\frac{x}{459.43}$$

The moles of MgO will equal the moles of Mg = $y/312.61$. Now we can write

$$\underbrace{\left(\frac{1}{2}\right)\underbrace{\frac{x}{459.43}}_{\text{mol Al}_2\text{O}_3}\underbrace{(101.96)}_{\frac{\text{g Al}_2\text{O}_3}{\text{mol Al}_2\text{O}_3}}}_{\text{Mass of Al}_2\text{O}_3} + \underbrace{\underbrace{\frac{y}{312.61}}_{\text{mol MgO}}\underbrace{(40.304)}_{\frac{\text{g MgO}}{\text{mol MgO}}}}_{\text{Mass of MgO}} = 0.134\ 4\ \text{g}$$

Substituting $y = 1.084\ 3 - x$ into the preceding equation gives

$$\left(\frac{1}{2}\right)\left(\frac{x}{459.43}\right)(101.96) + \left(\frac{1.084\ 3 - x}{312.61}\right)(40.304) = 0.134\ 4\ \text{g}$$

from which we find $x = 0.300\ 3$ g, which is 27.70% of the original mixture.

TEST YOURSELF If reproducibility is ±0.5 mg, product mass might have been 0.133 9 to 0.134 9 g. Find wt% $Al(C_9H_6NO)_3$ if the product weighed 0.133 9 g. (***Answer:*** 30.27%).

Please appreciate the huge uncertainty: A 0.5-mg difference in mass of product gives a 9% difference in calculated composition of the mixture.

27-4 Combustion Analysis

A historically important form of gravimetric analysis was *combustion analysis* to determine C and H in organic compounds burned in excess O_2. Instead of weighing combustion products, modern instruments use thermal conductivity, infrared absorption, flame photometry (for S), and coulometry (for halogens) to measure products.

Gravimetric Combustion Analysis

In gravimetric combustion analysis in Figure 27-7, partially combusted product is passed through catalysts such as Pt gauze, CuO, PbO_2, or MnO_2 at elevated temperature to complete the oxidation to CO_2 and H_2O. The combustion products are flushed through a chamber containing P_4O_{10} ("phosphorus pentoxide"), which absorbs water, and then through a chamber of Ascarite,* which absorbs CO_2. The increase in mass of each chamber tells how much H and C were initially present. A guard tube prevents atmospheric H_2O or CO_2 from entering the chambers.

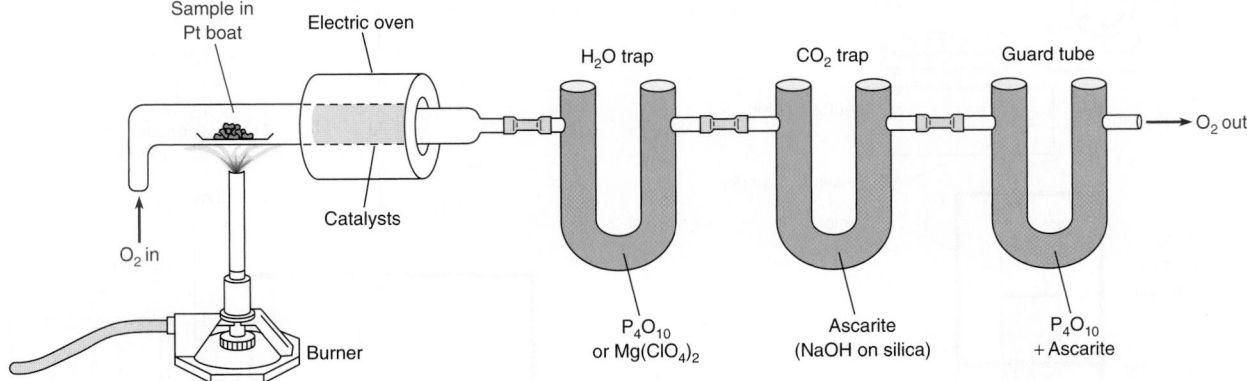

FIGURE 27-7 Gravimetric combustion analysis for carbon and hydrogen.

---

*The original Ascarite is NaOH coated on asbestos. Asbestos is no longer used because breathing particles of asbestos can cause fatal lung disease. In Ascarite II®, asbestos is replaced by an inert silica (SiO_2) carrier.

EXAMPLE Combustion Analysis Calculations

A compound weighing 5.714 mg produced 14.414 mg of CO_2 and 2.529 mg of H_2O upon combustion. Find the wt% of C and H in the sample.

Solution One mole of CO_2 contains 1 mol of carbon. Therefore,

Moles of C in sample $=$ moles of CO_2 produced

$$= \frac{14.414 \times 10^{-3} \text{ g } CO_2}{44.009 \text{ g/mol } CO_2} = 3.275 \times 10^{-4} \text{ mol}$$

Mass of C in sample $= (3.275 \times 10^{-4} \text{ mol C})(12.010\,6 \text{ g/mol C}) = 3.934 \text{ mg}$

$$\text{wt\% C} = \frac{3.934 \text{ mg C}}{5.714 \text{ mg sample}} \times 100 = 68.84\%$$

One mole of H_2O contains 2 mol of H. Therefore,

Moles of H in sample $= 2(\text{moles of } H_2O \text{ produced})$

$$= 2\left(\frac{2.529 \times 10^{-3} \text{ g } H_2O}{18.015 \text{ g/mol } H_2O}\right) = 2.808 \times 10^{-4} \text{ mol}$$

Mass of H in sample $= (2.808 \times 10^{-4} \text{ mol H})(1.008 \text{ g/mol H}) = 2.830 \times 10^{-4} \text{ g}$

$$\text{wt\% H} = \frac{0.283\,0 \text{ mg H}}{5.714 \text{ mg sample}} \times 100 = 4.95\%$$

TEST YOURSELF A 6.234-mg sample produced 12.123 mg CO_2 and 2.529 mg of H_2O. Find the wt% of C and H in the sample. (***Answer:*** 53.07, 4.54%)

Combustion Analysis Today[13]

Figure 27-8 shows an instrument that measures C, H, N, and S in a single operation. First, a ~2-mg sample is accurately weighed and sealed in a tin or silver capsule. The analyzer is swept with He gas that has been treated to remove traces of O_2, H_2O, and CO_2. At the start of

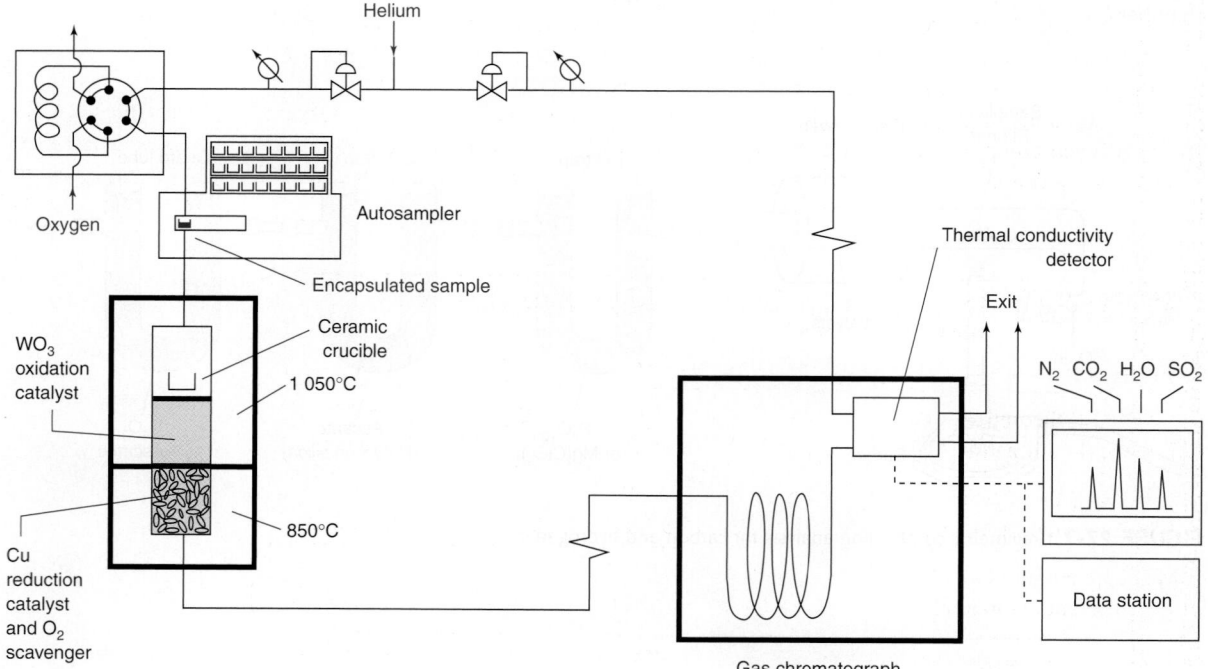

FIGURE 27-8 C, H, N, S elemental analyzer uses gas chromatography with thermal conductivity detection to measure N_2, CO_2, H_2O, and SO_2 combustion products. [Information from E. Pella, "Elemental Organic Analysis. 2. State of the Art," *Am. Lab.*, August 1990, p. 28.]

a run, a measured excess of O_2 is added to the He stream. Then the capsule is dropped into a preheated ceramic crucible, where the capsule melts and sample is rapidly oxidized.

$$C, H, N, S \xrightarrow{1\,050°C\,/\,O_2} CO_2(g) + H_2O(g) + N_2(g) + \underbrace{SO_2(g) + SO_3(g)}_{95\%\ SO_2}$$

Products pass through hot WO_3 oxidation catalyst to complete the combustion of C to CO_2. In the next zone, metallic Cu at 850°C reduces SO_3 to SO_2 and removes excess O_2:

$$Cu + SO_3 \xrightarrow{850°C} SO_2 + CuO(s)$$

$$Cu + \frac{1}{2}O_2 \xrightarrow{850°C} CuO(s)$$

The mixture of CO_2, H_2O, N_2, and SO_2 is separated by gas chromatography, and each component is measured with a thermal conductivity detector (Figure 24-19). Alternatively, CO_2, H_2O, and SO_2 can be measured by infrared absorbance.

A key to elemental analysis is *dynamic flash combustion*, which creates a short burst of gaseous products, instead of slowly bleeding products out over several minutes. Chromatographic analysis requires that the whole sample be injected at once. Otherwise, the injection zone is so broad that the products cannot be separated.

In dynamic flash combustion, the tin-encapsulated sample is dropped into the preheated furnace shortly after the flow of a 50 vol% O_2/50 vol% He mixture is started (Figure 27-9). The Sn capsule melts at 235°C and is instantly oxidized to SnO_2, thereby liberating 594 kJ/mol, and heating the sample to 1 700–1 800°C. By dropping the sample in before much O_2 is present, decomposition (cracking) occurs prior to oxidation, which minimizes the formation of nitrogen oxides. (Flammable liquid samples would be admitted prior to any O_2 to prevent explosions.)

Analyzers that measure C, H, and N, but not S, use better catalysts. The oxidation catalyst is Cr_2O_3. The gas then passes through hot Co_3O_4 coated with Ag to absorb halogens and sulfur. A hot Cu column scavenges excess O_2.

Oxygen analysis requires a different strategy. The sample is thermally decomposed (a process called **pyrolysis**) in the absence of added O_2. Gaseous products are passed through nickelized carbon at 1 075°C to convert oxygen from the compound into CO (not CO_2). Other products include N_2, H_2, CH_4, and hydrogen halides. Acidic products are absorbed by NaOH, and the remaining gases are separated and measured by gas chromatography with a thermal conductivity detector.

For halogenated compounds, combustion gives CO_2, H_2O, N_2, and HX (X = halogen). The HX is trapped in aqueous solution and titrated with Ag^+ ions in a coulometer (Section 17-3). This instrument counts the electrons produced (one e^- for each Ag^+) during complete reaction with HX.

Table 27-4 shows representative results for two of seven compounds sent to more than 35 laboratories to compare their performance in combustion analysis. The accuracy for all seven compounds is excellent: Mean values of wt% C, H, N, and S for ~150 measurements of each compound are almost always within 0.1 wt% of theoretical values. Precision for all seven compounds is summarized at the bottom of the table. The mean 95%

An **oxidation catalyst** completes the oxidation of sample, and a **reduction catalyst** carries out any required reduction and removes excess O_2.

The Sn capsule is oxidized to SnO_2, which

1. liberates heat to vaporize and crack (decompose) the sample
2. uses available oxygen immediately
3. ensures that sample oxidation occurs in gas phase
4. acts as an oxidation catalyst

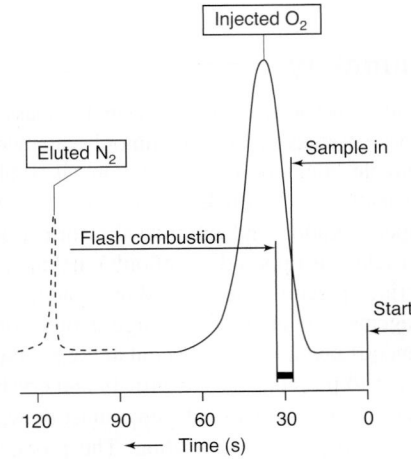

FIGURE 27-9 Sequence of events in dynamic flash combustion. [Information from E. Pella, "Elemental Organic Analysis. 1. Historical Developments," *Am. Lab.*, February 1990, p. 116.]

TABLE 27-4 Accuracy and precision of combustion analysis of pure compounds[a]				
Substance	C	H	N	S
$C_7H_9NO_2S$ theoretical wt%	49.10	5.30	8.18	18.73
Toluene-4-sulfonamide	49.1±0.63	5.3±0.31	8.2±0.38	18.7±0.89
$C_4H_7NO_2S$ theoretical wt%	36.07	5.30	10.52	24.08
4-Thiazolidinecarboxylic acid	36.0±0.33	5.3±0.16	10.5±0.16	24.0±0.53
Mean uncertainty (wt%) for 7 different compounds	±0.47	±0.24	±0.31	±0.76

a. Results for two of seven pure compounds that were analyzed by 33–45 laboratories each year over six years. Each lab analyzed each compound at least five times during at least two days. For each substance, first row gives theoretical wt% and second row gives measured wt%. Uncertainties are 95% confidence intervals computed for all results after rejecting outliers at the 1% significance level.

SOURCE: R. Companyó, R. Rubio, A. Sahuquillo, R. Boqué, A. Maroto, and J. Riu, "Uncertainty Estimation in Organic Elemental Analysis Using Information from Proficiency Tests," Anal. Bioanal. Chem. **2008**, 392, 1497.

confidence interval for C is ± 0.47 wt%. For H, N, and S, 95% confidence intervals are ± 0.24, ± 0.31, and ± 0.76 wt%, respectively. Chemists consider a result within ± 0.3 wt% of theoretical to be good evidence that the compound has the expected formula. This criterion can be difficult to meet for C and S with a single analysis because the 95% confidence intervals are larger than ± 0.3.

Silicon compounds, such as SiC, Si_3N_4, and silicates (rocks) can be analyzed by reaction with elemental fluorine (F_2) in a Ni vessel to produce SiF_4 and fluorinated products of every element in the periodic table except O, N, He, Ne, Ar, and Kr.[14] Products can be measured by mass spectrometry. Nitrogen in Si_3N_4 and some other metal nitrides can be analyzed by heating to 3 000°C in an inert atmosphere to liberate nitrogen as N_2, which is measured by thermal conductivity.

> F_2 is exceedingly reactive and dangerous. It must be handled only in systems designed for its use.

Terms to Understand

absorption	digestion	ignition	pyrolysis
adsorption	electric double layer	masking agent	supersaturated solution
colloid	gathering	mother liquor	thermogravimetric analysis
combustion analysis	gravimetric analysis	nucleation	
coprecipitation	homogeneous precipitation	peptization	
dialysis	hygroscopic substance	precipitant	

Summary

In gravimetric analysis, we relate the mass of a known product to the mass of analyte. Most commonly, analyte ion is precipitated by a suitable counterion. Precipitation takes place in two stages called nucleation and particle (crystal) growth. Measures taken to reduce supersaturation and promote the formation of large, easily filtered particles (as opposed to colloids) include (1) raising the temperature during precipitation, (2) slowly adding and vigorously mixing reagents, (3) maintaining a large sample volume, and (4) using homogeneous precipitation. Precipitates are usually digested in hot mother liquor to promote particle growth and crystallization. Precipitates are then filtered and washed; some must be washed with a volatile electrolyte to prevent peptization. The product is heated to dryness or ignited to achieve a reproducible, stable composition. Gravimetric calculations relate moles of product to moles of analyte.

In combustion analysis for C, H, N, S, and halogens, an organic compound in a tin capsule is rapidly heated with excess O_2 to give predominantly CO_2, H_2O, N_2, SO_2, and HX (hydrogen halides). A hot oxidation catalyst completes the process, and hot Cu scavenges excess O_2. For sulfur analysis, hot copper also converts SO_3 into SO_2. Products may be separated by gas chromatography and measured by their thermal conductivity. Some instruments use infrared absorption to measure CO_2, H_2O, and SO_2. HX is trapped in aqueous solution and measured by coulometric titration (counting electrons) with electrolytically generated Ag^+. Oxygen in organic compounds can be measured by pyrolysis in the absence of added O_2 to convert oxygen from the compound into CO.

Exercises

27-A. An organic compound with a formula mass of 417 g/mol was analyzed for ethoxyl (CH_3CH_2O-) groups by the reactions

$$ROCH_2CH_3 + HI \rightarrow ROH + CH_3CH_2I$$
$$CH_3CH_2I + Ag^+ + OH^- \rightarrow AgI(s) + CH_3CH_2OH$$

A 25.42-mg sample of compound produced 29.03 mg of AgI. How many ethoxyl groups are there in each molecule?

27-B. A 0.649-g sample containing only K_2SO_4 (FM 174.27) and $(NH_4)_2SO_4$ (FM 132.14) was dissolved in water and treated with $Ba(NO_3)_2$ to precipitate all SO_4^{2-} as $BaSO_4$ (FM 233.39). Find the weight percent of K_2SO_4 in the sample if 0.977 g of precipitate was formed.

27-C. Consider a mixture of the two solids, $BaCl_2 \cdot 2H_2O$ (FM 244.26) and KCl (FM 74.551), in an unknown ratio. (The notation

$BaCl_2 \cdot 2H_2O$ means that a crystal is formed with two water molecules for each $BaCl_2$.) When the unknown is heated to 160°C for 1 h, the water of crystallization is driven off:

$$BaCl_2 \cdot 2H_2O(s) \xrightarrow{160°C} BaCl_2(s) + 2H_2O(g)$$

A sample originally weighing 1.783 9 g weighed 1.562 3 g after heating. Calculate the weight percent of Ba, K, and Cl in the original sample.

27-D. A mixture containing only aluminum tetrafluoroborate, $Al(BF_4)_3$ (FM 287.39), and magnesium nitrate, $Mg(NO_3)_2$ (FM 148.31), weighed 0.282 8 g. It was dissolved in 1 wt% HF(aq) and treated with nitron solution to precipitate a mixture of nitron tetrafluoroborate and nitron nitrate weighing 1.322 g. Find the wt% Mg in the original solid mixture.

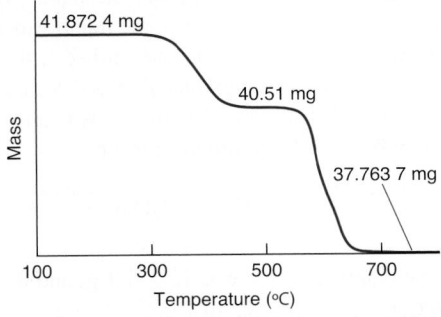

(d) Write the balanced reaction of $LaCoO_{3+x}$ with H_2O to produce La_2O_3, Co, and H_2O. If 41.872 4 mg of $LaCoO_{3+x}$ react completely, what will be the mass of solid product? Your answer will be an expression with x in it.

(e) From the observed product mass (37.763 7 mg) at 700°C, find x in $LaCoO_{3+x}$. Write the formula of the starting solid.

(f) In part (c), the mass lost from the ideal formula $LaCoO_3$, is 4.087 7 mg. What is the product near 500°C where the mass is ~40.51 mg?

Thermogravimetric analysis of $LaCoO_{3+x}$ in 5 vol% H_2 in Ar. [Data from O. Haas, Chr. Ludwig, and A. Wokaun, "Determination of the Bulk Cobalt Valence State of Co-Perovskites Containing Surface-Adsorbed Impurities," *Anal. Chem.* **2006**, *78*, 7273.]

27-E. $LaCoO_{3\pm x}$ has a perovskite structure (Box 19-1) with variable oxygen content. The figure shows the thermogravimetric curve observed when 41.872 4 mg were heated in 5 vol% H_2 in Ar. La(III) does not react, but cobalt is reduced to Co(*s*).

(a) What is the oxidation state of cobalt in the ideal formula $LaCoO_3$?

(b) Write the reaction of $LaCoO_3$ with H_2 to produce $La_2O_3(s)$, Co(*s*), and $H_2O(g)$.

(c) If 41.872 4 mg of $LaCoO_3$ react completely, what will be the mass of product (La_2O_3 + Co)?

Problems

Gravimetric Analysis

27-1.(a) What is the difference between absorption and adsorption?
(b) How is an inclusion different from an occlusion?

27-2. State four desirable properties of a gravimetric precipitate.

27-3. Why is high relative supersaturation undesirable in a gravimetric precipitation?

27-4. What measures can be taken to decrease the relative supersaturation during a precipitation?

27-5. Why are many ionic precipitates washed with electrolyte solution instead of pure water?

27-6. Why is it less desirable to wash AgCl precipitate with aqueous $NaNO_3$ than with HNO_3 solution?

27-7. Why would a reprecipitation be employed in a gravimetric analysis?

27-8. Explain what is done in thermogravimetric analysis.

27-9. Explain how the quartz crystal microbalance at the opening of Chapter 2 measures small masses.

27-10. A 50.00-mL solution containing NaBr was treated with excess $AgNO_3$ to precipitate 0.214 6 g of AgBr (FM 187.772). What was the molarity of NaBr in the solution?

27-11. To find the Ce^{4+} content of a solid, 4.37 g were dissolved and treated with excess iodate to precipitate $Ce(IO_3)_4$. The precipitate was collected, washed well, dried, and ignited to produce 0.104 g of CeO_2 (FM 172.114). What was the weight percent of Ce in the original solid?

27-12. Marie Curie dissolved 0.091 92 g of $RaCl_2$ and treated it with excess $AgNO_3$ to precipitate 0.088 90 g of AgCl. In her time (1900), the atomic mass of Ag was known to be 107.8 and that of Cl was 35.4. From these values, find the atomic mass of Ra that Marie Curie would have calculated.

27-13. A 0.050 02-g sample of impure piperazine contained 71.29 wt% piperazine (FM 86.136). How many grams of product (FM 206.240) will be formed when this sample is analyzed by Reaction 27-6?

27-14. A 1.000-g sample of unknown gave 2.500 g of bis(dimethylglyoximate)nickel(II) (FM 288.91) when analyzed by Reaction 27-7. Find the weight percent of Ni in the unknown.

27-15. Name the products obtained in Figure 27-6 when calcium salicylate monohydrate is heated to 550°C and to 1 000°C. Using the formula masses of these products, calculate what mass is expected to remain when 0.635 6 g of calcium salicylate monohydrate is heated to 550°C and to 1 000°C.

27-16. A method to measure soluble organic carbon in seawater includes oxidation of the organic materials to CO_2 with $K_2S_2O_8$, followed by gravimetric determination of the CO_2 trapped by a column of Ascarite. A water sample weighing 6.234 g produced 2.378 mg of CO_2 (FM 44.009). Calculate the ppm carbon in the seawater.

27-17. How many milliliters of 2.15% alcoholic dimethylglyoxime should be used to provide a 50.0% excess for Reaction 27-7 with 0.998 4 g of steel containing 2.07 wt% Ni? Assume that the density of the dimethylglyoxime solution is 0.790 g/mL.

27-18. Twenty dietary iron tablets with a total mass of 22.131 g were ground and mixed thoroughly. Then 2.998 g of the powder were dissolved in HNO_3 and heated to convert all iron into Fe^{3+}. Addition of NH_3 precipitated $Fe_2O_3 \cdot xH_2O$, which was ignited to give 0.264 g of Fe_2O_3 (FM 159.69). What is the average mass of $FeSO_4 \cdot 7H_2O$ (FM 278.01) in each tablet?

27-19. Finely ground mineral (0.632 4 g) was dissolved in 25 mL of boiling 4 M HCl and diluted with 175 mL H_2O containing two drops of methyl red indicator. The solution was heated to 100°C, and 50 mL of warm solution containing 2.0 g $(NH_4)_2C_2O_4$ were slowly added to precipitate CaC_2O_4. Then 6 M NH_3 was added until the indicator changed from red to yellow, showing that the liquid was neutral or slightly basic. After slow cooling for 1 h, the liquid was decanted and the solid transferred to a filter crucible and washed with cold 0.1 wt% $(NH_4)_2C_2O_4$ solution five times until no Cl^- was detected in the filtrate upon addition of $AgNO_3$ solution. The crucible was dried at 105°C for 1 h and then at 500 ± 25°C in a furnace for 2 h.

$$Ca^{2+} + C_2O_4^{2-} \xrightarrow{105°C} CaC_2O_4 \cdot H_2O(s) \xrightarrow{500°C} CaCO_3(s)$$
$$\text{FM 40.078} \qquad\qquad\qquad\qquad\qquad \text{FM 100.087}$$

The mass of the empty crucible was 18.231 1 g, and the mass of the crucible with $CaCO_3(s)$ was 18.546 7 g.

(a) Find the wt% Ca in the mineral.

(b) Why is the unknown solution heated to boiling and the precipitant solution, $(NH_4)_2C_2O_4$, also heated before slowly mixing the two?

(c) What is the purpose of washing the precipitate with 0.1 wt% $(NH_4)_2C_2O_4$?

(d) What is the purpose of testing the filtrate with $AgNO_3$ solution?

27-20. *Man in the vat problem.*[15] Long ago, a workman at a dye factory fell into a vat containing hot, concentrated sulfuric and nitric acids. He dissolved completely! Because nobody witnessed the accident, it was necessary to prove that he fell in so that the man's wife could collect his insurance money. The man weighed 70 kg, and a human body contains ~6.3 parts per thousand (mg/g) phosphorus. The acid in the vat was analyzed for phosphorus to see whether it contained a dissolved human.

(a) The vat contained 8.00×10^3 L of liquid, and a 100.0-mL sample was analyzed. If the man did fall into the vat, what is the expected quantity of phosphorus in 100.0 mL?

(b) The 100.0-mL sample was treated with a molybdate reagent that precipitated ammonium phosphomolybdate, $(NH_4)_3[P(Mo_{12}O_{40})] \cdot 12H_2O$. This substance was dried at 110°C to remove waters of hydration and heated to 400°C until it reached the constant composition $P_2O_5 \cdot 24MoO_3$, which weighed 0.371 8 g. When a fresh mixture of the same acids (not from the vat) was treated in the same manner, 0.033 1 g of $P_2O_5 \cdot 24MoO_3$ (FM 3 596.46) was produced. This *blank determination* gives the amount of phosphorus in the starting reagents. The $P_2O_5 \cdot 24MoO_3$ that could have come from the dissolved man is therefore $0.371\ 8 - 0.033\ 1 = 0.338\ 7$ g. How much phosphorus was present in the 100.0-mL sample? Is this quantity consistent with a dissolved man?

27-21. A 1.475-g sample containing NH_4Cl (FM 53.491), K_2CO_3 (FM 138.21), and inert ingredients was dissolved to give 0.100 L of solution. A 25.0-mL aliquot was acidified and treated with excess sodium tetraphenylborate, $Na^+B(C_6H_5)_4^-$, to precipitate K^+ and NH_4^+ ions completely:

$$(C_6H_5)_4B^- + K^+ \rightarrow (C_6H_5)_4BK(s)$$
$$\text{FM 358.33}$$

$$(C_6H_5)_4B^- + NH_4^+ \rightarrow (C_6H_5)_4BNH_4(s)$$
$$\text{FM 337.27}$$

The resulting precipitate amounted to 0.617 g. A fresh 50.0-mL aliquot of the original solution was made alkaline and heated to drive off all the NH_3:

$$NH_4^+ + OH^- \rightarrow NH_3(g) + H_2O$$

It was then acidified and treated with sodium tetraphenylborate to give 0.554 g of precipitate. Find the weight percent of NH_4Cl and K_2CO_3 in the original solid.

27-22. A mixture containing only Al_2O_3 (FM 101.96) and Fe_2O_3 (FM 159.69) weighs 2.019 g. When heated under a stream of H_2, Al_2O_3 is unchanged, but Fe_2O_3 is converted into metallic Fe plus $H_2O(g)$. If the residue weighs 1.774 g, what is the weight percent of Fe_2O_3 in the original mixture?

27-23. A solid mixture weighing 0.548 5 g contained only ferrous ammonium sulfate hexahydrate and ferrous chloride hexahydrate. The sample was dissolved in 1 M H_2SO_4, oxidized to Fe^{3+} with H_2O_2, and precipitated with cupferron. The ferric cupferron complex was ignited to produce 0.167 8 g of ferric oxide, Fe_2O_3 (FM 159.69). Calculate the weight percent of Cl in the original sample.

$FeSO_4 \cdot (NH_4)_2SO_4 \cdot 6H_2O$	$FeCl_2 \cdot 6H_2O$
Ferrous ammonium sulfate hexahydrate	Ferrous chloride hexahydrate
FM 392.13	FM 234.84

Cupferron
FM 155.16

27-24. *Propagation of error.* A mixture containing only silver nitrate and mercurous nitrate was dissolved in water and treated with excess sodium cobalticyanide, $Na_3[Co(CN)_6]$ to precipitate both cobalticyanide salts:

$AgNO_3$	FM 169.873
$Ag_3[Co(CN)_6]$	FM 538.643
$Hg_2(NO_3)_2$	FM 525.19
$(Hg_2)_3[Co(CN)_6]_2$	FM 1 633.62

(a) The unknown weighed 0.432 1 g and the product weighed 0.451 5 g. Find wt% $AgNO_3$ in the unknown. *Caution:* Keep all the digits in your calculator or else serious rounding errors may occur. Do not round off until the end.

(b) Even a skilled analyst is not likely to have less than a 0.3% error in isolating the precipitate. Suppose that there is negligible error in all quantities, except the mass of product, which has an uncertainty of 0.30%. Calculate the relative uncertainty in the mass of $AgNO_3$ in the unknown.

27-25. The thermogravimetric trace on the next page shows mass loss by $Y_2(OH)_5Cl \cdot xH_2O$ upon heating. In the first step, waters of hydration are lost to give ~8.1% mass loss. After a decomposition step, 19.2% of the original mass is lost. Finally, the composition stabilizes at Y_2O_3 above 800°C.

(a) Find x in the formula $Y_2(OH)_5Cl \cdot xH_2O$. Because the 8.1% mass loss is not accurately defined in the experiment, use the 31.8% total mass loss for your calculation.

(b) Suggest a formula for the material remaining at the 19.2% plateau. Be sure that the charges of all ions in your formula sum to 0. The cation is Y^{3+}.

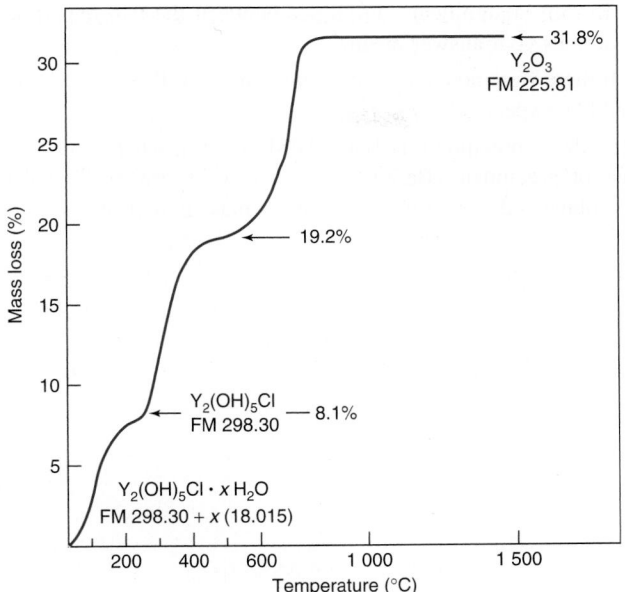

Thermogravimetric analysis of $Y_2(OH)_5Cl \cdot xH_2O$ [Data from T. Hours, P. Bergez, J. Charpin, A. Larbot, C. Guizard, and L. Cot, "Preparation and Characterization of Yttrium Oxide by a Sol-Gel Process," *Ceramic Bull.* **1992**, *71*, 200.]

27-26. *Thermogravimetric analysis and propagation of error.*[16] Crystals of deuterated potassium dihydrogen phosphate, $K(D_xH_{1-x})_2PO_4$, are used in optics as a light valve, as a light deflector, and for frequency doubling of lasers. The optical properties are sensitive to the fraction of deuterium in the material. A publication states that deuterium content can be determined by measuring the mass lost by dehydration of the crystal after slow heating to 450°C in a Pt crucible under N_2.

$$K(D_xH_{1-x})_2PO_4(s) \xrightarrow{450°C} KPO_3(s) + (D_xH_{1-x})_2O(g)$$
FM 136.085 3 + 2.012 55x $\qquad\qquad$ FM 118.070 3

(a) Let α be the mass of product divided by the mass of reactant:

$$\alpha = \frac{\text{mass of } KPO_3}{\text{mass of } K(D_xH_{1-x})_2PO_4}$$

Show that the coefficient x in $K(D_xH_{1-x})_2PO_4$ is related to α by the equation

$$x = \frac{58.667\ 0}{\alpha} - 67.618\ 3$$

What would be the value of α if starting material were 100% deuterated?

(b) One crystal measured in triplicate gave an average value of $\alpha = 0.856\ 7_7$. Find x for this crystal.

(c) From equation B.1 in Appendix B, show that the uncertainty in x (e_x) is related to the uncertainty in α (e_α) by the equation

$$e_x = \frac{58.667\ 0\ e_\alpha}{\alpha^2}$$

(d) The uncertainty in deuterium:hydrogen stoichiometry is e_x. The authors estimate that their uncertainty in α is $e_\alpha = 0.000\ 1$. From e_α, compute e_x. Write the stoichiometry in the form $x \pm e_x$. If e_α were 0.001 (which is perfectly reasonable), what would e_x be?

27-27. When the *high-temperature superconductor* yttrium barium copper oxide (see Chapter 16 opening and Box 16-3) is heated under flowing H_2, the solid remaining at 1 000°C is a mixture of Y_2O_3, BaO, and Cu. The starting material has the formula $YBa_2Cu_3O_{7-x}$, in which the oxygen stoichiometry varies between 7 and 6.5 ($x = 0$ to 0.5).

$$YBa_2Cu_3O_{7-x}(s) + (3.5-x)H_2(g) \xrightarrow{1\ 000°C}$$
$$\text{FM}$$
$$666.19 - 16.00x$$

$$\underbrace{\tfrac{1}{2}Y_2O_3(s) + 2BaO(s) + 3Cu(s)}_{YBa_2Cu_3O_{3.5}} + (3.5-x)H_2O(g)$$

(a) *Thermogravimetric analysis.* When 34.397 mg of $YBa_2Cu_3O_{7-x}$ were subjected to this analysis, 31.661 mg of solid remained after heating to 1 000°C. Find the value of x in $YBa_2Cu_3O_{7-x}$.

(b) *Propagation of error.* Suppose that the uncertainty in each mass in part (a) is ± 0.002 mg. Find the uncertainty in the value of x.

Combustion Analysis

27-28. What is the difference between combustion and pyrolysis?

27-29. What is the purpose of the WO_3 and Cu in Figure 27-8?

27-30. Why is tin used to encapsulate a sample for combustion analysis?

27-31. Why is sample dropped into the preheated furnace before the oxygen concentration reaches its peak in Figure 27-9?

27-32. Write a balanced equation for the combustion of benzoic acid, $C_6H_5CO_2H$, to give CO_2 and H_2O. How many milligrams of CO_2 and of H_2O will be produced by the combustion of 4.635 mg of benzoic acid?

27-33. Write a balanced equation for combustion of $C_8H_7NO_2SBrCl$ in a C,H,N,S elemental analyzer.

27-34. Combustion analysis of a compound known to contain just C, H, N, and O demonstrated that it is 46.21 wt% C, 9.02 wt% H, 13.74 wt% N, and, by difference, $100 - 46.21 - 9.02 - 13.74 = 31.03\%$ O. This means that 100 g of unknown would contain 46.21 g of C, 9.02 g of H, and so on. Find the atomic ratio C:H:N:O and express it as the lowest reasonable integer ratio.

27-35. A mixture weighing 7.290 mg contained only cyclohexane, C_6H_{12} (FM 84.159), and oxirane, C_2H_4O (FM 44.052). When the mixture was analyzed by combustion analysis, 21.999 mg of CO_2 (FM 44.009) were produced. Find the weight percent of oxirane in the mixture.

27-36. Combustion analysis of an organic compound gave the composition 71.17 ± 0.41 wt% C, 6.76 ± 0.12 wt% H, and 10.34 ± 0.08 wt% N. Find the coefficients h and n and their uncertainties x and y in the formula $C_8H_{h \pm x}N_{n \pm y}$.

27-37. Determination of sulfur by combustion analysis produces a mixture of SO_2 and SO_3 that can be passed through H_2O_2 to convert both into H_2SO_4, which is titrated with standard base. When 6.123 mg of a substance were burned, the H_2SO_4 required 3.01 mL of 0.015 76 M NaOH for titration. Find wt% sulfur in the sample.

27-38. *Statistics of coprecipitation.*[17] In Experiment 1, 200.0 mL of solution containing 10.0 mg of SO_4^{2-} (from Na_2SO_4) were treated with excess $BaCl_2$ solution to precipitate $BaSO_4$ containing some coprecipitated Cl^-. To find out how much coprecipitated Cl^- was present, the precipitate was dissolved in 35 mL of 98 wt% H_2SO_4 and boiled to liberate HCl, which was removed by bubbling N_2 gas through the H_2SO_4. The HCl/N_2 stream was passed into a reagent solution that reacted with Cl^- to give a color that was measured. Ten replicate trials gave values of 7.8, 9.8, 7.8, 7.8, 7.8, 7.8, 13.7, 12.7, 13.7, and 12.7 μmol Cl^-. Experiment 2 was identical to the first one, except that the 200.0-mL solution also contained 6.0 g of Cl^- (from NaCl). Ten replicate trials gave 7.8, 10.8, 8.8, 7.8, 6.9, 8.8, 15.7, 12.7, 13.7, and 14.7 μmol Cl^-.

(a) Find the mean, standard deviation, and 95% confidence interval for Cl^- in each experiment.

(b) Is there a significant difference between the two experiments? What does your answer mean?

(c) If there were no coprecipitate, what mass of $BaSO_4$ (FM 233.39) would be expected?

(d) If the coprecipitate is $BaCl_2$ (FM 208.23), what is the average mass of precipitate ($BaSO_4$ + $BaCl_2$) in Experiment 1. By what percentage is the mass greater than the mass in part (c)?

28 | Sample Preparation

Map of Italy, showing where the Po River was sampled in research by E. Zuccato, C. Chiabrando, S. Castiglioni, D. Calamari, R. Bagnati, S. Schiarea, and R. Fanelli, "Cocaine in Surface Waters: A New Evidence-Based Tool to Monitor Community Drug Use," *Environ. Health* **2005**, *4,* 14 available at www.ehjournal.net/content/4/1/14.

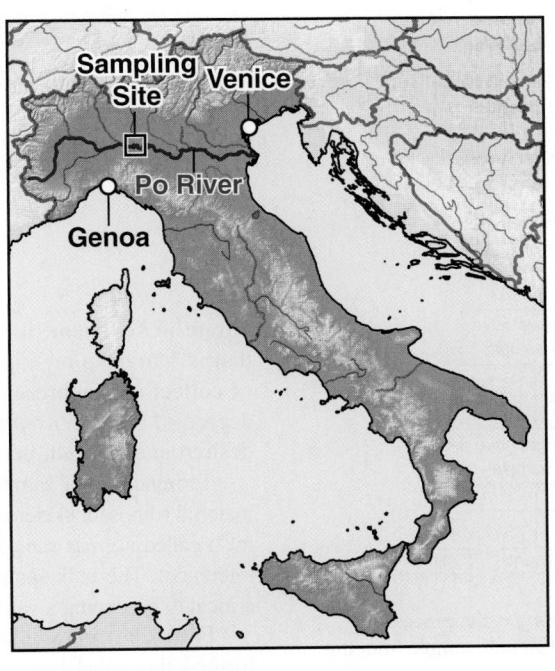

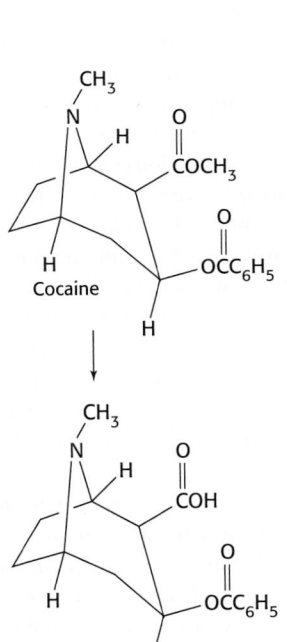

Cocaine

Benzoylecgonine

How honest do you expect people to be when questioned about illegal drug use? In Italy in 2001, 1.1% of people aged 15 to 34 years old acknowledged using cocaine "at least once in the preceding month." Researchers studying the occurrence of therapeutic drugs in sewage realized that they had a tool to measure illegal drug use.

After ingestion, cocaine is largely converted to benzoylecgonine before being excreted in urine. Scientists collected representative composite samples of water from the Po River and samples of wastewater entering treatment plants serving four Italian cities. They **preconcentrated** minute quantities of benzoylecgonine from large volumes of water by solid-phase extraction, which is described in this chapter. Extracted chemicals were washed from the solid phase by a small quantity of solvent, separated by liquid chromatography, and measured by mass spectrometry. Cocaine use was estimated from the concentration of benzoylecgonine, the volume of water flowing in the river, and the population of 5.4 million people living upstream of the collection site.

Benzoylecgonine in the Po River corresponded to 27 ± 5 100-mg doses of cocaine per 1 000 people per day by the 15- to 34-year-old population. Similar results were observed in water from four treatment plants. Cocaine use is much higher than people admit in a survey.

Heterogeneous: different composition from place to place in a material

Homogeneous: same composition everywhere

A chemical analysis is meaningless unless you begin with a meaningful sample. To measure cholesterol in a dinosaur skeleton or herbicide in a truckload of oranges, you must have a strategy for selecting a *representative sample* from a *heterogeneous* material. Figure 28-1 shows that the concentration of nitrate in sediment beneath a lake drops by two orders of magnitude in the first 3 mm below the surface. If you want to measure

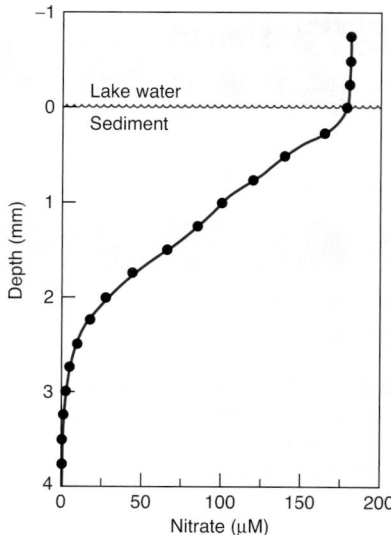

FIGURE 28-1 Depth profile of nitrate in sediment from freshwater Lake Søbygård in Denmark. A similar profile was observed in saltwater sediment. Measurements were made with a *biosensor* containing live bacteria that convert NO_3^- into N_2O, which was measured amperometrically by reduction at a silver cathode. [Data from L. H. Larsen, T. Kjær, and N. P. Revsbech, "A Microscale NO_3^- Biosensor for Environmental Applications," *Anal. Chem.* **1997,** *69,* 3527.]

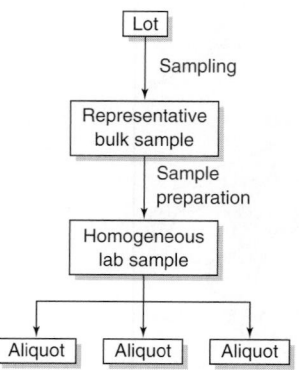

FIGURE 28-2 *Sampling* is the process of selecting a representative bulk sample from the lot. *Sample preparation* is the process that converts a bulk sample into a homogeneous laboratory sample. Sample preparation also refers to steps that eliminate interfering species or that concentrate the analyte.

TABLE 28-1	**Manganese concentration of serum stored in washed and unwashed polyethylene containers**
Container[a]	**Mn (ng/mL)**
Unwashed	0.85
Unwashed	0.55
Unwashed	0.20
Unwashed	0.67
Average	0.57 ± 0.27
Washed	0.096
Washed	0.018
Washed	0.12
Washed	0.10
Average	0.084 ± 0.045

a. Washed containers were rinsed with water distilled twice from fused-silica vessels, which introduce less contamination into water than does glass.

SOURCE: *J. Versieck,* Trends Anal. Chem. **1983,** *2, 110.*

nitrate in sediment, it makes an enormous difference whether you select a core sample that is 1 m deep or skim the top 2 mm of sediment for analysis. **Sampling** is the process of collecting a representative sample for analysis.[1] Real samples generally require some degree of *sample preparation* to remove substances that interfere in the analysis of the desired analyte and, perhaps, to convert analyte into a form suitable for analysis.[2]

Terminology of sampling and sample preparation is shown in Figure 28-2. A *lot* is the total material (dinosaur skeleton or truckload of oranges) from which samples are taken. A *bulk sample* (also called a *gross sample*) is taken from the lot for analysis or for *archiving* (storing for future reference). The bulk sample must be representative of the lot, and the choice of bulk sample is critical to producing a valid analysis. Box 0-1 gave a strategy for sampling heterogeneous material.

From the representative bulk sample, a smaller, homogeneous *laboratory sample* is formed that must have the same composition as the bulk sample. For example, we might obtain a laboratory sample by grinding the entire solid bulk sample to fine powder, mixing thoroughly, and keeping one bottle of powder for testing. Small portions (called *aliquots*) of the laboratory sample are used for individual analyses. *Sample preparation* is the series of steps that convert a representative bulk sample into a form suitable for analysis.

Besides choosing a sample judiciously, we must be careful about storing the sample. The composition may change with time after collection because of chemical changes, reaction with air, or interaction of the sample with its container. Glass is a notorious ion exchanger that alters the concentrations of trace ions and proteins[3] in solution. Therefore, plastic (especially Teflon) collection bottles are frequently employed. Even these materials can absorb trace levels of analytes. For example, a 0.2 μM $HgCl_2$ solution lost 40–95% of its concentration in 4 h in polyethylene bottles. A 2 μM Ag^+ solution in a Teflon bottle lost 2% of its concentration in a day and 28% in a month.[4]

Plastic containers must be washed before use. Table 28-1 shows that manganese in blood serum samples increased by a factor of seven when stored in unwashed polyethylene containers prior to analysis. In the most demanding *trace analysis* of lead at 1 pg/g in polar ice cores, polyethylene containers contributed a measurable flux of 1 fg of lead per cm^2 per day even after they had been soaked in acid for seven months.[5] Steel needles are an avoidable source of metal contamination in biochemical analysis.

A study of mercury in Lake Michigan found levels near 1.6 pM (1.6×10^{-12} M), which is *two orders of magnitude* below concentrations observed in earlier studies.[6] Previous investigators apparently unknowingly contaminated their samples. A study of handling techniques for the analysis of lead in rivers investigated variations in sample collection, sample containers, protection during transportation, filtration, preservatives, and preconcentration procedures.[7] Each step that deviated from best practice *doubled* the apparent concentration of lead in stream water. Clean rooms with filtered air supplies are essential in trace analysis. Even with the best precautions, the precision of trace analysis becomes poorer as the concentration of analyte decreases (Box 5-3).

"Unless the complete history of any sample is known with certainty, the analyst is well advised not to spend his [or her] time in analyzing it."[8] Your laboratory notebook should describe how a sample was collected and stored and exactly how it was handled, as well as stating how it was analyzed.

28-1 Statistics of Sampling[9]

For random errors, the overall *variance*, s_o^2, is the sum of the variance of the analytical procedure, s_a^2, and the variance of the sampling operation, s_s^2:

Additivity of variance: $$s_o^2 = s_a^2 + s_s^2 \qquad (28\text{-}1)$$

If either s_a or s_s is sufficiently smaller than the other, there is little point in trying to reduce the smaller one. For example, if s_s is 10% and s_a is 5%, the overall standard deviation is $11\% (\sqrt{0.10^2 + 0.05^2} = 0.11)$. A more expensive and time-consuming analytical procedure that reduces s_a to 1% only improves s_o from 11 to $10\% (\sqrt{0.10^2 + 0.01^2} = 0.10)$.

Variance = (standard deviation)²

$$\frac{\text{total}}{\text{variance}} = \frac{\text{analytical}}{\text{variance}} + \frac{\text{sampling}}{\text{variance}}$$

Origin of Sampling Variance

To understand the nature of the uncertainty in selecting a sample for analysis, consider a random mixture of two kinds of solid particles. The theory of probability allows us to state the likelihood that a randomly drawn sample has the same composition as the bulk sample. It might surprise you to learn how large a sample is required for accurate sampling.[10]

Suppose that the mixture contains n_A particles of type A and n_B particles of type B. The probabilities of drawing A or B from the mixture are

$$p = \text{probability of drawing A} = \frac{n_A}{n_A + n_B} \qquad (28\text{-}2)$$

$$q = \text{probability of drawing B} = \frac{n_B}{n_A + n_B} = 1 - p \qquad (28\text{-}3)$$

If n particles are drawn at random, the expected number of particles of type A are np and the standard deviation of many drawings is known from the binomial distribution to be

Standard deviation in sampling operation: $$s_n = \sqrt{npq} \qquad (28\text{-}4)$$

EXAMPLE Statistics of Drawing Particles

A mixture contains 1% KCl particles and 99% KNO_3 particles. If 10^4 particles are taken, what is the expected number of KCl particles, and what will be the standard deviation if the experiment is repeated many times?

Solution The expected number is

$$\text{Expected number of KCl particles} = np = (10^4)(0.01) = 100 \text{ particles}$$

and the standard deviation will be

$$\text{Standard deviation} = \sqrt{npq} = \sqrt{(10^4)(0.01)(0.99)} = 9.9$$

The standard deviation $\sqrt{npq}$ applies to both kinds of particles. The standard deviation is 9.9% of the expected number of KCl particles, but only 0.1% of the expected number of KNO_3 particles ($nq = 9\,900$). To measure nitrate, this sample is probably sufficient. For chloride, 9.9% uncertainty may not be acceptable.

TEST YOURSELF If 10^5 particles are taken, what is the relative standard deviation of each measurement? (***Answer:*** 3% for KCl and 0.03% for KNO_3)

How much sample corresponds to 10^4 particles? Suppose that the particles are 1-mm-diameter spheres. The volume of a 1-mm-diameter sphere is $\frac{4}{3}\pi(0.5 \text{ mm})^3 = 0.524 \text{ }\mu\text{L}$. The density of KCl is 1.984 g/mL and that of KNO_3 is 2.109 g/mL, so the average density of the mixture is $(0.01)(1.984) + (0.99)(2.109) = 2.108$ g/mL. The mass of mixture containing 10^4 particles is $(10^4)(0.524 \times 10^{-3} \text{ mL})(2.108 \text{ g/mL}) = 11.0$ g. *If you take 11.0-g test portions from a larger laboratory sample, the expected sampling standard deviation for chloride is 9.9%.* The sampling standard deviation for nitrate will be only 0.1%.

Wow! I'm surprised that an 11.0-g sample has such a large standard deviation.

How can you prepare a mixture of 1-mm-diameter particles? You might make such a mixture by grinding larger particles and passing them through a 16 mesh sieve, whose screen openings are squares with sides of length 1.18 mm (Table 28-2). Particles that pass through the screen are then passed through a 20 mesh sieve, whose openings are 0.85 mm, and material that

TABLE 28-2	Standard test sieves		
Sieve number	Screen opening (mm)	Sieve number	Screen opening (mm)
5	4.00	45	0.355
6	3.35	50	0.300
7	2.80	60	0.250
8	2.36	70	0.212
10	2.00	80	0.180
12	1.70	100	0.150
14	1.40	120	0.125
16	1.18	140	0.106
18	1.00	170	0.090
20	0.850	200	0.075
25	0.710	230	0.063
30	0.600	270	0.053
35	0.500	325	0.045
40	0.425	400	0.038

EXAMPLE: *Particles designated 50/100 mesh pass through a 50 mesh sieve but are retained by a 100 mesh sieve. Their size is in the range 0.150–0.300 mm.*

does not pass is retained for your sample. This procedure gives particles whose diameters are in the range 0.85–1.18 mm. We refer to the size range as 16/20 *mesh*.

Suppose that much finer particles of 80/120 mesh size (average diameter = 152 μm, average volume = 1.84 nL) were used instead. Now the mass containing 10^4 particles is reduced from 11.0 to 0.038 8 g. We could analyze a larger sample to reduce the sampling uncertainty for chloride.

EXAMPLE Reducing Sample Uncertainty with a Larger Test Portion

How many grams of 80/120 mesh sample are required to reduce the chloride sampling uncertainty to 1%?

Solution We are looking for a standard deviation of 1% of the number of KCl particles (= 1% of np):

$$\sigma_n = \sqrt{npq} = (0.01)np$$

Using $p = 0.01$ and $q = 0.99$, we find $n = 9.9 \times 10^5$ particles. With a particle volume of 1.84 nL and an average density of 2.108 g/mL, the mass required for 1% chloride sampling uncertainty is

$$\text{Mass} = (9.9 \times 10^5 \text{ particles})\left(1.84 \times 10^{-6}\frac{\text{mL}}{\text{particle}}\right)\left(2.108\frac{\text{g}}{\text{mL}}\right) = 3.84 \text{ g}$$

Even with an average particle diameter of 152 μm, we must analyze 3.84 g to reduce the sampling uncertainty to 1%. There is no point using an expensive analytical method with a precision of 0.1%, because the overall uncertainty will still be 1% from sampling.

TEST YOURSELF What mass of 170/200 mesh particles reduces the KCl sampling uncertainty to 1%? (**Answer:** particle diameter = 0.082_5 mm, 9.9×10^5 particles, 0.61 g)

There is no advantage to reducing the analytical uncertainty if the sampling uncertainty is high, and vice versa.

TABLE 28-3	Nickel content of crushed ore
Particle mesh size	Ni content (wt%)
>230	13.52 ± 0.69
120/230	13.20 ± 0.74
25/120	13.22 ± 0.49
10/25	10.54 ± 0.84
>10	9.08 ± 0.69

NOTE: *Uncertainty is ±1 standard deviation.*

SOURCE: *J. G. Dunn, D. N. Phillips, and W. van Bronswijk. "An Exercise to Illustrate the Importance of Sample Preparation in Analytical Chemistry," J. Chem. Ed. **1997**, 74, 1188.*

Sampling uncertainty arises from the random nature of drawing particles from a mixture. If the mixture is a liquid and the particles are molecules, there are about 10^{22} particles/mL. It will not require much volume of homogeneous liquid solution to reduce the sampling error to a negligible value. Solids, however, must be ground to very fine dimensions, and large quantities must be used to ensure a small sampling variance. Grinding invariably contaminates the sample with material from the grinding apparatus.

Table 28-3 illustrates another problem with heterogeneous materials. Nickel ore was crushed into small particles that were sieved and analyzed. Parts of the ore that are deficient in nickel are relatively resistant to fracture, so the larger particles do not have the same chemical composition as the smaller particles. It is necessary to grind the entire ore to a fine powder to have any hope of obtaining a representative sample.

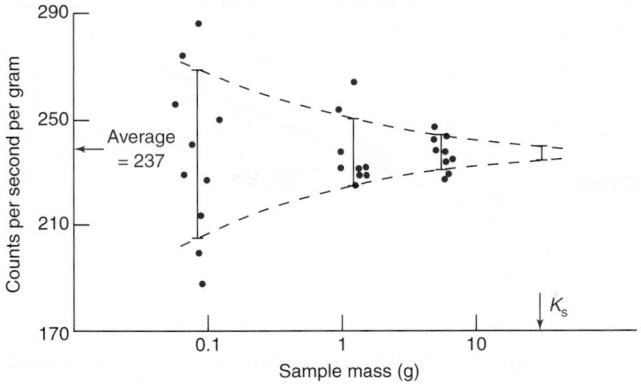

FIGURE 28-3 Sampling diagram of experimental results for ^{24}Na in liver homogenate. Dots are experimental points, and error bars extend ±1 standard deviation about the mean. Abscissa is logarithmic. [Data from B. Kratochvil and J. K. Taylor, "Sampling for Chemical Analysis," *Anal. Chem.* **1981,** *53,* 925A; National Bureau of Standards Internal Report 80-2164, 1980, p. 66.]

Choosing a Sample Size

A well-mixed powder containing KCl and KNO$_3$ is an example of a heterogeneous material in which the variation from place to place is random. *How much of a random mixture should be analyzed to reduce the sampling variance for one analysis to a desired level?*

To answer this question, consider Figure 28-3, which shows results for sampling the radioisotope ^{24}Na in human liver. The tissue was "homogenized" in a blender but was not truly homogeneous because it was a suspension of small particles in water. The average number of radioactive counts per second per gram of sample was about 237. When the mass of sample for each analysis was about 0.09 g, the standard deviation (shown by the error bar at the left in the diagram) was ±31 counts per second per gram of homogenate, which is ±13.1% of the mean value (237). When the sample size was increased to about 1.3 g, the standard deviation decreased to ±13 counts/s/g, or ±5.5% of the mean. For a sample size near 5.8 g, the standard deviation was reduced to ±5.7 counts/s/g, or ±2.4% of the mean.

Equation 28-4 told us that, when *n* particles are drawn from a mixture of two kinds of particles (such as liver tissue particles and droplets of water), the sampling standard deviation will be $\sigma_n = \sqrt{npq}$, where *p* and *q* are the fraction of each kind of particle present. The relative standard deviation is $\sigma_n/n = \sqrt{npq}/n = \sqrt{pq/n}$. The relative variance is

$$\text{Relative variance} \equiv R^2 = \left(\frac{\sigma_n}{n}\right)^2 = \frac{pq}{n} \quad \Rightarrow nR^2 = pq \quad \textbf{(28-5)}$$

Noting that the mass of sample drawn, *m*, is proportional to the number of particles drawn, we can rewrite Equation 28-5 in the form

Sampling constant: $\qquad\qquad mR^2 = K_s \qquad\qquad\qquad$ **(28-6)**

in which *R* is the relative standard deviation (expressed as a percentage) due to sampling and K_s is called the *sampling constant*. K_s is the mass of sample producing a relative sampling standard deviation of 1%.

Let's see if Equation 28-6 describes Figure 28-3. Table 28-4 shows that mR^2 is approximately constant for large samples, but agreement is poor for the smallest sample. Attributing the poor agreement at low mass to random sampling variation, we assign $K_s \approx 36$ g in Equation 28-6. This is the average from the 1.3- and 5.8-g samples in Table 28-4.

TABLE 28-4	Calculation of sampling constant for Figure 28-3	
Sample mass, *m* (g)	Relative standard deviation (%)	mR^2 (g)
0.09	13.1	15.4
1.3	5.5	39.3
5.8	2.4	33.4

EXAMPLE **Mass of Sample Required to Produce a Given Sampling Variance**

What mass in Figure 28-3 will give a sampling standard deviation of ±7%?

Solution With the sampling constant $K_s \approx 36$ g, the answer is

$$m = \frac{K_s}{R^2} = \frac{36\ \text{g}}{7^2} = 0.73\ \text{g}$$

A 0.7-g sample should give ~7% sampling standard deviation. This is strictly a sampling standard deviation. The net variance will be the sum of variances from sampling and from analysis (Equation 28-1).

TEST YOURSELF By what factor must the mass increase to reduce the sampling standard deviation by a factor of 2? (*Answer:* four times greater)

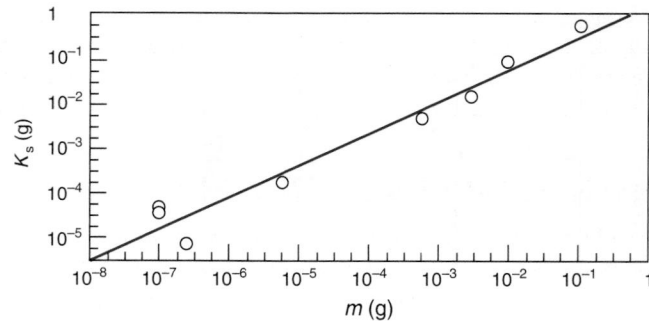

FIGURE 28-4 Data for Mn in powdered algae show that the sampling constant, K_s, in Equation 28-6 is approximately proportional to sample mass, m, over six orders of magnitude of mass. [Data from M. Rossbach and E. Zeiller, "Assessment of Element-Specific Homogeneity in Reference Materials Using Microanalytical Techniques," *Anal. Bioanal. Chem.* **2003,** *377,* 334.]

Figure 28-4 shows an example in which Equation 28-6 is approximately valid over six orders of magnitude of sample mass.

Choosing the Number of Replicate Analyses

We just saw that a single 0.7-g sample is expected to give a sampling standard deviation of $\pm 7\%$. *How many 0.7-g samples must be analyzed to give 95% confidence that the mean is known to within $\pm 4\%$?* The meaning of 95% confidence is that there is only a 5% chance that the true mean lies more than $\pm 4\%$ away from the measured mean. The question we just asked refers only to sampling uncertainty and assumes that analytical uncertainty is negligible.

Rearranging Student's t equation 4-7 allows us to answer the question:

The sampling contribution to the overall uncertainty can be reduced by analyzing more samples.

Required number of replicate analyses:
$$\underbrace{\mu - \bar{x}}_{e} = \frac{t s_s}{\sqrt{n}} \Rightarrow n = \frac{t^2 s_s^2}{e^2} \tag{28-7}$$

in which μ is the true population mean, $\bar{x}$ is the measured mean, n is the number of samples needed, s_s^2 is the sampling variance, and e is the uncertainty we seek. Both s_s and e must be expressed as absolute uncertainties or both must be expressed as relative uncertainties. Student's t is taken from Table 4-4 for 95% confidence at $n - 1$ degrees of freedom. Because n is not yet known, the value of t for $n = \infty$ can be used to estimate n. After a value of n is calculated, repeat the process a few times until n becomes constant.

> **EXAMPLE** **Sampling a Random Bulk Material**
>
> How many 0.7-g samples must be analyzed to give 95% confidence that the mean is known to within $\pm 4\%$?
>
> **Solution** A 0.7-g sample gives $s_s = 7\%$, and we are seeking $e = 4\%$. We will express both uncertainties in relative form. Taking $t = 1.960$ (from Table 4-4 for 95% confidence and ∞ degrees of freedom) as a starting value, we find
>
> $$n \approx \frac{(1.960)^2 (0.07)^2}{(0.04)^2} = 11.8 \approx 12$$
>
> For $n = 12$, there are 11 degrees of freedom, so a second trial value of Student's t (interpolated from Table 4-4) is 2.209. A second cycle of calculation gives
>
> $$n \approx \frac{(2.209)^2 (0.07)^2}{(0.04)^2} = 14.9 \approx 15$$
>
> For $n = 15$, there are 14 degrees of freedom and $t = 2.150$, which gives
>
> $$n \approx \frac{(2.150)^2 (0.07)^2}{(0.04)^2} = 14.2 \approx 14$$
>
> For $n = 14$, there are 13 degrees of freedom and $t = 2.170$, which gives
>
> $$n \approx \frac{(2.170)^2 (0.07)^2}{(0.04)^2} = 14.4 \approx 14$$
>
> The calculations reach a constant value near $n \approx 14$, so we need 14 samples of 0.7-g size to determine the mean value to within 4% with 95% confidence.
>
> **TEST YOURSELF** How many 2.8-g samples must be analyzed to give 95% confidence that the mean is known to within $\pm 4\%$? (***Answer:*** 6)

For the preceding calculations, we needed prior knowledge of the standard deviation. A preliminary study of the sample must be made before the remainder of the analysis can be planned. If there are many similar samples to be analyzed, a thorough analysis of one sample might allow you to plan less thorough—but adequate—analyses of the remainder.

28-2 Dissolving Samples for Analysis[11]

Once a *bulk sample* is selected, a *laboratory sample* must be prepared for analysis (Figure 28-2). A coarse solid sample should be ground and mixed so that the lab sample has the same composition as the bulk. Solids are typically dried at 110°C at atmospheric pressure to remove adsorbed water prior to analysis. Temperature-sensitive samples may simply be stored in an environment that brings them to a constant, reproducible moisture level.

The laboratory sample is usually dissolved for analysis. It is important to dissolve the entire sample, or else we cannot be sure that all of the analyte dissolved. If the sample does not dissolve under mild conditions, *acid digestion* or *fusion* may be used. Organic material may be destroyed by *combustion* (also called *dry ashing*) or *wet ashing* (oxidation with liquid reagents) to place inorganic elements in suitable form for analysis.

Grinding

Solids can be ground in a **mortar and pestle** like those shown in Figure 28-5. The steel mortar (also called a percussion mortar or "diamond" mortar) is a hardened steel tool into which the sleeve and pestle fit snugly. Ores and minerals can be crushed by striking the pestle lightly with a hammer. The *agate* mortar (or similar ones made of porcelain, mullite, or alumina) is designed for grinding small particles into a fine powder. Less expensive mortars tend to be more porous and more easily scratched, which leads to contamination of the sample with mortar material or portions of previously ground samples. A ceramic mortar can be cleaned by wiping with a wet tissue and washing with distilled water. Difficult residues can be removed by grinding with 4 M HCl in the mortar or by grinding an abrasive household scouring powder, followed by washing with HCl and water. A *boron carbide* mortar and pestle is five times harder than agate and less likely to contaminate the sample.

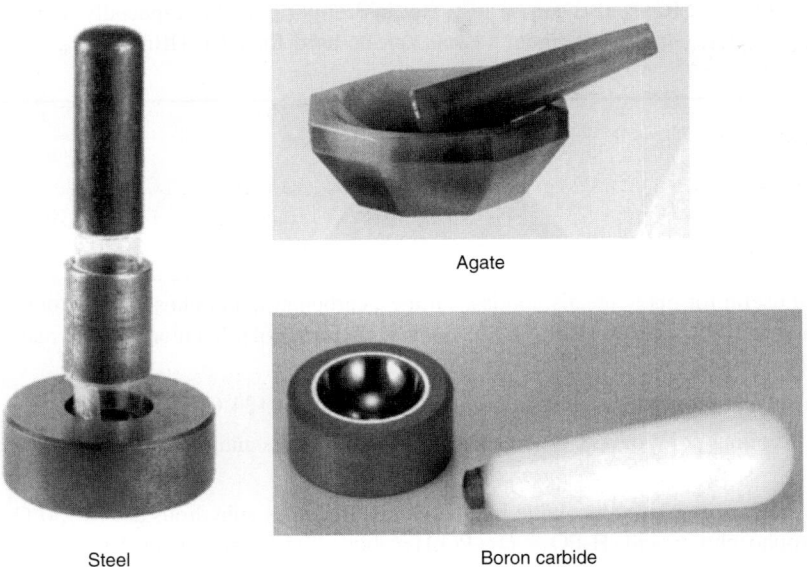

Agate

Steel

Boron carbide

FIGURE 28-5 Steel, agate, and boron carbide mortars and pestles. The mortar is the bowl and the pestle is the grinding tool. The boron carbide mortar is a hemispheric shell enclosed in a plastic or aluminum body. The pestle has a boron carbide button at the tip of a plastic handle. [Courtesy Thomas Scientific, Swedesboro, NJ.]

A **ball mill** is a grinding device in which steel or ceramic balls are rotated inside a drum to crush the sample to a fine powder. A Wig-L-Bug® pulverizes a sample by shaking it in a vial with a ball that moves back and forth. For soft materials, plastic vials and balls are appropriate. For harder materials, steel, agate, and tungsten carbide are used. A Shatterbox laboratory mill spins a puck and ring inside a grinding container at 825 revolutions per minute to pulverize up to 100 g of material (Figure 28-6). Tungsten carbide and zirconia containers are used for very hard samples.

Shatterbox grinding action

FIGURE 28-6 Shatterbox laboratory mill spins a puck and ring inside a container at high speed to grind up to 100 mL of sample to a fine powder. [SPEX SamplePrep LLC.]

Dissolving Inorganic Materials with Acids

Table 28-5 lists acids commonly used for dissolving inorganic materials. The nonoxidizing acids HCl, HBr, HF, H_3PO_4, dilute H_2SO_4, and dilute $HClO_4$ dissolve metals by the redox reaction

$$M + nH^+ \rightarrow M^{n+} + \frac{n}{2}H_2 \qquad (28\text{-}8)$$

Metals with negative reduction potentials should dissolve, although some, such as Al, form a protective oxide coat that inhibits dissolution. Volatile species formed by protonation of anions such as carbonate ($CO_3^{2-} \rightarrow H_2CO_3 \rightarrow CO_2$), sulfide ($S^{2-} \rightarrow H_2S$), phosphide ($P^{3-} \rightarrow PH_3$), fluoride ($F^- \rightarrow HF$), and borate ($BO_3^{3-} \rightarrow H_3BO_3$) will be lost from hot acids in open vessels. Volatile metal halides such as $SnCl_4$ and $HgCl_2$ and some molecular oxides such as OsO_4 and RuO_4 also can be lost. Hot hydrofluoric acid is especially useful for dissolving silicates. Glass or platinum vessels can be used for HCl, HBr, H_2SO_4, H_3PO_4, and

HF causes excruciating burns. **Exposure of just 2% of your body to concentrated (48 wt%) HF can kill you.** Flood the affected area with water for 5 min and then coat the skin with 2.5% calcium gluconate gel kept in the lab for this purpose, and seek medical help. If the gel is not available, use whatever calcium salt is handy. HF damage can continue to develop days after exposure.

TABLE 28-5	Acids for sample dissolution	
Acid	**Typical composition (wt% and density)**	**Notes**
HCl	37% 1.19 g/mL	Nonoxidizing acid useful for many metals, oxides, sulfides, carbonates, and phosphates. Constant boiling composition at 109°C is 20% HCl. As, Sb, Ge, and Pb form volatile chlorides that may be lost from an open vessel.
HBr	48–65%	Similar to HCl in solvent properties. Constant boiling composition at 124°C is 48% HBr.
H_2SO_4	95–98% 1.84 g/mL	Good solvent at its boiling point of 338°C. Attacks metals. Dehydrates and oxidizes organic compounds.
H_3PO_4	85% 1.70 g/mL	Hot acid dissolves refractory oxides insoluble in other acids. Becomes anhydrous above 150°C. Dehydrates to pyrophosphoric acid (H_2PO_3—O—PO_3H_2) above 200°C and dehydrates further to metaphosphoric acid ($[HPO_3]_n$) above 300°C.
HF	50% 1.16 g/mL	Used primarily to dissolve silicates, making volatile SiF_4. This product and excess HF are removed by adding H_2SO_4 or $HClO_4$ and heating. Constant boiling composition at 112°C is 38% HF. Used in Teflon, silver, or platinum containers. Extremely harmful upon contact or inhalation. Fluorides of As, B, Ge, Se, Ta, Nb, Ti, and Te are volatile. LaF_3, CaF_2, and YF_3 precipitate. F^- is removed by adding H_3BO_3 and taking to dryness with H_2SO_4 present.
$HClO_4$	60–72% 1.54–1.67 g/mL	Cold and dilute acid are not oxidizing, but hot, concentrated acid is an extremely powerful, explosive oxidant, especially useful for organic matter that has already been partially oxidized by hot HNO_3. Constant boiling composition at 203°C is 72%. **Before using $HClO_4$, evaporate the sample to near dryness several times with hot HNO_3 to destroy as much organic material as possible.**

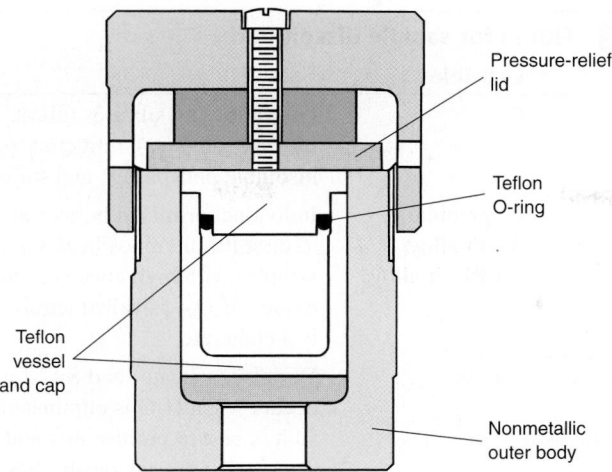

FIGURE 28-7 Microwave digestion bomb lined with Teflon. The outer container retains strength to 150°C but rarely reaches 50°C.
[Information from Parr Instrument Co., Moline, IL.]

$HClO_4$. HF should be used in Teflon, polyethylene, silver, or platinum vessels. The highest quality acids must be used to minimize contamination by the concentrated reagent.

Substances that do not dissolve in nonoxidizing acids may dissolve in the oxidizing acids HNO_3, hot, concentrated H_2SO_4, or hot, concentrated $HClO_4$. Nitric acid attacks most metals, but not Au and Pt, which dissolve in the 3:1 (vol/vol) mixture of $HCl:HNO_3$ called **aqua regia.** Strong oxidants such as Cl_2 or $HClO_4$ in HCl dissolve difficult materials such as Ir at elevated temperature. A mixture of HNO_3 and HF attacks refractory carbides, nitrides, and borides of Ti, Zr, Ta, and W. A powerful oxidizing solution known as "piranha solution" is a mixture of concentrated (30 wt%) H_2O_2 plus concentrated (98 wt%) H_2SO_4 variously described as 1:1 or 3:7 (vol/vol) mixtures. Hot, concentrated $HClO_4$ (described later for organic substances) is a **DANGEROUS**, powerful oxidant whose oxidizing power is increased by adding concentrated H_2SO_4 and catalysts such as V_2O_5 or CrO_3.

Digestion is conveniently carried out in a Teflon-lined **bomb** (a sealed vessel) heated in a microwave oven.[12] The vessel in Figure 28-7 has a volume of 23 mL and digests up to 1 g of inorganic material in up to 15 mL of concentrated acid or digests 0.1 g of organic material, which releases a great deal of $CO_2(g)$. Microwave energy heats the contents to 200°C in a minute. To prevent explosions, the lid releases gas from the vessel if the internal pressure exceeds 8 MPa (80 bar). The bomb cannot be made of metal, which absorbs microwaves. Cool the bomb prior to opening to prevent loss of volatile products.

An example of a complex sample that can be digested by microwave heating is mixed electronic waste.[13] Circuit boards are cut into small pieces and mixed thoroughly. Then 0.1 g is digested with 6 mL of 70 wt% HNO_3/2 mL 30 wt% H_2O_2/1 mL 49 wt% HF. Raw plastics can be digested with 9 mL of 70 wt% HNO_3. Power was ramped from 0 to 600 W over 5 min, held for 15 min, ramped to 1 400 W over 5 min, and held for 20 min.

Dissolving Inorganic Materials by Fusion

Substances that will not dissolve in acid can usually be dissolved by a hot, molten inorganic **flux** (Table 28-6). Finely powdered unknown is mixed with 2–20 times its mass of solid flux, and **fusion** (melting) is carried out in a platinum-gold alloy crucible at 300–1 200°C in a furnace or over a burner. Apparatus in Figure 28-8 fuses three samples at once over propane burners while rotating the crucibles to homogenize the melt. Then the molten flux is poured into beakers containing 10 wt% aqueous HNO_3 to dissolve the product.

Most fusions use lithium tetraborate ($Li_2B_4O_7$, m.p. 930°C), lithium metaborate ($LiBO_2$, m.p. 845°C), or a mixture of the two. A nonwetting agent such as KI can be added to prevent the flux from sticking to the crucible. For example, 0.2 g of cement might be fused with 2 g of $Li_2B_4O_7$ and 30 mg of KI.

A disadvantage of a flux is that impurities are introduced by the large mass of solid reagent. If part of the unknown can be dissolved with acid prior to fusion, it should be dissolved. Then the insoluble component is dissolved with less flux and the two portions are combined for analysis.

Basic fluxes in Table 28-6 ($LiBO_2$, Na_2CO_3, NaOH, KOH, and Na_2O_2) are best used to dissolve acidic oxides of Si and P. Acidic fluxes ($Li_2B_4O_7$, $Na_2B_4O_7$, $K_2S_2O_7$, and B_2O_3) are most suitable for basic oxides (including cements and ores) of the alkali metals, alkaline earths, lanthanides, and Al. KHF_2 is useful for lanthanide oxides. Sulfides and some oxides,

FIGURE 28-8 Automated apparatus that fuses three samples at once over propane burners. The Pt/Au crucibles rotate while they are inclined. [Courtesy Claisse, Quebec.]

Fusion is a last resort because the flux can introduce impurities.

TABLE 28-6 Fluxes for sample dissolution

Flux	Crucible	Uses
Na_2CO_3	Pt	For dissolving silicates (clays, rocks, minerals, glasses), refractory oxides, insoluble phosphates, and sulfates.
$Li_2B_4O_7$ or $LiBO_2$ or $Na_2B_4O_7$	Pt, graphite Au-Pt alloy, Au-Rh-Pt alloy	Individual or mixed borates are used to dissolve aluminosilicates, carbonates, and samples with high concentrations of basic oxides. $B_4O_7^{2-}$ is called tetraborate and BO_2^- is metaborate.
NaOH or KOH	Au, Ag	Dissolves silicates and SiC. Frothing occurs when H_2O is eliminated from flux, so it is best to prefuse flux and then add sample. Analytical capabilities are limited by impurities in NaOH and KOH.
Na_2O_2	Zr, Ni	Strong base and powerful oxidant good for silicates not dissolved by Na_2CO_3. Useful for iron and chromium alloys. Because it slowly attacks crucibles, a good procedure is to coat the inside of a Ni crucible with molten Na_2CO_3, cool, and add Na_2O_2. The peroxide melts at lower temperature than the carbonate, which shields the crucible from the melt.
$K_2S_2O_7$	Porcelain, SiO_2, Au, Pt	Potassium pyrosulfate ($K_2S_2O_7$) is prepared by heating $KHSO_4$ until all water is lost and foaming ceases. Alternatively, potassium persulfate ($K_2S_2O_8$) decomposes to $K_2S_2O_7$ upon heating. Good for refractory oxides, not silicates.
B_2O_3	Pt	Useful for oxides and silicates. Principal advantage is that flux can be completely removed as volatile methyl borate ($[CH_3O]_3B$) by several treatments with HCl in methanol.
$Li_2B_4O_7$ + Li_2SO_4 (2 : 1 wt/wt)	Pt	Example of a powerful mixture for dissolving refractory silicates and oxides in 10–20 min at 1 000°C. One gram of flux dissolves 0.1 g of sample. The solidified melt dissolves readily in 20 mL of hot 1.2 M HCl.

some iron and platinum alloys, and some silicates require an oxidizing flux for dissolution. For this purpose, pure Na_2O_2 may be suitable or oxidants such as KNO_3, $KClO_3$, or Na_2O_2 can be added to Na_2CO_3. Boric oxide can be converted into $B(OCH_3)_3$ after fusion and completely evaporated. Treat the solidified flux with 100 mL of CH_3OH saturated with HCl gas and heat gently. Repeat the procedure several times to remove all of the boron.

Platinum crucibles are expensive and should be heated in air, not a reducing atmosphere. Hot Pt should only be handled with Pt-tipped tongs. You can put Pt foil on the tip of ordinary tongs to handle Pt crucibles. The hot crucible should only be placed on a clean, inert surface, or in a Pt/Ir triangle. Carbon from smoky flames can embrittle Pt. Other elements including Sb, As, Pb, P, Se, Te, and Zn also embrittle Pt. Molten Ag, Au, and most base metals dissolve Pt. Oxides of Fe and Pb harm Pt above 1 000°C, as will silicates under reducing conditions.

Decomposition of Organic Substances

Digestion of organic material is classified as either **dry ashing,** when the procedure does not include liquid, or **wet ashing,** when liquid is used. Occasionally, fusion with Na_2O_2 (called

Parr oxidation) or alkali metals may be carried out in a sealed bomb. Section 27-4 discussed *combustion analysis*, in which C, H, N, S, and halogens are measured.

One form of *dry ashing* is microwave-induced combustion, examples of which are the analysis of halogens in coal[14] and metals in plant matter.[15] For example, 50- to 500-mg pellets of coal were wrapped in low-ash filter paper and placed on a quartz holder in a quartz vessel containing 6 mL of 50 mM $(NH_4)_2CO_3$ solution at the bottom. After wetting the filter paper with 50 mL of 6 M NH_4NO_3 (an oxidizer), the vessel was capped and pressurized with 20 bar of O_2. Application of 1 400 W of microwave power initiated combustion in which the coal reached a temperature of 1 400°C. Halides released by combustion dissolved in the $(NH_4)_2CO_3$ solution and were measured by ion chromatography.

Convenient *wet ashing* procedures include microwave digestion with acid in a Teflon bomb (Figure 28-7). For example, 0.25 g of animal tissue can be digested for metal analysis by placing the sample in a 60-mL Teflon vessel containing 1.5 mL of high-purity 70% HNO_3 plus 1.5 mL of high-purity 96% H_2SO_4 and heating it in a 700-W kitchen microwave oven for 1 min.[16] Teflon bombs fitted with temperature and pressure sensors allow safe, programmable control of digestion conditions. An important wet-ashing process is *Kjeldahl digestion* with H_2SO_4 for nitrogen analysis (Section 11-8).

In the *Carius method*, digestion is performed with fuming HNO_3 (which contains excess dissolved NO_2) in a sealed, heavy-walled glass tube at 200–300° C. For safety, the glass Carius tube should be contained in a steel vessel pressurized to approximately the same pressure expected inside the glass tube.[17] For trace analysis, sample should be placed inside a fused-silica tube inside the glass tube. Silica provides as little as 1–10% as much extractable metal as glass.[18]

Figure 28-9 shows microwave wet-ashing apparatus. Sulfuric acid or a mixture of H_2SO_4 and HNO_3 (~15 mL of acid per gram of unknown) is added to an organic substance in a glass digestion tube fitted with the reflux cap. In the first step, sample is *carbonized* for 10–20 min at gentle reflux until all particles have dissolved and the solution has a uniform black appearance. Power is turned off and the sample is allowed to cool for 1–2 min. Next, *oxidation* is carried out by adding H_2O_2 or HNO_3 through the reflux cap until the color disappears or the solution is just barely tinted. If the solution is not homogeneous, the power is turned up and the sample is heated to bring all solids into solution. Repeated cycles of oxidation and dissolution may be required. Once conditions for a particular type of material are found, the procedure is automated, with power levels and reagent delivery (by the peristaltic pump) programmed into the controller.

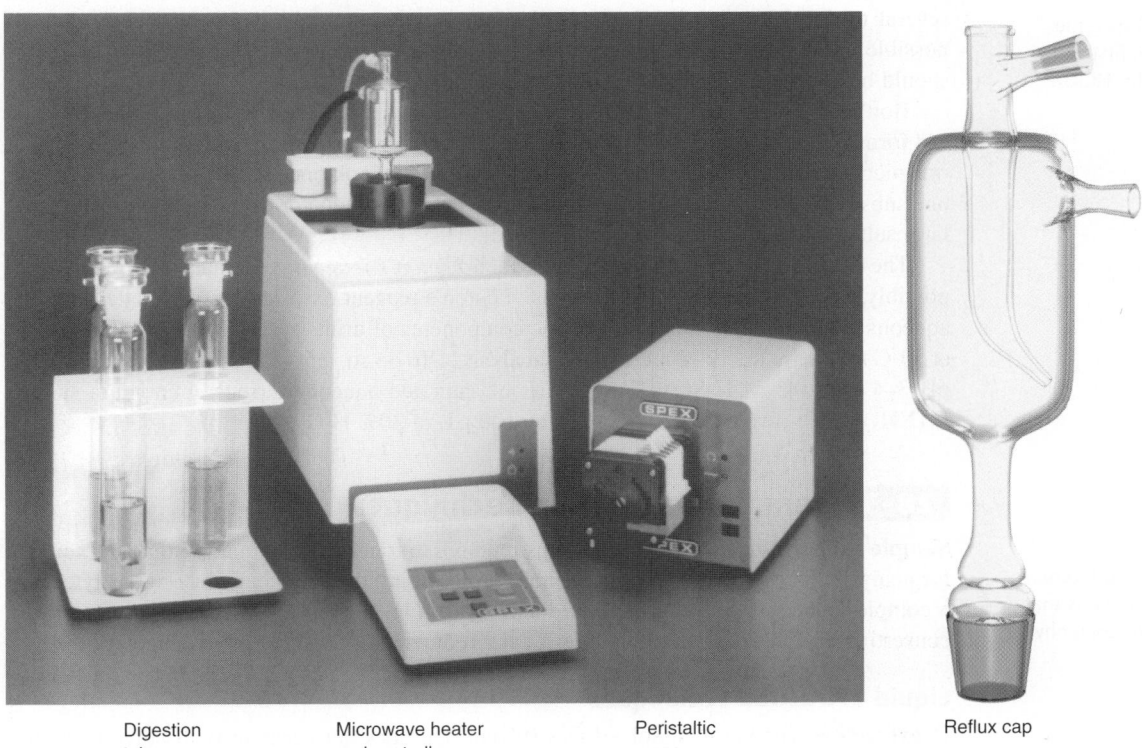

| Digestion tubes | Microwave heater and controller | Peristaltic pump | Reflux cap |

FIGURE 28-9 Microwave apparatus for digesting organic materials by wet ashing. [SPEX SamplePrep LLC.]

FIGURE 28-10 High-pressure autoclave allows digestion up to 270°C without H_2SO_4 in open vessels inside autoclave. [Information from B. Maichin, M. Zischka, and G. Knapp, "Pressurized Wet Digestion in Open Vessels," *Anal. Bioanal. Chem.* **2003,** *376,* 715.]

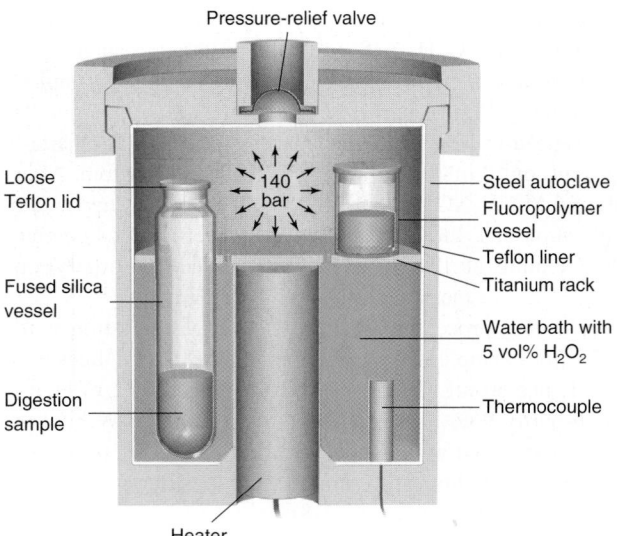

Pressure-relief valve

140 bar

Loose Teflon lid

Steel autoclave

Fluoropolymer vessel

Teflon liner

Titanium rack

Fused silica vessel

Water bath with 5 vol% H_2O_2

Digestion sample

Thermocouple

Heater

Lip

Hole

Spout

FIGURE 28-11 Reflux cap for wet ashing in an Erlenmeyer flask. The hole allows vapor to escape. The curved spout touches the inside of the flask. [Information from D. D. Siemer and H. G. Brinkley, "Erlenmeyer Flask-Reflux Cap for Acid Sample Decomposition," *Anal. Chem.* **1981,** *53,* 750.]

$HClO_4$ with organic material is an extreme **explosion hazard.** Always oxidize first with HNO_3. Always use a blast shield for $HClO_4$.

Section 24-4 described *solid-phase microextraction, purge and trap,* and *thermal desorption*—sample preparation methods that are especially useful for gas chromatography.

The high-pressure asher in Figure 28-10 uses a resistive heating element inside a sealed chamber for digestion at temperatures up to 270°C under a pressure up to 140 bar. High pressure allows acids to be heated to high temperature without boiling. At high temperature, HNO_3 oxidizes organic matter without assistance from H_2SO_4, which is not as pure as HNO_3 and is therefore less suitable for trace analysis. Silica or fluoropolymer vessels inside the sealed chamber are loosely sealed by Teflon caps that permit evolved gases to escape. The bottom of the vessel is filled with 5 vol% H_2O_2 in H_2O. Hydrogen peroxide reduces nitrogen oxides generated by digestion of organic matter. As an example, 1 g of animal tissue could be digested in a 50-mL fused-silica vessel containing 5 mL of high-purity 70 vol% HNO_3 plus 0.2 mL of high-purity 37 vol% HCl. Metals in the digestion solution could be measured at part-per-billion to part-per-million levels by inductively coupled plasma–atomic emission.

Wet ashing with refluxing HNO_3-$HClO_4$ (Figure 28-11) is a widely applicable, but hazardous, procedure.[19] **Perchloric acid has caused numerous explosions.** Use a good blast shield in a metal-lined fume hood designed for $HClO_4$. First, heat the sample slowly to boiling with HNO_3 but *without* $HClO_4$. Boil to near dryness to destroy easily oxidized material that might explode in the presence of $HClO_4$. Add fresh HNO_3 and repeat the evaporation several times. After the sample cools to room temperature, add $HClO_4$ and heat again. If possible, HNO_3 should be present during the $HClO_4$ treatment. A large excess of HNO_3 should be present when oxidizing organic materials.

Bottles of $HClO_4$ should not be stored on wooden shelves, because acid spilled on wood can form explosive cellulose perchlorate esters. Perchloric acid also should not be stored near organic reagents or reducing agents. A reviewer of this book once wrote, "I have seen someone substitute perchloric acid for sulfuric acid in a Jones reductor experiment with spectacular results—no explosion but the tube melted!"

The combination of Fe^{2+} and H_2O_2, called *Fenton's reagent*, generates $OH^•$ radical and, possibly, $Fe^{II}OOH$ as powerful oxidants.[20] Fenton's reagent oxidizes organic matter in dilute aqueous solution.[21] For example, organic components of urine could be destroyed in 30 min at 50°C to release traces of mercury for analysis.[22] To do so, a 50-mL sample was adjusted to pH 3–4 with 0.5 M H_2SO_4. Then 50 μL of saturated aqueous ferrous ammonium sulfate, $Fe(NH_4)_2(SO_4)_2$, were added, followed by 100 μL of 30% H_2O_2.

28-3 Sample Preparation Techniques

Sample preparation is the series of steps required to transform a sample so that it is suitable for analysis. Sample preparation may include dissolving the sample, extracting analyte from a complex matrix, concentrating dilute analyte to a level that can be measured, chemically converting analyte into a detectable form, and removing or masking interfering species.

Liquid Extraction Techniques

In **extraction,** analyte is dissolved in a solvent that does not necessarily dissolve the entire sample and does not decompose the analyte. In a typical *microwave-assisted extraction* of pesticides from soil, a mixture of soil plus acetone and hexane is placed in a Teflon-lined

FIGURE 28-12 Microwave oven with 1.8 kW of power for solvent extraction or acid digestion of up to 40 samples in 75-mL vessels contained in Kevlar explosion-proof sleeves. Different vessels are designed to handle temperatures up to 260–300°C and pressures up to 30–100 bar. Rupture disks release pressure if it exceeds a critical value. Vapors are vented to a fume hood. The temperature inside each vessel is monitored by infrared sensors and used to control microwave power. [© Courtesy CEM Corp., Matthews, NC.]

bomb (Figures 28-7 and 28-12) and heated by microwaves to 150°C. This temperature is 50–100° higher than the boiling points of solvents at atmospheric pressure. Pesticides dissolve, but soil remains behind. The liquid is then analyzed by chromatography.

Acetone absorbs microwaves, so it can be heated in a microwave oven. Hexane does not absorb microwaves. To perform an extraction with pure hexane, the liquid is placed in a fluoropolymer insert inside the Teflon vessel in Figure 28-7.[23] The walls of the insert or a special stir bar contain carbon black, which absorbs microwaves and heats the solvent.

Supercritical fluid extraction uses a *supercritical fluid* (Box 25-3) as the extraction solvent.[24] Carbon dioxide is the most common supercritical fluid because it is inexpensive and eliminates the need for costly disposal of waste organic solvents. Addition of a second solvent such as methanol increases the solubility of polar analytes. Nonpolar substances, such as petroleum hydrocarbons, can be extracted with supercritical argon.[26] Extraction can be monitored by infrared spectroscopy because Ar has no infrared absorption.

Figure 28-13a shows how a supercritical fluid extraction can be carried out. Pressurized fluid is pumped through a heated extraction vessel. Fluid can be left in contact with the sample for some time or it can be pumped through continuously. At the outlet of the extraction vessel, the fluid flows through a capillary tube to release pressure. Exiting CO_2 evaporates, leaving extracted analyte in the collection vessel. Alternatively, the CO_2 can be bubbled through a solvent in the collection vessel to leave a solution of analyte.

Figure 28-13b shows the extraction of organic compounds from dust collected with a vacuum cleaner from doormats at the chemistry building of Ohio State University. The chromatogram of

Some chelators can extract metal ions into supercritical CO_2 (containing small quantities of methanol or water). The ligand below dissolves lanthanides and actinides:[25]

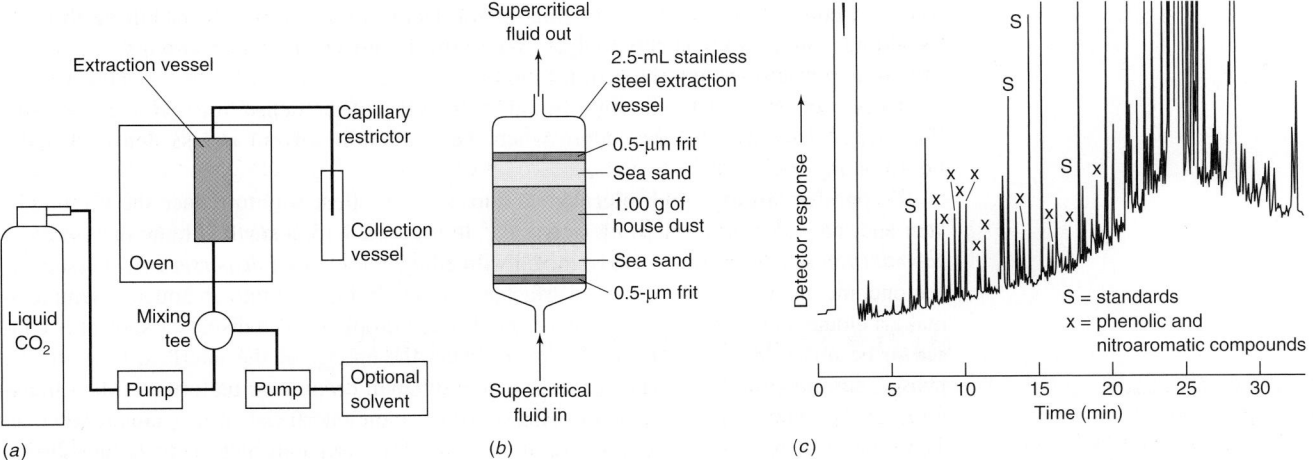

FIGURE 28-13 (a) Apparatus for supercritical fluid extraction. (b) Vessel for extracting house dust at 50°C with 20 mol% methanol/80 mol% CO_2 at 24.0 MPa (240 bar). (c) Gas chromatogram of CH_2Cl_2 solution of extract using a 30 m × 0.25 mm diphenyl$_{0.05}$dimethyl$_{0.95}$siloxane column (1-μm film thickness) with a temperature ranging from 40° to 280°C and flame ionization detection. [Information from T. S. Reighard and S. V. Olesik, "Comparison of Supercritical Fluids and Enhanced-Fluidity Liquids for the Extraction of Phenolic Pollutants from House Dust," *Anal. Chem.* **1996,** *68,* 3612.]

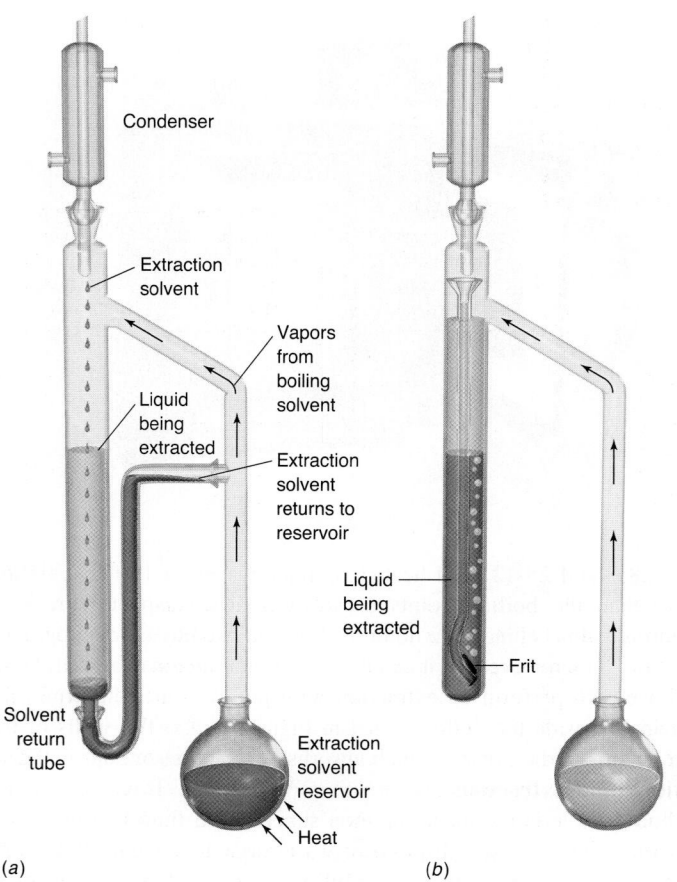

FIGURE 28-14 Continuous liquid-liquid extraction apparatus used when extraction solvent is (*a*) denser than the liquid being extracted or (*b*) lighter than the liquid being extracted.

Condenser

Extraction solvent

Vapors from boiling solvent

Liquid being extracted

Extraction solvent returns to reservoir

Solvent return tube

Extraction solvent reservoir

Heat

(a)

Liquid being extracted

Frit

(b)

the extract in Figure 28-13c exhibits myriad organic compounds that you and I inhale in every breath. In another study, polybrominated diphenyl ether flame retardants were found in house dust.[27] Levels in the U.S. were an order of magnitude higher than levels in Europe. Ingested dust was estimated to provide 0.1–6 μg of flame retardants per day to toddlers aged 1–4.

Figure 28-14 shows glassware for continuous **liquid-liquid extraction** of a nonvolatile analyte, typically from an aqueous solution into an organic solvent. In Figure 28-14a, the extracting solvent is denser than the liquid being extracted. Solvent boils from the flask and condenses into the extraction vessel. Dense droplets of solvent falling through the liquid column extract the analyte. When the liquid level is high enough, extraction solvent is pushed through the return tube to the solvent reservoir. By this means, analyte is slowly transferred from the light liquid at the left into the dense liquid in the reservoir. Figure 28-14b shows the procedure when the extraction solvent is less dense than the liquid being extracted.

Dispersive liquid-liquid microextraction is an excellent way to reduce the use of solvent and the generation of hazardous waste.[28] In Figure 28-15, a small volume of *extraction solvent* such as chloroform ($CHCl_3$) mixed with a larger volume of *disperser solvent* (such as acetone, methanol, acetonitrile, or a surfactant) is rapidly injected into an aqueous sample to make a cloudy *emulsion* (a dispersion of small liquid droplets). The disperser solvent is chosen to be miscible with both phases, so it lowers the energy of the interface between the phases, permitting a high-surface-area emulsion of small droplets to be formed. Mass transfer of analyte between the two phases is rapid because the interfacial surface area is so large. The tube is then centrifuged to break the emulsion into separate layers. In Figure 28-15, extraction solvent is denser than water. A sample can be withdrawn with a microsyringe and injected into a gas or liquid chromatograph for analysis. In one example, microorganism-derived ochratoxin A in wine can be extracted by injecting a mixture of 100 μL of $CHCl_3$ with 1.0 mL of acetone dispersant. The ratio of extraction solvent to dispersant must be controlled to obtain a cloudy emulsion. In general, dispersant is chosen to be miscible with water and with the extraction solvent. A vessel with a very narrow neck can be used to collect

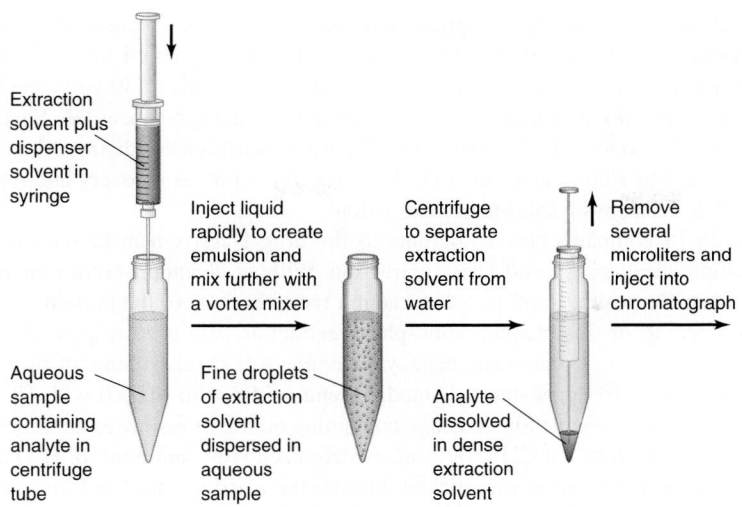

FIGURE 28-15 Dispersive liquid-liquid microextraction.

extraction solvents such as dodecanol or cyclohexane that float on water. Water is added to the aqueous phase until the extraction solvent is raised into the narrow neck.

Another way to reduce solvent use in liquid-liquid extraction is by **solid-supported liquid-liquid extraction,** in which the aqueous phase is suspended in a microporous medium through which organic solvent is passed to extract analytes (Figure 28-16).[29] This technique is conveniently carried out with a 96-well plate shown later in Figure 28-20. In a representative procedure for extracting prescription drugs from blood, human plasma was diluted with an equal volume of 0.5 M NH_3. Then 200 µL of diluted plasma were applied to a small column of microporous diatomaceous earth with suction for 10 s. After allowing 5 min for the aqueous phase to disperse through the solid phase, the column was washed with 1 mL of immiscible organic solvent (hexane : 2-methyl-1-butanol, 98:2 vol/vol), which flowed for 5 min by gravity. Suction was applied for 2 min to complete the elution. Solvent was evaporated to dryness and the residue was dissolved in mobile phase for liquid chromatography.

Solid-Phase Extraction[30]

Solid-phase extraction uses a small volume of a chromatographic stationary phase or molecularly imprinted polymer[31] (Box 26-2) to isolate analytes from the sample matrix to simplify analysis. Solid-phase extraction uses little solvent and generates little waste.

Figure 28-17 shows typical steps in the solid-phase extraction of 10 ng/mL of steroids from urine. First, a syringe containing 1 mL of C_{18}-silica is conditioned with 2 mL of methanol to remove adsorbed organic material. Then the column is washed with 2 mL of water (a).

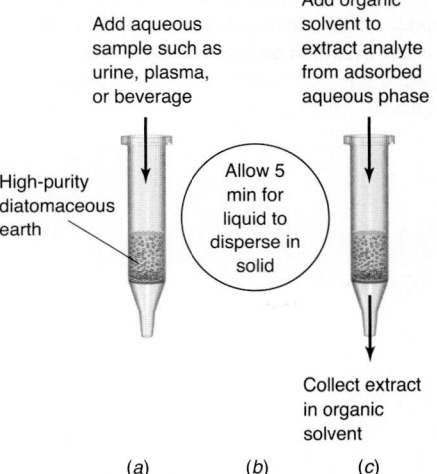

FIGURE 28-16 Solid-supported liquid-liquid extraction.

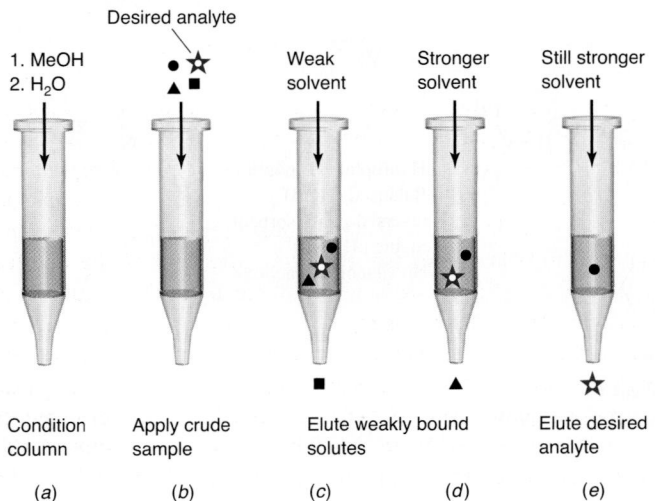

FIGURE 28-17 Solid-phase extraction.

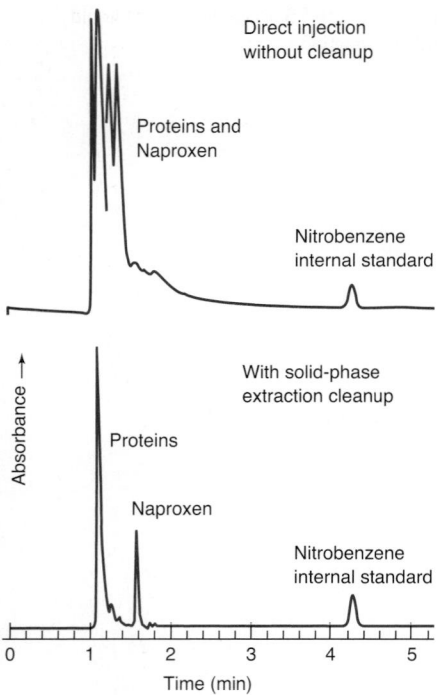

FIGURE 28-18 HPLC of Naproxen in blood serum with no cleanup (upper trace) or with prior sample cleanup (lower trace) by solid-phase extraction on C$_8$-silica. [Data from R. E. Majors and A. D. Broske, "New Directions in Solid-Phase Extraction Particle Design," *Am. Lab.*, February 2002, p. 22.]

In Figure 28-19, the notation 1 meq/g means 1 milliequivalent of ion-exchange sites per gram of resin. A milliequivalent is one millimole of charged sites.

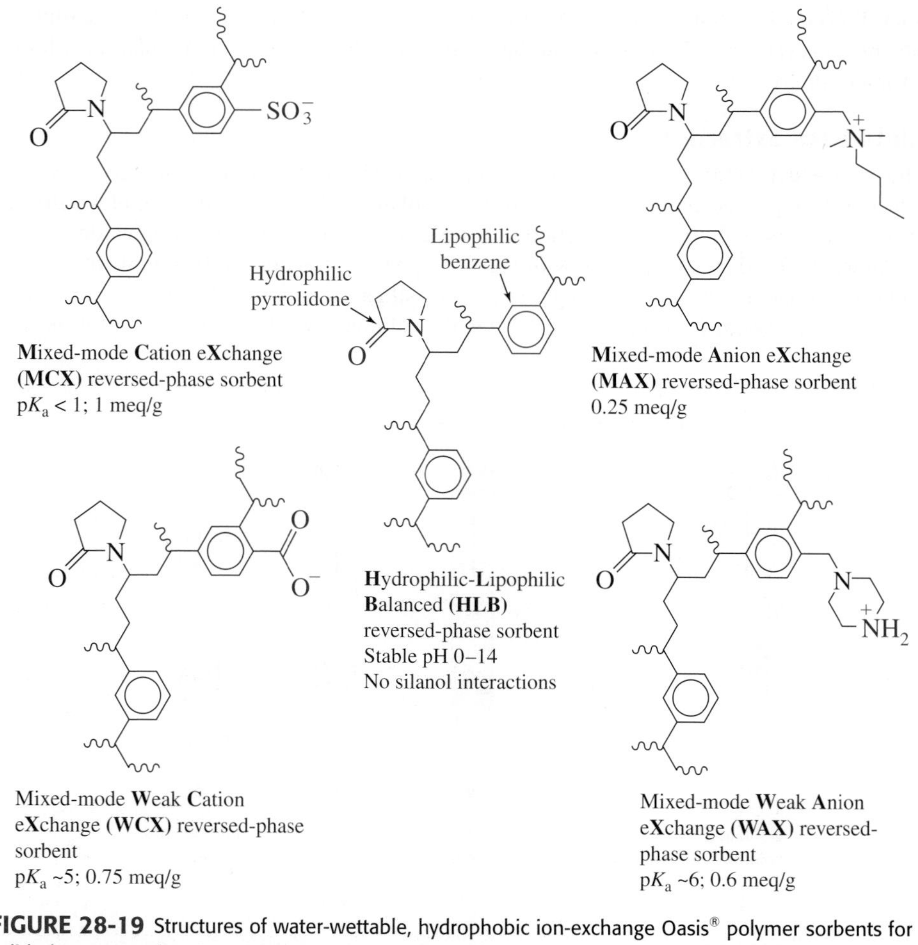

When the 10-mL urine sample is applied, nonpolar components adhere to the C$_{18}$-silica and polar components pass through (*b*). The column is then rinsed with 4 mL of 25 mM borate buffer at pH 8 to remove polar substances (*c*). Rinses with 4 mL of 40 vol% methanol/60% water and 4 mL of 20% acetone/80% water remove less polar substances (*d*). Finally, elution with two 0.5-mL aliquots of 73% methanol/27% water washes steroids from the column (*e*). Sample loading and elution at a slow rate (1–2 drops/s) improves recovery and reproducibility. Color Plate 36 shows a solid-phase extraction.

Figure 28-18 compares chromatograms of the drug Naproxen in blood serum with or without sample cleanup by solid-phase extraction. Without cleanup, serum proteins obscure the signal from Naproxen. Solid-phase extraction removes most of the protein.

At the opening of this chapter, solid-phase extraction was used to *preconcentrate* and partially purify traces of cocaine and benzoylecgonine. A 500-mL volume of river water was filtered, spiked with 10 ng of internal standard, and acidified to pH 2.0 with HCl. A solid-phase cation-exchange extraction cartridge containing 60 mg of resin was conditioned before use by washing with 6 mL of CH$_3$OH, 3 mL of deionized H$_2$O, and 3 mL of H$_2$O acidified to pH 2.0 with HCl. River water was sucked through the cartridge at 20 mL/min. Liquid was removed from the cartridge by suction for 5 min. Analytes were then eluted from the cartridge with 2 mL of CH$_3$OH followed by 2 mL of 2% NH$_3$ in CH$_3$OH. This process preconcentrates the sample by a factor of 500 mL/4 mL = 125.

The solid-phase extraction cartridge used to preconcentrate cocaine from river water is the mixed-mode cation-exchange reversed-phase sorbent at the upper left in Figure 28-19. This is one of a family of resins whose backbone contains lipophilic benzene rings and hydrophilic pyrrolidone rings. The resins are wetted by water and have affinity for both polar and nonpolar substances. The four ion-exchange derivatives are useful for retaining and then releasing different types of analytes when conditions such as pH, solvent, and ionic strength are changed.

Preconcentration of cocaine from 500 mL of river water was done with just 60 mg of resin in a syringe. For multiple samples or exploratory research, a *96-well plate* such as those shown in Figure 28-20 can be used. The conventional plate at the upper right has syringe-like wells,

Mixed-mode Cation eXchange (MCX) reversed-phase sorbent pK_a < 1; 1 meq/g

Mixed-mode Anion eXchange (MAX) reversed-phase sorbent 0.25 meq/g

Hydrophilic pyrrolidone

Lipophilic benzene

Hydrophilic-Lipophilic Balanced (HLB) reversed-phase sorbent Stable pH 0–14 No silanol interactions

Mixed-mode Weak Cation eXchange (WCX) reversed-phase sorbent pK_a ~5; 0.75 meq/g

Mixed-mode Weak Anion eXchange (WAX) reversed-phase sorbent pK_a ~6; 0.6 meq/g

FIGURE 28-19 Structures of water-wettable, hydrophobic ion-exchange Oasis® polymer sorbents for solid-phase extraction. [Data from Waters Corp., Milford, MA].

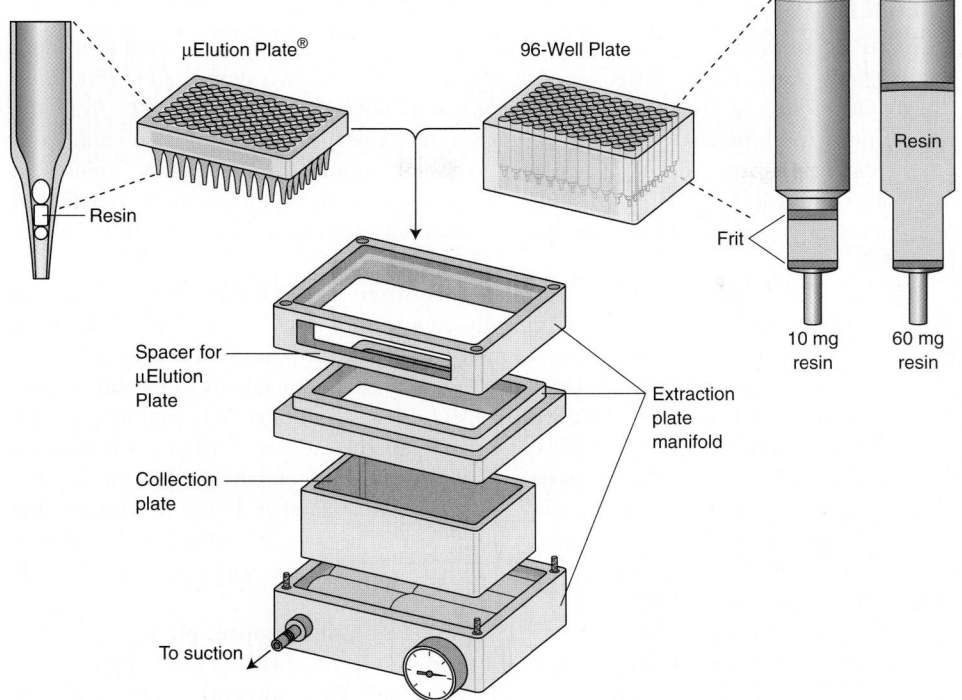

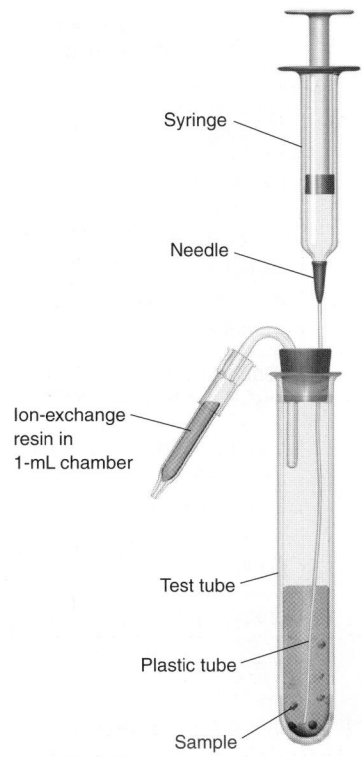

FIGURE 28-20 96-Well Plate and 96-well μElution Plate® for solid-phase extraction. [Information from Waters Corporation, Milford, MA.]

FIGURE 28-21 Apparatus for trapping basic or acidic gases by ion exchange. [Information from D. D. Siemer, "Ion Exchange Resins for Trapping Gases: Carbonate Determination," *Anal. Chem.* **1987,** *59,* 2439.]

each of which can contain 5 to 60 mg of resin. The μElution® plate at the upper left has Pasteur-pipet-like wells with a small volume which can be eluted by 25–50 μL of solvent.

One standard procedure for analysis of pesticides in 1 L of wastewater required 200 mL of dichloromethane for liquid-liquid extraction. The same analytes can be isolated by solid-phase extraction on a C_{18}-silica disk supported by a porous glass frit, such as that in Figure 2-17. Pesticides are recovered from the disk by supercritical fluid extraction with CO_2 that is finally vented into a small volume of hexane. This one kind of analysis avoids using 10^5 kg of CH_2Cl_2 per year.[32] Solid-phase extraction disks use sorbents such a C_{18}, ion exchangers, chelates, and activated carbon.

Ion-exchange resins can capture basic or acidic gases. Carbonate liberated as CO_2 from $(ZrO)_2CO_3(OH)_2 \cdot xH_2O$ used in nuclear fuel reprocessing can be measured by placing a known amount of powdered solid in the test tube in Figure 28-21 and adding 3 M HNO_3. When the solution is purged with N_2, CO_2 is captured by moist anion-exchange resin in the sidearm:

$$CO_2 + H_2O \rightarrow H_2CO_3$$
$$2\,\text{resin}^+OH^- + H_2CO_3 \rightarrow (\text{resin}^+)_2CO_3^{2-} + 2H_2O$$

Carbonate is eluted from the resin with 1 M $NaNO_3$ and measured by titration with acid. Table 28-7 gives other applications of this technique.

TABLE 28-7 Use of ion-exchange resin for trapping gases

Gas	Species trapped	Eluent	Analytical method
CO_2	CO_3^{2-}	1 M $NaNO_3$	Titrate with acid
H_2S	S^{2-}	0.5 M Na_2CO_3 + H_2O_2	S^{2-} is oxidized to SO_4^{2-} by H_2O_2. The sulfate is measured by ion chromatography.
SO_2	SO_3^{2-}	0.5 M Na_2CO_3 + H_2O_2	SO_3^{2-} is oxidized to SO_4^{2-} by H_2O_2. The sulfate is measured by ion chromatography.
HCN	CN^-	1 M Na_2SO_4	Titration of CN^- with hypobromite: $$CN^- + OBr^- \rightarrow CNO^- + Br^-$$
NH_3	NH_4^+	1 M $NaNO_3$	Colorimetric assay with Nessler's reagent: $$\underset{\substack{\text{Nessler's}\\\text{reagent}}}{2K_2HgI_4} + 2NH_3 \rightarrow \underset{\substack{\text{Absorbs strongly}\\\text{at 400–425 nm}}}{NH_2Hg_2I_3} + 4KI + NH_4I$$

SOURCE: D. D. Siemer, "Ion Exchange Resins for Trapping Gases: Carbonate Determination," Anal. Chem. *1987, 59, 2439.*

QuEChERS[33]

QuEChERS (pronounced "kĕ-chers") is an acronym for a sample preparation procedure that is *Qu*ick, *E*asy, *Ch*eap, *E*ffective, *R*ugged, and *S*afe. The two steps are (1) *extraction* of analytes followed by (2) *sample cleanup* to remove matrix components from the extract prior to chromatography. QuEChERS extracts pesticides and other organic contaminants from fruits, vegetables, plants, soil, and sediments. The method applies to antibiotics in meat. Little organic solvent is required, many samples can be handled at once, and good recovery of many analytes is observed. Vendors sell complete kits for this procedure.

In one variant of the procedure, chopped fruit is frozen at −20°C and homogenized with Dry Ice to form free-flowing powder, typically containing >80% water. For *extraction* of organic residues, 10 mL of acetonitrile are mixed with 10 g of powder in a capped, 50-mL plastic centrifuge tube. An internal standard and quality control standards to measure recovery are added. After 1 min of shaking well, salts (1 g NaCl + 4 g MgSO₄) and near-neutral buffer (1 g Na₃citrate · 2H₂O + 0.5 g Na₂Hcitrate · 1.5H₂O) are added. High salt concentration creates aqueous and organic phases and drives organic compounds into the organic phase. Buffer protects some base-sensitive analytes. After another 1 min of shaking, the sample is centrifuged for 5 min and the acetonitrile liquid phase is withdrawn. (To extract organic compounds from 5 g of dry, ground soil, you could mix the soil well with ~3–8 mL of H₂O prior to the QuEChERS procedure.)

For *sample cleanup*, 1 mL of extract is placed in a 2-mL capped, plastic centrifuge tube containing 150 mg of anhydrous MgSO₄ (to absorb residual H₂O), 25 mg of weak anion exchanger (called "primary secondary amine" PSA sorbent), and ~5 mg of optional solid sorbents such as C₁₈-silica and graphitized carbon black. Sample cleanup removes matrix substances such as fatty acid anions and carbohydrates. After 2 min of shaking and 5 min of centrifugation, supernatant liquid is withdrawn for gas or liquid chromatography.

Alternatively, acetonitrile extract is passed through a syringe-mounted cartridge containing sorbents. The cartridge requires 50 mL of 3 : 1 (vol/vol) acetonitrile : toluene for elution and eluate must then be evaporated to 0.5 mL for chromatography. The residue might need to be evaporated to dryness and reconstituted with low-boiling solvent for gas chromatography or a water-based solvent for liquid chromatography.

Figure 28-22 shows a small section of the gas chromatogram–mass spectrum of a pesticide extracted from a dietary supplement by QuEChERS. Both traces come from one sample. The upper trace with many peaks (going up to ~10⁶ counts) is the total ion

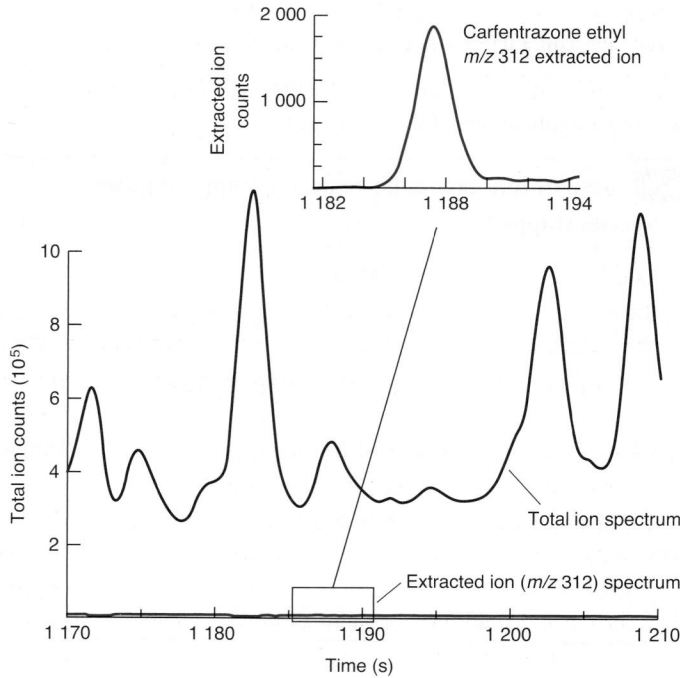

FIGURE 28-22 Partial gas chromatogram–time-of-flight mass spectrum of a pesticide extracted from a dietary supplement by QuEChERS. Column: 30 m × 0.25 mm Restek Rxi-5Sil MS (0.25 μm thick), He 1.5 mL/min, 90–340°C at 8°/min, 70 eV electron ionization, 1 μL splitless injection. Main spectrum is total ion count. Inset is *m/z* 312 extracted ion. [Data from J. Kowalski, M. Misselwitz, J. Thomas, and J. Cochran, "Evaluation of QuEChERS, Cartridge SPE Cleanup, and Gas Chromatography Time-of-Flight Mass Spectrometry for the Analysis of Pesticides in Dietary Supplements," *LCGC North Am.* **2010,** *28,* 972.]

chromatogram representing all components in the extract. Even though the sample has been "cleaned up," it is a mixture of many components. The lower trace displays only the mass spectral extracted ion signal at m/z 312, which comes from the pesticide of interest. The ion at m/z 312 with 2 000 counts is invisible in the total ion spectrum, but has a signal-to-noise ratio of 105 in the extracted ion spectrum shown in the inset at the top of the figure. The specificity of mass spectrometry is able to highlight one analyte in a complex chromatogram.

Derivatization

Derivatization is a procedure in which analyte is chemically modified to make it easier to detect or separate. For example, formaldehyde and other aldehydes and ketones in air, breath, or cigarette smoke[34] can be trapped and derivatized by passing air through a cartridge containing 0.35 g of silica coated with 0.3 wt% 2,4-dinitrophenylhydrazine. Carbonyls are converted into the 2,4-dinitrophenylhydrazone derivative, which is eluted with 5 mL of acetonitrile and analyzed by HPLC. Products are detected by their strong ultraviolet absorbance near 360 nm.

Terms to Understand

aqua regia
ball mill
bomb
derivatization
dispersive liquid-liquid
  microextraction

dry ashing
extraction
flux
fusion
liquid-liquid extraction
mortar and pestle

preconcentration
QuEChERS
sample preparation
sampling
solid-phase extraction

solid-supported liquid-liquid
  extraction
supercritical fluid extraction
variance
wet ashing

Summary

The variance of an analysis is the sum of the sampling variance and the analytical variance. Sampling variance can be understood in terms of the statistics of selecting particles from a heterogeneous mixture. If the probabilities of selecting two kinds of particles from a two-particle mixture are p and q, the standard deviation in selecting n particles is $\sqrt{npq}$. You should be able to use this relation to estimate how large a sample is required to reduce the sampling variance to a desired level. Student's t can be used to estimate how many repetitions of the analysis are required to reach a level of confidence in the final result.

Many inorganic materials can be dissolved in strong acids with heating. Glass vessels are often useful, but Teflon, platinum, or silver are required for HF, which dissolves silicates. If a non-

oxidizing acid is insufficient, aqua regia or other oxidizing acids may do the job. A Teflon-lined bomb heated in a microwave oven is a convenient means of dissolving difficult samples. If acid digestion fails, fusion in a molten salt will usually work, but the large quantity of flux adds trace impurities. Organic materials are decomposed by wet ashing with hot concentrated acids or by dry ashing with heat.

Analytes can be separated from complex matrices by sample preparation techniques that include liquid extraction, supercritical fluid extraction, dispersive liquid-liquid microextraction, solid-supported liquid-liquid extraction, solid-phase extraction, and QuEChERS. Derivatization transforms analyte into a more easily detected or separated form.

Exercises

28-A. A box contains 120 000 red marbles and 880 000 yellow marbles.

(a) If you draw a random sample of 1 000 marbles from the box, what are the expected numbers of red and yellow marbles?

(b) Now put those marbles back in the box and repeat the experiment. What will be the absolute and relative standard deviations for the numbers in part **(a)** after many drawings of 1 000 marbles?

(c) What will be the absolute and relative standard deviations after many drawings of 4 000 marbles?

(d) If you quadruple the size of the sample, you decrease the sampling standard deviation by a factor of _____. If you increase the sample size by a factor of n, you decrease the sampling standard deviation by a factor of _____.

(e) What sample size is required to reduce the sampling standard deviation of red marbles to ±2%?

28-B. (a) What mass of sample in Figure 28-3 is expected to give a sampling standard deviation of ±10%?

(b) With the mass from part **(a)**, how many samples should be taken to assure 95% confidence that the mean is known to within ±20 counts per second per gram?

28-C. A soil sample contains some acid-soluble inorganic matter, some organic material, and some minerals that do not dissolve in any combination of hot acids that you try.

(a) Suggest a procedure for dissolving the entire sample to measure inorganic constituents.

(b) Suggest a procedure for measuring pesticides and herbicides in the soil.

Problems

Statistics of Sampling

28-1. Explain what is meant by the statement, "Unless the complete history of any sample is known with certainty, the analyst is well advised not to spend his or her time in analyzing it."

28-2. Explain what is meant by "analytical quality" and "data quality" in the following quotation: "We need to update the environmental data quality model to explicitly distinguish *analytical quality* from *data quality*. We need to begin spending as much effort ensuring sample and subsample representativeness for heterogeneous matrices as we spend overseeing the analysis of an extract. We need to stop acting like data variability stemming from laboratory analysis is all-important while variability stemming from the sampling process can be ignored...."[35]

28-3. (a) In the analysis of a barrel of powder, the standard deviation of the sampling operation is ±4% and the standard deviation of the analytical procedure is ±3%. What is the overall standard deviation?

(b) To what value must the sampling standard deviation be reduced so that the overall standard deviation is ±4%?

28-4. What mass of sample in Figure 28-3 is expected to give a sampling standard deviation of ±6%?

28-5. Explain how to prepare a powder with an average particle diameter near 100 μm by using sieves from Table 28-2. How would such a particle mesh size be designated?

28-6. An example of a mixture of 1-mm-diameter particles of KCl and KNO_3 in a number ratio 1:99 follows Equation 28-4. A sample containing 10^4 particles weighs 11.0 g. What is the expected number and relative standard deviation of KCl particles in a sample weighing 11.0×10^2 g?

28-7. When you flip a coin, the probability of its landing on each side is $p = q = \frac{1}{2}$ in Equations 28-2 and 28-3. If you flip it n times, the expected number of heads equals the expected number of tails = $np = nq = \frac{1}{2}n$. The expected standard deviation for n flips is $\sigma_n = \sqrt{npq}$. From Table 4-1, we expect that 68.3% of the results will lie within ±$1\sigma_n$ and 95.5% of the results will lie within ±$2\sigma_n$.

(a) Find the expected standard deviation for the number of heads in 1 000 coin flips.

(b) By interpolation in Table 4-1, find the value of z that includes 90% of the area of the Gaussian curve. We expect that 90% of the results will lie within this number of standard deviations from the mean.

(c) If you repeat the 1 000 coin flips many times, what is the expected range for the number of heads that includes 90% of the results? (For example, your answer might be, "The range 490 to 510 will be observed 90% of the time.")

28-8. In analyzing a lot with random sample variation, you find a sampling standard deviation of ±5%. Assuming negligible error in the analytical procedure, how many samples must be analyzed to give 95% confidence that the error in the mean is within ±4% of the true value? Answer the same question for a confidence level of 90%.

28-9. In an experiment analogous to that in Figure 28-3, the sampling constant is found to be $K_s = 20$ g.

(a) What mass of sample is required for a ±2% sampling standard deviation?

(b) How many samples of the size in part **(a)** are required to produce 90% confidence that the mean is known to within 1.5%?

28-10. $^{87}Sr/^{86}Sr$ isotope ratios were measured in polar ice cores of varying sample size. The 95% confidence interval, expressed as a percentage of the mean $^{87}Sr/^{86}Sr$ isotope ratio, decreased with increasing quantity of Sr measured:

pg Sr	Confidence interval	pg Sr	Confidence interval
57	±0.057%	506	±0.035%
68	±0.069%	515	±0.027%
110	±0.049%	916	±0.018%
110	±0.045%	955	±0.022%

SOURCE: G. R. Burton, V. I. Morgan, C. F. Boutron, and K. J. R. Rosman, "High-Sensitivity Measurements of Strontium Isotopes in Polar Ice," Anal. Chim. Acta **2002**, *469*, 225.

We postulate that the confidence interval is related to the mass of sample by the relation $mR^2 = K_s$, where m is the mass of Sr in picograms, R is the confidence interval expressed as a percentage of the isotope ratio ($R = 0.022$ for the 955-pg sample), and K_s is a constant with units of picograms. Find the average value of K_s and its standard deviation. Justify why we expect $mR^2 = K_s$ to hold if all measurements are made with the same number of replications.

28-11. Consider a random mixture containing 4.00 g of Na_2CO_3 (density 2.532 g/mL) and 96.00 g of K_2CO_3 (density 2.428 g/mL) with a uniform spherical particle radius of 0.075 mm.

(a) Calculate the mass of a single particle of Na_2CO_3 and the number of particles of Na_2CO_3 in the mixture. Do the same for K_2CO_3.

(b) What is the expected number of particles in 0.100 g of the mixture?

(c) Calculate the relative sampling standard deviation in the number of particles of each type in a 0.100-g sample of the mixture.

Sample Preparation

28-12. From their standard reduction potentials, which of the following metals would you expect to dissolve in HCl by the reaction $M + nH^+ \rightarrow M^{n+} + \frac{n}{2} H_2$: Zn, Fe, Co, Al, Hg, Cu, Pt, Au? (When the potential predicts that the element will not dissolve, it probably will not. If it is expected to dissolve, it may dissolve if some other process does not interfere. Predictions based on standard reduction potentials at 25°C are only tentative, because the potentials and activities in hot, concentrated solutions vary widely from those in the table of standard potentials.)

28-13. The following wet-ashing procedure was used to measure arsenic in organic soil samples by atomic absorption spectroscopy: A 0.1- to 0.5-g sample was heated in a 150-mL Teflon bomb in a microwave oven for 2.5 min with 3.5 mL of 70% HNO_3. After the sample cooled, a mixture containing 3.5 mL of 70% HNO_3, 1.5 mL of 70% $HClO_4$, and 1.0 mL of H_2SO_4 was added and the sample was reheated for three 2.5-min intervals with 2-min unheated periods in between. The final solution was diluted with 0.2 M HCl for analysis. Why was $HClO_4$ not introduced until the second heating?

28-14. Barbital can be isolated from urine by solid-phase extraction with C_{18}-silica. The barbital is then eluted with 1 : 1 vol/vol acetone : chloroform. Explain how this procedure works.

$$\underset{\text{Barbital}}{\begin{array}{c}CH_3CH_2\\CH_3CH_2\end{array}\text{...}}$$

Barbital

28-15. To preconcentrate cocaine and benzoylecgonine from river water described at the opening of this chapter, solid-phase extraction was carried out at pH 2 using the mixed-mode cation-exchange resin in Figure 28-19. After passing 500 mL of river water through 60 mg of resin, the retained analytes were eluted first with 2 mL of CH_3OH and then with 2 mL of 2% ammonia solution in CH_3OH. Explain the purpose of using pH 2 for retention and dilute ammonia for elution.

28-16. Referring to Table 28-7, explain how an anion-exchange resin can be used for absorption and analysis of SO_2 released by combustion.

28-17. Why is it advantageous to use large particles (50 μm) for solid-phase extraction, but small particles (5 μm) for chromatography?

28-18. In 2002, workers at the Swedish National Food Administration discovered that heated, carbohydrate-rich foods, such as french fries, potato chips, and bread, contain alarming levels (0.1 to 4 μg/g) of acrylamide, a known carcinogen.[36]

$$\underset{\text{H}}{\overset{\text{O}}{H_2C=C}}—NH_2 \qquad \text{Acrylamide}$$

After the discovery, simplified methods were developed to measure ppm levels of acrylamide in food. In one procedure, 10 g of pulverized, frozen french fries were mixed for 20 min with 50 mL of H_2O to extract acrylamide, which is very soluble in water (216 g/100 mL). The liquid was decanted and centrifuged to remove suspended matter. The internal standard $^2H_3$-acrylamide was added to 1 mL of extract. A solid-phase extraction column containing 100 mg of cation-exchange polymer with protonated sulfonic acid groups ($—SO_3H$) was washed twice with 1-mL portions of methanol and twice with 1-mL portions of water. The aqueous food extract (1 mL) was then passed through the column to bind protonated acrylamide ($—NH_3^+$) to sulfonate ($—SO_3^-$) on the column. The column was dried for 30 s at 0.3 bar and then acrylamide was eluted with 1 mL of H_2O. Eluate was analyzed by liquid chromatography with a polar bonded phase. The chromatograms show the results moni-

tored by ultraviolet absorbance or by mass spectrometry. The retention time of acrylamide is different on the two columns because they have different dimensions and different flow rates.

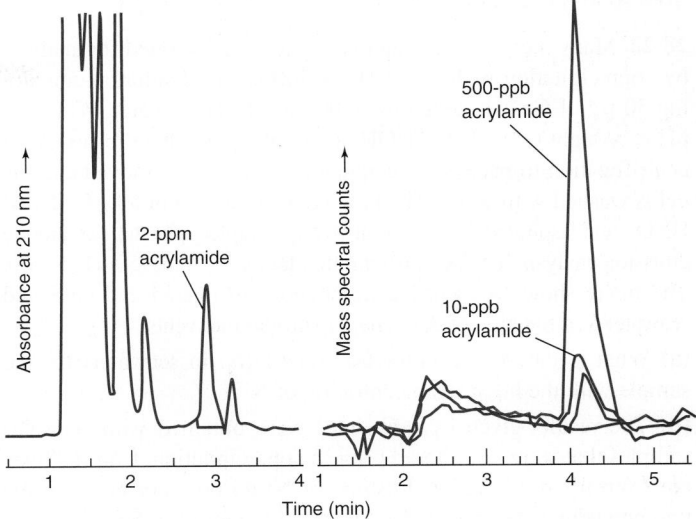

Chromatograms of acrylamide extract after passage through solid-phase extraction column. Left: Phenomenex Synergi Polar-RP 4-μm column eluted with 96:4 (vol/vol) $H_2O:CH_3CN$. Right: Phenomenex Synergi Hydro-RP 4-μm column eluted with 96:4:0.1 (vol/vol/vol) $H_2O:CH_3OH:HCO_2H$. [Data from L. Peng, T. Farkas, L. Loo, J. Teuscher, and K. Kallury, "Rapid and Reproducible Extraction of Acrylamide in French Fries Using a Single Solid-Phase Sorbent," *Am. Lab. News Ed.,* October 2003, p. 10.]

(a) What is the purpose of solid-phase extraction prior to chromatography? How does the ion-exchange sorbent retain acrylamide?

(b) Why are there many peaks when chromatography is monitored by ultraviolet absorbance?

(c) Mass spectral detection used selected reaction monitoring (Figure 22-33) with the m/z 72 → 55 transition for acrylamide and 75 → 58 for $^2H_3$-acrylamide. Explain how this detection method works and suggest structures for the ions with m/z 72 and 55 from acrylamide.

(d) Why does mass spectral detection give just one major peak?

(e) How is the internal standard used for quantitation with mass spectral detection?

(f) Where does $^2H_3$-acrylamide appear with ultraviolet absorbance? With mass spectral selected reaction monitoring?

(g) Why does the mass spectral method give quantitative results even though retention of acrylamide by the ion-exchange sorbent is not quantitative and elution of acrylamide from the sorbent by 1 mL of water might not be quantitative?

28-19. (a) Describe the steps in QuEChERS and explain their purpose.

(b) Why is an internal standard used in QuEChERS?

(c) What is displayed in the total ion chromatogram in Figure 28-22?

(d) What is displayed in the extracted ion chromatogram in Figure 28-22? What is the difference between an extracted ion chromatogram and a selected ion chromatogram? Which would have greater signal-to-noise ratio?

(e) What mass spectrometric method could be used to obtain even greater signal-to-noise ratio from the same QuEChERS extract?

28-20. (a) Explain how dispersive liquid-liquid microextraction reduces the use of solvent in comparison with liquid-liquid extraction.

(b) What is the purpose of the disperser solvent, which is used in much greater volume than the extraction solvent?

28-21. How does solid-supported liquid-liquid extraction differ from solid-phase extraction?

28-22. Many metals in seawater can be preconcentrated for analysis by coprecipitation with $Ga(OH)_3$. A 200-μL HCl solution containing 50 μg of Ga^{3+} is added to 10.00 mL of the seawater. When the pH is brought to 9.1 with NaOH, a jellylike precipitate forms. After centrifugation to pack the precipitate, the water is removed and the gel is washed with water. Then the gel is dissolved in 50 μL of 1 M HNO_3 and aspirated into an inductively coupled plasma for atomic emission analysis. The preconcentration factor is 10 mL/50 μL = 200. The figure shows elemental concentrations in filtered and unfiltered seawater as a function of depth near hydrothermal vents.

(a) What is the atomic ratio (Ga added):(Ni in seawater) for the sample with the highest concentration of Ni?

(b) The results given by gray lines were obtained with seawater samples that were not filtered prior to coprecipitation. Colored lines are from filtered samples. Results for Ni do not vary between the two procedures, but results for Fe vary. Explain what this means.

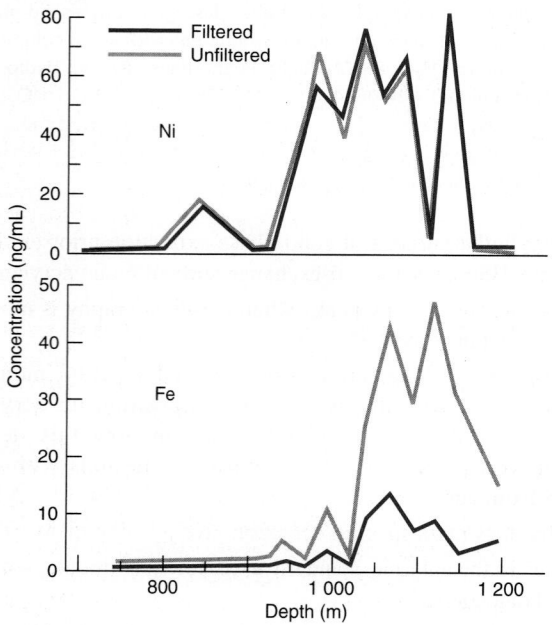

Depth profile of elements from filtered and unfiltered seawater near hydrothermal vents. [Data from T. Akagi and H. Haraguchi, "Simultaneous Multielement Determination of Trace Metals Using 10 mL of Seawater by Inductively Coupled Plasma Atomic Emission Spectrometry with Gallium Coprecipitation and Microsampling Technique," *Anal. Chem.* **1990**, *62*, 81.]

28-23. Barium titanate, a ceramic used in electronics, was analyzed by the following procedure: Into a Pt crucible was placed 1.2 g of Na_2CO_3 and 0.8 g of $Na_2B_4O_7$ plus 0.314 6 g of unknown. After fusion at 1 000°C in a furnace for 30 min, the cooled solid was extracted with 50 mL of 6 M HCl, transferred to a 100-mL volumetric flask, and diluted to the mark. A 25.00-mL aliquot was treated with 5 mL of 15% tartaric acid (which complexes Ti^{4+} and keeps it in aqueous solution) and 25 mL of ammonia buffer, pH 9.5. The solution was treated with organic reagents that complex Ba^{2+}, and the Ba complex was extracted into CCl_4. After acidification (to release the Ba^{2+} from its organic complex), the Ba^{2+} was back-extracted into 0.1 M HCl. The final aqueous sample was treated with ammonia buffer and methylthymol blue (a metal ion indicator) and

titrated with 32.49 mL of 0.011 44 M EDTA. Find the weight percent of Ba in the ceramic.

28-24. Acid-base equilibria of Cr(III) were summarized in Problem 10-36. Cr(VI) in aqueous solution above pH 6 exists as the yellow tetrahedral chromate ion, CrO_4^{2-}. Between pH 2 and 6, Cr(VI) exists as an equilibrium mixture of $HCrO_4^-$ and orange-red dichromate, $Cr_2O_7^{2-}$. Cr(VI) is a carcinogen, but Cr(III) is not considered to be as harmful. The following procedure was used to measure Cr(VI) in airborne particulate matter in workplaces.

1. Particles were collected by drawing a known volume of air through a polyvinyl chloride filter with 5-μM pore size.

2. The filter was placed in a centrifuge tube and 10 mL of 0.05 M $(NH_4)_2SO_4$/0.05 M NH_3 buffer, pH 8, were added. The immersed filter was agitated by ultrasonic vibration for 30 min at 35°C to extract all Cr(III) and Cr(VI) into solution.

3. A measured volume of extract was passed through a "strongly basic" anion exchanger (Table 26-1) in the Cl^- form. Then the resin was washed with distilled water. Liquid containing Cr(III) from the extract and the wash was discarded.

4. Cr(VI) was then eluted from the column with 0.5 M $(NH_4)_2SO_4$/0.05 M NH_3 buffer, pH 8, and collected in a vial.

5. The eluted Cr(VI) solution was acidified with HCl and treated with a solution of 1,5-diphenylcarbazide, a reagent that forms a colored complex with Cr(VI). The concentration of the complex was measured by its visible absorbance.

(a) What are the dominant species of Cr(VI) and Cr(III) at pH 8?

(b) What is the purpose of the anion exchanger in step 3?

(c) Why is a "strongly basic" anion exchanger used instead of a "weakly basic" exchanger?

(d) Why is Cr(VI) eluted in step 4 but not step 3?

28-25. The county landfill in the diagram was monitored to verify that toxic compounds were not leaching into the local water supply. Wells drilled at 21 locations were monitored over a year and pollutants were observed only at sites 8, 11, 12, and 13. Monitoring all 21 sites each month is very expensive. Suggest a strategy to use *composite samples* (Box 0-1) made from more than one well at a time to reduce the cost of routine monitoring. How will your scheme affect the minimum detectable level for pollutants at a particular site?

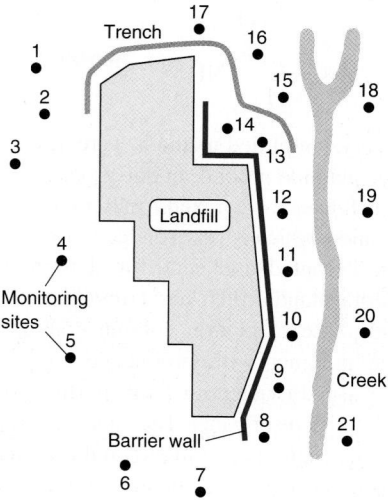

Diagram of county landfill, showing the location of wells used to monitor groundwater. [Information from P.-C. Li and R. Rajagopal, *Am. Environ. Lab.,* October 1994, p. 37.]

NOTES AND REFERENCES

Chapter 0

1. S. P. Beckett, *The Science of Chocolate*, 2nd ed. (Cambridge: Royal Society of Chemistry, 2008); G. Tannenbaum, "Chocolate: A Marvelous Natural Product of Chemistry," *J. Chem. Ed.* **2004**, *81*, 1131. This notation for a journal reference means *The Journal of Chemical Education* in the year **2004**, volume *81*, page 1131.

2. T. J. Wenzel, "A New Approach to Undergraduate Analytical Chemistry," *Anal. Chem.* **1995**, *67*, 470A. See also T. J. Wenzel, "The Lecture as a Learning Device," *Anal. Chem.* **1999**, *71*, 817A; T. J. Wenzel, "Cooperative Student Activities as Learning Devices," *Anal. Chem.* **2000**, *72*, 293A; T. J. Wenzel, "Practical Tips for Cooperative Learning," *Anal. Chem.* **2000**, *72*, 359A; T. J. Wenzel, "Undergraduate Research as a Capstone Learning Experience," *Anal. Chem.* **2000**, 72, 547A; T. J. Wenzel, "Active Learning Materials for Equilibrium Chemistry and Separation Science," *Anal. Bioanal. Chem.* **2011**, *400*, 637.

3. W. R. Kreiser and R. A. Martin, Jr., "High-Pressure Liquid Chromatographic Determination of Theobromine and Caffeine in Cocoa and Chocolate Products," *J. Assoc. Off. Anal. Chem.* **1978**, *61*, 1424; W. R. Kreiser and R. A. Martin, Jr., "High-Pressure Liquid Chromatographic Determination of Theobromine and Caffeine in Cocoa and Chocolate Products," *J. Assoc. Off. Anal. Chem.* **1980**, *63*, 591.

4. A good source for many well-tested analytical procedures is G. Latimer, Jr., ed., *Official Methods of Analysis of AOAC International*, 19th ed. (Gaithersburg, MD: AOAC International, 2012).

5. A. Carlin-Sinclair, I. Marc, L. Menguy, and D. Prim, "The Determination of Methylxanthines in Chocolate and Cocoa by Different Separation Techniques: HPLC, Instrumental TLC, and MECC," *J. Chem. Ed.* **2009**, *86*, 1307; S. E. Stitzel and R. E. Sours, "High-Performance Liquid Chromatography Analysis of Single-Origin Chocolates for Methylxanthine Composition and Provenance Determination," *J. Chem. Ed.* **2013**, *90*, 1227.

6. W. Fresenius, "The Position of the Analyst as Expert: Yesterday and Today," *Fresenius J. Anal. Chem.* **2000**, *368*, 548.

Chapter 1

1. A. M. Pollard and C. Heron, *Archaeological Chemistry*, 2nd ed. (Cambridge: Royal Society of Chemistry, 2008).

2. S. L. Gerstenberger, A. Martinson, and J. L. Kramer, "An Evaluation of Mercury Concentrations in Three Brands of Canned Tuna," *Environ. Toxicol. Chem.* **2010**, *29*, 237.

3. U. Shahin, S.-M. Yi, R. D. Paode, and T. M. Holsen, "Long-Term Elemental Dry Deposition Fluxes Measured Around Lake Michigan," *Environ. Sci. Tech.* **2000**, *34*, 1887.

Chapter 2

1. V. Tsionsky, "The Quartz-Crystal Microbalance in an Undergraduate Laboratory Experiment," *J. Chem. Ed.* **2007**, *84*, 1334, 1337, 1340; J. Janata, *Principles of Chemical Sensors* (Dordrecht: Springer, 2009).

2. A vibrating cantilever is 10^7 times more sensitive than a quartz crystal microbalance and can measure femtogram (10^{-15} g) quantities of analyte. (W. Tan, Y. Huang, T. Nan, C. Xue, Z. Li, Q. Zhang, and B. Wang, "Development of Protein A Functionalized Microcantilever Immunosensors for the Analyses of Small Molecules at Parts per Trillion Levels," *Anal. Chem.* **2010**, *82*, 615; H. Sone, S. Ichikawa, Y. Matsubara, M. Suzuki, H. Okano, T. Izumi, and S. Hosaka, "Prototype of Frame-Type Cantilever for Biosensor and Femtogram Detection," *Key Engineering Mater.* **2011**, 459, 134.)

3. For small (<2%) changes in resonance frequency (Δf) of a piezoelectric crystal oscillating in air, Δf is related to the change in mass (Δm) bound to the surface of the electrode by the Sauerbrey equation:

$$\Delta f = -\frac{f_o^2}{A\sqrt{\rho_q\mu_q}}\Delta m$$

where f_o is the resonance frequency, A is the area between the electrodes, ρ_q is the density of quartz (2 648 kg/m^3), and μ_q is the shear modulus of quartz (2.947×10^{10} kg m^{-1} s^{-2} for the commonly used "AT cut" of quartz).

A resonance curve can be constructed by plotting the electric conductance of the driving circuit versus frequency. Conductance reaches a peak at the resonance frequency. The width of the resonance peak changes in proportion to the *viscosity* of solute on the gold electrode surface. (*Viscosity* is a measure of the resistance of a liquid to flow.) So, for example, different conformations of DNA bound to a quartz crystal microbalance can be distinguished by the way they affect the width of the resonance curve. (A. Tsortos, G. Papadakis, and E. Gizeli, "Shear Acoustic Wave Biosensor for Detecting DNA Intrinsic Viscosity and Conformation: A Study with QCM-D," *Biosensors Bioelectronics* **2008**, *24*, 836; G. Papadakis, A. Tsortos, and E. Gizeli, "Acoustic Characterization of Nanoswitch Structures: Application to the DNA Holliday Junction," *Nano Lett.* **2010**, *10*, 5093.)

4. Training in basic laboratory techniques is available from www.jce.divched. org/ and from www.academysavant.com. A "living" resource for teaching and learning analytical chemistry techniques, including topics in instrumentation, is the *Analytical Sciences Digital Library* at http://www.asdlib.org/.

5. R. H. Hill and D. Finster, *Laboratory Safety for Chemistry Students* (Hoboken, NJ: Wiley, 2010). Free safety training videos are available at www. safety.dow.com.

6. *Prudent Practices in the Laboratory: Handling and Management of Chemical Hazards* (Washington: National Academies Press, 2011); R. J. Lewis, Sr., *Hazardous Chemicals Desk Reference*, 6th ed. (New York: Wiley, 2008); P. Patnaik, *A Comprehensive Guide to the Hazardous Properties of Chemical Substances*, 3rd ed. (New York: Wiley, 2007); G. Lunn and E. B. Sansone, *Destruction of Hazardous Chemicals in the Laboratory*, 3rd ed. (New York: Wiley, 2012); and M. A. Armour, *Hazardous Laboratory Chemical Disposal Guide*, 3rd ed. (Boca Raton, FL: CRC Press, 2003).

7. To recover gold from electronics, see J. W. Hill and T. A. Lear, "Recovery of Gold from Electronic Scrap," *J. Chem. Ed.* **1988**, *65*, 802. To remove Hg from gold, soak it in a 1:1 mixture of 0.01 M $(NH_4)_2S_2O_8$ and 0.01 M HNO_3; see T. Nomura and M. Fujisawa, "Electrolytic Determination of Mercury(II) in Water with a Piezoelectric Quartz Crystal," *Anal. Chim. Acta* **1986**, *182*, 267.

8. P. T. Anastas and J. C. Warner, *Green Chemistry: Theory and Practice* (New York: Oxford University Press, 1998); M. de la Guardia and S. Armenta, *Green Analytical Chemistry* (Oxford: Elsevier, 2011); M. Koel and M. Kaljurand, *Green Analytical Chemistry* (Cambridge: Royal Society of Chemistry, 2010); M. Tobiszewski, A. Mechlińska, and J. Namieśnik, "Green Analytical Chemistry—Theory and Practice," *Chem. Soc. Rev.* **2010**, *39*, 2869; B. Braun, R. Charney, A. Clarens, J. Farrugia, C. Kitchens, C. Lisowski, D. Naistat, and A. O'Neil, "Green Chemistry in the Curriculum," *J. Chem. Ed.* **2006**, *83*, 1126.

9. J. M. Bonicamp, "Weigh This Way," *J. Chem. Ed.* **2002**, *79*, 476.

10. B. B. Johnson and J. D. Wells, "Cautions Concerning Electronic Analytical Balances," *J. Chem. Ed.* **1986**, *63*, 86.

11. For a demonstration of buoyancy, see K. D. Pinkerton, "Sink or Swim: The Cartesian Diver," *J. Chem. Ed.* **2001**, *78*, 200A.

12. R. Batting and A. G. Williamson, "Single-Pan Balances, Buoyancy, and Gravity or 'A Mass of Confusion,'" *J. Chem. Ed.* **1984**, *61*, 51; J. E. Lewis and L. A. Woolf, "Air Buoyancy Corrections for Single-Pan Balances," *J. Chem. Ed.* **1971**, *48*, 639; F. F. Cantwell, B. Kratochvil, and W. E. Harris, "Air Buoyancy Errors and the Optical Scale of a Constant-Load Balance," *Anal. Chem.* **1978**, *50*, 1010; G. D. Chapman, "Weighing with Electronic Balances," National Research Council of Canada, Report NRCC 38659 (1996).

13. Air density (g/L) = (0.003 485 B − 0.001 318 v)/T, where B is barometric pressure (Pa), v is the vapor pressure of water in the air (Pa), and T is air temperature (K).

14. U. Henriksson and J. C. Eriksson, "Thermodynamics of Capillary Rise: Why Is the Meniscus Curved?" *J. Chem. Ed.* **2004**, *81*, 150.

15. Cleaning solution is prepared by dissolving 36 g of ammonium peroxydisulfate, $(NH_4)_2S_2O_8$, in a *loosely stoppered* 2.2-L bottle of 98 wt% sulfuric acid (H. M. Stahr, W. Hyde, and L. Sigler, "Oxidizing Acid Baths—without Chromate Hazards," *Anal. Chem.* **1982**, *54*, 1456A). Add $(NH_4)_2S_2O_8$ every few weeks to maintain oxidizing power. Keep it loosely stoppered to prevent gas buildup (P. S. Surdhar, "Laboratory Hazard," *Anal. Chem.* **1992**, *64*, 310A). Another powerful, oxidizing cleaning solution, called "piranha solution," is variously described as a 3 : 7 (vol/vol) or 1 : 1 mixture of 30 wt% H_2O_2 plus 98 wt% H_2SO_4. Far less hazardous cleaners that do not generate toxic waste are commercially available. For example, see the website for International Products Corp. http://www.ipcol.com/.

16. W. B. Guenther, "Supertitrations: High-Precision Methods," *J. Chem. Ed.* **1988**, *65*, 1097; D. D. Siemer, S. D. Reeder, and M. A. Wade, "Syringe Buret Adaptor," *J. Chem. Ed.* **1988**, *65*, 467.

17. M. M. Singh, C. McGowan, Z. Szafran, and R. M. Pike, "A Modified Microburet for Microscale Titration," *J. Chem. Ed.* **1998**, *75*, 371; "A Comparative Study of Microscale and Standard Burets," *J. Chem. Ed.* **2000**, *77*, 625.

18. D. R. Burfield and G. Hefter, "Oven Drying of Volumetric Glassware," *J. Chem. Ed.* **1987**, *64*, 1054.

19. R. H. Obenauf and N. Kocherlakota, "Identifying Contamination in Trace Metal Laboratories," *Spectroscopy Applications Supplement*, March 2006, p. 12.

20. B J. Vanderford, D. B. Mawhinney, R. A. Trenholm, J. C. Ziegler-Holady, and S. A. Snyder, "Assessment of Sample Preservation Techniques for Pharmaceuticals, Personal Care Products, and Steroids in Surface and Drinking Water," *Anal. Bioanal. Chem.* **2011**, *399*, 2227.

21. W. Vaccaro, "Minimizing Liquid Delivery Risk: Operators as Sources of Error," *Am. Lab. News Ed.* September 2007, p. 16; A. B. Carle, "Minimizing Liquid Delivery Risk: Barometric Pressure and Thermal Disequilibrium," *Am. Lab. News Ed.* January 2008, p. 8.

22. I. Suominen and S. Koivisto, "Increasing Precision When Pipetting Protein Samples: Assessing Reliability of the Reverse Pipetting Technique," *Am. Lab.* February 2011, p. 50.

23. M. Connors and R. Curtis, "Pipetting Error," *Am. Lab. News Ed.* June 1999, p. 20; ibid. December 1999, p. 12; R. H. Curtis and G. Rodrigues, ibid. February 2004, p. 12.

24. R. Curtis, "Minimizing Liquid Delivery Risk: Pipets as Sources of Error," *Am. Lab. News Ed.* March 2007, p. 8.

25. B. Kratochvil and N. Motkosky, "Precision and Accuracy of Mechanical-Action Micropipets," *Anal. Chem.* **1987**, *59*, 1064. A colorimetric calibration kit is available from Artel, Inc., Westbrook, ME, www.artel-usa.com/.

26. S. R. Crouch and F. J. Holler, *Applications of Microsoft® Excel in Analytical Chemistry*, 2nd ed. (Independence, Kentucky: Brooks/Cole, Cengage Learning, 2013); E. J. Billo, *Microsoft Excel for Chemists*, 2nd ed. (New York: Wiley, 2001); R. de Levie, *How to Use Excel® in Analytical Chemistry and in General Scientific Data Analysis* (Cambridge: Cambridge University Press, 2001); E. J. Billo, *Excel for Scientists and Engineers: Numerical Methods* (New York: Wiley, 2007); R. de Levie, *Advanced Excel for Scientific Data Analysis*, 3rd ed. (Orr's Island, Harpswell ME: Atlantic Academic, 2012).

27. D. Bohrer, P. Cícero do Nascimento, P. Martins, and R. Binotto, "Availability of Aluminum from Glass on an Al Form Ion Exchanger in the Presence of Complexing Agents and Amino Acids," *Anal. Chim. Acta* **2002**, *459*, 267.

Chapter 3

1. K. L. Wilson and J. W. Birks, "Mechanism and Elimination of a Water Vapor Interference in the Measurement of Ozone by UV Absorbance," *Environ. Sci. Technol.* **2006**, *40*, 6361; P. C. Andersen, C. J. Williford, and J. W. Birks, "Miniature Personal Ozone Monitor Based on UV Absorbance," *Anal. Chem.* **2010**, *82*, 7924.

2. U.S. Standard Reference Materials are available from SRMINFO@enh.nist.gov. European Certified Reference Materials are available from http://irmm.jrc.ec.europa.eu/reference_materials_catalogue/Pages/index.aspx.

3. J. R. Taylor, *An Introduction to Error Analysis*, 2nd ed. (Sausalito, CA: University Science Books, 1997). An outstandingly readable book.

4. W. A. Brand, "Atomic Weights: Not So Constant After All," *Anal. Bioanal. Chem.* **2013**, *405*, 2755.

5. P. De Bièvre, S. Valkiers, and P. D. P. Taylor, "The Importance of Avogadro's Constant for Amount-of-Substance Measurements," *Fresenius J. Anal. Chem.* **1998**, *361*, 227; L. Yang, Z. Mester, R. E. Sturgeon, and J. Meija, "Determination of the Atomic Weight of ^{28}Si-Enriched Silicon for a Revised Estimate of the Avogadro Constant," *Anal. Chem.* **2012**, *84*, 2321.

Chapter 4

1. Excellent, readable sources on statistics are D. B. Hibbert and J. J. Gooding, *Data Analysis for Chemistry* (Oxford: Oxford University Press, 2006); J. C. Miller and J. N. Miller, *Statistics and Chemometrics for Analytical Chemistry*, 6th ed. (Harlow, UK: Pearson Prentice Hall, 2010); R. Pearson, *Exploring Data in Engineering, the Sciences, and Medicine* (Oxford: Oxford University Press, 2011); S. L. R. Ellison, V. J. Barwick, and T. J. D. Farrand, *Practical Statistics for the Analytical Scientist* (Cambridge: RCS Publishing, 2009); D. Lucy, *Introduction to Statistics for Forensic Scientists* (Chichester: Wiley, 2004); C. G. G. Aitken and F. Taroni, *Statistics and the Evaluation of Evidence for Forensic Scientists*, 2nd ed. (Chichester: Wiley, 2004); and P. C. Meier and R. E. Zünd, *Statistical Methods in Analytical Chemistry*, 2nd ed. (New York: Wiley, 2000).

2. S. A. Lee, R. K. Ross, and M. C. Pike, "An Overview of Menopausal Oestrogen–Progestin Hormone Therapy and Breast Cancer Risk," *Br. J. Cancer* **2005**, *92*, 2049.

3. L. H. Keith, W. Crummett, J. Deegan, Jr., R. A. Libby, J. K. Taylor, and G. Wentler, "Principles of Environmental Analysis," *Anal. Chem.* **1983**, *55*, 2210.

4. When $t_{calculated}$ from Equation 4-9 is less than t_{table}, we conclude that two means are not significantly different at a chosen confidence level. This test does not provide the same confidence that two means *are equal*. The two one-sided t test (TOST) provides a way to show that two means are equivalent. See S. E. Lewis and J. E. Lewis, "The Same or Not the Same: Equivalence as an Issue in Educational Research," *J. Chem. Ed.* **2005**, *82*, 1408; G. B. Limentani, M. C. Ringo, F. Ye, M. L. Bergquist, and E. O. McSorley, "Beyond the *t*-Test: Statistical Equivalence Testing," *Anal. Chem.* **2005**, *77*, 221A; and M. J. Chatfield and P. J. Borman, "Acceptance Criteria for Method Equivalency Assessments," *Anal. Chem.* **2009**, *81*, 9841.

5. For a comprehensive approach to least-squares fitting of nonlinear curves, including analysis of uncertainty, see J. Tellinghuisen, "Understanding Least Squares through Monte Carlo Calculations," *J. Chem. Ed.* **2005**, *82*, 157; P. Ogren, B. Davis, and N. Guy, "Curve Fitting, Confidence Intervals and Correlations, and Monte Carlo Visualizations for Multilinear Problems in Chemistry: A General Spreadsheet Approach," *J. Chem. Ed.* **2001**, *78*, 827; R. de Levie, "Collinearity in Least-Squares Analysis," *J. Chem. Ed.* **2012**, *89*, 68; see also D. C. Harris, "Nonlinear Least-Squares Curve Fitting with Microsoft Excel Solver," *J. Chem. Ed.* **1998**, *75*, 119; C. Salter and R. de Levie, "Nonlinear Fits of Standard Curves: A Simple Route to Uncertainties in Unknowns," *J. Chem. Ed.* **2002**, *79*, 268; R. de Levie, "Estimating Parameter Precision in Nonlinear Least Squares with Excel's Solver," *J. Chem. Ed.* **1999**, *76*, 1594; S. E. Feller and C. F. Blaich, "Error Estimates for Fitted Parameters," *J. Chem. Ed.* **2001**, *78*, 409; R. de Levie, "When, Why, and How to Use Weighted Least Squares," *J. Chem. Ed.* **1986**, *63*, 10; P. J. Ogren and J. R. Norton, "Applying a Simple Linear Least-Squares Algorithm to Data with Uncertainties in Both Variables," *J. Chem. Ed.* **1992**, *69*, A130.

6. In this book, we plot analytical response on the y-axis versus concentration on the x-axis. The inverse calibration (y = concentration, x = response) provides a more precise estimate of concentration from a measured response. The advantage of the inverse calibration increases as the noise in the response increases. There are cases, such as some spectrophotometric measurements, in which uncertainty in response (absorbance) is smaller than uncertainty in concentration. In such cases, you should plot response on the x-axis and concentration on the y-axis. See J. Tellinghuisen, "Inverse vs Classical Calibration for Small Data Sets," *Fresenius J. Anal. Chem.* **2000**, *368*, 585; V. Centner, D. L. Massart, and S. de Jong, "Inverse Calibration Predicts Better Than Classical Calibration," *Fresenius J. Anal. Chem.* **1998**, *361*, 2; D. Grientschnig, "Relation Between Prediction Errors of Inverse and Classical Calibration," *Fresenius J. Anal. Chem.* **2000**, *367*, 497.

7. Equations for the perpendicular fit, including uncertainties and covariance, are given by J. V. de Julián-Ortiz, L. Pogliani, and E. Besalú, "Two-Variable Linear Regression: Modeling with Orthogonal Least-Squares Analysis," *J. Chem. Ed.* **2010**, *87*, 994.

8. W. Hyk and Z. Stojek, "Quantifying Uncertainty of Determination by Standard Additions and Serial Dilutions Methods Taking into Account Standard Uncertainties in Both Axes," *Anal. Chem.* **2013**, *85*, 5933. This article gives equations for the least-squares line when there is uncertainty in both *x* and *y*. It gives equations for uncertainty when using a calibration curve and the method of standard addition.

9. K. Danzer and L. A. Currie, "Guidelines for Calibration in Analytical Chemistry," *Pure Appl. Chem.* **1998**, *70*, 993.

10. C. Salter, "Error Analysis Using the Variance-Covariance Matrix," *J. Chem. Ed.* **2000**, *77*, 1239. Salter's Equation 8 is equivalent to Equation 4-27, although this equivalence is not obvious.

11. N. J. Lawryk and C. P. Weisel, "Concentration of Volatile Organic Compounds in the Passenger Compartments of Automobiles," *Environ. Sci. Tech.* **1996**, *30*, 810.

Chapter 5

1. C. Hogue, "Ferreting Out Erroneous Data," *Chem. Eng. News*, 1 April 2002, p. 49.

2. D. B. Hibbert, *Quality Assurance for the Analytical Chemistry Laboratory* (Oxford: Oxford University Press, 2007); W. Funk, V. Dammann, and G. Donnevert, *Quality Assurance in Analytical Chemistry* (Hoboken, NJ: Wiley, 2006); B. W. Wenclawiak, M. Koch, and E. Hadjiscostas, eds., *Quality Assurance in Analytical Chemistry* (Heidelberg: Springer-Verlag, 2004); E. Mullins, *Statistics for the Quality Control Chemistry Laboratory* (Cambridge: Royal Society of Chemistry, 2003); P. Quevauviller, *Quality Assurance for Water Analysis* (Chichester: Wiley, 2002); M. Valcárcel, *Principles of Analytical Chemistry* (Berlin: Springer-Verlag, 2000).

3. J. A. Paulos, "Weighting the Positives," *Scientific American*, January 2012, p. 20; J. G. McCully, "Screening Stats," *Scientific American*, May 2012, p. 8.

4. K. M. Phillips, K. Y. Patterson, A. S. Rasor, J. Exler, D. B. Haytowitz, J. M. Holden, and P. R. Pehrsson, "Quality-Control Material in the USDA National Food and Nutrient Analysis Program," *Anal. Bioanal. Chem.* **2006**, *384*, 1341.

5. C. C. Chan, H. Lam, Y. C. Lee, and X.-M. Zhang, eds., *Analytical Method Validation and Instrument Performance Verification* (New York: Wiley, 2004); J. M. Green, "A Practical Guide to Analytical Method Validation," *Anal. Chem.* **1996**, *68*, 305A; M. Swartz and I. S. Krull, "Validation of Bioanalytical Methods—Highlights of FDA's Guidance," *LCGC North Am.* **2003**, *21*, 136; J. D. Orr, I. S. Krull, and M. E. Swartz, "Validation of Impurity Methods," *LCGC North Am.* **2003**, *21*, 626 and 1146.

6. R. de Levie, "Two Linear Correlation Coefficients," *J. Chem. Ed.* **2003**, *80*, 1030.

7. E. Stottmeister et al., "Interlaboratory Trial on the Analysis of Alkylphenols, Alkylphenol Ethoxylates, and Bisphenol A in Water Samples According to ISO/CD 18857-2," *Anal. Chem.* **2009**, *81*, 6765.

8. W. Horwitz, L. R. Kamps, and K. W. Boyer, *J. Assoc. Off. Anal. Chem.* **1980**, *63*, 1344; W. Horwitz, "Evaluation of Analytical Methods Used for Regulation of Foods and Drugs," *Anal. Chem.* **1982**, *54*, 67A; P. Hall and B. Selinger, "A Statistical Justification Relating Interlaboratory Coefficients of Variation with Concentration Levels," *Anal. Chem.* **1989**, *61*, 1465; R. Albert and W. Horwitz, "A Heuristic Derivation of the Horwitz Curve," *Anal. Chem.* **1997**, *69*, 789.

9. J. Vial and A. Jardy, "Experimental Comparison of the Different Approaches to Estimate LOD and LOQ of an HPLC Method," *Anal. Chem.* **1999**, *71*, 2672; G. L. Long and J. D. Winefordner, "Limit of Detection," *Anal. Chem.* **1983**, *55*, 713A; W. R. Porter, "Proper Statistical Evaluation of Calibration Data," *Anal. Chem.* **1983**, *55*, 1290A; S. Geiß and J. W. Einmax, "Comparison of Detection Limits in Environmental Analysis," *Fresenius J. Anal. Chem.* **2001**, *370*, 673; M. E. Zorn, R. D. Gibbons, and W. C. Sonzogni, "Evaluation of Approximate Methods for Calculating the Limit of Detection and Limit of Quantitation," *Environ. Sci. Technol.* **1999**, *33*, 2291; J. D. Burdge, D. L. MacTaggart, and S. O. Farwell, "Realistic Detection Limits from Confidence Bands," *J. Chem. Ed.* **1999**, *76*, 434.

10. The procedure in the text leading to Equation 5-5 is the one most recommended for determining the detection limit. If you do not have replicate determinations of the blank and a low-concentration sample, but you do have a linear calibration curve such as the one in Figure 4-13, you can use the

least-squares parameters to estimate a limit of detection for analyte at a desired confidence level. The formula below comes from ISO 11843-2:2000 (International Organization for Standardization, Geneva, www.iso.org). Suppose you measure I calibration standards (including the blank), with J replicates at each concentration. Then you make K replicate measurements of unknown. The detection limit is

$$\text{Detection limit} = \frac{2ts_y}{m}\sqrt{\frac{1}{K} + \frac{1}{I \times J} + \frac{\bar{x}^2}{J\sum(x_i - \bar{x})^2}} \qquad \text{(A)}$$

where s_y is the standard deviation of y (Equation 4-20), m is the slope (Equation 4-16), and $\bar{x}$ is the mean value of x for the standards (including the blank). Student's t is selected from Table 4-4 for $(I \times J) - 2$ degrees of freedom. Column headings in Table 4-4 are for a two-tailed distribution. The required value of t in Equation A is for a one-tailed distribution. The equation gives the concentration of analyte that will lead with a probability $(1 - \beta)$ to the conclusion that the concentration of analyte in the unknown is larger than that of a blank. For 95% confidence, $\beta = 0.05$. In this case select t from the column labeled 90% confidence. For 99% confidence, $\beta = 0.01$, and you select t from the column labeled 98% confidence.

Example: Consider the calibration data in Problem 4-35, for which $m = 869.1$ mV/vol%, $s_y = 18.05$ mV, $\bar{x} = 0.544$ vol%, and $\sum(\bar{x}_i - \bar{x})^2 = 2.878$ vol%^{2}. There are seven calibration points, including the blank, so $I = 7$ and degrees of freedom = $7 - 2 = 5$. There is one measurement at each calibration concentration, so $J = 1$. There are four replicate measurements of unknown, so $K = 4$. Suppose we want a 99% confidence level detection limit. Therefore, select $t = 3.365$ from Table 4-4 for 98% confidence and 5 degrees of freedom.

$$\text{Detection limit} = \frac{2(3.365)(18.05\text{ mV})}{(869.1\text{ mV/vol\%})}\sqrt{\frac{1}{4} + \frac{1}{7 \times 1} + \frac{(0.544\text{ vol\%})^2}{(1)(2.878\text{ vol\%})^2}}$$
$$= (0.140)\sqrt{0.250 + 0.143 + 0.0357} = 0.092\text{ vol\%}$$

If more replicate measurements of unknown are made, the first term in the square root becomes smaller and the detection limit decreases.

11. M. Bader, "A Systematic Approach to Standard Addition Methods in Instrumental Analysis," *J. Chem. Ed.* **1980**, *57*, 703.

12. W. Hyk and Z. Stojek, "Quantifying Uncertainty of Determination by Standard Additions and Serial Dilutions Methods Taking into Account Standard Uncertainties in Both Axes," *Anal. Chem.* **2013**, *85*, 5933. This article gives equations for the least-squares line when there is uncertainty in both *x* and *y*. It gives equations for uncertainty when using a calibration curve and the method of standard addition.

13. W. R. Kelly, B. S. MacDonald, and W. F. Guthrie, "Gravimetric Approach to the Standard Addition Method in Instrumental Analysis," *Anal. Chem.* **2008**, *80*, 6154.

14. G. R. Bruce and P. S. Gill, "Estimates of Precision in a Standard Additions Analysis," *J. Chem. Ed.* **1999**, *76*, 805.

15. J. A. Day, M. Montes-Bayón, A. P. Vonderheide, and J. A. Caruso, "A Study of Method Robustness for Arsenic Speciation in Drinking Water Samples by Anion Exchange HPLC-ICP-MS," *Anal. Bioanal. Chem.* **2002**, *373*, 664.

16. X. Zhao and C. D. Metcalf, "Characterizing and Compensating for Matrix Effects Using Atmospheric Pressure Chemical Ionization Liquid Chromatography–Tandem Mass Spectrometry: Analysis of Neutral Pharmaceuticals in Municipal Wastewater," *Anal. Chem.* **2008**, *80*, 2010.

Chapter 6

1. D. P. Sheer and D. C. Harris, "Acidity Control in the North Branch Potomac," *J. Water Pollution Control Federation* **1982**, *54*, 1441.

2. R. E. Weston, Jr., "Climate Change and Its Effect on Coral Reefs," *J. Chem. Ed.* **2000**, *77*, 1574; R. Martin and A. Quigg, "Tiny Plants That Once Ruled the Seas," *Scientific American*, June 2013, p. 40.

3. P. D. Thacker, "Global Warming's Other Effects on the Oceans," *Environ. Sci. Technol.* **2005**, *39*, 10A.

4. J. K. Baird, "A Generalized Statement of the Law of Mass Action," *J. Chem. Ed.* **1999**, *76*, 1146; R. de Levie, "What's in a Name?" *J. Chem. Ed.* **2000**, *77*, 610.

5. For thermodynamic data, see N. Jacobson, "Use of Tabulated Thermochemical Data for Pure Compounds," *J. Chem. Ed.* **2001**, *78*, 814; http://webbook.nist.gov/chemistry/; M. W. Chase, Jr., *NIST-JANAF Thermochemical Tables*, 4th ed; *J. Phys. Chem. Ref. Data*: Monograph 9 (New York: American Chemical Society and American Physical Society, 1998).

6. L. M. Raff, "Spontaneity and Equilibrium: Why 'ΔG < 0 Denotes a Spontaneous Process' and 'ΔG = 0 Means the System is at Equilibrium' are Incorrect," *J. Chem. Ed.* **2014**, *91*, 386.

7. The solubility of most ionic compounds increases with temperature, despite the fact that the standard heat of solution (ΔH°) is negative for about half of them. Discussions of this seeming contradiction can be found in G. M. Bodner, "On the Misuse of Le Châtelier's Principle for the Prediction of the Temperature Dependence of the Solubility of Salts," *J. Chem. Ed.* **1980**, *57*, 117, and R. S. Treptow, "Le Châtelier's Principle Applied to the Temperature Dependence of Solubility," *J. Chem. Ed.* **1984**, *61*, 499.

8. A. K. Sawyer, "Solubility and K_{sp} of Calcium Sulfate: A General Chemistry Laboratory Experiment," *J. Chem. Ed.* **1983**, *60*, 416; J. Shukla, V. P. Mohandas, and A. Kumar, "Effect of pH on the Solubility of $CaSO_4 \cdot 2H_2O$ in Aqueous NaCl Solutions and Physicochemical Solution Properties at 35°C," *J. Chem. Eng. Data* **2008**, *53*, 2797.

9. A really good book to read about solubility and all types of equilibrium calculations is W. B. Guenther, *Unified Equilibrium Calculations* (New York: Wiley, 1991).

10. E. Koubek, "Demonstration of the Common Ion Effect," *J. Chem. Ed.* **1993**, *70*, 155.

11. For many *great* chemical demonstrations, see B. Z. Shakhashiri, *Chemical Demonstrations: A Handbook for Teachers of Chemistry* (Madison, WI: University of Wisconsin Press, 1983–2011), 5 volumes. See also L. E. Summerlin and J. L. Ealy, Jr., *Chemical Demonstrations: A Sourcebook for Teachers*, 2nd ed. (Washington, DC: American Chemical Society, 1988).

12. A demonstration of selective precipitation by addition of Pb^{2+} to a solution containing CO_3^{2-} and I^- is described by T. P. Chirpich, "A Simple, Vivid Demonstration of Selective Precipitation," *J. Chem. Ed.* **1988**, *65*, 359.

13. Classroom demonstration of complex equilibria: A. R. Johnson, T. M. McQueen, and K. T. Rodolfa, "Species Distribution Diagrams in the Copper-Ammonia System," *J. Chem. Ed.* **2005**, *82*, 408.

14. A computer database of critically selected equilibrium constants is found in R. M. Smith, A. E. Martell, and R. J. Motekaitis, *NIST Critical Stability Constants of Metal Complexes Database 46* (Gaithersburg, MD: National Institute of Standards and Technology, 2001). Measurement of equilibrium constants is described in A. Martell and R. Motekaitis, *Determination and Use of Stability Constants* (New York: VCH Publishers, 1992); K. A. Conners, *Binding Constants: The Measurement of Molecular Complex Stability* (New York: Wiley, 1987); and D. J. Leggett, ed., *Computational Methods for the Determination of Formation Constants* (New York: Plenum Press, 1985).

15. P. A. Giguère, "The Great Fallacy of the H^+ Ion," *J. Chem. Ed.* **1979**, *56*, 571; P. A. Giguère and S. Turrell, "The Nature of Hydrofluoric Acid: A Spectroscopic Study of the Proton-Transfer Complex, $H_3O^+ \cdot F^-$," *J. Am. Chem. Soc.* **1980**, *102*, 5473.

16. Z. Xie, R. Bau, and C. A. Reed, "A Crystalline $[H_9O_4]^+$ Hydronium Salt with a Weakly Coordinating Anion," *Inorg. Chem.* **1995**, *34*, 5403.

17. F. A. Cotton, C. K. Fair, G. E. Lewis, G. N. Mott, K. K. Ross, A. J. Schultz, and J. M. Williams, "X-Ray and Neutron Diffraction Studies of $[V(H_2O)_6][H_5O_2][CF_3SO_3]_4$," *J. Am. Chem. Soc.* **1984**, *106*, 5319.

18. J. M. Headrick, E. G. Diken, R. W. Walters, N. I. Hammer, R. A. Christie, J. Cui, E. M. Myshakin, M. A. Duncan, M. A. Johnson, and K. D. Jordan, "Spectral Signatures of Hydrated Proton Vibrations in Water Clusters," *Science* **2005**, *308*, 1765.

19. S. Wei, Z. Shi, and A. W. Castleman, Jr., "Mixed Cluster Ions as a Structure Probe: Experimental Evidence for Clathrate Structure of $(H_2O)_{20}H^+$ and $(H_2O)_{21}H^+$," *J. Chem. Phys.* **1991**, *94*, 3268.

20. K. Abu-Dari, K. N. Raymond, and D. P. Freyberg, "The Bihydroxide $(H_3O_2^-)$ Anion," *J. Am. Chem. Soc.* **1979**, *101*, 3688.

21. W. B. Jensen, "The Symbol for pH," *J. Chem. Ed.* **2004**, *81*, 21.

22. V. Buch, A. Milet, R. Vácha, P. Jungwirth, and J. P. Devlin, "Water Surface Is Acidic," *Proc. Natl. Acad. Sci. USA* **2007**, *104*, 7342.

23. D. K. Nordstrom, C. N. Alpers, C. J. Ptacek, and D. W. Blowes, "Negative pH and Extremely Acidic Mine Waters from Iron Mountain, California," *Environ. Sci. Technol.* **2000**, *34*, 254.

24. For a CO_2 fountain, see S.-J. Kang and E.-H. Ryu, "Carbon Dioxide Fountain," *J. Chem. Ed.* **2007**, *84*, 1671. For NH_3 fountains, see N. C. Thomas, S. Faulk, and R. Sullivan, "A Hand-Held Ammonia Fountain," *J. Chem. Ed.* **2008**, *85*, 1063; M. D. Alexander, "The Ammonia Smoke Fountain," *J. Chem. Ed.* **1999**, *76*, 210; N. C. Thomas, "A Chemiluminescent Ammonia Fountain," *J. Chem. Ed.* **1990**, *67*, 339; N. Steadman, "Ammonia Fountain Improvements," *J. Chem. Ed.* **1992**, *66*, 764.

25. L. M. Schwartz, "Ion-pair Complexation in Moderately Strong Aqueous Acids," *J. Chem. Ed.* **1995**, *72*, 823. Even though it is not "free," H_3O^+ in ion pairs with certain anions such as $CF_3CO_2^-$ and $CCl_3CO_2^-$ appears to participate in ionic conductance. (R. I. Gelb and J. S. Alper, "Anomalous Conductance in Electrolyte Solutions," *Anal. Chem.* **2000**, *72*, 1322.)

26. M. I. Stojanovska, V. M. Petruševski, and B. T. Šoptrajanov, "On the Existence of Hydrogen Salts of Monoprotic Acids," *J. Chem. Ed.* **2012**, 89, 1168; J. Emsley, "Very Strong Hydrogen Bonding," *Chem. Soc. Rev.* **1980**, *9*, 91; J. Roziere, M.-T. Roziere-Bories, and J. Williams, "Unusual Hydrogen Bonds. A Neutron Diffraction Study of the Hydrogen Dinitrate Ion in Cesium Hydrogen Dinitrate," *Inorg. Chem.* **1976**, *15*, 2490; J. C. Speakman and H. H. Mills, "The Crystal Structures of the Acid Salts of Some Monobasic Acids. Part VI. Sodium Hydrogen Diacetate," *J. Chem. Soc.* **1961**, 1164.

27. Z. Tian, B. Chan, M. B. Sullivan, L. Radom, and S. R. Kass, "Lithium Monoxide Anion: A Ground-State Triplet with the Strongest Base to Date," *Proc. Natl. Acad. Sci. USA* **2008**, *105*, 7647.

28. S. J. Hawkes, "All Positive Ions Give Acid Solutions in Water," *J. Chem. Ed.* **1996**, *73*, 516.

29. M. Kern, "The Hydration of Carbon Dioxide," *J. Chem. Ed.* **1960**, *37*, 14. Great demonstrations with CO_2, including one with carbonic anhydrase, are described by J. A. Bell, "Every Year Begins a Millennium," *J. Chem. Ed.* **2000**, *77*, 1098.

30. T. Loerting, C. Tautermann, R. T. Kroemer, I. Kohl, A. Hallbrucker, E. Mayer, and K. R. Liedl, "On the Surprising Kinetic Stability of Carbonic Acid," *Angew. Chem. Int. Ed.* **2000**, *39*, 891; R. Ludwig and A. Kornath, "In Spite of Chemist's Belief: Carbonic Acid is Surprisingly Stable," *Angew. Chem. Int. Ed.* **2000**, *39*, 1421.

31. J. A. Tossell, "H_2CO_3 and Its Oligomers: Structures, Stabilities, Vibrational and NMR Spectra, and Acidities," *Inorg. Chem.* **2006**, *45*, 5961.

32. I. Kohl, K. Winkel, M. Bauer, K. R. Liedl, T. Loerting, and E. Mayer, "Raman Spectroscopic Study of the Phase Transition of Amorphous to Crystalline β-Carbonic Acid," *Angew. Chem. Int. Ed.* **2009**, *48*, 2690; H. P. Reisenaur, J. P. Wagner, and P. R. Schreiner, "Gas-Phase Preparation of Carbonic Acid and Its Monomethyl Ester," *Angew. Chem. Int. Ed.* **2014**, *53*, 11766.

Chapter 7

1. S. P. Kounaves, M. H. Hecht, J. Kapit, R. C. Quinn, D. C. Catling, B. C. Clark, D. W. Ming, K. Gospodinova, P. Hredzak, K. McElhoney, and J. Shusterman, *Geophys. Res. Lett.* **2010**, *37*, L09201.

2. American Chemical Society, *Reagent Chemicals*, 10th ed. (New York: Oxford University Press, 2006).

3. W. B. Guenther, "Supertitrations: High-Precision Methods," *J. Chem. Ed.* **1988**, *65*, 1097; E. A. Butler and E. H. Swift, "Gravimetric Titrimetry: A Neglected Technique," *J. Chem. Ed.* **1972**, *49*, 425.

4. R. W. Ramette, "In Support of Weight Titrations," *J. Chem. Ed.* **2004**, *81*, 1715.

5. G. Grguric, "Denitrification as a Model Chemical Process," *J. Chem Ed.* **2002**, *79*, 179.

6. M. L. Ware, M. D. Argentine, and G. W. Rice, "Potentiometric Determination of Halogen Content in Organic Compounds Using Dispersed Sodium Reduction," *Anal. Chem.* **1988**, *60*, 383.

Chapter 8

1. H. Ohtaki and T. Radnai, "Structure and Dynamics of Hydrated Ions," *Chem. Rev.* **1993**, *93*, 1157.

2. E. Galbis, J. Hernández-Cobos, C. den Auwer, C. Le Naour, D. Guillaumont, E. Simoni, R. R. Pappalardo, and E. Sánchez-Marcos, "Solving the Hydration Structure of the Heaviest Actinide Aqua Ion Known: The Californium(III) Case," *Angew. Chem. Int. Ed.* **2010**, *49*, 3811.

3. A. G. Sharpe, "The Solvation of Halide Ions and Its Chemical Significance," *J. Chem. Ed.* **1990**, *67*, 309.

4. E. R. Nightingale, Jr., "Phenomenological Theory of Ion Solvation. Effective Radii of Hydrated Ions," *J. Phys. Chem.* **1959**, *63*, 1381.

5. K. H. Stern and E. S. Amis, "Ionic Size," *Chem. Rev.* **1959**, *59*, 1.

6. D. R. Driscol, "Invitation to Enquiry: The Fe^{3+}/CNS^- Equilibrium," *J. Chem. Ed.* **1979**, *56*, 603.

7. S. J. Hawkes, "Salts are Mostly NOT Ionized," *J. Chem. Ed.* **1996**, *73*, 421; S. O. Russo and G. I. H. Hanania, "Ion Association, Solubilities, and Reduction Potentials in Aqueous Solution," *J. Chem. Ed.* **1989**, *66*, 148.

8. K. S. Pitzer, *Activity Coefficients in Electrolyte Solutions*, 2nd ed. (Boca Raton, FL: CRC Press, 1991); B. S. Krumgalz, R. Pogorelskii, A. Sokolov, and K. S. Pitzer, "Volumetric Ion Interaction Parameters for Single-Solute Aqueous Electrolyte Solutions at Various Temperatures," *J. Phys. Chem. Ref. Data* **2000**, *29*, 1123.

9. J. Kielland, "Individual Activity Coefficients of Ions in Aqueous Solutions," *J. Am. Chem. Soc.* **1937**, *59*, 1675.

10. R. E. Weston, Jr., "Climate Change and Its Effect on Coral Reefs," *J. Chem. Ed.* **2000**, *77*, 1574.

11. R. A. Feely, C. L. Sabine, K. Lee, W. Berelson, J. Kleypas, V. J. Fabry, and F. J. Millero, "Impact of Anthropogenic CO_2 on the $CaCO_3$ System in the Oceans," *Science* **2004**, *305*, 362.

12. For more about equilibrium calculations, see W. B. Guenther, *Unified Equilibrium Calculations* (New York: Wiley, 1991); J. N. Butler, *Ionic Equilibrium: Solubility and pH Calculations* (New York: Wiley, 1998); and M. Meloun, *Computation of Solution Equilibria* (New York: Wiley, 1988). For equilibrium calculation software, see http://www.micromath.com/ and http://www.acadsoft.co.uk/.

13. J. J. Baeza-Baeza and M. C. Garcia-Álvarez-Coque, "Systematic Approach to Calculate the Concentration of Chemical Species in Multi-Equilibrium Problems," *J. Chem. Ed.* **2011**, *88*, 169.

14. J. J. Baeza-Baeza and M. C. Garcia-Álvarez-Coque, "Systematic Approach for Calculating the Concentration of Chemical Species in Multi-Equilibrium Problems: Inclusion of the Ionic Strength Effects," *J. Chem. Ed.* **2012**, *89*, 900. We modified the published approach to use a circular reference and to avoid asking Solver to find the ionic strength, which was not a reliable process for us.

15. E. Koort, P. Gans, K. Herodes, V. Pihl, and I. Leito, "Acidity Constants in Different Media (I = 0 and I = 0.1 M KCl) from the Uncertainty Perspective," *Anal. Bioanal. Chem.* **2006**, *385*, 1124.

Chapter 9

1. R. Schmid and A. M. Miah, "The Strength of the Hydrohalic Acids," *J. Chem. Ed.* **2001**, *78*, 116.

2. T. F. Young, L. F. Maranville, and H. M. Smith, "Raman Spectral Investigations of Ionic Equilibria in Solutions of Strong Electrolytes," in W. J. Hamer, ed., *The Structure of Electrolytic Solutions* (New York: Wiley, 1959).

3. E. S. Shamay, V. Buch, M. Parrinello, and G. L. Richmond, "At the Water's Edge: Nitric Acid as a Weak Acid," *J. Am. Chem. Soc.* **2007**, *129*, 12910.

4. Acid dissociation constants do not tell us which protons dissociate in each step. Assignments for pyridoxal phosphate come from nuclear magnetic resonance spectroscopy (B. Szpoganicz and A. E. Martell, "Thermodynamic and Microscopic Equilibrium Constants of Pyridoxal 5'-Phosphate," *J. Am. Chem. Soc.* **1984**, *106*, 5513).

5. For an alternative treatment approach, see H. L. Pardue, I. N. Odeh, and T. M. Tesfai, "Unified Approximations: A New Approach for Monoprotic Weak Acid-Base Equilibria," *J. Chem. Ed.* **2004**, *81*, 1367.

6. M. C. Bonneau, "The Chemistry of Fabric Reactive Dyes," *J. Chem. Ed.* **1995**, *72*, 724.

7. H. N. Po and N. M. Senozan, "The Henderson-Hasselbalch Equation: Its History and Limitations," *J. Chem. Ed.* **2001**, *78*, 1499; R. de Levie, "The Henderson-Hasselbalch Equation: Its History and Limitations," *J. Chem. Ed.* **2003**, *80*, 146.

8. F. B. Dutton and G. Gordon in H. N. Alyea and F. B. Dutton, eds., *Tested Demonstrations in Chemistry*, 6th ed. (Easton, PA: Journal of Chemical Education, 1965), p. 147; R. L. Barrett, "The Formaldehyde Clock Reaction," *J. Chem. Ed.* **1955**, *32*, 78. See also J. J. Fortman and J. A. Schreier, "Some Modified Two-Color Formaldehyde Clock Salutes for Schools with Colors of Gold and Green or Gold and Red," *J. Chem. Ed.* **1991**, *68*, 324; M. G. Burnett, "The Mechanism of the Formaldehyde Clock Reaction," *J. Chem. Ed.* **1982**, *59*, 160; and P. Warneck, "The Formaldehyde-Sulfite Clock Reaction Revisited," *J. Chem. Ed.* **1989**, *66*, 334.

9. Many other clock reactions are described in the literature. For a summary, see A. P. Oliveira and R. B. Faria, "The Chlorate-Iodine Clock Reaction," *J. Am. Chem. Soc.* **2005**, *127*, 18022.

10. The chemical we call sodium bisulfite ($NaHSO_3$) is apparently not the solid in the reagent bottle, which is reported to be sodium metabisulfite ($Na_2S_2O_5$) (D. Tudela, "Solid $NaHSO_3$ Does Not Exist," *J. Chem. Ed.* **2000**, *77*, 830; see also H. D. B. Jenkins and D. Tudela, "New Methods to Estimate Lattice Energies: Application to Bisulfite and Metabisulfite," *J. Chem. Ed.* **2003**, *80*, 1482). $NaHSO_3$ is produced when $Na_2S_2O_5$ reacts with H_2O. A reagent bottle I use for the formaldehyde clock reaction is labeled "sodium bisulfite," but no formula is given. The label gives the reagent assay "as SO_2: minimum 58.5%." Pure $NaHSO_3$ is equivalent to 61.56 wt% SO_2 and pure $Na_2S_2O_5$ is equivalent to 67.40 wt% SO_2.

11. J. B. Early, A. R. Negron, J. Stephens, R. Stauffer, and S. D. Furrow, "The Glyoxal Clock Reaction," *J. Chem. Ed.* **2007**, *84*, 1965.

12. E. T. Urbansky and M. R. Schock, "Understanding, Deriving, and Computing Buffer Capacity," *J. Chem. Ed.* **2000**, *77*, 1640.

Chapter 10

1. International Energy Agency data cited in *Chemical and Engineering News*, 4 June 2012, p 8.

2. The mass of dry air in the atmosphere is 5.14×10^{21} g. (K. E. Trenberth and L. Smith, "The Mass of the Atmosphere: A Constraint on Global Analyses," *J. Climate*, **2005**, *18*, 864 [http://dx.doi.org/10.1175/JCLI-3299.1].) Main constituents of dry air are N_2 (78.09 vol%), O_2 (20.95 vol%), Ar (0.93 vol%), and CO_2 (0.04 vol%). Volume weighted average molecular mass of air = 28.968 g/mol. Moles of gas in air = (5.14×10^{21} g)/(28.968 g/mol) = 1.77×10^{20} mol. CO_2 from fossil fuel burning in 2011 = (3.16×10^{16} g/44.010 g/mol) = 7.18×10^{14} mol. Treating the atmosphere as an ideal gas, ppm by volume is the same as ppm by moles. CO_2 added = 7.18×10^{14} mol/1.77×10^{20} mol = 4.05 ppm.

3. D. S. Arndt, M. O. Baringer, and M. R. Johnson, eds., "State of the Climate in 2009," Special Supplement to *Bull. Am. Meteorological Soc.* **2010**, *91*, No. 6.

4. B. J. Bozlee, M. Janebo, and G. Jahn, "A Simplified Model to Predict the Effect of Increasing Atmospheric CO_2 on Carbonate Chemistry in the Ocean," *J. Chem. Ed.* **2008**, *85*, 213.

5. P. D. Thacker, "Global Warming's Other Effects on the Oceans," *Environ. Sci. Technol.* **2005**, *39*, 10A.

6. C. Turley, J. Blackford, S. Widdicombe, D. Lowe, P. D. Nightingale, and A. P. Rees, "Reviewing the Impact of Increased Atmospheric on Oceanic pH and the Marine Ecosystem," in *Avoiding Dangerous Climate Change*, H. J. Schellnhuber, W. Cramer, N. Nakicenovic, T. Wigley, G. Yohe, eds. (Cambridge: Cambridge University Press, 2006).

7. R. Albright, B. Mason, M. Miller, and C. Langdon, "Ocean Acidification Compromises Recruitment Success of the Threatened Caribbean Coral Acropora palmata," *Proc. Natl. Acad. Sci. USA*, **2010**, *107*, 20400; R. E. Weston, Jr., "Climate Change and Its Effect on Coral Reefs," *J. Chem. Ed.* **2000**, *77*, 1574.

8. J. C. Orr, V. J. Fabry, O. Aumont, L. Bopp, S. C. Doney, R. A. Feely, A. Gnanadesikan, N. Gruber, A. Ishida, F. Joos, R. M. Key, K. Lindsay, E. Maier-Reimer, R. Matear, P. Monfray, A. Mouchet, R. G. Najjar, G.-K. Plattner, K. B. Rodgers, C. L. Sabine, J. L. Sarmiento, R. Schlitzer, R. D. Slater, I. J. Totterdell, M.-F. Weirig, Y. Yamanaka, and A. Yool, "Anthropogenic Ocean Acidification over the Twenty-first Century and Its Impact on Calcifying Organisms," *Nature* **2005**, *437*, 681.

9. M. D. Iglesias-Rodriguez, P. R. Halloran, R. E. M. Rickaby, I. R. Hall, E. Colmenero-Hidalgo, J. R. Gittins, D. R. H. Green, T. Tyrrell, S. J. Gibbs,

P. von Dassow, E. Rehm, E. V. Armbrust, and K. P. Boessenkool, "Phytoplankton Calcification in a High-CO_2 World," *Science*, **2008**, *320*, 336.

10. P. G. Daniele, A. De Robertis, C. De Stefano, S. Sammartano, and C. Rigano, "Na^+, K^+, and Ca^{2+} Complexes of Low Molecular Weight Ligands in Aqueous Solution," *J. Chem. Soc. Dalton Trans.* **1985**, 2353.

11. P. A. Sims, "Use of a Spreadsheet to Calculate the Net Charge of Peptides and Proteins as a Function of pH," *J. Chem. Ed.* **2010**, *87*, 803; R. H. Singiser, "Spreadsheet Enhancement to Calculating the pI of Proteins," *J. Chem. Ed.* **2011**, *88*, 142.

12. Experiment on surface acidity of a solid: L. Tribe and B. C. Barja, "Adsorption of Phosphate on Goethite," *J. Chem. Ed.* **2004**, *81*, 1624.

13. Experiment on pH of zero charge: M. Davranche, S. Lacour, F. Bordas, and J.-C. Bollinger, "Determination of the Surface Chemical Properties of Natural Solids," *J. Chem. Ed.* **2003**, *80*, 76.

14. W. Stumm and J. J. Morgan, *Aquatic Chemistry*, 3rd ed. (New York: Wiley, 1996), pp. 343–348; F. J. Millero, "Thermodynamics of the Carbon Dioxide System in the Oceans," *Geochim. Cosmochim. Acta* **1995**, *59*, 661; Ocean carbon thermodynamics: http://cdiac.esd.ornl.gov/oceans/glodap/cther.htm.

Chapter 11

1. M. J. Fedor, "Structure and Function of the Hairpin Ribozyme," *J. Mol. Biol.* **2000**, *297*, 269.

2. A. G. Dickson, http://cdiac.ornl.gov/oceans/Handbook_2007.html.

3. T. R. Martz, A. G. Dickson, and M. D. DeGrandpre, "Tracer Monitored Titrations: Measurement of Total Alkalinity," *Anal. Chem.* **2006**, *78*, 1817.

4. K. R. Williams, "Automatic Titrators in the Analytical and Physical Chemistry Laboratories," *J. Chem. Ed.* **1998**, *75*, 1133; K. L. Headrick, T. K. Davies, and A. N. Haegele, "A Simple Laboratory-Constructed Automatic Titrator," *J. Chem. Ed.* **2000**, *77*, 389.

5. M. Inoue and Q. Fernando, "Effect of Dissolved CO_2 on Gran Plots," *J. Chem. Ed.* **2001**, *78*, 1132; G. Gran, "Equivalence Volumes in Potentiometric Titrations," *Anal. Chim. Acta* **1988**, *206*, 111; F. J. C. Rossotti and H. Rossotti, "Potentiometric Titrations Using Gran Plots," *J. Chem. Ed.* **1965**, *42*, 375; L. M. Schwartz, "Uncertainty of a Titration Equivalence Point," *J. Chem. Ed.* **1992**, *69*, 879; L. M. Schwartz, "Advances in Acid-Base Gran Plot Methodology," *J. Chem. Ed.* **1987**, *64*, 947.

6. M. Rigobello-Masini and J. C. Masini, "Application of Modified Gran Functions and Derivative Methods to Potentiometric Acid Titration Studies of the Distribution of Inorganic Carbon Species in Cultivation Medium of Marine Microalgae," *Anal. Chim. Acta* **2001**, *448*, 239.

7. G. Papanastasiou and I. Ziogas, "Simultaneous Determination of Equivalence Volumes and Acid Dissociation Constants from Potentiometric Titration Data," *Talanta* **1995**, *42*, 827.

8. G. Wittke, "Reactions of Phenolphthalein at Various pH Values," *J. Chem. Ed.* **1983**, *60*, 239.

9. Demonstrations with universal indicator (a mixed indicator with many color changes) are described in J. T. Riley, "Flashy Solutions," *J. Chem. Ed.* **1977**, *54*, 29.

10. T. A. Canada, L. R. Allain, D. B. Beach, and Z. Xue, "High-Acidity Determination in Salt-Containing Acids by Optical Sensors," *Anal. Chem.* **2002**, *74*, 2535.

11. D. Fárcasiu and A. Ghenciu, "Acidity Functions from ^{13}C-NMR," *J. Am. Chem. Soc.* **1993**, *115*, 10901.

12. B. Hammouti, H. Oudda, A. El Maslout, and A. Benayada, "A Sensor for the In Situ Determination of Acidity Levels in Concentrated Sulfuric Acid," *Fresenius J. Anal. Chem.* **1999**, *365*, 310. For use of glass electrodes to measure pH as low as −4, see D. K. Nordstrom, C. N. Alpers, C. J. Ptacek, and D. W. Blowes, "Negative pH and Extremely Acidic Mine Waters from Iron Mountain, California," *Environ. Sci. Technol.* **2000**, *34*, 254.

13. M. Juhasz, S. Hoffmann, E. Stoyanov, K.-C. Kim, and C. A. Reed, "The Strongest Isolable Acid," *Angew. Chem. Int. Ed.* **2004**, *43*, 5352; E. S. Stoyanov, S. P. Hoffmann, M. Juhasz, and C. A. Reed, "The Structure of the Strongest Brønsted Acid: The Carborane Acid $H(CHB_{11}C_{11})$," *J. Am. Chem. Soc.* **2006**, *128*, 3160; M. M. Meyer, X.-B. Wang, C. A. Reed, L.-S. Wang, and S. R. Kass, "Experimental Determination of the Electron Binding Energy of Carborane Anions and the Gas Phase Acidity of Carborane Acids," *J. Am. Chem. Soc.* **2009**, *131*, 18050; A. Avelar, F. S. Tham, and C. A. Reed, "Superacidity of Boron Acids $H_2(B_{12}X_{12})$ (X = Cl, Br)," *Angew. Chem. Int. Ed.* **2009**, *48*, 3491.

14. R. A. Butler and R. G. Bates, "Double Potassium Salt of Sulfosalicylic Acid in Acidimetry and pH Control," *Anal. Chem.* **1976**, *48*, 1669.

15. Borax goes down to the pentahydrate upon standing: R. Naumann, C. Alexander-Weber, and F. G. K. Baucke, "Limited Stability of the pH Reference Material Sodium Tetraborate Decahydrate (Borax)," *Fresenius J. Anal. Chem.* **1994**, *350*, 119.

16. Instructions for purifying and using primary standards can be found in the following books: J. A. Dean, *Analytical Chemistry Handbook* (New York: McGraw-Hill, 1995), pp. 3-28 to 3-30; J. Bassett, R. C. Denney, G. H. Jeffery, and J. Mendham, *Vogel's Textbook of Quantitative Inorganic Analysis*, 4th ed. (Essex: Longman, 1978), pp. 296–306; I. M. Kolthoff and V. A. Stenger, *Volumetric Analysis*, Vol. 2 (New York: Wiley-Interscience, 1947).

17. A. A. Smith, "Consumption of Base by Glassware," *J. Chem. Ed.* **1986**, *63*, 85; G. Perera and R. H. Doremus, "Dissolution Rates of Commercial Soda-Lime and Pyrex Borosilicate Glasses," *J. Am. Ceramic Soc.* **1991**, *74*, 1554.

18. R. E. Oesper, "Kjeldahl and the Determination of Nitrogen," *J. Chem. Ed.* **1934**, *11*, 457. A beautifully illustrated history and biography.

19. D. Lee, "Plant Linked to Pet Deaths Had History of Polluting," *Los Angeles Times*, 9 May 2007, p. C1; B. Puschner, R. H. Poppenga, L. J. Lowenstine, M. S. Filigenzi, and P. A. Pesavento, "Assessment of Melamine and Cyanuric Acid Toxicity in Cats," *J. Vet. Diagn. Invest.* **2007**, *19*, 616.

20. X. Zheng, A. Zhao, G. Xie, Y. Chi, L. Zhao, H. Li, C. Wang, Y. Bao, W. Jia, M. Luther, M. Su, J. K. Nicholson, W. Jia, "Melamine-Induced Renal Toxicity Is Mediated by the Gut Microbiota," *Sci. Transl. Med.* 2013, *5*, 172ra22.

21. D. Lee and A. Goldman, "Anguished Chinese Flood Hospitals," *Los Angeles Times*, 19 September 2008, p. A3; "Tainted Milk Powder Seized," *Los Angeles Times*, 9 July 2010; R. M. Baum, *Chem. Eng. News*, 13 October 2008, p. 3.

22. J. J. Urh, "Protein Testing Enters the 21st Century: Innovative Protein Analyzer Not Affected by Melamine," *Am. Lab.* October 2008, p. 18.

23. L. Zhu, G. Gamez, H. Chen, K. Chingin, and R. Zenobi, "Rapid Detection of Melamine in Untreated Milk and Wheat Gluten by Ultrasound-Assisted Extractive Electrospray Ionization Mass Spectrometry," *Chem. Commun.* **2009**, 559; G. Huang, Z. Ouyang, and R. G. Cooks, "High-Throughput Trace Melamine Analysis in Complex Mixtures," *Chem. Commun.* **2009**, 556; Q. Xu, H. P. Wei, S. Du, H. B. Li, Z. P. Ji, and X. Y. Hu, "Detection of Subnanomolar Melamine Based on Electrochemical Accumulation Coupled with Enzyme Colorimetric Assay," *J. Agric. Food Chem.* **2013**, *61*, 1810.

24. The Kjeldahl digestion captures amine ($—NR_2$) or amide ($—C[=O]NR_2$) nitrogens (where R can be H or an organic group), but not oxidized nitrogen such as nitro ($—NO_2$) or azo ($—N≡N—$) groups, which must first be reduced to amines or amides.

25. W. Maher, F. Krikowa, D. Wruck, H. Louie, T. Nguyen, and W. Y. Huang, "Determination of Total Phosphorus and Nitrogen in Turbid Waters by Oxidation with Alkaline Potassium Peroxodisulfate," *Anal. Chim. Acta* **2002**, *463*, 283.

26. G. Cruz, "Boric Acid in Kjeldahl Analysis," *J. Chem. Ed.* **2013**, *90*, 1645; F. M. Scales and A. P. Harrison, "Boric Acid Modification of the Kjeldahl Method for Crop and Soil Analysis," *J. Ind. Eng. Chem.* **1920**, *12*, 350; T. Michalowski, A. G. Asuero, and S. Wybraniec, "The Titration in the Kjeldahl Method of Nitrogen Determination: Base or Acid as Titrant?" *J. Chem. Ed.* **2013**, *90*, 191.

27. http:// www.umass.edu/tei/mwwp/acrobat/epa351_3Norg.pdf; http://www.flowinjection.com/methods/tkn.aspx.

28. J. S. Fritz, *Acid-Base Titrations in Nonaqueous Solvents* (Boston: Allyn and Bacon, 1973); J. Kucharsky and L. Safarik, *Titrations in Non-Aqueous Solvents* (New York: Elsevier, 1963); W. Huber, *Titrations in Nonaqueous Solvents* (New York: Academic Press, 1967); I. Gyenes, *Titration in Non-Aqueous Media* (Princeton, NJ: Van Nostrand, 1967).

29. S. P. Porras, "Capillary Zone Electrophoresis of Some Extremely Weak Bases in Acetonitrile," *Anal. Chem.* **2006**, *78*, 5061.

30. R. de Levie, "A General Simulator for Acid-Base Titrations," *J. Chem. Ed.* **1999**, *76*, 987; R. de Levie, "Explicit Expressions of the General Form of the Titration Curve in Terms of Concentration," *J. Chem. Ed.* **1993**, *70*, 209; R. de Levie, "General Expressions for Acid-Base Titrations of Arbitrary Mixtures," *Anal. Chem.* **1996**, *68*, 585; R. de Levie, *Principles of Quantitative Chemical Analysis* (New York: McGraw-Hill, 1997); J. Burnett and W. A. Burns, "Using a Spreadsheet to Fit Experimental pH Titration Data to a Theoretical Expression: Estimation of Analyte Concentration and K_a," *J. Chem. Ed.* **2006**, *83*, 1190.

31. C. Salter and D. L. Langhus, "The Chemistry of Swimming Pool Maintenance," *J. Chem. Ed.* **2007**, *84*, 1124.

32. P. Ballinger and F. A. Long, "Acid Ionization Constants of Alcohols," *J. Am. Chem. Soc.* **1960**, *82*, 795.

Chapter 12

1. Z. Hou, K. N. Raymond, B. O'Sullivan, T. W. Esker, and T. Nishio, "Microbial Macrocyclic Dihydroxamate Chelating Agents," *Inorg. Chem.* **1998**, *37*, 6630. Ferrioxiamines found in the ocean at concentrations of 0.1–10 pM are presumably excreted by microorganisms to enable them to accumulate scarce iron from the ocean. (E. Mawji, M. Gledhill, J. A. Milton, G. A. Tarran, S. Ussher, A. Thompson, G. A. Wolff, P. J. Worsfold, and E. P. Achterberg, "Hydroxamate Siderophores: Occurrence and Importance in the Atlantic Ocean," *Environ. Sci. Technol.* **2008**, *42*, 8675.)

2. N. F. Olivieri and G. M. Brittenham, "Iron-Chelating Therapy and the Treatment of Thalassemia," *Blood* **1997**, *89*, 739.

3. E. J. Neufeld, "Oral Chelators Deferasirox and Deferiprone for Transfusional Iron Overload in Thalassemia Major: New Data, New Questions," *Blood* **2006**, *107*, 3436; K. Farmaki, "Reversal of Complications Following Intensive Combined Chelation in Beta Thalassemia Major Patients," Abstract LB4, 49th American Society of Hematology Annual Meeting, Atlanta, GA, December 2007.

4. D. T. Haworth, "Some Linguistic Details on Chelation," *J. Chem. Ed.* **1998**, *75*, 47.

5. Classroom demonstration: D. C. Bowman, "A Colorful Look at the Chelate Effect," *J. Chem. Ed.* **2006**, *83*, 1158.

6. The chelate effect is often attributed to a favorable entropy change for multidentate binding. Recent work does not support this explanation: V. Vallet, U. Wahlgren, and I. Grenthe, "Chelate Effect and Thermodynamics of Metal Complex Formation in Solution: A Quantum Chemical Study," *J. Am. Chem. Soc.* **2003**, *125*, 14941.

7. R. J. Abergel, E. G. Moore, R. K. Strong, and K. N. Raymond, "Microbial Evasion of the Immune System: Structural Modifications of Enterobactin Impair Siderocalin Recognition," *J. Am. Chem. Soc.* **2006**, *128*, 10998.

8. J. Künnemeyer, L. Terborg, S. Nowak, L. Telgmann, F. Tokmak, B. K. Krämer, A. Günsel, G. A. Wiesmüller, J. Waldeck, C. Bremer, and U. Karst, "Analysis of the Contrast Agent Magnevist and Its Transmetalation Products in Blood Plasma by Capillary Electrophoresis/Electrospray Ionization Time-of-Flight Mass Spectrometry," *Anal. Chem.* **2009**, *81*, 3600.

9. U. Lindner, J. Lingott, S. Richter, N. Jakubowski, and U. Panne, "Speciation of Gadolinium in Surface Water Samples and Plants," *Anal. Bioanal. Chem.* **2013**, *405*, 1865.

10. W. J. Blaedel and H. T. Knight, "Purification and Properties of Disodium Salt of Ethylenediaminetetraacetic Acid as a Primary Standard," *Anal. Chem.* **1954**, *26*, 741.

11. R. L. Barnett and V. A. Uchtman, "Crystal Structures of Ca(CaEDTA) · 7H$_2$O and NaCaNTA," *Inorg. Chem.* **1979**, *18*, 2674.

12. P. Lindqvist-Reis, C. Apostolidis, J. Rebizant, A. Morgenstern, R. Klenze, O. Walter, T. Fanghänel, and R. G. Haire, "The Structures and Optical Spectra of Hydrated Transplutonium Ions in the Solid State and in Solution," *Angew. Chem. Int. Ed.* **2007**, *46*, 919; S. Skanthakumar, M. R. Antonio, R. E. Wilson, and L. Soderholm, "The Curium Aqua Ion," *Inorg. Chem.* **2007**, *46*, 3485.

13. J. N. Mathur, P. Thakur, C. J. Dodge, A. J. Francis, and G. R. Choppin, "Coordination Modes in the Formation of the Ternary Am(III), Cm(III), and Eu(III) Complexes with EDTA and NTA," *Inorg. Chem.* **2006**, *456*, 8026.

14. A definitive reference for the theory of EDTA titration curves is A. Ringbom, *Complexation in Analytical Chemistry* (New York: Wiley, 1963).

15. For discussion of metal-ligand equilibria with numerous examples, see P. Letkeman, "Computer-Modeling of Metal Speciation in Human Blood Serum," *J. Chem. Ed.* **1996**, *73*, 165; A. Rojas-Hernández, M. T. Ramírez, I. González, and J. G. Ibanez, "Predominance-Zone Diagrams in Solution Chemistry," *J. Chem. Ed.* **1995**, *72*, 1099; and A. Bianchi and E. Garcia-España, "Use of Calculated Species Distribution Diagrams to Analyze Thermodynamic Selectivity," *J. Chem. Ed.* **1999**, *76*, 1727.

16. W. N. Perara and G. Hefter, "Mononuclear Cyano- and Hydroxo-Complexes of Iron(III)," *Inorg. Chem.* **2003**, *42*, 5917.

17. G. Schwarzenbach and H. Flaschka, *Complexometric Titrations*, H. M. N. H. Irving, trans. (London: Methuen, 1969); H. A. Flaschka, *EDTA Titrations* (New York: Pergamon Press, 1959); J. A. Dean, *Analytical Chemistry Handbook* (New York: McGraw-Hill, 1995); A. E. Martell and R. D. Hancock, *Metal Complexes in Aqueous Solution* (New York: Plenum Press, 1996).

18. S. Tandy, K. Bossart, R. Mueller, J. Ritschel, L. Hauser, R. Schulin, and B. Nowack, "Extraction of Heavy Metals from Soils Using Biodegradable Chelating Agents," *Environ. Sci. Technol.* **2004**, *38*, 937; B. Kos and D. Leštan, "Induced Phytoextraction/Soil Washing of Lead Using Biodegradable Chelate and Permeable Barriers," *Environ. Sci. Technol.* **2003**, *37*, 624; S. V. Sahi, N. L. Bryant, N. C. Sharma, and S. R. Singh, "Characterization of a Lead Hyperaccumulator Shrub," *Environ. Sci. Technol.* **2002**, *36*, 4676.

19. B. Nowack, R. Schulin, and B. H. Robinson, "Critical Assessment of Chelant-Enhanced Metal Phytoextraction," *Environ. Sci. Technol.* **2006**, *40*, 5225.

20. Indirect determinations of monovalent *cations* are described by I. M. Yurist, M. M. Talmud, and P. M. Zaitsev, "Complexometric Determination of Monovalent Metals," *J. Anal. Chem. USSR* **1987**, *42*, 911.

21. D. P. S. Rathore, P. K. Bhargava, M. Kumar, and R. K. Talra, "Indicators for the Titrimetric Determination of Ca and Total Ca + Mg with EDTA," *Anal. Chim. Acta* **1993**, *281*, 173.

22. H. Bao, "Purifying Barite for Oxygen Isotope Measurement by Dissolution and Reprecipitation in a Chelating Solution," *Anal. Chem.* **2006**, *78*, 304.

23. T. Darjaa, K. Yamada, N. Sato, T. Fujino, and Y. Waseda, "Determination of Sulfur in Metal Sulfides by Bromine Water-CCl$_4$ Oxidative Dissolution and Modified EDTA Titration," *Fresenius J. Anal. Chem.* **1998**, *361*, 442.

Chapter 13

1. J. Gorman in *Science News*, 9 September 2000, p. 165.

2. Books about equilibrium calculations: W. B. Guenther, *Unified Equilibrium Calculations* (New York: Wiley, 1991); J. N. Butler, *Ionic Equilibrium: Solubility and pH Calculations* (New York: Wiley, 1998); and M. Meloun, *Computation of Solution Equilibria* (New York: Wiley, 1988). Software for equilibrium calculations: http://www.micromath.com/ and http://www.acadsoft.co.uk/.

3. R. G. Bates, *Determination of pH*, 2nd ed. (New York: Wiley, 1973), p. 86, is the authoritative reference on pH. The pH uncertainty of primary standards could be greater than ±0.006 at temperatures other than 25°C.

4. R. B. Martin, "Aluminum: A Neurotoxic Product of Acid Rain," *Acc. Chem. Res.* **1994**, *27*, 204.

5. Our approach is similar to one by J. L. Guiñón, J. Garcia-Antón, and V. Pérez-Herranz, "Spreadsheet Techniques for Evaluating the Solubility of Sparingly Soluble Salts of Weak Acids," *J. Chem. Ed.* **1999**, *76*, 1157.

6. K. H. Weber, F. J. Morales, and F.-M. Tao, "Theoretical Study on the Structure and Stabilities of Molecular Clusters of Oxalic Acid with Water," *J. Phys. Chem. A* **2012**, *116*, 11601.

7. P. A. W. Dean, "The Oxalate Dianion, C$_2$O$_4^{2-}$: Planar or Nonplanar?" *J. Chem. Ed.* **2012**, *89*, 417.

8. A. Kraft, "Determination of the pK_a of Multiprotic, Weak Acids by Analyzing Potentiometric Acid-Base Titration Data with Difference Plots," *J. Chem. Ed.* **2003**, *80*, 554.

9. G. B. Kauffman, "Niels Bjerrum: A Centennial Evaluation," *J. Chem. Ed.* **1980**, *57*, 779, 863.

10. Table 6-1 gives $pK_w = 13.995$ at $\mu = 0$ at 25°C. The expression of K_w to which this value applies is given in terms of molalities, m:

$$K_w = \frac{m_{H^+}\gamma_{H^+} m_{OH^-}\gamma_{OH^-}}{\mathcal{A}_{H_2O}} = 10^{-13.995}$$

We want to evaluate K'_w for 0.1 M KCl. The factor for converting molality into molarity in 0.1 M K is 0.994 in Table 12-1-1A of H. S. Harned and B. B. Owen, *Physical Chemistry of Electrolyte Solutions,* 3rd ed. (New York: Reinhold, 1958), p. 725. The factor $\gamma_{H^+}\gamma_{OH^-}/\mathcal{A}_{H_2O}$ is 0.626 in 0.10 M KCl, interpolated from Table 15-2-1A of Harned and Owen, p. 752. K'_w is the product of concentrations $[H^+][OH^-]$:

$$[H^+][OH^-] = \frac{m_{H^+}(0.994)\gamma_{H^+} m_{OH^-}(0.994)\gamma_{OH^-}}{\mathcal{A}_{H_2O}}\frac{\mathcal{A}_{H_2O}}{\gamma_{H^+}\gamma_{OH^-}}$$

$$= 10^{-13.995}(0.994^2)\left(\frac{1}{0.626}\right) = 10^{-13.797}$$

Chapter 14

1. Some treatments of electrochemistry: C. H. Hamann, A. Hamnett, and W. Vielstich, *Electrochemistry,* 2nd ed. (Weinheim: Wiley-VCH, 2007); R. Holze, *Experimental Electrochemistry: A Laboratory Textbook* (Weinheim: Wiley-VCH, 2009); and H. B. Oldham, J. C. Myland, and A. M. Bond, *Electrochemical Science and Technology: Fundamentals and Applications* (Chichester: Wiley, 2012).

2. Quotation from Lady Pollock cited in J. Kendall, *Great Discoveries by Young Chemists* (New York: Thomas Y. Crowell Co., 1953, p. 63).

3. N. J. Tao, "Measurement and Control of Single Molecule Conductance," *J. Mater. Chem.* **2005,** *15,* 3260; N. Tao, "Electrochemical Fabrication of Metallic Quantum Wires," *J. Chem. Ed.* **2005,** *82,* 720; S. Lindsay, "Single-Molecule Electronic Measurements with Metal Electrodes," *J. Chem. Ed.* **2005,** *82,* 727; R. A. Wassel and C. B. Gorman, "Establishing the Molecular Basis for Molecular Electronics," *Angew. Chem. Int. Ed.* **2004,** *43,* 5120.

4. T. Morita and S. Lindsay, "Determination of Single Molecule Conductances of Alkanedithiols by Conducting-Atomic Force Microscopy with Large Gold Nanoparticles," *J. Am. Chem. Soc.* **2007,** *129,* 7262.

5. S. Weinberg, *The Discovery of Subatomic Particles* (Cambridge: Cambridge University Press, 2003), pp. 13–16. A wonderful book by a Nobel Prize winner.

6. P. Krause and J. Manion, "A Novel Approach to Teaching Electrochemical Principles," *J. Chem. Ed.* **1996,** *73,* 354.

7. L. P. Silverman and B. B. Bunn, "The World's Longest Human Salt Bridge," *J. Chem. Ed.* **1992,** *69,* 309.

8. Classroom demonstrations: J. D. Ciparick, "Half Cell Reactions: Do Students Ever See Them?" *J. Chem. Ed.* **1991,** *68,* 247; P.-O. Eggen, T. Grønneberg, and L. Kvittengen, "Small-Scale and Low-Cost Galvanic Cells," *J. Chem. Ed.* **2006,** *83,* 1201.

9. G. C. Smith, Md. M. Hossain, and P. MacCarthy, "Why Batteries Deliver a Fairly Constant Voltage until Dead," *J. Chem. Ed.* **2012,** *89,* 1416; M. J. Smith, A. M. Fonseca, and M. M. Silva, "The Lead-Lead Oxide Secondary Cell as a Teaching Resource," *J. Chem. Ed.* **2009,** *86,* 357; M. J. Smith and C. A. Vincent, "Structure and Content of Some Primary Batteries," *J. Chem. Ed.* **2001,** *78,* 519; M. J. Smith and C. A. Vincent, "Why Do Some Batteries Last Longer Than Others?" *J. Chem. Ed.* **2002,** *79,* 851; M. Tamez and J. H. Yu, "Aluminum-Air Battery," *J. Chem. Ed.* **2007,** *84,* 1936A; H. Goto, H. Yoneyama, F. Togashi, R. Ohta, A. Tsujimoto, E. Kita, K. Ohshima, and D. Rosenberg, "Preparation of Conducting Polymers by Electrochemical Methods and Demonstration of a Polymer Battery," *J. Chem. Ed.* **2008,** *85,* 1067.

10. R. S. Treptow, "The Lead-Acid Battery: Its Voltage in Theory and in Practice," *J. Chem. Ed.* **2002,** *79,* 334.

11. J. Ge, R. Schirhagl, and R. N. Zare, "Glucose-Driven Fuel Cell Constructed from Enzymes and Filter Paper," *J. Chem. Ed.* **2011,** *88,* 1283.

12. K. Klara, N. Hou, A. Lawman, L. Wu, D. Morrill, A. Tente, and L.-Q. Wang, "Developing and Implementing a Simple, Affordable Hydrogen Fuel Cell Laboratory in Introductory Chemistry," *J. Chem. Ed.* **2014,** *91,* 1924; M. Shirkhanzadeh, "Thin-Layer Fuel Cell for Teaching and Classroom Demonstrations," *J. Chem. Ed.* **2009,** *86,* 324; O. Zerbinati, "A Direct Methanol Fuel Cell," *J. Chem. Ed.* **2002,** *79,* 829.

13. L. Deng, C. Chen, M. Zhou, S. Guo, E. Wang, and S. Dong, "Integrated Self-Powered Microchip Biosensor for Endogenous Biological Cyanide," *Anal. Chem.* **2010,** *82,* 4283.

14. A. M. Feltham and M. Spiro, "Platinized Platinum Electrodes," *Chem. Rev.* **1971,** *71,* 177.

15. A. W. von Smolinski, C. E. Moore, and B. Jaselskis, "The Choice of the Hydrogen Electrode as the Base for the Electromotive Series" in *Electrochemistry, Past and Present,* ACS Symposium Series 390, J. T. Stock and M. V. Orna, eds. (Washington, DC, American Chemical Society, 1989), Chap. 9.

16. K. Rajeshwar and J. G. Ibanez, *Environmental Electrochemistry* (San Diego: Academic Press, 1997).

17. H. Frieser, "Enhanced Latimer Potential Diagrams Via Spreadsheets," *J. Chem. Ed.* **1994,** *71,* 786.

18. A. Arévalo and G. Pastor, "Verification of the Nernst Equation and Determination of a Standard Electrode Potential," *J. Chem. Ed.* **1985,** *62,* 882.

19. For a classroom demonstration using a cell as a chemical probe, see R. H. Anderson, "An Expanded Silver Ion Equilibria Demonstration: Including Use of the Nernst Equation and Calculation of Nine Equilibrium Constants," *J. Chem. Ed.* **1993,** *70,* 940. See also J. L. Brosmer and D. G. Peters, "Galvanic Cells and the Determination of Equilibrium Constants," *J. Chem. Ed.* **2012,** *89,* 763.

20. Structure of dehydroascorbic acid: R. C. Kerber, "'As Simple as Possible, But Not Simpler'—The Case of Dehydroascorbic Acid," *J. Chem. Ed.* **2008,** *85,* 1237.

21. J. E. Walker, "ATP Synthesis by Rotary Catalysis (Nobel Lecture)," *Angew. Chem. Int. Ed.* **1998,** *37,* 2309; P. D. Boyer, "Energy, Life, and ATP (Nobel Lecture)," *Angew. Chem. Int. Ed.* **1998,** *37,* 2297; W. S. Allison, "F_1-ATPase," *Acc. Chem. Res.* **1998,** *31,* 819.

22. For an advanced equilibrium problem based on the bromine Latimer diagram, see T. Michalowski, "Calculation of pH and Potential E for Bromine Aqueous Solution," *J. Chem. Ed.* **1994,** *71,* 560.

23. B. R. Staples, "Activity and Osmotic Coefficients of Aqueous Sulfuric Acid at 298.15 K," *J. Phys. Chem. Ref. Data* **1981,** *10,* 779.

24. K. T. Jacob, K. P. Jayadevan, and Y. Waseda, "Electrochemical Determination of the Gibbs Energy of Formation of $MgAl_2O_4$," *J. Am. Ceram. Soc.* **1998,** *81,* 209.

25. J. T. Stock, "Einar Biilmann (1873–1946): pH Determination Made Easy," *J. Chem. Ed.* **1989,** *66,* 910.

26. The cell in this problem would not give an accurate result because of the *junction potential* at each liquid junction (Section 15-3). A cell without liquid junctions is described by P. A. Rock, "Electrochemical Double Cells," *J. Chem. Ed.* **1975,** *52,* 787.

Chapter 15

1. G. Inzelt, A. Lewenstam, and F. Scholz, eds., *Handbook of Reference Electrodes* (Heidelberg: Springer-Verlag, 2013).

2. Practical aspects of electrode fabrication are discussed in D. T. Sawyer, A. Sobkowiak, and J. L. Roberts, Jr., *Electrochemistry for Chemists,* 2nd ed. (New York: Wiley, 1995); G. A. East and M. A. del Valle, "Easy-to-Make Ag/AgCl Reference Electrode," *J. Chem. Ed.* **2000,** *77,* 97.

3. A demonstration of potentiometry with a silver electrode (or a microscale experiment for general chemistry) is described by D. W. Brooks, D. Epp, and H. B. Brooks, "Small-Scale Potentiometry and Silver One-Pot Reactions," *J. Chem. Ed.* **1995,** *72,* A162.

4. D. Dobčnik, J. Stergulec, and S. Gomišček, "Preparation of an Iodide Ion-Selective Electrode by Chemical Treatment of a Silver Wire," *Fresenius J. Anal. Chem.* **1996,** *354,* 494.

5. I. R. Epstein and J. A. Pojman, *An Introduction to Nonlinear Chemical Dynamics: Oscillations, Waves, Patterns, and Chaos* (New York: Oxford University Press, 1998); I. R. Epstein, K. Kustin, P. De Kepper, and M. Orbán, *Scientific American,* March 1983, p. 112; H. Degn, "Oscillating Chemical Reactions in Homogeneous Phase," *J. Chem. Ed.* **1972,** *49,* 302.

6. Mechanisms of oscillating reactions are discussed by J. Miller, "A Nonbiological System Offers Insight into Biological Synchronization," *Physics*

Today, April 2009, p. 14; M. A. Pellitero, C. A. Lamsfus, and J. Borge, "The Belousov-Zhabotinskii Reaction: Improving the Oregonator Model with the Arrhenius Equation," *J. Chem. Ed.* **2013**, *90*, 82; O. Benini, R. Cervellati, and P. Fetto, "The BZ Reaction: Experimental and Model Studies in the Physical Chemistry Laboratory," *J. Chem. Ed.* **1996**, *73*, 865; R. J. Field and F. W. Schneider, "Oscillating Chemical Reactions and Nonlinear Dynamics," *J. Chem. Ed.* **1989**, *66*, 195; R. M. Noyes, "Some Models of Chemical Oscillators," *J. Chem. Ed.* **1989**, *66*, 190; P. Ruoff, M. Varga, and E. Körös, "How Bromate Oscillators are Controlled," *Acc. Chem. Res.* **1988**, *21*, 326; M. M. C. Ferriera, W. C. Ferriera, Jr., A. C. S. Lino, and M. E. G. Porto, "Uncovering Oscillations, Complexity, and Chaos in Chemical Kinetics Using *Mathematica*," *J. Chem. Ed.* **1999**, *76*, 861; G. Schmitz, L. Kolar-Anić, S. Anić, and Z. Čupić, "The Illustration of Multistability," *J. Chem. Ed.* **2000**, *77*, 1502.

7. C. D. Baird, H. B. Reynolds, and M. A. Crawford, "A Microscale Method to Demonstrate the Belousov–Zhabotinskii Reaction," *J. Chem. Ed.* **2011**, *88*, 960; E. Poros, V. Horváth, K. Curin-Csörgei, I. R. Epstein, and M. Orbán, "Generation of pH-Oscillations in Closed Chemical Systems: Method and Applications," *J. Am. Chem. Soc.* **2011**, *133*, 7174; H. E. Prypsztejn, "Chemiluminescent Oscillating Demonstrations: The Chemical Buoy, The Lighting Wave, and the Ghostly Cylinder," *J. Chem. Ed.* **2005**, *82*, 53; D. Kolb, "Overhead Projector Demonstrations," *J. Chem. Ed.* **1988**, *65*, 1004; R. J. Field, "A Reaction Periodic in Time and Space," *J. Chem. Ed.* **1972**, *49*, 308; J. N. Demas and D. Diemente, "An Oscillating Chemical Reaction with a Luminescent Indicator," *J. Chem. Ed.* **1973**, *50*, 357; J. F. Lefelhocz, "The Color Blind Traffic Light," *J. Chem. Ed.* **1972**, *49*, 312: P. Aroca, Jr., and R. Aroca, "Chemical Oscillations: A Microcomputer-Controlled Experiment," *J. Chem. Ed.* **1987**, *64*, 1017; J. Amrehn, P. Resch, and F. W. Schneider, "Oscillating Chemiluminescence with Luminol in the Continuous Flow Stirred Tank Reactor," *J. Phys. Chem.* **1988**, *92*, 3318; D. Avnir, "Chemically Induced Pulsations of Interfaces: The Mercury Beating Heart," *J. Chem. Ed.* **1989**, *66*, 211; K. Yoshikawa, S. Nakata, M. Yamanaka, and T. Waki, "Amusement with a Salt-Water Oscillator," *J. Chem. Ed.* **1989**, *66*, 205; L. J. Soltzberg, M. M. Boucher, D. M. Crane, and S. S. Pazar, "Far from Equilibrium—The Flashback Oscillator," *J. Chem. Ed.* **1987**, *64*, 1043; S. M. Kaushik, Z. Yuan, and R. M. Noyes, "A Simple Demonstration of a Gas Evolution Oscillator," *J. Chem. Ed.* **1986**, *63*, 76; R. F. Melka, G. Olsen, L. Beavers, and J. A. Draeger, "The Kinetics of Oscillating Reactions," *J. Chem. Ed.* **1992**, *69*, 596; J. M. Merino, "A Simple, Continuous-Flow Stirred-Tank Reactor for the Demonstration and Investigation of Oscillating Reactions," *J. Chem. Ed.* **1992**, *69*, 754.

8. T. Kappes and P. C. Hauser, "A Simple Supplementary Offset Device for Data Acquisition Systems," *J. Chem. Ed.* **1999**, *76*, 1429.

9. [Br⁻] also oscillates in this experiment. For an [I⁻] oscillator, see T. S. Briggs and W. C. Rauscher, "An Oscillating Iodine Clock," *J. Chem. Ed.* **1973**, *50*, 496; S. D. Furrow, "A Modified Recipe and Variations for the Briggs–Rauscher Oscillating Reaction," *J. Chem. Ed.* **2012**, *89*, 1421.

10. M. Shibata, H. Sakaida, and T. Kakiuchi, "Determination of the Activity of Hydrogen Ions in Dilute Sulfuric Acids by Use of an Ionic Liquid Salt Bridge Sandwiched by Two Hydrogen Electrodes," *Anal. Chem.* **2011**, *83*, 164.

11. M. P. S. Mousavi and P. Bühlmann, "Reference Electrodes with Salt Bridges Contained in Nanoporous Glass: An Unappreciated Source of Error," *Anal. Chem.* **2013**, *85*, 8895.

12. E. Bakker, P. Bühlmann, and E. Pretsch, "Carrier-Based Ion-Selective Electrodes and Bulk Optodes," *Chem. Rev.* **1997**, *97*, 3083; ibid. **1998**, *98*, 1593.

13. D. J. Graham, B. Jaselskis, and C. E. Moore, "Development of the Glass Electrode and the pH Response," *J. Chem. Ed.* **2013**, *90*, 345; C. E. Moore, B. Jaselskis, and A. von Smolinski, "Development of the Glass Electrode" in *Electrochemistry, Past and Present*, ACS Symposium Series 390, J. T. Stock and M. V. Orna, eds. (Washington, DC: American Chemical Society, 1989), Chap. 19.

14. W. G. Hines and R. de Levie, "The Early Development of Electronic pH Meters," *J. Chem. Ed.* **2010**, *87*, 1145; B. Jaselskis, C. E. Moore, and A. von Smolinski, "Development of the pH Meter" in *Electrochemistry, Past and Present*, ACS Symposium Series 390, J. T. Stock and M. V. Orna, eds. (Washington, DC: American Chemical Society, 1989), Chap. 18; A. Thackray and M. Myers, Jr., *Arnold O. Beckman: One Hundred Years of Excellence* (Philadelphia: Chemical Heritage Foundation, 2000).

15. R. P. Buck, S. Rondinini, A. K. Covington, F. G. K. Baucke, C. M. A. Brett, M. F. Camoes, M. J. T. Milton, T. Mussini, R. Naumann, K. W. Pratt, P. Spitzer, and G. S. Wilson, "Measurement of pH. Definitions, Standards, and Procedures," *Pure Appl. Chem.* **2002**, *74*, 2169; B. Lunelli and F. Scagnolari, "pH Basics," *J. Chem. Ed.* **2009**, *86*, 246.

16. R. de Levie, "Potentiometric pH Measurements of Acidity Are Approximations, Some More Useful than Others," *J. Chem. Ed.* **2010**, *87*, 1188.

17. F. S. Lopes, L. H. G. Coelho, and I. G. R. Gutz, "Unraveling the Role of Sulfur Compounds in Acid Rain Formation: Experiments on a Wetted Glass pH Electrode," *J. Chem. Ed.* **2010**, *87*, 157; L. M. Goss, "A Demonstration of Acid Rain and Lake Acidification: Wet Deposition of Sulfur Dioxide," *J. Chem. Ed.* **2003**, *80*, 39.

18. J. A. Lynch, V. C. Bowersox, and J. W. Grimm, "Acid Rain Reduced in Eastern United States," *Environ. Sci. Technol.* **2000**, *34*, 940; R. E. Baumgardner, Jr., T. F. Lavery, C. M. Rogers, and S. S. Isil, "Estimates of the Atmospheric Deposition of Sulfur and Nitrogen Species: Clean Air Status and Trends Network, 1990–2000," *Environ. Sci. Technol.* **2002**, *36*, 2614; www.epa.gov/acidrain.

19. W. F. Koch, G. Marinenko, and R. C. Paule, "An Interlaboratory Test of pH Measurements in Rainwater," *J. Res. Natl. Bur. Stand.* **1986**, *91*, 23.

20. *Free diffusion junction* electrodes are designed to minimize junction potentials. The junction is a Teflon capillary tube containing electrolyte that is periodically renewed by syringe.

21. Spectrophotometry with acid-base indictors is another means for measuring the pH of low ionic strength natural waters. (C. R. French, J. J. Carr, E. M. Dougherty, L. A. K. Eidson, J. C. Reynolds, and M. D. DeGrandpre, "Spectrophotometric pH Measurements of Freshwater," *Anal. Chim. Acta* **2002**, *453*, 13.)

22. The reference electrode in the Ross combination electrode is Pt│I₂, I⁻. This electrode is claimed to give improved precision and accuracy over conventional pH electrodes. (R. C. Metcalf, *Analyst* **1987**, *112*, 1573.)

23. C.-E. Lue, T.-C. Yu, C.-M. Yang, D. G. Pijanowska, and C.-S. Lai, "Optimization of Urea-EnFET Based on Ta_2O_5 Layer with Post Annealing," *Sensors* **2011**, *11*, 4562.

24. A. N. Bezbaruah and T. C. Zhang, "Fabrication of Anodically Electrodeposited Iridium Oxide Film pH Microelectrodes for Microenvironmental Studies," *Anal. Chem.* **2002**, *74*, 5726. Iridium oxide electrodes are commercially available or can be prepared by methods such as those described by J.-P. Ndobo-Epoy, E. Lesniewska, and J.-P. Guicquero, "Nano-pH Sensor for the Study of Reactive Materials," *Anal. Chem.* **2007**, *79*, 7560 or R.-G. Du, R.-G. Hu, R.-S. Huang, and C.-J. Lin, "In Situ Measurement of Cl⁻ Concentrations and pH at the Reinforcing Steel/Concrete Interface by Combination Sensors," *Anal. Chem.* **2006**, *78*, 3179.

25. L. W. Niedrach, "Electrodes for Potential Measurements in Aqueous Systems at High Temperatures and Pressures," *Angew. Chem.* **1987**, *26*, 161.

26. History of ion-selective electrodes: M. S. Frant, "Where Did Ion Selective Electrodes Come From?" *J. Chem. Ed.* **1997**, *74*, 159; J. Ruzicka, "The Seventies: Golden Age for Ion Selective Electrodes," *J. Chem. Ed.* **1997**, *74*, 167; T. S. Light, "Industrial Use and Applications of Ion Selective Electrodes," *J. Chem. Ed.* **1997**, *74*, 171; C. C. Young, "Evolution of Blood Chemistry Analyzers Based on Ion Selective Electrodes," *J. Chem. Ed.* **1997**, *74*, 177; R. P. Buck and E. Lindner, "Tracing the History of Selective Ion Sensors," *Anal. Chem.* **2001**, *73*, 88A.

27. E. Bakker and E. Pretsch, "Modern Potentiometry," *Angew. Chem. Int. Ed.* **2007**, *46*, 5660.

28. Y. Chen, R. N. Liang, and W. Qin, "Potentiometric Sensor for Sensitive and Selective Detection of Heparin," *Chinese Chem. Lett.* **2012**, *23*, 233.

29. E. Bakker and E. Pretsch, "The New Wave of Ion-Selective Electrodes," *Anal. Chem.* **2002**, *74*, 420A.

30. For interfering ions X with charge different from the primary ion A, you will find the incorrect, empirical Nicolsky-Eisenman equation in the literature:

$$E = \text{constant} \pm \frac{0.059\,16}{z_A}\log\left[\mathcal{A}_A + \sum_X K_{A,X}^{\text{Pot}}\, \mathcal{A}_X(z_A/z_X) \right]$$

where z_A is the charge of primary ion A and z_X is the charge of interfering ion X. This equation should not be used. (Y. Umezawa, K. Umezawa, and H. Sato, "Selectivity Coefficients for Ion-Selective Electrodes: Recommended Methods for Reporting $K_{A,X}^{\text{Pot}}$ Values," *Pure Appl. Chem.* **1995**, *67*, 507.)

Correct equations for interfering ions with charges different from that of the primary ion are complicated. You can find them in E. Bakker, R. Meruva, E. Pretsch, and M. Meyerhoff, "Selectivity of Polymer Membrane-Based Ion-Selective Electrodes: Self-Consistent Model Describing the Potentiometric Response in Mixed Ion Solutions of Different Charge," *Anal. Chem.* **1994**, *66*, 3021, and N. Nägele, E. Bakker, and E. Pretsch, "General Description of the Simultaneous Response of Potentiometric Ionophore-Based Sensors to Ions of Different Charge," *Anal. Chem.* **1999**, *71*, 1041.

31. E. Bakker, E. Pretsch, and P. Bühlmann, "Selectivity of Potentiometric Ion Sensors," *Anal. Chem.* **2000**, *72*, 1127; E. Bakker, "Determination of Unbiased Selectivity Coefficients of Neutral Carrier-Based Cation-Selective Electrodes," *Anal. Chem.* **1997**, *69*, 1061.

32. Y. Umezawa, K. Umezawa, and H. Sato, "Selectivity Coefficients for Ion-Selective Electrodes: Recommended Methods for Reporting $K_{A,X}^{Pot}$ Values," *Pure Appl. Chem.* **1995**, *67*, 507.

33. K. Ren, "Selectivity Problems of Membrane Ion-Selective Electrodes," *Fresenius J. Anal. Chem.* **1999**, *365*, 389.

34. M. H. Hecht, S. P. Kounaves, R. C. Quinn, S. J. West, S. M. M. Young, D. W. Ming, D. C. Catling, B. C. Clark, W. V. Boynton, J. Hoffman, L. P. DeFlores, K. Gospodinova, J. Kapit, and P. H. Smith, "Detection of Perchlorate and the Soluble Chemistry of Martian Soil at the Phoenix Lander Site," *Science* **2009**, *325*, 64; S. P. Kounaves, M. H. Hecht, S. J. West, J.-M. Morookian, S. M. M. Young, R. Quinn, P. Grunthaner, X. Wen, M. Weilert, C. A. Cable, A. Fisher, K. Gospodinova, J. Kapit, S. Stroble, P.-C. Hsu, B. C. Clark, D. W. Ming, and P. H. Smith, "The 2007 Mars Scout Lander MECA Wet Chemistry Laboratory," *J. Geophys. Res.* **2009**, *113*, E00A19.

35. S. P. Kounaves et al., "Discovery of Natural Perchlorate in the Antarctic Dry Valleys and Its Global Implications," *Environ. Sci. Technol.* **2010**, *44*, 2360.

36. J. D. Schuttlefield, J. B. Sambur, M. Gelwicks, C. M. Eggleston, and B. A. Parkinson, "Photooxidation of Chloride by Oxide Minerals: Implications for Perchlorate on Mars," *J. Am. Chem. Soc.* **2011**, *133*, 17521.

37. D. P.Glavin et al., "Evidence for Perchlorates and the Origin of Chlorinated Hydrocarbons Detected by SAM at the Rocknest Aeolian Deposit in Gale Crater," *J. Geophys. Res. Planets*, **2013**, *118*, 1.

38. J. W. Severinghaus, "First Electrodes for Blood P_{O_2} and P_{CO_2} Determination," *J. Appl. Physiol.* **2004**, *97*, 1599. A brief history paper freely available at http://jap.physiology.org/content/by/year.

39. Y. S. Choi, L. Lvova, J. H. Shin, S. H. Oh, C. S. Lee, B. H. Kim, G. S. Cha, and H. Nam, "Determination of Oceanic Carbon Dioxide Using a Carbonate-Selective Electrode," *Anal. Chem.* **2002**, *74*, 2435; ibid., **2000**, *72*, 4468.

40. M. Umemoto, W. Tani, K. Kuwa, and Y. Ujihira, "Measuring Calcium in Plasma," *Anal. Chem.* **1994**, *66*, 352A.

41. *Resistivity*, ρ, measures how well a substance retards the flow of electric current when an electric field is applied: $J = E/\rho$, where J is current density (current flowing through a unit cross section of the material, A/m^2) and E is electric field (V/m). Units of resistivity are V · m/A or Ω · m, because Ω = V/A, where Ω = ohm. Conductors have resistivities near 10^{-8} Ω · m, semiconductors have resistivities of 10^{-4} to 10^7 Ω · m, and insulators have resistivities of 10^{12} to 10^{20} Ω · m. The reciprocal of resistivity is *conductivity*. Resistivity does not depend on the dimensions of the substance. Resistance, R, is related to resistivity by the equation $R = \rho l/A$, where l is the length and A is the cross-sectional area of the conducting substance.

42. N. M. Milović, J. R. Behr, M. Godin, C.-S. J. Johnson Hou, K. R. Payer, A. Chandrasekaran, P. R. Russo, R. Sasisekharan, and S. R. Manalis, "Monitoring of Heparin and its Low-Molecular-Weight Analogs by Silicon Field Effect," *Proc. Natl. Acad. Sci. U S A* **2006**, *103*, 13374.

43. R. Stine, S. P. Mulvaney, J. T. Robinson, C. R. Tamanaha, and P. E. Sheehan, "Fabrication, Optimization, and Use of Graphene Field Effect Sensors," *Anal. Chem.* **2012**, *85*, 509.

44. Y. Ohno, K. Maehashi, and K. Matsumoto, "Label-Free Biosensors Based on Aptamer-Modified Graphene Field-Effect Transistors," *J. Am. Chem. Soc.* **2010**, *132*, 18012.

45. A. Düzgün, G. A. Zelada-Guillén, G. A. Crespo, S. Macho, J. Riu, and F. X. Rius, "Nanostructured Materials in Potentiometry," *Anal. Bioanal. Chem.* **2011**, *399*, 171.

46. B. Tian, T. Cohen-Karni, Q. Qing, X. Duan, P. Zie, and C. M. Lieber, "Three-Dimensional, Flexible Nanoscale Field-Effect Transistors as Localized Bioprobes," *Science* **2010**, *329*, 830.

47. C.-P. Chen, A. Ganguly, C.-Y. Lu, T.-Y. Chen, C.-C. Kuo, R.-S. Chen, W.-H. Tu, W. B. Fischer, K.-H. Chen, and L.-C. Chen, "Ultrasensitive in Situ Label-Free DNA Detection Using a GaN Nanowire-Based Extended-Gate Field-Effect Transistor Sensor," *Anal. Chem.* **2011**, *83*, 1938.

48. G.-J. Zhang, M. J. Huang, J. J. Ang, Q. Yao, and Y. Ning, "Label-Free Detection of Carbohydrate-Protein Interactions Using Nanoscale Field-Effect Transistor Biosensors," *Anal. Chem.* **2013**, *85*, 4392.

49. P. Estrela, D. Paul, Q. Song, L. K. J. Stadler, L. Wang, E. Huq, J. J. Davis, P. K. Ferrigno, and P. Migliorato, "Label-Free Sub-picomolar Protein Detection with Field-Effect Transistors," *Anal. Chem.* **2010**, *82*, 3531.

50. D. C. Jackman, "A Recipe for the Preparation of a pH 7.00 Calibration Buffer," *J. Chem. Ed.* **1993**, *70*, 853.

51. R. G. Bates, *Determination of pH: Theory and Practice* (New York: Wiley, 1964).

52. W. J. Hamer and Y.-C. Wu, "Osmotic Coefficient and Mean Activity Coefficients of Uni-univalent Electrolytes in Water at 25 °C," *J. Phys. Chem. Ref. Data* **1972**, *1*, 1047.

53. Equation 46 in E. Bakker, R. Meruva, E. Pretsch, and M. Meyerhoff, "Selectivity of Polymer Membrane-Based Ion-Selective Electrodes: Self-Consistent Model Describing the Potentiometric Response in Mixed Ion Solutions of Different Charge," *Anal. Chem.* **1994**, *66*, 3021.

Chapter 16

1. T. Astrup, S. L. S. Stipp, and T. H. Christensen, "Immobilization of Chromate from Coal Fly Ash Leachate Using an Attenuating Barrier Containing Zero-Valent Iron," *Environ. Sci. Technol.* **2000**, *34*, 4163; S. H. Joo, A. J. Feitz, and T. D. Waite, "Oxidative Degradation of the Carbothiolate Herbicide, Molinate, Using Nanoscale Zero-Valent Iron," *Environ. Sci. Technol.* **2004**, *38*, 2242; R. Miehr, P. G. Tratnyek, J. Z. Bandstra, M. M. Scherer, M. J. Alowitz, and E. U. Bylaska, "Diversity of Contaminant Reduction Reactions by Zerovalent Iron: Role of the Reductant," *Environ. Sci. Technol.* **2004**, *38*, 139; V. K. Sharma, C. R. Burnett, D. B. O'Connor, and D. Cabelli, "Iron(VI) and Iron(V) Oxidation of Thiocyanate," *Environ. Sci. Technol.* **2002**, *36*, 4182; C. F. Palomar-Ramírez, J. A. Bazan-Martínez, M. E. Palomar-Pardavé, M. A. Romero-Romo, and T. Ramírez-Silva, "Taking Advantage of a Corrosion Problem to Solve a Pollution Problem," *J. Chem. Ed.* **2011**, *88*, 1109.

2. J. B. Gruber, C. A. Morrison, D. C. Harris, M. D. Seltzer, T. H. Allik, J. A. Hutchinson and M. P. Scripsick, "Spectra of Tetravalent Chromium in Calcium Fluorophosphate," *J. Appl. Phys.*, **1995**, *77*, 2116.

3. Information on redox titrations: J. Bassett, R. C. Denney, G. H. Jeffery, and J. Mendham, *Vogel's Textbook of Inorganic Analysis*, 4th ed. (Essex, UK: Longman, 1978); H. A. Laitinen and W. E. Harris, *Chemical Analysis*, 2nd ed. (New York: McGraw-Hill, 1975); I. M. Kolthoff, R. Belcher, V. A. Stenger, and G. Matsuyama, *Volumetric Analysis*, Vol. 3 (New York: Wiley, 1957); A. Berka, J. Vulterin, and J. Zýka, *Newer Redox Titrants*, H. Weisz, trans. (Oxford: Pergamon, 1965).

4. J. Ermírio, F. Moraes, F. H. Quina, C. A. O. Nascimento, D. N. Silva, and O. Chiavone-Filho, "Treatment of Saline Wastewater Contaminated with Hydrocarbons by the Photo-Fenton Process," *Environ. Sci. Technol.* **2004**, *38*, 1183; B. Gözmen, M. A. Oturan, N. Oturan, and O. Erbatur, "Indirect Electrochemical Treatment of Bisphenol A in Water via Electrochemically Generated Fenton's Reagent," *Environ. Sci. Technol.* **2003**, *37*, 3716. $(H_2O)_5Fe^{IV}=O^{2+}$ is not the Fenton intermediate in acidic and neutral aqueous solution: O. Pestovsky, S. Stoian, E. L. Bominaar, X. Shan, E. Münck, L. Que, Jr., and A. Bakac, "Aqueous $Fe^{IV}=O$: Spectroscopic Identification and Oxo-Group Exchange," *Angew Chem. Int. Ed.* **2005**, *44*, 6871.

5. R. D. Webster, "New Insights into the Oxidative Electrochemistry of Vitamin E," *Acc. Chem. Res.* **2007**, *40*, 251.

6. D. T. Sawyer, "Conceptual Considerations in Molecular Science," *J. Chem. Ed.* **2005**, *82*, 985.

7. Equations 16-9 and 16-10 are analogous to the Henderson-Hasselbalch equation of acid-base buffers. Prior to the equivalence point, the redox titration is *buffered* to a potential near E_+ = formal potential for Fe³⁺|Fe²⁺ by the presence of Fe^{3+} and Fe^{2+}. After the equivalence point, the reaction is *buffered* to a potential near E_+ = formal potential for Ce⁴⁺|Ce³⁺. (R. de Levie "Redox Buffer Strength," *J. Chem. Ed.* **1999**, *76*, 574.)

8. D. W. King, "A General Approach for Calculating Speciation and Poising Capacity of Redox Systems with Multiple Oxidation States: Application to Redox Titrations and the Generation of pϵ-pH Diagrams," *J. Chem. Ed.* **2002**, *79*, 1135.

9. T. J. MacDonald, B. J. Barker, and J. A. Caruso, "Computer Evaluation of Titrations by Gran's Method," *J. Chem. Ed.* **1972**, *49*, 200.

10. M. da Conceição Silva Barreto, L. de Lucena Medieros, and P. C. de Holanda Furtado, "Indirect Potentiometric Titration of Fe(III) with Ce(IV) by Gran's Method," *J. Chem. Ed.* **2001**, *78*, 91.

11. R. D. Hancock and B. J. Tarbet, "The Other Double Helix: The Fascinating Chemistry of Starch," *J. Chem. Ed.* **2000**, *77*, 988.

12. $Ag^IAg^{III}O_2$ can be synthesized from $AgNO_3$ plus $K_2S_2O_8$ in hot aqueous NaOH (R. N. Hammer and J. Kleinberg, *Inorg. Synth.* **1953**, *4*, 12). The crystal structure is described by K. Yvon, A. Bezinge, P. Tissot, and P. Fischer, "Structure and Magnetic Properties of Tetragonal Silver(I,III) Oxide, AgO," *J. Solid State Chem.* **1986**, *65*, 225.

13. H. Van Ryswyk, E. W. Hall, S. J. Petesch, and A. E. Wiedeman, "Extending the Marine Microcosm Laboratory," *J. Chem. Ed.* **2007**, *84*, 306.

14. E. T. Urbansky, "Total Organic Carbon Analyzers as Tools for Measuring Carbonaceous Matter in Natural Waters," *J. Environ. Monit.* **2001**, *3*, 102.

15. L. J. Stolzberg and V. Brown, "Note on Photocatalytic Destruction of Organic Wastes: Methyl Red as a Substrate," *J. Chem. Ed.* **2005**, *82*, 526; J. A. Poce-Fatou, M. L. A. Gil, R. Alcántara, C. Botella, and J. Martin, "Photochemical Reactor for the Study of Kinetics and Adsorption Phenomena," *J. Chem. Ed.* **2004**, *81*, 537; J. C. Yu and L. Y. L. Chan, "Photocatalytic Degradation of a Gaseous Organic Pollutant," *J. Chem. Ed.* **1998**, *75*, 750.

16. S. Sakthivel and H. Kisch, "Daylight Photocatalysis by Carbon-Modified Titanium Dioxide," *Angew. Chem. Int. Ed.* **2003**, *42*, 4908.

17. S. Horikoshi, N. Serpone, Y. Hisamatsu, and H. Hidaka, "Photocatalyzed Degradation of Polymers in Aqueous Semiconductor Suspensions," *Environ. Sci. Technol.* **1998**, *32*, 4010.

18. *Standard Test Method for Total Oxygen Demand in Water*, ASTM D6238.

19. M. Riehl, "Determination of Biochemical Oxygen Demand of Area Waters: A Bioassay Procedure for Environmental Monitoring," *J. Chem. Ed.* **2012**, *89*, 807.

20. BOD and COD procedures are described in *Standard Methods for the Examination of Wastewater*, 21st ed. (Washington, DC: American Public Health Association, 2005), which is the standard reference for water analysis.

21. K. Catterall, H. Zhao, N. Pasco, and R. John, "Development of a Rapid Ferricyanide-Mediated Assay for Biochemical Oxygen Demand Using a Mixed Microbial Consortium," *Anal. Chem.* **2003**, *75*, 2584.

22. B. Wallace and M. Purcell, "The Benefits of Nitrogen and Total Organic Carbon Determination by High-Temperature Combustion," *Am. Lab. News Ed.*, February 2003, p. 58.

23. W. Gottardi, "Redox-Potentiometric/Titrimetric Analysis of Aqueous Iodine Solutions," *Fresenius J. Anal. Chem.* **1998**, *362*, 263.

24. S. C. Petrovic and G. M. Bodner, "An Alternative to Halogenated Solvents for Halogen/Halide Extractions," *J. Chem. Ed.* **1991**, *68*, 509.

25. G. L. Hatch, "Effect of Temperature on the Starch-Iodine Spectrophotometric Calibration Line," *Anal. Chem.* **1982**, *54*, 2002.

26. Y. Xie, M. R. McDonald, and D. W. Margerum, "Mechanism of the Reaction Between Iodate and Iodide Ions in Acid Solutions," *Inorg. Chem.* **1999**, *38*, 3938.

27. Prepare anhydrous $Na_2S_2O_3$ by refluxing 21 g of $Na_2S_2O_3 \cdot 5H_2O$ with 100 mL of methanol for 20 min. Then filter the anhydrous salt, wash with 20 mL of methanol, and dry at 70°C for 30 min. (A. A. Woolf, "Anhydrous Sodium Thiosulfate as a Primary Iodometric Standard," *Anal. Chem.* **1982**, *54*, 2134.)

28. J. Hvoslef and B. Pedersen, "The Structure of Dehydroascorbic Acid in Solution," *Acta Chem. Scand.* **1979**, *B33*, 503; D. T. Sawyer, G. Chiericato, Jr., and T. Tsuchiya, "Oxidation of Ascorbic Acid and Dehydroascorbic Acid by Superoxide in Aprotic Media," *J. Am. Chem. Soc.* **1982**, *104*, 6273;

R. C. Kerber, "As Simple As Possible, But Not Simpler—The Case of Dehydroascorbic Acid," *J. Chem. Ed.* **2008**, *85*, 1237.

29. R. J. Cava, "Oxide Superconductors," *J. Am. Ceram. Soc.* **2000**, *83*, 5.

30. D. C. Harris, M. E. Hills. and T. A. Hewston, "Preparation, Iodometric Analysis, and Classroom Demonstration of Superconductivity in $YBa_2Cu_3O_{8-x}$," *J. Chem. Ed.* **1987**, *64*, 847; D. C. Harris, "Oxidation State Chemical Analysis," in T. A. Vanderah, ed., *Chemistry of Superconductor Materials* (Park Ridge, NJ: Noyes, 1992); B. D. Fahlman, "Superconductor Synthesis: An Improvement," *J. Chem. Ed.* **2001**, *78*, 1182. Browse for "superconductor demonstration kit" to find commercial kits.

31. Experiments with an ¹⁸O-enriched superconductor show that the O_2 evolved in Reaction 1 is all derived from the solid, not from the solvent (M. W. Shafer, R. A. de Groot, M. M. Plechaty, G. J. Scilla, B. L. Olson, and E. I. Cooper, "Evolution and Chemical State of Oxygen Upon Acid Dissolution of $YBa_2Cu_3O_{6.98}$," *Mater. Res. Bull.* **1989**, *24*, 687; P. Salvador, E. Fernandez-Sanchez, J. A. Garcia Dominguez, J. Amdor, C. Cascales, and I. Rasines, "Spontaneous O_2 Release from $SmBa_2Cu_3O_{7-x}$ High T_c Superconductor in Contact with Water," *Solid State Commun.* **1989**, *70*, 71).

32. A more sensitive and elegant iodometric procedure is described by E. H. Appelman, L. R. Morss, A. M. Kini, U. Geiser, A. Umezawa, G. W. Crabtree, and K. D. Carlson, "Oxygen Content of Superconducting Perovskites $La_{2-x}Sr_xCuO_y$ and $YBa_2Cu_3O_y$," *Inorg. Chem.* **1987**, *26*, 3237. This method can be modified by adding standard Br_2 to analyze superconductors with oxygen in the range 6.0–6.5, in which there is formally Cu^+ and Cu^{2+}. The use of electrodes instead of starch to find the end point in iodometric titrations of superconductors is recommended. (P. Phinyocheep and I. M. Tang, "Determination of the Hole Concentration (Copper Valency) in the High T_c Superconductors," *J. Chem. Ed.* **1994**, *71*, A115.)

33. C. L. Copper and E. Koubek, "Analysis of an Oxygen Bleach," *J. Chem. Ed.* **2001**, *78*, 652.

34. S. E. Winter et al., "Gut Inflammation Provides a Respiratory Electron Acceptor for *Salmonella*," *Nature* **2010**, *467*, 426.

35. W. R. Stag, "Dissolved Oxygen," *J. Chem. Ed.* **1972**, *49*, 427; T. Martz, Y. Takeshita, R. Rolph, and P. Bresnahan, "Tracer Monitored Titrations: Measurement of Dissolved Oxygen," *Anal. Chem.* **2011**, *84*, 290.

36. M. T. Garrett, Jr., and J. F. Stehlik, "Classical Analysis," *Anal. Chem.* **1992**, *64*, 310A.

37. C. S. Pundir and R. Rawal, "Determination of Sulfite with Emphasis on Biosensing Methods: A Review," *Anal. Bioanal. Chem.* **2013**, *405*, 3049.

38. K. Peitola, K. Fujinami, H. Karppinen, H. Yamauchi, and L. Niiniströ, "Stoichiometry and Copper Valence in the $Ba_{1-y}CuO_{2+\delta}$ System," *J. Mater. Chem.* **1999**, *9*, 465.

39. S. Scaccia and M. Carewska, "Determination of Stoichiometry of $Li_{1+y}CoO_2$ Materials by Flame Atomic Absorption Spectrometry and Automated Potentiometric Titration," *Anal. Chim. Acta* **2002**, *453*, 35.

40. M. Karppinen, A. Fukuoka, J. Wang, S. Takano, M. Wakata, T. Ikemachi, and H. Yamauchi, "Valence Studies on Various Superconducting Bismuth and Lead Cuprates and Related Materials," *Physica* **1993**, *C208*, 130.

41. The oxygen liberated in Reaction 4 of this problem may be derived from the superconductor, not from solvent water. In any case, BiO_3^- reacts with Fe^{2+}, and Cu^{3+} does not, when the sample is dissolved in acid.

Chapter 17

1. T. R. I. Cataldi, C. Campa, and G. E. De Benedetto, "Carbohydrate Analysis by High-Performance Anion-Exchange Chromatography with Pulsed Amperometric Detection," *Fresenius J. Anal. Chem.* **2000**, *368*, 739.

2. J. M. Cullen and J. M. Allwood, "Mapping the Global Flow of Aluminum: From Liquid Aluminum to End-Use Goods," *Environ. Sci. Technol.* **2013**, *47*, 3057.

3. W. E. Haupin, "Electrochemistry of the Hall-Heroult Process for Aluminum Smelting," *J. Chem. Ed.* **1983**, *60*, 279; N. C. Craig, "Charles Martin Hall: The Young Man, His Mentor, and His Metal," *J. Chem. Ed.* **1986**, *63*, 557.

4. A. J. Bard and L. R. Faulkner, *Electrochemical Methods and Applications*, 2nd ed. (New York: Wiley, 2001); J. Wang, *Analytical Electrochemistry*, 3rd ed. (New York: Wiley-VCH, 2006); J. O'M. Bockris and A. K. N. Reddy, *Modern Electrochemistry*, 2nd ed. (Dordrecht, Netherlands: P Kluwer, 1998–2001,

3 vols.); F. Scholz, ed., *Electroanalytical Methods* (Berlin: Springer-Verlag, 2002).

5. E. C. Gilbert in H. N. Alyea and F. B. Dutton, eds., *Tested Demonstrations in Chemistry* (Easton, PA: Journal of Chemical Education, 1965), p. 145.

6. B. Naiman, "Preservation of Starch Indicator," *J. Chem. Ed.* **1937**, *14*, 138.

7. J. O'M. Bockris, "Overpotential: A Lacuna in Scientific Knowledge," *J. Chem. Ed.* **1971**, *48*, 352.

8. Z. Qiao, W. Shang, X. Zhang, and C. Wang, "Underpotential Deposition of Tin(II) on a Gold Disk Electrode and Determination of Tin in a Tin Plate Sample," *Anal. Bioanal. Chem.* **2005**, *381*, 1467.

9. D. N. Craig, J. I. Hoffman, C. A. Law, and W. J. Hamer, "Determination of the Value of the Faraday with a Silver-Perchloric Acid Coulometer," *J. Res. Natl. Bur. Stand.*, **1960**, *64A*, 381; H. Diehl, "High-Precision Coulometry and the Value of the Faraday," *Anal. Chem.* **1979**, *51*, 318A.

10. J. Greyson and S. Zeller, "Analytical Coulometry in Monier-Williams Sulfite-in-Food Determinations," *Am. Lab.*, July 1987, p. 44; D. T. Pierce, M. S. Applebee, C. Lacher, and J. Bessie, "Low Parts Per Billion Determination of Sulfide by Coulometric Argentometry," *Environ. Sci. Technol.* **1998**, *32*, 1734; R. B. Dabke, Z. Gebeyehu, and R. Thor, " Coulometric Analysis Experiments for the Undergraduate Laboratory," *J. Chem. Ed.* **2011**, *88*, 1707.

11. L. C. Clark, R. Wolf, D. Granger, and A. Taylor, "Continuous Recording of Blood Oxygen Tension by Polarography," *J. Appl. Physiol.* **1953**, *6*, 189.

12. G. C. Jensen, Z. Zheng, and M. E. Meyerhoff, "Amperometric Nitric Oxide Sensors with Enhanced Selectivity over Carbon Monoxide via Platinum Oxide Formation Under Alkaline Conditions," *Anal. Chem.* **2013**, *85*, 10057; S. S. Park, J. Kim, and Y. Lee, "Improved Electrochemical Microsensor for Real-Time Simultaneous Analysis of Endogenous Nitric Oxide and Carbon Monoxide Generation," *Anal. Chem.* **2012**, *84*, 1792; F. Schreiber, L. Polerecky, and D. De Beer, "Nitric Oxide Microsensor for High Spatial Resolution Measurements in Biofilms and Sediments," *Anal. Chem.* **2008**, *80*, 1152.

13. P. Miao, M. Shen, L. Ning, G. Chen, and Y. Yin, "Functionalization of Platinum Nanoparticles for Electrochemical Detection of Nitrite," *Anal. Bioanal. Chem.* **2011**, *399*, 2407.

14. J. Yinon, "Detection of Explosives by Electronic Noses," *Anal. Chem.* **2003**, *75*, 99A; M. C. C. Oliveros, J. L. P. Pavón, C. G. Pinto, M. E. F. Laespada, B. M. Cordero, and M. Forina, "Electronic Nose Based on Metal Oxide Semiconductor Sensors as a Fast Alternative for the Detection of Adulteration of Virgin Olive Oils," *Anal. Chim. Acta* **2002**, *459*, 219; C. L. Honeybourne, "Organic Vapor Sensors for Food Quality Assessment," *J. Chem. Ed.* **2000**, *77*, 338; E. Zubritsky, "E-Noses Keep an Eye on the Future," *Anal. Chem.* **2000**, *72*, 421A.

15. M. A. Bangar, D. J. Shirale, W. Chen, N. V. Myung, and A. Mulchandani, "Single Conducting Polymer Nanowire Chemiresistive Label-Free Immunosensor for Cancer Biomarker," *Anal. Chem.* **2009**, *81*, 2168.

16. L. Feng, C. J. Musto, J. W. Kemling, S. H. Lim, W. Zhong, and K. S. Suslick, "Colorimetric Sensor Array for Determination and Identification of Toxic Industrial Chemicals," *Anal. Chem.* **2010**, *82*, 9433.

17. F.-G. Banica, *Chemical Sensors and Biosensors* (Chichester: Wiley, 2012); B. Wang and E. V. Anslyn, eds., *"Chemosensors" Principles, Strategies, and Applications* (Hoboken: Wiley, 2011); J. Janata, *Principles of Chemical Sensors*, 2nd ed. (Heidelberg: Springer, 2009); R. Renneberg and F. Lisdat, eds., *Biosensing for the 21st Century* (Heidelberg: Springer, 2008); J. M. Cooper and A. E. G. Cass, eds., Biosensors, 2nd ed. (Oxford: Oxford University Press, 2004).

18. W. Jia, A. J. Bandodkar, G. Valdés-Ramírez, J. R. Windmiller, Z. Yang, J. Ramírez, G. Chan, and J. Wang, "Electrochemical Tatoo Biosensors for Real-Time Noninvasive Lactate Monitoring in Human Perspiration," *Anal. Chem.* **2013**, *85*, 6553; A. Guiseppi-Elie, "An Implantable Biochip to Influence Patient Outcomes Following Trauma-Induced Hemorrhage," *Anal. Bioanal. Chem.* **2011**, *399*, 4034.

19. Z. Matharu, J. Enomoto, and A. Revzin, "Miniature Enzyme-Based Electrodes for Detection of Hydrogen Peroxide Release from Alcohol-Injured Hepatocytes," *Anal. Chem.* **2013**, *85*, 932.

20. Y. Zhu, P. Chandra, and Y.-B. Shim, "Ultrasensitive and Selective Electrochemical Diagnosis of Breast Cancer Based on Hydrazine-Au Nanoparticle-Aptamer Bioconjugate," *Anal. Chem.* **2013**, *85*, 1058.

21. A. A. Lubin and K. W. Plaxco, "Folding-Based Electrochemical Biosensors: The Case for Responsive Nucleic Acid Architectures," *Acc. Chem. Res.* **2010**, *43*, 496; H. Li, H. Xie, Y. Cao, X. Ding, Y. Yin, and G. Li, "A General Way to Assay Protein by Coupling Peptide with Signal Reporter via Supermolecule Formation," *Anal. Chem.* **2013**, *85*, 1047.

22. J. Das, K. B. Cederquist, A. A. Zaragoza, P. E. Lee, E. H. Sargent, and S. O. Kelley, "An Ultrasensitive Universal Detector Based on Neutralizer Displacement," *Nat. Chem.* **2012**, *4*, 642.

23. Glucose electrodes that do not use enzymes have been described. (R. Ahmad, M. Vaseem, N. Tripathy, and Y.-B. Hahn, "Wide Linear-Range Detecting Nonenzymatic Glucose Biosensor Based on CuO Nanoparticles Inkjet-Printed on Electrodes," *Anal. Chem.* **2013**, *85*, 10448; X. Niu, M. Lan, H. Zho, and C. Chen, "Highly Sensitive and Selective Nonenzymatic Detection of Glucose Using Three-Dimensional Porous Nickel Nanostructures," *Anal. Chem.* **2013**, *85*, 3561.) These electrodes are not limited by the shelf life of enzymes and are said to be selective for glucose.

24. N. Mano and A. Heller, "Detection of Glucose at 2 fM Concentration," *Anal. Chem.* **2005**, *77*, 729.

25. Y. Xiang and Y. Lu, "Using Personal Glucose Meters and Functional DNA Sensors to Quantify a Variety of Analytical Targets," *Nat. Chem.* **2011**, *3*, 697; Y. Xiang and Y. Lu, "Portable and Quantitative Detection of Protein Biomarkers and Small Molecular Toxins Using Antibodies and Ubiquitous Personal Glucose Meters," *Anal. Chem.* **2012**, *84*, 4174.

26. A. Heller and B. Feldman, "Electrochemical Glucose Sensors and Their Applications in Diabetes Management," *Chem. Rev.* **2008**, *108*, 2482.

27. J. Nikolic, E. Expósito, J. Iniesta, J. González-Garcia, and V. Montiel, "Theoretical Concepts and Applications of a Rotating Disk Electrode," *J. Chem. Ed.* **2000**, *77*, 1191.

28. D. T. Miles, "Run-D.M.C.: A Mnemonic Aid for Explaining Mass Transfer in Electrochemical Systems," *J. Chem. Ed.* **2013**, *90*, 1649.

29. J. Lagrange and P. Lagrange, "Voltammetric Method for the Determination of H_2O_2 in Rainwater," *Fresenius J. Anal. Chem.* **1991**, *339*, 452.

30. R. G. Compton and C. E. Banks, *Understanding Voltammetry*, 2nd ed. (Singapore: World Scientific Press, 2011); R. G. Compton, C. Batchelor-McAuley, and E. J. F. Dickinson, *Understanding Voltammetry: Problems and Solutions* (London: Imperial College Press, 2012); A. J. Bard, "The Rise of Voltammetry: From Polarography to the Scanning Electrochemical Microscope," *J. Chem. Ed.* **2007**, *84*, 644; A. J. Bard and C. G. Zoski, "Voltammetry Retrospective," *Anal. Chem.* **2000**, *72*, 346A; A. M. Bond, *Broadening Electrochemical Horizons* (Oxford: Oxford University Press, 2002).

31. Low background current in diamond is attributed to (i) the absence of redox-active surface groups and (ii) low capacitance due to the absence of ionizable surface groups and low internal charge carrier concentration. (A. E. Fischer, Y. Show, and G. M. Swain, "Electrochemical Performance of Diamond Thin-Film Electrodes from Different Commercial Sources," *Anal. Chem.* **2004**, *76*, 2553.)

32. J. Xu, M. C. Granger, Q. Chen, J. W. Strojek, T. E. Lister, and G. M. Swain, "Boron-Doped Diamond Thin Film Electrodes," *Anal. Chem.* **1997**, *69*, 591A; J. A. Birbeck and T. A. Mathews, "Simultaneous Detection of Monoamine and Purine Molecules Using High-Performance Liquid Chromatography with a Boron-Doped Diamond Electrode," *Anal. Chem.* **2013**, *85*, 7398.

33. C. X. Lim, H. Y. Hoh, P. K. Ang, and K. P. Loh, "Direct Voltammetric Detection of DNA and pH Sensing on Epitaxial Graphene," *Anal. Chem.* **2010**, *82*, 7387.

34. To clean a Hg spill, consolidate the droplets with a piece of cardboard. Use an evacuated filter flask to suck up Hg with a Pasteur pipet attached to a hose. To remove residual Hg, sprinkle elemental zinc powder on the surface and dampen the powder with 5% aqueous H_2SO_4 to make a paste. Mercury dissolves in the zinc. After working the paste into contaminated areas with a sponge or brush, allow the paste to dry and sweep it up. Discard the powder as contaminated Hg waste. This procedure is better than sprinkling sulfur on the spill. Sulfur coats Hg but does not react with the bulk of the droplet. (D. N. Easton, "Management and Control of Hg Exposure," *Am. Lab.*, July 1988, p. 66.)

35. To remove traces of O_2 from N_2, bubble the gas through two consecutive columns of liquid. The first column removes O_2 by reaction with V^{2+}, and the second saturates the gas stream with water at the same vapor pressure as that in the voltammetry cell. Fill the second column with the same supporting electrolyte solution used for voltammetry. Prepare the first column by boiling 2 g NH_4VO_3 (ammonium metavanadate) with 25 mL 12 M HCl and reduce VO_3^- to V^{2+} with zinc amalgam. (Amalgam is prepared by covering granulated Zn with 2 wt% $HgCl_2$ solution and stirring for 10 min to reduce Hg^{2+} to Hg, which coats the Zn. The liquid is decanted, and the amalgam is washed three times with water by decantation. Amalgamation increases the overpotential for H^+ reduction at the Zn surface, so the Zn is not wasted by reaction with acid.) The blue or green oxidized vanadium solution turns violet upon reduction. When the violet color is exhausted through use, it can be regenerated by adding more zinc amalgam or HCl or both. Two V^{2+} bubble tubes can be used in series (in addition to a third tube with supporting electrolyte). When V^{2+} in the first tube is expended, the second tube is still effective.

36. J. G. Osteryoung and R. A. Osteryoung, "Square Wave Voltammetry," *Anal. Chem.* **1985**, *57*, 101A; J. G. Osteryoung, "Voltammetry for the Future," *Acc. Chem. Res.* **1993**, *26*, 77.

37. M. Mascini, *Aptamers in Bioanalysis* (Hoboken: Wiley, 2009); T. Mairal, V. C. Özlap, P. L. Sanchez, M. Mir, I. Katakis, and C. K. O'Sullivan, "Aptamers: Molecular Tools for Analytical Applications," *Anal. Bioanal. Chem.* **2008**, *390*, 989; G. Mayer, "The Chemical Biology of Aptamers," *Angew. Chem. Int. Ed.* **2009**, *48*, 2672; R. Mukhopadhyay, "Aptamers are Ready for the Spotlight," *Anal. Chem.* **2005**, *77*, 114A.

38. S. Song, L. Wang, J. Li, J. Zhao, and C. Fan, "Aptamer-Based Biosensors," *Trends Anal. Chem.* **2008**, *27*, 108.

39. A. Alberich, N. Serrano, J. M. Díaz-Cruz, C.Ariño, and M. Esteban, "Substitution of Mercury Electrodes by Bismuth-Coated Screen-Printed Electrodes in the Determination of Quinine in Tonic Water," *J. Chem. Ed.* **2013**, *90*, 1681.

40. M. J. Goldcamp, M. N. Underwood, J. L. Cloud, and S. Harshman, "An Environmentally Friendly, Cost-Effective Determination of Lead in Environmental Samples," *J. Chem. Ed.* **2008**, *85*, 976.

41. C. Kokkinos, A. Economou, P. S. Petrou, and S. E. Kakabakos, "Microfabricated Tin-Film Electrodes for Protein and DNA Sensing Based on Stripping Voltammetric Detection of Cd(II) Released from Quantum Dot Labels," *Anal. Chem.* **2013**, *85*, 10686.

42. L. M. Laglera, J. Santos-Echeandia, S. Caprara, and D. Monticelli, "Quantification of Iron in Seawater at the Low Picomolar Range Based on Optimization of Bromate/Ammonia/Dihydroxynaphthalene System by Catalytic Adsorptive Cathodic Stripping Voltammetry," *Anal. Chem.* **2013**, *85*, 2486; H. Obata and C. M. G. van den Berg, "Determination of Picomolar Levels of Iron in Seawater Using Catalytic Cathodic Stripping Voltammetry," *Anal. Chem.* **2001**, *73*, 2522; C. M. G. van den Berg, "Chemical Speciation of Iron in Seawater by Cathodic Stripping Voltammetry with Dihydroxynaphthalene," *Anal. Chem.* **2006**, *78*, 156.

43. R. G. Compton and C. E. Banks, *Understanding Cyclic Voltammetry* (Singapore: World Scientific Press, 2007); P. Zanello, *Inorganic Electrochemistry: Theory, Practice and Application* (Cambridge: Royal Society of Chemistry, 2003); G. A. Mabbott, "An Introduction to Cyclic Voltammetry," *J. Chem. Ed.* **1983**, *60*, 697; P. T. Kissinger and W. R. Heineman, "Cyclic Voltammetry," *J. Chem. Ed.* **1983**, *60*, 702; D. H. Evans, K. M. O'Connell, R. A. Petersen, and M. J. Kelly, "Cyclic Voltammetry," *J. Chem. Ed.* **1983**, *60*, 290; H. H. Thorp, "Electrochemistry of Proton-Coupled Redox Reactions," *J. Chem. Ed.* **1992**, *69*, 251.

44. S. M. Oja, M. Wood, and B. Zhang, "Nanoscale Electrochemistry," *Anal. Chem.* **2013**, *85*, 473; X. Li, D. Bergman, and B. Zhang, "Preparation and Electrochemical Response of 1–3 nm Pt Disk Electrodes," *Anal. Chem.* **2009**, *81*, 5496; J. J. Watkins, B. Zhang, and H. S. White, "Electrochemistry at Nanometer-Scaled Electrodes," *J. Chem. Ed.* **2005**, *82*, 713; S. Ching, R. Dudek, and E. Tabet, "Cyclic Voltammetry with Ultramicroelectrodes," *J. Chem. Ed.* **1994**, *71*, 602; E. Howard and J. Cassidy, "Analysis with Microelectrodes Using Microsoft Excel Solver," *J. Chem. Ed.* **2000**, *77*, 409.

45. For a detailed molecular description of a well characterized gold nanoparticle, see J. F. Parker, C. A. Fields-Zinna, and R. W. Murray, "The Story of a Monodisperse Gold Nanoparticle: $Au_{25}L_{18}$," *Acc. Chem. Res.* **2010**, *43*, 1289.

46. T. K. Chen, Y. Y. Lau, D. K. Y. Wong, and A. G. Ewing, "Pulse Voltammetry in Single Cells Using Platinum Microelectrodes," *Anal. Chem.* **1992**, *64*, 1264.

47. A. J. Cunningham and J. B. Justice, Jr., "Approaches to Voltammetric and Chromatographic Monitoring of Neurochemicals in Vivo," *J. Chem. Ed.* **1987**, *64*, A34.

48. S. K. MacLeod, "Moisture Determination Using Karl Fischer Titrations," *Anal. Chem.* **1991**, *63*, 557A.

49. S. Grünke and G. Wünsch, "Kinetics and Stoichiometry in the Karl Fischer Solution," *Fresenius J. Anal. Chem.* **2000**, *368*, 139.

50. A. Cedergren and S. Jonsson, "Progress in Karl Fischer Coulometry Using Diaphragm-Free Cells," *Anal. Chem.* **2001**, *73*, 5611.

51. S. A. Margolis and J. B. Angelo, "Interlaboratory Assessment of Measurement Precision and Bias in the Coulometric Karl Fischer Determination of Water," *Anal. Bioanal. Chem.* **2002**, *374*, 505.

52. D. A. Jayawardhana, R. M. Woods, Y. Zhang, C. Wang, and D. W. Armstrong, "Rapid, Efficient Quantification of Water in Solvents and Solvents in Water Using an Ionic Liquid-Based GC Column," *LCGC North Am.* **2012**, *30*, 142.

53. C. Zhao, A. M. Bond, and X. Lu, "Determination of Water in Room Temperature Ionic Liquids by Cathodic Stripping Voltammetry at a Gold Electrode," *Anal. Chem.* **2012**, *84*, 2784.

54. C. M. Sánchez-Sánchez, E. Expósito, A. Frías-Ferrer, J. González-García, V. Montiel, and A. Aldaz, "Chlor-Alkali Industry: A Laboratory Scale Approach," *J. Chem. Ed.* **2004**, *81*, 698; D. J. Wink, "The Conversion of Chemical Energy," *J. Chem. Ed.* **1992**, *69*, 108; S. Venkatesh and B. V. Tilak, "Chlor-Alkali Technology," *J. Chem. Ed.* **1983**, *60*, 276. Nafion is a trademark of DuPont Co.

55. R. S. Treptow, "The Lead-Acid Battery: Its Voltage in Theory and Practice," *J. Chem. Ed.* **2002**, *79*, 334. Includes *activity coefficients* of electrolyte in the battery.

56. Y. Chyan and O. Chyan, "Metal Electrodeposition on an Integrated, Screen-Printed Electrode Assembly," *J. Chem. Ed.* **2008**, *85*, 565.

57. D. Meziane, A. Barras, A. Kroma, J. Houdkova, R. Boukherroub, and S. Szunerits, "Thiol-yne Reaction on Boron-Doped Diamond Electrodes: Application for the Electrochemical Detection of DNA-DNA Hybridization Events," *Anal. Chem.* **2012**, *84*, 194.

58. D. Lowinsohn and M. Bertotti, "Coulometric Titrations in Wine Samples: Determination of S(IV) and the Formation of Adducts," *J. Chem. Ed.* **2002**, *79*, 103. Some species in wine in addition to sulfite react with I_3^-. A blank titration to correct for such reactions is described in this article.

59. M. E. Gomez and A. E. Kaifer, "Voltammetric Behavior of a Ferrocene Derivative," *J. Chem. Ed.* **1992**, *69*, 502.

Chapter 18

1. R. S. Stolarski, "The Antarctic Ozone Hole," *Scientific American*, January 1988. The 1995 Nobel Prize in Chemistry was shared by Paul Crutzen, Mario Molina, and F. Sherwood Rowland for "their work in atmospheric chemistry, particularly concerning the formation and decomposition of ozone." Their Nobel lectures can be found in P. J. Crutzen, "My Life With O_3, NO_x, and Other YZO_x Compounds," *Angew. Chem. Int. Ed. Engl.* **1996**, *35*, 1759; M. J. Molina, "Polar Ozone Depletion," ibid., 1779; F. S. Rowland, "Stratospheric Ozone Depletion by Chlorofluorocarbons," ibid., 1787.

2. J. H. Butler, M. Battle, M. L. Bender, S. A. Montzka, A. D. Clarke, E. S. Saltzman, C. M. Sucher, J. P. Severinghaus, and J. W. Elkins, "A Record of Atmospheric Halocarbons During the Twentieth Century from Polar Firn Air," *Nature* **1999**, *399*, 749.

3. O. B. Toon and R. P. Turco, "Polar Stratospheric Clouds and Ozone Depletion," *Scientific American,* June 1991; A. J. Prenni and M. A. Tolbert, "Studies of Polar Stratospheric Cloud Formation," *Acc. Chem. Res.* **2001**, *34*, 545.

4. C. G. Camara, J. V. Escobar, J. R. Hird, and S. J. Putterman, "Correlation Between Nanosecond X-ray Flashes and Stick-Slip Friction in Peeling Tape," *Nature* **2008**, *455*, 1089; E. Constable, J. Horvat, and R. A. Lewis, "Mechanism of X-ray Emission from Peeling Adhesive Tape," *Appl. Phys. Lett.* **2010**, *97*, 131502.

5. Classroom exercise to "derive" Beer's law: R. W. Ricci, M. A. Ditzler, and L. P. Nestor, "Discovering the Beer-Lambert Law," *J. Chem. Ed.* **1994**, *71*,

983. An alternative derivation: W. D. Bare, "A More Pedagogically Sound Treatment of Beer's Law: A Derivation Based on a Corpuscular-Probability Model," *J. Chem. Ed.* **2000**, *77*, 929.

6. D. R. Malinin and J. H. Yoe, "Development of the Laws of Colorimetry: A Historical Sketch," *J. Chem. Ed.*, **1961**, *38*, 129. The equation that we call "Beer's Law" embodies contributions by P. Bouguer (1698–1758), J. H. Lambert (1728–1777), and A. Beer (1825–1863). Beer published his work in 1852. F. Bernard independently reached similar conclusions and published them within a few months of Beer.

7. D. H. Alman and F. W. Billmeyer, Jr., "A Simple System for Demonstrations in Spectroscopy," *J. Chem. Ed.* **1976**, *53*, 166. For another approach, see F. H. Juergens, "Spectroscopy in Large Lecture Halls," *J. Chem. Ed.* **1988**, *65*, 266.

8. B. K. Niece, "Simultaneous Display of Spectral Images and Graphs Using a Web Camera and Fiber-Optic Spectrophotometer," *J. Chem. Ed.* **2006**, *83*, 761.

9. D. J. Williams, T. J. Flaherty, C. L. Jupe, S. A. Coleman, K. A. Marquez, and J. H. Stanton, "Beyond λ_{max}: Transforming Visible Spectra into 24-Bit Color Values," *J. Chem. Ed.* **2007**, *84*, 1873.

10. D. J. Williams, T. J. Flaherty, and B. K. Alnasleh, "Beyond λ_{max} Part 2: Predicting Molecular Color," *J. Chem. Ed.* **2009**, *86*, 333.

11. W. E. Wentworth, "Dependence of the Beer-Lambert Absorption Law on Monochromatic Radiation," *J. Chem. Ed.* **1966**, *43*, 262.

12. D. C. Harris, "Serum Iron Determination: A Sensitive Colorimetric Experiment," *J. Chem. Ed.* **1978**, *55*, 539.

13. A pictorial description of the dynamics of the $n \rightarrow \pi^*$ and $\pi \rightarrow \pi^*$ transitions is given by G. Henderson, "A New Look at Carbonyl Electronic Transitions," *J. Chem. Ed.* **1990**, *67*, 392.

14. R. B. Weinberg, "An Iodine Fluorescence Quenching Clock Reaction," *J. Chem. Ed.* **2007**, *84*, 797; R. B. Weinberg, "How Does Your Laundry Glow?" *J. Chem. Ed.* **2007**, *84*, 800A; J. P. Blitz, D. J. Sheeran, and T. L. Becker, "Classroom Demonstrations of Concepts in Molecular Fluorescence," *J. Chem. Ed.* **2006**, *83*, 758; J. W. Bozzelli, "A Fluorescence Lecture Demonstration," *J. Chem. Ed.* **1982**, *59*, 787; G. L. Goe, "A Phosphorescence Demonstration," *J. Chem. Ed.* **1972**, *49*, 412; E. M. Schulman, "Room Temperature Phosphorescence," *J. Chem. Ed.* **1976**, *53*, 522; F. B. Bramwell and M. L. Spinner, "Phosphorescence: A Demonstration," *J. Chem. Ed.* **1977**, *54*, 167; S. Roalstad, C. Rue, C. B. LeMaster, and C. Lasko, "A Room-Temperature Emission Lifetime Experiment for the Physical Chemistry Laboratory," *J. Chem. Ed.* **1997**, *74*, 853; A. MacCormac, E. O'Brien, and R. O'Kennedy, "Lessons from Fluorescence" (of foods), *J. Chem. Ed.* **2010**, *87*, 685.

15. M. J. Eckelman, P. T. Anastas, and J. B. Zimmerman, "Spatial Assessment of Net Mercury Emissions from the Use of Fluorescent Bulbs," *Environ. Sci. Technol.* **2008**, *42*, 8564; E. Engelhaupt, "Do Compact Fluorescent Bulbs Reduce Mercury Pollution," *Environ. Sci. Technol.* **2008**, *42*, 8176.

16. J.-S. Filhol, D. Zitoun, L. Bernaud, and A. Manteghetti, "Microwave Synthesis of a Long-Lasting Phosphor," *J. Chem. Ed.* **2009**, *86*, 72.

17. C. Gell, D. Brockwell, and A. Smith, *Handbook of Single Molecule Fluorescence Spectroscopy* (Oxford: Oxford University Press, 2013); C. Zander, J. Enderlein, and R. Keller, eds., *Single Molecule Detection in Solution* (New York: Wiley, 2002); R. A. Keller, W. P. Ambrose, A. A. Arias, H. Cai, S. R. Emory, P. M. Goodwin, and J. H. Jett, "Analytical Applications of Single-Molecule Detection," *Anal. Chem.* **2002**, *74*, 317A; J. Zimmermann, A. van Dorp, and A. Renn, "Fluorescence Microscopy of Single Molecules," *J. Chem. Ed.* **2004**, *81*, 553; T. A. Byassee, W. C. W. Chan, and S. Nie, "Probing Single Molecules in Single Living Cells," *Anal. Chem.* **2000**, *72*, 5606.

18. B. Fanget, O. Devos, and M. Draye, "Correction of Inner Filter Effect in Mirror Coating Cells for Trace Level Fluorescence Measurements," *Anal. Chem.* **2003**, *75*, 2790.

19. Q. Gu and J. E. Kenny, "Improvement of Inner Filter Effect Correction Based on Determination of Effective Geometric Parameters Using a Conventional Fluorimeter," *Anal. Chem.* **2009**, *81*, 420.

20. J. R. Albani, *Principles and Applications of Fluorescence Spectroscopy* (Oxford: Blackwell, 2007).

21. D. S. Chatellier and H. B. White, III, "What Color Is Egg White? A Biochemical Demonstration of the Formation of a Vitamin-Protein Complex Using Fluorescence Quenching," *J. Chem. Ed.* **1988**, *65*, 814.

22. F. Loe-Mie, G. Marchand, J. Berthier, N. Sarrut, M. Pucheault, M. Blanchard-Desce, F. Vinet, and M. Vaultier, "Toward an Efficient Microsystem for the Real-Time Detection and Quantification of Mercury in Water Based on a Specifically Designed Fluorogenic Binary Task-Specific Ionic Liquid," *Angew. Chem. Int. Ed.* **2010**, *49*, 424; L. Deng, X. Ouyang, J. Jin, C. Ma, Y. Jiang, J. Zheng, J. Li, Y. Li, W. Tan, and R. Yang, "Exploiting the Higher Specificity of Silver Amalgamation: Selective Detection of Mercury(II) by Forming Ag/Hg Amalgam," *Anal. Chem.* **2013**, *85*, 8594; A. Thakur, D. Mandal, and S. Ghosh, "Sensitive and Selective Redox, Chromogenic, and 'Turn-On' Fluorescent Probe for Pb(II) in Aqueous Environment," *Anal. Chem.* **2013**, *85*, 1665.

23. Y. Kurishita, T. Kohira, A. Ojida, and I. Hamachi, "Rational Design of FRET-Based Ratiometric Chemosensors for in Vitro and in Cell Fluorescence Analyses of Nucleoside Polyphosphates," *J. Am. Chem. Soc.* **2010**, *132*, 13290.

24. A. Prasanna de Silva, H. Q. N. Gunaratne, T. Gunnlaugsson, A. J. M. Huxley, C. P. McCoy, J. T. Rademacher, and T. E. Rice, "Signalling Recognition Events with Fluorescent Sensors and Switches," *Chem. Rev.* **1997**, *97*, 1515.

25. J. B. Rampal, ed., *DNA Arrays: Methods and Protocols* (Totowa, NJ: Humana Press, 2001); D. Gerion, F. Chen, B. Kannan, A. Fu, W. J. Parak, D. J. Chen, A. Majumdar, and A. P. Alivisatos, "Room-Temperature Single-Nucleotide Polymorphism and Multiallele DNA Detection Using Fluorescent Nanocrystals and Microarrays," *Anal. Chem.* **2003**, *75*, 4766.

26. T. S. Kuntzleman, K. Rohrer, and E. Schultz, "The Chemistry of Lightsticks: Demonstrations to Illustrate Chemical Processes," *J. Chem. Ed.* **2012**, *89*, 910; O. Jilani, T. M. Donahue, and M. O. Mitchell, "A Greener Chemiluminescence Demonstration," *J. Chem. Ed.* **2011**, *88*, 786; T. S. Kuntzleman, A. E. Comfort, and B. W. Baldwin, "Glowmatography," *J. Chem. Ed.* **2009**, *86*, 65; C. Salter, K. Range, and G. Salter, "Laser-Induced Fluorescence of Lightsticks," *J. Chem. Ed.* **1999**, *76*, 84; E. Wilson, "Light Sticks," *Chem. Eng. News*, 18 January 1999, p. 65.

27. A. Roda, G. Zomer, J. W. Hastings, and F. Berthold, *Chemiluminescence and Bioluminescence* (Cambridge: RCS Publishing, 2010).

28. J. K. Robinson, M. J. Bollinger, and J. W. Birks, "Luminol/H_2O_2 Chemiluminescence Detector for the Analysis of NO in Exhaled Breath," *Anal. Chem.* **1999**, *71*, 5131. Many substances can be analyzed by coupling their chemistry to luminol oxidation. See, for example, O. V. Zui and J. W. Birks, "Trace Analysis of Phosphorus in Water by Sorption Preconcentration and Luminol Chemiluminescence," *Anal. Chem.* **2000**, *72*, 1699.

29. A. Struss, P. Pasini, C. M. Ensor, N. Raut, and S. Daunert, "Paper Strip Whole Cell Biosensors: A Portable Test for the Semiquantitative Detection of Bacterial Quorum Signaling Molecules," *Anal. Chem.* **2010**, *82*, 4457.

30. T. Nagahata, H. Kajiwara, S.-I. Ohira, and K. Toda, "Simple Field Device for Measurement of Dimethyl Sulfide and Dimethylsulfoniopropionate in Natural Waters, Based on Vapor Generation and Chemiluminescence Detection," *Anal. Chem.* **2013**, *85*, 4461.

31. J. G. Ibanez, D. Zavala-Araiza, B. S.-M. Barranco, J. Torres-Perez, C. Camacho-Zuniga, C. Bohrmann-Linde, and M. W. Tausch, "A Demonstration of Simultaneous Electrochemiluminescence," *J. Chem. Ed.* **2013**, *90*, 470.

32. C. N. Pace, F. Vajdos, L. Fee, G. Grimsley, and T. Gray, "How to Measure and Predict the Molar Absorption Coefficient of a Protein," *Protein Science* **1995**, *4*, 2411; P. A. Sims, "Us of a Spreadsheet to Help Students Understand the Origin of the Empirical Equation that Allows Estimation of the Extinction Coefficients of Proteins," *J. Chem. Ed.* **2012**, *89*, 738.

33. R. T. A. MacGillivray, E. Mendez, S. K. Sinha, M. R. Sutton, J. Lineback-Zins, and K. Brew, "The Complete Amino Acid Sequence of Human Serum Transferrin," *Proc. Nat. Acad. Sci. USA* **1982**, *79*, 2504.

Chapter 19

1. F. S. Ligler and C. A. R. Taitt, eds., *Optical Biosensors: Present and Future* (Saint Louis: Elsevier, 2002); F. S. Ligler, "Perspective on Optical Biosensors and Integrated Sensor Systems," *Anal. Chem.* **2009**, *81*, 519.

2. L. Stryer, "Fluorescence Energy Transfer as a Spectroscopic Ruler," *Annu. Rev. Biochem.* **1978**, *47*, 819; C. Berney and G. Danuser, "FRET or No FRET: A Quantitative Comparison," *Biophys. J.* **2003**, *84*, 3992; http://www.probes.com/handbook/.

3. K. M. Sanchez, D. E. Schlamadinger, J. E. Gable, and J. E. Kim, "Förster Resonance Energy Transfer and Conformational Stability of Proteins," *J. Chem. Ed.* **2008**, *85*, 1253.

4. X. Liu, Z. A. Wang, R. H. Byrne, E. A. Kaltenbacher, and R. E. Bernstein, "Spectrophotometric Measurements of pH in Situ: Laboratory and Field Evaluations of Instrumental Performance," *Environ. Sci. Tech.* **2006**, *40*, 5036; T. R. Martz, J. J. Carr, C. R. French, and M. D. DeGrandpre, "A Submersible Autonomous Sensor for Spectrophotometric pH Measurements of Natural Waters," *Anal. Chem.* **2003**, *75*, 1844; W. Yao and R. H. Byrne, "Spectrophotometric Determination of Freshwater pH Using Bromocresol Purple and Phenol Red," *Environ. Sci. Technol.* **2001**, *35*, 1197; H. Yamazaki, R. P. Sperline, and H. Freiser, "Spectrophotometric Determination of pH and Its Applications to Determination of Thermodynamic Equilibrium Constants," *Anal. Chem.* **1992**, *64*, 2720.

5. To multiply a matrix times a vector, multiply each row of the matrix times each element of the vector as follows:

$$\begin{bmatrix} a_1 & a_2 \\ b_1 & b_2 \end{bmatrix} \begin{bmatrix} X \\ Y \end{bmatrix} = \begin{bmatrix} row1 \times column1 \\ row2 \times column2 \end{bmatrix} = \begin{bmatrix} a_1X + a_2Y \\ b_1X + b_2Y \end{bmatrix}$$

The product is a vector. The product of a matrix times a matrix is another matrix obtained by multiplying rows times columns:

$$\begin{bmatrix} a_1 & a_2 \\ b_1 & b_2 \end{bmatrix} \begin{bmatrix} c_1 & c_2 \\ d_1 & d_2 \end{bmatrix} = \begin{bmatrix} row\ 1 \times column\ 1 & row\ 1 \times column\ 2 \\ row\ 2 \times column\ 1 & row\ 2 \times column\ 2 \end{bmatrix}$$

$$= \begin{bmatrix} a_1c_1 + a_2d_1 & a_1c_2 + a_2d_2 \\ b_1c_1 + b_2d_1 & b_1c_2 + b_2d_2 \end{bmatrix}$$

The matrix **B** below is the inverse of **A** because their product is the unit matrix:

$$\underset{\textbf{A}}{\begin{bmatrix} 1 & 2 \\ 3 & 4 \end{bmatrix}} \underset{\textbf{B}}{\begin{bmatrix} -2 & 1 \\ \frac{3}{2} & -\frac{1}{2} \end{bmatrix}} = \begin{bmatrix} 1 \cdot -2 + 2 \cdot \frac{3}{2} & 1 \cdot 1 + 2 \cdot -\frac{1}{2} \\ 3 \cdot -2 + 4 \cdot \frac{3}{2} & 3 \cdot 1 + 4 \cdot -\frac{1}{2} \end{bmatrix} = \underset{\text{Unit matrix}}{\begin{bmatrix} 1 & 0 \\ 0 & 1 \end{bmatrix}}$$

6. Under certain conditions, it is possible for a solution with more than two principal species to exhibit an isosbestic point. See D. V. Stynes, "Misinterpretation of Isosbestic Points: Ambident Properties of Imidazole," *Inorg. Chem.* **1975**, *14*, 453.

7. G. Scatchard, *Ann. N. Y. Acad. Sci.* **1949**, *51*, 660.

8. D. A. Deranleau, "Theory of the Measurement of Weak Molecular Complexes," *J. Am. Chem. Soc.* **1969**, *91*, 4044.

9. J. Job, "Formation and Stability of Inorganic Complexes in Solution," *Ann. Chim.* **1928**, *9*, 113.

10. E. Bruneau, D. Lavabre, G. Levy, and J. C. Micheau, "Quantitative Analysis of Continuous-Variation Plots with a Comparison of Several Methods," *J. Chem. Ed.* **1992**, *69*, 833; V. M. S. Gil and N. C. Oliveira, "On the Use of the Method of Continuous Variation," *J. Chem. Ed.* **1990**, *67*, 473; Z. D. Hill and P. MacCarthy, "Novel Approach to Job's Method," *J. Chem. Ed.* **1986**, *63*, 162.

11. J. Ruzicka, *Flow Injection Analysis*, 4th ed., 2009. (Tutorial available free of charge from www.flowinjection.com.)

12. J. Ruzicka and E. H. Hansen, *Flow Injection Analysis*, 2nd ed. (New York: Wiley, 1988); J. Ruzicka and E. H. Hansen, "Flow Injection Analysis: From Beaker to Microfluidics," *Anal. Chem.* **2000**, *72*, 212A; J. Ruzicka and L. Scampavia, "From Flow Injection to Bead Injection," *Anal. Chem.* **1999**, *71*, 257A.

13. M. Valcarcel and M. D. Luque de Castro, *Flow Injection Analysis: Principles and Applications* (Chichester: Ellis Horwood, 1987).

14. M. Trojanowicz, *Flow Injection Analysis: Instrumentation and Applications* (Singapore: World Scientific, 2000).

15. S. Kolev and Ian McKelvie, eds., *Advances in Flow Injection Analysis and Related Techniques* (New York: Elsevier Science, 2008).

16. J. Shah, M. Rasul Jan, and N. Bashir, "Flow Injection Spectrophotometric Determination of Acetochlor in Food Samples," *Am. Lab. News Ed.*, March 2008, p. 12.

17. C. E. Lenehan, N. W. Barnett, and S. W. Lewis, "Sequential Injection Analysis," *Analyst* **2002**, *127*, 997.

18. R. S. Yalow, "Development and Proliferation of Radioimmunoassay Technology," *J. Chem. Ed.* **1999**, *76*, 767; R. P. Ekins, "Immunoassay, DNA Analysis, and Other Ligand Binding Assay Techniques," *J. Chem. Ed.* **1999**, *76*, 769; E. F. Ullman, "Homogeneous Immunoassays," *J. Chem. Ed.* **1999**, *76*, 781; E. Straus, "Radioimmunoassay of Gastrointestinal Hormones," *J. Chem. Ed.* **1999**, *76*, 788.

19. X.-B. Yin, Y.-Y. Xin, and Y. Zhao, "Label-Free Electrochemiluminescent Aptasensor with Attomolar Mass Detection Limits Based on a $Ru(phen)_3^{2+}$-Double-Strand DNA Composite Film Electrode," *Anal. Chem.* **2009**, *81*, 9299.

20. R. Petkewich, "Sleuthing Out Contamination," *Chem. Eng. News*, 1 September 2008, p. 49; J. Sanchís, L. Kantiani, M. Llorca, F. Rubio, A. Ginebreda, J. Fraile, T. Garrido, and M. Farré, "Determination of Glyphosate in Groundwater Samples," *Anal. Bioanal. Chem.* **2012**, *402*, 2335; W. Jiang, Z. Wang, R. C. Beier, H. Jiang, Y. Wu, and J. Shen, "Simultaneous Determination of 13 Fluoroquinolone and 22 Sulfonamide Residues in Milk by Dual-Colorimetric Enzyme-Linked Immunosorbent Assay," *Anal. Chem.* **2013**, *85*, 1995.

21. E. P. Diamandis and T. K. Christopoulos, "Europium Chelate Labels in Time-Resolved Fluorescence Immunoassays and DNA Hybridization Assays," *Anal. Chem.* **1990**, *62*, 1149A.

22. A. Scott, "The Shade Is Hot," *Chem. Eng. News*, 4 June 2012, p. 20.

23. A. Yella, H.-W. Lee, H. N. Tsao, C. Yi, A. K. Chandiran, Md. K. Nazeeruddin, E. W.-G. Diau, C.-Y. Yeh, S. M. Zakeeruddin, and M. Grätzel, "Porphyrin-Sensitized Solar Cells with Cobalt(II/III)-Based Redox Electrolyte Exceed 12% Efficiency," *Science*, **2011**, *334*, 629.

24. G. J. Meyer, "Efficient Light-to-Electrical Energy Conversion: Nanocrystalline TiO_2 Films Modified with Inorganic Sensitizers," *J. Chem. Ed.* **1997**, *74*, 652. Student experiment to construct a photocell: G. P. Smestad and M. Grätzel, "Demonstrating Electron Transfer and Nanotechnology: A Natural Dye-Sensitized Nanocrystalline Energy Converter," *J. Chem. Ed.* **1998**, *75*, 752. Solar cell kits for students are available from http://ice.chem.wisc.edu/ Catalog.html. Another demonstration that illustrates principles of photocells is J. G. Ibanez, M. W. Tausch, C. Bohrmann-Linde, I. Fernandez-Gallardo, A. Robles-Leyzaola, S. Krees, N. Meuter, and M. Tennior, "The Basis for Photocatalytic Writing," *J. Chem. Ed.* **2011**, *88*, 1116.

25. H.-S. Kim, C.-R. Lee, J.-H Im, K.-B. Lee, T. Moehl, A. Marchioro, S.-J. Moon, R. Humphry-Baker, J.-H. Yum, J. E. Moser, M. Grätzel, and N.-G. Park, "Lead Iodide Perovskite Sensitized All-Solid-State Submicron Thin Film Mesoscopic Solar Cell with Efficiency Exceeding 9%," *Scientific Reports*, **2012**, *2*, 591. DOI: 10.1038/srep00591.

26. M. G. Walter, E. L. Warren, J. R. McKone, S. W. Boettcher, Q. Mi, E. A. Santori, and N. S. Lewis, "Solar Water Splitting Cells," *Chem. Rev.* **2010**, *110*, 6446.

27. F. A. Settle, "Uranium to Electricity: The Chemistry of the Nuclear Fuel Cycle," *J. Chem. Ed.* **2009**, *86*, 316.

28. K. A. Kneas, W. Xu, J. N. Demas, and B. A. DeGraff, "Dramatic Demonstration of Oxygen Sensing by Luminescence Quenching," *J. Chem. Ed.* **1997**, *74*, 696.

29. C. Beleizão, S. Nagl, M. Schäferling, M. N. Berberan-Santos, and O. S. Wolfbeis, "Dual Fluorescence Sensor for Trace Oxygen and Temperature with Unmatched Range and Sensitivity," *Anal. Chem.* **2008**, *80*, 6449; S. Kochmann, C. Beleizão, M. N. Berberan-Santos, and O. S. Wolfbeis, "Sensing and Imaging of Oxygen with Parts per Billion Limits of Detection Based on the Quenching of the Delayed Fluorescence of $^{13}C_{70}$ Fullerene in Polymer Hosts," *Anal. Chem.* **2013**, *85*, 1300.

30. R. R. Islangulov, D. V. Kozlov, and F. N. Castellano, "Low Power Upconversion Using MLCT Sensitizers," *Chem. Commun.* **2005**, 3776; B. M. Wilke and F. N. Castellano, "Photochemical Upconversion: A Physical or Inorganic Chemistry Experiment for Undergraduates using a Conventional Fluorimeter," *J. Chem. Ed.* **2013**, *90*, 786. For red-to-yellow and red-to-green upconversion, see T. N. Singh-Rachford, A. Haefele, R. Ziessel, and F. N. Castellano, *J. Am. Chem. Soc.* **2008**, *130*, 16164.

31. A. L. Ouelette, J. J. Li, D. E. Cooper, A. J. Ricco, and G. T. A. Kovacs, "Evolving Point-of-Care Diagnostics Using Up-Converting Phosphor Bioanalytical Systems," *Anal. Chem.* **2009**, *81*, 3216.

32. X. Dong, C. H. Heo, S. Chen., H. M. Kim, and Z. Liu, "Quinoline-Based Two-Photon Fluorescent Probe for Nitric Oxide in Live Cells and Tissues," *Anal. Chem.* **2014**, *86*, 308.

33. M. D. DeGrandpre, M. M. Baehr, and T. R. Hammar, "Calibration-Free Optical Chemical Sensors," *Anal. Chem.* **1999**, *71*, 1152.

34. A Poisson distribution is valid when (a) all possible outcomes are random and independent of one another, (b) the maximum possible value of *n* is a large number, and (c) the average value of *n* is a small fraction of the maximum possible value.

Chapter 20

1. L. S. Berden, *Cavity Ringdown Spectroscopy—Techniques and Applications* (Chichester: Wiley, 2009); K. L. Bechtel, R. N. Zare, A. A. Kachanov, S. S. Sanders, and B. A. Paldus, "Cavity Ring-Down Spectroscopy for HPLC," *Anal. Chem.* **2005**, *77*, 1177; B. Bahnev, L. van der Sneppen, A. E. Wiskerke, F. Ariese, C. Gooijer, and W. Ubachs, "Miniaturized Cavity Ring-Down Detection in a Liquid Flow Cell," *Anal. Chem.* **2005**, *77*, 1188; L. N. Seetohul, Z. Ali, and M. Islam, "Broadband Cavity Enhanced Absorption Spectroscopy as a Detector for HPLC," *Anal. Chem.* **2009**, *81*, 4106.

2. www.picarro.com.

3. H. Sevian, S. Müller, H. Rudmann, and M. F. Rubner, "Using Organic Light-Emitting Electrochemical Thin-Film Devices to Teach Materials Science," *J. Chem. Ed.* **2004**, *81*, 1620.

4. J. M. Kauffman, "Water in the Atmosphere," *J. Chem. Ed.* **2004**, *81*, 1229.

5. S. K. Lower, "Thermal Physics (and Some Chemistry) of the Atmosphere," *J. Chem. Ed.* **1998**, *75*, 837; W. C. Trogler, "Environmental Chemistry of Trace Atmospheric Gases," *J. Chem. Ed.* **1995**, *72*, 973.

6. D. L. Hartmann, A. M. G. Klein Tank, M. Rusticucci, L. V. Alexander, S. Brönnimann, Y. Charabi, F. J. Dentener, E. J. Dlugokencky, D. R. Easterling, A. Kaplan, B. J. Soden, P. W. Thorne, M. Wild, and P. M. Zhai, 2013: Observations: Atmosphere and Surface, in: *Climate Change 2013: The Physical Science Basis. Contribution of Working Group I to the Fifth Assessment Report of the Intergovernmental Panel on Climate Change*, T. F. Stocker, D. Qin, G.-K. Plattner, M. Tignor, S. K. Allen, J. Boschung, A. Nauels, Y. Xia, V. Bex, and P. M. Midgley, eds. (Cambridge: Cambridge University Press, 2014).

7. M. G. D. Baumann, J. C. Wright, A. B. Ellis, T. Kuech, and G. C. Lisensky, "Diode Lasers," *J. Chem. Ed.* **1992**, *69*, 89; T. Imasaka and N. Ishibashi, "Diode Lasers and Practical Trace Analysis," *Anal. Chem.* **1990**, *62*, 363A.

8. S. Nakamura, S. Pearton, and G. Fasol, *The Blue Laser Diode: The Complete Story* (Heidelberg: Springer, 2000).

9. W. E. L. Grossman, "The Optical Characteristics and Production of Diffraction Gratings," *J. Chem. Ed.* **1993**, *70*, 741.

10. G. C.-Y. Chan and W. T. Chan, "Beer's Law Measurements Using Non-Monochromatic Light Sources—A Computer Simulation," *J. Chem. Ed.* **2001**, *78*, 1285.

11. J. C. Travis et al., "Intrinsic Wavelength Standard Absorption Bands in Holmium Oxide Solution for UV/Visible Molecular Absorption Spectrophotometry," *J. Phys. Chem. Ref. Data* **2005**, *34*, 41.

12. www.oceanoptics.com. Look at spectrometers and then standards.

13. S. C. Denson, C. J. S. Pommier, and M. B. Denton, "The Impact of Array Detectors on Raman Spectroscopy," *J. Chem. Ed.* **2007**, *84*, 67; J. M. Harnly and R. E. Fields, "Solid-State Array Detectors for Analytical Spectrometry," *Appl. Spectros.* **1997**, *51*, 334A; Q. S. Hanley, C. W. Earle, F. M. Pennebaker, S. P. Madden, and M. B. Denton, "Charge-Transfer Devices in Analytical Instrumentation," *Anal. Chem.* **1996**, *68*, 661A; J. V. Sweedler, K. L. Ratzlaff, and M. B. Denton, eds., *Charge Transfer Devices in Spectroscopy* (New York: VCH, 1994).

14. D. C. Harris, "Charles David Keeling and the Story of Atmospheric CO_2 Measurements," *Anal. Chem.* **2010**, *82*, 7865.

15. J. C. Pales and C. D. Keeling, "The Concentration of Atmospheric Carbon Dioxide in Hawaii," *J. Geophys. Res.* **1965**, *70*, 6053; V. N. Smith, "A Recording Infrared Analyzer," *Instruments* **1953**, *9*, 421.

16. U. Resch-Genger, D. Pfeifer, C. Monte, W. Pilz, A. Hoffmann, M. Spieles, K. Rurack, J. Hollandt, D. Taubert, B. Schönenberger, and P. Nording, "Traceability in Fluorometry: Part II. Spectral Fluorescence Standards," *J. Fluoresc.* **2005**, *15*, 314.

17. F. Baldini, A. N. Chester, J. Homola, and S. Martellucci, *Optical Chemical Sensors* (Dordrecht, Netherlands: Springer, 2006); F. S. Ligler and C. R. Taitt,

eds., *Optical Biosensors: Today and Tomorrow* (Amsterdam: Elsevier, 2008); X.-D. Wang and O. S. Wolfbeis, "Fiber-Optic Chemical Sensors and Biosensors (2008–2012)," *Anal. Chem.* **2013**, *85*, 487; F. Deiss, N. Sojic, D. J. White, and P. R. Stoddart, "Nanostructured Optical Fibre Arrays for High-Density Biochemical Sensing and Remote Imaging," *Anal. Bioanal. Chem.* **2010**, *396*, 53.

18. A. S. Kocincova, S. M. Borisov, C. Krause, and O. S. Wolfbeis, "Fiber-Optic Microsensors for Simultaneous Sensing of Oxygen and pH, and of Oxygen and Temperature," *Anal. Chem.* **2007**, *79*, 8486; I. Kasik, J. Mrazek, T. Martan, M. Pospisilova, O. Podrazky, V. Matejec, K. Hoyerova, and M. Kaminek, "Fiber-Optic pH Detection in Small Volumes of Biosamples," *Anal. Bioanal. Chem.* **2010**, *398*, 1883; D. Citterio, J. Takeda, M. Kosugi, H. Hisamoto, S.-I. Sasaki, H. Komatsu, and K. Suzuki, "pH-Independent Fluorescent Chemosensor for Highly Selective Lithium Ion Sensing," *Anal. Chem.* **2007**, *79*, 1237; W. Tan, R. Kopenman, S. L. R. Barker, and M. T. Miller, "Ultrasmall Optical Sensors for Cellular Measurements," *Anal. Chem.* **1999**, *71*, 606A; S. L. R. Barker, Y. Zhao, M. A. Marletta, and R. Kopelman, "Cellular Applications of a Fiber-Optic Biosensor Based on a Dye-Labeled Guanylate Cyclase," *Anal. Chem.* **1999**, *71*, 2071; M. Kuratli and E. Pretsch, "SO_2-Selective Optodes," *Anal. Chem.* **1994**, *66*, 85; S.-I. Ohira, P. K. Dasgupta, and K. A. Schug, "Fiber Optic Sensor for Simultaneous Determination of Atmospheric Nitrogen Dioxide, Ozone, and Relative Humidity," *Anal. Chem.* **2009**, *81*, 4183.

19. Z. Rosenzweig and R. Kopelman, "Analytical Properties and Sensor Size Effects of a Micrometer-Sized Optical Fiber Glucose Biosensor," *Anal. Chem.* **1996**, *68*, 1408.

20. C. Preininger, I. Klimant, and O. S. Wolfbeis, "Optical Fiber Sensor for Biological Oxygen Demand," *Anal. Chem.* **1994**, *66*, 1841.

21. J. T. Bradshawl, S. B. Mendes, and S. S. Saavedra, "Planar Integrated Optical Waveguide Spectroscopy," *Anal. Chem.* **2005**, *77*, 28A.

22. J. Homola, "Surface Plasmon Resonance Sensors for Detection of Chemical and Biological Species," *Chem. Rev.* **2008**, *108*, 462; R. Bakhtiar, "Surface Plasmon Resonance Spectroscopy: A Versatile Technique in a Biochemist's Toolbox," *J. Chem. Ed.* **2013**, *90*, 203; D. J. Campbell and Y. Xia, "Plasmons: Why Should We Care?" *J. Chem. Ed.* **2007**, *84*, 91; P. N. Abadian, C. P. Kelley, and E. D. Goluch, "Cellular Analysis and Detection Using Surface Plasmon Resonance Techniques," *Anal. Chem.* **2014**, *86*, 2799.

23. P. R. Griffiths and J. A. de Haseth, *Fourier Transform Infrared Spectrometry*, 2nd ed. (Hoboken, NJ: Wiley, 2007); W. D. Perkins, "Fourier Transform-Infrared Spectroscopy," *J. Chem. Ed.* **1986**, *63*, A5; ibid. **1987**, *64*, A269, A296; Q. S. Hanley, "Fourier Transforms Simplified: Computing an Infrared Spectrum from an Interferogram," *J. Chem. Ed.* **2012**, *89*, 391; B. Shepherd and M. K. Bellamy, "A Spreadsheet Exercise to Teach the Fourier Transform in FTIR Spectrometry," *J. Chem. Ed.* **2012**, *89*, 681.

24. Digital and electronic techniques for improving signal-to-noise ratio: T. C. O'Haver, "An Introduction to Signal Processing in Chemical Measurement," *J. Chem. Ed.* **1991**, *68*, A147; M. G. Prais, "Spreadsheet Exercises for Instrumental Analysis," *J. Chem. Ed.* **1992**, *69*, 488; R. Q. Thompson, "Experiments in Software Data Handling," *J. Chem. Ed.* **1985**, *62*, 866; B. H. Vassos and L. López, "Signal-to-Noise Improvement," *J. Chem. Ed.* **1985**, *62*, 542. K. Overway, "FT Digital Filtering: Simulating Fourier Transform Apodization via Excel," *J. Chem. Ed.* **2008**, *85*, 1151; M. P. Eastman, G. Kostal, and T. Mayhew, "An Introduction to Fast Fourier Transforms Through the Study of Oscillating Reactions," *J. Chem. Ed.* **1986**, *63*, 453.

25. G. M. Hieftje, "Signal-to-Noise Enhancement Through Instrumental Techniques," *Anal. Chem.* **1972**, *44*, 81A [No. 6], 69A [No. 7]; D. C. Tardy, "Signal Averaging," *J. Chem. Ed.* **1986**, *63*, 648.

26. N. N. Sesi, M. W. Borer, T. K. Starn, and G. M. Hieftje, "A Standard Approach to Collecting and Calculating Noise Amplitude Spectra," *J. Chem. Ed.* **1998**, *75*, 788.

27. M. L. Salit and G. C. Turk, "A Drift Correction Procedure," *Anal. Chem.* **1998**, *70*, 3184.

28. L. D. Rothman, S. R. Crouch, and J. D. Ingle, Jr., "Theoretical and Experimental Investigation of Factors Affecting Precision in Molecular Absorption Spectrophotometry," *Anal. Chem.* **1975**, *47*, 1226.

29. Let voltages V_{dif}^0 and V_{ref}^0 be recorded with pure solvent in both cells. Let voltages V_{dif} and V_{ref} be recorded with sample in one cell and solvent in the other cell. Absorbance is related to these voltages by $A = -\log[(1 + V_{\text{dif}}/V_{\text{ref}})/(1 + V_{\text{dif}}^0/V_{\text{ref}}^0)]$.

30. A. Savitzky and M. J. E. Golay, "Smoothing and Differentiation of Data by Simplified Least Squares Procedures," *Anal. Chem.* **1964**, *36*, 1627. This paper gives tables of coefficients for multipoint smoothing of 5 to 25 points by fitting polynomials of order 2, 3, 4, and 5, and for finding smoothed derivative curves.

31. Polynomial smoothing of spectroscopic data is applied in a student experiment described by R. R. de Oliveira, L. S. das Neves, and K. M. G. de Lima, "Experimental Design, Near-Infrared Spectroscopy, and Multivariate Calibration," *J. Chem. Ed.* **2012**, *89*, 1566.

Chapter 21

1. A. Sanz-Medel and R. Pereiro, *Atomic Absorption Spectrometry: An Introduction* (Oxfordshire: Coxmoor, 2008); L. H. J. Lajunen and P. Perämäki, *Spectrochemical Analysis by Atomic Absorption* (Cambridge: Royal Society of Chemistry, 2004); M. Cullen, *Atomic Spectroscopy in Elemental Analysis* (Oxford: Blackwell, 2003); J. R. Dean and D. J. Ando, *Atomic Absorption and Plasma Spectroscopy* (New York: Wiley, 2002); J. A. C. Broekaert, *Analytical and Atomic Spectrometry with Flames and Plasmas* (Weinheim: Wiley-VCH, 2002); L. Ebdon, E. H. Evans, A. S. Fisher, and S. J. Hill, *An Introduction to Analytical Atomic Spectrometry* (Chichester: Wiley, 1998).

2. Excellent training modules are available from www.academysavant.com.

3. M. J. Holden, S. A. Rabb, Y. B. Tewari, and M. R. Winchester, "Traceable Phosphorus Measurements by ICP-OES and HPLC for the Quantitation of DNA," *Anal. Chem.* **2007**, *79*, 1536.

4. History: A. Walsh, "The Development of Atomic Absorption Methods of Elemental Analysis 1952–1962," *Anal. Chem.* **1991**, *63*, 933A. B. V. L'vov, "Graphite Furnace Atomic Absorption Spectrometry," *Anal. Chem.* **1991**, *63*, 924A.

5. B. K. Niece and J. F. Hauri, "Determination of Mercury in Fish: A Low-Cost Implementation of Cold-Vapor Atomic Absorbance for the Undergraduate Environmental Chemistry Laboratory," *J. Chem. Ed.* **2013**, *90*, 487; J. V. Cizdziel, "Mercury in Environmental and Biological Samples Using Online Combustion with Sequential Atomic Absorption and Fluorescence Measurements," *J. Chem. Ed.* **2011**, *88*, 209; S. Armenta and M. de la Guardia, "Determination of Mercury in Milk by Cold Vapor Atomic Fluorescence," *J. Chem. Ed.* **2011**, *88*, 488; L. Beaudin, S. C. Johannessen, and R. W. Macdonald, "Coupling Laser Ablation and Atomic Fluorescence Spectrophotometery: Mercury Analysis of Small Sections of Fish Scales," *Anal. Chem.* **2010**, *82*, 8785.

6. J. Gray, "Simple, Environmentally Friendly Mercury Analysis by U.S. EPA Method 1631," *Am. Lab.*, March 2011, p. 22; K. Scoffin, "Mercury Analyzers in the Laboratory," *Am. Lab.*, April 2013, p. 20; H. Fleischer and K. Thurow, "Determination of Total Mercury Content in Wood Materials, Part 2: ICP-MS—A Multielement Method," *Am. Lab.*, September 2013, p. 6.

7. K. Leopold, M. Foulkes, and P. J. Worsfold, "Gold-Coated Silica as a Preconcentration Phase for the Determination of Total Dissolved Mercury in Natural Waters Using Atomic Fluorescence Spectrometry," *Anal. Chem.* **2009**, *81*, 3421.

8. N. Patel-Sorrentino, J.-Y. Benaim, D. Cossa, and Y. Lucas, "Synthesis of Hydrochloric Acid Solution for Total Mercury Determination in Natural Waters," *Anal. Bioanal. Chem.* **2011**, *399*, 1389.

9. History: R. F. Jarrell, "A Brief History of Atomic Emission Spectrochemical Analysis, 1666–1950," *J. Chem. Ed.* **2000**, *77*, 573; R. F. Jarrell, F. Brech, and M. J. Gustafson, "A History of Thermo Jarrell Ash Corporation and Spectroscopist Richard F. Jarrell," *J. Chem. Ed.* **2000**, *77*, 592; G. M. Hieftje, "Atomic Emission Spectroscopy—It Lasts and Lasts and Lasts," *J. Chem. Ed.* **2000**, *77*, 577.

10. R. J. Stolzberg, "Optimizing Signal-to-Noise Ratio in Flame Atomic Absorption Using Sequential Simplex Optimization," *J. Chem. Ed.* **1999**, *76*, 834; C. R. Dockery, M. J. Blew, and S. R. Goode, "Visualizing the Solute Vaporization Interference in Flame Atomic Absorption Spectroscopy," *J. Chem. Ed.* **2008**, *85*, 854.

11. G. Schlemmer and B. Radziuk, *Analytical Graphite Furnace Atomic Absorption Spectrometry: A Laboratory Guide* (Basel: Birkhäuser, 1999); D. J. Butcher and J. Sneddon, *A Practical Guide to Graphite Furnace Atomic Absorption Spectrometry* (New York: Wiley, 1998).

12. J. B. Voit, "Low-Level Determination of Arsenic in Drinking Water," *Am. Lab. News Ed.*, February 2002, p. 62.

13. M. Hornung and V. Krivan, "Determination of Trace Impurities in Tungsten by Direct Solid Sampling Using a Transversely Heated Graphite Tube," *Anal. Chem.* **1998**, *70*, 3444.

14. U. Schäffer and V. Krivan, "Analysis of High Purity Graphite and Silicon Carbide by Direct Solid Sampling Electrothermal Atomic Absorption Spectrometry," *Fresenius J. Anal. Chem.* **2001**, *371*, 859; R. Nowka and H. Müller, "Direct Analysis of Solid Samples by Graphite Furnace Atomic Absorption Spectrometry," *Fresenius J. Anal. Chem.* **1997**, *359*, 132; J. Štupar and F. Dolinšek, "Determination of Chromium, Manganese, Lead, and Cadmium in Biological Samples Including Hair Using Direct Electrothermal Atomic Absorption Spectrometry," *Spectrochim. Acta* **1996**, *B51*, 665.

15. J. Y. Cabon, "Influence of Experimental Parameters on the Determination of Antimony in Seawater by Atomic Absorption Spectrometry," *Anal. Bioanal. Chem.* **2003**, *374*, 1282.

16. D. L. Styris and D. A. Redfield, "Mechanisms of Graphite Furnace Atomization of Aluminum by Molecular Beam Sampling Mass Spectrometry," *Anal. Chem.* **1987**, *59*, 2891.

17. S. J. Hill, *Inductively Coupled Plasma Spectrometry and Its Applications* (Chichester: Wiley, 2007); J. R. Dean, *Practical Inductively Coupled Plasma Spectroscopy* (Chichester: Wiley, 2005); J. Noelte, ICP Emission Spectrometry—A Practical Guide (Weinheim: Wiley-VCH, 2003); S. Greenfield, "Invention of the Annular Inductively Coupled Plasma as a Spectroscopic Source," *J. Chem. Ed.* **2000**, *77*, 584.

18. S. Weagant, V. Chen, and V. Karanassios, "Battery-Operated, Argon-Hydrogen Microplasma on Hybrid, Postage Stamp-Sized Plastic-Quartz Chips for Elemental Analysis of Liquid Microsamples Using a Portable Optical Emission Spectrometer," *Anal. Bioanal. Chem.* **2011**, *401*, 2865.

19. N H. Tung, M. Chikae, Y. Ukita, P. H. Viet, and Y. Takamura, "Sensing Technique of Silver Nanoparticles as Labels for Immunoassay Using Liquid Electrode Plasma Atomic Emission Spectrometry," *Anal. Chem.* **2012**, *84*, 1210; A. Kitano, A. Iiduka, T. Yamamoto, Y. Ukita, E. Tamiya, and Y. Takamura, "Highly Sensitive Elemental Analysis for Cd and Pb by Liquid Electrode Plasma Atomic Emission Spectrometry with Quartz Glass Chip and Sample Flow," *Anal. Chem.* **2011**, *83*, 9424.

20. V. B. E. Thomsen, G. J. Roberts, and D. A. Tsourides, "Vacuumless Spectrochemistry in the Vacuum Ultraviolet," *Am. Lab.*, August 1997, p. 18H.

21. V. B. E. Thomsen, "Why Do Spectral Lines Have a Linewidth?" *J. Chem. Ed.* **1995**, *72*, 616.

22. Sigma-Aldrich data. M. Weber and J. Wüthrich, www.sigma-aldrich.com/tracecert.

23. J.-M. Mermet, A. Cosnier, S. Vélasquez, and S. Lebouil, "Efficient Use of Spectral Information Through Multiline Analysis in ICP-AES," *Am. Lab. News Ed.*, August 2007, p. 16.

24. M. R. Winchester, G. C. Turk, T. A. Butler, T. J. Oatts, C. Coleman, D. Nadrotowski, R. Sud, M. D. Hoover, and A. B. Stefaniak, "Certification of Beryllium Mass Fraction in SRM 1877 Beryllium Oxide Powder Using High-Performance Inductively Coupled Plasma Optical Emission Spectrometry with Exact Matching," *Anal. Chem.* **2009**, *81*, 2208.

25. A diagnostic for whether emission from an Ar plasma will be sensitive to matrix effects is the ratio of emission from Mg^+ and Mg atoms (at 280.270 and 285.213 nm, respectively). When the relative intensity is above 10, the plasma is not sensitive to variations in the sample matrix. When the ratio is below 4, there is high sensitivity to matrix effects. See J.-M. Mermet, "Mg as a Test Element for ICP Atomic Emission Diagnostics," *Anal. Chim. Acta* **1991**, *250*, 85 and J.-M. Mermet and E. Poussel, "ICP Emission Spectrometers: Analytical Figures of Merit," *Appl. Spectros.* **1995**, *49*[10], 12A. Intensities of the two lines must be corrected for different grating diffraction efficiencies at the two wavelengths. (J. W. Olesik, J. A. Kinser, and B. Harkleroad, "Inductively Coupled Plasma Optical Emission Spectrometry Using Nebulizers with Widely Different Sample Consumption Rates," *Anal. Chem.* **1994**, *66*, 2022.)

26. R. E. Russo, X. Mao, J. J. Gonzalez, V. Zorba, and J. Yoo, "Laser Ablation in Analytical Chemistry," *Anal. Chem.* **2013**, *85*, 6162; B. Hattendorf, C. Latkoczy, and D. Günther, "Laser Ablation–ICPMS," *Anal. Chem.* **2003**, *75*,

341A; R. E. Russo, X. Mao, and S. S. Mao, "The Physics of Laser Ablation in Microchemical Analysis," *Anal. Chem.* **2002**, *74*, 71A.

27. J. Frydenvang, K. M. Kinch, S. Husted, and M. B. Madsen, "An Optimized Calibration Procedure for Determining Elemental Ratios Using Laser-Induced Breakdown Spectroscopy," *Anal. Chem.* 2013, *85*, 1492; F. Claverie, J. Malherbe, N. Bier, J. L. Molloy, and S. E. Long, "Standard Addition Method for Laser Ablaton ICPMS Using a Spinning Platform," *Anal. Chem.* 2013, *85*, 3584; R. C. Chinni, "A Simple LIBS (Laser-Induced Breakdown Spectroscopy) Laboratory Experiment to Introduce Undergraduates to Calibration Functions and Atomic Spectroscopy," *J. Chem. Ed.* **2012**, *89*, 678.

28. F. Vanhaecke and P. Degryse, eds., Isotopic Analysis. Fundamentals and Applications using ICP–MS (Weinheim: Wiley-VCH, 2012); J. S. Becker, *Inorganic Mass Spectrometry: Principles and Applications* (Chichester: Wiley, 2008); R. Thomas, *Practical Guide to ICP-MS: A Tutorial for Beginners*, 2nd ed. (Boca Raton, Florida: CRC Press, 2008); H. E. Taylor, *Inductively Coupled Plasma-Mass Spectrometry* (San Diego: Academic Press, 2001); C. M. Barshick, D. C. Duckworth, and D. H. Smith, eds., *Inorganic Mass Spectrometry* (New York: Marcel Dekker, 2000); S. J. Hill, ed. *Inductively Coupled Plasma Spectrometry and Its Applications* (Sheffield, England: Sheffield Academic Press, 1999); A. Montaser, ed., *Inductively Coupled Plasma Mass Spectrometry* (New York: Wiley, 1998); G. Holland and S. D. Tanner, eds., *Plasma Source Mass Spectrometry* (Cambridge: Royal Society of Chemistry, 1999).

29. M. J. Felton, "Plasma Opens New Doors in Isotope Ratio MS," *Anal. Chem.* **2003**, *75*, 119A.

30. I. Rodushkin, E. Engström, and D. C. Baxter, "Sources of Contamination and Remedial Strategies in the Multi-Elemental Trace Analysis Laboratory," *Anal. Bioanal. Chem.* **2010**, *396*, 365.

31. K. Kawabata, Y. Kishi, and R. Thomas, "The Benefits of Dynamic Reaction Cell ICP-MS Technology to Determine Ultratrace Metal Contamination Levels in High-Purity Phosphoric and Sulfuric Acid," *Spectroscopy*, January 2003, p. 16.

32. R. Wahlen, L. Evans, J. Turner, and R. Hearn, "The Use of Collision/ Reaction Cell ICP-MS for the Determination of Elements in Blood and Serum Samples," *Spectroscopy*, December 2005, p. 84.

33. D. R. Bandura, V. I. Baranov, and S. D. Tanner, "Reaction Chemistry and Collisional Processes in Multipole Devices for Resolving Isobaric Interferences in ICP-MS," *Fresenius J. Anal. Chem.* **2001**, *370*, 454; D. R. Bandura, V. I. Baranov, and S. D. Tanner, "Inductively Coupled Plasma Mass Spectrometer with Axial Field in a Quadrupole Reaction Cell," *J. Am. Soc. Mass Spec.* **2002**, *13*, 1176.

34. P. T. Palmer, "Energy-Dispersive X-Ray Fluorescence Spectrometry: A Long Overdue Addition to the Chemistry Curriculum," *J. Chem. Ed.* **2011**, *88*, 868.

35. P. T. Palmer, "Introduction to Energy-Dispersive X-Ray Fluorescence (XRF) – An Analytical Chemistry Perspective," Analytical Sciences Digital Library, http://collection.asdlib.org/?p=1602.

36. Quotation from Frederick Soddy cited by S. Weinberg, *The Discovery of Subatomic Particles* (Cambridge: Cambridge University Press, 2003), p. 126.

37. L. Wittmers, Jr., A. Aufderheide, G. Rapp, and A. Alich, "Archaeological Contributions of Skeletal Lead Analysis," *Acc. Chem. Res.* **2002**, *35*, 669.

38. V. Cheam, G. Lawson, I. Lechner, and R. Desrosiers, "Recent Metal Pollution in Agassiz Ice Cap," *Environ. Sci. Technol.* **1998**, *32*, 3974.

39. A. Bazzi, B. Kreuz, and J. Fischer, "Determination of Calcium in Cereal with Flame Atomic Absorption Spectroscopy," *J. Chem. Ed.* **2004**, *81*, 1042.

40. L. Perring and M. Basic-Dvorzak, "Determination of Total Tin in Canned Food Using Inductively Coupled Plasma Atomic Emission Spectroscopy," *Anal. Bioanal. Chem.* **2002**, *374*, 235.

41. R. B. Saper, R. S. Phillips, A. Sehgal, N. Khouri, R. B. Davis, J. Paquin, V. Thuppil, and S. N. Kales, "Lead, Mercury, and Arsenic in US- and Indian-Manufactured Ayurvedic Medicines Sold via the Internet," *J. Am. Med. Assoc.*, **2008**, *300*, 915.

Chapter 22

1. J. T. Watson and O. D. Sparkman, *Introduction to Mass Spectrometry*, 4th ed. (Chichester: Wiley, 2007); F. Klink, *Introduction to Protein and Peptide Analysis with Mass Spectrometry* (Fullerton, CA: Academy Savant, 2004), computer training Program CMSP-10. Example of sequencing short peptides:

M. Langsdorf, A. Ghassempour, A. Römpp, and B. Spengler, "Isolation and Sequence Analysis of Peptides from the Skin Secretion of the Middle East Tree Frog *Hyla Savignyi*," *Anal. Bioanal. Chem.* **2010**, *398*, 2853.

2. W.-P. Peng, Y.-C. Yang, M.-W. Kang, Y. T. Lee, and H.-C. Chang, "Measuring Masses of Single Bacterial Whole Cells with a Quadrupole Ion Trap," *J. Am. Chem. Soc.* **2004**, *126*, 11766; J. J. Jones, M. J. Stump, R. C. Fleming, J. O. Lay, Jr., and C. L. Wilkins, "Investigation of MALDI-TOF and FT-MS Techniques for Analysis of *Escherichia coli* Whole Cells," *Anal. Chem.* **2003**, *75*, 1340.

3. H.-C. Lin, J.-L. Lin, J.-H. Lin, S.-W. Tsai, A. L. Yu, R. L. Chen, and C.-H. Chen, "High-Speed Mass Measurement of Nanoparticle and Virus," *Anal. Chem.* **2012**, *84*, 4965; Z. Nie, Y.-K. Tzeng, H.-C. Chang, C.-C. Chiu, C.-Y. Chang, C.-M. Chang, and M.-H. Tao, "Microscopy-Based Mass Measurement of a Single Whole Virus in a Cylindrical Ion Trap," *Angew. Chem. Int. Ed.* **2006**, *45*, 8131.

4. J. H. Gross, *Mass Spectrometry: A Textbook*, 2nd ed. (Heidelberg: Springer-Verlag, 2011); E. de Hoffmann and V. Stroobant, *Mass Spectrometry: Principles and Applications*, 3rd ed. (Chichester: Wiley, 2007); C. Dass, *Fundamentals of Contemporary Mass Spectrometry* (Wenheim: Wiley-VCH, 2007); K. Doward, *Mass Spectrometry: A Foundation Course* (Cambridge: Royal Society of Chemistry, 2004); C. G. Herbert and R. A. W. Johnstone, *Mass Spectrometry Basics* (Boca Raton, FL: CRC Press, 2002); R. K. Boyd, C. Basic, and R. A. Bethem, *Trace Quantitative Analysis by Mass Spectrometry* (Chichester: Wiley, 2008); C. Dass, *Principles and Practice of Biological Mass Spectrometry* (New York: Wiley, 2001).

5. For a demonstration of mass spectrometry, see N. C. Grim and J. L. Sarquis, "Mass Spectrometry Analogy on the Overhead Projector," *J. Chem. Ed.* **1995**, *72*, 930.

6. D. W. Koppenaal, C. J. Barinaga, M. B. Denton, R. P. Sperline, G. M. Hieftje, G. D. Schilling, F. J. Andrade, and J. H. Barnes, IV, "MS Detectors," *Anal. Chem.* **2005**, *77*, 419A.

7. Electron ionization mass spectra of many compounds are freely available at http://webbook.nist.gov/chemistry. The commercial NIST library of 250 000 compounds is accessed from http://www.sisweb.com/software/ms/nist.htm.

8. A. B. Fialkov, U. Steiner, L. Jones, and A. Amirav, "A New Type of GC–MS with Advanced Capabilities," *Int. J. Mass Spectrom.* **2006**, *251*, 47.

9. A. B. McCoy, B. J. Braams, A. Brown, X. Huang, Z. Jin, and J. M. Bowman "Ab Initio Diffusion Monte Carlo Calculations of the Quantum Behavior of CH_5^+ in Full Dimensionality," *J. Phys. Chem. A* **2004**, *108*, 4991; O. Asvany, P. Kumar P, B. Redlich, I. Hegemann, S. Schlemmer, and D. Marx, "Understanding the Infrared Spectrum of Bare CH_5^+," *Science* **2005**, *309*, 1219; X. Huang, A. B. McCoy, J. M. Bowman, L. M. Johnson, C. Savage, F. Dong, and D. J. Nesbitt, "Quantum Deconstruction of the Infrared Spectrum of CH_5^+," *Science* **2006**, *311*, 60.

10. J. D. Hearn and G. D. Smith, "A Chemical Ionization Mass Spectrometry Method for the Online Analysis of Organic Aerosols," *Anal. Chem.* **2004**, *76*, 2820.

11. C. Sandy, J.-F. Garnier, B. Rothweiler, F. Feyerherm, S. B. Mahmoud, and H. F. Prest, "Automated Chemical Ionization in GC-MS," *Am. Lab.* January 2007, p. 22.

12. V.S. Sevastyanov, ed. *Isotope Ratio Mass Spectrometry of Light Gas-Forming Elements* (Boca Raton, FL: CRC Press, 2014); I. T. Platzner, *Modern Isotope Ratio Mass Spectrometry* (New York: Wiley, 1997). For geochemical applications, see D. J. Weiss, C. Harris, K. Maher, and T. Bullen, "A Teaching Exercise to Introduce Stable Isotope Fractionation of Metals into Geochemistry Courses," *J. Chem. Ed.* **2013**, *90*, 1014, and M. Tanimizu, Y. Sohrin, and T. Hirata, "Heavy Element Stable Isotope Ratios: Analytical Approaches and Applications," *Anal. Bioanal. Chem.* **2013**, *405*, 2771.

13. W. Chen and M. V. Orna, "Recent Advances in Archaeological Chemistry," *J. Chem. Ed.* **1996**, *73*, 485.

14. A. M. Pollard and C. Heron, *Archaeological Chemistry*, 2nd ed. (Cambridge: Royal Society of Chemistry, 2008).

15. R. A. Gross, Jr., "A Mass Spectral Chlorine Rule for Use in Structure Determinations in Sophomore Organic Chemistry," *J. Chem. Ed.* **2004**, *81*, 1161.

16. J. T. Watson and K. Biemann, "High-Resolution Mass Spectra of Compounds Emerging from a Gas Chromatograph," *Anal. Chem.* **1964**, *36*, 1135. A classic paper on gas chromatography/mass spectrometry.

17. A. Kaufmann, "The Current Role of High-Resolution Mass Spectrometry in Food Analysis," *Anal. Bioanal. Chem.* **2012**, *403*, 1233.

18. R. M. Smith, *Understanding Mass Spectra: A Basic Approach*, 2nd ed. (Hoboken, NJ: Wiley, 2004).

19. P. E. Miller and M. B. Denton, "The Quadrupole Mass Filter: Basic Operating Concepts," *J. Chem. Ed.* **1986**, *63*, 617; M. Henchman and C. Steel, "Design and Operation of a Portable Quadrupole Mass Spectrometer for the Undergraduate Curriculum," *J. Chem. Ed.* **1998**, *75*, 1042; C. Steel and M. Henchman, "Understanding the Quadrupole Mass Filter through Computer Simulation," *J. Chem. Ed.* **1998**, *75*, 1049; J. J. Leary and R. L. Schmidt, "Quadrupole Mass Spectrometers: An Intuitive Look at the Math," *J. Chem. Ed.* **1996**, *73*, 1142.

20. I. Ferrer and E. M. Thurman, eds., *Liquid Chromatography Time-of-Flight Mass Spectrometry: Principles, Tools, and Applications* (Hoboken, NJ: Wiley, 2009).

21. R. J. Cotter, *Time-of-Flight Mass Spectrometry* (Washington, DC: American Chemical Society, 1997); R. J. Cotter, "The New Time-of-Flight Mass Spectrometry," *Anal. Chem.* **1999**, *71*, 445A.

22. R. E. March and J. F. J. Todd, *Quadrupole Ion Trap Mass Spectrometry*, 2nd ed. (Hoboken, NJ: Wiley, 2005); Z. Ziegler, "Ion Traps Come of Age," *Anal. Chem.* **2002**, *74*, 489A; C. M. Henry, "The Incredible Shrinking Mass Spectrometers," *Anal. Chem.* **1999**, *71*, 264A; R. E. March, "An Introduction to Quadrupole Ion Trap Mass Spectrometry," *J. Mass Spectrom.* **1997**, *32*, 351; R. G. Cooks and R. E. Kaiser, Jr., "Quadrupole Ion Trap Mass Spectrometry," *Acc. Chem. Res.* **1990**, *23*, 213.

23. O. D. Sparkman, Z. Penton, and F. G. Kitson, *Gas Chromatography and Mass Spectrometry: A Practical Guide*, 2nd ed. (Burlington, MA: Academic Press, 2011).

24. W. M. A. Niessen, *Liquid Chromatography–Mass Spectrometry*, 3rd ed. (Boca Raton, FL: Taylor & Francis, 2006); B. Ardrey, *Liquid Chromatography–Mass Spectrometry: An Introduction* (Chichester, UK: Wiley, 2003); I. Ferrer and E. M. Thurman, eds., *Liquid Chromatography Time-of-Flight Mass Spectrometry. Principles, Tools, and Applications for Accurate Mass Analysis* (Hoboken, NJ: Wiley, 2009); J. Abian, "Historical Feature: The Coupling of Gas and Liquid Chromatography with Mass Spectrometry," *J. Mass Spectrom.* **1999**, *34*, 157.

25. M. M. Vestling, "Using Mass Spectrometry for Proteins," *J. Chem. Ed.* **2002**, *80*, 122; C. M. Henry, "Winning Ways," *Chem. Eng. News*, 18 November 2002, p. 62.

26. R. B. Cole, ed., *Electrospray and MALDI Mass Spectrometry: Fundamentals, Instrumentation, and Applications* (Hoboken, NJ: Wiley, 2010).

27. G. J. van Berkel and V. Kertesz, "Using the Electrochemistry of the Electrospray Ion Source," *Anal. Chem.* **2007**, *79*, 5510; S. Liu, W. J. Griffiths, and J. Sjövall, "On-Column Electrochemical Reactions Accompanying the Electrospray Process," *Anal. Chem.* **2003**, *75*, 1022.

28. S. R. Mabbett, L. W. Zilch, J. T. Maze, J. W. Smith, and M. F. Jarrold, "Pulsed Acceleration Charge Detection Mass Spectrometry: Application to Weighing Electrosprayed Droplets," *Anal. Chem.* **2007**, *79*, 8431.

29. A. Cappiello, G. Famiglini, F. Mangani, and P. Palma, "A Simple Approach for Coupling Liquid Chromatography and Electron Ionization Mass Spectrometry," *J. Am. Soc. Mass Spectrom.* **2002**, *13*, 265; A. Cappiello, G. Famiglini, P. Palma, and A. Siviero, "Liquid Chromatography–Electron Ionization Mass Spectrometry: Fields of Application and Evaluation of the Performance of a Direct-EI Interface," *Mass Spectrom. Rev.* **2005**, *24*, 978; A. Cappiello, G. Famiglini, P. Palma, and H. Trufelli, "Advanced Liquid Chromatography–Mass Spectrometry Interface Based on Electron Ionization," *Anal. Chem.* **2007**, *79*, 5364.

30. A. Cappiello, G. Famiglini, V. Termopoli, H. Trufelli, R. Zazzeroni, S. Jacquoilleot, L. Radici, and O. Saib, "Application of Liquid Chromatography–Direct-Electron Ionization–MS in an in Vitro Dermal Absorption Study: Quantitative Determination of *trans*-Cinnamaldehyde," *Anal. Chem.* **2011**, *83*, 8537.

31. L. Hanley and R. Zimmermann, "Light and Molecular Ions: The Emergence of Vacuum UV Single-Photon Ionization in MS," *Anal. Chem.* **2009**, *81*, 4174.

32. R. A. Yost, C. G. Enke, D. C. McGilvery, D. Smith, and J. D. Morrison, "High Efficiency Collision-Induced Dissociation in an RF-Only Quadrupole," *Intl. J. Mass Spectrom. Ion Phys.* **1979**, *30*, 127.

33. A. R. Batt, M. S. Kostich, and J. M. Lazorchak, "Analysis of Ecologically Relevant Pharmaceuticals in Wastewater and Surface Water Using Selective Solid-Phase Extraction and UPLC-MS/MS," *Anal. Chem.* **2008**, *80*, 5021; M. Schultz and E. T. Furlong, "Trace Analysis of Antidepressant Pharmaceuticals and Their Select Degradates in Aquatic Matrixes by LC/ESI/MS/MS," *Anal. Chem.* **2008**, *80*, 1756; M. J. M. Bueno, A. Agüera, M. J. Gómez, M. D. Hernando, J. F. García-Reyes, and A. R. Fernández-Alba, "Application of Liquid Chromatography/Quadrupole-Linear Ion Trap Mass Spectrometry and Time-of-Flight Mass Spectrometry to the Determination of Pharmaceuticals and Related Contaminants in Wastewater," *Anal. Chem.* **2007**, *79*, 9372; C. Postigo, M. J. Lopez de Alda, and D. Barceló, "Fully Automated Determination in the Low Nanogram per Liter Level of Different Classes of Drugs of Abuse in Sewage Water by On-Line Solid-Phase Extraction-Liquid Chromatography–Electrospray-Tandem Mass Spectrometry," *Anal. Chem.* **2008**, *80*, 3123.

34. I. González-Mariño, J. B. Quintana, I. Rodríguez, M. González-Díez, and R. Cela, "Screening and Selective Quantification of Illicit Drugs in Wastewater by Mixed-Mode Solid-Phase Extraction and Quadrupole-Time of Flight Liquid Chromatography–Mass Spectrometry," *Anal. Chem.* **2012**, *84*, 1708.

35. T. Malchi, Y. Maor, G. Tadmor, M. Shenker, and B. Chefetz, "Irrigation of Root Vegetables with Treated Wastewater: Evaluating Uptake of Pharmaceuticals and the Associated Human Health Risks," *Environ. Sci. Technol.* **2014**, *48*, 9325.

36. D. Remane, D. Wetzel, and F. T. Peters, "Development and Validation of a Liquid Chromatography–Tandem Mass Spectrometry (LC–MS/MS) Procedure for Screening of Urine Specimens for 100 Analytes Relevant in Drug-Facilitated Crime," *Anal. Bioanal. Chem.* **2014**, *406*, 4411.

37. S. C. Nanita, A. M. Pentz, and F. Q.Bramble, "High-Throughput Pesticide Residue Quantitative Analysis Achieved by Tandem Mass Spectrometry with Automated Flow Injection," *Anal. Chem.* **2009**, *81*, 3134.

38. S. Dulaurent, C. Moesch, P. Marquet, J.-M. Gaulier, and G. Lachâtre, "Screening of Pesticides in Blood with Liquid Chromatography–Linear Ion Trap Mass Spectrometry," *Anal. Bioanal. Chem.* **2010**, *396*, 2235.

39. R. B. Cole, *Electrospray and MALDI Mass Spectrometry: Fundamentals, Instrumentation, Practicalities, and Biological Applications* (Hoboken, NJ: Wiley, 2010).

40. S. A. Hofstadler, R. Bakhtiar, and R. D. Smith, "Electrospray Ionization Mass Spectrometry: Instrumentation and Spectral Interpretation," *J. Chem. Ed.* **1996**, *73*, A82.

41. R. Bakhtiar, R. Hofstadler, and R. D. Smith, "Electrospray Ionization Mass Spectrometry: Characterization of Peptides and Proteins," *J. Chem. Ed.* **1996**, *73*, A118; C. E. C. A. Hop and R. Bakhtiar, "Electrospray Ionization Mass Spectrometry: Applications in Inorganic Chemistry and Synthetic Polymer Chemistry," *J. Chem. Ed.* **1996**, *73*, A162.

42. F. Hillenkamp and J. Peter-Katalinic, eds., 2nd ed., *MALDI MS: A Practical Guide to Instrumentation, Methods, and Applications* (Weinheim: Wiley-VCH, 2014); S. C. Moyer and R. J. Cotter, "Atmospheric Pressure MALDI," *Anal. Chem.* **2002**, *74*, 469A.

43. S. S. Rubakhin, W. T. Greenough, and J. V. Sweedler, "Spatial Profiling with MALDI MS: Distribution of Neuropeptides Within Single Neurons," *Anal. Chem.* **2003**, *75*, 5374.

44. R. L. Edwards, A. J. Creese, M. Baumert, P. Griffiths, J. Bunch, and H. J. Cooper, "Hemoglobin Variant Analysis via Direct Surface Sampling of Dried Blood Spots Coupled with High-Resolution Mass Spectrometry," *Anal. Chem.* **2011**, *83*, 2265.

45. Singly charged proteins can be produced by electrospray onto the inside surface of an electrically grounded metal tube prior to the inlet of the mass spectrometer. Protein mass up to at least 200 kDa can be measured with a precision of ~1 Da by a time-of-flight spectrometer. J. Lee, H. Chen, T. Liu, C. E. Berkman, and P. T. A. Reilly, "High Resolution Time-of-Flight Mass Analysis of the Entire Range of Intact Singly-Charged Proteins," *Anal. Chem.* **2011**, *83*, 9406.

46. J. J. Coon, "Collisions or Electrons? Protein Sequence Analysis in the 21st Century," *Anal. Chem.* **2009**, *81*, 3208. An alternative to electron transfer dissociation is *electron capture dissociation*, initiated by photoionization of a compound such as acetone in the source of the mass spectrometer: D. B. Robb, J. M. Brown, M. Morris, and M. W. Blades, "Tandem Mass Spectrometry Using the Atmospheric Pressure Electron Capture Ion Source," *Anal. Chem.* **2014**, *86*, 4439.

47. G. C. McAlister, J. D. Russell, N. G. Rumachik, A. S. Hebert, J. E. P. Syka, L. Y. Geer, M. S. Westphall, D. J. Pagliarini, and J. J. Coon, "Analysis of the Acidic Proteome with Negative Electron-Transfer Dissociation Mass Spectrometry," *Anal. Chem.* **2012**, *84*, 2875.

48. G. C. McAlister, D. Phanstiel, D. M. Good, W. T. Berggren, and J. J. Coon, "Implementation of Electron-Transfer Dissociation on a Hybrid Linear Ion Trap–Orbitrap Mass Spectrometer," *Anal. Chem.* **2007**, *79*, 3525. An alternative way to use the linear ion trap for electron-transfer dissociation is described by X. Liang, J. W. Hager, and S. A. McLuckey, "Transmission Mode Ion-Ion Electron-Transfer Dissociation in a Linear Ion Trap," *Anal. Chem.* **2007**, *79*, 3363.

49. S. L. Hubler, A. Jue, J. Keith, G. C. McAlister, G. Craciun, and J. J. Coon, "Valence Parity Renders *z*-Type Ions Chemically Distinct," *J. Am. Chem. Soc.* **2008**, *130*, 6388.

50. R. M. Alberici, R. C. Simas, G. B. Sanvido, W. Romão, P. M. Lalli, M. Benassi, I. B. S. Cunha, and M. N. Eberlin, "Ambient Mass Spectrometry: Bringing MS into the 'Real World'," *Anal. Bioanal. Chem.* **2010**, *398*, 265.

51. Poppy seed bagels: *JEOL DART Applications Notebook*, www.jeolusa.com. Examples of other applications of DART are E. Crawford and B. Musselman, "Evaluating a Direct Swabbing Method for Screening Pesticides on Fruit and Vegetable Surfaces Using Direct Analysis in Real Time Coupled to an Exactive Benchtop Orbitrap Mass Spectrometer," *Anal. Bioanal. Chem.* **2012**, *403*, 2807, and H. Danhelova, J. Hradecky, S. Prinosilova, T. Cajka, K. Riddellova, L. Vaclavik, and J. Hajslova, "Rapid Analysis of Caffeine in Various Coffee Samples Employing Direct Analysis in Real-Time Ionization–High Resolution Mass Spectrometry," *Anal. Bioanal. Chem.* **2012**, *403*, 2883.

52. J. S. Wiley, J. T. Shelley, and R. G. Cooks, "Heldheld Low-Temperature Plasma Probe for Portable 'Point and Shoot' Ambient Ionization Mass Spectrometry," *Anal. Chem.* **2013**, *85*, 6545.

53. Z. Tokáts, J. M. Wiseman, B. Gologan, and R. G. Cooks, "Mass Spectrometry Sampling Under Ambient Conditions with Desorption Electrospray Ionization," *Science* **2004**, *306*, 471.

54. G. A. Eiceman, Z. Karpas, and H. H. Hill, Jr., *Ion Mobility Spectrometry*, 3rd ed. (Boca Raton, FL: CRC Press, 2014); A. A. Shvartsburg, *Differential Ion Mobility Spectrometry: Nonlinear Ion Transport and Fundamentals of FAIMS* (Boca Raton, FL: CRC Press, 2009); C. L. Wilkins and S. Trimpin, eds., *Ion Mobility-Mass Spectrometry: Theory and Applications* (Boca Raton, FL: CRC Press, 2011).

55. C. S. Kaddis and J. A. Loo, "Native Protein MS and Ion Mobility: Large Flying Proteins with ESI," *Anal. Chem.* **2007**, *79*, 1778.

56. E. J. Davis, P. Dwivedi, M. Tam, W. F. Siems, and H. H. Hill, "High-Pressure Ion Mobility Spectrometry," *Anal. Chem.* **2009**, *81*, 3270.

57. S. D. Pringle, K. Giles, J. L. Wildgoose, J. P. Williams, S. E. Slade, K. Thalassino, R. H. Bateman, M. T. Bowers, and J. H. Scrivens, "An Investigation of the Mobility Separation of Some Peptide Protein Ions Using a New Hybrid Quadrupole/Travelling Wave IMS/oa-ToF Instrument," *Intl. J. Mass Spectrom.* **2007**, *261*, 1.

58. R. Guevremont,, "High-Field Asymmetric Waveform Ion Mobility Spectrometry: A New Tool for Mass Spectrometry," *J. Chromatogr. A.* **2004**, *1058*, 3.

59. S. D. Maleknia and K. M. Downard, "Charge Ratio Analysis Method: Approach for the Deconvolution of Electrospray Mass Spectra," *Anal. Chem.* **2005**, *77*, 111.

60. S. G. Valeja, N. K. Kaiser, F. Xian, C. L. Hendrickson, J. C. Rouse, and A. G. Marshall, "Unit Mass Baseline Resolution for an Intact 148 kDa Therapeutic Monoclonal Antibody by Fourier Transform Ion Cyclotron Resonance Mass Spectrometry," *Anal. Chem.* **2011**, *83*, 8391.

61. E. N. Nikolaev, R. Jertz, A. Grigoryev, and G. Baykut, "Fine Structure in Isotopic Peak Distributions Measured Using a Dynamically Harmonized Fourier Transform Ion Cyclotron Resonance Cell at 7 T," *Anal. Chem.* **2012**, *84*, 2275; J. J. Savory, N. K. Kaiser, A. M. McKenna, F. Xian, G. T. Blakney, R. P. Rodgers, C. L. Hendrickson, and A. G. Marshall, "Parts-Per-Billion Fourier Transform Ion Cyclotron Resonance Mass Measurement Accuracy with a 'Walking' Calibration Equation," *Anal. Chem.* **2011**, *83*, 1732.

62. F. He, C. L. Hendrickson, and A. G. Marshall, "Baseline Mass Resolution of Peptide Isobars: A Record for Molecular Mass Resolution," *Anal. Chem.* **2001**, *73*, 647.

63. K. Kumara and A. Dass, "Au$_{329}$(SR)$_{84}$ Nanomolecules: Compositional Assignment of the 76.3 kDa Plasmonic Faradaurates," *Anal. Chem.* **2014**, *86*, 4227.

64. R. Chaler, J. O. Grimalt, C. Pelejero, and E. Calvo, "Sensitivity Effects in U$_{37}^{k'}$ Paleotemperature Estimation by Chemical Ionization Mass Spectrometry," *Anal. Chem.* **2000**, *72*, 5892.

65. L. Charles and D. Pépin, "Electrospray Ion Chromatography–Tandem Mass Spectrometry of Oxyhalides at Sub-ppb Levels," *Anal. Chem.* **1998**, *70*, 353.

66. J. I. G. Alonso and P. Rodríguez-González, *Isotope Dilution Mass Spectrometry* (Cambridge: Royal Society of Chemistry, 2013).

67. J. D. Fassett and P. J. Paulsen, "Isotope Dilution Mass Spectrometry for Accurate Elemental Analysis," *Anal. Chem.* **1989**, *61*, 643A.

Chapter 23

1. T. Michalowski, "Effect of Mutual Solubility of Solvents in Multiple Liquid-Liquid Extraction," *J. Chem. Ed.* **2002**, *79*, 1267.

2. Derivation from Joshua Erickson, a student at Utah Valley State College.

3. K. Ueno, T. Imamura, and K. L. Cheng, *CRC Handbook of Organic Analytical Reagents*, 2nd ed. (Boca Raton, FL: CRC Press, 1992), pp. 431–444.

4. R. P. Paradkar and R. R. Williams, "Micellar Colorimetric Determination of Dithizone Metal Chelates," *Anal. Chem.* **1994**, *66*, 2752.

5. S. Maji, S. Lahiri, B. Wierczinski, and G. Korschinek, "Separation of Trace Level Hafnium from Tungsten: A Step Toward Solving an Astronomical Puzzle," *Anal. Chem.* **2006**, *78*, 2303.

6. H. H. Strain and J. Sherma, "M. Tswett: Adsorption Analysis and Chromatographic Methods," *J. Chem. Ed.* **1967**, *44*, 238 (a translation of Tswett's original article); H. H. Strain and J. Sherma, "Michael Tswett's Contributions to Sixty Years of Chromatography," *J. Chem. Ed.* **1967**, *44*, 235; L. S. Ettre, "M. S. Tswett and the Invention of Chromatography," *LCGC North Am.* **2003**, *21*, 458; L. S. Ettre, "The Birth of Partition Chromatography," *LCGC North Am.* **2001**, *19*, 506; L. S. Ettre, "The Story of Thin-Layer Chromatography," *LCGC North Am.* **2001**, *19*, 712; L. S. Ettre, "A. A. Zhuykhovitskii—A Russian Pioneer of Gas Chromatography," *LCGC North Am.* **2000**, *18*, 1148; V. R. Meyer, "Michael Tswett and His Method," *Anal. Chem.* **1997**, *69*, 284A; Ya. I. Yashin, "History of Chromatography (1903–1993)," *Russ. J. Anal. Chem.* **1994**, *49*, 939; K. I. Sakodynskii, "Discovery of Chromatography by M. S. Tsvet," *Russ. J. Anal. Chem.* **1994**, *48*, 897.

7. L. S. Ettre, "The Rebirth of Chromatography 75 Years Ago," *LCGC North Am.* **2007**, *25*, 640.

8. C. F. Poole, *The Essence of Chromatography* (Amsterdam: Elsevier, 2003) (graduate-level treatment of theory and specific techniques); J. M. Miller, *Chromatography: Concepts and Contrasts*, 2nd ed. (Hoboken, NJ: Wiley, 2005).

9. L. V. Heumann, "Colorful Column Chromatography: A Classroom Demonstration of a Three-Component Separation," *J. Chem. Ed.* **2008**, *85*, 524; M. J. Samide, "Separation Anxiety: An In-Class Game Designed to Help Students Discover Chromatography," *J. Chem. Ed.* **2008**, *85*, 1512; C. A. Smith and F. W. Villaescusa, "Simulating Chromatographic Separations in the Classroom," *J. Chem. Ed.* **2003**, *80*, 1023; M. L. Nagel, "Visualizing Separations: How Shopping Can Be Useful for Introducing Chromatography," *J. Chem. Ed.* **2013**, *90*, 93.

10. P. Sun and D. W. Armstrong, "Ionic Liquids in Analytical Chemistry," *Anal. Chim. Acta* **2011**, *661*, 1; T. D. Ho, C. Zhang, L. W. Hantao, and J. L Anderson, "Ionic Liquids in Analytical Chemistry: Fundamentals, Advances, and Perspectives," *Anal. Chem.* **2014**, *86*, 262.

11. G. Guiochon, A. Felinger, D. G. Shirazi, and A. M. Katti, *Fundamentals of Preparative and Nonlinear Chromatography* (Amsterdam: Elsevier, 2006); H. Schmidt-Traub, ed., *Preparative Chromatography of Fine Chemicals and Pharmaceutical Agents* (Weinheim: Wiley-VCH, 2005); S. Miller, "Prep LC Systems for Chemical Separations," *Anal. Chem.* **2003**, *75*, 477A; R. E. Majors, "The Role of the Column in Preparative HPLC," *LCGC North Am.* **2004**, *22*, 416; H. L. Zuo, F. Q. Yang, W. H. Huang, and Z. N. Xia, "Preparative Gas Chromatography and Its Applications," *J. Chromatogr. Sci.* **2013**, *51*, 704; G. Guiochon and A. Tarafder, "Fundamental Challenges and Opportunities for Preparative Supercritical Fluid Chromatography," *J. Chromatogr. A* **2011**, *1218*, 1037.

12. L. R. Snyder and J. W. Dolan, *High-Performance Gradient Elution* (Hoboken, NJ: Wiley, 2007), Equation V.15.

13. J. P. Foley and J. G. Dorsey, "Equations for Calculation of Chromatographic Figures of Merit for Ideal and Skewed Peaks," *Anal. Chem.* **1983**, *55*, 730; B. A. Bidlingmeyer and F. V. Warren, Jr., "Column Efficiency Measurement," *Anal. Chem.* **1984**, *56*, 1583A.

14. More accurate expressions are discussed in J. P. Foley, "Resolution Equations for Column Chromatography," *Analyst* **1991**, *116*, 1275.

15. F. Gritti and G. Guiochon, "The van Deemter Equation: Assumptions, Limits, and Adjustment to Modern High Performance Liquid Chromatography," *J. Chromatogr. A* **2013**, *1302*, 1; J. C. Giddings, *Unified Separation Science* (New York: Wiley, 1991); S. J. Hawkes, "Modernization of the van Deemter Equation for Chromatographic Zone Dispersion," *J. Chem. Ed.* **1983**, *60*, 393.

16. C. A. Lucy, L. L. M. Glavina, and F. F. Cantwell, "A Laboratory Experiment on Extracolumn Band Broadening in Liquid Chromatography," *J. Chem. Ed.* **1995**, *72*, 367; J. W. Dolan, "Extracolumn Effects," *LCGC North Am.* **2008**, *26*, 1092; J. W. Dolan, "Why Does an Improvement Make Things Worse?" *LCGC North Am.* **2012**, *30*, 216; S. Fekete, I. Kohler, S. Rudaz, and D. Guillarme, "Importance of Instrumentation for Fast Liquid Chromatography in Pharmaceutical Analysis," *J. Pharm. Biomed. Anal.* **2014**, *87*, 105.

17. F. Gritti and G. Guiochon, "Mass Transfer Kinetics, Band Broadening and Column Efficiency," *J. Chromatogr. A* **2012,** *1221*, 2; G. Desmet and K. Broeckhoven, "Equivalance of the Different C_m- and C_s-Term Expressions Used in Liquid Chromatography and a Geometrical Model Uniting Them," *Anal. Chem.* **2008**, *80*, 8076.

18. T. Teutenberg, "Potential of High Temperature Liquid Chromatography for the Improvement of Separation Efficiency," *Anal. Chim. Acta* **2009**, *643*, 1; R. Berta, M. Babják, and M. Gazdag, "A Study of Some Practical Aspects of High Temperature Liquid Chromatography in Pharmaceutical Applications," *J. Pharm. Biomed. Anal.* **2011**, *54*, 458.

19. J. C. Giddings and R. A. Robison, "Failure of Eddy Diffusion Concept of Gas Chromatography," *Anal. Chem.* **1962**, *34*, 885; J. H. Knox, "Band Dispersion in Chromatography – A Universal Expression for the Contribution from the Mobile Phase," *J. Chromatogr. A* **2002**, *960*, 7; S. Khirevich, A. Höltzel, A. Seidel-Morgenstern, and U. Tallarek, "Time and Length Scales of Eddy Dispersion in Chromatographic Beds," *Anal. Chem.* **2009**, *81*, 7057; G. Desmet, "A Finite Parallel Zone Model to Interpret and Extend Giddings' Coupling Theory for the Eddy-Dispersion in Porous Chromatographic Media," *J. Chromatogr. A* **2013**, *1314*, 124.

20. For numerical simulation of skewed bandshapes, see S. Sugata and Y. Abe, "An Analogue Column Model for Nonlinear Isotherms: The Test Tube Model," *J. Chem. Ed.* **1997**, *74*, 406 and B. R. Sundheim, "Column Operations: A Spreadsheet Model," *J. Chem. Ed.* **1992**, *69*, 1003.

21. A. Felinger, "Molecular Movement in an HPLC Column: A Stochastic Analysis," *LCGC North Am.* **2004**, *22*, 642; J. C. Giddings, *Dynamics of Chromatography* (New York: Marcel Dekker, 1965).

Chapter 24

1. D. Thieme and P. Hemmersbach, eds., *Doping in Sports* (Berlin: Springer-Verlag, 2010); T. C. Werner and C. K. Hatton, "Performance-Enhancing Drugs in Sports: How Chemists Catch Users," *J. Chem. Ed.* **2011**, *88*, 34; T. C. Werner, "The 'Anatomy' of a Performance-Enhancing Drug Test in Sports," *J. Chem. Ed.* **2012**, *89*, 624.

2. D. H. Catlin, M. H. Sekera, B. D. Ahrens, B. Starcevic, Y. C. Chang, and C. K. Hatton, "Tetrahydrogestrinone: Discovery, Synthesis, and Detection in Urine," *Rapid Commun. Mass Spectrom.* **2004**, *18*, 1245; E. M. Brun, R. Puchades, and Á. Maquieira, *Trends Anal. Chem.* **2011**, *30*, 771; P. Aqai, E. Cevik, A. Gerssen, W. Haasnoot, and M. W. F. Nielen, "High-Throughput Bioaffinity Mass Spectrometry for Screening and Identification of Designer Anabolic Steroids in Dietary Supplements," *Anal. Chem.* **2013**, *85*, 3255.

3. Training modules are available from www.academysavant.com. For exercises to complement lab experience in chromatography, see D. C. Stone, "Teaching Chromatography Using Virtual Laboratory Exercises," *J. Chem. Ed.* **2007**, *84*, 1488.

4. J. V. Hinshaw and L. S. Ettre, *Introduction to Open Tubular Gas Chromatography* (Cleveland, OH: Advantar Communications, 1994); H. M. McNair and J. M. Miller, *Basic Gas Chromatography*, 2nd ed. (New York: Wiley, 2009); R. L. Grob and E. F. Barry, eds., *Modern Practice of Gas Chromatography* (Hoboken, NJ: Wiley, 2004); E. F. Barry and R. L. Grob, *Columns for Gas Chromatography* (Hoboken, NJ: Wiley, 2007); M. C. McMaster, *GC/MS: A Practical User's Guide*, 2nd ed. (Hoboken, NJ: Wiley, 2008).

5. E. Cremer and F. Prior first explored gas-solid adsorption chromatography in the mid-1940s at the University of Innsbruck in Austria. Gas-liquid partition chromatography is attributed to A. J. P. Martin and A. T. James in 1950–1952 at the British National Institute of Medical Research. See L. S. Ettre, "The Beginnings of Gas Adsorption Chromatography 60 Years Ago," *LCGC North Am.* **2008**, *26*, 48.

6. A. Wollrab, "Lecture Experiments in Gas-Liquid Chromatography," *J. Chem. Ed.* **1982**, *59*, 1042; C. E. Bricker, M. A. Taylor, and K. E. Kolb, "Simple Classroom Demonstration of Gas Chromatography," *J. Chem. Ed.* **1981**, *58*, 41.

7. The Royal Society of Chemistry (RSC) provides a video tour of a gas chromatograph at www.youtube.com/watch?v=08YWhLTjlfo. Agilent Technologies also provides an instrument tour in their "Fundamentals of GC – Introduction and Overview" video at www.youtube.com/watch?v=piGSGkcwFAw. A free online simulation of gas chromatography is available at www.gcsimulator.org.

8. A. K. Vickers, D. Decker, and R. E. Majors, "The Art and Science of GC Capillary Column Production," *LCGC North Am.* **2007**, *25*, 616. L. S. Ettre, "Evolution of Capillary Columns for Gas Chromatography," *LCGC North Am.* **2001**, *19*, 48.

9. V. R. Meyer, "Amino Acid Racemization: A Tool for Fossil Dating," *Chemtech* **1992**, *22*, 412.

10. A. M. Pollard and C. Heron, *Archaeological Chemistry,* 2nd ed. (Cambridge: Royal Society of Chemistry, 2008)—a very good book.

11. P. A. Levkin, A. Levkina, and V. Schurig, "Combining the Enantioselectivities of L-Valine Diamide and Permethylated β-Cyclodextrin in One Gas Chromatographic Chiral Stationary Phase," *Anal. Chem.* **2006**, *78*, 5143; I. Molnár-Perl, ed., *Quantitation of Amino Acids and Amines by Chromatography* (New York: Elsevier Science, 2005).

12. T. Cserháti and E. Forgács, *Cyclodextrins in Chromatography* (Cambridge: Royal Society of Chemistry, 2003); B. Haldar, A. Mallick, and N. Chattopadhyay, "Supramolecular Inclusion in Cyclodextrins: A Pictorial Spectroscopic Demonstration," *J. Chem. Ed.* **2008**, *85*, 429; J. Hernández-Benito, M. P. Garcia-Santos, E. O'Brien, E. Calle, and J. Casado, "A Practical Integrated Approach to Supramolecular Chemistry III. Thermodynamics of Inclusion Phenomena," *J. Chem. Ed.* **2004**, *81*, 540; B. D. Wagner, P. J. MacDonald, and M. Wagner, "Visual Demonstration of Supramolecular Chemistry: Fluorescence Enhancement upon Host-Guest Inclusion," *J. Chem. Ed.* **2000**, *77*, 178; D. Díaz, I. Vargas-Baca, and J. Graci-Mora, "β-Cyclodextrin Inclusion Complexes with Iodine," *J. Chem. Ed.* **1994**, *71*, 708.

13. T. Beesley, "The State of the Art in Chiral Capillary Gas Chromatography," *LCGC North Am.* **2011**, *29*, 642; L. Szente and J. Szemán, "Cyclodextrins in Analytical Chemistry: Host-Guest Type Molecular Recognition," *Anal. Chem.* **2013**, *85*, 8024; A. Berthod, "Chiral Recognition Mechanisms," *Anal. Chem.* **2006**, *78*, 2093; T. J. Ward, "Chiral Separations," *Anal. Chem.* **2002**, *74*, 2863; A. Berthod, W. Li, and D. W. Armstrong, "Multiple Enantioselective Retention Mechanisms on Derivatized Cyclodextrin Gas Chromatographic Chiral Stationary Phases," *Anal. Chem.* **1992**, *64*, 873; K. Bester, "Chiral Analysis for Environmental Applications," *Anal. Bioanal. Chem.* **2003**, *376*, 302.

14. P. Sun and D. W. Armstrong, "Ionic Liquids in Analytical Chemistry," *Anal. Chim. Acta* **2011**, *661*, 1; T. D. Ho, C. Zhang, L. W. Hantao, and J. L. Anderson, "Ionic Liquids in Analytical Chemistry: Fundamentals, Advances, and Perspectives," *Anal. Chem.* **2014**, *86*, 262.

15. E. N. Coker, P. J. Davis, H. van Bekkum, and A. Kerkstra, "Experiments with Zeolites at the Secondary-School Level," *J. Chem. Ed.* **1999**, *76*, 1417.

16. R. P. W. Scott, *Introduction to Analytical Gas Chromatography*, 2nd ed. (New York; Marcel Dekker, 1998); N. H. Snow, "Determination of Free-Energy Relationships using Gas Chromatography," *J. Chem. Ed.* **1996**, *73*, 592; H. R. Ellison, "Enthalpy of Vaporization by Gas Chromatography," *J. Chem. Ed.* **2005**, *82*, 1086.

17. J. R. Conder and C. L. Young, *Physicochemical Measurement by Gas Chromatography* (Chichester: Wiley, 1979); R. Lebrón-Aguilar, J. E. Quintanilla-López, and J. M. Santiuste, "Solvation Molar Enthalpies and Heat Capacities of *n*-Alkanes and *n*-Alkylbenzenes on Stationary Phases of Wide-Ranging Polarity," *J. Chromatogr. A* **2010**, *1217*, 7767.

18. L. K. Nash, "Trouton and T-H-E Rule," *J. Chem. Ed.* **1984**, *61*, 981; M. MacLeod, M. Scheringer, and K. Hungerbühler, "Estimating Enthalpy of Vaporization from Vapor Pressure using Trouton's Rule," *Env. Sci. Tech.* **2007**, *41*, 2827.

19. In programmed temperature runs with a constant inlet pressure, the flow rate decreases during the run because the viscosity of the carrier gas increases as temperature increases. The effect can be significant (e.g., 30% decrease in linear velocity for a 200° temperature increase), so it is a good idea to set the initial linear velocity *above* the optimum value so that it does not decrease too far below the optimum. Equations for calculating flow rates as a function of temperature and pressure are given by J. V. Hinshaw and L. S. Ettre, *Introduction to Open Tubular Gas Chromatography* (Cleveland, OH: Advanstar Communications, 1994) and L. S. Ettre and J. V. Hinshaw, *Basic Relationships of Gas Chromatography* (Cleveland, OH: Advanstar Communications, 1993).

20. *Mythbusters* demonstrated the hazard potential of pressurized gas cylinders in their *Air Cylinder Rocket* episode, first aired on Oct. 18, 2006; J. V. Hinshaw, "The Storage and Use of Gases for Gas Chromatography," *LCGC North Am.* **2014**, *32*, 194; J. V. Hinshaw, "Gas Cylinder Setup and Use," *LCGC North Am.* **2014**, *32*, 488; Dow Chemical Company provides safety videos (http://safety.dow.com) on topics including safe handling of gas cylinders (http://safety.dow.com/en/safety-courses/specialized-topics/gas-cylinder-user).

21. R. J. Bartram and P. Froehlich, "Considerations on Switching from Helium to Hydrogen," *LCGC North Am.* **2010**, *28*, 890.

22. J. V. Hinshaw, "The Retention Gap Effect," *LCGC North Am.* **2004**, *22*, 624.

23. K. Grob, *Split and Splitless Injection for Quantitative Gas Chromatography* (New York: Wiley, 2001).

24. J. V. Hinshaw, "Syringes for Gas Chromatography," *LCGC North Am.* **2006**, *24*, 278.

25. Techniques and hardware for *large volume injections* for trace analysis are described by S. M. Song, P. Marriott, D. Ryan, and P. Wynne, "Analytical Limbo: How Low Can You Go?" *LCGC North Am.* **2006**, *24*, 1012.

26. A. N. Papas and M. F. Delaney, "Evaluation of Chromatographic Integrators and Data Systems," *Anal. Chem.* **1987**, *59*, 54A; M. K. L. Bicking, "Integration Errors in Chromatographic Analysis," *LCGC North Am.* **2006**, *24*, 402; J. W. Dolan, "Integration Problems," *LCGC North Am.* **2009**, *27*, 892.

27. J. V. Hinshaw, "The Flame Ionization Detector," *LCGC North Am.* **2005**, *23*, 1262.

28. If standards are not available, the relative response to different organic compounds can be approximated as being proportional to the number of carbon atoms not bound to one or more heteroatoms or halogens. (J. T. Scanlon and D. E. Willis, "Calculation of Flame Ionization Detector Relative Response Factors Using the Effective Carbon Number Concept," *J. Chromatogr. Sci.* **1985**, *23*, 333; E. Tissot, S. Rochat, C. Debonneville, and A. Chaintreau, "Rapid GC-FID Quantification Technique Without Authentic Samples Using Predicted Response Factors," *Flavour Frag. J.* **2012**, *27*, 290.

29. B. Plutowska and W. Wardencki, "Application of Gas Chromatography-Olfactometry (GC-O) in Analysis and Quality Assessment of Alcoholic Beverages—A Review," *Food Chem.* **2008**, *107*, 449.

30. B. Erickson, "Measuring Nitrogen and Phosphorus in the Presence of Hydrocarbons," *Anal. Chem.* **1998**, *70*, 599A.

31. K. A. Schug, I. Sawicki, D. D. Carlton, Jr., H. Fan, H. M. McNair, J. P. Nimmo, P. Kroll, J. Smuts, P. Walsh, and D. Harrison, "Vacuum Ultraviolet Detector for Gas Chromatography," *Anal. Chem.* **2014**, *86*, 8329.

32. X. W. Yan, "Unique Selective Detectors for Gas Chromatography: Nitrogen and Sulfur Chemiluminescence Detectors," *J. Sep. Sci.* **2006**, *29*, 1931.

33. R. N. Easter, J. A. Caruso, and A. P. Vonderheide, "Recent Developments and Novel Applications in GC-ICPMS," *J. Anal. Atom. Spec.* **2010**, *25*, 493.

34. S. Risticevic, V. H. Niri, D. Vuckovic, and J. Pawliszyn, "Recent Developments in Solid-Phase Microextraction," *Anal. Bioanal. Chem.* **2009**, *393*, 781; F. Pragst, "Application of Solid-Phase Microextraction in Analytical Toxicology," *Anal. Bioanal. Chem.* **2007**, *388*, 1393; J. B. Quintana and I. Rodríguez, "Strategies for the Microextraction of Polar Organic Contaminants in Water Samples," *Anal. Bioanal. Chem.* **2006**, *384*, 1447; P. Mayer, J. Tolls, J. C. M. Hermens, and D. MacKay, "Equilibrium Sampling Devices," *Environ. Sci. Technol.* **2003**, *37*, 185A; www.sigmaaldrich.com/analytical-chromatography/video/spme-video.html.

35. G. F. Ouyang and J. Pawliszyn, "A Critical Review in Calibration Methods for Solid-Phase Microextraction," *Anal. Chim. Acta* **2008**, *627*, 184.

36. B. Kolb and L. S. Ettre, *Static Headspace Gas Chromatography: Theory and Practice*, 2nd ed. (Hoboken, NJ: Wiley, 2006).

37. C. Bicchi, C. Iori, P. Rubiolo, and P. Sandra, "Headspace Sorptive Extraction, Stir Bar Sorptive Extraction, and Solid Phase Microextraction Applied to the Analysis of Roasted Arabica Coffee and Coffee Brew," *J. Agric. Food Chem.* **2002**, *50*, 449.

38. F. J. Camino-Sánchez, R. Rodríguez-Gómez, A. Zafra-Gómez, A. Santos-Fandila, and J. L. Vílchez, "Stir Bar Sorptive Extraction: Recent Applications, Limitations and Future Trends," *Talanta* **2014**, *130*, 388; B. Sgorbini, P. Rubiolo, C. Bicchi, E. Liberto, and C. Cordero, "Stir-Bar Sorptive Extraction and Headspace Sorptive Extraction: An Overview," *LCGC North Am.* **2009**, *27*, 376.

39. K. Dettmer and W. Engewald, "Absorbent Materials Commonly Used in Air Analysis for Adsorptive Enrichment and Thermal Desorption of Volatile Organic Compounds," *Anal. Bioanal. Chem.* **2002**, *373*, 490; M. Schneider and K.-U. Goss, "Systematic Investigation of the Sorption Properties of Tenax TA, Chromosorb 106, Porapak N, and Carbopak F," *Anal. Chem.* **2009**, *81*, 3017.

40. S. Cram, "How to Develop, Validate, and Troubleshoot Capillary GC Methods," American Chemical Society Short Course, 1996.

41. J. de Zeeuw and M. Barnes, "Fast, Practical GC and GC-MS," *Am. Lab. News Ed.*, January 2007, p. 26.

42. J. V. Seeley and S. K. Seeley, "Multidimensional Gas Chromatography: Fundamental Advances and New Applications," *Anal. Chem.* **2013**, *85*, 557; S-T. Chin and P. J. Marriott, "Multidimensional Gas Chromatography Beyond Simple Volatiles Separation," *Chem. Commun.* **2014**, *50*, 8819.

43. For software to optimize liquid and gas chromatography, see www.acdlabs.com. J. V. Hinshaw, "Strategies for GC Optimization: Software," *LCGC North Am.* **2000**, *18*, 1040; L. Wool and D. Decker, "Practical Fast Gas Chromatography for Contract Laboratory Program Pesticide Analyses," *J. Chromatogr. Sci.* **2002**, *40*, 434.

Chapter 25

1. C. E. Burgess, P. N. Pearson, C. H. Lear, H. E. G. Morgans, L. Handley, R. D. Pancost, and S. Schouten, "Middle Eocene Climate Cyclicity in the Southern Pacific: Implications for Global Ice Volume," *Geology* **2008**, *36*, 651; P. N. Pearson, B. E. van Dongen, C. J. Nicholas, R. D. Pancost, S. Schouten, J. M. Singano, and B. S. Wade, "Stable Warm Tropical Climate Through the Eocene Epoch," *Geology* **2007**, *35*, 211; C. Huguet, J.-H. Kim, J. S. Sinninghe Damsté, and S. Schouten, "Reconstruction of Sea Surface Temperature Variations in the Arabian Sea of the Last 23 kyr using Organic Proxies TEX_{86} and $U_{37}^{K'}$," *Paleooceanography* **2006**, *21*, PA3003.

2. L. R. Snyder, J. J. Kirkland, and J. W. Dolan, *Introduction to Modern Liquid Chromatography*, 3rd ed. (Hoboken, NJ: Wiley, 2010); L. R. Snyder, J. J. Kirkland, and J. L. Glajch, *Practical HPLC Method Development* (New York: Wiley, 1997). References 2 and 3 are definitive resources for liquid chromatography.

3. L. R. Snyder and J. W. Dolan, *High-Performance Gradient Elution* (Hoboken, NJ: Wiley, 2007).

4. M. W. Dong, *Modern HPLC for Practicing Scientists* (Hoboken, NJ: Wiley, 2006); S. Kromidas, *HPLC Made to Measure—A Practical Handbook for Optimization* (Weinheim: Wiley-VCH, 2006); M. C. McMaster, *HPLC: A Practical User's Guide*, 2nd ed. (Hoboken, NJ: Wiley, 2008); M. C. McMaster, *LC/MS: A Practical User's Guide* (Hoboken, NJ: Wiley, 2005); S. Kromidas, *More Practical Problem Solving in HPLC* (Weinheim: Wiley-VCH, 2005); S. Kromidas, *Practical Problem Solving in HPLC* (New York: Wiley, 2000); V. R. Meyer, *Practical High-Performance Liquid Chromatography*, 5th ed. (Chichester: Wiley, 2010); V. R. Meyer, *Pitfalls and Errors of HPLC in Pictures*, 3rd ed. (New York: Wiley, 2013); U. D. Neue, *HPLC Columns: Theory, Technology, and Practice* (New York: Wiley, 1997).

5. A wealth of practical tips can be found in *LCGC* magazine at www.chromatographyonline.com. Specific references are noted throughout the chapter. A compendium of the *LC Troubleshooting* columns is at www.lcresources.com/tsbible.

6. D. C. Stone, "Teaching Chromatography Using Virtual Laboratory Exercises," *J. Chem. Ed.* **2007**, *84*, 1488; A. Kadjo and P. K. Dasgupta, "Tutorial: Simulating Chromatography with Microsoft Excel Macros," *Anal. Chim. Acta* **2013**, *773*, 1.

7. P. G. Boswell, D. R. Stoll, P. W. Carr, M. L. Nagel, M. F. Vitha, and G. A. Mabbott, "An Advanced, Interactive, High-Performance Liquid Chromatography Simulator and Instructor Resources," *J. Chem. Ed.,* **2013**, *90*, 198. Exercises are provided in the supplementary information at the Journal website. The simulator can be downloaded at www.hplcsimulator.org.

8. For training modules, see www.academysavant.com and www.chromacademy.com.

9. Royal Society of Chemistry (RSC) provides a video tour of a high performance liquid chromatograph at www.youtube.com/watch?v=kz_egMtdnL4.

10. J. W. Dolan, "LC Method Scaling, Part I: Isocratic Separations," *LCGC North Am.* **2014**, *32*, 98.

11. J. W. Dolan, "Honoring Readers," *LCGC North Am.* **2003**, *21*, 888; F. Osmani and J. W. Dolan, "Evaluation of an LC Method," *LCGC North Am.* **2012**, *30*, 972.

12. F. Gritti, M. Martin, and G. Guiochon, "Influence of Viscous Friction Heating on the Efficiency of Columns Operated under Very High Pressures," *Anal. Chem.* **2009**, *81*, 3365; F. Gritti and G. Guiochon, "Heat Exchanges in Fast, High-Performance Liquid Chromatography. A Complete Thermodynamic Study," *Anal. Chem.* **2008**, *80*, 6488; F. Gritti and G. Guiochon, "Complete Temperature Profiles in Ultra-High-Pressure Liquid Chromatography Columns," *Anal. Chem.* **2008**, *80*, 5009; J. R. Mazzeo, U. D. Neue, M. Kele, and R. S. Plumb, "Advancing LC Performance with Smaller Particles and Higher Pressure," *Anal. Chem.* **2005**, *77*, 460A.

13. J. W. Dolan, "The Importance of Temperature," *LCGC North Am.* **2002**, *20*, 524; Y. Yang and D. R. Lynch, Jr., "Stationary Phases for High-Temperature LC Separations," *LCGC North Am. Supplement*, June 2004, p. S34.

14. For high-pH operation of any silica-based stationary phase, temperature should not exceed 40°C, organic buffers should be used instead of phosphate or carbonate, and methanol instead of acetonitrile should be the organic solvent. (J. J. Kirkland, J. D. Martosella, J. W. Henderson, C. H. Dilks, Jr., and J. B. Adams, Jr., "HPLC of Basic Compounds at High pH with a Silica-Based Bidentate-C18 Bonded-Phase Column," *Am. Lab.*, November 1999, p. 22; E. D. Neue, "HPLC Troubleshooting," *Am. Lab.*, February 2002, p. 72.)

15. A. Giaquinto, Z. X. Liu, A. Bach, and Y. Kazakevich, "Surface Area of Reversed-Phase HPLC Columns," *Anal. Chem.* **2008**, *80*, 6358.

16. F. Gritti and G. Guiochon, "Facts and Legends About Columns Packed with Sub-3-μm Core-Shell Particles," *LCGC North Am.* **2012**, *30*, 586.

17. B. A. Rogers, Z. Wu, B. C. Wei, X. Zhang, X. Cao, O. Alabi, and M. J. Wirth, "Submicrometer Particles and Slip Flow in Liquid Chromatography," *Anal. Chem.* **2015**, *87*, 2520; B. J. Rogers and M. J. Wirth, "Slip Flow Through Colloidal Crystals of Varying Particle Diameter," *ACS Nano* **2013**, *7*, 725; B. C. Wei, B. J. Rogers, and M. J. Wirth, "Slip Flow in Colloidal Crystals for Ultraefficient Chromatography," *J. Am. Chem. Soc.* **2012**, *134*, 10780.

18. C. West, C. Elfakir, and M. Lafosse, "Porous Graphitic Carbon: A Versatile Stationary Phase for Liquid Chromatography," *J. Chromatogr. A* **2010**, *1217*, 3201. For other robust stationary phases, see: J. Nawrocki, C. Dunlap, J. Zhao, C. V. McNeff, A. McCormick, and P. W. Carr, "Chromatography Using Ultra-Stable Metal Oxide-Based Stationary Phases for HPLC," *J. Chromatogr. A* **2004**, *1028*, 31; C. Paek, A. V. McCormick, and P. W. Carr, "New Method for Development of Carbon Clad Silica Phases for Liquid Chromatography," *J. Chromatogr. A* **2011**, *1218*, 1359.

19. A. Cavazzini, L. Pasti, A. Massi, N. Marchetti, and F. Dondi, "Recent Applications in Chiral High Performance Liquid Chromatography: A Review," *Anal. Chim. Acta* **2011**, *706*, 205; M. L. Tang, J. Zhang, S. L. Zhuang, and W. R. Liu, "Development of Chiral Stationary Phases for High-Performance Liquid Chromatographic Separation," *Trends Anal. Chem.* **2012**, *39*, 180; S. Ahuja, "A Strategy for Developing HPLC Methods for Chiral Drugs," *LCGC North Am.* **2007**, *25*, 1112.

20. A. P. Schellinger and P. W. Carr, "Solubility of Buffers in Aqueous-Organic Eluents for Reversed-Phase Liquid Chromatography," *LCGC North Am.* **2004**, *22*, 544; D. Sykora, E. Tesarova, and D. W. Armstrong, "Practical Considerations of the Influence of Organic Modifiers on the Ionization of Analytes and Buffers in Reversed-Phase LC," *LCGC North Am.* **2002**, *20*, 974; G. W. Tindall, "Mobile-Phase Buffers. I. The Interpretation of pH in Partially Aqueous Mobile Phases," *LCGC North Am.* **2002**, *20*, 1028; S. Espinosa, E. Bosch, and M. Rosés, "Acid-Base Constants of Neutral Bases in Acetonitrile-Water Mixtures," *Anal. Chim. Acta* **2002**, *454*, 157.

21. SF_6 demonstrations: P. Licence, D. Litchfield, M. P. Dellar, and M. Poliakoff, "'Supercriticality'; a Dramatic but Safe Demonstration of the Critical Point," *Green Chem.* **2004**, *6*, 352; R. Chang and J. F. Skinner, "A Lecture Demonstration of the Critical Phenomenon," *J. Chem. Ed.* **1992**, *69*, 158. CO_2 demonstration: V. T. Lieu, "Simple Experiment for Demonstration of Phase Diagram of Carbon Dioxide," *J. Chem. Ed.* **1996**, *73*, 837. H_2O + isobutyric acid: M. R. Johnson, "A Demonstration of the Continuous Phase (Second-Order) Transition of a Binary Liquid System in the Region Around Its Critical Point," *J. Chem. Ed.* **2006**, *83*, 1014.

22. K. De Klerck, D. Mangelings, and Y. Vander Heyden, "Supercritical Fluid Chromatography for the Enantioseparation of Pharmaceuticals," *J. Pharm. Biomed. Anal.* **2012**, *69*, 77; L. Nováková, A. Grand-Guillaume Perrenoud, I. Francois, C. West, E. Lesellier, and D. Guillarme, "Modern Analytical Supercritical Fluid Chromatography using Columns Packed with Sub-2-μm Particles: A Tutorial," *Anal. Chim. Acta* **2014**, *824*, 18; G. K. Webster, ed., *Supercritical Fluid Chromatography—Advances and Applications in Pharmaceutical Analysis* (Boca Raton: CRC Press, 2014).

23. G. B. Hansen, "Ultraviolet to Near-Infrared Absorption Spectrum of Carbon Dioxide Ice from 0.174 to 1.8 μm," *J. Geophys. Res. Planets* **2005**, *110*, E11003 (DOI: 10.1029/2005JE002531).

24. B. Olsen and B. Pack, ed., *Hydrophilic Interaction Chromatography—A Guide for Practitioners* (Hoboken, NJ; Wiley, 2013); P. Jandera, "Stationary and Mobile Phases in Hydrophilic Interaction Chromatography A Review," *Anal. Chim. Acta* **2011**, *692*, 1; M. R. Gama, R. G. da Costa Silva, C. H. Collins, and C. B. G. Bottoli, "Hydrophilic Interaction Chromatography," *Trends Anal. Chem.* **2012**, *37*, 48.

25. J. W. Dolan, "Mobile-Phase Degassing: What, Why, and How," *LCGC North Am.* **2014**, *32*, 482.

26. J. W. Dolan, "Column Care," *LCGC North Am.* **2008**, *26*, 692.

27. *Washing normal-phase column:* Remove the guard column so that impurities from the guard column are not washed into the analytical column. Bare silica and diol-bonded phases can be washed (in order) with heptane, chloroform, ethyl acetate, acetone, ethanol, and water. Then the order is reversed, using dried solvents, to reactivate the column. Use 15 mobile phase volumes ($15V_m$) of each solvent. Amino-bonded phases are washed in the same manner as silica, but 0.5 M ammonia is used after water. (F. Rabel and K. Palmer, *Am. Lab.*, August 1992, p. 65.) Between uses, normal-phase columns can be stored in 2-propanol or hexane.

28. J. W. Dolan, "Troubleshooting Basics. Peak Shape Problems," *LCGC North Am.* **2012**, *30*, 564.

29. J. W. Dolan, "Peak Shape Problems," *LCGC North Am.* **2008**, *26*, 610.

30. J. W. Dolan, "How Much Is Too Much?" *LCGC North Am.* **1999**, *17*, 508.

31. J. W. Dolan, "Extra-column Effects," *LCGC North Am.* **2008**, *26*, 1092.

32. J. W. Dolan, "LC Troubleshooting Strategies," *LCGC North Am.* **2009**, *27*, 1040.

33. J. W. Dolan, "LC Pumps," *LCGC North Am.* **2008**, *26*, 1168. See animation of pump action at www.lcresources.com/resources/getstart/2b01.htm.

34. K. G. Kehl and V. R. Meyer, "Argentometric Titration for the Determination of Liquid Chromatographic Injection Reproducibility," *Anal. Chem.* **2001**, *73*, 131.

35. M. Swartz, "Seeing Is Believing: Detectors for HPLC," *LCGC North Am.* **2010**, *28*, 880; M. Swartz, "HPLC Detectors: A Brief Review," *J. Liq. Chromatogr.* **2010**, *33*, 1130.

36. L. N. Seetohul, Z. Ali, and M. Islam, "Broadband Cavity Enhanced Absorption Spectroscopy as a Detector for HPLC," *Anal. Chem.* **2009**, *81*, 4106.

37. C. K. Zacharis and P. D. Tzanavaras, "Liquid Chromatography Coupled to On-line Post Column Derivatization for the Determination of Organic

Compounds: A Review on Instrumentation and Chemistries," *Anal. Chim. Acta* **2013**, *798*, 1; P. G. Rigas, "Post-Column Labeling Techniques in Amino Acid Analysis by Liquid Chromatography," *Anal. Bioanal. Chem.* **2013**, *405*, 7957.

38. S. Iijima, Y. Sato, M. Bounoshita, T. Miyaji, D. J. Tognarelli, and M. Saito, "Optimization of an Online Post-Column Derivatization System for Ultra High Performance Liquid Chromatography (UHPLC) and Its Applications to Analysis of Biogenic Amines," *Anal. Sci.* **2013**, *29*, 539.

39. J. A. Koropchak, S. Sadain, X. H. Yang, L.-E. Magnusson, M. Heybroek, M. Anisimov, and S. L. Kaufman, "Nanoparticle Detection Technology," *Anal. Chem.* **1999**, *71*, 386A; N. C. Megoulas and M. A. Koupparis, "Twenty Years of Evaporative Light Scattering Detection," *Crit. Rev. Anal. Chem.* **2005**, *35*, 301; L. Lucena, S. Cárdenas, and M. Valcárcel, "Evaporative Light Scattering Detection: Trends in Its Analytical Uses," *Anal. Bioanal. Chem.* **2007**, *388*, 1663.

40. P. H. Gamache, R. S. McCarthy, S. M. Freeto, D. J. Asa, M. J. Woodcock, K. Laws, and R. O. Cole, "HPLC Analysis of Nonvolatile Analytes Using Charged Aerosol Detection," *LCGC North Am.* **2005**, *23*, 150; M. Ligor, S. Studzińska, A. Horna, and B. Buszewski, "Corona-Charged Aerosol Detection: An Analytical Approach," *Crit. Rev. Anal. Chem.* **2013**, *43*, 64.

41. The effect of the gradient on the response of the charged aerosol detector can be cancelled by delivering an exact inverse gradient with a second pump and mixing the inverse gradient with eluate prior to the detector, as was done in Box 25-3. (T. Górecki, F. Lynen, R. Szucs, and P. Sandra, "Universal Response in Liquid Chromatography Using Charged Aerosol Detection," *Anal. Chem.* **2006**, *78*, 3186.)

42. M. Trojanowicz, "Recent Developments in Electrochemical Flow Detections—A Review Part II. Liquid Chromatography," *Anal. Chim. Acta* **2011**, *688*, 8.

43. W. R. LaCourse, *Pulsed Electrochemical Detection in High-Performance Liquid Chromatography* (New York: Wiley, 1997); W. R. LaCourse, "Pulsed Electrochemical Detection: Waveform Evolution," *LCGC North Am.* **2011**, *29*, 584.

44. J. W. Dolan, "Avoiding Refractive Index Detector Problems," *LCGC North Am.* **2012**, *30*, 1032.

45. For a concise, readable, and expert guide to method development for reversed-phase separations, see J. W. Dolan, "The Perfect Method," *LCGC North Am.* **2007**, *25*, 546, 632, 704, 944, 1014, 1094, 1178.

46. Acetonitrile can be hydrolyzed to sodium acetate and poured down the drain for disposal: $CH_3CN + NaOH + H_2O \rightarrow CH_3CO_2Na + NH_3$. Dilute the chromatography eluate to 10 vol% CH_3CN with water. To 1.0 L of 10 vol% CH_3CN, add 475 mL of 10 M NaOH. The solution can be left at 20°C for 25 days in a hood or heated to 80°C for 70 min to reduce the CH_3CN concentration to 0.025 vol%. Mix the hydrolysate with waste acid so that it is approximately neutral before disposal. (K. Gilomen, H. P. Stauffer, and V. R. Meyer, "Management and Detoxification of Acetonitrile Wastes from Liquid Chromatography," *LCGC* **1996**, *14*, 56.)

47. Tetrahydrofuran can be stored for at least half a year without oxidation by adding 25 vol% H_2O. (J. H. Zhao and P. W. Carr, "The Magic of Water in Tetrahydrofuran—Preventing Peroxide Formation," *LCGC* **1999**, *17*, 346.)

48. J. W. Dolan, "Selectivity in Reversed-Phase LC Separations, Part II: Solvent-Strength Selectivity," *LCGC North Am.* **2011**, *29*, 28.

49. J. D. Thompson and P. W. Carr, "High-Speed Liquid Chromatography by Simultaneous Optimization of Temperature and Eluent Composition," *Anal. Chem.* **2002**, *74*, 4150.

50. J. W. Dolan, "Selectivity in Reversed-Phase LC Separations, Part I: Solvent-Type Selectivity," *LCGC North Am.* **2010**, *28*, 1022.

51. E. M. Borges, "How to Select Equivalent and Complementary Reversed-Phase Liquid Chromatography Columns from Column Characterization Databases," *Anal. Chim. Acta* **2014**, *807*, 143; see www.hplccolumns.org for online column selection tools based on the hydrophobic subtraction model, L. R. Snyder, J. W. Dolan, and P. W. Carr, "A New Look at the Selectivity of RPC Columns," *Anal. Chem.* **2007**, *79*, 3254; see www.acdlabs.com/resources/freeware/colsel for data based on M. R. Euerby and P. Petersson, "Chromatographic Classification and Comparison of Commercially Available Reversed-Phase Liquid Chromatographic Columns Using Principal Component Analysis," *J. Chromatogr. A* **2003**, *994*, 13.

52. J. W. Dolan, "Starting Out Right, Part V—Changing Column Conditions," *LCGC North Am.* **2000**, *18*, 376; J. W. Dolan, "The Perfect Method, Part VI: Make It Faster," *LCGC North Am.* **2007**, *25*, 1094.

53. For a spreadsheet for method transfer calculations in isocratic and gradient modes, see www.unige.ch/sciences/pharm/fanal/lcap/telechargement-en.htm. Based on D. Guillarme, D. T.-T. Nguyen, S. Rudaz, and J.-L. Veuthey, "Method Transfer for Fast Liquid Chromatography in Pharmaceutical Analysis: Application to Short Columns Packed with Small Particle: Part I: Isocratic Separation and Part II: Gradient Experiments," *Eur. J. Pharm. Biopharm.* **2007**, *66*, 475 and **2008**, *68*, 430.

54. For more on developing gradient separations, see Reference 3 and J. W. Dolan, "The Scouting Gradient Alternative," *LCGC North Am.* **2000**, *18*, 478; J.W. Dolan, "Making the Most of a Gradient Scouting Run," *LCGC North Am.* **2013**, *31*, 30.

55. M. R. Euerby, P. Petersson, and M. A. James, "Translations Between Differing Liquid Chromatography Formats: Advantages, Principles, and Possible Pitfalls," *LCGC North Am.* **2014**, *32*, 558.

56. A. P. Schellinger, D. R. Stoll, and P. W. Carr, "High-Speed Gradient Elution Reversed-Phase Liquid Chromatography of Bases in Buffered Eluents. Part I: Retention Repeatability and Column Re-Equilibration," *J. Chromatogr. A* **2008**, *1192*, 41.

57. J. W. Dolan, "Gradient Elution, Part IV: Dwell-Volume Problems," *LCGC North Am.* **2013**, *31*, 456.

58. Available software includes: DryLab, Molnar Institute, Berlin (www.molnar-institut.com/drylab/); ChromSword (www.chromsword.com); LC & GC Simulator, Advanced Chemistry Development, Toronto (www.acdlabs.com/products/com_iden/meth_dev/).

59. J. M. Cintrón and L. A. Colón, "Organo-Silica Nano-Particles Used in Ultrahigh-Pressure Liquid Chromatography," *Analyst* **2002**, *127*, 701.

60. N. Tanaka, H. Kobayashi, K. Nakanishi, H. Minakuchi, and N. Ishizuka, "Monolithic LC Columns," *Anal. Chem.* **2001**, *73*, 420A; G. Guiochon, "Monolithic Columns in High Performance Liquid Chromatography," *J. Chromatogr. A* **2007**, *1168*, 101; J. Urban and P. Jandera, "Recent Advances in the Design of Organic Polymer Monoliths for Reversed-Phase and Hydrophilic Interaction Chromatography Separations of Small Molecules," *Anal. Bioanal. Chem.* **2013**, *405*, 2123.

61. E. S. Grumbach, D. M. Wagrowski-Diehl, J. R. Mazzeo, B. Alden, and P. C. Iraneta, "Hydrophilic Interaction Chromatography Using Silica Columns for the Retention of Polar Analytes and Enhanced ESI-MS Sensitivity," *LCGC North Am.* **2004**, *22*, 1010.

Chapter 26

1. J. M. Butler, *Forensic DNA Typing* (Amsterdam: Elsevier, 2005).

2. A. A. Zagorodni, *Ion Exchange Materials* (Amsterdam: Elsevier, 2007); F. de Dardel and T. V. Arden, "Ion Exchangers" in *Ullmann's Encyclopedia of Industrial Chemistry*, 7th ed. (New York: Wiley-VCH, 2010); J. Kammerer, R. Carle, and D. R. Kammerer, "Adsorption and Ion Exchange: Basic Principles and Their Application in Food Processing," *J. Agric. Food Chem.* **2011**, *59*, 22.

3. Y.-H. Chen, J.-Y. Lin, L.-P. Lin, H. Liang, and J.-F. Yaung, "Sequestration of Divalent Metal Ion by Superabsorbent Polymer in Diapers," *J. Chem. Ed.* **2010**, *87*, 920.

4. J. T. Way, "On the Power of Soils to Absorb Manure," *J. Royal Agric. Soc. Eng.* **1852**, *13*, 123; ibid., **1850**, *11*, 313.

5. F. R. Mansour, C. L. Kirkpatrick, and N. D. Danielson, "Ion Exclusion Chromatography of Aromatic Acids," *J. Chromatogr. Sci.* **2013**, *51*, 655.

6. B. M. Stewart, "The Production of High-Purity Water in the Clinical Laboratory," *Lab. Med.* **2000**, *31*, 605; S. Mabic, C. Regnault, and J. Krol, "The Misunderstood Laboratory Solvent: Reagent Water for HPLC," *LCGC North Am.* **2005**, *23*, 74; *A Guide to Laboratory Water Purification* (Labconco Corp., 2006); E. Riché, "Best Practices for Using a Water Purification System," www.youtube.com/watch?v=D89zoLy8fiE.

7. M. Cantuel, G. Bernardinelli, G. Muller, J. P. Riehl, and C. Piguet, "The First Enantiomerically Pure Helical Noncovalent Tripod for Assembling Nine-Coordinate Lanthanide(III) Podates," *Inorg. Chem.* **2004**, *43*, 1840;

Y. Yoshikawa and K. Yamasaki, "Chromatographic Resolution of Metal Complexes on Sephadex Ion Exchangers," *Coord. Chem. Rev.* **1979**, *28*, 205.

8. D. P. Elder, "Pharmaceutical Applications of Ion-Exchange Resins," *J. Chem. Ed.* **2005**, *82*, 575.

9. J. S. Fritz and D. T. Gjerde, *Ion Chromatography*, 4th ed. (Weinheim: Wiley-VCH, 2009); J. Weiss, *Handbook of Ion Chromatography*, 3rd ed. (Weinheim: Wiley-VCH, 2004); P. R. Haddad and P. E. Jackson, *Ion Chromatography: Principles and Applications* (New York: Elsevier, 1990); H. Small, "Landmarks in the Evolution of Ion Chromatography," *LCGC North Am.* **2013**, *32* (April Supplement 4B), 8; H. Small, "Ion Chromatography: An Account of Its Conception and Early Development," *J. Chem. Ed.* **2004**, *81*, 1277; B. Evans, "The History of Ion Chromatography: The Engineering Perspective," *J. Chem. Ed.* **2004**, *81*, 1285; P. R. Haddad, P. N. Nesterenko, and W. Buchberger, "Recent Developments and Emerging Directions in Ion Chromatography," *J. Chromatogr. A* **2008**, *1184*, 456.

10. Elevated thiocyanate and benzoate are detectable in sweat from smokers. E. Gilchrist, N. Smith, and L. Barron, "Probing Gunshot Residue, Sweat and Latent Human Fingerprints with Capillary-Scale Ion Chromatography and Suppressed Conductivity Detection," *Analyst* **2012**, *137*, 1576.

11. C. Pohl, "Recent Developments in Ion-Exchange Columns for Ion Chromatography," *LCGC North Am.* **2013**, *31* (April Supplement 4B), 16.

12. B. K. Ng, G. W. Dicinoski, R. A. Shellie, and P. R. Haddad, "Tools for Simulation and Optimization of Separations in Ion Chromatography," *LCGC North Am.* **2013**, *31* (April Supplement 4B), 33; J. E. Madden, M. J. Shaw, G. W. Dicinoski, N. Avdalovic, and P. R. Haddad, "Simulation and Optimization of Retention in Ion Chromatography Using Virtual Column 2 Software," *Anal. Chem.* **2002**, *74*, 6023.

13. T. Ehmann, C. Mantler, D. Jensen, and R. Neufang, "Monitoring the Quality of Ultra-Pure Water in Semiconductor Industry by Online Ion Chromatography," *Microchim. Acta* **2006**, *154*, 15; D. Z. Živojinović and L. V. Rajaković, "Application and Validation of Ion Chromatography for the Analysis of Power Plants Water: Analysis of Corrosion Anions in Conditioned Water-Steam Cycles," *Desalination* **2011**, *275*, 17.

14. T. Brinkmann, C. H. Specht, and F. H. Frimmel, "Non-Linear Calibration Functions in Ion Chromatography with Suppressed Conductivity Detection Using Hydroxide Eluents," *J. Chromatogr. A* **2002**, *957*, 99.

15. R. G. Brennan, T. A. Butler, and M. R. Winchester, "Achieving 0.2% Relative Expanded Uncertainty in Ion Chromatography Analysis Using a High-Performance Methodology," *Anal. Chem.* **2011**, *83*, 3801.

16. W. W. Buchberger, "Detection Techniques for Ion Chromatography of Inorganic Anions," *Trends Anal. Chem.* **2001**, *20*, 296; P. K. Dasgupta, H. Z. Liao, and C. P. Shelor, "Ion Chromatography Yesterday and Today: Detection," *LCGC North Am.* **2013**, *31* (April Supplement 4B), 23.

17. B. C. Yang, Y. J. Chen, M. Mori, S.-I. Ohira, A. K. Azad, P. K. Dasgupta, and K. Srinivasan, "Charge Detector for the Measurement of Ionic Solutes," *Anal. Chem.* **2010**, *82*, 951.

18. B. Buszewski, *Ion-Pair Chromatography and Related Techniques* (London: CRC Press, 2009); J. W. Dolan, "Ion Pairing—Blessing or Curse?" *LCGC North Am.* **2008**, *26*, 170.

19. M. A. Hervas and C. E. Fabara, "A Simple Demonstration of the Ion-Pairing Effect on the Solubility of Charged Molecules," *J. Chem. Ed.* **1995**, *72*, 437.

20. M. R. Bresler and J. P. Hogen, "Surfactant Adsorption: A Revised Physical Chemistry Lab," *J. Chem. Ed.* **2008**, *85*, 269; P. C. Schulz and D. Clausse, "An Undergraduate Physical Chemistry Experiment on Surfactants: Electrochemical Study of a Commercial Soap," *J. Chem. Ed.* **2003**, *80*, 1053; A. Domínguez, A. Fernández, N. González, E. Iglesias, and L. Montenegro, "Determination of Critical Micelle Concentration of Some Surfactants by Three Techniques," *J. Chem. Ed.* **1997**, *74*, 1227; K. R. Williams and L. H. Tennant, "Micelles in the Physical/Analytical Chemistry Laboratory. Acid Dissociation of Neutral Red Indicator," *J. Chem. Ed.* **2001**, *78*, 349; S. A. Tucker, V. L. Amszi, and W. E. Acree, Jr., "Studying Acid-Base Equilibria in Two-Phase Solvent Media," *J. Chem. Ed.* **1993**, *70*, 80.

21. L. R. Snyder, J. J. Kirkland, and J. L. Glajch, *Practical HPLC Method Development* (New York: Wiley, 1997); J. W. Dolan, "Improving an Ion-Pairing Method," *LCGC North Am.* **1996**, *14*, 768.

22. Agilent Technologies, *An Introduction to Gel Permeation Chromatography and Size Exclusion Chromatography* (publication no. 5990-6969EN, 2014), available online at no charge; GE Healthcare, *Gel Filtration: Principles and Methods* (publication no. 18-1022-18 AK, 2010); A. M. Striegel, W. W. Yau, J. J. Kirkland, and D. D. Bly, *Modern Size-Exclusion Liquid Chromatography*, 2nd ed. (Hoboken, NJ: Wiley, 2009); G. A. Marson and B. Baptista, "Principles of Gel Permeation Chromatography: Interactive Software," *J. Chem. Ed.* **2006**, *83*, 1567.

23. H. G. Barth and G. D. Saunders, "State of the Art and Future Trends of Size-Exclusion Chromatography Packings and Columns," *LCGC North Am.* **2012**, *30*, 544; C.-S. Wu, ed. *Handbook of Size Exclusion Chromatography*, 2nd ed. (New York: Marcel Dekker, 2004).

24. GE Healthcare, *Affinity Chromatography: Principles and Methods* (publication no. 18-1022-29 AE, 2007), available online at no charge.

25. X. M. Fang and W.-W. Zhang, "Affinity Separation and Enrichment Methods in Proteomic Analysis," *J. Proteomics* **2008**, *71*, 284.

26. T. K. Nevanen, H. Simolin, T. Suortti, A. Koivula, and H. Söderlund, "Development of a High-Throughput Format for Solid-Phase Extraction of Enantiomers Using an Immunosorbent in 384-Well Plates," *Anal. Chem.* **2005**, *77*, 3038.

27. Y. L. Hu, J. L. Pan, K. G. Zhang, H. X. Lian, and G. K. Li, "Novel Applications of Molecularly-Imprinted Polymers in Sample Preparation," *Trends Anal. Chem.* **2013**, *43*, 37.

28. Q. Zhao, X.-F. Li, and X. C. Le, "Aptamer-Modified Monolithic Capillary Chromatography for Protein Separation and Detection," *Anal. Chem.* **2008**, *80*, 3915.

29. K. K. S. Lau, J. Bico, K. B. K. Teo, M. Chhowalla, G. A. J. Amaratunga, W. I. Milne, G. H. McKinley, and K. K. Gleason, "Superhydrophobic Carbon Nanotube Forests," *Nano Lett.* **2003**, *3*, 1701.

30. GE Healthcare, *Hydrophobic Interaction and Reversed Phase Chromatography: Principles and Methods* (publication no. 11-0012-69 AA, 2006), available online at no charge.

31. H. H. Lauer and G. P. Rozing, *High Performance Capillary Electrophoresis*, 2nd ed. (publication no. 5990-3777EN, Agilent, 2010), available online at no charge; R. Weinberger, *Practical Capillary Electrophoresis*, 2nd ed. (San Diego: Academic Press, 2000); P. Landers, ed., *Handbook of Capillary and Microchip Electrophoresis and Associated Microtechniques*, 3rd ed. (Boca Raton, FL: CRC Press, 2008).

32. L. A. Holland, "Capillary Electrophoresis: Focus on Undergraduate Laboratory Experiments," *J. Chem. Ed.* **2011**, *88*, 254; M. M. Bushey, "Capillary Electrophoresis and High-Performance Liquid Chromatography Experiments Throughout the Undergraduate Curriculum," *J. Chem. Ed.* **2009**, *86*, 332 (and Supplementary Materials); C. L. Cooper, "Capillary Electrophoresis: Theoretical and Experimental Background," *J. Chem. Ed.* **1998**, *75*, 343.

33. J. Kemsley, "Heparin Undone," *Chem. Eng. News*, 12 May 2008, p. 38.

34. M. Guerrini, D. Beccati, Z. Shriver, A. Naggi, K. Viswanathan, A. Bisio, I. Capila, J. C. Lansing, S. Guglieri, B. Fraser, A. Al-Hakim, N. S. Gunay, Z. Q. Zhang, L. Robinson, L. Buhse, M. Nasr, J. Woodcock, R. Langer, G. Venkataraman, R. J. Linhardt, B. Casu, G. Torri, and R. Sasisekharan, "Oversulfated Chondroitin Sulfate Is a Contaminant in Heparin Associated with Adverse Clinical Events," *Nature Biotech.* **2008**, *26*, 669. Liquid chromatographic methods for routine heparin analysis have since been perfected: S. Beni, J. F. K. Limtiaco, and C. K. Larive, "Analysis and Characterization of Heparin Impurities," *Anal. Bioanal. Chem.* **2011**, *399*, 527.

35. J. W. Jorgenson and K. D. Lukacs, "Zone Electrophoresis in Open Tubular Glass Capillaries," *Anal. Chem.* **1981**, *53*, 1298.

36. For a demonstration of electrophoresis, see J. G. Ibanez, M. M. Singh, R. M. Pike, and Z. Szafran, "Microscale Electrokinetic Processing of Soils," *J. Chem. Ed.* **1998**, *75*, 634.

37. A. L. Hogerton and M. T. Bowser, "Monitoring Neurochemical Release from Astrocytes Using In Vitro Microdialysis Coupled with High-Speed Capillary Electrophoresis," *Anal. Chem.* **2013**, *85*, 9070; F.-M. Matysik, "Advances in Fast Electrophoretic Separations Based on Short Capillaries," *Anal. Bioanal. Chem.* **2010**, *397*, 961.

38. C. Cecala and J. V. Sweedler, "Sampling Techniques for Single-Cell Electrophoresis," *Analyst* **2012**, *137*, 2922; G. G. Wolken and E. A. Arriaga,

"Simultaneous Measurement of Individual Mitochondrial Membrane Potential and Electrophoretic Mobility by Capillary Electrophoresis," *Anal. Chem.* **2014**, *86*, 4217; A. J. Dickinson, P. M. Armistead, and N. L. Allbritton, "Automated Capillary Electrophoresis System for Fast-Single Cell Analysis," *Anal. Chem.* **2013**, *85*, 4797.

39. J. N. Rauch, J. Nie, T. J. Buchholz, J. E. Gestwicki, and R. T. Kennedy, "Development of a Capillary Electrophoresis Platform for Identifying Inhibitors of Protein-Protein Interactions," *Anal. Chem.* **2013**, *85*, 9824; D. B. Craig and L. N. Chase, "Arrhenius Plot for a Reaction Catalyzed by a Single Molecule of β-Galactosidase," *Anal. Chem.* **2012**, *84*, 2044; J. Y. Bao, S. M. Krylova, D. J. Wilson, O. Reinstein, P. E. Johnson, and S. N. Krylov, "Kinetic Capillary Electrophoresis with Mass-Spectrometry Detection Facilitates Label-Free Solution-Based Kinetic Analysis of Protein-Small Molecule Binding," *ChemBioChem* **2011**, *12*, 2551; J. Yang and M. T. Bowser, "Capillary Electrophoresis–SELEX Selection of Catalytic DNA Aptamers for a Small-Molecule Porphyrin Target," *Anal. Chem.* **2013**, *85*, 1525.

40. A. Rogacs and J. G. Santiago, "Temperature Effects on Electrophoresis," *Anal. Chem.* **2013**, *85*, 5103; C. J. Evenhuis, M. U. Musheev, and S. N. Krylov, "Universal Method for Determining Electrolyte Temperatures in Capillary Electrophoresis," *Anal. Chem.* **2011**, *83*, 1808; C. J. Evenhuis, R. M. Guijt, M. Macka, P. J. Marriott, and P. R. Haddad, "Temperature Profiles and Heat Dissipation in Capillary Electrophoresis," *Anal. Chem.* **2006**, *78*, 2684.

41. E. Fuguet, C. Ràfols, E. Bosch, and M. Rosés, "Fast High-Throughput Method for Determination of Acidity Constants by Capillary Electrophoresis. I. Monoprotic Weak Acids and Bases," *J. Chromatogr. A* **2009**, *1216*, 3646.

42. C. T. Culbertson and J. W. Jorgenson, "Flow Counterbalanced Capillary Electrophoresis," *Anal. Chem.* **1994**, *66*, 955.

43. J. E. Gómez and J. E. Sandoval, "The Effect of Conditioning of Fused-Silica Capillaries on their Electrophoretic Performance," *Electrophoresis* **2008**, *29*, 381; R. Weinberger, "Capillary Conditioning Revisited," *Am. Lab.*, August 2008, p. 28.

44. R. Weinberger, "Capillary Electrophoresis," *Am. Lab.*, April 2005, p. 25.

45. H. Whatley, "Making CE Work—Points to Consider," *LCGC* **1999**, *17*, 426.

46. M. Macka, P. Andersson, and P. R. Haddad, "Changes in Electrolyte pH Due to Electrolysis During Capillary Zone Electrophoresis," *Anal. Chem.* **1998**, *70*, 743.

47. N. E. Baryla, J. E. Melanson, M. T. McDermott, and C. A. Lucy, "Characterization of Surfactant Coatings in Capillary Electrophoresis by Atomic Force Microscopy," *Anal. Chem.* **2001**, *73*, 4558; M. M. Yassine and C. A. Lucy, "Enhanced Stability Self-Assembled Coatings for Protein Separations by Capillary Zone Electrophoresis Through the Use of Long-Chained Surfactants," *Anal. Chem.* **2005**, *77*, 620.

48. S. L. Simpson Jr., J. P. Quirino, and S. Terabe, "On-Line Preconcentration in Capillary Electrophoresis: Fundamentals and Applications," *J. Chromatogr. A* **2008**, *1184*, 504.

49. At a wavelength of 190 nm, phosphate at pH 7.2 has about three times the absorbance of borate at pH 9.2. Glycine, citrate, HEPES, and TRIS buffers have significant absorption near 210 nm. Borate buffer should be prepared from sodium tetraborate (borax, $Na_2B_4O_7 \cdot 10H_2O$), not from boric acid ($B(OH)_3$), which has somewhat different acid-base chemistry. An ultraviolet-absorbing impurity in borate buffer can be removed by passage through a C_{18} solid-phase extraction column.

50. B. J. de Kort, G. J. de Jong, and G. W. Somsen, "Native Fluorescence Detection of Biomolecular and Pharmaceutical Compounds in Capillary Electrophoresis: Detector Designs, Performance and Applications: A Review," *Anal. Chim. Acta* **2013**, *766*, 13.

51. O. O. Dada, D. C. Essaka, O. Hindsgaul, M. M. Palcic, J. Prendergast, R. L. Schnaar, and N. J. Dovichi, "Nine Orders of Magnitude Dynamic Range: Picomolar to Millimolar Concentration Measurement in Capillary Electrophoresis with Laser Induced Fluorescence Detection Employing Cascaded Avalanche Photodiode Photon Counters," *Anal. Chem.* **2011**, *83*, 2748.

52. J. J. P. Mark, R. Scholz, and F.-M. Matysik, "Electrochemical Methods in Conjunction with Capillary and Microchip Electrophoresis," *J. Chromatogr. A* **2012**, *1267*, 45.

53. E. J. Maxwell and D. D. Y. Chen, "Twenty Years of Interface Development for Capillary Electrophoresis-Electrospray Ionization–Mass Spectrometry," *Anal. Chim. Acta* **2008**, *627*, 25.

54. S. Terabe, "Capillary Separation: Micellar Electrokinetic Chromatography," *Ann. Rev. Anal. Chem.* **2009**, *2*, 99.

55. S. Terabe, K. Otsuka, K. Ichikawa, A. Tsuchiya, and T. Ando, "Electrokinetic Separations with Micellar Solutions and Open Tubular Capillaries," *Anal. Chem.* **1984**, *56*, 111; ibid., **1985**, *57*, 834.

56. Demonstrations: N. Gani and J. Khanam, "Are Surfactant Molecules Really Oriented in the Interface?" *J. Chem. Ed.* **2002**, *79*, 332; C. J. Marzzacco, "The Effect of SDS Micelle on the Rate of a Reaction," *J. Chem. Ed.* **1992**, *69*, 1024; C. P. Palmer, "Demonstrating Chemical and Analytical Concepts Using Electrophoresis and Micellar Electrokinetic Chromatography," *J. Chem. Ed.* **1999**, *76*, 1542.

57. Polar organic solvents with electrolytes such as sodium *p*-toluenesulfonate are compatible with capillary electrophoresis. Background electrolyte need not be an aqueous solution: E. Kenndler, "A Critical Overview of Non-Aqueous Capillary Electrophoresis," *J. Chromatogr. A* **2014**, *1335*, 16; ibid., **2014**, *1335*, 31.

58. G. K. E. Scriba, "Fundamental Aspects of Chiral Electromigration Techniques and Application in Pharmaceutical and Biomedical Analysis," *J. Pharm. Biomed. Anal.* **2011**, *55*, 688; H. A. Lu and G. N. Chen, "Recent Advances of Enantioseparations in Capillary Electrophoresis and Capillary Electrochromatography," *Anal. Methods* **2011**, *3*, 488; G. K. E. Scriba and K. Altria, "Using Cyclodextrins to Achieve Chiral and Non-chiral Separations in Capillary Electrophoresis," *LC GC Eur.* **2009**, *22*, 420; S. Conradi, C. Vogt, and E. Rohde, "Separation of Enantiomeric Barbiturates by Capillary Electrophoresis Using a Cyclodextrin-Containing Run Buffer," *J. Chem. Ed.* **1997**, *74*, 1122.

59. E. Zubritsky, "How Analytical Chemists Saved the Human Genome Project," *Anal. Chem.* **2002**, *74*, 22A.

60. J. M. Butler, E. Buel, F. Crivellente, and B. R. McCord, "Forensic DNA Typing by Capillary Electrophoresis Using the ABI Prism 310 and 3100 Genetic Analyzers for STR Analysis," *Electrophoresis* **2004**, *25*, 1397.

61. Z. F. Zhu, J. J. Lu, and S. R. Liu, "Protein Separation by Capillary Gel Electrophoresis: A Review," *Anal. Chim. Acta* **2012**, *709*, 21.

62. C. Sänger-van de Griend, "Revival of Capillary Electrophoretic Techniques in the Pharmaceutical Industry," *LCGC North Am.* **2012**, *30*, 954.

63. R. Weinberger, "An Interview with Ira Lurie of the DEA," *Am. Lab.*, January 2005, p. 6; I. S. Lurie, P. A. Hays, A. E. Garcia, and S. Panicker, "Use of Dynamically Coated Capillaries for the Determination of Heroin, Basic Impurities, and Adulterants with Capillary Electrophoresis," *J. Chromatogr. A* **2004**, *1034*, 227.

64. R. Weinberger, "Method Development for Capillary Electrophoresis," *Am. Lab.*, March 2003, p. 54.

65. S. P. Porras, M. L. Riekkola, and E. Kenndler, "The Principles of Migration and Dispersion in Capillary Zone Electrophoresis in Nonaqueous Solvents," *Electrophoresis* **2003**, *24*, 1485; M. L. Riekkola, "Recent Advances in Nonaqueous Capillary Electrophoresis," *Electrophoresis* **2002**, *23*, 3865; F. Steiner and M. Hassel, "Non-aqueous Capillary Electrophoresis. A Versatile Completion of Electrophoretic Separation Techniques," *Electrophoresis* **2000**, *21*, 3994.

66. S. Orlandini, R. Gotti, and S. Furlanetto, "Multivariate Optimization of Capillary Electrophoresis Methods: A Critical Review," *J. Pharm. Biomed. Anal.* **2014**, *87*, 290.

67. T. N. Chiesl, W. K. Chu, A. M. Stockton, X. Amashukeli, F. Grunthaner, and R. A. Mathies, "Enhanced Amine and Amino Acid Analysis Using Pacific Blue and the Mars Organic Analyzer Microchip Capillary Electrophoresis System," *Anal. Chem.* **2009**, *81*, 2537.

68. P. C. H. Li, *Fundamentals of Microfluidics and Lab on a Chip for Biological Analysis and Discovery*, 2nd ed. (Boca Raton: CRC Press, 2010).

69. M. Hemling, J. A. Crooks, P. M. Oliver, K. Brenner, J. Gilbertson, G. C. Lisensky, and D. B. Weibel, "Microfluidics for High School Students," *J. Chem. Ed.* **2014**, *91*, 112; P. A. E. Piunno, A. Zetina, N. Chu, A. J. Tavares, M. O. Noor, E. Petryayeva, U. Uddayasandar, and A. Veglio, "A Comprehensive Microfluidics Device Construction and Characterization Module for the Advanced Undergraduate Analytical Chemistry Laboratory," *J. Chem. Ed.* **2014**, *91*, 902; C. W. T. Yang, E. Ouellet, and E. T. Lagally, "Using Inexpensive Jell-O Chips for Hands-On Microfluidics Education," *Anal. Chem.* **2010**, *82*, 5408.

70. G. M. Whitesides, "The Origins and the Future of Microfluidics," *Nature* **2006**, *442*, 368.

71. K. M. Elkins, "Designing Polymerase Chain Reaction (PCR) Primer Multiplexes in the Forensic Laboratory," *J. Chem. Ed.* **2011**, *88*, 1422.

72. C. Hurth, J. Yang, M. Barrett, C. Brooks, A. Nordquist, S. Smith, and F. Zenhausern, "A Miniature Quantitative PCR Device for Directly Monitoring a Sample Processing on a Microfluidic Rapid DNA System," *Biomed. Microdevices* **2014**, *16*, 905.

73. W. Jung, J. Yang, M. Barrett, B. Duane, C. Brooks, C. Hurth, A. Nordquist, S. Smith, and F. Zenhausern, "Recent Improvement in Miniaturization and Integration of a DNA Analysis System for Rapid Forensic Analysis (MiDAS)," *J. Forensic Invest.* **2014**, *2*, 7.

74. N. D. Chasteen and P. M. Harrison, "Mineralization in Ferritin: An Efficient Means of Iron Storage," *J. Struct. Biol.* **1999**, *126*, 182.

75. Data from A. W. Moore, Jr., S. C. Jacobson, and J. M. Ramsey, "Microchip Separations of Neutral Species via Micellar Electrokinetic Capillary Chromatography," *Anal. Chem.* **1995**, *67*, 4184.

76. G. R. Asbury and H. H. Hill, Jr., "Evaluation of Ultrahigh Resolution Ion Mobility Spectrometry as an Analytical Separation Device in Chromatographic Terms," *J. Microcolumn Sep.* **2000**, *12*, 172; H. E. Revercomb and E. A. Mason, "Theory of Plasma Chromatography/Gaseous Electrophoresis," *Anal. Chem.* **1975**, *47*, 970.

Chapter 27

1. T. W. Richards, "Atomic Weights and Isotopes," *Chem. Rev.* **1925**, *1*, 1; H. Hartley, "Theodore William Richards Memorial Lecture," *J. Chem. Soc.* **1930**, 1937; C. M. Beck, II, "Classical Analysis: A Look at the Past, Present, and Future," *Anal. Chem.* **1994**, *66*, 225A; I. M. Kolthoff, "Analytical Chemistry in the U.S.A. in the First Quarter of This Century," *Anal. Chem.* **1994**, *66*, 241A; D. T. Burns, "Highlights in the History of Quantitation in Chemistry," *Fresenius J. Anal. Chem.* **1990**, *337*, 205; L. Niiniströ, "Analytical Instrumentation in the 18th Century," *Fresenius J. Anal. Chem.* **1990**, *337*, 213.

2. W. Wacławek and M. Wacławek, "Marie Składowska-Curie and Her Contributions of Chemistry, Radiochemistry, and Radiotherapy," *Anal. Bioanal. Chem.* **2011**, *400*, 1567.

3. Colloid experiment: C. D. Keating, M. D. Musick, M. H. Keefe, and M. J. Natan, "Kinetics and Thermodynamics of Au Colloid Monolayer Self Assembly," *J. Chem. Ed.* **1999**, *76*, 949.

4. Cellulose tubing, such as catalog number 3787 sold by Thomas Scientific, www.thomassci.com, is adequate for this demonstration.

5. M. Suzuki, "The Movement of Molecules and Heat Energy: Two Demonstrative Experiments," *J. Chem. Ed.* **1993**, *70*, 821.

6. D. Erdemir, A. Y. Lee, and A. S. Myerson, "Nucleation of Crystals from Solution: Classical and Two-Step Models," *Acc. Chem. Res.* **2009**, *42*, 621.

7. A. Abou-Hassan, O. Sandre, S. Neveu, and V. Cabuil, "Synthesis of Goethite by Separation of the Nucleation and Growth Processes of Ferrihydrite Nanoparticles Using Microfluidics," *Angew. Chem. Int. Ed.* **2009**, *48*, 2342.

8. D. Li, M. H. Nielsen, J. R. I. Lee, C. Frandsen, J. F. Banfield, and J. J. De Yoreo, "Direction-Specific Interactions Control Crystal Growth by Oriented Attachment," *Science* **2012**, *336*, 1014.

9. M. H. Nielsen, S. Aloni, and J. J. De Yoreo, "In Situ TEM Imaging of $CaCO_3$ Nucleation Reveals Coexistence of Direct and Indirect Pathways," *Science* **2014**, *345*, 1158. Video clips of nucleation in the supplementary material are available at no charge once you have registered at www.sciencemag.org. Movie S3 shows spontaneous transformation of a spherical particle into aragonite crystals.

10. M. Luo, X. Hou, C. He, Q. Liu, and Y. Fan, "Speciation Analysis of ^{129}I in Seawater by Carrier-Free AgI-AgCl Coprecipitation and Accelerator Mass Spectrometric Measurement," *Anal. Chem.* **2013**, *85*, 3715; J. Hou and P. Roos, "Method for Determination of Neptunium in Large-Sized Urine Samples Using Manganese Dioxide Coprecipitation and ^{242}Pu as Yield Tracer," *Anal. Chem.* **2013**, *85*, 1889; N. Guerin and X. Dai, "Rapid Preparation of Polonium Counting Sources for Alpha Spectrometry Using Copper Sulfide Microprecipitation," *Anal. Chem.* **2013**, *85*, 6524.

11. D. VanDorn, M. T. Ravalli, M. M. Small, B. Hillery, and S. Andreescu, "Adsorption of Arsenic by Iron Oxide Nanoparticles: A Versatile, Inquiry-Based Laboratory for a High School or College Science Course," *J. Chem. Ed.* **2011**, *88*, 1119; L. C. Roberts, S. J. Hug, T. Ruettimann, M. Billah, A. W. Khan, and M. T. Rahman, "Arsenic Removal with Iron(II) and Iron(III) Waters with High Silicate and Phosphate Concentrations," *Environ. Sci. Technol.* **2004**, *38*, 307. A more convenient method to remove arsenite and arsenate from drinking water is to pass the water through diatomaceous earth (silica skeletons of algae with 80–90% voids) containing hydrous ferric oxide. (M. Jang, S.-H. Min, T.-H. Kim, and J. K. Park, "Removal of Arsenite and Arsenate Using Hydrous Ferric Oxide Incorporated in Naturally Occurring Porous Diatomite," *Environ. Sci. Technol.* **2006**, *40*, 1636.)

12. G. W. Latimer, Jr., "Piperazine as the Diacetate," *J. Chem. Ed.* **1966**, *43*, 148; G. R. Bond, "Determination of Piperazine as Piperazine Diacetate," *Anal. Chem.* **1960**, *32*, 1332.

13. E. Pella, "Elemental Organic Analysis. 1. Historical Developments," *Am. Lab.*, February 1990, p. 116; "Elemental Organic Analysis. 2. State of the Art," *Am. Lab.*, August 1990, p. 28.

14. K. Russe, H. Kipphardt, and J. A. C. Broekaert, "Determination of Main and Minor Components of Silicon Based Materials by Combustion with F_2," *Anal. Chem.* **2000**, *72*, 3875.

15. R. W. Ramette, "Stoichiometry to the Rescue (A Calculation Challenge)," *J. Chem. Ed.* **1988**, *65*, 800.

16. G. Li, G. Su, X. Zhuang, Z. Li, and Y. He, "A New Method to Determine the Deuterium Content of DKDP Crystal with Thermo-Gravimetric Apparatus," *Opt. Mater.* **2006**, *29*, 220.

17. F. Torrades and M. Castellvi, "Spectrophotometric Determination of Cl^- in $BaSO_4$ Precipitate," *Fresenius J. Anal. Chem.* **1994**, *349*, 734.

Chapter 28

1. E. P. Popek, *Sampling and Analysis of Environmental Chemical Pollutants* (Amsterdam: Academic Press, 2003); P. Gy, *Sampling for Analytical Purposes* (Chichester, UK: Wiley, 1998); B. B. Kebbekus and S. Mitra, *Environmental Chemical Analysis* (London: Blackie, 1998); N. T. Crosby and I. Patel, *General Principles of Good Sampling Practice* (Cambridge: Royal Society of Chemistry, 1995).

2. J. Pawliszyn and H. L. Lord, eds., *Handbook of Sample Preparation* (Hoboken, NJ: Wiley, 2010); S. Mitra, ed., *Sample Preparation Techniques in Analytical Chemistry* (Hoboken, NJ: Wiley, 2003).

3. H. M. D. R. Herath, R.-R. Kim, P. J. Cabot, P. N. Shaw, and A. K. Hewavitharana, "Inaccuracies in the Quantification of Peptides—A Case Study Using β-Endorphin Assay," *LCGC North Am.* **2013**, *31*, 58.

4. D. T. Sawyer, A. Sobkowiak, and J. L. Roberts, Jr., *Electrochemistry for Chemists*, 2nd ed. (New York: Wiley, 1995), p. 262.

5. P. Vallelonga, K. Van de Velde, J. P. Candelone, C. Ly, K. J. R. Rosman, C. F. Boutron, V. I. Morgan, and D. J. Mackey, "Recent Advances in Measurement of Pb Isotopes in Polar Ice and Snow at Sub-Picogram per Gram Concentrations Using Thermal Ionisation Mass Spectrometry," *Anal. Chim. Acta* **2002**, *453*, 1.

6. R. P. Mason and K. A. Sullivan, "Mercury in Lake Michigan," *Environ. Sci. Technol.* **1997**, *31*, 942.

7. G. Benoit, K. S. Hunter, and T. F. Rozan, "Sources of Trace Metal Contamination Artifacts During Collection, Handling, and Analysis of Freshwaters," *Anal. Chem.* **1997**, *69*, 1006.

8. R. E. Thiers, *Methods of Biochemical Analysis* (D. Glick, ed.), Vol. 5 (New York: Interscience, 1957), p. 274.

9. B. Kratochvil and J. K. Taylor, "Sampling for Chemical Analysis," *Anal. Chem.* **1981**, *53*, 924A; S. L. Lohr, *Sampling: Design and Analysis*, 2nd ed. (Boston: Brooks/Cole, 2010); W. A. Fuller, *Sampling Statistics* (Hoboken, NJ: Wiley, 2009); H. A. Laitinen and W. E. Harris, *Chemical Analysis*, 2nd ed. (New York: McGraw-Hill, 1975), Chap. 27; S. K. Thompson, *Sampling* (New York: Wiley, 1992).

10. J. E. Vitt and R. C. Engstrom, "Effect of Sample Size on Sampling Error," *J. Chem. Ed.* **1999**, *76*, 99; R. D. Guy, L. Ramaley, and P. D. Wentzell, "Experiment in the Sampling of Solids for Chemical Analysis," *J. Chem. Ed.* **1998**, *75*, 1028. Demonstrations: L. S. Canaes, M. L. Brancalion,

A. V. Rossi, and S. Rath, "Using Candy Samples to Learn About Sampling Techniques and Statistical Data Evaluation," *J. Chem. Ed.* **2008**, *85*, 1083; M. R. Ross, "A Classroom Exercise in Sampling Technique," *J. Chem. Ed.* **2000**, *77*, 1015; J. R. Hartman, "In-Class Experiment on the Importance of Sampling Techniques and Statistical Analysis of Data," *J. Chem. Ed.* **2000**, *77*, 1017.

11. D. C. Bogen, *Treatise on Analytical Chemistry*, 2nd ed. (P. J. Elving, E. Grushka, and I. M. Kolthoff, eds.), Part I, Vol. 5 (New York: Wiley, 1982), Chap. 1.

12. H. M. Kingston and S. J. Haswell, *Microwave-Enhanced Chemistry* (Washington, DC: American Chemical Society, 1997); B. D. Zehr, "Development of Inorganic Microwave Dissolutions," *Am. Lab.*, December 1992, p. 24; B. D. Zehr, J. P. VanKuren, and H. M. McMahon, "Inorganic Microwave Digestions Incorporating Bases," *Anal. Chem.* **1994**, *66*, 2194.

13. Z. Grosser, L. Thompson, and L. Davidowski, "Inorganic Analysis for Environmental RoHS Compliance," *Am. Lab.*, October 2007, p. 30.

14. E. M. M. Flores, M. F. Mesko, D. P. Moraes, J. S. F. Pereira, P. A. Mello, J. S. Barin, and G. Knapp, "Determination of Halogens in Coal after Digestion using the Microwave-Induced Combustion Technique," *Anal. Chem.* **2008**, *80*, 1865.

15. M. F. Mesko, J. S. F. Pereira, D. P. Moraes, J. S. Barin, P. A. Mello, J. N. G. Paniz, J. A. Nóbrega, M. G. A. Korn, and E. M. M. Flores, "Focused Microwave-Induced Combustion: A New Technique for Sample Digestion," *Anal. Chem.* **2010**, *82*, 2155.

16. P. Aysola, P. Anderson, and C. H. Langford, "Wet Ashing in Biological Samples in a Microwave Oven Under Pressure Using Poly(tetrafluoroethylene) Vessels," *Anal. Chem.* **1987**, *59*, 1582.

17. S. E. Long and W. R. Kelly, "Determination of Mercury in Coal by Isotope Dilution Cold-Vapor Generation Inductively Coupled Plasma Mass Spectrometry," *Anal. Chem.* **2002**, *74*, 1477.

18. M. Rehkämper, A. N. Halliday, and R. F. Wentz, "Low-Blank Digestion of Geological Samples for Pt-Group Analysis Using a Modified Carius Tube Design," *Fresenius J. Anal. Chem.* **1998**, *361*, 217.

19. A. A. Schilt, *Perchloric Acid and Perchlorates*, 2nd ed. (Powell, OH: GFS Chemicals, 2003).

20. C. Walling, "Intermediates in the Reactions of Fenton Type Reagents," *Acc. Chem. Res.* **1998**, *31*, 155; P. A. MacFaul, D. D. M. Wayner, and K. U. Ingold, "A Radical Account of 'Oxygenated Fenton Chemistry,'" *Acc. Chem. Res.* **1998**, *31*, 159; D. T. Sawyer, A. Sobkowiak, and T. Matsushita, "Oxygenated Fenton Chemistry," *Acc. Chem. Res.* **1996**, *29*, 409; O. Pestovsky, S. Stoian, E. L. Bominaar, X. Shan, E. Münck, L. Que, Jr., and A. Bakac, "Aqueous Fe^{IV}=O: Spectroscopic Identification and Oxo-Group Exchange," *Angew. Chem. Int. Ed.* **2005**, *44*, 6871.

21. D. C. Luehrs and A. E. Roher, "Demonstration of the Fenton Reaction," *J. Chem. Ed.* **2007**, *84*, 1290.

22. L. Ping and P. K. Dasgupta, "Determination of Total Mercury in Water and Urine by a Gold Film Sensor Following Fenton's Reagent Digestion," *Anal. Chem.* **1989**, *61*, 1230.

23. G. LeBlanc, "Microwave-Accelerated Techniques for Solid Sample Extraction," *LCGC* **1999**, *17*, S30 (June 1999 supplement).

24. L. T. Taylor, *Supercritical Fluid Extraction* (New York: Wiley, 1996); C. L. Phelps, N. G. Smart, and C. M. Wai, "Past, Present, and Possible Future Applications of Supercritical Fluid Extraction Technology," *J. Chem. Ed.* **1996**, *73*, 1163; M. E. P. McNally, "Advances in Environmental SFE," *Anal. Chem.* **1995**, *67*, 308A; L. T. Taylor, "Strategies for Analytical SFE," *Anal. Chem.* **1995**, *67*, 364A.

25. Y. Lin and C. M. Wai, "Supercritical Fluid Extraction of Lanthanides with Fluorinated β-Diketones and Tributyl Phosphate," *Anal. Chem.* **1994**, *66*, 1971.

26. S. Liang and D. C. Tilotta, "Extraction of Petroleum Hydrocarbons from Soil Using Supercritical Argon," *Anal. Chem.* **1998**, *70*, 616.

27. H. M. Stapleton, N. G. Dodder, J. H. Offenberg, M. M. Schantz, and S. A. Wise, "Polybrominated Diphenyl Ethers in House Dust and Clothes Dryer Lint," *Environ. Sci. Technol.* **2005**, *39*, 925. "Best Papers of 2005," *Environ. Sci. Technol.* **2006**, *40*, 2087.

28. M. Saraji and M. K. Boroujeni, "Recent Developments in Dispersive Liquid-Liquid Microextraction," *Anal. Bioanal. Chem.* **2014**, *406*, 2027; P. Viñas, N. Campillo, I. López-Garcia, and M. Hernández-Córdoba, "Dispersive Liquid-Liquid Microextraction in Food Analysis. A Critical Review," *Anal. Bioanal. Chem.* **2014**, *406*, 2067.

29. www.biotage.com; L. Williams, H. Lodder, S. Merriman, A. Howells, S. Jordan, J. Labadie, M. Cleeve, C. Desbrow, R. Calverley, and M. Burke, "Extraction of Drugs from Plasma using ISOLUTE SLE+ Supported Liquid Extraction Plates," *LCGC Application Notebook*, February 2006, p. 12.

30. J. C. Arsenault, *Beginner's Guide to SPE: Solid-Phase Extraction* (Milford, MA: Waters Corp., 2012); M. J. Telepchak, T. F. August, and G. Chaney, *Forensic and Clinical Applications of Solid Phase Extraction* (Totowa, NJ: Humana Press, 2004).

31. J.-H. Zhang, M. Jiang, L. Zou, D. Shi, S.-R. Mei, Y.-X. Zhu, Y. Shi, K. Dai, and B. Lu, "Selective Solid-Phase Extraction of Bisphenol A Using Molecularly Imprinted Polymers and Its Application to Biological and Environmental Samples," *Anal. Bioanal. Chem.* **2006**, *385*, 780.

32. R. Hites and V. S. Ong, "Determination of Pesticides and Polychlorinated Biphenyls in Water: A Low-Solvent Method," *Environ. Sci. Technol.* **1995**, *29*, 1259.

33. M. C. Bruzzoniti, L. Checchini, R. M. De Carlo, S. Orlandini, L. Rivoira, and M. Del Bubba, "QuEChERS Sample Preparation for the Determination of Pesticides and Other Organic Residues in Environmental Matrices: A Critical Review," *Anal. Bioanal. Chem.* **2014**, *406*, 4089.

34. J. W. Wong, K. K. Ngim, T. Shibamoto, S. A. Mabury, J. P. Eiserich, and H. C. H. Yeo, "Determination of Formaldehyde in Cigarette Smoke," *J. Chem. Ed.* **1997**, *74*, 1100. As another example, an analysis of nitrite (NO_2^-) in natural waters is based on its reaction with 2,4-dinitrophenylhydrazine to give an azide ($R-N_3$) that is measured by HPLC with ultraviolet detection at 307 nm. (R. J. Kieber and P. J. Seaton, "Determination of Subnanomolar Concentrations of Nitrite in Natural Waters," *Anal. Chem.* **1995**, *67*, 3261.)

35. D. M. Crumbling, Letter to Editor, *C&E News,* 12 August 2002, p. 4.

36. E. Tareke, P. Rydberg, P. Karlsson, S. Eriksson, and M. Tornqvist, "Analysis of Acrylamide, A Carcinogen Formed in Heated Foodstuffs," *J. Agric. Food Chem.* **2002**, *50*, 4998; B. E. Erickson, "Finding Acrylamide," *Anal. Chem.* **2004**, *76*, 247A. See collection of papers in *J. AOAC Int.*, **2005**, *88*, 227.

GLOSSARY

ablation Vaporization of a small volume of material by a laser pulse.

abscissa Horizontal (x) axis of a graph.

absolute uncertainty An expression of the margin of uncertainty associated with a measurement. Absolute error also could refer to the difference between a measured value and the "true" value.

absorbance, A Defined as $A = \log(P_0/P)$, where P_0 is the radiant power of light (power per unit area) striking the sample on one side and P is the radiant power emerging from the other side. Also called *optical density*.

absorptance, a Fraction of incident radiant power absorbed by a sample.

absorption Occurs when a substance is taken up *inside* another. See also *adsorption*.

absorption coefficient Light absorbed by a sample is attenuated at the rate $P_2/P_1 = e^{-\alpha b}$, where P_1 is the initial radiant power, P_2 is the power after traversing a pathlength b, and α is called the absorption coefficient.

absorption spectrum A graph of absorbance of light versus wavelength, frequency, or wavenumber.

accuracy A measure of how close a measured value is to the "true" value.

acid A substance that increases the concentration of H^+ when added to water.

acid-base titration One in which the reaction between analyte and titrant is an acid-base reaction.

acid dissociation constant, K_a Equilibrium constant for the reaction of an acid, HA, with H_2O:

$$HA + H_2O \overset{K_a}{\rightleftharpoons} A^- + H_3O^+ \qquad K_a = \frac{\mathcal{A}_{A^-} \cdot \mathcal{A}_{H_3O^+}}{\mathcal{A}_{HA}}$$

acid error Systematic error that occurs in measuring the pH of strongly acidic solutions, where glass electrodes tend to indicate a value of pH that is too high.

acidic solution One in which the activity of H^+ is greater than the activity of OH^-.

acidity In natural waters, the quantity of carbonic acid and other dissolved acids that react with strong base when the pH of the sample is raised to 8.3. Expressed as mmol OH^- needed to raise the pH of 1 L to pH 8.3.

acid wash Procedure in which glassware is soaked in 3–6 M HCl for >1 h (followed by rinsing well with distilled water and soaking in distilled water) to remove traces of cations adsorbed on the surface of the glass and to replace them with H^+.

activation energy, E_a Energy needed for a process to overcome a barrier that otherwise prevents the process from occurring.

activity, $\mathcal{A}$ The value that replaces concentration in a thermodynamically correct equilibrium expression. The activity of X is given by $\mathcal{A}_X = [X]\gamma_X$, where γ_X is the activity coefficient and $[X]$ is the concentration.

activity coefficient, γ The number by which the concentration must be multiplied to give activity.

adduct Product formed when a Lewis base combines with a Lewis acid.

adjusted retention time t_r' In chromatography, this parameter is $t_r' = t_r - t_m$, where t_r is the retention time of a solute and t_m is the time needed for mobile phase to travel the length of the column.

adsorption Occurs when a substance becomes attached to the *surface* of another substance. See also *absorption*.

adsorption chromatography A technique in which solute equilibrates between the mobile phase and adsorption sites on the stationary phase.

adsorption indicator Used for precipitation titrations, it becomes attached to a precipitate and changes color when the surface charge of the precipitate changes sign at the equivalence point.

aerosol A suspension of very small liquid or solid particles in air or gas. Examples include fog and smoke.

affinity chromatography A technique in which a particular solute is retained by a column by virtue of a specific interaction with a molecule covalently bound to the stationary phase.

aliquot Portion.

alkali flame detector Modified flame ionization detector that responds to N and P, which produce ions when they contact a Rb_2SO_4-containing glass bead in the flame. Also called *nitrogen-phosphorus detector*.

alkalimetric titration With reference to EDTA, this technique involves titration of the protons liberated from EDTA upon binding to a metal.

alkaline error See *sodium error*.

alkalinity In natural water, the quantity of base (mainly HCO_3^-, CO_3^{2-}, and OH^-) that reacts with strong acid when the pH of the sample is lowered to 4.5. Expressed as mmol H^+ needed to lower the pH of 1 L to pH 4.5.

amalgam A solution of anything in mercury.

ambient temperature The temperature of the surroundings (such as room temperature).

amine A compound with the general formula RNH_2, R_2NH, or R_3N, where R is any group of atoms.

amino acid One of 20 building blocks of proteins, having the general structure

$$H_3\overset{+}{N}-\overset{\overset{\displaystyle R}{|}}{C}H-CO_2^-$$

where R is a different substituent for each acid.

ammeter Device for measuring electric current.

ammonium ion *The* ammonium ion is NH_4^+. *An* ammonium ion is any ion of the type RNH_3^+, $R_2NH_2^+$, R_3NH^+, or R_4N^+, where R is an organic substituent.

ampere, A SI unit of electric current. One ampere will produce a force of exactly 2×10^{-7} N/m when that current flows through two "infinitely" long, parallel conductors of negligible cross section, with a spacing of 1 m, in a vacuum. One ampere is a flow of one coulomb per second.

amperometric detector See *electrochemical detector*.

amperometric titration One in which the end point is determined by monitoring the current passing between two electrodes immersed in the sample solution and maintained at a constant potential difference.

amperometry Measurement of electric current for analytical purposes.

amphiprotic molecule One that can act as both a proton donor and a proton acceptor. The intermediate species of polyprotic acids are amphiprotic.

analysis of variance A statistical tool used to compare means and to dissect the overall random error in a process into contributions from its component sources.

analyte Substance being measured or detected.

analytical chromatography Chromatography of small quantities of material conducted for the purpose of qualitative or quantitative analysis or both.

analytical concentration See *formal concentration*.

anhydrous Adjective describing a substance from which all water has been removed.

anion A negatively charged ion.

anion exchanger An ion exchanger with positively charged groups covalently attached to the support. It can reversibly bind anions.

anode Electrode at which oxidation occurs. In electrophoresis, it is the positively charged electrode.

anodic depolarizer A molecule that is easily oxidized, thereby preventing the anode potential of an electrochemical cell from becoming too positive.

anodic wave   In voltammetry, a flow of current due to oxidation at the working electrode.

anolyte   Solution present in the anode chamber of an electrochemical cell.

antibody   A protein manufactured by an organism to sequester foreign molecules and mark them for destruction.

antigen   A molecule that is foreign to an organism and causes the organism to make antibodies.

antilogarithm   The antilogarithm of a is b if $10^a = b$.

apparent mobility   Constant of proportionality, μ_{app}, between the net speed, u_{net}, of an ion in solution and the applied electric field, E: $u_{net} = \mu_{app}E$. Apparent mobility is the sum of the *electrophoretic* and *electroosmotic mobilities*: $\mu_{app} = \mu_{ep} + \mu_{eo}$.

aprotic solvent   One that cannot donate protons (hydrogen ions) in an acid-base reaction.

aptamer   A short (15–40 bases) length of single- or double-stranded DNA (deoxyribonucleic acid) or RNA (ribonucleic acid) that strongly binds to a selected molecule.

aqua regia   A $3:1$ (vol/vol) mixture of concentrated (37 wt%) HCl and concentrated (70 wt%) HNO_3.

aqueous   In water (as an *aqueous* solution).

aquo ion   The species $M(H_2O)_n^{m+}$, containing just the cation M and its tightly bound water ligands.

argentometric titration   One using Ag^+ ion.

ashless filter paper   Specially treated paper that leaves negligible residue (such as 0.01 wt%) after ignition. It is used for gravimetric analysis. The filter is made from refined pulp and and is acid washed to remove impurities.

assessment   In quality assurance, the process of (1) collecting data to show that analytical procedures are operating within specified limits and (2) verifying that final results meet use objectives.

asymmetry factor, *B/A*   In chromatography, the parameter describing the shape of a peak. B is the distance from the peak apex to the back of the peak measured at 10% of peak height. A is the distance from the peak apex to the front of the peak at 10% of peak height. $B/A = 1$ is a symmetrical peak. $B/A > 1$ is tailing and $B/A < 1$ is fronting

asymmetry potential   When the activity of analyte is the same on the inside and outside of an ion-selective electrode, there should be no voltage across the membrane. In fact, the two surfaces are never identical, and some voltage (called the asymmetry potential) is usually observed.

atmosphere, atm   One atm is defined as a pressure of 101 325 Pa. It is equal to the pressure exerted by a column of Hg 760 mm in height at Earth's surface.

atmospheric pressure chemical ionization   A method to interface liquid chromatography to mass spectrometry. Liquid is nebulized into a fine aerosol by a coaxial gas flow and the application of heat. Electrons from a high-voltage corona discharge create cations and anions from analyte exiting the chromatography column. The most common species observed with this interface is MH^+, the protonated analyte, with little fragmentation.

atomic absorption spectroscopy   A technique in which the absorption of light by free gaseous atoms in a flame or furnace is used to measure the concentration of atoms.

atomic emission spectroscopy   A technique in which the emission of light by thermally excited atoms in a plasma, flame, or furnace is used to measure the concentration of atoms.

atomic fluorescence spectroscopy   A technique in which electronic transitions of atoms in a flame, furnace, or plasma are excited by light, and the fluorescence is observed at a right angle to the incident beam.

atomic mass   Number of grams of an element containing Avogadro's number of atoms. For elements with multiple stable isotopes, the atomic mass is a weighted average reflecting the abundance of the different isotopes.

atomization   Process in which a compound is decomposed into its atoms at high temperature.

attenuated total reflection   An analytical technique based on passage of light through a waveguide or optical fiber by total internal reflection. The absorption of the cladding is sensitive to the presence of analyte. Some of the evanescent wave is absorbed in the cladding during each reflection in the presence of analyte. The more analyte, the more signal is lost.

autoprotolysis   The reaction in which two molecules of the same species transfer a proton from one to the other; e.g., $CH_3OH + CH_3OH \rightleftharpoons CH_3OH_2^+ + CH_3O^-$. Also called *self-ionization*.

autoprotolysis constant   The equilibrium constant for an autoprotolysis reaction.

autotitrator   A device that dispenses measured amounts of titrant into a solution and monitors a property such as pH or electrode potential after each addition. The instrument performs the titration automatically and can determine the end point automatically. Data can be transferred to a spreadsheet for further interpretation.

auxiliary complexing agent   A species, such as ammonia, used to stabilize a metal ion and keep that metal in solution. It binds loosely enough to be displaced by a titrant.

auxiliary electrode   Current-carrying partner of the working electrode in an electrolysis. Also called *counter electrode*.

average   The sum of several values divided by the number of values. Also called *mean*.

Avogadro's number   The number of atoms in exactly 0.012 kg of ^{12}C, approximately 6.022×10^{23}.

azeotrope   The constant-boiling distillate produced by two liquids. It is of constant composition, containing both substances.

background correction   In atomic spectroscopy, a means of distinguishing signal due to analyte from signal due to absorption, emission, or scattering by the flame, furnace, plasma, or sample matrix.

background electrolyte   In capillary electrophoresis, the buffer in which separation is carried out. Also called *run buffer*.

back titration   One in which an excess of standard reagent is added to react with analyte. Then the excess reagent is titrated with a second reagent or with a standard solution of analyte.

ball mill   A drum in which solid sample is ground to fine powder by tumbling with hard ceramic balls.

band gap   Energy separating the valence band and conduction band in a semiconductor.

band-pass filter   A filter that allows a band of wavelengths to pass through it while absorbing or reflecting other wavelengths.

bandwidth   Usually, the range of wavelengths or frequencies of an absorption or emission band, typically measured at a height equal to half of the peak height. Also, the width of radiation emerging from the exit slit of a monochromator.

base   A substance that decreases the concentration of H^+ when added to water.

base "dissociation" constant   A misnomer for *base hydrolysis constant, K_b*.

base hydrolysis constant, K_b   The equilibrium constant for the reaction of a base, B, with H_2O:

$$B + H_2O \overset{K_b}{\rightleftharpoons} BH^+ + OH^- \qquad K_b = \frac{\mathcal{A}_{BH^+}\mathcal{A}_{OH^-}}{\mathcal{A}_B}$$

base peak   Most intense peak in a mass spectrum.

baseline   In chromatography, the background signal in absence of solute is the baseline signal. Peak heights and peak area are measured relative to a baseline projected underneath the peak.

baseline resolution   Occurs in chromatography when two adjacent peaks are sufficiently resolved that the signal between the peaks returns to the baseline (resolution > 1.5). Allows determination of accurate peak areas.

basic solution   One in which the activity of OH^- is greater than the activity of H^+.

battery   A glavanic cell used to produce electricity by the chemical reaction between reagents contained within the battery. When the reagents are consumed, no further electricity can be produced.

beam chopper   A rotating mirror that directs light alternately through the sample and reference cells of a double-beam spectrophotometer.

beam chopping A technique using a rotating *beam chopper* to modulate the signal in a spectrophotometer at a frequency at which noise is reduced. In atomic absorption, periodic blocking of the beam allows a distinction to be made between light from the source and light from the flame.

beamsplitter A partially reflective, partially transparent plate that reflects some light and transmits some light.

Beer's law Relates the absorbance, A, of a sample to its concentration, c, pathlength, b, and molar absorptivity, ε: $A = \varepsilon bc$. More correctly called *Beer-Lambert-Bouguer law*.

biamperometric titration An amperometric titration conducted with two polarizable electrodes held at a constant potential difference.

bilayer Formed by a surfactant, a two-dimensional membrane structure in which polar or ionic headgroups are pointing outward and nonpolar tails are pointing inward.

biochemical oxygen demand (BOD) In a water sample, the quantity of dissolved oxygen consumed by microorganisms during a 5-day incubation in a sealed vessel at 20°C. Oxygen consumption is limited by organic nutrients, so BOD is a measure of pollutant concentration.

biosensor Device that uses biological components such as enzymes, antibodies, or DNA, in combination with electric, optical, or other signals, to achieve a highly selective response to one analyte.

bipotentiometric titration A potentiometric titration in which a constant current is passed between two polarizable electrodes immersed in the sample solution. An abrupt change in potential characterizes the end point.

Bjerrum plot See *difference plot*.

blackbody An ideal surface that absorbs all photons striking it. If the blackbody is at constant temperature, it must emit as much radiant energy as it absorbs.

blackbody radiation Radiation emitted by a blackbody. The energy and spectral distribution of the emission depend only on the temperature of the blackbody.

blank A sample not intended to contain analyte. See also *field blank, method blank, reagent blank*.

blank titration One in which a solution containing all reagents except analyte is titrated. The volume of titrant needed in the blank titration should be subtracted from the volume needed to titrate unknown.

bleed In gas chromatography, slow loss of stationary phase from evaporation, thermal decomposition, or oxidation.

blind sample See *performance test sample*.

blocking Occurs when a metal ion binds tightly to a metal ion indicator. A blocked indicator is unsuitable for a titration because no color change is observed at the end point.

Boltzmann distribution Relative population of two states at thermal equilibrium:

$$\frac{N_2}{N_1} = \frac{g_2}{g_1} e^{-(E_2 - E_1)/kT}$$

where N_i is the population of the state, g_i is the degeneracy of the state, E_i is the energy of the state, k is Boltzmann's constant, and T is temperature in kelvins. *Degeneracy* refers to the number of states with the same energy.

bomb Sealed vessel for conducting high-temperature, high-pressure reactions.

bonded stationary phase In high-performance liquid chromatography, a stationary liquid phase covalently attached to the solid support.

Brønsted-Lowry acid A proton (hydrogen ion) donor.

Brønsted-Lowry base A proton (hydrogen ion) acceptor.

Brownian motion Random motion of a small particle in a liquid caused by collisions with molecules moving with random speeds in random directions.

buffer A mixture of a weak acid and its conjugate base. A buffered solution is one that resists changes in pH when acids or bases are added.

buffer capacity, β A measure of the ability of a buffer to resist changes in pH. The larger the buffer capacity, the greater the resistance to pH change. The definition of buffer capacity is $\beta = dC_b/dpH = -dC_a/dpH$,

where C_a and C_b are the number of moles of strong acid or base per liter needed to produce a unit change in pH. Also called *buffer intensity*.

buffer intensity See *buffer capacity*.

bulk sample Material taken from lot being analyzed—usually chosen to be representative of the entire lot. Also called *gross sample*.

bulk solution Chemists' jargon referring to the main body of a solution. In electrochemistry, we distinguish properties of the bulk solution from properties that may be different in the immediate vicinity of an electrode.

buoyancy Upward force exerted on an object in a liquid or gaseous fluid. An object weighed in air appears lighter than its actual mass by an amount equal to the mass of air that it displaces.

buret A calibrated glass tube with a stopcock at the bottom. Used to deliver known volumes of liquid.

calibration Process of relating the actual quantity (such as mass, volume, force, or electric current) to the quantity indicated on the scale of an instrument.

calibration check In a series of analytical measurements, a calibration check is an analysis of a solution formulated by the analyst to contain a known concentration of analyte. It is the analyst's own check that procedures and instruments are functioning correctly.

calibration curve A graph showing the value of some property versus concentration of analyte. When the corresponding property of an unknown is measured, its concentration can be determined from the graph. Also called *standard curve*.

calomel electrode A common reference electrode based on the half-reaction $Hg_2Cl_2(s) + 2e^- \rightleftharpoons 2Hg(l) + 2Cl^-$. See also *saturated calomel electrode*.

candela, cd Luminous intensity, in a given direction, of a source that emits monochromatic radiation of frequency 540 THz and that has a radiant intensity of 1/683 W/sr in that direction.

capacitance The electric capacitance of two parallel, charged surfaces is the charge on each surface divided by the electric potential difference (volts) between the two surfaces.

capacitor current See *charging current*.

capacity factor See *retention factor*.

capacity ratio See *retention factor*.

capillary electrochromatography A version of high-performance liquid chromatography in which mobile phase is driven by electroosmosis instead of a pressure gradient. Solutes are separated by partitioning between the mobile and stationary phases.

capillary electrophoresis Separation of a mixture into its components by a strong electric field imposed between the two ends of a narrow capillary tube filled with electrolyte solution. Unlike chromatography, there is no stationary phase in electrophoreses. Solutes are separated by differences in mobility.

capillary gel electrophoresis A form of capillary electrophoresis in which the tube is filled with a polymer gel that serves as a sieve for macromolecules. The largest molecules have the slowest migration through the gel. Also called *capillary sieving electrophoresis*.

capillary zone electrophoresis A form of capillary electrophoresis in which ionic solutes are separated because of differences in their electrophoretic mobility.

carboxylate anion Conjugate base (RCO_2^-) of a carboxylic acid.

carboxylic acid A molecule with the general structure RCO_2H, where R is any group of atoms.

carcinogen A cancer-causing agent.

carrier gas Mobile phase gas in gas chromatography.

catalyst A substance that increases the rate of a selected chemical reaction, but does not appear in the net chemical equation.

cathode Electrode at which reduction occurs. In electrophoresis, it is the negatively charged electrode.

cathodic depolarizer A molecule that is easily reduced, thereby preventing the cathode potential of an electrochemical cell from becoming very low.

cathodic wave In voltammetry, a flow of current due to reduction at the working electrode.

catholyte   Solution present in the cathode chamber of an electrochemical cell.

cation   A positively charged ion.

cation exchanger   An ion exchanger with negatively charged groups covalently attached to the support. It can reversibly bind cations.

certified reference material   Samples sold by national measurement institutes and containing known quantities of analytes to test accuracy of analytical procedures. The U.S. National Institute of Standards and Technology calls its certified materials *Standard Reference Materials*.

chain of custody   Trail followed by a sample from the time it is collected to the time it is analyzed and, possibly, archived.

characteristic   The part of a logarithm at the left of the decimal point.

charge balance   A statement that the sum of all positive charge in solution equals the magnitude of the sum of all negative charge in solution.

charge coupled device   An extremely sensitive detector in which light creates electrons and holes in a semiconductor material. The electrons are attracted to regions near positive electrodes, where the electrons are "stored" until they are ready to be counted. The number of electrons in each pixel (picture element) is proportional to the number of photons striking the pixel.

charged aerosol detector   Sensitive, nearly universal detector for liquid chromatography in which solvent is evaporated from eluate to leave an aerosol of fine particles of nonvolatile solute. Aerosol particles are charged by adsorption of N_2^+ ions and flow to a collector that measures total charge reaching the detector versus time.

charging current   Electric current arising from migration of ions and electrons to or from an electrode when the potential of the electrode is changed. There is no redox reaction of the ions at the electrode. Charging current charges or discharges the electric double layer at the electrode-solution interface. Also called *capacitor current* or *condenser current*.

charring   In a gravimetric analysis, the precipitate and filter paper are first *dried* gently. Then the filter paper is *charred* at intermediate temperature to destroy the paper without letting it inflame. Finally, precipitate is *ignited* at high temperature to convert it into its analytical form.

chelate effect   The observation that a single multidentate ligand forms metal complexes that are more stable than those formed by several individual ligands with the same ligand atoms.

chelating ligand   A ligand that binds to a metal through more than one atom.

chemical interference   In atomic spectroscopy, any chemical reaction that decreases the efficiency of atomization.

chemical ionization   A gentle method of producing ions for a mass spectrometer without extensive fragmentation of the analyte molecule, M. A reagent gas such as CH_4 is bombarded with electrons to make CH_5^+, which transfers H^+ to M, giving MH^+.

chemical oxygen demand (COD)   In a natural water or industrial effluent sample, the quantity of O_2 equivalent to the quantity of $K_2Cr_2O_7$ consumed by refluxing the sample with a standard dichromate–sulfuric acid solution containing Ag^+ catalyst. Because 1 mol of $K_2Cr_2O_7$ consumes $6e^-$ ($Cr^{6+} \rightarrow Cr^{3+}$), it is equivalent to 1.5 mol of O_2 ($O \rightarrow O^{2-}$). See also *oxidizability*.

chemiluminescence   Emission of light by an excited-state product of a chemical reaction.

chiral molecule   One that is not superimposable on its mirror image in any accessible conformation. Also called an *optically active molecule*, a chiral molecule rotates the plane of polarization of light.

chromatogram   A graph showing chromatography detector response as a function of elution time or volume.

chromatograph   An instrument used to perform chromatography.

chromatography   A technique in which molecules in a mobile phase are separated because of their different affinities for a stationary phase. The greater the affinity for the stationary phase, the longer a molecule is retained.

chromophore   The part of a molecule responsible for absorption of light of a particular frequency.

chronoamperometry   A technique in which the potential of a working electrode in an unstirred solution is varied rapidly while the current between the working and the auxiliary electrodes is measured. Suppose that the analyte is reducible and that the potential of the working electrode is made more negative. Initially, no reduction occurs. At a certain potential, the analyte begins to be reduced and the current increases. As the potential becomes more negative, the current increases further until the concentration of analyte at the surface of the electrode is sufficiently depleted. Then the current decreases, even though the potential becomes more negative. The maximum current is proportional to the concentration of analyte in bulk solution.

chronopotentiometry   A technique in which a constant current is forced to flow between two electrodes. The voltage remains fairly steady until the concentration of an electroactive species becomes depleted. Then the voltage changes rapidly as a new redox reaction assumes the burden of current flow. The elapsed time when the voltage suddenly changes is proportional to the concentration of the initial electroactive species in bulk solution.

cladding   Covering; overlay. Layer surrounding the core of an optical fiber.

Clark electrode   One that measures the activity of dissolved oxygen by amperometry.

coagulation   With respect to gravimetric analysis, small crystallites coming together to form larger crystals.

co-chromatography   In chromatography, addition of a known compound to an unknown to see if it is eluted at the same time as a component of the unknown. See *spike*.

coefficient of variation   The standard deviation, s, expressed as a percentage of the mean value $\bar{x}$: coefficient of variation $= 100 \times s/\bar{x}$. Also called *relative standard deviation*.

cofactor   A small, nonprotein molecule that is bound to an enzyme and is necessary for the function of the enzyme. Also called a *prosthetic group*.

coherence   Degree to which electromagnetic waves are in phase with one another. Laser light is highly coherent.

co-ion   An ion with the same charge as the ion of interest.

cold trapping   Splitless gas chromatography injection technique in which solute is condensed far below its boiling point in a narrow band at the start of the column.

collimated light   Light in which all rays travel in parallel paths.

collimation   Process of making light rays travel parallel to one another.

collisionally activated dissociation   Fragmentation of an ion in a mass spectrometer by high-energy collisions with gas molecules. In atmospheric pressure chemical ionization or electrospray interfaces, collisionally activated dissociation at the inlet to the mass filter can be promoted by varying the cone voltage. In tandem mass spectrometry, dissociation occurs in a collision cell between the two mass separators. Also called *collision-induced dissociation*.

collision cell   Middle stage of a tandem mass spectrometer in which the precursor ion selected by the first stage is fragmented by collisions with gas molecules.

collision-induced dissociation   See *collisionally activated dissociation*.

colloid   A dissolved particle with a diameter in the approximate range 1–500 nm. It is too large to be considered one molecule but too small to simply precipitate.

combination electrode   A glass electrode with a concentric reference electrode built on the same body.

combustion analysis   A technique in which a sample is heated in an atmosphere of O_2 to oxidize it to CO_2 and H_2O, which are collected and weighed or measured by gas chromatography. Modifications permit the simultaneous analysis of N, S, and halogens.

common ion effect   Occurs when a salt is dissolved in a solution already containing one of the ions of the salt. The salt is less soluble than it would be in a solution without that surplus ion. An application of Le Châtelier's principle.

complex ion   Historical name for any ion containing two or more ions or molecules that are each stable by themselves; e.g., $CuCl_3^-$ contains $Cu^+ + 3Cl^-$.

complexometric titration   One in which the reaction between analyte and titrant involves complex formation.

composite sample   A representative sample prepared from a heterogeneous material. If the material consists of distinct regions, the composite is made of portions of each region, with relative amounts proportional to the size of each region.

compound electrode   An ion-selective electrode consisting of a conventional electrode surrounded by a barrier that is selectively permeable to the analyte of interest. Alternatively, the barrier region might convert external analyte into a different species, to which the inner electrode is sensitive.

concentration   An expression of the quantity per unit volume or unit mass of a substance. Common measures of concentration are molarity (mol/L) and molality (mol/kg of solvent).

concentration cell   A galvanic cell in which both half-reactions are the same, but the concentrations in each half-cell are not identical. The cell reaction increases the concentration of species in one half-cell and decreases the concentration in the other.

concentration polarization   Occurs when an electrode reaction occurs so rapidly that the concentration of solute near the surface of the electrode is not the same as the concentration in bulk solution.

condenser current   See *charging current*.

conditional formation constant, K_f'   Equilibrium constant for formation of a complex under a particular, stated set of conditions, such as pH, ionic strength, and concentration of auxiliary complexing species. Also called *effective formation constant*.

conduction band   Energy levels containing conduction electrons in a semiconductor.

conduction electron   An electron free to move about within a solid and carry electric current. In a semiconductor, the energies of the conduction electrons are above those of the valence electrons that are localized in chemical bonds. The energy separating the valence and conduction bands is called the *band gap*.

conductivity, σ   Proportionality constant between electric current density, J (A/m^2), and electric field, E (V/m): $J = \sigma E$. Units are Ω^{-1} m^{-1}. Conductivity is the reciprocal of *resistivity*.

cone voltage   Voltage applied between the *skimmer cone* and a nearby orifice through which gaseous ions flow into the mass separator of a mass spectrometer. The magnitude of the voltage can be increased to promote collisionally activated dissociation of ions prior to mass separation.

confidence interval   Range of values within which there is a specified probability that the "true" value lies.

conjugate acid-base pair   An acid and a base that differ only through the gain or loss of a single proton.

constant-current electrolysis   Electrolysis in which a constant current flows between working and auxiliary electrodes. As reactants are consumed, increasing voltage is required to keep the current flowing, so this is the least selective form of electrolysis.

constant mass   In gravimetric analysis, the product is heated and cooled to room temperature in a desiccator until successive weighings are "constant." There is no standard definition of "constant mass"; but, for ordinary work, it is usually taken to be about ±0.3 mg. Constancy is usually limited by the irreproducible regain of moisture during cooling and weighing.

constant-voltage electrolysis   Electrolysis in which a constant voltage is maintained between working and auxiliary electrodes. This is less selective than controlled-potential electrolysis because the potential of the working electrode becomes more extreme as ohmic potential and overpotential change.

continuous-dynode electron multiplier   An electron detector that works like a photomultiplier tube. An electron striking the lead-doped glass wall of a horn-shaped tube liberates several electrons which are accelerated into the horn by increasingly positive potential. After many bounces, ~10^5 electrons reach the narrow end of the horn for each incident electron.

control chart   A graph in which successive observations of a process are recorded to determine whether the process is within specified control limits.

controlled-potential electrolysis   A technique for selective reduction (or oxidation), in which the voltage between working and reference electrodes is held constant.

convection   Process in which solute is carried from one place to another by bulk motion of the solution.

coprecipitation   Occurs when a substance whose solubility is not exceeded precipitates along with one whose solubility is exceeded.

correlation coefficient   The square of the correlation coefficient, R^2, is a measure of goodness of fit of data points to a straight line. The closer R^2 is to 1, the better the fit.

coulomb, C   Amount of charge per second that flows past any point in a circuit when the current is 1 ampere. There are approximately 96 485 coulombs in a mole of electrons.

coulometer   A device that generates a redox reagent for quantitative reaction with analyte and measures the number of electrons required to generate the redox reagent.

coulometric titration   One conducted with a constant current for a measured time.

coulometry   A technique in which the quantity of analyte is determined by measuring the number of coulombs needed for complete electrolysis.

counter electrode   See *auxiliary electrode*.

counterion   An ion with a charge opposite that of the ion of interest.

coupled equilibria   Reversible chemical reactions that have a species in common. For example, the product of one reaction could be a reactant in another.

critical point   Critical temperature and pressure of a substance.

critical pressure   Pressure above which a fluid cannot be condensed to two phases (liquid and gas), no matter how low the temperature.

critical temperature   Temperature above which a fluid cannot be condensed to two phases (liquid and gas), no matter how great a pressure is applied.

cross-linking   Covalent linkage between different strands of a polymer.

cryogenic focusing   In gas chromatography, cold trapping of solutes below ambient temperature at the beginning of a column. A *cryogen* is a cold fluid such as liquid nitrogen used for cooling.

crystallization   Process in which a substance comes out of solution slowly to form a solid with a regular arrangement of atoms.

cumulative formation constant, β_n   Equilibrium constant for a reaction of the type M + nX $\rightleftharpoons$ MX$_n$. Also called *overall formation constant*.

current, I   Amount of charge flowing through a circuit per unit time (C/s).

current density   Electric current per unit area (A/m^2).

cuvet   A cell with transparent walls used to hold samples for spectrophotometric measurements.

cyclic voltammetry   A polarographic technique with a triangular waveform. Both cathodic and anodic currents are observed for reversible reactions.

dalton, Da   Unit of atomic mass defined as 1/12 of the mass of ^{12}C.

dark current   Small current produced by a photodetector in the absence of light.

DART   See *direct analysis in real time*.

data quality objectives   Accuracy, precision, and sampling requirements for an analytical method.

dead volume, V_m   Volume of mobile phase within a chromatography column, equal to the entire space accessible to solvent and small solutes. Includes pores in the particles and volume between particles. If time required for solvent or unretained solute to travel through the column is t_m, the dead volume is $V_m = t_m F$, where F is the volume flow rate.

Debye-Hückel equation   Gives the activity coefficient, γ, of an ion as a function of ionic strength, μ. The *extended Debye-Hückel equation*, applicable to $\mu \leq 0.1$ M, is log $\gamma = [-0.51z^2\sqrt{\mu}]/[1 + (\alpha\sqrt{\mu}/305)]$, where z is the ionic charge and α is the effective hydrated size in picometers.

decant   To pour liquid off a solid or, perhaps, a denser liquid. The denser phase is left behind.

decomposition potential   In an electrolysis, that voltage at which rapid reaction first begins.

degrees of freedom   In statistics, the number of observations minus the number of parameters estimated from the observations.

deionized water   Water that has been passed through a cation exchanger (in the H^+ form) and an anion exchanger (in the OH^- form) to remove ions from the solution.

deliquescent substance   A *hygroscopic* solid that spontaneously picks up so much water from the air that the substance completely dissolves.

demasking   Removal of a masking agent from the species protected by the masking agent.

density   Mass per unit volume.

depolarizer   A molecule that is oxidized or reduced at a modest potential. It is added to an electrolytic cell to prevent the cathode or anode potential from becoming too extreme.

derivatization   Chemical alteration to attach a group to a molecule so that the molecule can be detected conveniently. Alternatively, treatment can alter volatility or solubility to allow easier separation.

desalting   Removal of salts (or any small molecules) from a solution of macromolecules. Gel filtration or dialysis are used for desalting.

DESI   See *desorption electrospray ionization*.

desiccant   A drying agent.

desiccator   A sealed chamber in which samples can be dried in the presence of a desiccant or by vacuum pumping or both.

desorption   The release of an adsorbed substance from a surface.

desorption electrospray ionization, DESI   A solvent is electrosprayed onto a surface to dissolve analyte from the surface into aerosol microdroplets, which can be analyzed with a mas spectrometer.

detection limit   The smallest quantity of analyte that is "significantly different" from a blank. The detection limit is often taken as the concentration of analyte that gives a signal equal to three times the standard deviation of signal from a blank. Also called *lower limit of detection*.

determinant   The value of the two-dimensional determinant $\begin{vmatrix} a & b \\ c & d \end{vmatrix}$ is the difference $ad - bc$.

determinate error   See *systematic error*.

deuterium arc lamp   Source of broadband ultraviolet radiation. An electric discharge (a spark) in deuterium gas causes D_2 molecules to dissociate and emit many wavelengths of radiation.

dialysis   A technique in which solutions are placed on either side of a semipermeable membrane that allows small molecules, but not large molecules, to cross. Small molecules in the two solutions diffuse across and equilibrate between the two sides. Large molecules are retained on their original side.

dielectric constant, ε   The electrostatic force, F, between two charged particles is given by $F = kq_1q_2/\varepsilon r^2$, where k is a constant, q_1 and q_2 are the charges, r is the separation between particles, and ε is the dielectric constant of the medium. The higher the dielectric constant, the less force is exerted by one charged particle on another.

difference plot   A graph of the mean fraction of protons bound to an acid versus pH. For complex formation, the difference plot gives the mean number of ligands bound to a metal versus pL ($= -\log[\text{ligand concentration}]$). Also called *Bjerrum plot*.

diffraction   The bending of light rays by a grating. This process occurs when electromagnetic radiation passes through or is reflected from slits with a spacing comparable to the wavelength. Interference of waves from adjacent slits produces a spectrum of radiation, with each wavelength emerging at a different angle.

diffuse part of the double layer   Region of solution near a charged surface in which excess counterions are attracted to the charge. The thickness of this layer is 0.3–10 nm.

diffuse reflection   Occurs when a rough surface reflects light in all directions.

diffusion   Net transport of a solute from a region of high concentration to a region of low concentration caused by the random movement of molecules in a liquid or gas (or, very slowly, in a solid).

diffusion coefficient, D   Defined by Fick's first law of diffusion: $J = -D(dc/dx)$, where J is the rate at which molecules diffuse across a plane of unit area and dc/dx is the concentration gradient in the direction of diffusion.

diffusion current   Current observed when the rate of electrolysis is limited by the rate of diffusion of analyte to the electrode. In polarography, diffusion current = limiting current – residual current.

diffusion layer   Region near an electrode containing excess product or decreased reactant involved in the electrode reaction. The thickness of this layer can be hundreds of micrometers.

digestion   Process in which a precipitate is left (usually warm) in the presence of mother liquor to promote particle recrystallization and growth. Purer, more easily filterable crystals result. Also used to describe any chemical treatment in which a substance is decomposed to transform the analyte into a form suitable for analysis.

dilution factor   Factor (initial volume of reagent)/(total volume of solution) used to multiply the initial concentration of reagent to find the diluted concentration.

dimer   A molecule made from two identical units.

diode   A semiconductor device consisting of a *pn* junction through which current can pass in only one direction. Current flows when the *n*-type material is made negative and the *p*-type material is made positive. A voltage sufficient to overcome the activation energy for carrier movement must be supplied before any current flows. For silicon diodes, this voltage is ~0.6 V. If a sufficiently large voltage, called the *breakdown voltage*, is applied in the reverse direction, current will flow in the wrong direction through the diode.

diprotic acids and bases   Compounds that can donate or accept two protons.

direct analysis in real time, DART   A DART source produces excited He or N_2, which is directed at the surface of an object to be sampled in ambient atmosphere. The excited species react with ambient moisture to create protonated water clusters that react with analyte M to make MH^+. The MH^+ is measured by mass spectrometry.

direct current polarography   Classical form of polarography, in which a linear voltage ramp is applied to the working electrode.

direct titration   One in which the analyte is treated with titrant and the volume of titrant required for complete reaction is measured.

dispersion   A measure of the ability of a monochromator to separate wavelengths differing by $\Delta\lambda$ through the angle $\Delta\phi$. The greater the dispersion, the greater the angle separating two closely spaced wavelengths. For a prism, dispersion refers to the rate of change of refractive index with wavelength, $dn/d\lambda$.

displacement titration   An EDTA titration procedure in which analyte is treated with excess $MgEDTA^{2-}$ to displace Mg^{2+}: $M^{n+} + MgEDTA^{2-} \rightleftharpoons MEDTA^{n-4} + Mg^{2+}$. The liberated Mg^{2+} is then titrated with EDTA. This procedure is useful if there is no suitable indicator for direct titration of M^{n+}.

disproportionation   A reaction in which an element in one oxidation state gives products containing that element in both higher and lower oxidation states; e.g., $2Cu^+ \rightleftharpoons Cu^{2+} + Cu(s)$.

distilled water   Water that has been boiled to separate it from less volatile impurities. The vapor is condensed back to liquid that is collected in a clean vessel.

distribution coefficient, D   For a solute partitioned between two phases, the distribution coefficient is the total concentration of all forms of solute in phase 2 divided by the total concentration in phase 1.

Donnan exclusion   Ions of the same charge as those fixed on an ion exchange resin are repelled from the resin. Thus, anions do not readily penetrate a cation-exchange resin, and cations are repelled from an anion-exchange resin.

dopant   When small amounts of substance B are added to substance A, we call B a dopant and say that A is doped with B. Doping is done to alter the properties of A.

Doppler effect   A molecule moving toward a source of radiation experiences a higher frequency than one moving away from the source.

double-focusing mass spectrometer A spectrometer that uses electric and magnetic sectors in series to obtain high resolution.

double-junction electrode An electrode with inner and outer compartments designed to minimize contact between analyte solution and the contents of the inner electrode. The outer compartment serves as a salt bridge with ions that are chemically compatible with the analyte.

double layer See *electric double layer*.

drift Slow change in the response of an instrument due to various causes such as changes in electrical components with temperature, variation in power-line voltage to an instrument, and aging of components within instruments. Same as *1/f noise* or *flicker noise*.

dropping-mercury electrode One that delivers fresh drops of Hg to a polarographic cell.

dry ashing Oxidation of organic matter with O_2 at high temperature to leave behind inorganic components for analysis.

dwell time In chromatography, time between mixing of solvents and when they reach the start of the column.

dwell volume Volume in chromatography between the point of mixing solvents and the start of the column.

dynamic range Range of analyte concentration over which a change in concentration gives a change in detector response.

dynode A metal surface that easily emits several electrons each time it is struck by one accelerated electron in a photomultiplier tube or an electron multiplier.

E^o Standard reduction potential.

$E^{o\prime}$ Effective standard reduction potential at pH 7 (or at some other specified conditions).

EDTA (ethylenediaminetetraacetic acid) $(HO_2CCH_2)_2NCH_2CH_2N-(CH_2CO_2H)_2$, the most widely used reagent for complexometric titrations. It forms 1:1 complexes with virtually all cations with a charge of 2 or more.

effective formation constant, K_f' See *conditional formation constant*.

effervescence Rapid release of gas with bubbling and hissing.

efflorescence Property by which the outer surface or entire mass of a substance turns into powder from loss of water of crystallization.

effluent See *eluate*.

einstein A mole of photons. The symbol of this unit is the same as its name, einstein.

electric charge, q Quantity of electricity, measured in coulombs.

electric discharge emission spectroscopy A technique in which atomization and excitation are stimulated by an electric arc, a spark, or microwave radiation.

electric double layer Region comprising the charged surface of an electrode or a particle plus the oppositely charged region of solution adjacent to the surface. Also called *double layer*.

electric potential, E The electric potential (in volts) at a point is the energy (in joules) needed to bring one coulomb of positive charge from infinity to that point. The *potential difference* between two points is the energy needed to transport one coulomb of positive charge from the negative point to the positive point.

electroactive species Any species that can be oxidized or reduced at an electrode.

electrochemical detector Liquid chromatography detector that measures current when an electroactive solute emerges from the column and passes over a working electrode held at a fixed potential with respect to a reference electrode. Also called *amperometric detector*.

electrochemiluminescence Light emitted by a chemical reaction at an electrode.

electrochemistry Use of electrical measurements on a chemical system for analytical purposes. Also refers to use of electricity to drive a chemical reaction or use of a chemical reaction to produce electricity.

electrode An electrical conductor through which electrons flow into or out of chemical species involved in a redox reaction.

electroendosmosis See *electroosmosis*.

electrogravimetric analysis A technique in which the mass of an electrolytic deposit is used to quantify the analyte.

electrokinetic injection In capillary electrophoresis, the use of an electric field to inject sample into the capillary. Because different species have different mobilities, the injected sample does not have the same composition as the original sample.

electrolysis A chemical reaction driven by passage of electric current.

electrolyte A substance that produces ions when dissolved.

electrolytic cell A cell in which a chemical reaction that would not otherwise occur is driven by a voltage applied between two electrodes.

electromagnetic spectrum The whole range of electromagnetic radiation, including visible light, radio waves, X-rays, etc.

electron capture detector Gas chromatography detector that is particularly sensitive to compounds with halogen atoms, nitro groups, and other groups with high electron affinity. Makeup gas (N_2 or 5% CH_4 in Ar) is ionized by β-rays from ^{63}Ni to liberate electrons that produce a small, steady current. High-electron-affinity analytes capture some of the electrons and reduce the detector current.

electronic balance A weighing device that uses an electromagnetic servomotor to balance the load on the pan. The mass of the load is proportional to the current needed to balance it.

electronic transition One in which an electron goes from one energy level to another.

electron ionization Method used to create ions from gaseous molecules in the inlet of a mass spectrometer by bombardment of the gas with high-energy electrons.

electron multiplier An ion detector that works like a photomultiplier tube. Cations striking a cathode liberate electrons. A series of *dynodes* multiplies the number of electrons by $\sim 10^5$ before they reach the anode.

electron-transfer dissociation Breaking of a chemical bond by the exothermic transfer of an electron from one species to another. This process is used in mass spectral sequencing of polypeptides because it cleaves peptide bonds without breaking other bonds in the molecule.

electroosmosis Bulk flow of fluid in a capillary tube induced by an electric field. Mobile ions in the diffuse part of the double layer at the wall of the capillary serve as the "pump." Also called *electroendosmosis*.

electroosmotic flow Uniform, pluglike flow of fluid in a capillary tube under the influence of an electric field. The greater the charge on the wall of the capillary, the greater the number of counterions in the double layer and the stronger the electroosmotic flow.

electroosmotic mobility, μ_{eo} Constant of proportionality between the electroosmotic speed, u_{eo}, of a fluid in a capillary and the applied electric field, E: $u_{eo} = \mu_{eo}E$. Also equal to the speed of a neutral species, $u_{neutral}$, divided by the electric field, E. See also *apparent mobility*.

electroosmotic velocity Speed with which solvent flows through a capillary electrophoresis column. It is measured by adding a detectable neutral molecule to the sample. Electroosmotic velocity is the distance from injector to detector divided by the time required for the neutral molecule to reach the detector.

electropherogram A graph of detector response versus time for electrophoresis.

electrophoresis Migration of ions in solution in an electric field. Cations move toward the cathode and anions move toward the anode. Ions can be separated from one another by their differing rates of migration in a strong electric field.

electrophoretic mobility, μ_{ep} Constant of proportionality between the electrophoretic speed, u_{ep}, of an ion in solution and the applied electric field, E: $u_{ep} = \mu_{ep}E$. See also *apparent mobility*.

electrospray ionization A method for interfacing liquid chromatography to mass spectrometry. A high potential applied to the liquid at the column exit creates charged droplets in a fine aerosol. Gaseous ions are derived from ions that were already in the mobile phase on the column. It is common to observe protonated bases (BH^+), ionized acids (A^-), and complexes formed between analyte, M (which could be neutral or charged), and stable ions such as NH_4^+, Na^+, HCO_2^-, or $CH_3CO_2^-$ that were already in solution.

eluate   What comes out of a chromatography column. Also called *effluent*.

eluent   Solvent applied to the beginning of a chromatography column.

eluent strength   A measure of the ability of a solvent to elute solutes from a normal-phase chromatography column. It is a measure of the solvent adsorption energy on bare silica.

eluotropic series   Ranks solvents according to their ability to displace solutes from the stationary phase in adsorption chromatography.

elution   Process of passing a liquid or a gas through a chromatography column.

emission spectrum   A graph of luminescence intensity versus luminescence wavelength (or frequency or wavenumber), obtained with a fixed excitation wavelength.

emissivity   A quotient given by the radiant emission from a real object divided by the radiant emission of a blackbody at the same temperature.

emulsion   A stable dispersion of immiscible liquids, which might be made by vigorous shaking. Homogenized milk is an emulsion of cream in an aqueous phase. Emulsions usually require an emulsifying agent (a surfactant) for stability. The emulsifying agent stabilizes the interface between the two phases by its affinity for both phases.

enantiomers   Mirror image isomers that cannot be superimposed on each other. Also called *optical isomers*.

endothermic reaction   One for which ΔH is positive; heat must be supplied to reactants for them to react.

end point   Point in a titration at which there is a sudden change in a physical property, such as indicator color, pH, conductivity, or absorbance. Used as a measure of the equivalence point.

energy   The product force $\times$ distance.

enthalpy change, ΔH   The heat absorbed or released when a reaction occurs at constant pressure.

enthalpy of hydration   Heat liberated when a gaseous species is transferred to water.

entropy   A measure of the "disorder" of a substance.

enzyme   A protein that catalyzes a chemical reaction.

enzyme-linked immunosorbent assay (ELISA)   A biochemical assay incorporating an enzyme attached to an antibody. After antibody binds to the intended analyte and excess antibody is washed away, a reagent is added that is converted to a detectable product (such as a colored or fluorescent substance) by the enzyme. The quantity of detectable product is proportional to the amount of analyte.

equilibrium   State in which the forward and reverse rates of all reactions are equal, so the concentrations of all species remain constant.

equilibrium constant, K   For the reaction $a\text{A} + b\text{B} \rightleftharpoons c\text{C} + d\text{D}$, $K = \mathcal{A}_C^c \mathcal{A}_D^d / \mathcal{A}_A^a \mathcal{A}_B^b$, where $\mathcal{A}_i$ is the activity of the ith species.

equimolar mixture of compounds   One that contains an equal number of moles of each compound.

equivalence point   Point in a titration at which the quantity of titrant is exactly sufficient for stoichiometric reaction with the analyte.

equivalent   For a redox reaction, the amount of reagent that can donate or accept one mole of electrons. For an acid-base reaction, the amount of reagent that can donate or accept one mole of protons. For ion exchange, the amount of reagent that will exchange with one mole of monovalent cation or anion.

equivalent weight   The mass of substance containing one equivalent.

error bar   Graphical depiction of the uncertainty in a measurement.

evanescent wave   Light that penetrates the walls of an optical fiber or waveguide in which the light travels by total internal reflection.

evaporative light-scattering detector   A liquid chromatography detector that makes a fine mist of eluate and evaporates solvent from the mist in a heated zone. The remaining particles of liquid or solid solute flow past a laser beam and are detected by their ability to scatter the light.

excitation spectrum   A graph of luminescence (measured at a fixed wavelength) versus excitation frequency or wavelength. It closely corresponds to an absorption spectrum because the luminescence is generally proportional to the absorbance.

excited state   Any state of an atom or a molecule having more than the minimum possible energy.

exitance, M   Power per unit area radiating from the surface of an object.

exothermic reaction   One for which ΔH is negative; heat is liberated when products are formed.

extended Debye-Hückel equation   See *Debye-Hückel equation*.

external standard   A known solution of analyte, distinct from the unknown solution, used to prepare a calibration curve showing analytical response versus analyte concentration.

extinction coefficient   See *molar absorptivity*.

extra-column volume   Volume of a chromatography system (not including the column) between the point of injection and the point of detection.

extracted ion chromatogram   Chromatogram made by collecting consecutive full-range mass spectra, but selecting just one value of m/z for display. Most time is spent monitoring values of m/z that will not be displayed. A *selected ion chromatogram* provides greater signal-to-noise ratio than an extracted ion chromatogram because all of the time is taken to monitor just one or a few values of m/z in the selected ion chromatogram.

extraction   Process in which a solute is transferred from one phase to another. Analyte is sometimes removed from a sample by extraction into a solvent that dissolves the analyte.

extrapolation   Estimation of a value that lies beyond the range of measured data.

1/f noise   See *drift*.

F test   For two variances, s_1^2 and s_2^2 (with s_1 chosen to be the larger of the two), the statistic F is defined as $F = s_1^2/s_2^2$. To decide whether s_1 is significantly greater than s_2, we compare F with the critical values in a table based on a certain confidence level. If the calculated value of F is greater than the value in the table, the difference is significant.

Fajans titration   A precipitation titration in which the end point is signaled by adsorption of a colored indicator on the precipitate.

false negative   A conclusion that the concentration of analyte is below a certain limit when, in fact, the concentration is above the limit.

false positive   A conclusion that the concentration of analyte exceeds a certain limit when, in fact, the concentration is below the limit.

farad, F   Unit of electrical capacitance; 1 farad of capacitance will store 1 coulomb of charge in a potential difference of 1 volt.

faradaic current   That component of current in an electrochemical cell due to oxidation and reduction reactions.

Faraday constant, F   The number of coulombs in a mole of elementary charges, approximately $9.648\ 5 \times 10^4$ C/mol of charge.

Faraday cup   A mass spectrometric ion detector in which each arriving cation is neutralized by an electron. The current required to neutralize the ions is proportional to the number of cations arriving at the Faraday cup.

Faraday's laws   Two laws stating that the extent of an electrochemical reaction is directly proportional to the quantity of electricity that has passed through the cell. The mass of substance that reacts is proportional to its formula mass and inversely proportional to the number of electrons required in its half-reaction.

ferroelectric material   A solid with a permanent electric polarization (dipole) in the absence of an external electric field. The polarization results from alignment of molecules within the solid.

Fick's first law of diffusion   See *diffusion coefficient*.

field blank   A blank sample exposed to the environment at the sample collection site and transported in the same manner as other samples between the lab and the field.

field effect transistor   A semiconductor device in which the electric field between gate and base governs the flow of current between source and drain.

filtrate   Liquid that passes through a filter.

flame ionization detector   A gas chromatography detector in which solute is burned in an H_2-air flame to produce CHO^+ ions. The current carried through the flame by these ions is proportional to the concentration of susceptible species in the eluate.

flame photometer A device that uses flame atomic emission and a filter photometer to quantify Li, Na, K, and Ca in liquid samples.

flame photometric detector Gas chromatography detector that measures optical emission from S, P, Pb, Sn, or other elements in a H_2-O_2 flame.

flicker noise See *drift*.

flocculate To cause particles of a dispersion, such as a colloid, to aggregate into larger particles.

flow injection analysis Analytical technique in which sample is injected into flowing liquid carrier containing a reagent that reacts with the analyte. Additional reagents might be added farther downstream. As sample flows from injector to detector, the sample zone broadens and reacts with reagent to form a product to which the detector responds.

fluorescence Process in which a molecule emits a photon 10^{-8} to 10^{-4} s after absorbing a photon. It results from a transition between states of the same spin multiplicity (e.g., singlet → singlet).

fluorescence detector Liquid chromatography detector that uses a strong light or laser to irradiate eluate emerging from a column and detects radiant emission from fluorescent solutes.

fluorophore The part of a molecule responsible for emission of light.

flux In sample preparation, flux is the agent used as the medium for a fusion. In transport phenomena, flux is the quantity of whatever you like crossing each unit area in one unit of time. For example, the flux of diffusing molecules could be mol/(m$^2 \cdot$ s). Heat flux would be J/(m$^2 \cdot$ s).

force The product mass × acceleration.

formal concentration, F The molarity of a substance if it did not change its chemical form on being dissolved. It represents the total number of moles of substance dissolved in a liter of solution, regardless of any reactions that take place when the solute is dissolved. Also called *analytical concentration* or *formality*.

formality, F See *formal concentration*.

formal potential Potential of a half-reaction (relative to a standard hydrogen electrode) when the formal concentrations of reactants and products are unity. Any other conditions (such as pH, ionic strength, and concentrations of ligands) also must be specified.

formation constant, K_f Equilibrium constant for the reaction of a metal with its ligands to form a metal-ligand complex. Also called *stability constant*. See also *effective formation constant*.

formula mass, FM The mass containing one mole of the indicated chemical formula of a substance. For example, the formula mass of $CuSO_4 \cdot 5H_2O$ is the sum of the masses of copper, sulfate, and five water molecules.

fortification Same as *spike*—a deliberate addition of analyte to a sample.

Fourier analysis Process of decomposing a function into an infinite series of sine and cosine terms. Because each term represents a certain frequency or wavelength, Fourier analysis decomposes a function into its component frequencies or wavelengths.

Fourier series Infinite sum of sine and cosine terms that add to give a particular function in a particular interval.

fraction of association, α For the reaction of a base (B) with H_2O, the fraction of base in the form BH^+.

fraction of dissociation, α For the dissociation of an acid (HA), the fraction of acid in the form A^-.

free energy See *Gibbs free energy*.

frequency, n The number of cycles per unit time for a repetitive event.

friction coefficient A molecule moving through a solution (from thermal motion or an electric field) is retarded by a force that is proportional to its speed. The constant of proportionality is the friction coefficient.

fronting Asymmetric chromatographic band where there is a gradual increase in concentration followed by an abrupt return to baseline. The peak maximum moves to longer retention time than for an ideal peak. Fronting results from injection of excess solute which alters the stationary phase polarity such that it becomes increasingly retentive.

fuel cell A glavanic cell used to produce electricity by the chemical reaction between reagents that flow into the cell. Products are pumped out of the cell. A fuel cell produces electricity as long as fuel is available.

fugacity The activity of a gas.

fugacity coefficient Activity coefficient for a gas.

fused-core particles See *superficially porous particles*.

fusion Process in which an otherwise insoluble substance is dissolved in a molten salt such as Na_2CO_3, Na_2O_2, or KOH. Once the substance has dissolved, the melt is cooled, dissolved in aqueous solution, and analyzed.

galvanic cell One that produces electricity by means of a spontaneous chemical reaction. Also called a *voltaic cell*.

gas chromatography A form of chromatography in which the mobile phase is a gas.

gathering A process in which a trace constituent of a solution is intentionally coprecipitated with a major constituent.

Gaussian distribution Theoretical bell-shaped distribution of measurements when all error is random. The center of the curve is the mean, μ, and the width is characterized by the standard deviation, σ. A *normalized* Gaussian distribution, also called the *normal error curve*, has an area of unity and is given by

$$y = \frac{1}{\sigma\sqrt{2\pi}} e^{-(x-\mu)^2/2\sigma^2}$$

Gaussian noise See *white noise*.

gel Chromatographic stationary phase particles, such as Sephadex or polyacrylamide, which are soft and pliable.

gel filtration chromatography See *molecular exclusion chromatography*.

gel permeation chromatography See *molecular exclusion chromatography*.

general elution problem In chromatography, the inability of a single isothermal or isocratic condition to provide adequate separation in reasonable time for solutes with a wide range of retention.

geometric mean For a series of n measurements with the values x_i, geometric mean $= \sqrt[n]{x_1 x_2 \dots x_n}$.

Gibbs free energy, G The change in Gibbs free energy, ΔG, for any process at constant temperature is related to the change in enthalpy, ΔH, and entropy, ΔS, by the equation $\Delta G = \Delta H - T \Delta S$, where T is temperature in kelvins. A process is spontaneous (thermodynamically favorable) if ΔG is negative.

glass electrode One that has a thin glass membrane across which a pH-dependent voltage develops. The voltage (and hence pH) is measured by a pair of reference electrodes on either side of the membrane.

glassy carbon electrode An inert carbon electrode, impermeable to gas, and especially well suited as an anode. The isotropic structure (same in all directions) is thought to consist of tangled ribbons of graphitelike sheets, with some cross-linking.

globar An infrared radiation source made of a ceramic such as silicon carbide heated by passage of electricity.

Gooch filter crucible A short, cup-shaped container with holes at the bottom, used for filtration and ignition of precipitates. For ignition, the crucible is made of porcelain or platinum and lined with a mat of ceramic fibers to retain the precipitate. For precipitates that do not need ignition, the crucible is made of glass and has a porous glass disk instead of holes at the bottom.

gradient elution Chromatography in which the composition of the mobile phase is progressively changed to increase the eluent strength of the solvent.

graduated cylinder A tube with volume calibrations along its length.

gram-atom The amount of an element containing Avogadro's number of atoms; it is the same as a mole of the element.

Gran plot A graph such as the plot of $V_b \cdot 10^{-pH}$ versus V_b used to find the end point of a titration. V_b is the volume of base (titrant) added to an acid being titrated. The slope of the linear portion of the graph is related to the dissociation constant of the acid.

graphite furnace A graphite tube that can be heated electrically to about 2 500 K to decompose and atomize a sample for atomic spectroscopy.

grating Either a reflective or a transmitting surface etched with closely spaced lines; used to disperse light into its component wavelengths.

gravimetric analysis Any analytical method that relies on measuring the mass of a substance (such as a precipitate) to complete the analysis.

gravimetric titration A titration in which the mass of titrant is measured, instead of the volume. Titrant concentration is conveniently expressed as mol reagent/kg titrant solution. Gravimetric titrations can be more accurate and precise than volumetric titrations.

green chemistry Principles intended to change our behavior in a manner that will help sustain the habitability of Earth. Green chemistry seeks to design chemical products and processes to reduce the use of resources and energy and the generation of hazardous waste.

greenhouse gas A component of Earth's atmosphere that absorbs infrared radiation from the ground and reradiates some of it back to the ground, thereby keeping Earth warmer than it would be in the absence of the greenhouse gas.

gross sample See *bulk sample*.

ground state State of an atom or a molecule with the minimum possible energy.

Grubbs test Statistical test used to decide whether to discard a datum that appears discrepant.

guard column In high-performance liquid chromatography, a short column packed with the same material as the main column and placed between the injector and the main column. The guard column removes impurities that might irreversibly bind to the main column and degrade it. Also called *precolumn*. In gas chromatography, the guard column is a length of empty, silanized capillary ahead of the chromatography column. Nonvolatile residues accumulate in the guard column.

half-cell Part of an electrochemical cell in which half of an electrochemical reaction (either the oxidation or the reduction reaction) occurs.

half-height Half of the maximum amplitude of a signal.

half-reaction Any redox reaction can be conceptually broken into two half-reactions, one involving only oxidation and one involving only reduction.

half-wave potential Potential at the midpoint of the rise in the current of a polarographic wave.

half-width Width of a signal at its half-height.

Hall-Héroult process Electrolytic production of aluminum metal from a molten solution of Al_2O_3 and cryolite (Na_3AlF_6).

Hammett acidity function The acidity of a solvent that protonates the weak base, B, is called the Hammett acidity function, H_0, and is given by

$$H_0 = pK_a \text{ (for BH}^+) + \log \frac{[B]}{[BH^+]}$$

For dilute aqueous solutions, H_0 approaches pH.

hanging-drop electrode One with a stationary drop of Hg that is used for stripping analysis.

hardness Total concentration of alkaline earth ions in natural water expressed as mg $CaCO_3$ per liter of water as if all of the alkaline earths present were $CaCO_3$. See also *permanent hardness* and *temporary hardness*.

headspace Gas phase above a solid or liquid sample within a sealed vial. The concentration of a volatile component in the headspace is proportional to its concentration in the solid or liquid.

Heisenberg uncertainty principle Certain pairs of physical quantities cannot be known simultaneously with arbitrary accuracy. If δE is the uncertainty in the energy difference between two atomic states and δt is the lifetime of the excited state, their product cannot be known more accurately than $\delta E \delta t \geq h/(4\pi)$, where h is Planck's constant. A similar relation holds between the position and the momentum of a particle. If position is known very accurately, then the uncertainty in momentum is large, and vice versa.

Henderson-Hasselbalch equation A logarithmic rearranged form of the acid dissociation equilibrium equation:

$$pH = pK_a + \log \frac{[A^-]}{[HA]}$$

Henry's law The concentration of a gas dissolved in solution is proportional to the partial pressure (P) of that gas in the gas phase: [dissolved gas] $= K_h P$, where K_h is the *Henry's law constant*. It is a function of the gas, the liquid, and the temperature.

hertz, Hz SI unit of frequency, s^{-1}, also called *reciprocal seconds*.

heterogeneous Having a composition that is not uniform throughout.

HETP, height equivalent to a theoretical plate The length of a chromatography column divided by the number of theoretical plates in the column.

hexadentate ligand One that binds to a metal atom through six ligand atoms.

high-performance liquid chromatography, HPLC A chromatographic technique using very small stationary phase particles and high pressure to force solvent through the column.

HILIC See *hydrophilic interaction chromatography*.

hole Absence of an electron in a semiconductor. When a neighboring electron moves into the hole, a new hole is created where the electron came from. By this means, a hole can move through a solid just as an electron can move through a solid.

hollow-cathode lamp One that emits sharp atomic lines characteristic of the element from which the cathode is made.

homogeneous Having the same composition everywhere.

homogeneous precipitation A technique in which a precipitating agent is generated slowly by a reaction in homogeneous solution, effecting a slow crystallization instead of a rapid precipitation of product.

HPLC See *high-performance liquid chromatography*.

hydrated radius Effective size of an ion or a molecule plus its associated water molecules in solution.

hydrodynamic flow Motion of fluid through a tube, driven by a pressure difference. Hydrodynamic flow is usually laminar, in which there is a parabolic profile of velocity vectors, with the highest velocity at the center of the stream and zero velocity at the walls.

hydrodynamic injection In capillary electrophoresis, the use of a pressure difference between the two ends of the capillary to inject sample into the capillary. Injection is achieved by applying pressure on one end, by applying suction on one end, or by siphoning.

hydrodynamic radius Effective radius of a molecule migrating through a fluid. It is defined by the *Stokes equation*, in which the friction coefficient is $6\pi\eta r$, where η is the viscosity of the fluid and r is the hydrodynamic radius of the molecule.

hydrolysis "Reaction with water." The reaction $B + H_2O \rightleftharpoons BH^+ + OH^-$ is often called hydrolysis of a base.

hydronium ion, H_3O^+ What we really mean when we write $H^+(aq)$.

hydrophilic interaction chromatography Chromatographic separation of polar solutes with a hydrophilic stationary phase using mixed organic-aqueous eluent. Eluent strength increases with decreasing organic solvent. Commonly called HILIC.

hydrophilic substance One that is soluble in water or attracts water to its surface.

hydrophobic interaction chromatography Chromatographic separation based on the interaction of a hydrophobic solute with a hydrophobic stationary phase.

hydrophobic substance One that is insoluble in water or repels water from its surface.

hygroscopic substance One that readily picks up water from the atmosphere.

hypothesis test Statistical test based on the probability that the *null hypothesis* is false. We usually reject the null hypothesis if its probability of being true is greater than 5%.

ignition The heating to high temperature of some gravimetric precipitates to convert them into a known, constant composition that can be weighed.

immiscible liquids Two liquids that do not form a single phase when mixed together.

immunoassay An analytical measurement using antibodies.

inclusion An impurity that occupies lattice sites in a crystal.

indeterminate error See *random error*.

indicator A compound having a physical property (usually color) that changes abruptly near the equivalence point of a chemical reaction.

indicator electrode One that develops a potential whose magnitude depends on the activity of one or more species in contact with the electrode. Same as *working electrode*.

indicator error Difference between the indicator end point of a titration and the true equivalence point.

indirect detection Chromatographic detection based on the *absence* of signal from a background species. For example, in ion chromatography, a light-absorbing ionic species can be added to the eluent. Nonabsorbing analyte replaces an equivalent amount of light-absorbing eluent when analyte emerges from the column, thereby decreasing the absorbance of eluate.

indirect titration One that is used when the analyte cannot be directly titrated. For example, analyte A may be precipitated with excess reagent R. The product is filtered, and the excess R washed away. Then AR is dissolved in a new solution, and R can be titrated.

inductively coupled plasma A high-temperature plasma that derives its energy from an oscillating radio-frequency field. It is used to atomize a sample for atomic emission spectroscopy.

inflection point One at which the derivative of the slope is 0: $d(dy/dx)/dx = d^2y/dx^2 = 0$. That is, the slope reaches a maximum or minimum value.

inner filter effect See *self-absorption*.

inorganic carbon In a natural water or industrial effluent sample, the quantity of dissolved carbonate and bicarbonate.

instrument precision Reproducibility observed when the same quantity of one sample is repeatedly introduced into an instrument.

intensity Power per unit area of a beam of electromagnetic radiation (W/m^2). Also called *radiant power* or *irradiance*.

intercalation Binding of a flat, aromatic molecule between the flat, hydrogen-bonded base pairs in DNA or RNA.

intercept For a straight line whose equation is $y = mx + b$, b is the intercept. It is the value of y when $x = 0$.

interference A phenomenon in which the presence of one substance changes the response in the analysis of another substance.

interference filter A filter that transmits a particular band of wavelengths and reflects others. Transmitted light interferes constructively within the filter, whereas light that is reflected interferes destructively.

interferogram A graph of light intensity versus retardation (or time) for the radiation emerging from an *interferometer*.

interferometer A device with a beamsplitter, fixed mirror, and moving mirror that breaks input light into two beams that interfere with each other. The degree of interference depends on the difference in pathlength of the two beams.

interlaboratory precision The reproducibility observed when aliquots of the same sample are analyzed by different people in different laboratories.

intermediate precision Precision observed when an assay is performed by different people on different instruments on different days in the same lab. Also called *ruggedness*.

internal conversion A nonradiative electronic transition between states of the same energy and same electron-spin multiplicity.

internal standard A known quantity of a compound other than analyte added to a solution containing an unknown quantity of analyte. The concentration of analyte is then measured relative to that of the internal standard.

interpolation Estimation of the value of a quantity that lies between two known values.

interstitial volume, V_0 The volume of mobile phase *outside* of the stationary phase in a molecular exclusion chromatography column. The total volume of mobile phase is the interstitial volume plus the volume of solvent inside the stationary phase particles. Formerly called *void volume*.

intersystem crossing A nonradiative, isoenergetic, electronic transition between states of different electron-spin multiplicity.

intra-assay precision Precision observed when analyzing aliquots of a homogeneous material several times by one person on one day with the same equipment.

iodimetry Use of triiodide (or iodine) as a titrant.

iodometry A technique in which an oxidant is treated with I^- to produce I_3^-, which is then titrated (usually with thiosulfate).

ion chromatography High-performance liquid chromatography ion-exchange separation of ions. See also *suppressed-ion chromatography* and *single-column ion chromatography*.

ion-exchange chromatography A technique in which solute ions are retained by oppositely charged sites in the stationary phase.

ion-exchange equilibrium An equilibrium involving replacement of a cation by a different cation or replacement of an anion by a different anion. Usually the ions in these reactions are bound by electrostatic forces.

ion-exchange membrane Membrane containing covalently bound charged groups. Oppositely charged ions in solution penetrate the membrane freely, but similarly charged ions tend to be excluded from the membrane by the bound charges.

ion-exclusion chromatography A technique in which electrolytes are separated from nonelectrolytes by an ion-exchange resin.

ion guide A symmetric, parallel arrangement of four, six, or eight electrically conductive rods to which a radio frequency signal is applied to confine ions to the center of the guide. Neutral molecules can escape between the rods and are removed by a vacuum pump. The guide transports a collimated beam of ions from a region of high pressure to a region of low pressure.

ionic atmosphere The region of solution around an ion or a charged particle. It contains an excess of oppositely charged ions.

ionic liquid A salt that melts near or below room temperature and has a large temperature range over which it remains liquid.

ionic radius Effective size of an ion in a crystal.

ionic strength, μ A measure of the total concentration of ions in solution, given by $\mu = \frac{1}{2}\sum c_i z_i^2$, where c_i is the concentration of the ith ion in solution and z_i is the charge on that ion. The sum extends over all ions in solution, including the ions whose activity coefficients are being calculated.

ionization interference In atomic spectroscopy, loss of signal intensity as a result of ionization of analyte atoms.

ionization suppressor An element used in atomic spectroscopy to decrease the extent of ionization of the analyte.

ion mobility spectrometer An instrument that measures the drift time of gaseous ions migrating in an electric field against a flow of gas. The "spectrum" of detector current versus drift time is really an electropherogram of a gas.

ionophore A molecule with a hydrophobic outside and a polar inside that can engulf an ion and carry the ion through a hydrophobic phase (such as a cell membrane).

ion pair A closely associated anion and cation, held together by electrostatic attraction. In solvents less polar than water, ions are usually found as ion pairs.

ion-pair chromatography Separation of ions on reversed-phase high-performance liquid chromatography column by adding to the eluent a hydrophobic counterion that pairs with analyte ion and is attracted to stationary phase.

ion-selective electrode One whose potential is selectively dependent on the concentration of one particular ion in solution.

ion spray See *electrospray ionization*.

irradiance Power per unit area (W/m^2) of a beam of electromagnetic radiation. Also called *radiant power* or *intensity*.

isobaric interference In mass spectrometry, overlap of two peaks with nearly the same mass. For example, $^{41}K^+$ and $^{40}ArH^+$ differ by 0.01 atomic mass unit and appear as a single peak unless the spectrometer resolution is great enough to separate them.

isocratic elution Chromatography using a single solvent (or single solvent mixture) for the mobile phase.

isoelectric buffer   A neutral, polyprotic acid occasionally used as a low-conductivity "buffer" for capillary zone electrophoresis. For example, a solution of pure aspartic acid ($pK_1 = 1.99$, $pK_2 = 3.90$, $pK_3 = 10.00$) has $pH = \frac{1}{2}(pK_1 + pK_2) = 2.94$. Calling pure aspartic acid a "buffer" is an oxymoron, because the buffer capacity is a *minimum* at pH 2.94 and increases to maxima at pH 1.99 and 3.90. However, as the pH drifts away from 2.94, the solution gains significant buffer capacity. When electrophoresis is conducted in a background electrolyte of aspartic acid, the pH stays near 2.94 and conductivity remains very low, permitting a high electric field to be used, thus enabling rapid separations.

isoelectric focusing   A technique in which a sample containing polyprotic molecules is subjected to a strong electric field in a medium with a pH gradient. Each species migrates until it reaches the region of its isoelectric pH. In that region, the molecule has no net charge, ceases to migrate, and remains focused in a narrow band.

isoelectric point   That pH at which the average charge of a polyprotic species is 0. Same as *isoelectric pH*.

isoionic point   The pH of a pure solution of a neutral, polyprotic molecule. The only ions present are H^+, OH^-, and those derived from the polyprotic species. Same as *isoionic pH*.

isosbestic point   A wavelength at which the absorbance spectra of two species cross each other. The appearance of isosbestic points in a solution in which a chemical reaction is occurring is evidence that there are only two components present, with a constant total concentration.

isotherm   Graph of C_s (mass transfer in stationary phase) versus C_m (mass transfer in mobile phase) at a given temperature.

isotope ratio mass spectrometry   A mass spectrometric technique designed to provide accurate measurements of the ratio of different ions of a selected element. The instrument has one detector dedicated to each isotope.

isotopologues   Molecules or ions with different numbers of isotopes of one or more elements. Example: $^{12}C_2H_4O$ and $^{12}C^{13}CH_4O$.

isotopomers   Molecules or ions with the same isotopic composition, but different locations of the isotopes. Example: $^{13}CH_3{}^{12}CH{=}O$ and $^{12}CH_3{}^{13}CH{=}O$

Job's method   See *method of continuous variation*.

Johnson noise   A form of white noise (Gaussian noise) arising from random fluctuations of electrons in an electronic device. Lowering the temperature lowers the Johnson noise. Also called *Nyquist noise*.

Jones reductor   A column packed with zinc amalgam. An oxidized analyte is passed through to reduce the analyte, which is then titrated with an oxidizing agent.

joule, J   SI unit of energy. One joule is expended when a force of 1 N acts over a distance of 1 m. This energy is equivalent to that required to raise 102 g (about $\frac{1}{4}$ pound) by 1 m at sea level.

Joule heating   Heat produced in an electric circuit by the flow of electricity. Power (J/s) $= I^2R$, where I is the current (A) and R is the resistance (ohms).

junction potential   An electric potential that exists at the junction (interface) between two different electrolyte solutions or substances. It arises in solutions as a result of unequal rates of diffusion of different ions.

Karl Fischer titration   A sensitive technique for measuring traces of water, based on the reaction of H_2O with an amine, I_2, SO_2, and an alcohol.

kelvin, K   SI unit of temperature defined such that the temperature of water at its triple point (where water, ice, and water vapor are at equilibrium) is 273.16 K and the absolute zero of temperature is 0 K. Also called *absolute temperature*.

kilogram, kg   SI unit of mass equal to the mass of a particular Pt-Ir cylinder kept at the International Bureau of Weights and Measures, Sèvres, France.

Kjeldahl nitrogen analysis   Procedure for the analysis of nitrogen in organic compounds. The compound is digested with boiling H_2SO_4 to convert nitrogen into NH_4^+, which is treated with base and distilled as NH_3 into a standard acid solution. The moles of acid consumed equal the moles of NH_3 liberated from the compound.

Kovats index   See *retention index*.

lab-on-a-chip   See *microfluidic chip*.

laboratory sample   Portion of bulk sample taken to the lab for analysis. Must have the same composition as the bulk sample.

laminar flow   Motion with a parabolic velocity profile of fluid through a tube. Motion is fastest at the center and zero at the walls.

laser   Source of intense, coherent monochromatic radiation. Light is produced by stimulated emission of radiation from a medium in which an excited state has been pumped to a high population. Coherence means that all light exiting the laser has the same phase.

laser-induced breakdown spectroscopy   Semiquantitative measurement of elements in a surface by vaporizing a small patch with a short laser pulse and measuring atomic emission from the plasma above the surface.

Latimer diagram   One that shows the reduction potentials connecting a series of species containing an element in different oxidation states.

law of mass action   For the chemical reaction $aA + bB \rightleftharpoons cC + dD$, the condition at equilibrium is $K = \mathcal{A}_C^c \mathcal{A}_D^d / \mathcal{A}_A^a \mathcal{A}_B^b$, where A_i is the activity of the ith species. The law is usually used in approximate form, in which activities are replaced by concentrations.

least squares   Process of fitting a mathematical function to a set of measured points by minimizing the sum of the squares of the distances from the points to the curve.

Le Châtelier's principle   If a system at equilibrium is disturbed, the direction in which it proceeds back to equilibrium is such that the disturbance is partly offset.

leveling effect   The strongest acid that can exist in solution is the protonated form of the solvent. A stronger acid will donate its proton to the solvent and be leveled to the acid strength of the protonated solvent. Similarly, the strongest base that can exist in a solvent is the deprotonated form of the solvent.

Lewis acid   One that can form a chemical bond by sharing a pair of electrons donated by another species.

Lewis base   One that can form a chemical bond by sharing a pair of its electrons with another species.

lifetime   In spectroscopy, the lifetime for an excited state is the time required for the population to decrease to 1/e times its initial value, where e is the base of the natural logarithm.

ligand   An atom or a group attached to a central atom in a molecule. The term is often used to mean any group attached to anything else of interest.

limiting current   In a polarographic experiment, the current that is reached at the plateau of a polarographic wave. See also *diffusion current*.

limiting reagent   The reactant that is used up first in a chemical reaction. After the limiting reagent has been consumed, the reaction cannot continue.

limit of quantitation   The minimum signal that can be measured "accurately," often taken as the mean signal for blanks plus 10 times the standard deviation of a low-concentration sample.

linear flow rate   In chromatography, the distance per unit time traveled by the mobile phase. Same as *linear velocity*.

linear interpolation   A form of interpolation in which the variation in some quantity is assumed to be linear. For example, to find the value of b when $a = 32.4$ in the following table,

a:	32	32.4	33
b:	12.85	x	17.96

you can set up the proportion

$$\frac{32.4 - 32}{33 - 32} = \frac{x - 12.85}{17.96 - 12.85}$$

which gives $x = 14.89$.

linear quadrupole ion-trap mass spectrometer   An instrument that separates gaseous ions by trapping them in stable trajectories inside a linear quadrupole by use of radio-frequency fields. Ions can be expelled from the trap in order of increasing m/z for mass spectrometry.

linear range   Concentration range over which the change in detector response is proportional to the change in analyte concentration.

linear response   The case in which the analytical signal is directly proportional to the concentration of analyte.

linearity   A measure of how well data in a graph follow a straight line, showing that response is proportional to the quantity of analyte.

linear-solvent-strength model   In liquid chromatography, a model in which the retention factor k is related to mobile phase composition Φ by the empirical equation, $\log k \approx \log k_w - S\Phi$, where $\log k_w$ and S are constants.

linear velocity, u_x   In chromatography, the distance per unit time traveled by the mobile phase. Also called *linear flow rate*, in contrast to *volume flow rate*.

linear voltage ramp   The linearly increasing potential that is applied to the working electrode in polarography.

line noise   Noise concentrated at discrete frequencies that come from sources external to an intended measuring system. Common sources include radiation emanating from the 60-Hz power line, vacuum-pump motors, and radio-frequency devices. Same as *whistle noise*.

lipid bilayer   Double layer formed by molecules containing hydrophilic headgroup and hydrophobic tail. The tails of the two layers associate with each other and the headgroups face the aqueous solvent.

liquid-based ion-selective electrode   One that has a hydrophobic membrane separating an inner reference electrode from the analyte solution. The membrane is saturated with a liquid ion exchanger dissolved in a nonpolar solvent. The ion-exchange equilibrium of analyte between the liquid ion exchanger and the aqueous solution gives rise to the electrode potential.

liquid chromatography   A form of chromatography in which the mobile phase is a liquid.

liquid-liquid extraction   Extraction of a solute from one liquid phase to another liquid phase.

liter, L   Common unit of volume equal to exactly 1 000 cm³.

logarithm   The base 10 logarithm of n is a if $10^a = n$ (which means log $n = a$). The natural logarithm of n is a if $e^a = n$ (which means ln $n = a$). The number e (= 2.718 28 . . .) is called the base of the natural logarithm.

longitudinal diffusion   Diffusion of solute molecules parallel to the direction of travel through a chromatography column.

lot   Entire material that is to be analyzed. Examples are a bottle of reagent, a lake, or a truckload of gravel.

lower limit of detection   See *detection limit*.

lower limit of quantitation   Smallest amount of analyte that can be measured with reasonable accuracy. Usually taken as 10 times the standard deviation of a low-concentration sample. Also called *quantitation limit*.

Luggin capillary   Used to measure electric potential near the opening of a capillary. A reference electrode is dipped in a reservoir of electrolyte solution connected to a capillary tube. With negligible current between the capillary opening and the reference electrode, the reference electric potential is established at the capillary outlet.

luminescence   Any emission of light by a molecule.

L'vov platform   Platform on which sample is placed in a graphite-tube furnace for atomic spectroscopy to prevent sample vaporization before the walls reach constant temperature.

macroporous resin   Cross-linked ion exchange resins that have molecular scale micropores and also macropores of ≥100 nm. These large pores enable large molecules such as proteins to access the large internal surface area of the particle. Also called *macroreticular resin*.

macroreticular resin   See *macroporous resin*.

magnetic sector mass spectrometer   A device that separates gaseous ions that have the same kinetic energy by passing them through a magnetic field perpendicular to their velocity. Trajectories of ions with a certain mass-to-charge ratio are bent exactly enough to reach the detector. Other ions are deflected too much or too little.

makeup gas   Gas added to the exit stream from a gas chromatography column for the purpose of changing flow rate or gas composition to optimize detection of analyte.

MALDI   See *matrix-assisted laser desorption/ionization*.

mantissa   The part of a logarithm to the right of the decimal point.

masking   Process of adding a chemical substance (a *masking agent*) to a sample to prevent one or more components from interfering in a chemical analysis.

masking agent   A reagent that selectively reacts with one (or more) component(s) of a solution to prevent the component(s) from interfering in a chemical analysis.

mass balance   A statement that the sum of the moles of any element in all of its forms in a solution must equal the moles of that element delivered to the solution.

mass chromatogram   See *selected ion chromatogram*.

mass spectrometer   An instrument that converts gaseous molecules into ions, accelerates them in an electric field, separates them according to their mass-to-charge ratio, and detects the amount of each species.

mass spectrometry   A technique in which gaseous molecules are ionized, accelerated by an electric field, and then separated according to their mass-to-charge ratio.

mass spectrometry–mass spectrometry, MS–MS   See *selected reaction monitoring*.

mass spectrum   In mass spectrometry, a graph showing the relative abundance of each ion as a function of its mass-to-charge ratio.

mass titration   One in which the mass of titrant, instead of the volume, is measured.

mass transfer   Net movement of a molecule from one location to another by mechanisms such as diffusion, convection, and deliberate mixing. In chromatography, mass transfer refers to movement of solute between the mobile and stationary phases.

mass-to-charge ratio, m/z   The mass of an ion in daltons divided by the charge of the ion measured in multiples of the elementary charge. For $^{23}Na^+$, for example, $m/z \approx 23/1 = 23$.

matrix   The medium containing analyte (that is, everything in the sample other than analyte). For many analyses, it is important that standards be prepared in the same matrix as the unknown.

matrix-assisted laser desorption/ionization, MALDI   A gentle technique for introducing predominantly singly charged, intact macromolecular ions into the gas phase. An intimate solid mixture of analyte plus a large excess of a small, ultraviolet-absorbing molecule is irradiated by a pulse from an ultraviolet laser. The small molecule (the matrix) absorbs the radiation, becomes ionized, evaporates, and expands in a supersonic jet that carries analyte into the gas phase. Matrix ions apparently transfer charge to the analyte.

matrix effect   A change in analytical signal caused by anything in the sample other than analyte.

matrix modifier   Substance added to sample for atomic spectroscopy to make the matrix more volatile or the analyte less volatile so that the matrix evaporates before analyte does.

mean   The sum of a set of results divided by the number of values in the set. Also called *average*.

mean activity coefficient   For the salt $(cation)_m(anion)_n$, the mean activity coefficient, $\gamma_\pm$, is related to the individual ion activity coefficients (γ_+ and γ_-) by the equation $\gamma_\pm = (\gamma_+^m \gamma_-^n)^{1/(m+n)}$.

mechanical balance   A balance having a beam that pivots on a fulcrum. Standard masses are used to measure the mass of an unknown.

median   For a set of data, that value above and below which there are equal numbers of data.

mediator   In electrolysis, a molecule that carries electrons between the electrode and the intended analyte. Used when the analyte cannot react directly at the electrode or when analyte concentration is so low that other reagents react instead. Mediator is recycled indefinitely by oxidation or reduction at the counter electrode.

memory effect   Interference in a later analysis caused by substances retained in the instrument or apparatus from an earlier analysis.

meniscus   Curved surface of a liquid.

mesh size   The number of openings per linear inch in a standard screen used to sort particles. A 100/200 mesh particle passes through a 100 mesh screen but not through a 200 mesh screen.

metal ion buffer Consists of a metal-ligand complex plus excess free ligand. The two serve to fix the concentration of free metal ion through the reaction $M + nL \rightleftharpoons ML_n$.

metal ion indicator A compound that changes color when it binds to a metal ion.

meter, m SI unit of length defined as the distance that light travels in a vacuum during $\frac{1}{299\ 792\ 458}$ of a second.

method blank A sample without deliberately added analyte. The method blank is taken through all steps of a chemical analysis, including sample preparation. See also *reagent blank*.

method of continuous variation Procedure for finding the stoichiometry of a complex by preparing a series of solutions with different metal-to-ligand ratios. The ratio at which the extreme response (such as spectrophotometric absorbance) occurs corresponds to the stoichiometry of the complex. Also called *Job's method*.

method of least squares Process of fitting a mathematical function to a set of measured points by minimizing the sum of the squares of the distances from the points to the curve.

method validation Process of proving that an analytical method is acceptable for its intended purpose.

micellar electrokinetic capillary chromatography A form of capillary electrophoresis in which a micelle-forming surfactant is present. Migration times of solutes depend on the fraction of time spent in the micelles.

micelle An aggregate of molecules with ionic headgroups and long, nonpolar tails. The inside of the micelle resembles hydrocarbon solvent, whereas the outside interacts strongly with aqueous solution.

microelectrode An electrode with a diameter on the order of 10 μm (or less). Microelectrodes fit into small places, such as living cells. Their small current gives rise to little ohmic loss, so they can be used in resistive, nonaqueous media. Small double-layer capacitance allows their voltage to be changed rapidly, permitting short-lived species to be studied.

microequilibrium constant An equilibrium constant that describes the reaction of a chemically distinct site in a molecule. For example, a base may be protonated at two distinct sites, each of which has a different equilibrium constant.

microfluidic chip Also called "lab-on-a-chip," a glass or plastic or other "chip" with features such as channels, valves, pumps, and reaction chambers designed to carry out chemical separations, analysis, or manipulations.

microporous particles Chromatographic stationary phase consisting of porous particles 1.5–10 μm in diameter, with high efficiency and high capacity for solute.

migration Induced motion of ions in a solution under the influence of an electric field.

migration time Time required for a solute to reach the detector in capillary electrophoresis.

miscible liquids Two liquids that form a single phase when mixed in any ratio.

mobile phase In chromatography, the phase that travels through the column.

mobile phase strength Refers to the ability of a mobile phase to elute a particular compound from a liquid chromatography column. Also called *solvent strength*.

mobility The terminal velocity that an ion reaches in a field of 1 V/m. Velocity = mobility × field. See also *apparent mobility* and *electrophoretic mobility*.

modulation amplitude In polarography, the magnitude of the voltage pulse applied to the working electrode.

Mohr titration Argentometric titration conducted in the presence of chromate. The end point is signaled by the formation of red $Ag_2CrO_4(s)$.

molality, m A measure of concentration equal to the number of moles of solute per kilogram of solvent.

molar absorptivity, ε Constant of proportionality in Beer's law: $A = \varepsilon bc$, where A is absorbance, b is pathlength, and c is the molarity of the absorbing species. Molar absorptivity tells how much light is absorbed at a particular wavelength by a particular substance. Also called *extinction coefficient*.

molarity, M A measure of concentration equal to the number of moles of solute per liter of solution.

mole, mol SI unit for the amount of substance that contains as many molecules as there are atoms in 12 g of ^{12}C. There are approximately 6.022×10^{23} molecules per mole.

molecular exclusion chromatography A technique in which the stationary phase has a porous structure into which small molecules can enter but large molecules cannot. Molecules are separated by size, with larger molecules moving faster than smaller ones. Also called *size exclusion, gel filtration, gel permeation,* or *size exclusion chromatography*.

molecular ion, M⁺· In mass spectrometry, an ion that has not lost or gained any atoms during ionization.

molecularly imprinted polymer A polymer synthesized in the presence of a template molecule. After the template is removed, the polymer has a void with the right shape to hold the template or closely related molecules. Polymer functional groups are positioned correctly to bind to template functional groups.

molecular mass The number of grams of a substance that contains Avogadro's number of molecules.

molecular orbital Describes the distribution of an electron within a molecule.

molecular sieve A crystalline solid particle with pores the size of small molecules. Zeolites (sodium aluminosilicates) are a common type.

mole fraction The number of moles of a substance in a mixture divided by the total number of moles of all components present.

monochromatic light Light with a very narrow range of wavelengths ("one color").

monochromator A device (usually a grating or prism) that disperses light into its component wavelengths and selects a narrow band of wavelengths to pass through the exit slit.

monoclonal antibody Identical antibody molecules produced by a single type of cell.

monodentate ligand One that binds to a metal ion through only one atom.

monolithic column Chromatographic column in which polymerization is conducted inside the column to fill the column with porous stationary phase. Monolithic columns allow faster flow rates because the pore structure is maintained at high pressure.

monoprotic acids and bases Compounds that can donate or accept one proton.

mortar and pestle A mortar is a hard ceramic or steel vessel in which a solid sample is ground with a hard tool called a pestle.

mother liquor Solution from which a substance has crystallized or precipitated or crystallized.

MSⁿ Successive cycles of selected reaction monitoring. The product ion from one cycle becomes the precursor ion for the next cycle. This experiment can be conducted in a single three-dimensional quadrupole ion-trap mass spectrometer under software control.

mull A fine dispersion of a solid in an oil.

multidentate ligand One that binds to a metal ion through more than one atom.

m/z See *mass-to-charge ratio*.

natural logarithm The natural logarithm (ln) of a is b if $e^b = a$, where e = 2.718 28…. See also *logarithm*.

nebulization Process of breaking a liquid into a mist of fine droplets.

nebulizer In atomic spectroscopy, this device breaks the liquid sample into a mist of fine droplets.

needle valve A valve with a tapered plunger that fits into a small orifice to constrict flow.

nephelometry A technique in which the intensity of light scattered at 90° by a suspension is measured to determine the concentration of suspended particles. In a precipitation titration, the scattering increases until the equivalence point is reached, and then remains constant.

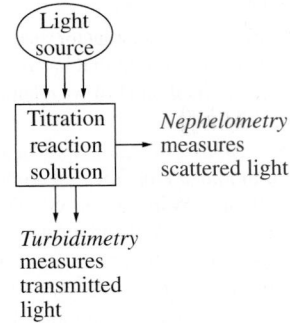

Nernst equation   Relates the voltage of a cell, E, to the activities of reactants and products:

$$E = E° - \frac{RT}{nF} \ln Q = E° - \frac{0.059\,16\text{ V}}{n} \log Q \text{ (at 298.15 K)}$$

where R is the gas constant, T is temperature in kelvins, F is the Faraday constant, Q is the reaction quotient, and n is the number of electrons transferred in the balanced reaction. $E°$ is the cell voltage when all activities are unity.

neutralization   Process in which a stoichiometric equivalent of acid is added to a base (or *vice versa*).

neutron-activation analysis   A technique in which radiation is observed from a sample bombarded by slow neutrons. The radiation gives both qualitative and quantitative information about the sample composition.

newton, N   SI unit of force. One newton will accelerate a mass of 1 kg by 1 m/s^2.

nitrogen chemiluminescence detector   Gas chromatography detector in which combustion of eluate at 1 800°C converts nitrogen into NO, which reacts with O_3 to create a chemiluminescent product. Response to N is 10^7 times greater than response to C.

nitrogen-phosphorus detector   See *alkali flame detector*.

nitrogen rule   A compound with an odd number of nitrogen atoms—in addition to C, H, halogens, O, S, Si, and P—will have an odd nominal mass. A compound with an even number of nitrogen atoms (0, 2, 4, …) will have an even nominal mass.

noise   Signals originating from sources other than those intended to be measured. See, for example, *line noise* and *white noise*.

nominal mass   *Integer* mass of the species with the most abundant isotope of each of the constituent atoms. For C, H, and Br, the most abundant isotopes are ^{12}C, ^{1}H, and ^{79}Br. Therefore, the nominal mass of C_2H_5Br is $(2 \times 12) + (5 \times 1) + (1 \times 79) = 108$.

nonelectrolyte   A substance that does not dissociate into ions when dissolved.

nonpolar compound   A compound, such as a hydrocarbon, with little charge separation within the molecule and no net ionic charge. Nonpolar compounds interact with other substances by weak van der Waals forces and are generally not soluble in water.

nonpolarizable electrode   One whose potential remains nearly constant, even when current flows; e.g., a saturated calomel electrode.

normal error curve   See *Gaussian distribution*.

normal hydrogen electrode, N.H.E.   Same as *standard hydrogen electrode*.

normality n times the molarity of a redox reagent, where n is the number of electrons donated or accepted by that species in a particular chemical reaction. For acids and bases, it is also n times the molarity, but n is the number of protons donated or accepted by the species.

normalized peak area   In capillary electrophoresis, the measured peak area divided by the migration time. Used in electrophoresis because different solutes migrate at different rates and pass through the detector at different rates. In chromatography, all solutes pass through the detector at the same rate, so normalization is not necessary.

normal-phase chromatography   A chromatographic separation utilizing a polar stationary phase and a less polar mobile phase.

nucleation   Process whereby molecules in solution come together randomly to form small crystalline aggregates that can grow into larger crystals.

null hypothesis   In statistics, the statement that two sets of data are drawn from populations with the same properties such as standard deviation σ (F test) or mean μ (t test).

Nyquist noise   See *Johnson noise*.

occlusion   An impurity that becomes trapped (sometimes with solvent) in a pocket within a growing crystal.

ohm, Ω   SI unit of electrical resistance. A current of 1 A flows across a potential difference of 1 V if the resistance of the circuit is 1 Ω.

ohmic potential   Voltage required to overcome the electrical resistance of an electrochemical cell.

Ohm's law   States that the current, I, in a circuit is proportional to voltage, E, and inversely proportional to resistance, R: $I = E/R$.

Ohm's law plot   In capillary electrophoresis, a graph of current versus applied voltage. The graph deviates from a straight line when Joule heating becomes significant.

olfactometric detection   Gas chromatography detector that identifies and quantifies odor-active chemicals. Eluate is sniffed by a trained individual.

on-column injection   Technique used in gas chromatography to place a thermally unstable sample directly on the column without excessive heating in an injection port. Solute is condensed at the start of the column by low temperature, and then the temperature is raised to initiate chromatography.

1/f noise   See *drift*.

open tubular column   In chromatography, a hollow capillary column whose walls are coated with stationary phase.

optical density, OD   See *absorbance*.

optical fiber   Fiber that carries light by total internal reflection because the transparent core has a higher refractive index than the surrounding cladding.

optical isomers   See *enantiomers*.

optically active molecule   Same as *chiral molecule*.

optode   A sensor based on an optical fiber. Also called *optrode*.

orbitrap mass spectrometer   A device that traps ions in stable orbits around a central electrode. Ions oscillate from one end of the trap to the other, inducing image currents in the outer electrodes. Fourier analysis of the image currents gives m/z for the oscillating ions.

ordinate   Vertical (y) axis of a graph.

order of magnitude   A power of 10.

orthogonal injection   In time-of-flight mass spectrometry, use of an ion guide and collimating ion optics to introduce ions with a small spread of kinetic energy into the spectrometer. Othogonal injection improves the attainable m/z resolution.

osmolarity   An expression of concentration that gives the total number of particles (ions and molecules) per liter of solution. For nonelectrolytes such as glucose, the osmolarity equals the molarity. For the strong electrolyte $CaCl_2$, the osmolarity is three times the molarity, because each mole of $CaCl_2$ provides 3 mol of ions ($Ca^{2+} + 2Cl^-$).

Ostwald ripening   Thermodynamically favorable process in which small particles (with typical sizes in the nanometer range) in a suspension dissolve and larger particles (micron sizes) crystallize.

outlier   A datum that lies far from the other data in a set of measurements.

overall formation constant, β_n   Same as *cumulative formation constant*.

overpotential   Voltage required to overcome the activation energy for a reaction at an electrode. The electrode potential is above that expected from the equilibrium potential, concentration polarization, and ohmic potential. Overpotential is 0 for a reversible reaction.

oxidant   See *oxidizing agent*.

oxidation   A loss of electrons or a raising of the oxidation state.

oxidation number   See *oxidation state*.

oxidation state   A bookkeeping device used to tell how many electrons have been gained or lost by a neutral atom when it forms a compound. Also called *oxidation number*.

oxidizability   In a natural water or industrial effluent sample, the quantity of O_2 equivalent to the quantity of $KMnO_4$ consumed by refluxing the sample with standard permanganate. Each $KMnO_4$ consumes five electrons and is chemically equivalent to 1.25 mol of O_2. See also *chemical oxygen demand*.

oxidizing agent   A substance that takes electrons in a chemical reaction. Also called *oxidant*.

p function   The negative logarithm (base 10) of a quantity: $pX = -\log X$.

packed column   A chromatography column filled with stationary phase particles.

parallax   Apparent displacement of an object when the observer changes position. Occurs when the scale of an instrument is viewed from a position that is not perpendicular to the scale. The apparent reading is not correct.

particle growth   Process in which molecules become attached to a crystal to form a larger crystal.

partition chromatography   A technique in which separation is achieved by equilibration of solute between two phases.

partition coefficient, K   The equilibrium constant for the reaction in which a solute is partitioned between two phases: solute (in phase 1) $\rightleftharpoons$ solute (in phase 2).

partition ratio   See *retention factor*.

parts per billion (ppb)   An expression of concentration denoting nanograms (10^{-9} g) of solute per gram of solution.

parts per million (ppm)   An expression of concentration denoting micrograms (10^{-6} g) of solute per gram of solution.

pascal, Pa   SI unit of pressure equal to 1 N/m^2. There are 10^5 Pa in 1 bar and 101 325 Pa in 1 atm.

pellicular particles   A type of stationary phase used in liquid chromatography. Contains a thin layer of liquid coated on a spherical bead. It has high efficiency (low plate height) but low capacity.

Penning ionization   Process in which an electronically excited atom such as He ionizes a molecule.

peptization   Occurs when washing some ionic precipitates with distilled water causes the ions that neutralize the charges of individual particles to be washed away. The particles then repel one another, disintegrate, and pass through the filter with the wash liquid.

performance test sample   In a series of analytical measurements, a performance test sample is inserted to see whether the procedure gives correct results when the analyst does not know the right answer. The performance test sample is formulated by someone other than the analyst to contain a known concentration of analyte. Also called a *quality control sample* or *blind sample*.

permanent hardness   Component of water hardness not due to dissolved alkaline earth bicarbonates. This hardness remains in the water after boiling. See also *hardness*.

pH   Defined as $pH = -\log \mathcal{A}_{H+}$ where $\mathcal{A}_{H+}$ is the activity of H^+. In most approximate applications, the pH is taken as $-\log[H^+]$.

phase ratio, β_m   In chromatography, the relative volume of mobile phase to stationary phase in the column. In older literature, the phase ratio may be given as ϕ which is the reciprocal of β_m.

phase transfer agent   A compound such as a crown ether or a salt of a hydrophobic ion used to extract an ionic species from water into an organic solvent.

pH meter   A potentiometer that can measure voltage when extremely little current is flowing. It is used with a glass electrode to measure pH.

pH of zero charge   The pH at which the net charge on the surface of a solid is zero.

phospholipid   A molecule with a phosphate-containing polar headgroup and long hydrocarbon (lipid) tail.

phosphorescence   Emission of light during a transition between states of different spin multiplicity (e.g., triplet → singlet). Phosphorescence is slower than fluorescence, with emission occurring ~10^{-4} to 10^2 s after absorption of a photon.

photochemistry   Chemical reaction initiated by absorption of a photon.

photoconductive detector   A detector whose conductivity changes when light is absorbed by the detector material.

photodiode array   An array of semiconductor diodes used to detect light. The array is normally used to detect light that has been spread into its component wavelengths. One small band of wavelengths falls on each detector.

photoioniztion detector   A gas chromatography detector that uses vacuum ultraviolet radiation to ionize aromatic and unsaturated compounds; it has little response to saturated hydrocarbons or halocarbons. Electrons produced by the ionization are collected and measured.

photomultiplier tube   One in which the cathode emits electrons when struck by light. The electrons then strike a series of dynodes (plates that are positive with respect to the cathode), and more electrons are released each time a dynode is struck. As a result, more than 10^6 electrons may reach the anode for every photon striking the cathode.

photon   A "particle" of light with energy $h\nu$, where h is Planck's constant and ν is the frequency of the light.

phototube   A vacuum tube with a photoemissive cathode. The electric current flowing between the cathode and the anode is proportional to the intensity of light striking the cathode.

photovoltaic detector   A photodetector with a junction across which the voltage changes when light is absorbed by the detector material.

pH-stat   A device that maintains a constant pH in a solution by continually injecting (or electrochemically generating) acid or base to counteract pH changes.

piezoelectric crystal   A crystal that deforms when an electric field is applied.

piezoelectric effect   Development of electric charge on the surface of certain crystals when subjected to pressure. Conversely, application of an electric field deforms the crystal.

pipet   A glass tube calibrated to deliver a fixed or variable volume of liquid.

pK   The negative logarithm (base 10) of an equilibrium constant: $pK = -\log K$.

Planck distribution   Equation giving the spectral distribution of blackbody radiation:

$$M_\lambda = \frac{2\pi hc^2}{\lambda^5}\left(\frac{1}{e^{hc/\lambda kT} - 1}\right)$$

where h is Planck's constant, c is the speed of light, λ is the wavelength of light, k is Boltzmann's constant, and T is temperature in kelvins. M_λ is the power (watts) per square meter of surface per meter of wavelength radiating from the surface. The integral $\int_{\lambda_1}^{\lambda_2} M_\lambda \, d\lambda$ gives the power emitted per unit area in the wavelength interval from λ_1 to λ_2.

Planck's constant   Fundamental constant of nature equal to the energy of light divided by its frequency: $h = E/\nu \approx 6.626 \times 10^{-34}$ J·s.

plane-polarized light   Light whose electric field oscillates in one plane.

plasma   A gas that is hot enough to contain free ions and electrons, as well as neutral molecules.

plasmon   A collective oscillation of the free electrons in a metal.

plate height, H   Length of a chromatography column divided by the number of theoretical plates in the column. Calculated as the variance, σ^2, of the analyte band divided by the distance, x, it has traveled: $H = \sigma^2/x$.

polar compound   A compound, such as an alcohol, that has positive and negative regions that attract neighboring molecules by electrostatic forces. Polar compounds tend to be soluble in water and insoluble in nonpolar solvents such as hydrocarbons.

polarity index, P'   In partition liquid chromatography, refers to the ability of a solvent to dissolve polar compounds. Polarity arises from dipole, hydrogen-bond-donating, and hydrogen-bond-accepting interactions.

polarizability   Proportionality constant relating the induced dipole to the strength of the electric field. When a molecule is placed in an electric field, a dipole is induced in the molecule by attraction of the electrons toward the positive pole and attraction of the nuclei toward the negative pole.

polarizable electrode One whose potential can change readily when a small current flows. Examples are Pt or Ag wires used as indicator electrodes.

polarogram A graph showing the relation between current and potential during a polarographic experiment.

polarograph An instrument used to obtain and record a polarogram.

polarographic wave S-shaped increase in current during a redox reaction in polarography.

polarography A voltammetry experiment using a dropping-mercury electrode.

polychromatic light Light of many wavelengths ("many colors").

polychromator A device that spreads light into its component wavelengths and directs each small band of wavelengths to a different region where it is detected by a photodiode array.

polyprotic acid or base Compound that can donate or accept more than one proton.

population inversion A necessary condition for laser operation in which the population of an excited energy level is greater than that of a lower energy level.

population mean Mean value for an infinite population of data. Same as *true mean*.

population standard deviation Standard deviation for an infinite population of data. Same as *true standard deviation*.

porous-layer column Gas chromatography column containing an adsorptive solid phase coated on the inside surface of its wall.

postprecipitation Adsorption of otherwise soluble impurities on the surface of a precipitate after the precipitation is over.

potential See *electric potential*.

potential difference See *electric potential*.

potentiometer A device that measures electric potential difference by balancing it with a known potential of the opposite sign. A potentiometer measures the same quantity as that measured by a voltmeter, but the potentiometer is designed to draw much less current from the circuit being measured.

potentiometry An analytical method in which an electric potential difference (a voltage) of a cell is measured.

potentiostat An electronic device that maintains a constant voltage between a pair of electrodes.

power Energy expended (work done) per unit time. SI units are $J/s = $ watts, W.

ppb See *parts per billion*.

ppm See *parts per million*.

precipitant A substance that precipitates a species from solution.

precipitation Occurs when a substance leaves solution rapidly (to form either microcrystalline or amorphous solid).

precipitation titration One in which the analyte forms a precipitate with the titrant.

precision How well replicate measurements agree with one another.

precolumn See *guard column*.

preconcentration Process of concentrating trace components of a mixture prior to their analysis.

precursor ion In tandem mass spectrometry (selected reaction monitoring), the ion selected by the first mass separator for fragmentation in the collision cell.

premix burner In atomic spectroscopy, one in which the sample is nebulized and simultaneously mixed with fuel and oxidant before being fed into the flame.

preoxidation Process of oxidizing analyte prior to titrating it with a reducing agent.

preparative chromatography Chromatography of large quantities of material conducted for the purpose of isolating pure material.

prereduction Process of reducing an analyte to a lower oxidation state prior to performing a titration with an oxidizing agent.

pressure Force per unit area, commonly measured in pascals (N/m^2) or bars.

pressure broadening In spectroscopy, line broadening due to collisions between molecules.

primary standard A reagent that is pure enough and stable enough to be used directly after weighing. The entire mass is considered to be pure reagent.

prism A transparent, triangular solid. Each wavelength of light passing through the prism is bent (refracted) at a different angle.

product The species created in a chemical reaction. Products appear on the right side of the chemical equation.

product ion In tandem mass spectrometry (selected reaction monitoring), the fragment ion from the collision cell selected by the final mass separator for passage through to the detector.

protic solvent One with an acidic hydrogen atom.

protocol In quality assurance, written directions stating what must be documented and how the documentation is to be done.

proton The ion H^+.

proton acceptor A Brønsted-Lowry base: a molecule that combines with H^+.

proton donor A Brønsted-Lowry acid: a molecule that can provide H^+ to another molecule.

protonated molecule In mass spectrometry, the ion MH^+ resulting from addition of H^+ to analyte M.

purge To force a fluid (usually gas) to flow through a substance or a chamber, usually to extract something from the substance being purged or to replace the fluid in the chamber with the purging fluid.

purge and trap A method for removing volatile analytes from liquids or solids, concentrating the analytes, and introducing them into a gas chromatograph. A carrier gas bubbled through a liquid or solid extracts volatile analytes, which are then trapped in a tube containing adsorbent. After analyte has been collected, the adsorbent tube is heated and purged in the reverse direction to desorb the analytes, which are collected by cold trapping in a gas chromatography column.

pyroelectric effect Variation with temperature in the electric polarization of a ferroelectric material.

pyrolysis Thermal decomposition of a substance.

Q test Statistical test used to decide whether to discard a datum that appears discrepant.

quadrupole ion-trap mass spectrometer See *three-dimensional ion-trap mass spectrometer* and *linear quadrupole ion-trap mass spectrometer*.

qualitative analysis Process of determining the identity of the constituents of a substance.

quality assurance Quantitative indications that demonstrate whether data requirements have been met. Also refers to the broader process that includes quality control, quality assessment, and documentation of procedures and results designed to ensure adequate data quality.

quality control Active measures taken to ensure the required accuracy and precision of a chemical analysis.

quality control sample See *performance test sample*.

quantitation limit See *lower limit of quantitation*.

quantitative analysis Process of measuring how much of a constituent is present in a substance.

quantitative transfer Transfer of the entire contents from one vessel to another. This process is usually accomplished by rinsing the first vessel several times with fresh liquid and pouring each rinse into the receiving vessel.

quantum yield In photochemistry, the fraction of absorbed photons that produce a particular result. For example, if a molecule can isomerize from a *cis* to a *trans* isomer when light is absorbed, the quantum yield for isomerization is the number of molecules that isomerize divided by the number that absorb photons. Quantum yield is in the range 0 to 1.

quaternary ammonium ion A cation containing four substituents attached to a nitrogen atom; e.g., $(CH_3CH_2)_4N^+$, the tetraethylammonium ion.

quenching Process in which emission from an excited molecule is decreased by energy transfer to another molecule called a *quencher*.

radian, rad SI unit of plane angle. There are 2π radians in a complete circle.

radiant power Power per unit area (W/m^2) of a beam of electromagnetic radiation. Also called *irradiance, intensity,* or *radiant flux.*

Raleigh scattering Scattering of light in all directions by molecules or particles that are much smaller than the wavelength of the light. The wavelength of scattered light is the same as that of incident light. The intensity of scattered light increases as $1/\lambda^4$, where λ is the wavelength.

Raman scattering Scattering of light in which the wavelength of scattered light is changed from that of incident light by an energy corresponding to vibrational energy of the molecule responsible for scattering. In Stokes Raman scattering, the molecule gains vibrational energy and scattered light has less energy than incident light. In anti-Stokes Raman scattering, an excited molecule loses vibrational energy and scattered light has more energy than incident light.

random error A type of error, which can be either positive or negative and cannot be eliminated, based on the ultimate limitations on a physical measurement. Also called *indeterminate error.*

random heterogeneous material A material in which there are fine-scale differences in composition with no pattern or predictability. When you collect a portion of the material for analysis, you obtain some of each of the different compositions.

random sample Bulk sample constructed by taking portions of the entire lot at random.

range The difference between the highest and lowest values in a set of data. Also called *spread.* With respect to an analytical method, range is the concentration interval over which linearity, accuracy, and precision are all acceptable.

raw data Individual values of a measured quantity, such as peak areas from a chromatogram or volumes from a buret.

reactant The species consumed in a chemical reaction. It appears on the left side of a chemical equation.

reaction quotient, Q Expression having the same form as the equilibrium constant for a reaction. However, the reaction quotient is evaluated for a particular set of existing activities (concentrations), which are generally not the equilibrium values. At equilibrium, $Q = K$.

reagent blank A solution prepared from all of the reagents except analyte. The blank measures the response of the analytical method to impurities in the reagents or any other effects caused by any component other than the analyte. The reagent blank, unlike the method blank, is not subjected to all sample preparation steps before analysis.

reagent gas In a chemical ionization source for mass spectrometry, reagent gas (normally methane, isobutane, or ammonia at ~1 mbar) is converted into strongly proton donating species such as CH_5^+ by a process beginning with electron ionization. Protonated reagent gas reacts with analyte to create protonated analyte.

reagent grade chemical A high-purity chemical generally suitable for use in quantitative analysis and meeting purity requirements set by organizations such as the American Chemical Society.

reciprocal centimeter, cm^{-1} The most common unit of wavenumber, $1/\lambda$, where λ is wavelength in cm.

reconstructed total ion chromatogram In chromatography, a graph of the sum of intensities of all ions detected at all masses (above a selected cutoff) versus time.

rectangular distribution A probability distribution in which the likelihood of observing a value is uniform over a certain range and zero elsewhere.

redox couple A pair of reagents related by electron transfer; e.g., $Fe^{3+} \mid Fe^{2+}$ or $MnO_4^- \mid Mn^{2+}$.

redox indicator A compound used to find the end point of a redox titration because its various oxidation states have different colors. The standard potential of the indicator must be such that its color changes near the equivalence point of the titration.

redox reaction A chemical reaction in which electrons are transferred from one element to another.

redox titration One in which the reaction between analyte and titrant is an oxidation-reduction reaction.

reduced plate height In chromatography, the quotient plate height/d, where the numerator is the height equivalent to a theoretical plate and the denominator is the diameter of stationary phase particles.

reducing agent A substance that donates electrons in a chemical reaction. Also called *reductant.*

reductant See *reducing agent.*

reduction A gain of electrons or a lowering of the oxidation state.

reference electrode One that maintains a constant potential against which the potential of another half-cell may be measured.

reflectance Fraction of incident radiant power reflected by an object.

refraction Bending of light when it passes between media with different refractive indexes.

refractive index, n The speed of light in any medium is c/n, where c is the speed of light in vacuum and n is the refractive index of the medium. The refractive index also measures the angle at which a light ray is bent when it passes from one medium into another. Snell's law states that $n_1 \sin \theta_1 = n_2 \sin \theta_2$, where n_i is the refractive index for each medium and θ_i is the angle of the ray with respect to the normal between the two media.

refractive index detector Liquid chromatography detector that measures the change in refractive index of eluate as solutes emerge from the column.

regeneration Process by which a chromatography column or capillary is returned to its initial state. For ion exchange, regeneration involves replacing ions taken up in the exchange process with the original ion.

relative retention, α In chromatography, the ratio of adjusted retention times for two components. If component 1 has an adjusted retention time of t'_{r1} and component 2 has an adjusted retention time of $t'_{r2} (> t'_{r1})$, the relative retention is $\alpha = t'_{r2}/t'_{r1}$. Also called *separation factor.* See also *unadjusted relative retention, γ.*

relative standard deviation See *coefficient of variation.*

relative supersaturation Defined as $(Q - S)/S$, where S is the concentration of solute in a saturated solution and Q is the concentration in a particular supersaturated solution.

relative uncertainty Uncertainty of a quantity divided by the value of the quantity. It is usually expressed as a percentage of the measured quantity.

releasing agent In atomic spectroscopy, a substance that prevents chemical interference.

repeatability Describes the spread in results when one person uses one procedure to analyze the same sample by the same method with the same equipment multiple times.

replicate measurements Repeated measurements of the same quantity.

reporting limit Concentration below which regulations dictate that an analyte is reported as "not detected." The reporting limit is typically set 5 to 10 times higher than the detection limit.

reprecipitation Sometimes a gravimetric precipitate can be freed of impurities by redissolving it in fresh solvent and reprecipitating it. The impurities are present at lower concentration during the second precipitation and are less likely to coprecipitate.

reproducibility Describes the spread in results when different people in different labs using different equipment each try to follow the same procedure with the same kind of sample.

residual current The small current that is observed prior to the decomposition potential in an electrolysis.

resin Small, hard particles of an ion exchanger, such as polystyrene with ionic substituents.

resistance, R A measure of the retarding force opposing the flow of electric current. SI unit is ohm, Ω.

resistivity, ρ A measure of the ability of a material to retard the flow of electric current. $J = E/\rho$, where J is the current density (A/m^2) and E is electric field (V/m). Units of resistivity are $V \cdot m/A = \text{ohm} \cdot m = \Omega \cdot m$. The resistance, R, of a conductor with a given length and cross-sectional area is given by $R = \rho \cdot \text{length/area}$. Resistivity is the reciprocal of *conductivity.*

resolution How close two bands in a spectrum or a chromatogram can be to each other and still be seen as two peaks. In chromatography, it is defined as the difference in retention times of adjacent peaks divided by their width. In mass spectrometry, resolution is the smallest difference in *m/z* values that can be detected as separate peaks. Report the *m/z* value at which resolution is measured.

resolving power   In mass spectrometry, resolving power can be defined as $m/\Delta m$, where Δm is the separation of two peaks when the overlap at the base is 10% of the peak height and m is the smaller of the two *m/z* values. Alternatively, resolving power can be taken as $m/m_{1/2}$, where $m_{1/2}$ is the width of a peak at half the maximum height. In this case, the dip between two barely resolved peaks is 8% below the peak heights.

response factor, *F*   Relative response of a detector to analyte (X) and internal standard (S): (signal from X)/[X] = *F*(signal from S)/[S]. Once you have measured *F* with a standard mixture, you can use it to find [X] in an unknown if you know [S] and the quotient (signal from X)/(signal from S).

results   What we ultimately report after applying statistics to treated data.

retardation, δ   Difference in pathlength between light striking the stationary and moving mirrors of an interferometer.

retention factor, *k*   In chromatography, the adjusted retention time for a peak divided by the time for the mobile phase to travel through the column. Retention factor is also equal to the ratio of the time spent by the solute in the stationary phase to the time spent in the mobile phase. Also called *capacity factor*, *capacity ratio*, and *partition ratio*.

retention gap   In gas chromatography, a 3- to 10-m length of empty, silanized capillary ahead of the chromatography column. The retention gap improves the peak shape of solutes that elute close to solvent when large volumes of solvent are injected or when the solvent has a very different polarity from that of the stationary phase.

retention index, *I*   In gas chromatography, the Kovats retention index is a logarithmic scale that relates the retention time of a compound to those of linear alkanes. Pentane would be given an index of 500, hexane 600, heptane 700, and so on.

retention ratio   In chromatography, the time required for solvent to pass through the column divided by the time required for solute to pass through the column.

retention time, *t*_r   The time from injection for an individual solute to reach the detector of a chromatography column.

retention volume, *V*_r   The volume of solvent needed to elute a solute from a chromatography column.

reversed-phase chromatography   Liquid chromatography in which the stationary phase is less polar than the mobile phase.

rings + double bonds formula   The number of rings + double bonds in a molecule with the formula $C_cH_hN_nO_x$ is $c - h/2 + n/2 + 1$, where c includes all Group 14 atoms (C, Si, Ge, Sn, Pb, which all make four bonds), h includes H + halogens (which make one bond), and n is the number of Group 15 atoms (N, P, As, Sb, Bi, which make three bonds). Group 16 atoms (which make two bonds) do not affect the result.

robustness   Ability of an analytical method to be unaffected by small, deliberate changes in operating parameters.

root-mean-square (rms) noise   Standard deviation of the noise in a region where the signal is flat:

$$\text{rms noise} = \sqrt{\frac{\sum\limits_i (A_i - \overline{A})^2}{n}}$$

where A_i is the measured signal for the *i*th data point, $\overline{A}$ is the mean signal, and n is the number of data points.

rotating disk electrode   A motor-driven electrode with a smooth flat face in contact with the solution. Rapid convection created by rotation brings fresh analyte to the surface of the electrode. A Pt electrode is especially suitable for studying anodic processes, in which a mercury electrode would be too easily oxidized.

rotational transition   Occurs when a molecule changes its rotation energy.

rubber policeman   A glass rod with a flattened piece of rubber on the tip. The rubber is used to scrape solid particles from glass surfaces in gravimetric analysis.

ruggedness   See *intermediate precision*.

run buffer   See *background electrolyte*.

salt   An ionic solid.

salt bridge   A conducting ionic medium in contact with two electrolyte solutions. It allows ions to flow without allowing immediate diffusion of one electrolyte solution into the other. It usually contains a salt that minimizes the junction potential at each end of the bridge.

sample cleanup   Removal of portions of the sample that do not contain analyte and may interfere with analysis.

sample preparation   Transforming a sample into a state that is suitable for analysis. This process can include concentrating a dilute analyte and removing or masking interfering species.

sampling   The process of collecting a representative sample for analysis.

sampling variance   The square of the standard deviation arising from heterogeneity of the sample, not from the analytical procedure. For inhomogeneous materials, it is necessary to take larger portions or more portions to reduce the uncertainty of composition due to variation from one region to another. Total variance is the sum of variances from sampling and from analysis.

saturated calomel electrode (S.C.E.)   A calomel electrode saturated with KCl. The electrode half-reaction is $Hg_2Cl_2(s) + 2e^- \rightleftharpoons 2Hg(l) + 2Cl^-$.

saturated solution   One that contains the maximum amount of a compound that can dissolve at equilibrium.

Scatchard plot   A graph used to find the equilibrium constant for a reaction such as $X + P \rightleftharpoons PX$. It is a graph of [PX]/[X] versus [PX] or any functions proportional to these quantities. The magnitude of the slope of the graph is the equilibrium constant.

S.C.E.   See *saturated calomel electrode*.

schlieren   Streaks in a liquid mixture observed before the two phases have mixed. Streaks arise from regions that refract light differently.

Schottky noise   See *white noise*.

second, s   SI unit of time equal to the duration of 9 192 631 770 periods of the radiation corresponding to the transition between two hyperfine levels of the ground state of ^{133}Cs.

segregated heterogeneous material   A material in which differences in composition are on a large scale. Different regions have obviously different composition.

selected ion chromatogram   A graph of detector response versus time when a mass spectrometer monitors just one or a few species of selected mass-to-charge ratio, *m/z*, emerging from a chromatograph. Also called *mass chromatogram*.

selected ion monitoring   Use of a mass spectrometer to monitor species with just one or a few mass-to-charge ratios, *m/z*.

selected reaction monitoring   A technique in which the precursor ion selected by one mass separator passes through a collision cell in which the precursor breaks into several fragment ions (product ions). A second mass separator then selects one (or a few) of these ions for detection. Selected reaction monitoring improves chromatographic signal-to-noise ratio because it is insensitive to almost everything other than the intended analyte. Also called *mass spectrometry–mass spectrometry (MS–MS)* or *tandem mass spectrometry*.

selectivity   See *specificity*.

selectivity coefficient   With respect to an ion-selective electrode, a measure of the relative response of the electrode to two different ions. In ion-exchange chromatography, the selectivity coefficient is the equilibrium constant for displacement of one ion by another from the resin.

self-absorption   In a luminescence measurement, a high concentration of analyte molecules can absorb energy from excited analyte. Also called *inner filter effect*. If the absorbed energy is dissipated as heat instead of light, fluorescence does not increase in proportion to analyte concentration. Analyte concentration can be so high that fluorescence *decreases* with increasing concentration. In flame emission atomic spectroscopy, there is a lower concentration of excited-state atoms in the cool, outer part of the flame than in the hot, inner flame. The cool atoms can absorb emission from the hot ones and thereby decrease the observed signal.

self-ionization   See *autoprotolysis*.

semiconductor   A material whose conductivity (10^{-7} to $10^4 \; \Omega^{-1} \cdot m^{-1}$) is intermediate between that of good conductors ($10^8 \; \Omega^{-1} \cdot m^{-1}$) and that of insulators (10^{-20} to $10^{-12} \; \Omega^{-1} \cdot m^{-1}$).

semipermeable membrane   A thin layer of material that allows some substances, but not others, to pass across the material. A dialysis membrane allows small molecules to pass, but not large molecules.

sensitivity   Capability of responding reliably and measurably to changes in analyte concentration. In quantitative terms, sensitivity is the amount of instrument response per unit change in concentration of analyte.

separation factor   See *relative retention*.

separator column   Ion-exchange column used to separate analyte species in ion chromatography.

septum   A disk, usually made of silicone rubber, covering the injection port of a gas chromatograph. The sample is injected by syringe through the septum.

sequential injection analysis   Analytical technique related to flow injection. Sample and reagents are taken into a holding coil through a multiport valve. After a suitable reaction time, flow is reversed and the zones of reagent, product, and sample are pushed through a detector to measure the amount of product. Flow is not continuous, so sequential injection consumes less reagents than does flow injection.

serial dilution   The process of making successive dilutions to obtain a desired concentration. The purpose is to transfer accurately small amounts of material that are too little to weigh accurately.

shot noise   A form of white noise (Gaussian noise) arising from the quantized nature of charge carriers and photons. At low signal levels, shot noise arises from random variation in the small number of photons reaching a detector or the small number of electrons and holes generated in a semiconductor.

SI units   International system of units based on the meter, kilogram, second, ampere, kelvin, candela, mole, radian, and steradian.

sieving   In electrophoresis, the separation of macromolecules by migration through a polymer gel. Movement of the smallest molecules is fastest and that of the largest is slowest.

signal averaging   Improvement of a signal by averaging successive scans. The signal increases in proportion to the number of scans accumulated. The noise increases in proportion to the square root of the number of scans. Therefore, the signal-to-noise ratio improves in proportion to the square root of the number of scans collected.

signal-to-noise ratio   The height of a signal divided by the noise in the baseline around the signal. Noise is commonly expressed as root-mean-square noise. The higher the signal-to-noise ratio, the less uncertainty there is in the signal.

significant figure   The number of significant digits in a quantity is the minimum number of digits needed to express the quantity in scientific notation without loss of precision. In experimental data, the first uncertain figure is the last significant figure.

silanization   Treatment of a chromatographic solid support or glass column with hydrophobic silicon compounds that bind to the most reactive Si—OH groups. It reduces irreversible adsorption and tailing of polar solutes.

silver-silver chloride electrode   A common reference electrode containing a silver wire coated with AgCl paste and dipped in a solution saturated with AgCl and (usually) KCl. The half-reaction is $AgCl(s) + e^- \rightleftharpoons Ag(s) + Cl^-$.

single-column ion chromatography   Separation of ions on a low-capacity ion-exchange column, using low-ionic-strength eluent.

single-electrode potential   Voltage measured when the electrode of interest is connected to the positive terminal of a potentiometer and a standard hydrogen electrode is connected to the negative terminal.

singlet state   One in which all electron spins are paired.

size exclusion chromatography   See *molecular exclusion chromatography*.

skimmer cone voltage   See *cone voltage*.

slope   For a straight line whose equation is $y = mx + b$, the value of m is the slope. It is the ratio $\Delta y / \Delta x$ for any segment of the line.

slurry   A suspension of a solid in a solvent.

Smith-Hieftje background correction   In atomic absorption spectroscopy, a method of distinguishing analyte signal from background signal, based on applying a periodic pulse of high current to the hollow-cathode lamp to distort the lamp signal. Signal detected during the current pulse is subtracted from signal detected without the pulse to obtain the corrected response.

smoothing   Use of a mathematical procedure or electrical filtering to improve the quality of a signal.

Snell's law   Relates angle of refraction, θ_2, to angle of incidence, θ_1, for light passing from a medium with refractive index n_1 to a medium of refractive index n_2: $n_1 \sin \theta_1 = n_2 \sin \theta_2$. Angles are measured with respect to the normal to the surface between the two media.

sodium error   Systematic error that ccurs when a glass pH electrode is placed in a strongly basic solution containing very little H^+ and a high concentration of Na^+. The electrode begins to respond to Na^+ as if it were H^+, so the pH reading is lower than the actual pH. Also called *alkaline error*.

solid-phase extraction   Preconcentration procedure in which a solution is passed through a short column of chromatographic stationary phase, such as C_{18} on silica. Trace solutes adsorbed on the column can be eluted with a small volume of solvent of high eluent strength.

solid-phase microextraction   Extraction of compounds from liquids or gases into a coated fiber dispensed from a syringe needle. After extraction, the fiber is withdrawn into the needle and the needle is inserted through the septum of a chromatograph. The fiber is extended inside the injection port and adsorbed solutes are desorbed by heating (for gas chromatography) or solvent (for liquid chromatography).

solid-state ion-selective electrode   An ion-selective electrode that has a solid membrane made of an inorganic salt crystal. Ion-exchange equilibria between the solution and the surface of the crystal account for the electrode potential.

solid-supported liquid-liquid extraction   A form of liquid-liquid extraction in which the liquid to be extracted is applied to a porous solid which retains the liquid. The second liquid is then passed through the porous solid to extract solute from the first liquid.

solubility product, K_{sp}   Equilibrium constant for the dissociation of a solid salt to give its ions in solution. For the reaction $M_m N_n(s) \rightleftharpoons mM^{n+} + nN^{m-}$, $K_{sp} = \mathcal{A}_{M^{n+}}^m \mathcal{A}_{N^{m-}}^n$ where A is the activity of each species.

solute   A minor component of a solution.

solution   A homogeneous mixture of two or more substances.

solvation   Interaction of solvent molecules with solute. Solvent molecules orient themselves around solute to minimize the energy through dipole and van der Waals forces.

solvent   Major constituent of a solution.

solvent extraction   A method in which a chemical species is transferred from one liquid phase to another. It is used to separate components of a mixture.

solvent strength   See *eluent strength* and *mobile phase strength*.

solvent trapping   Splitless gas chromatography injection technique in which solvent is condensed below its boiling point at the start of the column. Solutes dissolve in a narrow band in the condensed solvent.

speciation   Describes the distribution of an element or compound among different chemical forms.

species   Chemists refer to any element, compound, or ion of interest as a *species*. The word *species* is both singular and plural.

specifications   In quality assurance, specifications are written statements describing how good analytical results need to be and what precautions are required in an analytical method.

specific gravity   A dimensionless quantity equal to the mass of a substance divided by the mass of an equal volume of water at 4°C. Specific gravity is virtually identical with density in g/mL.

specificity   Capability of an analytical method to distinguish analyte from other species in the sample. Also called *selectivity*.

spectral interference   In atomic spectroscopy, any physical process that affects light intensity at the analytical wavelength. Created by substances that absorb, scatter, or emit light of the analytical wavelength.

spectrophotometer A device used to measure absorption of light. It includes a source of light, a wavelength selector (monochromator), and an electrical means of detecting light.

spectrophotometric analysis Any method in which light absorption, emission, reflection, or scattering is used to measure chemical concentrations.

spectrophotometric titration One in which absorption or emission of light is used to monitor the progress of the titration reaction and to find the equivalence point.

spectrophotometry In a broad sense, any method using light to measure chemical concentrations.

specular reflection Reflection of light at an angle equal to the angle of incidence.

spike Addition of a known compound (usually at a known concentration) to an unknown. In isotope dilution mass spectrometry, the spike is the added, unusual isotope. *Spike* is a noun and a verb. In *co-chromatography*, spiking is the simultaneous chromatography of a known compound with an unknown. If a known and an unknown have the same retention time on several columns, they are probably identical. Also called *fortification*.

spike recovery The fraction of a spike eventually found by chemical analysis.

split injection Technique used in capillary gas chromatography to inject a small fraction of sample onto the column; the rest of the sample is blown out to waste.

splitless injection Technique used in capillary gas chromatography for trace analysis and quantitative analysis. The entire sample in a low-boiling solvent is directed to the column, where the sample is concentrated by *solvent trapping* (condensing the solvent below its boiling point) or *cold trapping* (condensing solutes far below their boiling range). The column is then warmed to initiate separation.

spontaneous process One that is energetically favorable. It will eventually occur, but thermodynamics makes no prediction about how long it will take.

spread See *range*.

square wave voltammetry A form of *voltammetry* (measurement of current versus potential in an electrochemical cell) in which the potential waveform consists of a square wave superimposed on a voltage staircase. The technique is faster and more sensitive than voltammetry with other waveforms.

stability constant See *formation constant*.

stacking In electrophoresis, the process of concentrating ions into a narrow band at the interface of electrolytes of low conductivity and high conductivity. Stacking occurs because the electric field in low-conductivity electrolyte is stronger than the field in high-conductivity electrolyte. Ions in the low-conductivity region migrate rapidly until they reach the interface, where the electric field is much smaller.

standard addition A technique in which an analytical signal due to an unknown is first measured. Then a known quantity of analyte is added, and the increase in signal is recorded. From the response, it is possible to calculate what quantity of analyte was in the unknown.

standard curve A graph showing the response of an analytical technique to known quantities of analyte.

standard deviation A statistic measuring how closely data are clustered about the mean value. For a finite set of data, the standard deviation, s, is computed from the formula

$$s = \sqrt{\frac{\sum_i (x_i - \bar{x})^2}{n-1}} = \sqrt{\frac{\sum_i (x_i^2)}{n-1} - \frac{\left(\sum_i x_i\right)^2}{n(n-1)}}$$

where n is the number of results, x_i is an individual result, and $\bar{x}$ is the mean result. For a large number of measurements, s approaches σ, the true standard deviation of the population, and $\bar{x}$ approaches μ, the true population mean.

standard deviation of the mean The standard deviation of a set of measurements (σ) divided by the square root of the number of measurements (n) in the set: $\sigma/\sqrt{n}$.

standard error When we obtain parameters such as slope and intercept by curve fitting, their uncertainties are called standard errors, rather than standard deviations. Standard errors behave like the standard deviation of the mean: they decrease by $1/\sqrt{n}$, where n is the number of points in the curve fit. By contrast, standard deviation does not depend on n.

standard hydrogen electrode, S.H.E. One that contains $H_2(g)$ bubbling over a catalytic Pt surface immersed in aqueous H^+. The activities of H_2 and H^+ are both unity in the hypothetical standard electrode. The reaction is $H^+ + e^- \rightleftharpoons \frac{1}{2} H_2(g)$. Also called *normal hydrogen electrode (N.H.E.)*.

standardization Process of determining the concentration of a reagent by reaction with a known quantity of a second reagent.

standard operating procedure A written procedure that must be rigorously followed to ensure the quality of a chemical analysis.

standard reduction potential, E^o The voltage that would be measured when a hypothetical cell containing the desired half-reaction (with all species present at unit activity) is connected to a standard hydrogen electrode anode.

Standard Reference Material Certified reference material from the U.S. National Institute of Standards and Technology.

standard solution A solution whose composition is known by virtue of the way that it was made from a reagent of known purity or by virtue of its reaction with a known quantity of a standard reagent.

standard state The standard state of a solute is 1 M and the standard state of a gas is 1 bar. Pure solids and liquids are considered to be in their standard states. In equilibrium constants, dimensionless concentrations are expressed as a ratio of the concentration of each species to its concentration in its standard state.

standard uncertainty Same as the standard deviation of the mean.

stationary phase In chromatography, the phase that does not move through the column.

stepwise formation constant, K_n Equilibrium constant for a reaction of the type $ML_{n-1} + L \rightleftharpoons ML_n$.

steradian, sr SI unit of solid angle. There are 4π steradians in a complete sphere.

stimulated emission Emission of a photon induced by the passage of another photon of the same wavelength.

stir-bar sorptive extraction Sample preparation method similar to solid-phase microextraction, except the sorptive layer is coated on the outside of a stirring bar. Coating volume is greater than the fiber volume in solid-phase microextraction, so it provides ~10^2 times greater sensitivity for traces of analyte. Analyte is removed from the coating by thermal desorption for chromatography.

stoichiometry Ratios of substances participating in a chemical reaction.

Stokes equation The friction coefficient for a molecule migrating through solution is $6\pi\eta r$, where η is the viscosity of the fluid and r is the hydrodynamic radius (equivalent spherical radius) of the molecule.

stray light In spectrophotometry, unintentional light that reaches the detector but has not passed through the sample or is not part of the narrow set of wavelengths expected from the monochromator.

stripping analysis A sensitive polarographic technique in which analyte is concentrated from dilute solution by reduction into a drop (or a film) of Hg. It is then analyzed polarographically during an anodic redissolution process. Some analytes can be oxidatively concentrated onto an electrode other than Hg and stripped in a reductive process.

strong acids and bases Those that are completely dissociated (to H^+ or OH^-) in water.

strong electrolyte One that mostly dissociates into ions in solution.

Student's t A statistical tool used to express confidence intervals and to compare results from different experiments.

sulfur chemiluminescence detector Gas chromatography detector for the element sulfur. Exhaust from a flame ionization detector is mixed with O_3 to form an excited state of SO_2 that emits light, which is detected. Response to S is 10^7 times greater than response to C.

superconductor A material that loses all electrical resistance when cooled below a critical temperature.

supercritical fluid   A fluid whose temperature is above its critical temperature and whose pressure is above its critical pressure. It has properties of both a liquid and a gas.

supercritical fluid chromatography   Chromatography using supercritical fluid as the mobile phase. Capable of highly efficient separations of nonvolatile solutes and able to use detectors suitable for gas or liquid.

supercritical fluid extraction   Extraction of compounds (usually from solids) with a supercritical fluid solvent.

superficially porous particle   A stationary phase particle for liquid chromatography containing a thin, porous outer layer and a dense, nonporous core. Mass transfer is faster in the superfically porous particle than in a fully porous particle of the same diameter. Also called *fused-core particle*.

supernatant liquid   Liquid remaining above the solid after a precipitation. Also called *supernate*.

supersaturated solution   One that contains more dissolved solute than would be present at equilibrium.

support-coated column   Open tubular gas chromatography column in which the stationary phase is coated on solid support particles attached to the inside wall of the column.

supporting electrolyte   An unreactive salt added in high concentration to solutions for voltammetric measurements (such as polarography). The supporting electrolyte carries most of the ion-migration current and therefore decreases the coulombic migration (drift of ions in the electric field) of electroactive species to a negligible level. The electrolyte also decreases the resistance of the solution.

suppressed-ion chromatography   Separation of ions by using an ion-exchange column followed by a suppressor (membrane or column) to remove ionic eluent.

suppressor   In ion chromatography, a device that transforms ionic eluent into a nonionic form.

surface plasmon resonance   A sensitive means to measure the binding of molecules to a thin gold layer on the underside of a prism. Light directed through the prism is reflected from the gold surface. There is one narrow range of angles at which reflection is nearly 0 because the gold absorbs the light to set up oscillations (called *plasmons*) of the electron cloud in the metal. When a thin layer of material (such as protein or DNA) binds to the side of the gold away from the prism, the electrical properties of the gold are changed and the reflectivity changes.

surface tension   The work required to increase the surface area of a substance (J/m^2).

surfactant   A molecule with an ionic or polar headgroup and a long, nonpolar tail. Surfactants aggregate in aqueous solution to form micelles. Surfactants derive their name from the fact that they accumulate at boundaries between polar and nonpolar phases and modify the surface tension, which is the free energy of formation of the surface. Soaps are surfactants.

sweeping   In capillary electrophoresis, migration of a collector species such as a micelle or chelator to concentrate analyte into a narrow region at the front of the migrating collector species.

syringe   A device having a calibrated barrel into which liquid is sucked by a plunger. The liquid is expelled through a needle by pushing on the plunger.

systematic error   Error due to procedural or instrumental factors that cause a measurement to be consistently too large or too small. The error can, in principle, be discovered and corrected. Also called *determinate error*.

systematic treatment of equilibrium   A method that uses the charge balance, mass balance(s), and equilibria to completely specify a system's composition.

t **test**   Statistical test used to decide whether the results of two experiments are within experimental uncertainty of each other. The uncertainty must be specified to within a certain probability.

tailing   Asymmetric chromatographic band in which the peak maximum moves to shorter retention time than the ideal retention time and is followed by a gradual decrease in concentration. Tailing results either from saturation of stationary phase by solute or adsorption of a solute onto a few active sites on the stationary phase.

tandem mass spectrometry   See *selected reaction monitoring*.

tare   As a noun, *tare* is the mass of an empty vessel used to receive a substance to be weighed. As a verb, *tare* means setting the balance reading to 0 when an empty vessel or weighing paper is placed on the pan.

temperature programming   Raising the temperature of a gas chromatography column during a separation to reduce the retention time of late-eluting components.

temporary hardness   Component of water hardness due to dissolved alkaline earth bicarbonates. It is temporary because boiling causes precipitation of the carbonates.

test portion   Part of the laboratory sample used for one analysis. Also called *aliquot*.

theoretical plate   An imaginary construct in chromatography denoting a segment of a column in which one equilibration of solute occurs between stationary and mobile phases. The number of theoretical plates on a column with Gaussian bandshapes is defined as $N = t_r^2/\sigma^2$, where t_r is the retention time of a peak and σ is the standard deviation of the band. See also *plate height*.

thermal conductivity, κ   Rate at which a substance transports heat (energy per unit time per unit area) through a temperature gradient (degrees per unit distance). Energy flow $[J/(s \cdot m^2)] = -\kappa(dT/dx)$, where κ is the thermal conductivity $[W/(m \cdot K)]$ and dT/dx is the temperature gradient (K/m).

thermal conductivity detector   A device that detects substances eluted from a gas chromatography column by measuring changes in the thermal conductivity of the gas stream.

thermal desorption   A sample preparation technique used in gas chromatography to release volatile substances from a solid sample by heating.

thermistor   A device whose electrical resistance changes markedly with changes in temperature.

thermocouple   An electrical junction across which a temperature-dependent voltage exists. Thermocouples are calibrated for measurement of temperature and usually consist of two dissimilar metals in contact with each other.

thermogravimetric analysis   A technique in which the mass of a substance is measured as the substance is heated. Changes in mass indicate decomposition of the substance, often to well-defined products.

thermometric titration   One in which the temperature is measured to determine the end point. Most titration reactions are exothermic, so the temperature rises during the reaction and suddenly stops rising when the equivalence point is reached.

thin-layer chromatography   Liquid chromatography in which the stationary phase is coated on a flat glass or plastic plate. Solute is spotted near the bottom of the plate. The bottom edge of the plate is placed in contact with solvent, which creeps up the plate by capillary action.

three-dimensional quadrupole ion-trap mass spectrometer   An instrument that separates gaseous ions by trapping them in stable three-dimensional trajectories inside a metal chamber to which a radio-frequency electric field is applied. Application of an oscillating electric field between the ends of the chamber destabilizes the trajectories of ions with a particular mass-to-charge ratio, expelling them from the cavity and into a detector.

time-of-flight mass spectrometer   Ions of different mass accelerated through the same electric field have different velocities. The lighter ions move faster than the heavier ions. The time-of-flight spectrometer finds the mass-to-charge ratio by measuring the time that each group of ions requires to travel a fixed distance to the detector.

titer   A measure of concentration, usually defined as how many milligrams of reagent B will react with 1 mL of reagent A. One milliliter of $AgNO_3$ solution with a titer of 1.28 mg NaCl/mL will be consumed by 1.28 mg NaCl in the reaction $Ag^+ + Cl^- \rightarrow AgCl(s)$. The same solution of $AgNO_3$ has a titer of 0.993 mg of KH_2PO_4/mL, because 1 mL of $AgNO_3$ solution will be consumed by 0.993 mg KH_2PO_4 to precipitate Ag_3PO_4.

titrant   Substance added to the analyte in a titration.

titration   A procedure in which one substance (titrant) is carefully added to another (analyte) until complete reaction has occurred. The quantity of titrant required for complete reaction tells how much analyte is present.

titration curve A graph showing how the concentration of a reactant or a physical property of the solution varies as one reactant (the titrant) is added to another (the analyte).

titration error Difference between the observed end point and the true equivalence point in a titration.

tolerance Manufacturer's stated uncertainty in the accuracy of a device such as a buret or volumetric flask. A 100-mL flask with a tolerance of ± 0.08 mL may contain 99.92 to 100.08 mL.

total carbon In a natural water or industrial effluent sample, the quantity of CO_2 produced when the sample is completely oxidized by oxygen at $900°C$ in the presence of a catalyst.

total ion chromatogram A graph of detector response versus time when a mass spectrometer monitors all ions above a selected m/z ratio emerging from a chromatograph.

total organic carbon In a natural water or industrial effluent sample, the quantity of CO_2 produced when the sample is first acidified and purged to remove carbonate and bicarbonate and then completely oxidized by oxygen at $900°C$ in the presence of a catalyst.

total oxygen demand In a natural water or industrial effluent sample, the quantity of O_2 required for complete oxidation of species in the water at $900°$ C in the presence of a catalyst.

trace Tiny amount of a substance.

trace analysis Chemical analysis of very low levels of analyte, typically ppm and lower.

transition range For an acid-base indicator, the pH range over which the color change occurs. For a redox indicator, the potential range over which the color change occurs.

transmission quadrupole mass spectrometer A mass spectrometer that separates ions by passing them between four metallic cylinders to which are applied direct current and oscillating electric fields. Resonant ions with the right mass-to-charge ratio pass through the chamber to the detector while nonresonant ions are deflected into the cylinders and are lost.

transmittance, T The fraction of incident radiation that passes through a sample. Defined as $T = P/P_0$, where P_0 is the radiant power of light striking the sample on one side and P is the radiant power of light emerging from the other side of the sample.

treated data Concentrations or amounts of analyte found from raw data with a calibration curve or some other calibration method.

triangular distribution A probability distribution shaped like an isosceles triangle in which the likelihood of observing a value is maximum at the center and decreases linearly to zero at two symmetrically spaced limits. Outside the limits, the probability is zero.

triple point The one temperature and pressure at which the solid, liquid, and gaseous forms of a substance are in equilibrium with one another.

triplet state An electronic state in which there are two unpaired electrons.

tungsten lamp An ordinary light bulb in which electricity passing through a tungsten filament heats the wire and causes it to emit visible light.

turbidimetry A technique in which the decrease in intensity of light traveling through a turbid solution (a solution containing suspended particles) is measured. The greater the concentration of suspended particles, the less light is transmitted. See diagram under *nephelometry*.

turbidity Light-scattering property associated with suspended particles in a liquid. A turbid solution appears cloudy.

turbidity coefficient The transmittance of a turbid solution is given by $P/P_0 = e^{-\tau b}$, where P is the transmitted radiant power, P_0 is the incident radiant power, b is the pathlength, and τ is the turbidity coefficient.

ultra-high-performance liquid chromatography (UHPLC) Liquid chromatography with stationary phase particles in the 1.5–2 μm range. Higher pressure is required to force liquid through a UHPLC column than through columns with larger particle size.

ultraviolet detector Liquid chromatography detector that measures ultraviolet absorbance of solutes emerging from the column.

unadjusted relative retention, γ For components A and B separated by chromatography or electrophoresis, the unadjusted relative retention γ is the quotient of linear velocities or the quotient of unadjusted retention times: $\gamma = u_A/u_B = t_B/t_A$, where u is linear velocity and t is retention time.

uncertainty Variability within a set of measurements of a quantity.

underpotential depositon Reduction of M^{n+} to make an atomic monolayer of M on the surface of another material, such as gold. If the reduction potential to make the monolayer on gold is less negative than the potential required to reduce M^{n+} onto bulk metal M, we say that underpotential deposition occurs.

upconversion A process in which a higher energy photon is created by the combination of energy of several lower energy photons.

use objectives In quality assurance, use objectives are a written statement of how results will be used. Use objectives are required before specifications can be written for the method.

valence band Energy levels containing valence electrons in a semiconductor. The electrons in these levels are localized in chemical bonds.

van Deemter equation Describes the dependence of chromatographic plate height, H, on linear velocity, u_x: $H = A + B/u_x + Cu_x$. The constant A depends on band-broadening processes such as multiple flow paths that are independent of flow rate. B depends on the rate of diffusion of solute in the mobile phase. C depends on the rate of mass transfer (that is, the equilibration time) between the stationary and mobile phases.

van der Waals force Attractive force between any two molecules or surfaces that occurs at distances of ~1–5 nm. Movement of electrons in molecule A gives rise to instantaneous separation of charges. Positive regions of A attract electrons of molecule B, giving rise to charge separation in B and attraction between the two molecules.

variance, σ^2 The square of the standard deviation.

vibrational transition Occurs when a molecule changes its vibrational energy.

viscosity Resistance to flow in a fluid.

void volume, V_0 See *interstitial volume*.

volatile Easily vaporized.

volatilization Selective removal of a component from a mixture by transforming the component into a volatile (low-boiling) species and removing it by heating, pumping, or bubbling a gas through the mixture.

Volhard titration Titration of Ag^+ with SCN^- in the presence of Fe^{3+}. Formation of red $Fe(SCN)^{2+}$ marks the end point.

volt, V SI unit of electric potential. If the potential difference between two points is one volt, then one joule of energy is required to move one coulomb of charge between the two points.

voltaic cell See *galvanic cell*.

voltammetry An analytical method in which the relation between current and voltage is observed during an electrochemical reaction.

voltammogram A graph of current versus electrode potential in an electrochemical cell.

voltmeter Device for measuring electric potential difference. See also *potentiometer*.

volume flow rate In chromatography, the volume of mobile phase per unit time eluted from the column.

volume percent, vol% (Volume of solute/volume of solution) $\times$ 100.

volumetric analysis A technique in which the volume of material needed to react with the analyte is measured.

volumetric flask One having a tall, thin neck with a calibration mark. When the liquid level is at the calibration mark, the flask contains its specified volume of liquid at a specified temperature.

Walden reductor A column packed with silver and eluted with HCl. Analyte is reduced during passage through the column. The reduced product is titrated with an oxidizing agent.

wall-coated column Hollow chromatographic column in which the stationary phase is coated on the inside surface of the wall.

watt, W SI unit of power equal to an energy flow of one joule per second. When an electric current of one ampere flows through a potential difference of one volt, the power is one watt.

waveguide   A thin layer or hollow structure in which electromagnetic radiation is totally reflected.

wavelength, λ   Distance between consecutive crests of a wave.

wavenumber, $\tilde{\nu}$ Reciprocal of the wavelength, $1/\lambda$, usually expressed in units of cm^{-1}.

weak acids and bases   Those whose dissociation constants are not large.

weak electrolyte   One that only partly dissociates into ions when it dissolves.

weighing paper   Paper on which to place a solid reagent on a balance. Weighing paper has a very smooth surface, from which solids fall easily for transfer to a vessel.

weight percent, wt%   (Mass of solute/mass of solution) × 100.

weight/volume percent   [(Mass of solute, g)/(volume of solution, mL)] × 100.

Weston cell   A stable voltage source based on the reaction $Cd(s) + HgSO_4(aq) \rightleftharpoons CdSO_4(aq) + Hg(l)$. It was formerly used to standardize a potentiometer.

wet ashing   Destruction of organic matter in a sample by a liquid reagent (such as boiling aqueous $HClO_4$) prior to analysis of an inorganic component.

whistle noise   See *line noise*.

white light   Light comprising all wavelengths.

white noise   Random noise, also called *Gaussian noise*, due to random movement of charge carriers in an electric circuit (called *thermal noise, Johnson noise,* or *Nyquist noise*) or from random arrival of photons or charge carriers at a detector (called *shot noise* or *Schottky noise*).

width at half-height, $w_{1/2}$ Width of a signal at half of its maximum height. Also called *half-width*.

work   Energy required or released when an object is moved from one point to another. Units of work are joules, J.

working electrode   One at which the reaction of interest occurs. Same as *indicator electrode*.

X-ray fluorescence   Emission of X-rays from a material following the absorption of X-rays by the material.

y-intercept   The value of *y* at which a line crosses the *y*-axis.

Zeeman background correction   Technique used in atomic spectroscopy in which analyte signals are shifted outside the detector monochromator range by applying a strong magnetic field to the sample. Signal that remains is the background.

Zeeman effect   Shifting of atomic energy levels in a magnetic field.

zwitterion   A neutral molecule with a positive charge localized at one position and a negative charge localized at another position.

APPENDIX A   Logarithms and Exponents and Graphs of Straight Lines

If a is the base 10 logarithm of n ($a = \log n$), then $n = 10^a$. On a calculator, you find the logarithm of a number by pressing the "log" button. If you know $a = \log n$ and you wish to find n, use the "antilog" button or raise 10 to the power a:

$$a = \log n$$

$$10^a = 10^{\log n} = n \; (\Rightarrow n = \text{antilog } a)$$

Natural logarithms (ln) are based on the number e ($= 2.718\ 281\ \ldots$) instead of 10:

$$b = \ln n$$

$$e^b = e^{\ln n} = n$$

On a calculator, you find the ln of n with the "ln" button. To find n when you know $b = \ln n$, use the e^x key.

Here are some useful properties to know:

$\log (a \cdot b) = \log a + \log b \qquad \log 10^a = a$

$\log \left(\dfrac{a}{b}\right) = \log a - \log b \qquad a^b \cdot a^c = a^{(b+c)}$

$\log (a^b) = b \log a \qquad\qquad \dfrac{a^b}{a^c} = a^{(b-c)}$

Problems

Test yourself by simplifying each expression as much as possible:

(a) $e^{\ln a}$

(b) $10^{\log a}$

(c) $\log 10^a$

(d) $10^{-\log a}$

(e) $e^{-\ln a^3}$

(f) $e^{\ln a^{-3}}$

(g) $\log (10^{1/a^3})$

(h) $\log (10^{-a^2})$

(i) $\log (10^{a^2-b})$

(j) $\log (2a^3\,10^{b^2})$

(k) $e^{(a + \ln b)}$

(l) $10^{[(\log 3) - (4 \log 2)]}$

Solving a logarithmic equation: In working with the Nernst and Henderson-Hasselbalch equations, we will need to solve equations such as

$$a = b - c \log \dfrac{d}{gx}$$

for the variable, x. First isolate the log term:

$$\log \dfrac{d}{gx} = \dfrac{(b - a)}{c}$$

Then raise 10 to the value of each side of the equation:

$$10^{\log(d/gx)} = 10^{(b-a)/c}$$

But $10^{\log (d/gx)}$ is just d/gx, so

$$\dfrac{d}{gx} = 10^{(b-a)/c} \Rightarrow x = \dfrac{d}{g10^{(b-a)/c}}$$

Converting between ln x and log x: The relation between them is derived by writing $x = 10^{\log x}$ and taking the ln of both sides:

$$\ln x = \ln(10^{\log x}) = (\log x)(\ln 10)$$

because $\ln a^b = b \ln a$.

Answers

(a) a

(b) a

(c) a

(d) $1/a$

(e) $1/a^3$

(f) $1/a^3$

(g) $1/a^3$

(h) $-a^2$

(i) $a^2 - b$

(j) $b^2 + \log(2a^3)$

(k) be^a

(l) $3/16$

The general form of the equation of a straight line is

$$y = mx + b$$

where $m = \text{slope} = \dfrac{\Delta y}{\Delta x} = \dfrac{y_2 - y_1}{x_2 - x_1}$

$b = \text{intercept on } y\text{-axis}$

The meanings of slope and intercept are illustrated in Figure A-1.

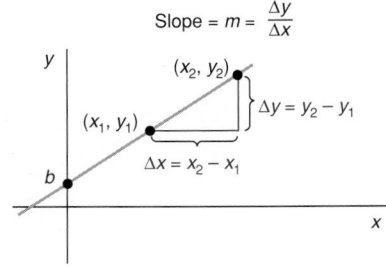

Slope $= m = \dfrac{\Delta y}{\Delta x}$

(x_2, y_2)

(x_1, y_1)

$\Delta y = y_2 - y_1$

$\Delta x = x_2 - x_1$

FIGURE A-1 Parameters of a straight line.

If you know two points $[(x_1, y_1)$ and $(x_2, y_2)]$ that lie on the line, you can generate the equation of the line by noting that the slope is the same for every pair of points on the line. Calling some general point on the line (x, y), we can write

$$\dfrac{y - y_1}{x - x_1} = \dfrac{y_2 - y_1}{x_2 - x_1} = m \qquad\qquad \text{(A-1)}$$

which can be rearranged to the form

$$y - y_1 = \left(\dfrac{y_2 - y_1}{x_2 - x_1}\right)(x - x_1)$$

$$y = \underbrace{\left(\dfrac{y_2 - y_1}{x_2 - x_1}\right)}_{m} x + \underbrace{y_1 - \left(\dfrac{y_2 - y_1}{x_2 - x_1}\right)x_1}_{b}$$

When you have a series of experimental points that should lie on a line, the best line is generally obtained by the method of least squares, described in Chapter 4. This method gives the slope and the intercept directly. If, instead, you wish to draw the "best" line by eye, you can derive the equation of the line by selecting two points *that lie on the line* and applying Equation A-1.

Sometimes you are presented with a linear plot in which x or y or both are nonlinear functions. An example is shown in Figure A-2, in which the potential of an electrode is expressed as a function of the activity of analyte. Given that the slope is 29.6 mV and the line passes through the point ($\mathcal{A} = 10^{-4}, E = -10.2$), find the equation of the line. To do this, first note that the y-axis is linear but the x-axis is *logarithmic*. That is, the function E versus $\mathcal{A}$ is *not* linear, but E versus log $\mathcal{A}$ is linear. The form of the straight line should therefore be

$$E = (29.6) \underbrace{\log \mathcal{A}}_{} + b$$
$$\begin{matrix} \uparrow & \uparrow & \uparrow \\ y & m & x \end{matrix}$$

To find b, we can use the coordinates of the one known point in Equation A-1:

$$\frac{y - y_1}{x - x_1} = \frac{E - E_1}{\log \mathcal{A} - \log \mathcal{A}_1} = \frac{E - (-10.2)}{\log \mathcal{A} - \log(10^{-4})} = m = 29.6$$

or

$$E + 10.2 = 29.6 \log \mathcal{A} + (29.6)(4)$$
$$E \text{ (mV)} = 29.6 \text{ (mV)} \log \mathcal{A} + 108.2 \text{ (mV)}$$

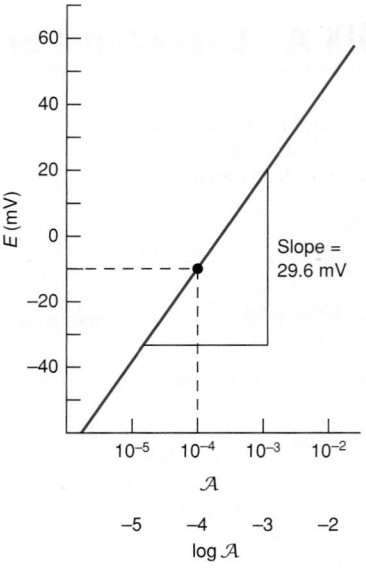

FIGURE A-2 A linear graph in which one axis is a logarithmic function.

The *standard uncertainties* (u) of these measured quantities are:

$$\text{Standard uncertainty: } u_v = \frac{s_v}{\sqrt{n_v}} \qquad u_w = \frac{s_w}{\sqrt{n_w}} \qquad \text{(B-5)}$$

Cells B10 and C10 in Figure B-1 give the standard deviations s_v and s_w and cells B12 and C12 give standard uncertainties u_v and u_w.

- Standard deviation s is a measure of uncertainty of individual measurements. *As you make more measurements, s approaches the population standard deviation σ.*
- Standard uncertainty $u = s/\sqrt{n}$ is a measure of the uncertainty of the mean. *As you make more measurements, uncertainty of the mean value approaches 0.*

Chapter 3 told us that the uncertainty of a sum or difference is the square root of the sum of squares of uncertainties of the individual terms. For $\bar{z} = \bar{v} - \bar{w}$, the standard uncertainty of $\bar{z}$ is

$$\text{Standard uncertainty of a sum or difference:} \qquad u_z = \sqrt{u_v^2 + u_w^2} \qquad \text{(B-6)}$$

The standard uncertainty in cell D14 is $u_z = \sqrt{0.062^2 + 0.022^2} = 0.066$.

Equation 4-7 told us that the confidence interval is

$$\text{Confidence interval for } \bar{z} = \bar{z} \pm ts_z/\sqrt{n} = \bar{z} \pm tu_z \qquad \text{(B-7)}$$

where t is Student's t, and we rewrite the standard deviation of the mean, $s_z/\sqrt{n}$, as the standard uncertainty, u_z. We know u_z from Equation B-6. We must decide how many degrees of freedom are associated with $\bar{z}$ so we can select the value of Student's t to find the confidence interval $\bar{z} \pm tu_z$. There are 3 degrees of freedom for $\bar{v}$ and 4 degrees of freedom for $\bar{w}$. The crux of the problem—and a key new idea of this section—is to arrive at the number of degrees of freedom for $\bar{z}$ so we can find the confidence interval for $\bar{z}$.

The number of degrees of freedom (df) for a sum or difference is given by

$$\text{Degrees of freedom for a sum or difference (Welch-Satterthwaite approximation):} \qquad df_z = \frac{u_z^4}{\dfrac{u_v^4}{df_v} + \dfrac{u_w^4}{df_w}} \qquad \text{(B-8)}$$

where u is the standard uncertainty of each quantity and df is the degrees of freedom for each quantity. Equation B-8 is a weighted average for the degrees of freedom. You have seen this equation for degrees of freedom in the t-test (Equation 4-10b). Equations B-8 and B-9 (below) apply when u_v and u_w are independent of one another.

Expression B-8 is evaluated in cell D15 in Figure B-1:

$$df_z = \frac{u_z^4}{\dfrac{u_v^4}{df_v} + \dfrac{u_w^4}{df_w}} = \frac{0.066^4}{\dfrac{0.062^4}{4-1} + \dfrac{0.022^4}{5-1}} = 3.771$$

which we truncate to $df_z = 3$ to find the value of Student's $t = 3.182$ in cell D18 for 95% confidence. The degrees of freedom for $\bar{z}$ depend on standard uncertainties and degrees of freedom for $\bar{v}$ and $\bar{w}$. Degrees of freedom are truncated to the smaller integer to provide a more conservative (wider) confidence interval. The Excel function TINV(0.05,D15) in cell D18 automatically truncates the degrees of freedom in cell D15.

Knowing the degrees of freedom and Student's t, we finally compute the 95% confidence interval in cell D19:

$$\text{Confidence interval} = \bar{z} \pm tu_z = 2.054 \pm (3.182)(0.066) = 2.054 \pm 0.210$$

which could sensibly be rounded to 2.05 (±0.21) wt%.

Confidence Intervals for Multiplication and Division

For multiplication and division, use relative uncertainties rather than absolute uncertainties. For the product $\bar{z} = \bar{v}\,\bar{w}$ or the quotient $\bar{z} = \bar{v}/\bar{w}$, the degrees of freedom are

$$\text{Degrees of freedom for a product or quotient uses relative uncertainty (Welch-Satterthwaite approximation):} \qquad df_z = \frac{(u_z/\bar{z})^4}{\dfrac{(u_v/\bar{v})^4}{df_v} + \dfrac{(u_w/\bar{w})^4}{df_w}} \qquad \text{(B-9)}$$

Confidence Interval for a Quotient

Suppose that you determine the density of a mineral by measuring its mass (m) four times and its volume (V) six times:

mean mass ($\bar{m}$) = 4.635 g $\qquad$ $\bar{V}$ = 1.13 mL

standard deviation (s_m) = 0.002_2 g $\qquad$ $s_V = 0.04_7$ mL

n_m = 4 measurements $\qquad$ n_V = 6 measurements

standard uncertainty $\qquad\qquad\quad u_V = (0.04_7)/\sqrt{6} = 0.01_9$ mL
$(u_m) = 0.002_2/\sqrt{4} = 0.001_1$ g

Density is mass/volume = 4.635 g/1.13 mL = 4.101 8 g/mL. Find the 95% confidence interval for the computed density.

Solution The arithmetic is best done with a spreadsheet to avoid errors. Using standard uncertainties as the estimate of error, convert absolute uncertainty to relative uncertainty for division and proceed just as you learned in Chapter 3:

$$\frac{\bar{m}\,(\pm u_m)}{\bar{V}\,(\pm u_V)} = \frac{4.635\,(\pm 0.001_1)\text{g}}{1.13\,(\pm 0.01_9)\text{ mL}} = \frac{4.635\,(\pm 0.02_{37}\%)\text{g}}{1.13\,(\pm 1._{698}\%)\text{ mL}}$$

$$= 4.101\ 8\,(\pm 1.698\%) = 4.101\ 8\,(\pm 0.069\ 7)\text{ g/mL}$$

because $\sqrt{(0.023\ 7\%)^2 + (1.698\%)^2} = 1.698\%$. The standard uncertainty of the density is 1.698% of 4.101 8 g/mL = $\pm$0.069 7 g/mL.

Degrees of freedom for the quotient are obtained from Equation B-9:

$$df_d = \frac{(u_d/d)^4}{\dfrac{(u_m/\bar{m})^4}{4-1} + \dfrac{(u_V/\bar{V})^4}{6-1}} = \frac{(0.069\ 7/4.101\ 8)^4}{\dfrac{(0.001\ 1/4.635)^4}{4-1} + \dfrac{(0.019/1.13)^4}{6-1}} = 5.002$$

For 95% confidence and 5 degrees of freedom, Student's t is 2.571 in Table 4-4. The 95% confidence interval for the density is 4.101 8 $\pm$ (2.571)(0.069 7) = 4.101 8 $\pm$ 0.179 1 g/mL. Finally, round to a reasonable number of significant digits:

$$95\% \text{ confidence interval} = 4.10 \pm 0.18 \text{ g/mL}$$

TEST YOURSELF Suppose that volume could be measured 100 times more precisely: V = 1.130 00 mL with s_V = 0.000 47 mL. Find the degrees of freedom and the 95% confidence interval for the density. (*Answer:* df_d = 5.93, which is truncated to $5 \Rightarrow t = 2.571$ and d = 4.101 8 $\pm$ 0.003 1 g/mL)

Mixed operations can be broken into sequences of addition and subtraction or multiplication and division. For a function of more than two variables, add another term to the denominators of Equations B-8 and B-9 for each variable.

In summary, here is how to find a confidence interval for $\bar{z}$ computed from several experimental quantities:

- The standard uncertainty (u) for each experimental quantity is its standard deviation of the mean.
- Apply the rules for propagation of uncertainty from Chapter 3 to the standard uncertainties to find the standard uncertainty u_z.
- Find the degrees of freedom for $\bar{z}$ by using Equation B-8 for sums and differences or Equation B-9 for products and quotients. Add one more term to the denominators of B-8 and B-9 for each additional variable. Truncate degrees of freedom to an integer.
- From the degrees of freedom, find Student's t and the confidence interval $\bar{z} + tu_z$.

Type B Uncertainty: Rectangular and Triangular Distributions

So far, we have dealt only with Type A standard uncertainty, which we *measure*. Uncertainty of atomic mass in the periodic table is a Type B uncertainty. People have measured the abundance of isotopes of elements in a variety of substances (Box 3-3) and concluded that the average atomic mass of an element lies in a certain range. Atomic mass does not follow a normal distribution and is not evenly distributed in that range. Different substances have atomic mass that lies in different parts of the range listed in the periodic table.

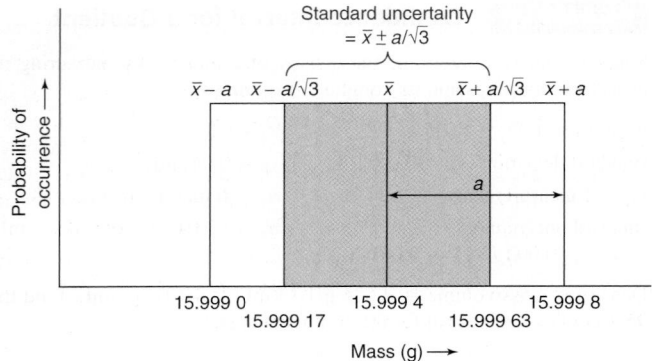

FIGURE B-2 *Rectangular distribution* has equal probability of finding a value anywhere in the range from $\bar{x} - a$ to $\bar{x} + a$, and zero probability of finding a value outside this range. The standard uncertainty (standard deviation) shown in color is $\pm a/\sqrt{3}$. Numbers on the abscissa apply to the atomic mass of oxygen.

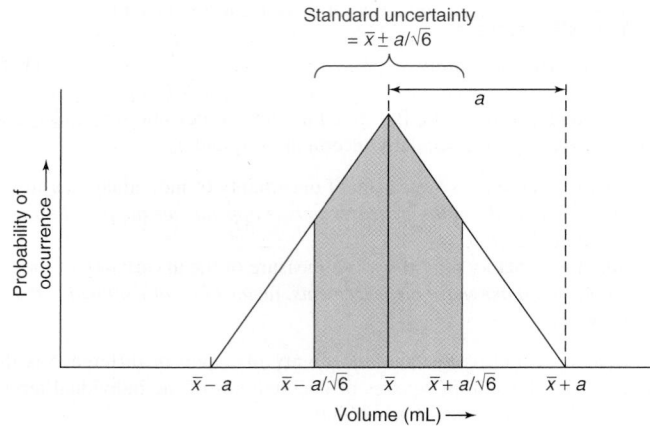

FIGURE B-3 *Triangular distribution* for volumetric glassware, including volumetric flasks and transfer pipets. The standard uncertainty interval (standard deviation) shown in color is $\pm a/\sqrt{6}$.

When we need uncertainty of atomic mass for chemical calculations, it is not practical to measure the isotopes in each bottle of reagent. It is common to suppose that the distribution of atomic mass follows a **rectangular distribution** (Figure B-2). In this ideal distribution, atomic mass is equally likely to lie anywhere in the interval from $\bar{x} - a$ to $\bar{x} + a$, where a is the uncertainty in the periodic table inside the cover of this book. For oxygen (Box 3-3), the atomic mass is 15.999 4 ± 0.000 4 g/mol. That is $\bar{x}$ = 15.999 4 g/mol and a = 0.000 4 g/mol. The standard uncertainty[6] for a rectangular distribution is $a/\sqrt{3}$.

Standard uncertainty for rectangular distribution
$$u_x = \pm a/\sqrt{3} \qquad \text{(B-10)}$$

As an example, let's find the 95% confidence interval for the molecular mass of C_2H_4. The uncertainty in atomic mass of carbon in the periodic table is 0.001 0, which is the number a in the rectangular distribution for carbon. For hydrogen, the uncertainty is a = 0.000 14. First, convert uncertainty (a) to standard uncertainty ($u_x = \pm a/\sqrt{3}$):

Atomic mass of C
$$= 12.010\ 6 \pm (0.001\ 0)/\sqrt{3} = 12.010\ 6 \pm 0.000\ 577 \leftarrow u_x$$

Atomic mass of H
$$= 1.007\ 98 \pm (0.000\ 14)/\sqrt{3} = 1.007\ 98 \pm (0.000\ 080\ 8) \leftarrow u_x$$

Now multiply u_x by the number of atoms for each element:

2C: $2(12.010\ 6 \pm 0.000\ 577) = 24.021\ 2 \pm \mathbf{0.001\ 15} \leftarrow 2 \times 0.000\ 577$

4H: $\underline{4(1.007\ 98 \pm (0.000\ 080\ 8) = 4.031\ 92 \pm \mathbf{0.000\ 323}} \leftarrow 4 \times 0.000\ 080\ 8$
$$28.053\ 12 \pm ?$$

Then compute the standard uncertainty for molecular mass by taking the square root of the sum of squares of u_x:

$$28.053\ 12 \pm \sqrt{\mathbf{0.001\ 15^2 + 0.000\ 323^2}}$$
$$28.053\ 12 \pm 0.001\ 19 \leftarrow u_x \text{ for } C_2H_4$$

We wish to express uncertainty as a 95% confidence interval. The rectangular distribution represents a large body of measurements. In Table 4-4, Student's t for 95% confidence approaches ~2 as the degrees of freedom become large. We use $t = 2$ to estimate the 95% confidence limit for molecular mass. In the literature of metrology (the science of measurement), the product tu_z is called the *expanded uncertainty* and is written ku_z instead of tu_z. The multiplier k, called the *coverage factor*, is commonly taken as 2.

To estimate a 95% confidence interval for molecular mass, multiply u_x = 0.001 19 for C_2H_4 by the coverage factor $k = 2$:

$$ku_x = 2(0.001\ 19\ \text{g/mol}) = 0.002\ 38$$

95% confidence interval: molecular mass = $28.053_1 \pm 0.002_4$ g/mol

In Chapter 3, when we knew nothing about statistics, we found that the uncertainty in molecular mass of C_2H_4 is $\pm 0.002_1$ g/mol. The procedure in Chapter 3 gives approximately the 95% confidence interval for molecular mass.

The uncertainty in volume held by a transfer pipet or volumetric flask is a Type B uncertainty that is conventionally modeled by the **triangular distribution** in Figure B-3. The tolerance for a Class A 10-mL volumetric flask is ±0.02 mL. The manufacturer tries hard to put the calibration line as close as possible to 10.00 mL. We model the uncertainty as a triangular distribution with a most probable value of 10.00 mL and a linear decrease to zero at 9.98 mL and 10.02 mL. The value of a in Figure B-3 is 0.02 mL for this flask. The standard uncertainty is

Standard uncertainty for triangular distribution
$$u_x = \pm a/\sqrt{6} \qquad \text{(B-11)}$$

which is smaller than u_x for the rectangular distribution because the probability in the triangular distribution is greatest at the mean value. For a 10-mL flask, the standard uncertainty is $u_x = \pm 0.02/\sqrt{6} = 0.008_{16}$ mL.

If you have the fortitude, you can calibrate a volumetric flask by weighing the water it contains when filled to the mark. If you repeat the calibration several times, you will *know* how much the flask contains and the standard deviation of the mean. Calibration changes uncertainty from Type B to Type A. Type A is a measured uncertainty and Type B is an estimated uncertainty.

Combining Type A and Type B Uncertainties

Suppose that you measure the density of ethanol by weighing the liquid required to fill an uncalibrated 10-mL volumetric flask to the mark at 20°C. You repeat the procedure five times and find a mean mass $\bar{m} = 7.888_1$ g with a standard deviation of 0.004_3 g and a Type A standard uncertainty of $u_{mass} = (0.004_3)/\sqrt{5} = 0.001_{92}$ g. You did not calibrate the flask, so its Type B uncertainty is a triangular distribution: $u_{volume} = \pm 0.02/\sqrt{6} = 0.008_{16}$ mL.

Standard uncertainties for mass and volume can be combined with the rules for propagation of uncertainty in Chapter 3:

$$\text{Density} = \frac{\bar{m}\ (\pm u_{mass})}{\bar{V}\ (\pm u_{volume})} = \frac{7.888_1\ (\pm 0.001_{92})\ \text{g}}{10.00\ (\pm 0.008_{16})\ \text{mL}} = \frac{7.888_1\ (\pm 0.02_{43}\%)\ \text{g}}{10.00\ (\pm 0.08_{16}\%)\ \text{mL}}$$
$$= 0.788\ 81\ (\pm 0.08_{51}\%)$$

because $\sqrt{(0.02_{43}\%)^2 + (0.08_{16}\%)^2} = 0.08_{51}\%$. The standard uncertainty of the density is $0.08_{51}\%$ of 0.788 81 g/mL = $\pm 0.000\ 67$ g/mL. *Express uncertainty as a ~95% confidence interval by multiplying the combined standard uncertainty by the coverage factor $k = 2$.* This *expanded uncertainty* is $(2)(0.000\ 67) = 0.001_{34}$ g/mL. The final answer is

$$\text{Density} = 0.788_8 \pm 0.001_3\ \text{g/mL}$$

which could be accompanied by the statement "uncertainty is an estimate of the 95% confidence interval obtained by multiplying standard uncertainty by a coverage factor of 2."

Exercises

B-3. *Propagation of uncertainty.* Create a spreadsheet like Figure B-1 for this problem. In isotope ratio mass spectrometry, individual calibrated detectors measure each isotope of interest. Consider data for the ratio $^{13}C/^{12}C$ for an unknown sample and a standard.

	$x = {}^{13}C/{}^{12}C$ in unknown	$y = {}^{13}C/{}^{12}C$ in standard
$(\bar{x} \text{ or } \bar{y}) = $ mean $^{13}C/^{12}C$	0.010 853	0.011 197
$s = $ standard deviation	0.000 017	0.000 010
$n = $ number of observations	12	12

The difference between $^{13}C/^{12}C$ for an unknown compound and a standard compound, called $\delta^{13}C$, measured in parts per thousand (‰), is

$$\delta^{13}C = \left[\frac{({}^{13}C/{}^{12}C)_{\text{sample}}}{({}^{13}C/{}^{12}C)_{\text{standard}}} - 1 \right] \times 1\,000$$

For data in the table $\delta^{13}C = [(0.010\,853/0.011\,197) - 1] \times 1\,000 = [0.969\,28 - 1] \times 1\,000 = -30.7‰$.

(a) Find the standard uncertainties (standard deviation of the mean) of $\bar{x} = ({}^{13}C/{}^{12}C)_{\text{sample}}$ and $\bar{y} = ({}^{13}C/{}^{12}C)_{\text{standard}}$.

(b) Find the standard uncertainty of the quotient $\bar{z} = \bar{x}/\bar{y} = ({}^{13}C/{}^{12}C)_{\text{sample}} / ({}^{13}C/{}^{12}C)_{\text{standard}}$. Express the answer as a relative uncertainty $u_z/\bar{z}$.

(c) Find the degrees of freedom for $\bar{z}$.

(d) Find the 95% confidence interval for $\bar{z}$.

(e) Find the 95% confidence interval for $\delta^{13}C$.

B-4. *Propagation of uncertainty.* In a spectrophotometric analysis, six replicate samples had absorbance values of 0.216, 0.214, 0.207, 0.220, 0.205, and 0.213. Six replicate reagent blanks had absorbance values of 0.032, 0.030, 0.029, 0.034, 0.035, and 0.030. Find the 95% confidence interval for the corrected absorbance, which is the average absorbance minus the average blank absorbance.

B-5. *Type A (measured) uncertainties.* You prepared an ammonia solution by diluting concentrated NH_3. The concentrated solution was titrated five times

to measure its concentration, with the result $[NH_3] = 27.63$ (± 0.14) wt% ($n = 5$), where ± 0.14 is the standard deviation and $n = 5$ is the number of replicate determinations. The density of concentrated NH_3 solution was measured four times and is $0.904_0 \pm 0.002_3$ g/mL ($n = 4$). A 1-mL transfer pipet was calibrated and found to deliver $1.004_2 \pm 0.002_0$ mL in $n = 6$ trials. The concentrated ammonia was delivered with the 1-mL pipet into a calibrated 500-mL volumetric flask whose volume is 499.86 ± 0.08 mL ($n = 4$).

(a) Find the standard uncertainty ($u_x = s_x/\sqrt{n}$) for each quantity x described above and find the percent relative standard uncertainty $= 100\,u_x/x$.

(b) Find the molecular mass of NH_3 and its standard uncertainty and percent relative standard uncertainty. Because the relative uncertainty is significantly smaller than other uncertainties in the procedure, we will take it to be 0.

(c) Find the molarity of NH_3 in the 500-mL solution. When you are done, write out all the arithmetic in one line of multiplication and division. The example on pages 54–55 provides a guide for you.

(d) The arithmetic in **(c)** has the form molarity $(M) = (a \times b \times c)/(d \times e)$. The standard uncertainty has the form $u_M/M = \sqrt{(u_a/a)^2 + (u_b/b)^2 + (u_c/c)^2 + (u_d/d)^2 + (u_e/e)^2}$. Find u_M.

(e) The degrees of freedom for molarity (df_M) computed from five variables combined in multiplication and division takes the form below. Find df_M.

$$df_M = \frac{(u_M/M)^4}{\dfrac{(u_a/\bar{a})^4}{df_a} + \dfrac{(u_b/\bar{b})^4}{df_b} + \dfrac{(u_c/\bar{c})^4}{df_c} + \dfrac{(u_d/\bar{d})^4}{df_d} = \dfrac{(u_e/\bar{e})^4}{df_e}}$$

(f) Find the 95% confidence interval for the molarity of NH_3 in the 500-mL flask. What percentage of the molarity is the 95% confidence interval?

B-6. *Mixed Type A (measured) and Type B (estimated) uncertainties.* Work the previous problem with the following changes: The 1-mL pipet and the 500-mL flask are not calibrated. Use the triangular distribution for their estimated uncertainties. The uncertainty in molecular mass of NH_3 remains negligible. Find the expanded uncertainty ($\approx 95\%$ confidence interval) with a coverage factor of $k = 2$.

Solutions to Exercises

	A	B	C	D	E
1	Isotope ratio propagation of uncertainty				
2		x =	y =	z =	
3		$^{13}C/^{12}C_{\text{sample}}$	$^{13}C/^{12}C_{\text{standard}}$	x/y	
4	mean	0.010853	0.011197		
5	s (stdev)	0.000017	0.000010		
6	n (observations)	12	12		
7	Standard uncertainty u = s/sqrt(n)	4.907E-06	2.887E-06		B7 = B5/SQRT(B6)
8	mean z = (mean x)/(mean y)			0.96928	D8 = B4/C4
9	u_z/z = sqrt($(u_x/x)^2 + (u_y/y)^2$)			0.00052	D9 = SQRT((B7/B4)^2+(C7/C4)^2)
10	degrees of freedom (df_z) for z =	$\dfrac{(u_z/\bar{z})^4}{\dfrac{(u_x/\bar{x})^4}{df_x} + \dfrac{(u_y/\bar{y})^4}{df_y}}$		17.46832	D10 = (D9)^4/((B7/B4)^4/(B6−1)
11					+(C7/C4)^4/(C6−1))
12					
13	Student's t			2.110	D13 = TINV(0.05,D10)
14	Confidence interval = t*u_z =			0.00106	D14 = D13*D9*D8

Spreadsheet for solution to Exercise B-3.

B-3. (a) Standard uncertainties are computed in cells B7 and C7. The calculation in B7 is $u_x = s_x/\sqrt{n_x} = 0.000\,017/\sqrt{12} = 4.907 \times 10^{-6}$.

(b) Relative standard uncertainty for $\bar{z} = \bar{x}/\bar{y}$ is computed in cell D9:
$$u_z/\bar{z} = \sqrt{(u_x/\bar{x})^2 + (u_y/\bar{y})^2} =$$
$$\sqrt{(4.907 \times 10^{-6}/0.010\,853)^2 + (2.887 \times 10^{-6}/0.011\,197)^2} = 0.000\,52.$$

(c) Degrees of freedom for $\bar{z}$ computed in cell D10:
$$df_z = \frac{(u_z/\bar{z})^4}{\dfrac{(u_x/\bar{x})^4}{df_x} + \dfrac{(u_y/\bar{y})^4}{df_y}}$$
$$= \frac{(0.000\,52)^4}{\dfrac{(4.907 \times 10^{-6}/0.010\,853)^4}{12-1} + \dfrac{(2.887 \times 10^{-6}/0.011\,197)^4}{12-1}} = 17.47$$

(d) To find the 95% confidence interval for $\bar{z}$, truncate degrees of freedom to 17 and find Student's $t = 2.110$ for 17 degrees of freedom with the statement TINV(0.05,D10) in cell D13. The 95% confidence interval for $\bar{z}$ in cell D14 is $\bar{z} \pm tu_z = \bar{z} \pm t(u_z/\bar{z})(\bar{z}) = 0.969\,28 \pm (2.110)(0.000\,52)(0.969\,28)$ $= 0.969\,28 \pm 0.001\,06$.

(e) We are looking for absolute uncertainty in the expression
$$\delta^{13}C = \left[\frac{(^{13}C/^{12}C)_{\text{sample}}}{(^{13}C/^{12}C)_{\text{standard}}} - 1\right] \times 1\,000.$$

The quotient in brackets is $\bar{x}/\bar{y} = \bar{z}$. Its 95% confidence limit in **(d)** is $0.969\,28 \pm 0.001\,06$. The difference in brackets is $(0.969\,28 \pm 0.001\,06) - 1 = -0.030\,72 \pm 0.001\,06$ because the uncertainty in the difference is $\sqrt{(0.001\,06)^2 + 0^2} = 0.001\,06$. Multiplying by 1 000 gives $\delta^{13}C = -(30.72 \pm 1.06)$‰ which might also be rounded to $-(30.7 \pm 1.1)$‰ or $-(31 \pm 1)$‰.

B-4. Averages of the absorbance ($\bar{a}$) and blank ($\bar{b}$) are computed in cells B11 and C11. Standard deviations s_a and s_b are found in cells B12 and C12.

Standard uncertainty (= standard deviation of the mean, $u = s/\sqrt{n}$) is computed in cells B14 and C14. By the rule for propagation of uncertainty of a sum or difference, the standard uncertainty for the corrected absorbance ($\bar{c}$) is computed as follows:

$\bar{c} = \bar{a} - \bar{b} = 0.212\,50 - 0.031\,67 = 0.180\,33$ (cell D15)
$u_c = \sqrt{u_a^2 + u_b^2} = \sqrt{0.002\,29^2 + 0.000\,99^2} = 0.002\,50$ (cell D16)

The degrees of freedom for the corrected absorbance are

$$df_c = \frac{u_c^4}{\dfrac{u_a^4}{df_a} + \dfrac{u_b^4}{df_b}} = \frac{0.002\,50^4}{\dfrac{0.002\,29^4}{6-1} + \dfrac{0.000\,99^4}{6-1}} = 6.800 \text{ (cell D17)}$$

which we truncate to 6 degrees of freedom, for which Student's t for 95% confidence is 2.447 (cell D20). The confidence interval is

$\bar{c} \pm tu_c = 0.180\,83 \pm (2.447)(0.002\,50) = 0.180\,83 \pm 0.006\,11$
$= 0.181 \pm 0.006$

B-5. (a) NH_3 wt%: $u_{NH_3} = 0.14/\sqrt{5} = 0.062\,6$ wt%

relative uncertainty $= (100)(0.062\,6 \text{ wt\%})/(27.63 \text{ wt\%}) = 0.226\,6\%$

density: $u_{\text{density}} = 0.002_3/\sqrt{4} = 0.001\,15$ g/mL

relative uncertainty $= (100)(0.001\,15 \text{ g/mL})/(0.904_0 \text{ g/mL}) = 0.127\,2\%$

pipet: $u_{\text{pipet}} = 0.002_0/\sqrt{6} = 0.000\,82$ mL

relative uncertainty $= (100)(0.000\,82 \text{ mL})/(1.004_2 \text{ mL}) = 0.081\,3\%$

flask: $u_{\text{flask}} = 0.08/\sqrt{4} = 0.04$ mL

relative uncertainty $= (100)(0.04 \text{ mL})/(499.86 \text{ mL}) = 0.008\,0\%$

	A	B	C	D	E
1	Corrected absorbance				
2				corrected	
3		absorbance	blank	absorbance	
4		a		c	
5		0.216	0.032		
6		0.214	0.030		
7		0.207	0.029		
8		0.220	0.034		
9		0.205	0.035		
10		0.213	0.030		
11	mean	0.21250	0.03167		B11 = AVERAGE(B5:B10)
12	s (stdev)	0.00561	0.00242		B12 = STDEV(B5:B10)
13	n (observations)	6	6		B13 = COUNT(B5:B10)
14	Standard uncertainty $u = s/\text{sqrt}(/n)$	0.00229	0.00099		B14 = B12/SQRT(B13)
15	mean c = mean a – mean b			0.18083	D15 = B11–C11
16	$u_c = \text{sqrt}(u_a^2 + u_b^2)$			0.00250	D16 = SQRT((B14^2+C14^2)
17	degrees of freedom (df) for c =	$\dfrac{u_c^4}{\dfrac{u_a^4}{df_a} + \dfrac{u_b^4}{df_b}}$		6.800	D17 = B16^4/
18					(B14^4/(B13–1)+C14^4/(C13–1))
19					
20	Student's t (95% confidence)			2.447	D20 = TINV(0.05,D17)
21	Confidence interval = $t \cdot u_c$ =			0.00611	D21 = D20*D16

Spreadsheet for solution to Exercise B-4.

(b) NH_3 molecular mass:

N atomic mass $= 14.006\ 8\ \pm\ (0.000\ 4)/\sqrt{3}$
$= 14\ 006\ 8\ \pm\ 0.000\ 23 \leftarrow u_x$

H atomic mass $= 1.007\ 98\ \pm\ (0.000\ 14)/\sqrt{3}$
$= 1.007\ 98\ \pm\ 0.000\ 080\ 8 \leftarrow u_x$

Multiply mass and uncertainty by the number of atoms for each element:

N: $\qquad\qquad\qquad\qquad\qquad$ $14.006\ 8 \pm 0.000\ 23 \leftarrow u_x$

$3H: 3[1.007\ 98 \pm 0.000\ 080\ 8] = 3.023\ 94 \pm 0.000\ 242 \leftarrow u_x$

NH_3 molecular mass: $\qquad$ $17.030\ 7\ \pm\ \sqrt{0.000\ 23^2 + 0.000\ 242^2}$
$= 17.030\ 7\ \pm\ 0.000\ 334$ g/mol

Relative uncertainty $= (100)(0.000\ 334)/(17.030\ 7) = 0.002\ 0\%$

$0.002\ 0\%$ uncertainty is negligible in comparison with the uncertainties in **(a)**.

(c)

$$\frac{\left(0.904\ 0\ \dfrac{\text{g solution}}{\text{mL solution}}\right)\left(0.276\ 3\ \dfrac{\text{g NH}_3}{\text{g solution}}\right)(1.004_2\ \text{mL solution})}{\left(17.030\ 7\ \dfrac{\text{g NH}_3}{\text{mol NH}_3}\right)(0.499\ 86\ \text{L})}$$

$$= 0.029\ 46\ \frac{\text{mol}}{\text{L}}$$

(d) $\quad \dfrac{u_M}{M} = \sqrt{\left(\dfrac{u_a}{a}\right)^2 + \left(\dfrac{u_b}{b}\right)^2 + \left(\dfrac{u_c}{c}\right)^2 + \left(\dfrac{u_d}{d}\right)^2 + \left(\dfrac{u_e}{e}\right)^2}$

where $\quad M =$ molarity of NH_3
$a = NH_3$ wt%
$b = NH_3$ solution density
$c =$ pipet volume
$d = NH_3$ molecular mass $(u_d \approx 0)$
$e =$ flask volume

You can express relative uncertainties as absolute numbers or percentages. I choose to use percent, which gives an answer u_M/M in percent:

$u_M/M = \sqrt{(0.226\ 6\%)^2 + (0.127\ 2\%)^2 + (0.081\ 3\%)^2 + (0)^2 + (0.008\ 0\%)^2}$
$= 0.272\%$. Now convert 0.272% to the decimal $0.002\ 72$ and find u_M:

$\qquad u_M = (0.002\ 72)(M) = (0.002\ 72)(0.029\ 46\ M) = 0.000\ 08_{02}\ M$

(e) $\quad df_M = \dfrac{(u_M/M)^4}{\dfrac{(u_a/\bar{a})^4}{df_a} + \dfrac{(u_b/\bar{b})^4}{df_b} + \dfrac{(u_c/\bar{c})^4}{df_c} + \dfrac{(u_d/\bar{d})^4}{df_d} + \dfrac{(u_e/\bar{e})^4}{df_e}}$

I choose to express the relative uncertainties as percentages in the arithmetic below. You can use absolute numbers instead of percentages, but the answer is the same either way.

$$df_M = \frac{(0.272\%)^4}{\dfrac{(0.226\ 6\%)^4}{5-1} + \dfrac{(0.127\ 2\%)^4}{4-1} + \dfrac{(0.081\ 3\%)^4}{6-1} + \dfrac{(0\%)^4}{df_d} + \dfrac{(0.008\ 0\%)^4}{4-1}}$$

$df_M = 7.29$, which we truncate to 7 degrees of freedom

(f) Student's t for 95% confidence and 7 degrees of freedom $= 2.365$

The 95% confidence interval for M is $M \pm t \cdot u_M =$
$0.029\ 46 \pm (2.365)(0.000\ 08_{02}) = 0.029\ 46 \pm 0.000\ 19$ mol/L.

The 95% confidence interval is $100 \times (0.000\ 19)/(0.029\ 46) = 0.64\%$ of the molarity.

B-6. (a) NH_3 wt% relative uncertainty $= 0.226\ 6\%$ (unchanged)

$\qquad\qquad$ density relative uncertainty $= 0.127\ 2\%$ (unchanged)

pipet: tolerance $= 0.006$ mL $=$ range a in triangular distribution

$\qquad u_{\text{pipet}} = a/\sqrt{6} = 0.006/\sqrt{6} = 0.002\ 45$ mL
relative uncertainty $= (100)(0.002\ 45\ \text{mL})/(1.000\ \text{mL}) = 0.245\%$

flask: tolerance $= 0.20$ mL $=$ range a in triangular distribution

$\qquad u_{\text{flask}} = a/\sqrt{6} = 0.20/\sqrt{6} = 0.081\ 6$ mL
relative uncertainty $= (100)(0.081\ 6\ \text{mL})/(500\ \text{mL}) = 0.016\%$

(b) NH_3 molecular mass $= 17.030\ 7 \pm 0.002\ 0\%$ (unchanged)

(c)

$$\frac{\left(0.904\ 0\ \dfrac{\text{g solution}}{\text{mL solution}}\right)\left(0.276\ 3\ \dfrac{\text{g NH}_3}{\text{g solution}}\right)(1.000\ \text{mL solution})}{\left(17.030\ 7\ \dfrac{\text{g NH}_3}{\text{mol NH}_3}\right)(0.500\ 0\ \text{L})}$$

$$= 0.029\ 33\ \frac{\text{mol}}{\text{L}}$$

(d)

$\dfrac{u_M}{M} = \sqrt{(0.226\ 6\%)^2 + (0.127\ 2\%)^2 + (0.245\%)^2 + (0)^2 + (0.016\%)^2}$
$= 0.358\%$
$u_M = (0.003\ 58)(M) = (0.003\ 58)(0.029\ 33\ M) = 0.000\ 10_5\ M$

(e) Degrees of freedom is undefined for Type B uncertainties

(f) Expanded uncertainty $= k \cdot u_M = 2(0.000\ 10_5) = 0.000\ 21$

Answer: $0.029\ 33 \pm 0.000\ 21$ M or $\pm 0.72\%$

The increased uncertainty in pipet volume only brings it up to a similar size to the uncertainties in NH_3 wt% and density, so the overall uncertainty is not changed very much.

References

1. A numerical recipe for evaluating Equation B-1 with a spreadsheet is given by R. de Levie, "Spreadsheet Calculation of the Propagation of Experimental Imprecision," *J. Chem. Ed.* **2000**, *77*, 534.

2. E. F. Meyer, "A Note on Covariance in Propagation of Uncertainty," *J. Chem. Ed.* **1997**, *74*, 1339.

3. C. Salter, "Error Analysis Using the Variance-Covariance Matrix," *J. Chem. Ed.* **2000**, *77*, 1239.

4. B. Wampfler, M. Rösslein, and H. Felber, "The New Measurement Concept Explained by Using an Introductory Example," *J. Chem. Ed.* **2006**, *83*, 1382; *Evaluation of Measurement Data—Guide to the Expression of Uncertainty in Measurement*, JCGM 100:2008 available at http://www.bipm.org/utils/common/documents/jcgm/JCGM_100_2008.E.pdf; S. L. R. Ellison and A. Williams, eds., *Eurachem/CITAC guide: Quantifying Uncertainty in Analytical Measurement*, 3rd ed., (2012) available at http://eurachem.org/index.php/publications/guides/quam; *NIST Uncertainty of Measurement Results*, http://physics.nist.gov/cuu/uncertainty/index.html.

5. A different approach to propagation of uncertainty is the Monte Carlo method, which simulates a large number of events with a spreadsheet. Each input variable for the simulation is given random variation that follows the assumed distribution of uncertainty for that variable. See G. Chew and T. Walczyk, "A Monte Carlo Approach for Estimating Measurement Uncertainty Using Standard Spreadsheet Software," *Anal. Bioanal. Chem.* **2012**, *402*, 2463.

6. Variance (σ^2) for a probability function $p(x)$ is *defined* as

$$\sigma^2 = \int_{-\infty}^{\infty} (x - \mu)^2\, p(x)\ dx$$

where μ is the population mean. For example, $p(x)$ could be the Gaussian distribution function 4-3. An equivalent definition of σ^2 is the (mean value of x^2) minus the (mean value of x)^2. For a Gaussian distribution, 68.3% of the population lies in the interval $\mu \pm \sigma$. For a rectangular distribution extending from $-a$ to $+a$, the standard deviation is $\sigma = a/\sqrt{3}$ and 57.7% of the population lies in the interval $\mu \pm \sigma$. For a triangular distribution extending from $-a$ to $+a$, the standard deviation is $\sigma = a/\sqrt{6}$ and 65.0% of the population lies in the interval $\mu \pm \sigma$.

C-1 Analysis of Variance

Analysis of variance (ANOVA) is a statistical tool widely used to compare means (as is the *t* test) and to apportion the overall variance of experimental measurements to different sources of error. Recall that *variance* is the square of the standard deviation. Variance is additive, but standard deviation is not. If a procedure consists of a sampling operation and an analysis operation, the overall variance is the sum

$$s_{overall}^2 = s_{sampling}^2 + s_{analysis}^2 \qquad \text{(C-1)}$$

where $s_{sampling}^2$ is the variance due to sampling and $s_{analysis}^2$ is the variance due to the analysis.

If one source of variance is much smaller than another, there is little value in further reducing the smaller source of error because it will have little effect on the overall variance. For example, if $s_{analysis} = 0.5$ and $s_{sampling} = 1.0$, then $s_{overall} = \sqrt{0.5^2 + 1.0^2} = 1.1_2$. If you reduce the analytical standard deviation from 0.5 to 0.25, the overall standard deviation is only reduced from 1.1_2 to 1.0_3, which might not have been worth the effort. To improve the precision of a procedure, you should attack the largest source of variance first. If you reduce the sampling standard deviation from 1.0 to 0.5, the overall standard deviation is reduced from 1.1_2 to $\sqrt{0.5^2 + 0.5^2} = 0.7_1$.

Analysis of Variance with a Single Factor

Consider the experiment in Figure C-1 in which the salt content (expressed as wt% Na) of potato chips is measured. Four chips are selected at random from a bag. Each is weighed and then ground into small pieces, and salt from the chip is extracted into water. The water is filtered into a volumetric flask. The solid residue is washed several times, and the washes are filtered into the volumetric flask. Three aliquots from each flask are then analyzed, with results shown in Figure C-2 and listed at the top of Table C-1.

There are at least two sources of random error in Figure C-1. Sampling error arises because each chip probably has a different amount of salt. If we had selected different chips, we would get different answers. Analytical error occurs because we cannot perfectly reproduce the analytical process on each replicate from one chip. How much variation is attributable to sampling error and how much to measurement error?

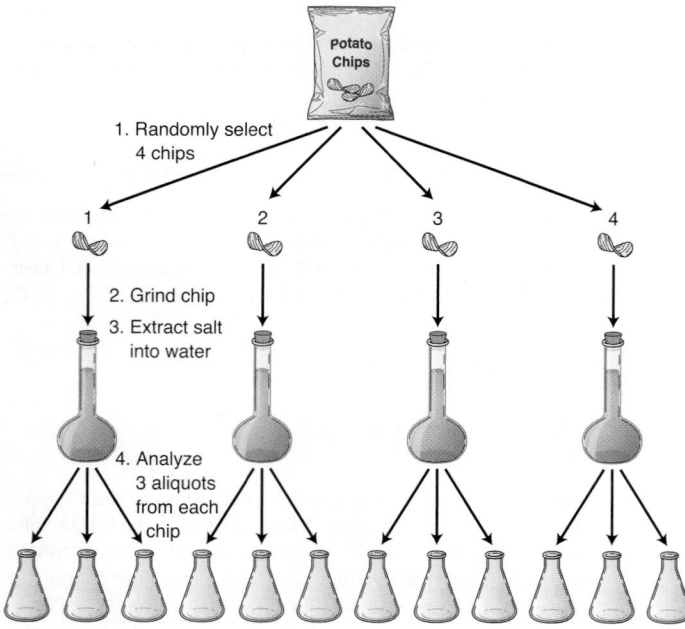

FIGURE C-1 Steps in measuring salt in potato chips. Step 1 introduces sampling variance, $s_{sampling}^2$. Steps 2 to 4 introduce analytical variance, $s_{analysis}^2$.

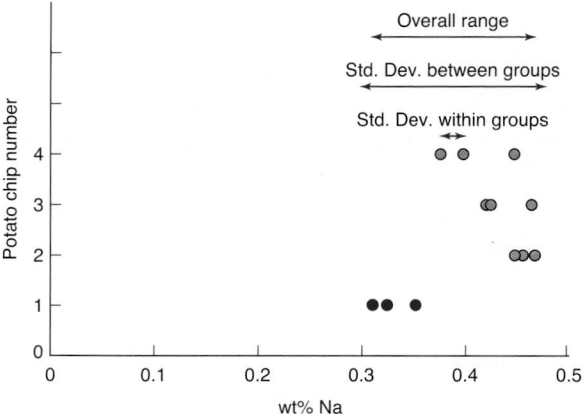

FIGURE C-2 Sodium content of four potato chips.

We demonstrate analysis of variance with tedious manual computations that you can understand. Once we have gone through the process, we will switch to using simple procedures built into Excel. In the jargon of analysis of variance, the potato chip is called the **factor,** which is the experimental variable under study. In Figure C-1 there are three repetitions of the analysis for each potato chip.

Analytical error leads to variable results from each individual potato chip at the top of Table C-1. If there were no measurement error, replicate values from one chip would be identical, which they are not. For example, the three values for chip 1 are 0.324, 0.311, and 0.352 wt% Na. If all chips were identical, and if measurement error were small, the average value of sodium in each chip would be the same. The averages of the four chips in Table C-1 are 0.329_0, 0.456_7, 0.435_7, and 0.407_3 wt% Na. Variation arises from real difference between chips and random measurement error for each chip.

Let the true sodium content of chip 1 be μ_1. If we made a large number of individual measurements of sodium in chip 1, the measurements would be normally distributed around μ_1 with a standard deviation $\sigma_{analysis}$ because they differ from the true mean by the random errors associated with the analytical procedure. From the $n = 3$ replicate measurements of chip 1, we observe a mean value $\bar{x}_1$. If we repeated the measurement of three replicates from chip 1 many times, we would obtain many average values for chip 1, which would be distributed about the true mean (μ_1) for chip 1 with a *standard deviation of the mean* equal to $\sigma_{analysis}/\sqrt{n} = \sigma_{analysis}/\sqrt{3}$.

Variance *Within* Samples

For chip 1 in Table C-1, the mean wt% Na is $\bar{x}_1 = 0.329_0$ wt% Na with a standard deviation of $s_1 = 0.020_{95}$ wt%, giving a variance of $s_1^2 = 4.39_0 \times 10^{-4}$ (wt%)^2. For simplicity, we will stop writing wt%.

The average variance of the four chips is $\frac{1}{4}(s_1^2 + s_2^2 + s_3^2 + s_4^2) = 5.96_5 \times 10^{-4}$, computed in Equation C-2 in Table C-1. We call this the average variance *within* samples, s_{within}^2, because it is the average from $n = 3$ replicate values from each chip. s_{within}^2 is an estimate of the analytical variance, $s_{analysis}^2$, because it represents the variability of results from individual chips.

The standard deviation of n measurements has $n - 1$ degree of freedom, because 1 degree of freedom is lost in computing the average. If we know the average and $n - 1$ values, we could calculate the nth value. In Table C-1, we have $n = 3$ replicate measurements from each of $h = 4$ samples. The total number of measurements is $nh = 3 \times 4 = 12$. Because we computed four averages to find the variance between averages, the degrees of freedom for s_{within}^2 is $12 - 4 = 8$.

Variance *Between* Samples

Next, we compute variance from the difference between means for four potato chips. If there were no variance due to sampling, the variance between means would only arise from random error in the analysis. If all samples

TABLE C-1　Analysis of sodium (wt%) in potato chips

Sample	Chip 1	Chip 2	Chip 3	Chip 4
	0.324	0.455	0.420	0.447
	0.311	0.467	0.463	0.377
	0.352	0.448	0.424	0.398
Average within sample	0.329_0	0.456_7	0.435_7	0.407_3
	$\bar{x}_1$	$\bar{x}_2$	$\bar{x}_3$	$\bar{x}_4$
Standard deviation within sample	0.020_{95}	0.009_{61}	0.023_{76}	0.035_{92}
	s_1	s_2	s_3	s_4

h = number of samples = 4　　　n = number of replicates = 3　　　nh = total measurements = 12

Finding variance *within* samples:

Variance within samples	$4.39_0 \times 10^{-4}$	$9.23_3 \times 10^{-5}$	$5.64_3 \times 10^{-4}$	$1.29_0 \times 10^{-3}$
	s_1^2	s_2^2	s_3^2	s_4^2

$$\text{Average variance within samples} = s_{\text{within}}^2 = \frac{1}{h}\sum(s_i^2) \tag{C-2}$$

$$= \frac{1}{4}(4.39_0 \times 10^{-4} + 9.23_3 \times 10^{-5} + 5.64_3 \times 10^{-4} + 1.29_0 \times 10^{-3})$$

$$s_{\text{within}}^2 = \mathbf{5.96_5 \times 10^{-4}}$$

Degrees of freedom for variance within samples = $nh - h = 12 - 4 = \mathbf{8}$

Finding variance *between* samples:

$$\text{Overall average} = \bar{x} = \frac{1}{nh}\sum(\text{all measurements}) \tag{C-3}$$

$$= \frac{1}{12}(0.324 + 0.311 + 0.352 + 0.455 + \ldots + 0.377 + 0.398)$$

$$= 0.407_2$$

$$\text{Variance of average values from overall mean value} = s_{\text{means}}^2 = \frac{1}{h-1}\sum(\text{sample mean} - \bar{x})^2 \tag{C-4}$$

$$= \frac{1}{3}[(0.329_0 - 0.407_2)^2 + (0.456_7 - 0.407_2)^2 + (0.435_7 - 0.407_2)^2 + (0.407_3 - 0.407_2)^2] = 3.12_4 \times 10^{-3}$$

Degrees of freedom for variance of average values from overall mean = $h - 1 = 4 - 1 = \mathbf{3}$

$$s_{\text{means}} = s_{\text{between}}/\sqrt{n} \tag{C-5}$$

$$\Rightarrow s_{\text{between}}^2 = ns_{\text{means}}^2 = 3(3.12_4 \times 10^{-3}) = \mathbf{9.37_3 \times 10^{-3}} = s_{\text{between}}^2 \tag{C-6}$$

Hypothesis testing with F test:

$h - 1 = 3$ degrees of freedom　↘

$$F = \frac{s_{\text{between}}^2}{s_{\text{within}}^2} = \frac{9.37_3 \times 10^{-3}}{5.96_5 \times 10^{-4}} = \mathbf{15.7} > 4.07 \text{ in Table 4-3} \tag{C-7}$$

$nh - h = 8$ degrees of freedom　↗

Assigning sources of variance:

$$s_{\text{between}}^2 = s_{\text{within}}^2 + ns_{\text{sampling}}^2 \tag{C-8}$$

$$\Rightarrow s_{\text{sampling}}^2 = \frac{1}{n}(s_{\text{between}}^2 - s_{\text{within}}^2) = \frac{1}{3}(9.37_3 \times 10^{-3} - 5.96_5 \times 10^{-4}) = 2.92_6 \times 10^{-3}$$

Standard deviations:

$$s_{\text{sampling}} = \sqrt{s_{\text{sampling}}^2} = \sqrt{2.92_6 \times 10^{-3}} = 0.0541 \tag{C-9}$$

$$s_{\text{analysis}} = s_{\text{within}} = \sqrt{s_{\text{within}}^2} = \sqrt{5.96_5 \times 10^{-4}} = 0.024\,4 \tag{C-10}$$

$$s_{\text{overall}} = \sqrt{s_{\text{analysis}}^2 + s_{\text{sampling}}^2} = \sqrt{(0.024_4)^2 + (0.054_1)^2} = 0.059 \tag{C-11}$$

Data from F. A. Settle and M. Pleva, "The Weakest Link Exercise," Anal. Chem. **1999**, *71, 538A. Two entries in each column are real and the third is a fictitious value added for the sake of this example. The article cited here takes the analysis one step further and shows how to break the overall error down into components from sampling, sample preparation, and analysis. See also D. Harvey, "Two Experiments Illustrating the Importance of Sampling in Quantitative Chemical Analysis," J. Chem. Ed.* **2002**, *79, 360.*

represent the same population whose only variance is s^2_{analysis}, then the standard deviation of the mean of sets of $n = 3$ replicate measurements would be $s_{\text{analysis}}/\sqrt{n} = s_{\text{analysis}}/\sqrt{3}$. If the variance *between* means is significantly larger than our estimate of s^2_{analysis} from the variance *within* each sample, then we conclude that there is additional variance from the sampling operation contributing to the overall variance between means.

The overall average of all $nh = 12$ measurements in Table C-1, calculated in Equation C-3 in Table C-1, is 0.407_2. The four mean values—one for each potato chip—are $\bar{x}_1 = 0.329_0$, $\bar{x}_2 = 0.456_7$, $\bar{x}_3 = 0.435_7$, and $\bar{x}_4 = 0.407_3$. The variance between these $h = 4$ mean values, which we call s^2_{means} in Equation C-4 in Table C-1, is 3.124×10^{-3}. It has $h - 1 = 3$ degrees of freedom because it is computed from h mean values.

If there were no sampling variance, then the variance between means would simply be another measure of analytical variance. The standard deviation of the mean is equal to the population standard deviation divided by $\sqrt{n}$: $s_{\text{means}} = s_{\text{between}}/\sqrt{n}$. Equation C-6 in Table C-1 gives the variance: $s^2_{\text{between}} = n s^2_{\text{means}} = 3(3.124 \times 10^{-3}) = 9.373 \times 10^{-3}$. If there were no error from sampling, s^2_{between} would be equivalent to s^2_{within} for analytical variance.

Testing the Null Hypothesis

The null hypothesis is that the two estimates of variance, s^2_{between} and s^2_{within}, *are not* significantly different. If we find that the variances *are* significantly different, then we conclude that sampling adds to the differences among chips.

The tool for comparing variance is the F test:

$$F_{\text{calculated}} = \frac{s^2_{\text{between}}}{s^2_{\text{within}}} = \frac{9.373 \times 10^{-3}}{5.965 \times 10^{-4}} = \mathbf{15.7} > 4.07 \text{ in Table 4-3} \quad \text{(C-7)}$$

where the upper arrow indicates $h - 1 = 3$ degrees of freedom and the lower arrow indicates $nh - h = 8$ degrees of freedom.

The calculated value of F exceeds the critical value in Table 4-3, so we conclude that the two variances are *not* from the same population.

Assigning Variance to Each Source

To recap, our estimate of variance from differences between the mean values for four potato chips, s^2_{between}, is significantly larger than the average variance for replicate values from single potato chips, s^2_{within}. We infer that both sampling and the analytical procedure contribute to the overall variance.

The variance s^2_{between} can be attributed to *both* analytical variance ($s^2_{\text{analysis}} = s^2_{\text{within}}$) and sampling variance, s^2_{sampling}. The relation is

$$s^2_{\text{between}} = s^2_{\text{within}} + n s^2_{\text{sampling}} \Rightarrow s^2_{\text{sampling}} = \frac{1}{n}(s^2_{\text{between}} - s^2_{\text{within}}) \quad \text{(C-8)}$$

from which we find $s^2_{\text{sampling}} = \frac{1}{3}(9.373 \times 10^{-3} - 5.965 \times 10^{-4}) = 2.926 \times 10^{-3}$.
The final results computed at the bottom of Table C-1 are

$$s_{\text{analysis}} = s_{\text{within}} = \sqrt{s^2_{\text{within}}} = \sqrt{5.965 \times 10^{-4}} = 0.024_4 \quad \text{(C-10)}$$

$$s_{\text{sampling}} = \sqrt{s^2_{\text{sampling}}} = \sqrt{2.926 \times 10^{-3}} = 0.054_1 \quad \text{(C-9)}$$

The sampling operation introduces about twice as much standard deviation as the analytical procedure. Overall variance has components from both sampling and analysis:

$$s^2_{\text{overall}} = s^2_{\text{analysis}} + s^2_{\text{sampling}}$$

$$\Rightarrow s_{\text{overall}} = \sqrt{(0.024_4)^2 + (0.054_1)^2} = 0.059 \quad \text{(C-11)}$$

Inspection of Figure C-2 shows that the Na content of chip 1 is different from that of chips 2, 3, and 4. The difference between chip 1 and the other three chips is the principal source of sampling variance in this experiment.

Why Did We Do All This Work?

Analysis of variance shows that sampling contributes about twice as much as the analytical procedure to the overall uncertainty of the result. To improve the quality of the analysis, we should focus on improving the sampling operation for the greatest effect.

One possible way to improve sampling would be to take several potato chips and combine them for each sample. For example, we could randomly select four potato chips, grind them together, and take one quarter of the thoroughly mixed solid for extraction and analysis. We could repeat this process three more times and end up with the same kind of data as in Table C-1. However, each sample would be a composite of four chips instead of one. We predict that the standard deviation of the mean will be reduced from that in Table C-1 by a factor of $\sim\sqrt{4} = 2$. This procedure would reduce the sampling uncertainty to approximately the same magnitude as the analytical uncertainty.

Spreadsheet for Analysis of Variance

Excel has a procedure for analysis of variance. In Figure C-3, enter data from Table C-1 into cells B4:E7. Excel calls the quantity in row 4 the *factor*. In our case, the factor is the potato chip selected for analysis. In other cases, the factor might be different methods of analysis or experimental variables such as solvent, pH, temperature, or flow rate.

From this point, in Excel 2010, go to Data and select Data Analysis. In the Data Analysis window, select Anova: Single Factor. Another window appears. For Input Range, enter B5:E7. For Output Range, write B9. Excel then prints rows 9 to 24.

Excel provides sample averages ($\bar{x}_1$ to $\bar{x}_4$) in cells E13:E16 and variances (s^2_1 to s^2_4) in cells F13:F16. The variance s^2_{within} (0.000 597) appears in cell E22 (under the heading MS) and its associated degrees of freedom (8) appears in cell D22. The variance s^2_{between} (0.009 37) appears in cell E21 and its associated degrees of freedom (3) appears in cell D21. The value $F_{\text{calculated}} = 15.7$ appears in cell F21. For comparison, the critical value $F = 4.07$ for 95% confidence is given in cell H21. Because $F_{\text{calculated}} > F_{\text{critical}}$, the difference between s^2_{between} and s^2_{within} is significant at the 95% confidence level. Cell G21 tells us that the chance of observing $F_{\text{calculated}} = 15.7$ by random chance if results came from populations with the same population standard deviation is 0.001 0. That is, the difference in variance is significant at the $100 \times (1 - 0.001\ 0) = 99.90\%$ confidence level.

Row 24 is the last output from Excel's ANOVA routine. But it is not the last information we desire. Type in the words and formulas below row 24 to find s_{analysis}, s_{sampling}, and s_{overall}.

Analysis of Variance for Two Factors with Replication

So far in our discussion of ANOVA, we have only considered repetitions of a single experimental factor. In Figures C-1 and C-3, the single factor is the selection of individual potato chips, which is sampling. Excel also allows us to carry out analysis of variance when there are two experimental factors.

In an extension of the potato chip experiment, sodium was measured by two different methods. Salt from four chips labeled A, B, C, and D was extracted into water as in Figure C-1. Aliquots of solution were then analyzed by atomic emission and by a titration. Results appear in cells B4:D12 of Figure C-4. For example, cells D7 and D8 show that chip B was found by titration to contain 0.455 and 0.467 wt% Na in duplicate determinations. Factor 1 in this experiment is the analytical method (emission or titration). Factor 2 is the selection of the chip (sampling). Factor 2 is replicated to obtain an estimate of repeatability.

To analyze variance in Figure C-4 with Excel 2010, go to Data and select Data Analysis. Then select Anova: Two-Factor with Replication. In the next window, enter B4:D12 for the Input Range. This range includes the labels of the columns and rows. In Rows per sample, enter 2 to indicate that each sample is analyzed twice. For Output Range, enter F3. Excel fills in most of the rest of Figure C-4. (We rearranged the Excel output so it would fit on a page.) Cells F3:I28 summarize results for each chip and for the total data set.

The important ANOVA results are in cells C30:I36. Variance within sets of identical treatments is $s^2_{\text{within}} = 0.000\ 663$ in cell F34 (under the heading MS). Variance for different samples analyzed by the same method is $s^2_{\text{sample}} = 0.010\ 018$ in cell F31. Variance between the two experimental

	A	B	C	D	E	F	G	H
1	ANOVA single factor - Analysis of variance of salt in potato chips							
2								
3		Raw data (wt% Na)						
4	Repetition	Chip 1	Chip 2	Chip 3	Chip 4	← Factor		
5	↓	0.324	0.455	0.420	0.447			
6		0.311	0.467	0.463	0.377			
7		0.352	0.448	0.424	0.398			
8								
9		Anova: Single Factor						
10								
11		SUMMARY						
12		Groups	Count	Sum	Average	Variance		
13		Column 1	3	0.987	0.329	0.000439		
14		Column 2	3	1.37	0.456667	9.23E-05		
15		Column 3	3	1.307	0.435667	0.000564		
16		Column 4	3	1.222	0.407333	0.00129		
17								
18								
19		ANOVA				↓ F = MS$_{between}$/MS$_{within}$		
20	Source of Variation	SS	df	MS	F	P-value	F crit	
21	Between Groups	0.028118	3	0.009373	15.71258	0.001025	4.066181	
22	Within Groups	0.004772	8	0.000597				
23				↑ Variance = SS/df				
24	Total	0.03289	11					
25								
26		Assigning sources of variance:						
27	Repeatability: s$_{within}$ = s$_{analysis}$ =	0.024423	= SQRT(E22)					
28	Stdev between factors = s$_{sampling}$ =	0.054087	= SQRT((E21−E22)/D21)					
29	Overall stdev = s$_{overall}$ =	0.059345	= SQRT(D27^2+D28^2)					

FIGURE C-3 Analysis of variance (ANOVA) with a single factor, which is the selection of the potato chip to analyze.

methods is $s^2_{columns} = 0.007\,744$ in cell F32. We will explain the variance labeled "interaction" in cell F33 later.

We want to know if variance from different factors is significantly different from the variance due to repeatability of the measurements, which is s^2_{within} in cell F34. Three F statistics are calculated. Cell G31 gives the quotient $F = s^2_{sample}/s^2_{within} = 0.010\,018/0.000\,663 = 15.1$. This value of F exceeds the critical value 4.07 for 95% confidence in cell I31. Therefore, the difference between samples (chips) is significant. The probability of the observed value of F is $p = 0.001\,17$ in cell H31, which means that there is only a 0.1% chance of obtaining this value if measurements come from sets with the same population mean and population standard deviation.

Cell G32 gives the quotient $F = s^2_{columns}/s^2_{within} = 0.007\,744/0.000\,663 = 11.7$. This value exceeds the critical value 5.32 in cell I32 ($p = 0.009\,1$). Therefore, the difference between columns (the two methods of analysis) is significant at the $1 - 0.009 = 99.1\%$ confidence level. That is, the two methods give different answers for identical samples. This conclusion is apparent because every result from titration in cells D5:D12 is lower than the results from emission in cells C5:C12. The paired t test in Section 4-3, applied to results from the two methods of analysis, produces the same conclusion.

What is an "interaction"? Suppose that you were optimizing a liquid chromatographic separation by varying the organic component of solvent (acetonitrile or methanol, which we will call factor 1) and the pH of the aqueous component of solvent (which we will call factor 2). Suppose that the result you measure is resolution of the two closest peaks in the chromatogram. The optimum combination of solvent and pH will give the greatest resolution. If you tabulate resolution as a function of solvent and pH in a spreadsheet such as Figure C-4, you can test whether solvent and pH each lead to significant differences in resolution.

"Interaction" in row 33 of the spreadsheet tells us if changing one factor changes the response to the other factor. For example, maybe at low pH the resolution varies with organic solvent composition. Perhaps at high pH, resolution becomes independent of organic solvent composition. In such case, we say that there is an *interaction* between the two factors. Changing pH changes the response to solvent.

In Figure C-4, the two factors are method of analysis and sampling. Intuitively, there is no reason to suspect that the performance of the analytical method will depend on which potato chips are drawn from the bag. The value $F = s^2_{interaction}/s^2_{within} = 1.21$ in cell G33 is less than the critical value of 4.07 in cell I33. We conclude that there is *no interaction* between factors in this experiment.

	A	B	C	D	E	F	G	H	I
1	ANOVA with two factors and repetition								
2									
3		Factor 2	Factor 1 analytical method			SUMMARY	Emission	Titration	Total
4		Chip #	Emission	Titration		A			
5	Repetition	A	0.411	0.324		Count	2	2	4
6		A	0.394	0.310		Sum	0.805	0.634	1.439
7	Repetition	B	0.485	0.455		Average	0.4025	0.317	0.35975
8		B	0.493	0.467		Variance	0.000144	9.8E-05	0.002518
9	Repetition	C	0.450	0.420		B			
10		C	0.481	0.463		Count	2	2	4
11	Repetition	D	0.474	0.447		Sum	0.978	0.922	1.9
12		D	0.427	0.377		Average	0.489	0.461	0.475
13						Variance	3.2E-05	7.2E-05	0.000296
14						C			
15						Count	2	2	4
16						Sum	0.931	0.883	1.814
17						Average	0.4655	0.4415	0.4535
18						Variance	0.00048	0.000925	0.00066
19						D			
20						Count	2	2	4
21						Sum	0.901	0.824	1.725
22						Average	0.4505	0.412	0.43125
23						Variance	0.001104	0.00245	0.001679
24						Total			
25						Count	8	8	
26						Sum	3.615	3.263	
27						Average	0.451875	0.407875	
28						Variance	0.001396	0.004	
29		ANOVA					$\downarrow F = MS/MS_{within}$		
30		Source of Variation	SS	df	MS	F	P-value	F crit	
31		Sample	0.030055	3	0.01001842	15.10504	0.00117	4.066181	
32		Columns	0.007744	1	0.007744	11.67584	0.009128	5.317655	
33		Interaction	0.002408	3	0.00080283	1.210454	0.366689	4.066181	
34		Within	0.005306	8	0.00066325				
35					$\uparrow$ Variance = SS/df				
36		Total	0.045514	15					

FIGURE C-4 Analysis of variance (ANOVA) with two factors, which are the selection of the potato chip and the method of analysis. [Data from F. A. Settle and M. Pleva, "The Weakest Link Exercise," *Anal. Chem.* **1999**, *71*, 538A.]

C-2 Efficiency in Experimental Design[1]

Operating parameters usually need to be optimized when we develop an analytical method. The least efficient way to do this is to vary one parameter at a time while keeping everything else constant. More efficient procedures are called *fractional factorial experimental design*[2] and *simplex optimization*.[3] We discuss one example of experimental design intended to provide maximum information in the fewest number of trials.

Suppose we have three different unknown solutions of acid, designated A, B, and C. If we titrate each one once with base, we find its concentration,

but we have no estimate of uncertainty. If we titrate each solution three times, for a total of nine measurements, we would find each concentration and its standard deviation.

A more efficient experimental design provides concentrations and standard deviations in fewer than nine experiments. One of many efficient designs is given in the spreadsheet in Figure C-5. Instead of titrating each acid by itself, we titrate mixtures of the acids. For example, in row 5 of the spreadsheet, a mixture containing 2 mL A, 2 mL B, and 2 mL C required 23.29 mL of 0.120 4 M NaOH, which amounts to 2.804 mmol of OH^-.

1. P. de B. Harrington, E. Kolbrich, and J. Cline, "Experimental Design and Multiplexed Modeling Using Titrimetry and Spreadsheets," *J. Chem. Ed.* **2002**, *79*, 863.

2. G. Hanrahan, *Environmental Chemometrics* (Boca Raton, FL: CRC Press, 2009); R. G. Brereton, *Applied Chemometrics for Scientists* (Chichester: Wiley, 2007); M. Otto, *Chemometrics* (Wenheim: Wiley-VCH, 2007); D. Montgomery, *Design and Analysis of Experiments*, 5th ed., (New York: Wiley, 2001); C. F. Wu and M. Hamada, *Experiments: Planning, Analysis, and Parameter Design Optimization* (New York: Wiley, 2000); M. Anderson and P. Whitcomb, *DoE Simplified: Practical Tools for Effective Experimentation* (Portland, OR: Productivity, Inc., 2000); G. E. P. Box, J. S. Hunter, and W. G. Hunter, *Statistics for Experimenters: Design, Innovation, and Discovery*, 2nd ed. (Hoboken, NJ: Wiley, 2005); R. S. Strange,

"Introduction to Experimental Design for Chemists," *J. Chem. Ed.* **1990**, *67*, 113; J. M. Gozálvez and J. C. García-Díaz, "Mixture Design Experiments Applied to the Formulation of Colorant Solutions," *J. Chem. Ed.* **2006**, *83*, 647.

3. S. N. Deming and S. L. Morgan, "Simplex Optimization of Variables in Analytical Chemistry," *Anal. Chem.* **1973**, *45*, 278A; D. J. Leggett, "Instrumental Simplex Optimization," *J. Chem. Ed.* **1983**, *60*, 707; S. Srijaranai, R. Burakham, T. Khammeng, and R. L. Deming, "Use of the Simplex Method to Optimize the Mobile Phase for the Micellar Chromatographic Separation of Inorganic Anions," *Anal. Bioanal. Chem.* **2002**, *374*, 145; D. Betteridge, A. P. Wade, and A. G. Howard, "Reflections on the Modified Simplex," *Talanta* **1985**, *32*, 709, 723.

	A	B	C	D	E
1	Experimental Design				
2					
3	Volumes of unknown acids (mL)			mL NaOH	mmol
4	A	B	C	(0.1204 M)	NaOH
5	2	2	2	23.29	2.804
6	2	3	1	20.01	2.409
7	3	1	2	21.72	2.615
8	1	2	3	28.51	3.433
9	2	2	2	23.26	2.801
10					
11			[C]	[B]	[A]
12		Molarity (M)	0.8099	0.4001	0.1962
13		Std. uncertainty (u_M)	0.0062	0.0062	0.0062
14			1.0000	0.0130	#N/A
15			R^2	S_y	
16	Highlight cells C12:E14				
17	Type "= LINEST(E5:E9,A5:C9,FALSE,TRUE)"				
18	Press CTRL+SHIFT+ENTER (on PC)				
19	Press COMMAND(⌘)+RETURN (on Mac)				

FIGURE C-5 Spreadsheet for efficient experimental design uses Excel LINEST routine to fit the function $y = m_A x_A + m_B x_B + m_C x_C$ to experimental data by a least-squares procedure.

(Acids could be delivered by transfer pipets whose tolerances are given in Table 2-4. So 2 mL means 2.000 mL with uncertainty in the third decimal place.) In row 6, the mixture contained 2 mL A, 3 mL B, and 1 mL C. Other permutations are titrated in rows 7 and 8. Then row 5 is repeated independently in row 9. Column E gives mmol of base for each run.

For each experiment, mmol of base consumed equals mmol of acid in the mixture:

$$\underbrace{\text{mmol OH}^-}_{y} = \underbrace{[A]V_A}_{m_A x_A} + \underbrace{[B]V_B}_{m_B x_B} + \underbrace{[C]V_C}_{m_C x_C} \qquad \text{(C-12)}$$

where [A] is the concentration of acid A (mol/L) and V_A is the volume of A in mL (mol/L $\times$ mL = mmol). Rows 5 through 9 of the spreadsheet tell us that

$$\left.\begin{array}{l} 2.804 = [A]\cdot 2 + [B]\cdot 2 + [C]\cdot 2 \\ 2.409 = [A]\cdot 2 + [B]\cdot 3 + [C]\cdot 1 \\ 2.615 = [A]\cdot 3 + [B]\cdot 1 + [C]\cdot 2 \\ 3.433 = [A]\cdot 1 + [B]\cdot 2 + [C]\cdot 3 \\ 2.801 = [A]\cdot 2 + [B]\cdot 2 + [C]\cdot 2 \end{array}\right\} \qquad \text{(C-13)}$$

Our problem is to find the best values for the molarities [A], [B], and [C].

As luck would have it, Excel will find these values for us by the least-squares procedure LINEST. In Chapter 4, we used LINEST to find the slope and intercept in the equation $y = mx + b$. In Figure C-5, we use LINEST to find the slopes for $y = m_A x_A + m_B x_B + m_C x_C + b$ (where the intercept, b, is 0). To execute LINEST, highlight cells C12:E14 and type "=LINEST(E5:E9, A5:C9,FALSE,TRUE)". Then press CTRL+SHIFT+ENTER on a PC or COMMAND(⌘)+RETURN on a Mac. The first argument of LINEST, E5:E9, contains values of y (= mmol OH$^-$). The second argument, A5:C9, contains values of x (= volumes of acid). The third argument (FALSE) tells the computer to set the intercept (b) to zero and the fourth argument (TRUE) says that we want statistics to be computed.

Excel finds the least-squares slopes in row 12 and their standard uncertainties (u_M = standard deviation of the mean) in row 13. These slopes are the molarities [C], [B], and [A]. For example, cell C12 tells us that [C] = 0.809 9 M.

We require at least n equations to solve for n unknowns. In this example, we have five equations (C-13), but just three unknowns ([A], [B], [C]). The two extra equations allow us to estimate uncertainty in the unknowns. If you do more experiments, you will generally decrease the uncertainty in concentration. An efficient experimental design with five experiments leaves us with more uncertainty than if we had done nine experiments. However, we have cut our effort nearly in half with the efficient design.

Standard uncertainties in cells C13:E13 depend on the quality of the least-squares fit to Equations C-13, which is related to uncertainties in volumes and estimation of the equivalence point of the titrations. To find the 95% confidence interval for concentration, multiply the standard uncertainty by Student's t for $5 - 3 = 2$ degrees of freedom (because we have 5 equations and determined 3 unknowns). In Table 4-4, $t = 4.303$ for 95% confidence and 2 degrees of freedom. Therefore, for each component (A, B, and C), the 95% uncertainty is $tu_M = (4.303)(0.006\ 2) = 0.027$. A reasonable expression of concentration is [C] = 0.810 ± 0.027 M or 0.81 ± 0.03 M.

Exercises

C-1. *Analysis of variance.* Chlorophylls a and b are plant pigments that absorb sunlight and transfer the energy into photosynthesis of carbohydrates from CO_2 and H_2O, releasing O_2 in the process. Chlorophylls were extracted from chopped up grass and measured by spectrophotometry. The table shows results for chlorophyll a from four separate analyses of five blades of grass.

Chlorophyll a (g/L)

Site 1:	Blade 1	Blade 2	Blade 3	Blade 4	Blade 5
	1.09	1.26	1.19	1.23	0.85
	0.86	0.96	1.21	1.30	0.65
	0.93	0.80	1.27	0.97	0.86
	0.99	0.73	1.12	0.97	1.03

Data from J. Marcos, A. Ríos, and M. Valcárcel, "Practicing Quality Control in a Bioanalytical Experiment," J. Chem. Ed. 1995, 72, 947.

Four replications from each blade of grass tell us the precision of the analytical procedure. Differences between mean values for each of the five blades of grass are a measure of variation due to sampling. (That is, not all blades of grass have the same composition.) Determine whether sampling variance is significantly different from analytical variance. Find the standard deviations attributable to sampling and to analysis, as well as the overall standard deviation arising from both sources.

C-2. Let's apply analysis of variance to Rayleigh's data in Figure 4-10 to see if the density of nitrogen from air and from chemical decomposition are significantly different. The single experimental *factor* is the choice of how nitrogen was prepared (from air or by chemical decomposition). Copy cells A1:C17 of Figure 4-10 into a fresh spreadsheet. In Excel 2010, go to the Data tab and select Data Analysis Anova: Single Factor. For Input Range, select B5:C12. (Excel ignores blank cell B12.) The ANOVA window should show that data are grouped by columns and alpha = 0.05 (95% confidence). For Output Range, type E6 and click OK. In Excel's output, identify the variance for each method of preparing nitrogen.

(a) Identify the variance in mass of nitrogen from air and the variance from decomposition. Identify the variance between groups (between methods of preparing nitrogen) and the variance within groups (repetitions of each method).

(b) Identify F = (variance between methods)/(variance within method). What is the critical value of F? Is the difference between methods of preparing nitrogen significant?

C-3. (a) Apply the paired t test to the emission and titration analyses in cells C5:C12 and D5:D12 of Figure C-4. Is there a significant difference between results produced by the two methods of analysis? Did ANOVA in Figure C-4 reach the same conclusion as the t test?

(b) You should have found in **(a)** that atomic emission gave a significantly higher value of wt% Na in potato chips than does titration. Now we ask you to *suggest a reason why one method gives systematically higher results.* Here is some background. In atomic emission analysis, the dissolved sample is fed into a flame or plasma, which excites gaseous Na atoms to a high energy level from which they emit light. Emission intensity is proportional to the concentration of sodium in solution. In the titration, the dissolved sample is titrated with a known concentration of $AgNO_3(aq)$. The reaction is $Ag^+ + Cl^- \rightarrow AgCl(s)$. The titration tells us how much Cl^- (and therefore NaCl) was in solution. Why might the two methods give systematically different values of wt% Na?

C-4. A mixture of potassium nitrate and sodium chloride was shipped in a train car. To determine what fraction was potassium nitrate, samples were withdrawn at random from five locations, and each sample was subjected to four replicate analyses, with the results shown below:

Sample	Weight percent potassium	Mean	Standard deviation
A	12.42, 12.28, 12.33, 12.36	12.34_{75}	0.05_{85}
B	12.27, 12.24, 12.19, 12.19	12.22_{25}	0.03_{95}
C	12.41, 12.48, 12.51, 12.39	12.44_{75}	0.05_{68}
D	12.42, 12.43, 12.47, 12.40	12.43_{00}	0.02_{94}
E	12.19, 12.28, 12.20, 12.32	12.24_{75}	0.06_{29}

Show that the variance among the five mean values is significantly greater than the variance within samples. What standard deviation can be ascribed to variation in the sample composition (s_{sampling}) and what standard deviation can be ascribed to irreproducibility (s_{analysis}) in the replicate analyses of each sample?

C-5. Acid-base titrations similar to those in Figure C-5 had volumes and results shown in the table.[1] Use Excel LINEST to find the concentrations of acids A, B, and C and their standard uncertainties and 95% confidence intervals.

Acid volume (mL)			mmol of NaOH required
A	**B**	**C**	
2	2	2	3.015
0	2	2	1.385
2	0	2	2.180
2	2	0	2.548
2	2	2	3.140

Solutions to Exercises

C-1. (a) Spreadsheet for site 1 analysis of variance:

	A	B	C	D	E	F	G
1	ANOVA: Chlorophyll a at Site 1						
2							
3	Blade 1	Blade 2	Blade 3	Blade 4	Blade 5		
4	1.09	1.26	1.19	1.23	0.85		
5	0.86	0.96	1.21	1.30	0.65		
6	0.93	0.80	1.27	0.97	0.86		
7	0.99	0.73	1.12	0.97	1.03		
8							
9	Anova: Single Factor						
10							
11	SUMMARY						
12	Groups	Count	Sum	Average	Variance		
13	Column 1	4	3.87	0.9675	0.009492		
14	Column 2	4	3.75	0.9375	0.055492		
15	Column 3	4	4.79	1.1975	0.003825		
16	Column 4	4	4.47	1.1175	0.029825		
17	Column 5	4	3.39	0.8475	0.024158		
18							
19	ANOVA						
20	Source of						
21	Variation	SS	df	MS	F	P-value	F crit
22	Between Groups	0.32048	4	0.08012	3.262436	0.0411	3.0556
23	Within Groups	0.36838	15	0.02456			
24							
25	Total	0.68886	19				

The calculated value of $F = (s_{\text{between}}^2)/(s_{\text{within}}^2) = 3.26$ in cell E22 is greater than the critical value of 3.05 in cell G22, so the difference is significant. $p = 0.04$ in cell F22 tells us that the difference is significant at the 96% level.

Assigning sources of variance:

$$s_{\text{sampling}}^2 = \frac{1}{n}(s_{\text{between}}^2 - s_{\text{within}}^2) = \frac{1}{4}(0.080_{12} - 0.024_{56}) = 0.013_{89}$$

where n = number of replicates of each grass blade, and the variances are found under the heading MS in cells D22 and D23.

Standard deviations:

$$s_{\text{sampling}} = \sqrt{s_{\text{sampling}}^2} = \sqrt{0.013_{89}} = 0.11_8$$

$$s_{\text{analysis}} = s_{\text{within}} = \sqrt{s_{\text{within}}^2} = \sqrt{0.024_{56}} = 0.15_7$$

$$s_{\text{overall}}^2 = s_{\text{analysis}}^2 + s_{\text{sampling}}^2 \Rightarrow s_{\text{overall}} = \sqrt{(0.011_8)^2 + (0.015_7)^2} = 0.19_6$$

C-2. Analysis of variance for Rayleigh's nitrogen density (ANOVA: Single Factor)

	A	B	C	D	E	F	G	H	I	J	K
1	Analysis of Rayleigh's Data										
2											
3	Mass of gas (g) collected from										
4		air	chemical		←Factor						
5	Repetition	2.31017	2.30143								
6	↓	2.30986	2.29890		Anova: Single Factor						
7		2.31010	2.29816								
8		2.31001	2.30182		SUMMARY						
9		2.31024	2.29869		Groups	Count	Sum	Average	Variance		
10		2.31010	2.29940		Column 1	7	16.1708	2.310109	2.035E-08		
11		2.31028	2.29849		Column 2	8	18.3958	2.299473	1.902E-06		
12			2.29889								
13	Average	2.31011	2.29947		ANOVA				↓ F = MS$_{between}$/MS$_{within}$		
14	Std Dev	0.00014	0.00138		Source of Variation	SS	df	MS	F	P-value	F crit
15					Between Groups	0.00042	1	0.000422	408.59465	3.32E-11	4.667193
16	B13 = AVERAGE(B5:B12)				Within Groups	1.3E-05	13	1.03E-06			
17	B14 = STDEV(B5:B12)							↑ Variance = SS/df			
18					Total	0.00044	14				

(a) Variance of mass of nitrogen from air $= 2.035 \times 10^{-8}$ (cell I10)

Variance of mass of nitrogen from decomposition $= 1.902 \times 10^{-6}$ (cell I11)

Variance between groups $= 0.0004\ 22$ (cell H15)

Variance within groups $= 1.03 \times 10^{-6}$ (cell H16)

(b) $F = $ H15/H16 $= 408$ (cell I15)

Critical value of $F = 4.67$ (cell K15)

$F_{observed} > F_{critical}$, so difference is significant (highly significant with $p = 3 \times 10^{-11}$ in cell J15)

C-3. (a) The spreadsheet uses the paired t test to find that $t_{calculated}$ in cell F11 is greater than the critical value of t in cell F15, so the difference between methods is significant. Cell F14 tells us that the difference is significant at the $(1 - 0.002\ 5) = 99.75\%$ level.

(b) Atomic emission measures dissolved Na^+ directly. Titration measures Cl^- directly. To find Na^+ from Cl^-, we assume that all Cl^- arises from NaCl extracted from the potato chip. If Na^+ in the sample arises from other salts in addition to NaCl, there will be more Na^+ than Cl^- in the solution. If all Cl^- is not in the form of NaCl, there could be more Cl^- than Na^+ in the solution.

	A	B	C	D	E	F	G
1	Paired t test						
2					t-Test: Paired Two Sample for Means		
3	Chip #	Emission	Titration				
4	A	0.411	0.324			Variable 1	Variable 2
5	A	0.394	0.310		Mean	0.4519	0.4079
6	B	0.485	0.455		Variance	0.0014	0.0040
7	B	0.493	0.467		Observations	8	8
8	C	0.450	0.420		Pearson Correlation	0.9857	
9	C	0.481	0.463		Hypothesized Mean Difference	0	
10	D	0.474	0.447		df	7	
11	D	0.427	0.377		t Stat	4.5811	
12	Calculated t statistic in cell F11 is				P(T<=t) one-tail	0.0013	
13	greater than critical t in cell F15.				t Critical one-tail	1.8946	
14	Therefore difference in the two methods				P(T<=t) two-tail	0.0025	
15	is significant.				t Critical two-tail	2.3646	

Spreadsheet for Exercise C-3.

	A	B	C	D	E	F	G
1	Potassium nitrate in train car						
2							
3	Sample A	B	C	D	E		
4	12.42	12.27	12.41	12.42	12.19		
5	12.28	12.24	12.48	12.43	12.28		
6	12.33	12.19	12.51	12.47	12.20		
7	12.36	12.19	12.39	12.40	12.32		
8							
9	Anova: Single Factor						
10							
11	SUMMARY						
12	Groups	Count	Sum	Average	Variance		
13	Column 1	4	49.39	12.3475	0.00342		
14	Column 2	4	48.89	12.2225	0.00156		
15	Column 3	4	49.79	12.4475	0.00323		
16	Column 4	4	49.72	12.43	0.00087		
17	Column 5	4	48.99	12.2475	0.00396		
18							
19	ANOVA						
20	Source of						
21	Variation	SS	df	MS	F	P-value	F crit
22	Between Groups	0.16828	4	0.04207	16.1394	3E-05	3.0556
23	Within Groups	0.0391	15	0.00261			
24							
25	Total	0.20738	19				

Spreadsheet for Exercise C-4.

C-4. In the spreadsheet,

$s_{within}^2 = 0.002\ 61$ in cell D23 with $20 - 5 = 15$ degrees of freedom in cell C23.

$s_{between}^2 = 0.042\ 07$ in cell D22 with $5 - 1 = 4$ degrees of freedom in cell C22.

$$F = \frac{s_{between}^2}{s_{within}^2} = 0.042\ 07/0.002\ 61 = 16.1 \text{ (cell E22)} > 3.06 \text{ (cell G22)}$$

(with $p = 3 \times 10^{-5}$ in cell F 22)

Assigning sources of variance:

$$s_{sampling}^2 = \frac{1}{n}(s_{between}^2 - s_{within}^2) = \frac{1}{4}(0.042\ 07 - 0.002\ 61) = 0.009\ 866$$

$$s_{sampling} = \sqrt{s_{sampling}^2} = \sqrt{0.009\ 866} = 0.09_9 \text{ wt\%}$$

$$s_{analysis} = s_{within} = \sqrt{s_{within}^2} = \sqrt{0.002\ 61} = 0.05_1 \text{ wt\%}$$

C-5. Molarities computed by LINEST are given in row 12 of the spreadsheet and standard uncertainties (u_M) are given in row 13. Student's t for 95% confidence and $5 - 3 = 2$ degrees of freedom is calculated from $t = $ TINV(0.05,2) = 4.303 in cell C21. The 95% confidence interval for the concentrations is $tu_M = (4.303)(0.0267) = 0.12$. The 95% confidence interval for each of the three components is ± 0.12 M.

	A	B	C	D	E
1	Experimental Design				
2					
3	Volumes of unknown acids (mL)			mmol	
4	A	B	C	NaOH	
5	2	2	2	3.015	
6	0	2	2	1.385	
7	2	0	2	2.180	
8	2	2	0	2.548	
9	2	2	2	3.140	
10					
11			[C]	[B]	[A]
12		Molarity (M)	0.25635	0.44035	0.83785
13		Std. uncertainty (u_M)	0.026729	0.026729	0.026729
14			0.999746	0.063896	#N/A
15			R^2	S_y	
16	Highlight cells C12:E14				
17	Type "= LINEST(D5:D9,A5:C9,FALSE,TRUE)"				
18	Press CTRL+SHIFT+ENTER (on PC)				
19	Press COMMAND(⌘)+ENTER (on Mac)				
20					
21		Student's t =	4.302653	C21 = TINV(0.05,2)	
22		95% confidence interval for concentrations = tu_M			0.115008

Spreadsheet for Exercise C-5.

APPENDIX C Analysis of Variance and Efficiency in Experimental Design

The *oxidation number*, or *oxidation state*, is a bookkeeping device used to keep track of the number of electrons formally associated with a particular element. The oxidation number is meant to tell how many electrons have been lost or gained by a neutral atom when it forms a compound. Because oxidation numbers have no real physical meaning, they are somewhat arbitrary, and not all chemists will assign the same oxidation number to a given element in an unusual compound. However, there are some ground rules that provide a useful start.

1. The oxidation number of an element by itself—e.g., $Cu(s)$ or $Cl_2(g)$—is 0.
2. The oxidation number of H is almost always $+1$, except in metal hydrides—e.g., NaH—in which H is -1.
3. The oxidation number of oxygen is almost always -2. The only common exceptions are peroxides, in which two oxygen atoms are connected and each has an oxidation number of -1. Two examples are hydrogen peroxide (H—O—O—H) and its anion (H—O—O$^-$). The oxidation number of oxygen in gaseous O_2 is, of course, 0.
4. The alkali metals (Li, Na, K, Rb, Cs, Fr) almost always have an oxidation number of $+1$. The alkaline earth metals (Be, Mg, Ca, Sr, Ba, Ra) are almost always in the $+2$ oxidation state.
5. The halogens (F, Cl, Br, I) are usually in the -1 oxidation state. Exceptions occur when two different halogens are bound to each other or when a halogen is bound to more than one atom. When different halogens are bound to each other, we assign the oxidation number -1 to the more electronegative halogen.

The sum of the oxidation numbers of each atom in a molecule must equal the charge of the molecule. In H_2O, for example, we have

$$\begin{aligned} 2 \text{ hydrogen} &= 2(+1) = +2 \\ \text{oxygen} &= \underline{-2} \\ \text{net charge} & \qquad\quad 0 \end{aligned}$$

In SO_4^{2-}, sulfur must have an oxidation number of $+6$ so that the sum of the oxidation numbers will be -2:

$$\begin{aligned} \text{oxygen} &= 4(-2) = \quad -8 \\ \text{sulfur} &= \qquad\quad \underline{+6} \\ \text{net charge} & \qquad\qquad -2 \end{aligned}$$

In benzene (C_6H_6), the oxidation number of each carbon must be -1 if hydrogen is assigned the number $+1$. In cyclohexane (C_6H_{12}), the oxidation number of each carbon must be -2, for the same reason. The carbons in benzene are in a higher oxidation state than those in cyclohexane.

The oxidation number of iodine in ICl_2^- is $+1$. This is unusual, because halogens are usually -1. However, because chlorine is more electronegative than iodine, we assign Cl as -1, thereby forcing I to be $+1$.

The oxidation number of As in As_2S_3 is $+3$, and the value for S is -2. This is arbitrary but reasonable. Because S is more electronegative than As, we make S negative and As positive; and, because S is in the same family as oxygen, which is usually -2, we assign S as -2, thus leaving As as $+3$.

The oxidation number of S in $S_4O_6^{2-}$ (tetrathionate) is $+2.5$. The *fractional oxidation state* comes about because six O atoms contribute -12. Because the charge is -2, the four S atoms must contribute $+10$. The average oxidation number of S must be $+\frac{10}{4} = 2.5$.

The oxidation number of Fe in $K_3Fe(CN)_6$ is $+3$. To make this assignment, we first recognize cyanide (CN^-) as a common ion that carries a charge of -1. Six cyanide ions give -6, and three potassium ions (K^+) give $+3$. Therefore, Fe should have an oxidation number of $+3$ for the whole formula to be neutral. In this approach, it is not necessary to assign individual oxidation numbers to carbon and nitrogen, as long as we recognize that the charge of CN is -1.

Problems

Answers are given at the end of this appendix.

D-1. Write the oxidation state of the boldface atom in each of the following species.

(a) **Ag**Br
(b) **S**$_2$O$_3^{2-}$
(c) **Se**F$_6$
(d) H**S**$_2$O$_3^-$
(e) H**O**$_2$
(f) **N**O
(g) **Cr**$^{3+}$
(h) **Mn**O$_2$
(i) **Pb**(OH)$_3^-$
(j) **Fe**(OH)$_3$
(k) **Cl**O$^-$
(l) K$_4$**Fe**(CN)$_6$
(m) **Cl**O$_2$
(n) **Cl**O$_2^-$
(o) **Mn**(CN)$_6^{4-}$

(p) **N**$_2$
(q) **N**H$_4^+$
(r) **N**$_2$H$_5^+$
(s) H**As**O$_3^{2-}$
(t) **Co**$_2$(CO)$_8$ (CO group is neutral)
(u) (CH$_3$)$_4$**Li**$_4$
(v) **P**$_4$O$_{10}$
(w) **C**$_2$H$_6$O (ethanol, CH$_3$CH$_2$OH)
(x) **V**O(SO$_4$)
(y) **Fe**$_3$O$_4$
(z) **C**$_3$H$_3^+$

Structure: $H — C{\oplus}\overset{\displaystyle H}{\underset{\displaystyle H}{\overset{\diagup C}{\diagdown_{C}}}}$

D-2. Identify the oxidizing agent and the reducing agent on the left side of each of the following reactions.

(a) $Cr_2O_7^{2-} + 3Sn^{2+} + 14H^+ \rightarrow 2Cr^{3+} + 3Sn^{4+} + 7H_2O$
(b) $4I^- + O_2 + 4H^+ \rightarrow 2I_2 + 2H_2O$
(c) $5CH_3\overset{\displaystyle O}{\overset{\|}{C}}H + 2MnO_4^- + 6H^+ \rightarrow 5CH_3\overset{\displaystyle O}{\overset{\|}{C}}OH + 2Mn^{2+} + 3H_2O$
(d) $\underset{\text{Glycerol}}{HOCH_2CHOHCH_2OH} + 2IO_4^- \rightarrow$
     $2H_2C{=}O + HCO_2H + 2IO_3^- + H_2O$
     $\qquad\qquad\underset{\text{Formaldehyde}}{} \; \underset{\text{Formic acid}}{}$
(e) $C_8H_8 + 2Na \rightarrow C_8H_8^{2-} + 2Na^+$
     C_8H_8 is cyclooctatetraene with the structure ⬡
(f) $I_2 + OH^- \rightarrow HOI + I^-$
     $\qquad\quad\underset{\substack{\text{Hypoiodous} \\ \text{acid}}}{}$

Balancing Redox Reactions

To balance a reaction involving oxidation and reduction, we must first identify which element is oxidized and which is reduced. We then break the net reaction into two imaginary *half-reactions*, one of which involves only oxidation and the other only reduction. Although free electrons never appear in a balanced net reaction, they do appear in balanced half-reactions. If we are dealing with aqueous solutions, we proceed to balance each half-reaction, using H_2O and either H^+ or OH^-, as necessary. *A reaction is balanced when the number of atoms of each element is the same on both sides and the net charge is the same on both sides.*[*]

---

[*]A completely different method for balancing complex redox equations by inspection has been described by D. Kolb, *J. Chem. Ed.* **1981**, *58*, 642. For some challenging problems in balancing redox equations, see R. Stout, *J. Chem. Ed.* **1995**, *72*, 1125.

(Continued)

Acidic Solutions

Here are the steps we will follow:

1. Assign oxidation numbers to the elements that are oxidized or reduced.
2. Break the reaction into two half-reactions, one involving oxidation and the other reduction.
3. For each half-reaction, balance the number of atoms that are oxidized or reduced.
4. Balance the electrons to account for the change in oxidation number by adding electrons to one side of each half-reaction.
5. Balance oxygen atoms by adding H_2O to one side of each half-reaction.
6. Balance the H atoms by adding H^+ to one side of each half-reaction.
7. Multiply each half-reaction by the number of electrons in the other half-reaction so that the number of electrons on each side of the total reaction will cancel. Then add the two half-reactions and simplify to the smallest integral coefficients.

EXAMPLE Balancing a Redox Equation

Balance the following equation using H^+, but not OH^-:

$$\underset{+2}{Fe^{2+}} + \underset{+7}{\underset{\text{Permanganate}}{MnO_4^-}} \rightleftharpoons \underset{+3}{Fe^{3+}} + \underset{+2}{Mn^{2+}}$$

Solution

1. *Assign oxidation numbers.* They are assigned for Fe and Mn in each species in the above reaction.
2. *Break the reaction into two half-reactions.*

   Oxidation half-reaction: $\underset{+2}{Fe^{2+}} \rightleftharpoons \underset{+3}{Fe^{3+}}$

   Reduction half-reaction: $\underset{+7}{MnO_4^-} \rightleftharpoons \underset{+2}{Mn^{2+}}$

3. *Balance the atoms that are oxidized or reduced.* Because there is only one Fe or Mn in each species on each side of the equation, the atoms of Fe and Mn are already balanced.
4. *Balance electrons.* Electrons are added to account for the change in each oxidation state.

   $$Fe^{2+} \rightleftharpoons Fe^{3+} + e^-$$
   $$MnO_4^- + 5e^- \rightleftharpoons Mn^{2+}$$

   In the second case, we need $5e^-$ on the left side to take Mn from $+7$ to $+2$.
5. *Balance oxygen atoms.* There are no oxygen atoms in the Fe half-reaction. There are four oxygen atoms on the left side of the Mn reaction, so we add four molecules of H_2O to the right side:

   $$MnO_4^- + 5e^- \rightleftharpoons Mn^{2+} + 4H_2O$$

6. *Balance hydrogen atoms.* The Fe equation is already balanced. The Mn equation needs $8H^+$ on the left.

   $$MnO_4^- + 5e^- + 8H^+ \rightarrow Mn^{2+} + 4H_2O$$

   At this point, each half-reaction must be completely balanced (the same number of atoms and charge on each side) *or you have made a mistake.*
7. *Multiply and add the reactions.* We multiply the Fe equation by 5 and the Mn equation by 1 and add:

   $$5Fe^{2+} \rightleftharpoons 5Fe^{3+} + 5e^-$$
   $$\underline{MnO_4^- + 5e^- + 8H^+ \rightleftharpoons Mn^{2+} + 4H_2O}$$
   $$5Fe^{2+} + MnO_4^- + 8H^+ \rightleftharpoons 5Fe^{3+} + Mn^{2+} + 4H_2O$$

The total charge on each side is $+17$, and we find the same number of atoms of each element on each side. The equation is balanced.

EXAMPLE A Reverse Disproportionation

Now try the next reaction, which represents the reverse of a *disproportionation*. (In a disproportionation, an element in one oxidation state reacts to give the same element in higher and lower oxidation states.)

$$\underset{0}{\underset{\text{Iodine}}{I_2}} + \underset{+5}{\underset{\text{Iodate}}{IO_3^-}} + \underset{-1}{Cl^-} \rightleftharpoons \underset{+1-1}{ICl_2^-}$$

Solution

1. The oxidation numbers are assigned above. Note that chlorine has an oxidation number of -1 on both sides of the equation. Only iodine is involved in electron transfer.
2. Oxidation half-reaction: $\underset{0}{I_2} \rightleftharpoons \underset{+1}{ICl_2^-}$

   Reduction half-reaction: $\underset{+5}{IO_3^-} \rightleftharpoons \underset{+1}{ICl_2^-}$

3. We need to balance I atoms in the first reaction and add Cl^- to each reaction to balance Cl.

   $$I_2 + 4Cl^- \rightleftharpoons 2ICl_2^-$$
   $$IO_3^- + 2Cl^- \rightleftharpoons ICl_2^-$$

4. Now add electrons to each.

   $$I_2 + 4Cl^- \rightleftharpoons 2ICl_2^- + 2e^-$$
   $$IO_3^- + 2Cl^- + 4e^- \rightleftharpoons ICl_2^-$$

   The first reaction needs $2e^-$ because there are two I atoms, each of which changes from 0 to $+1$.
5. The second reaction needs $3H_2O$ on the right side to balance oxygen atoms.

   $$IO_3^- + 2Cl^- + 4e^- \rightleftharpoons ICl_2^- + 3H_2O$$

6. The first reaction is balanced, but the second needs $6H^+$ on the left.

   $$IO_3^- + 2Cl^- + 4e^- + 6H^+ \rightleftharpoons ICl_2^- + 3H_2O$$

   As a check, the charge on each side of this half-reaction is -1, and all atoms are balanced.
7. Multiply and add.

   $$2(I_2 + 4Cl^- \rightleftharpoons 2ICl_2^- + 2e^-)$$
   $$\underline{IO_3^- + 2Cl^- + 4e^- + 6H^+ \rightleftharpoons ICl_2^- + 3H_2O}$$
   $$2I_2 + IO_3^- + 10Cl^- + 6H^+ \rightleftharpoons 5ICl_2^- + 3H_2O \qquad \text{(D-1)}$$

We multiplied the first reaction by 2 so that there would be the same number of electrons in each half-reaction. You could have multiplied the first reaction by 4 and the second by 2, but then all coefficients would simply be doubled. We customarily write the smallest coefficients.

Basic Solutions

The method many people prefer for basic solutions is to balance the equation first with H^+. The answer can then be converted into one in which OH^- is used instead. This is done by adding to each side of the equation a number of hydroxide ions equal to the number of H^+ ions appearing in the equation. For example, to balance Equation D-1 with OH^- instead of H^+, proceed as follows:

$$2I_2 + IO_3^- + 10Cl^- + 6H^+ \rightleftharpoons 5ICl_2 + 3H_2O$$
$$+ 6OH^- \qquad\qquad\qquad + 6OH^-$$
$$\overline{2I_2 + IO_3^- + 10Cl^- + \underbrace{6H^+ + 6OH^-}_{} \rightleftharpoons 5ICl_2 + 3H_2O + 6OH^-}$$

$$6H_2O$$
$$\Downarrow$$
$$3H_2O$$

Realizing that $6H^+ + 6OH^- = 6H_2O$, and canceling $3H_2O$ on each side, gives the final result:

$$2I_2 + IO_3^- + 10Cl^- + 3H_2O \rightleftharpoons 5ICl_2 + 6OH^-$$

Problems

D-3. Balance the following reactions by using H^+ but not OH^-.

(a) $Fe^{3+} + Hg_2^{2+} \rightleftharpoons Fe^{2+} + Hg^{2+}$

(b) $Ag + NO_3^- \rightleftharpoons Ag^+ + NO$

(c) $VO^{2+} + Sn^{2+} \rightleftharpoons V^{3+} + Sn^{4+}$

(d) $SeO_4^{2-} + Hg + Cl^- \rightleftharpoons SeO_3^{2-} + Hg_2Cl_2$

(e) $CuS + NO_3^- \rightleftharpoons Cu^{2+} + SO_4^{2-} + NO$

(f) $S_2O_3^{2-} + I_2 \rightleftharpoons I^- + S_4O_6^{2-}$

(g) $ClO_3^- + As_2S_3 \rightleftharpoons Cl^- + H_2AsO_4^- + SO_4^{2-}$

(h) $Cr_2O_7^{2-} + CH_3\overset{\text{O}}{\overset{\|}{C}}H \rightleftharpoons CH_3\overset{\text{O}}{\overset{\|}{C}}OH + Cr^{3+}$

(i) $MnO_4^{2-} \rightleftharpoons MnO_2 + MnO_4^-$

(j) $Hg_2SO_4 + Ca^{2+} + S_8 \rightleftharpoons Hg_2^{2+} + CaS_2O_3$

(k) $ClO_3^- \rightleftharpoons Cl_2 + O_2$

D-4. Balance the following equations by using OH^- but not H^+.

(a) $PbO_2 + Cl^- \rightleftharpoons ClO^- + Pb(OH)_3^-$

(b) $HNO_2 + SbO^+ \rightleftharpoons NO + Sb_2O_5$

(c) $Ag_2S + CN^- + O_2 \rightleftharpoons S + Ag(CN)_2^- + OH^-$

(d) $HO_2^- + Cr(OH)_3^- \rightleftharpoons CrO_4^{2-} + OH^-$

(e) $ClO_2 + OH^- \rightleftharpoons ClO_2^- + ClO_3^-$

(f) $WO_3^- + O_2 \rightleftharpoons HW_6O_{21}^{5-} + OH^-$

(g) $Mn_2O_3 + CN^- \rightleftharpoons Mn(CN)_6^{4-} + (CN)_2$

(h) $Cu^{2+} + H_2 \rightleftharpoons Cu + H_2O$

(i) $BH_4^- + H_2O \rightleftharpoons H_3BO_3 + H_2$

(j) $Mn_2O_3 + Hg + CN^- \rightleftharpoons Mn(CN)_6^{4-} + Hg(CN)_2$

(k) $MnO_4^- + HC\overset{\text{O}}{\overset{\|}{C}}CH_2CH_2OH \rightleftharpoons CH_2(CO_2^-)_2 + MnO_2$

(l) $K_3V_5O_{14} + HOCH_2CHOHCH_2OH \rightleftharpoons VO(OH)_2 + HCO_2^- + K^+$

Answers

D-1.

(a) $+1$	(j) $+3$	(s) $+3$
(b) $+2$	(k) $+1$	(t) 0
(c) $+6$	(l) $+2$	(u) -4
(d) $+2$	(m) $+4$	(v) $+5$
(e) $-\frac{1}{2}$	(n) $+3$	(w) -2
(f) $+2$	(o) $+2$	(x) $+4$
(g) $+3$	(p) 0	(y) $+8/3$
(h) $+4$	(q) -3	(z) $-2/3$
(i) $+2$	(r) -2	

D-2.

	Oxidizing agent	Reducing agent
(a)	$Cr_2O_7^{2-}$	Sn^{2+}
(b)	O_2	I^-
(c)	MnO_4^-	CH_3CHO
(d)	IO_4^-	Glycerol
(e)	C_8H_8	Na
(f)	I_2	I_2

Reaction (**f**) is called a *disproportionation*, because an element in one oxidation state is transformed into two different oxidation states—one higher and one lower than the original oxidation state.

D-3. (a) $2Fe^{3+} + Hg_2^{2+} \rightleftharpoons 2Fe^{2+} + 2Hg^{2+}$

(b) $3Ag + NO_3^- + 4H^+ \rightleftharpoons 3Ag^+ + NO + 2H_2O$

(c) $4H^+ + 2VO^{2+} + Sn^{2+} \rightleftharpoons 2V^{3+} + Sn^{4+} + 2H_2O$

(d) $2Hg + 2Cl^- + SeO_4^{2-} + 2H^+ \rightleftharpoons Hg_2Cl_2 + SeO_3^{2-} + H_2O$

(e) $3CuS + 8NO_3^- + 8H^+ \rightleftharpoons 3Cu^{2+} + 3SO_4^{2-} + 8NO + 4H_2O$

(f) $2S_2O_3^{2-} + I_2 \rightleftharpoons S_4O_6^{2-} + 2I^-$

(g) $14ClO_3^- + 3As_2S_3 + 18H_2O \rightleftharpoons$
$$14Cl^- + 6H_2AsO_4^- + 9SO_4^{2-} + 24H^+$$

(h) $Cr_2O_7^{2-} + 3CH_3CHO + 8H^+ \rightleftharpoons 2Cr^{3+} + 3CH_3CO_2H + 4H_2O$

(i) $4H^+ + 3MnO_4^{2-} \rightleftharpoons MnO_2 + 2MnO_4^- + 2H_2O$

(j) $2Hg_2SO_4 + 3Ca^{2+} + \frac{1}{2}S_8 + H_2O \rightleftharpoons 2Hg_2^{2+} + 3CaS_2O_3 + 2H^+$

(k) $2H^+ + 2ClO_3^- \rightleftharpoons Cl_2 + \frac{5}{2}O_2 + H_2O$

The balanced half-reaction for As_2S_3 in (g) is

$$As_2S_3 + 20H_2O \rightleftharpoons 2H_2AsO_4^- + 3SO_4^{2-} + 28e^- + 36H^+$$
$$+3 -2 \qquad\qquad +5 \qquad\quad +6$$

Because As_2S_3 is a single compound, we must consider the $As_2S_3 \rightarrow H_2AsO_4^-$ and $As_2S_3 \rightarrow SO_4^{2-}$ reactions together. The net change in oxidation number for the *two* As atoms is $2(5 - 3) = +4$. The net change in oxidation number for the *three* S atoms is $3[6 - (-2)] = +24$. Therefore, $24 + 4 = 28e^-$ are involved in the half-reaction.

D-4. (a) $H_2O + OH^- + PbO_2 + Cl^- \rightleftharpoons Pb(OH)_3^- + ClO^-$

(b) $4HNO_2 + 2SbO^+ + 2OH^- \rightleftharpoons 4NO + Sb_2O_5 + 3H_2O$

(c) $Ag_2S + 4CN^- + \frac{1}{2}O_2 + H_2O \rightleftharpoons S + 2Ag(CN)_2^- + 2OH^-$

(d) $2HO_2^- + Cr(OH)_3^- \rightleftharpoons CrO_4^{2-} + OH^- + 2H_2O$

(e) $2ClO_2 + 2OH^- \rightleftharpoons ClO_2^- + ClO_3^- + H_2O$

(f) $12WO_3^- + 3O_2 + 2H_2O \rightleftharpoons 2HW_6O_{21}^{5-} + 2OH^-$

(g) $Mn_2O_3 + 14CN^- + 3H_2O \rightleftharpoons 2Mn(CN)_6^{4-} + (CN)_2 + 6OH^-$

(h) $Cu^{2+} + H_2 + 2OH^- \rightleftharpoons Cu + 2H_2O$

(i) $BH_4^- + 4H_2O \rightleftharpoons H_3BO_3 + 4H_2 + OH^-$

(j) $3H_2O + Mn_2O_3 + Hg + 14CN^- \rightleftharpoons$
$$2Mn(CN)_6^{4-} + Hg(CN)_2 + 6OH^-$$

(k) $2MnO_4^- + HC\overset{\text{O}}{\overset{\|}{C}}CH_2CH_2OH \rightleftharpoons 2MnO_2 + 2H_2O + CH_2(CO_2^-)_2$

For (**k**), the organic half-reaction is $8OH^- + C_3H_6O_2 \rightleftharpoons$
$$C_3H_2O_4^{2-} + 6e^- + 6H_2O.$$

(l) $32H_2O + 8K_3V_5O_{14} + 5HOCH_2CHOHCH_2OH \rightleftharpoons$
$$40VO(OH)_2 + 15HCO_2^- + 9OH^- + 24K^+$$

For (**l**), the two half-reactions are $K_3V_5O_{14} + 9H_2O + 5e^- \rightleftharpoons$ $5VO(OH)_2 + 8OH^- + 3K^+$ and $C_3H_8O_3 + 11OH^- \rightleftharpoons 3HCO_2^- + 8e^- + 8H_2O.$

The *normality*, N, of a redox reagent is n times the molarity, where n is the number of electrons donated or accepted by that species in a chemical reaction.

$$N = nM \qquad \text{(E-1)}$$

For example, in the half-reaction

$$MnO_4^- + 8H^+ + 5e^- \rightleftharpoons Mn^{2+} + 4H_2O \qquad \text{(E-2)}$$

the normality of permanganate ion is five times its molarity, because each MnO_4^- accepts $5e^-$. If the molarity of permanganate is 0.1 M, the normality for the reaction

$$MnO_4^- + 5Fe^{2+} + 8H^+ \rightleftharpoons Mn^{2+} + 5Fe^{3+} + 4H_2O \qquad \text{(E-3)}$$

is $5 \times 0.1 = 0.5$ N (read "0.5 normal"). In this reaction, each Fe^{2+} ion donates one electron. The normality of ferrous ion *equals* the molarity of ferrous ion, even though it takes five ferrous ions to balance the reaction.

In the half-reaction

$$MnO_4^- + 4H^+ + 3e^- \rightleftharpoons MnO_2 + 2H_2O \qquad \text{(E-4)}$$

each MnO_4^- ion accepts only *three* electrons. The normality of permanganate for this reaction is equal to three times the molarity of permanganate. A 0.06 N permanganate solution for this reaction contains 0.02 M MnO_4^-.

The normality of a solution is a statement of the moles of "reacting units" per liter. One mole of reacting units is called one *equivalent*. Therefore, the units of normality are equivalents per liter (equiv/L). For redox reagents, *one equivalent is the amount of substance that can donate or accept one mole of electrons*. It is possible to speak of equivalents only with respect to a particular half-reaction. For example, in Reaction E-2, there are five equivalents per mole of MnO_4^-; but, in Reaction E-4, there are only three equivalents per mole of MnO_4^-. The mass of substance containing one equivalent is called the *equivalent mass*. The formula mass of $KMnO_4$ is 158.033 9. The equivalent mass of $KMnO_4$ for Reaction E-2 is $(158.033\ 9)/5 = 31.606\ 8$ g/equiv. The equivalent mass of $KMnO_4$ for Reaction E-4 is $(158.033\ 9)/3 = 52.678\ 0$ g/equiv.

EXAMPLE **Finding Normality**

Find the normality of a solution containing 6.34 g of ascorbic acid in 250.0 mL if the relevant half-reaction is

Ascorbic acid (vitamin C)   Dehydroascorbic acid

Solution The formula mass of ascorbic acid ($C_6H_8O_6$) is 176.124. In 6.34 g, there are $(6.34\ \text{g})/(176.124\ \text{g/mol}) = 3.60 \times 10^{-2}$ mol. Because each mole contains 2 equivalents in this example, $6.34\ \text{g} = (2\ \text{equiv/mol})(3.60 \times 10^{-2}\ \text{mol}) = 7.20 \times 10^{-2}$ equivalent. The normality is $(7.20 \times 10^{-2}\ \text{equiv})/(0.250\ 0\ \text{L}) = 0.288$ N.

EXAMPLE **Using Normality**

How many grams of potassium oxalate should be dissolved in 500.0 mL to make a 0.100 N solution for titration of MnO_4^-?

$$5H_2C_2O_4 + 2MnO_4^- + 6H^+ \rightleftharpoons 2Mn^{2+} + 10CO_2 + 8H_2O \quad \text{(E-5)}$$

Solution It is first necessary to write the oxalic acid half-reaction:

$$H_2C_2O_4 \rightleftharpoons 2CO_2 + 2H^+ + 2e^-$$

It is apparent that there are two equivalents per mole of oxalic acid. Hence, a 0.100 N solution will be 0.050 0 M:

$$\frac{0.100\ \text{equiv/L}}{2\ \text{equiv/mol}} = 0.050\ 0\ \text{mol/L} = 0.050\ 0\ \text{M}$$

Therefore, we must dissolve $(0.050\ 0\ \text{mol/L})(0.500\ 0\ \text{L}) = 0.025\ 0$ mol in 500.0 mL. Because the formula mass of $K_2C_2O_4$ is 166.216, we should use $(0.025\ 0\ \text{mol}) \times (166.216\ \text{g/mol}) = 4.15$ g of potassium oxalate.

The utility of normality in volumetric analysis lies in the equation

$$N_1V_1 = N_2V_2 \qquad \text{(E-6)}$$

where N_1 is the normality of reagent 1, V_1 is the volume of reagent 1, N_2 is the normality of reagent 2, and V_2 is the volume of reagent 2. V_1 and V_2 may be expressed in any units, as long as the same units are used for both.

EXAMPLE **Finding Normality**

A solution containing 25.0 mL of oxalic acid required 13.78 mL of 0.041 62 N $KMnO_4$ for titration, according to Reaction E-5. Find the normality and molarity of the oxalic acid.

Solution Setting up Equation E-6, we write

$$N_1(25.0\ \text{mL}) = (0.041\ 62\ \text{N})(13.78\ \text{mL})$$
$$\quad H_2C_2O_4 \qquad\qquad KMnO_4$$
$$N_1 = 0.022\ 94\ \text{equiv/L}$$

Because there are two equivalents per mole of oxalic acid in Reaction E-5,

$$M = \frac{N}{n} = \frac{0.022\ 94}{2} = 0.011\ 47\ \text{M}$$

Normality is sometimes used in acid-base or ion-exchange chemistry. With respect to acids and bases, the equivalent mass of a reagent is the amount that can donate or accept one mole of H^+. With respect to ion exchange, the equivalent mass is the mass of reagent containing one mole of charge.

APPENDIX F Solubility Products*

Formula	pK_{sp}	K_{sp}	Formula	pK_{sp}	K_{sp}
Azides: L = N_3^-			Chromates: L = CrO_4^{2-}		
CuL	8.31	4.9×10^{-9}	BaL	9.67	2.1×10^{-10}
AgL	8.56	2.8×10^{-9}	CuL	5.44	3.6×10^{-6}
Hg_2L_2	9.15	7.1×10^{-10}	Ag_2L	11.92	1.2×10^{-12}
TlL	3.66	2.2×10^{-4}	Hg_2L	8.70	2.0×10^{-9}
$PdL_2(\alpha)$	8.57	2.7×10^{-9}	Tl_2L	12.01	9.8×10^{-13}
Bromates: L = BrO_3^-			Cobalticyanides: L = $Co(CN)_6^{3-}$		
BaL · H_2O (f)	5.11	7.8×10^{-6}	Ag_3L	25.41	3.9×10^{-26}
AgL	4.26	5.5×10^{-5}	$(Hg_2)_3L_2$	36.72	1.9×10^{-37}
TlL	3.78	1.7×10^{-4}			
PbL_2	5.10	7.9×10^{-6}	Cyanides: L = CN^-		
			AgL	15.66	2.2×10^{-16}
			Hg_2L_2	39.3	5×10^{-40}
Bromides: L = Br^-			ZnL_2 (h)	15.5	3×10^{-16}
CuL	8.3	5×10^{-9}			
AgL	12.30	5.0×10^{-13}			
Hg_2L_2	22.25	5.6×10^{-23}	Ferrocyanides: L = $Fe(CN)_6^{4-}$		
TlL	5.44	3.6×10^{-6}	Ag_4L	44.07	8.5×10^{-45}
HgL_2 (f)	18.9	1.3×10^{-19}	Zn_2L	15.68	2.1×10^{-16}
PbL_2	5.68	2.1×10^{-6}	Cd_2L	17.38	4.2×10^{-18}
			Pb_2L	18.02	9.5×10^{-19}
Carbonates: L = CO_3^{2-}					
MgL	7.46	3.5×10^{-8}	Fluorides: L = F^-		
CaL (calcite)	8.35	4.5×10^{-9}	LiL	2.77	1.7×10^{-3}
CaL (aragonite)	8.22	6.0×10^{-9}	MgL_2	8.13	7.4×10^{-9}
SrL	9.03	9.3×10^{-10}	CaL_2	10.50	3.2×10^{-11}
BaL	8.30	5.0×10^{-9}	SrL_2	8.58	2.6×10^{-9}
Y_2L_3	30.6	2.5×10^{-31}	BaL_2	5.82	1.5×10^{-6}
La_2L_3	33.4	4.0×10^{-34}	LaL_3	18.7	2×10^{-19}
MnL	9.30	5.0×10^{-10}	ThL_4	28.3	5×10^{-29}
FeL	10.68	2.1×10^{-11}	PbL_2	7.44	3.6×10^{-8}
CoL	9.98	1.0×10^{-10}			
NiL	6.87	1.3×10^{-7}	Hydroxides: L = OH^-		
CuL	9.63	2.3×10^{-10}	MgL_2 (amorphous)	9.2	6×10^{-10}
Ag_2L	11.09	8.1×10^{-12}	MgL_2 (brucite crystal)	11.15	7.1×10^{-12}
Hg_2L	16.05	8.9×10^{-17}	CaL_2	5.19	6.5×10^{-6}
ZnL	10.00	1.0×10^{-10}	$BaL_2 \cdot 8H_2O$	3.6	3×10^{-4}
CdL	13.74	1.8×10^{-14}	YL_3	23.2	6×10^{-24}
PbL	13.13	7.4×10^{-14}	LaL_3	20.7	2×10^{-21}
			CeL_3	21.2	6×10^{-22}
			UO_2 ($\rightleftharpoons U^{4+} + 4OH^-$)	56.2	6×10^{-57}
Chlorides: L = Cl^-			UO_2L_2 ($\rightleftharpoons UO_2^{2+} + 2OH^-$)	22.4	4×10^{-23}
CuL	6.73	1.9×10^{-7}	MnL_2	12.8	1.6×10^{-13}
AgL	9.74	1.8×10^{-10}	FeL_2	15.1	7.9×10^{-16}
Hg_2L_2	17.91	1.2×10^{-18}	CoL_2	14.9	1.3×10^{-15}
TlL	3.74	1.8×10^{-4}	NiL_2	15.2	6×10^{-16}
PbL_2	4.78	1.7×10^{-5}			

*The designations α, β, or γ after some formulas refer to particular crystalline forms (which are customarily identified by Greek letters). Data for salts except oxalates are taken mainly from A. E. Martell and R. M. Smith, Critical Stability Constants, Vol. 4 (New York: Plenum Press, 1976). Data for oxalates are from L. G. Sillén and A. E. Martell, Stability Constants of Metal-Ion Complexes, Supplement No. 1 (London: The Chemical Society, Special Publication No. 25, 1971). Another source: R. M. H. Verbeeck et al., Inorg. Chem. **1984**, 23, 1922.

Conditions are 25°C and zero ionic strength unless otherwise indicated: (a) 19°C; (b) 20°C; (c) 38°C; (d) 0.1 M; (e) 0.2 M; (f) 0.5 M; (g) 1 M; (h) 3 M; (i) 4 M; (j) 5 M.

(Continued)

Formula	pK_{sp}	K_{sp}
Hydroxides: L = OH⁻—(Continued)		
CuL_2	19.32	4.8×10^{-20}
VL_3	34.4	4.0×10^{-35}
CrL_3 (d)	29.8	1.6×10^{-30}
FeL_3	38.8	1.6×10^{-39}
CoL_3 (a)	44.5	3×10^{-45}
VOL_2 ($\rightleftharpoons VO^{2+} + 2OH^-$)	23.5	3×10^{-24}
PdL_2	28.5	3×10^{-29}
ZnL_2 (amorphous)	15.52	3.0×10^{-16}
CdL_2 (β)	14.35	4.5×10^{-15}
HgO (red) ($\rightleftharpoons Hg^{2+} + 2OH^-$)	25.44	3.6×10^{-26}
Cu_2O ($\rightleftharpoons 2Cu^+ + 2OH^-$)	29.4	4×10^{-30}
Ag_2O ($\rightleftharpoons 2Ag^+ + 2OH^-$)	15.42	3.8×10^{-16}
AuL_3	5.5	3×10^{-6}
AlL_3 (α)	33.5	3×10^{-34}
GaL_3 (amorphous)	37	10^{-37}
InL_3	36.9	1.3×10^{-37}
SnO ($\rightleftharpoons Sn^{2+} + 2OH^-$)	26.2	6×10^{-27}
PbO (yellow) ($\rightleftharpoons Pb^{2+} + 2OH^-$)	15.1	8×10^{-16}
PbO (red) ($\rightleftharpoons Pb^{2+} + 2OH^-$)	15.3	5×10^{-16}
Iodates: L = IO_3^-		
CaL_2	6.15	7.1×10^{-7}
SrL_2	6.48	3.3×10^{-7}
BaL_2	8.81	1.5×10^{-9}
YL_3	10.15	7.1×10^{-11}
LaL_3	10.99	1.0×10^{-11}
CeL_3	10.86	1.4×10^{-11}
ThL_4 (f)	14.62	2.4×10^{-15}
UO_2L_2 ($\rightleftharpoons UO_2^{2+} + 2IO_3^-$)(e)	7.01	9.8×10^{-8}
CrL_3 (f)	5.3	5×10^{-6}
AgL	7.51	3.1×10^{-8}
Hg_2L_2	17.89	1.3×10^{-18}
TlL	5.51	3.1×10^{-6}
ZnL_2	5.41	3.9×10^{-6}
CdL_2	7.64	2.3×10^{-8}
PbL_2	12.61	2.5×10^{-13}
Iodides: L = I^-		
CuL	12.0	1×10^{-12}
AgL	16.08	8.3×10^{-17}
CH_3HgL ($\rightleftharpoons CH_3Hg^+ + I^-$) (b, g)	11.46	3.5×10^{-12}
CH_3CH_2HgL ($\rightleftharpoons CH_3CH_2Hg^+ + I^-$)	4.11	7.8×10^{-5}
TlL	7.23	5.9×10^{-8}
Hg_2L_2	28.34	4.6×10^{-29}
SnL_2 (i)	5.08	8.3×10^{-6}
PbL_2	8.10	7.9×10^{-9}
Oxalates: L = $C_2O_4^{2-}$		
CaL (b, d)	7.9	1.3×10^{-8}
SrL (b, d)	6.4	4×10^{-7}
BaL (b, d)	6.0	1×10^{-6}
La_2L_3 (b, d)	25.0	1×10^{-25}
ThL_2 (g)	21.38	4.2×10^{-22}
UO_2L ($\rightleftharpoons UO_2^{2+} + C_2O_4^{2-}$) (b, d)	8.66	2.2×10^{-9}

Formula	pK_{sp}	K_{sp}
Phosphates: L = PO_4^{3-}		
$MgHL \cdot 3H_2O$ ($\rightleftharpoons Mg^{2+} + HL^{2-}$)	5.78	1.7×10^{-6}
$CaHL \cdot 2H_2O$ ($\rightleftharpoons Ca^{2+} + HL^{2-}$)	6.58	2.6×10^{-7}
$SrHL$ ($\rightleftharpoons Sr^{2+} + HL^{2-}$) (b)	6.92	1.2×10^{-7}
$BaHL$ ($\rightleftharpoons Ba^{2+} + HL^{2-}$) (b)	7.40	4.0×10^{-8}
LaL (f)	22.43	3.7×10^{-23}
$Fe_3L_2 \cdot 8H_2O$	36.0	1×10^{-36}
$FeL \cdot 2H_2O$	26.4	4×10^{-27}
$(VO)_3L_2$ ($\rightleftharpoons 3VO^{2+} + 2L^{3-}$)	25.1	8×10^{-26}
Ag_3L	17.55	2.8×10^{-18}
Hg_2HL ($\rightleftharpoons Hg_2^{2+} + HL^{2-}$)	12.40	4.0×10^{-13}
$Zn_3L_2 \cdot 4H_2O$	35.3	5×10^{-36}
Pb_3L_2 (c)	43.53	3.0×10^{-44}
GaL (g)	21.0	1×10^{-21}
InL (g)	21.63	2.3×10^{-22}
Sulfates: L = SO_4^{2-}		
CaL	4.62	2.4×10^{-5}
SrL	6.50	3.2×10^{-7}
BaL	9.96	1.1×10^{-10}
RaL (b)	10.37	4.3×10^{-11}
Ag_2L	4.83	1.5×10^{-5}
Hg_2L	6.13	7.4×10^{-7}
PbL	6.20	6.3×10^{-7}
Sulfides: L = S^{2-}		
MnL (pink)	10.5	3×10^{-11}
MnL (green)	13.5	3×10^{-14}
FeL	18.1	8×10^{-19}
CoL (α)	21.3	5×10^{-22}
CoL (β)	25.6	3×10^{-26}
NiL (α)	19.4	4×10^{-20}
NiL (β)	24.9	1.3×10^{-25}
NiL (γ)	26.6	3×10^{-27}
CuL	36.1	8×10^{-37}
Cu_2L	48.5	3×10^{-49}
Ag_2L	50.1	8×10^{-51}
Tl_2L	21.2	6×10^{-22}
ZnL (α)	24.7	2×10^{-25}
ZnL (β)	22.5	3×10^{-23}
CdL	27.0	1×10^{-27}
HgL (black)	52.7	2×10^{-53}
HgL (red)	53.3	5×10^{-54}
SnL	25.9	1.3×10^{-26}
PbL	27.5	3×10^{-28}
In_2L_3	69.4	4×10^{-70}
Thiocyanates: L = SCN^-		
CuL (j)	13.40	4.0×10^{-14}
AgL	11.97	1.1×10^{-12}
Hg_2L_2	19.52	3.0×10^{-20}
TlL	3.79	1.6×10^{-4}
HgL_2	19.56	2.8×10^{-20}

APPENDIX G Acid Dissociation Constants

Name	Structure*	Ionic strength (μ) = 0		$\mu = 0.1\ M^{\S}$
		$pK_a{}^{\dagger}$	$K_a{}^{\ddagger}$	$pK_a{}^{\dagger}$
Acetic acid (ethanoic acid)	$CH_3CO_2\mathbf{H}$	4.756	1.75×10^{-5}	4.56
Alanine		2.344 (CO_2H) 9.868 (NH_3)	4.53×10^{-3} 1.36×10^{-10}	2.33 9.71
Aminobenzene (aniline)		4.601	2.51×10^{-5}	4.64
4-Aminobenzenesulfonic acid (sulfanilic acid)		3.232	5.86×10^{-4}	3.01
2-Aminobenzoic acid (anthranilic acid)		2.08 (CO_2H) 4.96 (NH_3)	8.3×10^{-3} 1.10×10^{-5}	2.01 4.78
2-Aminoethanethiol (2-mercaptoethylamine)	$\mathbf{H}SCH_2CH_2N\mathbf{H}_3^+$	—— ——		8.21 (SH) 10.73 (NH_3)
2-Aminoethanol (ethanolamine)	$HOCH_2CH_2N\mathbf{H}_3^+$	9.498	3.18×10^{-10}	9.52
2-Aminophenol		4.70 (NH_3) (20°) 9.97 (OH) (20°)	2.0×10^{-5} 1.05×10^{-10}	4.74 9.87
Ammonia	$N\mathbf{H}_4^+$	9.245	5.69×10^{-10}	9.26
Arginine		1.823 (CO_2H) 8.991 (NH_3) —— (NH_2)	1.50×10^{-2} 1.02×10^{-9} ——	2.03 9.00 (12.1)
Arsenic acid (hydrogen arsenate)		2.24 6.96 (11.50)	5.8×10^{-3} 1.10×10^{-7} 3.2×10^{-12}	2.15 6.65 (11.18)
Arsenious acid (hydrogen arsenite)	$As(O\mathbf{H})_3$	9.29	5.1×10^{-10}	9.14
Asparagine		—— ——	—— ——	2.16 (CO_2H) 8.73 (NH_3)

*Each acid is written in its protonated form. The acidic protons are indicated in **bold** type.

$^{\dagger}pK_a$ values refer to 25°C unless otherwise indicated. Values in parentheses are considered to be less reliable. Data are from A. E. Martell, R. M. Smith, and R. J. Motekaitis, NIST Database 46 (Gaithersburg, MD: National Institute of Standards and Technology, 2001).

‡The accurate way to calculate K_b for the conjugate base is $pK_b = 13.995 - pK_a$ and $K_b = 10^{-pK_b}$.

§See marginal note on page 191 for distinction between pK_a at $\mu = 0$ and at $\mu = 0.1$ M.

(Continued)

Name	Structure	Ionic strength (μ) = 0		μ = 0.1 M
		pK_a	K_a	pK_a
Aspartic acid	NH$_3^+$ $\|$ CHCH$_2$CO$_2$H β $\|$ α CO$_2$H	1.990 (α-CO$_2$H) 3.900 (β-CO$_2$H) 10.002 (NH$_3$)	1.02×10^{-2} 1.26×10^{-4} 9.95×10^{-11}	1.95 3.71 9.96
Aziridine (dimethyleneimine)	$\triangleright$NH$_2^+$	8.04	9.1×10^{-9}	——
Benzene-1,2,3-tricarboxylic acid (hemimellitic acid)	CO$_2$H CO$_2$H CO$_2$H	2.86 4.30 6.28	1.38×10^{-3} 5.0×10^{-5} 5.2×10^{-7}	2.67 3.91 5.50
Benzoic acid	CO$_2$H	4.202	6.28×10^{-5}	4.01
Benzylamine	CH$_2$NH$_3^+$	9.35	4.5×10^{-10}	9.40
2,2'-Bipyridine	N N H$^+$ $^+$H	—— 4.34	—— 4.6×10^{-5}	(1.3) 4.41
Boric acid (hydrogen borate)	B(OH)$_3$	9.237 (12.74) (20°) (13.80) (20°)	5.79×10^{-10} 1.82×10^{-13} 1.58×10^{-14}	8.98 —— ——
Bromoacetic acid	BrCH$_2$CO$_2$H	2.902	1.25×10^{-3}	2.71
Butane-2,3-dione dioxime (dimethylglyoxime)	HON NOH CH$_3$ CH$_3$	10.66 (12.0)	2.2×10^{-11} 1×10^{-12}	10.45 (11.9)
Butanoic acid	CH$_3$CH$_2$CH$_2$CO$_2$H	4.818	1.52×10^{-5}	4.62
cis-Butenedioic acid (maleic acid)	CO$_2$H CO$_2$H	1.92 6.27	1.20×10^{-2} 5.37×10^{-7}	1.75 5.84
trans-Butenedioic acid (fumaric acid)	CO$_2$H HO$_2$C	3.02 4.48	9.5×10^{-4} 3.3×10^{-5}	2.84 4.09
Butylamine	CH$_3$CH$_2$CH$_2$CH$_2$NH$_3^+$	10.640	2.29×10^{-11}	10.66
Carbonic acid* (hydrogen carbonate)	O $\|\|$ HO—C—OH	6.351 10.329	4.46×10^{-7} 4.69×10^{-11}	6.13 9.91
Chloroacetic acid	ClCH$_2$CO$_2$H	2.865	1.36×10^{-3}	2.69
3-Chloropropanoic acid	ClCH$_2$CH$_2$CO$_2$H	4.11	7.8×10^{-5}	3.92
Chlorous acid (hydrogen chlorite)	HOCl=O	1.96	1.10×10^{-2}	——

*The concentration of "carbonic acid" is considered to be the sum $[H_2CO_3] + [CO_2(aq)]$. See Box 6-4.

Name	Structure	Ionic strength (μ) = 0		$\mu = 0.1\,M$
		pK_a	K_a	pK_a
Chromic acid (hydrogen chromate)	HO—Cr—OH (with O double-bonded above and below Cr)	(−0.2) (20°) 6.51	1.6 3.1×10^{-7}	(−0.6) (20°C) 6.05
Citric acid (2-hydroxypropane-1,2,3-tricarboxylic acid)	$HO_2CCH_2CCH_2CO_2H$ with CO_2H and OH on central carbon	3.128 4.761 6.396	7.44×10^{-4} 1.73×10^{-5} 4.02×10^{-7}	2.90 4.35 5.70
Cyanoacetic acid	$NCCH_2CO_2H$	2.472	3.37×10^{-3}	——
Cyclohexylamine	cyclohexyl—NH_3^+	10.567	2.71×10^{-11}	10.62
Cysteine	NH_3^+ / CHCH$_2$SH / CO_2H	(1.7) (CO$_2$H) 8.36 (SH) 10.74 (NH$_3$)	2×10^{-2} 4.4×10^{-9} 1.82×10^{-11}	(1.90) 8.18 10.30
Dichloroacetic acid	Cl_2CHCO_2H	(1.1)	8×10^{-2}	(0.9)
Diethylamine	$(CH_3CH_2)_2NH_2^+$	11.00	1.0×10^{-11}	11.04
1,2-Dihydroxybenzene (catechol)	benzene with two OH	9.45 ——	3.5×10^{-10} ——	9.26 (13.3)
1,3-Dihydroxybenzene (resorcinol)	benzene with two OH	—— ——	—— ——	9.30 11.06
D-2,3-Dihydroxybutanedioic acid (D-tartaric acid)	$HO_2CCHCHCO_2H$ with two OH	3.036 4.366	9.20×10^{-4} 4.31×10^{-5}	2.82 3.97
2,3-Dimercaptopropanol	$HOCH_2CHCH_2SH$ / SH	—— ——	—— ——	8.63 10.65
Dimethylamine	$(CH_3)_2NH_2^+$	10.774	1.68×10^{-11}	10.81
2,4-Dinitrophenol	benzene with NO$_2$, O$_2$N, OH	4.114	7.69×10^{-5}	3.92
Ethane-1,2-dithiol	$HSCH_2CH_2SH$	—— ——	—— ——	8.85 (30°C) 10.43 (30°C)
Ethylamine	$CH_3CH_2NH_3^+$	10.673	2.12×10^{-11}	10.69
Ethylenediamine (1,2-diaminoethane)	$H_3\overset{+}{N}CH_2CH_2\overset{+}{N}H_3$	6.848 9.928	1.42×10^{-7} 1.18×10^{-10}	7.11 9.92

(Continued)

Name	Structure	Ionic strength (μ) = 0		μ = 0.1 M
		pK_a	K_a	pK_a
Ethylenedinitrilotetraacetic acid (EDTA)	$(HO_2CCH_2)_2\overset{+}{N}HCH_2CH_2\overset{+}{N}H(CH_2CO_2H)_2$	—— (CO_2H) —— (CO_2H) —— (CO_2H) —— (CO_2H) 6.273 (NH) 10.948 (NH)	—— —— —— —— 5.3×10^{-7} 1.13×10^{-11}	(0.0) (CO_2H) (μ = 1 M) (1.5) (CO_2H) 2.00 (CO_2H) 2.69 (CO_2H) 6.13 (NH) 10.37 (NH)
Formic acid (methanoic acid)	HCO_2H	3.744	1.80×10^{-4}	3.57
Glutamic acid	$\overset{\overset{NH_3^+}{\mid}}{CHCH_2CH_2\overset{\gamma}{CO_2H}}$; $\alpha\,CO_2H$	2.160 (α-CO_2H) 4.30 (γ-CO_2H) 9.96 (NH_3)	6.92×10^{-3} 5.0×10^{-5} 1.10×10^{-10}	2.16 4.15 9.58
Glutamine	$\overset{\overset{NH_3^+}{\mid}}{CHCH_2CH_2\overset{\overset{O}{\parallel}}{C}NH_2}$; CO_2H	—— ——	—— ——	2.19 (CO_2H) 9.00 (NH_3)
Glycine (aminoacetic acid)	$\overset{\overset{NH_3^+}{\mid}}{\underset{\underset{CO_2H}{\mid}}{CH_2}}$	2.350 (CO_2H) 9.778 (NH_3)	4.47×10^{-3} 1.67×10^{-10}	2.33 9.57
Guanidine	$H_2N-\overset{\overset{+NH_2}{\parallel}}{C}-NH_2$	——	——	(13.5) (μ = 1 M)
1,6-Hexanedioic acid (adipic acid)	$HO_2CCH_2CH_2CH_2CH_2CO_2H$	4.424 5.420	3.77×10^{-5} 3.80×10^{-6}	4.26 5.04
Hexane-2,4-dione	$CH_3\overset{\overset{O}{\parallel}}{C}CH_2\overset{\overset{O}{\parallel}}{C}CH_2CH_3$	9.38	4.2×10^{-10}	9.11 (20°C)
Histidine	$\overset{\overset{NH_3^+}{\mid}}{\underset{\underset{CO_2H}{\mid}}{CHCH_2}}$ (imidazole ring with NH)	(1.6) (CO_2H) 5.97 (NH) 9.28 (NH_3)	3×10^{-2} 1.07×10^{-6} 5.2×10^{-10}	(1.7) 6.05 9.10
Hydrazine	$H_3\overset{+}{N}-\overset{+}{N}H_3$	−0.99 7.98	1.0×10^{1} 1.05×10^{-8}	(−0.21) (μ = 0.5 M) 8.07
Hydrazoic acid (hydrogen azide)	$HN=\overset{+}{N}=\overset{-}{N}$	4.65	2.2×10^{-5}	4.45
Hydrogen cyanate	$HOC{\equiv}N$	3.48	3.3×10^{-4}	——
Hydrogen cyanide	$HC{\equiv}N$	9.21	6.2×10^{-10}	9.04
Hydrogen fluoride	HF	3.17	6.8×10^{-4}	2.94
Hydrogen peroxide	$HOOH$	11.65	2.2×10^{-12}	——
Hydrogen sulfide	H_2S	7.02 14.0*	9.5×10^{-8} 1.0×10^{-14}*	6.82 ——
Hydrogen thiocyanate	$HSC{\equiv}N$	(−1.1) (20°C)	1.3×10^{1}	——
Hydroxyacetic acid (glycolic acid)	$HOCH_2CO_2H$	3.832	1.48×10^{-4}	3.62

*D. J. Phillips and S. L. Phillips. "High Temperature Dissociation Constants of HS⁻ and the Standard Thermodynamic Values for S²⁻," J. Chem. Eng. Data **2000**, 45, 981.

| Name | Structure | Ionic strength (μ) = 0 | | μ = 0.1 M |
		pK_a	K_a	pK_a
Hydroxybenzene (phenol)	⬡—OH	9.997	1.01×10^{-10}	9.78
2-Hydroxybenzoic acid (salicylic acid)	CO₂H / OH (structure)	2.972 (CO₂H) (13.7) (OH)	1.07×10^{-3} 2×10^{-14}	2.80 (13.4)
L-Hydroxybutanedioic acid (malic acid)	$HO_2CCH_2CHCO_2H$ with OH	3.459 5.097	3.48×10^{-4} 8.00×10^{-6}	3.24 4.68
Hydroxylamine	$HON\overset{+}{H_3}$	5.96 (NH) (13.74) (OH)	1.10×10^{-6} 1.8×10^{-14}	5.96 ——
8-Hydroxyquinoline (oxine)	(structure)	4.94 (NH) 9.82 (OH)	1.15×10^{-5} 1.51×10^{-10}	4.97 9.65
Hypobromous acid (hydrogen hypobromite)	HOBr	8.63	2.3×10^{-9}	——
Hypochlorous acid (hydrogen hypochlorite)	HOCl	7.53	3.0×10^{-8}	——
Hypoiodous acid (hydrogen hypoiodite)	HOI	10.64	2.3×10^{-11}	——
Hypophosphorous acid (hydrogen hypophosphite)	H_2POH with O	(1.3)	5×10^{-2}	(1.1)
Imidazole (1,3-diazole)	(structure)	6.993 (14.5)	1.02×10^{-7} 3×10^{-15}	7.00 ——
Iminodiacetic acid	$H_2\overset{+}{N}(CH_2CO_2H)_2$	(1.85) (CO₂H) 2.84 (CO₂H) 9.79 (NH₂)	1.41×10^{-2} 1.45×10^{-3} 1.62×10^{-10}	(1.77) 2.62 9.34
Iodic acid (hydrogen iodate)	$HOI=O$ with O	0.77	0.17	——
Iodoacetic acid	ICH_2CO_2H	3.175	6.68×10^{-4}	2.98
Isoleucine	NH_3^+ / $CHCH(CH_3)CH_2CH_3$ / CO_2H	2.318 (CO₂H) 9.758 (NH₃)	4.81×10^{-3} 1.75×10^{-10}	2.26 9.60
Leucine	NH_3^+ / $CHCH_2CH(CH_3)_2$ / CO_2H	2.328 (CO₂H) 9.744 (NH₃)	4.70×10^{-3} 1.80×10^{-10}	2.32 9.58
Lysine	$\alpha\ NH_3^+$ / $CHCH_2CH_2CH_2CH_2\overset{\epsilon}{NH_3^+}$ / CO_2H	(1.77) (CO₂H) 9.07 (α-NH₃) 10.82 (ϵ-NH₃)	1.7×10^{-2} 8.5×10^{-10} 1.51×10^{-11}	2.15 9.15 10.66

(Continued)

Name	Structure	Ionic strength (μ) = 0		μ = 0.1 M
		pK_a	K_a	pK_a
Malonic acid (propanedioic acid)	$HO_2CCH_2CO_2H$	2.847 / 5.696	1.42×10^{-3} / 2.01×10^{-6}	2.65 / 5.27
Mercaptoacetic acid (thioglycolic acid)	$HSCH_2CO_2H$	3.64 (CO_2H) / 10.61 (SH)	2.3×10^{-4} / 2.5×10^{-11}	3.48 / 10.11
2-Mercaptoethanol	$HSCH_2CH_2OH$	9.72	1.9×10^{-10}	9.40
Methionine	NH_3^+ | $CHCH_2CH_2SCH_3$ | CO_2H	—— / ——	—— / ——	2.18 (CO_2H) / 9.08 (NH_3)
2-Methoxyaniline (o-anisidine)		4.526	2.98×10^{-5}	——
4-Methoxyaniline (p-anisidine)		5.357	4.40×10^{-6}	5.33
Methylamine	$CH_3\overset{+}{N}H_3$	10.632	2.33×10^{-11}	10.65
2-Methylaniline (o-toluidine)		4.447	3.57×10^{-5}	——
4-Methylaniline (p-toluidine)		5.080	8.32×10^{-6}	5.09
2-Methylphenol (o-cresol)		10.31	4.9×10^{-11}	10.09
4-Methylphenol (p-cresol)		10.269	5.38×10^{-11}	10.04
Morpholine (perhydro-1,4-oxazine)		8.492	3.22×10^{-9}	——
1-Naphthoic acid		3.67	2.1×10^{-4}	——
2-Naphthoic acid		4.16	6.9×10^{-5}	——
1-Naphthol		9.416	3.84×10^{-10}	9.14
2-Naphthol		9.573	2.67×10^{-10}	9.31
Nitrilotriacetic acid	$H\overset{+}{N}(CH_2CO_2H)_3$	—— (CO_2H) / 2.0 (CO_2H) (25°) / 2.940 (CO_2H) (20°) / 10.334 (NH) (20°)	—— / 0.010 / 1.15×10^{-3} / 4.63×10^{-11}	(1.0) / 1.81 / 2.52 / 9.46

Name	Structure	pK_a	K_a	pK_a
		Ionic strength (μ) = 0		μ = 0.1M
2-Nitrobenzoic acid		2.185	6.53×10^{-3}	——
3-Nitrobenzoic acid		3.449	3.56×10^{-4}	3.28
4-Nitrobenzoic acid	O_2N——CO_2H	3.442	3.61×10^{-4}	3.28
Nitroethane	$CH_3CH_2NO_2$	8.57	2.7×10^{-9}	——
2-Nitrophenol		7.230	5.89×10^{-8}	7.04
3-Nitrophenol		8.37	4.3×10^{-9}	8.16
4-Nitrophenol	O_2N——OH	7.149	7.10×10^{-8}	6.96
N-Nitrosophenylhydroxylamine (cupferron)		——	——	4.16
Nitrous acid	$HON{=}O$	3.15	7.1×10^{-4}	——
Oxalic acid (ethanedioic acid)	HO_2CCO_2H	1.250 4.266	5.62×10^{-2} 5.42×10^{-5}	(1.2) 3.80
Oxoacetic acid (glyoxylic acid)	$HCCO_2H$	3.46	3.5×10^{-4}	3.05
Oxobutanedioic acid (oxaloacetic acid)	$HO_2CCH_2CCO_2H$	2.56 4.37	2.8×10^{-3} 4.3×10^{-5}	2.26 3.90
2-Oxopentanedioic (α-ketoglutaric acid)	$HO_2CCH_2CH_2CCO_2H$	—— ——	—— ——	(1.9) (μ = 0.5 M) 4.44 (μ = 0.5 M)
2-Oxopropanoic acid (pyruvic acid)	CH_3CCO_2H	2.48	3.3×10^{-3}	2.26
1,5-Pentanedioic acid (glutaric acid)	$HO_2CCH_2CH_2CH_2CO_2H$	4.345 5.422	4.52×10^{-5} 3.78×10^{-6}	4.19 5.06
Pentanoic acid (valeric acid)	$CH_3CH_2CH_2CH_2CO_2H$	4.843	1.44×10^{-5}	4.63 (18°C)
1,10-Phenanthroline		4.91	1.23×10^{-5}	(1.8) 4.92
Phenylacetic acid	——CH_2CO_2H	4.310	4.90×10^{-5}	4.11

(Continued)

| Name | Structure | Ionic strength (μ) = 0 | | μ = 0.1 M |
		pK_a	K_a	pK_a
Phenylalanine	NH$_3^+$ CHCH$_2$—⬡ CO$_2$H	2.20 (CO$_2$H) 9.31 (NH$_3$)	6.3×10^{-3} 4.9×10^{-10}	2.18 9.09
Phosphoric acid* (hydrogen phosphate)	O ‖ HO—P—OH OH	2.148 7.198 12.375	7.11×10^{-3} 6.34×10^{-8} 4.22×10^{-13}	1.92 6.71 11.52
Phosphorous acid (hydrogen phosphite)	O ‖ HP—OH OH	(1.5) 6.78	3×10^{-2} 1.66×10^{-7}	—— ——
Phthalic acid (benzene-1,2-dicarboxylic acid)	⬡CO$_2$H CO$_2$H	2.950 5.408	1.12×10^{-3} 3.90×10^{-6}	2.76 4.92
Piperazine (perhydro-1,4-diazine)	H$_2$N⁺—⬡—⁺NH$_2$	5.333 9.731	4.65×10^{-6} 1.86×10^{-10}	5.64 9.74
Piperidine	⬡NH$_2^+$	11.125	7.50×10^{-12}	11.08
Proline	⬠—CO$_2$H N ⁺H$_2$	1.952 (CO$_2$H) 10.640 (NH$_2$)	1.12×10^{-2} 2.29×10^{-11}	1.89 10.46
Propanoic acid	CH$_3$CH$_2$CO$_2$H	4.874	1.34×10^{-5}	4.69
Propenoic acid (acrylic acid)	H$_2$C=CHCO$_2$H	4.258	5.52×10^{-5}	——
Propylamine	CH$_3$CH$_2$CH$_2$NH$_3^+$	10.566	2.72×10^{-11}	10.64
Pyridine (azine)	⬡NH$^+$	5.20	6.3×10^{-6}	5.24
Pyridine-2-carboxylic acid (picolinic acid)	⬡NH$^+$ CO$_2$H	(1.01) (CO$_2$H) 5.39 (NH)	9.8×10^{-2} 4.1×10^{-6}	(0.95) 5.21
Pyridine-3-carboxylic acid (nicotinic acid)	HO$_2$C ⬡NH$^+$	2.03 (CO$_2$H) 4.82 (NH)	9.3×10^{-3} 1.51×10^{-5}	2.08 4.69
Pyridoxal-5-phosphate	O O=CH ‖ HO—POCH$_2$—⬡—OH HO N ⁺H CH$_3$	—— —— —— ——	—— —— —— ——	(1.4) (POH) 3.51 (OH) 6.04 (POH) 8.25 (NH)
Pyrophosphoric acid (hydrogen diphosphate)	O O ‖ ‖ (HO)$_2$POP(OH)$_2$	(0.9) 2.28 6.70 9.40	0.13 5.2×10^{-3} 2.0×10^{-7} 4.0×10^{-10}	(0.8) (1.9) 5.94 8.25

*pK₃ from A. G. Miller and J. W. Macklin, Anal. Chem. **1983**, 55, 684.

AP32   APPENDIX G Acid Dissociation Constants

Name	Structure	Ionic strength (μ) = 0		$\mu = 0.1M$
		pK_a	K_a	pK_a
Pyrrolidine	(structure) NH_2^+	11.305	4.95×10^{-12}	11.3
Serine	NH_3^+ | $CHCH_2OH$ | CO_2H	2.187 (CO_2H) 9.209 (NH_3)	6.50×10^{-3} 6.18×10^{-10}	2.16 9.05
Succinic acid (butanedioic acid)	$HO_2CCH_2CH_2CO_2H$	4.207 5.636	6.21×10^{-5} 2.31×10^{-6}	3.99 5.24
Sulfuric acid (hydrogen sulfate)	$HO-\overset{\overset{O}{\|\|}}{\underset{\underset{O}{\|\|}}{S}}-OH$	1.987 (pK_2)	1.03×10^{-2}	1.54
Sulfurous acid (hydrogen sulfite)	$HO\overset{\overset{O}{\|\|}}{S}OH$	1.857 7.172	1.39×10^{-2} 6.73×10^{-8}	1.66 6.85
Thiosulfuric acid (hydrogen thiosulfate)	$HO\overset{\overset{O}{\|\|}}{\underset{\underset{O}{\|\|}}{S}}SH$	(0.6) (1.6)	0.3 0.03	—— (1.3)
Threonine	NH_3^+ | $CHCHOHCH_3$ | CO_2H	2.088 (CO_2H) 9.100 (NH_3)	8.17×10^{-3} 7.94×10^{-10}	2.20 8.94
Trichloroacetic acid	Cl_3CCO_2H	(−0.5)	3	——
Triethanolamine	$(HOCH_2CH_2)_3NH^+$	7.762	1.73×10^{-8}	7.85
Triethylamine	$(CH_3CH_2)_3NH^+$	10.72	1.9×10^{-11}	10.76
1,2,3-Trihydroxybenzene (pyrogallol)	(structure with OH, OH, OH)	—— —— ——	—— —— ——	8.96 11.00 (14.0) (20°C)
Trimethylamine	$(CH_3)_3NH^+$	9.799	1.59×10^{-10}	9.82
Tris(hydroxymethyl)amino- methane (tris or tham)	$(HOCH_2)_3CNH_3^+$	8.072	8.47×10^{-9}	8.10
Tryptophan	NH_3^+ | $CHCH_2$—(indole) | CO_2H	—— ——	—— ——	2.37 (CO_2H) 9.33 (NH_3)
Tyrosine	NH_3^+ | $CHCH_2$—(benzene)—OH | CO_2H	—— —— ——	—— —— ——	2.41 (CO_2H) 8.67 (NH_3) 11.01 (OH)
Valine	NH_3^+ | $CHCH(CH_3)_2$ | CO_2H	2.286 (CO_2H) 9.719 (NH_3)	5.18×10^{-3} 1.91×10^{-10}	2.27 9.52
Water*	H_2O	13.997	1.01×10^{-14}	——

*The constant given for water is K_w.

Reaction	$E°$ (volts)	$dE°/dT$ (mV/K)
Aluminum		
$Al^{3+} + 3e^- \rightleftharpoons Al(s)$	-1.677	0.533
$AlCl^{2+} + 3e^- \rightleftharpoons Al(s) + Cl^-$	-1.802	
$AlF_6^{3-} + 3e^- \rightleftharpoons Al(s) + 6F^-$	-2.069	
$Al(OH)_4^- + 3e^- \rightleftharpoons Al(s) + 4OH^-$	-2.328	-1.13
Antimony		
$SbO^+ + 2H^+ + 3e^- \rightleftharpoons Sb(s) + H_2O$	0.208	
$Sb_2O_3(s) + 6H^+ + 6e^- \rightleftharpoons 2Sb(s) + 3H_2O$	0.147	-0.369
$Sb(s) + 3H^+ + 3e^- \rightleftharpoons SbH_3(g)$	-0.510	-0.030
Arsenic		
$H_3AsO_4 + 2H^+ + 2e^- \rightleftharpoons H_3AsO_3 + H_2O$	0.575	-0.257
$H_3AsO_3 + 3H^+ + 3e^- \rightleftharpoons As(s) + 3H_2O$	$0.247\ 5$	-0.505
$As(s) + 3H^+ + 3e^- \rightleftharpoons AsH_3(g)$	-0.238	-0.029
Barium		
$Ba^{2+} + 2e^- + Hg \rightleftharpoons Ba(in\ Hg)$	-1.717	
$Ba^{2+} + 2e^- \rightleftharpoons Ba(s)$	-2.906	-0.401
Beryllium		
$Be^{2+} + 2e^- \rightleftharpoons Be(s)$	-1.968	0.60
Bismuth		
$Bi^{3+} + 3e^- \rightleftharpoons Bi(s)$	0.308	0.18
$BiCl_4^- + 3e^- \rightleftharpoons Bi(s) + 4Cl^-$	0.16	
$BiOCl(s) + 2H^+ + 3e^- \rightleftharpoons Bi(s) + H_2O + Cl^-$	0.160	
Boron		
$2B(s) + 6H^+ + 6e^- \rightleftharpoons B_2H_6(g)$	-0.150	-0.296
$B_4O_7^{2-} + 14H^+ + 12e^- \rightleftharpoons 4B(s) + 7H_2O$	-0.792	
$B(OH)_3 + 3H^+ + 3e^- \rightleftharpoons B(s) + 3H_2O$	-0.889	-0.492
Bromine		
$BrO_4^- + 2H^+ + 2e^- \rightleftharpoons BrO_3^- + H_2O$	1.745	-0.511
$HOBr + H^+ + e^- \rightleftharpoons \frac{1}{2}Br_2(l) + H_2O$	1.584	-0.75
$BrO_3^- + 6H^+ + 5e^- \rightleftharpoons \frac{1}{2}Br_2(l) + 3H_2O$	1.513	-0.419
$Br_2(aq) + 2e^- \rightleftharpoons 2Br^-$	1.098	-0.499
$Br_2(l) + 2e^- \rightleftharpoons 2Br^-$	1.078	-0.611
$Br_3^- + 2e^- \rightleftharpoons 3Br^-$	1.062	-0.512
$BrO^- + H_2O + 2e^- \rightleftharpoons Br^- + 2OH^-$	0.766	-0.94
$BrO_3^- + 3H_2O + 6e^- \rightleftharpoons Br^- + 6OH^-$	0.613	-1.287
Cadmium		
$Cd^{2+} + 2e^- + Hg \rightleftharpoons Cd(in\ Hg)$	-0.380	
$Cd^{2+} + 2e^- \rightleftharpoons Cd(s)$	-0.402	-0.029
$Cd(C_2O_4)(s) + 2e^- \rightleftharpoons Cd(s) + C_2O_4^{2-}$	-0.522	
$Cd(C_2O_4)_2^{2-} + 2e^- \rightleftharpoons Cd(s) + 2C_2O_4^{2-}$	-0.572	
$Cd(NH_3)_4^{2+} + 2e^- \rightleftharpoons Cd(s) + 4NH_3$	-0.613	
$CdS(s) + 2e^- \rightleftharpoons Cd(s) + S^{2-}$	-1.175	
Calcium		
$Ca(s) + 2H^+ + 2e^- \rightleftharpoons CaH_2(s)$	0.776	
$Ca^{2+} + 2e^- + Hg \rightleftharpoons Ca(in\ Hg)$	-2.003	
$Ca^{2+} + 2e^- \rightleftharpoons Ca(s)$	-2.868	-0.186

*All species are aqueous unless otherwise indicated. The reference state for amalgams is an infinitely dilute solution of the element in Hg. The temperature coefficient, $dE°/dT$, allows us to calculate the standard potential, $E°(T)$, at temperature T: $E°(T) = E° + (dE°/dT)\Delta T$, where ΔT is $T - 298.15$ K. Note the units mV/K for $dE°/dT$. Once you know $E°$ for a net cell reaction at temperature T, you can find the equilibrium constant, K, for the reaction from the formula $K = 10^{nFE°/RT \ln 10}$, where n is the number of electrons in each half-reaction, F is the Faraday constant, and R is the gas constant.

SOURCES: *The most authoritative source is S. G. Bratsch, J. Phys. Chem. Ref. Data **1989**, 18, 1. Additional data come from L. G. Sillén and A. E. Martell, Stability Constants of Metal-Ion Complexes (London: The Chemical Society, Special Publications Nos. 17 and 25. 1964 and 1971); G. Milazzo and S. Caroli, Tables of Standard Electrode Potentials (New York: Wiley, 1978); T. Mussini, P. Longhi, and S. Rondinini, Pure Appl. Chem. **1985**, 57, 169. Another good source is A. J. Bard, R. Parsons, and J. Jordan. Standard Potentials in Aqueous Solution (New York: Marcel Dekker, 1985). Reduction potentials for 1 200 free radical reactions are given by P. Wardman, J. Phys. Chem. Ref. Data **1989**, 18, 1637.*

Reaction	$E°$ (volts)	$dE°/dT$ (mV/K)
$Ca(acetate)^+ + 2e^- \rightleftharpoons Ca(s) + acetate^-$	-2.891	
$CaSO_4(s) + 2e^- \rightleftharpoons Ca(s) + SO_4^{2-}$	-2.936	
$Ca(malonate)(s) + 2e^- \rightleftharpoons Ca(s) + malonate^{2-}$	-3.608	

Carbon

Reaction	$E°$ (volts)	$dE°/dT$ (mV/K)
$C_2H_2(g) + 2H^+ + 2e^- \rightleftharpoons C_2H_4(g)$	0.731	
$O{=}\bigcirc{=}O + 2H^+ + 2e^- \rightleftharpoons HO{-}\bigcirc{-}OH$	0.700	
$CH_3OH + 2H^+ + 2e^- \rightleftharpoons CH_4(g) + H_2O$	0.583	-0.039
Dehydroascorbic acid $+ 2H^+ + 2e^- \rightleftharpoons$ ascorbic acid $+ H_2O$	0.390	
$(CN)_2(g) + 2H^+ + 2e^- \rightleftharpoons 2HCN(aq)$	0.373	
$H_2CO + 2H^+ + 2e^- \rightleftharpoons CH_3OH$	0.237	-0.51
$C(s) + 4H^+ + 4e^- \rightleftharpoons CH_4(g)$	$0.131\ 5$	$-0.209\ 2$
$HCO_2H + 2H^+ + 2e^- \rightleftharpoons H_2CO + H_2O$	-0.029	-0.63
$CO_2(g) + 2H^+ + 2e^- \rightleftharpoons CO(g) + H_2O$	$-0.103\ 8$	$-0.397\ 7$
$CO_2(g) + 2H^+ + 2e^- \rightleftharpoons HCO_2H$	-0.114	-0.94
$2CO_2(g) + 2H^+ + 2e^- \rightleftharpoons H_2C_2O_4$	-0.432	-1.76

Cerium

Reaction	$E°$ (volts)		$dE°/dT$ (mV/K)
	1.72		1.54
	1.70	1 F $HClO_4$	
$Ce^{4+} + e^- \rightleftharpoons Ce^{3+}$	1.44	1 F H_2SO_4	
	1.61	1 F HNO_3	
	1.47	1 F HCl	
$Ce^{3+} + 3e^- \rightleftharpoons Ce(s)$	-2.336		0.280

Cesium

Reaction	$E°$ (volts)	$dE°/dT$ (mV/K)
$Cs^+ + e^- + Hg \rightleftharpoons Cs(in\ Hg)$	-1.950	
$Cs^+ + e^- \rightleftharpoons Cs(s)$	-3.026	-1.172

Chlorine

Reaction	$E°$ (volts)	$dE°/dT$ (mV/K)
$HClO_2 + 2H^+ + 2e^- \rightleftharpoons HOCl + H_2O$	1.674	0.55
$HClO + H^+ + e^- \rightleftharpoons \frac{1}{2}Cl_2(g) + H_2O$	1.630	-0.27
$ClO_3^- + 6H^+ + 5e^- \rightleftharpoons \frac{1}{2}Cl_2(g) + 3H_2O$	1.458	-0.347
$Cl_2(aq) + 2e^- \rightleftharpoons 2Cl^-$	1.396	-0.72
$Cl_2(g) + 2e^- \rightleftharpoons 2Cl^-$	$1.360\ 4$	-1.248
$ClO_4^- + 2H^+ + 2e^- \rightleftharpoons ClO_3^- + H_2O$	1.226	-0.416
$ClO_3^- + 3H^+ + 2e^- \rightleftharpoons HClO_2 + H_2O$	1.157	-0.180
$ClO_3^- + 2H^+ + e^- \rightleftharpoons ClO_2 + H_2O$	1.130	0.074
$ClO_2 + e^- \rightleftharpoons ClO_2^-$	1.068	-1.335

Chromium

Reaction	$E°$ (volts)	$dE°/dT$ (mV/K)
$Cr_2O_7^{2-} + 14H^+ + 6e^- \rightleftharpoons 2Cr^{3+} + 7H_2O$	1.36	-1.32
$CrO_4^{2-} + 4H_2O + 3e^- \rightleftharpoons Cr(OH)_3\ (s, hydrated) + 5OH^-$	-0.12	-1.62
$Cr^{3+} + e^- \rightleftharpoons Cr^{2+}$	-0.42	1.4
$Cr^{3+} + 3e^- \rightleftharpoons Cr(s)$	-0.74	0.44
$Cr^{2+} + 2e^- \rightleftharpoons Cr(s)$	-0.89	-0.04

Cobalt

Reaction	$E°$ (volts)		$dE°/dT$ (mV/K)
	1.92		1.23
$Co^{3+} + e^- \rightleftharpoons Co^{2+}$	1.817	8 F H_2SO_4	
	1.850	4 F HNO_3	
$Co(NH_3)_5(H_2O)^{3+} + e^- \rightleftharpoons Co(NH_3)_5(H_2O)^{2+}$	0.37	1 F NH_4NO_3	
$Co(NH_3)_6^{3+} + e^- \rightleftharpoons Co(NH_3)_6^{2+}$	0.1		
$CoOH^+ + H^+ + 2e^- \rightleftharpoons Co(s) + H_2O$	0.003		-0.04
$Co^{2+} + 2e^- \rightleftharpoons Co(s)$	-0.282		0.065
$Co(OH)_2(s) + 2e^- \rightleftharpoons Co(s) + 2OH^-$	-0.746		-1.02

Copper

Reaction	$E°$ (volts)	$dE°/dT$ (mV/K)
$Cu^+ + e^- \rightleftharpoons Cu(s)$	0.518	-0.754
$Cu^{2+} + 2e^- \rightleftharpoons Cu(s)$	0.339	0.011
$Cu^{2+} + e^- \rightleftharpoons Cu^+$	0.161	0.776
$CuCl(s) + e^- \rightleftharpoons Cu(s) + Cl^-$	0.137	
$Cu(IO_3)_2(s) + 2e^- \rightleftharpoons Cu(s) + 2IO_3^-$	-0.079	
$Cu(ethylenediamine)_2^+ + e^- \rightleftharpoons Cu(s) + 2$ ethylenediamine	-0.119	
$CuI(s) + e^- \rightleftharpoons Cu(s) + I^-$	-0.185	
$Cu(EDTA)^{2-} + 2e^- \rightleftharpoons Cu(s) + EDTA^{4-}$	-0.216	
$Cu(OH)_2(s) + 2e^- \rightleftharpoons Cu(s) + 2OH^-$	-0.222	
$Cu(CN)_2^- + e^- \rightleftharpoons Cu(s) + 2CN^-$	-0.429	
$CuCN(s) + e^- \rightleftharpoons Cu(s) + CN^-$	-0.639	

(Continued)

Reaction	$E°$ (volts)		$dE°/dT$ (mV/K)
Dysprosium			
$Dy^{3+} + 3e^- \rightleftharpoons Dy(s)$	-2.295		0.373
Erbium			
$Er^{3+} + 3e^- \rightleftharpoons Er(s)$	-2.331		0.388
Europium			
$Eu^{3+} + e^- \rightleftharpoons Eu^{2+}$	-0.35		1.53
$Eu^{3+} + 3e^- \rightleftharpoons Eu(s)$	-1.991		0.338
$Eu^{2+} + 2e^- \rightleftharpoons Eu(s)$	-2.812		-0.26
Fluorine			
$F_2(g) + 2e^- \rightleftharpoons 2F^-$	2.890		-1.870
$F_2O(g) + 2H^+ + 4e^- \rightleftharpoons 2F^- + H_2O$	2.168		-1.208
Gadolinium			
$Gd^{3+} + 3e^- \rightleftharpoons Gd(s)$	-2.279		0.315
Gallium			
$Ga^{3+} + 3e^- \rightleftharpoons Ga(s)$	-0.549		0.61
$GaOOH(s) + H_2O + 3e^- \rightleftharpoons Ga(s) + 3OH^-$	-1.320		-1.08
Germanium			
$Ge^{2+} + 2e^- \rightleftharpoons Ge(s)$	0.1		
$H_4GeO_4 + 4H^+ + 4e^- \rightleftharpoons Ge(s) + 4H_2O$	-0.039		-0.429
Gold			
$Au^+ + e^- \rightleftharpoons Au(s)$	1.69		-1.1
$Au^{3+} + 2e^- \rightleftharpoons Au^+$	1.41		
$AuCl_2^- + e^- \rightleftharpoons Au(s) + 2Cl^-$	1.154		
$AuCl_4^- + 2e^- \rightleftharpoons AuCl_2^- + 2Cl^-$	0.926		
Hafnium			
$Hf^{4+} + 4e^- \rightleftharpoons Hf(s)$	-1.55		0.68
$HfO_2(s) + 4H^+ + 4e^- \rightleftharpoons Hf(s) + 2H_2O$	-1.591		-0.355
Holmium			
$Ho^{3+} + 3e^- \rightleftharpoons Ho(s)$	-2.33		0.371
Hydrogen			
$2H^+ + 2e^- \rightleftharpoons H_2(g)$	$0.000\,0$		0
$H_2O + e^- \rightleftharpoons \frac{1}{2}H_2(g) + OH^-$	$-0.828\,0$		$-0.836\,0$
Indium			
$In^{3+} + 3e^- + Hg \rightleftharpoons In(in\ Hg)$	-0.313		
$In^{3+} + 3e^- \rightleftharpoons In(s)$	-0.338		0.42
$In^{3+} + 2e^- \rightleftharpoons In^+$	-0.444		
$In(OH)_3(s) + 3e^- \rightleftharpoons In(s) + 3OH^-$	-0.99		-0.95
Iodine			
$IO_4^- + 2H^+ + 2e^- \rightleftharpoons IO_3^- + H_2O$	1.589		-0.85
$H_5IO_6 + 2H^+ + 2e^- \rightleftharpoons HIO_3 + 3H_2O$	1.567		-0.12
$HOI + H^+ + e^- \rightleftharpoons \frac{1}{2}I_2(s) + H_2O$	1.430		-0.339
$ICl_3(s) + 3e^- \rightleftharpoons \frac{1}{2}I_2(s) + 3Cl^-$	1.28		
$ICl(s) + e^- \rightleftharpoons \frac{1}{2}I_2(s) + Cl^-$	1.22		
$IO_3^- + 6H^+ + 5e^- \rightleftharpoons \frac{1}{2}I_2(s) + 3H_2O$	1.210		-0.367
$IO_3^- + 5H^+ + 4e^- \rightleftharpoons HOI + 2H_2O$	1.154		-0.374
$I_2(aq) + 2e^- \rightleftharpoons 2I^-$	0.620		-0.234
$I_2(s) + 2e^- \rightleftharpoons 2I^-$	0.535		-0.125
$I_3^- + 2e^- \rightleftharpoons 3I^-$	0.535		-0.186
$IO_3^- + 3H_2O + 6e^- \rightleftharpoons I^- + 6OH^-$	0.269		-1.163
Iridium			
$IrCl_6^{2-} + e^- \rightleftharpoons IrCl_6^{3-}$	1.026	1 F HCl	
$IrBr_6^{2-} + e^- \rightleftharpoons IrBr_6^{3-}$	0.947	2 F NaBr	
$IrCl_6^{2-} + 4e^- \rightleftharpoons Ir(s) + 6Cl^-$	0.835		
$IrO_2(s) + 4H^+ + 4e^- \rightleftharpoons Ir(s) + 2H_2O$	0.73		-0.36
$IrI_6^{2-} + e^- \rightleftharpoons IrI_6^{3-}$	0.485	1 F KI	
Iron			
$Fe(phenanthroline)_3^{3+} + e^- \rightleftharpoons Fe(phenanthroline)_3^{2+}$	1.147		
$Fe(bipyridyl)_3^{3+} + e^- \rightleftharpoons Fe(bipyridyl)_3^{2+}$	1.120		
$FeOH^{2+} + H^+ + e^- \rightleftharpoons Fe^{2+} + H_2O$	0.900		0.096
$FeO_4^{2-} + 3H_2O + 3e^- \rightleftharpoons FeOOH(s) + 5OH^-$	0.80		-1.59
$Fe^{3+} + e^- \rightleftharpoons Fe^{2+}$	$\begin{cases} 0.771 \\ 0.732 \\ 0.767 \\ 0.746 \\ 0.68 \end{cases}$	$\begin{matrix} \\ \text{1 F HCl} \\ \text{1 F HClO}_4 \\ \text{1 F HNO}_3 \\ \text{1 F H}_2\text{SO}_4 \end{matrix}$	1.175

Reaction	$E°$ (volts)	$dE°/dT$ (mV/K)
$FeOOH(s) + 3H^+ + e^- \rightleftharpoons Fe^{2+} + 2H_2O$	0.74	−1.05
$Ferricinium^+ + e^- \rightleftharpoons ferrocene$	0.400	
$Fe(CN)_6^{3-} + e^- \rightleftharpoons Fe(CN)_6^{4-}$	0.356	
$Fe(glutamate)^{3+} + e^- \rightleftharpoons Fe(glutamate)^{2+}$	0.240	
$FeOH^+ + H^+ + 2e^- \rightleftharpoons Fe(s) + H_2O$	−0.16	0.07
$Fe^{2+} + 2e^- \rightleftharpoons Fe(s)$	−0.44	0.07
$FeCO_3(s) + 2e^- \rightleftharpoons Fe(s) + CO_3^{2-}$	−0.756	−1.293

Lanthanum

Reaction	$E°$ (volts)	$dE°/dT$ (mV/K)
$La^{3+} + 3e^- \rightleftharpoons La(s)$	−2.379	0.242
$La(succinate)^+ + 3e^- \rightleftharpoons La(s) + succinate^{2-}$	−2.601	

Lead

Reaction	$E°$ (volts)	$dE°/dT$ (mV/K)
$Pb^{4+} + 2e^- \rightleftharpoons Pb^{2+}$	1.69 1 F HNO$_3$	
$PbO_2(s) + 4H^+ + SO_4^{2-} + 2e^- \rightleftharpoons PbSO_4(s) + 2H_2O$	1.685	
$PbO_2(s) + 4H^+ + 2e^- \rightleftharpoons Pb^{2+} + 2H_2O$	1.458	−0.253
$3PbO_2(s) + 2H_2O + 4e^- \rightleftharpoons Pb_3O_4(s) + 4OH^-$	0.269	−1.136
$Pb_3O_4(s) + H_2O + 2e^- \rightleftharpoons 3PbO(s, red) + 2OH^-$	0.224	−1.211
$Pb_3O_4(s) + H_2O + 2e^- \rightleftharpoons 3PbO(s, yellow) + 2OH^-$	0.207	−1.177
$Pb^{2+} + 2e^- \rightleftharpoons Pb(s)$	−0.126	−0.395
$PbF_2(s) + 2e^- \rightleftharpoons Pb(s) + 2F^-$	−0.350	
$PbSO_4(s) + 2e^- \rightleftharpoons Pb(s) + SO_4^{2-}$	−0.355	

Lithium

Reaction	$E°$ (volts)	$dE°/dT$ (mV/K)
$Li^+ + e^- + Hg \rightleftharpoons Li(in\ Hg)$	−2.195	
$Li^+ + e^- \rightleftharpoons Li(s)$	−3.040	−0.514

Lutetium

Reaction	$E°$ (volts)	$dE°/dT$ (mV/K)
$Lu^{3+} + 3e^- \rightleftharpoons Lu(s)$	−2.28	0.412

Magnesium

Reaction	$E°$ (volts)	$dE°/dT$ (mV/K)
$Mg^{2+} + 2e^- + Hg \rightleftharpoons Mg(in\ Hg)$	−1.980	
$Mg(OH)^+ + H^+ + 2e^- \rightleftharpoons Mg(s) + H_2O$	−2.022	0.25
$Mg^{2+} + 2e^- \rightleftharpoons Mg(s)$	−2.360	0.199
$Mg(C_2O_4)(s) + 2e^- \rightleftharpoons Mg(s) + C_2O_4^{2-}$	−2.493	
$Mg(OH)_2(s) + 2e^- \rightleftharpoons Mg(s) + 2OH^-$	−2.690	−0.946

Manganese

Reaction	$E°$ (volts)	$dE°/dT$ (mV/K)
$MnO_4^- + 4H^+ + 3e^- \rightleftharpoons MnO_2(s) + 2H_2O$	1.692	−0.671
$Mn^{3+} + e^- \rightleftharpoons Mn^{2+}$	1.56	1.8
$MnO_4^- + 8H^+ + 5e^- \rightleftharpoons Mn^{2+} + 4H_2O$	1.507	−0.646
$Mn_2O_3(s) + 6H^+ + 2e^- \rightleftharpoons 2Mn^{2+} + 3H_2O$	1.485	−0.926
$MnO_2(s) + 4H^+ + 2e^- \rightleftharpoons Mn^{2+} + 2H_2O$	1.230	−0.609
$Mn(EDTA)^- + e^- \rightleftharpoons Mn(EDTA)^{2-}$	0.825	−1.10
$MnO_4^- + e^- \rightleftharpoons MnO_4^{2-}$	0.56	−2.05
$3Mn_2O_3(s) + H_2O + 2e^- \rightleftharpoons 2Mn_3O_4(s) + 2OH^-$	0.002	−1.256
$Mn_3O_4(s) + 4H_2O + 2e^- \rightleftharpoons 3Mn(OH)_2(s) + 2OH^-$	−0.352	−1.61
$Mn^{2+} + 2e^- \rightleftharpoons Mn(s)$	−1.182	−1.129
$Mn(OH)_2(s) + 2e^- \rightleftharpoons Mn(s) + 2OH^-$	−1.565	−1.10

Mercury

Reaction	$E°$ (volts)	$dE°/dT$ (mV/K)
$2Hg^{2+} + 2e^- \rightleftharpoons Hg_2^{2+}$	0.908	0.095
$Hg^{2+} + 2e^- \rightleftharpoons Hg(l)$	0.852	−0.116
$Hg_2^{2+} + 2e^- \rightleftharpoons 2Hg(l)$	0.796	−0.327
$Hg_2SO_4(s) + 2e^- \rightleftharpoons 2Hg(l) + SO_4^{2-}$	0.614	
$Hg_2Cl_2(s) + 2e^- \rightleftharpoons 2Hg(l) + 2Cl^-$	$\begin{cases} 0.268 \\ 0.241\ (saturated\ calomel\ electrode) \end{cases}$	
$Hg(OH)_3^- + 2e^- \rightleftharpoons Hg(l) + 3OH^-$	0.231	
$Hg(OH)_2 + 2e^- \rightleftharpoons Hg(l) + 2OH^-$	0.206	−1.24
$Hg_2Br_2(s) + 2e^- \rightleftharpoons 2Hg(l) + 2Br^-$	0.140	
$HgO(s, yellow) + H_2O + 2e^- \rightleftharpoons Hg(l) + 2OH^-$	0.098 3	−1.125
$HgO(s, red) + H_2O + 2e^- \rightleftharpoons Hg(l) + 2OH^-$	0.097 7	−1.120 6

Molybdenum

Reaction	$E°$ (volts)	$dE°/dT$ (mV/K)
$MoO_4^{2-} + 2H_2O + 2e^- \rightleftharpoons MoO_2(s) + 4OH^-$	−0.818	−1.69
$MoO_4^{2-} + 4H_2O + 6e^- \rightleftharpoons Mo(s) + 8OH^-$	−0.926	−1.36
$MoO_2(s) + 2H_2O + 4e^- \rightleftharpoons Mo(s) + 4OH^-$	−0.980	−1.196

Neodymium

Reaction	$E°$ (volts)	$dE°/dT$ (mV/K)
$Nd^{3+} + 3e^- \rightleftharpoons Nd(s)$	−2.323	0.282

Neptunium

Reaction	$E°$ (volts)	$dE°/dT$ (mV/K)
$NpO_3^+ + 2H^+ + e^- \rightleftharpoons NpO_2^{2+} + H_2O$	2.04	
$NpO_2^{2+} + e^- \rightleftharpoons NpO_2^+$	1.236	0.058

(Continued)

Reaction	$E°$ (volts)	$dE°/dT$ (mV/K)
$NpO_2^+ + 4H^+ + e^- \rightleftharpoons Np^{4+} + 2H_2O$	0.567	-3.30
$Np^{4+} + e^- \rightleftharpoons Np^{3+}$	0.157	1.53
$Np^{3+} + 3e^- \rightleftharpoons Np(s)$	-1.768	0.18

Nickel

Reaction	$E°$ (volts)	$dE°/dT$ (mV/K)
$NiOOH(s) + 3H^+ + e^- \rightleftharpoons Ni^{2+} + 2H_2O$	2.05	-1.17
$Ni^{2+} + 2e^- \rightleftharpoons Ni(s)$	-0.236	0.146
$Ni(CN)_4^{2-} + e^- \rightleftharpoons Ni(CN)_3^{2-} + CN^-$	-0.401	
$Ni(OH)_2(s) + 2e^- \rightleftharpoons Ni(s) + 2OH^-$	-0.714	-1.02

Niobium

Reaction	$E°$ (volts)	$dE°/dT$ (mV/K)
$\frac{1}{2}Nb_2O_5(s) + H^+ + e^- \rightleftharpoons NbO_2(s) + \frac{1}{2}H_2O$	-0.248	-0.460
$\frac{1}{2}Nb_2O_5(s) + 5H^+ + 5e^- \rightleftharpoons Nb(s) + \frac{5}{2}H_2O$	-0.601	-0.381
$NbO_2(s) + 2H^+ + 2e^- \rightleftharpoons NbO(s) + H_2O$	-0.646	-0.347
$NbO_2(s) + 4H^+ + 4e^- \rightleftharpoons Nb(s) + 2H_2O$	-0.690	-0.361

Nitrogen

Reaction	$E°$ (volts)	$dE°/dT$ (mV/K)
$HN_3 + 3H^+ + 2e^- \rightleftharpoons N_2(g) + NH_4^+$	2.079	0.147
$N_2O(g) + 2H^+ + 2e^- \rightleftharpoons N_2(g) + H_2O$	1.769	-0.461
$2NO(g) + 2H^+ + 2e^- \rightleftharpoons N_2O(g) + H_2O$	1.587	-1.359
$NO^+ + e^- \rightleftharpoons NO(g)$	1.46	
$2NH_3OH^+ + H^+ + 2e^- \rightleftharpoons N_2H_5^+ + 2H_2O$	1.40	-0.60
$NH_3OH^+ + 2H^+ + 2e^- \rightleftharpoons NH_4^+ + H_2O$	1.33	-0.44
$N_2H_5^+ + 3H^+ + 2e^- \rightleftharpoons 2NH_4^+$	1.250	-0.28
$HNO_2 + H^+ + e^- \rightleftharpoons NO(g) + H_2O$	0.984	0.649
$NO_3^- + 4H^+ + 3e^- \rightleftharpoons NO(g) + 2H_2O$	0.955	0.028
$NO_3^- + 3H^+ + 2e^- \rightleftharpoons HNO_2 + H_2O$	0.940	-0.282
$NO_3^- + 2H^+ + e^- \rightleftharpoons \frac{1}{2}N_2O_4(g) + H_2O$	0.798	0.107
$N_2(g) + 8H^+ + 6e^- \rightleftharpoons 2NH_4^+$	0.274	-0.616
$N_2(g) + 5H^+ + 4e^- \rightleftharpoons N_2H_5^+$	-0.214	-0.78
$N_2(g) + 2H_2O + 4H^+ + 2e^- \rightleftharpoons 2NH_3OH^+$	-1.83	-0.96
$\frac{3}{2}N_2(g) + H^+ + e^- \rightleftharpoons HN_3$	-3.334	-2.141

Osmium

Reaction	$E°$ (volts)	$dE°/dT$ (mV/K)
$OsO_4(s) + 8H^+ + 8e^- \rightleftharpoons Os(s) + 4H_2O$	0.834	-0.458
$OsCl_6^{2-} + e^- \rightleftharpoons OsCl_6^{3-}$	0.85 1 F HCl	

Oxygen

Reaction	$E°$ (volts)	$dE°/dT$ (mV/K)
$OH + H^+ + e^- \rightleftharpoons H_2O$	2.56	-1.0
$O(g) + 2H^+ + 2e^- \rightleftharpoons H_2O$	2.430 1	$-1.148\ 4$
$O_3(g) + 2H^+ + 2e^- \rightleftharpoons O_2(g) + H_2O$	2.075	-0.489
$H_2O_2 + 2H^+ + 2e^- \rightleftharpoons 2H_2O$	1.763	-0.698
$HO_2 + H^+ + e^- \rightleftharpoons H_2O_2$	1.44	-0.7
$\frac{1}{2}O_2(g) + 2H^+ + 2e^- \rightleftharpoons H_2O$	1.229 1	$-0.845\ 6$
$O_2(g) + 2H^+ + 2e^- \rightleftharpoons H_2O_2$	0.695	-0.993
$O_2(g) + H^+ + e^- \rightleftharpoons HO_2$	-0.05	-1.3

Palladium

Reaction	$E°$ (volts)	$dE°/dT$ (mV/K)
$Pd^{2+} + 2e^- \rightleftharpoons Pd(s)$	0.915	0.12
$PdO(s) + 2H^+ + 2e^- \rightleftharpoons Pd(s) + H_2O$	0.79	-0.33
$PdCl_6^{4-} + 2e^- \rightleftharpoons Pd(s) + 6Cl^-$	0.615	
$PdO_2(s) + H_2O + 2e^- \rightleftharpoons PdO(s) + 2OH^-$	0.64	-1.2

Phosphorus

Reaction	$E°$ (volts)	$dE°/dT$ (mV/K)
$\frac{1}{4}P_4(s, \text{white}) + 3H^+ + 3e^- \rightleftharpoons PH_3(g)$	-0.046	-0.093
$\frac{1}{4}P_4(s, \text{red}) + 3H^+ + 3e^- \rightleftharpoons PH_3(g)$	-0.088	-0.030
$H_3PO_4 + 2H^+ + 2e^- \rightleftharpoons H_3PO_3 + H_2O$	-0.30	-0.36
$H_3PO_4 + 5H^+ + 5e^- \rightleftharpoons \frac{1}{4}P_4(s, \text{white}) + 4H_2O$	-0.402	-0.340
$H_3PO_3 + 2H^+ + 2e^- \rightleftharpoons H_3PO_2 + H_2O$	-0.48	-0.37
$H_3PO_2 + H^+ + e^- \rightleftharpoons \frac{1}{4}P_4(s) + 2H_2O$	-0.51	

Platinum

Reaction	$E°$ (volts)	$dE°/dT$ (mV/K)
$Pt^{2+} + 2e^- \rightleftharpoons Pt(s)$	1.18	-0.05
$PtO_2(s) + 4H^+ + 4e^- \rightleftharpoons Pt(s) + 2H_2O$	0.92	-0.36
$PtCl_4^{2-} + 2e^- \rightleftharpoons Pt(s) + 4Cl^-$	0.755	
$PtCl_6^{2-} + 2e^- \rightleftharpoons PtCl_4^{2-} + 2Cl^-$	0.68	

Plutonium

Reaction	$E°$ (volts)	$dE°/dT$ (mV/K)
$PuO_2^+ + e^- \rightleftharpoons PuO_2(s)$	1.585	0.39
$PuO_2^{2+} + 4H^+ + 2e^- \rightleftharpoons Pu^{4+} + 2H_2O$	1.000	$-1.615\ 1$
$Pu^{4+} + e^- \rightleftharpoons Pu^{3+}$	1.006	1.441
$PuO_2^{2+} + e^- \rightleftharpoons PuO_2^+$	0.966	0.03
$PuO_2(s) + 4H^+ + 4e^- \rightleftharpoons Pu(s) + 2H_2O$	-1.369	-0.38
$Pu^{3+} + 3e^- \rightleftharpoons Pu(s)$	-1.978	0.23

Reaction	$E°$ (volts)		$dE°/dT$ (mV/K)
Potassium			
$K^+ + e^- + Hg \rightleftharpoons K(in\ Hg)$	−1.975		
$K^+ + e^- \rightleftharpoons K(s)$	−2.936		−1.074
Praseodymium			
$Pr^{4+} + e^- \rightleftharpoons Pr^{3+}$	3.2		1.4
$Pr^{3+} + 3e^- \rightleftharpoons Pr(s)$	−2.353		0.291
Promethium			
$Pm^{3+} + 3e^- \rightleftharpoons Pm(s)$	−2.30		0.29
Radium			
$Ra^{2+} + 2e^- \rightleftharpoons Ra(s)$	−2.80		−0.44
Rhenium			
$ReO_4^- + 2H^+ + e^- \rightleftharpoons ReO_3(s) + H_2O$	0.72		−1.17
$ReO_4^- + 4H^+ + 3e^- \rightleftharpoons ReO_2(s) + 2H_2O$	0.510		−0.70
Rhodium			
$Rh^{6+} + 3e^- \rightleftharpoons Rh^{3+}$	1.48	1 F HClO$_4$	
$Rh^{4+} + e^- \rightleftharpoons Rh^{3+}$	1.44	3 F H$_2$SO$_4$	
$RhCl_6^{2-} + e^- \rightleftharpoons RhCl_6^{3-}$	1.2		
$Rh^{3+} + 3e^- \rightleftharpoons Rh(s)$	0.76		0.4
$2Rh^{3+} + 2e^- \rightleftharpoons Rh_2^{4+}$	0.7		
$RhCl_6^{3-} + 3e^- \rightleftharpoons Rh(s) + 6Cl^-$	0.44		
Rubidium			
$Rb^+ + e^- + Hg \rightleftharpoons Rb(in\ Hg)$	−1.970		
$Rb^+ + e^- \rightleftharpoons Rb(s)$	−2.943		−1.140
Ruthenium			
$RuO_4^- + 6H^+ + 3e^- \rightleftharpoons Ru(OH)_2^{2+} + 2H_2O$	1.53		
$Ru(dipyridyl)_3^{3+} + e^- \rightleftharpoons Ru(dipyridyl)_3^{2+}$	1.29		
$RuO_4(s) + 8H^+ + 8e^- \rightleftharpoons Ru(s) + 4H_2O$	1.032		−0.467
$Ru^{2+} + 2e^- \rightleftharpoons Ru(s)$	0.8		
$Ru^{3+} + 3e^- \rightleftharpoons Ru(s)$	0.60		
$Ru^{3+} + e^- \rightleftharpoons Ru^{2+}$	0.24		
$Ru(NH_3)_6^{3+} + e^- \rightleftharpoons Ru(NH_3)_6^{2+}$	0.214		
Samarium			
$Sm^{3+} + 3e^- \rightleftharpoons Sm(s)$	−2.304		0.279
$Sm^{2+} + 2e^- \rightleftharpoons Sm(s)$	−2.68		−0.28
Scandium			
$Sc^{3+} + 3e^- \rightleftharpoons Sc(s)$	−2.09		0.41
Selenium			
$SeO_4^{2-} + 4H^+ + 2e^- \rightleftharpoons H_2SeO_3 + H_2O$	1.150		0.483
$H_2SeO_3 + 4H^+ + 4e^- \rightleftharpoons Se(s) + 3H_2O$	0.739		−0.562
$Se(s) + 2H^+ + 2e^- \rightleftharpoons H_2Se(g)$	−0.082		0.238
$Se(s) + 2e^- \rightleftharpoons Se^{2-}$	−0.67		−1.2
Silicon			
$Si(s) + 4H^+ + 4e^- \rightleftharpoons SiH_4(g)$	−0.147		−0.196
$SiO_2(s, quartz) + 4H^+ + 4e^- \rightleftharpoons Si(s) + 2H_2O$	−0.990		−0.374
$SiF_6^{2-} + 4e^- \rightleftharpoons Si(s) + 6F^-$	−1.24		
Silver			
$Ag^{2+} + e^- \rightleftharpoons Ag^+$	2.000	4 F HClO$_4$	
	1.989		0.99
	1.929	4 F HNO$_3$	
$Ag^{3+} + 2e^- \rightleftharpoons Ag^+$	1.9		
$AgO(s) + H^+ + e^- \rightleftharpoons \frac{1}{2}Ag_2O(s) + \frac{1}{2}H_2O$	1.40		
$Ag^+ + e^- \rightleftharpoons Ag(s)$	0.799 3		−0.989
$Ag_2C_2O_4(s) + 2e^- \rightleftharpoons 2Ag(s) + C_2O_4^{2-}$	0.465		
$AgN_3(s) + e^- \rightleftharpoons Ag(s) + N_3^-$	0.293		
$AgCl(s) + e^- \rightleftharpoons Ag(s) + Cl^-$	0.222		
	0.197	saturated KCl	
$AgBr(s) + e^- \rightleftharpoons Ag(s) + Br^-$	0.071		
$Ag(S_2O_3)_2^{3-} + e^- \rightleftharpoons Ag(s) + 2S_2O_3^{2-}$	0.017		
$AgI(s) + e^- \rightleftharpoons Ag(s) + I^-$	−0.152		
$Ag_2S(s) + H^+ + 2e^- \rightleftharpoons 2Ag(s) + SH^-$	−0.272		
Sodium			
$Na^+ + e^- + Hg \rightleftharpoons Na(in\ Hg)$	−1.959		
$Na^+ + \frac{1}{2}H_2(g) + e^- \rightleftharpoons NaH(s)$	−2.367		−1.550
$Na^+ + e^- \rightleftharpoons Na(s)$	−2.714 3		−0.757

(Continued)

Reaction	$E°$ (volts)		$dE°/dT$ (mV/K)
Strontium			
$Sr^{2+} + 2e^- \rightleftharpoons Sr(s)$	-2.889		-0.237
Sulfur			
$S_2O_8^{2-} + 2e^- \rightleftharpoons 2SO_4^{2-}$	2.01		
$S_2O_6^{2-} + 4H^+ + 2e^- \rightleftharpoons 2H_2SO_3$	0.57		
$4SO_2 + 4H^+ + 6e^- \rightleftharpoons S_4O_6^{2-} + 2H_2O$	0.539		-1.11
$SO_2 + 4H^+ + 4e^- \rightleftharpoons S(s) + 2H_2O$	0.450		-0.652
$2H_2SO_3 + 2H^+ + 4e^- \rightleftharpoons S_2O_3^{2-} + 3H_2O$	0.40		
$S(s) + 2H^+ + 2e^- \rightleftharpoons H_2S(g)$	0.174		0.224
$S(s) + 2H^+ + 2e^- \rightleftharpoons H_2S(aq)$	0.144		-0.21
$S_4O_6^{2-} + 2H^+ + 2e^- \rightleftharpoons 2HS_2O_3^-$	0.10		-0.23
$5S(s) + 2e^- \rightleftharpoons S_5^{2-}$	-0.340		
$S(s) + 2e^- \rightleftharpoons S^{2-}$	-0.476		-0.925
$2S(s) + 2e^- \rightleftharpoons S_2^{2-}$	-0.50		-1.16
$2SO_3^{2-} + 3H_2O + 4e^- \rightleftharpoons S_2O_3^{2-} + 6OH^-$	-0.566		-1.06
$SO_3^{2-} + 3H_2O + 4e^- \rightleftharpoons S(s) + 6OH^-$	-0.659		-1.23
$SO_4^{2-} + 4H_2O + 6e^- \rightleftharpoons S(s) + 8OH^-$	-0.751		-1.288
$SO_4^{2-} + H_2O + 2e^- \rightleftharpoons SO_3^{2-} + 2OH^-$	-0.936		-1.41
$2SO_3^{2-} + 2H_2O + 2e^- \rightleftharpoons S_2O_4^{2-} + 4OH^-$	-1.130		-0.85
$2SO_4^{2-} + 2H_2O + 2e^- \rightleftharpoons S_2O_6^{2-} + 4OH^-$	-1.71		-1.00
Tantalum			
$Ta_2O_5(s) + 10H^+ + 10e^- \rightleftharpoons 2Ta(s) + 5H_2O$	-0.752		-0.377
Technetium			
$TcO_4^- + 2H_2O + 3e^- \rightleftharpoons TcO_2(s) + 4OH^-$	-0.366		-1.82
$TcO_4^- + 4H_2O + 7e^- \rightleftharpoons Tc(s) + 8OH^-$	-0.474		-1.46
Tellurium			
$TeO_3^{2-} + 3H_2O + 4e^- \rightleftharpoons Te(s) + 6OH^-$	-0.47		-1.39
$2Te(s) + 2e^- \rightleftharpoons Te_2^{2-}$	-0.84		
$Te(s) + 2e^- \rightleftharpoons Te^{2-}$	-0.90		-1.0
Terbium			
$Tb^{4+} + e^- \rightleftharpoons Tb^{3+}$	3.1		1.5
$Tb^{3+} + 3e^- \rightleftharpoons Tb(s)$	-2.28		0.350
Thallium			
	1.280		0.97
	0.77	1 F HCl	
$Tl^{3+} + 2e^- \rightleftharpoons Tl^+$	1.22	1 F H_2SO_4	
	1.23	1 F HNO_3	
	1.26	1 F HClO_4	
$Tl^+ + e^- + Hg \rightleftharpoons Tl(in\ Hg)$	-0.294		
$Tl^+ + e^- \rightleftharpoons Tl(s)$	-0.336		-1.312
$TlCl(s) + e^- \rightleftharpoons Tl(s) + Cl^-$	-0.557		
Thorium			
$Th^{4+} + 4e^- \rightleftharpoons Th(s)$	-1.826		0.557
Thulium			
$Tm^{3+} + 3e^- \rightleftharpoons Tm(s)$	-2.319		0.394
Tin			
$Sn(OH)_3^+ + 3H^+ + 2e^- \rightleftharpoons Sn^{2+} + 3H_2O$	0.142		
$Sn^{4+} + 2e^- \rightleftharpoons Sn^{2+}$	0.139	1 F HCl	
$SnO_2(s) + 4H^+ + 2e^- \rightleftharpoons Sn^{2+} + 2H_2O$	-0.094		-0.31
$Sn^{2+} + 2e^- \rightleftharpoons Sn(s)$	-0.141		-0.32
$SnF_6^{2-} + 4e^- \rightleftharpoons Sn(s) + 6F^-$	-0.25		
$Sn(OH)_6^{2-} + 2e^- \rightleftharpoons Sn(OH)_3^- + 3OH^-$	-0.93		
$Sn(s) + 4H_2O + 4e^- \rightleftharpoons SnH_4(g) + 4OH^-$	-1.316		-1.057
$SnO_2(s) + H_2O + 2e^- \rightleftharpoons SnO(s) + 2OH^-$	-0.961		-1.129
Titanium			
$TiO^{2+} + 2H^+ + e^- \rightleftharpoons Ti^{3+} + H_2O$	0.1		-0.6
$Ti^{3+} + e^- \rightleftharpoons Ti^{2+}$	-0.9		1.5
$TiO_2(s) + 4H^+ + 4e^- \rightleftharpoons Ti(s) + 2H_2O$	-1.076		0.365
$TiF_6^{2-} + 4e^- \rightleftharpoons Ti(s) + 6F^-$	-1.191		
$Ti^{2+} + 2e^- \rightleftharpoons Ti(s)$	-1.60		-0.16
Tungsten			
$W(CN)_8^{3-} + e^- \rightleftharpoons W(CN)_8^{4-}$	0.457		
$W^{6+} + e^- \rightleftharpoons W^{5+}$	0.26	12 F HCl	
$WO_3(s) + 6H^+ + 6e^- \rightleftharpoons W(s) + 3H_2O$	-0.091		-0.389

Reaction	$E°$ (volts)		$dE°/dT$ (mV/K)
$W^{5+} + e^- \rightleftharpoons W^{4+}$	-0.3	12 F HCl	
$WO_2(s) + 2H_2O + 4e^- \rightleftharpoons W(s) + 4OH^-$	-0.982		-1.197
$WO_4^{2-} + 4H_2O + 6e^- \rightleftharpoons W(s) + 8OH^-$	-1.060		-1.36
Uranium			
$UO_2^+ + 4H^+ + e^- \rightleftharpoons U^{4+} + 2H_2O$	0.39		-3.4
$UO_2^{2+} + 4H^+ + 2e^- \rightleftharpoons U^{4+} + 2H_2O$	0.273		-1.582
$UO_2^{2+} + e^- \rightleftharpoons UO_2^+$	0.16		0.2
$U^{4+} + e^- \rightleftharpoons U^{3+}$	-0.577		1.61
$U^{3+} + 3e^- \rightleftharpoons U(s)$	-1.642		0.16
Vanadium			
$VO_2^+ + 2H^+ + e^- \rightleftharpoons VO^{2+} + H_2O$	1.001		-0.901
$VO^{2+} + 2H^+ + e^- \rightleftharpoons V^{3+} + H_2O$	0.337		-1.6
$V^{3+} + e^- \rightleftharpoons V^{2+}$	-0.255		1.5
$V^{2+} + 2e^- \rightleftharpoons V(s)$	-1.125		-0.11
Xenon			
$H_4XeO_6 + 2H^+ + 2e^- \rightleftharpoons XeO_3 + 3H_2O$	2.38		0.0
$XeF_2 + 2H^+ + 2e^- \rightleftharpoons Xe(g) + 2HF$	2.2		
$XeO_3 + 6H^+ + 6e^- \rightleftharpoons Xe(g) + 3H_2O$	2.1		-0.34
Ytterbium			
$Yb^{3+} + 3e^- \rightleftharpoons Yb(s)$	-2.19		0.363
$Yb^{2+} + 2e^- \rightleftharpoons Yb(s)$	-2.76		-0.16
Yttrium			
$Y^{3+} + 3e^- \rightleftharpoons Y(s)$	-2.38		0.034
Zinc			
$ZnOH^+ + H^+ + 2e^- \rightleftharpoons Zn(s) + H_2O$	-0.497		0.03
$Zn^{2+} + 2e^- \rightleftharpoons Zn(s)$	-0.762		0.119
$Zn^{2+} + 2e^- + Hg \rightleftharpoons Zn(in\ Hg)$	-0.801		
$Zn(NH_3)_4^{2+} + 2e^- \rightleftharpoons Zn(s) + 4NH_3$	-1.04		
$ZnCO_3(s) + 2e^- \rightleftharpoons Zn(s) + CO_3^{2-}$	-1.06		
$Zn(OH)_3^- + 2e^- \rightleftharpoons Zn(s) + 3OH^-$	-1.183		
$Zn(OH)_4^{2-} + 2e^- \rightleftharpoons Zn(s) + 4OH^-$	-1.199		
$Zn(OH)_2(s) + 2e^- \rightleftharpoons Zn(s) + 2OH^-$	-1.249		-0.999
$ZnO(s) + H_2O + 2e^- \rightleftharpoons Zn(s) + 2OH^-$	-1.260		-1.160
$ZnS(s) + 2e^- \rightleftharpoons Zn(s) + S^{2-}$	-1.405		
Zirconium			
$Zr^{4+} + 4e^- \rightleftharpoons Zr(s)$	-1.45		0.67
$ZrO_2(s) + 4H^+ + 4e^- \rightleftharpoons Zr(s) + 2H_2O$	-1.473		-0.344

Reacting ions	$\log \beta_1$	$\log \beta_2$	$\log \beta_3$	$\log \beta_4$	$\log \beta_5$	$\log \beta_6$	Temperature (°C)	Ionic strength (μ, M)
Acetate, $CH_3CO_2^-$								
Ag^+	0.73	0.64					25	0
Ca^{2+}	1.24						25	0
Cd^{2+}	1.93	3.15					25	0
Cu^{2+}	2.23	3.63					25	0
Fe^{2+}	1.82						25	0.5
Fe^{3+}	3.38	7.1	9.7				20	0.1
Mg^{2+}	1.25						25	0
Mn^{2+}	1.40						25	0
Na^+	−0.18						25	0
Ni^{2+}	1.43						25	0
Zn^{2+}	1.28	2.09					20	0.1
Ammonia, NH_3								
Ag^+	3.31	7.23					25	0
Cd^{2+}	2.51	4.47	5.77	6.56			30	0
Co^{2+}	1.99	3.50	4.43	5.07	5.13	4.39	30	0
Cu^{2+}	3.99	7.33	10.06	12.03			30	0
Hg^{2+}	8.8	17.5	18.50	19.28			22	2
Ni^{2+}	2.67	4.79	6.40	7.47	8.10	8.01	30	0
Zn^{2+}	2.18	4.43	6.74	8.70			30	0
Cyanide, CN^-								
Ag^+		20	21				20	0
Cd^{2+}	5.18	9.60	13.92	17.11			25	?
Cu^+		24	28.6	30.3			25	0
Ni^{2+}			30				25	0
Tl^{3+}	13.21	26.50	35.17	42.61			25	4
Zn^{2+}		11.07	16.05	19.62			25	0
Ethylenediamine (1,2-diaminoethane), $H_2NCH_2CH_2NH_2$								
Ag^+	4.70	7.70	9.7				20	0.1
Cd^{2+}	5.69	10.36	12.80				25	0.5
Cu^{2+}	10.66	19.99					20	0
Hg^{2+}	14.3	23.3	23.2				25	0.1
Ni^{2+}	7.52	13.84	18.33				20	0
Zn^{2+}	5.77	10.83	14.11				20	0
Hydroxide, OH^-								
Ag^+	2.0	3.99					25	0
Al^{3+}	9.00	17.9	25.2	33.3			25	0
	$\log \beta_{22} = 20.3 \quad \log \beta_{43} = 42.1$							
Ba^{2+}	0.64						25	0
Bi^{3+}	12.9	23.5	33.0	34.8			25	0
	$\log \beta_{126} = 165.3 \ (\mu = 1)$							
Be^{2+}	8.6	14.4	18.8	18.6			25	0
	$\log \beta_{12} = 10.82 \ (\mu = 0.1) \quad \log \beta_{33} = 32.54 \ (\mu = 0.1) \quad \log \beta_{65} = 66.24 \ (\mu = 3) \quad \log \beta_{86} = 85 \ (\mu = 0)$							
Ca^{2+}	1.30						25	0
Cd^{2+}	3.9	7.7	10.3 ($\mu = 3$)	12.0 ($\mu = 3$)			25	0
	$\log \beta_{12} = 4.6 \quad \log \beta_{44} = 23.2$							
Ce^{3+}	4.9						25	3
	$\log \beta_{22} = 12.4 \quad \log \beta_{53} = 35.1$							
Co^{2+}	4.3	9.2	10.5	9.7			25	0
	$\log \beta_{12} = 3 \quad \log \beta_{44} = 25.5$							
Co^{3+}	13.52						25	3

*The overall (cumulative) formation constant, β_n, is the equilibrium constant for the reaction $M + nL \rightleftharpoons ML_n$: $\beta_n = [ML_n]/([M][L]^n)$. β_n is related to stepwise formation constants (K_i) by $\beta_n = K_1K_2 \ldots K_n$ (Box 6-2). β_{nm} is the cumulative formation constant for the reaction $mM + nL \rightleftharpoons M_mL_n$: $\beta_{nm} = [M_mL_n]/([M]^m[L]^n)$. The subscript n refers to the ligand and m refers to the metal. Data from L. G. Sillén and A. E. Martell, *Stability Constants of Metal-Ion Complexes* (London: The Chemical Society. Special Publications No. 17 and 25, 1964 and 1971); and A. E. Martell, R. M. Smith, and R. J. Motekaitis, *NIST Critical Stability Constants of Metal Complexes Database 46* (Gaithersburg, MD: National Institute of Standards and Technology, 2001).

Reacting ions	$\log \beta_1$	$\log \beta_2$	$\log \beta_3$	$\log \beta_4$	$\log \beta_5$	$\log \beta_6$	Temperature (°C)	Ionic strength (μ, M)
Cr^{2+}	8.5						25	1
Cr^{3+}	10.34	17.3 ($\mu = 0.1$)					25	0
	$\log \beta_{22} = 24.0$ ($\mu = 1$)	$\log \beta_{43} = 37.0$ ($\mu = 1$)	$\log \beta_{44} = 50.7$ ($\mu = 2$)					
Cu^{2+}	6.5	11.8	14.5 ($\mu = 1$)	15.6 ($\mu = 1$)			25	0
	$\log \beta_{12} = 8.2$ ($\mu = 3$)	$\log \beta_{22} = 17.4$	$\log \beta_{43} = 35.2$					
Fe^{2+}	4.6	7.5	13	10			25	0
Fe^{3+}	11.81	23.4		34.4			25	0
	$\log \beta_{22} = 25.14$	$\log \beta_{43} = 49.7$						
Ga^{3+}	11.4	22.1	31.7	39.4			25	0
Gd^{3+}	4.9						25	3
	$\log \beta_{22} = 14.14$							
Hf^{4+}	13.7				52.8		25	0
Hg_2^{2+}	8.7						25	0.5
	$\log \beta_{12} = 11.5$ ($\mu = 3$)	$\log \beta_{45} = 48.24$ ($\mu = 3$)						
Hg^{2+}	10.60	21.8	20.9				25	0
	$\log \beta_{12} = 10.7$	$\log \beta_{33} = 35.6$						
In^{3+}	10.1	20.2	29.5	33.8			25	0
	$\log \beta_{22} = 23.2$ ($\mu = 3$)	$\log \beta_{44} = 47.8$ ($\mu = 0.1$)	$\log \beta_{64} = 43.1$ ($\mu = 0.1$)					
La^{3+}	5.5						25	0
	$\log \beta_{22} = 10.7$ ($\mu = 3$)	$\log \beta_{95} = 38.4$						
Li^+	0.36						25	0
Mg^{2+}	2.6	−0.3 ($\mu = 3$)					25	0
	$\log \beta_{44} = 18.1$ ($\mu = 3$)							
Mn^{2+}	3.4			7.7			25	0
	$\log \beta_{12} = 6.8$	$\log \beta_{32} = 18.1$						
Na^+	0.1						25	0
Ni^{2+}	4.1	9	12				25	0
	$\log \beta_{12} = 4.7$ ($\mu = 1$)	$\log \beta_{44} = 28.3$						
Pb^{2+}	6.4	10.9	13.9				25	0
	$\log \beta_{12} = 7.6$	$\log \beta_{43} = 32.1$	$\log \beta_{44} = 36.0$	$\log \beta_{86} = 68.4$				
Pd^{2+}	13.0	25.8					25	0
Rh^{3+}	10.67						25	2.5
Sc^{3+}	9.7	18.3	25.9	30			25	0
	$\log \beta_{22} = 22.0$	$\log \beta_{53} = 53.8$						
Sn^{2+}	10.6	20.9	25.4				25	0
	$\log \beta_{22} = 23.2$	$\log \beta_{43} = 49.1$						
Sr^{2+}	0.82						25	0
Th^{4+}	10.8	21.1		41.1 ($\mu = 3$)			25	0
	$\log \beta_{22} = 23.6$ ($\mu = 3$)	$\log \beta_{32} = 33.8$ ($\mu = 3$)	$\log \beta_{53} = 53.7$ ($\mu = 3$)					
Ti^{3+}	12.7						25	0
	$\log \beta_{22} = 24.6$ ($\mu = 1$)							
Tl^+	0.79	−0.8 ($\mu = 3$)					25	0
Tl^{3+}	13.4	26.6	38.7	41.0			25	0
U^{4+}	13.4						25	0
VO^{2+}	8.3						25	0
	$\log \beta_{22} = 21.3$							
Y^{3+}	6.3						25	0
	$\log \beta_{22} = 13.8$	$\log \beta_{53} = 38.4$						
Zn^{2+}	5.0	10.2	13.9	15.5			25	0
	$\log \beta_{12} = 5.5$ ($\mu = 3$)	$\log \beta_{44} = 27.9$ ($\mu = 3$)						
Zr^{4+}	14.3				54.0		25	0
	$\log \beta_{43} = 55.4$	$\log \beta_{84} = 106.0$						
Nitrilotriacetate, $N(CH_2CO_2^-)_3$								
Ag^+	5.16						20	0.1
Al^{3+}	9.5						20	0.1
Ba^{2+}	4.83						20	0.1
Ca^{2+}	6.46						20	0.1
Cd^{2+}	10.0	14.6					20	0.1

(Continued)

Reacting ions	$\log \beta_1$	$\log \beta_2$	$\log \beta_3$	$\log \beta_4$	$\log \beta_5$	$\log \beta_6$	Temperature (°C)	Ionic strength (μ, M)
Nitrilotriacetate, $N(CH_2CO_2^-)_3$—*(continued)*								
Co^{2+}	10.0	13.9					20	0.1
Cu^{2+}	11.5	14.8					20	0.1
Fe^{3+}	15.91	24.61					20	0.1
Ga^{3+}	13.6	21.8					20	0.1
In^{3+}	16.9						20	0.1
Mg^{2+}	5.46						20	0.1
Mn^{2+}	7.4						20	0.1
Ni^{2+}	11.54						20	0.1
Pb^{2+}	11.47						20	0.1
Tl^+	4.75						20	0.1
Zn^{2+}	10.44						20	0.1
Oxalate, $^-O_2CCO_2^-$								
Al^{3+}			15.60				20	0.1
Ba^{2+}	2.31						18	0
Ca^{2+}	1.66	2.69					25	1
Cd^{2+}	3.71						20	0.1
Co^{2+}	4.69	7.15					25	0
Cu^{2+}	6.23	10.27					25	0
Fe^{3+}	7.54	14.59	20.00				?	0.5
Ni^{2+}	5.16	6.5					25	0
Zn^{2+}	4.85	7.6					25	0
1,10-Phenanthroline,								
Ag^+	5.02	12.07					25	0.1
Ca^{2+}	0.7						20	0.1
Cd^{2+}	5.17	10.00	14.25				25	0.1
Co^{2+}	7.02	13.72	20.10				25	0.1
Cu^{2+}	8.82	15.39	20.41				25	0.1
Fe^{2+}	5.86	11.11	21.14				25	0.1
Fe^{3+}			14.10				25	0.1
Hg^{2+}		19.65	23.4				20	0.1
Mn^{2+}	4.50	8.65	12.70				25	0.1
Ni^{2+}	8.0	16.0	23.9				25	0.1
Zn^{2+}	6.30	11.95	17.05				25	0.1

APPENDIX J Logarithm of the Formation Constant for the Reaction $M(aq) + L(aq) \rightleftharpoons ML(aq)$*

M	F^-	Cl^-	Br^-	I^-	NO_3^-	ClO_4^-	IO_3^-	SCN^-	SO_4^{2-}	CO_3^{2-}
Li^+	0.23	—	—	—	—	—	—	—	0.64	—
Na^+	−0.2	−0.5	—	—	−0.55	−0.7	−0.4	—	0.72	1.27
K^+	−1.2[a]	−0.5	—	−0.4	−0.19	−0.03	−0.27	—	0.85	—
Rb^+	—	−0.4	—	0.04	−0.08	0.15	−0.19	—	0.60	—
Cs^+	—	−0.2	0.03	−0.03	−0.02	0.23	−0.11	—	0.3	—
Ag^+	0.4	3.31	4.6	6.6	−0.1	−0.1	0.63	4.8	1.3	—
$(CH_3)_4N^+$	—	0.04	0.16	0.31	—	0.27	—	—	—	—
Mg^{2+}	2.05	0.6	−1.4[d]	—	—	—	0.72	−0.9[d]	2.23	2.92
Ca^{2+}	0.63	0.2[b]	—	—	0.5	—	0.89	—	2.36	3.20
Sr^{2+}	0.14	−0.22[a]	—	—	0.6	—	1.00	—	2.2	2.81
Ba^{2+}	−0.20	−0.44[a]	—	—	0.7	—	1.10	—	2.2	2.71
Zn^{2+}	1.3	0.4	−0.07	−1.5[d]	0.4	—	—	1.33	2.34	4.76
Cd^{2+}	1.2	1.98	2.15	2.28	0.5	—	0.51[a]	1.98	2.46	3.49[b]
Hg_2^{2+}	—	—	—	—	0.08[f]	—	—	—	1.30[f]	—
Hg^{2+}	1.03[f]	7.30	9.07[f]	12.87[f]	0.11[d]	—	—	9.64	1.34[f]	11.0[f]
Sn^{2+}	—	1.64	1.16	0.70[e]	0.44[a]	—	—	0.83[a]	—	—
Y^{3+}	4.81	−0.1[a]	−0.15[a]	—	—	—	—	−0.07[f]	3.47	8.2
La^{3+}	3.60	−0.1[a]	—	—	0.1[a]	—	—	0.12[a]	3.64	5.6[d]
In^{3+}	4.65	2.32[c]	2.01[c]	1.64[c]	0.18	—	—	3.15	1.85[a]	—

*Unless otherwise indicated, conditions are 25°C and $\mu = 0$.

a. $\mu = 1\ M$; b. $\mu = 0.1\ M$; c. $\mu = 0.7\ M$; d. $\mu = 3\ M$; e. $\mu = 4\ M$; f. $\mu = 0.5\ M$.

SOURCE: A. E. Martell, R. M. Smith, and R. J. Motekaitis, *NIST Critical Stability Constants of Metal Complexes Database 46* (Gaithersburg, MD: National Institute of Standards and Technology, 2001).

The table in this appendix recommends primary standards for many elements. An *elemental assay standard* must contain a known amount of the desired element. A *matrix matching standard* must contain extremely low concentrations of undesired impurities, such as the analyte. If you want to prepare 10 ppm Fe in 10% aqueous NaCl, the NaCl must not contain significant Fe impurity, because the impurity would then have a higher concentration than the deliberately added Fe.

Rather than using compounds in the table, many people purchase certified solutions whose concentrations are traceable to standards from the National Institute of Standards and Technology (NIST) or other national institutes of standards. By *NIST traceable*, we mean that the solution has been prepared from a standard material certified by NIST or that it has been compared with a NIST standard by a reliable analytical procedure.

Manufacturers frequently indicate elemental purity by some number of 9s. This deceptive nomenclature is based on the measurement of certain impurities. For example, 99.999% (five 9s) pure Al is certified to contain ≤0.001% *metallic* impurities, based on the analysis of other metals present. However, C, H, N, and O are not measured. The Al might contain 0.1% Al_2O_3 and still be "five 9s pure." For the most accurate work, the dissolved gas content in solid elements may also be a source of error.

Carbonates, oxides, and other compounds may not possess the expected stoichiometry. For example, TbO_2 will have a high Tb content if some Tb_4O_7 is present. Ignition in an O_2 atmosphere may be helpful, but the final stoichiometry is never guaranteed. Carbonates may contain traces of bicarbonate, oxide, and hydroxide. Firing in a CO_2 atmosphere may improve the stoichiometry. Sulfates may contain some HSO_4^-. Some chemical analysis may be required to ensure that you know what you are really working with.

Most metal standards dissolve in 6 M HCl or HNO_3 or a mixture of the two, possibly with heating. Frothing accompanies dissolution of metals or carbonates in acid, so vessels should be loosely covered by a watchglass or Teflon lid to prevent loss of material. Concentrated HNO_3 (16 M) may *passivate* some metals, forming an insoluble oxide coat that prevents dissolution. If you have a choice between using a bulk element or a powder as standards, the bulk form is preferred because it has a smaller surface area on which oxides can form and impurities can be adsorbed. After a pure metal to be used as a standard is cut, it should be etched ("pickled") in a dilute solution of the acid in which it will be dissolved to remove surface oxides and contamination from the cutter. The metal is then washed well with water and dried in a vacuum desiccator.

Dilute solutions of metals are best prepared in Teflon or plastic vessels, because glass is an ion exchanger that can replace analyte species. Specially cleaned glass vials are commercially available for trace organic analysis. Volumetric dilutions are rarely more accurate than 0.1%, so gravimetric dilutions are required for greater accuracy. Of course, weights should be corrected for buoyancy with Equation 2-1. Evaporation of standard solutions is a source of error that is prevented if the mass of the reagent bottle is recorded after each use. If the mass changes between uses, the contents are evaporating.

Calibration standards

Element	Source[a]	Purity	Comments[b]
Li	SRM 924 (Li_2CO_3)	100.05 ± 0.02%	E; dry at 200°C for 4 h.
	Li_2CO_3	five–six 9s	M; purity calculated from impurities. Stoichiometry unknown.
Na	SRM 919 or 2201 (NaCl)	99.9%	E; dry for 24 h over $Mg(ClO_4)_2$.
	Na_2CO_3	three 9s	M; purity based on metallic impurities.
K	SRM 918 (KCl)	99.9%	E; dry for 24 h over $Mg(ClO_4)_2$.
	SRM 999 (KCl)	52.435 ± 0.004% K	E; ignite at 500°C for 4 h.
	K_2CO_3	five–six 9s	M; purity based on metallic impurities.
Rb	SRM 984 (RbCl)	99.90 ± 0.02%	E; hygroscopic. Dry for 24 h over $Mg(ClO_4)_2$.
	Rb_2CO_3		M
Cs	Cs_2CO_3		M
Be	metal	three 9s	E, M; purity based on metallic impurities.
Mg	SRM 929	100.1 ± 0.4%	E; magnesium gluconate clinical standard.
		5.403 ± 0.022% Mg	Dry for 24 h over $Mg(ClO_4)_2$.
	metal	five 9s	E; purity based on metallic impurities.
Ca	SRM 915 ($CaCO_3$)	three 9s	E; use without drying.
	$CaCO_3$	five 9s	E, M; dry at 200°C for 4 h in CO_2. User must determine stoichiometry.
Sr	SRM 987 ($SrCO_3$)	99.8%	E; ignite to establish stoichiometry. Dry at 110°C for 1 h.
	$SrCO_3$	five 9s	M; up to 1% off stoichiometry. Ignite to establish stoichiometry. Dry at 200°C for 4 h.
Ba	$BaCO_3$	four–five 9s	M; dry at 200°C for 4 h.

Transition metals: Use pure metals (usually ≥four 9s) for elemental and matrix standards. Assays are based on impurities and do not include dissolved gases.

Lanthanides: Use pure metals (usually ≥four 9s) for elemental standards and oxides as matrix standards. Oxides may be difficult to dry and stoichiometry is not certain.

a. SRM is the National Institute of Standards and Technology designation for a Standard Reference Material.

b. E means elemental assay standard; M means matrix matching standard.

SOURCE: J. R. Moody, R. R. Greenberg, K. W. Pratt, and T. C. Rains, "Recommended Inorganic Chemicals for Calibration," Anal. Chem. **1988**, 60, 1203A.

Calibration standards (*continued*)

Element	Source[a]	Purity	Comments[b]
B	SRM 951 (H_3BO_3)	100.00 ± 0.01	E; expose to room humidity (~35%) for 30 min before use.
Al	metal	five 9s	E, M; SRM 1257 Al metal available.
Ga	metal	five 9s	E, M; SRM 994 Ga metal available.
In	metal	five 9s	E, M
Tl	metal	five 9s	E, M; SRM 997 Tl metal available.
C			No recommendation.
Si	metal	six 9s	E, M; SRM 990 SiO_2 available.
Ge	metal	five 9s	E, M
Sn	metal	six 9s	E, M; SRM 741 Sn metal available.
Pb	metal	five 9s	E, M; several SRMs available.
N	NH_4Cl	six 9s	E; can be prepared from $HCl + NH_3$.
	N_2	>three 9s	E
	HNO_3	six 9s	M; contaminated with NO_x. Purity based on impurities.
P	SRM 194 ($NH_4H_2PO_4$)	three 9s	E
	P_2O_5	five 9s	E, M; difficult to keep dry.
	H_3PO_4	four 9s	E; must titrate 2 hydrogens to be certain of stoichiometry.
As	metal	five 9s	E, M
	SRM 83d (As_2O_3)	99.9926 ± 0.0030 %	Redox standard. As assay not ensured.
Sb	metal	four 9s	E, M
Bi	metal	five 9s	E, M
O	H_2O	eight 9s	E, M; contains dissolved gases.
	O_2	>four 9s	E
S	element	six 9s	E, M; difficult to dry. Other sources are H_2SO_4, Na_2SO_4, and K_2SO_4. Stoichiometry must be proved (e.g., no SO_3^{2-} present).
Se	metal	five 9s	E, M; SRM 726 Se metal available.
Te	metal	five 9s	E, M
F	NaF	four 9s	E, M; no good directions for drying.
Cl	NaCl	four 9s	E, M; dry for 24 h over $Mg(ClO_4)_2$. Several SRMs (NaCl and KCl) available.
Br	KBr	four 9s	E, M; need to dry and demonstrate stoichiometry.
	Br_2	four 9s	E
I	sublimed I_2	six 9s	E
	KI	three 9s	E, M
	KIO_3	three 9s	Stoichiometry not ensured.

APPENDIX L   DNA and RNA

Genetic information is coded in the sequence of nucleotides in deoxyribonucleic acid (DNA), which is made of two long strands wrapped around each other to form a helix (Figure L-1). The backbone of each strand consists of alternating units of phosphate and the sugar, deoxyribose. One of four different nucleotide bases designated A (adenine), T (thymine), C (cytosine), and G (guanine) is attached to each sugar. The two strands of DNA are connected by hydrogen bonds between the nucleotide bases. A and T form two hydrogen bonds with each other. C and G form three hydrogen bonds. Figure L-2 shows that the two strands of DNA wind around each other to form a double helix.

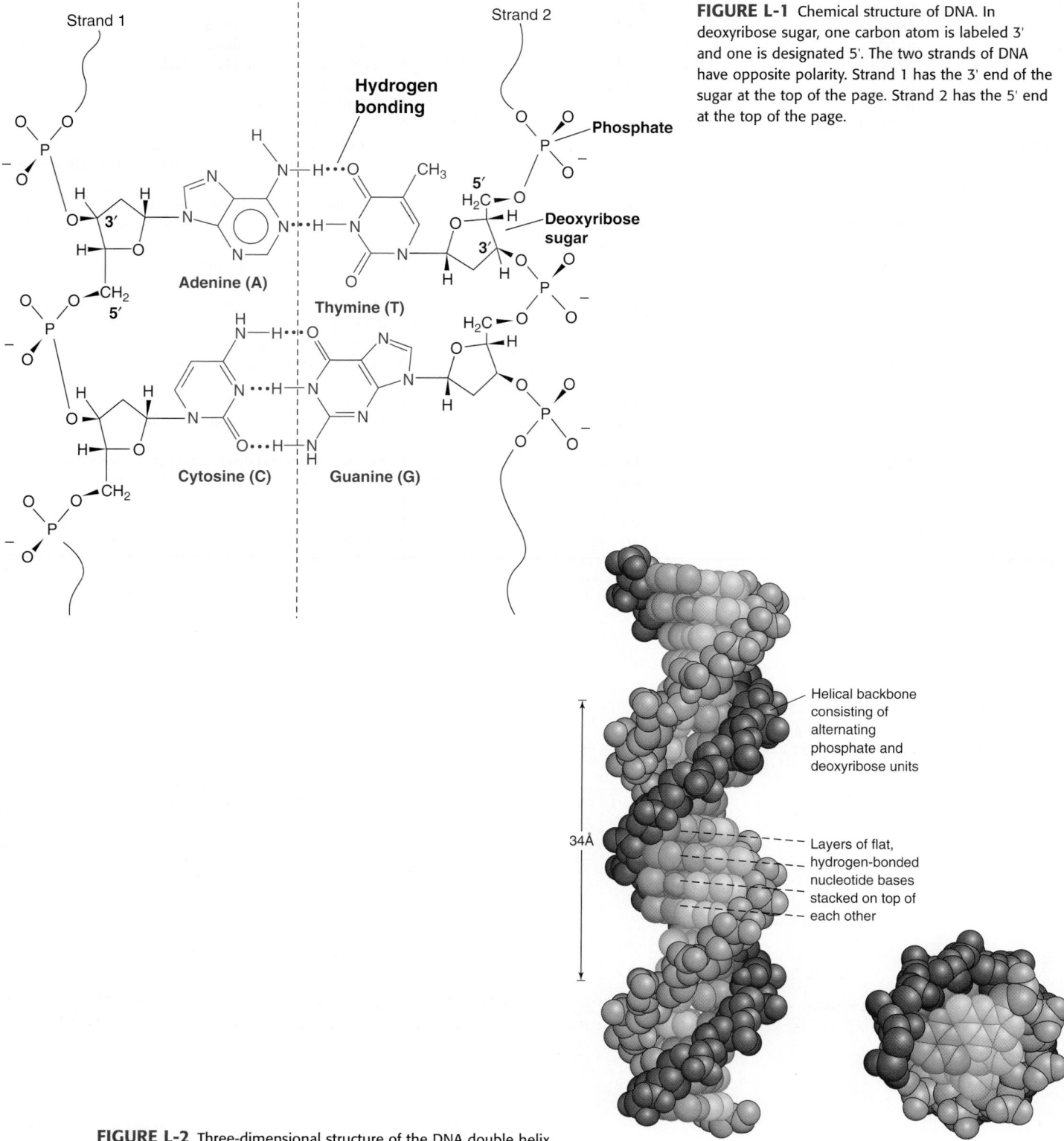

FIGURE L-1 Chemical structure of DNA. In deoxyribose sugar, one carbon atom is labeled 3′ and one is designated 5′. The two strands of DNA have opposite polarity. Strand 1 has the 3′ end of the sugar at the top of the page. Strand 2 has the 5′ end at the top of the page.

FIGURE L-2 Three-dimensional structure of the DNA double helix.
[J. M. Berg, J. L. Tymoczko, and L. Stryer, Biochemistry, 5th ed. (New York: W. H. Freeman & Co, 2002).]

View down the axis of the double helix

Before cell division, the two strands of DNA separate and each serves as a *template* for constructing a new strand with a complementary sequence (Figure L-3). For example, base A on the original strand will select base T for the complementary strand, and G on the original strand will select C for the complementary strand. The enzyme *DNA polymerase* synthesizes the new strand using one strand of DNA as a template and the four deoxyribonucleotide triphosphate building blocks, designated dATP, dTTP, dCTP, and dGTP, to create a complementary strand. During cell division, each daughter cell receives one full set of DNA from the parent cell.

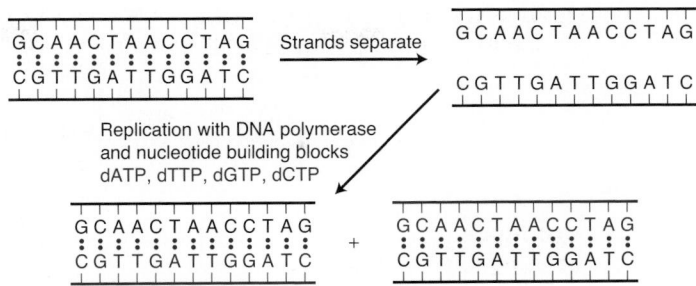

FIGURE L-3 For replication, DNA double helix separates and each strand is used as a template to synthesize a new strand shown in color.

The structure of ribonucleic acid (RNA) is similar to that of DNA, except that thymine (T) is replaced by uracil (U) and the sugar deoxyribose is replaced by ribose.

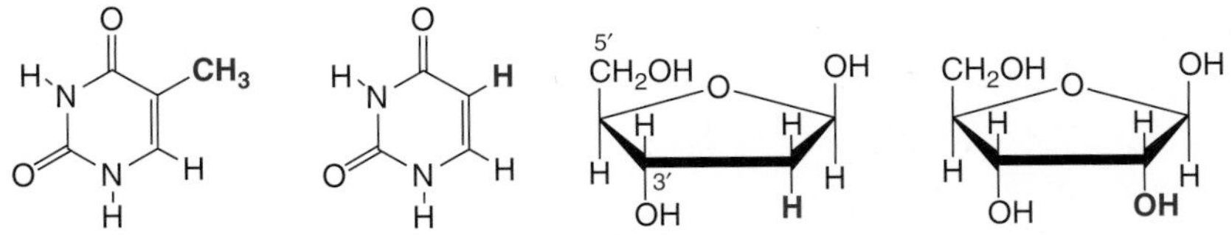

Thymine (T)
in DNA

Uracil (U)
in RNA

Deoxyribose
in DNA

Ribose
in RNA

DNA sequencing described at the opening of Chapter 15 is based on the release of H^+ when each nucleotide base is added to the 3′ end of the growing chain. The reaction drawn in Figure L-4 shows one H^+ from the 3′-OH of deoxyribose being produced for each nucleotide incorporated into the growing DNA strand.

Not exactly $1H^+$ per nucleotide base is released in the synthesis of DNA because of acid-base equilibria in Figure L-4. pK_a for release of the fourth (last) proton from the triphosphate group is ~7.0 at an ionic strength of $\mu = 0.1$ M (R. C. Phillips, S. J. Philip George, and R. J. Rutman, *J. Am. Chem. Soc.* **1966**, *88*, 2631). pK_a for release of the fourth (last) proton from pyrophosphate is 8.25 at $\mu = 0.1$ M (Appendix G). At pH 7.5 in the DNA sequencing chip at the opening of Chapter 15, the reactant nucleotide triphosphate binds ~$0.24H^+$ and the product pyrophosphate binds ~$0.85H^+$. Therefore $(0.85 - 0.24) \approx 60\%$ of the $1H^+$ released in the chemical reaction in Figure L-4 is taken up by the product pyrophosphate. Net release of H^+ depends on pH and ionic strength.

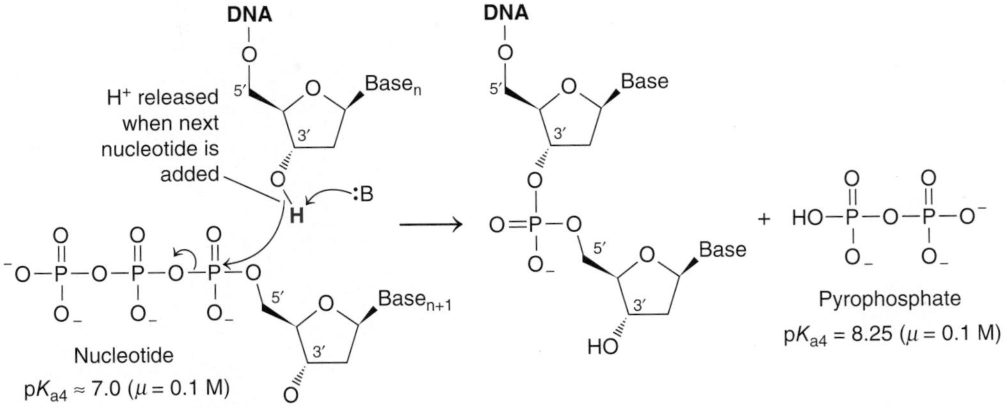

FIGURE L-4 Addition of one nucleotide to the growing chain of DNA releases H^+ from deoxyribose at the end of the existing chain. Base ":B" that removes the H^+ atom is part of DNA polymerase. Some H^+ released from the deoxyribose reactant binds to the pyrophosphate product, so net production is less than $1H^+$ per nucleotide.

SOLUTIONS TO EXERCISES

Chapter 1

1-A. **(a)** $\dfrac{(25.00 \text{ mL})(0.791\ 4 \text{ g/mL})/(32.042 \text{ g/mol})}{0.500\ 0 \text{ L}} = 1.235 \text{ M}$

(b) 500.0 mL of solution weighs (1.454 g/mL) × (500.0 mL) = 727.0 g and contains 25.00 mL (= 19.78 g) of methanol. The mass of chloroform in 500.0 mL must be 727.0 − 19.78 = 707.2 g. The molality of methanol is

$$\text{Molality} = \frac{\text{mol methanol}}{\text{kg chloroform}}$$

$$= \frac{(19.78 \text{ g})/(32.042 \text{ g/mol})}{0.707\ 2 \text{ kg}} = 0.872\ 9\ m$$

If you keep all the numbers in your calculator, the answer is 0.873 1 *m*. You will find small discrepancies due to intermediate round-offs in many answers in this book.

1-B. **(a)**

$$\left(\frac{48.0 \text{ g HBr}}{100.0 \text{ g solution}}\right)\left(1.50\ \frac{\text{g solution}}{\text{mL solution}}\right) = \left(\frac{0.720 \text{ g HBr}}{\text{mL solution}}\right) = \left(\frac{720 \text{ g HBr}}{\text{L solution}}\right)$$

$$\text{Formal concentration} = \frac{720 \text{ g HBr/L}}{80.912 \text{ g/mol}} = 8.90 \text{ M}$$

(b) $\dfrac{36.0 \text{ g HBr}}{0.480 \text{ g HBr/g solution}} = 75.0 \text{ g solution}$

(c) 233 mmol = 0.233 mol

$\dfrac{0.233 \text{ mol}}{8.90 \text{ mol/L}} = 0.026\ 2 \text{ L} = 26.2 \text{ mL}$

(d) $M_{conc} \cdot V_{conc} = M_{dil} \cdot V_{dil}$

$(8.90 \text{ M}) \cdot (x \text{ mL}) = (0.160 \text{ M}) \cdot (250 \text{ mL}) \Rightarrow x = 4.49 \text{ mL}$

1-C. **(a)** Each mol of $Ca(NO_3)_2$ (FM 164.088) contains 2 mol NO_3^- (FM 62.005), so the fraction of mass that is nitrate is

$$\left(\frac{2 \text{ mol } NO_3^-}{\text{mol } Ca(NO_3^-)_2}\right)\left(\frac{62.005 \text{ g } NO_3^-/\text{mol } NO_3^-}{164.088 \text{ g } Ca(NO_3^-)_2/\text{mol } Ca(NO_3^-)_2}\right)$$

$$= 0.755\ 8\ \frac{\text{g } NO_3^-}{\text{g } Ca(NO_3)_2}$$

If $Ca(NO_3)_2 = 12.6$ ppm (= 12.6 μg $Ca(NO_3)_2$/g solution), $NO_3^- =$ (0.755 8)(12.6 ppm) = 9.52 ppm because 1 μg $Ca(NO_3)_2$ contains 0.755 8 μg NO_3^-.

(b) 0.144 mM $Ca(NO_3)_2$

$= \left(1.44 \times 10^{-4}\ \dfrac{\text{mol } Ca(NO_3)_2}{\text{L}}\right)\left(164.088\ \dfrac{\text{g } Ca(NO_3)_2}{\text{mol } Ca(NO_3)_2}\right) = 0.023\ 63 \text{ g/L}$

Assuming that the density of solution is close to 1.00 g/mL, the concentration of $Ca(NO_3)_2$ is 0.023 63 g/(1 000 g solution).

$$\text{ppm} = \frac{\text{mass of substance}}{\text{mass of sample}} \times 10^6 = \frac{0.023\ 63 \text{ g } Ca(NO_3)_2}{1\ 000 \text{ g solution}} \times 10^6 = 23.6 \text{ ppm}$$

(c) We found in **(a)** that the mass fraction of nitrate in calcium nitrate is 0.755 8. Therefore, if a solution contains 23.6 ppm $Ca(NO_3)_2$, it contains (0.755 8)(23.6 ppm) = 17.9 ppm NO_3^-.

1-D. mol $OBr^- = (0.005\ 00 \text{ L})(0.623 \text{ M}) = 3.11_5$ mmol
mol $NH_3 = (183 \times 10^{-6} \text{ L})(14.8 \text{ M}) = 2.70_8$ mmol

The reaction requires 2 mol NH_3 for 3 mol OBr^-. Therefore, 3.11_5 mmol OBr^- require $\left(\dfrac{2 \text{ mmol } NH_3}{3 \text{ mmol } OBr^-}\right)(3.11_5 \text{ mmol } OBr^-) = 2.077$ mmol NH_3

There is more than enough NH_3 for the reaction, so NaOBr is the limiting reagent. NH_3 left over $= 2.70_8$ mmol $− 2.07_7$ mmol $= 0.63_1$ mmol.

Chapter 2

2-A. **(a)** At 15°C, water density = 0.999 102 6 g/mL.

$$m = \frac{(5.397\ 4 \text{ g})\left(1 - \dfrac{0.001\ 2 \text{ g/mL}}{8.0 \text{ g/mL}}\right)}{\left(1 - \dfrac{0.001\ 2 \text{ g/mL}}{0.999\ 102\ 6 \text{ g/mL}}\right)} = 5.403\ 1 \text{ g}$$

(b) At 25°C, water density = 0.997 047 9 g/mL and $m = 5.403\ 1$ g.

2-B. Use Equation 2-1 with $m' = 0.296\ 1$ g, $d_a = 0.001\ 2$ g/mL, $d_w = 8.0$ g/mL, and $d = 5.24$ g/mL $\Rightarrow m = 0.296\ 3$ g.

2-C. $\dfrac{c'}{d'} = \dfrac{c}{d}$

Let the primes stand for 16°C:

$$\Rightarrow \frac{c' \text{ at } 16°C}{0.998\ 946\ 0 \text{ g/mL}} = \frac{0.051\ 38 \text{ M}}{0.997\ 299\ 5 \text{ g/mL}}$$

$\Rightarrow c'$ at 16°C = 0.051 46 M

2-D. The stock solution is approximately 50 mM. To obtain ~1 mM requires a 1/50 dilution. This can be done by diluting 2 mL up to 100 mL. Pipet 2 mL of stock solution into a 100 mL volumetric flask and dilute to volume with water. The exact molarity will be 51.38 mM/50 = 1.028 mM. If the molarity calculation is not obvious, you could use the dilution formula 1-3:

$$M_{conc} \cdot V_{conc} = M_{dil} \cdot V_{dil}$$

$(51.38 \text{ mM})(2 \text{ mL}) = (x \text{ mM})(100 \text{ mL}) \Rightarrow x = 1.028 \text{ mM}$

A slightly better procedure is to dilute 5 mL of stock solution up to 250 mL because the 5 mL pipet has less relative uncertainty than the 2-mL pipet and the 250-mL flask has less relative uncertainty than the 100-mL flask.

To obtain ~2 mM solution requires twice as much stock solution, so dilute 4 mL up to 100 mL or dilute 10 mL up to 250 mL. Molarity = (4 mL)(51.38 mM)/(100 mL) = 2.055 mM.

To obtain ~3 mM solution requires three times as much stock solution, so dilute 6 mL (2 deliveries of 3 mL) up to 100 mL or dilute 15 mL up to 250 mL. Molarity = (6 mL)(51.38 mM)/(100 mL) = 3.083 mM.

To obtain ~4 mM solution requires four times as much stock solution, so dilute 8 mL (2 deliveries of 4 mL) up to 100 mL or dilute 20 mL up to 250 mL. Molarity = (8 mL)(51.38 mM)/(100 mL) = 4.110 mM.

2-E. Column 3 of Table 2-7 tells us that water occupies 1.003 3 mL/g at 22°C. Therefore, (15.569 g) × (1.003 3 mL/g) = 15.620 mL.

Chapter 3

3-A. **(a)** 12.529 6 ± 0.000 3 g
$\underline{-12.437\ 2 \pm 0.000\ 3 \text{ g}}$
0.092 4 g ← 3 significant digits

(b) uncertainty $= \sqrt{0.000\ 3^2 + 0.000\ 3^2} = 0.000\ 4_2$

Mass of precipitate = 0.092 4 ± 0.000 4 (±0.5%) g

3-B. **(a)** $[12.41\ (\pm 0.09) \div 4.16\ (\pm 0.01)] \times 7.068\ 2\ (\pm 0.000\ 4)$

$$= \frac{12.41(\pm 0.725\%) \times 7.068\ 2(\pm 0.005\ 7\%)}{4.16(\pm 0.240\%)}$$

$= 21.086(\pm 0.764\%)$ (because $\sqrt{0.725^2 + 0.005\ 7^2 + 0.240^2} = 0.764$)

$= 21.0_9\ (\pm 0.1_6)$ or $21.1\ (\pm 0.2)$

$$\text{Relative uncertainty} = \frac{0.1_6}{21.0_9} \times 100 = 0.8\%$$

(b) $[3.26\ (\pm 0.10) \times 8.47\ (\pm 0.05)] - 0.18\ (\pm 0.06)$

$= [3.26\ (\pm 3.07\%) \times 8.47\ (\pm 0.59\%)] - 0.18\ (\pm 0.06)$

$= [27.612\ (\pm 3.12\%)] - 0.18\ (\pm 0.06)$

$= [27.612\ (\pm 0.863)] - 0.18\ (\pm 0.06)$

$= [27.4_3\ (\pm 0.8_6)]$ or $27.4\ (\pm 0.9)$; relative uncertainty = $3._2\%$

(c) $6.843 (\pm 0.008) \times 10^4 \div \underbrace{[2.09 (\pm 0.04) - 1.63 (\pm 0.01)]}_{\text{Combine absolute uncertainties}}$

$= \underbrace{6.843 (\pm 0.008) \times 10^4 \div [0.46 (\pm 0.041\ 2)]}_{\text{Combine relative uncertainties}}$

$= 6.843 (\pm 0.117\%) \times 10^4 \div [0.46 (\pm 8.96\%)] = 1.49 (\pm 8.96\%) \times 10^5$
$= 1.49 (\pm 0.1_3) \times 10^5$; relative uncertainty $= 9._0\%$

(d) $\%e_y = \frac{1}{2}\%e_x = \frac{1}{2}\left(\frac{0.08}{3.24} \times 100\right) = 1.235\%$

$(3.24 \pm 0.08)^{1/2} = 1.80 \pm 1.235\%$
$\qquad\qquad = 1.80 \pm 0.02_2 \ (\pm 1._2\%)$

(e) $\%e_y = 4\%e_x = 4\left(\frac{0.08}{3.24} \times 100\right) = 9.877\%$

$(3.24 \pm 0.08)^4 = 110.20 \pm 9.877\%$
$\qquad\qquad = 1.1_0 \ (\pm 0.1_1) \times 10^2 \ (\pm 9._9\%)$

(f) $e_y = 0.434\ 29\ \dfrac{e_x}{x} = 0.434\ 29\left(\dfrac{0.08}{3.24}\right) = 0.010\ 7$

$\log (3.24 \pm 0.08) = 0.510\ 5 \pm 0.010\ 7$
$\qquad\qquad\qquad = 0.51 \pm 0.01\ (\pm 2._1\%)$

(g) $\dfrac{e_y}{y} = 2.302\ 6\ e_x = 2.302\ 6\ (0.08) = 0.184$

$10^{3.24 \pm 0.08} = 1.74 \times 10^3 \pm 18.4\%$
$\qquad\qquad = 1.7_4\ (\pm 0.3_2) \times 10^3\ (\pm 18\%)$

3-C. **(a)** 2.000 L of 0.169 M NaOH (FM = 39.997) requires
0.338 mol = 13.52 g NaOH.

$\dfrac{13.52 \text{ g NaOH}}{0.534 \text{ g NaOH/g solution}} = 25.32$ g solution

$\dfrac{25.32 \text{ g solution}}{1.52 \text{ g solution/mL solution}} = 16.6_6$ mL

(b) Molarity =

$$\dfrac{[16.66\,(\pm 0.10)\text{ mL}]\left[1.52(\pm 0.01)\dfrac{\text{g solution}}{\text{mL}}\right] \times \left[0.534(\pm 0.004)\dfrac{\text{g NaOH}}{\text{g solution}}\right]}{\left(39.997\ 1\dfrac{\text{g NaOH}}{\text{mol}}\right)(2.000 \text{ L})}$$

Because the relative errors in formula mass and final volume are negligible
(≈ 0), we can write

Relative error in molarity $= \sqrt{\left(\dfrac{0.10}{16.66}\right)^2 + \left(\dfrac{0.01}{1.52}\right)^2 + \left(\dfrac{0.004}{0.534}\right)^2} = 1.16\%$

Molarity $= 0.169\ (\pm 0.002)$

3-D. Use the function $y = 10^x$, in which $y = [\text{H}^+]$ and $x = -\text{pH}$. The
uncertainty is $e_y/y = 2.302\ 6\ e_x$.

$[\text{H}^+] = 10^{-\text{pH}} = 10^{-4.44} = 3.63 \times 10^{-5}\text{ M}$
$\dfrac{e_{[\text{H}^+]}}{[\text{H}^+]} = 2.302\ 6\ e_{\text{pH}} = (2.3026)(0.04) = 0.092\ 1$

$e_{[\text{H}^+]} = (0.092\ 1)[\text{H}^+] = (0.092\ 1)[3.63 \times 10^{-5}\text{ M}] = 3.34 \times 10^{-6}\text{ M}$
$[\text{H}^+] = 3.63\ (\pm 0.334) \times 10^{-5}\text{ M} = 3.6\ (\pm 0.3) \times 10^{-5}\text{ M}$

3-E. $0.050\ 0(\pm 2\%)$ mol =

$$\dfrac{[4.18(\pm x)\text{ mL}]\left[1.18(\pm 0.01)\dfrac{\text{g solution}}{\text{mL}}\right]\left[0.370(\pm 0.005)\dfrac{\text{g HCl}}{\text{g solution}}\right]}{36.461\dfrac{\text{g HCl}}{\text{mol}}}$$

Error analysis:

$$(0.02)^2 = \left(\dfrac{x}{4.18}\right)^2 + \left(\dfrac{0.01}{1.18}\right)^2 + \left(\dfrac{0.005}{0.370}\right)^2$$
$$x = 0.05 \text{ mL}$$

3-F. Atomic masses from periodic table:
N: $14.006\ 8 \pm 0.000\ 4$
H: $1.007\ 98 \pm 0.000\ 14$

$n \times$ (atomic mass $\pm$ uncertainty):
N: $14.006\ 8 \pm 0.000\ 4 = 14.006\ 8 \pm 0.000\ 4$
3H: $3(1.007\ 98 \pm 0.000\ 14) = 3.023\ 94 \pm 0.000\ 42$

NH_3: $17.030\ 74 \pm \sqrt{0.000\ 4^2 + 0.000\ 42^2}$
$= 17.030\ 7_4 \pm 0.000\ 5_8$ g/mol

Percent relative uncertainty $= 100 \times \dfrac{0.000\ 5_8}{17.030\ 7_4} = 0.003\ 4\%$

Chapter 4

4-A. Mean $= \frac{1}{5}(116.0 + 97.9 + 114.2 + 106.8 + 108.3)$
$\qquad = 108.6_4$

Standard deviation $= \sqrt{\dfrac{(116.0 - 108.6_4)^2 + \cdots + (108.3 - 108.6_4)^2}{5 - 1}}$
$\qquad\qquad = 7.1_4$

Standard uncertainty $=$ standard deviation of the mean $= \dfrac{7.1_4}{\sqrt{5}} = 3.1_9$

Range $= 116.0 - 97.9 = 18.1$

90% confidence interval $= 108.6_4 \pm \dfrac{(2.132)(7.1_4)}{\sqrt{5}} = 108.6_4 \pm 6.8_1$

$G_{\text{calculated}} = |97.9 - 108.6_4|\ /\ 7.1_4 = 1.50$
$G_{\text{table}} = 1.672$ for five measurements

Because $G_{\text{calculated}} < G_{\text{table}}$, we retain 97.9.

4-B.

	A	B	C	D
1	Computing standard deviation			
2				
3		Data = x	x − mean	(x-mean)^2
4		17.4	−0.44	0.1936
5		18.1	0.26	0.0676
6		18.2	0.36	0.1296
7		17.9	0.06	0.0036
8		17.6	−0.24	0.0576
9	sum =	89.2		0.452
10	mean =	17.84		
11	std dev =	0.3362		
12				
13	Formulas:	B9 = B4+B5+B6+B7+B8		
14		B10 = B9/5		
15		B11 = SQRT(D9/(5−1))		
16		C4 = B4−$B$10		
17		D4 = C4^2		
18		D9 = D4+D5+D6+D7+D8		
19				
20	Calculations using built-in functions:			.
21	sum =	89.2		
22	mean =	17.84		
23	std dev =	0.3362		
24				
25	Formulas:	B21 = SUM(B4:B8)		
26		B22 = AVERAGE(B4:B8)		
27		B23 = STDEV(B4:B8)		

4-C. **(a)** We need to find the fraction of the area of the Gaussian curve between $x = -\infty$ and $x = 40\,860$ h. When $x = 40\,860$, $z = (40\,860 - 62\,700)/10\,400 = -2.100\,0$. The Gaussian curve is symmetric, so the area from $-\infty$ to $-2.100\,0$ is the same as the area from $2.100\,0$ to $+\infty$. Table 4-1 tells us that the area between $z = 0$ and $z = 2.1$ is 0.482 1. Because the area from $z = 0$ to $z = \infty$ is 0.500 0, the area from $z = 2.100\,0$ to $z = \infty$ is $0.500\,0 - 0.482\,1 = 0.017\,9$. The fraction of brakes expected to be 80% worn in less than 40 860 miles is 0.017 9, or 1.79%.

(b) At 57 500 miles, $z = (57\,500 - 62\,700)/10\,400 = -0.500\,0$. At 71 020 miles, $z = (71\,020 - 62\,700)/10\,400 = +0.800\,0$. The area under the Gaussian curve from $z = -0.500\,0$ to $z = 0$ is the same as the area from $z = 0$ to $z = +0.500\,0$, which is 0.191 5 in Table 4-1. The area from $z = 0$ to $z = +0.800\,0$ is 0.288 1. The total area from $z = -0.500\,0$ to $z = +0.800\,0$ is $0.191\,5 + 0.288\,1 = 0.479\,6$. The fraction of brakes expected to be 80% worn between 57 500 and 71 020 miles is 0.479 6, or 47.96%.

4-D. The answers in cells C4 and C9 of the following spreadsheet are **(a)** 0.052 and **(b)** 0.361.

	A	B	C
1	Mean =	Std dev =	
2	62700	10400	
3			
4	Area from $-\infty$ to 45800 =		0.052081
5	Area from $-\infty$ to 60000 =		0.397580
6	Area from $-\infty$ to 70000 =		0.758637
7			
8	Area from 60000 to 70000		
9		= C6−C5 =	0.361056
10			
11	Formula:		
12	C4 = NORMDIST(45800,A2,B2,TRUE)		

4-E. **(a)** $\bar{x}_1 = \dfrac{31.40 + 31.24 + 31.18 + 31.43}{4}$

$= 31.31_2$ mM

$s_1 = \sqrt{\dfrac{(31.40 - 31.31_2)^2 + (31.24 - 31.31_2)^2 + \cdots + (31.43 - 31.31_2)^2}{4 - 1}}$

$= 0.12_1$ mM

$u_1 = s_1/\sqrt{n} = (0.12_1 \text{ mM})/\sqrt{4} = 0.06_1$

$\bar{x}_2 = \dfrac{30.70 + 29.49 + 30.01 + 30.15}{4} = 30.08_8$ mM

$s_2 = \sqrt{\dfrac{(30.70 - 30.08_8)^2 + (29.49 - 30.08_8)^2 + \cdots + (30.15 - 30.08_8)^2}{4 - 1}}$

$= 0.49_7$ mM

$u_2 = s_2/\sqrt{n} = (0.49_7 \text{ mM})/\sqrt{4} = 0.24_9$

(b) $F_{\text{calculated}} = \dfrac{s_1^2}{s_2^2} = \dfrac{(0.49_7)^2}{(0.12_1)^2} = 16._8$

For 3 degrees of freedom in both standard deviations, $F_{\text{table}} = 9.28$. $F_{\text{calculated}} > F_{\text{table}}$, so standard deviations are significantly different.

4-F. For 117, 119, 111, 115, 120 μmol/100 mL, $\bar{x} = 116._4$ and $s = 3._{58}$.

The 95% confidence interval for 4 degrees of freedom is

$$\bar{x} \pm \dfrac{ts}{\sqrt{n}} = 116._4 \pm \dfrac{(2.776)(3._{58})}{\sqrt{5}} = 116._4 \pm 4._4$$

$$= 112._0 \text{ to } 120._8 \text{ } \mu\text{mol/100 mL}$$

The 95% confidence interval does not include the accepted value of 111 μmol/100 mL, so the *difference is significant*.

4-G. **(a)** pg/g corresponds to 10^{-12} g/g, which is parts per trillion.

(b) $F_{\text{calculated}} = 4.6^2/3.6^2 = 1.6_3 < F_{\text{table}} = 5.05$ (for 5 degrees of freedom in both numerator and denominator). Standard deviations are not significantly different at 95% confidence level.

(c) Because $F_{\text{calculated}} < F_{\text{table}}$, we use Equations 4-10a and 4-9a.

$$s_{\text{pooled}} = \sqrt{\dfrac{s_1^2(n_1 - 1) + s_2^2(n_2 - 1)}{n_1 + n_2 - 2}}$$

$$= \sqrt{\dfrac{4.6^2(6 - 1) + 3.6^2(6 - 1)}{6 + 6 - 2}} = 4.1_3$$

$$t_{\text{calculated}} = \dfrac{|\bar{x}_1 - \bar{x}_2|}{s_{\text{pooled}}} \sqrt{\dfrac{n_1 n_2}{n_1 + n_2}} = \dfrac{|51.1 - 34.4|}{4.1_3} \sqrt{\dfrac{6 \cdot 6}{6 + 6}} = 7.0_0$$

Because $t_{\text{calculated}}\,(7.0_0) > t_{\text{table}}\,(= 2.228$ for 10 degrees of freedom), the difference is significant at the 95% confidence level.

(d) $F_{\text{calculated}} = 3.6^2/1.2^2 = 9.0_0 > F_{\text{table}} = 5.05$. The standard deviations are significantly different at the 95% confidence level. Therefore, we use Equations 4-10b and 4-9b to compare the means:

$$\text{Degrees of freedom} = \dfrac{(s_1^2/n_1 + s_2^2/n_2)^2}{\dfrac{(s_1^2/n_1)^2}{n_1 - 1} + \dfrac{(s_2^2/n_2)^2}{n_2 - 1}}$$

$$= \dfrac{(3.6^2/6 + 1.2^2/6)^2}{\dfrac{(3.6^2/6)^2}{6 - 1} + \dfrac{(1.2^2/6)^2}{6 - 1}} = 6.10 \approx 6$$

$$t_{\text{calculated}} = \dfrac{|\bar{x}_1 - \bar{x}_2|}{\sqrt{(s_1^2/n_1) + (s_2^2/n_2)}} = \dfrac{|34.4 - 42.9|}{\sqrt{3.6^2/6 + 1.2^2/6}} = 5.4_9$$

Because $t_{\text{calculated}}\,(= 5.4_9) > t_{\text{table}}\,(= 2.447$ for 6 degrees of freedom), the difference is significant at the 95% confidence level.

4-H. **(a)**

x_i	y_i	$x_i y_i$	x_i^2	d_i	d_i^2
0.00	0.466	0	0	−0.004 6	2.12×10^{-5}
9.36	0.676	6.327	87.61	+0.001 6	2.56×10^{-6}
18.72	0.883	16.530	350.44	+0.004 8	2.30×10^{-5}
28.08	1.086	30.495	788.49	+0.004 0	1.60×10^{-5}
37.44	1.280	47.923	1 401.75	−0.005 8	3.36×10^{-5}
Sum: 93.60	4.391	101.275	2 628.29		9.64×10^{-5}

$$D = \begin{vmatrix} \Sigma(x_i^2) & \Sigma x_i \\ \Sigma x_i & n \end{vmatrix}$$

$$= (2\,628.29)(5) - (93.60)(93.60) = 4\,380.5$$

$$m = \begin{vmatrix} \Sigma(x_i y_i) & \Sigma x_i \\ \Sigma y_i & n \end{vmatrix} \div D$$

$$= \dfrac{(101.275)(5) - (93.60)(4.391)}{D}$$

$$= 95.377 \div 4\,380.5 = 0.021\,773$$

$$b = \begin{vmatrix} \Sigma(x_i^2) & \Sigma(x_i y_i) \\ \Sigma x_i & \Sigma y_i \end{vmatrix} \div D$$

$$= \dfrac{(2\,628.29)(4.391) - (101.275)(93.60)}{D}$$

$$= 2\,061.48 \div 4\,380.5 = 0.470\,60$$

$$s_y^2 = \dfrac{\Sigma(d_i^2)}{n - 2} = \dfrac{9.64 \times 10^{-5}}{3}$$

$$= 3.21 \times 10^{-5}; s_y = 0.005\,67$$

$$u_m = \sqrt{\dfrac{s_y^2 n}{D}} = \sqrt{\dfrac{(3.21 \times 10^{-5})5}{4\,380.5}} = 0.000\,191$$

$$u_b = \sqrt{\dfrac{s_y^2 \Sigma(x_i^2)}{D}} = \sqrt{\dfrac{(3.21 \times 10^{-5})(2\,628.29)}{4\,380.5}}$$

$$= 0.004\,39$$

Equation of the best line:

$$y = [0.021\,8\,(\pm 0.000\,2)]x + [0.471\,(\pm 0.004)]$$

	A	B	C	D	E	F	G	H	I
1	Least-Squares Spreadsheet								
2									
3		x	y						
4		0	0.466						
5		9.36	0.676						
6		18.72	0.883						
7		28.08	1.086						
8		37.44	1.280						
9									
10			LINEST output:						
11	m	0.02177	0.47060	b					
12	u_m	0.00019	0.00439	u_b					
13	R^2	0.99977	0.00567	s_y					
14									
15	n =	5	B15 = COUNT(B4:B8)						
16	Mean y =	0.878	B16 = AVERAGE(C4:C8)						
17	$(x_i\text{-mean } x)^2 =$	876.096	B17 = DEVSQ(B4:B8)						
18									
19	Measured y =	0.973	Input						
20	Number of replicate measurements of y (k) =	1	Input						
21	Derived x	23.0739	B21 = (B19-C11)/B11						
22	$u_x =$	0.2878	B22 = (C13/ABS(B11))*SQRT((1/B20)+(1/B15)+((B19-B16)^2)/(B11^2*B17))						

Chart: Absorbance at 595 nm vs Protein (μg); $y = 0.0218x + 0.4706$

Spreadsheet for Exercise 4-C

(c) $x = \dfrac{y - b}{m} = \dfrac{0.973 - 0.471}{0.021\,8} = 23.0\ \mu g.$

If you keep more digits for m and b, $x = 23.07\ \mu g$.

Uncertainty in x (s_x)

$$u_x = \frac{s_y}{|m|}\sqrt{\frac{1}{k} + \frac{1}{n} + \frac{(y - \bar{y})^2}{m^2\,\Sigma(x_i - \bar{x})^2}}$$

$$= \frac{0.005\,67}{|0.021\,77|}\sqrt{\frac{1}{1} + \frac{1}{5} + \frac{(0.973 - 0.878\,2)^2}{(0.021\,77)^2(876.1)}} = 0.29\ \mu g.$$

Final answer is $23.1 \pm 0.3\ \mu g$.

Chapter 5

5-A. **(a)** Standard deviation of 9 samples = $s = 0.000\,6_{44}$

Mean blank = $y_{blank} = 0.001\,1_{89}$

$y_{dl} = y_{blank} + 3s = 0.001\,1_8 + (3)(0.000\,6_{44}) = 0.003\,1_{12}$

(b) Minimum detectable concentration $= \dfrac{3s}{m} = \dfrac{(3)(0.000\,6_{44})}{2.24 \times 10^4\ M^{-1}}$

$= 8._6 \times 10^{-8}\ M$

(c) Lower limit of quantitation $= \dfrac{10s}{m} = \dfrac{(10)(0.000\,6_{44})}{2.24 \times 10^4\ M^{-1}} = 2._9 \times 10^{-7}\ M$

5-B. **(a)** $[Ni^{2+}]_f = [Ni^{2+}]_i \dfrac{V_i}{V_f} = [Ni^{2+}]_i\left(\dfrac{25.0}{25.5}\right) = 0.980_4[Ni^{2+}]_i$

(b) $[S]_f = (0.028\,7\ M)\left(\dfrac{0.500}{25.5}\right) = 0.000\,562_7\ M$

(c) $\dfrac{[Ni^{2+}]_i}{0.000\,562\,7 + 0.980\,4[Ni^{2+}]_i} = \dfrac{2.36\ \mu A}{3.79\ \mu A}$

$\Rightarrow [Ni^{2+}]_i = 9.00 \times 10^{-4}\ M$

5-C. There are 9 points, so there are $9 - 2 = 7$ degrees of freedom. For 90% confidence, $t = 1.895$, so the 90% confidence interval is $\pm(1.895)(0.098\ mM) = \pm0.19\ mM$. For 99% confidence, $t = 3.500$, and the 99% confidence interval is $\pm (3.500)(0.098\ mM) = \pm0.34\ mM$.

5-D. Use the standard mixture to find the response factor. We know that, when $[X] = [S]$, the ratio of signals A_X/A_S is 1.31.

$$\frac{A_X}{[X]} = F\left(\frac{A_S}{[S]}\right) \Rightarrow F = \frac{A_X/A_S}{[X]/[S]} = \frac{1.31}{1} = 1.31$$

In the mixture of unknown plus standard, the concentration of S is

$$[S] = \underbrace{(4.13\ \mu g/mL)}_{\substack{\text{Initial} \\ \text{concentration}}}\underbrace{\left(\frac{2.00}{10.0}\right)}_{\substack{\text{Dilution} \\ \text{factor}}} = 0.826\ \mu g/mL$$

For the unknown mixture: $F = \dfrac{A_X/A_S}{[X]/[S]}$

$1.31 = \dfrac{0.808}{[X]/[0.826\ \mu g/mL]} \Rightarrow [X] = 0.509\ \mu g/mL$

Because X was diluted from 5.00 to 10.0 mL in the mixture with S, the original concentration of X was $(10.0/5.0)(0.509\ \mu g/mL) = 1.02\ \mu g/mL.$

5-E. **(a)** Cells B10:C11 of the spreadsheet give slope (m) = 1.075 6, $u_m = 0.051\ 7$; intercept (b) = 0.008 4, $u_b = 0.033\ 5$. The theoretical value of the intercept is 0. The observed value (0.008 4) is less than one standard uncertainty (0.033 5) away from 0, which is within experimental error of 0.

(b) Cells B19 and B20 in the spreadsheet give $[C_{10}H_8]/[C_{10}D_8] = x = 0.598$ with a standard uncertainty $u_x = 0.035\ 5.$

(c) Student's t for 95% confidence and 1 degree of freedom is 12.706. The 95% confidence interval for $[C_{10}H_8]/[C_{10}D_8]$ is $0.598 \pm (12.706) \times (0.035\ 5) = 0.598 \pm 0.451$. Relative uncertainty = $0.451/0.598 = 75\%$. The uncertainty from a 3-point calibration is huge (75%) because Student's

	A	B	C	D	E	F	G	H	I	J
1	$C_{10}H_8/D_{10}H_8$ internal standard calibration curve									
2	[X]/[S]	A_x/A_s								
3	0.100	0.101								
4	0.500	0.573								
5	1.000	1.072								
6	Highlight cells B10:C12									
7	Type "= LINEST(B3:B5,A3:A5,TRUE,TRUE)"									
8	PC: CTRL + SHIFT + ENTER									
9	Mac: COMMAND + RETURN									
10	m	1.0756	0.0084	b						
11	u_m	0.0517	0.0335	u_b						
12	R^2	0.9977	0.0330	s_y						
13	n =	3	B13 = COUNT(B3:B5)							
14	Mean y =	0.582	B14 = AVERAGE(B3:B5)							
15	$\sum(x_1 - \text{mean } x)^2 =$	0.406667	B15 = DEVSQ(A3:B5)							
16										
17	Measured y =	0.652	Input							
18	k = Number of replicate measurements of y =	1	Input							
19	Derived x =	0.598	B19 = (B17–C10)/B10							
20	$u_x =$	0.0355	B20 = (C12/ABS(B10))*							
21	SQRT((1/B18)+(1/B13)+((B17–B14)^2)/(B10^2*B15))									

Graph inset (cells G2:J15): plot of A_x/A_s vs [X]/[S] with line $y = 1.0756x + 0.0084$.

Spreadsheet for Exercise 5-E

---

t is 12.706 when there is only 1 degree of freedom. If we had just one more data point (giving 2 degrees of freedom), the 95% confidence interval would decrease by a factor of 3.

5-F. For the data in this problem, mean = 0.84_{11} and standard deviation = 0.18_{88}. Stability criteria are

- There should be no observations outside the action lines—One observation (day 101) lies above the upper action line.
- There are not 2 out of 3 consecutive measurements between warning and action lines—OK.
- There are not 7 consecutive measurements all above or all below the center line—OK.
- There are not 6 consecutive measurements all steadily increasing or all steadily decreasing, wherever they are located—OK.
- There are not 14 consecutive points alternating up and down, regardless of where they are located—OK.
- There is no obvious nonrandom pattern—OK.

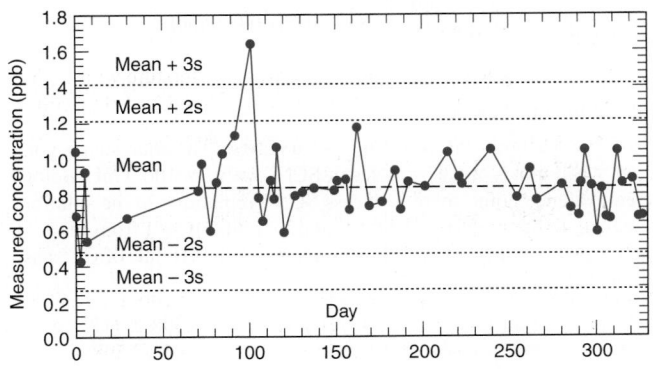

Chapter 6

6-A. **(a)** $Ag^+ + Cl^- \rightleftharpoons AgCl(aq)$
$\underline{AgCl(s) \rightleftharpoons Ag^+ + Cl^-}$
$AgCl(s) \rightleftharpoons AgCl(aq)$

$K_1 = 2.0 \times 10^3$
$\underline{K_2 = 1.8 \times 10^{-10}}$
$K_3 = K_1 K_2 = 3.6 \times 10^{-7}$

(b) The answer to **(a)** tells us $[AgCl(aq)] = 3.6 \times 10^{-7}$ M

(c) $AgCl_2^- \rightleftharpoons AgCl(aq) + Cl^-$
$Ag^+ + Cl^- \rightleftharpoons AgCl(s)$
$\underline{AgCl(aq) \rightleftharpoons Ag^+ + Cl^-}$
$AgCl_2^- \rightleftharpoons AgCl(s) + Cl^-$

$K_1 = 1/(9.3 \times 10^1)$
$K_2 = 1/(1.8 \times 10^{-10})$
$\underline{K_3 = 1/(2.0 \times 10^3)}$
$K_4 = K_1 K_2 K_3 = 3.0 \times 10^4$

6-B. **(a)** $\dfrac{(x)(x)(1.00 + 8x)^8}{(0.010\ 0 - x)(0.010\ 0 - 2x)^2} = 1 \times 10^{11}$

(b) $[Br^-]$ and $[Cr_2O_7^{2-}]$ will both be 0.005 00 M because Cr^{3+} is the *limiting reagent*. The reaction requires 2 moles of Cr^{3+} per mole of BrO_3^-. The Cr^{3+} will be used up first, making 1 mol Br^- and 1 mol $Cr_2O_7^{2-}$ per 2 mol Cr^{3+} consumed. To solve the preceding equation, we set $x = 0.005\ 00$ M in all terms except $[Cr^{3+}]$. The concentration of Cr^{3+} will be a small, unknown quantity.

$$\frac{(0.005\ 00)(0.005\ 00)[1.00 + 8(0.005\ 00)]^8}{(0.010\ 0 - 0.005\ 00)[Cr^{3+}]^2} = 1 \times 10^{11}$$

$[Cr^{3+}] = 2._6 \times 10^{-7}$ M

$[BrO_3^-] = 0.010\ 0 - 0.005\ 00 = 0.005\ 00$ M

6-C. K_{sp} for $La(IO_3)_3$ is small (1.0×10^{-11}), so we presume that the concentration of iodate will not be altered by the small amount of $La(IO_3)_3$ that dissolves.

$$[La^{3+}] = \frac{K_{sp}}{[IO_3^-]^3} = \frac{1.0 \times 10^{-11}}{(0.050)^3} = 8.0 \times 10^{-8} \text{ M}$$

The answer agrees with the assumption that iodate from $La(IO_3)_3 \ll 0.050$ M.

6-D. We expect $Ca(IO_3)_2$ to be more soluble because its K_{sp} is larger and the two salts have the same stoichiometry. If the stoichiometry were not the same, we could not directly compare values of K_{sp}. Our prediction could be wrong if, for example, the barium salt formed a great deal of the ion pairs $Ba(IO_3)^+$ or $Ba(IO_3)_2(aq)$ and the calcium salt did not form ion pairs.

6-E. $[Fe^{3+}][OH^-]^3 = (10^{-10})[OH^-]^3 = 1.6 \times 10^{-39}$
$\Rightarrow [OH^-] = 2.5 \times 10^{-10} \text{ M}$
$[Fe^{2+}][OH^-]^2 = (10^{-10})[OH^-]^2 = 7.9 \times 10^{-16}$
$\Rightarrow [OH^-] = 2.8 \times 10^{-3} \text{ M}$

6-F. We want to reduce $[Ce^{3+}]$ to 1.0% of 0.010 M = 0.000 10 M. The concentration of oxalate in equilibrium with 0.000 10 M Ce^{3+} is computed as follows:
$[Ce^{3+}]^2 [C_2O_4^{2-}]^3 = K_{sp} = 5.9 \times 10^{-30}$
$(0.000\ 10)^2 [C_2O_4^{2-}]^3 = 5.9 \times 10^{-30}$

$$[C_2O_4^{2-}] = \left(\frac{5.9 \times 10^{-30}}{(0.000\ 10)^2}\right)^{1/3} = 8.4 \times 10^{-8} \text{ M}$$

To see if 8.4×10^{-8} M $C_2O_4^{2-}$ will precipitate 0.010 M Ca^{2+}, evaluate Q for CaC_2O_4:
$$Q = [Ca^{2+}][C_2O_4^{2-}] = (0.010)(8.4 \times 10^{-8}) = 8.4 \times 10^{-10}$$
Because $Q < K_{sp}$ for CaC_2O_4 $(=1.3 \times 10^{-8})$, Ca^{2+} will not precipitate.

6-G. Assuming that all nickel is in the form $Ni(en)_3^{2+}$, $[Ni(en)_3^{2+}] = 1.00 \times 10^{-5}$ M. This uses up just 3×10^{-5} mol of en, which leaves the en concentration at 0.100 M. Adding the three equations gives
$$Ni^{2+} + 3en \rightleftharpoons Ni(en)_3^{2+}$$
$$K = K_1 K_2 K_3 = 2.1_4 \times 10^{18}$$
$$[Ni^{2+}] = \frac{[Ni(en)_3^{2+}]}{K[en]^3}$$
$$= \frac{(1.00 \times 10^{-5})}{(2.1_4 \times 10^{18})(0.100)^3} = 4.7 \times 10^{-21} \text{ M}$$

Now we verify that $[Ni(en)^{2+}]$ and $[Ni(en)_2^{2+}] \ll 10^{-5}$ M:
$[Ni(en)^{2+}] = K_1[Ni^{2+}][en] = 1.5 \times 10^{-14}$ M
$[Ni(en)_2^{2+}] = K_2[Ni(en)^{2+}][en] = 3.2 \times 10^{-9}$ M

6-H. (a) Neutral—Neither Na^+ nor Br^- has any acidic or basic properties.
(b) Basic—$CH_3CO_2^-$ is the conjugate base of acetic acid, and Na^+ is neither acidic nor basic.
(c) Acidic—NH_4^+ is the conjugate acid of NH_3, and Cl^- is neither acidic nor basic.
(d) Basic—PO_4^{3-} is a base, and K^+ is neither acidic nor basic.
(e) Neutral—Neither ion is acidic nor basic.
(f) Basic—The quaternary ammonium ion is neither acidic nor basic, and the $C_6H_5CO_2^-$ anion is the conjugate base of benzoic acid.
(g) Acidic—Fe^{3+} is acidic, and nitrate is neither acidic nor basic.

6-I. $K_{b1} = K_w/K_{a2} = 4.3 \times 10^{-9}$
$K_{b2} = K_w/K_{a1} = 1.6 \times 10^{-10}$

6-J. $K \equiv K_{b2} = K_w/K_{a2} = 1.2 \times 10^{-8}$

6-K. (a)
$[H^+][OH^-] = x^2 = K_w \Rightarrow x = \sqrt{K_w} \Rightarrow pH = -\log\sqrt{K_w} = 7.469$ at 0°C, 7.082 at 20°C, and 6.770 at 40°C.
(b) Because $[D^+]=[OD^-]$ in pure D_2O, $K = 1.35 \times 10^{-15} = [D^+][OD^-]$
$= [D^+]^2 \Rightarrow [D^+] = 3.67 \times 10^{-8}$ M $\Rightarrow pD = 7.435$.

Chapter 7

7-A.
(a)

Ascorbic acid
$C_6H_8O_6$

Dehydroascorbic acid
$C_6H_8O_7$

Formula mass of ascorbic acid = 6(atomic mass of C) + 8(atomic mass of H) + 6(atomic mass of O) = 6(12.010 6) + 8(1.007 98) + 6(15.999 4) = 176.124 g/mol

(b) $\dfrac{0.197\ 0 \text{ g of ascorbic acid}}{176.124 \text{ g/mol}} = 1.118\ 5$ mmol.

Molarity of I_3^- = 1.118 5 mmol/29.41 mL
$\qquad = 0.038\ 03$ M

(c) 31.63 mL of I_3^- = 1.203 mmol of I_3^-
$\qquad\qquad = 1.203$ mmol of ascorbic acid
$\qquad\qquad = 0.211\ 9$ g = 49.94% of the tablet

7-B. (a)

$C_8H_5O_4K$

(b) $\dfrac{0.824 \text{ g/acid}}{204.22 \text{ g/mol}} = 4.03_{49}$ mmol.

This many mmol of NaOH is contained in 0.038 314 kg of NaOH solution

$\Rightarrow$ concentration $= \dfrac{4.03_{49} \times 10^{-3} \text{ mol NaOH}}{0.038\ 314 \text{ kg solution}}$
$\qquad\qquad\qquad\qquad = 0.105_{31}$ mol/kg solution

(c) mol NaOH = (0.057 911 kg solution)(0.105$_{31}$ mol/kg solution) = 6.09$_{86}$ mmol

Because 2 mol NaOH react with 1 mol H_2SO_4,

$[H_2SO_4] = \dfrac{\frac{1}{2}(6.09_{86} \text{ mmol})}{10.00 \text{ mL}} = 0.305\ \dfrac{\text{mmol}}{\text{mL}} = 0.305$ M

7-C. 34.02 mL of 0.087 71 M NaOH = 2.983 9 mmol of OH^-. Let x be the mass of malonic acid and y be the mass of anilinium chloride.

Then $x + y = 0.237\ 6$ g and
(moles of anilinium chloride) + 2(moles of malonic acid) = 0.002 983 9

$\dfrac{y \text{ g}}{129.59 \text{ g/mol}} + 2\left(\dfrac{x \text{ g}}{104.06 \text{ g/mol}}\right) = 0.002\ 983\ 9$ mol

Substituting $y = 0.237\ 6 - x$ gives $x = 0.100\ 01$ g = 42.09% malonic acid. Anilinium chloride = 57.91%

7-D. The reaction is $SCN + Cu^+ \rightarrow CuSCN(s)$. The equivalence point occurs when moles of Cu^+=moles of $SCN^- \Rightarrow V_e = 100.0$ mL. Before the equivalence point, there is excess SCN^- remaining in the solution. We calculate the molarity of SCN^- and then find $[Cu^+]$ from the relation $[Cu^+] = K_{sp}/[SCN^-]$. For example, when 0.10 mL of Cu^+ has been added,

$$[SCN^-] = \left(\frac{100.0 \text{ mL} - 0.10 \text{ mL}}{100.0 \text{ mL}}\right)(0.080\ 0 \text{ M})\left(\frac{50.0 \text{ mL}}{50.1 \text{ mL}}\right)$$

$$= 7.98 \times 10^{-2} \text{ M}$$

$$[Cu^+] = 4.8 \times 10^{-15}/7.98 \times 10^{-2}$$

$$= 6.0 \times 10^{-14}$$

$$pCu^+ = 13.22$$

At the equivalence point, $[Cu^+][SCN^-] = x^2 = K_{sp} \Rightarrow x = [Cu^+] = 6.9 \times 10^{-8} \Rightarrow pCu^+ = 7.16$.

Past the equivalence point, there is excess Cu^+. For example, when $V = 101.0$ mL,

$$[Cu^+] = (0.040\ 0 \text{ M})\left(\frac{101.0 \text{ mL} - 100.0 \text{ mL}}{151.0 \text{ mL}}\right) = 2.6 \times 10^{-4} \text{ M}$$

$$pCu^+ = 3.58$$

mL	pCu	mL	pCu	mL	pCu
0.10	13.22	75.0	12.22	100.0	7.16
10.0	13.10	95.0	11.46	100.1	4.57
25.0	12.92	99.0	10.75	101.0	3.58
50.0	12.62	99.9	9.75	110.0	2.60

7-E. $V_e = 23.66$ mL for AgBr. At 2.00, 10.00, 22.00, and 23.00 mL, AgBr is partially precipitated and excess Br^- remains.

At 2.00 mL

$$[Ag^+] = \frac{K_{sp}(\text{for AgBr})}{[Br^-]}$$

$$= \frac{5.0 \times 10^{-13}}{\underbrace{\left(\dfrac{23.66 \text{ mL} - 2.00 \text{ mL}}{23.66 \text{ mL}}\right)}_{\substack{\text{Fraction} \\ \text{remaining}}}\underbrace{(0.0500\ 0 \text{ mL})}_{\substack{\text{Original molarity} \\ \text{of Br}^-}}\underbrace{\left(\dfrac{40.00 \text{ mL}}{42.00 \text{ mL}}\right)}_{\substack{\text{Dilution} \\ \text{factor}}}}$$

$$= 1.15 \times 10^{-11} \text{ M} \Rightarrow pAg^+ = 10.94$$

By similar reasoning, we find
at 10.00 mL $pAg^+ = 19.66$
at 22.00 mL $pAg^+ = 9.66$
at 23.00 mL $pAg^+ = 9.25$

At 24.00, 30.00, and 40.00 mL, AgCl is precipitating and excess Cl^- remains in solution.

At 24.00 mL:

$$[Ag^+] = \frac{K_{sp}(\text{for AgCl})}{[Cl^-]}$$

$$= \frac{1.8 \times 10^{-10}}{\left(\dfrac{47.32 \text{ mL} - 24.00 \text{ mL}}{23.66 \text{ mL}}\right)(0.050\ 00 \text{ M})\left(\dfrac{40.00 \text{ mL}}{64.00 \text{ mL}}\right)}$$

$$= 5.8 \times 10^{-9} \text{ M} \Rightarrow pAg^+ = 8.23$$

By similar reasoning, we find
at 30.00 mL $pAg^+ = 8.07$
at 40.00 mL $pAg^+ = 7.63$

At the second equivalence point (47.32 mL), $[Ag^+] = [Cl^-]$, and we can write
$$[Ag^+][Cl^-] = x^2 = K_{sp}(\text{for AgCl})$$

$$\Rightarrow [Ag^+] = 1.34 \times 10^{-5} \text{ M} \Rightarrow pAg^+ = 4.87$$

At 50.00 mL, there is an excess of $(60.00 - 47.32) = 2.68$ mL of Ag^+

$$[Ag^+] = \left(\frac{2.68 \text{ mL}}{90.00 \text{ mL}}\right)(0.084\ 54 \text{ M}) = 2.5 \times 10^{-3} \text{ M}$$

$$pAg^+ = 2.60$$

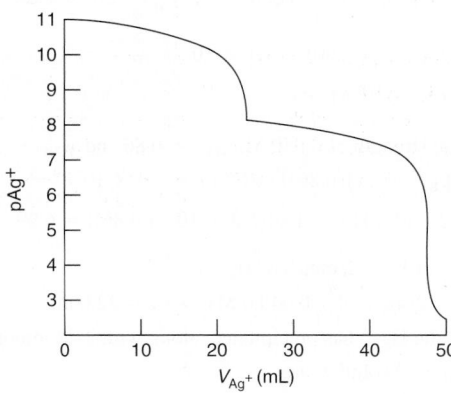

7-F. **(a)** 12.6 mL of Ag^+ are required to precipitate I^-, then $(27.7 - 12.6) = 15.1$ mL are required to precipitate SCN^-.

$$[SCN^-] = \frac{\text{moles of } Ag^+ \text{ needed to react with } SCN^-}{\text{original volume of } SCN^-}$$

$$= \frac{[27.7(\pm0.3) - 12.6(\pm0.4) \text{ mL}][0.068\ 3\ (\pm0.000\ 1) \text{ M}]}{50.00\ (\pm0.05) \text{ mL}} \qquad \text{(a)}$$

$$= \frac{[15.1(\pm0.5)][0.068\ 3\ (\pm0.000\ 1)]}{50.00\ (\pm0.05)}$$

$$= \frac{[15.1(\pm3.31\%)][0.068\ 3(\pm0.146\%)]}{50.00\ (\pm0.100\%)} = 0.020\ 6(\pm0.000\ 7) \text{ M}$$

In expression **(a)**, I chose to keep the volumes in mL rather than L. As long as the units are consistent and cancel out, this practice is allowed. You could have written all volumes in L.

(b) $[SCN^-](\pm4.0\%) =$

$$\frac{[27.7\ (\pm0.3) - 12.6\ (\pm?)][0.068\ 3\ (\pm0.000\ 1)]}{50.00\ (\pm0.05)}$$

Let the error in 15.1 mL be $y\%$:

$$(4.0\%)^2 = (y\%)^2 + (0.146\%)^2 + (0.100\%)^2$$

$$\Rightarrow y = 4.00\% = 0.603 \text{ mL}$$

$$27.7\ (\pm0.3) - 12.6\ (\pm?) = 15.1\ (\pm0.603)$$

$$\Rightarrow 0.3^2 + ?^2 = 0.603^2 \Rightarrow ? = 0.5 \text{ mL}$$

Chapter 8

8-A. **(a)** $\mu = \frac{1}{2}([K^+] \cdot 1^2 + [NO_3^-] \cdot (-1)^2) = 0.2 \text{ mM}$

(b) $\mu = \frac{1}{2}([Cs^+] \cdot 1^2 + [CrO_4^{2-}] \cdot (-2)^2)$

$$= \frac{1}{2}([0.4] \cdot 1 + [0.2] \cdot 4) = 0.6 \text{ mM}$$

(c) $\mu = \frac{1}{2}([Mg^{2+}] \cdot 2^2 + [Cl^-] \cdot (-1)^2 + [Al^{3+}] \cdot 3^2)$

$$= \frac{1}{2}([0.2] \cdot 4 + [0.4 \quad + \quad 0.9] \cdot 1 + [0.3] \cdot 9) = 2.4 \text{ mM}$$

$$\qquad\qquad\qquad \uparrow \qquad\qquad \uparrow$$

$$\qquad\qquad \text{From MgCl}_2 \quad \text{From AlCl}_3$$

8-B. For 0.005 0 M $(CH_3CH_2CH_2)_4N^+Br^-$ plus 0.005 0 M $(CH_3)_4N^+Cl^-$, $\mu = 0.010$ M. The size of the ion $(CH_3CH_2CH_2)_4N^+$ is 800 pm. At $\mu = 0.01$ M, $\gamma = 0.912$ for an ion of charge ±1 with $\alpha = 800$ pm. $\mathcal{A} = (0.005\ 0)(0.912) = 0.004\ 6$.

8-C. $\mu = 0.060$ M from KSCN, assuming negligible solubility of AgSCN.

$K_{sp} = [Ag^+]\gamma_{Ag^+}[SCN^-]\gamma_{SCN^-} = 1.1 \times 10^{-12}$

The activity coefficients at $\mu = 0.060$ M are $\gamma_{Ag^+} = 0.79$ and $\gamma_{SCN^-} = 0.80$.

$K_{sp} = [Ag^+](0.79)[0.060](0.80) = 1.1 \times 10^{-12}$

$\Rightarrow [Ag^+] = 2.9 \times 10^{-11}$ M.

8-D. At an ionic strength of 0.050 M, $\gamma_{H^+} = 0.86$ and $\gamma_{OH^-} = 0.81$.

$[H^+]\gamma_{H^+}[OH^-]_{OH^-} = (x)(0.86)(x)(0.81) = 1.0 \times 10^{-14} \Rightarrow x =$

$[H^+] = 1.2 \times 10^{-7}$ M. pH $= -\log[(1.2 \times 10^{-7})(0.86)] = 6.99$.

8-E. (a) Moles of $I^- = 2($ moles of $Hg_2^{2+})$

$(V_e)(0.100$ M$) = 2(40.0$ mL$)(0.040\ 0$ M$) \Rightarrow V_e = 32.0$ mL

(b) Virtually all the Hg_2^{2+} has precipitated, along with 3.20 mmol of I^-.
The ions remaining in solution are

$[NO_3^-] = \dfrac{3.20 \text{ mmol}}{100.0 \text{ mL}} = 0.032\ 0$ M

$[I^-] = \dfrac{2.80 \text{ mmol}}{100.0 \text{ mL}} = 0.028\ 0$ M

$[K^+] = \dfrac{6.00 \text{ mmol}}{100.0 \text{ mL}} = 0.060\ 0$ M

$\mu = \dfrac{1}{2}\Sigma c_i z_i^2 = 0.060\ 0$ M

(c) $\mathcal{A}_{Hg_2^{2+}} = K_{sp}/\mathcal{A}_{I^-}^2 = K_{sp}/[I^-]^2\gamma_{I^-}^2$
$= 4.6 \times 10^{-29}/(0.028\ 0)^2(0.795)^2 = 9.3 \times 10^{-26}$

$\Rightarrow pHg_2^{2+} = -\log \mathcal{A}_{Hg_2^{2+}} = 25.03$

8-F. (a) $[Cl^-] = 2[Ca^{2+}]$

(b) $\underbrace{[Cl^-] + [CaCl^+]}_{\text{Species containing Cl}^-} = 2\{\underbrace{[Ca^{2+}] + [CaCl^+] + [CaOH^+]}_{\text{Species containing Ca}^{2+}}\}$

(c) $[Cl^-] + [OH^-] = 2[Ca^{2+}] + [CaCl^+] + [CaOH^+] + [H^+]$

8-G. Charge balance:

$[F^-] + [HF_2^-] + [OH^-] = 2[Ca^{2+}] + [CaOH^+] + [CaF^+] + [H^+]$

Mass balance: CaF_2 gives 2 mol F for each mol Ca.

$\underbrace{[F^-] + [CaF^+] + 2[CaF_2(aq)] + [HF] + 2[HF_2^-]}_{\text{Species containing F}^-}$

$= 2\{\underbrace{[Ca^{2+}] + [CaOH^+] + [CaF^+] + [CaF_2(aq)]}_{\text{Species containing Ca}^{2+}}\}$

8-H. Charge balance:
$2[Ca^{2+}] + [CaOH^+] + [H^+]$
$\quad = [CaPO_4^-] + 3[PO_4^{3-}] + 2[HPO_4^{2-}] + [H_2PO_4^-] + [OH^-]$

Mass balance: Equate 2(calcium species) = 3(phosphate species)
$2\{\underbrace{[Ca^{2+}] + [CaOH^+] + [CaPO_4^-]}_{\text{Species containing calcium}}\}$

$= 3\{\underbrace{[CaPO_4^-] + [PO_4^{3-}] + [HPO_4^{2-}] + [H_2PO_4^-] + [H_3PO_4]}_{\text{Species containing phosphate}}\}$

8-I. (a) Pertinent reactions:

$Mn(OH)_2(s) \underset{}{\overset{K_{sp}}{\rightleftharpoons}} Mn^{2+} + 2OH^- \qquad K_{sp} = 10^{-12.8}$

$Mn^{2+} + 2OH^- \overset{K_1}{\rightleftharpoons} MnOH^+ \qquad K_1 = 10^{3.4}$

$H_2O \overset{K_w}{\rightleftharpoons} H^+ + OH^- \qquad K_w = 10^{-14.00}$

Charge balance: $2[Mn^{2+}] + [MnOH^+] + [H^+] = [OH^-]$
Mass balance: $\underbrace{[OH^-] + [MnOH^+]}_{\text{Species containing OH}^-} = 2\{\underbrace{[Mn^{2+}] + [MnOH^+]}_{\text{Species containing Mn}^{2+}}\} + [H^+]$

(Mass balance is equivalent to charge balance.)
Equilibrium constant expressions:
$K_{sp} = [Mn^{2+}]\gamma_{Mn^{2+}}[OH^-]^2\gamma_{OH^-}^2$

$K_1 = \dfrac{[MnOH^+]\gamma_{MnOH^+}}{[Mn^{2+}]\gamma_{Mn^{2+}}[OH^-]\gamma_{OH^-}}$

$K_w = [H^+]\gamma_{H^+}[OH^-]\gamma_{OH^-}$

	A	B	C	D	E	F	G	H
1	Manganese hydroxide equilibria							
2	1. *Estimate* pMn in cell B8							
3	2. Use Solver to adjust B8 to minimize sum in cell H15							
4	Ionic strength					Extended		
5	μ	0.1	fixed			Debye-	Activity	
6				Size		Hückel	coefficient	
7	Species	pC	C (M)	α (pm)	Charge	log γ	γ	
8	Mn²⁺	4.271737	5.349E-05	600	2	−3.977E-01	4.002E-01	
9	OH⁻		1.130E-04	350	−1	−1.183E-01	7.615E-01	
10	MnOH⁺		6.016E-06	400	1	−1.140E-01	7.691E-01	
11	H⁺		1.408E-10	900	1	−8.343E-02	8.252E-01	
12								
13	pK$_{sp}$ =	12.80	K$_{sp}$ =	1.58E−13		Charge balance:		b$_i$
14	pK$_1$ =	−3.40	K$_1$ =	2.51E+03	b$_1$ = 0 = 2[Mn²⁺] + [MnOH⁺] + [H⁺] − [OH⁻] =			−3.66E−15
15	pK$_w$ =	14.00	K$_w$ =	1.00E−14			Σb$_i$² =	1.34E-29
16						H14 = 2*C8+C10+C11−C9		
17	Ion size estimates:					H15 = H14^2		
18	MnOH⁺ size ≈ 400 pm					C8 = 10^−B8		
19						C9 = SQRT(D13/(C8*G8))/G9		
20	Initial values:					C10 = D14*C8*G8*C9*G9/G10		
21	pMn =	4				C11 = D15/(C9*G9*G11)		
22						F8 = −0.51*E8^2*SQRT($B$5)/(1+D8*SQRT($B$5)/305)		
23						G8 = 10^−F8		

Spreadsheet for Exercise 8-I(a).

We need to estimate (number of unknowns) − (number of equilibria) = 4 − 3 = 1 concentration, for which I choose $pMn^{2+} = 4$ from the K_{sp} equilibrium. The spreadsheet has the same form as for $Mg(OH)_2$ solubility in the text, but with μ fixed at 0.1 M in cell B5. Concentrations after executing Solver to minimize cell H15 by varying cell B8 are shown in cells C8:C11.

(b) The spreadsheet is the same as (a), except that cell B5 contains the formula "= 0.5*(E8^2*C8+E9^2*C9+E10^2*C10+E11^2*C11)". Solver finds $[Mn^{2+}] = 3.30 \times 10^{-5}$, $[MnOH^+] = 5.68 \times 10^{-6}$, $[OH^-] = 7.18 \times 10^{-5}$, $[H^+] = 1.43 \times 10^{-10}$, and $\mu = 1.05 \times 10^{-4}$ M.

(c) Na^+ and ClO_4^- ions create ionic atmospheres that lessen the attraction of Mn^{2+} and OH^-, thereby increasing the solubility of $Mn(OH)_2$ by a factor of $(5.35 \times 10^{-5}$ M$)/((3.30 \times 10^{-5}$ M$) = 1.6$.

Chapter 9

9-A. $pH = -\log \mathcal{A}_{H^+}$. But $\mathcal{A}_{H^+} \mathcal{A}_{OH^-} = K_w \Rightarrow \mathcal{A}_{H^+} = K_w / \mathcal{A}_{OH^-}$. For 1.0×10^{-2} M NaOH, $[OH^-] = 1.0 \times 10^{-2}$ M and $\gamma_{OH^-} = 0.900$ (Table 8-1, with $\mu = 0.010$ M).

$$\mathcal{A}_{H^+} = \frac{K_w}{[OH^-]\gamma_{OH^-}} = \frac{1.0 \times 10^{-14}}{(1.0 \times 10^{-2})(0.900)}$$
$$= 1.11 \times 10^{-12} \Rightarrow pH = -\log \mathcal{A}_{H^+} = 11.95$$

9-B. (a) Charge balance: $[H^+] = [OH^-] + [Br^-]$

Mass balance: $[Br^-] = 1.0 \times 10^{-8}$ M
Equilibrium: $[H^+][OH^-] = K_w$
Setting $[H^+] = x$ and $[Br^-] = 1.0 \times 10^{-8}$ M, the charge balance tells us that $[OH^-] = x - 1.0 \times 10^{-8}$. Putting this into the K_w equilibrium gives
$(x)(x - 1.0 \times 10^{-8}) = 1.0 \times 10^{-14}$
$\Rightarrow x = 1.0_5 \times 10^{-7}$ M $\Rightarrow pH = 6.98$

(b) Charge balance: $[H^+] = [OH^-] + 2[SO_4^{2-}]$

Mass balance: $[SO_4^{2-}] = 1.0 \times 10^{-8}$ M
Equilibrium: $[H^+][OH^-] = K_w$
As before, writing $[H^+] = x$ and $[SO_4^{2-}] = 1.0 \times 10^{-8}$ M gives $[OH^-] = x - 2.0 \times 10^{-8}$ M and $[H^+][OH^-] = (x)[x - (2.0 \times 10^{-8})] = 1.0 \times 10^{-14} \Rightarrow x = 1.10 \times 10^{-7}$ M $\Rightarrow pH = 6.96$.

9-C.

2-Nitrophenol
$C_6H_5NO_3$
FM = 139.11
$K_a = 5.89 \times 10^{-8}$

$$F_{HA} \text{ (formal concentration)} = \frac{1.23 \text{ g}/(139.11 \text{ g/mol})}{0.250 \text{ L}}$$
$$= 0.035\ 4 \text{ M}$$
$$HA \rightleftharpoons H^+ + A^-$$
$$\underset{F-x}{\quad} \underset{x}{\quad} \underset{x}{\quad}$$

$$\frac{x^2}{0.035\ 4 - x} = 5.89 \times 10^{-8} \text{ M}$$
$$\Rightarrow x \approx \sqrt{(0.035\ 4)(5.89 \times 10^{-8})} = 4.57 \times 10^{-5} \text{ M}$$
$$\Rightarrow pH = -\log x = 4.34$$

9-D.

But $[H^+] = 10^{-pH} = 6.9 \times 10^{-7}$ M $\Rightarrow [A^-] = 6.9 \times 10^{-7}$ M and $[HA] = 0.010 - [H^+] = 0.010$.

$$K_a = \frac{[H^+][A^-]}{[HA]} = \frac{(6.9 \times 10^{-7})^2}{0.010} = 4.8 \times 10^{-11} \Rightarrow pK_a = 10.32$$

9-E. As $[HA] \to 0$, $pH \to 7$. If $pH = 7$,

$$\frac{[H^+][A^-]}{[HA]} = K_a \Rightarrow [A^-] = \frac{K_a}{[H^+]}[HA]$$
$$= \frac{10^{-5.00}}{10^{-7.00}}[HA] = 100[HA]$$

$$\alpha = \frac{[A^-]}{[HA] + [A^-]} = \frac{100[HA]}{[HA] + 100[HA]} = \frac{100}{101} = 99\%$$

If $pK_a = 9.00$, we find $\alpha = 0.99\%$.

9-F. $CH_3CH_2CH_2CO_2^- + H_2O \rightleftharpoons CH_3CH_2CH_2CO_2H + OH^-$
$\underset{F-x}{\quad} \underset{\quad}{\quad} \underset{x}{\quad} \underset{x}{\quad}$

$$K_b = \frac{K_w}{K_a} = 6.58 \times 10^{-10}$$

$$\frac{x^2}{F - x} = \frac{x^2}{0.050 - x} = K_b$$
$$\Rightarrow x \approx \sqrt{(0.050)(6.58 \times 10^{-10})} = 5.7_4 \times 10^{-6} \text{ M}$$

$$pH = -\log\left(\frac{K_w}{x}\right) = 8.76$$

9-G. (a) $CH_3CH_2NH_2 + H_2O \rightleftharpoons CH_3CH_2NH_3^+ + OH^-$
$\underset{F-x}{\quad} \underset{\quad}{\quad} \underset{x}{\quad} \underset{x}{\quad}$

Because $pH = 11.82$, $[OH^-] = K_w/10^{-pH} = 6.6 \times 10^{-3}$ M $= [BH^+]$.
$[B] = F - x = F - [OH^-] = 0.093$ M.

$$K_b = \frac{[BH^+][OH^-]}{[B]} = \frac{(6.6 \times 10^{-3})^2}{0.093} = 4.7 \times 10^{-4}$$

(b) $CH_3CH_2NH_3^+ \rightleftharpoons CH_3CH_2NH_2 + H^+$
$\underset{F-x}{\quad} \underset{x}{\quad} \underset{x}{\quad}$

$$K_a = \frac{K_w}{K_b} = 2.1 \times 10^{-11}$$

$$\frac{x^2}{F - x} = K_a \Rightarrow x = 1.4_5 \times 10^{-6} \text{ M} \Rightarrow pH = 5.84$$

9-H.

Compound	pK_a (for conjugate acid)	
Ammonia	9.24	← Most suitable, because pK_a is closest to pH 9.00
Aniline	4.60	
Hydrazine	8.02	
Pyridine	5.20	

There would be problems using the other compounds even if they had the right pK_a. Hydrazine is highly toxic, dangerously unstable, and very reactive as a strong reducing agent. Pyridine is highly flammable, smells bad, and is somewhat toxic.

9-I. $pH = 4.25 + \log 0.75 = 4.13$

9-J. (a) $pH = pK_a + \log\dfrac{[B]}{[BH^+]}$

$$= 8.04 + \log\frac{[(1.00 \text{ g})/(74.08 \text{ g/mol})]}{[(1.00 \text{ g})/(110.54 \text{ g/mol})]} = 8.21$$

(b) $pH = pK_a + \log\dfrac{\text{mol B}}{\text{mol BH}^+}$

$$8.00 = 8.04 + \log\frac{\text{mol B}}{(1.00 \text{ g})/(110.54 \text{ g/mol})}$$
$$\Rightarrow \text{mol B} = 0.008\ 25 = 0.611 \text{ g of glycine amide}$$

(c)

	B	+	H^+	→	BH^+
Initial moles:	0.013 499		0.000 500		0.009 046
Final moles:	0.012 999		—		0.009 546

$$pH = 8.04 + \log\left(\frac{0.012\ 999}{0.009\ 546}\right) = 8.17$$

(d)

	BH^+	+	OH^-	$\rightarrow$	B
Initial moles:	0.009 546		0.001 000		0.012 999
Final moles:	0.008 546		—		0.013 999

$$pH = 8.04 + \log\left(\frac{0.013\ 999}{0.008\ 546}\right) = 8.25$$

(e) The solution in **(a)** contains 9.046 mmol glycine amide hydrochloride and 13.499 mmol glycine amide. Now we are adding 9.046 mmol of OH^-, which will convert all of the glycine amide hydrochloride into glycine amide. The new solution contains $9.046 + 13.499 = 22.545$ mmol of glycine amide in 190.46 mL. The concentration of glycine amide is $(22.545\ \text{mmol})/(190.46\ \text{mL}) = 0.118_4$ M. The pH is determined by hydrolysis of glycine amide:

$$H_2N\overset{O}{\underset{NH_2}{\diagdown}} + H_2O \rightleftharpoons H_3\overset{+}{N}\overset{O}{\underset{NH_2}{\diagdown}} + OH^-$$

$$\underset{0.118_4 - x}{} \qquad\qquad\qquad \underset{x}{} \qquad \underset{x}{}$$

$$\frac{x^2}{0.118_4 - x} = K_b = \frac{K_w}{K_a} = \frac{10^{-14.00}}{10^{-8.04}} = 1.10 \times 10^{-6}\ \text{M}$$

$$\Rightarrow x = 3.60 \times 10^{-4}\ \text{M}$$
$$pH = -\log(K_w/x) = 10.56$$

9-K. Two candidates are MES (HA, $pK_a = 6.27$) and citric acid (H_3A, $pK_3 = 6.40$). Both compounds are weak acids, so add base to give A^- to make a buffer. Weigh out $(0.25\ \text{L})(0.2\ \text{mol/L}) = 0.05$ mol of solid HA and dissolve it in ~200 mL of H_2O. Add concentrated NaOH (such as 10 M) dropwise while stirring and monitoring pH with an electrode. It will take ~2.5 mL of 10 M NaOH to neutralize half of the MES or ~12.5 mL to remove the first, second, and half of the third H^+ of citric acid. Dilute the solution with H_2O to 250 mL after adjusting the pH.

9-L. The reaction of phenylhydrazine with water is

$$B + H_2O \rightleftharpoons BH^+ + OH^- \qquad K_b$$

We know that pH = 8.13, so we can find $[OH^-]$.

$$[OH^-] = \frac{\mathcal{A}_{OH^-}}{\gamma_{OH^-}} = \frac{K_w/10^{-pH}}{\gamma_{OH^-}} = 1.78 \times 10^{-6}\ \text{M}$$

(using $\gamma_{OH^-} = 0.76$ for $\mu = 0.10$ M)

$$K_b = \frac{[BH^+]\gamma_{BH^+}[OH^-]\gamma_{OH^-}}{[B]\gamma_B}$$

$$= \frac{(1.78 \times 10^{-6})(0.80)(1.78 \times 10^{-6})(0.76)}{[0.010 - (1.78 \times 10^{-6})](1.00)}$$

$$= 1.93 \times 10^{-10}$$

$$K_a = \frac{K_w}{K_b} = 5.19 \times 10^{-5} \Rightarrow pK_a = 4.28$$

To find K_b, we made use of the equality $[BH^+] = [OH^-]$.

9-M. We use Goal Seek to vary cell B5 until cell D4 is equal to K_a. The spreadsheet shows that $[H^+] = 4.236 \times 10^{-3}$ (cell B5) and pH = 2.37 in cell B7.

	A	B	C	D	E
1	Ka = 10^–pKa =	0.00316228		Reaction quotient	
2	Kw =	1.00E-14		for Ka =	
3	FHA =	0.03		[H+][A–]/[HA] =	
4	FA =	0.015		0.0031623	
5	H =	4.236E-03	<–Goal Seek solution		
6	OH = Kw/H =	2.3609E-12		D4 = H*(FA+H–OH)/(FHA–H+OH)	
7	pH = –logH =	2.37			

If we were doing this problem by hand with the approximation that what we mix is what we get, $[H^+] = K_a[HA]/[A^-] = 10^{-2.50}[0.030]/[0.015] = 0.006\ 32$ M $\Rightarrow$ pH = 2.20.

Chapter 10

10-A. (a) $H_2SO_3 \rightleftharpoons HSO_3^- + H^+$

$$\underset{0.050 - x}{} \qquad \underset{x}{} \quad \underset{x}{}$$

$$\frac{x^2}{0.050 - x} = K_1 = 1.39 \times 10^{-2} \Rightarrow x = 2.03 \times 10^{-2}$$
$$[HSO_3^-] = [H^+] = 2.03 \times 10^{-2}\ \text{M} \Rightarrow pH = 1.69$$
$$[H_2SO_3] = 0.050 - x = 0.030\ \text{M}$$

$$[SO_3^{2-}] = \frac{K_2[HSO_3^-]}{[H^+]} = K_2 = 6.7 \times 10^{-8}\ \text{M}$$

(b) $[H^+] = \sqrt{\dfrac{K_1K_2(0.050) + K_1K_w}{K_1 + (0.050)}}$

$$= 2.71 \times 10^{-5}\ \text{M} \Rightarrow pH = 4.57$$

$$[H_2SO_3] = \frac{[H^+][HSO_3^-]}{K_1} = \frac{(2.71 \times 10^{-5})(0.050)}{1.39 \times 10^{-2}} = 9.7 \times 10^{-5}\ \text{M}$$

$$[SO_3^{2-}] = \frac{K_2[HSO_3^-]}{[H^+]} = 1.2 \times 10^{-4}\ \text{M}$$

$$[HSO_3^-] = 0.050\ \text{M}$$

(c) $SO_3^{2-} + H_2O \rightleftharpoons HSO_3^- + OH^-$

$$\underset{0.050 - x}{} \qquad\qquad \underset{x}{} \quad \underset{x}{}$$

$$\frac{x^2}{0.050 - x} = K_{b1} = \frac{K_w}{K_{a2}} = 1.49 \times 10^{-7}$$

$$[HSO_3^-] = x = 8.6 \times 10^{-5}\ \text{M}$$

$$[H^+] = \frac{K_w}{x} = 1.16 \times 10^{-10}\ \text{M} \Rightarrow pH = 9.94$$

$$[SO_3^{2-}] = 0.050 - x = 0.050\ \text{M}$$

$$[H_2SO_3] = \frac{[H^+][HSO_3^-]}{K_1} = 7.2 \times 10^{-13}\ \text{M}$$

10-B. (a) $pH = pK_2\ (\text{for}\ H_2CO_3) + \log\dfrac{[CO_3^{2-}]}{[HCO_3^-]}$

$$10.80 = 10.329 + \log\frac{(4.00\ \text{g})/(138.21\ \text{g/mol})}{x/(84.01\ \text{g/mol})}$$

$$\Rightarrow x = 0.822\ \text{g}$$

(b)

	CO_3^{2-}	+	H^+	$\rightarrow$	HCO_3^-
Initial moles:	0.028 9$_4$		0.010 0		0.009 78
Final moles:	0.018 9$_4$		—		0.019 7$_8$

$$pH = 10.329 + \log\frac{0.018\ 9_4}{0.019\ 7_8} = 10.31$$

(c)

	CO_3^{2-}	+	H^+	$\rightarrow$	HCO_3^-
Initial moles:	0.028 9$_4$		x		—
Final moles:	0.028 9$_4 - x$		—		x

$$10.00 = 10.329 + \log\frac{0.028\ 9_4 - x}{x} \Rightarrow x = 0.019\ 7\ \text{mol}$$

$$\Rightarrow \text{volume} = \frac{0.019\ 7\ \text{mol}}{0.320\ \text{M}} = 61.6\ \text{mL}$$

10-C.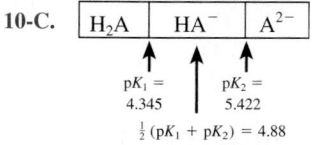

pH 4.40 is above pK_1. At pH = pK_1, there would be a 1:1 mixture of H_2A and HA^-. We must add enough KOH to convert some H_2A to HA^- to create a mixture with pH = 4.40. 5.02 g $H_2A/(132.11\ \text{g/mol}) = 0.038\ 0$ mol H_2A.

$$H_2A \quad + \quad OH^- \quad \rightarrow \quad HA^-$$

Initial moles:	0.038 0	x	—
Final moles:	$0.038\ 0 - x$	—	x

$$pH = pK_1 + \log\frac{[HA^-]}{[H_2A]}$$

$$4.40 = 4.345 + \log\frac{x}{0.038\ 0 - x} \Rightarrow x = 0.020\ 2 \text{ mol}$$

Volume of KOH = (0.020 2 mol)/(0.800 M) = 25.2 mL.

10-D. (a) Call the three forms of glutamine H_2G^+, HG, and G^-. The form shown is HG.

$$[H^+] = \sqrt{\frac{K_1K_2(0.010) + K_1K_w}{K_1 + 0.010}} = 1.9_8 \times 10^{-6}\ M \Rightarrow pH = 5.70$$

(b) Call the four forms of cysteine H_3C^+, H_2C, HC^-, and C^{2-}. The form shown is HC^-.

$$[H^+] = \sqrt{\frac{K_2K_3(0.010) + K_2K_w}{K_2 + 0.010}} = 2.8_9 \times 10^{-10}\ M \Rightarrow pH = 9.54$$

(c) Call the four forms of arginine H_3A^{2+}, H_2A^+, HA, and A^-. The form shown is HA.

$$[H^+] = \sqrt{\frac{K_2K_3(0.010) + K_2K_w}{K_2 + 0.010}} = 4.2_8 \times 10^{-11}\ M \Rightarrow pH = 10.37$$

10-E.

	pH 9.00	pH 11.00
Principal species:		
Secondary species:		
Percentage in major form:	66.5%	52.9%

The percentage in the major form was calculated with the formulas for α_{H_2A} (Equation 10-19 at pH 9.00) and α_{HA^-} (Equation 10-20 at pH 11.00).

10-F.

	pH 9.0	pH 10.0
Predominant form:	NH_3^+ $CHCH_2CH_2CO_2^-$ CO_2^-	NH_2 $CHCH_2CH_2CO_2^-$ CO_2^-
Secondary form:	NH_2 $CHCH_2CH_2CO_2^-$ CO_2^-	NH_3^+ $CHCH_2CH_2CO_2^-$ CO_2^-

	pH 9.0	pH 10.0
Predominant form:	NH_2 $CHCH_2$—⬡—OH CO_2^-	NH_2 $CHCH_2$—⬡—OH CO_2^-
Secondary form:	NH_3^+ $CHCH_2$—⬡—OH CO_2^-	NH_2 $CHCH_2$—⬡—O^- CO_2^-

10-G. The isoionic pH is the pH of a solution of pure neutral lysine, which is

$$NH_2$$
$$CHCH_2CH_2CH_2CH_2NH_3^+$$
$$CO_2^-$$

$$[H^+] = \sqrt{\frac{K_2K_3F + K_2K_w}{K_2 + F}} \Rightarrow pH = 9.93$$

10-H. We know that the isoelectric point will be near $\frac{1}{2}(pK_2 + pK_3) \approx 9.95$. At this pH, the fraction of lysine in the form H_3L^{2+} is negligible. Therefore, the electroneutrality condition reduces to $[H_2L^+] = [L^-]$, for which the expression isoelectric pH = $\frac{1}{2}(pK_2 + pK_3)$ = 9.95 applies.

Chapter 11

11-A. The titration reaction is $H^+ + OH^- \rightarrow H_2O$ and $V_e = 5.00$ mL. Three representative calculations are given:

At 1.00 mL: $[OH^-] = \left(\frac{4.00\ mL}{5.00\ mL}\right)(0.010\ 0\ M)\left(\frac{50.00\ mL}{51.00\ mL}\right) = 0.007\ 84\ M$

$$pH = -\log\left(\frac{K_w}{[OH^-]}\right) = 11.89$$

At 5.00 mL: $H_2O \rightleftharpoons H^+ + OH^-$
$$\qquad\qquad\qquad x \quad\ x$$

$$x^2 = K_w \Rightarrow x = 1.0 \times 10^{-7}\ M$$
$$pH = -\log x = 7.00$$

At 5.01 mL: $[H^+] = \left(\frac{0.01\ mL}{55.01\ mL}\right)(0.100\ M)$

$$= 1.82 \times 10^{-5}\ M \Rightarrow pH = 4.74$$

V_e (mL)	pH	V_e	pH	V_e	pH
0.00	12.00	4.50	10.96	5.10	3.74
1.00	11.89	4.90	10.26	5.50	3.05
2.00	11.76	4.99	9.26	6.00	2.75
3.00	11.58	5.00	7.00	8.00	2.29
4.00	11.27	5.01	4.74	10.00	2.08

11-B. The titration reaction is $HCO_2H + OH^- \rightarrow HCO_2^- + H_2O$ and $V_e = 50.0$ mL. For formic acid, $K_a = 1.80 \times 10^{-4}$. Four representative calculations are given:

At 0.0 mL: $HA \rightleftharpoons H^+ + A^-$
$$\qquad\qquad\qquad x \quad\ x$$

$$\frac{x^2}{0.005\ 0 - x} = K_a \Rightarrow x = 2.91 \times 10^{-3}\ M \Rightarrow pH = 2.54$$

At 48.0 mL:

	HA	+	OH^-	$\rightarrow$	A^-	+	H_2O
Initial:	50		48		—		—
Final:	2		—		48		48

$$pH = pK_a + \log\frac{[A^-]}{[HA]} = 3.744 + \log\frac{48.0}{2.0} = 5.12$$

At 50.0 mL: $A^- + H_2O \overset{K_b}{\rightleftharpoons} HA + OH^-$
$$\qquad\qquad\quad F-x \qquad\qquad x \quad\ x$$

$$K_b = \frac{K_w}{K_a} \quad\text{and}\quad F = \left(\frac{50\ mL}{100\ mL}\right)(0.05\ M)$$

$$\frac{x^2}{0.025\ 0 - x} = 5.56 \times 10^{-11} \Rightarrow x = 1.18 \times 10^{-6}\ M$$

$$pH = -\log\left(\frac{K_w}{x}\right) = 8.07$$

At 60.0 mL: $[OH^-] = \left(\dfrac{10.0\text{ mL}}{110.0\text{ mL}}\right)(0.050\ 0\text{ M})$

$= 4.55 \times 10^{-3}\text{ M} \Rightarrow pH = 11.66$

V_b (mL)	pH	V_b	pH	V_b	pH
0.0	2.54	45.0	4.70	50.5	10.40
10.0	3.14	48.0	5.12	51.0	10.69
20.0	3.57	49.0	5.43	52.0	10.99
25.0	3.74	49.5	5.74	55.0	11.38
30.0	3.92	50.0	8.07	60.0	11.66
40.0	4.35				

11-C. The titration reaction is $B + H^+ \rightarrow BH^+$ and $V_e = 50.0$ mL.
Representative calculations:

At $V_a = 0.0$ mL: $\underset{0.100-x}{B} + H_2O \rightleftharpoons \underset{x}{BH^+} + \underset{x}{OH^-}$

$\dfrac{x^2}{0.100-x} = 2.6 \times 10^{-6} \Rightarrow x = 5.09 \times 10^{-4}$ M

$pH = -\log\left(\dfrac{K_w}{x}\right) = 10.71$

At $V_a = 20.0$ mL:

	B	+	H^+	$\rightarrow$	BH^+
Initial:	50.0		20.0		—
Final:	30.0		—		20.0

$pH = pK_a \text{ (for } BH^+\text{)} + \log\dfrac{[B]}{[BH^+]} = 8.41 + \log\dfrac{30.0}{20.0} = 8.59$

At $V_a = V_e = 50.0$ mL: All B has been converted into the conjugate acid, BH^+. The formal concentration of BH^+ is $\left(\frac{100\text{ mL}}{150\text{ mL}}\right)(0.100\text{ M}) = 0.066\ 7$ M. The pH is determined by the reaction

$\underset{0.066\ 7-x}{BH^+} \rightleftharpoons \underset{x}{B} + \underset{x}{H^+}$

$\dfrac{x^2}{0.066\ 7-x} = K_a = \dfrac{K_w}{K_b} \Rightarrow x = 1.60 \times 10^{-5}\text{ M} \Rightarrow pH = 4.80$

At $V_a = 51.0$ mL: There is excess H^+:

$[H^+] = \left(\dfrac{1.0\text{ mL}}{151.0\text{ mL}}\right)(0.200\text{ M}) = 1.32 \times 10^{-3}\text{ M} \Rightarrow pH = 2.88$

V_a (mL)	pH	V_a	pH	V_a	pH
0.0	10.71	30.0	8.23	50.0	4.80
10.0	9.01⁻	40.0	7.81	50.1	3.88
20.0	8.59	49.0	6.72	51.0	2.88
25.0	8.41	49.9	5.71	60.0	1.90

11-D. The titration reactions are
$HO_2CCH_2CO_2H + OH^- \rightarrow {}^-O_2CCH_2CO_2H + H_2O$
${}^-O_2CCH_2CO_2H + OH^- \rightarrow {}^-O_2CCH_2CO_2^- + H_2O$

and the equivalence points are 25.0 and 50.0 mL. Let's designate malonic acid as H_2M. Also, $K_{a1} = 1.42 \times 10^{-3}$ and $K_{a2} = 2.01 \times 10^{-6}$.

At 0.0 mL: $\underset{0.050\ 0-x}{H_2M} \rightleftharpoons \underset{x}{H^+} + \underset{x}{HM^-}$

$\dfrac{x^2}{0.050\ 0-x} = K_1 \Rightarrow x = 7.75 \times 10^{-3}\text{ M} \Rightarrow pH = 2.11$

At 8.0 mL:

	H_2M	+	OH^-	$\rightarrow$	HM^-	+	H_2O
Initial:	25		8		—		—
Final:	17		—		8		—

$pH = pK_1 + \log\dfrac{[HM^-]}{[H_2M]} = 2.847 + \log\dfrac{8}{17} = 2.52$

At 12.5 mL: $V_b = \frac{1}{2}V_e \Rightarrow pH = pK_1 = 2.85$

At 19.3 mL:

	H_2M	+	OH^-	$\rightarrow$	HM^-	+	H_2O
Initial:	25		19.3		—		—
Final:	5.7		—		19.3		—

$pH = pK_1 + \log\dfrac{19.3}{5.7} = 3.38$

At 25.0 mL: At V_e, H_2M has been converted into HM^-.

$[H^+] = \sqrt{\dfrac{K_1K_2F + K_1K_w}{K_1 + F}}$

where $F = \left(\dfrac{50}{75}\right)(0.050\ 0\text{ M}) = 0.033\ 3$ M.

$[H^+] = 5.23 \times 10^{-5}\text{ M} \Rightarrow pH = 4.28$

At 37.5 mL: $V_b = \frac{3}{2}V_e \Rightarrow pH = pK_2 = 5.70$

At 50.0 mL: At $2V_e$, H_2M has been converted into M^{2-}:

$$M^{2-} + H_2O \rightleftharpoons HM^- + OH^-$$

$\underset{(\frac{50\text{ mL}}{100\text{ mL}})(0.050\ 0)-x}{} \quad \underset{x}{} \quad \underset{x}{}$

$\dfrac{x^2}{0.025\ 0-x} = K_{b1} = \dfrac{K_w}{K_{a2}} \Rightarrow x = 1.12 \times 10^{-5}$ M

$\Rightarrow pH = -\log\left(\dfrac{K_w}{x}\right) = 9.05$

At 56.3 mL: There are 6.3 mL of excess NaOH.

$[OH^-] = \left(\dfrac{6.3\text{ mL}}{106.3\text{ mL}}\right)(0.100\text{ M}) = 5.93 \times 10^{-3}\text{ M} \Rightarrow pH = 11.77$

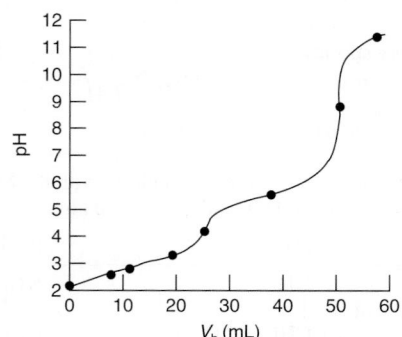

11-E.

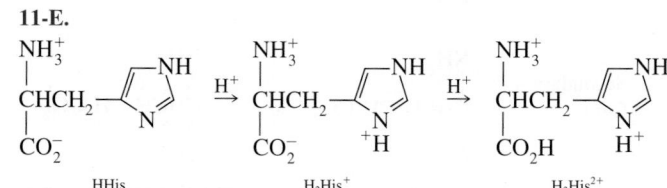

The equivalence points occur at 25.0 and 50.0 mL.

At 0 mL: HHis is the second intermediate form derived from the triprotic acid H_3His^{2+}.

$[H^+] = \sqrt{\dfrac{K_2K_3(0.050\ 0) + K_2K_w}{K_2 + (0.050\ 0)}} = 2.37 \times 10^{-8}\text{ M} \Rightarrow pH = 7.62$

At 4.0 mL:

	HHis	+	H^+	$\rightarrow$	H_2His^+
Initial:	25		4		—
Final:	21		—		4

$$\mathrm{pH} = \mathrm{p}K_2 + \log\frac{21}{4} = 6.69$$

At 12.5 mL: $\mathrm{pH} = \mathrm{p}K_2 = 5.97$

At 25.0 mL: The histidine has been converted into H_2His^+ at the formal

concentration $F = \left(\dfrac{25\ \text{mL}}{50\ \text{mL}}\right)(0.050\ 0\ M) = 0.025\ 0\ M$.

$$[H^+] = \sqrt{\frac{K_1 K_2 F + K_1 K_w}{K_1 + F}} = 1.16 \times 10^{-4}\ M \Rightarrow \mathrm{pH} = 3.94$$

At 26.0 mL:

	H_2His^+	+	H^+	$\rightarrow$	H_3His^{2+}
Initial:	25		1		—
Final:	24		—		1

$$\mathrm{pH} = \mathrm{p}K_1 + \log\frac{24}{1} = 2.98$$

The approximation that histidine reacts completely with HCl breaks down between 25 and 50 mL. If you use titration equations in Table 11-5, you would find $\mathrm{pH} = 3.28$, instead of 2.98, at $V_a = 26.0$ mL.

At 50.0 mL: The histidine has been converted into H_3His^{2+} at the formal

concentration $F = \left(\dfrac{25\ \text{mL}}{75\ \text{mL}}\right)(0.050\ 0\ M) = 0.016\ 7\ M$.

$$\begin{array}{ccccc} H_3His^{2+} & \rightleftharpoons & H_2His^+ & + & H^+ \\ 0.016\ 7 - x & & x & & x \end{array}$$

$$\frac{x^2}{0.016\ 7 - x} = K_1 = 10^{-1.6} \Rightarrow 0.011\ 5\ M \Rightarrow \mathrm{pH} = 1.94$$

Alternatively, if you use $K_1 = 0.03$ instead of $10^{-1.6}$, you would find $x = 0.011\ 9\ M$ and $\mathrm{pH} = 1.92$. Both ways to do this problem are reasonable and the answers are within the uncertainty of K_1.

11-F. Figure 10-1: bromothymol blue: blue $\rightarrow$ yellow
Figure 10-2: thymol blue: yellow $\rightarrow$ blue
Figure 10-3: thymolphthalein: colorless $\rightarrow$ blue

11-G. The titration reaction is $HA + OH^- \rightarrow A^- + H_2O$. It requires 1 mol of NaOH to react with 1 mol of HA. Therefore, the formal concentration of A^- at the equivalence point is

$$\underbrace{\left(\frac{27.63}{127.63}\right)}_{\text{Dilution factor for NaOH}} \times \underbrace{(0.093\ 81)}_{\text{Initial concentration of NaOH}} = 0.020\ 31\ M$$

Because the pH is 10.99, $[OH^-] = 9.77 \times 10^{-4}$ M, and we can write
$$A^- + H_2O \rightleftharpoons HA + OH^-$$

$$K_b = \frac{[HA][OH^-]}{[A^-]} = \frac{(9.77 \times 10^{-4})^2}{0.020\ 31 - (9.77 \times 10^{-4})} = 4.94 \times 10^{-5}$$

$$K_a = \frac{K_w}{K_b} = 2.03 \times 10^{-10} \Rightarrow \mathrm{p}K_a = 9.69$$

For the 19.47-mL point, we have

	HA	+	OH^-	$\rightarrow$	A^-	+	H_2O
Initial:	27.63		19.47		—		—
Final:	8.16		—		19.47		—

$$\mathrm{pH} = \mathrm{p}K_a + \log\frac{[A^-]}{[HA]} = 9.69 + \log\frac{19.47}{8.16} = 10.07$$

11-H. When $V_b = \frac{1}{2}V_e$, $[HA] = [A^-] = 0.033\ 3\ M$ (using a correction for dilution by NaOH). $[Na^+] = 0.033\ 3\ M$ as well. Ionic strength = 0.033 3 M.

$$\mathrm{p}K_a = \mathrm{pH} - \log\frac{[A^-]\gamma_{A^-}}{[HA]\gamma_{HA}}$$

$$= 4.62 - \log\frac{(0.033\ 3)(0.854)}{(0.033\ 3)(1.00)} = 4.69$$

The activity coefficient of A^- was found by interpolation in Table 8-1.

11-I. **(a)** The derivatives are shown in the spreadsheet below. In the first derivative graph, the maximum is near 119 mL. In Figure 11-7, the second derivative graph gives an end point of 118.9 μL.

(b) Column G in the spreadsheet gives $V_b(10^{-pH})$. In a graph of $V_b(10^{-pH})$ vs. V_b, the points from 113 to 117 μL give a straight line whose slope is -1.178×10^6 and whose intercept (end point) is 118.7 μL.

11-J. **(a)** pH 9.6 is past the equivalence point, so excess volume, V, is given by

$$[OH^-] = 10^{-4.4} = (0.100\ 0\ M)\frac{V}{50.00 + 10.00 + V}$$

$$\Rightarrow V = 0.024\ \text{mL}$$

(b) pH 8.8 is before the equivalence point:

$$8.8 = 6.27 + \log\frac{[A^-]}{[HA]} \Rightarrow \frac{[A^-]}{[HA]} = 339$$

	A	B	C	D	E	F	G
1	Derivatives of titration curve						
2							
3				First Derivative		Second Derivative	
4	μL NaOH	pH	μL	Derivative	μL	Derivative	Vb*10^-pH
5	107	6.921					
6	110	7.117	108.5	6.533E-02			
7	113	7.359	111.5	8.067E-02	110	5.11E-03	4.94E-06
8	114	7.457	113.5	9.800E-02	112.5	8.67E-03	3.98E-06
9	115	7.569	114.5	1.120E-01	114	1.40E-02	3.10E-06
10	116	7.705	115.5	1.360E-01	115	2.40E-02	2.29E-06
11	117	7.878	116.5	1.730E-01	116	3.70E-02	1.55E-06
12	118	8.090	117.5	2.120E-01	117	3.90E-02	9.59E-07
13	119	8.343	118.5	2.530E-01	118	4.10E-02	5.40E-07
14	120	8.591	119.5	2.480E-01	119	-5.00E-03	3.08E-07
15	121	8.794	120.5	2.030E-01	120	-4.50E-02	1.94E-07
16	122	8.952	121.5	1.580E-01	121	-4.50E-02	1.36E-07
17							
18	C6 = (A6+A5)/2			E7 = (C7+C6)/2		G7 = A7*10^-B7	
19	D6 = (B6-B5)/(A6-A5)			F7 = (D7-D6)/(C7-C6)			

Spreadsheet for Exercise 11-I(a).

Titration reaction:

	HA	+	OH⁻	→	A⁻	+	H₂O

	HA	OH⁻	A⁻	H₂O
Relative initial quantities:	10	V	—	—
Relative final quantities:	$10 - V$	—	V	—

To attain a ratio $[A^-]/[HA] = 339$, we need $V/(10 - V) = 339 \Rightarrow V = 9.97$ mL. The indicator error is $10 - 9.97 = 0.03$ mL.

11-K. (a) $A = 2\,080[HIn] + 14\,200[In^-]$

(b) $[HIn] = x$; $[In^-] = 1.84 \times 10^{-4} - x$
$A = 0.868 = 2\,080x + 14\,200(1.84 \times 10^{-4} - x)$
$\Rightarrow x = 1.44 \times 10^{-4}$ M

$$pK_a = pH - \log\frac{[In^-]}{[HIn]}$$

$$= 6.23 - \log\frac{(1.84 \times 10^{-4}) - (1.44 \times 10^{-4})}{(1.44 \times 10^{-4})} = 6.79$$

Chapter 12

12-A. For every mole of K^+ entering the first reaction, 4 moles of EDTA are produced in the second reaction.
Moles of EDTA = moles of Zn^{2+} used in titration

$$[K^+] = \frac{\frac{1}{4}(\text{moles of } Zn^{2+})}{\text{volume of original sample}}$$

$$= \frac{\frac{1}{4}[28.73\ (\pm0.03)\ \text{mL}][0.043\ 7\ (\pm0.000\ 1)\ \text{M}]}{250.0\ (\pm0.1)\ \text{mL}}$$

$$= \frac{[\frac{1}{4}(\pm0\%)][28.73\ (\pm0.104\%)][0.043\ 7\ (\pm0.229\%)]}{250.0\ (\pm0.040\ 0\%)}$$

$$= 1.256\ (\pm0.255\%) \times 10^{-3}\ \text{M} = 1.256\ (\pm0.003)\ \text{mM}$$

12-B. Total $Fe^{3+} + Cu^{2+}$ in 25.00 mL = $(16.06\ \text{mL}) \times (0.050\ 83\ \text{M}) = 0.816\ 3$ mmol.

Second titration:

millimoles EDTA used: $(25.00\ \text{mL})(0.050\ 83\ \text{M}) = 1.270\ 8$
millimoles Pb^{2+} needed: $(19.77\ \text{mL})(0.018\ 83\ \text{M}) = \underline{0.372\ 3}$
millimoles Fe^{3+} present: (difference) $\qquad 0.898\ 5$

Because 50.00 mL of unknown were used in the second titration, Fe^{3+} in 25.00 mL $= \frac{1}{2}(0.898\ 5\ \text{mmol}) = 0.449\ 2$ mmol. The millimoles of Cu^{2+} in 25.00 mL are $0.816\ 3 - 0.449\ 2 = 0.367\ 1$ mmol/25.00 mL $= 0.014\ 68$ M.

12-C. Designating the total concentration of free EDTA as [EDTA], we can write

$$K_f' = \frac{[CuY^{2-}]}{[Cu^{2+}][EDTA]} = \alpha_{Y^{4-}} \cdot K_f = (2.9 \times 10^{-7})(10^{18.78})$$

$$= 1.7_4 \times 10^{12}$$

Representative calculations:

At 0.1 mL:

$$[EDTA] = \left(\frac{25.0\ \text{mL} - 0.1\ \text{mL}}{25.0\ \text{mL}}\right)(0.040\ 0\ \text{M})\left(\frac{50.0\ \text{mL}}{50.1\ \text{mL}}\right) = 0.039\ 8\ \text{M}$$

$$[CuY^{2-}] = \left(\frac{0.1\ \text{mL}}{50.1\ \text{mL}}\right)(0.080\ 0\ \text{M}) = 1.60 \times 10^{-4}\ \text{M}$$

$$[Cu^{2+}] = \frac{[CuY^{2-}]}{K_f'[EDTA]} = \frac{(1.60 \times 10^{-4}\ \text{M})}{(1.7_4 \times 10^{12})(0.039\ 8\ \text{M})}$$

$$= 2.3 \times 10^{-15}\ \text{M} \Rightarrow pCu^{2+} = 14.64$$

At 25.0 mL:

Formal concentration of $CuY^{2-} = \left(\frac{25.0\ \text{mL}}{75.0\ \text{mL}}\right)(0.080\ 0\ \text{M}) = 0.026\ 7\ \text{M}$

	Cu²⁺	+	EDTA	⇌	CuY²⁻
Initial concentration:	—		—		0.026 7
Final concentration:	x		x		$0.026\ 7 - x$

$$\frac{0.026\ 7 - x}{x^2} = 1.7_4 \times 10^{12} \Rightarrow [Cu^{2+}] = 1.2_4 \times 10^{-7}\ \text{M} \Rightarrow pCu^{2+} = 6.91$$

At 26.0 mL:

$$[Cu^{2+}] = \left(\frac{1.0\ \text{mL}}{76.0\ \text{mL}}\right)(0.080\ 0\ \text{M}) = 1.05 \times 10^{-3}\ \text{M} \Rightarrow pCu^{2+} = 2.98$$

Volume (mL)	pCu²⁺	Volume	pCu²⁺	Volume	pCu²⁺
0.1	14.64	15.0	12.07	25.0	6.91
5.0	12.84	20.0	11.64	26.0	2.98
10.0	12.42	24.0	10.86	30.0	2.30

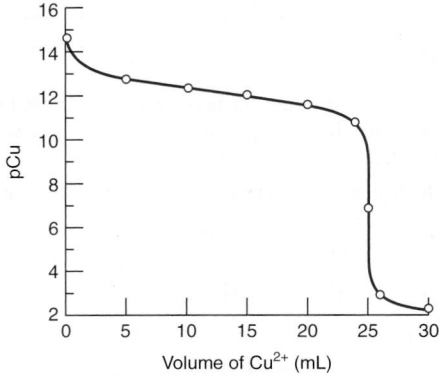

12-D. We seek a relationship between $[H_2Y^{2-}]$ and $[Y^{4-}]$, which we can relate to total EDTA:

$HY^{3-} \rightleftharpoons H^+ + Y^{4-}$	K_6
$H_2Y^{2-} \rightleftharpoons H^+ + HY^{3-}$	K_5

$$H_2Y^{2-} \rightleftharpoons 2H^+ + Y^{4-} \qquad K = K_5K_6 = \frac{[H^+]^2[Y^{4-}]}{[H_2Y^{2-}]}$$

$$[H_2Y^{2-}] = \frac{[H^+]^2[Y^{4-}]}{K_5K_6} = \frac{[H^+]^2\alpha_{Y^{4-}}[EDTA]}{K_5K_6}$$

Using the values $[H^+] = 10^{-5.00}$ M, $\alpha_{Y^{4-}} = 2.9 \times 10^{-7}$, and $[EDTA] = 1.2_4 \times 10^{-7}$ M gives $[H_2Y^{2-}] = 1.1 \times 10^{-7}$ M.

12-E. (a) One volume of Mn^{2+} requires two volumes of EDTA to reach the equivalence point. The formal concentration of MnY^{2-} at the equivalence point is $(\frac{1}{3})(0.010\ 0\ \text{M}) = 0.003\ 33$ M.

$$\begin{array}{ccccc} Mn^{2+} & + & EDTA & \rightleftharpoons & MnY^{2-} \\ x & & x & & 0.003\ 33 - x \end{array}$$

$$\frac{0.003\ 33 - x}{x^2} = \alpha_{Y^{4-}} \cdot K_f = (3.8 \times 10^{-4})10^{13.89} = 2.9 \times 10^{10}$$

$$\Rightarrow x = [Mn^{2+}] = 3.4 \times 10^{-7}\ \text{M}$$

(b) pH is constant, so the *quotient* $[H_3Y^-]/[H_2Y^{2-}]$ is constant throughout the *entire* titration.

$$\frac{[H_2Y^{2-}][H^+]}{[H_3Y^-]} = K_4 \Rightarrow \frac{[H_3Y^-]}{[H_2Y^{2-}]} = \frac{[H^+]}{K_4} = \frac{10^{-7.00}}{10^{-2.69}} = 4.9 \times 10^{-5}$$

12-F. K_f for $CoY^{2-} = 10^{16.45} = 2.8 \times 10^{16}$
$\alpha_{Y^{4-}} = 0.041$ at pH 9.00

$$\alpha_{Co^{2+}} = \frac{1}{1 + \beta_1[C_2O_4^{2-}] + \beta_2[C_2O_4^{2-}]^2} = 6.8 \times 10^{-6}$$

(using $\beta_1 = K_1 = 10^{4.69}$ and $\beta_2 = K_1K_2 = 10^{7.15}$)

$K_f' = \alpha_{Y^{4-}} \cdot K_f = 1.1_6 \times 10^{15}$
$K_f'' = \alpha_{Co^{2+}}\alpha_{Y^{4-}} \cdot K_f = 7.9 \times 10^9$
At 0 mL:
$\quad [Co^{2+}] = \alpha_{Co^{2+}}(1.00 \times 10^{-3}\ \text{M}) = 6.8 \times 10^{-9}\ \text{M} \Rightarrow pCo^{2+} = 8.17$

At 1.00 mL:

$$C_{Co^{2+}} = \left(\frac{1.00 \text{ mL}}{2.00 \text{ mL}}\right) \underset{\substack{\text{Initial}\\\text{concentration}}}{(1.00 \times 10^{-3} \text{ M})} \underset{\substack{\text{Dilution}\\\text{factor}}}{\left(\frac{20.00 \text{ mL}}{21.00 \text{ mL}}\right)}$$

Fraction remaining

$$= 4.76 \times 10^{-4} \text{ M}$$

$$[Co^{2+}] = \alpha_{Co^{2+}}C_{Co^{2+}} = 3.2 \times 10^{-9} \text{ M} \Rightarrow pCo^{2+} = 8.49$$

At 2.00 mL: This is the equivalence point.

$$\underset{x}{C_{Co^{2+}}} + \underset{x}{\text{EDTA}} \overset{K''_f}{\rightleftharpoons} \underset{\left(\frac{20.00 \text{ mL}}{22.00 \text{ mL}}\right)(1.00 \times 10^{-3} \text{ M}) - x}{CoY^{2-}}$$

$$K''_f = \frac{9.09 \times 10^{-4} - x}{x^2} \Rightarrow x = 3.4 \times 10^{-7} \text{ M} = C_{Co^{2+}}$$

$$[Co^{2+}] = \alpha_{Co^{2+}}C_{Co^{2+}} = 2.3 \times 10^{-12} \text{ M} \Rightarrow pCo^{2+} = 11.64$$

At 3.00 mL:

$$[\text{excess EDTA}] = \frac{1.00 \text{ mL}}{23.00 \text{ mL}}(1.00 \times 10^{-2} \text{ M}) = 4.35 \times 10^{-4} \text{ M}$$

$$[CoY^{2-}] = \frac{20.00 \text{ mL}}{23.00 \text{ mL}}(1.00 \times 10^{-3} \text{ M}) = 8.70 \times 10^{-4} \text{ M}$$

Knowing [EDTA] and [CoY^{2-}], we use K'_f to find [Co^{2+}]:

$$K'_f = \frac{[CoY^{2-}]}{[Co^{2+}][\text{EDTA}]} = \frac{[8.70 \times 10^{-4} \text{ M}]}{[Co^{2+}][4.35 \times 10^{-4} \text{ M}]}$$

$$\Rightarrow [Co^{2+}] = 1.7 \times 10^{-15} \text{ M} \Rightarrow pCo^{2+} = 14.76$$

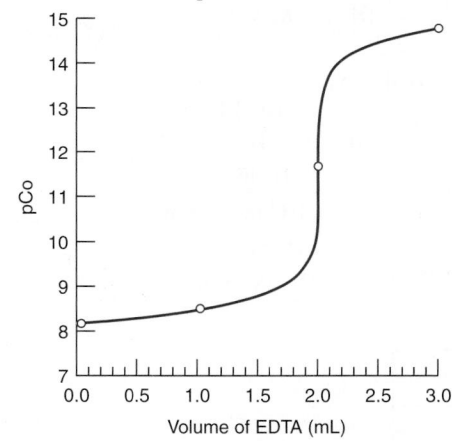

12-G.   25.0 mL of 0.120 M iminodiacetic acid = 3.00 mmol
25.0 mL of 0.050 0 M Cu^{2+} = 1.25 mmol

	Cu^{2+}	+	2 iminodiacetic acid	$\rightleftharpoons$	CuX_2^{2-}
Initial mmol:	1.25		3.00		—
Final mmol:	—		0.50		1.25

$$\frac{[CuX_2^{2-}]}{[Cu^{2+}][X^{2-}]^2} = K_f$$

$$\frac{[1.25 \text{ mmol}/50.0 \text{ mL}]}{[Cu^{2+}][(0.50 \text{ mmol}/50.0 \text{ mL})(4.6 \times 10^{-3})]^2} = 3.5 \times 10^{16}$$

$$\Rightarrow [Cu^{2+}] = 3.4 \times 10^{-10} \text{ M}$$

Chapter 13

13-A.   Hydroxybenzene = HA with pK_{HA} = 9.997

Dimethylamine = B from monoprotic BH^+ with pK_{BH^+} = 10.774

Mixture contains 0.010 mol HA, 0.030 mol B, and 0.015 mol HCl in 1.00 L

Chemical reactions:

$$\begin{aligned}
HA &\rightleftharpoons A^- + H^+ & K_{HA} &= 10^{-9.997}\\
BH^+ &\rightleftharpoons B + H^+ & K_{BH^+} &= 10^{-10.774}\\
H_2O &\rightleftharpoons H^+ + OH^- & K_w &= 10^{-14.00}
\end{aligned}$$

Charge balance: $\quad [H^+] + [BH^+] = [OH^-] + [A^-] + [Cl^-]$

Mass balances: $\quad [Cl^-] = 0.015 \text{ M}$

$$[BH^+] + [B] = 0.030 \text{ M} \equiv F_B$$

$$[HA] + [A^-] = 0.010 \text{ M} \equiv F_A$$

We have seven equations and seven chemical species.

Fractional composition equations:

$$[BH^+] = \alpha_{BH^+}F_B = \frac{[H^+]F_B}{[H^+] + K_{BH^+}}$$

$$[B] = \alpha_B F_B = \frac{K_{BH^+}F_B}{[H^+] + K_{BH^+}}$$

$$[HA] = \alpha_{HA}F_A = \frac{[H^+]F_A}{[H^+] + K_{HA}}$$

$$[A^-] = \alpha_{A^-}F_A = \frac{K_{HA}F_A}{[H^+] + K_{HA}}$$

Substitute into charge balance:

$$[H^+] + \alpha_{BH^+}F_B = K_w/[H^+] + \alpha_{A^-}F_A + [0.015 \text{ M}] \tag{A}$$

	A	B	C	D	E	F	G	H	I
1	Mixture of 0.010 M HA, 0.030 M B, and 0.015 M HCl								
2									
3	F_A =	0.010		F_B =	0.030		[Cl⁻] =	0.015	
4	pK_{HA} =	9.997		pK_{BH+} =	10.774		pK_w =	14.000	
5	K_{HA} =	1.01E-10		K_{BH+} =	1.68E-11		K_w =	1.00E-14	
6									
7	Species in charge balance:						Other concentrations:		
8	[H⁺] =	4.67E-11		[A⁻] =	6.83E-03		[HA] =	3.17E-03	
9	[BH⁺] =	2.20E-02		[Cl⁻] =	0.015		[B] =	7.95E-03	
10				[OH⁻] =	2.14E-04		pH =	10.331	
11							↑ initial value is a guess		
12	Positive charge minus negative charge				-4.92E-17		= B8+B9-E8-E9-E10		
13	Formulas:								
14	B5 = 10^-B4			B8 = 10^-H10			H5 = 10^-H4		
15	E5 = 10^-E4			E10 = H5/B8			E9 = H3		
16	B9 = B8*E3/(B8+E5)						E8 = B5*B3/(B8+B5)		
17	H9 = E5*E3/(B8+E5)						H8 = B8*B3/(B8+B5)		

Spreadsheet for Exercise 13-A.

We solve Equation A for $[H^+]$ by using Solver in the spreadsheet, with an initial guess of pH = 10 in cell H10. In Solver Options, select the tab for All Methods. Set Constraint Precision = 1E-15, use Automatic Scaling, Max Time = 100, and Iterations = 200. Then click OK. In the Solver window, Set Objective E12 Equal To Value of 0 By Changing Variable Cells H10. Click Solve and Solver finds pH = 10.33 in cell H10, giving a net charge near 0 in cell E12.

13-B. We use effective equilibrium constants, K', as follows:

$$K_{HA} = \frac{[A^-]\gamma_{A^-}[H^+]\gamma_{H^+}}{[HA]\gamma_{HA}} = 10^{-9.997}$$

$$K'_{HA} = K_{HA}\left(\frac{\gamma_{HA}}{\gamma_{A^-}\gamma_{H^+}}\right) = \frac{[A^-][H^+]}{[HA]}$$

$$[HA] = \alpha_{HA}F_A = \frac{[H^+]F_A}{[H^+] + K'_{HA}}$$

$$[A^-] = \alpha_{A^-}F_A = \frac{K'_{HA}F_A}{[H^+] + K'_{HA}}$$

$$K_{BH^+} = \frac{[B]\gamma_B[H^+]\gamma_{H^+}}{[BH^+]\gamma_{BH^+}} = 10^{-10.774}$$

$$K'_{BH^+} = K_{BH^+}\left(\frac{\gamma_{BH^+}}{\gamma_B\gamma_{H^+}}\right) = \frac{[B][H^+]}{[BH^+]}$$

$$[BH^+] = \alpha_{BH^+}F_B = \frac{[H^+]F_B}{[H^+] + K'_{BH^+}}$$

$$[B] = \alpha_B F_B = \frac{K'_{BH^+}F_B}{[H^+] + K'_{BH^+}}$$

$$K_w = [H^+]\gamma_{H^+}[OH^-]\gamma_{OH^-} = 10^{-13.995}$$

$$K'_w = \frac{K_w}{\gamma_{H^+}\gamma_{OH^-}} = [H^+][OH^-]$$

$$[OH^-] = K'_w/[H^+]$$
$$pH = -\log([H^+]\gamma_{H^+})$$
$$[H^+] = (10^{-pH})/\gamma_{H^+}$$

Modify the spreadsheet in Exercise 13-A. To handle the circular reference in Excel 2010, select the File tab and select Options. In the Options window, select Formulas. In Calculation Options, check "Enable iterative

calculation" and set Maximum Change to 1e-15. Click OK. Add activity coefficients in cells A8:H9. Compute effective equilibrium constants in row 5. Because we are going to the trouble of using activity coefficients, we use $pK_w = 13.995$ in cell H4 instead of the less accurate value of 14.00. With an initial ionic strength of 0 in cell E17 and a guess of pH = 10 in cell H14, the net charge in cell E16 is 0.005 56 m. Write the formula for ionic strength in cell E17. Execute Solver to find the pH in cell H14 that reduces the net charge in cell E16 to near 0. The final pH in cell H14 is 10.358, giving an ionic strength of 0.022 5 m in cell E17.

13-C. **(a)** 2-Aminobenzoic acid = HA from diprotic H_2A^+, $pK_1 = 2.08$, $pK_2 = 4.96$

Dimethylamine = B from monoprotic BH^+, $pK_a = 10.774$
Mixture contains 0.040 mol HA, 0.020 mol B, and 0.015 mol HCl in 1.00 L.

Chemical reactions:
$$H_2A^+ \rightleftharpoons HA + H^+ \qquad K_1 = 10^{-2.08}$$
$$HA \rightleftharpoons A^- + H^+ \qquad K_2 = 10^{-4.96}$$
$$BH^+ \rightleftharpoons B + H^+ \qquad K_a = 10^{-10.774}$$
$$H_2O \rightleftharpoons H^+ + OH^- \qquad K_w = 10^{-14.00}$$

Charge balance: $\quad [H^+] + [H_2A^+] + [BH^+] = [OH^-] + [A^-] + [Cl^-]$

Mass balances: $\quad [Cl^-] = 0.015$ M
$$[BH^+] + [B] = 0.020 \text{ M} \equiv F_B$$
$$[H_2A^+] + [HA] + [A^-] = 0.040 \text{ M} \equiv F_A$$

We have eight equations and eight chemical species.

Fractional composition equations:

$$[BH^+] = \alpha_{BH^+}F_B = \frac{[H^+]F_B}{[H^+] + K_a}$$

$$[B] = \alpha_B F_B = \frac{K_a F_B}{[H^+] + K_a}$$

$$[H_2A^+] = \alpha_{H_2A^+}F_A = \frac{[H^+]^2 F_A}{[H^+]^2 + [H^+]K_1 + K_1K_2}$$

$$[HA] = \alpha_{HA}F_A = \frac{K_1[H^+]F_A}{[H^+]^2 + [H^+]K_1 + K_1K_2}$$

$$[A^-] = \alpha_{A^-}F_A = \frac{K_1K_2F_A}{[H^+]^2 + [H^+]K_1 + K_1K_2}$$

	A	B	C	D	E	F	G	H	I
1	Mixture of 0.010 M HA, 0.030 M B, and 0.015 M HCl								
2	With activities								
3	$F_A =$	0.010		$F_B =$	0.030		$[Cl^-] =$	0.015	
4	$pK_{HA} =$	9.997		$pK_{BH+} =$	10.774		$pK_w =$	13.995	
5	$K_{HA}' =$	1.35E-10		$K_{BH+}' =$	1.68E-11		$K_w' =$	1.35E-14	
6									
7	Activity coefficients:								
8	$H^+ =$	0.86		A^-	0.86		HA	1.00	
9	$OH^- =$	0.86		BH^+	0.86		B	1.00	
10									
11	Species in charge balance:						Other concentrations:		
12	$[H^+] =$	5.07E-11		$[A^-] =$	7.26E-03		$[HA] =$	2.74E-03	
13	$[BH^+] =$	2.25E-02		$[Cl^-] =$	0.015		$[B] =$	7.47E-03	
14				$[OH^-] =$	2.67E-04		pH =	10.358	
15							↑ initial value is a guess		
16	Positive charge minus negative charge =				7.67E-17	= B12+B13–E12–E13–E14			
17				Ionic strength =	2.25E-02	= 0.5*(B12+B13+E12+E13+E14)			
18	Formulas:								
19	B5 = (10^–B4)*H8/(E8*B8)				H8 = H9 = 1				
20	E5 = (10^–E4)*E9/(H9*B8)				E13 = H3				
21	H5 = (10^–H4)/(B8*B9)								
22	B8 = B9 = E8 = E9 = 10^(–0.51*1^2*(SQRT($E$17)/(1+SQRT($E$17))–0.3*$E$17))								
23	B12 = (10^–H14)/B8				E14 = H5/B12				
24	B13 = B12*E3/(B12+E5)				H13 = E5*E3/(B12+E5)				
25	E12 = B5*B3/(B12+B5)				H12 = B12*B3/(B12+B5)				

Spreadsheet for Exercise 13-B

	A	B	C	D	E	F	G	H	I
1	Mixture of 0.040 M HA, 0.020 M B, and 0.015 M HCl								
2									
3	F_A =	0.040		F_B =	0.020		[Cl⁻] =	0.015	
4	pK_1 =	2.080		pK_a =	10.774		K_w =	1.00E-14	
5	pK_2 =	4.960		K_a =	1.68E-11				
6	K_1 =	8.32E-03							
7	K_2 =	1.10E-05							
8									
9	Species in charge balance:						Other concentrations:		
10	[H⁺] =	7.03E-05		[H₂A⁺] =	2.90E-04		[HA] =	3.43E-02	
11	[BH⁺] =	2.00E-02		[A⁻] =	5.36E-03		[B] =	4.79E-09	
12	[OH⁻] =	1.42E-10		[Cl⁻] =	0.015		pH =	4.153	
13							↑ initial value is a guess		
14	Positive charge minus negative charge				0.00E+00		= B10+B11+E10-B12-E11-E12		
15	Formulas:								
16	B16 = 10^-B4		B7 = 10^-B5			E5 = 10^-E4			
17	B10 =10^-H12		B12 = H4/B10			E12 = H3			
18	B11 = B10*E3/(B10+E5)								
19	E10 = B10^2*B3/(B10^2+B10*B6+B6*B7)								
20	E11 = B6*B7*B3/(B10^2+B10*B6+B6*B7)								
21	H10 = B10*B6*B3/(B10^2+B10*B6+B6*B7)								
22	H11 = E5*E3/(B10+E5)								

Spreadsheet for Exercise 13-C

Substitute into charge balance:

$$[H^+] + \alpha_{H_2A^+} F_A + \alpha_{BH^+} F_B = K_w/[H^+] + \alpha_{A^-} F_A + [0.015\ M] \quad \text{(A)}$$

Solve Equation A for $[H^+]$ by using Solver with an initial guess of pH = 7 in cell H12. In the Solver window, Set Target Cell <u>E14</u> Equal To Value of <u>0</u> By Changing Variable Cells <u>H12</u>. Click Solve to find pH = 4.15 in cell H12, giving a net charge near 0 in cell E14.

(b) From the concentrations in the spreadsheet, we find the following fractions of 2-aminobenzoic acid: H_2A^+ = 0.7%, HA = 85.9%, and A^- = 13.4%. The fractions of dimethylamine are BH^+ = 100.0% and B = 0.0%. The simple prediction is that HCl would consume B, giving 100% BH^+. The remaining 5 mmol B consumes 5 mmol HA, making 5 mmol A^- and leaving 35 mmol HA.

Predicted fractions: A^- = 5/40 = 12.5%, HA = 35/40 = 87.5%

Estimated pH = pK_2 + log([A^-]/[HA]) = 4.96 + log(5/35) = 4.11

13-D. Effective equilibrium constants:

$$H_2T \overset{pK_1 = 3.036}{\rightleftharpoons} HT^- + H^+ \qquad K_1' = K_1\left(\frac{\gamma_{H_2T}}{\gamma_{HT^-}\gamma_{H^+}}\right) = \frac{[HT^-][H^+]}{[H_2T]}$$

$$HT^- \overset{pK_2 = 4.366}{\rightleftharpoons} T^{2-} + H^+ \qquad K_2' = K_2\left(\frac{\gamma_{HT^-}}{\gamma_{T^{2-}}\gamma_{H^+}}\right) = \frac{[T^{2-}][H^+]}{[HT^-]}$$

$$PyH^+ \overset{pK_a = 5.20}{\rightleftharpoons} Py + H^+ \qquad K_a' = K_a\left(\frac{\gamma_{PyH^+}}{\gamma_{Py}\gamma_{H^+}}\right) = \frac{[Py][H^+]}{[PyH^+]}$$

$$H_2O \overset{pK_w = 13.995}{\rightleftharpoons} H^+ + OH^- \qquad K_w' = \frac{K_w}{\gamma_{H^+}\gamma_{OH^-}} = [H^+][OH^-]$$

$$[OH^-] = K_w'/[H^+] \qquad pH = -\log([H^+]\gamma_{H^+})$$

The spreadsheet is modified from the one in the text by addition of activity coefficients in cells A10:H11. The ionic strength for computing activity coefficients with the Davies equation is in cell E20. Use Solver to find the pH in cell H17 that makes the net charge in cell E19 nearly 0. In the spreadsheet, a pH of 4.114 gives an ionic strength of 0.054 0 m.

13-E. $AgCN(s) \rightleftharpoons Ag^+ + CN^- \qquad pK_{sp} = 15.66$

$[CN^-] = K_{sp}/[Ag^+]$

$HCN(aq) \rightleftharpoons CN^- + H^+ \qquad pK_{HCN} = 9.21$

$$[HCN(aq)] = \frac{[H^+][CN^-]}{K_{HCN}} = \frac{[H^+]K_{sp}}{K_{HCN}[Ag^+]}$$

$Ag^+ + H_2O \rightleftharpoons AgOH(aq) + H^+ \qquad pK_{Ag} = 12.0$

$$[AgOH(aq)] = \frac{K_{Ag}[Ag^+]}{[H^+]}$$

Mass balance: total silver = total cyanide

$[Ag^+] + [AgOH(aq)] = [CN^-] + [HCN(aq)]$

Substitute expressions for concentrations into the mass balance:

$$[Ag^+] + \frac{K_{Ag}[Ag^+]}{[H^+]} = \frac{K_{sp}}{[Ag^+]} + \frac{[H^+]K_{sp}}{K_{HCN}[Ag^+]} \quad \text{(A)}$$

Rearrange Equation A to solve for $[Ag^+]$ as a function of $[H^+]$ or use Solver to find $[Ag^+]$ as a function of $[H^+]$. We use the algebraic solution, which is easy for this exercise. Multiply both sides by $[Ag^+]$ and solve:

$$[Ag^+]^2 + \frac{K_{Ag}[Ag^+]^2}{[H^+]} = K_{sp} + \frac{[H^+]K_{sp}}{K_{HCN}}$$

$$[Ag^+]^2\left(\frac{[H^+] + K_{Ag}}{[H^+]}\right) = K_{sp}\left(\frac{K_{HCN} + [H^+]}{K_{HCN}}\right)$$

$$[Ag^+] = \sqrt{\frac{K_{sp}(K_{HCN} + [H^+])[H^+]}{K_{HCN}([H^+] + K_{Ag})}} \quad \text{(B)}$$

The spreadsheet uses Equation B to find $[Ag^+]$ in column C. pH is input in column A. To find the pH of unbuffered solution, we find the pH at which the net charge in column H is zero. We used Solver to find that pH = 7.28 in cell A12 makes the net charge in cell H12 equal to 0.

To see if the solubility of Ag_2O is exceeded, evaluate the solubility product at pH 7.28: $[Ag^+]^2[OH^-]^2 = (1.4 \times 10^{-7})^2(K_w/10^{-7.28})^2 = 5 \times 10^{-29} < K_{Ag_2O} = 10^{15.42}$. We predict that Ag_2O will not precipitate from the unbuffered, saturated solution of AgCN.

Spreadsheet for Exercise 13-D

	A	B	C	D	E	F	G	H
1	Mixture of 0.020 M Na⁺HT⁻, 0.015 M PyH⁺Cl⁻, and 0.010 M KOH = with activity							
2								
3	F_{H2T} =	0.020		F_{PyH^+} =	0.015		$[K^+]$ =	0.010
4	pK_1 =	3.036		pK_a =	5.20		pK_w =	13.995
5	pK_2 =	4.366		K_a' =	6.31E-06		K_w' =	1.52E-14
6	K_1' =	1.38E-03						
7	K_2' =	9.67E-05						
8								
9	Activity coefficients from Davies equation:							
10	H^+ =	0.817		HT^- =	0.817		OH^- =	0.817
11	PyH^+ =	0.817		T^{2-} =	0.445			
12								
13	Species in charge balance:						Other concentrations:	
14	$[H^+]$ =	9.42E-05		$[OH^-]$ =	1.61E-10		$[H_2T]$ =	6.51E-04
15	$[PyH^+]$ =	1.41E-02		$[HT^-]$ =	9.54E-03		$[Py]$ =	9.42E-04
16	$[Na^+]$ =	0.020		$[T^{2-}]$ =	9.80E-03			
17	$[K^+]$ =	0.010		$[Cl^-]$ =	0.015		pH =	4.114
18							↑initial value is a guess	
19	Positive charge minus negative charge =				0.00E+00			
20	Ionic strength =				5.40E-02			
21	Formulas:							
22	B6 = 10^-B4*(1/(E10*B10))			B14 = (10^-H17)/B10			E14 = H5/B14	
23	B7 = 10^-B5*(E10/(E11*B10))			B16 = B3			E17 = E3	
24	E5 = 10^-E4*(B11/B10)			B17 = H3				
25	H5 = (10^-E4)/(B10*H10)							
26	B10=B11=E10=H10 = 10^(-0.51*1^2*(SQRT($C$20)/(1+SQRT($C$20))-0.3*$C$20))							
27	E11 = 10^(-0.51*2^2*(SQRT($C$20)/(1+SQRT($C$20))-0.3*$C$20))							
28	B15 = B14*E3/(B14+E5)							
29	E15 = B14*B6*B3/(B14^2+B14*B6+B6*B7)							
30	E16 = B6*B7*B3/(B14^2+B14*B6+B6*B7)							
31	H14 = B14^2*B3/(B14^2+B14*B6+B6*B7)							
32	H15 = E5*E3/(B14+E5)							
33	C21 = 0.5*(B14+B15+B16+B17+E14+E15+4*E16+E17)							
34	E19 = B14+B15+B16+B17-E14-E15-2*E16-E17							
35	E20 = 0.5*(B14+B15+B16+B17+E14+E15+4*E16+E17)							

Spreadsheet for Exercise 13-D

Spreadsheet for Exercise 13-E

	A	B	C	D	E	F	G	H
1	Solubility of AgCN							
2								
3	pK_{sp} =	15.66		K_{sp} =	2.2E-16		K_w =	1.00E-14
4	pK_{HCN} =	9.21		K_{HCN} =	6.2E-10			
5	pK_{Ag} =	12.00		K_{Ag} =	1.0E-12			
6								Net
7	pH	$[H^+]$	$[Ag^+]$	$[CN^-]$	[HCN]	[AgOH]	$[OH^-]$	charge
8	0	1.0E+00	6.0E-04	3.7E-13	6.0E-04	6.0E-16	1.00E-14	1.0E+00
9	2	1.0E-02	6.0E-05	3.7E-12	6.0E-05	6.0E-15	1.00E-12	1.0E-02
10	4	1.0E-04	6.0E-06	3.7E-11	6.0E-06	6.0E-14	1.00E-10	1.1E-04
11	6	1.0E-06	6.0E-07	3.7E-10	6.0E-07	6.0E-13	1.00E-08	1.6E-06
12	7.28	5.3E-08	1.4E-07	1.6E-09	1.4E-07	2.6E-12	1.89E-07	3.7E-21
13	8	1.0E-08	6.1E-08	3.6E-09	5.8E-08	6.1E-12	1.00E-06	-9.3E-07
14	10	1.0E-10	1.6E-08	1.4E-08	2.2E-09	1.6E-10	1.00E-04	-1.0E-04
15	12	1.0E-12	1.0E-08	2.1E-08	3.4E-11	1.0E-08	1.00E-02	-1.0E-02
16	14	1.0E-14	1.5E-09	1.5E-07	2.4E-12	1.5E-07	1.00E+00	-1.0E+00
17								
18	B8 = 10^-A8			D8 = $E$3/C8		E8 = B8*D8/$E$4		F8 = $E$5*C8/B8
19	C8 = SQRT($E$3*($E$4+B8)*(B8)/($E$4*(B8+$E$5)))							
20	G8 = $H$3/B8			H8 = B8+C8-D8-G8				

Spreadsheet for Exercise 13-E

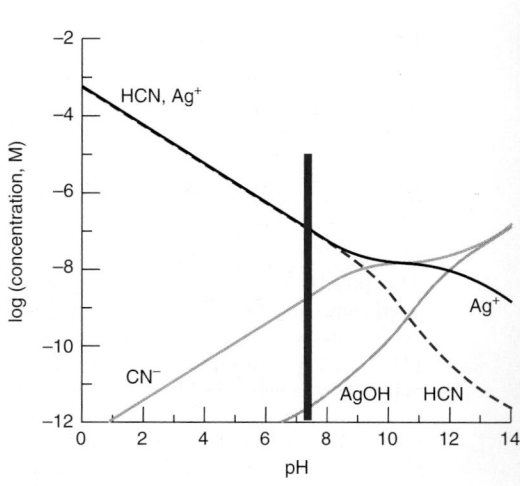

Graph for Exercise 13-E

	A	B	C	D	E	F	G
1	Difference plot for acetic acid						
2				C15 = 10^-B15/$B$8			
3	Titrant NaOH =	0.4905	C_b (M)	D15 = 10^-$B$9/C15			
4	Initial volume =	200	V_o (mL)	E15 = $B$7+($B$6-$B$3*A15)			
5	Acetic acid =	3.96	L (mmol)	-(C15-D15)*($B$4+A15))/$B$5			
6	HCl added =	0.484	A (mmol)	F15 = $C15/($C15+$E$10)			
7	Number of H+ =	1	n	G15 = (E15-F15)^2			
8	Activity coeff =	0.78	γ_H				
9	pK_w' =	13.869					
10	pK_a =	4.726		K_a =	1.881E-05	= 10^-B10	
11	$\Sigma(resid)^2$ =	0.0045	= sum of column G				
12							
13	v	pH	[H+] =	[OH-] =	Measured	Theoretical	(residuals)^2 =
14	mL NaOH		$(10^{-pH})/\gamma_H$	$(10^{-pKw})/[H^+]$	n_H	$n_H = \alpha_{HA}$	$(n_{meas} - n_{theor})^2$
15	0.00	2.79	2.08E-03	6.50E-12	1.017	0.991	0.000685
16	0.30	2.89	1.65E-03	8.19E-12	1.002	0.989	0.000163
17	:						
18	4.80	4.78	2.13E-05	6.36E-10	0.527	0.531	0.000018
19	5.10	4.85	1.81E-05	7.47E-10	0.490	0.491	0.000001
20	:						
21	10.20	11.39	5.22E-12	2.59E-03	-0.004	0.000	0.000014
22	10.50	11.54	3.70E-12	3.66E-03	0.016	0.000	0.000259

Spreadsheet for Exercise 13-F

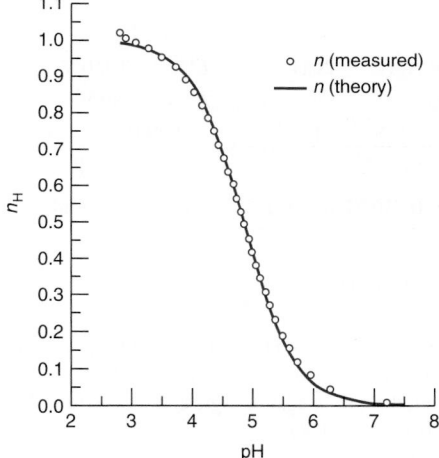

Graph for Exercise 13-F

13-F. (a) $\bar{n}_H = \dfrac{\text{moles of bound } H^+}{\text{total moles of weak acid}} = \dfrac{[HA]}{[HA] + [A^-]}$

$= \dfrac{[HA]}{F_{HA}} = \dfrac{F_{HA} - [A^-]}{F_{HA}}$ (A)

Charge balance: $[H^+] + [Na^+] = [OH^-] + [Cl^-]_{HCl} + [A^-]$

$\Rightarrow -[A^-] = [OH^-] + [Cl^-]_{HCl} - [H^+] - [Na^+]$

Put this expression for $-[A^-]$ into numerator of (A):

$\bar{n}_H(\text{measured}) = \dfrac{F_{HA} + [OH^-] + [Cl^-]_{HCl} - [H^+] - [Na^+]}{F_{HA}}$

$= 1 + \dfrac{[OH^-] + [Cl^-]_{HCl} - [H^+] - [Na^+]}{F_{HA}}$

Making the same substitutions used in Section 13-4 produces Equation 13-62 with $n = 1$. The theoretical expression is $\bar{n}_H(\text{theoretical}) = \alpha_{HA} = [H^+]/([H^+] + K_a)$.

(b) Optimized values of pK_w' and pK_a in cells B9 and B10 in the spreadsheet are 13.869 and 4.726. These were obtained from initial guesses of $pK_w' = 14$ and $pK_a = 5$ after executing Solver to minimize the sum of squared residuals in cell B11. The NIST database lists $pK_a = 4.757$ at $\mu = 0$ and $pK_a = 4.56$ at $\mu = 0.1$ M. Our value of 4.726 at $\mu = 0.1$ M suggests that the titration experiment was not very accurate.

Chapter 14

14-A. The cell voltage is $E° = 1.35$ V because all activities are unity.

$I = P/E = 0.010\ 0\ W/1.35\ V = 7.41 \times 10^{-3}$ C/s

mol e^-/s = $(7.41 \times 10^{-3}$ C/s$)/(9.649 \times 10^4$ C/mol$)$

$= 7.68 \times 10^{-8}$ mol e^-/s = 2.42 mol e^-/365 days

HgO accepts $2e^-$ when it is reduced from Hg(II) to Hg(0), so the consumption rate is = (2.42 mol e^-/365 days)(1 mol HgO/2 mol e^-) = 1.21 mol HgO/365 days = 0.262 kg HgO = 0.578 lb

14-B.

(a) $5Br_2(aq) + 10e^- \rightleftharpoons 10Br^-$

$2IO_3^- + 12H^+ + 10e^- \rightleftharpoons I_2(s) + 6H_2O$

$I_2(s) + 5Br_2(aq) + 6H_2O \rightleftharpoons 2IO_3^- + 10Br^- + 12H^+$

$E_+° = 1.098$ V

$E_-° = 1.210$ V

$E° = 1.098 - 1.210 = -0.112$ V

$K = 10^{10(-0.112)/0.059\ 16} = 1 \times 10^{-19}$

(b) $Cr^{2+} + 2e^- \rightleftharpoons Cr(s)$

$Fe^{2+} + 2e^- \rightleftharpoons Fe(s)$

$Cr^{2+} + Fe(s) \rightleftharpoons Cr(s) + Fe^{2+}$

$E_+° = -0.89$ V

$E_-° = -0.44$ V

$E° = -0.89 - (-0.44) = -0.45$ V

$K = 10^{2(-0.45)/0.059\ 16} = 6 \times 10^{-16}$

(c)

$$\underline{\begin{array}{l} Cl_2(g) + 2e^- \rightleftharpoons 2Cl^- \\ Mg^{2+} + 2e^- \rightleftharpoons Mg(s) \end{array}}$$

$$Mg(s) + Cl_2(g) \rightleftharpoons Mg^{2+} + 2Cl^-$$

$$E_+^\circ = 1.360 \text{ V}$$

$$\underline{E_-^\circ = -2.360 \text{ V}}$$

$$E^\circ = 1.360 - (-2.360) = 3.720 \text{ V}$$

$$K = 10^{2(3.720)/0.059\,16} = 6 \times 10^{125}$$

(d)

$$\underline{\begin{array}{l} 3[MnO_2(s) + 4H^+ + 2e^- \rightleftharpoons Mn^{2+} + 2H_2O] \\ 2[MnO_4^- + 4H^+ + 3e^- \rightleftharpoons MnO_2(s) + 2H_2O] \end{array}}$$

$$5MnO_2(s) + 4H^+ \rightleftharpoons 2MnO_4^- + 3Mn^{2+} + 2H_2O$$

$$E_+^\circ = 1.230 \text{ V}$$

$$\underline{E_-^\circ = 1.692 \text{ V}}$$

$$E^\circ = 1.230 - 1.692 = -0.462 \text{ V}$$

$$K = 10^{6(-0.462)/0.059\,16} = 1 \times 10^{-47}$$

An alternate way to answer **(d)** is

$$\underline{\begin{array}{l} 5[MnO_2(s) + 4H^+ + 2e^- \rightleftharpoons Mn^{2+} + 2H_2O] \\ 2[MnO_4^- + 8H^+ + 5e^- \rightleftharpoons Mn^{2+} + 4H_2O] \end{array}}$$

$$5MnO_2(s) + 4H^+ \rightleftharpoons 2MnO_4^- + 3Mn^{2+} + 2H_2O$$

$$E_+^\circ = 1.230 \text{ V}$$

$$\underline{E_-^\circ = 1.507 \text{ V}}$$

$$E^\circ = 1.230 - 1.507 = -0.277 \text{ V}$$

$$K = 10^{10(-0.277)/0.059\,16} = 2 \times 10^{-47}$$

(e)

$$\underline{\begin{array}{l} Ag^+ + e^- \rightleftharpoons Ag(s) \\ Ag(S_2O_3)_2^{3-} + e^- \rightleftharpoons Ag(s) + 2S_2O_3^{2-} \end{array}}$$

$$Ag^+ + 2S_2O_3^{2-} \rightleftharpoons Ag(S_2O_3)_2^{3-}$$

$$E_+^\circ = 0.799 \text{ V}$$

$$\underline{E_-^\circ = 0.017 \text{ V}}$$

$$E^\circ = 0.799 - 0.017 = 0.782 \text{ V}$$

$$K = 10^{0.782/0.059\,16} = 2 \times 10^{13}$$

(f)

$$\underline{\begin{array}{l} CuI(s) + e^- \rightleftharpoons Cu(s) + I^- \\ Cu^+ + e^- \rightleftharpoons Cu(s) \end{array}}$$

$$CuI(s) \rightleftharpoons Cu^+ + I^-$$

$$E_+^\circ = -0.185 \text{ V}$$

$$\underline{E_-^\circ = 0.518 \text{ V}}$$

$$E^\circ = -0.185 - 0.518 = -0.703 \text{ V}$$

$$K = 10^{-0.703/0.059\,16} = 1 \times 10^{-12}$$

14-C. (a)

$$\underline{\begin{array}{ll} Br_2(l) + 2e^- \rightleftharpoons 2Br^- & E_+^\circ = 1.078 \text{ V} \\ Fe^{2+} + 2e^- \rightleftharpoons Fe(s) & E_-^\circ = -0.44 \text{ V} \end{array}}$$

$$Br_2(l) + Fe(s) \rightleftharpoons 2Br^- + Fe^{2+}$$

$$E = \left\{1.078 - \frac{0.059\,16}{2}\log(0.050)^2\right\} - \left\{-0.44 - \frac{0.059\,16}{2}\log\frac{1}{0.010}\right\}$$

$$= 1.155 - (-0.50) = 1.65 \text{ V}$$

Electrons flow from the more negative Fe electrode (−0.50 V) to the more positive Pt electrode (1.155 V) through the circuit.

(b)

$$\underline{\begin{array}{ll} Fe^{2+} + 2e^- \rightleftharpoons Fe(s) & E_+^\circ = -0.44 \text{ V} \\ Cu^{2+} + 2e^- \rightleftharpoons Cu(s) & E_-^\circ = 0.339 \text{ V} \end{array}}$$

$$Fe^{2+} + Cu(s) \rightleftharpoons Fe(s) + Cu^{2+}$$

$$E = \left\{-0.44 - \frac{0.059\,16}{2}\log\frac{1}{0.050}\right\} - \left\{0.339 - \frac{0.059\,16}{2}\log\frac{1}{0.020}\right\}$$

$$= -0.48 - (0.289) = -0.77 \text{ V}$$

Electrons flow from the more negative Fe electrode (−0.48 V) to the more positive Cu electrode (0.289 V) through the circuit.

(c)

$$\underline{\begin{array}{ll} Cl_2(g) + 2e^- \rightleftharpoons 2Cl^- & E_+^\circ = 1.360 \text{ V} \\ Hg_2Cl_2(s) + 2e^- \rightleftharpoons 2Hg(l) + 2Cl^- & E_-^\circ = 0.268 \text{ V} \end{array}}$$

$$Cl_2(g) + 2Hg(l) \rightleftharpoons Hg_2Cl_2(s)$$

$$E = \left\{1.360 - \frac{0.059\,16}{2}\log\frac{(0.040)^2}{0.50}\right\} - \left\{0.268 - \frac{0.059\,16}{2}\log(0.060)^2\right\}$$

$$= 1.434 - (0.340) = 1.094 \text{ V}$$

Electrons flow from the more negative Hg electrode (0.340 V) to the more positive Pt electrode (1.434 V) through the circuit.

14-D. (a)

$$\begin{array}{ll} H^+ + e^- \rightleftharpoons \tfrac{1}{2}H_2(g) & E_+^\circ = 0 \text{ V} \\ Ag^+ + e^- \rightleftharpoons Ag(s) & E_-^\circ = 0.799 \text{ V} \end{array}$$

$$E^\circ = E_+^\circ - E_-^\circ = -0.799 \text{ V}$$

$$E = \left\{0 - 0.059\,16\log\frac{P_{H_2}^{1/2}}{[H^+]}\right\} - \left\{0.799 - 0.059\,16\log\frac{1}{[Ag^+]}\right\}$$

(b) $[Ag^+] = \dfrac{K_{sp}}{[I^-]} = \dfrac{8.3 \times 10^{-17}}{0.10} = 8.3 \times 10^{-16} \text{ M}$

$$E = \left\{0 - 0.059\,16\log\frac{\sqrt{0.20}}{0.10}\right\} - \left\{0.799 - 0.059\,16\log\frac{1}{8.3 \times 10^{-16}}\right\}$$

$$= -0.038 - (-0.093) = 0.055 \text{ V}$$

Electrons flow from the more negative Ag electrode (−0.093 V) to the more positive Pt electrode (−0.038 V) through the circuit.

(c)

$$\begin{array}{ll} H^+ + e^- \rightleftharpoons \tfrac{1}{2}H_2(g) & E_+^\circ = 0 \text{ V} \\ AgI(s) + e^- \rightleftharpoons Ag(s) + I^- & E_-^\circ = ? \end{array}$$

$$0.055 = \left\{0 - 0.059\,16\log\frac{\sqrt{0.20}}{0.10}\right\} - \{E_-^\circ - 0.059\,16\log(0.10)\}$$

$$\Rightarrow E_-^\circ = -0.153 \text{ V}$$

(Appendix H gives $E_-^\circ = -0.152$ V.)

14-E.

$$\begin{array}{ll} Ag(CN)_2^- + e^- \rightleftharpoons Ag(s) + 2CN^- & E_+^\circ = -0.310 \text{ V} \\ Cu^{2+} + 2e^- \rightleftharpoons Cu(s) & E_-^\circ = 0.339 \text{ V} \end{array}$$

$$E = \left\{-0.310 - 0.059\,16\log\frac{[CN^-]^2}{[Ag(CN)_2^-]}\right\} - \left\{0.339 - \frac{0.059\,16}{2}\log\frac{1}{[Cu^{2+}]}\right\}$$

We know that $[Ag(CN)_2^-] = 0.010$ M and $[Cu^{2+}] = 0.030$ M. To find $[CN^-]$ at pH 8.21, we write

$$\frac{[CN^-]}{[HCN]} = \frac{K_a}{[H^+]} \Rightarrow [CN^-] = 0.10\,[HCN]$$

But, because $[CN^-] + [HCN] = 0.10$ M, $[CN^-] = 0.009\,1$ M. Putting this concentration into the Nernst equation gives $E = -0.187 - (0.294) = -0.481$ V. Electrons flow from the more negative Ag electrode (−0.187 V) to the more positive Cu electrode (0.294 V).

14-F. (a)

$$PuO_2^+ + e^- + 4H^+ \rightleftharpoons Pu^{4+} + 2H_2O$$

$$\begin{array}{ll} PuO_2^{2+} \rightarrow PuO_2^+ & \Delta G = -1F(0.966) \\ PuO_2^+ \rightarrow Pu^{4+} & \Delta G = -1F\,E^\circ \\ Pu^{4+} \rightarrow Pu^{3+} & \Delta G = -1F(1.006) \\ \hline PuO_2^{2+} \rightarrow Pu^{3+} & \Delta G = -3F(1.021) \end{array}$$

$$-3F(1.021) = -1F(0.966) - 1F\,E^\circ - 1F(1.006) \Rightarrow E^\circ = 1.091 \text{ V}$$

(b)

$$\underline{\begin{array}{ll} 2PuO_2^{2+} + 2e^- \rightleftharpoons 2PuO_2^+ & E_+^\circ = 0.966 \text{ V} \\ \tfrac{1}{2}O_2(g) + 2H^+ + 2e^- \rightleftharpoons H_2O & E_-^\circ = 1.229 \text{ V} \end{array}}$$

$$2PuO_2^{2+} + H_2O \rightleftharpoons 2PuO_2^+ + \tfrac{1}{2}O_2(g) + 2H^+$$

$$E = \left\{0.966 - \frac{0.059\,16}{2}\log\frac{[PuO_2^+]^2}{[PuO_2^{2+}]^2}\right\} - \left\{1.229 - \frac{0.059\,16}{2}\log\frac{1}{P_{O_2}^{1/2}[H^+]^2}\right\}$$

At pH 2.00: $E = \{0.966\} - \{1.100\} = -0.134$ V

At pH 7.00: $E = \{0.966\} - \{0.805\} = +0.161$ V

$[PuO_2^+]$ cancels $[PuO_2^{2+}]$ because they are equal. At pH 2.00, we insert $[H^+] = 10^{-2.00}$ M and $P_{O_2} = 0.20$ bar to find $E = -0.134$ V. Because $E < 0$, the reaction is not spontaneous and water is not oxidized. At pH 7.00, we find $E = +0.161$ V, so water will be oxidized.

14-G.
$$2H^+ + 2e^- \rightleftharpoons H_2(g) \qquad\qquad E_+^\circ = 0 \text{ V}$$
$$Hg_2Cl_2(s) + 2e^- \rightleftharpoons 2Hg(l) + 2Cl^- \qquad E_-^\circ = 0.268 \text{ V}$$

$$E = \left\{-\frac{0.059\ 16}{2}\log\frac{P_{H_2}}{[H^+]^2}\right\} - \left\{0.268 - \frac{0.059\ 16}{2}\log\,[Cl^-]^2\right\}$$

We find $[H^+]$ in the right half-cell by considering the acid-base chemistry of KHP, the intermediate form of a diprotic acid:

$$[H^+] = \sqrt{\frac{K_1K_2(0.050) + K_1K_w}{K_1 + 0.050}} = 6.5 \times 10^{-5}\text{ M}$$

$$E = \left\{-\frac{0.059\ 16}{2}\log\frac{1}{(6.5\times 10^{-5})^2}\right\} - \left\{0.268 - \frac{0.059\ 16}{2}\log\,(0.10)^2\right\}$$

$$= -0.247_7 - 0.327_2 = -0.575\text{ V}$$

Electrons flow from the more negative Pt electrode (-0.247_7 V) to the more positive Hg electrode (0.327_2 V).

14-H.
$$2H^+ + 2e^- \rightleftharpoons H_2(g) \qquad E_+^\circ = 0 \text{ V}$$
$$Hg^{2+} + 2e^- \rightleftharpoons Hg(l) \qquad E_-^\circ = 0.852 \text{ V}$$

$$E = \left\{-\frac{0.059\ 16}{2}\log\frac{P_{H_2}}{[H^+]^2}\right\} - \left\{0.852 - \frac{0.059\ 16}{2}\log\frac{1}{[Hg^{2+}]}\right\}$$

$$0.083 = \left\{-\frac{0.059\ 16}{2}\log\frac{1}{1^2}\right\} - \left\{0.852 - \frac{0.059\ 16}{2}\log\frac{1}{[Hg^{2+}]}\right\}$$

$$\Rightarrow [Hg^{2+}] = 2.5 \times 10^{-32}\text{ M}$$

$[HgI_4^{2-}] = 0.001\ 0$ M. To make this much HgI_4^{2-}, the concentration of I^- must have been reduced from 0.500 M to 0.496 M, because one Hg^{2+} ion reacts with four I^- ions.

$$K = \frac{[HgI_4^{2-}]}{[Hg^{2+}][I^-]^4} = \frac{(0.001\ 0)}{(2.5\times 10^{-32})(0.496)^4} = 7 \times 10^{29}$$

14-I.
$$-\quad CuY^{2-} + 2e^- \rightleftharpoons Cu(s) + Y^{4-} \qquad E_+^\circ = ?$$
$$\underline{\quad Cu^{2+} + 2e^- \rightleftharpoons Cu(s) \qquad\qquad E_-^\circ = 0.339\text{ V}}$$
$$CuY^{2-} \overset{1/K_f}{\rightleftharpoons} Cu^{2+} + Y^{4-} \qquad\qquad E^\circ$$

$$E^\circ = \frac{0.059\ 16}{2}\log\frac{1}{K_f} = -0.556\text{ V}$$

$$E_+^\circ = E^\circ + E_-^\circ = -0.556 + 0.339 = -0.217\text{ V}$$

14-J. To compare glucose and H_2 at pH = 0, we need to know E° for each. For H_2, $E^\circ = 0$ V. For glucose, we find E° from $E^{\circ\prime}$:

$$HA + 2H^+ + 2e^- \rightleftharpoons G + H_2O$$
$$\text{Gluconic acid} \qquad\qquad\qquad \text{Glucose}$$

$$E = E^\circ - \frac{0.059\ 16}{2}\log\frac{[G]}{[HA][H^+]^2}$$

But $F_G = [G]$ and $[HA] = \dfrac{[H^+]F_{HA}}{[H^+] + K_a}$. Putting these into the Nernst equation gives

$$E = E^\circ - \frac{0.059\ 16}{2}\log\frac{F_G}{\left(\dfrac{[H^+]F_{HA}}{[H^+] + K_a}\right)[H^+]^2}$$

$$= \underbrace{E^\circ - \frac{0.059\ 16}{2}\log\frac{[H^+] + K_a}{[H^+]^3}}_{\text{This is } E^{\circ\prime} = -0.45\text{ V when }[H^+] = 10^{-7}} - \frac{0.059\ 16}{2}\log\frac{F_G}{F_{HA}}$$

$$-0.45\text{ V} = E^\circ - \frac{0.059\ 16}{2}\log\frac{10^{-7.00} + 10^{-3.56}}{(10^{-7.00})^3}$$

$$\Rightarrow E^\circ = +0.066_6\text{ V for glucose}$$

Because E° for H_2 is more negative than E° for glucose, H_2 is the stronger reducing agent at pH 0.00.

14-K. **(a)** Each H^+ must provide $\frac{1}{2}(34.5$ kJ$)$ when it passes from outside to inside.

$$\Delta G = -\frac{1}{2}(34.5 \times 10^3\text{ J}) = -RT\ln\frac{\mathcal{A}_{high}}{\mathcal{A}_{low}}$$

$$\frac{\mathcal{A}_{high}}{\mathcal{A}_{low}} = 1.05 \times 10^3 \Rightarrow \Delta pH = \log(1.05 \times 10^3) = 3.02\text{ pH units}$$

(b) $\Delta G = -nNFE$ (where n = charge of H^+ = 1 and N = 1 mol H^+)

$$-\frac{1}{2}(34.5 \times 10^3\text{ J}) = -(1\text{ mol})(F\text{ C/mol})(E\text{ V}) \Rightarrow E = 0.179\text{ J/C} = 0.179\text{ V}$$

(c) If $\Delta pH = 1.00$, $\mathcal{A}_{high}/\mathcal{A}_{low} = 10$.

$$\Delta G(pH) = -RT\ln 10 = -5.7 \times 10^3\text{ J}$$

$$\Delta G(\text{electric}) = \left[\frac{1}{2}(34.5) - 5.7\right]\text{kJ} = 11.5_5\text{ kJ}$$

$$E = \frac{\Delta G(\text{electric})}{nNF} = \frac{11.5_5 \times 10^3\text{ J}}{(1)(1\text{ mol})(96\ 485\text{ C/mol})} = 0.120\text{ V}$$

Chapter 15

15-A. The reaction at the silver electrode (written as a reduction) is $Ag^+ + e^- \rightleftharpoons Ag(s)$, and the cell voltage is written as

$$E = E_+ - E_- = E_+ - E(\text{S.C.E.}) = E_+ - 0.241$$
$$= \left(0.799 - 0.059\ 16\log\frac{1}{[Ag^+]}\right) - 0.241$$
$$= 0.558 + 0.059\ 16\log\,[Ag^+]$$

Titration reaction: $Br^- + Ag^+ \rightarrow AgBr(s) \qquad K_{sp} = 5.0 \times 10^{-13}$

The equivalence point is $V_e = 25.0$. Between 0 and 25 mL, there is unreacted Ag^+ in the solution.

$$1.0\text{ mL:}\quad [Ag^+] = \left(\frac{24.0\text{ mL}}{25.0\text{ mL}}\right)\!(0.100\text{ M})\!\left(\frac{50.0\text{ mL}}{51.0\text{ mL}}\right) = 0.094\ 1\text{ M}$$

$$\underset{\substack{\text{Fraction of }Ag^+ \\ \text{remaining}}}{\nearrow} \qquad \underset{\substack{\text{Initial concentration} \\ \text{of }Ag^+}}{} \qquad \underset{\substack{\text{Dilution} \\ \text{factor}}}{\nwarrow}$$

$$\Rightarrow E = \ = 0.558 + 0.059\ 16\log[0.094\ 1] = 0.497\text{ V}$$

$$12.5\text{ mL:}\quad [Ag^+] = \left(\frac{12.5}{25.0}\right)\!(0.100\text{ M})\!\left(\frac{50.0}{62.5}\right) = 0.040\ 0\text{ M}$$
$$\Rightarrow E = 0.475\text{ V}$$

$$24.0\text{ mL:}\quad [Ag^+] = \left(\frac{1.0}{25.0}\right)\!(0.100\text{ M})\!\left(\frac{50.0}{74.0}\right) = 0.002\ 70\text{ M}$$
$$\Rightarrow E = 0.406\text{ V}$$

$$24.9\text{ mL:}\quad [Ag^+] = \left(\frac{0.10}{25.0}\right)\!(0.100\text{ M})\!\left(\frac{50.0}{74.9}\right) = 2.67 \times 10^{-4}\text{ M}$$
$$\Rightarrow E = 0.347\text{ V}$$

Beyond 25 mL, all AgBr has precipitated and there is excess Br^- in solution.

$$25.1\text{ mL:}\quad [Br^-] = \left(\frac{0.1\text{ mL}}{75.1\text{ mL}}\right)\!(0.200\text{ M}) = 2._{67} \times 10^{-4}\text{ M}$$
$$\Rightarrow [Ag^+] = K_{sp}/[Br^-] = (5.0 \times 10^{-13})/(2._{67} \times 10^{-4}) = 1._{88} \times 10^{-9}\text{ M}$$
$$\Rightarrow E = +0.042\text{ V}$$

$$26.0\text{ mL:}\quad [Br^-] = \left(\frac{1.0}{76.0}\right)\!(0.200\text{ M}) = 2.6_6 \times 10^{-4}\text{ M}$$
$$\Rightarrow [Ag^+] = 1.9_0 \times 10^{-10}\text{ M} \Rightarrow E = -0.017\text{ V}$$

$$\text{At } 35.0\text{ mL:}\quad [Br^-] = \left(\frac{10.0}{85.0}\right)\!(0.200\text{ M}) = 0.023\ 5\text{ M}$$
$$\Rightarrow [Ag^+] = 2.1_2 \times 10^{-11}\text{ M} \Rightarrow E = -0.073\text{ V}$$

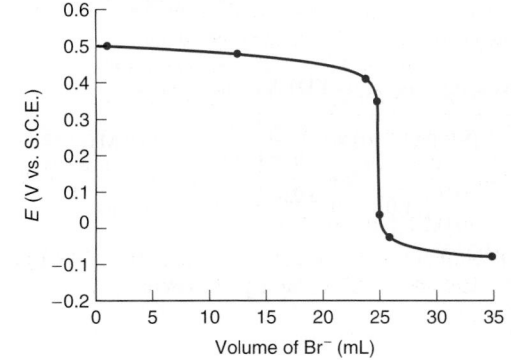

15-B. The cell voltage is given by Equation C, in which K_f is the formation constant for $Hg(EDTA)^{2-}$ ($= 10^{21.5}$). To find the voltage, we must calculate $[HgY^{2-}]$ and $[Y^{4-}]$ at each point. The concentration of HgY^{2-} is 1.0×10^{-4} M when $V = 0$ and is thereafter affected only by dilution because $K_f(HgY^{2-}) \gg K_f(MgY^{2-})$. The concentration of Y^{4-} is found from the Mg-EDTA equilibrium at all but the first point. At $V = 0$ mL, the Hg-EDTA equilibrium determines $[Y^{4-}]$.

$$0 \text{ mL: } \frac{[HgY^{2-}]}{[Hg^{2+}][EDTA]} = \alpha_{Y^{4-}}K_f \text{ (for } HgY^{4-}\text{)} = (0.30)(10^{21.5})$$

$$\frac{1.0 \times 10^{-4} - x}{(x)(x)} = 9.5 \times 10^{20} \Rightarrow x = [EDTA] = 3.2 \times 10^{-13} \text{ M}$$

$$[Y^{4-}] = \alpha_{Y^{4-}}[EDTA] = 9.7 \times 10^{-14} \text{ M}$$

Using Equation C, we write

$$E = 0.852 - 0.241 - \frac{0.059\,16}{2} \log \frac{10^{21.5}}{1.0 \times 10^{-4}}$$

$$-\frac{0.059\,16}{2} \log (9.7 \times 10^{-14}) = 0.242 \text{ V}$$

10.0 mL: $V_e = 25.0$ mL, so $\frac{10}{25}$ of Mg^{2+} is in the form MgY^{2-}, and $\frac{15}{25}$ is in the form Mg^{2+}.

$$[Y^{4-}] = \frac{[MgY^{2-}]}{[Mg^{2+}]}/K_f \text{ (for } MgY^{2-}\text{)} = \left(\frac{10}{15}\right)/6.2 \times 10^8 = 1.08 \times 10^{-9} \text{ M}$$

$$[HgY^{2-}] = \left(\frac{50.0 \text{ mL}}{60.0 \text{ mL}}\right)(1.0 \times 10^{-4} \text{ M}) = 8.33 \times 10^{-5} \text{ M}$$
$$\underbrace{}_{\text{Dilution factor}}$$

$$E = 0.852 - 0.241 - \frac{0.059\,16}{2} \log \frac{10^{21.5}}{8.33 \times 10^{-5}}$$

$$-\frac{0.059\,16}{2} \log (1.08 \times 10^{-9}) = 0.120 \text{ V}$$

20.0 mL: $[Y^{4-}] = \left(\frac{20}{5}\right)/6.2 \times 10^8 = 6.45 \times 10^{-9}$ M

$$[HgY^{2-}] = \left(\frac{50.0}{70.0}\right)(1.0 \times 10^{-4} \text{ M}) = 7.14 \times 10^{-5} \text{ M}$$

$$\Rightarrow E = 0.095 \text{ V}$$

24.9 mL: $[Y^{4-}] = \left(\frac{24.9}{0.1}\right)/6.2 \times 10^8 = 4.02 \times 10^{-7}$ M

$$[HgY^{2-}] = \left(\frac{50.0}{74.9}\right)(1.0 \times 10^{-4} \text{ M}) = 6.68 \times 10^{-5} \text{ M}$$

$$\Rightarrow E = 0.041 \text{ V}$$

25.0 mL: This is the equivalence point, at which $[Mg^{2+}] = [EDTA]$.

$$\frac{[MgY^{2-}]}{[Mg^{2+}][EDTA]} = \alpha_{Y^{4-}}K_f \text{ (for } MgY^{2-}\text{)}$$

$$\frac{\left(\frac{50.0 \text{ mL}}{75.0 \text{ mL}}\right)(0.010\,0) - x}{x^2} = 1.85 \times 10^8 \Rightarrow x = 6.0 \times 10^{-6} \text{ M}$$

$$[Y^{4-}] = \alpha_{Y^{4-}}(6.0 \times 10^{-6} \text{ M}) = 1.80 \times 10^{-6} \text{ M}$$

$$[HgY^{2-}] = \left(\frac{50.0}{75.0}\right)(1.0 \times 10^{-4} \text{ M}) = 6.67 \times 10^{-5} \text{ M}$$

$$\Rightarrow E = 0.021 \text{ V}$$

26.0 mL: Now there is excess EDTA in the solution:

$$[Y^{4-}] = \alpha_{Y^{4-}}[EDTA] = (0.30)\left[\left(\frac{1.0 \text{ mL}}{76.0 \text{ mL}}\right)(0.020\,0 \text{ M})\right] = 7.89 \times 10^{-5} \text{ M}$$

$$[HgY^{2-}] = \left(\frac{50.0}{76.0}\right)(1.0 \times 10^{-4} \text{ M}) = 6.58 \times 10^{-5} \text{ M}$$

$$\Rightarrow E = -0.027 \text{ V}$$

15-C. At intermediate pH, the voltage will be constant at 100 mV. When $[OH^-] \approx [F^-]/10 = 10^{-6}$ M (pH = 8), the electrode begins to respond to OH^- and the voltage will decrease (i.e., the electrode potential will change in the same direction as if more F^- were being added). Near pH = 3.17 ($= pK_a$ for HF), F^- reacts with H^+ and the concentration of free F^- decreases. At pH = 1.17, $[F^-] \approx 1\%$ of 10^{-5} M $= 10^{-7}$ M, and $E \approx 100 + 2(59) = 218$ mV. A qualitative sketch of this behavior is shown here.

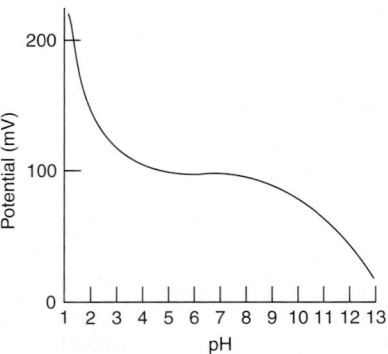

15-D. **(a)** For 1.00 mM Na^+ at pH 8.00, we can write
$E = \text{constant} + 0.059\,16 \log ([Na^+] + 36[H^+])$
$-0.038 = \text{constant} + 0.059\,16 \log[(1.00 \times 10^{-3}) + (36 \times 10^{-8})]$
$\Rightarrow \text{constant} = +0.139$ V
For 5.00 mM Na^+ at pH 8.00, we have
$E = +0.139 + 0.059\,16 \log [(5.00 \times 10^{-3}) + (36 \times 10^{-8})]$
$= 0.003$ V

(b) For 1.00 mM Na^+ at pH 3.87, we have
$E = +0.139 + 0.059\,16 \log [(1.00 \times 10^{-3}) + (36 \times 10^{-3.87})]$
$= 0.007$ V

15-E. A graph of E(mV) versus $\log[NH_3(M)]$ gives a straight line whose equation is $E = 563.4 + 59.05 \times \log [NH_3]$. For $E = 339.3$ mV, $[NH_3] = 1.60 \times 10^{-4}$ M. The sample analyzed contains (100 mL) $\times$ $(1.60 \times 10^{-4}$ M$) = 0.016\,0$ mmol of nitrogen. But this sample represents just 2.00% (20.0 mL/1.00 L) of the food sample. Therefore, the food contains (0.016 mmol N)/0.020 0 = 0.800 mmol N = 11.2 mg of N = 3.59 wt% N.

15-F. The function to plot on the y-axis is $(V_0 + V_S)10^{E/S}$, where $S = (\beta RT/nF) \ln 10$. β is 0.985. Putting in $R = 8.314\,5$ J/mol·K), $F = 96\,485$ C/mol, $T = 298.15$ K, and $n = -2$ gives $S = -0.029\,136$ J/C $= -0.029\,136$ V. (Recall that joule/coulomb = volt.)

V_S (mL)	E (V)	y
0	0.046 5	0.633 8
1.00	0.040 7	1.042 5
2.00	0.034 4	1.781 1
3.00	0.030 0	2.615 2
4.00	0.026 5	3.571 7

The data are plotted in Figure 15-34, which has a slope of $m = 0.744\,84$ and an intercept of $b = 0.439\,19$, giving an x-intercept of $-b/m = -0.58_{965}$ mL. The concentration of original unknown is

$$c_X = \frac{(x\text{-intercept})c_S}{V_0} = -\frac{(-0.58_{965} \text{ mL})(1.78 \text{ mM})}{25.0 \text{ mL}}$$
$$= 4.2 \times 10^{-5} \text{ M}$$

(We decided that the last significant digit in the x-intercept was the 0.01 decimal place because the original data were only measured to the 0.01 decimal place.)

Chapter 16

16-A. Titration: $Sn^{2+} + 2Ce^{4+} \rightarrow Sn^{4+} + 2Ce^{3+}$ $V_e = 10.0$ mL
Representative calculations:

0.100 mL: Fraction of the way to V_e = 0.100 mL/10.00 mL

$$E_+ = 0.139 - \frac{0.059\ 16}{2}\log\frac{[Sn^{2+}]}{[Sn^{4+}]}$$

$$= 0.139 - \frac{0.059\ 16}{2}\log\frac{9.90}{0.100} = 0.080\ V$$

$$E = E_+ - E_- = 0.080 - 0.241 = -0.161\ V$$

10.00 mL: $\quad 2E_+ = 2(0.139) - 0.059\ 16\log\frac{[Sn^{2+}]}{[Sn^{4+}]}$

$$E_+ = 1.47 - 0.059\ 16\log\frac{[Ce^{3+}]}{[Ce^{4+}]}$$

$$\overline{\quad 3E_+ = 1.748 - 0.059\ 16\log\frac{[Sn^{2+}][Ce^{3+}]}{[Sn^{4+}][Ce^{4+}]}\quad}$$

At the equivalence point, $[Sn^{4+}] = \frac{1}{2}[Ce^{3+}]$ and $[Sn^{2+}] = \frac{1}{2}[Ce^{4+}]$, which makes the log term 0. Therefore, $3E_+ = 1.748$ and $E_+ = 0.583$ V.

$$E = E_+ - E_- = 0.583 - 0.241 = 0.342\ V$$

10.10 mL: The first 10.00 mL went into making Ce^{3+}. The next 0.10 mL left unreacted Ce^{4+}:

$$E_+ = 1.47 - 0.059\ 16\log\frac{[Ce^{3+}]}{[Ce^{4+}]}$$

$$= 1.47 - 0.059\ 16\log\frac{10.0}{0.10} = 1.35_2\ V$$

$$E = E_+ - E_- = 1.35_2 - 0.241 = 1.11\ V$$

mL	E (V)	mL	E (V)
0.100	−0.161	10.00	0.342
1.00	−0.130	10.10	1.11
5.00	−0.102	12.00	1.19
9.50	−0.064		

16-B. Standard potentials: indigo tetrasulfonate, 0.36 V; $Fe[CN]_6^{3-}\,|\,Fe[CN]_6^{4-}$, 0.356 V; $Tl^{3+}\,|\,Tl^+$, 0.77 V. The end-point potential will be between 0.356 and 0.77 V. Indigo tetrasulfonate changes color near 0.36 V. Therefore, it will not be a useful indicator for this titration.

16-C. Titration: $MnO_4^- + 5Fe^{2+} + 8H^+ \rightarrow Mn^{2+} + 5Fe^{3+} + 4H_2O$
$Fe^{3+} + e^- \rightleftharpoons Fe^{2+}$ $E° = 0.68$ V in 1 M H_2SO_4
$MnO_4^- + 8H^+ + 5e^- \rightarrow Mn^{2+} + 4H_2O$ $E° = 1.507$ V

The equivalence point comes at 15.0 mL. Before the equivalence point:

$$E = E_+ - E_- = \left(0.68 - 0.059\ 16\log\frac{[Fe^{2+}]}{[Fe^{3+}]}\right) - 0.241$$

1.0 mL: $[Fe^{2+}]/[Fe^{3+}] = 14.0/1.0 \Rightarrow E = 0.371$ V
7.5 mL: $[Fe^{2+}]/[Fe^{3+}] = 7.5/7.5 \Rightarrow E = 0.439$ V
14.0 mL: $[Fe^{2+}]/[Fe^{3+}] = 1.0/14.0 \Rightarrow E = 0.507$ V

At the equivalence point, use Equation E of Demonstration 16-1:

$$6E_+ = 8.215 - 0.059\ 16\log\frac{1}{[H^+]^8} \overset{pH=0}{\Rightarrow} E_+ = 1.369\ V$$

$$E = E_+ - E_- = 1.369 - 0.241 = 1.128\ V$$

After the equivalence point:

$$E = E_+ - E_- = \left(1.507 - \frac{0.059\ 16}{5}\log\frac{[Mn^{2+}]}{[MnO_4^-][H^+]^8}\right) - 0.241$$

16.0 mL: $[Mn^{2+}]/[MnO_4^-] = 15.0/1.0$ and $[H^+] = 1$ M
$\Rightarrow E = 1.252$ V

30.0 mL: $[Mn^{2+}]/[MnO_4^-] = 15.0/15.0$ and $[H^+] = 1$ M
$\Rightarrow E = 1.266$ V

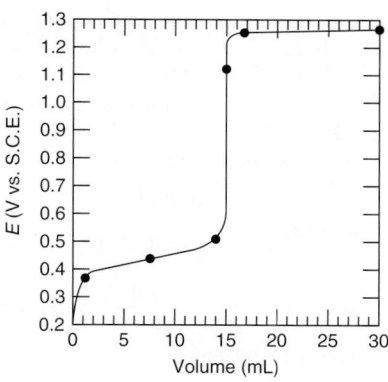

16-D. The Gran plot of $V \cdot 10^{-E/0.059\ 16}$ versus V is shown in Figure 16-4. The data from 8.5 to 12.5 mL appear to be on a straight line. The least-squares line through these four points has a slope $m = -1.567\ 3 \times 10^{-11}$ and an intercept $b = 2.170\ 2 \times 10^{-10}$. The x-intercept is $-b/m = 13.85$ mL. The amount of Ce^{4+} required to reach the equivalence point is $(0.100\ \text{mmol/mL})(13.85\ \text{mL}) = 1.385$ mmol, and the concentration of unknown Fe^{2+} is 1.385 mmol/50.0 mL = 0.027 7 M.

Titrant volume, V (mL)	E (volts)	$V \cdot 10^{-E/0.059\ 16}$
6.50	0.635	$1.200\ 3 \times 10^{-10}$
8.50	0.651	$8.421\ 0 \times 10^{-11}$
10.50	0.669	$5.162\ 6 \times 10^{-11}$
11.50	0.680	$3.685\ 1 \times 10^{-11}$
12.50	0.696	$2.148\ 8 \times 10^{-11}$

16-E. Let x = mg of $FeSO_4 \cdot (NH_4)_2SO_4 \cdot 6H_2O$ and $(54.85 - x)$ = mg of $FeCl_2 \cdot 6H_2O$
mmol of Ce^{4+} = mmol $FeSO_4 \cdot (NH_4)_2SO_4 \cdot 6H_2O$ + mmol $FeCl_2 \cdot 6H_2O$

$$(13.39\ \text{mL})(0.012\ 34\ \text{M}) = \frac{x\ \text{mg}}{392.13\ \text{mg/mmol}} + \frac{(54.85 - x)}{234.84\ \text{mg/mmol}}$$

$\Rightarrow x = 40.01$ mg $FeSO_4 \cdot (NH_4)_2SO_4 \cdot 6H_2O$
mass of $FeCl_2 \cdot 6H_2O$ = 14.84 mg = 0.063 19 mmol = 4.48 mg Cl

$$\text{wt\% Cl} = \frac{4.48\ \text{mg}}{54.85\ \text{mg}} \times 100 = 8.17\%$$

Chapter 17

17-A. Cathode: $2H^+ + 2e^- \rightleftharpoons H_2(g)$ $\quad E° = 0$ V

Anode (written as a reduction):
$\frac{1}{2}O_2(g) + 2H^+ + 2e^- \rightleftharpoons H_2O$ $\quad E = 1.229$ V

$$E(\text{cathode}) = 0 - \frac{0.059\ 16}{2}\log\frac{P_{H_2}}{[H^+]^2}$$

$$E(\text{anode}) = 1.229 - \frac{0.059\ 16}{2}\log\frac{1}{[H^+]^2 P_{O_2}^{1/2}}$$

$$E(\text{cell}) = E(\text{cathode}) - E(\text{anode})$$

$$= -1.229 - \frac{0.059\ 16}{2}\log P_{H_2}P_{O_2}^{1/2} = -1.229\ V$$

$$E = E(\text{cell}) - I \cdot R - \text{overpotentials}$$
$$= -1.229 - (0.100\ \text{A})(2.00\ \Omega)$$
$$\underbrace{-0.85\ V}_{\substack{\text{Anode}\\\text{overpotential}}} \underbrace{- 0.068\ V}_{\substack{\text{Cathode}\\\text{overpotential}}} = -2.35\ V$$

From Table 17-1

For Au electrodes, overpotentials of 0.963 and 0.390 V give $E = -2.78$ V.

17-B. (a) To electrolyze 0.010 M SbO^+ requires a potential of

$$E(\text{cathode}) = 0.208 - \frac{0.059\ 16}{3}\log\frac{1}{[SbO^+][H^+]^2}$$

$$= 0.208 - \frac{0.059\ 16}{3}\log\frac{1}{(0.010)(1.0)^2} = 0.169\ V$$

$E(\text{cathode versus Ag}|\text{AgCl}) = E(\text{versus S.H.E.}) - E(\text{Ag}|\text{AgCl})$

$= 0.169 - 0.197 = -0.028 \text{ V}$

(b) The concentration of Cu^{2+} in equilibrium with $Cu(s)$ at 0.169 V is
$Cu^{2+} + 2e^- \rightleftharpoons Cu(s)$ $E° = 0.339$

$E(\text{cathode}) = 0.339 - \dfrac{0.059\ 16}{2} \log \dfrac{1}{[Cu^{2+}]}$

$0.169 = 0.339 - \dfrac{0.059\ 16}{2} \log \dfrac{1}{[Cu^{2+}]} \Rightarrow [Cu^{2+}] = 1.8 \times 10^{-6} \text{ M}$

Percentage of Cu^{2+} not reduced $= \dfrac{1.8 \times 10^{-6}}{0.10} \times 100 = 1.8 \times 10^{-3}\%$

Percentage of Cu^{2+} reduced $= 99.998\%$

17-C. **(a)** $Co^{2+} + 2e^- \rightleftharpoons Co(s)$ $E° = -0.282 \text{ V}$
$E(\text{cathode versus S.H.E.}) = -0.282 - \dfrac{0.059\ 16}{2} \log \dfrac{1}{[Co^{2+}]}$
Putting in $[Co^{2+}] = 1.0 \times 10^{-6} \text{ M}$ gives $E = -0.459 \text{ V}$ and
$E(\text{cathode versus S.C.E.}) = -0.459 - \underbrace{0.241}_{E(\text{S.C.E.})} = -0.700 \text{ V}$

(b) $Co(C_2O_4)_2^{2-} + 2e^- \rightleftharpoons Co(s) + 2C_2O_4^{2-}$ $E° = -0.474 \text{ V}$
$E(\text{cathode versus S.C.E.})$

$= -0.474 - \dfrac{0.059\ 16}{2} \log \dfrac{[C_2O_4^{2-}]^2}{[Co(C_2O_4)_2^{2-}]} - 0.241$

For $[C_2O_4^{2-}] = 0.10 \text{ M}$ and $[Co(C_2O_4)_2^{2-}] = 1.0 \times 10^{-6} \text{ M}$, $E = -0.833 \text{ V}$.
(c) We can think of the reduction as $Co^{2+} + 2e^- \rightleftharpoons Co(s)$, for which
$E° = -0.282 \text{ V}$. But $[Co^{2+}]$ is the tiny amount in equilibrium with 0.10 M
EDTA plus $1.0 \times 10^{-6} \text{ M Co(EDTA)}^{2-}$. In Table 12-2, we find that the
formation constant for $Co(EDTA)^{2-}$ is $10^{16.45} = 2.8 \times 10^{16}$.

$K_f = \dfrac{[Co(EDTA)^{2-}]}{[Co^{2+}][EDTA^{4-}]} = \dfrac{[Co(EDTA)^{2-}]}{[Co^{2+}]\alpha_{Y^{4-}}F}$

where F is the formal concentration of EDTA ($= 0.10 \text{ M}$) and
$\alpha_{Y^{4-}} = 3.8 \times 10^{-4}$ at pH 7.00 (Table 12-1). Putting in $[Co(EDTA)^{2-}] =$
$1.0 \times 10^{-6} \text{ M}$ and solving for $[Co^{2+}]$ gives $[Co^{2+}] = 9.4 \times 10^{-19} \text{ M}$.

$E = -0.282 - \dfrac{0.059\ 16}{2} \log \dfrac{1}{9.4 \times 10^{-19}} - 0.241 = -1.056 \text{ V}$

17-D. **(a)** 75.00 mL of 0.023 80 M KSCN = 1.785 mmol of SCN^-,
which gives 1.785 mmol of AgSCN, containing 0.103 7 g of SCN. Final
mass $= 12.463\ 8 + 0.103\ 7 = 12.567\ 5 \text{ g}$.
(b) Anode: $AgBr(s) + e^- \rightleftharpoons Ag(s) + Br^-$ $E° = 0.071 \text{ V}$
$E(\text{anode}) = 0.071 - 0.059\ 16 \log[Br^-]$
$= 0.071 - 0.059\ 16 \log [0.10] = 0.130 \text{ V}$
$E(\text{cathode}) = E(\text{S.C.E.}) = 0.241 \text{ V}$
$E = E(\text{cathode}) - E(\text{anode}) = 0.111 \text{ V}$

(c) Removal of 99.99% of 0.10 M KI will leave $[I^-] = 1.0 \times 10^{-5} \text{ M}$.
The concentration of Ag^+ in equilibrium with this much I^- is
$[Ag^+] = K_{sp}/[I^-] = (8.3 \times 10^{-17})/(1.0 \times 10^{-5}) = 8.3 \times 10^{-12} \text{ M}$.
The concentration of Ag^+ in equilibrium with 0.10 M Br^- is $[Ag^+] = K_{sp}/$
$[Br^-] = (5.0 \times 10^{-13})/(0.10) = 5.0 \times 10^{-12} \text{ M}$. Therefore, $8.3 \times 10^{-12} \text{ M}$
Ag^+ will begin to precipitate 0.10 M Br^-. The separation is not possible.

17-E. Corrected coulometric titration time $= 387 - 6 = 381 \text{ s}$.
$\text{Moles } X^- \text{ reacted} = \dfrac{It}{nF} = \dfrac{(4.23 \text{ mA})(381 \text{ s})}{(1)(96\ 485 \text{ C/mol})} = 16.7 \ \mu\text{mol}$.
$[\text{organohalide}] = 16.7 \ \mu\text{M}$. If all halogen is Cl, 16.7 μmol = 592 μg Cl/L.

17-F. **(a)** Use the internal standard equation with $X = Pb^{2+}$ and
$S = Cd^{2+}$. From the standard mixture, we find the response factor, F:

$$\dfrac{\text{Signal}_X}{[X]} = F\left(\dfrac{\text{signal}_S}{[S]}\right)$$

$\dfrac{1.58 \ \mu\text{A}}{[41.8 \ \mu\text{M}]} = F\left(\dfrac{1.64 \ \mu\text{A}}{[32.3 \ \mu\text{M}]}\right) \Rightarrow F = 0.744_5$

$[Cd^{2+}]$ standard added to unknown
$= \left(\dfrac{10.00 \text{ mL}}{50.00 \text{ mL}}\right)(3.23 \times 10^{-4} \text{ M}) = 6.46 \times 10^{-5} \text{ M}$
For the unknown mixture, we can now say
$\dfrac{\text{Signal}_X}{[X]} = F\left(\dfrac{\text{signal}_S}{[S]}\right)$

$\dfrac{3.00 \ \mu\text{A}}{[Pb^{2+}]} = 0.744_5\left(\dfrac{2.00 \ \mu\text{A}}{[64.6 \ \mu\text{M}]}\right) \Rightarrow [Pb^{2+}] = 130._2 \ \mu\text{M}$
$[Pb^{2+}]$ in diluted unknown $= 130._2 \ \mu\text{M}$. In undiluted unknown,
$[Pb^{2+}] = \left(\dfrac{50.00 \text{ mL}}{25.00 \text{ mL}}\right)(130._2 \ \mu\text{M}) = 2.60 \times 10^{-4} \text{ M}$.

(b) First find the relative uncertainty in the response factor:
$F = \dfrac{(1.58 \pm 0.03 \ \mu\text{A})(32.3 \pm 0.1 \ \mu\text{M})}{(1.64 \pm 0.03 \ \mu\text{A})(41.8 \pm 0.1 \ \mu\text{M})} \Rightarrow F = 0.744\ 5 \pm 0.019\ 9\ (\pm 2.67\%)$
Find $[Pb^{2+}]$ uncertainty by writing calculation from **(a)** in one string:
$[Pb^{2+}] =$

$(3.00 \pm 0.03 \ \mu\text{A}) \dfrac{(10.00 \pm 0.05 \text{ mL})\ \cancel{(50.00 \pm 0.05 \text{ mL})}}{\cancel{(50.00 \pm 0.05 \text{ mL})}(25.00 \pm 0.05 \text{ mL})}(3.23(\pm 0.01) \times 10^{-4} \text{ M})$

$\overline{\qquad\qquad\qquad (2.00 \pm 0.03 \ \mu\text{A})(0.744\ 5 \pm 0.019\ 9) \qquad\qquad\qquad}$

$\Rightarrow [Pb^{2+}] = 2.60\ (\pm 0.08) \times 10^{-4} \text{ M}$

17-G. We see two consecutive reductions. From $E_{pa} - E_{pc} = 60 \text{ mV}$, we
conclude that $n = 1 \text{ e}^-$ is involved in each reduction (using Equation
17-20). A possible sequence of reaction is
$Co(III)(B_9C_2H_{11})_2^- \rightarrow Co(II)(B_9C_2H_{11})_2^{2-} \rightarrow Co(I)(B_9C_2H_{11})_2^{3-}$
The equality of the anodic and cathodic peak heights suggests that the
reactions are reversible. The expected sampled current (panel a) and square
wave (panel b) polarograms are sketched below.

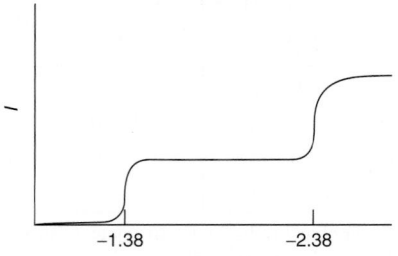

(a)

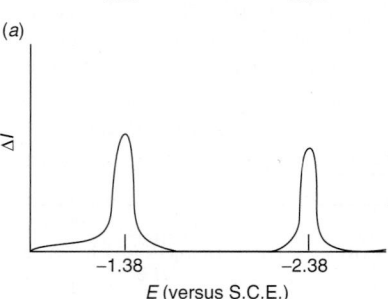

(b)

17-H. Electricity required for H_2O in 0.847 6 g of polymer =
$(63.16 \text{ C} - 4.23 \text{ C}) = 58.93 \text{ C}$.

$\dfrac{58.93 \text{ C}}{96\ 485 \text{ C/mol}} = 0.610\ 8 \text{ mmol of e}^-$
which corresponds to $\frac{1}{2}(0.610\ 8 \text{ mmol}) = 0.305\ 4 \text{ mmol of}$
$I_2 = 0.305\ 4 \text{ mmol of } H_2O = 5.502 \text{ mg } H_2O$
Water content $= 100 \times \dfrac{5.502 \text{ mg } H_2O}{847.6 \text{ mg polymer}} = 0.649\ 1 \text{ wt}\%$

Chapter 18

18-A. **(a)** $A = -\log P/P_0 = -\log T = -\log(0.45) = 0.347$
(b) Absorbance is proportional to concentration, so the absorbance will
double to 0.694, giving $T = 10^{-A} = 10^{-0.694} = 0.202 \Rightarrow \%T = 20.2\%$.

18-B. (a) $\varepsilon = \dfrac{A}{cb} = \dfrac{0.624 - 0.029}{(3.96 \times 10^{-4}\ \text{M})(1.000\ \text{cm})} = 1.50 \times 10^3\ \text{M}^{-1}\ \text{cm}^{-1}$

(b) $c = \dfrac{A}{\varepsilon b} = \dfrac{0.375 - 0.029}{(1.50 \times 10^3\ \text{M}^{-1}\ \text{cm}^{-1})(1.000\ \text{cm})} = 2.31 \times 10^{-4}\ \text{M}$

(c) $c = \left(\dfrac{25.00\ \text{mL}}{\underbrace{2.00\ \text{mL}}_{\text{Dilution factor}}}\right) \dfrac{0.733 - 0.029}{(1.50 \times 10^3\ \text{M}^{-1}\ \text{cm}^{-1})(1.000\ \text{cm})} = 5.87 \times 10^{-3}\ \text{M}$

18-C. (a) 1.00×10^{-2} g of NH_4Cl in $1.00\ L = 1.869 \times 10^{-4}$ M. In the colored solution, the concentration is $(\frac{10\ \text{mL}}{50\ \text{mL}})(1.869 \times 10^{-4}\ \text{M}) = 3.739 \times 10^{-5}$ M. $\varepsilon = A/bc = (0.308 - 0.140)/[(1.00)(3.739 \times 10^{-5})] = 4.49_3 \times 10^3\ \text{M}^{-1}\ \text{cm}^{-1}$.

(b) $\dfrac{\text{Absorbance of unknown}}{\text{Absorbance of reference}}$

$= \dfrac{0.592 - 0.140}{0.308 - 0.140} = \dfrac{\text{concentration of unknown}}{\text{concentration of reference}}$

$\Rightarrow$ concentration of NH_3 in unknown

$= \left(\dfrac{0.452}{0.168}\right)(1.869 \times 10^{-4}\ \text{M}) = 5.028 \times 10^{-4}\ \text{M}$

100.00 mL of unknown $= 5.028 \times 10^{-5}$ mol of N $= 0.704\ 3$ mg of N

$\Rightarrow$ wt% N $= 100 \times (0.704\ 3\ \text{mg})/(4.37\ \text{mg}) = 16.1\%$

18-D. (a) Cu in flask C $= (1.00\ \text{mg})\left(\dfrac{10\ \text{mL}}{250\ \text{mL}}\right)\left(\dfrac{15\ \text{mL}}{30\ \text{mL}}\right) = 0.020\ 0$ mg.

This entire quantity is in the isoamyl alcohol (20.00 mL), so the concentration is $(2.00 \times 10^{-5}\ \text{g})/[(0.020\ 0\ \text{L})(63.546\ \text{g/mol})] = 1.57_4 \times 10^{-5}$ M.

(b) Observed absorbance

$=$ absorbance due to Cu in rock $+$ blank absorbance $= \varepsilon bc + 0.056$

$= (7.90 \times 10^3\ \text{M}^{-1}\text{cm}^{-1})(1.00\ \text{cm})(1.57_4 \times 10^{-5}\ \text{M}) + 0.056 = 0.180$

Note that the observed absorbance is equal to the absorbance from Cu in the rock *plus* the blank absorbance. In the lab, we measure the observed absorbance and subtract the blank absorbance from it to find the absorbance due to copper.

(c) $\dfrac{\text{Cu in unknown}}{\text{Cu in known}} = \dfrac{A \text{ of unknown}}{A \text{ of known}}$

$\dfrac{x\ \text{mg}}{1.00\ \text{mg}} = \dfrac{0.874 - 0.056}{0.180 - 0.056} \Rightarrow x = 6.60\ \text{mg Cu}$

18-E. Absorbance is corrected by multiplying observed absorbance by (total volume/initial volume). For example, at 36.0 μL, A(corrected) $= (0.399) \times [(2\ 025 + 36)/2\ 025] = 0.406$. A graph of corrected absorbance versus volume of Pb^{2+} has two linear regions intersecting at 46.7 μL. Mol Pb^{2+} in this volume $= (46.7 \times 10^{-6}\ \text{L})(7.515 \times 10^{-4}\ \text{M}) = 3.510 \times 10^{-8}$ mol. [semi-xylenol orange] $= (3.510 \times 10^{-8}\ \text{mol})/(2.025 \times 10^{-3}\ \text{L}) = 1.73 \times 10^{-5}$ M.

Chapter 19

19-A. (a) $c = A/\varepsilon b = 0.463/[(4\ 170)(1.00)] = 1.110 \times 10^{-4}$ M $= 8.99$ g/L $= 8.99$ mg of transferrin/mL. The Fe concentration is 2.220×10^{-4} M $= 0.012\ 4$ g/L $= 12.4$ μg/mL.

(b) $A_\lambda = \Sigma \varepsilon bc$

At 470 nm: $0.424 = 4\ 170[\text{T}] + 2\ 290[\text{D}]$

At 428 nm: $0.401 = 3\ 540[\text{T}] + 2\ 730[\text{D}]$

where [T] and [D] are the concentrations of transferrin and desferrioxamine, respectively. Solving the simultaneous equations gives [T] $= 7.30 \times 10^{-5}$ M and [D] $= 5.22 \times 10^{-5}$ M. The fraction of iron in transferrin (which binds two ferric ions) is $2[\text{T}]/(2[\text{T}] + [\text{D}]) = 73.7\%$. The fraction in desferrioxamine is 26.3%.

The spreadsheet solution looks like this:

	A	B	C	D	E	F	G
1	Transferrin/Desferrioxamine mixture						
2							
3	Wavelength	Coefficient matrix		Absorbance		Concentrations	
4				of unknown		in mixture	
5	428	3540	2730	0.401		7.2992E-05	← [TRF]
6	470	4170	2290	0.424		5.2238E-05	← [DFO]
7		K		A		C	

Spreadsheet for Exercise 19-A

19-B.

	A	B	C	D	E	F	G
1	Analysis of a mixture when you have more data points than components of the mixture						
2					Measured		
3			Molar Absorptivity		Absorbance	Calculated	
4	Wavelength		$(\text{M}^{-1}\ \text{cm}^{-1})$		of Mixture	Absorbance	
5	(nm)	Tartrazine	Sunset Yellow	Ponceau 4R	A_m	A_{calc}	$[A_{calc} - A_m]^2$
6	350	6.229E+03	2.019E+03	4.172E+03	0.557	0.5363	4.269E-04
7	375	1.324E+04	4.474E+03	2.313E+03	0.853	0.8375	2.411E-04
8	400	2.144E+04	7.403E+03	3.310E+03	1.332	1.3437	1.378E-04
9	425	2.514E+04	8.551E+03	4.534E+03	1.603	1.5999	9.635E-06
10	450	2.200E+04	1.275E+04	6.575E+03	1.792	1.8018	9.524E-05
11	475	1.055E+04	1.940E+04	1.229E+04	2.006	1.9992	4.633E-05
12	500	1.403E+03	1.869E+04	1.673E+04	1.821	1.8336	1.590E-04
13	525	0.000E+00	7.641E+03	1.528E+04	1.155	1.1303	6.109E-04
14	550	0.000E+00	3.959E+02	9.522E+03	0.445	0.4743	8.597E-04
15	575	0.000E+00	0.000E+00	1.814E+03	0.084	0.0864	5.698E-06
16						sum =	2.592E-03
17	Pathlength =	1.000	cm				
18	Concentrations in the mixture		F6 = B6*$B$17*$B$20+C6*$B$17*$B$21+D6*$B$17*$B$22				
19	(to be found by Solver)						
20	Tartrazine =	3.7129E-05	M				
21	Sunset yellow =	5.2691E-05	M				
22	Ponceau 4R =	4.7622E-05	M				

19-C.

	A	B	C	D	E	F	G	H	I	J
1	Finding K for Binding of Methyl Orange (X) to Bovine Serum Albumin (P)									
2	Aliquots of protein were added to 1.00 mL of 5.7 μM methyl orange									
3	X_o = total concentration of methyl orange									
4	P_o = total concentration of protein									
5	K = [PX]/([P][X])									
6	[PX] = [K(P_o+X_o)+1 ± sqrt{[K(P_o+X_o)+1]^2 − 4K^2$X_o$$P_o$}]/(2K)						← (Eq. A)			
7	Change in absorbance at 490 nm = ΔA = [PX]Δε for 1 cm pathlength						← (Eq. B)			
8	Δε = $ε_{PX}$ − $ε_X$									
9	Parameters to find with Solver:									
10	K =	5.59E+04		initial estimate =	2.7E+04					
11	Δε =	7.58E+03		initial estimate =	1.0E+04M^{-1}cm^{-1}					
12										
13					[PX] (M)					[PX]/X_o
14	Protein	X_o	P_o	$ΔA_{obs}$	Eq. A using	[X] =	[P] =	$ΔA_{calc}$	$(A_{obs} − A_{calc})^2$	fraction of
15	addition	(M)	(M)	at 490 nm	minus sign	X_o − [PX]	P_o − [PX]	Eq. B		X reacted
16	1	5.7E-06	8.0E-06	0.0118	1.517E-06	4.183E-06	6.483E-06	0.0115	8.7864E-08	0.266
17	2	5.7E-06	1.14E-05	0.0148	1.969E-06	3.731E-06	9.431E-06	0.0149	1.6442E-08	0.345
18	3	5.7E-06	1.63E-05	0.0187	2.485E-06	3.215E-06	1.381E-05	0.0188	2.0176E-08	0.436
19	4	5.7E-06	3.28E-05	0.0268	3.539E-06	2.161E-06	2.926E-05	0.0268	9.3629E-10	0.621
20	5	5.7E-06	4.04E-05	0.0291	3.829E-06	1.871E-06	3.657E-05	0.0290	4.7689E-09	0.672
21	Formulas							sum =	1.3019E-07	
22	E16 = (($B$10*(B16+C16)+1)−SQRT(($B$10*(B16+C16)+1)^2 − 4*$B$10^2*B16*C16))/(2*$B$10)									
23	F16 = B16−E16				I16 = (D16−H16)^2					
24	G16 = C16−E16				J16 = E16/B16					
25	H16 = $B$11*E16				I21 = SUM(I16:I20)					

Spreadsheet for Exercise 19-C

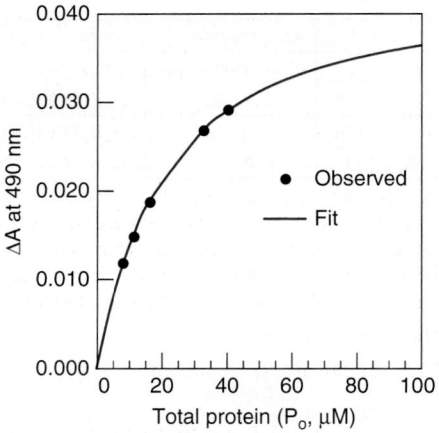

19-D.

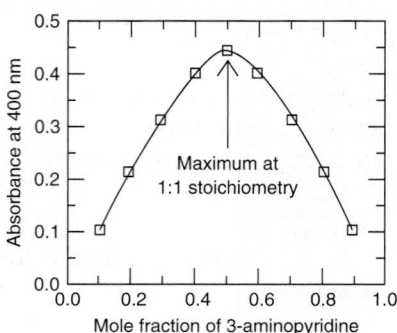

Chapter 20

20-A. **(a)** For $λ = 10.00$ μm and $Δλ = 0.01$ μm, $λ/Δλ = 10.00/0.01 = 10^3$. The resolution is 10^4, so these lines will be resolved.

(b) $λ = \dfrac{1}{\tilde{ν}} = \dfrac{1}{(1\,000\ \text{cm}^{-1})(10^{-4}\ \text{cm/μm})} = 10$ μm

$Δλ = \dfrac{λ}{10^4} = 10^{-3}$ μm

⇒ 10.001 μm could be resolved from 10.000 μm

$\left.\begin{array}{l} 10.000\ \text{μm} = 1000.0\ \text{cm}^{-1} \\ 10.001\ \text{μm} = 999.9\ \text{cm}^{-1} \end{array}\right\}$ Difference = 0.1 cm^{-1}

(c) 5.0 cm × 2 500 lines/cm = 12 500 lines
Resolution = 1 · 12 500 = 12 500 for $n = 1$
= 10 · 12 500 = 125 000 for $n = 10$

(d) $\dfrac{Δϕ}{Δλ} = \dfrac{n}{d\cos ϕ} = \dfrac{2}{\left(\dfrac{1\ \text{mm}}{250}\right)\cos 30°}$

$= 577\dfrac{\text{radians}}{\text{mm}} = 0.577\dfrac{\text{radians}}{\text{μm}}$

Convert radians to degrees $= \dfrac{\text{radians}}{π} × 180$

$⇒ \dfrac{Δϕ}{Δλ} = 33.1$ degrees/μm

The two wavelengths are 1 000 cm^{-1} = 10.00 μm and 1 001 cm^{-1} = 9.99 μm ⇒ $Δλ = 0.01$ μm.

$Δϕ = 0.577\dfrac{\text{radians}}{\text{μm}} × 0.01$ μm

$= 6 × 10^{-3}$ radian = 0.3°

20-B. True transmittance $= 10^{-1.000} = 0.100$. With 1.0% stray light, the apparent transmittance is

$$\text{Apparent transmittance} = \frac{P + S}{P_0 + S} = \frac{0.100 + 0.010}{1 + 0.010} = 0.109$$

The apparent absorbance is $-\log T = -\log 0.109 = 0.963$.

Apparent concentration $= 96.3\%$ of true concentration $\Rightarrow$ error $= -3.7\%$

20-C. **(a)** $\Delta \tilde{v} = 1/2\delta = 1/(2 \cdot 1.266\,0 \times 10^{-4}\,\text{cm}) = 3\,949\,\text{cm}^{-1}$
(b) Each interval is $1.266\,0 \times 10^{-4}$ cm. 4 096 intervals $= (4\,096)(1.266\,0 \times 10^{-4}\,\text{cm}) = 0.518\,6$ cm. This is a range of $\pm\Delta$, so $\Delta = 0.259\,3$ cm.
(c) Resolution $\approx 1/\Delta = 1/(0.259\,3\,\text{cm}) = 3.86\,\text{cm}^{-1}$
(d) Mirror velocity $= 0.693$ cm/s

$$\text{Interval} = \frac{1.266\,0 \times 10^{-4}\,\text{cm}}{0.693\,\text{cm/s}} = 183\,\mu\text{s}$$

(e) (4 096 points) (183 μs/point) $= 0.748$ s
(f) The beamsplitter is germanium on KBr. KBr absorbs light below $400\,\text{cm}^{-1}$, which the background transform shows clearly.

20-D. The graphs show that signal-to-noise ratio is proportional to $\sqrt{n}$. The confidence interval is $\pm ts/\sqrt{n}$, where s is the standard deviation, n is the number of experiments, and t is Student's t from Table 4-4 for 95% confidence and $n - 1$ degrees of freedom. For the first line of the table, $n = 8$, $s = 1.9$, and $t = 2.365$ for 7 degrees of freedom. 95% confidence interval $= \pm(2.365)(1.9)/\sqrt{8} = \pm 1.6$. For remaining rows, 95% confidence interval $= 3.9, 5.1, 5.9, 9.0, 11.2, 14.0, 24.0, 23.9,$ and 27.2.

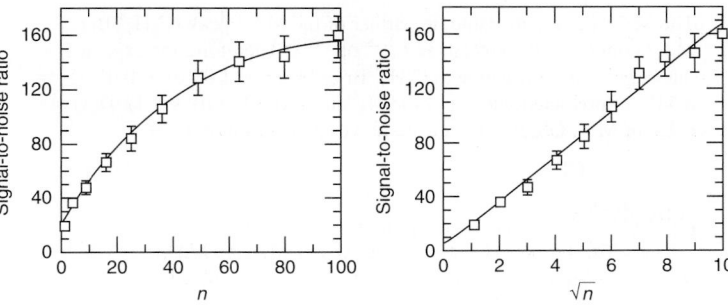

Chapter 21

21-A. A graph of intensity versus concentration of added standard has an x-intercept of -0.164 ± 0.005 μg/mL. Because the sample was diluted by a factor of 10, the original concentration is 1.64 ± 0.05 μg/mL.

	A	B	C	D
1	Standard Addition Constant Volume Least-Squares			
2	x = Added Li	y		
3	(μg/mL)	Signal		
4	0.000	309		
5	0.081	452		
6	0.162	600		
7	0.243	765		
8	0.324	906		
9	B11:C13 = LINEST(B4:B8,A4:A8,TRUE,TRUE)			
10		LINEST output:		
11	m	1860.5	305.0	b
12	u_m	26.3	5.2	u_b
13	R^2	0.9994	6.7	s_y
14	x-intercept = –b/m =	–0.164		
15	n =	5	= COUNT (A4:A8)	
16	Mean y =	606.400	= AVERAGE(B4:B8)	
17	$\Sigma(x_i - \text{mean } x)^2$ =	0.06561	= DEVSQ(A4:A8)	
18	Std uncertainty of			
19	x-intercept (u_x) =	0.0049		
20	B19=(C13/ABS(B11))*SQRT((1/B15) + B16^2/(B11^2*B17))			

21-B. The concentration of Mn in the unknown mixture is (13.5 μg/mL) $(1.00/6.00) = 2.25$ μg/mL.

Standard mixture:

$$\frac{A_X}{[X]} = F\left(\frac{A_S}{[S]}\right)$$

$$\frac{1.05}{[2.50\,\mu\text{g/mL}]} = F\left(\frac{1.00}{[2.00\,\mu\text{g/mL}]}\right) \Rightarrow F = 0.840$$

Unknown mixture:

$$\frac{A_X}{[X]} = F\left(\frac{A_S}{[S]}\right)$$

$$\frac{0.185}{[\text{Fe}]} = 0.840\left(\frac{0.128}{[2.25\,\mu\text{g/mL}]}\right) \Rightarrow [\text{Fe}] = 3.87\,\mu\text{g/mL}$$

The original concentration of Fe must have been

$$\frac{6.00}{5.00}(3.87\,\mu\text{g/mL}) = 4.65\,\mu\text{g/mL} = 8.33 \times 10^{-5}\,\text{M}$$

21-C. **(a)** I estimate the Fe signal height to be 21.2 vertical units. The root-mean-square noise level is given as $s = 0.30$ vertical units. A height of $3s$ is 0.90. The concentration of Fe that would give a signal height of 0.90 units is

$$\left(\frac{0.90}{21.2}\right)(0.048\,5\,\mu\text{g/mL}) = 0.002\,1\,\mu\text{g/mL}(= 2.1\,\text{ppb}).$$

(b) Standard deviation of 7 standards $= 0.22_{48}$ ng/L $\equiv s/m$
Detection limit $= 3s/m = 0.67$ ng/L
Quantitation limit $= 10s/m = 2.2$ ng/L

21-D. **(a)** The higher result in experiment 2 compared with experiment 1 is probably the effect of diluting interfering species, so they do not interfere as much in experiment 2 as in experiment 1. Dilution lowers the concentration of species that might react with Li or make smoke that scatters light. In experiment 3, interference is present to the same extent as in experiment 2, but the standard addition procedure corrects for the interference. The whole point of standard addition is to measure the effect of the complex interfering matrix on the response to known quantities of analyte.

(b) Experiments 4–6 use a hotter flame than experiments 1–3. High temperature appears to eliminate most of the interference observed at lower temperature. Dilution has only a tiny effect on the results.

(c) Because it appears from experiments 1–3 that standard addition gives a true result, we surmise that experiments 3 and 6, and possibly 5, are within experimental error of each other. I would probably report the "true" value as the mean of experiments 3 and 6 (81.4 ppm). It might also be reasonable to take the average of experiments 3, 5, and 6 (80.8 ppm).

21-E. Possible assignments

Energy (keV)	Assignment and comments
2.32	S K_α
	K_β at 2.46 keV is a shoulder on the right side of K_α peak
	Notice the asymmetry of the peak on the right side
2.97	Ar K_α (K_β at 3.19 keV is a ripple in the baseline)
3.32	K K_α (K_β at 3.59 keV is a ripple is under 3.69 peak)
3.69	Ca K_α
4.03	Ca K_β
4.64	Fe K_α – Si K_α = 6.40 – 1.74 = 4.66 keV
	Silicon is from the detector
5.90	Mn K_α (K_β at 6.49 keV is under 6.41 peak)
6.41	Fe K_α
7.06	Fe K_β
8.05	Cu K_α (K_β at 8.90 keV is tail on 8.60 peak)
8.63	Zn K_α
9.20	?
9.58	Zn K_β

22-A. **(a)** Resolving power $= \dfrac{m}{m_{1/2}} = \dfrac{53}{0.6_0} \approx 88$

We should be able to barely distinguish two peaks differing by 1 Da at a mass of 88 Da. We will probably not be able to distinguish two peaks at 100 and 101 Da.

22-B. $C_2H_5^{+\bullet}$

	$2 \times 12.000\ 00$
	$+5 \times 1.007\ 825$
$-e^-$ mass	$-1 \times 0.000\ 55$
	$29.038\ 58$

$HCO^{+\bullet}$

	$1 \times 12.000\ 00$
	$1 \times 1.007\ 825$
	$+1 \times 15.994\ 91$
$-e^-$ mass	$-1 \times 0.000\ 55$
	$29.002\ 18$

We need to distinguish a mass difference of $29.038\ 58 - 29.002\ 18 = 0.036\ 40$. The required resolving power is $m/\Delta m = 29.0/(0.036\ 40) \approx 800$.

22-C. ^{35}Cl abundance $\equiv a = 0.757\ 8$

^{37}Cl abundance $\equiv b = 0.242\ 2$

Relative abundance of $C_6H_4{}^{35}Cl_2 = a^2 = 0.574\ 2_6$

Relative abundance of $C_6H_4{}^{35}Cl^{37}Cl = 2ab = 0.367\ 0_8$

Relative abundance of $C_6H_4{}^{37}Cl_2 = b^2 = 0.058\ 66_1$

Relative abundances: $^{35}Cl_2 : {}^{35}Cl^{37}Cl : {}^{37}Cl_2 = 1 : 0.639\ 2 : 0.102\ 2$

Figure 22-7 shows the stick diagram.

22-D. **(a)** $C_{14}H_{12}$

$R + DB = c - h/2 + n/2 + 1 = 14 - 12/2 + 0/2 + 1 = 9$

A molecule with two rings + seven double bonds is *trans*-stilbene:

(b) $C_4H_{10}NO^+$

$R + DB = c - h/2 + n/2 + 1 = 4 - 10/2 + 1/2 + 1 = \dfrac{1}{2}$ Huh?

We came out with a fraction because the species is an ion in which at least one atom does not make its usual number of bonds. In the following structure, N makes four bonds instead of three:

A fragment in a mass spectrum

22-E. **(a)** The principal difference between the two spectra is the appearance of a significant peak at m/z 72 in A that is missing in B. This peak represents loss of a neutral molecule with an even mass of 28 Da from the molecular ion. The McLafferty rearrangement can split C_2H_4 from 3-methyl-2-pentanone but not from 3,3-dimethyl-2-butanone, which lacks a γ-CH group.

3-Methyl-2-pentanone

Ethylene
28 Da

$m/z = 72$

Spectrum A must be from 3-methyl-2-pentanone and spectrum B is from 3,3-dimethyl-2-butanone.

(b) Expected intensity of M + 1 relative to $M^{+\bullet}$ for $C_6H_{12}O$:

Intensity $= \underbrace{6 \times 1.08\%}_{^{13}C} + \underbrace{12 \times 0.012\%}_{^{2}H} + \underbrace{1 \times 0.038\%}_{^{17}O} = 6.7\%$ of $M^{+\bullet}$

22-F. **(a)** —OH C_6H_6O $M^{+\bullet} = 94$

Rings + double bonds $= c - h/2 + n/2 + 1 = 6 - 6/2 + 0/2 + 1 = 4$

Expected intensity of M+1 from Table 21-2:

$$\underset{\text{Carbon}}{1.08(6)} + \underset{\text{Hydrogen}}{0.012(6)} + \underset{\text{Oxygen}}{0.038(1)} = 6.59\%$$

Observed intensity of M+1 $= 68/999 = 6.8\%$

Expected intensity of M+2 $= (0.005\ 8)(6)(5) + 0.205(1) = 0.38\%$

Observed intensity of M+2 $= 0.3\%$

(b) —Br C_6H_5Br $M^{+\bullet} = 156$

The two nearly equal peaks at m/z 156 and 158 scream out "bromine!"

Rings + double bonds $= c - \underset{\substack{\uparrow \\ h \text{ includes H + Br}}}{h/2} + n/2 + 1 = 6 - 6/2 + 0/2 + 1 = 4$

Expected intensity of M+1 $= \underset{\text{Carbon}}{1.08(6)} + \underset{\text{Hydrogen}}{0.012(5)} = 6.54\%$

Observed intensity of M+1 $= 46/566 = 8.1\%$

Expected intensity of M+2 $= \underset{\text{Carbon}}{(0.005\ 8)(6)(5)} + \underset{\text{Bromine}}{97.3(1)} = 97.5\%$

Observed intensity of M+2 $= 520/566 = 91.9\%$

The M+3 peak is the isotopic partner of the M+2 peak ($C_6H_5{}^{81}Br$). M+3 contains ^{81}Br plus either 1 ^{13}C or 1 ^{2}H. Therefore, the expected intensity of M+3 (relative to $C_6H_5{}^{81}Br$ at M+2) is $1.08(6) + 0.012(5) = 6.54\%$ of predicted intensity of $C_6H_5{}^{81}Br$ at M+2 $= (0.065\ 4)(97.3) = 6.4\%$ of $M^{+\bullet}$. Observed intensity of M+3 is $35/566 = 6.2\%$.

(c) —CO_2H $C_7H_3O_2Cl_3$ $M^{+\bullet} = 224$

From Figure 22-7, the M:M+2:M+4 pattern looks like a molecule containing three chlorine atoms. The correct structure is shown here, but there is no way you could assign the isomeric structure from the data.

Rings + double bonds $= c - h/2 + n/2 + 1 = 7 - 6/2 + 0/2 + 1 = 5$

Expected intensity of M+1 from Table 22-2:

$$\underset{\text{Carbon}}{1.08(7)} + \underset{\text{Hydrogen}}{0.012(3)} + \underset{\text{Oxygen}}{0.038(2)} = 7.67\%$$

Observed intensity of M+1 $= 63/791 = 8.0\%$

Expected intensity of M+2 $= (0.005\ 8)(7)(6) + 0.205(2) + 32.0(3) = 96.7\%$

Observed intensity of M+2 $= 754/791 = 95.4\%$

The M+3 peak is the isotopic partner of $C_7H_3O_2{}^{35}Cl_2{}^{37}Cl$ at M+2. M+3 contains one ^{37}Cl plus either 1 ^{13}C or 1 ^{2}H or 1 ^{17}O. Expected intensity of M+3 (relative to $C_7H_3O_2{}^{35}Cl_2{}^{37}Cl$ at M+2) $= 1.08(7) + 0.012(3) + 0.038(2) = 7.67\%$ of predicted intensity of $C_7H_3O_2{}^{35}Cl_2{}^{37}Cl$ at M+2. The predicted intensity of $C_7H_3O_2{}^{35}Cl_2{}^{37}Cl$ is $32.0(3) = 96.0\%$ of $M^{+\bullet}$.

Expected intensity of M+3 $= 7.67\%$ of $96.0\% = 7.4\%$ of $M^{+\bullet}$. Observed intensity $= 60/791 = 7.6\%$.

M+4 is composed mainly of $C_7H_3O_2{}^{35}Cl^{37}Cl_2$ plus a small amount of $C_7H_3{}^{16}O^{18}O^{35}Cl_2{}^{37}Cl$. Other formulas such as $^{12}C_6{}^{13}CH_3{}^{16}O^{17}O^{35}Cl_2{}^{37}Cl$ also add up to M+4, but they are even less likely to occur because they have two minor isotopes (^{13}C and ^{17}O). Expected intensity of M+4 from $C_7H_3O_2{}^{35}Cl^{37}Cl_2$ is $5.11(3)(2) = 30.7\%$ of $M^{+\bullet}$. The contribution from $C_7H_3{}^{16}O^{18}O^{35}Cl_2{}^{37}Cl$ is based on the predicted intensity of $C_7H_3O_2{}^{35}Cl_2{}^{37}Cl$ at M+2. The predicted intensity of $C_7H_3O_2{}^{35}Cl_2{}^{37}Cl$ is $32.0(3) = 96.0\%$ of $M^{+\bullet}$. The predicted intensity from $C_7H_3{}^{16}O^{18}O^{35}Cl_2{}^{37}Cl$ at M+4 is $0.205(2) = 0.410\%$ of $96.0\% = 0.4\%$. Total expected intensity of M+4 is $30.7\% + 0.4\% = 31.1\%$ of $M^{+\bullet}$. Observed intensity $= 264/791 = 33.4\%$.

Expected intensity of M+5 from $^{12}C_6{}^{13}CH_3O_2{}^{35}Cl^{37}Cl_2$ and $^{12}C_7H_2{}^2HO_2{}^{35}Cl^{37}Cl_2$ and $C_7H_3{}^{16}O^{17}O^{35}Cl^{37}Cl_2$ is based on the predicted intensity of $C_7H_3O_2{}^{35}Cl^{37}Cl_2$ at M+4. M+5 should have 1.08(7) + 0.012(3) + 0.038(2) = 7.7% of $C_7H_3O_2{}^{35}Cl^{37}Cl_2$ at M+4 = 7.7% of 30.7% = 2.4%. Observed intensity = 19/791 = 2.4%.

Expected intensity of M+6 from $C_7H_3O_2{}^{37}Cl_3$ is 0.544(3)(2)(1) = 3.26% of $M^{+\bullet}$. There will also be a small contribution from $C_7H_3{}^{16}O^{18}O^{35}Cl^{37}Cl_2$, which will be 0.205(2) = 0.410% of predicted intensity of $C_7H_3{}^{16}O^{35}Cl^{37}Cl_2$ = 0.410% of 30.7% of $M^{+\bullet}$ = 0.13% of $M^{+\bullet}$. The total expected intensity at M+6 is therefore 3.26 + 0.13 = 3.4% of $M^{+\bullet}$.

Observed intensity = 29/791 = 3.7%.

(d) 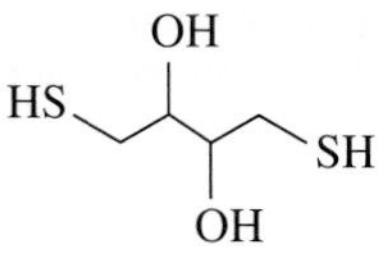

$C_4H_{10}O_2S_2$ $M^{+\bullet} = 154$

Sulfur gives a significant M+2 peak (4.52% of M per sulfur). The observed M+2 is 12/122 = 9.8%, which could represent two sulfur atoms. The composition $C_4H_{10}O_2S_2$ has two sulfur atoms and has a molecular mass of 154. The known structure is shown here, but you could not deduce the structure from the composition.

Rings + double bonds = $c - h/2 + n/2 + 1$
$= 4 - 10/2 + 0/2 + 1 = 0$

Expected intensity of M+1:

$$\underset{\text{Carbon}}{1.08(4)} + \underset{\text{Hydrogen}}{0.012(10)} + \underset{\text{Oxygen}}{0.038(2)} + \underset{\text{Sulfur}}{0.801(2)} = 6.12\%$$

Observed intensity of M+1 = 9/122 = 7.4%

Expected intensity of M+2 = (0.005 8)(4)(3) + 0.205(2) + 4.52(2) = 9.52%

Observed intensity of M+2 = 12/122 = 9.8%

22-G.

Analysis of electrospray mass spectrum of lysozyme

Observed $m/z \equiv m_n$	$m_{n+1} - 1.008$	$m_n - m_{n+1}$	Charge = n = $\dfrac{m_{n+1} - 1.008}{m_n - m_{n+1}}$	Molecular mass = $n \times (m_n - 1.008)$
1 789.1	1 589.39	198.7	7.99 ≈ 8	14 304.7
1 590.4	1 430.49	158.9	9.00 ≈ 9	14 304.5
1 431.5	1 300.49	130.0	10.00 ≈ 10	14 304.9
1 301.5	1 192.09	108.4	11.00 ≈ 11	14 305.4
1 193.1	—	—	12	14 305.1
				Mean = 14 304.9 (±0.3)

22-H. **(a)** To find the response factor, we insert values from the first line of the table into the equation:

$$\frac{\text{Area of analyte signal}}{\text{Area of standard signal}} = F\left(\frac{\text{concentration of analyte}}{\text{concentration of standard}}\right)$$

$$\frac{11\,438}{2\,992} = F\left(\frac{13.60 \times 10^2}{3.70 \times 10^2}\right) = F = 1.04_0$$

For the next two sets of data, we find $F = 1.02_0$ and 1.06_4, giving a mean value $F = 1.04_1$.

(b) The concentration of internal standard in the mixture of caffeine-D_3 plus cola is

$$(1.11\text{ g/L}) \times \frac{0.050\,0\text{ mL}}{1.050\text{ mL}} = 52.8_6\text{ mg/L}$$

The concentration of caffeine in the chromatography solution is

$$\frac{\text{Area of analyte signal}}{\text{Area of standard signal}} = F\left(\frac{\text{concentration of analyte}}{\text{concentration of standard}}\right)$$

$$\frac{1\,733}{1\,144} = 1.04_1\left(\frac{[\text{caffeine}]}{52.8_6\text{ mg/L}}\right) \Rightarrow [\text{caffeine}] = 76.9\text{ mg/L}$$

The unknown beverage had been diluted from 1.000 to 1.050 mL when the standard was added, so the concentration of caffeine in the original beverage was $\left(\frac{1.050\,0\text{ mL}}{1.00\text{ mL}}\right)(76.9\text{ mg/L}) = 80.8$ mg/L.

Chapter 23

23-A. **(a)** Fraction remaining $= q = \dfrac{V_1}{V_1 + KV_2}$

$$0.01 = \frac{10}{10 + 4.0V_2} \Rightarrow V_2 = 248\text{ mL}$$

(b) $q^3 = 0.01 = \left(\dfrac{10}{10 + 4.0\,V_2}\right)^3 \Rightarrow V_2 = 9.1$ mL and total volume = 27.3 mL.

23-B. **(a)** $k_1 = \dfrac{t_{r1} - t_m}{t_m}$

$$\Rightarrow t_m = \frac{t_{r1}}{k_1 + 1} = \frac{10.0\text{ min}}{5.00} = 2.00\text{ min}$$

$t_{r2} = t_m(k_2 + 1) = (2.00\text{ min})(5.00 + 1) = 12.0$ min

$$\sigma_1 = \frac{t_{r1}}{\sqrt{N}} = \frac{10.0\text{ min}}{\sqrt{1\,000}} = 0.316\text{ min}$$

$\Rightarrow w_{1/2}$ (peak 1) $= 2.35\sigma_1 = 0.74$ min

$w_1 = 4\sigma_1 = 1.26$ min

$$\sigma_2 = \frac{t_{r2}}{\sqrt{N}} = \frac{12.0\text{ min}}{\sqrt{1\,000}} = 0.379\text{ min}$$

$\Rightarrow w_{1/2}$ (peak 2) $= 2.35\sigma_2 = 0.89$ min

$w_2 = 4\sigma_2 = 1.52$ min

(b)

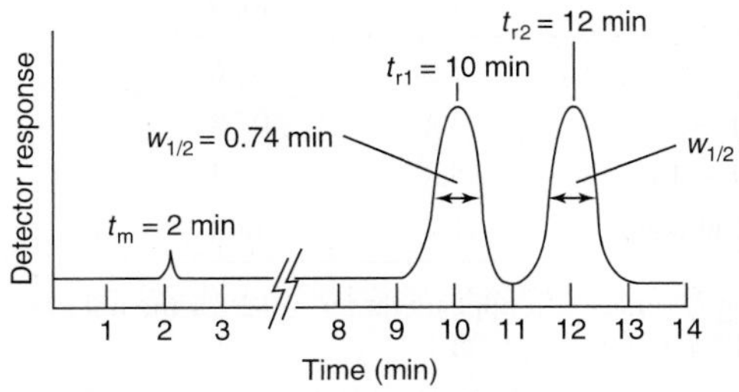

(c) Resolution $= \dfrac{\Delta t_r}{w_{av}} = \dfrac{2\text{ min}}{(1.26\text{ min} + 1.52\text{ min})/2} = 1.44$

23-C. **(a)** Relative distances measured from the figure:
$t_m = 10._4$ $t'_r = 39._8$ for octane $t'_r = 76._0$ for nonane
$k = t'_r/t_m = 3.8_3$ for octane and 7.3_1 for nonane

(b) Let t_s = time in stationary phase, t_m = time in mobile phase, and t = total time on column. We know that $k = t_s/t_m$. But

$$t = t_s + t_m = t_s + \frac{t_s}{k} = t_s\left(1 + \frac{1}{k}\right) = t_s\left(\frac{k+1}{k}\right)$$

Therefore, $t_s/t = \dfrac{k}{k+1} = 3.8_3/4.8_3 = 0.79_3$

(c) $\alpha = t'_r\,(\text{nonane})/\,t'_r\,(\text{octane}) = 76._0/39._8 = 1.9_1$

(d) $K = kV_m/V_s = 3.8_3(V_m/\frac{1}{2}V_m) = 7.6_6$

23-D. **(a)** For ethyl acetate, I measure $t_r = 20.3$ mm and $w = 3.0$ mm. (You will have different numbers depending on the size of the figure in your book.) Therefore, $N = 16t_r^2/w^2 = 730$ plates. For toluene, the values are $t_r = 66.8$ mm, $w = 8.3$ mm, and $N = 1\,040$ plates.

(b) We expect $w_{1/2} = (2.35/4)w$. The measured value of $w_{1/2}$ is in good agreement with the calculated value.

23-E. The column is overloaded, causing a gradual rise and an abrupt fall of the peak. As the sample size is decreased, the overloading decreases and the peak becomes more symmetric.

23-F. **(a)** We know that $\alpha = 1.068$ and $k_1 = 5.16$. Since $\alpha = k_2/k_1$, $k_2 = \alpha k_1 = (1.068)(5.16) = 5.51_1$.

(b) Resolution $= \dfrac{\sqrt{N}}{4}\left(\dfrac{\alpha - 1}{\alpha}\right)\left(\dfrac{k_2}{1 + k_2}\right)$

$$1.50 = \frac{\sqrt{N}}{4}\left(\frac{1.068 - 1}{1.068}\right)\left(\frac{5.51_1}{1 + 5.51_1}\right) \Rightarrow N = 1.24_0 \times 10^4$$

Required length = $(1.24_0 \times 10^4$ plates$)(0.520$ mm/plate$) = 6.45$ m

(c) The relation between retention factor and retention time is
$$k_1 = (t_1 - t_m)/t_m = t_1/t_m - 1 \Rightarrow$$
$$t_1 = t_m(k_1 + 1) = t_m(6.16) = (2.00 \text{ min})(6.16) = 12.32 \text{ min}$$
$$t_2 = t_m(k_2 + 1) = t_m(6.51_1) = (2.00 \text{ min})(6.51_1) = 13.02 \text{ min}$$

$$w_{1/2} = \sqrt{\frac{5.55}{N}}\, t_r = \sqrt{\frac{5.55}{1.24_0 \times 10^4}}(12.32 \text{ min}) = 0.26_1 \text{ min for component 1}$$

$$w_{1/2} = \sqrt{\frac{5.55}{1.24_0 \times 10^4}}(13.02 \text{ min}) = 0.27_5 \text{ min for component 2}$$

(d) We are told that $V_s/V_m = 0.30$. The relation between retention factor and partition coefficient for each solute is $k = KV_s/V_m \Rightarrow K = kV_m/V_s$. For component 1, $K_1 = k_1 V_m/V_s = (5.16)(1/0.30) = 17._2$

Chapter 24

24-A. **(a)** Between nonane (C_9H_{20}) and decane ($C_{10}H_{22}$), but very close to decane.
(b) Adjusted retention times are 13.83 (C_{10}) and 15.42 min (C_{11})

$$1\,050 = 100\left[10 + (11 - 10)\frac{\log t_r'(\text{unknown}) - \log 13.83}{\log 15.42 - \log 13.83}\right]$$

$$\Rightarrow t_r'(\text{unknown}) = 14.60 \text{ min} \Rightarrow t_r(\text{unknown}) = 16.40 \text{ min}$$

24-B. **(a)** A plot of $\log t_r'$ versus (number of carbon atoms) should be a fairly straight line for a homologous series of compounds.

Peak	t_r'	$\log t_r'$
$n = 7$	2.9	0.46
$n = 8$	5.4	0.73
$n = 14$	85.8	1.93
Unknown	41.4	1.62

From a graph of $\log t_r'$ versus n, it appears that $n = 12$ for the unknown.
(b) $k = t_r'/t_m = (42.5 - 1.1)/1.1 = 38$

24-C. **(a)** Plate number (N) is proportional to column length. Resolution is proportional to $\sqrt{N}$. If everything is the same except for length, we can say that

$$\frac{1.5}{1.0} = \frac{R_2}{R_1} = \frac{\sqrt{N_2}}{\sqrt{N_1}} \Rightarrow N_2 = 2.25N_1$$

The column must be 2.25 times longer to achieve the desired resolution, and the elution time will be 2.25 times longer.
(b) If N and k are constant, the relation between resolution and relative retention is

$$\frac{1.5}{1.0} = \frac{R_2}{R_1} = \frac{(\alpha_2 - 1)/\alpha_2}{(\alpha_1 - 1)/\alpha_1} = \frac{(\alpha_2 - 1)/\alpha_2}{(1.016 - 1)/1.016} \Rightarrow \alpha_2 = 1.024_2$$

Alcohols are polar, so we could probably increase the relative retention by choosing a more polar stationary phase. (Diphenyl)$_{0.05}$(dimethyl)$_{0.95}$polysiloxane is listed as nonpolar. We could try an intermediate polarity phase such as (diphenyl)$_{0.35}$(dimethyl)$_{0.65}$polysiloxane. The more polar phase will probably retain alcohols more strongly and increase the retention time.

24-D. **(a)** $S = [\text{butanol}] = \dfrac{234 \text{ mg}/(74.12 \text{ g/mol})}{10.0 \text{ mL}} = 0.315_7 \text{ M}$

$X = [\text{hexanol}] = \dfrac{312 \text{ mg}/(102.17 \text{ g/mol})}{10.0 \text{ mL}} = 0.305_4 \text{ M}$

$\dfrac{A_X}{[X]} = F\left(\dfrac{A_S}{[S]}\right) \Rightarrow \dfrac{1.45}{[0.305_4 \text{ M}]} = F\left(\dfrac{1.00}{[0.315_7 \text{ M}]}\right) \Rightarrow F = 1.49_9$

(b) I estimate the areas by measuring peak height and $w_{1/2}$ in millimeters. Your answer will be different from mine if the figure size in your book is different from that in my manuscript. However, the relative peak areas should be the same.

Butanol: Height = 41.3 mm; $w_{1/2}$ = 2.2 mm;
  Area = 1.064 × peak height × $w_{1/2}$ = 96._7 mm^2

Hexanol: Height = 21.9 mm; $w_{1/2}$ = 6.9 mm;
  Area = 161 mm^2

(c) The volume of solution is not stated, but the concentration is directly proportional to the number of moles. We can substitute moles for concentrations in the internal standard equation:

$$\frac{A_X}{[X]} = F\left(\frac{A_S}{[S]}\right) \Rightarrow \frac{161 \text{ mm}^2}{\text{mg hexanol}/(102.17 \text{ mg/mmol})}$$

$$= 1.49_9\left(\frac{96.7 \text{ mm}^2}{112 \text{ mg}/(74.12 \text{ mg/mmol})}\right)$$

$$\Rightarrow \text{hexanol} = 171 \text{ mg}$$

(d) The greatest uncertainty is in the width of the fairly narrow butanol peak. The uncertainty in width is ~5–10%.

24-E. $S = [\text{pentanol}]$; $X = [\text{hexanol}]$. We will substitute mmol for concentrations, because the volume is unknown and concentrations are proportional to mmol.
For the standard mixture, we can write

$$\frac{A_X}{[X]} = F\left(\frac{A_S}{[S]}\right) \Rightarrow \frac{1\,570}{[1.53]} = F\left(\frac{922}{[1.06]}\right) \Rightarrow F = 1.18_0$$

For the unknown mixture,

$$\frac{816}{[X]} = 1.18_0\left(\frac{843}{[0.57]}\right) \Rightarrow [X] = 0.47 \text{ mmol}$$

Chapter 25

25-A. $\dfrac{\text{Area}_A}{[A]} = F\left(\dfrac{\text{Area}_B}{[B]}\right) \Rightarrow \dfrac{10.86}{[1.03]} = F\left(\dfrac{4.37}{[1.16]}\right) \Rightarrow F = 2.79_9$

The concentration of internal standard (B) mixed with unknown (A) is 12.49 mg/25.00 mL = 0.499 6 mg/mL

$$\frac{5.97}{[A]} = 2.79_9\left(\frac{6.38}{[0.499\,6]}\right) \Rightarrow [A] = 0.167_0 \text{ mg/mL}$$

$$[A] \text{ in original unknown} = \frac{25.00}{10.00}(0.167_0 \text{ mg/mL})$$

$$= 0.418 \text{ mg/mL}$$

25-B. **(a)** Equation 23-16: $k = \dfrac{t_r - t_m}{t_m} \Rightarrow \dfrac{t_1 - 1.00}{1.00} = 1.35$

$$\Rightarrow t_1 = 2.35 \text{ min}$$

Equation 23-15: $\alpha = \dfrac{t_{r2}'}{t_{r1}'} \Rightarrow 4.53 = \dfrac{t_2 - 1.00}{t_1 - 1.00} \Rightarrow t_2 = 7.12 \text{ min}$

Equation 23-23: $\text{Resolution} = \dfrac{\Delta t_r}{w_{av}}$

$$\Rightarrow 7.7 = \frac{7.12 - 2.35}{w_{av}} \Rightarrow w_{av} = 0.62 \text{ min}$$

(b) From Equation 23-31, we know that $w_{1/2}$ is proportional to t_r if N is constant. Therefore, $\dfrac{w_{1/2}(\text{peak 1})}{w_{1/2}(\text{peak 2})} = \dfrac{t_1}{t_2} = \dfrac{2.35}{7.12} = 0.330$. We know that

w_{av}, the average width at the base, is 0.62 min. For each peak, w = 4σ and $w_{1/2}$ = 2.35σ , so w = 1.70$w_{1/2}$.
$w_{av} = 0.62 = \frac{1}{2}(w_1 + w_2) = \frac{1}{2}[1.70w_{1/2} (\text{peak 1}) + 1.70w_{1/2} (\text{peak 2})]$

Substituting $w_{1/2}$ (peak 1) = 0.330$w_{1/2}$ (peak 2) into the previous equation gives $w_{1/2}$ (peak 2) = 0.54$_8$ min. Then $w_{1/2}$ (peak 1) = 0.330$w_{1/2}$ (peak 2) = 0.18$_1$ min.
(c) Because the areas are equal, we can say

$$\text{Height}_R \times w_R = \text{height}_S \times w_S \Rightarrow \frac{\text{Height}_R}{\text{Height}_S} = \frac{w_S}{w_R} = \frac{0.54_8}{0.18_1} = 3.0$$

25-C.

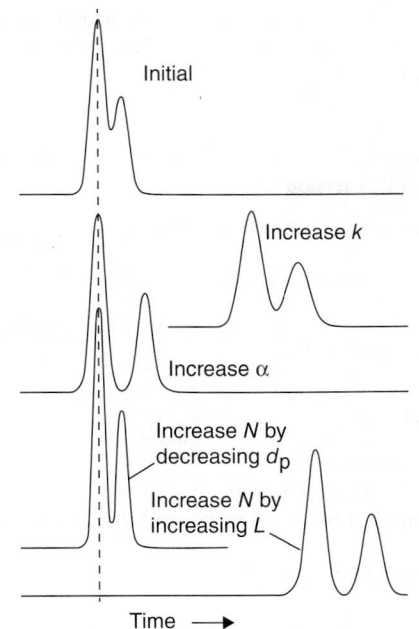

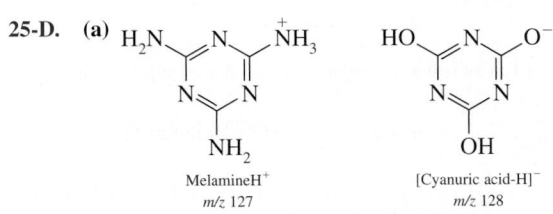

Time ⟶

25-D. **(a)**

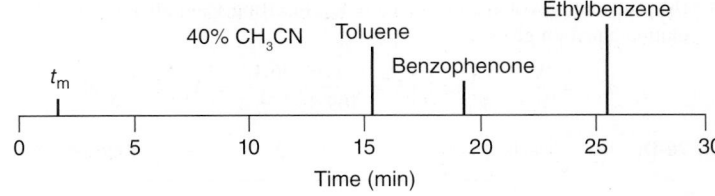

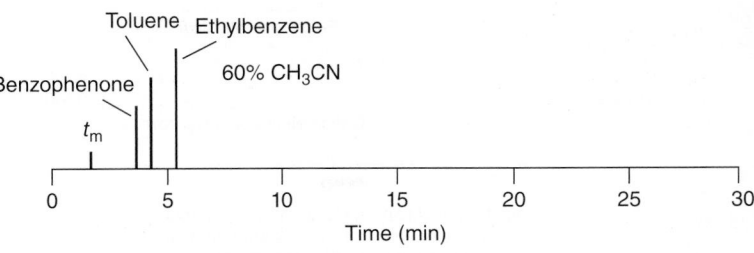

MelamineH⁺
m/z 127

[Cyanuric acid-H]⁻
m/z 128

When melamineH⁺ decomposes to *m/z* 85, it loses a mass of 42 Da, which is probably CN_2H_2. The remaining cation is $C_2N_4H_5^+$, which might have the structure

$$H_2N-C \underset{N}{\overset{||}{=}} C \underset{NH}{\overset{+NH_2}{<}}$$

[Cyanuric acid-H]⁻ decomposes to *m/z* 42, which is most likely cyanate, $N\equiv C-O^-$.

(b) In selected reaction monitoring for melamine, the mass spectrometer isolates *m/z* 127⁺ from the milk, and this ion then undergoes collisionally activated dissociation to *m/z* 85⁺. There are not many other molecular species in the milk that provide the same two ions. For cyanuric acid, the spectrometer isolates *m/z* 128⁻, and this ion dissociates to *m/z* 42⁻. Again, there are not many other species in the milk that provide the same two ions.

25-E. **(a)** C_8-silica is a reversed-phase column. In reversed-phase chromatography mobile phase strength increases with increasing organic solvent. Greatest retention is observed with 40% acetonitrile and least retention at 60% acetonitrile. Toluene and benzophenone are co-eluted at 51.6% acetonitrile.

(b) Convert the log *k* from the graph to *k* and then to retention time t_r. For example, for 40% CH_3CN, log *k* of ethyl benzene is 1.18 from the graph, so *k* = 15.1. Retention time is obtained from *k* by rearranging Equation 23-16

$$k = \frac{t_r - t_m}{t_m} \Rightarrow t_r = t_m \times (1 + k).$$

t_m can be estimated with Equation 25-5.

$$t_m \approx \frac{Ld_c^2}{2F} \approx \frac{15 \text{ cm} \times (0.46 \text{ cm})^2}{2 \times 1.0 \text{ mL/min}} = 1.6 \text{ min}$$

$t_r = 1.6 \text{ min} \times (1 + 15.1) = 25.8 \text{ min}$

(c) Equation 23-33 shows how resolution depends on plate number, relative retention, and retention factor.

$$\text{Resolution} = \frac{\sqrt{N}}{4}\left(\frac{\alpha - 1}{\alpha}\right)\left(\frac{k_2}{1 + k_2}\right)$$

Benzophenone and toluene are the closest two peaks, and so would have least resolution. Retention factors of benzophenone and toluene are $10^{0.13} = 1.35$ and $10^{0.21} = 1.62$, respectively.

Equation 23-20: Relative retention is $\alpha = k_2/k_1 = 1.62/1.35 = 1.20$

Equation 25-1: $N \approx \dfrac{3\,000 \cdot L(\text{cm})}{d_p(\mu m)} \approx \dfrac{3\,000 \times 15}{5} \approx 9\,000$

$$\text{Resolution} = \frac{\sqrt{9\,000}}{4}\frac{(1.20 - 1)}{1.20}\left(\frac{1.62}{1 + 1.62}\right) = 2.4$$

(d) Attributes of a good separation are (1) $0.5 \le k \le 20$, (2) resolution $\ge$ 2, (3) operating pressure $\le$ 15 MPa, (4) $0.9 \le$ asymmetry factor ≤ 1.5. With 60% acetonitrile, *k* ranges from 1.35 to 2.40. Resolution in **(c)** is >2. Pressure in figure caption is 7–8 MPa. Gaussian peak has asymmetry factor = 1. Yes: separation at 60% acetonitrile has all attributes of good separation. Short separation time (< 6 minutes) is an added advantage.

Chapter 26

26-A. **(a)** Each K^+ applied to the column releases one H^+ from the column. The moles of NaOH titrant required will equal the moles of H^+ exchanged from the column.

mol NaOH = $0.045\,8$ M $\times 0.010\,00$ L = 4.58×10^{-4} mol

volume NaOH = $(4.58 \times 10^{-4}$ moles$)/(0.023\,1$ M$) = 19.8_3$ mL

(b) Equivalents of cation in sample = mol OH^- titrant required

= $0.139\,6$ M $\times 0.030\,64$ L = 4.277 mmol OH^- = 4.277 meq

Milliequivalents per gram of sample = $\dfrac{4.277 \text{ meq}}{0.269\,2 \text{ g}} = 15.8_9$ meq/g

(c) Equivalent mass = mass of a substance containing one equivalent. That is, it equals the mass of sample loaded onto the ion exchange column divided by the equivalents determined by titration.

Equivalent mass = $0.269\,2$ g$/4.277 \times 10^{-3}$ eq = 62.94 g/eq

26-B. 13.03 mL of 0.022 74 M NaOH = 0.296 3 mmol of OH^-, which must equal the total cation charge (= $2[VO^{2+}] + 2[H_2SO_4]$) in the 5.00 mL aliquot. 50.0 mL therefore contains 2.963 mmol of cation charge. The VO^{2+} content is (50.0 mL)(0.024 3 M) = 1.215 mmol = 2.43 mmol of charge. The H_2SO_4 must therefore be $(2.963 - 2.43)/2 = 0.266_5$ mmol.

1.215 mmol $VOSO_4$ = 0.198 g $VOSO_4$ in 0.244 7 g sample = 80.9 wt%

0.266_5 mmol H_2SO_4 = 0.026 1 g H_2SO_4 in 0.244 7 g sample = 10.7 wt%

H_2O (by difference) = 8.4 wt%

26-C. **(a)** Fractionation range of Sephadex G-50 = 1 500–30 000, so hemoglobin should not be retained and ought to be eluted in a volume of 36.4 mL.

(b) The elution volume of $^{22}NaCl$ is V_m. Inserting $V_o = 36.4$ mL into the elution equation gives

$$K_{av} = \frac{V_r - V_o}{V_m - V_o} \Rightarrow 0.65 = \frac{V_r - 36.4}{109.8 - 36.4} \Rightarrow V_r = 84.1 \text{ mL}$$

26-D. **(a)**

Net flow of cations and anions is to the right, because electroosmotic flow is stronger than electrophoretic flow at high pH

(b) I^- has a greater mobility than Cl^-. Therefore, I^- swims upstream faster than Cl^- (because electrophoresis opposes electroosmosis) and is eluted later than Cl^-. The mobility of Br^- is greater than that of I^- in Table 14-1. Therefore, Br^- will have a longer migration time than I^-.
(c) Bare I^- is a larger ion than bare Cl^-, so the charge density in I^- is lower than the charge density in Cl^-. Therefore, I^- should have a smaller hydrated radius than Cl^-. This means that I^- has less friction than Cl^- and a greater mobility than Cl^-.

Chapter 27

27-A. 1 mol ethoxyl groups produces 1 mol AgI. 29.03 mg AgI = $0.123\ 6_5$ mmol. The amount of compound analyzed is 25.42 mg/(417 g/mol) = $0.060\ 9_6$ mmol. There are

$$\frac{0.123\ 6_5 \text{ mmol ethoxyl groups}}{0.060\ 9_6 \text{ mmol compound}} = 2.03\ (= 2) \text{ ethoxyl groups/molecule}$$

27-B. There is 1 mol SO_4^{2-} in each mole of each reactant and of the product. Let x = g of K_2SO_4 and y = g of $(NH_4)_2SO_4$.

$$x + y = 0.649 \text{ g} \tag{1}$$

$$\underset{\substack{\text{Moles of}\\K_2SO_4}}{\frac{x}{174.27}} + \underset{\substack{\text{Moles of}\\(NH_4)_2SO_4}}{\frac{y}{132.14}} = \underset{\substack{\text{Moles of}\\BaSO_4}}{\frac{0.977}{233.39}} \tag{2}$$

Making the substitution $y = 0.649 - x$ in Equation 2 gives $x = 0.397$ g = 61.1% of the sample.

27-C. Formula and atomic masses: Ba(137.327), Cl(35.452), K(39.098), H_2O(18.015), KCl(74.550), $BaCl_2 \cdot 2H_2O$(244.26), H_2O lost = $1.783\ 9 - 1.562\ 3 = 0.221\ 6$ g = $1.230\ 1 \times 10^{-2}$ mol of H_2O. For 2 mol H_2O lost, 1 mol $BaCl_2 \cdot 2H_2O$ must have been present. $\frac{1}{2}(1.230\ 1 \times 10^{-2}$ mol H_2O lost) = $6.150\ 4 \times 10^{-3}$ mol $BaCl_2 \cdot 2H_2O = 1.502\ 3$ g. The Ba and Cl contents of the $BaCl_2 \cdot 2H_2O$ are

$$Ba = \left(\frac{137.327}{244.26}\right)(1.502\ 3 \text{ g}) = 0.844\ 62 \text{ g}$$

$$Cl = \left(\frac{2(35.452)}{244.26}\right)(1.502\ 3 \text{ g}) = 0.436\ 09 \text{ g}$$

Because the total sample weighs $1.783\ 9$ g and contains $1.502\ 3$ g of $BaCl_2 \cdot 2H_2O$, the sample must contain $1.783\ 9 - 1.502\ 3 = 0.281\ 6$ g of KCl, which contains

$$K = \left(\frac{39.098}{74.550}\right)(0.281_6) = 0.147\ 69 \text{ g}$$

$$Cl = \left(\frac{35.452}{74.550}\right)(0.281_6) = 0.133\ 91 \text{ g}$$

Weight percent of each element:

$$Ba = \frac{0.844\ 62}{1.783\ 9} = 47.35\%$$

$$K = \frac{0.147\ 69}{1.783\ 9} = 8.28\%$$

$$Cl = \frac{0.436\ 09 + 0.133\ 91}{1.783\ 9} = 31.95\%$$

27-D. Let x = mass of $Al(BF_4)_3$ and y = mass of $Mg(NO_3)_2$. We can say that $x + y = 0.282\ 8$ g. We also know that

$$\text{Moles of nitron tetrafluoroborate} = 3(\text{moles of } Al(BF_4)_3) = \frac{3x}{287.39}$$

$$\text{Moles of nitron nitrate} = 2(\text{moles of } Mg(NO_3)_2) = \frac{2y}{148.31}$$

Equating the mass of product to the mass of nitron tetrafluoroborate plus the mass of nitron nitrate, we can write

$$1.322 = \underset{\substack{\text{Mass of}\\\text{product}}}{} \underset{\substack{\text{Mass of nitron}\\\text{tetrafluoroborate}}}{\left(\frac{3x}{287.39}\right)(400.18)} + \underset{\substack{\text{Mass of nitron}\\\text{nitrate}}}{\left(\frac{2y}{148.31}\right)(375.39)}$$

Making the substitution $x = 0.282\ 8 - y$ allows us to find $y = 0.158\ 9$ g of $Mg(NO_3)_2 = 1.072$ mmol of Mg = $0.026\ 05$ g of Mg = 9.210% of the original solid sample.

27-E. **(a)** Co(III) ($La^{3+}Co^{3+}(O^{2-})_3$)

(b) $LaCoO_3(s) + \frac{3}{2}H_2(g) \rightarrow \underbrace{\frac{1}{2}La_2O_3(s) + Co(s)}_{} + \frac{3}{2}H_2O(g)$

FM 245.836 9   FM 221.837 8   FM 18.015 4

(c) 245.836 9 g of $LaCoO_3$ would create 221.837 8 g of solid product

$$\frac{\text{product from 41.872 4 mg}}{41.872\ 4 \text{ mg}} = \frac{221.837\ 8 \text{ g}}{245.836\ 9 \text{ g}} \Rightarrow \text{product} = 37.784\ 7_3 \text{ mg}$$

(d) $LaCoO_{3+x}(s) + \frac{3}{2}H_2(g) \rightarrow \underbrace{\frac{1}{2}La_2O_3(s) + Co(s)}_{} + \left(\frac{3}{2} + x\right)H_2O(g)$

FM (245.836 9 + 15.999 4x)   FM 221.837 8

$$\frac{\text{final mass}}{\text{initial mass}} = \frac{\text{final mass}}{41.872\ 4 \text{ mg}} = \frac{221.837\ 8}{245.836\ 9 + (15.999\ 4)x}$$

$$\Rightarrow \text{final mass} = \frac{(221.837\ 8)(41.872\ 4 \text{ mg})}{245.836\ 9 + (15.999\ 4)x}$$

(e) Observed final mass = 37.763 7 mg. Substituting this mass into the equation derived in **(d)** gives

$$\text{final mass} = 37.763\ 7 \text{ mg} = \frac{(221.837\ 8)(41.872\ 4 \text{ mg})}{245.836\ 9 + (15.999\ 4)x} \Rightarrow x = 0.008\ 6$$

Formula of the starting solid is $LaCoO_{3.008\ 6}$
(f) Mass loss at 500°C $\approx 41.872\ 4$ mg $- 40.51$ mg = 1.36 mg, which is 1/3 of the theoretical complete mass loss from $LaCoO_3$ (1/3 of 4.087 7 mg = 1.362 6 mg). The plateau at 500°C corresponds to the change Co(III) $\rightarrow$ Co(II), which is the same as $LaCoO_3 \rightarrow LaCoO_{2.5}$.

Chapter 28

28-A. **(a)** Expected number of red marbles = np_{red} = (1 000)(0.12) = 120. Expected number of yellow = nq_{yellow} = (1 000)(0.88) = 880.

(b) Absolute: $\sigma_{red} = \sigma_{yellow} = \sqrt{npq}$

$$= \sqrt{(1\ 000)(0.12)(0.88)} = 10.28$$

Relative: σ_{red}/n_{red} = 10.28/120 = 8.56%
$\sigma_{yellow}/n_{yellow}$ = 10.28/880 = 1.17%
(c) For 4 000 marbles, n_{red} = 480 and n_{yellow} = 3 520.
$\sigma_{red} = \sigma_{yellow} = \sqrt{npq} = \sqrt{(4\ 000)(0.12)(0.88)} = 20.55$
σ_{red}/n_{red} = 4.28% $\sigma_{yellow}/n_{yellow}$ = 0.58%

(d) 2, $\sqrt{n}$

(e) $\frac{\sigma_{red}}{n_{red}} = 0.02 = \frac{\sqrt{n(0.12)(0.88)}}{(0.12)n}$

$\Rightarrow n = 1.83 \times 10^4$
28-B. **(a)** $mR^2 = K_s \Rightarrow m(10)^2 = 36 \Rightarrow m = 0.36$ g

(b) An uncertainty of ± 20 counts per second per gram is $100 \times 20/237 = 8.4\%$.

$$n = \frac{t^2 s_s^2}{e^2} = \frac{(1.96)^2 (0.10)^2}{(0.084)^2} = 5.4 \approx 5$$

$\Rightarrow t = 2.776$ (4 degrees of freedom)

$$n \approx \frac{(2.776)^2 (0.10)^2}{(0.084)^2} = 10.9 \approx 11 \Rightarrow t = 2.228$$

$$n \approx \frac{(2.228)^2 (0.10)^2}{(0.084)^2} = 7.0 \approx 7 \Rightarrow t = 2.447$$

$$n \approx \frac{(2.447)^2 (0.10)^2}{(0.084)^2} = 8.5 \approx 8 \Rightarrow t = 2.365$$

$$n \approx \frac{(2.365)^2 (0.10)^2}{(0.084)^2} = 7.9 \approx 8$$

28-C. **(a)** Acid-soluble inorganic matter and organic material can probably be dissolved (and oxidized) together by wet ashing with HNO_3 + H_2SO_4 in a Teflon-lined bomb in a microwave oven. Insoluble residue should be washed well with water and the washings combined with the acid solution. After the residue has been dried, it can be fused with one of the fluxes in Table 28-6, dissolved in dilute acid, and combined with the previous solution.

(b) Grind the soil well with a mortar and pestle or ball mill or a Shatterbox mill. Mix 5 g of dry powder well with ~3–8 mL of H_2O. Then conduct the QuEChERS procedure followed by chromatography of the extract.

Answers to Problems

Chapter 0

1. Mn^{2+} interferes with method A, but not with method B, whose result (0.009 mg/L) is more reliable.

Chapter 1

1. **(a)** meter (m), kilogram (kg), second (s), ampere (A), kelvin (K), mole (mol)   **(b)** hertz (Hz), newton (N), pascal (Pa), joule (J), watt (W)

2. Table 1-3

3. **(a)** milliwatt $= 10^{-3}$ watt   **(b)** picometer $= 10^{-12}$ meter **(c)** kiloohm $= 10^3$ ohm   **(d)** microfarad $= 10^{-6}$ farad   **(e)** terajoule $= 10^{12}$ joule   **(f)** nanosecond $= 10^{-9}$ second   **(g)** femtogram $= 10^{-15}$ gram   **(h)** decipascal $= 10^{-1}$ pascal

4. **(a)** 100 fJ or 0.1 pJ   **(b)** 43.172 8 nF   **(c)** 299.79 THz or 0.299 79 PHz   **(d)** 0.1 nm or 100 pm   **(e)** 21 TW   **(f)** 0.483 amol or 483 zmol

5. **(a)** 8×10^{12} kg of C   **(b)** 2.9×10^{13} kg CO_2 **(c)** 4 tons CO_2 per person

6. 15 days for chunk *white* tuna; 3.5 days for chunk *light* tuna

7. 7.457×10^4 J/s, 6.416×10^7 cal/h

8. **(a)** 2.0 W/kg and 3.0 W/kg   **(b)** The person consumes 1.1×10^2 W.

9. 1.47×10^3 J/s, 1.47×10^3 W

10. **(a)** 0.621 37 miles/km   **(b)** 51 miles per gallon   **(c)** diesel produces 5.38 metric tons and gasoline produces 6.42 metric tons of CO_2

11. See Table 1-2.

12. 6 tons/year

13. **(a)** molarity = moles of solute/liter of solution   **(b)** molality = moles of solute/kilogram of solvent   **(c)** density = grams of substance/milliliter of substance   **(d)** weight percent $= 100 \times$ (mass of substance/mass of solution or mixture)   **(e)** volume percent $= 100 \times$ (volume of substance/volume of solution or mixture)   **(f)** parts per million $= 10^6 \times$ (grams of substance/grams of sample)   **(g)** parts per billion $= 10^9 \times$ (grams of substance/grams of sample)   **(h)** formal concentration = moles of formula/liter of solution

14. CH_3CO_2H is a weak electrolyte that is partially dissociated. $[CH_3CO_2H] + [CH_3CO_2^-] = 0.01$ M. $[CH_3CO_2H]$ alone is <0.01 M.

15. 1.10 M

16. 5.48 g

17. **(a)** 39 ppb = 3.9 mPa at the surface of the Earth where atmospheric pressure = 1 bar   **(b)** 19 mPa at an altitude of 16 km = 2 000 ppb where atmospheric pressure = 9.6 kPa

18. **(a)** 1.9×10^{-7} bar   **(b)** 11 nM

19. **(a)** 2.11×10^{-7} M   **(b)** Ar: 3.77×10^{-4} M; Kr: 4.60×10^{-8} M; Xe: 3.5×10^{-9} M

20. 10^{-3} g/L, 10^3 μg/L, 1 μg/mL, 1 mg/L

21. 7×10^{-10} M

22. 26.5 g $HClO_4$, 11.1 g H_2O

23. **(a)** 1 670 g solution   **(b)** 1.18×10^3 g $HClO_4$ **(c)** 11.7 mol

24. 1.51 *m*

25. **(a)** 6.0 amol/vesicle   **(b)** 3.6×10^6 molecules **(c)** 3.35×10^{-20} m^3, 3.35×10^{-17} L   **(d)** 0.30 M

26. 4.4×10^{-3} M, 6.7×10^{-3} M

27. **(a)** 1 046 g, 376.6 g/L   **(b)** 9.07 *m*

28. Cal/g, Cal/ounce: Shredded Wheat (3.6, 102); doughnut (3.9, 111); hamburger (2.8, 79); apple (0.48, 14)

29. 2.5×10^6 g F$^-$, 3.2×10^6 g H_2SiF_6

30. 6.18 g in a 2-L volumetric flask

31. Dissolve 6.18 g $B(OH)_3$ in 2.00 kg H_2O.

32. 3.2 L

33. 8.0 g

34. **(a)** 55.6 mL   **(b)** 1.80 g/mL

35. 1.52 g/mL

36. 1.29 mL

37. 14.4 g

38. **(a)** 60.05 g/mol, 84.01 g/mol   **(b)** 3.57 g   **(c)** 72 mL **(d)** vinegar   **(e)** 3 bar

Chapter 2

1. Familiarize yourself with hazards of what you are about to do and do not carry out a dangerous procedure without adequate precautions.

2. Nonpolar organic liquids might penetrate rubber gloves. Concentrated HCl acid is a polar aqueous solution that is not likely to penetrate through rubber gloves.

3. Reducing Cr(VI) to Cr(III) decreases the toxicity of the metal. Converting Cr(III)(aq) to Cr(OH)$_3$(s) decreases solubility. Evaporation minimizes waste volume.

4. Green chemistry is a set of principles intended to help sustain the habitability of Earth. Green chemistry seeks to design chemical products and processes to reduce the use of resources and the generation of hazardous waste.

5. The lab notebook must (1) state what was done; (2) state what was observed; and (3) be understandable to a stranger.

7. The buoyancy correction is 1 when the substance being weighed has the same density as the weight used to calibrate the balance.

8. 14.85 g

9. smallest: PbO_2; largest: Li

10. 4.239 1 g, lower by 0.06%

11. 1.266_8 g—the mass in air is negligibly different from the mass in vacuum because the density of CsCl is high.

12. **(a)** 0.000 164 g/mL   **(b)** 0.823 g

13. **(a)** 979 Pa   **(b)** 0.001 1 g/mL   **(c)** 1.001 0 g

14. 99.999 1 g

15. **(a)** 11 pg   **(b)** -4.4 Hz frequency change comes from binding of 14 pg, which corresponds to 1.3 cytosine nucleotides per DNA molecule.

16. TD means "to deliver" and TC means "to contain."

17. Dissolve $(0.250\ 0\ L)(0.150\ 0\ mol/L) = 0.037\ 50$ mol K_2SO_4 (= 6.535 g, FM 174.26 g/mol) in <250 mL H_2O in a 250-mL volumetric flask. Add H_2O and mix. Dilute to mark and mix well by inverting flask many times.

18. Plastic flask is for trace analysis of ppb analytes that might be adsorbed on glass.

19. **(a)** See Section 2.6.   **(b)** transfer pipet

20. **(a)** Use forward mode described in text.   **(b)** Use reverse mode described in text.

21. Trap prevents liquid from being sucked into vacuum system. Watchglass keeps dust out of the sample.

22. phosphorus pentoxide

23. **(a)** 3.053_0 g   **(b)** 3.050_6 g   **(c)** You need a 2 000-fold dilution, which could be done with a 20-fold dilution followed by a 100-fold dilution. Dilute 5.00 mL up to 100.0 mL. Then dilute 10.00 mL of the resulting solution up to 1 000.0 mL.

24. 9.979 9 mL

25. 0.2%; 0.499 0 M

26. 49.947 g in vacuum; 49.892 g in air

27. true mass = 50.506 g; mass in air = 50.484 g

28. Procedure (ii) has smaller relative uncertainty. Calibrate the pipet and volumetric flask to reduce the uncertainty further.

29. (b) 54 days

30. 0.70%

Chapter 3

1. (a) 5   (b) 4   (c) 3

2. (a) 1.237   (b) 1.238   (c) 0.135   (d) 2.1   (e) 2.00

3. (a) 0.217   (b) 0.216   (c) 0.217

4. (b) 1.18 (three significant figures)   (c) 0.71 (two significant figures)

5. (a) 3.71   (b) 10.7   (c) 4.0×10^1 (d) 2.85×10^{-6} (e) 12.625 1   (f) 6.0×10^{-4} (g) 242

6. (a) 175.324   (b) 140.094

7. (a) 12.3   (b) 75.5   (c) 5.520×10^3 (d) 3.04 (e) 3.04×10^{-10} (f) 11.9   (g) 4.600   (h) 4.9×10^{-7}

11. low; systematic

12. (a) 25.031 mL is systematic error; ±0.009 mL is random error (b) 1.98 and 2.03 mL are systematic errors; ±0.01 and ±0.02 mL are random errors   (c) random error   (d) random error

13. (a) Carmen   (b) Cynthia   (c) Chastity   (d) Cheryl

14. 3.124 (±0.005), 3.124 (±0.2%)

15. (a) 2.1 (±0.2 or ±11%)   (b) 0.151 (±0.009 or ±6%) (c) 0.22_3 (±0.02_4 or ±11%)   (d) 0.097_1 (±0.002_2 or ±$2._2$%)

16. (a) 10.18 (±0.07 or ±0.7%)   (b) 174 (±3 or ±2%) or 174.4 (±2.7 or ±1.5%)   (c) 0.147 (±0.003 or ±2%) or 0.147_4 (±0.002_8) (±1.9%)   (d) 7.86 (±0.01 or ±0.1%)   (e) 2 185.8 (±0.8 or ±0.04%)   (f) 1.464_3 (±0.007_8 or ±0.5_3%) (g) 0.496_9 (±0.006_9 or ±1.3_9%)

18. (b) 0.450 7 (±0.000 5) M

19. 1.035 7 (±0.000 2) g

20. (a) 105.988_3 ± 0.001_6 g/mol   (b) 0.667 ± 0.001 M (c) uncertainty increases by 5 in the fifth decimal place, so answer is still 0.667 ± 0.001 M

21. 255.184 ± 0.009

22. (a) 1.500_0 ± 0.005_5 μg/mL   (b) 1.500_0 ± 0.009_6 μg/mL; uncertainty in the 1-mL pipet is 3 times larger than the next highest uncertainty.

23. 6.022 136 9 (48) $\times 10^{23}$

Chapter 4

1. Smaller standard deviation means better precision. There is no necessary relationship between standard deviation and accuracy.

2. (a) 0.682 6   (b) 0.954 6   (c) 0.341 3   (d) 0.191 5 (e) 0.149 8

3. (a) 1.527 67   (b) 0.001 26   (c) 1.59×10^{-6} (d) 0.000 45 (e) 1.527_7 ± 0.001_3

4. (a) $0.890 \; 2_0$ g, $0.896 \; 4_9$ g   (b) $0.027 \; 8_5$ g, $0.011 \; 9_5$ g (c) Predicted quotient $\sigma_{16}/\sigma_4 = 0.5$, observed quotient = 0.429

5. (a) 0.044 6   (b) 0.417 3   (c) 0.404 0

6. (a) 0.5   (b) 0.8%   (c) 8.7%

9. Confidence interval is region around measured mean in which true mean is likely to lie.

10. 50%. The 90% bars must be longer because 90% of them must reach the mean.

11. Case 1: Compare measured result to known value with Equation 4-8.

Case 2: Compare replicate measurements with Equations 4-9 and 4-10 after F test.

Case 3: Compare individual differences with Equations 4-11 and 4-12.

12. 90%: 0.14_8 ±0.02_8; 99%: 0.14_8 ±0.05_6

13. $\bar{x}$ ± 0.000 10 (1.527 83 to 1.528 03)

14. (a) dL = deciliter = 0.1 L   (b) yes ($t_{\text{calculated}} = 2.12 < t_{\text{table}} = 2.262$)

15. Difference is <u>not</u> significant ($t_{\text{calculated}} = 0.99 < t_{\text{table}} = 2.57$).

17. Differences are <u>not</u> significant ($F_{\text{calculated}} = 2.43 < F_{\text{table}} = 9.28$ and $t_{\text{calculated}} = 1.55 < t_{\text{table}} = 2.447$).

18. Difference <u>is</u> significant ($F_{\text{calculated}} = 92.7 > F_{\text{table}} = 6.26$, so we use Equations 4-9b and 4-10b. $t_{\text{calculated}} = 11.3 > t_{\text{table}} = 2.57$ for 5 degrees of freedom).

19. yes (90%: $\bar{x}$ ± 1.1_8%)

20. 1–2 difference <u>is</u> significant ($F_{\text{calculated}} = 5.3 > F_{\text{table}} \approx 2.2$, so we use Equations 4-9b and 4-10b. $t_{\text{calculated}} = 18.2 > t_{\text{table}} = 2.02$ for 40 degrees of freedom); 2–3 difference is <u>not</u> significant ($F_{\text{calculated}} = 1.3 < F_{\text{table}} \approx 2.2$, so we use Equations 4-9a and 4-10a. $t_{\text{calculated}} = 1.39 < t_{\text{table}} \approx 2.02$ for 43 degrees of freedom).

21. Difference <u>is</u> significant at 95% and 99% levels ($t_{\text{calculated}} = 2.88$).

22. Difference <u>is</u> significant in both cases. First case 95% confidence interval = 94.9_4 to 99.0_6. Second case 95% confidence interval = 94.6_9 to 98.4_7.

23. (a) Differences are <u>not</u> significant. Rainwater $t_{\text{calculated}} = 1._{61} < t_{\text{table}} = 2.228$. Drinking water $t_{\text{calculated}} = 1._{89} < t_{\text{table}} = 2.306$ (b) Yes. Chromatography $t_{\text{calculated}} = 2._{61} > t_{\text{table}} = 2.228$. Spectrophotometry $t_{\text{calculated}} = 4._{74} > t_{\text{table}} = 2.306$.

24. Retain 216. $G_{\text{calculated}} = 1.52 < G_{\text{table}} = 1.672$.

25. Statement (i) is true.

26. $m = -1.299$ (±0.001) $\times 10^4$ or -1.298_7 (±0.001_3) $\times 10^4$; $b = 3$ (±3) $\times 10^2$

27. $m = 0.6_4 \pm 0.1_2$; $b = 0.9_3 \pm 0.2_6$; $s_y = 0.27$

29. $m = -0.1379 \pm 0.0066$; $b = 0.195 \pm 0.163$; $s_y = 0.198$

30. We must measure the response of an analytical procedure to known quantities before the procedure can be used for an unknown.

31. If negative value is within experimental error of 0, the value is 0. If negative value is beyond experimental error, something is wrong with the analysis.

32. $15.2_2 \pm 0.8_6$ μg, $15._2 \pm 1._5$ μg

33. (a) $2.0_0 \pm 0.3_8$ (b) 0.2_6 (c) ±1.6 for (a) and ±1.1 for (b)

34. (a) 10.1 μg   (b) 10.08 ± 0.45 μg ($u_x = 0.204\,5$ μg, $t = 2.179$)

35. (a) $m = 869 \pm 11$, $b = -22._1 \pm 8._9$ (b) 145.0 mV (c) 0.19_2 (±0.01_4) vol%, 95% confidence 0.19_2 (±0.03_5) vol%

36. 21.9 μg

37. (a) Entire range is linear.   (b) log(current) = 0.969 2 log (concentration, μg/mL) + 1.338 9   (c) 4.80 μg/mL (d) ±0.49 μg/mL

Chapter 5

5. Calibration check sample is made by the analyst; performance test sample is made by someone else and the analyst does not know the expected answer.

8. False positive concludes that analyte exceeds a limit when it is really below the limit. False negative concludes that analyte is below a limit when it is really above the limit.

9. ~1% of samples with no analyte give signal above detection limit. 50% of samples with analyte at detection limit produce signal below detection limit.

11. iii

14. yes: 7 consecutive measurements all above or all below the center line

15. $R^2 = 0.993\,2$, y error bar = ±ts_y = ±162

16. (a) 22.2 ng/mL: precision = 23.8%, accuracy = 6.6% 88.2 ng/mL: precision = 13.9%, accuracy = –6.5% 314 ng/mL: precision = 7.8%, accuracy = –3.6% (b) signal detection limit = $129._6$; detection limit = 4.8×10^{-8} M; quantitation limit = 1.6×10^{-7} M

17. (a) 4%, 128% **(b)** 1.4%

18. recovery = 96%; concentration detection limit = 0.064 μg/L

19. detection limits: 0.086, 0.102, 0.096, and 0.114 μg/mL; mean = 0.10 μg/mL

20. Testing 2 independent samples drawn at the same time from each athlete would reduce the rate of false positives to $0.01 \times 0.01 = 0.000\ 1$.

21. Lab C vs. Lab A: $F_{calculated} = 31._0 > F_{table} = 3.88$ (2 degrees of freedom for s_C and 12 degrees of freedom for s_A). $t_{calculated} = 2.4_1 < t_{table} = 4.303$ for 95% confidence and 2 degrees of freedom $\Rightarrow$ difference <u>is not</u> significant.

Lab C vs. Lab B: $F_{calculated} = 1.9_4 < F_{table} = 4.74$ (2 degrees of freedom for s_C and 7 degrees of freedom for s_A). $s_{pooled} = 0.61_6$. $t_{calculated} = 2.4_7 > t_{table} = 2.262$ for 95% confidence and 9 degrees of freedom $\Rightarrow$ difference <u>is</u> significant. The conclusion that C is greater than B but C is not greater than A makes no sense. The problem is $s_C \gg s_A$ and number of replicates for C $\ll$ number of replicates for A. I suggest that C > A and C > B. I would ask for more replicates from C.

22. Addition of a small volume keeps the matrix nearly constant by not diluting the sample.

23. (c) 1.04 ppm

24. (a) 8.72 ± 0.43 ppb **(b)** 116 ppm **(c)** ± 6 ppm **(d)** ± 18 ppm

25. (a) tap water, 0.091 ng/mL; pond water, 22.2 ng/mL **(b)** This is a matrix effect. Something in pond water decreases the Eu(III) emission.

26. (a) 0.140 M **(b)** standard deviation = ± 0.005 M; 95% confidence = ± 0.015 M

27. (a) 8.2 ± 1.7 mg alliin/g garlic **(b)** 3.8 ± 0.8 mg allicin/g garlic

28. (a) Standard addition line: $y = 42.852x + 1.088\ 8$; Pb in 4.60 mL = 0.025 4 ppm. Pb in 1.00 mL = 0.117 ppm. **(b)** Standard uncertainty of intercept = 0.000 98 ppm; 95% confidence interval for intercept = 0.004 2 ppm. Relative uncertainty of intercept = 16.6%. Uncertainty in concentration = 0.019 ppm.

29. Standard addition is appropriate when sample matrix is unknown or complex and hard to duplicate. Internal standard is appropriate when (1) there will be uncontrolled losses of sample during the analysis, or (2) instrument conditions vary from run to run.

30. (a) 0.168_4 **(b)** 0.847 mM **(c)** 6.16 mM **(d)** 12.3 mM

31. 9.09 mM

32. (b) $m = 3.47$, $u_m = 0.15$, $b = 0.038$, $u_b = 0.057$, $s_y = 0.072$ **(c)** [X]/[S] = 0.560 with $u_x = 0.024\ 6$, $t = 2.776$, $\pm tu_x = (2.776)(0.024\ 6) = 0.068$, [X]/[S] = 0.56 ± 0.07 **(d)** 95% confidence interval for intercept = -0.12 to $+0.20$, which includes 0

Chapter 6

4. (a) $K = 1/[Ag^+]^3 [PO_4^{3-}]$ **(b)** $K = P_{CO_2}^6/P_{O_2}^{15/2}$

5. 1.2×10^{10}

6. 2.0×10^{-9}

7. (a) decrease **(b)** give off **(c)** negative

8. 5×10^{-11}

9. (a) right **(b)** right **(c)** neither **(d)** right **(e)** smaller

10. (a) 4.7×10^{-4} bar **(b)** 153°C

11. (a) 7.82 kJ/mol **(b)** A graph of ln K vs. $1/T$ will have a slope of $-\Delta H°/R$.

12. (a) right **(b)** $P_{H_2} = 1\ 366$ Pa, $P_{Br_2} = 3\ 306$ Pa, $P_{HBr} = 57.0$ Pa **(c)** neither **(d)** formed

13. 0.663 mbar

14. 5×10^{-8} M

15. 8.5 zM

16. 3.9×10^{-7} M

17. (a) 2.1×10^{-8} M **(b)** 8.4×10^{-4} M

18. BX_2 coprecipitates with AX_3.

19. no, 0.001 4 M

20. no

21. $I^- < Br^- < Cl^- < CrO_4^{2-}$

23. (a) BF_3 **(b)** AsF_5

24. 0.096 M

25. $[Zn^{2+}] = 2._{93} \times 10^{-3}$ M, $[ZnOH^+] = 9 \times 10^{-6}$ M, $[Zn(OH)_2(aq)] = 6 \times 10^{-6}$ M, $[Zn(OH)_3^-] = 8 \times 10^{-9}$ M, $[Zn(OH)_4^{2-}] = 9 \times 10^{-14}$ M

26. 15%

27. 1.1×10^{-5} M

29. (a) an adduct **(b)** dative or coordinate covalent **(c)** conjugate **(d)** $[H^+] > [OH^-]$, $[H^+] < [OH^-]$

33. (a) HI **(b)** H_2O

34. $2H_2SO_4 \rightleftharpoons HSO_4^- + H_3SO_4^+$

35. (a) (H_3O^+, H_2O); $(H_3\overset{+}{N}CH_2CH_2\overset{+}{N}H_3, H_3\overset{+}{N}CH_2CH_2NH_2)$ **(b)** $(C_6H_5CO_2H, C_6H_5CO_2^-)$; $(C_5H_5NH^+, C_5H_5N)$

36. (a) 2.00 **(b)** 12.54 **(c)** 1.52 **(d)** -0.48 **(e)** 12.00

37. (a) 6.998 **(b)** 6.132

38. 1.0×10^{-56}

39. 7.8

40. (a) endothermic **(b)** endothermic **(c)** exothermic

44. $Cl_3CCO_2H \rightleftharpoons Cl_3CCO_2^- + H^+$

$\bigcirc\!\!\!-\overset{+}{N}H_3 \rightleftharpoons \bigcirc\!\!\!-NH_2 + H^+$

$La^{3+} + H_2O \rightleftharpoons LaOH^{2+} + H^+$

45. $\bigcirc\!\!N + H_2O \rightleftharpoons \bigcirc\!\!\overset{+}{N}H + OH^-$

$HOCH_2CH_2S^- + H_2O \rightleftharpoons HOCH_2CH_2SH + OH^-$

46. K_a: $HCO_3^- \rightleftharpoons H^+ + CO_3^{2-}$
K_b: $HCO_3^- + H_2O \rightleftharpoons H_2CO_3 + OH^-$

47. (a) $H_3\overset{+}{N}CH_2CH_2\overset{+}{N}H_3 \overset{K_{a1}}{\rightleftharpoons} H_2NCH_2CH_2\overset{+}{N}H_3 + H^+$
$H_2NCH_2CH_2\overset{+}{N}H_3 \overset{K_{a2}}{\rightleftharpoons} H_2NCH_2CH_2NH_2 + H^+$ **(b)** $^-O_2CCH_2CO_2^- + H_2O \overset{K_{b1}}{\rightleftharpoons} HO_2CCH_2CO_2^- + OH^-$
$HO_2CCH_2CO_2^- + H_2O \overset{K_{b2}}{\rightleftharpoons} HO_2CCH_2CO_2H + OH^-$

48. a, c

49. $CN^- + H_2O \rightleftharpoons HCN + OH^-$; $K_b = 1.6 \times 10^{-5}$

50. $H_2PO_4^- \overset{K_{a2}}{\rightleftharpoons} HPO_4^{2-} + H^+$
$HC_2O_4^- + H_2O \overset{K_{b2}}{\rightleftharpoons} H_2C_2O_4 + OH^-$

51. $K_{a1} = 7.04 \times 10^{-3}$, $K_{a2} = 6.25 \times 10^{-8}$, $K_{a3} = 4.3 \times 10^{-3}$

52. 3.0×10^{-6}

53. (a) 1.2×10^{-2} M **(b)** Solubility will be greater.

54. 0.22 g

Chapter 7

7. 32.0 mL

8. 43.20 mL $KMnO_4$, 270.0 mL $H_2C_2O_4$

9. 0.149 M

10. 0.100 3 M

11. 92.0 wt%

12. (a) 0.020 34 M **(b)** 0.125 7 g **(c)** 0.019 83 M

13. 56.28 wt%

14. 8.17 wt%

15. 0.092 54 M

16. (a) 17 L **(b)** 793 L **(c)** 1.05×10^3 L

18. (a) 13.08 **(b)** 8.04 **(c)** 2.53

19. (a) 6.06 **(b)** 3.94 **(c)** 2.69

20. $[AgCl(aq)] = 370$ nM, $[AgBr(aq)] = 20$ nM, $[AgI(aq)] = 0.32$ nM

21. (a) SO_4^{2-} [from soil] + Ba^{2+} [from $BaCl_2(s)$] $\rightarrow BaSO_4(s)$

   (b) 0.11_8 mmol   (c) 0.11_8 mmol   (d) 1.1 wt%

23. $[Ag^+] = 9.1 \times 10^{-9}$ M; $Q = [Ag^+][Cl^-] = 2.8 \times 10^{-10} > K_{sp}$ for AgCl

24. $V_{e1} = 18.76$ mL, $V_{e2} = 37.52$ mL

25. (a) 14.45   (b) 13.80   (c) 8.07   (d) 4.87   (e) 2.61

26. (a) 19.00, 18.85, 18.65, 17.76, 14.17, 13.81, 7.83, 1.95   (b) no

27. $V_X = V_M^\circ (C_M^\circ - [M^+] + [X^-])/(C_X^\circ + [M^+] - [X^-])$

32. negative

34. 0.574 0 M, 1 376 mg

35. $CO_2(g)$ would bubble out of solution when HNO_3 is added.

Chapter 8

2. (a) true   (b) true   (c) true

3. Increasing ion concentration promotes dissociation of yellow HBG^- to blue BG^{2-}.

4. (a) 0.008 7 M   (b) 0.001_2 M

5. (a) 0.660   (b) 0.54   (c) 0.18   (d) 0.83

6. 0.88_7

7. (a) 0.42_2 (b) 0.43_2

8. 0.20_2

9. increase

10. 7.0×10^{-17} M

11. 6.6×10^{-7} M

12. $\gamma_{H^+} = 0.86$, pH = 2.07

13. 11.94, 12.00

14. 0.329

15. 0.63

18. $[H^+] + 2[Ca^{2+}] + [Ca(HCO_3)^+] + [Ca(OH)^+] + [K^+] =$
   $[OH^-] + [HCO_3^-] + 2[CO_3^{2-}] + [ClO_4^-]$

19. $[H^+] = [OH^-] + [HSO_4^-] + 2[SO_4^{2-}]$

20. $[H^+] = [OH^-] + [H_2AsO_4^-] + 2[HAsO_4^{2-}] + 3[AsO_4^{3-}]$

21. (a) charge: $2[Mg^{2+}] + [H^+] + [MgBr^+] + [MgOH^+] = [Br^-] + [OH^-]$;
   mass: $[MgBr^+] + [Br^-] = 2\{[Mg^{2+}] + [MgBr^+] + [MgOH^+]\}$

   (b) $[Mg^{2+}] + [MgBr^+] + [MgOH^+] = 0.2$ M; $[MgBr^+] + [Br^-] = 0.4$ M

22. 2.3×10^6 N, 5.2×10^5 pounds, no

23. $[CH_3CO_2^-] + [CH_3CO_2H] = 0.1$ M

24. $[Y^{2-}] = [X_2Y_2^{2+}] + 2[X_2Y^{4+}]$

25. $3\{[Fe^{3+}] + [Fe(OH)^{2+}] + [Fe(OH)_2^+] + 2[Fe_2(OH)_2^{4+}] + [FeSO_4^+]\}$
   $= 2\{[FeSO_4^+] + [SO_4^{2-}] + [HSO_4^-]\}$

26. $[H^+] = 1.08 \times 10^{-11}$ M, $[NH_4^+] = 9.29 \times 10^{-4}$ M, $[OH^-] = 9.29 \times 10^{-4}$ M, and $[NH_3] = 4.91 \times 10^{-2}$ M; pH = 10.97 and fraction of hydrolysis = 1.86%

27. Same results as found by Goal Seek in Figure 8-8

28. (b) $[A^-] = 1.00 \times 10^{-2}$ M, $[OH^-] = 2.39 \times 10^{-6}$ M, $[HA] = 2.39 \times 10^{-6}$ M, $[H^+] = 5.08 \times 10^{-9}$ M, $\mu = 0.010$ 0 M, pH = 8.33, fraction of hydrolysis = 0.024%

29. (b) $[Ca^{2+}] = 0.015$ 6 M, $[CaOH^+] = 0.005$ 3 M, $[OH^-] = 0.036$ 4 M, $[H^+] = 4 \times 10^{-13}$ M; solubility = 1.1 g/L, fraction of hydrolysis = $[CaOH^+]/\{[Ca^{2+}] + [CaOH^+]\} = 25\%$, solubility = 1.5_5 g/L

30. $[Na^+] = [Cl^-] = 0.024$ 86 M, $[NaCl(aq)] = 0.000$ 143 M, ionic strength = 0.024 86 M, ion pair fraction = 0.57%

31. $[Na^+] = 0.047$ 75 M, $[SO_4^{2-}] = 0.022$ 75M, $[NaSO_4^- (aq)] = 0.002$ 246 M ionic strength = 0.070 51 M, ion pair fraction = 9.0%

32. (a) $[Mg^{2+}] = [SO_4^{2-}] = 0.016$ 16 M, $[MgSO_4 (aq)] = 0.008$ 844 M, ionic strength = 0.064 63 M, ion pairing fraction = 35.4%

   (b) $Mg^{2+} + OH^- \rightleftharpoons MgOH^+$ $K_1 = 10^{2.6}$

   $[MgOH^+] \approx K_1[Mg^{2+}][OH^-] = 10^{2.6}[0.016]10^{-7} = 6 \times 10^{-7}$ M, which is negligible in comparison with $[Mg^{2+}] = 0.016$ M

   $SO_4^{2-} + H_2O \rightleftharpoons HSO_4^- + OH^-$ $pK_b = 12.01$

   $[HSO_4^-] \approx K_b[SO_4^{2-}]/[OH^-] = 10^{-12.01}[0.016]/10^{-7} = 2 \times 10^{-7}$ M, which is negligible in comparison with $[SO_4^{2-}] = 0.016$ M

33. $[Li^+] = [F^-] = 0.050$ 1 M, $[LiF(aq)] = 0.002$ 88 M, $[HF] = 8.5 \times 10^{-7}$ M, $[OH^-] = 8.7 \times 10^{-7}$ M, $[H^+] = 1.7 \times 10^{-8}$ M, $\mu = 0.050$ 1 M

34. (a) 4.3×10^{-5} (b) 5.2×10^{-4} M = 21 mg/L   (c) 0.023 bar

Chapter 9

1. Added H^+ suppresses H_2O ionization (Le Châtelier's principle).

2. (a) 3.00   (b) 12.00

3. 6.89, 0.61

4. (a) 0.809   (b) 0.791   (c) Activity coefficient depends slightly on counterion.

5. (a)

6. pH = 3.00, $\alpha = 0.995\%$

7. 5.50

8. 5.51, 3.1×10^{-6} M, 0.060 M

10. 99% dissociation when F = $(0.010$ $2)K_a$

11. 4.20

12. 5.79

13. (a) 3.03, 9.4%   (b) 7.00, 99.9%

14. 5.64, 0.005 3%

15. 2.86, 14%

16. 99.6%, 96.5%

18. Acid in lemon juice converts volatile RNH_2 into less volatile RNH_3^+.

19. 11.00, 0.995%

20. 11.28, [B] = 0.058 M, $[BH^+] = 1.9 \times 10^{-3}$ M

21. 10.95

22. 0.007 6%, 0.024%, 0.57%

23. 3.6×10^{-9}

24. 4.1×10^{-5}

25. 0.999, 0.000 999

26. Weigh 0.020 0 mol acetic acid (=1.201 g) into a beaker with ~75 mL of water. Monitor pH with an electrode while adding 3 M NaOH (~4 mL is required) until pH = 5.00. Quantitatively transfer the liquid to a 100-mL volumetric flask with many small washings of the beaker and dilute to volume.

27. Place 16.9 mL of 28 wt% NH_3 into ~160 mL H_2O in a beaker. Add ~9 mL of 37.2 wt% HCl. Monitor pH with an electrode while adding ~3 more mL HCl dropwise until the pH is exactly 9.00. Quantitatively transfer the liquid to a 250-mL volumetric flask with many small washings and dilute to volume.

28. Buffer contains ~5 mmol $[H_3BO_3]$ and ~5 mmol $[H_2BO_3^-]$. If generated acid consumes half of $[H_2BO_3^-]$, pH is reduced from 9.24 to pH = pK_a +

   $\log [H_2BO_3^-]/[H_3BO_3] = 9.24 + \log (2.5$ mmol/7.5 mmol) = 8.76.

29. When volume changes, the ratio $[A^-]/[HA]$ does not change.

30. More concentrated buffer has more A^- and HA to consume added acid or base.

31. At low or high pH, additional acid or base is much less than the amount of H^+ or OH^- already present.

32. When pH = pK_a, a given increment of added acid or base has the least effect on the ratio $[A^-]/[HA]$.

33. Henderson-Hasselbalch equation is equivalent to K_a equilibrium expression, which is always true. It is an approximation that [HA] and $[A^-]$ are unchanged from what was placed in solution.

34. 4-aminobenzenesulfonic acid

35. 4.70

36. (a) 0.180   (b) 1.00   (c) 1.80

37. 1.5

38. (a) 14   (b) 1.4×10^{-7}

39. (a) NaOH   (b) 1. Weigh out $(0.250 \text{ L})(0.050\ 0 \text{ M}) = 0.012\ 5$ mol of HEPES and dissolve in ~200 mL. 2. Adjust pH to 7.45 with NaOH. 3. Dilute to 250 mL.

40. 3.38 mL

41. (b) 7.18   (c) 7.00   (d) 6.86 mL

42. (a) 2.56   (b) 2.61   (c) 2.86

43. 16.2 mL

44. (a) pH = 5.06, [HA] = 0.001 99 M, $[A^-]$ = 0.004 01 M

45. (a) approximate pH = 11.70, more accurate pH = 11.48

46. 6.86

47. $pK_a \approx 5.2$. Chemical shift at low pH (8.67 ppm) is for $C_5H_5NH^+$. Chemical shift at high pH (7.89) is for C_5H_5N. At pH = pK_a, there will be equal amounts of both species and the chemical shift will be ½(8.87 + 7.89) = 8.28 ppm. This chemical shift intercepts the curve through the data points at pH ≈ 5.2.

Chapter 10

1. H^+ produced by acid dissociation reacts with OH^- from base hydrolysis.

2. $H_3\overset{+}{N}\text{—CHR—}CO_2^-$; pK values apply to $-NH_3^+$, $-CO_2H$, and, in some cases, R.

3. 4.37×10^{-4}, 8.93×10^{-13}

4. (a) pH = 2.51, $[H_2A]$ = 0.096 9 M, $[HA^-]$ = 3.11×10^{-3} M, $[A^{2-}]$ = 1.00×10^{-8} M

(b) 6.00, 1.00×10^{-3} M, 1.00×10^{-1} M, 1.00×10^{-3} M

(c) 10.50, 1.00×10^{-10} M, 3.16×10^{-4} M, 9.97×10^{-2} M

5. (a) pH = 1.95, $[H_2M]$ = 0.089 M, $[HM^-]$ = 1.12×10^{-2} M, $[M^{2-}]$ = 2.01×10^{-6} M

(b) pH = 4.28, $[H_2M]$ = 3.7×10^{-3} M, $[HM^-]$ ≈ 0.100 M, $[M^{2-}]$ = 3.8×10^{-3} M

(c) pH = 9.35, $[H_2M]$ = 7.04×10^{-12} M, $[HM^-]$ = 2.23×10^{-5} M, $[M^{2-}]$ = 0.100 M

6. pH = 11.60, [B] = 0.296 M, $[BH^+]$ = 3.99×10^{-3} M, $[BH_2^{2+}]$ = 2.15×10^{-9} M

7. pH = 3.69, $[H_2A]$ = 2.9×10^{-6} M, $[HA^-]$ = 7.9×10^{-4} M, $[A^{2-}]$ = 2.1×10^{-4} M

8. 4.03

9. (a) pH = 6.002, $[HA^-]$ = 0.009 8 M, $[H_2A]$ = 0.000 098 M, $[A^{2-}]$ = 0.000 099 M   (b) pH = 4.50, $[HA^-]$ = 0.006 1 M, $[H_2A]$ = 0.001 9 M, $[A^{2-}]$ = 0.002 0 M

10. $[CO_2(aq)]$ = $10^{-4.9}$ M, pH = 5.67

11. (a) $[CO_3^{2-}] = K_{a2}K_{a1}K_H P_{CO_2}/[H^+]^2$
(b) 0°C: 6.6×10^{-5} mol kg^{-1}; 30°C: 1.8×10^{-4} mol kg^{-1}
(c) 0°C: $[Ca^{2+}][CO_3^{2-}]$ = 6.6×10^{-7} mol² kg^{-2} (aragonite dissolves,

calcite does not); 30°C: $[Ca^{2+}][CO_3^{2-}]$ = 1.8×10^{-6} mol² kg^{-2} (neither dissolves)

12. 2.96 g

13. 2.22 mL

14. Procedure: Dissolve 10.0 mmol (1.23 g) picolinic acid in ~75 mL H_2O in a beaker. Add NaOH (~5.63 mL) until the measured pH is 5.50. Transfer to a 100-mL volumetric flask and use small portions of H_2O to rinse the beaker into the flask. Dilute to 100.0 mL and mix well.

15. 26.5 g Na_2SO_4 + 1.31 g H_2SO_4

16. no

17.

Glutamic acid

Tyrosine

18. (a) 2.8×10^{-3} (b) 2.8×10^{-8}

19. (a) NaH_2PO_4 and Na_2HPO_4 would be simplest, but other combinations such as H_3PO_4 and Na_3PO_4 or H_3PO_4 and Na_2HPO_4 would work just fine. (b) 4.55 g Na_2HPO_4 + 2.15 g NaH_2PO_4 (c) One of several ways: Weigh out 0.050 0 mol Na_2HPO_4 and dissolve it in 900 mL of water. Add HCl while monitoring the pH with a pH electrode. When the pH is 7.45, stop adding HCl and dilute to exactly 1 L with H_2O.

20. pH = 5.64, $[H_2L^+]$ = 0.010 0 M, $[H_3L^{2+}]$ = 1.36×10^{-6} M, [HL] = 3.68×10^{-6} M, $[L^-]$ = 2.40×10^{-11} M

21. 78.9 mL

22. (a) 5.88   (b) 5.59

23. (a) HA   (b) A^- (c) 1.0, 0.10

24. (a) 4.00   (b) 8.00   (c) H_2A (d) HA^- (e) A^{2-}

25. (a) 9.00   (b) 9.00   (c) BH^+ (d) 1.0×10^3

26.

27. α_{HA} = 0.091, α_{A^-} = 0.909, $[A^-]/[HA]$ = 10

28. 0.91

29. α_{H_2A} = 0.876, 0.049 1; α_{HA^-} = 0.124, 0.693; $\alpha_{A^{2-}}$ = 4.60×10^{-4}, 0.258

30. $\alpha_{H_2A^-} = 0.893, 0.500, 5.4 \times 10^{-5}, 2.2 \times 10^{-5}, 1.55 \times 10^{-12}$
$\alpha_{HA^-} = 0.107, 0.500, 0.651, 0.500, 1.86 \times 10^{-4}$
$\alpha_{A^{2-}} = 5.8 \times 10^{-7}, 2.2 \times 10^{-5}, 0.349, 0.500, 0.9998$

31. **(b)** $8.6 \times 10^{-6}, 0.61, 0.39, 1.6 \times 10^{-6}$

32. 0.36

33. 96%

35. At pH 10: $\alpha_{H_3A} = 1.05 \times 10^{-9}, \alpha_{H_2A^-} = 0.0409, \alpha_{HA^{2-}} = 0.874,$
$\alpha_{A^{3-}} = 0.0854$

36. **(b)** $[Cr(OH)_3(aq)] = 10^{-6.84}$ M

(c) $[Cr(OH)_2^+] = 10^{-4.44}$ M, $[Cr(OH)^{2+}] = 10^{-2.04}$ M

37. Acidic substituents: aspartic acid, cysteine, glutamic acid, tyrosine
Basic substituents: arginine, histidine, lysine

38. Isoelectric pH: protein has no net charge, even though it has many positive and negative sites. Isoionic pH is the pH of a solution containing only protein, H^+, and OH^-.

39. The _average_ charge is 0. There is no pH at which _all_ molecules have zero charge.

40. isoelectric pH 5.59, isoionic pH 5.72

Chapter 11

1. Equivalence point: quantity of titrant is the exact amount needed for reaction with analyte. End point: marked by abrupt change in physical property such as pH or indicator color.

2. 13.00, 12.95, 12.68, 11.96, 10.96, 7.00, 3.04, 1.75

3. $pH = -\log[H^+]$. Even though $[H^+]$ hardly changes near V_e, its logarithm changes rapidly near V_e because $[H^+]$ decreases by orders of magnitude with tiny additions of OH^- when there is hardly any H^+ present.

4. Sketch like Figure 11-2. Initial pH is determined by acid dissociation of HA. Between initial point and V_e, added OH^- converts equivalent quantity of HA into A^-, giving buffer (HA and A^-). At V_e, HA has been converted to A^- whose pH is controlled by hydrolysis of A^-. After V_e, the pH is determined by excess OH^-.

5. If analyte is too weak or too dilute, there is little change in pH at equivalence point.

6. 3.00, 4.05, 5.00, 5.95, 7.00, 8.98, 10.96, 12.25

7. $V_e/11$; $10V_e/11$; $V_e = 0$, pH = 2.80; $V_e/11$, pH = 3.60; $V_e/2$, pH = 4.60; $10V_e/11$, pH = 5.60; V_e, pH = 8.65; $1.2V_e$, pH = 11.96

8. 8.18

9. 5.4×10^7

10. 0.107 M

11. 9.72

12. Sketch like Figure 11-9. Initial pH is determined by
$B + H_2O \rightleftharpoons BH^+ + OH^-$. Between the initial point and V_e, added H^+ converts equivalent quantity of B into BH^+, giving buffer (B and BH^+). At V_e, B has been converted to BH^+ whose pH is controlled by acid dissociation of BH^+. After V_e, pH is determined by excess H^+.

13. At V_e, B is converted to BH^+, which is an acid.

14. 11.00, 9.95, 9.00, 8.05, 7.00, 5.02, 3.04, 1.75

15. $V_e/2$

16. 2.2×10^9

17. 10.92, 9.57, 9.35, 8.15, 5.53, 2.74

18. **(a)** 9.45 **(b)** 2.55 **(c)** 5.15

19. Initial pH determined by H_2A acid dissociation. $V_0 < V_b < V_{e2}$: buffer mixture of H_2A and HA^-. V_{e1}: H_2A converted to HA^-, whose pH is determined by acid-base reactions of HA^-. $V_{e1} < V < V_{e2}$: buffer mixture of HA^- and A^{2-}. V_{e2}: HA^- converted into A^{2-}, whose base hydrolysis determines pH. After V_{e2}: excess OH^- determines pH.

20. positive (Average charge = 0 at isoelectric point. When H^+ is added to reach isoionic point, some basic groups become protonated.)

21. isoionic (Point H could be attained by mixing HA plus NaCl. pH is equivalent to that of pure HA, which is isoionic pH.)

22. Upper curve: $\frac{3}{2}V_e$ is at pH = pK_2 (1:1 mixture of HA^- and A^{2-}) Lower curve: "pK_2" ($= pK_{BH^+}$) occurs at 1:1 mixture of B and BH^+. To create this mixture, B is first transformed into BH^+ at V_e by reaction with HA. At $2V_e$ one more equivalent of B has been added, giving a 1:1 mole ratio B : BH^+.

23. 11.49, 10.95, 10.00, 9.05, 8.00, 6.95, 6.00, 5.05, 3.54, 1.79

24. 2.51, 3.05, 4.00, 4.95, 6.00, 7.05, 8.00, 8.95, 10.46, 12.21

25. 11.36, 10.21, 9.73, 9.25, 7.53, 5.81, 5.33, 4.86, 3.41, 2.11, 1.85

26. 5.01

27. **(a)** 1.99

28. **(b)** 7.13

29. 2.72

30. **(a)** 9.54 **(b)** 7.9×10^{-10}

31. 6.28 g

32. $pK_2 = 9.84$

33. Gran plot finds V_e by extrapolating from points prior to V_e.

34. end point = 23.39 mL

35. end point = 10.727 mL

36. $[HIn]/[In^-]$ changes from 10:1 at pH = $pK_{HIn} - 1$ to 1:10 at pH = $pK_{HIn} + 1$

37. Properly chosen indicator pH transition range coincides with steep part of titration curve, which includes equivalence point.

38. If we know pK_{HIn} and we measure $[In^-]/[HIn]$ spectroscopically, we can calculate pH from Henderson-Hasselbalch equation.

39. H_2SO_4, HCl, HNO_3, or $HClO_4$

40. yellow, green, blue

41. **(a)** red **(b)** orange **(c)** yellow

42. **(a)** red **(b)** orange **(c)** yellow **(d)** red

43. no (end-point pH must be >7)

44. **(a)** 2.47

45. **(a)** violet **(b)** blue **(c)** yellow

46. **(a)** 5.62 **(b)** methyl red using yellow end point

47. 2.859 wt%

48. _Alkalinity_ = mol of H^+ needed to reach pH 4.5, which is pH of H_2CO_3. Alkalinity measures $[OH^-] + [CO_3^{2-}] + [HCO_3^-]$ plus other bases present. Bromocresol green is blue above pH 5.4 and yellow below pH 3.8. Green color range includes pH 4.5.

49. HCl standardization: any base in Table 11-4; NaOH: any acid in Table 11-4

50. Higher equivalent mass requires more mass of primary standard with less relative error in weighing the reagent.

51. Dry potassium acid phthalate at 105° and weigh accurately into a flask. Titrate with NaOH, using pH electrode or phenolphthalein to observe end point.

52. 0.079 34 mol/kg

53. 1.023_8 g, systematic error = 0.08%, calculated HCl molarity is low

54. 0.100 0 M

55. 0.31 g

56. **(a)** 20.254 wt% **(b)** 17.985 g

57. **(a)** 204.221 ± 0.005 g/mol **(b)** 1.00000 ± 0.00003

58. 15.1 wt%

59. **(a)** 15.3 wt% **(b)** 8.40 **(c)** 13% **(d)** 1.02

60. Acid stronger than H_3O^+ is leveled to H_3O^+ in water. Bases stronger than OH^- are leveled to OH^-.

61. Reactions of CH_3O^- and $CH_3CH_2O^-$ (RO^-) are driven to the right because of the high concentration of H_2O: $RO^- + H_2O \rightarrow CH_3OH + OH^-$.

62. (a) acetic acid **(b)** pyridine

63. Each reacts with H_2O to give OH^-:
$$NH_2^- + H_2O \rightarrow NH_3 + OH^-$$

64. CH_3OH is less polar than H_2O. If CH_3OH is added to the aqueous solution, the neutral pyridine molecule will tend to be favored over the protonated pyridinium cation. It takes higher concentration of acid to protonate pyridine in aqueous CH_3OH than in H_2O.

69. (b) $K = 0.279$, pH $= 4.16$

74. 0.139 M

75. 0.815

Chapter 12

1. Multidentate ligands form more stable complexes than do similar, monodentate ligands.

2. (a) 2.7×10^{-10} **(b)** 0.57

3. (a) 2.5×10^7 **(b)** 4.5×10^{-5} M

4. 5.60 g

5. Neutral H_5DTPA is $DTPA(CO_2H)_2(CO_2^-)_3(NH^+)_3$. The predominant species at pH 14 is $DTPA^{5-}$. By analogy with EDTA, all carboxyl groups are probably ionized at pH 3–4, so the predominant species is H_3DTPA^{2-}, which is $DTPA(CO_2^-)_5(NH^+)_3$. At pH 14 and at pH 3, sulfate is in the form SO_4^{2-}. 10^{-3} M H^+ displaces Ba^{2+} from DTPA, but 10^{-14} M H^+ does not.

6. (a) 100.0 mL **(b)** 0.016 7 M **(c)** 0.041 **(d)** 4.1×10^{10}
(e) 7.8×10^{-7} M **(f)** 2.4×10^{-10} M

7. (a) 2.93 **(b)** 6.79 **(c)** 10.52

8. (a) 1.70 **(b)** 2.18 **(c)** 2.81 **(d)** 3.87 **(e)** 4.87
(f) 6.85 **(g)** 8.82 **(h)** 10.51 **(i)** 10.82

9. ∞, 10.30, 9.52, 8.44, 7.43, 6.15, 4.88, 3.20, 2.93

10. 4.6×10^{-11} M

14. Auxiliary complexing agent keeps analyte in solution, but gives up analyte to EDTA.

15. (a) 25 **(b)** 0.016

16. (a) 15.03 **(b)** 15.05 **(c)** 16.30 **(d)** 17.02 **(e)** 17.69

17. (b) $\alpha_{ML} = 0.28$, $\alpha_{ML_2} = 0.70$

18. (a) $K_1 = [FeT]/([Fe^{3+}][T])$, $K_2 = [Fe_2T]/([Fe^{3+}][FeT])$
(d) $[T] = 0.27_7$; $[Fe_aT] = 0.55_3$; $[Fe_bT] = 0.09_2$; $[Fe_2T] = 0.07_7$

20. (b) 1.34 mL, pNi $= 7.00$; 21.70 mL, pNi $= 8.00$; 26.23 mL, pNi $= 17.00$

23. Most Mg^{2+} is not bound to the small amount of indicator. Free Mg^{2+} reacts with EDTA before MgIn reacts. [MgIn] is constant until all of Mg^{2+} has been consumed. When MgIn begins to react, the color changes.

24. 1. metal ion indicator; 2. Hg electrode; 3. ion-selective electrode; 4. glass electrode

25. HIn^{2-}, wine-red, blue

26. Buffer (i): yellow $\rightarrow$ blue; other buffers: violet $\rightarrow$ blue, which is harder to see

27. Analyte precipitates without EDTA, reacts slowly with EDTA, or blocks the indicator.

28. Analyte displaces a metal ion from a complex.

30. Hardness $\approx [Ca^{2+}] + [Mg^{2+}]$. Temporary hardness from $Ca(HCO_3)_2$ is lost by heating. Permanent hardness (e.g., $CaSO_4$) is not affected by heat.

31. 10.0 mL, 10.0 mL

32. 0.020 0 M

33. 0.995 mg

34. 0.092 54 M

35. 21.45 mL

36. $[Ni^{2+}] = 0.012\ 4$ M, $[Zn^{2+}] = 0.007\ 18$ M

37. 0.024 30 M

38. 0.092 28 M

39. observed: 32.7 wt%; theoretical: 32.90 wt%

Chapter 13

1. $PbS(s) + H^+ \rightleftharpoons Pb^{2+} + HS^-$
$PbCO_3(s) + H^+ \rightleftharpoons Pb^{2+} + HCO_3^-$

2. (a) pH $= 9.98$ **(b)** pH $= 10.00$ **(c)** pH $= 9.45$

3. pH $= 9.95$

4. predicted values: $pK_1' = 2.350$, $pK_2' = 9.562$

5. pH $= 10.194$ from spreadsheet and 10.197 by hand method

6. pH $= 4.52$

7. pH $= 5.00$

8. ionic strength $= 0.025$ M, pH $= 4.94$

9. (a) pH $= 7.420$ **(b)** pH $= 7.403$

10. pH $= 4.44$

11. (e) $[Fe^{3+}] = 4.20$ mM, $[SCN^-] = 2.03$ μM, $[H^+] = 15.8$ mM, $[Fe(SCN)^{2+}] = 2.97$ μM, $[Fe(SCN)_2^+] = 106$ pM, $[FeOH^{2+}] = 0.802$ mM, $\mu = 0.043\ 4$ M **(f)** Hydrolysis of Fe(III) produces $0.000\ 8$ M H^+ **(g)** Calculated quotient $= 293$, graph gives 270
(h) $[Fe^{3+}] = 4.45$ mM, $[SCN^-] = 2.81$ μM, $[H^+] = 15.6$ mM, $[Fe(SCN)^{2+}] = 2.19$ μM, $[Fe(SCN)_2^+] = 68.2$ pM, $[FeOH^{2+}] = 0.546$ mM, $[OH^-] = 1.18$ pM, $\mu = 0.244$ M; Calculated quotient $= 156$, graph gives 150

12. (a) $[SO_4^{2-}] = 1.50$ mM, $[La^{3+}] = 0.57$ mM, $[H^+] = 1.14$ μM, $[La(SO_4)^+] = 1.36$ mM, $[La(SO_4)_2^-] = 67$ μM, $[LaOH^{2+}] = 1.13$ μM, $[OH^-] = 10.5$ nM, ionic strength $= 0.006\ 29$ M, pH $= 5.98$ **(b)** ionic strength of strong electrolyte $= 15.0$ mM; actual ionic strength $= 6.3$ mM
(c) 28.5% **(d)** pK_a for HSO_4^- is 1.99 and we expect the solution to be near neutral pH. **(e)** no; $[La^{3+}][OH^-]^3\gamma_{La^{3+}}\gamma_{OH^-}^3 = 2.4 \times 10^{-28} < K_{sp}$ for $La(OH)_3 = 2 \times 10^{-21}$

13. $[CN^-] = 1.51$ μM, $[H^+] = 1.29 \times 10^{-12}$ M, $[OH^-] = 0.012\ 9$ M, $[Ag^+] = 0.241$ nM, $[AgOH] = 0.187$ nM, $[Ag(OH)(CN)^-] = 46.9$ μM, $[Ag(CN)_2^-] = 0.100$ M, $[Ag(CN)_3^{2-}] = 4.19$ μM, $[HCN] = 1.90$ nM, $[Na^+] = 0.013\ 0$ M, $[K^+] = 0.100$ M, ionic strength $= 0.113$ M; principal form of silver is $Ag(CN)_2^-$ $= 99.95\%$ of Ag

14. $[Fe^{2+}] = 1.74$ mM, $[G^-] = 0.954$ mM, $[H^+] = 3.67$ nM, $[FeG^+] = 18.7$ mM, $[FeG_2] = 29.0$ mM, $[FeG_3^-] = 0.459$ mM, $[FeOH^+] = 0.121$ mM, $[HG] = 21.0$ mM, $[H_2G^+] = 12.8$ nM, $[OH^-] = 3.67$ μM, $[Cl^-] = 20.9$ mM; fraction of Fe in each form: $[Fe^{2+}]$, 3.49%; $[FeG^+]$, 37.40%; $[FeG_2]$, 57.95%; $[FeG_3^-]$, 0.92%; $[FeOH^+]$, 0.24%; fraction of glycine in each form: $[G^-]$, 0.95%; $[HG]$, 21.02%; $[H_2G^+]$, 0.00%; $[FeG^+]$, 18.70%; $2[FeG_2]$, 57.95%; $3[FeG_3^-]$, 1.38%; HCl added $= 20.9$ mmol; ionic strength $= 24.1$ mM; chemistry: $FeG_2 \rightleftharpoons FeG^+ + G^-$ followed by $G^- + H^+ \rightleftharpoons HG$. G^- released when FeG_2 dissolves requires HCl to lower the pH to 8.50.

15. (b) Fixing pK_w' at 13.797 causes n_H(measured) to deviate systematically above $\bar{n}_H$(theoretical) at the end of the titration when n_H should approach 0.

16. (a) $\bar{n}_H$(experimental) $= 3 + \dfrac{[OH^-] + [Cl^-]_{HCl} - [H^+] - [Na^+]}{F_{H_3A}}$

$\bar{n}_H$(theoretical) $= 3\alpha_{H_3A} + 2\alpha_{H_2A} + \alpha_{HA}$
(b) $pK_w' = 13.819$, $pK_1 = 8.33$, $pK_2 = 9.48$, $pK_3 = 10.19$

17. (b) $[T^{2-}] = \dfrac{F_{H_2T}}{\dfrac{[H^+]^2}{K_1K_2} + \dfrac{[H^+]}{K_2} + 1 + K_{NaT^-}[Na^+] + K_{NaHT}[Na^+]\dfrac{[H^+]}{K_2}}$

(c) $[HT^-] = \dfrac{F_{H_2T}}{\dfrac{[H^+]}{K_1} + 1 + \dfrac{K_2}{[H^+]} + K_{NaT^-}[Na^+]\dfrac{K_2}{[H^+]} + K_{NaHT}[Na^+]}$

$[H_2T] = \dfrac{F_{H_2T}}{1 + \dfrac{K_1}{[H^+]} + \dfrac{K_1K_2}{[H^+]^2} + K_{NaT^-}[Na^+]\dfrac{K_1K_2}{[H^+]^2} + K_{NaHT}[Na^+]\dfrac{K_1}{[H^+]}}$

(d) pH = 4.264, $[PyH^+]$ = 0.013 4, $[Na^+]$ = 0.018 5, $[K^+]$ = 0.010 0, $[OH^-]$ = 1.84 × 10^{-10}, $[HT^-]$ = 0.010 0, $[T^{2-}]$ = 0.007 92, $[Cl^-]$ = 0.015 0, $[NaT^-]$ = 0.001 17, $[H_2T]$ = 5.93 × 10^{-4}, $[Py]$ = 0.001 56, $[NaHT]$ = 2.97 × 10^{-4} M

Chapter 14

2. (a) 6.241 509 48 × 10^{18} e$^-$/C   **(b)** 96 485.338 3 C/mol

3. (a) 71.$_5$ A   **(b)** 4.35 A   **(c)** 79 W

4. (a) 3.00 mA, 1.87 × 10^{16} e$^-$/s   **(b)** 9.63 × 10^{-19} J/e$^-$
(c) 5.60 × 10^{-5} mol   **(d)** 447 V

5. (a) I_2 **(b)** $S_2O_3^{2-}$ **(c)** 861 C   **(d)** 14.3 A

6. (a) NH_4^+ and Al, reducing agents; ClO_4^-, oxidizing agent
(b) 9.576 kJ/g

8. (a) $Fe(s)|FeO(s)|KOH(aq)|Ag_2O(s)|Ag(s)$;
$FeO(s) + H_2O + 2e^- \rightleftharpoons Fe(s) + 2OH^-$;
$Ag_2O(s) + H_2O + 2e^- \rightleftharpoons 2Ag(s) + 2OH^-$
(b) $Pb(s)|PbSO_4(s)|K_2SO_4(aq)\|H_2SO_4(aq)|PbSO_4(s)|PbO_2(s)|Pb(s)$;
$PbSO_4(s) + 2e^- \rightleftharpoons Pb(s) + SO_4^{2-}$;
$PbO_2(s) + 4H^+ + SO_4^{2-} + 2e^- \rightleftharpoons PbSO_4(s) + 2H_2O$

9. $Fe^{3+} + e^- \rightleftharpoons Fe^{2+}$; $Cr_2O_7^{2-} + 14H^+ + 6e^- \rightleftharpoons 2Cr^{3+} + 7H_2O$

10. (a) Electrons flow from Zn to C.   **(b)** 1.32 kg

11. (a) anode: $C_6Li \rightleftharpoons C_6 + Li^+ + e^-$; cathode: $2Li_{0.5}CoO_2 + Li^+ + e^- \rightleftharpoons 2LiCoO_2$; $2Li_{0.5}CoO_2 + LiC_6 = 267.863$ g/formula mass
(b) 3 600 C; 0.037 311 mol e$^-$ **(d)** 370 W · h/kg

12. Cl_2 has the most positive $E°$.

13. (a) Fe(III)   **(b)** Fe(II)

14. At equilibrium, E = 0. $E°$ is constant

15. (a) e$^-$ moves from Zn to Cu   **(b)** Zn^{2+}

16. −0.356 V

17. (a) $Pt(s)|Br_2(l)|HBr(aq, 0.10$ M$)\|Al(NO_3)_3(aq, 0.010$ M$)|Al(s)$
(b) E_+ = −1.716$_4$ V, E_- = 1.137$_2$ V, E = −2.854 V, electrons flow from Al to Pt, $\frac{3}{2}Br_2(l) + Al(s) \rightleftharpoons 3Br^- + Al^{3+}$ **(c)** Br_2 **(d)** 1.31 kJ
(e) 2.69 × 10^{-8} g/s

18. Activities of solids do not change until they are used up. $OH^-(aq)$ is created at cathode and consumed at the anode, so its concentration remains constant.

19. (a) 1.219 V   **(b)** 4.88 g/h   **(c)** 26.8 horsepower

20. (b) Cathode: $2MnO_2(s) + H_2O(l) + 2e^- \rightleftharpoons Mn_2O_3(s) + 2OH^-$
$E°$ = +0.147 V
Anode: $ZnO(s) + H_2O(l) + 2e^- \rightleftharpoons Zn(s) + 2OH^-$ $E°$ = −1.260 V
Net: $2MnO_2(s) + Zn(s) \rightleftharpoons Mn_2O_3(s) + ZnO(s)$ $E°$ = −1.407 V
(c) KOH(aq) can leak, forming $K_2CO_3(s)$ with CO_2 from air
(d) $E°$ = +0.147 V

21. (a) 0.572 V   **(b)** e$^-$ flow from left to right   **(c)** 0.568 V

22. 0.799 2 V

23. $HOBr + 2e^- + H^+ \rightleftharpoons Br^- + H_2O$; 1.341 V

24. $3X^+ \rightleftharpoons X^{3+} + 2X(s)$; $E_2° > E_1°$

25. 0.580 V, electrons flow from Ni to Cu

26. (a) Cathode: $PbO_2(s) + SO_4^{2-} + 4H^+ + 2e^- \rightleftharpoons PbSO_4(s) + 2H_2O$
$E°$ = −0.355 V
Anode: $PbSO_4(s) + 2e^- \rightleftharpoons Pb(s) + SO_4^{2-}$ $E°$ = 1.685 V
Net: $Pb(s) + PbO_2(s) + SO_4^{2-} + 4H^+ \rightleftharpoons$
$2PbSO_4(s) + 2H_2O$ $E°$ = 2.040 V
(b) $Pb(s)|PbSO_4(s)|H_2SO_4(aq)|PbSO_4(s)|PbO_2(s)|Pb(s)$
(c) Anode: $PbSO_4(s) + 2H_2O \rightleftharpoons PbO_2(s) + SO_4^{2-} + 4H^+ + 2e^-$
Cathode: $PbSO_4(s) + 2e^- \rightleftharpoons Pb(s) + SO_4^{2-}$
(d) E = 2.040 − $\dfrac{0.059\ 16}{2}$ log $\dfrac{m_{H_2O}^2 \gamma_{H_2O}^2}{m_{SO_4^{2-}}^2 \gamma_{SO_4^{2-}}^2 m_{H^+}^4 \gamma_{H^+}^4}$

(e) E_{net} = 2.040 − $\dfrac{0.059\ 16}{2}$ log $\dfrac{(0.66)^2}{(11.0)^4(5.5)^2(0.22)^3}$ = 2.159 V

27. $2LiOH(s) + CO_2(g) \rightleftharpoons Li_2CO_3(s) + H_2O(l)$; LiOH weighs less than the same number of moles of NaOH or KOH.

28. (a) 1.33 V; 1 × 10^{45}

29. (a) K = 10^{47} **(b)** K = 1.9 × 10^{-6}

30. (b) K = 2 × 10^{16} **(c)** −0.02$_0$ V   **(d)** 10 kJ   **(e)** 0.21

31. K = 1.0 × 10^{-9}

32. 0.101 V

33. 34 g/L

34. (a) 0.063 M   **(b)** 0.030 M, solubility decreases as temperature increases

35. 0.117 V

36. −1.664 V

37. K = 3 × 10^5

38. (a) $Al_2O_3(s) + MgO(s) \rightleftharpoons MgAl_2O_4(s)$
(b) −29.51 kJ/mol   **(c)** $\Delta H°$ = −23.60 kJ/mol, $\Delta S°$ = 5.90 J/(K · mol)

39. In right half-cell, $Hg^{2+} + Y^{4-} \rightleftharpoons HgY^{2-}$ is at equilibrium, even though net cell reaction $Hg^{2+} + H_2 \rightleftharpoons Hg(l) + 2H^+$ is not at equilibrium.

40. (b) 0.14$_3$ M

41. (b) A = −0.414 V, B = 0.059 16 V   **(c)** Hg → Pt

42. 9.6 × 10^{-7}

43. 7.1 × 10^{14}

44. 0.76

45. 7.5 × 10^{-8}

46. $E°'$ is reduction potential at pH 7, instead of pH 0. Living systems have pH much closer to 7 than to 0.

47. (c) 0.317 V

48. −0.041 V

49. −0.268 V

50. −0.036 V

51. 7.2 × 10^{-4}

52. −0.447 V

53. (a) [Ox] = 3.82 × 10^{-5} M, [Red] = 1.88 × 10^{-5} M
(b) $[S^-]$ = [Ox], [S] = [Red]   **(c)** −0.092 V

Chapter 15

1. (b) 0.044 V

2. (a) 0.326 V   **(b)** 0.086 V   **(c)** 0.019 V   **(d)** −0.021 V
(e) 0.021 V

3. 0.684 V

4. 0.627

5. 0.243 V

6. (c) 0.068 V

7. Ag electrode responds to Ag^+. If silver halide is present, $[Ag^+]$ = K_{sp}/[halide], so changing [halide] changes electrode potential.

8. 0.481 V; −0.039 V

9. 3 × 10^{21}

10. (a) $Fe^{3+} + e^- \rightleftharpoons Fe^{2+}$ **(b)** 1 × 10^{11} **(c)** 6 × 10^{10}

11. $[CN^-]$ = 0.847 mM; [KOH] = 0.29$_6$ M

12. Junction potential arises because different ions diffuse at different rates across a liquid junction, leading to separation of charge. Figure 14-4 has no liquid junction.

13. H^+ diffuses into KCl faster than K^+ diffuses into HCl. K^+ has a greater mobility than Na^+, so NaCl|KCl has opposite sign. HCl|KCl voltage > NaCl|KCl voltage because mobility difference between H^+ and K^+ > mobility difference between K^+ and Na^+.

14. left

15. Junction potentials are dominated by the high mobility of H^+ and OH^-. 0.1 M NaOH|KCl (saturated) junction potential is dominated by high concentrations of K^+ and Cl^-, which have nearly equal mobility.

16. Both half-cells contain saturated KCl so it makes sense to use saturated KCl in the salt bridge.

17. (a) 42.4 s   (b) 208 s

18. (a) $3._2 \times 10^{13}$ (b) 8%   (c) 49.0, 8%

19. Both half-cell reactions are the same. Ideally, cell voltage = 0 if there were no junction potential. Measured voltage can be attributed to junction potential.

20. (c) 0.1 M HCl|1 mM KCl, 93.6 mV; 0.1 M HCl|4 M KCl, 4.7 mV

21. Use MOPSO and HEPES buffers and calibrate at 37°C.

22. Uncertainty in pH of standard buffers, junction potential, junction potential drift, sodium or acid errors at extreme pH values, equilibration time, hydration of glass, temperature of measurement and calibration, and cleaning of electrode

23. 10.67

24. Potassium hydrogen tartrate and potassium hydrogen phthalate

25. Na^+ competes with H^+ for cation exchange sites on glass, which responds as if H^+ were present.

26. +0.10 pH unit

27. (a) 274 mV   (b) 285 mV

28. pH = 5.686; slope = -57.17_3 mV/pH unit; theoretical slope = -58.17 mV/pH unit; $\beta = 0.983$

29. (b) 0.465   (c) $Na_2HPO_4 = 0.026\ 8\ m$ and $KH_2PO_4 = 0.019\ 6\ m$

30. (b) $p(a_H\gamma_{Cl})° = 6.972$, $\gamma_{Cl} = 0.777$, $a_H = 1.37 \times 10^{-7}$, pH = 6.862

31. (a) Analyte ions equilibrate between outer solution and membrane ligand L, creating a slight charge imbalance because other ions are not free to cross the solution-membrane interface. Changes in analyte ion concentration in outer solution change the potential difference across the solution-membrane interface.   (b) Compound electrodes contain a conventional electrode surrounded by a membrane that isolates (or generates) the analyte to which the electrode responds.

32. Smaller $K_{A,X}^{Pot}$ is more selective for the intended ion.

33. A mobile molecule dissolved in the membrane liquid phase binds tightly to the ion of interest and weakly to interfering ions.

34. Metal ion buffer maintains low concentration of metal ion (M) from large reservoir of metal complex (ML) and free ligand (L). Without buffer, M could bind to container wall or to a ligand in solution and be lost.

35. Electrodes respond to *activity*. If the ionic strength is constant, the activity coefficient of analyte will be constant in all standards and unknowns.

36. (a) -0.407 V   (b) $1.5_5 \times 10^{-2}$ M   (c) $1.5_2 \times 10^{-2}$ M

37. +0.029 6 V

38. 0.211 mg/L

39. Group 1: K^+; Group 2: Sr^{2+} and Ba^{2+}; $[K^+] \approx 100[Li^+]$

40. 3.8×10^{-9} M

41. (a) $E = 51.10\ (\pm0.24) + 28.14\ (\pm0.08_5)\ log[Ca^{2+}]\ (s_y = 0.2_7)$
(b) 0.951   (c) $2.43\ (\pm0.04) \times 10^{-3}$ M

42. -0.331 V

43. 3.0×10^{-5} M

44. (a) 0.36 ± 0.15 ppm   (b) There is too much added standard.

45. $log\ K_{Na^+,Mg^{2+}}^{Pot} = -8.09, -8.15$; $log\ K_{Na^+,K^+}^{Pot} = -4.87$

46. Na^+ error = 0.25%; Ca^{2+} error = 2.5%

47. $E = 120.2 + 28.80\ log([Ca^{2+}] + 6.0 \times 10^{-4}\ [Mg^{2+}])$

49. $[Hg^{2+}]$ is computed from equilibrium constants. Equilibrium constants for the $HgCl_2$ system might be in error. When we make a buffer by mixing *calculated* quantities of reagents, we are at the mercy of the quality of tabulated equilibrium constants.

50. (a) 1.13×10^{-4} (b) 4.8×10^4

51. Analyte on the surface of gate changes electric potential of gate, thereby regulating current between the source and drain. The key to ion-specific response is to have a chemical on the gate that selectively binds one analyte.

Chapter 16

1. (d) 0.490, 0.526, 0.626, 0.99, 1.36, 1.42, 1.46 V

2. (d) 1.58, 1.50, 1.40, 0.733, 0.065, 0.005, -0.036 V

3. (d) $-0.120, -0.102, -0.052, 0.21, 0.48, 0.53$ V

4. (b) 0.570, 0.307, 0.184 V

5. (d) $-0.143, -0.102, -0.061, 0.096, 0.408, 0.450$

6. diphenylamine sulfonic acid: colorless → red-violet;

diphenylbenzidine sulfonic acid: colorless → violet;

tris(2,2′-bipyridine)iron: red → pale blue;

ferroin: red → pale blue

7. no

8. In preoxidation and prereduction, oxidation state of analyte is made suitable for titration. Preoxidation or prereduction reagent must be destroyed so it does not react with titrant.

9. $2S_2O_8^{2-} + 2H_2O \xrightarrow{boiling} 4SO_4^{2-} + O_2 + 4H^+$

$Ag^{3+} + H_2O \xrightarrow{boiling} Ag^+ + \frac{1}{2}O_2 + 2H^+$

$2H_2O_2 \xrightarrow{boiling} O_2 + 2H_2O$

10. A column of Zn granules coated with Zn amalgam reduces analyte passed through the column.

11. Ag is not a strong enough reducing agent to reduce Cr^{3+} and TiO^{2+}.

12. A weighed amount of solid mixture is added to excess standard aqueous Fe^{2+} plus H_3PO_4. Excess Fe^{2+} is then titrated with standard $KMnO_4$ to find out how much Fe^{2+} was consumed by $(NH_4)_2S_2O_8$. H_3PO_4 masks yellow color of Fe^{3+}.

13. (a) $MnO_4^- + 8H^+ + 5e^- \rightleftharpoons Mn^{2+} + 4H_2O$
   (b) $MnO_4^- + 4H^+ + 3e^- \rightleftharpoons MnO_2(s) + 2H_2O$
   (c) $MnO_4^- + e^- \rightleftharpoons MnO_4^{2-}$

14. $3MnO_4^- + 5Mo^{3+} + 4H^+ \rightarrow 3Mn^{2+} + 5MoO_2^{2+} + 2H_2O$; 0.011 29 M

15. $2MnO_4^- + 5H_2O_2 + 6H^+ \rightarrow 2Mn^{2+} + 5O_2 + 8H_2O$; 0.586 4 M

16. (a) Scheme 1: $6H^+ + 2MnO_4^- + 5H_2O_2 \rightarrow 2Mn^{2+} + 5O_2 + 8H_2O$

Scheme 2: $6H^+ + 2MnO_4^- + 3H_2O_2 \rightarrow 2Mn^{2+} + 4O_2 + 6H_2O$

(b) Scheme 1: 25.43 mL; Scheme 2: 42.38 mL

17. $2MnO_4^- + 5H_2C_2O_4 + 6H^+ \rightarrow 2Mn^{2+} + 10CO_2 + 8H_2O$; 3.826 mM

18. $C_3H_8O_3 + 8Ce^{4+} + 3H_2O \rightleftharpoons 3HCO_2H + 8Ce^{3+} + 8H^+$; 41.9 wt%

19. $Fe(NH_4)_2(SO_4)_2 \cdot 6H_2O$; 78.67 wt%

20. oxidation number = 3.761; 217 μg/g

21. (a) 0.020 34 M   (b) 0.125 7 g   (c) 0.019 82 M

22. I^- reacts with I_2 to give I_3^-. This reaction increases the solubility of I_2 and decreases its volatility.

23. Standard I_3^- can be prepared from a weighed amount of KIO_3 with H^+ plus I^-. Alternatively, I_3^- can be standardized by reaction with standard $S_2O_3^{2-}$ prepared from anhydrous $Na_2S_2O_3$.

24. Starch is not added early in iodometry, so it does not irreversibly bind to I_2.

25. $S_4O_6^{2-} + 2e^- \rightleftharpoons 2S_2O_3^{2-}$ or $S_4O_6^{2-} + 4H^+ + 2e^- \rightleftharpoons 2H_2SO_3$
$E°$ for the second half-reaction above is 0.57 V. $E°$ for the half-reaction $\frac{1}{2}O_2(g) + 2H^+ + 2e^- \rightleftharpoons H_2O$ is 1.23 V. O_2 is a stronger oxidant than tetrathionate.

26. (a) 1.433 mmol   (b) 0.076 09 M   (c) 12.8 wt%
(d) Do not add starch until right before the end point.

27. 11.43 wt%; just before the end point

28. (a) 98.66% (b) 97.98% (c) 196.0 ML (d) $1O_2$ makes $4Mn(OH)_3$ which makes $2I_3^-$ (e) 11.7 mg O_2/L (f) 80%
(g) $2HNO_2 + 2H^+ + 3I^- \rightarrow 2NO + I_3^- + 2H_2O$

29. 0.007 744 M; just before the end point

30. (a) 7×10^2 (b) 1.0 (c) 0.34 g/L

31. mol NH_3 = 2(initial mol H_2SO_4) – mol thiosulfate

32. (a) no, no (b) $I_3^- + SO_3^{2-} + H_2O \rightarrow 3I^- + SO_4^{2-} + 2H^+$
(c) 5.07 9×10^{-3} M, 406.6 mg/L (d) no: $t_{calculated} = 2.56 < t_{table} = 2.776$

33. 5.730 mg

34. (a) 0.125 (b) 6.875 ± 0.038

36. (a) 0.191 5 mmol (b) 2.80 (c) 0.20 (d) 0.141 3, difference is experimental error

37. Bi oxidation state = +3.200 0 (±0.003 3)

Cu oxidation state = +2.200 1 (±0.004 6)

formula = $Bi_2Sr_2CaCu_2O_{8.400\ 1\ (±0.005\ 7)}$

Chapter 17

1. Difference is due to overpotential.

2. 2.68 h

3. −1.228 8 V

4. (a) −1.906 V (b) 0.20 V (c) −2.71 V (d) −2.82 V

5. (a) V_2

6. (a) Cu: E_+ = 0.339 V; Ag|AgCl: E_- = 0.197 V;
$E_{predicted} = E_+ - E_- = 0.142$ V
Difference between $E_{observed}$ and $E_{predicted}$ is from neglect of Cu^{2+} activity coefficient and 3 M KCl instead of saturated KCl in reference electrode.
(b) Increased current increases overpotentials, resulting in anode intercept >122 mV and cathode intercept < 85 mV.

7. (a) 6.64×10^3 J (b) 0.012 4 g/h

8. OH^- generated at cathode and Cl^- in anode compartment cannot cross Nafion membrane. Na^+ crosses the membrane freely to preserve charge balance.

9. Electrical losses (ohmic, overpotentials, and concentration polarization) decrease the magnitude of the voltage that can be delivered by a cell and increase the magnitude of the voltage required to reverse the spontaneous cell reaction.

10. anode, 54.77 wt%

11. −0.619 V, negative

12. $[Cd^{2+}] = 2.8 \times 10^{-12}$ M $\Rightarrow E(cathode) = -0.744$ V

13. 94%

14. When Br_2 appears at end point, current flows at low voltage by oxidation of Br^- at one electrode and reduction of Br_2 at the other electrode.

15. Mediator shuttles electrons between analyte and electrode. After being oxidized or reduced by analyte, mediator is regenerated at the electrode.

16. (a) $5._2 \times 10^{-9}$ mol (b) 0.000 2_6 mL

17. (a) 5.32×10^{-5} mol (b) 2.66×10^{-5} mol (c) 5.32×10^{-3} M

18. 151 μg/mL

19. 2.00 nmol fructose → 4.00 nmol e^- = 0.386 mC, which agrees with integrated area

20. (a) p-type (b) 11%

21. (a) current density = 1.00×10^2 A/m², overpotential = 0.85 V
(b) −0.036 V (c) 1.160 V (d) −2.57 V

22. 96 486.6_7 ± 0.2_8 C/mol

23. (a) $H_2SO_3 <$ pH 1.86; pH 1.86 $< HSO_3^- <$ pH 7.17; $SO_3^{2-} >$ pH 7.17
(b) cathode: $H_2O + e^- \rightarrow \frac{1}{2}H_2(g) + OH^-$; anode: $3I^- \rightarrow I_3^- + 2e^-$
(c) $I_3^- + HSO_3^- + H_2O \rightarrow 3I^- + SO_4^{2-} + 3H^+$;
$I_3^- + 2S_2O_3^{2-} \rightleftharpoons 3I^- + S_4O_6^{2-}$ (d) 3.64 mM

24. (a) $B = c$; $C = x$; $D = n$
$o + A = 2B \Rightarrow o + A = 2c \Rightarrow A = 2c - o$
$h + 2A = 3D + E \Rightarrow h + 2(2c - o) = 3n + E \Rightarrow E = h + 4c - 2o - 3n$
$F = E - C = h + 4c - 2o - 3n - c = h - c/2 + o - 3n$
(b) $F/4$ (c) $2.44_3 \times 10^{-8}$ mol (d) 57.9 mg O_2/L (e) 2.26×10^{-4} M

26. (e) 321 μC

27. At lower rotational speed, the magnitude of the current on both ends of the curves would decrease.

28. 15 μm, 7.8×10^2 A/m²

31. 0.12%

32. 0.096 mM

34. (a) $Cu^{2+} + 2e^- \rightarrow Cu(s)$ (b) $Cu(s) \rightarrow Cu^{2+} + 2e^-$ (c) 313 ppb

35. Estimated relative peak heights are 1, 1.5_6, and 1.9_8. Fe(III) in seawater = 1.0×10^2 pM

37. Peak C: $RNO + 2H^+ + 2e^- \rightarrow RNHOH$. There was no RNO present before the initial scan.

38. 7.8×10^{-10} m²/s

41. 5.2 μm

42. $ROH + SO_2 + B \rightarrow BH^+ + ROSO_2^-$
$H_2O + I_2 + ROSO_2^- + 2B \rightarrow ROSO_3^- + 2BH^+I^-$

Chapter 18

1. (a) double (b) halve (c) double

2. (a) 184 kJ/mol (b) 299 kJ/mol

3. 5.33×10^{14} Hz, 1.78×10^4 cm⁻¹, 3.53×10^{-19} J/photon, 213 kJ/mol

4. rotation, vibration, electronic excitation, electronic excitation, and bond breaking

5. ν = 5.088 491 0 and 5.083 335 8×10^{14} Hz, λ = 588.985 54 and 589.582 86 nm, $\tilde{\nu}$ = 1.697 834 5 and 1.696 114 4×10^4 cm⁻¹

6. (a) Hg vapor lamp (b) photodiode (c) 254 nm filter
(d) b = lengths of two side legs + bottom leg (e) $A_{254} \propto P_{O_3}$

7. T = fraction of light transmitted; A = −log T; ε = constant of proportionality between A and (concentration × pathlength)

8. graph of absorbance vs. wavelength

9. Color transmitted is complement of color absorbed.

10.

Curve	Absorption peak (nm)	Predicted color (Table 18-1)	Observed color
A	760	green	green
B	700	green	blue green
C	600	blue	blue
D	530	violet	violet
E	500	red or purple red	red
F	410	green-yellow	yellow

11. High absorbance: too little light reaches detector; low absorbance: too little difference between sample and reference

12. 3.56×10^4 M⁻¹ cm⁻¹

13. violet-blue

14. 2.19×10^{-4} M

15. (a) 0.613 (b) $1.22_3 \times 10^{-6}$ M (c) 1.67×10^5 M⁻¹ cm⁻¹

16. (a) 325 nm: T = 0.90, A = 0.045; 300 nm: T = 0.061, A = 1.22
(b) 2.0% (c) T_{winter} = 0.142; T_{summer} = 0.095; 49%

17. neocuproine masks Cu(I)

18. (a) 6.97×10^{-5} M (b) 6.97×10^{-4} M (c) 1.02 mg

19. yes

20. (a) 7.87×10^4 M⁻¹ cm⁻¹ (b) 1.98×10^{-6} M

21. (a) 8.512×10^4 M⁻¹ cm⁻¹ (b) 10.7 (c) 0.14 wt% = 1.40 mg/mL

22. (a) $4.97 \times 10^4 \ M^{-1} \ cm^{-1}$ (b) $4.69 \ \mu g$

23. (a) 1.711 g   (b) 475.5 µg Fe/mL   (c) Dilute 1, 2, 3, 4, and 5 mL of stock solution to 500 mL with 0.1 M H_2SO_4 to obtain 0.951, 1.90, 2.85, 3.80, and 4.76 µg Fe/mL   (d) Dilute 10 mL of stock solution to 50 mL with 0.1 M H_2SO_4 to obtain ~50 µg Fe/mL. Then dilute 1, 2, 3, 4, or 5 mL of ~50 µg Fe/mL stock solution to 50 mL with 0.1 M H_2SO_4 to obtain ~1, 2, 3, 4, or 5 µg Fe/mL.

24. (a) 3.511 g   (b) $1\,033._0$ µg Fe/mL   (c) Dilute 10 mL of $1\,033._0$ µg Fe/mL to 250 ml. Then dilute 5 mL of resulting solution to 250 mL $\Rightarrow$ 0.826 µg Fe/mL.

Volume for first dilution (mL)	Volume for second dilution (mL)	Final µg Fe/mL
10	5	0.826
10	10	1.653
10	15	2.479
15	15	3.719
15	20	4.958
20	20	6.611
20	25	8.264
20	30	9.917

25. $6.516 \pm 0.020 \ [\pm 0.30\%]$ µg Fe/mL

26. (b) 2.33×10^{-7} mol Fe(III)   (c) 5.83×10^{-5} M

27. theoretical equivalence point = 13.3 µL; observed end point = 12.2 µL = 1.83 Ga/transferrin. Ga does not appear to bind in the absence of oxalate.

28. (a) end point = 21.4 µL, 2.29 mmol Au(0)/g   (b) 1.24 mmol $C_{12}H_{25}S$/g   (c) 1.52 mmol Au(I)/g, mole ratio Au(I):$C_{12}H_{25}S$ = 1.23

29. (a) 4 biotins per streptavidin   (b) When adding BF to SA, fluorescence increases slowly until SA is saturated, and then fluorescence of excess BF increase rapidly. When adding SA to BF, fluorescence of BF decreases until it is all bound to SA, and then fluorescence has constant, low value.

30. 630 kg CO_2, 11 kg SO_2

31. $\Delta E(S_1 - T_1)$ = 36 kJ/mol

32. Following absorption of a photon, fluorescence is prompt emission with no change in electronic spin state. Phosphorescence is slower and electronic spin changes during emission. Phosphorescence also comes at lower energy than fluorescence.

33. Luminescence: light emitted after molecule absorbs light. Chemiluminscence: light emitted by a molecule created in an excited state by a chemical reaction.

34. Rayleigh scattering: electrons in molecules oscillate at the frequency of incoming radiation and emit that same frequency in all directions. The time scale is ~10^{-15} s for visible light. Raman scattering: molecules extract vibrational energy from incoming light and scatter light with less than incoming energy. Rayleigh and Raman time scale $\approx 10^{-15}$ s for visible light. Fluorescence occurs in 10^{-8} to 10^{-4} s.

35. wavelength: absorption < fluorescence < phosphorescence

36. Excitation spectrum resembles absorption spectrum.

37. Fluorescence is proportional to concentration up to 5 µM (within 5%); yes.

38. $3.56 \ (\pm 0.07) \times 10^{-4}$ wt%; 95% confidence interval: $3.56 \ (\pm 0.22) \times 10^{-4}$ wt%

Chapter 19

1. [X] = 8.03×10^{-5} M, [Y] = 2.62×10^{-4} M

2. $[Cr_2O_7^{2-}] = 1.78 \times 10^{-4}$ M, $[MnO_4^-] = 8.36 \times 10^{-5}$ M

3. If spectra of two compounds with constant total concentration cross at any wavelength, all mixtures with the same concentration go through that point.

4. M + L(λ_{max} 439 nm) $\rightarrow$ ML(λ_{max} 485 nm) followed by ML + M $\rightarrow$ $M_2L(\lambda_{max}$ 566 nm)

5. [A] = 9.11×10^{-3} M, [B] = 4.68×10^{-3} M

6. [TB] = 1.22×10^{-5} M, [STB] = 9.30×10^{-6} M, [MTB] = 1.32×10^{-5} M

7. [p-xylene] = 0.062 7 M, [m-xylene] = 0.079 5 M, [o-xylene] = 0.075 9 M, [ethylbenzene] = 0.076 1 M

8. (a) [In^-] = 3.28 µM, [HIn] = 6.91 µM   (b) pH = 6.78   (c) 3.31 µM, 6.97 µM; procedure (a) is probably more accurate because it uses more data points.

9. [In^-] = 0.794 µM, [HIn] = 0.436 µM, pK_a = 4.00

10. (f) [$CO_2(aq)$] = 3.0 µM   (g) $\mu \approx 10^{-4}$ M, yes

11. $K = 4.0 \times 10^9 \ M^{-1}$, S = 0.29 to 0.84

12. (a) $K = (4.3 \pm 0.8) \times 10^4 \ M^{-1}$ (b) fraction bound = 0.21 and 0.77

13. (b) K = 0.464, $\varepsilon = 1.074 \times 10^4 \ M^{-1} \ cm^{-1}$

14. (b) K = 0.464, $\varepsilon = 1.073 \times 10^4 \ M^{-1} \ cm^{-1}$

15. (a) 1:1   (b) K must not be very large.   (c) to maintain constant ionic strength

16. (xylenol orange)$_2$Zr$_3$ has a xylenol orange mole fraction of 0.4.

17. Maximum is at mole fraction of A = 1/3. $K = 10^8$ gives high extent of reaction and useable Job's plot. $K = 10^7$ gives rounded Job's plot with less well defined peak. For $K = 10^6$, extent of reaction is too little for useful Job's plot.

18. Job's plot has broad peak at mole fraction = 0.50, consistent with 1:1 complex.

20. Each enzyme molecule bound to antibody 2 catalyzes many cycles of reaction in which a colored or fluorescent product is created. Many product molecules are created for every analyte molecule.

21. Background fluorescence decays to near zero prior to recording emission from lanthanide ion.

22. Few biological molecules absorb 800–1 000 nm near-infrared radiation, so little matrix fluorescence is stimulated.

24. Dimethylphenol is a weak acid HA. Interpretation: A^- is strong quencher with $K_{sv} \approx 1\,350$. HA is weak quencher with $K_{sv} \approx 100$. $pK_a \approx 10.8$ divides the two regions.

25. (b) N_{av} = 55.9   (c) [M] = 0.227 mM; Q = 0.881 molecules per micelle   (d) P_0 = 0.414; P_1 = 0.365; P_2 = 0.161

Chapter 20

3. D_2, silicon carbide globar

4. Resolution increases with number of illuminated grooves and diffraction order selected by blaze angle. Dispersion is proportional to diffraction order and inversely proportional to the grating line spacing. Blaze angle is selected to give specular reflection at desired diffraction angle.

5. Filter removes higher order diffraction (different wavelengths) at same angle as desired diffraction.

6. Advantage–increased ability to resolve closely spaced spectral peaks Disadvantage–more noise because less light reaches detector

8. When temperature of DTGS crystal changes by absorption of infrared light, the voltage difference between the two faces changes.

9. (a) 2.38×10^3 lines/cm   (b) 143

11. (a) 1.7×10^4 (b) 0.05 nm   (c) 5.9×10^4 (d) 0.000 43°, 0.013°

12. (a) T = 0.036 4, A = 1.439   (b) 0.002 3%   (c) 0.000 022, 0.000 22

13. 0.124 2 mm

14. 77 K: 1.99 W/m²; 298 K: 447 W/m²

15. (a) $M_\lambda = 8.79 \times 10^9$ W/m³ at 2.00 µm; $M_\lambda = 1.164 \times 10^9$ W/m³ at 10.00 µm   (b) 1.8×10^2 W/m²   (c) 2.3×10^1 W/m²   (d) $M_{2.00 \ \mu m}/M_{10.00 \ mm}$ = 7.55 at 1 000 K; $M_{2.00 \ \mu m}/M_{10.00 \ mm}$ = 3.17×10^{-22} at 100 K

16. (a) 2.51×10^{-6} (b) (i) Ordinate is ringdown lifetime. The greater the absorbance, the shorter the lifetime. Spectrum is analogous to transmittance spectrum. (ii) 1 653.725 00 nm (iii) infrared

17. (a) $34°$ (b) $0°$

18. Light in the fiber strikes the wall at an angle greater than the critical angle for total reflection. If bending angle is not too great, angle of incidence exceeds critical angle.

19. $\theta_{critical}$ (solvent/silica) = $76.7°$, $\theta_{critical}$ (silica/air) = $43.2°$. Total reflection is from silica/air interface.

21. For same angle of incidence, number of reflections increases as thickness of waveguide decreases.

22. (a) $80.7°$ (b) 0.955

23. (a) $61.04°$ (b) $51.06°$

24. $n_{prism} > \sqrt{2}$

25. (b) $76°$ (c) 0.020

26. (a) 0.964 (b) 343 nm, 5.83×10^{14} Hz

27. (b) blue

28. (a) ± 2 cm (c) 0.5 cm^{-1} (d) 2.5 μm

29. Background transform gives P_0. Sample transform gives P. Transmittance = P/P_0, not $P - P_0$.

30. White noise such as random motion of electrons is independent of frequency. $1/f$ noise such as drift and flicker of lamp decreases with increasing frequency. Line noise such as 60 Hz wall frequency results from disturbances at discrete frequencies.

31. Beam is alternately directed through sample and reference. Chopping moves analytical signal from zero frequency to frequency of chopper, which can be selected so that $1/f$ noise and line noise are minimal.

32. V_{diff} is ideally 0 if sample and reference are identical because same lamp intensity goes through both. If sample absorbs some radiation, V_{diff} responds to absorption with little noise from source flicker.

33. 7

35. Expected detection limit after 10 s of averaging is $37/\sqrt{10} = 11.7$; Expected detection limit after 40 s = $37/\sqrt{10} = 5.8_5$

36. (300 cycles, 32.9) (100 cycles, 19.0) (1 cycle, 1.90)

37. y_{12}(5-pt cubic smoothing) = $\dfrac{-3y_{10} + 12y_{11} + 17y_{12} + 12y_{13} - 3y_{14}}{35}$

$= \dfrac{-3(0.869) + 12(1.956) + 17(6.956) + 12(5.000) - 3(6.086)}{35} = 5.167$

y_{12}(9-pt cubic smoothing) = 4.228

Chapter 21

1. Emission, because population of the excited state varies with temperature

2. Advantages: sensitivity and small sample size. Disadvantage: poor reproducibility with manual sample injection

3. Drying removes water; ashing removes as much matrix as possible without evaporating analyte; atomization vaporizes analyte.

4. Advantages: less chemical interference and self absorption; lamps are not required; simultaneous multi-element analysis is possible. Disadvantage: cost

5. An atom moving toward radiation source sees higher frequency than one moving away from source. Increasing temperature gives increased speeds (more broadening) and increased mass gives decreased speeds (less broadening).

7. Spectral interference: overlap of analyte signal with other signals. Chemical interference: matrix component decreases atomization of analyte. Isobaric interference: overlap of different species with same mass-to-charge ratio.

8. La^{3+} acts as releasing agent by binding PO_4^{3-} and freeing Pb^{2+}.

9. (a) Dynamic reaction cell contains reactive gas and its electric field limits mass range of ions passing through. Interfering species are eliminated by chemical reaction or analyte signal is moved to a mass where there is

no interference. (b) $^{87}Sr^+ \rightarrow {}^{87}Sr^{19}\dot{F}^+$ (m/z 106) no longer overlaps $^{87}Rb^+$ (m/z 87)

10. Matrix-matching is intended to match extent to which analyte is ablated, transported to the plasma, and atomized.

11. Excitation spectrum: sample absorbs light when narrow laser frequency coincides with atomic frequency. Emission spectrum: sample excited by laser emits broad radiation. Monochromator bandwidth is not narrow enough to discriminate between emission at different wavelengths, so broad envelope is observed.

12. Pb: 1.2 ± 0.2; Tl: 0.005 ± 0.001; Cd: 0.04 ± 0.01; Zn: 2.0 ± 0.3; Al: $7 (\pm 2) \times 10^1$ ng/cm^2

13. 589.3 nm

14. 0.025

15. Na: 0.003_8 nm; Hg: $0.000 5_6$ nm

16. (a) 283.0 kJ/mol (b) 3.67×10^{-6} (c) $+8.5\%$ (d) 1.03×10^{-2}

17.

wavelength (nm):	591	328	154
N*/N$_0$ at 2 600 K in flame:	2.6×10^{-4}	1.4×10^{-7}	1.8×10^{-16}
N*/N$_0$ at 6 000 K in plasma:	5.2×10^{-2}	2.0×10^{-3}	1.2×10^{-7}

Br is not readily observed by atomic absorption because its lowest excited state requires far-ultraviolet radiation absorbed by N_2 and O_2 in the air. Br is not readily observed in emission because the excited state is not sufficiently populated.

18. Y takes C from BaC, so BaC + Y $\rightleftharpoons$ Ba + YC is driven to right.

19. 20 GW/cm^2, 5 ng

20. Analyte and standard are lost in equal proportions, so their ratio remains constant.

21. 0.429 ± 0.012 wt%

22. (a) 7.49 μg/mL (b) 25.6 μg/mL

23. 17.4 ± 0.3 μg/mL

24. (a) CsCl inhibits ionization of Sn. (b) $m = 0.782 \pm 0.018$; $b = 0.86 \pm 1.55$; $R^2 = 0.997$ (c) There is little interference at 189.927 nm, which is the better choice of wavelengths. At 235.485 nm, there is interference from Fe, Cu, Mn, Zn, Cr, and, perhaps, Mg. (d) limit of detection = 9 μg/L; limit of quantitation = 31 μg/L (e) 0.8 mg/kg

25. [Ti] = 0.224 6 mM, [S] = 4.273 mM, [transferrin] = 0.109 6 mM, Ti/transferrin = 2.05

26. When an electron is removed from an inner shell of an atom by absorption of an X-ray, an electron from an outer shell falls into the vacancy. The excess energy of the electron making the transition is emitted as an X-ray. Electronic energy levels are different for every element, so the signature (energies of emitted X-rays) is different for each element.

27. Ti K_β could be the weak peak near 4.9 keV. Se K_β at 12.50 keV comes under the strong Pb L_β peak. Zr K_β at 17.67 keV is a weak peak.

28. $K_\alpha = 74.97$ and $K_\beta = 84.94$ keV are beyond the range of the spectrum. The K energies for Pb exceed the energy available from common X-ray tubes.

29. 0.392 keV = 3.78×10^4 kJ/mol = 40 times greater than N≡N bond energy

30. 6.40 keV Fe K_α, 7.05 Fe K_β, 7.50 Ni K_α, 8.07 Cu K_α, 8.62 Zn K_α, 10.57 Pb L_α, 12.60 Pb L_β, ~14.1 Sr K_α?, ~15.78 Zr K_α, 17.50 Mo K_α, 19.59 Mo K_β

31. 3.70 keV Ca K_α, 4.01 Ca K_α, 6.40 keV Fe K_α, 7.06 Fe K_β, 9.99 Hg L_α, 10.55 Pb L_α, 11.84 Hg L_β, 12.63 Pb L_β, 25.25 Sn K_α, 28.48 Sn K_β

Chapter 22

2. Electron ionization: 70 eV electron striking pentobarbital creates M$^{+\bullet}$ ($m/z = 226$) with enough energy to break into fragments. Chemical ionization: pentobarbital reacts with proton donor CH_5^+ to make MH$^+$ ($m/z = 227$). Some fragmentation occurs also.

3. 1 Da$\equiv$1/12 mass of ^{12}C = 1.660 54 $\times$ 10^{-24} g, 83.5 ($\pm$2.3) fg

4. 58.5 from the spectrum in the text

5. 1.5 $\times$ 10^4, yes

6. $\sim$3 100

7. 2.0 $\times$ 10^6, 3.4 $\times$ 10^6

8. C$_6$H$_{11}^+$

9. ^{31}P$^+$ = 30.973 21, ^{15}N^{16}O$^+$ = 30.994 47, ^{14}N^{16}OH$^+$ = 31.005 25

10. (a) m/z 141: ^{13}C^{12}C$_6$^{1}H$_{10}$^{14}N^{16}O$_2^+$ and ^{12}C$_7$^{2}H^{1}H$_9$^{14}N^{16}O$_2^+$ and ^{12}C$_7$^{1}H$_{10}$^{15}N^{16}O$_2^+$ and ^{12}C$_7$^{1}H$_{10}$^{14}N^{17}O^{16}O$^+$, intensity = 8.1%

(b) ^{12}C$_7$^{1}H$_{10}$^{14}N^{18}O^{16}O$^+$ and ^{13}C$_2$^{12}C$_5$^{1}H$_{10}$^{14}N^{16}O$_2^+$ and ^{13}C^{12}C$_6$^{1}H$_{10}$^{15}N^{16}O$_2^+$

11. 2.4% not counting the predominant species, which is [^{12}C$_{12}$^{1}H$_{18}$^{16}O$_8$^{35}Cl$_2$^{37}Cl]$^-$

12. Calculated exact mass = 395.007 23, 397.004 28, 399.001 33, 400.998 38

Difference = 0.2, 0.6, 1.2 ppm

13. (a) m/z 141 m/z 279

(b) m/z 277 m/z 221

14. 1 : 1.946 : 0.946 3

15. 1 : 8.05 : 16.20

16. 0.342 7 : 1 : 0.972 8 : 0.315 4

17. (a) 1 010 Da   **(b)** 1 011.14 Da   **(c)** 1 : 0.778 7

18. (a) 4   **(b)** 6   **(c)** 1$\frac{1}{2}$ could be [H$_2$C$=$C(H)$-$CH$_2$]$^+$

19. (a) Cl, C$_6$H$_5$Cl: M$^{+\bullet}$ = 112

Predicted intensities M+1 : M+2 : M+3 = 6.54 : 32.2 : 2.11%

(b) Cl, C$_6$H$_4$Cl$_2$: M$^{+\bullet}$ = 146

Predicted intensities M+1 : M+2 : M+3 : M+4 : M+5 = 6.53 : 64.2 : 4.19 : 10.33 : 0.67%

(c) NH$_2$, C$_6$H$_7$N: M$^{+\bullet}$ = 93

Predicted intensities M+1 : M+2 = 6.93 : 0.17%

(d) (CH$_3$)$_2$Hg: M$^{+\bullet}$ = 228
Predicted intensities M+1 : M+2 = 169.2 : 231.7%

(e) CH$_2$Br$_2$: M$^{+\bullet}$ = 172
Predicted intensities M+2 : M+4 = 194.6 : 94.6%

(f) , 1,10-phenanthroline, C$_{12}$H$_8$N$_2$: M$^{+\bullet}$ = 180

Predicted intensities M+1 : M+2 = 13.8 : 0.8%
(C$_{13}$H$_8$O also fits the observed intensities)

(g) , ferrocene, C$_{10}$H$_{10}$Fe: M$^{+\bullet}$ = 186

Predicted intensities M−2 : M+1 : M+2 = 6.37 : 13.23 : 0.83%

20. M$^{+\bullet}$ = 206, CH^{79}Br$_2$^{35}Cl

21. (a) 6   **(b)** 350   **(c)** 350, 315, 280, 245, 210 = M$^+$, (M − Cl)$^+$, (M − 2Cl)$^+$, (M − 3Cl)$^+$, (M − 4Cl)$^+$, (M − 5Cl)$^+$

22. U.S. diet might contain more C$_4$ and CAM plants and European diet might contain more C$_3$ plants

23. (a) mass of p$^+$ + e$^-$ = mass of ^{1}H
(b) mass of p$^+$ + n + e$^-$ = 2.016 489 963 Da; mass of ^{2}H = 2.014 10 Da
(c) 2.15 $\times$ 10^8 kJ/mol   **(d)** 1.31 $\times$ 10^3 kJ/mol   **(e)** 5 $\times$ 10^5

24.

Mass:	84	85	86	87	88	89	90
Intensity:	1	0.152	0.108	0.010 3	0.003 62	0.000 171	0.000 037

25. Electric sector narrows the range of ion kinetic energy prior to separation by the magnetic sector.

26. Reflectron improves resolving power by allowing slower ions to catch up to faster ions of the same mass.

27. 0.000 2 Da at m/z 100 and 0.04 Da at m/z 20 000

28. 4.39 $\times$ 10^4 m/s; 45.6 μs; 2.20 $\times$ 10^4 spectra/s; 1.56 $\times$ 10^4 spectra/s

29. 10.2 μs for 100 Da and 1.02 ms for 1 000 000 Da

30. (a) 93 m   **(b)** 93 km

31. Electrospray sees ions already in solution. Atmospheric pressure chemical ionization creates ions in the corona discharge.

32. Gas chromatography flow rates do not overwhelm the vacuum pumping capacity of a mass spectrometer. Liquid chromatography flow rates overwhelm the vacuum pump unless the liquid flow rate is very small ($\leqslant$500 nL/min).

33. Photons energy $\approx$10 eV is just enough for ionization without extra energy for fragmentation. A 70 eV electron has more energy than required for ionization.

34. Ions accelerated by an electric field beak into fragments when they collide with N$_2$ or Ar. This process can be conducted at the entrance to the spectrometer or in a collision cell in the middle of tandem mass spectrometry.

35. Reconstructed total ion chromatogram shows current from all ions above a selected mass as a function of time. Extracted ion chromatogram shows current from one or a few selected ions from the full mass spectrum. Selected ion chromatogram monitors only the current from one ion to increase signal-to-noise ratio.

36. An ion of one m/z selected by the first mass separator is directed to a collision cell to produce fragment ions. One selected fragment ion is monitored. This name MS/MS refers to two consecutive mass separations. Signal/noise is improved because there are few sources of the precursor ion other than analyte, and it is unlikely that other precursor ions can decompose to give the same product ion.

37. (a) negative ion mode, neutral solution   **(b)** 14.32

38. n_A = 12 and n_I = 20; mean molecular mass (disregarding peak G) is 15 126 Da

39. charge = 4; molecular mass = 7 848.48 Da

40. Au$_{329}$(SR)$_{84}$

41.

37:3:	[MNH$_4$]$^+$ = C$_{37}$H$_{72}$ON	[MH]$^+$ = C$_{37}$H$_{69}$O
M+1:	41.2% predicted	40.8% predicted
	35.8% observed	23.0% observed
M+2:	7.9% predicted	7.9% predicted
	7.0% observed	8.0% observed
37:2:	[MNH$_4$]$^+$ = C$_{37}$H$_{74}$ON	[MH]$^+$ = C$_{37}$H$_{71}$O
M+1:	41.3% predicted	40.8% predicted
	40.8% observed	33.4% observed
M+2:	7.9% predicted	7.9% predicted
	3.7% observed	8.4% observed

42. Q1 selects ^{35}ClO$_3^-$ and Q3 selects ^{35}ClO$_2^-$ from the collision cell.

43. (d) 7.63$_9$ μmol V/g

Chapter 23

1. three

2. **(a)** no **(b)** Shaking forms small droplets of one phase in the other, shortening the distance that solute must diffuse to transfer from one phase to the other.

3. 3

4. **(a)** AlY^- is an anion, but AlL_3 is neutral. **(b)** hydrophobic cation

5. $mHL + M^{m+} \rightleftharpoons ML_m + mH^+$ is driven right at high pH, increasing fraction of metal in the form ML_m, which is extracted into organic solvent.

6. Neutral ML_n is extracted into organic solvent. ML_n formation is favored by increasing β and increasing K_a. Increasing K_L decreases [L] in aqueous phase, thereby decreasing $[ML_n]$. Increasing $[H^+]$ decreases $[L^-]$ available for complexation.

7. When pH $> pK_{BH}+$, predominant form is B, which is extracted into organic phase. When pH $> pK_{HA}$, predominant form is A^-, which is extracted into water.

8. **(a)** 0.080 M **(b)** 0.50

9. 0.088

10. **(c)** 4.5 **(d)** greater

12. **(a)** 0.16 M in benzene **(b)** 2×10^{-6} M in benzene

13. 2 pH units

14. **(a)** 2.6×10^4 at pH 1 and 2.6×10^{10} at pH 4 **(b)** 3.8×10^{-4}

15. **(a)** % extracted $= \left(100\,D\dfrac{V_{aq}}{V_{org}}\right)\bigg/\left(1 + D\dfrac{V_{aq}}{V_{org}}\right)$

16. $(n, \text{fraction extracted}) = (1, 0.667), (2, 0.750), (3, 0.784), (10, 0.838)$
$q_{limit} = 0.865$; 95% of q_{limit} is attained in 6 extractions

17. 1-C, 2-D, 3-A, 4-E, 5-B

18. Increased partition coefficient $\Rightarrow$ increased fraction of solute in stationary phase $\Rightarrow$ decreased fraction of solute moving through the column.

19. **(a)** $k =$ (time solute is in stationary phase)/(time solute is in mobile phase) **(b)** fraction of time in mobile phase $= 1/(1 + k)$

20. **(a)** 17.4 cm/min **(b)** 0.592 min **(c)** 6.51 min

21. **(a)** 13.9 m/min, 3.00 mL/min **(b)** $k = 7.02$, fraction of time $= 0.875$ **(c)** 295

22. **(a)** 40 cm long $\times$ 4.25 cm diameter **(b)** 5.5 mL/min **(c)** 1.11 cm/min for both

23. **(a)** 2.0 **(b)** 0.33 **(c)** 20

24. 19 cm/min

25. 0.6, 6

26. $k = 3.59$, $K = 4.69$

27. 603, 0.854

28. **(a)** Column 2 **(b)** Column 3 **(c)** Column 3 **(d)** Column 4

30. **(a)** Smaller molecules diffuse faster than larger molecules. **(b)** $\sim 10^{-5}$ m^2/s

31. **(a)** $1.2_6 \times 10^4$ **(b)** 20 μm **(c)** 0.72

32. **(a)** 1 **(b)** 2 **(c)** 1 **(d)** neither **(e)** B **(f)** B **(g)** 5.2 **(h)** 6.7 **(i)** 1.3

33. **(a)** increase flow rate **(b)** decrease flow rate **(c)** no effect

34. Linear velocity determines the time available for longitudinal diffusion and mass transfer to occur.

35. 0.1 mm

36. Longitudinal diffusion occurs much faster in gas chromatography than in liquid chromatography.

37. Smaller particles enable more rapid mass transfer between mobile and stationary phases.

38. 33 mL/min

39. 1-C_m, 2-A, 3-EC, 4-B, 5-C_s

40. Silanization caps hydroxyl groups to which strong hydrogen bonding can occur.

41. Overload gives fronting in gas chromatography and tailing in liquid chromatography.

42. **(a)** Broader injected band $\Rightarrow$ broader eluted peaks **(b)** Higher sample concentration might overload the column.

43. 2.65 mm

44. **(a)** 1.1×10^2 **(b)** 0.89 mm

45. **(a)** 2.1 **(b)** 5.2×10^3 **(c)** Eq. 23-30: $N = 7.72 \times 10^3$; Equation 23-32: $N = 7.75 \times 10^3$

46. **(a)** resolution $= 0.83$ **(b)** resolution > 1.5

47. 10.4 mL

48. $0.013\,3$ mL^2, $0.005\,2$ mL^2, $0.002\,2$ mL^2. 25.0 s

49. 1.8×10^3

50. **(a)** $N = 41\,000$ plates **(b)** $N = 11\,000$ plates **(c)** $N = 34\,000$ plates

51. **(a)** $k = 11.25, 11.45$ **(b)** 1.018 **(c)** C_6HF_5: 60 800 plates, height = 0.493 mm; C_6H_6: 66 000 plates, height = 0.455 mm **(d)** C_6HF_5: 55 700 plates, C_6H_6: 48 800 plates **(e)** 0.96 **(f)** 0.93

Chapter 24

1. **(a)** Packed columns: high sample capacity; open tubular columns: better separation efficiency (smaller plate height), shorter analysis time, increased sensitivity **(b)** wall-coated: liquid stationary phase bonded to wall of column; porous-layer: solid stationary phase on wall of column **(c)** reduced tendency for stationary phase to bleed from the column

2. Band broadening from multiple path term (A) in van Deemter equation is eliminated.

3. **(a)** Advantage: higher resolution; disadvantage: lower sample capacity requires more sensitive detector **(b)** Advantage: more plates and greater resolution; disadvantage: longer run time **(c)** Advantage: greater retention and greater sample capacity; disadvantage: lower resolution, longer run time, increased bleeding

4. **(a)** nonpolar solutes **(b)** polar solutes **(c)** light gasses

5. **(a)** Advantages: compounds with wide range of retention characteristics can be resolved and separated in reasonable time; low initial temperature permits splitless or on-column injection. Disadvantage: sloping baseline from column bleed, long cooling time between runs **(b)** Reduced retention time for late-eluting peaks and reduced thermal decomposition

6. **(a)** Inert, low viscosity, compatible with detector **(b)** Diffusion is more rapid in H_2 and He.

7. **(a)** Split injection for high concentrations, high resolution, and dirty samples; splitless injection for trace analysis; on-column injection for quantitative analysis and thermally sensitive samples **(b)** Solvent trapping condenses solute in narrow band of solvent at beginning of column. Cold trapping condenses solute in a narrow band at the start of the column before temperature is raised.

8. **(a)** all analytes **(b)** C atoms bearing H **(c)** halogens, CN, NO_2, conjugated C=O **(d)** P and S and other elements selected by wavelength **(e)** P and N (and much less to hydrocarbons) **(f)** Aromatic and unsaturated compounds **(g)** S **(h)** Most elements (selected individually by wavelength) **(i)** all analytes **(j)** all analytes

9. Any substance other than carrier gas changes conductivity of gas stream.

10. Reconstructed total ion chromatogram (least selective) sees everything. Selected ion monitoring only observes selected m/z. Selected reaction monitoring (most selective) takes one ion from analyte, breaks it into fragments, and monitors just one of the fragments.

11. Derivatization converts analyte to a form that is easier to separate or detect.

12. **(a)** Analyte is extracted from sample into thin coating on silica fiber; cold trapping condenses analyte at start of column during slow evaporation

from fiber; no.   **(b)** Stir-bar sorptive extraction is 100x more sensitive because analyte is extracted into thick coating on stir bar.

13. Purge and trap collects all analyte from the unknown. Splitless injection is required so analyte is not lost during injection.

14. (1) goal of the analysis, (2) sample preparation method, (3) detector, (4) column, and (5) injection method

15. **(a)** $t'_r = 4.7$ min, $k = 1.3$ **(b)** 80   **(c)** 104   **(d)** 6.1 min

16. **(a)** 836   **(b)** no   **(c)** no

17. **(a)** hexane < butanol = benzene < 2-pentanone < heptane < octane

(b) hexane < heptane < butanol < benzene < 2-pentanone < octane

(c) hexane < heptane < octane < benzene < 2-pentanone < butanol

18. **(a)** 3, 1, 2, 4, 5, 6   **(b)** 3, 4, 1, 2, 5, 6   **(c)** 3, 4, 5, 6, 2, 1

19. **(a)** 35 kJ/mol   **(b)** 0.18 bar   **(c)** 1.05 bar   **(d)** lower vapor pressure ⇒ greater retention   **(e)** gaseous mobile phase

20. $t'_r = 27.1$ min

21. t_r decreases with increasing temperature, approaching t_m at some temperature above the boiling point of butanol.

22. **(a)** $N = 4.0 \times 10^4$, $H = 0.75$ mm   **(b)** 1.7

23. **(a)** $k = 6.4$ **(b)** $N = 4.0 \times 10^4$, $H = 0.50$ mm   **(c)** 1.1

24. **(a)** Thin stationary phase gives rapid mass transfer in stationary phase. Wide bore column gives slow mass transfer in mobile phase. **(b)** Injections: 0.16 ng (narrow bore), 56 ng (wide bore)

25. **(a)** Air does not require a gas tank, but O_2 in air could degrade column   **(b)** air: 9.3 cm/s, 0.036 cm; H_2: 17.6 cm/s, 0.051 **(c)** air: 8 300; H_2: 5 900   **(d)** air: 32 s; H_2: 17 s   **(e)** C_s becomes negligible   **(f)** solutes diffuse in H_2 faster than in air

26. **(a)** Increase column length and decrease diameter; change stationary phase; optimize flow rate; possibly increase film thickness and decrease temperature to increase retention factor   **(b)** flow rate and temperature

27. **(a)** 1.25_3 **(c)** 77.6 mM

28. 0.41 μM

29. 932

30. **(a)** Multiple small extractions are more efficient than one large extraction   **(b)** reduced likelihood of thermal decomposition and larger quantity of molecules to smell   **(c)** polar column for polar analytes **(d)** 132   **(e)** enables larger injection so components can be smelled when eluted

31. **(a)** lower solubility of MTBE in groundwater   **(b)** m/z 73, only 3 components have this ion   **(c)** m/z 73 is $CH_3OC^+(CH_3)_2$

32. **(a)** increase vapor pressure of analyte   **(b)** low temperature cold traps analytes during thermal desorption from fiber   **(d)** nonsmoker: 78 ± 5 μg/L; nonsmoker whose parents smoke: 192 ± 6 μg/L

33. $[^{14}NO_2^-] = 1.8$ μM; $[^{14}NO_3^-] = 384$ μM

34. **(a)** $A = 0$ **(b)** $B = 2D_m$

(c) $C = C_s + C_m = \dfrac{2k}{3(k+1)^2}\dfrac{d^2}{D_s} + \dfrac{1 + 6k + 11k^2}{24(k+1)^2}\dfrac{r^2}{D_m}$

(d) $u_x(\text{optimum}) = \sqrt{\dfrac{B}{C}}$; $H_{min} = 2\sqrt{B(C_s + C_m)}$

35. **(a)** 0.58, 1.9   **(b)** 0.058 mm, 0.19 mm   **(c)** 3.0×10^5 **(d)** $\beta = r/2d_f$ **(e)** 4.0

36. $D_m = 3.0 \times 10^{-5}$ m²/s, $D_s = 5.0 \times 10^{-10}$ m²/s (diffusion in gas is 6.0×10^4 times faster than diffusion in stationary phase)

37. **(b)** limiting $m = KV_fC_o = 6.9$ ng for $K = 100$ and 690 ng for $K = 10\,000$ **(c)** 0.69%, 41%

Chapter 25

1. **(a)** Reversed phase: nonpolar solutes are more soluble in nonpolar solvent. Normal phase: polar solutes are more soluble in polar solvent. **(b)** pressure gradient (= density gradient)

2. Solvent competes with solute for adsorption sites. Solvent-adsorbent interaction is independent of solute.

3. Solute equilibrates between mobile phase and aqueous surface layer. Water in eluent competes with the surface layer to dissolve polar solute.

4. **(a)** Small particles increase resistance to flow.   **(b)** Mass transfer rate increases if diffusion distance is smaller ⇒ decreased C term in van Deemter equation. Flow paths with small particles are more uniform ⇒ smaller A term.   **(c)** Covalently attached to solid support

5. **(a)** $L = 33, 17, 10, 5$ cm   **(b)** 10, 6, 3 min   **(c)** 9, 15, 30 MPa **(d)** 1 600, 820, 490, 240 μL   **(e)** 1.5 μm requires UHPLC

6. **(a)** Pores increase surface area, which increases sample capacity. **(b)** Pores must be large enough for solute to enter. Large pores decrease available surface area, so pores should not be larger than necessary to admit solute.

7. 0.14 min, 0.30 min

8. **(a)** $N = 1\,500$, $H = 33$ μm, 19 particles/plate   **(b)** 2.4 particles/ plate, maximum speed

9. SiO_2 dissolves in base. Stationary phase bond to SiO_2 hydrolyzes in acid. Bulky groups hinder approach of H_3O^+ to Si–O–Si bond.

10. Additive binds to the sites that cause tailing.

11. **(b)** Reduce sample concentration, unclog frit with reverse flow, use Type B silica, use additives in solvent

12. **(b)** C term approaches zero and A term is small for small particles. **(c)** A and C terms are small; particle has flow resistance of 2.7-μm-diameter particle.

13. **(a)** 1 560 for L enantiomer and 1 310 for D enantiomer **(b)** 1.25   **(c)** 1.19

14. **(a)** 18   **(b)** 10   **(c)** Mass transfer rate and optimum velocity both increase.

15. **(a)** normal-phase   **(b)** bonded reversed-phase **(c)** bonded reversed-phase with buffered mobile phase   **(d)** HILIC **(e)** ion-exchange or ion chromatography   **(f)** molecular-exclusion **(g)** ion-exchange with wide pore stationary phase   **(h)** molecular-exclusion

16. 0.27 m²

17. **(a)** shorter   **(b)** amine   **(c)** greater   **(d)** shorter

18. **(a)** 1.76 mm/s   **(b)** 3.51   **(c)** 192 000, 22.9 μm   **(d)** 0.59 **(e)** 1.006_6 **(f)** $5.5_3 \times 10^5$, 12.7 m   **(g)** decrease flow, decrease mobile phase strength, or change solvent   **(h)** 0.90

19. **(a)** (R)-Gimatecan would be eluted at 6.96 min from (R,R)-stationary phase and 6.10 min from (S,S)-stationary phase.   **(b)** (R)-Gimatecan at 6.96 min is lost beneath tail of (S)-gimatecan   **(c)** $\alpha = 1.23$ **(d)** 2.3 (better than baseline separation)

20. 126 mm

21. **(a)** $t_m = 0.38$ min for column A and 0.26 min for column B **(b)** More polar glucuronide is less retained by reversed-phase column. **(c)** 1.3, 3.3   **(d)** Polar, hydrophilic silica retains glucuronide more than less polar morphine. Gradient goes to increasing polarity to remove more polar solute.   **(e)** 6.2

22. **(b)** 5.8 W

23. **(a)** Universal <210 nm, selective >210 nm   **(b)** universal but insensitive   **(c)** and **(d)** near universal for non-volatile compounds **(e)** redox-active compounds   **(f)** fluorescent molecules   **(g)** molecules with N   **(h)** ionic or ionizable compounds

24. **(a)** Many chromophores absorb <210 nm but only some absorb at 254 nm   **(b)** increasing gradient of CH_3OH absorbs weakly at 210 nm **(c)** measured $t_m = 1.23$ min, predicted $t_m = 1.32$ min

25. **(a)** m/z 304 is BH^+ (cocaine protonated at N)   **(b)** loss of $C_6H_5CO_2H$ **(c)** m/z 304 selected by Q1 is isotopically pure, so there is no ^{13}C partner **(d)** Q1 selects only m/z 304. Q3 selects only the fragment m/z 182. Few components in plasma that give m/z 304 also break into m/z 282. **(e)** phenyl group   **(f)** Measure cocaine/D_5-cocaine response factor. Add D_5-cocaine internal standard to plasma and measure peak areas of cocaine and D_5-cocaine.

26. (a) m/z 234 is MH^+, m/z 84 is $C_5H_{10}N^+$ (b) $237 \rightarrow 84$

27. (b) (i) Run uracil or sodium nitrate with ultraviolet detector (ii) observe first baseline disturbance (iii) $t_m \approx Ld_c^2/(2F)$ (c) Run toluene with ultraviolet detector. (ii) and (iii) from part (b) also work in HILIC. (d) 1.1 min for both

28. Extra-column volume: from injection to detection not including column. Dwell volume: from point of mixing solvents to start of column. Extra-column volume causes peak broadening. Dwell volume delays start of gradient elution.

29. Rugged procedure is not seriously affected by small changes in conditions.

30. $0.5 \leq k \leq 20$, resolution ≥ 2, pressure ≤ 15 MPa, $0.9 \leq$ asymmetry factor ≤ 1.5

31. Run wide gradient selected to produce $k^* \approx 5$. Measure Δt between first and last peaks. Use gradient if $\Delta t/t_G > 0.40$. Use isocratic elution if $\Delta t/t_G < 0.25$.

32. (i) Determine goal (ii) Sample preparation (iii) Choose detector (iv) Run wide gradient to select isocratic or gradient elution (v) For isocratic separation vary %B until $0.5 \leq k \leq 20$. For adequate resolution, try minor adjustment in %B, different organic solvent of equivalent mobile phase strength, or different column. Select column length or particle size to increase resolution or shorter separation time.

33. (a) 53% tetrahydrofuran in H_2O (b) Mix 530 mL tetrahydrofuran + 470 mL H_2O. Do not adjust volume. (c) Tetrahydrofuran has ultraviolet absorption and attacks polyether ether ketone plastic components.

34. Run A: high %B, low T; Run B: high %B, high T; Run C: low %B, high T; Run D: low %B, low T. From appearance of chromatograms, explore conditions between Runs A, B, C, and D to improve separation.

35. k (50% B) = 3.1. k (40% B): 27.8 min predicted, 20.2 min observed

36. (a) ~36 min (b) 43.0 min using points from $\Phi = 0.35$ to 0.6
(c) k is high at start of gradient and decrease exponentially until compounds are almost unretained at end of gradient (d) By the time a compound is eluted, it is almost unretained and has the peak width of a weakly retained compound.

37. 38%

38. (a) lower (b) higher

39. (a) A weak acid, $pK_a = 7.7$, $k_{HA} = 6.5$; B weak acid, $pK_a = 3.7$, $k_{HA} = 5.9$; C weak base, $pK_a = 4.7$, $k_B = 3.2$ (b) pH $\approx$ 3 to 6.5
(c) different buffers are required to cover different ranges of pH

40. (a) pH 2.0 provides retention within $0.5 \leq k \leq 20$ (b) pH $\approx$ 4
(c) pH 7.5

41. Acetophenone – neutral and retained at pH 3–7; salicylic acid – 50% HA at pH 3 with some retention, A^- with little retention at pH 5–7; nicotine 50% B with some retention at pH 7, BH^+ or BH_2^{2+} at pH 3–7 with little retention

42. (a) $\Delta t/t_G = 0.32 > 0.25 \Rightarrow$ gradient elution (b) 40 to 70% acetonitrile in 60 min

43. (a) Change solvent strength, temperature, or pH. Use a different solvent or a different kind of stationary phase. (b) Slower flow rate, different temperature, longer column, smaller particle size

44. (a) ~29 min (b) $k^* = 12.9$, $F = 4.7$ mL/min, $m = 4.7$ mg, $t_G = 11.5$ min

Chapter 26

1. Decreased swelling, increased capacity and selectivity, longer equilibration time

2. Wash column with NaOH to convert resin to OH^- form. Elute all bound OH^- with excess Cl^-. Then titrate OH^- in eluate with HCl.

3. 38–75 μm; 200/400 mesh

4. (a) pyruvate -1, 2-oxovalerate -1, and maleate -2
(b) I^- is highly polarizable

5. (a) Anionic proteins are protonated as pH decreases, and so less retained. (b) Increased anion concentration displaces proteins.

6. Possibly $NH_3 < (CH_3)_3N < CH_3NH_2 < (CH_3)_2NH$ due to pK_a

7. Cations and anions have been removed. Nonionic impurities are not removed.

8. (a) 1 000-fold (b) 0.057 nM (c) $\leq$340 000 nM $(6 \times 10^6 \times [Fe^{3+}]_{seawater})$

9. Cation charge = 0.002 02 M, anion charge = -0.001 59 M. Some concentrations are inaccurate or some ionic material was not detected.

10. Elution order is from most dissociated to least dissociated.

11. (a) Ions partition into the polar water layer formed on the zwitterionic surface. Decreasing CH_3CN increases the polarity of the eluent to elute polar solutes. (b) Aerosol from nebulized effluent gains charge proportional to mass of nonvolatile analyte.

12. Separator column separates ions by ion exchange. Suppressor neutralizes eluent, exchanging Cl^- for OH^- to convert H^+Cl^- into H_2O.

13. (a) 3.8×10^2 h (b) 8.0 mA, 0.16 A

15. (a) Acids are formed by suppression. Strong acids like HCl fully dissociate, weak acids like HCN barely dissociate. (b) Suppression product is H_2CO_3 which partially dissociates.

16. Indirect detection, where a non-absorbing analyte anion replaces an absorbing eluent anion in the eluate

17. (a) Octyl sulfonate binds to nonpolar stationary phase to make it a cation exchanger. (b) 29 ng/mL

18. (a) affinity (b) molecular exclusion (c) molecular exclusion
(d) hydrophobic interaction

19. (a) Compare elution volume to that of mass standards (b) 25 nm

20. (a) 40.2 mL (b) 0.53

21. 0.16

22. (a) 10^6 Da (b) 10^4 Da (c) 10^4 Da

23. (a) 5.7 mL (b) 11.5 mL (c) solutes must be adsorbed

24. 320 000

25. Decreasing salt makes protein more soluble in aqueous mobile phase.

26. (a) μ_{ep} depends inversely on analyte size

(b) phenol $-1.53 \times 10^{-8} \dfrac{m^2}{V \cdot s}$ and trichlorophenol $-2.85 \times 10^{-8} \dfrac{m^2}{V \cdot s}$; only a fraction of HA is ionized, so $\mu_{predicted}$ is only a fraction of μ_{A^-}
(c) ethylphenol $\approx$ methylphenol < phenol < trichlorophenol

27. Bulk flow of fluid caused by electromigration of excess ions in double layer

28. Negative wall $\Rightarrow$ flow toward cathode; positive wall $\Rightarrow$ flow toward anode; speed depends on protonation of silanol and amines.

29. cation; $\mu_{app} = 9.30 \times 10^{-8}$ m^2/(V $\cdot$ s); $\mu_{ep} = +5.04 \times 10^{-8}$ m^2/(V $\cdot$ s)

30. Arginine's positive side chain makes its derivative the least negative.

31. longitudinal diffusion

32. (a) In absence of electroosmotic flow, anionic analyte migrates to anode. (b) Higher conductivity than sample to cause stacking
(c) Narrow capillary dissipates heat better. (d) Li_3PO_4 has lower conductivity than Na_3PO_4. Lower conductivity requries higher field to generate same current. Higher field yields more plates.

33. Lower pH, add amine, and coat capillary. Reverses wall charge.

34. When partitioned into a micelle, neutral analyte moves with velocity of micelle. Migration time depends on fraction of analyte in micelle.

35. (a) 1.15×10^4 Pa (b) 1.17 m height not possible, so use pressure of 11.5 kPa = 0.114 atm.

36. (a) 29.5 fmol (b) 3.00×10^3 V

37. 9.2×10^4 plates, 4.1×10^3 plates (My measurements are about 1/3 lower than the values labeled in the figure from the original source.)

38. (a) maleate (b) fumarate is eluted first. (c) maleate is eluted first.

39. (a) pH 2: 920 s; pH 12: 150 s (b) pH 2: never; pH 12: 180 s

40. (a) $t_{120 kV}/t_{28 kV} = 4.3$ (observed ratio = 3.9) (b) $N_{120 kV}/N_{28 kV} = 4.3$
(c) $\sigma_{120 kV}/\sigma_{28 kV} = 0.48$ (d) Increasing voltage decreases migration time, giving bands less time to broaden by diffusion.

41. Linear to 25 000 V, then broadening due to capillary heating

42. $1.3_5 \times 10^4$ plates, 30 μm

43. 20.5 min

44. light chain = 17 300 Da, heavy chain = 23 500 Da

45. 2.0×10^5 plates

46. Thiamine < (niacinamide + riboflavin) < niacin; thiamine is most soluble.

47. Cyclobarbital and thiopental each separate into two peaks because each has a chiral carbon atom.

48. **(a)** Increases at low velocity due to longitudinal diffusion and at high velocity due to finite time solute needs to equilibrate with micelle **(b)** extra-column effects due to injection and detection.

49. 5.55

50. **(c)** diffusion broadening decreases **(d)** $N_{obs} = 1.45 \times 10^5$, $N_{theory} = 2.06 \times 10^5$ **(e)** 1.6

Chapter 27

1. **(a)** adsorption on surface, absorption inside **(b)** inclusion occupies lattice site, occlusion is a pocket of impurity

2. insoluble, filterable, pure, known composition

3. High supersaturation could give impure, colloidal product.

4. increase temperature, mix during addition, use dilute reagents, homogeneous precipitation

5. Electrolyte preserves electric double layer to prevent peptization.

6. HNO_3 evaporates during drying; $NaNO_3$ does not evaporate.

7. Reprecipitation gives purer product.

8. Mass of a sample is measured as sample is heated.

9. Adding mass to electrode lowers the oscillation frequency.

10. 0.022 86 M

11. 1.94 wt%

12. 225.3 g/mol

13. 0.085 38 g

14. 50.79 wt%

15. 0.191 4 g calcium carbonate, 0.107 3 g calcium oxide

16. 104.1 ppm

17. 7.22 mL

18. 0.339 g

19. **(a)** 19.98% **(b)** Heating reduces supersaturation to make larger particles. **(c)** $(NH_4)_2C_2O_4$ provides oxalate to reduce solubility of CaC_2O_4 and electrolyte to prevent peptization. **(d)** Negative test for Cl^- assures that original solution has been washed away and there is no dissolved solid that will increase mass of dry product.

20. **(a)** 5.5 mg/100 mL **(b)** 5.834 mg, yes

21. 14.5 wt% K_2CO_3, 14.6 wt% NH_4Cl

22. 40.4 wt%

23. 22.65 wt%

24. **(a)** 40.05 wt% **(b)** 39%

25. **(a)** 1.82 **(b)** $Y_2O_2(OH)Cl$ or $Y_2O(OH)_4$

26. **(a)** 0.854 976 **(b)** 0.856 3 **(d)** 0.856 ± 0.008, 0.86 ± 0.08

27. **(b)** 0.204 (± 0.004)

28. Combustion: heat in excess O_2 to convert $C \rightarrow CO_2$ and $H \rightarrow H_2O$ Pyrolysis: decompose with heat without O_2. $O \rightarrow CO$

29. WO_3 catalyzes $C \rightarrow CO_2$ with excess O_2 Cu converts $SO_3 \rightarrow SO_2$ and removes excess O_2

30. Capsule melts and is oxidized to SnO_2 to liberate heat and crack sample. Sn consumes O_2, ensuring that sample oxidation occurs in gas phase. Sn is oxidation catalyst.

31. Pyrolysis gives gaseous products prior to oxidation to minimize formation of NO_x.

32. 11.69 mg CO_2, 2.051 mg H_2O

33.

$$C_8H_7NO_2SBrCl + 9\tfrac{1}{4}O_2 \rightarrow 8CO_2 + \tfrac{5}{2}H_2O + \tfrac{1}{2}N_2 + SO_2 + HBr + HCl$$

34. $C_4H_9NO_2$

35. 10.5 wt%

36. $C_8H_{9.06 \pm 0.17}N_{0.997 \pm 0.010}$

37. 12.4 wt%

38. **(a)** 95% confidence: $10.16_0 \pm 1.93_6$ μmol Cl^- (Experiment 1), $10.77_0 \pm 2.29_3$ μmol Cl^- (Experiment 2) **(b)** difference is not significant **(c)** 24.2_{96} mg $BaSO_4$ **(d)** 4.35%

Chapter 28

1. You must know how a sample was collected and stored.

2. Analytical quality means accuracy and precision of analytical method. Data quality means sample is representative and analytical procedure is appropriate for intended purpose.

3. **(a)** 5% **(b)** 2.6%

4. 1.0 g

5. 120/170 mesh

6. $10^4 \pm 0.99\%$

7. **(a)** 15.8 **(b)** 1.647 **(c)** 474–526

8. 95%: 8; 90%: 6

9. **(a)** 5.0 g **(b)** 7

10. 0.34 ± 0.14 pg

11. **(a)** Na_2CO_3: 4.47 μg, 8.94×10^5 particles; K_2CO_3: 4.29 μg, 2.24×10^7 particles **(b)** 2.33×10^4 **(c)** Na_2CO_3: 3.28%; K_2CO_3: 0.131%

12. Zn, Fe, Co, Al

13. prevents possible explosion

14. Barbital from aqueous solution is retained by column, but dissolves in acetone/chloroform, which elutes it from the column.

15. pH 2: amine is protonated and carboxylic acid is neutral, so analytes are retained by cation exchanger. High pH: amine is neutral and carboxylate is negative, so analytes are not retained by cation-exchanger.

16. $SO_2 + H_2O \rightarrow H_2SO_3$; $2Resin^+OH^- + H_2SO_3 \rightarrow (Resin^+)_2SO_3^{2-} + H_2O$; SO_3^{2-} eluted with Na_2CO_3 and oxidized by H_2O_2 to SO_4^{2-} measured by ion chromatography

17. Extraction: large particles allow rapid flow without pressure. Chromatography: small particles increase separation efficiency, but require high pressure.

18. **(a)** Solid-phase extraction retains CH_2=$CHCONH_3^+$ while passing many other components. **(b)** Many components absorb UV. **(c)** Acrylamide, m/z 72 selected by Q1 and dissociates by collisions in Q2. Product m/z 55 selected in Q3 for detection; CH_2=$CHCONH_3^+$ (m/z 72) $\rightarrow CH_2$=$CHC\equiv O^+$ (m/z 55); CD_2=$CDCONH_3^+$ (m/z 75) $\rightarrow CD_2$=$CDC\equiv O^+$ (m/z 58); **(d)** Acrylamide is the only m/z 72 that produces m/z 55; **(e)** [acrylamide]/[internal standard] = [area of m/z 72$\rightarrow$55]/ [area of m/z 75$\rightarrow$58]; **(f)** UV: same time as acrylamide with similar signal; SRM: detector sees either acrylamide or the internal standard, with no interference from each other. **(g)** Fraction of acrylamide and fraction of internal standard that bind to and are recovered from solid-phase extraction column are essentially equal.

19. **(b)** standard suffers same losses as analyte; **(c)** total signal from all ions at any moment; **(d)** extracted ion chromatogram shows m/z 312 taken from the full spectrum; selected ion chromatogram shows m/z 312 recorded continuously and has greater signal : noise ratio; **(e)** selected reaction monitoring

20. **(a)** Liquid-liquid extraction uses ≥ 100 mL of organic solvent to extract aqueous phase by continuous distillation. Dispersive liquid-liquid microextraction uses ~ 10–100 μL of immiscible organic solvent plus ~ 0.5–1 mL of dispersant solvent to make a cloudy emulsion for extraction. **(b)** Disperser solvent is miscible with both aqueous and organic phases, lowering interfacial energy and permitting high-surface-area emulsion with rapid mass transfer to be formed.

21. Solid-supported liquid-liquid extraction: aqueous phase is suspended in a microporous medium through which organic solvent is passed to extract analytes. Solid-phase extraction: aqueous sample is passed through small column of stationary phase that retains analytes. Impurities and analytes are eluted by a series of washes with small volumes of solvent with increasing solvent strength.

22. **(a)** 53 **(b)** Ni is in solution or in particles too fine to be removed by filtration. Fe is present as a suspension of solid particles that are removed by filtration.

23. 64.90 wt%

24. **(a)** Cr(III) is $Cr(OH)_2^+$ and $Cr(OH)_3(aq)$; Cr(VI) is CrO_4^{2-}; **(b)** Anion exchanger retains CrO_4^{2-}, but not $Cr(OH)_2^+$ and $Cr(OH)_3(aq)$; **(c)** "Weakly basic" anion exchanger contains $(-^+NHR_2)$ that might lose its positive change in basic solution. "Strongly basic" anion exchanger $(-^+NR_3)$ remains cationic in basic solution; **(d)** CrO_4^{2-} is eluted when $[SO_4^{2-}]$ increases from 0.05 M to 0.5 M.

25. Monitor wells 8, 11, 12, and 13 individually, but pool samples from other sites. If no warning level of analyte is found in composite sample, assume composite is OK. If analyte is found in composite, then analyze each well separately. Pooling samples from n wells reduces sensitivity for analyte in any one well by $1/n$.

INDEX

Physical Constants (2010)

Term	Symbol	Value	
Elementary charge	e	1.602 176 655 (35)*	$\times 10^{-19}$ C
		4.803 204 78 (10)	$\times 10^{-10}$ esu
Speed of light in vacuum	c	2.997 924 58	$\times 10^{8}$ m/s
			$\times 10^{10}$ cm/s
Planck's constant	h	6.626 069 57 (29)	$\times 10^{-34}$ J·s
			$\times 10^{-27}$ erg·s
$h/2\pi$	$\hbar$	1.054 571 726 (47)	$\times 10^{-34}$ J·s
			$\times 10^{-27}$ erg·s
Avogadro's number	N_A	6.022 141 29 (27)	$\times 10^{23}$ mol^{-1}
Gas constant	R	8.314 462 1 (75)	J/(mol·K)
			V·C/(mol·K)
			$\times 10^{-2}$ L·bar/(mol·K)
			$\times 10^{7}$ erg/(mol·K)
		8.205 736 1 (74)	$\times 10^{-5}$ m^3·atm/(mol·K)
			$\times 10^{-2}$ L·atm/(mol·K)
		1.987 204 1 (18)	cal/(mol·K)
Faraday constant (= Ne)	F	9.648 533 65 (21)	$\times 10^{4}$ C/mol
Boltzmann's constant (= R/N)	k	1.380 648 8 (13)	$\times 10^{-23}$ J/K
			$\times 10^{-16}$ erg/K
Electron rest mass	m_e	9.109 382 91 (40)	$\times 10^{-31}$ kg
			$\times 10^{-28}$ g
Proton rest mass	m_p	1.672 621 777 (74)	$\times 10^{-27}$ kg
			$\times 10^{-24}$ g
Dielectric constant (permittivity) of free space	ε_0	8.854 187 817	$\times 10^{-12}$ C^2/(N·m^2)
Gravitational constant	G	6.673 84 (80)	$\times 10^{-11}$ m^3/(s^2·kg)

Numbers in parentheses are the one-standard-deviation uncertainties in the last digits.

SOURCE: *2010 CODATA Values from http://physics.nist.gov/cuu/constants/index.html (August 2011).*

Concentrated Acids and Bases

Name	Approximate weight percent	Molecular mass	Approximate molarity	Approximate density (g/mL)	mL of reagent needed to prepare 1 L of ~ 1.0 M solution
Acid					
Acetic	99.8	60.05	17.4	1.05	57.3
Hydrochloric	37.2	36.46	12.1	1.19	82.4
Hydrofluoric	49.0	20.01	28.4	1.16	35.2
Nitric	70.4	63.01	15.8	1.41	63.5
Perchloric	70.5	100.46	11.7	1.67	85.3
Phosphoric	85.5	97.99	14.7	1.69	67.8
Sulfuric	96.0	98.08	18.0	1.84	55.5
Base					
Ammonia[†]	28.0	17.03	14.8	0.90	67.6
Sodium hydroxide	50.5	40.00	19.3	1.53	51.8
Potassium hydroxide	45.0	56.11	11.5	1.44	86.6

[†]*28.0 wt% ammonia is the same as 56.6 wt% ammonium hydroxide.*